JURAN'S
QUALITY
HANDBOOK

Other McGraw-Hill Handbooks of Interest

Avallone & Baumeister • MARKS' STANDARD HANDBOOK FOR MECHANICAL ENGINEERS

Bleier • FAN HANDBOOK

Brady et al. • MATERIALS HANDBOOK

Bralla • DESIGN FOR MANUFACTURABILITY HANDBOOK

Brink • HANDBOOK OF FLUID SEALING

Chironis & Sclater • MECHANISMS AND MECHANCIAL DEVICES SOURCEBOOK

Cleland & Gareis • GLOBAL PROJECT MANAGEMENT HANDBOOK

Considine • PROCESS/INDUSTRIAL INSTRUMENTS AND CONTROLS HANDBOOK

Czernik • GASKET: DESIGN, SELECTION, AND TESTING

Elliot et al. • STANDARD HANDBOOK OF POWERPLANT ENGINEERING

Frankel • FACILITY PIPING SYSTEMS HANDBOOK

Haines & Wilson • HVAC SYSTEMS DESIGN HANDBOOK

Harris • SHOCK AND VIBRATION HANDBOOK

Hicks • HANDBOOK OF MECHANICAL ENGINEERING CALCULATIONS

Hicks • STANDARD HANDBOOK OF ENGINEERING CALCULATIONS

Higgins • MAINTENANCE ENGINEERING HANDBOOK

Hodson • MAYNARD'S INDUSTRIAL ENGINEERING HANDBOOK

Ireson et al. • HANDBOOK OF RELIABILITY ENGINEERING AND MANAGEMENT

Karassik et al. • PUMP HANDBOOK

Kohan • PLANT OPERATIONS AND SERVICES HANDBOOK

Lancaster • HANDBOOK OF STRUCTURAL WELDING

Lewis • FACILITY MANAGER'S OPERATION AND MAINTENANCE HANDBOOK

Lingaiah • MACHINE DESIGN DATA HANDBOOK

Modern Plastics Magazine • PLASTICS HANDBOOK

Mulcahy • MATERIALS HANDLING HANDBOOK

Mulcahy • WAREHOUSE DISTRIBUTION AND OPERATIONS HANDBOOK

Nayyar • THE PIPING HANDBOOK

Parmley • STANDARD HANDBOOK OF FASTENING AND JOINING

Rohsenow • HANDBOOK OF HEAT TRANSFER

Rosaler • STANDARD HANDBOOK OF PLANT ENGINEERING

Rothbart • MECHANICAL DESIGN HANDBOOK

Shigley & Mischke • STANDARD HANDBOOK OF MACHINE DESIGN

Skousen • VALVE HANDBOOK

Solomon • SENSORS HANDBOOK

Stoecker • INDUSTRIAL REFRIGERATION HANDBOOK

Suchy • HANDBOOK OF DIE DESIGN

Turner & Taylor • THE HANDBOOK OF PROJECT-BASED MANAGEMENT

Walsh • MCGRAW-HILL MACHINING AND METALWORKING HANDBOOK

Walsh • ELECTROMECHANICAL DESIGN HANDBOOK

Wang • HANDBOOK OF AIR CONDITIONING AND REFRIGERATION

Woodson et al. • HUMAN FACTORS DESIGN HANDBOOK

Wrennall & Lee • HANDBOOK OF COMMERCIAL AND INDUSTRIAL FACILITIES MANAGEMENT

Ziu • HANDBOOK OF DOUBLE CONTAINMENT PIPING SYSTEMS

JURAN'S QUALITY HANDBOOK

Joseph M. Juran   Co-Editor-in-Chief

A. Blanton Godfrey   Co-Editor-in-Chief

Robert E. Hoogstoel   Associate Editor

Edward G. Schilling   Associate Editor

Fifth Edition

McGraw-Hill

New York   San Francisco   Washington, D.C.   Auckland   Bogotá
Caracas   Lisbon   London   Madrid   Mexico City   Milan
Montreal   New Delhi   San Juan   Singapore
Sydney   Tokyo   Toronto

Library of Congress Cataloging-in-Publication Data

Juran's quality handbook / Joseph M. Juran, co-editor-in-chief,
  A. Blanton Godfrey, co-editor-in-chief. — 5th ed.
      p.   cm.
   Previous eds. published under title: Juran's quality control handbook.
   Includes indexes.
   ISBN 0-07-034003-X
    1. Quality control—Handbooks, manuals, etc. I. Juran, J. M.
(Joseph M.), date. II. Godfrey, A. Blanton. III. Title:
Quality handbook. IV. Title: Quality control handbook.
TS156.Q3618   1998
  658.5'62—dc21
                                   98-43311
                                     CIP

McGraw-Hill

A Division of The McGraw·Hill Companies

6 7 8 9 0 DOC/DOC 0 3 2 1

ISBN 0-07-034003-X

The sponsoring editor of this book was Linda Ludewig. The editing supervisor was David E. Fogarty, and the production supervisor was Sherri Souffrance. This book was set in the HB1 design in Times Roman by Joanne Morbit and Michele Pridmore of McGraw-Hill's Professional Book Group Hightstown composition unit.

Printed and bound by R. R. Donnelley & Sons Company.

This book was printed on acid-free paper.

McGraw-Hill books are available at special quantity discounts to use as premiums and sales promotions, or for use in corporate training programs. For more information, please write to the Director of Special Sales, McGraw-Hill, Professional Publishing, Two Penn Plaza, New York, NY 10121-2298. Or contact your local bookstore.

CONTENTS

CONTRIBUTORS

Charles A. Aubrey II *Vice President, American Express, New York, NY* (SECTION 33, FINANCIAL SERVICES INDUSTRIES)

Lawrence Bernstein *Vice President, Operations Systems, AT&T Network Systems (retired); Have Laptop-Will Travel, Short Hills, NJ* (SECTION 20, SOFTWARE DEVELOPMENT)

Marcos E. J. Bertin *Chairman of the Board, Firmenich S. A., Buenos Aires, Argentina* (SECTION 43, QUALITY IN LATIN AMERICA)

Donald M. Berwick, M.D. *President and Chief Executive Officer, Institute for Healthcare Improvement, Boston, MA* (SECTION 32, HEALTH CARE SERVICES)

Maureen Bisognano *Executive Vice President and Chief Operating Officer, Institute for Healthcare Improvement, Boston, MA* (SECTION 32, HEALTH CARE SERVICES)

Robert C. Camp, Ph.D., PE *Best Practice Institute™, Rochester, NY* (SECTION 12, BENCHMARKING)

O. John Coletti *Manager, Special Vehicle Engineering, Ford Motor Company, Danou Technical Center, Allen Park, MI* (SECTION 3, THE QUALITY PLANNING PROCESS)

Dr. Tito Conti *President, OAM s.r.l., Ivrea, Italy* (SECTION 38, QUALITY IN WESTERN EUROPE)

Joseph A. DeFeo *Executive Vice President and Chief Operating Officer, Juran Institute, Inc., Wilton, CT* (SECTION 13, STRATEGIC DEPLOYMENT)

Irving J. DeToro *The Quality Network, Palm Harbor, FL* (SECTION 12, BENCHMARKING)

Ms. M. K. Detrano *Quality Director, Network Systems Product Realization, Lucent Technologies (retired); Detrano Consulting, Somerville, NJ* (SECTION 28, QUALITY IN A HIGH TECH INDUSTRY)

Dr. Necip Doganaksoy *Research and Development, General Electric Corporate, Schenectady, NY* (SECTION 48, RELIABILITY CONCEPTS AND DATA ANALYSIS)

John A. Donovan *Director PARIS Associates, Fairfax, VA* (SECTION 21, SUPPLIER RELATIONS)

Dr. Edward J. Dudewicz *Professor and Consultant, Department of Mathematics, Syracuse University, Syracuse, NY* (SECTION 44, BASIC STATISTICAL METHODS)

John F. Early *Vice President, Planning and Strategic Support, Empire Blue Cross and Blue Shield, New York, NY* (SECTION 3, THE QUALITY PLANNING PROCESS)

Dr. Al C. Endres *Director, Center for Quality, and Chair, Dept. of Management, University of Tampa, Tampa, FL* (SECTION 19, QUALITY IN RESEARCH AND DEVELOPMENT)

Professor Luis A. Escobar *Dept. of Experimental Statistics, Louisiana State University, Baton Rouge, LA* (SECTION 48, RELIABILITY CONCEPTS AND DATA ANALYSIS)

Edward Fuchs *Chapel Hill, NC* (SECTION 25, CUSTOMER SERVICE)

William R. Garwood *President, Eastman Chemical Co., Europe, Middle East, and Africa Region (retired), Hilton Head, SC* (SECTION 15, HUMAN RESOURCES AND QUALITY)

Dr. A. Blanton Godfrey *Chairman and Chief Executive Officer, Juran Institute, Inc., Wilton, CT* (CO-EDITOR-IN-CHIEF; SECTION 4, THE QUALITY CONTROL PROCESS; SECTION 14, TOTAL QUALITY MANAGEMENT)

E. F. "Bud" Gookins *Managing Partner, Strategic Quality Consulting, Elyria, OH* (SECTION 23, INSPECTION AND TEST)

Vice President Al Gore *Washington, DC* (SECTION 31, GOVERNMENT SERVICES)

Dr. Frank M. Gryna   *Distinguished Univ. Professor of Management, The Center for Quality, College of Business, The University of Tampa, Tampa, FL* (SECTION 8, QUALITY AND COSTS; SECTION 18, MARKET RESEARCH AND MARKETING; SECTION 22, OPERATIONS)

Dr. Gerald J. Hahn   *Research and Development, General Electric Corporate, Schenectady, NY* (SECTION 48, RELIABILITY CONCEPTS AND DATA ANALYSIS)

Gary L. Hallen   *Eastman Chemical Company, Kingsport, TN* (SECTION 15, HUMAN RESOURCES AND QUALITY)

Robert E. Hoogstoel   *Vice President, Juran Institute (retired), Pittsboro, NC* (ASSOCIATE EDITOR; SECTION 33, FINANCIAL SERVICES INDUSTRIES; SECTION 39, QUALITY IN CENTRAL AND EASTERN EUROPE; APPENDIX IV)

Dr. J. Stuart Hunter   *Professor Emeritus, Princeton University, Princeton, NJ* (SECTION 47, DESIGN AND ANALYSIS OF EXPERIMENTS)

Yoshio Ishizaka   *President and Chief Executive Officer, Toyota Motor Sales USA, Torrance, CA* (SECTION 29, AUTOMOTIVE INDUSTRY)

Dr. Joseph M. Juran   *Founder, Juran Institute, Wilton, CT* (CO-EDITOR-IN-CHIEF; SECTION 1, HOW TO USE THE HANDBOOK; SECTION 2, HOW TO THINK ABOUT QUALITY; SECTION 4, THE QUALITY CONTROL PROCESS; SECTION 5, THE QUALITY IMPROVEMENT PROCESS; SECTION 7, QUALITY AND INCOME; SECTION 35, QUALITY AND SOCIETY; SECTION 36, QUALITY AND THE NATIONAL CULTURE; SECTION 40, QUALITY IN THE UNITED STATES)

Dr. Noriaki Kano   *Professor, Dept. of Management Science, Faculty of Engineering, Science University of Tokyo, Tokyo, Japan* (SECTION 41, QUALITY IN JAPAN)

Professor Yoshio Kondo   *Professor Emeritus, Kyoto University, Kyoto, Japan* (SECTION 41, QUALITY IN JAPAN)

Professor Yuanzhang Liu   *Institute of Systems Science, Academia Sinica, Beijing, China* (SECTION 42, QUALITY IN THE PEOPLE'S REPUBLIC OF CHINA)

Frank P. Maresca   *Manager Quality Improvement, Mobil Oil Corporation (retired), Fairfax, VA* (SECTION 21, SUPPLIER RELATIONS)

Donald W. Marquardt   *Donald W. Marquardt and Associates, Wilmington, DE* (SECTION 11, THE ISO 9000 FAMILY OF INTERNATIONAL STANDARDS; SECTION 27, PROCESS INDUSTRIES) Don Marquardt passed away a matter of days after submitting his final manuscripts for this handbook. He was the victim of a heart attack. The editors are grateful for his contribution to this book. Consistent with his performance in a long career in quality, Don's submissions were crisp, complete, and timely.

Dr. William Q. Meeker   *Professor of Statistics, and Distinguished Professor of Liberal Arts and Sciences, Department of Statistics, Iowa State University, Ames, IA* (SECTION 48, RELIABILITY CONCEPTS AND DATA ANALYSIS)

Patrick Mene   *Vice President, Quality, The Ritz Carlton Hotel Company, Atlanta, GA* (SECTION 30, TRAVEL AND HOSPITALITY INDUSTRIES)

August Mundel   *Professional Engineer, White Plains, NY* (APPENDIX III)

Daniel P. Olivier   *President, Certified Software Solutions, San Diego, CA* (SECTION 10, COMPUTER APPLICATIONS IN QUALITY SYSTEMS)

Fredric I. Orkin   *President, Frederic I. Orkin and Associates, Inc., Highland Park, IL* (SECTION 10, COMPUTER APPLICATIONS IN QUALITY SYSTEMS)

Gabriel A. Pall   *Senior Vice President, Juran Institute, Wilton, CT* (SECTION 16, TRAINING FOR QUALITY)

Gerard T. Paul   *Management Consultant, Weston, CT* (SECTION 17, PROJECT MANAGEMENT AND PRODUCT DEVELOPMENT)

Dr. Thomas C. Redman   *President, Navesink Consulting Group, Rumson, NJ* (SECTION 9, MEASUREMENT, INFORMATION, AND DECISION-MAKING; SECTION 34, SECOND-GENERATION DATA QUALITY SYSTEMS)

James F. Riley, Jr.   *Vice President Emeritus, Juran Institute, Inc., University Park, FL* (SECTION 6, PROCESS MANAGEMENT)

Peter J. Robustelli   *Senior Vice President, Juran Institute, Inc., Wilton, CT* (SECTION 16, TRAINING FOR QUALITY)

Professor Dr. Lennart Sandholm   *President, Sandholm Associates AB, Djursholm, Sweden* (SECTION 37, QUALITY IN DEVELOPING COUNTRIES)

Dr. Edward G. Schilling *Professor of Statistics, Rochester Institute of Technology, Center for Quality and Applied Statistics, Rochester, NY,* (ASSOCIATE EDITOR; SECTION 46, ACCEPTANCE SAMPLING; APPENDIX I)

Leonard A. Seder *Quality Consultant (retired), Lexington, MA* (SECTION 24, JOB SHOP INDUSTRIES)

Professor James A. F. Stoner *Graduate School of Business Administration, Fordham University, New York, NY* (SECTION 26, ADMINISTRATIVE AND SUPPORT OPERATIONS)

Dr. Harrison M. Wadsworth *Professor Emeritus, School of Industrial and Systems Engineering, Georgia Institute of Technology, Atlanta, GA* (SECTION 45, STATISTICAL PROCESS CONTROL)

Professor Charles B. Wankel *College of Business, St. John's University, Jamaica, NY* (SECTION 26, ADMINISTRATIVE AND SUPPORT OPERATIONS)

Professor Frank M. Werner *Associate Dean, Graduate School of Business Administration, Fordham University at Tarrytown, Tarrytown, NY* (SECTION 26, ADMINISTRATIVE AND SUPPORT OPERATIONS)

Ms. Josette H. Williams *Juran Institute, Inc., Wilton, CT* (APPENDIX V)

Ms. C. M. Yuhas *Have Laptop-Will Travel, Short Hills, NJ* (SECTION 20, SOFTWARE DEVELOPMENT)

PREFACE TO THE FIFTH EDITION

In the preface to the Fourth Edition of this handbook, Dr. Juran commented on the events of the four decades between signing the contract for the First Edition of this handbook (1945) and the publication of the Fourth Edition (1988). He noted the growth of the handbook itself—in circulation and in status—and the parallel growth of importance of quality in society generally. The growth was attributable to the increasing complexity of products and the systems in which they participate, and, because of our increasing dependence on these systems, to the unprecedented potential for disruption when these products fail. This threat (and its occasional frightening fulfillment) is what he long ago identified as "life behind the quality dikes."

In the decade that has passed since the Fourth Edition, the importance of quality has continued to grow rapidly. To some extent, that growth is due in part to the continuing growth in complexity of products and systems, society's growing dependence on them, and, thus, society's growing dependence on those "quality dikes." But the main impetus for the growing importance of quality in the past decade has been the realization of the critical role quality plays as the key to competitive success in the increasingly globalized business environment. Upper managers now understand much more clearly the importance of quality—convinced by the threat of the consequences of product failure, by the rapid shift of power to the buyers and by the demands of global competition in costs, performance, and service.

As the importance of achieving quality has sunk in, the quest to learn how to achieve it has grown also. The emergence in the United States of America of the Malcolm Baldrige National Quality Award, and its many offspring at the state level, have promoted the development of quality by providing a comprehensive, home-grown organizational model for the achievement of quality, and by opening to view organizations that have applied this model successfully. It is difficult to overstate the importance of these models of excellence in the promotion of quality practice over the past decade. They have provided managers at all levels with evidence that "it can be done here," and, more important, they have provided in unusual detail, roadmaps of how it was done. In Europe, the European Quality Award and its offspring have provided much the same motive power to the quality movement that the Baldrige Award has provided in the United States.

The mounting success of quality in the industrial sector has caused recognition of the importance of quality to spread throughout manufacturing industries, the traditional home ground of quality ideas and applications, and beyond to the service sector, government, and non-profit enterprises. In this regard, we are especially pleased to welcome the contribution on quality in government of Vice President of the United States Al Gore.

In recognition of these changes, the editors have made some fundamental changes in this handbook.

1. We have changed the name from *Juran's Quality Control Handbook*, to *Juran's Quality Handbook*. The new name signals the change in emphasis from quality control, traditionally the concern of those working on the manufacturing floor, to an emphasis on the management of quality generally, a concern of managers throughout an organization.

2. We have changed the structure to reflect the new emphasis on managing quality. The Fifth Edition has 48 sections, arranged in five groups: Managerial, Functional, Industry, International, and Statistical.

The revision has not consisted merely of rearrangement. Once again, as in the Fourth Edition, the content of this edition has has undergone extensive editing and updating. There are many entirely new sections on new subjects. There are total rewrites of other sections. And there are many new additions of case studies, examples and other material even to the few "classic sections." An editorial undertaking of this scope and magnitude would be unthinkable without the help and support of a number of our colleagues and friends.

The founding editor of the handbook, Joseph M. Juran, has placed his unmistakable stamp of vision and clarity on this new edition—the fifth in which he has played a guiding role—by his contributions to its planning and, more directly, in the six major sections that he authored. My association with him since I joined Juran Institute in 1987 has provided a deep and rewarding exploration of the evolving field of quality management. Sharing the position of Editor-in-Chief of the present volume has been a part of that experience.

Our Associate Editors, Edward Schilling and Robert Hoogstoel, shared the major literary and diplomatic burden of helping the contributors create handbook sections that would at once reveal their individual subject-matter expertise and would mesh smoothly with the other sections to make a coherent and useful desk reference, in the long tradition of this book. Ed Schilling edited Sections 44 through 48, those concerned with mathematical statistics and related applications; Bob Hoogstoel edited most of the remaining sections and provided overall coordination of the editorial effort.

The grounding in practical experience which has characterized earlier editions of this book is strengthened further in this edition by the examples provided by the numerous managers who have shared their experiences on the quality journey through their presentations at Juran Institute's annual IMPRO conferences, workshops and seminars. We also wish to acknowledge the generous support of Juran Institute, Inc. throughout this endeavor. Many of the figures and charts come straight from Juran Institute publications and files, many others were created with support from people and facilities within the Institute.

Among the many colleagues at Juran Institute who have made major exertions on behalf of this book, Josette Williams stands out. Her own editorial and publishing experience have sharpened her sense of what goes and what doesn't, a sense she shared willingly. Jo provided a comforting presence as she managed the flow of correspondence with the contributors, and helped the editors enormously by performing calmly and expertly as liaison with the publisher astride the flow of manuscripts, the counterflow of page proofs, and the publisher's myriad last-minute questions of detail and the manuscript tweakings by contributors. Jo went far beyond the usual bounds of the responsibilities of an assistant editor. She worked closely with authors, editors, the publisher, and others in making this edition happen. Her style and grasp of language and clarity of expression are present in almost every section. This handbook owes much to her dedication, focus, and thousands of hours of hard work. Fran Milberg played a major role in preparing the manuscript for submission. My Executive Assistant, Jenny Edwards, frequently found her considerable workload in that job added to by the sudden, often unpredictable demands associated with the preparation of the manuscript, answering authors' questions, and keeping me on track. It was too much to ask of a normal person, but Jenny, as always, rose to the occasion, for which I am most grateful. Many others among the Juran Institute support staff helped at various stages of manuscript preparation, including: Laura Sutherland, Jane Gallagher, Marilyn Maher, and Carole Wesolowski. In the early stages of organizing for this effort we were grateful for the assistance of Sharon Davis and Rosalie Kaye. Special thanks go to Hank Williams who spent hours at the copier and many other hours helping Josette make sure manuscripts were sent on time to all the right places.

It would be unfair (and unwise) to omit mention of those closest to the contributors and editors of this book, the wives and husbands whose personal plans had occasionally to be put on hold in favor of work on the book. Larry Bernstein and C.M.Yuhas sidestepped the problem by making Section 20, Software Development, a family project, as is their joint consultancy. Other contributors no doubt were faced with dealing with the inevitable impingement on family life in their own ways. As for the editors, we unite to thank our wives for their support in this endeavor: Dr. Juran's wife of 73 years, known to him as "Babs," and to the rest of us as a gracious inspiration and

Editorial Assistant Emerita of the first three editions of this book and numerous of his earlier books and papers; Judy Godfrey, now a survivor of three books; Jean Schilling, a veteran editor of her husband's earlier publications and who has been patient and supportive in this effort; and Jewel Hoogstoel, for whom the answer to her persistent question is "It is done." We hope they will share the editors' mutual sense of accomplishment.

A. BLANTON GODFREY
Co-Editor-in-Chief

JURAN'S
QUALITY
HANDBOOK

SECTION 1
HOW TO USE
THE HANDBOOK

J. M. Juran

INTRODUCTION

This is a reference book for all who are involved with quality of products, services, and processes. Experience with the first four editions has shown that "all who are involved" include:

- The various *industries* that make up the international economy: manufacture, construction, services of all kinds—transportation, communication, utilities, financial, health care, hospitality, government, and so on.

- The various *functions* engaged in producing products (goods and services) such as research and development, market research, finance, operations, marketing and sales, human resources, supplier relations, customer service, and the administration and support activities.

- The various *levels in the hierarchy*—from the chief executives to the work force. It is a mistake to assume that the sole purpose of the book is to serve the needs of quality managers and quality specialists. The purpose of the book is to serve the entire quality function, and this includes participation from every department of the organization and people in all levels of the organization.

- The various staff *specialists* associated with the processes for planning, controlling, and improving quality.

While there is a great deal of know-how in this book, it takes skill and a bit of determination to learn how to find and make use of it. This first section of the handbook has therefore been designed to help the reader to find and apply those contents which relate to the problem at hand.

The handbook is also an aid to certain "stakeholders" who, though not directly involved in producing and marketing products and services, nevertheless have "a need to know" about the qualities produced and the associated side effects. These stakeholders include the customer chain, the public, the owners, the media, and government regulators.

USES OF THE HANDBOOK

Practitioners make a wide variety of uses of *Juran's Quality Handbook.* Experience has shown that usage is dominated by the following principal motives:

- To study the narrative material as an aid to solving problems
- To find structured answers in tables, charts, formulas, and so on
- To review for specific self-training
- To find special tools or methods needed for reliability engineering, design of experiments, or statistical quality control
- To secure material for the teaching or training of others
- To use as the basic reference in quality training, quality management operations, and even in designing and managing fundamental work functions

Beyond these most frequent uses, there is a longer list of less frequent uses such as:

- To review for personal briefing prior to attending a meeting
- To cross-check one's approach to tackling a problem
- As a reference for instructors and students during training courses
- To indoctrinate the boss
- To train new employees
- To help sell ideas to others, based on: (1) the information in the handbook and (2) the authoritative status of the handbook

Usage appears to be more frequent during times of change, for example, while developing new initiatives, working on new contracts and projects, reassigning functions, or trying out new ideas.

Irrespective of intended use, the information is very likely available. The problem for the practitioner becomes one of: (1) knowing where to find it and (2) adapting the information to his or her specific needs.

ORGANIZATION OF THE HANDBOOK

Knowing "where to find it" starts with understanding how the handbook is structured. The handbook consists of several broad groupings, outlined as follows.

The Managerial Group (Sections 2 through 17).   This group deals with basic concepts and with the processes by which quality is managed (planning, control, and improvement) along with sections of a coordinating nature such as human resources and project management.

The Functional Group (Sections 18 through 26).   This group roughly follows the sequence of activities through which product concepts are converted into marketable goods and services.

The Industry Group (Sections 27 through 33).   This group illustrates how quality is attained and maintained in selected leading industries.

The International Group (Sections 35 through 43).   This group deals with how quality is managed in selected geographical areas throughout the world.

Collectively the previous four groups comprise about 70 percent of the handbook.

The Statistical Group (Sections 44 through 48).   This group shows how to use the principal statistical tools as aids to managing for quality. This group also includes the glossary of symbols (Appendix I) as well as the supplemental tables and charts (Appendix II).

Collectively the statistical group comprises about 25 percent of the handbook.

HOW TO FIND IT

There are three main roads for locating information in the handbook:

1. Table of contents

2. Index

3. Cross-references

In addition, there are supplemental devices to aid in securing elaboration. Note that the handbook follows a *system of dual numbering,* consisting of the section number followed by page number, figure number, or table number. For example, page number 16.7 is the seventh page in Section 16. Figure or table number 12.4 is in Section 12, and is the fourth figure or table in that section.

Table of Contents.   There are several tables of contents. At the beginning of the book is the list of *section and subsection headings,* each of which describes, in the broadest terms, the contents of that section.

Next there is the *list of contents* that appears on the first page of each section. Each item in any section's list of contents becomes a *major heading* within that section.

Next, under each of these major headings, there may be one or more *minor headings,* each descriptive of its bundle of contents. Some of these bundles may be broken down still further by alphabetic or numeric lists of subtopics.

In a good many cases, it will suffice merely to follow the hierarchy of tables of contents to find the information sought. In many other cases it will not. For such cases, an alternative approach is to use the Index.

Use of the Index.   A great deal of effort has gone into preparing the index so that, through it, the reader can locate all the handbook material bearing on a subject. For example, the topic "Pareto analysis" is found in several sections. The index entry for "Pareto analysis" assembles *all* uses of the term Pareto analysis and shows the numbers of the pages on which they may be found.

The fact that information about a single topic is found in more than one section (and even in many sections) gives rise to criticisms of the organization of the handbook, that is, Why can't all the information on one topic be brought together in one place? The answer is that we require multiple and interconnected uses of knowledge, and hence these multiple appearances cannot be avoided. In fact, what must be done to minimize duplication is to make one and only one exhaustive explanation at some logical place and then to use cross-referencing elsewhere. In a sense, all the information on one topic *is* brought together—in the index.

Some key words and phrases may be explained in several places in the handbook. However, there is always one passage which constitutes the major explanation or definition. In the index, the word "defined" is used to identify this major definition, for example, "Evolutionary operation, defined."

The index also serves to assemble all case examples or applications under one heading for easy reference. For example, Section 5 deals with the general approach to quality improvement and includes examples of the application of this approach. However, additional examples are found in other sections. The index enables the reader to find these additional examples readily, since the page numbers are given.

Cross-References.   The handbook makes extensive use of cross-references in the text in order to: (1) guide the reader to further information on a subject and (2) avoid duplicate explanations of the same subject matter. The reader should regard these cross-references, wherever they occur, as extensions of the text. Cross-referencing is to either: (1) specific major headings in various sections or (2) specific figure numbers or table numbers. Study of the referenced material will provide further illumination.

A Note on Abbreviations. Abbreviations of names or organizations are usually used only after the full name has previously been spelled out, for example, American Society for Quality (ASQ). In any case, all such abbreviations are listed and defined in the index.

MAIN ROAD AND SIDE ROAD

The text of the handbook emphasizes the "main road" of quality management know-how, that is, the comparatively limited number of usual situations which nevertheless occupy the bulk of the time and attention of practitioners. Beyond the main road are numerous "side roads," that is, less usual situations which are quite diverse and which require special solutions.

(The term "side road" is not used in any derogatory sense. The practitioner who faces an unusual problem must nevertheless find a solution for it.)

As to these side roads, the handbook text, while not complete, nevertheless points the reader to available solutions. This is done in several ways.

Citations.   The handbook cites numerous papers, books, and other bibliographic references. In most cases these citations also indicate the nature of the special contribution made by the work cited in order to help the reader to decide whether to go to the original source for elaboration.

Special Bibliographies.   Some sections provide supplemental lists of bibliographical material for further reference. The editors have attempted to restrict the contents of these lists to items which: (1) bear directly on the subject matter discussed in the text or (2) are of uncommon interest to the practitioner. A special bibliography in Appendix III lists quality standards and specifications.

Literature Search.   Papers, books, and other references cited in the handbook contain further references which can be hunted up for further study. Use can be made of available abstracting and indexing services. A broad abstracting service in the engineering field is Engineering Index, Inc., 345 E. 47 St., New York, NY 10017. A specialized abstracting service in quality control and applied statistics is Executive Sciences Institute, Inc., 1005 Mississippi Ave., Davenport, IA 52803. In addition, various other specialized abstracting services are available on such subjects as reliability, statistical methods, research and development, and so on.

In searching the literature, the practitioner is well advised to make use of librarians. To an astonishing degree, library specialists have devised tools for locating literature on any designated subject: special bibliographies, abstracting services, indexes by subject and author, and so on. Librarians are trained in the use of these tools, and they maintain an effective communication network among themselves.

The Internet.   In the past few years an amazing new tool has appeared on the scene, the Internet. It is now possible to find almost any book in print and many that are out of print in just a few minutes using some of the well-designed web sites. Using the new search engines, one can find hundreds (or even thousands) of articles on numerous topics or by selected authors. Many special sites have been developed that focus on quality management in its broadest sense. A simple e-mail contact with an author may bring forth even more unpublished works or research in progress. Sites developed by university departments doing research in quality are especially useful for searching for specific examples and new methods and tools.

Author Contact.   The written book or paper is usually a condensation of the authors's knowledge; that is, what he or she wrote is derived from material which is one or two orders of magnitude more voluminous than the published work. In some cases it is worthwhile to contact the author for further elaboration. Most authors have no objection to being contacted, and some of these contacts lead not only to more information but also to visits and enduring collaboration.

Other Sources.   Resourceful people are able to find still other sources of information relating to the problem at hand. They contact the editors of journals to discover which companies have faced similar problems, so that they may contact these companies. They contact suppliers and customers to learn if competitors have found solutions. They attend meetings—such as courses, seminars, and conferences of professional societies—at which there is discussion of the problem. There is hardly a problem faced by any practitioner which has not already been actively studied by others.

ADAPTING TO USE

In many cases a practitioner is faced with adapting, to a special situation, knowledge derived from a totally different technology, that is, industry, product, or process. Making this transition requires that he or she identify the commonality, that is, the common principle to which both the special situation and the derived knowledge correspond.

Often the commonality is managerial in nature and is comparatively easy to grasp. For example, the concept of self-control is a management universal and is applicable to any person in the company.

Commonality of a statistical nature is even easier to grasp, since so much information is reduced to formulas which are indifferent to the nature of the technology involved.

Even in technological matters, it is possible to identify commonalities despite great outward differences. For example, concepts such as process capability apply not only to manufacturing processes, but to health care, services and administrative, and support processes as well. In like manner, the approaches used to make quality improvements by discovering the causes of defects have been classified into specific categories which exhibit a great deal of commonality despite wide differences in technology.

In all these situations, the challenge to practitioners is to establish a linkage between their own situations and those from which the know-how was derived. This linkage is established by discovering the commonality which makes them both members of one species.

SECTION 2
HOW TO THINK ABOUT QUALITY

J. M. Juran

This section deals with the fundamental concepts that underlie the subject of managing for quality. It defines key terms and makes critical distinctions. It identifies the key processes through which quality is managed. It demonstrates that while managing for quality is a timeless concept, it has undergone frequent revolution in response to the endless procession of changes and crises faced by human societies.

WHAT IS QUALITY?

The Meanings of "Quality."   Of the many meanings of the word "quality," two are of critical importance to managing for quality:

1. "Quality" means those *features of products* which meet customer needs and thereby provide customer satisfaction. In this sense, the meaning of quality is oriented to income. The purpose of such higher quality is to provide greater customer satisfaction and, one hopes, to increase income. However, providing more and/or better quality features usually requires an investment and hence usually involves increases in costs. Higher quality in this sense usually "costs more."

Product features that meet customer needs	Freedom from deficiencies
Higher quality enables companies to:	Higher quality enables companies to:
Increase customer satisfaction Make products salable Meet competition Increase market share Provide sales income Secure premium prices	Reduce error rates Reduce rework, waste Reduce field failures, warranty charges Reduce customer dissatisfaction Reduce inspection, test Shorten time to put new products on the market Increase yields, capacity Improve delivery performance
The major effect is on sales.	
Usually, higher quality costs more.	Major effect is on costs. Usually, higher quality costs less.

FIGURE 2.1   The meanings of quality. [*Planning for Quality, 2d ed. (1990). Juran Institute, Inc., Wilton, CT, pp. 1–10.*]

2. "Quality" means *freedom from deficiencies*—freedom from errors that require doing work over again (rework) or that result in field failures, customer dissatisfaction, customer claims, and so on. In this sense, the meaning of quality is oriented to costs, and higher quality usually "costs less." Figure 2.1 elaborates on these two definitions.

Figure 2.1 helps to explain why some meetings on managing for quality end in confusion.

A meeting of managers is discussing, "Does higher quality cost more, or does it cost less?" Seemingly they disagree, but in fact some of them literally do not know what the others are talking about. The culprit is the word "quality," spelled the same way and pronounced the same way, but with two meanings.

At one bank the upper managers would not support a proposal to reduce waste because it had the name "quality improvement." In their view, higher quality also meant higher cost. The subordinates were forced to relabel the proposal "productivity improvement" in order to secure approval.

Such confusion can be reduced if training programs and procedures manuals make clear the distinction between the two meanings of the word "quality." However, some confusion is inevitable as long as we use a single word to convey two very different meanings. There have been efforts to clarify matters by adding supplemental words, such as "positive" quality and "negative" quality. To date, none of these efforts has gained broad acceptance.

There also have been efforts to coin a short phrase that would clearly and simultaneously define both the major meanings of the word "quality." A popular example is "fitness for use." However, it is unlikely that any short phrase can provide the depth of meaning needed by managers who are faced with choosing a course of action. The need is to understand the distinctions set out in Figure 2.1.

Customer Needs and Conformance to Specification.   For most quality departments, the long-standing definition of quality was "conformance to specification." In effect, they assumed that products that conformed to specifications also would meet customer needs. This assumption was logical, since these departments seldom had direct contact with customers. However, the assumption can be seriously in error. Customer needs include many things not found in product specifications: service explanations in simple language, confidentiality, freedom from burdensome paperwork, "one-stop shopping," and so on. (For elaboration and discussion, see AT&T 1990.)

The new emphasis on customer focus has caused the quality departments to revise their definition of "quality" to include customer needs that are not a part of the product specification.

Definitions of Other Key Words.   The definitions of "quality" include certain key words that themselves require definition.

Product: The output of any process. To many economists, products include both goods and services. However, under popular usage, "product" often means goods only.

Product feature: A property possessed by goods or services that is intended to meet customer needs.

Customer: Anyone who is affected by the product or by the process used to produce the product. Customers may be external or internal.

Customer satisfaction: A state of affairs in which customers feel that their expectations have been met by the product features.

Deficiency: Any fault (defect or error) that impairs a product's fitness for use. Deficiencies take such forms as office errors, factory scrap, power outages, failures to meet delivery dates, and inoperable goods.

Customer dissatisfaction: A state of affairs in which deficiencies (in goods or services) result in customer annoyance, complaints, claims, and so on.

In the world of managing for quality, there is still a notable lack of standardization of the meanings of key words. However, any organization can do much to minimize internal confusion by standardizing the definitions of key words and phrases. The basic tool for this purpose is a *glossary.* The glossary then becomes a reference source for communication of all sorts: reports, manuals, training texts, and so on.

Satisfaction and Dissatisfaction Are Not Opposites.   Customer *satisfaction* comes from those features which induce customers to buy the product. *Dissatisfaction* has its origin in deficiencies and is why customers complain. Some products give little or no dissatisfaction; they do what the producer said they would do. Yet they are not salable because some competing product has features that provide greater customer satisfaction.

The early automated telephone exchanges employed electromagnetic analog switching methods. Recently, there was a shift to digital switching methods, owing to their superior product features. As a result, analog switching systems, even if absolutely free from product deficiencies, were no longer salable.

Big Q And Little Q.   Definitions of words do not remain static. Sometimes they undergo extensive change. Such a change emerged during the 1980s. It originated in the growing quality crisis and is called the concept of "Big Q."

Table 2.1 shows how the quality "umbrella" has been broadening dramatically. In turn, this broadening has changed the meanings of some key words. Adoption of Big Q grew during the 1980s, and the trend is probably irreversible. Those most willing to accept the concept of Big Q have been the quality managers and the upper managers. Those most reluctant have been managers in the technological areas and in certain staff functions.

QUALITY: THE FINANCIAL EFFECTS

The Effect on Income.   Income may consist of sales of an industrial company, taxes collected by a government body, appropriations received by a government agency, tuitions received by a school, and donations received by a charity. Whatever the source, the amount of the income relates

TABLE 2.1   Contrast, Big Q and Little Q

Topic	Content of little Q	Content of big Q
Products	Manufactured goods	All products, goods, and services, whether for sale or not
Processes	Processes directly related to manufacture of goods	All process manufacturing support; business, etc.
Industries	Manufacturing	All industries, manufacturing, service, government, etc., whether for profit or not
Quality is viewed as:	A technological problem	A business problem
Customer	Clients who buy the products	All who are affected, external and internal
How to think about quality	Based on culture of functional departments	Based on the universal trilogy
Quality goals are included:	Among factory goals	In company business plan
Cost of poor quality	Costs associated with deficient manufactured goods	All costs that would disappear if everything were perfect
Evaluation of quality is based mainly on:	Conformance to factory specifications, procedures, standards	Responsiveness to customer needs
Improvement is directed at:	Departmental performance	Company performance
Training in managing for quality is:	Concentrated in the quality department	Companywide
Coordination is by:	The quality manager	A quality council of upper managers

Source: *Planning for Quality,* 2d ed. (1990). Juran Institute, Inc., Wilton, CT, pp. 1–12.

in varying degrees to the features of the product produced by the recipient. In many markets, products with superior features are able to secure superior income, whether through higher share of market or through premium prices. Products that are not competitive in features often must be sold at below-market prices.

Product deficiencies also can have an effect on income. The customer who encounters a deficiency may take action of a cost-related nature: file a complaint, return the product, make a claim, or file a lawsuit. The customer also may elect instead (or in addition) to stop buying from the guilty producer, as well as to publicize the deficiency and its source. Such actions by multiple customers can do serious damage to a producer's income. Section 7, Quality and Income, is devoted to the ways in which quality can influence income.

The Effect on Costs.   The cost of poor quality consists of all costs that would disappear if there were no deficiencies—no errors, no rework, no field failures, and so on. This cost of poor quality is shockingly high. In the early 1980s, I estimated that within the U.S. manufacturing industries, about a third of the work done consisted of redoing what had already been done. Since then, estimates from a sample of service industries suggest that a similar situation prevails in service industries generally.

Deficiencies that occur prior to sale obviously add to producers' costs. Deficiencies that occur after sale add to customers' costs as well as to producers' costs. In addition, they reduce producers' repeat sales. Section 8, Quality and Costs, is devoted to the ways in which quality can influence costs.

HOW TO MANAGE FOR QUALITY: THE JURAN TRILOGY

To attain quality, it is well to begin by establishing the "vision" for the organization, along with policies and goals. (These matters are treated elsewhere in this handbook, especially in Section 13, Strategic Deployment.) Conversion of goals into results (making quality happen) is then done through managerial processes—sequences of activities that produce the intended results. Managing for quality makes extensive use of three such managerial processes:

- Quality planning
- Quality control
- Quality improvement

These processes are now known as the "Juran trilogy." They parallel the processes long used to manage for finance. These financial processes consist of

Financial planning: This process prepares the annual financial budget. It defines the deeds to be done in the year ahead. It translates those deeds into money. It determines the financial consequences of doing all those deeds. The final result establishes the financial goals for the organization and its various divisions and units.

Financial control: This process consists of evaluating actual financial performance, comparing this with the financial goals, and taking action on the difference—the accountant's "variance." There are numerous subprocesses for financial control: cost control, expense control, inventory control, and so on.

Financial improvement: This process aims to improve financial results. It takes many forms: cost-reduction projects, new facilities to improve productivity, new product development to increase sales, acquisitions, joint ventures, and so on.

These processes are universal—they provide the basis for financial management, no matter what the type of enterprise is.

The financial analogy helps managers realize that they can manage for *quality* by using the same processes of planning, control, and improvement. Since the *concept* of the trilogy is identical to that used in managing for finance, managers are not required to change their conceptual approach. Much of their previous training and experience in managing for finance is applicable to managing for quality.

While the conceptual approach does not change, the procedural steps differ. Figure 2.2 shows that each of these three managerial processes has its own unique sequence of activities.

Each of the three processes is also a universal—it follows an unvarying sequence of steps. Each sequence is applicable in its respective area, no matter what is the industry, function, culture, or whatever.

Figure 2.2 shows these unvarying sequences in abbreviated form. Extensive detail is provided in other sections of this handbook as follows:

Section 3, The Quality Planning Process

Section 4, The Quality Control Process

Section 5, The Quality Improvement Process

Quality planning	Quality control	Quality improvement
Establish quality goals	Evaluate actual performance	Prove the need
Identify who the customers are	Compare actual performance with quality goals	Establish the infrastructure
Determine the needs of the customers	Act on the difference	Identify the improvement projects
Develop product features that respond to customers' needs		Establish project teams
Develop processes able to produce the product features		Provide the teams with resources, training, and motivation to: Diagnose the causes Stimulate remedies
Establish process controls; transfer the plans to the operating forces		Establish controls to hold the gains

FIGURE 2.2   The three universal processes of managing for quality. [*Adapted from Juran, J. M. (1989). The Quality Trilogy: A Universal Approach to Managing for Quality. Juran Institute, Inc., Wilton, CT.*]

The Juran Trilogy Diagram.   The three processes of the Juran trilogy are interrelated. Figure 2.3 shows this interrelationship.

The Juran trilogy diagram is a graph with time on the horizontal axis and cost of poor quality on the vertical axis. The initial activity is quality planning. The planners determine who the customers are and what their needs are. The planners then develop product and process designs to respond to those needs. Finally, the planners turn the plans over to the operating forces: "You run the process, produce the product features, and meet the customers' needs."

Chronic and Sporadic.   As operations proceed, it soon emerges that the process is unable to produce 100 percent good work. Figure 2.3 shows that over 20 percent of the work must be redone due to quality deficiencies. This waste is *chronic*—it goes on and on. Why do we have this chronic waste? Because *the operating process was planned that way.*

Under conventional responsibility patterns, the operating forces are unable to get rid of this planned chronic waste. What they can do is to carry out *quality control*—to prevent things from getting worse. Figure 2.3 also shows a sudden *sporadic* spike that has raised the defect level to over 40 percent. This spike resulted from some *unplanned* event such as a power failure, process breakdown, or human error. As a part of their job of quality control, the operating forces converge on the scene and take action to restore the status quo. This is often called "corrective action," "troubleshooting," "putting out the fire," and so on. The end result is to restore the error level back to the planned chronic level of about 20 percent.

The chart also shows that in due course the chronic waste was driven down to a level far below the original level. This gain came from the third process in the trilogy—*quality improvement.* In effect, it was seen that the chronic waste was an opportunity for improvement, and steps were taken to make that improvement.

The Trilogy Diagram and Product Deficiencies.   The trilogy diagram (Figure 2.3) relates to *product deficiencies.* The vertical scale therefore exhibits units of measure such as cost of poor quality, error rate, percent defective, service call rate, and so on. On this same scale, perfection is at zero, and *what goes up is bad.* The results of reducing deficiencies are to reduce the cost of poor quality, meet more delivery promises, reduce customer dissatisfaction, and so on.

The Trilogy Diagram and Product Features.   When the trilogy diagram is applied to product features, the vertical scale changes. Now the scale may exhibit units of measure such as millions of instructions per second, mean time between failures, percent on-time deliveries, and so on. For such diagrams, *what goes up is good,* and a logical, generic vertical scale is "product salability." (For elaboration on the Juran trilogy, see Juran 1986.)

Allocation of Time within the Trilogy.   An interesting question for managers is, "How do people allocate their time relative to the processes of the trilogy?" Figure 2.4 is a model designed to show this interrelationship in a Japanese company (Itoh 1978).

In Figure 2.4 the horizontal scale represents the percentage allocation of any person's time and runs from zero to 100 percent. The vertical scale represents levels in the hierarchy. The diagram shows that the upper managers spend the great majority of their time on planning and improvement. They spend a substantial amount of time on strategic planning. The time they spend on control is small and is focused on major control subjects.

At progressively lower levels of the hierarchy, the time spent on strategic planning declines, whereas the time spent on control and maintenance grows rapidly. At the lowest levels, the time is dominated by control and maintenance, but some time is still spent on planning and improvement.

QUALITY: A CONTINUING REVOLUTION

A young recruit who joins an organization soon learns that it has in place numerous processes (systems) to manage its affairs, including managing for quality. The recruit might assume that humans have always used those processes to manage for quality and will continue to so in the future. Such

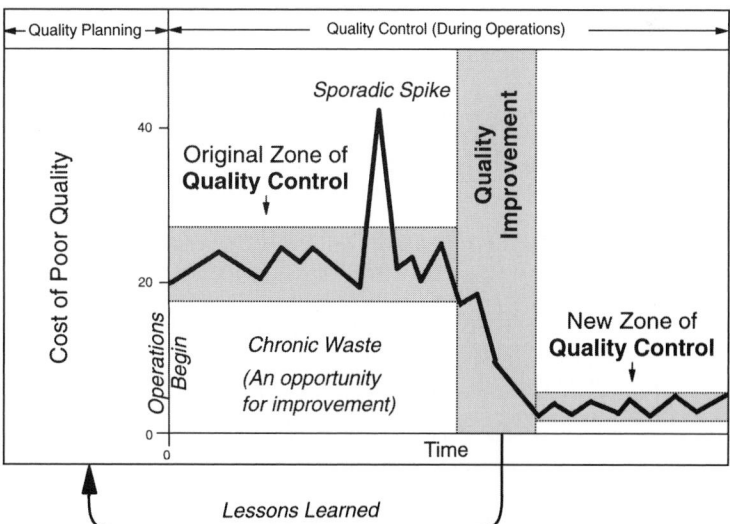

FIGURE 2.3   The Juran trilogy diagram. [*Adapted from Juran, J. M. (1989). The Quality Trilogy: A Universal Approach to Managing for Quality. Juran Institute, Inc., Wilton, CT.*]

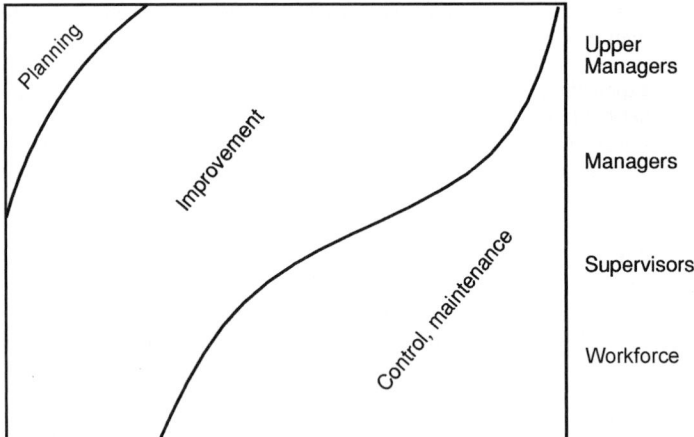

FIGURE 2.4   The Itoh model. [*Adapted from Management for Quality, 4th ed. (1987). Juran Institute, Inc., Wilton, CT, p. 18.*]

assumptions would be grossly in error. The processes used to manage for quality have undergone extensive change over the millennia, and there is no end in sight.

Primitive Societies

The Family.   Quality is a timeless concept. The origins of ways to manage for quality are hidden in the mists of the ancient past. Yet we can be sure that humans have always faced problems of quality. Primitive food-gatherers had to learn which fruits were edible and which were poisonous. Primitive hunters had to learn which trees supplied the best wood for making bows or arrows. The resulting know-how was then passed down from generation to generation.

The nuclear human organizational unit was the family. Isolated families were forced to create self-sufficiency—to meet their own needs for food, clothing, and shelter. There was division of work among family members. Production was for self-use, so the design, production, and use of a product were all carried out by the same persons. Whereas the technology was primitive, the coordination was superb. The same human beings received all inputs and took all remedial action. The limiting factor for achieving quality was the primitive state of the technology.

The Village: Division of Labor.   Villages were created to serve other essential human requirements such as mutual defense and social needs. The village stimulated additional division of labor and development of specialized skills. There emerged farmers, hunters, fishermen, and artisans of all sorts—weavers, potters, shoemakers. By going through the same work cycle over and over again, the artisans became intimately familiar with the materials used, the tools, the steps in the process, and the finished product. The cycle included selling the product to users and receiving their feedback on product performance. The experience derived from this intimate familiarity then enabled human ingenuity to take the first steps toward the evolution of technology.

The Village Marketplace: Caveat Emptor.   As villages grew, the village marketplace appeared, where artisans and buyers met on scheduled market days. In this setting, producer and user met face to face with the goods between them. The goods typically were natural products or were made from natural materials. The producers and purchasers had long familiarity with the products, and the quality of the products could to a high degree be judged by the unaided human senses.

Under such a state of affairs, the village magistrates tended to avoid being drawn into quality disputes between seller and buyer. This forced buyers to be vigilant so as to protect themselves against poor quality. In effect, the seller was responsible for supplying the goods, but the buyer became responsible for supplying the quality "assurance." This arrangement was known as *caveat emptor*— "let the buyer beware." Thus buyers learned to beware by use of product inspection and test. They looked closely at the cloth, smelled the fish, thumped the melon, tasted a grape. Their failure to beware was at their own peril. In the village marketplace, *caveat emptor* was quite a sensible doctrine. It is widely applied to this day in villages all over the world.

A further force in the village marketplace was the fact of common residence. Producer and buyer both lived in the same village. Each was subject to scrutiny and character evaluation by the villagers. Each also was subject to village discipline. For the artisan, the stakes were high. His status and livelihood (and those of his family) were closely tied to his reputation as a competent and honest artisan. In this way, the concept of craftsmanship became a quiet yet powerful stimulus to maintain a high level of quality.

Effects of the Growth of Commerce.   In due course villages expanded into towns and cities, and improved transport opened the way to trade among regions.

> A famous example of organized multiregional trade was the Hanseatic League which was centered among the cities of northern Europe from about the 1200s to the 1600s. Its influence extended into Scandinavia and Russia as well as to the Mediterranean and Black seas (von der Porten 1994).

Under trade among regions, producer and user could no longer meet face to face in the marketplace. Products were now made by chains of suppliers and processors. Marketing was now done by chains of marketers. The buyers' direct point of contact was now with some merchant rather than with the producer. All this reduced the quality protections inherent in the village marketplace to a point requiring invention of new forms of quality assurance. One such invention was the quality warranty.

Quality Warranties.   Early quality warranties were no doubt in oral form. Such warranties were inherently difficult to enforce. Memories differed as to what was said and meant. The duration of the warranty might extend beyond the life of the parties. Thus the written warranty was invented.

> An early example was on a clay tablet found amid the ruins of Nippur in ancient Babylon. It involved a gold ring set with an emerald. The seller guaranteed that for twenty years the emerald would not fall out of the gold ring. If it did fall out of the gold ring before the end of twenty years, the seller agreed to pay to the buyer an indemnity of ten *mana* of silver. The date is the equivalent of 429 B.C. (Bursk et al. 1962, vol. I, p. 71).

Quality warranties are now widely used in all forms of trade and commerce. They stimulate producers to give priority to quality and stimulate sellers to seek out reliable sources of supply. So great is their importance that recent legislation has imposed standards to ensure that the wording of warranties does not mislead the buyers.

Quality Specifications.   Sellers need to be able to communicate to buyers the nature of what they have to sell. Buyers need to be able to communicate to sellers the nature of what they want to buy. In the village marketplace, oral communication could take place directly between producer and buyer. With the growth of commerce, communication expanded to include chains of producers and chains of merchants who often were widely separated. New forms of communications were needed, and a major invention was the written quality specification. Now quality information could be communicated directly between designer and producer or between seller and buyer no matter how great the distance between them and how complex the nature of the product.

Like warranties, written specifications are of ancient origin. Examples have been found in Egyptian papyrus scrolls over 3500 years old (Durant 1954). Early specifications focused on defining products and the processes for producing them. In due course the concept was extended to defining the materials from which the products were made. Then, as conflicts arose because sellers and buyers used different methods of test, it became necessary to establish inspection and test specifications as well.

Measurement.   The emergence of inspection and test specifications led to the evolution of measuring instruments. Instruments for measuring length, volume, and time evolved thousands of years ago. Instruments have continued to proliferate, with ever-increasing precision. In recent centuries, the precision of measurement of time has increased by over ten orders of magnitude (Juran 1995, Chapter 10).

Artisans and Guilds.

The artisan's possession of the skills of a trade was a source of income and status as well as self-respect and respect from the community. However, as villages grew into towns and cities, the numbers of artisans grew as well. The resulting competition became destructive and threatened the benefits derived from craftsmanship.

To perpetuate their benefits, the artisans within a trade organized trade unions—guilds. Each guild then petitioned the city authorities to confer on the guild members a monopoly on practicing their trade.

Guilds flourished for centuries during the Middle Ages until the Industrial Revolution reduced their influence. They used their monopolistic powers chiefly to provide a livelihood and security for their members. The guilds also provided extensive social services to their members. (For elaboration, see Bursk et al. 1962, vol. III, pp. 1656–1678.)

The Guild Hierarchy.   Each guild maintained a hierarchy of (usually) three categories of workers: the apprentice, the journeyman, and the master. Considerable formality surrounded the entry into each category.

At the bottom was the apprentice or novice, whose entry was through an indenture—a formal contract that bound the apprentice to serve a master for a specified period of years. In turn, the master became responsible for teaching the trade to the apprentice.

To qualify for promotion, the apprentice was obliged to serve out the full term of the indenture. In addition, he was required to pass an *examination* by a committee of masters. Beyond the oral part of the examination, the apprentice was required to produce a perfect piece of work—a *masterpiece*—that was then inspected by the examination committee. Success in the examination led to a ceremonial admission to the status of journeyman.

The journeyman's right to practice the trade was limited. He could become an employee of a master, usually by the day. He also could *journey* to other towns, seeking employment in his trade. Only after admission to the rank of master could he set up shop on his own.

Admission to the rank of master required first that there be an opening. Guilds imposed limits on the numbers of masters in their areas. On the death or retirement of an active master, the guild would decide whether to fill that opening. If so, a journeyman would be selected and admitted, again through a formal ceremony.

Guilds and Quality Planning.   Guilds were active in managing for quality, including quality planning. They established specifications for input materials, manufacturing processes, and finished products, as well as for methods of inspection and test.

Guilds and Quality Control.   Guild involvement in quality control was extensive. They maintained inspections and audits to ensure that artisans followed the quality specifications. They established means of "traceability" to identify the producer. In addition, some applied their "mark" to finished products as added assurance to consumers that quality met guild standards.

Control by the guilds also extended to sales. The sale of poor-quality goods was forbidden, and offenders suffered a range of punishments—all the way from fines to expulsion from membership. The guilds also established prices and terms of sale and enforced them.

Guilds and Quality Improvement.   An overriding guild policy was solidarity—to maintain equality of opportunity among members. To this end, internal competition among members was limited to "honest" competition. Quality improvement through product or process innovation was *not* considered to be "honest" competition. This limitation on quality improvement did indeed help to maintain equality among members, but it also made the guild increasingly vulnerable to competition from other cities that did evolve superior products and processes.

Guilds and External Forces.   The guilds were able to control internal competition, but external competition was something else. Some external competition came in the form of jurisdictional disputes with other guilds, which consumed endless hours of negotiation. More ominous was competition from other cities, which could be in quality as well as in price and value.

The policy of solidarity stifled quality improvement and thereby became a handicap to remaining competitive. Thus the guilds urged the authorities to restrict imports of foreign goods. They also imposed strict rules to prevent their trade secrets from falling into the hands of foreign competitors. (The Venetian glass industry threatened capital punishment to those who betrayed such secrets.)

Inspection and Inspectors.   The concepts of inspection and inspectors are of ancient origin. Wall paintings and reliefs in Egyptian tombs show the inspections used during stone construction projects. The measuring instruments included the square, level, and plumb bob for alignment control. Surface flatness of stones was checked by "boning rods" and by threads stretched across the faces of the stone blocks:

> As shops grew in size, the function of inspection gave rise to the full-time job of inspector. In due course, inspectors multiplied in numbers to become the basis for inspection departments, which in turn gave birth to modern quality departments. (Singer et al. 1954, vol. I, p. 481).

Government Involvement in Managing for Quality.   Governments have long involved themselves in managing for quality. Their purposes have included protecting the safety and health of citizens, defending and improving the economics of the state, and protecting consumers against fraud. Each of these purposes includes some aspect of managing for quality.

Safety and Health of the Citizens.   Early forms of protection of safety and health were after-the-fact measures. The Code of Hammurabi (c. 2000 B.C.) prescribed the death penalty for any builder of a house that later collapsed and killed the owner. In medieval times, the same fate awaited the baker who inadvertently had mixed rat poison with the flour.

Economics of the State.   With the growth of trade between cities, the quality reputation of a city could be an asset or a liability. Many cities took steps to protect their reputation by imposing quality controls on exported goods. They appointed inspectors to inspect finished products and affix a seal to certify as to quality. This concept was widely applied to high-volume goods such as textiles.

Continued growth of commerce then created competition among nations, including competition in quality. Guilds tended to stifle quality improvement, but governments favored improving the quality of domestic goods in order to reduce imports and increase exports. For example, in the late sixteenth century, James VI of Scotland imported craftsmen from the Low Countries to set up a textile factory and to teach their trade secrets to Scottish workers (Bursk et al. 1962, vol. IV, pp. 2283–2285).

Consumer Protection.   Many states recognized that as to some domestic trade practices, the rule of *caveat emptor* did not apply. One such practice related to measurement. The states designed official standard tools for measuring length, weight, volume, and so on. Use of these tools was then mandated, and inspectors were appointed to ensure compliance. (See, for example, Juran 1995, chap. 1.) The twentieth century witnessed a considerable expansion in consumer protection legislation. (For elaboration, see Juran 1995, chap. 17.)

The Mark or Seal.   A mark or seal has been applied to products over the centuries to serve multiple purposes. Marks have been used to

Identify the producer, whether artisan, factory, town, merchant, packager, or still others: Such identification may serve to fix responsibility, protect the innocent against unwarranted blame, enable buyers to choose from among multiple makers, advertise the name of the maker, and so on.

Provide traceability: In mass production, use of lot numbers helps to maintain uniformity of product in subsequent processing, designate expiration dates, make selective product recalls, and so on.

Provide product information, such as type and quantities of ingredients used, date when made, expiration dates, model number, ratings (such as voltage, current), and so on.

Provide quality assurance: This was the major purpose served by the marks of the guilds and towns. It was their way of telling buyers, "This product has been independently inspected and has good quality."

An aura of romance surrounds the use of seals. The seals of some medieval cities are masterpieces of artistic design. Some seals have become world-renowned. An example is the British "hallmark" that is applied to products made of precious metals.

The Industrial Revolution.   The Industrial Revolution began in Europe during the mid-eighteenth century. Its origin was the simultaneous development of power-driven machinery and sources of mechanical power. It gave birth to factories that soon outperformed the artisans and small shops and made them largely obsolete.

The Factory System: Destruction of Crafts.   The goals of the factories were to raise productivity and reduce costs. Under the craft system, productivity had been low due to primitive technology, whereas costs had been high due to the high wages of skilled artisans. To reach their goals, the factories reengineered the manufacturing processes. Under the craft system, an artisan performed every one of the numerous tasks needed to produce the final product—pins, shoes, barrels, and so on. Under the factory system, the tasks within a craft were divided up among several or many factory workers. Special tools were designed to simplify each task down to a short time cycle. A worker then could, in a few hours, carry out enough cycles of his or her task to reach high productivity.

Adam Smith, in his book, *The Wealth of Nations,* was one of the first to publish an explanation of the striking difference between manufacture under the craft system versus the factory system. He noted that pin making had been a distinct craft, consisting of 18 separate tasks. When these tasks were divided among 10 factory workers, production rose to a per-worker equivalent of 4800 pins a day, which was orders of magnitude higher than would be achieved if each worker were to produce pins by performing all 18 tasks (Smith 1776). For other types of processes, such as spinning or weaving, power-driven machinery could outproduce hand artisans while employing semiskilled or unskilled workers to reduce labor costs.

The broad economic result of the factory system was mass production at low costs. This made the resulting products more affordable and contributed to economic growth in industrialized countries, as well as to the associated rise of a large "middle class."

Quality Control under the Factory System.   The factory system required associated changes in the system of quality control. When craft tasks were divided among many workers, those workers were no longer their own customers, over and over again. The responsibility of workers was no longer to provide satisfaction to the buyer (also customer, user). Few factory workers had contact with buyers. Instead, the responsibility became one of "make it like the sample" (or specification).

Mass production also brought new technological problems. Products involving assemblies of bits and pieces demanded interchangeability of those bits and pieces. Then, with the growth of technology and of interstate commerce, there emerged the need for standardization as well. All this required greater precision throughout—machinery, tools, measurement. (Under the craft system, the artisan fitted and adjusted the pieces as needed).

In theory, such quality problems could be avoided during the original planning of the manufacturing processes. Here the limitation rested with the planners—the "master mechanics" and shop supervisors. They had extensive, practical experience, but their ways were empirical, being rooted in craft practices handed down through the generations. They had little understanding of the nature of process variation and the resulting product variation. They were unschooled in how to collect and analyze data to ensure that their processes had "process capability" to enable the production workers to meet the specifications. Use of such new concepts had to await the coming of the twentieth century.

Given the limitations of quality planning, what emerged was an expansion of inspection by departmental supervisors supplemented by full-time inspectors. Where inspectors were used, they were made responsible to the respective departmental production supervisors. The concept of a special department to coordinate quality activities broadly also had to await the coming of the twentieth century.

Quality Improvement. The Industrial Revolution provided a climate favorable for continuous quality improvement through product and process development. For example, progressive improvements in the design of steam engines increased their thermal efficiency from 0.5 percent in 1718 to 23.0 percent in 1906 (Singer et al. 1958, vol. IV). Inventors and entrepreneurs emerged to lead many countries into the new world of technology and industrialization. In due course, some companies created internal sources of inventors—research laboratories to carry out product and process development. Some created market research departments to carry out the functions of entrepreneurship.

In contrast, the concept of continuous quality improvement to reduce chronic waste made little headway. One likely reason is that most industrial managers give higher priority to increasing income than to reducing chronic waste. The guilds' policy of solidarity, which stifled quality improvement, also may have been a factor. In any event, the concept of quality improvement to reduce chronic waste did not find full application until the Japanese quality revolution of the twentieth century.

The Taylor System of Scientific Management. A further blow to the craft system came from F. W. Taylor's system of "scientific management." This originated in the late nineteenth century when Taylor, an American manager, wanted to increase production and productivity by improving manufacturing planning. His solution was to separate planning from execution. He brought in engineers to do the planning, leaving the shop supervisors and the work force with the narrow responsibility of carrying out the plans.

Taylor's system was stunningly successful in raising productivity. It was widely adopted in the United States but not so widely adopted elsewhere. It had negative side effects in human relations, which most American managers chose to ignore. It also had negative effects on quality. The American managers responded by taking the inspectors out of the production departments and placing them in newly created inspection departments. In due course, these departments took on added functions to become the broad-based quality departments of today. (For elaboration, see Juran 1995, chap. 17.)

The Rise of Quality Assurance.
The anatomy of "quality assurance" is very similar to that of quality control. Each evaluates actual quality. Each compares actual quality with the quality goal. Each stimulates corrective action as needed. What differs is the prime purpose to be served.

Under quality control, the prime purpose is to serve those who are directly responsible for conducting operations—to help them regulate current operations. Under quality assurance, the prime purpose is to serve those who are not directly responsible for conducting operations but who have a need to know—to be informed as to the state of affairs and, hopefully, to be assured that all is well.

In this sense, quality assurance has a similarity to insurance. Each involves spending a small sum to secure protection against a large loss. In the case of quality assurance, the protection consists of an early warning that may avoid the large loss. In the case of insurance, the protection consists of compensation after the loss.

Quality Assurance in the Village Marketplace.   In the village marketplace, the buyers provided much of the quality assurance through their vigilance—through inspection and test before buying the product. Added quality assurance came from the craft system—producers were trained as apprentices and were then required to pass an examination before they could practice their trade.

Quality Assurance through Audits.   The growth of commerce introduced chains of suppliers and merchants that separated consumers from the producers. This required new forms of quality assurance, one being quality warranties. The guilds created a form of quality assurance by establishing product and process standards and then auditing to ensure compliance by the artisans. In addition, some political authorities established independent product inspections to protect their quality reputations as exporters.

Audit of Suppliers' Quality Control Systems.   The Industrial Revolution stimulated the rise of large industrial companies. These bought equipment, materials, and products on a large scale. Their early forms of quality assurance were mainly through inspection and test. Then, during the twentieth century, there emerged a new concept under which customers defined and mandated *quality control systems*. These systems were to be instituted and followed by suppliers as a condition for becoming and remaining suppliers. This concept was then enforced by audits, both before and during the life of the supply contracts.

At first, this concept created severe problems for suppliers. One was the lack of standardization. Each buying company had its own idea of what was a proper quality control system, so each supplier was faced with designing its system to satisfy multiple customers. Another problem was that of multiple audits. Each supplier was subject to being audited by each customer. There was no provision for pooling the results of audits into some common data bank, and customers generally were unwilling to accept the findings of audits conducted by personnel other than their own. The resulting multiple audits were especially burdensome to small suppliers.

In recent decades, steps have been taken toward standardization by professional societies, by national standardization bodies, and most recently, by the International Standards Organization (ISO). ISO's 9000 series of standards for quality control systems is now widely accepted among European companies. There is no legal requirement for compliance, but as a marketing matter, companies are reluctant to be in a position in which their competitors are certified as complying to ISO 9000 standards but they themselves are not.

There remains the problem of multiple audits. In theory, it is feasible for one audit to provide information that would be acceptable to all buyers. This is already the case in quality audits conducted by Underwriters' Laboratories and in financial audits conducted by Dun & Bradstreet. Single audits may in the future become feasible under the emerging process for certification to the ISO 9000 series.

Extension to Military Procurement.   Governments have always been large buyers, especially for defense purposes. Their early systems of quality assurance consisted of inspection and test. During the twentieth century, there was a notable shift to mandating quality control systems and then using audits to ensure conformance to the mandated systems. The North Atlantic Treaty Organization (NATO) evolved an international standard—the Allied Quality Assurance Publications (AQAP)—that includes provisions to minimize multiple audits. (For elaboration, see Juran 1977.)

Resistance to Mandated Quality Control Systems.   At the outset, suppliers resisted the mandated quality control systems imposed by their customers. None of this could stop the movement toward quality assurance. The economic power of the buyers was decisive. Then, as suppliers gained experience with the new approach, they realized that many of its provisions were simply good business practice. Thus the concept of mandated quality control systems seems destined to become a permanent feature of managing for quality.

Shift of Responsibility.   It should be noted that the concept of mandating quality control systems involves a major change of responsibility for quality assurance. In the village marketplace, the producer supplies the product, but the buyer has much of the responsibility for supplying the quality assurance. Under mandated quality control systems, the producer becomes responsible for supplying both the product and the quality assurance. The producer supplies the quality assurance by

- Adopting the mandated system for controlling quality
- Submitting the data that prove that the system is being followed

The buyers' audits then consist of seeing to it that the mandated system is in place and that the system is indeed being followed.

The Twentieth Century and Quality.   The twentieth century witnessed the emergence of some massive new forces that required responsive action. These forces included an explosive growth in science and technology, threats to human safety and health and to the environment, the rise of the consumerism movement, and intensified international competition in quality.

An Explosive Growth in Science and Technology.   This growth made possible an outpouring of numerous benefits to human societies: longer life spans, superior communication and transport, reduced household drudgery, new forms of education and entertainment, and so on. Huge new industries emerged to translate the new technology into these benefits. Nations that accepted industrialization found it possible to improve their economies and the well-being of their citizenry.

The new technologies required complex designs and precise execution. The empirical methods of earlier centuries were unable to provide appropriate product and process designs, so process yields were low and field failures were high. Companies tried to deal with low yields by adding inspections to separate the good from the bad. They tried to deal with field failures through warranties and customer service. These solutions were costly, and they did not reduce customer dissatisfaction. The need was to prevent defects and field failures from happening in the first place.

Threats to Human Safety and Health and to the Environment.   With benefits from technology came uninvited guests. To accept the benefits required changes in lifestyle, which, in turn, made quality of life dependent on continuity of service. However, many products were failure-prone, resulting in many service interruptions. Most of these were minor, but some were serious and even frightening—threats to human safety and health, as well as to the environment.

Thus the critical need became quality. Continuity of the benefits of technology depended on the quality of the goods and services that provided those benefits. The frequency and severity of the interruptions also depended on quality—on the continuing performance and good behavior of the products of technology. This dependence came to be known as "life behind the quality dikes." (For elaboration, see Juran 1970.)

Expansion of Government Regulation of Quality.   Government regulation of quality is of ancient origin. At the outset, it focused mainly on human safety and was conducted "after the fact"—laws provided for punishing those whose poor quality caused death or injury. Over the centuries, there emerged a trend to regulation "before the fact"—to become preventive in nature.

This trend was intensified during the twentieth century. In the field of human health, laws were enacted to ensure the quality of food, pharmaceuticals, and medical devices. Licensing of practitioners was expanded. Other laws were enacted relating to product safety, highway safety, occupational safety, consumer protection, and so on.

Growth of government regulation was a response to twentieth-century forces as well as a force in its own right. The rise of technology placed complex and dangerous products in the hands of amateurs—the public. Government regulation then demanded product designs that avoided these dangers. To the companies, this intervention then became a force to be reckoned with. (For elaboration, see Juran 1995, chap. 17.)

The Rise of the Consumerism Movement.   Consumers welcomed the features offered by the new products but not the associated new quality problems. The new products were unfamiliar—most consumers lacked expertise in technology. Their senses were unable to judge which of the competing products to buy, and the claims of competing companies often were contradictory.

When products failed in service, consumers were frustrated by vague warranties and poor service. "The system" seemed unable to provide recourse when things failed. Individual consumers were unable to fight the system, but collectively they were numerous and hence potentially powerful, both economically and politically. During the twentieth century, a "consumerism" movement emerged to make this potential a reality and to help consumers deal more effectively with these problems. This same movement also was successful in stimulating new government legislation for consumer protection. (For elaboration, see Juran 1995, chap. 17.)

Intensified International Competition in Quality.   Cities and countries have competed for centuries. The oldest form of such competition was probably in military weaponry. This competition then intensified during the twentieth century under the pressures of two world wars. It led to the development of new and terrible weapons of mass destruction.

A further stimulus to competition came from the rise of multinational companies. Large companies had found that foreign trade barriers were obstacles to export of their products. To get around these barriers, many set up foreign subsidiaries that then became their bases for competing in foreign markets, including competition in quality.

The most spectacular twentieth-century demonstration of the power of competition in quality came from the Japanese. Following World War II, Japanese companies discovered that the West was unwilling to buy their products—Japan had acquired a reputation for making and exporting shoddy goods. The inability to sell became an alarm signal and a stimulus for launching the Japanese quality revolution during the 1950s. Within a few decades, that revolution propelled Japan into a position of world leadership in quality. This quality leadership in turn enabled Japan to become an economic superpower. It was a phenomenon without precedent in industrial history.

QUALITY TO CENTER STAGE

The cumulative effect of these massive forces has been to "move quality to center stage." Such a massive move logically should have stimulated a corresponding response—a revolution in managing for quality. However, it was difficult for companies to recognize the need for such a revolution—they lacked the necessary alarm signals. Technological measures of quality did exist on the shop floors, but managerial measures of quality did not exist in the boardrooms. Thus, except for Japan, the needed quality revolution did not start until very late in the twentieth century. To make this revolution effective throughout the world, economies will require many decades—the entire twenty-first century. Thus, while the twentieth century has been the "century of productivity," the twenty-first century will be known as the "century of quality."

The failure of the West to respond promptly to the need for a revolution in quality led to a widespread crisis. The 1980s then witnessed quality initiatives being taken by large numbers of companies. Most of these initiatives fell far short of their goals. However, a few were stunningly successful and produced the lessons learned and role models that will serve as guides for the West in the decades ahead.

Lessons Learned.   Companies that were successful in their quality initiatives made use of numerous strategies. Analysis shows that despite differences among the companies, there was much commonality—a lengthy list of strategies was common to most of the successful companies. These common strategies included

Customer focus: Providing customer satisfaction became the chief operating goal.

Quality has top priority: This was written into corporate policies.

Strategic quality planning: The business plan was opened up to include planning for quality.

Benchmarking: This approach was adopted in order to set goals based on superior results already achieved by others.

Continuous improvement: The business plan was opened up to include goals for quality improvement. It was recognized that quality is a moving target.

Training in managing for quality: Training was extended beyond the quality department to all functions and levels, including upper managers.

Big Q was adopted to replace little Q.

Partnering: Through cross-functional teams, partnering was adopted to give priority to company results rather than to functional goals. Partnering was extended to include suppliers and customers.

Employee empowerment: This was introduced by training and empowering the work force to participate in planning and improvement, including the concept of self-directed teams.

Motivation: This was supplied through extending the use of recognition and rewards for responding to the changes demanded by the quality revolution.

Measurements were developed to enable upper managers to follow progress toward providing customer satisfaction, meeting competition, improving quality, and so on.

Upper managers took charge of managing for quality by recognizing that certain responsibilities were *not delegable*—they were to be carried out by the upper managers, personally.

These responsibilities included

- Serve on the quality council
- Establish the quality goals
- Provide the needed resources
- Provide quality-oriented training
- Stimulate quality improvement
- Review progress
- Give recognition
- Revise the reward system

Inventions Yet to Come.   Many of the strategies adopted by the successful companies are without precedent in industrial history. As such, they must be regarded as experimental. They did achieve results for the role model companies, but they have yet to demonstrate that they can achieve comparable results in a broader spectrum of industries and cultures. It is to be expected that the efforts to make such adaptations will generate new inventions, new experiments, and new lessons learned. There is no end in sight.

ACKNOWLEDGMENTS

This section of the handbook has drawn extensively from the following two books:

Juran, J. M. (ed.) (1995). A History of Managing for Quality. Sponsored by Juran Foundation, Inc. Quality Press, Milwaukee Press, WI.

Upper Management and Quality: Making Quality Happen, 6th ed. (1993) Juran Institute, Inc., Wilton, CT.

The author is grateful to Juran Foundation, Inc., and Juran Institute, Inc., for permission to quote from these works.

REFERENCES

AT&T Quality Library (1990). *Achieving Customer Satisfaction.* Bell Laboratories Quality Information Center, Indianapolis, IN.

Bursk, Edward C., Clark, Donald T., and Hidy, Ralph W. (1962). *The World of Business,* vol. I, p. 71, vol. III, pp. 1656–1678, vol. IV, pp. 2283–2285. Simon and Shuster, New York.

Davies, Norman De G. (1943). *The Tomb of Rekh-mi-re at Thebes,* vol. 2, Plate LXII. Metropolitan Museum of Art, New York.

Durant, Will (1954). *The Story of Civilization,* Part I: *Our Oriental Heritage,* pp. 182–183. Simon and Schuster, New York.

Itoh, Yasuro (1978). *Upbringing of Component Suppliers Surrounding Toyota.* International Conference on Quality Control, Tokyo.

Juran, J. M. (1970). "Consumerism and Product Quality." *Quality Progress,* July, pp. 18–27.

Juran, J. M. (1977). *Quality and Its Assurance—An Overview.* Second NATO Symposium on Quality and Its Assurance, London.

Juran, J. M. (1986). "The Quality Trilogy: A Universal Approach to Managing for Quality." *Quality Progress,* August, pp. 19–24.

Juran, J. M. (1995) ed. *A History of Managing for Quality.* Sponsored by Juran Foundation, Inc. Quality Press, Milwaukee, WI.

Singer, Charles, Holmyard, E. J., and Hall, A. R. (eds.) (1954). *A History of Technology,* vol. I, Fig. 313, p. 481. Oxford University Press, New York.

Singer, Charles, Holmyard, E. J., and Hall, A. R. (eds.) (1958). *A History of Technology,* vol. IV, p. 164. Oxford University Press, New York.

Smith, Adam (1776). *The Wealth of Nations.* Random House, New York, published in 1937.

Upper Management and Quality: Making Quality Happen, 6th ed. Juran Institute, Inc., Wilton, CT.

von der Porten, Edward (1994). "The Hanseatic League, Europe's First Common Market." *National Geographic,* October, pp. 56–79.

SECTION 3
THE QUALITY PLANNING PROCESS*

John F. Early and O. John Coletti

*In the Fourth Edition, material covered by this section was supplied by Joseph M. Juran and Frank M. Gryna in sections on Companywide Planning for Quality, Product Development, and Manufacturing Planning.

DEFINITION OF QUALITY PLANNING

"Quality planning," as used here, is a structured process for developing products (both goods and services) that ensures that customer needs are met by the final result. The tools and methods of quality planning are incorporated along with the technological tools for the particular product being developed and delivered. Designing a new automobile requires automotive engineering and related disciplines, developing an effective care path for juvenile diabetes will draw on the expert methods of specialized physicians, and planning a new approach for guest services at a resort will require the techniques of an experienced hotelier. All three need the process, methods, tools, and techniques of quality planning to ensure that the final designs for the automobile, diabetic care, and resort services not only fulfill the best technical requirements of the relevant disciplines but also meet the needs of the customers who will purchase and benefit from the products.

THE QUALITY PLANNING PROBLEM

The quality planning process and its associated methods, tools, and techniques have been developed because in the history of modern society, organizations have rather universally demonstrated a consistent failure to produce the goods and services that unerringly delight their customers. As a customer, everyone has been dismayed time and time again when flights are delayed, radioactive contamination spreads, medical treatment is not consistent with best practices, a child's toy fails to function, a new piece of software is not as fast or user-friendly as anticipated, government responds with glacial speed (if at all), or a home washing machine with the latest high-tech gadget delivers at higher cost clothes that are no cleaner than before. These frequent, large quality gaps are really the compound result of a number of smaller gaps illustrated in Figure 3.1.

The first component of the quality gap is the *understanding gap,* that is, lack of understanding of what the customer needs. Sometimes this gap opens up because the producer simply fails to consider who the customers are and what they need. More often the gap is there because the supplying organization has erroneous confidence in its ability to understand exactly what the customer really needs. The final perception gap in Figure 3.1 also arises from a failure to understand the customer and the customer needs. Customers do not experience a new suit of clothes or the continuity in service from a local utility simply based on the technical merits of the product. Customers react to how they *perceive* the good or service provides them with a benefit.

FIGURE 3.1   The quality gap and its constituent gaps. [*Inspired by A. Parasuraman, Valarie A. Zeithami, and Leonard L. Berry (1985). "A Conceptual Model for Service Quality and Its Implications for Further Research." Journal of Marketing, Fall, pp. 41–50.*]

The second constituent of the quality gap is a *design gap.* Even if there were perfect knowledge about customer needs and perceptions, many organizations would fail to create designs for their goods and services that are fully consistent with that understanding. Some of this failure arises from the fact that the people who understand customers and the disciplines they use for understanding customer needs are often systematically isolated from those who actually create the designs. In addition, designers—whether they design sophisticated equipment or delicate human services—often lack the simple tools that would enable them to combine their technical expertise with an understanding of the customer needs to create a truly superior product.

The third gap is the *process gap.* Many splendid designs fail because the process by which the physical product is created or the service is delivered is not capable of conforming to the design consistently time after time. This lack of process capability is one of the most persistent and bedeviling failures in the total quality gap.

The fourth gap is the *operations gap.* The means by which the process is operated and controlled may create additional deficiencies in the delivery of the final good or service.

• Establish the project
• Identify the customers
• Discover the customer needs
• Develop the product
• Develop the process
• Develop the controls and
    transfer to operations

FIGURE 3.2 Quality planning steps. (*Juran Institute, Inc., Copyright 1994. Used by permission.*)

THE QUALITY PLANNING SOLUTION

Quality planning provides the process, methods, tools, and techniques for closing each of these component gaps and thereby ensuring that the final quality gap is at a minimum. Figure 3.2 summarizes at a high level the basic steps of quality planning. The remainder of this section will provide the details and examples for each of these steps.

The first step, establish the project, provides the clear goals, direction, and infrastructure required if the constituent quality gaps are to be closed. The next step provides for systematic identification of all the customers. It is impossible to close the understanding gap if there is the least bit of uncertainty, fuzziness, or ignorance about who all the customers are.

The discovery of customer needs in the third step provides the full and complete understanding required for a successful product design to meet those needs. It also evaluates customer perceptions explicitly so that the final perception gap can be avoided.

The develop product step uses both quality planning tools and the technology of the particular industry to create a design that is effective in meeting the customer needs, thereby closing the design gap. The process gap is closed in the next step, develop process. Quality planning techniques ensure that the process is capable of delivering the product as it was designed, consistently, time after time.

Finally, the operations gap is closed by developing process controls that keep the process operating at its full capability. Successful elimination of the operations gap also depends on an effective transfer of the plans to the operating forces. A strong transfer plan, executed well, will provide operations with all the processes, techniques, materials, equipment, skills, and so on to delight customers on a continuing basis.

The remainder of this section will provide details, practical guidance, and examples for each of these steps. Many detailed examples will be included of how Ford Motor Company applied the principles of quality planning to develop its 1994 Ford Mustang.

STEP 1: ESTABLISH THE PROJECT

A quality planning project is the organized work needed to prepare an organization to deliver a new or revised product, following the steps associated with quality planning. Generally speaking, the following activities are associated with establishing a quality planning project:

- Identify which projects are required to fulfill the organization's strategy.
- Prepare a mission statement for each project.
- Establish a team to carry out the project.
- Plan the project.

Identification of Projects. Deciding which projects to undertake is usually the outgrowth of the strategic and business planning of an organization. (See Section 13, Strategic Deployment for a discussion of how specific projects are deployed from an organization's vision, strategies, and goals.) Typically, quality planning projects create new or updated products that are needed to reach specific strategic goals, to meet new or changing customer needs, to fulfill legal or customer mandates, or to take advantage of a new or emerging technology.

Upper management must take the leadership in identifying and supporting the critical quality planning projects. Acting as a quality council or similar body, management needs to fulfill the following key roles.

Setting Quality Goals. Top management identifies opportunities and needs to improve quality and sets strategic goals for the organization.

Nominating and Selecting Projects. The quality council selects those major quality planning projects critical to meeting strategic quality goals.

Selecting Teams. Once a project has been identified, the quality council appoints a team to see the project through the remaining steps of the quality planning process.

Supporting Project Team. New techniques and processes are generally required to meet quality goals. It is up to the quality council to see that each quality planning team is well prepared and equipped to carry out its mission. The quality council's support may include

- Providing education and training in quality planning tools and techniques
- Providing a trained facilitator to help the team work effectively and learn the quality planning process
- Reviewing team progress
- Approving revision of the project mission
- Identifying/helping with any problems
- Coordinating related quality planning projects
- Helping with logistics, such as a meeting site
- Providing expertise in data analysis and survey design
- Furnishing resources for unusually demanding data collection
- Communicating project results

Monitoring Progress. The quality council is generally responsible for keeping the quality planning process on track, evaluating progress, and making midcourse corrections to improve the effectiveness of the entire process. Once the quality council has reviewed the sources for potential projects, it will select one or more for immediate attention. Next, it must prepare a mission statement for the project.

Prepare Mission Statement. Once the quality council has identified the need for a project, it should prepare a mission statement that incorporates the specific goal(s) of the project. The mission statement is the written instruction for the team that describes the intent or purpose of the project. The team mission describes

- The scope of the planning project, that is, the product and markets to be addressed
- The goals of the project, that is, the results to be achieved

Writing mission statements requires a firm understanding of the driving force behind the project. The mission helps to answer the following questions:

- Why does the organization want to do the project?
- What will it accomplish once it is implemented?

A mission statement also fosters a consensus among those who either will be affected by the project or will contribute the time and resources necessary to plan and implement the project goal.

Examples

- The team mission is to deliver to market a new low-energy, fluorocarbon-free refrigerator.
- The team will create accurate control and minimum cost for the inventory of all stores.

While these mission statements describe what will be done, they are still incomplete. They lack the clarity and specificity that is required of a complete quality planning mission statement that incorporates the goal(s) of a project. Well-written and effective mission statements define the scope of the project by including one or more of the following.

Inherent performance: How the final product will perform on one or more dimensions, e.g., 24-hour response time.

Comparative performance: How the final product will perform vis-à-vis the competition, e.g., the fastest response time in the metropolitan area.

Customer reaction: How customers will rate the product compared with others available, e.g., one company is rated as having a better on-time delivery service compared with its closest rival.

Market: Who are or will be the customers or target audience for this product, and what share of the market or market niche will it capture, e.g., to become the "preferred" source by all business travelers within the continental United States.

Performance deficiencies: How will the product perform with respect to product failure, e.g., failure rate of less than 200 for every million hours of use.

Avoidance of unnecessary constraints: Avoid overspecifying the product for the team, e.g., if the product is intended for airline carryon, specifying the precise dimensions in the mission may be too restrictive. There may be several ways to meet the carryon market.

Basis for Establishing Quality Goals. In addition to the scope of the project, a mission statement also must include the goal(s) of the project. An important consideration in establishing quality goals is the choice of the basis for which the goal(s) are set.

Technology as a Basis. In many organizations, it has been the tradition to establish the quality goals on a technological basis. Most of the goals are published in specifications and procedures that define the quality targets for the supervisory and nonsupervisory levels.

The Market as a Basis. Quality goals that affect product salability should be based primarily on meeting or exceeding market quality. Because the market and the competition undoubtedly will be changing while the quality planning project is under way, goals should be set so as to meet or beat the competition estimated to be prevailing when the project is completed. Some internal suppliers are internal monopolies. Common examples include payroll preparation, facilities maintenance, cafeteria service, and internal transportation. However, most internal monopolies have potential competitors. There are outside suppliers who offer to sell the same service. Thus the performance of the internal supplier can be compared with the proposals offered by an outside supplier.

Benchmarking as a Basis.   "Benchmarking" is a recent label for the concept of setting goals based on knowing what has been achieved by others. (See Section 12.) A common goal is the requirement that the reliability of a new product be at least equal to that of the product it replaces and at least equal to that of the most reliable competing product. Implicit in the use of benchmarking is the concept that the resulting goals are attainable because they have already been attained by others.

History as a Basis.   A fourth and widely used basis for setting quality goals has been historical performance; i.e., goals are based on past performance. Sometimes this is tightened up to stimulate improvement. For some products and processes, the historical basis is an aid to needed stability. In other cases, notably those involving chronically high costs of poor quality, the historical basis helps to perpetuate a chronically wasteful performance. During the goal-setting process, the management team should be on the alert for such misuse of the historical basis.

Quality Goals Are a Moving Target.   It is widely recognized that quality goals must keep shifting to respond to the changes that keep coming over the horizon: new technology, new competition, threats, and opportunities. While organizations that have adopted quality management methods practice this concept, they may not do as well on providing the means to evaluate the impact of those changes and revise the goals accordingly.

Project Goals.   Specific goals of the project, i.e., what the project team is to accomplish, are part of an effective mission statement. In getting the job done, the team must mentally start at the finish. The more focused it is on what the end result will look like, the easier it will be to achieve a successful conclusion.

Measurement of the Goal.   In addition to stating what will be done and by when, a project goal must show how the team will measure whether or not it has achieved its stated goals. It is important to spend some time defining how success is measured. Listed below are the four things that can be measured:

1. Quality
2. Quantity
3. Cost
4. Time

An effective quality planning project goal must have five characteristics for it to provide a team with enough information to guide the planning process. The goal must be

- Specific
- Measurable
- Agreed to by those affected
- Realistic—It can be a stretch, but it must be plausible.
- Time specific—when it will be done

An example of a poorly written goal might look something like this: "To design a new car that is best in class." Contrast this with the following example: "To design, and put into production within 3 years, a new, midsized car that is best in class and priced, for the public, at under $20,000 (at time of introduction). The design also should allow the company to sell the car and still have an average return of between 4 and 6 percent."

The second example is much more detailed, measurable, and time-specific compared with the first. The target or end result is clearly stated and provides enough direction for the team to plan the product features and processes to achieve the goal.

The Ford Mustang—Mission and Goals. Before moving ahead with any product development, Ford agreed to a clear mission for the Mustang. The short version was "The Car The Star." Whenever a large group of people from various functional organizations is brought together to work on a project, there is a natural tendency to bring a lot of their "home office priorities and objectives" to the team. Unattended, these home office priorities can diffuse the team's focus and create a significant amount of conflict within the team.

To address this issue, the team chose the statement "The Car is the Star" to align the efforts of all team members and to focus them on a single objective. This statement was a simple and effective way to galvanize the team around the fact that there was one common purpose and a single superordinate objective.

Specifically, this meant that all team members could adjudicate their daily decisions and actions consistent with the overall team objective of making the car a reality and success. Program goals were established for 18 separate parameters. These "18 panel charts" enabled the project to focus on very specific success factors. (See Figure 3.3.)

	Topic	Description
Panel 1:	Quality	Things gone wrong/1000, things gone right/1000, repairs/1000 and customer satisfaction @ 3 months in service, plus things gone wrong/1000 @ 4 years in service.
Panel 2:	Timing	Summary of the major milestones from the total program workplan.
Panel 3:	Vehicle hardpoints	Summary of the architectural hardpoints such as wheelbase, tread, length, height, width, interior room, leg room, etc.
Panel 4:	Vehicle dynamics	Subjective targets for performance feel, ride, handling, noise/vibration/harshness, brake performance, seat performance, etc.
Panel 5:	Weight	Curb weight and emission test weight for all models.
Panel 6:	Fuel economy	Metro-highway fuel consumption is declared for all models. Avoidance of gas guzzler is also declared.
Panel 7:	Performance	0–60 mph elapsed time is declared for all models.
Panel 8:	Complexity	Number of discrete customer decisions and buildable combinations is declared.
Panel 9:	Serviceability	Projection for the number of total service hours through 50,000 miles is declared.
Panel 10:	Damageability	Projected repair cost for a typical collision is declared.
Panel 11:	Safety emissions	Compliance with all applicable Federal Motor Vehicle Safety Standards and Federal Clean Air Standards is declared.
Panel 12:	Variable cost	Variable cost versus prior model is declared.
Panel 13:	Program investment	Total investment for tooling, facilities, launch, and engineering is declared.
Panel 14:	Pricing	Wholesale delivery price is declared for all models.
Panel 15:	Volumes	Five year projection for trend volume is declared.
Panel 16:	Profitability	Program profitability is declared in terms of fully accounted profits.
Panel 17:	Features	All product standard features and option by model are declared.
Panel 18:	Export	Export markets and volumes are declared.

FIGURE 3.3 Ford Mustang 18 panel chart goals.

New Product Policies. Companies need to have very clear policy guidance with respect to quality and product development. Most of these should relate to all new products, but specific policies may relate to individual products, product lines, or groups. Four of the most critical policies are as follows.

Deficiencies in New and Carryover Designs. Many organizations have established the clear policy that no new product or component of a product will have a higher rate of deficiencies than the old product or component that it is replacing. In addition, they often require that any carryover design must have a certain level of performance; otherwise, it must be replaced with a more reliable design. The minimum carryover reliability may be set by one or more of the following criteria: (1) competitor or benchmark reliability, (2) customer requirements, or (3) a stretch goal beyond benchmark or customer requirements.

Intended versus Unintended Use. Should stepladders be designed so that the user can stand on the top step without damage, even though the step is clearly labeled "Do Not Step Here?" Should a hospital design its emergency room to handle volumes of routine, nonemergency patients who show up at its doors? These are policy questions that need to be settled before the project begins. The answers can have a significant impact on the final product, and the answers need to be developed with reference to the organization's strategy and the environment within which its products are used.

Requirement of Formal Quality Planning Process. A structured, formal process is required to ensure that the product planners identify their customers and design products and processes that will meet those customer needs with minimum deficiencies. Structured formality is sometimes eschewed as a barrier to creativity. Nothing could be more misguided. Formal quality planning identifies the points at which creativity is demanded and then encourages, supports, and enables that creativity. Formal planning also ensures that the creativity is focused on the customers and that creative designs ultimately are delivered to the customer free of the destructive influences of deficiencies.

Custody of Designs and Change Control. Specific provision must be made to ensure that approved designs are documented and accessible. Any changes to designs must be validated, receive appropriate approvals, be documented, and be unerringly incorporated into the product or process. Specific individuals must have the assigned authority, responsibility, and resources to maintain the final designs and administer change control.

Ford Policies with Respect to the Mustang. Ford had three specific policies with respect to carryover and new designs. New designs were required to be more reliable than the old. They also were required to provide demonstrated cost-benefit contributions to the final product. Finally, major features were expected to exceed the performance of the chief competitor—Camaro/Firebird.

Because Mustang needed to maintain its reputation as a reliable performance car, a more stringent testing policy was established. In addition to the safety, economy, reliability, durability, and other tests that all Ford cars must pass, the Mustang was required to pass tests such as 500 drag starts without a failure and repeated hard braking without fading.

Establish Team. The cross-functional approach to quality planning is effective for several reasons:

Team involvement promotes sharing of ideas, experiences, and a sense of commitment to being a part of and helping "our" organization achieve its goal.

The diversity of team members brings a more complete working knowledge of the product and processes to be planned. Planning a product requires a thorough understanding of how things get done in many parts of the organization.

Representation from various departments or functions promotes the acceptance and implementation of the new plan throughout the organization. Products or processes designed with the active participation of the affected areas tend to be technically superior and accepted more readily by those who must implement them.

Guidelines for Team Selection.   When selecting a team, the quality council identifies those parts of the organization which have a stake in the outcome. There are several places to look:

- Those who will be most affected by the result of the project
- Departments or functions responsible for various steps in the process
- Those with special knowledge, information, or skill in the design of the project
- Areas that can be helpful in implementing the plan

The Mustang Team.   Ford established a dedicated team of individuals from all key parts of the company, including

Stakeholder organizations	*Role*
Vehicle Engineering	Team leader
Product and Business Planning	Prepare the product assumptions and business case.
Body and Chassis Engineering	Define the product upgrades to satisfy the customer for comfort, convenience, and safety and to meet the forecasted federal motor vehicle safety standards.
Power Train Engineering	Define the power train upgrades to satisfy the customer wants for power and drivability and meet the clean air and gas guzzler requirements.
Vehicle Design	Provide excitement with a highly styled design.
Manufacturing and Assembly	Ensure that there is a feasible, high-quality product.
Sales and Marketing	Provide the "voice of the customer" and ensure that the product hits the target market.
Purchasing	Bring the suppliers to the party.
Finance	Help the team develop a financially attractive business package.

Besides their professional expertise for the job, many of the team members also were Mustang enthusiasts who brought an understanding and passion for the customer needs that would be hard to duplicate.

Because an automobile is a highly complex consumer product, the work was divided into five chunk teams: body exterior and structure, body interior and electrical, power train, chassis, and vehicle. Each chunk team had members from all the major stakeholders and was able to manage its portion of the program with autonomy within the overall design and 18 panel chart parameters.

The overall coordination and direction of the work among the chunk teams was managed with a weekly "major matters" meeting among the chunk team leaders and the program manager. These meetings focused on the major program metrics incorporated in the 18 panel charts.

STEP 2: IDENTIFY THE CUSTOMERS

This step may seem unnecessary; of course, the planners and designers know who their customers are: the driver of the automobile, the depositor in the bank account, the patient who takes the medication. But these are not the only customers—not even necessarily the most important customers. Customers comprise an entire cast of characters that needs to be understood fully.

Generally, there are two primary groups of customers: the *external customers*—those outside the producing organization; and the *internal customers*—those inside the producing organization.

Types of External Customers.   The term "customer" is often used loosely; it can refer to an entire organization, a unit of a larger organization, or a person. There are many types of customers—some obvious, others hidden. Below is a listing of the major categories to help guide complete customer identification.

The purchaser	Someone who buys the product for himself or herself or for someone else, e.g., anyone who purchases food for his or her family.
The end user/ultimate customer	Someone who finally benefits from the product, e.g., the patient who goes to health care facility for diagnostic testing.
Merchants	People who purchase products for resale, wholesalers, distributors, travel agents and brokers, and anyone who handles the product, such as a supermarket employee who places the product on the shelf.
Processors	Organizations and people who use the product or output as an input for producing their own product, e.g., a refinery that receives crude oil and processes it into different products for a variety of customers.
Suppliers	Those who provide input to the process, e.g., the manufacturer of the spark plugs for an automobile or the law firm that provides advice on the company's environmental law matters. Suppliers are also customers. They have information needs with respect to product specification, feedback on deficiencies, predictability of orders, and so on.
Original equipment manufacturers (OEMs)	Purchasers of a product to incorporate into their own, e.g., a computer manufacturer using another producer's disk drives for its computers.
Potential customers	Those not currently using the product but capable of becoming customers; e.g., a business traveler renting a car may purchase a similar automobile when the time comes to buy one for personal use.
Hidden customers	An assortment of different customers who are easily overlooked because they may not come to mind readily. They can exert great influence over the product design: regulators, critics, opinion leaders, testing services, payers, the media, the public at large, those directly or potentially threatened by the product, corporate policymakers, labor unions, professional associations.

Internal Customers. Everyone inside an organization plays three roles: supplier, processor, and customer. Each individual receives something from someone, does something with it, and passes it to a third individual. Effectiveness in meeting the needs of these internal customers can have a major impact on serving the external customers. Identifying the internal customers will require some analysis because many of these relationships tend to be informal, resulting in a hazy perception of who the customers are and how they will be affected. For example, if a company decides to introduce just-in-time manufacturing to one of its plants, this will have significant effect on purchasing, shipping, sales, operations, and so on.

Most organizations try to set up a mechanism that will allow seemingly competing functions to negotiate and resolve differences based on the higher goal of satisfying customer needs. This might include conducting weekly meetings of department heads or publishing procedure manuals. However, these mechanisms often do not work because the needs of internal customers are not fully understood, and communication among the functions breaks down. This is why a major goal in the quality planning process is to identify who the internal customers are, discover their needs, and plan how those needs will be satisfied. This is also another reason to have a multifunctional team involved in the planning; these are people who are likely to recognize the vested interests of internal customers.

Identifying Customers. In addition to the general guidance just laid out, it is most often helpful to draw a relatively high-level flow diagram of the processes related to the product being planned. Careful analysis of this flow diagram often will provide new insight, identifying customers that might have been missed and refining understanding of how the customers interact with the process. Figure 3.4 is an example of such a diagram. A review of this diagram reveals that the role of "customer" is really two different roles—placing the order and using the product. These may or may not be played by the same individuals, but they are two distinct roles, and each needs to be understood in terms of its needs.

Customers for Ford Mustang. The most critical customer for Ford Mustang was the "ultimate customer," namely, the person who would purchase the vehicle. The prime demographic target group was the Mustang GT buyer. (The GT model is the image setter and the model that must appeal to the enthusiast if the car line is to be successful.)

Other vital few external customers included

Dealers

Suppliers

Government

• Environmental Protection Agency (EPA)
• The National Highway and Traffic Safety Administration (NHTSA)
• The Federal Trade Commission (FTC)

Media (refers to both the internal and external sources)

• Internal media

Ford Communication Network (FCN)

Employee publications

• External media

Enthusiast magazines such as *Motor Trend Magazine, Car and Driver, Road & Track, Automobile Magazine, AutoWeek,* etc.

Trade magazines such as *Ward's AutoWorld, Automotive News, Automotive Industries, Automotive Engineering.*

National television programs such as "Motorweek," "MotorTrend."

Local and regional television programs

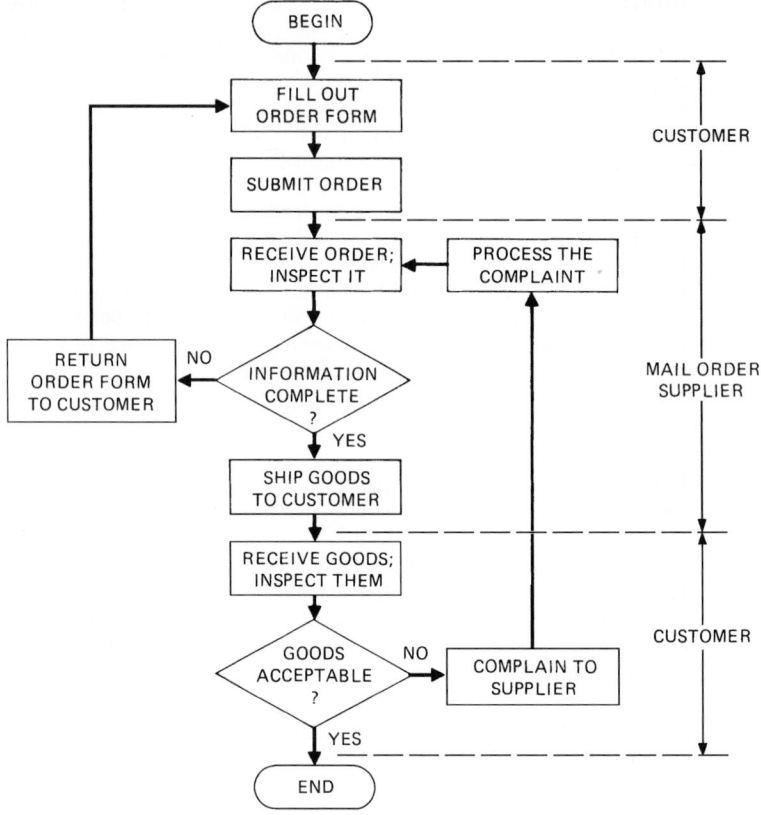

FIGURE 3.4 The flow diagram and customers. [*From Juran, J. M. (1988), Quality Control Handbook, 4th ed. McGraw-Hill, New York, p. 6.6.*]

Local and regional radio programs

Local and regional newspaper coverage

Internal Customers

Corporate management

Product Engineering Office

Assembly plant

Ford Customer Service Division

Marketing

Sales

Purchasing

Public affairs

Internal media

- Ford Communication Network (FCN)
- Employee publications

STEP 3: DISCOVER CUSTOMER NEEDS

The third step of quality planning is to discover the needs of both internal and external customers for the product. Some of the key activities required for effective discovery of customer needs include

- Plan to collect customers' needs.
- Collect a list of customers' needs in their language.
- Analyze and prioritize customers' needs.
- Translate their needs into "our" language.
- Establish units of measurement and sensors.

Our own experience tells us that the needs of human beings are both varied and complex. This can be particularly challenging to a planning team because the actions of customers are not always consistent with what they *say* they want. The challenge for quality planning is to identify the most important needs from the full array of those needs expressed or assumed by the customer. Only then can the product delight the customers.

When designing a product, there are actually two related but distinct aspects of what is being developed: the *technology* elements of what the product's features *will actually do or how it will function* and the *human* elements of the *benefits customers will receive from using the product.* The two must be considered together.

Discovering customer needs is a complex task. Experience shows that customers usually do not state, in simple terms, exactly what they want; often they do not even mention some of their most basic needs. Accuracy of bank statements, competence of a physician, reliability of a computer, and grammatical correctness of a publication may be assumed and never stated without probing.

One of the ways customers express their needs is in terms of problems they experience and their expectation that a product will solve their problems. For example, a customer may state, "I cannot always answer my telephone personally, but I do not want callers to be either inconvenienced or disgusted with nonresponsive answering systems," or, " My mother's personal dignity and love of people are very important to me. I want to find an extended care facility that treats her like a person, not a patient." Even when the need is not *expressed* in such terms, the art and science of discovering needs are to understand exactly the benefit that the customer expects.

When a product's features meet a customer's need, it gives the customer a feeling of satisfaction. If it fails to deliver the promised feature defect-free, the customer feels dissatisfaction. Even if a product functions the way it has been designed, a competing product, by virtue of superior service or performance, may provide customers with greater satisfaction.

Stated Needs and Real Needs.   Customers commonly state their needs as seen from their viewpoint and in their language. Customers may state their needs in terms of the goods or services they wish to buy. However, their real needs are the benefits they believe they will receive. To illustrate:

Customer wishes to buy...	Benefit customer needs might include...
Fresh pasta	Nourishment and taste
Newest personal computer	Write reports quickly and easily Find information on the Web Help children learn math
Health insurance	Security against financial disaster Access to high quality health care Choice in health care providers
Airline ticket	Transportation, Comfort, Safety, and Convenience

Failure to grasp the difference between stated needs and real needs can undermine a quality planning project. Understanding the real needs does not mean that the planners can dismiss the customers' statements and substitute their own superior technical understanding as being the customers' real needs. Understanding the real needs means asking and answering such questions as

Why is the customer buying this product?

What service does he or she expect from it?

How will the customer benefit from it?

How does the customer use it?

What has created customer complaints in the past?

Why have customers selected competitor products over ours?

Perceived Needs. Customers understandably state their needs based on their *perceptions.* These may differ entirely from the supplier's perceptions of what constitutes product quality. Planners can mislead themselves by considering whether the customers' perceptions are wrong or right rather than focusing on how these perceptions influence their buying habits. While such differences between customers and suppliers are potential troublemakers, they also can be an opportunity. Superior understanding of customer perceptions can lead to competitive advantage.

Cultural Needs. The needs of customers, especially internal customers, go beyond products and processes. They include primary needs for job security, self-respect, respect of others, continuity of habit patterns, and still other elements of what we broadly call the "cultural values"; these are seldom stated openly. Any proposed change becomes a threat to these important values and hence will be resisted until the nature of the threat is understood.

Needs Traceable to Unintended Use. Many quality failures arise because a customer uses the product in a manner different from that intended by the supplier. This practice takes many forms. Patients visit emergency rooms for nonemergency care. Untrained workers are assigned to processes requiring trained workers. Equipment does nor receive specified preventive maintenance.

Factors such as safety may add to the cost, yet they may well result in a reduced overall cost by helping to avoid the higher cost arising from misuse of the product. What is essential is to learn the following:

What will be the actual use (and misuse)?

What are the associated costs?

What are the consequences of adhering only to intended use?

Human Safety. Technology places dangerous products into the hands of amateurs who do not always possess the requisite skills to handle them without accidents. It also creates dangerous byproducts that threaten human health, safety, and the environment. The extent of all this is so great that much of the effort of product and process planning must be directed at reducing these risks to an acceptable level. Numerous laws, criminal and civil, mandate such efforts.

"User Friendly." The amateur status of many users has given rise to the term "user friendly" to describe that product feature that enables amateurs to make ready use of technological products. For example, the language of published information should be

Simple

Unambiguous

Readily understood (Notorious offenders have included legal documents, owners' operating manuals, administrative forms, etc. Widely used forms such as governmental tax returns should be field tested on a sample of the very people who will later be faced with filling out such forms.)

Broadly compatible   (For example, new releases of software should be "upward compatible with earlier releases.")

Promptness of Service.   Services should be prompt. In our culture, a major element of competition is promptness of service. Interlocking schedules (as in mail delivery or airline travel) are another source of a growing demand for promptness. Still another example is the growing use of just-in-time manufacturing, which requires dependable deliveries of materials to minimize inventories. All such examples demonstrate the need to include the element of promptness in planning to meet customer needs.

Customer Needs Related to Deficiencies.   In the event of product failure, a new set of customer needs emerges—how to get service restored, and how to get compensated for the associated losses and inconvenience. Clearly, the ideal solution to all this is to plan quality so that there will be no failures. At this point, we will look at what customers need when failures do occur.

Warranties.   The laws governing sales imply that there are certain warranties given by the supplier. However, in our complex society, it has become necessary to provide specific, written contracts to define just what is covered by the warranty and for how long a time. In addition, it should be clear who has what responsibilities.

Effect of Complaint Handling on Sales.   While complaints deal primarily with product dissatisfaction, there is a side effect on salability. Research in this area has pointed out the following: Of the customers who were dissatisfied with products, nearly 70 percent did not complain. The proportions of these who did complain varied depending on the type of product involved. The reasons for not complaining were principally (1) the effort to complain was not worth it, (2) the belief that complaining would do no good, and (3) lack of knowledge about how to complain. More than 40 percent of the complaining customers were unhappy with the responsive action taken by the suppliers. Again, percentages varied depending on the type of product.

Future salability is strongly influenced by the action taken on complaints. This strong influence also extends to brand loyalty. Even customers of popular brands of "large ticket" items, such as durable goods, financial services, and automobile services, will reduce their intent to buy when they perceive that their complaints are not addressed.

This same research concluded that an organized approach to complaint handling provides a high return on investment. The elements of such an organized approach may include

- A response center staffed to provide 24-hour access by consumers and/or a toll-free telephone number
- Special training for the employees who answer the telephones
- Active solicitation of complaints to minimize loss of customers in the future

Keeping Customers Informed.   Customers are quite sensitive to being victimized by secret actions of a supplier, as the phrase "Let the buyer beware!" implies. When such secrets are later discovered and publicized, the damage to the supplier's quality image can be considerable. In a great many cases, the products are fit for use despite some nonconformances. In other cases, the matter may be debatable. In still other cases, the act of shipment is at the least unethical and at the worst illegal.

Customers also have a need to be kept informed in many cases involving product failures. There are many situations in which an interruption in service will force customers to wait for an indefinite period until service is restored. Obvious examples are power outages and delays in public transportation. In all such cases, the customers become restive. They are unable to solve the problem—they must leave that to the supplier. Yet they want to be kept informed as to the nature of the problem and especially as to the likely time of solution. Many suppliers are derelict in keeping customers informed and thereby suffer a decline in their quality image. In contrast, some airlines go to great pains to keep their customers informed of the reasons for a delay and of the progress being made in providing a remedy.

Plan to Collect Customers' Needs.   Customer needs keep changing. There is no such thing as a final list of customer needs. While it can be frustrating, planning teams must realize that even while they are in the middle of the planning process, forces such as technology, competition, social change, and so on, can create new customer needs or may change the priority given to existing needs. It becomes extremely important to check with customers frequently and monitor the marketplace.

Some of the most common ways to collect customer needs include

- Customer surveys, focus groups, and market research programs and studies
- Routine communications, such as sales and service calls and reports, management reviews, house publications
- Tracking customer complaints, incident reports, letters, and telephone contacts
- Simulated-use experiments and planning processes that involve the customer
- Employees with special knowledge of the customer: sales, service, clerical, secretarial, and supervisory who come into contact with customers
- Customer meetings
- User conferences for the end user
- Information on competitors' products
- Personal visits to customer locations; observe and discuss

  How product is used

  Unintended uses

  Service failures by others

  What current or new features will relieve onerous tasks

  Changes in habits and culture

  Changes in sales

  Existence of or changes in premium prices

  Sale of spare parts or after-sale service

  Purchase of options

- Government or independent laboratory data
- Changes in federal, state, and local regulations that will identify current need or new opportunity
- Competitive analysis and field intelligence comparing products with those of competitors
- Personal experience dealing with the customer and the product (However, it is important to be cautious about giving personal experience too much weight without direct verification by customers. The analysts must remember that looking at customer needs and requirements from a personal viewpoint can be a trap.)

Often customers do not express their needs in terms of the benefits they wish to receive from purchasing and using the product.

Discovering Mustang Customer Needs. In designing the Mustang, Ford relied heavily on the following sources for customer needs:

- Quantitative market research
- Qualitative market research
- Inspection studies
- Observational research
- Dealer input
- Customer surveys
- Direct interaction with customers
- Media feedback
- Product evaluation reports—internal and external
- Competitive evaluations

Collect List of Customers' Needs in Their Language. For a list of customers' needs to have significant meaning in planning a new product, they must be stated in terms of benefits sought. Another way of saying this is to capture needs in the customer's voice. By focusing on the benefits sought by the customer rather than on the means of delivering the benefit, designers will gain a better understanding of what the customer needs and how the customer will be using the product. Stating needs in terms of the benefits sought also can reveal opportunities for improved quality that often cannot be seen when concentrating on the product features alone.

Analyze and Prioritize Customer Needs. The information actually collected from customers is often too broad, too vague, and too voluminous to be used directly in designing a product. Both specificity and priority are needed to ensure that the design really meets the needs and that time is spent on designing for those needs which are truly the most important. The following activities help provide this precision and focus:

- Organizing, consolidating, and prioritizing the list of needs for both internal and external customers
- Determining the importance of each need for both internal and external customers
- Breaking down each need into precise terms so that a specific design response can be identified
- Translating these needs into the supplying organization's language
- Establishing specific measurements and measurement methods for each need

One of the best planning tools to analyze and organize customers' needs is the "quality planning spreadsheet."

Quality Planning Spreadsheets. Quality planning generates a large amount of information that is both useful and necessary, but without a systematic way to approach the organization and analysis of this information, the planning team may be overwhelmed by the volume and miss the message it contains.

   Although planners have developed various approaches for organizing all this information, the most convenient and basic planning tool is the *quality planning spreadsheet*. The spreadsheet is a highly versatile tool that can be adapted to a number of situations. The quality planning process makes use of several kinds of spreadsheets, such as

- Customer needs spreadsheet
- Needs analysis spreadsheet

- Product design spreadsheet
- Process design spreadsheet
- Process control spreadsheet

Besides recording information, these tools are particularly useful in analyzing relationships among the data that have been collected and in facilitating the stepwise conversion of customer needs into product features and then product features into process characteristics and plans. This conversion is illustrated in Figure 3.5. Analysis of customers and their needs provides the basis for designing the product. The summary of that design feeds the process design, which feeds the control spreadsheet.

For most planning projects, simple matrix spreadsheets will suffice. For other projects, more complex quality functional deployment spreadsheets are helpful in computing design trade-offs. All these spreadsheets are designed to allow the team to record and compare the relationships among many variables at the same time. We will illustrate some of these spreadsheets at the appropriate point in the planning process. Figure 3.6 illustrates the generic layout of any one of these spreadsheets. In general, the row headings are the "what's" of the analysis—the customers to be satisfied, the needs to be met, and so on. The columns are the "how's"—the needs that, when met, will satisfy the customer, the product features that will meet the needs, and so on. The bottom row of the spreadsheet generally contains specific measurable goals for the "how" at the top. The body of the spreadsheet expresses with symbols or numerics the impact of the "how" on the "what"—e.g., none, moderate, strong, very strong. Other columns can be added to give specific measures of the importance of the respective rows, benchmarks, and so on.

Customer Needs Spreadsheet.   Figure 3.7 provides a simple example of a customer needs spreadsheet. The left column lists, in priority order, all the external and internal customers. The column headings are the various needs that have been discovered. By either checking or entering a designation for importance, it is possible to create a simple but comprehensive picture of the importance of meeting each need. All product development must operate within a budget. Prioritizing the customers and their needs ensures that the budget is focused on what is most important.

Precise Customer Needs.   Once the needs that must be met have been prioritized, they must be described in sufficiently precise terms to design a product based on them. A customer needs spreadsheet helps assemble this analysis. At this point, customer needs are probably a mixture of relatively broad expectations such as "ease of use" and more specific requests such as "access on Saturday." Figure 3.8 illustrates how broad needs (called "primary") are broken into succeeding levels of specificity ("secondary," "tertiary," etc.) Note that primary and secondary do not mean more and less important, they mean, respectively, less specific and more specific. Each need must be broken down to the level at which it can (1) be measured and (2) serve as an unambiguous guide for product design. In some cases two levels of detail may suffice, in others four or five may be required. Figure 3.8 illustrates how this might be done for the primary need "convenience" associated with a group medical practice.

Mustang Customer Needs.   In addition to basic transportation, the following needs surfaced as most important for Mustang customers:

Safety
Performance
Image
Comfort

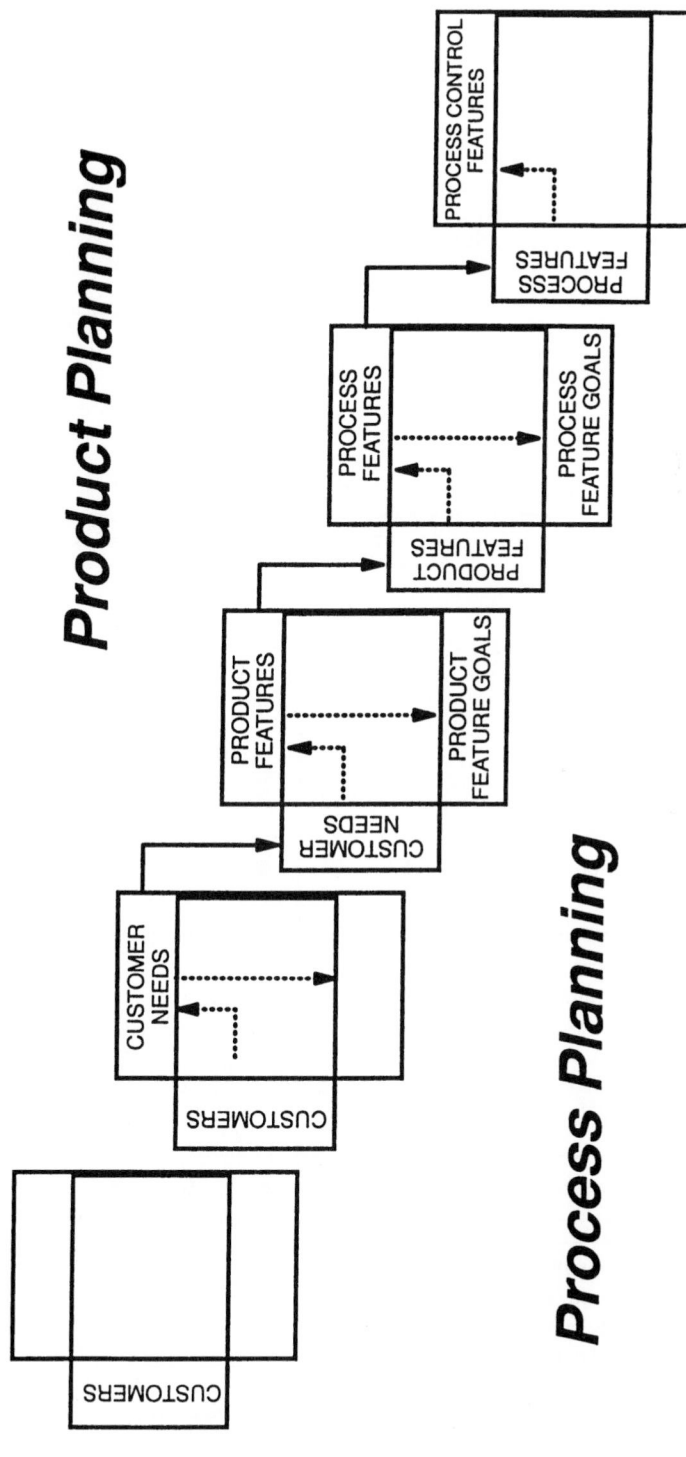

FIGURE 3.5 Spreadsheets in quality planning. *(Juran Institute, Inc. Copyright 1994. Used by permission.)*

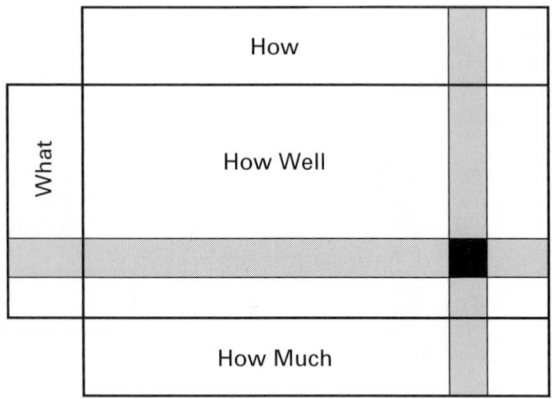

FIGURE 3.6   Generic planning spreadsheet. (*Juran Institute, Inc. Copyright 1994. Used by permission.*)

Customers	Customer Needs							
	Attractive	Informative/well-written articles	Catchy cover lines	Stable circulation	It sellls	Enough time	Material complete	No last minute changes
Readers	●	●	●					
Advertisers	●	○	●	●	●			
Printers						●	●	●
Typesetters						●	●	●
Color separators						●	●	●
Newsstand	●	○	●	●	●			

Legend
● Very Strong
○ Strong
△ Weak

FIGURE 3.7   Customer needs spreadsheet for a magazine. (*Juran Institute, Inc. Copyright 1994. Used by permission.*)

Primary need	Secondary need	Tertiary need
Convenience	Hours of operation	Open between 5:00 and 9:00 p.m.
		Saturday hours
	Transportation access	Within three blocks of bus stop
		Ample parking
	Short wait	Urgent appointment within 24 hours
		Routine appointment within 14 days
		Waiting time at appointment less than 15 minutes
	Complementary available services	Pharmacy on site
		Lab on site

FIGURE 3.8 Needs analysis spreadsheet for a medical office. (*Juran Institute, Inc. Copyright 1994. Used by permission.*)

Convenience

Entertainment

Reliability/dependability

Durability

Low cost of ownership

Ergonomics

Fuel economy

Handling—in all types of weather

Ride quality—regardless of road surface

Translate Their Needs into "Our" Language. The precise customer needs that have been identified may be stated in any of several languages, including

The customer's language

The supplier's ("our") language

A common language

An old aphorism claims that the British and Americans are separated by a common language. The appearance of a common language or dialect can be an invitation to trouble because both parties believe that they understand each other and expect to be understood. Failure to communicate because of the unrecognized differences can build additional misunderstanding that only compounds the difficulty. It is imperative, therefore, for planners to take extraordinary steps to ensure that they properly understand customer needs by systematically translating them. The need to translate applies to both internal and external customers. Various company functions employ local dialects that are often not understood by other functions.

Vague terminology constitutes one special case for translation that can arise even (and often especially) between customers and suppliers that believe they are speaking the same dialect. Identical words have multiple meanings. Descriptive words do not describe with technological precision.

Aids to Translation. Numerous aids are available to clear up vagueness and create a bridge across languages and dialects. The most usual are listed following:

A *glossary* is a list of terms and their definitions. It is a published agreement on the precise meanings of key terms. The publication may be embellished by other forms of communication, such as sketches, photographs, and videotapes.

Samples can take many forms, such as physical goods (e.g., textile swatches, color chips, audio cassettes) or services (e.g., video recordings to demonstrate "samples" of good service—courtesy, thoughtfulness, etc.) They serve as specifications for product features. They make use of human senses beyond those associated with word images.

A *special organization* to translate communications with external customers may be required because of the high volume of translation. A common example is the order-editing department, which receives orders from clients. Some elements of these orders are in client language. Order editing translates these elements into supplier language, e.g., product code numbers, supplier acronyms, and so on.

Standardization is used by many mature industries for the mutual benefit of customers and suppliers. This standardization extends to language, products, processes, and so on. All organizations make use of short designations for their products, such as code numbers, acronyms, words, phrases, and so on. Such standardized nomenclature makes it easy to communicate with internal customers.

Measurement is the most effective remedy for vagueness and multiple dialects—"Say it in numbers." This is the first, but not the last, point in the planning process where measurement is critical. Quality planning also requires measurement of product features, process features, process capability, control subjects, and so on.

Establish Units of Measurement and Sensors.

Sound quality planning requires precise communication between customers and suppliers. Some of the essential information can be conveyed adequately by words. However, an increasingly complex and specialized society demands higher precision for communicating quality-related information. The higher precision is best attained when we say it in numbers.

Quantification requires a system of measurement. Such a system consists of

A *unit of measurement,* which is a defined amount of some quality feature, permits evaluation of that feature in numbers, e.g., hours of time to provide service, kilowatts of electric power, or concentration of a medication.

A *sensor,* which is a method or instrument of measurement, carries out the evaluation and states the findings in numbers in terms of the unit of measure, e.g., a clock for telling time, a thermometer for measuring temperature, or an x-ray to measure bone density.

By measuring customer needs, one has established an objective criterion for whether or not the needs are met. In addition, only with measurement can one answer questions such as, Is our quality getting better or worse? Are we competitive with others? Which one of our operations provides the best quality? How can we bring all operations up to the level of the best?

Units of Measure for Product Features.   The first task in measurement is to identify the appropriate unit of measurement for each customer need. For product features, we know of no simple, convenient, generic formula that is the source of many units of measure. The number and variety of product features are simply enormous. In practice, each product feature requires its own unique unit of measure. A good starting point is to ask the customers what their units of measure are for evaluating product quality. If the supplier's units of measure are different, the stage is set for customer dissatisfaction, and the team will need to come up with a unit of measure acceptable to both parties. Even if the customers have not developed an explicit unit of measure, ask them how they would know whether the need were met. Their response may carry with it an implicit unit of measure.

Application to Goods.   Units of measure for quality features of goods make extensive use of "hard" technological units. Some of these are well known to the public: time in minutes, temperature in degrees, or electric current in amperes. Many others are known only to the specialists.

There are also "soft" areas of quality for goods. Food technologists need units of measure for flavor, tenderness, and still other properties of food. Household appliances must be "handsome" in appearance. Packaging must be "attractive." To develop units of measure for such features involves much effort and ingenuity.

Application to Services.   Evaluation of service quality includes some technological units of measure. A widespread example is promptness, which is measured in days, hours, and so on. Environmental pollutants (e.g., noise, radiation, etc.) generated by service companies are likewise measured using technological units of measure.

Service quality also involves features such as courtesy of service personnel, decor of surroundings, and readability of reports. Since these features are judged by human beings, the units of measure (and the associated sensors) must be shown to correlate with a jury of customer opinion.

The Ideal Unit of Measure.   The criteria for an ideal unit of measure are summarized below. An ideal unit of measure

- Is understandable
- Provides an agreed basis for decision making
- Is conducive to uniform interpretation
- Is economical to apply
- Is compatible with existing designs of sensors, if other criteria also can be met

Measuring Abstractions.   Some quality features seem to stand apart from the world of physical things. Quality of service often includes courtesy as a significant quality feature. Even in the case of physical goods, we have quality features, such as beauty, taste, aroma, feel, or sound. The challenge is to establish units of measure for such abstractions.

The approach to dealing with abstractions is to break them up into identifiable pieces. Once again, the customer may be the best source to start identifying these components. For example, hotel room appearance is certainly a quality feature, but it also seems like an abstraction. However, we can divide the feature into observable parts and identify those specifics which collectively constitute "appearance," e.g., the absence of spots or bare patches on the carpet, clean lavatory, linens free from discoloration and folded to specified sizes, windows free of streaks, bedspreads free of wrinkles and hanging to within specific distances from the floor, and so on. Once units of measure have been established for each piece or component, they should be summarized into an index, e.g., number of soiled or damaged carpets to total number of hotel rooms, number of rooms with missing linens to total number of rooms, or number of customer complaints.

Establish the Sensor.   To say it in numbers, we need not only a unit of measure, but we also need to evaluate quality in terms of that unit of measure. A key element in making the evaluation is the sensor.

A "sensor" is a specialized detecting device or measurement tool. It is designed to recognize the presence and intensity of certain phenomena and to convert this sense knowledge into information. In turn, the resulting information becomes an input to decision making because it enables us to evaluate actual performance.

Technological instruments are obviously sensors. So are the senses of human beings. Trends in some data series are used as sensors. Shewhart control charts are sensors.

Precision and Accuracy of Sensors.   The "precision" of a sensor is a measure of the ability of the sensor to reproduce its results over and over on repeated tests. For most technological sensors, this reproducibility is high and is also easy to quantify.

At the other end of the spectrum are the cases in which we use human beings as sensors: inspectors, auditors, supervisors, and appraisers. Human sensors are notoriously less precise than technological sensors. Such being the case, planners are well advised to understand the limitations inherent in human sensing before making decisions based on the resulting data.

The "accuracy" of a sensor is the degree to which the sensor tells the truth—the extent to which its evaluation of some phenomenon agrees with the "true" value as judged by an established standard. The difference between the observed evaluation and the true value is the "error," which can be positive or negative.

For technological sensors, it is usually easy to adjust for accuracy by recalibrating. A simple example is a clock or watch. The owner can listen to the time signals provided over the radio. In contrast, the precision of a sensor is not easy to adjust. The upper limit of precision is usually inherent in the basic design of the sensor. To improve precision beyond its upper limit requires a redesign. The sensor may be operating at a level of precision below that of its capability owing to misuse, inadequate maintenance, and so on. For this reason, when choosing the appropriate sensor for each need, planners will want to consider building in appropriate maintenance schedules along with checklists on actions to be taken during the check.

Translating and Measuring Mustang Customer Needs.   The customer need for performance illustrates how high-level needs breakdown into a myriad of detailed needs. Performance included all the following detailed, precise needs:

Performance feel off the line

Wide-open throttle (WOT) 0 to 60 mi/h elapsed time

WOT $1/4$-mile elapsed time

WOT 40 to 60 mi/h passing time

WOT 30 to 70 mi/h passing time

Part-throttle response

Seat-of-the-pants feel that can only be measured by a jury of customers

Competitor performance was used as a minimum benchmark, but Ford knew that its competitors also were working on new models and had to stretch the needs analysis to include "what-if" scenarios that were tested with panels of consumers and automotive experts.

Product Design Spreadsheet.   All the information on the translation and measurement of a customer need must be recorded and organized. Experience recommends placing these data so that they will be close at hand during product design. The example in Figure 3.9 shows a few needs all prepared for use in product design. The needs, their translation, and their measurement are all placed to the left of the spreadsheet. The remainder of the spreadsheet will be discussed in the next section.

STEP 4: DEVELOP PRODUCT

Once the customers and their needs are fully understood, we are ready to design the product that will meet those needs best. Product development is not a new function for a company. Most companies have some process for designing and bringing new products to market. In this step of the quality planning process, we will focus on the role of quality in product development and how that role combines with the technical aspects of development and design appropriate for a particular industry. Within product development, product design is a creative process based largely on technological or functional expertise.

The designers of products traditionally have been engineers, systems analysts, operating managers, and many other professionals. In the quality arena, designers can include any whose experience, position, and expertise can contribute to the design process. The outputs of product design are detailed designs, drawings, models, procedures, specifications, and so on.

The overall quality objectives for this step are two:

1. Determine which product features and goals will provide the optimal benefit for the customer

2. Identify what is needed so that the designs can be delivered without deficiencies.

Product Features

Needs	Translation	Units of Measure	Sensors	Cross resource checking	Auto search for open times	Check resource constraints	FAX information to scheduling source	Mail instructions to patient
No double bookings	Double bookings	Yes/No	Review by scheduler	●				
Pt. comes prepared	Pt. followed MD's instructions	Yes/No/Partial	Review by person doing procedure				△	●
All appointments used	No "holds" used	Yes/No	Review by scheduler		●	○		
All info. easy to find	Do not have to "search"	Yes/No	Review by scheduler	○	○			
Quick confirmation	Quick confirmation	Minutes	Software/Review by scheduler		○			
Product Feature Goals				100% of time for all information entered	One key stroke	Cannot change appt. w/o author from source	Reminder always generated for receiver	For all appointments

Legend

● Very Strong
○ Strong
△ Weak

FIGURE 3.9 Product design spreadsheet for outpatient appointment function. (*Juran Institute, Inc. Copyright 1994. Used by permission.*)

In the case of designing services, the scope of this activity is sometimes puzzling. For example, in delivering health care, where does the *product* of diagnosing and treating end and the *processes* of laboratory testing, chart reviews, and so on begin? One useful way to think about the distinction is that the *product* is the "face to the customer." It is what the customer sees and experiences. The patient sees and experiences the physician interaction, waiting time, clarity of information, and so on. The effectiveness and efficiency of moving blood samples to and around the laboratory have an effect on these product features but are really features of the process that delivers the ultimate product to the customer.

Those who are designing physical products also can benefit from thinking about the scope of product design. Remembering that the customer's needs are the benefits that the customer wants from the product, the design of a piece of consumer electronics includes not only the contents of the box itself but also the instructions for installation and use and the "help line" for assistance.

There are six major activities in this step:

- Group together related customer needs.
- Determine methods for identifying product features.
- Select high-level product features and goals.
- Develop detailed product features and goals.
- Optimize product features and goals.
- Set and publish final product design.

Group Together Related Customer Needs.
Most quality planning projects will be confronted with a large number of customer needs. Based on the data developed in the preceding steps, the team can prioritize and group together those needs which relate to similar functionality. This activity does not require much time, but it can save a lot of time later. Prioritization ensures that the scarce resources of product development are spent most effectively on those items which are most important to the customer. Grouping related needs together allows the planning team to "divide and conquer," with subteams working on different parts of the design. Such subsystem or component approaches to design, of course, have been common for years. What may be different here is that the initial focus is on the *components of the customers' needs, not the components of the product.* The component design for the product will come during the later activities in this step.

Determine Methods for Identifying Product Features.
There are many complementary approaches for identifying the best product design for meeting customers' needs. Most design projects do not use all of them. Before starting to design, however, a team should develop a systematic plan for the methods it will use in its own design. Here are some of the options.

Benchmarking.   This approach identifies the best in class and the methods behind it that make it best. See Section 12 for details.

Basic Research.   One aspect of research might be a new innovation for the product that does not currently exist in the market or with competitors. Another aspect of basic research looks at exploring the feasibility of the product and product features. While both these aspects are important, be careful that fascination with the technological abilities of the product do not overwhelm the primary concern of its benefits to the customer.

Market Experiments.   Introducing and testing ideas for product features in the market allow one to analyze and evaluate concepts. The focus group is one technique that can be used to measure customer reactions and determine whether the product features actually will meet customer needs. Some organizations also try out their ideas, on an informal basis, with customers at trade shows and association meetings. Still others conduct limited test marketing with a prototype product.

Creativity.   Developing product features allows one to dream about a whole range of possibilities without being hampered by any restrictions or preconceived notions. Quality planning is a proven, structured, data-based approach to meeting customers' needs. But this does not mean it is rigid and uncreative. At this point in the process, the participants in planning must be encouraged and given the tools they need to be creative so as to develop alternatives for design. After they have selected a number of promising alternatives, then they will use hard analysis and data to design the final product.

Planning teams can take advantage of how individuals view the world: from their own perspective. Every employee potentially sees other ways of doing things. The team can encourage people to suggest new ideas and take risks. Team members should avoid getting "stuck" or take too much time to debate one particular idea or issue. They can put it aside and come back to it later with a fresh viewpoint. They can apply new methods of thinking about customers' needs or problems, such as the following:

- *Changing key words or phrases.* For example, call a "need" or "problem" an "opportunity." Instead of saying, "deliver on time," say, "deliver exactly when needed."
- *Random association.* For example, take a common word such as "apple" or "circus" and describe your business, product, or problem as the word. For example, "Our product is like a circus because…"
- *Central idea.* Shift your thinking away from one central idea to a different one. For example, shift the focus from the product to the customer by saying, "What harm might a child suffer, and how can we avoid it?" rather than, "How can we make the toy safer?"
- *Putting yourself in the other person's shoes.* Examine the question from the viewpoint of the other person, your competitor, your customer—and build their case before you build your own.
- *Dreaming.* Imagine that you had a magic wand that you could wave to remove all obstacles to achieving your objectives. What would it look like? What would you do first? How would it change your approach?
- *The spaghetti principle.* When you have difficulty considering a new concept or how to respond to a particular need, allow your team to be comfortable enough to throw out a new idea, as if you were throwing spaghetti against the wall, and see what sticks. Often even "wild" ideas can lead to workable solutions.

The initial design decisions are kept as simple as possible at this point. For example, the idea of placing the control panel for the radio on the steering wheel would be considered a high-level product feature. Its exact location, which controls, and how they function can be analyzed later in more detail. It may become the subject of more detailed product features as the planning project progresses.

Standards, Regulations, and Policies.

This is also the time to be certain that all relevant standards, regulations, and policies have been identified and addressed. While some of these requirements are guidelines for how a particular product or product feature can perform, others mandate how they must perform. These may come from inside the organization, and others may come from specific federal, state, or local governments, regulatory agencies, or industry associations. All product features and product feature goals must be analyzed against these requirements before making the final selection of product features to be included in the design.

It is important to note that if there is a conflict when evaluating product features against any standards, policies, or regulations, it is not always a reason to give up. Sometimes one can work to gain acceptance for a change when it will do a better job of meeting customer needs. This is especially true when it comes to internal policies. However, an advocate for change must be prepared to back the arguments up with the appropriate data.

Criteria for Design.

As part of the preparation for high level design, the design team must agree on the explicit criteria to be used in evaluating alternative designs and design features. All designs must fulfill the following general criteria:

- Meet the customers' needs
- Meet the suppliers' and producers' needs
- Meet (or beat) the competition
- Optimize the combined costs of the customers and suppliers

In addition to the preceding four general criteria, the team members should agree explicitly on the criteria that it will use to make its selection. (If the choices are relatively complex, the team should consider using the formal discipline of a selection matrix.) One source for these criteria will be the team's mission statement and goals. Some other types of criteria the team may develop include

- The impact of the feature on the needs
- The relative importance of the needs being served
- The relative importance of the customers whose needs are affected
- The feasibility and risks of the proposed feature
- The impact on product cost
- The relationship to competitive features uncovered in benchmarking
- The requirements of standards, policies, regulations, mandates, and so on

As part of the decision on how to proceed with design, teams also must consider a number of other important issues regarding what type of product feature will be the best response to customers' needs. When selecting product features, they need to consider whether to

- Develop an entirely new functionality
- Replace selected old features with new ones
- Improve or modify existing features
- Eliminate the unnecessary

Regulations and Standards for Ford's Mustang.   The Federal Motor Vehicle Safety Standards (FMVSS) are, of course, a prime concern for designing any automobile. Ford had established its own safety standards that were more extensive than the federal mandates and included a significant margin of additional safety on all quantitative standards.

Select High-Level Product Features and Goals.   This phase of quality planning will stimulate the team to consider a whole array of potential product features and how each would respond to the needs of the customer. This activity should be performed without being constrained by prior assumptions or notions as to what worked or did not work in the past. A response that previously failed to address a customer need or solve a customer problem might be ready to be considered again because of changes in technology or the market.

The team begins by executing its plan for identifying the possible product features. It should then apply its explicit selection criteria to identify the most promising product features.

The product design spreadsheet in Figure 3.9 is a good guide for this effort. Use the right side of the spreadsheet to determine and document the following:

- Which product features contribute to meeting which customer needs
- That each priority customer need is addressed by at least one product feature
- That the total impact of the product features associated with a customer need is likely to be *sufficient* for meeting that need
- That every product feature contributes to meeting at least one significant customer need
- That every product feature is *necessary* for meeting at least one significant customer need (i.e., removing that feature would leave a significant need unmet)

Now the team must set goals for each feature. In quality terms, a goal is an aimed-at quality target (such as aimed-at values and specification limits). As discussed earlier, this differs from quality standards in that the standard is a mandated model to be followed that typically comes from an external source. While these standards serve as "requirements" that usually dictate uniformity or how the product is to function, product feature goals are often voluntary or negotiated. Therefore, the quality planning process must provide the means for meeting both quality standards and quality goals.

Criteria for Setting Product Feature Goals.   As with all goals, product feature goals must meet certain criteria. While the criteria for establishing product feature goals differ slightly from the criteria for project goals verified in step 1, there are many similarities. Product feature goals should encompass all the important cases and be

- Measurable
- Optimal
- Legitimate
- Understandable
- Applicable
- Attainable

Measuring Product Features Goals.   Establishing the measurement for a product feature goal requires the following tasks:

- Determine the unit of measure: meters, seconds, days, percentages, and so on.
- Determine how to measure the goal (i.e., determine what is the sensor).
- Set the value for the goal.

The work done in measuring customer needs should be applied now. The two sets of measurements may be related in one of the following ways:

- Measurement for the need and for the product feature goal may use the same units and sensors. For example, if the customer need relates to timeliness measured in hours, one or more product features normally also will be measured in hours, with their combined effects meeting the customer need.
- Measurement for the product feature may be derived in a technical manner from the need measurement. For example, a customer need for transporting specified sizes and weights of loads may be translated into specific engineering measurements of the transport system.
- Measurement for the product feature may be derived from a customer behavioral relationship with the product feature measure. For example, automobile manufacturers have developed the specific parameters for the dimensions and structure of an automobile seat that translate into the customer rating it "comfortable."

Since we can now measure both the customer need and the related product feature goals, it is possible for the quality planning team to ensure that the product design will go a long way toward meeting the customers' needs, even before building any prototypes or conducting any test marketing.

For large or complex projects, the work of developing product features is often divided among a number of different individuals and work groups. After all these groups have completed their work, the overall quality planning team will need to integrate the results. Integration includes

- Combining product features when the same features have been identified for more than one cluster
- Identifying and resolving conflicting or competing features and goals for different clusters
- Validating that the combined design meets the criteria established by the team

Develop Detailed Product Features and Goals.   For large and highly complex products, it will usually be necessary to divide the product into a number of components and even subcomponents for detailed design. Each component will typically have its own design team that will complete the detailed design described below. In order to ensure that the overall design remains integrated, consistent, and effective in meeting customer needs, these large, decentralized project require

- A steering or core team that provides overall direction and integration
- Explicit charters with quantified goals for each component
- Regular integrated design reviews for all components
- Explicit integration of designs before completion of the product design phase

Once the initial detailed product features and goals have been developed, then the technical designers will prepare a preliminary design, with detailed specifications. This is a necessary step before a team can optimize models of product features using a number of quality planning tools and ultimately set and publish the final product features and goals.

It is not uncommon for quality planning teams to select product features at so high a level that they are not specific enough to respond to precise customer needs. Just as in the identification of customers' primary needs, high-level product features need to be broken down further into terms that are clearly defined and which can be measured.

Optimize Product Features and Goals.   Once the preliminary design is complete, it must be optimized. That is, the design must be adjusted so that it meets the needs of both customer and supplier while minimizing their combined costs and meeting or beating the competition.

Finding the optimum can be a complicated matter unless it is approached in an organized fashion and follows quality disciplines. For example, there are many designs in which numerous variables converge to produce a final result. Some of these designs are of a business nature, such as design of an information system involving optimal use of facilities, personnel, energy, capital, and so on. Other such designs are technological in nature, involving optimizing the performance of hardware. Either way, finding the optimum is made easier through the use of certain quality disciplines.

Finding the optimum involves balancing the needs, whether they are multicompany needs or within-company needs. Ideally, the search for the optimum should be done through the participation of suppliers and customers alike. There are several techniques that help achieve this optimum.

Design Review.   Under this concept, those who will be affected by the product are given the opportunity to review the design during various formative stages. This allows them to use their experience and expertise to make such contributions as

- Early warning of upcoming problems
- Data to aid in finding the optimum
- Challenge to theories and assumptions

Design reviews can take place at different stages of development of the new product. They can be used to review conclusions about customer needs and hence the product specifications (characteristics of product output). Design reviews also can take place at the time of selecting the optimal product design. Typical characteristics of design reviews include the following:

- Participation is mandatory.
- Reviews are conducted by specialists, external to the planning team.
- Ultimate decisions for changes remain with the planning team.
- Reviews are formal, scheduled, and prepared for with agendas.
- Reviews will be based on clear criteria and predetermined parameters.
- Reviews can be held at various stages of the project.

Ground rules for good design reviews include

- Adequate advance planning of review agenda and documents
- Clearly defined meeting structure and roles
- Recognition of interdepartmental conflicts in advance
- Emphasis on constructive, not critical, inputs
- Avoidance of competitive design during review
- Realistic timing and schedules for the reviews
- Sufficient skills and resources provided for the review
- Discussion focus on untried/unproved design ideas
- Participation directed by management

Joint Planning.   Planning teams should include all those who have a vested interest in the outcome of the design of the product along with individuals skilled in product design. Under this concept, the team, rather than just the product designers, bears responsibility for the final design.

Structured Negotiation.   Customers and suppliers are tugged by powerful local forces to an extent that can easily lead to a result other than the optimum. To ensure that these negotiating sessions proceed in as productive a fashion as possible, it is recommended that ground rules be established before the meetings. Here are some examples:

- The team should be guided by a spirit of cooperation, not competition, toward the achievement of a common goal.
- Differences of opinion can be healthy and can lead to a more efficient and effective solution.
- Everyone should have a chance to contribute, and every idea should be considered.
- Everyone's opinions should be heard and respected without interruptions.
- Avoid getting personal; weigh pros and cons of each idea, looking at its advantages before its disadvantages.
- Challenge conjecture; look at the facts.
- Whenever the discussion bogs down, go back and define areas of agreement before discussing areas of disagreement.
- If no consensus can be reached on a particular issue, it should be tabled and returned to later on in the discussion.

Create New Options.   Often teams approach a product design with a history of how things were done in the past. Optimization allows a team to take a fresh look at the product and create new options. Some of the most common and useful quality tools for optimizing the design include the following:

Competitive analysis provides feature-by-feature comparison with competitors' products. (See the following for an example.)

Salability analysis evaluates which product features stimulate customers to be willing to buy the product and the price they are willing to pay. (See the following for an example.)

Value analysis calculates not only the incremental cost of specific features of the product but also the cost of meeting specific customer needs and compares the costs of alternative designs. (See the following for an example.)

Criticality analysis identifies the "vital few" features that are vulnerable in the design so that they can receive priority for attention and resources.

Failure mode and effect analysis (*FMEA*) calculates the combined impact of the probability of a particular failure, the effects of that failure, and the probability that the failure can be detected and corrected, thereby establishing a priority ranking for designing in failure-prevention countermeasures. (See Section 19 under Reliability Analysis.)

Fault-tree analysis aids in the design of preventive countermeasures by tracing all possible combinations of causes that could lead to a particular failure. (See Section 19 under Reliability Analysis; also see Section 48.)

Design for manufacture and assembly evaluates the complexity and potential for problems during manufacture to make assembly as simple and error-free as possible. Design for maintainability evaluates particular designs for the ease and cost of maintaining them during their useful life.

Competitive Analysis.   Figure 3.10 is an example of how a competitive analysis might be displayed. The data for a competitive analysis may require a combination of different approaches such as laboratory analysis of the competitors' products, field testing of those products, or in-depth interviews and on-site inspections where willing customers are using a competitor's product.

Note that by reviewing this analysis, the planning team can identify those areas in which the design is vulnerable to the competition, as well as those in which the team has developed an advantage. Based on this analysis, the team will then need to make optimization choices about whether to upgrade the product or not. The team may need to apply a value analysis to make some of these choices.

Salability Analysis.   An example of salability analysis is shown in Figure 3.11. This analysis is similar to a competitive analysis, except that the reference point is the response of customers to the proposed design rather than a comparison with the features of the competitors' designs. Note, however, that elements of competitive and salability analyses can be combined, with the salability analysis incorporating customer evaluation of both the proposed new design and existing competitive designs.

Complex products, such as automobiles, with multiple optional features and optional configurations offer a unique opportunity to evaluate salability. Observed installation rates of options on both the existing car line and competitors' cars provide intelligence on both the level of market demand for the feature and the additional price that some segments of the market will pay for the feature—although the other segments of the market may place little or no value on it.

Value Analysis.   Value analysis has been quite common in architectural design and the development of custom-engineered products, but it also can be applied successfully to other environments as well, as illustrated in Figure 3.12. By comparing the costs for meeting different customer needs, the design team can make a number of significant optimization decisions. If the cost for meeting low-priority needs is high, the team must explore alternative ways to meet those needs and even consider not addressing them at all if the product is highly price sensitive. If very important needs have not consumed much of the expense, the team will want to make certain that it has met those needs fully and completely. While low expense for meeting a high-priority need is not necessarily inappropriate, it does present the designers with the challenge of making certain that lower-priority needs are not being met using resources that could be better directed toward the higher-priority needs. It is not uncommon for products to be overloaded with "bells and whistles" at the expense of the fundamental functionality and performance.

Mustang's Performance Features.   One of the critical challenges for engineering the performance of Mustang was to develop the ideal power-to-weight ratio that would meet the performance needs of the customers. However, as is the case with most ratios of this sort, the ratio can be improved either by reducing the weight or by increasing the power. What is more, fuel economy is also affected by the weight.

Design trade-offs among weight, power, and fuel economy involved not only detailed engineering calculations but also careful assessments of customer reactions and the competing views of different functions within the company. Reaching a final design that met the customer needs and fulfilled sound engineering principles required strong project leadership in addition to the data.

Product Feature & Goal	Check if Product Feature is Present			Feature Performance vs. Goal (*)			Identify if Significant Risk or Opportunity
	Product A	Product B	Ours	Product A	Product B	Ours	
Retreive messages from all touch tone phones easily	Yes	Yes	Yes	4	5	4	—
Change message from any remote location	Yes	No	Yes	3	—	5	O
2 lines built in	No	No	Yes	—	—	4	O

Below Add Features in Competitors Product Not Included in Ours	Check if Product Feature is Present			Feature Performance vs. Goal (*)			Identify if Significant Risk or Opportunity
	Product A	Product B	Ours	Product A	Product B	Ours	
No cassette used to record message	Yes	Yes		4	—		R
Telephone and answering machine in one unit	Yes	Yes		3	4		R

FIGURE 3.10 Competitive analysis. (*Juran Institute, Inc. Copyright 1994. Used by permission.*)

Name of Product Car Repair Service—Tune-up	How Do Customers Rate Product? Poor Fair Satisfactory Good Excellent	Basis for Rating Prior Use vs. Opinion	How Do Customers See Differences Between Our Products and Competing Products? Positively (+) Negatively (−) No Difference	Would Customers Buy If Price Were Not Important? Yes No	Would Customer Buy If Price Were Important? Price	Would Customer Buy If Price Were Important? Yes No	Of All Products Listed, Prioritize Which Would Customers Buy and Its Basis? Price Features	Identify if Significant Risk or Opportunity
Ours—	E	U		Y	$175	Y	2-F	
Competitor A—	G	O	+	N	$145	Y	3-P	O
Competitor B—	E	U	O	Y	$175	Y	1-F	R

Name of Product Pick-up and delivery of car to be repaired Product Feature Goal: Same Day Service	How Do Customers Rate Product? Poor Fair Satisfactory Good Excellent	Basis for Rating Prior Use vs. Opinion	How Do Customers See Differences Between Our Features Against Competing Features? Positively (+) Negatively (−) No Difference	Does the Addition of the Feature Make the Product: More Salable Less Salable No Difference	Identify if Significant Risk or Opportunity
Ours—Offered	G	U		O	
Competitor A— Not Offered	S	O	+	−	O
Competitor B— Offered. Also provides loaner car to customer	E	U	−	+	R

FIGURE 3.11 Salability analysis for automobile maintenance service. (*Juran Institute, Inc. Copyright 1994. Used by permission.*)

Product: Store Front Prenatal Clinic

Customer Need (listed in priority order)	Product Feature & Goals						Cost of Meeting Need
	Walk in appointments handled by Nurse, 5 days a week	Board Certified Obstetrician, 2 days a week	Social Worker, 5 days a week	Nutritional Counselor, 5 days a week	On-site Billing Clerk takes Medicaid insurance from all eligible patients	On-site laboratory most results under 1 hour	
Convenient to use	60,000	30,000	10,000	10,000	20,000	40,000	170,000
Confidence in staff		70,000	10,000	15,000			95,000
Reasonable cost						25,000	25,000
Sensitivity			5,000	5,000			20,000
Informed choices			15,000	15,000			20,000
Cost for Feature	60,000	100,000	40,000	45,000	20,000	65,000	330,000

FIGURE 3.12   Value analysis for prenatal clinic. (*Juran Institute, Inc. Copyright 1994. Used by permission.*)

Set and Publish Final Product Design.   After the design has been optimized and tested, it is time to select the product features and goals to be included in the final design. This is also the stage where the results of product development are officially transmitted to other functions through various forms of documentation. These include the specifications for the product features and product feature goals, as well as the spreadsheets and other supporting documents. All this is supplemented by instructions, both oral and written. To complete this activity, the team must first determine the process for authorizing and publishing product features and product feature goals. Along with the features and goals, the team should include any procedures, specifications, flow diagrams, and other spreadsheets that relate to the final product design. The team should pass along results of experiments, field testing, prototypes, and so on, that are appropriate. If an organization has an existing process for authorizing product goals, it should be reexamined in light of recent experience. Ask these questions, Does the authorization process guarantee input from key customers—both internal and external? Does it provide for optimization of the design? If an organization has no existing goal authorization process, now is a good time to initiate one.

STEP 5: DEVELOP PROCESS

Once the product is developed, it is necessary to determine the means by which the product will be created and delivered on a continuing basis. These means are, collectively, the "process." "Process development" is the set of activities for defining the specific means to be used by operating personnel for meeting product quality goals. Some related concepts include

Subprocesses: Large processes may be decomposed into these smaller units for both the development and operation of the process.

Activities: The steps in a process or subprocess.

Tasks: The detailed step-by-step description for execution of an activity.

In order for a process to be effective, it must be goal oriented, with specific measurable outcomes; systematic, with the sequence of activities and tasks fully and clearly defined and all inputs and outputs fully specified; and capable, i.e., able to meet product quality goals under operating conditions and legitimate, with clear authority and accountability for its operation.

The eleven major activities involved in developing a process are

- Review product goals.
- Identify operating conditions.
- Collect known information on alternate processes.
- Select general process design.
- Identify process features and goals.
- Identify detailed process features and goals.
- Design for critical factors and human error.
- Optimize process features and goals.
- Establish process capability.
- Set and publish final process features and goals.
- Set and publish final process design.

Review Product Goals. Ideally, this review will be relatively simple. Product quality goals should have been validated with the prior participation of those who would be affected. In many companies, however, product and process design often are executed by different teams. There is no real joint participation on either group's part to contribute to the results that both the teams are expected to produce. This lack of participation usually reduces the number of alternative designs that could have been readily adopted in earlier stages but become more difficult and more expensive to incorporate later. In addition, those who set the product goals have a vested interest in their own decisions and exhibit cultural resistance to proposals by the process design team to make changes to the product design. If the product and process design efforts are being performed by different groups, then review and confirmation of the product quality goals are absolutely critical.

Review of product quality goals ensures that they are understood by those most affected by the process design. The review helps achieve the optimum. Process designers are able to present product designers with some realities relative to the costs of meeting the quality goals. The review process should provide a legitimate, unobstructed path for challenging costly goals.

Identify Operating Conditions. Seeking to understand operating conditions requires investigation of a number of dimensions.

User's Understanding of the Process. By "users," we mean those who either contribute to the processes in order to meet product goals or those who employ the process to meet their own needs. Users consist, in part, of internal customers (organization units or persons) responsible for running the processes to meet the quality goals. Operators or other workers are users. Process planners need to know how these people will understand the work to be done. The process must be designed either to accommodate this level of understanding or to improve the level of understanding.

How the Process Will be Used. Designers always know the *intended* use of the process they develop. However, they may not necessarily know how the process is *actually* used (and misused) by the end user. Designers can draw on their own experiences but usually must supplement these with direct observation and interviews with those affected.

The Environments of Use.   Planners are well aware that their designs must take account of environments that can influence process performance. Planners of physical processes usually do take account of such environmental factors as temperature, vibration, noise level, and so on. Planners who depend heavily on human responses, particularly those in the service areas, should address the impact of the environment on human performance in their process designs. For example, a team designing the process for handling customer inquiries should consider how environmental stress can influence the performance of the customer service representatives. This stress can result from large numbers of customer complaints, abusive customers, lack of current product information, and so on.

Collect Known Information on Alternative Processes.   Once the goals and environment are clear, the planning team needs reliable information on alternative processes available for meeting those goals in the anticipated environment.

Process Anatomy.   At the highest level, there are some basic process anatomies that have specific characteristics that planners should be aware of. A "process anatomy" is a coherent structure that binds or holds the process together. This structure supports the creation of the goods or the delivery of the service. The selection of a particular anatomy also will have a profound influence on how the product is created and the ability of the organization to respond to customers' needs. Figure 3.13 illustrates these.

The Autonomous Department.   The "autonomous process" is defined as a group of related activities that are usually performed by one department or a single group of individuals. In this process form, the department or group of individuals receives inputs from suppliers, such as raw materials, parts, information, or other data, and converts them into finished goods and services, all within a single self-contained department.

An example of an autonomous process is the self-employed professional, e.g., a physician, consultant, or artisan. In financial services, it might be the loan-approval department. In manufacturing, a well-known example is a tool room. It starts with tool steel and engineering drawings and creates punches, dies, fixtures, and gauges to be used on the manufacturing floor. Even though we refer to this kind of process anatomy as "autonomous," outputs or deliverables from other processes are still required from outside sources that serve as inputs into this process. The self-employed physician, for example, may purchase equipment and materials from supply houses, pharmaceutical companies, and so on.

The Assembly Tree.   The "assembly tree" is a familiar process that incorporates the outputs of several subprocesses. Many of these are performed concurrently and are required for final assembly or to achieve an end result at or near the end of the process. This kind of process anatomy is widely used by the great mechanical and electronic industries that build automotive vehicles, household appliances, electronic apparatus, and so on. It is also used to define many processes in a hospital, such as in the case of performing surgery in the operating room. The branches or leaves of the tree represent numerous suppliers or in-house departments making parts and components. The elements are assembled by still other departments.

In the office, certain processes of data collection and summary also exhibit features of the assembly tree. Preparation of major accounting reports (e.g., balance sheet, profit statement) requires assembly of many bits of data into progressively broader summaries that finally converge into the consolidated reports. The assembly-tree design has been used at both the multifunctional and departmental levels. In large operations, it is virtually mandatory to use staff specialists who contribute different outputs at various multifunctional levels. An example of this is the budget process. While it is not mandatory to use staff specialists for large departmental processes, this is often the case. This can be illustrated by the design department, where various design engineers contribute drawings of a project that contribute to the overall design.

Process Anatomies

Flow of basic materials

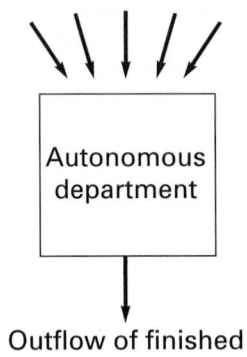

Autonomous
department

Outflow of finished
and tested product

Autonomous Department

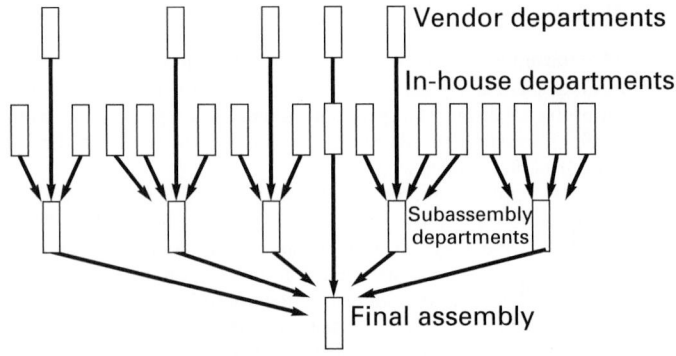

Vendor departments

In-house departments

Subassembly
departments

Final assembly

To test and usage

Assembly Tree

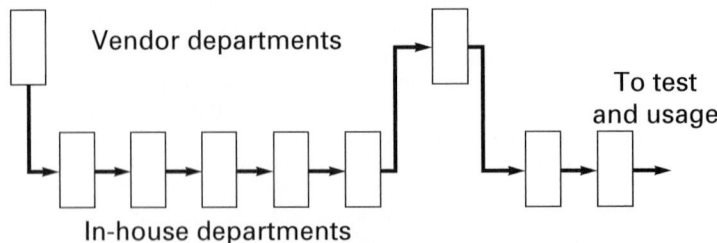

Vendor departments

To test
and usage

In-house departments

Procession

FIGURE 3.13   Process anatomies. (*Juran Institute, Inc. Copyright 1994. Used by permission.*)

The Procession. Another familiar form, the "procession process," uses a sequential approach as the basis for the process. This differs from the assembly tree, in which many of the activities are performed concurrently. The procession approach tends to take a more linear approach, whereby the lower-level processes are performed sequentially. It mandates that certain activities must be completed before others can begin because the outputs of each of the subprocesses serve as the inputs for each succeeding subprocess.

The intent of selecting a process anatomy is to determine the overall structure or architecture of the process that produces the product features and meets product feature goals. It does not necessarily follow that choosing one process anatomy over another locks the team into using that same architecture exclusively throughout the entire system. Quite the contrary, the team may select the assembly-tree process as the structure for the overall system but use a combination of autonomous and procession anatomies as the basis for subprocesses at the functional, departmental, or unit level.

Process Quality Management. Increasingly, many planners are applying a fourth, less traditional form of management known as "process quality management" to their major processes. This new, alternative management form has come about in response to an increased realization that many of today's business goals and objectives are becoming even more heavily dependent on large, complex, cross-functional business processes. Process quality management emphasizes that there are several critical processes that are crucial to an organization if it is to maintain and grow its business. (See Section 6 for a full discussion.)

Measuring the Process. In selecting a specific process design, the team will need to acquire information on the effectiveness and efficiency of alternative designs, including

- Deficiency rates
- Cycle time
- Unit cost
- Output rate

To acquire the needed data, the planners must typically use a number of different approaches, including

- Analyzing the existing process
- Analyzing similar or related processes
- Testing alternative processes
- Analyzing new technology
- Acquiring information from customers
- Simulating and estimating
- Benchmarking

Select General Process Design. Just as product design began with a high-level description expanded to the details, process design should begin by describing the overall process flow with a high-level process-flow diagram. From this diagram it will be possible to identify the subprocesses and major activities that can then be designed at a more detailed level. In developing the high-level flow, as well as the greater detail later, the team should ensure that it meets the following criteria:

- Will deliver the quality goals for the product
- Incorporates the countermeasures for criticality analysis, FMEA, and fault-tree analysis
- Meets the project goals
- Accounts for actual, not only intended, use
- Is efficient in consumption of resources
- Demands no investments that are greater than planned

While some process designs will largely repeat existing designs and some others will represent "green field" or "blank sheet" redesigns, most effective process redesigns are a combination of the tried and true existing processes with some significant quantum changes in some parts of the process. The preceding criteria should be the guides for whether a particular part of the process should be incorporated as it is, improved, or replaced with a fundamentally different approach.

This is the point in process design to think as creatively as possible, using some of the same techniques discussed under product development. Consider the impact of radically different anatomies. Would the customer be served better with dedicated, multispecialty units or with highly specialized expert functionality accessed as needed? What approach is mostly likely to reduce deficiencies? How can cycle time by cut dramatically? Is there a new technology that would allow us to do it differently? Can we develop such a technology?

Once the high-level flow is completed, each activity and decision within the flow diagram needs to be fully documented with a specification of the following for each:

- Inputs
- Outputs
- Goals for outputs
- Cycle time
- Cost
- General description of the conversion of inputs to outputs

Clear specification of these factors makes it possible to divide up the work of detailed design later and still be confident that the final design will be consistent and coordinated.

Once the initial new process flow is completed, it should be reviewed for opportunities to improve it, such as

- Eliminate sources of error that lead to rework loops.
- Eliminate or reduce redundant subprocesses, activities, or tasks.
- Decrease the number of handoffs.
- Reduce cycle time.
- Replace tasks, activities, or processes that have outputs with defects.
- Correct sequencing issues in the process to reduce the amount of activity or rework.

Carryover of Process Designs.   For each subprocess or major activity, one of the following questions must be answered in the affirmative:

If it is a carryover design, is the process capable of meeting the product quality goals?

If it is a new design, can we demonstrate that it is at least as effective at meeting product quality goals while also maintaining or improving cost and cycle time?

Testing Selected Processes.   One of the key factors for a successful design is incorporating the lessons learned from testing the product, the product features, and the overall process and subprocesses to ensure that they meet quality goals. Testing should be conducted throughout the entire quality planning process to allow for changes, modifications, and improvements to the plan before it is transferred to operations. Testing is performed at various points to analyze and evaluate alternate designs of the overall process and subprocesses.

There are a number of options for testing the efficiency and effectiveness of a process prior to full-scale implementation. They include the following:

Pilot test: A pilot test tests the overall process on a small scale or with a small segment of the total population. The segment to receive testing will vary depending on the process itself. Testing may be limited to a particular location, department, or function.

Modular test: Sometimes it is not possible to test the entire process at one time, but it may be possible to test crucial elements of the process separately. A modular test is a test of individual segments of the process. Generally, the outputs of certain subprocesses influence the ability of other processes to perform efficiently and effectively. These critical processes require their own tests to isolate problems that may occur and allow improvements to be made.

Simulation: This design technique observes and manipulates a mathematical or physical model that represents a real-world process for which, for technical or economic reasons, direct experimentation is not possible. Different circumstances can be applied to test how the process will perform under varying conditions, inputs, and worst-case scenarios.

Dry run: A dry run is a walk-through of the new process, with the planning team playing a dominant operating role in the process. This is a test of the process under operating conditions. The purpose is to test the process. Any resulting product is not sent to customers. Usually the team has worked so closely with designing the process that it can lose sight of how the various pieces actually fit together. The dry run gives the team one last opportunity to step back and see, from a conceptual standpoint, whether the process can work as designed before other tests are performed or before the process is transferred to operations.

Acceptance test: This is a highly structured form of testing common in complex systems, such as computer systems. A test plan is designed by a special team not directly involved in the design of the process being tested. The test plan sets up the proper environmental conditions, inputs, relevant interventions, and operating conditions. The test is intended to stress, in relevant ways, the important functional and other features in which the process could fail. In some cases, it is vital that the new process design be tested under operating conditions by the people who will actually operate it—assuming they are different from the planning team. The team may not have understood problem operating conditions; there may be unforeseen problems or resistance that cannot be overcome. Without such a test, these factors could contribute to a very costly mistake. Therefore, in such cases, acceptance testing under real conditions is essential.

Comparisons or benchmarks. Other units inside and outside the organization may already be using a process similar to the one designed. The process can be validated by comparing it with existing similar processes.

Test Limitations. All tests have some limitations. The following are common limitations that should be understood and addressed.

Differences in operating conditions: Dry runs and modular testing obviously differ from operating conditions. Even pilot tests and benchmarks will differ in some details from the actual, full implementation. Some common differences between conditions for testing and conditions for full-scale use include

- People operating the process
- Customers of the process
- Extreme values and unusual conditions
- Interactions with other processes and other parts of the organization.

Differences in size: Especially with critical failures, such as breakdown of equipment, loss of key personnel, or any other potential failure, as in the case of complications in a surgical procedure, a test might not be large enough to allow these rare failures to occur with any high degree of certainty.

Cultural resistance: Cultural reactions to tests differ from reactions to permanent changes. Such reactions might be either more or less favorable than full-scale implementation. Tests may go well because they lack the cultural impact of full implementation. They may go poorly because participants will not give the test the same careful attention they would give the "real" work.

Other effects. Sometimes designing a new process or redesigning an existing process may create or exacerbate problems in other processes. For example, improved turnaround time in approving home loans may create a backlog for the closing department. Such interactions among processes might not occur in an isolated test.

Identify Process Features and Goals. A "process feature" is any property, attribute, and so on that is needed to create the goods or deliver the service and achieve the product feature goals that will satisfy a customer need. A "process goal" is the numeric target for one of the features.

Whereas product features answer the question, "What characteristics of the product do we need to meet customers needs?" process features answer the question, "What mechanisms do we need to create or deliver those characteristics (and meet quality goals) over and over again without deficiencies?" Collectively, process features define a process. The flow diagram is the source of many, but not all, of these features and goals.

As the process design progresses from the macro level down into details, a long list of specific process features emerges. Each of these is aimed directly at producing one or more product features. For example:

- Creating an invoice requires a process feature that can perform arithmetic calculations so that accurate information can be added.
- Manufacturing a gear wheel requires a process feature that can bore precise holes into the center of the gear blank.
- Selling a credit card through telemarketing requires a process feature that accurately collects customer information

Most process features fall into one of the following categories:

- *Procedures*—a series of steps followed in a regular, definite order
- *Methods*—an orderly arrangement of a series of tasks, activities, or procedures
- *Equipment and supplies*—"physical" devices and other hard goods that will be needed to perform the process
- *Materials*—tangible elements, data, facts, figures, or information (these, along with equipment and supplies, also may make up inputs required as well as what is to be done to them)
- *People*—numbers of individuals, skills they will require, goals, and tasks they will perform
- *Training*—skills and knowledge required to complete the process
- *Other resources*—additional resources that may be needed
- *Support processes*—can include secretarial support, occasionally other support, such as outsources of printing services, copying services, temporary help, and so on.

Just as in the case of product design, process design is easier to manage and optimize if the process features and goals are organized into a spreadsheet indicating how the process delivers the product features and goals. Figure 3.14 illustrates such a spreadsheet.

The spreadsheet serves not only as a convenient summary of the key attributes of the process, it also facilitates answering two key questions that are necessary for effective and efficient process design. First, will every product feature and goal be attained by the process? Second, is each process feature absolutely necessary for at least one product feature; i.e., are there any unnecessary or redundant process features? Also, verify that one of the other process features cannot be used to create the same effect on the product.

Often high-level process designs will identify features and goals that are required from companywide macro processes. Examples might include cycle times from the purchasing process, specific data from financial systems, and new skills training. Because the new process will depend on these macro processes for support, now is the time to verify that they are capable of meeting the goals. If they are not, the macro processes will need to be improved as part of the process design, or they will need to be replaced with an alternative delivery method.

Identify Detailed Process Features and Goals. In most cases, it will be most efficient and effective for individual subteams to carry out the detailed designs of subprocesses and major

Product Feature	Product Feature Goal	Process Features			
		Spray delivery capacity	Crew Size	Certified materials	Scheduling forecast on P.C. to determine to/from and work needed
Time to perform job	Less than one hour 100 percent of time	○	●		●
Guaranteed appointment time	99 percent of jobs within 15 minutes of appointment				●
All materials environmentally safe	All naturally occuring/no synthetics			●	
Legend ● Very Strong ○ Strong △ Weak		10 gallons per minute	One person per 10,000 sq. ft. of yd.	100% approved by State Dept. of Agriculture	Forecast time always within 10 percent of actual
		Process Feature Goals			

FIGURE 3.14 Process design spreadsheet for a lawn care service. (*Juran Institute, Inc. Copyright 1994. Used by permission.*)

activities. These detailed designs will have the process features and goals as their objectives and criteria. Each subprocess team will develop the design to the level at which standard operating procedures can be developed, software coded, equipment produced or purchased, and materials acquired.

Design for Critical Factors and Human Error. One key element of process design is determining the effect that critical factors will have on the design. "Critical factors" are those aspects which present serious danger to human life, health, and the environment or risk the loss of very large sums of money. Some examples of such factors involve massive scales of operations: airport traffic control systems, huge construction projects, systems of patient care in hospital, and even the process for managing the stock market. Planning for such factors should obviously include ample margins of safety as to structural integrity, fail-safe provisions, redundancy systems, multiple alarms, and so on. Criticality analysis and failure-mode and effect analysis (see Section 19) are helpful tools in identifying those factors which require special attention at this point.

Workers vary in their capabilities to perform specific tasks and activities. Some workers perform well, whereas others do not perform nearly as well. What is consistent about all workers is that they are a part of the human family, and human beings are fallible. Collectively, the extent of human errors is large enough to require that the process design provides for means to reduce and control human error. Begin by analyzing the data on human errors, and then apply the Pareto principle. The vital few error types individually become candidates for special process design. The human errors that can be addressed by process design fall into these major classes:

- Technique errors arising from individuals lacking specific, needed skills
- Errors aggravated by lack of feedback
- Errors arising from the fact that humans cannot remain indefinitely in a state of complete, ready attention

Technique Errors.   Some workers consistently outperform others on specific quality tasks. The likely reason is possession of a special "knack." In such cases, designers should study the methods used by the respective workers to discover the methodologic differences. These differences usually include the knack—a small difference in method that produces a big difference in performance. Once the knack is discovered, the process designers can arrange to include the knack in the technology. Alternatively, the knack can be brought into the workers' training program so that all workers are brought up to the level of the best.

Lack of Instant Feedback.   A useful principle in designing human tasks is to provide instant feedback to the worker so that the performance of the work conveys a message about the work to the worker. For example, a worker at a control panel pushes a switch and receives three feedbacks: the feel of the shape of the switch handle, the sound of an audible click signaling that the switch went all the way, and the sight of a visual illumination of a specific color and shape. Providing such feedback is part of self-control and allows the worker to modify his or her performance to keep the process within its quality goals.

Human Inattention Errors.   A technique for designing human work is to require human attention as a prerequisite for completing the work; i.e., the task cannot be performed unless the person doing it devotes attention to it and to nothing else. A widespread case in point is inspection of documents, products, or whatever. Human checking can be done in two very different ways.

> *By passive deeds:*   Listening, looking, reading. Such deeds are notoriously subject to lapses in human attention. Also, such deeds leave no trail behind them. We have no way of knowing whether the human being in question is really paying attention or is in a state of inattention. For example, a person providing visual inspection of a product moving along an assembly line or someone proofreading a report may become fatigued. They can easily experience a momentary lapse in their attention, causing them to miss spotting a defect or to fail to notice that a column of numbers does not add up correctly.

> *By active deeds:*   Operating a keyboard, writing, spelling. Such deeds cannot be performed at all without paying attention to the task at hand and to the exclusion of all else. These active deeds do leave a trail behind them. They are therefore far less error-prone than passive checking. An example would be someone having to attach the leads of a voltage meter to a circuit board to check its resistance or a blood bank technician retesting each sample to verify blood type.

Inadvertent human errors and other types of errors can also be reduced by "errorproofing"— building processes so that the error either cannot happen or is unlikely to happen.

Principles of Errorproofing.   Research has indicated that there are a number of different classifications of errorproofing methods, and these are spelled out below.

> *Elimination:*   This consists of changing the technology to eliminate operations that are error-prone. For example, in some materials handling operations, the worker should insert a protective pad between the lifting wire and the product so that the wire will not damage the product. Elimination could consist of using nylon bands to do the lifting.

> *Replacement:*   This method retains the error-prone operation but replaces the human worker with a nonhuman operator. For example, a human worker may install the wrong component into an assembly. A properly designed robot avoids such errors. Nonhuman processes, so long as they are properly maintained, do not have lapses in attention, do not become weary, do not lose their memory, and so on.

> *Facilitation:*   Under this method, the error-prone operation is retained, and so is the human worker. However, the human worker is provided with a means to reduce any tendency toward errors. Color coding of parts is an example.

> *Detection:*   This method does nothing to prevent the human error from happening. Instead, it aims to find the error at the earliest opportunity so as to minimize the damage done. A widespread example is automated testing between steps in a process.

Mitigation: Here again, the method does nothing to prevent the human error from happening. However, means are provided to avoid serious damage done. A common example is providing a fuse to avoid damage to electrical equipment.

Optimize Process Features and Goals. After the planners have designed for critical factors and made modifications to the plan for ways of reducing human error, the next activity is to optimize first the subprocesses and then the overall process design. In step 4, develop product, the concept of optimization was introduced. The same activities performed for optimizing product features and product feature goals also apply to process planning. Optimization applies to both the design of the overall process and the design of individual subprocesses.

Establish Process Capability. Before a process begins operation, it must be demonstrated to be capable of meeting its quality goals. The concepts and methods for establishing process capability are discussed in detail in Section 22, under Process Capability. Any planning project must measure the capability of its process with respect to the key quality goals. Failure to achieve process capability should be followed by systematic diagnosis of the root causes of the failure and improvement of the process to eliminate those root causes before the process becomes operational.

Reduction in Cycle Time. Process capability relates to the effectiveness of the process in meeting customer needs. One special class of needs may relate to subprocess cycle time—the total time elapsed from the beginning of a process to the end. Reducing cycle time has almost become an obsession for many organizations. Pressures from customers, increasing costs, and competitive forces are driving companies to discover faster ways of performing their processes. Often these targeted processes include launching new products, providing service to customers, recruiting new employees, responding to customer complaints, and so on. For existing processes, designers follow the well-known quality-improvement process to reduce cycle time. Diagnosis identifies causes for excessive time consumption. Specific remedies are then developed to alleviate these causes. (See Section 5, The Quality Improvement Process.)

Set and Publish Final Process Features and Goals. After the planning team has established the flow of the process, identified initial process features and goals, designed for critical processes and human error, optimized process features and goals, and established process capabilities, it is ready to define all the detailed process features and goals to be included in the final design. This is also the stage where the results of process development are officially transmitted to other functions through various forms of documentation. These include the specifications for the product features and product feature goals as well as the spreadsheets and other supporting documents. All this is supplemented by instructions, both oral and written.

Filling out the process design spreadsheet is an ongoing process throughout process development. The spreadsheet should have been continually updated to reflect design revisions from such activities as reviewing alternative options, designing for critical factors and human error, optimizing, testing process capability, and so on. After making the last revision to the process design spreadsheet, it should be checked once more to verify the following:

• That each product feature has one or more process features with strong or very strong relation. This will ensure the effective delivery of the product feature without significant defects. Each product feature goal will be met if each process goal is met.

• That each process feature is important to the delivery of one or more product features. Process features with no strong relationship to other product features are unnecessary and should be discarded.

The completed process design spreadsheet and detailed flow diagrams are the common information needed by managers, supervisors, and workers throughout the process. In addition, the planning team must ensure that the following are also specified for each task within the process:

• Who is responsible for doing it

- How the task is to be competed
- Its inputs
- Its outputs
- Problems that can arise during operations and how to deal with them
- Specification of equipment and materials to be used
- Information required by the task
- Information generated by the task
- Training, standard operating procedures, job aids that are needed

STEP 6: DEVELOP PROCESS CONTROLS/ TRANSFER TO OPERATIONS

In this step, planners develop controls for the processes, arrange to transfer the entire product plan to operational forces, and validate the implementation of the transfer. There are seven major activities in this step.

- Identify controls needed.
- Design feedback loop.
- Optimize self-control and self-inspection.
- Establish audit.
- Demonstrate process capability and controllability.
- Plan for transfer to operations.
- Implement plan and validate transfer.

Once planning is complete, these plans are placed in the hands of the operating departments. It then becomes the responsibility of the operational personnel to manufacture the goods or deliver the service and to ensure that quality goals are met precisely and accurately. They do this through a planned system of quality control. Control is largely directed toward continuously meeting goals and preventing adverse changes from affecting the quality of the product. Another way of saying this is that no matter what takes place during production (change or loss of personnel, equipment or electrical failure, changes in suppliers, etc.), workers will be able to adjust or adapt the process to these changes or variations to ensure that quality goals can be achieved.

Identify Controls Needed.   Process control consists of three basic activities:

- Evaluate the actual performance of the process.
- Compare actual performance with the goals.
- Take action on the difference.

Detailed discussions of these activities in the context of the feedback loop are contained in Section 4, The Quality Control Process.

Control begins with choosing quality goals. Each quality goal becomes the target at which the team directs its efforts. All control is centered around specific things to be controlled. We will call these things "control subjects." Each control subject is the focal point of a feedback loop. Control subjects are a mixture of

Product features: Some control is carried out by evaluating features of the product itself (e.g., the invoice, the gear wheel, the research report, etc.) Product controls are associated with the decision: Does this product conform to specifications or goals? Inspection is the major activity for answering this question. This inspection is usually performed at points where the inspection results make it possible to determine where breakdowns may have occurred in the production process.

Process features: Much control consists of evaluating those process features which most directly affect the product features, e.g., the state of the toner cartridge in the printer, the temperature of the furnace for smelting iron, or the validity of the formulas used in the researcher's report. Some features become candidates for control subjects as a means of avoiding or reducing failures. These control subjects typically are chosen from previously identified critical factors or from conducting FMEA, FTA, and criticality analysis. Process controls are associated with the decision: Should the process run or stop?

Side-effect features: These features do not affect the product, but they may create troublesome side effects, such as irritations to employees, offense to the neighborhood, threats to the environment, and so on.

These three types of control subjects may be found at several different stages of the process:

- Setup/startup
- During operations, including

  Running control
  Product control
  Supporting operations control
  Facility and equipment control

Design Feedback Loop. Once the control subjects are selected, it is time to design the remainder of the feedback loop by

- Setting the standards for control—i.e., the levels at which the process is out of control and the tools, such as control charts, that will be used to make the determination
- Deciding what action is needed when those standards are not met, e.g., troubleshooting.
- Designating who will take those actions

A detailed process flow diagram should be used to identify and document the points at which control measurements and actions will be taken. Then each control point should be documented on a control spreadsheet similar to Figure 3.15.

Optimize Self-Control and Self-Inspection. As discussed in more detail in Section 22, Operations, self-control takes place when workers know what they are supposed to do. Goals and targets are clearly spelled out and visible.

- Workers know what they are doing. Their output is measured, and they receive immediate feedback on their performance.
- Workers have the ability and the means to regulate the outcomes of the process. They need a capable process along with the tools, training, and authority to regulate it.

In addition to providing the optimal conditions for process operation and control, establishing self-control has a significant, positive impact on the working environment and the individuals in it. Whenever possible, the design of the quality control system should stress self-control by the operating forces. Such a design provides the shortest feedback loop but also requires the designers to ensure that the process capability is adequate to meet the product quality goals.

Once self-control is established, self-inspection should be developed. Self-inspection permits the worker to check that the product adheres to quality standards before it is passed on to the next

				PROCESS CONTROLS			
PROCESS FEATURE	CONTROL SUBJECT	SENSOR	GOAL	MEASURE-MENT FREQUENCY	SAMPLE SIZE	CRITERION	RESPONS-IBILITY
PROCESS FEATURE 1							
PROCESS FEATURE 2							
⋮							
WAVE SOLDER	SOLDER TEMPER-ATURE	THERMO-COUPLE	505°F	CONTIN-UOUS	N/A	⩾510°F, DECREASE HEAT; 500°F, INCREASE HEAT	OPERATOR
	CONVEYOR SPEED	FT/MIN METER	4.5 FT/MIN	1/HOUR	N/A	⩾5 FT/MIN, REDUCE SPEED; ⩽4 FT/MIN, INCREASE SPEED	OPERATOR
	ALLOY PURITY	LAB. CHEM. ANALYSIS	1.5% MAX TOTAL CONTAMIN-ANTS	1/MONTH	15 GRAMS	⩾1.5%, DRAIN BATH, REPLACE SOLDER	PROCESS ENGINEER

FIGURE 3.15   Control spreadsheet. [*From Juran, J. M. (1988), Quality Control Handbook, 4th ed. McGraw-Hill, New York, 6.9.*]

station in the production cycle. Production and front-line workers are made to feel more responsible for the quality of their work. Feedback on performance is immediate, thereby facilitating process adjustments. Traditional inspection also has the psychological disadvantage of using an "outsider" to report the defects to the worker. The costs of a separate inspection department can be reduced.

However, some prerequisite criteria must first be established:

Quality is number one: Quality must undoubtedly be made the highest priority. If this is not clear, the workers succumb to schedule and cost pressures and classify as acceptable products that should be rejected.

Mutual confidence: Managers must trust the workers enough to be willing to delegate the responsibility and the authority to carry out the work. Workers must also have enough confidence in managers to be willing to accept this responsibility and authority.

Training: Workers should be trained to make the product conformance decisions and should also be tested to ensure that they make good decisions.

Specifications must be unequivocally clear.

The quality audit and audit of control systems are treated elsewhere in detail—see, for example, Section 22, under Audit of Operations Quality. While the audit of a control system is a function independent of the planning team, the planning team does have the responsibility for ensuring that adequate documentation is available to make an effective audit possible and that there are provisions of resources and time for conducting the audit on an ongoing basis.

Demonstrate Process Capability and Controllability.   While process capability must be addressed during the design of the process, it is during implementation that initial findings of process capability and controllability must be verified.

Plan for Transfer to Operations.   In many organizations, receipt of the process by operations is structured and formalized. An information package is prepared consisting of certain standardized essentials: goals to be met, facilities to be used, procedures to be followed, instructions, cautions, and so on. There are also supplements unique to the project. In addition, provision is made for briefing and training the operating forces in such areas as maintenance, dealing with crisis, and so on. The package is accompanied by a formal document of transfer of responsibility. In some organizations, this transfer takes place in a near-ceremonial atmosphere.

The structured approach has value. It tends to evolve checklists and countdowns that help ensure that the transfer is orderly and complete. If the organization already has a structure for transfer, project information may be adapted to conform with established practice. If the company has a loose structure or none at all, the following material will aid in planning the transfer of the project.

Regardless of whether the organization has a structure or not, the team should not let go of the responsibility of the project until it has been validated that the transfer has taken place and everyone affected has all the information, processes, and procedures needed to produce the final product.

Transfer of Know-How.   During process design, the planners acquire a great deal of know-how about the process. The operating personnel could benefit from this know-how if it were transferred. There are various ways of making this transfer, and most effective transfers make use of several complementary channels of communication, including

Process specifications

Briefings

On-the-job training

Formal training courses

Prior participation

Audit Plan for the Transfer.   As part of the plan for formal transfer, a separate audit plan should also be developed as a vehicle for validating the transfer of the plan. This kind of audit is different from the control audits described previously. The purpose of this audit is to evaluate how successful the transfer was. For the audit to have real meaning, specific goals should be established during the planning phase of the transfer. Generally, these goals relate to the quality goals established during the development of the product, product features, and process features. The team may decide to add other goals inherent to the transfer or to modify newly planned quality goals during the first series of operations. For example, during the first trial runs for producing the product, total cycle time may exceed expected goals by 15 percent. This modification takes into account that workers may need time to adjust to the plan. As they become more skilled, gain experience with the process, and get more comfortable with their new set of responsibilities, cycle time will move closer to targeted quality goals.

The audit plan for the transfer should include the following:

- Goals to meet
- How meeting the goals will be measured
- The time phasing for goals, measurement, and analysis
- Who will audit
- What reports will be generated
- Who will have responsibility for corrective action for failure to meet specific goals

Implement Plan and Validate Transfer.   The final activity of the quality planning process is to implement the plan and validate that the transfer has occurred. A great deal of time and effort has gone into creating the product plan, and validating that it all works is well worth the effort.

REFERENCES

Designs for World Class Quality (1995). Juran Institute, Wilton, CT.

Juran, Joseph M. (1992). *Quality by Design.* Free Press, New York.

Parasuraman, A., Zeithami, Valarie A., and Berry, Leonard L. (1985). "A Conceptual Model for Service Quality and Its Implications for Further Research." *Journal of Marketing,* Fall, pp. 41–50.

Veraldi, L. C. (1985). "The Team Taurus Story." MIT Conference paper, Chicago, Aug. 22. Center for Advanced Engineering Study, MIT, Cambridge, MA.

SECTION 4
THE QUALITY CONTROL PROCESS

J. M. Juran
A. Blanton Godfrey

INTRODUCTION

Quality Control Defined.   This section describes the quality control process. "Quality control" is a universal managerial process for conducting operations so as to provide stability—to prevent adverse change and to "maintain the status quo."

To maintain stability, the quality control process evaluates actual performance, compares actual performance to goals, and takes action on the difference.

Quality control is one of the three basic managerial processes through which quality can be managed. The others are quality planning and quality improvement, which are discussed in Sections 3 and 5, respectively. The Juran trilogy diagram (Figure 4.1) shows the interrelation of these processes.

Figure 4.1 is used in several other sections in this handbook to describe the relationships between quality planning, quality improvement, and quality control and the fundamental managerial processes in total quality management. What is important for this section is to concentrate on the two "zones of control." In Figure 4.1 we can easily see that although the process is in control in the middle of the chart, we are running the process at an unacceptable level of waste. What is necessary here is not more control but improvement—actions to change the level of performance.

After the improvements have been made, a new level of performance has been achieved. Now it is important to establish new controls at this level to prevent the performance level from deteriorating to the previous level or even worse. This is indicated by the second zone of control.

The term "control of quality" emerged early in the twentieth century (Radford 1917, 1922). The concept was to broaden the approach to achieving quality, from the then-prevailing after-the-fact inspection, to what we now call "defect prevention." For a few decades, the word "control" had a broad meaning which included the concept of quality planning. Then came events which narrowed the meaning of "quality control." The "statistical quality control" movement gave the impression that quality control consisted of using statistical methods. The "reliability" movement claimed that quality control applied only to quality at the time of test but not during service life.

In the United States, the term "quality control" now often has the narrow meaning defined previously. The term "total quality management" (TQM) is now used as the all-embracing term. In

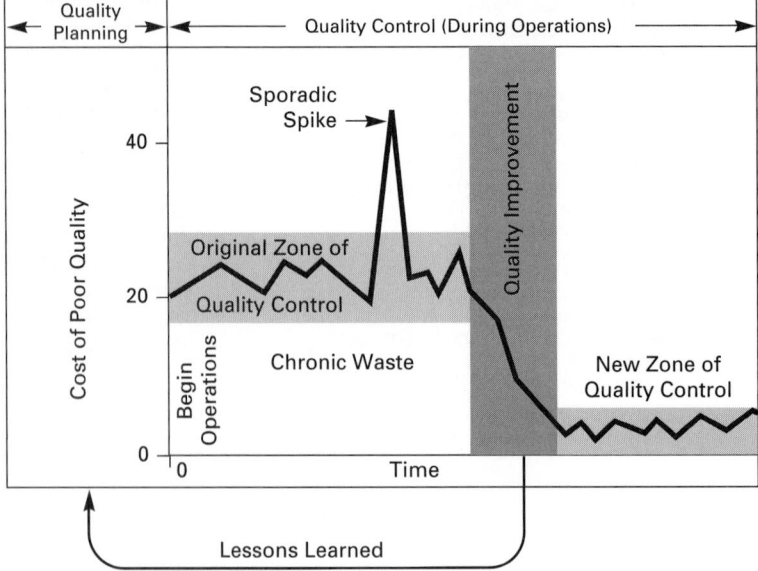

FIGURE 4.1   The Juran trilogy diagram. (*Juran Institute, Inc., Wilton, CT.*)

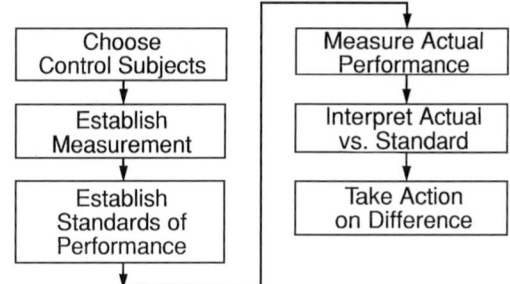

FIGURE 4.2   The input-output diagram for the quality control process.

Europe, the term "quality control" is also acquiring a narrower meaning. Recently, the European umbrella quality organization changed its name from European Organization for Quality Control to European Organization for Quality. In Japan, the term "quality control" retains a broad meaning. Their "total quality control" is roughly equivalent to our term "total quality management." In 1997 the Union of Japanese Scientists and Engineers (JUSE) adopted the term total quality management (TQM) to replace total quality control (TQC) to more closely align themselves with the more common terminology used in the rest of the world.

The quality control process is one of the steps in the overall quality planning sequence described in Section 3, The Quality Planning Process, and briefly again in Section 14, Total Quality Management. Figure 4.2 shows the input-output features of this step.

In Figure 4.2 the input is operating process features developed to produce the product features required to meet customer needs. The output consists of a system of product and process controls which can provide stability to the operating process.

The Relation to Quality Assurance.   Quality control and quality assurance have much in common. Each evaluates performance. Each compares performance to goals. Each acts on the difference. However they also differ from each other. Quality control has as its primary purpose to maintain control. Performance is evaluated during operations, and performance is compared to goals during operations. The resulting information is received and used by the operating forces.

Quality assurance's main purpose is to verify that control is being maintained. Performance is evaluated after operations, and the resulting information is provided to both the operating forces and others who have a need to know. Others may include plant, functional, or senior management; corporate staffs; regulatory bodies; customers; and the general public.

The Feedback Loop.   Quality control takes place by use of the feedback loop. A generic form of the feedback loop is shown in Figure 4.3.

The progression of steps in Figure 4.3 is as follows:

1. A *sensor* is "plugged in" to evaluate the actual quality of the *control subject*—the product or process feature in question. The performance of a process may be determined directly by evaluation of the process feature, or indirectly by evaluation of the product feature—the product "tells" on the process.
2. The sensor reports the performance to an *umpire.*
3. The umpire also receives information on what is the quality *goal* or standard.
4. The umpire compares actual performance to standard. If the difference is too great, the umpire energizes an *actuator.*
5. The actuator stimulates the *process* (whether human or technological) to change the performance so as to bring quality into line with the quality goal.

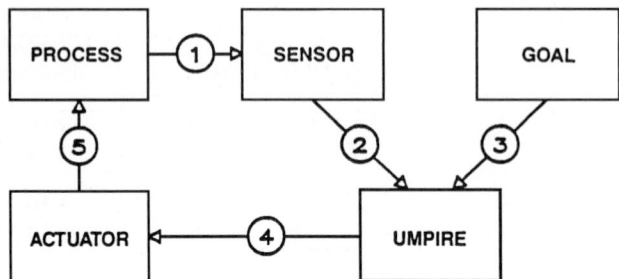

FIGURE 4.3   The generic feedback loop. (*Making Quality Happen, Juran Institute, Inc., senior executive workshop, p. F-3, Wilton, CT.*)

6. The process responds by restoring conformance.

Note that in Figure 4.3 the elements of the feedback loop are functions. These functions are universal for all applications, but responsibility for carrying out these functions can vary widely. Much control is carried out through automated feedback loops. No human beings are involved. Common examples are the thermostat used to control temperature and the cruise control used in automobiles to control speed.

Another frequent form of control is self-control carried out by a human being. An example of such self-control is the village artisan who performs every one of the steps of the feedback loop. The artisan chooses the control subjects, sets the quality goals, senses what is the actual quality performance, judges conformance, and becomes the actuator in the event of nonconformance. For a case example involving numerous artisans producing Steinway pianos, see Lenehan (1982). Self-directing work teams also perform self-control as is meant here. See Section 15 for a further discussion of this concept.

This concept of self-control is illustrated in Figure 4.4. The essential elements here are the need for the worker or work-force team to know what they are expected to do, to know how they are actually doing, and to have the means to adjust their performance. This implies they have a capable process and have the tools, skills, and knowledge necessary to make the adjustments and the authority to do so.

A further common form of feedback loop involves office clerks or factory workers whose work is reviewed by umpires in the form of inspectors. This design of a feedback loop is largely the result of the Taylor system of separating planning from execution. The Taylor system emerged a

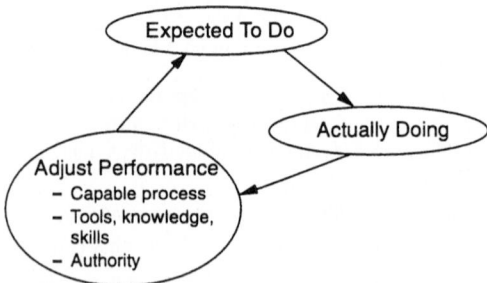

FIGURE 4.4   Self-control. (*"Quality Control," Leadership for the Quality Century, Juran Institute, Inc., senior executive workshop, p. 5, Wilton, CT.*)

century ago and contributed greatly to increasing productivity. However, the effect on quality control was negative.

THE ELEMENTS OF THE FEEDBACK LOOP

The feedback loop is a universal. It is fundamental to any problem in quality control. It applies to all types of operations, whether in service industries or manufacturing industries, whether for profit or not. It applies to all levels in the hierarchy, from the chief executive officer to the work force, inclusive. However, there is wide variation in the nature of the elements of the feedback loop.

In Figure 4.5 a simple flowchart is shown describing the quality control process with the simple universal feedback loop imbedded.

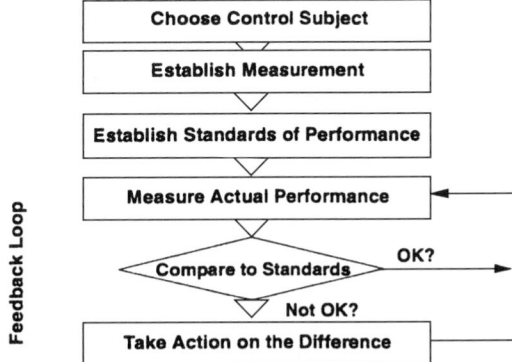

FIGURE 4.5 The quality control process. (*"Quality Control," Leadership for the Quality Century, Juran Institute, Inc., senior executive workshop, p. 2, Wilton, CT.*)

Choose the Control Subject. Each feature of the product (goods and services) or process becomes a *control subject*—a center around which the feedback loop is built. The critical first step is to choose the control subject. Control subjects are derived from multiple sources which include:

Stated customer needs for product features

Technological analysis to translate customer needs into product and process features

Process features which directly impact the product features

Industry and government standards

Needs to protect human safety and the environment

Needs to avoid side effects such as irritations to employees or offense to the neighboring community

At the worker level, control subjects consist mainly of product and process features set out in specifications and procedures manuals. At managerial levels the control subjects are broader and increasingly business-oriented. Emphasis shifts to customer needs and to competition in the marketplace. This shift in emphasis then demands added, broader control subjects which, in turn, have an influence on the remaining steps of the feedback loop.

Establish Measurement.   After choosing the control subject, the next step is to establish the means of measuring the actual performance of the process or the quality level of the goods or services. Measurement is one of the most difficult tasks in quality management and is discussed in almost every section of this handbook, especially in the industry sections. In establishing the measurement we need to clearly specify the means of measurement (the sensor), the frequency of measurement, the way the data will be recorded, the format for reporting the data, the analysis to be made on the data to convert the data to usable information, and who will make the measurement. See Section 9, Measurement, Information, and Decision-Making, for a thorough discussion of this subject.

Establish Standards of Performance: Product Goals and Process Goals.   For each control subject it is necessary to establish a standard of performance—a quality goal (also called targets, objectives, etc.). A standard of performance is an aimed-at achievement toward which effort is expended. Table 4.1 gives some examples of control subjects and the associated goals.

The prime goal for *products* is to meet customer needs. Industrial customers often specify their needs with some degree of precision. Such specified needs then become quality goals for the producing company. In contrast, consumers tend to state their needs in vague terms. Such statements must then be translated into the language of the producer in order to become product goals.

Other goals for products which are also important are those for reliability and durability. Whether the products meet these goals can have a critical impact on customer satisfaction and loyalty and on overall costs. The failures of products under warranty can seriously impact the profitability of a company through both direct costs and indirect costs (loss of repeat sales, word of mouth, etc.).

The *processes* which produce products have two sets of quality goals:

1. To produce products which do meet customer needs. Ideally, each and every unit of product should meet customer needs.

2. To operate in a stable and predictable manner. In the dialect of the quality specialist, each process should be "under control." We will later elaborate on this, under the heading Process Conformance. These goals may be directly related to the costs of producing the goods or services.

Quality goals may also be established for departments or persons. Performance against such goals then becomes an input to the company's reward system. Ideally such goals should be:

Legitimate: They should have undoubted official status.

Measurable: So that they can be communicated with precision.

Attainable: As evidenced by the fact that they have already been attained by others.

Equitable: Attainability should be reasonably alike for individuals with comparable responsibilities.

TABLE 4.1   Examples of Control Subjects and Associated Quality Goals

Control subject	Goal
Vehicle mileage	Minimum of 25 mi/gal highway driving
Overnight delivery	99.5% delivered prior to 10:30 a.m. next morning
Reliability	Fewer than three failures in 25 years of service
Temperature	Minimum 505°F; maximum 515°F
Purchase-order error rate	No more than 3 errors/1000 purchase orders
Competitive performance	Equal or better than top three competitors on six factors
Customer satisfaction	90% or better rate, service outstanding or excellent
Customer retention	95% retention of key customers from year to year
Customer loyalty	100% of market share of over 80% of customers

Quality goals may be set from a combination of the following bases:

Goals for product features and process features are largely based on *technological* analysis.

Goals for departments and persons should be based on *benchmarking* rather than historical performance. For elaboration, see Section 12, Benchmarking.

Quality goals at the highest levels are in the early stages of development. The emerging practice is to establish goals on matters such as meeting customers' changing needs, meeting competition, maintaining a high rate of quality improvement, improving the effectiveness of business processes, and revising the planning process so as to avoid creating new failure-prone products and processes.

Measure Actual Performance.   The critical step in quality control is to measure the actual performance of the product or the process. To make this measurement we need a sensor, a device to make the actual measurement.

The Sensor.   A "sensor" is a specialized detecting device. It is designed to recognize the presence and intensity of certain phenomena, and to convert the resulting data into "information." This information then becomes the basis of decision making. At lower levels of organization the information is often on a real-time basis and is used for current control. At higher levels the information is summarized in various ways to provide broader measures, detect trends, and identify the vital few problems.

The wide variety of control subjects requires a wide variety of sensors. A major category is the numerous technological instruments used to measure product features and process features. Familiar examples are thermometers, clocks, yardsticks, and weight scales. Another major category of sensors is the data systems and associated reports which supply summarized information to the managerial hierarchy. Yet another category involves the use of human beings as sensors. Questionnaires and interviews are also forms of sensors.

Sensing for control is done on a huge scale. This has led to the use of computers to aid in the sensing and in conversion of the resulting data into information. For an example in an office environment (monitoring in telephone answering centers), see Bylinsky (1991). For an example in a factory environment (plastic molding), see Umscheid (1991).

Most sensors provide their evaluations in terms of a *unit of measure*—a defined amount of some quality feature—which permits evaluation of that feature in numbers. Familiar examples of units of measure are degrees of temperature, hours, inches, and tons. For a discussion of units of measure, see Section 9, Measurement, Information, and Decision-Making. A considerable amount of sensing is done by human beings. Such sensing is subject to numerous sources of error.

Compare to Standards.   The act of comparing to standards is often seen as the role of an umpire. The umpire may be a human being or a technological device. Either way, the umpire may be called on to carry out any or all of the following activities:

1. Compare the actual quality performance to the quality goal.
2. Interpret the observed difference; determine if there is conformance to the goal.
3. Decide on the action to be taken.
4. Stimulate corrective action.

These activities require elaboration and will shortly be examined more closely.

Take Action on the Difference.   In any well-functioning quality control system we need a means of taking action on the difference between desired standards of performance and actual performance. We need an actuator. This device (human or technological or both) is the means for stimulating action to restore conformance. At the worker level it may be a keyboard for giving orders to an office computer or a calibrated knob for adjusting a machine tool. At the management level it may be a memorandum to subordinates.

The Process.   In all of the preceding discussion we have assumed a process. This may also be human or technological or both. It is the means for producing the product features, each of which is a control subject. All work is done by a process which consists of an input, labor, technology, procedures, energy, materials, and output. For a more complete discussion of process, see Section 6, Process Management.

The PDCA Cycle.   There are many ways of dividing the feedback loop into elements and steps. Some of them employ more than six elements; others employ fewer than six. A popular example of the latter is the so-called PDCA cycle (also the Deming wheel) as shown in Figure 4.6. Deming (1986) referred to this as the Shewhart cycle, which is the name many still use when describing this version of the feedback loop.

Study the results.
What did we learn?
What can we predict?

What could be the most important accomplishments of this team? What changes might be desirable? What data are available? Are new observations needed? If yes, plan a change or test. Decide how to use the observations.

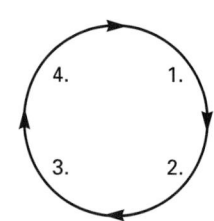

Observe the effects of the change or test.

Carry out the change or test decided upon, preferably on a small scale.

Step 5.   Repeat Step 1, with knowledge accumulated.
Step 6.   Repeat Step 2, and onward.

FIGURE 4.6   The PDCA cycle. (*Deming, 1986.*)

In this example the feedback loop is divided into four steps labeled Plan, Do, Check, and Act. These steps correspond roughly to the six steps discussed previously:

"Plan" includes choosing control subjects and setting goals.

"Do" includes running the process.

"Check" includes sensing and umpiring.

"Act" includes stimulating the actuator to take corrective action.

An early version of the PDCA cycle was included in W. Edwards Deming's first lectures in Japan (Deming 1950). Since then, additional versions have been devised and published. For elaboration, see Koura (1991).

Some of these versions have attempted to label the PDCA cycle in ways which make it serve as a universal series of steps for both quality control and quality improvement. The authors feel that this confuses matters, since two very different processes are involved. (The process for quality improvement is discussed in Section 5.)

THE PYRAMID OF CONTROL

Control subjects run to large numbers, but the number of "things" to be controlled is far larger. These things include the published catalogs and price lists sent out, multiplied by the number of items in each; the sales made, multiplied by the number of items in each sale; the units of product produced, multiplied by the associated numbers of quality features; and so on for the numbers of items associated with employee relations, supplier relations, cost control, inventory control, product and process development, etc.

A study in one small company employing about 350 people found that there were *over a billion things to be controlled* (Juran 1964, pp. 181–182).

There is no possibility for upper managers to control huge numbers of control subjects. Instead, they divide up the work of control, using a plan of delegation somewhat as depicted in Figure 4.7.

This division of work establishes three areas of responsibility for control: control by nonhuman means, control by the work force, and control by the managerial hierarchy.

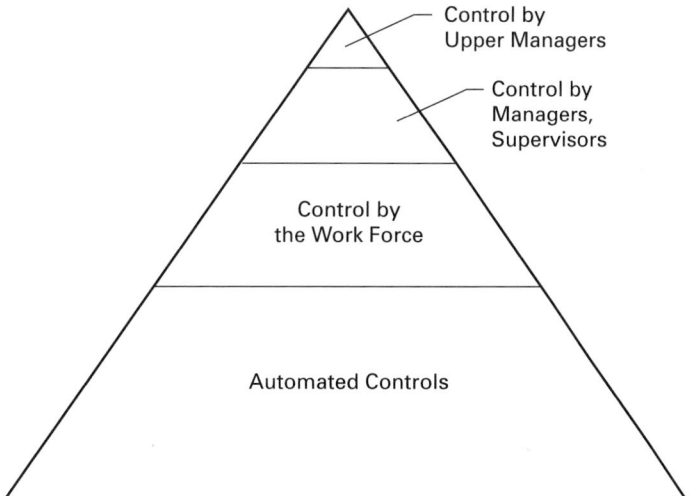

FIGURE 4.7   The pyramid of control. (*Making Quality Happen, Juran Institute, Inc., senior executive workshop, p. F-5, Wilton, CT.*)

Control by Nonhuman Means.   At the base of the pyramid are the automated feedback loops and error-proofed processes which operate with no human intervention other than maintenance of facilities (which, however, is critical), These nonhuman methods provide control over the great majority of things. The control subjects are exclusively technological, and control takes place on a real-time basis.

The remaining controls in the pyramid require human intervention. By a wide margin, the most amazing achievement in quality control takes place during a biological process which is millions of years old—the growth of the fertilized egg into an animal organism. In human beings the genetic instructions which program this growth consist of a sequence of about three billion "letters." This sequence—the human genome—is contained in two strands of DNA (the double helix) which "unzip" and replicate about a million billion times during the growth process from fertilized egg to birth of the human being.

Given such huge numbers, the opportunities for error are enormous. (Some errors are harmless, but others are damaging and even lethal.) Yet the actual error rate is of the order of about one in 10 billion. This incredibly low error rate is achieved through a feedback loop involving three processes (Radman and Wagner 1988):

A high-fidelity selection process for attaching the right "letters," using chemical lock-and-key combinations

A proofreading process for reading the most recent letter, and removing it if incorrect

A corrective action process to rectify the errors which are detected

Control by the Work Force.   Delegating such decisions to the work force yields important benefits in human relations and in conduct of operations. These benefits include shortening the feedback loop; providing the work force with a greater sense of ownership of the operating processes, often referred to as "empowerment"; and liberating supervisors and managers to devote more of their time to planning and improvement.

It is feasible to delegate most quality control decisions to the work force. Many companies already do. However, to delegate *process control* decisions requires meeting the criteria of "self-control." To delegate *product control* decisions requires meeting the criteria for "self-inspection." (See later in this section under Self-Control and Self-Inspection, respectively.)

Control by the Managerial Hierarchy.   The peak of the pyramid of control consists of the "vital few" control subjects. These are delegated to the various levels in the managerial hierarchy, including the upper managers.

Managers should avoid getting deeply into making decisions on quality control. Instead, they should:

Make the vital few decisions.

Provide criteria to distinguish the vital few decisions from the rest. For an example of providing such criteria see Table 4.3 under the heading: The Fitness for Use Decision.

Delegate the rest under a decision making process which provides the essential tools and training.

The distinction between vital few matters and others originates with the control subjects. Table 4.2 shows how control subjects at two levels—work force and upper management—affect the elements of the feedback loop.

PLANNING FOR QUALITY CONTROL

Planning for control is the activity which provides the system—the concepts, methodology, and tools—through which company personnel can keep the operating processes stable and thereby produce the product features required to meet customer needs. The input-output features of this system (also plan, process) were depicted in Figure 4.2.

The Customers and Their Needs.   The principal customers of quality control systems are the company personnel engaged in control—those who carry out the steps which form the feedback loop. Such personnel require (1) an understanding of customers' quality needs and (2) a definition

TABLE 4.2   Contrast of Quality Control at Two Levels—Work Force and Upper Management

	At work force levels	At managerial levels
Control goals	Product and process features in specifications and procedures	Business oriented, product salability, competitiveness
Sensors	Technological	Data systems
Decisions to be made	Conformance or not?	Meet customer needs or not?

Source: Making Quality Happen, Juran Institute, Inc., senior executive workshop, p. F-4, Wilton, Ct.

of their own role in meeting those needs. However, most of them lack direct contact with customers. Planning for quality control helps to bridge that gap by supplying a translation of what are customers' needs, along with defining responsibility for meeting those needs. In this way, planning for quality control includes providing operating personnel with information on customer needs (whether direct or translated) and definition of the related control responsibilities of the operating personnel. Planning for quality control can run into extensive detail. See, for example, Duyck (1989) and Goble (1987).

Who Plans?   Planning for quality control has in the past been assigned variously to

Staff planners who also plan the operating processes

Staff quality specialists

Multifunctional teams of planners and operating personnel

Departmental managers and supervisors

The work force

Planning for quality control of critical processes has traditionally been the responsibility of those who plan the operating process. For noncritical processes the responsibility was usually assigned to quality specialists from the Quality Department. Their draft plans were then submitted to the operating heads for approval.

Recent trends have been to increase the use of the team concept. The team membership includes the operating forces and may also include suppliers and customers of the operating process. The recent trend has also been to increase participation by the work force. For elaboration, see Juran (1992, pp. 290–291). The concept of self-directing work teams has been greatly expanded in recent years and includes many of these ideas. See Section 15 for more details on this topic.

Quality Control Concepts.   The methodologies of Quality Control are built around various concepts such as the feedback loop, process capability, self-control, etc. Some of these concepts are of ancient origin; others have evolved in this century. During the discussion of planning for quality control, we will elaborate on some of the more widely used concepts.

The Flow Diagram.   The usual first step in planning for quality control is to map out the flow of the operating process. (Design of that process is discussed in Section 3, The Quality Planning Process.) The tool for mapping is the "flow diagram." Figure 4.8 is an example of a flow diagram. (For more examples of this tool, see Appendix V.)

The flow diagram is widely used during planning of quality controls. It helps the planning team to

Understand the overall operating process. Each team member is quite knowledgable about his/her segment of the process, but less so about other segments and about the interrelationships.

Identify the control subjects around which the feedback loops are to be built. [For an example, see Siff (1984).] The nature of these control subjects was discussed previously under the heading, The Control Subject.

Design control stations. (See the following section.)

Control Stations.   A "control station" is an area in which quality control takes place. In the lower levels of organization, a control station is usually confined to a limited physical area. Alternatively, the control station can take such forms as a patrol beat or a "control tower." At higher levels, control stations may be widely dispersed geographically, as is the scope of a manager's responsibility.

A review of numerous control stations shows that they are usually designed to provide evaluations and/or early warnings in the following ways:

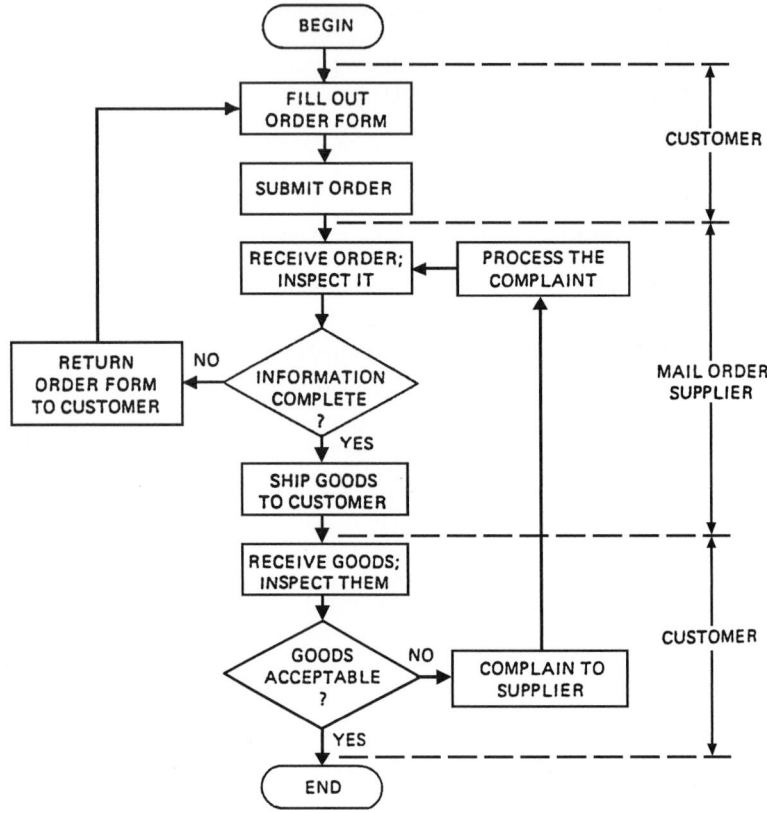

FIGURE 4.8   The flow diagram.

At changes of jurisdiction, where responsibility is transferred from one organization to another

Before embarking on some significant irreversible activity such as signing a contract

After creation of a critical quality feature

At the site of dominant process variables

At areas ("windows") which allow economical evaluation to be made

STAGES OF PROCESS CONTROL

The flow diagram not only discloses the progression of events in the operating process, it also suggests which stages should become the centers of control activity. Several of these stages apply to the majority of operating processes.

Setup (Startup) Control.   The end result of this form of control is the decision of whether or not to "push the start button." Typically this control involves

A *countdown* listing the preparatory steps needed to get the process ready to produce. Such countdowns sometime come from suppliers. Airlines provide checklists to help travelers plan their trips; electric power companies provide checklists to help householders prepare the house for winter weather.

Evaluation of process and/or product features to determine whether, if started, the process will meet the goals.

Criteria to be met by the evaluations.

Verification that the criteria have been met.

Assignment of responsibility. This assignment varies, depending largely on the criticality of the quality goals. The greater the criticality, the greater is the tendency to assign the verification to specialists, supervisors and "independent" verifiers rather than to nonsupervisory workers.

Running Control.   This form of control takes place periodically during the operation of the process. The purpose is to make the "run or stop" decision—whether the process should continue to produce product or whether it should stop.

Running control consists of closing the feedback loop, over and over again. The process and/or product performance is evaluated and compared with goals. If the product and/or process conforms to goals, and if the process has not undergone some significant adverse change, the decision is "continue to run." If there is nonconformance or if there has been a significant change, then corrective action is in order.

The term "significant" has meanings beyond those in the dictionary. One of these meanings relates to whether an indicated change is a real change or is a false alarm due to chance variation. The design for process control should provide the tools needed to help the operating forces distinguish between real changes and false alarms. Statistical process control (SPC) methodology is aimed at providing such tools (see Section 45).

Product Control.   This form of control takes place after some amount of product has been produced. The purpose of the control is to decide whether or not the product conforms to the product quality goals. Assignment of responsibility for this decision differs from company to company. However, in all cases those who are to make the decision must be provided with the facilities and training which will enable them to understand the product quality goals, evaluate the actual product quality, and decide whether there is conformance.

Since all this involves making a factual decision, it can in theory be delegated to anyone, including members of the work force. In practice, this delegation is not made to those whose assigned priorities might bias their judgment. In such cases the delegation is usually to those whose responsibilities are free from such biases, for example, "independent" inspectors. Statistical quality control (SQC) is a methodology frequently employed to yield freedom from biases.

Facilities Control.   Most operating processes employ physical facilities: equipment, instruments, and tools. Increasingly the trend has been to use automated processes, computers, robots, etc. This same trend makes product quality more and more dependent on maintenance of the facilities.

The elements of design for facilities control are well known:

Establish a schedule for conducting facilities maintenance.

Establish a checklist—a list of tasks to be performed during a maintenance action.

Train the maintenance forces to perform the tasks.

Assign clear responsibility for adherence to schedule.

The weakest link in facilities control has been adherence to schedule. To ensure strict adherence to schedule requires an independent audit.

In cases involving introduction of new technology, a further weak link is training the maintenance forces (White 1988).

During the 1980s the auto makers began to introduce computers and other electronics into their vehicles. It soon emerged that many repair shop technicians lacked the technological education base needed to diagnose and remedy the associated field failures. To make matters worse, the auto makers did not give high priority to standardizing the computers. As a result a massive training backlog developed.

For an excellent treatise on facilities maintenance, see Nowlan and Heap (1978).

Concept of Dominance.   Control subjects are so numerous that planners are well advised to identify the vital few control subjects so that they will receive appropriate priority. One tool for identifying the vital few is the concept of dominance.

Operating processes are influenced by many variables, but often one variable is more important than all the rest combined. Such a variable is said to be the "dominant variable." Knowledge of which process variable is dominant helps planners during allocation of resources and priorities. The more usual dominant variables include:

1. *Set-up dominant:*   Some processes exhibit high stability and reproducibility of results, over many cycles of operation. A common example is the printing process. The design for control should provide the operating forces with the means for precise setup and validation before operations proceed.

2. *Time-dominant:*   Here the process is known to change progressively with time, for example, depletion of consumable supplies, heating up, and wear of tools. The design for control should provide means for periodic evaluation of the effect of progressive change and for convenient readjustment.

3. *Component-dominant:*   Here the main variable is the quality of the input materials and components. An example is the assembly of electronic or mechanical equipments. The design for control should be directed at supplier relations, including joint planning with suppliers to upgrade the quality of the inputs.

4. *Worker-dominant:*   In these processes, quality depends mainly on the skill and knack possessed by the workers. The skilled trades are well-known examples. The design for control should emphasize aptitude testing of workers, training and certification, quality rating of workers, and error-proofing to reduce worker errors.

5. *Information-dominant:*   Here the processes are of a "job-shop" nature, so that there is frequent change in what product is to be produced. As a result, the job information changes frequently. The design for control should concentrate on providing an information system which can deliver accurate, up-to-date information on just how this job differs from its predecessors.

Seriousness Classification.   Another way of identifying the vital few control subjects is through "seriousness classification." Under this concept each product feature is classified into one of several defined classes such as critical, major, and minor. These classifications then guide the planners in allocation of resources, assignment of priorities, choice of facilities, frequency of inspection and test, etc.

For elaboration, see Section 22, Operations, under Classification of Defects.

Process Capability.   One of the most important concepts in the quality planning process is "process capability." The prime application of this concept is during planning of the operating processes. This application is treated in more depth in Section 22, Operations.

This same concept also has application in quality control. To explain this, a brief review is in order. All operating processes have an inherent uniformity for producing product. This uniformity can often be quantified, even during the planning stages. The process planners can use the resulting information for making decisions on adequacy of processes, choice of alternative processes, need for revision of processes, and so forth, with respect to the inherent uniformity and its relationship to process goals.

Applied to planning for quality control, the state of process capability becomes a major factor in decisions on frequency of measuring process performance, scheduling maintenance of facilities, etc.

The greater the stability and uniformity of the process, the less the need for frequent measurement and maintenance.

Those who plan for quality control should have a thorough understanding of the concept of process capability and its application to both areas of planning—planning the operating processes as well as planning the controls.

THE CONTROL SPREADSHEET

The work of the planners is usually summarized on a control spreadsheet. This spreadsheet is a major planning tool. An example can be seen in Figure 4.9.

In this spreadsheet the horizontal rows are the various control subjects. The vertical columns consist of elements of the feedback loop plus other features needed by the operating forces to exercise control so as to meet the quality goals.

Some of the contents of the vertical columns are unique to specific control subjects. However, certain vertical columns apply widely to many control subjects. These include unit of measure, type of sensor, quality goal, frequency of measurement, sample size, criteria for decision-making, and responsibility for decision making.

Who Does What?   The feedback loop involves multiple tasks, each of which requires a clear assignment of responsibility. At any control station there may be multiple people available to perform those tasks. For example, at the work-force level, a control station may include setup specialists, operators, maintenance personnel, inspectors, etc. In such cases it is necessary to agree on who should make which decisions and who should take which actions. An aid to reaching such agreement is a special spreadsheet similar to Figure 4.9.

In this spreadsheet the essential decisions and actions are listed in the left-hand column. The remaining columns are headed up by the names of the job categories associated with the control station. Then, through discussion among the cognizant personnel, agreement is reached on who is to do what.

The spreadsheet (Figure 4.9) is a proven way to find answers to the long-standing, but vague, question, "Who is responsible for quality?" This question has never been answered because it is

CONTROL SUBJECT / PROCESS CONTROL FEATURES	UNIT OF MEASURE	TYPE OF SENSOR	GOAL	FREQUENCY OF MEASUREMENT	SAMPLE SIZE	CRITERIA FOR DECISION MAKING	RESPONSIBILITY FOR DECISION MAKING	. . .
Wave solder conditions Solder temperature	Degree F (°F)	Thermo-couple	505 °F	Continuous	N/A	510 °F reduce heat 500 °F increase heat	Operator	. . .
Conveyor speed	Feet per minute (ft/min)	ft/min	4.5 ft/min	1/hour	N/A	5 ft/min reduce speed 4 ft/min increase speed	Operator	. . .
Alloy purity	% Total contaminates	lab chemical analysis	1.5% max	1/month	15 grams	At 1.5%, drain bath, replace solder	Process engineer	. .
⋮	⋮	⋮	⋮	⋮	⋮	⋮	⋮	

FIGURE 4.9   Spreadsheet for "Who does what?" (*Making Quality Happen, Juran Institute, Inc., senior executive workshop, p. F-8, Wilton, CT.*)

inherently unanswerable. However if the question is restated in terms of decisions and actions, the way is open to agree on the answers. This clears up the vagueness.

PROCESS CONFORMANCE

Does the process conform to its quality goals? The umpire answers this question by interpreting the observed difference between process performance and process goals. When current performance does differ from the quality goals, the question arises: What is the cause of this difference?

Special and Common Causes of Variation.   Observed differences usually originate in one of two ways: (1) the observed change is caused by the behavior of a major variable in the process (or by the entry of a new major variable) or (2) the observed change is caused by the interplay of multiple minor variables in the process.

Shewhart called (1) and (2) "assignable" and "nonassignable" causes of variation, respectively (Shewhart 1931). Deming later coined the terms "special" and "common" causes of variation (Deming 1986). In what follows we will use Deming's terminology.

"Special" causes are typically sporadic, and often have their origin in single variables. For such cases it is comparatively easy to conduct a diagnosis and provide remedies. "Common" causes are typically chronic and usually have their origin in the interplay among multiple minor variables, As a result, it is difficult to diagnose them and to provide remedies. This contrast makes clear the importance of distinguishing special causes from common causes when interpreting differences. The need for making such distinctions is widespread. Special causes are the subject of quality control; common causes are the subject of quality improvement.

The Shewhart Control Chart.   It is most desirable to provide umpires with tools which can help to distinguish between special causes and common causes. An elegant tool for this purpose is the Shewhart control chart (or just control chart) shown in Figure 4.10.

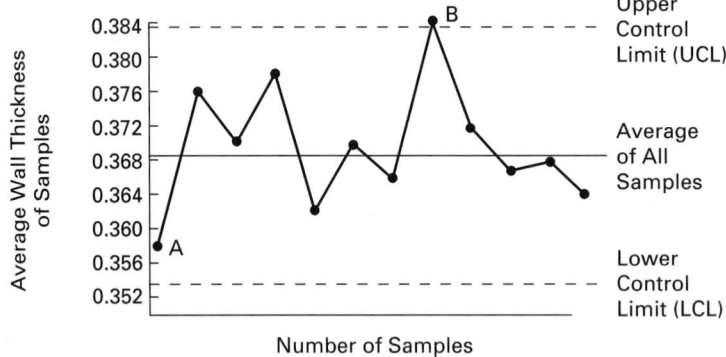

FIGURE 4.10   The Shewhart control chart. (*"Quality Control," Leadership for the Quality Century, Juran Institute, Inc., senior executive workshop, p. 4, Wilton, CT.*)

In Figure 4.10 the horizontal scale is time, and the the vertical scale is quality performance. The plotted points show quality performance as time progresses.

The chart also exhibits three horizontal lines. The middle line is the average of past performance and is therefore the expected level of performance. The other two lines are statistical "limit lines."

They are intended to separate special causes from common causes, based on some chosen level of odds, such as 20 to 1.

Points Within Control Limits.

Point A on the chart differs from the historical average. However, since point A is within the limit lines, this difference could be due to common causes (at odds of less than 20 to 1.) Hence we assume that there is no special cause.

In the absence of special causes, the prevailing assumptions include:

Only common causes are present.

The process is in a state of "statistical control."

The process is doing the best it can.

The variations must be endured.

No action need be taken—taking action may make matters worse (a phenomenon known as "hunting" or "tampering."

The preceding assumptions are being challenged by a broad movement to improve process uniformity. Some processes exhibit no points outside of control chart limits, yet the interplay of minor variables produces some defects.

In one example, a process in statistical control was nevertheless improved by an order of magnitude. The improvement was by a multifunctional improvement team which identified and addressed some of the minor variables. This example is a challenge to the traditional assumption that variations due to common causes must be endured (Pyzdek 1990).

In other cases the challenge is more subtle. There are again no points outside the control limits, but in addition, *no defects are being produced.* Nevertheless the customers demand greater and greater uniformity. Examples are found in business processes (precision of estimating), as well as in manufacture (batch-to-batch uniformity of chemicals, uniformity of components going into random assembly). Such customer demands are on the increase, and they force suppliers to undertake projects to improve the uniformity of even the minor variables in the process. There are many types of control charts. See Section 45, Statistical Process Control, for a more detailed discussion of this important tool.

Points Outside of Control Limits.

Point B also differs from the historical average, but is outside of the limit lines. Now the odds are heavily against this being due to common causes—over 20 to 1. Hence we assume that point B is the result of special causes. Traditionally such "out-of-control" points became *nominations* for corrective action.

Ideally all such nominations should stimulate prompt corrective action to restore the status quo. In practice many out-of-control changes do not result in corrective action. The usual reason is that the changes involving special causes are too numerous—the available personnel cannot deal with all of them. Hence priorities are established based on economic significance or on other criteria of importance. Corrective action is taken for the high-priority cases; the rest must wait their turn. Some changes at low levels of priority may wait a long time for corrective action.

A further reason for failure to take corrective action is a lingering confusion between statistical control limits and quality tolerances. It is easy to be carried away by the elegance and sensitivity of the control chart. This happened on a large scale during the 1940s and 1950s. Here are two examples from the personal experience of the one of the authors:

A large automotive components factory placed a control chart at every machine.

A viscose yarn factory created a "war room" of over 400 control charts.

In virtually all such cases the charts were maintained by the quality departments but ignored by the operating personnel. Experience with such excesses has led managers and planners to be wary of employing control charts just because they are sensitive detectors of change. Instead, the charts should be justified based on value added. Such justifications include:

Customer needs are directly involved.

There is risk to human safety or the environment.

Substantial economics are at stake.

The added precision is needed for control.

Statistical Control Limits and Quality Tolerances.   For most of human history quality goals consisted of product features or process features, usually defined in words. The growth of technology then stimulated the growth of measurement plus a trend to define quality goals in numbers. In addition, there emerged the concept of limits or "tolerances" around the goals. For example:

At least 95 percent of the shipments shall meet the scheduled delivery date.

The length of the bar shall be within 1 mm of the specified number.

Such quality goals had official status. They were set by product or process designers, and published as official specifications. The designers were the official quality legislators—they enacted the laws. Operating personnel were responsible for obeying the quality laws—meeting the specified goals and tolerances.

Statistical control limits in the form of control charts were virtually unknown until the 1940s. At that time, these charts lacked official status. They were prepared and published by quality specialists from the Quality Department. To the operating forces, control charts were a mysterious, alien concept. In addition, the charts threatened to create added work in the form of unnecessary corrective action. The operating personnel reasoned as follows: It has always been our responsibility to take corrective action whenever the product becomes nonconforming. These charts are so sensitive that they detect process changes which do not result in nonconforming product. We are then asked to take corrective action even when the products meet the quality goals and tolerances.

So there emerged a confusion of responsibility. The quality specialists were convinced that the control charts provided useful early-warning signals which should not be ignored. Yet the quality departments failed to recognize that the operating forces were now faced with a confusion of responsibility. The latter felt that so long as the products met the quality goals there was no need for corrective action. The upper managers of those days were of no help—they did not involve themselves in such matters. Since the control charts lacked official status, the operating forces solved their problem by ignoring the charts. This contributed to the collapse, in the 1950s, of the movement known as "statistical quality control."

The 1980s created a new wave of interest in applying the tools of statistics to the control of quality. Many operating personnel underwent training in "statistical process control." This training helped to reduce the confusion, but some confusion remains. To get rid of the confusion, managers should:

Clarify the responsibility for corrective action on points outside the control limits. Is this action mandated or is it discretionary?

Establish guidelines on action to be taken when points are outside the statistical control limits but the product still meets the quality tolerances.

The need for guidelines for decision making is evident from Figure 4.11. The guidelines for quadrants A and C are obvious. If both process and product conform to their respective goals, the process may continue to run. If neither process nor product conform to their respective goals, the process should be stopped, and remedial action should be taken. The guidelines for quadrants B and D are often vague, and this vagueness has been the source of a good deal of confusion. If the choice of action is delegated to the work force, the managers should establish clear guidelines.

Numerous efforts have been made to design control chart limits in ways which help operating personnel to detect whether product quality is threatening to exceed the product quality limits. For a recent example, see Carr (1989). Another approach, based on product quality related to product quality limits, is "PRE-Control." See Juran (1988, pp. 24.31–24.38).

		PRODUCT	
		CONFORMS	DOES NOT CONFORM
PROCESS	DOES NOT CONFORM	B VAGUE	C CLEAR
	CONFORMS	A CLEAR	D VAGUE

FIGURE 4.11   Example of areas of decision making. (*Making Quality Happen, Juran Institute, Inc., senior executive workshop, p.F-21, Wilton, CT.*)

Self-Control; Controllability.   Workers are in a state of self-control when they have been provided with all the essentials for doing good work. These essentials include:

Means of knowing what are the quality goals.

Means of knowing what is their actual performance.

Means for changing their performance in the event that performance does not conform to goals. To meet this criterion requires an operating process which (1) is inherently capable of meeting the goals and (2) is provided with features which make it possible for the operating forces to adjust the process as needed to bring it into conformance with the goals.

These criteria for self-control are applicable to processes in all functions and all levels, from general manager to nonsupervisory worker.

It is all too easy for managers to conclude that the above criteria have been met. In practice, there are many details to be worked out before the criteria can be met. The nature of these details is evident from checklists which have been prepared for specific processes in order to ensure meeting the criteria for self-control. Examples of these checklists include those designed for product designers, production workers, and administrative and support personnel. Examples of such checklists can be found by referring to the subject index of this handbook.

If all the criteria for self-control have been met at the worker level, any resulting product nonconformances are said to be *worker-controllable*. If any of the criteria for self-control have not been met, then management's planning has been incomplete—the planning has not fully provided the means for carrying out the activities within the feedback loop. The nonconforming products resulting from such deficient planning are then said to be *management-controllable*. In such cases it is risky for managers to hold the workers responsible for quality.

Responsibility for results should, of course, be keyed to controllability. However, in the past many managers were not aware of the extent of controllability as it prevailed at the worker level. Studies conducted by Juran during the 1930s and 1940s showed that at the worker level the proportion of management-controllable to worker-controllable nonconformances was of the order of 80 to 20. These findings were confirmed by other studies during the 1950s and 1960s. That ratio of 80 to 20 helps to explain the failure of so many efforts to solve the companies' quality problems solely by motivating the work force.

Effect on the Process Conformance Decision.   Ideally the decision of whether the process conforms to process quality goals should be made by the work force. There is no shorter feedback loop. For many processes this is the actual arrangement. In other cases the process conformance decision is assigned to nonoperating personnel—independent checkers or inspectors. The reasons include:

The worker is not in a state of self-control.

The process is critical to human safety or to the environment.

Quality does not have top priority.

There is lack of mutual trust between the managers and the work force.

PRODUCT CONFORMANCE; FITNESS FOR USE

There are two levels of product features, and they serve different purposes. One of these levels serves such purposes as:

Meeting customer needs

Protecting human safety

Protecting the environment

Product features are said to possess "fitness for use" if they are able to serve the above purposes. The second level of product features serves purposes such as:

Providing working criteria to those who lack knowledge of fitness for use

Creating an atmosphere of law and order

Protecting innocents from unwarranted blame

Such product features are typically contained in internal specifications, procedures, standards, etc. Product features which are able to serve the second list of purposes are said to possess *conformance to specifications,* etc. We will use the shorter label "conformance."

The presence of two levels of product features results in two levels of decision making: Is the product in conformance? Is the product fit for use? Figure 4.12 shows the interrelation of these decisions to the flow diagram.

The Product Conformance Decision.   Under prevailing policies, products which conform to specification are sent on to the next destination or customer. The assumption is that products which conform to specification are also fit for use. This assumption is valid in the great majority of cases.

The combination of large numbers of product features when multiplied by large volumes of product creates huge numbers of product conformance decisions to be made. Ideally these decisions should be delegated to the lowest levels of organization—to the automated devices and the operating work force. Delegation of this decision to the work force creates what is called "self-inspection."

Self-Inspection.   We define "self-inspection" as a state in which decisions on the *product* are delegated to the work force. The delegated decisions consist mainly of: Does product quality conform to the quality goals? What disposition is to be made of the product?

Note that self-inspection is very different from self-control, which involves decisions on the *process.*

The merits of self-inspection are considerable:

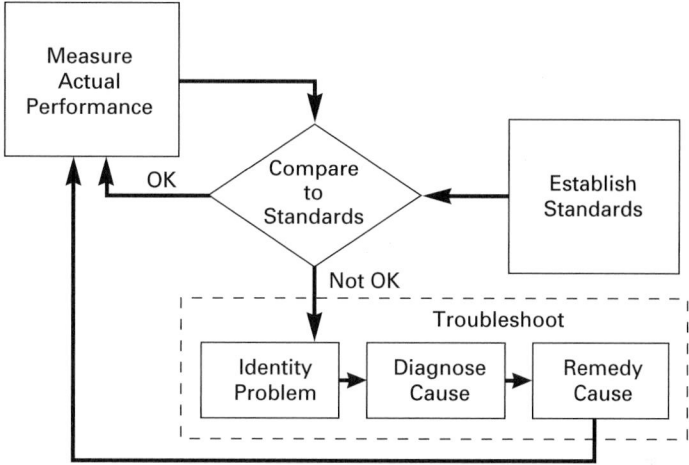

FIGURE 4.12   Flow diagram of decisions on conformance and fitness for use.

The feedback loop is short; the feedback often goes directly to the actuator—the energizer for corrective action.

Self-inspection enlarges the job of the work force—it confers a greater sense of job ownership.

Self-inspection removes the police atmosphere created by use of inspectors, checkers, etc.

However, to make use of self-inspection requires meeting several essential criteria:

Quality is number one: Quality must have undoubted top priority.

Mutual confidence: The managers must have enough trust in the work force to be willing to make the delegation, and the work force must have enough confidence in the managers to be willing to accept the responsibility.

Self-control: The conditions for self-control should be in place so that the work force has all the means necessary to do good work.

Training: The workers should be trained to make the product conformance decisions.

Certification: The recent trend is to include a certification procedure. Workers who are candidates for self-inspection undergo examinations to ensure that they are qualified to make good decisions. The successful candidates are certified and may be subject to audit of decisions thereafter. For examples, see Nowak (1991, Military Airlift Command) and Pearl (1988, Corning Glass Works).

In many companies these criteria are not fully met, especially the criterion of priority. If some parameter other than quality has top priority, there is a real risk that evaluation of product conformance will be biased. This problem happens frequently when personal performance goals are in conflict with overall quality goals. For example, a chemical company found that it was rewarding sales personnel on revenue targets without regard to product availability or even profitability. The sales people were making all their goals, but the company was struggling.

The Fitness for Use Decision.   The great majority of products do conform to specifications. For the nonconforming products there arises a new question: Is the nonconforming product nevertheless fit for use?

A complete basis for making this decision requires answers to questions such as:

Who will be the user(s)?

How will this product be used?

Are there risks to structural integrity, human safety, or the environment?

What is the urgency for delivery?

How do the alternatives affect the producer's and the user's economics?

To answer such questions can involve considerable effort. Companies have tried to minimize the effort through procedural guidelines. The methods in use include:

Treat all nonconforming product as unfit for use: This approach is widely used for products which can pose risks to human safety or the environment—products such as pharmaceuticals or nuclear energy.

Create a mechanism for decision making: An example is the Material Review Board so widely used in the defense industry. This device is practical for matters of importance, but is rather elaborate for the more numerous cases in which little is at stake.

Create a system of multiple delegation: Under such a system, the "vital few" decisions are reserved for a formal decision-making body such as a Material Review Board. The rest are delegated to other people.

Table 4.3 is an example of a table of delegation used by a specific company. (Personal communication to one of the authors.)

For additional discussion on the fitness-for-use decision, see Juran (1988, pp. 18.32–18.36).

Disposition of Unfit Product.
Unfit product is disposed of in various ways: scrap, sort, rework, return to supplier, sell at a discount, etc. The internal costs can be estimated to arrive at an economic optimum. However, the effects go beyond money: schedules are disrupted, people are blamed, etc. To minimize the resulting human abrasion, some companies have established rules of conduct such as:

TABLE 4.3 Multiple Delegations of Decision Making on Fitness for Use*

Effect of nonconformance is on	Amount of product or money at stake is	
	Small	Large
Internal economics only	Department head directly involved, quality engineer	Plant managers involved, quality manager
Economic relations with supplier	Supplier, purchasing agent, quality engineer	Supplier, manager
Economic relations with client	Client, salesperson, quality engineer	Client: for Marketing, Manufacturing, Technical, Quality
Field performance of the product	Product designer, salesperson, quality engineer	Client: managers for Technical, Manufacturing, Marketing Quality
Risk of damage to society or of nonconformance to government regulations	Product design manager, compliance officer, lawyer, quality managers	General manager and team of upper managers

*For those industries whose quality mission is really one of conformance to specification (for example, atomic energy, space), the real decision maker on fitness for use is the client or the government regulator.

Choose that alternative which minimizes the total loss to all parties involved. Now there is less to argue about, and it becomes easier to agree on how to share the loss.

Avoid looking for blame. Instead, treat the loss as an opportunity for quality improvement.

Use "charge backs" sparingly. Charging the vital few losses to the departments responsible has merit from an accounting viewpoint. However, when applied to the numerous minor losses, this is often uneconomic as well as detrimental to efforts to improve quality.

Failure to use products which meet customer needs is a waste. Sending out products which do not meet customer needs is worse. Personnel who are assigned to make product conformance decisions should be provided with clear definitions of responsibility as well as guidelines for decision making. Managers should, as part of their audit, ensure that the processes for making product conformance decisions are appropriate to company needs.

Corrective Action.   The final step in closing the feedback loop is to actuate a change which restores conformance with quality goals. This step is popularly known as "troubleshooting" or "fire-fighting."

Note that the term "corrective action" has been applied loosely to two very different situations, as shown in Figure 4.1. The feedback loop is well designed to eliminate *sporadic* nonconformance like that "spike" in Figure 4.1; the feedback loop is *not* well designed to deal with the area of chronic waste shown in the figure. Instead, the need is to employ the quality improvement process of Section 5.

We will use the term "corrective action" in the sense of troubleshooting—eliminating sporadic nonconformance.

Corrective action requires the journeys of diagnosis and remedy. These journeys are simpler than for quality improvement. Sporadic problems are the result of adverse change, so the diagnostic journey aims to discover what has changed. The remedial journey aims to remove the adverse change and restore conformance.

Diagnosis of Sporadic Change.   During the diagnostic journey the focus is on "What has changed." Sometimes the causes are not obvious, so the main obstacle to corrective action is diagnosis. The diagnosis makes use of methods and tools such as:

Autopsies to determine with precision the symptoms exhibited by the product and process.

Comparison of products made before and after the trouble began to see what has changed; also comparison of good and bad products made since the trouble began.

Comparison of process data before and after the problem began to see what process conditions have changed.

Reconstruction of the chronology, which consists of logging on a time scale (of hours, days, etc.): (1) the events which took place in the process before and after the sporadic change, that is, rotation of shifts, new employees on the job, maintenance actions, etc., and (2) the time-related product information, that is, date codes, cycle time for processing, waiting time, move dates, etc.

Analysis of the resulting data usually sheds a good deal of light on the validity of the various theories of causes. Certain theories are denied. Other theories survive to be tested further.

Operating personnel who lack the training needed to conduct such diagnoses may be forced to shut down the process and request assistance from specialists, the maintenance department, etc. They may also run the process "as is" in order to meet schedules and thereby risk failure to meet the quality goals.

Corrective Action—Remedy.   Once the cause(s) of the sporadic change is known, the worst is over. Most remedies consist of going back to what was done before. This is a return to the familiar, not a journey into the unknown (as is the case with chronic problems). The local personnel are usually able to take the necessary action to restore the status quo.

Process designs should provide means to adjust the process as required to attain conformance with quality goals. Such adjustments are needed at start-up and during running of the process. This aspect of design for process control ideally should meet the following criteria:

There should be a known relationship between the process variables and the product results.

Means should be provided for ready adjustment of the process settings for the key process variables.

A predictable relationship should exist between the amount of change in the process settings and the amount of effect on the product features.

If such criteria are not met, the operating personnel will, in due course, be forced to cut and try in order to carry out remedial action. The resulting frustrations become a disincentive to putting high priority on quality. Burgam (1985) found:

> In one foundry an automated process design for controlling the amount of metal poured failed to provide adequate regulation. As a result, human regulation took over. The workers then played safe by overpouring, since underpoured castings had to be scrapped. The result was much waste until a new technology solved the problem.

Some companies provide systematic procedures for dealing with sporadic changes. See, for example, Sandorf and Bassett (1993).

For added discussion on troubleshooting, see Section 22.

THE ROLE OF STATISTICAL METHODS

An essential activity within the feedback loop is the collection and analysis of data. This activity falls within the scientific discipline known as "statistics." The methods and tools used are often called "statistical methods." These methods have long been used to aid in data collection and analysis in many fields: biology, government, economics, finance, management, etc. Section 44 contains a thorough discussion of the basic statistical methods, while Section 45 contains a good discussion on those methods used in statistical process control.

During this century, much has happened to apply statistical methodology to quality-oriented problems. This has included development of special tools such as the Shewhart control chart. An early wave of such application took place during the 1920s, largely within the Bell System. A second and broader wave was generated during the 1940s and 1950s. It came to be known as statistical quality control. A third wave, broader still, emerged during the 1980s, and came to be widely known as statistical process control. This is covered in Section 45.

Statistical Process Control (SPC).   The term has multiple meanings, but in most companies it is considered to include basic data collection; analysis through such tools as frequency distributions, Pareto principle, Ishikawa (fish bone) diagram, Shewhart control chart, etc.; and application of the concept of process capability.

Advanced tools, such as design of experiments and analysis of variance (see Section 47), are a part of statistical methods but are not normally considered to be a part of statistical process control.

The Merits.   These statistical methods and tools have contributed in an important way to quality control and also to the other processes of the Juran trilogy—quality improvement and quality planning. For some types of quality problems the statistical tools are more than useful—the problems cannot be solved at all without using the appropriate statistical tools.

The SPC movement has succeeded in training a great many supervisors and workers in basic statistical tools. The resulting increase in statistical literacy has made it possible for them to improve their grasp of the behavior of processes and products. In addition, many have learned that decisions based on data collection and analysis yield superior results.

The Risks.   There is danger in taking a tool-oriented approach to quality instead of a problem-oriented or results-oriented approach. During the 1950s this preoccupation became so extensive that the entire statistical quality control movement collapsed; the word "statistical" had to be eliminated from the names of the departments.

The proper sequence in managing is first to establish goals and then to plan how to meet those goals, including choice of the appropriate tools. Similarly, when dealing with problems—threats or opportunities—experienced managers start by first identifying the problems. They then try to solve those problems by various means, including choice of the proper tools.

During the 1980s, numerous companies did, in fact, try a tool-oriented approach by training large numbers of their personnel in the use of statistical tools. However, there was no significant effect on the "bottom line." The reason was that no infrastructure had been created to identify which projects to tackle, to assign clear responsibility for tackling those projects, to provide needed resources, to review progress, etc.

Managers should ensure that training in statistical tools does not become an end in itself. One form of such assurance is through measures of progress. These measures should be designed to evaluate the effect on operations, such as improvement in customer satisfaction or product performance, reduction in cost of poor quality, etc. Measures such as numbers of courses held, or numbers of people trained, do *not* evaluate the effect on operations and hence should be regarded as subsidiary in nature.

Information for Decision Making.   Quality control requires extensive decision-making. These decisions cover a wide variety of subject matter and take place at all levels of the hierarchy. The planning for quality control should provide an information network which can serve all decision makers. At some levels of the hierarchy, a major need is for real-time information to permit prompt detection and correction of nonconformance to goals. At other levels, the emphasis is on summaries which enable managers to exercise control over the vital few control subjects (see Sections 9 and 34). In addition the network should provide information as needed to detect major trends, identify threats and opportunities, and evaluate performance of organization units and personnel.

In some companies the quality information system is designed to go beyond control of product features and process features; the system is also used to control the quality performance of organizations and individuals, for example, departments and department heads. For example, many companies prepare and regularly publish scoreboards showing summarized quality performance data for various market areas, product lines, operating functions, etc. These performance data are often used as indicators of the quality performance of the personnel in charge.

To provide information which can serve all those purposes requires planning which is directed specifically to the information system. Such planning is best done by a multifunctional team whose mission is focused on the quality information system. That team properly includes the customers as well as the suppliers of information. The management audit of the quality control system should include assurance that the quality information system meets the needs of the various customers. (For additional discussion relating to the quality information system, see Section 9, Measurement, Information, and Decision Making.)

THE QUALITY CONTROL MANUAL

A great deal of quality planning is done through "procedures" which are really repetitive-use plans. Such procedures are thought out, written out, and approved formally. Once published, they become the authorized ways of conducting the company's affairs. It is quite common for the procedures relating to managing for quality to be published collectively in a "quality manual" (or similar title). A significant part of the manual relates to quality control.

Quality manuals add to the usefulness of procedures in several ways:

Legitimacy: The manuals are approved at the highest levels of organization.

Readily findable: The procedures are assembled into a well-known reference source rather than being scattered among many memoranda, oral agreements, reports, minutes, etc.

Stable: The procedures survive despite lapses in memory and employee turnover.

Study of company quality manuals shows that most of them contain a core content which is quite similar from company to company. Relative to quality control, this core content includes procedures for:

Application of the feedback loop to process and product control

Ensuring that operating processes are capable of meeting the quality goals

Maintenance of facilities and calibration of measuring instruments

Relations with suppliers on quality matters

Collection and analysis of the data required for the quality information system

Training the personnel to carry out the provisions of the manual

Audit to ensure adherence to procedures

The need for repetitive-use quality control systems has led to evolution of standards at industry, national, and international levels. For elaboration, see Section 11, The ISO 9000 Family of International Standards. For an example of developing standard operating procedures, including the use of videocassettes, see Murphy and McNealey (1990). Work-force participation during preparation of procedures helps to ensure that the procedures will be followed. See, in this connection, Gass (1993).

Format of Quality Manuals.   Here again, there is much commonality. The general sections of the manual include:

1. An official statement by the general manager. It includes the signatures which confer legitimacy.

2. The purpose of the manual and how to use it.

3 The pertinent company (or divisional, etc.) quality policies.

4 The organizational charts and tables of responsibility relative to the quality function.

5. Provision for audit of performance against the mandates of the manual.

Additional sections of the manual deal with applications to functional departments, technological products and processes, business processes, etc. For elaboration, see Juran (1988, pp. 6.40–6.47).

Managers are able to influence the adequacy of the Quality Control manual in several ways:

Participate in defining the criteria to be met by the manual.

Approve the final draft of the manual to make it official.

Periodically audit the up-to-dateness of the manual as well as conformance to the manual.

CONTROL THROUGH THE REWARD SYSTEM

An important influence on Quality Control is the extent to which the reward system (merit rating, etc.) emphasizes quality in relation to other parameters. This aspect of quality control is discussed throughout Section 15. See also Section 40, under Motivation, Recognition, and Reward.

PROVISION FOR AUDIT

Experience has shown that control systems are subject to "slippage" of all sorts. Personnel turnover may result in loss of essential knowledge. Entry of unanticipated changes may result in obsolescence. Shortcuts and misuse may gradually undermine the system until it is no longer effective.

The major tool for guarding against deterioration of a control system has been the audit. Under the audit concept a periodic, independent review is established to provide answers to the following questions: Is the control system still adequate for the job? Is the system is being followed?

The answers are obviously useful to the operating managers. However, that is not the only purpose of the audit. A further purpose is to provide those answers to people who, though not directly involved in operations, nevertheless have a need to know. If quality is to have top priority, those who have a need to know include the upper managers.

It follows that one of the responsibilities of managers is to mandate establishment of a periodic audit of the quality control system.

QUALITY CONTROL: WHAT IS NEW?

Recent decades have witnessed a growing trend to improve the effectiveness of quality control by formal adoption of modern concepts, methodologies, and tools. These have included:

Systematic planning for quality control, with extensive participation by the operating personnel

Formal application of the feedback loop, and establishment of clear responsibility for the associated decisions and actions

Delegation of decisions to the work force through self-control and self-inspection

Wide application of statistical process control and the associated training of the operating personnel

A structured information network to provide a factual basis for decision making

A systematic process for corrective action in the event of sporadic adverse change

Formal company manuals for quality control, with periodic audits to ensure up-to-dateness and conformance

SUMMARY

The quality control process is a universal managerial process for conducting operations so as to provide stability—to prevent adverse change and to "maintain the status quo." Quality control takes place by use of the feedback loop. Each feature of the product or process becomes a *control subject*—a center around which the feedback loop is built. As much as possible, human control should be done by the work force—the office clerical force, factory workers, salespersons, etc. The flow diagram is widely used during the planning of quality controls. The weakest link in facilities control has been adherence to schedule. To ensure strict adherence to schedule requires an independent audit. Knowing which process variable is dominant helps planners during allocation of resources and priorities. The work of the planners is usually summarized on a control spreadsheet. This spreadsheet is a major planning tool.

The question "Who is responsible for quality?" is inherently unanswerable. However, if the question is restated in terms of decisions and actions, the way is open to agree on the answers. The design for process control should provide the tools needed to help the operating forces distinguish between real changes and false alarms. It is most desirable to provide umpires with tools which can help to distinguish between special causes and common causes. An elegant tool for this purpose is the

Shewhart control chart (or just control chart). The criteria for self-control are applicable to processes in all functions, and all levels, from general manager to nonsupervisory worker. Responsibility for results should be keyed to controllability. Ideally the decision of whether the process conforms to process quality goals should be made by the work force. There is no shorter feedback loop.

To make use of self-inspection requires meeting several essential criteria: quality is number one; mutual confidence, self-control, training, and certification are the others. Personnel who are assigned to make product conformance decisions should be provided with clear definitions of responsibility as well as guidelines for decision making. The proper sequence in managing is first to establish goals and then to plan how to meet those goals, including the choice of the appropriate tools. The planning for quality control should provide an information network which can serve all decision makers.

TASKS FOR MANAGERS

Managers should avoid getting deeply into making decisions on quality control. They should make the vital few decisions, provide criteria to distinguish the vital few from the rest, and delegate the rest under a decision-making process.

To eliminate the confusion relative to control limits and product quality tolerance, managers should clarify the responsibility for corrective action on points outside the control limits and establish guidelines on action to be taken when points are outside the statistical control limits but the product still meets the quality tolerances.

Managers should, as part of their audit, ensure that the processes for making product conformance decisions are appropriate to company needs. They should also ensure that training in statistical tools does not become an end in itself. The management audit of the quality control system should include assurance that the quality information system meets the needs of the various customers.

Managers are able to influence the adequacy of the quality control manual in several ways: participate in defining the criteria to be met, approve the final draft to make it official, and periodically audit the up-to-dateness of the manual as well as the state of conformance.

REFERENCES

Burgam, Patrick M. (1985). "Application: Reducing Foundry Waste." *Manufacturing Engineering,* March.

Bylinsky, Gene (1991). "How Companies Spy on Employees." *Fortune,* November, pp. 131–140.

Carr, Wendell E. (1989). "Modified Control Limits." *Quality Progress,* January, pp. 44–48.

Deming, W. Edwards (1986). *Out of the Crisis,* MIT Center for Advanced Engineering Study. Cambridge, MA.

Deming, W. Edwards (1950). *Elementary Principles of the Statistical Control of Quality,* Nippon Kagaku Gijutsu Renmei (Japanese Union of Scientists and Engineers), Tokyo.

Duyck, T. O. (1989). "Product Control Through Process Control and Process Design." *1989 ASQC Quality Congress Transactions,* pp. 676–681.

Gass, Kenneth C. (1993). "Getting the Most Out of Procedures." *Quality Engineering,* June.

Goble, Joann (1987). "A Systematic Approach to Implementing SPC," *1987 ASQC Quality Congress Transactions,* pp. 154–164.

Juran, J. M. (1992). *Juran on Quality by Design,* The Free Press, A Division of Macmillan, Inc., New York.

Juran, J. M., ed. (1988). *Juran's Quality Control Handbook,* McGraw-Hill, New York.

Juran, J. M. (1964). *Managerial Breakthrough,* McGraw-Hill, New York.

Koura, Kozo (1991). "Deming Cycle to Management Cycle." *Societas Qualitas,* Japanese Union of Scientists and Engineers, Tokyo, May–June.

Lenehan, Michael (1982). "The Quality of the Instrument, Building Steinway Grand Piano K 2571." *The Atlantic Monthly,* August, pp. 32–58.

Murphy, Robert W., and McNealey, James E. (1990). "A Technique for Developing Standard Operating Procedures to Provide Consistent Quality." *1990 Juran IMPRO Conference Proceedings,* pp. 3D1–3D6.

Nowlan, F. Stanley, and Heap, Howard F. *Reliability Centered Maintenance.* Catalog number ADA 066579, National Technical Information Service, Springfield, Va.

Nowak, Major General John M. (1991). "Transitioning from Correction to Prevention." *1991 Juran IMPRO Conference.,* pp. 1-15–1-20.

Pearl, Daniel H. (1988). "Operator Owned SPC, How to Get it and Keep It." *1988 ASQC Congress Transactions,* pp. 817–821.

Pyzdek, Thomas (1990). "There's No Such Thing as a Common Cause." *ASQC Quality Congress Transactions,* pp. 102–108.

Radford, G. S. (1922). *The Control of Quality of Manufacturing,* Ronald Press Company, New York.

Radford, G. S. (1917). "The Control of Quality." *Industrial Management,* vol. 54, p. 100.

Radman, Miroslav, and Wagner, Robert (1988). "The High Fidelity of DNA Duplication." *Scientific American,* August, pp. 40–46.

Sandorf, John P., and Bassett, A. Thomas III (1993). "The OCAP: Predetermined Responses to Out-of-Control Conditions." *Quality Progress,* May, pp. 91–95.

Shainin, Dorian, and Shainin, Peter D. (1988). "PRE-Control." *Juran's Quality Control Handbook,* 4th ed., McGraw-Hill, New York.

Siff, Walter C. (1984). "The Strategic Plan of Control—A Tool For Participative Management." *1984 ASQC Quality Congress Transactions,* pp. 384–390.

Shewhart, W. A. (1931). *Economic Control of Quality of Manufactured Product,* Van Nostrand, New York, 1931. Reprinted by ASQC, Milwaukee, 1980.

Umscheid, Thomas E. (1991). "Production Monitoring Enhances SPC." *Quality Progress,* December, pp. 58–61.

White, Joseph B. (1988). "Auto Mechanics Struggle to Cope With Technology in Today's Cars." *The Wall Street Journal,* July 26, p. 37.

SECTION 5

THE QUALITY IMPROVEMENT PROCESS

J. M. Juran[1]

---

[1]In the fourth edition, the section on quality improvement was prepared by Frank M. Gryna.

THE PURPOSE OF THIS SECTION

The purpose of this section is to explain the nature of quality improvement and its relation to managing for quality, show how to establish quality improvement as a continuing process that goes on year after year, and define the action plan and the roles to be played, including those of upper management.

WHAT IS IMPROVEMENT?

As used here, "improvement" means "the organized creation of beneficial change; the attainment of unprecedented levels of performance." A synonym is "breakthrough."

Two Kinds of Beneficial Change.   Better quality is a form of beneficial change. It is applicable to both the kinds of quality that are summarized in Section 2, Figure 2.1:

Product features: These can increase customer satisfaction. To the producing company, they are income-oriented.

Freedom from deficiencies: These can create customer dissatisfaction and chronic waste. To the producing company, they are cost-oriented.

Quality improvement to increase income may consist of such actions as

Product development to create new features that provide greater customer satisfaction and hence may increase income

Business process improvement to reduce the cycle time for providing better service to customers

Creation of "one-stop shopping" to reduce customer frustration over having to deal with multiple personnel to get service

Quality improvement to reduce deficiencies that create chronic waste may consist of such actions as

Increase of the yield of factory processes

Reduction of the error rates in offices

Reduction of field failures

The end results in both cases are called "quality improvement." However, the processes used to secure these results are fundamentally different, and for a subtle reason.

Quality improvement to increase income starts by setting new goals, such as new product features, shorter cycle times, and one-stop shopping. Meeting such new goals requires several kinds of planning, including quality planning. This quality planning is done through a universal series of steps: identify the "customers" who will be affected if the goal is met, determine the needs of those customers, develop the product features required to meet those needs, and so on. Collectively, this series of steps is the "quality planning roadmap," which is the subject matter of Section 3, The Quality Planning Process.

In the case of chronic waste, the product goals are already in place; so are the processes for meeting those goals. However, the resulting products (goods and services) do not all meet the goals. Some do and some do not. As a consequence, the approach to reducing chronic waste is different from the quality planning roadmap. Instead, the approach consists of (1) discovering the causes—why do some products meet the goal and others do not—and (2) applying remedies to remove the causes. *It is this approach to quality improvement that is the subject of this section.*

Continuing improvement is needed for both kinds of quality, since competitive pressures apply to each. Customer needs are a moving target. Competitive costs are also a moving target. However, improvement for these two kinds of quality has in the past progressed at very different rates. The chief reason is that many upper managers, perhaps most, give higher priority to increasing sales than to reducing costs. This difference in priority is usually reflected in the respective organization structures. An example is seen in the approach to new product development.

Structured Product Development.   Many companies maintain an organized approach for evolving new models of products, year after year. Under this organized approach:

Product development projects are a part of the business plan.

A New Products Committee maintains a business surveillance over these projects.

Full-time product and process development departments are equipped with personnel, laboratories, and other resources to carry out the technological work.

There is clear responsibility for carrying out the essential technological work.

A structured procedure is used to progress the new developments through the functional departments.

The continuing existence of this structure favors new product development on a year-to-year basis.

This special organization structure, while necessary, is not sufficient to ensure good results. In some companies, the cycle time for getting new products to market is lengthy, the new models compete poorly in the market, or new chronic wastes are created. Such weaknesses usually are traceable to weaknesses in the quality planning process, as discussed in Section 3, The Quality Planning Process.

Unstructured Reduction of Chronic Waste.   In most companies, the urge to reduce chronic waste has been much lower than the urge to increase sales. As a result:

The business plan has not included goals for reduction of chronic waste.

Responsibility for such quality improvement has been vague. It has been left to volunteers to initiate action.

The needed resources have not been provided, since such improvement has not been a part of the business plan.

The lack of priority by upper managers is traceable in large part to two factors that influence the thinking processes of many upper managers:

1. Not only do many upper managers give top priority to increasing sales, but some of them even regard cost reduction as a form of lower-caste work that is not worthy of the time of upper managers. This is especially the case in high-tech industries.

2. Upper managers have not been aware of the size of the chronic waste, nor of the associated potential for high return on investment. The "instrument panel" available to upper managers has stressed performance measures such as sales, profit, cash flow, and so on but not the size of chronic waste and the associated opportunities. The quality managers have contributed to this unawareness by presenting their reports in the language of quality specialists rather than in the language of management—the language of money.

The major focus of this section of the handbook is to show how companies can mobilize their resources to deal with this neglected opportunity.

THE GROWTH OF CHRONIC WASTE

Chronic waste does not seem to have been a major problem during the early centuries of artisanship. The artisan typically carried out many tasks to complete a unit of product. *During each of these tasks, he was his own customer.* His client lived in the same village, so the feedback loops were short and prompt.

The Industrial Revolution of the mid-eighteenth century greatly reduced the role of artisans while creating large factories and complex organizational structures that became breeding grounds for chronic waste. The Taylor system of the early twentieth century improved productivity but had a negative effect on quality. To minimize the damage, the companies expanded product inspection. This helped to shield customers from receiving defective products but encouraged the resulting chronic waste, which became huge.

The widespread practice of relying on inspection was shattered by the Japanese quality revolution that followed World War II. That revolution greatly reduced chronic waste, improved product features, and contributed to making Japan an economic superpower. In addition, it greatly intensified international competition in quality. This competition soon created a growing crisis in Western countries, reaching alarming proportions by the 1980s.

THE INITIATIVES OF THE 1980S

In response to the crisis, many companies, especially in the United States, undertook initiatives to improve their quality. For various reasons, most of these initiatives fell far short of their goals. However, a relatively few companies made stunning improvements in quality and thereby became the role models. The methods used by these role models have been analyzed and have become the lessons learned—what actions are needed to attain quality leadership and what processes must be devised to enable those actions to be taken.

Lessons Learned.   Analysis of the actions taken by the successful companies shows that most of them carried out many or all of the strategies set out below:

They enlarged the business plan at all levels to include goals for quality improvement.

They designed a process for making improvements and set up special organizational machinery to carry out that process.

They adopted the big Q concept—they applied the improvement process to business processes as well as to manufacturing processes.

They trained all levels of personnel, including upper management, in how to carry out their respective missions of managing for quality.

They empowered the work force to participate in making improvements.

They established measures to evaluate progress against the improvement goals.

The managers, including the upper managers, reviewed progress against the improvement goals.

They expanded use of recognition for superior quality performance.

They revised the reward system to recognize the changes in job responsibilities.

The Rate of Improvement Is Decisive.   The central lesson learned was that the annual rate of quality improvement determines which companies emerge as quality leaders. Figure 5.1 shows the effect of differing rates of quality improvement.

In this figure, the vertical scale represents product saleability, so what goes up is good. The upper line shows the performance of company A, which at the outset was the industry quality leader. Company A kept getting better, year after year. In addition, company A was profitable. Seemingly, Company A faced a bright future.

The lower line shows that company B, a competitor, was at the outset not the quality leader. However, company B has improved at a rate much faster than that of company A. Company A is now threatened with loss of its quality leadership. The lesson is clear:

The most decisive factor in the competition for quality leadership is the rate of quality improvement.

The sloping lines of Figure 5.1 help to explain why Japanese goods attained quality leadership in so many product lines. The major reason is that the Japanese rate of quality improvement was for decades *revolutionary* when compared with the *evolutionary* rate of the West.

Figure 5.2 shows my estimate of the rates of quality improvement in the automobile industry. [For elaboration, see Juran (1993).] There are also lessons to be learned from the numerous initiatives during the 1980s that failed to produce useful results. These have not been well analyzed, but one

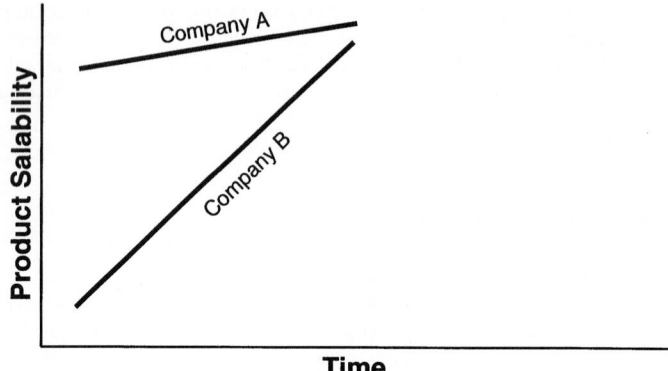

FIGURE 5.1 Two contrasting rates of improvement. (*From Making Quality Happen, 1988, Juran Institute, Wilton, CT, p. D4.*)

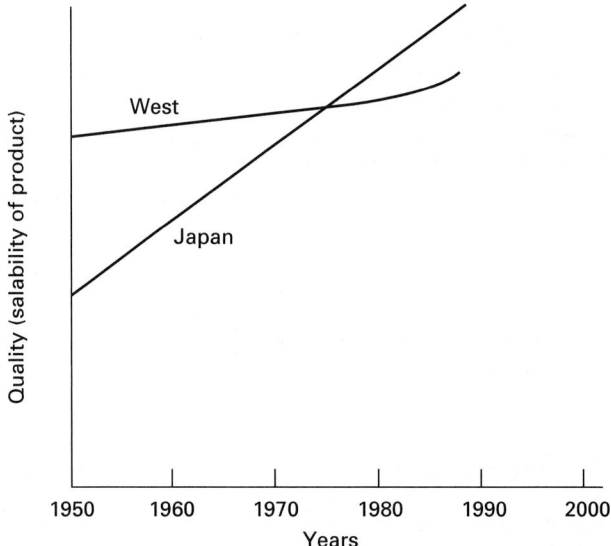

FIGURE 5.2 Estimate of rates of quality improvement in the automobile industry. (*From Making Quality Happen, 1988, Juran Institute, Wilton, CT, p. D5.*)

lesson does stand out. Collectively, those failed initiatives show that attaining a revolutionary rate of quality improvement is *not simple at all*. There are numerous obstacles and much cultural resistance, as will be discussed throughout this section.

THE OUTLOOK FOR THE TWENTY-FIRST CENTURY

By the late 1990s, the efforts to meet competition in quality were proceeding along two lines based on two very different philosophies:

1. Political leaders focused on traditional political solutions—import quotas, tariffs, legislation on "fair trade," and so on.

2. Industrial leaders increasingly became convinced that the necessary response to competition was to become more competitive. This approach required applying the lessons learned from the role models across the entire national economy. Such a massive scaling up likely would extend well into the twenty-first century.

The Emerging Consensus.   The experience of recent decades and the lessons learned have led to an emerging consensus as to the status of managing for quality, the resulting threats and opportunities, and the actions that need to be taken. As related to quality improvement, the high points of this consensus include the following:

> Competition in quality has intensified and has become a permanent fact of life. A major needed response is a high rate of quality improvement, year after year.

> Customers are increasingly demanding improved quality from their suppliers. These demands are then transmitted through the entire supplier chain. The demands may go beyond product improvement and extend to improving the system of managing for quality. (For an example, see Krantz 1989. In that case, a company used product inspection to shield its customers from receiving defective products. Nevertheless, a large customer required the company to revise its system of managing for quality as a condition for continuing to be a supplier.)

> The chronic wastes are known to be huge. In the United States during the early 1980s, about a third of what was done consisted of redoing what was done previously, due to quality deficiencies (estimate by the author). The emerging consensus is that such wastes should not continue on and on, since they reduce competitiveness in costs.

> Quality improvement should be directed at all areas that influence company performance—business processes as well as factory processes.

> Quality improvement should not be left solely to voluntary initiatives; it should be built into the system.

> Attainment of quality leadership requires that the upper managers personally take charge of managing for quality. In companies that did attain quality leadership, the upper managers personally guided the initiative. I am not aware of any exceptions.

THE REMAINDER OF THIS SECTION

The remainder of this section focuses on "how to do it." The division of the subject includes

> The basic concepts that underlie quality improvement

> How to mobilize a company's resources so as to make quality improvement an integral part of managing the company

> The improvement process itself—the universal sequence of steps for making any improvement

> How to "institutionalize" improvement so that it goes on and on, year after year

QUALITY IMPROVEMENT: THE BASIC CONCEPTS

The quality improvement process rests on a base of certain fundamental concepts. For most companies and managers, annual quality improvement is not only a new responsibility, it is also a radical change in the style of management—a change in company culture. Therefore, it is important to grasp the basic concepts before getting into the improvement process itself.

Improvement Distinguished from Control.   Improvement differs from control. The trilogy diagram (Figure 5.3) shows this difference. (Note that Figure 5.3 is identical with Figure 2.4 in Section 2.) In this figure, the chronic waste level (the cost of poor quality) was originally about 23 percent of the amount produced. This chronic waste was built into the process—"It was planned that way." Later, a quality improvement project reduced this waste to about 5 percent. Under my definition, this reduction in chronic waste is an improvement—it attained an unprecedented level of performance.

Figure 5.3 also shows a "sporadic spike"—a sudden increase in waste to about 40 percent. Such spikes are unplanned—they arise from various unexpected sources. The personnel promptly got rid of that spike and restored the previous chronic level of about 23 percent. This action did not meet the definition of an improvement—it did not attain an unprecedented level of performance. Usual names for such actions are "troubleshooting", "corrective action", or "firefighting."

All Improvement Takes Place Project by Project.   There is no such thing as improvement generally. All improvement takes place project by project and in no other way.

As used here, "improvement project" means "a chronic problem scheduled for solution." Since improvement project has multiple meanings, the company glossary and training manuals should define it. The definition is helped by including some case examples that were carried out successfully in that company.

Quality Improvement Is Applicable Universally.   The huge numbers of projects carried out during the 1980s and 1990s demonstrated that quality improvement is applicable to

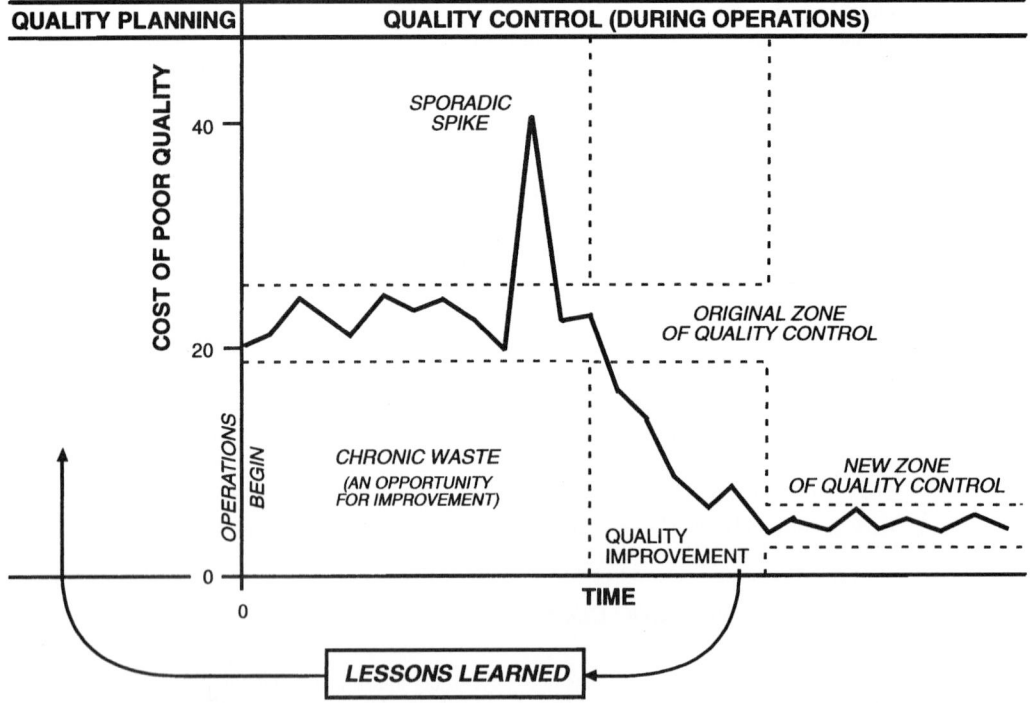

FIGURE 5.3   The Juran trilogy diagram. (*Adapted from Juran, J. M, 1989, The Quality Trilogy: A Universal Approach to Managing for Quality, Juran Institute, Inc., Wilton, CT.*)

Service industries as well as manufacturing industries

Business processes as well as manufacturing processes

Support activities as well as operations

Software as well as hardware

During the 1980s and 1990s, quality improvement was applied to virtually all industries, including government, education, and health. [For a seminal book related to the health industry, see Berwick et al. (1990).]

In addition, quality improvement has been applied successfully to the entire spectrum of company functions: finance, product development, marketing, legal, and so on.

In one company, the legal vice-president doubted that quality improvement could be applied to legal work. Yet within 2 years he reduced by more that 50 percent the cycle time of filing for a patent. (Private communication to the author.)

(For elaboration and many case examples, see the Proceedings of the Juran Institute's Annual IMPRO Conferences on Quality Management.)

Quality Improvement Extends to All Parameters.

Published reports of quality improvements show that the effects have extended to all parameters:

Productivity: The output per person-hour

Cycle time: The time required to carry out processes, especially those which involve many steps performed sequentially in various departments. Section 6, Process Management, elaborates on improvement as applied to such processes.

Human safety: Many projects improve human safety through errorproofing, fail-safe designs, and so on.

The environment: Similarly, many projects have been directed at protecting the environment by reducing toxic emissions and so on.

Some projects provide benefits across multiple parameters. A classic example was the color television set (Juran 1979). The Japanese Matsushita Company had purchased an American color television factory (Quasar). Matsushita then made various improvements, including

Product redesign to reduce field failures

Process redesign to reduce internal defect rates

Joint action with suppliers to improve quality of purchased components

The results of these and other changes are set out in the before and after data:

	1974	1977
Fall-off rate, i.e., defects (on assembled sets) requiring repair	150 per 100 sets	4 per 100 sets
Number of repair and inspection personnel	120	15
Failure rate during the warranty period	70%	10%
Cost of service calls	$22 million	$4 million

The manufacturer benefited in multiple ways: lower costs, higher productivity, more reliable deliveries, and greater saleability. The ultimate users also benefited—the field failure rate was reduced by over 80 percent.

The Backlog of Improvement Projects Is Huge. The existence of a huge backlog is evident from the numbers of improvements actually made by companies that carried out successful initiatives during the 1980s and 1990s. Some reported making improvements by the thousands, year after year. In very large companies, the numbers are higher still, by orders of magnitude.

The backlog of improvement projects exists in part because the planning of new products and processes has long been deficient. In effect, the planning process has been a dual hatchery. It hatched out new plans. It also hatched out new chronic wastes, and these accumulated year after year. Each such chronic waste then became a potential improvement project.

A further reason for a huge backlog is the nature of human ingenuity—it seems to have no limit. Toyota Motor Company has reported that its 80,000 employees offered 4 million suggestions for improvement during a single year—an average of 50 suggestions per person per year (Sakai 1994).

Quality Improvement Does Not Come Free. Reduction of chronic waste does not come free—it requires expenditure of effort in several forms. It is necessary to create an infrastructure to mobilize the company's resources toward the end of annual quality improvement. This involves setting specific goals to be reached, choosing projects to be tackled, assigning responsibilities, following progress, and so on.

There is also a need to conduct extensive training in the nature of the improvement process, how to serve on improvement teams, how to use the tools, and so on.

In addition to all this preparatory effort, each improvement project requires added effort to conduct diagnoses to discover the causes of the chronic waste and provide remedies to eliminate the causes.

The preceding adds up to a significant front-end outlay, but the results can be stunning. They *have* been stunning in the successful companies—the role models. Detailed accounts of such results have been widely published, notably in the proceedings of the annual conferences held by the U.S. National Institute for Standards and Technology (NIST), which administers the Malcolm Baldrige National Quality Award.

Reduction in Chronic Waste Is Not Capital-Intensive. Reduction in chronic waste seldom requires capital expenditures. Diagnosis to discover the causes usually consists of the time of the quality improvement project teams. Remedies to remove the causes usually involve fine-tuning the process. In most cases, a process that is already producing over 80 percent good work can be raised to the high 90s without capital investment. Such avoidance of capital investment is a major reason why reduction of chronic waste has a high return on investment (ROI).

In contrast, projects for product development to increase sales involve outlays to discover customer needs, design products and processes, build facilities, and so on. Such outlays are largely classified as capital expenditures and thereby lower the ROI estimates.

The Return on Investment Is Among the Highest. This is evident from results publicly reported by national award winners in Japan (Deming Prize), the United States (Baldrige Award), and elsewhere. More and more companies have been publishing reports describing their quality improvements, including the gains made. [For examples, see the Proceedings of the Juran Institute's Annual IMPRO Conferences on Quality Management for 1983 and subsequent years. See especially, Kearns and Nadler (1995).]

While these and other published case examples abound, the actual return on investment from quality improvement projects has not been well researched. I once examined 18 papers published by companies and found that the average quality improvement project had yielded about $100,000 of cost reduction (Juran 1985). The companies were large—sales in the range of over $1 billion (milliard) per year.

I have also estimated that for projects at the $100,000 level, the investment in diagnosis and remedy combined runs to about $15,000. The resulting ROI is among the highest available to managers. It has caused some managers to quip: "The best business to be in is quality improvement."

The Major Gains Come from the Vital Few Projects.   The bulk of the measurable gains comes from a minority of the quality improvement projects—the "vital few." These are multifunctional in nature, so they need multifunctional teams to carry them out. In contrast, the majority of the projects are in the "useful many" category and are carried out by local departmental teams. Such projects typically produce results that are orders of magnitude smaller than those of the vital few.

While the useful many projects contribute only a minor part of the measurable gains, they provide an opportunity for the lower levels of the hierarchy, including the work force, to participate in quality improvement. In the minds of many managers, the resulting gain in quality of work life is quite as important as the tangible gains in operating performance.

QUALITY IMPROVEMENT—SOME INHIBITORS

While the role-model companies achieved stunning results through quality improvement, most companies did not. Some of these failures were due to honest ignorance of how to mobilize for improvement, but there are also some inherent inhibitors to establishing improvement on a year-to-year basis. It is useful to understand the nature of some of the principal inhibitors before setting out.

Disillusioned by the Failures.   The lack of results mentioned earlier has led some influential journals to conclude that improvement initiatives are inherently doomed to failure. Such conclusions ignore the stunning results achieved by the role-model companies. (Their results prove that such results are achievable.) In addition, the role models have explained how they got those results, thereby providing lessons learned for other companies to follow. Nevertheless, the conclusions of the media have made some upper managers wary about going into quality improvement.

"Higher Quality Costs More."   Some managers hold to a mindset that "higher quality costs more." This mindset may be based on the outmoded belief that the way to improve quality is to increase inspection so that fewer defects escape to the customer. It also may be based on the confusion caused by the two meanings of the word "quality."

Higher quality in the sense of improved product features (through product development) usually requires capital investment. In this sense, it does *cost* more. However, higher quality in the sense of lower chronic waste usually costs less—a lot less. Those who are responsible for preparing proposals for management approval should be careful to define the key words—Which kind of quality are they talking about?

The Illusion of Delegation.   Managers are busy people, yet they are constantly bombarded with new demands on their time. They try to keep their workload in balance through delegation. The principle that "a good manager is a good delegator" has wide application, but it has been overdone as applied to quality improvement. The lessons learned from the role-model companies show that going into annual quality improvement adds minimally about 10 percent to the workload of the entire management team, *including the upper managers.*

Most upper managers have tried to avoid this added workload through sweeping delegation. Some established vague goals and then exhorted everyone to do better—"Do it right the first time." In the role-model companies, it was different. In every such company, the upper managers took charge of the initiative and personally carried out certain nondelegable roles. (See below, under The Nondelegable Roles of Upper Managers.)

Employee Apprehensions.   Going into quality improvement involves profound changes in a company's way of life—far more than is evident on the surface. It adds new roles to the job descriptions and more work to the job holders. It requires accepting the concept of teams for tackling projects—a concept that is alien to many companies and which invades the jurisdictions of the functional departments. It raises the priority of quality, with damaging effects on other priorities. It requires training on how to do all this. Collectively, it is a megachange that disturbs the peace and breeds many unwanted side effects.

To the employees, the most frightening effect of this profound set of changes is the threat to jobs and/or status. Reduction of chronic waste reduces the need for redoing prior work and hence the jobs of people engaged in such redoing. Elimination of such jobs then becomes a threat to the status and/or jobs of the associated supervision. It should come as no surprise if the efforts to reduce waste are resisted by the work force, the union, the supervision, and others.

Nevertheless, quality improvement is essential to remaining competitive. Failure to go forward puts all jobs at risk. Therefore, the company should go into improvement while realizing that employee apprehension is a very logical reaction of worried people to worrisome proposals. The need is to open a communication link to explain the why, understand the worries, and search for optimal solutions. In the absence of forthright communication, the informal channels take over, breeding suspicions and rumors. For added discussion, see below, under the Quality Council: Anticipating the Questions.

Additional apprehension has its origin in cultural patterns. (See below, under Resistance to Change, Cultural Patterns.) (The preceding apprehensions do not apply to improvement of product features to increase sales. These are welcomed as having the potential to provide new opportunities and greater job security.)

SECURING UPPER MANAGEMENT APPROVAL AND PARTICIPATION

The lessons learned during the 1980s and 1990s include a major finding: Personal participation by upper managers is indispensable to getting a high rate of annual quality improvement. This finding suggests that advocates for quality initiatives should take positive steps to convince the upper managers of

The merits of annual quality improvement

The need for active upper management participation

The precise nature of the needed upper management participation

Awareness: Proof of the Need.   Upper managers respond best when they are shown a major threat or opportunity. An example of a major threat is seen in the case of company G, a maker of household appliances. Company G and its competitors R and T were all suppliers to a major customer involving four models of appliances. (See Table 5.1.) This table shows that in 1980, company G was a supplier for two of the four models. Company G was competitive in price, on-time delivery, and product features, but it was definitely inferior in the customer's perception of quality, the chief problem being field failures. By 1982, lack of response had cost company G the business on model number 1. By 1983, company G also had lost the business on model number 3.

TABLE 5.1   Suppliers to a Major Customer

Model number	1980	1981	1982	1983
1	G	G	R	R
2	R	R	R	R
3	G	G	G	R
4	T	R	R	R

Awareness also can be created by showing upper managers other opportunities, such as cost reduction through cutting chronic waste.

The Size of the Chronic Waste. A widespread major opportunity for upper managers is to reduce the cost of poor quality. In most cases, this cost is greater than the company's annual profit, often much greater. Quantifying this cost can go far toward proving the need for a radical change in the approach to quality improvement. An example is shown in Table 5.2. This table shows the estimated cost of poor quality for a company in a process industry using the traditional accounting classifications. The table brings out several matters of importance to upper managers:

TABLE 5.2   Analysis of Cost of Poor Quality

Category	Amount, $	Percent of total
Internal failures	7,279,000	79.4
External failures	283,000	3.1
Appraisal	1,430,000	15.6
Prevention	170,000	1.9
	9,162,000	100.0

The order of magnitude: The total of the costs is estimated at $9.2 million per year. For this company, this sum represented a major opportunity. (When such costs have never before been brought together, the total is usually much larger than anyone would have expected.)

The areas of concentration: The table is dominated by the costs of internal failures—they are 79.4 percent of the total. Clearly, any major cost reduction must come from the internal failures.

The limited efforts for prevention: The figure of 1.9 percent for prevention suggests that greater investment in prevention would be cost-effective.

(For elaboration, see Section 8, Quality and Costs.)

The Potential Return on Investment. A major responsibility of upper managers is to make the best use of the company's assets. A key measure of judging what is best is *return on investment (ROI)*. In general terms, ROI is the ratio of (1) the estimated gain to (2) the estimated resources needed. Computing ROI for projects to reduce chronic waste requires assembling estimates such as

The costs of chronic waste associated with the projects

The potential cost reductions if the projects are successful

The costs of the needed diagnosis and remedy

Many proposals to go into quality improvement have failed to gain management support because no one has quantified the ROI. Such an omission is a handicap to the upper managers—they are unable to compare (1) the potential ROI from quality improvement with (2) the potential ROI from other opportunities for investment.

Quality managers and others who prepare such proposals are well advised to prepare the information on ROI in collaboration with those who have expertise in the intricacies of ROI. Computation of ROI gets complicated because two kinds of money are involved—capital and expenses. Each is money, but in some countries (including the United States) they are taxed differently. Capital expenditures are made from after-tax money, whereas expenses are paid out of pretax money.

This difference in taxation is reflected in the rules of accounting. Expenses are written off promptly, thereby reducing the stated earnings and hence the income taxes on earnings. Capital expenditures are written off gradually—usually over a period of years. This increases the stated earnings and hence the income taxes on those earnings. All this is advantageous to proposals to go into quality improvement because quality improvement is seldom capital intensive. (Some upper managers tend to use the word *investment* as applying only to capital investment.)

Use of Bellwether Projects.   Presentation of the cost figures becomes even more effective if it is accompanied by a "bellwether project"—a case example of a successful quality improvement actually carried out within the company. Such was the approach used in the ABC company, a large maker of electronic instruments.

Historically, ABC's cost of poor quality ran to about $200 million annually. A notorious part of this was the $9 million of annual cost of scrap for a certain electronic component. The principal defect type was defect X. It had been costing about $3 million per year.

The company had launched a project to reduce the frequency of defect X. The project was a stunning success—it had cut the cost of defect X from $3 million to $1 million—an annual improvement of $2 million. The investment needed to make this improvement was modest—about one-fourth of a million—to fine-tune the process and its controls. The gain during the first year of application had been eight times the investment.

This bellwether project was then used to convince the upper managers that expansion of quality improvement could greatly reduce the company's cost of poor quality and do so at a high return on the investment.

In most companies, the previously successful quality improvements can serve collectively as a bellwether project. The methodology is as follows:

Identify the quality improvement projects completed within the last year or two.

For each such project, estimate (1) what was gained and (2) what was the associated expenditure.

Summarize, and determine the composite ROI.

Compare this composite with the returns being earned from other company activities. (Such comparisons usually show that quality improvement provides the highest rate of return.)

Getting the Cost Figures.   Company accounting systems typically quantify only a minority of the costs of poor quality. The majority are scattered throughout the various overheads. As a result, quality specialists have looked for ways to supply what is missing. Their main efforts toward solution have been as follows:

1. *Make estimates:*   This is the "quick and dirty" approach. It is usually done by sampling and involves only a modest amount of effort. It can, in a few days or weeks, provide (a) an evaluation of the approximate cost of chronic waste and (b) indicate where this is concentrated.

2. *Expand the accounting system:*   This is much more elaborate. It requires a lot of work from various departments, especially Accounting and Quality. It runs into a lot of calendar time, often two or three years. (For elaboration, see Section 8, Quality and Costs.)

In my experience, estimates involve much less work, can be prepared in far less time, and yet are adequate for managerial decision making.

Note that the demand for "accuracy" of the cost figures depends on the use to which the figures will be put. Balancing the books demands a high degree of accuracy. Making managerial decisions sometimes can tolerate a margin of error. For example, a potential improvement project has been estimated to incur about $300,000 in annual cost of poor quality. This figure is challenged. The contesting estimates range from $240,000 to $360,000—quite a wide range. Then someone makes an incisive observation: "It doesn't matter which estimate is correct. Even at the lowest figure, this is a good opportunity for improvement, so let's tackle it." In other words, the managerial decision to tackle the project is identical despite a wide range of estimate.

Languages in the Hierarchy.   A subtle aspect of securing upper management approval is *choice of language.* Industrial companies make use of two standard languages—the language of money and the language of things. (There are also local dialects, each peculiar to a specific function.) However, as seen in Figure 5.4, use of the standard languages is not uniform.

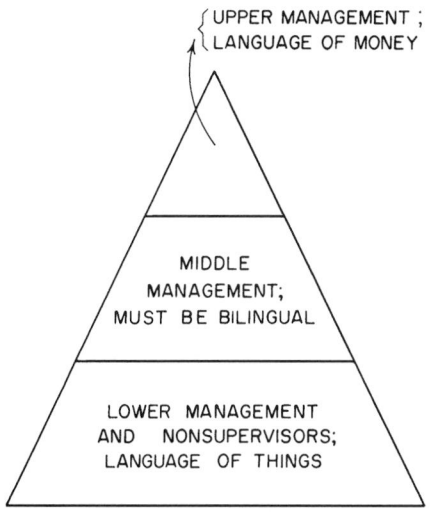

FIGURE 5.4 shows the use of standard languages in different levels of a typical hierarchy. At the apex, the principal language of the top management team is the language of money. At the base, the principal language of the first-line supervisors and the work force is the language of things. In between, the middle managers and the specialists need to understand both the principal languages—the *middle managers should be bilingual.*

It is quite common for chronic waste to be measured in the language of things: percent errors, process yields, hours of rework, and so on. Converting these measures into the language of money enables upper managers to relate them to the financial measures that have long dominated the management "instrument panel."

Years ago, I was invited to visit a major British manufacturer to study its approach to managing for quality, and to provide a critique. I found that the company's cost of poor quality was huge, that it was feasible to cut this in two in 5 years, and that the resulting

FIGURE 5.4   Common languages in the hierarchy.

return on investment would be much greater than that of making and selling the company's products. When I explained this to the managing director, he was most impressed—it was the first time that the problem of chronic waste had been explained to him in the language of return on investment. He promptly convened his directors (vice presidents) to discuss what to do about this opportunity.

Presentations to Upper Managers.   Presentations to upper managers should focus on the goals of the upper managers, not on the goals of the advocates. Upper managers are faced with meeting the needs of various stakeholders: customers, owners, employees, suppliers, the public (e.g., safety, health, the environment), and so on. It helps if the proposals identify specific problems of stakeholders and estimate the benefits to be gained.

Upper managers receive numerous proposals for allocating the company's resources: invade foreign markets, develop new products, buy new equipment to increase productivity, make acquisitions, enter joint ventures, and so on. These proposals compete with each other for priority, and a major test is return on investment (ROI). It helps if the proposal to go into quality improvement includes estimates of ROI.

Explanation of proposals is sometimes helped by converting the supporting data into units of measure that are already familiar to upper managers. For example:

Last year's cost of poor quality was five times last year's profit of $1.5 million.

Cutting the cost of poor quality in half would increase earnings by 13 cents per share of stock.

Thirteen percent of last year's sales orders were canceled due to poor quality.

Thirty-two percent of engineering time was spent in finding and correcting design weaknesses.

Twenty-five percent of manufacturing capacity is devoted to correcting quality problems.

Seventy percent of the inventory carried is traceable to poor quality.

Twenty-five percent of all manufacturing hours were spent in finding and correcting defects.

Last year's cost of poor quality was the equivalent of the X factory making 100 percent defective work during the entire year.

Experience in making presentations to upper management has evolved some useful do's and don'ts.

Do summarize the total of the estimated costs of poor quality. The total will be big enough to command upper management attention.

Do show where these costs are concentrated. A common grouping is in the form of Table 5.2. Typically (as in that case), most of the costs are associated with failures, internal and external. Table 5.2 also shows the fallacy of trying to start by reducing inspection and test. The failure costs should be reduced first. After the defect levels come down, inspection costs can be reduced as well.

Do describe the principal projects that are at the heart of the proposal.

Do estimate the potential gains, as well as the return on investment. If the company has never before undertaken an organized approach to reducing quality-related costs, then *a reasonable goal is to cut these costs in two* within a space of 5 years.

Do have the figures reviewed in advance by those people in finance (and elsewhere) to whom upper management looks for checking the validity of financial figures.

Don't inflate the present costs by including debatable or borderline items. The risk is that the decisive review meetings will get bogged down in debating the validity of the figures without ever getting to discuss the merits of the proposals.

Don't imply that the total costs will be reduced to zero. Any such implication will likewise divert attention from the merits of the proposals.

Don't force the first few projects on managers who are not really sold on them or on unions who are strongly opposed. Instead, start in areas that show a climate of receptivity. The results obtained in these areas will determine whether the overall initiative will expand or die out.

The needs for quality improvement go beyond satisfying customers or making cost reductions. New forces keep coming over the horizon. Recent examples have included growth in product liability, the consumerism movement, foreign competition, and legislation of all sorts. Quality improvement has provided much of the response to such forces.

Similarly, the means of convincing upper managers of the need for quality improvement go beyond reports from advocates. Conviction also may be supplied by visits to successful companies, hearing papers presented at conferences, reading reports published by successful companies, and listening to the experts, both internal and external. However, none of these is as persuasive as results achieved within one's own company.

A final element of presentation to upper managers is to explain their personal responsibilities in launching and perpetuating quality improvement. (See below, under The Nondelegable Roles of Upper Managers.)

MOBILIZING FOR QUALITY IMPROVEMENT

Until the 1980s, quality improvement in the West was not mandated—it was not a part of the business plan or a part of the job descriptions. Some quality improvement did take place, but on a voluntary basis. Here and there a manager or a nonmanager, for whatever reason, elected to tackle some improvement project. He or she might persuade others to join an informal team. The result might be favorable, or it might not. This voluntary, informal approach yielded few improvements. The emphasis remained on inspection, control, and firefighting.

The Need for Formality.   The quality crisis that followed the Japanese quality revolution called for new strategies, one of which was a much higher rate of quality improvement. It then became evident that an informal approach would not produce thousands (or more) improvements year after year. This led to experiments with structured approaches that in due course helped some companies to become the role models.

Some upper managers protested the need for formality. "Why don't we just do it?" The answer depends on how many improvements are needed. For just a few projects each year, informality is adequate; there is no need to mobilize. However, making improvements by the hundreds or the thousands does require a formal structure. (For some published accounts of company experiences in mobilizing for quality improvement, see under References, Some Accounts of Mobilizing for Quality Improvement.)

As it has turned out, mobilizing for improvement requires two levels of activity, as shown in Figure 5.5. The figure shows the two levels of activity. One of these mobilizes the company's resources to deal with the improvement projects *collectively*. This becomes the responsibility of management. The other activity is needed to carry out the projects *individually*. This becomes the responsibility of the quality improvement teams.

THE QUALITY COUNCIL

The first step in mobilizing for quality improvement is to establish the company's quality council (or similar name). The basic responsibility of this council is to launch, coordinate, and "institutionalize" annual quality improvement. Such councils have been established in many companies. Their experiences provide useful guide lines.

Membership.   Council membership is typically drawn from the ranks of senior managers. Often the senior management committee is also the quality council. Experience has shown that quality councils are most effective when upper managers are personally the leaders and members of the senior quality councils.

In large companies, it is common to establish councils at the divisional level as well as at the corporate level. In addition, some individual facilities may be so large as to warrant establishing a local quality council. When multiple councils are established, they are usually linked together—members of high-level councils serve as chairpersons of lower-level councils. Figure 5.6 is an example of such linkage.

Experience has shown that organizing quality councils solely in the lower levels of management is ineffective. Such organization limits quality improvement projects to the "useful many" while neglecting the "vital few" projects—those which can produce the greatest results. In addition, quality councils solely at lower levels send a message to all: "Quality improvement is not high on upper management's agenda."

Responsibilities.   It is important for each council to define and publish its responsibilities so that (1) the members agree on what is their mission, and (2) the rest of the organization can become informed relative to upcoming events.

Activities by management	Activities by teams
Establish quality councils	Analyze symptoms
Select projects; write mission statements	Theorize as to causes
Assign teams	Test theories
Review progress	Establish causes
Provide recognition and rewards	Stimulate remedies and controls

FIGURE 5.5   Mobilizing for quality improvement.

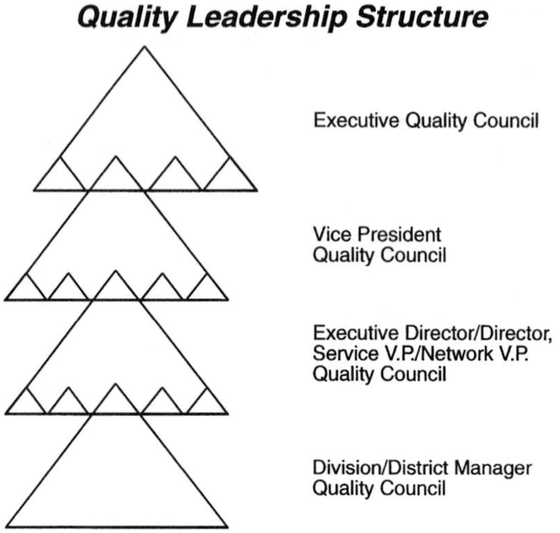

Quality Leadership Structure

Executive Quality Council

Vice President
Quality Council

Executive Director/Director,
Service V.P./Network V.P.
Quality Council

Division/District Manager
Quality Council

FIGURE 5.6 How quality councils are linked together. (*From Making Quality Happen, 1988, Juran Institute, Wilton, CT, p. D17.*)

Many quality councils have published their statements of responsibility. Major common elements have included the following:

Formulate the quality policies, such as focus on the customer, quality has top priority, quality improvement must go on year after year, participation should be universal, or the reward system should reflect performance on improvement.

Estimate the major dimensions, such as status of quality compared with competitors, extent of chronic waste, adequacy of major business processes, or results achieved by prior improvements.

Establish processes for selecting projects, such as soliciting and screening nominations, choosing projects to be tackled, preparing mission statements, or creating a favorable climate for quality improvement.

Establish processes for carrying out the projects, such as selecting team leaders and members or defining the role of project teams.

Provide support for the project teams, such as training (see Section 16, Training for Quality), time for working on projects, diagnostic support, facilitator support, or access to facilities for tests and tryouts.

Establish measures of progress, such as effect on customer satisfaction, effect on financial performance, or extent of participation by teams.

Review progress, assist teams in the event of obstacles, and ensure that remedies are implemented.

Provide for public recognition of teams.

Revise the reward system to reflect the changes demanded by introducing annual quality improvement.

Anticipating the Questions. Announcement of a company's intention to go into annual quality improvement always stimulates questions from subordinate levels, questions such as

What is the purpose of this new activity?

How does it relate to other ongoing efforts to make improvements?

How will it affect other quality-oriented activities?

What jobs will be affected, and how?

What actions will be taken, and in what sequence?

In view of this new activity, what should we do that is different from what we have been doing?

Quality councils should anticipate the troublesome questions and, to the extent feasible, provide answers at the time of announcing the intention to go into annual quality improvement. Some senior managers have gone to the extent of creating a videotape to enable a wide audience to hear the identical message from a source of undoubted authority.

Apprehensions about Elimination of Jobs.   Employees not only want answers to such questions, they also want assurance relative to their apprehensions, notably the risk of job loss due to quality improvement. Most upper managers have been reluctant to face up to these apprehensions. Such reluctance is understandable. It is risky to provide assurances when the future is uncertain.

Nevertheless, some managers have estimated in some depth the two pertinent rates of change:

1. The rate of creation of job openings due to attrition: retirements, offers of early retirement, resignation, and so on. This rate can be estimated with a fair degree of accuracy.

2. The rate of elimination of jobs due to reduction of chronic waste. This estimate is more speculative—it is difficult to predict how soon the improvement rate will get up to speed. In practice, companies have been overly optimistic in their estimates.

Analysis of these estimates can help managers to judge what assurances they can provide, if any. It also can shed light on choice of alternatives for action: retrain for jobs that have opened up, reassign to areas that do have job openings, offer early retirement, assist in finding jobs in other companies, and/or provide assistance in the event of termination.

Assistance from the Quality Department.   Many quality councils secure the assistance of the Quality Department to

Provide inputs needed by the council for planning to introduce quality improvement

Draft proposals and procedures

Carry out essential details such as screening nominations for projects

Develop training materials

Develop new measures for quality

Prepare reports on progress

It is also usual, but not invariable, for the quality manager to serve as secretary of the quality council.

QUALITY IMPROVEMENT GOALS IN THE BUSINESS PLAN

Companies that have become the quality leaders—the role models—all adopted the practice of enlarging their business plan to include quality-oriented goals. In effect, they translated the threats and opportunities faced by their companies into quality goals such as

Increase on-time deliveries from 83 to 98 percent over the next 2 years.

Reduce the cost of poor quality by 50 percent over the next 5 years.

Such goals are clear—each is quantified, and each has a timetable. Convincing upper managers to establish such goals is a big step, but it is only the first step.

Deployment of Goals.   Goals are merely a wish list until they are *deployed*—until they are broken down into specific projects to be carried out and assigned to specific individuals or teams who are then provided with the resources needed to take action. Figure 5.7 shows the anatomy of the deployment process. In the figure, the broad (strategic) quality goals are established by the quality council and become a part of the company business plan. These goals are then divided and allocated to lower levels to be translated into action. In large organizations there may be further subdivision before the action levels are reached. The final action level may consist of individuals or teams.

In response, the action levels select improvement *projects* that collectively will meet the goals. These projects are then proposed to the upper levels along with estimates of the resources needed. The proposals and estimates are discussed and revised until final decisions are reached. The end result is an agreement on which projects to tackle, what resources to provide, and who will be responsible for carrying out the projects.

This approach of starting at the top with strategic quality goals may seem like purely a top-down activity. However, the deployment process aims to provide open discussion in both directions before final decisions are made, and such is the way it usually works out.

The concept of strategic quality goals involves the vital few matters, but it is not limited to the corporate level. Quality goals also may be included in the business plans of divisions, profit centers, field offices, and still other facilities. The deployment process is applicable to all of these. (For added discussion of the deployment process, see Section 13, Strategic Planning.)

The Project Concept.   As used here, a *project* is a chronic problem scheduled for solution. The project is the focus of actions for quality improvement. All improvement takes place project by project and in no other way.

Some projects are derived from the quality goals that are in the company business plan. These are relatively few in number, but each is quite important. Collectively, these are among the vital few projects (see below, under Use of the Pareto principle). However, most projects are derived not from the company business plan but from the nomination-selection process, as discussed below.

Use of the Pareto Principle.   A valuable aid to selection of projects during the deployment process is the *Pareto Principle*. This principle states that in any population that contributes to a com-

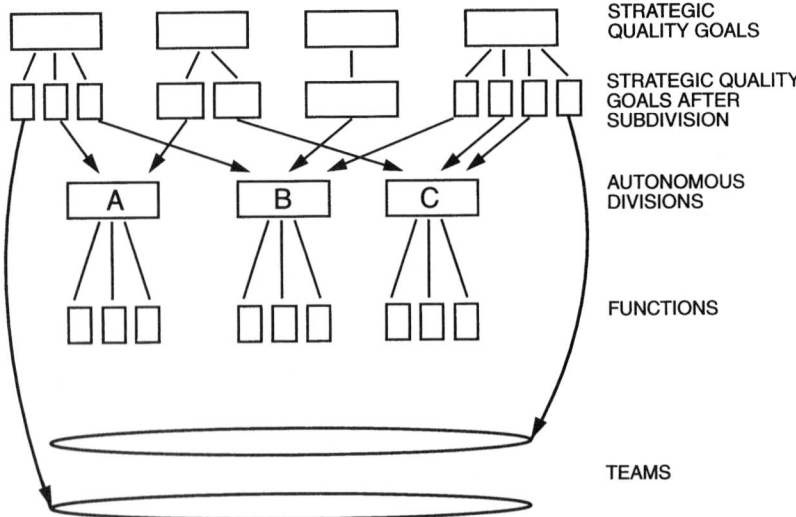

FIGURE 5.7   Anatomy of the deployment process. (*From Visual OPQ9-2, Juran Institute, Inc., Wilton, CT.*)

mon effect, a relative few of the contributors—the vital few—account for the bulk of the effect. The principle applies widely in human affairs. Relatively small percentages of the individuals write most of the books, commit most of the crimes, own most of the wealth, and so on.

An example of using the Pareto principle to select projects is seen in a paper mill's goal of reducing its cost of poor quality. The estimated total was $9,070,000 per year, divided among seven accounting categories. (See Table 5.3.) One of these seven categories is called "broke." It amounts to $5,560,000, or 61 percent of total. Clearly, there will be no major improvement in the total unless there is a successful attack on broke—this is where the money is concentrated. (Broke is paper mill dialect for paper so defective that it must be returned to the beaters for reprocessing.)

This paper mill makes 53 types of paper. When the broke is analyzed by type of paper, the Pareto principle is again in evidence. (See Table 5.4.) Six of the 53 product types account for $4,480,000, which is 80 percent of the $5,560,000. There will be no major improvement in broke unless there is a successful attack on these six types of paper. Note that studying 12 percent of the product types results in attacking 80 percent of the cost of broke.

Finally, the analysis is extended to the defect types that result in the major cost of broke. There are numerous defect types, but five of them dominate. (See Table 5.5.) The largest number is $612,000 for "tear" on paper type B. Next comes $430,000 for "porosity" on paper type A, and so on. Each such large number has a high likelihood of being nominated for an improvement project.

Identification of the vital few (in this case, accounting categories, product types, and defect types) is made easier when the tabular data are presented in graphic form. Figures 5.8, 5.9, and 5.10 present the paper mill data graphically. Like their tabular counterparts, each of these graphs contains three elements:

1. The contributors to the total effect, ranked by the magnitude of their contribution
2. The magnitude of the contribution of each expressed numerically and as a percentage of total
3. The cumulative percentage of total contribution of the ranked contributors

TABLE 5.3   Pareto Analysis by Accounts

Accounting category	Annual quality loss,* $thousands	Percent of total quality loss	
		This category	Cumulative
Broke	5560	61	61
Customer claim	1220	14	75
Odd lot	780	9	84
High material cost	670	7	91
Downtime	370	4	95
Excess inspection	280	3	98
High testing cost	190	2	100
TOTAL	9070		

*Adjusted for estimated inflation since time of original study.

TABLE 5.4   Pareto Analysis by Products

Product type	Annual broke loss,* $thousands	Percent of broke loss	Cumulative percent broke loss
A	1320	24	24
B	960	17	41
C	720	13	54
D	680	12	66
E	470	8	74
F	330 (4480)	6	80
47 other types	1080	20	100
TOTAL 53 types	5560	100	

*Adjusted for estimated inflation since time of original study.

TABLE 5.5   Matrix of Quality Costs*

Type	Trim, $thousands	Visual defects,† $thousands	Caliper, $thousands	Tear, $thousands	Porosity, $thousands	All other causes, $thousands	Total, $thousands
A	270	94	None‡	162	430	364	1320
B	120	33	None‡	612	58	137	960
C	95	78	380	31	74	62	720
D	82	103	None‡	90	297	108	680
E	54	108	None‡	246	None‡	62	470
F	51	49	39	16	33	142	330
TOTAL	672	465	419	1157	892	875	4480

*Adjusted for estimated inflation since time of original study.

†Slime spots, holes, wrinkles, etc.

‡Not a specified requirement for this type.

Key:	Loss	Cum. Loss	% Category	% Cum.
BK	5,560,000	5,560,000	61	61
CC	1,220,000	6,780,000	13	75
OL	780,000	7,560,000	9	83
MC	670,000	8,230,000	7	91
DT	370,000	8,600,000	4	95
XI	280,000	8,880,000	3	98
TC	190,000	9,070,000	2	100
Total	9,070,000		100	

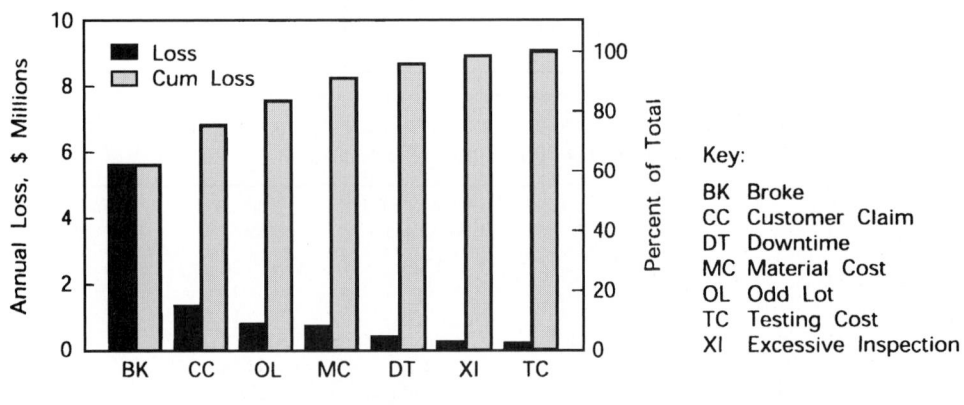

FIGURE 5.8   Pareto analysis: annual loss by category.

Key:	Loss	Cum. Loss	Type	% Loss	Cum. %		
A	1,320,000	1,320,000	A	24	24	1320	1320
B	960,000	2,280,000	B	17	41	960	2280
C	720,000	3,000,000	C	13	54	720	3000
D	680,000	3,680,000	D	12	66	680	3680
E	470,000	4,150,000	E	8	75	470	4150
F	330,000	4,480,000	F	6	81	330	4480
Other	1,080,000	5,560,000	Other	19	100	1080	5560
TOTAL	5,560,000			100		5560	

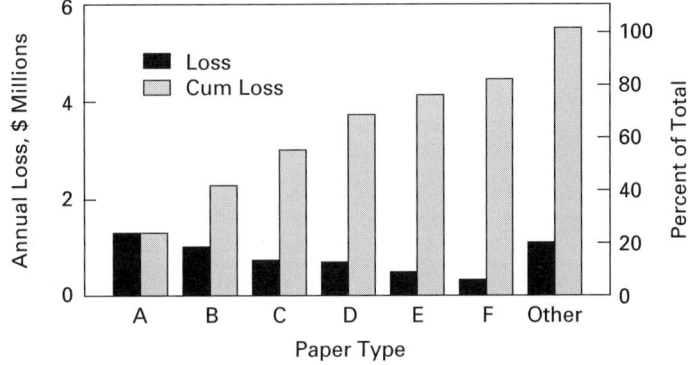

FIGURE 5.9   Pareto analysis: annual loss by paper type.

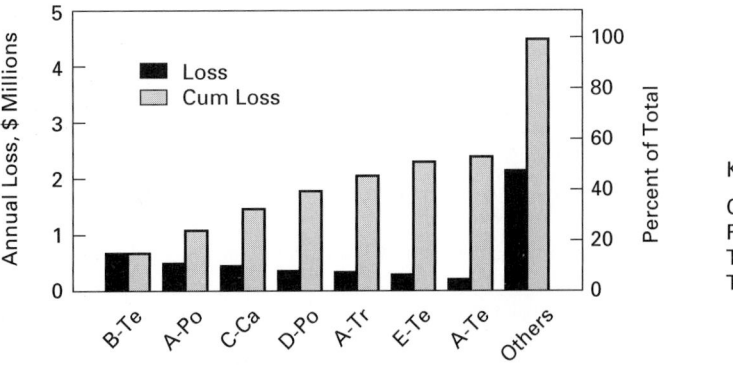

Key:

Ca   Caliper
Po   Porosity
Te   Teat
Tr   Trim

FIGURE 5.10   Pareto analysis: annual loss by defect type.

In addition to facilitating analysis, presentation of the data in the form of a Pareto diagram greatly enhances communication of the information, most notably in convincing upper management of the source of a problem and gaining support for a proposed course of action to remedy the problem. [For an account of how I came to misname the Pareto principle, see Juran (1975).]

The Useful Many Projects.   Under the Pareto principle, the vital few projects provide the bulk of the improvement, so they receive top priority. Beyond the vital few are the useful many projects. Collectively they contribute only a minority of the improvement, but they provide most of the opportunity for employee participation. Choice of these projects is made through the nomination-selection process.

THE NOMINATION AND SELECTION PROCESS

Most projects are chosen through the nomination and selection process, involving several steps:

Project nomination

Project screening and selection

Preparation and publication of project mission statements

Sources of Nominations.   Nominations for projects can come from all levels of the organization. At the higher levels, the nominations tend to be extensive in size (the vital few) and multifunctional in their scope. At lower levels, the nominations are smaller in size (the useful many) and tend to be limited in scope to the boundaries of a single department.

Nominations come from many sources. These include

Formal data systems such as field reports on product performance, customer complaints, claims, returns, and so on; accounting reports on warranty charges and on internal costs of poor quality; and service call reports. (Some of these data systems provide for analyzing the data to identify problem areas.) [For an example of project nomination based on customer complaints, see Rassi (1991).]

Special studies such as customer surveys, employee surveys, audits, assessments, benchmarking against competitive quality, and so on.

Reactions from customers who have run into product dissatisfactions are often vocal and insistent. In contrast, customers who judge product features to be not competitive may simply (and quietly) become ex-customers.

Field intelligence derived from visits to customers, suppliers, and others; actions taken by competitors; and stories published in the media (as reported by sales, customer service, technical service, and others).

The impact of quality on society, such as new legislation, extension of government regulation, and growth of product liability lawsuits.

The managerial hierarchy, such as the quality council, managers, supervisors, professional specialists, and project teams.

The work force through informal ideas presented to supervisors, formal suggestions, ideas from quality circles, and so on.

Proposals relating to *business processes.*

Effect of the Big Q Concept.   Beginning in the 1980s, the scope of nominations for projects broadened considerably under the big Q concept. (For details relative to the big Q concept, see Section 2, Figure 2.1.)

The breadth of the big Q concept is evident from the wide variety of projects that have already been tackled:

Improve the precision of the sales forecast.

Reduce the cycle time for developing new products.

Increase the success rate in bidding for business.

Reduce the time required to fill customers' orders.

Reduce the number of sales cancellations.

Reduce the errors in invoices.

Reduce the number of delinquent accounts.

Reduce the time required to recruit new employees.

Improve the on-time arrival rate (for transportation services).

Reduce the time required to file for patents.

(For examples from many industries, see proceedings of IMPRO conferences. See also *The Juran Report.*) (For elaboration on projects in business processes, see Section 6, Process Management.)

The Nomination Processes.   Nominations must come from human beings. Data systems are impersonal—they make no nominations. Various means are used to stimulate nominations for quality improvement projects:

Call for nominations: Letters or bulletin boards are used to invite all personnel to submit nominations, either through the chain of command or to a designated recipient such as the secretary of the quality council.

Make the rounds: In this approach, specialists (such as quality engineers) are assigned to visit the various departments, talk with the key people, and secure their views and nominations.

The council members themselves: They become a focal point for extensive data analyses and proposals.

Brainstorming meetings: These are organized for the specific purpose of making nominations.

Whatever the method used, it will produce the most nominations if it urges use of the big Q concept— the entire spectrum of activities, products, and processes.

Nominations from the Work Force.   The work force is potentially a source of numerous nominations. Workers have extensive residence in the workplace. They are exposed to many local cycles of activity. Through this exposure, they are well poised to identify the existence of quality problems and to theorize about their causes. As to the details of goings on in the workplace, no one is better informed than the work force. "That machine hasn't seen a maintenance man for the last 6 months." In addition, many workers are well poised to identify opportunities and to propose new ways.

Work force nominations consist mainly of local useful many projects along with proposals of a human relations nature. For such nominations, workers can supply useful theories of causes as well as practical proposals for remedies. For projects of a multifunctional nature, most workers are handicapped by their limited knowledge of the overall process and of the interactions among the steps that collectively make up the overall.

In some companies, the solicitation of nominations from the work force has implied that such nominations would receive top priority. The effect was that the work force was deciding which projects the managers should tackle first. It should have been made clear that workers' nominations must compete for priority with nominations from other sources.

Joint Projects with Suppliers and Customers.   All companies buy goods and services from suppliers; over half the content of the finished product may come from suppliers. In earlier decades, it was common for customers to contend that "the supplier should solve his quality problems." Now there is growing awareness that these problems require a partnership approach based on

Establishing mutual trust

Defining quality in terms of customer needs as well as specifications

Exchanging essential data

Direct communication at the technical level as well as the commercial level

This approach gains momentum from joint projects between suppliers and customers. Published examples include

Alcoa and Kodak, involving photographic plates (Kegarise and Miller 1985).

Alcoa and Nalco, involving lubricants for rolling mills (Boley and Petska 1990).

Alcoa and Phifer, involving aluminum wire (Kelly et al. 1990).

NCR and its customers, establishing a universal code for tracking product failures as they progress through the customer chain (Daughton 1987).

Efforts to serve customers are sometimes delayed by actions of the customers themselves.

A maker of technological instruments encountered delays when installing the instruments in customers' premises, due to lack of site preparation. When the installers arrived at the site, the foundation was not yet in place, supply lines such as compressed air were not yet in place, and so on. The company analyzed a number of these delays and then created a videotape on site preparation. The company sent this videotape to customers at the time of signing the contract. Once the site was ready, the customers sent back a certificate to this effect. The result was a sharp drop in installation time, improved delivery to customers, as well as a cost reduction (communication to the author).

For further information on Quality Councils, see Section 13, Strategic Planning, and Section 14, Total Quality Management.

PROJECT SCREENING

A call for nominations can produce large numbers of responses—numbers that are beyond the digestive capacity of the organization. In such cases, an essential further step is *screening* to identify those nominations which promise the most benefits for the effort expended.

To start with a long list of nominations and end up with a list of agreed projects requires an organized approach—an infrastructure and a methodology. The screening process is time-consuming, so the quality council usually delegates it to a secretariat, often the Quality Department. The secretariat screens the nominations—it judges the extent to which the nominations meet the criteria set out below. These judgments result in some preliminary decision making. Some nominations are rejected. Others are deferred. The remainder are analyzed in greater depth to estimate potential benefits, resources needed, and so on.

The quality councils and/or the secretariats have found it useful to establish criteria to be used during the screening process. Experience has shown that there is need for two sets of criteria:

1. Criteria for choosing the first projects to be tackled by any of the project teams

2. Criteria for choosing projects thereafter

Criteria for the First Projects.   During the beginning stages of project-by-project improvement, everyone is in a learning state. Projects are assigned to project teams who are in training. Completing a project is a part of that training. Experience with such teams has evolved a broad criterion: *The first project should be a winner.* More specifically:

The project should deal with a *chronic* problem—one that has been awaiting solution for a long time.

The project should be *feasible.* There should be a good likelihood of completing it within a few months. Feedback from companies suggests that the most frequent reason for failure of the first project has been failure to meet the criterion of feasibility.

The project should be *significant.* The end result should be sufficiently useful to merit attention and recognition.

The results should be *measurable,* whether in money or in other significant terms.

Criteria for Projects Thereafter.   These criteria aim to select projects that will do the company the most good:

Return on investment: This factor has great weight and is decisive, all other things being equal. Projects that do not lend themselves to computing return on investment must rely for their priority on managerial judgment.

The amount of potential improvement: One large project will take priority over several small ones.

Urgency: There may be a need to respond promptly to pressures associated with product safety, employee morale, and customer service.

Ease of technological solution: Projects for which the technology is well developed will take precedence over projects that require research to discover the needed technology.

Health of the product line: Projects involving thriving product lines will take precedence over projects involving obsolescent product lines.

Probable resistance to change: Projects that will meet a favorable reception take precedence over projects that may meet strong resistance, such as from the labor union or from a manager set in his or her ways.

Some companies use a systematic approach to evaluate nominations relative to these criteria. This yields a composite evaluation that then becomes an indication of the relative priorities of the nominations. [For an example, see Hartman (1983); also see DeWollf et al. (1987).]

PROJECT SELECTION

The end result of the screening process is a list of recommended projects in their order of priority. Each recommendation is supported by the available information on compatibility with the criteria and potential benefits, resources required, and so on. This list is commonly limited to matters in which the quality council has a direct interest.

The quality council reviews the recommendations and makes the final determination on which projects are to be tackled. These projects then become an official part of the company's business. Other recommended projects are outside the scope of the direct interest of the quality council. Such projects are recommended to appropriate subcouncils, managers, and so on. None of the preceding prevents projects from being undertaken at local levels by supervisors or by the work force.

Vital Few and Useful Many.   During the 1980s, some companies completed many quality improvement projects. Then, when questions were raised—"What have we gotten for all this effort?"—they were dismayed to learn that there was no noticeable effect on the "bottom line." Investigation then showed that the reason was traceable to the process used for project selection. The projects actually selected had consisted of

Firefighting projects: These are special projects for getting rid of sporadic "spikes." Such projects did not attack the chronic waste and hence could not improve financial performance.

Useful many projects: By definition, these have only a minor effect on financial performance.

Projects for improving human relations: These can be quite effective in their field, but the financial results are usually not measurable.

To achieve a significant effect on the bottom line requires selecting the "vital few" projects as well as the "useful many." It is feasible to work on both, since different people are assigned to each.

There is a school of thought that contends that the key to quality leadership is "tiny improvements in a thousand places"—in other words, the useful many (Gross 1989). Another school urges focus on the vital few. In my experience, neither of these schools has the complete answer.

The vital few projects are the major contributors to quality leadership and to the bottom line. The useful many projects are the major contributors to employee participation and to the quality of work life. Each is necessary; neither is sufficient.

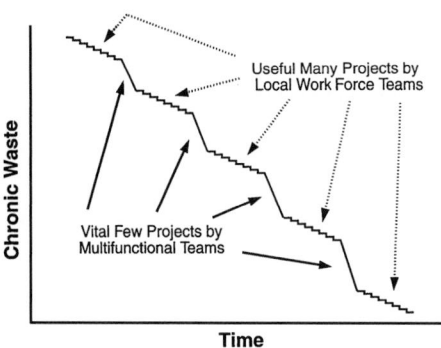

FIGURE 5.11   Interrelation of projects, vital few and useful many.

The vital few and useful many projects can be carried out simultaneously. Successful companies have done just that. They did so by recognizing that while there are these two types of projects, they require the time of different categories of company personnel.

The interrelation of these two types of projects is shown in Figure 5.11. In this figure, the horizontal scale is time. The vertical scale is chronic waste. What goes up is bad. The useful many improvements collectively create a gradually sloping line. The vital few improvements, though less frequent, contribute the bulk of the total improvement.

Cost Figures for Projects.   To meet the preceding criteria (especially that of return on investment) requires information on various costs:

The cost of chronic waste associated with a given nomination

The potential cost reduction if the project is successful

The cost of the needed diagnosis and remedy

For the methodology of providing the cost figures, see above, under Getting the Cost Figures.

Costs versus Percent Deficiencies.   It is risky to judge priorities based solely on the percentage of deficiencies (errors, defects, and so on). On the face of it, when this percentage is low, the priority of the nomination also should be low. In some cases this is true, but in others it can be seriously misleading.

In a large electronics company the percentage of invoices protested by customers was 2.4 percent. While this was uncomfortable, it was below the average for similar processes in the industry. Then a study in-depth showed that nearly half the time of the sales force was spent placating the protesting customers and getting the invoices straightened out. During that time, the sales people were not selling anything (communication to the author).

Even more dramatic was the case of the invoices in Florida Power and Light Company. Protested invoices ran to about 60,000 per year, which figured to about 0.2 percent of all invoices. The cost to straighten these out came to about $2.1 million annually. A quality improvement project then cut the percent errors to about 0.05 percent, at an annual saving of over $1 million. Even more important was the improvement in customer relations and the related reduction of complaints to the Public Utility Commission (Florida Power and Light Company 1984).

Elephant-Sized and Bite-Sized Projects.   Some projects are "elephant-sized"; i.e., they cover so broad an area of activity that they must be subdivided into multiple "bite-sized" projects. In such cases, one project team can be assigned to "cut up the elephant." Other teams are then assigned to tackle the resulting bite-sized projects. This approach shortens the time to complete the project, since the teams work concurrently. In contrast, use of a single team stretches the time out to several years. Frustration sets in, team membership changes due to attrition, the project drags, and morale declines.

In the Florida Power and Light Company invoice case, the project required several teams, each assigned to a segment of the invoicing process.

In Honeywell, Inc., a project to improve the information security system required creation of seven teams, involving 50 team members. (See Parvey 1990.)

A most useful tool for cutting up the elephant is the Pareto analysis. For an application, see the paper mill example earlier, under Use of the Pareto Principle.

For elephant-sized projects, separate mission statements (see below) are prepared for the broad coordinating team and for each team assigned to a bite-sized project.

Cloning.   Some companies consist of multiple autonomous units that exhibit much commonality. A widespread example is the chains of retail stores, repair shops, hospitals, and so on. In such companies, a quality improvement project that is carried out successfully in one operating unit logically becomes a nomination for application to other units. This is called *cloning the project.*

It is quite common for the other units to resist applying the improvement to their operation. Some of this resistance is cultural in nature (not invented here, and so on). Other resistance may be due to real differences in operating conditions. For example, telephone exchanges perform similar functions for their customers. However, some serve mainly industrial customers, whereas others serve mainly residential customers.

Upper managers are wary of ordering autonomous units to clone improvements that originated elsewhere. Yet cloning has advantages. Where feasible, it provides additional quality improvements without the need to duplicate the prior work of diagnosis and design of remedy. What has emerged is a process as follows:

Project teams are asked to include in their final report their suggestions as to sites that may be opportunities for cloning.

Copies of such final reports go to those sites.

The decision of whether to clone is made by the sites.

However, the sites are required to make a response as to their disposition of the matter. This response is typically in one of three forms:

1. We have adopted the improvement.
2. We will adopt the improvement, but we must first adapt it to our conditions.
3. We are not able to adopt the improvement for the following reasons.

In effect, this process requires the units to adopt the improvement or give reasons for not doing so. The units cannot just quietly ignore the recommendation.

A more subtle but familiar form of cloning is done through projects that have repetitive application over a wide variety of subject matter.

A project team develops computer software to find errors in spelling. Another team evolves an improved procedure for processing customer orders through the company. A third team works up a procedure for conducting design reviews.

What is common about such projects is that the end result permits repetitive application of the same process to a wide variety of subject matter: many different misspelled words, many different customer orders, and many different designs.

MISSION STATEMENTS FOR PROJECTS

Each project selected should be accompanied by a written mission statement that sets out the intended end result of the project. On approval, this statement defines the mission of the team assigned to carry out the project.

Purpose of Mission Statements.   The mission statement serves a number of essential purposes:

It defines the intended end result and so helps the team to know when it has completed the project.

It establishes clear responsibility—the mission becomes an addition to each team member's job description.

It provides legitimacy—the project becomes official company business. The team members are authorized to spend the time needed to carry out the mission.

It confers rights—the team has the right to hold meetings, to ask people to attend and assist the team, and to request data and other services germane to the project.

The Numerical Goal.   The ideal mission statement quantifies two critical elements: (1) the intended amount of improvement and (2) the timetable.

Examples of such mission statements follow:

During the coming fiscal year, reduce the time to fill customer orders to an average of 1.5 days.

Reduce the field failure rate of product X by 50 percent over the next 3 years.

The numbers that enter the goals have their origin in various sources. They may originate in

Demands from customers who have their own goals to meet.

Actions taken by competitors, with associated threats to share of market.

Benchmarking to find the best results now being achieved. (The fact that they are being achieved proves that they are achievable.)

In some cases, the available information is not enough to support a scientific approach to goal setting. Hence the goal is set by consensus—by a "jury of opinion."

Perfection as a Goal.   There is universal agreement that perfection is the ideal goal—complete freedom from errors, defects, failures, and so on. The reality is that the absence of perfection is due to many kinds of such deficiencies and that each requires its own improvement project. If a company tries to eliminate all of them, the Pareto principle applies:

The vital few kinds of deficiencies cause most of the trouble but also readily justify the resources needed to root them out. Hence they receive high priority during the screening process and become projects to be tackled.

The remaining many types of deficiencies cause only a small minority of the trouble. As one comes closer and closer to perfection, each remaining kind of deficiency becomes rarer and rarer and hence receives lower and lower priority during the screening process.

All companies tackle those rare types of failure which threaten human life or which risk significant economic loss. In addition, companies that make improvements by the thousands year after year tackle even the mild, rare kinds of deficiency. To do so they enlist the creativity of the work force through such means as quality circles.

Some critics contend that publication of any goal other than perfection is proof of a misguided policy—a willingness to tolerate defects. Such contentions arise from lack of experience with the realities. It is easy to set goals that demand perfection now. Such goals, however, require companies to tackle failure types so rare that they do not survive the screening process.

Nevertheless, there has been progress. During the twentieth century there was a remarkable revision in the unit of measure for deficiencies. In the first half of the century, the usual measure was in percent defective, or defects per hundred units. By the 1990s, many industries had adopted a measure of defects per million units. The leading companies now do make thousands of improvements year after year. They keep coming closer to perfection, but it is a never-ending process.

While many nominated projects cannot be justified solely on their return on investment, they may provide the means for employee participation in the improvement process, which has value in its own right.

Publication of Mission Statements.   Publication of the mission statement makes a project an official part of company business. However, the quality council cannot predict precisely what the project team will encounter as it tackles the project. Experience with numerous projects has provided guidelines as to what to include (and exclude) from mission statements.

What to include: A mission statement may include information about the importance of the problem. It may include data about the present level of performance as well as stating the intended goal. It may include other factual information such as known symptoms of the problem.

What not to include: The mission statement should not include anything that may bias the approach of the project team, such as theories of causes of the problem or leading questions. The statement also should avoid use of broad terms (people problems, communication, and so on) for which there are no agreed definitions.

(The preceding are derived from the training materials of Aluminum Company of America.)

Some companies separate the statement of the problem from the mission statement. In one Dutch company, the quality council published a problem statement as follows:

The lead time of project-related components, from arrival to availability in the production departments, is too long and leads to delays and interruptions in production.

The subsequent mission statement was as follows:

Investigate the causes of this problem and recommend remedies that would lead to a 50 percent reduction in production delays within 3 months after implementation. A preliminary calculation estimated the cost savings potential to be approximately 800,000 Dutch guilders ($400,000). (Smidt and Doesema 1991)

Revision of Mission Statements.   As work on the project progresses, the emerging new information may suggest needed changes in the mission statement, changes such as the following:

The project is bigger than anticipated; it should be subdivided.

The project should be deferred because there is a prerequisite to be carried out first.

The project should change direction because an alternative is more attractive.

The project should be aborted because any remedy will be blocked.

Project teams generally have been reluctant to come back to the quality council for a revision of the mission statement. There seems to be a fear that such action may be interpreted as a failure to carry out the mission or as an admission of defeat. The result can be a dogged pursuit of a mission that is doomed to failure.

The quality council should make clear to all project teams that they have the duty as well as the right to recommend revision of mission statements if revision is needed. This same point also should be emphasized during the training of project teams.

THE PROJECT TEAM

For each selected project, a team is assigned. This team then becomes responsible for completing the project.

Why a Team?   The most important projects are the vital few, and they are almost invariably multifunctional in nature. The symptoms typically show up in one department, but there is no agreement on where the causes lie, what the causes are, or what the remedies should be. Experience has shown that the most effective organizational mechanisms for dealing with such multifunctional problems are multifunctional teams.

Some managers prefer to assign problems to individuals rather than to teams. ("A camel is a horse designed by a committee.") The concept of individual responsibility is in fact quite appropriate if applied to quality control. ("The best form of control is self-control.") However, improvement, certainly for multifunctional problems, inherently requires teams. For such problems, assignment to individuals runs severe risks of departmental biases in the diagnosis and remedy.

A process engineer was assigned to reduce the number of defects coming from a wave soldering process. His diagnosis concluded that a new process was needed. Management rejected this conclusion, on the ground of excess investment. A multifunctional team was then appointed to restudy the problem. The team found a way to solve the problem by refining the existing process (Betker 1983).

Individual biases also show up as cultural resistance to proposed remedies. However, such resistance is minimal if the remedial department has been represented on the project team.

Appointment of Teams; Sponsors.   Project teams are not attached to the chain of command on the organization chart. This can be a handicap in the event that teams encounter an impasse. For this reason, some companies assign council members or other upper managers to be sponsors (or "champions") for specific projects. These sponsors follow team progress (or lack of progress). If the team does run into an impasse, the sponsor may be able to help the team get access to the proper person in the hierarchy.

Teams are appointed by sponsors of the projects, by process owners, by local managers, or by others. In some companies, work force members are authorized to form teams (quality circles, and so on) to work on improvement projects. Whatever the origin, the team is empowered to make the improvement as defined in the mission statement.

Most teams are organized for a specific project and are disbanded on completion of the project. Such teams are called *ad hoc,* meaning "for this purpose." During their next project, the members will be scattered among several different teams. There are also "standing" teams that have continuity—the members remain together as a team and tackle project after project.

Responsibilities and Rights.   A project team has rights and responsibilities that are coextensive with the mission statement. The basic responsibilities are to carry out the assigned mission and to follow the universal improvement process (see below). In addition, the responsibilities include

Proposing revisions to the mission statement

Developing measurement as needed

Communicating progress and results to all who have a need to know

The rights of the teams were set out earlier, under Purpose of Mission Statements: convene meetings, ask people for assistance, and request data and other services needed for the project.

Membership.   The team is selected by the sponsor after consulting with the managers who are affected. The selection process includes consideration of (1) which departments should be represented on the team, (2) what level in the hierarchy team members should come from, and (3) which individuals in that level.

The departments to be represented should include

The ailing department: The symptoms show up in this department, and it endures the effects.

Suspect departments: They are suspected of harboring the causes. (They do not necessarily agree that they are suspect.)

Remedial departments: They will likely provide the remedies. This is speculative, since in many cases the causes and remedies come as surprises.

Diagnostic departments: They are needed in projects that require extensive data collection and analysis.

On-call departments: They are invited in as needed to provide special knowledge or other services required by the team (Black and Stump 1987).

This list includes the usual sources of members. However, there is need for flexibility.

In one company, once the team had gotten under way, it was realized that the internal customer—a "sister facility"—was not represented. Steps were taken to invite the facility in, to avoid an "us versus them" relationship (Black and Stump 1987).

Choice of level in the hierarchy depends on the subject matter of the project. Some projects relate strongly to the technological and procedural aspects of the products and processes. Such projects require team membership from the lower levels of the hierarchy. Other projects relate to broad business and managerial matters. For such projects, the team members should have appropriate business and managerial experience.

Finally comes the selection of individuals. This is negotiated with the respective supervisors, giving due consideration to workloads, competing priorities, and so on. The focus is on the individual's ability to contribute to the team project. The individuals need

Time to attend the team meetings and to carry out assignments outside the meetings—"the homework."

A *knowledge base* that enables the individual to contribute theories, insights, and ideas, as well as job information based on his or her hands-on experience.

Training in the quality improvement process and the associated tools. During the first projects, this training can and should be done concurrently with carrying out the projects.

Most teams consist of six to eight members. Larger numbers tend to make the team unwieldy as well as costly. (A convoy travels only as fast as the slowest ship.)

Should team members all come from the same level in the hierarchy? Behind this question is the fear that the biases of high-ranking members will dominate the meeting. Some of this no doubt takes place, especially during the first few meetings. However, it declines as the group dynamics take over and as members learn to distinguish between theory and fact.

Once the team is selected, the members' names are published, along with their project mission. The act of publication officially assigns responsibility to the individuals as well as to the team. In effect, serving on the project team becomes a part of the individuals' job descriptions. This same publication also gives the team the legitimacy and rights discussed earlier.

Membership from the Work Force.   During the early years of using quality improvement teams, companies tended to maintain a strict separation of team membership. Teams for multifunctional projects consisted exclusively of members from the managerial hierarchy plus professional specialists. Teams for local departmental projects (such as quality circles) consisted exclusively of members from the work force. Figure 5.12 compares the usual features of quality circles with those of multifunctional teams.

Experience then showed that as to the details of operating conditions, no one is better informed than the work force. Through residence in the workplace, workers can observe local changes and recall the chronology of events. This has led to a growing practice of securing such information by interviewing the workers. The workers become "on call" team members.

These same interviews have disclosed that many workers can contribute much more than knowledge of workplace conditions. They can theorize about causes. They have ideas for remedies. In addition, it has become evident that such participation improves human relations by contributing to job satisfaction.

One result of all this experience has been a growing interest in broadening worker participation generally. This has led to experimenting with project teams that make no distinction as to rank in the hierarchy. These teams may become the rule rather than the exception. (For further discussion on the trends in work force participation, see Section 15, Human Resources and Quality.)

Upper Managers on Teams.   Some projects by their nature require that the team include members from the ranks of upper management. Here are some examples of quality improvement projects actually tackled by teams that included upper managers:

Shorten the time to put new products on the market.

Improve the accuracy of the sales forecast.

Reduce the carryover of prior failure-prone features into new product models.

Establish a teamwork relationship with suppliers.

Feature	Quality circles	Project teams
Primary purpose	To improve human relations	To improve quality
Secondary purpose	To improve quality	To improve participation
Scope of project	Within a single department	Multidepartmental
Size of project	One of the useful many	One of the vital few
Membership	From a single department	From multiple departments
Basis of membership	Voluntary	Mandatory
Hierarchical status of members	Typically in the workforce	Typically managerial or professional
Continuity	Circle remains intact, project after project	Team is ad hoc, disbands after project is completed.

FIGURE 5.12   Contrast, quality circles, and multifunctional teams. (*From Making Quality Happen, 1988, Juran Institute, Wilton, CT, p. D30.*)

Develop the new measures of quality needed for strategic quality planning.

Revise the system of recognition and rewards for quality improvement.

There are some persuasive reasons urging that all upper managers personally serve on some project teams. Personal participation on project teams is an act of leadership by example. This is the highest form of leadership. Personal participation on project teams also enables upper managers to understand what they are asking their subordinates to do, what kind of training is needed, how many hours per week are demanded, how many months does it take to complete the project, and what kinds of resources are needed. Lack of upper management understanding of such realities has contributed to the failure of some well-intentioned efforts to establish annual quality improvement.

In one company, out of 150 quality improvement projects tackled, 12 involved teams composed of senior directors (Egan 1985).

[For one upper manager's account of his experience when serving on a project team, see Pelletier (1990).]

Model of the Infrastructure.   There are several ways to show in graphic form the infrastructure for quality improvement—the elements of the organization, how they relate to each other, and the flow of events. Figure 5.13 shows the elements of infrastructure in pyramid form. The pyramid depicts a hierarchy consisting of top management, the autonomous operating units, and the major staff functions. At the top of the pyramid is the corporate quality council and the subsidiary councils, if any. Below these levels are the multifunctional quality improvement teams. (There may be a committee structure between the quality councils and the teams).

At the *intra*department level are teams from the work force—quality circles or other forms. This infrastructure permits employees in all levels of organization to participate in quality improvement projects, the useful many as well as the vital few.

TEAM ORGANIZATION

Quality improvement teams do not appear on the organization chart. Each "floats"—it has no personal boss. Instead, the team is supervised *impersonally* by its mission statement and by the quality improvement roadmap.

The team does have its own internal organizational structure. This structure invariably includes a team *leader* (chairperson and so on) and a team *secretary*. In addition, there is usually a *facilitator.*

FIGURE 5.13   Model of the infrastructure for quality improvement. (*From Visual GMQH15, Juran Institute, Inc., Wilton, CT.*)

The Team Leader.   The leader is usually appointed by the sponsor—the quality council or other supervising group. Alternatively, the team may be authorized to elect its leader.

The leader has several responsibilities. As a team member, the leader *shares* in the responsibility for completing the team's mission. In addition, the leader has administrative duties. These are *unshared* and include

Ensuring that meetings start and finish on time

Helping the members to attend the team meetings

Ensuring that the agendas, minutes, reports, and so on are prepared and published

Maintaining contact with the sponsoring body

Finally, the leader has the responsibility of *oversight*. This is met not through the power of command—the leader is not the boss of the team. It is met through the power of leadership. The responsibilities include

Orchestrating the team activities

Stimulating all members to contribute

Helping to resolve conflicts among members

Assigning the homework to be done between meetings

To meet such responsibilities requires multiple skills, which include

A trained capability for leading people

Familiarity with the subject matter of the mission

A firm grasp of the quality improvement process and the associated tools

The Team Secretary.   The team secretary is appointed by the sponsor or, more usually, by the team leader. Either way, the secretary is usually a member of the project team. As such, he or she shares in the responsibility for carrying out the team mission.

In addition, the secretary has unshared administrative responsibilities, chiefly preparing the agendas, minutes, reports, and so on. These documents are important. They are the team's chief means of communication with the rest of the organization. They also become the chief reference source for team members and others. All of which suggests that a major qualification for appointment to the job of secretary is the ability to write with precision.

The Team Members.   "Team members" as used here includes the team leader and secretary. The responsibilities of any team member consist mainly of the following:

Arranging to attend the team meetings

Representing his or her department

Contributing job knowledge and expertise

Proposing theories of causes and ideas for remedy

Constructively challenging the theories and ideas of other team members

Volunteering for or accepting assignments for homework

Finding the Time to Work on Projects.   Work on project teams is time-consuming. Assigning someone to a project team adds about 10 percent to that person's workload. This added time is needed to attend team meetings, perform the assigned homework, and so on. Finding the time to do all this is a problem to be solved, since this added work is thrust on people who are already fully occupied.

No upper manager known to me has been willing to solve the problem by hiring new people to make up for the time demanded by the improvement projects. Instead, it has been left to each team member to solve the problem in his or her own way. In turn, the team members have adopted such strategies as

Delegating more activities to subordinates

Slowing down the work on lower-priority activities

Improving time management on the traditional responsibilities

Looking for ongoing activities that can be terminated. (In several companies, there has been a specific drive to clear out unneeded work to provide time for improvement projects.)

As projects begin to demonstrate high returns on investment, the climate changes. Upper managers become more receptive to providing resources. In addition, the successful projects begin to reduce workloads that previously were inflated by the presence of chronic wastes. [Relative to team organization, see AT&T Quality Library, Quality Improvement Cycle (1988, pp. 7–12). Relative to team meetings, see also AT&T Quality Improvement Team Helper (1990, pp. 17–21).]

FACILITATORS

Most companies make use of internal consultants, usually called "facilitators", to assist quality improvement teams, mainly teams that are working on their first projects. A facilitator is not a member of the team and has no responsibility for carrying out the team mission. (The literal meaning of the word *facilitate* is "to make things easy.") The prime role of the facilitator is to help the team to carry out its mission.

The Roles.   The usual roles of facilitators consist of a selection from the following:

Explain the company's intentions: The facilitator usually has attended briefing sessions that explain what the company is trying to accomplish. Much of this briefing is of interest to the project teams.

Assist in team building: The facilitator helps the team members to learn to contribute to the team effort: propose theories, challenge theories of others, and/or propose lines of investigation. Where the team concept is new to a company, this role may require working directly with individuals to stimulate those who are unsure about how to contribute and to restrain the overenthusiastic ones. The facilitator also may evaluate the progress in team building and provide feedback to the team.

Assist in training: Most facilitators have undergone training in team building and in the quality improvement process. They usually have served as facilitators for other teams. Such experiences qualify them to help train project teams in several areas: team building, the quality improvement roadmap, and/or use of the tools.

Relate experiences from other projects: Facilitators have multiple sources of such experiences:

- Project teams previously served
- Meetings with other facilitators to share experiences in facilitating project teams
- Final published reports of project teams
- Projects reported in the literature

Assist in redirecting the project: The facilitator maintains a detached view that helps to sense when the team is getting bogged down. As the team gets into the project, it may find itself getting deeper and deeper into a swamp. The project mission may turn out to be too broad, vaguely defined, or not doable. The facilitator usually can sense such situations earlier than the team and can help guide it to a redirection of the project.

Assist the team leader: Facilitators provide such assistance in various ways:

- Assist in planning the team meetings. This may be done with the team leader before each meeting.
- Stimulate attendance. Most nonattendance is due to conflicting demands made on a team member's time. The remedy often must come from the member's boss.
- Improve human relations. Some teams include members who have not been on good terms with each other or who develop friction as the project moves along. As an "outsider," the facilitator can help to direct the energies of such members into constructive channels. Such action usually takes place outside the team meetings. (Sometimes the leader is part of the problem. In such cases the facilitator may be in the best position to help out.)
- Assist on matters outside the team's sphere of activity. Projects sometimes require decisions or actions from sources that are outside the easy reach of the team. Facilitators may be helpful due to their wider range of contacts.

Support the team members: Such support is provided in multiple ways:

- Keep the team focused on the mission by raising questions when the focus drifts.
- Challenge opinionated assertions by questions such as "Are there facts to support that theory?"
- Provide feedback to the team based on perceptions from seeing the team in action.

Report progress to the councils: In this role the facilitator is a part of the process of reporting on progress of the projects collectively. Each project team issues minutes of its meetings. In due course each also issues its final report, often including an oral presentation to the council. However, reports on the projects *collectively* require an added process. The facilitators are often a part of this added reporting network.

The Qualifications.   Facilitators undergo special training to qualify them for the preceding roles. The training includes skills in team building, resolving conflicts, communication, and management of change; knowledge relative to the quality improvement processes, e.g., the improvement roadmap and the tools and techniques; and knowledge of the relationship of quality improvement to the company's policies and goals. In addition, facilitators acquire maturity through having served on project teams and having provided facilitation to teams.

This prerequisite training and experience are essential assets to the facilitator. Without them, he or she has great difficulty winning the respect and confidence of the project's team.

Sources and Tenure.   Most companies are aware that to go into a high rate of quality improvement requires extensive facilitation. In turn, this requires a buildup of trained facilitators. However, facilitation is needed mainly during the startup phase. Then, as team leaders and members acquire training and experience, there is less need for facilitator support. The buildup job becomes a maintenance job.

This phased rise and decline has caused most companies to avoid creating full-time facilitators or a facilitator career concept. Facilitation is done on a part-time basis. Facilitators spend most of their time on their regular job. [For an interesting example of a company's thinking process on the question of full-time versus part-time facilitators, see Kinosz and Ice (1991). See also Sterett (1987).]

A major source of facilitators is line supervisors. There is a growing awareness that service as a facilitator provides a breadth of experience that becomes an aid on the regular job. In some companies, this concept is put to deliberate use. Assignment to facilitation serves also as a source of training in managing for quality. A second major source of facilitators is specialists. These are drawn from the Human Relations Department or from the Quality Department. All undergo the needed training discussed earlier.

A minority of large companies use a category of full-time specialists called "quality improvement manager" (or similar title). Following intensive training in the quality improvement process, these managers devote full time to the quality improvement activity. Their responsibilities go beyond facilitating project teams and may include

Assisting in project nomination and screening

Conducting training courses in the quality improvement process

Coordinating the activities of the project team with those of other activities in the company

Assisting in the preparation of summarized reports for upper managers

(For elaboration on facilitators and their roles, see "Quality Improvement Team Helper," a part of AT&T's Quality Library.)

THE UNIVERSAL SEQUENCE FOR QUALITY IMPROVEMENT

A quality improvement team has no personal boss. Instead, the team is supervised *impersonally.* Its responsibilities are defined in

The project mission statement: This mission statement is unique to each team.

The universal sequence[2] (or roadmap) for quality improvement: This is identical for all teams. It defines the actions to be taken by the team to accomplish its mission.

Some of the steps in the universal sequence have already been discussed in this section: proof of the need, project nomination and selection, and appointment of project teams. The project team has the principal responsibility for the steps that now follow—taking the two "journeys."

The Two Journeys.   The universal sequence includes a series of steps that are grouped into two journeys:

1. The *diagnostic journey* from symptom to cause. It includes analyzing the symptoms, theorizing as to the causes, testing the theories, and establishing the causes.
2. The *remedial journey* from cause to remedy. It includes developing the remedies, testing and proving the remedies under operating conditions, dealing with resistance to change, and establishing controls to hold the gains.

Diagnosis is based on the factual approach and requires a firm grasp of the meanings of key words. It is helpful to define some of these key words at the outset.

Definition of Key Words

A "defect" is any state of unfitness for use or nonconformance to specification. Examples are illegible invoice, oversizing, and low mean time between failures. Other names include "error", "discrepancy", and "nonconformance."

A "symptom" is the outward evidence of a defect. A defect may have multiple symptoms. The same word may serve as a description of both defect and symptom.

A "theory" is an unproved assertion as to reasons for the existence of defects and symptoms. Usually, multiple theories are advanced to explain the presence of defects.

A "cause" is a proved reason for the existence of a defect. Often there are multiple causes, in which case they follow the Pareto principle—the vital few causes will dominate all the rest.

---

[2] The concept of a universal sequence evolved from my experience first in Western Electric Company (1924–1941) and later during my years as an independent consultant, starting in 1945. Following a few preliminary published papers, a universal sequence was published in book form (Juran 1964). This sequence then continued to evolve based on experience gained from applications by operating managers.

The creation of the Juran Institute (1979) led to the publication of the videocassette series *Juran on Quality Improvement* (Juran 1981). This series was widely received and became influential in launching quality improvement initiatives in many companies. These companies then developed internal training programs and spelled out their own versions of a universal sequence. All these have much in common with the original sequence published in 1964. In some cases, the companies have come up with welcome revisions or additions.

A "dominant cause" is a major contributor to the existence of defects and one that must be remedied before there can be an adequate improvement.

"Diagnosis" is the process of studying symptoms, theorizing as to causes, testing theories, and discovering causes.

A "remedy" is a change that can eliminate or neutralize a cause of defects.

Diagnosis Should Precede Remedy.   It may seem obvious that diagnosis should precede remedy, yet biases or outdated beliefs can get in the way.

For example, during the twentieth century many upper managers held deep-seated beliefs that most defects were due to work force errors. The facts seldom bore this out, but the belief persisted. As a result, during the 1980s, many of these managers tried to solve their quality problems by exhorting the work force to make no defects. (In fact, defects are generally over 80 percent management-controllable and under 20 percent worker-controllable.)

Untrained teams often try to apply remedies before the causes are known. ("Ready, fire, aim.") For example:

An insistent team member "knows" the cause and pressures the team to apply a remedy for that cause.

The team is briefed as to the technology by an acknowledged expert. The expert has a firm opinion about what is the cause of the symptom, and the team does not question the expert's opinion.

As team members acquire experience, they also acquire confidence in their diagnostic skills. This confidence then enables them to challenge unproved assertions.
Where deep-seated beliefs are widespread, special research may be needed.

In a classic study, Greenridge (1953) examined 850 failures of electronic products supplied by various companies. The data showed that 43 percent of the failures were traceable to product design, 30 percent to field operation conditions, 20 percent to manufacture, and the rest to miscellaneous causes.

THE DIAGNOSTIC JOURNEY

The diagnostic journey starts with analyzing the symptoms of the chronic quality problem. Evidence of defects and errors comes in two forms:

The *words* used in written or oral descriptions
The *autopsies* conducted to examine the defects in-depth

Understanding the Symptoms.   Symptoms are often communicated in words such as incorrect invoices, machine produces poor copies, or "I don't feel well." Understanding such expressions is often hindered because key words have multiple or vague meanings. In such cases, the person who prepared the report becomes an essential source of information.

An inspection report persistently showed a high percentage of defects due to "contamination." Various remedies were tried to reduce contamination. All were unsuccessful. In desperation, the investigators spoke with the inspectors to learn about the meaning of contamination. The inspectors explained that there were 12 defect categories on the inspection form. If the observed defect did not fit any of the categories, they would report the defect as contamination.

A frequent source of misunderstanding is the use of generic words to describe multiple subspecies of defects.

In a plant making rubber products by the latex dip process, the word *tears* was used on the data sheets to describe torn products. One important manager regarded tears as due to workers' errors and urged a remedy through motivational and disciplinary measures. Actually, there were three species of tears: *strip tears* from a stripping operation, *click tears* from a press operation, and *assembly tears* from an assembly operation. Only strip tears were due to worker errors, and their frequency was only 15 percent. Revising the manager's belief became possible only after clearing up the meaning of the terminology and quantifying the relative frequencies of the subspecies of tears.

A useful tool for reducing semantic confusion is the "glossary." A team is assigned to think out the meanings of key words. The resulting agreements are then published as part of the official company glossary.

Autopsies. An important aid to understanding the meanings behind the words is the "autopsy" (to see with one's own eyes). Scientific autopsies can furnish extensive objective knowledge about symptoms and thereby can supplement or override the information contained in the written reports.

The report on tests of a product may include a category of "electrical" defects. Autopsies of a sample of such defects may show that there are multiple subspecies: open circuit, short circuit, dead battery, and so on.

[For a case example of using autopsies, see Black and Stump (1987).]

FORMULATION OF THEORIES

All progress in diagnosis is made theory by theory—by affirming or denying the validity of the theories about causes. The process consists of three steps: generating theories, arranging theories in some order, and choosing theories to be tested.

Generating Theories. Securing theories should be done systematically. Theories should be sought from all potential contributors—line managers and supervisors, technologists, the work force, customers, suppliers, and so on. Normally, the list of theories is extensive, 20 or more. If only 3 or 4 theories have emerged, it usually means that the theorizing has been inadequate.

One systematic way of generating theories is called "brainstorming." Potential contributors are assembled for the specific purpose of generating theories. Creative thinking is encouraged by asking each person, in turn, to propose a theory. No criticism or discussion is allowed until all theories are recorded. The end result is a list of theories that are then subjected to discussion.

Experience has shown that brainstorming can have a useful effect on team members who carry strong opinions. Such members may feel that their views should be accepted as facts. "I know this is so." However, other members regard these views as theories—unproved assertions. It all leads to a growing awareness of the difference between theory and fact.

Another systematic approach—"nominal group technique"—is similar to brainstorming. Participants generate their theories silently, in writing. Each then offers one theory at a time, in rotation. After all ideas have been recorded, they are discussed and then prioritized by vote.

Theories should not be limited to those which relate to errors on specific products or processes. In some cases, the cause may lie in some broader *system* that affects multiple products.

A manager observes, "In the last 6 weeks, we have lost four needed batches of unrelated products due to four different instruments being out of calibration. This shows that we should review our *system* for maintaining the accuracy of instruments."

Arranging Theories. The brainstorming process provides a helter-skelter list of theories. Orderly arrangement of such a list helps the improvement team to visualize the interrelation of the theories. In addition, an orderly arrangement is an essential aid to choosing which theories to test. The orderly arrangement can be made in several ways:

Storyboarding: A supplement to brainstorming, this is a form of orderly arrangement of theories. As each theory is proposed, it is recorded on an index card. The cards are then appropriately arranged on a board to form a visual display of the theories. [See Betker (1985) for an example of use of storyboarding in an electronics company.]

Tabular arrangement: Another form of arrangement is a table showing a logical hierarchy: theories, subtheories, sub-subtheories, and so on. Table 5.6 is an example as applied to yield of fine powder chemicals.

Cause-and-effect diagram: This popular diagram (also known as an *Ishikawa diagram* or *fishbone diagram* was developed in 1950 by the late Professor Kaoru Ishikawa. An example is shown in Figure 5.14.

To create the diagram, the effect (symptom) is written at the head of the arrow. Potential causes (theories) are then added to complete the diagram. A common set of major categories of causes consists of personnel, work methods, materials, and equipment. Figure 5.14 shows the cause-and-effect diagram as prepared for the same list of theories as was arranged in Table 5.6. Note how the diagram aids in identifying interrelationships among theories.

Cause-and-effect diagrams were first applied to manufacturing problems. They have since demonstrated that they are applicable to all manner of industries, processes, and problems. As a result, they are now in universal use in every conceivable application.

A cause-and-effect diagram can be combined with a *force-field analysis.* The team identifies the situations and events that contribute to the problem (these are the "restraining forces"). The actions necessary to counter the restraining forces are then identified (these actions are the "driving forces"). Finally, a diagram combining the restraining and driving forces is prepared to assist in diagnosis. [For example, see Stratton (1987).]

Choosing Theories to Be Tested. Theories are numerous, yet most turn out to be invalid. As a result, project teams have learned to discuss priorities for testing theories and to arrive at a con-

TABLE 5.6 Orderly Arrangement of Theories

Raw material	Moisture content
Shortage of weight	Charging speed of wet powder
Method of discharge	Dryer, rpm
Catalyzer	Temperature
Types	Steam pressure
Quantity	Steam flow
Quality	Overweight of package
Reaction	Type of balance
Solution and concentration	Accuracy of balance
B solution temperature	Maintenance of balance
Solution and pouring speed	Method of weighing
pH	Operator
Stirrer, rpm	Transportation
Time	Road
Crystallization	Cover
Temperature	Spill
Time	Container
Concentration	
Mother crystal	
Weight	
Size	

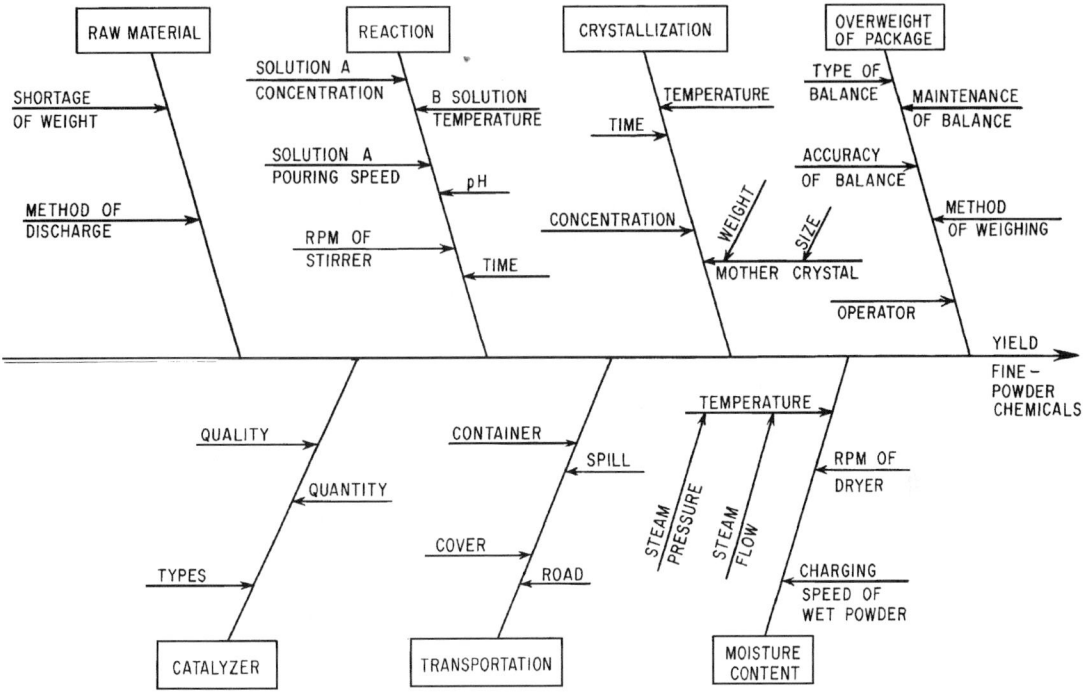

FIGURE 5.14 Ishikawa cause-and-effect diagram.

sensus. This approach has proved to be effective in reducing the teams' time and effort, as well as in minimizing the frustration of pursuing invalid theories.

Here and there companies have evolved structured matrixes for arriving at a quantitative score for each theory. A simple method is to ask each team member to rank all theories in his or her order of importance. The totals of the rank numbers then become an input to the final consensus on priorities.

TEST OF THEORIES

There are many strategies for testing theories, and they follow the Pareto principle—a relative few of them are applicable to most problems. What follows is a brief description of some vital few strategies along with their principal areas of application.

A critical question is whether to test one theory at a time, one group of interrelated theories at a time, or all theories simultaneously. To make a proper choice requires an understanding of the methods of data collection and analysis (see below). The team may be need to secure the advice of specialists in data analysis.

The Factual Approach.   The basic concept behind diagnosis is the factual approach—to make decisions based on fact rather than on theory. This concept enables amateurs in the technology nevertheless to contribute usefully to the project. Thus the teams must learn to distinguish theory from fact. Facts are supported by suitable evidence. Theories are unsupported assertions. Sometimes the distinction is subtle.

In one team, the engineering member asserted that changing the temperature of the solder bath would reduce the frequency of the defect under study but would create a new defect that would make matters worse. His belief was based on data *collected over 10 years earlier on different equipment.* The team challenged his assertion, conducted a new trial, and found that the higher temperature caused no such adverse effect (Betker 1983).

Flow Diagrams.   For many products, the anatomy of the producing process is a "procession"—a sequential series of steps, each performing a specific task. Most team members are familiar with some of the steps, but few are familiar with the entire procession. Note that the steps in the procession may include those within the external supplier chain as well as those taken during marketing, use, and customer service.

Preparing a flow diagram helps all members to better understand the progression and the relation of each step to the whole. [See, for example, Engle and Ball (1985).] (For details on constructing flow diagrams, see Section 3, The Quality Planning Process.)

Process Capability Analysis.   One of the most frequent questions raised by improvement team members refers to "process capability." Some members contend that "this process is inherently unable to meet the specifications." The opposing contention is that "the process is capable but it isn't being run right." In recent decades, tools have been devised to test these assertions, especially as applied to manufacturing processes.

A common test of process capability uses the "Shewart control chart." Data are take from the process at (usually) equal chronological intervals. *Having established by control chart analysis that the process is inherently stable,* the data are then compared with the terms of the specification. *This comparison provides a measure of the ability of the process to consistently produce output within specified limits.* (For elaboration on the Shewart control chart, see Section 45.)

While evaluation of process capability originally was applied to manufacturing processes, it has since been applied increasingly to administrative and business processes in all industries. A common example has been the application to cycle time of such processes.

Many of these processes consist of a procession in which the work is performed in a sequence of steps as it moves from department to department. It may take days (weeks, or even months) to complete a cycle, yet the time required to do the work has taken only a few hours. The remaining time has consisted of waiting for its turn at each step, redoing, and so on.

For such processes, the theoretical process capability is the cumulative work time. A person who is trained to perform all the steps and has access to all the database might meet this theoretical number. Some companies have set a target of cutting the cycle time to about twice the theoretical capability.

Process Dissection.   A common test of why a capable process isn't being run right is "process dissection." This strategy tries to trace defects back to their origins in the process. There are multiple forms of such process dissection.

Test at Intermediate Stages.   When defects are found at the end of a procession, it is not known which operational step did the damage. In such cases, a useful strategy may be to inspect or test the product at intermediate steps to discover at which step the defect first appears. Such discovery, if successful, can drastically reduce the effort of testing theories.

Stream-to-Stream Analysis.   High-volume products often require multiple sources ("streams") of production—multiple suppliers, machines, shifts, workers, and so on. The streams may seem to be identical, but the resulting products may not be. Stream-to-stream analysis consists of separating the production into streams of origin and testing for stream-to-stream differences in an effort to find the guilty stream, if any.

Time-to-Time Analysis.   Another form of process dissection is time-to-time analysis. The purpose is to discover if production of defects is concentrated in specific spans of time. This type of analysis has been used to study time between abnormalities, effect of change of work shifts, influence of the seasons of the year, and many other such potential causes.

A frequent example of time-to-time analysis is the Shewhart control chart, which also can show whether the variability in a process is at random or is due to assignable causes. (See Section 45.)

A special case of time-to-time changes is *drift*—a continuing deterioration of some aspect of the process. For example, in factory operations, the chemical solution gradually may become more dilute, the tools gradually may wear, or the workers may become fatigued.

In time-to-time analysis, the process (or product) is measured (usually) at equal time intervals. Graphic presentation of the data is an aid to interpretation. Presentation in cumulative form (cumulative sum charts) is an aid to detecting drift.

There are also "piece-to-piece" and "within-piece" variations.

An example of piece-to-piece variation is seen in foundry processes that produce castings in "stacks." In such cases, the quality of the castings may depend on their location in the stack. An example of within-piece variation is in lathe operations, where the diameter of a cylindrical piece is not uniform.

Simultaneous Dissection.   Some forms of process dissection can test multiple theories simultaneously. A classic example is the Multi-Vari[3] chart. See Figure 5.15. In this figure, a vertical line depicts the range of variation within a single unit of product, as compared with specification tolerance limits. In the left-hand example, the within-piece variation alone is too great in relation to the tolerance. Hence no improvement is possible unless within-piece variation is reduced. The middle example is one in which within-piece variation is comfortable; the problem is piece-to-piece variation. In the right-hand example, the problem is excess time-to-time variability. Traver (1983) presents additional examples of Multi-Vari charts.

Defect Concentration Analysis.   In "defect concentration analysis", the purpose is to discover concentrations that may point to causes. This method has been used in widely varied applications.

During one of the London cholera epidemics of the mid-nineteenth century, Dr. John Snow secured the addresses of those in the Soho district who had died of cholera. He then plotted the

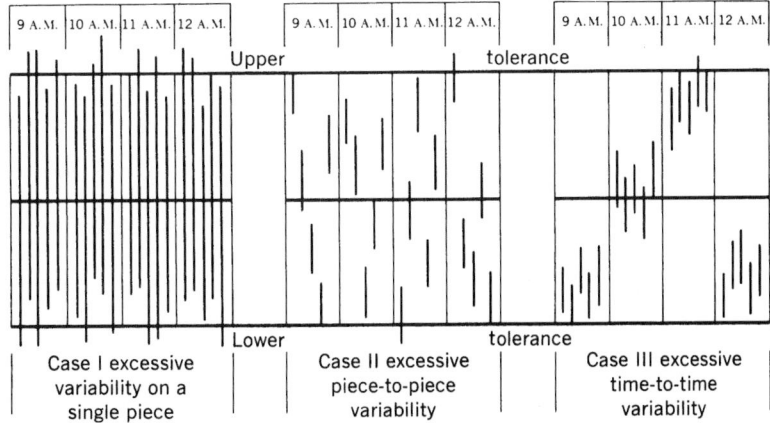

FIGURE 5.15 Multi-Vari chart.

[3]The name Multi-Vari was given to this form of analysis by L. A. Seder in his classic paper, "Diagnosis with Diagrams," in *Industrial Quality Control,* (January 1950 and March 1950). The concept of the vertical line had been used by J. M. Juran, who derived it from the method used in financial papers to show the ranges of stock prices.

addresses on a map of that district. (See Figure 5.16.) The addresses were concentrated around the Broad Street pump, which supplied drinking water for the Soho district. In those days, no one knew what caused cholera, but a remedy was provided by removing the handle from the pump.

In the case of manufactured products, it is common to plot defect locations on a drawing of the product. See Figure 5.17. This concentration diagram shows the location of defects on an office copier. The circled numbers show various locations on the equipment. The numbers adjacent to the circles show how many defects were found in the sample of machines under study. It is seen that locations 24 and 2 account for about 40 percent of the defects.

Concentration analysis has been applied to military operations.

During World War II, the United States Air Force studied the damage done to aircraft returning from combat missions. One form of analysis was to prepare diagrams to show where enemy bullet holes

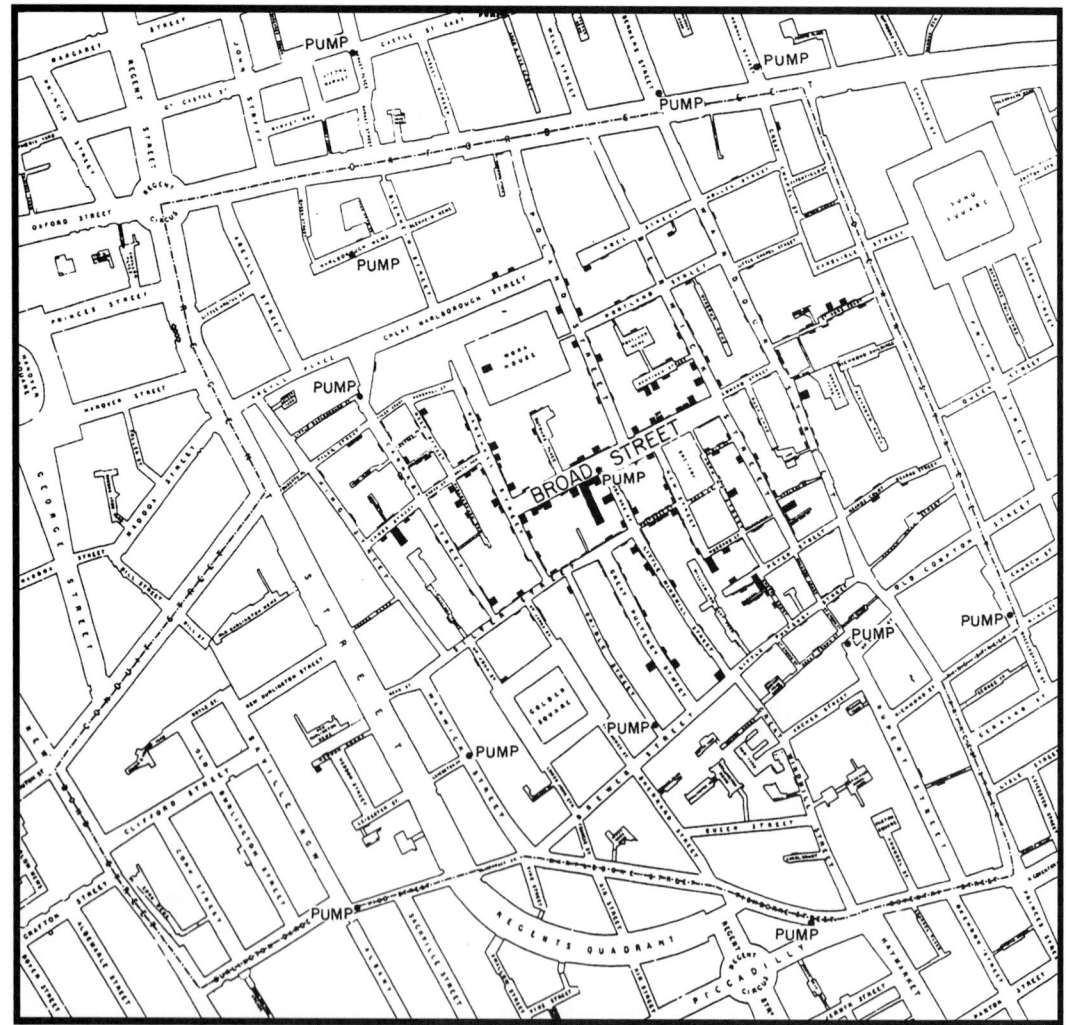

FIGURE 5.16 Dr. John Snow's concentration analysis.

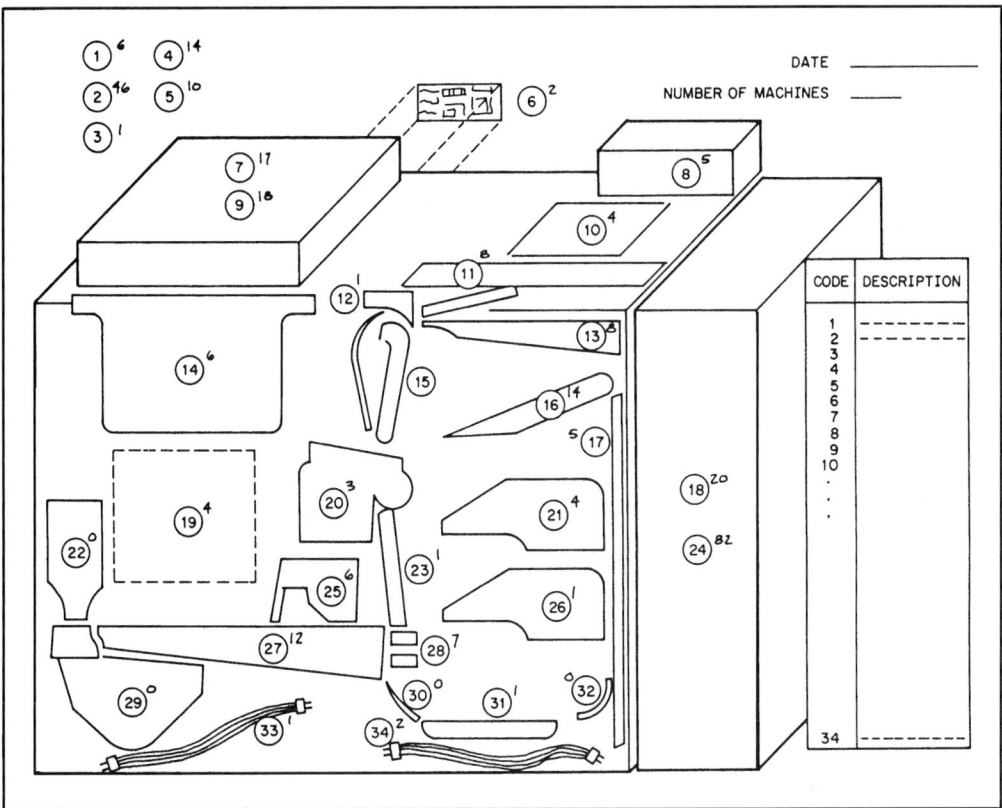

FIGURE 5.17   Concentration diagram: defects on copiers.

and other forms of damage were concentrated. The diagrams also seemed to show that some areas of the aircraft *never received damage.* The conclusion was that damage to those areas had destroyed the planes and that redesign was needed to reduce the vulnerability of those areas.

Association Searches.   Some diagnosis consists of relating data on symptoms to some theory of causation such as design, process, worker, and so on. Possible relationships are examined using various statistical tools such as correlation, ranking, and matrixes.

Correlation: In this approach, data on frequency of symptoms are plotted against data on the suspected cause. Figure 5.18 is an example in which the frequency of pitted castings was related to the "choke" thickness in the molds.

Ranking: In this approach, the data on defects are ranked in their order of frequency. This ranking is then compared with the incidence of the suspected cause.

Table 5.7 shows the frequency of the defect "dynamic unbalance" for 23 types of automotive torque tubes. The suspected cause was a swaging operation that was performed on some of the product types. The table shows which types had undergone swaging. It is clear that swaged product types were much worse than the unswaged types.

In some cases, it is feasible to study data on multiple variables using a structured cookbook method of analysis. An early published example is the SPAN plan (Seder and Cowan 1956). This

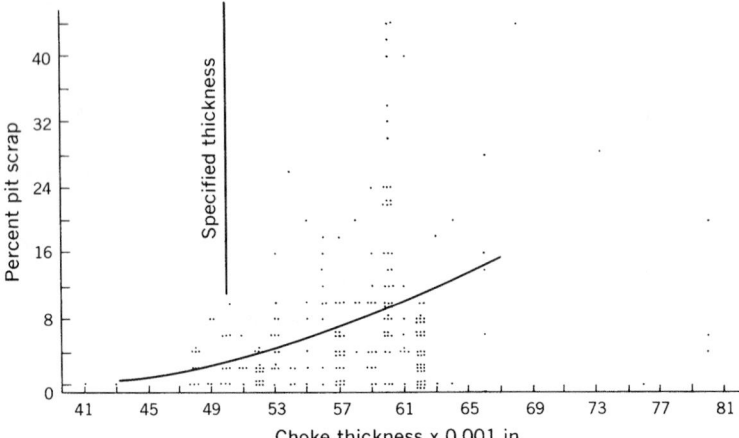

FIGURE 5.18   Test of theories by correlation.

TABLE 5.7   Test of Theories by Ranking

Type	% defective	Swaged (marked X)	Type	% defective	Swaged (marked X)
A	52.3	X	M	19.2	X
B	36.7	X	N	18.0	X
C	30.8	X	O	17.3	
D	29.9	X	P	16.9	X
E	25.3	X	Q	15.8	
F	23.3	X	R	15.3	
G	23.1	X	S	14.9	
H	22.5		T	14.7	
I	21.8	X	U	14.2	
J	21.7	X	V	13.5	
K	20.7	X	W	12.3	
L	20.3				

approach uses standardized data collection and analysis forms to permit successive separation of observed total product variability into five stages: lot-to-lot, stream-to-stream, time-to-time, within-piece (or positional), and error of measurement. Other forms of search for association are set out in the statistical group of Sections 44 to 48 of this handbook.

Cutting New Windows.   In some cases, the data available from operations are not able to test certain of the theories. In such cases, it may be necessary to create new data specifically for the purpose of testing theories. This is called "cutting new windows" and takes several forms.

Measurement at Intermediate Stages.   A common example is seen in products made by a procession of steps but tested only after completion of all steps. (See preceding, under Process Dissection.) In such cases, cutting new windows may consist of making measurements at intermediate stages of the procession.

In a project to reduce the time required to recruit new employees, data were available on the total time elapsed. Test of the theories required cutting new windows by measuring the time elapsed for each of the six steps in the recruitment process.

In a process for welding of large joints in critical pressure vessels, all finished joints were x-rayed to find any voids in the welds. The process could be dissected to study some sources of variation: worker-to-worker, time-to-time, and joint-to-joint. However, data were not available to study other sources of variations: layer-to-layer, bead-to-bead, and within bead. Cutting new windows involved x-raying some welds after each bead was laid down.

Creation of New Measuring Devices.   Some theories cannot be tested with the measuring devices used during operations. In such cases, it may be necessary to create new devices.

In a project to reduce defects in automotive radiators, some theories focused on the heat-treating and drying operations that occurred inside a closed brazing oven. To measure what was happening inside the oven, an insulated box—about the size of a radiator—was equipped with thermocouples and designed to log time and temperatures within the oven. The box was placed on the assembly line along with the radiators and sent through the oven on a normal brazing cycle. The resulting data were used to modify the temperature profile inside the oven. Down went the failure rate (Mizell and Strattner 1981).

Nondissectable Features.   A "dissectable" product feature is one that can be measured during various stages of processing. A "nondissectible" feature cannot be measured during processing; many nondissectible features do not even come into existence until all steps in the process have been completed. A common example is the performance of a television set. In such cases, a major form of test of theories is through design of experiments (see below).

Design of Experiments.
Test of theories through experiment usually involves producing trial samples of product under specially selected conditions. The experiment may be conducted either in a laboratory or in the real world of offices, factories, warehouses, users' premises, and so on.

It is easy enough to state the "minimal criteria" to be met by an experiment. It should

Test the theories under study without being confused by extraneous variables

Discover the existence of major causes even if these were not advanced as theories

Be economic in relation to the amounts at stake

Provide reliable answers

To meet these criteria requires inputs from several sources:

The *managers* identify the questions to which answers are needed.

The *technologists* select and set priorities on the proper variables to be investigated.

The *diagnosticians* provide the statistical methods for planning the experimental design and analyzing the resulting data.

Designs of experiments range from simple rifleshot cases to the complex unbridled cases, and most of them are not matters to be left to amateurs. In its simplest form, the "rifleshot experiment" uses a split-lot method to identify which of two suspects is the cause. For example, if processes A and B are suspects, a batch of homogeneous material is split. Half goes through process A; half goes through process B. If two types of material are also suspects, each is sent through both processes, A and B, creating a two-by-two design of experiment. As more variables get involved, more combinations are needed, but now the science of design of experiments enters to simplify matters.

In the "unbridled experiment", a sample (or samples) of product are followed through the various processes under a plan that provides for measuring values of the selected suspects at each stage. The resulting product features are also measured. The hope is that analysis of the resulting data will find the significant relationships between causes and effects.

The unbridled experiment should be defined in writing to ensure that it is understood and that it represents a meeting of the minds. Carefully planned experiments have a high probability of identifying the guilty suspects. The disadvantage is the associated cost and the time interval needed to get answers.

Statisticians have developed remarkably useful tools: to get rid of unwanted variables through "randomizing"; to minimize the amount of experimentation through skillful use of factorial, blocked, nested, and other designs; to read the meaning out of complex data. (See Section 47, Design and Analysis of Experiments.)

Measurement for Diagnosis.   A frequent roadblock to diagnosis is the use of shop instruments to make the measurements. These instruments were never intended to be used for diagnosis. They were provided for other purposes such as process regulation and product testing. There are several principal categories of cases in which measurement for diagnosis differs from measurement for operations:

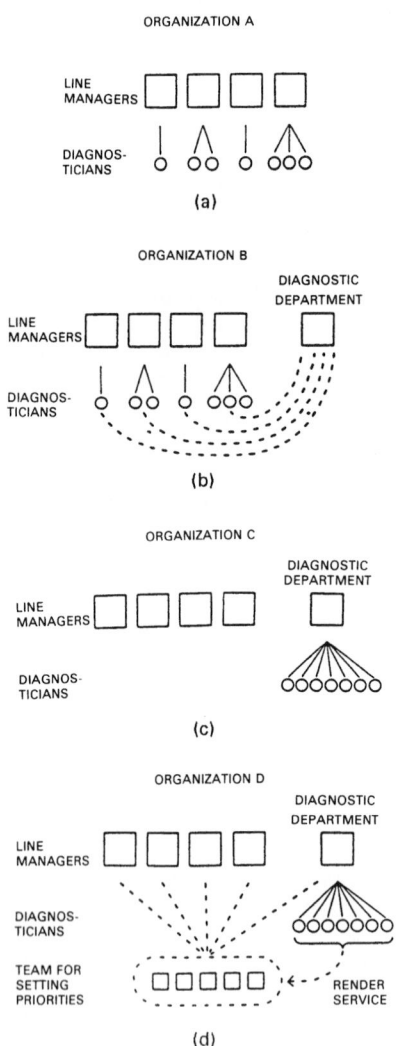

FIGURE 5.19   Alternatives for organization of diagnosticians.

Measurement by variables instead of attributes. Process capability studies usually demand variables measurements.

Measurement with a precision superior to that of the shop instruments. In some cases, the instruments provided for operations lack adequate precision and hence are a dominant cause of the quality problem.

Creation of new instruments to cut new windows or to deal with nondissectible processes.

Measurement to test suspected variables that are not controlled or even mentioned by the specifications.

Responsibility for Diagnosis.   Some of the work of diagnosis consists of the discussions that take place during project team meetings: analyzing symptoms, theorizing about causes, selecting theories for test, and so on. In addition, the work of diagnosis involves test of theories, which consists mainly of data collection and analysis and is done largely as homework outside of team meetings.

For some projects, the homework consists of data collection and analysis on a small scale. In such cases, the project team members themselves may be able to do the homework. Other projects require extensive data collection and analysis. In such cases, the project team may delegate or subcontract much or all of the work to *diagnosticians*—persons who have the needed time, skills, and objectivity. Despite such delegation, the project team remains responsible for getting the work done.

In large organizations working on many improvement projects, the work of diagnosis occupies the full-time equivalent of numerous diagnosticians. In response, many companies create full-time job categories for diagnosis, under such titles as quality engineer. Where to locate these on the organizational chart has led to several alternatives. (See Figure 5.19.)

1. The diagnosticians are assigned to line managers in proportion to the needs of their departments. (See Figure 5.19*a*). This arrangement is preferred by line managers. In practice, these arrangements tend to end up with the diagnosticians being assigned to help the line managers meet current goals, fight fires, and so on. Such assignments then take priority over the chronic problems.

2. The diagnosticians are assigned to the various line managers (as above) but with a "dotted line" running to a central diagnostic department such as Quality Engineering. (See Figure 5.19*b*). This arrangement is better from the standpoint of training diagnosticians, offering them an obvious career path and providing them with consulting assistance. However, the arrangement runs into conflicts on the problem of priorities—on which projects should the diagnosticians be working.

3. The diagnosticians are assigned to a central diagnostic department such as Quality Engineering. (See Figure 5.19*c*). This arrangement increases the likelihood that chronic projects will have adequate priority. In addition, it simplifies the job of providing training and consulting assistance for diagnosticians. However, it makes no specific provision for line manager participation in choice of projects or in setting priorities. Such an omission can be fatal to results.

4. The diagnosticians are assigned to a central department but with a structured participation by the line managers. (See Figure 5.19*d*). In effect, the line managers choose the projects and establish priorities. The diagnostic department assigns the diagnosticians in response to these priorities. It also provides training, consulting services, and other assistance to the diagnosticians. This arrangement is used widely and has demonstrated its ability to adapt to a wide variety of company situations.

The choice among these (and other) alternatives depends on many factors that differ from one company to another.

RETROSPECTIVE ANALYSIS; LESSONS LEARNED

Lessons learned are based on experience that is derived from prior historical events. These events become lessons learned only after analysis—"retrospective analysis."

An enormous amount of diagnosis is done by analysis of historical events. A common example is seen in quality control of an industrial process. It is done by measuring a sample of units of product as they emerge from the process. Production of each unit is a historical event. Production of multiple units becomes multiple historical events. Analysis of the measurements is analysis of historical events and thereby an example of retrospective analysis.

The Santayana Review.   A short name is needed as a convenient label for this process of retrospective analysis. I have proposed calling it the *Santayana review*. The philosopher George Santayana once observed that "Those who cannot remember the past are condemned to repeat it." This is a terse and accurate expression of the concept of lessons learned through retrospective analysis. The definition becomes:

> The Santayana review is the process of deriving lessons learned from retrospective analysis of historical events.

The Influence of Cycle Time and Frequency.   Use of the Santayana review has depended largely on

> The cycle time of the historical events

> The frequency of these same events, which is closely correlated with their cycle time

The influence of these two factors, cycle time and frequency, is best understood by looking at a few examples.

Application to High-Frequency Cycles.   High-frequency events abound in companies of all kinds. The associated processes are of a mass production nature, and they process various products:

Industry	Mass Processing of
Utilities	Invoices
Factories	Goods
All industries	Payroll checks

The resulting cycles can number millions and even billions annually. Nevertheless, many companies manage to run these processes at extremely low levels of error. They do so by analysis of samples from the processes—by analyzing data from historical events.

It is fairly easy to apply the Santayana review in such mass production cases. The data are available in large numbers—sampling is a necessity to avoid drowning in data. The data analysis is often simple enough to be done locally by personnel trained in basic statistics. The effort involved is modest, so there is seldom any need to secure prior approval from higher levels. As a result, the Santayana review is widely applied. Of course, those who make such applications seldom consider that they are engaged in a study of prior historical events. Yet this is precisely what they are doing.

Application to Intermediate-Frequency Cycles.   As used here, "intermediate frequency" is an order of magnitude of tens or hundreds of cycles per year—a few per month or week. Compared with mass production, these cycles are longer, each involves more functions, each requires more effort, and more is at stake. Examples within this range of frequency include recruitment of employees or bids for business.

Applications of the Santayana review to intermediate-frequency cycles have been comparatively few in number, but the opportunities abound. It is obviously desirable to reduce the time needed to recruit employees. It is also desirable to increase the percentage of successful bids. (In some industries, the percentage is below 10 percent). The low level of retrospective analysis is traceable to some realities of the Santayana review as it applies to intermediate-frequency cycles:

The application is to a multifunctional process, usually requiring a team effort.

It can require a lot of work now, for benefits to come later, and with no ready way of computing return on investment.

There is rarely a clear responsibility for doing the work.

The urge to volunteer to do the work is minimal, since the improvement will benefit the organization generally but not necessarily the volunteer's department.

(These realities do not preclude application of the Santayana review to high-frequency cycles, since usually the application is to departmental processes, the amount of work is small, and the urge to volunteer is present because the results will benefit the volunteer's department).

Application to Low-Frequency Cycles.   As used here, "low frequency" refers to a range of several cycles per year down to one cycle in several years. Examples on an annual schedule include the sales forecast and the budget. Examples on an irregular schedule include new product launches, major construction projects, and acquisitions.

Application of the Santayana review to low-frequency cycles has been rare. Each such cycle is a sizable event; some are massive. A review of multiple cycles becomes a correspondingly sizable undertaking.

An example is the historical reviews conducted by a team of historians in British Petroleum Company. This team reviews large business undertakings: joint ventures, acquisitions, and major construction projects. The reviews concern matters of business strategy rather than conformance to functional goals. Each review consumes months of time and requires about 40 interviews to supply what is not in the documented history. The conclusions and recommendations are presented to the highest levels (Gulliver 1987).

A widespread low-frequency process that desperately needs application of the Santayana review is the launching of new products. Such launchings are carried out through an elaborate multifunctional process. Each product launched has a degree of uniqueness, but the overall process is quite similar from one cycle to another. Such being the case, it is entirely feasible to apply the Santayana review.

Much of the time required during the launch cycle consists of redoing what was done previously. Extra work is imposed on internal and external customers. The extent and cost of these delays can be estimated from a study of prior cycles. Retrospective analysis can shed light on what worked and what did not and thereby can improve decision making.

Note that the bulk of this delay and cost does *not* take place within the product development department. An example is seen in the launch of product X that incurred expenses as follows (in $millions):

Market research	0.5
Product development	6.0
Manufacturing facilities	22.0
Marketing planning	2.0
Total	30.5

All this was lost because a competitor captured the market by introducing a similar product 2 years before the launch of product X. The bulk of the loss—80 percent—took place *outside* the product development department.

Some Famous Case Examples.   The potential of the Santayana review can best be seen from some famous historical case examples.

Sky watchers and calendars: One of the astounding achievements of ancient civilizations was the development of precise calendars. These calendars were derived from numerous observations of the motions of celestial bodies, cycle after cycle. Some of these cycles were many years in length. The calendars derived from the data analysis were vital to the survival of ancient societies. For example, they told when to plant crops.

Prince Henry's think tank: During the voyages of discovery in the fifteenth and sixteenth centuries, Portuguese navigators were regarded as leaders in guiding ships to their destinations and bringing them back safely. As a result, Portuguese navigators were preferred and demanded by ship owners, governments, and insurers. The source of this leadership was an initiative by a Portuguese prince—Prince Henry the Navigator (1394–1460.) In the early 1400s, Prince Henry established (at Sagres, Portugal) a center for marine navigation—a unique, unprecedented think tank. The facilities included an astronomical observatory, a fortress, a school for navigators, living quarters, a hospital, and a chapel. To this center, Prince Henry brought cartographers, instrument makers, astronomers, mathematicians, shipwrights, and drafters. He also established a data bank— a depository of logs of marine voyages describing prevailing winds, ocean currents, landmarks, and so on. Lessons learned from these logs contributed to Portuguese successes during the voyages of discovery around the coast of Africa, through the Indian Ocean, and across the Atlantic.

Mathew Maury's navigation charts: In the mid-nineteenth century, Mathew Maury, a U.S. Navy lieutenant, analyzed the logs of thousands of naval voyages. He then entered the findings (current speeds, wind directions, and so on) on the navigation charts using standardized graphics and terminology. One of the first ships to use Maury's charts was the famous *Flying Cloud.* In 1851 it sailed from New York to San Francisco in 89 days. The previous record was 119 days (Whipple 1984). The new record then endured for 138 years!

Research on recurring disasters: Some individual disasters are so notorious that the resulting glare of publicity forces the creation of a formal board of inquiry. However, the most damage is done by repetitive disasters that, although less than notorious individually, are notorious collectively. Some institutions exist to study these disasters *collectively.* At their best, these institutions have contributed mightily to the wars against diseases, to reduction of accidents, and to making buildings fireproof. A fascinating example is a multinational study to shed light on the relation of diet to cancer. Figure 5.20 shows the resulting correlation (Cohen 1987).

The Potential for Long-Cycle Events.   The usefulness of the Santayana review has been amply demonstrated in the case of short-cycle, high-frequency activities. As a result, the Santayana review is widely applied to such cases and with good effect. The opportunities for application to long-cycle, low-frequency activities are enormous. However, the actual applications have been comparatively rare due to some severe realities.

Sponsorship requires a consensus among multiple managers rather than an initiative by one manager.

The associated work of the diagnostician is usually extensive and intrudes on the time of others.

The resulting lessons learned do not benefit current operations. The benefits apply to future operations.

The results do not necessarily benefit the departmental performances of participating managers.

There is no ready way to compute return on investment.

It is understandable that projects facing such realities have trouble in securing priorities. As matters stand, an initiative by upper managers is needed to apply the Santayana review to long-cycle

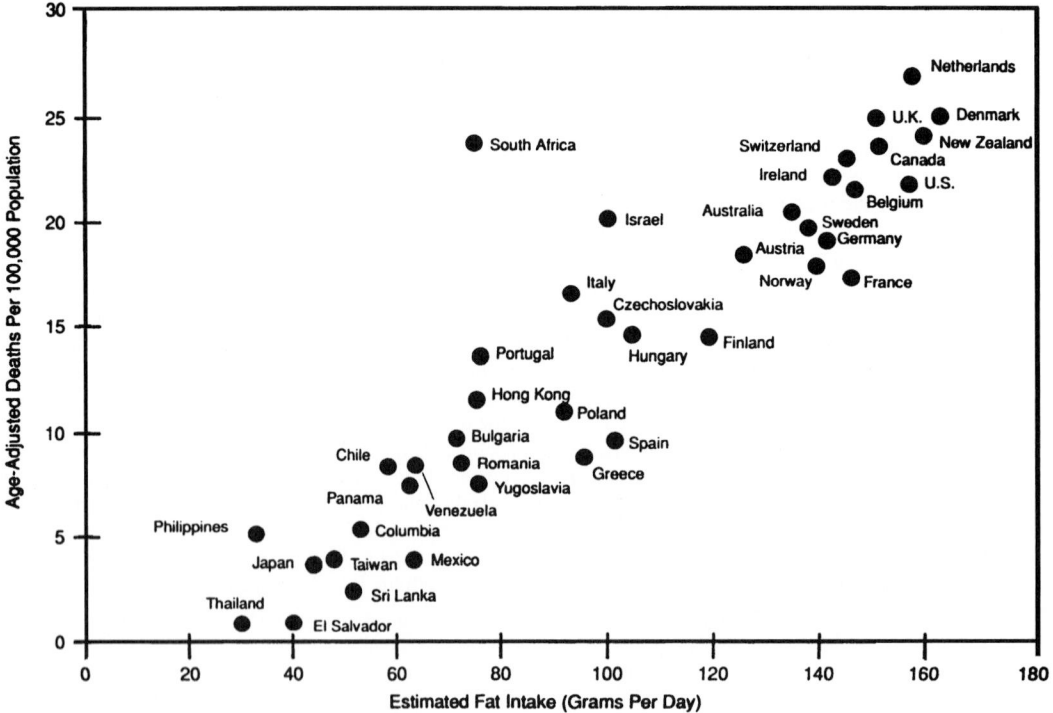

FIGURE 5.20   Correlation, diet and cancer.

activities. To date, such initiatives have been few, and published papers have been rare. The paper relative to the experience at British Petroleum is decidedly an exception (Gulliver 1987).

Will the pace of application accelerate? I doubt it. My prognosis is that the pace will remain evolutionary until some spectacular result is achieved and widely publicized. This is a discouraging forecast, the more so in the light of the quotation from Santayana: "Those who cannot remember the past are condemned to repeat it." (For extensive additional discussion and case examples, see Juran 1992.)

THE REMEDIAL JOURNEY

Once the causes are established, the diagnostic journey is over, and the remedial journey begins. While each remedy is unique to its project, the managerial approach to selecting and applying remedies is common to all projects.

Choice of Alternatives.   For most projects, there are multiple proposals for remedy. Choice of remedy then depends on the extent to which the proposals meet certain essential criteria. The proposed remedies should

Remove or neutralize the cause(s)

Optimize the costs

Be acceptable to those who have the last word

Remedies: Removing the Causes.   Proposed remedies typically must clear three hurdles before becoming effective:

1. The project team accepts the proposal based on logical reasoning—on its belief that the proposed remedy will meet the preceding criteria.
2. The proposal is tested out on a small scale, whether in operations or in the laboratory.
3. The proposal is tested full scale during operations.

In many companies a fourth hurdle has existed; the responsibility of project teams is vague, or limited to *recommending* remedies, with no responsibility to follow through. In such cases, many recommendations are simply not acted on. Results are much better in companies that make the teams responsible for ensuring that the remedies are in fact applied and that they are effective under operating conditions.

Many remedies consist of technological changes. These encounter the biases of some remedial departments, such as favoring remedies that involve buying new facilities. In many cases, however, the optimal remedy is through making better use of existing facilities. [For examples, see Black and Stump (1987); also see Bigelow and Floyd (1990).]

Actually, the remedies with the highest return on investment have involved managerial changes rather than technological changes. Dramatic evidence of this was seen when teams from the United States visited their Japanese counterparts to learn why Japanese quality was superior. Such visits were made to plants making steel, rubber tires, die castings, large-scale integrated circuits, automobiles, and so on. The Americans were astonished to find that the Japanese facilities (machinery, tools, instruments, and so on) were identical to those used in the American plants—they had even been bought from the same suppliers. The difference in quality had resulted from making better use of the existing facilities (personal experience of the author).

Still other remedies consist of revising matters of a broad managerial nature—policies, plans, organization, standards, procedures. Such remedies have effects that extend well beyond the specific project under study. Getting such remedies accepted requires special skills in dealing with cultural resistance (see below).

Occasionally, remedies can be remarkably imaginative. In a plant making chips for integrated circuits, a vibration problem caused by a nearby railroad was solved by constructing a swimming pool between the plant and the railroad. Another problem was due to cement dust from an adjacent concrete mixing plant. The remedy: Buy the plant and demolish it.

For some chronic quality problems, the remedy consists of *replanning* some aspect of the process or product in question. (For the methodology, see Section 3, The Quality Planning Process.)

Remedies: Optimizing the Costs.
In complex processes it is easy to apply a remedy that reduces costs in department A, only to learn that this cost reduction is more than offset by increased costs in department B. The cure can be worse than the disease. The project team should check out the side effects of the remedy to ensure that the costs are optimal for the company. This same check should extend to the effect on external customers' costs.

A well-chosen project team can do much to optimize costs because the membership is multifunctional. However, the team should look beyond the functions of its members. It also should enlist the aid of staff personnel from departments such as finance to assist in reviewing the figures and estimates. (For details on quantifying quality-related costs, see Section 8, Quality and Costs.)

Remedies: Acceptability.
Any remedy involves a change of some kind—redesign the product or process, revise the tool, and/or retrain the worker. Each such change falls within the jurisdiction of some functional department that then becomes the remedial department for the project in question. Normally the jurisdictional lines are respected, so the responsibility for making the change lies with the remedial department, not with the project team.

All this is simplified if someone from the remedial department is a member of the project team, which is usually the case. Such a member keeps his or her superiors informed and thereby helps to ensure that the proposed remedy will be adopted forthwith.

Matters are more complex if the remedial department has not been represented on the project team. Now the team must recommend that the remedial department adopt the remedy. This recommendation may encounter resistance for cultural reasons, including possible resentment at not having been represented. The project team is then faced with trying to convince the remedial department of the merits of the change. In the event of an impasse, the team may appeal through its sponsor or in other ways, such as through channels in the hierarchy.

Ideally, the remedial department is represented on the team from the outset. This is not always feasible—at the outset it is not known what will turn out to be the causes and hence the remedy. However, once the nature of the remedy becomes evident, the corresponding remedial department should be invited to join the team.

The concept of anticipating resistance applies to other sources as well—the union, the local community, and so on. The team is well advised to look for ways to establish a dialogue with those who are potentially serious opponents of the remedy. (For discussion of cultural resistance, see below under Resistance to Change.)

The Remedy for Rare but Critical Defects.
Some defects, while rare, can result in catastrophic damage to life or property. For such defects, there are special remedies.

Increase the factor of safety through additional structural material, use of exotic materials, design for misuse as well as intended use, fail-safe design, and so on. Virtually all of these involve an increase in costs.

Increase the amount and severity of test. Correlation of data on severe tests versus normal tests then provides a prediction of failure rates.

Reduce the process variability. This applies when the defects have their origin in manufacture.

Use automated 100 percent test. This concept has been supported recently by a remarkable growth in the technology: nondestructive test methods, automated testing devices, and computerized controls.

Use redundant 100 percent inspection. Inspection by human beings can be notoriously fallible. To find rare but critical defects, use can be made of multiple 100 percent inspections.

Remedy through Replication.
One form of replication of remedies is *cloning,* as discussed earlier in this section under Project Selection, Cloning. Through cloning, a remedy developed in one project may have application elsewhere in the same company. Replication also may be achieved through a generic remedy that applies to an assortment of error types.

Office work has long had the annoying problem of misspelled words. These misspellings are scattered among numerous different words. Now, word processing programs include a dictionary in their memory as a means of detecting misspelled words. The planners found a way to deal with numerous error types, each of which is comparatively rare.

Test under Operating Conditions.
Remedies are often tested in the laboratory before being adopted. A common approach is to develop a theoretical model and then construct and test some prototypes. This is a valuable step that can screen out inadequate remedies. Yet it is limited as a predictor of results in the real world of operations.

The theoretical model is based on assumptions that are never fully met.

The prototypes are constructed in a laboratory environment rather than in the operating environment.

The testing is done on a small sample size and under closely controlled test conditions.

The testing is done by trained technicians under the guidance of supervisors and engineers.

These and other limitations create the risk that the remedy, despite having passed its laboratory examination with flying colors, will not prove adequate under operating conditions. This has led some companies to require that the project team remain attached to the project until the remedy has been proved under operating conditions.

Control at the New Level; Holding the Gains.
To enable the operating forces to hold the gains requires (1) a successful transfer of the remedy from the laboratory to operations and (2) a systematic means of holding the gains—the control process. Ideally, the remedial change should be irreversible. Failing this, it may be necessary to conduct periodic audits to ensure that the change remains in place.

In a famous foundry project, one change involved the replacement of old ladle spouts with larger-diameter spouts. To make the change irreversible, the old spouts were destroyed. A different remedy required the melters to use scales to weigh accurately the amount of metal to be poured. This change could be reversed—some melters did not use the scales; they went right back to estimating by eye and feel.

Transfer to operations should include the revisions in operating standards, procedures, and so on needed to serve as a basis for training, control, and audit. These matters tend to be well defined with respect to the technology. In contrast, standards and procedures are often vague or silent on matters such as why the criteria should be met, what can happen if they are not met, equipment maintenance, and work methods. Failure to deal with these latter areas can be a threat to holding the gains.

Transfer to operations should include transfer of information related to the change. This transfer may require formal training in the use of the new processes and methods. It helps if the training also extends to the reasons behind the change, the resulting new responsibilities for decisions and actions, and the significant findings that emerged during the project.

The final step is establishing controls to hold the gains. This is done through the feedback loop—a cyclic process of evaluating actual performance, comparing this with the standard, and taking action on the difference. (Various aspects of the control process are discussed in Section 4, The Quality Control Process; Section 45, Statistical Process Control; and Section 11, ISO 9000 Family of Standards.)

HUMAN ERROR: DIAGNOSIS AND REMEDY

In some projects, the contributing causes include human error. Such errors are committed by all human beings—managers, supervisors, professional specialists, and the work force. Except for work force errors, the subject has received very little research, so the database is small. In view of this, what follows focuses on work force errors.

Extent of Work Force Errors.   Most errors are controllable by management. Errors are controllable by workers only if the criteria for self-control have all been met—if the worker has the means of

Knowing what he or she is supposed to do

Knowing what is his or her actual performance

Regulating his or her performance

Investigators in many countries have conducted studies on controllability. As reported to me, these generally confirm my own conclusion that in industry, by and large, controllability prevails as follows:

Management-controllable: over 80 percent

Worker-controllable: under 20 percent

Species of Work Force Error.   It has long been a widely held belief by managers that work force errors are due to lack of motivation. However, recent research has shown that there are multiple species of work force errors and that only a minority of such errors have their origin in lack of motivation.

Table 5.8 shows the distribution of 80 errors made by six office workers engaged in preparing insurance policy contracts. There are 29 types of errors, and they are of multiple origins.

Error type 3 was made 19 times, but worker B made 16 of them. Yet, except for error type 3, worker B makes few errors. Seemingly, there is nothing wrong with worker B, except on defect type 3. Seemingly also, there is nothing wrong with the job instructions, since no one else had trouble with error type 3. It appears that worker B and no one else is misinterpreting some instruction, resulting in that clump of 16 errors of type 3.

TABLE 5.8   Matrix of Errors by Insurance Policy Writers

Error type	Policy writer						Total
	A	B	C	D	E	F	
1	0	0	1	0	2	1	4
2	1	0	0	0	1	0	2
3	0	(16)	1	0	2	0	(19)
4	0	0	0	0	1	0	1
5	2	1	3	1	4	2	(13)
6	0	0	0	0	3	0	3
⋮	⋮	⋮	⋮	⋮	⋮	⋮	⋮	⋮
27							
28							
29							
TOTAL	6	(20)	8	3	(36)	7	80

Error type 5 is of a different species. There are 13 of these, and every worker makes this error, more or less uniformly. This suggests a difference in approach between all the workers on the one hand and the inspector on the other. Such a difference is usually of management origin, but the realities can be established by interviews with the respective employees.

A third phenomenon is the column of numbers associated with worker E. The total is 36 errors—worker E made nearly half the errors, and he or she made them in virtually all error type categories. Why did worker E make so many errors? It might be any of several reasons, such as inadequate training, lack of capability to do exacting work, and so on. Further study is needed, but some managers might prefer to go from symptom directly to remedy—find a less demanding job for that worker.

This single table of data demonstrates the existence of multiple species of worker error. The remedy is not as simplistic as "motivate the worker." Analysis of many such tables, plus discovery of the causes, has identified four principal species of work force error: inadvertent, technique, conscious, and communication. Table 5.9 shows the interrelations among the error patterns, the likely subspecies, and the likely remedies. The error species are examined below.

Inadvertent Errors. "Inadvertent" means "caused by lack of attention." Inadvertent errors are made because of human inability to maintain attention. (Ancient generals and admirals limited the length of the sentry's watch because of the risk of lack of attention.) (If not paying attention is deliberate, then the resulting errors are conscious rather than inadvertent.)

Diagnosis to identify errors as inadvertent is aided by understanding their distinguishing features. They are

Unintentional: The worker does not want to make errors.

Unwitting: At the time of making an error, the worker is unaware of having made it.

Unpredictable: There is nothing systematic as to when the next error will be made, what type of error will be made, or which worker will make the error. Due to this unpredictability, the error pattern exhibits *randomness.* Conversely, data that show a random pattern of worker error

TABLE 5.9 Interrelation among Human Error Patterns

Pattern disclosed by analysis of worker error	Likely subspecies of error causing this pattern	Likely solution
On certain defects, no one is error-prone; defect pattern is random.	Errors are due to inadvertence.	Error-proof the process.
On certain defects, some workers are consistently error-prone, while others are consistently "good."	Errors are due to lack of technique (ability, know-how, etc.). Lack of technique may take the form of secret ignorance. Technique may consist of known knack or of secret knowledge.	Discovery and propagation of knack. Discovery and elimination of secret ignorance.
Some workers are consistently error-prone over a wide range of defects.	There are several potential causes: Conscious failure to comply to standards. Inherent incapacity to perform this task. Lack of training.	Solution follows the cause: Motivation. Transfer worker. Supply training.
On certain defects, all workers are error-prone.	Errors are management controllable.	Meet the criteria for self-control. Standardize the language; provide translation, glossaries.

suggest that the errors are due to inadvertence. The randomness may apply to the types of error, to the workers who make the errors, and to the time when the errors are made.

The cause of inadvertent errors is inattention. But what causes inattention? The search for an answer leads into the complexities of psychological (e.g., monotony) and physiologic (e.g., fatigue) phenomena. These are not fully understood, even by experts. To explore these complexities in-depth means going deeper and deeper into an endless swamp. Practical managers prefer to go around the swamp—to go directly from symptom to remedy.

Remedies for Inadvertent Errors.   Remedies for inadvertent errors involve two main approaches:

1. Reduce the dependence on human attention through error-proofing: fail-safe designs, countdowns, redundant verification, cutoffs, interlocks, alarm signals, automation, and robots. (Use of bar codes has greatly reduced errors in identifying goods).
2. Make it easier for workers to remain attentive. Reorganize work to reduce fatigue and monotony by use of job rotation, sense multipliers, templates, masks, overlays, and so on.

[For an uncommonly useful paper on error-proofing, with numerous examples, especially as applied to service industries, see Chase and Stewart (1994).]

Technique Errors.   Technique errors are made because workers lack some "knack"—some essential technique, skill, or knowledge needed to prevent errors from happening. Technique errors exhibit certain outward features. They are

Unintentional: The worker does not want to make errors.

Specific: Technique errors are unique to certain defect types—those types for which the missing technique is essential.

Consistent: Workers who lack the essential technique consistently make more defects than workers who possess the technique. This consistency is readily evident from data on worker errors.

Unavoidable: The inferior workers are unable to match the performance of the superior workers because they (the inferior workers) do not know "what to do different."

An example of technique errors is seen in the gun assembly case. Guns were assembled by 22 skilled artisans, each of whom assembled complete guns from bits and pieces. After the safety test, about 10 percent of the guns could not be opened up to remove the spent cartridge—a defect known as "open hard after fire." For this defect it was necessary to disassemble the gun and then reassemble, requiring about 2 hours per defective gun—a significant chronic waste.

Following much discussion, a table like Table 5.10 was prepared to show the performance of the assemblers. This table shows the frequency of "open hard after fire" by assembler and by month over a 6-month period. Analysis of the table brings out some significant findings.

The *departmental* defect rate varied widely from month to month, ranging from a low of 1.8 percent in January to a high of 22.6 percent in February. Since all workers seemed to be affected, this variation had its cause outside the department. (Subsequent analysis confirmed this.)

The ratio of the five best worker performances to the five worst showed a *stunning consistency.* In each of the 6 months, the five worst performances add up to an error rate that is at least 10 times as great as the sum of the five best performances. There must be a reason for such a consistent difference, and it can be found by studying the work methods—the techniques used by the respective workers.

The knack: The study of work methods showed that the superior performers used a file to cut down one of the dimensions on a complex component; the inferior performers did not file the file. This filing constituted the knack—a small difference in method that accounts for a large differ-

TABLE 5.10   Matrix Analysis to Identify Technique Errors

Assembly operator rank	Nov.	Dec.	Jan.	Feb.	Mar.	Apr.	Total
1	4	1	0	0	0	0	5
2	1	2	0	5	1	0	9
3	3	1	0	3	0	3	10
4	1	1	0	2	2	4	10
5	0	1	0	10	2	1	14
6	2	1	0	2	2	15	22
⋮	⋮	⋮	⋮	⋮	⋮	⋮	⋮
17	18	8	3	37	9	23	98
18	16	17	0	22	36	11	102
19	27	13	4	62	4	14	124
20	6	5	2	61	22	29	125
21	39	10	2	45	20	14	130
22	26	17	4	75	31	35	188
TOTAL	234	146	34	496	239	241	1390
% Defective	10.6	6.6	1.8	22.6	10.9	11.0	10.5
5 best	9	6	0	20	5	8	48
5 worst	114	62	12	265	113	103	669
Ratio	13	10	∞	13	23	13	14

ence in results. (Until the diagnosis was made, the superior assemblers did not realize that the filing greatly reduced the incidence of "open hard after fire.")

Usually the difference in worker performance is traceable to some superior knack used by the successful performers to benefit the product. In the case of the gun assemblers, the knack consisted of filing the appropriate component. In other cases, the difference in worker performance is due to unwitting *damage done to the product by the inferior performers*—sort of "negative knack."

There is a useful rule for predicting whether the difference in worker performance is due to a beneficial knack or to a negative knack. If the superior performers are in the minority, the difference is probably due to a beneficial knack. If the inferior performers are in the minority, then the difference in performance is likely due to a negative knack.

In an aircraft assembly operation, data analysis by individual workers revealed that one worker met the production quota consistently, whereas the others did not. The worker explained that he had taken his powered screwdriver home and rebuilt the motor. The company replaced all the motors, with a resulting increase in quality and productivity.

Analysis of data on damage to crankshafts showed that only one worker's product was damaged. Study in the shop then revealed that this worker sometimes bumped a crankshaft into a nearby conveyor. Why? Because the worker was left-handed and the workplace layout was too inconvenient for a left-handed person.

The gun assembly case shows the dangers of assuming that differences in worker performance are due to a lack of motivation. Such an assumption is invalid as applied to technique errors. Technique errors are doomed to go on and on until ways are found to provide the inferior workers with an answer to the question, "What should I do different than I am doing now?"

How are such questions to be answered? Worker improvement teams sometimes can provide answers. Failing this, they will keep on doing what they have been doing (and keep on making the same defects) until the answers are provided by management.

Remedies for Technique Errors.   Solution of numerous cases of technique errors has yielded a structured generic approach:

1. Collect data on individual worker performances.
2. Analyze the data for consistent worker-to-worker differences.
3. For cases of consistent differences, study the work methods used by the best and worst performers to identify their differences in technique.
4. Study these differences further to discover the beneficial knack that produces superior results (or the negative knack that damages the product).
5. Bring everyone up to the level of the best through appropriate remedial actions such as
   - Train the inferior performers in use of the knack or in avoidance of damage.
   - Change the technology so that the process embodies the knack.
   - Error-proof the process in ways that force use of the knack or that prohibit use of the negative knack.
   - Institute controls and audits to hold the gains.

Conscious Errors.   Conscious errors involve distinctive psychological elements. Conscious errors are

Witting: At the time of making an error, the worker is aware of it.

Intentional: The error is the result of a deliberate decision on the part of the worker.

Persistent: The worker who makes the error usually intends to keep it up.

Conscious errors also exhibit some unique outward evidences. Whereas inadvertent errors exhibit randomness, conscious errors exhibit consistency—some workers consistently make more errors than others. However, whereas technique errors typically are restricted to one or a few defect types, conscious errors tend to cover a wide spectrum of defect types.

On the face of it, workers who commit conscious errors deserve to be disciplined, but this principle has only partial validity. Many such errors are actually *initiated by management.*

A major source of conscious errors is an atmosphere of blame. In such an atmosphere, workers defend themselves by violating company rules. They omit making out the rework tickets, they hide the scrap, and so on.

Another widespread source of conscious errors is conflict in priorities. For example, in a sellers' market, priority on delivery schedules can prevail over some quality standards. The pressures on the managers are transmitted down through the hierarchy and can result in conscious violation of quality standards to meet the schedules.

In addition, some well-intentioned actions by management can have a negative effect. For example, the managers launch a poster campaign to urge everyone to do better work. However, the campaign makes no provision to solve some quality problems well known to the workers: poor quality from suppliers, incapable processes, inadequate maintenance of facilities, and so on. Thus management loses credibility—the workers conclude that the real message of the managers is "Do as we say, not as we do."

Some conscious errors are initiated by the workers. Workers may have real or imagined grievances against the boss or the company. They may get revenge by not meeting standards. Some become rebels against the whole social system and use sabotage to show their resentment. Some of the instances encountered are so obviously antisocial that no one—not the fellow employees, not the union—will defend the actions.

To a degree, conscious worker errors can be dealt with through the disciplinary process. However, managers also have access to a wide range of constructive remedies for conscious worker errors.

Remedies for Conscious Errors.   Generally, the remedies listed here emphasize securing changes in behavior but without necessarily changing attitudes. The remedies may be directed toward the persons or the "system"—the managerial and technological processes.

Depersonalize the order: In one textile plant, the spinners were failing to tie the correct knots ("weaver's knots") when joining two ends of yarn together. The pleas and threats of the supervisor were of no avail. The spinners disliked the supervisor, and they resented the company's poor responses to their grievances. The problem was solved when the personnel manager took the informal leader of the spinners to the Weaving Department to show her how the weavers were having trouble due to wrong knots. Despite their unsolved grievances, once they learned about the events in the weaving room, the spinners were unwilling to continue making trouble for their fellow workers. The principle involved here is *the law of the situation*—one *person* should not give orders to another *person*; both should take their orders from the situation. *The law of the situation* is a phrase coined by Mary Parker Follett. [See Metcalf and Urwick (1941).] The situation in the Weaving Department requires that weaver's knots be tied. Hence this situation is binding on the president, the managers, the supervisors, and the spinners.

Establish accountability: To illustrate, in one company the final product was packaged in bulky bales that were transported by conventional forklift trucks. Periodically, a prong of a fork would pierce a bale and do a lot of damage. Yet there was no way of knowing which trucker moved which bale. When the company introduced a simple means of identifying which trucker moved which bale, the amount of damage dropped dramatically.

Provide balanced emphasis: Workers discover the company's real priorities on multiple standards (quality, productivity, delivery) from the behavior of management. For example, scoreboards on productivity and delivery rates should be supplemented with a scoreboard on quality to provide evidence of balanced emphasis.

Conduct periodic quality audits: Systems of continuing traceability or scorekeeping are not always cost-effective. Quality audits can be designed to provide, on a sampling basis, information of an accountability and scorekeeping nature.

Provide assistance to workers: Visual aids to help prevent defects can be useful. Some companies have used wall posters listing the four or five principal defects in the department, along with a narrative and graphic description of the knack that can be used to avoid each defect.

Create competition, incentives: These devices have potential value if they are not misused. Competition among workers and teams should be designed to be in good humor and on a friendly level, such as prevails among departmental sports teams. Financial incentives are deceptively attractive. They look good while pay is going up—during that part of the cycle there are "bonuses" for good work. However, during a spate of poor work, removal of the bonuses converts the incentives into penalties, with all the associated arguments about who is responsible. Nonfinancial incentives avoid the pitfall of bonuses becoming penalties, but they should be kept above the gimmickry level.

Error-proof the operation: Error-proofing has wide application to conscious errors. (See Section 22, Operations, under Error-Proofing the Process.)

Reassign the work: An option usually available to managers is selective assignment, i.e., assign the most demanding work to workers with the best quality record. Application of this remedy may require redesign of jobs—separation of critical work from the rest so that selective assignment becomes feasible.

Use the tools of motivation: This subject is discussed in Section 15, Human Resources and Quality.

This list of remedies helps to solve many conscious errors. However, prior study of the symptoms and surrounding circumstances is essential to choosing the most effective remedy.

Communication Errors.   A fourth important source of human error is traceable to errors in communication. There are numerous subspecies of these, but a few of them are especially troublesome.

Communication omitted: Some omissions are by the managers. There are situations in which managers take actions that on their face seem antagonistic to quality but without informing the workers why. For example, three product batches fail to conform to quality feature X. In each case, the inspector places a hold on the batch. In each case, a material review board concludes that the batch is fit for use and releases it for delivery. However, neither the production worker nor the inspector is told why. Not knowing the reason, these workers may (with some logic) conclude that feature X is unimportant. This sets the stage for future unauthorized actions. In this type of case and in many others, company procedures largely assume that the workers have no need to know. (The release forms of material review boards contain no blank to be filled in requiring the members to face the question: "What shall we communicate to the work force?" Lacking such a provision, the question is rarely faced, so by default there is no communication.

Communication inhibited: In most hierarchies, the prevailing atmosphere historically has inhibited communication from the bottom up. The Taylor system of the late nineteenth century made matters worse by separating planning from execution. More recently, managers have tried to use this potential source of information through specific concepts such as suggestion systems, employee improvement teams, and most recently, self-directed teams of workers. The twentieth century rise in education levels has greatly increased the workers' potential for participating usefully in planning and improvement of operations. It is a huge underemployed asset. Managers are well advised to take steps to make greater use of this asset.

Transmission errors: These errors are not conscious. They arise from limitations in human communication. Identical words have multiple meanings, so the transmitter may have one meaning in mind, but the receiver has a different meaning in mind. Dialects differ between companies and even within companies. (The chief language in the upper levels is money, whereas in the lower levels it is things.)

A critical category of terminology contains the words used to transmit broad concepts and matters of a managerial nature: policies, objectives, plans, organization structure, orders (commands), advice, reviews, incentives, and audits. The recipients (receivers) are mainly internal, across all functions and all levels. The problem is to ensure that receivers interpret the words in ways intended by transmitters. There is also the problem of ensuring that the responses are interpreted as intended.

In other cases, the intention of the transmitter is clear, but what reaches the receiver is something else. A misplaced comma can radically change the meaning of a sentence. In oral communications, background noise can confuse the receiver.

In an important football game, on a key play near the end of the game, amid the deafening noise of the crowd, the defensive signal was *three*, which called for a man-to-man defense. One defensive player thought he heard *green*, which called for a zone defense. The error resulted in loss of the game (Anderson 1982).

Remedies for Communication Errors.   Communication errors are sufficiently extensive and serious to demand remedial action. The variety of error types has required corresponding variety in the remedies.

Translation: For some errors, the remedy is to create ways to translate the transmitters' communications into the receivers' language. A common example is the Order Editing Department, which receives orders from clients. Some elements of these orders are in client language. Order Editing translates these elements into the supplier's language, through product code numbers, acronyms, and other means. The translated version is then issued as an internal document within the supplier's company. A second example is the specialists in the Technical Service Department. The specialists in this department are trained to be knowledgeable about their company's products. Through their contacts with customers, they learn of customer needs. This combined knowledge enables them to assist both companies to communicate, including assistance in translation.

The glossary: This useful remedy requires reaching agreement on definitions for the meanings of key words and phrases. These definitions are then published in the form of a glossary—a list of terms and their definitions. The publication may be embellished by other forms of communication: sketches, photographs, and/or videotapes.

Standardization: As companies and industries mature, they adopt standardization for the mutual benefit of customers and suppliers. This extends to language, products, processes, and so on. In the case of physical goods, standardization is very widely used. Without it, a technological society would be a perpetual tower of Babel. All organizations make use of short designations for their products: code numbers, acronyms and so on. Such standardized nomenclature makes it easier to communicate internally. If external customers also adopt the nomenclature, the problem of multiple dialects is greatly reduced. The *Airline Flight Guide* publishes flight information for multiple airlines. This information is well standardized. Some clients learn how to read the flight guide. For such clients, communication with the airlines is greatly simplified.

Measurement: Saying it in numbers is an effective remedy for some communication problems, e.g., those in which adjectives (such as *roomy, warm, quick,* and so on) are used to describe product features). (For elaboration, see Section 9, Measurement, Information, and Decision-Making.)

A role for upper managers: Companies endure extensive costs and delays due to poor communication. The remedies are known, but they do not emerge from day-to-day operations. Instead, they are the result of specific projects set up to create them. In addition, they evolve slowly because they share the common feature of "invest now for rewards later." Upper managers are in a position to speed up this evolution by creating project teams with missions to provide the needed remedies.

RESISTANCE TO CHANGE

On the face of it, once a remedy has been determined, all that remains is to apply it. Not so. Instead, obstacles are raised by various sources. There may be delaying tactics or rejection by a manager, the work force, or the union. "Resistance to change" is the popular name for these obstacles.

Cultural Patterns. An understanding of resistance to change starts with the realization that every change actually involves two changes:

1. The *intended* change
2. The *social consequence* of the intended change

The social consequence is the troublemaker. It consists of the impact of the intended change on the *cultural pattern* of the human beings involved—on their pattern of beliefs, habits, traditions, practices, status symbols, and so on. This social consequence is the root source of the resistance to change. Dealing with this resistance requires an understanding of the nature of cultural patterns.

Ideally, advocates of change should be aware that all human societies evolve cultural patterns and that these are fiercely defended as a part of "our way of life." In addition, the advocates should try to discover precisely what their proposals will threaten—which habits, whose status, what beliefs. Unfortunately, too many advocates are *not even aware of the existence of cultural patterns,* let alone their detailed makeup.

To make matters more complex, those who resist the change often state their reasons as objections to the merits of the intended change, whereas their real reasons relate to the social consequences. As a result, the advocates of the intended change are confused because the stated reasons are not the real reasons for the resistance.

To illustrate, companies that first tried to introduce computer-aided design (CAD) ran into resistance from the older designers, who claimed that the new technology was not as effective as design analysis by a human being. Interviews then found that the real reasons included the fear of losing status because the younger engineers could adapt more readily to the change.

Rules of the Road. Behavioral scientists have evolved some specific rules of the road for dealing with cultural resistance (Mead 1951). These rules are widely applicable to industrial and other organizational entities (Juran 1964).

Provide participation: This is the single most important rule for introducing change. Those who will be affected by the change should participate in the planning as well as in the execution. Lack of participation leads to resentment, which can harden into a rock of resistance.

Provide enough time: How long does it take for members of a culture to accept a change? They need enough time to evaluate the impact of the change. Even if the change seems beneficial, they need to learn what price they must pay in cultural values.

Start small: Conducting a small-scale tryout before going all out reduces the risks for the advocates as well as for members of the culture.

Avoid surprises: A major benefit of the cultural pattern is its predictability. A surprise is a shock to this predictability and a disturber of the peace.

Choose the right year: There are right and wrong years—even decades—for timing a change.

Keep the proposals free of excess baggage: Avoid cluttering the proposals with extraneous matters not closely related to getting the results. The risk is that the debates will get off the main subject and into side issues.

Work with the recognized leadership of the culture: The culture is best understood by its members. They have their own leadership, and this is sometimes informal. Convincing the leadership is a significant step in getting the change accepted.

Treat the people with dignity: The classic example was the relay assemblers in the Hawthorne experiments. Their productivity kept rising, under good illumination or poor, because in the laboratory they were being treated with dignity.

Reverse the positions: Ask the question: "What position would I take if I were a member of the culture?" It is even useful to go into role playing to stimulate understanding of the other person's position. [For a structured approach, see Ackoff (1978).]

Deal directly with the resistance: There are many ways of dealing directly with resistance to change.

- Try a program of persuasion.
- Offer a *quid pro quo*—something for something.
- Change the proposals to meet specific objections.
- Change the social climate in ways that will make the change more acceptable.
- Forget it. There are cases in which the correct alternative is to drop the proposal. Human beings do not know how to plan so as to be 100 percent successful.

[For added discussion, see Schein (1993). See also Stewart (1994) for a discussion of self-rating one's resistance to change.]

Resolving Differences. Sometimes resistance to change reaches an impasse. Coonley and Agnew (1941) once described a structured process used for breaking an impasse on the effort to establish quality standards on cast iron pipe. Three conditions were imposed on the contesting parties:

1. They must identify their areas of agreement and their areas of disagreement. "That is, they must first agree on the exact point at which the road began to fork." When this was done, it was found that a major point of disagreement concerned the validity of a certain formula.

2. "They must agree on why they disagreed." They concluded that the known facts were inadequate to decide whether the formula was valid or not.

3. "They must decide what they were going to do about it." The decision was to raise a fund to conduct the research needed to establish the necessary facts. "With the facts at hand, the controversies disappeared."

THE LIFE CYCLE OF A PROJECT: SUMMARY

The universal sequence for improvement sets up a common pattern for the life cycle of projects. Following project selection, the project is defined in a mission statement and is assigned to a project team.

The team then meets, usually once a week for an hour or so. During each meeting, the team

Reviews the progress made since the previous meeting

Agrees on the actions to be taken prior to the next meeting (the homework)

Assigns responsibility for those actions

Gradually, the team works its way through the universal sequence. The diagnostic journey establishes the causes. The remedial journey provides the remedies and establishes the controls to hold the gains.

During all this time, the team issues minutes of its meetings as well as periodic progress reports. These reports are distributed to team members and also to nonmembers who have a need to know. Such reports form the basis for progress review by upper managers.

The final report contains a summary of the results achieved, along with a narrative of the activities that led to the results. With experience, the teams learn to identify lessons learned that can be applied elsewhere in the company. [Relative to the life cycle of a project, see AT&T Quality Library, *Quality Improvement Cycle* (1988, pp. 13–17).]

INSTITUTIONALIZING QUALITY IMPROVEMENT

Numerous companies have initiated quality improvement, but few have succeeded in institutionalizing it so that it goes on year after year. Yet many of these companies have a long history of annually conducting product development, cost reduction, productivity improvement, and so on. The methods they used to achieve such annual improvement are well known and can be applied to quality improvement.

Enlarge the annual business plan to include goals for quality improvement.

Make quality improvement a part of the job description. In most companies, the activity of quality improvement has been regarded as incidental to the regular job of meeting the goals for quality, cost, delivery, and so on. The need is to make quality improvement a part of the regular job.

Establish upper management audits that include review of progress on quality improvement.

Revise the merit rating and reward system to include a new parameter—performance on quality improvement—and give it proper weight.

Create well-publicized occasions to provide recognition for performance on improvement.

THE NONDELEGABLE ROLES OF UPPER MANAGERS

The upper managers must participate extensively in the quality initiative. It is not enough to create awareness, establish goals, and then leave all else to subordinates. This has been tried and has failed over and over again. I know of no company that became a quality leader without extensive participation by upper managers.

It is also essential to define just what is meant by "participation." It consists of a list of roles to be played by the upper managers, *personally*. What follows is a list of roles actually played by upper managers in companies that have become quality leaders. These roles can be regarded as "nondelegable."

Serve on the quality council: This is fundamental to upper managers' participation. It also becomes an indicator of priorities to the rest of the organization.

Acquire training in managing for quality: Sources of such training include visits to successful companies. Training is also available at courses specially designed for upper managers and through attending conferences. (Upper managers risk losing credibility if they try to lead while lacking training in managing for quality.)

Approve the quality vision and policies: A growing number of companies have been defining their quality vision and policies. Invariably, these require upper management approval before they may be published.

Approve the major quality goals: The quality goals that enter the business plan must be deployed to lower levels to identify the deeds to be done and the resources needed. The upper managers become essential parties to the deployment process.

Establish the infrastructure: The infrastructure includes the means for nominating and selecting projects, preparing mission statements, appointing team leaders and members, training teams and facilitators, reporting progress, and so on. Lacking such an infrastructure, quality improvement will take place only in local areas and with no noticeable effect on the *bottom line.*

Provide resources: During the 1980s, many upper managers provided extensive resources for training their personnel, chiefly in awareness and in statistical tools. In contrast, only modest resources were provided for training in managing for quality and for setting up the infrastructure for quality improvement.

Review progress: A major shortcoming in personal participation by upper managers has been the failure to maintain a regular review of progress in making quality improvements. During the 1980s, this failure helped to ensure lack of progress—quality improvement could not compete with the traditional activities that did receive progress reviews from upper managers.

Give recognition: Recognition usually involves ceremonial events that offer highly visible opportunities for upper managers to show their support for quality improvement. Upper managers should seize these opportunities; most upper managers do so. (See below, under Recognition.)

Revise the reward system: Traditional reward systems provide rewards for meeting traditional goals. These systems must now be opened up to give proper weight to performance on quality improvement. Upper managers become involved because any changes in the reward system require their approval. (See below, under Rewards.)

Serve on project teams: There are some persuasive reasons behind this role. See preceding, under The Project Team, Upper Managers on Teams.

Face up to employee apprehensions: See preceding, under The Quality Council, Apprehensions about Elimination of Jobs.

Such is a list of the nondelegable roles of upper managers. In companies that have become quality leaders, the upper managers carry out most, if not all, of these roles. No company known to me has attained quality leadership without the upper managers carrying out those nondelegable roles.

PROGRESS REVIEW

Scheduled, periodic review of progress by upper managers is an essential part of maintaining annual quality improvement. Activities that do not receive such review cannot compete for priority with activities that do receive such review. Subordinates understandably give top priority to matters that are reviewed regularly by their superiors.

Review of Results.   Results take multiple forms, and these are reflected in the design of the review process. Certain projects are of such importance *individually* that the upper managers want to follow them closely. The remaining projects receive their reviews at lower levels. However, for the purpose of upper management review, they are summarized to be reviewed *collectively* by upper management.

There is also a need for regular review of the quality improvement *process.* This is done through audits that may extend to all aspects of managing for quality. (Refer to Section 11, ISO 9000 Family of Standards.)

Inputs to Progress Review.   Much of the database for progress review comes from the reports issued by the project teams. However, it takes added work to analyze these reports and to prepare the summaries needed by upper managers. Usually this added work is done by the secretary of the quality council with the aid of the facilitators, the team leaders, and other sources such as finance.

As companies gain experience, they design standardized reporting formats to make it easy to summarize reports by groups of projects, by product lines, by business units, by divisions, and for the corporation. One such format, used by a large European company, determines for each project

The original estimated amount of chronic waste

The original estimated reduction in cost if the project were to be successful

The actual cost reduction achieved

The capital investment

The net cost reduction

The summaries are reviewed at various levels. The corporate summary is reviewed quarterly at the chairman's staff meeting (personal communication to the author).

Evaluation of Performance.   One of the objectives of progress review is evaluation of performance. This evaluation extends to individuals as well as to projects. Evaluation of individual performance on improvement projects runs into the complication that the results are achieved by teams. The problem then becomes one of evaluating individual contribution to team efforts. This new problem has as yet no scientific solution. Thus each supervisor is left to judge subordinates' contributions based on inputs from all available sources.

At higher levels of organization, the evaluations extend to judging the performance of supervisors and managers. Such evaluations necessarily must consider results achieved on multiple projects. This has led to evolution of measurement (metrics) to evaluate managers' performance on projects collectively. These metrics include

Numbers of improvement projects: initiated, in progress, completed, and aborted

Value of completed projects in terms of improvement in product performance, reduction in costs, and return on investment

Percentage of subordinates active on project teams

Superiors then judge their subordinates based on these and other inputs.

RECOGNITION

"Recognition" as used here means "public acknowledgment of superior performance." (Superior performance deserves public acknowledgment.) Recognition tells recipients that their efforts are appreciated. It adds to their self-respect and to the respect received from others.

Most companies are quite effective at providing recognition. They enlist the ingenuity of those with special skills in communication—Human Relations, Marketing, Advertising—as well as the line managers. The numerous forms of recognition reflect this ingenuity:

Certificates, plaques, and such are awarded for serving on project teams, serving as facilitator, and completing training courses.

Project teams present their final report in the office of the ranking local manager.

Project summaries are published in the company news media, along with team pictures. Some companies create news supplements or special newsletters devoted to quality improvement. Published accounts of successful projects not only provide recognition, they also serve as case materials for training purposes and as powerful stimulators to all.

Dinners are held to honor project teams.

Medals or prizes may be awarded to teams judged to have completed the best projects during some designated time period. The measure of success always includes the extent of results achieved and sometimes includes the methods used to achieve the results. [For an account of the annual competition sponsored by Motorola, see Feder (1993); see also Motorola's Team Competition (1992).]

REWARDS

As used here, "rewards" refers to salaries, salary increases, bonuses, promotions, and so on resulting from the annual review of employee performance. This review has in the past focused on meeting goals for traditional parameters: costs, productivity, schedule, and quality. Now a new parameter—quality improvement—must be added to recognize that quality improvement is to become a part of the job description.

Note that reward differs sharply from recognition. The crucial difference lies in whether the work is voluntary or mandatory.

Recognition is given for superior performance, which is *voluntary*. (People can hold their jobs by giving adequate performance.)

Rewards are given for *mandated* performance—doing the work defined in the job description. Willful failure to do this work is a violation of the employment contract and is a form of insubordination.

The new parameter—quality improvement—is time-consuming. It adds a new function. It invades the cultural pattern. Yet it is critical to the company's ability to remain competitive. This is why the parameter of quality improvement must enter the job descriptions and the reward system. Failing this, employees will continue to be judged on their performance against traditional goals, and quality improvement will suffer due to lack of priority.

One well-known company had added the parameter "performance on quality improvement" to its annual review system. All personnel in the managerial hierarchy are then rated into one of three classes: more than satisfactory, satisfactory, or less than satisfactory. Those who fall into the lowest class are barred from advancement for the following 12 months (personal communication to the author).

(For additional discussion, see Section 15, Human Resources and Quality.)

TRAINING

Throughout this section there have been numerous observations on the needs for training. These needs are extensive because quality improvement is a new function in the company that assigns new responsibility to all. To carry out these new responsibilities requires extensive training. Some details of this training have been discussed here and there in this section. (For additional discussion, see Section 16, Training for Quality.)

SUMMARY OF RESPONSIBILITIES FOR QUALITY IMPROVEMENT

Quality improvement requires action at all levels of the organization, as summarized in Figure 5.21. (For more on the planning and coordination of these activities on multiple levels, see Section 13, Strategic Deployment, and Section 14, Total Quality Management.)

ACKNOWLEDGMENTS

This section of the handbook has drawn extensively from various training materials published by the Juran Institute, Inc., and Juran, J. M., ed. (1955), *A History of Managing for Quality,* sponsored by the Juran Foundation, Inc. Quality Press, Milwaukee. I am also grateful to the Juran Foundation, Inc., and the Juran Institute, Inc., for permission to extract from those works.

REFERENCES

Ackoff, Russell L. (1978). *The Art of Problem Solving.* John Wiley & Sons, New York, pp. 45–47.

AT&T Quality Library. For Information, call AT&T Customer Information Center, 1-800-432-6600.

Anderson, Dave (1982). "When 3 Meant 6." *New York Times,* December 27, pp. C1, C3.

Berwick, Donald M., Godfrey, A. Blanton, and Roessner, Jane (1990). *Curing Health Care.* Jossey-Bass, San Francisco.

Betker, Harry A. (1983). "Quality Improvement Program: Reducing Solder Defects on Printed Circuit Board Assembly." *The Juran Report,* No. 2, November, pp. 53–58.

Betker, Harry A. (1985). "Storyboarding: It's No Mickey Mouse Technique." *The Juran Report,* No. 5, Summer, pp. 25–30.

Bigelow, James S., and Floyd, Raymond C. (1990). "Quality Improvement through Root Cause Analysis." *IMPRO 1990 Conference Proceedings,* Juran Institute, Inc., Wilton, CT.

Black, Dennis A., and Stump, Jerry R. (1987). "Developing AQI Team 'Attitudes for Progress.' " *IMPRO 1987 Conference Proceedings,* Juran Institute, Inc., Wilton, CT.

Boley, Daniel C., and Petska, Joseph P. (1990). "The Genesis of a Partnership: A Tale of Two Plants." *IMPRO 1990 Conference Proceedings,* Juran Institute, Inc., Wilton, CT.

Chase, Richard B., and Stewart, Douglas M. (1994). "Make Your Service Fail-Safe." *Sloan Management Review,* Spring, pp. 35–44.

Upper management	Quality improvement teams	Operating departments
Organize quality councils	Receive and review mission statements	Implement remedies
Secure and screen project nominations	Conduct diagnostic journey	Implement controls
Select projects	Conduct remedial journey	
Prepare mission statements	Deal with cultural resistance	
Assign teams; establish training	Establish controls to hold gains	
Review progress	Report results	

FIGURE 5.21 Responsibility for quality improvement.

Cohen, Leonard A. (1987). "Diet and Cancer." *Scientific American,* November, pp. 42–48.

Coonley, Howard, and Agnew, P. G. (1941). *The Role of Standards in the System of Free Enterprise.* American National Standards Institute, New York.

Daughton, William J. (1987). "Achieving Customer Satisfaction with a Major Automobile Manufacturer." *IMPRO 1987 Conference Proceedings,* Juran Institute, Inc., Wilton, CT.

DeWolff, Thomas P., Anderson, Ralph J., and Stout, Gordon L. (1987). "Quality Improvement Process and Cultural Change." *IMPRO 1987 Conference Proceedings,* Juran Institute, Inc., Wilton, CT.

Egan, John (1985). "The Jaguar Obsession." *Quality (EOQC),* January, pp. 3–4.

Engle, David, and Ball, David (1985). "Improving Customer Service for Special Orders." *IMPRO 1985 Conference Proceedings,* Juran Institute, Inc., Wilton, CT.

Feder, Barnaby J. (1993). "At Motorola, Quality Is a Team Sport." *New York Times,* January 21.

Florida Power and Light Company (1984). *American Productivity Center,* Case Study 39, September.

Greenridge, R. M. C. (1953). "The Case of Reliability vs. Defective Components et al." *Electronic Applications Reliability Review,* No. 1, p. 12.

Gross, Neil (1989). "A Wave of Ideas, Drop by Drop." *Business Week,* Innovation, pp. 22, 28, and 30.

Gulliver, Frank R. (1987). "Post-Project Appraisals Pay." *Harvard Business Review,* March-April, pp. 128–132.

Hartman, Bob (1983). "Implementing Quality Improvement." *The Juran Report,* No. 2, November, pp. 124–131.

IMPRO Proceedings, Annual Conference on Quality Management, Juran Institute, Inc., Wilton, CT.

Juran, J. M. (1964). *Managerial Breakthrough.* McGraw-Hill, New York. Revised edition, 1995.

Juran, J. M. (1975). "The Non-Pareto Principle; Mea Culpa." *Quality Progress,* May, pp. 8–9.

Juran, J. M. (1978). "Japanese and Western Quality—A Contrast." From a paper originally presented at the International Conference on Quality Control, Tokyo, Oct. 17–20.

Juran, J. M. (1981). *Juran on Quality Improvement,* a series of 16 videocassettes on the subject. Juran Institute Inc., Wilton, CT.

Juran, J. M. (1985). "A Prescription for the West—Four Years Later." European Organization for Quality, 29th Annual Conference. Reprinted in *The Juran Report,* No. 5, Summer 1985.

Juran, J. M. (1992). *Juran on Quality by Design.* Free Press, New York, pp. 407–425.

Juran, J. M. (1993), "Made in USA, a Renaissance in Quality." *Harvard Business Review,* July-August, pp. 42–50.

Juran, J. M., ed. (1995). *A History of Managing for Quality.* Sponsored by Juran Foundation, Inc., Quality Press, ASQC, Milwaukee.

Juran Report, The. Issued periodically by Juran Institute, Inc., Wilton, CT.

Kearns, David T., and Nadler, David A. (1995). *How Xerox Reinvented Itself and Beat Back the Japanese.* Harper Business, New York.

Kelly, Bob, Howard, Tom, and Gambel, Anthony (1990). "Quality Forces Alcoa/Phirer Partnership." *IMPRO 1990 Conference Proceedings,* Juran Institute, Inc., Wilton, CT.

Kegarise, Ronald J., and Miller, George D. (1985). "An Alcoa-Kodak Joint Team." *IMPRO 1985 Conference Proceedings,* Juran Institute, Inc., Wilton, CT.

Kinosz, Donald L., and Ice, James W. (1991). "Facilitation of Problem Solving Teams." *IMPRO 1991 Conference Proceedings,* Juran institute, Inc., Wilton, CT.

Krantz, K. Theodor (1989). "How Velcro Got Hooked on Quality." *Harvard Business Review,* September-October, pp. 34–40.

Mead, Margaret, ed. (1951). *Cultural Patterns and Technical Change.* UNESCO, Paris; also published by Mentor Books, New American Library of World Literature, New York, 1955.

Metcalf, Henry C., and Urwick, L., eds. (1941). *Dynamic Administration.* Harper and Row, New York.

Mizell, Michael, and Strattner, Lawrence (1981). "Diagnostic Measuring in Manufacturing." *Quality,* September, pp. 29–32.

Motorola's Team Competition (1992). *Productivity Views,* January/February, pp. 3–4.

NIST, National Institute of Standards and Technology, Gaithersburg, MD 20899. (It manages the U.S. Malcolm Baldrige National Quality Award.)

Parvey, Dale E. (1990). "The Juran Improvement Methodology Applied to Information Security Systems." *IMPRO 1990 Conference Proceedings,* Juran Institute, Inc., Wilton, CT.

Pelletier, Don (1990). "Quality Improvement and the Executive Employee." *IMPRO 1990 Conference Proceedings,* Juran Institute, Inc., Wilton, CT.

Rassi, Alan J. (1991). "A Three-Phase Approach to Exceptional Customer Service." *IMPRO 1991 Conference Proceedings,* Juran Institute, Inc., Wilton, CT.

Sakai, Shinji (1994). "Rediscovering Quality—The Toyota Way." *IMPRO 1994 Conference Proceedings,* Juran Institute, Inc., Wilton, CT.

Schein, Edgar H. (1993). "How Can Organizations Learn Faster? The Challenge of Entering the Green Room." *Sloan Management Review,* Winter, pp. 85–92.

Seder, L. A., and Cowan, D. (1956). "The Span Plan of Process Capability Analysis. American Society for Quality Control," *General Publication 3,* September. (It was originally called "S.P.A.N. capability method," in which the letters stood for "systematic procedure for attaining necessary capability.")

Seder, L. A.(1950). "Diagnosis with Diagrams." *Industrial Quality Control,* Vol. 6, Nos. 4 and 5, January and March.

Smidt, Harm, and Doesema, Tjeerd (1991). "Lead Time Reduction of Project Related Purchased Components." *IMPRO 1991 Conference Proceedings,* Juran Institute, Inc., Wilton, CT.

Sterett, Kent W. (1987). "Quality at Florida Power and Light. (The Role of Facilitators and Performance Reviews)." *IMPRO 1987 Conference Proceedings,* Juran Institute, Inc., Wilton, CT.

Stewart, Thomas A. (1994). "Rate Your Readiness to Change." *Fortune,* February 7, pp. 106–110.

Stratton, Donald A. (1987). "Force Field Analysis: A Powerful Tool for Facilitators." *The Juran Report,* No. 8, pp. 105–111.

Traver, Robert W. (1983). "Locating the Key Variables." *ASQC Quality Congress Transactions,* Milwaukee, pp. 231–237.

Whipple, A. B. C. (1984). "Stranded Navy Man Who Charted the World's Seas." *Smithsonian,* March, pp. 171–186.

Some Accounts of Mobilizing for Quality Improvement

Hodgson, Alan (1991). "Starting the Amersham Quality Revolution" (at Amersham International). *IMPRO 1991 Conference Proceedings,* Juran Institute, Inc., Wilton, CT.

Lewicki, Thomas A., and Romo, Robert A. (1987). "Quality Through Collaboration" (at General Motors Central Foundry). *IMPRO 1987 Conference Proceedings,* Juran Institute, Inc., Wilton, CT.

McBride, P. S. (1993). "Short's Total Quality Management Programme: Quality Management in Practice" (at Short Brothers, Belfast, Northern Ireland). *Quality Forum,* June 1993.

Payne, Bernard J. (1987). "Overall QI/JQI: From 1% AQL to 10 ppm" (at Mullard Blackburn). *IMPRO 1987 Conference Proceedings,* Juran Institute, Inc., Wilton, CT.

Scanlon, Robert J. (1991). "Installing a Quality Process in Record Time" (at Southern Pacific Transportation Company). *IMPRO 1991 Conference Proceedings,* Juran Institute, Inc., Wilton, CT.

Tenner, Arthur R., and De Toro, Irving J. (1992). *Total Quality Management: Three Steps to Continuous Improvement* (at Banc One, based on the work of Charles A. Aubrey II). Addison-Wesley Publishing Company, Reading, MA.

Young, John (1990). "Enabling Quality Improvement in British Telecom." *IMPRO 1991 Conference Proceedings,* Juran Institute, Inc., Wilton, CT.

SECTION 6
PROCESS MANAGEMENT

James F. Riley, Jr.

INTRODUCTION

Why Process Quality Management?   The dynamic environment in which business is conducted today is characterized by what has been referred to as "the six c's:" change, complexity, customer demands, competitive pressure, cost impacts, and constraints. All have a great impact on an organization's ability to meet its stated business goals and objectives. Traditionally, organizations have responded to these factors with new products and services. Rarely have they made changes in the processes that support the new goods and services.

Experience shows that success in achieving business goals and objectives depends heavily on large, complex, cross-functional business processes, such as product planning, product development, invoicing, patient care, purchasing, materials procurement, parts distribution, and the like. In the absence of management attention over time, many of these processes become obsolete, overextended, redundant, excessively costly, ill-defined, and not adaptable to the demands of a constantly changing environment. For processes that have suffered this neglect (and this includes a very large number of processes for reasons that will be discussed later in this section) quality of output falls far short of the quality required for competitive performance.

A business process is the logical organization of people, materials, energy, equipment, and information into work activities designed to produce a required end result (product or service) (Pall 1986).

There are three principal dimensions for measuring process quality: effectiveness, efficiency, and adaptability. The process is *effective* if the output meets customer needs. It is *efficient* when it is effective at the least cost. The process is *adaptable* when it remains effective and efficient in the face of the many changes that occur over time. A process orientation is vital if management is to meet customer needs and ensure organizational health.

On the face of it, the need to maintain high quality of processes would seem obvious. To understand why good process quality is the exception, not the rule, requires a close look at how processes are designed and what happens to them over time.

First, the design. The western business organization model, for reasons of history, has evolved into a hierarchy of functionally specialized departments. Management direction, goals, and measurements are deployed from the top downward through this vertical hierarchy. However, the processes which yield the products of work, in particular those products which customers buy (and which justify the existence of the organization), flow horizontally across the organization through functional departments (Figure 6.1). Traditionally, each functional piece of a process is the responsibility of a department, whose manager is held accountable for the performance of that piece. However, no one is accountable for the entire process. Many problems arise from the conflict between the demands of the departments and the demands of the overall major processes.

In a competition with functional goals, functional resources, and functional careers, the cross-functional processes are starved for attention. As a result, the processes as operated are often neither effective nor efficient, and they are certainly not adaptable.

A second source of poor process performance is the natural deterioration to which all processes are subject in the course of their evolution. For example, at one railroad company, the company telephone directory revealed that there were more employees with the title "rework clerk" than with the title "clerk." Each of the rework clerks had been put in place to guard against the recurrence of some serious problem that arose. Over time, the imbalance in titles was the outward evidence of processes which had established rework as the organization's norm.

The rapidity of technological evolution, in combination with rising customer expectations, has created global competitive pressures on costs and quality. These pressures have stimulated an exploration of cross-functional processes—to identify and understand them and to improve their performance. There is now much evidence that within the total product cycle a major problem of poor process performance lies with process management technologies. Functional objectives frequently conflict with customer needs, served as they must be by cross-functional processes. Further, the processes generate a variety of waste (missed deadlines, factory scrap, etc.). It is not difficult to identify products, such as invoice generation, preparation of an insurance policy, or paying a claim, that take over 20 days to accomplish less than 20 min of actual work. They are also not easily changed in response to the continuously changing environment. To better serve customer needs there is a need to restore these processes to effectiveness, efficiency, and adaptability.

The Origins of PQM.   IBM Corporation was among the first American companies to see the benefits of identifying and managing business processes. The spirit of IBM's first efforts in manag-

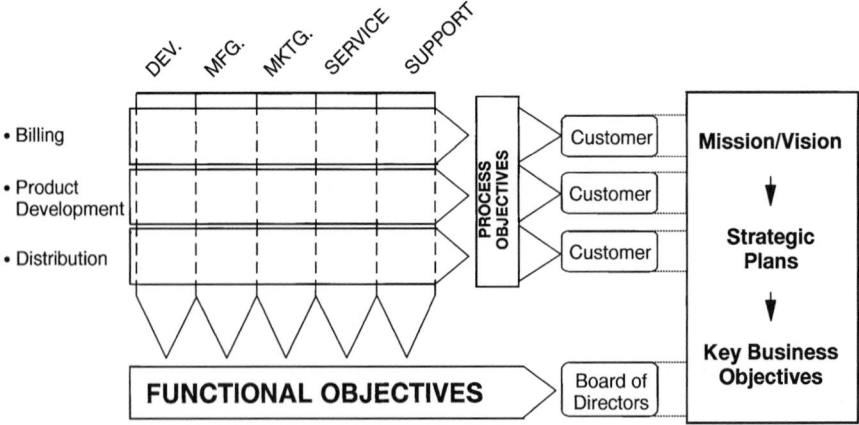

FIGURE 6.1   Workflow in a functional organization. (*Source: Juran Institute, Wilton, CT.*)

ing business processes in the early 1980s was expressed in the words of one executive: "Focus for improvement must be on the job process" (Kane 1986). Process Management has long been practiced in manufacturing. In product manufacturing, the plant manager "owns" a large part of the manufacturing process. This manager has complete responsibility for operating this part of the manufacturing process and is accountable for the results. As owner, the manager is expected to control, improve, and optimize the manufacturing process to meet customer needs and business needs (cost, cycle time, waste elimination, value creation, etc.). In pursuit of these targets, managers of the manufacturing process have developed some indispensable concepts and tools, including definition of process requirements, step-by-step process documentation, establishment of process measurements, removal of process defects, and assurance of process optimization. In fact, much of the science of industrial engineering is concerned with these tasks. Recognizing the value of these tools in manufacturing and their applicability to business processes, the IBM senior management committee directed that process management methodology be applied to all major business processes (such as product development, business planning, distribution, billing, market planning, etc.), and not just to the manufacturing process.

Around the same time, a number of other North American companies, including AT&T, Ford Motor Company, Motorola, Corning, and Hewlett-Packard, also began applying process management concepts to their business processes. In all of these companies, the emphasis was placed on cross-functional and cross-organizational processes. Application of process management methodology resulted in breaking down the functional barriers within the processes. In each case, a new, permanent managerial structure was established for the targeted process.

By mid-1985, many organizations and industries were managing selected major business processes with the same attention commonly devoted to functions, departments, and other organizational entities. Early efforts bore such names as Business Process Management, Continuous Process Improvement, and Business Process Quality Improvement.

Business Process Reengineering (BPR) should be mentioned as part of this family of methodologies. Like the methodologies mentioned previously in this section, BPR accomplishes a shift of managerial orientation from function to process. According to the consultants who first described BPR and gave it its name, BPR departs from the other methodologies in its emphasis on radical change of processes rather than on incremental change. Furthermore, BPR frequently seeks to change more than one process at the same time. Because of the economic climate of the early 1990s, and the outstanding payback that some writers attribute to BPR, its popularity grew rapidly for a time.

However, there is evidence, including the testimony of Michael Hammer, one of the most widely read writers on BPR, that in many early applications, the lure of rapid improvement caused some managers (and their consultants), who ignored human limitations, to impose too much change in too short a time, with a devastating effect on long-term organization performance. Furthermore, in many early applications, users became so fascinated by the promise of radical change that they changed everything, overlooking elements of the existing process design that worked perfectly well and would have been better carried over as part of the new design. Such a carryover would have saved time, reduced demand on the designers, and produced a better result.

Much has been published on process management. AT&T (1988), Black (1985), Gibson (1991–92), Hammer and Champy (1993), Kane (1986 and 1992), Pall (1987), Riley (1989), Rummler (1992), Schlesiona (1988), and Zachman (1990) have all proposed similar methodological approaches that differ from one another in minor details. The specific details of the methodology presented in this section were developed by consultants at the Juran Institute, Inc. [Gibson et al. (1990); Riley et al. (1994)], based on years of collective experience in a variety of industries.

Process Quality Management (PQM) Defined.   The methodology described in this section is one which has been introduced with increasing success by a number of prominent corporations, including the ones already mentioned. While it may vary in name and details from company to company, the methodology possesses a core of common features which distinguishes it from other approaches to managing quality. That core of features includes: a conscious orientation toward customers and their needs; a specific focus on managing a few key cross-functional processes which most affect satisfaction of customer needs; a pattern of clear ownership—accountability for each key process; a cross-functional team responsible for operating the process; application at the process level of quality-management processes—quality control, quality improvement, and quality planning. In this section, the methodology will be referred to as process quality management, or PQM.

AN APPLICATION EXAMPLE: THE CONTRACT MANAGEMENT PROCESS

Before discussing the details of PQM, an example will illustrate how a process, operating in a traditional functional hierarchy, may respond poorly to a seemingly minor change in an operating environment, and how the effects of that change can stimulate dramatic improvements, made possible by applying the process management approach. It also illustrates how the potential for dramatic improvement offered by a new technology (information technology, in this case) is more easily recognized when there is accountability for making those improvements.

In the early 1980s, a major multinational manufacturer of information processing systems decided to change its traditional rent-or-lease product pricing policy to include outright purchase. This strategic change led to a complete revision of the company's contracting policies, including terms and conditions. Instead of firm list prices, published discounts were now available; for especially large procurements, the company offered unpublished discounts with a number of financing options. A new contract management process evolved out of the new policy, an incremental modification of the existing process. The new process had to accommodate special contracts with a variety of nonstandard terms and conditions.

Within 2 years, more than 10 percent of the company's revenue was generated by "special contracts." However, as the percentage of this revenue increased, the ratio of sales closed to proposals made plummeted to fewer than 1 out of 5—a process yield of 20 percent. Both customers and field marketing representatives complained about the long turnaround time (the time elapsed from receipt of a request for a proposal until delivery of the proposal to the customer), which averaged 14 weeks. The process was simply unresponsive to customer business needs.

Facing lost business opportunities and a barrage of complaints from field marketing representatives, the executive quality council targeted this process for the application of process quality management. The director of contract management was designated as the process owner, and formed a process management team comprising representatives from field marketing, field administration, business systems, product development, finance, marketing practices, and legal services.

Originally, the contract management process was a serial set of steps (Figure 6.2). The process started with the field marketing representative, who received a request for a special contract proposal from the customer. A draft of the contract proposal was then prepared in the branch office with the help of branch administration and reviewed by the branch manager. Subsequently, it was submitted for regional management review (usually in another geographic location) and finally for a comprehensive evaluation at headquarters by large-account marketing, marketing practices, finance, and business systems. If the proposal was in order, a contract was prepared. The contract management department then arranged for up to 28 sequential approvals of the contract at the executive level, involving various functions, such as product development, finance, legal services, and the like.

Having successfully passed all these hurdles, the contract was then approved and returned to marketing division headquarters for further refinement and processing. Eventually, some 3 to 4 months later, the proposed contract returned to the branch office for presentation to the customer. In many instances, it arrived too late. The customer had taken the business to a competitor.

The process management team flow-charted the process, and validated a number of hypotheses. These included: manual processing was slow; there were postal service delays; the serial approval process, consisting of up to 28 high-level executive signoffs, took too long; the memo-generated guidelines for the format and content of the contract proposal were vague, conflicting, and difficult to access; and there was substantial resistance to changing to the new, purchase-only strategy, especially by the finance, business practices, and legal functions.

Branch Office	Regional Office	HQ Large-Account Marketing	HQ Special-Contract Management	Assorted High-Level Managers
Prepare special-contract proposal→	Review, approve, forward→	Log in, review, and assign for approval→	Review, provide price, terms, and conditions; schedule approvals→	28 sequential approvals ↓
Plan presentation			← Consolidate changes and approvals; finalize special contract	←

FIGURE 6.2 Special-contract management process (before application of process-management principles). (*Source: Juran Institute, Wilton, CT.*)

After several months of process redesign and test, the team launched the new contract management process, shown in Figure 6.3. The new process addressed all causes of delay that the team had discovered. It made especially good use of new information technology support, which was unavailable years before when the contract management process began operation.

In designing the new process, the team incorporated a number of important new features:

- The team wrote new guidelines for contract proposal preparation and installed them on-line, using a national electronic mail system. They established a system to keep the guidelines continuously updated. This measure accelerated both preparation and transmission of contract proposals.

- They arranged for approval authority for simple contracts—those of lower dollar volume or having no special engineering requirements—to be delegated to the regional marketing manager.

- They established two review boards for concurrent review and approval of special contracts. The concurrent processes replaced the serial process.

- They found that the teamwork required by the cross-functional arrangement had the added effect of reducing interfunctional rivalry and resistance to the new marketing policy.

- They established turnaround time requirements for each activity in the process, then measured and tracked actual performance against the standards. Whenever they experienced delays beyond the specified time targets, they initiated corrective action. For example, each review board had 5 business days to review proposals, approve them, and pass them on. With the target established, it was a relatively simple matter for the board to monitor its own performance against the standard.

This new management approach resulted in an 83 percent improvement in average turnaround time (from 14 weeks to 17 days), and an increase in process yield of 180 percent (from 20 to 56 percent). Still, the team was not satisfied. They implemented two more process redesigns in the next 3 years. After 5 years of PQM focus, the special-contract management process was performing at a 60 percent yield. For simple contracts, which account for 92 percent of the process volume, the turnaround time is 24 hours.

Before it was redesigned, this process consumed the equivalent of 117 full-time people; as of 1995, after the several redesigns, it required fewer than 60. Special-contract revenue now exceeds 30

Branch Office	Regional Office	HQ Large-Account Marketing	HQ Special-Contract Management	Assorted High-Level Managers	Review Board #1	Review Board #2
Prepare special-contract proposal→	Review, approve→		Screen, log in, and follow review board actions→	[Sequential approvals eliminated]→	Evaluate and approve→	Evaluate and approve ↓
Plan presentation			← Finalize special contract			←

FIGURE 6.3 Special-contract management process (after application of process-management principles). (*Source: Juran Institute, Wilton, CT.*)

percent of total U.S. revenue—an all-time high. Customers and company management agree that the present process performance may be judged effective and efficient. As additional redesigns are required to respond to the inevitable changes in environment, the managers believe that the process will also prove to be adaptable.

THE PQM METHODOLOGY

Overview.   A PQM effort is initiated when executive management selects key processes, identifies owners and teams, and provides them with process mission statements and goals. After the owners and team are trained in process methodology, they work through the three phases of PQM methodology: planning, transfer, and operational management.

The *planning phase,* in which the process design (or redesign) takes place, involves five steps:

1. Defining the present process.
2. Determining customer needs and process flow.
3. Establishing process measurements.
4. Conducting analyses of measurement and other data.
5. Designing the new process. The output is the new process plan.

Planning is the most time-consuming of the three phases.

The *transfer phase* is the second phase, in which the plans developed in the first phase are handed off from the process team to the operating forces and put into operation.

Operational management is the third phase of PQM. Here, the working owner and team first monitor new process performance, focusing on process effectiveness and efficiency measurements. They apply quality control techniques, as appropriate, to maintain process performance. They use quality improvement techniques to rid the process of chronic deficiencies. Finally, they conduct a periodic executive management review and assessment to ensure that the process continues to meet customer needs and business needs, and remains competitive.

Replanning, the cycling back to the first phase, is invoked when indicated. PQM is not a one-time event; it is itself a continuous process.

Initiating PQM Activity

Selecting the Key Process(es).   Organizations operate dozens of major cross-functional business processes. From these a few key processes are selected as the PQM focus. The organization's Strategic Plan provides guidance in the selection of the key processes. (See Section 13, Strategic Deployment.)

There are several approaches to selecting key business processes:

- The Critical Success Factor approach holds that for any organization relatively few (no more than eight) factors can be identified as "necessary and sufficient" for the attainment of its mission and vision. Once identified, these factors are used to select the key business processes and rank them by priority (Hardaker and Ward 1987).
- The Balanced Business Scorecard (Kaplan and Norton 1992) measures business performance in four dimensions: financial performance, performance in the eyes of the customer, internal process performance, and performance in organization learning and innovation. For each dimension, performance measures are created and performance targets are set. Using these measures to track performance provides a "balanced" assessment of business performance. The processes which create imbalances in the scorecard are identified as the processes that most need attention—the key processes.
- Another approach is to invite upper management to identify a few (four to six) organization-specific critical selection criteria to use in evaluating the processes. Examples of such criteria are:

effect on business success, effect on customer satisfaction, significance of problems associated with the process, amount of resources currently committed to the process, potential for improvement, affordability of adopting process management, and effect of process on schedule. Using the criteria and some simple scoring system (such as "low, medium, or high"), the managers evaluate the many processes from the long list of the organization's major business processes (10 to 25 of them) and, by comparing the evaluations, identify the key processes. (The long list may be prepared in advance in a process identification study conducted separately, often by the chief quality officer, and often with the support of a consultant.

Whatever approach is used to identify key processes, the process map can be used to display the results. The "process map" is a graphic tool for describing an organization in terms of its business processes and their relationships to the organization's principal stakeholders. The traditional organization chart answers the question: "Who reports to whom?" The process map answers the question: "How does the organization's work get done?"

Figure 6.4 describes the work of the Educational Testing Service (ETS), the organization that prepares and administers educational entrance examinations in the United States. In this process map, organizations and departments are represented by shaded blocks labeled in bold type. The key operational units of ETS, including external units designated "partners" by ETS, are located within a boundary line labeled "ETS." The important business processes of ETS are listed within that boundary and marked by an asterisk (*). These are the processes eligible for the PQM focus, shown in their relationship to various parts of the organization. Picturing the organization from a process perspective provides upper management with a useful tool in thinking about and discussing the organization in terms of its work and the processes it employs to get the work done.

Organizing: Assigning Ownership, Selecting the Team, and PQM Infrastructure.   Because certain major cross-functional business processes, the *key processes,* are critical to business success, the

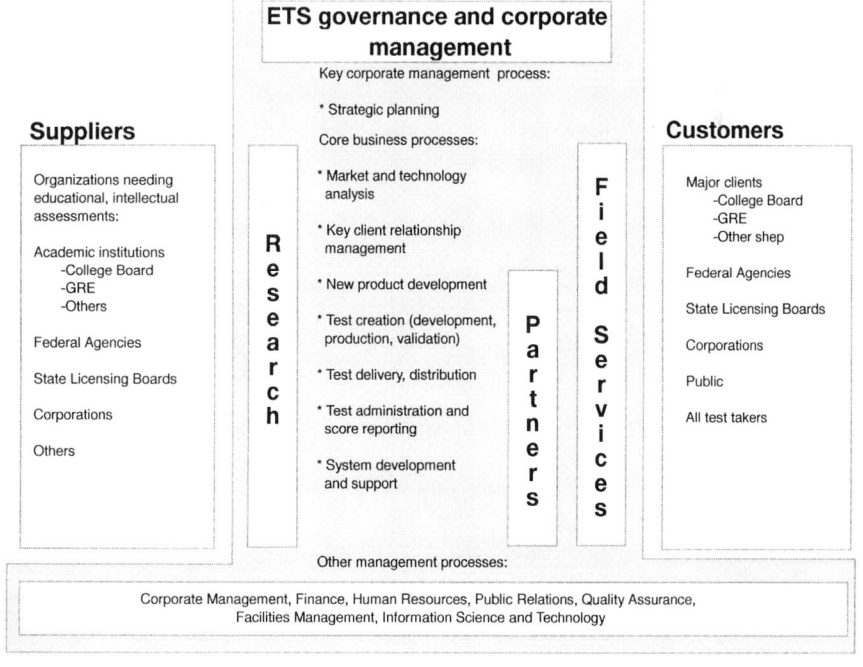

FIGURE 6.4   Process map of major business processes at Educational Testing Services. (*Source: Juran Institute, Wilton, CT.*)

quality council sees to it that those processes are organized in a special way. After selecting key processes, the quality council appoints a process owner, who is responsible for making the process effective, efficient, and adaptable, and is accountable for its performance (Riley, 1989 and 1994).

For large complex processes, especially in large companies, a two-tier ownership arrangement is most often used. An appointed executive owner operates as a sponsor, champion, and supporter at the upper management level, and is accountable for process results. At the operating level, a working owner, usually a first- or second-level manager, leads the process-management team responsible for day-to-day operation. The owner assignments—executive owner and working owner—are ongoing. The major advantages of this structure are that there is at the same time "hands on" involvement and support of upper management and adequate management of the process details.

The process-management team is a peer-level group which includes a manager or supervisor from each major function within the process. Each member is an expert in a segment of the process. Ideally, process management teams have no more than eight members, and the individuals chosen should be proven leaders. The team is responsible for the management and continuous improvement of the process. The team shares with the owner the responsibilities for effectiveness and efficiency. Most commonly, the team assignments are ongoing.

From time to time a process owner creates an ad hoc team to address some special issue (human resources, information technology, activity-based costing, etc.). The mission of such a project-oriented team is limited, and the team disbands when the mission is complete. The ad hoc team is different from the process-management team.

Figure 6.5 is a simplified diagram of a multifunctional organization and one of its major processes. The shaded portions include: the executive owner, the working owner, the process management team, and the stakeholders—functional heads at the executive level who have work activities of the business process operating within their function. Customarily, the stakeholders are members of the quality council, along with the executive owner. Taken together, this shaded portion is referred to as the PQM Infrastructure.

Establishing the Team's Mission and Goals. The preliminary process mission and improvement goals for the process are communicated to the owners (executive and working levels) and team by the quality council. To do their jobs most effectively, the owners and team must make the mission and goals their own. They do this in the first step of the planning phase: defining the process.

The Planning Phase: Planning the New Process. The first phase of PQM is Planning, which consists of five steps: (1) defining the process, (2) discovering customer needs and flow-

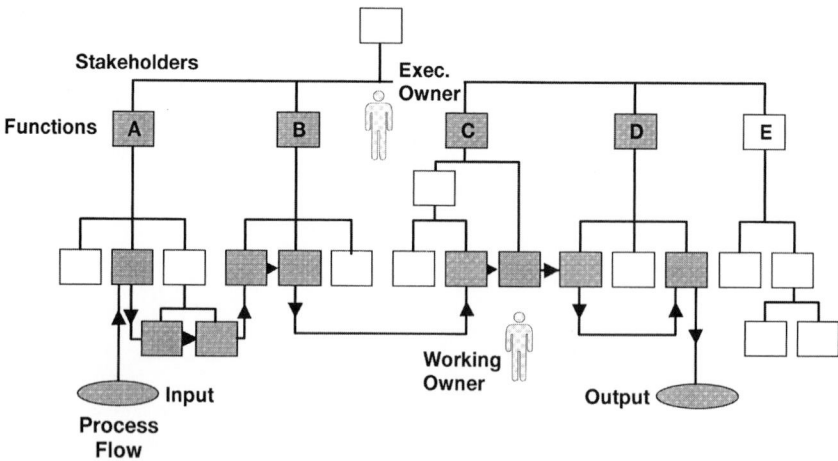

FIGURE 6.5 Organization infrastructure for process management in multifunctional organizations. (*Source: Juran Institute, Wilton, CT.*)

charting the process, (3) establishing measurements of the process, (4) analyzing process measurements and other data, and (5) designing (or redesigning) the process. The output of the Planning Phase is the new process plan.

Defining the Current Process.   The owner(s) and team collaborate to define the process precisely. In accomplishing this, their starting point and principal reference is the process documentation developed by the quality council during the selection of key processes and identification of owners and teams. This documentation includes preliminary statements of mission and goals.

Effective mission and goal statements explicitly declare:

- The purpose and scope of the process
- "Stretch" targets for customer needs and business needs

(The purpose of the stretch target is to motivate aggressive process improvement activity.)

As an example, a mission statement for the Special-Contract Management Process is: Provide competitive special pricing and supportive terms and conditions for large information systems procurements that meet customer needs for value, contractual support, and timeliness at affordable cost.

The goals for the same process are:

1. Deliver approved price and contract support document within 30 days of date of customer's letter of intent.
2. Achieve a yield of special-contract proposals (percent of proposals closed as sales) of not less than 50 percent.

The team must reach consensus on the suitability of these statements, propose modifications for the quality council's approval, if necessary, and also document the scope, objectives, and content. Based on available data and collective team experience, the team will document process flow, the process strengths and weaknesses, performance history, measures, costs, complaints, environment, resources, and so on. This will probably involve narrative documentation and will certainly require the use of flow diagrams.

Bounding the business process starts with an inventory of the major subprocesses—six to eight of them is typical—that the business process comprises. The inventory must include the "starts-with" subprocess (the first subprocess executed), the "ends-with" subprocess (the last executed), and the major subprocesses in between. If they have significant effect on the quality of the process output, activities upstream of the process are included within the process boundary. To provide focus and avoid ambiguity, it is also helpful to list subprocesses which are explicitly excluded from the business process. The accumulating information on the process components is represented in diagram form, which evolves, as the steps of the planning phase are completed, from a collection of subprocesses to a flow diagram.

Figure 6.6 shows the high-level diagram of the special-contract process that resulted from process analysis but before the process was redesigned. At the end of the process definition step such a diagram is not yet a flow diagram, as there is no indication of the sequence in which the subprocesses occur. Establishing those relationships as they presently exist is the work of Step 2.

Discovering Customer Needs and Flowcharting the Process.   For the process to do its work well, the team must identify all of the customers, determine their needs, and prioritize the information. Priorities enable the team to focus its attention and spend its energies where they will be most effective. (The subject of identifying customers and their needs is covered in detail in Section 3, The Quality Planning Process.)

Determining customer needs and expectations requires ongoing, disciplined activity. Process owners must ensure that this activity is incorporated in the day-to-day conduct of the business process as the customer requirements subprocess and assign accountability for its performance. The output of this vital activity is a continually updated customer requirement statement.

On the process flow chart it is usual to indicate the key suppliers and customers and their roles in the process, as providers or receivers of materials, product, information, and the like. Although the

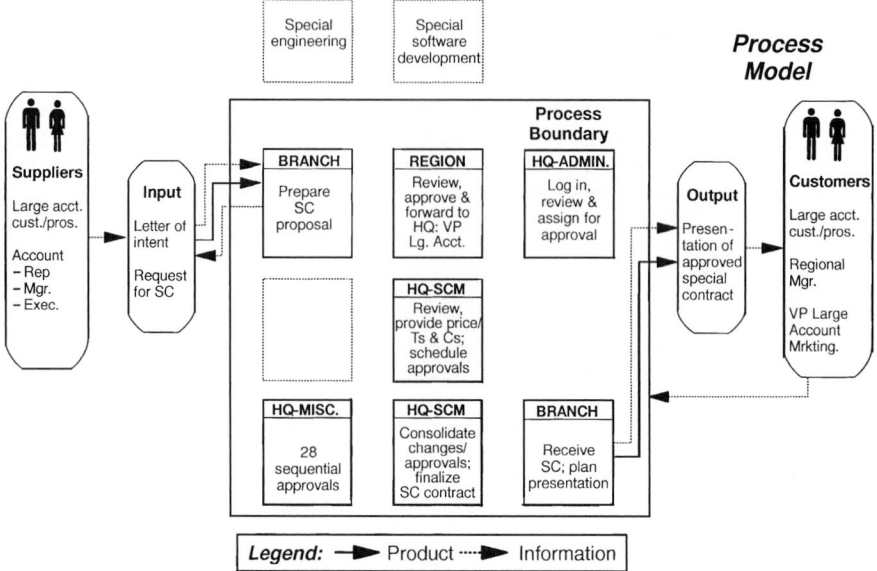

FIGURE 6.6   High-level diagram of the special-contract process, an output of process analysis. (*Source: Juran Institute, Wilton, CT.*)

diagram can serve a number of specialized purposes, the most important here is to create a common, high-level understanding among the owner and team members of how the process works—how the subprocesses relate to each other and to the customers and suppliers and how information and product move around and through the process. In creating the process flow chart, the team will also verify the list of customers and may, as understanding of the process deepens, add to the list of customers.

The process flow chart is the team's primary tool for analyzing the process to determine whether it can satisfy customer needs. By walking through the chart together, step by step, sharing questions and collective experience, the team determines whether the process is correctly represented, making adjustments to the diagram as necessary to reflect the process as it presently operates.

When the step is complete, the team has a starting point for analysis and improvement of the process. In Figure 6.8, product flow is shown by solid lines and information flow by dotted lines.

Establishing Process Measurements.   What gets measured, gets done. Establishing, collecting, and using the correct measures is critical in managing business process quality. "Process capability," "process performance," and other process measures have no practical significance if the process they purport to describe is not managed. To be managed, the process must fulfill certain minimum conditions:

a. It has an owner.

b. It is defined.

c. Its management infrastructure is in place.

d. Its requirements are established.

e. Its measurements and control points are established.

f. It demonstrates stable, predictable, and repeatable performance.

A process which fulfills these minimum conditions is said to be *manageable.* Manageability is the precondition for all further work in PQM.

Of these criteria, (a) through (d) have already been addressed in this section. Criteria (e) and (f) are addressed in the following.

Process Measurements (See also Section 9).   In deciding what aspects of the process to measure, we look for guidance to the process mission and to our list of customer needs. Process measures based on customer needs provide a way of measuring process effectiveness. For example, if the customer requires delivery of an order within 24 hours of order placement, we incorporate into our order-fulfillment process a measure such as "time elapsed between receipt of order and delivery of order," and a system for collecting, processing, summarizing, and reporting information from the data generated. The statistic reported to the executive owner will be one such as "percent of orders delivered within 24 hours," a statistic which summarizes on-time performance. The team will also need data on which to base analysis and correction of problems and continuous improvement of the process. For this purpose, the team needs data from which they can compute such descriptive statistics as distribution of delivery times by product type, and so on. The uses to which the data will be put must be thought through carefully at the time of process design to minimize the redesign of the measures and measurement systems.

Process measures based on cost, cycle time, labor productivity, process yield, and the like are measures of process efficiency. Suppose that a goal for our order-fulfillment process is to reduce order-picking errors to one error per thousand order lines. Managing to that goal requires identification of order-picking errors in relation to the number of order lines picked. For order-picking errors that are inadvertent—that is, when they happen, the picker is unaware of them—measuring them requires a separate inspection to identify errors. In a random audit on a sample of picked orders, an inspector identifies errors and records them. As with delivery-time measurement, the team must think through all the uses it will make of these measurements. For a report of estimated error rate, the data needed are: number of errors and number of order lines inspected. To improve process performance in this category, the data must help the team identify error sources and determine root cause. For that to occur, each error must be associated with time of day, shift, product type, size of package, etc., so that the data can be stratified to test various theories of root cause.

While not a measurement category, process adaptability is an important consideration for process owners and teams. Adaptability will be discussed later in this section.

Process measurements must be linked to business performance. If certain key processes must run exceptionally well to ensure organization success, it follows that collective success of the key processes is good for the organization's performance. Process owners must take care to select process measures that are strongly correlated with traditional business indicators, such as revenue, profit, ROI, earnings per share, productivity per employee, and so on. In high-level business plan reviews, managers are motivated and rewarded for maintaining this linkage between process and organization performance measures because of the two values which PQM supports: organization success is good, and process management is the way we will achieve organization success.

Figure 6.7 shows some typical process measurements and the traditional business indicators with which they are linked. To illustrate, "percent of sales quota achieved" is a traditional business indicator relating to the business objective of improving revenue. The special-contract management process has a major impact on the indicator, since more than 30 percent of U.S. revenue comes from that process. Therefore, the contract close rate (ratio of the value of firm contracts to the total value of proposals submitted) of the special-contract management process is linked to percent of sales quota and other traditional revenue measures, and is therefore a measure of great importance to management. Measurement points appear on the process flow diagram.

Control Points.   Process measurement is also a part of the control mechanisms established to maintain planned performance in the new process. To control the process requires that each of a few selected process variables be the control subjects of a feedback control loop. Typically, there will be five to six control points at the macroprocess level for variables associated with: external output, external input, key intermediate products, and other high-leverage process points.

The control points in the special-contract management process are represented graphically in Figure 6.8. Feedback loop design and other issues surrounding process control are covered in detail in Section 4, The Quality Control Process.

The Traditional Business View		The Process View	
Business Objective	Business Indicator	Key Process	Process Measure
Higher revenue	Percent of sales quota achieved	Contract management	Contract close rate
	Percent of revenue plan achieved	Product development	Development cycle time
	Value of orders canceled after shipment	Account management	Backlog management and system assurance timeliness
	Receivable days outstanding		Billing quality index
Reduce costs	Inventory turns	Manufacturing	Manufacturing cycle time

FIGURE 6.7   Linkages among business objectives, traditional business indicators, and process measures generated by the process-management approach—a few examples. (*Source: Juran Institute, Wilton, CT.*)

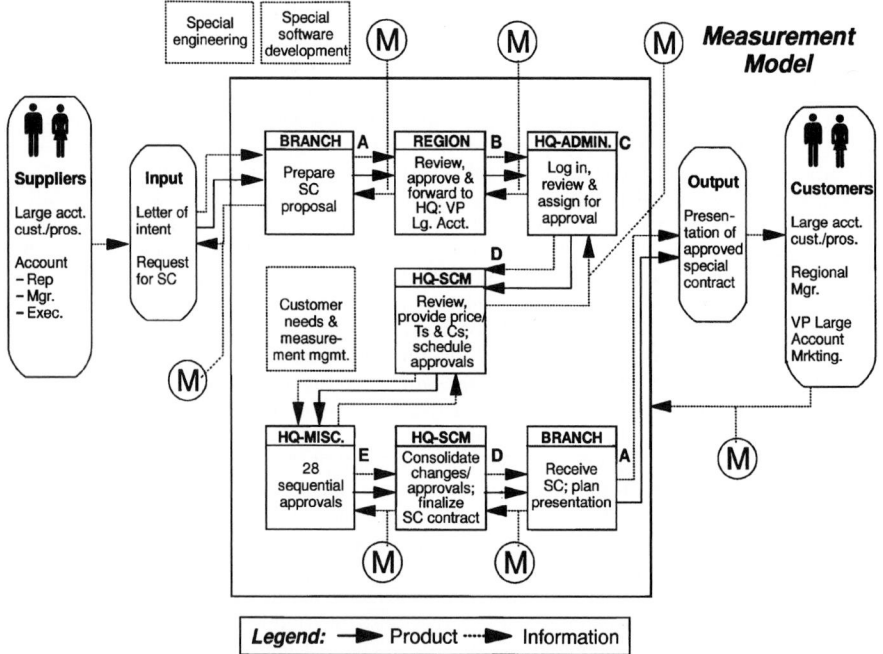

FIGURE 6.8   Flowchart of the special-contract management process, including process control points. (*Source: Juran Institute, Wilton, CT.*)

Process Variability, Stability, and Capability.   As in all processes, business processes exhibit variability. The tools of statistical process control such as Shewhart charts (see Section 45, Statistical Process Control) help the team to minimize process variation and assess process stability.

Evaluation of process capability is an important step in process quality improvement. Process capability is a measure of variation in a process operating under stable conditions. "Under stable conditions" means that all variation in the process is attributable to random causes. The usual criterion for stability is that the process, as plotted and interpreted on a Shewhart control chart, is "in control."

Statistical process control, process capability, and associated tools are useful components of the process team's tool kit. They are covered in detail in Section 44, Basic Statistical Methods.

The output of the measurement step is a measurement plan, a list of process measurements to be made and the details of making each one—who will make it, how it will be made, on what schedule, and so on.

Analyzing the Process.   Process Analysis is performed for the following purposes:

- Assess the current process for its effectiveness and efficiency.
- Identify the underlying causes of any performance inadequacy.
- Identify opportunities for improvement.
- Make the improvements.

First, referring to the process flowchart, the team breaks the process into its component activities using a procedure called "process decomposition," which consists of progressively breaking apart the process, level by level, starting with the macrolevel. As decomposition proceeds, the process is described in ever finer detail.

As the strengths and weaknesses of the process are understood at one level, the process management team's interim theories and conclusions will help decide where to go next with the analysis. The team will discover that certain subprocesses have more influence on the performance of the overall business process than others (an example of the Pareto principle). These more significant subprocesses become the target for the next level of analysis.

Decomposition is complete when the process parts are small enough to judge as to their effectiveness and efficiency. Figure 6.9 gives examples from three levels of decomposition (subprocess, activity, and task) of three typical business processes (procurement, development engineering, and office administration).

Measurement data are collected according to the measurement plan to determine process effectiveness and efficiency. The data are analyzed for effectiveness (conformance to customer needs) and long-term capability to meet current and future customer requirements.

The goal for process efficiency is that all key business processes operate at minimum total process cost and cycle time, while still meeting customer requirements.

Process *effectiveness* and *efficiency* are analyzed concurrently. Maximizing effectiveness and efficiency together means that the process produces high quality at low cost; in other words, it can provide the most *value* to the customer.

"Business process adaptability" is the ability of a process to readily accommodate changes both in the requirements and the environment, while maintaining its effectiveness and efficiency over time.

To analyze the business process, the flow diagram is examined in four steps and modified as necessary. The steps are:

- Examine each decision symbol

  Is this a checking activity?

Business Process	Subprocess	Activity	Task
Procurement	Vendor selection	Vendor survey	Documentation of outside vendor
Development engineering	Hardware design	Engineering change	Convening the Change Board
Office administration	Providing secretarial services	Calendar management	Making a change to existing calendar

FIGURE 6.9 Process decomposition—examples of process elements disclosed within typical business processes. (*Source: Juran Institute, Wilton, CT.*)

If so, is this a complete check, or do some types of errors go undetected?

Is this a redundant check?

- Examine each rework loop

Would we need to perform these activities if we had no failures?

How "long" is this rework loop (as measured in number of steps, time lost, resources consumed, etc.)?

Does this rework loop prevent the problem from recurring?

- Examine each activity symbol

Is this a redundant activity?

What is the value of this activity relative to its cost?

How have we prevented errors in this activity?

- Examine each document and database symbol

Is this necessary?

How is this kept up to date?

Is there a single source for this information?

How can we use this information to monitor and improve the process?

The "Process Analysis Summary Report" is the culmination and key output of this process analysis step. It includes the findings from the analysis, that is, the reasons for inadequate process performance and potential solutions that have been proposed and recorded by owner and team as analysis progressed. The completion of this report is an opportune time for an executive owner/stakeholder review.

The owner/stakeholder reviews can be highly motivational to owners, teams, stakeholders, and the Quality Council. Of particular interest is the presentation of potential solutions for improved process operation. These have been collected throughout the planning phase and stored in an idea bin. These design suggestions are now documented and organized for executive review as part of the process analysis summary report presentation.

In reviewing the potential solutions, the executive owner and quality council provide the selection criteria for acceptable process design alternatives. Knowing upper management's criteria for proposed solutions helps to focus the process-management team's design efforts and makes a favorable reception for the reengineered new process plan more likely.

Designing (or Redesigning) the Process.   In Process Design, the team defines the specific operational means for meeting stated product goals. The result is a newly developed Process Plan. Design changes fall into five broad categories: workflow, technology, people and organization, physical infrastructure, and policy and regulations.

In the design step, the owner and team must decide whether to create a new process design or to redesign the existing process. Creating a new design might mean radical change; redesign generally means incremental change with some carryover of existing design features.

The team will generate many design alternatives, with input from both internal and external sources. One approach to generating these design alternatives from internal sources is to train task-level performers to apply creative thinking to the redesign of their process.

Ideas generated in these sessions are documented and added to the idea bin. Benchmarking can provide a rich source of ideas from external sources, including ideas for radical change. Benchmarking is discussed in detail in Section 12.

In designing for process effectiveness, the variable of most interest is usually process cycle time. In service-oriented competition, lowest process cycle time is often the decisive feature. Furthermore, cycle-time reduction usually translates to efficiency gains as well. For many processes, the most promising source of cycle-time reduction is the introduction of new technology, especially information technology.

Designing for speed creates surprising competitive benefits: growth of market share and reduction of inventory requirements. Hewlett-Packard, Brunswick Corp., GE's Electrical Distribution and Control Division, AT&T, and Benetton are among the companies who have reported stunning achievements in cycle-time reduction for both product development and manufacturing (Dumaine, 1989). In each of the companies, the gains resulted from efforts based on a focus on major processes. Other common features of these efforts included:

- Stretch objectives proposed by top management
- Absolute adherence to schedule, once agreed to
- Application of state-of-the art information technology
- Reduction of management levels in favor of empowered employees and self-directed work teams
- Putting speed in the culture

In designing for speed, successful redesigns frequently originate from a few relatively simple guidelines: eliminate handoffs in the process, eliminate problems caused upstream of activity, remove delays or errors during handoffs between functional areas, and combine steps that span businesses or functions. A few illustrations are provided:

- *Eliminate handoffs in the process:*   A "handoff" is a transfer of material or information from one person to another, especially across departmental boundaries. In any process involving more than a single person, handoffs are inevitable. It must be recognized, however, that the handoff is time-consuming and full of peril for process integrity—the missed instruction, the confused part identification, the obsolete specification, the miscommunicated customer request.

  In the special-contract management process, discussed previously in this section, the use of concurrent review boards eliminated the 28 sequential executive approvals and associated handoffs.

- *Eliminate problems caused upstream of activity.*   Errors in order entry at a U.S. computer company were caused when sales representatives incorrectly configured systems. As a result, the cost of the sales-and-order process was 30 percent higher than that of competitors, and the error rates for some products were as high as 100 percent. The cross-functional redesign fixed both the configurations problem and sales-force skills so that on-time delivery improved at significant cost savings (Hall, Rosenthal, and Wade 1993).

- *Remove delays or errors during handoffs between functional areas:*   The processing of a new policy at a U.K. insurance company involved 10 handoffs and took at least 40 days to complete. The company implemented a case-manager approach by which only one handoff occurred and the policy was processed in less than 7 days (Hall, Rosenthal, and Wade 1993).

- *Combine steps that span businesses or functions:*   At a U.S. electronics equipment manufacturer, as many as seven job titles in three different functions were involved in the nine steps required to design, produce, install, and maintain hardware. The company eliminated all but two job titles, leaving one job in sales and one job in manufacturing (Hall, Rosenthal, and Wade 1993).

The Ford accounts payable process provides a classic example of process redesign. Details are given by Hammer and Champy (1993). Process Quality Management is successful when the design step involved radical change. Hammer and Champy propose the following principles for such radical change of a process:

- Organize the process around outcomes, not tasks.
- Have those who use the output of the process perform the process.
- Incorporate information-processing work into the real work that produces the information.
- Treat geographically dispersed resources as though they were centralized.
- Coordinate parallel functions within the process, not in subsequent steps.
- Put the decision point where the work is performed and build control into the process.
- Capture information only once and at the source.

Before the new design is put into place, a design review is in order. Its purpose is to temper the enthusiasm of the team with the objectivity of experienced outsiders. Typically, the process owner assembles a panel of experts from within the organization (but outside the process) to provide the evaluation of design alternatives.

Process design testing is performed to determine whether the process design alternative will work under operating conditions. Design testing may include trials, pilots, dry runs, simulations, etc. The results are used to predict new process performance and cost/benefit feasibility.

Successful process design requires employee participation and involvement. To overlook such participation creates a lost opportunity and a barrier to significant improvement. The creativity of the first-line work force in generating new designs can be significant.

Byrne (1993) reports that many companies are adopting "horizontal organizational design," which features the use of self-directed work teams organized around the process. Eastman Chemical has over 1000 teams; increasing reliance on self-directed teams has enabled the company to eliminate senior VP positions for administration, manufacturing, and R&D. (See also Section 15, Human Resources and Quality.)

Lexmark International, a former IBM division, abolished 60 percent of the management jobs in manufacturing and support services. Instead, they organized around cross-functional teams worldwide.

Creating the New Process Plan.   After we have redefined a key process, we must document the new process and carefully explain the new steps. The new process plan now includes the new process design and its control plan for maintaining the new level of process performance. The new process plan for the special-contract management process, shown as a high-level process schematic, is shown in Figure 6.10.

The Transfer Phase: Transferring the New Process Plan to Operations.   The transfer phase consists of three steps: (1) planning for implementation problems, (2) planning for implementation action, and (3) deploying the new process plan.

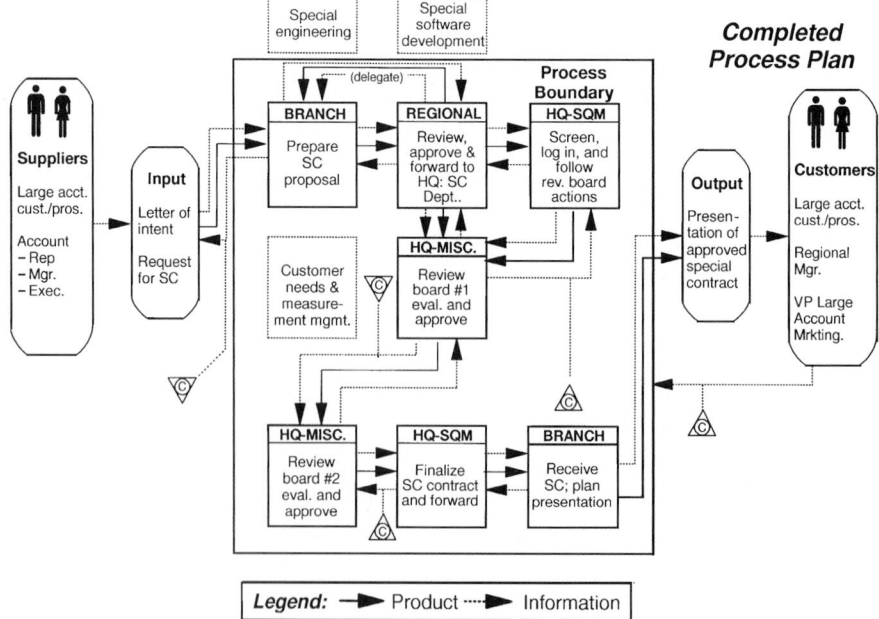

FIGURE 6.10   Completed process plan diagram for the special-contract management process. (*Source: Juran Institute, Wilton, CT.*)

Planning for Implementation Problems. A major PQM effort may involve huge expenditures and precipitate fundamental change in an organization, affecting thousands of jobs. All of this poses major management challenges. All of the many changes must be planned, scheduled, and completed so that the new process may be deployed to operational management. Figure 6.11 identifies specific categories of problems to be addressed and the key elements that are included.

Of the five categories listed in Figure 6.11, People and Organization is usually the source of the most challenging change issues in any PQM effort. Implementation issues in the people and organizational design category include: new jobs, which are usually bigger; new job descriptions; training people in the new jobs; new performance plans and objectives; new compensation systems (incentive pay, gainsharing, and the like); new recognition and reward mechanisms; new labor contracts with unions; introduction of teamwork and team-building concepts essential to a process orientation; formation of self-directed work teams; team education; reduction of management layers; new reporting relationships; development and management of severance plans for those whose jobs are eliminated; temporary continuation of benefits; out-placement programs; and new career paths based on knowledge and contribution, rather than on promotion within a hierarchy. The list goes on. Additionally, there are changes in technology, policy, physical infrastructure, etc., to be dealt with.

The importance of change management skills becomes clear. Deploying a new process can be a threat to those affected. The owner and team must be skilled in overcoming resistance to change.

Creating Readiness for Change: A Model for Change. Change happens when four conditions are combined. First, the current state must be seen as unsatisfactory, even painful; it must constitute a tension for change. Second, there must be a satisfactory alternative, a vision of how things can be better. Third, some practical steps must be available to reach the satisfactory state, including instruction in how to take the steps, and support during the journey. Fourth, to maintain the change, the organization and individuals must acquire skills and reach a state of self-efficacy.

These four conditions reinforce the intent to change. Progress toward that change must be monitored continuously so as to make the change a permanent one. In the operational management phase, operational controls, continuous improvement activity, and ongoing review and assessment all contribute to ensuring that the new process plan will continue to perform as planned. (See also Resistance to Change and how to deal with it in Section 5, The Quality Improvement Process.)

Planning for Implementation Action. The output of this step is a complex work plan, to be carried out by the Owner and Process Management Team. They will benefit from skills in the techniques of Project Management. (See Section 17, Project Management and Product Development.)

Deploying the New Process Plan. Before actually implementing the new process, the team tests the process plan. They test selected components of the process and may carry out computer simulations. The purpose is to predict the performance of the new process and determine feasibility. Also, the tests help the team refine the "roll out" of the process and decide whether to conduct parallel

Category	Key Elements Included
Workflow	Process anatomy (macro/micro, cross-functional, intrafunctional, inter-departmental, and intradepartmental)
Technology	Information technology and automation
People and organization	Jobs, job description, training and development, performance management, compensation (incentive-based or not), recognition/reward, union involvement, teams, self-directed work teams, reporting relationships and delayering
Infrastructure (physical)	Location, space, layout, equipment, tools, and furnishings
Policy/regulations	Government, community, industry, company, standards, and culture
New-process design issues	

FIGURE 6.11 Design categories. (*Source: Juran Institute, Wilton, CT.*)

operation (old process and new process running concurrently). The team must decide how to deploy the new process. There are several options:

- *Horizontal deployment,* function by function.
- *Vertical deployment,* top down, all functions at once.
- *Modularized deployment,* activity by activity, until all are deployed.
- *Priority deployment,* subprocesses and activities in priority sequence, those having the highest potential for improvement going first.
- *Trial deployment,* a small-scale pilot of the entire process, then expansion for complete implementation. This technique was used in the first redesign of the Special-Contract Management process, that is, a regional trial preceded national expansion. The insurance company USAA conducts all pilot tests of new process designs in their Great Lakes region. In addition to "working the bugs out of the new design before going national," USAA uses this approach as a "career-broadening experience for promising managers," and to "roll out the new design to the rest of the organization with much less resistance" (Garvin 1995).

Full deployment of the new process includes the development and deployment of an updated control plan. Figure 6.12 lists the contents of a new process plan.

Operational Management Phase: Managing the New Process.   The Operational Management Phase begins when the process is put into operation. The major activities in operational management are: (1) process quality control, (2) process quality improvement, and (3) periodic process review and assessment.

Process Quality Control.   "Process control" is an ongoing managerial process, in which the actual performance of the operating process is evaluated by measurements taken at the control points, comparing the measurements to the quality targets, and taking action on the difference. The goal of process control is to maintain performance of the business process at its planned level. (See Section 4, The Quality Control Process).

Process Quality Improvement.   By monitoring process performance with respect to customer requirements, the process owner can identify gaps between what the process is delivering and what is required for full customer satisfaction. These gaps are targets for process quality improvement

---

- Process mission
- Process goals
- Process management infrastructure (that is, owner/team/stakeholders)
- Process contract
- Process description/model
- Customer requirements (that is, customer list, customer needs, and requirements statement)
- Process flow
- Measurement plan
- Process analysis summary report
- Control plan
- Implementation action plan
- Resource plan
- Schedules/timeline

FIGURE 6.12   Contents of complete process plan (*Source: Juran Institute, Wilton, CT.*)

efforts. They are signaled by defects, complaints, high costs of poor quality, and other deficiencies. (See Section 5, The Quality Improvement Process.)

Periodic Process Review and Assessment.   The owner conducts reviews and assessments of current process performance to ensure that the process is performing according to plan. The review should include review and assessment of the process design itself to protect against changes in the design assumptions and anticipated future changes such as changes in customer needs, new technology or competitive process designs. It is worthwhile for the process owner to establish a schedule for reviewing the needs of customers and evaluating and benchmarking the present process.

As customer needs change, process measures must be refined to reflect these changes. This continuous refinement is the subject of a measurement management subprocess, which is established by the owners and team and complements the customer needs subprocess. The two processes go hand in hand.

The process management category in the Malcolm Baldrige National Quality Award criteria (1998) provides a basis for management review and assessment of process performance.

·Other external award criteria from worldwide sources, as well as many national and international standards, serve as inspiration and guidance for owners and teams contemplating process reviews. (See Section 14, Total Quality Management, and Section 11, The ISO 9000 Family of International Standards.)

THE INTEGRATION OF PQM WITH TQM

The criteria of the Malcolm Baldrige National Quality Award have come to be regarded as the de facto definition of TQM. (See Section 14.) Process quality management is an important concept within the TQM framework.

Organizations have learned not to limit managerial attention to the financial dimension. They have gained experience in defining, identifying, and managing the quality dimension. They are accustomed to thinking strategically—setting a vision, mission, and goals, all in alignment. And they will have experience reviewing progress against those goals.

The quality improvement process, which began in Japan in the 1950s and was widely deployed in the United States in the early 1980s, was an important step beyond functional management. Organizations found that quality improvement required two new pieces of organization machinery— the quality council and the cross-functional project team. The Quality Council usually consists of the senior management team; to its traditional responsibility for management of finance the responsibility for the management of quality is added. The project team recognizes that, in a functional organization, responsibility for reduction of chronic deficiencies has to be assigned to a cross-functional team.

PQM is a natural extension of many of the lessons learned in early quality improvement activities. It requires a conceptual change—from reliance on functional specialization to an understanding of the advantages of focusing on major business processes. It also requires an additional piece of organization machinery: an infrastructure for each of the major processes.

SUMMARY AND CRITICAL SUCCESS FACTORS FOR PQM IMPLEMENTATION

Key Points.   PQM is distinguished by the following:

- A strategic orientation, that is

  A clear mission, values, and vision for the organization

  Strategic goals tied to the organization vision, which are shared by executive leadership and deployed throughout the organization in the form of key business objectives

  Alignment and linkage of the organization's processes to its vision, strategic goals, and objectives

- A cross-functional orientation in place of the hierarchical organization.
- Cross-functional process teams, supported by the management system (education, communication, performance management, recognition and reward, compensation, new career path structures, etc.). The mission of each team is to dramatically improve the effectiveness, efficiency, and adaptability of each major business process to which it is assigned.
- Prime organizational focus on the needs of customers, external and internal, and business needs such as cost, cycle time, waste elimination.
- The driving of all work processes by quality of products and services and overall value creation.

Critical Success Factors for PQM Implementation.   The following factors are important to the success of a PQM initiative:

- Leadership from the top of the organization
- Management which communicates the vision, strategic goals, and key business objectives throughout the organization
- Vision shared by all in the organization
- Employees empowered and accountable to act in support of these key business objectives
- Expertise in change management available throughout the organization to facilitate dramatic change
- Continuous improvement
- Widespread skills in project management to enable the many PQM teams to manage schedules, costs, and work plans being coordinated and implemented throughout the organization
- Executive management promotion of the importance, impact, progress, and success of the PQM effort throughout the organization, and to external stakeholders
- Upper management's obligation is to enable and promote three principal objectives: customer focus, process orientation, and empowered employees at all levels

    Leaders of those organizations who have adopted PQM as a management tool know that Process Quality Management is a continuous managerial focus, not a single event or a quick fix. They also know that a constant focus on business processes is essential to the long-term success of their organization.

REFERENCES

AT&T Quality Steering Committee (1988). *Process Quality Management & Improvement Guidelines,* AT&T Information Center, Indianapolis, IN.

Black, John (1985). *Business Process Analysis—Guide to Total Quality,* Boeing Commercial Airplane Company, Seattle, WA, revised 1987.Byrne, John A. (1993). "The Horizontal Corporation," *Business Week,* Dec. 20, pp. 76–81.

Dumaine, Brian (1989). "How Managers Can Succeed Through Speed," *Fortune,* Feb. 13, pp. 54–60.

Garvin, David A. (1995). "Leveraging Processes for Strategic Advantage," *Harvard Business Review,* Sept./Oct., vol. 73, no. 5, pp. 77–90.

Gibson, Michael J. W. (1991–92). "The Quality Process: Business Process Quality Management," *International Manufacturing Strategy Resource Book,* pp. 167–179.

Gibson, Michael J. W., Gabriel, A., and Riley, James F., Jr. (1990). *Managing Business Process Quality (MBPQ),* 1st ed., Juran Institute, Inc., Wilton, CT.

Hall, Gene, Rosenthal, Jim, and Wade, Judy (1993). "How to Make Reengineering Really Work," *Harvard Business Review,* Nov./Dec., vol. 71, no. 6, pp. 199–130.

Hammer, Michael, and Champy, James (1993). *Reengineering the Corporation,* Harper Collins, New York.

Hardaker, Maurice, and Ward, Bryan K. (1987). "Getting Things Done: How to Make a Team Work," *Harvard Business Review,* Nov./Dec., vol. 65, pp. 112–119.

Kane, Edward J. (1986). "IBM's Focus on the Business Process," *Quality Progress,* April, p. 26.

Kane, Edward J. (1992). "Process Management Methodology Brings Uniformity to DBS," *Quality Progress,* June, vol. 25, no. 6, pp. 41–46.

Kaplan, Robert S., and Norton, David P. (1992). "The Balanced Scorecard—Measures that Drive Performance," *Harvard Business Review,* Jan./Feb., vol. 7, no. 1, pp. 71–79, reprint #92105.

Pall, Gabriel A. (1987). *Quality Process Management,* Prentice-Hall, Inc., Englewood Cliffs, NJ.

Riley, James F., Jr. (1989). *Executive Quality Focus: Discussion Leader's Guide,* Science Research Associates, Inc., Chicago.

Riley, James F., Jr., Pall, Gabriel A., and Harshbarger, Richard W. (1994). *Reengineering Processes for Competitive Advantage: Business Process Quality Management* (*BPQM*), 2nd ed., Juran Institute, Inc., Wilton, CT.

Rummler, Geary (1992). "Managing the White Space: The Work of Geary Rummler," *Training and Development,* Special Report, August, pp. 26–30.

Schlesiona, Peter (1988). *Business Process Management,* Science Research Associates, Inc., Chicago.

Zachman, James W. (1990). "Developing and Executing Business Strategies Using Process Quality Management," *IMPRO Conference Proceedings,* Juran Institute, Wilton, CT., pp. 2a-9–21.

SECTION 7
QUALITY AND INCOME

J. M. Juran

QUALITY AND COMPANY ECONOMICS

Quality affects company economics in two principal ways:

The effect of quality on costs: In this case "quality" means freedom from troubles traceable to office errors, factory defects, field failures, and so on. Higher "quality" means fewer errors, fewer defects, and fewer field failures. It takes effort to reduce the numbers of such deficiencies, but in the great majority of cases, the end result is cost reduction. This type of effect of quality on company economics is discussed in Section 8, Quality and Costs.

The effect of quality on income: In this case "quality" means those features of the product which respond to customer needs. Such features make the product salable and provide "product satisfaction" to customers. Higher quality means better and/or more features which provide greater satisfaction to customers.

This section focuses on the relationship between product features and company income. ("Company" includes any operating institution—an industrial company, a government agency, a school, and so on. "Income" means gross receipts, whether from sales, appropriations, tuitions, and so on.) The section discusses the forces through which quality affects income and the methods in use for studying the cause-effect relationships. Closely related to this subject of quality and income are two other sections of this handbook:

Market Research and Marketing (Section 18)

Customer Service (Section 25)

The above two effects of quality—on costs and on income—interact with each other. Product deficiencies not only add to suppliers' and customers' costs, they also discourage repeat sales. Customers who are affected by field failures are, of course, less willing to buy again from the guilty supplier. In addition, such customers do not keep this information to themselves—they publicize it so that it becomes an input to other potential buyers, with negative effects on the sales income of the supplier.

In recent decades there has been much study of the effect of poor quality on company economics. (See generally, Section 8, Quality and Costs.) In contrast, study of the effect of quality on income has lagged. This imbalance is all the more surprising since most upper managers give higher priority to increasing income than to reducing costs. This same imbalance presents an opportunity for improving company economics through better understanding of the effect of quality on income.

MAJOR ECONOMIC INFLUENCES

The ability of an industrial company to secure income is strongly influenced by the economic climate and by the cultural habits which the various economies have evolved. These overriding influences affect product quality as well as other elements of commerce.

National Affluence and Organization.   The form of a nation's economy and its degree of affluence strongly influence the approach to its quality problems.

Subsistence Economies.   In such economies the numerous impoverished users have little choice but to devote their income to basic human needs. Their protection against poor quality is derived more from their collective political power than from their collective economic power. Most of the world's population remains in a state of subsistence economy.

Planned Economies.   In all countries there are some socialized industries—government monopolies for some products or services. In some countries the entire economy is so organized. These monopolies limit the choice of the user to those qualities which result from the national planning and its execution. For elaboration, see Section 36, Quality and the National Culture.

Shortages and Surpluses.   In all economies, a shortage of goods (a "sellers' market") results in a relaxing of quality standards. The demand for goods exceeds the supply, so users must take what they can get (and bid up the price to boot). In contrast, a buyers' market results in a tightening of quality standards.

Life Behind the Quality Dikes.
As societies industrialize, they revise their lifestyle in order to secure the benefits of technology. Collectively, these benefits have greatly improved the quality of life, but they have also created a new dependence. In the industrial societies, great masses of human beings place their safety, health, and even their daily well-being behind numerous "quality dikes." For elaboration, see Section 35, Quality and Society, under the heading Life Behind the Quality Dikes.

Voluntary Obsolescence.
As customers acquire affluence, the industrial companies increasingly bring out new products (and new models of old products) which they urge prospective users to buy. Many of the users who buy these new models do so while possessing older models which are still in working order. This practice is regarded by some economists and reformers as a reprehensible economic waste.

In their efforts to put an end to this asserted waste, the reformers have attacked the industrial companies who bring out these new models and who promote their sale. Using the term "planned obsolescence," the reformers imply (and state outright) that the large companies, by their clever new models and their powerful sales promotions, break down the resistance of the users. Under this theory, the responsibility for the waste lies with the industrial companies who create the new models.

In the experience and judgment of the author, this theory of planned obsolescence is mostly nonsense. The simple fact, obvious both to manufacturers and consumers, is that *the consumer makes the decision* (of whether to discard the old product and buy the new). Periodically, this fact is dramatized by some massive marketing failure.

A few decades ago E.I. DuPont de Nemours & Co., Inc. (DuPont) brought out the product Corfam, a synthetic material invented to compete with leather for shoe uppers (and for other applications). Corfam was a technological triumph. Though costly, it possessed excellent properties for shoe uppers: durability, ease of care, shape retention, scuff resistance, water repellency, and ability to "breathe." DuPont became a major supplier of shoe uppers materials, but in 1971 it withdrew from the business because Corfam "never attained sufficient sales volume to show a profit."

Industry observers felt that the high durability of Corfam was an irrelevant property due to rapid style obsolescence; i.e., the life of the shoes was determined not by the inherent durability of Corfam, but by style obsolescence. In essence, a large corporation undertook a program which was antagonistic to obsolescence, but the users decided against it. DuPont's investment in Corfam may have exceeded $100 million.

In a case involving an even larger investment, the Ford Motor Company's Edsel automobile failed to gain consumer acceptance despite possessing numerous product innovations and being promoted by an extensive marketing campaign.

Involuntary Obsolescence.
A very different category of obsolescence consists of cases in which long-life products contain failure-prone components which will not last for the life of the product. The life of these components is determined by the manufacturer's design. As a result, even though the user decides to have the failed component replaced (to keep the product in service), *the manufacturer has made the real decision* because the design determined the life of the component.

This situation is at its worst when the original manufacturer has designed the product in such a way that the supplies, spare parts, and so on are nonstandard, so that the sole source is the original manufacturer. In such a situation, the user is locked into a single source of supply. Collectively, such cases have lent themselves to a good deal of abuse and have contributed to the consumerism movement. (For elaboration, see Juran 1970.)

CONTRAST IN VIEWS: CUSTOMER AND SUPPLIER

Industrial companies derive their income from the sale of their products. These sales are made to "customers," but customers vary in their functions. Customers may be merchants, processors, ultimate users, and so on, with resulting variations in customer needs. Response to customer needs requires a clear understanding of just what those needs are.

Human needs are complex and extend beyond technology into social, artistic, status, and other seemingly intangible areas. Suppliers are nevertheless obliged to understand these intangibles in order to be able to provide products which respond to such needs.

The Spectrum of Affluence.   In all economies the affluence of the population varies across a wide spectrum. Suppliers respond to this spectrum through variations in product features. These variations are often called "grades."

For example, all hotels provide overnight sleeping accommodations. Beyond this basic service, hotels vary remarkably in their offerings, and the grades (deluxe, four star, and so on) reflect this variation. In like manner, any model of automobile provides the basic service of point-to-point transportation. However, there are multiple grades of automobiles. The higher grades supply services beyond pure transportation—higher levels of safety, comfort, appearance, status, and so on.

Fitness for Use and Conformance to Specification.   Customers and suppliers sometimes differ in their definition of what is quality. Such differences are an invitation to trouble. To most customers, quality means those features of the product which respond to customer needs. In addition, quality includes freedom from failures, plus good customer service if failures do occur. One comprehensive definition for the above is "fitness for use."

In contrast, many suppliers had for years defined quality as conformance to specification at the time of final product test. This definition fails to consider numerous factors which influence quality as defined by customers: packaging, storage, transport, installation, reliability, maintainability, customer service, and so on.

Table 7.1 tabulates some of the differences in viewpoint as applied to long-life goods.

The ongoing revolution in quality has consisted in part of revising the suppliers' definition of quality to conform more nearly with the customers' definition.

Cost of Use.   For consumable products, the purchase price paid by the customer is quite close to the cost of using (consuming) the product. However, for long-lived products, the cost of use can diverge considerably from the purchase price because of added factors such as operating costs, maintenance costs, downtime, depreciation, and so on.

The centuries-old emphasis on purchase price has tended to obscure the subsequent costs of use. One result has been suboptimization; i.e., suppliers optimize their costs rather than the combined costs of suppliers and customers.

The concept of life-cycle costing offers a solution to this problem, and progress is being made in adopting this concept. (See Life Cycle Costing, below.)

Degrees of User Knowledge.   In a competitive market, customers have multiple sources of supply. In making a choice, product quality is an obvious consideration. However, customers vary greatly in their ability to evaluate quality, especially prior to purchase.

Table 7.2 summarizes the extent of customer knowledge and strength in the marketplace as related to quality matters.

TABLE 7.1 Contrasting Views: Customer and Suppliers

	Principal views	
Aspects	Of customers	Of manufacturers
What is bought	A service needed by the customer	Goods made by the manufacturer
Definition of quality	Fitness for use during the life of the product	Conformance to specification on final test
Cost	Cost of use, including Purchase price Operating costs Maintenance Downtime Depreciation Loss on resale	Cost of manufacture
Responsibility for keeping in service	Over the entire useful life	During the warranty period
Spare parts	A necessary evil	A profitable business

Source: *Juran's Quality Control Handbook,* 4th ed., McGraw-Hill, New York, 1988, p. 3.7.

TABLE 7.2 Customer Influences on Quality

Aspects of the problem	Original equipment manufacturers (OEMs)	Dealers and repair shops	Consumers
Makeup of the market	A few, very large customers	Some large customers plus many smaller ones	Very many, very small customers
Economic strength of any one customer	Very large, cannot be ignored	Modest or low	Negligible
Technological strength of customer	Very high; has engineers and laboratories	Low or nil	Nil (requires technical assistance)
Political strength of customer	Modest or low	Low to nil	Variable, but can be very great collectively
Fitness for use is judged mainly by:	Qualification testing	Absence of consumer complaints	Successful usage
Quality specifications dominated by:	Customers	Manufacturer	Manufacturer
Use of incoming inspection	Extensive test for conformance to specification	Low or nil for dealers; in-use tests by repair shops	In-use test
Collection and analysis of failure data	Good to fair	Poor to nil	Poor to nil

Source: *Juran's Quality Control Handbook,* 4th ed., McGraw-Hill, New York, 1988, p. 3.8.

The broad conclusions which can be drawn from Table 7.2 are as follows:

• Original equipment manufacturers (OEMs) can protect themselves through their technological and/or economic power as much as through contract provisions. Merchants and repair shops must rely mainly on contract provisions supplemented by some economic power.

• Small users have very limited knowledge and protection. The situation of the small user requires some elaboration.

With some exceptions, small users do not fully understand the technological nature of the product. The user does have sensory recognition of some aspects of fitness for use: the bread smells fresh-baked, the radio set has clear reception, the shoes are good-looking. Beyond such sensory judgments, and especially concerning the long-life performance of the product, the small user must rely mainly on prior personal experience with the supplier or merchant. Lacking such prior experience, the small user must choose from the propaganda of competing suppliers plus other available inputs (neighbors, merchants, independent laboratories, and so on).

To the extent that the user does understand fitness for use, the effect on the supplier's income is somewhat as follows:

As seen by the user, the product or service is	The resulting income to the supplier is
Not fit for use	None, or in immediate jeopardy
Fit for use but noticeably inferior to competitive products	Low due to loss of market share or need to lower prices
Fit for use and competitive	At market prices
Noticeably superior to competitive products	High due to premium prices or greater share of market

In the foregoing, the terms "fitness for use," "inferior," "competitive," and "superior" all relate to the situation as *seen by the user.* (The foregoing table is valid as applied to both large customers and small users.)

Stated Needs and Real Needs.
Customers state their needs as they see them, and in their language. Suppliers are faced with understanding the real needs behind the stated needs and translating those needs into suppliers' language.

It is quite common for customers to state their needs in the form of goods, when their real needs are for the services provided by those goods. For example:

Stated needs	Real needs
Food	Nourishment, pleasant taste
Automobile	Transportation, safety, comfort, etc.
Color TV	Entertainment, news, etc.
Toothpaste	Clean teeth, sweet breath, etc.

Preoccupation with selling *goods* can divert attention from the real needs of customers.

> Two hair net manufacturers were in competition. They devoted much effort to improving the qualities of the product and to strengthening their marketing techniques. But hair nets became extinct when someone developed a hair spray which gave the user a better way of providing the basic service—holding her hair in place. (Private communication to J. M. Juran.)

In a classic, widely read paper, "Marketing Myopia," Levitt (1960), stressed service orientation as distinguished from product orientation. In his view, the railroads missed an opportunity for expansion due to focus on railroading rather than on transportation. In like manner, the motion picture industry missed an opportunity to participate in the growing television industry due to focus on movies rather than on entertainment. (Levitt 1960.)

To understand the real needs of customers requires answers to questions such as: Why are you buying this product? What service do you expect from it?

Psychological Needs.
For many products, customer needs extend beyond the technological features of the product; the needs also include matters of a psychological nature. Such needs apply to both goods and services.

A man in need of a haircut has the option of going to (1) a "shop" inhabited by "barbers" or (2) a "salon" inhabited by "hair stylists." Either way, he is shorn by a skilled artisan. Either way, his resulting outward appearance is essentially the same. What differs is his remaining assets and his sense of well-being.

What applies to services also applies to physical goods. There are factories in which chocolate-coated candies are conveyed by a belt to the packaging department. At the end of the belt are two teams of packers. One team packs the chocolates into modest cardboard boxes destined for budget-priced merchant shops. The other team packs the chocolates into satin-lined wooden boxes destined to be sold in deluxe shops. The resulting price for a like amount of chocolate can differ by severalfold. The respective purchasers encounter other differences as well: the shop decor, level of courtesy, promptness of service, sense of importance, and so on. However, the goods are identical. Any chocolate on that conveyer belt has not the faintest idea of whether it will end up in a budget shop or in a deluxe shop.

Technologists may wonder why consumers are willing to pay such price premiums when the goods are identical. However, to many consumers, the psychological needs are perceived as real needs, and the consumers act on their perceptions. Most suppliers design their marketing strategies to respond to customers' perceived needs.

"User-Friendly" Needs. The "amateur" status of many users has given rise to the term "user friendly" to describe a condition which enables amateurs to use technological and other complex products with confidence. For example:

The language of published information should be simple, nonambiguous, and readily understood. Notorious offenders have included legal documents, owners' operating manuals, forms to be filled out, and so on. Widely used forms (such as Federal tax returns) should be field tested on a sample of the very people who will later be faced with filling out the forms.

Products should be broadly compatible. Much of this has been done through standardization committees or through natural monopolies. An example of lack of such compatibility during the 1980s was the personal computer—many personal computers were able to "talk" to computers made by the same manufacturer but not to computers made by other manufacturers.

The Need to Be Kept Informed. Customers sometimes find themselves in a state of uncertainty: Their train is late, and they don't know when to expect it; there is a power outage, and they don't know when power will be restored. In many such cases, the supplier company has not established the policies and processes needed to keep customers informed. In actuality, the customers, even if kept informed, usually have no choice but to wait it out. Nevertheless, being kept informed reduces the anxiety—it provides a degree of assurance that human beings are aware of the problem and that it is in the process of being solved.

The New York subway system rules require conductors to explain all delays lasting two minutes or more. One survey reported that this rule was followed only about 40 percent of the time. A City Hall report concluded that "shortage of information is a significant source of public antagonism toward the Transit Authority" (Levine 1987).

In contrast, some airlines go to pains to keep their customers informed of the reasons for a delay and of the progress being made in providing a remedy.

A different category of cases involves companies secretly taking actions adverse to quality but without informing the customer. The most frequent are those in which products not conforming to specification are shipped to unwary customers. In the great majority of such cases, the products are fit for use despite the nonconformances. In other cases, the matter may be debatable. In still other cases, the act of shipment is at the least unethical and at the worst illegal.

In a highly publicized case, Oldsmobile cars were being delivered containing Chevrolet engines. Yet the Oldsmobile sales promotion had emphasized the quality of its engines. In due course the manufacturer made restitution but not before suffering adverse publicity.

Once discovered, any secretive actions tend to arouse suspicions, even if the product is fit for customer use. The customers wonder, "What else has been done secretly without our being informed?"

The usual reason for not informing the customer is a failure to raise the question: What shall we tell the customers? It would help if every nonconformance document included a blank space headed "What is to be communicated to the customers?" The decision may be to communicate nothing, but at least the question has been faced.

Cultural Needs.   The needs of customers, especially internal customers, include cultural needs—preservation of status, continuity of habit patterns, and still other elements of what is broadly called the cultural pattern. Some of the inability to discover customer needs is traceable to failure to understand the nature and even the existence of the cultural pattern.

Cultural needs are seldom stated openly—mostly they are stated in disguised form. A proposed change which may reduce the status of some employee will be resisted by that employee. The stated reasons for the resistance will be on plausible grounds, such as the effect on costs. The real reason will not emerge. No one will say, "I am against this because it will reduce my status." Discovery of the real needs behind the stated needs is an important step toward a meeting of the minds.

(For elaboration on the nature of cultural patterns and the "rules of the road," see Section 5, The Quality Improvement Process, under Resistance to Change; see also Juran 1964, Chapter 9.)

Needs Traceable to Unintended Use.   Many quality failures arise because the customer uses the product in a manner different from that intended by the supplier. This practice takes many forms:

Untrained workers are assigned to processes requiring trained workers.

Equipment is overloaded or is allowed to run without adherence to maintenance schedules.

The product is used in ways never intended by the supplier.

All this influences the relationship between quality and income. The critical question is whether the quality planning should be based on *intended use* or *actual use*. The latter often requires adding a factor of safety during the planning. For example:

Fuses and circuit breakers are designed into electrical circuits for protection against overloads.

Software is written to detect spelling errors.

Public utility invoicing may include a check of customers' prior usage to guard against errors in reading the meters.

Such factors of safety may add to the cost. Yet they may well result in an optimal overall cost by helping to avoid the higher cost arising from actual use or misuse.

NEEDS RELATED TO PRODUCT DISSATISFACTION

When products fail, a new set of customer needs arises—how to restore service and get compensated for the associated losses and inconvenience. These new needs are communicated through customer complaints, which then are acted on by special departments such as Customer Service.

Inadequate company response to consumer complaints and to the terms of warranties has contributed importantly to the rise of the "consumerism" movement. (See Section 35, Quality and Society, under The Growth of Consumerism.)

Studies of how to respond to customer complaints have identified the key features of a response system which meets customer needs. (For elaboration, see Section 25, Customer Service; see also, United States Office of Consumer Affairs, 1985–86.)

Complaints also affect product salability. This has been researched in studies commissioned by the United States Office of Consumer Affairs. The findings may be summarized as follows:

Of customers who were dissatisfied with products, nearly 70 percent did not complain. The proportions varied with the type of product involved. The reasons for not complaining included: the effort to complain was not worth it; the belief that complaining would do no good; lack of knowledge of how to complain.

Over 40 percent of the complaining customers were unhappy with the responsive action taken by the suppliers. Here again the percentage varied depending on the type of product involved.

Future salability is strongly influenced by the action taken on complaints. Figure 7.1 shows broadly the nature of consumer behavior following product dissatisfaction. This strong influence extends to brand loyalty. Figure 7.2 shows the extent of this influence as applied to "large ticket" durable goods, financial services, and automobile services, respectively. A similar, strong influence extends also to product line loyalty.

That same research concluded that an organized approach to complaint handling provides a high return on investment. The elements of such an organized approach may include:

A response center staffed to provide 24-h access by consumers

A toll-free telephone number

A computerized database

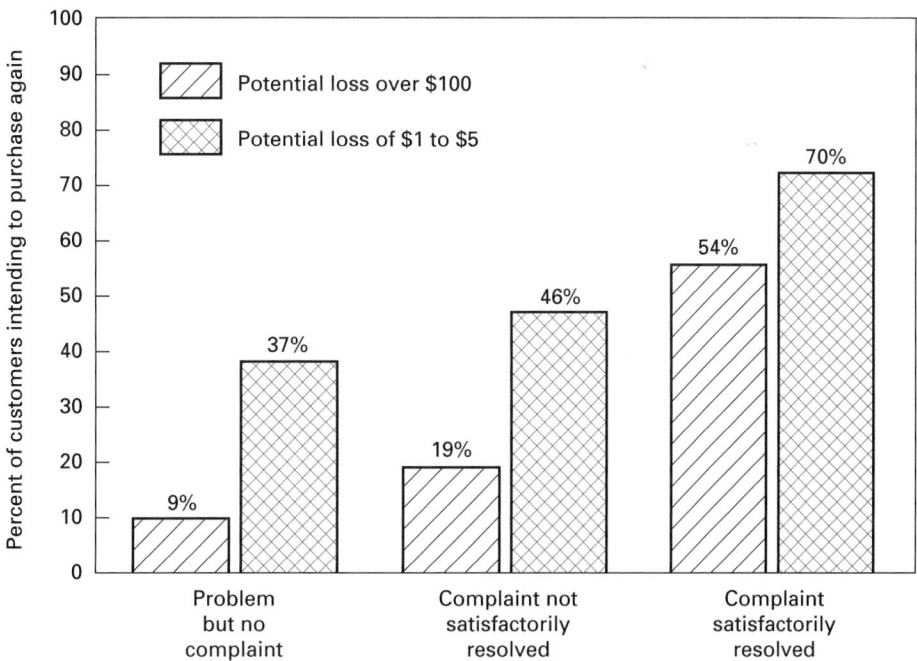

FIGURE 7.1   Consumer behavior after experiencing product dissatisfaction. [*Planning for Quality, 2nd ed. (1990), Juran Institute Inc., Wilton, CT, pp. 4–12.*]

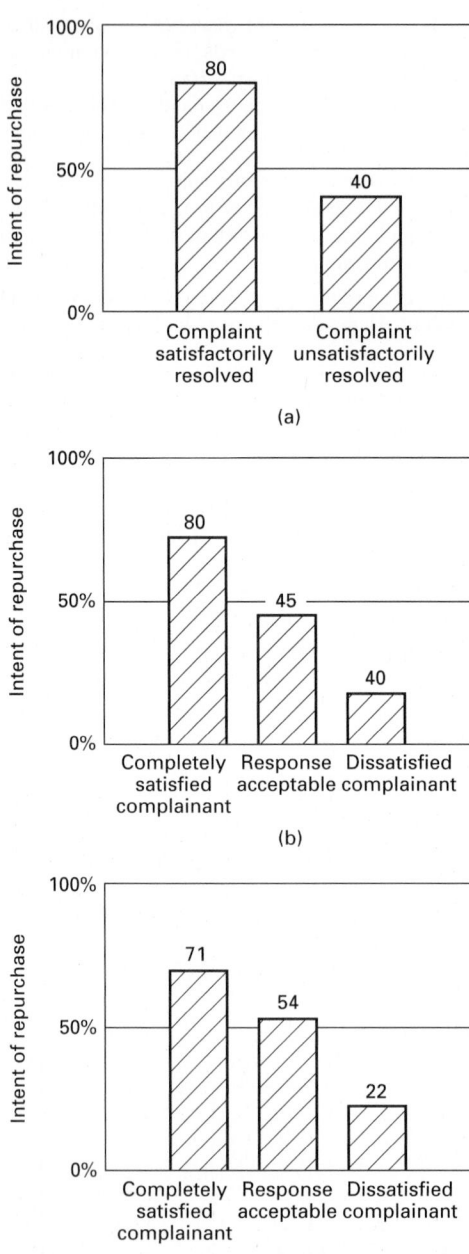

FIGURE 7.2   Consumer loyalty versus complaint resolution. Large-ticket durable goods; financial services; automotive services. (*Planning for Quality, 2nd ed. (1990),Juran Institute, Inc., Wilton, CT, pp. 4–14.*)

Special training for the personnel who answer the telephones

Active solicitation of complaints to minimize loss of customers in the future

(For added detail, see the full report, United States Office of Consumer Affairs, 1985–86.)

SOURCES OF CUSTOMER NEEDS

The most simplistic assumption is that customers are completely knowledgeable as to their needs and that market research can be used to extract this information from them. In practice, customer knowledge can be quite incomplete. In some cases the customer may be the last person to find out. It is unlikely that any customer ever expressed the need for a Walkman (a miniature, portable audiotape player) before such devices came on the market. However, once they became available, many customers discovered that they needed one.

These gaps in customer knowledge are filled in mainly by the forces of the competitive market and by the actions of entrepreneurs.

Inadequate Available Products.   When available products are perceived as inadequate, a vacuum waiting to be filled emerges. Human ingenuity then finds ways to fill that vacuum:

The number of licensed New York taxicabs has remained frozen for years. The resulting shortage has been filled by unlicensed cabs, limousines, and so on.

Government instructions for filling out tax forms have been confusing to many taxpayers. One result has been the publication of some best-selling books on how to prepare tax returns.

The service provided by tradesmen has been widely regarded as expensive and untimely. One result has been the growth of a large do-it-yourself industry.

Relief from Onerous Chores.   There seems to be no end to the willingness of affluent people to pay someone else to do onerous chores. Much former kitchen work is now being done in factories (soluble coffee, canned foods, and more). The prices of the processed foods are often several times the prices of the raw foods. Yet to do the processing at home involves working for a very low hourly wage. Cleaning chores have been extensively transferred to household appliances. The end is not in sight. The same kinds of transfer have taken place on a massive scale with respect to industrial chores (data processing, materials handling, etc.)

Reduction of Time for Service.   Some cultures exhibit an urge to "get it over with." In such cultures, those who can serve customers in the shortest time are rewarded by a higher share of market. A spectacular example of this urge is the growth of the "fast food" industry. In other industries, a major factor in choosing suppliers is the time spent to get service. An example is choice of gasoline filling stations. [See Ackoff (1978), Fable 5.4, p. 108.] This same need for prompt service is an essential element in the urge to go to "just-in-time" manufacture.

Changes in Customer Habits.   Customer habits can be notoriously fickle. Obvious examples are fashions in clothing and concerns over health that have reduced the consumption of beef and increased that of poultry. Such shifts are not limited to consumers. Industrial companies often launch "drives," most of which briefly take center stage and then fade away. The associated "buzz words" similarly come and go.

Role of the Entrepreneur.   The entrepreneur plays a vital role in providing customers with new versions of existing products. In addition, the entrepreneur identifies new products, some of them unheard of, which might create customer needs where none have existed previously. Those new products have a shocking rate of mortality, but the rewards can be remarkably high, and that is what attracts the independent entrepreneur. Moreover, the entrepreneurs can make use of the power of advertising and promotion, which some do very effectively. The legendary Charles Revson, founder of Revlon, stated it somewhat as follows: "In our factory we make lipstick; in our advertising we sell hope."

CUSTOMER NEEDS AND THE QUALITY PLANNING PROCESS

Discovery of customer needs is critical to generating income; it is also one of the major steps on the "quality planning road map." That road map includes other steps which, in varying degrees, influence the relationship between quality and income. (For elaboration, see Section 3, The Quality Planning Process.)

QUALITY AND PRICE

There is general awareness that product price bears some rational relationship to product quality. However, researchers on the subject have often reported confused relationships, some of which appear to run contrary to logical reasoning. To interpret this research, it is useful to separate the subject into consumer products and industrial products.

Consumer Products.   Numerous researchers have tried to quantify the correlation between product quality and product price. (See, for example, Riesz 1979; also Morris and Bronson 1969.) A major database for this research has been the journal *Consumer Reports,* a publication of Consumers Union, a nonprofit supplier of information and advice to consumers. The specific information used in the research consisted of *Consumer Reports'* published quality ratings of products, along with the associated prevailing market prices.

The research generally concluded that there is little positive correlation between quality ratings and market prices. For a significant minority of products, the correlation was negative. Such conclusions were reached as to foods, both convenience and nonconvenience (Riesz 1979). Similar conclusions were reached for other consumable products, household appliances, tools, and other long-life products (Morris and Bronson 1969).

Researchers offer various theories to explain why so many consumers seem to be acting contrary to their own best interests:

- The quality ratings are based solely on evaluations of the functional features of the products—the inherent quality of design. The ratings do not evaluate various factors which are known to influence consumer behavior. These factors include: *service* in such forms as attention, courtesy, promptness; also *decor* in such forms as pleasant surroundings, attractive packaging.
- Consumers generally possess only limited technological literacy, and most are unaware of the quality ratings.
- Lacking objective quality information, consumers give weight to the image projected by manufacturers and merchants through their promotion and advertising.
- The price itself is perceived by many consumers as a quality rating. There appears to be a widespread belief that a higher-priced product is also a higher-quality product. Some companies have exploited this belief as a part of their marketing and pricing strategy ("Pricing of Products Is Still an Art" 1981).

Price Differences.   Premium-priced products usually run about 10 to 20 percent higher than other products. For example, branded products often are priced in this range relative to generic products. However, there are many instances of much greater price differences.

Haircuts given in some "salons" sell at several times the price prevailing in "barber shops."

Chocolates packaged in elegant boxes and sold in deluxe shops may sell for several times the price of the identical chocolates packaged in simple boxes and sold in budget shops.

The spectrum of restaurant meal prices exceeds an order of magnitude.

Branded pharmaceuticals may sell for several times the price of generic drugs which are asserted to be therapeutically equivalent.

What emerges is that for many consumers, perception of the quality-price relationship is derived from unique interpretations of the terms used:

Quality is interpreted as including factors which go beyond the functional features of the product.

Price is interpreted as relating to "value" and is paid for those added factors, along with the inherent functional features.

Price premiums are conspicuous, and are often resisted fiercely by buyers, even when there are clear quality differences. In contrast, buyers are usually willing to reward superior quality with higher share of market in lieu of price differences. In many cases, the supplier can gain more from higher share of market than from price premiums, because of the arithmetic of the break-even chart. (See Figure 7.3, below.) For an interesting case example involving the risks of price premiums based on superior quality, see Smith (1995).

Efforts to Quantify Value.   Efforts to quantify value have largely been limited to functional properties of products. For example, the U.S. Department of Transportation evaluated the perfor-

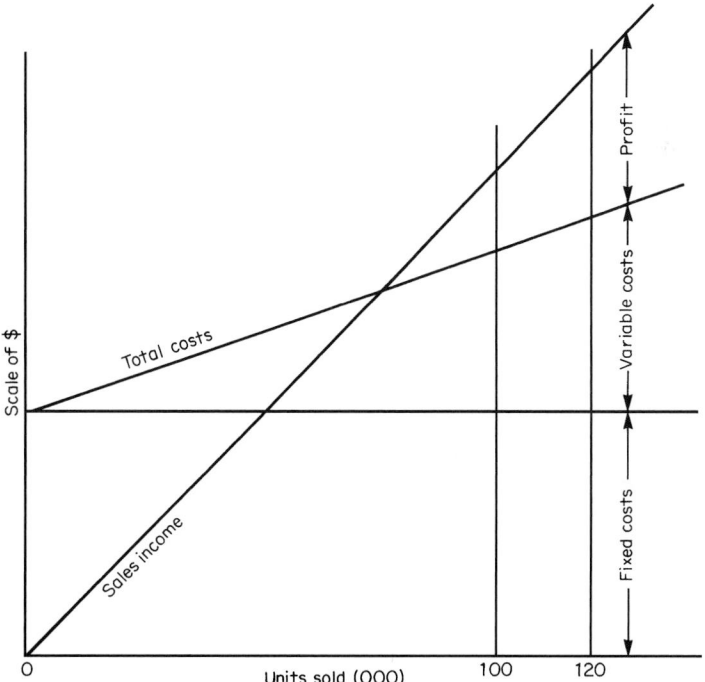

FIGURE 7.3   Break-even chart. (*Juran's Quality Control Handbook, 4th ed., McGraw-Hill, New York, pp. 3.13.*)

mance of automobile tires for several qualities, notably tread wear. That made it possible to estimate the cost per unit of distance traveled (under standard test conditions). (See "Consumer Guide on Tires" 1980.) Consumers Union sometimes makes such evaluations for consumer products. (See, for example, "Dishwashing Liquids" 1984.)

Such evaluations can be useful to consumers. However, the impact of such information is limited because of consumer unawareness and because consumer perceptions are based on broader concepts of quality and value.

Industrial Products.   Industrial products also employ the concepts of quality, price, and value. However, industrial buyers are generally much better informed of the significance of these concepts. In addition, industrial buyers are better provided with the technological and economic information needed to make rational decisions.

The principle of "standard product, market price" can be difficult to apply due to product quality differences.

> A company making standard power tools improved the reliability of the tools, but the marketing manager resisted increasing the price on the ground that since they were standard tools he would lose share of market if he raised prices. A field study then disclosed that the high-reliability tools greatly reduced the costs of the (industrial) users in maintenance and especially in downtime. This information then became the means of convincing users to accept a price increase (of $500,000 per year). (Consulting experience of J. M. Juran.)

Commodity versus Specialty or System.   An important question in much industrial buying is whether the product being bought is a commodity or something broader. The greater breadth may involve a specialty or a system of which the commodity is a part, but which includes other attributes of special value to the buyer.

Commodities are typically bought at market prices, and the price strongly influences the purchasing decisions. However, a perceived quality superiority is nevertheless an asset which may be translated into higher share of market or into a price premium. Many companies have opted for price premiums despite the fact that customers resist accepting price premiums more strongly than awarding higher market share.

The report entitled *Pricing High Quality Products* (PIMS 1978) raises questions concerning this strategy. According to the report, the market is willing to pay premium prices for high-quality products. However, if the premium price is not demanded, the market responds by awarding so high an increase in market share that the supplier ends up with a return on investment greater than that resulting solely from premium pricing.

Perceived quality superiority takes many forms: predictable uniformity of product, promptness of delivery, technological advice and service, assistance in training customer personnel, prompt assistance in troubleshooting, product innovation, sharing of information, joint quality planning, and joint projects for quality improvement. For a case example of a joint quality improvement project involving Aluminum Company of America and Eastman Kodak, see Kegarise and Miller (1986). (See also Kegarise et al. 1987.)

Specialties; the "Bundled" Price.   Specialties are standard products which are tailored specifically for use by specific customers. The product is "special" because of added nonstandard features and services which become the basis for "bundled" prices. Bundled prices provide no breakdown of price between the goods (commodities) and the associated additional features and services.

Bundled prices are an advantage to the supplier as long as the product remains a specialty and requires the added features and services. However, if wide use of the specialty results in standardization, the need for the added services diminishes. In such cases it is common for competitors to offer the standard product at lower prices but without the technical services. This is a form of "unbundling" the price. (For an interesting research on pricing in the chemicals industry, along with an approach to evaluation of the "additional attributes," see Gross 1978.)

QUALITY AND SHARE OF MARKET

Growth in share of market is among the highest goals of upper managers. Greater market share means higher sales volume. In turn, higher sales volume accelerates return on investment disproportionally due to the workings of the break-even chart (Figure 7.3).

In Figure 7.3, to the right of the break-even line, an increase of 20 percent in sales creates an increase of 50 percent in profit, since the "constant" costs do not increase. (Actually, constant costs do vary with volume, but not at all in proportion.) The risks involved in increasing market share are modest, since the technology, production facilities, market, and so on are already in existence and of proved effectiveness.

Effect of Quality Superiority.   Quality superiority can often be translated into higher share of market, but it may require special effort to do so. Much depends on the nature and degree of superiority and especially on the ability of the buyer to perceive the difference and its significance.

Quality Superiority Obvious to the Buyer.   In such cases, the obvious superiority can be translated into higher share of market. This concept is fully understood by marketers, and they have from time immemorial urged product developers to come up with product features which can then be propagandized to secure higher share of market. Examples of such cases are legion.

Quality Superiority Translatable into Users' Economics.   Some products are outwardly "alike" but have unlike performances. An obvious example is the electric power consumption of an appliance. In this and similar examples, it is feasible to translate the technological difference into the language of money. Such translation makes it easier for amateurs in technology to understand the significance of the quality superiority.

The power tool case (above) realized the same effect. The superior reliability was translated into the language of money to secure a price premium. It could instead have been used to secure higher share of market. In the tire wear case (above) there was a translation into cost per unit of distance traveled.

The initiative to translate may also be taken by the buyer. Some users of grinding wheels keep records on wheel life. This is then translated into money—grinding wheel costs per 1000 pieces processed. Such a unit of measure makes it unnecessary for the buyer to become expert in the technology of abrasives.

Collectively, cases such as the above can be generalized as follows:

There is in fact a quality difference among competing products.

This difference is technological in nature so that its significance is not understood by many users.

It is often possible to translate the difference into the language of money or into other forms within the users' systems of values.

Quality Superiority Minor but Demonstrable.   In some cases, quality superiority can secure added share of market even though the "inferior" product is fit for use.

A manufacturer of antifriction bearings refined his processes to such an extent that his products were clearly more precise than those of his competitors. However, competitors' products were fit for use, so no price differential was feasible. Nevertheless, the fact of greater precision impressed the clients' engineers and secured increased share of market. (Consulting experience of J. M. Juran.)

In consumer products, even a seemingly small product difference may be translated into increased market share if the consumers are adequately sensitized.

A manufacturer of candy-coated chocolates seized on the fact that his product did not create chocolate smudge marks on consumers' hands. He dramatized this in television advertisements by contrasting the appearance of children's hands after eating his and competitors' (uncoated) chocolate. His share of market rose dramatically.

Quality Superiority Accepted on Faith.   Consumers can be persuaded to accept, on faith, assertions of product superiority which they themselves are unable to verify. An example was an ingenious market research on electric razors. The sponsoring company (Schick) employed an independent laboratory to conduct the tests. During the research, panelists shaved themselves twice, using two electric razors one after the other. On one day the Schick razor was used first and a competing razor immediately after. On the next day the sequence was reversed. In all tests the contents of the second razor were weighed precisely. The data assertedly showed that when the Schick was the second razor, its contents weighed more than those of competitors. The implication was that Schick razors gave a cleaner shave. Within a few months the Schick share of market rose as follows:

September   8.3 percent
December   16.4 percent

In this case, the consumers had *no way to verify* the accuracy of the asserted superiority. They had the choice of accepting it on faith, or not. Many accepted it on faith.

No Quality Superiority.   If there is no demonstrable quality superiority, then share of market is determined by marketing skills. These take such forms as persuasive propaganda, attractive packaging, and so on. Price reductions in various forms can provide increases in share of market, but this is usually temporary. Competitors move promptly to take similar action.

Consumer Preference and Share of Market.   Consumers rely heavily on their own senses to aid them in judging quality. This fact has stimulated research to design means for measuring quality by using human senses as measuring instruments. This research has led to development of objective methods for measuring consumer preference and other forms of consumer response. A large body of literature is now available, setting out the types of sensory tests and the methods for conducting them. (For elaboration, see Section 23, Inspection and Test, under Sensory Tests.)

At first these methods were applied to making process control and product acceptance decisions. More recently the applications have been extended into areas such as consumer preference testing, new product development, advertising, and marketing.

For some products it is easy to secure a measure of consumer preference through "forced choice" testing. For example, a table is set up in a department store and passers-by are invited to taste two cups of coffee, A and B, and to express their preference. Pairs of swatches of carpet may be shown to panels of potential buyers with the request that they indicate their preferences. For comparatively simple consumer products, such tests can secure good data on consumer preference.

The value of consumer preference data is greatly multiplied through correlation with data on share of market. Figure 7.4 shows such a correlation for 41 different packaged consumer food products. This was an uncommonly useful analysis and deserves careful study.

Each dot on Figure 7.4 represents a food product sold on supermarket shelves. Each product has competitors for the available shelf space. The competing products sell for identical prices and are packaged in identically sized boxes containing identical amounts of product. What may influence the consumer are

- The contents of the package, as judged by senses and usage, which may cause the consumer to prefer product A over product B.
- The marketing features such as attractiveness of the package, appeal of prior advertising, and reputation of the manufacturer.

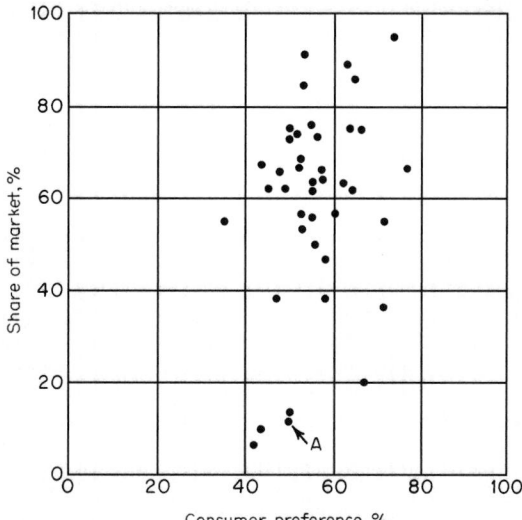

FIGURE 7.4   Consumer preference versus share of market.
(*Juran's Quality Control Handbook, 4th ed., McGraw-Hill, New York, p. 3.15.*)

On Figure 7.4 the horizontal scale shows consumer preference over the leading competitor as determined by statistically sound preference testing. The vertical scale shows the share of market versus the leading competitor, considering the two as constituting 100 percent.

In Figure 7.4 no product showed a consumer preference below 25 percent or above 75 percent. Such preference levels would mean that the product is so superior (or inferior) that three users out of four can detect the difference. Since all other factors are essentially equal, a product which is so overwhelmingly preferred takes over the entire market, and its competition disappears.

In contrast to the vacant areas on the horizontal scale of consumer preference, the vertical scale of share of market has data along the entire spectrum. One product (marked A on Figure 7.4) lies squarely on the 50 percent consumer preference line, which probably means (under forced-choice testing) that the users are guessing as to whether they prefer that product or its competitor. Yet product A has only 10 percent share of market, and its competitor 90 percent. Not only that, this inequality in share of market has persisted for years. The reason is that the 90 percent company was the first to bring that product to market. As a result it acquired a "prior franchise" and has retained its position through good promotion.

The conclusion is that when competing products are quite similar in consumer preference, any effect of such small quality differentials is obscured by the effect of the marketing skills. In consequence, it is logical to conclude that when quality preferences are clearly evident to the user, such quality differences are decisive in share of market, all other things being equal. When quality differences are slight, the decisive factor in share of market is the marketing skills.

As a corollary, it appears that companies are well advised to undertake quality improvements which will result in either (1) bringing them from a clearly weak to an acceptable preference or (2) bringing them from an acceptable preference to a clearly dominant preference. However, companies are not well advised to undertake quality improvements which will merely move them from one acceptable level to another, since the dominant role in share of market in such cases is played by the marketing skills. [For elaboration, see Juran (1959)].

It is easy for technologists to conclude that what they regard as important in the product is also of prime concern to the user. In the carpet industry, the engineers devote much effort to improving wear qualities and other technological aspects of fitness for use. However, a market survey established that consumers' reasons for selecting carpets were primarily sensory:

Color	56 percent
Pattern	20 percent
Other sensory qualities	6 percent
Nonsensory qualities	18 percent

For more complex consumer products it is feasible, in theory, to study the relation of quality to market share by securing quantitative data on (1) actual changes in buying patterns of consumers and (2) actions of suppliers which may have created these changes. In practice, such information is difficult to acquire. It is also difficult to conclude, in any one instance, why the purchase was of model A rather than B. What does emerge are "demographic" patterns, i.e., age of buyers, size of family, and so on, which favor model A rather than B. (For elaboration, see Section 18, Market Research and Marketing.) For products sold through merchants, broad consumer dissatisfaction with quality can translate into "merchant preference," with extensive damage to share of market.

A maker of household appliances was competitive with respect to product features, price, and promptness of delivery. However, it was not competitive with respect to field failure, and this became a major source of complaints from consumers to the merchants. Within several years the maker (B) lost all of its leadership in share of market, as shown in the table below. This table stimulated the upper managers of company B to take action to improve product reliability.

	Leaders in market share during:			
Model price	Base year	Base year plus 1	Base year plus 2	Base year plus 3
High	A	C	C	C
Medium	B	B	C	C
Low	C	C	C	C
Special	B	B	B	C

Industrial Products and Share of Market.   Industrial products are sold more on technological performance than on sensory qualities. However, the principle of customer preference applies, as does the need to relate quality differences to customer preference and to share of market. The methodology is discussed in Section 18, Market Research and Marketing.

Quality and Competitive Bidding.   Many industrial products and, especially, large systems, are bought through competitive bidding. Most government agencies are required by law to secure competitive bids before awarding large contracts. Industrial companies require their purchasing managers to do the same. The invitations to bid usually include the parameter of quality, which may be specified in detail or though performance specifications.

To prospective suppliers the ratio of awards received to bids made is of great significance. The volume of sales and profit depends importantly on this ratio. In addition, the cost of preparing bids is substantial; for large systems, the cost of bid preparation is itself large. Finally, the ratio affects the morale of the people involved. (Members of a winning team fight with their competitors; members of a losing team fight with each other.) It is feasible to analyze the record of prior bids in order to improve the percent of successful bids. Table 7.3 shows such an analysis involving 20 *unsuccessful* bids.

To create Table 7.3, a multifunctional team analyzed 20 unsuccessful bids. It identified the main and contributing reasons for failure to win the contract. The team's conclusions show that the installation price was the most influential factor—it was a contributing cause in 10 of the 14 cases which

TABLE 7.3   Analysis of Unsuccessful Bids*

Contract proposal	Quality of design	Product price	Installation price†	Reciprocal buying	Other
			Bid not accepted due to		
A1		×	×		×
A2			××		
A3	××	×			
A4	××		×		
A5	××				
A6	××				
A7		××			
A8		××			
A9			××		
A10			××		
B1	×		×		
B2				××	
B3				××	
B4				××	
B5		×	×		
B6		×	××		
B7	××				
B8		×	×		
B9				×	
B10	×	×	×		
Totals	7	8	10 (of 14)	4	1

*×=Contributing reason; ××=main reason
†Only 14 bids were made for installation.
Source: *Juran's Quality Control Handbook,* 4th ed., McGraw-Hill, New York, 1988, p. 3.7.

included bids for installation. This finding resulted in a revision of the process for estimating the installation price and to an improvement in the bidding/success ratio.

QUALITY LEADERSHIP AND BUSINESS STRATEGY

Among marketers there has always been a school of thought which gives quality the greatest weight among the factors which determine marketability. A survey by Hopkins and Bailey (1971) of 125 senior marketing executives as to their preference for their own product superiority showed the following:

Form of product superiority	Percent of marketing executives giving first preference to this form
Superior quality	40
Lower price (or better value)	17
More features, options, or uses	12
All others	31

Such opinions are supported by the "Profit Impact of Market Strategy" (PIMS) study, (Schoeffler, Buzzell, and Heany 1974). The PIMS study, involving 521 businesses, undertook (among other things) to relate (1) quality competitiveness to (2) share of market. The findings can be expressed as follows:

Quality versus competitors	Number of businesses in these zones of share of market		
	Under 12%	12–26%	27%+
Inferior	79	58	35
Average	51	63	53
Superior	39	55	88
Total	169	176	176

Building Quality Leadership.   Quality leadership is often the result of an original quality superiority which gains what marketers call a "prior franchise." Once gained, this franchise can be maintained through continuing product improvement and effective promotion.

Companies which have attained quality leadership have usually done so on the basis of one of two principal strategies:

- Let nature take its course. In this approach, companies apply their best efforts, hoping that in time these efforts will be recognized.

- Help nature out by adopting a positive policy—establish leadership as a formal goal and then set out to reach that goal.

Those who decide to make quality leadership a formal goal soon find that they must also answer the question: Leadership in what? Quality leadership can exist in any of the multiple aspects of fitness for use, but the focus of the company will differ depending on which aspect is chosen.

If quality leadership is to consist of:	*The company must focus on:*
Superior quality of design	Product development, systems development
Superior quality of conformance	Manufacturing quality controls
Availability	Reliability and maintainability programs
Guarantees, field services	Customer service capability

Once attained, quality leadership endures until there is clear cumulative evidence that some competitor has overtaken the leader. Lacking such evidence, the leadership can endure for decades and even centuries. However, quality leadership can also be lost through some catastrophic change.

A brewery reportedly changed its formulation in an effort to reduce costs. Within several years, its share of market declined sharply. The original formula was then restored but market share did not recover. (See "The Perils of Cutting Quality" 1982.)

In some cases, the quality reputation is built not around a specific company but around an association of companies. In that event, this association adopts and publicizes some mark or symbol. The quality reputation becomes identified with this mark, and the association goes to great lengths to protect its quality reputation.

The medieval guilds imposed strict specifications and quality controls on their members. Many medieval cities imposed "export controls" on selected finished goods in order to protect the quality reputation of the city (Juran 1995, Chapter 7).

The growth of competition in quality has stimulated the expansion of strategic business planning to include planning for quality and quality leadership. (For elaboration, see Section 13, Strategic Deployment.)

The "Market Leader" Concept.   One approach to quality leadership is through product development in collaboration with the leading user of such products—a user who is influential in the market and hence is likely to be followed. For example, in the medical field, an individual is "internationally renowned; a chairman of several scientific societies; is invited to congresses as speaker or chairman; writes numerous scientific papers" (Ollson 1986).

Determining who is the leading user requires some analysis. (In some respects the situation is similar to the marketer's problem of discovering who within the client company is the most influential in the decision to buy.) Ollson lists 10 leader types, each playing a different role.

Carryover of Failure-Prone Features.   Quality leadership can be lost by perpetuating failure-prone features of predecessor models. The guilty features are well known, since the resulting field failures keep the field service force busy restoring service. Nevertheless, there has been much carryover of failure-prone features into new models. At the least, such carryover perpetuates a sales detriment and a cost burden. At its worst, it is a cancer which can destroy seemingly healthy product lines.

A notorious example was the original xerographic copier. In that case the "top 10" list of field failure modes remained essentially identical, model after model. A similar phenomenon existed for years in the automobile industry.

The reasons behind this carryover have much in common with the chronic internal wastes which abound in so many companies:

- The alarm signals are disconnected. When wastes go on, year after year, the accountants incorporate them into the budgets. That disconnects the alarm signals—no alarms ring as long as actual waste does not exceed budgeted waste.

- There is no clear responsibility to get rid of the wastes. There are other reasons as well. The technologists have the capability to eliminate much of the carryover. However, those technologists are usually under intense pressure from the marketers to develop new product and process features in order to increase sales. In addition, they share a distaste for spending their time cleaning up old problems. In their culture, the greatest prestige comes from developing the new.

The surprising result can be that each department is carrying out its assigned responsibilities, and yet the product line is dying. Seemingly nothing short of upper management intervention—setting goals for getting rid of the carryover—can break up the impasse.

QUALITY, EARNINGS, AND STOCK MARKET PRICES

At the highest levels of management, and among boards of directors, there is keen interest in financial measures such as net income and share prices on the stock markets. It is known that quality influences these measures, but so do other variables. Separating out the effect of quality has as yet not been feasible other than through broad correlation studies.

During the early 1990s, some of the financial press published articles questioning the merits of the Malcolm Baldrige National Quality Award, Total Quality Management (TQM), and other quality initiatives. These articles were challenged, and one result was analysis of the stock price performance of Baldrige Award winners compared with that of industrial companies generally. The results were striking. From the dates of receiving the Award, the stock price of the Baldrige winners had advanced 89 percent, as compared with 33 percent for the broad Standard & Poor's index of 500 stocks (*Business Week* 1993, p. 8.)

In 1991 the General Accounting Office (GAO) published the results of a study of 20 "finalist" applicants for the Baldrige Award (companies which were site-visited). The report concluded that "In nearly all cases, companies that used total quality management practices achieved better employee relations, higher productivity, greater customer satisfaction, increased market share, and improved profitability" (General Accounting Office 1991).

LIFE CYCLE COSTING

In its simplest form, a sales contract sets out an agreed price for a specific product (goods or services), e.g., *X* cents for a glass of milk; *Y* dollars for a bus ticket. For such consumable products, the purchase price is also the cost of using the product. Drinking the milk or riding the bus normally involves no added cost for the user, beyond the original purchase price. Expressed in equation form,

$$\text{Purchase price} = \text{cost of use}$$

For long-life products, this simple equation is no longer valid. Purchase price expands to include such factors as cost of capital invested, installation cost, and deductions for resale value. Cost of use expands to include costs of operation and maintenance. It is true even for "simple" consumer products. For some articles of clothing, the cumulative costs of cleaning and maintenance can exceed the original purchase price.

The famous comedian Ed Wynn is said to have worn the same $3.50 shoes throughout his long career. They cost him $3000 in repairs.

Concept of the Optimum.   The basic concept of life-cycle costing is one of finding the optimum—finding that set of conditions which (1) meets the needs of both supplier and customer and (2) minimizes their combined costs. (Life cycle cost is only one of several names given to this concept of an optimum. Other names include: cost of ownership, cost of use or usage, mission cost, lifetime cost.)

The life cycle cost concept is widely applicable, but application has lagged. The concept can be defined in models which identify the factors to be considered, the data to be acquired, and the equations to be used in arriving at the optimum. The lag in application is not due to difficulty in setting up the models. Instead, the lag is due to inadequacies in acquiring the needed data, and especially to cultural resistance. (See below.)

Steps in Life Cycle Cost Analysis.   The literature has gone far to organize life cycle cost analyses. The steps set out below represent a typical organized approach. For elaboration on various organized approaches, see: Brook and Barasia (1977); Ebenfelt and Ogren (1974); Stokes and Stehle (1968); Toohey and Calvo (1980); Wynholds and Skratt (1977).

Identify the Life Cycle Phases.   Optimizing requires striking a balance among numerous costs, some of which are antagonistic to others. The starting point is to identify the phases or activities through which the product goes during its life cycle. These phases are mapped out in a flow diagram as an aid to the team doing the analysis. Typical phases include: product research; product development; product design; manufacturing planning; production; installation; provision of spares; operation; maintenance; support services; modifications; disposal.

Identify the Cost Elements.   The next step is to identify the cost elements for each phase. For example, operating costs for civilian aircraft include: maintenance labor and material, spares holding, delay/flight interruptions, administrative, insurance, training, flight operation, crew, aircraft and traffic servicing, fuel and oil (Rose and Phelps 1979). For an example from the Tennessee Valley Authority, see Duhan and Catlin 1973.

Acquire the Cost Data.   This step can be a formidable obstacle. The prevailing accounting systems provide only part of the essential cost information. The rest must be acquired by special study—by estimate or by enlarging the accounting system. The work involved can be reduced by concentrating on the vital few cost categories—those which involve most of the money. Attention must also be given to those categories which are highly sensitive, i.e., they are leveraged to respond to small changes in other factors—the "cost drivers."

Analyze the Relationships.   This step quantifies the interrelationship among the cost factors. For example, a comparatively simple analysis establishes that for automotive vehicles, tire wear correlates mainly with distance traveled and speed of travel. For aircraft, tire wear correlates mainly with number of landings and takeoffs.

However, many analyses are far more complex. A common example is the relationship of (1) designed mean time between failures (MTBF) and mean time to repair (MTTR) to (2) the subsequent costs of operation and maintenance Repair and Maintenance (R&M). For some products (such as certain military categories) repair and maintenance costs over the life of the product run to multiples of the original purchase price. These R&M costs are highly sensitive to the designed MTBF and MTTR. Efforts to quantify the interrelationship run into complex estimates bounded by a wide range of error. For a case example involving military avionics, see Toohey and Calvo (1980).

Formulate Aids to Decision Making.   The purpose of these analyses is to aid decision making. Typically the decision maker first establishes which categories of cost are to be included in the decision-making process. Then, on the basis of the analysis, equations are set up to arrive at the life cycle cost in terms of those same established cost categories. For example, the state of Virginia arrived at the following equation for estimating cost per hour for a certain class of highway machinery:

> Cost per hour equals initial price, plus repair parts, plus foregone interest, less resale value, all divided by operating hours (Doom 1969).

Breadth of Application.
Ideally the life cycle cost analysis should provide aid to making strategic decisions on optimizing costs. In practice this is feasible only for simple products or for problems of limited scope: state government purchase of room air conditioners (Doom 1969); optimum inventory levels (Dushman 1970); repair level strategy, i.e., discard, base repair, or depot repair (Henderson 1979); effect of test system requirements of operation and support costs (Gleason 1981); optimization of number of thermal cycles (Shumaker and DuBuisson 1976).

Probably the widest application has been in the area of industrial products. (See below, under Application to Industrial Products.)

Irrespective of the area of application, most investigators have concluded that the decisions which determine life cycle cost are concentrated in the early stages of the product life cycle. Figure 7.5 is a typical model (Björklund 1981).

Figure 7.5 shows that life cycle cost is determined mainly by decisions made during the very early phases of the life cycle. Such a concentration makes clear the need for providing the product researchers, developers, and designers with a good database on the subsequent costs of production, installation, operation, and maintenance.

Application to Consumer Products.
In a classic study, Gryna (1970) found that for various household appliances and television sets, the ratio of life cycle costs to original price ranged from 1.9 to 4.8. (See Table 7.4.)

A study, Consumer Appliances: The Real Cost (M.I.T. 1974), found the following proportions of life cycle costs to prevail during the year 1972 for color TV sets and household refrigerators:

Elements of life cycle costs	Color TV sets	Refrigerators
Purchase price	53	36
Power	12	58
Service	35	6
Total	100	100

Lund (1978) provides some supplemental information based on a follow-up study.

Fody (1977) reported on how a U.S. government agency made its first application of the life cycle cost concept to procurement of room air conditioners. The suppliers made their bids on the basis of

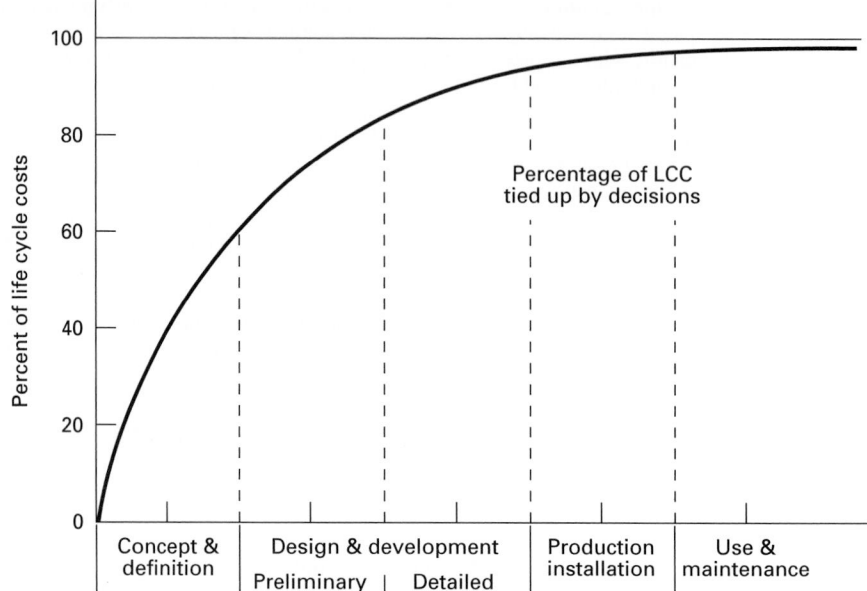

FIGURE 7.5 Phases affecting life cycle cost adapted from Björklund, 1981, p. 3. (*Juran's Quality Control Handbook, 4th ed., McGraw-Hill, New York, pp. 3.23.*)

TABLE 7.4 Life cycle Costs: Consumer Products

Product	Original price, $	Cost of operation plus maintenance, $	Total cost, $	Ratio to (life cycle cost to original price)
Room air conditioner	200	465	665	3.3
Dishwasher	245	372	617	2.5
Freezer	165	628	793	4.8
Range, electric	175	591	766	4.4
Range, gas	180	150	330	1.9
Refrigerator	230	561	791	3.5
TV (black and white)	200	305	505	2.5
TV (color)	560	526	1086	1.9
Washing machine	235	617	852	3.6

Source: *Juran's Quality Control Handbook,* 4th ed., McGraw-Hill, New York, p. 3.23.

original price. However, the agency considered in addition the expected electric power cost based on certified energy efficiency ratings. The basis for awarding the contracts then became the lowest life cycle costs rather than the lowest bid price.

Life cycle costs for automobiles have been studied in depth. Table 7.5 (Federal Highway Administration 1984) shows life cycle costs for intermediate size cars driven 120,000 mi (192,000 km) in 12 years.

Although data on life cycle costs of consumer products have become increasingly available, consumer use of such data has lagged. The major reasons include:

TABLE 7.5 Life-Cycle Costs, Automobiles

Original price	$10,320
Additional "ownership" costs	
Accessories	198
Registration	240
Titling	516
Insurance	6,691
Scheduled maintenance	1,169
Nonoperating taxes	33
Subtotal	$8,847
Operation and maintenance costs	
Gasoline	$6,651
Unscheduled maintenance	4,254
Tires	638
Oil	161
Gasoline tax, federal	514
Gasoline tax, other	771
Sales taxes	130
Parking, tolls	1,129
Subtotal	$14,248
Grand total	$33,415

Source: *Juran's Quality Control Handbook,* 4th ed.,
McGraw-Hill, New York, p. 3.24.

1. Cultural resistance (see below)
2. The economics of administering numerous small long-life contracts
3. The complexities created by multiple ownership

The most notable example of multiple ownership is passenger automobiles. In the United States, these often go through multiple ownership before being scrapped. Even under short-term warrantees, transfer of ownership creates problems of administering warrantee contracts. Existing practice usually imposes a charge for such transfer between successive owners. For contracts over the useful life of the product, this problem becomes considerably more complicated.

Application to Industrial Products.   Application to industrial products has probably been the area of greatest progress. A major example is seen in the airlines' evolution of life cycle costing strategy for aircraft maintenance. A critical element was the creation of an adequate database relative to field operation and maintenance. Data analysis then resulted in a change in grand strategy for maintenance, from the overhaul concept to the concept of on-condition maintenance. In addition, the data analysis resulted in a superior feedback to product designers and manufacturers. For an uncommonly well-documented explanation, see Nowlan and Heap (1978).

Part of the competition to sell industrial equipment consists of convincing prospective buyers that their operating and maintenance costs will be low. In some cases this conviction is created by guaranteeing the operating costs or by offering low-cost maintenance contracts. Some manufacturers provide record-keeping aids to enable users to accumulate data on competitive products as an aid to future purchasing decisions. Some industrial users build up data banks on cost of downtime for various types of industrial equipment as an input to future decision making.

The approach to making decisions to acquire capital equipment follows generally the steps set out above under the heading Steps in Life-Cycle Cost Analysis. Kaufman (1969) gives an explanation of methodology along with case examples of application.

Application to Defense Industries.   During the twentieth century many governments great-ly expanded their acquisition of military weaponry, both in volume and in complexity. Mostly the gov-ernments acquired these weapons by purchase rather than by expansion of government arsenals and shipyards. It was most desirable that the life cycle cost concept be applied to such weapons. However, a major obstacle was the deeply rooted practice of buying on the basis of the lowest bid price.

Starting in about the 1960s the U.S. Department of Defense organizations stepped up their efforts to make the life cycle cost concept effective in procurement contracts. Directives were issued to define the new emphasis and to clear away old obstacles. However, as events unfolded, it became evident that to apply the concept to government procurement was more difficult than for compara-ble situations in civilian procurement. The differences have their origin in such factors as the nature of the respective missions, the system of priorities, the organization for decision making and the extent of public scrutiny. [For a more detailed discussion of these differences, see Gansler (1974), Pedrick (1968), and Bryan (1981).]

The urge for applying the life cycle concept to military products has stimulated an extensive lit-erature. Most of the published papers relate to division of the subject and to the structure of models. [See, for example, Barasia and Kiang (1978), Peratino (1968), and Ryan (1968).]

There are also numerous papers on application. These are mainly directed at subsystems, e.g., optimizing inventory levels. Alternatively, the applications are directed at lower-level components. A published example relates to standardization of electronic modules (Laskin and Smithhisler 1979). Another example deals with standardization of test equipment (Rosenberg and Witt 1976). [See also Eustis (1977) and Gallagher and Knobloch (1971).] Application to subsystems or lower-level com-ponents obviously runs the risk of suboptimizing unless care is taken to examine the impact of any proposed change on related subsystems or components.

Cultural Resistance.   Cultural resistance is a major force holding back the application of the life cycle cost concept. Purchase based on original price has dominated commercial practice for thousands of years. The skills, habit patterns, and status of many persons—product designers, pur-chasing managers, marketers—have long been built around the original purchase price concept. Changing to life cycle costing demands a change in habit patterns, with associated risks of damage to long-standing skills and status.

The most deeply rooted habits are probably those of consumers—small buyers for personal use. They keep few records on costs of operation and maintenance and tend to underestimate the amounts. For less-than-affluent consumers, the purchase of a costly product is obscured by the fact that they may lack the capital needed even for the original price and hence must borrow part of it. In addition, the laws of sales are well worked out as applied to original price contracts but are still in evolution as applied to life cycle cost contracts.

Obviously, makers of consumer goods cannot abandon marketing on original price when such is the cultural pattern. What they can do is to experiment by offering some optional models designed for lower cost of usage as a means of gaining experience and time for the day when life cycle cost-ing comes into wider use.

Makers of industrial products also face cultural resistance in trying to use life cycle costing as a business opportunity. However, with good data they can make a persuasive case and strike respon-sive chords in buyers who see in these data a way to further the interests of their companies and themselves.

Contracts Based on Amount of Use.   An alternative approach to life cycle costing is through sales contracts which are based on the amount of use. Such contracts shift all the life cycle costs to the supplier, who then tends to redesign the system in a way which optimizes the cost of pro-viding service.

The public utilities—e.g., telephone, power—are long-standing examples. These utilities neither sell a product nor do they often even lease a product; they sell only the service (e.g., watt-hours of electricity, message units of telephone service). In such cases, the ownership of the equipment remains with the utility, which also has the responsibility for keeping the equipment maintained and repaired. As a result, the income of the utility is directly bound up with keeping the equipment in service. There are numerous other instances; e.g., the rental car is often rented based on the actual mileage driven; laundromat machines are rented on the basis of hours of use.

Sale of goods can sometimes be converted into a sale of use. It is common practice for vehicle fleets to "buy" tires based on mileage. Airlines buy engines based on hours of use. There is much opportunity for innovation in the use of this concept.

For consumer products, the metering of actual use adds many complications. Common practice is therefore to use elapsed time as an approximation of amount of use.

PERFECTIONISM

The human being exhibits an instinctive drive for precision, beauty, and perfection. When unrestrained by economics, this drive has created the art treasures of the ages. In the arts and in esthetics, this timeless human instinct still prevails.

In the industrial society, there are many situations in which this urge for perfection coincides with human needs. In food and drug preparation, certain organisms must be completely eliminated or they will multiply and create health hazards. Nuclear reactors, underground mines, aircraft, and other structures susceptible to catastrophic destruction of life require a determined pursuit of perfection to minimize dangers to human safety. So does the mass production of hazardous products.

However, there are numerous other situations in which the pursuit of perfection is antagonistic to society, since it consumes materials and energy without adding to fitness for use, either technologically or esthetically. This wasteful activity is termed "perfectionism" because it adds cost without adding value.

Perfectionism in Quality of Design.   This is often called "overdesign." Common examples include:

Long-life designs for products which will become obsolete before they wear out.

Costly finishes on nonvisible surfaces.

Tolerances or features added beyond the needs of fitness for use. (The military budget reviewers call this "gold-plating.")

Some cases of overdesign are not simple matters of yes or no. For example, in television reception there are "fringe areas" which give poor reception with conventional circuit design. For such areas, supplemental circuitry is needed to attain good quality of image. However, this extra circuitry is for many areas an overdesign and a waste. The alternative of designing an attachment to be used only in fringe areas creates other problems, since these attachments must be installed under nonfactory conditions.

Overdesign can also take place in the areas of reliability and maintainability. Examples include:

"Worst case" designs that guard against failures resulting from a highly unlikely combination of adverse conditions. Such designs can be justified in critical situations but seldom otherwise.

Use of unduly large factors of safety.

Use of heavy-duty or high-precision components for products which will be subjected to conventional usage.

Defense against overdesign is best done during design review, when the design is still fluid. The design review team commonly includes members from the functions of production, marketing, use, and customer service. Such a team can estimate the economic effects of the design. It can then challenge those design features which do not contribute to fitness for use and which therefore will add costs without adding value. Some forms of design review classify the characteristics, e.g., essential, desirable, unessential. The unessential then become prime candidates for removal.

Perfectionism in Quality of Conformance.   Typical examples include:

Insistence on conformance to specification despite long-standing successful use of nonconforming product

Setting appearance standards at levels beyond those sensed by users

One defense against this type of perfectionism is to separate two decisions which are often confused: (1) the decision on whether product conforms to specification and (2) the decision on whether nonconforming product is fit for use. Decision 1 may be relegated to the bottom of the hierarchy. Decision 2 should be made only by people who have knowledge of the conditions of use.

A further defense is to quantify the costs and then shift the burden of proof.

A marketing head insisted on overfilling the product packages beyond the label weight. The production head computed the cost of the overfill and then challenged the marketer to prove that this cost would be recovered through added sales.

The Perfectionists.   Those who advocate perfectionism often do so with the best intentions and always for reasons which seem logical to them. The resulting proposals are nevertheless of no benefit to users for one of several common reasons:

The added perfection has no value to the user. (The advocate is not aware of this.)

The added perfection has value to the user but not enough to make up for the added cost. (The advocate is unaware of the extent of the costs involved.)

The added perfection is proposed not to benefit the user but to protect the personal position of the advocate who has some functional interest in perfection but no responsibility for the associated cost.

The weaknesses of such proposals all relate back to costs: ignorance of the costs; no responsibility for the costs; indifference to costs due to preoccupation with something else. Those who do have responsibility for the costs should quantify them and then dramatize the results in order to provide the best challenge.

ACKNOWLEDGMENTS

This section has drawn extensively from the following:

Juran, J. M. (1991). *Juran on Quality by Design.* The Free Press, a division of Macmillan, New York.

Juran, J. M., ed. (1995). *A History of Managing for Quality.* Sponsored by Juran Foundation, Inc. Quality Press, Milwaukee.

Leadership for the Quality Century: A Reference Guide, 6th ed. (1996). Juran Institute, Inc., Wilton, CT.

Planning for Quality, 2nd ed. (1990). Juran Institute, Inc., Wilton, CT.

The author is grateful to the copyright holders for permission to quote from these works.

REFERENCES

Ackoff, Russell L. (1978). *The Art of Problem Solving.* John Wiley & Sons, New York.

Barasia, R. K. and Kiang, T. D. (1978). "Development of a Life Cycle Management Cost Model." *Proceedings, Annual Reliability and Maintainability Symposium.* IEEE, New York, pp. 254–260.

Björklund, O. (1981). "Life Cycle Costs; An Analysis Tool for Design, Marketing, and Service." *Q-Bulletin of Alfa-Laval.*

Brook, Cyril, and Barasia, Ramesh (1977). "A Support System Life Cycle Cost Model." *Proceedings, Annual Reliability and Maintainability Symposium.* IEEE, New York, pp. 297–302.

Bryan, N. S. (1981). "Contracting for Life Cycle Cost to Improve System Affordability." *Proceedings, Annual Reliability and Maintainability Symposium.* IEEE, New York, pp. 342–345.

Business Week (1993). "Betting to Win on the Baldie Winners." October 18.

Consumer Appliances: The Real Cost (1974). Prepared jointly by the Massachusetts Institute of Technology Center for Policy Alternatives with the Charles Stark Draper Laboratory, Inc. Sponsored by the National Science Foundation. Available from RANN Document Center, National Science Foundation, Washington, D.C. 20550.

"Consumer Guide on Tires Is Issued by U.S. Agency" (1980). *New York Times,* December 31, p. 6.

"Dishwashing Liquids" (1984). *Consumer Reports,* July, pp. 412–414.

Doom, I. F. (1969). "Total Cost Purchasing Applied to Heavy Equipment Procurement." Virginia Highway Research Council, Charlottesville.

Duhan, Stanley, and Catlin, John C., Sr. (1973). "Total Life Cost and the 'Ilities'." *Proceedings, Annual Reliability and Maintainability Symposium.* IEEE, New York, pp. 491–495.

Dushman, Allan (1970). "Effect of Reliability on Life Cycle Inventory Cost." *Proceedings, Annual Reliability and Maintainability Symposium.* IEEE, New York, pp. 549–561.

Ebenfelt, Hans, and Ogren, Stig (1974). "Some Experiences from the Use of an LCC Approach." *Proceedings, Annual Reliability and Maintainability Symposium.* IEEE, New York, pp. 142–146.

Eustis, G. E. (1977). "Reduced Support Costs for Shipboard Electronic Systems." *Proceedings, Annual Reliability and Maintainability Symposium.* IEEE, New York, pp. 316–319.

Federal Highway Administration, Office of Highway Planning, Highway Statistics Division (1984). "Cost of Owning and Operating Automobiles and Vans." Washington, D.C.

Fody, Theodore, J. (1977). "The Procurement of Window Air Conditioners Using Life-Cycle Costing." *Proceedings, Annual Reliability and Maintainability Symposium.* IEEE, New York, pp. 81–88.

Gallagher, B. M., and Knobloch, W. H. (1971). "Helicopter Auxiliary Power Unit Cost of Ownership." *Proceedings, Annual Reliability and Maintainability Symposium.* IEEE, New York, pp. 285–291.

Gansler, J. S. (1974). "Application of Life-Cycle Costing to the DOD System Acquisition Decision Process." *Proceedings, Annual Reliability and Maintainability Symposium.* IEEE, New York, pp. 147–148.

General Accounting Office (1991). "Management Practices: U.S. Companies Improve Performance Through Quality Efforts." Washington, DC.

Gleason, Daniel (1981). "The Cost of Test System Requirements." *Proceedings, Annual Reliability and Maintainability Symposium.* IEEE, New York, pp. 108–113.

Gross, Irwin (1978). "Insights from Pricing Research." In *Pricing Practices and Strategies.* E. L. Bailey (ed.). The Conference Board, New York, pp. 34–39.

Gryna, F. M., Jr. (1970). "User Costs of Poor Product Quality." (Doctoral dissertation, University of Iowa, Iowa City.) For additional data and discussion, see Gryna, F. M., Jr. (1977). "Quality Costs: User vs. Manufacturer." *Quality Progress,* June, pp. 10–13.

Henderson, J. T., P. E. (1979). "A Computerized LCC/ORLA Methodology." *Proceedings Annual Reliability and Maintainability Symposium.* IEEE, New York, pp. 51–55.

Hopkins, David S., and Bailey, Earl (1971). "New Product Pressures." *The Conference Record,* June, pp. 16–24.

Juran, J. M. (1959). "A Note on Economics of Quality." *Industrial Quality Control,* February, pp. 20–23.

Juran, J. M. (1964, 1995). *Managerial Breakthrough* (first published 1964). McGraw-Hill, New York.

Juran, J. M. (1970). "Consumerism and Product Quality." *Quality Progress,* July, pp. 18–27.

Juran, J. M. (1992). *Juran on Quality by Design.* The Free Press, a division of Macmillan, New York.

Juran, J. M. (ed.) (1995). *A History of Managing for Quality.* Quality Press, Milwaukee.

Kaufman, R. J. (1969). "Life-Cycle Costing: Decision-Making Tool for Capital Equipment Acquisitions." *Journal of Purchasing,* August, pp. 16–31.

Kegarise, R. J., and Miller, G. D. (1986). "An Alcoa-Kodak Joint Team." *The Juran Report,* no. 6, pp. 29–34.

Kegarise, R. J., Heil, M., Miller, G. D., and Miller, G. (1987). "A Supplier/Purchaser Project: From Fear to Trust." *The Juran Report,* no. 8, pp. 248–252.

Laskin, R., and Smithhisler, W. L. (1979). "The Economics of Standard Electronic Packaging." *Proceedings, Annual Reliability and Maintainability Symposium.* IEEE, New York, pp. 67–72.

Levine, Richard (1987). "Breaking Routine: Voice of the Subway." *New York Times,* January 15.

Levitt, Theodore (1960). "Marketing Myopia." *Harvard Business Review,* July-August, pp. 26–28ff.

Lund, Robert T. (1978). "Life Cycle Costing: A Business and Societal Instrument." *Management Review,* April, pp. 17–23.

M.I.T. (1974). "The Real Cost." Massachusetts Institute of Technology Study. Cambridge, MA.

Morris, Ruby Turner, and Bronson, Claire Sekulski (1969). "The Chaos of Competition Indicated by Consumer Reports." *Journal of Marketing,* July, pp. 26–34.

Nowlan, F. S., and Heap, H. F. (1978). *Reliability Centered Maintenance.* Document ADA066579. Defense Documentation Center, Alexandria, VA.

Ollson, John Ryding (1986). "The Market-Leader Method; User-Oriented Development." *Proceedings 30th EOQC Annual Conference,* pp. 59–68.

Pedrick, P. C. (1968). "Survey of Life-Cycle Costing Practices of Non-Defense Industry." *Proceedings, Annual Reliability and Maintainability Symposium.* IEEE, New York, pp. 188–192.

Peratino, G. S. (1968). "Air Force Approach to Life Cycle Costing." *Proceedings, Annual Reliability and Maintainability Symposium.* IEEE, New York, pp. 184–187.

"The Perils of Cutting Quality." (1982). *New York Times,* August 22.

PIMS (1978). *Pricing High-Quality Products.* The Strategic Planning Institute, Cambridge, MA.

"Pricing of Products Is Still an Art." (1981). *Wall Street Journal,* November 25, pp. 25, 33.

Riesz, Peter C. (1979). "Price-Quality Correlations for Packaged Food Products." *Journal of Consumer Affairs.*

Rose, John, and Phelps, E. L. (1979). "Cost of Ownership Application to Airplane Design." *Proceedings, Annual Reliability and Maintainability Symposium,* IEEE, New York, pp. 47–50.

Rosenberg, H., and Witt, J. H. (1976). "Effects on LCC of Test Equipment Standardization." *Proceedings, Annual Reliability and Maintainability Symposium.* IEEE, New York, pp. 287–292.

Ryan, W. J. (1968). "Procurement Views of Life-Cycle Costing." *Proceedings, Annual Reliability and Maintainability Symposium.* IEEE, New York, pp. 164–168.

Schoeffler, Sidney, Buzzell, Robert D., and Heany, Donald F. (1974). "Impact of Strategic Planning on Profit Performance." *Harvard Business Review,* March-April, pp. 137–145.

Shumaker, M. J., and DuBuisson, J. C. (1976). "Tradeoff of Thermal Cycling vs. Life Cycle Costs." *Proceedings, Annual Reliability and Maintainability Symposium.* IEEE, New York, pp. 300–305.

Smith, Lee (1995). "Rubbermaid Goes Thump." *Fortune,* October 2, pp. 90–104.

Stokes, R. G., and Stehle, F. N. (1968). "Some Life-Cycle Cost Estimates for Electronic Equipment: Methods and Results." *Proceedings, Annual Reliability and Maintainability Symposium,* IEEE, New York, pp. 169–183.

Toohey, Edward F., and Calvo, Alberto B. (1980). "Cost Analyses for Avionics Acquisition." *Proceedings, Annual Reliability and Maintainability Symposium.* IEEE, New York, pp. 85–90.

United States Office of Consumer Affairs (1985–86). "Consumer Complaint Handling in America; An Update Study." Washington, DC, 20201.

Wynholds, Hans W., and Skratt, John P. (1977). "Weapon System Parametric Life-Cycle Cost Analysis." *Proceedings, Annual Reliability and Maintainability Symposium.* IEEE, New York, pp. 303–309.

SECTION 8
QUALITY AND COSTS

Frank M. Gryna

INTRODUCTION

This section discusses how quality has an impact on the costs of goods and services in an organization. Section 7, Quality and Income, addresses the issue of quality and sales revenue. Thus, the two sections provide a framework of how quality is related to the total financial picture of an organization.

We identify and measure the costs associated with poor quality for three reasons: to quantify the size of the quality problem to help justify an improvement effort, to guide the development of that effort, and to track progress in improvement activities. Among the concepts and methodologies covered are traditional categories of quality costs, a broadened concept of categories including lost revenue and process capability costs, activity-based costing, data collection methods, return on quality, presentation of findings, gaining approval for an improvement effort, using cost data to support continuous improvement, optimum quality level, and reporting cost data. The underlying theme in the section is the use of quality-related costs to support a quality improvement effort rather than as a system of reporting quality costs.

We will follow the convention of using the term "product" to denote goods or services.

EVOLUTION OF QUALITY AND COSTS

During the 1950s there evolved numerous quality-oriented staff departments. The heads of these new departments were faced with "selling" their activities to the company managers. Because the main language of those managers was money, the concept of studying quality-related costs provided the vocabulary to communicate between the quality staff departments and the company managers.

Over the decades, as the staff quality specialists extended their studies, some surprises emerged:

1. The quality-related costs were much larger than had been shown in the accounting reports. For most companies, these costs ran in the range of 10 to 30 percent of sales or 25 to 40 percent of operating expenses. Some of these costs were visible, some of them were hidden.

2. The quality costs were not simply the result of factory operation, the support operations were also major contributors.

3. The bulk of the costs were the result of poor quality. Such costs had been buried in the standards, but they were in fact avoidable.

4. While these quality costs were avoidable, there was no clear responsibility for action to reduce them, neither was there any structured approach for doing so.

Quality specialists used the data to help justify quality improvement proposals and to track the cost data over time.

Those early decades of experience led to some useful lessons learned.

Lessons Learned.   These lessons, discussed below, can help us to formulate objectives for tracking and analyzing the impact of quality on costs.

The Language of Money Is Essential.   Money is the basic language of upper management. Despite the prevalence of estimates, the figures provide upper managers with information showing the overall size of the quality costs, their prevalence in areas beyond manufacture, and the major areas for potential improvement.

Without the quality cost figures, the communication of such information to upper managers is slower and less effective.

The Meaning of "Quality Costs."   The term "quality costs" has different meanings to different people. Some equate "quality costs" with the costs of poor quality (mainly the costs of finding and correcting defective work); others equate the term with the costs to attain quality; still others use the term to mean the costs of running the Quality department. In this handbook, the term "quality costs" means the cost of poor quality.

Quality Cost Measurement and Publication Does Not Solve Quality Problems.   Some organizations evaluate the cost of poor quality and publish it in the form of a scoreboard in the belief that publication alone will stimulate the responsible managers to take action to reduce the costs. These efforts have failed. The realities are that publication alone is not enough. It makes no provision to identify projects, establish clear responsibilities, provide resources to diagnose and remove causes of problems, or take other essential steps. New organization machinery is needed to attack and reduce the high costs of poor quality (see, generally, Section 5, The Quality Improvement Process).

Scoreboards, if properly designed, can be a healthy stimulus to competition among departments, plants, and divisions. To work effectively, the scoreboard must be supplemented by a structured improvement program. In addition, scoreboards must be designed to take into account inherent differences in operations among various organizational units. Otherwise, comparisons made will become a source of friction.

Scope of Quality Costs Is Too Limited.   Traditionally, the measurement of quality cost focuses on the cost of nonconformities, i.e., defects in the goods or services delivered to external and internal customers. These are often called external and internal failure costs. An important cost that is not measured is lost sales due to poor quality (this is called a "hidden cost" because it is not easily measured). Another omitted cost is the extra cost in processes that were producing conforming output but which are inefficient. These inefficiencies are due to excess product or process variability (even though within specification limits) or inefficiencies due to redundant or non-value-added process steps.

Traditional Categories of Quality Costs Have Had a Remarkable Longevity. About 1945, a pioneering effort proposed that quality-related costs be assigned to one of three categories: failure costs, appraisal costs, and prevention costs. The pioneers emphasized that these categories were not the only way to organize quality costs; the important point was to obtain a credible estimate of the total quality cost. But many practitioners found the categories useful and even found ingenious ways to adapt the categories to special applications such as engineering design.

The experience that led to these lessons learned also included some changes in the quality movement:

1. An explosion in the acceptance of the concept of continuous improvement in all sectors—profit, nonprofit, and public.

2. Progress in understanding and in quantifying the impact of quality on sales revenue.

3. Emphasis on examining cross-functional processes to reduce errors and cycle time and improve process capability to increase customer satisfaction. These analyses confirm the benefits of (*a*) diagnosis of causes to reduce errors and (*b*) process analysis to identify redundant work steps and other forms of non-value-added activities.

From the lessons learned and the changes in the quality movement, we can identify some objectives for evaluating quality and costs.

Objectives of Evaluation. The primary objectives are

1. Quantify the size of the quality problem in language that will have an impact on upper management. The language of money improves communication between middle managers and upper managers. Some managers say: "We don't need to spend time to translate the defects into dollars. We realize that quality is important, and we already know what the major problems are." Typically when the study is made, these managers are surprised by two results. First, the quality costs turn out to be much higher than had been realized. In many industries they are in excess of 20 percent of sales. Second, while the distribution of the quality costs confirms some of the known problem areas, it also reveals other problem areas that had not previously been recognized.

2. Identify major opportunities for reduction in cost of poor quality throughout all activities in an organization. Costs of poor quality do not exist as a homogeneous mass. Instead, they occur in specific segments, each traceable to some specific cause. These segments are unequal in size, and a relative few of the segments account for the bulk of the costs. A major byproduct of evaluation of costs of poor quality is identification of these vital few segments. This results in setting priorities to assure the effective use of resources. We need to collect data on the cost of poor quality, analyze the data, and plan an improvement strategy that attacks chunks of the glacier rather than ice chips.

3. Identify opportunities for reducing customer dissatisfaction and associated threats to sales revenues. Some costs of poor quality are the result of customer dissatisfaction with the goods or service provided. This dissatisfaction results in a loss of current customers—"customer defections"—and an inability to attract new customers (for elaboration see Section 18 under Linking Customer Satisfaction Results to Customer Loyalty and to Processes). Addressing the areas of dissatisfaction helps to improve retention of current customers and create new customers.

4. Provide a means of measuring the result of quality improvement activities instituted to achieve the opportunities in 2 and 3 above. Measuring progress helps to keep a focus on improvement and also spotlights conditions that require removal of obstacles to improvements.

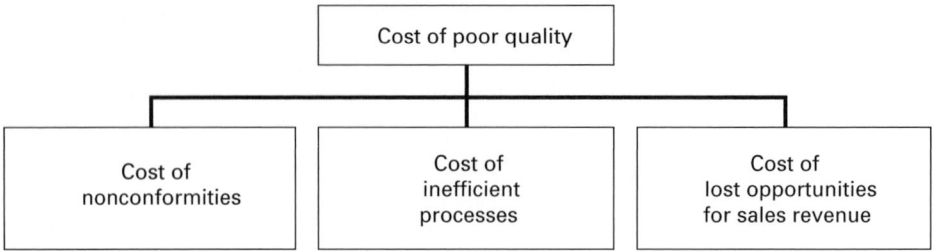

FIGURE 8.1   Components of the cost of poor quality.

5. Align quality goals with organization goals. Measuring the cost of poor quality is one of four key inputs for assessing the current status of quality (the others are market standing on quality relative to competition, the organization quality culture, and the activities composing the quality system). Knowing the cost of poor quality (and the other elements) leads to the development of a strategic quality plan consistent with overall organization goals.

Collectively, these objectives strive to increase the value of product and process output and enhance customer satisfaction. This section uses the framework shown in Figure 8.1. Note that this framework extends the traditional concept of quality costs to reflect not only the costs of nonconformities but also process inefficiencies and the impact of quality on sales revenue. Sometimes, the term "economics of quality" is employed to describe the broader concept and differentiate it from the traditional concept of "quality cost."

We must emphasize the main objective in collecting this data, i.e., to energize and support quality improvement activities. This is summarized in Figure 8.2. The term "cost of quality" used in this figure includes the prevention, appraisal, and failure categories which are discussed below.

CATEGORIES OF QUALITY COSTS

Many companies summarize these costs into four categories. Some practitioners also call these categories the "cost of quality." These categories and examples of typical subcategories are discussed below.

Internal Failure Costs.   These are costs of deficiencies discovered before delivery which are associated with the failure (nonconformities) to meet explicit requirements or implicit needs of external or internal customers. Also included are avoidable process losses and inefficiencies that occur even when requirements and needs are met. These are costs that would disappear if no deficiencies existed.

Failure to Meet Customer Requirements and Needs.   Examples of subcategories are costs associated with:

Scrap: The labor, material, and (usually) overhead on defective product that cannot economically be repaired. The titles are numerous—scrap, spoilage, defectives, etc.

Rework: Correcting defectives in physical products or errors in service products.

Lost or missing information: Retrieving information that should have been supplied.

Failure analysis: Analyzing nonconforming goods or services to determine causes.

Scrap and rework—supplier: Scrap and rework due to nonconforming product received from suppliers. This also includes the costs to the buyer of resolving supplier quality problems.

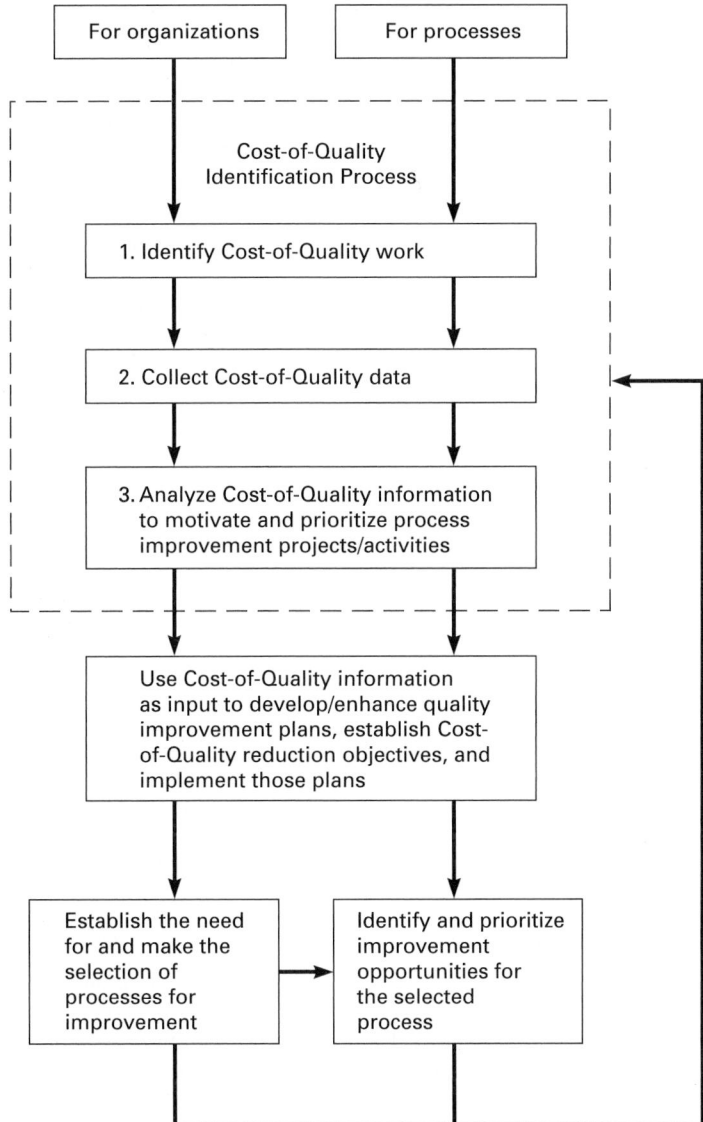

FIGURE 8.2   Cost of quality and quality improvement. (*AT&T 1990, p. 16.*)

One hundred percent sorting inspection: Finding defective units in product lots which contain unacceptably high levels of defectives.

Reinspection, retest: Reinspection and retest of products that have undergone rework or other revision.

Changing processes: Modifying manufacturing or service processes to correct deficiencies.

Redesign of hardware: Changing designs of hardware to correct deficiencies.

Redesign of software: Changing designs of software to correct deficiencies.

Scrapping of obsolete product: Disposing of products that have been superseded.

Scrap in support operations: Defective items in indirect operations.

Rework in internal support operations: Correcting defective items in indirect operations.

Downgrading: The difference between the normal selling price and the reduced price due to quality reasons.

Cost of Inefficient Processes.   Examples of subcategories are

Variability of product characteristics: Losses that occur even with conforming product (e.g., overfill of packages due to variability of filling and measuring equipment).

Unplanned downtime of equipment: Loss of capacity of equipment due to failures.

Inventory shrinkage: Loss due to the difference between actual and recorded inventory amounts.

Variation of process characteristics from "best practice":   Losses due to cycle time and costs of processes as compared to best practices in providing the same output. The best-practice process may be internal or external to the organization.

Non-value-added activities: Redundant operations, sorting inspections, and other non-value-added activities. A value-added activity increases the usefulness of a product to the customer; a non-value-added activity does not. (The concept is similar to the 1950s concept of value engineering and value analysis.)

For elaboration on variation and the cost of quality, see Reeve (1991). For a discussion of waste in "white collar" processes, see Quevedo (1991).

External Failure Costs.   These are costs associated with deficiencies that are found after product is received by the customer. Also included are lost opportunities for sales revenue. These costs also would disappear if there were no deficiencies.

Failure to Meet Customer Requirements and Needs.   Examples of subcategories are

Warranty charges: The costs involved in replacing or making repairs to products that are still within the warranty period.

Complaint adjustment: The costs of investigation and adjustment of justified complaints attributable to defective product or installation.

Returned material: The costs associated with receipt and replacement of defective product received from the field.

Allowances: The costs of concessions made to customers due to substandard products accepted by the customer as is or to conforming product that does not meet customer needs.

Penalties due to poor quality: This applies to goods or services delivered or to internal processes such as late payment of an invoice resulting in a lost discount for paying on time.

Rework on support operations: Correcting errors on billing and other external processes.

Revenue losses in support operations: An example is the failure to collect on receivables from some customers.

Lost Opportunities for Sales Revenue.   Examples are

Customer defections: Profit margin on current revenue lost due to customers who switch for reasons of quality. An important example of this category is current contracts that are canceled due to quality.

New customers lost because of quality: Profit on potential customers lost because of poor quality.

New customers lost because of lack of capability to meet customer needs: Profit on potential revenue lost because of inadequate processes to meet customer needs.

Appraisal Costs.   These are the costs incurred to determine the degree of conformance to quality requirements. Examples are

Incoming inspection and test: Determining the quality of purchased product, whether by inspection on receipt, by inspection at the source, or by surveillance.

In-process inspection and test: In-process evaluation of conformance to requirements.

Final inspection and test: Evaluation of conformance to requirements for product acceptance.

Document review: Examination of paperwork to be sent to customer.

Balancing: Examination of various accounts to assure internal consistency.

Product quality audits: Performing quality audits on in-process or finished products.

Maintaining accuracy of test equipment: Keeping measuring instruments and equipment in calibration.

Inspection and test materials and services: Materials and supplies in inspection and test work (e.g., x-ray film) and services (e.g., electric power) where significant.

Evaluation of stocks: Testing products in field storage or in stock to evaluate degradation.

In collecting appraisal costs, what is decisive is the kind of work done and not the department name (the work may be done by chemists in the laboratory, by sorters in Operations, by testers in Inspection, or by an external firm engaged for the purpose of testing). Also note that industries use a variety of terms for "appraisal," e.g., checking, balancing, reconciliation, review.

Prevention Costs.   These are costs incurred to keep failure and appraisal costs to a minimum. Examples are

Quality planning: This includes the broad array of activities which collectively create the overall quality plan and the numerous specialized plans. It includes also the preparation of procedures needed to communicate these plans to all concerned.

New-products review: Reliability engineering and other quality-related activities associated with the launching of new design.

Process planning: Process capability studies, inspection planning, and other activities associated with the manufacturing and service processes.

Process control: In-process inspection and test to determine the status of the process (rather than for product acceptance).

Quality audits: Evaluating the execution of activities in the overall quality plan.

Supplier quality evaluation: Evaluating supplier quality activities prior to supplier selection, auditing the activities during the contract, and associated effort with suppliers.

Training: Preparing and conducting quality-related training programs. As in the case of appraisal costs, some of this work may be done by personnel who are not on the payroll of the Quality department. The decisive criterion is again the type of work, not the name of the department performing the work.

Note that prevention costs are costs of special planning, review, and analysis activities for quality. Prevention costs do *not* include basic activities such as product design, process design, process maintenance, and customer service.

The compilation of prevention costs is initially important because it highlights the small investment made in prevention activities and suggests the potential for an increase in prevention costs with the aim of reducing failure costs. The author has often observed that upper management immediately grasps this point and takes action to initiate an improvement effort. Experience also suggests, however, that continuing measurement of prevention costs can usually be excluded in order to (1) focus on the major opportunity, i.e., failure costs, and (2) avoid the time spent discussing what should be counted as prevention costs.

This part of the section focuses on the question "How much is it costing our organization by not doing a good job on quality?" Thus we will use the term "cost of poor quality." Most (but not all) of the total of the four categories is the cost of poor quality (clearly, prevention costs are not a cost of poor quality.) Strictly defined, the cost of poor quality is the sum of internal and external failure costs categories. But this assumes that those elements of appraisal costs—e.g., 100 percent sorting inspection or review—necessitated by inadequate processes are classified under internal failures. This emphasis on the cost of poor quality is related to a later focus in the section, i.e., quality improvement, rather than just quality cost measurement.

A useful reference on definitions, categories, and other aspects is Campanella (1999). For an exhaustive listing of elements within the four categories see Atkinson, Hamburg, and Ittner (1994). Winchell (1991) presents a method for defining quality cost terms directly in the language used by an organization.

Example from Manufacturing Sector.   An example of a study for a tire manufacturer is shown in Table 8.1.

Some conclusions are typical for these studies:

1. The total of almost $900,000 per year is large.
2. Most (79.1 percent) of the total is concentrated in failure costs, specifically in "waste-scrap" and consumer adjustments.
3. Failure costs are about 5 times the appraisal costs. Failure costs must be attacked first.
4. A small amount (4.3 percent) is spent on prevention.

TABLE 8.1   Annual Quality Cost—Tire manufacturer

1. Cost of quality failures—losses		
a. Defective stock	$ 3,276	0.37
b. Repairs to product	73,229	8.31
c. Collect scrap	2,288	0.26
d. Waste-scrap	187,428	21.26
e. Consumer adjustments	408,200	46.31
f. Downgrading products	22,838	2.59
g. Customer ill will	Not counted	
h. Customer policy adjustment	Not counted	
Total	$697,259	79.10%
2. Cost of appraisal		
a. Incoming inspection	$ 23,655	2.68
b. Inspection 1	32,582	3.70
c. Inspection 2	25,200	2.86
d. Spot-check inspection	65,910	7.37
Total	$147,347	16.61%
3. Cost of prevention		
a. Local plant quality control engineering	$ 7,848	0.89
b. Corporate quality control engineering	30,000	3.40
Total	$ 37,848	4.29%
Grand total	$882,454	100.00%

5. Some consequences of poor quality could not be quantified, e.g., "customer ill will" and "customer policy adjustment." Here, the factors were listed as a reminder of their existence.

As a result of this study, management decided to increase the budget for prevention activities. Three engineers were assigned to identify and pursue specific quality improvement projects.

Example from Service Sector. Table 8.2 shows a monthly quality cost report for an installment loan process at one bank. Only those activities that fall into the four categories of quality cost are shown. The monthly cost of quality of about $13,000 for this *one* process is equivalent to about $160,000 per year for this bank. Table 8.2 quantifies loan costs throughout a typical consumer loan life cycle (including loan payoff). Note that the internal and external failure costs account for about half of total quality costs. These failure costs—which are preventable—are now tracked and managed for reduction or elimination so that they do not become unintentionally built into the operating structure. From the moment the customer contacts the bank with a problem, all costs related to resolving the problem are external failure costs. (Also note the significant amount of appraisal cost.)

These two examples illustrate studies at the plant level and the process level, but studies can be conducted at other levels, e.g., corporate, division, plant, department, process, product, component, or on a specific problem. Studies made at higher levels are typically infrequent, perhaps annual. Increasingly, studies are conducted as part of quality improvement activities on one process or one problem and then the frequency is guided by the needs of the improvement effort. For further discussion, see below under Using Cost of Poor Quality Concept to Support Quality Improvement.

The concept of cost of poor quality applies to a gamut of activities. For examples from manufacturing see Finnegan and Schottmiller (1990) and O'Neill (1988). In the service sector, useful discussions are provided as applied to hotels (Bohan and Horney 1991) and for educational testing (Wild and Kovacs

TABLE 8.2 Quality Cost Report—Installment Loans

Operation	Prevention	Appraisal	Internal failure	External failure
Making a loan:				
Run credit check	0	0	26	0
Process GL tickets and I/L input sheets	0	0	248	0
Review documents	0	3014	8	0
Make document corrections	0	0	1014	0
Follow up on titles, etc.	0	157	0	0
Review all output	0	2244	0	0
Correct rejects and incorrect output	0	0	426	0
Correct incomplete collateral report	0	0	0	78
Work with dealer on problems	0	0	0	2482
I/L system downtime	0	0	520	0
Time spent training on I/L	1366	0	0	0
Loan payment:				
Receive and process payments	0	261	784	0
Respond to inquiries when no coupon is presented with payments	0	0	784	0
Loan payoff:				
Process payoff and release document	0	0	13	0
Research payoff problems	0	0	13	0
Total cost of quality (COQ)	1366	5676	3836	2560
COQ as % of total quality cost	10.2	42.2	28.5	19.1
COQ as % of reported salary expense (25.6%)	2.6	10.8	7.3	4.9

Source: Adapted from Aubrey (1988).

1994). Applications have also been made to functional areas. For marketing, see Carr (1992) and Nickell (1985); for engineering see Schrader (1986); for "white collar" see Hou (1992) and Keaton et al. (1988). The Conference Board (1989) presents the results of a survey of 111 companies (manufacturing and service) on current practices in measuring quality costs.

Finalizing the Definitions.   Although many organizations have found it useful to divide the overall cost into the categories of internal failure, external failure, appraisal, and prevention, the structure may not apply in all cases. Clearly, the practitioner should choose a structure that suits company need. In defining the cost of poor quality for a given organization, the following points should be kept in mind.

1. The definitions should be tailor-made for each organization. The usual approach is to review the literature and select those detailed categories which apply to the organization. The titles used should meet the needs of the organization, not the literature. This selected list is then discussed with the various functions to identify additional categories, refine the wording, and decide on broad groupings, if any, for the costs. The resulting definitions are "right" for the organization.

2. The key categories are the failure cost elements because these provide the major opportunity for reduction in costs and for removal of the causes of customer dissatisfaction. These costs should be attacked first. Appraisal costs are also an area for reduction, especially if the causes of the failures are identified and removed so as to reduce the need for appraisal.

3. Agreement should be reached on the categories of cost to include before any data are collected. Upper management should be a party to this agreement. Initially, summarized data on scrap and rework can gain management's attention and stimulate the need for a full study. Such summaries can be an impetus for management to become personally involved, for example, by calling and chairing the meetings to finalize the definition of the cost of poor quality. The quality specialist and the accountant both have key roles.

4. Certain costs routinely incurred may have been accepted as inevitable but are really part of the cost of poor quality. Examples are the costs of redesigning the product made necessary by deficiencies in fitness for use and the costs of changing the manufacturing process because of an inability to meet product specifications. If the original design and original manufacturing plans had been adequate, these costs would not have occurred. Typically, these costs have been accepted as normal operating costs, but should be viewed as opportunities for improvement and subsequent cost reduction.

5. As the detailed categories of the cost of poor quality are identified, some categories will be controversial. Much of the controversy centers around the point: "These are not quality-related costs but costs that are part of normal operating expenses and therefore should not be included." Examples are inclusion of full overhead in calculating scrap costs, preventive maintenance, and loss in morale.

In most companies, the cost of poor quality is a large sum, frequently larger than the company's profits. This is true even when the controversial categories are not included, so it is prudent to omit these categories and avoid the controversy in order to focus attention on the major areas of potential cost reduction. Some efforts to quantify quality costs have failed because of tenacious insistence by some specialists that certain controversial categories be included. A useful guide is to ask: "Suppose all defects disappeared. Would the cost in question also disappear?" A "yes" answer means that the cost is associated with quality problems and therefore should be included. A "no" answer means that the category should not be included in the cost of poor quality.

At the minimum, controversial categories should be separated out of the totals so that attention will be directed to the main issues, i.e., the failure costs.

Hidden Costs.   The cost of poor quality may be understated because of costs which are difficult to estimate. The "hidden" costs occur in both manufacturing and service industries and include:

1. Potential lost sales (see above under External Failure Costs).
2. Costs of redesign of products due to poor quality.

3. Costs of changing processes due to inability to meet quality requirements for products.
4. Costs of software changes due to quality reasons.
5. Costs of downtime of equipment and systems including computer information systems.
6. Costs included in standards because history shows that a certain level of defects is inevitable and allowances should be included in standards:
   a. Extra material purchased: The purchasing buyer orders 6 percent more than the production quantity needed.
   b. Allowances for scrap and rework during production: History shows that 3 percent is "normal" and accountants have built this into the cost standards. One accountant said, "Our scrap cost is zero. The production departments are able to stay within the 3 percent that we have added in the standard cost and therefore the scrap cost is zero." Ah, for the make-believe "numbers game."
   c. Allowances in time standards for scrap and rework: One manufacturer allows 9.6 percent in the time standard for certain operations to cover scrap and rework.
   d. Extra process equipment capacity: One manufacturer plans for 5 percent unscheduled downtime of equipment and provides extra equipment to cover the downtime. In such cases, the alarm signals ring only when the standard value is exceeded. Even when operating within those standards, however, the costs should be a part of the cost of poor quality. They represent opportunities for improvement.
7. Extra indirect costs due to defects and errors. Examples are space charges and inventory charges.
8. Scrap and errors not reported. One example is scrap that is never reported because of fear of reprisals, or scrap that is charged to a general ledger account without an identification as scrap.
9. Extra process costs due to excessive product variability (even though within specification limits): For example, a process for filling packages with a dry soap mix meets requirements for label weight on the contents. The process aim, however, is set above label weight to account for variability in the filling process. See Cost of Inefficient Processes above under Internal Failure Costs.
10. Cost of errors made in support operations, e.g., order filling, shipping, customer service, billing.
11. Cost of poor quality within a supplier's company. Such costs are included in the purchase price.

These hidden costs can accumulate to a large amount—sometimes three or four times the reported failure cost. Where agreement can be reached to include some of these costs, and where credible data or estimates are available, then they should be included in the study. Otherwise, they should be left for future exploration.

Progress has been made in quantifying certain hidden costs, and therefore some of them have been included in the four categories discussed above. Obvious costs of poor quality are the tip of the iceberg.

Atkinson et al. (1991) trace the evolution of the cost of quality, present research results from four organizations (manufacturing and service), and explain how cost of quality data is applied in continuous improvement programs.

International Standards and Quality Costs.   The issue of quality costs is addressed in ISO 9004-1 (1994), *Quality Management and Quality System Elements—Guidelines,* Section 6, "Financial Considerations of Quality Systems." This standard is advisory rather than mandatory. Three approaches to data collection and reporting are identified (but others are not excluded):

1. *Quality costing approach:*   This is the failure, appraisal, and prevention approach described above.
2. *Process cost approach.*   This approach collects data for a process rather than a product. All process costs are divided into cost of conformity and cost of nonconformity. The cost of conformity includes *all* costs incurred to meet stated and implied need of customers. Note that this is the cost incurred when a process is running without failure, i.e., material, labor, and overhead including prevention and process control activities. This cost includes process inefficiencies. The cost of nonconformity is the traditional cost of internal and external failures. The focus is to reduce both the cost of conformity and the cost of nonconformity.

3. *Quality loss approach:*   This approach includes, but goes beyond, internal and external failure costs. Conceptually it tries to collect data on many of the "hidden" costs such as loss of sales revenue due to poor quality, process inefficiencies, and losses when a quality characteristic deviates from a target value even though it is within specification limits. Under this approach the costs can be estimated by using the Taguchi quality loss function.

For a comparison of these three approaches, see Schottmiller (1996). To provide further guidance, Technical Committee 176 of the International Organization for Standardization is developing a document, ISO/CD 10014, *Guideline for Managing the Economics of Quality.* This document will address both costs and customer satisfaction and will apply to "for profit" and "not for profit" organizations. Shepherd (1998) reviews the experiences with quality costs of over 50 organizations that successfully implemented ISO 9000.

MAKING THE INITIAL COST STUDY

A study of the cost of poor quality is logically made by the accountant, but the usual approach follows a different scenario. A quality manager learns about the quality cost concept and speaks with the accountant about making a study. The accountant responds that "the books are not kept that way." The accountant does provide numbers on scrap, rework, or certain other categories, but is not persuaded to define a complete list of categories and collect the data. The quality manager then follows one of two routes: (1) unilaterally prepares a definition of the categories and collects data or (2) presents to upper management the limited data provided by the accountant, and recommends that a full study be made using the resources of Accounting, Quality, and other functions. The second approach is more likely to achieve acceptance of the results of the study.

Sequence of Events.   The following sequence applies to most organizations.

1. Review the literature on quality costs. Consult others in similar industries who have had experience with applying quality cost concepts.
2. Select one organizational unit of the company to serve as a pilot site. This unit may be one plant, one large department, one product line, etc.
3. Discuss the objectives of the study with the key people in the organization, particularly those in the accounting function. Two objectives are paramount: determine the size of the quality problem and identify specific projects for improvement.
4. Collect whatever cost data are conveniently available from the accounting system and use this information to gain management support to make a full cost study.
5. Make a proposal to management for a full study. The proposal should provide for a task force of all concerned parties to identify the work activities that contribute to the cost of poor quality. Work records, job descriptions, flowcharts, interviews, and brainstorming can be used to identify the activities.
6. Publish a draft of the categories defining the cost of poor quality. Secure comments and revise.
7. Finalize the definitions and secure management approval.
8. Secure agreement on responsibility for data collection and report preparation.
9. Collect and summarize the data. Ideally, this should be done by Accounting.
10. Present the cost results to management along with the results of a demonstration quality improvement project (if available). Request authorization to proceed with a broader company-wide program of measuring the costs and pursuing projects. See below under Gaining Approval for the Quality Improvement Program.

Clearly, the sequence must be tailored for each organization.

The costs associated with poor quality typically span a variety of departments (see Figure 8.3), and thus it is important to plan for this in data collection.

Data Collection.   The initial study collects cost data by several approaches:

1. *Established accounts:*   Examples are appraisal activities conducted by an Inspection department and warranty expenses to respond to customer problems.
2. *Analysis of ingredients of established accounts:*   For example, suppose an account called "customer returns" reports the cost of all goods returned. Some of the goods are returned because they are defective. Costs associated with these are properly categorized as "cost of poor quality." Other goods may be returned because the customer is reducing inventory. To distinguish the quality costs from the others requires a study of the basic return documents.
3. *Basic accounting documents:*   For example, some product inspection is done by Production department employees. By securing their names and the associated payroll data, we can quantify these quality costs.
4. *Estimates:*   Input from knowledgeable personnel is clearly important. In addition, several approaches may be needed.
    a. *Temporary records:*   For example, some production workers spend part of their time repairing defective product. It may be feasible to arrange with their supervisor to create a temporary record to determine the repair time and thereby the repair cost. This cost can then be projected for the time period to be covered by the study.
    b. *Work sampling:*   Here, random observations of activities are taken and the percent of time spent in each of a number of predefined categories can then be estimated (see Esterby 1984). In one approach, employees are asked to record the observation as prevention, appraisal, failure, or first time work (AT&T 1990, p. 35).
    c. *Allocation of total resources:*   For example, in one of the engineering departments, some of the engineers are engaged part time in making product failure analyses. The department, however, makes no provision for charging engineering time to multiple accounts. Ask each engineer to make an estimate of time spent on product failure analysis by keeping a temporary activity log for several representative weeks. As the time spent is due to a product failure, the cost is categorized as a failure cost.
    d. *Unit cost data:*   Here, the cost of correcting one error is estimated and multiplied by the number of errors per year. Examples include billing errors and scrap. Note that the unit cost per error may consist of costs from several departments.
    e. *Market research data:*   Lost sales revenue due to poor quality is part of the cost of poor quality. Although this revenue is difficult to estimate, market research studies on customer satisfaction and loyalty can provide input data on dissatisfied customers and customer defections.

Cost-of-poor-quality activity	Location in organization				
	Dept. A	Dept. B	Dept. C	Dept. D	Etc.
Discover status of late order	•	•			
Correct erroneous bills	•		•		
Expedite installation of late shipment	•			•	
Troubleshoot failures on installation		•	•	•	
Perform warranty repairs	•		•	•	
Dispose of scrap		•			
Replace unacceptable installation	•		•	•	
Etc.					

FIGURE 8.3   Costs of poor quality across departments. (*Romagnole and Williams 1995.*)

Calculations can then be made to estimate the lost revenue. Table 8.3 shows a sample case from the banking industry. Note that this approach starts with annual revenue and does not take into account the loss over the duration of years that the customer would have been loyal to the company. For a more comprehensive calculation, see Section 18 under Linking Customer Satisfaction Results to Customer Loyalty Analysis and to Processes. Also note that the calculations in Table 8.3 do not consider the portion of satisfied customers who will be wooed away by competition.

A special problem is whether all of the defective product has been reported. An approach to estimate the total amount of scrap is the input-output analysis. A manufacturer of molded plastic parts followed this approach:

1. Determine (from inventory records) the pounds of raw material placed in the manufacturing process.
2. Determine the pounds of finished goods shipped. If necessary, convert shipments from units to pounds.
3. Calculate the overall loss as step 1 minus step 2.
4. Make subtractions for work in process and finished goods inventory.

The result is the amount of raw material unaccounted for and presumably due to defective product (or other unknown reasons). A comparison of this result and the recorded amount of defectives provides a practical check on the recorded number of defectives.

Another special problem is the rare but large cost, e.g., a product liability cost. Such costs can be handled in two ways: (1) report the cost in a special category separating it from the total for the other categories or (2) calculate an expected cost by multiplying the probability of occurrence of the unlikely event by the cost if the event does occur.

A common mistake in data collection is the pursuit of great precision. This is not necessary—it is a waste of resources and time. We determine the cost of poor quality in order to justify and support quality improvement activities and identify key problem areas. For that purpose, a precision of ±20 percent is adequate in determining the cost of poor quality.

Briscoe and Gryna (1996) discuss the categories and data collection as applied to small business.

COST OF POOR QUALITY AND ACTIVITY-BASED COSTING

One of the issues in calculating the costs of poor quality is how to handle overhead costs. Three approaches are used in practice: include total overhead using direct labor or some other base, include

TABLE 8.3   Revenue Lost through Poor Quality

$10,000,000	Annual customer service revenue
1,000	Number of customers
× 25%	Percent dissatisfied
250	Number dissatisfied
× 75%	Percent of switchers (60–90% of dissatisfied)
188	Number of switchers
× $10,000	Average revenue per customer
$1,880,000	Revenue lost through poor quality

Source: The University of Tampa (1990).

variable overhead only (the usual approach), or do not include overhead at all. The allocation of overhead can, of course, have a significant impact on calculating the total cost of poor quality and also on determining the distribution of the total over the various departments. Activity-based costing (ABC) can help by providing a realistic allocation of overhead costs.

Traditionally, manufacturing overhead costs are allocated to functional departments and to products based on direct labor hours, direct labor dollars, or machine hours. This method works fine but only if the single base used (e.g., direct labor hours) accounts for most of the total operational costs, as has been true in the past. Times have changed.

During the past 20 years, many manufacturing and service firms have experienced significant changes in their cost structure and the way their products are made. Direct labor accounted for about 50 to 60 percent of the total cost of the product or service, with overhead constituting about 20 percent of the cost. As companies became more automated through robotics and other means, direct labor costs have declined to about 5 to 25 percent and overhead costs have increased to about 75 percent in the total cost mix of the product.

Activity-based costing (ABC) is an accounting method that aims to improve cost-effectiveness through a focus on key cost elements. In doing this, ABC allocates overhead based on the factors (activities) which cause overhead cost elements to be incurred. These causal factors—called cost drivers—are measurable activities that increase overhead costs. ABC refines the way product costs are determined by using computer technology to economically track overhead costs in smaller categories with many cost drivers. These cost drivers are analogous to an allocation base such as direct labor hours in traditional allocation of overhead. But instead of just one cost driver (e.g., direct labor hours) there are many cost drivers such as machine setups, purchase orders, shipments, maintenance requests, etc. For each cost driver, an overhead rate is determined by dividing total costs for the driver (e.g., total cost for all machine setups) by the number of driver events (e.g., number of setups). The results might be for example, $90 per machine setup, $40 per purchase order, $120 per shipment, $80 per maintenance request. These overhead rates can then be applied to specific products, thus recognizing that every product or service does not utilize every component of overhead at exactly the same intensity, or in some cases, may not even use a given component at all. This more precise allocation of overhead can change the total cost of poor quality and the distribution over departments, thus influencing the priorities for improvement efforts.

Activity-based costing is more than an exercise in allocating overhead. This broader viewpoint of "activity-based cost management" emphasizes improvement in terms of cost reduction to be applied where it is most needed.

A basic accounting text that explains both activity-based costing and quality costs is Garrison and Noreen (1994). Other references relating activity-based costing and quality costs are Dawes and Siff (1993), Hester (1993), and Krause and Gryna (1995).

RETURN ON QUALITY

Improvement requires an investment of resources, and the investment must be justified by the blossoming benefits of improvement. The long-term effect of applying the cost of poor quality concept is shown in Figure 8.4. We will call the comparison of benefits to investment the "return on quality" (ROQ). Thus ROQ is really a return of investment (ROI) in the same sense as other investments such as equipment or an advertising program.

Using the expanded scope of cost of poor quality (see above under Categories of Quality Costs), the benefits of an improvement effort involve both reductions in cost and increases in sales revenue.

Some of the issues involved in estimating the benefits are

Reduced cost of errors: Expected savings, of course, must be based on specific plans for improvement. Often, such plans have a goal of cutting these costs by 50 percent within 5 years, but such a potential benefit should not be assumed unless the problem areas for improvement have been explicitly identified and an action plan with resources has been developed. In estimating present costs, don't inflate the present costs by including debatable or borderline items.

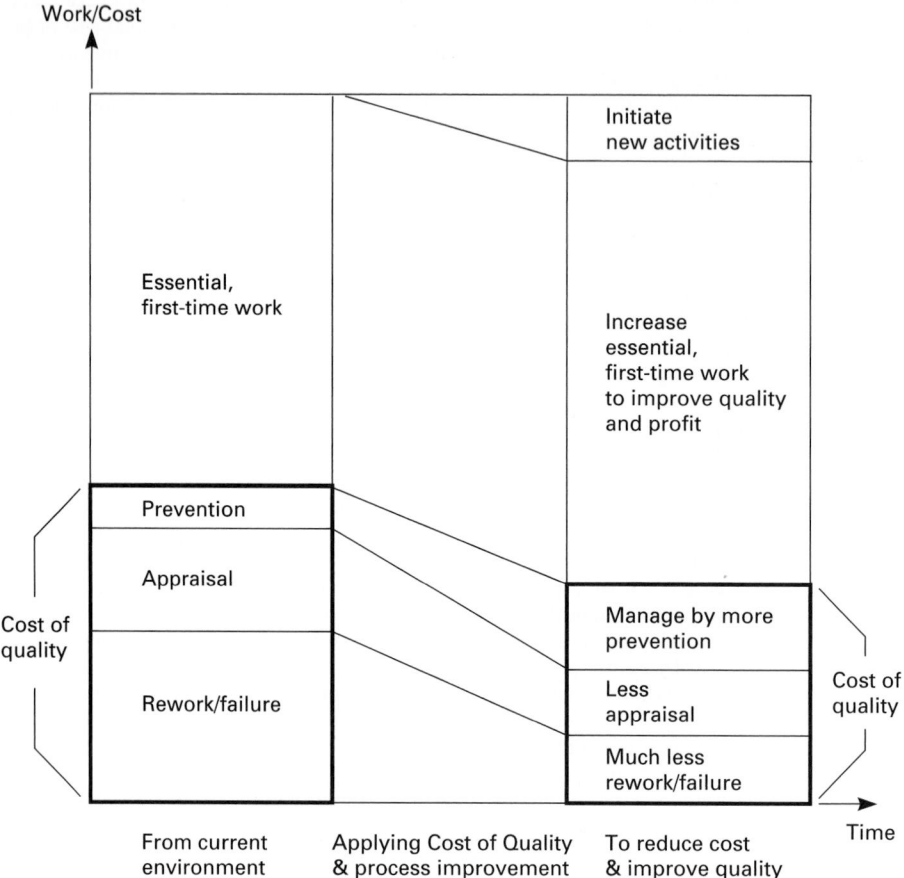

FIGURE 8.4   Effects of identifying cost of quality. (*AT&T 1990, p. 9.*)

Decisive review meetings will get bogged down in debating the validity of the figures instead of discussing the merits of the proposals for improvement.

Improved process capability: Expected savings can come from a reduction in variability (of product characteristics or process characteristics) and other process losses such as redundant operations, sorting inspections, retrieving missing information, and other non-value-added activities. As with other benefits, these expected savings must be based on improvement plans.

Reduced customer defections: One early indicator of defections can be responses to the market research question, "Would you purchase this product again?" In an early application by the author, 10.5 percent of a sample of current customers of washing machines said they would not repurchase; the reason was dissatisfaction with the machine, not with the dealer or price. At $50 profit per machine, the lost profit due to likely customer defections was then estimated. Progress has been made in quantifying the benefits of an effort to reduce defections. The parameters include the economic effect of losing customers over the "customer life," the level of quality to retain present customers (see Section 18 under Customer Satisfaction versus Customer Loyalty), and the effect on retention of the quality of handling customer complaints (see Section 18 under Linking Customer Satisfaction Results to Customer Loyalty and to Processes). Additional discussion is provided by Rust, Zahorik, and Keiningham (1994),

Chapter 6. Again, potential benefits should not be projected unless problem areas have been identified and an action plan has been formulated.

Increase in new customers: This is a most difficult benefit to quantify and predict. Quality improvements that make goods or services attractive to new customers will increase sales revenue but the amount and the timing depend on many internal actions and external market forces. Note that as the cost of poor quality is reduced, additional resources become available to finance new features for the goods and services-without increasing the price. The result can be a dramatic increase in market share.

The investments required to achieve the benefits may include diagnosis and other forms of analysis, training, redesign of products and processes, testing and experimentation, and equipment. Surprisingly, many improvement projects require little in costly equipment or facilities. The investment is mainly in analysis work.

An issue in calculating an ROQ is the matter of assumptions and estimates. Both must be realistic for the ROQ to be viewed as credible. Avoid assumptions which really represent ideal conditions; such conditions never occur. Also, the impact of numerical estimates on the ROQ can be evaluated (if necessary) using "sensitivity analysis." This involves changing the estimates (say by ±20 percent) and recalculating the ROQ to see the effect of the accuracy of the estimate on the ROQ. The ROQ will change but if the amount is small, it adds credibility to the estimates.

Note that one source of benefits (reducing the cost of errors) is based on relatively "hard" data of costs already incurred that will continue unless improvement action is instituted. A second source (reducing the cost of defections) also represents a loss (of sales revenue) already incurred. Other sources (improvement of process capability and gaining new customers) are not based on current losses but do represent important opportunity costs. In the past, because of the difficulty of quantifying other sources of benefits, the cost savings of a quality improvement program have been based primarily on the cost of errors. Advances made in quantifying the impact of quality on sales revenue, however, are making it possible to add the revenue impact to the return on quality calculation. At a minimum, the ROQ calculation can be based on savings in the cost of errors. When additional data are available (e.g., process information or market research data), then estimates for one or more of the other three sources should be included to calculate the total benefits. A note of caution: this expanded view of the cost of poor quality could mean that "traditional" quality improvement efforts (reducing cost of errors) will become entangled with other efforts (increasing sales revenue), leading to a blurring of the traditional efforts on reducing errors. We need both efforts—just as much as we need the sun and the moon.

The rate of return on an investment in quality activities translates into the ratio of average annual benefits to the initial investment. (Note that the reciprocal—investment divided by annual savings—represents the time required for savings to pay back the investment, i.e., the "payback period.") But this calculation of ROQ provides an approximate rate of return because it neglects the number of years involved for the savings and also the time value of money. A more refined approach involves calculating the "net present value" of the benefits over time. This means using the mathematics of finance to calculate the amount today which is equivalent to the savings achieved during future years (see Grant, Ireson, and Leavenworth1990, Chapter 6). Rust, Zahorik, and Keinubghan (1994) describe an approach for calculating the ROQ incorporating savings in traditional losses due to errors, sales revenue enhancement using market research information for customer retention, and the time value of money.

Wolf and Bechert (1994) describe a method to determine the payback of a reduction in failure costs when prevention and appraisal expenditures are made. Bester (1993) discusses the concept of net value productivity which addresses the cost of quality and the value of quality.

GAINING APPROVAL FOR THE QUALITY IMPROVEMENT PROGRAM

Those presenting the results of the cost study should be prepared to answer this question from management: "What action must we take to reduce the cost of poor quality?"

The control of quality in many companies follows a recognizable pattern—as defects increase, we take action in the form of more inspection. This approach fails because it does not remove the causes of defects; i.e., it is detection but not prevention. To achieve a significant and lasting reduction in defects and costs requires a structured process for attacking the main sources of loss—the failure costs. Such an attack requires proceeding on a project-by-project basis. These projects in turn require resources of various types (see Section 5, The Quality Improvement Process). The resources must be justified by the expected benefits. For every hour spent to identify one of the vital few problems, we often spend 20 hours to diagnose and solve the problem.

To gain approval from upper management for a quality improvement effort, we recommend the following steps.

1. Establish that the costs are large enough to justify action (see, for example, Tables 8.1 and 8.2).

   a. Use the grand total to demonstrate the need for quality improvement. This is the most significant figure in a quality cost study. Usually, managers are stunned by the size of the total—they had no idea the amount was so big. One memorable example was a leading manufacturer of aircraft engines. When the total quality costs were made known to the managing director, he promptly convened his senior executives to discuss a broad plan of action. Those presenting the report should be prepared for the report to be greeted with skepticism. The cost may be such that it will not be believed. This can be avoided if management has previously agreed to the definition of the cost of poor quality and if the accounting function has collected the data or has been a party to the data collection process. Also, don't inflate the present costs by including debatable or borderline items.

   b. Relate the grand total to business measures. Interpretation of the total is aided by relating total quality costs to other figures with which managers are familiar. Two universal languages are spoken in the company. At the "bottom," the language is that of objects and deeds: square meters of floor space, output of 400 tons per week, rejection rates of 3.6 percent, completion of 9000 service transactions per week. At the "top," the language is that of money: sales, profit, taxes, investment. The middle managers and the technical specialists must be *bilingual.* They must be able to talk to the "bottom" in the language of objects and to the "top" in the language of money. Table 8.4 shows actual examples of the annual cost of poor quality related to various business measures. In one company which was preoccupied with meeting delivery schedules, the quality costs were translated into equivalent added production. Since this coincided with the chief current goals of the managers, their interest was aroused. In another company, the total quality costs of $176 million per year for the company were shown to be equivalent to one of the company plants employing 2900 people, occupying 1.1 million ft^2 of space and requiring $6 million of in-process inventory. These latter three figures in turn meant the equivalent of one of their major plants making 100 percent defective work every working day of the year. This company is the quality leader in its industry. Similarly, in an airplane manufacturing company, it was found useful to translate the time spent on rework to the backlog of delivery of airplanes, i.e., reducing the rework time made more hours available for producing the airplanes.

   c. Show the subtotals for the broad major groupings of quality costs, when these are available. A helpful grouping is by the four categories discussed above under Categories of Quality Costs. Typically, most of the quality costs are associated with failures, internal and external. The proper sequence is to reduce the failure costs first, not to start by reducing inspection costs. Then as the defect levels come down, we can follow through and cut the inspection costs as well.

2. Estimate the savings and other benefits:

   a. If the company has never before undertaken an organized program to reduce quality-related costs, then a reasonable goal is to cut these costs in two, within a space of 5 years.

   b. Don't imply that the quality costs can be reduced to zero.

   c. For any benefits that cannot be quantified as part of the return on quality, present these benefits as intangible factors to help justify the improvement program. Sometimes, benefits can be related to problems of high priority to upper management such as meeting delivery

TABLE 8.4 Languages of Management

Money (annual cost of poor quality)
24% of sales revenue
15% of manufacturing cost
13 cents per share of common stock
$7.5 million per year for scrap and rework compared to a profit of $1.5 million per year
$176 million per year
40% of the operating cost of a department

Other languages
The equivalent of one plant in the company making 100% defective work all year
32% of engineering resources spent in finding and correcting quality problems
25% of manufacturing capacity devoted to correcting quality problems
13% of sales orders canceled
70% of inventory carried attributed to poor quality levels
25% of manufacturing personnel assigned to correcting quality problems

schedules, controlling capital expenditures, or reducing a delivery backlog. In a chemical company, a key factor in justifying an improvement program was the ability to reduce significantly a major capital expenditure to expand plant capacity. A large part of the cost of poor quality was due to having to rework 40 percent of the batches every year. The improvement effort was expected to reduce the rework from 40 percent to 10 percent, thus making available production capacity that was no longer needed for rework.

3. Calculate the return on investment resulting from improvement in quality. Where possible, this return should reflect savings in the traditional cost of poor quality, savings in process capability improvement, and increases in sales revenue due to a reduction in customer defections and increases in new customers. See above under Return on Quality.
4. Use a successful case history (a "bellwether" project) of quality improvement in the company to justify a broader program.
5. Identify the initial specific improvement projects. An important tool is the Pareto analysis which distinguishes between the "vital few" and the "useful many" elements of the cost of poor quality. This concept and other aids in identifying projects is covered in Section 5 under Use of the Pareto Principle. An unusual application of the concept is presented in Figure 8.5. Here, in this "double Pareto" diagram, the horizontal axis depicts (in rank order) within each major category the cost of poor quality and the vertical axis shows the subcategories (in rank order) of the major category. Thus, the first two major categories (patient care–related and facility-related) account for 80 percent of the cost. Also, of the 64 percent which is patient care related about 54 percent is related to variation in hospital practice and 46 percent to outpatient thruput optimization. This graph has proven to be a powerful driver for action.
6. Propose the structure of the improvement program including organization, problem selection, training, review of progress, and schedule. See Section 5, The Quality Improvement Process. Justification is essential for an effective program of quality improvement. Approaches to justification are discussed in more detail in Section 5 under Securing Upper Management Approval and Participation.

USING COST OF POOR QUALITY CONCEPT TO SUPPORT CONTINUING IMPROVEMENT

As formal quality improvement continues with projects addressed to specific problems, the measurement of the cost of poor quality has several roles.

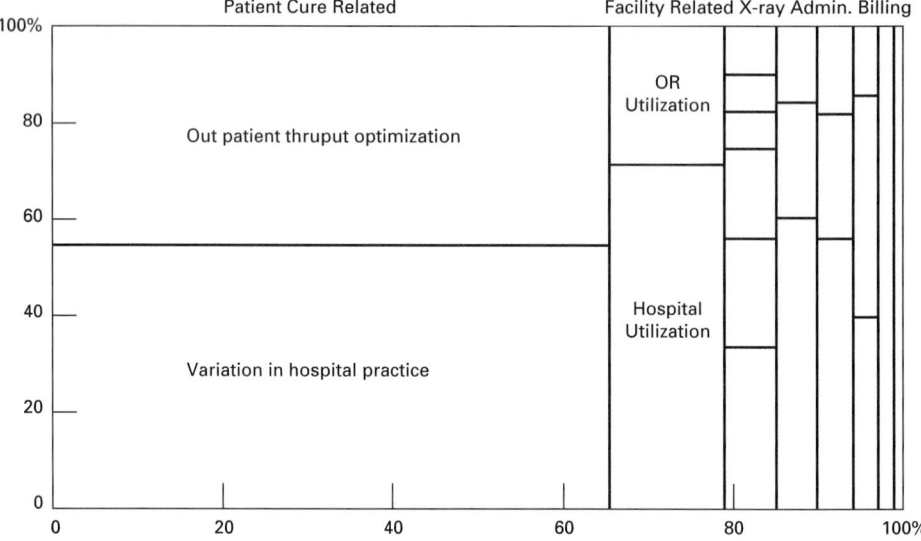

FIGURE 8.5   Mayo Rochester improvement opportunity. (*Adapted from Rider 1995.*)

Roles to Support Improvement.   These include:

1. Identify the most significant losses for an individual problem and the specific costs to be eliminated. This helps to focus the diagnostic effort on root causes. Figure 8.6 shows an example of a special report that dissects one element of cost ("penalties") and summarizes data on the cost of poor quality to identify "activity cost drivers" and "root causes."

2. Provide a measure of effectiveness of the remedies instituted on a specific project. Thus, a project quality improvement team should provide for measuring the costs to confirm that the remedies have worked.

3. Provide a periodic report on specific quality costs. Such a report might be issued quarterly or semiannually.

4. Repeat the full cost of poor quality study. This study could be conducted annually to assess overall status and help to identify future projects.

5. Identify future improvement projects by analyzing the full study (see item 4 above) using Pareto analysis and other techniques for problem selection.

Note that the emphasis is on using the cost of poor quality to identify improvement projects and support improvement team efforts rather than focusing on the gloomy cost reporting.

Optimum Cost of Poor Quality.   When cost summaries on quality are first presented to managers, one of the usual questions is: "What are the right costs?" The managers are looking for a standard ("par") against which to compare their actual costs so that they can make a judgment on whether there is a need for action.

Unfortunately, few credible data are available because (1) companies almost never publish such data and (2) the definition of cost of poor quality varies by company. [In one published study, Ittner (1992) summarizes data on the four categories for 72 manufacturing units of 23 companies in 5 industry sectors.] But three conclusions on cost data do stand out: The total costs are higher for complex industries, failure costs are the largest percent of the total, and prevention costs are a small percent of the total.

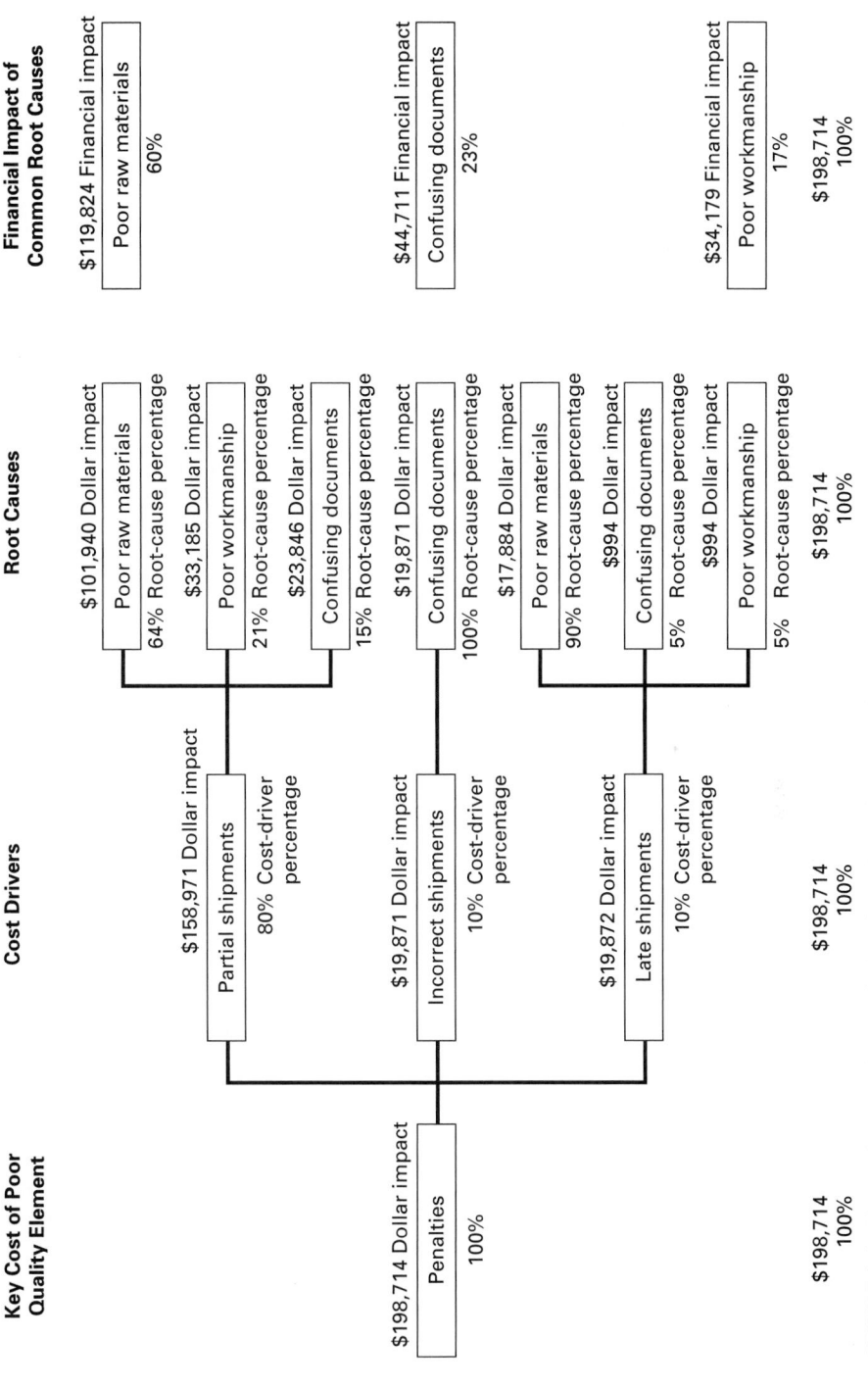

FIGURE 8.6 The completed cost-driver analysis. *(Adapted from Atkinson, Hamburg, and Itner, 1994, p. 142.)*

The study of the distribution of quality costs over the major categories can be further explored using the model shown in Figure 8.7. The model shows three curves:

1. *The failure costs:*   These equal zero when the product is 100 percent good, and rise to infinity when the product is 100 percent defective. (Note that the vertical scale is cost per good unit of product. At 100 percent defective, the number of good units is zero, and hence the cost per good unit is infinity.)

2. *The costs of appraisal plus prevention:*   These costs are zero at 100 percent defective, and rise as perfection is approached.

3. *The sum of curves 1 and 2:*   This third curve is marked "total quality costs" and represents the total cost of quality per good unit of product.

Figure 8.7 suggests that the minimum level of total quality costs occurs when the quality of conformance is 100 percent, i.e., perfection. This has not always been the case. During most of the twentieth century the predominant role of (fallible) human beings limited the efforts to attain perfection at finite costs. Also, the inability to quantify the impact of quality failures on sales revenue resulted in underestimating the failure costs. The result was to view the optimum value of quality of conformance as less than 100 percent.

While perfection is obviously the goal for the long run, it does not follow that perfection is the most economic goal for the short run, or for every situation. Industries, however, are facing increasing pressure to reach for perfection. Examples include:

1. *Industries producing goods and services that have a critical impact on human safety and well-being:*   The manufacture of pharmaceuticals and the generation of mutual fund statements provide illustrations.

2. *Highly automated industries:*   Here, it is often possible to achieve a low level of defects by proper planning of the manufacturing process to assure that processes are capable of meeting

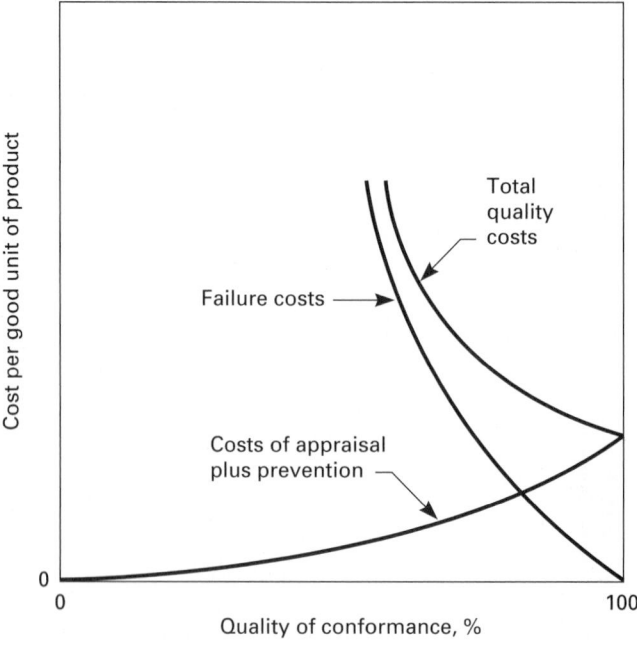

FIGURE 8.7   Model for optimum quality costs.

specifications. In addition, automated inspection often makes it economically feasible to perform 100 percent inspection to find all the defects.

3. *Companies selling to affluent clients:*   These customers are often willing to pay a premium price for perfect quality to avoid even a small risk of a defect.

4. *Companies striving to optimize the user's cost:*   The model depicted in Figure 8.7 shows the concept of an optimum from the viewpoint of the producer. When the user's costs due to product failure are added to such models, those costs add further fuel to the conclusion that the optimum point is perfection. The same result occurs if lost sales income to the manufacturer is included in the failure cost.

The prospect is that the trend to 100 percent conformance will extend to more and more goods and services of greater and greater complexity.

To evaluate whether quality improvement has reached the economic limit, we need to compare the benefits possible from specific projects with the costs involved in achieving these benefits. When no justifiable projects can be found, the optimum has been reached.

REPORTING ON THE COST OF POOR QUALITY

As structured quality improvement teams following the project-by-project approach have emerged as a strong force, reporting on the cost of poor quality has focused on supporting these team activities. These reports provide information which helps to diagnose the problem and to track the change in costs as a remedy is implemented to solve the problem. Regularly issued reports with a fixed format usually do not meet the needs of improvement teams. Teams may need narrowly focused information and they may need it only once. What data are needed is determined by the team, and the team often collects its own data (Winchell 1993).

Some companies use periodic reporting on the cost of poor quality in the form of a scoreboard. Such a scoreboard can be put to certain constructive uses. To create and maintain the scoreboard, however, requires a considerable expenditure of time and effort. Before undertaking such an expenditure, the company should look beyond the assertions of the advocates; it should look also at the realities derived from experience. (Many companies have constructed quality cost scoreboards and have then abandoned them for not achieving the results promised by the advocates.) After finalizing the categories for a cost scoreboard, the planning must include collecting and summarizing the data, establishing bases for comparison, and reporting the results.

Summarizing the Data.   The most basic ways are

1. By product, process, component, defect type, or other likely defect concentration pattern

2. By organizational unit

3. By category cost of poor quality

4. By time

Often, the published summaries involve combinations of these different ways.

Bases for Comparison.   When managers use a scoreboard on the cost of poor quality, they are not content to look at the gross dollar figures. They want, in addition, to compare the costs with some base which is an index of the opportunity for creating these costs. A summary of some widely used bases, along with the advantages and disadvantages of each, is presented in Table 8.5. The base used can greatly influence the interpretation of the cost data.

It is best to start with several bases and then, as managers gain experience with the reports, retain only the most meaningful. The literature stresses that quality costs be stated as percent of sales income. This is a useful base for some, but not all, purposes.

TABLE 8.5   Measurement Bases for Quality Costs

Base	Advantages	Disadvantages
Direct labor hour	Readily available and understood	Can be drastically influenced by automation
Direct labor dollars	Available and understood; tends to balance any inflation effect	Can be drastically influenced by automation
Standard manufacturing cost dollars	More stability than above	Includes overhead costs both fixed and variable
Value-added dollars	Useful when processing costs are important	Not useful for comparing different types of manufacturing departments
Sales dollars	Appeals to higher management	Sales dollars can be influenced by changes in prices, marketing costs, demand, etc.
Product units	Simplicity	Not appropriate when different products are made unless "equivalent" item can be defined

Reporting the Results.   The specific matters are the same as for those for other reports—format, frequency, distribution, responsibility for publication. Atkinson, Hamburg, and Ittner (1994) describe how reporting can help promote a cultural change for quality, Dobbins and Brown (1989) provide "tips" for creating reports, and Onnias (1985) describes a system used at Texas Instruments.

The likely trend is for cost of poor quality and other quality-related information to become integrated into the overall performance reporting system of organizations. Kaplan and Norton (1996) propose that such a system provide a "balanced scorecard." Such a scorecard allows managers to view an organization from four perspectives:

1. How do customers see us? (Customer perspective.)

2. What must we excel at? (Internal perspective.)

3. Can we continue to improve and create value? (Innovation and learning perspective.)

4. How do we look to shareholders? (Financial perspective.) The scorecard would include a limited number of measures—both the financial result measures and the operational measures that drive future financial performance.

Periodically (say annually), a comprehensive report on the cost of poor quality is useful to summarize and consolidate results of project teams and other quality improvement activities. The format for this report need not be identical to the initial cost of poor quality study but should (1) reflect results of improvement efforts and (2) provide guidance to identify major areas for future improvement efforts.

ECONOMIC MODELS AND THE COST OF POOR QUALITY

The costs of poor quality affect two parties—the provider of the goods or service and the user. This section discusses the impact on the provider, i.e., a manufacturer or a service firm. Poor quality also increases the costs of the user of the product in the form of repair costs after the warranty period, various losses due to downtime, etc. Gryna (1977) presents a methodology with case examples of user costs of poor quality. The extent of these user costs clearly affects future purchasing decisions of the user and thereby influences the sales income of the provider. This section stresses the potential for profit improvement by reducing provider costs and by reducing loss of sales revenue due to poor quality.

An extension of this thinking is provided by applying concepts from the economics discipline. Cole and Mogab (1995) apply economic concepts to analyze the difference between the "mass production/scientific management firm" and the "continuous improvement firm" (CIF). A defining feature of the CIF is the ability to add to the net customer value of the marketed product. The net customer value is defined as the total value realized by the customer from the purchase and use of the goods or service less that which must be sacrificed to obtain and use it. In terms of the economy of countries, Brust and Gryna (1997) discuss five links between quality and macroeconomics.

From the birth of the cost of poor quality with the emphasis on the cost of errors in manufacturing, the concept is now extended in the scope of cost elements and applies to manufacturing and service industries in both the profit and nonprofit sectors.

REFERENCES

AT&T (1990). *AT&T Cost of Quality Guidelines.* Document 500-746, AT&T's Customer Information Center, Indianapolis, IN.

Atkinson, Hawley, Hamburg, John, and Ittner, Christopher (1994). *Linking Quality to Profits.* ASQ Quality Press, Milwaukee, and Institute of Management Accountants, Montvale, NJ.

Atkinson, John Hawley, et al. (1991). *Current Trends in Cost of Quality: Linking the Cost of Quality and Continuous Improvement.* Institute of Management Accountants, Montvale, NJ.

Aubrey, Charles A., II (1988). "Effective Use of Quality Cost Applied to Service." *ASQC Quality Congress Transactions,* Milwaukee, pp. 735–739.

Bester, Yogi (1993). "Net-Value Productivity: Rethinking the Cost of Quality Approach." *Quality Management Journal,* October, pp. 71–76.

Bohan, George P., and Horney, Nicholas F. (1991). "Pinpointing the Real Cost of Quality in a Service Company." *National Productivity Review,* Summer, pp. 309–317.

Briscoe, Nathaniel R., and Gryna, Frank M. (1996). *Assessing the Cost of Poor Quality in a Small Business.* Report No. 902, College of Business Research Paper Series, The University of Tampa, Tampa, FL.

Brust, Peter J., and Gryna, Frank M. (1997). *Product Quality and Macroeconomics—Five Links.* Report No. 904, College of Business Research Paper Series, The University of Tampa, Tampa, FL.

Campanella, Jack, Ed. (1999). *Principles of Quality Costs,* 3rd ed., ASQ, Milwaukee.

Carr, Lawrence P. (1992). "Applying Cost of Quality to a Service Business." *Sloan Management Review,* Summer, pp. 72–77.

Cole, William E., and Mogab, John W. (1995). *The Economics of Total Quality Management: Clashing Paradigms in the Global Market.* Blackwell, Cambridge, MA.

Dawes, Edgar W., and Siff, Walter (1993). "Using Quality Costs for Continuous Improvement." *ASQC Annual Quality Congress Transactions,* Milwaukee, pp. 810–816.

Dobbins, Richard K., and Brown, F. X. (1989). "Quality Cost Analysis—Q.A. Versus Accounting." *ASQC Annual Quality Congress Transactions,* pp. 444–452.

Esterby, L. James (1984). "Measuring Quality Costs by Work Sampling." *Quality Costs: Ideas and Applications.* Andrew F. Grimm, Louis A. Pasteelnick, and James F. Zerfas, eds., ASQC Quality Press, Milwaukee, pp. 487–491.

Finnegan, John T., and Schottmiller, John C. (1990). "Cost of Quality—Multinational and Multifunctional." *ASQC Quality Congress Transactions,* Milwaukee, pp. 746–751.

Garrison, Ray H., and Noreen, Eric W. (1994). *Managerial Accounting.* Irwin, Boston.

Grant, Eugene L., Ireson, W. Grant, and Leavenworth, Richard S. (1990). *Principles of Engineering Economy,* 8th ed., John Wiley and Sons, New York.

Gryna, Frank M. (1977). "Quality Costs: User vs. Manufacturer." *Quality Progress,* June, pp. 10–15.

Harrington, H. James (1986). *Poor Quality Cost.* ASQC Quality Press, Milwaukee, and Marcel Dekker, New York.

Hester, William F. (1993). "True Quality Cost With Activity Based Costing." *ASQC Annual Quality Congress Transactions,* Milwaukee, pp. 446–454.

Hou,Tien-fang (1992). "Cost-of-Quality Techniques for Business Processes." *ASQC Annual Quality Congress Transactions,* Milwaukee, pp. 1131–1137.

ISO 9004-1 (1994). *Quality Management and Quality System Elements—Guidelines,* Section 6, "Financial Considerations of Quality Systems." ISO, Geneva.

Ittner, Christopher Dean (1992). *The Economics and Measurement of Quality Costs: An Empirical Investigation.* Doctoral dissertation, Harvard University, Cambridge, MA.

Kaplan, Robert S., and Norton, David P. (1996). *The Balanced Scorecard,* Harvard Business School Press, Boston.

Keaton, Willie E., et al. (1988). "White Collar Quality Pinpointing Areas for Breakthrough." *IMPRO 1988 Conference Proceedings,* Juran Institute, Inc., Wilton, CT, pp. 3C-19 to 3C-24.

Krause, James D., and Gryna, Frank M. (1995). *Activity Based Costing and Cost of Poor Quality—a Partnership.* Report No. 109, College of Business Research Paper Series, The University of Tampa, Tampa, FL.

Nickell, Warren L. (1985). "Quality Improvement in Marketing." *The Juran Report,* no. 4 (Winter), pp. 29–35.O'Neill, Ann R. (1988). "Quality Costs In = Continuous Improvement Out." *ASQC Annual Quality Congress Transactions,* Milwaukee, pp. 180–187.

Onnias, Arturo (1985). "The Quality Blue Book." *IMPRO 1985 Conference Proceedings,* Juran Institute, Inc., Wilton, CT, pp. 127–134.

Quevedo, Rene (1991). "Quality, Waste, and Value in White-Collar Environments." *Quality Progress,* January, pp. 33–37.

Reeve, J. M. (1991). "Variation and the Cost of Quality." *Quality Engineering,* vol. 4, no. 1, pp. 41–55.

Rider, C. T. (1995). "Aligning Your Improvement Strategy for the Biggest Payback." *IMPRO 1995 Conference Proceedings,* Juran Institute Inc., Wilton, CT, pp. 7E. 1-1 to 7E.1-11.

Romagnole, Anthony J., and Williams, Mary A. (1995). "Costs of Poor Quality." *IMPRO 1995 Conference Handout,* Juran Institute, Inc., Wilton, CT.

Rust, Roland T., Zahorik, Anthony J., and Keiningham, Timothy L. (1994). *Return on Quality.* Probus, Chicago.

Schottmiller, John C. (1996). "ISO 9000 and Quality Costs." *ASQC Annual Quality Congress Proceedings,* Milwaukee, pp. 194–199.

Schrader, Lawrence, J. (1986). "An Engineering Organization's Cost of Quality Program." *Quality Progress,* January, pp. 29–34.

Shepherd, Nick A. (1998). "Cost of Quality and ISO Implementation Using Cost of Quality as a Driver for Effective Use of ISO 9000 and 14000." *ASQ Annual Quality Congress Transactions,* pp. 776–783.

The Conference Board (1989). *Current Practices in Measuring Quality.* Research Bulletin No. 234, New York.

The University of Tampa (1990). "Costly Mistakes." *Visions,* Fall, p. 7.

Wasserman, Gary S., and Lindland, John L. (1994). "Minimizing the Cost of Quality Over Time: A Dynamic Quality Cost Model." *ASQC Annual Quality Congress Proceedings,* pp. 73–81.

Wild, Cheryl, and Kovacs, Carol (1994). "Cost-of-Quality Case Study." *IMPRO 1994 Conference Proceedings,* Juran Institute, Inc., Wilton, CT, pp. 3B-19 to 3B-29.

Winchell, William O. (1991). "Implementation of Quality Costs—Making It Easier." *ASQC Annual Quality Congress Transactions,* Milwaukee, pp. 592–597.

Winchell, William O. (1993). "New Quality Cost Approach Gives Proven Results." *ASQC Annual Quality Congress Transactions,* Milwaukee, pp. 486–492.

Wolf, C., and Bechert, J. (1994). "Justifying Prevention and Appraisal Quality Cost Expenditures: A Benefit/Cost Decision Model." *Quality Engineering,* vol. 7, no. 1, pp. 59–70.

MEASUREMENT, INFORMATION, AND DECISION MAKING

Thomas C. Redman

INTRODUCTION

Managers Need Information.   The general business leader and quality manager share an eternal lament first voiced by Alexander the Great: "Data, data, data! I'm surrounded by data. Can't anyone get me the information I really need?" Alexander needed answers to some rather basic questions. "Where is the enemy? How large are his forces? How well will they fight? When will our supplies arrive?" His spies, lieutenants, and others gathered data. But they could not always satisfy Alexander's need for information.

Today's business manager also has a voracious appetite for information (see also Drucker 1995). Obtaining needed information is time-consuming, expensive, and fraught with difficulty. And in the end, the manager is often less informed than Alexander the Great. Basic questions like "What do customers really want? How well are we meeting their needs? What is the competitor going to do next?" are not easier to answer. This, despite an amazing proliferation of measuring devices, customer surveys, statistical methods, and database and networking technology. Just as quality management is a never-ending journey, so too is the task of learning of, obtaining, sorting through, synthesizing, and understanding all the data and information that could productively be used. It seems that the manager's appetite for information will never be satisfied.

Systems of Measurement.   We usually consider information in light of decisions that managers must make and the actions they take. Information plays a similar role in science. The scientific method is essentially a process by which hypotheses are proposed, experiments designed to test aspects of those hypotheses, data collected and analyzed, and the hypotheses either advanced, discarded, or modified. Time-tested and rigid standards apply. Most businesses cannot afford the required rigor. The system of measurement advanced here parallels the scientific method, but the standards are different. "Caution" is the watchword in science. "Success" is the watchword in business. (See Section 47 for further reference to the scientific method and experimental design.)

In some cases, the manager is presented with well-defined choices. A simple example is the question: "Is this process in a state of control?" The manager can either decide that the process is in control and take no action, or decide that the process is not in control and take action to find and eliminate a special cause. (See Section 4: The Quality Control Process.) In other situations, the range of options is ill-defined and/or unbounded. In many cases (perhaps too many), the manager may even choose to gather more data. In almost all cases, it appears axiomatic that the better the information, the better the decision. Here *better information* may have any number of attributes, including more complete, more accurate, more relevant, more current, from a more reliable source, more precise, organized in a more convincing fashion, presented in a more appealing format, and so forth.

A critical step in obtaining needed information is measurement. To measure is "to compute, estimate, or ascertain the extent, dimensions, or capacity of, especially by a certain rule or standard" (Webster 1979). Measurement, then, involves the collection of raw data. For many types of measurements, specialized fields have grown up and there is a considerable body of expertise in making measurements. Chemical assays and consumer preference testing are two such areas. Data collection may involve less formal means—searching a library, obtaining data originally gathered for other purposes, talking to customers, and the like. For our purposes, all such data collection shall be considered measurement.

Of course there is more to developing the information the manager needs then simply collecting data (indeed, therein lies Alexander's lament). For examples we cited "more relevant" and "presented in a more appealing format" as attributes of better information. It is evident that the choice of what to measure and the analysis, synthesis, and presentation of the resultant information are just as important as the act of measurement itself. High-quality information results only from high-caliber and integrated design, data collection, and analysis/synthesis/presentation. Thus we distinguish between the *act of measurement,* or data collection, and the *measurement process,* which includes design, data collection, and analysis/synthesis/presentation. This process is presented in Figure 9.1 and described throughout this section.

But good information is not a decision. So Figure 9.1 goes further. The measurement process is preceded by a step "Understand framework" and followed by a step "Make decision/take action." These steps put the measurement process in its proper context and represent the suppliers to and customers of the measurement process. Finally, Figure 9.1 features one further important subtlety. Decision makers in most organizations are often too busy to carefully consider all the data and evaluate alternatives. What they require are recommendations, not just clearly presented information. So the analysis/synthesis/presentation step is better described as analysis/synthesis/recommendations/presentation of results and recommendations. It could be

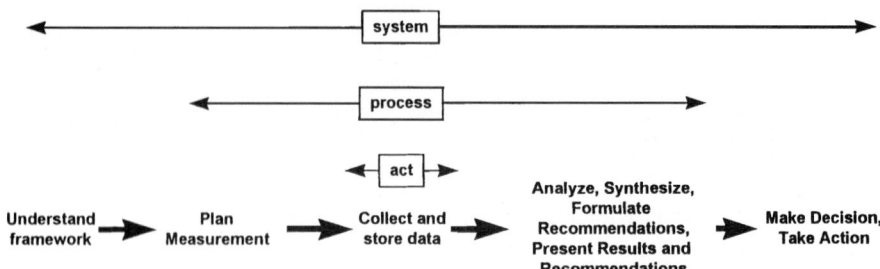

FIGURE 9.1   The act of measurement is but one step in a larger measurement system. Herein we consider the measurement process as consisting of steps needed to collect data and present results. The larger measurement system also embraces the decisions that are made and the framework in which the process operates.

argued that formulation of results is better thought of as a part of decision making. But it is more often the case that those analyzing data also are expected to make recommendations. Indeed, they may be petitioning decision makers to adopt their recommendations.

As will be discussed, decision making is often a complex political process, in and of itself. Thus the framework in which information is produced is critical. At a high level, the framework defines the overall context in which decisions are to be made, including such diverse considerations as the organization's business goals, its competitive position, and its available resources; customer requirements; the goals and biases of decision makers; and any relevant constraints on the measurement process or decision making. For any particular decision, the framework includes the specific issues to be addressed. Taken together, these five elements (framework, design, data collection, analysis/synthesis/recommendations/presentation, and decision/action) compose a "measurement system".

Ultimately, the goal is to help the organization make consistently better decisions and take better actions. Here, *better decisions* are defined in terms of results in achieving organizational objectives. Good measurement systems support a number organizational goals, not just a few (Kaplan and Norton 1992, 1993; Meyer 1994). It is usually true that "what gets measured, gets managed." Most organizations intend to pursue a variety of goals, satisfying customers, providing a healthy and fulfilling workplace, meeting financial objectives, and so forth. In many organizations financial measurements are the most fully developed and deployed. It is little wonder that financial considerations dominate decision making in such organizations (Eccles 1991).

The most important point of this section is that those who define and operate a measurement process (i.e., those who gather or analyze data or who recommend what the organization should do) must consider the system as a whole, including the environment in which it operates. It is not sufficient to be technically excellent or for one or two elements to be outstanding (see Futrell 1994 for a review of the failures of customer satisfaction surveys). Overall effectiveness is as much determined by how well the elements relate to one another and to other systems in the enterprise as by excellence in any area. The following personal anecdote illustrates this point.

Very early in my career, I was asked to recommend which color graphics terminal our department should buy. I spent a lot of time and effort on the measurement process. I talked to several users about their needs, called a number of vendors, arranged several demonstrations, and read the relevant literature. At the time (early 1980s) the underlying technology was in its infancy—it had many problems with it and a terminal cost about $15,000. Further, many people anticipated dramatic and near-term improvements to the technology and substantial price reductions. So I recommended that we wait a year and then reconsider the purchase. I was proud of my work and recommendation and presented it to my manager. He promptly informed me I had misunderstood his question. The question was not "Should we buy a terminal?" but "Which terminal should we buy?" (and none was not a permitted answer). In retrospect, I could have: explicitly defined the possible decisions in advance, or thought through the framework. My manager was a forward thinker. He had seen the potential for personal computing and clearly wanted to experiment. Even casual consideration of his objectives would have made it clear that "wait a year" was an unacceptable recommendation.

About This Section.   The primary audiences for this section are persons interested in helping their organizations make better decisions. I have already noted that obtaining relevant information can be time-consuming, expensive, and fraught with difficulty. On the other hand, there are many cost-effective practices that everybody can take to reduce the gap between the information they desire and the information they have at their disposal.

Readers may have several interests:

- Those whose measurement and decision processes are well defined may be especially interested in comparing their processes to the "ideals" described here.

- Those who make measurements and/or decisions routinely, but whose processes are ad hoc, may be especially interested in applying the techniques of process management to improve their measurements and decisions.

- Those whose interests involve only a single "project" may be especially interested in learning about the steps their project should involve.

This section does not consider many technical details. Entire volumes have been written on technical details of measurement making (see Finkelstein and Leaning 1984 and Roberts 1979, for example), statistical analysis (see Sections 4, 44, 45, and 47 of this Handbook, for example), graphical data presentation, and the like.

In the next section, we consider measurement systems in greater detail. Outputs of each step are defined and the activities needed to produce those outputs are described. Then, we consider more complex measurement systems, involving hierarchies of measurements. The fourth section provides practical advice for starting and evolving a measurement system. The final section summarizes good practice in a list of the "Top 10 Measurement System Principles."

Three examples, of escalating complexity, are used to illustrate the main points. At the so-called "operational level," we consider the measurement system for a single step for the billing process summarized in Figure 9.2. At the tactical level, we consider the measurement system needed to support changes to the feature set associated with the invoice (the result of the billing process). We also consider measurement systems needed to support strategic decisions. Virtually everyone is involved in some way or another at all three levels of decision making.

A middle manager may find himself or herself playing the following roles:

- A designer of the measurement system used at the operational level
- A decision maker at the tactical level
- A supplier of data to strategic decisions

MEASUREMENT SYSTEMS AND THEIR ELEMENTS

Figure 9.3 expands on Figure 9.1 in two ways: It lists the work products produced at each step and describes the work activities in more detail. The first two steps (understand framework and plan measurement) are planning steps. There is no substitute for careful planning, so most of the discussion is on planning. The most important work products of the system are decisions and actions. Other work products aim to produce better decisions and actions. So we begin the discussion with the final step: Make decision/take action.

Make Decisions/Take Action.   The first step in defining a measurement system is to understand who will make the decisions and how. Many decisions, and virtually all decisions of consequence, are not made by an individual, but by a committee or other group. In some cases, this helps build support for implementation. In others, it is more a vehicle for diffusing accountability. Some groups decide by majority rule, others by consensus. Most groups have a few key individuals. Some are thoughtful leaders, others are self-centered, domineering bullies. Few decision makers are completely unbiased. Most are concerned, at least to some degree, with their careers. Individuals intuitively make decisions based on different criteria. Some are risk takers, others are risk-averse. Some are concerned only with the near-

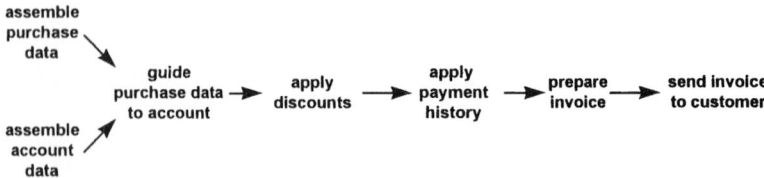

FIGURE 9.2   A hypothetical billing process.

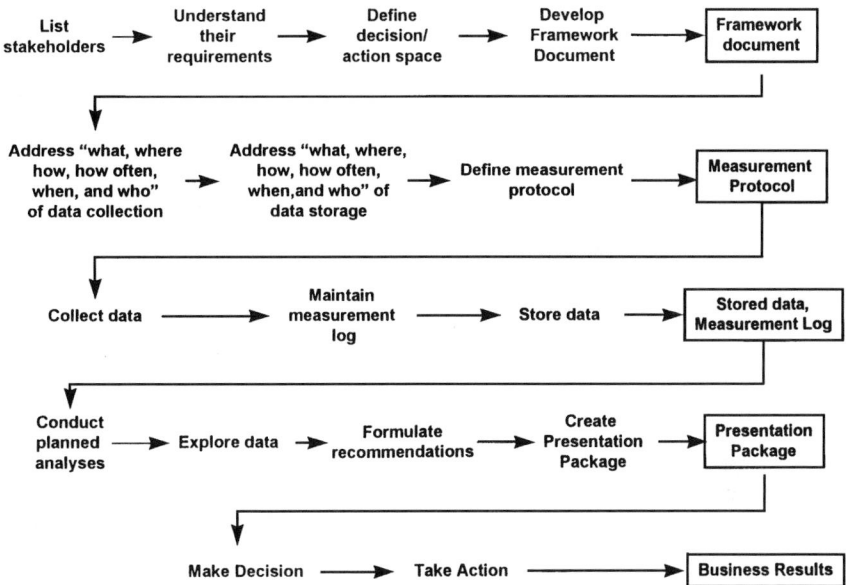

FIGURE 9.3 Further detail about subprocesses of the measurement system. For each subprocess, the principal work product and several steps are given.

term financial impacts, others consider the long-term. And those who may be saddled with responsibility for implementation have other perspectives. Decision making is thus also a political process that the designer of the measurement system is well advised to understand.

Understand Framework.   Prior to determining what to measure and how to measure it, it is important to understand the overall framework in which the measurement system operates. We've already noted the political nature of decision making. Those who make decisions and take actions are members of organizations, and all organizations have their own politics and cultures that define acceptable directions, risks, behaviors, and policies that must be followed. These features of the organization form much of the context or framework for the measurement system. Good measurement systems usually work in concert with the organizational culture. But they are also capable of signaling need for fundamental changes in the culture.

Defining the framework is somewhat akin to stakeholder analysis in strategic planning. Stakeholders include at least three groups: customers, owners (and perhaps society), and employees. Each has valid, and sometimes conflicting, goals for the organization. These impact the organization's business model and in turn, the measurement system. We consider each in turn. See Figure 9.4.

1. *Customers:*   One goal of most organizations is to maintain and improve customer satisfaction, retention, and loyalty. Customer needs are usually stated in subjective terms. At the operational level, a consumer may simply want "the bill to be correct." At the tactical level, an important requirement may be that the invoice feed the customer's invoice payment system. Finally, at the strategic level, business customers may want to establish single sources of supply with companies they trust. It is important to recognize that there is an element of subjectivity in each customer requirement. Technicians are often dismayed by customers' lack of ability to give clear, objective requirements. But customers and their communications abilities are explicitly part of the overall framework.

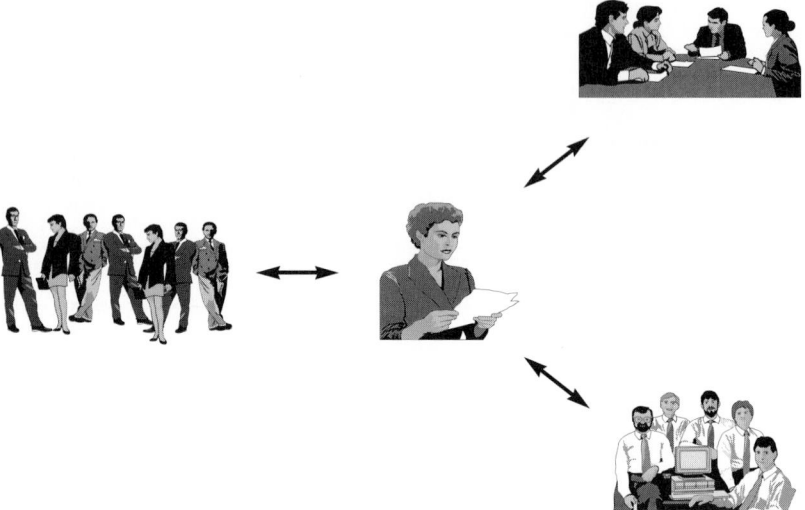

FIGURE 9.4 The measurement system is impacted by and impacts virtually all systems within an organization.

2. *Owners:* Owners of the business are usually less concerned about day-to-day considerations and more concerned about the economic viability, both long- and short-term, of the business. Their interests are reflected in corporate direction, competitive position, and financial performance. They usually wish, implicitly at least, to see costs kept to the lowest levels.

The impact of their interests on the measurement system is that certain things are more important to measure than others. Consider a company that wishes to pursue a strategy of price leadership. It wishes to be perceived as "just as good" as a major competitor, but wants to keeps costs as low as possible. Such a company will design its measurement system around competitive intelligence. It will not, for example, invest to learn of customers' preferred bill formats. In contrast, a company pursuing customer value and intimacy will design its measurement system to learn more about customers, how they use the organization's products and services, and how to use these measurements in defect prevention. It will invest to learn how to make its invoices a source of advantage.

3. *Employees:* Insofar as measurements systems are concerned, employees are stakeholders because they depend on the organization for their livelihood. Many are also stakeholders because they make decisions and take actions, others because they collect data, and so forth. Employees may view the measurement system as a device of management control, but good measurement systems are also empowering. In our operational example, the day-to-day decision maker could be owner of the guiding function, the billing process owner, or even a product manager. There may be good reasons for the process owner—local custom, required skill, union rules—to be the decision maker. But it is usually preferable that decisions be made as close to the action as possible. So unless there is a compelling reason otherwise, the owner of the guiding function is the preferred decision maker. In contrast, poor measurement systems require much additional time and are of dubious value.

A second aspect of understanding the framework involves the range of possible decisions and actions. A list of such decisions and actions is called the "decision/action space." In some cases creating this list is a straightforward exercise and in others it is nearly impossible. At the lowest, or operational, level it is usually possible to describe the decision space completely. Thus, in our first example, there are only a few possible decisions:

- The process is in control and performing at an acceptable level and should be left alone,
- The process is out of control and must be brought under control.
- The process is in control but not performing at an acceptable level. It must be improved.

At the tactical and strategic levels, the exercise of defining the decision/action space becomes more difficult. The range of possible decisions may be enormous, many possible decisions may be difficult to specify beforehand, and some decisions may be enormously complex. In our second (tactical) example, one possible decision is to leave the invoice alone. But the invoice can also be improved, possibly in a virtually unlimited number of ways. There are any number of incremental improvements to the formatting and accounting codes. Or the invoice can be wholly redesigned. A paper invoice may be replaced with an electronic one, for example. Finally the invoice may even be replaced with a superior invoice based on electronic commerce on the Internet.

Experience suggests that the more carefully the decision/action space is defined, the better the resultant decisions. This is just as true at the strategic level as it is at the operational level.

Unfortunately there are no complete templates for defining the framework. Any number of other considerations may be pertinent. For examples, legal obligations, safety rules, or technical limitations may be very important.

Framework Document. The end result should be a framework document that captures the major points of the work conducted here. It should describe major business goals and strategies and customer requirements, define the decision/action space and decision makers (by name in some cases, by job classification in others), note important constraints (financial and other), and reference more detailed business plans and customer requirements. And, as the business grows and changes, so too should the framework document.

Plan Measurements.
Once the planner understands decision space and the framework in which the measurement system operates, plans for the remaining steps of the process are made. The output of this step is a "measurement protocol", a document that describes the "whats, whens, wheres, hows, and how often" of data collection, storage, and planned analyses. Perhaps most importantly, the protocol should also describe the "whos"—who is responsible for each step. Figure 9.5 portrays the landscape to be covered. The most important issues to be addressed are discussed below.

Data Collection: What to Measure. Above, we noted that most customer requirements are stated in subjective terms. These requirements have to be translated into a set of objective measurements. A good example involves the early days of telephony. Customers' most basic requirement was "to hear and be heard." An almost unlimited number of problems can thwart this basic requirement. And, except for the actual speech into the mouthpiece and sound emanating from the earpiece, a phone call is carried electrically. A remarkable series of experiments helped determine that three parameters, loss, noise, and echo, each measurable on any telephone circuit or portion thereof,

	Data collection	Data storage	Analysis, synthesis, recommendations presentation
What			
Where			
When			
How			
How often			
Who			

FIGURE 9.5   The landscape to be covered by a measurement protocol.

largely determined whether the customer could hear and be heard (Cavanaugh, Hatch, and Sullivan 1976; AT&T 1982).

In recent years, Quality Function Deployment (Hauser and Clausing 1988) has proven to be an invaluable tool in helping map subjective user requirements into objective criteria for process performance. Figure 9.6 illustrates an ideal scenario in which the user requirement for a "correct bill" is first translated into a small number of objective parameters that are further translated into requirements on steps in the billing process (and, in particular, on guiding).

Naturally, many requirements will never lend themselves to objective measurement. Our second example, involving changes to a feature set, is such an example. Here the primary sources of data will be customer satisfaction surveys, customer feedback, observation of competitors' features, and views of likely technological innovations.

In some cases, it is pretty clear what you would like to measure, but you simply can't measure it. A famous story involves the vulnerability to enemy fire in World War II planes. The goal was to have more aircraft complete their missions and return safely. And, ideally, one would like to determine where aircraft that didn't return were hit. But these aircraft were not available. Good surrogate measurements are needed in such cases. The problem with World War II aircraft was addressed by examining where planes that did return were hit and assuming that those that didn't return were hit elsewhere.

In almost all cases, literally dozens of possible measurements are possible. The planner is usually well advised to select "a critical few" measurements. There are decreasing returns as measurements are added, and too many measurements can overwhelm the measurement system. The planner should list possible measurements and rank-order them. There will be an essential few that he/she must select. Other measurements should only grudgingly be admitted. Reference to the framework is usually most helpful in making the necessary selections.

Many planners fall into a trap by concentrating on getting a few good measures for each step of a process and giving insufficient attention to overall measurement. There is a compelling logic that, in billing for example, if each step performs correctly, then the overall process will perform correctly. Unfortunately this logic is often incorrect. Too many problems can arise between steps, where accountability is not clear. An overall measure of bill correctness and measures of correctness at each step are needed. The principles of business process management are covered in Section 6.

Precise definitions of what is measured are essential, as slight changes in definition can produce very different results, a fact that advertisers may exploit (see Schlesinger 1988 for one example).

Data Collection: Where.   The planner must simultaneously determine where to make measurements. In quality management, the usual evolution is from inspection at the end of a production process to measurement of in-process performance. Immature systems place greater weight on inspection, more mature ones on in-process measurement (see Ishiakawa 1990).

Data Collection: When, How, How Often.   In some cases, no new data are collected, but rather existing data from "customer accounts" or other databases are used. The planner is still advised

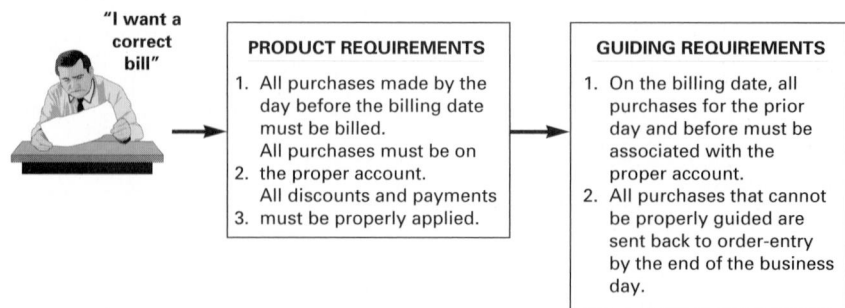

FIGURE 9.6   Customer requirements are usually subjective. They need to be translated into objective measurable parameters. Here it is done for the requirement "I want a correct bill" and the guiding step of the billing process.

to learn about the intricacies of data collection, as data collected for one purpose may not suit another.

The planner next specifies how, when, and how often measurements are to be made. Each should be spelled out in full detail. "How" involves not only how a particular measurement is to be made, but also how the measurement equipment is to be calibrated and maintained and how accurate data are to be obtained. "When" and "how often" must be addressed to ensure that ample data are available. The interested reader is referred to Sections 44, 45, and 47.

Data Storage and Access. Perhaps nothing is more frustrating than knowing "the data are in the computer" but being unable to get them. Planners too often give insufficient attention to this activity, and data storage and retrieval becomes the Achilles' heel of the system. Suffice it to note that, despite an explosion in database technology, especially in ease of use, data storage and retrieval are not easy subjects and should be carefully planned.

Data Analysis, Synthesis, Recommendations, and Presentation. Finally, the planner must consider the analysis, synthesis, formulation of recommendations, and presentation step. While all the analyses that will be carried out cannot be planned, certain basic ones should be. Thus, in our operational example, the addition of a point to a control chart at specified intervals should be planned in advance.

Who. Just as important as what, where, when, and how is who. Who collects data, who stores them, who plots points on control charts, who looks at data in other ways. All should be specified.

Measurement Protocol. The output of this step is a measurement protocol that documents plans for data collection and storage, analysis/synthesis and presentation. In effect, the measurement protocol defines the sub-processes to be followed in subsequent steps. Written protocols appear to be common in many manufacturing, service, and health-care settings. In many other areas, particularly service areas, written protocols are less common. This is poor and dangerous practice. Protocols should be written and widely circulated with those who must follow them. There are simply too many ways to interpret and/or bypass verbal instructions. The protocol should be carefully maintained. Like many useful documents it will be in constant revision.

Collect Data. When all goes well, data collection involves nothing more than following the measurement protocol. All going well seems to be the exception rather than the rule, however. For this reason, those making measurements should maintain careful logs. Good discipline in maintaining logs is important. Logs should be kept even when calibration and measurement procedures go as planned. It is most important that any exceptions be carefully documented. One topic, data quality, deserves special attention (see also Redman 1996 and Section 34). Unfortunately, measuring devices do not always work as planned. Operators may repeat measurements when, for example, values are out of range. Or data analysts may delete suspect values. Detecting and correcting erred data goes by many names: data scrubbing, data cleanup, data editing, and so forth. It is better to detect and correct readings as they are made, rather than later on. And naturally it is best to control data collection so that errors are prevented in the first place. But wherever errors are caught, changes in data must be carefully logged.

Protocols for tactical and strategic systems often call for literature scans, attendance at professional conferences, discussion with consultants, and so forth. When data are gathered in this manner, it is important that sources be documented. It is best to determine original sources.

Analyze, Synthesize, Formulate Results, and Present Results and Recommendations. Once data are collected, they must be summarized and presented in a form that is understandable to decision makers. This step is often called "data analysis". But that term is a misnomer. "Analysis" is defined as "separating or breaking up of any whole into its parts so as to find out their nature, proportion, function, relationship, etc." Analysis is absolutely essential, but it is only one-fourth of the required activity. The other three-fourths are "synthesis", "formulation of results", and "presentation". Synthesis is "composition; the putting of two or more things together so as to form a whole: opposed to analysis." Alexander's henchmen (and many others) seem not to have heard of synthesis. Next, specific "recommendations" for decision/action are developed. Finally, presentation involves putting the most important results and recommendations into an easily understood format.

That said, we use "analysis" as a shorthand for analysis, synthesis, formulation of results, and presentation. There are four steps:

- Completing planned analysis
- Exploratory data analysis (when appropriate)
- Formulation of results and recommendations
- Presentation of results and recommendations

For the operations example, the planned data analysis and presentation involves nothing more than computing an average and control limits and plotting them and requirements lines on a chart. Such a chart is presented in Figure 9.7. Simple as it is, the control chart is ideal:

1. It prescribes the proper decision for the manager.
2. It is graphical (and most people more readily interpret graphs) and visually appealing.
3. It is easy to create, as it does not require extensive calculation (indeed, much of its utility on the factory floor stems from this point).
4. It empowers those who create or use it.
5. Perhaps most important, the control chart provides a basis for predicting the future, not just explaining the past. And the whole point of decision making is to make a positive impact on the future.

Unfortunately, in tactical and strategic situations, this step is not so simple. We have already noted that many analyses should be planned in advance. Time should also be allotted for data exploration (also called data analysis, exploratory data analysis, data visualization, etc.). There are often "hidden treasures" data, waiting to be discovered. Indeed explosions in database and graphical exploration technologies can increase any organization's ability to explore data. More and more, all data, including details of all customer transactions and operations, are available. The most critical element, though, is not tools, but inquisitive people who enjoy the detective work needed to uncover the treasures.

The output of this step is a "presentation package." It may be nothing more than the control chart. The new tools also make it possible for every organization to present results in a clear,

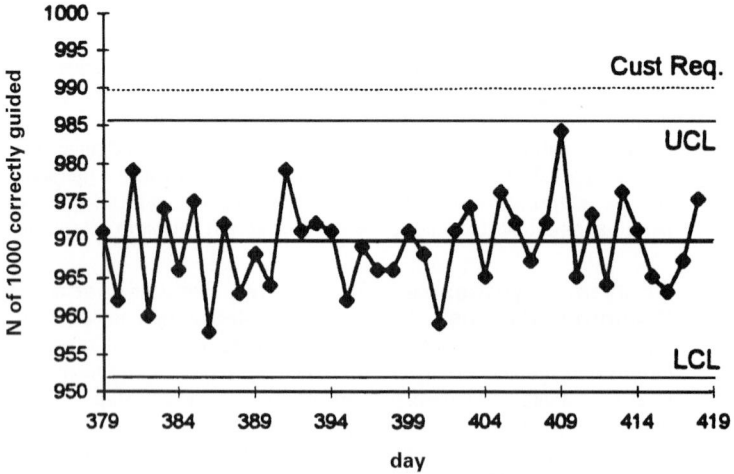

FIGURE 9.7   The control chart, quite possibly the best presentation vehicle in quality management.

understandable, and engaging manner to every organization. Experience suggests that good presentation is

- *Comprehensive:* The presentation covers the points of view of all pertinent stakeholders. It is, in effect, a "balanced scorecard" (Kaplan and Norton 1992; Eccles 1991).
- *Presented in layers:* High-level summaries that cover the landscape are presented in overview and layers of additional detail are available, as needed.
- *Graphical:* Almost everyone prefers well-conceived graphical summaries to other forms of presentation. See Tukey (1976), Tufte (1983), and Chambers et al. (1983) for good examples and practice.
- *Fair and unbiased.*
- *To the point:* Recommendations should be specific and clear.

We conclude with a quotation: "We also find that common data displays, when applied carefully, are often sufficient for even complex analyses…" (Hoaglin and Velleman 1995).

Data Quality and Measurement Assurance.   Clearly, decisions are no better than the data on which they are based. And a data quality program can help ensure that data are of the highest possible quality. One component of a data quality program is measurement assurance. The National Institute of Standards and Technology (NIST) defines a measurement assurance program as "a quality assurance program for a measurement process that quantifies the total uncertainty of measurements (both random and systematic components of error) with respect to national or other standards and demonstrates that the total uncertainty is sufficiently small to meet the user's requirements." (Carey 1994, quoting Belanger 1984). Other definitions (Eisenhart 1969; Speitel 1982) expanded, contracted, or refocused this definition slightly. All definitions explicitly recognize that "data collection," as used here, is generally itself a repeatable process. So the full range of quality methods is applicable. Clearly measurement assurance is a component of a good measurement system.

But the data quality program should be extended to the entire system, not just data collection. The measurement system can be corrupted at any point, not just at data collection. Consider how easy it is for a manager to make an inappropriate decision. Figure 9.8 illustrates a simple yet classic situation. Note that the control chart clearly indicates a stable process that is meeting customer needs. However, an inexperienced manager, at the point in time indicated on the chart, notes "deteriorated performance in the previous three periods." The manager may decide that corrective action is needed and take one, even though none is indicated. Such a manager is "tampering." Unless saved by dumb luck, the best he or she can accomplish is to waste time and money. At worst, he or she causes the process to go out of control with deteriorated performance.

In many organizations it is common to "manage the measurements, not the process." Departmental managers urged to reduce their travel costs may be facile at moving such costs to their training budgets, for example. In good measurement systems, the measurement assurance program helps all components function as planned.

Checklist.   Figure 9.9 summarizes the elements of a good measurement system for the billing example. The planner, moving from left to right on the figure, first defines the desired decision/action space. Next, the overall context is defined. It consists of three components: the customer's overall requirements and specific requirements on this step, the choice of operator as the decision maker and his/her understanding of control charts, and the budget allotted for quality control on this assembly line. The measurement protocol specifies the data collection plan, and raw data is collected accordingly. The control chart is the presentation vehicle.

To conclude this section, Figure 9.10 presents a measurement system checklist. It may be used to plan, evaluate, or improve a system. Not all items on the list are of equal importance in all systems.

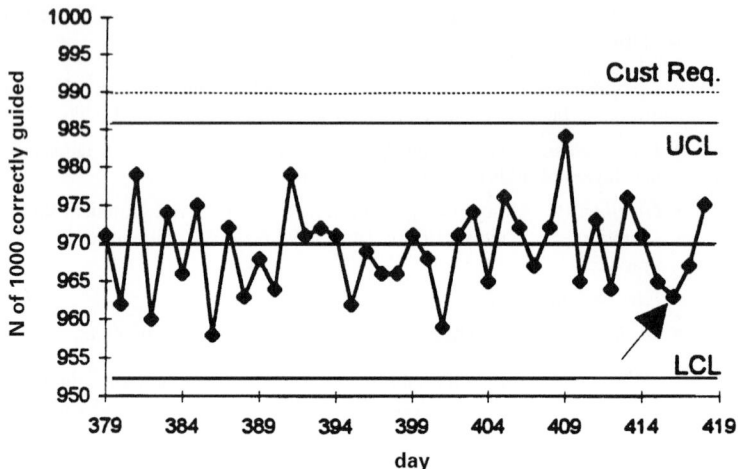

FIGURE 9.8 The control chart misinterpreted. At point 417, the overzealous manager sees evidence of degraded performance.

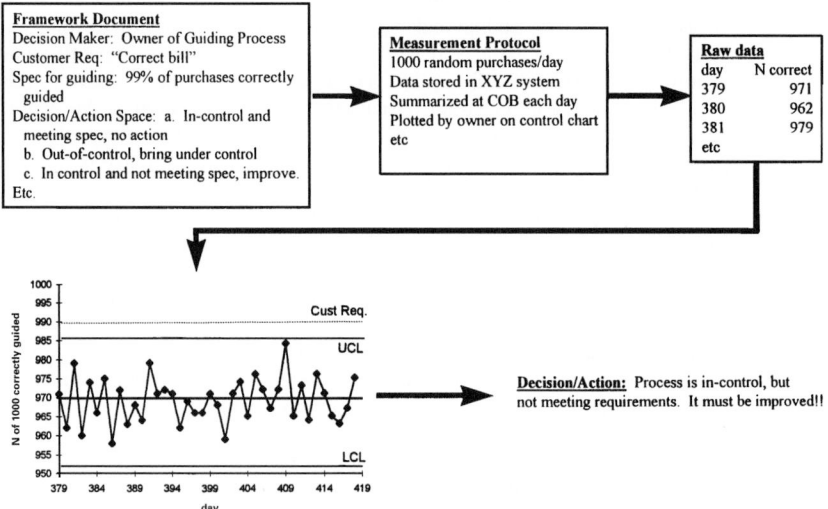

FIGURE 9.9 The major elements of a measurement system for the guiding step are summarized.

HIERARCHIES OF MEASUREMENTS

Objective Measurement versus Subjective Measurement. Texts on science, management, and statistics often extol the virtues of objective measurement, made on a nominal scale, over subjective "opinions," especially those of the "Yes/No" variety. The reasons are simple: Opinions are imprecise and subject to change.

These disadvantages aside, subjective measurements have certain advantages that should not be dismissed. First and foremost, customer opinion is the final arbiter in quality management. Second,

---

PLANNER'S CHECKLIST

Make decision/take action
- Is the decision/action space specified as clearly as possible?
- Are the planned actions consistent with the decision and the data that lead to them?

Define framework
- Are customer requirements clearly defined?
- Are the requirements of owners clearly defined?
- Are employee requirements clearly defined?
- Are any other major stakeholders identified? Are their requirements clearly defined?
- Are decision makers named?
- Is the framework documented?

Plan measurements
- Are plans for making measurement clearly laid out, including:
  - What is to be measured?
  - When are measurements to be made?
  - Where are measurements made?
  - How are measurements to be made, including calibration routines, data editing, and how a measurement log is to be used?
  - How often are measurements to be made?
  - Who is responsible for measurement, including calibration?
- Are plans for data storage clearly laid out, including:
  - What data are to be stored?
  - When are they to be stored?
  - Where are they to be stored?
  - How is storage to be accomplished?
  - How often are data stored? How often are data backed up?
  - Who is responsible for data storage and data management?
- Are planned analyses and data presentations clearly defined, including:
  - What analyses are planned?
  - When are planned analyses conducted?
  - Where are analyses conducted (i.e., which analytic environment)?
  - How are planned analyses carried out?
  - How often are routine analyses made?
  - Who conducts the planned analyses?
- Is the measurement protocol written?
- Are those who make measurements familiar with the protocol?
- Is the protocol under change management?

Collect and store data
- Is the measurement protocol followed?
- Do data collection plans provide high-quality data?
- Is a measurement log of exceptions maintained?

Analysis, synthesis, present results
- Is the measurement protocol followed?
- Are presentation packages clear and understandable?
- Are results presented in a comprehensive and fair manner?
- Is sufficient time allotted for data exploration?

Data quality program
- Is the data quality program comprehensive? Does it cover all aspects of the measurement system?
- Is the data quality program documented?

FIGURE 9.10   A measurement system designer's checklist.

there are always requirements that customers are not able to state in objective terms. A fairly typical customer requirement is: "I want to trust my suppliers." "Trust" simply does not translate well into objective terms. Of course, the customer may be able to give examples that illustrate how a supplier can gain trust, but these examples do not form a solid basis for a measurement system. Finally, there is a richness of emotion in customer opinion that does not translate well into objective parameters. If the customer discussing trust above adds: "If this critical component fails, our product will fail. We'll lose customers and jobs if that happens," there can be no mistaking this customer's priorities and the sense of urgency associated with such components.

While it will still be necessary to make objective measurements, the designer of the measurement system should also work to ensure that everyone hears the "voice of the customer" (Davis 1987).

Systems of Systems. So far, we have treated our operational, tactical, and strategic examples as if they were independent of one another. Obviously this is not the case. Ideally, measurements made at a lower level can be integrated into higher-level measurements, and higher-level measurements provide context (framework) to help interpret lower-level measurements. We call this property "integrability." The customer requirement for a "correct bill" should lead to overall measures of billing process performance and of the performance of each step. It is possible to achieve a certain amount of integrability. But as a practical matter, integrability is harder than one might expect, even with financial measurements.

Manager's Intuition. Measurement systems can take a manager only so far. Even the best measurement systems cannot eliminate all risk in decision making. And Lloyd Nelson has noted that "the most important figures are unknown and unknowable" (Deming 1986). Corporate traditions are made of the fiercely independent leader who defied the numbers to take the organization into new and successful directions (although I suspect that many organizations are also long gone because of similar decisions). Morita's decision to invest in the Walkman is a famous example of such a tradition. So clearly, there are times when the decision maker's intuition must take over. Managers and organizations must recognize this situation. Certainly decisions supported by data are preferred, but organizations must not require that all decisions be fully supported. The danger that an opportunity will pass or a potential crisis will escalate while diligent managers seek data to support a clear direction is simply too great. Organizations must support individual managers who take prudent risks. And individual managers should train their intuition.

STARTING AND EVOLVING MEASUREMENT SYSTEMS

Just as organizations are adaptive, so too are their systems. Measurement systems must be among the most adaptive. Day-to-day, they are integral to the functioning of the organization at all levels. It is not enough for measurement systems to keep pace with change. They also must signal the needs for more fundamental change. In this section we provide some practical advice for starting and evolving measurement systems.

1. It is usually better to build on the existing system than to try to start from scratch. All existing organizations have embedded measurement systems. Even start-ups have rudimentary financial systems. So one almost never begins with a "clean sheet of paper." Fundamental business changes will require new types of measurements (indeed the quality revolution forced many organizations to expand their focus to include customers) but they will only rarely eliminate all established measurements or measurement practices. In addition, in times of great change, the measurement system is an old friend that can be counted on to provide a certain amount of security.

2. Experiment with new measurements, analyses, and presentations. Learn what others are doing and incorporate the best ideas into your system.

3. Prototype. It is difficult to introduce new measures. Prototyping helps separate good ideas from bad, provides an environment in which technical details can safely be worked out, and gives people needed time to learn how to use them.

4. Actively eliminate measures that are no longer useful. This can be very difficult in some organizations. But we have already noted that good measurement systems are not oppressive. Similarly, we have noted the need to create new measures. So those that have outlived their usefulness must be eliminated.

5. Expect conflicts. We noted in the previous section that it is not usually possible to fully integrate measurements. Conflicts will arise. Properly embraced, they are a source of good ideas for adapting a measurement system.

6. Actively train people about new measures, their meaning, and how to use them.

SUMMARY OF PRINCIPLES

We conclude this section with the Top 10 Measurement System Principles:

1. Manage measurement as an overall system, including its relationships with other systems of the organization.

2. Understand who makes decisions and how they make them.

3. Make decisions and measurements as close to the activities they impact as possible.

4. Select a parsimonious set of measurements and ensure it covers what goes on "between functions."

5. Define plans for data storage and analyses/syntheses/recommendations/presentations in advance.

6. Seek simplicity in measurement, recommendation, and presentation.

7. Define and document the measurement protocol and the data quality program.

8. Continually evolve and improve the measurement system.

9. Help decision makers learn to manage their processes and areas of responsibility instead of the measurement system.

10. Recognize that all measurement systems have limitations.

ACKNOWLEDGMENTS

The author wishes to thank Anany Levitin of Villanova University and Charles Redman of Nashville, TN, whose comments on an early draft of this section led to substantial improvement.

REFERENCES

AT&T Engineering and Operations in the Bell System. Members of Technical Staff, AT&T Bell Laboratories (1982), Holmdel, NJ.

Belanger, B. (1984). *Measurement Assurance Programs: Part 1: General Introduction.* National Bureau of Standards, NBS Special Publication 676-1, Washington, D.C.

Carey, M. B. (1994) "Measurement Assurance: Role of Statistics and Support from International Statistical Standards." *International Statistical Review,* vol. 61, pp. 27–40.

Cavanaugh, J. R., R. W. Hatch, and J. L. Sullivan (1976). "Models for the Subjective Effects of Loss, Noise and Talker Echo on Telephone Connections." *Bell System Technical Journal,* vol. 55, November, pp. 1319–1371.

Chambers, J. M., W. S. Cleveland, B. Kleiner, and P. A. Tukey (1983). *Graphical Methods for Data Analysis.* Wadsworth International Group, Belmont, CA.

Davis, S. M. (1987). *Future Perfect.* Addison-Wesley, Reading, MA.

Deming, W. E. (1986). *Out of the Crisis.* Massachusetts Institute of Technology Center for Advanced Engineering Study, Cambridge, MA.

Drucker, P. F. (1995). "The Information Executives Truly Need." *Harvard Business Review,* vol. 73, no. 1, January–February, pp. 54–63.

Eccles, R. G. (1991). "The Performance Measurement Manifesto." *Harvard Business Review,* vol. 69, no 1, January–February, pp. 131–137.

Eisenhart, C. (1969). "Realistic Evaluation of the Precision and Accuracy of Instrument Calibration Systems," in Ku, H. K., (ed.). *Precision Measurements and Calibration—Statistical Concepts and Procedures.* National Bureau of Standards, NBS Special Publication 330-1, Washington, D.C.

Finkelstein, L., and M. S. Leaning (1984). "A review of fundamental concepts of measurement." *Measurement.* vol. 2, January–March, pp. 25–34.

Futrell, D. (1994). "Ten Reasons Why Surveys Fail." *Quality Progress,* vol. 27, April, pp. 65–69.

Hauser, R., and D. Clausing (1988). "The House of Quality." *Harvard Business Review,* vol. 66, no. 3, May–June, pp. 17–23.

Hoaglin, D. C., and P. F. Velleman (1995). "A Critical Look at Some Analyses of Major League Baseball Salaries." *The American Statistician,* vol. 49, August, pp. 277–285.

Ishikawa, K. (1990). *Introduction to Quality Control.* 3A Corporation, Tokyo.

Kaplan, R. S., and D. P. Norton (1992). "The Balanced Scorecard—Measures that Drive Performance." *Harvard Business Review,* vol. 70, no. 1, January–February, pp. 71–79.

Kaplan, R. S., and D. P. Norton (1993). "Putting the Balanced Scorecard to Work." *Harvard Business Review,* vol. 71, no. 5, September–October, pp. 134–149.

Meyer, C. (1994). "How the Right Measures Help Teams Excel." *Harvard Business Review,* vol. 72, no. 3, May–June, pp. 95–103.

Redman, T. C. (1996). *Data Quality for the Information Age.* Artech, Norwood, MA.

Roberts, F. S. (1979). *Measurement Theory with Applications to Decisionmaking, Utility and the Social Sciences.* Addison-Wesley, Reading, MA.

Schlesinger, J. M. (1988). "Ford's Claims About Quality Show Difficulty of Identifying Top Autos." *Wall Street Journal,* March 14.

Speitel, K. F. (1982). "Measurement Assurance," in G. Salvendy, *Handbook of Industrial Engineering.* Wiley, New York.

Tufte, E. R. (1983). *The Visual Display of Quantitative Information.* Graphics Press, Cheshire, CT.

Tukey, J. W. (1976). *Exploratory Data Analysis.* Addison-Wesley, Reading, MA.

Webster (1979). *Webster's New Twentieth Century Dictionary,* 2nd ed. William Collins Publishers, Simon & Schuster, New York.

SECTION 10

COMPUTER APPLICATIONS TO QUALITY SYSTEMS[1]

Fredric I. Orkin
Daniel Olivier

INTRODUCTION

Everyone dealing with the quality function by now has either faced the computer or retired. The exponential growth of computer systems will continue unabated for the foreseeable future and invade all aspects of our jobs. Management and practitioners alike must devote significant effort supplementing their computer skills and evaluating the risks to their companies associated with change. The computer revolution has also been a primary force in the expansion of quality activity into service businesses and functions. Breakthrough technology in communications and data storage has already begun the transition from powerful stand-alone machines back to large, virtual machines of unlimited potential. The Internet and its derivatives will spawn this revolution.

This section will include segments of the quality program affected by the computer: (1) communications and computer systems, (2) design control, (3) testing and validation, (4) software quality, (5) statistical application, (6) computer-aided inspection, (7) reliability and safety, and (8) future trends.

Examples and checklists are provided to assist the reader in the application of computer and related technology. The contributions of computers to quality-related activities are presented throughout this Handbook. See particularly Section 28, Quality in a High Tech Industry.

Communications and Computer Systems. For purposes of this section, a computer system is defined as a group of interacting hardware equipment and software that performs an independent function. Computer systems can be as small as appliance- or automobile-type microprocessor

---

[1]In the Fourth Edition, material for the section on computers was supplied by Fredric I. Orkin.

chips with self-contained software or as large as mainframe computers running specialized database management software. Personal computers running commercial word processing and spread sheet software programs are generally considered as productivity tools rather than computer systems unless the commercial software is specifically developed and tested to perform a unique function. Examples are spread sheet programs that analyze financial, statistical, or engineering data, and word processing programs that present forms for entry of specialized information on data. Those commercial software applications are called "templates" and "macros."

Computer systems have experienced a rapid evolution in recent years, from stand-alone processing systems dominated by central mainframe processors to widely distributed architectures that provide almost unlimited communication capabilities. The explosion in popularity of systems such as the Internet has opened up tremendous new opportunities for information access. Almost any subject can be researched through the use of the Internet search engines. Systems such as the Internet also present new quality challenges.

Figure 10.1 lists some of the well-known quality societies and information sources that are currently accessible via the Internet. Figure 10.2 is a copy of the logo on The World Wide Web site home page for the American Society for Quality.

Of increasing popularity are "Intranet" applications. Intranet is a private corporate network using Internet technologies. The Internet and Intranet, combined with the increasing power of on-line document management systems, have transformed the way that companies communicate and perform work. The future will be characterized by automated on-line services that provide unlimited access to and rapid retrieval of information from huge databases of specifications and reference materials. These new capabilities have several quality implications:

- Increased privacy concerns about access to personal information, such as salary and evaluation data
- Increased security concerns about external access to corporate information, such as financial and product development data
- Accuracy and currency of information available

Also of concern is the potential for the introduction through these channels of programs that may cause corruption or destruction of data. The most familiar corrupter is called a "virus." A virus is computer code that executes when an infected program is run. This has traditionally meant that only stand-alone programs (executable files) can be infected. Recently, however, viruses have been inserted into programs that can allow nonexecutable files to propagate a virus.

The threat of viruses is real. There were about 4000 known viruses at the end of 1994, and the number of new viruses is growing at a steady pace (McAfee 1996). Viruses can infect data on computers from mainframes to laptops. Viruses can be distributed across phone lines from bulletin boards or can enter computers via floppy disks. A virus infection can destroy files; it can cost countless hours to remove the virus and recover lost data from affected programs.

Quality practices must address the growing virus risk, which is a consequence of the increasing interconnection of computer systems. Available protection techniques are shown in Figure 10.3.

DESIGN CONTROLS

ISO 9001 is the model for quality assurance in design/development, production, installation, and servicing promulgated by the International Organization for Standardization. Subclause 4.4, Design Control, spells out the elements which make a quality system acceptable from the standpoint of assurance of quality in design and development. These elements of design control also apply to the development of computer systems and are used by successful companies to assure efficient and high-quality development. Design control elements include classical engineering activities but are addressed in the process-oriented terminology consistent with ISO 9000. Design control phases are discussed in terms of process inputs and outputs for each of the design and development phases: requirements definition; design; code and debug; unit, integration, and acceptance testing; and operation and maintenance.

American National Standards Institute (ANSI)
American Society for Quality (ASQ)
Association for Computing Machinery (ACM)
Deming Electronic Network
Electronic Industries Association (EIA)
European Standards Institute
Institute of Electrical and Electronics Engineers (IEEE)
International Electrotechnical Commission (IEC)
International Organization for Standardization (ISO)
The Juran Institute
National Institute of Standards and Technology (NIST)
Software Engineering Institute (SEI)
Underwriters Laboratories, Inc. (UL)

FIGURE 10.1   Quality resources available via the Internet.

FIGURE 10.2   Logo for the World Wide Web site home page for the American Society for Quality (http://www.asq.org).

Password controls to restrict system access only to authorized users

Password procedures that force password uniqueness and periodic changes

Training of personnel on how a virus may be detected and on how controls are used for checking the integrity of files and disks

Internal procedures to limit the programs that can be entered onto company workstations and networks

Use of virus checking programs to detect the presence of a virus

Diskless workstations to prevent the introduction of foreign files into critical computer systems

Periodic scanning of systems to detect the presence of viruses

Audits of system access and file activity to detect the presence of viruses

Regular system backups using different tapes (an infected tape if reused may be useless) to support file recovery if data is lost

FIGURE 10.3   Techniques to prevent virus infections.

Design and Development Plan.   The first element of design control is an organized plan for establishing design and development activities. The plan defines the activities to be conducted in support of the design effort and the parties responsible for each.

Organizational and Technical Interfaces.   Further, the plan defines the relationship between departments, technical disciplines, and subcontractors who are contributing to the design. It also establishes the documents to be produced by each group and responsibilities regarding review.

Design Input.   During the design input phase, the specification of requirements is developed. This document defines the essential elements of the product and becomes the basis for evaluating when the product design and development is complete.

Design Output.   The design output is the aggregate of all documents, design descriptions, analyses, test procedures, and test results that are produced during the actual design activities. For hardware development this may include functional block diagrams, schematics, drawings, and detailed component specifications. For software the design output may include design descriptions, module hierarchy charts, object-oriented diagrams, and other representations of the program. (See Section 20 for a discussion of software development.)

Design Review.   Informal reviews conducted by the project team provide quick feedback on specification compliance, misunderstandings, and errors. Formal reviews that include independent peers are scheduled milestone events that are often used to separate design phases. Reviews have been shown to be a more effective method for improving product quality than relying on testing (Wheeler, Brykczynski, and Meeson 1996). Reviews are important factors in increasing product quality and improving the efficiency of the design and development process. The most significant benefit from reviews is the early identification of errors. The earlier in the design process that errors are identified, the less costly they are to correct (Boehm 1985).

Design Verification.   Verification is the incremental checking performed during each phase of the design and development process to ensure that the process is being followed and that the low-level derived requirements have been correctly implemented. Verification activities include unit level testing, analysis studies, integration level testing, and reviews.

Design Validation.   Validation is assurance that the product conforms to the defined input requirements and needs of the customer. Validation must also encompass stress testing in the actual or simulated customer-use environment.

Design Changes.   Changes to the approved design are accomplished with controls similar to those applied for the initial development process. The changed design must be subjected to the same steps in the process.

Benefits of Design Control.   The addition of reviews during the early project stages will require scheduled time. Some will complain that valuable code and prototype development time will be lost. However, history has demonstrated that the investment in reviews is repaid in customer satisfaction and in reduced rework during testing and reduced errors in maintenance, as shown in Figure 10.4 (Olivier 1996). The chart, developed by the authors, shows two critical characteristics of design control reviews.

First is a savings in the effort invested in the development project. The added time required up front for reviews is represented by the shaded area on the left of the graph. The additional time spent in rework and refining the product during later stages if reviews are not conducted is shown by the shaded area on the right. It can be readily seen that the area on the left is smaller than the area on the right, indicating less effort and, therefore, reduced total cost attributable to design reviews. Errors found early in the development process are much easier to correct. Studies referenced by Humphrey

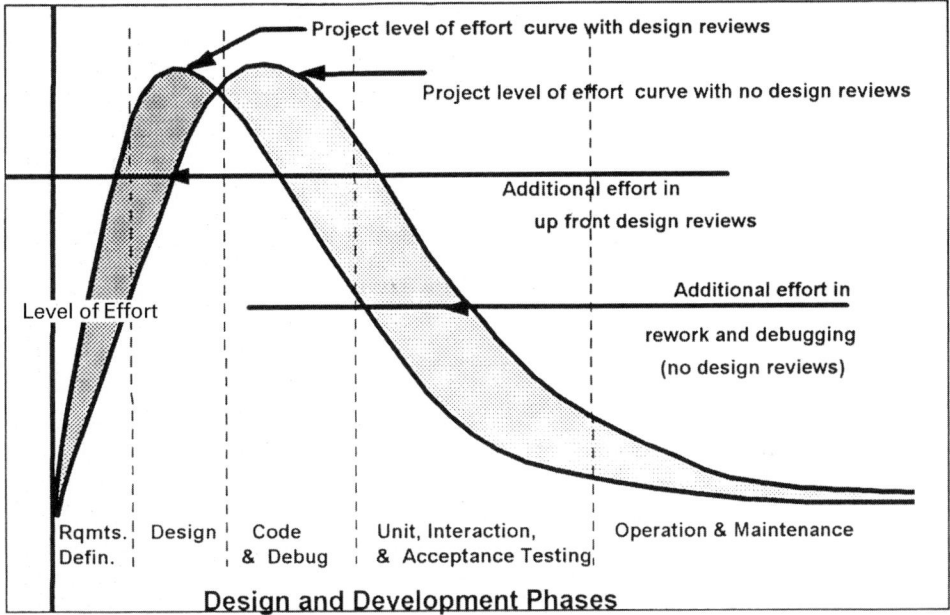

FIGURE 10.4   Impact of design review effort investment on projects. (*Source: Olivier 1996.*)

(1989) show that AT&T Bell Laboratories found third-party code review inspections to be 20 times more effective in finding errors than testing for them.

A second benefit of design review as shown in the figure is an increased level of quality. The quality improvement is demonstrated by the area under the curve shown for the maintenance phase. It is smaller when design control reviews are conducted, representing fewer errors and less rework. Freedman and Weinberg (1990) reference a 10-times reduction in the number of errors reaching each phase of testing for projects with a full system of reviews.

TESTING AND VALIDATION

With the increasing complexity and criticality of computer programs, traditional ad hoc methods for testing have become inadequate. Current computer system design and development models incorporate validation tests to ensure that system requirements have been satisfied. These models also include verification review and test techniques as part of the development process (IEEE 1986). Figure 10.5 illustrates the application of verification and validation to design and development phases (Olivier 1993).

Testing Environment.   Testing must ensure that the system operates correctly in the actual environment or, where such testing is not possible, in an environment that simulates the conditions of actual use. Stress testing in the actual-use environment is very effective in identifying errors that may otherwise remain undetected until after product release. Effective techniques to assure correct operation in the user environment must include "beta"-type testing, where early product versions are provided for customer-use testing to assure that the system functionality is consistent with the actual use environment.

Requirements-based testing conditions might not be representative of operational scenarios, and the requirements themselves might not be error free (Collins et al. 1994). The customers for most

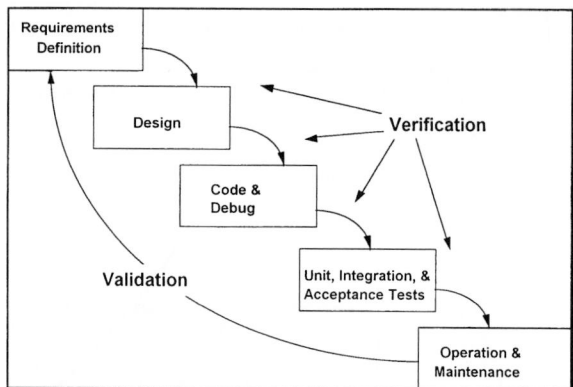

FIGURE 10.5   Verification and validation during design and development phases. (*Source: Olivier 1993.*)

computer systems are not knowledgeable enough, nor can they define requirements specifications precisely enough for validation (Kaner et al. 1993). Empirical data show that the majority of system failures arise from errors in the requirements themselves, not from failure to satisfy the defined requirements (Leveson 1995). Systems ultimately may be used in ways not envisioned during the development process. Errors are experienced as a result of user entry of unexpected values, the use of unanticipated equipment interfaces, and misinterpretation of reports and displays. These conditions create a marketing challenge for purveyors of software, who may ignore the challenge and resign themselves to providing less than perfect product.

Buyer Beware!   In purchasing software, the buyer would do well to keep a few caveats in mind:

- Testing for satisfaction of requirements supports the software quality model. However, this approach alone is no longer practical. Each significant piece of new off-the-shelf commercial software can be assumed to contain errors, even after thousands of millions of executions (Collins et al. 1994).
- While fitness for use recognizes the importance of the satisfaction of defined requirements, there may be errors in the defined requirements themselves.
- Freedom from errors is only one of many attributes which add up to fitness for use.

The success of the Microsoft Corporation and other software developers is based on the realization that customers will tolerate bugs in *software* as a trade-off for timely product enhancements, increased functionality, and low cost. This is in direct contrast to the reactions of the same customers to errors experienced in *hardware* quality where zero-defect performance is expected.

For many software developers, a result of these customer attitudes is that structured design is subordinated to early coding and arduous testing. At Microsoft, there is an attempt to accelerate the release of new products by investing tremendous hours in early coding and testing and subsequent debugging. A reference to the Windows NT development process during the last week of April 1993 showed the fixing of 1132 bugs while 713 new bugs serious enough to warrant testing were also discovered (*Wall Street Journal* 1993).

The unending pressure to increase the complexity of emerging computer and software systems is forcing the industry to rethink the cost and risk of this "brute force" strategy. In the authors' opinion, this complexity will dictate the gradual development of new software development tools and a shift by successful developers back to more structured design techniques.

Techniques available today that can be used to increase the effectiveness of the testing program for computer systems include:

- Training development personnel on effective review and testing techniques. Testers must understand that the goal of testing is to identify as many errors as possible and not to test a program to show that it works (Meyers 1979).

- Analyzing errors to identify the root cause. Error analysis leads to more effective review techniques and checklists that focus on avoiding historical error causes.

- Designing systems to support testability. This means including diagnostic routines and error-conditions logging, as well as designing tools in conjunction with the design and development of the core system functions.

- Tracking the number and severity of errors found per unit test time.

SOFTWARE ATTRIBUTES

Quality software programs exhibit certain attributes across programming languages and applications. A list of some of these key attributes is provided in Table 10.1 (McCall et al. 1977). These attributes include both subjective and objective measures and are only a subset of the total attributes of software programs. The applicable elements should be selected by the quality professional for application to each specific program.

Many application programs today are purchased off the shelf. For these programs quality assessment is different than for programs developed internally. It is often worthwhile to contact the vendor who developed the off-the-shelf program and current users to gain insight into the quality of the program. The authors have prepared a list of suggested steps for the evaluation and selection of off-the-shelf software programs (see Figure 10.6).

If the purchased programs are to be modified by in-house personnel, other information should also be requested, including: design documentation, source code, test procedures, and support utilities used for development and testing.

If software programs are developed internally, the developers should prepare a quality assurance plan to identify specific activities that are to be performed. Such actions should include those shown in Figure 10.6, as well as:

TABLE 10.1 Software Quality Attributes

Quality attribute	Description
Correctness	Extent to which a program satisfies its specifications and fulfills the user's mission objectives
Reliability	Extent to which a program can be expected to perform its intended function with required precision
Efficiency	Amount of computing resources and code required by a program to perform a function
Integrity	Extent to which access to software or data by unauthorized persons can be controlled
Usability	Effort required to learn how to operate, prepare input, and interpret output of a program
Maintainability	Effort required to locate and fix an error in an operational program
Testability	Effort required to test a program to ensure that it performs its intended function
Flexibility	Effort required to modify an operational program
Portability	Effort required to transfer a program from one hardware configuration and/or software system environment to another
Reusability	Extent to which a program can be used in other application—related to the packaging and scope of the functions that programs perform
Interoperability	Effort required to couple one system with another

Source: McCall, Richards, and Walters (1977).

1. Identify present and future requirements.
2. Survey available packages.
3. Examine documentation and manuals.
4. Determine data and communication interoperability with existing programs.
5. Survey existing users as to product acceptability. (Internet queries provide an optimal way to obtain this information.)
6. Request quality assurance and testing information from the developer.
7. Request a list of known bugs.
8. Review copyright and licensing requirements.
9. Obtain programs for internal trial execution.
10. Negotiate contract to include maintenance services and upgrades.

FIGURE 10.6   Steps in acquiring an application package.

- Internal audits to assure that internal design procedures are being followed
- Process and product quality measures
- Oversight of subcontractor activities
- Management of the configuration of the code and documentation
- Tracking of open program bugs and corrective actions
- Analyzing program bugs to identify process improvement opportunities
- Final release approval

Simulation and Fuzzy Logic Quality Tools.   Simulation modeling as a computer tool is gaining acceptance in many fields. It involves the use of mathematics to replace and predict physical test results. "By altering the input variables of a simulation model, different configurations can be created and, through successive experiments, each can be evaluated by analyzing the output variables designed to measure performance" (Smith 1994).

Finite element analysis software, combined with design of experiments (DOE), can result in a powerful quality engineering tool. Rizzo reports that simulation models were used to reveal a nonlinear buckling behavior of a short segment of the power conductor of an electronic system. Engineers were fascinated to find that their initial judgment was proven wrong by the model (Rizzo 1994). Chalsma also reports that "computer modeling and simulation have pushed much automobile testing back into the design stage" (Chalsma 1994).

"The effort to make computer-based systems more intelligent and more human certainly is centered on the use of fuzzy logic" (Chen 1996). "Fuzzy logic" is the application of mathematics (set theory) to represent and manipulate data that possess nonstatistical uncertainty.

Fuzzy logic represents a new way to deal with the vagueness of everyday life and has a major future role in product and process development (Bezdek 1996).

Zadeh (1965) observed that, as a system becomes more complex, the need to describe it with precision becomes less important. Subjective descriptors, such as hot, cold, near, far, soon, and late enter the design equations of control theory via fuzzy logic.

Traditional control systems are open- or closed-loop (Figure 10.7). They collect input data, perform computation, and provide an output. Closed-loop systems use feedback of the output function in the computation. With fuzzy systems, the input is "fuzzified" by placing it into overlapping groups or sets—for example, cool, warm, or hot—rather than into the set characterized by a specific temperature or range of temperatures. A feedback control system then uses a controller made up of fuzzy-logic components (if-then rules, variables, sets, and translation mechanisms) that describe the desired control. The results are then "defuzzified" and processed to the output (Zadeh 1965).

Figure 10.8, adapted from Miller (1996), shows the block diagram of a typical fuzzy-logic control system. Specialized electronic hardware can handle from 800 to 8,000,000 rules per second.

Fuzzy-logic controllers are found in such wide-ranging applications as home appliances, nondestructive testing, machine tool control, and inspection systems (see Figure 10.9).

Industrial application of fuzzy logic is on an exponential growth curve. Applications include feed-forward controls for industrial machine tools, robotic systems, medical devices, and process equipment. Kierstan (1995) reports of application of fuzzy logic systems to temperature control in automated food processing clean rooms with extreme temperatures, environment control in microbe hostile atmospheres, and monitoring the size and shape of pizzas!

STATISTICS

One of the first, and still important, uses of the computer in quality control is for statistical analysis (Besterfield 1979). Most statistical techniques discussed in this handbook may be programmed by using one of the standard programming languages. "The Shewhart Chart is a child of the pre-computer age, when all record-keeping was done manually. With today's off-the-shelf

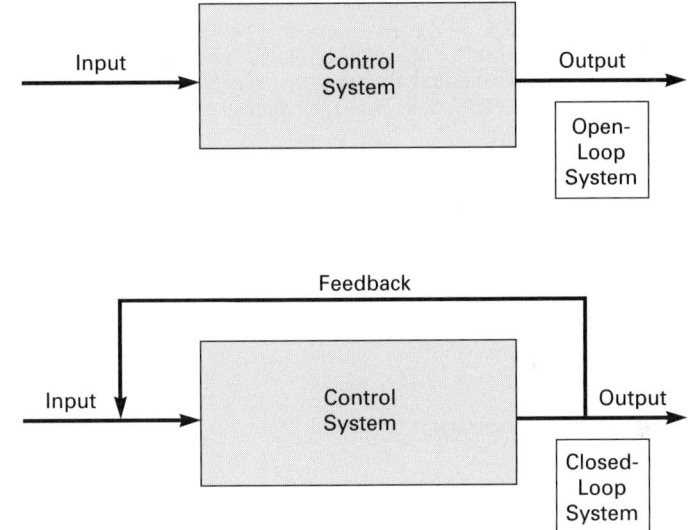

FIGURE 10.7   Traditional control system. (*Source: Zadeh 1965.*)

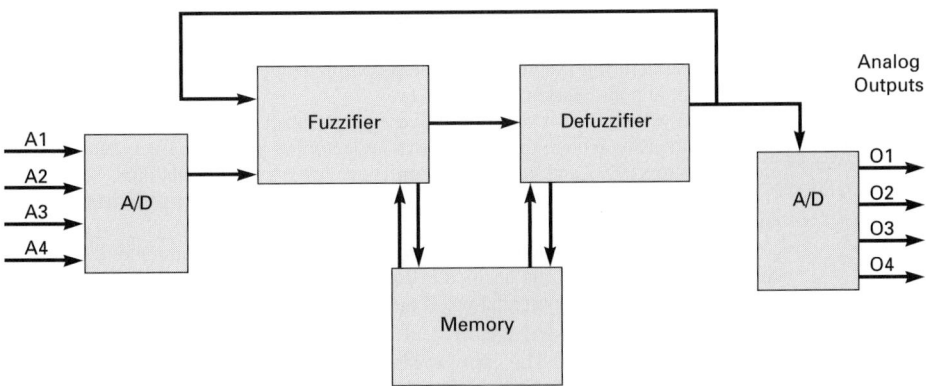

FIGURE 10.8   Fuzzy logic block diagram. (*Reprinted with permission from Electronics Now magazine, May 1996 issue. © Gernsback Publications, Inc., 1998.*)

> Washing machines that sense the degree of dirt in the wash water and optimize the cycle
>
> Video cameras with electronic image stabilizers that remove jitter caused by hand movement
>
> Air conditioners that sense the number of occupants and thermal load as well as environmental conditions of temperature and humidity
>
> Refrigerators that reduce heat shock when the door is opened and new food added
>
> Vacuum cleaners that sense the dirt load and control the cleaning cycle
>
> Wind power rotor pitch control
>
> Transit system speed control
>
> Soda can printing defect inspection
>
> Automobile brake and transmission control

FIGURE 10.9   Applications of fuzzy logic spark the imagination.

computing capabilities, the total technology system needed can readily be put together." (Marquardt 1984.)

Sources of Statistical Software.   *Quality Progress* annually publishes commercial sources for software. The 1996 issue lists 183 companies that supply statistical software products covering (Struebing 1996):

- Capability studies
- Design of experiments
- Sampling
- Simulation
- Statistical methods
- Statistical process control

COMPUTER-AIDED INSPECTION

Rapid acceptance of computer-aided design (CAD), computer-aided engineering (CAE), and computer-aided manufacturing (CAM) provides the database and information structure for the introduction of computer-aided inspection and computer-aided test. Cost incentives are also expected to accelerate the development of inspection and test automation to keep pace with major gains in manufacturing automation.

Reduction of defect levels to the parts-per-million range often requires computer-aided technology. Many industries are increasingly accepting inspection systems that are integrated with automated manufacturing systems. "This step completes the computer-integrated manufacturing (CIM) loop" (Reimann and Sarkis 1993).

Generally, automatic inspection will couple a transducer to a computer. Transducers can take the form of dimensional position indicators or indicators of physical effects such as force, flow, vibration, electrical properties, and magnetic properties. An American National Standards Institute (ANSI) standard for integrating the CAD and dimensional measuring instruments was published in 1990 (ANSI/CAM-I 1990).

The multitude of potential applications for automated inspection and the equipment and computer functions related to them are detailed in Table 10.2. This table, developed by the authors, should prove useful as a checklist for potential project ideas.

In-Cycle Gauging.   The use of touch-trigger probes to take measurements during a numerically controlled machine-tool cycle is called "in-cycle gauging." As reported by Reimann and Sarkis (1993), recent developments in this technology have led to the availability of new techniques that can compensate for machine variances and develop parametric probe-path data.

TABLE 10.2 Potential Applications for Automated Inspection

Industry applications	Equipment type	Transducer type	Computer function
Dimensional gauging	Automatic high-speed, noncontact video inspection, and optical comparators	Optical, laser, video, solid-state camera	Video image processing; autofocus; mass storage for uninterrupted cycle execution; part and table multiple-axis servo positioning; inspection of unaligned parts
	Coordinate measurement machine	Touch probe	Geometrical tolerance programming, tolerance analysis, data handling, multiple probe calibration, laser calibration, math processing, contouring, operator prompting, editing, feedback, accept/reject decision
	Computer-assisted gauging (lab)	Touch probe, electronic, air	Supervised prompting, automatic mastering, magnification set, zeroing display, statistics, counting, spec comparison, diagnostic testing
	Electronic gauges and measuring systems with computer interface	Calipers, micrometers, snap gauges, bore gauges, indicator probes, height gauges, air gauges, ultrasonic gauges, magnetic gauges, etc.	Direct digital output, gauges to host computer through interface
	In-cycle gauging on numerical control (NC) machines	Touch probe	On machine measurements, tool wear compensation, temperature compensation automatic check of tool offset, work location, table and spindle relationship
	Bench laser micrometer	Laser	Automatic laser scan, data handling, statistical dimension calculations, part sorting, accept/reject decision
	Holography	Laser	Automatic stress, strain, displacement, image processing

TABLE 10.2 Potential Applications for Automated Inspection *(Continued)*

Industry applications	Equipment type	Transducer type	Computer function
	Laser interferometer	Laser	Automatic temperature and humidity compensation data handling and storage, math processing
	3-D theodolite, coordinate, measurement	Optical	Interactive operator prompting, automatic angular measurement, data handling
	Scanning laser acoustic microscope (SLAM)	Laser, acoustic	Beam scanning, data processing
Electrical and electronic instrumentation	Temperature measurement	Thermocouple, thermistor, resistance temperature detector (RTD)	Calibration; data acquisition, analysis, and processing
	Robotic-printed circuit board test	Electronic	Robot control, fully automatic board test
	Weight and balance, filling and packaging, inspection	Electronic	Automatic tare, statistical processing, data recording
	Circuit analyzers	Electronic	Special-purpose test systems
	Automatic test equipment functional testers	All	Special-purpose test systems with complete real-time input, processing and output data
	Cable testers	Electrical	Automated harness continuity and high-potential testing
	Sem/iconductor testers		Automated test of standard and special-purpose chips
Lab devices and equipment	Chromatographs	Optical	Fully automatic preprogrammed sampling and data recording
	Strength of materials	Probe, force, displacement, strain gauge	Preprogrammed cycle operation: data, chart, and graphic output records; multichannel recording; on-line data processing

	Hardness testing	Probe	Robotic, fully automatic testing and recording, results analysis, and prediction
	Analyzers	All	Automatic calibration, testing, and recording
	Electron microscopes	Electromagnetic	Processing and materials analysis, preprogrammed for failure analysis
Optical imaging	Video borescope, fiber-optic inspection	Optical	Digital data image processing documentation
	Photographic	Optical	Fully automatic strobe, photographic sequencing and processing
	Video microscopes	Optical	Video image processing data documentation
	High-speed video recording	Optical	Automatic 200–12,000 frames per second stop-motion recording of machine and manual processes; motion analysis; data processing
Environmental and functional test equipment	Test chamber controls	Temperature, humidity, altitude	Preprogrammed cycle controls, time and data records
	Leak detection	Vacuum, gas, acoustic	Automatic zeroing, built-in calibration, automatic sequencing, tolerance checking, data processing and display
	Shock and vibration testing	Accelerometer	Automatic cycle control, built-in calibration, data logging and display
	Built-in equipment	Electrical, electronic	Preprogrammed part and system functional and environmental cycling, recording
	EMI measurement	Electronic, magnetic	Data processing, math analysis, recording
Materials testing equipment	Surface and roughness measurement	Stylus follower, air flow	Operator prompting, data analysis
	Coating thickness, sheeting thickness	Electronic, video, ultrasonic, beta backscatter	Calculation and math processing; display: self-calibration; automatic filter changing and positioning; prompting self-diagnostics; feedback; accept/reject decision

TABLE 10.2 Potential Applications for Automated Inspection (*Continued*)

Industry applications	Equipment type	Transducer type	Computer function
	Paper, plastic, and coated product process inspection for holes, particulates, streaks, thickness	Laser	Automatic high-speed processing, feedback controls, data analysis, and alarms
Nondestructive test equipment	Magnetic particle, eddy current	Probe	Self-regulation, calibration, data handling, defect recognition
	Ultrasonic flaw detection	Sonic, vibration	Automated quantitative analysis, curve matching, automated procedures, graphics data acquisition and storage
	Scanning laser acoustic microscope (SLAM) flaw detection	Laser, acoustic	Beam scanning, data processing, flow detection
	X-ray, fluoroscopic	Optical, electronic	Automatic calibration, operator prompting, data handling, statistics, stored programming, defect recognition
	Acoustic emission	Acoustic	Independent channel monitoring and display, linear, zone location, tolerance comparison, preprogrammed tests, graphics output, triangulation, source location
	Infrared test systems	Optical, video	Calibration, system control
	Radiographic, gamma	Optical, gamma	Programmable, automatic, self-diagnostic, safety malfunction interrupts, automatic defect recognition, robotic part handling, automatic detection of missing parts
	Computer-aided tomography (CAT)	X-ray	Data acquisition, processing, interpretation and imaging
	Nuclear magnetic resonance (NMR) scanner	Magnetic	Data acquisition, processing, interpretation and imaging

RELIABILITY AND SAFETY

Product Reliability and Safety.   The incorporation of microprocessors in a product under development changes the way in which the quality, reliability, and safety design objectives are defined, managed, and analyzed. Products that utilize microprocessors require an additional category consisting of program software, which adds a unique dimension to the problem of design analysis. While methods have evolved to analyze, predict, and control the reliability of conventional systems, software as an entity has an altogether different character, presents greater difficulties, and must be treated separately.

Even though the software aspect of the microprocessor is more difficult to analyze, the product objectives of a fail-safe design and self-testing system are easier to achieve when the design incorporates microprocessors. Some products, because of the type of application for which they will be used, have higher criticality than do others. The level of design effort that goes into each type of product must be commensurate with the application. The following discussion focuses on the reliability and quality of the microprocessor itself as an electronic device, the reliability of the software program associated with the processor, and the design advantages that the microprocessor allows.

Microprocessor Reliability.   Microprocessors and peripheral devices are supplied by many manufacturers. Typical device costs vary with the number of instructions, execution time, and underlying technology. The quality level of these hardware devices has a direct effect on the reliability of the computer system. They can be procured at specified levels of quality as defined by MIL-M-38510D (1977) and MIL-STD-883C (1983), which are keyed to the models for predicting reliability and failure rate that are given in MIL-HDBK-217F (1991). These specifications define levels of qualification testing, screening tests, and burn-in.

Commercial microprocessors have also demonstrated excellent performance, but testing is not documented. However, as a minimum, they generally undergo incoming inspection, visual inspection, electrical parameter testing, and burn-in.

Software Reliability.   Built-in software is an element of the total product that governs computational or control functions. Each step of the program development process is inherently error-prone. Errors can be introduced from misinterpretation, mistakes in translation, or mistakes in coding. Unless special attention is given to each step, a large number of errors can be introduced. It is most cost-effective to detect and eliminate errors as early in the program development process as possible.

Figure 10.10 shows the cost of correcting an error relative to the phase in which the error is detected. The figure indicates that an error corrected during operation can cost over 75 times as much as if it had been corrected during preliminary design (Anderson and Martin 1982).

Software reliability is more difficult to measure than the reliability of hardware. Hardware reliability tends to follow a "bathtub curve" that represents a high initial failure rate and then a constant low rate until reaching the end of product life. Software does not wear out and reliability is a function of error removal. Software errors in the program may be coding errors or the result of misunderstanding of the requirements. Reliability prediction measures include tracking of the rate at which discrepancies are found and the number of discrepancies (bugs) that remain open. As long as the rate of bug removal exceeds the rate of bug introduction, the rate of new bugs detected will decrease over time. This relationship has led to prediction of software reliability through tracking of the error detection rate to predict the level of software reliability. This relationship is graphically shown in Figure 10.11.

Although zero errors in software prior to release may be a desirable goal, for many large software development efforts this may not be achievable with current tools. If all errors cannot be eliminated prior to release, it is essential that safety and critical performance errors be corrected.

System Fail-Safe Design Using Microprocessors.   The incorporation of a microprocessor into the design of a product permits greater freedom in designing the system. Failures of single circuit components can be detected by the microprocessor before the effects of those failures produce a hazard to the user. Techniques such as watchdog timers, drift analysis, and redundant voting circuits can place the system in a predicted, safe configuration before shutdown.

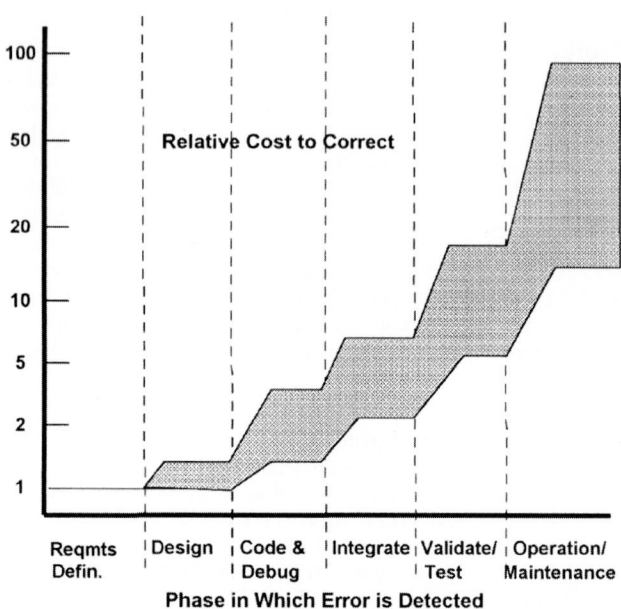

FIGURE 10.10   Relative cost of error correction. (*Source: Anderson and Martin 1982.*)

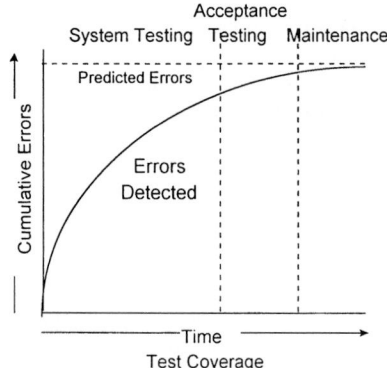

FIGURE 10.11   Predicting software reliability on the basis of errors discovered.

Quality analysis techniques such as hazards analysis and failure mode, effect, and criticality analysis (FMECA) are important "preventive" tools (see Section 3, The Quality Planning Process; Section 19, Quality in Research and Development).

Fault Tree Analysis Input.   Fault tree analysis (FTA) is another technique that can be used to identify potential safety risks. An FTA is a top-down analysis technique aimed at identifying causes or combinations of causes that can lead to an unsafe condition. The analysis is primarily qualitative, which makes FTA an excellent complement to FMECA. While FMECA is especially effective in identifying potential failure conditions and predicting system reliability, the primary strength of the

FTA is the ability to predict potential unsafe conditions that may result due to operator errors, misuse, or abuse conditions. Often, unsafe conditions may result from unforeseen use of the system rather that from a system failure.

In addition to established analysis techniques such as FMECA and FTA, there are several expected safety risk mitigation controls that should be implemented. These controls include the steps shown in Figure 10.12 (Olivier 1995).

Application of these and other controls is not only part of good engineering practices but is also an essential element of an effective safety program.

Several regulatory standards bodies, such as the International Electrotechnical Commission (IEC) and United Laboratories (UL), have developed testing laboratories and published standards regarding procedures and controls necessary to reduce personnel and property hazards. [See IEC 601-1 (1988), IEC 1025 (1990), and UL 544 (1993).] These standards take into consideration a survey of known existing standards and are based on the inputs from a wide variety of users and professional organizations. The standards are frequently updated to remain current with social and technological advances. The standards cover a wide range of conditions that can affect product safety ranging from electrical standards and electromagnetic compatibility requirements to cautions and warning notices. Both IEC and UL publish a list of current standards and can be contacted via the Internet.

Compliance with IEC and UL standards presents a baseline for establishing levels of safeness for products. Requirements for electrical systems, safe levels of radiation, and practices for labeling as defined in these standards should be adopted as a minimum level for products such as medical devices. Verification that the requirements defined by these standards have been realized can be tested through laboratories.

The safety risk analysis program should recognize the dynamic nature of the market. A system that was intended for a very specific application may be used for more general purposes, and new potential safety risk conditions may be experienced. As a consequence of the evolution of a system, the safety risk analysis program must also evolve and continue throughout the life of the system.

Testing.   Although somewhat tedious, it is also possible to verify, through test of the final design, that certain undesirable failure modes cannot produce a hazardous result. Such test methods involve the systematic simulation of an internal fault and observation of the response of the system.

Electronic systems including microprocessors are susceptible to *environmental* threats, which require different types of protection. Electromagnetic interference (EMI) and radio frequency interference (RFI) are present in most operating environments. These effects are not always easy to measure and trace to their source. Both EMI and RFI can produce transient, unpredictable behavior, which even the most thorough fail-safe circuit design cannot manage. Forethought must be given to the methods by which the system will be shielded and protected from these effects.

Because of the often nonreproducible nature of EMI/RFI-induced failures, such problems could easily be thought to exist in the software logic of the microprocessor. Therefore, EMI/RFI must be ruled out before an unrewarding search through the software is begun. Electrostatic discharge, temperature, humidity, and vibration also need to be considered.

Checking the status of hardware on start-up

Monitoring hardware equipment during run time

Data range checking to reduce the likelihood of operator entry errors

Defining system fail-safe states in the case of failures

Installing security controls to restrict access to authorized users

Providing safety margins for critical tolerances

Low-level testing and review of safety related functions

Complying with regulatory standards (UL and IEC)

External laboratory testing

FIGURE 10.12   Safety check controls. (*Source: Olivier 1995.*)

The Reliability and Fail-Safe Design Program.   The amount of effort that goes into a product to ensure that it is fail-safe and reliable depends on the criticality of its use and its inherent complexity. A product can be graded into four categories, as shown in Figure 10.13.

Low Complexity and Noncritical.   At the low end of the product scale are devices that are relatively simple and noncritical in terms of their application. Many simple commercial products fall into this category.

Low Complexity and Critical.   Also at the low-complexity end of the product scale are those devices that are relatively simple but absolutely critical. An example is a medical syringe, which must be sterile and must be sharp; these are not complex characteristics, but they are critical. Other examples are an electrical fuse or a lug nut for an automobile wheel.

High Complexity and Noncritical.   For systems that have high complexity and low criticality, emphasis must be placed on ensuring that the reliability is sufficient to the needs of the application. The reliability emphasis for complex systems is due to the inverse relationship between complexity and reliability. Unless adequate reliability procedures and programs are followed, complex systems are likely to perform disappointingly with respect to mean time between failure.

High Complexity and Critical.   At the high end of the scale are microprocessor systems used in aircraft control systems and patient-connected medical devices. These types of device are both complex and critical, and their development requires a great deal more attention to the details of design and the procedures by which the designs are brought forward. For products that are both complex and critical in operation, the use of a microprocessor simplifies some problems but introduces others in the validation of the software.

The destruction of a European Space Agency Ariane 5 rocket only 39 seconds after launch on its maiden flight illustrates the rigor required for this category. One line of code, without benefit of error trapping and graceful recovery, shut down both of the guidance system computers. "One bug, one crash. Of all the careless lines of code recorded in the annals of computer science, this one may stand as the most efficient" (Gleick 1996).

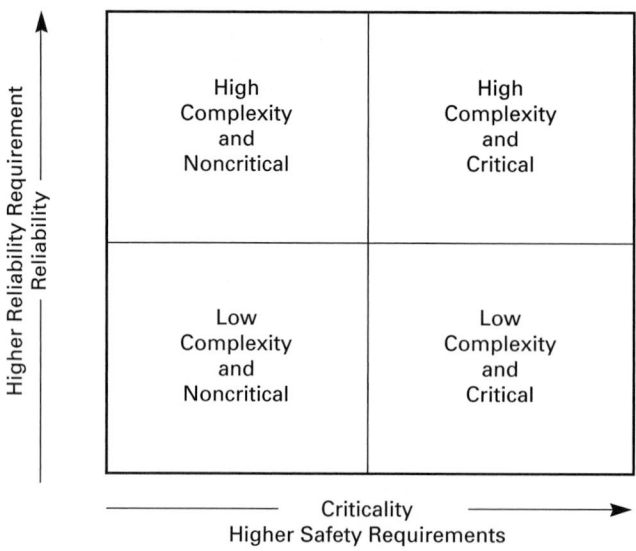

FIGURE 10.13   Evaluation matrix: complexity versus criticality.

TABLE 10.3 Recommended Tasks for Product's Complexity and Criticality

		Criticality	Low	Low	High	High
		Complexity	Low	High	Low	High
	Requirements Activities					
1.	Requirements definition		▓	▓	▓	▓
2.	Requirements traceability		▓	▓	▓	▓
3.	Environmental hardening			▓	▒	▓
4.	Built in diagnosis			▓	▓	▓
5.	High reliability components			▓	▒	▓
6.	Derating of components				▓	▓
7.	Requirements inspections				▓	▓
8.	Fail safe design				▓	▓
9.	Preliminary hazard analysis				▓	▓
10.	Fault Tree analysis				▒	▓
	Implementation and Design					
11.	Design reviews		▓	▓	▓	▓
12.	Reliability analysis models			▒	▓	▓
13.	Formal component testing			▒	▓	▓
14.	Prototype demonstrations			▓	▓	▓
15.	Formal integration testing			▓	▓	▓
16.	Formal software inspections				▓	▓
17.	Program proofs				▓	▓
18.	FMECA				▓	▓
19.	Sneak path circuit analysis				▓	▓
20.	Operations and support risk analysis				▓	▓
	System Testing					
21.	Functional requirements		▓	▓	▓	▓
22.	Manufacturing process validation test			▓	▓	▓
23.	Environmental tests			▓	▓	▓
24.	Accelerated life test			▓	▓	▓
25.	Field (Beta) test			▓	▒	▓
26.	Independent validation			▓	▒	▓
27.	Actual/simulated use test			▓	▒	▓
28.	Stress test			▓	▒	▓
29.	Fail safe test				▓	▓
30.	Regulatory standards compliance (UL, IEC, FDA, ISO 9000)		▓	▓		▓

▓ Mandatory Activity   ▒ Recommended Activity   ☐ Optional Activity

Matrix of Categories. Table 10.3 displays a matrix of program approaches and procedures that can be applied to enhance reliability for the four categories of complexity and criticality. (Note that Table 10.3 is divided into the same four categories as Figure 10.13). These procedures can be applied more or less stringently and in appropriate combinations in order to meet the development effort objectives.

FUTURE TRENDS

Although the future is impossible to predict precisely, one thing is certain: Computer systems will continue to revolutionize the definition of quality practices. Some current trends include:

- Data from disparate quality tracking systems will be increasingly integrated to provide systemwide measures.

- With the speed of new computer systems doubling every 18 months for the foreseeable future, the availability of this steadily increasing computer system capability will lead to more extensive automation within businesses of every type (Scientific American 1995).

- The cost of scrap, rework, warranties, and product liability will impart continuing importance to monitoring of the system, the process, and the machines that assure quality of output (McKee 1983).

- Paperless factories and automated document management systems will emphasize automated analysis tools and document control. [The cost of copying paper is approximately $0.05 per page as against $0.00021 per page for electronic copies (Gates 1995).]

- Evaluation of the effectiveness of software quality systems will become an increasing responsibility of the quality professional.

- "As documents become more flexible, richer in multimedia content, and less tethered to paper, the ways that people collaborate and communicate will become richer and less tied to location" (Gates 1995).

- Company networks will reach beyond employees into the world of suppliers, consultants, and customers (Gates 1995).

A 1995 survey by IEEE asked experts in academia and industry to share their vision of software's future. Significant to the quality professional are predictions of:

- "Parallel processing and networked computing," which will continue the exponential expansion of computer power into the foreseeable future. Factory automation will follow a parallel path.

- "Object-oriented technology integrated into a `Business Information Navigator' application," which will tie internal and worldwide external databases together. Fuzzy logic search routines will retrieve data and answer questions posed in sentence form.

- "Conferencing of data and video as the work force moves to telecommuting." Paperless systems and video communications will allow much of your work to be done at home.

- "Information overload."

ACKNOWLEDGMENTS

The reliability and safety engineering work of James Wingfield, Ph.D., is recognized and appreciated.

REFERENCES

Anderson, R. T., and Martin, T. L. (1982). *Guidebook for Management of Software Quality.* Reliability Technology Associates, Lockport, IL, P. 8.

ANSI/CAM-I (1990). Standard 101-1990, *Dimensional Measuring Interface Specification.* Computer Aided Manufacturing—International, Inc., Arlington, TX.

Besterfield, Dale H. (1979). *Quality Control.* Prentice-Hall, Englewood Cliffs, NJ.

Bezdek, James C. (1996). "A Review of Probabilistic, Fuzzy, and Neural Models for Pattern Recognition," in *Fuzzy Logic and Neural Network Handbook,* Chen, ed. McGraw-Hill, New York, pp. 2.1–2.33.

Boehm, Barry W. (1985). "Verifying and Validating Software Requirements and Design Specifications." *IEEE Software,* January, pp. 471–472.

Chalsma, Jennifer K. (1994). "Better tests mean better cars." *Machine Design,* vol. 66, no. 1, January 10, pp. 36–42.

Chen, ed. (1996). *Fuzzy Logic and Neural Network Handbook.* McGraw-Hill, New York, pp. 2.1–2.33.

Collins, W. Robert, Miller, Keith W., Spielman, Bethany J., Wherry, Phillip (1994). *Communications of the ACM,* vol. 37, no. 1, pp. 81–91.

Freedman, Daniel P., and Weinberg, Gerald M. (1990). *Handbook of Walkthroughs, Inspections, and Technical Reviews: Evaluating Programs, Projects, and Products.* Dorset House, New York.

Gates, Bill (1995). *The Road Ahead.* Viking Penguin, pp. 63, 119.

Gleick, James (1996). "Little Bug, Big Bang," *New York Times Magazine,* December 1.

Humphrey, Watts S. (1989). *Managing the Software Process.* Addison-Wesley, Reading, MA, p. 187.

IEC 601-1 (1988). IEC 601-1: *1988 Medical Electrical Equipment,* Part 1: *General Requirements for Safety,* 2nd ed. International Electrotechnical Commission, Geneva.

IEC 1025 (1990). *Fault Tree Analysis,* 1st ed., 1990-10, p. 11. International Electrotechnical Commission, Geneva.

IEEE (1986). IEEE Standard 1012-1986, *Software Verification and Validation Plans.* Institute of Electrical and Electronic Engineers, New York, pp. 9–12.

Kaner, Cem, Falk, Jack, and Nguyen, Hung Quoc (1993). *Testing Computer Software,* 2nd ed. International Thomson Computer Press, p. 59.

Kierstan, Mark (1995). "Food Hygiene, Quality and Safety: Toward the Year 2000." *British Food Journal,* vol. 97, no. 10, pp. 8–10.

Leveson, Nancy G. (1995). *Safeware: System Safety and Computers.* Addison Wesley, Reading, MA, p. 157.

Marquardt, D. (1984). "New Technical and Educational Directions for Managing Product Quality." *The American Statistician,* vol. 38, no. 1, February, pp. 8–13.

McAfee (1996). "An Introduction to Computer Viruses and other Destructive Programs." McAfee, www.mcafee.com.

McCall, J., Richards, P., and Walters, G. (1977). "Factors in Software Quality—Concepts and Definitions of Software Quality." Joint General Electric and U.S. Air Force Report No. RADC-TR-77-369, vol. 1, November, pp. 3–5.

McKee, Keith E. (1983). "Quality in the 21st Century." *Quality Progress,* vol. 16, no. 6, June, pp. 16–20.

Meyers, Glenford J. (1979). *The Art of Software Testing,* Wiley Interscience, New York, p. 5.

MIL-HDBK-217F (1991). *Reliability Prediction for Electronic Equipment.* U.S. Department of Defense, Washington, DC. Available from Department of Defense and NASA, Washington, DC 20301.

Miller, Byron, (1996). "Fuzzy Logic." *Electronics Now,* vol. 67, no. 5, pp. 29–30.

MIL-M-38510D (1977). *General Specifications for Microcircuits.* U.S. Department of Defense, Washington, DC. Available from Department of Defense and NASA, Washington, DC 20301.

MIL-STD-883C (1983). *Test Methods and Procedures for Microcircuits.* U.S. Department of Defense, Washington, DC. Available from Department of Defense and NASA, Washington, DC 20301.

Olivier, Daniel (1993). "Required Documentation for Software Validation." *Medical Device and Diagnostic Industry,* July.

Olivier, Daniel P. (1995). "Software Safety: Historical Problems and Proposed Solutions." *Medical Device and Diagnostic Industry,* July.

Olivier, Daniel P. (1996). "Implementation of Design Controls Offers Practical Benefits." *Medical Device and Diagnostic Industry,* July.

Reimann, Michael D., and Sarkis, Joseph (1993). "An Architecture for Integrated Automated Quality Control." *Journal of Manufacturing Systems,* vol. 12, no. 4, pp. 341–355.

Rizzo, Anthony (1994) "Diagrams." *Mechanical Engineering,* vol. 116, no. 5, pp. 76–78, May.

Scientific American (1995). "Microprocessors in 2020." *Scientific American,* September, pp. 62–67.

Smith, D. L. (1994). "The Leisure Industry." *Service Industries Journal,* vol. 14, no. 3, pp. 395–408, July (ISSN 0264-2069).

Struebing, Laura (1996). "Quality Progress' 13th Annual QA/QC Software Directory." *Quality Progress,* vol. 29, no. 4, pp. 31–59, April.

UL 544 (1993). *Standard for Medical and Dental Equipment,* UL 544, 3rd ed. Underwriters Laboratories, Inc., Northbrook, IL, August 31.

Wall Street Journal (1993). "Climbing the Peak: 200 Code Writers Badgered by a Perfectionist Guru, Bring Forth Windows NT." *Wall Street Journal,* May 26, p. A12.

Wheeler, David A., Brykczynski, Bill, and Meeson, Reginald N., Jr., (1996). *Software Inspections: An Industry Best Practice.* IEEE Computer Society Press, pp. 10–11.

Zadeh, L. A. (1965). *Information Control,* vol. 8, pp. 338–352.

SECTION 11
THE ISO 9000 FAMILY OF INTERNATIONAL STANDARDS

Donald W. Marquardt

OVERVIEW

Role in Facilitating International Trade. The ISO 9000 standards exist principally to facilitate international trade. In the pre-ISO 9000 era there were various national and multinational quality system standards. These were developed for military and nuclear power industry needs, and, to a lesser extent, for commercial and industrial use. These various standards had commonalities and historical linkages. However, they were not sufficiently consistent in terminology or content for widespread use in international trade.

The ISO 9000 standards have had great impact on international trade and quality systems implementation by organizations worldwide. These international standards have been adopted as national standards by over 100 countries and regional groups of countries. They are applied in a wide range of industry/economic sectors and government regulatory areas. The ISO 9000 standards deal with the management systems used by organizations to design, produce, deliver, and support their products. The standards apply to all generic product categories: hardware, software, processed materials, and services. Specific ISO 9000 family standards provide quality management guidance, or quality assurance requirements, or supporting technology for an organization's management system. The standards provide guidelines or requirements on *what* features are to be present in the management system of an organization but do not prescribe *how* the features are to be implemented. This nonprescriptive character gives the standards their wide applicability for various products and situations. The ISO 9000 family does not deal with any technical specifications for a product. The ISO 9000 standards for an organization's management system are complementary to any technical specifications, standards, or regulations applicable to the organization's products or to its operations.

The standards in the ISO 9000 family are produced and maintained by Technical Committee 176 of the International Organization for Standardization (ISO). The first meeting of ISO/TC176 was held in 1980. ISO 8402, the vocabulary standard, was first published in 1986. The initial ISO 9000 series was published in 1987, consisting of:

- The fundamental concepts and road map guideline standard ISO 9000
- Three alternative requirements standards for quality assurance (ISO 9001, ISO 9002, or ISO 9003)
- The quality management guideline standard ISO 9004

Since 1987, additional standards have been published. The ISO 9000 family now contains a variety of standards supplementary to the original series, some numbered in the ISO 10000 range. In particular, revisions of the basic ISO 9000 series, ISO 9000 through ISO 9004, were published in 1994. This section is written in relation to the 1994 revisions. Table 11.1 lists the standards published as of the beginning of 1996. Additional standards are under development.

ISO 9001, ISO 9002, and ISO 9003 have been adopted and implemented worldwide for quality assurance purposes in both two-party contractual situations and third-party certification/registration situations. ISO 9001 and ISO 9002 together have predominant market share in this segment. Their use continues to grow, as does the infrastructure of certification/registration bodies, accreditation bodies, course providers, consultants, and auditors trained and certified for auditing to these standards. Mutual recognition arrangements between and among nations continue to develop, with the likelihood of ISO-sponsored quality system accreditation recognition in the near future. The number of quality systems that have been certified/registered worldwide now exceeds 100,000 and continues to grow.

The periodic surveillance audits that are part of the third-party certification/registration arrangements worldwide provide continuing motivation for supplier organizations to maintain their quality systems in complete conformance and to improve the systems to continually meet their objectives for quality.

The market for quality management and quality assurance standards is itself growing, partly in response to trade agreements such as European Union (EU), General Agreement on Tariffs and Trade (GATT), and North American Free Trade Association (NAFTA). These agreements all are dependent upon standards that implement the reduction of nontariff trade barriers. The ISO 9000 family occupies a key role in the implementation of such agreements.

Certain industry/economic sectors are developing industry-wide quality system standards, based upon the verbatim adoption of ISO 9001, together with industry-wide supplemental requirements. The automotive industry, the medical devices industry, government regulatory agencies, and military procurement agencies are adopting this approach in many places worldwide.

External Driving Forces.   The driving forces that have resulted in widespread implementation of the ISO 9000 standards can be summed up in one phrase: the globalization of business. Expressions such as the "post-industrial economy" and "the global village" reflect profound changes during recent decades. These changes include:

TABLE 11.1 The ISO 9000 Family of International Standards

ISO 8402	Quality Vocabulary (1994)
ISO 9000	Quality Management and Quality Assurance standards
	Part 1: Guidelines for Selection and Use (1994)
	Part 2: Generic Guidelines for the Application of ISO 9001, ISO 9002, and ISO 9003 (1993)
	Part 3: Guidelines for the Application of ISO 9001 to the Development, Supply, and Maintenance of Software (1991, reissue 1993)
	Part 4: Application for Dependability Management (1993)
ISO 9001	Quality Systems—Model for Quality Assurance in Design, Development, Production, Installation and Servicing (1994)
ISO 9002	Quality Systems—Model for Quality Assurance in Production, Installation, and Servicing (1994)
ISO 9003	Quality Systems—Model for Quality Assurance in Final Inspection and Test (1994)
ISO 9004	Quality Management and Quality System Elements
	Part 1: Guidelines (1994)
	Part 2: Guidelines for Services (1991, reissue 1993)
	Part 3: Guidelines for Processed Materials (1993)
	Part 4: Guidelines for Quality Improvement (1993)
ISO 10005	Quality Management—Guidelines for Quality Plans (1995)
ISO 10007	Guidelines for Configuration Management (1994)
ISO 10011	Guidelines for Auditing Quality Systems
	Part 1: Auditing (1990, reissue 1993)
	Part 2: Qualification Criteria for Quality Systems Auditors (1991, reissue 1993)
	Part 3: Management of Audit Programs (1991, reissue 1993)
ISO 10012	Quality Assurance Requirements for Measuring Equipment
	Part 1: Management of Measuring Equipment (1992)
ISO 10013	Guidelines for Developing Quality Manuals (1994)

Source: Marquardt, D. W., et al. (1991). "Vision 2000: The Strategy for the ISO 9000 Series Standards in the '90s," *Quality Progress,* May, pp. 25–31.

- New technology in virtually all industry/economic sectors
- Worldwide electronic communication networks
- Widespread worldwide travel
- Dramatic increase in world population
- Depletion of natural resource reserves

  Arable land, fishing grounds, fossil fuels

- More intensive use of land, water, energy, air

  Widespread environmental problems/concerns

- Downsizing of large companies and other organizations

  Flattened organizational structure

  Outsourcing of functions outside the core functions of the organization

- Number and complexity of language, culture, legal, and social frameworks encountered in the global economy

  Diversity a permanent key factor

- Developing countries becoming a larger proportion of the total global economy

  New kinds of competitors and new markets

These changes have led to increased economic competition, increased customer expectations for quality, and increased demands upon organizations to meet more stringent requirements for quality of their products.

The globalization of business is a reality even for many small- and medium-size companies. These smaller companies, as well as large companies, now find that some of their prime competitors are likely to be based in another country. Fewer and fewer businesses are able to survive by considering only the competition within the local community. This affects the strategic approach and the product planning of companies of all sizes.

Internal Response to the External Forces. Companies everywhere are dealing with the need to change. There is greater focus on human resources and organizational culture, on empowering and enabling people in their jobs. Dr. W. Edwards Deming often said that many workers do not know what their job is. ISO 9000 implementation involves establishing policy, setting objectives for quality, designing management systems, documenting procedures, and training for job skills. All of these are parts of clarifying what people's jobs are.

Companies are adopting the process perspective. This concept is emphasized in the 1994 revision of the ISO 9000 standards. In implementing the ISO 9000 standards, companies are using flowcharts and other devices to emphasize work-process diagnosis and to find opportunities for process simplification and improvement. Metrics are being used increasingly to characterize product quality and customer satisfaction more effectively.

Companies are implementing better product design and work-process design procedures, and improved production strategies. Benchmarking and competitive assessment are used increasingly. Enterprise models, electronic data exchange, and other information technology approaches are growing in scope and impact.

It may be asked: In this world of rapid change, how can a single family of standards, ISO 9000, apply to all industry/economic sectors, all products, and all sizes of organizations?

The "Separate and Complementary" Concept. The ISO 9000 standards are founded on the concept that the assurance of consistent product quality is best achieved by simultaneous application of two kinds of standards:

- Product Standards (technical specifications)
- Quality system (management system) Standards

I call this the "separate and complementary" concept because the two types of standards are separate from each other and they are complementary. The two types of standards are needed to provide confidence that products will meet consistently the requirements for quality.

Product standards provide the technical specifications that apply to the characteristics of the product and, often, the characteristics of the process by which the product is produced. Product standards are specific to the particular product: both its intended functionality and the end-use situations the product may encounter.

The management system is the domain of the ISO 9000 standards. It is by means of the distinction between product specifications and management system features that the ISO 9000 standards apply to all industry/economic sectors, all products, and all sizes of organizations.

The standards in the ISO 9000 family, both guidance and requirements, are written in terms of *what* features are to be present in the management system of an organization but do not prescribe *how* the features are to be implemented. The technology selected by an organization determines how the relevant features will be incorporated in its own management system. Likewise, an organization is free to determine its own management structure.

Comment. In regard to terminology, three terms are in current use, all of them having the same essential meaning. *Quality system* is the formal term currently defined internationally in ISO 8402, the ISO/TC176 vocabulary standard. *Management system* is the term frequently used in the daily language of business. *Quality management system* is the term coming into increasing use for discussing an organization's management system when the focus is upon the overall performance of the organization and its results in relation to the organization's objectives for quality. A benefit of the term "quality management system" is its effectiveness in emphasizing both:

- The commonalities in management system features
- The differences in the objectives for the results of an organization's management system, for various areas of application (e.g., quality management systems and environmental management systems)

Characteristics of ISO 9000 Standards.   Some of the ISO 9000 family standards contain *requirements,* while others contain *guidelines.*

ISO 9001, ISO 9002, and ISO 9003 are *requirements* standards. They are quality system models to be used for quality assurance purposes for providing confidence in product quality. A requirements standard becomes binding upon a company or organization wherever:

- It is explicitly called up in a contract between the organization and its customer
- The organization seeks and earns third-party certification/registration

The text of a requirements standard is phrased in terms of the verb "shall," with the meaning that the stated requirements are mandatory.

ISO 9004 is an example of a *guideline* standard. Guideline standards are advisory documents. They are phrased in terms of the word "should," with the meaning that they are recommendations. The scope of ISO 9004 is broader than the scope of ISO 9001, because it covers not only quality system features necessary to provide customer confidence in product quality, but also quality system features for organizational effectiveness.

All of the ISO 9000 family standards are *generic,* in the sense that they apply to any product or any organization. All of the ISO 9000 family standards are *nonprescriptive* in the sense that they describe what management system functions shall or should be in place; but they do not prescribe how to carry out those functions.

The Clauses of ISO 9001 and Their Typical Structure.   The ISO 9000 family is best known for ISO 9001, the most comprehensive of the quality assurance requirements standards. As indicated by their titles listed in Table 11.1, ISO 9002 is identical to ISO 9001 except that ISO 9002 does not contain requirements for the design function (clause 4.4). Between them, ISO 9001 and ISO 9002 account for the largest current market share of use of the ISO 9000 family documents. The third quality assurance requirements standard, ISO 9003, is much less comprehensive and is based on final product inspection only. Its current market share is very small, less than 2 percent in most parts of the world.

The clause titles of ISO 9001 are shown in Table 11.2. The actual quality management system requirements are spelled out in clause 4, specifically in subclauses 4.1 through 4.20. The scope of ISO 9001 is focused on management system features that directly affect product quality. This emphasis is consistent with the most fundamental purpose of ISO standards: to facilitate international trade.

To illustrate the structure, content, and style of ISO 9001, two brief subclauses are quoted below:

Quality systems, general (clause 4.2.1)

> The supplier shall establish, document, and maintain a quality system as a means of ensuring that product conforms to specified requirements. The supplier shall prepare a quality manual covering the requirements of this International Standard. The quality manual shall include or make reference to the quality system procedures and outline the structure of the documentation used in the quality system.

Document and data control, general (clause 4.5.1)

> The supplier shall establish and maintain documented procedures to control all documents and data that relate to the requirements of this International Standard, including, to the extent applicable, documents of external origin such as standards and customer drawings.

TABLE 11.2   International Standard ISO 9001:1994(E)

Quality systems—Model for quality assurance in design, development, production, installation, and servicing

Clause titles	
1	Scope
2	Normative reference
3	Definitions
4	Quality system requirements
4.1	Management responsibility
4.2	Quality system
4.3	Contract review
4.4	Design control
4.5	Document and data control
4.6	Purchasing
4.7	Control of customer-supplied product
4.8	Product identification and traceability
4.9	Process control
4.10	Inspection and testing
4.11	Control of inspection, measuring, and test equipment
4.12	Inspection and test status
4.13	Control of nonconforming product
4.14	Corrective and preventive action
4.15	Handling, storage, packaging, preservation, and delivery
4.16	Control of quality records
4.17	Internal quality audits
4.18	Training
4.19	Servicing
4.20	Statistical techniques

Some key words and their meanings are

Supplier: The organization to which the standard is addressed; namely the organization that will supply the products to the customer organization

Establish: To institute permanently

Document: To record in readable form

Maintain: To keep up-to-date at all times

Documents: Examples are overall quality manual, quality system procedures, work instructions for specific jobs, etc. (as distinct from *records* of actions completed, measurements made, etc.)

As a result of these clauses not being prescriptive as to how the requirements are to be implemented, it is expected that there may be wide variations from one supplier to another. The appropriate method of implementation will depend upon such characteristics as the type of product, its complexity, regulatory requirements that must be satisfied for legal reasons, and size of supplier organization.

One important benefit of the nonprescriptive character of the ISO 9000 standards—in particular, ISO 9001—is across-the-board applicability to *all organizational structures.* The requirements of ISO 9001 are equally relevant whether the supplier organization is large or small; has one site or many; is downsized, bereft of middle management; is heavily networked and/or based on joint ventures; uses contract subsuppliers, part-time and/or temporary personnel; or has multinational legal and/or economic arrangements. The only organizational requirement is that the organization shall have "management with executive responsibility" and that the executive management "appoint a member of the supplier's own management" to be the management representative with responsibility for the establishment, implementation, maintenance, and reporting on the performance of the quality system.

THE FACETS OF PRODUCT QUALITY

The guideline standard ISO 9000-1:1994 explains many concepts that are fundamental to the ISO 9000 family. Among these is the concept of the four facets of product quality:

1. Quality due to definition of needs for the product

   • Defining and updating the product to meet marketplace requirements and opportunities

2. Quality due to product design

   • Designing into the product the characteristics that enable it to meet marketplace requirements and opportunities:

   Features that influence intended functionality

   Features that influence the robustness of product performance under variable conditions of production and use

3. Quality due to conformance to product design

4. Quality due to product support throughout the product life cycle

Facets 1, 2, 3, and 4 encompass all stages of the product life cycle.

The publication of the ISO 9000 series in 1987 brought necessary harmonization on an international scale. As expected, the initial emphasis of ISO 9000 standards application was primarily on facet 3, eliminating nonconformities in product supplied to customers. But, to many people's surprise there was little use of ISO 9003. By the late 1980s suppliers were recognizing that the preventive approach of ISO 9001 and ISO 9002 was more effective than final-inspection alone as the means to achieve quality due to conformance to product design. In the years before the ISO 9000 standards were developed and put into commercial and industrial uses, the national predecessor counterparts of ISO 9003 had the predominant market share; they focused on a final-inspection-only approach to facet 3.

ISO 9002 implementation now has the largest market share of the three requirements standards. With the growing worldwide emphasis on quality, suppliers involved in international trade have continued to gain maturity of understanding about quality systems. Consequently, the market share of ISO 9001 has increased, reflecting the widening appreciation of facet 2 by customers and suppliers.

The 1994 revisions of the ISO 9000 standards include significant changes in many features of the requirements and guidelines for a quality management system. These changes tend to strengthen the quality contributions from all of facets 1, 2, 3, and 4. However, primary emphasis still remains on facet 3 and the first group of features under facet 2. The next revision is likely again to have some broadening of the emphasis, reflecting the continually changing needs of international trade.

THE COMMONALITIES AND DISTINCTIONS BETWEEN QUALITY MANAGEMENT AND QUALITY ASSURANCE

One of the most pressing needs in the early years of ISO/TC176 work was to harmonize internationally the meanings of terms such as "quality control" and "quality assurance." These two terms, in particular, were used with diametrically different meanings among various nations, and even within nations. In my role as convener of the working group that wrote the ISO 9000:1987 standard, I proposed early in the 1980s that the term "quality management" be introduced into the ISO 9000 standards as the umbrella term for quality control and quality assurance. The term "quality management" was defined, included in ISO 8402, adopted internationally, and is now used worldwide. This, in turn, enabled agreement on harmonized definitions of the meanings of each of the terms "quality control" and "quality assurance."

However, discussions in TC176 during 1995 revealed that the essential commonalities and distinctions between quality management and quality assurance are still not universally understood. This may be a result of the expansion of ISO 9000 standards use to many more countries than participated in the early 1980s, or lack of widespread reference to ISO 8402, or deficiencies in the ISO 8402 definitions. Undoubtedly, all these reasons have contributed. In any event, the meanings of the terms "quality management" and "quality assurance" need careful articulation to achieve clarity. Table 11.3 describes the essence and is the same as the meanings intended in ISO 8402:1986 and ISO 8402:1994. The quality control aspects of the umbrella term "quality management" are focused on the word "achieving," but all bullet points in the left-hand column of Table 11.3 relate at least indirectly to quality control. The right-hand column of Table 11.3 shows that the quality assurance aspects of the umbrella term "quality management" have primary focus on the notions of *demonstrating and providing confidence* through objective evidence.

VISION 2000

"Vision 2000" refers to the report of the ISO/TC176 Ad Hoc Task Force (Marquardt et al. 1991). It outlines the strategy adopted by TC176 for the ISO 9000 standards in the 1990s. Several key concepts and strategies from that report are essential to any discussion of the ISO 9000 standards.

Generic Product Categories.   The task force identified four generic product categories:

- Hardware
- Software
- Processed materials
- Services

Table 11.4, from Marquardt et al. (1991), provides descriptors of the four generic product categories. Several of these now have formal definitions in ISO 8402:1994. These categories encompass

TABLE 11.3   The Prime Focus of Quality Management and Quality Assurance

The Prime Focus of	
Quality management	Quality assurance
• *Achieving* results that satisfy the requirements for quality	• *Demonstrating* that the requirements for quality have been (and can be) achieved
• Motivated by stakeholders *internal* to the organization, especially the organization's management	• Motivated by stakeholders, especially customers, *external* to the organization
• Goal is to satisfy *all stakeholders*	• Goal is to satisfy all *customers*
• Effective, efficient, and continually improving, overall quality-related *performance* is the intended result	• *Confidence* in the organization's products is the intended result
• Scope covers all activities that affect the total quality-related *business results* of the organization	• Scope of demonstration covers activities that directly affect quality-related *process and product results*

TABLE 11.4   Generic Product Categories*

Generic product category	Kinds of product
Hardware	Products consisting of manufactured pieces, parts, or assemblies thereof
Software	Products such as computer software, consisting of written or otherwise recordable information, concepts, transactions, or procedures
Processed materials[†]	Products (final or intermediate) consisting of solids, liquids, gases, or combinations thereof, including particulate materials, ingots, filaments, or sheet structures
Services	Intangible products which may be the entire or principal offering or incorporated features of the offering, relating to activities such as planning, selling, directing, delivering, improving, evaluating, training, operating, or servicing a tangible product

*All generic product categories provide value to the customer only at the times and places the customer interfaces with and perceives benefits from the product. However, the value from a service often is provided primarily by activities at a particular time and place of interface with the customer.

†Processed materials typically are delivered (packaged) in containers such as drums, bags, tanks, cans, pipelines, or rolls

Source: Marquardt et al. (1991).

all the kinds of product that need explicit attention in quality management and quality assurance standardization. The initial 1987 standards were acknowledged to have inherited some of the hardware bias of the predecessor standards. To remedy this, supplemental standards for each of the other three generic product categories were developed and published (ISO 9000-3; ISO 9004-2; ISO 9004-3); see Table 11.1.

One of the principal strategies in Vision 2000 was stated as follows:

> We envision that, by the year 2000, there will be an intermingling, a growing together, of the terminology, concepts, and technology used in all four generic product categories. This vision implies that, by the year 2000, the need for separate documents for the four generic product categories will have diminished. Terminology and procedures for all generic product categories will be widely understood and used by practitioners, whatever industry/economic sector they might be operating in.
>
> Consequently, our Vision 2000 for TC176 is to develop a single quality management standard (an updated ISO 9004 that includes new topics as appropriate) and an external quality assurance requirements standard (an updated ISO 9001) tied together by a road map standard (an updated ISO 9000). There would be a high degree of commonality in the concepts and architecture of ISO 9004 and ISO 9001. The requirements in ISO 9001 would continue to be based upon a selection of the guidance elements in ISO 9004. Supplementary standards that provide expanded guidance could be provided by TC176 as needed.

This strategy continues to guide TC176 in its work on the next revisions.

Acceptance, Compatibility, and Flexibility.   Vision 2000 proposed four goals that relate to maintaining the ISO 9000 standards so that they continually meet the needs of the marketplace. These goals are *universal acceptance,* being adopted and used worldwide; *current compatibility,* facilitating combined used without conflicting requirements; *forward compatibility,* with successive revisions being accepted by users; and *forward flexibility,* using architecture that allows new features to be incorporated readily.

TC176 continues to use these goals as guides, recognizing as in Vision 2000 that "Proposals that are beneficial to one of the goals might be detrimental to another goal. As in all standardization, compromises and paradoxes might be needed in specific situations."

Avoiding Proliferation.   Vision 2000 recognized that the role of the ISO 9000 standards to facilitate international trade could be maintained only if the remarkable, rapid, worldwide success in replacing national standards with harmonized ISO 9000 international standards did not itself lead to new rounds of proliferation. The issue was stated as follows:

If the ISO 9000 series were to become only the nucleus of a proliferation of localized standards derived from, but varying in content and architecture from, the ISO 9000 series, then there would be little worldwide standardization. The growth of many localized certification schemes would present further complications. Once again, there could be worldwide restraint of trade because of proliferation of standards and inconsistent requirements.

and

Vision 2000 emphatically discourages the production of industry/economic-sector-specific generic quality standards supplemental to, or derived from, the ISO 9000 series. We believe such proliferation would constrain international trade and impede progress in quality achievements. A primary purpose of the widespread publication of this article is to prevent the proliferation of supplemental or derivative standards.

It is, however, well understood that product-specific standards containing technical requirements for specific products or processes or describing specific product test methods are necessary and have to be developed within the industry/economic sector.

Proliferation has been virtually eliminated worldwide in terms of national standards because of the withdrawal of prior national standards and adoption of the ISO 9000 standards. Moreover, the fundamental role of the ISO 9000 standards in relation to other areas of international standardization has been incorporated into the ISO/IEC Directives, which govern the operations of all Technical Committees of ISO and IEC. (ISO and IEC together coordinate and publish international voluntary consensus standards for all sectors of the economy and all technical fields. IEC, the International Electrotechnical Commission, deals with standards in industries related to electrical and electronic engineering; ISO deals with all other areas of standardization.) Clause 6.6.4 of the ISO/IEC Directives, Part 2, reads:

6.6.4   When a technical committee or sub-committee wishes to incorporate quality systems requirements in a standard for a product, process, or service, the standards shall include a reference to the relevant quality systems standard (ISO 9001, ISO 9002 or ISO 9003). It shall not add to, delete, change or interpret the requirements in the quality systems standard.

Any requests for additions, deletions, changes or interpretations shall be submitted to the secretariat of ISO/TC176/SC2: *Quality systems*.

When the industry or sector terminology is sufficiently different, a document explaining the relationship between the quality assurance terminology and the sector terminology may be prepared.

This clause may be viewed as an operational definition of avoiding proliferation. It is being applied to good effect within ISO/IEC, and as a result a number of other ISO TCs have not prepared proposed new standards that would have represented unnecessary proliferation within specific industry/economic sectors. However, in one area of application, the ISO Technical Management Board has ruled that a new Technical Committee, TC207, on Environmental Management Systems should be set up. This is discussed further in this section (see Other Areas of Application), and should be viewed as one of the necessary "compromises and paradoxes" quoted above from Vision 2000.

BEYOND VISION 2000

Even before publication of the 1994 "Phase 1" revisions, TC176 and its subcommittees began explicit planning for the next revisions, which had been referred to in Vision 2000 as the Phase 2 revisions. In addition, TC176 appointed a Strategic Planning Advisory Group (SPAG). The SPAG study included a formal strategic planning effort examining the TC176 products, markets, benefits to users, beliefs about the value of such standardization, external trends, competitive factors, and unmet market needs. From these examinations emerged a number of strategic opportunities and, ultimately, strategic goals.

The essential concepts and strategies of Vision 2000 were reaffirmed. However, the study concluded that ISO 9004 should and could have more impact in guiding practitioners of quality management. To do so requires an expansion of scope and change of approach. During 1995 TC176 also reexamined, in various meetings and study groups, the developing plans for the next revisions of the ISO 9000 standards. At the Durban, South Africa, meeting in November 1995, work was completed on specifications for the revision of ISO 9000, ISO 9001, ISO 9004, and a proposed new document on quality management principles. The specifications were prepared for formal comment by the member bodies representing the various nations. Also TC176 achieved tentative consensus on the architecture and content of the ISO 9000 family for the year 2000. One of the guiding themes is to avoid unnecessary proliferation of standards within the ISO 9000 family itself, as well as external to TC176. The leaders of the delegations of the more than 40 countries represented at Durban spent several days on these issues, including the detailed texts of vision and mission statements and of key strategies for TC176 activities and products.

QUALITY SYSTEM CERTIFICATION/REGISTRATION

Origin of the Need.   The earliest users of quality assurance requirements standards were large customer organizations such as electric power providers and military organizations. These customers often purchase complex products to specific functional design. In such situations the quality assurance requirements are called up in a two-party contract, where the providing organization (i.e., the supplier) is referred to as the "first party" and the customer organization is referred to as the "second party." Such quality assurance requirements typically include provisions for the providing organization to have internal audits sponsored by its management to verify that its quality system meets the contract requirements. These are first-party audits. Such contracts typically also include provisions to have external audits sponsored by the management of the customer organization to verify that the supplier organization's quality system meets the contract requirements. These are second-party audits. Within a contractual arrangement between two such parties, it is possible to tailor the requirements, as appropriate, and to maintain an ongoing dialogue between customer and supplier.

When such assurance arrangements become a widespread practice throughout the economy, the two-party, individual-contract approach becomes burdensome. There develops a situation where each organization in the supply chain is subject to periodic management system audits by many customers and is itself subjecting many of its subsuppliers to such audits. There is a lot of redundant effort throughout the supply chain because each organization is audited multiple times for essentially the same requirements. The conduct of audits becomes a significant cost element for both the auditor organizations and auditee organizations.

Certification/Registration-Level Activities.   The development of quality system certification/registration is a means to reduce the redundant, non-value-adding effort of these multiple audits. A third-party organization, which is called a "certification body" in some countries, or a "registrar" in other countries (including the United States), conducts a formal audit of a supplier organization to assess conformance to the appropriate quality system standard, say, ISO 9001 or ISO 9002. When the supplier organization is judged to be in complete conformance, the third party issues a *certificate* to the supplying organization and *registers* the organization's quality system in a publicly available register. Thus, the terms "certification" and "registration" carry the same marketplace meaning because they are two successive steps signifying successful completion of the same process.

To maintain its registered status, the supplier organization must pass periodic surveillance audits by the registrar. Surveillance audits are often conducted semiannually. They may be less comprehensive than the full audit. If so, a full audit is performed every few years.

In the world today, there are hundreds of certification bodies/registrars. Most of them are private, for-profit companies. Their services are valued by the supplier organizations they register, and by the customer organizations of the supplier organizations, because the registration service adds value in the supply chain. It is critical that the registrars do their work competently and objectively and that all registrars meet standard requirements for their business activities. They are, in fact, supplier organizations that provide a needed service product in the economy.

Accreditation-Level Activities.   To assure competence and objectivity of the registrars, systems of registrar *accreditation* have been set up worldwide. Accreditation bodies audit the registrars for conformity to standard international guides for the operation of certification bodies. The quality system of the registrar comes under scrutiny by the accreditation body through audits that cover the registrar's documented quality management system, the qualifications and certification of auditors used by the registrar, the record keeping, and other features of the office operations. In addition, the accreditation body witnesses selected audits done by the registrar's auditors at a client supplier organization's facility.

Accreditation and Registration Flowchart, Including Related Activities.   This process as it operates in the United States is depicted graphically in Figure 11.1. The three columns in the figure depict the three areas of activity of the Registrar Accreditation Board: accreditation of registrar companies, certification of individuals to be auditors, and accreditation of the training courses which are part of the requirements for an auditor to be certified. The relevant ISO/IEC international standards and guides that apply in each of these activities are shown in Figure 11.1. The ISO criteria documents for auditing quality systems are the ISO 10011 standard, Parts 1, 2, and 3. See Table 11.1. Part 2 of ISO 10011 deals specifically with auditor qualifications.

In the United States the registrar accreditation is carried out by the Registrar Accreditation Board (RAB) under a joint program with the American National Standards Institute (ANSI). This joint program is called the American National Accreditation Program for Registrars of Quality Systems. The auditor certification and training course accreditation portions of the entire scheme shown in Figure 11.1 also are carried out by RAB.

Materials governing accreditation procedures for the American National Accreditation Program for Registrars of Quality Systems are available from the Registrar Accreditation Board, P.O. Box 3005, Milwaukee, WI 53201-3005.

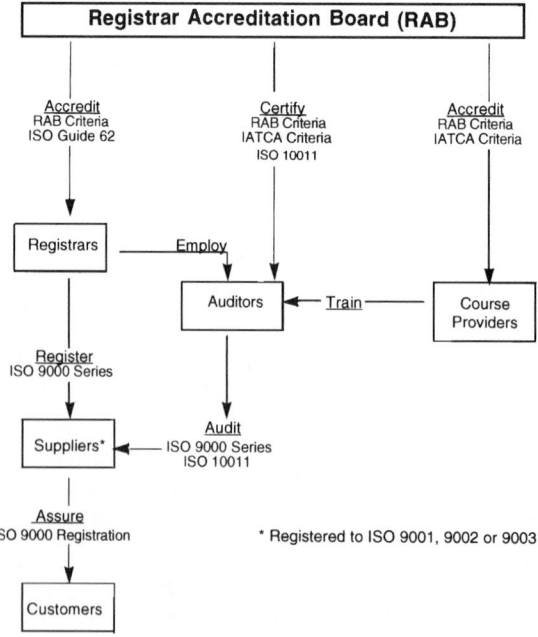

FIGURE 11.1   Accreditation and registration process.

Mutual International Acceptance. Various other countries have implemented these three areas of activity, too:

- Accreditation of certification bodies/registrars
- Certification of auditors
- Accreditation of auditor training courses

The systems in the Netherlands and in the United Kingdom have been in place longer than most. At this time, various bilateral mutual recognition agreements are in place between certain countries whereby, for example, the certification of an auditor in one country carries over into automatic recognition of that certification in another country. In other situations, a memorandum of understanding has been negotiated between, say, the accreditation bodies in two countries, whereby they enter into a cooperative mode of operation preliminary to entering into a formal mutual recognition agreement. Under a memorandum of understanding, the accreditation bodies may conduct jointly the audit of a registrar, and the auditors may jointly document the results of the audit. However, each of the accreditation bodies would make its own decision whether to grant or continue, as the case may be, the accreditation.

In principle, there should be no need for a supplier organization to obtain more than one certification/registration. A certificate from a registrar accredited anywhere in the world should, in principle, be accepted by customer organizations anywhere else in the world. In practice, it takes time to build infrastructure comparable to Figure 11.1 in any country. It takes additional time (measured in years) for that infrastructure to mature in its operation and for confidence to build in other countries. Of course, not all countries decide to set up their own infrastructure but may choose to have their supplier organizations who wish to become registered do so by employing the services of an accredited registrar from another country.

Indeed, many registrar companies have established operations internationally and provide services in many countries. Such registrars often seek accreditation in multiple countries because their customers (the supplier organizations) look for accreditation under a system with which they are familiar and have developed confidence.

At the present time, there are a multiplicity of arrangements involving single or multiple accreditations of registrars, single or multiple certifications of auditors, and single or multiple accreditations of training courses. The overall system is moving toward widespread mutual recognition, but the ultimate test of credibility is the marketplace willingness to accept a single certification and a single accreditation.

The ISO/IEC are themselves sponsoring development and implementation of a truly international system of mutual recognition for the accreditation of certification bodies/registrars. Called QSAR, it has been worked out in detail through international negotiations and is starting to be implemented. It is expected that the QSAR arrangement will accelerate the process of mutual recognition internationally.

The current status where registrars and course providers may have multiple accreditations, and auditors may have multiple certifications, may seem to have more redundancy than necessary. If we step back and compare the current situation to the alternative of widespread second-party auditing of supplier organizations' quality systems, it must be acknowledged that the present situation is better in the following ways:

- Much less redundancy of auditing
- Much improved consistency of auditing
- Potential for even less redundancy and further improved consistency through the use of international standards and guides as criteria and through mutual harmonization efforts driven by the marketplace.

Formal International Mutual Recognition. For the United States, there is one further complication. Almost alone among the countries of the world, the U.S. standards system is a private-sector activity. American National Standards Institute (ANSI), a private sector organization, is the coordinating body for standards in the United States. Under the ANSI umbrella many organizations produce and maintain numbers of American National Standards. Most of these standards relate to product technical specifications. Among the largest U.S. producers of standards are organizations such as the American Society of Testing and Materials (ASTM), the American Society of Mechanical Engineers (ASME), and the Institute of Electrical and Electronics Engineers (IEEE), but there are many other organizations that produce American National Standards applicable to specific products or fields of activity. The ANSI system provides a consistent standards development process that is open, fair, and provides access to all parties that may be materially affected by a standard. The success of the U.S. system is attested to by the predominance of the U.S. economy internationally and the widespread adoption of U.S. standards for multinational or international use.

However, there are three levels of activities and infrastructure in relation to conformity assessment in international trade. Two of these levels have already been discussed: the certification/registration level and the accreditation level. The third level is the *recognition* level. At the recognition level, the national government of country A affirms to the government of country B that A's certification and accreditation infrastructure conforms to international standards and guides. In most countries of the world, where the standards system is run by a government or semigovernment agency and the accreditation activities are carried out by that agency, the recognition level is virtually automatic. In the United States, various government agencies may be called upon to provide the formal recognition.

For example, in dealing with the European Union (EU) on products that fall under one of the EU Directives that regulate products that have health, safety, and environmental risks, the EU insists upon dealing through designated government channels. The relevant U.S. government agency varies from one EU Directive to another. In many areas, the recognition responsibility will come under the recently authorized National Voluntary Conformity Assessment System Evaluation (NVCASE) program to be run by the Department of Commerce, through the National Institute of Standards and Technology. The NVCASE program had not come into operation at the time of this writing.

CONFORMITY ASSESSMENT AND INTERNATIONAL TRADE

The conformity assessment approach of the European Union typifies what is happening in many parts of the world. For a regulated product to be sold in any EU country, it must bear the "CE" mark. Under the Modular Approach of the EU, to qualify for use of the mark the supplier organization must produce evidence of conformity in four areas:

- Technical documentation of product design
- Type testing
- Product surveillance (by samples, or by each product)
- Quality assurance surveillance.

Depending on the directive, the EU will offer suppliers various routes (modules) to satisfy the requirements. These routes range from "Internal Control of Production," which focuses on the product surveillance aspects, to "Full Quality Assurance," which typically focuses on certification/registration to ISO 9001 and relies upon the ISO 9001 requirements for capability in product design. In most modules the manufacturer must submit product units, and/or product design technical information, and/or quality system information to a certification body that has been designated by the government as a "notified body." The notified body must, in some modules, also provide for product tests where required. Several modules involve certification to ISO 9001, ISO 9002, or ISO 9003.

The implementation of this modular approach to conformity assessment for regulated products by the European Union (then called the European Community) was the largest, single, early impetus to the rapid spread of certification/registration to ISO 9001 or ISO 9002 worldwide. For example, about half of the dollar volume of U.S. trade with Europe is in regulated products.

Nevertheless, the global trends in technology and in requirements for quality, and the cost savings of third-party versus widespread second-party auditing, as discussed previously in this section, are powerful additional incentives and staying power for sustained international use and growth of third-party quality system certification/registration.

Moreover, for a supplier organization it is not effective to attempt to have two quality management systems, one for regulated products and another for nonregulated products. Consequently, there are multiple incentives for large numbers of supplier organizations, engaged directly or indirectly in international trade, to operate a quality management system that conforms to ISO 9001 or ISO 9002, as appropriate.

STANDARDIZATION PROBLEMS AND OPPORTUNITIES FOR THE ISO 9000 FAMILY

The rapid worldwide adoption and implementation of the ISO 9000 standards, and the rapid growth of the infrastructure of certification/registration bodies, accreditation bodies, course providers, consultants, auditors, trade magazines, books, and ISO 9000 journalists is virtually unprecedented in any field of standardization. This can appropriately be regarded as a remarkable success story for the ISO 9000 standards. Accompanying this success story are several problems and opportunities that need careful attention.

Problems and Opportunities.   In all important endeavors, "problems" and "opportunities" abound. It is important to understand that these are two sides of the same coin. Every problem is a doorway to opportunities; and every opportunity carries problems with it. This universal truth certainly applies in relation to the ISO 9000 family. Coming to grips with these problem-opportunities deepens our understanding of the ISO 9000 standards themselves. These problem-opportunities will be with us for a long time because they come from fundamental economic issues that are always present in national and international trade. Some, or all, of them could be make-or-break issues for the success of ISO 9001, and with it the ISO 9000 family. As a preliminary to discussing a number of problem-opportunities, it is important to understand what is implied by the concept of standardization.

The Scope of Standardization.   "Standardization" encompasses activities in two interrelated areas:

• The conception, planning, production, promotion, and selling of *standards*
• The conception, planning, establishment, control, promotion, and maintenance of standards *implementation*

In practice, the standardization problems and opportunities for the ISO 9000 family relate to *implementation* activities at least as much as to the *standards* themselves. The topics discussed in the following illustrate the predominance of problem-opportunities that arise from implementation activities.

The Concept of Continuous Improvement.   The philosophy of using quality assurance standards has changed over the years. In the 1960s the virtually universal perspective of business managers was "If it ain't broke, don't fix it." In that philosophical environment of maintaining the status quo, the characteristics shown at the top of Table 11.5 prevailed in most of industry and commerce. Today, as illustrated by the minimal use of ISO 9003, the philosophy of 30 years ago is giving way to a philosophy of continuous improvement. Continuous improvement is increasingly necessary for economic survival in the global economy and is becoming a widely pursued goal. It is the only reliable route to sustaining marketplace advantage for both customer and supplier.

As shown at the bottom of Table 11.5, the focus of quality assurance standardization is now on prevention of nonconformities. This requires a "process" focus, which is reflected in the documentation and many other features.

TABLE 11.5 Philosophy of Using Quality Assurance Standards

30 years ago	
• Quality goal:	Maintain status quo.
• Business goal:	The best deal for this contract.
• Methods:	Final-inspection oriented; sort good product from bad. "Records" paperwork was featured.
• Customer/supplier relationship:	Adversarial.
Today	
• Quality goal:	Continuous improvement.
• Business goal:	Mutual marketplace advantage.
• Methods:	Prevention oriented; don't make bad product. "Process" documentation is featured.
• Customer/supplier relationship:	Partnership.

The Role of Continuous Improvement in ISO 9001.   In the first years of use since 1987 an unfortunate mind-set has been adopted by many registrars/certifiers and their auditors. This mind-set can be called the "status quo mind-set." Reminiscent of the 1960s, it is characterized by the ditty:

> "Say what you do.
> Do what you say."

This simple ditty is correct as far as it goes, but is *far short of the requirements in the ISO 9000 standards set in 1987 and revised in 1994.* It focuses only on adherence to established procedures. For example, the status quo mind-set ignores the *linked requirements* to demonstrate continuing adequacy of the quality system:

* For business objectives
* For customer satisfaction

Intrinsic to the 1994 revision of ISO 9001 is a reshaped mind-set as summarized in Table 11.6. The reshaped mind-set is *a cycle of continuous improvement,* which can be depicted in terms of the classic plan-do-check-act management cycle. The P-D-C-A cycle is expected to be used explicitly in the next revisions of the ISO 9000 standards.

Comment.   Various people, in both the United States and Japan (e.g., Shewhart, Deming, Mizuno), have been associated with the early evolution of the P-D-C-A management cycle. The history is traced by Kolesar (1994, pp. 14–16). In its form and application to depict the management cycle, "P-D-C-A plays a central role in Japanese thought" (Kolesar 1994, p. 16). In recent years, P-D-C-A has been widely accepted in the Western world as a simple but robust embodiment of management as an activity. The P-D-C-A management cycle is compatible, in particular, with the contemporary concept that all work is accomplished by a process (ISO 9000-1:1994, clause 4.6).

Continuous improvement is a necessary consequence of implementing ISO 9001 (and ISO 9002). There are two groupings of linked clauses in ISO 9001 that work together to ensure continuous improvement.

The linkages among the clauses are really quite clear in ISO 9001 if your mind-set does not block them out. The intention of the ISO 9000 standards has always been that the clauses are elements of an integrated quality system. In implementing any system, the interrelationships among the elements, that is, the linkages, are as important as the elements themselves.

The linkages among clauses in ISO 9001 can be recognized in three ways:

- Explicit cross references between linked clauses shown by parenthetic expressions "(see *z.zz*)" in the text of certain clauses
- Use of key words or phrases in clauses that have linked requirements, for example, "objectives" for quality in clauses 4.1.1 (Quality Policy) and 4.1.3 (Management Review)
- Content interrelationships which cause linkages among the activities required by two or more clauses, for example, clauses 4.1.1 (Quality Policy), 4.1.3 (Management Review), 4.14 (Corrective Action), and 4.17 (Internal Quality Audits).

The two groupings of linked clauses that work together to ensure continuous improvement are

1. *Continuous improvement via objectives for quality and ensuring the effectiveness of the quality system:*   The 1994 revision of ISO 9001 expands and strengthens the requirements for executive management functions and links a number of clauses by requirements to define "objectives" for quality [clauses 4.1.1 (Quality Policy) and 4.1.3 (Management Review)] and to "ensure the effectiveness of the quality system" [clauses 4.1.3 (Management Review), 4.2.2 (Quality System), 4.16 (Quality Records), 4.17 (Internal Quality Audits)]. In today's competitive economy, the objectives for quality must continually be more stringent in order to maintain a healthy business position. More stringent objectives for quality translate inevitably into the need for an increasingly effective quality system.

2. *Continuous improvement via internal audits and management review:*   The 1994 revision of ISO 9001 expands and strengthens the requirements of four clauses in the 1987 standard that link together for continuous improvement in the supplier organization. These are internal quality audits (clause 4.17), corrective and preventive action (clause 4.14), management representative (clause 4.1.2.3), and management review (clause 4.1.3). The interplay of activities required by these four clauses provides a mechanism to institutionalize the pursuit of continuous improvement.

It is instructive to display simultaneously the similarities of the three managing processes that have had great influence on the practice of quality management. These are the P-D-C-A cycle, the ISO 9001/ISO 9002 requirements, and the Juran Trilogy. These are summarized in Table 11.7. Viewed from this perspective, the similarities are striking.

If the ISO 9000 implementation community fails to embrace adequately the continuous improvement requirements in the standards, the competitiveness of the standards themselves will be seriously compromised in the global marketplace.

TABLE 11.6   The Reshaped Mind-Set

A cycle of continuous improvement (built into ISO 9001 by the linked requirements)

- *Plan* your objectives for quality and the processes to achieve them.
- *Do* the appropriate resource allocation, implementation, training, and documentation.
- *Check* to see if
    You are implementing as planned
    Your quality system is effective
    You are meeting your objectives for quality
    Your objectives for quality are relevant to the expectations and needs of customers
- *Act* to improve the system as needed.

TABLE 11.7   Similarities of Three Managing Processes

P-D-C-A cycle	ISO 9001/9002 requirements	Juran Trilogy
Plan	Say what you will do • "Plan, define, establish, document" • "Objectives for quality" • "Provide resources"	Planning • Determine needs • Establish product • Establish process • Develop process • Set goals
Do	Do what you say • "Implement, maintain" Record what you did • "Quality records"	Control (remove sporadic deficiencies) • Run process • Evaluate performance • Compare to goals • Act on differences
Check	Check results versus expectations • Management review • Internal audits • External audits	Improvement (remove chronic deficiencies) • Nominate projects • Establish teams • Use improvement process • Provide resources
Act	Act on any deficiencies • Quality system revision • Preventive action • Corrective action	

Source: Kolesar (1994).

The Role of Statistical Techniques.   From the earliest days of the quality movement, statistical techniques have been recognized as having an important role. In fact, during the 1940s and 1950s, statistical techniques were viewed as the predominant aspect of quality control. During succeeding decades, the management system increasingly took center stage. In the 1987 version of ISO 9001, clause 4.20 on statistical techniques paid only lip service to its subject. The implementation of quality assurance standards worldwide has reinforced the deterioration of emphasis on statistical techniques to the point that important technical advances in statistical techniques for quality have been deprecated by many practitioners who have claimed that only the simplest, most primitive, and most ancient statistical techniques are needed, provided they are conscientiously implemented. Then they neglect to implement even those!

It is critical that this situation be remedied and that statistical techniques and management systems each be given an important place in quality.

Fortunately, the 1994 version of ISO 9001 contains explicit, meaningful requirements relating to statistical techniques (clause 4.20, Statistical Techniques), citing their relation to "process capability" and "product characteristics." Recalling the ISO 9001 mechanisms to link clauses, there is direct linkage between clause 4.20 and process control (clause 4.9, especially 4.9g), as well as indirect linkages to other clauses. These requirements, if conscientiously implemented, would guarantee that statistical techniques are enabled to have their important place in quality under the ISO 9001 umbrella.

Unfortunately, a large majority of personnel in the existing infrastructure of auditors, registrars, and accreditation bodies—and their supporting consultants, training course providers, etc.—have little knowledge or experience in statistical techniques. Consequently, despite the requirements in clause 4.20 of ISO 9001:1994, the clause still is receiving little emphasis.

Like all problems, this creates opportunities for entrepreneurial supplier companies, registrars, auditors, consultants, and course providers. I hope many will seize these opportunities (which really are obligations) in the near future. Moreover, this problem places responsibilities on accreditation bodies. They have the responsibility to ensure that the third-party registration system is implemented in conformance to the applicable international standards and guides. In particular, this includes ISO 9001 itself, and clause 4.20.

In the United States, the Registrar Accreditation Board (RAB) has taken some initiatives in this direction. The RAB has, among these initiatives, issued to all ANSI/RAB-accredited registrars a bulletin stating RAB's intention to monitor the operations of registrars with respect to the linked requirements and the implementation of clause 4.20. I have emphasized this issue in various national and international presentations and publications (e.g., Marquardt 1995, 1996a), encouraging the accreditation bodies and certification/registration bodies in other countries to take similar steps. There are opportunities for ISO/TC176, too. TC176 is cooperating with ISO/TC69 (Application of Statistical Methods) to provide guidance documentation on the use of statistical techniques when implementing the standards in the ISO 9000 family.

Interpretations of the Standards.   In actual application ISO standards are published by ISO in English and French (and often in Russian), which are the official ISO languages. ISO requires that "the texts in the different official language versions shall be technically equivalent and structurally identical" (ISO/IEC Directives, Part 1, 1.5, 1989). Sometimes ISO itself publishes standards in languages other than the official languages; then "each is regarded as an original-language version" (ISO/IEC Directives, Part 1, F.3, 1995). "However, only the terms and definitions given in the official languages can be considered as ISO terms and definitions" (ISO/IEC Directives, Part 3, B2.2, 1989).

When a nation "adopts" an ISO standard, the standard is first *translated* by the national body into the national language, and processed through the official national procedures for adoption. In the United States, the translation issue is minimal, consisting, when deemed necessary, of replacement of British English (the ISO official English) with American English spellings or other stylistic editorial details. In the United States, the adoption process is under the American National Standards Institute procedures, which ensure the objectivity, fairness, and lack of bias that might favor any constituency; these are requirements for all American National Standards.

In situations where the national language is not one of the ISO official languages, ISO has, at present, no formal procedure for validating the accuracy of the national body translation. Translation from one language to another always presents challenges when great accuracy of meaning should be preserved. There are many ways in which the meaning may be changed by a translation. These changes can be a troublesome source of nontariff trade barriers in international trade.

The problem-opportunity relating to interpretations of the ISO 9000 standards goes beyond problems of translation into languages other than the official ISO languages. In the global economy many *situations of use* are encountered; the intended meaning of the standard is not always clear to those applying the standard in some situations of use. For such situations each member body of ISO is expected to set up interpretation procedures. There will, nevertheless, be cases where an official ISO interpretation is required. ISO has, at present, no formal procedure for developing and promulgating such official interpretations. ISO/TC176 has taken the initiative with ISO Central Secretariat to establish an official procedure; ultimately ISO/TC176 should be the point of the final interpretation of the ISO 9000 standards which it is responsible to prepare and maintain.

When the situation of use is a *two-party* contractual situation between the supplier organization and the customer organization, differences of interpretation should normally be revealed and mutually resolved at an early stage (e.g., during contract negotiation and contract review). Official international interpretations become more necessary in third-party certification/registration situations. In *third-party* situations negotiations between supplier and customer tend to focus on the technical specifications for the product, plus only those quality system requirements, if any, that go beyond the scope of the relevant ISO 9000 requirements standard.

Defining the Scope of Certification/Registration

Background.   There is great variability in the documented definitions of scope of registration of suppliers' quality systems. This variability is observed from supplier to supplier for a given registrar, from registrar to registrar in a given nation, and from one nation to another. Greater consistency in defining and documenting this scope is an essential prerequisite for:

- Establishing marketplace credibility of quality system certification/registration to ISO 9001, ISO 9002, or ISO 9003
- Negotiating meaningful mutual recognition arrangements among nations

Beyond the benefits of marketplace credibility, there are important benefits to registrars if the ground rules for defining scope are consistent for all parties. This topic is an important problem-opportunity for the ISO 9000 standards.

To describe adequately the scope of certification/registration of a supplier's quality system, four questions must be asked:

- Which standard?
- Which geographic sites or operating units?
- Which products?
- Which portions of the supply chain?

The first three elements of scope are dealt with in ISO/IEC Guides in generic terms. The last (supply-chain boundaries) is not dealt with in the Guides but is equally important.

Certificates of registration are the original records from which other records (e.g., lists of registered quality systems) are derived. Examination of samples of certificates and registers shows that even the first three elements are not universally or uniformly documented today.

Selection of Standard.   The procedure should provide confidence to the customer that the selection of the appropriate standard jointly by the supplier and the registrar has taken adequately into consideration the amount and nature of product design activity that is involved in the products produced by the supplier, as well as the nature of the production processes through which the supplier adds value to the product. In some cases ISO 9003 or ISO 9002 has been selected when it appears that a more comprehensive quality assurance model would be more appropriate. In some cases, the mismatch may not be readily apparent to the supplier's customer.

An example where clarity is important is distributor operations. Many are registered to ISO 9003 under the rationale that the distributor does not produce the (tangible) products themselves. However, a distributor's product is the service products of acquiring, stocking, preserving, order fulfilling, and delivery. Hence ISO 9002 is appropriate to cover the production of these services. Distributors who design their service products should be registered to ISO 9001.

Specification of Boundaries in Terms of Geographic Location or Operating Unit.   The procedure should inform the customer whether the product the customer receives is processed within the registered quality system, even in situations where the supplier may have multiple sites or operating units dealing with the same product, not all of which may be registered. The lack of consistent procedures for scope description in regard to geographic locations or operating units included sets the stage for misrepresentation.

Specification of Boundaries in Terms of Product Processed.   The procedure should inform the customer whether the product the customer receives is processed within the registered quality system, even in situations where the supplier may deal with multiple products at the same site or operating unit and not all of the products may be processed within the registered quality system. The lack of consistent procedures in regard to product processed sets the stage for misrepresentation.

Specification of Boundaries in Terms of Supply-Chain Criteria.   The procedure should inform the customer regarding:

- The starting points of the supplier's registered operations (e.g., the raw materials, parts, components, services, and intermediate products that are provided by subsuppliers)
- The ending points of the supplier's registered operations (i.e., the remaining steps on the way to the ultimate consumer that are excluded from the supplier's registered operations)

- The nature of the value that has been added by the supplier's registered operations

Where the registered quality system represents only a fraction of the supplier's operations, or a fraction of the total value added in the product, this should be stated in registration documentation so that, as a consequence, customers may be aware of this fact.

The procedures should not invite suppliers who wish to be registered, but want to exclude portions of their operations from scrutiny by the registrar, to declare the excluded portions to be subcontractor operations. It does not matter whether the excluded portions are in another nation, elsewhere in the same nation, or simply another part of the same production site.

Procedures for this element of scope would apply also to support functions that are critical to product quality, such as a test laboratory, which may not be included in the supplier's quality system as registered to ISO 9001 or ISO 9002.

Guiding Principle.   There are many registrars; each is registering many supplier quality systems. Each supplier is dealing with many customers. It is impractical to monitor adequately the operations of such a system solely by periodic audits conducted by an accreditation body. Consequently the guiding principle should be

Primary reliance must be placed on the concept of "truth in labeling," by means of which every customer has routine, ready access to the information upon which to judge all four elements of scope of a supplier's registered quality system.

Alternate Routes to Certification/Registration.   Organizations differ in regard to the status of their quality management efforts. Some are at an advanced state of maturity and effectiveness. Others have hardly begun. Most are at some intermediate state. The ISO 9000 standards have as their primary purpose the facilitation of international trade. They are, therefore, positioned to ensure a level of maturity and effectiveness that meets the needs for reducing nontariff trade barriers in international trade. This required level of maturity and effectiveness will change with the passage of time (compare the minimal use of ISO 9003 and the growth of use of ISO 9001).

At any point in time, there will be some organizations that have well-established, advanced quality management systems based on an approach that may go beyond the requirements of ISO 9001. For such organizations, the cost of registration/certification by the usual third-party route is perceived to be high compared to the incremental value added to their quality management system. This is, and will continue to be, a significant problem-opportunity.

In the United States a number of such companies that have major international presence, especially ones in the electronics and computer industry, have been working with organizations involved in the implementation of third-party certification/registration to devise an approach that would gain international acceptance. The approach would have to take cognizance of their existing quality management maturity and reduce the cost of certification/registration, while supporting their international trade by providing the assurance conferred by certification/registration.

Industry-Specific Adoptions and Extensions of ISO 9000 Standards

Industry-Specific Situations.   In some sectors of the global economy there are industry-specific adoptions and extensions of the ISO 9000 standards. These situations are a classic example of a problem-opportunity. As problems, such adaptations and extensions strain the goal of nonproliferation. As opportunities, they have been found effective in a *very few industries* where there are special circumstances and where appropriate ground rules can be developed and implemented consistently. These special circumstances have been characterized by

1. Industries where the product impact on the health, safety, or environmental aspects is potentially severe; as a consequence most nations have regulatory requirements regarding the quality management system of a supplier
2. Industries that have had well-established, internationally deployed industry-specific or supplier-specific quality system requirements documents prior to publication of the ISO 9000 standards

Fortunately, in the very few instances so far, the operational nonproliferation criteria of the ISO/IEC Directives have been followed.

Medical Device Industry.   Circumstance 1 relates to the medical device manufacturing industry. For example, in the United States, the Food and Drug Administration (FDA) developed and promulgated the Good Manufacturing Practice (GMP) regulations. The GMP operates under the legal imprimatur of the FDA regulations, which predate the ISO 9000 standards. The FDA regularly inspects medical device manufacturers for compliance with the GMP requirements. Many of these requirements are quality management system requirements that parallel the subsequently published ISO 9002:1987 requirements. Other GMP regulatory requirements relate more specifically to health, safety, or environmental aspects. Many other nations have similar regulatory requirements for such products.

In the United States, the FDA is in late stages of developing and promulgating revised GMPs that parallel closely the ISO 9001:1994 standard, plus specific regulatory requirements related to health, safety, or environment. The expansion of scope to include quality system requirements related to product design reflects the recognition of the importance of product design and the greater maturity of quality management practices in the medical device industry worldwide. Similar trends are taking place in other nations, many of which are adopting ISO 9001 verbatim for their equivalent of the GMP regulations.

In ISO, a new technical committee, ISO/TC210, has been formed specifically for medical device systems. TC210 has developed standards that provide supplements to ISO 9001 clauses. These supplements primarily reflect the health, safety, and environment aspects of medical devices and tend to parallel the regulatory requirements in various nations. These standards are in late stages of development and international approval at this time.

Automotive Industry.   Circumstance 2 relates to the automotive industry. In the years preceding publication of the 1987 ISO 9000 standards, various original equipment manufacturers (OEMs) in the automotive industry had developed company-specific proprietary quality system requirements documents. These requirements were part of OEM contract arrangements for purchasing parts, materials, and subassemblies from the thousands of companies in their supply chain. The OEMs had large staffs of second-party auditors to verify that these OEM-specific requirements were being met.

Upon publication of ISO 9001:1994, the major U.S. OEMs began implementation of an industry-wide common standard, labeled QS-9000, that incorporates ISO 9001 verbatim plus industry-specific supplementary requirements. Some of the supplementary requirements are really prescriptive approaches to some of the generic ISO 9001 requirements; others are additional quality system requirements which have been agreed on by the major OEMs; a few are OEM-specific.

QS-9000 is being deployed by these OEMs in their worldwide operations. Part of this deployment involves separate registrations to QS-9000, through existing registrars who have been accredited specifically to the QS-9000 system. These QS-9000 registrars must use auditors who have had specific accredited training in those QS-9000 requirements which are more prescriptive than, or go beyond, ISO 9001 requirements. Accreditations are provided through specially designated accreditation bodies, including RAB in the United States.

The QS-9000 system, by removing the redundancy of multiple second-party audits to multiple requirements documents, is providing cost reductions for both the OEMs and the large number of organizations in their supply chain. Assuming that credibility is maintained by continuous improvement to meet marketplace needs and requirements, the goals of improved quality industry-wide and worldwide, together with reduced costs, can be attained.

Computer Software.   The global economy has become permeated with electronic information technology (IT). The IT industry now plays a major role in shaping and driving the global economy. As in past major technological advances, the world seems fundamentally very different, and paradoxically, fundamentally the same. Computer software development occupies a central position in this paradox.

First, it should be noted that computer software development is not so much an industry as it is a *discipline.*

Second, many IT practitioners emphasize that computer software issues are complicated by the multiplicity of ways that computer software quality may be critical in a supplier organization's business. For example:

- The supplier's product may be complex software whose functional design requirements are specified by the customer.
- The supplier may actually write most of its software product, or may integrate off-the-shelf packaged software from subsuppliers.
- The supplier may incorporate computer software/firmware into its product, which may be primarily hardware and/or services.
- The supplier may develop and/or purchase from subsuppliers software that will be used in the supplier's own design and/or production processes of its product.

However, it is important to acknowledge that hardware, processed materials, and services often are involved in a supplier organization's business in these same multiple ways, too.

What, then, are the issues in applying ISO 9001 to computer software development? There is general consensus worldwide that:

- The generic quality management system activities and associated requirements in ISO 9001 are relevant to computer software, just as they are relevant in other generic product categories (hardware, other forms of software, processed materials, and services).
- There are some things that are *different* in applying ISO 9001 to computer software.

There is at this time no worldwide consensus as to *which* things, if any, are different enough to make a difference and what to do about any things that are different enough to make a difference.

ISO/TC176 developed and published ISO 9000-3:1991 as a means of dealing with this important, paradoxical issue. ISO 9000-3 provides guidelines for applying ISO 9001 to the development, supply, and maintenance of (computer) software. ISO 9000-3 has been useful and widely used. ISO 9000-3 offers guidance that goes beyond the requirements of ISO 9001, and it makes some assumptions about the life cycle model for software development, supply, and maintenance. In the United Kingdom a separate certification scheme (TickIT) for software development has been operated for several years, using the combination of ISO 9001 and ISO 9000-3. The scheme has received both praise and criticism from various constituencies worldwide. Those who praise the scheme claim that it

- Addresses an important need in the economy to provide assurance for customer organizations that the requirements for quality in software they purchase (as a separate product, or incorporated in a hardware product) will be satisfied
- Includes explicit provisions beyond those for conventional certification to ISO 9001 to assure competency of software auditors, their training, and audit program administration by the certification body
- Provides a separate certification scheme and logo to exhibit this status publicly

Those who criticize the scheme claim that it

- Is inflexible and attempts to prescribe a particular life cycle approach to computer software development which is out of tune with current best practices for developing many types of computer software
- Includes unrealistically stringent auditor qualifications in the technology aspects of software development, qualifications whose technical depth is not necessary for effective auditing of management systems for software development
- Is almost totally redundant with conventional third-party certification to ISO 9001, under which the certification body/registrar already is responsible for competency of auditors, and accreditation bodies verify the competency as part of accreditation procedures

- Adds substantial cost beyond conventional certification to ISO 9001 and provides little added value to the supply chain

In the United States a proposal to adopt a TickIT-like software scheme was presented to the ANSI/RAB accreditation program. The proposal was rejected, primarily on the basis that there was not consensus and support in the IT industry and the IT-user community.
At this writing:

- ISO/TC176 is revising ISO 9000-3 for the short term to bring it up-to-date with ISO 9001:1994 and to remedy some technical deficiencies.
- ISO/TC176 is planning the next revision of ISO 9001 with the long-term intention of incorporating ISO 9001 quality assurance requirements stated in a way that will meet the needs of all four generic product categories without supplementary application guideline standards such as ISO 9000-3.
- Various national and international groups, conferences, and organizations are discussing whether there is enough of a difference to warrant a special program, and if so, what such a program should look like.

The one thing that is currently clear is that no worldwide consensus exists.

Other Areas of Application.   The special case of environmental management systems and their relation to quality management systems has been discussed earlier in this section. This situation, too, is a classic example of a problem-opportunity from the perspective of the ISO 9000 standards. Companies are likely to have to do business under both sets of requirements: the ISO 9000 standards from ISO/TC176 and the ISO 14000 standards from ISO/TC207. The opportunity for mutually beneficial consistency promises important benefits. These benefits relate to the operational effectiveness of having one consistent management approach in both areas of the business activities and can translate also into cost benefits of such a single approach. The ISO Technical Management Board has mandated that TC176 and TC207 achieve compatibility of their standards.
In the United States and other nations, the compatibility of the ISO 9000 standards and the ISO 14000 standards is one part of the standardization job. The implementation part requires that similar harmonization and compatibility be established in each nation in the infrastructure of accreditation bodies, certification/registration bodies, and auditor certification bodies, operating under internationally harmonized guidelines. At this writing the ISO 14000 infrastructure is in its infancy.

RELATION OF ISO 9000 STANDARDS TO NATIONAL QUALITY AWARDS AND TO TOTAL QUALITY MANAGEMENT (TQM)

Various nations and regional bodies have established quality awards. The most widely known of these are the Deming Award in Japan; the Malcolm Baldrige National Quality Award (MBNQA) in the United States; and the European Quality Award, a European regional award. These awards incorporate concepts and principles of Total Quality Management (TQM).
TQM means different things to different people. ISO 8402:1994 defines TQM as follows:

> …management approach of an organization centered on quality, based on the participation of all its members and aiming at long-term success through customer satisfaction, and benefits to all members of the organization and to society.

For purposes of this section the criteria of the Baldrige Award or the Deming Award or the European Quality Award can be considered to be an operational definition of full-scale TQM implementation.

Questions often are asked about the relationships between the criteria upon which these awards are based and the content of the ISO 9000 standards. This discussion is in two parts: the relationship to ISO 9001 and the relationship to ISO 9004.

Overall, it is important to understand that the purpose of the ISO 9000 standards is to facilitate international trade. To achieve that purpose, the ISO 9000 standards focus on the supplier organization functions that most directly affect product quality. The ISO 9000 standards are intended for implementation by the large majority of supplier organizations. By contrast, the purposes of the award criteria are (1) to select, from among all the supplier organizations in a nation or region, those few organizations that exemplify the very best level of achievement in quality management and (2) to provide criteria and guidelines for other organizations that may wish to improve in the direction of becoming best of the best and are willing to make the substantial investment to achieve that lofty level of quality performance.

Relationship to ISO 9001. ISO 9001 is a requirements standard for two-party contractual or third-party registration use in support of international trade. Commensurate with this role, ISO 9001 focuses only on the functions that most directly affect product quality. It does not, therefore, deal with questions of economic effectiveness and cost efficiency. It deals only with specific personnel aspects and specific sales and marketing aspects that directly affect product quality. Thus, the *scope* of ISO 9001 is narrower than the scopes of the cited national awards. For example, the MBNQA criteria examine many specific items in seven broad categories of an organization's activities. These seven categories are: leadership, information and analysis, strategic planning, human resource development and management, process management, business results, and customer focus and satisfaction. The ISO 9001 requirements give greatest emphasis to the process management category of MBNQA, and have lesser emphasis on the other categories.

In view of the differing purposes of ISO 9001 and the award criteria, there is a difference also in the *depth* of examination of the supplier organization's quality management system. The MBNQA and ISO 9001 both embrace the concept that all work is accomplished by a process and that an organization's activities can be viewed as a network of processes. Both MBNQA and ISO 9001 recognize the need to examine the approach, the deployment, and the results (Marquardt 1996b) in the examination of a process.

In the late 1980s the author developed Table 11.8 to describe the relative depth of expectations in terms of the appropriate assessment questions at various levels, including ISO 9001 and award criteria. Questions are shown for approach, deployment, and results.

It is instructive to compare ISO 9000 registration (specifically ISO 9001 or ISO 9002) to the achievement of an MBNQA award. As described in Table 11.9, ISO 9000 registration has in many ways more modest requirements, but it does emphasize to a greater degree the necessity of a consistent, disciplined, documented quality system, with periodic internal and external audits that serve to hold the gains and institutionalize continuous improvement.

Relationship to ISO 9004. ISO 9004 is the standard in the ISO 9000 family that provides to organizations quality management guidelines that cover a *wider scope* and *greater depth* than the requirements in ISO 9001. The additional scope and depth go part way toward the scope and depth of award criteria such as the MBNQA. In keeping with the purpose of the ISO 9000 standards, the scope and depth is at a level that is achievable by a large proportion of organizations in the global economy. Thus, ISO 9004 is deliberately positioned in an intermediate range, between ISO 9001 and the award criteria. The marketplace uniqueness of ISO 9004 is that it offers to organizations a framework for building a quality management system that will be effective and efficient and that will focus on features that have a direct effect on product quality, features that are fully consistent with ISO 9001, ISO 8402, and other standards in the ISO 9000 family. This enables the organization to use one consistent set of terminology that is internationally standardized and one consistent framework.

Many organizations worldwide have adopted the strategic perspective that ISO 9001 provides for them minimum adequate criteria for effective operation and for meeting the marketplace

TABLE 11.8   Assessment Questions (for Any Process in a Quality Management System)

Approach	Deployment	Results
First level of depth		
Is there a defined process?	Is the process fully deployed?	Are the results meeting requirements?
Second level of depth (e.g., ISO 9001 requirements)		
Is the process appropriate to the needs of the function? Is there documentation appropriate to each person's needs at each organizational level? Is the documentation controlled (accurate, up-to-date, available when and where needed)?	Is a process used wherever this generic function is implemented in the organization? Is every involved person trained (understands the process, why it exists, how to use it)? Does every person have value for the function?	Have quantitative metrics been defined? Are the metrics understood and used by those involved? Do the values of the metrics show that the process is appropriate to the function for both quality management and quality assurance purposes?
Third level of depth (National or International Award Criteria)		
Is the process state of the art, world class, best of the best for this function? Is there innovation, technology advantage, cost advantage, functional superiority in this process? Does the excellence in this process provide superior value that is perceived by the customer as exceeding expectations?	Is the process deployed consistently and universally to accomplish this function in a standard way with consistent, transferable training everywhere? Does the consistency of the results of this process provide superior value that is perceived by the customer as exceeding expectations?	Are the metrics designed and implemented in a way that elicits value-adding behavior responses? Do the values of the metrics clearly portray continuous improvement and world-class status? Do the financial results demonstrate that customers perceive superior value?

TABLE 11.9   ISO 9000 Registration in Relation to Malcolm Baldrige Award

ISO 9000 requires:
• Adequate quality systems
• Objective evidence for every requirement
• Complete, controlled, up-to-date documentation
• Periodic surveillance audits that verify continuing compliance to requirements

MBNQA looks for:
• Best-of-the-best quality systems
• Clear evidence of product quality superiority
• Clear evidence of customer perception of superiority
• Historic trends that lend credence to one-time audit

requirements for quality. ISO 9004 provides a guideline to enrich and enhance the ISO 9001 baseline, and to take deliberate, planned initiatives that build upon the baseline in the direction of TQM. The award criteria, such as those of MBNQA, provide a comprehensive operational statement of full-scale TQM. With this insight, the standards and the award criteria are compatible and complementary. Both are necessary in the global economy.

REFERENCES

Kolesar, P. J. (1994). "What Deming Told the Japanese in 1950." *Quality Management Journal,* vol. 2, issue 1, pp. 9–24.

Marquardt, D. W. (1994). "Credibility of Quality Systems Certification: How to Deal with Scopes of Certification, Conflicts of Interest and Codes of Conduct." In Peach, R. W. (ed.), *ISO 9000 Handbook,* 2d ed. Irwin Professional Publishing, Fairfax, VA, chap. 17.

Marquardt, D. W.(1995). "The Missing Linkage in ISO 9001: What's Being Done About It?" *ASQC 49th Annual Quality Congress Proceedings,* American Society for Quality Control, Milwaukee, pp. 1056–1061.

Marquardt, D. W. (1996a). "The Importance of Linkages in ISO 9001." *ISO 9000 News,* vol. 5, no. 1, pp. 11–13.

Marquardt, D. W. (1996b). "The Functions of Quality Management in Relation to the P-D-C-A Management Cycle." Manuscript submitted for publication.

Marquardt, D. W., et al. (1991). "Vision 2000: The Strategy for the ISO 9000 Series Standards in the '90s." *Quality Progress,* May, pp. 25–31. Reprinted in *ISO 9000 Quality Management* (an ISO standards compendium containing all standards in the ISO 9000 family, plus the Vision 2000 paper), 5th ed. (1994), ISO Central Secretariat, Geneva. Also reprinted in a number of other countries and languages, and in the *ISO 9000 Handbook,* op cit., chap. 11.

SECTION 12
BENCHMARKING

Robert C. Camp
Irving J. DeToro

This section defines benchmarking and outlines the 10-step benchmarking process as developed in Camp (1989 and 1994). It summarizes the activities of typical benchmarking teams, including their objectives, tasks, and responsibilities.

INTRODUCTION

The hottest and least understood new term in the quality field is "benchmarking." Xerox does it. Ford does it. GTE, AT&T, DEC, TI, duPont, HP, J&J, IBM, and Motorola do it. Just what is it?

Benchmarking is an ongoing investigation and learning experience. It ensures that the best practices are uncovered, adopted, and implemented. Benchmarking is a process of industrial research that enables managers to perform company-to-company comparisons of processes and practices to identify the "best of the best" and to attain a level of superiority or competitive advantage.

Benchmarking is a method of establishing performance goals and quality improvement projects based on industry best practices. It is one of the most exciting new tools in the quality field. Searching out and emulating the best can fuel the motivation of everyone involved, often producing breakthrough results.

The Japanese word *dantotsu*—striving to be the best of the best—captures the essence of benchmarking. It is a positive, proactive process to change operations in a structured fashion to achieve superior performance. The purpose of benchmarking is to gain competitive advantage.

Benchmarking: Definition.   The formal definition of benchmarking is "The continuous process of measuring products, services, and practices against the company's toughest competitors or those companies renowned as industry leaders." (Camp 1994).

Benchmarking Objectives.   The purpose of benchmarking is derived primarily from the need to establish credible goals and pursue continuous improvement. It is a direction-setting process, but more important, it is a means by which the practices needed to reach new goals are discovered and understood.

Benchmarking legitimizes goals based on an external orientation instead of extrapolating from internal practices and past trends. Because the external environment changes so rapidly, goal setting, which is internally focused, often fails to meet what customers expect from their suppliers.

Customer expectations are driven by the standards set by the best suppliers in the industry as well as by great experiences with suppliers in other industries. Thus, the ultimate benefit of benchmarking is to help achieve the leadership performance levels that fully satisfy these ever-increasing customer expectations.

Benchmarking is an important ingredient in strategic planning and operational improvement. To remain competitive, long-range strategies require organizations to adapt continuously to the changing marketplace. To energize and motivate its people, an organization must:

- Establish that there is a need for change
- Identify what should be changed
- Create a picture of how the organization should look after the change

Benchmarking achieves all three. By identifying gaps between the organization and the competition, benchmarking establishes that there is a need. By helping understand how industry leaders do things, benchmarking helps identify what must be changed. And by showing what is possible and what other companies have done, benchmarking creates a picture of how the organization should look after the change.

BENCHMARKING FUNDAMENTALS

Embarking on a benchmarking activity requires acceptance of the following fundamentals:

- Know the operation. Assess strengths and weaknesses. This should involve documentation of work process steps and practices as well as a definition of the critical performance measurements used.
- Know industry leaders and competitors. Capabilities can be differentiated only by knowing the strengths and weaknesses of the leaders.
- Incorporate the best and gain superiority. Adapt and integrate these best practices to achieve a leadership position.

Practices and Performance Levels.   Benchmarking can be divided into two parts: practices and performance levels. From experience, most managers now understand that benchmarking should first focus on industry best practices. The performance levels that result from these practices can be analyzed and synthesized later. Having identified the best practices of several companies, the lessons learned can be integrated to create world-class work processes. At that stage, the expected performance from these work processes can be determined so that service levels that are superior to the best of the competitors' can be delivered.

When preparing for benchmarking, it is important to engage line management so that the findings are understood and accepted and result in a commitment to take action. This requires concerted management involvement and carefully designed communications to the organization that must implement the action plans.

THE 10-STEP BENCHMARKING PROCESS

The 10-step process for conducting a benchmarking investigation consists of the following five essential phases (see Figure 12.1).

Phase 1:   Planning

- Decide what to benchmark. All functions have a product or output. These are priority candidates to benchmark for opportunities to improve performance.
- Identify whom to benchmark. World-class leadership companies or functions with superior work practices, wherever they exist, are the appropriate comparisons.
- Plan the investigation, and conduct it. Collect data sources. A wide array of sources exists, and a good starting point is a business library. An electronic search of recently published information on an area of interest can be requested. Begin collecting. Observe best practices.

Phase 2:   Analysis

- It is important to have a full understanding of internal business processes before comparing them to external organizations. After this, examine the best practices of other organizations. Then measure the gap.
- Project the future performance levels. Comparing the performance levels provides an objective basis on which to act and helps to determine how to achieve a performance edge.

Phase 3:   Integration

- Redefine goals and incorporate them into the planning process.
- Communicate benchmarking findings and gain acceptance from upper management.
- Revise performance goals.
- Remember, the competition will not stand still while organizations improve. Thus, goals that reflect projected improvement are necessary.
- On the basis of the benchmarking findings, the targets and strategies should be integrated into business plans and operational reviews and updated as needed.

Phase 4:   Action

- Best practices are implemented and periodically recalibrated as needed.
- Develop and implement action plans.

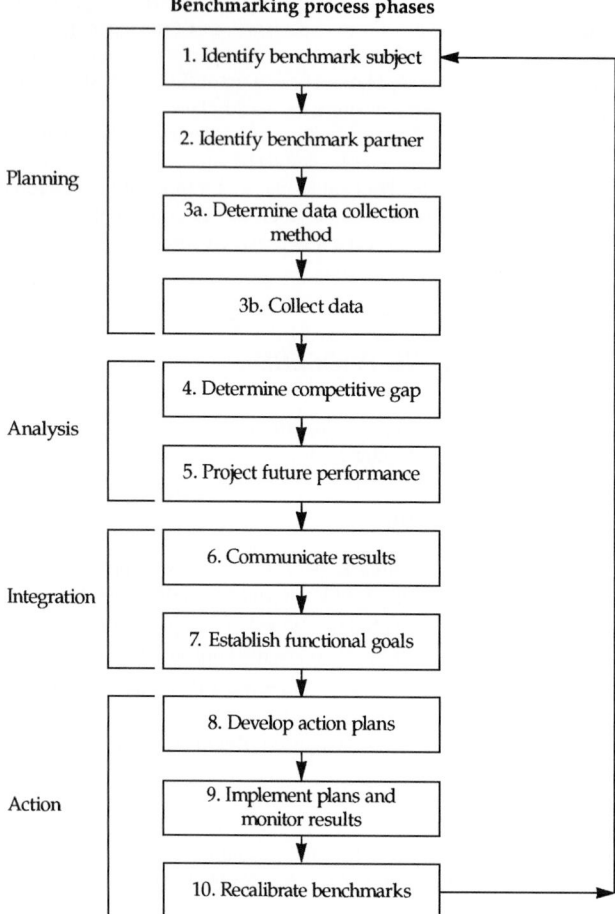

FIGURE 12.1 The formal 10-step benchmarking process. (*Quality Resources, a division of The Kraus Organization Limited, White Plains, NY, through ASQC Quarterly Press.*)

- Monitor progress.
- Recalibrate the benchmarks.

Phase 5: Maturity

- Determine when a leadership position is attained. Maturity is achieved when best practices are incorporated in all business processes; when benchmarking becomes a standard part of guiding work; and when performance levels are continually improving toward a leadership position. Assess benchmarking as an ongoing process.

Benchmarking Triggers.   Events that cause a benchmarking project to be initiated usually fall into one of three groups: problem, innovation, or policy.

Problem. If a crisis occurs within an organization, such as a major cost overrun or a major customer threatening to cancel an existing contract, the topic for an improvement project may be apparent.

Innovation. If the organization becomes aware of some innovative technology, practice, or process employed by another organization, this information may well cause an organization to commission a benchmarking study.

Policy. If the organization does not have a significant problem or need to understand an innovative practice, then selecting a benchmarking project may be difficult. This is especially true for an organization employing a total quality management (TQM) philosophy. However, a well-established quality management effort includes a strategic planning process. In organizations with such efforts, it is common for one of the outputs of the planning process to be a list of nominations for appropriate benchmarking projects.

Benchmarking Teams. Benchmarking is conducted by teams consisting of individuals with direct operational experience and knowledge of the process. Members should possess analytical, research, process documentation, and team facilitation skills. These requirements favor candidates with engineering or technical backgrounds, and those with research experience. Benchmarking teams are typically commissioned by the process champion. Teams rarely function effectively if they consist of more than 9 to 12 members. Team size of 3 to 6 is preferred. Large teams can be considered but will most likely break down into small subgroups to do their work.

STEP 1: WHAT TO BENCHMARK

The first step in determining what to benchmark is identifying the product or output of the business process or function. Fundamental to this is the development of a clear mission statement detailing the reason for the organization's existence, including key outputs expected by its customers and critical to fulfilling the mission successfully. Next, each function's broad purposes should be broken down into specific outputs to be benchmarked. Outputs should be documented to a level of detail necessary for analyses of key tasks, handoffs, and both in-process and end results measurements, and for quality, cost, and delivery analyses.

One good way to determine which outputs are most in need of benchmarking is to pose a set of questions that might reveal current issues facing the function. Questions might focus on customer care (including service), cost, or perception of product offerings. Another way to identify key outputs is to convert the problems, issues, and challenges faced by the function into problem statements and then to develop these into a cause-and-effect Ishikawa diagram. The causals in the diagram are candidates for benchmarking.

Successfully completing a benchmarking project is dependent on selecting a worthwhile topic. It should not be too large, too trivial, or one that would not secure a performance advantage. To avoid these problems, the topic should be selected after some analysis to ensure that the organization's resources are being justifiably expended. The intent of step 1 is to confirm a topic already selected, to narrow or broaden the scope of a project, or to select a topic that could best contribute to the organization's success.

If resources are constrained, consideration should be given to improving the earliest possible area in a process since, if that is improved, there may be a beneficial effect on all subsequent activities.

The candidate process should be tested for reasonableness by asking such questions as: Is this the area customers complain about the most? Are there areas with major cost overruns that need attention? Is there something that if not fixed immediately will be affected in the marketplace?

Documentation. The team must describe how the work is currently performed by preparing detailed flowcharts. This is essential because it helps the team gain consensus on how the work is actually performed, the time it takes to perform the work, the cost, and the errors created in the current work flow. This understanding is essential because comparison to a superior system will not reveal deficiencies in the current system unless such understanding and documentation exists.

A brief, two- or three-page, description of the benchmarking project should be prepared and circulated among sponsors, managers, stakeholders, and other interested parties. This document captures all the thinking that has gone into selecting the project, the potential resources required, and the expected outcome. As more information is gathered and the team completes other steps in the 10-step benchmarking model, the project description can be updated and used as a means of keeping sponsors informed.

STEP 2:   WHOM TO BENCHMARK

The difficulty is in identifying which leading-edge companies possess processes that truly have best practices. Determining whom to benchmark against is a search process that starts with consideration of, in broad terms, an operation's primary competitors and then extends to leading companies that are not competitors. While the process is one of comparison, the goal is to identify and understand where things done differently can produce breakthrough results.

A successful approach encompasses internal, competitive, and functional benchmarking. *Internal* benchmarking is the comparison of practices among similar operations within a firm. One distinct benefit of internal benchmarking is that it forces documentation and allows easy comparison of the work process to uncover the best practices. *Competitive* benchmarking is the comparison to the best direct competitors, and serves to prevent complacency. However, it is *functional* benchmarking—the comparison of functional activities, even in dissimilar industries—that holds the most potential for discovering and stimulating innovative practices.

Internal Benchmarking.   Within every organization exists another department that may be performing identical or similar work to that of the benchmarking team. If this is true, then the first benchmarking partners may exist within the same organization. Technical training, for example, usually is not the only training activity within an organization. Typically there is also management training, sales training, computer skills training, diversity training, and specific job training.

Competitive Benchmarking.   Benchmarking competitors is an essential part of any external comparison. However, benchmarking is not industrial espionage, but industrial information gathering, and so a few cautions are needed to study competition in an ethical and legal manner. One approach is to collect information without directly contacting the competitor. So much information exists in the public domain that it may be possible to determine a competitor's best practices without having to talk directly to that competitor. Some of these public sources are discussed in step 3, and detailed search techniques are available in Camp (1994).

If a direct benchmarking exchange with a competitor is possible, the technique to avoid legal and ethical problems in comparing best practices is to focus on *process* information and not company data. For example, it is possible to discuss the rate of billing errors, cost or cycle time per thousand invoices, or to compare process steps or computer technology employed. It is not appropriate to ask for or discuss the number of products sold per invoice, the revenue value of each invoice, or the number of such invoices per accounting period. The benchmarking team should avoid asking for information that its own organization would be unwilling to provide.

Competitors sometimes minimize the difficulties of benchmarking other industry members by cooperating in a joint study conducted by a third party. In this case, a consultant may be retained by a group of competitors to collect and analyze data, protecting the sources of information, and publishing industry-wide information from the participants.

Formal benchmarking has been ethically managed since the early 1980s, in part because of the existence of the Benchmarking Code of Contact developed by the International Benchmarking Clearing House (Camp 1989). This code describes the nature of a benchmarking exchange that should take place between partners, and covers such items as the legality, exchange, confidentiality, and use of information.

Functional Benchmarking and World-Class Leaders.   A benchmarking study should not be restricted to one industry when an organization in a different industry is achieving superior results in a similar function. This source of best practices is where the large process gains are possible. The search here is not restricted to a common application, but to a method or practice within a process that can be adopted and adapted to a specific process.

For example, a low-cost airline in southwest Texas undertook a project to maximize its revenue by benchmarking the time a plane spends on the ground. After all, revenues are generated only when the plane is in the air. The project was to determine how ground crews could safely clean and service the aircraft, refuel it, change tires, and provide food service in the most efficient manner to minimize ground time. The benchmarking team decided to meet with individuals who have perfected such techniques—race pit crews at the Indianapolis Speedway. Racing professionals certainly are experts at these very practices because they are faced with the same issues, and have developed an expertise in rapid service.

Other examples of functional and innovative benchmarking are found at Xerox. It has worked with some of America's largest corporations including the following: American Express (billing and collection); American Hospital Supply (automated inventory control); Ford Motor Company (manufacturing floor layout); IBM and General Electric (customer service support centers); L. L. Bean, Hershey Foods, and Mary Kay Cosmetics (warehousing and distribution); Westinghouse (National Quality Award application process, warehouse controls, bar coding); and Florida Power and Light (quality process).

Having identified the sources of best practices, they should now be used in the team's specific benchmarking project. That is, for each major portion of the targeted process to be benchmarked, the team needs to identify who are comparative organizations internally, competitively, functionally, and innovatively. Typically, a team can brainstorm about 20 internal and competitive organizations to benchmark. Functional and innovative comparisons require more work to identify. Part of the work begun here will be completed in step 3A. The key point here is that the team should try to identify at least 100 prospective benchmarking partners from which it will select three or four to study in detail. Limiting the candidates at the outset may result in selecting inappropriate partners later.

Partnering.   The organization with which one would benchmark is known as a "partner." The concept of partnership is important because it conveys the notion that there must be something in a benchmarking exchange for both parties. There may a common interest in a particular process, or the benchmarking partner may ask for assistance in arranging a benchmarking visit with another department of the requesting organization. Without such mutual benefit, the likelihood of a successful, long-term benchmarking exchange is low.

STEP 3A: COLLECT DATA

The benchmarking team should tap the following information sources.

Internal.   Some organizations have a company library, but everyone has access to a public library. Using online computer capability, a search of everything published on the team's topic during the last 5 years can be conducted. This effort usually reveals a wealth of information that can be expanded by directly contacting the authors of the most pertinent articles. Other internal sources include reviewing internal market research or competitive studies.

External.   So much organizational information exists in the public domain that the problem is one of sifting the mountains of data. Sources such as professional associations, public seminars, lectures, trade shows, and speeches before public audiences are all available.

Original.   It may be necessary to contact directly some potential benchmarking partners through phone or mail surveys. This approach uses a series of increasingly detailed questionnaires. For example, if 100 organizations have been identified as candidates, then the benchmarking team can send out a simple questionnaire to each one asking about the performance of the subject process and the organization's willingness to benchmark. The team should provide its performance data on the same questions being asked. This sets the right tone of openness, sharing, and possible collaboration. Once the responses are received, more detailed questionnaires can be mailed until the benchmarking team believes it has found the select few organizations it wants to visit. If these selected organizations agree to a benchmarking exchange, then the team needs to prepare for the site visit.

STEP 3B: *CONDUCT THE SITE VISIT*

Conducting a site visit is not a trivial activity. Extensive preparation is required to ensure the visit is mutually useful and productive. Simply visiting with an other organization may be socially pleasing, but it is not apt to yield any significant learning. In addition, world-class organizations that are overwhelmed with requests for benchmarking visits will not allow one to occur without extensive assurances that the visit will be productive.

Questions that will be asked during the site visit should be prepared and sent to the benchmarking partners before the visit. This is usually the first test the prospective partner uses to understand the degree to which the requesting team is prepared.

Preparing good penetrating questions takes some thought. The team should prepare a prioritized list of the topics about which it wants to ask. There are several question forms, such as open-ended, multiple choice, forced choice, and scaled questions. Each has a useful purpose and all forms should be used to extract the required information.

The questions should be tested on an internal benchmarking partner. Any first attempt to develop succinct, pertinent questions is difficult, and using them internally will reveal any design deficiencies. Once perfected, these questions are asked during the site visit.

Another reason for extensively preparing questions is that it is not useful to ask one partner one set of questions, and another partner a different set of questions. When that occurs, answers are impossible to correlate and summarize.

An agenda for the site visit should be prepared and exchanged, and the participants identified. As part of this site visit preparation, a benchmarking protocol should be agreed to so that questions about what information will be exchanged and what documents will be available are answered before the visit. It is less than professional for a team to request information that is denied.

Typically, a benchmarking site visit team consists of a small number of individuals, perhaps four, each of whom plays a specific role in the meeting. One individual is designated as the team presenter, and he or she should be prepared to present professionally the project team's current flowcharts and related process information. Another team member should be prepared to ask the questions that have been forwarded to the partner. A third team member is the scribe and should have a laptop computer to record the responses to all questions.

A fourth member can fill a number of support roles. If some specialized knowledge is required to understand the information presented, the team may invite such a specialist to attend. Some project teams regularly invite a senior manager to attend, because this affords the manager the opportunity to see a world-class performer in action. This exposure has the effect of enabling senior managers to rethink what their organization needs to accomplish. All team members should be organized, should rehearse, and should prepare themselves thoroughly to conduct a professional site visit.

In turn, the benchmarking partner will likely have a team of two to four participants with an agenda of their own, who will follow a similar meeting format. Thus, each team might take a half day to review and answer the questions and the remaining time to visit the operation to observe it firsthand. The time required to accomplish the agenda of both teams may not be sufficient where a process is complex. The need to satisfy both agendas provides another argument to support the need for thorough preparation to ensure a successful exchange of information.

Once the site visit is complete, the benchmarking team should write a site visit report. It captures the answers recorded by the scribe, as well as impressions and information from the other team members. It is essential that this report be prepared the same day as the meeting, in the hotel lobby, airport lounge, or wherever the team members can debrief while their impressions are still fresh in memory. This report, and all other site visit reports, become the basis for the best practices report (Figure 12.2). It is this report on which the team will base its recommendations to management.

At this point some will argue that there is so much work in conducting a benchmarking study, and so much normal work to be completed, that an outside consultant should be hired to perform the task. When time pressures are extreme, when external sources have more creditability, or where specialized skills are required such as questionnaire design, hiring an outside consultant to perform the work may be appropriate. However, when an external source is used there is little organizational learning, and, perhaps, little ownership in the findings. The rule of thumb followed by many organizations familiar with benchmarking is 70/30; 70 percent of the studies done internally, 30 percent done externally.

STEP 4: ANALYZE THE PERFORMANCE GAP

The team must now analyze all the information collected. The specific task here is to analyze the data to determine if the processes benchmarked are at parity, ahead of, or behind others. Said differently, the team needs to identify gaps or differences in performance that exist between the team's process and that of the best-in-class and that of world-class organizations. This analysis should include which inputs, outputs, processes, or steps within a process are superior, and by what *measure* each of these components is superior.

CONTENTS
Introduction
  Project Purpose
    Study Background
    Problems Cited
    Company Business Objectives
    Critical Performance Metrics
    The Focus of Our Benchmarking Study
Method of Investigation
  What to Benchmark
  Whom to Benchmark
  Project Scope
    Study Duration
    Benchmark Team
    Study Costs
    Conversion Costs
Summary of Benchmark Findings
  Comparative Work Flowcharts
  Best Practices Descriptive Study
Analysis of Performance Gaps and Gains
  Current Performance Gap
    Overall Gap
    Historical Trends
    Impact of Superior Training
  Projected Performance Gap
    Impact of Implementing Benchmark Practices
  Projected Performance Advantages
Appendix
  Benchmark Study Questionnaire
  Site Visits Reports
  Quantitative Cost Analysis
  Comparison of Industry Best Practices

FIGURE 12.2 Best practices report. (*DeToro 1995, Table 14.5.*)

"By what measure?" is an important question, because summary data that seem comparable can be quite different if calculated differently and if different elements or assumptions are used in the calculation. Quantitative measures are often quite unreliable, and different methods of deriving data make any meaningful comparisons difficult.

Another important part of analyzing performance is the development of a summary statistic. This data element is usually a measure of overall performance that is useful in making preliminary comparisons between organizations and for projecting data, as will be seen in step 5. Computing a summary statistic—a quick measure of productivity—may be as simple as dividing an organization's output by the number of people involved. In warehousing, for example, dividing the number of orders picked and packed by the number of individuals involved is a rough measure of performance.

In addition to analyzing the performance gap between the team's process and that of its partners, the team also needs to understand how the targeted operation would perform if all the best practices observed were adopted. This is important because if the combination of all best practices from all benchmarking partners is implemented, that may allow the benchmarking team to surpass the performance of every partner studied.

For many reasons, however, it may be impossible to adopt all the best practices. Transition cost, disruption, cultural issues, and conflicting values are all reasons organizations usually offer for making minimal change. But as a result of the comparison of best practices, the full extent of the opportunity to achieve superior performance is now known, as are the specific work steps and methodology in achieving this improvement. *A more compelling case for dramatic process improvement and change is hard to make.*

The team needs to compare each of the steps of its process and determine the results achieved by the benchmarking partners. That is, what is the value, performance advantage, or cost advantage in each of the key steps of the subject process if the project team were to adopt the best practices of the organizations studied? An example of an analysis of performance gaps is shown for a technical training group in Table 12.1.

Another important activity in assessing benchmarking data is visually comparing the flowcharts employed by each of the benchmarked firms. This side-by-side comparison reveals important improvement opportunities. Typical observations and implications are listed in Table 12.2.

The team should compile a "best-of-best practices" chart for the benchmarked process. This chart includes, for each step in the targeted work process, a description of the best practice, the performance advantage associated with that best practice, and the value or gain possible if this best practice were employed. This information is the basis on which recommendations will be made in step 6.

STEP 5: *PROJECT PERFORMANCE LEVELS*

In the analysis step, the relative position today is defined, but industry practices are not static. Both industry and competitors continue to pursue improvement. Therefore one must not only analyze the gap as it exists at the time of measurement but also project where the benchmark and gap are likely to be in the future (see Camp 1989, p. 152 for a graphic portrayal).

At this point in the project, management has the right to ask the team the following:

- How does the organization compare today with the industry's best?
- How will this organization gain a performance advantage?
- What will it mean to the operation? The organization?
- How much will it cost to convert?

To answer these questions, the benchmarking team needs to ask the following:

- What were the historical performance trends?
- What is the current performance gap?

TABLE 12.1   Performance Gap Analysis

Group: Technical Training
Process: Curriculum Development
Performance Analysis Summary

Current method	Best method	Performance advantage	$ value gain	Conversation considerations
		Work step 1: Needs analysis		
Discuss deficiencies with trainers	Ask client to identify current performance deficiencies, set program goals and set customer requirements	Save one full day in the field observing incumbents	$350	Develop new survey questionnaire
	Conduct task analysis to identify job duties and create a task map	Improve accuracy of learner objectives and save half day of rewrites	$175	Train curriculum development team
Observe incumbents on the job	Observe incumbent performers on the job	Same	None	$2500 travel costs/year
Survey managers and top performers	Observe top and average performers on the job to measure performance differences to compute potential for improvement	Set measurable targets for improvement against which to assess success	Save 1 day	
Create learner objectives	Convert into learner objectives	Same	None	

Source: DeToro 1995, Table 14.6.

- How will industry performance change?
- Will the performance gap widen, narrow, or remain the same?
- What are the implications for the subject business?
- How can the organization gain a significant performance advantage?

The team can begin to answer these questions by choosing a summary statistic and using it as an overall measure to understand changes that have and will take place. Historical data (from the organization and from the overall industry) and projected improvements can be compared and analyzed to determine:

TABLE 12.2   Flowchart Comparisons

Observations	Implications
Fewer steps	Fewer people, lower cost
Different sequences	Better work flow, reduced cycle time
Different handoffs	Short cuts, reduced errors
Design differences	Better asset utilization, lower cost
Automated steps	Streamlined processes, reduced errors
Steps outsourced	Specialized skills utilized, lower costs

Source: DeToro 1995.

Recent trends

Size of the performance gap

Why the performance gap exists

If the gap is widening or narrowing and why

The next activity for the team is to decide what improvements the industry is capable of achieving in the future and plot those industry trends along with improvements the team expects to secure during the next 2 to 5 years. Actually, the team should develop two projections: one showing the effects of no changes in the current process and a second showing the effects of implementing the best practices discovered through benchmarking.

The last activity in step 5 is assessing the operational implications of the changes and the financial impact or value in closing the gap. For instance, the projected changes may have operational effects on suppliers, customers, and staff, as well as methods of working. Each of these impacts needs to be identified and the implications of the proposed changes shared with the affected groups. This is necessary to ensure that those affected have ample opportunity to absorb and adjust. Similarly, the financial effect needs to be calculated so that the value of implementing the best practices is identified and conveyed to the affected groups. A powerful technique to capture the minds and hearts of senior managers is to translate the operational and financial benefits of implementing best practices into the contribution they will make to the organization's goals and objectives, and the contribution they will make toward realizing the organization's vision.

STEP 6: COMMUNICATE FINDINGS

Collecting and analyzing the best practices data and projecting the operational implications are important and necessary, but are not enough to attain improved organizational performance. The benchmarking team has the responsibility to secure management's approval for the recommendations. Management skepticism, unwillingness, or outright resistance to accept the team's findings may exist. Thus, the team's task is to communicate its findings in such a way as to obtain acceptance. To accomplish this, the team must ensure that management understands the findings, thinks the team is creditable, and accepts its recommendations. This is achieved as follows.

Decide Who Needs to Know.   There are both formal and informal decision makers in every organization, and there are customers, suppliers, staff, and associates with whom the teams work. From among these constituencies the team must decide who (by name) needs to know about its findings, what they need to know, and why they need to know it.

Select the Best Presentation Vehicle.   Cultures vary, and what is accepted in one organization may not be viable in another. In some cases written reports are required with detailed supporting documentation. In other cultures, an oral presentation with overhead transparencies that is preceded by prebriefings and handouts is required. In other organizations, a one-page executive summary and informal discussion is all that is needed. The team must tailor the delivery of its report to its audience.

Organize Findings.   The team should capture its findings in the best practices report (Figure 12.2). This document represents the accumulation of all the work the team has completed and is an integral part of the communication effort. The report is also important because it captures work that may be useful to other benchmarking teams. Lastly, the discipline of developing a document like the best practices report helps ensure the team's findings will be thoroughly prepared.

Best practice reports should include descriptive and quantitative data; present facts, not opinions; stress performance gains, not methods of investigation; and include a preliminary estimate of the cost to implement the recommendations.

Present Recommendations.   The presentation of best practices should be made with the complete benchmarking team in attendance. Various portions of the presentation can be delegated so that responsibilities are shared. This approach represents a developmental opportunity and is one of the highest forms of recognition.

The team's recommendations should always be accepted. That is not to say that every benchmarking team's recommendations should be automatically accepted without challenge; however, questions should be raised and answered and issues should be identified and resolved well before the final presentation. If the team has kept its management sponsor appraised throughout the benchmarking study, and if senior management has reviewed the team's work and provided guidance and support, it is highly unlikely that recommendations would be rejected.

STEP 7:   ESTABLISH FUNCTIONAL GOALS

After management approves the recommendations, the impacts of practice changes must be identified and communicated to the affected individuals. Management normally does not concern itself with the day-to-day details of process improvements; however, management does need to know what these recommended changes mean to the work unit, the department, and the organization as a whole. An important related question is: What does implementing these recommendations mean to the organization this year? To answer these questions the team must complete the following three tasks:

Revise Operational Goals.   Every organization and the units within it have clear, definite direction regarding the short- and long-term objectives that must be accomplished. The anticipated changes are reflected in a revised set of goals and objectives. Failure to revise direction after a benchmarking recommendation is accepted indicates that the organization is not serious on following through with the recommended actions. This may reflect less on the team, and more on management's unwillingness to improve performance aggressively.

Analyze the Impact on Others.   Obviously, changes are not implemented in a vacuum, and significant changes can reverberate throughout an organization. Groups normally affected include customers, suppliers, management, staff, and associates. An example of *not* considering the impact of changes on others occurred at Xerox Distribution. The picking operation became more efficient as a result of benchmarking the L. L. Bean operation. Xerox, however, did not consider the impact of its improved productivity, and the packing department was unprepared to handle the increased volume. A sample of the impacts that a technical training group identified are listed in Table 12.3.

Changes, and the effects that result from those changes, are important. The types of changes that can be considered significant include but are not limited to the following:

Changes	Effects
Layoffs	Reorganization
Redeployment	Reduction in allocated funds
Altered customer requirements	Revised reporting relationships
Altered priorities	Supplier changes
New job descriptions	New input requirements

Secure Management Approval.   Once again, management must endorse, support, approve, and to some extent implement the changes brought about by the benchmarking effort. Clearly, this applies if the changes involve redirecting the goals and objectives of the organization, or if there is a significant effect on individuals.

At this point the team is prepared to implement their recommendations.

TABLE 12.3   Benchmarking Impact on Other Groups

Impact on other organization's operating plans as the result of implementing the recommendations of a benchmarking team at Xerox Distribution

Suppliers:
- Technical engineers will have to provide operating plan manuals and hands-on training to curriculum designers.
- Graphics department will have to create all artwork on PCs.
- Outside cleaning service will have to clean two additional classrooms per day.

Internal support groups:
- Cafeteria will have to adjust for 50 additional students per week.
- Corporate travel department must book 25 additional rooms to accommodate larger classes.
- Lunch schedule will have to be staggered.

Field organization:
- Field technicians will have to complete a preschool self-study program prior to attending classroom training.
- All technicians will have to travel to Headquarters training center. Regional training centers will be closed.
- Travel budgets must be increased 10 percent to reflect longer distances traveled.
- Field managers will be required to certify technicians' preparation for class.

Company:
- Company-based erosion may be revised downward.
- Market share erosion projection may be lowered.

Source: DeToro 1995.

STEP 8:   DEVELOP ACTION PLAN

The benchmarking team must now assess implementation priorities, develop an action plan, and when approval for the action has been secured, proceed to implementation. Taking the following steps enhances the team's opportunity for success.

Set Implementation Priorities.   Not all benchmarking practices will yield the same payoffs, and some may be more difficult to implement than others. Some changes may be more costly than others, and still others will require staff changes. Some will have significant consequences upstream for suppliers, and downstream for customers.

Given these considerations, implementing best practices, most likely, will be done in stages. But, it may not be so obvious which are the best practices to implement first. One approach to dealing with this issue is not to choose the practice to be implemented, but to choose the criteria by which a best practice should be selected. Once the criteria have been defined, the selection of the practice to implement may be straightforward. Priority criteria are shown in Table 12.4.

Teams can develop a worksheet to assess the various values of each criterion. This way decisions can be based on information and analysis, not opinion. The example shown in Table 12.5 demonstrates how a technical training team decided which best practice to implement. This analysis should be completed for each practice under consideration.

After this is completed, a distillation of information results in a summary that rank-orders the practices to be implemented. This is shown in Table 12.6.

Show Revisions to the Performance Gap.   The benchmarking team now needs to reflect on the staging of implementation and its effect on the performance gap. The logic here is that the projection of the gap prepared earlier must be modified to reflect the phased implementation. A graph, such as the one in Figures 12.3a and b, illustrates, to the team and management, the expected gains within the implementation period.

TABLE 12.4 Implementation Priority Criteria

Criteria	Consideration
Performance improvement	How much will this contribute to the work unit output?
Time to implement	How soon can this process be installed and see results?
Cost to implement	What will it cost to implement?
Training required	Who needs to be trained on what, where, when, for how long?
Success probability	What are the risks? What is the certainty of results?
Controls needed	What inspectors, measurements, and monitoring systems are needed?
Staff changes	What changes in people are required?
Impacts	Who is affected by the changes internally? Externally?
	Do facilities have to be enlarged?
	Does the company have to relocate?

Source: DeToro 1995.

TABLE 12.5 Implementation Priority Criteria for a Technical Training Team

Criteria	+/−Considerations
Performance improvement	+Saves several weeks
	+Improves test scores
	−None
Time to implement	+Three weeks
	+No impact on current work
	−First quarter too busy
Resources required	+No additional people required
Cost to implement	+Estimate $2500
	+No new equipment required
	−Requires a budget allocation
	−Requires new software $4500
	−Requires new laser printer $10,000
Training required	+Five designers attend two weeks of training
	−Training costs $8000
Success probability	+Very high, no new technology involved
	+Staff likes new approach
Controls needed	+Monitoring and retraining
Staff changes	+Hire curriculum design specialist
	+Promote one editor to designer
	−Add $45,000 to overhead
Impact	+No direct impact

Note: Not all these criteria are required every time. Perhaps only two or three are necessary for a specific best practice.
Source: DeToro 1995.

Develop Action Plans.   The team needs to complete the following:

- Describe the specific tasks that must be completed and the results expected.
- Sequence the tasks chronologically and designate a targeted completion date for each.
- Assess the resources required to implement the best practices. This covers items such as budget, people, equipment, and materials.

TABLE 12.6 Implementation and Strategy

Work practice	Priority/strategy
1. Needs analysis	Implement immediately
2. Task analysis	Implement within 30 days
3. Program blueprinting	Implement immediately
4. Write curriculum	Implement 10 days after designers are trained
5. Pilot program methods	Wait until all practices are installed
6. Evaluation methods	Wait until first pilot is conducted

Source: DeToro 1995.

- Assign responsibilities to specific, named individuals for each action item. Explain how to complete the task.
- Establish a monitoring system to track progress and alert the team when corrective action is required. This includes setting up a reporting mechanism to keep the process owners informed of the project status at all times.

An example of a simplified action plan is illustrated in Figure 12.4.

STEP 9:   IMPLEMENT PLAN AND MONITOR RESULTS

The implementation plan has been approved, and the team now moves toward instituting best practices. A few considerations are appropriate here to help the benchmarking team understand the progress of its implementation efforts. The team should first pilot one best practice before a wholesale implementation is begun. That pilot may be as simple as a simulation of the new process, in which paper is passed around to represent the various steps in order to understand the flow of information and the activities that would be performed. In a more complex process, an actual trial run would be conducted in which data or material are subjected to the new work methods. In the latter case, it is always a good idea to push the pilot hard to determine if the new process can withstand the pressure of high-volume, complex transactions.

The team needs to establish measurements to gauge implementation progress. These process measures, also called "efficiency measures," include cost, time, quality (no defects), and a series of effectiveness measures that determine customer satisfaction.

Many quality tools are available to help the team during the implementation phase. The seven basic and seven management tools (see Camp 1994, p. 137) should be employed to ensure that data are captured and analyzed to reflect the performance of the new best practices.

STEP 10:   RECALIBRATE BENCHMARKS

To ensure success and effectiveness, benchmarks must be planned and recalibrated. There are no hard and fast rules on the frequency and method for recalibration.

Several approaches can be pursued. Specific, targeted studies can be done to fill information gaps. A complete reassessment of all critical benchmark targets and best practice findings can be done. Or a new, more productive investigation can be pursued. It is important to remember that at some point a complete reassessment must be done to ensure that information remains relevant and timely.

One approach is to recalibrate benchmarks annually. More frequently is usually not worthwhile since practices generally do not change that rapidly. Recalibration after 3 years is a massive undertaking since almost all processes will be affected. (Only some processes may be affected by annual

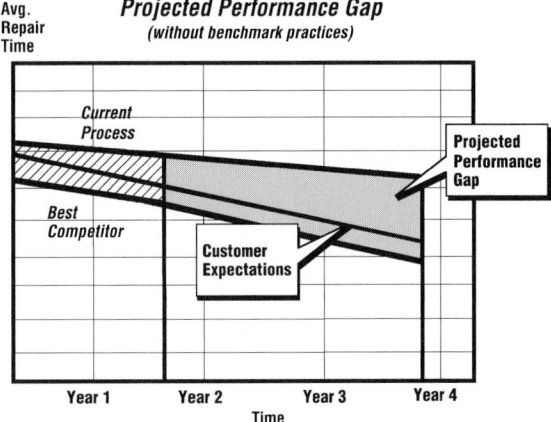

Description of Projected Trends

• Our investigations reveal that the competition is investing in computer-aided instruction materials, which would further reduce their training time from 3 days to 2.5 days, and improve the average time to repair equipment to 0.8 hours within two years.

• During the same period, we would expect, barring no strategic changes in our training methods, to achieve 1.2 hours per service call.

• Without strategic changes, our competition would widen its advantage to 0.4 hrs. (0.8 vs. 1.2) within 2 years.

FIGURE 12.3a   Impact of projected performances—current. (*DeToro 1995, Figure 14.6.*)

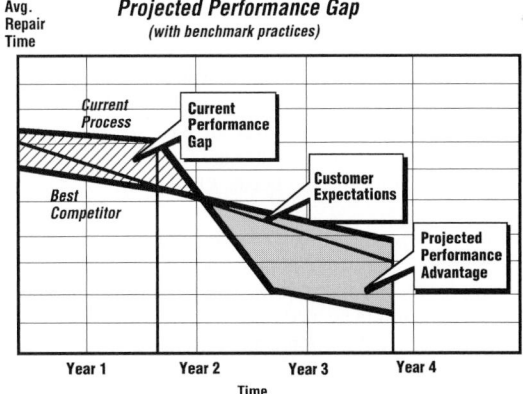

Impact of Benchmark Practices

• With the adoption of best industry practices, we could within six months achieve an average test score of 0.7 hours, giving us a performance advantage of +0.4 over our best competitor.

• Even allowing for competitive improvements, we would still maintain a performance advantage of ≥ 0.1 for several years.

FIGURE 12.3b   Impact of projected performance—with changes. (*DeToro 1995, Figure 14.7.*)

New practice	Action items	Assigned to	Resources required	Budget required	Completion date
Revise needs and analyze needs	1. Write new procedure manual	Fred Kennedy	PC	$1500	10-6
	2. Set approval points	Fred Kennedy	Schedule calendar	—	10-8
	3. Create a progress chart	Carol Jenkins	PERT	$75	10-30
	4. Train designers	Bill Peterson	Classroom facilities	$500	11-15
	5. Explain procedures to suppliers	Carol Jenkins	Procedure manual	$250	12-1
	6. Coordinate staff training	Fred Kennedy	Curriculum	$400	1-4

FIGURE 12.4   Action plan. (*DeToro 1995, Table 14.15.*)

recalibration.) Some level of insightful maintenance is probably most productive. Each business unit and benchmarking team needs to determine the frequency of recalibration based on its own industry and needs. Obviously, rapid external change argues for faster benchmarking updates. However, a useful rule is to examine the need to recalibrate within 3 years of the original study.

For example, Company A examines its accounts receivable process every 3 years. It has made only minor adjustments in the 6 years since an early benchmarking project resulted in a major redesign of the process. The customer service process is a different story. Because the technology available to support the process is changing so rapidly, the process team has been conducting benchmark studies almost continuously over the past 5 years.

Each business unit and benchmarking team needs to determine the frequency of recalibration on the basis of its own industry and needs. Obviously, rapid external change argues for more frequent benchmarking updates.

Recalibration is performed by reexercising the 10-step benchmarking process. It is imperative that all steps are reexamined. None should be skipped, because assuming nothing has changed is a dangerous approach. Feedback from internal sources should readily reveal deficiencies and areas where new information is needed.

The full value of recalibration is not only in refining the output of the benchmarking process, but also making the process more efficient and responsive to benchmarking needs.

SUCCESS FACTORS AND MANAGEMENT CONSIDERATIONS

Having process owners conduct their own benchmarking project is fundamental to its success. Those who work the process know it best. They are almost always the most qualified to analyze it. When the process owners conduct their own benchmarking, they develop a commitment to the process and resulting best practices.

Successful benchmarking is not done by separate staffs. There are, however, individuals who act in a competency capacity to help ensure that the benchmarking process is followed. But the actual benchmarking is done by process owners or process representatives, with assistance.

There are several considerations for managing benchmarking activities. Among them is the way benchmarking is communicated. It can make a big difference. When continuous improvement is used, it means that all must work together to improve how things are done so that a stronger organization results. Benchmarking is the process to find the best practices that are implemented in work processes that will lead to continuous improvement.

How resources are found to conduct benchmarking must also be considered. While it is true that some incremental resources are needed for new benchmarking, there is often another option. Somewhere, in nearly all organizations, there are resources devoted to continuous improvement. Some of these should be devoted to benchmarking. Benchmarking should be looked on as part of the ongoing effort to improve, and as part of the process owner's job. It should not be seen as extra work. There are an unlimited number of best practices, each offering significant potential for improvement, for increased results, and for superior performance. The process operators should be urged to go and find them.

To get the most out of benchmarking, what is not needed is blindly copying other institutions. That will not get superior performance. Creatively adapting best practices will. The implementation phase of benchmarking should be the creative phase. Benchmarking means combining the creative talents of the people running the processes with best practices. Benchmarking should uncover the best practices. But their innovative implementation should be the way the organization goes beyond and establishes a competitive advantage.

For example, Xerox is most concerned with customer satisfaction. When benchmarking L. L. Bean, Xerox became patently aware of Bean's customer satisfaction policy. If a customer is unhappy with a product, Bean will take back the product and return the customer's money. Xerox believed that to be a best practice. But Xerox went beyond just copying the practice. Xerox asked its customers for feedback about the best practice and found some interesting results. Customers did not want their money back; they wanted the device to work. So Xerox had to adapt the best practice to work for its customers.

The result is Xerox's Total Satisfaction Guarantee, which says that if the customer is dissatisfied with a Xerox product within a stated time frame after purchase, Xerox will replace the product, at the customer's request, until he or she is satisfied with it. So Xerox did not just copy the L. L. Bean best practice. Xerox went a step further to adapt that best practice to its specific customer needs.

Behavioral Benefits.   Benchmarking is essentially a learning experience. It helps an organization focus and drive for consensus on what needs to be done and how to achieve it, not argue over what should be done. Benchmarking can provide the stimulus for improvement by people at all levels through an externally focused, competitive situation to achieve world-class performance with increased customer satisfaction. Very few people are willing to settle for second place once they are aware of what needs to be done and know how to do it.

Competitiveness.   The bottom-line benefit of benchmarking is improved competitiveness and increased value in the eyes of customers. Effective use of benchmarking to develop and implement improvement actions can help organizations achieve superior customer service levels. This, in turn, will lead to increased market share and improved financial results.

REFERENCES

Biesada, Alexandra (1991). "Benchmarking." *Financial World,* 17 September, pp. 28–54.

Camp, Robert C. (1989). *Benchmarking: The Search for Industry Best Practices That Lead to Superior Performance.* ASQC Quality Press, Milwaukee.

Camp, Robert C. (1994). *Business Process Benchmarking: Finding and Implementing Best Practices.* ASQC Quality Press, Milwaukee.

DeToro, Irving J. (1995). *Business Process Benchmarking Workshop.* The Quality Network Inc., Rochester, NY, August.

DeToro, Irving J. and Tenner, Arthur R. (1997). *Process Redesign: The Implementation Guide for Managers.*

Addison Wesley, Reading, MA.

Ishikawa, Kaoru (1980). *Guide to Quality Control.* Quality Resources, White Plains, NY.

Kelsch, John E. (1982). "Benchmarking: Shrewd Way to Keep Your Company Ahead of its Competition." *Boardroom Report,* December, pp. 3–5.

Port, Otis, and Smith, Geoffrey (1992). "Beg, Borrow, and Benchmark." *Business Week,* 30 November, pp. 74–75.

Rummler, G. A., and Brache, A. P. (1990). *Improving Performance: How to Manage the White Space on the Organizational Chart.* Jossey-Bass, San Francisco.

Tucker, Francis G., Zivan, Seymour M., and Camp, Robert C. (1987). "How to Measure Yourself Against the Best." *Harvard Business Review,* January-February, pp. 8–10.

"World-class Organizations: Xerox" (1990). *Industry Week,* 9 March, pp. 14, 16.

SECTION 13
STRATEGIC DEPLOYMENT

Joseph A. DeFeo

INTRODUCTION

Strategic Planning (SP) is a systematic approach to defining long-term business goals and identifying the means to achieve them. Once an organization has established its long-term goals, effective strategic planning enables it, year by year, to create an annual business plan which includes the necessary annual goals, resources, and actions needed to move toward that future.

To institute organization-wide change efforts of any kind (a program of annual quality improvement, for example), an organization must incorporate the effort into the strategic planning process and into the annual business plan. This will ensure that the effort will become part of the plan and not compete with the well-established priorities for resources. Otherwise, the best-intended change effort will fail.

In recent years, total quality management (TQM) has become a pervasive change process and a natural candidate for inclusion in the strategic plan of many organizations. The integration of TQM and strategic planning is so natural, in fact, that the combination of TQM and strategic planning has become known by its own separate term. Unfortunately, different organizations have chosen different terms for this process. Some have used a Japanese term, "hoshin kanri." Others have partially translated the term and called it "hoshin planning." Still others have used a rough translation of the terms and called it "policy deployment." In an earlier version of the Malcolm Baldrige National Quality Award, this process was called "strategic quality planning." Later this award criterion was renamed "strategic planning." Although the criteria in the Award guidelines clearly define the deployment nature of the concept, the term strategic planning is often misunderstood to be the creation of the strategic plan and not the careful deployment of strategic goals, subgoals, and annual goals and the assignment of the resources and actions to achieve them. We will try to highlight this difference and use the term strategic deployment throughout this section. Many organizations have overcome failures of change programs and have achieved long-lasting results through strategic deployment.

This section describes the strategic deployment process and explains how it is managed within organizations. It addresses such important issues as: how to align strategic goals with the organization's vision and mission; how to deploy those goals throughout the organization; and how to derive the benefits of Strategic Deployment.

To this end, this section

1. Defines strategic quality deployment
2. Describes the benefits of strategic quality deployment
3. Describes the systematic approach to strategic quality deployment
4. Describes some of the issues surrounding the introduction of strategic quality deployment into an organization
5. Explains the specific roles of senior management in implementing and ensuring the success of strategic quality deployment

What Is Strategic Deployment?   Strategic deployment is a systematic approach to integrating customer-focused organization-wide improvement efforts with the strategic plan of an organization. More specifically, strategic deployment is a systematic process by which an organization defines its long-term goals with respect to quality, and integrates them—on an equal basis—with financial, human resources, marketing, and research and development goals into one cohesive business plan. The plan is then deployed throughout the entire organization.

As a component of a total quality management system, strategic deployment enables an organization to plan and execute strategic organizational breakthroughs. Over the long term, the intended collective effect of such breakthroughs is to achieve competitive advantage.

Strategic deployment has evolved during the 1990s as an integral part of many organizational change processes, especially total quality management. Strategic deployment is part of the foundation that supports the broader system of managing total quality throughout an organization. The relationship between strategic quality deployment and the broader system is shown in Figure 13.1.

Strategic deployment also is a key element of the Malcolm Baldrige National Quality Award (see Section 4) and the European Foundation for Quality Management (EFQM) Award, as well as other international and state awards. The criteria for these awards stress that customer-driven quality and operational performance excellence are key strategic business issues which need to be an integral part of overall business planning. A critical assessment of the Malcolm Baldrige National Quality Award winners demonstrates that those companies which won the Award out-performed those that did not (Figure 13.2). For the fourth year quality paid off—and big. The "Baldrige Index" outperformed the Standard & Poor's 500 stock index by almost 3 to 1. The index shows the composite growth of companies that have won the U.S. Malcolm Baldrige National Quality Award since 1988.

Godfrey (1997) has observed that to be effective strategic deployment should be used as a tool, a means to an end, not as the goal itself. It should be an endeavor that involves people throughout the organization. It must capture existing activities, not just add to already overflowing plates. It must help senior managers face difficult decisions, set priorities, and not just start new initiatives but eliminate many current activities which add no value.

History

Strategic Planning: The Past.   Until recently, strategic plans typically consisted only of financial goals or market goals. The approach can be described as "organization-wide financial planning" and

FOUNDATION

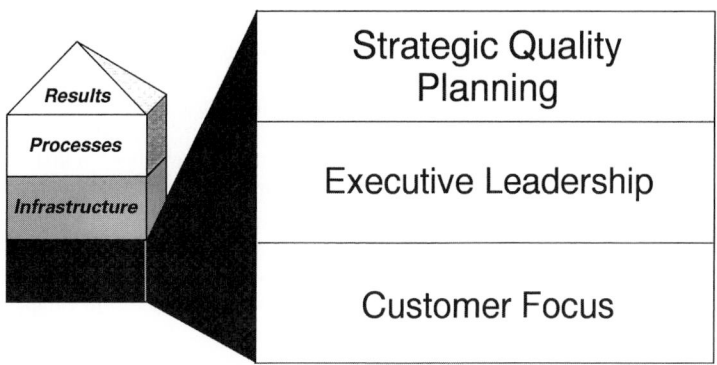

FIGURE 13.1 Strategic Quality Planning management relationship. (*Juran Institute, Wilton, CT.*)

	1988–1996 Investments	Value on 12/1/97	Percent change
All recipients	$7496.54	$33,185.69	362
Standard & Poor's 500	$7496.54	$18,613.28	148

Data: National Institute of Standards and Technology.

FIGURE 13.2 Malcolm Baldrige National Quality Award winner performance. (*Business Week 1998, March 16, p. 60.*)

it formed the basis of most strategic planning. This approach consisted of: establishing financial goals, developing plans to meet the goals, providing the needed resources, establishing measures of actual performance, reviewing performance against goals, and providing rewards based on results. It resulted in the annual business plan and the budget. This plan became the driver of all activity within the organization. Where such a plan covered a period of 5 or more years it was usually referred to as a "strategic business plan." The results of this effort enabled an organization to focus all employees on the financial goals and the means to achieve them.

The major components of this strategic planning process are

A hierarchy of goals: This includes but is not limited to financial goals supported by financial goals at lower levels such as divisional and departmental budgets, sales quotas, cost standards, project cost estimates, etc.

A formalized methodology: A methodology for establishing the goals (an annual budgeting process) and for providing the needed resources to achieve the strategic plan and annual business plan.

An infrastructure: The infrastructure (usually) includes a Finance Committee; a full-time Controller and supporting personnel; and all top management, meeting regularly to review and adjust the plans when needed.

A control process: The control process includes organization-wide financial metrics; systems for data collection and analysis; financial reports; reviews of financial performance against goals; and adjustment, when needed, of the plan itself.

Provision of rewards: Performance against financial goals is given substantial weight in the system of employee performance management and recognition and reward of key employees.

Universal participation: The financial goals, reports, reviews, etc., are designed hierarchically to parallel the company's organization hierarchy. These hierarchically deployed goals make it possible for managers at all levels to support the upper managers' goals.

A common language: The planning process typically focuses on major, common metrics—revenues and profits—expressed in a common unit of measure—a currency unit, such as the U.S. dollar. There are also other common metrics which are widely used; ratios such as return on investment and return on sales are examples. In addition, such key words as "budget," "expense," "profit," etc., acquire standardized meanings, so that communication becomes more and more precise. Hence, the organization creates a language it can understand.

Training: In successful organizations, it is common for employees at all levels to undergo training in various financial concepts, processes, methods, tools, etc.—in other words—to learn to speak and understand the same language. Companies which have so trained their employees in all functions and at all levels are well poised to outperform companies in which such training is confined to the finance department.

Strategic Planning: Today. The approach used to establish organization-wide financial goals has evolved into a more robust strategic plan. To be effective at achieving rapid change in a global environment, many organizations incorporated establishment of organization-wide change efforts, such as total quality management, into the strategic plan. The generic steps and features inherent in managing for the annual business plan are likewise applicable to managing for quality. It also makes it easier to incorporate organization-wide improvement programs into one cohesive plan. In earlier versions of the Malcolm Baldrige National Quality Award this was referred to as the *strategic* quality plan (SQP). The strategic quality plan should include:

Quality goals: The major quality goals get incorporated and are supported by a hierarchy of goals at lower levels: subgoals, projects, etc. Improvement goals are goals aimed at creating a breakthrough in performance of a product, serving process, or people by focusing on the needs of customers, suppliers, and shareholders. The strategic quality plan incorporates the voice of the customer with quality goals and integrates them throughout the plan. This integration enables the goals to be legitimate and balance the financial goals (which are important to shareholders) with those of importance to the customers. It also eliminates the concern that there are two plans, one for finance and one for quality.

A formalized methodology: A systematic, structured process for establishing improvement goals and providing resources:

- *A new infrastructure* is created which includes the establishment of an upper-management team or "Executive Council," a quality office and supporting personnel to review all goals.
- *A review and control process* which includes systems for collection and analysis of customer data, reports of key quality performance indicators, and reviews to monitor performance against goals.

Provision of rewards: Performance against improvement goals is given substantial weight in the system of merit rating and recognition. A change in the structure that includes rewarding the right behaviors is required.

Universal participation: The goals, reports, reviews, etc., are designed to gain participation from within the organization's hierarchy. This participation involves every employee at every level, providing support for the change initiative and helping achieve the desired results.

A common language: Key terms, such as quality, benchmarking, and strategic quality deployment, acquire standard meanings so that communication becomes more and more precise.

Training: It is common for all employees to undergo training in various change concepts, processes, methods, tools, etc. Companies which have so trained their workforce, in all functions, at all levels, and at the right time, are well poised to outperform companies in which such training has been confined to the quality department or managers.

These required changes seem numerous and extensive. Prior to the 1980s the asserted benefits of establishing improvement goals were generally not persuasive to upper managers. Most of the reasons are implied in that same list of changes:

- Going into total quality management or expanding strategic planning is a lot of work.
- It adds to the workload of upper managers as well as managers at lower levels.
- It is quite disturbing to the established cultural pattern.
- "We've already tried it and it failed."

However, to compete globally, organizations have needed to get the most out of their assets and resources. Strategic deployment provides the means to accomplish this.

Why Do Strategic Deployment? The Benefits.   The first question that often arises in the beginning stages of strategic deployment in an organization is: Why do it? Can it help us become a global competitor? To answer these questions requires a look at the benefits that other organizations have realized from strategic deployment. They report that strategic deployment

1. Focuses the organization's resources on the activities that are essential to increasing customer satisfaction, lowering costs, and increasing shareholder value (see Figure 13.2).
2. Creates a planning and implementation system that is responsive, flexible, and disciplined.
3. Encourages interdepartmental cooperation.
4. Provides a process to execute breakthroughs year after year.
5. Empowers managers and employees by providing them with the authority to carry out the planned activities.
6. Eliminates unnecessary and wasteful team activities that are not in the plan.
7. Eliminates the existence of many potentially conflicting plans—the finance plan, the marketing plan, the technology plan, and the improvement plan.
8. Focuses resources to ensure financial plans are achievable.

Different organizations have tried to implement total quality management systems as well as other change management systems. Some organizations have achieved stunning results; others have been disappointed by their results, often achieving little in the way of bottom-line savings or increased customer satisfaction. Some of these efforts have been classified as failures. One of the primary causes of these disappointments has been the inability to incorporate these "quality programs" into the business plans of the organization. Other reasons for failure were that

1. Strategic planning was assigned to planning departments, not to the upper managers themselves. These planners lacked training in concepts and methods and were not among the decision makers in the organization. This led to a strategic plan which did not include improvement goals aimed at customer satisfaction, process improvement, etc.

2. Individual departments had been pursuing their own departmental goals, failing to integrate them with the overall organizational goals.

3. New products or services continued to be designed with failures from prior designs that were carried over into new models, year after year. The new designs were not evaluated or improved and hence were not customer-driven.

4. Multifunctional "re-engineering" projects have suffered delays and waste due to inadequate participation and to lack of early warnings by upper management, and have ended before positive business results were achieved.

5. There has been no clear responsibility for reducing cycle times or waste associated with major business processes. Clear responsibilities were limited to local (intradepartmental) processes.

6. Improvement goals were assumed to apply only to manufactured goods and manufacturing processes. Customers became irritated not only by receipt of defective goods; they were also irritated by receiving incorrect invoices and late deliveries. The business processes which produce invoices and deliveries were not subject to modern quality planning and improvement because there were no such goals in the annual plan to do so.

The deficiencies of the past strategic planning processes had their origin in the lack of a systematic, structured approach to integrate programs into one plan. As more companies became familiar with strategic quality deployment, many adopted its techniques which treat managing for change on the same organization-wide basis as managing for finance. The remedy is what we call strategic quality deployment.

LAUNCHING STRATEGIC DEPLOYMENT

Creating a strategic plan that is customer-focused requires that leaders become coaches and teachers, personally involved, consistent, eliminate the atmosphere of blame, and make their decisions on the best available data. Juran (1988) has stated:

> You need participation by the people that are going to be impacted, not just in the execution of the plan but in the planning itself. You have to be able to go slow, no surprises, use test sites in order to get an understanding of what are some things that are damaging and correct them.

The Strategic Deployment Process.   The strategic quality deployment process requires that an organization incorporate customer focus into the organization's vision, mission, values, policies, strategies, and long- and short-term goals and projects. Projects are the day-to-day, month-to-month activities that link quality improvement activities, re-engineering efforts, and quality planning teams to the organization's business objectives.

The elements needed to establish strategic deployment are generally alike for all organizations. However, each organization's uniqueness will determine the sequence and pace of application and the extent to which additional elements must be provided.

There exists an abundance of jargon used to communicate the strategic deployment process. Depending on the organization, one may use different terms to describe similar concepts. For example, what one organization calls a vision, another organization may call a mission (see Figure 13.3).

The following definitions are in widespread use and are used in this section:

Vision: A desired future state of the organization. Imagination and inspiration are important components of a vision. Typically, a vision can be viewed as the ultimate goal of the organization, one that may take 5 or even 10 years to achieve.

Mission: The purpose or reason for the organization's existence, i.e., what we do and whom we serve.

Selected definitions	
Mission:	What business we are in
Vision:	Desired future state of organization
Values:	Principles to be observed to meet vision or
	Principle to be served by meeting vision
Policy:	Commitment to customer

FIGURE 13.3   Organizational vision/mission. (*Juran Institute, Wilton, CT.*)

Strategies: Means to achieve the vision. Strategies are few and define the key success factors such as price, value, technology, market share, and culture that the organization must pursue. Strategies are sometimes referred to as "key objectives" or "long-term goals."

Goals: What the organization must achieve over a 1- to 3-year period; the aim or end to which work effort is directed. Goals are referred to as "long term" (2 to 3 years) and "short term" (1 to 2 years). Achievement of goals signals the successful execution of a strategy.

Values: What the organization stands for and believes in.

Policies: A guide to managerial action. An organization may have policies in a number of areas: quality, environment, safety, human resources, etc. These policies guide day-to-day decision making.

Project: An activity of duration as long as 3 to 9 months that addresses a deployed goal, and whose successful completion contributes to assurance that the strategic goals are achieved. A project most usually implies assignment of selected individuals to a team which is given the responsibility and authority to achieve the specific goal.

Deployment plan: To turn a vision into action, the vision must be broken apart and translated into successively smaller and more specific parts—key strategies, strategic goals, etc.—to the level of projects and even departmental actions. The detailed plan for decomposition and distribution throughout the organization is called the "deployment plan." It includes the assignment of roles and responsibilities and identification of resources needed to implement and achieve the project goals (Figure 13.4).

Key performance indicators: Measurements that are visible throughout the organization for evaluating the degree to which the strategic plan is being achieved.

THE ELEMENTS OF STRATEGIC DEPLOYMENT

Establish the Vision. Strategic deployment begins with a vision that is customer-focused:

In the companies we know that are successfully making the transition to a more collaborative organization, the key to success is developing and living by a common strategic vision. When you agree on an overall direction, you can be flexible about the means to achieve it… (Tregoe and Tobia 1990)

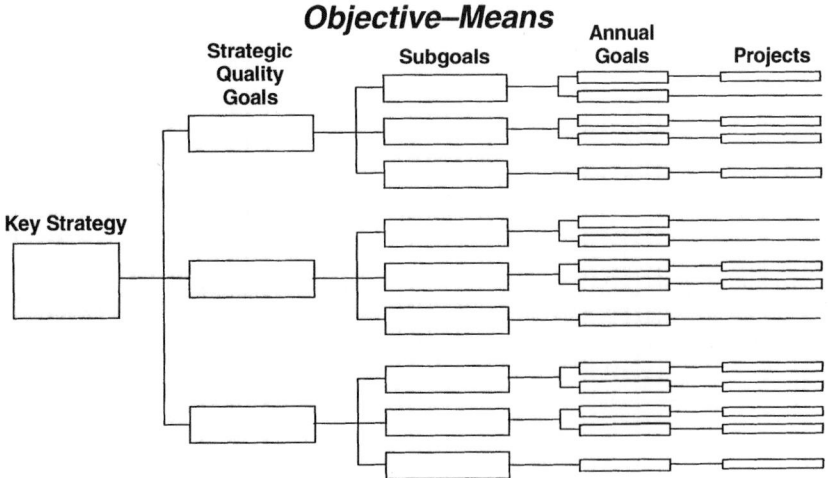

FIGURE 13.4 Deploying the vision. (*Juran Institute, Wilton, CT.*)

Really powerful visions are simply told. The Ten Commandments, the Declaration of Independence, a Winston Churchill World War II speech—all present messages that are so simple and direct you can almost touch them. Our corporate strategies should be equally compelling. (Roberts 1987)

A vision should define the benefits a customer, an employee, a shareholder, or society at large can expect from the organization:

> We will lead in delivering affordable, quality health care that exceeds the service and value of our customers' expectations. (Kaiser-Permanente)
> To be the leading consumer battery company in the world. (Duracell International)
> To engineer, produce, and market the world's finest automobiles. (Cadillac Motor Car Division)
> To be the best producers of manufactured housing and the easiest to do business with. (Schult Homes)
> To be the number one provider of orthopaedic medical devices in the world. (Howmedica)

Each of the preceding visions offers a very different view of the direction and character of the organization. Each conveys a general image to customers and employees of where the organization is headed. For the organization, the vision provides, often for the first time in its history, a clear picture of where it is headed and why it is going there.

Good vision statements should also be compelling and shared throughout the organization. It is often a good idea to make the vision a stretch for the organization but possible of achievement within 3 to 5 years, and to state a measurable achievement (e.g., being the best). In creating the vision, organizations should take into account its customers, the markets in which it wants to compete, the environment within which the organization operates, and the current state of the organization's culture.

Vision statements, by themselves, are little more than words. Publication of such a statement does not inform the members of an organization what they should do differently from what they have done in the past. The strategic deployment process and the strategic plan become the basis for making the vision a reality. The words of the vision are just a reminder of what the organization is pursuing. The vision must be carried out through deeds and action.

Some common pitfalls in forming a vision are

1. Focusing the vision exclusively on shareholders as customers.
2. Thinking that once a strategic plan is written it will be carried out with no further work.
3. Failing to explain the vision as a benefit to customers, employees, suppliers, and other stakeholders.
4. Creating a vision that is either too easy or too difficult to achieve.
5. Failing to consider the effects that the rapid changes taking place in the global economy will have 3 to 5 years in the future.
6. Failing to involve key employees at all levels in creating the vision.
7. Failing to benchmark competitors or to consider all possible sources of information on future needs, internal capabilities, and external trends.

Agree on a Mission.   Most organizations also have a mission statement. A mission statement is designed to address the question, "What business(es) are we in?" A mission is often confused with a vision and even published as one. A mission statement should clarify the organization's purpose or reason for existence. That's all.

The following are some examples:

> The Ritz-Carlton Hotel is a place where the genuine care and comfort of our guests is our highest mission. (Ritz-Carlton Hotel)

> We exist to create, make, and market useful products and services to satisfy the needs of our customers throughout the world. (Texas Instruments)
>
> Our mission is to be a leader in meeting the present and future health care needs of the people of our communities through a network of high-quality services, teaching and research programs which share common goals and values. (Sentara Health System)

In the Sentara example, the references to leadership and the future may lead the reader to confuse this mission statement (what business we are in) with a vision statement (what we aim to become). Only the organization itself can decide whether these words belong in its mission statement. It is in debating such points that an organization comes to consensus on its vision and mission.

Together, a vision and a mission provide a common agreed-upon direction for the entire organization. This direction can be used as a basis for daily decision making.

Develop Key Strategies. The first step in converting the vision into an achievable plan is to break the vision into a small number (usually four or five) key strategies. Key strategies represent the most fundamental choices that the organization will make about how it will go about reaching its vision. Each strategy must contribute significantly to the overall vision. For example:

> Supporting leadership through quality platforms: customer orientation, employee involvement, benchmarking, use of quality tools, planning for quality, and customer focus. (Xerox)
>
> Three critical strategies implemented to transform Cadillac: A cultural change where teamwork and employee involvement are considered a competitive advantage, a focus on the customer with customer satisfaction in the master plan, and a more disciplined approach to planning that focuses all employees on the quality objectives. (Cadillac Motor Car Division)

Responsibility for executing these key strategies is distributed (or deployed) to key executives within the organization, the first step in a succession of subdivisions and deployments by which the vision is converted to action.

In order to determine what the key strategies should be, one needs to assess five areas of the organization and obtain the necessary data on

1. Customer loyalty, customer satisfaction
2. Costs related to poor quality
3. Organization culture (satisfaction)
4. Internal business process (including suppliers)
5. Competitive benchmarking

Each of these assessments can form the basis for a balanced business scorecard (see The Scorecard later in this section). Setting key strategies requires specific data on the quality position and environment. These data must be analyzed to discover specific strengths, weaknesses, opportunities, and threats as they relate to customers, quality, and costs. Once complete, the key strategies can be created or modified to reflect measurable and observable long-term goals.

Develop Strategic Goals

The Nature of Strategic Goals. Next an organization sets specific, measurable strategic goals that must be achieved for the broad strategy to be a success. These quantitative goals will guide the organization's efforts toward achieving each strategy. As used here, a goal is an aimed-at target. A goal must be specific. It must be quantifiable (measurable) and is to be met within a specific period of time. At first, an organization may not know how specific the goal should be. Over time the measurement systems will improve and the goal setting will become more specific and more measurable.

Despite the uniqueness of specific industries and organizations, certain subjects for goals are widely applicable. There are seven areas that are minimally required to assure that the proper goals are established. They are

Product performance: Goals in this area relate to product features which determine response to customer needs, e.g., promptness of service, fuel consumption, mean time between failures, and courteousness. These product features directly influence product salability and impact revenues when they are met.

Competitive performance: This has always been a goal in market-based economies, but seldom a part of the business plan. The trend to make competitive performance a long-term business goal is recent but irreversible. It differs from other goals in that it sets the target relative to the competition, which, in a global economy, is a rapidly moving target. For example: All of our products will be considered the "best in class" within 1 year of introduction as compared to products of the top five competitors.

Quality improvement: Goals in this area may be aimed at improving product deficiencies or process failures or reducing the cost-of-poor-quality waste in the system. Improvement goals are deployed through a formal structure of quality improvement projects with assignment of associated responsibilities. Collectively, these projects focus on reducing deficiencies in the organization, thereby leading to improved performance.

Cost of poor quality: Goals related to quality improvement usually include a goal of reducing the costs due to poor quality or waste in the processes. These costs are not known with precision, though they are estimated to be very high. Nevertheless, it is feasible, through estimates, to bring this goal into the business plan and to deploy it successfully to lower levels. A typical cost-of-poor-quality goal is to reduce the cost of poor quality 50 percent each year for 3 years.

Performance of business processes: Goals in this area have only recently entered the strategic business plan. The goals relate to the performance of major processes which are multifunctional in nature, e.g., new product development, supply-chain management, and information technology, and subprocesses such as accounts receivable and purchasing. For such macroprocesses, a special problem is to decide who should have the responsibility for meeting the goal? We discuss this later under Deployment to Whom?

Customer satisfaction: Setting specific goals for customer satisfaction helps keep the organization focused on the customer. Clearly, deployment of these goals requires a good deal of sound data on the current level of satisfaction/dissatisfaction and what factors will contribute to increasing satisfaction and removing dissatisfaction. If the customers' most important needs are known, the organization's strategies can be altered to meet those needs most effectively.

Customer loyalty and retention: Beyond direct measurement of customer satisfaction, it is even more useful to understand the concept of customer loyalty. Customer loyalty is a measure of customer purchasing behavior vis a vis a given supplier. A customer whose needs for product offered by supplier A who buys solely from that supplier is said to display a loyalty with respect to A of 100 percent. A study of loyalty opens the organization to a better understanding of product salability from the customer's viewpoint and provides the incentive to determine how to better satisfy customer needs. The organization can benchmark to discover the competition's performance, then set goals to exceed that performance (see Figure 13.5).

The goals selected for the annual business plan are chosen from a list of nominations made by all levels of the hierarchy. Only a few of these nominations will survive the screening process and end up as part of the organization-wide business plan. Other nominations may instead enter the business plans at lower levels in the organization. Many nominations will be deferred because they fail to attract the necessary priority and therefore will get no organization resources.

Upper managers should become an important source of nominations for strategic goals, since they receive important inputs from sources such as membership on the executive council, contacts with customers, periodic reviews of business performance, contacts with upper managers in other organizations, shareholders, and employee complaints.

Product performance (*customer focus*): This relates to performance features which determine response to customer needs such as promptness of service, fuel consumption, MTBF, and courtesy. (Product includes goods and services.)

Competitive performance: Meeting or exceeding competitive performance has always been a goal. What is new is putting it into the business plan.

Quality improvement: This is a new goal. It is mandated by the fact that the rate of quality improvement decides who will be the quality leader of the future.

Reducing the cost of poor quality: The goal here relates to being competitive as to costs. The measures of cost of poor quality must be based on estimates.

Performance of macroprocesses: This relates to the performance of major multifunctional processes such as billing, purchasing, and launching new products.

FIGURE 13.5 Quality goals in the business plan. (*Juran Institute, Wilton, CT.*)

Goals which affect product salability and revenue generation should be based primarily on meeting or exceeding marketplace quality. Some of these goals relate to projects which have a long lead time, e.g., a new product development involving a cycle time of several years, computerizing a major business process, a large construction project which will not be commissioned for several years. In such cases the goal should be set so as to meet the competition estimated to be prevailing when these projects are completed, thereby "leapfrogging" the competition.

In industries which are natural monopolies (e.g., certain utilities) the organizations often are able to make comparisons through use of industry data banks. In some organizations there is internal competition as well—the performances of regional branches are compared with each other.

Some internal departments may also be internal monopolies. However, most internal monopolies have potential competitors—outside suppliers who offer the same services. The performance of the internal supplier can be compared with the proposals offered by an outside supplier.

A third and widely used basis for setting goals has been historical performance. For some products and processes the historical basis is an aid to needed stability. For other cases, notably those involving high chronic costs of poor quality, the historical basis has done a lot of damage by helping to perpetuate a chronically wasteful performance. During the goal-setting process, upper managers should be on the alert for such misuse of the historical data. Goals for chronically high cost of poor quality should be based on planned breakthroughs using the quality improvement process described in Section 5.

Establish Values. Some organizations create value statements to further define themselves. Values are what an organization stands for and believes in. A list of values must be supported with actions and deeds from management, lest its publication create cynicism in the organization. Training and communication of values for all employees becomes a prerequisite to participation in the planning process. Organization policies must be changed to support the values of the organization. Some examples of published values are:

Constant respect for people, uncompromising integrity. (Motorola)

Living the values we have established, a culture that supports customer focus, positive morale, empowerment, and job satisfaction. Values that guide us are: customer delight, commitment, teamwork, continuous improvement, trust and integrity, and mutual respect. (AT&T)

Value statements are gaining popularity in many organizations. They provide a reminder of what is important when carrying out the strategic plan.

Communicate Company Policies.   "Policy" as used here is a guide to managerial action. Published policy statements are the result of a good deal of deliberation by management, followed by approval at the highest level. The senior executive team or quality council plays a prominent role in this process.

Policy declarations are a necessity during a period of major change, and organizations have acted accordingly. Since the 1980s we have seen an unprecedented surge of activity in publishing "quality policies." While the details vary, the published policies have much in common from company to company. For instance, most published quality policies declare the intention to meet the needs of customers. The wording often includes identification of specific needs to be met, e.g., "The company's products should provide customer satisfaction."

Most published policies include language relative to competitiveness in quality, e.g., "Our company's products shall equal or exceed the competition."

A third frequent area of published quality policy relates to quality improvement, declaring, for example, the intention to conduct improvement annually.

Some quality policy statements include specific reference to internal customers or indicate that the improvement effort should extend to all phases of the business. For example:

> Reilly Industries is dedicated to meeting and exceeding the requirements of all our customers—both internal and external—with our products and services. To achieve customer satisfaction through continuous improvement in the quality or our products, processes, and services requires the total commitment of all our employees. We shall ensure that the necessary environment, training, and tools are available to support this commitment.

The quality policy of Chrysler Corporation is

> To be the best. This policy requires that every individual and operating unit fully understand the requirements of their customers, and deliver products and services that satisfy these requirements at a defect-free level. (Chrysler Corporation)

Enforcement of policies is a new problem due to the relative newness of documented quality policies. In some organizations provision is made for independent review of adherence to policies. ISO 9000, the international standard for quality assurance, requires a quality policy as a declaration of intent to meet needs of customers. An audit process is mandated to ensure the policy is carried out.

Upper Management Leadership.   A fundamental step in the establishment of any strategic plan is the participation of upper management acting as an executive council. Membership typically consists of the key executives. Top-level management must come together as a team to determine and agree upon the strategic direction of the organization. The council is formed to oversee and coordinate all strategic activities aimed at achieving the strategic plan. The council is responsible for executing the strategic business plan and monitoring the key performance indicators. At the highest level of the organization, an executive council should meet monthly or quarterly.

The executive council is also responsible for ensuring that other business units have a similar council at the subordinate levels of the organization. In such cases the councils are interlocked, i.e., members of upper-level councils serve as chairpersons for lower-level councils (see Figure 13.6).

If a council or something similar to it is not in place, the organization will have to create one. In a global organization processes are too complex to be managed functionally. A council ensures a multifunctional team working together to maximize process efficiency and effectiveness. Although this may sound easy, in practice it is not. The senior management team members may not want to give up the monopolies they have enjoyed in the past. For instance, the manager of sales and marketing is accustomed to defining customer needs, the manager of engineering is accustomed to sole responsibility for creating products, and the manager of manufacturing has enjoyed free rein in producing products. In the short run, these managers may not easily give up their monopolies to become team players.

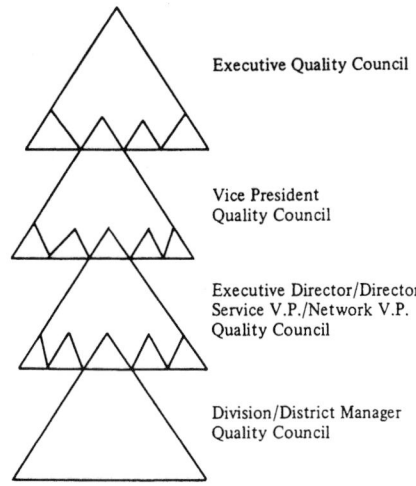

Executive Quality Council

Vice President
Quality Council

Executive Director/Director
Service V.P./Network V.P.
Quality Council

Division/District Manager
Quality Council

FIGURE 13.6 Interlocking councils. (*Juran Institute, Wilton, CT.*)

Deploy Goals. The deployment of long- and short-term goals is the conversion of goals into operational plans and projects. "Deployment" as used here means subdividing the goals and allocating the subgoals to lower levels. This conversion requires careful attention to such details as the actions needed to meet these goals, who is to take these actions, the resources needed, and the planned timetables and milestones. Successful deployment requires establishment of an infrastructure for managing the plan. Goals are deployed to multifunctional teams, functions, and individuals (see Figure 13.7).

Subdividing the Goals. Once the strategic goals have been agreed to, they must be subdivided and communicated to lower levels. The deployment process also includes dividing up broad goals into manageable pieces (short-term goals or projects). For example:

1. An airline goal of attaining 95 percent on-time arrivals may require specific short-term (8 to 12 months) projects to deal with such matters as

   • The policy of delaying departures in order to accommodate delayed connecting flights
   • The organization for decision making at departure gates
   • The availability of equipment to clean the plane
   • The need for revisions in departmental procedures
   • The state of employee behavior and awareness

2. A hospital's goal of improving the health status of the communities they serve may require projects that

   • Reduce incidence of preventable disease and illness
   • Improve patient access to care
   • Improve the management of chronic conditions
   • Develop new services and programs in response to community needs

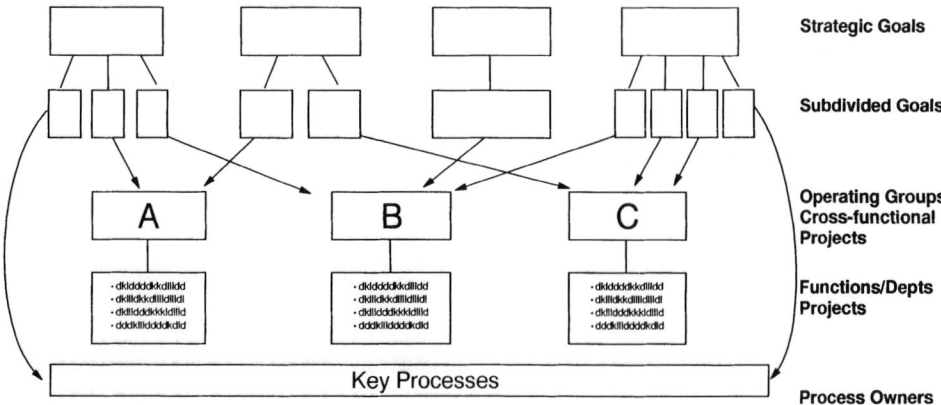

FIGURE 13.7 Deployment of strategic goals. (*Juran Institute, Wilton, CT.*)

Such deployment accomplishes some essential purposes:

- The subdivision continues until it identifies specific deeds to be done.
- The allocation continues until it assigns specific responsibility for doing the specific deeds.

Those who are assigned responsibility respond by determining the resources needed and communicating this to higher levels. Many times the council must define specific projects, complete with team charters and team members, to ensure goals are met (see Figure 13.8). (For more on the improvement process, see Section 5, The Quality Improvement Process.)

Deployment to Whom?   The deployment process starts with the identification of needs of the organization and the upper managers. Those needs determine what deeds are required. The deployment process leads to an optimum set of goals through consideration of the resources required. The specific projects to be carried out address the subdivided goals. For example: In the early 1980s, the newly designed Ford Taurus/Sable goal of becoming "Best in Class" was divided into more than 400 specific subgoals, each related to a specific product feature. The total planning effort was enormous and required over 1500 project teams.

To some degree, deployment can follow hierarchical lines, such as corporate to division and division to function. However, this simple arrangement fails when goals relate to cross-functional business processes and problems that affect customers.

Major activities of organizations are carried out by use of interconnecting networks of business processes. Each business process is a multifunctional system consisting of a series of sequential operations. Since it is multifunctional, the process has no single "owner," hence no obvious answer to the question: Deployment to whom? Deployment is thus made to multifunctional teams. At the conclusion of the team project an owner is identified. The owner (who may be more than one person) then monitors and maintains this business process. (See Section 6, Process Management.)

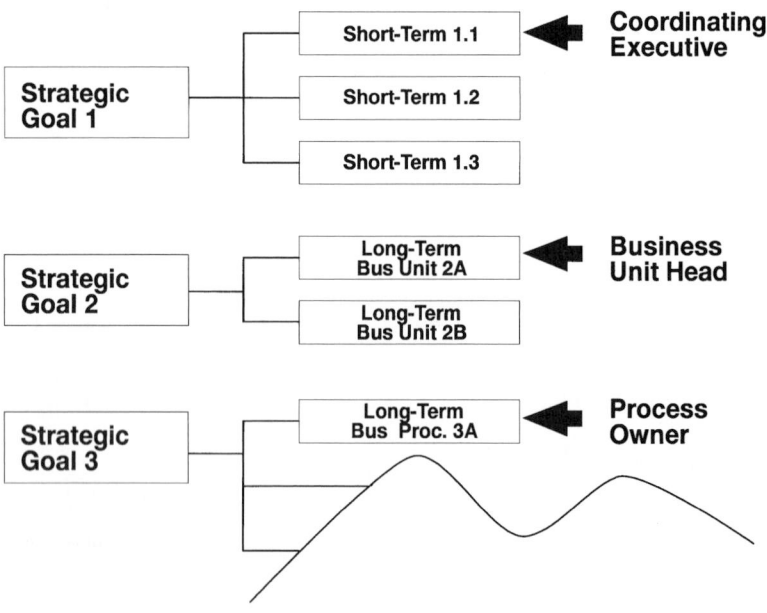

FIGURE 13.8   Subgoals. (*Juran Institute, Wilton, CT.*)

Communicating the Plan: "Catch Ball."   Once the goals have been established, the goals are communicated to the appropriate organization units. In effect, the executive leadership asks their top management, "What do you need to support this goal?" The next level managers discuss the goal and ask their subordinates a similar question, and so on. The responses are summarized and passed back up to the executives. This process may be repeated several times until there is general satisfaction with the final plan.

This two-way communication process is called "catch ball," a term coined by the Japanese. Catch ball includes the following:

1. Clear communication of what top management proposes as the key focus areas of the strategic plan for the coming business year
2. Identification and nomination by managers at various lower levels of other areas for organization attention
3. Decisions as to what departments and functions should do about the areas that have been identified in the plan

This two-way communication requires that the recipients be trained in how to respond. The most useful training is prior experience in quality improvement. Feedback from organizations using catch ball suggests that it outperforms the process of unilateral goal setting by upper managers. For example: Fannie Mae, a financial services company, has been very successful in introducing its strategic quality plan. During a 6- to 9-month period, Fannie Mae developed its strategic quality plan, with its mission, vision, key strategies, and strategic goals. They used catch ball as a team-building exercise and as an opportunity for the senior management to state clearly what they wanted as direction and goals for Fannie Mae. After the senior executives drafted the original vision, mission, and key strategies, they asked the directors and middle-level managers to provide comments during highly interactive working-group sessions. They also developed draft strategic goals during these sessions. The almost 100 directors and managers created a wealth of ideas. Next, the senior executives refined these strategic goals and incorporated many of the comments about vision, mission, and key strategies into the next plan. They next involved over 600 managers and supervisors in more interactive work sessions to gain further comments and ideas for deploying the goals to subgoals and specific projects. In the end, senior management came up with a final version of the key strategies and goals for the next 5 years. This vision was then presented to everyone who was involved to agree to and sign.

Union Electric Company (UE) formed lead teams in each department in order to effectively deploy objectives down to the lowest organizational level. (At UE this is the department level.) This allowed more people than ever before at these lower levels to be involved in the business planning process (Weigel 1990).

A Useful Tool for Deployment.   The tree diagram is a graphic tool that aids in the deployment process (see Figure 13.4). It displays the hierarchical relationship of the vision, key strategies, strategic goals, long-term goals, short-term goals, and projects, and indicates where each is assigned in the organization. A tree diagram is useful in visualizing the relationship between goals and objectives or teams and goals. It also provides a visual way to determine if all strategies are supported.

Measure Progress with Key Performance Indicators

Why Is Measurement Necessary?   There are several reasons why measurement of performance is necessary and why there should be an organized approach to it:

1. Performance measures indicate the degree of accomplishment of objectives and, therefore, quantify progress toward the attainment of goals.
2. Performance measures are needed to monitor the continuous improvement process, which is central to the changes required to become competitive.

3. Measures of individual, team, and business unit performance are required for periodic performance reviews by management.

Once goals have been set and broken down into subgoals, key measures (performance indicators) need to be established. A measurement system that clearly monitors performance against plans has the following properties:

1. Indicators that link strongly to strategic goals and to the vision and mission of the organization

2. Indicators that include customer concerns; that is, the measures focus on the needs and requirements of internal and external customers

3. A small number of key measures of key processes that can be easily obtained on a timely basis for executive decision making

4. The identification of chronic waste or cost of poor quality

For example: MetPath, Inc. established measures of their processes early in the implementation of their business plan and were able to monitor and quantify the following:

1. A tenfold reduction of the process errors responsible for a patient's specimen being lost or broken before it can be tested

2. Significant cost savings due to decreased errors in proficiency testing

3. An eightfold reduction in turnaround time, the time it takes to deliver a specific health care service

The best measures of the implementation of the strategic planning process are simple, quantitative, and graphical. A basic spread sheet which describes the key measures and how they will be implemented is shown in Figure 13.9. It is simply a method to monitor the measures.

As goals are set and deployed, the means to achieve them at each level must be analyzed to ensure that they satisfy the objective that they support. Then the proposed resource expenditure must be compared with the proposed result and the benefit/cost ratio assessed. Examples of such measures are

- Financial results:

  Gains

  Investment

  Return on investment

- Human resources:

  Trained

  Active on project teams

- Number of projects:

  Undertaken

  In process

  Completed

  Aborted

- New product development:

  Number or percentage of successful product launches

  Return on investment of new product development effort

  Cost of developing a product versus the cost of the product it replaces

  Percent of revenue attributable to new products

Annual quality goals	Specific measurements	Frequency	Format	Data source	Name

FIGURE 13.9   Measurement of quality goals. (*Juran Institute, Wilton, CT.*)

Percent of market share gain attributable to products launched during the last 2 years

Percent of on-time product launches

Cost of poor quality associated with new product development

Number of engineering changes in the first 12 months of introduction

- Supply-chain management:

Manufacturing lead times—fill rates

Inventory turnover

Percent on-time delivery

First-pass yield

Cost of poor quality

The following is an example of measures that one bank used to monitor teller quality:

- Speed:
  1. Number of customers in the queue
  2. Amount of time in the queue

- Timeliness:
  1. Time per transaction
  2. Turnaround time for no-wait or mail transactions

- Accuracy:
  1. Teller differences
  2. Amount charged off/amount handled

Once the measurement system is in place, it must be reviewed periodically to ensure goals are being met.

Reviewing Progress.   A formal, efficient review process will increase the probability of reaching the goals. When planning actions, an organization should look at the gaps between measurement of the current state and the target it is seeking. The review process looks at gaps between what has been achieved and the target (see Figure 13.10).

Frequent measurements of strategic deployment progress displayed in graphic form help identify the gaps in need of attention. Success in closing those gaps depends on a formal feedback loop with clear responsibility and authority for acting on those differences. In addition to the review of results, progress reviews are need for projects under way to identify potential problems before it is too late to take effective action. Every project should have specific, planned review points, much like those in Figure 13.11.

Organizations today include key performance indicators on the following:

Product Performance.   A product's features may be very numerous. For the great majority of product features, there exist performance metrics and technological sensors to provide objective product evaluation.

Competitive Quality.   These metrics relate to those qualities which influence product salability, e.g., promptness of service, responsiveness, courtesy of pre-sale and after-sale service, and order fulfillment accuracy. For automobiles, qualities include top-speed, acceleration, braking distance, and safety. For some product features, the needed data must be acquired from customers, through negotiation, persuasion, or purchase. For other product features, it is feasible to secure the data through laboratory tests. In still other cases, it is necessary to conduct market research.

Trends must now be studied so that goals for new products can be set to correspond to the state of competition anticipated at the time of launch.

Some organizations operate as natural monopolies, e.g, regional public utilities. In certain of such cases, the industry association gathers and publishes performance data. In the case of internal monopolies, (e.g., payroll preparation, transportation) it is sometimes feasible to secure competitive information from organizations which offer similar services for sale.

Performance on Quality Improvement.   This evaluation is important to organizations which go into quality improvement on a project-by-project basis. Due to lack of commonality among the projects, collective evaluation is limited to the summary of such features as

- *Number of projects:*   Undertaken, in-process, completed, aborted.
- *Financial results:*   Amounts gained, amounts invested, returns on investment.
- *Persons involved as project team members:*   Note that a key measure is the proportion of the organization's management team which is actually involved in improvement projects. Ideally, this proportion should be over 90 percent. In the great majority of organizations the actual proportion has been less than 10 percent.

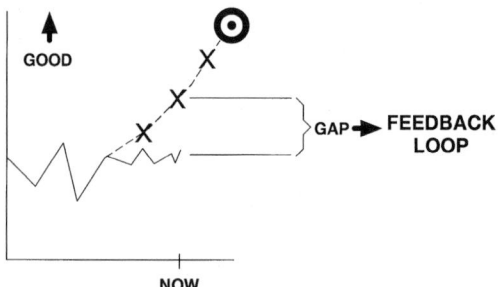

FIGURE 13.10   Review. (*Juran Institute, Wilton, CT.*)

Projects	Project leaders	Baseline measurements	Targets	Initial plan	Review points				Review leader
					Resources	Analysis	Plan	Results	

FIGURE 13.11   Progress review plan. (*Juran Institute, Wilton, CT.*)

Cost of Poor Quality.   We define "cost of poor quality" as those costs which would disappear if our products and processes were perfect and generated no waste. Those costs are huge. As of the 1980s, about a third of the work in the economy of the United States consisted of redoing prior work because products and processes were not perfect.

The costs are not known with precision. In most organizations the accounting system provides only a minority of the information needed to quantify this cost of poor quality. It takes a great deal of time and effort to extend the accounting system so as to provide full coverage. Most organizations have concluded that such effort is not cost effective.

What can be done is to fill the gap by estimates which provide upper managers with approximate information as to the total cost of poor quality and as to which are the major areas of concentration. These areas of concentration then become the target for quality improvement projects. Thereafter the completed projects provide fairly precise figures on quality costs before and after the improvements.

Product and Process Deficiencies.   Even though the accounting system does not provide for evaluating the cost of poor quality, much evaluation is available through measures of product and process deficiencies, either in natural units of measure or in money equivalents; for example, cost of poor quality per dollar of sales, dollar of cost of sales, hour of work, or unit shipped. Most measures lend themselves to summation at progressively higher levels. This feature enables goals in identical units of measure to be set at multiple levels: corporate, division, department.

Performance of Business Processes.   Despite the wide prevalence and importance of business processes, they have been only recently controlled as to performance. A contributing factor is their multifunctional nature. There is no obvious owner and hence no clear, sole responsibility for their performance. Responsibility is clear only for the subordinate microprocesses. The system of upper management controls must include control of the macroprocesses. That requires establishing goals in terms of cycle times, deficiencies, etc., and means for evaluating performances against those goals.

The Scorecard.   To enable upper managers to "know the score" relative to achieving strategic quality deployment, it is necessary to design a report package, or scorecard. In effect, the strategic plan dictates the choice of subjects and identifies the measures needed on the upper management scorecard.

The scorecard should consist of several conventional components:

- Key performance indicators (at the highest levels of the organization)
- Quantitative reports on performance, based on data
- Narrative reports on such matters as threats, opportunities, pertinent events
- Audits conducted (see Business Audits later in this section)

These conventional components are supplemented as required to deal with the fact that each organization is different. The end result should be a report package which assists upper managers to meet the quality goals in much the same way as the financial report package assists the upper managers to meet the financial goals.

The council has the ultimate responsibility for design of such a scorecard. In large organizations, design of such a report package requires inputs from the corporate offices and divisional offices alike. At the division level the inputs should be from multifunctional sources.

The report package should be specially designed to be read at a glance and to permit easy concentration on those exceptional matters which call for attention and action. Reports in tabular form should present the three essentials: goals, actual performances, and variances. Reports in graphic form should, at the least, show the trends of performances against goals. The choice of format should be made only after learning what are the preferences of the customers, i.e., the upper managers.

Managerial reports are usually published monthly or quarterly. The schedule is established to coincide with the meetings schedule of the council or other key reviewing body. The editor of the scorecard is usually the Director of Quality (Quality Manager, etc.), who is usually also the secretary of the council.

At Texas Instruments, Inc., the scorecard is a quality report package (the "Quality Blue Book"), deliberately designed to parallel the company's financial reporting system, down to the color of the cover (blue). The report is organized into

1. Leading indicators, e.g., quality of purchased components
2. Concurrent indicators, e.g., product test results, process conditions, and service to customers
3. Lagging indicators, e.g., data feedback from customers and returns
4. Data on cost of poor quality

The report is issued monthly and is the basis for annual performance appraisal of managers' contributions to quality (Onnias 1985).

The scorecard should be reviewed formally on a regular schedule. Formality adds legitimacy and status to the reports. Scheduling the reviews adds visibility. The fact that upper managers personally participate in the reviews indicates to the rest of the organization that the reviews are of great importance.

In the past few years many organizations have combined their measurements from financial, customer, operational, and human resource areas into "instrument panels" or "balanced business scorecards." [See Kaplan and Norton (1992) or Godfrey (1998) for more details.]

Business Audits.   An essential tool for upper managers is the audit. By "audit," we mean an independent review of performance. "Independent" signifies that the auditors have no direct responsibility for the adequacy of the performance being audited.

The purpose of the audit is to provide independent, unbiased information to the operating managers and others who have a need to know. For certain aspects of performance, those who have a need to know include the upper managers.

To ensure quality, upper management must confirm that

1. The systems are in place and operating properly
2. The desired results are being achieved

Duracell International, Inc. performed what they called a "worldwide quality audit" to review the progress they had made toward realizing the vision to be the best. According to C.R. Kidder, Duracell's former chairman and CEO, the idea was to test Duracell products that had been bought anonymously from retail outlets around the world against competitor products acquired in the same way. Buying the Duracell product at retail instead of simply gathering samples from Duracell manufacturing facilities ensured that the product tested was representative of product purchased by consumers and ensured comparability with the competitive products. The products were tested and the results shared with Duracell executives, in the expectation that doing so would raise the visibility of the competitive status of the product and create pressure to make improvements necessary to close any competitive gaps revealed in the testing. The test information was organized to compare Duracell product against competitive product on two dimensions: quality (leakage, labeling, size, etc.) and performance (number of service hours). In 1985, the audit showed Duracell was about even with its competitors. By 1993, Duracell had the longest-lasting, highest-quality product in the world.

These audits may be based on externally developed criteria, on specific internal objectives, or on some combination of both. Three well-known external sets of criteria to audit company performance are those of the United States' Malcolm Baldrige National Quality Award (MBNQA), the European Quality Award (EQA), and Japan's Deming Prize. All provide similar criteria for assessing business excellence throughout the entire organization.

Traditionally, quality audits have been used to provide assurance that products conform to specifications and that operations conform to procedures. At upper-management levels, the subject matter of quality audits expands to provide answers to such questions as

- Are our policies and goals appropriate to our company's mission?
- Does our quality provide product satisfaction to our clients?
- Is our quality competitive with the moving target of the marketplace?
- Are we making progress in reducing the cost of poor quality?
- Is the collaboration among our functional departments adequate to ensure optimizing company performance?
- Are we meeting our responsibilities to society?

Questions such as these are not answered by conventional technological audits. Moreover, the auditors who conduct technological audits seldom have the managerial experience and training needed to conduct business-oriented quality audits. As a consequence, organizations that wish to carry out quality audits oriented to business matters usually do so by using upper managers or outside consultants as auditors.

Juran (1998) has stated:

> One of the things the upper managers should do is maintain an audit of how the processes of managing for achieving the plan is being carried out. Now, when you go into an audit, you have three things to do. One is to identify what are the questions to which we need answers. That's non-delegable, the upper managers have to participate in identifying these questions. Then you have to put together the information that's needed to give the answers to those questions. That can be delegated and that's most of the work, collecting and analyzing the data. And there's the decisions of what to do in light of those answers, that's non-delegable. That's something the upper managers must participate in.

Audits conducted by executives at the highest levels of the organization where the president personally participates are usually called "The President's Audit" (Kondo 1988). Such audits can have major impacts throughout the organization. The subject matter is so fundamental in nature that the audits reach into every major function. The personal participation of the upper managers simplifies the problem of communicating to the upper levels, and increases the likelihood that action will be forthcoming. (See Section 41 under Company-Wide Quality Control Education and Training.) The very fact that the upper managers participate in person sends a message to the entire organization relative to the priority placed on quality and to the kind of leadership being provided by the upper managers—leading, not cheerleading (Shimoyamada 1987). (For elaboration on quality audits, see Section 14, Total Quality Management.)

STRATEGIC DEPLOYMENT OR NOT: THE DECISIVE ELEMENT

Benefits of Implementing Strategic Deployment.   Whether the upper managers should adopt strategic deployment is a decision unique to each organization. What is decisive is the importance of integrating major change initiatives or quality programs into the strategic plan. The potential benefits of strategic deployment are clear:

1. The goals become clear—the planning process forces clarification of any vagueness.
2. The planning process then makes the goals achievable.
3. The control process helps to ensure that the goals are reached.
4. Chronic wastes are reduced through the quality improvement process.
5. Creation of new wastes is reduced through revision of the business planning process.

Strategic Deployment Implementation: Risks and Lessons Learned.   There are also some important lessons learned about the risks in implementing strategic deployment.

1. Pursuing too many objectives, long term and short term, at the same time will dilute the results and blur the focus of the organization.
2. Excessive planning and paper work will drive out the needed activities and demotivate managers.
3. Trying to plan strategically without adequate data about customers, competitors, and internal employees can create an unachievable plan or a plan with targets so easy to achieve that the financial improvements are not significant enough.
4. If the executive leadership delegates too much of the responsibility, there will be a real and perceived loss of leadership and direction.
5. For an organization to elevate quality and customer focus to top priority creates the impression that it is reducing the importance of finance, which formerly occupied that priority. This perceived downgrading is particularly disruptive to those who have been associated with the former top-priority financial goals.

EMBARKING ON STRATEGIC DEPLOYMENT

Probably the biggest disruption is created by imposing a structured approach on those who prefer not to have it. Resistance to the structured approach is evident at the very outset.

> The single most important prerequisite for embarking on a long-term, effective company-wide quality improvement effort is the creation of an environment conducive to the many changes that are necessary for success. We've aggressively sought to eliminate barriers that have taken years or decades to establish. The process of change takes time, however, and change will occur only as an evolutionary process. (Delaplane 1987)

(See Figure 13.12.)

Technology   Culture

FIGURE 13.12   Cultural changes. (*Juran Institute, Wilton, CT.*)

HIGHLIGHTS

Strategic deployment is a systematic approach for integrating customer focus and company-wide change programs (such as quality improvement) with the strategic plans throughout the entire organization. The strategic deployment process provides focus and enables organizations to align improvement goals and actions with their vision, mission, and key strategies. Strategic deployment provides the basis for senior management to make sound strategic choices and prioritize the organization's improvement and other change activities. Activities not aligned with the organization's strategic goals should be terminated or eliminated.

REFERENCES

Delaplane, Gary W. (1987). "Integrating Quality Into Strategic Planning." *IMPRO 1987 Conference Proceedings,* Juran Institute, Inc., Wilton, CT, pp. 21–29.

Godfrey, A. Blanton (1997). "A Short History of Managing Quality in Health Care," In Chip Caldwell, ed., *The Handbook For Managing Change in Health Care.* ASQ Quality Press, Milwaukee, WI.

Godfrey, A. Blanton (1998). "Hidden Costs to Society," *Quality Digest,* vol. 18, no.6.

Juran, J. M. (1988). *Juran on Planning for Quality,* Free Press, New York.

Kaplan, Robert S. and Norton, David P. (1992). "The Balanced Scorecard—Measures That Drive Performance." *Harvard Business Review,* January–February, vol. 70, no.1.

Kondo, Yoshio (1988). "Quality In Japan." In J. M. Juran, ed., *Juran's Quality Control Handbook,* 4th ed., McGraw-Hill, New York. (Kondo provides a detailed discussion of quality audits by Japanese top managements, including the president's audit. See Section 35F, "Quality in Japan", under Internal QC Audit by Top Management.)

Onnias, Arturo (1985). "The Quality Blue Book." *IMPRO 1985 Conference Proceedings,* Juran Institute, Inc., Wilton, CT, pp. 127–131.

Roberts, Michael (1987). *The Strategic CEO.* New York, p. 334.

Shimoyamada, Kaoru (1987). "The President's Audit: QC Audits at Komatsu." *Quality Progress,* January, pp. 44–49. (Special Audit Issue).

Treqoe, Benjamin and Tobia, Peter (1990). "Strategy and the New American Organization," *Industry Week.* August 6.

Weigel, Peter J. (1990). "Applying Policy Deployment Below the Corporate Level." *IMPRO 1990 Conference Proceedings,* Juran Institute, Inc., Wilton, CT.

SECTION 14
TOTAL QUALITY MANAGEMENT

A. Blanton Godfrey

GENERAL INTRODUCTION

In the past 10 or 20 years a few companies have radically transformed their business performance. Many of the concepts and methods they have used are now collectively called "total quality" or "total quality management." Many other terms have also been used. These include "business transformation, performance excellence, business excellence, and six sigma." The successes of these companies have dramatically changed how they and others see both quality and business management today. They are rethinking how they are organized, how they manage themselves, and even what businesses they should be in.

INTRODUCTION TO TOTAL QUALITY MANAGEMENT

In the past two decades many organizations throughout the world have been under tremendous pressure. Some have been battered by international competition, others by new entrepreneurial companies that redefined businesses, and yet others were seriously challenged by new technologies which created formidable alternatives to their products and services. Some leading companies have changed rapidly. While some of the new companies have now become major players, other companies are still engaged in daily battles for survival, and many other companies have disappeared.

Many companies have found that all of their radical restructuring, reengineering, downsizing, and numerous quality programs may have helped them survive, but they still do not have a distinctive quality advantage. Their future will be determined by three key areas: alignment, linkage, and replication. Combined with the fundamental concepts of quality management (continuous improvement, customer focus, and the value of every member of the organization), their work in these three key areas is transforming the way they are managing the entire organization.

During these years there has been an increasing global emphasis on quality management. In global competitive markets, *quality* has become the most important single factor for success. Quality management has become the competitive issue for many organizations. Juran has gone so far as to state that, "Just as the twentieth century was the century of productivity, the twenty-first century will be the quality century."

Reimann (1992*a*), then Director for Quality Programs, National Institute of Standards and Technology, U.S. Department of Commerce, in testimony to the U.S. Congress, stated this clearly, "There is now far clearer perception that quality is central to company competitiveness and to national competitiveness."

In the United States, the President and the Secretary of Commerce have given their personal support and attention to quality, thus elevating quality on the national agenda. Their efforts have helped the American public understand that quality is a main component in national competitiveness. In other countries, such as Argentina, Brazil, France, Greece, Malaysia, Mexico, and Singapore, there has also been leadership from the top levels of government and business, creating national programs of awareness, training, and awards.

In October 1991 a leading international business magazine, *Business Week,* published a bonus issue devoted entirely to the subject of quality. The editor-in-chief, Stephen Shepard, called this bonus issue "the most ambitious single project" in *Business Week*'s 62-year history. Shepard further commented that quality "may be the biggest competitive issue of the late twentieth and early twenty-first centuries."

This issue was sold out in a matter of days. The demand was so high in the United States and throughout the world that *Business Week* had to make two additional printings of tens of thousands of magazines. At the end of the year, the magazine editors of the United States named this issue the "Magazine of the Year," the top honor for magazines in the United States.

During 1991, the U.S. General Accounting Office (GAO) completed a study of Malcolm Baldrige National Quality Award winners and site-visited companies. The GAO studied carefully the relationship between quality management activity and success and profitability. This report, GAO Report 91-190, became GAO's all-time best selling report. In early 1995 the National Institute of Standards and Technology of the U.S. Department of Commerce issued a new report contrasting the stock market success of the Malcolm Baldrige National Quality Award–winning companies (companies with divisional winners and site-visited companies) with average companies. The results were convincing. The National Quality Award Program in the United States does not maintain information on an individual organization's financial results, but for the fourth year in a row a special stock comparison study has shown significant differences (Port 1998). The Malcolm Baldrige National Quality Award recipients as a group have outperformed the Standard & Poor 500 by nearly a 2.5 to 1 margin (see Figure 14.1). The recipients achieved a 362-percent rate of growth versus a 148-percent rate of growth for average companies (Port 1998, p. 113).

In Europe, the creation of the European Foundation for Quality Management in 1988 has already had a significant impact on the understanding of quality management as a leadership issue and as a competitive tool. The introduction in 1992 of the European Quality Award has had a major impact in raising senior executive awareness and understanding of quality management concepts and methods. The oldest award is, of course, the Deming Application Prize, which was started in 1951 by the Union of Japanese Scientists and Engineers (JUSE). This prize stimulated the adoption of quality control in virtually every sector of Japanese industry. Over time the prize criteria evolved into the concept of company-wide quality control (CWQC) and total quality control (TQC) (Kondo, Kume, and Schimizu 1995, p. 4).

We should mention here that we will use the generic term "total quality management" to mean the vast collection of philosophies, concepts, methods, and tools now being used throughout the

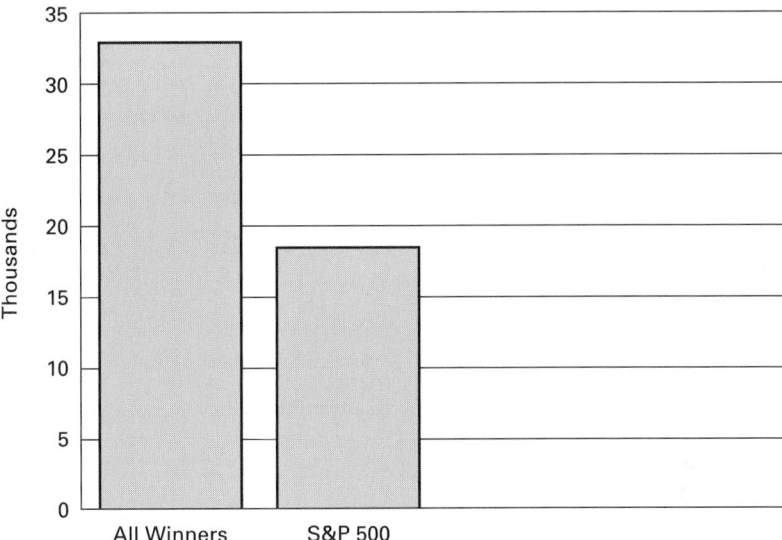

FIGURE 14.1 Corporate growth of investment of $7496.54 in MBNQA winners versus the S&P 500. (*Port 1998, p. 113.*)

world to manage quality. Other terms are frequently used. Total quality management (TQM) is probably the most frequently used term in the United States, while total quality control (TQC) was until recently most often used in Japan, although this may be changing. "The term TQC (total quality control) has begun to be replaced in Japan by the term TQM (total quality management)" (Kondo 1995, p. vi). Kondo himself uses the equivalent term "Companywide Quality Management" in his recent book (Kondo 1995). Another term sometimes encountered is "continuous quality improvement" (CQI). In 1997, JUSE announced a formal change from the term TQC (total quality control) to TQM (total quality management) (The TQM Committee 1997*a,* p. 1). This name change was made both to adopt a more internationally accepted term and to provide an opportunity to revisit the origin of quality control and rebuild the concept to meet new environmental challenges in business management. The TQM Committee of JUSE explained this change in four publications (The TQM Committee 1997*a,* 1997*b,* 1997*c,* and 1997*d*). A summary of their thinking is provided by the diagram in Figure 14.2.

In JUSE's view, TQM is a management approach that strives for the following in any business environment:

* Under strong top-management leadership, establish clear mid- and long-term vision and strategies.
* Properly utilize the concepts, values, and scientific methods of TQM.
* Regard human resources and information as vital organizational infrastructures.
* Under an appropriate management system, effectively operate a quality assurance system and other cross-functional management systems such as cost, delivery, environment, and safety.
* Supported by fundamental organizational powers, such as core technology, speed, and vitality, ensure sound relationships with customers, employees, society, suppliers, and stockholders.
* Continuously realize corporate objectives in the form of achieving an organization's mission, building an organization with a respectable presence, and continuously securing profits.

In any discussion of total quality it is useful to start with the basics: the results we expect, the three fundamental concepts, the three strong forces, the three critical processes, and the key elements of the total quality infrastructure.

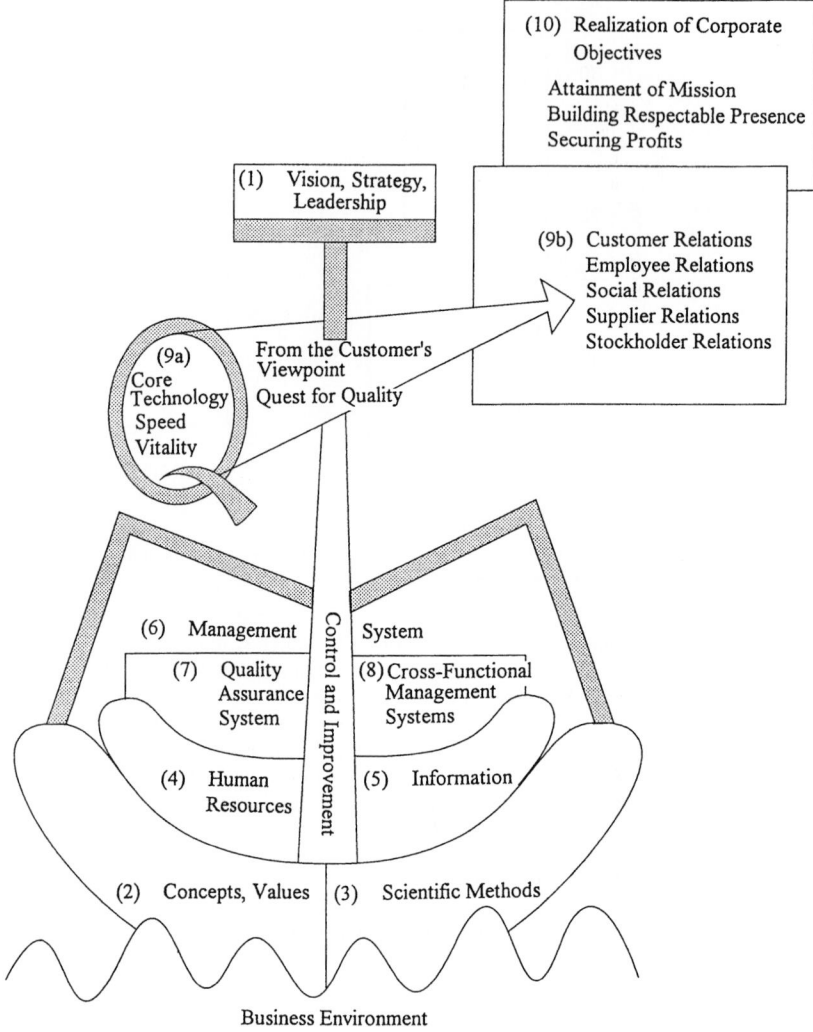

FIGURE 14.2   The overall picture of TQM. *The TQM Committee 1997a, p. 4.*)

The Results of Total Quality.   The almost universally accepted goals of total quality are lower costs, higher revenues, delighted customers, and empowered employees. These goals need little explanation. Figure 14.3 from Juran Institute's *Leadership for the Quality Century* workshop graphically illustrates this.

In the past few years we have moved quickly from believing that managing quality just means conformance to specifications and requirements. Quality also means meeting and even exceeding the needs and expectations of customers. Quality includes having the right features, correct documentation, and error-free invoices. It also includes the proper functioning of critical business processes—on-time delivery, friendly and accurate technical support, *and* no failures. Quality involves reducing all the costs of poor quality.

Lower Costs.   Higher quality can mean lowering costs by reducing errors, reducing rework, and reducing non-value-added work. In the past 15 or 20 years companies around the world have repeatedly demonstrated that higher quality frequently means lower costs. The costs associated with preventing errors during design are often far less than correcting the errors during production, the costs of preventing errors during production are far less than correcting the errors after final inspection, and the costs of finding and correcting errors during final inspection are far less than fixing the errors after the customer has received the goods or services. Our understanding of these costs has grown rapidly in the past decade (Godfrey 1998, p. 18). Sörqvist (1998, pp. 36–39) defines these costs in five basic categories: traditional poor-quality costs, hidden poor-quality costs, lost income, customers' costs, and socioeconomic costs. See Section 8 for a more detailed discussion of quality and costs. See Section 7 for a discussion of quality and income.

Higher Revenues.   Higher quality can mean better satisfied customers, increased market share, improved customer retention, more loyal customers, and even premium prices. Customers are increasingly beginning to expect and demand high-quality goods and services. By exceeding the levels of quality offered by competitors in the marketplace, organizations can add new customers, retain old customers, and move into new markets. Often, informed customers are willing to pay a price premium for higher levels of quality that provide new and useful features or that reduce total life-cycle costs.

Delighted Customers.   "Delighted" customers are customers who buy over and over again, customers who advertise your goods and services for you, customers who check you first when they are going to buy anything else to see if you also offer those goods or services. Loyal customers will frequently increase their purchases to the point of selecting sole suppliers for certain goods and services (Reichheld 1996).

Market studies have recently shown the dramatic impact of such delighted customers. In one study, customers giving satisfaction ratings of 5 (on a 1 to 5 scale) were 4 times less likely to leave during the next 12 months and 5 times as likely to purchase additional services than those giving satisfaction ratings of 4. In fact, those giving ratings of 2, 3, and 4 were remarkably similar, basically neutral.

Empowered Employees.   For many years organizations viewed empowered employees as a means for achieving lower costs, higher revenues, and delighted customers. Now most leading organizations realize that creating such employees is also a major goal of total quality management. These organizations not only aim to solve the problems of today, but they also want to create an organization that can solve, or even avoid, the problems of tomorrow.

The concept of empowered employees embraces many new ideas. Empowered employees are in self-control. They have the means to measure the quality of their own work processes, to interpret the measurements, and compare these measurements to goals and take action when the process is not on target.

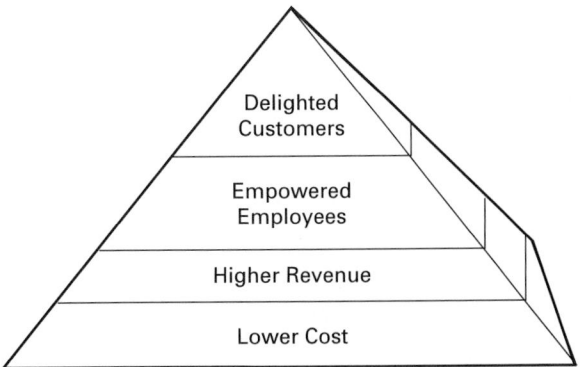

FIGURE 14.3   Results of total quality management. (*Leadership for the Quality Century, 1997, Juran Institute, Inc., Wilton, CT.*)

But the concept of empowered employees goes far beyond self-control. Employees also know how to change the process and to improve performance, improving both the effectiveness and the efficiency of the process.

They also understand how to plan for quality. They understand who their customers are; what the customers need, want, and expect; how to design new goods and services to meet these needs; how to develop the necessary work processes; how to develop and use the necessary quality measurements; and how to continuously improve these processes. The basic principles of empowerment are covered in Section 15.

The Three Fundamental Concepts.

In the past few years many leading companies throughout the world have begun to revisit the fundamental concepts of quality management: customer focus, continuous improvement, and the value of every individual.

Customer Focus.   That customer focus is a fundamental concept of quality management perhaps seems obvious. After all, organizations only exist to provide goods and services to customers. Aren't all organizations customer focused?

For most organizations the answer to this question is a resounding *no*. During the evolutionary progress of almost every industry, the first phase is a focus on quality of the new product in the most basic terms. The goal is to make the product work. The early automobiles, airplanes, and telephones are dramatic examples. There are no customers in the beginning, only wild-eyed inventors trying to make something no one has ever seen before. Later in this section we trace the basic evolution of quality in typical organizations and industries.

The customers provide little input at this stage. Most are not even sure they want these goods or services. They have little understanding of what they are, why they should buy them, or what they do, much less ideas on how to make them better. Unfortunately, many organizations do not progress far beyond this stage. The technology-driven companies and organizations providing health care are two highly visible examples.

One of the hottest trends in business today is the creation of custom products for mass markets. Bicycles, jeans, shoes, carpets, and numerous other items are being customized to individual taste. Financial services, hotel services, meals, and even health care are being designed and delivered to meet individual customer needs (Peppers and Rogers 1993).

The biggest challenge facing companies today is linking measurement of how well they meet customer needs to the actual behavior of the customers. Knowing that you have a 4.3 customer satisfaction rating means nothing. What truly matters is whether a 4.3 satisfaction rating is creating business or losing business.

More and more companies are finding that keeping customers (reducing the churn) is far more profitable than acquiring new ones. A Xerox study found that sales to current customers were over 20 percent more profitable than sales to new customers. The other critical factor is what percent of the customer's business you have. Becoming the dominant supplier can have stunning business results.

Reichheld (1996) documents many of the examples of how companies have gone beyond customer satisfaction and customer retention to customer loyalty. Building customer loyalty is becoming a bedrock of corporate strategic planning and process management.

Continuous Improvement.   Juran (1964) documented the structured approach that many companies use to achieve breakthrough improvements. In recent years rapid change has become a way of life. Many companies now employ this and similar approaches to create improvements by the hundreds and even thousands.

But this was not always the case. For thousands of years societies and governments have been organized to prevent change. In some societies doing something in a different way was punishable by death. In his novel, *The Egyptian,* Waltari (1949) describes how a physician in ancient Egypt was trained to perform 128 different procedures. Only these could be performed, and there was only one way to perform each. Even artists were trained carefully in the only way to draw a bird, a crocodile, or a person.

In medieval Europe the various trade guilds established rigid guidelines for the making of each object. Daring to experiment in the ways things were made or the materials used was grounds for expulsion from the guild. In the Byzantine language the word for change was the same as the word for danger. Change in societies, in production practices, in armies, in governments came slowly. Many societies endured half-witted rulers rather than risk changing the form of government.

Organizations and companies mirrored society. Companies were governed by thick policy manuals and corporate executive instructions in multivolume sets. Promotions were given to those "not rocking the boat." Strong hierarchies were created to control all operations and individual work. Scientific management, the so-called Taylor System was used to carefully define each step in the work process and each person's role. Job descriptions defined clearly what one did and what one did not do.

Things began to change rapidly in the years following World War II. The Japanese were so far behind in many areas of commercial production that they had to improve rapidly to survive. The continuous improvement methods they perfected worked well. Faced with severe competition, many U.S. firms started copying these ideas, some with great success.

The literature now abounds with examples of astonishing improvements. These improvements are being made in manufacturing companies, hospitals, telecommunications companies, government agencies at every level, all types of service companies, and in schools. The names of the means used to achieve these results have become quite familiar to all of us: cross-functional teams, quality control circles, re-engineering, quality action teams, creative idea suggestion systems, process improvement teams, quality in daily work, and many others.

Value of Every Associate.   The value of each associate in an organization is another idea that sounds simple on the surface. For years companies have published clear statements about the strength of their organizations being the people who work for them. But most of these are just hollow statements. The companies are still blindly following the Taylor system. A few planners, managers, or engineers are planning all the steps of every process, defining carefully worded job descriptions, and enforcing the unthinking following of instructions.

Even the most cursory review of history illuminates how radical an idea it is to have each person thinking, creating ideas, challenging authority, and making changes to the system. Entire armies marched side by side with spears pointed forward at exactly the same angle. Archers fired precisely when told. Musketeers marched in ranks, fired precisely timed volleys, reloaded, and fired again. But only on the orders of the commander.

Individuals were trained in long apprentice programs by demanding masters. Rows of clerks transcribed exactly what was written. The early factories contained rows and rows of workers each doing each task exactly the same way.

The average number of implemented ideas per employee per year in the United States is still only 0.16. That is one idea implemented for every six employees per year. In organizations truly valuing the ideas and personal contributions of each employee the number is dramatically higher. Already in the United States, Toyota is achieving eight implemented ideas per employee at its Georgetown manufacturing facilities. Overall, Toyota receives 4,000,000 ideas from its 80,000 employees. Since over 95 percent are implemented, this is over 46 implemented ideas per employee per year (Yasuda 1991).

Some companies in the United States have achieved similar results. Globe Metallurgical and Milliken have averaged over one implemented idea per employee per week. Milliken is now one of the country's leaders at 68 ideas implemented per associate per year. One employee in a Marriott hotel contributed 63 improvement suggestions in one month (Fromm and Schlesinger 1993, p. 8)!

But ideas contributed are just one measure of individual contributions. Other contributions may be even more important. These include participation on quality improvement and quality planning teams, membership on business process re-engineering teams, work on statistical quality control and self-control of their own work processes, and working as members of high-performance or self-directing work teams.

Eastman Chemical was already 6 years into its quality journey in 1985 when they began to recognize the strong connection between culture, values, and quality excellence. Their objective was to identify, understand, and emphasize the people elements of their quality policy. They now use their internally developed quality management process as a vehicle to bring all employees into the improvement efforts. They use interlocking teams of employees at every level to define how each work process links together with the next and with the customers' needs and expectations.

Eastman Chemical has also formally defined "empowerment" as the creation of a culture "where people have the knowledge, skills, authority, and desire to decide, act, and take responsibility for the results of their actions and for the contribution to the success of the company." They implement this clear, working definition of empowerment by providing just-in-time training where employees come to class with improvement projects already selected. Quality coaches (facilitators) provide direct support back on the job.

The Three Strong Forces.   There are three primary drivers of performance excellence: alignment, linkage, and replication. To achieve breakthrough results the organization must focus its efforts on the most important issues—it must have its strategy correct and the organization's goals, resources, and activities aligned with the strategy. The organization must also understand the cross-functional nature of work, the linkages across the organization. Sometimes called "systems thinking" or "process thinking," this understanding of the way work is done is crucial. Associates in the organization must also be able to replicate successes quickly. A simple improvement may be worth only a few thousand dollars. But replicated 100 times it may become a major contribution to the organization's success.

Alignment.   A recent study by the Association of Management Consulting Firms in the United States found that executives, consultants, and business school professors all agree that business strategy is now the single most important management issue and will remain so for at least the next 5 years (Byrne 1996, p. 46). In the past few years, there has been a new understanding of the importance of strategy. This strategy must include:

1. A clear vision of where the company is going—this must be clearly stated and communicated to every member of the organization in language he or she understands.

2. Clear definitions of the small number of key objectives that must be achieved if the company is to realize its vision.

3. Translation of these key objectives throughout the entire organization so that each person knows how performing his or her job helps the company achieve the objectives. This alignment of all associates with the top priorities of the company is absolutely critical (Sugiura 1992).

One of the biggest changes in the strategic planning process has been the inclusion of many layers of the workforce, customers, suppliers, and even competitors in the planning process. These changes are creating a whole new set of buzzwords: co-evolution, business ecosystems, strategic intent, business designs, core competencies, game theory, and white-space opportunities. The key differences include the creation of networks of new relationships with customers, suppliers, and rivals to gain new competitive advantages, new markets, and new opportunities.

The second big change has been the inclusion of numbers of employees of all ages, levels, and job functions in the planning process. Some years ago, Electronic Data Systems Corporation (EDS) launched a major strategy initiative involving 2500 of its 55,000 employees. A core group of 150 worked full time for a year coordinating the input from the larger group. Finland's Nokia Group recently involved 250 employees in a strategic review. Nokia's head of strategy development, Chris Jackson, reports that the involvement of more people not only makes their ability to implement the strategy more viable, but they also win a high degree of commitment by the process (Byrne 1996, p. 52).

To be effective, strategic quality planning must be used as a tool—a means to an end—and not as the goal itself. It must be an endeavor that involves people throughout the organization. It must capture existing activities, not just add more activities to already overflowing plates. Finally, it must help senior managers face difficult decisions, set priorities, and eliminate many current activities, not just start new ones.

The third change has been the extreme focus, perhaps even obsession, on customers. The new strategic planning starts with customers. Hewlett-Packard brings both customers and suppliers together with general managers from many different business units to work on strategies. For example, they brought together managers from divisions making service-bay diagnostic systems for Ford with those making workstations for auto plants and those developing electronic components for cars. Many of the ideas for new opportunities came directly from the customers.

Far too many companies have stopped with creating the strategic plan. Their plans are beautifully developed and packaged, but come to nothing. Somehow these companies assume that packaging and distributing the plans to a select number of managers is actually going to make things happen. Nothing could be further from the truth. To really get results these plans must be carefully deployed throughout the entire organization. Every associate must be clearly aligned with the key objectives of the company, every associate must understand the strategic goals and how he or she contributes. Every strategic goal must be broken into subgoals and these must be subdivided into annual goals. The organization must then clearly define the specific work projects which support the annual goals. They must assign clear priorities, establish specific measurements, and provide the resources to achieve the desired results for each project. Strategic deployment is covered in detail in Section 13.

Linkage (Process Management or Systems Thinking). In the past few years companies throughout the world have embraced the concept of re-engineering with a fervor that defies description. Pioneered in the early 1980s by companies such as IBM, Ford, AT&T, and NCR, and popularized in Michael Hammer's best-selling book, *Reengineering the Corporation,* re-engineering has become a common tool for corporations throughout the world (Hammer and Champy 1993). The definition of re-engineering by Hammer as "the radical redesign of business processes for dramatic improvement" captured and excited the imagination of managers around the world. More recently, Hammer has stated that "the key word in the definition of reengineering is 'process': a complete end-to-end set of activities that together create value for a customer" (Hammer 1996, p. xii).

As companies have rediscovered the importance of linking their activities across all functions and departments in the company, they have also rediscovered how critical it is to think of how many activities are actually in series. Unless we link our efforts across all parts of the company, we fail to achieve the results we so desperately need.

With this critical emphasis on linkage (or process management) the worlds of total quality management and re-engineering converge. A fundamental tenet of quality management since Shewhart in the 1920s (if not before) has been the importance of controlling the process. Deming later further developed Shewhart's ideas of statistical process control with the now famous PDCA cycle (Plan, Do, Check, Act), and Juran pioneered the concepts of process improvement with his text *Managerial Breakthrough* (Juran 1964). As leading companies moved into rapid improvement activities in the 1980s, the need for process management became clear. In the manufacturing plants the series nature of work was obvious. If any part of an assembly line failed or created a bottleneck, the whole line suffered. What wasn't so obvious was how many administrative processes were also series systems. With a mistake in the order entry step, there may be no way to complete the delivery of the product or service on time and correctly.

The steps to managing the critical linkages and making dramatic and continuous improvements to the key processes are now well defined. The first step is identifying the organization's key processes. There are numerous methods for doing this, but the essence of them all is narrowing down the list to the most important few and making sure everyone knows them. The next step is creating the necessary measurements. Many companies have long lists of measurements for almost every task in the organization. Most of these measurements are focused on departmental activities and many are related to the budget. But these same companies have few measurements on the critical processes that drive the success of the company. They cannot tell you how long it takes from the receipt of an order from a customer to the time the customer receives the goods or services, much less the time until they receive payment. They do not know the real cost of processing the order, delivering the product, or providing follow-up service.

The final step in managing the critical linkages is to actually get serious about managing these linkages. Without major changes in the structure of the organization, without assigned process owners, without realignment of authorities, responsibilities, and accountabilities, nothing much happens.

Although quality management has for many years been about process control, improvement, and planning, we have still not developed all of the needed understanding, tools, and measurements to manage in this critical new way. This is a major challenge for the future.

The single most important word in the definition of process is "customer." As many of us have discovered in the past decade, a company is a collection of processes, and the customer only sees the company in terms of the output of those processes. The customer does not care how the company is organized, who reports to whom, what the various titles are, or even where the different departments are located. The customer does not even care what parts of the goods or services are produced by the company, the company's suppliers, or the company's competitors. The customers request products, want them delivered exactly when promised, want the required service to be available when needed, and want the bills to be exactly as agreed upon.

If a process is not providing value to the customer, the process is producing waste. There are many subprocesses in a company that exist primarily as enablers for the company to produce value to the customer. Most key processes touch the customer directly, and these processes must add value for the customer.

The second key to managing processes is to determine exactly what value is added by each step in the process. When we see a purchase order for a $30 book with six signatures that has taken 6 weeks to process, we know there is a better way. What value has this process added? Organizations throughout the world have been stunned to learn how many steps they have in key processes, how many useless handoffs, and how much wasted time and effort.

Just focusing on cycle-time reductions can illuminate how unmanaged many of our key processes are. The Royal Leicester Infirmary in the U.K. reduced a neurological testing procedure from 40 days to 1 day and removed 40 percent of the administrative costs by redesigning the process in which 14 departments worked together. Motorola reduced from 6 weeks to under 100 minutes the process time required to take a pager order, produce the pager, and ship it.

The third critical area of managing the critical linkages is the realization that almost all key processes cut across many different areas of the company. To manage these processes successfully requires a team-based approach involving employees with new skills, new understanding of the company's strategy, goals, and competitors, and new tools for doing their rapidly expanding jobs.

The challenge for the future is to continue to identify these skills, tools, and understandings, and to know which are part of the essential core knowledge of the company (which must be taught to all employees) and which are the needed new skills and tools. Companies that a few years ago thought they could rush through a "quality training program" and be through with it are now finding that training has become a full-time activity.

Many of the ideas of process management, teamwork, and problem-solving skills are now finding their way into business and engineering schools, but companies need to quickly introduce all employees to the key processes, the measurements used, and the way the company continuously challenges and changes process performance. The topic of process management is covered in Section 6.

Replication.   Probably the most powerful and the least understood way to dramatically accelerate the results of quality and productivity improvement efforts is the third strong force, replication. An example from a leading international service company makes this clear. The CEO was justifiably proud of some of their accomplishments. In one location a true chronic problem had been solved—the savings were over $350,000/year. In another location, a different chronic problem had been reduced by 75 percent. The increased revenues were also in the hundreds of thousands of dollars.

It was not hard for the CEO to do the math. If each of the more than 250 locations could duplicate these results, the company would exceed its aggressive financial goals for the next year. But he knew how hard it would be to get each of the locations to understand what had been done in these two locations, to modify the approach to fit their situations, and to apply a similar problem-solving methodology and achieve similar results.

When we address replication we are learning first hand about resistance to change, the dreaded not-invented-here syndrome, the entrenched beliefs that every location is different, and even the reluctance of many corporations to "stifle innovation and creativity" by directing business units and branches to act. Problems remain unsolved, new solutions are invented and tried, opportunities are missed, and companies muddle along with slow rates of change and disappointments in results. The successful companies take action; they make things happen. They use passive means to encourage replication, they use active means to force replication, and they make replication an obligation not an option.

Passive systems include sharing, reward and recognition, newspaper articles, and team presentations. The results of quality improvement projects are made known widely throughout the organization. In these systems we assume that those with similar problems or opportunities will hear about the project, obtain the information they need, and act.

Active sharing systems force the issue. At Honda's annual facilitator network meetings (attended by over 3000 people worldwide), participants are expected to share one completed and well-documented project and to study thoroughly four others that could be used in their location. Upon returning to their location, they are expected to implement these four projects. The support structure is in place to assist them, and results are expected.

The Three Critical Processes for Quality Management.   These three management processes are not new. They are the same management processes we have used for years to manage finance. This commonality is helpful to managers. Their long experience in managing for finance becomes useful to them when they enter the world of managing for quality. These three processes are closely interconnected.

Quality Planning.   The logical place to start is quality planning. Quality planning consists of a universal sequence of events—a quality planning roadmap. We first identify the customers and their needs. We then design products (goods and services) which respond to those needs. We also design processes which can produce these goods and services. Finally, we turn the plan over to the operating forces. They then have the responsibility of conducting operations. They run the process, produce the goods and services, and satisfy the customers. The quality planning process is summarized in Figure 14.4.

The quality planning process is discussed in Section 3. In later sections we provide some in-depth coverage of some of the more technical tools used in quality planning (or as some call it, quality by design). These tools include experimental design (Section 47) and reliability prediction and reliability estimation (Section 48).

But no matter how well we apply our methods and tools of quality planning, most processes are not perfect. They have associated with them some chronic waste: time delays, errors, rework, non-value-added work, scrap. This waste is built into the plan; it goes on and on. We first have to provide the control systems necessary to maintain quality at the planned levels. And next we must search for opportunities to make dramatic improvements in the levels of quality achieved. Figure 14.5 makes these relationships clear.

FIGURE 14.4   The quality planning process. (*Leadership for the Quality Century, 1997, Juran Institute, Inc., Wilton, CT.*)

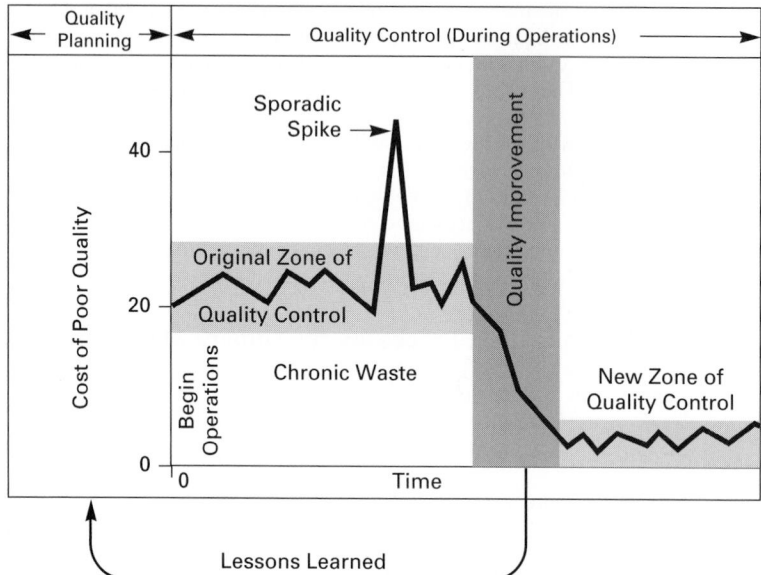

FIGURE 14.5   The Juran trilogy. (*Juran Institute, Inc., Wilton, CT.*)

In this figure we have plotted the cost of poor quality on the vertical scale, so what goes up is bad. These are the costs associated with imperfection. Despite our best efforts at planning, the costs in this example are about 20 percent. These costs could be from defects or they could be even harder to see. Examples of hard-to-detect costs are work-in-process inventory, non-value-added work, underutilized capacity, and unnecessary delays and hand-offs.

Our first job is to build the quality control system to ensure that our quality performance is at least as good as planned. On the diagram in Figure 14.5 we see a sporadic spike, a major deviation from our planned level of performance. In this example, the quality control system seems to be working well. Since this point is a spike, it indicates that the problem was detected quickly, the cause of the problem found quickly, and the cause was removed quickly. Little time elapsed before the quality performance was back at the planned levels.

In many real-life cases our quality control systems do not function this well. Several days or even weeks may go by before we realize we have a problem. Then we may spend more days or weeks investigating the possible causes of the problem and more days or weeks developing remedies. The new level of costs of poor quality persists during this time, causing much damage to the organization.

Quality Control.   What the operating forces can do is minimize this waste. They do this through quality control. Quality control relies on five basics: a clear definition of quality; a target, a clear goal; a sensor, a way to measure actual performance; a way to interpret the measurement and compare with the target; and a way to take action, to adjust the process if necessary. Quality control is discussed in Section 4. Statistical process control is covered thoroughly in Section 45.

Quality Improvement.   But all of this activity only keeps quality at the planned level. We must take deliberate, specific actions if we wish to change this level. As Deming pointed out some time ago, "Putting out the fires in a hotel doesn't make the hotel any better." As he states in *Out of the Crisis* (Deming 1982, p. 51), "Putting out fires is not improvement of the process. Neither is discovery and removal of a special cause detected by a point out of control" (our sporadic spike in Figure 14.5). "This only puts the process back to where it should have been in the first place (an insight of Dr. Joseph M. Juran, years ago)."

Juran (1964) describes the quality improvement process used by individuals and organizations to make "breakthrough" changes in levels of performance. The quality improvement process is directed at long-standing performance levels. The quality improvement process questions whether this is the best that can be attained. Juran describes the quality improvement process in Section 5.

The Total Quality Management Infrastructure.   Figure 14.6 shows the main elements of the total quality infrastructure. These elements include the quality system, customer-supplier partnerships, total organization involvement, measurement and information, and education and training.

The Quality System.   The total quality infrastructure consists of several key pieces. The first, and one of the most important, is the quality system. Best defined by ISO Standard 9004-1, the quality system is a critical building block for total quality management. The ISO Quality System standards are described in detail in Section 11.

A good quality system also contains customer supplier partnerships. Again the ISO 9000 series of standards provides a good starting point for contractual relationships by adding a solid quality management structure. But many companies are going far beyond contractual relationships. Many customer-supplier relationships in the leading U.S. companies are evolving quickly to resemble those pioneered by Toyota and other leading Japanese automotive companies.

To achieve quality improvement at a revolutionary pace, we must also have total organization involvement. In the words of of an interdisciplinary study group convened at Columbia University in 1988 to study global competitiveness (Starr 1988):

> ...we have collected some basic principles of what makes a firm competitive, the first of which is quality.
> The successful business no longer sees employees as a cost of production but as a resource for production. Although job uncertainty will never be eliminated, it must be recognized that long-term commitment of and to workers is at least as important as machinery or technology. Employee involvement in efforts to improve productivity and quality is vital, and they must also be able to share in the gains.

A key element of the infrastructure is measurement and information. Donald Peterson, former chairman of the Ford Motor Company, stresses how important having the right information is. When Ford benchmarked Mazda they were quite impressed with how well Mazda manages this part of the business. Peterson (1992) states,

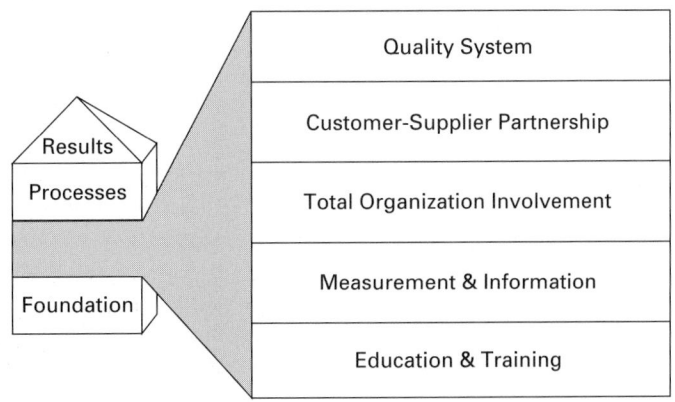

FIGURE 14.6   The total quality infrastructure. (*Leadership for the Quality Century, 1997, Juran Institute, Inc., Wilton, CT.*)

> Perhaps, most important, Mazda had been able to identify the types of information and records that were truly useful. It didn't bother with any other data. (At Ford) we were burdened with mountains of useless data and stifled by far too many levels of control over them.

The last, and perhaps most important, part of the infrastructure is education and training. Organizations must train the teams in how to work as teams and in how to diagnose problems and provide remedies. This type of training should be directed at changing behavior. The training should be just in time. The best learning comes by doing. Training in how to improve quality should be done during actual improvement projects. The training should be designed to help the teams complete these projects quickly and successfully. Training is covered in Section 16.

THE EVOLUTION OF TOTAL QUALITY

In many countries, industries, and companies TQM has appeared to evolve through several distinct steps or phases. These phases include a focus on product quality, on product process quality, service quality, service process quality, business planning, strategic quality planning, and integrated strategic quality planning.

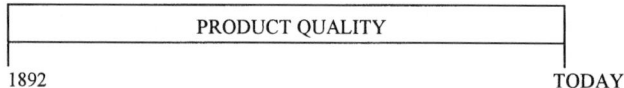

```
┌─────────────────────────────────────────────────────┐
│                  PRODUCT QUALITY                      │
└─────────────────────────────────────────────────────┘
 1892                                            TODAY
```

Product Quality. All organizations began their quality management efforts with a focus on product quality. At the first introduction of a product, this is necessarily a definition of product quality from the producer's point of view. Since the product is unknown to the customers, the customers have little input as to the definition of quality. They may be surveyed for needs and wants, but in the case of a truly new product their inputs are ambiguous and somewhat vague.

In a recent study of the evolution of quality in telecommunications, this was clearly the case (Endres and Godfrey 1994). The telephone was truly a new product. Potential customers were amazed it worked at all and had absolutely no idea how it worked. The driving forces for defining quality were the engineers trying to make it work well enough to be a salable product. As early as 1892, the Bell System was developing inspection procedures to ensure that the specifications and requirements developed by the engineers (the company's definitions of quality) were being met by the production personnel.

This was also the case in other industries such as health care. For many years in modern medicine the definitions of quality focused on outcomes. These were defined by the medical specialists, the doctors. Elaborate quality assurance procedures, usually based on inspection, were developed to review the outcomes and assign responsibility for less-than-perfect outcomes. In Section 29 on automotive quality, a similar evolution is traced.

For other industries (service or manufacturing), this also appears to be the case. The early airlines concentrated their entire efforts on product quality—providing quick transportation from point A to point B. This basic definition of product quality—safe, fast, reasonably on-time air travel—occupied all of their efforts for years.

We should also note that this focus on product quality has continued up to the present. Telecommunications companies extended their efforts beyond initial quality to reliability in the field, then to availability, usability, maintainability, and other definitions of product quality. Some of the methods used to manage these broadened definitions of product quality have become quite sophisticated.

In health care, much recent work on clinical outcomes would fall in this category. Researchers have extended the traditional definitions of outcome to include patient performance, lack of pain, and ability to work. This carries the traditional definition of outcome quality far beyond the walls of the hospital (Godfrey 1997).

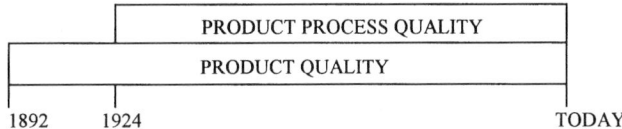

Product Process Quality.   The next phase of the evolution for telecommunications quality began in 1924 with the creation of the control chart. For some time it had been becoming evident that controlling product quality by final inspection was quite expensive. In the installation forces of the rapidly growing American Telephone & Telegraph company, the rallying cry had become, "Do it right the first time." Finding the wiring errors in complex switching machines after the machine had been assembled was a time-consuming, costly process. It was far more economical to ensure functioning parts and carefully control the assembly than to go back and try to find the problems.

Earlier examples abound. During World War I, there is evidence that the British developed rather sophisticated control procedures for ensuring proper tensioning of the wires between the wings of the biplane fighters.

But it was the creation of the control chart that made it clear how easily process control could be transferred to the operating forces. This would reduce the reliance on final inspection and free up numerous people for productive work. In this way product quality could be improved and costs driven down at the same time.

This stage of product process quality—focus on the processes producing the products—has also continued to the present. Many sophisticated methods have been added to the arsenal: engineering process control, experimental design, evolutionary operations, robust design, and more recently process simplification and reengineering.

In health care there are numerous examples: patient-focused care, care maps, clinical guidelines, protocols. Any methods that try to improve the outcomes of our work through improving the processes by which we create those outcomes are in the product process phase. For the most part, in other industries, these have focused on the cost side of producing the product.

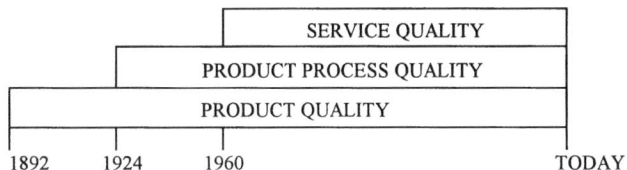

Service Quality.   The next phase for some industries began in the early 1960s. We suspect for other industries it had begun much earlier, for others not until the 1980s or even 1990s. This was the expansion of the traditional definition of product quality to include the services surrounding the product. For telecommunications this expansion includes repair and maintenance services, order entry, billing, and modular phones that the customer could easily self-install and maintain. In health care many new ideas emerged. These included patient-focused care and many other means of providing services beyond basic clinical ones. Many new ideas emerged concerning admissions, waiting times, bedside manner, housekeeping, laundry, room layouts and decor, phones, TVs, food (beyond basic nutrition), parking, and other services surrounding the basic product of the correct outcome. These are sometimes called the features, or the salability part of quality.

For many manufacturing companies the 1960s and 1970s were the wake-up calls for this aspect of quality. The customer was no longer just interested in the quality of the car. Service provided by the dealers, availability of parts, roadside assistance, the sales experience, financing, leasing, and many other aspects of the supplier/customer relationship became part of the competitive quality battleground. In the late 1980s we saw this accelerated with the introduction of the Japanese luxury cars (Acura, Infiniti, and Lexus) with their special dealerships, new service relationships, and new levels of support. General Motors has applied these concepts to the basic car in its Saturn Division.

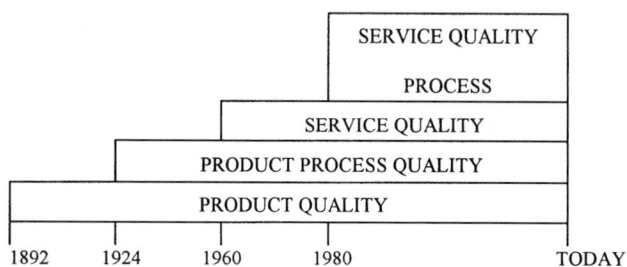

Service Quality Process.   In the 1980s a new focus on quality occurred. Pioneered by IBM, companies started focusing on the costs of providing the quality of these services—or business process quality management or improvement. Many of the same techniques (for the most part rather standard industrial engineering tools) used in product process quality were applied for the first time to the horizontal processes that cut across organizations and had been, for the most part, totally unmanaged.

Some new ideas also emerged. The concept of a process owner and a process team expanded the power of a quality council by continuously examining and identifying opportunities for team interventions in critical business processes. In many ways these process teams acted as focused councils, deploying improvement, control, and planning teams to the macro process or to micro processes within the macro process.

Again the focus was primarily on costs. These re-engineering or business process quality efforts were directed at reducing cycle times, reducing numbers of steps or handoffs, and improving efficiency overall. Many of these business process interventions also improved the quality of the output.

We began to see a cycle emerging. The evolutionary process of total quality management seems to alternate between a focus on quality and a focus on the costs to attain that quality.

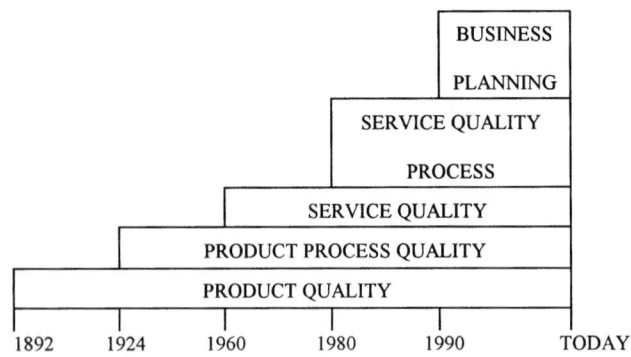

Business Planning.   In the past few years we have observed many companies starting to integrate quality management into their business planning cycles. This integration of the quality goals with the financial goals has been a major thrust of the leading companies. Recently this integration was listed as the major effort currently underway by the respondents to the Business Roundtable survey on TQM.

The setting of quality goals; the subdividing of goals into subgoals, annual goals, and projects; and the deployment throughout the organization (hoshin kanri, hoshin planning, policy deployment, or strategic quality planning) has become one of the major breakthroughs in total quality management for many companies. Some are actually going beyond the annual business planning cycle to incorporate these methods in their 5-year or even longer-term plans.

This naturally leads to the question: What are the next steps in the evolution of total quality? The immediate next step to the pyramid is fairly obvious—strategic quality planning. Some companies are beginning to go even further: they have implemented integrated strategic planning where they are involving customers and suppliers in joint strategic planning. The details of strategic quality planning and those of strategic deployment are given in Section 13.

THE IMPACT OF NATIONAL AND INTERNATIONAL QUALITY AWARDS ON TOTAL QUALITY MANAGEMENT

One of the most useful trends in the past decade has been the self-assessment activities of many companies throughout the world. Companies worldwide are using the criteria of the Malcolm Baldrige National Quality Award, the European Quality Award, the Deming Application Prize, and many other national quality awards to assess their current performance against a reasonable set of guidelines for total quality.

These assessments can provide senior managers with a clear baseline of current quality performance levels. When these managers are willing to take the time to understand the criteria, understand what their own assessment scores mean, and to understand what is necessary to improve these scores, they can develop meaningful and realistic action plans for improving their organizations.

A very important step in this process is to first understand one's own organization's performance level and compare it to the performance level of another organization.

One of the most striking benefits of these national and international quality awards has been the stunning increase in senior management contact with true leaders in total quality. For the first time in the history of the United States, senior managers are hearing what other companies have achieved in quality, how they obtained these results, and what the executive leadership's role in these achievements was.

This benchmarking is, both on a personal level and an organizational level, one of the most important trends in modern quality management. When asked by *Boardroom Reports* if there was a single most important thing a company could do to change the company culture and achieve remarkable results, the then-chairman of Ford Motor Company, Donald Peterson, answered: "There sure is. Each company must find out which other companies in the world are best in that industry. Then, each company must benchmark operations against the most efficient—and most profitable—foreign and domestic businesses....those that do—such as Xerox—have had incredible results."

Peterson went on to explain that at Ford, "we began by comparing our manufacturing processes, design, marketing, financial management and quality, with the best of the Japanese operations."

Peterson stated:

> Comparisons should be based on speed, capital investment, wasted effort, number of employees, and any other yardstick with which the company can measure both its own and other operations.
> The next step is to get managers in the key departments to acknowledge that another business is doing all or part of their job better. That becomes easier when the CEO says that he looks on the benchmarking as an opportunity, not as criticism. Good managers are energized by that challenge.
> Next, send groups of managers to visit the companies with the superior operations.

Ford included union representatives on the visiting teams. Later they assembled teams of key people in the affected departments to discuss the ideas they had seen. They then decided which ideas they could implement and how. Benchmarking is covered in Section 12.

The Creation of the Malcolm Baldrige National Quality Award.   During the 1980s there was a growing interest in the United States in promoting what is now called total quality. Many leaders in the United States felt that a national quality award, similar to the Deming Application Prize of the Union of Japanese Scientists and Engineers, would help stimulate the quality efforts of U.S. companies.

A number of individuals and organizations proposed such an award, leading to a series of hearings before the House of Representatives Subcommittee on Science, Research, and Technology. Finally, on January 6, 1987, the Malcolm Baldrige National Quality Improvement Act of 1987 was passed. The act was signed by President Ronald Reagan on August 20, 1987 and became Public Law 100-107. This act provided for the establishment of the Malcolm Baldrige National Quality Award Program. The purpose of this award program was to help improve quality and productivity by (House Resolution 812, U.S. Congress):

(A) helping stimulate American companies to improve quality and productivity for the pride of recognition while obtaining a competitive edge through increased profits;
(B) recognizing the achievements of those companies which improve the quality of their goods and services and provide an example to others;
(C) establishing guidelines and criteria that can be used by business, industrial, governmental and other organizations in evaluating their own quality improvement efforts; and
(D) providing specific guidance for other American organizations that wish to learn how to manage for high quality by making available detailed information on how winning organizations were able to change their cultures and achieve eminence.

The act provided that up to two awards could be presented to companies in each of three categories:

- Small businesses
- Companies or their subsidiaries
- Companies which primarily provide services

The act also stated that companies must apply for the award by submitting an application, in writing, for the award. And the company must permit a "rigorous evaluation of the way in which the business and other operations have contributed to improvements in the quality of goods and services."

In 1998 pilot examinations of health care and educational organizations were conducted. Recently the Malcolm Baldrige National Quality Award was expanded to include educational and health care organizations.

The act also called on the Director of the National Bureau of Standards (now the National Institute of Standards and Technology) to

rely upon an intensive evaluation by a competent board of examiners which shall review the evidence submitted by the organization and, through a site visit, verify the accuracy of the quality improvements claimed. The examination should encompass all aspects of the organization's current quality management in its future goals. The award shall be given only to organizations which have made outstanding improvements in the quality of their goods or services (or both) and which demonstrate effective quality management through the training and involvement of all levels of personnel in quality improvement.

In addition to the establishment of the Board of Examiners, the act also called for the establishment of a Board of Overseers consisting of at least five individuals who have demonstrated preeminence in the field of quality management.

The Malcolm Baldrige National Quality Award Development Strategy.   In creating the Malcolm Baldrige National Quality Award, the first step was to develop the criteria which would be used to evaluate the organizations applying. The Director of the National Bureau of

Standards selected Dr. Curt Reimann as Director of the Malcolm Baldrige National Quality Award. Dr. Reimann immediately began calling on individuals and organizations throughout the United States and the world for their suggestions and contributions to creating the criteria and the process by which these criteria would be evaluated. Dr. Reimann and his staff collected much information on other awards, such as the JUSE Deming Prize and the NASA quality award, as background information. They then selected a small team of volunteers to help create the first draft of the criteria. These draft criteria were then reviewed in intensive focus group sessions by selected experts from organizations throughout the United States (Reimann 1992*b*).

One of the most important actions taken by the director, his team, and the volunteers at this stage was to create a clear design strategy for the award program. The elements of the strategy were:

- To create a national value system for quality
- To provide a basis for diagnosis and information transfer
- To create a vehicle for cooperation across organizations
- To provide for a dynamic award system which would evolve through consensus and be continuously improved

The design strategy has been followed carefully. The award criteria have been changed and improved each year. This section does not try to capture the evolution of the criteria but rather describes the current, 1998 criteria. These criteria are presented in detail in a booklet available free from the National Institute of Standards and Technology (U.S. Department of Commerce 1998).

The Malcolm Baldrige National Quality Award Criteria are the basis for making awards and giving feedback to the applicants. The criteria also have three other important purposes:

- To help raise quality performance standards and expectations
- To facilitate communication and sharing among and within organizations of all types, based upon a common understanding of key quality and operational performance requirements
- To serve as a working tool for planning, training, assessment, and other uses

The Malcolm Baldrige National Quality Award Core Values.   There are 11 core values and concepts embodied in the award criteria. These core values and concepts are as follows:

Customer-Driven Quality.   Emphasis here is placed on product and service attributes that contribute value to the customer and lead to customer satisfaction and preference. The concept goes beyond just meeting basic customer requirements, including those that enhance the product and service attributes and differentiate them from competing offerings. Customer-driven quality is thus described as a strategic concept directed towards customer retention and market share gain.

This focus on the customer has even been emphasized by the President of the United States (George Bush, *1993 Award Criteria,* National Institute of Standards and Technology, Malcolm Baldrige National Quality Award):

> In business, there is only one definition of quality—the customer's definition. With the fierce competition of the international market, quality means survival.

The emphasis on quality management and the customer has crossed administrations and political parties in the United States, as shown by this statement from the President (William J. Clinton, *1998 Award Criteria,* National Institute of Standards and Technology, Malcolm Baldrige National Quality Award):

> Quality is one of the keys to the continued competitive success of U.S. businesses. The Malcolm Baldrige National Quality Award, which highlights customer satisfaction, workforce empowerment, and increased productivity, has come to symbolize America's commitment to excellence.

Leadership.   A key part of the MBNQA focus is on senior executive leadership. The leaders must create a customer orientation, clear and visible quality values, and high expectations. This concept stresses the personal involvement required of leaders. This involvement extends to areas of public responsibility and corporate citizenship as well as to areas of development of the entire work force. This concept also emphasizes such activities as planning, communications, review of company quality performance, recognition, and serving as a role model.

Continuous Improvement and Learning.   This concept includes both incremental and "breakthrough" improvement activities in every operation, function, and work process in the company. It stresses that improvements may be made through enhancing value to customers; reducing errors, defects, and waste; improving responsiveness and cycle-time performance; improving productivity and effectiveness in the use of all resources; and improving the company's performance and leadership position in fulfilling its public responsibilities and corporate citizenship.

The addition of learning to the core values in recent editions of the criteria refers to the adaptation to change leading to new goals and/or approaches. Learning and improvement need to be embedded in the way the organization operates.

Valuing Employees.   This concept stresses the fact that a company's success depends increasingly on the knowledge, skills, and motivation of its work force. Employee success depends increasingly on having opportunities to learn and to practice new skills. Companies need to invest in the development of the work force through education, training, and opportunities for continued growth. There is an increasing awareness in the United States that overall organization performance depends more and more on work force quality and involvement. Factors that bear upon the safety, health, well-being, and morale of employees need to be part of the company's continuous improvement objectives.

Fast Response.   The value of shortening time cycles is also emphasized. Faster and more flexible response to customers is becoming each year a more critical requirement of business management. Improvements in these areas often require redesigning work processes, eliminating unnecessary work steps, and making better use of technology. Measures of time performance should be among the quality indicators used by leading organizations. Objectives regarding response time, quality, and productivity should be integrated.

Design Quality and Prevention.   Throughout the criteria the importance of prevention-based quality systems are highlighted. Design quality is a primary driver of "downstream" quality. This concept includes fault-tolerant (robust) products and processes. It also includes concept-to-customer times, the entire time for the design, development, production, and delivery to customer of new goods and services.

The concept of continuous improvement and corrective action involving "upstream" interventions is also covered here. This concept stresses that changes should be made as far upstream as possible for the greatest savings. This value also recognizes that major success factors in competition include the design-to-introduction cycle times. To meet the demands on rapidly changing national and international markets, companies need to carry out "concurrent engineering" of activities from basic research to commercialization.

Long-Range Outlook.   This concept stresses the need to take a long-range view of the organization's future and consider all stakeholders: customers, employees, stockholders, and the community. Planning must take into account new technologies, the changing needs of customers, and the changing customer mix, new regulatory requirements, community/societal expectations, and competitors' strategies. Emphasis is also placed on long-term development of employees and suppliers, and on fulfilling public responsibilities and serving as a corporate citizenship role model.

Management by Fact.   This concept stresses the need to make decisions based on reliable data, information, and analyses. These data need to accurately reflect the needs, wants, expectations, and perceptions of the customers; to give accurate descriptions of the performance of goods and services sold; to reflect clearly the market situation; to portray accurately the offerings, performance levels, and satisfaction levels of competitors' goods and services; to provide clear findings of employee-rated issues; and to accurately portray the cost and financial matters. The role of analysis is stressed. Here, also, emphasis is placed on the role of benchmarking in comparing organizational quality performance with the performance of competitors or best-in-class organizations.

The need for organization-wide performance indicators is also stressed. These indicators are measurable characteristics of goods, services, processes, and company operations. They are used to evaluate, track, and improve performance. They should be clearly linked to show the relationships between strategic goals and all activities of the company.

Partnership Development.   The need to develop both internal and external partnerships to accomplish overall goals is also emphasized in the MBNQA core values. These partnerships may include labor-management relationships; relationships with key suppliers; working agreements with technical colleges, community colleges, and universities; and strategic alliances with other organizations.

Corporate Responsibility and Citizenship.   The core values and concepts also emphasize that the organization's quality system should address corporate responsibility and citizenship. This includes business ethics, protection of public health, public safety, and the environment. The company's day-to-day operations and the entire life cycle of the products sold should be considered as they impact health, safety, and environment. Quality planning should anticipate any adverse impacts from facilities management, production, distribution, transportation, use, and disposal of products.

Corporate responsibility also refers to leadership and support of such areas as education, resource conservation, community services, improving industry and business practices, and sharing of non-proprietary quality-related information, tools, and concepts.

Results Focus.   The Award criteria stress results throughout. They emphasize that performance measurements need to focus on key results. But these results should not be just financial. Results should be guided and balanced by the interests of all stakeholders—customers, employees, stock-holders, suppliers and partners, the public, and the community. Company strategy should explicitly include all stakeholder requirements. The use of a balanced composite of performance measurements offers an effective means to communicate short- and longer-term priorities, to monitor actual performance, and to marshal support for improving results.

The Malcolm Baldrige National Quality Award Criteria.   The core values and concepts described previously are embodied in seven categories:

1.0   Leadership

2.0   Strategic Planning

3.0   Customer and Market Focus

4.0   Information and Analysis

5.0   Human Resource Focus

6.0   Process Management

7.0   Business Results

The dynamic relationships among these seven categories are best described by Figure 14.7, as presented in the 1998 Award Criteria booklet.

Leadership, Strategic Planning, and Customer and Market Focus represent the leadership triad. These categories are placed together to emphasize the importance of a leadership focus on strategy and customers. Human Resource Focus, Process Management, and Business Results represent the results triad. A company's employees and its supplier partners through its key processes accomplish the work of the organization that yields the business results. All company actions point towards business results— a composite of customer, financial, and nonfinancial performance results, including

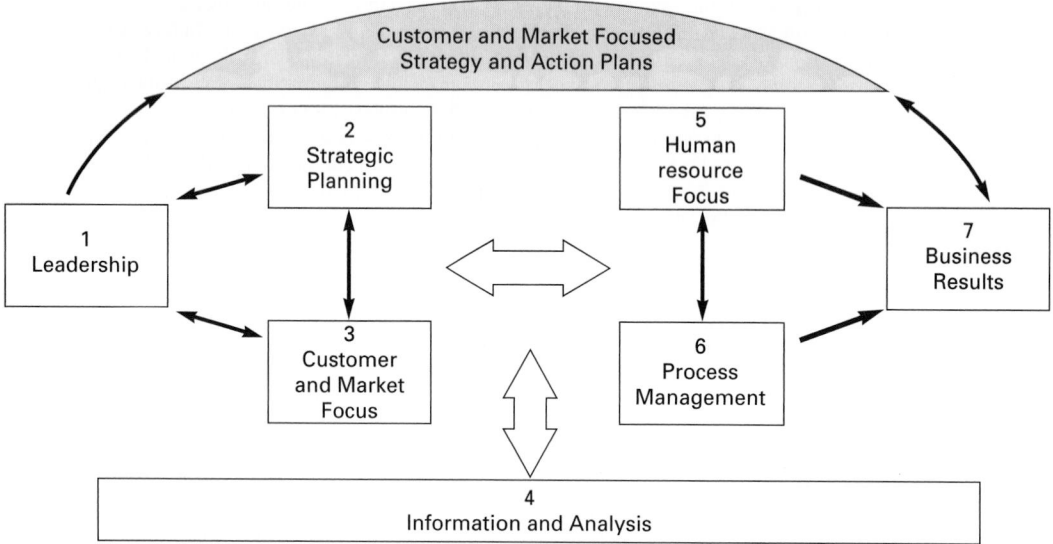

FIGURE 14.7   The Malcolm Baldrige Criteria for Performance Excellence Framework—a systems perspective. (*National Institute of Standards and Technology, 1998, p. 43.*)

human resource results and public responsibility. Information and Analysis is critical to effective management and to a fact-based system for improving company performance and competitiveness. Information and Analysis serve as the foundation for the performance management system.

The seven categories are further subdivided into 20 examination items, each focusing on a major requirement. Each item contains one or more areas to address. There are 29 areas to address. The seven categories, the 20 examination items, and the points for each category and examination item are shown in Table 14.1.

The areas to address give specific instructions as to what information should be contained in the application form. Notes supporting each section give further explanation and clarification. The notes also help the applicant understand where certain data should be reported when there seem to be several possibilities.

An example of an examination item and its four areas to address is provided as Figure 14.8.

The Emphasis on Results in the Malcolm Baldrige National Quality Award.

During the first years of the Malcolm Baldrige National Quality Award, some people felt that too much emphasis was placed on quality systems and too little emphasis was placed on quality results. Conti (1992) compared the strengths and weaknesses of the Deming Application Prize, the European Quality Award, and the Malcolm Baldrige National Quality Award in an incisive paper. Conti's criticisms of systems-based assessments are right on the mark. The proof of the effectiveness of any quality system *must* be in the results produced by the system.

Conti points out the MBNQA's apparent overfocus on systems and underfocus on actual results. In actual fact, the applicants for the Malcolm Baldrige National Quality Award have always emphasized results, sometimes even entering charts and data in inappropriate places in the application form. The examiners also looked for results in almost every area addressed.

However, the language in the application form was not clear in the early years, and it was possible to interpret the application process as only giving 10 percent weight to customer satisfaction results and 10 percent weight to internal results. Some companies, and many reviewers, read the guidelines this way.

The 1998 revision makes it absolutely clear that the focus is on results. Criteria 7, Business Results, is now worth 450 points out of 1000.

TABLE 14.1 Malcolm Baldrige National Quality Award: 1998 Criteria for Performance Excellence. U.S. Department of Commerce, National Institute of Standards and Technology, National Quality Program, Gaithersburg, MD.

1995 Examination Items and Point Values

1.0	Leadership (110 points)	
	1.1	Leadership System (80)
	1.2	Company Responsibility and Citizenship (30)
2.0	Strategic Planning (80 points)	
	2.1	Strategy Deployment Process (40)
	2.2	Company Strategy (40)
3.0	Customer and Market Focus (80 points)	
	3.1	Customer and Market Knowledge (40)
	3.2	Customer Satisfaction and Relationship Enhancement (40)
4.0	Information and Analysis (80 points)	
	4.1	Selection and Use of Information and Data (25)
	4.2	Selection and Use of Comparative Information and Data (15)
	4.3	Analysis and Review of Company Performance (40)
5.0	Human Resource Focus (100 points)	
	5.1	Work Systems (40)
	5.2	Employee Education (30)
	5.3	Employee Well-Being and Satisfaction (30)
6.0	Process Management (100 points)	
	6.1	Management of Product and Service Processes (60)
	6.2	Management of Support Processes (20)
	6.3	Management of Supplier and Partnering Processes (20)
7.0	Business Results (450 points)	
	7.1	Customer Satisfaction Results (125)
	7.2	Financial and Market Results (125)
	7.3	Human Resource Results (50)
	7.4	Supplier and Partner Results (25)
	7.5	Company-Specific Results (125)
Total Points 1000		

Source: National Institute of Standards and Technology, 1998 Criteria for Performance Excellence, *Malcolm Baldrige National Quality Award,* Gaithersburg, MD, 1998.

The actual applications are full of charts, graphs, tables, and other forms of results. The winning companies are well on the way to "management by fact," and it is not surprising that they report their activities in fact-rich documents. The examiners expect this and often refuse to score any examination item highly that doesn't have convincing data to support a statement. One of the most common statements on a scored application is, "Lack of evidence to support claim of...."

Another misconception about the scoring is a belief that the examiners and judges rely wholly on a total score in making their final decisions on applicants. This is not at all the case. The seven category scores are always highly visible to all examiners and judges, and individual category scores are discussed at length. It is highly unlikely that a company scoring poorly in any single category would ever be selected for an award.

The scores, individual categories and total, are mainly used in the early stages of the awards process. High-scoring applications are selected for the consensus review stage. High-scoring applications after consensus scoring are selected for site visits. After site visits, scores are *not* recalculated. The actual findings of the site-visit teams are submitted to the judges, and the judges get further information from the site-visit team leader or members. At this stage of judging, scores have become much less important and are rarely used. The site-visit teams concentrate a great deal of their activity on finding the evidence to support claims in the applications, verifying results, and examining supporting documents. These visits focus very much on results, not just approach or

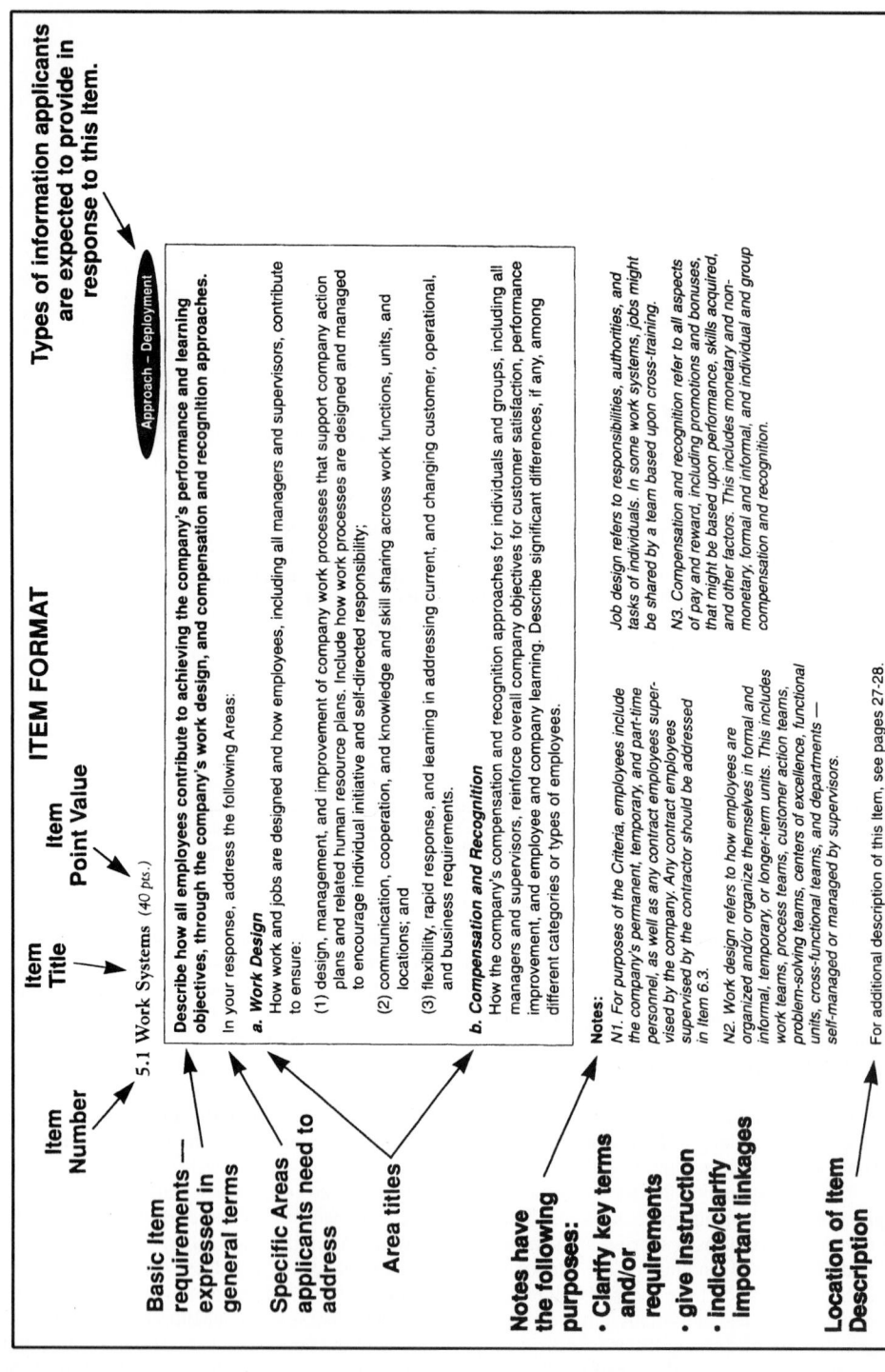

FIGURE 14.8 Example of areas to address. *(National Institute of Standards and Technology 1998, p. 37.)*

deployment. The focus is on whether the company's approach is *working* and is working across the company and across all functions. Examiners verify data, interview employees, and review actual operations and facilities.

During the site visit, examiners look for measurements of both internal and external quality. They look for measures of suppliers' quality levels. They interview employees and ascertain the results of the training, teamwork, and quality improvement processes. They look at customer satisfaction data, competitive evaluations, and benchmarks. They look for evidence of actual, sustained improvement and world-class performance.

Administration of the Malcolm Baldrige National Quality Award. The Malcolm Baldrige National Quality Award is administered through a complex set of processes under the management of the U.S. Department of Commerce, Technology Administration, National Institute of Standards and Technology. Administration for the Award is provided by the American Society for Quality. Most of the actual work of reviewing and scoring applications, site visits, judging, and developing the management processes is done by several hundred volunteers from U.S. companies, universities, government, consultants, and other organizations. These volunteers perform several key roles.

The Board of Overseers is a small group of people who have established preeminence in quality management. For example, the recent Chair of the Board of Overseers has been Robert Galvin, the chairman of the Executive Committee of Motorola. Motorola was one of the first winning companies of the MBNQA. Quality management experts Armand V. Feigenbaum, William Golomski, and Joseph M. Juran have all served as members of the Board of Overseers.

The overseers are concerned mostly with questions of process. They ensure that proper processes for managing the MBNQA are in place, are working, and are continuously improved. They review recommendations by the judges as to process improvements, but the overseers are not involved in the actual evaluation and judging of the applicants.

Issues of concern for the overseers include number of awards, award categories, changes to the Act, and technology sharing and transfer based on lessons learned.

The Board of Examiners consists of over 200 persons selected according to expertise, experience, and peer recognition. They do not represent companies or organizations, but serve as volunteers for the common good. All members of the Board of Examiners receive 3 days of rigorous training using case studies, scoring exercises, and team-building sessions. They become a powerful network for quality improvement throughout the United States.

The Board of Examiners consists of three distinct groups: Judges, Senior Examiners, and Examiners. There are nine Judges. The Judges oversee the entire process of administering the Award, help select examiners, review the scored applications, select the organizations to receive site visits, and review the results of the site visits. They then decide which, if any, organizations to recommend for the Malcolm Baldrige National Quality Award.

The final decision for the awards is made by the Secretary of Commerce after further background evaluations of the recommended organizations. These further evaluations are intended solely to determine if an organization is facing environmental charges, Justice Department action, or other problems. If these concerns are substantial, the Secretary may remove the organization from the recommended list. The Secretary may not add any organization to the list and has no other influence on the awards process.

The Judges are involved in oversight at every stage of the MBNQA process but only get involved in the review of actual applications after many hours of work by the examiners. These evaluations, screenings, and site visits provide the foundation on which the award process is built.

There are approximately 20 to 30 Senior Examiners, and they play a crucial role. They are selected for their experience and expertise. Many have been examiners for several years or directly involved in winning organizations' quality management. They score applications and manage the consensus review process.

There are almost 200 Examiners. The Examiners score all the applications, perform site visits with the Senior Examiners, and provide input each year on how to improve the application guidelines, the scoring process, and the entire awards process.

The MBNQA process follows several, carefully defined steps. The first is the annual improvement of the criteria, the guidelines, and the entire awards process. The next step is the completion of the eligibility determination form by the applicant company. Applicants must have their eligibility approved prior to applying for the award. Each applicant then completes and files the application. The award applications then go through four stages of review:

Stage 1:   Independent review by at least five members of the Board of Examiners

Stage 2:   Consensus review and evaluation for applications that score well in Stage 1

Stage 3:   Site visits to applicants that score well in Stage 2

Stage 4:   Judge's review and recommendations

The scoring system used by the Board of Examiners is described in the application guidelines. It is based on three evaluation dimensions: (1) approach, (2) deployment, and (3) results. All examination items require applicants to furnish information relating to one or more of these dimensions. The Scoring Guidelines are reproduced as Figure 14.9.

Each year, after the recommendations for the winning companies are forwarded to the Secretary of Commerce, the Judges review the entire MBNQA process. Feedback is solicited from all members of the Board of Examiners, applicant companies, the Administrator of the award process (ASQ), the staff of the National Quality Award Office, and other interested parties. The suggestions for improvement are carefully considered, and each year a number of changes are made to the award criteria, the application guidelines, and the award process. This constant improvement is one of the greatest strengths of the Malcolm Baldrige National Quality Award.

THE EUROPEAN QUALITY AWARD

Conti (1993) presents a comprehensive view of a total quality system in *Building Total Quality*. In this outstanding text Conti uses the European Quality Award as the fundamental model for total quality and gives many expansions of this model tied to business performance.

Conti's view of a total quality system is well worth understanding. He breaks the system down into five first-level subdivisions: the role of management, corporate values/culture, infrastructure, involvement/use/role of human resources, and the adequacy/use of technical resources. These are shown in Figure 14.10.

Conti also suggests a further deployment of this model from the first-level criteria to the second-level criteria. He admits that the choice of criteria is more subjective at this level and based on experience. He suggests that when used by different companies and/or different market sectors it is essential to assign appropriate weights to the different criteria. In fact, some of the assigned weights might be zero for certain companies. Conti's second-level criteria are shown in Figure 14.11.

Conti was a leading contributor in developing the European Quality Award, so it comes as no surprise that his model is closely connected to the underlying model of this award.

The European Quality Award (EQA) shares many concepts and criteria elements with the Malcolm Baldrige National Quality Award, but the two awards differ in some important ways. Conti illustrates the differences clearly and openly discusses strengths and weaknesses of both approaches as well as compares them both with the Deming Application Prize.

The logical model of the EQA is quite clear (Figure 14.12). The first element is leadership which drives people management, policy and strategy, and resources. These, in turn, drive all processes which drive people satisfaction, customer satisfaction, and impact of society. These three drive business results.

One of the major differences between the Malcolm Baldrige National Quality Award and the European Quality Award is the emphasis the EQA puts on self-assessment. The EQA makes the principle of self-assessment an entry requirement for companies applying for the award. Conti (1997) stresses this in a more recent text, *Organizational Self-Assessment.*

A second difference between the EQA and the MBNQA, which Conti feels is a weakness in the EQA, is the apparent absence of the fundamental internal results category. Some people argue that

RESULTS

SCORE	RESULTS
0%	■ no results or poor results in areas reported
10% to 30%	■ early stages of developing trends; some improvements *and/or* early good performance levels in a few areas ■ results not reported for many to most areas of importance to the applicant's key business requirements
40% to 60%	■ improvement trends *and/or* good performance levels reported for many to most areas of importance to the applicant's key business requirements ■ no pattern of adverse trends *and/or* poor performance levels in areas of importance to the applicant's key business requirements ■ some trends *and/or* current performance levels — evaluated against relevant comparisons *and/or* benchmarks — show areas of strength *and/or* good to very good relative performance levels
70% to 90%	■ current performance is good to excellent in most areas of importance to the applicant's key business requirements ■ most improvement trends *and/or* performance levels are sustained ■ many to most trends *and/or* current performance levels — evaluated against relevant comparisons *and/or* benchmarks — show areas of leadership and very good relative performance levels
100%	■ current performance is excellent in most areas of importance to the applicant's key business requirements ■ excellent improvement trends *and/or* sustained excellent performance levels in most areas ■ strong evidence of industry and benchmark leadership demonstrated in many areas

APPROACH/DEPLOYMENT

SCORE	APPROACH/DEPLOYMENT
0%	■ no systematic approach evident; anecdotal information
10% to 30%	■ beginning of a systematic approach to the primary purposes of the Item ■ early stages of a transition from reacting to problems to a general improvement orientation ■ major gaps exist in deployment that would inhibit progress in achieving the primary purposes of the Item
40% to 60%	■ a sound, systematic approach, responsive to the primary purposes of the Item ■ a fact-based improvement process in place in key areas; more emphasis is placed on improvement than on reaction to problems ■ no major gaps in deployment, though some areas or work units may be in very early stages of deployment
70% to 90%	■ a sound, systematic approach, responsive to the overall purposes of the Item ■ a fact-based improvement process and organizational learning/sharing are key management tools; clear evidence of refinement and improved integration as a result of improvement cycles and analysis ■ approach is well-deployed, with no major gaps; deployment may vary in some areas or work units
100%	■ a sound, systematic approach, fully responsive to all the requirements of the Item ■ a very strong, fact-based improvement process and extensive organizational learning/sharing are key management tools; strong refinement and integration — backed by excellent analysis ■ approach is fully deployed without any significant weaknesses or gaps in any areas or work units

FIGURE 14.9 Baldrige scoring guidelines. (*National Institute of Standards and Technology 1998, p. 35.*)

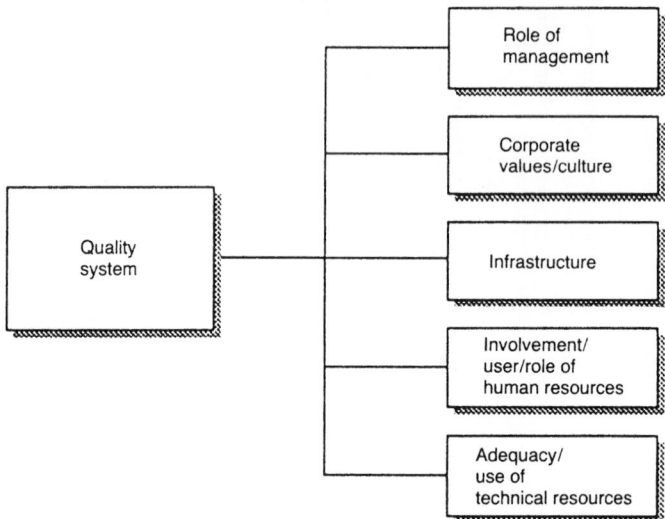

FIGURE 14.10   First-level deployment of a quality system. (*Conti 1993, p. 112. Used with permission.*)

internal results are implicit in other categories, but Conti feels that "it is inadmissible that such an important category should be absent or implicit in some other category." (Conti 1993, p. 289.)

Another difference between the MBNQA and EQA is the way the awards are administered. The MBNQA is competitive; it is given to a maximum of two companies in each of three categories: manufacturing, service, and small business. So far this has not been a problem, since the maximum number of companies has never been reached. The EQA is essentially noncompetitive, every company that reaches the pass mark receives a "prize." The Award is given to the best prize winner. In some ways this makes the EQA even more competitive, since companies have a great desire to win the award not just a "prize."

THE DEMING APPLICATION PRIZE

Another major contribution to the development of total quality has been the Union of Japanese Scientists and Engineers' Deming Application Prize. In his definitive book, *Companywide Quality Control,* Kondo describes the creation and evolution of the Deming Prize (Kondo 1995, pp. 37–42):

> In recognition of Deming's friendship and contributions to Japan, the Deming Prize was established in 1951 at JUSE's suggestion to encourage the development of QC in Japan. The prizes were originally funded with Deming's generous gift of the royalties from transcripts of his eight-day QC course lectures and the Japanese translation of his book, *Some Theory of Sampling,* along with other donations.

> There are two types of Deming Prize—the Deming Prize for individuals and the Deming Application Prize for companies and divisions (Kondo 1995, p. 38).

> Deming Application Prizes are awarded to companies or operating divisions that have achieved outstanding results through the skillful application of CWQC (companywide quality control) based on

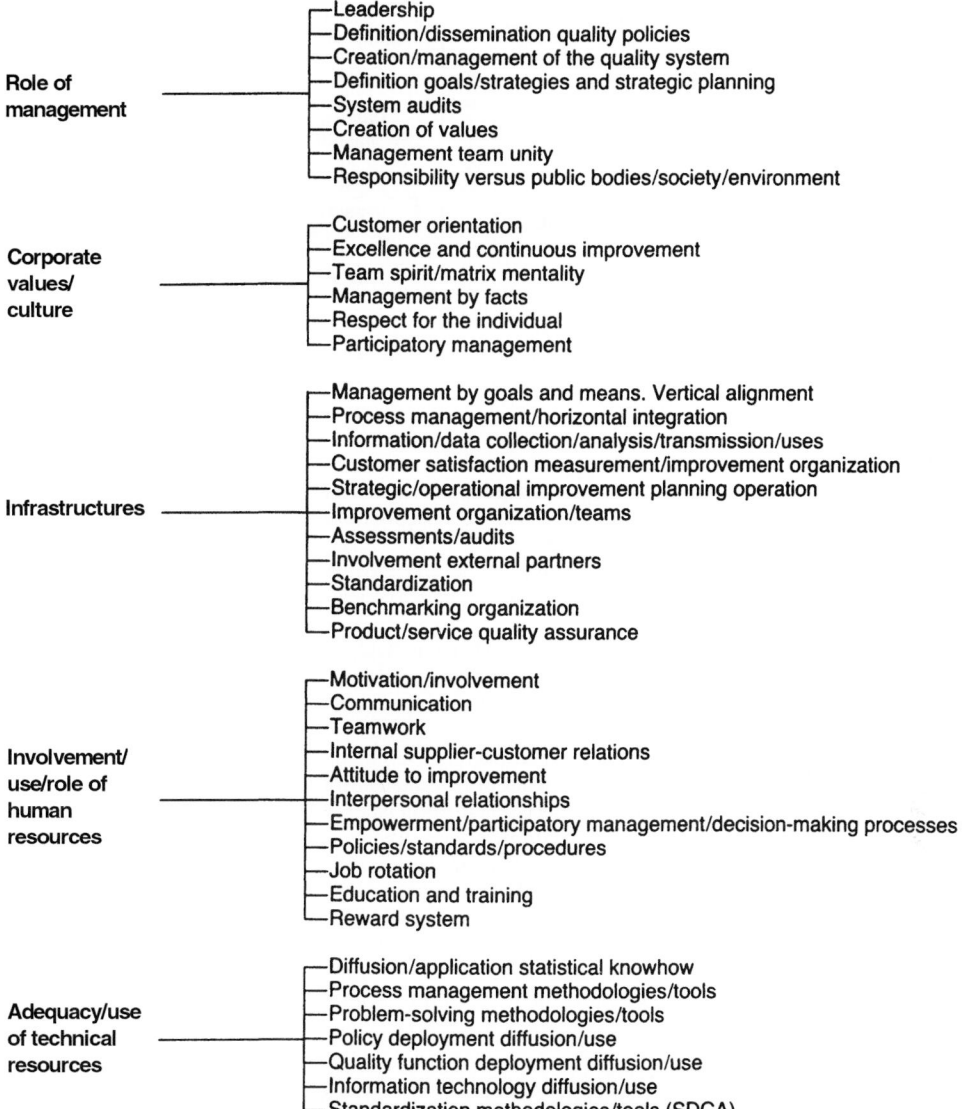

Role of management
- Leadership
- Definition/dissemination quality policies
- Creation/management of the quality system
- Definition goals/strategies and strategic planning
- System audits
- Creation of values
- Management team unity
- Responsibility versus public bodies/society/environment

Corporate values/ culture
- Customer orientation
- Excellence and continuous improvement
- Team spirit/matrix mentality
- Management by facts
- Respect for the individual
- Participatory management

Infrastructures
- Management by goals and means. Vertical alignment
- Process management/horizontal integration
- Information/data collection/analysis/transmission/uses
- Customer satisfaction measurement/improvement organization
- Strategic/operational improvement planning operation
- Improvement organization/teams
- Assessments/audits
- Involvement external partners
- Standardization
- Benchmarking organization
- Product/service quality assurance

Involvement/ use/role of human resources
- Motivation/involvement
- Communication
- Teamwork
- Internal supplier-customer relations
- Attitude to improvement
- Interpersonal relationships
- Empowerment/participatory management/decision-making processes
- Policies/standards/procedures
- Job rotation
- Education and training
- Reward system

Adequacy/use of technical resources
- Diffusion/application statistical knowhow
- Process management methodologies/tools
- Problem-solving methodologies/tools
- Policy deployment diffusion/use
- Quality function deployment diffusion/use
- Information technology diffusion/use
- Standardization methodologies/tools (SDCA)

FIGURE 14.11 Second-level deployment of a quality system. (*Conti 1993, p. 113. Used with permission.*)

statistical methods and are considered likely to continue to do so in the future, where CWQC is defined as "the activity of economically designing, producing, and supplying products and services of the quality demanded by customers, based on customer-focused principles and with full consideration of the public welfare."

In over the 40-year existence of the Deming Application Prize, there have been many modifications and improvements to the prize criteria and the administration of the prize. The Deming Application Prize is not competitive; every company whose application is accepted may win. The

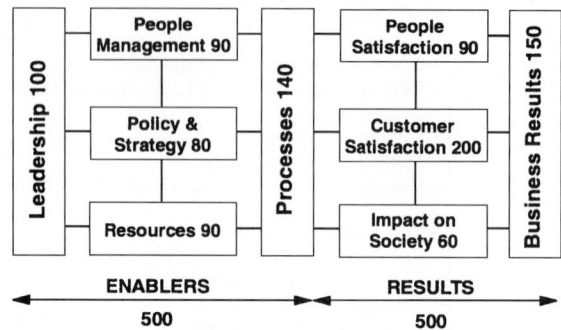

FIGURE 14.12   European Quality Award: the scoring process.
(*Leadership for the Quality Century, Juran Institute, Inc., Wilton, CT.
Used with permission.*)

examiners are selected by JUSE from a small group of scholars and other distinguished experts associated with not-for-profit organizations, who share a deeply rooted and basically uniform approach to quality management (Conti 1993, p. 286).

Conti gives a simple chart illustrating the first-level deployment and an example of second-level deployment for the assessment model of the Deming Application Prize. This model is given as Figure 14.13.

There are several differences between the Malcolm Baldrige National Quality Award and the Deming Application Prize. There is no limit to the number of companies that may receive a Deming Application Prize in any one year. There is a stronger emphasis on the use of statistical methods than in the Baldrige Award. The company decides itself when it is to receive an objective assessment of whether its activities have reached the level capable of passing the Deming Application Prize examination. Usually the company engages a team of consultants from JUSE to provide on-going consulting support during the 4 or 5 years preceding the official examination.

Kondo points out that one of the main differences between the Deming Application Prize and the Malcolm Baldrige National Quality Award is that the checklist of items applicants must satisfy to win a Baldrige Award is far more detailed, extending to 23 pages. Due to interest from around the world, the Deming Prize Committee created new regulations in 1984 making it possible for countries outside Japan to apply. In 1989 an American electric utility, Florida Power & Light, became the first overseas company to win. In 1991 Philips Taiwan became the second winner of the Deming Application Prize for Overseas Companies, and in 1994 the AT&T Power Systems division became the third.

COMPARISON OF NATIONAL/INTERNATIONAL QUALITY AWARDS AND INTERNATIONAL STANDARDS

Over the past few years there have been numerous attempts to compare the ISO 9000 series of standards with the Malcolm Baldrige National Quality Award, the Deming Application Prize, and the European Quality Award. Conti (1993, p. 283) provides one comparison (Figure 14.14). The ISO 9000 series of standards is covered in depth in Section 11.

In Conti's chart (Figure 14.14), the ISO 9000 series of standards provides a way of assessment and certification for the excellence of a quality system that is at the far left of the x axis. That is, the ISO system focuses on products and is the least comprehensive of the systems. It is also at the bottom of the y axis, indicating no assessment of excellence of results but just of system.

Conti sees the MBNQA, the Deming Application Prize, and the EQA as roughly equal in their focus on comprehensiveness of the quality system but the EQA as somewhat more focused on

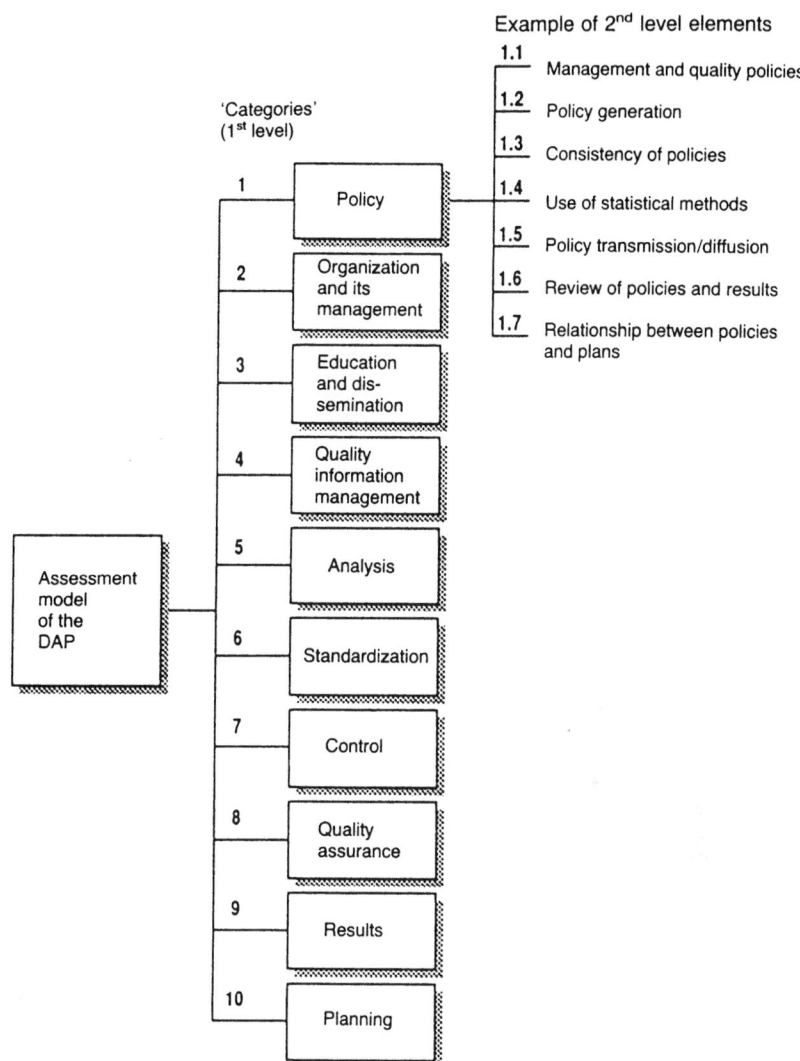

Example of 2nd level elements

1.1 — Management and quality policies

1.2 — Policy generation

1.3 — Consistency of policies

1.4 — Use of statistical methods

1.5 — Policy transmission/diffusion

1.6 — Review of policies and results

1.7 — Relationship between policies and plans

'Categories'
(1st level)

1 — Policy

2 — Organization and its management

3 — Education and dis- semination

4 — Quality information management

5 — Analysis

6 — Standardization

7 — Control

8 — Quality assurance

9 — Results

10 — Planning

Assessment model of the DAP

FIGURE 14.13 The Deming Application Prize. (*Conti 1993, p. 281. Used with permission.*)

results. The MBNQA has changed since Conti's comparisons, and it is now more focused on results, perhaps equal to the EQA.

RE-ENGINEERING, SIX SIGMA, AND OTHER EXTENSIONS OF TOTAL QUALITY MANAGEMENT

In the past few years there have been many redefinitions of total quality management. TQM has become an umbrella term for many different collections of concepts, methods, and tools. As new con- cepts are created, they are often added as extensions to the basic collection. Sometimes the creators

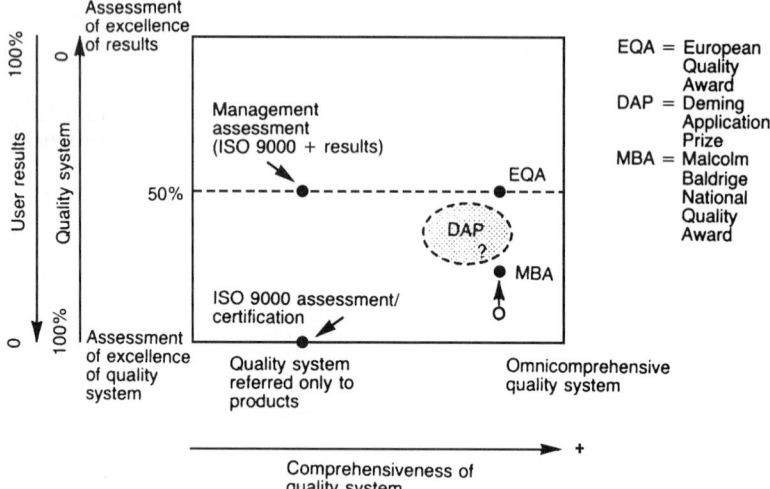

FIGURE 14.14 Comparison of awards, assessments, and ISO. (*Conti 1993, p. 283. Used with permission.*)

of these ideas and tools attempt to differentiate their ideas from TQM and energetically stake a separate place for their efforts. But most of the leading companies continuously integrate the new methods with the older successful methods and discard what is not working along the way.

A few years ago a great effort was made to stake a claim that re-engineering was somehow different from other methods considered part of TQM. Some companies actually created new departments of re-engineering separate from the quality departments or continuous improvement departments. Soon these efforts were merged. In Section 6, Process Management, the history of business process management, business process quality improvement, and re-engineering are covered thoroughly.

More recently there has been an emphasis on "six sigma" and "black belt" quality training. The concept of six sigma was introduced in the 1980s by Motorola in its efforts to reduce the defects in manufactured products to only a few parts per million. Later, Motorola extended the six sigma concept to business processes and service operations. Motorola allowed for a process average drift by as much as 1.5 standard deviations, so their term of six sigma actually sets the targets at 4.5 standard deviations—or 3.4 parts/million.

As other companies have taken ideas from Motorola and other leading companies and added their own variations, six sigma has come to be "a programme aimed at the near-elimination of defects from every product, process and transaction" (Tomkins 1997, p. 22). Six sigma has thus become a disciplined, quantitative approach for improving operations in all types of industries and business functions. The six sigma initiatives may sound quite familiar to many leading companies with successful total quality management systems but often sound quite new to those companies who have just dabbled in quality management in the past.

The basic steps of the six sigma process are quite similar to the quality improvement processes and quality control processes introduced in Section 5, Quality Improvement, and Section 4, Quality Control. The five basic steps are usually explained as define, measure, analyze, improve, and control. The implementation stresses leadership at the highest levels of the company. For many companies this has been the CEO, such as Jack Welch at General Electric, Bob Galvin at Motorola, and Larry Bossidy at Allied Signal. The implementation is then cascaded throughout every level of management, and clear responsibilities are understood (Hoerl 1998, p. 36).

The main focus of six sigma, like many other quality initiatives, is on cost and waste reduction, yield improvements, capacity improvements, and cycle-time reductions. Heavy emphasis is put on satisfying customer needs. Organizations try to estimate the financial impact of each operation. These companies also establish clear performance metrics for each improvement in costs, quality, yields, and capacity improvements. Financial figures are absolutely required. The projects undertaken are usually substantial with improvements commonly in the $50,000 to $100,000 range.

Another difference in the six sigma initiatives and many total quality management programs is the assignment of full-time staff. The team leaders and facilitators (often called black belts and master black belts) are chosen carefully and work 50 to 100 percent of their time on the improvement projects. The training for these people is also extensive, usually 4 or 5 weeks of intensive, highly quantitative training (Hoerl 1998, p. 36). Some companies have actually implemented training programs lasting up to 6 months for their new black belts.

Over the past few years many other concepts, methods, and tools have become part of the overall total quality management philosophy. Many of these concepts have become part of this handbook. For thousands of years organizations have tried to learn the secrets of others' successes, but in the past few years benchmarking has become a frequently used (and sometimes misused) management tool. Benchmarking is covered in Section 12. Quality assurance systems have been standardized and third-party assessments have become commonplace. These developments are covered in Section 11. Organizations have extended their quality systems into their suppliers, operations creating true customer/supplier relationships. These extensions are covered in Section 21. Customer service, customer satisfaction, and customer loyalty have become critical elements in all quality systems. Many of these new developments are discussed in Section 25. New methods and tools have been developed for extending quality management to information systems. These are covered in Section 34.

REFERENCES

Byrne, John A. (1996). "Strategic Planning." *Business Week,* New York, Aug. 26.

Conti, Tito (1993). *Building Total Quality,* Chapman and Hall, London.

Conti, Tito (1992). "A Critical Review of the Current Approach to Quality Awards." *Proceedings of the EOQ Conference,* Brussels, June, pp. 130–139.

Deming, W. Edwards (1993). *The New Economics For Industry, Government, Education,* Massachusetts Institute of Technology, Center for Advanced Engineering Study, Boston.

Deming, W. Edwards (1982). *Out of the Crisis,* Massachusetts Institute of Technology, Center for Advanced Engineering Study, Boston.

Endres, Al C., and Godfrey, A. Blanton (1994). "The Evolution of Quality Management Within Telecommunications." *IEEE Communications Society Journal,* October.

Fromm, Bill, and Schlesinger, Len (1993). *The Real Heroes of Business and Not a CEO Among Them,* Currency Doubleday, New York.

Garvin, David A. (1991). "How the Baldrige Award Really Works." *Harvard Business Review,* Nov.–Dec., pp. 80–93.

Godfrey, A. Blanton (1998). "Hidden Costs to Society." *Quality Digest,* vol. 18, no. 6, June, p. 18.

Godfrey, A. Blanton (1997). "A Short History of Managing for Quality in Health Care." *The Handbook for Managing Change in Health Care,* Chip Caldwell, ed., ASQ Quality Press, Milwaukee, pp. 1–9.

Godfrey, A. Blanton (1995). "The Malcolm Baldrige National Quality Award: Seven Years of Progress—7000 Lessons Learned." *Best on Quality,* vol. 5, Quality Press, Milwaukee, pp. 27–68.

Godfrey, A. Blanton (1992). *Testimony on the Malcolm Baldrige National Quality Award before the Sub-Committee on Technology and Competitiveness of the Committee on Science, Space and Technology.* U.S. House of Representatives, February 5, U.S. Government Printing Office.

Godfrey, A. Blanton (1992). "Malcolm Baldrige National Quality Award—The Spark of the Quality Revolution," *Electronics Buyers News,* January.

Godfrey, A. Blanton, and Myers, Dale H. (1994). "Self-Assessment Using the Malcolm Baldrige National Quality Award." *Proceedings of the First European Forum on Quality Self-Assessment,* March 3–4, Milano, Italy, pp. 67–77.

Hamel, Gary, and Prahalad, C. K. (1994). *Competing for the Future.* Harvard Business School Press, Boston.

Hammer, Michael (1996). *Beyond Reengineering,* Harper Business Division of Harper Collins Publishers, New York.

Hammer, Michael, and Champy, James (1993). *Reengineering the Corporation,* Harper Collins, Inc., New York.

Hoerl, Roger W. (1998). "Six Sigma and the Future of the Quality Profession." *Quality Progress,* vol. 31, no. 6, June, pp. 35–42.

Juran, J. M. (1995). *Managerial Breakthrough, 30th Anniversary Edition,* McGraw-Hill, Inc., New York.

Juran, J. M. (1992). *Juran on Quality By Design,* Free Press, New York.

Juran, J. M. (1989). *Juran on Leadership for Quality—An Executive Handbook,* Free Press, New York.

Juran, J. M. (1964). *Managerial Breakthrough,* McGraw-Hill, Inc., New York.

Juran, J. M., and A. Blanton Godfrey (1990). "Worker Participation—Developments in the USA." *Proceedings of the International Congress of Quality Control Circles,* Tokyo, November.

Juran, J. M., and Godfrey, A. Blanton (1990). "Total Quality Management (TQM)—Status in the U.S." *Proceedings on the Senior Management Conference on TQC,* JUSE, Tokyo, November.

Juran, Joseph M., and Gryna, Frank M. (1993). *Quality Planning and Analysis,* 3d ed., McGraw-Hill, New York.

Juran, Joseph M., and Gryna, Frank M., eds. (1988). *Juran's Quality Control Handbook,* 4th ed., McGraw-Hill, New York.

Kano, Noriaki, and Koura, Kozo (1991). "Development of Quality Control Seen Through Companies Awarded the Deming Prize." *Reports of Statistical Application Research,* Japanese Union of Scientists and Engineers, vol. 37, nos. 1–2, 1990–1991, pp. 79–105.

Kondo, Yoshio (1995). *Companywide Quality Control: Its Background and Development,* 3A Corporation, Tokyo.

Kondo, Yoshio, Kume, Hitoshi, and Schimizu, Shoichi (1995). "The Deming Prize." in *The Best on Quality: Targets, Improvement, Systems,* Quality Press, Milwaukee, vol. 5, pp. 3–19.

National Institute of Standards and Technology (1998). *1998 Criteria for Performance Excellence,* Malcolm Baldrige National Quality Award, Gaithersburg, MD.

National Institute of Standards and Technology (1992). *1992 Criteria for Performance Excellence.* Malcolm Baldrige National Quality Award, Gaithersburg, MD.

One Hundredth Congress of the United States of America (1987). "The Malcolm Baldrige Quality Improvement Act of 1987." *H.R. 812, Federal Register,* January 6.

Peppers, Don, and Rogers, Martha (1993). *The One to One Future: Building Relationships One Customer at a Time,* Currency Doubleday, New York.

Peterson, Donald (1992). "How Donald Peterson Turned Ford Around," Interview in *Boardroom Reports,* New York, June 15.

Peterson, Wayne (1992). *Testimony on the Malcolm Baldrige National Quality Award before the Sub-Committee on Technology and Competitiveness of the Committee on Science, Space and Technology.* U.S. House of Representatives, U.S. Government Printing Office, Feb. 5.

Port, Otis (1998). "Quality Claims Its Own Bull Market." *Business Week,* March 16, p. 113.

Reichheld, Frederick F. (1996). *The Loyalty Effect,* Harvard Business School Press, Boston.

Reimann, Curt W. (1992a). *Testimony on the Malcolm Baldrige National Quality Award before the Sub-Committee on Technology and Competitiveness of the Committee on Science, Space and Technology,* U.S. House of Representatives, U.S. Government Printing Office, Feb. 5.

Reimann, Curt W. (1992b). "The First Five Years of the Malcolm Baldrige National Quality Award." *Proceedings of IMPRO92,* Juran Institute, Inc., Wilton, CT, November 11–13.

Sörqvist, Lars (1998). "Poor Quality Costing." Doctoral Thesis No. 23, Royal Institute of Technology, Stockholm.

Starr, Martin K., ed. (1988). *Global Competitiveness—Getting the U.S. Back on Track,* W.W. Norton, New York.

Sugiura, Hideo (1992). "Productivity in the Work Place: The Honda Story." International Productivity Congress, Asian Productivity Organization, Tokyo.

The TQM Committee (1997a) "A Manifesto of TQM (1)—Quest for a Respectable Organization Presence." *Societas Qualitas,* vol. 10, no. 6, Jan./Feb.

The TQM Committee (1997b) "A Manifesto of TQM (2)—Quest for a Respectable Organization Presence." *Societas Qualitas,* vol. 11, no. 1, Mar./Apr.

The TQM Committee (1997*c*) "A Manifesto of TQM (3)—Quest for a Respectable Organization Presence." *Societas Qualitas,* vol. 11, no. 2, May/June.

The TQM Committee (1997*d*) "A Manifesto of TQM (4)—Quest for a Respectable Organization Presence." *Societas Qualitas,* vol. 11, no. 3, July/Aug.

Tomkins, Richard (1997). "GE Beats Expected 13% Rise." *Financial Times,* October 10, p. 22.

U.S. Department of Commerce (1998). *1998 Criteria for Performance Excellence.* National Institute of Standards and Technology, National Quality Program, Gaithersburg, MD.

Wadsworth, H. M., Stephens, K. S., and Godfrey, A. B. (1986). *Modern Methods for Quality Control and Improvement,* John Wiley and Sons, New York.

Waltari, Mika (1949). *The Egyptian,* Werner Söderström Osakeyhtiöm, Helsinki.

Yasuda, Yuzo (1991). *40 Years, 20 Million Ideas: The Toyota Suggestion System,* translated by Fredrich Czupryna, Productivity Press, Cambridge, MA.

SECTION 15

HUMAN RESOURCES AND QUALITY[1]

W. R. Garwood
Gary L. Hallen

---

[1] In the Fourth Edition, material for the section on human resources was supplied by Edward M. Baker.

HUMAN RESOURCES AND CUSTOMER SATISFACTION

Purpose of this Section: Definition of Human Resources.   The purpose of this section is to present concepts, structures, methods, and tools which have helped successful organizations manage human resources effectively in directing their efforts toward the pursuit of high-quality products (including services).

"Human resources" will be used in this section to denote the collection of people and all associated networks and structures within which they work together to make a collective contribution to quality. The term includes all human relationships throughout an organization. (We should note that human resources is commonly used as a name of what was once called the "personnel department." The term is also sometimes used to denote the focus of that department's work. These meanings are not used in this section.)

Human resources are differentiated from hard assets—money, machinery, etc.—by several characteristics, including a warm spirit which can be a multiplier for continuous improvement as people learn to work effectively.

Human Resources, Total Quality Management (TQM), and Business Success.
There is plenty of evidence that business success and quality go hand in hand, and that TQM is a sound business strategy for achieving them both. The General Accounting Office, official auditor of operations for the United States Government, performed an evaluation of 20 companies who were high-scoring applicants for the Malcolm Baldrige National Quality Award (U.S. General Accounting Office 1991). High-scoring performance was equated with successful introduction of TQM in the respective companies. The study showed that for a substantial number of the companies under study, increased quality (on a number of dimensions) and improved financial performance were associated with the introduction of TQM.

Major TQM elements (as embodied in the criteria of the Malcolm Baldrige National Quality Award and other major state, national, and regional quality awards around the world) which relate directly to human resources, and the Baldrige points associated with them are

4.1	Human resource planning and evaluation	20 of 1000
4.2	High-performance work systems	45 of 1000
4.3	Employee education, training, and development	50 of 1000
4.4	Employee well-being and satisfaction	25 of 1000
6.3	Human resource results	35 of 1000

Every other item in the Baldrige scheme relies heavily on well-managed human resources for its achievement, most notably the items under category 7.0, Customer focus and satisfaction. Customer satisfaction depends in large part on the customer's experience with organization contacts, service, and delivery.

Employee Involvement, Empowerment, and TQM.   Total Quality Management results in primary focus on improvement efforts. Employee involvement is a part of TQM. A survey of Fortune 1000 companies revealed that 80 percent of these companies see employee involvement as part of TQM, and only 20 percent see TQM as part of their employee involvement initiatives (Lawler et al. 1995).

Employee empowerment is an advanced form of employee involvement. Empowerment is a condition in which the employee has the knowledge, skills, authority, and desire to decide and act within prescribed limits. The employee takes responsibility for the consequences of the actions and for contribution to the success of the enterprise.

In an empowered organization, employees take action to respond to the needs and opportunities they face every day regarding: customer satisfaction; safe operations; quality and value of products and services; environmental protection; business results; and continuous improvement of processes, products, and people. The full potential of employee empowerment is realized in the empowered organization, when employees: align their goals with appropriate higher organization purpose; have the authority and opportunity to maximize their contribution; are capable of taking appropriate action; are committed to the organization's purpose; and have the means to achieve it. Thus, empowerment may be shown by the equation:

$$\text{Empowerment} = \text{alignment} \times \text{authority} \times \text{capability} \times \text{commitment}$$

Alignment.   For employees to be aligned with the organization's higher purpose, they must:

- Know the needs of customers and other stakeholders
- Know, concur in, and be prepared to contribute effort to organization strategies, goals, objectives, and plans

Authority and Opportunity.   For employees to have the authority and opportunity to maximize their contribution, the organization must so arrange affairs that:

- Individual authority, responsibility, and capability are consistent
- Barriers to successful exercise of authority have been removed
- The necessary tools and support are in place

The Ritz-Carlton Hotel Company authorizes every employee to spend up to $2000 on the spot to resolve a customer problem so as to satisfy the customer. (This example of empowerment is most frequently invoked by employees working at the front desk). At Walt Disney World and other Disney operations, cast members (Disney-talk for "employees") are authorized to replace lost tickets, spilled food, and damaged souvenirs, even if the damage was caused by the guest. In the Ritz-Carlton and Disney organizations, such empowerment is seen as a direct means to satisfy a customer and to strengthen the identification of the employee with the vision, mission, and values of the company. In this sense, when employees have the authority and ability to solve direct customer problems of this sort, they acquire a sense of ownership of the organization.

Capability.   Without capability, it can prove dangerous for employees to take some actions. Experiences at Eastman Chemical for example, have shown that peer feedback from fellow employees who have not been properly trained in giving feedback can be construed as harsh and not constructive. The organization objectives cannot be fulfilled if these employees do not know what actions to take or how to take them. Therefore, employees must have the capability to achieve appropriate goals. Empowered employees know how to do what needs doing, and have the skills and information to do it.

Training is a significant means of developing employee capability. At Ritz-Carlton, for example, all new employees undergo 48 hours of orientation training before they visit their workplace to begin work with customers face-to-face. After 21 days on the job, there is a 4-hour follow-up orientation, during which the first 21 days of experience are reviewed and discussed in light of the company's vision: "Ladies and gentlemen serving ladies and gentlemen."

Commitment.   Commitment is a state of mind which is in evidence when the employee assumes responsibility for creating success, and takes initiative to achieve that success. For example, a Federal Express package containing house-closing documents from Florida is due in Albany, New York by Saturday at 10 a.m. for signing and return for the Monday closing. The package is misplaced at the local Federal Express office and is not located until midafternoon. The driver, called from his home to make the delivery, makes it, then, on his own initiative, waits for the document to be signed and repackaged, and carries the returned document to the Federal Express office for its return journey in time for the closing (Hoogstoel 1996).

The organization must earn the commitment of employees by continuously demonstrating that the employees are valued members of the organization, and by appropriately recognizing and rewarding them.

Human Resources and Healthy Human Relations.   The immediate objective of human resource management is to achieve healthy human relations throughout the organization. Healthy human relations are relations which are open, positive, and efficient. Healthy human relations contribute to the employees' sense of ownership, to good business results, and even to what Deming (1986) called "joy in the workplace." In general, it is easier to achieve other organizational objectives in an environment of healthy human relations.

HEALTHY HUMAN RELATIONS: THE BUILDING BLOCKS

It took many organizations years to understand that a successful total quality system required a special culture within the organization to sustain continuous improvements. The literature of the 1970s and 1980s described the culture in which total quality systems flourished and the cultures where organizations were unable to sustain a successful total quality effort. Deming described these cultures in his writings, stating the lessons as the basis for his 14 points (Deming 1986). Juran wrote about the importance of diagnostic work in an open system (Juran 1989). Imai has stated that "In Total Quality Control (TQC), the first and foremost concern is with the quality of people. Instilling quality into people has always been fundamental to TQC. [An organization] able to build quality into its people is already halfway toward producing quality products." (Imai 1986) He states that workers should use their brains as well as their hands. Imai has discussed in some detail the subject of changing the corporate culture in such a way that everyone can participate positively in continuous improvements, and contribute to the competitiveness of the organization.

By putting these learnings into practice, organizations all over the world have accelerated the movement toward total quality systems. In the Malcolm Baldrige quality assessment criteria, considerable value is assigned to these cultural elements. Some cultural building blocks have been identified which support a total quality system: open communication, trust, employment stability, and performance appraisal and coaching.

Open Communications.   Employees must have unobstructed access to pertinent information. At Springfield Remanufacturing Company (Stack 1992), the emphasis is on "open book management," which includes frank and systematic disclosure of company financial performance. The idea is to let everyone see the scorecard that the company is using and to let everyone contribute to the performance on that scorecard.

Communications must be clear, timely, believable, and supported by data and facts. Employees must have information that once was thought not relevant to their jobs—"there is not a need to know." This includes information about cost of product, cost of energy, time cost of money, waste levels, cost of waste, levels of customer satisfaction/dissatisfaction, cost per employee, earnings pressures, etc. In a total quality system, employees are expected to be process managers, problem solvers, and decision makers. Open communications are needed because employees need information to make the day-to-day decisions. Without the information they cannot fulfill their roles.

Members of empowered organizations need to understand clearly the organization's vision, mission, and objectives. At Mt. Edgecumbe School in Sitka, AK, the entire faculty and student body participated in fashioning the school's vision. Similarly, 600 employees in Eastman's filter products division contributed over 4500 ideas to help establish the division's mission, vision, principles, and values, which are still used many years later. As a result of their participation, employees understand these, and act as owners in helping to see that actions are taken consistent with the organization's mission and vision (Hoogstoel 1996).

The organization's vision, mission, and objectives should be clearly defined by senior management, then clearly communicated throughout the organization. Pertinent issues should also be communicated, such as what will happen to future employment if a job becomes unnecessary due to improvements. Open communications are developed as the managers take steps to achieve freedom from fear, establish data-oriented decision processes, and establish the habit of sharing business goals and results. At Dana Corporation, the manager of each operating unit meets quarterly with all employees to review objectives and present the quarter's results against those objectives (Dana 1994). At Nissan in Smyrna, TN, the plant manager meets weekly with groups of employees at noon to talk about things that need to be communicated with employees. At Nissan, quarterly business meetings with all employees in various departments are held to review performance measures, costs, and other business targets.

Freedom from Fear.   Ideas and feedback from employees are essential. These can only come when employees feel they can give their comments without exposure to blame, reprisals, or other consequences administered by a capricious management. One of the reasons Eastman Chemical abandoned its traditional performance appraisal system was that some employees expressed the fear that they might no longer be given above-average ratings if they shared their ideas with others within the evolving team environment. They feared their ideas might be ridiculed or cause others to view them as "troublemakers." Whether well-founded or not, fear that management will view any negative comment as adverse is a powerful disincentive for employees to provide suggestions, challenge the status quo, or to offer accurate and honest feedback. This sort of fear in the organization inhibits employees from making improvement suggestions; employees fear that such suggestions will be viewed by management as criticism of managerial practices. Fear also can inhibit people from working toward improved efficiency; they may believe that such improvements will result in eliminating their jobs. Of course, fear of reprisal to an employee who makes a mistake could result in the employee covering up a mistake and, for example, shipping off-quality product to the customer.

Deming spoke of the importance of driving out fear so that everyone "could put in their best performance," unafraid of consequences (Deming 1986). Some organizations promote a proactive approach where employees are requested to talk back to the system to question results, challenge old procedures, and to question management practices.

It has been estimated that upward of 80 percent of quality problems are caused by management action (the remainder attributable to what may fairly be called worker error). The implication is that the search for root causes of problems will, in upward of 80 percent of problems studied, lead to the management realm—the systems, procedures, policies, equipment, etc., under managerial control. For many traditional managers, this probing of problems may be too close to home. The potential embarrassment may prove too great and the departure from conditions of the past intolerable. The managers' reaction may be, consciously or not, to resist the activities of the problem-solving teams to the point of causing the teams to fear going further. Thus, it requires substantial effort and cultural change to replace this fear with open communications.

Data Orientation.   Employees can best participate and work toward customer satisfaction and continuous improvement when they have knowledge of facts and data. Employees must know the facts and data regarding parameters such as costs, defect rates, and production and service capabilities, to be able to contribute to determining root causes, evaluating possible problem solutions, and making process improvements. Data orientation makes decisions objective and impersonal. Employees need easy access to these data. They also need training and coaching to help them understand the meaning of the data. To make proper interpretations and decisions, they need to understand and apply statistical concepts, such as the theory of variation.

At Eastman, every team has a set of key result measures that are endorsed by their management as being really the key measures of success. These measures are plotted, and employees have the ability to respond to the performance. Management information systems are in place so that all employees, including those in the lab and at the workstations, can access the data.

The concepts of variation, common causes, special causes, and root causes are at the very heart of total quality systems. Many organizations work hard and unsuccessfully at solving problems because they lack understanding of these concepts. Recognizing this, U.S. companies have accelerated training on statistical analysis skills and problem-solving skills. A 1993 survey indicates that training in these areas more than doubled between 1990 and 1993 within Fortune 1000 companies (Macy 1995).

Computer literacy plays a key role in these data orientations. In empowered organizations, computer literacy among employees is common. Employees' access to data and ability to analyze them are enhanced by electronic mail, immediate access to laboratory data, analytical software, and other computer-oriented communications.

Within well-developed TQM efforts, employee problem-solving teams methodically and continuously improve their processes. Their work requires asking questions of the process, gathering data, and analyzing data to answer the questions and eliminate defects. The teams require training in how to ask the questions, gather the data, and analyze them to answer the questions. Trained and otherwise empowered, such teams are able to produce quality that is the best in their industry. Whether in manufacturing industries, service industries, or government, the reliance on data is the same in all successful organizations.

At Eastman Chemical, workers have explained to Baldrige examiners and to hundreds of visitors how they routinely gather and analyze data to understand and improve their operations. There is often disbelief that ordinary people can learn to do such extraordinary things. "Ordinary people" can do this and more, but to enable such widespread and effective teamwork requires organizational change of the sort elaborated in this section.

Sharing Business Goals and Results.   Discretionary effort exerted toward improvement is an example of "ownership" behavior. To truly feel and behave as owners, employees need to know the goals of the business—objectives regarding sales, costs, earnings, customer satisfaction, etc. Employees also need to know clearly how their work can contribute to the accomplishment of these goals. Furthermore, employees need to know how the organization is performing regarding these goals—they need to know the results of the business if they are to sustain their focus on these goals.

Trust.   Trust is multidirectional (up, down, and sideways within the organization). Unilateral trust is too much to expect; trust must be bilateral. Management cannot expect to trust the message from subordinates unless the subordinates trust management not to punish the messenger. Here are three examples that will illustrate this.

At Frontera Grill and Topolobampo, two of Chicago's most popular restaurants, coowners Rick and Deann Groen Bayless apply the same standard to employee attendance as they apply to themselves: Absence to minister to a friend or family member in need is honored as a legitimate absence.

At Eastman Chemical, clock cards and gate guards were eliminated. Employees are trusted to arrive and depart on time, and to not carry out unauthorized company property. Eastman does occasionally audit people departing, and anyone caught stealing is terminated as not belonging in the system. Due to some problems a major customer in Minnesota was having with material from Eastman's Acid Division in Tennessee, some of Eastman's manufacturing people had to make a trip to the customer's plant. Rather than sending a group of engineers or managers to solve the problem, Eastman sent only one manager and the entire team of six operations people all via the company airplane. Eastman and the customer trusted the team to represent Eastman and solve the problem, which it did.

At the Nissan plant in Smyrna, the senior manager continually communicates both in person and through company newsletters that the importance of employees as thinking and creative persons is paramount. This is reinforced by a true trust culture in which self-supervision and good communications are the key. Instead of 10 to 12 levels of supervision, only five are needed at Smyrna. Employees in the daily assembly operations area are given a large amount of responsibility—quality and maintenance are entrusted to the technicians on the line—they are also empowered to stop the line, if necessary (Bernstein 1988).

Managerial behaviors that encourage trust include:

- Open communications
- Consistent communications—saying the same to all listeners
- Honesty—telling the truth, even when it is awkward to do so

- Fairness—maintaining the same policy for everyone, especially as regards promotion, vacation, pay scale, opportunity to contribute, and the like. At Lake Superior Paper Industries, every employee is salaried. Whether exempt or nonexempt, all share the same benefits package: paid sick days, no time clocks, no reserved parking spaces, no special lunch room, and minimized status differentials (Gilbert 1989).

- Respect for the opinion of others—listening to people to understand their needs, ideas, and concerns; being open to feedback, such as from employee attitude surveys

- Integrity—being guided by a clear, consistent set of principles; saying what will be done; doing what was promised.

Promises and agreements are demonstrated through actions. Aristotle said we are what we repeatedly do. Building trust takes lots of consistent actions over a long period of time.

Managerial behaviors that impede trust include:

- Dishonesty—telling untruths or half truths

- Fostering rumors—generating rumors; allowing rumors to persist; failing to provide information

- Isolating people—separating them physically, without adequate communication; separating them socially and psychologically by providing too little communication

- Breaking promises and agreements

Employees must respect and rely on each other. Fair treatment, honesty in relationships, and confidence in each other create trust. Trust is often referred to in Eastman Chemical as "the big T," absolutely necessary for healthy human relations.

Employment Stability.   Employment stability is a worthwhile objective in the TQM organization for many reasons. Principal among these is the protection of the organization's considerable training investment and preservation of the carefully developed atmosphere of trust on which TQM is built. Further, as employment stability becomes more rarely available in the job market place, its promise becomes the more attractive for many job seekers. Thus, as employment stability is a worthy strategic element, it is worth examining ways to address two issues which affect the organization's employment stability: the threat to individual jobs posed by improvement activity and cyclical employment fluctuation.

The Threat of Improving Jobs Away.   As employees work continuously to improve their work processes, some jobs will become unnecessary. Assurance of employment stability is essential before employees can be expected to work wholeheartedly toward continuous improvement. Management must make it very clear to employees that employment will not be terminated by the organization if their jobs are made unnecessary due to improvements. Retraining for other jobs will be provided if necessary. Alcatel Austria reports that through retraining and skill development programs, it has managed to retain its technically skilled workers, even when changing technology renders the old skills virtually obsolete (*European Foundation for Quality Management Special Report* 1995).

Cyclical Employment Fluctuation.   In many industries, the fluctuations in business activity imposed by the business cycle create a need to plan for employment stability of the permanent work force. The creation of "rings of defense" makes stable employment possible. The authors learned the term "rings of defense" during a visit to the Cummins Engine Company plant in Jamestown, NY. There, it was a recognition that employment stability was desirable and required a detailed strategy to achieve it. To cover peak activity and needs which cannot be met by the basic work force, alternative employment programs can be used to supplement the permanent work force. These alternatives include:

- Use of overtime

- Use of temporary employees
- Use of contract employees
- A coordinated plan to make this all work together

Think of the business case for providing employment stability. It is common for a company to spend thousands of dollars to train employees, then lose that training and experience when an improvement occurs which makes people redundant. Planning is needed to achieve employment stability. Achieving it is not easy. Not every company can, but it is highly desirable for successful long-term TQM.

Performance Appraisal and Coaching.   Performance appraisal and coaching can be a positive part of the human relations system or it can be highly destructive. The difference is in its purpose and execution. If the purpose of performance appraisal is to coach employees to a higher level of performance, it can be very helpful. If performance appraisal is done only to rank the work force for purposes of pay and advancement, it can be destructive.

Scholtes (1987) has proposed more positive performance appraisal systems, which emphasize the coaching potential. To implement such systems requires that the managers conducting the appraisals be skilled and trained how to coach. In a system of open communication, important appraisal input also comes from peers and subordinates. These inputs are useful in coaching for improved behaviors. Peer appraisal requires that team members be trained, then coached in peer appraisal skills. Prior to the initial experience of appraising peers, the team members may find it helpful to learn something about personality dynamics within the team. Instruments such as the Myers-Briggs Type Indicator can help them. (Myers-Briggs Type Indicator is a registered trademark of Consulting Psychologists Press, Inc., Palo Alto, CA).

Many organizations moving toward increased employee empowerment have found that the traditional performance appraisal system becomes an obstacle to progress. An annual employee rating that is tied to the employee's pay and selection opportunities inhibits teamwork and, therefore, continuous improvement. The individual performance of employees working in a team environment cannot be reliably determined. It is still common to conduct appraisals in empowered organizations. However, appraisals focus on employee development. Matters of selection and promotional opportunities and of pay are handled separately.

Employee Development System.   In the late 1980s, Eastman Chemical replaced its traditional performance appraisal system with an "employee development system," which focuses on employee development and coaching. This process is shown in Figure 15.1. Key principles of this process include:

- Continual improvement of employees' capabilities is important to achieve a competitive advantage.
- Employees want to do a good job.
- Employees will assume a high degree of participation and responsibility for their own development. Employee participation in the development process builds understanding and commitment to personal development.
- Expectations for the current job must be known and agreed upon with the key stakeholders (customers, supervision, etc.). Individual and team responsibilities should be clearly understood.
- Opportunities for development need to be identified so that effective, focused plans can be prepared to improve performance.
- Frequent feedback, coaching, positive reinforcement, and development planning are essential to assist employees in developing their capabilities in order to improve performance.
- A formal process is necessary to facilitate effective development and coaching.
- Development and coaching should be a positive experience for the employee and supervisor.

DESIGN PRINCIPLES OF WORK AND ORGANIZATION

Design Work for Optimum Satisfaction of Employee, Organization, and Customer.   Successful organizations are designed to achieve high employee commitment and organizational performance focused on satisfying, and even delighting, the customers. A proper work design allows people to take action regarding their day-to-day responsibilities for customer satisfaction and employee satisfaction.

Customer Satisfaction.   Employees must know the customer's needs. They must know whether and how these needs are met, and what improvements can be made to further delight the customer. Employees must have the opportunity to work to ensure that customer needs are satisfied and that performance in this regard is continuously improved. The team must have processes which provide this opportunity. The organization must be designed to ensure that customer needs are known, understood, and communicated, and that there is feedback on the team's performance in meeting the customer's needs. The organizational design should provide the employees the opportunity to work toward continuously improved performance. Improvement must not be expected to happen by accident. An example of such an organization is the Procter & Gamble plant (Buckeye Cellulose Division) located in Foley, FL. The employees at this plant work in self-regulating teams and participate in partnerships with key customers, improving their products and better satisfying customer needs (authors' visit to Foley ca. 1988).

Employee Satisfaction.   People naturally want to grow and learn. To enable this to happen, the environment should provide the employees with:

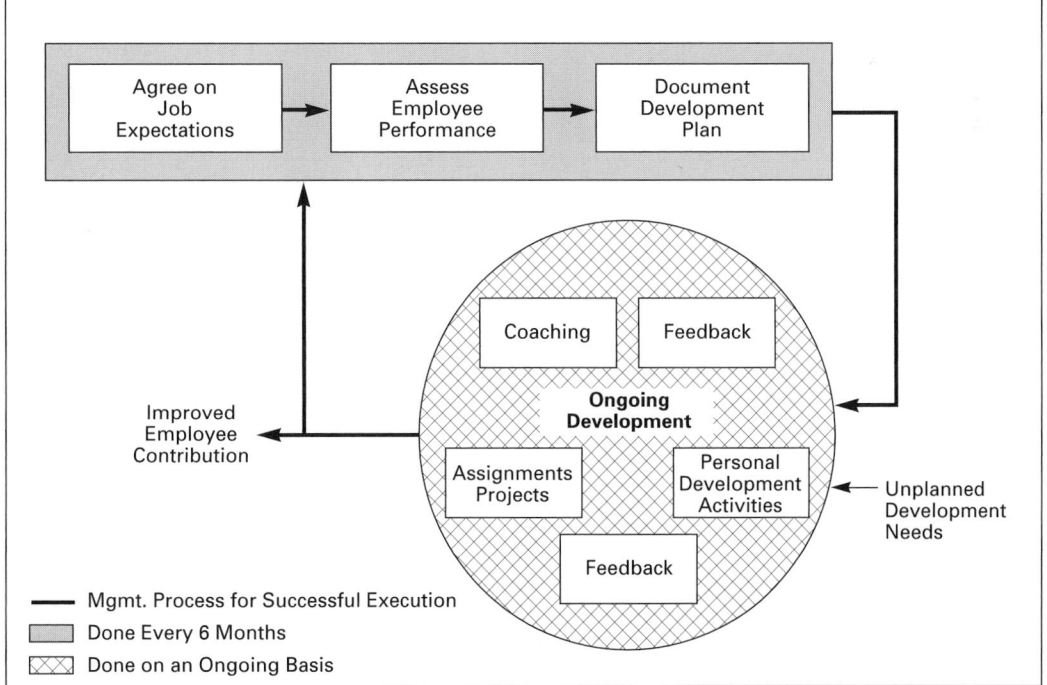

FIGURE 15.1   Eastman Chemical Company development and coaching process. (*Eastman Chemical Company.*)

- An understanding of the purpose of the work, strategies for accomplishing it, and the organization's expectations of them.
- Adequate pay
- Career growth opportunities
- Adequate authority for their jobs
- Sufficient training and tools
- A feeling of safety in the work environment

The organization must be designed to ensure that employee satisfaction is measured and fed back to the managers. The managers must routinely act on improvement opportunities presented by the employee satisfaction feedback.

Netas is a joint venture in Turkey between the Turkish PTT (the state telephone company) and Northern Telecom Limited of Canada. The Netas philosophy states that a satisfied work force is the key to a satisfied customer (*European Foundation for Quality Management* 1995). At Ritz-Carlton, data show that hotels with the highest employee satisfaction levels also have the highest customer satisfaction levels (Davis 1995). Honeywell and numerous other organizations also recognize the positive correlation between employee satisfaction and customer satisfaction.

In addition to customer satisfaction and employee satisfaction, the organization design must also provide for safe operations, quality and value of products and services, environmental protection, and continuous improvement of processes, products, and people.

Of course, the needs of the business must also be met. The organization must be properly designed to ensure that it meets financial and business objectives, including growth objectives. The design must also enable plants and facilities to operate in a manner that protects the environment and the health and safety of the employees and the public.

The many business objectives must be defined and communicated so they can be met. A common realistic vision is needed. Again, the organization must be designed to measure performance toward these objectives and provide feedback to the employees.

The way we get this knowledge is to measure the satisfaction of customers, employees, and the organization. Surveys can be very helpful here, as are face-to-face meetings. A balance of measures is needed.

The important point of this section is that organizations need to be designed with balance to satisfy customers, employees, and investors. Ignoring any one of the stakeholders results in an unbalanced design.

A last word on satisfaction: Miller argues that there is a healthy form of dissatisfaction, which he calls "creative dissatisfaction" (Miller 1984). It is a healthy thing for employees to be creatively dissatisfied. For example, in a TQM system, where people work together for continuous improvement, failure of the system to perform up to its best every day should cause a healthy level of dissatisfaction. The drive to make things better comes from this creative dissatisfaction with the status quo.

Design a System that Promotes High Levels of Employee Involvement at All Levels in Continuous Improvement. Traditional American management was based on Frederick Taylor's teachings of specialization. At the turn of the twentieth century, Taylor recommended that the best way to manage manufacturing organizations was to standardize the activity of general workers into simple, repetitive tasks and then closely supervise them (Taylor 1947). Workers were "doers"; managers were "planners." In the first half of the twentieth century, this specialized system resulted in large productivity increases and a very productive economy. As the century wore on, workers became more educated, and machinery and instruments more numerous and complicated. Many organizations realized the need for more interaction among employees. The training and experience of the work force was not being used. Experience in team systems, where employees worked together, began in the latter half of the twentieth century, though team systems did not seriously catch on until the mid-1970s as pressure mounted on many organizations to improve performance. Self-directed teams began to catch on in the mid-1980s (Wellins et al. 1991).

For maximum effectiveness, the work design should require a high level of employee involvement.

Empowerment and Commitment.   Workers who have been working under a directive-command management system, in which the boss gives orders and the worker carries them out, cannot be expected to adapt instantly to a highly participative, high-performance work system. There are too many new skills to learn, too many old habits to overcome. In the experience of the authors and according to reports from numerous organizations which have employed high-performance work systems, such systems must evolve. This evolution is carefully managed, step by step, to prepare team members for the many new skills and behaviors required of them. Figure 15.2 presents the steps in the evolutionary process experienced over the period of several years at Eastman Chemical. At each new step, the degree of required involvement increases, as does the degree of empowerment conferred on the worker. Many experienced managers have also reported that as involvement and empowerment increase, so, too, does employee commitment to the team, its work, and the long-term goals of the organization.

The *directive command* is the form most people learn in the military. A command is not to be questioned, but to be followed. It usually results in compliance.

The first stage of involvement is the *consultative environment,* in which the manager consults the people involved, asks their opinions, discussed their opinions, then takes unilateral action.

A more advanced state of involvement is the appointment of a *special team* or project team to work on a specific problem, such as improving the cleaning cycle on a reactor. This involvement often produces in team members pride, commitment, and a sense of ownership.

An example of special quality teams is the blitz team from St. Joseph's Hospital in Paterson, NJ. Teams had been working for about a year as a part of the TQM effort there. Teams were all making substantial progress but senior management was impatient because TQM was moving too slowly. Recognizing the need for the organization to produce quick results in the fast-paced marketplace, the team developed the "blitz team" method (from the German word for lightning).

The blitz team approach accelerated the standard team problem-solving approach by adding the services of a dedicated facilitator. The facilitator reduced elapsed time in three areas: problem-solving focus, data processing, and group dynamics. Because the facilitator was very experienced in the problem-solving process, the team asked the facilitator to use that experience to provide more guidance and direction than is normally the style on such teams. The result was that the team was more focused on results and took fewer detours than is usual. In the interest of speed, the facilitator took responsibility for the processing of data between meetings, thus enabling reduction of the time elapsed between team meetings. Further, the facilitator managed the team dynamics more skillfully than might be expected of an amateur in training within the company.

The team went from first meeting to documented root causes in one week. Some remedies were designed and implemented within the next few weeks. The team achieved the hospital's project

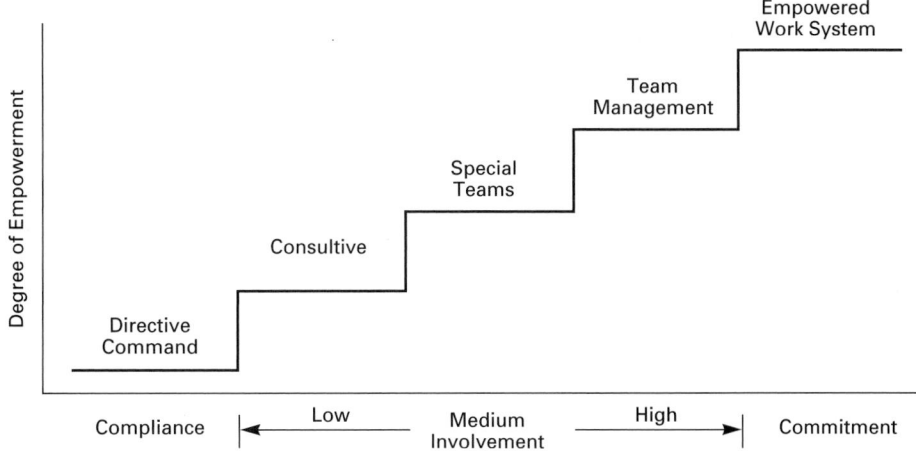

FIGURE 15.2   Relationship of commitment and empowerment. (*Eastman Chemical Company.*)

objectives by reducing throughput delays for emergency room (ER) patients. ER patients are treated more quickly, and worker frustrations have been reduced (Niedz 1995).

The special quality teams can focus sharply on specific problems. The success of such a team depends on assigning to the team people capable of implementing solutions quickly. In any TQM system, identifying and assigning the right people is a core part of the system design. Another part of the system design is provision for training teams in problem solving. For many teams, achieving maximum project speed requires a dedicated facilitator.

Team management is the use of teams to manage everyday business and the continuous improvement of the business. It represents an extension of the authority and responsibility of the natural team.

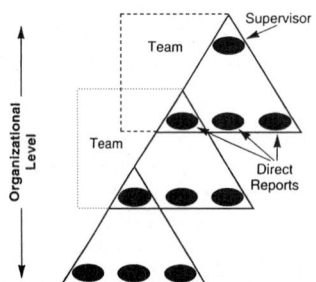

FIGURE 15.3   Interlocking team concept. (*Eastman Chemical Company.*)

Natural teams; interlocking team structure: An interlocking team structure creates an environment for sharing ideas and establishing common goals throughout the organization. The concept is that the teams are "natural teams," composed of employees who naturally work together. They are part of a work group, they work within a common work process, they share common goals, and they report to the same supervisor. The team is made up of a supervisor and those reporting directly to the supervisor (Figure 15.3). The team meets regularly, perhaps weekly or biweekly, to monitor its own performance, and to identify and work on opportunities for continuously improved performance of the team. The supervisor is leader of this team and is, in turn, a member of a team of peer supervisors led by the next-higher-level supervisor.

Every employee is therefore on at least one team. For example, at Texas Instruments Europe most managers belong to at least two quality steering teams, chairing one of them (*European Foundation for Quality Management Special Report* 1995).

An example of a very successful natural team is the Refrigeration Team (15 members) at Eastman Chemical. The team managed the maintenance of refrigeration equipment at a large chemical plan. In each year of a 3-year period, the team reduced annual refrigeration maintenance costs by $1 million. Savings came from improved processes, and in some cases developing completely different methods of maintaining and cleaning equipment. One critical activity was the annual shutdown (for maintenance) of scrubbers, which previously took a week per scrubber. As the team worked together to identify ways to perform annual scrubber maintenance more efficiently, the team eventually reduced the total time to 4 hours, which resulted in very large savings, as the production operations had been running at full capacity prior to the shutdown.

Implementing team management requires some additional design and training. The design needs to make clear who is on each natural team, when the teams will meet, whether participation is required, whether the meetings will require overtime, how the meetings will be organized, and how the team will measure its success.

Each team needs training in group dynamics, team problem solving, statistical thinking, statistical process control (SPC), and team leadership. Leadership is usually rotated every 6 to 12 months. Therefore, some training is needed to prepare people for their assignments. As teams mature, they will need training in providing feedback to each other.

Juran has predicted that the next improvements in TQM will be widespread introduction of empowered work systems including teams variously labeled as high-performance teams, self-managing teams, self-directed teams, empowered teams, natural work groups, and self-regulating teams, with self-directed work teams being the most common. A *self-regulating team* is a permanent team of employees empowered to manage themselves and their daily work. Team responsibilities may include: planning and scheduling of work, resource management, budgeting, quality control, process improvement, performance management, and other activities related to human resources. Thus, these teams perform some activities which have been traditionally performed by supervisors. The supervisor is still involved, but now serves as a coach. Self-regulating teams manage their business.

These are not ad hoc teams. They are permanent teams, frequently organized around a work process. They see themselves as "entrepreneurial business units," providing value-adding services, responsible for the satisfaction of their customers. For existing organizations which make the transition to empowered teams, some work system redesign is usually required.

These teams will take the organization to even higher levels of performance. Members of the empowered team support each other and the organization to achieve organizational and team goals (which, many times, they have helped establish). Peters states that the self-managing team should become the basic organizational building block (Peters 1987). Empowered work systems are designed so that the teams are empowered (having capability, authority, desire, and understanding of their purpose) and thus positioned to meet the needs of customers, business, and employees.

Empowered work team membership can make 80 to 90 percent of daily decisions regarding the team's business. The team is responsible for its own actions and team results, and thus for the overall product and service provided. Members accept that their job is expanded to include improving the work processes of the team. They also accept more responsibility for problem solving, leadership, and team development as the team matures. Administrative tasks such as scheduling and training are coordinated by the team itself. (Good coaching is required during the maturing process.)

The team receives cross training to make team members multiskilled and provide the flexibility needed to meet changing customer needs with high quality outputs. Through the cross training to provide "multiskills," team members can typically perform two to three different jobs. Multiskilling creates a more flexible organization, therefore a more efficient one. Associated with this is the ability to respond more readily to changing customer needs and to operate with a reduced number of people, compared to the number required in the absence of multiskilling. A survey of innovative organizations indicated new "greenfield" organizational start-ups or new plant designs usually exhibit "50 to 70 percent" better performance (profitability) than a "traditionally" designed organization and facility (Macy 1995).

Empowered, self-regulating teams monitor their progress and work on redesigning their work processes with regular redesign steps in 12- to 24-month intervals. Teams at the Procter & Gamble plant in Cape Girardeau, MO, conduct regular redesigns every 2 years to capture learnings of things that have worked and did not work in the previous work designs. These teams had been in operation for over 10 years and have done several redesigns for continual improvement. This continuous improvement keeps what works best and modifies things that do not work as well. Coaching for empowered teams is very important particularly during the first 2 to 3 years of operation.

Empowered teams distinguish themselves from traditional work groups in numerous ways. An example is the lengths to which a team goes to understand what satisfies and dissatisfies the team's customers. In one company, one team sent several team members on visits to learn customer reactions at first hand, directly from the workers who used their product. They reported their findings to the team, using video footage made in customer operations by a team member.

Empowerment principles are applied in the engineering environment, where employees work in cross-functional teams to design a new plant or product in minimum time and with maximum effectiveness. Sometimes these teams include suppliers, customers, and the necessary company functions as well as contractors. Toyota, Ford, Honda, and General Motors are among the companies which have effectively used such teams to design new automobiles in record times. Design activity within such a team permits rapid reaction throughout the organization to design proposals, including early warning of impacts up- and downstream from the proposed change. A proposed change—a new supplier, a tightened dimensional tolerance, a new paint material—can quickly be evaluated for its effect on other operations. This aspect is often referred to as "simultaneous engineering."

These principles are applicable at the highest levels of the company. In one company, a group of division presidents operate as a self-managed group of peer managers directing worldwide manufacturing efforts. These managers meet regularly to develop work processes and monitor measures of success. They rotate leadership. Ownership and commitment are evident here, just as in a team of operating technicians.

Other work processes being managed by such teams include writing insurance policies, making chemicals, and operating a bioclimatic zone in the San Diego Zoo. Some organizations which use empowered, self-regulating teams are AT&T Power Systems, Eastman Chemical, Harris Semiconductor, and Federal Express.

Behaviors.   Numerous positive behaviors may be observed among the members of empowered teams and are ascribed to the environment created by the team. Observations of such teams at Eastman Chemical have shown that members often:

- Focus on satisfying the purpose of the organization (satisfying the customer and achieving financial objectives) rather than merely trying to satisfy the supervisor. Team members talk about customer and business needs and the team's performance toward meeting those needs. Team members question management when reasons for objectives are not clear—members want to know the why as well as the what. Members are not afraid to question or offer suggestions to others (such as management) because the workers truly have the interests of the business and customers as top priorities.

- Behave more as owners of the organization. More discretionary effort is observed. An example comes from the fibers production area of Eastman Chemical. An operator who needed a wrench simply stopped by the hardware store on his way home and purchased one, and would not even let the company reimburse him for it; this was his special contribution to the work effort. Often, employees do tasks which are not in the job description. Team members eagerly monitor performance indicators because doing so helps them relate their actions to the success of the team and the business. Experienced people coach new employees by proactively sharing their ideas and suggestions to ensure that everyone on the team is pulling together 100 percent to achieve maximum team performance. Often discipline problems within the team—such as a member arriving late, being unnecessarily absent, or slacking on the job—are handled proactively by team members without outside supervision ever getting involved. Shift employees willingly come in earlier than necessary to ensure smooth exchange of knowledge regarding shift needs. Behaviors of this sort are evidence that team members feel responsible for the success of the organization, and prefer to take responsibility rather than abdicating that responsibility to the boss. In short, these behaviors are evidence of commitment to the organization's objectives.

- Are more entrepreneurial and innovative. If team members see a problem, they are observed to seek energetically to solve it. If they think of a better way of doing something, they are observed soliciting input from those around them to ensure the idea is workable and to spread the ownership for the idea (which will result in improved commitment of others for the idea). They freely seek to implement their ideas, and are willing to try new techniques which may yield improvements. Members are often observed constructively building on ideas from each other in search of solutions to problems.

- Communicate well within the team as well as outside the team. Open communications are key in all of the above examples.

Leadership Style.   Members of empowered teams share leadership responsibilities, sometimes willingly and sometimes reluctantly. Decision making is more collaborative, with consensus the objective. Teams work toward win-win agreements. Teamwork is encouraged. Emphasis is more on problem solution and prevention, rather than on blame. During the authors' visit to Procter & Gamble's plant in Foley, FL, the host employee commented that a few years before he would not have believed he would ever be capable of conducting this tour. His new leadership roles had given him confidence to relate to customers and other outsiders.

Citizenship.   Honesty, fairness, trust, and respect for others are more readily evident. In mature teams, members are concerned about each other's growth in the job—i.e., members reaching their full potential. Members more willingly share their experiences and coach each other, as their goal is focused on the team success, rather than on their personal success. Members more readily recognize and encourage each other's (and the team's) successes. Teams at Eastman Chemical drew up citizenship documents with their ideas about how to show respect and responsibility.

Reasons for High Commitment.   As previously stated, empowered team members have the authority, capability, desire, and understand the organization's direction. At Eastman, we believe that this makes members feel and behave as owners, and makes them more willing to accept greater responsibility. They also have greater knowledge, which further enhances their motivation and willingness to accept responsibility.

Means of Achieving High Performance.   We have observed that as employees accept more responsibility, have more motivation, and greater knowledge, they freely participate more toward the interests of the business. They begin to truly act like owners, displaying greater discretionary effort and initiative.

As previously stated, empowered team members have authority, capability, and desire and understand the organization's direction. Consequently, members feel and behave as owners, and are willing to accept greater responsibility. They also have greater knowledge, which further enhances their motivation and willingness to accept responsibility. An empowered organization is contrasted to a traditional organization in Figure 15.4.

Key Features of an Empowered, High-Performance Organization.   An outstanding example of an organization which exemplifies these characteristics is SOL, a Finnish building cleaning company.

Creating an Environment.   Liisa Joronen, head of SOL, believes that an organization must give its employees every opportunity to perform at its best. To achieve such a state, she has taken steps

Element	Traditional organization	Empowered organization
Guidance	Follow rules/procedures	Actions based on principles
Employee focuses on	Satisfying the supervisor	Satisfying the customer, and achieving the business objectives
Operator flexibility	One skill	Multiple skills
Participation	Limited	High Involvement
Empowerment	Follows instructions, asks permission	Takes initiative; a can-do attitude; discretionary effort
Employee viewed as	A *pair of hands* to do defined task	Human resource, with *head, heart, hands, and spirit*
Leadership for work processes	Managers only	Shared by managers and operators
Management communication style	Paternalistic	Adult to adult
Responsibility for continuous improvement	Management	Shared by managers, staff, and operators
Work unit defined by	Function (such as manufacturing or sales)	The work process, which may be cross-functional
Administrative decisions are made by	Management	Shared responsibility of team members and management (if self-directed team, the team may be responsible solely for certain administrative decisions)
Quality control is the responsibility of	Laboratory	Team capability for process control

FIGURE 15.4 Traditional versus empowered organization. (*Eastman Chemical Company.*)

to create an environment that gives SOL workers whatever they need to get the job done. This includes:

- Complete freedom to work when, where, and how an employee chooses, so long as customer needs are met. Workers are given cellular phones, voice mail, e-mail, laptop computers, and home computers as needed, resulting in much easier access to people than traditional setups allow.
- The absence of organization charts, job titles, and status symbols, including secretaries and company cars (Joronen often rides her bicycle to meetings with external customers). Each office worker has an area of primary responsibility, with the understanding that he or she will also contribute in other areas as vacations or other conditions warrant.
- An open-book policy on company performance. Each month, employees receive updated information on financial figures, absenteeism, turnover, and the like, right down to individual performance among the staff.
- A clean desk policy, which eliminates territorial office claims. Employees use whatever work space is available. When the work is done, the employee clears the area of his/her materials so that others may use it (Juran Institute 1996).

Over the past 20 years, enough progress has been made with various empowered organizations that we can now observe some key features of successful efforts. These have come from experiences of various consultants, visits by the authors to other companies, and published books and articles. These key features can help us learn how to design new organizations or redesign old ones to be more effective. The emphasis is on key features rather than a prescription of how each is to operate in detail. This list is not exhaustive but is a helpful checklist, useful for a variety of organizations.

Focus on External Customers.   The focus is on the external customers, their needs, and the products or services that satisfy those needs.

- The organization has the structure and job designs in place to reduce variation in process and product.
- The organizational layers are few.
- There is a focus on the business and customers.
- Boundaries are set to reduce variances at the source.
- Networks are strong.
- Communications are free-flowing and unobstructed.
- Employees understand who the critical customers are, what their needs are, and how to meet the customer needs with their own actions. Thus, all actions are based on satisfying the customer. The employees (operator, technicians, plant manager, etc.) understand that they work for the customer, rather than for the plant manager.
- Supplier and customer input are used for managing the business.

Guidance Is by Principles

- There is a common vision, which is shared with and understood by all in the organization.
- All actions and decisions are based on a stated philosophy, which refers to the organization's mission, values, and principles.

Dana Corporation provides a good example. In the early 1970s, Dana eliminated company-wide procedures, replacing them with a one-page statement of philosophies and policies. Dana states "We do not believe in company-wide procedures. If an organization requires procedures, it is the responsibility of the appropriate management to create them" (Dana Corporation 1994).

There Is a Relentless Pursuit of Continuous Improvement and Innovation.   Robert Galvin, former CEO of Motorola and Chairman of the Executive Committee, has said that the most important quality lesson he learned in his years at Motorola was that "perfection is attainable" (Galvin 1996). He did much to lead Motorola on the path to perfection, through continuous improvement and innovation.

The Baldrige Award is itself an example of continuous improvement:

- The organization and its supporting systems encourage all employees to improve products, processes, teams, and themselves.
- Continuous learning is part of the job.
- Stretch improvement goals are established. Examples include Motorola's "six sigma" and Xerox's "10x improvement goals.
- Informed risk taking is encouraged.
- A systematic means exists for periodic organization renewal.
- Coaching and development systems are in place for all teams and individuals.

Management can encourage improvement by giving appropriate recognition, such as by special awards and celebrations for achieving incremental performance goals, for reaching new records, etc.

Shared Leadership: There Are New Roles for Both Operators and Managers.   Supervisors in empowered, high-performing organizations find themselves in new roles, which include coaching and developing teams and individuals, clarifying business expectations and responsibilities, managing the interface between teams and their environment, allocating resources among teams, and ensuring that continuous improvements are occurring. All of this represents a span of responsibility greatly increased beyond that of the traditional supervisor.

Managers make provision for everyone's input to decisions affecting the larger organization, and for participative planning processes at the functional level.

Another perspective on leadership comes from Netas, and may be the harbinger of an emerging principle of management in empowered organizations. Managers at Netas believe that leaders can come from anywhere in the organization, that the organization has in it many leaders, and that a leader creates leaders. To foster leadership, top managers need to be visibly involved. Top managers at Netas contribute to leader development by leading teams, chairing all customer and distributor conferences, and attending presentations of continuous improvement teams (*European Foundation for Quality Management Special Report* 1995).

At Eastman Chemical, a manager (call him John) was a team manager in charge of 15 mechanics in a single area. As his empowerment grew, John soon became responsible for two teams, then three teams. John later went from the plant in Tennessee to the plants in England and Spain to teach other team managers how to coach their people.

Operators also find themselves in new jobs, such as scheduling work, hosting suppliers, visiting customers, working on cross-functional process teams, and leading the search for root causes.

The Team Concept Emerges.   The emerging team concept includes the following features:

- Work force empowerment
- Control of variances at the source
- Sharing of leadership and responsibilities by supervisor and team
- Agreed-on, well defined behavioral norms for individuals and team
- Constructive peer feedback for team and individual development
- Performance measures which enable the team to monitor its performance and take actions when needed
- Team responsibility for finished products or services
- Team participation in the process of hiring new team members

Empowered organizations have many support systems in common which enable them to function well. These include:

- Reward and recognition of desired skills and behaviors
- Business and accounting information to support decision making at the point of action
- Systems to obtain, maintain, and develop qualified personnel
- Leadership to enable teams to achieve their mission

The Role of Management in Supporting an Empowered Organization.   In empowered organizations, managers create an environment to make people great, rather than control them. Successful managers are said to "champion" employees and make them feel good about their jobs, their company, and themselves. Marvin Runyon, when head of the Nissan plant in Smyrna, TN, stressed that "management's job is to provide an environment in which people can do their work" (Bernstein 1988).

The role of management includes the following:

Create a vision of the business, and share it widely.   The Baldrige Award Category 1 states (and numerous leaders agree) that all employees must have a shared vision of the organization as an empowered, high-performance organization, which satisfies its customers, is efficient and effective, and works toward continuous improvement. Management's job is to create, share, and maintain this vision. Deming reminded us of the need for constancy of purpose in achieving the long-term objectives for the organization.

Set the organizational objectives and strategies, and share them.

Structure and align the organization to achieve the strategies.   Create a role for everyone in the business.

Allocate resources (including efforts such as research and education).

Communicate business information.

Listen for the needs of the organization.   Encouraging open, two-way communication which is essential for employees to contribute fully toward the organization's objectives is another management task. At the Dupont plant in Asturias, Spain, managers visit team meetings, and encourage team leaders.

Create the environment for sharing ideas and forming common goals throughout the organization.

- Create interlocking team structures.
- Identify and eliminate barriers to teamwork. Symbols which differentiate employee groups can be powerful barriers to teamwork. Such symbols include dress codes, car parking privileges, special dining rooms, clock cards, and special vacation policies, to name several. At the DuPont plant in Asturias, Spain, much attention has been devoted to eliminating symbols. Every employee has a business card, no matter what the job level is, but no titles or other indications are given to differentiate job level. Office furnishings are identical for each person with an office. Every employee dresses consistently, with no differentiation. The traditional performance rating system can also be a barrier to teamwork. Eastman Chemical has eliminated its traditional performance appraisal system, which had focused on annual ratings for employees, and replaced it with a system focusing on employee growth and development. The new system is designed to foster cooperation rather than competition among employees. Any disparities in benefits from one group to another can create a barrier to teamwork. Such disparities should be identified and removed. Employee suggestion systems which emphasize individual reward for suggestions create conflicting incentives for team members. Self-interest may drive a team member to reserve valuable suggestions for the channel which provides the greatest reward. Such conflicting incentives create insurmountable barriers to team success and must be addressed before launching team efforts. Figure 15.5 indicates major barriers to team empowerment, as reported in a survey of U.S. companies.

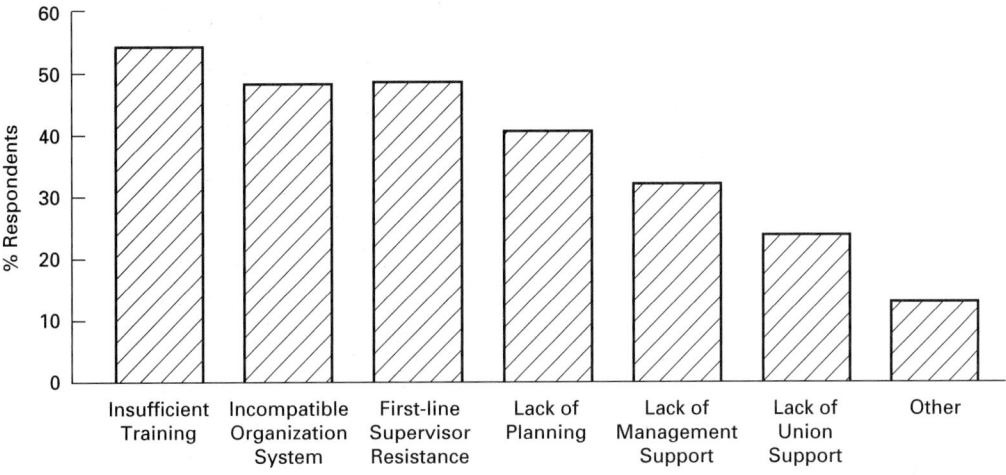

Barriers To Empowered Teams.
Survey of American Industry

FIGURE 15.5 Barriers to empowered teams. (*Industry Week.*)

- Emphasize and support training. When there is sufficient staffing, employees can receive needed training without requiring the team to work overtime. Some operations have designed their work system to include an extra crew, to provide work coverage while some personnel fulfill training needs.

- Pay for skills and knowledge. The pay system is carefully designed to recognize and encourage multiskilling. As employees learn additional skills and knowledge, they earn more pay. Compensation for team members may comprise up to four elements: base pay for the job; pay for individual performance, skill, and knowledge; pay based on the performance of the team; and pay based on the organization's overall performance. In the context of empowered teams, the simplest pay system is to pay a base level, and then achieve motivation through intrinsic factors of the job design and sharing gains in productivity.

- Assure job stability. When improved performance will render certain jobs unnecessary, employees may resist fully contributing in continuous improvement efforts. With its transition to empowered, self-regulating teams, Eastman eliminated over 50 first-line supervisory jobs. However, no one was laid off; former supervisors were eventually absorbed into other jobs within the company, often with the help of retraining. Many excess supervisors became team facilitators and coaches. During the transition, some excess supervisors became part of a "redeployment pool," assigned special projects, even for a few years, while waiting for appropriate job opportunities to open up, often through attrition. Most of these redeployed supervisors later admitted that life got better for them as a result of their new opportunities.

Reinforce positive behaviors. A critical role of management is to reward and reinforce behaviors and results which help achieve the overall goals set for the organization. This reinforcement reminds employees that these behaviors and results are truly important. Without this reinforcement, employees may come to believe that their actions and accomplishments are not really important since they resulted in no positive personal consequence. They may then wonder if something else has become more important, which leads to reduced focus on the organization's objectives.

The Role of the Empowered Manager versus the Traditional Management Role.
At Eastman Chemical, the supervisor responsible for the high-pressure cleaning team is recognized by his team members as being very effective at giving appropriate positive reinforcement, using employees to conduct job training, and encouraging innovative techniques. The team has achieved numerous internal quality awards. Team members have recognized the supervisor with their own award as "outstanding supervisor."

In another area of Eastman Chemical, a manager began to learn about empowerment. His interest in the subject led him to read over 300 articles and books on the subject. Eventually, he began teaching others about empowerment principles.

In the empowering, high-performing work system, managers exercise leadership in different ways from managers in traditional hierarchical organizations. Instead of making decisions for subordinates, managers prefer to empower people to make their own decisions. Instead of making job assignments themselves, managers teach others to make job assignments. Instead of closely monitoring and judging job performance, they coach employees to monitor and continuously improve their own job performance. And rather than managing day-to-day job needs, managers enable teams to manage those needs themselves.

Figure 15.6 is a summary of the shifting role of the supervisor.

Coaching is vital to empowerment, teamwork, and the development of employees. For those readers who are faced with the need to coach—and that includes virtually all who have a responsibility for supervision at any level—and who find themselves in unfamiliar territory, the following inventory of tasks, skills, and behaviors may serve as a useful checklist.

Effective coaching involves five elements: (1) leadership, (2) communication, (3) team development, (4) boundary management, (5) personal management.

Leadership.   Leadership involves four tasks:

- Clarify the vision. This requires much communication, understanding goals and objectives of the organization, ensuring that clear expectations for the team and individuals are set, and ensuring focus and constancy.
- Provide balance between intervention and freedom.
- Challenge and develop team leadership.
- Lead by example.

From	To
Directing	Empowering (giving authority, information, coaching, etc.)
Assigning	Teaching
Appraising	Coaching
Managing	Enabling
Minimal span of responsibility (5–10 people)	Broad span of responsibility (15–50 people)
Work vertically	Work horizontally
Capitalize on existing skills	Improve skills
Data reporting	Improvement of people and processes
Sell management's ideas	Sell team's ideas
Focus on doing things right	Focus on doing the right things
Task-oriented	Improvement, development-oriented
Concern with method	Concern with outcomes

FIGURE 15.6   Shifting supervisory role. (*Eastman Chemical Company.*)

Communication.   Roles and responsibilities for good communication include:

- Clarify principles found in vision, mission, and values.
- Clarify the goals and objectives of the larger organization.
- Communicate clear expectations and consequences.
- Transfer personal skills and knowledge to team members effectively.
- Listen to team members to understand their personal goals and interests.
- Provide encouragement and promote winning attitude.
- Give and receive sincere, specific, and timely feedback on current performance of individuals, team, and self.
- Facilitate flow of needed information.

Team Development.   This involves people and technology. The coach's roles and responsibilities regarding people include:

- Guide the team to maturity.
- Encourage and assist others in developing empowering habits.
- Promote continuous learning.
- Provide counsel.

The coach's roles and responsibilities regarding technology include:

- Increase the team's capability to understand manufacturing and service processes and to handle daily work assignments.
- Provide direction by focusing on process standardization.
- Stabilize and control processes.
- Ensure continuous improvement.

Boundary Management.   The coach's roles and responsibilities here include:

- Ensure that the team receives the information and resources needed to manage its operations.
- Help the team understand the scope and limits of its decision-making responsibilities.
- Help the team learn how to make its assigned decisions: Identify the information needed, from whom to get it, what to consider, whom to consult, and how to communicate decisions within and outside of the team.

   In doing these things, the coach helps the team learn to manage its own boundaries, and thus to be more effective.

Personal Management.   The coach here commits the time and energy to:

- Learn continuously
- Seek meaning and pleasure in the work
- Understand the new coaching role for managers
- Develop the technical skills necessary to manage the team
- Develop the people necessary to coach the team

EMPLOYEE SELECTION

Achieving an empowered, high-performance organization requires capable and motivated employees. Lake Superior Paper Industries states that its hiring process is the foundation for its success (Gilbert 1989). The employees should have the skills, education, intelligence, and dedication to participate fully toward the organization's goals. As Ritz-Carlton points out, employees must be selected, not simply hired. The selection process is critical, to ensure the right match of people and assignments.

What the Employees Will Be Asked to Do. An empowered, high-performance organization is organized around the relentless pursuit of continuous improvement. It operates in an environment of multiskilled teams and shared leadership. To operate effectively in this environment, employees must learn to think and communicate, become proficient at group problem solving and team leadership. To do their part toward integration of customer and supplier, they will need a good customer-based education. From that base, substantial training is required. Prospective employees must have the potential to become exceptional in these dimensions.

The Selection Process. In 1992, Eastman set about staffing a plant then under construction. The company interviewed 4000 people to fill 72 operations jobs. Prospective employees went through a nine-step process, which included three interviews and trial team activities to evaluate skill at problem solving and interfacing with others. On the whole, the work force hired was well-educated. Forty percent had a four-year college degree; 35 percent had managerial experience. After being hired, and before start-up of the operations, they received a year of training to enhance their technical and team-interaction skills. At IMPRO96, a team reported there had been no employee turnover in 4 years from this plant. World-class quality and productivity results were reported for their products.

A recently constructed DuPont plant in Spain used a five-phase selection process: (1) initial screening (a 2-year technical school diploma was the minimum requirement); (2) testing; (3) two interviews to identify ability and will; (4) interactive sessions, using team simulations; and (5) the offering process, including discussion of the benefits package, 10 elements of the work system, pay process, and other conditions of employment. Hiring is performed by a team that demonstrates the empowered work culture that DuPont is trying to create. New employees are involved in the subsequent hiring process.

Some important selection criteria are

- Good teamwork skills
- Demonstrated initiative
- Good oral communication skills
- Clear judgment and ability to solve problems
- Ability to learn quickly
- Ability to plan, organize, and manage work
- Appropriate level of professional/technical knowledge
- Good individual leadership and influencing skills
- Adaptability to changing conditions

As regards most of these criteria, candidates can be assessed by observing their behavior, especially collaborative interactions with peers during related work situations (for candidates who are presently employed), or during interviews and appropriately designed team problem-solving exercises. At Lake Superior Paper Industries, a candidate goes through a group interview of seven peers to determine whether the candidate will be a good "fit" with the team system. "The team members put a little stress on the interviewee. During the interview, the candidate is asked what makes a person mad or what makes a person upset. We probably turn away some superior craftsmen because they wouldn't fit. If any team members casts a negative note, then the candidate will not be hired. However, the team member must have defensible, legitimate reasons for saying no." (Gilbert 1989)

In the selection of candidates for certain specialized work, specialized selection criteria may be applied, such as orientation to providing customer service and ability to persuade and sell. Each criterion identified requires a corresponding practical and reliable means of candidate assessment.

The Transition to a Participative Work Environment.   For the most part, the above paragraphs address start-up operations. In existing operations, making the transition to an empowered work environment requires that existing employees receive different training from that given employees in a start-up operation. In addition to imparting technical and team skills, the training must aim at overcoming worker resistance to the change. Some employees will not take well to the participative work environment and may even refuse to accept it. In this case, the organization may need to start with employees who volunteer to participate. Those desiring not to participate initially may be able to "sit on the sidelines" for a while, in traditional roles, performing less participative jobs, while observing the participative work environment. Many people in this situation eventually decide to participate, persuaded by the success of the teams, and encouraged by the additional training provided.

TRAINING IN A TOTAL QUALITY ORGANIZATION

One of the largest barriers to successful empowered teams is insufficient training. An attribute that successful organizations have in common is commitment to extensive training of employees. A survey of innovative organizations shows that for a new ("green field") high-performance facility or a business start-up, training and retraining in the first 3 years, including on-the-job training, classroom training, and visits to customers and suppliers, occupies 20 to 25 percent of total available employee time per year. After this initial period, the training decreases to a constant 12 percent of employee time thereafter. By comparison, the number for the average Fortune 1000 company is 3.5 percent (Macy 1995).

Training Builds Capability.   At Motorola, the education strategy is part of the corporate strategy. Employees receive an average of 40 hours' training per year. Training curricula are in place to achieve corporate objectives. For example, to reduce manufacturing cycle time, elements include:

- Group technology/material flow
- Leveling production schedules
- Pull production systems
- Improve the production system
- Workplace organization
- Total productive maintenance
- Changeover time reduction

For Motorola's design of experiments (DOE) tools, a 17-day course is followed, which has evolved from lecture-oriented to project-oriented, work-oriented mode. Participants take home computer software to facilitate selection and application of the tools (Wiggenhorn 1987).

To empower workers successfully requires desire (on their part), authority (defined and conferred by management), and worker capability to achieve the results expected.

Multiskilled workers increase the organization's flexibility and facilitate teamwork. A multiskilled work force is a key feature of the desired organization and a key objective of the training activity.

The training process for developing multiple skills includes such steps as:

- Identify the skills needed
- Establish training technique

- Develop an assessment system
- Provide ongoing feedback and coaching
- Provide for rewards which encourage multiple skill development

Training for Effectiveness.   Imai (1986) states the Japanese axiom: Quality control starts with training and ends in training. In major Japanese companies, training is conducted regularly for top management, middle management, and workers.

Important training issues to be addressed are: pace, amount, and affordability. At Saturn, everyone is required to devote 5 percent of his or her time each year for training. Those who do so receive a 5 percent pay bonus (Garwood 1988). Wainright Industries invests the equivalent of 7 percent of its payroll in training—7 times national average in the U.S. (Garwood 1994). At Motorola, the training budget exceeds 10 percent of payroll.

At Motorola, the training may begin at the top of the organization. To ensure effectiveness of course on creativity, upper management receives training first. Then middle management is trained; and only after middle management is trained are the lower levels trained .

Measuring the Effectiveness of Training.   A training budget amounting to 5 to 10 percent of payroll is a large investment. Managers will quite naturally demand that the training be effective. To verify effectiveness requires measurement, testing, and assessment. Long experience with apprenticeship programs has proved that results of skills training can be reliably assessed using tests and hands-on demonstrations. In a multiskilled, high-performance system, skills are often required to be certified by a training team before an employee is allowed to work alone using that skill. Certification will include verbal testing and demonstration of the skill in the work area. In hazardous environments, where safety is a major concern, the certification process is very important.

Training for High-Performance Total Quality Environment.   There is evidence that the growing acceptance of TQM principles includes growth in related training of employees. Figure 15.7 presents the results of a survey of the training practices of approximately 300 companies over the period from 1985 to 1993. This figure reports the percentage of companies surveyed for which 60 percent or more of employees received training in the skill reported within the 3 years prior to the survey. The skills are frequently identified as being necessary for effective employee involvement and total quality management. Three of these skills are in the category of interpersonal and group skills. This is unsurprising, as so many employee-involvement and TQM processes involve meetings, interpersonal interactions, group problem solving, and influencing others. Also critical is knowledge in two technical areas: statistical analysis and business understanding. They are central to organizational improvement and TQM efforts (Lawler et al. 1995).

Training should focus on developing technical skills and social skills. Technical skills are the job-related skills to do the technical tasks of the job. Social skills are the skills of personal interaction and administration which, together, enable team members to work collaboratively to manage their business. Some of the team skills are communication skills, group dynamics, conflict management, holding efficient meetings, and decision making. They also include such technical skills as computer application, problem solving, and statistical analysis. The administrative skills include hiring, making work assignments, planning for vacation and relief, scheduling, and planning for training.

Cross training allows employees to understand and, when necessary, to perform the jobs of other team members. This provides each employee with a more complete understanding of the overall organization and how the various pieces of the organization fit together. With knowledge of the big picture, employees are more able to act like owners and take responsibility for customer delight and continuous improvement. Work is more interesting. This knowledge allows employees to address questions and issues more effectively and efficiently. Notable examples of extensive cross training are provided by the San Diego Zoo and Walt Disney World.

Types of training	1987 (N = 323)	1990 (N = 313)	1993 (N = 279)
Group decision-making/ problem-solving skills	5	6	16
Leadership skills	4	3	8
Skills in understanding the business (accounting, finance, etc.)	4	2	5
Quality/statistical analysis skills	6	9	22
Team-building skills	5	8	17
Job-skills training	N/A	35	48
Cross training	N/A	N/A	13

FIGURE 15.7   Percentage indicating that more than 60 percent of employees had training in past 3 years. (*From Lawler et al. 1995.*)

INTEGRATION OF THE COMMITTED PEOPLE WITH THE TOTAL QUALITY SYSTEM

The whole must be more than simply the sum of the parts. With a successful sports team, individual players alone do not make the team successful, no matter how capable the players may be individually. Similarly with a high-performance car—the individual components must be carefully matched and integrated into a properly designed system to achieve the intended results. It is not enough to have a good transmission, engine, wheels, etc.; these must be integrated into a smoothly operating system to make a world-class car. So, in a high-performance organization, the employees, structure, tasks, information, decision making, and rewards must be carefully integrated into a total system. Employees in high-performing organizations receive extensive training in both the technical skills and social skills. A balance between these skills must exist.

Building the Technical Skills.   The technical system is concerned with the production and service requirements of the work process. The system includes such elements as process operations, equipment, methods, instrumentation, procedures, knowledge, tools, and techniques, and provides for multiskilled operators.

Developing Social Skills for Better Personal Interactions.   To meet the needs of the people who are working within the process, the social system takes into account numerous factors, including individual attitudes and beliefs, employee-employer relationships, relationships between groups and among group members, human learning and growth, group norms, issues of power and politics, and personal motivation.

Communication skills, especially the listening skill, are among the most important social skills. A collaborative rather than competitive relationship with team members is important. These skills include how to conduct effective team meetings, how to resolve conflict, and how to make decisions.

Key Measures of Success.   Netas has 77 continuous improvement teams involving 310 employees. They are trained in value engineering, design of experiments, and SPC tools. They have reduced the number of defects per thousand units from 105 in 1989 to about 10 within 5 years (*European Foundation for Quality Management Special Report* 1995). This type of result is possible when the organization uses key measures of success to focus on results.

What to Measure and How to Measure It.   Measures of success should meet the following criteria:

- Team action significantly influences the key measure. Trends and changes in the data are traceable to the team's behaviors.
- The measures are important to the customer and the team. The team should be proud to tell its customers what it is measuring.
- The data which support the measure are simple to capture, analyze, and understand. Indexes which combine various measures into a single number are often too complicated to understand and not useful to the team. The team loses sight of how it influences a composite measure.
- The data are timely. The measure gives the team adequate warning of impending trouble.
- There are ample data to make the measure statistically significant.

Some commonly encountered measures which satisfy these criteria are percent of on-time deliveries, number of consecutive days worked without a lost-time injury, percent of customer survey ratings giving us a superior score, and percent of errorfree production batches.

At Federal Express, one of the primary measures is "percent of packages delivered by 10:00 a.m." with the goal of approximately 99.9 percent. Every employee has influence upon the achievement of this and has access to this data (Garwood ca. 1990).

At Eastman Chemical, customer loyalty and value, as well as over 20 specific related attributes, are measured by surveying over 2000 customers worldwide each year. Results are posted on company bulletin boards for all to see. Customer-complaint response time is another important measure, the goal being to acknowledge customer complaints within 24 hours. Performance has improved from an initial level in the 20 percent range to performance which is now over 90 percent. The improvement is due to numerous process improvements such as the assignment of advocates who are fitted with beepers and portable phones for faster communication.

Some Results Commonly Achieved

- Improved quality and operations (fewer defects and errors; increased production output per employee)
- Cost reduction (reduced operating and maintenance costs)
- Increased employee satisfaction (positive self-esteem, career path known)

POSITIVE REINFORCEMENT

As previously discussed in this section, and shown in the empowerment model (see Figure 15.1), empowerment requires four key elements: employees must know the purpose of the organization, and have the capability, the authority, and the desire to act. This part of the section addresses the desire of employees to act as owners of the organization. Positive reinforcement is essential toward creating this desire.

Definition and Key Elements of Performance Management.   Performance management focuses on achieving the desired motivation and behaviors. Daniels (1982) defines performance management as "a systematic, data-oriented approach to managing people at work that relies on positive reinforcement as the major way to maximize performance. The approach is based on the work of B. F. Skinner, whose studies revealed that behavior is a function of its consequences. As applied by numerous organizations, performance management follows five steps:

Pinpoint the Desired Results.   Measurement and goals are vital here. Ensure that everyone understands these. The pinpointed behavior should be measurable, controllable by the performer, and valuable to the performer and the customer.

Evaluate Need for Elements.   Evaluate whether desired performance can be achieved purely by motivation, or whether some additional empowerment elements (such as information, skill training, tools and equipment, authority, or knowing the purpose) is needed. Provide for them as needed.

Identify Reinforcers.   Reinforcement of desired behavior tells people that the behavior is important. When you quit reinforcing the behavior, people think it is no longer important. Though typically provided by supervision, reinforcement may also come from peers, the work group, self (intrinsic motivation), or subordinates. Good reinforcement should be positive, specific to the desired behavior, sincere, and provided immediately after the performance. Positive consequences should be a part of the job environment that encourages high performance.

Reinforcement should be planned. The refrigeration team at Eastman saved over $1 million a year, for 3 years in a row; in anticipation of this, the team planned its reinforcement which included an annual team dinner with spouses.

Measure and Communicate Performance.   Baseline historical performance should be known. Achievable stretch goals should be established which are clear and understandable. Successful teams usually find it better first to set an achievable goal rather than one large goal. Then when they reach that goal, they reset the goal, and so forth, as performance continues to improve over time. People can usually do more than expected, and thus the bar should continuously be raised whenever the earlier goals have been achieved. Performance should be visibly communicated by using scoreboards and graphs.

Deliver.   The reinforcement should be delivered when earned by the performer. Successful performances are frequently associated with lots of positive consequences along the way. This is true in sports, the arts, child rearing, and the business world. A key concept of positive reinforcement is to "catch people doing something right" (Blanchard and Johnson 1981). Consequences which have the strongest influence on behaviors are those which are positive, immediate, and certain. Such positive consequences enhance the desire to improve continuously.

Examples of Positive Reinforcement.
Successful teams celebrate their success. The sports world is filled with examples of how positive reinforcement drives continuous improvement: A football player who scores is immediately congratulated by fellow players; a baseball player who hits a home run is congratulated by fellow base runners who await him at home plate; coaches and parents cheer encouragingly for youngsters playing basketball in the youth leagues.

An athlete works hard to outdo past performances continually: winning streaks, consecutive games played, consecutive games in which the player gets a hit (in baseball), number of aces scored (in tennis), lowest golf score, or consecutive games in which the player scores (soccer).

Successful teams in work organizations also reach for new goals, and frequently celebrate their success when they achieve one. At Eastman Chemical, positive reinforcement is a common theme for team meetings. For example, an agenda item might be: "What tasks have we done well?" Celebrations of milestones or goal achievements are common and may be accompanied by tangible or intangible rewards (such as T shirts). Refreshments and mementos are often part of team celebrations. At Tennessee Eastman refreshments are served for over 20,000 people each month as part of team celebrations. An important part of team celebrations is the comments from management and team members to reinforce the behaviors and results achieved. Celebratory lunches and dinners are also popular. Additionally, hundreds of President's Awards have been issued at Tennessee Eastman for team performance, with plaques proudly displayed throughout the plant site.

Intrinsic Motivation.
A survey of companies that use TQM (Lawler et al. 1995) shows that people are motivated by success in making improvements and satisfying customers. In organizations where there is a very positive perception of TQM effectiveness, worker satisfaction is also positive. These natural rewards can be very powerful. These intrinsic motivators are things that make the job more interesting or stimulating, that build competence and pride, that promote self-control, and that result in a sense of purpose and achievement for the employee. For example, the Refrigeration Team

that made such gains in reducing the cost of maintaining refrigeration equipment exhibited the drive and accomplishment that grow from intrinsic motivation.

HUMAN RESOURCES CAN BE THE MOST EFFECTIVE SUSTAINABLE COMPETITIVE ADVANTAGE

Lester Thurow (1992) states in his book *Head to Head:* "The skills of the workforce are going to be the key competitive weapon in the twenty-first century. Brainpower will create new technologies, but skilled labor will be the arms and legs that allow one to employ—to be the low-cost masters of—the new product and process technologies that are being generated."

From discussion with many managers from the United States, Japan, Latin America, and Europe, we conclude that all have some consensus around the notion that Total Quality Management and business success depend upon long-term positive contribution from the people in their work systems. Those organizations that get the highest performance from employees who can work together effectively with the technology of their systems are projected to be long-term maximizers.

This is not easy to implement. If it were easy, every good company would be working to make itself a high-performing organization. The list of characteristics below might explain why successful implementation does not always occur. High-performance organizations require:

- Paradigm shifts
- New cultures
- Long-term vision
- Relinquishing management control
- Investment in training
- Careful selection of employees
- Effective communications

Competitive advantages are never easy to get and are sometimes hard to keep. The advantage of a high-performing work force is difficult to build but capable of sustaining a long-term advantage.

REFERENCES

Bernstein, Paul (1988). "The Trust Culture." *SAM Advanced Management Journal,* pp. 4–8.

Blanchard, Kenneth, and Johnson, Spencer (1981). *The One Minute Manager.* William Morrow, New York.

Dana Corporation (1994). *The Philosophies and Policies of Dana,* Toledo OH.

Daniels, Aubrey C, and Rosen, Theodore A. (1982). *Performance Management: Improving Quality and Productivity Through Positive Reinforcement.* Performance Management Publications, Atlanta.

Davis, Susan Musselman (1995). "Communicating the Quality Message at the Ritz-Carlton." *IMPRO95 Conference Proceedings,* Juran Institute, Inc., Wilton, CT.

Deming, W. Edwards (1986). *Out of the Crisis.* MIT Press, Cambridge, MA.

European Foundation for Quality Management Special Report (1995). European Quality Publications, Ltd. London, pp. 39, 42–46, 54–56.

Galvin, Robert (1996). Remarks made at satellite telecast, *J. M. Juran on Quality: Yesterday, Today, and Tomorrow.* October 10, 1996, produced by International Institute for Learning, Inc., and presented by PBS Business Channel, New York

Garwood, William (1988). Recollections of a visit to the Saturn automobile plant in Spring Hill, TN.

Garwood, William (ca. 1990). Recollections of a visit to Federal Express operations facility in Memphis, TN.

Garwood, William (1994). Recollections of remarks by the president of Wainwright Industries at the presentation ceremony for the Missouri Quality Awards.

Gilbert, Gaye E. (1989). *Framework for Success: Sociotechnical Systems at Lake Superior Paper Industries.* Case Study of the American Productivity & Quality Center, Houston, TX, July.

Hoogstoel, Robert E. (1996). Personal recollection related to the authors.

Imai, Masaaki (1986). *Kaizen: The Key to Japan's Competitive Success.* Random House Business Division, New York. 1986.

Juran, Joseph M. (1989). *Juran on Leadership for Quality: An Executive Handbook.* The Free Press, New York.

Juran Institute (1996) "Maximizing Employee Assets." *Quality Minute,* vol. 3, no. 9. Center for Video Education, White Plains, NY.

Lawler, Edward E., III, Mohrmann, Susan Albers, Ledford, Gerald E. Jr. (1995). *Creating High Performance Organizations: Practices and Results of Employee Involvement and Total Quality Management in Fortune 1000 Companies.* Jossey-Bass Publishers, San Francisco, pp. 13–14, 51–58, 76.

Macy, Barry A. (1995). *Survey of Innovative Organizations: New Organizational Start-Up/New Plant Design Characteristics and Site Selection Criteria.* The Texas Center for Productivity and Quality of Work Life, College of Business Administration, Texas Tech University, Lubbock, TX, December 5.

Miller, Lawrence M. (1984). *American Spirit.* William Morrow, New York.

Myers, Isabel, Briggs, Katharine (1987). Consulting Psychologists Press, Inc., Palo Alto, Cal.

Niedz, Barbara A. (1995) "The Blitz Team" *IMPRO95 Conference Proceedings,* Juran Institute, Inc., Wilton, CT.

Peters, Tom (1987). *Thriving on Chaos: Handbook for a Management Revolution.* Random House, New York.

Scholtes, Peter R. (1987). *A New View of Performance Appraisal.* Joiner Associates, Inc., Madison, WI.

Stuck, Jack (1992). *The Great Game of Business.* Springfield Remanufacturing Company, Springfield, MO.

Taylor, Frederick W. (1947). *The Principles of Scientific Management.* Harper and Row, New York.

Thurow, Lester (1992). *Head to Head: The Coming Economic Battle Among Japan, Europe, and America.* William Morrow, New York, pp. 51–52.

U.S. Department of Commerce, *Malcolm Baldrige National Quality Award Criteria.* Technology Administration, National Institute of Standards and Technology, Gaithersburg, MD.

U.S. General Accounting Office (1991). "Management Practices: U.S. Companies Improve Performance Through Quality Efforts." NSIAD-91-190, May.

Wellins, Richard S., Byham, William C., Wilson, Jeanne M. (1991). *Empowered Teams: Creating Self-Directed Work Groups That Improve Quality, Productivity, and Participation.* Jossey-Bass Publishers, San Francisco.

Wiggenhorn, William (1987). Remarks to visitors at Motorola University, November 12, 1987, as recorded by Robert Hoogstoel and shared with the authors.

Other suggested resources:

Ankarlo, Loren, and Callaway, Jennifer (1994). *Implementing Self-Directed Work Teams.* CareerTrack Publications, Boulder, CO.

Graham, Pauline (1995). *Mary Parker Follett: Prophet of Management.* Harvard Business School Press, Boston, p. 127.

Greenleaf, Robert (1977). *Servant Leadership: A Journey Into the Nature of Legitimate Power and Greatness.* Paulist Press, Mahwah, NJ.

Juran, Joseph M. (1988). *Quality Control Handbook,* 4th ed., sec. 15. McGraw-Hill, New York.

Ketchum, Lyman D., and Trist, Eric (1992). *All Teams are Not Created Equal: How Employee Empowerment Really Works.* Sage Publications, Newbury Park, CA.

Kinder, Mickey (1994). *Motivating People in Today's Workplace.* CareerTrack Publications, Boulder, CO.

Nadkarni, R. A. (1995). "A Not-So-Secret Recipe for Successful TQM." *Quality Progress,* November, pp. 91–96.

Quality Progress (1995). July, p. 18.

Senge, Peter M. (1990). "The Leader's New Work: Building Learning Organizations." *Sloan Management Review.* MIT Sloan School of Management, Fall, pp. 6–101.

Wellins, Richard S., Byham, William C., Dixon, George R. (1994). *Inside Teams,* Jossey-Bass Publishers, San Francisco.

SECTION 16
TRAINING FOR QUALITY[1]

Gabriel A. Pall
Peter J. Robustelli

INTRODUCTION

Traditionally, companies have provided training focused on the technical knowledge, skills, and abilities to complete specific tasks as they relate to job responsibilities. In the past decade, however, there has been a shift in this traditional pattern of training just for technical reasons as the need emerged to focus also on the holistic development of the professional. Professional development and the term "professional" have been a reflection of the overall change in educational distribution throughout the workforce. There has been a larger population of better-educated individuals entering the workforce with a greater need for ongoing education and development. There has also been great debate raging over the value of our national education system in preparing workforce candidates for the new work environment. These factors have placed a greater responsibility, and burden, on organizations to respond. Some have responded well, and many others are still struggling with the acknowledgment and the "how to" of this phenomenon. To complicate these matters further, there has been a virtual explosion in information technology advancements, which continue to change at a blinding pace. All of these combined have made a strong case for shifting our approach to training in general, and to Quality training in particular.

---

[1]In the Fourth Edition, material for the section on training for quality was supplied by Frank M. Gryna.

Methods of the Past.   Most organizations have an established manager of training or training department. This department is generally tasked with the training of all people in all things, everything from technical training on the use of customer service techniques to leadership training for the executives. Traditionally the training person or department would be consulted on the most effective way to meet a training need. This would involve assigning either an expert or group of experts from within one or more functional departments to craft a means to meet the training need. These means may be

- On-the-job training
- Classroom training
- Self-instruction through video cassettes, programmed instruction, home study, etc.
- Visits to other companies
- Membership in professional associations
- Computer-based learning

Most large organizations (Fortune 500 size) have a fairly sophisticated training infrastructure in place and are capable of handling the wide variety of requests they receive. They may either develop a training intervention to meet that need or contract with an external professional to provide that expertise in collaboration with the internal expert responsible for that discipline. However, training departments have taken a "hands-off" approach to quality-related subject matter. As a result, most organizations that chose to implement an organized Total Quality Management (TQM) system established a separate quality training organization. This was a component of the "Quality department" and the responsibility of the quality manager. The quality training organization was responsible for developing the curriculum and materials and implementing and evaluating the quality training independent of other corporate training efforts.

Major Issues.   Organizations in the 1990s have begun to realize the concept of "Big Q"; that is, quality is related to every department and process within the organization. This is a breakthrough in itself! As a result of this, the requirement for quality-focused training and education has extended throughout the organization, increasing the scope of activities and level of knowledge of the quality professionals. This increased knowledge among the quality professionals began to foster a barrier to collaboration; there was a distinct separation, even competition, between quality training and other training within organizations.

There has been a major downsizing (re-engineering) effort in the business community during this decade. It has driven many organizations into departmental consolidation and a general rethinking of how they do business. The quality and training departments were not exempted from this. The trend has been a merging of the quality, training, and human resources departments into a more consolidated "professional development" team, threatening each with a loss of identity.

Downsizing and consolidation has forced more work on fewer people, thereby making time an even more precious commodity than it was before. As a result, parts of the organization are in competition for whatever discretionary time is left from service- and production-oriented activities.

MANAGEMENT OF THE QUALITY TRAINING FUNCTION

In view of the major issues training for quality is facing, there has to be an organized and integrated approach to management of quality training in this environment. Some key components of this are

- A delineation of responsibilities for who contributes and in what ways
- A strong and unswerving focus on the customer—internal and external
- A plan established with clear strategies and tactics for quality training
- A budget to fund the plan

Responsibilities. Training for quality, and the TQM system the training supports, can succeed only if there is accountability and responsibility for its implementation and effectiveness. This accountability and responsibility lies with the same group that it does in any other key competitive or developmental strategy—with the leadership team. It is their responsibility to agree on the strategy and assure that it will support the other operational, cultural, and financial corporate strategies. They also have a responsibility to review and evaluate the results of the training strategy. They are not responsible for the planning, design, and execution of the quality strategy; this responsibility generally lies with a component of the human resources function, with technical support provided by key quality professionals.

The responsible parties are

- *Executive leadership:* The executive team (ET) bears the responsibility for creating a quality culture in the organization. A quality culture is a product of behaviors, skills, tools, and methods as they are applied to the work. These changes don't come about without showing people "how" to implement and sustain this culture. Therefore the ET must become educated in quality and stimulate their professional development team to offer options for training for quality. On the basis of these options, the ET will then develop and approve a strategy and strategic goals for the quality training effort. This effort may be organizationwide and long-term (3 to 5 years), or very narrowly focused on a particular segment of the organization or product/service line, and planned for a relatively short duration.

- *Human resources:* The human resources (HR) function (or a subfunction) bears the responsibility for implementing the quality training strategy. The implementation activities include the selection of subject matter, training design and delivery, and establishing an evaluation process. This is integrated into other corporate training activities and follows the same implementation process. The subject matter may be internally sourced, or may be outsourced to external quality training providers. The major difference between how this is approached now compared to the past is that there is a strong trend to seamlessly integrate the quality training into the professional development curriculum and to include a high degree of customization to reflect the organization's culture. This is especially true for organizations that have a mature Total Quality Management system in place.

- *The quality professionals:* The quality professionals (or Quality department) bear the responsibility to collaborate with the HR professionals to share their technical expertise on quality, much the same as key sales professional would share their expertise in developing the curriculum for sales training. This is also a departure from the past, when organizations had elaborate (and sometimes very large) Quality departments that identified, developed, and delivered the quality training, separate from the training department. This created barriers in the implementation of TQM as an integral part of all activities (Big Q) and contributed to the "quality versus real work" dilemma of the late 1980s and early 1990s.

Focusing on the Customer. An underlying principle of quality is to have an unswerving focus on the customer. Training for quality demands the same. A clear understanding of who the customers are, what their needs are, and what the features should be of a training strategy and the subsequent training subject matter that responds to those needs are critical components in training for quality. A contemporary and integrated training system for quality requires an organization to design the system using a process that incorporates all of the basics of quality planning.

A clear understanding of the customer means that all of those who will participate or benefit from the quality training must be considered in the design and delivery. Often organizations discover, too late, that the subject matter and delivery mechanisms of the quality training have been developed on the basis of how course designers *perceive* who the customers are and what their needs may be. Responsive organizations carefully approach this identification of customers and their objectives, and communicate how the quality training can help achieve those objectives. Many times, for lack of a clearly defined corporate training strategy, organizations waste huge amounts of time and money developing quality training or training associates on tools and techniques that they will never use. It was commonplace in the past for organizations to measure their success in quality in terms of the number of individuals they trained and the number of subjects they were trained in.

Developing a Training Plan for Quality. Developing the strategic training plan for quality is critical to the success of any TQM implementation. A strategic training plan addresses these key areas: quality awareness, executive education, management training, technical training, resources, budgeting, and staffing.

Quality Awareness. This addresses the foundation and principles of quality: the definition of quality; the quality processes of improvement, planning, and control (the Juran Trilogy); customer focus; measurement and data collection; reward and recognition; teamwork; and introduction to quality tools. The objective of quality awareness is to convey a basic understanding of

- Why quality is important
- What quality means in our environment
- How quality affects our daily work
- Where we can begin to apply quality concepts and techniques

Organizations have approached training for quality awareness in a variety of ways: (1) They have looked upon this awareness training as all that is necessary for a quality system implementation. (Just tell them and they will do it.) (2) They have implemented various types of technical quality training (statistical quality control, quality improvement process, benchmarking, etc.), ignoring the necessity to establish a foundation for understanding why this is important, a basic requirement for success in adult learning (Knowles 1980). (3) They have focused the quality awareness training on intermittent levels within the organization and ignored others, thereby creating an imbalance in basic knowledge within the organization.

Quality awareness training is critical to the success of any TQM implementation. The training should start at the top of the organization as the introductory component of the executive education. Our experience has shown us that, while many executives have become educated in the language and methods of quality, the diversity in their individual backgrounds has prevented them from finding the common threads to tie all of their collective knowledge together. Developing comprehensive quality awareness subject matter is an effective method to galvanize the thinking of the executive team toward quality and set the stage for initiating the quality training process for the rest of the organization.

Organizations that have won the Malcolm Baldrige National Quality Award (MBNQA), or have seriously competed for it, have developed clear and organized quality awareness training for all employees. Typically, the training is delivered to everyone in the organization, then to all new employees as part of their orientation. This material presents in very basic terms the organization's quality philosophy and how employees are expected to support it.

The United States Customs Service has instituted a quality effort it calls "People, Processes and Partnership." Vice President Al Gore has called the Customs Service the "vanguard" in the reinvention of the United States government to become more customer-focused and businesslike. The awareness training material in "People, Processes and Partnership" is focused on first-line supervisory personnel (above nonexempt and below first-line managers). Of particular interest is Day 3 of the training (Table 16.1), when the entire day is devoted to a description and discussion of the current union contract. Many companies have treated this subject as a separate issue, failing to understand the value of integrating it with quality training. One of the key strategies of the U.S. Customs Service in their reinvention efforts is to develop a strong working relationship with the union.

Executive Education. If quality awareness is the beginning of the training journey for the executive team then where does the journey end? It doesn't! Quality training is a systematic process that is continuously evolved and integrated into the organization's professional development process. Previously, executive education was treated in the same way as other quality training. Executives were asked to gather as a group, and a trainer or lecturer would impart knowledge to them. Given the changes the business environment has undergone these days, the approach has been modified to include a number of other options, both in delivery and content. The subject matter includes the primary subject matter of a TQM system:

TABLE 16.1 People, Processes and Partnership Detailed Agenda for Quality Awareness Training

Time	Description
Day 1	
8:00–9:00	Introduction: Welcome, participant introduction, course layout, logistics
9:00–10:00	Why change: Why customs is changing, what's changing
10:00–10:15	Break
10:15–11:00	Why change: Supervisor responsibilities
11:00–11:30	Change management: Discussion
11:30–12:30	Lunch
12:30–1:45	Change management: Exercise—Barriers to change, brainstorm
1:45–2:00	Break
2:00–2:30	Leadership skills: Coaching
2:30–3:15	Leadership skills: Effective teams, exercise—Toxic waste dump
3:15–3:45	Leadership skills: Effective meetings
3:45–4:15	Leadership skills: Making decisions, negotiating differences
4:15–4:30	Pluses and deltas
Day 2	
8:00–8:10	Review
8:10–9:10	Customer focus: Discussion
9:10–10:10	Business approaches: BPM Trilogy, process planning discussion
10:10–10:25	Break
10:25–11:15	Business approaches: Exercise—Build a paper airplane
11:15–11:30	Business approaches: Strategic problem solving
11:30–12:30	Lunch
12:30–1:10	Business approaches: Writing a mission statement
1:10–1:45	Measurement: Discussion
1:45–2:00	Break
2:00–2:25	Measurement: Variation
2:25–3:30	Measurement: Process control spread sheets, discussion
3:30–4:15	Empowerment: Video and debrief
4:15–4:30	Pluses and deltas
Day 3	
8:00–8:10	Review
8:10–9:45	Part/IBN/contract: Partnership, discussion
9:45–10:00	Break
10:00–11:15	Part/IBN/contract: Partnership, exercise—challenges
11:15–12:00	Part/IBN/contract: Interest-based negotiations, discussion
12:00–1:00	Lunch
1:00–1:30	Part/IBN/contract: Exercise—Part 2: Facts and myths
1:30–2:00	Part/IBN/contract: National agreement with NTEU
2:00–2:10	Break
2:10–3:20	Business analysis tools: Tools 1–4
3:20–3:30	Break
3:30–4:15	Business analysis tools: Tools (5), Pareto diagram discussion
4:15–4:30	Pluses and deltas
Day 4	
8:00–8:10	Review
8:10–9:30	Business analysis tools: Tools (6–10)
9:30–9:45	Break
9:45–10:45	Comparisons: Comparisons discussion, exercise: Scenarios
10:45–11:45	Why change: Linkages again, discussion
11:45–12:00	Pluses and deltas

Source: United States Customs Service.

- Awareness
- Quality leadership
- Roles and responsibilities
- Reward and recognition
- Team processes
- Strategic quality planning
- Customer satisfaction and loyalty
- Benchmarking
- Customer supplier relationships
- Business process quality
- Self-directed work teams

The delivery process has been enhanced. A critical goal for executive quality training is that the executives always be at the leading edge of the organizational learning curve. Executives do not necessarily need to know all of the technical how-to's of a particular subject. However, it is critical for them to be able to articulate what the training is and why it is important. Some of the contemporary delivery techniques that are particularly well suited for executives are

Modular training: Breaking down the training subject matter into bite-sized pieces of no more than 2 hours in length allows a module to be inserted into a scheduled meeting. This minimizes the impact on executive time, provides a break of pattern of a traditional meeting agenda, and makes the subject matter easier to present and digest. It also exposes the executives to a steady flow of quality information.

Just-in-time training: Any training is best delivered as it is required. This allows the recipient to immediately put the knowledge to use (80 percent of what is learned and immediately applied outside of the training environment is retained). For executives, just-in-time training means receiving training in time to support the organization's quality effort, at the leading edge of the implementation time line. The executives' responsibility is to know and understand what quality training is being given and why it is important for the organization to learn and use this subject matter. Their role is promoting the application of the knowledge by others within the organization as well as demonstrating their own application of it. Executives play this role in various ways: by participating in the executive reviews, by taking part in the rollout of the quality training (e.g., presenting an introductory portion of a training course), and by applying the subject matter in their everyday work.

Mentoring: A practical and effective approach to quality training for executives is to use a quality expert, on a regularly scheduled basis, as a personal mentor or coach. This allows the executive to receive professional one-on-one guidance. This will expose the executive to the quality subject matter while applying it to matters of executive concern. Through questions and challenges of concepts and techniques, the executive can develop a personal understanding and sense of confidence. This technique overcomes the challenges of executive time and just-in-time exposure that are posed by group training sessions.

Lecture by peers: Most executives feel more comfortable "hearing it from someone in a situation similar to mine." Peer lecture adds a sense of reality to the quality subject matter. This is best accomplished by having other executives, either internal or external, present their experiences and challenges. Hearing how others at similar levels of responsibility and in similar environments have learned, applied, and improved on quality management applications is an effective learning method for executives.

Self-study: Another personalized learning technique that is effective with executives, is to provide them with the materials, tools a plan to engage in a systematic quality education process. This is crafted jointly by the executive and an internal or external quality expert. There are scheduled review points where the quality expert will consult with the executive to assure conformance to the education plan and exploitation of opportunities to apply the knowledge. Self-study is usually applied in conjunction with the mentoring approach.

Conferences: Attending quality conferences and public seminars gives the executive the flexibility to pick and choose specific subject matter and applications which are the most appropriate. This method is most effectively executed when the executive is matching the learning experiences to the company-wide quality management implementation plan or a specific quality application, such as business process quality, strategic quality planning, self-directed work teams, etc.

Budget Rent-A-Car Company provides an example of a quality training curriculum for executives (Table 16.2). The executive team is called the Quality Committee. The committee designed a comprehensive package of quality subject matter to support a 3-year TQM implementation.

TABLE 16.2   Budget Rent-a-Car Quality Committee

Unit	Description
	Introduction
Section 1	Workshop objectives
Section 2	Why TQM at Budget
	Overview
	Objectives
	Opportunities and advantages
	Case examples of quality and definition
	Summary
Section 3	How to think about quality
	Overview
	Objectives
	Who are the customers?
	How to think about total quality
	The results
	What is total quality management?
	Measuring quality
	Process measurement
	Summary
Section 5	Budget's TQM implementation plan
	Overview
	Objectives
	The roadmap
	Budget's infrastructure
	Roles and responsibilities
	The quality-driven organization
	Upper managers' roles
	Integrating versus separating quality responsibility
	More overhead and paperwork?
	Team operations
	Project mission statements: publication
	The project team
	Contrasting thought processes
	Working together as a team
	Communications skills
	Summary

TABLE 16.2   Budget Rent-a-Car Quality Committee *(Continued)*

Unit	Description
Section 6	Quality improvement Overview Objectives Examples of quality improvement Revolutionary improvement Budget's structured quality improvement process Evaluate alternatives Quality tool: Remedy selection matrix Summary
Section 7	Quality planning Overview Objectives Quality planning versus quality improvement Examples of quality planning The process and tools But we do not have time Summary
Section 8	Quality control Overview Objectives The quality control process Summary
Section 9	Quality for work unit teams Overview Objectives Summary
Section 10	Selecting the correct quality process to help guide your project Overview Objectives Selecting the correct quality process Guidelines for identifying the type of project Relationships among the three quality processes Summary
Section 11	Role for suppliers Overview Objectives Return to the triple role model Summary
Section 12	Strategic quality planning Overview Objectives What is strategic quality planning? The Quality Committee's Role in Strategic Quality Planning Summary
Section 13	Quality Committee meetings Importance of planning the meeting Agenda: Initial quality committee meeting Quality tool: Opportunity selection matrix Agenda: Second quality committee meeting Planning for reviews

Source: Budget Rent-a-Car Company.

Management Training.   Management-level employees are usually the first candidates for quality training. This group includes the organization levels from first-line manager (just below the executive level) to supervisor. Individuals from these levels usually make up the first group of employees to "break the ice" in learning and using quality management concepts and techniques. They make up the membership of pilot quality improvement teams, business process quality management teams, quality planning teams, etc. Whereas the executives are trained in broad strategy and concepts, in preparation for their leadership roles, these individuals receive an abbreviated version of the executive subject matter, and more detailed information on the tools, techniques, and methods. They must understand the how-to's of a TQM system implementation.

Management training focuses on both the technical and human side of quality. Management's quality knowledge must go beyond the strategic quality plan to include quality improvement tools. Additionally, they must be trained to be sensitive to the organizational culture. Most TQM implementation failures are attributable to a lack of attention to this group of individuals. The lack of attention may take the form of failure to provide quality training or to consider their input in redefining the quality culture. Quality training subject matter for this group will fall in both the strategic and tactical categories. Some of the subjects, grouped by category, are shown in Table 16.3. Table 16.4 is a day plan of quality improvement training for managers offered by Juran Institute.

Technical Training.   This type of quality training consists of a wide variety of tools and techniques that enhance the employees' ability to collect and analyze data and present the resulting information for decision making. Because concern for quality has permeated virtually every industry and organization, these tools and techniques vary greatly in type and application. There are, however, a core group of them that are applicable in most industries. A prime example of this is a sample curriculum (Table 16.5) from Florida Power and Light Co. for "Application Expert," a series of training workshops to develop internal experts in statistical quality control (SQC). (FP&L was the recipient of the Deming Prize in 1989, the first non-Japanese company to accomplish this.)

Resources.   Every quality training program needs resources. There must be a purposeful effort to identify the staffing and materials funding necessary to achieve quality training goals. Organizations have begun to understand the value of an organized and focused quality training program. They also realize that there has to be a resource commitment made that is visible and actionable by those responsible for carrying out the training. There have been many instances

TABLE 16.3 Management Training Subjects

Strategic	Tactical
Developing strategic measures and goals	Quality processes
Deploying the strategic quality plan	• Improvement
Understanding business processes	• Planning
Quality systems	• Control
Quality culture	Quality tools
• Quality values	Facilitation skills
• Empowered employees	Communication skills
• Customer focus	Data collection and analysis
• Collaboration	Inspection and measurement
• Commitment	Assessments
• Creativity	Cost of quality
Reward and recognition	Statistical methods
Review and audit	Quality team roles and responsibilities
	Benchmarking
	Self-directed work teams

TABLE 16.4 Quality Improvement for Managers

Time	Description
Day 1	
8:00–8:30	**Introduction**
8:30–10:00	**Module 1: What is Quality?**—Overview, objectives, what is quality? Who are the customers?, external customers, triple role, total quality
10:00–10:15	**Break**
10:15–12:00	**Module 2: What is TQM?**—Overview, objectives, why have total quality management?, delighted customers, empowered employees, higher revenue, lower cost and the cost of poor quality, cost of poor quality categories, the results, what is total quality management?
12:00–1:00	**Lunch**
1:00–2:45	**Module 3: The Juran Trilogy**—Overview, objectives, a financial analogy, quality planning, quality control, quality improvement
2:45–3:00	**Break**
3:00–4:30	**Module 4: Organizing for Quality Improvement, Quality Councils and Project Teams**—Overview, objectives, quality council, cross-functional teams, roles and responsibilities
4:30–5:00	**Questions and Answers/Wrap-up**
5:00	**Adjourn**
Day 2	
8:30–9:00	**Review Day One**
9:00–10:30	**Module 5: The Diagnostic Journey**—Overview, objectives, analysis of symptoms, Pareto diagram, flow diagrams, formulate theories of causes
10:30–10:45	**Break**
10:45–12:00	**Module 5: Cont'd.**—Brainstorming: generating creative ideas, the cause-effect diagram, test theories, stratification, histograms, scatter diagram
12:00–1:00	**Lunch**
1:00–2:30	**Module 6: The Remedial Journey**—Overview, objectives, the remedial journey, consider alternatives, design-controls, provide a means to measure the process, establish the control standards, determine how actual performance compares to the standard, holding the gains
2:30–2:45	**Break**
2:45–4:00	**Module 7: Your Role in TQM**—Overview, objectives, change in results, reward and recognition
4:00–5:00	**Questions and Answers/Wrap-up**
5:00	**Adjourn**

where an aggressive training plan for quality has been developed and not implemented. The reasons stated are invariably lack of funds, time, or people to carry out the plan. Why haven't these resource requirements been considered? The common approach has been to develop these plans, then look into departmental budgets to fund the implementation. These funds are usually committed to other endeavors and are either refused or begrudgingly committed. Training for quality has to be budgeted, staffed and planned like any other business activity. If these resources are to be part of departmental budgets, the departments have to be part of the planning process (as customers) and see the benefits. World-class quality organizations, e.g., Malcolm Baldrige National Quality Award winners, budget a minimum of 40 hours per year for quality-related training. Many of them have a very small staff of trainers, depending instead on departmental volunteers to carry out the bulk of the training efforts. These volunteers are qualified by the quality-training professionals and bring an air of "credibility" to the training experience. The funding for quality training is planned for and budgeted by the Training department or individual departments. The Training department generally budgets for all developmental costs. The Quality department provides a technical consulting role to the training department in the development of materials and qualification of instructors.

TABLE 16.5 Technical Training

	Application Expert—Session I
Unit 1	Basic concepts of Total Quality Control
Unit 2	Basic data analysis
Unit 3	Histograms and frequency distributions
Unit 4	Overview of probability distributions
Unit 5	Discrete probability distributions
Unit 6	The normal distribution
Unit 7	Estimation of the mean
Unit 8	Introduction to hypothesis testing
Unit 9	Test of the mean with a known population variance
Unit 10	Test of the mean with an unknown population variance
Unit 11	Test of a single population variance
Unit 12	Estimation of the population variance
Unit 13	Test of two population variances
Unit 14	Tests of two population means

	Application Expert—Session II
Unit 1	Test of one proportion
Unit 2	Test of two proportions
Unit 3	Contingency tables
Unit 4	Review/update
Unit 5	Sampling
Unit 6	One-way analysis of variance (ANOVA)
Unit 7	Two-way ANOVA
Unit 8	Estimation of parameters
Unit 9	Project examples
Unit 10	Introduction to control charts
Unit 11	X-bar and R charts
Unit 12	X and R charts
Unit 13	X and R charts (individual observations)
Unit 14	Process evaluation
Unit 15	np chart
Unit 16	p chart
Unit 17	c and u charts

	Application Expert—Session III
Unit 1	Design of experiments
Unit 2	Correlation (I)
Unit 3	Correlation (II)
Unit 4	Regression (I)
Unit 5	Regression (II)
Unit 6	Regression (III)
Unit 7	Regression (IV)
Unit 8	Reliability overview
Unit 9	Reliability A type
Unit 10	Reliability—EMEA
Unit 11	Reliability—Weibull
Unit 12	Reliability—B type

Source: Florida Power and Light Co.

Budgeting.   Training for quality, as any other key activity, requires a dedicated financial commitment. These finances can be centralized or decentralized. However, they need to be committed specifically to the strategy and tactics that support training for quality. Many organizations have chosen a decentralized approach, in which each individual business unit designates a segment of its overall training budget to training for quality. The quality projects that are planned for the year are analyzed for training requirements and prioritized. The budget for quality training is then matrixed against these requirements and the executives make specific decisions whether to increase, realign, or decrease their quality training resources. Identifying and budgeting for what is to be achieved, rather than letting the amount of money available drive how much training will be done, gives the executives a different perspective on training for quality and budgeting. It allows them to make return-on-investment decisions that can dramatically change the organization's view of the value of training for quality.

Staffing.   The personnel requirements necessary to support training for quality have changed significantly over the past decade. Formerly, there was dedicated staffing for all aspects of training for quality, from development of materials to delivery. These functions were among the key roles of the "quality department." As the need for quality training increased, these departments tended to grow, sometimes exceeding 20 to 30 people. As organizations began to re-engineer themselves, these departments became targets to consolidate and streamline with the corporate training function. Currently there are many basic types of staffing options that organizations utilize. There is, however, one constant trend: quality departments and the training function for quality are leaner and multifunctional. Various staffing structures are described below.

Centralized: This is the traditional structure, in which there is a dedicated group of individuals that research, develop, instruct, and evaluate the quality-training curriculum. Many organizations have found that this structure tends to segregate the quality-training strategy from the general education strategies, creating a learning barrier. Participants view the quality training as something distinctly different from other professional development activities.

Hub and spoke: This is a more common arrangement for quality training. It consists of a training coordinator or manager at the headquarters unit with a "dotted line" relationship to training professionals in other divisions or business units. These individuals are usually part of the quality department infrastructure that is deployed throughout the organization. A good example of the hub-and-spoke structure is employed by the Golden Hope Plantations, S/B, in Kuala Lumpur, Malaysia. This organization has successfully applied the hub-and-spoke model to headquarters and each of its four major business units. These training professionals report directly to the division general manager and have a "dotted line" relationship to the director of Quality and Environment. The director has a strong collaborative relationship with the director of Human Resources, who has the overall training responsibility. This is a very efficient and effective professional development approach.

Decentralized: In this model, each department, division, or business unit has its own approach to training for quality. There is a common strategy. However each unit is charged with developing its own tactics and deploying them within their areas of responsibilities. This is not an effective approach to quality training and takes a much higher level of coordination.

Shared: This is similar to the hub-and-spoke model, with slight enhancements. The trainers draft volunteers from the organization to support the training plan. These volunteers are selected from those who have received quality training in the past and may be supporting the quality system implementation as facilitator, team leader, or team member.

External quality consultants: In the 1990s, there has been an exponential increase in the availability of external quality consultants. For many organizations this has provided an effective supplement to their training efforts and already lean quality departments. External quality consultants provide an immediate, qualified resource that can support the organization's planning, development, delivery, and evaluation of training for quality.

CURRICULUM DESIGN

Curriculum design is the mainstay of a successful quality training system. The process by which the subject matter is selected and shaped into a curriculum is of critical concern. Key elements of this process are

- Analysis of customer needs
- Instructional design
- Content development
- Pilot testing

In quality training, the designers of the training curriculum and subject matter must understand clearly the needs of the trainees. First among these needs are their business needs. In the past, training for quality was limited to a basic curriculum of tools and techniques, and it was left up to the customers (those benefiting from the training) to determine how to best apply the training to their business environment. Contemporary quality curriculum design focuses on those areas where quality can support the business objectives, through either some skill application or philosophical understanding.

Once the business needs are identified, the designers develop learning objectives (outcomes) and tie them to method of instruction (instructional design). Instructional design consists of developing specific "learning events" or "educational tactics" that can assist an individual in translating written word into knowledge. These learning events may take the form of problem-solving exercises, role playing, case study analysis, group activities, and other interactive learning techniques. The key objective is to minimize the traditional passive approaches to learning (lecture and reading) in favor of a more interactive, "hands-on" approach. An effective quality-training curriculum will string these learning events together, paying particular attention to the timing, content, sequence, and technique, to successfully bridge from one to another while continuing to focus on achieving the learning objectives.

Content development is also based on the learning objectives, which, in turn, is influenced by three factors: the performance needs, job requirements, and audience.

The *performance needs* represent the difference between the level of skills, knowledge, and abilities required to meet the business needs and the current level the target audience possesses.

The *job requirements* identify the specific learning that is missing and specify how the skill, knowledge, or ability will be used in the execution of the job. The knowledge level of the audience will determine the level of complexity appropriate to assure an effective learning experience. With all of these things considered, designers may then set to work at developing quality subject matter that will meet the learning objectives and create an effective and enjoyable learning experience.

Once the content has been developed, the quality training should be pilot-tested with a group of objective participants. This group can consist of trainers or a randomly selected group of the target participants. The key to a successful pilot is that the participants be given a specific set of instructions describing how to evaluate the content and design. The pilot testing is time-consuming. If the test is improperly planned, it will be ineffective. However, with a pilot audience that has been properly prepared, it can become a seamless step in the curriculum design process that pays off in a more effective educational product that clearly achieves the learning objectives.

Figure 16.1 is a flowchart depicting the product development process used by Juran Institute. The flowchart depicts some steps that have not been described above, specifically having to do with marketing and an annual plan. These steps are included in the Juran Institute process because its

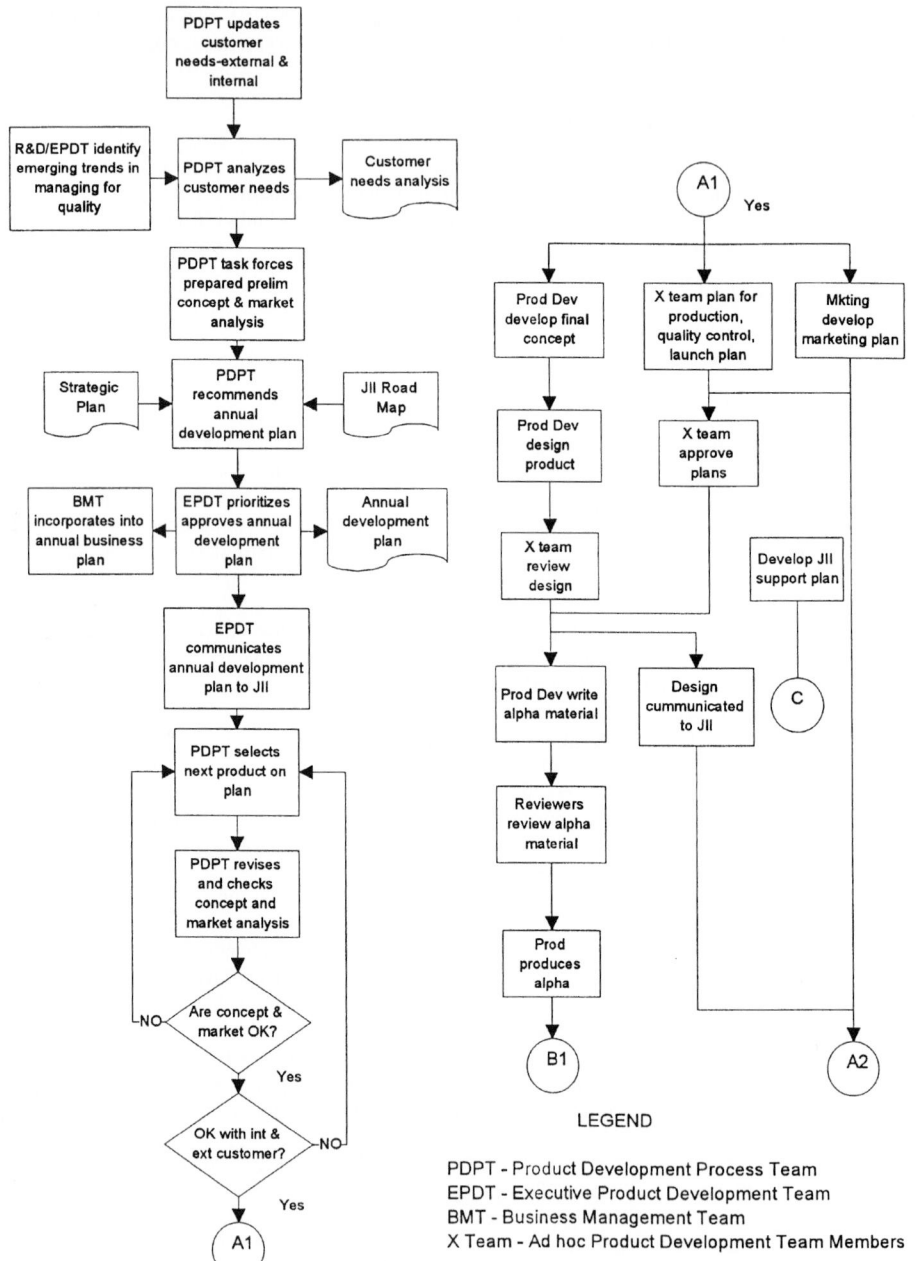

FIGURE 16.1   Product development process.

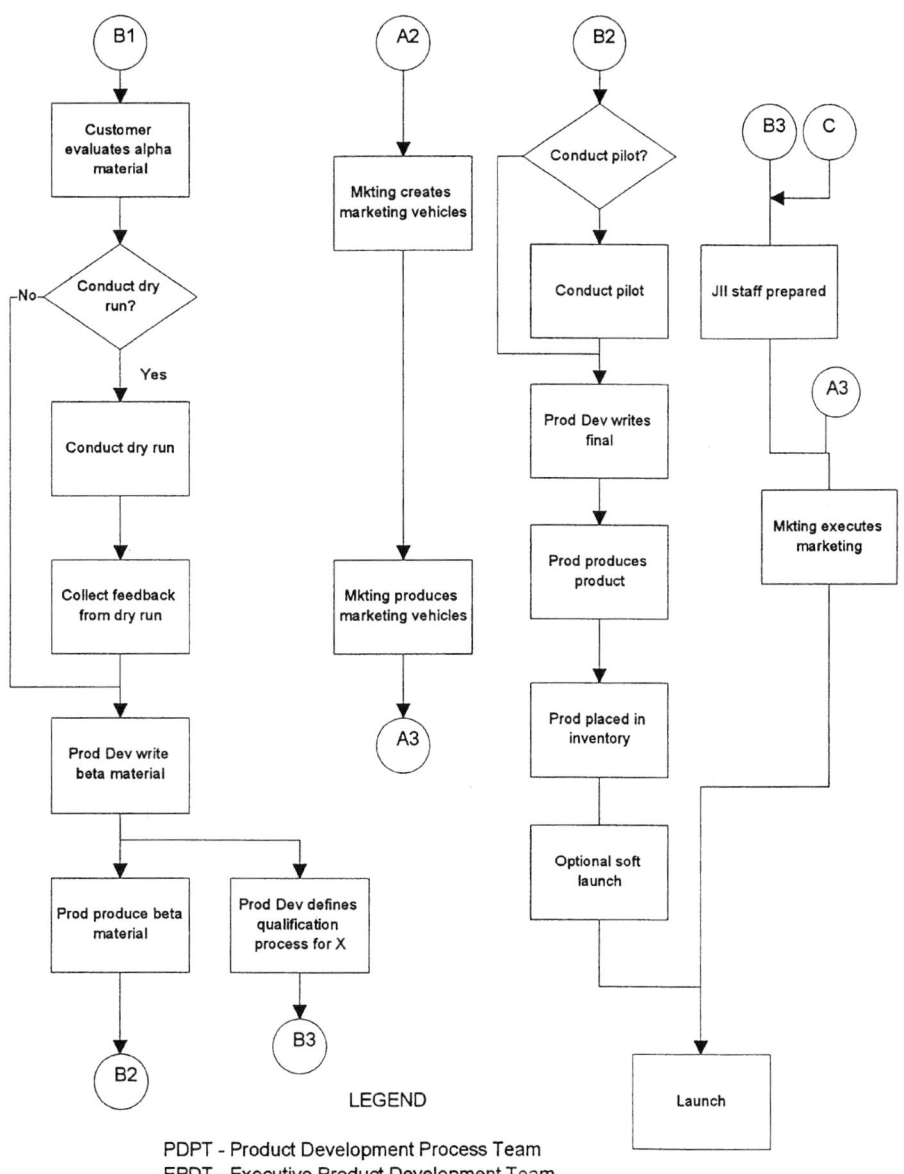

LEGEND

PDPT - Product Development Process Team
EPDT - Executive Product Development Team
BMT - Business Management Team
X Team - Ad hoc Product Development Team Members

FIGURE 16.1 *(Continued)*

quality-training products (text materials, videos, software, and seminars) are for public distribution as well as for internal use. A process for developing product only for internal use would normally exclude these steps.

IMPLEMENTATION AND DELIVERY

Training for quality is usually delivered in two steps:

1. *Implementation planning and preparation:* The assessment, selection, and deployment of delivery media, mechanisms, and facilities; the establishment of physical infrastructure and information technology support; assignment of support staff and other resources.
2. *Delivery:* Transfer of the training content (subject matter) to the students, in an individual or group setting.

While instructional design is usually optimized around a set of delivery methods and media, the implementation planner still has a wide choice of alternatives in arriving at the most effective and also most cost-efficient solution.

Instructional Techniques. Before selecting delivery media, the implementation designer must decide what instructional techniques will be most appropriate for delivering the training. Training will be most effective in transferring skills and enhancing knowledge when the following conditions exist:

Program material is easily available to the student.

The training is measurable on the basis of acquired competency.

Facilitation is provided (if delivery occurs in a group setting).

Materials possess high graphic clarity.

There is interactive content.

There is a high degree of learner control over the pace and receipt of training.

The training includes performance-based testing and scoring.

The training materials are reusable and revisable.

The training is supported by practice and exercise.

Some of these conditions conflict with the choice of delivery medium, requiring the designer to make trade-offs. For instance, if learner control is an important requirement, then classroom training delivered by a live trainer may be a poor choice.

Delivery Systems, Media, and Devices. The choice of instructional techniques will determine what delivery system, medium (or media, as appropriate), and devices will be the most effective and cost-efficient for a given training event. Delivery system options include: (1) Lecture-based, (2) computer-based, and (3) video-based delivery systems, each with its related set of delivery media and support technologies.

The available media options include:

- Live lecture
- Audience-response systems (ARS)
- Workbooks and case studies
- Audiotape
- Live video (TV)

- Recorded video: tape, videodisc or CD-ROM
- Multimedia computer (capable of delivering text, audio, motion video, and graphics)

The cost aspect of delivery system and media choice is critical. The normal range of development times for the most common delivery approaches varies greatly. For each delivery approach, development time is the time required to create 1 hour of delivered training (known as student contact time). Instructor-led delivery requires, on the average, 6 to 10 development hours per contact hour; workbook-based delivery requires 40 to 50 hours, textual computer-based training requires 100 to 200 hours, and interactive multimedia videodisk training could demand as many as 200 to 500 development hours per student contact hour.

Additional factors affecting the total cost of quality training include: students' salary, travel expense, and trainer-related costs (salary and travel for internal trainers, fees and travel for externally sourced consultants and facilitators).

Summary.   Instructor-led programs are appropriate when the training request requires quick response and the design calls for frequent interaction. Computer-based training is more efficient for a large audience across which to defray the costs. It requires adequate development time. Multimedia approaches are best when the training program includes multiple activities (information presentation, exercises, case studies, laboratory or practice sessions, etc.). While usually expensive, multimedia approaches help avoid bad compromises, like reading books from computer screens or using "talking heads" instead of graphic illustration.

External versus Internal Sourcing.   For maximum training effectiveness, the quality of the delivery system—and in particular that of the trainers—is critical. The most appropriate and best delivery system or trainers may not always be available within an organization. For this reason the sourcing of delivery systems and trainers (including facilitators and sometimes even support personnel) is an important decision during implementation planning.

This decision usually depends on two factors: (1) the size of the training budget and (2) training policy.

In many organizations the quality training budget allows only the most basic training skills to be established internally—usually limited to training on quality control, standardization, and basic quality assurance techniques. Specialized subject matter knowledge and training skills, such as related to business process design and re-engineering, benchmarking, cost of quality analysis, strategic quality planning, and the like, are considered unaffordable by many organizations. These sorts of training subjects and skills, and related delivery systems will probably be outsourced.

On the other hand, there are even large organizations which, as a matter of policy, do not internally source any of their job- or skills-related training—including training for quality. These organizations routinely outsource all their quality-related training using hired facilitators, trainers, and consultants. One favored source for such externally acquired training is the academic community: Training in a number of quality-related topics is available from faculty members of business and engineering schools.

Training Methods.   There continues to be a competition between time-tested, conventional training methods and technology-based knowledge transfer, which approaches the problem in another way. In each case, one can again focus on individualized methods and group training.

Conventional Training Methods.   *On-the-job training* (OJT) is still an effective training method in use today, specifically applied to the training of quality improvement and quality planning teams. In an OJT environment, the student (learner) performs the job about which the training is provided under the supervision of an experienced trainer or mentor. This approach is particularly effective for such subjects as quality control, quality assurance, process design or redesign, benchmarking, and general problem solving (e.g., quality improvement).

Independent study, by definition, is any method of studying alone. Independent study in the form of self-instruction (where students take responsibility for their own learning) is frequently used to acquire the knowledge and skills needed for quality management tools.

Self-directed learning is an instructional approach in which the highly motivated student takes the initiative to master preplanned training material. Self-directed learning may be completed by the individual trainee, using self-instructional packages, or conducted with the help of trainers, facilitators, mentors, etc. The student is provided a study plan and all the materials needed to execute it. These are usually text, less frequently graphics and visual aids. Such packages are known as *learning resource packages.*

Group learning methods generally do not require special technology, equipment, or facilities. Each method, however, requires a skilled instructor to plan, prepare, present, and facilitate the learning related activities.

Instructor-led group training methods are preferable when:

- Skilled instructors are available
- Knowledge must be transferred to a large number of students
- Training (content and instructional design) must be developed and delivered within a short time
- Students lack basic learning skills (such as reading and arithmetic)
- Subject matter is difficult to comprehend
- Students can be convened in one place, on a predetermined schedule

The particular methods of group training include: lecture, discussion, presentation and demonstration, case studies, role playing, and instructional games. Again, depending on the quality-related subject, each one of these methods finds effective use in training for quality.

Technology-Based Training.   Technology-based training (TBT) teaches, manages, and supports the instruction process. Most importantly, it is capable of making learning easier and more effective. TBT includes computer-supported learning resources, computer-managed instruction and computer-assisted instruction.

Computer-supported learning resources neither teach nor manage the instructional process; they represent the library or repository of resources from which a student may learn.

Computer-managed instruction (CMI) is the management of instruction by computers. In a CMI environment, learning resources do not have to be technology-based, and CMI can function independently as the manager of the instruction process. The key point is that CMI does not directly involve learning; however, it offers the power to make learning more efficient and effective.

As a management system, CMI operates in three modes: testing, study prescription, and record keeping. Of these, study prescription is the mode most directly related to the actual instructional process: the CMI system generates an instructional prescription for each unmastered learning objective. Accordingly, each student can receive individualized study prescriptions. This significantly reduces the time each student needs to study and is the basis for the instructional efficiencies associated with CMI.

Computer-assisted instruction (CAI) uses a computer in the instructional process for knowledge transfer to the student. CAI may operate in several modes of instruction, such as tutorial, exercise and practice, instructional games, modeling and simulation, and problem solving.

In training for quality, each of the modes of CAI can be effective, depending on the subject matter. Exercise and practice is used mainly in tools training; problem solving and tutorial are the key modalities for training in the basic quality management methodologies, such as quality planning, control, and improvement. Modeling and simulation are highly effective in training related to business process design, re-engineering, and analysis. Instructional games are primarily used in teaching training and facilitation skills, leadership skills, collaborative team building, and meeting-management skills.

Distance learning is an increasingly productive version of CAI: It supplies instruction to students dispersed over wide geographic areas. It enables the simultaneous study of advanced subjects by groups of individuals who could not otherwise attend a learning event in a single geographic location at the same time. Distance learning also allows the possibility of participation in learning activities at a time of the student's choice.

Distant students can be organized either as individuals (located in a variety of places), or groups (at one or several sites).

In practice, quality-related distance learning programs are based on one of two strategies: (1) video teleconferencing (group synchronous, i.e., all students participate simultaneously) or (2) computer conferencing (individual or group, synchronous or asynchronous).

The National Technology University (NTU) is an excellent example of video teleconferencing, a model for which is the regular classroom. Video teleconferencing courses emanate from some 30 participating universities with uplinks or broadcast stations. Direct phone lines from the receiving sites to the campus classroom provide for instructor-student interaction. Electronic mail and telephone response service supplement this interaction. NTU has broadcast a number of quality-related courses which have been reported as challenging and highly applicable to the students' work environment.

Learning Laboratories.   Most modern quality management processes and related tools require the introduction and use of new technologies—especially as they relate to the management of information flow and data. Many organizations, however, discover that their employees often do not have the skills to capitalize on these new technologies.

How does an organization integrate new technology into the way it manages quality? How does the organization get its employees to utilize the new technology? Who needs to be trained?

More than a classroom experience is required to answer these questions. Hands-on learning experiences must be included in the training. Learning laboratories provide the means to achieve this. They are specifically designed for education and training and not for technical research or production. The model of complete knowledge transfer by means of learning laboratories comprises:

- Classroom introduction to quality management concepts through lecture, discussion and team activities
- Training exercises and case studies
- Practice with real tools and applications in a realistic, workplacelike environment

The three types of learning laboratories most frequently used in training for quality are (1) computer-based, (2) programmable automation, and (3) manufacturing equipment laboratories.

Collaborative Training Systems.   Collaborative training systems are the most advanced versions of computer teleconferencing. Over the past decade, three major developments have contributed to the growth of participative, on-line training and collaborative work sharing. These include:

1. The significant increase in the internal training needs of organizations in both the public and private sectors
2. The advent of work-process-oriented relationships based on collaborative structures such as project teams, empowered cross-functional teams, and self-directed work teams
3. The development of "groupware"—the class of systems and software that enable efficient communications and the collaborative implementation of projects within and among work teams

Employee Training.   Over the past two decades, and more than ever before, internal employee training has become a key strategy and, in many cases a prerequisite for achieving business success. Many organizations in the United States, Japan, and Western Europe have launched massive training initiatives in the basics of business and technology, quality, and other strategic disciplines.

In an article in *Across the Board,* Steve Blickstein (1996) states that the education and training costs of U.S. businesses are now estimated at over $30 billion per year. The Saturn division of General Motors provides its 9200 workers an average of 90 hours of training per year. Saturn expects all employees to spend about 5 percent of their time in training and education each year. Intel University now has a training budget equal to about 5.7 percent of payroll. Motorola University has a budget of $120 million per year and 14 branches in the United States, with more branches in Europe, Asia, and Latin America. IBM has a standing policy requiring every employee to receive a minimum of 40 hours of training every year.

Many companies have developed their own in-house world-class education programs—with related professional training organizations and support systems—for their employees. Some claim that they now provide over 80 percent of the training of their employees in-house, whereas a few years ago much more training was done by schools, consultants, and other outside providers.

Collaborative Work Practices.   The application of collaborative computing technology to a variety of work shared by a number of individual or group participants results in collaborative work practices. These practices represent an increasingly useful, effective, and highly time- and cost-efficient way of carrying out shared work by participants who are geographically dispersed, or whose schedules conflict. Collaborative work practices allow specially trained experts (trainers, facilitators, team leaders, or process owners) to teach, and work on advanced solutions together with other professional participants, when a reasonable size group could not be assembled in a their own physical vicinity.

Collaborative work practices are based on the *team* as the basic work unit. The team may perform work in two modes: (1) *learning* as a team, and (2) *processing* as a team—executing tasks or activities learned, resulting in a specific *work product.* In either mode, participants can be organized, and geographically located either as individuals (located in many places) or as teams (the larger team may comprise several smaller subteams or squads, each located at a different place).

Participants—both team-member and individual—may work in one of two timing relationships: synchronous (all simultaneously) or asynchronous (at different times, usually of their own choosing). Each of these relationships offers specific advantages to the participants and to the organization implementing collaborative work practices. In practice, collaborative solutions are implemented by using a mix of these geographic and timing strategies.

"Groupware" and Collaborative Computing.   Groupware is the generic term for software designed to be used by groups on a shared basis. Collaborative computing is the system of communications network and appropriate collection of applications. While these systems enable asynchronous and geographically separated efforts, collaborative computing supports more traditional teams as well. The system should train team members on a given subject (e.g., process improvement, programming, project management), provide tools to support the subsequent team efforts, and allow archiving and retrieval for current and future projects.

Training and Performance Support.   Organizations face considerable challenges with their current training efforts:

- How to replicate success across the organization?
- How to measure the effectiveness of training?
- How to keep training, and work materials up-to-date, and relevant?
- How to reduce the costs of training, and teamwork?
- How to ensure the correct application of new skills and tools?
- How to provide "just-in-time" training?

The collaborative system should facilitate training, both in a classroom environment and at the desktop. Designing materials for both environments reduces development costs and provides continuity throughout the project cycle.

Measurement.   The success of training, and collaborative projects is not obvious from casual observation; it has to be measured in terms of specific, widely accepted and understood metrics. These metrics include:

- Learning metrics (such as comprehension, retention, on-the-job application)
- Cycle time
- Work-product quality
- Individual and team productivity
- Cost

These metrics are developed from the identified critical system requirements prior to building a collaborative computing system, and, to the degree possible, the system should track, analyze, and report on these metrics automatically. These empirical data can be combined with focus groups or other techniques to assure the system is optimized.

EVALUATION

Why Evaluate?   Evaluation is critical to effective training for quality. Evaluation is not just an afterthought but a necessary and systematic part of an effective training process for quality. Good training begins with an accurate assessment of training needs, and then proceeds through needs analysis, instructional design, and content development of training events, ending with more effective on-the-job and organizational performance of those trained.

Here we present practical guidance to assist training professionals to (1) select and design the evaluation approach, (2) evaluate the indicators of success, and (3) manage the evaluation process.

Systematic Approach to Training Evaluation.   The classic framework of training evaluation is based on the simple four-part evaluation process for training, proposed by Donald Kirkpatrick in 1976. It assesses:

1. Trainees' reactions
2. Trainees' learning
3. Whether and how trainees are using what they learned
4. Whether and how the use of learning has enhanced job performance

Over the years this basic model has been modified by a number of authors and training professionals, primarily in the area of establishing intensive front-end evaluation. This modification allows a more objective measurement of the *incremental* learning accomplished during the training session. Another aspect of training evaluation is related to *retention*—which can only be assessed sometime after the training event, say 6 to 12 months or more.

In the case of training for quality, the function of evaluation is broader than simply checking to see whether a particular training event has achieved learning results. The purpose of quality-related training is to enhance the value of the products and services of the organization, through the systematic improvement of the skills and knowledge of trainees who contribute to those products and services. Ultimately, the purpose of training for quality is to enhance customer satisfaction and loyalty.

Therefore, in designing an evaluation approach for quality-related training, the evaluation must fit the strategic context of improving product and service quality. The implications for evaluation design are

Objectives of training for quality should be derived from the organizations's quality strategy—established through strategic quality planning.

Training for quality should be conducted only after the organization receiving training has deployed its quality measurement system.

Training for quality should be delivered on a prioritized basis such that those employees with direct effect on customer satisfaction receive training first. Trainees should be ready for the training and have the prerequisite knowledge and preparation. Examples during training should include references to product and service quality; training sessions should also allow students to practice quality management skills, including appropriate tools. Students should leave training committed to the application of their newly acquired knowledge. Once back on their jobs, students need accurate and timely feedback on their impact on organizational performance, especially regarding customer satisfaction.

In designing evaluation of training for quality, the key question is how much and what sort of evaluation should be carried out. Evaluation design is a continuing process, since evaluations also must be continuously improved and adapted to changing needs. Evaluation will always work best when it has been part of the overall training strategy and training design. As with any reasonably complex process, managing the evaluation process should include the use of project management concepts, tools, and techniques, as well as professional quality management expertise as appropriate.

In summary, evaluation must be employed throughout the training process, not just during and after the training event itself. A good evaluation of training for quality will help assess:

1. Are the goals of training for quality linked to major business goals, and is the quality related training strategy driven by critical business needs?
2. Do training plans deliver the required amount of learning at the right time and in the most effective and efficient ways?
3. Are training outcomes (e.g., learning, retention, and application on the job to enhance organizational performance and customer satisfaction) being achieved?

As training for quality alone cannot meet business needs, the proper role for training evaluation is to help build strong partnerships between training professionals and their customers.

Why Training Fails.   Training for quality can fail for a variety of rather conventional inadequacies: in facilities, in training materials, in leaders, and in budgets. Such inadequacies are usually obvious enough to generate alarm signals to those directing the training program. The more subtle reasons for failure are also the more serious, since they may generate only subtle alarm signals, or no signals at all. Such subtle reasons (Juran 1989) include:

- Lack of prior participation by line managers
- Too narrow a base
- Failure to change behavior

Management has a role to play in heading off failures of training programs. That role consists of laying down the necessary policies and guidelines, which include the need for a strategic plan for training for quality and the requirement that trainees should apply their new learning to their jobs.

THE FUTURE OF TRAINING FOR QUALITY

In any organization, the prime role of training is to develop work-related skills, knowledge, and expertise. Training for quality occupies a special position in the spectrum of training activities, as it supplies quality-management expertise that can have a major impact on the organization's relationship with its customers.

Quality management practices straddle the entire spectrum of organizational performance. In an industrial and manufacturing environment, training for quality traditionally addresses levels of performance related to operations and troubleshooting—primarily through training in quality control.

After World War II, and in particular as a result of the research and practical work of such experts as Deming, Feigenbaum, Ishikawa, and Juran, quality management began to shift its focus to quality improvement and planning for quality—with an attendant shift in the focus of training for quality.

Another, more recent, development has been the extension of training for quality to include training in teamwork, team building, meeting management, and team facilitation—as necessary methods and techniques supporting the implementation of quality management through the empowerment of teams and individual employees.

The coming century will see a further expansion of the training for quality into the *technological domain,* where problem-solving methodologies will be complemented and enhanced by artificial intelligence and collaborative computing techniques. Another expected development is the application of systems theory to quality-related training, as a shift in *systems thinking* takes place from a closed system view (e.g., technological and chemical processes, and related quality issues) to an open system view (e.g., organizational and business processes, with quality, information, and human-resource issues). Finally, the increasing relevance of *economic and financial* theory to quality management will require an extension of the training for quality to include economic and financial subjects (e.g., cost-of-quality analysis, activity-based management, and the like).

RESOURCES

Examples of Training for Quality.   Examples of currently available, highly effective approaches and sources for quality-related training include:

American Society for Quality Control

American Society for Training and Development

Corporate universities (Motorola, Arthur Andersen, etc.)

IBM Quality Institute

Juran Institute

National Technological University (NTU)

References

Blickstein, Steve (1996). "Does Training Pay Off?" *Across the Board,* June.
Juran Institute (1992). *Total Quality Management: A Practical Guide.* Juran Institute, Wilton, CT.
Juran, J. M. (1989). *Juran on Leadership for Quality.* The Free Press, New York, pp. 342–343.
Kelly, L. (ed.) (1995).The ASTD Technical and Skills Training Handbook: McGraw-Hill, New York.
Knowles, M. S. (1980). *The Modern Practice of Adult Education,* 2nd ed. Cambridge Book Co., New York, 1980.
Senge, P. M. (1990). *The Fifth Discipline.* Doubleday, New York.
Swanson, R. A. (1994). *Analysis for Improving Performance.* Berrett-Koehler, San Francisco.

SECTION 17
PROJECT MANAGEMENT AND PRODUCT DEVELOPMENT

Gerard T. Paul

INTRODUCTION

Development is the translation of research findings or other knowledge into a plan or design for new, modified, or improved products, processes, and services, whether intended for sale or use. It includes the conceptual formulation, design, and testing of alternatives, the construction of prototypes, and the operation of initial, scaled-down systems or pilot plants (Industrial Research Institute 1996). The focus of the present section is the process of project management applied especially to the development of new physical products.

A *project* is a task which is undertaken in a structured manner. A project organizes the task and its proposed resolution into a structure in which there is clear definition of the undertaking and the corresponding plan for its execution. Moreover, there is visibility of the progress toward completion in terms of time and resources applied versus targets previously established. The project structure also permits managerial oversight and approval at key checkpoints along the way.

The project approach has long been favored for undertakings such as product development that involve a significant expenditure of personnel, time, and resources, especially when they are considered essential to the well-being of the enterprise. More recently, the project approach has been shown effective when applied to apparently lesser problems, especially when applied serially to accomplish incremental change and even to effect breakthrough, as in quality improvement. In this instance, Juran (1988) has described a project as "a problem scheduled for solution."

What follows draws on the author's experience in developing technical products for manufacturing. The management process outlined here is specific to product development, but is applicable generally to any project undertaking, whether to product or process development, or to product development in the service sector—as for example, restoring the grandeur of Michelangelo's ceiling in the Sistine Chapel or preparing this handbook for publication. In short, with some suitable modifications and fine tuning to fit the specific task at hand, the project approach works for just about any undertaking worth the effort.

The primary attraction of the project concept as a management tool is its focus on results and the means to achieve those results. It is structured; there is a beginning, a middle and an end. When a project has been completed successfully, something happens; a new product, a new service, an improved process, comes into being where it did not exist before.

PROJECT MANAGEMENT FRAMEWORK—THREE PHASES

There are three steps or phases in every project corresponding to the beginning, the middle and the end. These are the *Project Concept* phase, the *Project Discovery* phase, and the *Project Implementation* phase. There is often also a fourth step—the *Verification* phase—which continues beyond implementation. Each phase is a multicomponent undertaking, and some of those components specifically contribute to the quality of the process itself and of the resulting product. The means to achieve improved quality must be woven into the project management process from beginning to end.

The Project Concept Phase. Some projects start out as vague ideas. It is important to eliminate vagueness quickly. In every case, moreover, vagueness surrounding key project goals must be eliminated during the concept phase. There are ways to do it, as we shall see. The Concept stage consists of those thought processes, discussions, and activities that allow us to describe precisely and concisely what it is we intend to do, and then write it down. Proper completion of the concept phase consists in writing the *Project Concept Statement,* clarifying and refining the project concept statement, writing the *Product Requirements Document,* preparing the *Project Authorization* for the *Discovery Phase,* and gaining approval to proceed to the Discovery Phase.

The Project Concept Statement. Start with the *Project Concept Statement.* Keep it simple—the complications will arise to greet us soon enough. Clear focus will help us deal with them. The Concept Statement is also an important means to eliminate vagueness. Consider two examples:

When the order for Florence's majestic Cathedral of Santa Maria del Fiore (St. Mary of the Flower) was given over to the master builder Arnolfo di Cambio in 1296, a proclamation described the citizens' requirements: "The Florentine Republic, soaring ever above the conception of the most competent judges, desires an edifice shall be constructed so magnificent in its height and beauty that it shall surpass anything of its kind produced in the times of their greatest power by the Greeks and the Romans." (McCarthy 1963).

The cathedral was required to be *higher* and *more beautiful* than anything known in the ancient world. This was a big idea—the city fathers knew it was a big idea—but it is clear and easy to understand. Higher is simply higher. More beautiful is ambiguous to our ears, but it was much less so to the Florentines of that time who, by their own remarkable insights and efforts, had begun the process of creating beauty on the model of ancient Greece and Rome. Arnolfo di Cambio, a man standing at the threshold of the Renaissance, and a sculptor as well as an architect, knew very well what was required of him. Creating beauty was the easy part.

In May 1961 the young president John Kennedy stood before a joint session of Congress and focused the country's intellect and management skill outside of earthly bounds with this challenge: "I believe the nation should commit itself, before the decade is out, to landing a man on the moon and returning him safely to earth." (Chaikin 1994.)

Note that the President was quite specific with his big idea: Land a man on the moon, then bring him back safely, and do it in 9 years. In this instance, the president introduced the schedule as a key component of the goal. There were, however, no cost limits set in either case. That is most unusual. Cost was a secondary consideration in these titanic examples where the honor, the glory, and the prestige of the state and nation were at stake. That is not likely to be the case with the more humble efforts that come our way.

The *Project Concept Statement* is always short and to the point. Sometimes one sentence will do, as these examples attest. A single page will certainly be sufficient to contain the essentials of the concept statement. Be distrustful of a concept statement that is wordy and complex. This is usually a sign of fuzzy thinking which, in turn, leads to imprecise goal setting. The conduct of a project is an iterative process. The complex details may emerge only gradually.

The essentials are: the *statement of goals,* an indication of the *time allotted* (schedule) to achieve those goals, and the *resource constraints* (people, material, and money) under which the first two must be accomplished. It is very important to write these down in their simplest form and get agreement (sign-off) on these basic elements. It is often surprising how difficult it is to reduce an idea to a simple concept statement. It is tempting to skip this "trivial step" and start the project before the concept statement has been written down, understood, agreed to, and approved by senior management. It is well to remember Theodore Levitt's (1962) pithy admonition on this point, "Unless you know where you're going, any road will lead you there." In a survey of projects that had failed to meet schedule goals (Figure 17.1), "Poor definition of product requirements" is the reason most often cited as contributing to delay in bringing the product to market (70 percent) (Product Development Consulting 1990).

The following example is from the author's product development experience in the instrumentation field during the early 1990s. The goal and focus of the project is a device we have named here— *the application controller.* It is described in the Project Concept Statement (Figure 17.2). Note that the statement specifically addresses only the three elements required of the Concept Statement: goals, schedule, and resources. The details are left to follow.

Project Concept Statement Reality Check. Before moving beyond the subject of the Project Concept Statement, we must acknowledge a real-world consideration that is well understood by experienced project managers. It is the fact that one can run into great difficulty laying out and gaining approval for a truly innovative proposal, or for one that is merely out of the ordinary for our industry, or for our company, or even for our department. We must also concede that in many of those

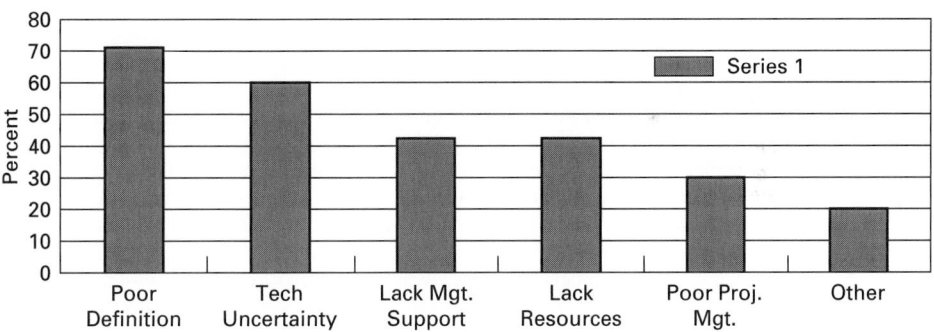

FIGURE 17.1 Reasons that products failed to meet schedule goals. (*Source: Product Development Consulting 1990.*)

Project Concept Statement
Project: Application Controller

Product goals:
1. Develop and bring to market an *application controller,* based on personal computer technology, to replace the current product line based on application-specific technology, and outpace competitors.
2. The resulting product must be fully featured, yet small and completely integrated, similar to current specialized units but with PC power and enhanced user features, including customer "turnkey" operation.
3. The cost target is $1250 for the basic configuration. At a cost of $1500 or more the product is not viable. The anticipated volume is 2000 units/year or more.

Project schedule: Complete the development and introduce the product in 2 years.

Project development expense: The development expense is not to exceed $2,000,000. Estimated payback time is 1 year, based on the expectation of cumulative gross margin to be achieved during the first full year of product shipments. Estimated product life is 5 years or more.

FIGURE 17.2 Project Concept Statement.

cases there is good reason for senior management to be wary and distrustful of ideas and proposals that depart significantly from the norm. Under these circumstances, there is a real, and often justi-fied, motivation on the part of the project manager to avoid presenting and seeking formal approval for the project concept until the environment can be properly prepared for its acceptance.

Peters and Waterman (1983) allude to this phenomenon in their description of the "skunk works" as an out-of-sight, underfunded, often bootlegged, idea incubator where, they believe, most of the really good new products are spawned in corporate America. The very existence of the uncontrolled skunk works incubator seems to be at odds with the clear-cut outline for writing out and gaining agreement on the Concept Statement before starting the project. Yet the existence of the skunk works suggests the need to be wary of premature disclosure of offbeat ideas. As is the case with most appar-ent anomalies, we must contend with shades of gray.

The truth is that if management would strive to track down and ruthlessly eliminate every new or unusual idea before it has gained headway, the effect would be to drive out the best and most cre-ative people and eventually to destroy the organization. On the other hand, if management has no mechanism in place for making decisions in an orderly way, for setting priorities and controlling events, the organization will surely founder and fall. The successful project manager must function effectively in the gray-shaded, fertile area between these pathological extremes.

Pressing Ahead—Making the Transition to the Next Phase.   The transition from one project phase to the next marks a milestone. It is, therefore, appropriate to obtain management approval as we complete the Concept Phase and before we begin the Discovery Phase. To gain approval we need to prepare two very simple documents in addition to the Project Concept Statement. These are the *Product Requirements Document* and the *Project Authorization for the Discovery Phase.* We then secure management sign-off on them.

The Product Requirements Document.   The purpose of this document is to establish the broad out-lines of the project. The "product" is the project's focus; it is the resulting deliverable. In that sense there is universal meaning for the word "product."

The Product Requirements Document is an expansion of the Concept Statement. It is still too early to completely describe the product; the description put forth at this point will be subject to refinement and outright change. That is what the Discovery Phase is about. Nevertheless, in the Product Requirements Document, those driving the project—the project leaders and the product champions—describe their mental picture of the product or service that they intend to produce. In addition to text, the document will often include graphical forms, drawings, etc. The Product Requirements Document need not be especially lengthy or detailed—three to five pages is usually enough. For example, a suitable outline for the Product Requirements Document for the application controller would cover the topics indicated in Figure 17.3.

The Project Authorization—Discovery Phase.   This is the companion to the Product Requirements Document, and it includes a concise description of the resources and an estimate of the time likely to be required to undertake and complete the Discovery Phase. It is also appropriate at this point to make a rough estimate of the total scope of the project. Note that there is overlap among the three Concept Phase documents. Overlap is to be expected, inasmuch as the documents all deal with same thing. But each document is produced from a somewhat different viewpoint. The Concept Statement is written primarily from the executive point of view; the Product Requirements Document is often written from the marketing and business management point of view; and the Project Authorization for the Discovery Phase is usually written from the viewpoint of those having responsibility for implementation.

In a corporation of medium or large size, the Project Concept Statement might come from top management, or from the head of a business unit, or it might bubble up from within the organization and be seized upon by a product champion. The Product Requirements Document might be written by the marketing department, product department, or the product champion. The Development Phase Proposal will usually come from development engineering. Whatever the case, it is advantageous to assign responsibilities ahead of time to those departments and people who will produce this project documentation.

Product Requirements Document Outline
Product: Application Controller
1. Target specifications
   a. Performance specifications
   b. Cost specifications
   c. Schedule required
   d. Those specification requirements that are vital to success

2. User interface style and hardware

3. Major software functions, i.e.:
   a. Operating system performance requirements
   b. Memory, disk space requirements
   c. Special calculations and reports
   d. Real-time data acquisition

4. Compatibility with other products and accessories

5. New technology planned—advantages versus risk

6. A block diagram or outline of the product

7. Place in product line—profit potential

8. Rough estimate of schedule and resources likely to be required

9. The list of deliverables

FIGURE 17.3   Product Requirements Document outline.

Continuing with the example of the application controller, Figure 17.4 is an example of a *Project Authorization Form.* It can be used to record management approval to proceed to the next project phase, in this case, The Discovery Phase.

With management approval of the Concept Statement, The Requirements Document, and the Project Authorization, the project activities expand. The rate of spending, although still modest at this point, will increase, and resources will be further committed to marketing fact-finding and technical studies, possibly including feasibility studies. These efforts move the project from the *Concept Phase* to the next phase—an expansive, intellectually stimulating, and dynamic period which we have labeled the *Discovery Phase.*

The Project Discovery Phase.   In the Product Requirements Document in Figure 17.3, we are faced with the problem of dealing simultaneously with several interrelated issues. We need more information, and it is necessary to begin to fill in details at this point. One useful technique for making significant headway under these circumstances is to analyze the key issues related to product feasibility and viability in the marketplace.

An analysis matrix aids in balancing the twin goals of low cost and small size. It presents the percentage contribution that each of the major hardware components makes to each goal.

The elements of cost relate to the dollar value of the components involved, while the elements of size relate to the volume of space required for each component and to the footprint of the final assembly. The use of percentages generalizes the presentation and also highlights important contributors. Of course in the very beginning of the project we cannot know these cost and size contributions precisely. But even an approximate valuation yields important insights as to where the key elements are with respect to cost and size. In this example the printed circuit board, the hard disk, and the display stood out with respect to cost, while the display, the printed circuit board and the power supply were key contributors to size. These were established by the matrix analysis as the "vital few" items that must be addressed and resolved in order to establish product viability. Thus, as a result of the matrix analysis many specific questions arose, for example:

- Will there be enough room on the printed circuit board?
- Will high-density devices be required?
- How can the unit be held to a reasonable size?

Project Authorization Form

Project title: _Application Controller_ _____

Phase authorization: _Discovery Phase_ _____

Effort and expense for this phase:

Start date: _____ Completion date:_____

Development staff

Engineering _____ Person-months _____

Electrical design _____ Person-months _____

Mechanical design _____ Person-months _____

Total staff _____ **Person-months** _____

Total cost: $ _____

Key goals of this phase: _____

_____

_____

_____

Appropriate signatures:

Engineering Manager _____

Product Manager _____

Vice President Operations _____

Comptroller _____

FIGURE 17.4 Project Authorization form.

The point here is that these questions flowed directly and in an orderly manner from the matrix analysis, even though the potential solutions were interactive and interdependent one with another. We knew from the matrix what was important. Thus, while the project leader and the marketing and product management members of the team continued the work of fleshing out the *Product Functional Specifications,* the technical specialist members of the team proceeded through a series of "what if" design alternatives, forcing the product configuration to evolve, to become simplified and more refined, while all the time striving to make the design process converge.

The Project Team.   The project team is at the center of the approach outlined here. The team is responsible for the management and it plays an important part in the execution of the entire project from concept to introduction. The composition of the team depends on the nature of the project, the makeup of the sponsoring organization, and the time phase of the undertaking at hand. In the case of a manufactured product, for example, the team usually consists of members representing:

- Marketing (i.e., product requirements, customer communication)
- Engineering (i.e., product development)
- Manufacturing
- Quality assurance
- Sales
- Service

This core team of five or six people is responsible for guiding the project and making sure that the needs of their respective constituencies are put forward and responded to appropriately. *The team manages the project.* The team must secure the necessary resources to accomplish the tasks at hand. Many, but not necessarily all, of the team members will be assigned to work on the project full time. The team will hold regular meetings. Everyone associated with the project will remain informed as to its progress through distribution of the minutes and periodic project reviews. These reviews are usually held monthly or quarterly, depending on the project's scope and phase. A project near completion often requires more frequent reviews.

The project leader has a special position on the team as the first among equals. The project leader calls the meetings, writes and distributes minutes, and sees to it that the project team carries out its tasks. The project leader also initiates and maintains a file of documentation related to the project. In the example, this file consists of the Project Concept Statement, the Product Requirements Document, the Development Plan, Functional Specifications, meeting minutes, pertinent memos, etc. The file establishes the documentation trail for the project.

The Practiced Team.   Probably nothing has more impact on the successful operation of the team (and therefore the project itself) than the effectiveness of the team leader. It is necessary to keep all the team members focused on the key objectives and the appropriate priorities as the effort progresses. This means calling the meetings, keeping them as short as possible consistent with the daily demands (try to keep regular staff meetings to an hour or less), then writing and distributing the minutes promptly. In fact, particularly in medium and large organizations, the minutes often have a much wider distribution than to the team members themselves. To that wider distribution, *the minutes are the project.* They are, accordingly, to be taken very seriously by the project leader. The meeting minutes are not something to be delegated.

Over time, this mode of operation develops a solid working relationship among team members characterized by forthright and effective communication, trust, mutual respect, and support. These bonds must be formed early and the team's working behavior must be fully in place and well exercised so that when (as inevitably will be the case) the serious, *even threatening,* problems occur, the team will be able to withstand the pressure and handle these effectively and successfully.

In fact it is of little use to call a meeting of virtual strangers when a crisis occurs. Even worse, under these circumstances, the team meeting can become associated with the stress and angst related to grappling with and trying to solve serious problems instead of the positive, sometimes even exhilarating, experience that these meetings ought to be. The *practiced team* is much better able to deal with and solve serious problems as they occur. They will do so because they will have experienced many small successes (progress) working together, thereby increasing by increments their skill and confidence so they are ready to take on the "big one" when it occurs.

Furthermore, the well-practiced team can often foresee and avoid serious or crippling problems before they burst forth, fully matured, to attack and undo their accomplishments. If undertaken properly and consistently, the cumulative effect of this methodology can be very salutary. By fostering a climate of shared purpose rather than one of adversarial relationship, the team is not only better able to deal effectively with serious problems as they arise but it will also be open to and poised to deal with change, even to the point of making the *innovative leap* when that action is required and appropriate. *Thus, the practiced team is well suited to fostering innovation.*

An Example of Team Innovation—Surface-Mount Technology.   The application controller ultimately required new, higher-density circuitry. This high density is achieved by a method of packaging called "surface-mount technology." The challenge was to bring a product design which uses surface-mount integrated circuit devices into a production environment built up exclusively around older, through-hole printed circuit board assembly techniques.

Key members of the application controller team represented Manufacturing, Quality Assurance, and Service, constituencies most affected by the proposed introduction of this new technology. It helped that everyone knew that this technology had to be adopted eventually. The questions were: Why now? Why does it have to start with us? Are we ready? Is the organization ready? Nevertheless, everyone on the team gained ownership of the goal to use surface-mount devices on the main PC board. Without these devices, the board layout was not feasible in the space available. There were potential cost savings and reliability improvements to be made as well.

Everyone on the team became intimate with the facts and details of this issue, and they went to work to make the change happen. In the end it was done using a combination of in-house and, with the effective participation of the Purchasing department, external vendor resources. The in-house development support staff and the manufacturing staff were trained and primed to deal with this new technology. The first production version of the application controller had about 15 percent surface-mount devices on the main PC board. However, an enhanced version which followed 3 years later had approximately 80 percent surface-mount devices installed. By that time the use of surface-mount technology had become routine for the company.

Dealing with change often engenders anticipation, excitement, enthusiasm, and the spontaneous release of creative energy. Just as often, however, the demands of change can bring on feelings of apprehension, of inadequacy, of cynicism (they'll never do it!), even of outright fear on the part of some participants. Which of these reactions prevails is often less related to the change itself then to manner in which the team has prepared for and presented its proposal to the organization at large. The application controller team was not only confidently successful in the application of surface-mount technology within its project, but also exhilarated and proud to have made an *innovative leap* in applying a new technology to the company's process and products.

The Feasibility Study.   The initiation of a project often requires a feasibility study. The objective of the feasibility study is to answer the question: Is it possible to successfully complete this project? The effort might be unprecedented: Is it possible, given the current state of technology, to land a man on the moon and bring him safely back to earth? Alternatively, the question might be more down to earth: Is the goal achievable given the skills and the resources that are available (or are likely to be available) to our organization?

Establishing Feasibility—IBM System 360, Model 50.   In the early 1960s, the author was an engineering manager for the IBM Model 50, one of the System 360 computers under development at that time at the IBM facility in Poughkeepsie, NY. Everything was new: the system architecture, the technology used in the digital circuitry, the high-density packaging, the interconnect scheme, power distribution, design automation support tools, everything. Several key questions were: How will the logic circuit set perform? What will be the cycle time? Are there any nasty surprises lurking just out of sight, for example, electrical noise or cross talk?

Senior management decided to build a feasibility model to get the answers to these questions. The development team was delighted with that assignment. The design was essentially complete. The approach was simple: Count the circuit logic levels from the beginning to the end of the cycle. And, because wiring length contributes to delay and cycle time, it was necessary to make a reasonable estimate of the wiring configuration likely to be encountered in the final package. Accordingly, circuit packaging, interconnect, and layout were made to conform as closely as possible to the final configuration, as we were able to envision it at the time.

The actual number of circuits and packaging units that were used in the feasibility model were a small fraction of those that were eventually used in the final Model 50. Probably no more than 1 percent of the final circuit count turned up in the feasibility model. Moreover, some important key elements were not addressed in this study; for example, the read-only memory used for instruction sequence control was not involved in this evaluation—that was left for a later time.

Nevertheless, in spite of these shortcomings—one might say because of these shortcomings—the feasibility study was a great success. The unit was designed, laid out, built and made to run in a relatively short time—a matter of several months. Top management and the development team got the answers to their questions, and those answers generally affirmed the design assumptions. In fact, the circuitry was, for the most part, faster than the conservative predictions being made by their designers, although we were surprised by the greater-than-expected contribution that the module-to-module interconnect made to the total delay. Overall, we were very encouraged by the performance that was predicted by the feasibility model, and that predicted performance became fact a couple of years later in the production version of the Model 50 (Pugh et al. 1991).

Looking back on this experience it is quite clear that there were many more questions—some of them only partially formulated or not articulated at all—that were addressed in this feasibility study. Among these were: Does the design support and implementation system work from beginning to end? Will some key component break down (for example the design automation tools or the packaging)? Can the team overcome problems? Can it get the feasibility model to work? Can it deliver on time? Yet, despite all of these questions, there is no substitute for a working unit—even one modeling just a tiny fraction of the final version—to instill confidence in top management, and to strengthen everyone's resolve going forward.

Establishing Feasibility—Brunelleschi's Breakthrough.   In the historical concept statement for the construction of the cathedral in Florence, the requirement to establish technical feasibility for the undertaking was built into the project goal itself. Di Cambio, the first architect, conceived the cathedral and the Duomo, even though the means to construct the dome itself were literally out of reach when he began construction at the beginning of the fourteenth century. In fact, over a hundred years elapsed between di Cambio's laying of the foundation for Santa Maria del Fiore and Filippo Brunelleschi's breakthrough which enabled construction of the great Duomo in the year 1420 (see Figure 17.5).

The problem was how to put a roof over the enormous expanse of the tribune, which stood unfinished and gaping while the republic cogitated. No precedent existed, for no dome of comparable size had been raised since ancient times and the methods used by the ancients were mysterious (McCarthy 1963).

To resolve the problem, the leaders of the Florentine Republic held a design competition in 1420 to which masters from every corner of Italy were invited. Filippo Brunelleschi, who had been studying architecture in Rome, and who had been specifically studying the methods used by the ancient builders, found a way of raising the dome without the use of external masonry supports, a feat everyone believed to be impossible.

Brunelleschi proposed an ingenious construction of total novelty and of unexpected benefits in economy and utility. Looking at it today, more than 500 years after its construction, it is easy to take the design's success for granted.

But the wool merchants and others of the city's leaders who funded the cathedral's construction did not leap at Brunelleschi's plan. Ever the cautious innovators, they chose instead to have Brunelleschi first try his theories and methodology for raising the dome in a much smaller church currently under construction. Only after the dome in the smaller church was shown to be feasible, was Brunelleschi given the authorization to proceed with the great cathedral.

Establishing Feasibility in the Marketplace.   Establishing feasibility applies to more than the technical issues as outlined in the examples above; it also applies to establishing feasibility according to the needs of the real world in which the product must compete. The best—perhaps the only—way in which this can be done is to be there to talk to, interact with, and respond to potential customers for the planned product. Visiting and talking to customers, especially potential users, is an essential part of establishing feasibility in the marketplace. One structured technique that has proven effective is the focus group.

The Focus Group.   This is a small group (rarely more than a dozen people) who are selected as representing a specific customer population and whose discussion is focused on a well-defined question. The objective of conducting a focus group is to secure customer feedback about the product. While the beta test is usually undertaken just prior to product release in order to be sure that the product as designed is functionally complete and robust in its operation, the focus group evaluation may be taken at any point in the product development cycle from the concept stage right up to the point of product release, and even afterward. Primary attention is paid here to the marketability of the product.

The focus session is usually arranged by a consultant. Potential product users are carefully selected and invited to join the focus group. The product is presented to the focus group in a scripted manner, demonstrations are made as appropriate, members of the group may actually use the proposed product, and a series of questions is asked to determine potential customer reaction. The session may be secretly witnessed by interested parties and even photographed for future analysis and for the record.

Sometimes the group will be interviewed as a whole, sometimes members will be shown the product alone, and interviewed singly as well. Questions are asked in such a way as to obtain a quantitative response to what is usually, in the focus member's mind, a qualitative impression.

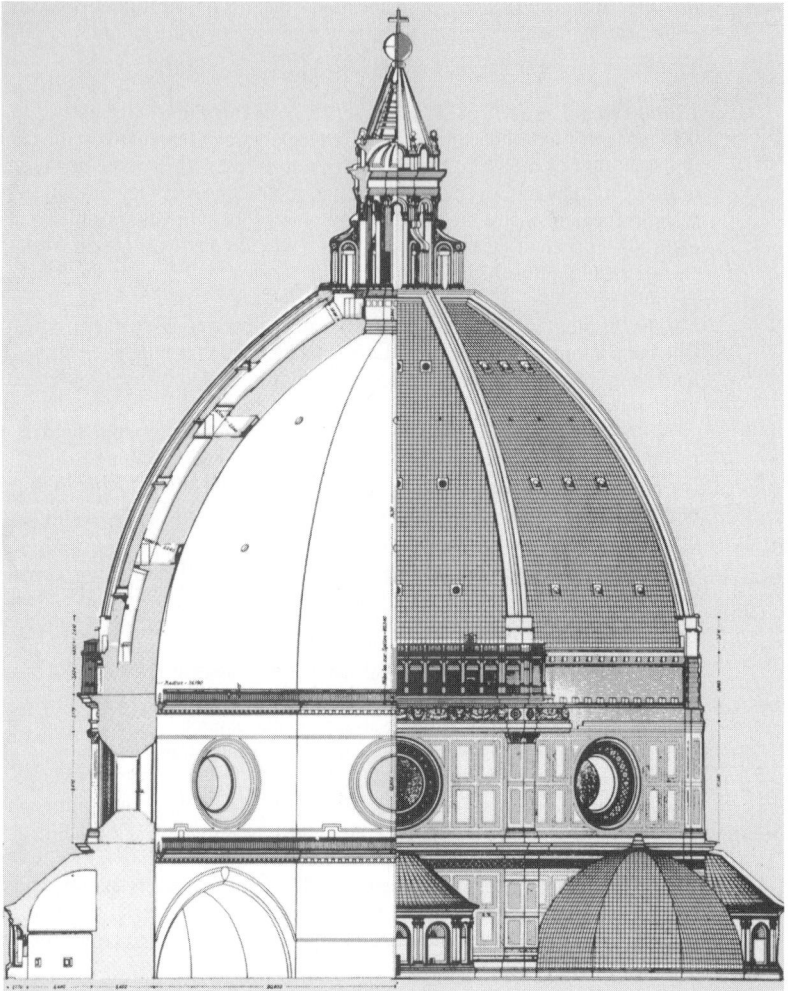

FIGURE 17.5   The cupola of the Duomo. (*Source: Used with permission of La Mandragora, Florence, Italy.*)

For example it might go something like this: (1) Do you need this product in your office, laboratory, etc.? (2) If you do need it, how badly do you want one: (*a*) Need it right now. (*b*) It is pretty important; need it soon. (*c*) Will take another look next year. (*d*) Do not need one in the foreseeable future.

How much would you be willing to pay for this box? Write down your answers and pass them to the referee. Or, would you be willing to pay $7000? How about $6000? How about $5000? How about $4000? And so forth.

What is the most important/valuable feature? Is anything important missing? What other features would you like to see in this unit? What might that feature be worth in additional cost?

There might be questions about distribution: local store, mail order, direct order, Internet, etc.? How about service, warranty, etc.?

Establish Feasibility—A Sacred Trust. In the final analysis, feasibility can address many questions: Does it work? Is it reliable? Can we afford the development and manufacturing start-up costs? Will we meet the target manufacturing cost? Is the schedule realistic? Can we make the schedule? Do we have a competitive edge? Will the potential customers buy it? These questions ought to be asked early and often during every significant project undertaking. The project manager should make every reasonable effort to get the best answers possible to the key questions related to project goals and performance.

Moving from the Discovery Phase to the Implementation Phase. To go or not to go, that is the question. But it is often not explicitly asked, let alone answered. Interestingly, at this point at the end of the Discovery Phase it is often more difficult to kill a project than to keep it going. Yet, if it ought to be stopped, now is the time to stop it.

Usually, a project will have had to pass over a multitude of hurdles in order to get through the Discovery Phase. Several project champions may have emerged in the decision-making structure; managers will have staked out positions for or against the project; careers are on the line; there may even be a working model. It is very difficult at this point to contemplate stopping the project. But now is the last time to stop the project without incurring the much greater expense and level of commitment associated with the Implementation Phase.

One generic failure mode that can trip up the project team is the emphasis on one aspect or one issue to the exclusion or de-emphasis of others. For example, a technically oriented company might have the tendency to work on establishing technical and scientific feasibility during the Discovery Phase, but not pay adequate attention to the marketing and business questions that always accompany the launch of a new product or service. Simply because the technical feasibility has been established does not mean that the product will be successful in the marketplace. Conversely, a marketing-oriented organization might get carried away with the business potential of their concept, but fail to do the technical feasibility and manufacturability work necessary to place the product or service on a solid technical footing prior to launch.

It is well, therefore, to go through a formal review at this point to make the go no-go decision and to have management formally sign off on the Implementation Phase. This review must be broadly based, taking into account all the technical, marketing, and business aspects of the proposed product or service before going forward.

The Project Proposal. The project proposal helps move a project from the Discovery Phase to the Implementation Phase. The following example, a composite from several product development projects, illustrates the document. It is about 30 pages in its full length. For each of the elements which will collectively define the move, it includes information on which to base the decision to make that move.

Product Concept, 1.5 Pages. The product concept is fleshed out beyond the few sentences suggested earlier in the development process. Still the concept is short and to the point.

Marketing Strategy, 2 Pages. Outlines the marketing approach. Is it direct or through dealers? What are the plans for marketing communication and sales? What are the customer benefits, etc.?

Target Specifications, 3 Pages. One product is small and simple to use, yet the target specifications are complex and detailed, especially details of the application and user interface specifications.

Pricing and Competition, 1 Page. Another product is unique and, at the present time at least, there is no direct competition for the unit. But there is plenty of indirect competition. Target pricing is based on a combination of competitive market factors and target manufacturing cost.

Engineering Approach, 9 Pages. This is a detailed description of the product, including several figures. In one of the products, there is a complex optical subassembly to describe, as well as the electronic PC board, user interface details, and system control logic.

Engineering Estimate of Manufacturing Cost, 1 Page. This is the initial cost estimate and it is usually made by the engineering staff on the basis of the cost of parts and an experienced estimate of what it will take to assemble and test the package for shipment. Later, during the Implementation Phase, the cost will be re-estimated and established by the manufacturing staff on the basis of detailed documentation produced by engineering.

Reliability and Installation and Estimated Warranty Cost, 1 Page.   This an estimate by engineering of the reliability to be expected and the installation and warranty costs to be anticipated. Later during the Implementation Phase, the reliability figures will be established by the QA department by calculation and testing methodology. Similarly, the Service organization will establish installation and warranty costs for the product.

Schedule, 4 Pages.   In this case, the schedule highlights are in the body of the report. Usually a detailed computerized schedule chart is also provided. The highlights are:

Year 1:

Begin effort in earnest	January 1
Complete Concept Document	February 1
Enter Discovery Phase (management approval)	February 15
Complete Discovery Phase	August 15
Project Proposal (this document)	September 15
Enter Implementation Phase (management approval)	October 1

Year 2:

Complete optical design	January 1
Complete mechanical design	January 1
Complete electrical design	January 1
Complete software design	February 1
Build 10 preproduction prototypes	March 1
Release drawings	April 1
Test prototypes	April 1–May 1
Evaluate prototypes internally and with potential customers—beta test	May 1–July 1
Set up first production run	July 1
Begin first shipments	August 1

These days, 18 months is just a moderately fast program. In the author's experience product development projects generally take from 1 to 3 years. One year or less is very short and more than 3 years is very long. However, in the retail PC computer industry, for example, some project development times have been shortened to 6 months or less. We have been told that in this industry more than one team may be operating concurrently and out of phase by six months. Each team has a development cycle of 1 year, but a new product is announced and shipped semiannually because each development team completes its efforts 6 months apart from the other. Now that's really fast! Under those circumstances there may be a very much shortened Discovery Phase, and management approval must be forthcoming in a day or two instead of 2 weeks as shown above.

Development Cost, 1 Page.   This includes an estimate of the cost of the engineering staff over the period of time outlined in the schedule. Thus, when the schedule stretches out, the costs go up as well. It includes outside consultants—for example, industrial design consultation—as well as the cost for the preproduction prototypes which, while usually assembled in the factory, are often expensed to the engineering development budget. That's one of the many reasons why very big, expensive, and complex products—for example, the modern commercial jet airplane—can be handled only by a few very large and well-established producers.

Capital Cost and Tooling, 1 Page.   This will include the estimate of expenses for capital equipment that will be required to undertake the project. This equipment might be used in engineering or in manufacturing or both. Specialized tooling is also included in this estimate, for example, the tools that are required to make the plastic covers for the instrument and for making several small parts in the optical assembly. Specialized tooling is also required to make and test the PC board in production.

Risk Factors and Risk Management, 5 Pages.   This is a thorough review of the risks that exist in a given development going forward. In some cases, the Discovery Phase will have eliminated many of the technical risks that exist in the undertaking. More often, however, there are still many risks that remain because there is some point at which the additional time that might be taken to resolve problems also delays that product's introduction to market. That delay can often entail more cost and risk than simply going forward in spite of the unknown factors that remain. This is always a matter of judgment, and some unknown factors simply cannot be resolved other than by going forward. At some point, therefore, management must make the conscious decision to move ahead and to take the risks involved with that decision. But first they must honestly examine those risks.

For example, the risks described in one Project Proposal report were (1) issues concerning the optics, (2) possible problems related to certification testing to specific published standards, (3) instrument temperature self-rise and possible cooling problems, (4) product testing and evaluation, and (5) production and distribution issues.

In the end, none of these became a serious problem. But they were all potentially serious, and it is better to acknowledge these before blindly plunging forward with implementation without having developed any contingency plans for dealing with them.

Summary and Change Record.   Finally, there is an executive summary, up front just after the index page, and a few pages right behind the index page for keeping track of changes to the Project Proposal document itself. This document ought to be kept up to date regularly—say quarterly—and filed either electronically or in hard copy in the project history file. In one particular case, the summary required just one page and the following items were included in bold type with the text on that single page:

The product is a completely new concept.

It is expected to sell in the quantity of 3000 units per year.

The cost is expected to be $2500.

The price will be set at $6000.

The development cost is estimated to be $1,000,000.

The tooling will be about $250,000.

First shipments are expected to begin August 1.

The major risks are optics and application software.

Thus in one quick read of that single page, management or anyone else can get the complete picture very quickly. They can read on for more detail, or go to a particular section that interests them. They have the big picture, the context, and the proposal details in front of them. And, the requirements for a go no-go decision are in hand. There should be no valid reason to ask for more information.

The Product Functional Specifications.   In order that everyone work to the product specifications, it is essential that the statement of requirements is agreed upon as early in the project as possible. The statement may be in the form of text—paper or electronic; it may include drawings or graphics; it may refer to existing systems or products. If there is a performance requirement, then it should be stated explicitly and quantitatively to the extent possible.

The list of *deliverables* is identified. In the case of a manufactured product this includes, for example: assembled and working prototypes for marketing, manufacturing, and quality testing and agency certification requirements; drawings and appropriate documentation for release to the factory; documentation for the methods of fabrication, assembly and test; components specifications and appropriate vendors; part numbers and bills of materials for customer deliverables, including user manuals.

Deliverables can include all that is necessary for the complete product introduction including, for example: the schedule of preproduction and early production units and their allocation to customers for evaluation; the plan and schedule for sales training and service training; provision for customer demonstration units; preparation and distribution of marketing introduction materials such as brochures, customer presentations, and advertising material.

Moreover, there must be a close linkage among the project product development staff, Manufacturing, QA, Service, and those who are responsible for introduction of the new product through initial buildup and eventual acceptance. All of the project's expenditure and the prestige of the sponsor is on the line at introduction. If there is a mistake or oversight at that point, it is most costly. So it is important that the plan carry the project right through product introduction to make sure that everything is in place for its success. When there is a handoff to be made—for example, from the product development team to the product marketing team—the transition planning should be so thorough and complete as to assure that the prospect of a slipup at this critical stage is minimized.

Project Implementation Phase.   The project implementation is usually the most time-consuming and the most costly phase of a project. This need not be the case. In fact, if the proper foundation has been laid for implementation, this can be the most straightforward phase. The key is careful attention paid to producing appropriate documentation, doing the groundwork of the Concept and Discovery Phases, and proceeding to implementation on that strong foundation.

There are many tools available today to help the management team with the mechanics of project management: the ubiquitous word-processing and spread sheet programs, project management and scheduling programs, e-mail and other communication methodologies, and presentation software; the list is long. There are also many books and articles that address this subject, and more become available each day. A short list of additional readings at the end of this section points the reader to detailed treatment of these issues, which is beyond the scope of the present discussion.

As useful as these tools are, they are no substitute for a disciplined analysis such as the one advocated here. Today's modern tools help get the job done more efficiently and expeditiously. But they don't help clarify the fundamental vision of the project goals and the means for achieving them. Remember that Arnolfo di Cambio and Filippo Brunelleschi didn't have our modern technology at their service, yet their accomplishment in project management is awesome, even by the standards of the late twentieth century.

The Three Critical Periods for Customer Acceptance of a New Product.   The development, manufacture, and introduction of industrial instrumentation and related computer control devices is typical of the sort of activity which requires project management. The applications for which these instruments are used are specialized, and the products themselves are quite complex. Moreover, we have found that there is a high degree of variability in the training, experience, and ability of potential customers. Often the customers are expert, or at least knowledgeable, in the target application, although they are not necessarily skilled in operating the instrument or especially adept using computer software, environments now so often used to control modern instrumentation systems.

Our further experience has been that the design team members—engineers and technical application specialists—tend to concentrate on specifications and functionality directed to the long-term use of the product, sometimes to the detriment of the potential customer's initial encounter with the product. On the other hand, sales and field specialists are very sensitive to the customer's initial reaction to a new product. However, their opinions are often not sought until the product is announced and on the verge of initial delivery. By that time it is too late to respond to criticisms that the product is "incomplete," "too complicated," "not user-friendly," etc. The reasons given for the failure to aggressively seek input from the field run generally along the following lines:

1. The new product is a secret.
2. If we start talking to field people about it before it is ready, they will stop selling the current model and orders will dry up!
3. It is difficult to talk in terms of concepts or about the future with people in the field (even our people) who are used to dealing in specifics and with the here and now.

4. Even if the field representatives don't like the new product's user interface, we don't have the time or resources now to make improvements. No product is perfect. Let them sell what they have. If these issues are opened up now we may as well throw the schedule out the window!

Of course, there is a measure of validity in each argument, especially the first one. But there is also a high price to pay if due consideration is not given to the three critical periods for product acceptance. The critical periods are

- The initial moments
- The first day
- The remainder of the product's life

The fact is that the amount of specification and development effort spent on the first two items invariably pales in comparison with the long-term considerations for product use. Yet, if the first moments with the product are not focused upon and handled correctly, we may not attract enough customers either to experience its depth and sophistication or to make the product viable. What is more, the price paid in development time and effort need not be high if these critical periods are addressed specifically and early enough.

The Initial Moments. It is during the initial moments of exposure to the product that the customer gains the critical first impressions. Exposure may be in a salesperson's demonstration; it may be in a self-directed operating session; it may come in the form of a box on the floor, from which the customer must take the contents, assemble them, and set them into operation. The initial moments may merge into the first day with the product.

Simplicity ought to be the focus when dealing with the initial moments. One useful technique is to build in operating support for the salesperson's demonstration. For example, as applied to instrumentation systems, operating methods and results can be "factory-installed." These previously stored methods and results can be used later in the field to demonstrate instrument functionality.

Furthermore, we have found that it is often much easier to run an instrument analysis than it is to set the system up and initialize an operating method. It is also quicker and easier to evolve specialized operating methods from an available suite of stored methods than to develop them from scratch. Accordingly, use of "factory-installed" methods helps the salesperson and customer during the initial encounter; these then form the basis for creating the specially tuned methods that the customer will ultimately require. This approach establishes a solid basis for the customers' understanding and acceptance of the product and for their eventual mastery of its operation.

Some products, especially computer-based products, are often set up for self-demonstration. Simply push a button or point to an icon on the screen and the "canned" demonstration takes off. This is standard practice, for example, with electronic musical keyboard instruments—and it can be very effective.

Product packaging and documentation are critically important for establishing first impressions. Opening that box on the floor can present a daunting challenge to the customer who is faced with getting the new purchase set up and running, especially for the first-time user. One useful technique is to separate installation and setup instructions from the User's Reference or Operating Guide, which is a more formidable document. It is also helpful to use lots of diagrams and other visual aids to help with the initial assembly and setup task.

An event from the author's experience exemplifies some of the first-moment issues and illustrates the means taken to deal with them. When the applications controller was first introduced to the sales force and field applications support specialists, there was genuine enthusiasm for the product concept and features. But there was also serious concern expressed about the apparent complexity of what should have been an easy-to-use product targeted to the low-priced end of the market. There was also concern about the lack of several important software features that the development staff said were to "come later." In fact, there was enough anxiety expressed by our own field sales and support staff that the decision was made to delay introduction for several months and address the deficiencies noted. We knew the effort would be worthwhile because the sales staff really liked the product. They said that it just wasn't ready yet.

A working group was formed, composed of inside development experts and several field specialists who where selected to represent their colleagues' needs and the customers' requirements. Speedy resolution of the problems was critical, and it was surprising how quickly a consensus developed about what had to be done to go successfully to market. The missing software features were agreed upon and developed more swiftly than we had imagined was possible.

Meanwhile, the working group specifically focused on the first moments and the first day with the product. They decided that two prespecified methods would be devised and stored in the unit and that the product would come out of the box ready to go. The user had only to turn on the power and a control screen would come up with the default method in place. A press of the start button would cause the unit to run the analysis, display the results on the screen, and automatically print out the analysis report.

Except for the few missing features, there was little of this "ready-to-go" functionality that did not exist already in the product. Only the streamlined start-up sequence was added. Yet, these simple adaptations, which focused specifically on the first moments with the product, completely changed the sales specialists' perception of the product's complexity. The product went from hard to use to easy to use. This was one of the most important elements of our response to the sales specialists' concerns, and it was the easiest and fastest to implement.

The product was announced and shipped on its revised schedule several months later. It was a very successful introduction, and sales of the applications controller accelerated rapidly. There is no question that we did the right thing by holding the product back and addressing the deficiencies pointed out by the field sales staff, and by focusing on the customers' first critical moments with the product.

The First Day.   The approach to dealing with the first day with a new and reasonably complex product is an extension of the methodology and techniques outlined for the first moments. In this instance the focus is on a longer period of customer interaction with the product and on the deeper understanding that builds up as the product is activated and exercised. Many of us share the common experience of learning to use computer hardware and software products. Sometimes these can take months to master. In fact many features of these products are left unlearned and unused by the average person. They learn what they need to know to get the current job done. When they need to know more, they will go back to learn the additional features needed to address those new application requirements.

It is important, therefore, to give the customers the means to control the level of information and learning that they must deal with, especially at the beginning. It is also important that they quickly get the new product running in a meaningful way, in the first moments if that is possible, but certainly during the first day. Initial success with the product quickly overcomes frustration and fear, and encourages the customer to go forward with the learning process.

One technique that can be helpful in this regard is the use of a "tutorial." The tutorial usually consists of a manual or flowchart and a prearranged operating sequence that takes the customer through the product function in an orderly way. Moreover, previously stored methods and data can be used with the tutorial to support customer training. Development of an effective tutorial is an excellent way to focus on the first day with a new product. In fact, the applications controller mentioned earlier did have a tutorial feature built into it, and a specific manual was prepared to go with the tutorial.

New technology, especially new computer technology, is having an important impact in this area. In the case of one brand new computer-controlled instrument product that the author is aware of, the manuals have been replaced by a CD-ROM. Stored on that CD-ROM are the control software, the "help" files for on-line user support, the manual that shows how to assemble the complete system, and the User's Reference Manual. That CD-ROM itself is packaged in an attractive cardboard foldout that gives straightforward, step-by-step instructions on how to install the CD-ROM and get the system up and running. It can be made easy! What is more, in this particular application there is a specialized technique for preparing samples to be analyzed. The CD-ROM contains a tutorial movie video that shows, on screen and up close, how to properly prepare the sample for analysis. That kind of customer first-day support with a new product is hard to beat.

The Remainder of the Product's Life.   The natural focus for a development team is the day-in, day-out use of the product. So that is where the attention is usually placed. In fact, the first moments, the first day and the rest of the customer's life with a new product form a continuum. The things done to improve the customer's first moments and first day yield benefits throughout the use of the product. For example, consider the improved product accessibility that the built-in tutorial provides when the experienced operator has to train someone new in the product operation.

We noted in the above-mentioned computer-controlled instrument product that there was extensive use of context-sensitive "help" files that, along with a good user's reference manual, form the basis of solid support for the user over the lifetime of the product.

Of course, no amount of focus on the first moments or the first-day operation can mask an inferior functionality. The product has to be logical, predictable, and solid in everyday use to properly benefit from improvements to aspects of the customer's initial encounter with the product.

Support for Multiple Languages. One additional point to be made in the area of customer acceptance is the possible requirement to translate a product or service into another language. Sometimes there is a dilemma with respect to adapting our product or service to language and usage other than our own. Is it worth the time and expense to translate the product or service in anticipation of increased sales or to fend off a possible drop in sales in foreign markets, or not? If so, in what languages and with what priority?

Sometimes the answer is clearly yes. The high-priority languages are usually those of industrialized countries. Of course, there are many special cases. Sometimes the answer is unclear until the translation is made and the potential market is tested. Sometimes it is just not worth the effort and expense. Whatever the case, to operate in global markets, it is wise to plan product and service for eventual translation into languages other than that of its original formulation.

There is more to this than meets the eye. In fact, the terminology in current use for support of additional languages is not "translation" but rather "localization," to account for this broader view of the subject. For example, word-for-word translations often don't fit into the space available. There may be alternative ways to ask a question and then deal with the response. The protocols for stating dates and addresses vary. Monetary units are different from country to country. The list of differences is long and vexing. Therefore, localization must be addressed early in the product planning process or the product may be locked out of some markets later. For example, if there are written legends on the product or service, it is worth planning ahead for changing those legends into another language. This is of special concern if expensive tooling is required. Finally—and this is very important—it is the author's experience that, while the adaptation of the product for multiple languages is an internal development responsibility, it is usually best to go to an outside, specialized, service vendor to implement the translation itself.

General Applicability of This Approach to Customer Product Acceptance. We believe that the approach outlined here—focusing on the first moments, the first day, and the remainder of product life—has general applicability. The author's experience and examples demonstrate this approach in the context of complex instrument system products. To adapt this methodology to a wide range of applications it will be necessary for project managers to pose these questions in their particular milieu. This will usually mean contacting people in the field, potential users of the product or service, describing the project—balancing somehow the need for secrecy against the need to obtain information—and considering carefully their responses.

The prudent project manager will put a line item into the Product Functional Specifications to specifically address the three critical periods for customer acceptance of a new product.

Verification Phase. Although the Verification Phase, by its nature, takes place after the preparation of the Product Proposal and the Functional specifications, the Product Test Plan ought to be prepared simultaneously with those two documents. Early consideration of the test plan can shed light on the product requirements. For example, suppose the product is not supposed to weigh more than $6^1/2$ pounds—3 kilograms—or that it must survive a fall from a specific height, or that it is to be exposed to outdoor use. It is essential to know that at the outset and not to be surprised by test requirements after the product has been designed and the manufacturing cycle committed.

Consider the example of testing for CE Mark compliance for generation of radio-frequency interference (RFI) and for susceptibility to RFI. (The CE Mark and its test protocols originated in the

European Community.) One may think that this applies only to high-technology electronic products. It does not. Once, in an independent RFI testing lab, the author observed that the staff was testing a lawn mower for RFI emissions and susceptibility. What was a lawn mower doing here? The newly designed lawnmower ignition system was controlled by a microprocessor. To qualify for export to Europe the lawn mower was required to qualify for the CE Mark.

There is much that can be done ahead of time. For example, filters can be added to circuits that have the potential to radiate. Specialty filter components can be placed on cable connectors coming in and going out of the electronic assemblies and PC boards. There is usually plenty of time to do a design review for whatever testing will be required in advance of undertaking the design, and certainly well in advance of freezing the design and undertaking the test itself. The people who do this type of testing on a regular basis will have many suggestions for those who are new to this process. An early design conference and product review for testing and verification can save lots of money, time, and angst later on.

Product Test Plan.   The product test plan must be prepared simultaneously with the Product Development Plan because it will be necessary to design in the features that will assure that the unit will satisfy the test criteria. In the author's experience, a typical test plan outline would be along the following lines:

1. CE Mark certification. This involves a safety review and test, designated IEC-1010-1, and a RFI radiation and susceptibility test, designated IEC-1010-2. When the unit has met the criteria required by this certification review and testing, the CE Mark label may be applied.

2. Additional certifications may include CSA (Canadian Standards Association) and UL (Underwriters Laboratory).

3. Test to product design requirements; for example, component analysis; fuse rating; power on-off cycling; line voltage sag, surge, dropout; drop test in shipping container; thermal profile; accelerated life test.

4. Test to application requirements; for example, product life test at 10,000 hours or more power-on operation without failure; mean time between failure analysis and prediction; performance at 35° at 80 percent relative humidity; performance at 10° at 50 percent relative humidity; operation at 50/60-Hz high line/low line; functionality with accessories and peripheral devices; lifetime predictions for critical components.

5. Maintainability/serviceability; for example, serviceability analysis; mean time to repair analysis and prediction; product performance verification; establishment of depot or on-site repair and spare parts strategy; assurance of parts interchangeability; simulation of installation procedure; and the like.

6. Documentation; for example, review all manuals and support material for completeness and accuracy; review advertising literature for claims; assess patent status prior to announcement; review factory test procedures and specifications; review product test and diagnostic procedures; etc.

The items outlined above focus primarily on the hardware component of the product. These days, however, many products are subject to highly complex and sophisticated control mechanisms that need to be tested very thoroughly as well. Many times these control systems are implemented via a personal computer that is connected to the product or by firmware embedded in a dedicated device that provides the user interface and controls the product at hand. This is often the case with today's instrumentation devices and with modern numerically controlled machine tools.

Testing software and firmware usually focuses on testing the many combinations of control inputs and product readout conditions that can occur, in contrast to physical stress testing of the software. But that is not always the case. A product—for example a database query system—might have dozens, hundreds, or even thousands of terminals connected. What will happen if the traffic on those terminals rises beyond the normal level of experience? Will the system slow down; will it become unstable; will it crash? It is sometimes advisable to set up a stress test, activate all of those terminals at once, and find out.

More typically, however, the focus will be the functionality of the complete range of product states and user inputs. To deal with this it is necessary to set up a complete check list of all the functions to be exercised and the conditions under which they must be tested. Often these test procedures can run to many tens or even hundreds of pages.

Repeating a test like this over and over again as features are added, changes are made, and bugs are fixed, can become very fatiguing for the test technician and induce errors. In the PC environment, some products are now available that remember the test sequence and repeat it again and again as required.

Still other products (e.g., Lotus Notes, from Lotus Development Corp.) help to keep track of bugs and problems as they are found. These are especially valuable as problems are turned up from multiple sources. In modern development environments, problem reporting can come from many sources: the developer in the next cubicle, the evaluator in the lab, the SQA engineer, the beta-site evaluator, the service technician, the customer. Today, a problem-reporting environment that spans several continents and many time zones is not unusual. Accordingly, it can be very difficult to report problems in a timely way, and to prioritize and deal with them in an orderly fashion. It is, therefore, important to take advantage of all the tools that are available to cope with and streamline this demanding and complex activity.

Beta Testing.   In its original meaning, beta testing usually involved the exercise and evaluation of a complete product working in the operating system environment. Beta testing would ordinarily precede announcement and release. Today, the idea of beta testing has been expanded to include customer evaluation and input. Sometimes it is done in a flexible and free-form evaluation: Try it out, see how you like it. Sometimes it is done in a very structured manner: Here is the unit for one week or one month, please use it in the manner outlined, and answer the following questions about its performance and your impressions.

We certainly recommend the latter approach. The beta test is important because it causes the developers to interact directly with the customer. It must be taken seriously. In fact, the beta test—and many other elements of a project development for that matter—ought themselves to be organized into a subproject with all of the elements of project management applied to it, though on an appropriately smaller scale. A concept statement is appropriate to answer such questions as: What is the test intended to accomplish? How many customers and testers will be contacted? How many units will be required? What questions will be asked? How are the results to be interpreted?

In contrast to the focus group, which may be established quite early in the project schedule—and, we might add, the earlier the better—the beta test must come at the end of the implementation phase, and usually just before announcement and release. Therefore, there is often a severe time constraint on the information that flows from the beta test, for on it depends whether the developers should (or can) make the changes and enhancements to the product that the beta test suggests. There is little point in undertaking this verification step if there will be no provision for dealing with the results.

Beta testing is difficult but worthwhile. When well conducted, in a projectlike manner, it can yield critically valuable information on which to base midcourse correction and fine tuning prior to final rollout.

Project Management—The Augmented Product.   Sometimes the focus of the team and the project manager is so intently on the product itself and its functionality that important supportive components and materials are often sadly neglected. This is particularly the case in organizations that produce technical products. We alluded to this tendency in the three critical periods for customer acceptance and their specific support needs.

There is always more to a product or service than the "product" itself. We call the larger entity the "augmented product." One classic example of augmentation is the brand name. A nationally known brand is often perceived as more valuable than a fledgling newcomer. Connecting one's product to an already accepted offering can add value and increase initial acceptance. Similarly, advertising, testimonials, and local connections can add value in this sense. Packaging of the product or service, including supporting material, is a feature of the augmented product. Official government or state acceptance [e.g., selection as a government contractor, or, in some European countries, approval to display the notice, "By Appointment to His (or Her) Majesty," etc.] are augmentation features which may aid product success.

Two examples of the effect of product augmentation on marketing goods are provided by Bose Corporation and the Saturn division of General Motors. Bose, through a direct-to-the-customer marketing approach of its Wave Radio, an approach as innovative as the patented design of the radio itself, has augmented the product, securing superior product acceptance and pricing. Saturn set out at its founding in 1985 to create a network of dealers known for superior service. Year after year, Saturn dealers have been ranked at the top in customer satisfaction. That service has augmented their product, the car itself, enabling the company to position itself strongly with first-time car buyers.

The wrapping really matters. Paying attention to it may spell the difference between success and failure for the project manager.

A FINAL ADMONITION

There is no formula or cookbook recipe that can assure success in these complex endeavors. We have suggested here a framework on which to proceed. In every case, the methodology will have to be adapted to the mechanisms, technologies, and the culture of the enterprise, which the project manager must successfully negotiate to carry out the project. While there is no guarantee of success, following the suggestions of this section will put the manager and his or her team on the right track. Having a track, recognizing what it is, and knowing where it is located is essential. There are hostile forces at work—some created by the project itself—that tend to push the team off course and threaten its well being. The known track will help the project manager and the team proceed, even in the face of those forces. Think of the outline presented here as the light on the miner's hard hat, its rays penetrating the darkness and showing the way forward. Use it wisely. Use it well.

REFERENCES

Chaikin, Andrew (1994). *A Man on the Moon—The Apollo Astronauts.* Viking, New York.

Industrial Research Institute (July, 1996). *Industrial Research and Development Facts,* Section 19, Industrial Research Institute, Washington, DC.

Juran, J. M. (1988). *Juran's Quality Control Handbook,* 4th ed. McGraw-Hill, New York, p. 22.19.

Levitt, Theodore (1962). *Innovation in Marketing.* McGraw-Hill, New York, p. 75.

McCarthy, Mary (1963). *The Stones of Florence.* Harcourt, Brace & World, New York, p. 68.

Peters, Thomas J., and Waterman, Jr., Robert H. (1983). *In Search of Excellence: Lessons from America's Best-Run Companies.* Macmillan, New York, p. 20.

Product Development Consulting (1990). Boston, MA.

Pugh, Emerson W., Johnson, Lyle R., and Palmer, John H. (1991). *IBM's 360 and Early 370 Systems.* The MIT Press, Cambridge, MA.

ADDITIONAL READING

Gouillart, Francis J., and Kelly, James N. (1995). *Transforming the Organization.* McGraw-Hill, New York.

Lewis, James P. (1995). *Fundamentals of Project Management.* American Management Association, New York.

Ludin, Ralph L., and Ludin, Irwin S. (1993). *The Noah Project: The Secrets of Practical Project Management.* Gower, Brookfield, VT.

Rummler, G. A., and Brache, A. P. (1990). *Improving Performance: How to Manage the White Space on the Organization Chart.* Jossey-Bass, San Francisco.

Shores, A. Richard (1994). *Reengineering the Factory.* Quality Press, American Society for Quality, Milwaukee.

SECTION 18

MARKET RESEARCH AND MARKETING

Frank M. Gryna

INTRODUCTION

This section addresses the quality-related issues associated with the marketing function. The discussion is divided into two parts: market research for quality and quality activities within the marketing function.

Among the concepts and methodologies covered are the distinction between customer satisfaction and customer loyalty; concept of field intelligence; difference between customer needs, expectations, satisfaction, and perception; tools and techniques of market research on current and new products; importance of measuring customer satisfaction relative to competition; significance of determining the relative importance of various product attributes; role of sales and marketing in quality-related activities; applying quality concepts to improve the effectiveness of the marketing function.

We will follow the convention of using the term "product" to denote goods or services.

QUALITY IN THE MARKETING FUNCTION OF LEADING ORGANIZATIONS

As an overview to this section, we will examine how some leading organizations in quality transformed their marketing function from the traditional to the quality-focused. Hurley (1994) collected information through in-depth interviews at seven organizations: Federal Express, Globe Metallurgical, Marriott, Texas Instruments Defense Systems and Electronics, Toyota Motor Sales USA, Xerox Corp., and Zytec Corp. Five of these organizations were Baldrige Award winners. The findings revealed that the companies: (1) learn, in detail, the customer view of quality; (2) know the company core processes and how they relate to customer satisfaction; and (3) develop a culture with a strong external orientation. The findings suggest that some fundamental changes in marketing occur when quality management is initiated; e.g., an integrated organizational approach to marketing and selling replaces the traditional functional approach. In achieving this transformation, two frameworks are helpful—one strategic and the other tactical.

The concepts of strategic quality planning (see Section 13, Strategic Deployment) apply to the marketing function. A marketing group within the Digital Equipment Corporation started with a mission and developed six strategies to achieve the mission (Kern 1993). Working through several phases, a tree diagram relates mission, strategies, customers, and demands (Figure 18.1). This diagram focuses on the strategy of "customer-driven marketing and quality" and the "Sales" customer. Thus, the "demands" column represents the needs of the sales function as a customer of the marketing function. Follow-up activity includes rating the importance of each demand, and evaluating Marketing's current performance in satisfying the demands, performance of competitors, and the impact that an improvement would have in enabling Sales to meet goals with the end customer.

Going to the tactical level, we can identify the steps the marketing function must take to improve customer satisfaction. Stowell (1989) proposes the following steps:

1. Understand customer requirements (e.g., identify customers, document customer decision-making processes).
2. Identify Marketing's products and processes (e.g., brochures, market research, information delivery).
3. Match customer requirements to Marketing's products and processes.
4. Eliminate ineffective products and processes (of the marketing function).
5. Improve the remaining processes (e.g., use process management concepts and tools to improve marketing processes).
6. Add new marketing processes as required (e.g., a process to provide customers with certain information to make decisions).
7. Review the processes for each new product (i.e., examine marketing products and processes when the organization brings a new product or service to the market).

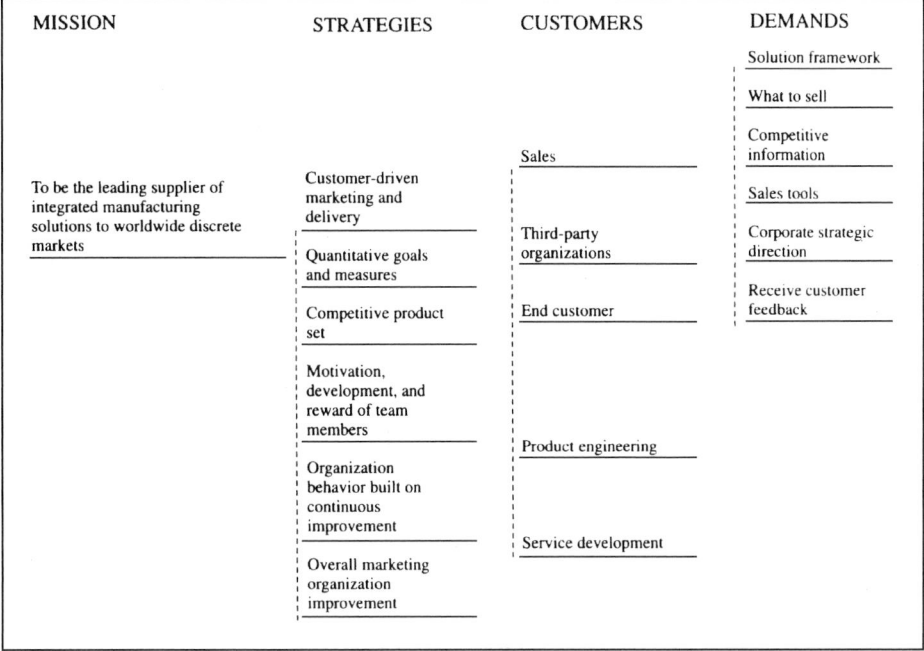

FIGURE 18.1   Final tree diagram. (*Kern 1993.*)

8. Improve the customer buying process (e.g., by making it easier for the customer to get and use information on products or services).

9. Involve employees in marketing quality (e.g., by participation in quality improvement teams as described in Section 5, The Quality Improvement Process).

From these frameworks for developing quality in marketing, we proceed to examine the role of market research and other roles of the marketing function to achieve both customer satisfaction and customer loyalty. Some books on marketing now discuss the integration of quality throughout marketing activities, e.g., see Churchill and Peter (1995).

CUSTOMER SATISFACTION VERSUS CUSTOMER LOYALTY

The importance of achieving customer satisfaction is a key theme of this handbook. That theme means not only meeting formal requirements on the product but also addressing other customer needs to achieve a satisfied customer. It is useful, however, to go further and make a distinction between customer satisfaction and customer loyalty (see Table 18.1) In brief, a satisfied customer will buy from our company but also from our competitors; a loyal customer will buy primarily (or exclusively) from our company. A dissatisfied customer is unlikely to be loyal but a satisfied customer is not necessarily loyal (see below).

Level of Satisfaction for Customer Loyalty.   Sometimes acceptable levels of customer satisfaction still result in a significant loss of new sales. Table 18.2 presents two examples from the telephone service industry. For example, although 92 percent of the customers who rated AT&T as

TABLE 18.1   Customer Satisfaction versus Customer Loyalty

Customer satisfaction	Customer loyalty
What customers say—opinions about a product	What customers do—buying decisions
Customer expects to buy from several suppliers in the future	Customer expects to buy primarily from one or two suppliers in the future
Company aims to satisfy a broad spectrum of customers	Company identifies key customers and "delights" them
Company measures satisfaction primarily with the product for spectrum of customers	Company measures satisfaction with all aspects of interaction with key customers and also their intention to repurchase.
Company measures satisfaction primarily for the current customers	Company also analyzes and learns the reasons for lost customers (defections)
Company emphasizes staying competitive on quality for a spectrum of customers	Company continuously adds value by creating new products based on evolving needs of key customers

"excellent" will probably repurchase, 8 percent probably will not. (These are "customer defections.") Note that even when the customer view of quality is "good," a quarter or more of the present customers may not return. Thus, customer satisfaction is a necessary but not a sufficient condition for customer retention and customer loyalty.

A similar story exists in the manufacturing sector (Burns and Smith 1991). In one company, the Harris Corporation, satisfaction with specific attributes of the product and service is measured on a scale of 1 to 10; loyalty is measured as the percentage of customers who will "continue to purchase" from Harris (Figure 18.2). Again note the high level of satisfaction required to achieve a high loyalty. (It was reported that the fit of the data was greatly improved when separate plots were made for Harris commercial divisions versus government and military divisions.) Also, the linkage between loyalty (and satisfaction) to financial results was validated through separate plots of satisfaction versus return on sales (ROS). These plots showed that a change of one point on satisfaction corresponded to about a 10 percent change in the ROS for commercial divisions and about a 1.5 percent change in ROS for government and military divisions.

To generalize, satisfaction can vary from disloyal customers who are so unhappy ("terrorists") that they speak out against a service or product to loyal customers ("apostles") so delighted that they try to convert others (Heskett et al. 1994)

Economic Worth of a Loyal Customer.   The sales revenue from a loyal customer measured over the period of repeat purchases can be dramatic, e.g., $5000 for a pizza eater, $25,000 for a business person staying at a favorite hotel chain, $300,000 for a loyalist to a brand of automobile.

To calculate the economic worth of the loyal customer, we can combine revenue projections with expenses over the expected life time of repeat purchases. (In highly dynamic high-tech industries, a limited period, say 5 years, may be appropriate to address uncertainties.) The economic worth is calculated as the net present value (NPV) of the net cash flow (profits) over the time period. The NPV is the value in today's dollars of the profits over time. This can be calculated by using a spread sheet that has an NPV function and applying an interest rate to the net cash flow estimates. With knowledge of the NPV, we can evaluate alternative marketing and other strategies to attract and retain customers.

Profits from loyal customers increase over time. Further, loyal customers represent a high customer "retention rate"; disloyal customers (those who switch to a competitor) contribute to the customer "defection rate." For example, a 60 percent retention rate corresponds to a 40 percent defection rate. Reichheld (1994) presents an example from the credit card industry. When a company is able to decrease the defection rate, the average customer life and the profits increase.

He also found that although retention/defection rates and NPV vary by industry, profits generally rise as defection rate decreases (retention rate increases). Figure 18.3 shows, for a variety of industries, the increase in NPV of the future customer profit stream resulting from an decrease of 5 per-

TABLE 18.2   Customer Satisfaction and Sales

Customer's view of quality	GTE business customers, of those who so rated quality who will recommend supplier, %	AT&T customers, of those who so rated quality, who are very willing to repurchase, %
Excellent	96	92
Good	76	63
Fair	35	18
Poor	3	0

Source: For GTE data, Gillett (1989); for AT&T data, Scanlan (1989).

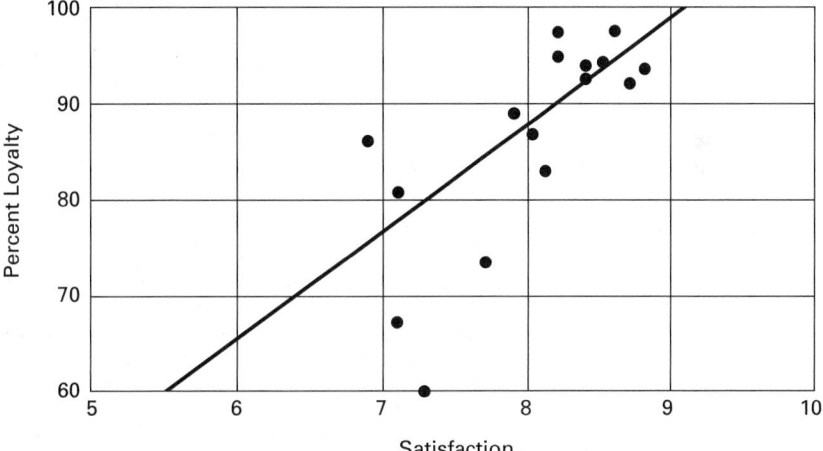

FIGURE 18.2   Relationship between loyalty and satisfaction rating (Harris divisions). (*Burns and Smith 1991.*)

centage points in defection rate (Reichheld 1996). He concludes that the 5-percentage-point decrease in defection rate can increase profits by 25 to 100 percent. In the same study, Reichheld also reports some sobering facts of the relationship between customer satisfaction ratings and customer loyalty. In some industries, more than 90 percent of customers report that they are "satisfied" or "very satisfied." Meanwhile repurchase rates are only about 35 percent. Also, about 70 percent of customers who defected had said on a survey just prior to defecting that they were satisfied or very satisfied.

Finally, Goodman (1991) points out other important marketplace phenomena that apply to many consumer and industrial products:

Most customers do not complain if a problem exists (50 percent encounter a problem but do not complain; 45 percent complain at the local level; 5 percent complain to top management).

On problems with loss of over a $100 and where the complaint has been resolved, only 45 percent of customers will purchase again (only 19 percent if the complaint has not been resolved).

Word-of-mouth behavior is significant. If a large problem is resolved to the customer's satisfaction, about 8 persons will be told about the experience; if the customer is dissatisfied with the resolution, 16 other persons will be told.

These business realities make it important to address customer satisfaction, customer retention, and customer loyalty by the collection of field intelligence through market research.

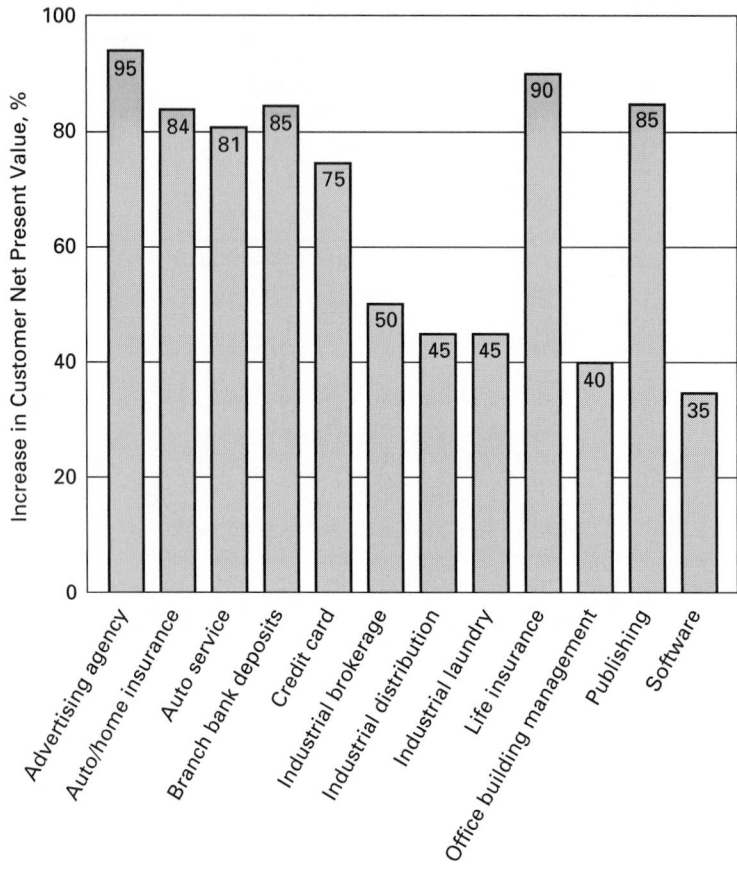

Industry

FIGURE 18.3   Profit impact of 5 percent increase in retention rate. [*From Reichheld (1996).*
Reprinted by permission of Harvard Business School Press.]

CONCEPT OF FIELD INTELLIGENCE

The term "field intelligence" will be the label for all information on product use whether relating to in-house use (other departments, sister divisions) or external use (merchants, processors, ultimate users). It includes information about product performance and the impact of that performance on all concerned—the merchant chain, users, service shops, society, government, the press, and the advocates.

Sources of Field Intelligence Information.   To develop and market its products, an organization requires information about quality in the marketplace. Some of this information is readily available in the form of feedback from customers. This feedback may include a chorus of complaints from customers or changes in market share due to product quality or other parameters. Such information helps to monitor the performance of current products but also serves as input to the development of future products.

The feedback on current products provides an important source of information both for designing future products and for monitoring the performance of current products. This feedback often provides alarm signals that must be acted on to maintain good customer relations. The main effect of an individual alarm signal (a complaint from one customer) is to stimulate action that provides satisfaction to the customer. Collectively, these signals supply one measure of overall customer dissatisfaction. They are, however, a poor measure of product quality since many quality failures do not result in alarm signals. Several areas can be completely silent:

1. *Dissatisfied customers who silently switch to a competitor's product:* Current customers are certainly a main source of business for the future, but those customers will switch to a competing product if they believe that the other product is superior. Often, the customer has no complaint about the product currently in use, but discovers a competing product that has superior performance or value.

2. *Noncustomers:* A major gap is information on the needs and attitudes of people who are not customers. Intelligence is needed to learn why they are not customers, and what it would take to cause them to purchase the product.

3. *Product dissatisfaction beyond the warranty period:* All companies have extensive systems for keeping track of product performance during the warranty period. The warranty period, however, is usually between 10 and 20 percent of the actual product life expected by the customer. Once the warranty period is passed, the feedback of information from the customer to the company becomes sporadic at best, and thus the company lacks information about the complete history of the product. As fitness for use applies for the life of the product, field intelligence information is needed over the lifetime.

Analysis of Available Field Intelligence. Included are

Decline in Sales. In practice, it is not a simple matter to draw conclusions about a decline in sales. The analyses require a joint effort among Marketing, Quality, and other departments to draw the proper conclusions. A discussion of the effect of quality on sales income is provided in Section 7, Quality and Income.

Field Complaints. Most organizations have systems of collecting and analyzing information on customer complaints. (For a discussion of these systems see Section 25, Customer Service, under Strategic Intent.) A Pareto analysis of field failures, complaints, product returns, etc., serves to identify the vital few quality problems to be addressed in both current and future products.(See Section 5, The Quality Improvement Process, under Quality Improvement Goals in the Business Plan.) This type of analysis is in wide use.

Salespersons' Reports. Salespeople, distributors, and other people in the marketing process are a natural, though imperfect, field intelligence system.

Sale of Spare Parts. Despite the fact that many users put up with these replacements as a necessary evil, the manufacturers should analyze the sales of spare parts to identify major field problems from the viewpoint of the customer.

Data from Customers. Information from customers consists not only of complaints but can include other information on quality such as warranty cards returned by customers upon receipt of the product. These cards can provide some information about the customer's perception of product quality upon delivery.

Also, purchasers of expensive equipment usually have detailed data systems for recording maintenance and repair work. Examples include construction, transportation, and manufacturing processing equipment. These data systems become a gold mine of information for the manufacturers of the equipment.

Government Reports. Governments are increasingly involved in product evaluation, mainly in their capacity as regulators. An example of such information is the reporting on tests conducted to evaluate the safety of automotive vehicles.

Independent Laboratories. Some of these laboratories make comparative tests for competing lines of products. They then publish the results for the benefit of their subscribers as an aid to making purchasing decisions.

The Internet. Increasingly, the wide variety of information available on the Internet provides another input on industry intelligence.

Reasons for Incomplete Use. The raw data for some (not all) types of field intelligence are readily available, but full use is not made of the data for two reasons: (1) the data are not recognized as valuable in planning for future products and (2) the data are not in a form useful to product development and other functions.

Planning for Field Intelligence. The main needs are to:

1. Discover current customer dissatisfactions not evident from complaints or alarm signals.
2. Discover the reasons for customer defections.
3. Discover the relative priority of quality versus other product parameters.
4. Discover the status of quality in relation to competitors.
5. Determine the specific fitness-for-use needs of customers.
6. Identify opportunities for improving income by improving fitness for use.
7. Identify opportunities in new markets or market segments.

Such broad needs make it necessary to plan carefully the approach to collecting the intelligence information. This is done through market research. The usual steps are (1) formulate the problem, (2) determine the research design, (3) design the data collection methods and forms, (4) design the sample and collect the data, (5) analyze and interpret the data, and (6) prepare the research report. For elaboration, see Churchill (1991). The most important users of market research results are design and operations people—not marketing people. The discussion below centers on designing the market research. It is essential that all users of the market research have input to the research design to assure that the content and depth of the research results meets their needs. Market research should never be left solely in the hands of market research specialists.

The Data Plan. Field intelligence can be useful to the manufacturer in various ways to improve product design, processes, tests, field service, marketing strategy. An interdepartmental team is able to come up with a list of needs as discussed below:

The vital few quality characteristics on which information is needed

Product performance with respect to those vital few characteristics, expressed in agreed units of measure

Information on users' costs and satisfaction as well as dissatisfaction

Failure information, including exact failure modes

Environments of use, both for successful use and for failure

The team can, in addition, usefully face up to the problem of data communications by preparation of a glossary of terms and a system of code numbers to facilitate data entry to computers, and other matters to improve field feedback. The data plan is the blueprint of what intelligence will be gathered and how it will be used.

Data Sources. In addition to using the available sources of field intelligence previously discussed, other approaches to data collection may be necessary.

1. *Natural field contacts:*   Every company has some employees who are in direct contact with the field: the sales force, complaint investigators, technical service specialists, service shop personnel. It is feasible to secure some field intelligence through these contacts, but only if a detailed data collection plan is defined and the time is provided for acquiring the information.

2. *Controlled use:*   In some cases a company makes enough use of its own products to provide a significant database on product performance. In other cases, the need may be to create an outside database by placing products in employee homes or with a consumer panel. In such cases it is feasible to design a data plan and to acquire useful field intelligence.

3. *Purchase of data:*   Contracts may be made to buy data from users of the product. The arrangements specify the data plan to be followed. They also provide for such associated features as training of personnel, audit of validity of data, and return of failed samples. For example, one automobile manufacturer agreed to pay all of the repair expenses for its cars at a car rental agency in return for detailed data on the repairs.

4. *Product monitoring:*   Technology is emerging which enables the "health" of products to be monitored during operation. In one case, instruments are now available to monitor the performance of computers. Users are asked to install the instruments as a means of collecting data on error rates. The data are then (*a*) used by clients to compare competing systems and (*b*) sold to manufacturers of the computers.

5. *Captive service center:*   Some consumer product companies maintain service centers throughout the country at which repair work on their products is performed. Thus, information on product performance, failure modes, and the cost of repairs is accumulated as part of the process of providing the repair service. Some companies also repair competing equipment and gather competitive information in the process.

6. *Maintaining the product for the customer:*   A variation of the captive service center is the mobile service center. In this form, a field force of repair people visits the customer's site and performs all preventive and corrective maintenance work on the product. Examples include computers and office equipment.

7. *Following progression of the product on site:*   In this approach a company "staples itself" to its product and observes how the customer uses the product. The idea is to identify opportunities for improving fitness for use (see below under Market Research for New and Modified Products).

8. *Use of "mystery shoppers":*   Some service industry firms employ an outside organization to pose as a customer and report back on the interaction with individuals in the service firm. For example, some banks use these "mystery shoppers" to obtain data on the interaction between bank customers and tellers.

9. *Continuing measurements to obtain customer perception of quality:*   In service industries, and increasingly in manufacturing industries, companies employ a variety of tools to evaluate how customers feel about quality. See the discussion below under Market Research—Tools and Techniques.

10. *Special surveys and studies.*   These are special market research studies conducted on an infrequent basis to obtain evaluations of quality versus the competition, answer specific questions about a product, or learn about customer perceptions of the relative priority of quality versus other product parameters. Examples of such studies are discussed below under Competitive Evaluations by Field Studies.

Clearly, it is important that field intelligence information and data be organized to make it easy for product development and other functions to easily and continuously access the information. This minimizes the need for special reports.

The Sampling Concept.   A common and fatal error in the pursuit of field intelligence is to go after 100 percent of the data. Normally a well-chosen sample will provide adequate field intelligence at a cost which is reasonable rather than prohibitive.

For example, a vehicle maker tried to secure complete field data from its 5000 dealers. The data quality was poor and the cost was high. The company then changed its approach. It concentrated on data from a sample of 35 dealers who were well distributed geographically and who accounted for 5 percent of the sales. The result was more prompt feedback for better decision making, and at a much lower cost.

Section 44, Basic Statistical Methods, discusses methodology for determining the sample size required. For a full discussion of reliability and validity in market research, see Churchill 1991, Chapter 9.

CUSTOMER BEHAVIOR CONCEPTS

The American Marketing Association defines marketing research as "the function which links the consumer, customer, and public to the marketer, through information—information used to: identify and define marketing opportunities and problems; generate, refine, and evaluate marketing actions; monitor marketing performance; and improve understanding of marketing as a process."

As applied to the quality of goods and services, market research is the systematic collection, recording, and analysis of data concerning quality as viewed by the customer. The research should address both components of quality, i.e., product features and freedom from deficiencies. In explaining the concepts of market research applicable to quality, we will first define certain terms concerning customer behavior. The terms are needs, expectations, satisfaction, and perception.

Customer Needs.   Customer needs are the basic physiological and psychological requirements and desires for survival and well-being. A. H. Maslow is a primary source of information on both physiological and psychological needs. He identifies a hierarchy of such needs as physiological, safety, social, ego, and self-fulfillment (for elaboration, see Section 15, Human Resources and Quality). Henry A. Murray developed a list of 20 physiological needs, e.g., achievement, order, autonomy (for a summary of the Murray study and a discussion of its marketing implications, see Onkvisit et al., 1994).

It is useful, also, to distinguish between stated needs and real needs. A consumer states a need for a "clothes dryer," but the real need is "to remove moisture"; a consumer wants a "lawn mower," but the real need is "to maintain height of lawn." In both cases, to express the need in terms of a basic verb and noun can spawn new product ideas. One historical example is the replacement of hair nets by hair spray to satisfy the basic need of "secure hair." Some needs are disguised or even unknown to the customer at the time of purchase. Such needs often lead to the customer using the product in a manner different from that intended by the supplier—a telephone number needed for emergencies is used for routine questions, a hair dryer is used in winter weather to thaw a lock, a tractor is used in unusual soil conditions. Designers view such applications as misuse of the product but might better view them as new applications and markets for their products.

Some of these applications are misuse, but such needs must be understood and, in some cases, alternative design concepts considered. Then there are other needs that go far beyond the utilitarian. Some needs may be perceptual (e.g., the now-classic example of Stew Leonard's supermarket customers who believed that only unwrapped fish on ice could be fresh); some needs may address social responsibility (e.g., the need for "green" products that protect the environment).

Customer needs may be clear or they may be disguised; they may be rational or less than rational. To create and retain customers, those needs must be discovered and served.

Discovering and understanding customer needs is necessary to define specific product attributes for subsequent market research and product development. Sometimes, a standard list of attributes is employed to obtain input on customer satisfaction. An example of this approach is Service Quality Gap analysis (ServQual). ServQual obtains input on customer service only and examines whether customer service is meeting customer expectations. The instrument covers five dimensions of service quality: tangibles (e.g., appearance of physical facilities and employees), reliability (dependability and accuracy of service), responsiveness (promptness), assurance (knowledge and courtesy of employees), and empathy (caring, individualized attention). A questionnaire uses 22 statements to obtain customer feedback on both their expectations and their perceptions with respect to the five dimensions. Customers respond on a 7-point scale ranging from "strongly disagree" to "strongly agree." For elaboration, see Zeithaml et al. (1990).

Customer Expectations. Customer expectations are the anticipated characteristics and performance of the goods or service.

Kano and Gitlow (1995) suggest there are three levels of customer expectation related to product attributes (see Section 3, The Quality Planning Process). The "expected" level of quality represents the minimum or "must be" attributes. We cannot drive satisfaction up with these attributes because they are taken for granted, but if performance of the basic attributes is poor then strong dissatisfaction will result. At the "unitary" (or desired) level, better performance leads to greater satisfaction but (in a limited time period) usually in small increments. For the "attractive" (or surprising) level, better performance results in delighted customers because the attributes or the level of performance are a pleasant surprise to the customers. Of course, these attributes must be translated into the product design.

On the basis of Kano's work, the Hospital Corporation of America finds it useful to identify several levels of customer expectation. Thus, at level I, a customer assumes that a basic need will be met; at level II, the customer will be satisfied; at the level III, the customer will be delighted with the service. For example, suppose a patient must receive 33 radiation treatments. Waiting time in the therapy area is one attribute of this outpatient service. At level I, the patient assumes that the radiation equipment will be functioning each day for use; at level II, the patient will be satisfied if the waiting time in the area is moderate, say 15 minutes. At level III, the patient will be delighted if the waiting time is short, say 1 minute. To achieve a unique competitive advantage, we must focus on level III; i.e., we must delight, not just satisfy. For additional examples and discussion of the three levels applied to different products see Hofmeister et al. (1996).

Rust et al. (1994) propose that there is a hierarchy of six levels of expectation ranging from the "ideal" (what would happen under the best of circumstances) to the "worst possible" (the worst outcome that can be imagined).

Customer Satisfaction. Customer satisfaction is the degree to which the customer believes that the expectations are met or exceeded by the benefits received. Satisfaction depends on many factors but Carlzon (1987) recommends focusing on "moments of truth." A moment of truth is the time during which a customer comes in contact with a company or its product and thereby forms either a positive or negative impression. These moments of truth can occur before, during, or after the purchase of a product.

Note that customer expectation has a strong influence on satisfaction. Suppose, for example, a customer stays at a luxury hotel and there is some minor inconvenience. Satisfaction will likely be low because the customer expects perfection at the luxury hotel. Contrast this with the customer staying at a budget motel. That hotel can have poor features but as long as the customer gets a reasonable night's sleep, the customer will have high satisfaction, because the expectation at the budget hotel is low.

Customer Perception. Customer perception is the impression made by the product. The perception occurs after a customer selects, organizes, and interprets information on the product. Customer perceptions are heavily based on previous experience. But other factors influence perception, and these factors can occur before the purchase, at the point of purchase, and after the purchase.

Spectrum of Customers. We define a customer as anyone who is affected by the product or process. Three categories of customers then emerge: (1) external customers, both current and potential; (2) internal customers; and (3) suppliers as customers. All three categories of customers have needs which must be understood and addressed during the planning of quality.

The paragraphs below describe how the tools and techniques of market research can collect intelligence on customer attitudes and behavior.

MARKET RESEARCH—TOOLS AND TECHNIQUES

The discussion below focuses on some basic tools and techniques. For further details on market research see Churchill (1991). Rhey and Gryna (1998) discuss applications to small business.

Telephone Calls to Customers.   Here, customers are called and asked for impressions on the quality of the item they purchased. The information gained can provide a general impression of quality, but it can also lead to specific action. Two examples will illustrate this tool.

On a quarterly basis, a hospital calls a sample of discharged patients (about one in three) and asks for opinions about the stay in the hospital. This phone interview consists of 36 questions or subquestions on nursing, physician care, hospital processes, and business processes. Separate questions on recommending the hospital and the physician to friends or relatives are included. From the information collected (and other research) it was learned that a major dissatisfier was the excessive waiting time for services within the hospital. The waiting time was in part due to the large number of personnel that a patient interacts with at the hospital. Further analysis resulted in the redesign of work and additional training for nursing personnel. This reduced the average number of personnel contacts for the patient from 60 to 24, for a typical stay of 3 days.

The Sales department in a high-technology company complained that "about 20 percent of the customers say quality is poor, and this is making it difficult for us to get repeat orders—no wonder we're having trouble meeting our sales goals." The Design and Manufacturing departments were amazed because of steps recently taken to achieve superior quality. The Quality department contacted some customers and heard a different story. Problems had occurred with early units of a new model, but the problems had been corrected and, said the customers, "Your overall quality record is really fine." Thus, the salespeople heard about problems but not about good performance and had therefore concluded that quality was poor.

Contacting the customers resulted in clarification of customer perception, but it also led to positive action. First, the salespeople were given information on product quality to make them believers of the superior quality level, and they highlighted this information in their selling efforts. Second, the positive reaction of customers resulted in Sales giving additional customer names to the Quality department and encouraging them to contact customers on a planned basis.

Visits to Individual Customers.   Another form of research is the periodic visit to major customers by a marketing or engineering representative of the company. These visits are not made in response to complaints but are designed to learn about customer experiences with the product and to provide answers to specific questions. (In practice, the visits are often not structured and are only partially successful in collecting useful information.)

In one case, a manufacturer of car wax held discussions with customers to learn how customers evaluated quality of wax. Within the company, high priority was attached to the "gloss" properties of the wax. The research revealed that the customers did not associate this property with car wax (although it was associated with house paint). Even when the term "gloss" was brought up for discussion it generated little reaction by the customer. What the customers did talk about was the "beading" property of the wax. Customers described beading as "when the water rolls off the surface." About 82 percent of the respondents indicated that they used "water beading" as a measure of continued wax performance. This study led to a change in priorities for the manufacturer and even influenced the selection of a name for the wax.

Special Arrangements with Individual Customers.   Establishing a special arrangement with a few customers is a simple and effective way to obtain information in depth. An electronics manufacturer does this with two or three key customers in each of several industries. The manufacturer offers, with the help of the customer (the "quality partner"), to maintain detailed records on the performance of the equipment in the customer's plant. The records benefit the manufacturer by relating performance to specific applications and environments rather than to assumed conditions. The customer benefits by having a direct link to the manufacturing and engineering operations of the manufacturer.

Focus Groups.   To better understand customer perceptions of its products, a food company sponsors meetings of small groups of customers to discuss product requirements. (The company holds an average of one such meeting per day.) The technique is called the "focus group" method. A focus group consists of about eight to ten current or potential customers who meet for a few hours to discuss a product. Here are some key features:

1. The discussion has a focus, hence the name.
2. The discussion can focus on current products, proposed products, or future products.
3. A moderator who is skilled in group dynamics guides the discussion.
4. The moderator has a clear goal on the information needed and a plan for guiding the discussion.
5. Often company personnel observe and listen in an adjacent room shielded by a one-way mirror.

Focus groups can discuss many facets of a product or can discuss quality only. A discussion on quality can be broad (e.g., obtaining views on what are the factors constituting fitness for use) or can have a narrower scope (e.g., determining customer sensitivity to various degrees of surface imperfections on silverware).

Depending on the goals, the participants in a focus group may be average customers, noncustomers, or special customers. Sometimes the participants are a special segment of society. For example, a toy manufacturer assembles a focus group of youngsters and provides them with a variety of toys for use in the group session. The children are observed to see which toys command the most attention and what kind of abuse the toys must take. The parents get their chance to talk in a separate focus group. A manufacturer of hospital supplies uses focus groups of nurses who apply products under simulated hospital conditions and offer comments while the product designers are observing and listening behind a one-way mirror. The designers translate the comments of nurses as wisdom based on experience, while the feedback of company marketing personnel is viewed as gossip.

A key to a successful focus group is the qualification of the focus group moderator. This is not a task for a well-intentioned amateur. Greenbaum (1988) presents the following criteria for a good focus group moderator:

1. *Quick learner:*   Absorbs and understands all inputs
2. *A "friendly" leader:*   Develops rapport with the group
3. *Knowledgeable but not all-knowing:*   Avoids being an expert
4. *Excellent memory:*   Ties together inputs during the meeting
5. *Good listener:*   Hears both content and implication
6. *A facilitator, not a performer:*   Secures information from the participants
7. *Flexible:*   Willing to adjust to the flow of information
8. *Empathic:*   Relates to the nervousness of some group members
9. *A "big picture" thinker:*   Separates the important observations from the less significant inputs
10. *Good writer:*   Writes clear, concise summaries with meaningful conclusions

Electronic meeting support (EMS) tools can help to record and summarize views during a focus group meeting. Typically, the participants meet together in a room, with each participant sitting at a computer. Using the computers, the participants—simultaneously and anonymously—

respond to the questions posed by the focus group moderator. Responses can be in many forms, e.g., numerically, yes/no, brainstorming phrases. The responses are collected by the computer network and displayed for all to see. After a review of the responses, further questions can be posed and the responses immediately summarized and displayed, e.g., the participants might be asked to rank the relative importance of brainstorming ideas collected in a previous step. EMS can have important advantages over face-to-face meetings: remove inhibitions of participants, save time through simultaneous collection of ideas, reduce the extraneous inputs ("noise") in the group communication process, and provide a transcript of the results. Anderson and Slater (1995) describe the benefits and present a case study applying EMS to a focus group.

A focus group is useful in exploratory research to pinpoint problems, obtain specific information on quality matters, and identify issues for further research. This caucus of customers requires a minimum of investment for significant potential benefits.

Mail Surveys.   Still another means of collecting field intelligence is the mail survey. Here, a questionnaire is developed listing various attributes of the product. The customer responds by using a satisfaction scale such as excellent, very good, good, fair, or poor. Space is also provided for other comments or suggestions. For example, a bank uses a questionnaire to probe 20 attributes of service by asking customers about the relative importance of the attributes and the degree of satisfaction.

Questionnaires are often lacking in two important areas of intelligence. First is the matter of the relative importance of each of the quality attributes. A list of attributes helps to be specific about the term "quality" but a failure to ask customers about the relative importance assumes that the attributes are of equal importance—a highly unlikely assumption. Second is the likelihood of the customer repurchasing or recommending the goods or service. Satisfaction does not necessarily mean loyalty in terms of repurchase (see above under Customer Satisfaction versus Customer Loyalty). A useful type of question to pose is the likelihood of repurchasing or recommending. Thus, intelligence information such as the percent of customers who are "highly likely to repurchase" provides data on customer loyalty that go beyond customer satisfaction data.

The following example illustrates the use of questionnaires and focus groups. An annual mail survey of automobile customers revealed a significant number of complaints on fit of doors. Within the company this term meant problems of "margin" and "flushness." Margin is the space between the front door and fender, or between the front and rear doors. Sometimes the margin was not uniform, e.g., a wider space at the top as compared to space at the bottom of the door. Flushness refers to the smoothness of fit of the door with the body of the car after the door is shut. The manufacturer took action, but it was doomed to failure. Steps were taken during processing and assembly to correct the margin and flushness problems, but the action didn't correct the problem—a later survey again reported a problem on fit of doors.

Fortunately, the company also held periodic focus group meetings in which a group of people were paid to attend a meeting where they answered a questionnaire and discussed issues in detail. The questionnaire listed fit of doors as one category of problems, and some customers in the focus group checked it. On the spot they were asked, "What do you mean by fit of doors?" Their answer was not margin or flushness but two other matters. First, fit meant the amount of effort required to close the door. They complained that they had to "slam the door hard in order to get it to close completely the first time." Second, fit meant sound. As the door was shut they wanted to hear a solid "businesslike" sound instead of the metallic, loose sound they heard (telling them the door was not closed) as they walked away from the car. The market research gave a better understanding of the symptom, and then the company pursued the right problem, which required changes in the product design and the manufacturing process. The company had long realized that fit of doors was important, but the first time the company acted, it solved the wrong problem.

Much experience is available to help in the design of questionnaires. Churchill (1991) suggests nine steps in preparing questionnaires along with 64 "dos and don'ts." For additional advice on preparing questionnaires, see Rust et al. 1994, Chapter 4.

Questionnaires should be viewed as only one of several means of gathering intelligence on quality. In practice, questionnaires need to be supplemented by contacts with individual customers, focus groups, and the other methods to furnish intelligence in sufficient depth to correct current problems and supply information for new product development. Willets (1989) provides an example of a comprehensive approach. Information is collected on a continuous basis through 11 methods including analysis of sales trends, analysis of complaints, and customer focus groups. Formal questioning is conducted at four levels of customers:

1. Those who approve the purchase, e.g., a senior executive
2. Those who influence the decision, e.g., a technical executive
3. Those who sign the purchase order, e.g., a buyer
4. Those who are users, e.g., a store manager who uses a computer

As part of field intelligence, organizations must learn where they stand on quality with respect to competition. For physical products, some knowledge can be acquired from laboratory testing. For physical and service products, however, it is also necessary to conduct field studies (see below).

COMPETITIVE EVALUATIONS BY FIELD STUDIES

The ultimate evaluation of quality is made by the user under the conditions of the marketplace. The field intelligence gathered on these evaluations must be based on factual inputs and not on hearsay. These evaluations aim to discover the users' viewpoints on fitness for use, and also to provide a comparison to competitors. The distinguishing feature of the field studies is that the user is the prime source of data.

Such studies should not be planned by any one department but by a team involving members from Marketing, Product Development, Quality, Manufacturing, and other areas as needed. This team must agree beforehand on what questions need to be answered by the field study. The types of questions which should be considered are:

1. What is the relative importance of various product qualities as seen by the user?
2. For each of the key qualities, how does our product compare with competitors' products, as seen by the users?
3. What is the effect of the quality differences on user costs, well-being, and other aspects of fitness for use?
4. What are users' problems about which they do not complain but which our product might nevertheless be able to remedy?
5. What ideas do users have that might be useful in new-product development?

Below are examples from the service and manufacturing sectors.

Examples of Field Studies.   GTE employs several approaches (including both survey and analytical) to measure and analyze customer satisfaction (Drew and Castrogiovanni 1995). In one approach, customers are asked to rate GTE on overall quality and on five attributes of telephone service. One of the summary statistics is the percent of "excellent" ratings. Other national local carriers serve as benchmarks. Table 18.3 shows the results for one franchise area. Note that the performance is mixed: scores are generally in the middle of the range, never at the top, but one (for installation) is at the bottom. For each attribute, a goal can be set and a schedule set for achieving the goal based on knowledge of proposed process improvements and customer satisfaction results. Separate data on the relative importance of the attributes resulted in setting the first priority on narrowing the ratings gap for local dial service.

In a case involving an industrial product, a field study was made as part of a strategic planning analysis (Utzig 1980). Although both failure costs and customer complaints were low, the product

TABLE 18.3 Benchmark Data for Local Telephone Service

Attribute	Company score	Competitor scores
Overall quality	38.6	36.6–46.3
Local dial quality	40.6	36.7–47.9
Billing quality	34.5	28.7–37.2
Installation quality	41.2	43.2–53.3
Long distance quality	47.5	40.9–55.3
Operator quality	41.5	35.0–47.1

Source: Drew and Castrogiovanni (1995).

had been losing market share for several years. Quality was one of five areas of strategic planning that was studied in order to improve market share. Table 18.4 shows a summary of the results. The six attributes studied were collectively a measure of quality. On a scale of 1 to 10, customers were asked for two responses—the relative importance of each attribute and a perceived performance rating versus two key competitors. The average score for General Electric was lowest—not just for one attribute but for all of them. These unpleasant findings were not believed. (After all, the failure costs were at an acceptable level and complaints were few.) The study was repeated—the results were still not believed. It was repeated again. Three geographical areas, covering all types of customers, were studied before the results were believed. The details revealed that customers not only wanted fewer failures; they also wanted to improve their own productivity, and this meant purchasing products with higher efficiency, durability, and better maintainability and serviceability. The desire for improved service was a surprise. The company thought they were the leader on service (because theirs was a comprehensive service organization). The customers, however, reported that the speed and quality of service were not equal to that of the competition. All of this led to actions in product design, the manufacturing processes, product service, and the quality assurance system.

A manufacturer of health products also uses a "multiattribute study" in which customers are asked to consider several product attributes and indicate both the relative importance and a competitive rating. An overall score is obtained for each manufacturer by multiplying the relative importance by the score for that attribute and then adding up these products. (For an extension to include costs, see Gryna 1983.) The Construction Industry Institute (Swartz 1995) recommends a similar approach to calculate a monthly index for specific construction jobs based on ratings of five attributes (cost, schedule, quality, safety, and management). Weights are assigned to each attribute to define the relative importance. The overall index is calculated as the sum of the weighted scores for the attributes.

Gale (1994) proposes that market research be performed not on quality alone but on value. Value has two components—quality and price. Table 18.5 shows the calculations for the "quality profile" for the Perdue Company versus the average of competitors in the chicken business. The ratio column shows the Perdue rating divided by the average of the competitor ratings. These ratios are then multiplied by the relative importance ("weight") assigned by customers to the various attributes. The "market perceived quality ratio" is the sum of the weighted ratios or 1.26 (126.1/100).

Now suppose Perdue chicken sold for 69 cents per pound and others sold for 59 cents. The quality rating can be combined with the price data to yield a customer value map (Figure 18.4). In this figure, the "fair value line" indicates where quality is balanced against price. The slope of the line is the relative weight assigned by the customer to quality versus the weight assigned to price in the buying decision (in this case, $1/3$ on quality and $2/3$ on price). Any company below and to the right of the line is providing better customer value; anyone above and to the left of the line is providing worse customer value. Gale discusses how such customer value analyses can help to plan cost-effective improvement actions in quality and competitiveness.

Woodruff and Gardial (1996) examine value as a tradeoff in positive consequences (e.g., benefits of a product) and negative consequences (e.g., initial price, repairs). They discuss concepts and techniques for obtaining and analyzing data on customer value and customer satisfaction.

TABLE 18.4   Customer-Based Measurements

Product attribute	Mean importance rating	Competitor performance ratings		
		GE	*A*	*B*
Reliable operation	9.7	8.1	9.3	9.1
Efficient performance	9.5	8.3	9.4	9.0
Durability/life	9.3	8.4	9.5	8.9
Easy to inspect and maintain	8.7	8.1	9.0	8.6
Easy to wire and install	8.8	8.3	9.2	8.8
Product service	8.8	8.9	9.4	9.2

Source: Utzig (1980), p. 150.

TABLE 18.5   Quality Profile: Chicken Business, after Frank Perdue Performance Scores

Quality attribute (1)	Weight (2)	Perdue (3)	Ave. competitor (4)	Ratio (5) = (3)/(4)	Weight times ratio (6) = (2) × (5)
Yellow bird	10	8.1	7.2	1.13	11.3
Meat-to-bone	20	9.0	7.3	1.23	24.6
No pinfeathers	20	9.2	6.5	1.42	28.4
Fresh	15	8.0	8.0	1.00	15.0
Availability	10	8.0	8.0	1.00	10.0
Brand image	25	9.4	6.4	1.47	36.8
	100				126.1
Customer satisfaction		8.8	7.1		

Source: Buzzell and Gale (1987).

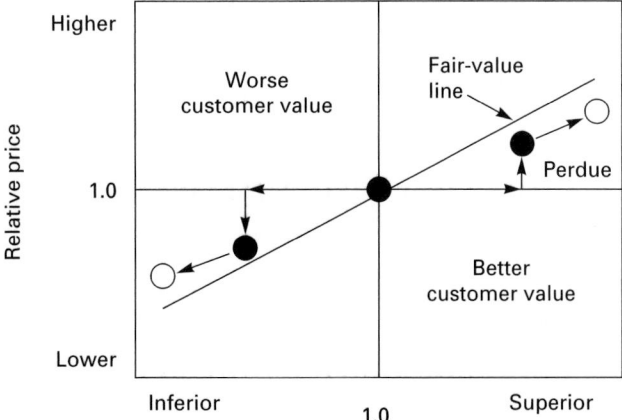

FIGURE 18.4   Customer value map: chicken business. [*From Gale (1994).
Reprinted with permission of The Free Press, a Division of Simon & Schuster.*]

The relative importance of the product attributes can be determined by several methods. In one approach, customers are asked to allocate 100 points over the various attributes. If desired, the allocation can be refined by using a systematic procedure involving questions to check the consistency of the numbers assigned. This systematic method was originally developed by C. W. Churchman, R. L. Ackoff, and E. L. Arnoff and is described in Barish and Kaplan (1978), Chapter 13. Finally, customers can be presented with combinations of product attributes and asked to indicate their preferences. The importance ratings can then be calculated. Churchill (1991), Appendix 9B, describes this method ("conjoint measurement").

AMERICAN CUSTOMER SATISFACTION INDEX

The American Customer Satisfaction Index (ACSI) is a national economic indicator of customer evaluations of goods and services. The index, calculated from about 50,000 telephone survey responses, covers 200 firms in seven sectors of the economy. For most companies, about 250 customers are contacted. A scale of 0 to 100 is used with 0 as the lowest rating and 100 as the highest. Although the ACSI is a satisfaction index, it goes beyond single measures such as the percent of respondents rating the product in the top one or two rating categories in a survey (Fornell et al. 1995). The basic model for the ACSI consists of a series of equations describing relations among six submodels—perceived quality, customer expectations, perceived value, customer satisfaction, customer complaints, and customer retention. Customers are asked questions directly related to each of the six elements of the model.

Important uses of the satisfaction data are

1. Compare industries.
2. Compare individual firms with the industry average.
3. Make comparisons over time.
4. Predict long-term performance. Empirical evidence is growing that customer satisfaction is related to internal company measures and also to stock market performance.
5. Answer specific questions. With appropriate assumptions, the model can be used to evaluate the impact of a specific quality improvement initiative, e.g., a training program for service personnel. Thus, the effect of such an initiative on satisfaction, retention, and future sales revenue (net present value) can be predicted in quantitative form. See Fornell et al. (1995) for elaboration.

As with all models, it is important to understand the assumptions and other elements of methodology. Additional information on the ACSI (including a methodologies report) can be acquired from the American Society for Quality.

LINKING CUSTOMER SATISFACTION RESULTS TO CUSTOMER LOYALTY ANALYSIS AND TO PROCESSES

Customer satisfaction and customer loyalty are distinct concepts (see above under Customer Satisfaction versus Customer Loyalty) but the two must be linked through action steps to achieve high customer loyalty and minimize customer defections. Some actions provide early warnings of defections and help to prevent the agony of further defections while other actions are "after the fact" of a defection. Action steps can include:

Ask about Repurchase Intention.   Incorporate in customer satisfaction measurement one or more questions on the likelihood that the customer will repurchase or recommend the product. The bottom line measure of customer satisfaction is the extent of repeat purchases. One service organization employs four measures: overall service quality, satisfaction with the price, was the service worth the price, and likelihood to repurchase. In addition, a study of a customer lifetime purchase pattern can determine loyalty in terms of repeat purchases. Winchell (1996) explains how to calculate the investment that would be justified for an improvement effort to increase the level of customer satisfaction. This approach combines data on customer satisfaction and willingness to repurchase (such as in Table 18.2) with a time profile of purchase patterns. The justifiable investment is calculated using the net present value concept.

Track Retention and Loyalty Information.   Track and distribute information on customer retention and customer loyalty. For example, an insurance company measures the percent of customers who do not allow a policy to lapse due to nonpayment of the annual premium. But that company is now examining a distinction between a retention measure and a loyalty measure. As a loyalty measure, the company is attempting to estimate what percent of insurance products being purchased by their customers are purchased from their company versus competitor insurance companies, i.e., the share of spending a firm earns from its customers. The initial estimate was low and identified a major market opportunity. Reichheld (1996, Chapter 8) discusses this "share of wallet" measure and other measures of customer loyalty.

Assure Understanding of Results.   Assure that market research results are understood and acted upon by line managers in product development, operations, and other areas. To accomplish this, one bank employs a guide to help each branch manager interpret and act on the market research results. The branch manager receives a separate customer satisfaction report, specific to the manager's branch, and uses the guide to identify the branch's strengths and weaknesses on the basis of those attributes the customers noted as most important. The branch manager and his/her staff complete the guide and communicate the results to all branch employees.

At this bank, 14 indicators of customer satisfaction are measured monthly. The results are reported to branch managers along with performance for the previous month, a comparison to other branches in the same state, and a comparison to the bank corporation as a whole. To assist in interpreting the results, the managers are provided with suggested questions. A brief plan is completed including action items to occur which will improve customer satisfaction.

A related issue is designing the satisfaction survey and other research to collect the right information in sufficient detail so that the line managers can take action on the research. This highlights the importance of obtaining input from the line managers for designing the satisfaction research (and also subsequent loyalty analysis).

Present Results for Action.   Present the results in a format that stimulates action. Graphing the market research results can be helpful. Rust et al. (1994) show an example of how the mapping of satisfaction and importance ratings can relate customer views and potential action (Figure 18.5). In this approach, attributes in which importance is high and satisfaction is poor represent the greatest potential for gain.

In Figure 18.5, the four quadrants are roughly defined by the averages on the two axes. Interpretation of the quadrants is typically as follows:

Upper left (satisfaction strong, importance low): maintain the status quo

Upper right (satisfaction strong, importance high): leverage this competitive strength through advertising and personal selling

Lower left (satisfaction weak, importance low): assign little or no priority on action

Lower right (satisfaction weak, importance high): add resources to achieve an improvement

Such maps could also be created for data on importance and percent of customers who are delighted, not just satisfied (see Rust et al. 1994). For additional examples of maps for customer retention modeling, see Lowenstein (1995, Chapter 9).

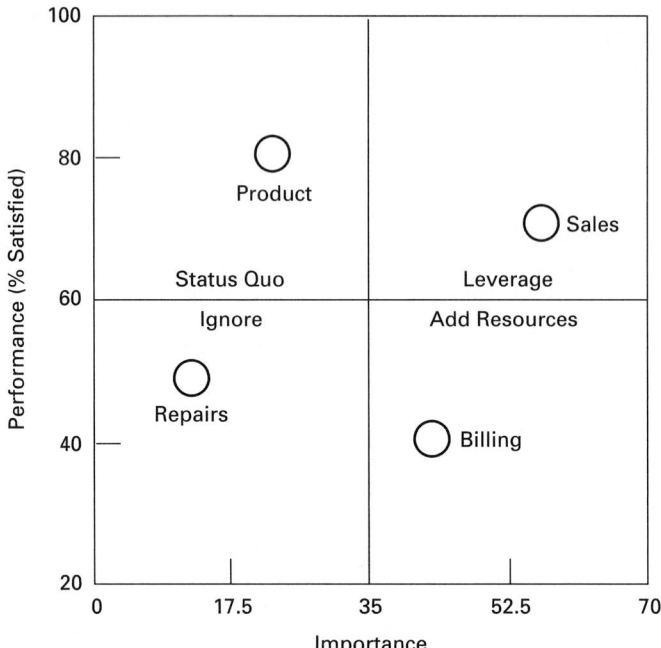

FIGURE 18.5 Performance versus importance in driving satisfaction: quadrant map. (*Rust et al. 1994.*)

Mikulak (1992) discusses the application of a "two-dimensional survey format" for satisfaction data on hotel guests. A battery manufacturer uses a two-dimensional diagram to depict the relation of its battery quality to that of its competition. The diagram applies the two components of quality; i.e., one axis has a scale for the key product feature (service life) and the other axis has a scale of freedom from deficiencies (a composite of various deficiencies). Each point on the diagram reflects both features and deficiencies. It is then possible to show both the company's and competitors' quality by plotting single points, each point representing quality for one company (or even one plant).

Link Results to Processes. Link the satisfaction results to operational processes or to specific activities. Kordupleski et al. (1993) show an example from the General Business Systems Division of AT&T.

Overall quality is made up of the combined quality of the business processes (Figure 18.6). In this example, note:

> 30 percent of the variation in overall quality is explained by product, 30 percent by sales, 10 percent by installation, 15 percent by repair and 15 percent by billing. Thus, service processes contribute 70 percent of the variation.
>
> Customer needs are identified for the five processes, and the relative importance of each need on the business process measure is shown. Note how the needs are stated in customer language, e.g., "accuracy, no surprises," "resolve on first call." (Customer needs are determined using focus groups and other methods discussed above under Market Research—Tools and Techniques.)
>
> To link customer needs to processes, internal metrics are tracked such that an improvement in the metric results in meeting a customer need.

This overall approach helps to ensure that improvement efforts have a strong customer emphasis.

Business Process	Customer Need	Internal Metric
Product (30%)	Reliability (40%)	% Repair call
	Easy to use (20%)	% Calls for help
	Features/functions (40%)	Function performance test
Sales (30%)	Knowledge (30%)	Supervisor observations
	Response (25%)	% Proposal made on time
	Follow-up (10%)	% Follow-up made
Installation (10%)	Delivery interval (30%)	Average order interval
	Does not break (25%)	% Repair reports
	Installed when promised (10%)	% Installed on due date
Repair (15%)	No repeat trouble (30%)	% Repeat reports
	Fixed fast (25%)	Average speed of repair
	Kept informed (10%)	% Customers informed
Billing (15%)	Accuracy, no surprises (45%)	% Billing inquiries
	Resolve on first call (35%)	% Resolved first call
	Easy to understand (10%)	% Billing inquiries

(Overall quality)

FIGURE 18.6   Strategic marketing information used to focus business processes. (*Kordupleski et al., 1993.*)

Analyze Complaints.   Use complaints as an early indicator of potential customer defections. Clearly, the frequency and nature of the complaints must be analyzed—see above under Concept of Field Intelligence. In addition, however, the effectiveness of the complaint-handling process and its impact on customer defections should be examined. Juran and Gryna (1993, pp. 518–520) explain a procedure for estimating the lost profit due to product problems coupled with ineffective handling of complaints. The language of economic impact can help to focus upper management attention on the complaints and the complaint-handling process.

Analyze Competitive Studies.   Analyze competitive studies on quality to identify differences in satisfaction that are likely to result in customer defections. Analysis of the multiattribute studies illustrated above can provide early warnings of reasons for customer defection and thereby trigger preventive action.

Determine the Reasons for Defections.   Conduct customer research to determine the reasons for defections. Basically, this means asking customers why they left. Experience suggests, however, that the reasons stated by customers are often not the real reasons; e.g., price is often mentioned as a key reason but probing usually reveals other reasons. The effort required to obtain actionable research results can be evaluated against the sales lost due to defections. Learning the reasons for defections can help to prevent a larceny—the loss of customers to competition.

Reichheld (1996, Chapter 7) discusses how "failure analysis" (a failure is a customer defection) can help to discover the causes of defections. Such an analysis includes applying the classical improvement technique of the "five whys"—i.e., asking why an event occurred at least five times to get to the root cause of a defection. For further discussion of the concept and techniques for customer loyalty and retention, see Lowenstein (1995).

MARKET RESEARCH FOR NEW AND MODIFIED PRODUCTS

In addressing quality for new and modified products, the broad steps for market research are: identify user needs and quantify user views, analyze the present use of the product, analyze the total system of use, discover market opportunities for new products, and link market research results to new product development. These steps can lead to a smash hit of success in the marketplace.

Identify User Needs and Quantify User Views.   Needs are the basic physiological and psychological body requirements for survival and well-being (see above under Customer Behavior Concepts).

Understanding, in sufficient depth, the needs of the customer requires listening to the "voice of the customer" by conducting special studies and also analyzing the present use of the product. In one example of a special study, a support department of AT&T Bell Laboratories conducted a survey of needs of internal customers (Weimer 1995). The result was a multilevel hierarchy of needs. Service quality was identified as one of the primary (strategic) needs. Customers defined service quality with 10 secondary (tactical) needs such as "no red tape" and "access to service information." These needs were further defined in terms of 42 tertiary (operational) needs such as "deal with person providing service directly" and "know who supplies a service or owns a process."

Sometimes, such studies require a team of people with special skills. For example, Philips NV sent a team of industrial designers, cognitive psychologists, anthropologists, and sociologists in mobile vans to conduct brainstorming with adults and children on ideas for new electronic products to meet changing needs (McKenna 1995). Also, companies must keep the dialogue flowing by making sure that all sources of customer contact, e.g., customer "hot line" phone calls, are converted to information for product designers (McKenna 1995). A further extension to gather information places the customer in the product development loop. This involves the concept of mass customization of products rather than mass production. Here, the individual customer designs the product to meet personal needs and simultaneously the company obtains information on customer needs. For examples involving products like greeting cards, windows, and clothing see Pine et al. (1995). The Ford Motor Company's planning for the Taurus model of automobile included extensive market research of customers. "Customers" included anyone affected by the automobile, e.g., potential purchasers, insurance companies, regulating agencies, manufacturing plants. Market research information, including competitor data, was obtained on 429 features of the automobile. A similar approach was employed for a major redesign of the Mustang automobile (see Section 3, The Quality Planning Process).

Multiattribute customer satisfaction studies, including those which include competitive data, are an important source of information on customer needs (see above under Competitive Evaluations by Field Studies). Such studies should include both satisfaction ratings and also the importance ratings on various attributes.

Analyze the Present Use of the Product.   This includes collecting and analyzing all available information on current product usage. See, generally, Section 25, Customer Service. Sometimes, a physical good or a service system can be designed to collect, filter, and interpret usage information. For example, Goodyear has developed a "smart tire" which contains a microchip that collects and analyzes air pressure data; the Ritz-Carlton Hotel chain tracks customer preferences (e.g., hypoallergenic pillows) and automatically transmits the information worldwide. For a discussion of the concept of "knowledge-based offerings" see Davis and Botkin (1994).

Visit the Scene of Action.   Go to the places where the product is used. We need to "staple ourselves to the product" to learn about customer problems and opportunities. To do this properly requires a high level of trust and "customer intimacy." It means watching for opportunities on the two components of quality, i.e., product features and freedom from deficiencies. Visits on site provide information about:

1. *Conditions of use:*   Actual conditions can differ markedly from those assumed by a product designer, but field conditions are the realities.
2. *Problems reported by the user:*   These problems can be of various types, e.g., difficult installation, breakdowns, inconvenience during use.
3. *Problems not reported by the user:*   Only a small portion of dissatisfied customers bother to complain; they just quietly walk away to a competitor. Thus, it is helpful to search out customers who defected and learn the reasons why. In addition, customers may report being satisfied but be unaware of problems or opportunities for improvement. Information on customer costs in using the product can be a source of ideas. Such information includes:
    a. Information on the costs to the user of operating and maintaining the product throughout its life. Performance data of this nature become targets for the product development function to exceed through future product designs.
    b. Information on the amount of employee training required to operate the product. To the extent that future designs can reduce the training, the design will be superior.
    c. Information on the amount of technical service required to support the product. An example includes the amount of diagnosis and maintenance work required to repair product failures.
4. *Remedial steps already taken by the user* (*or contemplated by the user*) *to improve the product:* This source of ideas for product changes even includes some built-in field testing.
5. *Needs for which the user sees no present solution:*   These needs represent an opportunity for the product development function to create a new and unique design for the marketplace.

Visits to the customer sites need not require a large amount of resources; a sample of customers is sufficient. Before the visits are made, however, a plan must spell out the questions to be answered and the procedures to be followed to collect the information. Some companies take unusual steps to conduct this deep, personalized research. For example, at the Weyerhaeuser Corp. some employees work for the customer—at the customer site—for a week. At one location, the Weyerhaeuser employee on site discovered that the bar-code label (attached by Weyerhaeuser to the newsprint rolls) was sticking to the printing presses. By relocating the bar code a few inches, the problem was solved (Jacob 1994). The Motorola Corp, as well as other organizations, encourages senior executives to visit customers. These visits, often scheduled on a quarterly basis, may address a customer complaint or provide executives with an eyewitness customer perspective to help identify new opportunities.

Analyze the Total System of Use of the Product.

The study of the user's operation can be aided by dissecting the total system of use. This is done by documenting all of the steps, analyzing them, and identifying opportunities for new product development. Saunders (1992) describes how the owners of different process steps may have different—and even conflicting—needs. Then we must sort, categorize, and prioritize the voices of these customers and even educate them on conflicting needs.

There have been a number of examples of the systems approach that enable us to identify some distinct categories:

Transfer of decentralized processing to a central process facility (e.g., frozen food, instant photography)

Modular concepts (e.g., prefabricated housing modules)

Elimination of user maintenance (e.g., self-lubricated bearings, aluminum exteriors)

New centralized service (e.g., centrally generated gas for heating, a central computer center into which various terminals feed)

Extending the shelf life of a product (e.g., by making a basic change in the manufacturing process, the shelf life of a brand of potato chips was increased from 2 months to 1 year)

Making a product compatible with other products (e.g., linking of products to a computer)

Input on customer needs and present usage leads to new products.

Discover Opportunities for New Products.   The opportunities for improvement span the full range of customer use from initial receipt through operation and maintenance. The primary stages and examples of ideas for improvements of an industrial product are given in Table 18.6.

In another approach, Bovee and Thill (1992) suggest five ways of adding value through enhanced customer service with an application to financial services and retailing (Table 18.7).

Taking a broader viewpoint, Treacy and Wiersema (1995) recommend that companies focus on one of three "value disciplines":

1. *Operational excellence:*   Provide customers with reliable products at competitive prices and minimal inconvenience.

2. *Product leadership:*   Provide customers with state-of-the-art products.

3. *Customer intimacy:*   Provide each customer with products that meet the specific customer needs.

Note how the focus on one of these disciplines not only influences the search for new product opportunities but also the market research process itself.

TABLE 18.6   Opportunities for Improving Fitness for Use of an Industrial Product

Stage	Opportunities
Receiving inspection	Provide data so incoming inspection can be eliminated
Material storage	Design product and packaging for ease of identification and handling
Processing	Do preprocessing of material (e.g., ready mixed concrete); design product to maximize productivity when it is used in customer's manufacturing operation
Finished goods storage warehouse and field	Design product and packaging for ease of identification and handling
Installation, alignment, and checkout	Use modular concepts and other means to facilitate setups by customer rather than manufacturer
Maintenance, preventive	Incorporate preventive maintenance in product (e.g., self-lubricated bearings)
Maintenance, corrective	Design product to permit self-diagnosis by user

TABLE 18.7   Adding Value Through Customer Service

Possible ways to add value (basic product)	Example from financial services (checking account)	Example from retailing (compact discs)
Be flexible	Let customers design their own checks	Accept returns of CDs that customers discovered they do not like
Tolerate customer errors	Cover overdrafts without charging a fee	Extend credit when customers forget to bring money
Give personal attention	Help customers with individual tax questions	Learn customers' musical tastes and suggest new CDs they might enjoy
Provide helpful information	Publish a brochure on financial planning	Distribute a newsletter that reviews new stereo equipment
Increase convenience	Install ATMs	Let customers order by phone

Source: Bovee and Thill (1992).

Linking Market Research Results to New Product Development.   In-depth market research results provide an essential input to new product development (NPD). Formal linking is accomplished using quality function deployment (QFD) and other approaches to convert customer needs to product and process requirements (see Section 3, The Quality Planning Process). The market research described in this Section to capture customer expectations, the relative importance of product attributes, and competitor positions is the direct link to QFD. To assure that the market research meets the needs of product development, it is important that product developers provide input and suggestions (content and depth) to a market research plan before the research is conducted.

Most organizations take steps to learn customer needs and use that information in new product development. But organizations striving for superiority probe customer needs in much more depth and scope than organizations who are satisfied with staying competitive. For superiority, what is needed is "customer intimacy," which goes beyond "customer satisfaction." Customer intimacy makes use of the techniques discussed above. But it also involves broader matters such as sharing responsibility for customer results, tailoring product solutions for *individual* (rather than group) customer needs, and developing trust with customers to share operating information and practice other forms of real partnerships. Such a profound approach requires much time and effort and may mean carefully choosing customers (and rejecting others) who have the technical capability and culture to form partnerships that have significant economic and other benefits to both partners. For elaboration, see Wiersema (1996).

To evaluate the benefits of conducting market research for current and new products, it is useful to conduct an audit of the usage of the research. This might include a review of the list of recipients of the research results, analysis of the knowledge of the recipients on how to interpret and use the results, and learning their satisfaction with the subject matter chosen and depth provided in the research results.

Next, we examine other quality-related activities within the marketing function.

QUALITY ACTIVITIES WITHIN THE MARKETING FUNCTION

Orsini (1994) presents an analysis of the seven Malcolm Baldrige National Quality Award categories and identifies the areas of marketing expertise that can foster results in specific Baldrige categories. The Baldrige categories of "Customer Focus and Satisfaction" and "Management of Process Quality" can benefit most from marketing expertise. A Pareto analysis of 12 areas of marketing expertise reveals that "research methods," "consumer behavior," and "market assessment" are the highest potential areas of marketing involvement in quality. Johnson and Chvala (1996) provide an overview of "total quality" and its application within the marketing function.

Albrecht (1992) describes how the Storage Technology Corp. uses four principles of deployment and five corporate change mechanisms to implement quality concepts in a regional marketing organization. The principles of deployment are

1. Utilize topics, subjects, processes that interest marketing representatives.
2. Start with interested parties. Leave those who resist until later.
3. Create opportunities for involvement of marketing people.
4. Provide value-added service for sales account management.

These principles are then related to five corporate change mechanisms: management network, education, quality improvement program, communication, and recognition.

Some marketing departments (particularly in the service industries) play a broader role in quality. Such a broader role includes taking the initiative to establish the need, do the initial planning, and coordinate the various quality activities throughout the company.

As with all functions in an organization, activities within the marketing function can benefit from the application of quality concepts. Table 18.8 shows some marketing activities and their related quality activities. Some of these are discussed in subsequent paragraphs.

TABLE 18.8   Marketing and Quality-Related Activities

Marketing activity	Quality-related activity
Launching new products	Conducting a test market to identify product weaknesses and weaknesses in the marketing plan.
Labeling	Ensuring that products conform to label claims.
	Ensuring that label information is accurate and complete.
Advertising	Identifying the product features that will persuade customers to purchase a product.
	Verifying the accuracy of quality claims included in advertising.
Assistance to customers in product selection	Presenting quality-related data to help customers evaluate alternative products.
Assistance to merchants	Providing merchants with quality-related information for use by salespeople.
	Providing merchants with technical advice on product storage, handling, and sales demonstrating.
Preparation of sales contract	Defining product requirements on performance, other technical requirements, and level of quality (e.g., defects per million).
	Defining requirements on execution of a contract, e.g., a quality plan, submission of specified documentation during contact.
	Defining warranty provisions.
	Defining incentive provisions on quality and reliability.
Order entry and filling	Applying quality improvement concepts to reduce lead time or reduce errors.

Quality in Sales Activities: Customer Relationship Process.   Every business has a process for working with customers, i.e., a "customer-relationship process" (CRP). Most organizations never define or analyze that process, but some are using quality principles and tools to define, study, and document their CRPs. Corcoran et al.(1995) explain how some organizations define the customer-relationship process in terms of five phases: establish the relationship, analyze the customer's requirements, recommend a solution and gain customer commitment, implement the recommendation, and maintain and expand the relationship. Managing an organization's CRP will help to:

Identify and analyze its people's interactions with customers.

Measure its people's performance against customer requirements and competitors' performance.

Involve customers in a dialogue about how the organization can change its CRP to better meet their needs.

Establish a common language that can be used within the organization and with customers to describe how the organization works with customers now—and how it would like to work with them in the future.

Clarify the roles, high-value activities, and competencies required of the front-line people whose daily interactions build the customer relationship.

Standardize and replicate the actions and behaviors that customers most value and that can differentiate the organizations from its competition.

Establish improvement priorities and allocate resources accordingly.

Better understand—and meet—the needs of various market segments by examining the CRPs that best meet their needs.

Based on CRP research, Corcoran et al. (1995) conclude that recent trends in many firms represent an evolution from traditional selling to consultative selling. Under consultative selling, salespeople have three roles in helping the customer—strategic orchestrator, business consultant, and long-term ally.

The concept of attracting, maintaining, and improving customer relationships is sometimes called "relationship marketing." Emphasis is on long-term retention of customers rather than making immediate sales. *The Journal of the Academy of Marketing Science* devoted a special issue to relationship marketing. One subprocess of this is "database marketing" which uses modern information technology to collect, assemble, and analyze information on customer characteristics (see *Business Week* 1994). This approach focuses on attracting the likeliest customers, targeting campaigns once they have become customers, and then structuring loyalty programs to reward them for continued patronage.

Thus, the process of personal selling has a quality-related dimension. But product promotion also involves advertising, public relations, and special sales promotions.

Advertising of Quality.   Customer satisfaction is the result of the comparison between customer expectations and the performance delivered (see above under Customer Behavior Concepts). Expectations evolve from a number of inputs, one of which is advertising. Thus advertising can influence customer satisfaction. Advertising enlarges sales income either through (1) product advertising, which aims to induce people to buy the product, or (2) institutional advertising, which aims to create a favorable image of the company. Advertising also provides information to help the customer make purchasing decisions. Overpromising in advertising (or by salespeople) can result in a dramatic loss of customer loyalty.

Objective Evidence.   Advertising based on objective product and quality data is widely used for industrial products and, to some degree, for consumer products as well. These objective presentations likewise take numerous forms:

Laboratory and inspection test results: tensile strength, frequency distribution, Weibull plots, safety ratings, tar and nicotine content

Usage data: mean time between failures; fuel consumption; cost per unit product, e.g., copies from a copier machine; lower frequency of adjustments, replacements, service

Listing of features possessed by the product (often to show that competitors lack these features)

Warranty provided (see below under Warranty)

Demonstrations of product usage by a series of still pictures or on television

Evidence of user satisfaction: testimonials from named users; data on share of market, e.g., "more than all other makes combined"

Results of tests by independent test laboratories: marks of certifications from such independent laboratories

Advertising which compares product features with those of competitors has a long history. Such advertising, however, has traditionally avoided naming the competitors in question. More recently there has been a dramatic increase in the frequency of naming the competitors. Some of these advertisements have stimulated competitors to complain to government regulators who then demand objective proof of the claims made in the advertisement. Alternatively, the competitors file legal actions to recover damages. In the face of such threats, advertisers are well advised to assure that the claims made in the advertising are based on objective test data.

Warranty: Industrial Products.   A warranty is a form of assurance that products are fit for use or, if defective, that the customer will receive some extent of compensation. Parties to a sale of industrial products are knowledgeable in contract relations, and they draft a purchase contract to embody their known needs and to cover contingencies experienced under previous contract arrangements. The resulting written contracts reflect mutual agreement on various pertinent matters including the warranty.

Warranties on industrial products are created uniquely for each product or type of product. For many commodity items, the warranty calls for compliance to specifications, and the warranty extends to survive "acceptance" of the item by the customer. Usually, no time period is specified. For other industrial products, the warranty often includes a time period and special provisions covering consequential damages. In some cases the warranty concept is extended to place predictable limits on users' costs. For example, a manufacturer of electrical generating equipment warrants to its industrial clients that the cost of power generated will be no higher than the cost of purchased power. If it is higher, the manufacturer agrees to pay the difference.

Warranty: Consumer Products.   In contrast to the tailor-made warranties which are written into large industrial contracts, warranties for consumer products are relatively standardized. Usually printed on good-quality paper with an artwork border, these warranties look like, and are, legal certificates. A preamble is included stating the good intentions and care supplied by the manufacturer. Specific statements of the manufacturer's written responsibility then follow.

Consumer product warranties are either "full" or "limited." The term "full warranty" refers to the consumer's rights, not to the portion of the physical product that is covered by the warranty, i.e., it does not have to cover the entire product. A full warranty means the following:

1. The manufacturer will fix or replace any defective product free of charge.
2. The warranty is not limited in time.
3. The warranty does not exclude or limit payment for consequential damages (see below).
4. If the manufacturer has been unable to make an adequate repair, the consumer may choose between a refund and a replacement.
5. The manufacturer cannot impose unreasonable duties on the consumer. For example, the warranty cannot require the consumer to ship a piano to the factory (one manufacturer listed such a condition).
6. The manufacturer is not responsible if the damage to the product was caused by unreasonable use.

The full warranty also provides that not only the original purchaser, but any subsequent owner of the product during the warranty period, is entitled to make claims.

A limited warranty is a warranty that does not meet the requirements for a full warranty. Typically, the limited warranty may exclude labor costs, may require the purchaser to pay for transportation charges, and may also be limited to the original purchaser of the product. As a practical matter, most warranties on consumer products are limited warranties and must be so labeled.

Some organizations provide an "unconditional service guarantee." Such a guarantee is not only without conditions, it is easy to understand, relevant, and easy to invoke. This type of guarantee can have a strong impact on sales, and it can also provide a strong internal company focus on customer satisfaction. For examples and discussion, see Hart (1988).

Role of the Marketing Function during the Launching of New Products.   A key factor in launching new products is the fit of the product with market needs.

Even with superior field intelligence and product development, the final test of the new product can be made only in the marketplace. Before mass producing and distributing the product on a nationwide basis, many companies use the concept of a "test market." In a test market, the product is sold on a limited basis for the purpose of:

1. Measuring the potential sales performance to decide whether or not to go ahead with full-scale marketing
2. Identifying weaknesses in the product
3. Identifying weaknesses in the marketing plan (product name, packaging, advertisements)

Conducting a test market can be expensive, and generally this step is taken only after pretesting has shown that the product will likely be a winner in the marketplace. A common method of pretesting involves a focus group of potential customers (see above under Market Research—Tools and Techniques).

As in the collection of field intelligence prior to the start of product development, it is essential that a team of people from marketing, product development, manufacturing, and other departments decide what questions need to be answered by the test market. After such agreement, the marketing department (with the aid of others, if necessary) collects the information in the marketplace. The mechanisms for collecting the information can include questionnaires, phone calls, focus groups, and other market research techniques.

The decision of whether or not to use a test market requires a comparison of the costs and risks. The costs include not only the expenses for data collection but also the income lost by delaying a full-scale market introduction. The key risk is introducing a product that fails. Even with prior market research, weaknesses in a new product are sometimes not revealed until the product lives in the marketplace environment. Examples of product weaknesses that did not show up until the marketplace follow:

Some bags of candy are almost impossible to open—the consumer resorts to biting the bag.

Because of their shape at the bottom, some tall, plastic bottles of soft drinks will not stand erect.

Some trash bags are difficult to use because they are a mystery to open and no instructions are provided.

Use of test marketing can help to prevent such disasters. The advent of three-dimensional computer graphics has led to the development of "virtual shopping simulation" for consumer products. In this market research technique, the atmosphere of an actual retail store is created on a computer screen, the shopper can "pick up" a package from a shelf, examine the package from all sides, and "purchase" the product by touching an image of a shopping cart. Product managers can test new products and marketing concepts without incurring manufacturing and other costs. For elaboration, see Burke (1996).

The process of launching new products is an example of an activity that is worthy of review by upper management (see Section 19, Quality in Research and Development, under Organizing for Research and Development Quality).

QUALITY IMPROVEMENT IN MARKETING

The marketing function has opportunities for identifying and acting upon chronic quality-related problems within marketing. For the approach and specific tools see Section 5, The Quality Improvement Process.

To cite just one area, Shapiro et al. (1992), studied the order management cycle (OMC) at 18 companies in different industries. In evaluating the effectiveness of the OMC, 10 steps (from order planning to postsales service) in the process were identified, the progress of individual orders traced, and the responsibilities of various departments analyzed. Further analysis revealed that four problems existed in a typical order management cycle:

1. Most companies never view the OMC as a process or system. No one person seems to understand the entire process.
2. Each step in the OMC has a mix of overlapping responsibilities.
3. Top management has little contact with the OMC.
4. The customer is as remote to the OMC as top management.

Improvement of the cycle in an individual company should start with charting the detailed process steps and applying appropriate quality improvement tools (see Section 5, The Quality Improvement Process).

QUALITY MEASUREMENT IN MARKETING

In all activities—including marketing—what gets measured gets done. Figure 18.6 shows examples of internal metrics for strategic marketing information. As in other applications of quality measurement, the measurements should have a customer focus, provide for both evaluation of performance and feedback for self-control, and include early, concurrent, and lagging indicators of performance.

BENCHMARKING FOR QUALITY IN MARKETING

Section 12, Benchmarking, explains the general concept and the steps in benchmarking. Johnson and Chvala (1996) identify 11 marketing areas to be benchmarked: sales management, sales-service support, direct marketing, industrial marketing, services marketing, new product development, advertising, sales promotion, distribution, supplier relationships, and retailing. Drew and Castrogiovanni (1995) describe how market research surveys can serve as a mechanism for benchmarking in a service industry. Identical questionnaires are sent to one's own customer and to customers of competitors. Data are gathered for various service processes (e.g., billing, installation) and the best rating for each becomes the benchmark for that process.

ACKNOWLEDGMENT

The author extends his thanks to Dr. Hemant Rustogi of The University of Tampa for his helpful comments on this section.

REFERENCES

Albrecht, Thomas J. (1992). "Quality Deployment in a Marketing Organization." *Proceedings IMPRO Conference,* Juran Institute Inc., Wilton, CT, pp. 6A-1 to 6A-9.Anderson, Elizabeth Scott, and Slater, Jill Smith (1995). "Electronic Meeting Software Makes Communicating Easier." *Quality Progress,* April, pp. 83–86.

Barish, Norman N., and Kaplan, Seymour (1978). *Economic Analysis for Engineering and Managerial Decision Making.* McGraw-Hill, New York.

Bovee, Courtland L., and Thill, John V. (1992). *Marketing.* McGraw-Hill, New York, p. 730.

Burke, Raymond R. (1996). "Virtual Shopping: Breakthrough in Marketing Research." *Harvard Business Review,* March-April, pp. 120–131.

Burns, Roger K., and Smith, Walter (1991). "Customer Satisfaction—Assessing Its Economic Value." *ASQC Annual Quality Congress Proceedings* pp. 316–321.

Business Week (1994). September 5, pp. 56–62.

Buzzell, Robert D., and Gale, Bradley T. (1987). *The Pims Principle.* The Free Press, New York.

Carlzon, Jan (1987). *Moments of Truth,* Harper and Row, New York.

Churchill, Gilbert A., Jr., (1991). *Marketing Research Methodological Foundations,* The Dryden Press, Chicago, pp. 399–400.

Churchill, Gilbert A., Jr., and Peter, J. Paul (1995). *Marketing: Creating Value for Customers.* Austen Press of Irwin, Burr Ridge, Ill.

Corcoran, Kevin J., Petersen, Laura K., Baitch, Daniel B., and Barrett, Mark F. (1995). *High Performance Sales Organizations.* Irwin, Chicago.

Davis, Stan, and Botkin, Jim (1994). "The Coming of Knowledge-Based Business." *Harvard Business Review,* September-October, pp. 165–170.

Drew, J. H., and Castrogiovanni, C.A. (1995). "Quality Management for Services: Issues in Using Customer Input." *Quality Engineering,* vol. 7, no. 3, pp. 551–566.

Fornell, Claes, Ittner, Christopher, and Larcker, David (1995). "Understanding and Using the ACSI: Assessing the Financial Implications of Quality Initiatives." *Proceedings, IMPRO Conference,* Juran Institute Inc., Wilton, CT.

Gale, Bradley T. (1994). *Managing Customer Value.* The Free Press, A Division of Simon & Schuster, New York.

Gillett, Tom F. (1989). "New Ways of Understanding Customers' Service Needs," a paper distributed by Charles Bultmann, in a session titled "How to Define Customer Needs and Expectations: An Overview," at the Customer Satisfaction Measurement Conference, American Marketing Association and ASQC.

Goodman, John A. (1991). "Measuring and Quantifying the Market Impact of Consumer Problems." Presentation on February 18 at the St.Petersburg-Tampa Section of the ASQC.

Greenbaum, Thomas L. (1988). *The Practical Handbook and Guide to Focus Group Research.* Lexington Books, D.C. Heath and Company, Lexington, MA, pp. 50–54.

Gryna, Frank M. (1983). "Marketing Research and Product Quality." *ASQC Quality Congress Proceedings,* Milwaukee, pp. 385–392.

Gryna, Frank M. (1995). "Exploratory Research on Quality Assessment in Financial Services." *College of Business Research Paper Series,* The University of Tampa.

Hart, Christopher W. L. (1988). "The Power of Unconditional Service Guarantees." *Harvard Business Review,* July-August, pp. 54–62.

Heskett, J. L., Jones, Thomas O., Loveman, Gary W., Sasser, W. Earl, Jr., and Schlesinger, Leonard A. (1994). "Putting the Service-Profit Chain to Work." *Harvard Business Review,* March-April, pp. 164–174.

Hofmeister, Kurt R., Walters,Christi, and Gongos, John (1996). Discovering Customer Wow's." *ASQC Annual Quality Congress Transactions,* Milwaukee, pp. 759–770.

Hurley, Robert F. (1994). "TQM and Marketing: How Marketing Operates in Quality Companies." *Quality Management Journal,* vol. 1, no. 4, July, pp. 42–51.

Jacob, Rahul (1994). "Why Some Customers Are More Equal Than Others." *Fortune,* September 19, pp. 215–224.

Johnson, William C., and Chvala, Richard J. (1996). *Total Quality in Marketing.* St. Lucie Press, Delray Beach, FL.

Journal of the Academy of Marketing Sciences (1995). Fall, vol. 23, no. 4.

Juran, J. M., and Gryna, Frank M. (1993). *Quality Planning and Analysis,* 3d ed. McGraw-Hill, New York.

Kano, Noriaki, and Gitlow, Howard (1995). "The Kano Program." Notes from seminar on May 4 and 5, University of Miami.

Kern, Jill Phelps (1993). "Toward Total Quality Marketing." *Quality Progress,* January, pp. 39–42.

Kordupleski, Raymond E., Rust, Roland T., and Zahorik, Anthony J. (1993). "Why Improving Quality Doesn't Improve Quality (Or Whatever Happened to Marketing?)." *California Management Review,* Spring, pp. 82–95.

Lowenstein, Michael W. L. (1995). *Customer Retention.* ASQC Quality Press, Milwaukee.

McKenna, Regis (1995). "Real-Time Marketing." *Harvard Business Review,* July-August, pp. 87–95.

Mikulak, Raymond J. (1992). "Are You Sure You're Meeting Your Customer's Needs?" *ASQC Annual Quality Congress Transactions,* Milwaukee, pp. 237–243.

Onkvisit, Sak, and Shaw, John J. (1994). *Consumer Behavior.* Macmillan, New York, pp. 41–42.

Orsini, Joseph L. (l994). "Make Marketing Part of the Quality Effort." *Quality Progress,* April, pp. 43–46.

Pine, B. Joseph, II, Peppers, Don, and Rogers, Martha (1995). "Do You Want to Keep Your Customers Forever?" *Harvard Business Review*, March-April, pp. 103–114.

Reichheld, Frederick F. (1994). "Loyalty and the Renaissance of Marketing." *Marketing Management,* vol. 2, no. 4, pp. 10–21.

Reichheld, Frederick F. (1996). *The Loyalty Effect.* Harvard Business School Press, Boston.

Rhey, William L., and Gryna, Frank M. (1998). "Market Research in Small Business: Observations and Recommendations." The University of Tampa, College of Business Research Paper Series, Report No. 906, Tampa, FL.

Rust, Roland T., Zahorik, Anthony J., and Keiningham, Timothy L. (1994). *Return on Quality.* Probus, Chicago.

Saunders, David M. (1992). "Hearing, Seeing, and Touching Your Customer's Process," *ASQC Annual Quality Congress Transactions,* Milwaukee, pp. 64–70.

Scanlan, Phillip M. (1989). "Integrating Quality and Customer Satisfaction Measurement." Handout at Customer Satisfaction Measurement Conference, American Marketing Association and ASQC.

Shapiro, Benson P., Rangan, V. K., and Sviokla, J.J. (1992). Staple Yourself to an Order." *Harvard Business Review,* July-August, pp. 113–122.

Stevens, Eric R. (1987). "Implementing an Internal Customer Satisfaction Improvement Process." *Juran Report,* no. 8, Juran Institute Inc., Wilton, CT, pp. 140–145.

Stowell, Daniel M. (1989). "Quality in the Marketing Process." *Quality Progress,* October, pp. 57–62.

Swartz, Kathleen (1995). "A Tool for Knowing Exactly What Your Customer Thinks." *ASQC Annual Quality Congress Transactions,* Milwaukee, pp. 490–494.

Treacy, Michael, and Wiersema, Fred (1995). *Discipline of Market Leaders.* Addison-Wesley, Reading, MA.

Utzig, Lawrence (1980). "Quality Reputation—A Precious Asset." *ASQC Technical Conference Transactions,* Milwaukee, pp. 145–154.

Weimer, Constance K. (1995). "Translating R&D Needs Into Support Measures." *ASQC Annual Quality Congress Proceedings,* Milwaukee, pp. 113–119.

Wiersema, Fred (1996). *Customer Intimacy.* Knowledge Exchange, Santa Monica, CA.

Willets, Gary G. (1989). "Internal and External Measures of Customer Satisfaction." *Customer Satisfaction Measurement Conference Notes,* ASQC and American Marketing Association, Atlanta.

Winchell, William O. (1996). "Driving Buyer Satisfaction by Quality Cost." *ASQC Annual Quality Congress Proceedings,* Milwaukee, pp. 205–211.

Woodruff, Robert B., and Gardial, Sarah F. (1996). *Know Your Customer.* Blackwell, Cambridge, MA.

Zeithaml, Valerie A., Parasuraman, A., and Berry, Leonard A. (1990). *Delivering Service Quality: Balancing Customer Perceptions and Expectations.* The Free Press, New York.

SECTION 19
QUALITY IN RESEARCH AND DEVELOPMENT

Al C. Endres

INTRODUCTION

This section discusses managing for quality in research organizations and in development processes. The material will focus on concepts, infrastructure, methods, and tools for simultaneously improving customer satisfaction and reducing costs associated with both these areas. (Managing for quality within the software development process is discussed in Section 20, Software Development.) Frequently the *combined* term "R&D" is used to describe cross-departmental processes, which integrate new knowledge and technology emanating from the research function with the subsequent development of new (or improved) processes and products. However, in this section, it will be useful to distinguish between managing for quality in research organizations and managing for quality in development processes.

Juran's original spiral of progress in quality (see Juran and Gryna 1988, p. 2.5) focused on (for a manufacturing organization) the cross-functional flow involved in the "development" of a new product. In the context of the original spiral, requirements for the new product emanated from marketing research. Marketing conducted research to define customers' needs, as well as to obtain customers' feedback on how well the organization had met those needs. Based upon customers' feedback, and changing customer needs, a new turn of the spiral began.

Marketing research is not the only possible origin of new technology and product ideas. Post-it notes actually resulted from the "failure" of an experiment that was recognized by a researcher as an opportunity for a new product. Furthermore, Roussel et al. (1991) emphasize the criticality of using exploratory research, conducted *proactively* to support an organization's strategic focus. Nussbaum (1997) quotes Thomson's vice president of consumer electronics for multimedia products as stating that design processes are being used "to address overall strategic business issues." Strategic research is being increasingly focused on the delivery of concepts and technologies which will drive new or improved technologies, such as lasers and photonics. These technologies are then used to generate breakthroughs for the organization's next generation of products. These respective origins (marketing research and strategy-directed research) for new technologies and product concepts can be characterized as "market pull" and "technology push," respectively.

Regardless of the means for identifying needs and opportunities, managing for quality in research organizations and development processes has become recognized as an increasingly critical activity. In addition to focusing on information, technology, goods, and services which are fit for use, there has been an increasing need to decrease R&D cycle times and costs. The chief executive of Hitachi Corporation's portable computer division has said (Markoff 1996) that "Speed is God, and time is the devil." The importance of speed in the automotive industry's new product design and development processes has also been emphasized (Reitman and Simpson 1995). Ford, Honda, and Toyota have all targeted approximately 33 percent reductions in their cycle times from concept approval to production. Clearly, managing for quality in the R&D processes can simultaneously reduce cycle times and costs. At a Shell Research center Jensen and Morgan (1990) found that a quality team's project for improving the project requirements process resulted in decreasing project cycle times by 12 months. At Corning Laboratories (Smith 1991) $21 million dollars of cost reductions were realized over a 4-year period while new products were pushed out faster, and with lower costs. An early project, which addressed reducing researchers' idle time during experiments, produced $1.2 million in "easy savings." Similarly Hutton and Boyer (1991) reported on a quality improvement project in Mitel Telecom's Semiconductor Division that resulted in custom prototype lead times being reduced from 22 weeks to 6 weeks.

The Missions of Research and Development.   In order to manage the research function and development processes, it is critical to define and understand their respective missions. To help distinguish among various types of research and development activities, the Industrial Research Institute (1996) has provided the following definitions:

- "Basic" (or "fundamental") research consists of original experimental and/or theoretical investigations conducted to advance human knowledge in scientific and engineering fields.

- "Directed basic" (or "exploratory") research is original scientific or technical work that advances knowledge in relevant (to corporate business strategies) scientific and engineering fields, or that creates useful concepts that can be subsequently developed into commercial materials, processes, or products and, thus, make a contribution to the company's profitability at some time in the foreseeable future. It may not respond directly to a specific problem or need, but it is selected and directed in those fields where advances will have a major impact on the company's future core businesses.

- "Applied" research is an investigation directed toward obtaining specific knowledge related to existing or planned commercial products, processes, systems, or services.

- "Development" is the translation of research findings or other knowledge into a plan or design for new, modified, or improved products/processes/services whether intended for sale or use. It includes the conceptual formulation, design, and testing of product/process/service alternatives, the construction of prototypes, and the operation of initial, scaled-down systems or pilot plants.

Building from Roussel et al. (1991), the following general definitions for the research and development processes are useful:

- *Research:*   The process used by an organization to acquire new knowledge and understanding.
- *Development:*   The process used by an organization to apply and connect scientific knowledge acquired from research for the provision of products and/or services commensurate with the organization's mission.

Although the latter definitions are broad, they are helpful. Both have been constructed to incorporate the word "process." One of the tenets of Total Quality Management (TQM) is to improve key processes which result in "products" which are "fit for use" by an organization's internal and external customers. In support of this perspective, Nussbaum (1997) has stated: "At the leading edge of design is the transformation of the industry to one that focuses on process as well as product." Similarly, Himmelfarb (1996a) has suggested that one key responsibility of senior managers is to ensure that the product development process is well defined (via flowcharts), documented, understood, monitored, and improved. It is therefore useful to define the "products" and "customers" of the research and development processes, which, in turn, can be used to define, measure, plan, control, and improve process quality.

Products of Research and Development Processes.   Juran (1992) has defined a product as "the output of any process," and noted that the word "product" can refer to either goods or services. For the purpose of this section, product will be used to denote the final or intermediary outputs of either the research organization or the development process. The primary "products" of a research organization are information, knowledge, and technology. The products of the development process are new or improved processes, goods, or services which result from the application of the knowledge and technology. For example, a likely output of a research project is a report containing the conclusions stemming from the project. Corresponding examples of final outputs of the product development process are designs and specifications released for production. Both the research and development processes also have intermediate or in-process outputs. Likely intermediate outputs of the research process are mathematical models, formulas, calculations, or the results from an experimental design. Correspondingly, likely intermediate outputs of the development process are physical models, prototypes, or minutes from design review meetings.

Processes of Research and Development.   Examples of key research processes identified and improved at Eastman Chemical Company have been provided by Holmes and McClaskey (1994): business unit organization interaction, needs validation and revalidation, concept development, technology transfer, and project management. Figure 19.1, from Holmes and McClaskey (1994), is a macrolevel process map of Eastman Chemical Company's "Innovation" process. Steps 1 to 4 represent the macrolevel research activities which generate the "new or improved product and process concept" stemming from step 4. The last step is the macrolevel development process which yields the processes and product designs for use in operations and markets, respectively.

Many organizations depict their product development processes through flowcharts reflecting their processes' major phases and "gates" (decision points). Altland (1995) has discussed the use of a "phase-gated" robust technology development process used by Kodak to help ensure that process and product technologies are "capable of manufacture and are compatible with intended product applications."

The flowchart in Figure 19.2, from Boath (1993), represents the results of "reengineering" of an organization's new product development process.

The new process led to a 25 percent increase in efficiency in "resource utilization." At IBM Rochester, Rocca (1991) reported that after the organization redesigned its product development process from a *sequential* progression of activities to *overlapping* planning, design, and development activities, it essentially halved development times. Wheelwright and Clark (1992) compare the phased development processes of Kodak, General Electric, and Motorola and relate them to the organizations' development *strategies*. Himmelfarb (1992) has named this overlapping multifunctional process concept "fast parallel new product development," and provides examples of use and benefits at Deere, Eaton, AT&T, Hewlett-Packard, Motorola, and NCR. Himmelfarb (1992) and Iizuka (1987) both stress the importance of understanding the "as is" development process and specific responsibilities within the process. Iizuka states: "The most important factor in building quality into a process is to define the process clearly...[and to show] what each department should do at every stage of the process."

Raven (1996) of Merrill Lynch's Insurance Group Services, Inc., provided an example of a project management process for product development in a financial service organization. The

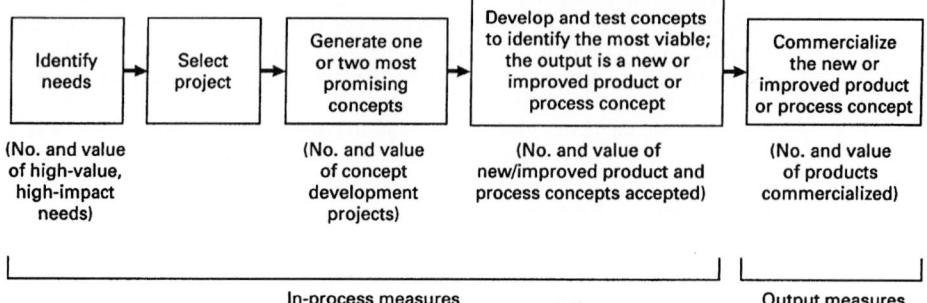

FIGURE 19.1   Eastman Chemical's innovation process (*Holmes and McClaskey 1994.*)

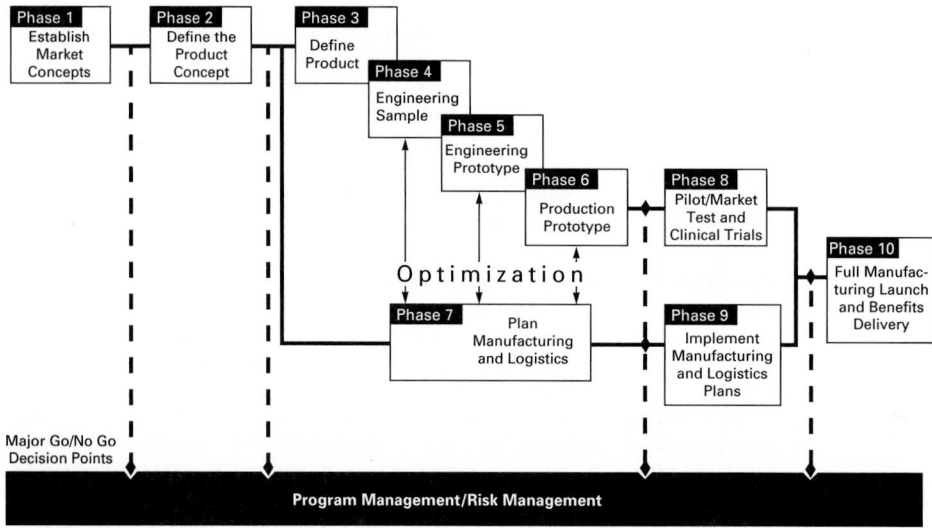

FIGURE 19.2   A new product development process. (*Boath 1993.*)

nine-step process depicted in Figure 19.3 was cited by Florida's Sterling (Quality) Award Examiners as being an example of a "…role model for excellence." The activities associated with each of the nine steps are listed in Table 19.1.

Another example of new service product development process improvement was presented by Swanson (1995) of the Educational Testing Service organization. He discusses dividing the business process reengineering project into three phases: data collection and assessment, best practices investigation, and process design. In the data collection and assessment phase, the reengineering team used past projects and customer perception data to identify and prioritize improvement opportunities. During investigation of best practices, the team used the prioritized problem list as a basis for uncovering "a wealth of data on sound product development practices and grounded the redesign in the current state of the art." During the design phase, the best practices for the current problems were integrated within the new process, and an implementation plan was developed. The total reengineering project spanned a period of $6^{1}/_{2}$ months and was expected to reduce cycle times by 70 percent and development cost by 60 percent. Himmelfarb (1996b) provides additional examples of new product development in service industries.

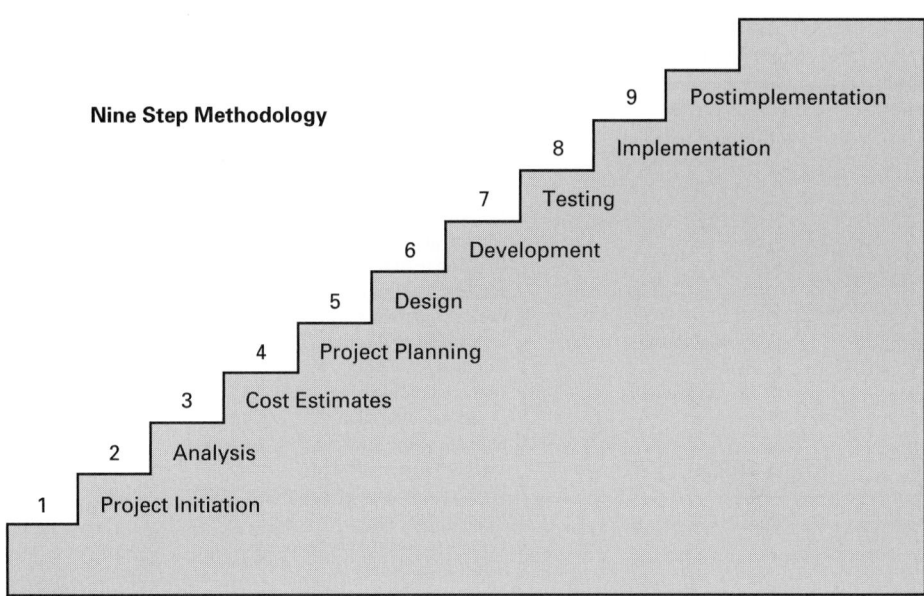

Nine Step Methodology

9 Postimplementation
8 Implementation
7 Testing
6 Development
5 Design
4 Project Planning
3 Cost Estimates
2 Analysis
1 Project Initiation

FIGURE 19.3   Merrill Lynch Insurance Group Services project planning and development process. (*Raven 1996.*)

DEFINING QUALITY FOR RESEARCH AND DEVELOPMENT

Defining Research Quality.   In this section, research quality will be defined from the perspective of both customer satisfaction (effectiveness of features) and costs (efficiency of providing the features). General Electric (Garfinkel 1990) defined several dimensions of research quality: technical quality of research, impact of research ("game changer" versus incremental), business relevance, and timeliness (early or late relative to the targeted market requirements). At DuPont, Darby (1990) defined R&D quality as "creating, anticipating, and meeting customer requirements" which required "continual improvement of knowledge, application, and alignment with business objectives."

The primary products of the exploratory and applied research process are information, knowledge, and technology. Godfrey (1991) provides a general discussion of information quality. Research product quality can therefore be defined both from the perspective of customers' satisfaction with the features of the information, and the absence of deficiencies of the information (which decreases costs and cycle times, and hence increases efficiency). Features of research information include timeliness, utility, accuracy, and costs. Research deficiencies can either occur during the research process or be reflected in the end products of the research. Possible deficiencies in research *products* may be that the knowledge is late, inaccurate, irrelevant, or of relative poor value for the investment. Deficiencies in research *processes* are associated with process "rework" or "scrap," e.g., having to reissue a section of a progress report because of using a wrong formula or having to redo an experiment because an audit revealed that reference samples had been contaminated.

Combining these perspectives, research quality is defined as *the extent to which the features of the information and knowledge provided by the research function meet users' requirements.*

Defining Development Process Quality.   The primary result of development is new or improved products and processes. The quality of a development process will be defined as *the extent to which the development process efficiently provides process and product features capable of repeatedly meeting their targeted design goals, e.g., for costs, safety, and performance.*

TABLE 19.1 Activities within Steps of Merrill Lynch Insurance Group Services Project Planning and Development Process

Step	Activities
1. Project initiation	*a.* Prepare recommendations *b.* Executive committee review *c.* Decide on approval (yes/no)
2. Analysis	*a.* Determine scope *b.* Obtain sign-off on scope *c.* Develop requirements *d.* Review requirements *e.* Conduct market research
3. Cost estimates	*a.* Determine cost estimates *b.* Conduct feasibility study
4. Project planning	*a.* Prepare timelines *b.* Develop action plans *c.* Schedule meetings
5. Design	*a.* Develop system design *b.* Develop business procedures
6. Development	*a.* Prepare SEC and state filings *b.* Complete system programming *c.* Develop test plan *d.* Develop work flows, policy and procedure bulletins *e.* Prepare training, marketing, and sales materials *f.* Determine purchasing and print requirements *g.* Obtain sign-off
7. Testing	*a.* Conduct program testing *b.* Conduct system testing *c.* Conduct user acceptance testing *d.* Conduct regression testing *e.* Conduct quality assurance tests *f.* Conduct branch office testing *g.* Obtain sign-off
8. Implementation	*a.* Distribute policy and procedure bulletins, training materials, marketing and sales materials *b.* Conduct operational training sessions *c.* Implement new systems, procedures, and processes
9. Postimplementation	*a.* Conduct postproject reviews and surveys

Source: Raven (1996).

Resultant product and process features must be thought of from the perspective of "big Q" thinking. Port (1996) discusses the growing importance of environmentally friendly products and processes. Regulators are compelling designers to address such issues as the German ordinance requiring manufacturers to assure the disposibility of all packaging used in product transport, and, in the Netherlands, the rule that manufacturers must accept old or broken appliances for recycling.

Deficiencies (and hence inefficiencies) in the development process are associated with process rework or scrap. Berezowitz and Chang (1997) cite a study at Ford Motor Company discussed by Hughes (1992) which concluded that while the work done in the product "design phase typically accounted for 5 percent of the ongoing total cost," it accounted for 70 percent of the influence on products' future quality. Boznak and Decker (1993) report that costs associated with deficiencies in product design and development processes can be very expensive. They reference one computer manufacturer whose costs "exceeded $21 million…(which) equated to 420,000 hours of non-value-added work…who lost nearly $55 million in gross margin opportunity on one product. Failure to effectively manage its product development processes put the company's entire $1.54

billion international business at risk." The authors suggest that the company's practices which caused this near catastrophe would have been precluded had those practices complied with the requirements of ISO 9000. (See Section 11 for discussion of the ISO 9000 standards.)

Examples of design "rework" include design changes necessitated by an outdated requirements package and partial redesigns necessitated by missing one or more design objectives (including schedules and costs). Perry and Westwood (1991) measured the quality of Blount's development process by the extent to which technical targets are met, e.g., "meeting specific process capability targets" and "the percent and degree of customer needs that are met, and the number of problems discovered at various stages of the product development process." At Motorola's Semiconductor Sector, Fiero and Birch (1989) reported that reducing development process deficiencies increased the percentage of fabricated prototypes passing all tests upon first submission from 25 percent to 65 percent. Furthermore, by involving 10 functional areas, Motorola was able to shorten development cycle times from 380 to 250 days. The reported investment of $150,000 resulted in potential additional revenues of $8 million per year.

PLANNING AND ORGANIZING FOR QUALITY IN RESEARCH AND DEVELOPMENT

Identifying and Addressing Barriers. To successfully plan for and utilize the concepts required to manage for quality in research or development, management must first understand and then address potential implementation pitfalls and barriers associated with developing and implementing quality initiatives within R&D environments. Hooper (1990) and Endres (1992, 1997) discuss cultural and organizational barriers that must be addressed. For example, researchers' fear that quality initiatives will stifle individual creativity, resulting in bureaucratic controls, can be addressed through the choice of pilot projects. A project can be chosen to demonstrate that improving research quality can provide researchers with better resources or processes for conducting more efficient research (e.g., reducing cycle times for obtaining reference articles; obtaining more information from fewer experiments using statistically designed experiments). Hooper (1990) identifies as an organizational barrier to improving R&D quality R&D's traditional isolation from customers and business. Oestmann (1990) discusses how Caterpillar addressed the problem of researchers being isolated from their customers by moving "...experienced research engineers into the field, close to high populations of customers. Their assignment is to understand the customer—how he used his machines today and how he will use them in the future, what drives the customer to make buying decisions now and in the future. The objective of this is to envision what *technologies* will be needed to produce superior future products." After research evolved the most promising technologies, Caterpillar used cross-disciplinary teams to develop the required product concepts. Teams comprising representatives from Marketing, Engineering, Manufacturing, and Research develop concepts for solving customers' needs "and then rate each idea based on its value to the customer."

For development personnel, Gryna (1988) discusses the importance of placing product developers in a state of "self-control." (See Section 22, Operations, under Concept of Controllability; Self-control.) Prior to holding designers responsible for the quality of their work products the three major criteria (I, II, III) provided in Table 19.2 must be met. Gryna, using input from designers, developed the specific items listed under each criterion. The table may be used as a checklist to identify opportunities for improving designers' work products, and subsequently, their motivation for quality improvement.

Leadership and Infrastructure Development. For upper managers to successfully lead a quality initiative, they must understand their respective roles and responsibilities in managing for quality. Holmes and McClaskey (1994) have stated that at Eastman Chemical:

Top Research Management Leadership was the most significant and essential success factor. Research management changed the way it managed research by focusing on the major output and by

TABLE 19.2   A Self-Control Checklist for Designers

I. Have designers been provided with the means of knowing what they should be doing?
   A. Do they know the variety of applications for the product?
      1. Do they have complete information on operating environments?
      2. Do they have access to the user to discuss applications?
      3. Do they know the potential field mususes of the product?
   B. Do they have a clear understanding of product requirements on performance, life, warranty period, reliability, maintainability, accessibility, availability, safety, operating costs, and other product features?
      1. Have nonquantitative features been defined in some manner?
      2. Do designers know the level of product sophistication suitable for the user involved?
   C. Are adequate design guidelines, standards, handbooks, and catalogs available?
   D. Do designers understand the interaction of their part of the design with the remainder of the design?
   E. Do they understand the consequences of a failure (or other inadequacy) of their design on: (1) the functioning of the total system? (2) warranty costs? (3) user costs?
   F. Do they know the relative importance of various components and characteristics within components?
   G. Do they know what are the manufacturing process capabilities relative to the design tolerances?
   H. Do they derive tolerances based on functional needs or just use standard tolerances?
   I. Do they know the shop and field costs incurred because of incomplete design specifications or designs requiring change?
II. Have designers been provided with the means for knowing what they are doing?
   A. Do the have the means of testing their design in regard to the following:
      1. Performance, reliability, and other tests?
      2. Tests for unknown design interactions or effects?
      3. Mock-up or pilot run?
   B. Is there an independent review of the design?
   C. Have the detail drawings been checked?
   D. Are designers required to record the analyses for the design?
   E. Do they receive adequate feedback from development tests, manufacturing tests, proving ground tests, acceptance tests, and user experience?
      1. Are the results quantified where possible, including severity and frequency of problems and costs to the manufacturer and user?
      2. Does failure information contain sufficient technical detail on causes?
      3. Have designers visited the user site when appropriate?
   F. Are designers aware of material substitutions, or process changes?
   G. Do they receive notice when their design specifications are not followed in practice?
III. Have designers been provided with the means of regulating the design process?
   A. Are they provided with information on new alternative materials or design approaches? Do they have a means of evaluating these alternatives?
   B. Have they been given performance information on previous designs?
   C. Are the results of research efforts on new products transmitted to designers?
   D. Are designers' approvals required to use products from new suppliers?
   E. Do designers participate in defining the criteria for shipment of products?
   F. May designers propose changes involving trade-offs between functional performance, reliability, and maintainability?
   G. Are designers told of changes to their designs before they are released?
   H. Have causes of design failures been determined by thorough analysis?
   I. Do designers have the authority to follow their designs through the prototype stage and make design changes where needed?
   J. May designers initiate design changes?
   K. Are field reports reviewed with designers before making decisions on design changes?
   L. Do designers understand the procedures and chain of command for changing a design?

---

personally leading the analysis and improvement of the key management processes which drive the output. Research management since 1990 has institutionalized QM (Quality Management) by making it the way Research is managed. The ECC Research success story is certainly another illustration of a quote by Dr. J. M. Juran (1992b): "To my knowledge no company has obtained world class quality without top managers taking charge."

A key responsibility of upper management in leading a quality initiative within research or development is to organize and develop an infrastructure for initiating, expanding, and perpetuating quality in both research organizations and development processes.

Organizing for Research and Development Quality. Several R&D organizations have developed structures which facilitate the attainment of their goals for improving customer satisfaction and reducing the costs of poor quality. Wood and McCamey (1993) discuss the use of a steering team at Procter & Gamble for "maintaining momentum," representing all levels of the organization, and from which subgroups were spun off "to manage areas such as communication, training, planning, measurement," and team support. "The role of the steering team was to keep the division focused on business results and setting clear, measurable targets." Taylor and Jule (1991) discuss the role of the quality council at Westinghouse's Savannah River Laboratory, consisting of the laboratory chairman, department heads, two senior research fellows, and the laboratory's TQM manger. The council was supported by department/section councils in developing, implementing, and tracking an annual Quality Improvement Plan (QIP). The QIP was developed by a team of laboratory managers chartered by the director to assess quality progress during the previous year and "select topical areas for improvement in the coming year based on employee input.…" Each department manager was assigned a topical area and required to develop an improvement plan. The separate improvement plans were then reviewed and integrated into a quality improvement plan for the entire laboratory. Menger (1993) has discussed the organization and activities of the World Class Quality (WCQ) Committee at Corning's Technology Group, consisting of representatives from five major groups reporting to Corning's vice-chairman. The WCQ identifies priorities and reviews progress in its group's members, establishing and improving key results indicators (KRIs) for cycle times, productivity, and customer and employee satisfaction. Figure 19.4, from Menger (1993), portrays the organization structure and process used to track and improve performance.

Figure 19.5, presented by Hildreth (1993), is a structure used to manage key business processes e.g., clinical research, development, product transfer in manufacturing, in R&D at Lederle-Praxis Biologicals. (See Section 6 for further discussion of managing key business process quality.) The Executive Quality Council is supported by a Business Process Quality Management (BPQM) Council and site-specific quality councils.

In addition to organization structure, other elements of infrastructure required to perpetuate R&D quality initiatives are training, councils, teams, facilitators, measurement, and rewards and recognition.

Training for Quality in Research and Development. Before managers or researchers can lead and implement quality concepts, processes, or tools, their needs for education and training must be identified and met. Wood and McCamey (1993) of Procter & Gamble discuss the importance of tailoring the training to the R&D environment:

> Our training had two key features: 1) it was focused on business needs and 2) it was tailored to the audience. These features reflected lessons we learned from other parts of the company; e.g., training that was not focused on real business issues lacked buy-in, and a training program developed for manufacturing could not be transplanted wholesale into an R&D organization.

Similarly, at Bell Laboratories Godfrey (1985) reported that a key ingredient for successfully training design engineers in experimental design and reliability statistics is the use of case studies

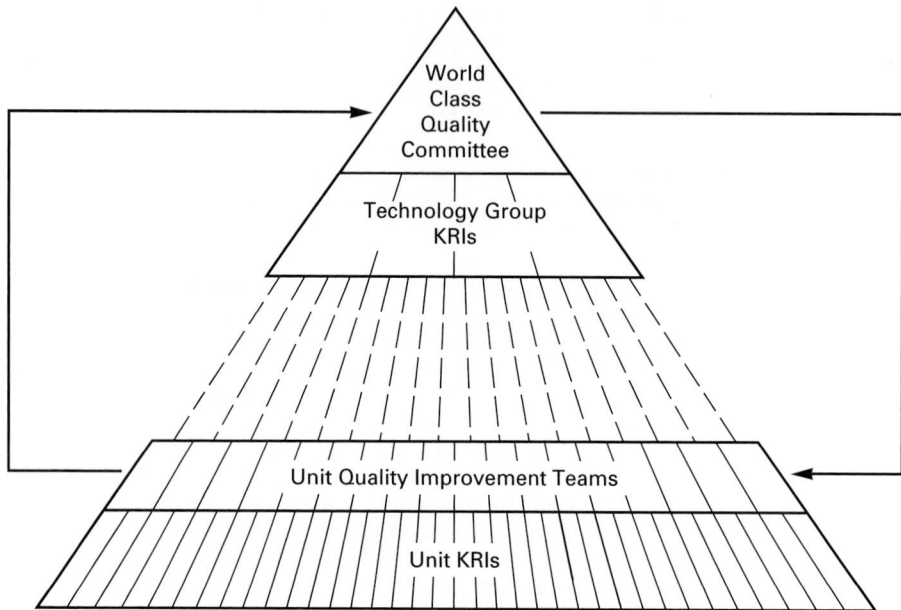

FIGURE 19.4 Corning's Technology Group quality organization and KRI improvement process. (*Menger 1993, p. 1-14.*)

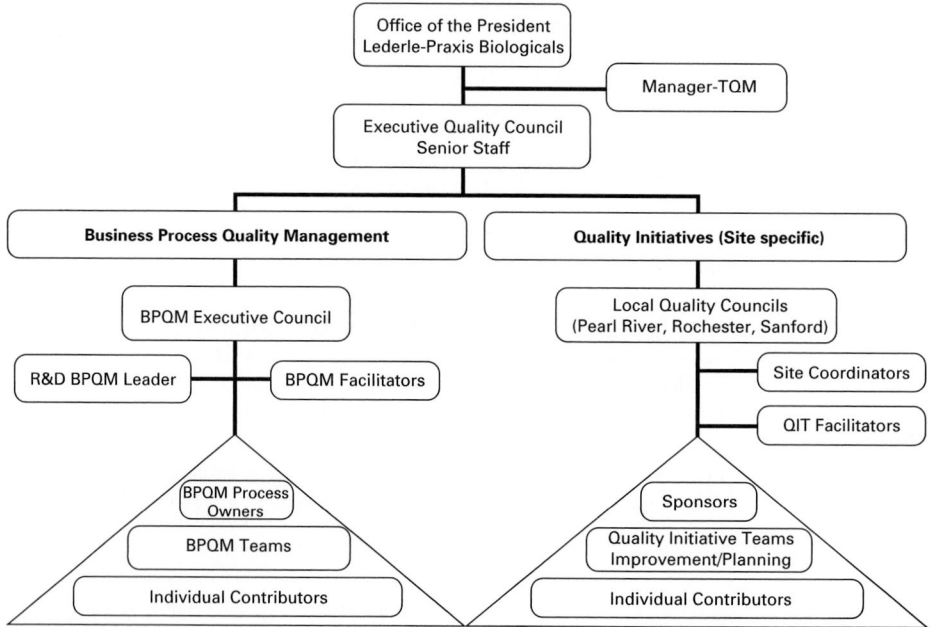

FIGURE 19.5 BPQM and site quality councils. (*Hildreth 1993, p. 2A-14.*)

based upon real problems that "Bell Labs engineers have had...." Training designers in modern technology can yield significant paybacks. At Perkin-Elmer, DeFeo (1987) reported that training design engineers in Boothroyd and Dewhurst's (1987, 1994) design for assembly (DFA) methodology resulted in "weighted average" decreases of 48 percent in assembly times and 103 percent increases in assembly efficiencies.

Yoest (1991), reporting on a study conducted by Sverdrup Technologies at Arnold Engineering Development Center, Arnold Air Force Base, concluded that teams whose facilitators and team leaders are specifically trained for their roles are more likely to successfully achieve their missions than teams whose leaders and facilitators did not receive training. Konosz and Ice (1991) at Alcoa's Technical Center have similarly stated that "The successful implementation of problem-solving teams and quality improvement processes requires three critical components: (1) management leadership and involvement, (2) team training and (3) process facilitation." They provide additional detail on the selection and training of team facilitators within an R&D environment.

Determining R&D Quality Status.
It has been said that in order to plan and improve, you must be able to control, and in order to control, you must be able to measure. Developing good measures for R&D quality *has* proven to be a key ingredient for improving the performance of research functions and development processes. To help distinguish among various types of measurements and measurement processes, it is useful to distinguish between measures used to *manage the quality of specific R&D processes and products,* and measures used to *assess overall R&D quality status.*

Measuring Quality in R&D Processes and Products. The utility and types of measures for R&D process and product quality can be viewed from several perspectives. Gendason and Brown (1993) have stated that for any metric to be "useful as a management tool, it must have three characteristics: it must be something that is countable; it must vary within a time frame that makes reaction to a `down trend' meaningful; and one must be able to define a goal value for the metric." Endres (1997) has classified measures with respect to *timeliness, application,* and *completeness.*

Measures: Timeliness. Traditional measures for research quality have been lagging indicators, in that they report on what the research organization has already accomplished. Mayo (1994) discusses Bell Labs' use of measures of new product revenues in a given year divided by total R&D costs in that year. Garfinkel (1990) at GE's Corporate R&D center has discussed GE's use of patents granted per million dollars invested in research as a benchmarking performance measurement.

Sekine and Arai (1994) provide tables of possible *design process* deficiency measures associated with management, lead times, costs, and quality. For example, a suggested measure for design quality is the ratio of the total costs of poor quality attributable to design problems to the total cost of poor quality caused by design, manufacture, or others. The authors state that, on the average, 60 percent of losses are attributable to design problems, 30 percent are attributable to manufacturing problems, and 10 percent to other areas, e.g., installation. Goldstein (1990) has suggested similar measures for design quality e.g., tracking the ratio of design corrective changes to the total number of drawings released for each new product.

Examples of *concurrent* indicators are the results of *peer reviews* and *design reviews.* Roberts (1990) discusses peer reviews used to verify progress by checking calculations, test data reduction, and research reports. Bodnarczuk (1991) provides insights into the nature of peer reviews in basic research at Fermi National Accelerator Laboratory.

Hiller (1986) at Electrolux AB in Stockholm, Sweden defines design review as "a documented review of a production project which is carried out at predetermined times and with participants who have backgrounds and experience different from those which the originator of the design could be expected to have." Hiller identifies, in the context of a phase-gated development process, four types of design reviews:

1. Preliminary (for specifications, drawings, early model)
2. Intermediate (for prototype test results)

3. Final (for prepilot lots and beta test results)

4. Production (for pilot lot products from production tools)

Gryna (1988) provides guidelines for structuring design reviews. Gryna provided Table 19.3 (adapted from Jacobs 1967) which summarizes design review team membership and responsibilities.

Kapur (1996) provides a similar design review responsibility matrix for a six-phase product design cycle. Concurrent indicators can also be used to help develop leading indicators for predicting, and in some cases, controlling R&D performance. The basic requirement is to identify coincident R&D process indicators that are demonstrably correlated, if not causative, with outcomes of research and development processes. For example, Cole (1990) of Kodak presented Figure 19.6, which demonstrates the relationship between compliance scores during the product development projects and the length of the development cycle. There is an obvious correlation, which may be useful in identifying the major contributing factors (within the scoring system) to protracted development cycles.

A similar approach has been discussed by Rajasekera (1990) at Bell Laboratories. Rajasekera has provided a list of what he has identified as key quality driver issues in an industrial research laboratory:

1. Project mission

2. Top management support

3. Project schedule/plan

4. Client consultation

5. Personnel involved

6. Technical tasks

7. Client acceptance

8. Monitoring and feedback

9. Communication

10. Troubleshooting

and also provides an associated scoring mechanism to monitor project quality during each stage of the project.

An additional leading indicator for research effectiveness used by Holmes and McClaskey (1994) at Eastman Chemical is the estimated net present value of new/improved concepts accepted (by business units for products, and manufacturing departments for processes) for commercialization. Figure 19.7 from Endres (1997) demonstrates that the effect of implementing TQM in Eastman Chemical Research virtually doubled research's productivity.

Measures: Applications.   In addition to viewing each R&D measure (or measurement process, e.g., peer review) with respect to timeliness, it is also helpful to examine each with respect to its intended application. That is, is the measure intended to address customer satisfaction levels (in which case it will relate to the key features of the goods and services provided by R&D), or is the measure intended to address customer dissatisfaction and organizational inefficiency (in which case it will relate to identification and quantification of key deficiencies of goods and services or of their R&D processes)? Juran (Section 2, How to Think about Quality) discusses the relative effects of features and deficiencies on customer satisfaction and organization performance.

PROCESS AND PRODUCT FEATURES.   Benchmarking the best practices of other R&D organizations is an important driver for measuring R&D quality. Lander et al. (1994) discuss the results of an industrial research organization benchmarking study of the best features of practices in R&D portfolio planning, development, and review. The study, by the Strategic Decisions Group, found that "best practice" companies exhibit common features, they:

1. Measure R&D's contribution to strategic objectives

2. Use decision-quality tools and techniques to evaluate proposed (and current) R&D portfolios

3. Coordinate long-range business and R&D plans

4. Agree on clear measurable goals for the projects

TABLE 19.3   Design Review Team Membership and Responsibility

Group member	Responsibilities	PDR	IDR	FDR
		Type of design review*		
Chairperson	Calls, conducts meetings of Group, and issues interim and final reports	X	X	X
Design Engineer(s) (of product)	Prepares and presents design and substantiates decisions with data from tests or calculations	X	X	X
Reliability Manager or Engineer	Evaluates design for optimum reliability consistent with goals	X	X	X
Quality Manager or Engineer	Ensures that the functions of inspection, control, and test can be efficiently carried out		X	X
Manufacturing Engineer	Ensures that the design is producible at minimum cost and schedule		X	X
Field Engineer	Ensures that installation, maintenance, and user considerations were included in the design		X	X
Procurement Representative	Assures that acceptable parts and materials are available to meet cost and delivery schedules		X	
Materials Engineer	Ensures that materials selected will perform as required		X	
Tooling Engineer	Evaluates design in terms of the tooling costs required to satisfy tolerance and functional requirements		X	
Packaging and Shipping Engineer	Assures that the product is capable of being handled without damage, etc.		X	X
Marketing Representative	Assures that requirements of customers are realistic and fully understood by all parties	X		
Design Engineers (not associated with unit under review)	Constructively reviews adequacy of design to meet all requirements of customer	X	X	X
Consultants, Specialists on components, value, human factors, etc. (as required)	Evaluates design for compliance with goals of performance, cost, and schedule	X	X	X
Customer Representative (optional)	Generally voices opinion as to acceptability of design and may request further investigation on specific items			X

* P = Preliminary; I = Intermediate; F = Final.
Source: Gryna (1988), adapted from Jacobs (1967).

The study also revealed that "companies which are excellent at the four best practices:

1. Have established an explicit decision process that focuses on aligning R&D with corporate strategy and creating economic value
2. Use metrics that measure this alignment and the creation of value
3. Maintain a fertile organizational setting that supports decision quality and the implementation of change efforts."

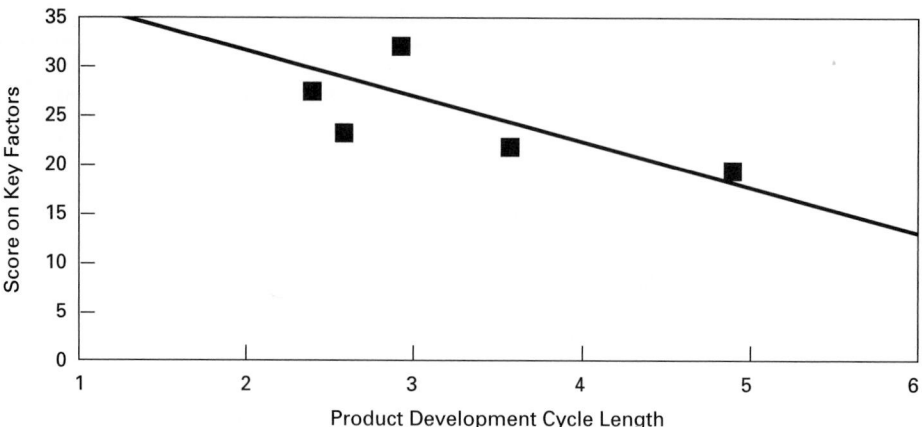

FIGURE 19.6   Correlation between development process compliance scores and cycle times at Kodak. (*Cole 1990.*)

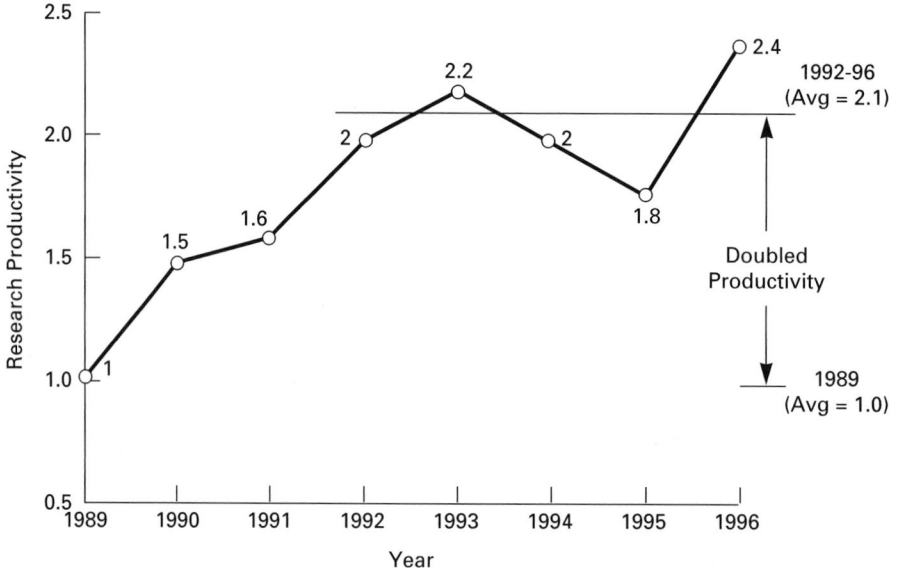

FIGURE 19.7   Eastman Chemical Research productivity as a ratio of 1989 NPV of improved concepts accepted and commercialized with major research input divided by total research expenditures. (*Endres 1997.*)

Figure 19.8 represents, at a macro level, the features of the process commonly used by the best-practice companies for R&D portfolio planning and review.

Among the organizations identified "for their exemplary R&D decision quality" practices were 3M, Merck, Hewlett-Packard, General Electric, Procter & Gamble, Microsoft, and Intel. Matheson et al. (1994) also provide examples of tools which organizations can use to identify their greatest opportunities for implementing and improving best practices in R&D planning and implementation. Hersh et al. (1993) discuss the use, in addition to the benchmarking for best practices, of internal customer surveys at ALCOA to identify and prioritize key R&D performance features at Alcoa's Technical Center. They used the survey results to establish four major categories of their customers' requirements:

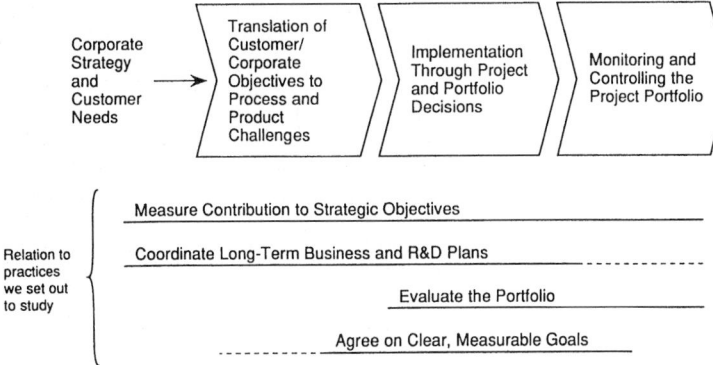

FIGURE 19.8 Common process for implementing best practices for R&D planning, implementation, and review. (*Lander et al. 1994, p. 3-14.*)

1. Manage technology effectively
2. Link technology and business strategies
3. Build strong customer relationships
4. Provide socially and legally acceptable solutions

Each of these feature categories contained activities, whose relative customer priority was also determined. For example, the first category, *manage technology effectively,* contained the highest-priority requirement to "Assume accountability for attaining mutually determined project objectives," and the second-highest-priority requirement to "Meet customer cost and performance expectations." Wasson (1995) also discusses the use of the survey data in developing customer-focused vision and mission statements for the Alcoa Technical Center. Endres (1997) provides additional details on the survey and its results.

PROCESS AND PRODUCT DEFICIENCIES. Identifying customers' requirements is necessary but not sufficient. R&D organizations must also define and implement methods for improving their customers' satisfaction levels and their process's efficiencies. Ferm et al. (1993) also discuss the use of business unit surveys at AlliedSignal's Corporate Research & Technology Laboratory to "create a broad, generic measure of customer satisfaction…and then use the feedback to identify improvement opportunities, to assess internal perceptions of quality, and to set a baseline for the level of…research conformance to customer requirements." (In addition to surveying its business-unit customers, the Laboratory management gave the same survey to Laboratory employees. The resulting data enabled comparison of employee perceptions of Laboratory performance to the perceptions of external customers.) One of the vital few needs identified for action was the need to convince the business units that the Laboratory was providing good value for project funding. Further analysis of the business units' responses revealed that the business units believed Laboratory results were not being commercialized rapidly enough. However, the Laboratory believed that the business units had accepted responsibility for the commercialization process. In response to this observation, a joint Laboratory and (one) business-unit team was formed to clearly define and communicate responsibilities *throughout* the research project and subsequent commercialization and development processes.

Wasson (1995) at Alcoa's Technical Center has also provided several explicit measures used to determine customer satisfaction:

1. Percentage of agreed-upon deliverables delivered
2. Percentage of technical results achieved
3. Results of customer satisfaction survey

Measures: Completeness.   Endres (1997) uses the word "completeness" to indicate the degree to which measures are simultaneously comprehensive (i.e., taken together, they provide answers to the question: "Is the R&D organization meeting its performance objectives?") and aligned (i.e., there is a direct linkage between each variable measured and one or more of those objectives). Juran (1964) and Boath (1992) have identified the need for a comprehensive hierarchy of measures. Figure 19.9, from Boath (1992), is an R&D performance measurement pyramid.

Although the concept of multiple levels of measures is useful, it is incomplete. To be complete, performance measures for research organizations and development processes must also be aligned. Menger (1993) discussed the development and use of key result indicators (KRIs) to drive progress in Corning's Technology Group, which contained research, development, and engineering. Corning's World-Class Quality Committee (WQC) defines the KRIs for the Technology Group. General areas for improvement and measurement used are

1. Cycle time

2. Productivity

3. Customer satisfaction

4. Employee satisfaction

The WQC then requires each of the 15 major units in the Technology Group to define explicit performance measures for each of the previous general areas for improvement. "Twice a year the committee spends the better part of two days visiting each of the 15 units…(to) review the quality of their KRIs, consistency of unit KRIs with those of the technology group, progress made on the KRIs, and plans for improvement.…"

Additional examples of linking R&D performance measures are provided by Rummler and Brache (1995) who provide a comprehensive example of linking organizational-, process-, and job/performer-level measures for a product development process.

Assessing Overall R&D Quality Status.   The previous discussions on measurement have focused on classifying and developing measures for Research organizations and Development processes. Juran

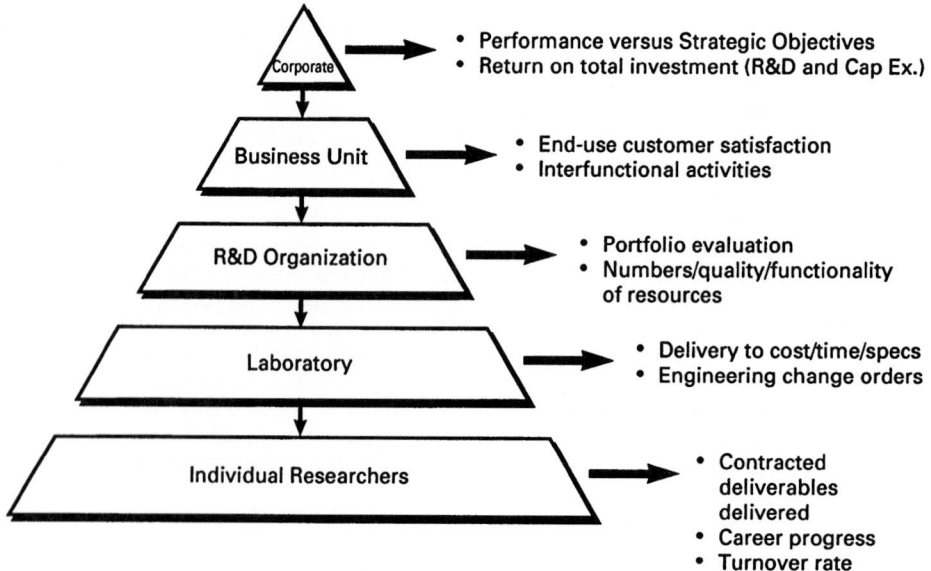

FIGURE 19.9   Boath's pyramid of R&D measures. (*Boath 1992, p. 2-4-4.*)

and Gryna (1993) have defined the benefits of determining the broad overall status of quality in organizations. This process has been defined as quality assessment. Quality assessment comprises:

1. Customer-based quality measurement
2. Quality culture review
3. Cost of poor quality determination
4. Quality system review

Examples of determining R&D customers' priorities and perspectives of performance have been discussed earlier. The assessment of some elements of quality culture in research has been discussed in an example presented by Holmes and McClaskey (1994). In 1989 Eastman Research had determined that though many elements of TQM had been installed (e.g., "Many processes had been studied and flow charted; some processes were being routinely measured and reviewed"), research output, as measured by the NPV of new/improved concepts accepted, had not improved. The authors conducted interviews with Research personnel that determined that although communications had improved:

1. Few process improvements had been implemented.
2. Most first-level managers and individual researchers saw nothing beneficial from the quality initiative.
3. Employees were confused as to what Research management wanted them to deliver ("What is Research's main output?").

As a result of the interviews, Eastman Chemical refocused its effort on improving the key processes that directly affected its primary deliverable category: new/improved concepts accepted for commercialization. The ultimate effect of shifting initiative focus from team activities and tools to mission and output is reflected in Figure 19.7.

Cost of poor quality has been discussed generally by Gryna (see Section 8, Quality and Costs). At Corning, Kozlowski (1993) discusses using quality cost data to identify high cost-of-poor-quality areas. For example, one primary contributor to internal failure costs was the "rework" associated with having to redo experiments. An improvement team assigned to reduce associated costs determined that an internal training program on experimental design was necessary to improve efficiency, and that it was necessary to improve communications with support groups through formally defining and sharing experimental objectives.

Quality System Assessments for R&D.   Quality Systems assessments may be conducted using the Baldrige criteria or the ISO 9000 standards. In Section 14, Total Quality Management, Godfrey provides insight into the use and benefits of the Baldrige National Quality Award. In Section 11, The ISO 9000 Family of International Standards, Marquardt provides similar perspectives of the use of ISO 9000 family of international standards for reviewing quality systems.

BALDRIGE ASSESSMENTS FOR R&D ORGANIZATIONS.   Within research organizations, Kozlowski (1993) has discussed using the Baldrige criteria to provide "outside focus to the quality process....This outside focus, specifically the emphasis on the customer, is the single biggest difference between where we started in 1985, and where we are today." Van der Hoeven (1993) has discussed the process used at IBM's Thomas J. Watson Research Center to organize a Baldrige assessment, and the importance of translating the Baldrige criteria into relevant interpretations for a research organization. Each Baldrige category was allocated to a senior research executive. For example, strategic planning and data collection and analysis were assigned to the VP of technical plans and controls; the director of quality coordinated work on training and writing the category assessments. Van der Hoeven reported that "it required a significant effort to interpret and formulate appropriate responses....this careful tailoring of responses to the Baldrige questions, in terms of existing division processes and management systems…is unique. And the assessment raises gaps in processes and practices to the surface." For example, the assessment revealed the need to improve processes for strategic planning, customer satisfaction, and capturing quality data in the divisionwide database.

McClaskey (1992) discusses how the Baldrige criteria were used at Eastman Chemical Company to accelerate the rate of performance in research, and provides guidelines effectively translating the criteria into action. One example is: "Give awards for both improvement as well as level [of quality]." Endres (1997) provides additional material from McClaskey's paper.

In Section 14, Total Quality Management, Godfrey provides additional insights into the way organizations use the Baldrige criteria.

ISO 9000 ASSESSMENTS FOR R&D ORGANIZATIONS.   Although the Baldrige criteria provide organizations with a comprehensive review mechanism for improving quality systems, some organizations perceive the criteria as being too complex for beginning their quality journey. The pervasive preference for the ISO 9000 quality system standards over the Baldrige criteria can be attributed to the fact that their scope is more limited, being focused on quality control and corrective action systems. Also, the ISO standards are frequently required by suppliers' customers. These drivers for the use of standards has led to the need to tailor and implement ISO standards for research and design organizations.

Fried (1993) discusses the process AT&T's Transmission Systems Business Unit (TSBU) used to pursue ISO 9001 registration. One consequence was the need for each of the TSBU design sites to support the decision by attaining ISO 9001 registration. Each TSBU design laboratory appointed an ISO coordinator; ISO managers were appointed in each of their two major geographical locations. A key initial decision was to review ISO 9001 and to identify those sections which were applicable to the design organizations. Each of the elements that were judged applicable were further categorized as "global" (where compliance could be most effectively addressed by a solution common to multiple organizations) or "local" (where compliance would require a site-by-site approach). Table 19.4 summarizes the results of the review process.

After holding ISO 9001 overview meetings with the design managers and engineers, the site coordinators and area managers coordinated self-assessments and subsequent improvement action planning. Communicating the needed changes to design procedures, coordinating planning with the manufacturing organizations, and coaching on audit participation were identified as being crucial activities in TSBU's successful registration process.

TABLE 19.4   ISO 9001 Elements for AT&T's TSBU R&D Units

ISO 9001 Element	Applicable?	Global/local
Management responsibility	Yes	Both
Quality system	Yes	Both
Contract review	No	
Design control	Yes	Local
Document control	Yes	Local
Purchasing	Yes	Local
Purchaser supplied product	No	
Product indentification and traceability	No	
Process control	No	
Inspection and testing	No	
Inspection, measuring, and test equipment	Yes	Global
Inspection and test status	No	
Control of nonconforming product	No	
Corrective action	Yes	Local
Handling, storage, packaging, and delivery	Yes	Local
Quality records	Yes	Local
Internal quality audits	Yes	Global
Training	Yes	Local
Servicing	No	
Statistical techniques	No	

Source: Fried (1993), p. 2B-25.

Endres (1997) includes materials from a presentation by Gibbard and Davis (1993) on pursuit of ISO 9001 registration by Duracell's Worldwide Technology Center (DWTC). An initial barrier identified was the belief of the technical managers and staff that formal procedures were unnecessary and would "stifle creativity." The authors suggest that the way to address this resistance is for upper management to drive registration via a "top-down effort," including required periodic progress reviews in which upper management participates. DWTC reported that two primary benefits of ISO registration were that it "forced us to identify precisely who our customers were for all projects carried out in our center…" and that ISO established "the foundation of a quality management system on which a program for quality improvement could be built."

OPERATIONAL QUALITY PLANNING FOR RESEARCH AND DEVELOPMENT

Quality Planning: Concepts and Tools for Design and Development.   The focus of the following materials is to provide examples of methodology and tools which support the implementation of Juran's operational quality planning process within the design and development process.

Operational Quality Planning Tools.   As discussed in Section 3, Juran's quality planning process is used to identify customers and their needs, develop product design features responding to those needs and process design features required to yield the product design features, and develop process control required to ensure that the processes repeatedly and economically yield the desired product features. Quality Function Deployment (QFD) is a tool for collecting and organizing the required information needed to complete the operational quality planning process. Zeidler (1993) provides examples of using customer focus groups, surveys, and QFD at Florida Power and Light to identify customers' needs and to determine design features for a new voice response unit. Zeidler concluded: QFD not only ensures customer satisfaction with a quality product or service, but reduces development time, start-up costs, and expensive after-the-fact design changes. It's also a useful political tool, since it guarantees that all affected parts of the organization are members of the QFD team.

Designing for Human Factors: Ergonomics and Errorproofing.   As a design feature, the design's ability to be built/delivered, and used by customers, must be considered from two perspectives: that of operations (manufacturing and service) and that of the customer. From the perspective of manufacturing or service operations, designers must consider the limitations of operators and delivery personnel. They must also consider the possible types of errors that may be committed during operations and use. Ergonomics or "human engineering" is used to address the needs and limitations of operators, service providers, and the customers. Thaler (1996) presents the results of an ergonomics improvement project for facilitating the assembly of aircraft doors. Originally operators "had to hold the doors in place with one hand while trimming or drilling with the other and carrying them for several feet." This job design resulted in a high incidence of worker back injuries. The job redesign included designing a universal clamp to hold the aircraft doors in any position and providing the operators with adjustable work chairs and transportation carts. These and other improvements resulted in a 75 percent reduction in OSHA lost workday incidents and dramatically decreased workers' compensation costs. Gross (1997) provides additional insights and guidance for improving manufacturability and customer usability by integrating ergonomics with the design process.

In contrast with planning for ease of assembly, installation, and use, poka-yoke (pronounced POH-kah YOH-kay) is a methodology for preventing, or correcting errors as soon as possible. The term's English translation is "prevent inadvertent mistake." Poka-yoke was developed by Shigeo Shingo, a Japanese manufacturing engineer. The "MfgNet" Internet newsletter (the WEB site address is http://www.mfgnet.com/poka-yoke.html) provides an example from Varian Associates, a semiconductor equipment manufacturer. Varian had previously placed blame for machine assembly and field installation and service problems on its assemblers and service personnel respectively. Using poka-yoke concepts, designs for new high-current ion-implanter equipment have been targeted so that they can be assembled, installed, and serviced in "only one way—the right way...." For example, in production, poka-yoke was used to ensure that correct alignment of a key assembly, "called a manipulator, which focuses an ion beam," is assured with the "use of holes tapped in the aluminum and graphite assembly." Similarly, "the design of an implanter's front door prevents it from being assembled in any way but the correct one." The *Mistake-Proofing Workshop Participant's Manual* (1995) provides a list of seven steps for developing poka-yoke devices. J. Grout provides multiple cases, examples (with illustrative photographs), and references for poka-yoke concepts. One example provided by Grout is the design of the 3.5-in computer "floppy disk." The disk's beveled corner design permits it to be inserted into a computer only by correctly orienting it. (The Web site address for accessing Grout's poka-yoke information is http://www.Cox.smu.edu/jgrout/pokayoke.html#read.) Kohoutek (1996b) also discusses "human-centered" design and presents approaches and references for predicting human error rates for given activities.

Designing for Reliability, Maintainability, and Availability

Designing for Reliability. A product feature that customers require for products is reliability. Juran and Gryna (1993) have defined reliability as the "chance that a product will work for the required time." Introducing the concept of operating environment, Ireson (1996) states that reliability is the "the ability or capability of the product to perform the specified function in the designated environment for a minimum length of time or minimum number of cycles or events," which also references specific operating conditions/environments. It is important to note that a precise, and agreed upon, definition of a "failure" is needed by customers, designers, and reliability engineers. MIL-STD-721C (1981), Notice 1, *Definition of Terms for Reliability and Maintainability,* and MIL-STD-2074 (1978), *Failure Classification for Reliability Testing,* provide additional definitions and classification information. An excellent source for many terms used in quality management is ISO 8402 (1994), *Quality Management and Quality Assurance—Vocabulary.* International sources for obtaining ISO Standards are listed on the International Organization for Standardization's Web site: http://www.hike.te.chiba-u.ac.jp/Acadia/ISO/home.html. Rees (1992) also discusses the importance of identifying and defining the intended purpose of the application and test procedure *prior to* defining failures.

The following materials will describe approaches and tools for "designing in" reliability. (Section 48 provides information on reliability concepts and the use of statistical tools for *analyzing* reliability data emanating from design tests and field failure data.) MIL-STD-785B, Notice 2, *Reliability Program for Systems and Equipment,* provides both general reliability program requirements and required specific tasks. Major program elements discussed by Juran and Gryna (1993) are

1. Setting reliability goals
2. Reliability modeling
3. Apportioning the reliability goals
4. Stress analysis
5. Reliability prediction
6. Failure mode and effects analysis
7. Identification of critical parts
8. Design review
9. Supplier selection
10. Control of reliability in manufacturing
11. Reliability testing
12. Failure reporting and corrective action

Section 18 discusses market research for identifying quality and reliability goals. Standinger (1990) discusses using competitive benchmarking and Weibull distributions for establishing reliability goals for products during the infant mortality, random failure, and wear-out phases for a new product's life cycle. Table 19.5, from Juran and Gryna (1993), provides typical indicators for reliability performance for which specific numerical goals may be established.

As seen earlier, design reviews can be used as concurrent indicators for a design's reliability. Therefore, one of the key requirements for design review meetings is to ensure that reliability goals have been established, and that intrinsic and actual reliability are being measured and improved during the design's evolution, manufacture, and use. Reliability of procured materials must be considered during supplier selection and control. Section 21 discusses the management of supplier performance. The effect of manufacturing processes on reliability must be addressed during process design selection and implementation. Section 22, Operations, provides guidance for controlling quality and reliability during manufacturing.

Juran and Gryna (1993) divide the process of reliability quantification into the three phases: apportionment, prediction, and analysis. Reliability apportionment is the process used to divide and allocate the design's overall reliability goal among its major subsystems and then to their components. Reliability prediction is the process of using reliability modeling and actual past performance data to predict reliability for expected operating conditions and duty cycles. Reliability analysis utilizes the results of reliability predictions to identify opportunities for improving either predicted or actual reliability performance.

Reliability Apportionment.   The top two sections in Table 19.6, from Juran and Gryna (1993), provide an example of reliability apportionment. A missile system's reliability goal of 95 percent for 1.45 hours must be apportioned among its subsystems and their components. The top section of the table demonstrates the first level apportionment of the 95 percent goal to the missile's six subsystems. The middle section of the table exemplifies the apportionment of the goal of one of those subsystems; the reliability goal of 0.995 for the missile's explosive subsystem is apportioned to its three components. The allocation for the fusing circuitry is 0.998 or, in terms of mean time between failures, 725 hours.

TABLE 19.5   Typical Reliability Indicators

Figure of merit	Meaning
Mean time between failures (MTBF)	Mean time between successive failures of a repairable product
Failure rate	Number of failures per unit time
Mean time to failure (MTTF)	Mean time to failure of a nonrepairable product or mean time to first failure of a repairable product
Mean life	Mean value of life ("life" may be related to major overhaul, wear-out time; etc.)
Mean time to first failure (MTFF)	Mean time to first failure of a repairable product
Mean time between maintenance (MTBM)	Mean time between a specified type of maintenance action
Longevity	Wear-out time for a product
Availability	Operating time expressed as a percentage of operating and repair time
System effectiveness	Extent to which a product achieves the requirements of the user
Probability of success	Same as reliability (but often used for "one-shot" or non-time-oriented products)
b_{10} life	Life during which 10% of the population would have failed
b_{50} life	Median life, or life during which 50% of the population would have failed
Repairs/100	Number of repairs per 100 operating hours

Source: Juran and Gryna (1993), p. 262.

TABLE 19.6 An Example of Reliability Apportionment And Prediction

System breakdown					
Subsystem	Type of operation	Reliability	Unreliability per hour	Failure rate objective*	Reliability
Air frame	Continuous	0.997	0.003	0.0021	483
Rocket motor	One-shot	0.995	0.005		1/200 operations
Transmitter	Continuous	0.982	0.018	0.0126	80.5 h
Receiver	Continuous	0.988	0.012	0.0084	121 h
Control system	Continuous	0.993	0.007	0.0049	207 h
Explosive system	One-shot	0.995	0.005		1/200 operations
System		0.95	0.05		

Explosive subsystem breakdown				
Unit	Operating mode	Reliability	Unreliability	Reliability objective
Fusing circuitry	Continuous	0.998	0.002	725 h
Safety and arming mechanism	One-shot	0.999	0.001	1/1000 operations
Warhead	One-shot	0.998	0.022	2/1000
Explosive subsystem		0.995	0.005	

Unit breakdown			
Fusing circuitry component part classification	Number used, n	Failure rate per part, λ, %/1000 h	Total part failure rate, $n\lambda$, %/1000 h
Transistors	93	0.30	27.90
Diodes	87	0.15	13.05
Film resistors	112	0.04	4.48
Wirewound resistors	29	0.20	5.80
Paper capacitors	63	0.04	2.52
Tantalum capacitors	17	0.50	8.50
Transformers	13	0.20	2.60
Inductors	11	0.14	1.54
Solder joints and wires	512	0.01	5.12
			71.51

$$\text{MTBF} = \frac{1}{\text{failure rate}} = \frac{1}{\Sigma n\lambda} = \frac{1}{0.0007151} = 1398 \text{ h}$$

*For a mission time of 1.45 h.

Source: Juran and Gryna (1993), adapted from G. N. Beaton (1959). "Putting the R&D Reliability Dollar to Work," *Proceedings of the Fifth National Symposium on Reliability and Quality Control,* IEEE, New York, p. 65.

Kohoutek (1996a) suggests that, in order to allow for design margins, only 90 percent of the system failure rate be apportioned to its subsystems and their components. He discusses five other methods for reliability apportionment. Kapur (1996) provides several examples of using alternative apportionment methods. Kohoutek also discusses the use of reliability policies to support goal setting and improvement for both individual products and product families.

Reliability Modeling, Prediction, Analysis, and Improvement. In general, before a prediction of reliability can be made, a model of the system must be constructed, stress levels for the model's components be determined, and, on the basis of the estimated stress levels, failure rates for the components be obtained and used to estimate the reliability of subsystems and systems. Turmel and Gartz (1997) provide a layout for an "item quality plan" which includes the part's critical characteristics and specification limits. It also includes the manufacturing process to be used and test and inspection procedures, with requirements for process stability and capability measures for these processes and procedures.

Reliability Modeling.   In order to construct a model for reliability prediction, the interrelationships among the system's subsystems and their components must be understood. Gryna (1988) suggests the following steps to developing reliability models and using them for reliability prediction:

1. *Define the product:*   The system, subsystems, and units must be precisely defined in terms of their functional configurations and boundaries. This precise definition is aided by preparation of a functional block diagram (Figure 19.10) which shows the subsystems and lower-level products, their interrelation, and the interfaces with other systems. For large systems it may be necessary to prepare functional block diagrams for several levels of the product hierarchy.

Given a functional block diagram and a well-defined statement of the functional requirements of the product, the conditions which constitute failure or unsatisfactory performance can be defined. The functional block diagram also makes it easier to define the boundaries of each unit and to assure that important items are neither neglected nor considered more than once. For example, a switch used to connect two units must be classified as belonging to one unit or the other (or as a separate unit.)

2. *Develop a reliability block diagram:*   The reliability block diagram (Figure 19.11) is similar to the functional block diagram, but it is modified to emphasize those aspects which influence reliability. The diagram shows, in sequence, those elements which must function for successful operation of each unit. Redundant paths and alternative modes should be clearly shown. Elements which are not essential to successful operation need not be included, e.g., decorative escutcheons. Also, because of the many thousands of individual parts that constitute a complex product, it is necessary to exclude from the calculation those classes of parts that are used in mild applications. The contribution of such parts to product unreliability is relatively small. Examples of items that can generally be disregarded are terminal strips, knobs, chassis, and panels.

3. *List factors relevant to reliability:*   These factors include part function, tolerances, part ratings, internal environments and stresses, and duty (on time) cycles. This detailed information makes

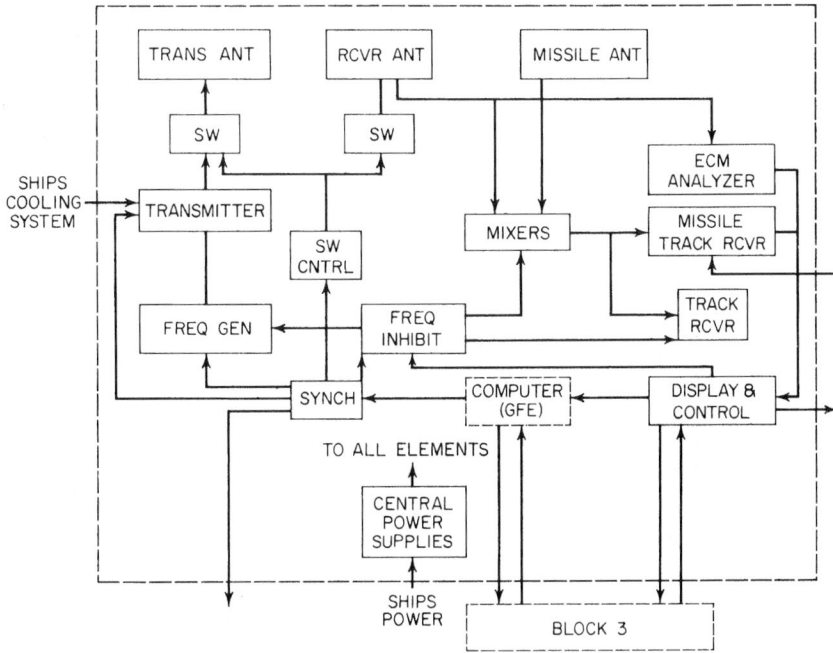

FIGURE 19.10   Functional block diagram. (*From Handbook of Reliability Engineering, NAVAIR 00-65-502, courtesy the Commander, Naval Air Systems Command.*)

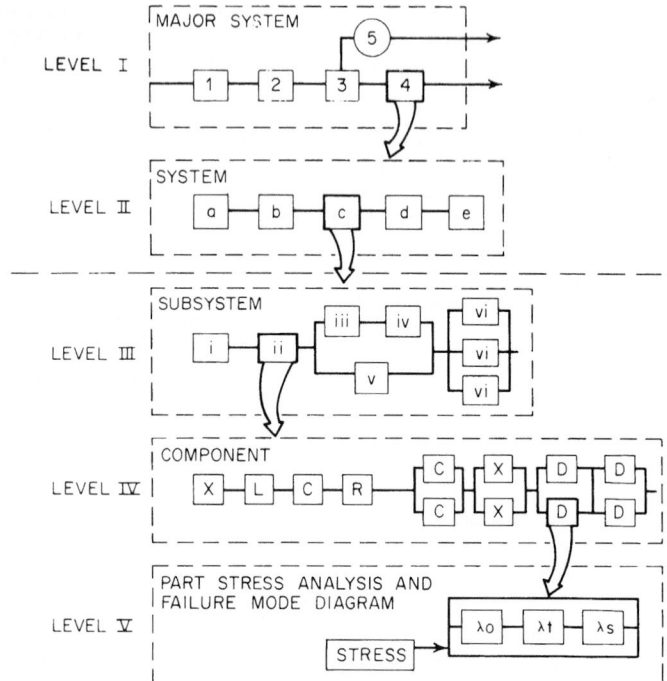

FIGURE 19.11   Reliability block diagram. (*From Handbook of Reliability Engineering, NAVAIR 00-65-502, courtesy the Commander, Naval Air Systems Command.*)

it possible to perform a stress analysis, which will not only provide information on the appropriate adjustments to standard input data but also serve to uncover weak or questionable areas in the design. (A methodology used by designers to improve part and product ability to perform in various environments is called "robust design." Phadke (1989) and Taguchi (1995) provide approaches and examples.) Parts with dependent failure probabilities should be grouped together into modules so that the assumptions upon which the prediction is based are satisfied.

4. *Determine part reliability data:*   The required part data consist of information on catastrophic failures and on tolerance variations with respect to time under known operating and environmental conditions. Acquiring these data is a major problem for the designer, since there is no single reliability data bank comparable to handbooks such as those which are available for *physical* properties of materials. Instead, the designer (or supporting technical staff) must either build up a data bank or use reliability data from a variety of sources:

Field performance studies conducted under controlled conditions

Specification life tests

Data from parts manufacturers or industry associations

Customers' part-qualification and inspection tests

Government agency data banks such as MIL-HDBK-217F, which contains component failure rate data and curves for various components' operating environments and stress levels. The handbook also provides examples of reliability prediction procedures appropriate for various stages of the design's evolution.

5. *Make estimates:* In the absence of basic reliability data, it may be feasible to make reasonably accurate estimates based upon past experience with similar part types. Lacking such experience, it becomes necessary to obtain the data via part evaluation testing.

6. *Determine block and subsystem failure rates:* The failure rate data obtained in step 4 or 5 are used to calculate failure rates for the higher-level systems and the total system. (Pertinent subsystem or assembly correction factors, such as those determined for the effects of preventive maintenance, should also be applied.)

7. *Determine the appropriate reliability unit of measure:* This is the choice of the reliability index or indicators as listed in Table 19.5

8. *Use the reliability model and predictions* to identify the design's "weak points" and the required actions and responsibilities for reliability improvement.

Reliability Prediction. The bottom portion of Table 19.6 provides an example of predicting, for known part counts, the failure rates for each component of the fusing circuitry. The prediction is based upon the assumptions of the statistical independence of the failure times of the components, conformance to an exponential failure distribution, and equal hours of operation. The estimated unit failure rate is of 0.7151 per/1000 hours of operation or 0.0007151 failures per hour. The reciprocal of the latter failure rate yields an estimated mean time between unit failures of 1398 hours, which exceeds the 725 hours requirement for the fusing circuitry. MIL-HDBK-217F (1991), Notice 2, Reliability Prediction of Electronic Equipment, provides formulas for estimating failure rates for classes of electronic parts and microcircuits in various operating environments. Additional sources of reliability data from Air Force databases, Navy databases, and Army databases are available in the Reliability Engineer's Toolkit (1993) available as ADA278215 from the National Technical Information Service in Springfield, VA. (The latter reference also provides sources for reliability prediction software programs.) The Government Industry Data Exchange Program (GIDEP) provides an on-line menu for accessing a data bank for reliability and maintainability data and information, and provides participants with alerts for known part problems. GIDEP may be contacted at: GIDEP Operations Center, Corona, CA 91718-8000 or on the World Wide Web at www.gidep.corona.navy.mil/data_inf/opscntr.htm.

Reliability Analysis

FAILURE MODE, EFFECT, AND CRITICALITY ANALYSIS; FAULT TREE ANALYSIS. In planning for reliability, the engineer's analysis of the expected effects of operating conditions on design reliability and safety are often enhanced by use of failure mode effect and criticality analysis (FMECA) and fault tree analysis (FTA). General introductions to failure mode effects analysis (FMEA), FMECA, and FTA are provided in Section 48. FMEA and FMECA are intended for use by product and process designers in identifying and addressing potential failure modes and their effects. Figure 19.12, from Gryna (1988), is an example of a FMECA for a traveling lawn sprinkler which includes for each part number its failure mode, result of the failure mode, cause of failure mode, estimated probability of failure mode, severity of the failure mode, and alternative countermeasures for preventing the failure. MIL-STD-1629A, (1984) Notice 2, Procedures for Performing a Failure Mode, Effects, and Criticality Analysis, provides, with examples, additional details on developing severity classifications and criticality numbers.

Whereas FMECA examines all possible failure modes from the component level upward, FTA focuses on particular known undesirable effects of a failure, e.g., fire and shock, and proceeds to identify all possible failure paths resulting in the specified undesirable outcome. Figure 19.13, from Hammer (1980), is a fault tree for a safety circuit. The failure outcome of concern is that x-rays will be emitted from a machine whose door has been left open. The spadelike symbol with a straight bottom is an "and gate," meaning the output occurs only if all input events below it happen. The spade symbol with the curved bottom is an "or gate," meaning the output occurs if any one or more of the input events below it happen. The probabilities of specific occurrences can be estimated by providing estimates of the probabilities of occurrence of each event in the fault tree. In Section 48, Reliability Concepts and Data Analysis, Meeker et al. cite Hoyland and Rausand (1994) and Lewis

1 = Very low (<1 in 1000)
2 = Low (3 in 1000)
3 = Medium (5 in 1000)
4 = High (7 in 1000)
5 = Very high (>9 in 1000)

T = Type of failure
P = Probability of occurrence
S = Seriousness of failure to system
H = Hydraulic failure
M = Mechanical failure
W = Wear failure
C = Customer abuse

Product   HRC-1
Date   Jan. 14, 1987
By   S.M.

Component part number	Possible failure	Cause of failure	T	P	S	Effect of failure on product	Alternatives
Worn bearing 4224	Bearing worn	Not aligned with bottom housing	M	1	4	Spray head wobble or slowing down	Improve inspection
Zytel 101		Excessive spray head wobble	M	1	3	DITTO	Improve worm bearing
Bearing stem 4225	Excessive wear	Poor bearing/material combination	M	5	4	Spray head wobbles and loses power	Change stem material
Brass		Dirty water in bearing area	M	5	4	DITTO	Improve worm seal area
		Excessive spray head wobble	M	2	3	DITTO	Improve operating instructions
Thrust washer 4226	Excessive wear	High water pressure	M	2	5	Spray head will stall out	Inform customer in instructions
Fulton 404		Dirty water in washers	M	5	5	DITTO	Improve worm seal design
Worm 4527	Excessive wear in bearing area	Poor bearing/material combination	M	5	4	Spray head wobbles and loses power	Change bearing stem material
Brass		Dirty water in bearing area	M	5	4	DITTO	Improve worm seal design
		Excessive spray head wobble	M	2	3	DITTO	Improve operating instructions

FIGURE 19.12   Failure mode, effect, and criticality analysis. (*Gryna 1988, from Hammer 1980.*)

(1996) as providing examples which include calculations for event probabilities. Lazor (1996) also provides examples and comparisons of FMECA and FTA analyses, with an interesting discussion on the relationship between fault trees and reliability block diagrams. In Section 48 Meeker et al. provide references for computer software for facilitating FMEA/FMECA and FTA analyses.

OTHER FAILURE ANALYSIS PREDICTION TECHNIQUES.   Other analytical techniques have been developed to aid in analyzing possible causes of product failures. The Transactions on Reliability of the Institute of Electrical and Electronics Engineers is a good source of information on such techniques. Worst-case analysis, statistical tolerancing, and sneak-circuit analysis will be highlighted.

"Worst-case" analysis, often facilitated via computer software, is a detailed environmental analysis. The purpose is to identify the conditions under which maximum stresses will be placed on components/circuits, and to verify the ability of the product to meet its goals when subjected to extremes, or highly probable combinations of electrical and physical conditions.

Although worst-case analysis is useful for identifying which combinations of conditions will produce the most severe environments (or interferences for mechanical assemblies), it does not consider the probability that these combinations will actually occur. Under varying sets of assumptions, statistical tolerancing can be used to estimate the actual probabilities of the worst-case conditions. These estimates can then be used by designers to decide on trade-offs of tolerance versus cost. Statistical tolerancing decisions generally result in allowing larger component tolerances. Dudewicz (1988) discusses statistical tolerancing and provides guidelines and examples for comparison with worst-case tolerance analysis. See Section 45 under "Statistical Estimation, Tolerance Intervals." MIL-STD-785B, (1988), Notice 2, provides additional information.

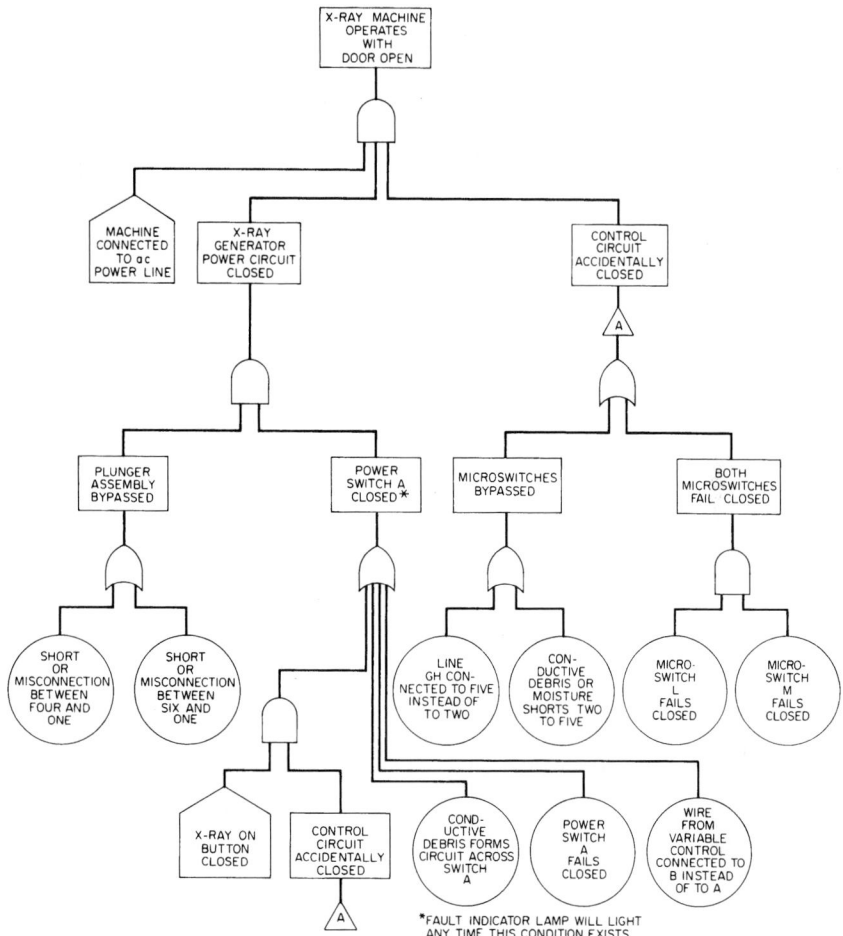

FIGURE 19.13 Fault-tree analysis of an interlock safety circuit. (*Gryna 1988.*)

In the analysis of electrical circuits, sneak-circuit analysis is similar to worst-case analysis and is a valuable supplement to it. Sneak-circuit analysis is usually performed by computer software to identify latent paths in a circuit which could cause the occurrence of unanticipated and unwanted functions which could prevent or degrade desired performance, *even with all components functioning properly*. Rome Laboratory's *Reliability Engineer's Toolkit* (1993) provides an example and identifies some available software.

Reliability Improvement. The general approach to quality improvement (see Section 5, The Quality Improvement Process) is widely applicable to reliability improvement as far as the economic analysis and the managerial tools are concerned. The differences are in the technological tools used for diagnosis and remedy. Projects can be identified through reliability prediction; design review; failure mode; effect, and criticality analysis; and other reliability evaluation techniques.

Action to improve reliability during the design phase is best taken by the designer. The reliability engineer can help by defining areas needing improvement and by assisting in the development of alternatives. The following actions indicate some approaches to improving a design:

1. Review the users' needs to see if the function of the unreliable parts is really necessary to the user. If not, eliminate those parts from the design. Alternatively, look to see if the reliability index (figure of merit) correctly reflects the real needs of the user. For example, availability is sometimes more meaningful than reliability. If so, a good maintenance program might improve availability and hence ease the reliability problem.

2. Consider trade-offs of reliability for other parameters, e.g., functional performance or weight. Here again it may be found that the customer's real needs may be better served by such a trade-off.

3. Use redundancy to provide more than one means for accomplishing a given task in such a way that all the means must fail before the system fails.

There are several types of redundancy, a common form being parallel redundancy. A familiar example is the multiengine aircraft, which is so designed that even if one engine fails, the aircraft will still be able to continue on to a safe landing.

Under conditions of independent failures, the overall reliability for parallel redundancy is expressed by the formula

$$P_s = 1 - (1 - P_i)^n$$

where P_s = reliability of the system
P_i = reliability of the individual elements in the redundancy
n = number of identical redundant elements

Figure 19.14 shows some simple examples of series-parallel and parallel-series redundancies and calculates the system reliability versus that prevailing for the case of no redundancy.

4. Review the selection of any parts that are relatively new and unproven. Use standard parts whose reliability has been proven by actual field use. (However, be sure that the conditions of previous use are applicable to the new product.)

5. Use derating to assure that the stresses applied to the parts are lower than the stresses the parts can normally withstand. Derating is one method that design engineers use to improve component reliability or provide additional reliability margins. Juran and Gryna (1993) define derating as the assignment of a product (component) to operate at stress levels *below* its normal rating, e.g., a capacitor rated at 300 V is used in a 200-V application.

Kohoutek also provides examples of derating graphs, to be used by design engineers for specific types of integrated circuits. Before using the graphs for a specific application, the design engineer first determines the expected operating temperatures, voltages, stresses, etc. of the component under study, then uses the graphs to select the appropriate derating factor.

6. Use "robust" design methods that enable a product to handle unexpected environments.

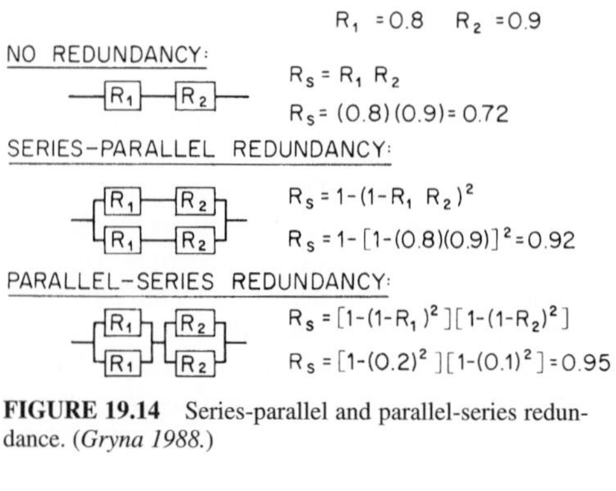

$$R_1 = 0.8 \quad R_2 = 0.9$$

NO REDUNDANCY:

$$R_s = R_1 R_2$$
$$R_s = (0.8)(0.9) = 0.72$$

SERIES–PARALLEL REDUNDANCY:

$$R_s = 1 - (1 - R_1 R_2)^2$$
$$R_s = 1 - [1 - (0.8)(0.9)]^2 = 0.92$$

PARALLEL–SERIES REDUNDANCY:

$$R_s = [1 - (1 - R_1)^2][1 - (1 - R_2)^2]$$
$$R_s = [1 - (0.2)^2][1 - (0.1)^2] = 0.95$$

FIGURE 19.14 Series-parallel and parallel-series redundance. (*Gryna 1988.*)

7. Control the operating environment to provide conditions that yield lower failure rates. Common examples are (*a*) potting electronic components to protect them against climate and shock, and (*b*) use of cooling systems to keep down ambient temperatures.

8. Specify replacement schedules to remove and replace low-reliability parts before they reach the wear-out stage. In many cases the replacement is made but is contingent on the results of checkouts or tests which determine whether degradation has reached a prescribed limit.

9. Prescribe screening tests to detect infant-mortality failures and to eliminate substandard components. The tests take various forms—bench tests, "burn in," accelerated life tests.

Jensen and Petersen (1982) provide a guide to the design of burn-in test procedures. Chien and Kuo (1995) offer further useful insight into maximizing burn-in effectiveness.

10. Conduct research and development to attain an improvement in the basic reliability of those components which contribute most of the unreliability. While such improvements avoid the need for subsequent trade-offs, they may require advancing the state of the art and hence an investment of unpredictable size. Research in failure mechanisms has created a body of knowledge called the "physics of failure" or "reliability physics." *Proceedings of the Annual Meeting on Reliability Physics,* sponsored by the Institute of Electrical and Electronic Engineers, Inc., is an excellent reference.

Although none of the foregoing actions provides a perfect solution, the range of choice is broad. In some instances the designer can arrive at a solution single-handedly. More usually it means collaboration with other company specialists. In still other cases the customer and/or the company management must concur because of the broader considerations involved.

Designing for Maintainability.   Although the design and development process may yield a product that is safe and reliable, it may still be unsatisfactory. Users want products to be available on demand. Designers must therefore also address ease of preventive maintenance and repair. Maintainability is the accepted term used to address and quantify the extent of need for *preventive* maintenance and the ease of repair.

A formal definition of maintainability is provided by MIL-STD-721C (1981):

> The measure of the ability of an item to be retained in or restored to specified condition when maintenance is performed by personnel having specified skill levels, using prescribed procedures and resources, at each prescribed level of maintenance and repair.

The definition emphasizes the distinction between maintainability, a *design* parameter, and maintenance, an *operational* activity.

Mean time to repair (MTTR) is an index used for quantifying maintainability, analogous to the term MTBF used as an index for reliability. Table 19.7, from MIL-STD-721C (1981), summarizes 11 possible indexes for maintainability.

MIL-HDBK-472 (1984), *Maintainability Prediction of Electronic Equipment,* may be used to estimate maintainability for various design alternatives. Kowalski (1996) provides an example of allocating a system's maintainability requirement among its subsystems. The allocation is analogous to the method by which reliability was apportioned (See above under Reliability Apportionment.). Kowalski also discusses the impact of *testability* on the ability to achieve maintainability goals. MIL-STD-2165A (1993), *Testability Program for Systems and Equipments,* defines testability as "a design characteristic which allows the status (operable, inoperable, or degraded) of an item to be determined and the isolation of faults within the item to be performed in a timely manner," and provides guidelines for testability planning and reviews. Turmel and Gartz (1997) of Eastman Kodak provide, for a *specific* test method, a test capability index (TCI) index for measuring the proportion of the specification range taken by the intrinsic variation of a test/measurement method. The reported guideline was to target test variation at less than 25 percent of the total tolerance range.

TABLE 19.7 Maintainability Figures of Merit

Figure of merit	Meaning
Mean time to repair (MTTR)	Mean time to correct a failure
Mean time to service	Mean time to perform an act to keep a product in operating condition
Mean preventive maintenance time	Mean time for scheduled preventive maintenance
Repair hours per 100 operating hours	Number of hours required for repairs per 100 product operating hours
Rate of preventive maintenance actions	Number of preventive maintenance actions required per period of operative or calendar hours
Downtime probability	Probability that a failed product is restored to operative condition in a specified downtime
Maintainability index	Score for a product design based on evaluation of defined maintainability features
Rate of maintenance cost	Cost of preventive and corrective maintenance per unit of operating or calendar time

Source: MIL-STD-721C (1981).

Designing for Availability.   Both design reliability and maintainability affect the probability of a product being available when required for use. Availability is calculated as the ratio of operating time to operating time plus downtime. However, downtime can be viewed in two ways:

1. *Total downtime:*   This includes the active repair time (diagnosis and repair), preventive maintenance time, and logistics time (time spent waiting for personnel, spare parts, etc.). When total downtime is used, the resulting ratio is called operational availability (A_o).

2. *Active repair time:*   When active repair time is used, the resulting ratio is called "intrinsic availability" (A_i).

Under certain conditions, "steady state" availability can be calculated as:

$$A_o = \frac{\text{MTBF}}{\text{MTBF} + \text{MDT}} \quad \text{and} \quad A_i = \frac{\text{MTBF}}{\text{MTBF} + \text{MTTR}}$$

where MTBF = mean time between failures
    MDT = mean total downtime
    MTTR = mean active time to repair

These formulas indicate that a specified product availability may be improved (increased) by increasing product reliability (MTBF), or by decreasing time to diagnose and repair failures (MDT or MTTR). Achieving any combination of these improved results requires an analysis of the trade-offs between the benefits of increasing reliability or maintainability. Gryna (1988) provides some specific trade-off decisions that should be considered by designers for increasing maintainability (decreasing diagnosis and repair times):

Modular versus nonmodular construction: Modular design requires added design effort but reduces the time required for diagnosis and remedy in the field. The fault need only be localized to the module level, after which the defective module is unplugged and replaced. This concept has been used by manufacturers of consumer products such as television sets.

Repair versus throwaway: For some products or modules, the cost of field repair exceeds the cost of making new units in the factory. In such cases, design for throwaway is an economic improvement in maintainability.

Built-in versus external test equipment: Built-in test capability reduces diagnostic time, but usually requires additional cost. However, the additional costs can also reduce overall repair costs by providing users with simple repair instructions for various failure modes diagnosed by the diagnostic equipment or software. For example, office copiers provide messages on where and how to remove paper jams.

Kowalski (1996) provides additional examples of criteria for maintainability design.

Formulas for steady-state availability have the advantage of simplicity. However, they are based upon the following assumptions:

1. The product is operating in the constant-failure-rate portion of its overall life, where time between failures is exponentially distributed.
2. Downtime and repair times are also exponentially distributed.
3. Attempts to locate system failures do not change failure rates.
4. No reliability growth occurs. (Such growth might be due to design improvements or removal of suspect parts.)
5. Preventive maintenance is scheduled outside the time frame included in the availability calculation.

For these conditions, O'Connor (1995) provides formulae and examples for various reliability block diagrams, e.g., series, parallel, and parallel-standby configurations. Malec (1996) provides general formulas and examples for calculating instantaneous availability and *mission interval availability,* the probability that a product will be available throughout the length of its mission.

Identifying and Controlling Critical Components. The design engineer will identify certain components as critically affecting reliability, availability, and maintainability (RAM) or for attaining cost objectives. These critical components are those which emerge from the various applicable analyses: the reliability block diagrams, stress analysis, FMEA/FMECA, FTA, and RAM studies. These components may be deemed critical because of their estimated effects on design RAM and cost, insufficient knowledge of their actual performance, or the uncertainty of their suppliers' performance. One approach to ensuring their performance and resolving uncertainties is to develop and manage a list of critical components. The critical components list (CCL) should be prepared early in the design effort. It is common practice to formalize these lists, showing, for each critical component, the nature of the critical features, and the plan for controlling and improving its performance. The CCL becomes the basic planning document for: (1) test programs to qualify parts; (2) design guidance in application studies and techniques; and (3) design guidance for application of redundant parts, circuits, or subsystems.

Configuration Management. Configuration management is the process used to define, identify, and control the composition and cost of a product. A configuration established at a specific point in time is called a "baseline." Baseline documents include drawings, specifications, test procedures, standards, and inspection or test reports. Configuration management begins during the design of the product, and continues throughout the remainder of the product's commercial life. As applied to the product's design phase, configuration management is analogous, at the level of total product, to the process described in the last paragraph for the identification and control of critical components. Gryna (1988) states that "configuration refers to the physical and functional characteristics of a product, including both hardware and software," and defines three principal activities which comprise configuration management:

Configuration identification: The process of defining and identifying every element of the product.

Configuration control: The process which manages a design change from the time of the original proposal for change through implementation of approved changes.

Configuration accounting: The process of recording the status of proposed changes and the implementation status of approved changes.

Configuration management is needed to help ensure:

1. All participants in the quality spiral know the current status of the product in service and the proposed status of the product in design or design change.
2. Prototypes, operations, and field service inventories reflect design changes
3. Design and product testing are conducted on the latest configurations.

Design Testing.   Once the foregoing tools and analyses of design quality have been invoked, it is necessary to assure that the resulting design can ultimately be manufactured, delivered, installed, and serviced to meet customers' requirements. To assure this, it is imperative to conduct actual tests on prototypes and pilot units prior to approval for full-scale manufacturing. Table 19.8 summarizes the various types and purposes of design evaluation tests.

In Section 48, Meeker et al. discuss the purpose and design of environmental stress tests, accelerated life tests, reliability growth tests, and reliability demonstration testing and analysis of the data from these tests. Graves and Menten (1996) and Schinner (1996) provide similar discussions on designing experiments for reliability measurement and improvement, and accelerated life testing respectively. *The Reliability Engineer's Toolkit* (1993) discusses the selection and use of reliability test plans from MIL-HDBK-781 (1987), *Reliability Test Methods, Plans and Environments for Engineering Development, Qualification and Production.*

Comparing Results of Field Failures with Accelerated Life Tests.   In order to verify design reliability within feasible time frames, it is often necessary to "accelerate" failure modes by use of various environmental stress factors. A key issue to address when introducing stress factors is to ensure that the failure modes that they produce are equal to those observed in actual use. Gryna (1988) provides an example of using plots on probability paper to compare and relate test results to "field" failures. Figure 19.15 contains plots of the estimated cumulative failure percentages versus number of accelerated test days and actual field usage days for two air conditioner models. Since the two lines are essentially parallel, it appears that the basic failure modes produced by the accelerated and field usage environments are equivalent. The test data are plotted in *tens of days*. The 5-year warranty period is represented by a heavy vertical line. Following the vertical line from where it intersects the field data line, and proceeding horizontally to the lines for the accelerated test data, the accelerated test time required to predict the percentage of field failures occurring during the 5-year warranty period is estimated at 135 days for one air conditioner model and 175 days for the other model.

TABLE 19.8   Summary of Tests Used for Design Evaluation

Type of test	Purpose
Performance	Determine ability of product to meet basic performance requirements
Environmental	Evaluate ability of product to withstand defined environmental levels; determine internal environments generated by product operation; verify environmental levels specified
Stress	Determine levels of stress that a product can withstand in order to determine the safety margin inherent in the design; determine modes of failure that are not associated with time
Reliability	Determine product reliability and compare to requirements; monitor for trends
Maintainability	Determine time required to make repairs and compare to requirements
Life	Determine wear-out time for a product, and failure modes associated with time or operating cycles
Pilot run	Determine if fabrication and assembly processes are capable of meeting design requirements; determine if reliability will be degraded.

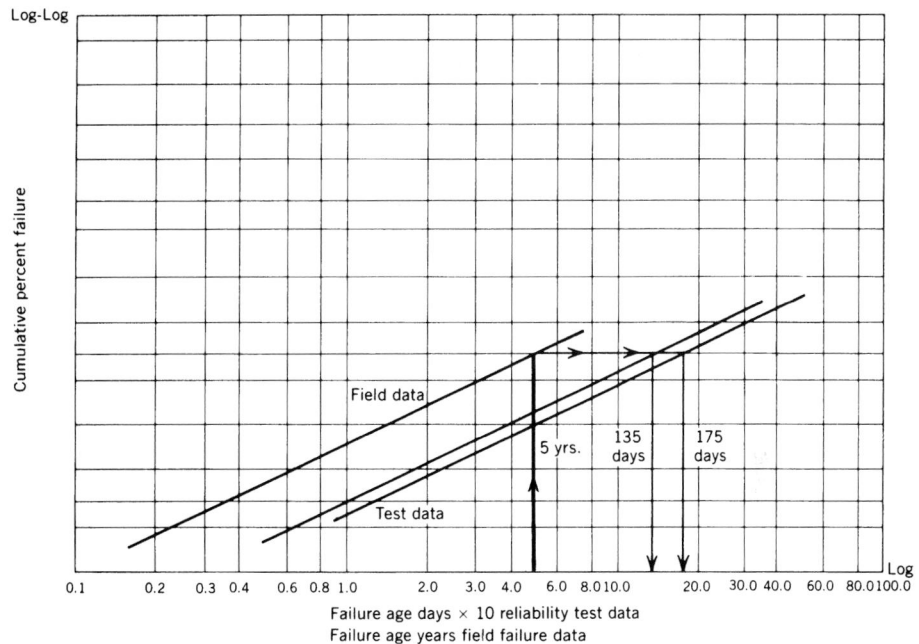

FIGURE 19.15   Weibull plot of accelerated test versus field failure data for two air conditioner models. (*Gryna 1988.*).

Failure Reporting and Corrective Action Systems.   In order to drive improvements in RAM and safety of designs, an organization must define and develop a formal process for reporting, analyzing, and improving these design parameters. Many organizations call this process "failure reporting and corrective action systems" (FRACAS). Figure 19.16, reproduced from the *Reliability Engineer's Toolkit* (1993), is a high-level flow diagram for a generic FRACAS process. In addition to the process steps, process-step responsibilities are identified by function. The same publication also provides a checklist for identifying gaps in existing FRACAS processes. Ireson (1996) provides additional guidance on reliability information collection and analysis, with discussion on data requirements at the various phases of design, development, production, and usage. Adams (1996) focuses on details of identifying the root causes of failures and driving corrective action, with an example of a "business plan" for justifying investment in the equipment and personnel required to support a failure analysis process.

REFERENCES

Adams, J. (1996). "Failure Analysis System—Root Cause and Corrective Action." *Handbook of Reliability Engineering And Management,* 2nd ed. Ireson, W., Coombs, C., and Moss, R., eds. McGraw-Hill, New York, chap. 13.

Altland, Henry (1995). "Robust Technology Development Process for Imaging Materials at Eastman Kodak." *Proceedings of Symposium on Managing for Quality in Research and Development,* Juran Institute, Wilton, CT, p. 2B-15.

Berezowitz, W., and Chang, T. (1997). "Assessing Design Quality During Product Development." *ASQC Quality Congress Proceedings,* Milwaukee, p. 908.

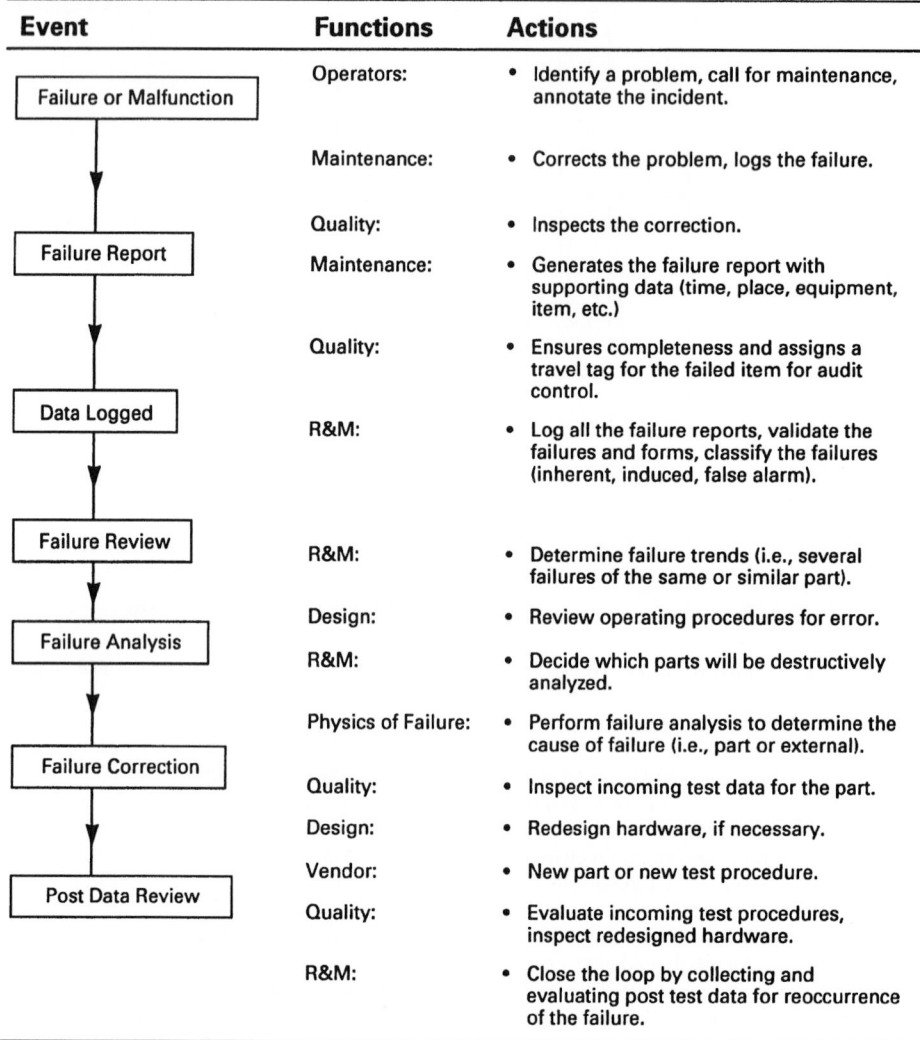

Event	Functions	Actions
Failure or Malfunction	Operators:	• Identify a problem, call for maintenance, annotate the incident.
	Maintenance:	• Corrects the problem, logs the failure.
	Quality:	• Inspects the correction.
Failure Report	Maintenance:	• Generates the failure report with supporting data (time, place, equipment, item, etc.)
	Quality:	• Ensures completeness and assigns a travel tag for the failed item for audit control.
Data Logged	R&M:	• Log all the failure reports, validate the failures and forms, classify the failures (inherent, induced, false alarm).
Failure Review	R&M:	• Determine failure trends (i.e., several failures of the same or similar part).
	Design:	• Review operating procedures for error.
Failure Analysis	R&M:	• Decide which parts will be destructively analyzed.
	Physics of Failure:	• Perform failure analysis to determine the cause of failure (i.e., part or external).
Failure Correction	Quality:	• Inspect incoming test data for the part.
	Design:	• Redesign hardware, if necessary.
	Vendor:	• New part or new test procedure.
Post Data Review	Quality:	• Evaluate incoming test procedures, inspect redesigned hardware.
	R&M:	• Close the loop by collecting and evaluating post test data for reoccurrence of the failure.

FIGURE 19.16   FRACAS flow diagram. (*Reliability Engineer's Toolkit 1993.*)

Boath, D. (1992), "Using Metrics to Guide the TQM Journey in R&D." *Proceedings of the Symposium on Managing for Quality in Research and Development,* Juran Institute, Wilton, CT.

Boath, D. (1993). "Reengineering The Product Development Process." *Proceedings of Symposium on Managing for Quality in Research And Development,* Juran Institute, Wilton, CT, p. 2B-36.

Bodnarczuk, M. (1991). "Peer Review, Basic Research, and Engineering: Defining a Role for QA Professionals in Basic Research Environments." *Proceedings of Symposium on Managing For Quality In Research And Development,* Juran Institute, Wilton, CT.

Boothroyd G., and Dewhurst, P. (1987). *Design for Assembly,* BDI, Wakefield, RI.

Boothroyd G., and Dewhurst, P. (1994). *Design for Manufacture and Assembly,* Marcel Dekker, New York.

Boznak, R., and Decker, A. (1993). *Competitive Product Development.* Copublished by ASQC Quality Press/Business One Irwin, Milwaukee, p. 53.

Chien, W., and Kuo, W. (1995). "Modelling and Maximizing Burn-in Effectiveness." *IEEE Transactions on Reliability,* vol. 44, no. 1, pp. 19–25.

Cole, R. (1990). " Quality in the Management of Research and Development." *Proceedings of Symposium on Managing for Quality in Research and Development,* Juran Institute, Wilton, CT.

Darby, Robert A. (1990). "R&D Quality in a Diversified Company." *Proceedings of Symposium on Managing for Quality in Research and Development,* Juran Institute, Wilton, CT.

DeFeo, J. (1987). "Quality Training: the key to successful Quality Improvement." *Proceedings, IMPRO Conference,* Juran Institute, Wilton, CT, pp. 4A-15 et. seq.

Dudewicz, E. (1988). "Basic Statistical Methods." Section 23, *Juran's Quality Control Handbook,* 4th ed., McGraw-Hill, New York.

Endres A. C. (1992). "Results and Conclusions from Applying TQM to Research." *ASQC Quality Congress Proceedings,* Milwaukee.

Endres, A. C. (1997). *Improving R&D Performance The Juran Way.* John Wiley & Sons, New York.

Ferm, P., Hacker, S., Izod, T., and Smith, G. (1993), "Developing a Customer Orientation in a Corporate Laboratory Environment," *Proceedings of Symposium on Managing for Quality in Research and Development,* Juran Institute, Wilton, CT, pp. A-13–A-22.

Fiero, J., and Birch, W. (1989). "Designing Cost-Effective Products." *ASQC Quality Congress Proceedings,* Milwaukee, pp. 725–730.

Fried, L. K. (1993). "AT&T Transmission Systems ISO 9001 Registration: The R&D Compliance Experience," *Proceedings of Symposium on Managing for Quality in Research and Development,* Juran Institute, Wilton, CT, pp. 2B-21–2B-27.

Garfinkel, M.(1990). "Quality In R&D." *Proceedings of Symposium on Managing for Quality in Research and Development,* Juran Institute, Wilton, CT.

Gendason, P., and Brown, E. (1993). "Measure of R&D Effectiveness: A Performance Evaluation Construct." *Proceedings of Symposium on Managing for Quality in Research and Development,* Juran Institute, Wilton, CT, pp. 2A-17–2A-25.

Gibbard, H. F., and Davis, C. (1993). "Implementation of ISO 9001 in an R&D Organization," *Proceedings IMPRO Conference,* Juran Institute, Wilton, CT, pp. 3B.3-1–3B.3-11.

Godfrey, A. B. (1985). "Training Design Engineers in Quality." *Proceedings, IMPRO Conference,* Juran Institute, Wilton, CT, p. 166 et seq.

Godfrey, A. B. (1991). "Information Quality: A Key Challenge for the 1990's." *The Best on Quality,* vol. 4, Hanser Publishers, Munich.

Goldstein, R. (1990). "The Cost Of Engineering Design Corrections." *ASQC Quality Congress Proceedings,* Milwaukee, pp. 549–554.

Graves, S., and Menten, T. (1996). "Designing Experiments to Measure and Improve Reliability," *Handbook of Reliability Engineering and Management,* 2nd ed. Ireson, W., Coombs, C., and Moss, R., eds. McGraw-Hill, New York, chap. 11.

Gross, C. (1997). "Ergonomic Quality: Using Biomechanics Technology To Create A Strategic Advantage In Product Design." *ASQC Quality Congress Proceedings,* Milwaukee, pp. 869–879.

Gryna, F. (1988). "Product Development." Section 13, *Juran's Quality Control Handbook,* 4th ed. McGraw-Hill, New York.

Hammer, W. (1980). *Product Safety Management and Engineering.* Prentice Hall, Englewood Cliffs, NJ.

Hersh, Jeff F., Backus, Marilyn C., Kinosz, Donald L., and Wasson, A. Robert (1993), "Understanding Customer Requirements for the Alcoa Technical Center," *Proceedings of Symposium on Managing for Quality in Research and Development,* Juran Institute, Wilton, CT.

Hildreth, S. (1993). "Rolling-Out BPQM in the Core R&D of Lederle-Praxis Biologicals, American Cyanamid." *Proceedings of Symposium on Managing for Quality in Research and Development,* Juran Institute, Wilton, CT, pp. 2A-9–2A-16.

Hiller, L. (1986). "Using Design Review to Improve Quality." *Proceedings, IMPRO Conference,* Juran Institute, Wilton, CT.

Himmelfarb, P. (1992). *Survival of the Fittest—New Product Development during the 90's.* Prentice Hall, Englewood Cliffs, NJ, p. 14.

Himmelfarb, P. (1996a). "Senior Managers' Role in New-Product Development." *Quality Progress,* October, pp. 31–33.

Himmelfarb, P. (1996b). "Fast New-Product Development at Service Sector Companies." *Quality Progress,* April, pp. 41–43.

Holmes, J., and McClaskey, D. (1994). "Doubling Research's Output Using TQM." *Proceedings of Symposium on Managing for Quality in Research and Development,* Juran Institute, Wilton, CT, pp. 4–7.

Hooper, J. (1990). "Quality Improvement In Research and Development." *Proceedings of Symposium on Managing for Quality in Research and Development,* Juran Institute, Wilton, CT.

Hoyland, A., and Rausand, M. (1994). *System Reliability Theory: Models and Statistics Methods,* John Wiley & Sons, New York.

Hughes, J. (1992). "Concurrent Engineering: A Designer's Perspective." Report SG-7-3, Motorola University Press, Schaumburg, IL.

Hutton, D., and Boyer, S. (1991). "Lead-Time Reduction In Development—A Case Study." *ASQC Quality Congress Proceedings,* Milwaukee, pp. 14–18.

Industrial Research Institute (1996). *Industrial Research and Development Facts,* Industrial Research Institute, Washington, DC, July, p. 7.

Iizuka, Y. (1987). "Principles of New Product Development Management." *Proceedings, International Congress on Quality Circles,* Tokyo, pp. 177–182.

Ireson, G. (1996). "Reliability Information Collection And Analysis." *Handbook of Reliability Engineering and Management,* 2nd ed. Ireson, W., Coombs, C., and Moss, R., eds. McGraw-Hill, New York, chap. 10.

ISO 8402 (1994), Quality Management and Quality Assurance—Vocabulary.

Jacobs, R. (1967). "Implementing Formal Design Review." *Industrial Quality Control,* February, pp. 398–404.

Jensen, Ronald P., and Morgan, Martha N. (1990). "Quality in R&D—Fit or Folly." *Proceedings of Symposium on Managing for Quality in Research and Development,* Juran Institute, Wilton, CT.

Jensen, F., and Petersen, N. (1982). *Burn-in: An Engineering Approach to the Design and Analysis of Burn-in Procedures.* John Wiley & Sons, New York.

Juran, J. M. (1964). *Managerial Breakthrough,* 1st ed., (2nd ed., 1995). McGraw-Hill, New York.

Juran, J. M.(1992a). *Juran On Quality By Design,* Free Press, New York, p. 5.

Juran, J. M. (1992b). Closing speech at Impro Conference, Juran Institute, Wilton, CT.

Juran, J. M., and Gryna, F. (1993). *Quality Planning and Analysis,* 1st ed. McGraw-Hill, New York.

Kapur, K. (1996). "Techniques of Estimating Reliability At Design Stage." *Handbook of Reliability Engineering and Management,* 2nd ed. Ireson, W., Coombs, C., and Moss, R., eds. McGraw-Hill, New York, p. 24.5.

Kohoutek, H. (1996a). "Reliability Specifications And Goal Setting." *Handbook of Reliability Engineering and Management,* 2nd ed. Ireson, W., Coombs, C., and Moss, R., eds. McGraw-Hill, New York, chap. 7.

Kohoutek, H. (1996b). "Human-Centered Design." *Handbook of Reliability Engineering and Management,* 2nd ed. Ireson, W., Coombs, C., and Moss, R., eds. McGraw-Hill, New York, chap. 9.

Konosz, D. L., and Ice, J. W. (1991). "Facilitation of Problem Solving Teams." *Proceedings of Symposium on Managing for Quality in Research and Development,* Juran Institute, Wilton, CT.

Kowalski, R. (1996). "Maintainability and Reliability." *Handbook of Reliability Engineering and Management,* 2nd ed. Ireson, W., Coombs, C., and Moss, R., eds. McGraw-Hill, New York, chap. 15.

Kozlowski, T. R. (1993), "Implementing a Total Quality Process into Research And Development: A Case Study," *Proceedings of Symposium on Managing for Quality in Research and Development,* Juran Institute, Wilton, CT.

Lander, L., Matheson, D., Ransley, D. (1994). "IRI's Quality Director's Network Takes R&D Decision Quality Benchmarking One Step Further." *Proceedings of Symposium on Managing for Quality in Research and Development,* Juran Institute, Wilton, CT, pp. 3-11 to 3-18.

Lazor, J.(1996). "Failure Mode and Effects Analysis (FMEA) and Fault Tree Analysis (FTA) (Success Tree Analysis—STA)." *Handbook of Reliability Engineering and Management,* 2nd ed. Ireson, W., Coombs, C., and Moss, R., eds. McGraw-Hill, New York, chap. 6.

Lewis, E. (1996). *Introduction to Reliability Engineering.* John Wiley & Sons, New York.

Malec, H. (1996). "System Reliability." *Handbook of Reliability Engineering and Management,* 2nd ed. Ireson, W., Coombs, C., and Moss, R., eds. McGraw-Hill, New York, chap. 21.

Markoff, J. (1996). "Quicker Pace Means No Peace in Silicon Valley." *New York Times,* June 3, 1996.

Matheson D., Matheson, J., Menke, M. (1994). "SDG's Benchmarking Study of R&D Decision Making Quality Provides Blueprint for Doing the Right R&D." *Proceedings of Symposium on Managing for Quality In Research and Development,* Juran Institute, Wilton, CT, pp. 3-1 to 3-9.

Mayo, J. (1994). "Total Quality Management at AT&T Bell Laboratories." *Proceedings of Symposium on Managing For Quality in Research and Development,* Juran Institute, Wilton CT, pp. 1-1 to 1-9.

McClaskey, D. J. (1992), "Using the Baldrige Criteria to Improve Research," *Proceedings of Symposium on Managing for Quality in Research and Development,* Juran Institute, Wilton, CT.

Menger, E., L. (1993). "Evolving Quality Practices at Corning Incorporated." *Proceedings of Symposium on Managing for Quality in Research and Development,* Juran Institute, Wilton CT, pp. 1-9 to 1-20.

MIL-HDBK-217F (1991). *Reliability Prediction of Electronic Equipment.* United States military handbooks, specifications, and standards documents are available from the Defense Automated Printing Service, 700 Robbins Avenue, Building 4, Section D, Philadelphia, PA, 19111-5094.

MIL-HDBK-472 (1984). *Maintainability Prediction of Electronic Equipment.*

MIL-HDBK-781 (1987). *Reliability Test Methods, Plans and Environments for Engineering Development, Qualification and Production.*

MIL-STD-721C (1981). *Definitions of Terms for Reliability and Maintainability.*

MIL-STD-785B (1988). *Reliability Program for Systems and Equipment Development and Production.*

MIL-STD-1629A (1984). *Procedures for Performing a Failure Mode, Effects, and Criticality Analysis.*

MIL-STD-2074 (1978). *Failure Classification for Reliability Testing.*

MIL-STD-2165A (1993). *Testability Program for Systems and Equipments.*

Mistake-Proofing Workshop Participant's Manual (1995). Productivity, Inc., Portland, OR.

Nussbaum, B.(1997). "Annual Design Award Winners." *Business Week,* June 2.

O'Connor, P. (1995). *Practical Reliability Engineering,* 3rd ed. John Wiley & Sons, New York.

Oestmann, E. (1990). "Research On Cat Research Quality." *Proceedings of Symposium on Managing for Quality in Research and Development,* Juran Institute, Wilton, CT.

Perry, W.; and Westwood, M. (1991). "Results from Integrating a Quality Assurance System with Blount's Product Development Process." *Proceedings of Symposium on Managing for Quality in Research and Development,* Juran Institute, Wilton, CT.

Phadke, M. (1989). *Quality Engineering Using Robust Design.* Prentice Hall, Englewood Cliffs, NJ.

Port, O. (1996), "'Green' Product Design." *Business Week,* June 10.

Rajasekera, J. (1990). "Outline of a quality plan for industrial research and development projects." *IEEE Transactions on Engineering Management,* vol. 37, no. 3, August, pp. 191–197.

Raven, J. (1996). "Merrill Lynch Insurance Group Services' Project Management Process." Handout distributed at Florida's Sterling Award Conference, May 29, 1996, Orlando.

Rees, R. (1992). "The Purpose of Failure." *Reliability Review,* vol. 12, March, pp. 6–7.

Reitman, V., and Simpson R. (1995). "Japanese Car Makers Speed Up Car Making." *The Wall Street Journal,* December 29.

Reliability Engineer's Toolkit: An Application Oriented Guide for the Practicing Reliability Engineer (1993). ADA278215. U.S. Department of Commerce, National Technical Information Service, Springfield, VA.

Roberts, G. (1990). "Managing Research Quality." *Proceedings of Symposium on Managing for Quality in Research and Development,* Juran Institute, Wilton, CT.

Rocca, C. J. (1991). "Rochester Excellence…Customer Satisfaction." *Proceedings of Symposium on Managing for Quality in Research and Development,* Juran Institute, Wilton, CT.

Roussel, P. A., Saad, N. K., and Erickson, T. J. (1991). *Third Generation R&D,* Harvard Business School Press, Boston, MA.

Rummler, G., and Brache, A. (1995). *Improving Performance—How to Manage the White Space on the Organization Chart,* 2nd ed. Jossey-Bass, San Francisco.

Schinner, C., 1996. "Accelerated Testing," *Handbook of Reliability Engineering and Management,* 2nd ed. Ireson, W., Coombs, C., and Moss, R., eds. McGraw-Hill, New York, chap. 12.

Sekine, K., and Arai, K. (1994). *Design Team Revolution.* Productivity Press, Portland, OR, p. 191.

Smith, Geoffrey (1991). "A Warm Feeling Inside." *Business Week: Bonus Issue,* October 25, p. 158.

Standinger, H. (1990). "Validating Marketing Specifications: A Foundation for the Design Process." *Proceedings of Symposium on Managing for Quality in Research and Development,* Juran Institute, Wilton, CT.

Swanson, L. (1995). "New Product Development: A Reengineered Process Design." *Proceedings, IMPRO Conference,* Juran Institute, Wilton, CT, pp. 3A.3-1–3A.3-12.

Taguchi, G. (1995). "Quality Engineering(Taguchi Methods) for the Development of Electronic Circuit Technology." *IEEE Transactions on Reliability,* vol. 44, pp. 225–229.

Taylor, D. H., and Jule, W. E. (1991). "Implementing Total Quality At Savannah River Laboratory." *Proceedings of Symposium on Managing for Quality in Research and Development,* Juran Institute, Wilton, CT.

Thaler, J. (1996). "The Sikorsky Success Story." *Workplace Ergonomics,* March/April.

Turmel, J., and Gartz, L. (1997). "Designing in Quality Improvement: A Systematic Approach to Designing for Six Sigma." *ASQC Annual Quality Congress Proceedings,* ASQC, Milwaukee, pp. 391–398.

Van der Hoeven, B. J. (1993). "Managing for Quality in IBM Research." Unpublished paper provided by author.

Wasson, A. (1995), "Developing and Implementing Performance Measures for an R&D Organization Using Quality Processes," *Proceedings of Symposium on Managing for Quality in Research and Development,* Juran Institute, Wilton CT.

Wheelwright, S., and Clark, K. (1992). *Revolutionizing Product Development,* Free Press, New York, pp. 151–161.

Wood, L., and McCamey, D. (1993). "Implementing Total Quality in R&D." *Research Technology Management,* July-August, pp. 39–41.

Yoest, D. T. (1991). "Comparison of Quality Improvement Team Training Methods and Results in a Research and Development Organization." *Proceedings of Symposium on Managing for Quality in Research and Development,* Juran Institute, Wilton, CT.

Zeidler, P. (1993). "Using Quality Function Deployment to Design and Implement a Voice Response Unit at Florida Power and Light Company." *Proceedings of Symposium on Managing for Quality in Research and Development,* Juran Institute, Wilton, CT, pp. A-45–A-56.

SECTION 20
SOFTWARE DEVELOPMENT[1]

Lawrence Bernstein
C. M. Yuhas

INTRODUCTION

The story of software development is as uniquely American as the westward expansion. It is rough-and-tumble, driven by a get-it-done attitude. It is studded with heroic figures who make brilliant leaps into new territory. Its products have amazed and astonished the population at large and created a mystique of invincibility. Only a moment ago have its parallels to the lone gunslinger, the paladin troubleshooter, and the flashy but inefficient pony express, come under some civilizing rule. It has been the place to be for the last three decades, and this is just the beginning. Every kid in an earlier time watched Westerns; now they plug into the Internet. The reason is simple. Creating and using the magic stuff called "software" is emotionally engaging, difficult, and wildly exciting.

But just as it is a long way from the fight at the OK Corral to the Supreme Court, it is a long way from a hacker's personal program to a system that is a true software product, appropriately tested and documented for its customers. This is a view of that work in progress.

---

[1] In the Fourth Edition the material on software development was prepared by Patrick J. Fortune.

PERSPECTIVES ON SOFTWARE PRODUCT DEVELOPMENT

The One Constant Objective.   At the risk of belaboring the wild west analogy, we might say that the Mark Twain of programming is Fred Brooks. He wrote *The Mythical Man-Month* in 1975, and recently reissued it with updates (Brooks 1995). He is a clear-eyed, astringent observer of both strengths and foibles in programming. He says, "A clean, elegant programming product must present to each of its users a coherent mental model of the application, of strategies for doing the application, and of the user-interface tactics to be used in specifying actions and parameters. The conceptual integrity of the product, as perceived by the user, is the most important factor in ease of use....[M]anaging large programming projects is qualitatively different from managing small ones, just because of the number of minds involved. Deliberate, and even heroic, management actions are necessary to achieve coherence."

The time and effort required to produce anything depends upon so many variables that systems have been chronically and universally plagued with slipped schedules and overrun budgets. Systems for large on-line data servers need much more engineering than personal programs, yet the techniques for estimating how much more engineering are relatively new. The communication, testing, and production involved raise the effort by at least a factor of ten. Big systems have been expensive, late, and not totally satisfying, even when they more or less work.

People Who Program.   In the concepts for the future, there are elements of coming full circle back to the much maligned intractable individualist. The "cottage industry" mentality used to be the bane of team programming, but prototyping calls for a small team with a coherent view. Boehm's (1988) spiral model of software development, Figure 20.1, starts with a kernel of an idea that expands.

The "mass customization" concept is intuitively attractive, promising reduced cost and improved quality by moving away from custom development to integrating off-the-shelf components into made-to-order systems. The Internet is the ultimately diversified large system made from small parts. This brings attention back to the master programmer. As Chen (1995) points out in "From Software Art to Software Engineering," this is an issue for software project managers. Quality theory presumes low variances in productivity, but, in fact, several studies have found wide variance in individual performance. Yourdon (1979) has found that the best person in one study was 28 times faster than the worst at coding and debugging an exercise and that the best software program was 10 times more efficient in terms of CPU and memory use than the worst. Their actual performance had no significant correlation with the programmers' years of experience.

Probably 1 percent of all programmers are virtuosi. The characteristics they share are as follows: broad competency coupled with depth and profound innovation in at least one area, ability to teach their colleagues, skill at bottom-up design, possession of a mental spatial map of a system, lively intellect, and the ability to iterate a system to make it, as Einstein remarked, "as simple as possible, but no simpler." Since the most promising methods of developing large systems depend on having virtuosi as leaders, managers need to learn to attract and reward such people with the quality work, corporate culture, and collegial esteem that they find significant.

Ethics.   One benefit of this circling back to an emphasis on the individual or small group is that it holds the potential of improving the ethical climate in which software development is conducted. A large project with hundreds of peer workers belongs to nobody. A large project conceived by a few has an owner, someone who is ultimately answerable for choices. Here are two examples of what results from a lack of ownership that should be dear to taxpayers' hearts.

The U.S. Army stopped the development of a system that was supposed to replace 3700 older systems by the year 2002. After $158 million and 3 years of effort, the new system was far behind schedule and well over budget. The method for choosing the developer encouraged low bidding to get the contract, then throwing money at a slipping schedule. The Army failed, having specified no

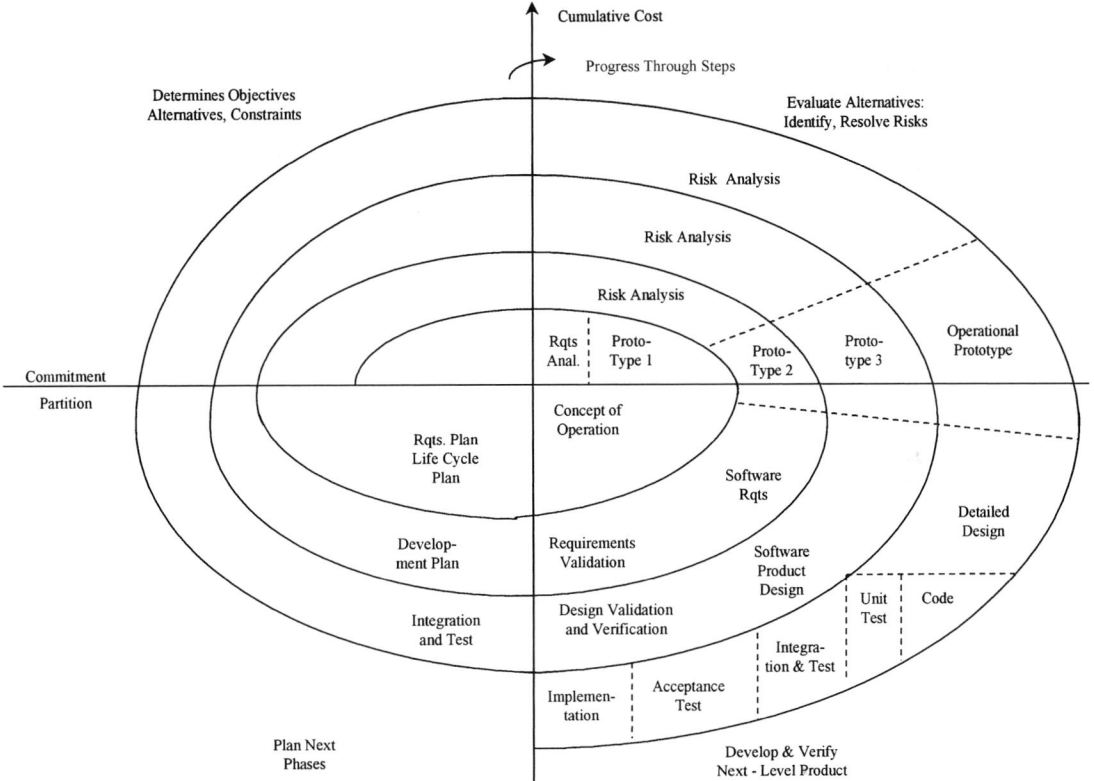

FIGURE 20.1   Boehm's spiral model of the software process. (*Boehm 1988.*)

way to know if the project was on track. The developer failed in not knowing the scope of the problem better than the customer. Both failed in that nobody was ultimately responsible. Appointing a project manager, splitting projects into pieces for which individuals are responsible, and enforcing risk management could produce a more logical approach.

The second example is worse because it is life-threatening. The Department of Energy (DOE) keeps a record of every ounce of plutonium, enriched uranium, and other potentially dangerous radioactive materials created, transported, or sold in the United States. This amounts to several hundred tons. The tracking system was obsolete and the DOE contracted for a replacement. When Congress asked the General Accounting Office (GAO) to check up on the project a year later, the GAO reported that the developer had started programming without adequately analyzing the problem, the user needs, or the final cost of operation. Furthermore, the developer could provide no specifications, no test results, and no status reports. The GAO recommended canceling the project because "the history of software development is littered with systems that failed under similar circumstances." Nevertheless, the DOE switched over to the new system without ever requiring that the software pass any acceptance tests. Clearly there are ethical issues involving all three organizations, but individuals, not organizations, practice ethics.

Challenges.   The software challenge, then, has many facets:

1. In high-risk, life-critical systems, a clear delineation of ethical responsibility is necessary.

2. The issue of defining minimum skill levels and enforcing proficiency in new technologies among software professionals has not yet been resolved.

3. The issue of personal privacy for all system users, including those whose lives are touched by systems without their full comprehension of the ramifications (think of the supermarket scanner or the video store rental), is scarcely addressed beyond some tentative forays into encryption.

4. Another major challenge is the integration of systems. The technological issues of component-based development and the creation of platforms to provide the infrastructure needed to develop applications must be addressed. Project managers need to be aware of the tools available for estimating the resources required to design and develop large systems.

5. The last great challenge, without whose solution all other discussion is moot, is reliability. How can large, integrated systems be tested sufficiently to warrant trust? Schemes such as software rejuvenation acknowledge that it is impossible to catch every bug. Testing methods that can guarantee a secure window of operation need to be linked with human factors analysis to determine what to do when those windows are breached. For example, it does no good to have the autopilot of an airplane compensate for tilting due to ice buildup on the wings if it then disengages without warning and hands the pilots a plane that is out of control and irretrievable. These problems must be resolved before computer technology can become sufficiently mature and trustworthy to be transparent, allowing focus on the task rather than the tool.

HISTORY

The Design Process.   Software development as a formal process began in the 1960s with a North Atlantic Treaty Organization (NATO) conference that set the stage for the early large-scale system development approach. Barry Boehm's waterfall model of linear development (Figure 20.2) came out of this conference. In Boehm's model, the requirements definition phase cascaded down into database design, which in turn cascaded into coding, then testing , then installation, then verification, and out tumbled live operation at the end. All software was customized for a particular site. When systems were used across many sites, on-site technical people tailored unique configurations. In the mid-1960s, the blue-suited IBMer was ubiquitous, tuning operating systems to meet the needs of the local development community. A decade later, it was obvious that software would have to run at multiple sites without such costly nursemaiding. Site-specific tests were added to the test suites, and software was pretested and installed at multiple sites. In that same decade of the 1970s, programming shops moved from punch card/batch processing to on-line time-sharing systems.

It was a milestone achievement when system developers were able to install the same software in several locations without the need for on-site configuration management. The principle outcome was the development of general purpose operating systems. Previously, customer input, output, and schedules were tailored to each application. The operating system provided program scheduling, memory management, I/O management and access to secondary storage devices. Generalized operating systems, plus the script languages like IBM's Job Control Language and UNIX's Shell allowed the loading and linking of programs independent of application.

Advanced software shops in 1975 adopted incremental development and formal configuration management. The integration of a program administration, control, and build function, called "software manufacturing," into the development environment allowed for managing different configurations. The idea of manufacturing multiple configurations spawned the concept of total software configuration management, which included field releases and updates from developers to system integration testers. This created a new level of reliability in managing the total development of a software product. The final verification and build could take place at the development site, leaving only a site-acceptance test at the customer's location.

This immediately created a new need. Freed from the turmoil of on-site builds, customers now had the luxury of focusing on what they actually got, and, not being emotionally tied to it by the blood-

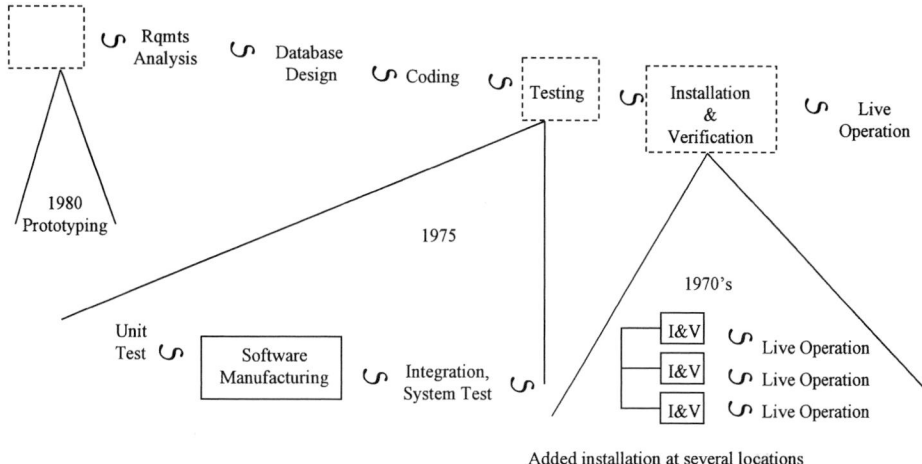

FIGURE 20.2   Early evolution of the software development cycle; the 1960s waterfall model of linear system development is overlaid with new ideas in 1970, 1975, 1980. (*Bernstein 1981.*)

shed of getting it working, they said it was not really what they had in mind, but maybe if just a few little things could be added .... Up-front prototyping was born (Figure 20-2). Software is most often created to do things in new ways, so neither customer nor developer could write a hard requirement at the beginning that would totally satisfy every desire at the end. Postdelivery changes, however, made scheduling and cost estimating very unreliable. The software had to evolve as the system developed. By 1980, the industry had accepted the idea that poor requirements analysis was at the root of much maintenance work. The practice of involving the customer in using prototypes first improved the effectiveness of the requirements analysis. The experiments of Barry Boehm et al. (1984) at the University of Southern California showed that a 40 percent cost and effort saving over the waterfall design method could be achieved by introducing a formal prototype step. Once again, Boehm brilliantly conceptualized the fragmented experiences of many developers when he articulated his spiral model for incremental development (Figure 20.1). The prototype is the foundation of that model. Feedback from the prototype experience directs the definition stage and informsevery subsequent stage. Sometimes development is set back several stages to fix problems on the basis of this feedback. The approach is used in successful systems in the 1990s and is the foundation of the definition of risk management as it is used today. Earlier assumptions about implementation being more important than design led managers to favor top-down decomposition. Managers in the 1990s have revised their assumptions and now favor a combination of top-down design with bottom-up prototyping.

Developmental Tools.   Concurrent with this evolution of the design process was a similar evolution in the supportive nature of the development environment. IBM and AT&T studies both showed that software developers could be 30 percent more productive if they had subsecond response time in an on-line system. The Programmer's Workbench, built on Ken Thompson and Dennis Ritchie's UNIX operating system (Ritchie 1984), was a great breakthrough in the mid-1980s that provided an editor, formatter, and photo composer. Debugging tools and higher-level languages exploded, leveraging the power of each bit of code. Some, like Bjarne Stroustrup's (1994) C++, supported a radical change in the architecture of software systems.

Data Management.   In the early 1970s, problem domain knowledge and data management knowledge were combined with program control in the application software. This required the development of a unique file management system and a unique data reference system for each application—in other words, totally customized, one-use-only software. Each application program had to know how to physically access the data from secondary storage, how to keep track of its own indices, and how to manage data files, including building its own backup and recovery systems. Database management systems at the end of that decade revolutionized this method. They isolated the physical access and management of the data files from the application programs. The program control and problem-solving knowledge were contained in the application, but the details of the data management and layout were universal. This approach spared developers the work of managing the files and disk space and allowed several applications to share some files easily. Database management was as important as the widespread use of an operating system in improving the efficiency of software development.

Database management systems first modeled the relationships among data in terms of hierarchies because hierarchies were easy to do. Airline reservations and payrolls fit nicely into this form, but there were many classes of problems such as inventories and work management that could not be modeled as hierarchies. Standards bodies came up with a network model that was more generic. Meanwhile, Date (1982) at IBM was working on a relational model to make inquiries easier. He had a more general model of relationships between data elements that became very popular. By 1985, relational databases were common and had created a new industry. Companies like Oracle, Ingress, Sybase, and Informix produced packaged software based on Date's idea.

A parallel development, expert systems, separated the rules for executing the program from both the data and the program control structure. By 1995, the use of stand-alone expert systems evolved into the practice of fully integrating those techniques in mainline development. The Japanese software industry made good use of fuzzy logic techniques in systems controlling hardware such as car engines. The potential of expert systems has not yet been realized because it is too complicated to keep the knowledge databases up to date and the computer power (available in 1998) is insufficient to execute complicated knowledge rules.

In the 1990s, everyone else began to understand the view that David Parnas (1979) had articulated 20 years earlier about information hiding as a programming principle. He taught that modules of code should be encapsulated with well-defined interfaces and that the interior of such a module should be the private property of its programmer, not discernible from outside. It was his opinion that programmers are most efficient if shielded from the inner workings of modules not their own. His information-hiding theory is the idea behind object-oriented design. This approach brings the data and the programs that operate on that data together in small programs called *methods*. This approach is robust under change because changes to the simplified component design are propagated easily throughout a system using a technique called *inheritance.* Most of the gain from object-oriented design derives from the fact that prefab libraries of modules or classes are designed and tested for reuse.

Although reuse and the inheritance concept for data updating were obvious benefits of object-oriented design, Michael Jackson (1983) pointed out that avoiding undesired interactions between modules was a critical part of making object-oriented design work. His idea was to normalize out all syntax and retain only the semantics, but to normalize to core values was expensive in terms of system performance. To that end, a significant development in support of Jackson's concept was the UNIX pipe to provide data normalization between modules, though only for data in text format. UNIX chose to use text as the interface medium. Developers could then build very complex software and make extensive use of libraries. Detailed specifications of how modules would work together did not need to be defined before a module was built. Another contribution to building complex systems was the "tag-value" approach. When the specification of a field was linked to the data itself, there was no need to know the order or sequence of data carried across the interface. The tag-value approach achieved Jackson's decoupling idea in a limited way. Object-oriented design is a generalized concept of interface between modules which carries in its methods the meaning of data items. If the tag-value is embedded in its methods and generalized, any data structure (not only text) can be passed between modules. The vehicle is a communication object, a specialized object created expressly for communicating among modules to preserve module isolation.

Architecture.   In 1757, Benjamin Franklin was sailing from New York to London and had lots of time to speculate on the common nautical wisdom that it could never be known how a ship would sail until she was built and could be tested. The trouble was, our philosopher decided, that "one man builds the hull, another rigs her, a third lades her and sails her. No one of these has the advantage of knowing all the ideas and experience of the others, and therefore cannot draw just conclusions from a combination of the whole." Franklin thought that a set of accurate experiments on all such matters, jointly undertaken and carried out to a common end, would be far more efficient than trial and error. He would have made a superb system architect. The industry is just beginning in the decade of the 1990s to act on the understanding that it is important to have structure before jumping into program requirements.

Kruchten (1995) describes a "4+1" model of software architecture using five concurrent views, each addressing a specific set of concerns of interest to different stakeholders in the system. The end users embody the first view, the logical, which creates the object model for object-oriented design. This view acts as a driver to help system integrators and programmers discover architectural elements during the architecture design. The end user describes the service that must be supplied. The system integrators embody the second view, the process, which considers concurrency and synchronization aspects. This view validates and illustrates the architectural design and is the starting point for tests of the prototype and drives the system engineers. How the software will execute on its target machine in terms of throughput, response time, and availability is its concern. System engineers embody the third view, the physical, which maps software onto hardware and considers distributed aspects. Programmers embody the fourth view, the execution, which is the software's static organization in its development environment. The software is organized into manageable chunks that can be created with well-defined interfaces.

The idea is to organize a description of architectural decisions around these four views, then illustrate the workings of the system with a few use cases, called scenarios, which are the "+1," or fifth, view. These scenarios are developed around the functions that are the most important, the most used, and represent the most significant technical risk. The experience of running the scenarios from all four views gives the basis for writing requirements shown in Figure 20.3.

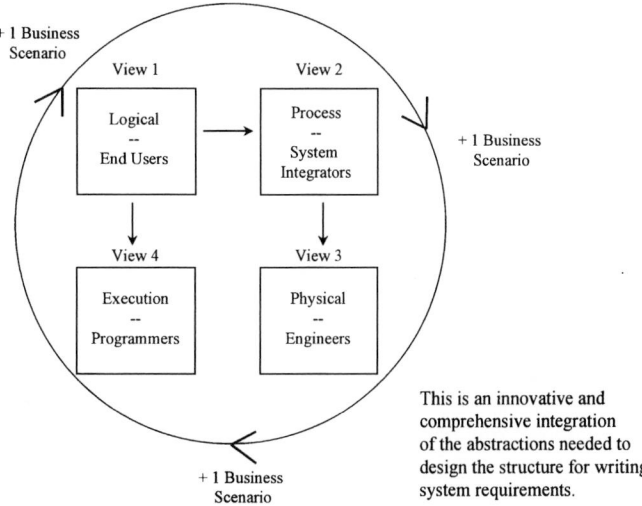

FIGURE 20.3   Kruchten's "4+1" model for developing software architecture. (*Kruchten 1995.*)

Linking Technology and Processes.   The Software Engineering Institute (SEI), a national laboratory affiliated with Carnegie-Mellon University, has established itself as an arbiter of good software practices. There is a consensus that technology must be exercised within the framework of mature methods and practices in order for productivity, reliability, speed, and coherence to be forthcoming. The SEI, in its People Capability Maturity Model, defined the following levels of organizational maturity concerning development processes (Software Engineering Institute 1995):

Level 1—Initial: Anarchy rules. Programmers see themselves as creative artists not subject to any rules or common procedures. Standards are ignored if they even exist. Project plans are viewed as suggestions. No one can specify a delivery date or a quality level for the system. The SEI has estimated that, unfortunately, 75 percent of the organizations it assessed are at this level.

Level 2—Repeatable: Development processes are stable and controlled. Project commitments, costs, schedules, and changes are rigorously managed. The organization adheres to "tribal folklore."

Level 3—Defined: Processes are codified and institutionalized, allowing for process improvement. A separate software engineering process group often exists.

Level 4—Managed: Software metrics beyond tracking costs and schedules are in place. Items measured might be time spent in specific development stages (design, coding, testing), time spent in defining requirements, time spent in code inspections, and number of defects found in each stage. Flaws are identified and remedied.

Level 5—Optimized: Characterized by formal emphasis on continuous process improvement using the data collected in level 4; feedback to the appropriate processes is automatic. The SEI found only two organizations at level 5 among those it assessed: Loral's (formerly IBM) Space Shuttle team in Houston and Motorola India Electronics' programming team in Bangalore.

These criteria for development processes have yet to be universally adopted. Beyond these are the refinements for making platforms and suites, like Microsoft Office, genuine software products in the sense that they would be characterized by tested reliability.

THE LIFE CYCLE OF SOFTWARE

Software is conceived, born, lives, and is replaced. There are different approaches to this life cycle. Boehm's spiral model is the most generally accepted, but any approach must include requirements analysis, preliminary design, detailed design, coding, testing, installation, and maintenance. Engineering discipline breaks down a system into its conceptual, functional, and technical operating components to reduce its complexity and support effectiveness and efficiency. Analysis of many systems concerning the allotment of resources to each life-cycle phase reveals two consistent facts: first, almost half of the development resources (excluding maintenance) are used for testing, and, second, almost 70 percent of the total resources over the life of the software is used in maintenance. Clearly, cost savings can be realized only with methods and tools that yield high-quality code initially and strictly control changes subsequently. If every error or change costs a single unit of effort/cost to fix in development, every error or change found during testing costs three units of effort/cost. Every error or change found in the field after installation costs 30 units of effort/cost to fix. Quality work from the very beginning of the project is the most important concern, before productivity, before financials, before schedules, because everything else flows from quality.

The complexity and high specialization of software development suggests that professional systems staffs will continue to produce systems on behalf of user groups. Large data processing, process control, and other real-time systems will continue to involve professionals even as the industry moves to personal computers.

Requirements Analysis.   The requirements analysis spells out what a system must do. Its end products are a clear and meaningful statement of the problem scope, a description of how the problem is presently addressed with deficiencies noted, a statement of constraints, and a list of new features that are needed. Royce (1975) describes the difficulty of this exercise and the many opportunities for misconception on the part of both the customer and the requirements engineers, which are then compounded by the developers. He says that merely writing requirements is insufficient; requirements need validation based on the developers' knowledge of the problem domain, or on simulation, or on prototyping. DeMarco (1978) describes methods for writing requirements and includes descriptions of data-flow diagrams and structure charts.

A cost-benefit analysis of each system feature is a useful tool for decision making by the customer. A process called "Quality Function Deployment" has merit, but it is very labor-intensive. Efforts to simplify and automate this method may make it practical for general use. Hewlett-Packard is very successful in using the break-even point from the business case as a figure of merit for system development.

The engineering effort varies, depending on the problem constraints of the environment (Figure 20-4). PCs need only vendor software and user programs. A workstation could add some reusable software and perhaps some object-oriented design. The proportions change as the size of the server increases. Application servers that have replaced departmental databases benefit from professional software engineering. The most software engineering is required by the large data servers. As servers handle more corporate data, the investment in software engineers is well spent. The most important corporate asset is data, so reliability, protection, and confidentiality are crucial.

Design, Preliminary and Detailed.   The design phase proceeds directly from the requirements. Functions specified from the requirements are assigned to logical software modules. These modules are examined from the user's viewpoint, as described in Kruchten's 4+1 model. In object-oriented design, the system architects synthesize object classes at this point.

The module descriptions contain explanations of the processing of the module inputs and outputs. Often, special processing concerns of reliability and performance are treated in these descriptions. This information is transcribed into the program's commentary and becomes the design record documentation. Other forms of documentation such as flowcharts and manuals have proved cumbersome and ultimately useless.

System integrators can now write detailed program specifications sufficient to develop the physical databases and files and, from those, write code for the application modules. This phase validates and illustrates the architecture from the computer's point of view. These modules will be put together to execute on the target machine for the system. A data dictionary made up of file descriptions, data element layouts, and security requirements can be the specification for construction of the databases and files.

Design Reviews.   Design reviews punctuate all the preceding phases. The design team can conduct its own reviews or can bring in outside reviewers drawn from customers, employees, problem-domain experts, software architects, or formal quality reviewers. The system is challenged with various scenarios to elicit responses to expected and unexpected stimuli, data for performance budgets, and indications of reliability. The design review is the most important element of static (not executing on a machine) testing. Formal static tests minimize design errors.

Design Simplification.   When someone asked Michaelangelo how he created his sculptures, he said he saw the figure in the stone and then just took away everything else. A successful design phase in the life cycle of a system should do that too, by eliminating redundant functions, making algorithms simpler, and reusing modules from common libraries. Vic Vyssotsky, an early pioneer in complex system development, used to say, "Make it work, then make it work better." But simulation, prototyping, and cost/benefit analysis have removed some of that seat-of-the-pants element. Software project managers will always face the reality, however, that customers are interested in price, functions, and schedule before they get the system, but after delivery they are interested in throughput, response time, and availability. Forty percent of changes introduced after deployment are due to changes in the requirements. The project

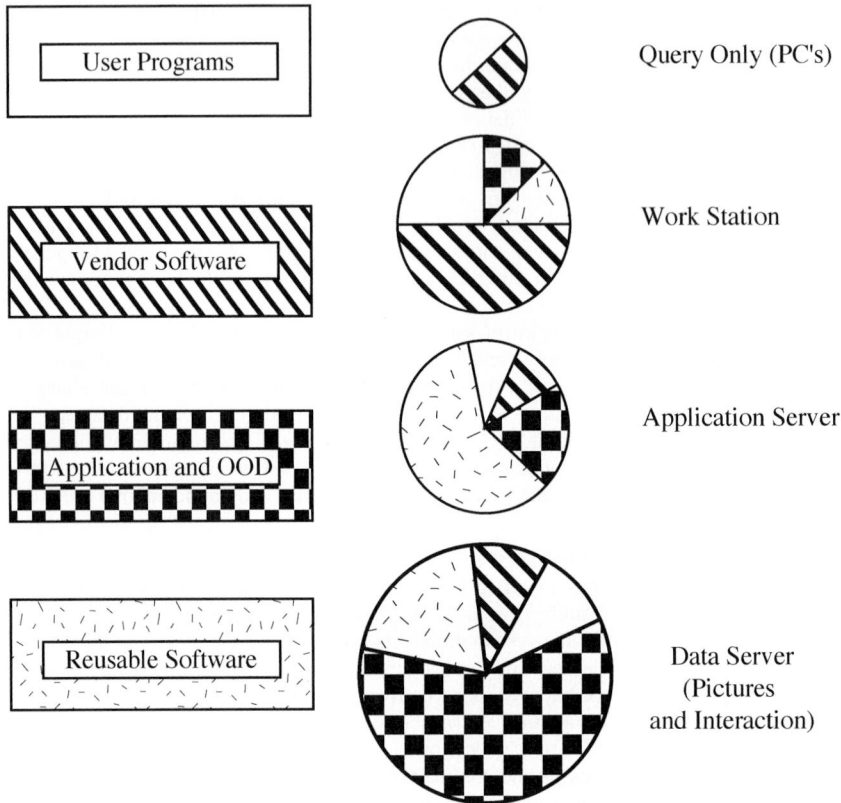

The relative amount of engineering is represented by the size of the circles.
The effort varies depending on the problem constraints of the environment.

FIGURE 20.4   Degree of software engineering needed under various circumstances. (*Bernstein 1981.*)

manager's job, defined by Buckle (1984) in his classic, *Managing Software Projects,* is to satisfy all six concerns early by forcing consideration of the inevitable trade-offs among schedules, resources, functions, and performance.

Coding.   Coding converts detailed design specifications into the form a computer can execute. Source language and shell script are developed. Code is compiled. Binding rules are put in place for linking all modules together into a single load module. Finally the code executes.

At this point, the design is placed under formal change control. Only changes that correct errors are tolerated. The importance of the design reviews becomes apparent because the cost of uncovering errors at this late point is twice that of finding them during static design reviews.

Keeping in mind the dual objectives of reducing the number of errors to uncover during testing and reducing maintenance, the development team can simplify coding in three ways: (1) The preceding phases in the life cycle should have been thoughtfully accomplished. (2) The internal structure of the programs should be streamlined and the control should flow cleanly from start to finish. (3) The level of the programming language should be as high as possible.

Structured programming combines top-down flow control with the use of explicitly defined data types and a restricted set of control primitives, just DO-WHILE and IF-THEN-ELSE. Restricting entry and exit points creates simple control flows, as does segregating program subfunctions into blocks from program start to end. Branching back to earlier statements is discouraged. Each restriction reduces complexity, improves clarity, and makes maintenance easy. Some, such as Knuth (1974) and Dahl (1972), have narrowly equated structured programming to the absence of GO-TO statements. Dijkstra's (1979) explanations are much more general. He says a structured approach involves thinking on successive levels of abstraction so that at any point,only features relevant to that level are considered. Irrelevant issues are not permitted to affect the speed of building or the quality of the software.

The last way developers can affect the quality of the coding is in the choice of language. The language options available to any particular project are determined by those that the selected hardware can support and by the parameters of the problem itself. The features of a particular language direct to some degree the data and logic structures. For example, object-oriented design could not exist before there were languages like C++, Smalltalk, Eiffel, and others to allow its articulation. The wisdom of using higher-level languages, those closest to English, was debated in the 1960s and continues today. Nicholas Wirth's (1968) landmark article on PL/360 pointed out that intermediate-level languages could provide concise syntax but still allow programmers to get at the architecture of the machine. Depending on one's viewpoint, high-level languages either protect or prevent programmers' access to the machine.

The language choice involves trade-offs. High-level languages work well when throughput and response time demands do not push the capability of the target machine. When projects run into performance problems, programmers have resorted to taking advantage of the power of the machine by dropping down into assembly language or machine object code. This poor practice should be avoided because it introduces discontinuity into the system source code and creates unmanageable complexity in the process of producing multiple versions of the software for several sites. The development system needs to contain all the tools for building and testing the system. The tools have to work together and be easy to change when they are upgraded. Therefore, projects that anticipate driving the target machine to 70 percent or more utilization should use an intermediate-level language. C and its offspring, C++, are the most popular intermediate languages. FORTRAN, COBOL, Ada, and Java are popular high-level languages.

Testing.   Testing consumes the most resources of all the development phases. At this point, the static testing of the earlier phases gives way to dynamic testing. Dynamic testing has four parts: unit testing, string testing, system testing, and stress testing.

Unit Testing.   Each module is tested individually with data chosen by examining its source code. The details of the program internals are critical, so unit testing (or "white-box" testing) is best performed by the programmer who wrote the code. The data are chosen to ensure that each branch of the program is executed. Additional stress is applied by introducing data outside the specified data range with at least one point well beyond the range and at least one point at or near the boundary condition. The module should operate as specified. It is a good practice to file the test data and test results with the source code for the module.

As programs become complex, there can be too many branches to test. The test data should then be selected by examining scenarios of the expected system use and by considering potential failure modes. Again, it is imperative to test within the specified data range, outside the range, and at the boundary. Unit testing also applies to purchased components, whose source code structure and comments should be inspected.

String Testing.   Several modules are run concurrently in string testing. Since a logical sequence may require the presence of modules that are still in development, dummy programs called "stubs" can be used to generate calls to the string being tested or to accept and check its results. As those missing modules become available in the project library, the stubs are replaced and string testing continues.

System Testing.   Hardware and software are integrated and challenged to determine if the requirements are satisfied during system testing. The test cases presented to the system are those that were developed in parallel with the program design, using data from the system requirements and scenarios based on the expected use of the system. Since this is a "black-box" process in that the code is not examined, system test is best done by an independent group. Documentation is delivered along with the hardware and software, so it is also reviewed during system test.

Stress Testing.   Software is stressed by offering more than the maximum anticipated loads, by offering no load at all, and by offering the appropriate load in a very short time frame. Software is also stressed by running it longer than the specified run times. Systems fail when arrays or files run out of allotted space or the physical capacities of tapes, disks, or buffers are exceeded. Short-term buffer capacities may be exceeded or temporary arrays may overflow. This type of failure can be a problem in real-time data acquisition and control systems or in batch systems that must turn around in a fixed time. Special reliability tests are based on expected operational scenarios.

Dijkstra voices the frustration of system testers when he remarks that "testing can show the presence of bugs, but not their absence." The number of tests needed for a finite-state machine depends on the number of states. The length of the test trajectory depends on the memory of the system. But at some point, it becomes impractical to do further testing. For this reason, a theory of software dynamics needs to be developed, as will be discussed more fully later.

Installation.
Putting a system into its first site, then installing subsequent versions at that site and additional sites can be an organizational mess. Software manufacturing is a systematic approach to building systems, producing user manuals and other deliverable documents, identifying the system configuration exactly, controlling changes, and packaging software for delivery, as seen in Figure 20.5.

The manufacturing process begins the moment a developer surrenders a completed functional unit. The manufacturing group becomes a single collection point for all the software and controls the source code for various environments. One manufacturing group may control several projects.

Whether or not the development machine and the target machine share the same physical hardware, the operating system used for development work is usually different from the one that will be used for the execution of the software product. The development machine should be optimized for maximum development productivity and software manufacturing. The target machine should be optimized for executing the product.

To decide whether a project warrants a discreet software manufacturing group, the following questions should be answered:

- Is the project or group of projects large enough to require the full attention of one second-level project manager—that is, are there approximately 25 designers and programmers?

- Will future enhancements to the software be made by building on the established, working base?

- Will the product be delivered to one or more sites distant from the development site, but be maintained from the development site?

- Will the customer's employees bear the primary responsibility of running and maintaining the system?

Maintenance.
System maintenance includes all the activities needed to make sure that the system continues to function properly after it is released to the customer. Maintenance consumes a whopping 70 percent of the total effort of owning a working system. Quality control during the developmental phases cannot be sufficiently emphasized as a means of reducing these costs.

Since both satisfying new requirements and fixing errors may require system modification, the process applied in the first phases of software development must be repeated as each need arises. As the system ages, its structure starts to deteriorate. Fully 20 percent of the maintenance staff must be devoted to improving the basic architecture to reduce memory and I/O usage and to speed up the programs.

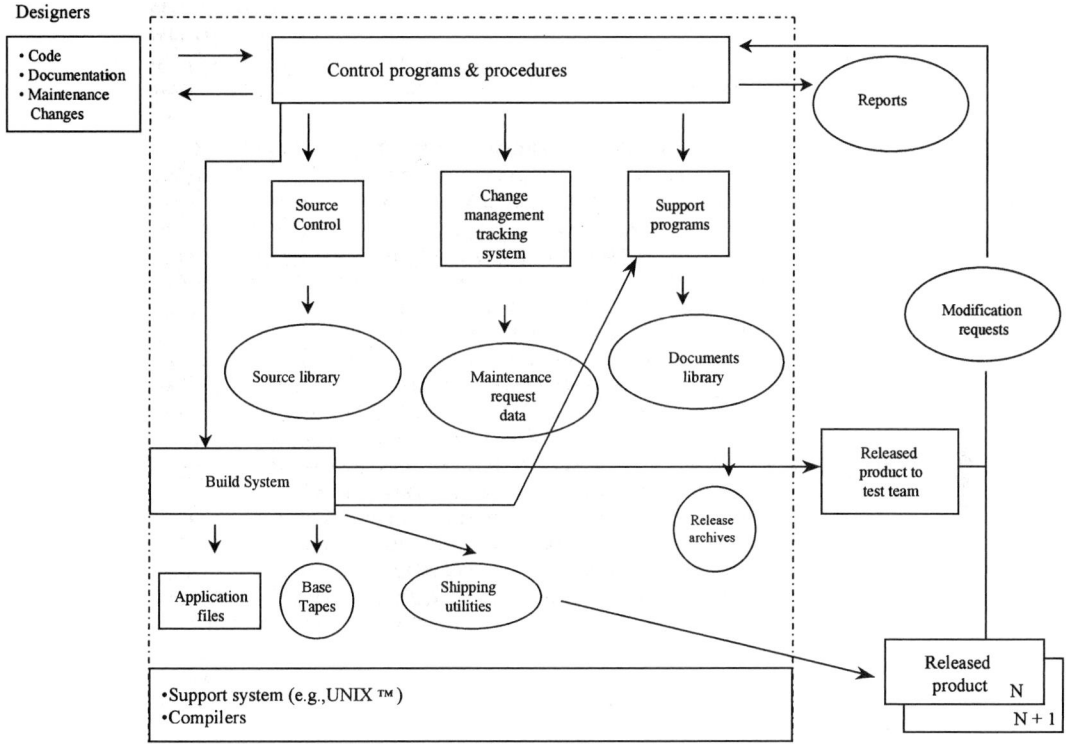

FIGURE 20.5   The process of software manufacturing. (*Bernstein and Yuhas 1989.*)

WHAT CAN GO WRONG DURING THE SOFTWARE LIFE CYCLE?

What are the signs and symptoms that indicate trouble in some phase of the software life cycle? Many a project limps along, obviously ill, but undiagnosed because its management is afraid to face the possibility that the illness is incurable. In fact, there are tools, techniques, and technology to treat many problems. There are also areas that are ripe for study to develop protocols that will help prevent problems.

Certain clues suggest project problems. The absence of a single project manager who has overall responsibility for the success of the work is almost a guarantee that the final product will not have the conceptual integrity that Brooks (1995) says is critical to user satisfaction. Disagreements among managers on facts and disagreements with the customer, particularly on delivery, indicate serious communication problems. Key people being shifted from crisis to crisis or, worse, leaving the project in frustration shows a poor use of the virtuosi. When commitments are an issue (too few being made, those made not being communicated to project members, commitments made before features are specified or made without a development planto produce them), it means that management is not committed to the project. When housekeeping details are lax, there is no good control method. For example, software should not ship before testing is finished, modules should have owners, and critical items should be tracked. Design problems can manifest themselves in several ways: program listings are useless because of too many or too few comments; there is a preoccupation with blindly enforcing standards without an understanding of their usefulness; bugs are found late in the software life cycle; there seem to be a variety of approaches to software design and no apparent correlation between design and programming.

A project audit by outside people can spot these warnings and encourage a project team to generate cures. Audits do not replace competent management; they provide a snapshot of project status. The purpose of an audit is to discover and define organizational problems and stimulate the organization to produce its own recovery plan. In a nonthreatening way, an audit can encourage the articulation of perceptions about the project from many viewpoints and synthesize common themes. In developing software systems there are many risks to which projects can fall prey, the following being the major ones:

Scanty Domain Knowledge.   The most important risk is the lack of a sufficient number of highly qualified people, particularly those having problem domain knowledge. Techniques for injecting quality processes or software technology into organizations include just-in-time training, jump starts with technology experts assigned to the development team for a few months, hiring consultants with experience, and building prototypes.

A look at the history of the tools, processes, and technology that have made a difference shows that each advance allows the programmer to expand the effect of a single line of written code in terms of the number of machine instructions that are executed as a result. That relative productivity, the expansion factor, is the explosion of a written line of code into its actual machine code, expressed as a ratio. The higher the expansion factor, the less the programmer has to write to complete a job (Table 20.1).

Each advance has to eventually be within the grasp of the average programmer, and organizations must make capital investments to keep up. The most recent big payoff came in the threefold increase in productivity due to object-oriented design. Though originally the province of virtuosi, by 1995 Java made this method available to programmers who were not in the top 1 percent of their field. Large-scale reuse of modules will be the next leap forward, but only when the problems of undesired interactions and inconsistent software behavior are more thoroughly investigated.

Unrealistic Schedules and Budgets.   As software shops struggle to win contracts, tension develops between the supplier and the customer. Neither has sufficient data on which to base an estimate of costs and schedules. The project manager has to employ deliberate techniques to counteract this lack.

Function prototyping and performance modeling from the project start, design reviews and weekly project meetings to track problem lists, and action item lists are reality checks for customer and designer alike. Useful metrics are availability of the system in the field, performance in terms of throughput and response time, trouble reports, repairs, customer use of the system, feature cost, and time to market.

TABLE 20.1   The Expansion Factor for Each Major Breakthrough in Software Technology

Event::Power	Date adopted	Expansion factor
Macro assembler: Machine instruction :: 3:1	1965	3
FORTRAN/COBOL: Macro assembler :: 5:1	1970	15
Database management: File manager :: 2:1	1975	30
Regression testing: Big bang testing :: 1.25:1	1978	37.5
On-line: Batch development :: 1.25:1	1980	47
Prototyping: Top-down design :: 1.6:1	1985	75
AGL: In-line reports :: 1.08:1	1988	81
Subsecond time-sharing: On-line development :: 1.4:1	1990	113
Reuse UNIX libraries: No reuse :: 1.35:1	1992	142
Object-oriented : Procedure programming :: 3:1	1995	475
Projection		
Large-scale reuse : Small-scale reuse :: 1.5:1		638

Source: L. Bernstein.

Slovenly Requirements Analysis.   When the wrong functions are developed initially, the lure of "feature creep" is seductive and insidious. Money spent on analyzing the problem difficulty pays off by a factor of 5 in radically compressed development costs. On-line software is 5 times more difficult to produce than basic report generation software. Communications or real-time software is 10 times as difficult. The study of the problem difficulty starts in the requirements analysis by computing the number of function points, the logical measure of software functions as seen by the user. While there is still no standardized method for counting function points (one of those areas which is ripe for protocol development), the National Computer Board in Singapore has instituted quick counting methods to arrive at an acceptable estimate in just 3 to 4 days for a system consisting of 1000 to 2000 function points. The power of function points comes from their emphasis on the external point of view. They show the essential value of what the software is and what it does. They are available early in the development to form a basis for effort and cost estimates.

Rapid prototyping and design analysis can reduce the number of function points that must be addressed by the new software by simplifying design and finding modules containing redundant functions. A careful and thorough look at this analysis shortens the final development interval and reduces the system cost.

Poor User Interfaces.   Software designers are slow to admit human performance engineers into a project. Resisting the human factors (often the tendency that brought them to programming in the first place) results in systems that are hard to understand and full of folklore. Customers want a system that is a tool to solve their business problems, not one that is in itself a major exercise in ferreting out secrets and idiosyncrasies. Human factors issues are present at all stages of development. The best results come when system designers and human factors professionals collaborate throughout a design project. Human factors engineering brings the following to the process:

- Research foundation and theory to generalize problems
- Models to predict human behavior and describe tasks
- Standards representing consensus on particular design parameters
- Principles for guiding the design process
- Methods of finding out about users, their tasks, and how they use systems
- Techniques for creating and testing system designs
- Tools for designing user interfaces and for evaluating designs

Human factors professionals can build bridges to the customer. Projects that enjoy a team relationship with customers are twice as productive as projects with merely a contractual one. Their productivity is directly linked to the quality of their product because frequent demonstrations on a system prototype keep both customer and developer on the same track, forestalling misunderstandings before they are irreparable. The investment of dedicating one person full time to creating a liaison with the end user is small compared to the benefits. This person educates the design group to the dimensions of the problem domain and is an information conduit to the customer on design details.

Shortfalls in Externally Furnished Components.   Sad experience with generic components can foster the desire to reinvent the wheel or the communication library, as the case may be. But producing standard items from scratch every time is wasteful. However, whether such components are off-the-shelf or made from scratch, there is a price to be paid in managerial energy for assuring their capability. Rigorous design review is required to be sure these components perform as expected in the environment of the solution system. Getting the supplier to respond rapidly to problems requires attention to managing the supplier. Some projects fail because of the "kid in the candy store" syndrome. They go to the other extreme; far from doing everything from scratch, they introduce new technology too fast, when it is incomplete, unstable, or exhausts computer system resources.

The question of reuse is murky in the absence of a working theory of software dynamics. Experienced software managers know to retest and reverify system operation through an exhaustive set of regression and functional tests when even the smallest segment of new code is added to a system. A programmer who did not design the reused module finds it difficult to fix problems with it and this often leads to redoing the software rather than reusing it. Data from NASA and AT&T Bell Laboratories suggest that reuse does not become profitable until the third use of a module and then only if the module is used essentially unchanged. The cost, relative to starting from scratch, of reusing a module with no change is 5 percent; the slightest change drives the cost up to 60 percent, as shown in Figure 20-6.

Abundant reuse will be the next high-payoff leap forward in technology when adequate module isolation becomes common practice and software dynamics are better understood. Until then, the following risks, though manageable, remain:

- Reuse is successful only when throughput and response time are not overriding concerns.
- An asset base of software modules is possible now only when they are in C libraries and when they are utility functions.
- Shifting to new technology is difficult with a library of reusable modules.

Unmanageable Team Size.   Small is back in fashion. There is a productivity advantage to small teams because quality control is much easier to enforce. Teams no larger than 50 developers are most effective because above that number, communication problems wash out any advantage of extra hands. When the size of the problem demands more people, it is best to break the system down into smaller subsystems and create well-defined interfaces among the modules. Putnam (1992) has presented data demonstrating that if the number of development people is sharply reduced, the development time is lengthened slightly but the development effort in staff months is greatly reduced. More important, fewer errors occur which means much greater reliability. It is ultimately reliability that is most significant to the customer.

In the 1990s, the industry tried organizing around product teams which were empowered with profit and loss responsibility. Products were subdivided into units small enough to be handled by comprehensive 50-person teams including development, support, and sales. Then products were grouped into product lines under a single management. This worked efficiently for a single product line. Each product line had its own financial bottom line, however, so there was no incentive to transfer technology or processes across product lines. There was no reuse nor company-wide efficiency.

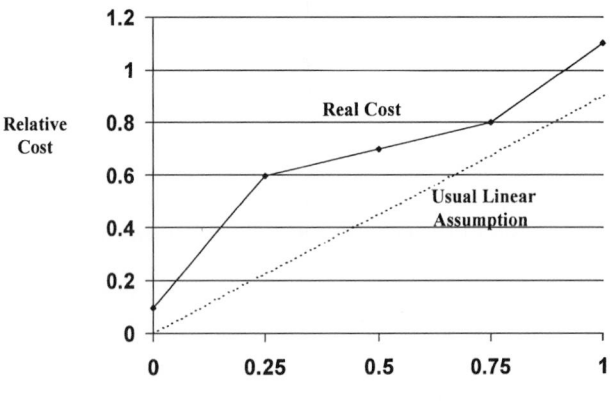

FIGURE 20.6   Cost of modifying modules for reuse—NASA data for 2954 modules. (*Adapted from Selby 1988.*)

Since the marketing people were essentially in competition within one company, there was redundant feature development. These problems were widespread through the industry. They caused the breakdown of the Japanese software factories. Human factors analysis of the original idea would have revealed its flaw.

Lack of Theory of Software Dynamics. The increasing role of reuse forces an investigation of software dynamics. The industrial revolution was fueled by interchangeable parts made possible by the acceptance of interface standards and design constraints based on solid analytical methods. Yet until researchers developed feedback-control theory, electronic systems had to be handcrafted and tuned to prevent them from failing intermittently. Much of today's software suffers from this same hypersensitivity. Minute inconsistencies among different sections of a program result in sporadic faults that even the most rigorous debugging cannot eliminate. A theory of software dynamics promises to ensure that software will be scalable, robust, and reliable. Batory (1993) argues that this is impossible with current methods, especially when a large, feature-rich collection of software components is being developed. Faults can appear to be transient since they may not recur when the software is run with the same input data. The internal states of the software may be different from run to run. This happens because timing and interleaving of messages, signals, interrupts, and shared variables will vary from run to run. Then residual bugs occasionally get triggered, leading to transient failures called "Heisen bugs."

There are reactive methods to recover from a problem after it has happened, but an anticipatory approach called "software rejuvenation" prevents failures from occurring by periodically and gracefully stopping a process. Reset and restarted, it kicks off with a clean set of desired states. This important aspect of software dynamics forces states to be periodic. Without software rejuvenation, processes are nonlinear and nonstationary, practically impossible to analyze. Under this scheme, instead of a system running for a year, it would run for 1 day, 365 times. Avoiding untested domains avoids the faults. A major challenge in software dynamics will be the development of analytically based design constraints to make software behavior periodic and stable in its operating domain.

STANDARDS

Those who manage public and private networks are daily challenged to maintain operations within parameters set by their employers, their customers, their suppliers, and the government. Standards are a way to prevent undue pressure from any one of those groups. For example, "open systems" became the watchword of the 1990s in direct response to IBM's proprietary systems. Standards also define approaches that suppliers are expected to follow so heterogeneous networks can be managed. An internationally standardized solution to this problem is based on the Open Systems Interconnection (OSI) Reference Model, shown in Table 20.2.

TABLE 20.2   Open Systems Interconnection Model—The Foundation for Platforms in Distributed Software Systems

Layer	Layer title	Functions
7	Application	User Interface and network
6	Presentation	Resolve data format differences
5	Session	Logival connections and session connection
4	Transport	Data flow and transmission errors
3	Network	Packetizing and routing
2	Data link	Sharing transport media and low-level error handling
1	Physical	Transport and channel

Source: L. Bernstein.

TABLE 20.3   European ISO Approach

All processes must be documented and have records of compliance
1. Management responsibility
2. Quality system
3. Contract review
4. Design control
5. Document and data control
6. Purchasing
7. Control of customer-supplied product
8. Product identification and traceability
9. Process control
10. Inspection and testing
11. Control of inspection
12. Inspection and test station
13. Control of nonconforming product
14. Correction action and prevention action
15. Handling, storage, packaging, and delivery
16. Control of quality records
17. Internal quality audits
18. Training
19. Servicing
20. Statistical technique

Source: ISO/IEC (1991).

When it is used between two computers, the OSI stack allows the separation of data from layer to layer and this stack, which is the physical embodiment of the layers, is contained in both computers. The stack then separates the application layer from the physical layer. This became the foundation technology for the platform approach to building distributed software systems. This model lets the transmitting computer separate its data in concert with its stack, separating the application layer from the physical layer, while the computer that receives the transmitted data reconstructs the data from the physical layer up the stack to the application layer. This separates details of the inter-machine communication from the function desired in the application. Most client/server applications use the Internet Protocol (IP) at layer 3 and the Transmission Control Protocol (TCP) for the transport, layer 4. Applications using TCP/IP for interfaces conform to layers 3 and 4. The power of this approach is that different transport media can be used simultaneously in the network without the individual network applications being aware of it. (See sidebar for Vinton Cerf's charming explanation of the Internet.) The client/server architecture became popular in the United States while the ISO approach was being standardized in Europe (Table 20.3).

CONCLUSION

The hot topics at the end of the twentieth century are reuse through object-oriented design, component selection and systems integration, and the Internet. Adopting the Internet for communications is the first step toward incorporating it within distributed computing systems. The Java language gives HotJava users the power to develop small, specialized software applications, or "applets," to distribute over the Internet. Large-scale networked computing becomes practical with the Internet as its backbone. Emerging programming approaches, including Java from Sun Microsystems, Object Linking Environment (OLE) from Microsoft, and OpenDOC standards, are competing to be the standard for distributed-component computing.

Forrester Research projects that Internet Web server software will grow threefold by 1997. On-line transactions will have the greatest growth. By the year 2000, nearly half of the applications in

THE INTERNET

Vinton Cerf, the inventor of the Internet, describes how it works:

The Internet is a huge collection of computer networks (about 100,000 of them in early 1996) that are interconnected around the world. Roughly 10M computers are "on" the Internet. These computers are sometimes called "hosts" because, historically, most applications were put up on large, central computers which "hosted" various services. Today, "hosts" can be clients OR servers or, sometimes, both. Networks are interconnected by special computers called "routers" whose job it is to "route" traffic from a source "host" to a destination "host" passing through some number of intervening networks.

The procedures used in the Internet to facilitate communication between computers are called, collectively, "protocols." These are conventions, formats and procedures that govern how information is organized and transported across the intervening networks between the source and destination computers.

Two of the key protocols used in the Internet are called "TCP" and "IP." Internet Protocol, or IP, is the most basic and on its foundation, the rest of the Internet sits. Transmission Control Protocol (TCP) is a "layer" above IP and provides services not available in IP. The two are usually referenced together as "TCP/IP."

"Traffic" on the Internet is made up of "packets" of data. Each packet has a finite, but not fixed length (content), a "from" address, and a "to" address. It may be helpful to think of these packets as "electronic postcards" with all the features you already know about postcards.

When you put a postcard into the mail slot, you have some expectation that the postcard will be delivered, eventually, to its destination. You don't know whether it will go by boat, by plane, by car or by train or perhaps all four. You are not really sure it will be delivered or, if it is, how long it will take. If you put in a number of postcards addressed to the same place, they may arrive in a different order than you sent them.

Internet Protocol packets are just like postcards, but about a hundred million times faster (assuming a postcard takes a day or two). When you send an Internet packet into the Internet, there is no guarantee that it will be delivered. If you send several of them, they may be delivered out of order, they may even be accidentally replicated (something that doesn't usually happen to a postcard!). Internet Protocol provides what is called a "best efforts" communication service.

It is often a surprise for people to learn that guaranteed delivery is NOT a part of the basic Internet system. However, the next layer of protocol, containing TCP, makes up for the potential shortcomings of the Internet Protocol layer. The best way to understand what TCP does is to imagine what you would have to do if you were to try to send a novel to a friend, but the only way you could was to send it as a series of postcards.

First, you would cut up the pages so they could fit on a postcard. Then you would notice that not all the resulting postcards had numbers, and since postcards often arrive out of order, you think to number each postcard so your friend could put them back in order to make it easier to read the novel. Second, since you know that postcards may be lost, you would keep copies of each one, in case you have to send a duplicate to make up for a lost one.

How will you know if one is lost? Well, it would be convenient if your friend would send YOU a postcard every so often to say he had received all postcards up to postcard number X (for some value of X). On receipt of that postcard, you could discard the duplicates up to postcard number X that you had been holding. Of course, your friend's postcard might be lost, so you also need some kind of "time out" after which you start resending copies of postcards that had not yet been acknowledged. Your friend may not have sent any confirmation because he was missing some postcards.

If your friend receives duplicate postcards, he can easily deal with that since the cards are numbered, so duplicates can be ignored or discarded. Finally, you might realize that your friend's mailbox has a finite size. If you sent all the postcards of the novel at once, they might, by some miracle, all be delivered at the same time and might not fit in the mailbox. Then some would fall on the floor, be eaten by the dog, and you'd have to resend them anyway. So it might be a good idea to agree not to send more than perhaps 100 at a time and await a postcard acknowledging successful receipt before sending others.

The Transmission Control Protocol does all these things to deliver data reliably over the basic Internet Protocol service. That's really all there is to it. Of course, I have left out some other important details such as routing (how do packets get to where they are supposed to go) and naming conventions, but I will leave that for the next installment."

© Vinton G. Cerf, 1995, 1996.

[Permission is granted to reproduce freely provided credit is given to the author.]

use will have been developed expressly for the Web and the presence of legacy systems will diminish. The trend will most likely be toward a hybrid solution of Web-ready applications and specially designed robust web servers.

Along with this enthusiasm for new areas, there is a disturbing mood of cynicism and fatalism among developers who are keenly aware that software needs more science and less art. Without a formal design discipline, software quality depends on a project manager's extreme attention to detail.

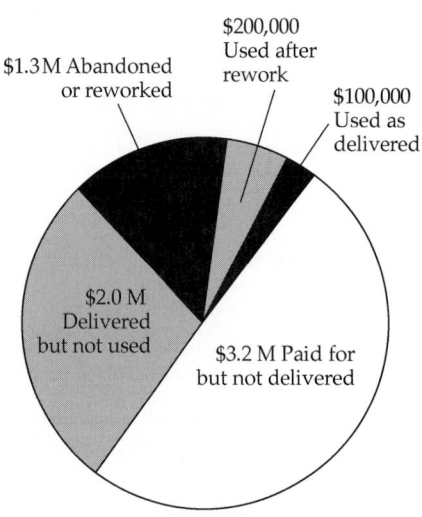

The National Software Council presentation to the United States Congress on November 2, 1995, bluntly stated that the United States continues to have major problems developing large and complex software systems (Figure 20.7).

Until the theoretical bases are developed, investment might be made profitably in four areas that have been shown to improve the quality of software: (1) developing programmer skills, (2) analysis and possible simplification of the problem difficulty, (3) improvement of the customer relationship, and (4) acquiring the best tools and technology available. Spending money on each factor can substantially improve the quality of the software produced.

These are ultimately stopgap measures until the major problem is addressed. The major problem is that there are far too many people in power who have neither the experience nor the education in practical strategies for managing software projects. The American competitive edge has depended on technological innovation approximately every 4 years. When executives do not understand how software development proceeds and put in place management processes with hard controls, cutting back on advanced applied research, the American advantage will surely evaporate. The rise of management ignorant of methods for handling software projects has created the stifling climate of fear that impedes progress. There is an ethical imperative to make software trustworthy that requires the cooperative efforts of industry, government, and universities. The behavior of software under load in the field must be studied so that principles of software dynamics can be developed. Executives must realize that nobody makes money for long when software fails, then educate themselves in ways to encourage the growth of the people who can advance this immature science.

FIGURE 20.7 Customers still find it difficult to get what they pay for. The National Software Council found this situation largely unchanged in 1995. A 1979 U.S. Government Accounting Office report (FGMSD-80-4), which breaks down results from $6.8 million in nine federal software projects shows that only 2 percent of the software was used as delivered. Information provided by *ACM SIGSOFT Software Engineering Notes,* vol. 10, no. 5, October 1985. (*Bernstein 1981.*)

REFERENCES

Arnold, Ken, and Gosling, James(1996). *The Java Programming Language,* Addison-Wesley, Reading, MA, May.

Arthur, Lowell Jay (1983). *Programmer Productivity: Myths, Methods, and Murphology, A Guide for Managers, Analysts, and Programmers.* John Wiley & Sons, New York, pp. 25–27.

Batory, D., Singhal, V., Sirkin, M., and Thomas, J. (1993). "Scalable Software Libraries," Department of Computer Sciences, The University of Texas at Austin, SIGSOFT '93, December.

Bernstein, L. (1981). "Software Project Management Audits." *Journal of Systems and Software,* vol. 2, no. 4, December, pp. 281–287.

Bernstein, L., and Yuhas, C. (1989). "Software Manufacturing." *UNIX Review,* vol. 7, no. 7, July, pp. 38–45.Boehm, Barry W., Gray, T. E., and Seewaldt, T. (1984). "Prototyping Versus Specifying: A Multiproject Experiment." *IEEE Transactions on Software Engineering,* vol. SE-10, no. 3, May.

Boehm, Barry W. (1988). "A Spiral Model of Software Development and Enhancement," *Software Management,* 4th ed. Donald J. Reifer, ed. IEEE Computer Society Press, Los Alamitos, CA, order number 3342-01. (Reifer's compendium is a first-rate collection of software papers.)

Boehm, Barry W. (1979). "Software Engineering," *Classics in Software Engineering.* Yourdin Press, New York, p. 325.

Brooks, Frederick P., Jr. (1995). *The Mythical Man-Month: Essays on Software Engineering Anniversary Edition.* Addison-Wesley, New York.

Brown, Eric, Colony, George F., O'Herron, Rich, and Smith, Nicole. (1995). *Software Strategy Service, Web Server Strategies,* vol. 6, no. 8, November. Forrester Research, Cambridge, MA.

Buckle, J. K. (1984). *Managing Software Projects.* Robert E. Krieger Publishing, Malabar, FL.

Buxton, J. N., and Randall, B. (1969). "Software Engineering Techniques," report on a conference sponsored by the NATO Science Committee, Rome, October 27–31. Scientific Affairs Division, NATO, Brussels 39, Belgium, April 1970.

Carnegie Mellon University/Software Engineering Institute (1995). *The Capability Maturity Model: Guidelines for Improving the Software Process.* Addison-Wesley, Reading, MA.

Chen, Stephen (1995). "From Software Art to Software Engineering." *Engineering Management Journal,* vol. 7, no. 4, December, pp. 23–27.

Cusumano, Michael A. (1991). *Japan's Software Factories: A Challenge to US Management.* Oxford University Press, New York.

Dahl, O. J., Dijsktra, E. W., and Hoare, C. A. R. (1972). *Structured Programming.* Academic Press, London.

Date, C. J. (1982). *An Introduction to Database Systems,* 3rd ed. Addison-Wesley, New York.

DeMarco, T. (1978) *Structured Analysis and System Specification.* Yourdan Press, New York.

Desmond, John (1994). "IBM's Workgroup Hides Repository," *Application Development Trends,* April, 25.

Dijkstra, E. (1979) "The Humble Programmer." 1972 ACM Turing Award Lecture in Classics in Software Engineering.

Duvall, Lorraine (1995). "A Study of Software Management: The State of the Practice in the United States and Japan." *Journal of Systems Software,* vol. 31, pp. 109–124. Elsevier Science, New York.

Fairly, Richard (1994). "Risk Management for Software Projects," *IEEE Software,* May.

Fayad, Mohamed E., Tsai, Wei-Tek, and Fulghum, Milton L. (1996). "Transition to Object-Oriented Software Development." *Communications of the ACM,* February, vol. 39, no. 2, pp. 108–121.

Fulton, N. D., et al. (1994). "A Reusable Module for Software." *Rejuvenation,* January 31. BL011267-940131-01TM.

Gibbs, W. Wayt (1996). "Battling the Enemy Within: A Billion Dollar Fiasco Is Just the Tip of the Military's Software Problem," *Scientific American,* vol. 274, no. 4, April, pp. 34–36.

Gibbs, W. Wayt (1996). "Systematic Errors," *Scientific American,* vol. 274, no. 5, May, pp. 30–31.

Gleick, James (1987). *Chaos: Making a New Science.* R.R. Donnelley & Sons, Harrisonburg, VA.

Gray, J., and Siewiorek, D. P. (1991). "High Availability Computer Systems." *IEEE Computer,* vol. 24, no. 9, September, pp. 39–48.

Hines, John R. (1996). "Java: Jive?" *IEEE Spectrum,* vol. 33, no. 3, March. World Wide Web http://java.sun.cor.

"HOTJ, The HotJava Browser," (1995). White paper available from Sun Microsystems, sunsoft@tbsmlet.com

Huang, Y., Kintala, C.M.R., Kolettis, N. and Fulton, N.D. (1995). "Software Rejuvenation: Analysis, Module and Applications." *Proceedings of the 25th International Symposium on Fault-Tolerant Computing* (FTC-95), Pasadena, CA, June, pp. 381–390.

ISO/IEC (1991). International Standard "Information technology–Software product evaluation—Quality characteristics and guidelines for their use," 1st ed. December 15. Reference number ISO/IEC 9126:1991(E), Case Postale 56, CH1221, Geneva 20, Switzerland.

Jackson, Michael A. (1975). *Principles of Program Design.* Academic Press, New York.

Jackson, Michael (1983). *System Development.* Prentice Hall International, Englewood Cliffs, NJ.

"Java, The Java Language," White paper available from Sun Microsystems, telephone (800) 433-4224.

Jones, Capers (1994). *Assessment and Control of Software Risks.* Prentice Hall, Englewood Cliffs, NJ.

Jones, Capers (1986). *Programming Productivity.* McGraw-Hill, New York, pp. 83–210.

Knuth, D. E. (1974). "Structured Programming with 6070 Statements," *Computing Surveys,* vol. 6, pp. 261–301.

Kruchten, Phillipe B. (1995). "The 4+1 View Model of Architecture," *IEEE Software,* vol. 12, no. 6, November, pp. 42–50.

Lutz, Robyn R. (1993). "Targeting Safety-Related Errors During Software Requirements Analysis," Jet Propulsion Laboratory, California Institute of Technology, *SIGSOFT,* December.

Mills, Harlan (1988). *Software Productivity.* Dorset House, New York, pp. 13–18.

Neumann, Peter G. (1995) *Computer-Related Risk,* ACM Press, New York.

O'Brien, Larry (1996). "Java Changes Everything," *Software Development,* vol. 4, no. 2, February.

Parnas, D. L. "On the Criteria to Be Used in Decomposing System into Modules," *Classics in Software Engineering.* Yourdin Press, New York.

Penzias, Arno (1995). *Harmony: Business, Technology, + Life after Paperwork.* HarperBusiness, New York.

Poulin, J. S., Aruso, J. M., and Hancock, D. R. (1993). "The Business Case for Software Reuse." *IBM Systems Journal,* vol. 32, no. 4, pp. 567–594.

Putnam, Lawrence H., and Meyers, Ware (1992). *Measures for Excellence, Reliable Software on Time, Within Budget.* Yourdin Press, Prentice Hall, Englewood Cliffs, NJ, p. 103.

Ritchie, Dennis (1984). "The Evolution of the UNIX Time-Sharing System," *AT&T Bell Laboratories Technical Journal,* vol. 63, no. 8, part 2, p. 1571, October.

Ross, Philip E. (1994). "The Day the Software Crashed," *Forbes,* April 25, 150.

Royce, Winston (1975). "Software Requirements Analysis: Sizing and Costing," *Practical Strategies for Developing Large Software Systems.* Ellis Horowitz, ed. Addison-Wesley, Reading, MA.

Selby, R. (1988). "Empirically Analyzing Software Reuse in a Production Environment." *Software Reuse: Emerging Technology.* W. Tracz, ed. IEEE Computer Society Press, New York, pp. 176–189.

Software Engineering Institute (1995). "Overview of the People Capability Maturity Model (P-CMM)," CMU/SEI-95-MM-02, September, pp. O-13 to O-32.

Stroustrup, Bjarne (1994). *The Design and Evolution of C++.* Addison-Wesley, Reading, MA.

Van Hoff, Arthur (1995). *Hooked on Java.* Addison Wesley, Reading, MA. (Discussion of applets.)

Van Hoff, Arthur, and Sami, Shaio (1996). "What Is This Thing Called Java?" *Datamation,* vol. 42, no. 5, March 1, p. 45.

Walston, C. E., and Felix, C. P. (1977). "A Method of Programming Measurement and Estimation," *IBM Systems Journal,* vol. 16, no. 1, pp. 54–60. Weinberg, G. (1971). *The Psychology of Computer Programming,* Van Nostrand Reinhold, New York.

Weinberg, Gerald M.(1988). *Understanding the Professional Programmer.* Dorset House, New York, pp. 69–79.

Wirth, Nicholas (1968). "PL/360, A Programming Language for the 360 Computers." *Journal of the ACM,* vol. 16, p. 37.

Yochum, Doreen S., Laws, Elaine P., and Barlow, Gretchen K. (1995). "An Integrated Human Resources Approach to Moving Information Technology Professionals Towards Best-In-Class." *AT&T Technical Journal,* 1995.

Yourdon, Edward Nash (1979). *Classics in Software Engineering,* Yourdon Press, New York, p. 122.

SECTION 21
SUPPLIER RELATIONS[1]

J. A. Donovan
F. P. Maresca

INTRODUCTION

> It is a tragedy that, in the West, relations between…buyer and seller have been confrontational and adversarial.…It is not uncommon for a supplier with a history of loyal service to be unceremoniously dumped when the buyer finds another supplier selling more cheaply. Nor is it uncommon for a supplier to gouge a customer during a seller's market and boom time.…In the long haul, this win-lose philosophy can turn both sides into losers. (K. R. Bhote 1987)

For many operations, purchased goods and services represent a significant component of cost. Typically, these goods and services are either: (1) those which are used or consumed or become a part of the end product (for example, raw materials and chemicals) or (2) those which support the production process itself or the personnel involved (for example, plant machinery and equipment, computer equipment, travel services, and office supplies).

---

[1] In the fourth edition, information on the section covering supplier relations was supplied by Frank M. Gryna.

The quality, or the fitness for use, of these purchased goods and services can heavily influence an operation's finished product quality. Furthermore, poor-quality suppliers can be a major contributor to an operation's overall cost of poor quality. It has been estimated that for the average American manufacturer, the cost of poor quality ranges from 10 to 30 percent of sales, an astounding and too often accepted cost leak (see Section 8, Quality and Costs). By ignoring this "pot of gold," manufacturers have unknowingly contributed to overall customer dissatisfaction and disloyalty, and adversely affected their own competitive positions.

Managing supplier relations has historically been the responsibility of an organization's purchasing department (also known as Procurement, Sourcing, or Materials Management). This section will review the historic roles and responsibilities of the purchasing department, and then describe how the quality revolution has redefined Purchasing's role—from a passive information transfer agent between requisitioner and supplier—to the facilitator of what we will define as the supply chain. We will then demonstrate how quality planning, quality control, and quality improvement can be applied to supplier relations to generate continuous improvement, customer satisfaction, value, and ultimately competitive advantage through the management of this supply chain.

The full value of supplier relations is achievable only if suppliers are viewed as partners with their customers in pursuit of mutual goals, rather than adversaries in a win-lose battle concerning price. The basis for building such supplier relations is cooperation, collaboration, and trust. Those not willing to build supplier relations on this foundation need not read any further.

TRADITIONAL ROLE OF PURCHASING

Following World War II, when growing demand for goods and services was satisfied by increasing plant capacity, Operations was identified as the strategic component of an organization. Purchasing was relegated to a staff support role. Purchasing's mission was to ensure that suppliers provided an uninterrupted supply of required goods and services, delivered on time and at the right price, where "right price" was usually interpreted as "lowest price."

Personnel in Purchasing departments developed competencies in supplier negotiations, bid evaluation and analysis, document administration, and market knowledge. Supplier negotiations were viewed as the major value-added activity of the Purchasing department, and supplier relations developed during these negotiations. This often resulted in adversarial supplier relations, which were focused on short-term performance. Availability and low price became the most important criteria for measuring supplier performance. As Carlisle and Parker wrote (1989): "This adversarial tendency…resulted in a great deal of management energy being spent on both sides in search of ways to capture some of the other's profit margin."

If a supplier change was made, little consideration was given to any resulting costs incurred. The new supplier's product or service might deviate slightly from that of the original supplier, translating into costs in other areas of the production process. This propensity to change suppliers resulted in many disadvantages to the purchaser, including:

- Excess inventory due to obsolescence
- Production shutdowns due to installation or operation requirements
- Transition costs such as training or maintenance testing disposal costs
- Production disruptions due to poor quality detected after the testing had been completed
- Increases in variation in the finished product
- Increases in scrap, product defect, or customer dissatisfaction

Rarely were these costs identified, aggregated, analyzed, and reduced. Furthermore, as Deming (1981) stated: "No one can outguess the future loss of business from a dissatisfied customer." In the adversarial climate that prevailed, little opportunity for collaborative root-cause analysis existed. The costs of "lowest price" purchasing became part of the Operations overhead, and was accepted by management as a cost of doing business. If action was required, additional supplier changes might occur, thus creating more hidden costs of ownership (see Figure 21.1).

Quality problems stemming from this price fixation are nothing new, as illustrated in a famous letter of September 1, 1865, from Vauban, the Fortifications Commissioner of Louis XIV of France, to his minister, Mr. Louvois, describing quality deficiencies experienced in the fortifications program:

> …There still remain a number of buildings of previous years which are not yet terminated, and which shall never be, if we are to believe the builders. All this is due, Monseigneur, to the confusion caused by the frequent reductions in price which are attributed in your construction contracts. It is a fact that all the broken contracts, agreements not kept, and renewal of adjudications only attract the people who know nothing about the business, rogues, and ignoramuses as contractors, while those who know what they are doing do not even attempt to sign such contracts. I say that in addition they increase the price and delay the construction of the buildings which is thereby much worse.…Pay the correct price. It will always in the long run be the cheapest deal you could make. (Dunaud 1995)

QUALITY REVOLUTION

In the late 1970s and early 1980s, companies in the United States were finally shocked into the realization that quality was vital to long-term success. Basic industries, such as steel and rubber, and producers of major products, such as automobiles, consumer electronics, and optical goods, lost market share to imported goods, especially from Japan. This market-share erosion could not be fully rationalized as resulting from lower prices made possible by a strong dollar, cheap foreign labor, or illegal "dumping" (selling at below manufacturer's cost to gain a foothold in a market).

The success of these imports was largely attributable to superior quality. Put more painfully, U.S. domestic goods had become inferior to imported goods. The automobiles of General Motors, Ford, and Chrysler had a purchased content as high as 70 percent. For these companies, the logical conclusion was that the quality of finished products was largely determined by the quality of purchased goods and services. Thus, senior management of these companies began paying closer attention to supplier quality, which became a critical differentiator in supplier selection.

As Purchasing departments began to focus on supplier quality, the fourth of Deming's 14 points, "End the practice of awarding business on the basis of price tag (alone)" (Deming 1986), became the

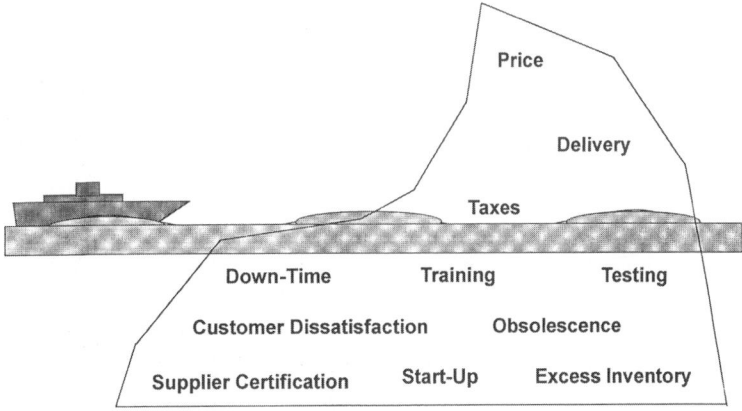

FIGURE 21.1 The total cost of changing suppliers.

framework by which purchasers approached the acquisition of goods and services. Supplier relations evolved from confrontation between adversaries to collaboration between partners trying to satisfy their common customer, the end user of the finished product or service. Because of the demonstrated benefits of this evolution, this new approach to supplier quality in the automobile industry set a pattern which began spreading throughout industry in the developed world.

SUPPLIER RELATIONS CONCEPTS DEFINED

As Purchasing's role has evolved—from passive information transfer agent between requisitioner and supplier to facilitator of the supply chain—the definitions applicable to the Purchasing function have evolved as well.

Purchasing.   The tasks, activities, events, and processes required to facilitate the acquisition and delivery of a good or service required by an end user.

Supplier Relations.   The tasks, activities, events, and processes required to facilitate the ongoing interface between suppliers of goods and services and the end users of those goods and services.

Supply Chain.   The tasks, activities, events, processes, and interactions undertaken by *all* suppliers and *all* end users in the development, procurement, production, delivery, and consumption of a specific good or service. The coordination, integration, and monitoring of this supply chain is referred to as "supply-chain management." The extended enterprise of the supply chain includes the end users, prime supplier or distributor of a product or service, prime manufacturer, and the multiple tiers of suppliers providing goods and services to these prime manufacturers and distributors, as illustrated in Figure 21.2.

Purchasing personnel find the scope of their job expanding. Purchasing is no longer expected simply to acquire goods and services, but to engage in the proactive management of supplier relations, searching for opportunities to add value throughout this supply chain. But what is value, and how can this value be articulated, identified, measured, and managed?

Definitions of Quality and Value.   "Quality" may be defined as fitness for use. The fitness for use of an acquisition can only be assessed based on a thorough understanding of the relevant customers and their needs. Value is the relative cost of acquiring quality. If two different supply chains are able to produce a product with identical fitness for use, the chain which can achieve the required fitness for use at the lower total cost of ownership is the one with the greater value. Therefore, the ability to provide a given level of quality at a reduced total cost of ownership [as a result, for example, of an initiative to reduce the cost of poor quality (COPQ)] will always result in value generation.

Total Cost of Ownership.   Sometimes referred to as life-cycle cost or total system cost, total cost of ownership is the sum of all costs associated with the acquisition, installation, operations, maintenance, and retirement of a good or service.

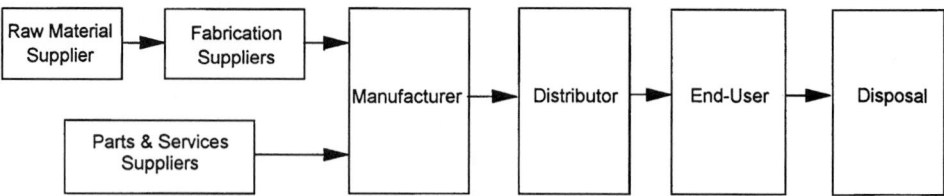

FIGURE 21.2   Elements of a supply chain.

QUALITY INCORPORATED INTO TRADITIONAL PURCHASING

Purchasing as a Strategic Process.   Consider the potential opportunity if time, resources, energy, and management priorities focus on the *processes* by which these goods and services were scheduled, designed, manufactured, and purchased, rather than simply focusing on the acquisition alone. Quality and cost reduction opportunities can be identified, measured, and managed. Where two firms compete in identical markets, the ability of one firm to identify, measure, and manage these opportunities faster than another firm creates a clear competitive advantage.

Therefore, purchasing, while traditionally thought of as a utility, nonvalue-added function, is increasingly being recognized as a strategic function—an opportunity for process management and improvement and a tool for achieving competitive advantage.

> On average, manufacturers shell out 55 cents of each dollar of revenue on goods and services, from material to mail. Shrinking that bill by 5 percent can add almost 3 percent to net profits. The same arithmetic applies to service businesses.
>
> Cutting purchasing costs has surprisingly little to do with browbeating suppliers. Purchasers at companies like AT&T and Chrysler aim to reduce the total cost—not just the price—of each part or service they buy. They form enduring partnerships with suppliers that let them chip away at key costs year after year. Companies are also packaging once fragmented purchases of goods and services into company-wide contracts for each.
>
> Allied-Signal's mastery of purchasing has led to an expected 21 percent surge in profits. When the company signs a supplier, it expects a steep price cut and also demands that the supplier commit to lowering the components total cost by 6 percent in real terms each year.
>
> One way purchasing will reshape U.S. business: Look for super-suppliers to emerge as customers buy more from fewer suppliers. (Tully 1995)

Importance of Supplier Quality.   To identify supplier-relations opportunities and to capitalize on them, an understanding of supplier's quality is of paramount importance. Consider the following:

- The costs associated with poor quality suppliers are high. For one home appliance manufacturer, 75 percent of all warranty claims were traced to poor quality of purchased items.
- The growing interdependency of suppliers and end users in identifying and implementing such opportunities as "just-in-time" delivery, electronic data interchange (EDI), electronic funds transfer (EFT), cycle-time reduction, and outsourcing initiatives.
- The trend to minimize incoming inspection.
- The growing trend of purchase decisions being made not on lowest price but on the total cost of ownership of the product or service.

These considerations require the purchasing function to abandon its traditional role of transaction-performance management. Expressions of this emerging approach are contained in statements from two eminent American companies.

From AT&T in 1995:

Mission: Provide worldwide professional procurement services that are a competitive advantage for AT&T and its customers.

Vision: Be THE benchmark for procurement excellence.

From Chrysler Corporation:

Mission: Manage and prepare the extended enterprise to the maximum benefit of Chrysler and its customers.

The implications of this role change are profound.

- Supplier selection is no longer the sole prerogative of the Purchasing department.
- Cooperation, collaboration, and joint problem solving among internal customers, purchasing, and suppliers is required.

- Purchasing personnel focus on process, abandoning the focus on transaction.
- Within the end user's firm, the purchasing function is elevated to a strategic level and its transaction activities and responsibilities minimized or eliminated.

A successful transition to a strategic approach to purchasing requires everyone in an organization to embrace a new belief system concerning purchasing. In the transition, senior management will find it necessary to aggressively promote the new view, which might be summarized as follows:

> Purchasing has become a key strategic process within our organization, requiring a staff of highly skilled professionals committed to working with our end users and suppliers, in a collaborative, problem-solving environment, facilitating quality and continuous improvement.

Shift to Strategic Purchasing.   The differences between the traditional view of purchasing and the strategic view are dramatic. They are summarized in Table 21.1. The differences require some significant changes in culture and behavior.

Total Cost of Ownership.   The most fundamental shift in the purchasing professional's behavior is to base purchase decisions on the total cost of ownership. Taking a total process approach (rather than a transactional approach) to quantifying the total cost of ownership will result in the identification of supplier, end-user, and joint costs which will need to be identified and measured. Many of these costs will be reduced through joint problem solving. Table 21.2 offers a sample list of elements of Total Cost of Ownership.

TABLE 21.1   Traditional Versus Strategic View of the Purchasing Process

Aspect in the purchasing process	Traditional view	Strategic view
Supplier/buyer relationship	Adversarial, competitive, distrusting	Cooperative, partnership, based on trust
Length of relationship	Short term	Long term, indefinite
Criteria for quality	Conformance to specifications	Fitness for use
Quality assurance	Inspection upon receipt	No incoming inspection necessary
Communications with suppliers	Infrequent, formal, focus on purchase orders, contracts, legal issues	Frequent, focus on the exchange of plans, ideas, and problem-solving opportunities
Inventory valuation	An asset	A liability
Supplier base	Many suppliers, managed in aggregate	Few suppliers, carefully selected and managed
Interface between suppliers and end users	Discouraged	Required
Purchasing's strategy	Manage transactions, troubleshoot	Manage processes and relationships
Purchasing business plans	Independent of end-user organization business plans	Integrated with end-user organization business plans
Geographic coverage of suppliers	As required to facilitate leverage	As required to facilitate problem solving and continuous improvement
Focus of Purchasing decisions	Price	Total cost of ownership
Key for Purchasing's success	Ability to negotiate	Ability to identify opportunities and collaborate on solutions

TABLE 21.2 Sample Checklist for Total Cost of Ownership Consideration

Category	Subcategory	Cost component
Preacquisition	Preprocurement cost	Engineering/design Supplier survey Supplier audit/site visits Product testing/technical review Regulatory compliance Market assessment Customer reviews/briefings
Acquisition	Material equipment cost	Price of material/equipment Cost of special features Shipping/handling/storage Spare parts Leased items Taxes
	New technology costs	Modification/retrofit Additional training
	Foreign acquisition costs	Foreign surtax Import duties Foreign currency risk Additional testing requirements
	Installation/start-up costs	Labor Subcontractor Special testing Construction equipment Required overhead Training Special tools Service engineering Inspection
Ownership	Operating/maintenance costs	Administration/overhead Ongoing labor Routine testing requirements Ongoing training Energy usage Preventative maintenance
	Inventory costs	Personnel required Inventory carrying costs
	Failure costs	Cost of expected down time Replacement parts
	Obsolescence costs	Energy efficiency Productivity loss
	Other costs of ownership	Environmental impact Licensing, permitting Environmental control equipment Conformance costs Standardization costs
Disposal	Disposition cost	Removal Salvage costs/value Disposal

SUPPLY-CHAIN OPTIMIZATION

The goal of a strategic purchasing function is to facilitate the performance of the supply chain. This process facilitation includes participation of the end users and suppliers. Supply-chain optimization is the *ongoing* management and continuous measurable improvement in the performance of this supply chain, generating value for *all* involved. The entire supply chain must be considered, including indirect suppliers, manufacturers, distributors, and end users. Note that the key words in this definition are:

- *Ongoing:* Supply-chain optimization is not an event, but an ongoing process
- *Measurable:* The results of supply-chain optimization are tangible benefits
- *Improvement:* The foundation of supply-chain optimization is continuous improvement
- *All:* True supply-chain optimization requires participation of all parties involved to share in the benefits

Axioms of Supply-Chain Optimization.   The Consortium for Advanced Manufacturing— International (CAM—I) group of Arlington, TX, is a not-for-profit organization that was formed in 1972 to further the cooperative research and development efforts of companies with common interests in competitive business practices and enabling technologies. In 1992, CAM—I's membership began to focus on the performance and optimization of the supply chain. They conducted extensive benchmarking on Best Practices in customer-supplier relations. They have developed axioms for successful supply-chain optimization (Figure 21.3).

The last axiom, concerning competition, was written to quell the fears of those who said that supply-chain relationships may cause suppliers to lose touch with the marketplace.

Goal of Supply-Chain Optimization.   The overriding goal of quality-focused supplier chain optimization is increased customer satisfaction through the joint (suppliers and end user) creation of value in the supply chain. On the supplier side, participation in such an initiative as supply chain optimization extends beyond the role of the account executive and includes the participation of those actually involved in the manufacturing and delivery of the product in question.

In addition, on the end user side, participation in such a venture extends beyond the Purchasing department, and includes participants from the core operating business units. In fact, while such a team effort is typically facilitated by a Purchasing individual, the team should be lead, and accountability of results assigned to, a member of the core business unit. More will be said about organizing for supplier relations later.

Supply-chain optimization creates value in the following six areas:

Quality Improvement: Continuous reduction in product variation and the ability to plan and build quality into each component and service, with measurable results.

---

1. There is a shared specific focus on satisfying their common end consumer.
2. There is an alignment of vision.
3. There is a fundamental level of cooperation, commitment to performance, and trust.
4. There is open and effective communication.
5. Decisions are made by maximizing the use of the competencies and knowledge on both sides of the relationship.
6. All stakeholders are committed to generating long-term mutual benefits.
7. There is a common view of how success is measured.
8. Both sides are committed to continuous improvement and breakthrough advancements.
9. Whenever competitive pressures exist in the environment they are allowed to exist in the extended enterprise.

FIGURE 21.3   Axioms for successful supply-chain optimization (Reprinted with permission of CAM—I. Zampino, Boykin, Doyle, Parker and CAM—1 1995.)

Cycle-Time Reduction: Continuous reduction in the time required to make and implement key decisions and perform various processes.

Cost of Poor Quality Reduction: Continuous measurement and reduction of costs associated with the prevention, inspection, and failure resulting from poor quality.

Total Cost of Ownership Reduction: Purchasing decisions based on total cost of ownership, including preprocurement, acquisition, operation, and disposal costs, rather than price alone. Continuously manage the ongoing acquisition based on the identification and elimination of root-cause cost drivers which contribute to total cost of ownership.

Technology/Innovation: Continuous identification and deployment of value-added technologies through joint planning and development.

Shared Risk: Continuous identification of opportunities to identify and share risk throughout the supply chain.

Successful supply chain optimization requires that the sourcing process operate as a single seamless entity, rather than a set of discrete processes. Members of the supply chain establish goals and work together toward these goals, which target the satisfaction of customer needs. As stated by Parker and Doyle (1995), "the goal of supply chain optimization is to have the three or more links function as one organism, where real-time decision making occurs throughout the supply chain."

ORGANIZING FOR SUPPLIER RELATIONS—INTERNAL

The following paragraphs will review the internal organizational characteristics required for the successful initiation, implementation, and ongoing nurturing of supplier relations which generate value within the supply chain. First, we will briefly review the sourcing process. Second, we will review two of the classic organization structures currently in use—Functional and Process-Based Organizations—and briefly outline the strengths and weaknesses of each. Third, we will identify 10 principles which facilitate the supply chain in highly effective purchasing organizations. Fourth, we will discuss the skill set required for strategic purchasing. Last, we will review the possible impact several contemporary organizational issues are having upon the successful management of the sourcing process.

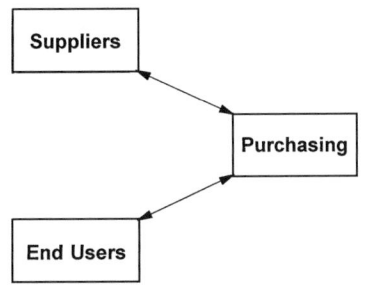

FIGURE 21.4 Traditional role of Purchasing department: managing the transaction between supplier and end users.

The Sourcing Process. As previously stated, supplier relations are defined as those tasks, activities, events, and processes required to facilitate the ongoing interface between suppliers of goods and services and the end users of those goods and services. Historically, these processes were thought to be incorporated in the buying process, and therefore Purchasing departments became collective groups of buyers, whose job was to buy the correct requisitioned item at as low a cost as possible. These relationships are represented in Figure 21.4.

Increasingly, organizations developed an understanding of the potential value-added activities in the ongoing management of the supply chain. More and more, companies are identifying the tasks, activities, events, and processes required to facilitate the ongoing interface between suppliers of goods and services and the end users of those goods and services as including:

- Quality planning
- Business planning and customer goal alignment

- Market assessments and analysis
- Customer identification
- Customer needs assessments
- Design specification determination and analysis
- Forecasting of Purchasing activity
- Consolidation of forecasted Purchasing activity
- Supplier evaluations and selection
- Establishment of supplier agreements
- Communication of supplier agreements
- Spot buying in response to emergency events
- Shipment and logistics planning and optimization
- Inventory control and optimization
- Quality control
- Accounts payable
- Value analysis
- Customer satisfaction assessments
- Quality improvement

Purchasing now becomes the facilitator of the sourcing process, requiring joint participation of purchasing, end users, and suppliers. The new purchasing relationships are represented by Figure 21.5.

The redefined role of purchasing, and the activities required to facilitate supplier relations, are such that the successful planning, control, and improvement of these processes will provide a higher value generated to the end user. Furthermore, an enterprise-wide commitment to a strategic approach to managing supplier relations would likely result in that firm gaining a competitive advantage vis-a-vis a competitor using its Purchasing department in a more traditional, nonstrategic role.

Process-Based Organizations Replacing Functional Organizations.   Given the strategic Purchasing role and the goal of supply chain optimization, how should Purchasing people be organized to enable this process to operate at an optimum level for an extended period of time? Businesses are constantly redrawing their lines within workgroups, departments, divisions, even entire companies, trying to enable productivity increases, cycle-time reduction, revenue enhance-

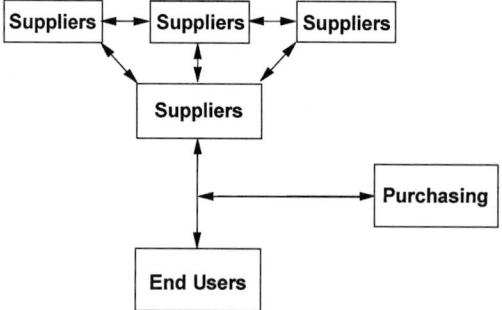

FIGURE 21.5   Revised role of Purchasing department: managing the transaction between supplier and end users.

ment, or an increase in customer satisfaction. The recent trend, often initiated by re-engineering projects or downsizing initiatives, is the shift from function-based to process-based organizations.

Function-Based Organizations. In a function-based organization, departments are established based on specialized expertise. Responsibility and accountability for process and results are usually distributed piecemeal among departments. Figure 21.6 depicts the organization of a function-based manufacturing facility.

A function-based organization typically develops and nurtures talent, and fosters expertise and excellence within the functions themselves. Therefore, it offers several long-term benefits. However, function-based organizations can result in a slow, bureaucratic decision-making apparatus, as well as the creation of functional business plans and objectives which may be inconsistent with overall strategic business unit plans and objectives. Many organizations are beginning to experiment with an alternative to the function-based organization in response to today's "make it happen fast" world. Increasingly, organizations are being rotated 90° into processed-based organizations.

Process-Based Organizations. In a process-based organization, reporting responsibilities are associated with a process and accountability is assigned to a process owner. In a process-based organization, each process is provided with the functionally specialized resources necessary. This has the effect of eliminating the barriers associated with the traditional function-based organization, making it easier to create cross-functional teams to manage the process on an ongoing basis. Figure 21.7 depicts such an organization. In this example, four processes are depicted: Sourcing; Product Development; Budgeting; and Recruiting.

Process-based organizations are usually accountable to the business unit or units which receive the benefits of the process under consideration. Therefore, process-based organizations are usually associated with responsiveness, efficiency, and customer focus.

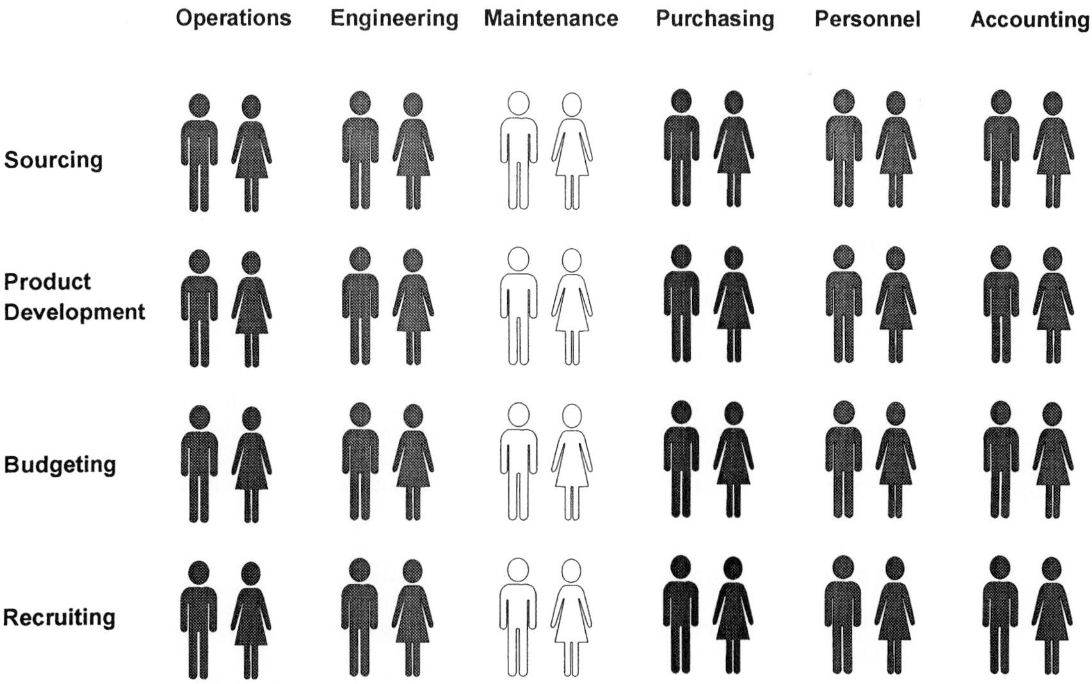

FIGURE 21.6 Function-based organization.

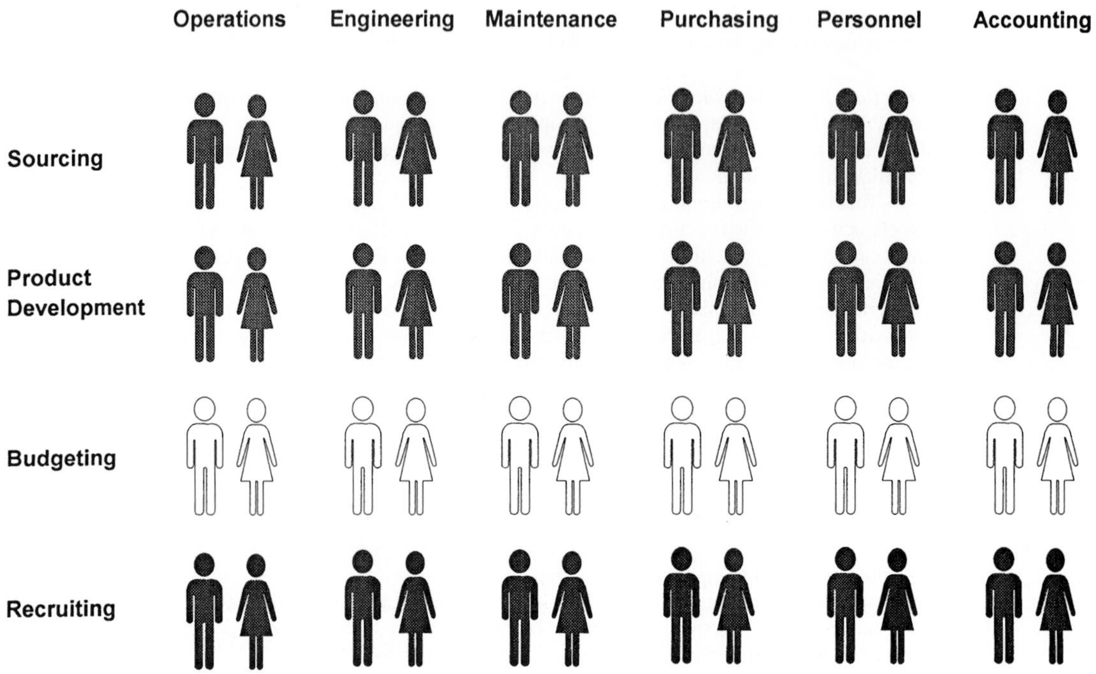

	Operations	Engineering	Maintenance	Purchasing	Personnel	Accounting
Sourcing						
Product Development						
Budgeting						
Recruiting						

FIGURE 21.7   Process-based organization.

However, over time, pure process-based organizations run the risk of diluting and diminishing the skill level within the various functions. Furthermore, a lack of process standardization can evolve, which can result in inefficiencies and organizational redundancies. Additionally, such organizations frequently require a matrix reporting structure, which can result in some confusion if the various business units have conflicting objectives.

Merging Functional Excellence with Process Orientation.   There is no exact way to organize internally a Purchasing department to nurture and develop supplier relations to an optimum level systematically. What is required, however, is an organization which identifies and captures the benefits of supply chain optimization in a responsive, customer-focused manner, while promoting and nurturing the expertise required to manage and improve continuously the Sourcing process on an ongoing basis.

This organization will likely be a hybrid of the functional and process-based organizations, with the business unit accountable for objectives, priorities, and results, and the functional department accountable for process management and improvement and resource development. Although the exact makeup of the organization will vary firm to firm, the following principles should be adhered to in organizing for supply relations.

10 Principles for Organizing for Supplier Relations

1. Recognize the Purchasing function as a strategic, highly value-added function within an organization, to be staffed by highly skilled professionals.

Purchasing value added is not in generating and managing transactions, but in facilitating ongoing relationships between commercial establishments in a way that is constantly generating value.

The Purchasing function should never be used, then, as a "dumping ground" for otherwise misplaced professionals.

2. Assign leadership within the Purchasing function to visionary, results-oriented individuals, who have both the full support of senior corporate management, as well as credibility at the operations plant level.

Purchasing management must establish and effectively communicate the sourcing vision, facilitating quantum leaps in performance of the sourcing process, and eliminating barriers, ensuring the continuous improvement of the sourcing process performance.

3. Develop purchasing strategies in alignment with business unit strategies.

The strategic value of a Purchasing department is recognized and captured only when these activities are driven by, and ultimately contribute to, a larger business unit strategy.

4. Hold the business unit management accountable for the successful implementation of the sourcing strategies.

The implementation of purchasing strategies can result in a competitive advantage. The likelihood of successful implementation is greatly enhanced when those held accountable for business strategy implementation are also the focal point for purchasing strategy implementation.

5. Hold the Purchasing management accountable for the performance and continuous improvement of the sourcing process. While the business unit is held accountable for the results of the purchasing strategy, functional management should be held accountable for the execution of the sourcing process, specifically that: the correct process is being adhered to, the process reflects industry best practices, the individuals are sufficiently trained to manage the sourcing process, and the process is measured, managed, and improved.

6. Organize cross-functional teams to manage the acquisition of goods and services.

A cross-functional team approach to strategy development helps facilitate a customer-driven, fact-based approach to the development and implementation of purchasing strategies which are consistent with overall business unit objectives.

7. Maintain an ongoing focus of the cross-functional team on the total supply chain performance, including the total cost of ownership, the identification of opportunities for increased value, and identifying and achieving competitive advantage.

The value added of a procurement strategy is realized when a total cost of ownership approach is used, as opposed to a low-price focus. A cross-functional team is best equipped to identify, measure, and manage this total cost of ownership of an acquisition.

8. Develop, implement, and manage purchasing strategies by consolidating and segmenting procurement activities across strategic business unit boundaries wherever feasible.

This principle is required if purchasing is to optimize adequately the total cost of ownership of the good or service to be acquired over an extended period of time, as well as to minimize redundant purchasing activity across an entire enterprise.

9. Maintain open, honest, and frequent communications with and between end users. This is critical for the ongoing successful identification and implementation of purchasing strategies.

Open, honest, and frequent communications are required to understand and consolidate customers' needs, as well as to measure customer satisfaction and identify and attack continuous improvement opportunities.

10. Base the development and implementation of purchasing strategies on collaboration and cooperation between business units, a decision-making process based on facts, and a measurement system whereby continuous improvement is built into the ongoing relationship between end user and supplier.

Analysis and decisions affecting the total cost of ownership of an acquisition, supplier selection, and continuous improvement initiatives must be based on fact and collaboration if the full, long-term benefits of enhancing supplier relations are to be realized.

Skill Sets Required for Purchasing Department Professionals.

As industry moves away from transaction-based Purchasing departments, the skill sets required of the

Purchasing professional have changed as well. The clerical and "win-lose" negotiating skills of the past are being replaced by leadership, facilitation, communication, consensus building, and creativity skills. Increasingly, Purchasing professionals are asked to lead teams, develop purchasing strategies which support an operation's business strategy, assist in the implementation of information technology related to purchasing, logistics, and accounts payable activities, or deliver presentations to councils of senior management. Clearly, the skill sets of the "historic purchasing model" are inadequate for these strategic activities.

Shaver (1993) offers the following set of skills and aptitudes for the Purchasing professional:

- Adaptability
- Results orientation
- Attention to detail
- Coaching and training
- Communication (nonverbal, oral, written)
- Listening
- Decision making
- Delegation
- Ethics
- Expert power (technical competencies)
- Information power
- Interpersonal
- Intellectual power (memory, formal education, creativity)
- Leadership
- Meeting management (facilitating and controlling)
- Management and supervision (monitoring individual and group behavior)
- Negotiation
- Persuasion skills
- Public speaking
- Reward power (ability to provide resources, outcomes)
- Risk orientation
- Service

Kolchin and Giunipero (1993), in a study commissioned by the Center for Advanced Purchasing Studies, identified the top 10 skills required by Purchasing professionals for the year 2000:

- Interpersonal communication
- Customer focus
- Ability to make decisions
- Negotiation
- Analytical skills
- Managing change
- Conflict resolution
- Problem solving
- Influencing and persuasion
- Computer literacy

A review of this list and Shaver's indicates that the Purchasing individual is viewed as more than just a "buyer." Facilitating ongoing relationships between and within commercial entities and identifying and capitalizing on value-added opportunities requires skills not easily found in today's marketplace for labor.

Empowerment, Outsourcing, Downsizing, Re-engineering.   Beginning in the early 1980s, corporate initiatives variously called restructuring, re-engineering, or downsizing, have resulted in sudden, and often dramatic, shifts, changes, and reductions in the work force. The causes, although somewhat complex, generally have to do with the rapid evolution in technology and the streamlining made possible by this automation and the cost pressures of a global marketplace which are forcing companies to rethink the services they provide, as well as who provides those services.

The most common themes of these changes are discussed next.

1. *Empowerment:*   Management is beginning to realize that it does not, and in fact cannot, have all the answers regarding successful supply chain management. Therefore, it is giving teams of individuals, as well as individuals themselves, greater accountability for their decision making and performance. Empowerment facilitates not only cycle-time reduction, but a reduction in the number of managers required to operate a business.

2. *Outsourcing:*   Defined as identifying and subcontracting to an outside supplier a process currently conducted in house, outsourcing is undertaken to cut costs, improve quality, or both. Generally, outsourcing is confined to utility processes (processes that are required but do not provide a competitive advantage, such as security, facility maintenance and repair, laboratory testing, income tax preparation, and legal services.

3. *Downsizing:*   A reduction in the work force, or elimination of entire departments within an organization, for the purpose of reducing costs is called "downsizing."

4. *Re-engineering:*   The fundamental change, and radical redesign, of a business process in order to achieve dramatic results is referred to as "re-engineering."

It is important to note that re-engineering and downsizing are two very different activities which are frequently confused. The focus of re-engineering is process and process redesign, while downsizing is purely a cost reduction initiative. So, although re-engineering might possibly produce downsizing as a result, the two are not the same. See Champy (1995) or Hammer and Champy (1993) for further clarification.

In organizing for supplier relations, the effects of these dramatic work force shifts can be significant, and, if not carefully implemented, can be a serious setback for supplier relations. Established strategic sourcing processes and relationships with suppliers can be significantly disrupted if care is not taken in implementing these organization changes. The focus of these restructuring initiatives should be on minimizing transactional activity, thus allowing for the reallocation of resources toward the efficient, effective delivery of strategic sourcing processes.

ORGANIZING FOR SUPPLIER RELATIONS—EXTERNAL

Once a firm is organized internally to capture the competitive advantage offered by supply-chain management, its next step is to organize the supplier base to capture these benefits. A Pareto analysis of the supply base will likely reveal that 80 percent of the potential benefits of strategic supply chain management are achievable by focusing on about 20 percent of the supply base itself. Supply-chain management requires that these vital few suppliers align and champion the process of managing and reducing total systems costs.

There will likely always be a need for some "spot" procurement, where end users need to buy an unanticipated item on a quick turnaround. Furthermore, there will likely be some acquisitions of an unusual, nonrepetitive nature, such as some engineered equipment. Not all commodities are, therefore, strong candidates for aggressive supply chain management. Supply-chain management is to be reserved for commodities deemed to be of strategic importance to the firm.

The Sourcing Strategy Model.   A recommended approach to initiating supply chain management is to analyze your historic commodity spend profile and isolate two factors of the commodities being purchased. The first factor should be the "criticality of the purchase." This will likely be a subjective assessment of a commodity's importance to the business. For example, raw materials or contract labor might be regarded as highly critical, while office supplies or tools might be assigned a lower criticality. The criticality rating of a commodity is a subjective ranking, and should therefore be assessed by a team of senior procurement management in close consultation with senior line management.

The second factor which needs to be captured is the amount of the spend for each commodity family. A petrochemical facility might have a significant expenditure on pipes, valves, and fittings, while its travel services expenditures might be relatively low. Conversely, a consulting company would likely have no amount of expenditure on pipes, valves, and fittings, but a significant spend on travel services.

After collecting this information, organize the information into four groups, as shown in Figure 21.8.

The following strategies should be applied to each quadrant to generate the optimum benefit from strategic supply chain management at a cost commensurate with value:

Quadrant 1 (low criticality/low expenditure):   Typically, these expenditures represent items of low strategic value and cost. Classic examples might include office supplies, books and publications, or food services. The acquisition of these goods and services offers little opportunity for generating competitive advantage, but can generate a high amount of nonvalue-added transaction work. This quadrant represents utility, rather than competitive advantage activities, and, therefore, is a good candidate for outsourcing. At the very least, these transactions could be, and should be, eliminated by end user direct buying through either an automated system or a mechanism such as a procurement card.

Quadrant 2 (low criticality/high expenditure):   These expenditures represent a higher strategic component than quadrant 1 and also a significant component of transaction activity within traditional purchasing functions. Classic examples of this quadrant could include pipes, valves, and fittings, miscellaneous electrical supplies, or contract labor. A total cost of ownership analysis of these commodities often reveals that significant cost components include nonstandardization of purchases, excess inventory, and insufficient planning. These commodities represent good candidates for sourcing teams to manage on an ongoing basis. However, their low complexity and

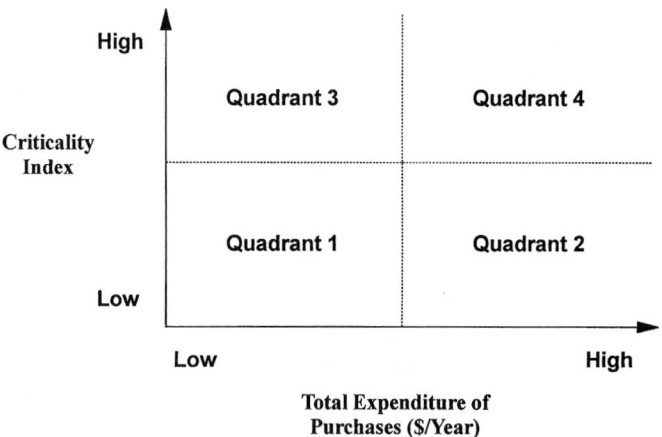

FIGURE 21.8   Classification of purchased commodities.

routine nature often limit the potential benefits of an aggressive supply chain management focus. At the very least, Material Requirement Planning (MRP) Information Systems, Electronic Data Interchange (EDI), Electronic Funds Transfer (EFT), Supply-Base Reduction, and end-user direct releases from established sourcing agreements should be used to minimize the transaction activities of these acquisitions.

Quadrant 3 (high criticality/low expenditure):   This quadrant might include highly engineered equipment, some fabricated materials, or specialized contract labor. These items can represent a high cost on an individual basis, but a low cost in aggregate due to the infrequency of their purchase. A thorough total cost of ownership analysis often reveals some cost reduction opportunities in this quadrant, especially when issues such as overspecification and nonstandardization of equipment and labor are considered. The infrequent, low aggregate value of these acquisitions makes the suppliers of these commodities poor alliance candidates. This is the one quadrant where some transaction workload may remain after an organization commits to a strategic procurement function.

Quadrant 4 (high criticality/high expenditure):   This quadrant contains the commodities which are the critical focus of a strategic procurement initiative. Traditionally, these purchases were made on a basis which prevented the supplier and the end user from interacting, thus the purchasing function served as a wedge to prevent collaboration. Strategic purchasing facilitates this end-user/supplier relationship so that potential breakthroughs of supply-chain management can be achieved. For these commodities, sourcing teams should be established, strategic suppliers should be identified and selected, and an ongoing team effort involving end users, purchasing, and the supplier should manage the commodity acquisition, and focus on creating value within the supply chain.

Benefits of the Sourcing Strategy Model.   Table 21.3 summarizes the strategies and implications of using this sourcing model.

JURAN TRILOGY AS IT APPLIES TO SUPPLIER RELATIONS

Once an organization has been established to facilitate quality supplier relations, the trilogy of Quality Planning, Quality Control, and Quality Improvement can be applied to the supply chain. The

TABLE 21.3   Implications of Sourcing Model

Quadrant	High-level strategy	Implication of strategy
I	Minimize transactions through automation, direct end-user buying, or outsourcing.	Fewer resources working on non-value added activities.
II	Focus on excess inventory to determine total cost of ownership cost drivers. Standardize, combine purchases across business units, establish long-term agreements with high-quality suppliers.	Transactions eliminated. Inventory reduced. End-user buying against established supplier agreements.
III	Focus on specifications and design to determine total cost of ownership cost drivers.	Reduced risk of overspecifying items.
IV	Focus on primary supply-chain management opportunities.	Look to capture value by working collaboratively with end users and suppliers to improve quality, reduce cycle time, and COPQ; minimize cost drivers impacting the total cost of ownership; identify technological, innovation, and risk-sharing opportunities.

relationship between the trilogy of quality processes and supplier relations is described and illustrated in Table 21.4.

Driving the managing of supplier relations is the identification and satisfying of customer needs. Application of the Trilogy processes to the supply chain fall easily into phases, executed in order, often overlapping: planning, control, improvement. The "planning phase" is concerned with identifying, understanding, and implementing a sourcing strategy which meets those customer needs; the "control phase" with managing process performance and the performance of the suppliers engaged in the process; and the "improvement phase" with identifying and capitalizing on value-added opportunities within the supply chain.

Following is a detailed explanation of the activities and deliverables of these three phases of managing for supplier relations.

Planning for Supplier Relations

> Honda's success on this continent [North America] is a direct result of the company's overall philosophy of manufacturing…manufacturing's success depends on two groups: the people who make the products, and the suppliers that provide the parts and raw material from which the products are made. (Kevin Fitzgerald 1995)

Planning for supplier relations is the activity of identifying customer needs and analyzing and developing a sourcing strategy to meet those needs. One of the key deliverables of the planning process is an initial model detailing the customer's total cost of ownership of the subject commodity. Thus, data collection and analysis will also be required throughout the planning process. The focus of this planning process is the identification of the appropriate customer and assessment of the current and future needs of these customers for the commodity in question (Table 21.5). Additionally, as the output of the planning process is a recommended sourcing process flow, a thorough understanding of the supply industry structure, dynamics, and trends is essential.

TABLE 21.4   Juran Trilogy Applied to Supplier Relations

Process	Definition	Process	Definition
Quality Planning	The activity of developing the products and processes required to meet customer needs.	Planning for supplier relations	The activity of identifying customer needs and analyzing and developing a sourcing strategy to meet those needs.
Quality Control	The activity of evaluating actual performance, comparing actual performance to goals, and taking action on the difference.	Control for supplier relations	The activity of evaluating suppliers' performances, selecting the vital few suppliers capable of optimizing performance, and the measurement of supplier performance.
Quality Improvement	The activity of raising quality performance to unprecedented levels.	Improvement for supplier relations	The activity of identifying and acting upon sourcing process improvement opportunities.

The planning process requires:

1. Early customer involvement to identify current and future sourcing needs
2. Extensive research and data collection regarding the alternative processes available to satisfy these needs.

Most successful sourcing planning has followed a methodology similar to the following:

Step 1.   Document the organization's historic, current, and future procurement activity.
In the absence of planning for supplier relations, it is assumed that purchasing is generally handled in a reactive business process which satisfies immediate, local operational needs. The documentation of the historic, current, and anticipated purchase activity across an organization's various business units enables that organization to take the first, necessary step toward achieving purchasing leverage; synergies within and between organization business units; and a strategic, collaborative, proactive approach to managing the sourcing process.

Available tools: data collection, trend analysis.

Step 2.   Identify a commodity from the procurement activity which represents both high expenditure and high criticality to the business (quadrant IV commodities).
A simple Pareto analysis will often reveal the vital few commodities which drive an organization's purchasing needs and costs. Focusing resources on these vital few commodities will enable an organization to begin to capture the value of supply-chain management early.

Available tools: pareto analysis, data histograms, stratification, management presentations

Step 3.   For this commodity, assemble a cross-functional team. The team includes representatives of the customer and of company functions—technical, purchasing, quality, and financial, for example. The team's mission is to define the customer's sourcing need for this commodity and to develop a sourcing strategy which will meet this need.

Available tools: brainstorming, team building, flow charting.

Step 4.   Determine the sourcing needs of the customer through data collection, survey, and other needs assessment activities. This is the critical step which, if not properly and thoroughly conducted, can derail any well-intentioned cross-functional team. It is often fatal to assume that the customer's needs are obvious. Extensive data collection through surveys, customer visits, and focus groups will pay off later on.

Available tools: brainstorming, data collection, flow charting, cause-and-effect diagrams, force field analysis, hypothesis formulation, and testing.

Step 5.   Analyze the supply industry's structure, capabilities, and trends.
Once the customer needs have been identified and validated, an industry analysis is required. It is the supply chain, and not the purchase itself, which will ultimately delight the customer with fitness for use and value. Thus, the various supply chains available, and their performance and cost structures, must be understood. This is an extensive research phase of the planning process, and might require the team to split temporarily into several subteams.

TABLE 21.5   Inputs/Outputs to the Planning Process

Customer needs	Recommended sourcing strategy
Industry data	Consolidated buying
Expenditure data	Supplier base reduction
Cost of ownership data	Total cost of ownership model defined

Available tools: industry data collection and analysis, flow charting, benchmarking, process capability analysis.

Step 6.   Analyze the cost components of the commodity's total cost of ownership.

This, too, will require extensive data collection and analysis and even benchmarking to identify how others have managed this commodity. This model of total cost of ownership will be redefined, refined, and optimized throughout the life of the commodity management team.

Available tools: data collection and analysis, brainstorming, flow charting, cause-and-effect diagrams, histograms, Pareto analysis.

Step 7.   Translate the customer needs into a sourcing process which will satisfy the customer and provide the opportunity to manage and optimize the total cost of ownership.

The customer needs as identified in step 4 will need to be mapped into the various alternative sourcing processes identified in step 5. An optimal sourcing strategy can be determined by optimizing the total cost of ownership, based on the results of step 6. Translation requires extensive dialog and feedback to identify and gauge fitness for use of the sourcing strategy.

Available tools: data collection and analysis, brainstorming, flow charting, cause-and-effect diagrams, histograms, Pareto analysis, force field analysis, customer and supplier visits.

Step 8.   Obtain management endorsement to transfer the sourcing strategy into operation. Implement it.

This strategy should now be transferred from the cross-functional team to operations management for implementation. The "selling job" which is often required to facilitate change is reduced by the ongoing involvement on the team of those affected. The strategy should include, at a minimum, the following: scope (for example, global, regional, local); terms and condition of agreement; and method of end-user release. A dry run or a pilot test should be conducted to demonstrate feasibility of concept. Once the pilot has been implemented and the feasibility of concept demonstrated, the revised process should proceed through a site-by-site acceptance test and implementation. Some training will be required.

Available tools: executive briefing, pilot testing, process debugging, acceptance testing, training.

The planning phase of the sourcing initiative in all likelihood resulted in some consolidation of the supplier base, where cross-divisional or multiple business units identified opportunities to exercise economies of scale by consolidating similar purchasing activity with fewer suppliers.

Here is an illustrative example of the planning process applied to the sourcing of personal computers at a financial institution.

> Data analysis indicates that most PCs are purchased at local computer stores from the winner of a three-bid competition. As a result, there is little standardization in the hardware and software used at the institution. PCs are historically purchased in small quantities, generating significant work for the Purchasing, Accounts Payable, and Information Technology support groups, who acquire, pay for, install, maintain, and manage the equipment.
>
> Analysis reveals that purchase price is actually a fraction of the total cost of ownership of the personal computer. Equipment support, software evaluation, training, and inventory control also represent significant, hidden costs.
>
> In this case, the sourcing process recommendation is to standardize the equipment and software, negotiate purchase and service agreements with a single computer distributor with wide geographical coverage, and limit purchases to semi-annual bulk acquisitions. Several local charities are identified for the donation of obsolete equipment. The supplier with the agreement now has specific key performance indicators by which its performance can be measured and monitored.

Control for Supplier Relations.   Control is applied to supplier relations in evaluating supplier performance and selecting the vital few suppliers capable of optimizing performance. As in planning, the focus of control must be the satisfaction of customer needs. However, as a result of the completed planning process, several criteria for performance evaluation and measurement are already in place. The purpose of control is to maintain acceptable performance. Applied to supplier relations, the purpose of control is to maintain the level of customer satisfaction at the level defined in the planning process.

The suppliers identified in the planning process are typically those suppliers which *can* perform the revised sourcing process. A thorough, ongoing evaluation conducted by a cross-functional team further narrows the supplier base and helps facilitate the selection of those few suppliers who *will* be able to optimize the total cost of ownership of the commodity. Therefore, it is in the application of control that the evolution begins from the traditional purchasing approach toward supply-chain management.

Control is a process requiring:

- Clearly defined supply-chain quality goals established in planning
- Extensive, ongoing data collection and evaluation of the performance of the suppliers against these supply-chain quality goals
- Corrective action where required

Most successful sourcing control processes follow a methodology similar to the following:

Step 1. Create a Cross-Functional Team.   The cross-functional control team includes customer, purchasing, and operation personnel. Its mission is the ongoing management, measurement, and evaluation of the performance of the supply-chain process established by the planning team during the planning phase. The team will initially need to identify quality goals and key performance indicators. Extensive customer involvement with the team should be expected.

Available tools: brainstorming, team building, flow charting, data collection, management presentation.

Step 2. Determine Critical Performance Metrics.   Performance metrics will have been proposed in the planning phase. However, the control team will need to identify and establish processes for capturing and reporting this information. Extensive supplier involvement should be expected in this step.

Available tools: data collection, flow charting, check sheet, run chart, scatter diagrams, process capability indices.

Step 3. Determine Minimum Standards of Performance.   In addition to critical performance metrics, the team establishes *minimum* standards for suppliers before they are considered for further strategic development. These standards would likely include several financial, legal, and environmental considerations. Some minimum acceptable quality standards might also be proposed such as percent defective, warranty performance, and delivery considerations. These minimum standards, along with the critical performance metrics established in step 2, are communicated to both the customer and supplier community.

Available tools: brainstorming, data collection and analysis; management, supplier, and customer presentation.

Step 4. Reduce the Supplier Base.   The team eliminates suppliers unable to achieve the minimum performance requirements, and shifts activity to suppliers who do achieve those performance standards. Through the application of the minimum standards of performance, the control process offers another opportunity for reducing the supplier base.

Available tools: data collection and analysis, management presentation.

Step 5. Assess Supplier Performance.   Based on actual supplier performance, begin the process of the ongoing evaluation and assessment of the performance of the remaining suppliers. This typically involves evaluations of supplier quality systems now in place, supplier capacity and capability, and fitness for use of the commodity being supplied.

Supplier assessment comprises three separate but interrelated assessments, undertaken by the cross-functional team. These assessments ensure conformance to quality and performance standards and establish a baseline for the improvement process.

Assessment 1. Supplier Quality Systems Assessment.   This assessment evaluates the quality systems the supplier currently has in place. It requires a visit to the supplier site by an evaluation team or by a third party who will certify the quality system as acceptable. This assessment should evaluate the supplier's:

1. Focus on customer's needs
2. Management commitment to Total Quality Management
3. Defined, documented, and fully implemented quality system
4. Employee empowerment in terms of monitoring their own work for defect
5. Use of fact-based, root-cause analysis to investigate and correct quality problems
6. Programs to encourage and evaluate quality improvement with their suppliers
7. Commitment to continuous improvement in all phases of its operation

Cost considerations may favor reliance on a third-party supplier certification instead of an evaluation by employees of the purchaser. Where this is done, it is important that the end-user organization clearly understand what this certification does and does not include.

The standards for supplier certification most often referred to are:

1. The ISO 9000 standards (ISO 9001, 9002, 9003), designed as models and guidelines of the minimum requirements for an effective quality system. (See Section 11, The ISO 9000 Family of International Standards.)
2. The ISO 14000 standards, designed as models and guidelines of the minimum requirements for an effective environmental system. (See Section 11, The ISO 9000 Family of International Standards.)
3. Quality System Requirements QS-9000, developed by the Chrysler/Ford/General Motors Supplier Quality Requirements Task Force. It is based on ISO 9000 standards, to which may be added automotive interpretations and further requirements (for example, continuous improvement and advanced product quality planning).
4. The Malcolm Baldrige Assessment, designed for applicants of the U.S. Malcolm Baldrige National Quality Award. It evaluates the Process Systems in place and the underlying organization and cultural issues of leadership, degree of empowerment, and utilization of information and information technology in place to facilitate quality planning, quality control, and quality improvement. (See Section 14, Total Quality Management.)

Assessment 2. Supplier Business Management.   This assessment evaluates the supplier's capability as an ongoing business entity to meet the end user's current and future business needs. This includes assessment of the supplier's current and future financial and operating performance. This assessment should evaluate the supplier with respect to:

• Research and development initiatives to ensure consistency with its customers' needs and future plans
• Cost structure to ensure financial health
• Production capacity to ensure ongoing capacity to produce and distribute the required goods and services
• Information technology to evaluate willingness and capability to initiate information-sharing initiatives such as Electronic Data Interface (EDI) and Electronic Funds Transfer (EFT)

The assessment includes measurement of such indicators as debt-to-equity ratio, percent of profit reinvested in the business, inventory-to-sales ratio, employee turnover statistics, and capacity utilization.

Assessment 3. Supplier Product Fitness for Use.   This assessment evaluates the fitness for use of the product or service being supplied. The focus is on quality, delivery, and service. Specifically, this assessment should evaluate:

- Conformance to customer requirements
- Process capability (Cpk) (see Section 6, Process Management)
- Key performance indicators

The assessment includes measurements of such indicators as the following:

- Percent of nonconforming products shipped
- Cycle times of key processes
- Customer satisfaction
- Identified and measured cost of poor quality

Available tools: supplier site visits, data collection and analysis, third-party evaluations.

Improvement for Supplier Relations.   The improvement phase includes:

- The management, measurement, and continuous improvement of the sourcing process
- The expansion of control and initiation of continuous improvement within the supply chain itself to ensure value creation

These improvement initiatives build on the foundations of quality, total cost of ownership, and supply-chain management already established in the planning and control phases. Fundamental to improvement in the performance of the entire supply chain is that trust has been established between all parties in the entire supply chain—from suppliers through end users. The objective of the improvement phase is to develop a supply chain which acts as a single entity, develops common goals, formulates real-time decision making, measures performance through a single set of key performance indicators, and is collectively responsive to the needs of the end user.

With trust as the foundation, supply-chain management and optimization can proceed. This sense of trust cannot be achieved by a single act of signing a long-term contract or by prominently displaying a banner indicating a commitment to quality. It must be demonstrated by behaviors and actions demonstrated over an extended period of time. As the climate of cooperation grows, the degree of trust between all supply-chain participants becomes deeper, and opportunities for value creation, joint problem solving, and innovation are identified and realized.

Five Tiers of Progression.   In the control phase, the end user and suppliers have identified and flow charted the entire supply chain. The continuous improvement phase generally progresses through five levels of cooperation: (1) joint team formation, (2) cost reduction, (3) value enhancement, (4) information sharing, and (5) resource sharing.

Level 1. Joint Team Formation.   The improvement phase begins with the establishment of a joint (end user/supplier) team. Although the team could have several objectives, the initial focus should be on:

- Alignment of goal
- Analysis of the supply-chain business process
- Identification and remediation of chronic problem

Goal alignment ensures that each link in the supply chain develops goals and objectives and proposes initiatives whose focus is the needs of the end user. Furthermore, goal alignment and the activities associated with it are a natural first step in developing the synergies and trust required for further supply-chain development.

In conducting the business process analysis of the supply chain, the team begins to identify the elements of the chain and to collect data to measure its performance. This data collection should focus on the areas of the supply chain which have a high probability of generating quality problems, such as excessive cycle time, rework, and scrap, or which are likely to create customer dissatisfaction.

Supply-chain business process analysis represents the initial steps of identifying the chain (typically using flow charting) and collecting data which describe the performance of this supply chain. This data collection phase should focus on the areas of high probability of quality problems, such as cycle time, rework, scrap, or customer dissatisfaction.

Chronic problem identification and remediation offers a preliminary opportunity to work collaboratively on problem solving in this joint-team environment. This offers a classic opportunity for a quality improvement team with membership from the various members of the supply chain. The team's efforts will likely result in near-term process improvement and enhanced customer satisfaction, and offer an opportunity for collaboration and trust to be nurtured within the chain itself.

See Section 5, The Quality Improvement Process, for further discussion of the quality improvement methodology.

Level 2. Cost Reduction.   Level 1 initiatives help create a culture of trust and collaboration between supplier and end user, especially as the result of the work of joint problem-solving teams. The teams were established to identify and gather the "low-hanging fruit," that is, reduce the occurrence of chronic problems in their joint business processes which are relatively easy to solve, once identified. Level 2 requires an approach to process improvement in more depth, often involving suppliers to the supplier or customers of the end users. Proactive managing of the supply chain begins at this point to replace the bilateral relationship between end user and supplier.

A COPQ study of the supply chain provides powerful guidance for organizations engaged in cost reduction. The costs are usually sorted into three categories:

- External failure costs (that is, warranty, customer dissatisfaction, recall costs)
- Internal failure costs (that is, scrap, rework, rejected raw material, downtime costs)
- Appraisal costs (that is, inspection, testing, verification costs)

For significant concentrations of COPQ revealed in the supply chain, joint teams are established to reduce those costs, project by project.

See Section 8, Quality and Costs, for further discussion of COPQ analysis.

As activities advance to a higher level, the activities of the lower levels continue. For example, as the chain moves into level 2 and begins measuring and managing cost reduction opportunities, the tools and initiatives of level 1 continue. This accumulating effect continues throughout the five levels.

Level 3. Value Enhancement.   As the teams begin reducing COPQ, the supply-chain itself begins to function as a single business process, rather than as a set of separate ones. At this point, the team needs to flow chart the activity of the supplier chain and evaluate the value added by each link in the chain. Two questions addressed at this stage are: "Does this step add value?" and "What would happen if we were to skip this step?" The nonvalue-added steps are identified and eliminated.

Level 4. Information Exchange.   At this point in the supply-chain improvement evolution, what was traditionally treated as confidential information is being routinely shared and more widely distributed throughout the chain. Furthermore, electronic commerce tools such as EDI, Internet and Intranet applications, and groupware applications such as Lotus Notes are facilitating the transfer of information, the collaboration of ideas, and real-time decision making.

Level 5. Resource Sharing.   In the latter stages of supply-chain management and improvement, the "walls" that traditionally separated departments, divisions, and companies have been eliminated. Fewer are working in corporate silos; the supply chain is beginning to function as a single process—involving personnel from several different suppliers within the chain, from the customer's organization and the end user. Personnel within the chain are routinely collaborating on ideas and improvement opportunities, and performance is continuously measured. Personnel from the various suppliers within the supply chain are often co-located with their customers to further facilitate this collaboration.

At the highest level of supply-chain management, the extent of data, resource, and risk sharing has increased to a dramatic level. Not only are personnel co-located with their customers, but technology plans and risk-taking initiatives and investments are shared throughout the supply chain, and benefits and losses are jointly apportioned. A seamless supply-chain process begins to emerge, generating value for customers as well as suppliers.

Agile Supply-Chain Implementation. *The Wheels: A Vehicle for Dialogue* is a comprehensive set of supply-chain assessment tools to help organizations, working with their suppliers, to identify, manage, and achieve breakthroughs in the performance of the supply chain. The tools, developed by a team at the Rochester Institute of Technology, are described by Graham (1996). A key requirement within the supply chain is identified as agility, implying the quick and resourceful manufacturing system required in today's highly competitive, global economy.

Assessments made by using these tools enable a dialogue between customers and suppliers which will facilitate performance improvement of the supply chain. The tool set is built around six questions. By examining customers, suppliers, and the supply chain in light of these six questions, we can develop a thorough picture of the supply chain and its component companies. The six questions are:

- Are we developing and producing the right things?
- Are we producing the right things well?
- Are we delivering the right things quickly enough?
- Are we creating the best operational climate?
- Are we collectively anticipating and improving?
- Are we all becoming more successful?

Further information regarding these tools is available from The Center for Integrated Manufacturing Studies, Rochester Institute of Technology, Rochester, NY.

Results of Supply-Chain Management. Parker and Doyle (1995) report the impressive results of some supply-chain optimization efforts selected from their experience:

- *Quality:* 20 to 70 percent reduction in variability
- *Cycle-time:* 30 to 90 percent reduction
- *Waste:* 15 to 30 percent reduction in cost of poor quality
- *Technology:* R&D resources increased by a factor of 3 or more by utilizing the entire supply chain
- *Risk:* overall reduction of hazards/obstacles through sharing

The Chrysler Experience. Dyer (1996) describes the experience and the significant benefits generated through sustained collaboration, trust, and joint problem solving between Chrysler and its suppliers. In 1989, Chrysler began a program to identify and develop supplier partners. At that time, their production supplier base was 2500 suppliers. The first step in supplier partnering required an aggressive supply-base reduction, and, between 1989 and 1994, they reduced their base by 54 percent. As of 1996, more than 90 percent of the remaining suppliers are assured of business for the life (and frequently beyond) of the automobile model for which they were supplying parts. The average term of a supplier's contract was reported to be 4.4 years, more than twice what it had been in 1989, when it was 2.1 years. Furthermore, Chrysler had replaced the detailed supply contracts with more flexible oral agreements.

According to Dyer, Chrysler collaborated with its remaining supply base to find ways to reduce the cost of making automobiles, while assuring suppliers that any savings would be shared among the Chrysler/supplier participants. For each of its five vehicle platforms (large cars, small cars, minivans, Jeeps, and trucks) Chrysler organized itself into cross-functional teams. Each team chose suppliers very early in the vehicle's concept-development phase, and gave the suppliers near-total responsibility for a given component or system's design.

Chrysler also used a concept called "target costing." Starting with a prediction of the market price of the vehicle, the team worked backwards to establish the allowable cost of every system, subsystem, and component. Using target costing, Chrysler managed to completely change its supplier relationships from adversarial price buying to collaborative cost reduction.

Chrysler also instituted a cost-reduction program called the "supplier cost reduction effort (SCORE)." SCORE enabled suppliers to identify and formally submit cost-improvement suggestions, which in turn would be reviewed and endorsed by Chrysler management. The results of this strategic shift in supplier relations are impressive.

- *Product development cycle time:*   The time Chrysler needs to develop a new vehicle is approaching 160 weeks, down from 234 weeks during the 1980s.
- *Cost reduction:*   Since its inception in 1990, Chrysler has implemented over 5000 suggestions from its suppliers as part of the SCORE Program, and generated savings in excess of $1.7 billion.
- *Reduced procurement transaction costs:*   Since 1988, Chrysler has reduced the number of buyers by 30 percent, and sharply increased the dollar value procured by each of the remaining buyers. This has been accomplished largely by supply-base reduction and the near elimination of the competitive bidding process.
- *Revenue and profit enhancement:*   Since 1989, Chrysler's U.S. market share has increased from 12.2 to 14.7 percent. Furthermore, their profit per vehicle produced has increased from $250 to $2,110 per vehicle.
- *Continuous improvement:*   The long-term supplier relationships, tied to measurable performance improvement, target costing, and cost reduction resulted in a developing culture of continuous improvement throughout the supply chain.

The next step for Chrysler and its suppliers is to roll out this supplier-relations culture to additional tiers in the supply chains. Suppliers to Chrysler will be expected to replicate programs such as early supplier involvement, target costing, cross-functional teams, and a SCORE-type program with their suppliers. Eventually, the entire supply chain will be involved in the proactive, collaborative effort.

Legal Issues in Supplier Relations.   "Law" consists of those principles, practices, rules, statutes, and requirements of behavior, formally adopted and enforced by a society as a whole, so that an orderly society can exist. Law can be subdivided into public law (those areas dealing with the relationship between individuals and the state as a whole) and private law (those areas dealing with the relationship between individuals or groups of individuals with each other). The laws of contracts are contained within private law.

The contract remains a primary facilitator of commercial transactions conducted within the business world. Therefore, a broad understanding of contract law is essential for any professional whose job is to manage and facilitate supplier relations. Within general contract law is the statutory law of the Uniform Commercial Code (UCC). The UCC governs the sale of goods, and is applicable in 49 of the 50 United States (Louisiana being the only exception).

International transactions, growing in volume and strategic importance, are governed by the Convention of the International Sale of Goods (CISG).

In the evolving world of supplier alliances and supply-chain management, there still remains the nagging question of how formal the relationships between supplier and customer should be. Several leaders in the application of supply-chain management, such as Honda of America, have effectively streamlined and simplified this relationship. Honda's typical agreement contains no mention of length of time of the agreement and no mention of quantity or dollar value. Basically, it is an agreement to buy from the supplier without restrictions regarding specific part or length of time of the agreement. Honda and its major suppliers have effectively created contracts which establish a relationship, without the restrictive or punitive language typical of the traditional contract.

However, other companies still cling to the traditional lengthy contracts containing indemnifications, prohibitive warranties, force majeure clauses, termination clauses, and other restrictive clauses which require extensive legal reviews, revisions, and negotiations.

To facilitate an ongoing relationship of collaboration, cooperation, and trust, the shorter, simplified contract is preferred as easier to establish and administer. This approach may not be practical in the early stages of a developing supplier relationship. However, as the relationship is developed, and trust within the supply chain grows, the parties will find that the contract review process can be streamlined, and lengthy contract documentation can also be simplified and, in some cases, eliminated.

Several leaders in supply-chain management have almost totally abandoned the use of lengthy contracts, and have opted for the use of their standard purchase order form, containing broad, simple language that allows the supply chain to be optimized with few restrictions imposed by legal requirements. The key is to establish the relationship and aggressively commit to and facilitate an environment of collaboration, cooperation, and trust. Increasingly simplified contracts reflect the growing mutual trust and sharing of risk and reward which characterize the successful supply chain.

REFERENCES

Bhote, K. R. (1987). *Supply Management,* AMA Publications Division, New York.

Carlisle, J. A. and Parker, R. C.(1989). *Beyond Negotiation,* John Wiley & Sons, Chichester, England.

Champy, James (1995). *Re-engineering Management,* Harper Business, New York.

Deming, W. Edwards (1981). Seminar notes for "Japanese Methods for Productivity and Quality," Course No. 617, W. Edwards Deming, Washington, DC.

Deming, W. Edwards (1986). *Out of the Crisis,* Massachusetts Institute of Technology, Center for Advanced Engineering Study, Cambridge, MA.

Dunaud, Michel (1995). "How the French Arms Industry Mastered Quality." J. M. Juran, ed., *A History of Managing for Quality,* Quality Press, Milwaukee, WI, p. 421.

Dyer, Jeffrey H. (1996). "How Chrysler Created An American Keiretsu," *Harvard Business Review,* July-August, vol. 74, no. 4, pp. 42–56.

Fitzgerald, Kevin (1995). "For Superb Supplier Development," *Purchasing Magazine,* September 21, pp. 32–40.

Graham, Robert (1996). *The Wheels: A Vehicle for Dialogue,* Rochester Institute of Technology, Rochester, NY.

Hammer, Michael, and Champy, James (1993). *Re-engineering the Corporation,* Harper Business, New York.

Juran, J. M. (1995). "A History of Managing for Quality," *Quality Progress,* August 1995, pp. 125–129.

Kolchin, Michael, and Giunipero, Larry (1993). *Purchasing Education and Training Requirements and Resources,* Center for Advanced Purchasing Studies, Tempe, AZ.

Parker and Doyle (1995). "Strategic supply-chain Management Program of the Consortium for Advanced Manufacturing—International," Rochester Institute of Technology, Rochester, NY.

Shaver, Robert (1993). "Potential Sources of Power, Influence and Control in Your Career," teaching notes. Madison School of Business, University Of Wisconsin, Madison, WI.

Tully, Shawn (1995). "Purchasing's New Muscle," *Fortune Magazine,* vol. 131, no. 3, Feb. 20, pp. 75–83.

Zampino, Boykin, Doyle, Parker & CAM-I (1995). *Axioms of Supply-chain Management,* Consortium for Advanced Manufacturing International, Arlington, TX.

SECTION 22
OPERATIONS

Frank M. Gryna

INTRODUCTION

The word *operations* as used in this handbook encompasses two areas: manufacture in the manufacturing sector and backroom activities in the service sector. In manufacturing industries, operations are those activities, typically carried out in a factory, which transform material into the final product. In service industries, operations are those activities which process customer transactions but which do not involve direct contact with external customers (e.g., backroom activities such as customer order preparation and payment processing). These two industry sectors have their own special needs. The discussion in this section covers both the planning and the execution of operations activities.

Activities that involve direct contact with external customers are clearly of high priority. In this handbook, such activities are discussed in Section 25, Customer Service, and in the group of industry sections, Sections 27 to 34.

I will use *product* to denote goods or services.

QUALITY IN THE OPERATIONS FUNCTION OF THE FUTURE

For many industries—manufacturing and service—emerging factors demand different approaches to quality in the twenty-first century. This galaxy of factors includes

1. *Demand for lower levels of defects and errors:* As products and processes have become more complex, new "world-class quality" levels are increasingly common. For many products, levels of 1 to 3 percent are being replaced by 1 to 10 parts per million. Also, many processes must meet "good manufacturing practices" and other forms of regulation.

2. *Emphasis on reduced inventory levels:* Under the "just-in-time" (JIT) production system, the concept of large lot sizes is challenged by reducing setup time, redesigning processes, and stan-

dardizing jobs. The results can be smaller lot sizes and substantial reductions in inventory. Such a system relies on a process that is capable of meeting quality requirements because little or no inventory exists to replace defective product. Thus JIT is not viable unless product quality is acceptable. Schonberger (1996) explains JIT and the impact on product quality.

3. *Time-based competition:* Performance is now measured not only by costs and quality but also by responsiveness to customer needs. This responsiveness means offering more products (i.e., product features) at lower cost and in less time. The time parameter puts pressure on the product development process, which can result in inadequate review of new designs. Increasingly, however, managers realize that quality problems can be on a critical path that will slow down the delivery process. Stalk and Hout (1990) examine a variety of issues, not only issues of quality, on the impact of time-based competition.

4. *Impact of technology:* Technology (including computer information systems) is clearly improving the quality of goods and services by providing (a) a wider variety of outputs and (b) more consistent output. One of the effects has been to reduce the emphasis on direct labor efficiency in operations. The infusion of technology makes some jobs more complex, thereby requiring extensive job skills and quality planning; technology also makes other jobs less complex but may contribute to job monotony.

5. *Agile competition:* This term refers to competition based on a group of correlated concepts that includes responding to constantly changing customer opportunities, being able to change over from one product to another quickly, manufacturing goods and producing services to customer order in arbitrary lot sizes, customizing goods and services for individual customers, and drawing on the expertise of people and facilities within a company or among groups of cooperating companies. Clearly, the impact on both product features and defect levels will be far reaching. Goldman, Nagel, and Preiss (1995) describe the concept and include examples. To cite one example from the apparel industry, a blouse and skirt were designed, cut to customer size order, printed, sewn, and distributed—at a trade show.

6. *Outsourcing:* Many organizations have reduced their total personnel by transferring complete functions to a supplier (outsourcing). In one survey, 86 percent of firms used outsourcing in 1995 versus 58 percent in 1992 (*Business Week,* April 1, 1996). In a financial services firm of about 8000 people, 74 percent are "contract" personnel, most of whom come from one supplier. Extensive steps are taken to ensure the quality of the services. Examples of activities for outsourcing include manufacturing operations, billing, service, and human resource tasks. Clearly, steps must be taken to ensure the quality of these tasks. Bettis, Bradley, and Hamel (1992) examine the implications of outsourcing on competitiveness and offer cautions and suggestions. The impact of contract workers and outsourcing on quality has been selected as a research project by the National Science Foundation.

These "lean manufacturing" factors, which are not independent, suggest that quality during operations can no longer focus on inspection and checking. We must recognize these factors as we pursue universal—and intoxicating—principles such as customer focus, continuous improvement, and employee empowerment in the operations function.

Schonberger (1996) explores the future of world-class manufacturing; Godfrey (1995) identifies critical issues in service quality.

PLANNING FOR QUALITY DURING OPERATIONS

Increasingly, planning for quality *before* the execution of operations is seen as essential. International standards such as the ISO 9000 and ISO 14000 series provide a minimum framework for planning (for elaboration, see Section 11, The ISO 9000 Family of International Standards). These standards cover important matters such as process control, inspection and testing, material control, product traceability, control of measuring equipment, control of nonconforming product, quality documentation, process environmental conditions, and the impact of processes on the external environment.

Responsibility for Planning.   The responsibility for this planning varies by industry. In the mechanical and electronics industries, the work is usually performed within the manufacturing function by a specialist department (e.g., manufacturing engineering, process engineering). For process industries, the work is usually divided into two parts. Broad planning (e.g., type of manufacturing process) is performed within the research and development function; detailed planning is executed within the manufacturing function. Similarly, the service industries show variety in assigning the planning responsibility. For example, in backoffice operations of the financial services industry the local operations manager handles the planning, whereas in the fast-foods industry planning for food preparation is usually handled by a corporate planning function.

The main factors influencing the decision on responsibility are the complexities of the products being made, the anatomy of the manufacturing process, the technological literacy of the work force, and the managerial philosophy of reliance on systems versus reliance on people.

Some industrialized countries delegate only a small amount of manufacturing planning to departmental supervision or to the work force. In the United States, this situation is largely a residue of the Taylor system of separating manufacturing planning from execution. This system gave rise to separate departments for manufacturing planning.

The Taylor system was proposed early in the twentieth century, at a time when the educational level of the work force was low, while at the same time products and manufacturing technology were becoming more complex. The system was so successful in improving productivity that it was widely adopted in the United States. It took firm root and remains as the dominant approach to manufacturing planning not only interdepartmentally but within departments as well.

Times have changed. A major premise of the Taylor system, i.e., technological illiteracy of the work force, is obsolete because of the dramatic increase in the educational levels of the work force. Many companies recognize that extensive job knowledge resides in the work force and are taking steps to use that knowledge. Manufacturing planning should be a collaborative effort in which the work force has the opportunity to contribute to the planning. In the United States, this collaboration is slow-moving because of the widespread adoption of the Taylor system and the vested interests that have been created by that approach.

Some companies are taking dramatic organizational steps to integrate quality matters into manufacturing planning. In one case, a separate quality department was eliminated, and the personnel and their activities were merged within the research and engineering department (Kearney 1984). A formal "manufacturing plan of control" was established for each operation by analyzing the material and process variables that affected key product properties. This document was prepared by a team of people from research and engineering (including quality professionals) and various areas of manufacturing. Each product grouping made use of such a team.

INITIAL PLANNING FOR QUALITY

Planning starts with evaluating emerging technologies for operations, a review of product designs, determining the importance of product characteristics, documenting processes with process diagrams, and correlating process variables with product results.

Emerging Technologies.   Sometimes an organization is faced with evaluating emerging operations technologies that it must develop concurrently with overall business planning. When this is the case, a number of issues arise. These include compatibility of the technology with existing operations, difficulties in launching new products, flexibility to accommodate volume and model mix changes, personnel requirements, and of course, the investment required. A four-step approach (Figure 22.1) is presented by Scharlacken (1992). This approach starts with a multidisciplinary team representing all the groups that will be affected by the new technology. The team develops a 3- to 5-year *technology profile* that reflects management's strategy to grow, maintain, or harvest the key product lines. In evaluating alternative technologies (task 2), note subtask 3. This subtask calls for modeling a proposed process on a computer using simulation software. The simulation reveals information about process characteristics such as output, reliability, bottlenecks, and downtime—before

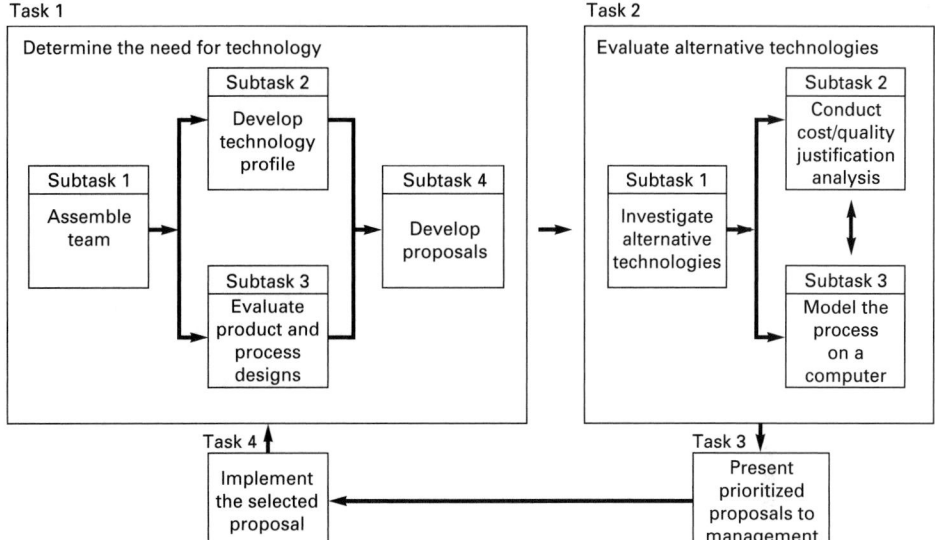

Task 1
Determine the need for technology

Subtask 1 — Assemble team

Subtask 2 — Develop technology profile

Subtask 3 — Evaluate product and process designs

Subtask 4 — Develop proposals

Task 2
Evaluate alternative technologies

Subtask 1 — Investigate alternative technologies

Subtask 2 — Conduct cost/quality justification analysis

Subtask 3 — Model the process on a computer

Task 4 — Implement the selected proposal

Task 3 — Present prioritized proposals to management

FIGURE 22.1   The approach to MTP. (*Adapted from Scharlacken 1992.*)

resources are committed to the process. This four-step approach, developed for manufacturing technology planning, also applies to the service sector.

Review of Product Designs Prior to Operations.

In both the manufacturing and service sectors, there is a clear advantage to having a new product design reviewed for feasibility by operations personnel before the design is finalized for the marketplace. In practice, the extent of such a review varies greatly—from essentially nothing ("tossing it over the wall" to the operations people) to a structured review (using formal criteria and follow-up on open issues). For physical goods, the design requirements are summarized in a product specification that is examined by a *design review* process (see also Section 19, under Designing for Reliability). The emphasis is on the evaluation of the product design for the adequacy of field performance. For backoffice operations in the service sector, requirements for a new service product may be described in a *service-level agreement.* This agreement summarizes the type and amount of service to be provided in the backoffice to satisfy the needs of the customer. Hart (1995) describes the concept of *internal guarantees,* e.g., a promise by one part of an organization to deliver a good or service to the complete satisfaction of an internal customer at the risk of incurring a monetary or other type of penalty.

Design review must include an evaluation of producibility to cover the following operational matters:

1. *Clarity of all requirements.*

2. *Relative importance of various product characteristics.*

3. *Design for manufacturability:*   This technique focuses on simplifying a design to make it more producible. The emphasis is on reducing the total number of parts, the number of different parts, and the total number of manufacturing operations. This type of analysis is not new—*value engineering* tools have been useful in achieving design simplification [see, for example, Cooper and Slagmulder (1997)]. What is new, however, is the computer software available for analyzing a design and identifying opportunities for simplifying assembly products. Such software dissects the assembly step by step, poses questions concerning parts and subassemblies, and provides a summary of the number of parts, the assembly time, and the theoretical minimum number of parts or subassemblies. Use of such software enables the designers to learn the principlesfor ease of manufacturing analogous to reliability, maintainability, and safety analyses. In

one example, the proposed design of a new electronic cash register was analyzed with design for manufacturabilty (DFM) software. As a result, the number of parts was reduced by 65 percent. A person using no screws or bolts can assemble the register in less than 2 minutes—blindfolded. This simplified terminal was put onto the marketplace in 24 months—a record. Such design simplification reduces other sources of quality problems during manufacture.

4. *Process robustness:*   A process is *robust* if it is flexible, easy to operate, and error-proof and its performance will tolerate uncontrollable variations in factors internal and external to the process. Such an ideal can be approached by careful process planning. Snee (1993) provides examples of actions that can be taken to create robust processes. See also Section 47, under Taguchi Off-Line Quality Control, for a discussion of the Taguchi approach for achieving robust processes to minimize process variation

5. *Availability of processes to meet requirements:*   In manufacturing industries, this means processes that have the capability to manufacture products with basic and special characteristics. Specification limits on these characteristics usually have important technical and economic aspects to evaluate. In service industries, backoffice processes are needed with the capability to produce accurate results often within a specified time.

6. *Identification of special needs,* e.g., handling, transportation, and storage during manufacture.

7. *Availability of measurement to evaluate requirements:*   In manufacturing, this may involve basic and special quality information equipment. The service sector often requires measurement of time, e.g., waiting time for service and elapsed time to complete the service.

8. *Special skills required of operations personnel.*

Specific criteria should be developed for each of these matters. This review of the product design must be supplemented by a review of the *process* design, which is discussed later in this section. These reviews provide an early warning to anticipate difficulties during operations.

Relative Importance of Product Characteristics.   Planners are better able to allocate available time and money where they will do the most good when they are well informed about the relative importance of the diverse characteristics of the product. Two useful techniques are the identification of critical items and the classification of characteristics of the product.

Identification of Critical Items.   *Critical items* are those features of a product which require a high level of attention to ensure that all requirements are achieved. At one company, part of a procedure to identify "quality-sensitive parts" uses specific criteria such as part complexity and high-failure-rate parts. For each such part, special planning for quality is undertaken, e.g., supplier involvement before and during the contract, process capability studies, reliability verification, and other activities.

Classification of Characteristics.   Under this system, the relative importance of individual features or properties of a product is determined and indicated on drawings and other documents. The classification can be simply "functional" or "nonfunctional" or can include several degrees of importance. An example of the latter is a system featuring four classes of seriousness: critical, major, minor, and incidental. The classification uses criteria based on the impact of the quality characteristic on safety, operating failure, performance, service, and manufacture. The input data are derived from study of the part and its application, field and test data, reliability design analysis, warranty experience, and past experience on similar designs. Shenoy (1994) explains how various customer needs relate to product control characteristics in terms of strong, medium, or weak relationships (see Table 22.1). Somerton and Mlinar (1996) explain how to obtain, organize, and prioritize customer-based data to determine key product and process characteristics. Many tools such as quality function deployment and failure mode, effects, and criticality analysis are employed in their process.

The classifications should be made by personnel with sufficient background in the functioning of the product. For most products, this must include technical personnel from the product development function. However, they frequently voice two objections to spending time on classifying characteristics:

TABLE 22.1 Relationship Matrix: Final Product Control Characteristics

Customer needs	GSM	Moisture	Brightness	Tensile index	Burst index	Tear index	Stretch	Acidity	Coating
Grade	@		0	Δ	Δ	Δ			0
Thickness	@								0
Smoothness			Δ						@
Whiteness				@					0
Ink drying property		0						@	Δ
Dimensional stability		0		Δ	Δ	0	@		
Folds				@	0	@			
Durability					@	@	@		

Note: @, strong; 0, medium; Δ, weak.
Source: Shenoy (1994).

1. "All the characteristics are critical." The realities, however, are that the multitude of characteristics inevitably requires setting priorities for manufacturing and inspection efforts. Setting these priorities requires knowledge of the relative importance of characteristics. If this knowledge is not provided by the development engineer, the decisions will, by default, be made by others who have less background in the design.

2. "The size of the tolerance range already provides a classification of relative importance." In fact, the criticality depends not so much on the allowable range of dimension as on the effect a substantial departure from that range might have on the function of the assembled system. For example, say the dimension under study is 2.000 cm ±. 010 cm. The design engineer may be asked to predict the functional effect on the system if the part dimension were 2.020 cm (a variation from target of *twice* the design tolerance). A factor of 2.0 provides a significant enough step for the engineer to form an opinion on the importance of the characteristic.

A classification approach also can be applied to the manufacturing process using the process capability index as a criterion (see below under Process Capability: The Concept). Product-process combinations can then be studied to identify high-risk areas, e.g., critical-critical, critical-major, and major-critical.

The classification of characteristics often leads to a useful dialogue between the Product Development and Manufacturing Planning Departments before the start of production. For example, a characteristic was classified by the designer as "incidental" but required a costly process to meet the specification range. After discussions, the design engineer increased the range and also reclassified the characteristics. The expanded range permitted the substitution of a less costly process. For further discussion of classifying characteristics, see Section 3, under Design for Critical Factors and Human Error.

In the service sector, an overnight delivery service has not only identified but also assigned importance weights to 12 service quality indicators (Table 22.2). These measures are tracked every day, both individually and in total.

Process Diagram.   Understanding the process can be aided by laying out the overall process in a flow diagram (similar diagrams use the terms *map, logic,* or *blueprint*). Several types are helpful.

One type of flow diagram shows the work paths followed by the materials through their progression into finished product. Planners use such a diagram to divide the flow into logical sections called *workstations.* For each workstation they prepare a formal document listing such items as operations to be performed, sequence of operations, facilities and instruments to be employed, and the process conditions to be maintained. This formal document becomes the plan to be carried out by production supervision and work force. It serves as the basis for control activities by the inspectors. It also becomes the standard against which the process audits are conducted. An example of a flow diagram for a coating process is shown in Figure 22.2.

TABLE 22.2 Federal Express Service Quality Indicators

Indicator	Weight	Indicator	Weight
Abandoned calls	1	Missed pickups	10
Complaints reopened	5	Missing proofs of delivery	1
Damaged packages	10	Overgoods (lost and found)	5
International	1	Right-day late deliveries	1
Invoice adjustments requested	1	Traces	1
Lost packages	10	Wrong-day late deliveries	5

Source: American Management Association (1992).

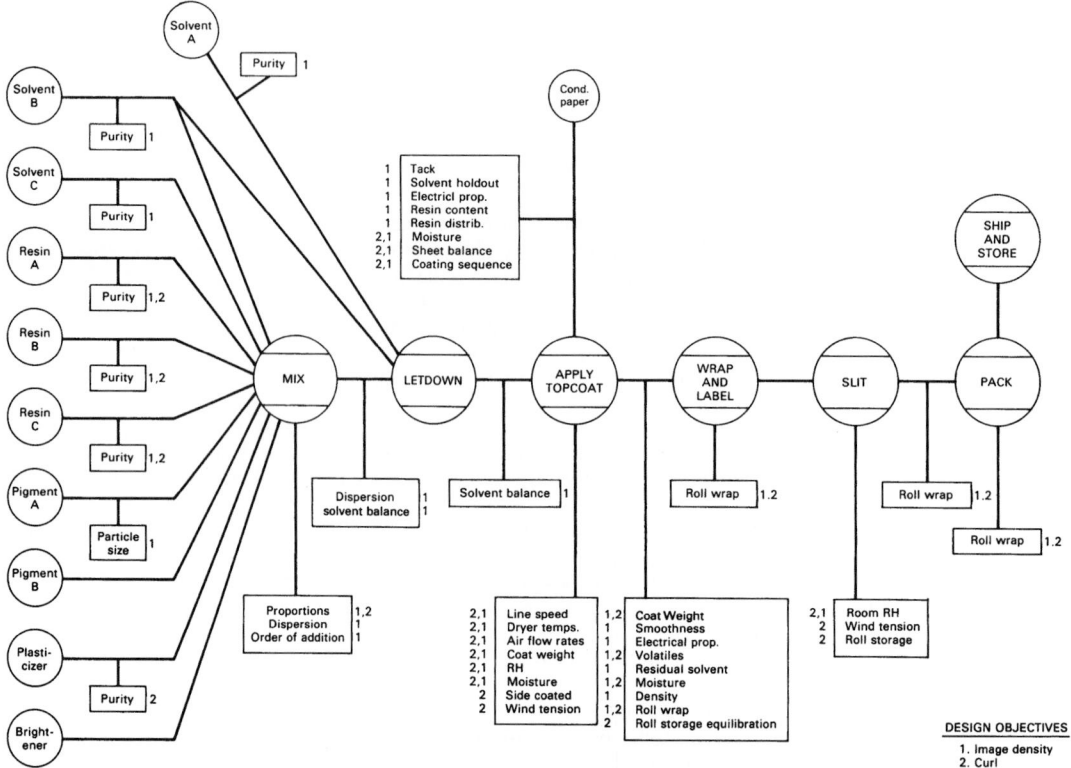

FIGURE 22.2 Strategic plan of control. Product and process analysis chart (P-PAC).

From the service sector, Figure 22.3 shows a flow diagram (called a *blueprint*) for processing a transaction at a discount brokerage house. Note the separation of customer contact activities ("tangible service evidence") from the backroom activities below the "line of visibility." The symbol *F* identifies "fail points"—those steps likely to cause problems and which require special attention through extra staffing, facility layout, or other means. Also note that service time standards are shown for selected activities.

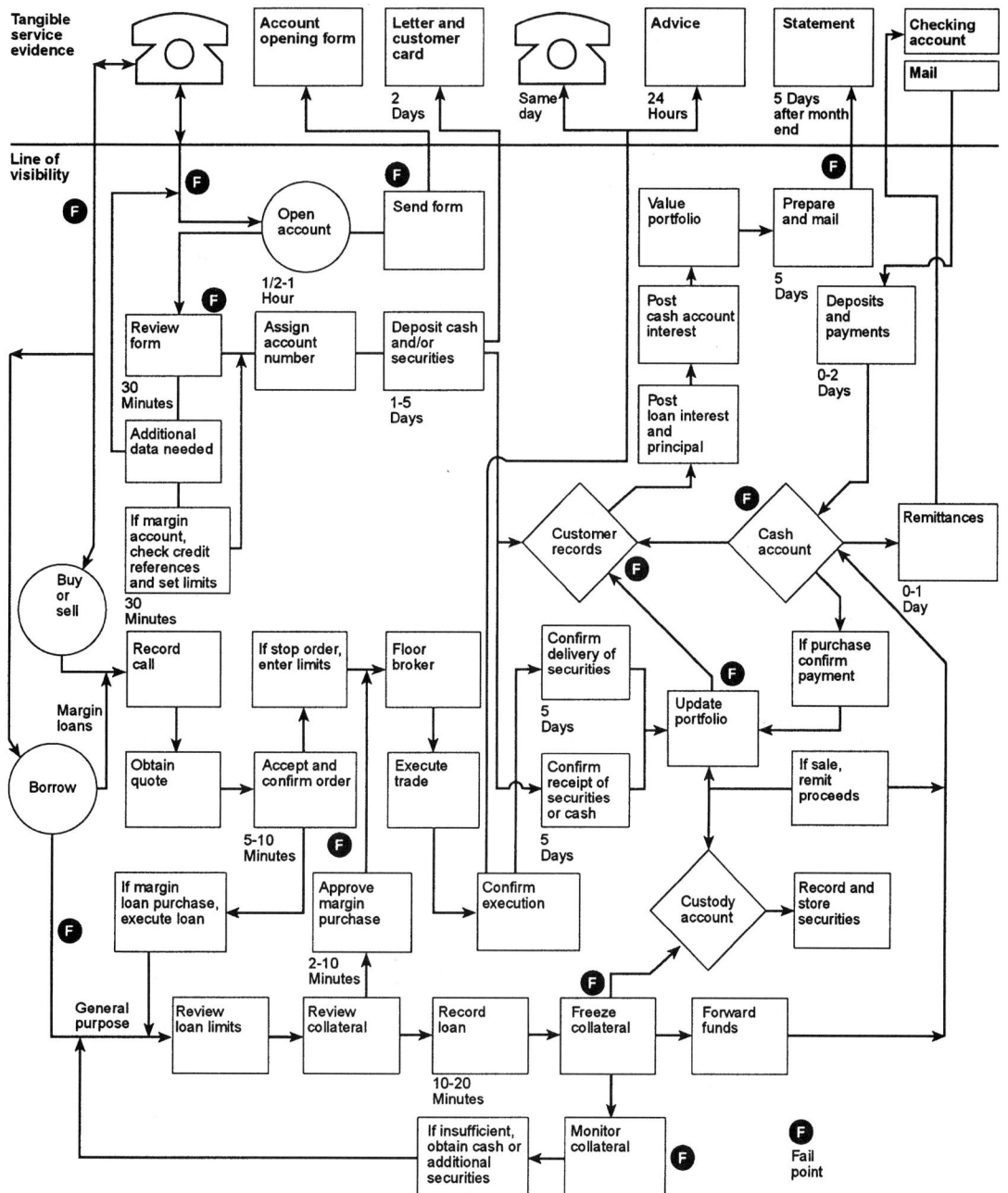

FIGURE 22.3   Blueprint for a service delivery system of a discount brokerage operation. (*Shostack 1984.*)

The flow diagram in Figure 22.4 shows the major functions and key activities in a billing process. The diagram shows not only the process flow in time (top to bottom) but also the flow across organizational boundaries (left to right). Note that each block in the flow diagram is numbered to reference more detailed flow diagrams or work instructions that describe the activity (Juran Institute 1995).

Correlation of Process Variables with Product Results.   A critical aspect of planning during manufacture is to discover, by data collection and analysis, the relationships between process variables or parameters and product results. Such knowledge enables the planner to specify various controls on the variables to achieve the specified product results. In Figure 22.2, process variables are shown in a rectangle attached to the circle representing the operation; product characteristics are listed in a rectangle between operations, at the point where conformance can be verified. Some characteristics (e.g., coat weight) are both a process variable and a product characteristic. For each control station in a process, designers identify the numerous control subjects over which control is to be exercised. Each control subject requires a feedback loop made up of multiple process control features. A process control spreadsheet helps to summarize the detail. An example is shown in Figure 22.5. For elaboration, see Juran (1992, p. 286). For a thorough discussion of planning for quality, including the identification of critical control points during manufacture, see Clark and Milligan (1994). They apply many useful quality tools to the manufacture of a simple product—honey.

Determining the optimal settings and tolerances for process variables sometimes requires much data collection and analysis. Carpenter (1982) discusses a case involving a statistical analysis of data on 33 parameters to pinpoint the key process variables in a copper ore roasting operation. Dodson (1993) presents a procedure, with tables, to determine the optimal target value for a process with upper and lower specification limits where the economic value of the product is considered.

The consequences of a lack of knowledge (of the relationship between process variables and product results) can be severe. In electronic component manufacturing, some yields are low and will likely remain that way until the process variables are studied in depth. In all industries, the imposition of new quality demands (e.g., reduction in weight of automotive components) can cause a sharp

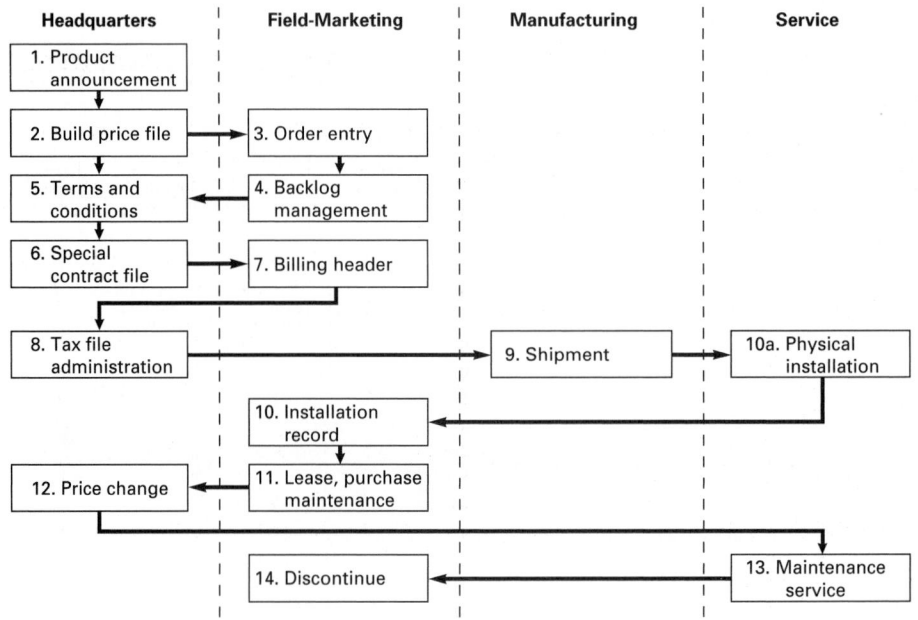

FIGURE 22.4   Billing process.

PROCESS CONTROL FEATURES/CONTROL SUBJECT	UNIT OF MEASURE	TYPE OF SENSOR	GOAL	FREQUENCY OF MEASUREMENT	SAMPLE SIZE	CRITERIA FOR DECISION MAKING	RESPONSIBILITY FOR DECISION MAKING	
Wave solder conditions: Solder temp.	°F	Thermo-couple	505°F	Continuous	N/A	510°F reduce heat 500°F increase heat	Operator	
Conveyor speed	Ft/Min (fpm)	fpm meter	4.5 fpm	1/hour	N/A	5 fpm reduce speed 4 fpm increase speed	Operator	
Alloy purity	% con-tamin-ants	Lab chemical analysis	1.5% max.	1/month	15 grams	At 1.5% drain bath, replace solder	Process Engineer	

FIGURE 22.5   A process control spreadsheet. (*Juran 1992.*)

rise in scrap (and hence in costs) because not enough is known about the process variables to adapt promptly to the new demands.

Only upper management can supply the missing essentials, which consist of

1. The budget for personnel needed full time to assist by analyzing existing data, determining the need for additional studies, designing the experiments, collecting the new data, analyzing, and so on.

2. The budget for training in the quality disciplines. The full-time analysts should, of course, have this training in depth. In addition, it is helpful for the process engineers to become knowledge-able as well. The necessary training programs are widely available.

The return on these investments is in the form of higher yields, higher productivity, lower costs, and better quality.

Some industries must meet explicit government regulations concerning manufacturing practices, and these must be recognized during manufacturing planning. An example is the good manufactur-ing practices (GMP) regulations in health-related industries.

PROCESS CAPABILITY: THE CONCEPT

Process capability is the measured, inherent reproducibility of the product turned out by a process.

Basic Definitions.   Experience has taught us that each key word in this definition must itself be clearly defined.

Process: This refers to some unique combination of machine, tools, methods, materials, *and peo-ple* engaged in production. The output of the process may be a physical good, such as an integrated circuit chip or a chemical; the output may be a service product, such as a credit card statement or answers provided on a consumer hot line.

Capability: This word is used in the sense of a competence, based on tested performance, to produce quality products.

Measured: Process capability is quantified from data which, in turn, are the results of measurement of work performed by the process. The measurement may be made on a physical property such as the pH value of a chemical. The measurement on a service product may be the time required to generate a credit card statement.

Inherent reproducibility: This refers to the product uniformity resulting from a process that is in a state of statistical control, e.g., in the absence of time-to-time "drift" or other assignable (special) causes of variation.

Product: The measurement is made on the product (goods or service) because it is product variation, which is the end result, that we use to quantify process capability.

Machine capability versus process capability: Some practitioners distinguish between these two terms. *Machine capability* refers to the reproducibility under one set of process conditions (e.g., one operator, homogeneous raw materials, uniform manufacturing practice). *Process capability* refers to the reproducibility over a long period of time with normal changes in workers, materials, and other process conditions.

Uses of Process Capability Information.   Process capability information serves multiple purposes:

1. Predicting the extent of variability that processes will exhibit. Such capability information, when provided to designers, provides important information in setting realistic specification limits.
2. Choosing, from among competing processes or equipment, that which is best to meet the specifications.
3. Planning the interrelationship of sequential processes. For example, one process may distort the precision achieved by a predecessor process, as in hardening of gear teeth. Quantifying the respective process capabilities often points the way to a solution.
4. Providing a quantified basis for establishing a schedule of periodic process control checks and readjustments.
5. Testing theories of causes of defects during quality improvement programs.
6. Serving as a basis of quality performance requirements for purchased product or equipment. In certifying suppliers, some organizations use a capability index (see below) as one element of certification criteria. In these applications, the value of the capability index desired from suppliers can be a function of the type of commodity being purchased.

These purposes account for the growing use of the process capability concept.

Process Patterns.   The concept of process capability can be better understood by an examination of the usual process patterns encountered. To make this examination, we can measure a sample, summarize the data in a histogram, and compare the result against the specification limits.

Typical histograms are shown in Figure 22.6. The examination has a three-part focus:

1. *Centering of the histogram:*   This defines the aim of the process.
2. *Width of the histogram:*   This defines the variability about the aim.
3. *Shape of the histogram:*   For most characteristics, a normal or bell-shaped curve is expected. Any significant deviation from the normal pattern has a cause that, once determined, can shed much light on the variability in the process. For example, histograms with two or more peaks reveal that multiple "populations" have been mixed together, e.g., different suppliers of material or services.

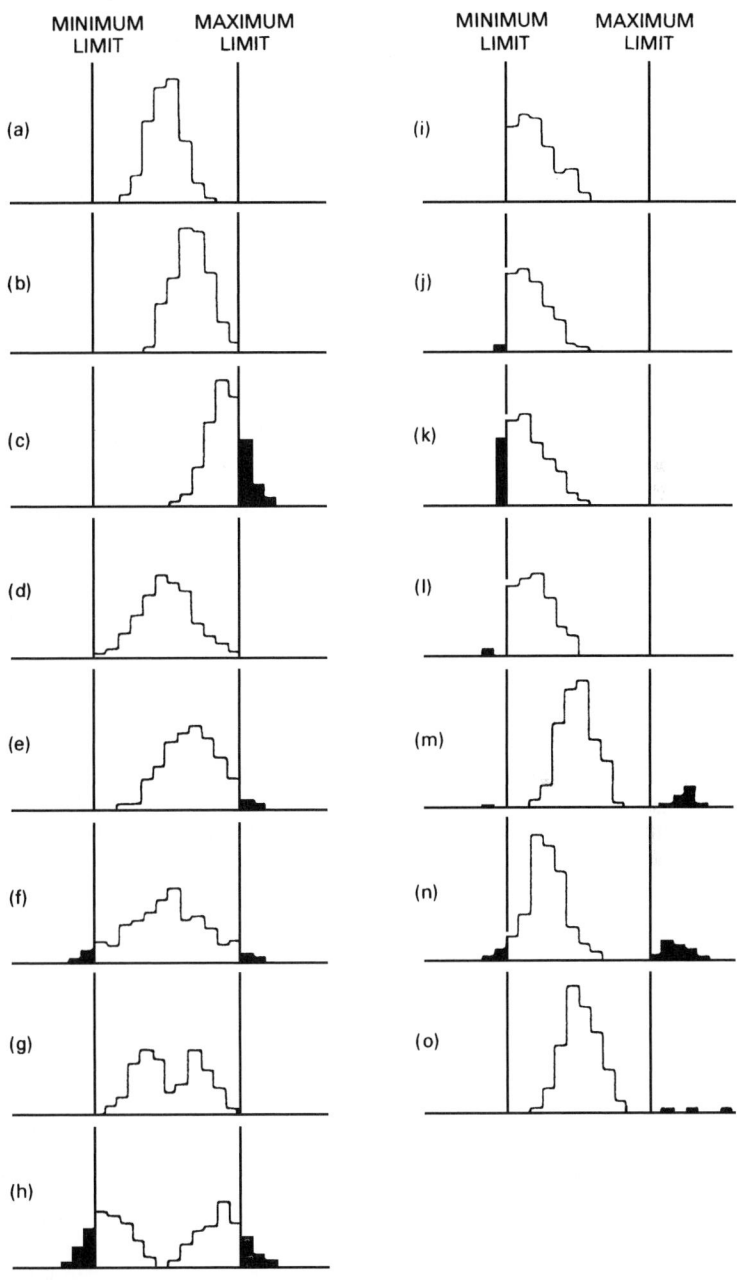

FIGURE 22.6 Histogram distribution patterns.

Histograms and chronological plots of process data indicate several reasons why some processes are not capable of holding specification limits. These are

1. *The inherent variability of the process is too large for the proposed specification limits:* Assuming that the process is in statistical control, the only courses of action are to revise the process, expand the limits, or live with a certain level of defectives.
2. *The process is misdirected:*   Here the planner must provide the work force with the means to evaluate the extent of misdirection and to make compensating adjustments in the process.
3. *The measurement process itself is inadequate.*
4. *There is process drift:*   Here the need is to quantify the amount of drift in a given period of time and to provide means for resetting the process to compensate for this drift.
5. *There are cyclic changes in the process:*   We must identify the underlying cause and either remove it or reduce the effect on the process.
6. *The process is erratic:*   Sudden changes can take place in processes. As the capability studies quantify the size of these changes and help to discover the reasons for them, appropriate planning action can be taken:
   a. Temporary phenomena (e.g., cold machine coming up to operating temperature) can be dealt with by scheduling warming periods plus checks at the predicted time of stability.
   b. More enduring phenomena (e.g., change due to new material supply) can be dealt with by specifying reverification at the time of introducing such change.

Note that the ability of the process to produce quality products consists of two different abilities:

1. The ability to achieve the desired average value (often called the *target* or *nominal specification*). This ability is evaluated by comparing the actual average with the target.
2. The ability to reproduce results consistently. This ability is evaluated by quantifying the width of the histogram (e.g., in terms of 6σ; see below). This "process capability" is compared with the specification in order to judge the adequacy of the process.

Process Mixture.   A common obstacle to using the inherent capability of a process is that for reasons of productivity, product data from several processes are combined. Examples of this are widespread: multicavity plastic molding, multiple-unit film deposition for electronic components, and multiple-head filling of containers. What these processes have in common is a multiplicity of "machines" mounted on a single frame. The multiple character of these producing sources super-imposes a stream-to-stream variation that materially affects the ability of the process to meet the specifications.

In such cases, any conventional sampling of product ends up with data that are a composite of two different sources of variation:

1. The stream-to-stream variation, traceable to differences in the mold cavities, spindles, heads, and so on.
2. The within-stream variation, which characterizes a single "pure" process.

To quantify the stream-to-stream variation requires that the product from different streams (e.g., cavities, spindles, molds, or heads) be segregated. Once segregated, the data for each stream can be treated in the conventional manner. Tarver (1984) presents procedures for process capability studies of multiple-stream processes.

An example of such mixture of data from a service industry is presented in Figure 22.7. Data were plotted to analyze turnover time in rooms in a laboratory at Brigham and Women's Hospital (Laffel and Plsek 1989). *Turnover time* was defined as the time between the moment all catheters and sheaths are removed from one patient and the time local anesthetic is injected into the next patient. Simply collecting data yielded some surprises: The mean time was 78 minutes (45 minutes had been the usual estimate), and the variation ranged from 20 to 150 minutes. In one part of the analysis, data were stratified by room, and histograms on turnover time were plotted by room (see Figure 22.7). Note what we learn when the total data are stratified by room: Room 1 had a shorter mean time and much less variation than

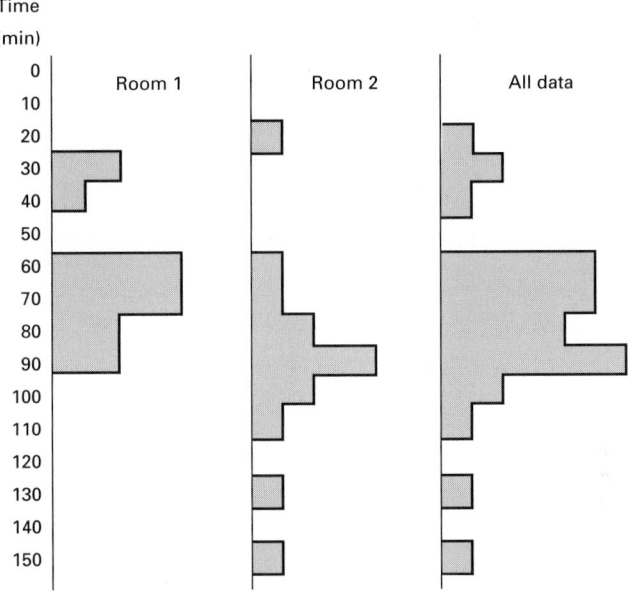

FIGURE 22.7 Room turnover times. (*Laffel and Plsek 1989.*)

room 2. Further analysis showed that when a nurse called for the next patient before the preceding case was completed, turnover time was relatively short. No one had been aware, until the data were recorded and the analysis made, that the timing of the call was a critical determinant of turnover time.

Standardized Formula. The most widely adopted formula for process capabilities is

$$\text{Process capability} = 6\sigma$$

where σ is the standard deviation of the process under a state of statistical control, i.e., under no drift and no sudden changes.

If the process is centered at the nominal specification and follows a normal probability distribution, 99.73 percent of production will fall within $\pm 3\sigma$ of the nominal specification.

Some industrial processes do operate under a state of statistical control. For such processes, the computed process capability of 6σ can be compared directly with specification tolerances, and judgments of adequacy can be made. The majority of industrial processes, however, do exhibit drift and do exhibit sudden changes. These departures from the ideal are a fact of life, and the practitioner must deal with them.

Nevertheless, there is great value in standardizing on a formula for process capability based on a state of statistical control. Under this state, the product variations are the result of numerous small variables (rather than being the effect of a single large variable) and hence have the character of random variation. It is most helpful for planners to have such limits in quantified form.

The standardized formula (process capability $= 6\sigma$) assumes a normal probability distribution. This is often the case, but it is not universally true. For example, dimensions that are close to a physical limit, such as the amount "out of round" (where a value of zero is desired), tend to show "skewed" distributions. In such cases, $\pm 3\sigma$ does not include 99.73 percent of the population. Whether the distribution is normal or not, it is useful to analyze capability graphically as a way to gain understanding of the distribution that is difficult to achieve with numerical analysis alone. (See Frequency Distribution and Histogram, below.)

Relation to Product Specification.   A major reason for quantifying process capability (i.e., process variation) is to be able to compute the ability of the process to hold product specifications. For processes that are in a state of statistical control (see below), a comparison of 6σ to the specification limits permits ready calculation of percentage defective by conventional statistical theory. See Section 44, under Continuous Probability Distributions. The comparison of process capability with specification limits leads to some broad plans of action (see Table 22.3).

Capability Index.   In most processes, not only are there departures from a state of statistical control but the process is not necessarily being operated to secure optimal yields; e.g., the average of the process is not centered between the upper and lower tolerance limits. To allow for these realities, planners try to select processes with the 6σ process capability well within the specification range. The two factors are expressed in a *capability index* C_p:

$$C_p = \frac{\text{specification range}}{\text{process capability}} = \frac{\text{USL} - \text{LSL}}{6\sigma}$$

where USL is the upper specification limit, and LSL is the lower specification limit

Figure 22.8 shows four of many possible relations between process variability and specification limits and the likely courses of action for each. Note that in all these cases the average of the process is at the midpoint between the specification limits.

Table 22.4 shows selected values of C_p and the corresponding level of defects assuming that the process average is midway between the specification limits. A process that is just meeting specification limits (specification range$=\pm3\sigma$) has a C_p of 1.0. The criticality of many applications and the reality that the process average will not remain at the midpoint of the specification range suggest that C_p should be a least 1.33. Note that the C_p index measures whether the process variability can fit within the specification range. It does not indicate if the process is actually running within the specification, because the index does not include a measure of the process average (this is addressed below under Process Performance Measurement).

Three capability indices commonly in use are shown in Table 22.5. Of these, the simplest is C_p. The higher the value of any of these indices, the lower will be the amount of product that is outside specification limits.

Pignatiello and Ramberg (1993) provide an excellent discussion of various capability indices. Bothe (1997) provides a comprehensive reference book that includes extensive discussion of mathematical aspects. These references explain how to calculate confidence bounds for various process

TABLE 22.3   Action to Be Taken

	Product meets specifications		Product does not meet specifications	
	Process variation small relative to specifications	Process variation large relative to specifications	Process variation small relative to specifications	Process variation large relative to specifications
Process is in control	Consider cost reduction through less precise process; consider value to designer of tighter specifications.	Closely monitor process setting.	Process is "misdirected" to wrong average. Generally easy to correct permanently.	Process may be misdirected and also too scattered. Correct misdirection. Consider economics of more precise process versus wider specifications versus sorting the product.
Process is out of control	Process is erratic and unpredictable. Investigate causes of lack of control. Decision to correct based on economics of corrective action.		Process is misdirected or erratic or both. Correct misdirection. Discover cause for lack of control. Consider economics of more precise process versus wider specifications versus sorting the product.	

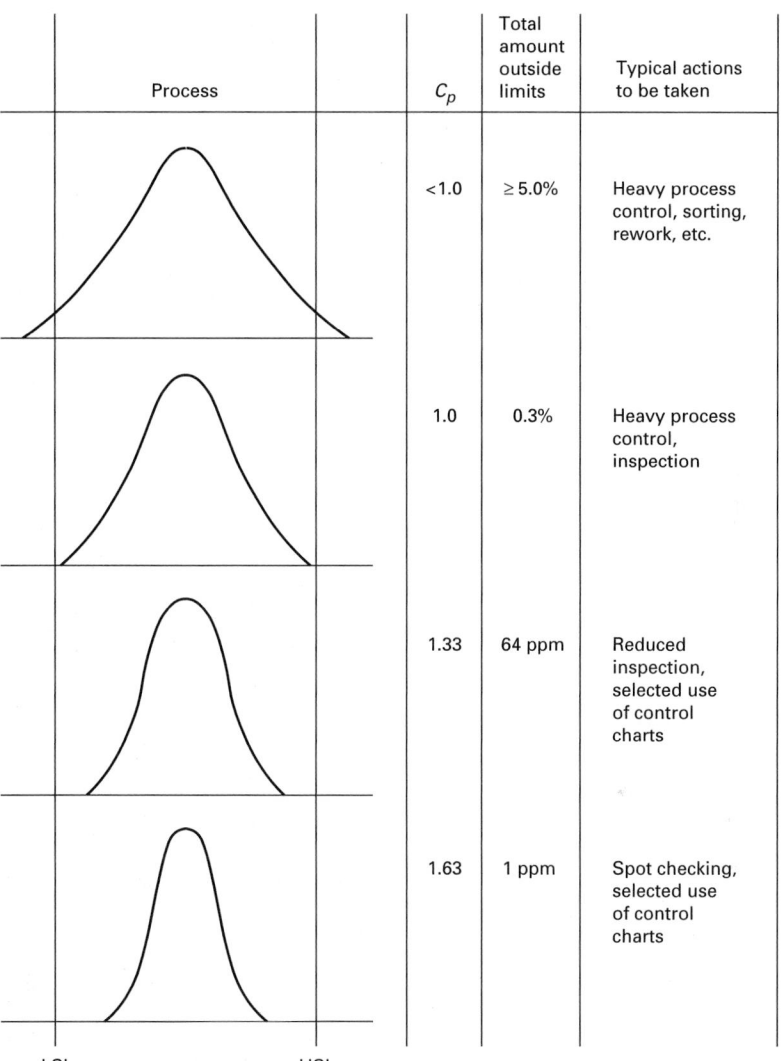

Process	C_p	Total amount outside limits	Typical actions to be taken
	<1.0	≥5.0%	Heavy process control, sorting, rework, etc.
	1.0	0.3%	Heavy process control, inspection
	1.33	64 ppm	Reduced inspection, selected use of control charts
	1.63	1 ppm	Spot checking, selected use of control charts

LSL   USL

FIGURE 22.8   Four examples of process variability.

capability indices. Herman (1989) cites important precautions in using capability indices in the process industries—variability among lots and measurement variability are relevant issues.

Capability and Performance Studies.   Two types of process studies can be identified:

1. Process capability studies that estimate the inherent or potential process capability, i.e., what the process can do under certain conditions. This type of study is discussed below.

2. Process performance studies that measure the present performance of the process, i.e., what the process *is* doing. This type of study is discussed later in this section under Process Performance Measurement.

TABLE 22.4   Process Capability Index C_p and Product Outside Specification Limits

Process capability index C_p	Total product outside two-sided specification limits*
0.5	13.36%
0.67	4.55%
1.00	0.3%
1.33	64 ppm
1.63	1 ppm
2.00	0

*Assuming the process is centered midway between the specification limits.

TABLE 22.5   Process Capability and Process Performance Indices

Process capability	Process performance
$C_p = \dfrac{USL - LSL}{6\sigma}$	$P_p = \hat{C}_p = \dfrac{USL - LSL}{6s}$
$C_{pk} = \min\left(\dfrac{USL - \mu}{3\sigma}, \dfrac{\mu - LSL}{3\sigma}\right)$	$P_p = \hat{C}_{pk} = \min\left\|\dfrac{USL - \overline{X}}{3s}, \dfrac{\overline{X} - LSL}{3s}\right\|$
$C_{pm} = \dfrac{USL - LSL}{6\sqrt{\sigma^2 + (\mu - T)^2}}$	$P_p = \hat{C}_{pm} = \dfrac{USL - LSL}{6\sqrt{s^2 + (\overline{X} - T)^2}}$

PROCESS CAPABILITY MEASUREMENT

Measuring the inherent or potential process capability requires that the process be stable. Stability is evaluated by a control chart.

The Control Chart.   A *control chart* is a graphic comparison of process performance data to statistical control limits, not specification limits. The performance data consist of groups of measurements (*rational subgroups*) selected in regular sequence of production while preserving the order. The statistical control limits help to evaluate capability by first evaluating whether the process is operating at its minimum inherent variation.

Process variations are traceable to two kinds of causes: (1) random, i.e., due solely to "common" or chance causes, and (2) assignable, i.e., due to findable "special" causes. Ideally, only random causes should be present in a process because this represents the minimum possible amount of variation with the given set of process conditions. A process that is operating without assignable causes of variation is said to be in a *state of statistical control.* A control chart analysis should be made and assignable causes eliminated from the process prior to calculating 6σ as a measure of process capability. When this is done, 6σ then represents the inherent process capability. If 6σ is calculated without first making a control chart analysis, the calculated value of 6σ probably will be inflated. Many control chart analyses reveal the presence of assignable causes even though production people profess that the process is operating with the minimum possible variation. A description of control chart methodology, including formulas and procedures, is given in Section 45, Statistical Process Control.

Determination of Process Capability from a Control Chart Analysis.   If, and only if, a process is in statistical control, the following relationship holds:

$$\text{Estimate of } \sigma = \frac{\overline{R}}{d_2}$$

Table A in Appendix II provides values of d_2. Knowing the standard deviation, process capability limits can be set at $\pm 3s$ and this used as an estimate of $\pm 3\sigma$. (This calculation converts $\overline{R}$ to a standard deviation of *individual* values. Control limits represent 3 standard deviations of sample *averages*.)

The Assumption of Statistical Control and Its Effect on Process Capability.

If a process is out of control and the causes cannot be eliminated economically, the standard deviation and process capability limits nevertheless can be computed (with the out-of-control points included). These limits will be inflated because the process will not be operating at its best. In addition, the instability of the process means that the prediction is approximate.

It is important to distinguish between a process in a state of statistical control and a process that is meeting specifications. A state of statistical control does *not* necessarily mean that the product from the process conforms to specifications. Statistical control limits on sample averages *cannot* be compared directly with specification limits because the specification limits refer to individual units. For some processes that are not in control, the specifications are being met, and no action is required; other processes are in control, but the specifications are not being met and action is needed. In summary, we need processes that are both stable (in statistical control) and capable (meet product specifications).

The increasing use of capability indices has brought with it the inevitable failure to understand and verify some important assumptions that are essential for statistical validity of the results. Four key assumptions are

1. *Process stability:*   This means a state of statistical control with no drift or oscillation (see above).
2. *Normality of the characteristic being measured:*   This is needed to draw statistical inferences about the population.
3. *Representativeness of samples:*   This includes random sampling.
4. *Independence of the measurements:*   This means that consecutive measurements cannot be correlated.

In practice, these assumptions are often not verified. Examination likely would reveal that one or more of the assumptions is not realistic. These assumptions are not theoretical refinements—they are important conditions for properly applying capability indices. Before applying capability indices, the reader is urged to read the paper by Pignatiello and Ramberg (1993). Also, the October 1992 issue of the *Journal of Quality Technology* is devoted to statistical issues concerning capability indices. McCoy (1991) summarizes the situation well—how effective the indices are depends on how they are used and understanding the risks involved. These risks can be minimized by statistically and visually comparing the indices with the full data versus specifications as depicted in a histogram.

PROCESS PERFORMANCE MEASUREMENT

If a process is in statistical control, then the measure of process performance results also in determining process capability. Several of the techniques described below for determining process performance use the same calculations as techniques described earlier for process capability (the difference involves the assumption of statistical control, as explained below). Mentch (1980) provides a further breakdown of process analysis into four categories and presents methods and examples for each.

Measuring Present Process Performance.

Specific tools for this type of study include process performance indices, the frequency distribution and histograms, probability paper, plot of individual measurements, and attributes data analysis. It is highly preferable to use variables rather than attributes data, i.e., numerical measurements rather than accept-reject information.

Process Performance Indices.   Table 22.5 presents process performance indices P corresponding to the process capability indices C_p discussed earlier. For example, Kane (1986) discusses the use of a *performance index* C_{pk} that reflects the current process mean's proximity to either the upper specification limit USL or lower specification limit LSL. C_{pk} is estimated by

$$C_{pk} = \min\left(\frac{\overline{X} - \text{LSL}}{3s}, \frac{\text{USL} - \overline{X}}{3s}\right)$$

For Kane's example where

$$\text{USL} = 20 \qquad \overline{X} = 16$$

$$\text{LSL} = 8 \qquad s = 2$$

the standard capability ratio is estimated as

$$C_p = \frac{\text{USL} - \text{LSL}}{6\sigma} = \frac{20 - 8}{12} = 1.0$$

which implies that if the process were centered between the specification limits (at 14), then only a small proportion (about 0.27 percent) of product would be defective.

However, when we calculate C_{pk}, we obtain

$$C_{pk} = \min\left(\frac{16 - 8}{6}, \frac{20 - 16}{6}\right) = 0.67$$

which alerts us that the process mean is *currently* nearer the USL. (Note that if the process were centered at 14, the value of C_{pk} would be 1.0.) An acceptable process will require reducing the standard deviation and/or centering the mean.

Interpretation of C_{pk}.   In using C_{pk} to evaluate a process, we must recognize that C_{pk} is an abbreviation of two parameters—the average and the standard deviation. Such an abbreviation can inadvertently mask important detail on these parameters; e.g., three extremely different processes can all have the same C_{pk} [for elaboration, see Juran and Gryna (1993, p. 402)].

Increasing the value of C_{pk} may require a change in the process average, the process standard deviation, or both. For some processes, it may be easier to increase the value of C_{pk} by changing the average value (perhaps through a simple adjustment of the process aim) than to reduce the standard deviation (by investigating the many causes of variability). The histogram of the process always should be reviewed to highlight both the average and the spread of the process.

Calculating and interpreting these performance indices do not require the assumptions of statistical control or normality of the distribution. Capability indices are useful in estimating future performance (based on certain assumptions); performance indices are useful as measures of past performance. In both cases, plotting data over time helps to identify trends and evaluate the success of improvement efforts.

Frequency Distribution and Histogram.   In this type of study, a sample of about 50 consecutive units is taken, during which time no adjustments are made on the machines or tools. The units are all measured, the data are tallied in frequency distribution form, and the standard deviation s is calculated and used as an estimate of σ. The characteristics are assumed to follow a normal probability distribution where $\pm 3\sigma$ standard deviations include 99.73 percent of the population. Process performance is then defined as $\pm 3\sigma$ or 6σ. For example, analysis of 60 measurements yielded

$$\overline{X} = 9.6 \qquad s = 2.5$$

Computer programs are available to calculate the average and standard deviation, develop and plot the histogram, and make various checks on the assumption of a normal probability distribution.

The process capability is calculated as ±3(2.5), or ±7.5, or a total of 15.0.

Normal probability paper (see Section 44, under The Normal Distribution) can be used to graphically determine process performance. Process data are plotted on the probability paper, and the mean value and 3 standard deviation values are then estimated graphically. In some normal probability paper, the upper and lower horizontal grid lines represent ±3σ, respectively. Weibull probability paper is available to handle nonnormal distributions.

An extensive discussion of determining process capability using probability paper is given by Lehrman (1991). The paper presents the detailed steps in making the plot, a discussion of normal and skewed distributions, and confidence limits on the capability index.

Plots of Individual Measurements.   A simple plot of individual measurements, in order of production, can be surprisingly revealing.

In a classic study of a machine process, watch parts were measured for each of five quality characteristics. The resulting measurements were plotted in chronological order on a chart that also showed the five sets of specification limits. The study demonstrated that the process was capable of meeting the specification limits. The study also showed that the poor performance (12 percent nonconforming product) was due to the inadequacy of the instruments provided to the work force. Provision of adequate instruments reduced the defect level to 2 percent and made possible a sharp reduction in the amount of gauging done by inspectors (consulting experience of J. M. Juran).

Limitations of Histograms and Probability Paper Analyses.   These methods of evaluating process performance do not evaluate the inherent capability of the process because they are usually performed without first evaluating the process for statistical control. The data may include measurements from several populations. There may be time-to-time changes such as solutions becoming dilute or tools becoming worn. Such process conditions result in observed dispersions that are wider than the inherent capability of the process. To evaluate the inherent capability requires use of a control chart (see above under Process Capability Measurement).

Six-Sigma Concept of Process Capability.   For some processes, shifts in the process average are so common that such shifts should be recognized in setting acceptable values of C_p. In some industries, shifts in the process average of ±1.5 standard deviations (of individual values) are not unusual. To allow for such shifts, high values of C_p are needed. For example, if specification limits are at ±6σ (not ±3σ), and if the mean shifts ±1.5σ, then only 3.4 ppm will be beyond specification limits. The Motorola Company's "six-sigma" approach recognizes the likelihood of these shifts in the process average and makes use of a variety of quality engineering techniques to change the product, the process, or both in order to achieve a C_p of at least 2.0. Craig (1993) describes a seven-step approach applied to electronics manufacturing.

Attributes Data Analysis.   The methods discussed earlier assume that numerical measurements are available from the process. This is the preferable type of data for a capability study. Sometimes, however, the only data available are in attribute form, i.e., the number defective and the number acceptable. Attributes data require large sample sizes and should be used only where variable measurement is impractical.

To illustrate, I will analyze data from a process for preparing insurance policies. Policy writers fill in blank policy forms with data from various inputs. The forms then go to a checker, who reviews them for errors. For a specified time period, the checker reported 80 errors from 6 policy writers and covering 29 types of errors (Table 22.6). Using errors as the unit of measure, the process performance can be calculated as 80/6, or 13.3 per writer. Note that none of the writers was close to the average.

The current *performance* of the process can be described as 13.3 errors per writer, but analysis revealed that this is not the *capability* of the process:

TABLE 22.6   Matrix of Errors by Insurance Policy Writers

Error type	Policy writers						Total
	A	B	C	D	E	F	
1	0	0	1	0	2	1	4
2	1	0	0	0	1	0	2
3	0	16	1	0	2	0	19
4	0	0	0	0	1	0	1
5	2	1	3	1	4	2	13
6	0	0	0	0	3	0	3
28							
29							
Totals	6	20	8	3	36	7	80

- For writer B, 16 of the 20 errors were due to a misunderstanding of a procedure. When this was clarified, error type 3 for writer B did not recur.

- In contrast, writer E made 36 errors in essentially all 29 categories. The writer was reassigned to other work.

- Error type 5 caused a problem for all the writers. Analysis revealed a difference in interpretation of the work instruction between the writers and the checker. When this was cleared up, error type 5 disappeared.

The process capability can now be calculated by excluding the preceding abnormal performances: type 3 errors by worker B, type 5 errors, and errors of worker E. The error data for the remaining 5 writers becomes 4, 3, 5, 2, and 5, with an average of 3.8 errors per writer. This process capability estimate of 3.8 compares with the original process performance estimate of 13.3.

Note that this example calculates process capability in terms of errors or mistakes rather than variability of a process parameter. Hinckley and Barkan (1995) point out that in many assembly processes, nonconforming product can be caused by excessive variability on one or more parameters or by mistakes (e.g., missing parts, wrong parts, or other processing errors). Mistakes are not included in a process capability calculation based on variability. For some processes, particularly complex processes, mistakes can be a major cause of failing to meet customer quality goals. The actions required to reduce mistakes are different from those required to reduce variability on a parameter.

OTHER ASPECTS OF PROCESS CAPABILITY

These aspects include complex processes, service industries, quality improvement, and planning for a study.

Process Capability in Service Industries.   The concept of process capability analysis grew up in the manufacturing industries. This concept focuses on evaluating process variability (6 standard deviations) as a measure of process capability. The concept, however, can apply to *any* process, including nonmanufacturing processes in the manufacturing industries and the spectrum of processes in the service industries. Little has been published on the application of process capability other than its application to manufacturing processes.

For certain parameters in service processes, process capability can be measured using 6σ and various capability indices. For example, in a loan association, the cycle time to complete the loan-approval process is critical and could be analyzed. Time data are readily available in quantitative form for calculating 6σ.

Other service processes may not have variables data available. For example, a firm provides a service of guaranteeing checks written by customers at retail establishments. The decision whether to guarantee is based on a process that employs an on-line evaluation of six factors. A percentage of checks guaranteed by the firm have insufficient funds, and the customer must be pursued for payment. The percentage of checks that default ("bounce") could be viewed as a measure of process capability. This approach uses discrete (attributes) data rather than the classic approach of calculating 6σ from variables data. The example given earlier on insurance policy writing illustrates the use of attributes data to calculate process capability for a service industry process.

With the emphasis on processes in quality management, evaluating the capability of processes requires not only evaluating capability based on variability (e.g., 6σ) but also a broader view. Juran (1992, pp. 240–256) describes the issues involved in developing a broader framework.

Process Capability and Quality Improvement.   Capability indices serve a role in quantifying the ability of a process to meet customer quality goals. The emphasis, however, should be on improving processes and not just determining a capability index for a product characteristic. Achieving customer quality goals (particularly for quality levels of 1 to 10 ppm) means meeting requirements on all variables and attributes characteristics. On variables characteristics, decreasing the amount of variability (even when specification limits are being met) has many advantages. Juran and Gryna (1993) discuss six of these advantages. Achieving decreased variability requires the use of basic and advanced improvement techniques. Sections 3, 4, and 47 cover many of these techniques. The Taguchi approach uses experimental design to determine the optimal values of process variables that will minimize the variation in a process while keeping a mean on target. Shina (1991) describes an application to a wave soldering process. The results were measured in terms of weekly solder defects in parts per million. Results, before and after the Taguchi application, were

	Mean	Standard deviation
Before	808.50	213.80
After	98.50	55.30

Because of this reduction in defects, follow-on projects such as computer control of the process and acquisition of additional process equipment were curtailed. The Taguchi approach is discussed in Section 47, under Orthogonal Arrays and Taguchi Off-Line Quality Control.

Planning for the Process Capability Study.   A capability study is made for different reasons, e.g., to respond to a customer request for a capability index number or to evaluate and improve product quality. Prior to data collection, clarify the purposes of making the study and the steps taken to ensure that the purpose is achieved.

In some cases, the capability study will focus on determining a histogram and capability index for a relatively simple process. Here, the planning should ensure that process conditions (e.g., feeds, speeds, temperature, and pressure) are completely defined and recorded. All other inputs clearly must be representative, i.e., specific equipment, material, and of course, personnel.

For more complex processes or where defect levels of 1 to 10 ppm are desired, the following steps are recommended:

1. Develop a process description including inputs, process steps, and output quality characteristics. This can range from simply identifying the equipment to developing a mathematical equation showing the effect of each process variable on the quality characteristics.

2. Define the process conditions for each process variable. In a simple case, this means stating the settings for temperature and pressure. For some processes, however, it means determining the optimal value or aim for each process variable. The statistical design of experiments provides the methodology (see Section 47, Design and Analysis of Experiments). Also, determine the operating

ranges of the process variables around the optimum because this will affect the variability of the product results.

3. Make sure that each quality characteristic has at least one process variable that can be used to adjust it.

4. Decide if measurement error is significant. This can be determined from a separate error of measurement study (see Section 23, under Error of Measurement). In some cases, the error of measurement can be evaluated as part of the overall study.

5. Decide if the capability study will focus on variability only (6σ) or whether it also will include mistakes or errors that cause quality problems.

6. Plan for the use of control charts to evaluate stability of the process.

7. Prepare a data collection plan that documents results on quality characteristics along with the process conditions (e.g., values of all process variables) and preserves information on the order of measurements so that trends can be evaluated.

8. Plan what methods will be used to analyze data from the study to ensure, before starting the study, that all necessary data for the analysis will be available. The analyses will include not only process capability calculations on variability but also analysis of attribute data on mistakes and analysis of data from statistically designed experiments built into the study.

9. Be prepared to spend time investigating interim results before process capability calculations can be made. These investigations can include analysis of optimal values and ranges of process variables, out-of-control points on control charts, or other unusual results. The investigations can lead to the ultimate objective, i.e., improvement of the process.

Note that these steps focus on improvement rather than just on determination of a capability index.

In a classic paper, Bemesderfer (1979) describes an eight-point program for evaluating new processes prior to production. Middleton (1992) presents a detailed example of a broad process capability study that incorporates a capability index, attributes measurement, and experimental design. Keenan (1995) discusses making a process capability study during product development, prior to regular production startup. Bothe (1992) describes an approach for making a capability study for an entire product by first determining the probability of each product characteristic being within specifications, calculating the combined probability of all characteristics being within specifications, and then expressing this combined probability as a C_{pk} value.

ERROR-PROOFING THE PROCESS

An important element of manufacturing planning is the concept of designing the process to be error-free through error-proofing. Where this type of design is economic, it can

Prevent defects or nonconformities that fallible human beings would otherwise make through inadvertence

Make effective a knack that would otherwise require retraining many workers

Prevent defects or nonconformities resulting from carelessness, indifference, and similar reasons

Bypass complex analysis for causes by finding a solution even though the cause of defects remains a mystery

Methods of Error-Proofing.   Some of the more usual forms are summarized below.

Fail-Safe Devices.   These consist of

1. *Interlocking sequences:*   For example, to ensure that operation A is performed, the subsequent operation B locates from a hole that only operation A creates.

2. *Alarms and cutoffs:*   These are used to signal depletion of material supply, broken threads, or other abnormalities. The alarms are also fail safe; i.e., they are silent only if all is well. If there is doubt, they sound anyhow.

3. *All-clear signal:*   These are designed to signal only if all remedial steps have been taken.

4. *Foolproof fixtures:*   These serve not only as fixtures but also as instruments to check the quality of work from preceding operations.

5. *Limiting mechanisms:*   For example, a slipping-type torque wrench to prevent overtightening.

Magnification of Senses.   Examples are

1. Locating indexes and fixtures to outperform human muscle in precision of position.

2. Optical magnification to improve visibility.

3. Remote-control viewing (closed-circuit television) to permit viewing of the process despite distance, heat, fumes, etc.

4. Multiple signals to improve likelihood of recognition and response, e.g., simultaneously ringing of bells and flashing of lights; audiovisual systems

5. Use of pictures in place of numbers (e.g., cards on the hood of a car undergoing assembly, to show pictorially the equipment needed for that car)

Redundancy.   This consists of extra work performed purely as a quality safeguard. Examples are

1. *Multiple-identity codings:*   These are intended to prevent product mixups, e.g., color codes or other recognition schemes on drug labels, tool steel, aluminum sheet, etc.

2. *Redundant actions and approvals:*   For example, the drug industry requires that formulation of recipes be prepared and approved by two registered pharmacists working independently.

3. *Audit review and checking procedures:*   These are widely used to ensure that the plans are being followed.

4. *Design for verification:*   The product may include specially designed provision for verification (holes for viewing, coupons for test, etc.). It also includes the rapidly growing use of nuclear tracers.

5. *Multiple test stations:*   For example, a can-filling line may provide checks for empty cans through height gauges, weighing scales, and air jets (for blowing empties off the conveyer).

Countdowns.   These are arranged by structuring sensing and information procedures to parallel the operating procedures so that the operational steps are checked against the sensing and informational needs. A dramatic example is the elaborate countdown for the launching of a space vehicle. Surgical operations require countdowns, accounting for all materials and tools used (e.g., sponges, surgical instruments, etc.). A useful principle is to use an active rather than passive form of countdown. For example, a welder counts all welds *aloud* in progressing from spot to spot. When the count reaches 17, the last weld has been made—just as called for by the specification.

Special Checking and Control Devices.   Examples from the service sector include

1. *Automatic dispensing devices:*   Examples are drink-filling and other portion-control devices in the fast-food sector.

2. *Software to detect incorrect information or data:*   This includes "spell check" in word processing and software to detect errors in data such as extreme charges on an invoice or too many digits in a data field.

3. *Software to detect missing information or data.*

4. *Hand-held devices to check or perform calculations,* e.g., on meter readings or rental car charges.

5. *Automatic recording of information:*   This includes the use of bar codes at the checkout counter of a supermarket and the scoping of a package with a wand to track the location of a package at each transfer during a delivery process.

6. *Automatic timing devices:*   Examples are those used for controlling cooking in fast-food operations.

Binroth (1992) presents research on three categories of errors in automotive manufacturing, i.e., missing components, incorrect processing, and wrong components.

Error-Proofing Principles.   A useful principle in error-proofing is that of providing feedback to the worker; i.e., *the performance of the work conveys a message to the worker.* For example, a worker at a control panel pushes a switch and receives three feedbacks: the *feel* of the shape of the switch handle, the *sound* of an audible click signaling that the switch went all the way, and the *sight* of a visual illumination of a specific color and shape.

In a classic study, Nakajo and Kume (1985) discuss five principles of error-proofing developed from an analysis of about 1000 examples collected mainly from assembly lines. The principles are elimination, replacement, facilitation, detection, and mitigation (see Table 22.7).

OTHER ELEMENTS OF EQUIPMENT AND WORK METHODS PLANNING

Planning for equipment goes beyond making a process capability study. Other factors include providing for process adjustments and for preventive maintenance.

Providing for Adjustments to Processes.   Many processes require periodic adjustments. Manufacturing planners should (1) identify the process variables that must be monitored for possible adjustment, (2) provide rules for determining when an adjustment is necessary, (3) provide instructions for determining the amount of adjustment, and (4) provide a convenient physical means for making the adjustment.

Each product characteristic should have a process variable that can be used to adjust it. As corollaries to this principle, Bemesderfer (1979) proposes

1. A single process variable should correspond to a single characteristic.
2. The degree of adjustment required during the process for a given change in the characteristic should be constant.
3. The range of possible adjustments must be consistent with the range of application need.
4. The setting accuracy must be consistent with the product tolerance requirements.
5. The controlling accuracy, once the process is set, must be consistent with the product tolerance requirements.

TABLE 22.7   Summary of Error-Proofing Principles

Principle	Objective	Example
Elimination	Eliminate the possibility of error	Redesign the process or product so that the task is no longer necessary
Replacement	Substitute a more reliable process for the worker	Use robotics (e.g., in welding or painting)
Facilitation	Make the work easier to perform	Color code parts
Detection	Detect the error before further processing	Develop computer software which notifies the worker when a wrong type of keyboard entry is made (e.g., alpha versus numeric)
Mitigation	Minimize the effect of the error	Utilize fuses for overloaded circuits

To the degree that these aims cannot be achieved, the process will be difficult for a worker to control.

Preventive Maintenance.   Maintenance of equipment is generally recognized as essential, but pressures for production can result in delaying the scheduled preventive maintenance. Sometimes the delay is indefinite, the equipment breaks down, and the maintenance becomes corrective instead of preventive.

The planning should determine how often preventive maintenance is necessary, what form it should take, and how processes should be audited to ensure that preventive maintenance schedules are followed.

In the event of objections to the proposed plan for preventive maintenance on the grounds of high cost, data on the cost of poor quality from the process can help to justify the maintenance plan.

The concept of *total productive maintenance* (TPM) aims to use equipment at its maximum effectiveness by eliminating waste and losses caused by equipment malfunctions. Shenoy (1994) identifies six major process losses in a paper mill and relates them to three measures of equipment effectiveness. The concept is shown in Figure 22.9.

This model provides a means of quantifying productivity and quality.

To quantify availability:

- Available hours: 4272 hours
- Downtime due to equipment failures, setups, and adjustment: 560 hours
- Availability = $(4272-560)/4272 = 0.869$

To quantify performance efficiency:

- Theoretical cycle time: 0.4 hours/ton
- Production amount: 7773 tons
- Operating time: 3712.5 hours
- Performance efficiency = $(0.4 \times 7773)/3712.5 = 0.837$

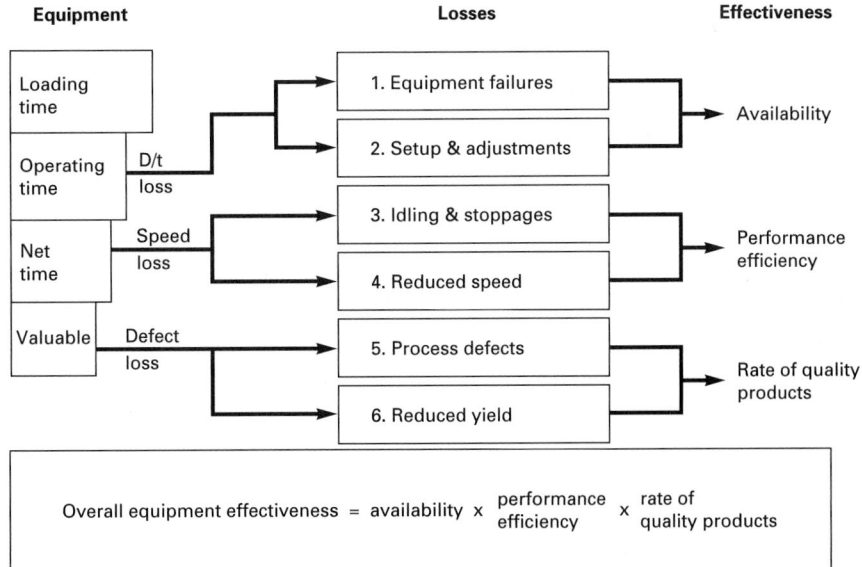

FIGURE 22.9   Six big losses and equipment effectiveness. (*Shenoy 1994.*)

To quantify the defect loss:

- Salable product: 7621 tons
- Rate of quality product: 7621/7773 = 0.98

The overall effectiveness is

$$0.869 \times 0.837 \times 0.98 \times 100 = 71.2\%$$

Note that downtime losses and cycle time (speed) losses were the major contributing factors to the low effectiveness of 71.2 percent. Particularly in backroom operations of service industries, computer downtime is emerging as a problem for operations.

OVERALL REVIEW OF MANUFACTURING PLANNING

Review of the proposed process can be accomplished most effectively through preproduction trials and runs. Techniques such as failure mode, effect, and criticality analysis can provide an even earlier warning before any product is made. Checklists for review of proposed processes also can be useful. These approaches are discussed below.

Preproduction Trials.   Because the manufacturing plan starts as a mental concept, it will be "scaled up" many orders of magnitude if it goes into large-scale production. There is great risk in going directly into production from the conceptual plan primarily because of the risk of quality failures. To reduce this risk, companies make use of trial production lots (called *pilot plant production, preproduction,* etc.) to discover deficiencies in the planning and to remedy them before going into full-scale production. In some industries this concept is formalized into regular phases of scaling up.

The scaling up of production is actually a continuation of the scaling up that takes place from product design concept to prototype or model construction and test. The adequacy of the full-scale manufacturing plan cannot be judged from the record of models made in the model shop. In the model shop the basic purpose is to prove engineering feasibility; in the production shop the purpose is to meet standards of quality, cost, and delivery. The model shop machinery, tools, personnel, supervision, motivation, etc. are all different from the corresponding situations in the production shop.

Tool Tryout.   At the workstation level, as new tools are completed, they are subjected to a tryout procedure that, in most companies, is highly formalized. The tryout consists of producing enough product from the new tool to demonstrate that it can meet quality standards under shop conditions.

These formalized tryouts conclude with the execution of a formal document backed up by supporting data, which always include the quality data. The release of the tool for full-scale production is contingent on the approval of this tryout document.

Limited Trial Lots.   Beyond the tryouts at individual workstations, there is a need for collective tryouts. These require trial production lots, which must be scheduled for the prime purpose of proving in the manufacturing process. The trial lot is usually made in the regular production shop and provides an extensive preview of the problems that will be encountered in large-scale production. In the process industries, the equivalent intermediate scaling up is the *pilot plant.* It is widely used to provide the essential information (on quality, costs, productivity, etc.) needed to determine whether and how to go into full-scale production.

Software Verification.   Software used with a process requires a tryout just like new tools—with the same degree of formality and approval process.

Experimental Lots.   The trial lot concept provides opportunities for planners to test out alternatives, and they often combine the concept of experimentation with that of proving in the nonexperimental portion of the trial.

Attainment of good process yields is one of the most important purposes of experimental lots. These experiments can make use of all the techniques discussed in Section 47, Design and Analysis of Experiments, and in the various statistical sections.

Preproduction Runs.   Ideally, product lots should be put through the entire system, with the deficiencies found and corrected before going into full-scale production. In practice, companies usually make some compromises with this ideal approach. The preproduction may be merely the first of the regular production, but with special provision for prompt feedback and correction of errors as found. Alternatively, the preproduction may be limited to those features of product and process design which are so new that prior experience cannot reliably provide a basis for good risk taking. While some companies do adhere to a strict rule of proving in the product and process through preproduction lots, the more usual approach is one of flexibility, in which the use of preproduction lots depends on

1. The extent to which the product embodies new or untested quality features
2. The extent to which the design of the manufacturing process embodies new or untried machines, tools, etc.
3. The amount and value of product which will be out in the field before there is conclusive evidence of the extent of process, product, and use difficulties

These trials sometimes include "production validation tests" to ensure that the full-scale process can meet the design intent. Figure 22.10 shows an example from Ford Motor Company.

Failure Mode, Effect, and Criticality Analysis for Processes   A failure mode, effect, and criticality analysis is useful in analyzing the proposed design of a product (see Sections 3 and 48). The same technique can dissect the potential failure modes and their effects on a proposed process. Ishiyama (1977) discusses the application of the failure mode and effects analysis technique to both product design and manufacturing processes in the automobile industry.

The fault-tree analysis technique is also useful in analyzing a design (see Sections 3 and 48). Proposed manufacturing processes can be analyzed with this same technique. Raheja (1982) discusses this approach. Fault-tree analysis can help to identify areas of a process that require error-proofing.

A supplier of telecommunications services was about to implement a complex process to deliver a new service to customers (Plsek 1989). A team from field operations was asked to evaluate the process prior to implementation. The team constructed a flow chart and an FMEA (Figure 22.11). The FMEA was used to set priorities for addressing potential failure modes. Three factors were considered in setting priorities: probability of occurrence, impact on end customers, and impact on internal costs. Each factor was scored using a scale of 1 to 5 ("very low" to "very high").
Then

$$\text{Priority score} = \text{probability of occurrence} \times (\text{customer impact} + \text{cost impact})$$

Priority scores could range from 2 to 50. Failure modes that were highly likely to occur and whose occurrence would seriously affect *both* the customer and internal cost received the highest priority scores. Using the FMEA, the team was able to identify dozens of potential failure modes that had not been addressed adequately in the process. Eight of these were critical and could have resulted in major customer dissatisfaction or high costs.

Using this type of analysis, we probably will not identify and prevent *all* potential failure modes, but we will identify and prevent *some* serious failures that traditional process design techniques might overlook.

The fault-tree analysis technique is also useful in analyzing a design because it traces *all possible combinations of causes that could lead to a particular failure.* Proposed manufacturing processes can be analyzed with this same technique. Raheja (1982) discusses this approach. Fault-tree analysis can help to identify areas of a process that require error-proofing.

PRODUCTION VALIDATION PLAN AND REPORT   #3

1.14-21812-A   PAGE 1 OF 3

SYSTEM 1.00 Body   ASSEMBLY Door Latch-Diab.   PROGRAM   DES. ENGR. S. Martin

SUBSYSTEM 1.14 Locks Handles & Mechs.   COMP N/A   DES LEVEL AA   CONC'R W.R.

TEST NAME SOURCE	ACCEPTANCE CRITERIA	TEST RESULTS	DES.	S/SIZE		REL. ACCEP. CRIT.		TIMING		REMARKS
				REQ.	TEST	REQ.	ACTUAL	SCH.	ACTUAL	
Life Cycle ES-Diab-652181 AB	35,000 cycles minimum at 650# rebound and 250#	21 completed 120,000 cycles	AA	21	21	$P_{.90} = .90$	$P_{.90} = .90$	7-15	7-20	
Static Strength FMVSS (ES)	Prim. Sec. longitude 3000 2250 transverse 4000 2700	-3 σ greater than min. strength req'd (see Problem Plots)	AA	60	60	-3 σ> req't.	-3 σ> req't.	7-24	7-20	
Corrosion Resistance (ES) 48-hour salt spray	Operating efforts must not increase 25% over drawing specifications	+ 3 σ less than max. effort allowed (see Problem Plots)	AA	40	40	+ 3 σ< req't.	+ 3 σ< req't.	7-30	8-15	

FIGURE 22.10 Production validation report. (*Ford Motor Company.*)

Process Step	Potential Failure Modes	Probability of Occurrence	Impact-On: Customer	Impact-On: Cost	Priority	Effect
	Presubscription order too early	4	5	5	40	Can't use 1+ 10-digit dialing
	Presubscription order too late	4	5	5	40	Billing errors
	Incorrect information about exchange access end office	1	1	2	3	Equipment ordered in wrong office; delay of service

FIGURE 22.11   Process FMEA for telecommunications service. (*Stampen and Stampen 1995.*)

The ultimate review of plans for manufacturing or service operations consists of an actual test. This may consist of a small-scale "pilot test" under laboratory conditions, a modular test of a portion of a process, a computer simulation, a "dry run" of the process using regular operating personnel, or a full trial of the process under typical process conditions and including acceptance or rejection of the process output.

Evaluation of Processes.   Processes should be evaluated for four parameters: effectiveness (of output), efficiency, adaptability, and cycle time. One approach for evaluating a process is the rating method developed by IBM in connection with its business process management activities. The method defines five levels of process maturity. The highest level, level 1, designates a business process that operates at maximum effectiveness and efficiency and serves as a benchmark or leader; the lowest level, level 5, suggests a process that is ineffective, may have major deficiencies, and the process management approach has not been instituted. Melan (1993) defines the specific criteria associated with each level. Criteria include both organizational matters (e.g., a process owner) and technical matters (e.g., measurements for effectiveness and efficiency).

Black (1993) describes a program for certification of a manufacturing process at Caterpillar, Inc. The requirements include a process that is significant, the product must meet specifications, all customers must be satisfied, and there must be evidence of continuous improvement. Recertification occurs annually.

A 12-step procedure for certification includes identify critical product characteristics, determine if the characteristics are in control, determine process capability, identify critical process parameters and their limits, determine if process parameters are in control and within limits, and develop an action plan to control process parameters. The procedure is at a sufficiently detailed level to achieve process control. As an example, for a foundry process making an engine block, limits on the temperature of the molten metal might be set at 2650 and 2675°F.

This internal certification procedure is particularly helpful when product is transferred to another "profit center." Under the profit center concept, a division chooses its suppliers, and they may be internal or external to Caterpillar. For example, the foundry profit center delivers an engine block to the engine assembly profit center. If the foundry engine block process is certified as meeting product requirements, the assembly division accepts the block without extensive incoming inspection. Thus the certification procedure reduces costs and maintains quality, thus helping the foundry profit center to retain the assembly division as an internal customer.

Evaluation and Reduction of Process Cycle Time.   Competitive pressures to reduce cycle time are now a galvanizing force to diagnose processes for improvement. Juran (1992) explains how a flow diagram can reveal

The number of functions that are affected

The extent to which the same macroprocess is used for the vital few customers and the useful many

The existence of redoing of prior work

The extent and location of bottlenecks such as numerous needs for signatures

Additional analysis is made of the vital few individual steps (microprocesses). Here the analysis focuses on

Is there a customer for the work done in this step?

Can this step be performed after serving the customer rather than before?

What can be done to reduce the time to perform this step?

Numerous ways have been found to shorten the cycle time for macroprocesses. These include

1. Provide a simplified process for the useful many applications.
2. Reduce the number of steps and handoffs.
3. Eliminate wasteful "loops."
4. Reduce changeover time.
5. Change from consecutive to concurrent processing.

These and other remedies, of course, can benefit from changes in technology. In any case, they have resulted in some stunning reductions in cycle time.

PLANNING PROCESS CONTROLS

The process specification, procedures, and instruction sheets prepared by the planners are the software of manufacturing planning. Their purpose is to inform the production people how to set up, run, and regulate the processes so that the result will be good product. Conversely, the production people should follow these plans. Otherwise, good product might not be the result.

Many companies institute process controls to provide assurance that the plans will in fact be followed. There are several kinds of these controls, and they are established by some combination of manufacturing engineers, quality engineers, production supervisors, and workers. The precise combination varies widely from company to company.

Process control is based on the feedback loop as discussed in Section 3, under Design Feedback Loop. The steps for planning manufacturing process controls follow closely the universal approach for use of the feedback loop.

Control Criteria.   While execution of the control plan is typically delegated to the work force, it is common to impose criteria to be met before the process is allowed to run. These criteria are imposed in three main areas:

1. *Setup criteria:*   For some processes the start of production must await meeting setup criteria (e.g., five pieces in a row must test "good"). In critical cases this form of early warning assurance may require that a supervisor or inspector independently approve the setup.
2. *Running criteria:*   For many processes there is a need to check the running periodically to decide whether the process should continue to run or should stop for readjustment. The criteria here relate to such things as frequency of check, size of sample, manner of sample selection, tests to be made, tolerances to be met.

3. *Equipment maintenance criteria:* In some processes, the equipment itself must be closely controlled if quality is to be maintained. This type of control is preventive in nature and is quite different in concept from repair of equipment breakdowns. This preventive form of equipment maintenance includes a carefully drawn set of criteria that define the essential performance characteristics of the equipment. Then, on a scheduled basis (strictly adhered to), the equipment is checked against these criteria. In the United States this aspect of equipment maintenance is not well developed, and there is need to take positive steps to strengthen it.

Relation to Product Controls. Process controls are sometimes confused with product controls, but there is a clear difference. Process controls are associated with the decision: Should the process run or stop? Product controls are associated with the decision: Does the product conform to specification? Usually both these decisions require input derived from sampling and measuring the product. (It is seldom feasible to measure the process directly.) However, the method of selecting the samples is often different. Production usually makes the "process run or stop" decision and tends to sample in ways which tell the most about the process. Inspection usually (in the United States) makes the "product conformance" decision and tends to sample in ways that tell the most about the product.

This difference in sampling can easily result in different conclusions on the "same" product. Production commonly does its sampling on a scheduled basis and at a time when the product is still traceable to specific streams of the process. Inspection often does its sampling on a random basis and at a time when traceability has begun to blur.

Despite the different purposes being served, it is feasible for the two departments to do joint planning. Usually they are able to establish their respective controls so that both purposes are well served and the respective data reinforce each other.

Control Systems and the Concept of Dominance. Specific systems for controlling characteristics can be related to the underlying factors that dominate a process. The main categories of dominance include those discussed below.

1. *Setup-dominant:* Such processes have high reproducibility and stability for the entire length of the batch to be made. Hence the control system emphasizes verification of the setup before production proceeds. Examples of such processes are drilling, labeling, heat sealing, printing, and presswork.

2. *Time-dominant:* Such a process is subject to progressive change with time (wear of tools, depletion of reagent, machine heating up). The associated control system will feature a schedule of process checks with feedback to enable the worker to make compensatory changes. Screw machining, volume filling, wood carding, and papermaking are examples of time-dominant processes.

3. *Component-dominant:* Here the quality of the input materials and components is the most influential. The control system is strongly oriented toward supplier relations along with incoming inspection and sorting of inferior lots. Many assembly operations and food formulation processes are component-dominant.

4. *Worker-dominant:* For such processes quality depends mainly on the skill and knack possessed by the production worker. The control system emphasizes such features as training courses and certification for workers, error-proofing, and worker and quality rating. Workers are dominant in processes such as welding, painting, and order picking.

5. *Information-dominant:* These are usually processes in which the job information undergoes frequent change. Hence the control system places emphasis on the accuracy and up-to-dateness of the information provided to the worker (and everyone else). Examples include order editing and "travelers" used in job shops.

The different types of dominance differ also in the tools used for process control. Table 22.8 lists the forms of process dominance along with the usual tools used for process control. Additional discussion of control tools as related to process dominance is included in Section 45.

Evaluation of Proposed Control Tools.   Proposed control tools need to be evaluated for both deficiencies and excesses. One health care manufacturer uses "process failure analysis" to analyze proposed control tools. A flowchart is first prepared to identify the elements of the manufacturing system and the output. Possible failure mechanisms are listed and the control system is analyzed in terms of

1. *The failure:*   probability of occurrence, criticality, effects, etc.
2. *The measurement:*   method, frequency, documentation, etc.
3. *The standard of comparison:*   selection, limits, etc.
4. *The feedback:*   method, content, speed

The proposed control for each failure mechanism is analyzed and classified as deficient, appropriate, or excessive.

PLANNING FOR EVALUATION OF PRODUCT

The planning must recognize the need for formal evaluation of product to determine its suitability for the marketplace. Three activities are involved:

1. Measuring the product for conformance to specifications
2. Taking action on the nonconforming product
3. Communicating information on the disposition of nonconforming product

These activities are discussed in Section 23 under Inspection and Test. However, these activities impinge on the manufacturing planning process. For example, several alternatives are possible for determining conformance, i.e., to have it done by production workers, by an independent inspection force, or by a combination of both. Second, the disposition of nonconforming product involves participation by production personnel, in such forms as segregation of product in the shop, and documentation. Finally, the communication of the decisions should include feedback to Production.

TABLE 22.8   Control Tools for Forms of Process Dominanace

Setup-dominant	Time-dominant	Component-dominant	Worker-dominant	Information-dominant
Inspection of process conditions	Periodic inspection $\bar{X}$ chart Median chart	Supplier rating Incoming inspection Prior operation	Acceptance inspection p chart	Computer-generated information "Active" checking of
First piece inspection	$\bar{X}$ and R chart Precontrol	control Acceptance inspection	c chart Operating	documentation Barcodes and electronic
Lot plot	Narrow-limit gauging	Mockup evaluation	scoring	entry
Precontrol	p chart		Recertification of	Process audits
Narrow limit gauging	Process variables check		workers Process audits	
Attribute visual inspection	Automatic recording Process audits			

AUTOMATED MANUFACTURING

In many manufacturing facilities, the computer is leading the march to automation. Several terms are important:

Computer-integrated manufacturing (CIM): This is the process of applying the computer in a planned fashion from design through manufacturing and shipping of the product. CIM has a broad scope.

Computer-aided manufacturing (CAM): This is the process in which the computer is used to plan and control the work of specific equipment.

Computer-aided design (CAD): This is the process by which the computer assists in the creation or modification of a design.

A basic reference for these areas is provided by Chang, Wysk, and Wang (1991).

Benefits to Product Quality. Automation may provide as large an increase in factory productivity as did the introduction of electric power. Product quality will benefit in several ways:

1. Automation can eliminate some of the monotonous or fatiguing tasks that cause errors by human beings. For example, when a manual seam welding operation was turned over to a robot, the scrap rate plunged from 15 percent to zero (Kegg 1985).

2. Process variation can be reduced by the automatic monitoring and continuous adjustment of process variables.

3. An important source of process troubles can be reduced, i.e., the number of machine setups.

4. Machines not only can measure product automatically but also can record, summarize, and display the data for line production operators and staff personnel. Feedback to the worker can be immediate, thus providing an early warning of impending troubles.

5. With cellular manufacture (see below), tracing a part to its origin is simplified, and this facilitates accountability for quality.

6. With CAD, the quality engineer can provide inputs early in the design stage. When a design is placed in the computer, the quality engineer can review that design over and over again and keep abreast of design changes.

Achieving these benefits requires a spectrum of concepts and techniques. Three of these are discussed below: the key functions of CIM, group technology, and flexible manufacturing systems.

With the emergence of an electronic information network provided by the Internet, a group of companies can operate as one virtual factory. This enables companies to exchange and act on information concerning inventory levels, delivery schedules, supplier lists, product specifications, and test data. It also means that CAD/CAM information and other manufacturing process information can be exchanged, data can be transferred to machines in a supplier's plant, and supplier software can be used to analyze producibility and to begin actual manufacturing.

Key Functions of Computer-Integrated Manufacturing. To integrate the computer from design through shipping involves a network of functions and associated computer systems. Willis and Sullivan (1984) describe this in terms of eight functions: design and drafting (CAD/CAM), production scheduling and control, process automation, process control, material handling and storage, maintenance scheduling and control, distribution management, and finance and accounting. Such a CIM system rests on a foundation of databases covering both manufacturing data and product data.

Lee (1995) discusses manufacturing initiatives in the perspective of global manufacturing competitiveness. CIM must integrate engineering and production with suppliers and customers globally to interactively design, plan, and process the manufacturing activities. A globalized CIM system includes the following technologies:

1. *Concurrent product design:*   Real-time design tools that will support innovation in a remote site to eliminate the time barriers to rapid transition of designs to production.

2. *Manufacturing planning:*   Tools for quick selection of resources and optimal process steps.

3. *Virtual manufacturing:*   A set of computer modeling and simulation tools to evaluate and predict the performance of products and processes, eliminating production delays and ensuring first-pass success.

4. *Remote performance monitoring, control, and diagnostics:*   Sensing and control tools for monitoring the machines and equipment remotely to control the behavior of the manufacturing process.

5. *Knowledge learning and acquisition:*   Intelligent tools for the acquisition and organization of process data to share with other manufacturing sites. The system allows global access to process data.

6. *Communications and integration:*   Multimedia information environment for information processing and transferring among geographically dispersed participants.

7. *Natural language translation:*   Automated translation of text between different languages.

Lee also describes research in progress to further develop these technologies.

Many manufacturing processes include automatic, self-calibrating systems with real-time closed-loop control. These processes achieve target values for product characteristics while minimizing variation around the target values.

Group Technology.   *Group technology* is the process of examining all items manufactured by a company to identify those with sufficient similarity that a common design or manufacturing plan can be used. The aim is to reduce the number of new designs or new manufacturing plans. In addition to the savings in resources, group technology can improve both the quality of design and the quality of conformance by using proven designs and manufacturing plans. In many companies, only 20 percent of the parts initially thought to require new design actually need it; of the remaining new parts, 40 percent could be built from an existing design, and the other 40 percent could be created by modifying an existing design.

Flexible Manufacturing System.   A *flexible manufacturing system* (FMS) is a group of several computer-controlled machine tools, linked by a materials handling system and a computer, to accommodate varying production requirements. The system can be reprogrammed to accommodate design changes or new parts. This system is in contrast to a fixed automation system, in which machinery, materials handling equipment, and controllers are organized and programmed for production of a single part or limited range of parts.

Quality Planning for Automated Processes.   Planning for automated processes requires special precautions:

1. Changes in the product design may be necessary to facilitate automated manufacture. For example, robots have difficulty picking up a randomly oriented part in a bin, but a redesign of the part may solve the problem.

2. Automated manufacturing equipment is complex and has the reliability and maintainability problems of most complex products. Design planning and evaluation tools (see Section 19, Research and Development) should be a part of the design process for automated equipment.

3. All software must be thoroughly tested (see Section 20, Software Development).

4. Knowledge of process capability, precise setup of equipment, and preventive maintenance are essential.

5. When feasible, on-line automatic inspection should be integrated with the operation. With manual operation of a process, the worker can observe a defect and take action. Automated processes can have mechanical, programming, or other problems that can create a disaster if not detected early.

6. Special provisions are necessary for measurement. These include the need for rugged gauges, cleaning of the measuring surfaces, reliability of gauges, and adherence to calibration schedules.

7. Some personnel will have greater responsibility under automated manufacture, particularly when computers are made available to workers for data entry and process control. All of this requires training.

The potential benefits of the automated factory will require significant time and resources for planning. However, automation will never be total. For example, there will never be robot plumbers in the factory. Therefore, in addition to ensuring that operating personnel have the new technical skills required of automated equipment, we also must give thought to the personnel requirements of the more conventional processes and plan for recruiting, training, and retaining people for these processes too.

PLANNING FOR SELECTION, TRAINING, AND RETENTION OF PERSONNEL

The principles of selection, training, and retention of personnel are known but are not always practiced with sufficient intensity in many functions, including operations. However, this is changing as organizations increasingly spend time and resources on these personnel matters to help achieve quality goals. First, we will consider the selection of personnel.

Selection of Personnel.   Norrell, a human resource company, provides client companies with traditional temporary help, managed staffing, and outsourcing services. A survey of over 1000 clients clarified the client definition of *quality* as excellence of personnel for a number of criteria. These criteria are shown in Table 22.9 for clerical and technical-industrial positions. These criteria are used in the selection and training of personnel assigned to the client companies. In another example based on a survey of five service organizations and nine manufacturers, Jeffrey (1995) identified 15 competencies that these organizations and their customers viewed as important in customer service activities by front-line employees. Many of these competencies apply to both front-line personnel and backoffice personnel.

To help in personnel selection, one human resource firm is developing a series of 50 to 100 questions to pose to prospective employees who would be assigned to client companies. The firm has data on client satisfaction with individual employees (from marketing research studies). Employees who are rated superior by clients answer certain questions differently than other employees who are not

TABLE 22.9   Criteria for Excellence

Clerical	Technical-Industrial
Punctuality	Being on time
Productivity	Showing up every day
Job skills	Having the right skills
Attitude	Keeping busy
Attire	Following safety procedures
Communication skills	Being productive
Employee preparation	Following safety procedures
Quick response by branch	Working together
	Norrell office responsiveness

superior. (Other questions result in the same response from most employees.) Responses to these "differentiating questions" will help to select new employees.

McDonald's Corporation uses an innovative job interview process for new "crew members." The interviewer asks "targeted questions" that probe the degree of customer satisfaction orientation of the applicant, teamwork orientation, work standards, and job fit. Applicants are asked to respond in terms of their own work experience. In addition, "targeted situations" are presented, and applicants are asked what action they would personally take in the situation. These situations cover interactions both with customers and with other team members.

Personality is one important attribute for many (but not all) positions in the operations function. This is increasingly the case as organizing by teams becomes more prevalent. One chemical manufacturer even places job applicants in a team problem-solving situation as part of the selection process.

One tool for evaluating personality types is the Myers-Briggs Type Indicator. This personality test describes 16 personality types that are based on four preference scales: extrovert or introvert, sensing or intuition, thinking or feeling, and judgment or perception. Thus one personality type is an extrovert, sensing, thinking, judgment person. Analyzing responses to test questions from prospective or current employees helps to determine the personality types of individuals. Organizations need many personality types, and the Myers-Briggs approach describes the contributions to the organization of each of the 16 types. By understanding the types and making job assignments accordingly, an organization can take advantage of all personality types to achieve high performance in the workplace. McDermott (1994) explains the 16 types and how the tool can help in recruiting new personnel and assigning current personnel.

Next we will consider some training aspects with respect to quality. It should be emphasized, however, that even well-planned training cannot make up for the lack of personal characteristics of people that are essential for certain positions. Thus intensive efforts in the selection process are justified.

Training.   The general subject of training for quality is treated in Section 16, Training for Quality. The training required for operations personnel depends on the responsibilities assigned. Major areas of training are

1. *Job skills:*   This is the minimal training. Such training must include provisions for updating, as knowledge on special knacks or other process information becomes available. Critical skills such as welding should have formal skills testing to certify that personnel can apply their training to make product that meets specifications. Passing these tests becomes a requirement for this job. Instances of falsification of tests have occurred, and steps must be taken to ensure valid results.

2. *Problem-solving tools:*   Depending on the responsibilities assigned to operations, training in problem solving may be needed. A notable example is the training provided to quality circles (data collection, cause-and-effect diagrams, Pareto analysis, histograms and other graphic techniques, etc.).

3. *Process control tools:*   Increasingly, production workers are receiving training in statistical control charts and other analysis techniques for routine control of a process.

4. *Importance of meeting specifications:*   It is useful periodically to reinforce the importance of meeting all specifications. In one chemical company, visits are made by workers to customer sites. Immediately after each visit, the workers hold a "reflections meeting" to discuss their observations and to decide how to get the message to the rest of the work force.

5. *Basic skills:*   To function effectively in a world of increasing technology and complex information, operations personnel must have basic skills in communication and mathematics. Communication skills include reading, writing, speaking, and listening; mathematical skills include arithmetic and recording and graphing of data. The 1993 National Adult Literacy Survey revealed that 47 percent of American adults have such poor literacy skills that they are unable to perform tasks that are more difficult than filling out a bank deposit slip or finding an intersection on a street map. Do not be surprised if you discover that some personnel do not understand that $8^3/4$ is the same as 8.75. We cannot assume that all personnel possess the basic skills required of specific positions. For a discussion of how basic skills can have an impact on quality, see Perkins (1994).

One organization sets aside 5 percent of working time for training, and when the training is completed, personnel receive a 5 percent bonus. This applies at all levels in the organization.

To ensure that training efforts meet the job needs of those being trained, it is useful to formalize the setting of priorities for the training. One approach is illustrated in Figure 22.12. The position is coordinating marketing and human resource activities in a department at a community bank. To relate job abilities to training priorities, the department manager and the employee jointly create the matrix in Figure 22.12. A scale of 1 to 10 is used, with 10 being the most important. The "Employee now" column shows the current proficiency of the employee in each ability; the "Plan for improvement" column shows the goal of an agreed-on plan to enhance the ability; the "Index of improvement" column is the difference between the previous two columns; the "Priority weight" column is the product of the "Relative importance" and "Index of improvement" columns. In the last column, the raw priority scores are converted to percentages for easier comparison. This approach for setting training priorities represents two steps of an eight-step process called *training function deployment,* analogous to quality function deployment. For elaboration, see Stampen and Stampen (1995).

A national chain of restaurants provides four levels of training for new employees. A formal test (written plus a "practicum") is required at each level. Employees decide when they are ready for the test. As a level is achieved, the employee receives a pay increase and other benefits.

Solectron, a winner of the Malcolm Baldrige National Quality Award, has a work force of about 3000 employees representing 20 nationalities and 40 languages. For a description of the company's training approach and results in skills improvement for quality, communication skills, managerial leadership, and lead worker interpersonal training, see Yee and Musselwhite (1993).

Some organizations, such as IBM, are extending formal certification of skills to various levels of management. This means that even experienced managers receive training in new managerial skills and then pass a certification examination. Such certification is entered in a database of skills used to select managers for new positions.

Retention. Investing increased resources in selection and training leads to stronger efforts to retain these skilled employees. Compensation, of course, is an essential contributor to employee retention. Other factors, however, are also essential, including

1. Career planning and development
2. Designing jobs for self-control (see below)
3. Providing sufficient empowerment and other means for personnel to excel
4. Removing the sources of job stress and burnout
5. Providing continuous coaching for personnel
6. Providing for participation in departmental planning
7. Providing the opportunity to interact with customers (both external and internal)
8. Providing a variety of forms of reward and recognition

Retaining superior operations personnel, particularly in the fast-paced operations environment, is clearly important to achieve quality goals. It means hire the best people, give them the tools they need, train them, and reward them in tangible and intangible ways.

ORGANIZATIONAL FORMS ON THE OPERATIONS FLOOR

Many firms organize around functional departments having a well-defined management hierarchy. This applies both to the major functions (e.g., Operations, Marketing, Product Development) and also to sections within a functional department such as Operations.

Step 1	Step 2					
Identify Abilities	**Establish Training Priorities**					
	RELATIVE IMPORTANCE	EMPLOYEE NOW	PLAN FOR IMPROVEMENT	INDEX OF IMPROVEMENT (PLAN - NOW)	PRIORITY WGHT. (Importance Index)	PRIORITY WGHT. AS A PERCENTAGE
Coordinate Educational Offerings	9	9	9	0	0	0%
Develop Presentation Materials	6	8	9	1	6	3%
Employee Benefit Tracking	6	1	7	6	36	19%
Team Participation & Support	7	4	6	2	14	7%
Future H.R. Projects	8	2	7	5	40	21%
Payroll Back-up	9	1	7	6	54	29%
Design & Layout of Communications	5	3	5	2	10	5%
Facilitate Marketing Efforts	10	6	8	2	20	11%
Dvlp rltnshp w/ printers, ad agent, hotels	8	7	8	1	8	4%

FIGURE 22.12   Training priorities. (*Stampen and Stampen 1995.*)

Organizing by function has certain advantages—clear responsibilities, efficiency of activities within a function, and so on. But this organizational form also creates "walls" between the departments. These walls—sometimes visible, sometimes invisible—often cause serious communication barriers. The outcome can be efficient operations *within* each department but with a less-than-optimal result delivered to external (and internal) customers.

Clearly, the participation of the work force in planning and improvement has become a way of life. It seems likely that self-managing teams will replace the Taylor system (see Section 15, under Empowerment and Commitment: Self-Regulating Team).

The "organization of the future" will be influenced by the interaction of two systems that are present in all organizations: the technical system (equipment, procedures, etc.) and the social system (people, roles, etc.)—thus the name *sociotechnical systems* (STSs).

Much of the research on sociotechnical systems has concentrated on designing new ways of organizing work, particularly at the work force level. For example, supervisors are emerging as "coaches"; they teach and empower rather than assign and direct. Operators are becoming "technicians"; they perform a multiskilled job with broad decision making rather than a narrow job with limited decision making. Team concepts play an important role in these new approaches. Some organizations now report that within a given year, 40 percent of their people participate on a team; some organizations have a goal of 80 percent. Permanent teams (e.g., process team, self-managing team) are responsible for all output parameters, including quality; ad hoc teams (e.g., a quality project team) are typically responsible for improvement in quality. A summary of the most common types of quality teams is given in Table 22.10.

The literature on organizational forms in operations and other functions is extensive and increases continuously. For a discussion of research conducted on teams, see Katzenbach and Smith (1993). Mann (1994) explains how managers in process-oriented operations will need to develop skills as coaches, developers, and "boundary managers."

CONCEPT OF CONTROLLABILITY; SELF-CONTROL

When work is organized in a way that enables a person to have full mastery over the attainment of planned results, that person is said to be in a state of *self-control* and therefore can be held responsible for the results. Self-control is a universal concept, applicable to a general manager responsible for running a company division at a profit, a plant manager responsible for meeting the various goals set for that plant, a technician running a chemical reactor, or a bank clerk processing checks. The concept also applies to work teams.

To achieve self-control, people must be provided with a means for

1. Knowing what they are supposed to do, e.g., the product specification or the work procedure.

2. Knowing what they are actually doing, e.g., instruments to measure process variables or the amount of output conforming to quality requirements.

3. Regulating the process, e.g., the authority and ability to regulate the work process.

TABLE 22.10 Summary of Types of Quality Teams

	Quality project team	Quality circle	Business process quality team	Self-managing team
Purpose	Solve cross-functional quality problems	Solve problems within a department	Plan, control, and improve the quality of a key cross-functional process	Plan, execute, and control work to achieve a defined output
Membership	Combination of managers, professionals, and work force from multiple departments	Primarily work force from one department	Primarily managers and professionals from multiple departments	Primarily work force from one work area
Basis of and size of membership	Mandatory; 4–8 members	Voluntary 6–12 members	Mandatory; 4–6 members	Mandatory; all members in the work area (6–18)
Continuity	Team disbands after project is completed	Team remains intact, project after project	Permanent	Permanent
Other names	Quality improvement team	Employee involvement group	Business process management team; process team	Self-supervising team; semiautonomous team

The three basic criteria for self-control make possible a separation of defects into categories of "controllability," of which the most important are

1. *Worker-controllable:*   A defect or nonconformity is worker-controllable if all three criteria for self-control have been met.

2. *Management-controllable:*   A defect or nonconformity is management-controllable if one or more of the criteria for self-control have not been met.

The theory behind these categories is that only the management can provide the means for meeting the criteria for self-control. Hence any failure to meet these criteria is a failure of management, and the resulting defects are therefore beyond the control of the workers. This theory is not 100 percent sound. Workers commonly have the duty to call management's attention to deficiencies in the system of control, and sometimes they do not do so. (Sometimes they do, and it is management who fails to act.) However, the theory is much more right than wrong.

Whether the defects or nonconformities in a plant are mainly management-controllable or worker-controllable is a fact of the highest order of importance. To reduce the former requires a program in which the main contributions must come from the managers, supervisors, and technical specialists. To reduce the latter requires a different kind of program in which much of the contribution comes from the workers. The great difference between these two kinds of programs suggests that managers should quantify their knowledge of the state of controllability before embarking on major programs.

An example of controllability study is given in Table 22.11. A diagnostic team was set up to study scrap and rework reports in six machine shop departments for 17 working days. The defect cause was entered on each report by a quality engineer who was assigned to collect the data. When the cause was not apparent, the team reviewed the defect and, when necessary, contacted other specialists (who had been alerted by management about the priority of the project) to identify the cause. The purpose of the study was to resolve a lack of agreement on the causes of chronically high scrap and rework. It did the job. The study was decisive in obtaining agreement on the focus of the improvement program. In less than 1 year over $2 million was saved, and important strides were made in reducing production backlogs.

Controllability also can be evaluated by posing specific questions for each of the three criteria of self-control. (typical questions that can be posed are presented below.) Although this approach does not yield a quantitative evaluation of management-controllable and worker-controllable defects, it does show whether the defects are primarily management-controllable or worker-controllable.

TABLE 22.11   Controllability Study in a Machine Shop, %

Management-controllable	
Inadequate training	15
Machine inadequate	8
Machine maintenance	8
Other process problems	8
Materials handling	7
Tool, fixture, gauge (TFG) maintenance	6
TFG inadequate	5
Wrong material	3
Operation run out of sequence	3
Miscellaneous	5
Total	68
Worker-controllable	
Failure to check work	11
Improperly operated	11
Other (e.g., piece mislocated)	10
Total	32

In my experience, defects are about 80 percent management-controllable. This figure does not vary much from industry to industry but varies greatly among processes. Other investigators, in Japan, Sweden, the Netherlands, and Czechoslovakia, have reach similar conclusions.

While the available quantitative studies make clear that defects are mainly management-controllable, many industrial managers do not know this or are unable to accept the data. Their long-standing beliefs are that most defects are the result of worker carelessness, indifference, and even sabotage. Such managers are easily persuaded to embark on worker-motivation schemes which, under the usual state of facts, aim at a small minority of the problems and hence are doomed to achieve minor results at best. The issue is not whether quality problems *in industry* are management-controllable. The need is to determine the answer *in a given plant.* This cannot be answered authoritatively by opinion but requires solid facts, preferably through a controllability study of actual defects, as in Table 22.11.

The concept of self-control draws attention to the importance of manufacturing planning. Manufacturing planning for quality is the means of *prevention* of both management- and worker-controllable defects on the manufacturing floor.

Collins and Collins (1993) provide six examples from the manufacturing and service sectors illustrating problems that were originally blamed on people but which really were management-controllable (often called *systems-controllable*). A similar situation is found in the service sector. Berry, Parasuramen, and Zeithmal (1994) identified 10 lessons learned in improving service quality. Three of these are service design ("The real culprit is poor service system design"), employee research (Ask employees why service problems occur and what they need to do their jobs), and servant leadership (Managers must serve by coaching, teaching, and listening to employees). Note that these three lessons are directly related to the concept of self-control.

Often in practice the three criteria are not fully met. For example, some specifications may be vague or disregarded (the first criterion); feedback of data may be insufficient, often vague, or too late (the second criterion); and people do not know how to correct a process (the third criterion).

The section on operations focuses on the factory floor and backroom operations in service firms. The concept of self-control, however, also applies to operations activities that involve extensive front-line customer contact.

The freedom provided to individuals working in a state of self-control inspires initiative, creativity, and a sense of well-being, all leading to self-development of the individual. Designing—and maintaining—work activities to meet the three criteria of self-control is a prerequisite to motivating personnel to achieve quality goals. Only management can create and maintain the conditions for self-control. If jobs are designed for self-control, management might hear a chorus of appreciation from the work force—followed by a smash hit of success on quality. Self-control is related to the broader concept of democracy in the workplace [see Rubinstein (1993) for elaboration]. The three criteria for self-control are discussed below.

KNOWLEDGE OF "SUPPOSED TO DO"

This knowledge commonly consists of the following:

1. The product standard, which may be a written specification, a product sample, or other definition of the end result to be attained.

2. The process standard, which may be a written process specification, written process instructions, an oral instruction, or other definition of "means to an end"

3. A definition of responsibility, i.e., what decisions to make and what actions to take (discussed earlier in this section)

Product Specifications. The ideal source of knowledge is the use required by the user. In most situations this is translated into a product specification. In developing these product specifications, some essential precautions must be observed.

Provide Unequivocal Information.   Two obstacles to proper knowledge can exist:

1. The specification may be vague. For example, when fiberglass tanks are transported in vehicles, the surface of the supporting cradles should be smooth. It was recognized that weld spatter would be deposited on the cradle surface, so an operation was specified to scrape the surface "smooth." However, there was no definition of "how smooth," and many rejections resulted.

2. There may be conflicting specifications. The supervisor's "black book" has had a long, durable career. Changes in specifications may fail to be communicated, especially when there is a constant parade of changes. In one instance, an inspector rejected product that lacked an angle cut needed for clearance in assembly. It was discovered that the inspector was using drawing revision D, the production floor had used revision B, and the design office had issued revision E just 3 days before.

Provide Information on Seriousness.   All specifications contain multiple characteristics, and these are not equally important. When workers are informed of the vital few characteristics, their emphasis is better placed.

Explain the "Why."   Explanation of the purposes served by the product and by the specification enlarges the knowledge of "supposed to do" and provides motivation through the resulting feeling of participation.

For example, a specification on weight called for a nominal value of 40.0 g with a tolerance of ±0.5 g. Although the total tolerance of 1.0 was being met, most of the tolerance range was being used up, and this created some problems later in assembly. A process capability study showed that the process capability was 0.10 g—far better than the tolerance of 1.0 g. But why was most of the tolerance range being used? Discussion revealed that (1) workers had not been told of the impact of inconsistent weights on later assembly, and (2) workers had not been instructed on centering the process to the nominal specification value.

Provide Standards.   In those cases where the specification cannot be quantitative, physical or photographic standards should be provided. There is an extensive array of needs here, especially on widely prevailing characteristics such as product appearance. (For years, enormous numbers of electrical connections were soldered in the absence of clear standards for an acceptable soldered connection.) If these standards are not provided by the managers and engineers, then, by default, the standards will be set by the inspectors and workers.

Process Specifications.   Work methods and process conditions (e.g., temperature, pressure, time cycles) must be unequivocally clear. A steel manufacturer uses a highly structured system of identifying key process variables, defining process control standards, communicating the information to the work force, monitoring performance, and performing diagnosis when problems arise. The process specification is a collection of process control standard procedures. A procedure is developed for controlling each of the key process variables (variables that must be controlled in order to meet specification limits on the product). The procedure answers the following questions:

1. What are the process standards?
2. Why is control needed?
3. Who is responsible for control?
4. How is measurement made?
5. When is measurement made?
6. How are routine data reported?
7. Who is responsible for data reporting?
8. How is audit conducted?

9. Who is responsible for audit?

10. What is done with product that is out of compliance?

11. Who developed the standard?

Often, detailed process instructions are not known until workers have experience with the process. Updating of process instructions based on job experience can be conveniently accomplished by posting a cause-and-effect diagram (see Section 5, under Formulation of Theories, Arrangement of Theories) in the Operations Department and inviting employees to attach index cards to the diagram. Each card recommends additional process instructions based on recent experience.

The Primester Division of Eastman Chemical immediately communicates changes in product and process specifications electronically to operations, and the software includes checks for understanding, assimilation, and retention of the changes.

Ford and Leader (1989) explain experiences in integrating group dynamics, communication skills, conflict management, and other "human dynamics" issues in a statistical process control activity. Newberg and Nielsen (1990) explain an approach to "operator control" in which operators participate to remove barriers (to operator control), develop process controls, and receive specific job training for a process producing soup. This participation includes operator use of flow diagrams to establish "critical control points" for the process.

Checklist for Manufacturing. The preceding discussion covers the first criterion of self-control; people must have the means for knowing what they are supposed to do. To evaluate adherence to this criterion, a checklist of questions can be created, including the following:

1. Are there written product specifications, process specifications, and work instructions? If written in more than one place, do they all agree? Are they legible? Are they conveniently accessible to the worker?

2. Does the specification define the relative importance of different quality characteristics? Are advisory tolerances on a process distinguished from mandatory tolerances on a product? If control charts or other control techniques are to be used, is it clear how these relate to product specifications?

3. Are standards for visual defects displayed in the work area?

4. Are the written specifications given to the worker the same as the criteria used by inspectors? Are deviations from the specification often allowed?

5. Does the worker know how the product is used?

6. Has the worker been adequately trained to understand the specification and perform the steps needed to meet the specification? Has the worker been evaluated by test or other means to see if he or she is qualified?

7. Does the worker know the effect on future operations and product performance if the specification is not met?

8. Does the worker receive specification changes automatically and promptly?

9. Does the worker know what to do with defective raw material and defective finished product?

10. Have the responsibilities in terms of decisions and actions been clearly defined?

(A checklist for self-control as applied to manufacturing operations was originally presented by L. A. Seder in the second edition of this handbook.)

The manufacturing sector has a long history of documenting quality and other requirements in the form of product and process specifications, work procedures, and other forms of written information. In the service sector, the formalization of quality requirements and associated documentation is now evolving. Highly detailed product specifications and process specifications are not yet common documents in service firms. Nevertheless, providing personnel in the service sector with the knowledge of what they are supposed to do is essential for self-control.

A framework starts with identifying control subjects. Control subjects are the features (or characteristics) that must be addressed to meet customer needs. Control subjects are a mixture of

Features of the product: Some control is carried out by evaluating features of the work product itself (e.g., the time to process an application or the completeness of a report). In manufacturing, these features are described in a product specification.

Features of the process: Much control consists of evaluating those features of the work process which directly affect the product features and therefore customer needs (e.g., the availability of equipment, the staffing levels for a service desk, the frequency of "out of stock" conditions, etc.). In manufacturing, a "process specification" describes these features. Such a specification is translated into procedures for use by operations personnel. Gass (1994) explains an approach for preparing and implementing procedures that encourages the use of the procedures on the operations floor. For service processes, Pyzdek (1994) describes a framework of service systems engineering.

Side effects: Some features that do not affect the work product directly but which may create troublesome side effects (e.g., irritations to employees, offense to the neighborhood, or threats to the environment) also can be control subjects.

Examples of control subjects and their relation to products and processes are shown in Table 22.12. To choose control subjects requires these steps: identify the major work process, identify the process objective, describe the work process, identify customers of the process, discover customer needs, and finally, select the control subjects. For elaboration, see Section 3, under Step 6: Develop Process Controls.

For self-control, these control subjects should be quantified and measured using appropriate units of measure and sensors. This quantification involves two kinds of indicators that must be made explicit for those running the process:

1. *Performance indicators:*   These measure the output of the process and its conformance to customer needs as defined by the unit of measure for the control subject.

2. *Process indicators:*   These measure activities or variation within the process that affect the performance indicators.

For clarity to personnel running a process, these indicators should have target values and maximum and minimum limits, where appropriate.

TABLE 22.12   Control Subjects

Major work product	Major work process	Control subjects
Photo developing	Film processing	Maintenance of chemicals
		Accuracy of placement of film on spool
Medical insurance	Claim processing	Accuracy of claim form
		Completeness of supporting documentation
Printing	Billing process	Accuracy of invoices
		Maintenance of customer information
Over-the-counter cold medications	Packaging of bottles	Safety seals
		Number of tablets per bottle
Catering services	Food preparation	Freshness of ingredients
		Oven temperature
Industrial tubing	Manufacture of tubing	Speed of intrusion machine
		Heat of machine
24-Hour banking services	Maintenance of ATM machines	Availability of cash
		Number of service people available

Source: Juran Institute (1995, pp. 1–45).

Checklist for Services.   Based on research (Shirley and Gryna 1998) with personnel in backroom operations of the financial services industry, the following questions can help to evaluate if personnel "know what they are supposed to do":

Work Procedures

1. Are job descriptions published, available, and up to date?
2. Do personnel know who their customers are? Have they ever met them?
3. Do personnel who perform the job have any impact on the formulation of the job procedure?
4. Are job techniques and terminologies consistent with the background and training of personnel?
5. Are there guides and aids (e.g., computer prompts) that lead personnel to the next step in a job?
6. Are there provisions to audit procedures periodically and make changes? Are changes communicated to all affected personnel?
7. Are there provisions for deviations from "home office" directives to meet local conditions?
8. Are procedures "reader friendly"?
9. Does supervision have a thorough knowledge of the operations to provide assistance when problems arise?
10. Do procedures given to personnel fully apply to the job they do in practice?
11. Have personnel responsibilities been clearly defined in terms of decisions and actions?
12. Do personnel know what happens to their output in the next stage of operations and understand the consequences of not doing the job correctly?
13. If appropriate, is job rotation used?

Performance Standards

14. Are formal job standards on quality and quantity needed? If "yes," do they exist. Are they in written form?
15. Have personnel been told about the relative priority of quality versus quantity of output? Do personnel really understand the explanation?
16. Are job standards reviewed and changed when more tasks are added to a job?
17. Do personnel feel accountable for their output, or do they believe that shortcomings are not under their control?
18. Does information from a supervisor about how to do a job always agree with information received from a higher level manager?

Training

19. Are personnel given an overview of the entire organization?
20. Is there regularly scheduled training to provide personnel with current information on customer needs and new technology?
21. Do personnel and their managers provide input to their training needs?
22. Does training include the "why," not just the "what"?
23. Does the design of the training program consider the background of those to be trained?
24. Do the people doing the training provide enough detail? Do they know how to do the job?
25. Where appropriate, are personnel who are new to a job provided with mentors?

KNOWLEDGE OF "IS DOING"

This is the second criterion for self-control. For self-control, people must have the means of knowing whether their performance conforms to standard. This conformance applies to

1. The product in the form of specifications on product characteristics

2. The process in the form of specifications on process variables

The knowledge is secured from three primary sources: measurements inherent in the process, measurements by production workers, and measurements by inspectors.

Measurement Inherent in the Process.
Many processes are engineered to include much instrumentation. The resulting information provides a feedback to enable the workers to close the loop. Even where the feedback is into an automated system, the data are usually available to human workers acting as monitors.

Measurements by Workers
Where the worker is to use the instruments, it is necessary to provide training in how to measure, what sampling criteria to use, how to record, how to chart, and what kinds of corrective action to take. The difficulty of motivating the workers to follow these instructions is so widespread a problem that many companies go to great lengths to minimize the need for worker action by providing instruments that require little or no human effort to measure, record, and control.

When these instruments are provided to workers, it is also necessary to ensure that these instruments are compatible with those used by inspectors and in other operations later in the progression of events.

On one construction project, the "form setters" were provided with carpenter levels and rulers to set the height of forms prior to the pouring of concrete. The inspectors were provided with a complex optical instrument. The differences in measurement led to many disputes.

Control of the process is strengthened if the worker is provided with the type of gauge that provides numerical measurements on a characteristic rather than providing accept-reject information.

A problem arises when the measurement necessary to control a process must be made in a laboratory off the production floor. The time required to send a sample to the laboratory, to have the analysis made, and to have the data relayed back to production can result in a delay to proper control of a process. One solution is the development of auxiliary measuring devices that can be used on the production floor by the worker and thereby provide immediate feedback. An example comes from a process used to control the concentration of chloride in a corn product derivative. Traditionally, the raw material undergoes centrifuging, a sample of the product is sent to a laboratory for analysis, the test results are forwarded to Production, and any necessary changes are then made in the centrifugal loads. (Chloride level, for the most part, is dependent on the load size of crystallized liquor being spun in the centrifugals.) Under the new setup, the worker takes a spot sample at the surge bin and analyzes the product for parts per million chloride on an ion analyzer and thereby directly regulates the process. Total time between processing a batch and obtaining a measurement plunges from 90 to 20 minutes. As a result, the amount of inferior product due to delayed process adjustment is greatly reduced.

Parker (1981) describes how the use of on-line gauges overcame problems in obtaining adequate sampling of product at a paper mill.

Measurements by Inspectors.
When an Inspection Department makes measurements that are to serve as a basis for action by Operations, the feedback usually goes to both workers and supervisors.

Criteria for Good Feedback to Workers.   The needs of production workers (as distinguished from supervisors or technical specialists) require that the data feedback read at a glance deals only with the few important defects, deals only with worker-controllable defects, provides prompt information about symptom and cause, and provides enough information to guide corrective actions. Criteria of good feedback are

1. *Read at a glance:*   The pace of events on the factory floor is swift. Workers should be able to review the feedback in stride.

Where the worker needs information about process performance over time, charts can provide an excellent form of feedback, provided they are designed to be consistent with the assigned responsibility of the worker (Table 22.13). It is useful to use visual displays to highlight recurrent problems. A problem described as "outer hopper switch installed backwards" displayed on a wall chart in large block letters has much more impact than the same message buried away as a marginal note in a work folder. Carlisle (1981) describes the effectiveness of such a system.

2. *Deal only with the few important defects:*   Overwhelming workers with data on all defects will result in diverting attention from the vital few.

3. *Deal only with worker-controllable defects:*   Any other course provides a basis for argument which will be unfruitful.

4. *Provide prompt information about symptom and cause:*   Timeliness is a basic test of good feedback the closer the system is to "real time" signaling, the better.

5. *Provide enough information to guide corrective action:*   The signal should be in terms that make it easy to decide on remedial action.

Software helps to collect, analyze, and display process data on a real-time basis. The control chart in Figure 22.13 shows an example of operator feedback in printed wiring assembly manufacture at Group Technologies. The unit of measure is defects per million opportunities (DPMO); the "Alarm line" is the computed average DPMO. Depending on the color of this line, the operator receives guidance on controlling the process. If the color is black, the process is acceptable, and the operator maintains the process under current conditions; a yellow color alerts the operator to exercise caution; a red color directs the operator to stop the process and seek help from a supervisor or a "reaction team." Limits for each of the zones are set by the process engineer based on customer specifications.

TABLE 22.13   Worker Responsibility versus Chart Design

Responsibility of the worker is to	Chart should be designed to show
1. Make individual units of product meet a product specification	The measurements of individual units of product compared to product specification limits
2. Hold process conditions to the requirements of a process specification	The measurements of the process conditions compared with the process specification limits
3. Hold averages and ranges to specified statistical control limits	The averages and ranges compared to the statistical control limits
4. Hold percent nonconforming below some prescribed level	Actual percent nonconforming compared to the limiting level

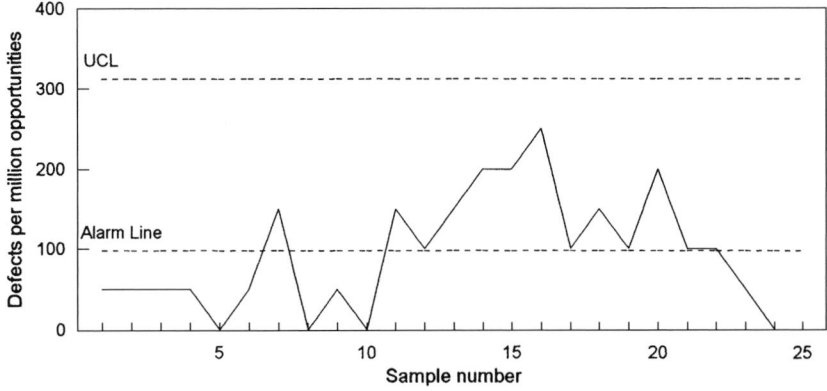

FIGURE 22.13   Operator feedback chart. (*Group Technologies Corporation internal document.*)

The control chart is just one of the many tools of statistical process control (SPC). SPC provides an arsenal of tools that operations people need to understand and apply. For elaboration, see Section 45, Statistical Process Control.

Feedback Related to Worker Action.   The worker needs to know what kind of process change to make to respond to a product deviation. Sources of this knowledge are

1. The process specification
2. Cut-and-try experience by the worker
3. The fact that the units of measure for product and process are identical

Lacking all these, the workers can only cut and try further or stop the process and sound the alarm.

Sometimes it is feasible for the data feedback to be supplemented with other graphic information that enables process personnel to decide on and take appropriate action on the process. Foster and Zirk (1992) explain how complex research and development information can be converted into "multiple curve plots" for use by process workers. An illustration is given in Figure 22.14. In a hydrocarbon cracking process, two of the operator-controlled variables are fuel capacity and burner air opening (BAO). The graphs serve as operating guides for workers to adjust the two variables.

Feedback to Supervisors.   Beyond the need for feedback at the workstations, there is need to provide supervisors with short-term summaries. These take several forms.

Matrix Summary.   A common form of matrix is workers versus defects; i.e., the vertical columns are headed by worker names and the horizontal rows by the names of defect types. The matrix makes clear which defect types predominate, which workers have the most defects, and what the interaction is. Other matrices include machine number versus defect type, defect type versus calendar week, and so on.

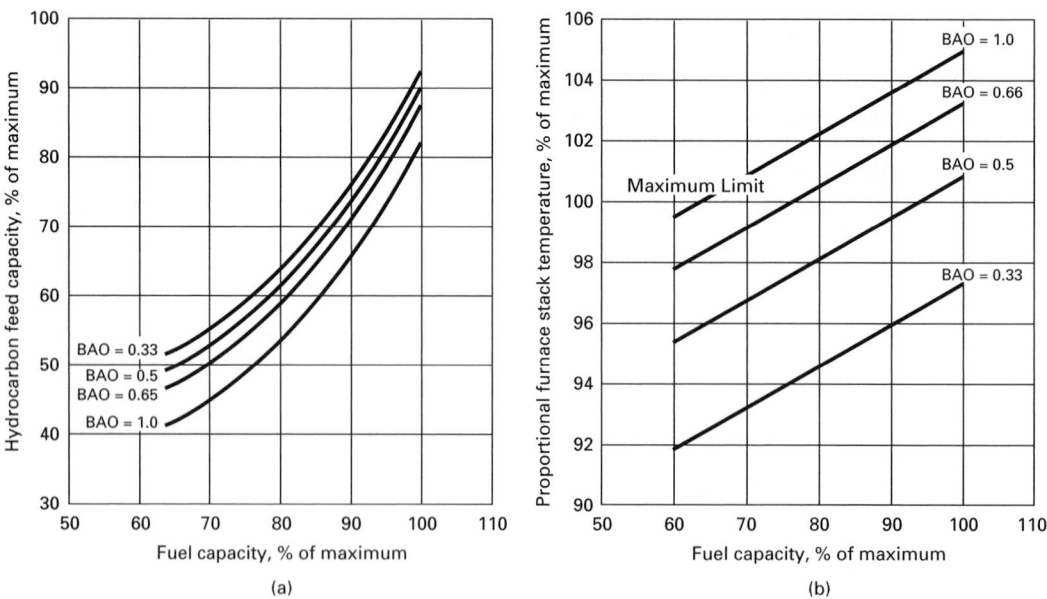

FIGURE 22.14   (*a*) Hydrocarbon feed capacity. (*b*) Proportional furnace stack temperature. [Parameter: burner air opening, area fraction; fixed condition: furnace draft = 0.065 inches H_2O.] (*Foster and Zirk 1992.*)

When the summary is published, it is usual to circle the matrix cells to highlight the vital few situations that call for attention. An elaboration of the matrix is to split the cell diagonally, permitting the entry of two numbers, e.g., number defective and number produced.

Pareto Analysis.   Some companies prefer to minimize the detail and provide information on the total defects for the day plus a list of the top three (or so) defects encountered and how many of each. Increasingly, supervisors can monitor processes by reviewing summarized and detailed process data on a personal computer (PC) in their office. One manufacturer even makes data available on a PC at the homes of people who are "on call" to assist on process problems.

Special Graphing of Multiple Parameters.   Traditionally, individual parameters are tracked in a process. The single-measure approach can lead to focusing on the numbers and not on the practices that drive desired performance. Madigan (1993) describes a chart that provides a "holistic approach to understanding operations." First, measures are identified that describe factors that are critical to the success of the operation. For example, in a building services operation, the measures relate to training, cleaning area, safety, wages, and absentee rate. Data are gathered from four plants. For each measure, the plant with the best result serves as a benchmark. The *measures matrix chart* consists of concentric circles that plot the data from all five measures and provide a profile of the average values and a profile of each plant. The approach is used in both service and manufacturing applications at Eastman Kodak. For further explanation, see Madigan (1993).

Automated Quality Information.   Production volume and complexity are important factors in determining the role of the computer. However, the role can be important even in relatively simple processes. For example, in a fast-food franchise, the elapsed time in filling a customer order is visibly shown (the goal is 45 seconds); the number of calls waiting to be answered in an insurance service center is clearly displayed. Section 10, Computer Applications in Quality Systems, explains the role of computers in analyzing and reporting data during production and other phases of the product life cycle.

Checklist for Manufacturing   A checklist to evaluate the second criterion of self-control includes questions such as

1. Are gauges provided to the worker? Do they provide numerical measurements rather than sort good from bad? Are they precise enough? Are they regularly checked for accuracy?

2. Is the worker told how often to sample work? Is sufficient time allowed?

3. Is the worker told how to evaluate measurements to decide when to adjust the process and when to leave it alone?

4. Is there a check to see that the worker does follow instructions on sampling work and making process adjustments?

5. Are inspection results provided to the worker, and are these results reviewed by the supervisor with the worker?

Examples of the second criterion of self-control also can be found in the service sector. These focus on identification and measurement of service indicators. A credit card provider has identified 18 key processes covering all activities: 2 business processes, 6 support processes, 10 product and service production and delivery processes. Two examples of production processes are credit screening and payment processing. For the total of 18 processes, over 100 internal and supplier process measures were defined. (In addition, external customer satisfaction research is conducted.) Daily and monthly performance results are available through video monitors and are also posted. Each morning, the head of operations meets with senior managers to discuss the latest results, identify problems, and propose solutions. Employees can access a summary of this meeting via telephone or electronic mail.

To emphasize the importance of quality, the measurement system is linked to compensation by a daily bonus system that provides up to 12 percent of base salary for nonmanagers and 8 to 12 percent for managers. For elaboration, see Davis et al. (1995).

Hestand (1991) explains how a bank uses a "report card" to measure the services provided to internal customers. Performance on specific quality measures is rated from A (excellent) to F (fails to meet requirements). At each participating branch, ratings are provided by a branch manager (BR MAN), an operations manager (OP MAN), and a customer service representative (CSR). A sample of the results for two measurements is presented in Table 22.14. Thus, for the accuracy of proof item encoding measurement, the total frequency of grades reported ranged from 2 A's, to 38 B's, to 0 F's. (The letter grades are also expressed in numerical form.) Further detail is automatically provided on D and F grades along with a procedure on follow-up toreport action taken.

One fast-food firm creates teams of "crew members" (workers at one location) who are trained to manage the site without a full-time manager (Harvard Business School 1994). Not only does this mean installing on-line technology such as the time to prepare an order, but it also means providing crew members with the same operating and financial information provided to a restaurant general manager to run the site. At these "team-managed units," the crew members make decisions on such matters as ordering food. Thus knowledge that long separated "brain workers" from "hand workers" now resides in a computer on the operations floor.

Lochner (1995) explains the use of a balanced set of measures in health care activities and makes the distinction between process measures (room occupancy rate) versus results measures (waiting time at admissions). Latzko (1993) explains the use of quality measurements to identify key potential "quality deviations" in a retail lending function at a bank. Early (1989) provides guidance on developing quality measures in the service sector.

Checklist for Services.   A checklist developed from discussions with personnel in backroom operations in financial service is provided below (Shirley and Gryna 1998):

Review of Work

1. Are personnel provided with the time and instructions for making self-review of their work?
2. Can errors be detected easily?
3. Are independent checks on quality needed? Are they performed? Are these checks performed by peer personnel or others?

TABLE 22.14   Examples of Quality Measurements from Internal Customers: Proof Department

72. Promptness of responses to special requests							
	A	B	C	D	F	N/A	?
Total	7	26	30	5	0	22	3	2.51
BR MAN	1	7	10	2	0	11	0	2.35
OP MAN	3	9	12	1	0	6	0	2.56
CSR	3	10	8	2	0	5	3	2.61

73. Accuracy of proof item encoding for transactions received by branch							
	A	B	C	D	F	N/A	?
Total	2	38	23	3	0	26	0	2.59
BR MAN	1	10	6	1	0	13	0	2.61
OP MAN	1	20	9	0	0	1	0	2.73
CSR	0	8	8	2	0	12	0	2.33

Source: Hestand (1991).

4. Is a review of work performed at various checkpoints in a process, not just when work is completed? Is the sample size sufficient?

5. Is there an independent audit of an entire process to ensure that individual work assignments are integrated to achieve process objectives?

6. Where appropriate, are detailed logs kept on customer contracts?

Feedback

7. Do upper management and supervision both provide the same message and actions on the importance of quality versus quantity?

8. If needed, do standards exist on making corrections to output?

9. Where appropriate, is feedback provided to both individuals and a group of personnel? Is time provided for discussion with the supervisor, and does the discussion occur?

10. Is feedback provided to those who need it? Is it timely? Is it personnel specific?

11. Does feedback provide the level of detail needed particularly to correct problem areas? Have personnel been asked what detail is needed in the feedback?

12. Is feedback provided from customers (external or internal) to show the importance of the output and its quality?

13. Does feedback include information on both quality and quantity?

14. Is positive in addition to negative (corrective) feedback provided?

15. Is negative (corrective) feedback given in private?

16. Do personnel receive a detailed report of errors by specific type of error?

17. Where appropriate, are reports prepared describing trends in quality (in terms of specific errors)? Is this done for individual personnel and for an entire process performed by a group of people?

18. Are there certain types of errors that are tracked with feedback from external customers? Could some of these be tracked with an internal early indicator?

ABILITY TO REGULATE

This is the third criterion for self-control. Regulating the process must always include both the authority to regulate and the ability to regulate. Regulation depends on a number of conditions, including

1. The process must be capable of meeting the specifications: (See above under Process Capability.)

2. The process must be responsive to regulating mechanisms in a predictable cause-and-effect relationship (this is essential to minimize variation): In a process for making polyethylene film, the workers were required to meet multiple product parameters. The equipment had various regulating devices, each of which could vary performance with respect to one or more parameters. The workers, however, could not "dial in" a predetermined list of settings that would meet all parameters. Instead, it was necessary to cut and try in order to meet all parameters simultaneously. During the period of cut and try, the machine produced nonconforming product to an extent that interfered with meeting standards for productivity and delivery. The workers were unable to predict how long the cut-and-try process would go on before full conformance was achieved. Consequently, it became the practice to stop cut and try after a reasonable amount of time and to let the process run, whether in conformance or not.

Skrabec (1991) describes how a cause-and-effect diagram can be combined with a process flow diagram to relate key input variables to key output variables. The result provides a guide for making process changes.

3. *The worker must be trained in how to use the regulating mechanisms and procedures:* This training should cover the entire spectrum of action: under what conditions to act, what kind and extent of changes to make, how to use the regulating devices, and why these changes need to be done.

Of three qualified workers on a food process, only one operated the process every week and became proficient. The other two workers were used when the primary worker was on vacation or was ill, and thus they never became proficient. Continuous training of the relief people was considered uneconomical, and agreements with the union prohibited their use except under the situations cited. This problem is management-controllable; i.e., additional training or a change in union agreements is necessary.

4. *The act of adjustment should not be personally distasteful to the worker, e.g., should not require undue physical exertion:* In a plant making glass bottles, one adjustment mechanism was located next to a furnace area. During the summer months, this area was so hot that workers tended to keep out as much as possible.

When the regulation consists of varying the human component of the operation, the question of process capability arises in a new form: Does the worker have the capability to regulate? This important question is discussed in Section 5, under Technique Errors, which includes some examples of discovering worker "knack."

5. *The process must be maintained sufficiently to retain its inherent capability:* Without adequate maintenance, equipment breaks down and requires frequent adjustments—often with an increase in both defects and variability around a nominal value. Clearly, such maintenance must be both preventive and corrective. The importance of maintenance has given rise to the concept of total productive maintenance (TPM). Under this approach, teams are formed to identify, analyze, and solve maintenance problems for the purpose of maximizing the uptime of process equipment. These teams consist of production line workers, maintenance personnel, process engineers, and others as needed. Problems are kept narrow in scope to encourage a steady stream of small improvements. Examples of improvement include a reduction in the number of tools lost and simplification of process adjustments.

Process Control Tools.   The tools selected are often related to one of five forms of dominance in a process: setup, time, component, worker, and information. For a listing of specific process control tools related to each form, see above under Planning Process Controls.

Checklist for Manufacturing.   A checklist to evaluate the third criterion of self-control typically includes such questions as the following:

1. Has the quality capability of the process been measured to include both inherent variability and variability due to time? Is the capability periodically checked?
2. Has the worker been told how often to reset the process or how to evaluate measurements to decide when the process should be reset?
3. Is there a process adjustment that the worker can make to eliminate defects? Under what conditions should the worker adjust the process? When should the worker shut down the machine and seek more help? Whose help?
4. Have the worker actions that cause defects, and the necessary preventive action, been given to the worker, preferably in the written form?
5. Is there an adequate preventive maintenance program on the process?
6. Is there a hidden knack possessed by some workers that needs to be discovered and transmitted to all workers?

Following a brief discussion of empowerment, a similar checklist will be provided for the service sector.

Empowerment. In providing sufficient authority for process regulation, the concept of empowerment plays an important part. *Empowerment* is the process of delegating decision-making authority to lower levels within the organization. Particularly dramatic is empowerment of the work force. But empowerment goes far beyond delegating authority and providing additional training. It means encouraging people to take the initiative and broaden their scope; it also means being supportive if they make mistakes.

As employees become more empowered in their work, the feeling of ownership and responsibility becomes more meaningful. Further, the act of empowering employees provides evidence of management's trust. Additional evidence is furnished when management shares confidential business information with employees. For many organizations, such steps are clearly a change in the culture.

The concept of empowerment applies both to individuals and to groups of workers. Self-managed teams (see Section 15, under Empowerment and Commitment: Self-Regulating Team) provide an illustration of empowerment for groups of workers. With empowerment comes the need to redefine the basic roles of upper management, middle management, and the work force. One model at a bank looks like this:

Upper managers act as shapers and coaches. As shapers, they create, communicate, and support the organization's mission. As coaches, they help when asked but avoid entering into the day-to-day problems of middle management.

Middle managers not only run their areas of responsibility but also work as a group to integrate all parts of the organization. In addition, they support the work force by eliminating obstacles to progress.

The *workers* are the primary producers of the output for customers. Their closeness to and knowledge about their work means that they uses their empowerment to determine how the work can best be done.

Dramatic illustrations of empowerment of the work force from authorizing a worker to stop the production line to authorizing clerks to make check-cashing decisions span all industries. Shay et al. (1991) trace the history, approach, and results of empowerment at an unusual firm—a century-old manufacturing plant with seven unions. The new system uses self-managing teams with new roles for supervisors, operators, and inspectors. Hayes (1994) shows how an employee questionnaire can be developed to learn the extent of empowerment within a firm. Simons (1995) explains how in an environment of empowering personnel, a firm may be exposed to some business risks that can be minimized by following four approaches, which he describes.

Checklist for Services. A checklist for the third criterion, developed from discussions with personnel in backroom operations of financial service, is provided below (Shirley and Gryna 1998).

Job Design
1. Is the process (including procedures, equipment, software, etc.) given to personnel capable of meeting standards on quality and quantity of output? Has this capability been verified by trial under normal operating conditions?
2. Has the design of the job made use of the principles of error-proofing?
3. Does the job design minimize monotonous or unpleasant tasks?
4. Have provisions been made in the job design to anticipate and minimize errors due to normal interruptions in the work cycle?
5. Can special checks be created (e.g., balancing of accounts) to detect errors?
6. Can steps be incorporated in data entry processes to reject incorrect entries?
7. Does the job design include provisions for action when wrong information is submitted or information is missing as an input to a job?
8. Is paperwork periodically examined and obsolete records destroyed to simplify working conditions?

9. When volume of work changes significantly, are there provisions for adjusting individual responsibilities or adding resources?

10. Are there external factors (e.g., no account number on a check, cash received instead of a check, etc.) that hinder the ability to perform a task?

11. Are some personnel cross-trained for different tasks to provide an adequate supply of experienced personnel for filling in when needed?

12. If appropriate, is a "productive hour" scheduled each day in which phone calls and other interruptions are not allowed, thus providing time to be away from the work location to attend to other tasks?

13. Has equipment, including any software, been designed to be compatible with the abilities and limitations of personnel?

14. Is there an adequate preventive maintenance program for computers and other equipment used by personnel?

15. Is there a hidden knack possessed by some personnel that needs to be discovered and explained to all personnel?

16. For a job requiring special skills, have personnel been selected to ensure the best match of personnel skills and job requirements?

Changes in Job Design

17. Are proposed changes limited by technology (e.g., address fields on forms)?

18. Can personnel institute changes in a job when they show that the change will provide benefits? Are personnel encouraged to suggest changes?

19. What levels of approval by management are required for proposed changes to be instituted? Could certain types of changes be identified as not needing any level of management approval?

20. Do management actions confirm that they are open to recommendations from all personnel?

Handling Problems

21. Have personnel been provided with the time and training to identify problems, analyze problems, and develop solutions? Does this include diagnostic training to look for patterns of errors and determine sources and causes?

22. Are personnel permitted to exceed permitted process limits (e.g., maximum time on a customer phone call) if they believe it is necessary?

23. When personnel encounter an obstacle on a job, do they know where to seek assistance? Is the assistance conveniently available?

Use of Checklists on Self-Control.   The checklists presented have several applications in operations:

1. *The design of new jobs and redesign of old jobs to assist in the prevention of errors and to place personnel in a state of self-control:*   The checklist can serve as a tool to evaluate all quality-related aspects of the job, e.g., clarity of formal requirements and job instructions, adequacy of feedback to control the job, capability of the process itself, and the means of regulating the process. When a re-engineering effort is in progress, the design of specific jobs within a process using self-control criteria ensures that the job requirements and the needs of personnel are in harmony.

2. *Analysis and diagnosis of current jobs that have quality problems:*   The search for root causes of current quality problems can be difficult and time-consuming. The checklist can help identify potential causes by providing an exhaustive list of candidate areas.

3. *Use by supervisors to discuss jobs with personnel:*   In reviewing both general job performance and current job problems, the checklist can focus on specific aspects of the job—some of which

are under control of the person and some of which are not—and help the supervisor to function as a coach. Similarly, the checklist can be beneficial to self-managing teams.

4. *Getting prepared for auditors on financial transactions:*   Auditors check output and procedures in many financial areas, e.g., moving money from one account to another. The checklist can help to prepare for the auditing process.

5. *Help to focus on a broad improvement strategy:*   Use of the checklists for reviewing current jobs may reveal that many job designs have basic weaknesses such as lack of clarity in quality requirements or job instructions, timely and relevant feedback on output, or capability of the process itself. Why not ask personnel to review the checklist and then prioritize the specific problem areas?

6. *Use in training classes in quality:*   The concept of self-control helps to plan new jobs and to analyze quality problems on current jobs. Exercises on the three elements of self-control can be made part of a training class. Checklists developed by the participants or the checklists in this section can serve as a basis for discussion by the participants.

You might wish to refine the checklists or develop your own to meet the needs of your organization. This could be done by circulating the checklist internally and asking for additions to the list, identification of critical items, or any input that would make the list more useful. Additional ideas also could be generated by using the list for an exercise and discussion in a training class on quality.

TROUBLESHOOTING

Chronic versus Sporadic Troubles.   Quality troubles exist in two different forms: chronic and sporadic. *Chronic* troubles go on and on because no way has ever been found to eliminate them. For example, a process has for years operated at 10 percent nonconforming. No one has succeeded in reducing this level, so we learn to live with it.

Sporadic troubles are the result of some sudden adverse change. For example, a process that is usually at 10 percent nonconforming suddenly goes to 25 percent. Such a change sets off a number of alarm signals that demand prompt action to restore the status quo (to go back to the usual 10 percent). *Troubleshooting* (also called *firefighting*) is the process of dealing with the sporadic and restoring quality to the original level. Section 5, The Quality Improvement Process, discusses a structured approach for dealing with chronic problems; Section 4, The Quality Control Process, presents a structured approach for sporadic problems.

For organizations that do not have a formal effort to reduce chronic and sporadic problems, operations managers often spend 30 percent of their time on troubleshooting; for the supervisors reporting to these managers, the time consumed frequently exceeds 60 percent.

Responsibility for Troubleshooting.   While Operations Department responsibility for troubleshooting is fairly clear, the ability to carry out this responsibility varies. The main variables are the *complexity* of the adverse change and the *extent* to which operations personnel are trained in the tools of diagnosis.

A complex adverse change can require an extent of data collection and analysis that goes beyond the training and experience of operations personnel. This same complexity also can require extensive time for data collection and analysis—time that is not available to Operations Department supervisors.

In such complex cases, a team approach may be needed. The team is usually drawn from the following:

Operations personnel to supply theories and authorize data collection

Technicians to carry out data collection

Diagnosticians to design the data plan and to analyze the subsequent data

"Outsiders" as the needs arise

The responsibility for creating such a team rests with the Operations Department.

The trend is to train operations personnel to become self-sufficient in troubleshooting. If non-supervisors are trained to do troubleshooting within their own work areas, the supervisors will have more time to participate on the cross-functional project teams that typically are needed for chronic troubles. The training needed is common to that needed by operations personnel for dealing with quality problems generally. The problem-solving training provided to quality teams is useful (see Section 5, under The Diagnostic Journey). At each step of the journey, the tools and techniques mentioned are candidates for training topics.

SELF-INSPECTION

Once goods or services have been produced, there arises the question: Do they conform to specification? In manufacturing industries, the activity to determine conformance is called *inspection* or *test* (see Section 23, Inspection and Test). In the service sector, typical terms used are *checking, examination, review,* and *reconciliation.*

In the manufacturing sector in the United States, the responsibility for making the conformance decision often rests with full-time inspectors in an independent Inspection Department but this is changing fast to in-process inspection by the worker with an audit inspection by an independent inspector. In the service sector, output is checked by the person creating the output, and typically no independent check occurs. Often the only independent check is by the customer (internal or external).

Under the concept of self-inspection, the worker who made the product also measures the product and decides whether it conforms to specifications. (Special or complex tests are performed by a separate department.)

Note that the worker is *not* given the responsibility for determining the disposition of any non-conforming product. Also, self-inspection does *not* involve transfer of full-time inspectors to the Production Department. It involves abolishing the jobs of full-time inspectors and having the inspection done on a part-time basis by the production workers. Provision is made for an audit (see below under Audit of Decisions).

Self-inspection has decided advantages over the traditional delegation of inspection to a separate department:

1. Production workers are made to feel more responsible for the quality of their work.
2. Feedback on performance is immediate, thereby facilitating process adjustments. Traditional inspection also has the psychological disadvantage of using an "outsider" to report the defects to a worker.
3. The costs of a separate Inspection Department can be reduced.
4. The job enlargement that takes place by adding inspection to the production activity of the worker helps to reduce the monotony and boredom that are inherent in many jobs.
5. Elimination of a specific station for inspecting all products reduces the total manufacturing cycle time.

The current emphasis on downsizing in organizations, coupled with the benefits of self-inspection, has resulted in pressures to reduce the size of inspection departments in manufacturing. Sometimes the reduction in independent inspection activities has been premature.

Criteria for Self-Inspection.   Before self-inspection can be adopted, some essential criteria must be met:

1. Quality is the number 1 priority within an organization. If this is not clear, the worker may succumb to schedule and cost pressures and classify as acceptable products that should be rejected.

2. There is mutual confidence between managers and workers. Managers must have sufficient confidence in the work force to be willing to entrust to them the responsibility of deciding whether the product conforms to specification. In turn, workers must have enough confidence in management to be willing to accept this responsibility.

3. The criteria for self-control are met. Failure to eliminate the management-controllable causes of defects suggests that management does not view quality as a high priority, and this may bias the workers during inspections.

4. Workers are trained to understand the specifications and perform the inspection. In some companies, "certification" (for making product conformance decisions) is issued only to those workers who demonstrate their competence.

5. Specifications are unequivocally clear.

6. Workers understand the use that will be made of the products (internally and externally) in order to grasp the importance of a conformance decision.

7. The process permits assignment of clear responsibility for decision making.

Several references provide elaboration on these criteria: Ziegler (1995) discusses six elements common to successful self-inspection; Whittingham (1987) explains some implementation details, including use of a "work conditions questionnaire" (to evaluate the criteria of self-control) and the importance of feedback to workers.

Sequence for Instituting the Self-Inspection Concept.   The many benefits of self-inspection suggest that steps be taken to successfully apply it broadly. However, the criteria listed above are not easy to meet. In practice, it is unlikely that the criteria could be met for all products, all operations, and all personnel. It is best to apply the concept only to products and processes that are stabilized and meet product specifications and to personnel who have demonstrated their competence.

This competence can be verified by a trial period during which workers make conformance decisions while duplicate decision making is done by inspectors. The purpose of this duplication is to discover, through data, which workers consistently make good product-conformance decisions.

Audit of Decisions.   During the trial period, the inspection is conducted for two purposes:

1. Product approval, lot by lot
2. Comparison of inspector results with worker results

As the comparison establishes validity of the worker's decisions, the duplicate inspections are reduced in frequency until their prime purpose is to determine whether the worker continues to make good decisions (hence the name *audit of decisions*). At this stage, any knowledge of the product is incidental. If an audit reveals that wrong decisions were made by the workers, then the product evaluated since the last audit is suspect and must be investigated.

Results of Self-Inspection.   In a coning operation of textile yarn, the traditional method of inspection often resulted in finished cones sitting for several days in the inspection department, thereby delaying any feedback to production. Under self-inspection, the worker received immediate feedback and could more promptly get machines repaired and setups improved. Overall, the program reduced nonconformities from 8 to 3 percent. An audit inspection of the products that were classified by the workers as "good" showed that virtually all of them were classified correctly. In this company, workers also can classify product as "doubtful." In one analysis, worker inspections classified 3 percent of the product as doubtful, after which an independent inspector reviewed the doubtful product and classified two-thirds of it as acceptable and one-third as nonconforming.

A pharmaceutical manufacturer employed a variety of tests and inspections before a capsule product was released for sale. These checks included chemical tests, weight checks, and visual inspections of the capsules. A 100 percent visual inspection traditionally had been conducted by an Inspection Department. Defects ranged from "critical" (e.g., an empty capsule) to "minor" (e.g., faulty print). This inspection was time-consuming and frequently caused delays in production flow. A trial experiment of self-inspection by machine operators was instituted. Operators performed a visual inspection on a sample of 500 capsules. If the sample was acceptable, the operator shipped the full container to the warehouse; if the sample was not acceptable, the full container was sent to the Inspection Department for 100 percent inspection. During the experiment, both the samples and the full containers were sent to the Inspection Department for 100 percent inspection with reinspection of the sample recorded separately. The experiment reached two conclusions: (1) the sample inspection by the operators gave consistent results with the sample inspection by the inspectors, and (2) the sample of 500 gave consistent results with the results of 100 percent inspection.

The experiment convinced all parties to switch to the sample inspection by operators. Under the new system, good product was released to the warehouse sooner, and marginal product received a highly focused 100 percent inspection. In addition, the level of defects *decreased*. The improved quality level was attributed to the stronger sense of responsibility by operators (they themselves decided if product was ready for sale) and the immediate feedback received by operators from self-inspection. However, there was another benefit—the inspection force was reduced by 50 people and these 50 people, were shifted to other types of work, including experimentation and analysis activities on the various types of defects. Inspectors became analysts.

Schilling (1994) provides a sobering explanation of the importance of inspection and its relationship to acceptance sampling, acceptance control, and process control.

In the service sector, relatively little use is made of full-time independent personnel who check work output. Workers check their own work, with perhaps a sampling check by a supervisor. Thus an operations supervisor in a bank money transfer department spends about 1.5 hours a day in 10-minute segments going from clerk to clerk and examining the last piece of work completed, and a housekeeping supervisor in a hotel samples a number of rooms to verify the quality of the work performed by the maids.

The lack of extensive checking and inspection activity in the service sector certainly does not mean that all output conforms to service requirements or goals. Particularly in backroom activities, service processes have many rework loops. One example of extensive rework is the department of 32 people in a regional office of a utility. The sole purpose of this department is to detect errors in internal billing charges received from other units of the company. The budget of this department can easily be justified by the savings achieved in detecting the errors. Two obvious questions are (1) Was there any form of inspection by those who generated the charges? and (2) Are there any plans to prevent these errors?

AUDIT OF OPERATIONS QUALITY

A *quality audit* is an independent evaluation of various aspects of quality performance for the purpose of providing information to those in need of assurance with respect to that performance. Application to manufacturing has been extensive and includes both audit of activities (systems audits) and audit of product (product audit). For products (e.g., medical, financial) that are subject to government regulations, audits are often concerned with compliance to these regulations.

Systems Audit.   Systems audits (sometimes called *process audits*) can be conducted for any activity that affects the final quality of goods or services. The audit is usually made of a specific activity against a specific document, such as process operating instructions, employee training manuals, certification of personnel for critical operations, and quality provisions in purchasing documents. The checklists presented earlier in this section for the three criteria of self-control can suggest useful specific subjects for audits. Priority is assigned to subjects that affect customer satisfaction. Adherence to existing procedures is often emphasized, but systems audits often uncover situations of inadequate or nonexistent procedures.

Peña (1990) explains an audit approach for processes. Two types of audits are employed: engineering and monitor. The *engineering process audit* is conducted by a quality assurance engineer and entails an intense review of all process steps, including equipment parameters, handling techniques, and statistical process control. Table 22.15 shows the audit checklist. The *monitor process audit* is conducted by a certified auditor; it covers a broad range of issues, e.g., whether specifications are correct and whether logs are filled in and maintained. Discrepancies (critical, major, or minor) are documented and corrective action is required in writing. Critical defects must be corrected immediately; majors and minors must be resolved within 5 working days.

McDonald's Corporation conducts a system evaluation of restaurants using visits (announced and unannounced) by a trained consultant. The evaluation includes quality, service, cleanliness, and sanitation. Highly detailed audit items include numerical standards on food-processing variables. Key questions for the overall systems evaluation cover training, ordering, scheduling, production control, equipment, and leadership. An overall grade (A, B, C, or F) encompasses operational standards and customer expectations.

A major airline employs audits to evaluate service in three areas: airport arrival and departure, aircraft interior and exterior, and airport facilities. Forty-seven specific activities are audited periodically, and then performance measurements are made and compared with numerical goals. Two examples on the aircraft are the condition (appearance) of carpets inside the planes and the adhesion of paint on the planes.

Dedhia (1985) describes an audit system for an electronics manufacturer. The audit consists of 14 subsystems each having an audit checklist. Routine audits are performed by quality audit personnel on a scheduled basis. For selected activities, annual audits are conducted by a team from

TABLE 22.15 Audit Checklist

1. Is the specification accessible to production staff?
2. Is the current revision on file?
3. Is the copy on file in good condition with all pages accounted for?
4. If referenced documents are posted on equipment, do they match the specification?
5. If the log sheet is referenced in specifications, is a sample included in the specification?
6. Is the operator completing the log sheet according to specifications?
7. Are lots with out-of-specification readings authorized and taken care of in writing by the engineering department or the proper supervisor?
8. Are corrections to paperwork made according to specification?
9. Are equipment time settings according to specifications?
10. Are equipment temperature settings according to specification?
11. Is the calibration sticker on equipment current?
12. Do chemicals or gases listed in the specification match usage on line?
13. Do quantities listed in the specification match the line setup?
14. Are changes of chemicals or gases made according to specification?
15. Is the production operator certified? If not, is this person authorized by the supervisor?
16. Is the production operating procedure according to specification?
17. Is the operator performing the written cleaning procedure according to specification?
18. If safety requirements are listed in the specification, are they being followed?
19. If process control procedures are written in the specification, are the actions performed by process control verifiable?
20. If equipment maintenance procedures are written in the specification, are the actions performed verifiable? according to specification?

Source: Peña (1990).

Manufacturing, Quality Engineering, Test Engineering, Purchasing, and other departments. The system includes a numerical audit rating based on classifying each discrepancy as major or minor. A rating below 90 percent requires an immediate corrective action response. Craner (1994) explains how managers (and others) conduct audits at a medical device firm. Lane (1989) relates how a micro-electronics firm reduced redundant inspection and shifted the resources to a defect-prevention effort that included audits of broad systems, individual process, and products.

Who Performs the Audits?   There are several categories of personnel to whom systems audits may be delegated:

Production Management: In this situation, the middle or upper operations managers undertake the audit of execution versus plan. Because most production activities are highly visible, skilled observers can learn much from shop tours. Generally, operations managers possess these skills and, in addition, put a high value on direct observation.

Inspectors: Some inspections are conducted not to measure the product but to observe the process (see, for example, Section 23, under Patrol Inspection). Such observations are themselves a review of execution versus plan. It is often feasible to extend these patrol inspections to review other aspects of execution.

Independent auditors: For critical work, the auditing preferably should be done by those who are not a part of the Inspection Department. Usually such auditors review the practices of inspectors as well as production workers.

The independent audit tends to be more completely planned than an audit by operations management or by inspectors. In addition, the entire concept of the independent audit has the support of upper management, which receives the audit reports for review. For further discussion, see Section 11, under Quality System Certification/Registration.

A self-audit and an independent audit can be combined to provide a two-tier audit each with an audit plan, execution, and report. Advantages include using the expertise of the person responsible for the activity, ensuring objectivity with an independent auditor, and minimizing some of the human relationship issues. The aim of both the self-audit and the independent audit is to build an atmosphere of trust based on the reputation of the auditors, the approach used during the audit, and an emphasis on being helpful to the activity audited.

There are two trends in auditing worth noting. First, the scope of audits often goes beyond determining compliance with specific procedures and requirements to include broader issues such as the effectiveness, efficiency, and adaptability of processes (see Section 6, Process Management). Second, audits increasingly emphasize helping operational areas to meet customer expectations and needs. Myers and Heller (1995) provide an illustration of how the broader scope (modeled after the Baldrige Award) helps to align business processes with customer needs and also recognize employees' efforts. Audits having a broad companywide scope are often called *assessments.*

The audit of decisions discussed previously requires a regular examination (product audit) of product conformance along with the associated documentation. This cannot be done readily by the independent auditors who are on the scene so infrequently. Instead, it is assigned to a special category of auditor created at the time of delegating conformance decisions to production workers.

Product Audit.   This form of audit provides information on the extent of product conformance to specification and fitness for use. For elaboration, see Section 11.

OVERALL QUALITY MEASUREMENT IN OPERATIONS

The management of key work processes must include provision for measurement. In developing units of measure, the reader should review the basics of quality measurement discussed in Section 9, Measurement, Information, and Decision Making.

Table 22.16 shows examples for manufacturing activities and for backroom operations in the service sector. Note that the measurements cover both output from operations and input to operations. Also note that the examples include early indicators, concurrent indicators, and lagging indicators of performance.

The units in Table 22.16 become candidates for data analysis using statistical techniques such as control charts. More important, the selection of the unit of measure and the periodic collection and reporting of data demonstrate to operating personnel that management regards quality as having priority importance. This helps to maintain a focus on improvement, which we will discuss below.

Many control subjects for quality measurement are forms of work output. In reviewing current units in use, a fruitful starting point is the measure of productivity. *Productivity* is usually defined as the amount of output related to input resources. Surprisingly, some organizations still mistakenly calculate only one measure of output—the total (acceptable *and* nonacceptable). Clearly, the pertinent output measure is that which is usable by customers, *acceptable output*).

MAINTAINING A FOCUS ON CONTINUOUS IMPROVEMENT

Historically, the operations function always has been involved in troubleshooting sporadic problems (see above under Troubleshooting). As chronic problems were identified, these were addressed using various approaches, such as quality improvement teams (see Section 5, The Quality Improvement Process). Often the remedies for improvement involve quality planning or replanning (see Section 3, The Quality Planning Process). These three types of actions are summarized in Table 22.17.

Global competitive pressures and other forces will result in an even stronger emphasis on improvement. Continuous improvement in the future will need to

1. Draw on many sources of information to identify improvement opportunities that go beyond nonconformance to specifications. These sources of information include studies on the cost of poor quality (see Section 8, Quality and Costs), market research on customer satisfaction and loyalty (see Section 18, Market Research and Marketing), assessments of quality culture (see Section 15, Human Resources and Quality), and broad assessments using, for example, Baldrige Award criteria (see Section 14, Total Quality Management).

2. Address process improvement in terms of effectiveness, efficiency, adaptability, and cycle time (see Section 6, Process Management).

TABLE 22.16 Examples of Quality Measurements in Operations

Quality of output from operations
Percentage of output meeting specifications at initial inspection ("first-time yield")
Percentage of output meeting specifications at intermediate and final inspections
Amount of scrap, rework (quantity, cost, percentage, etc.)
Percentage of invoices returned due to errors
Average cycle time to fill customer orders for products or documentation
Warranty and adjustment costs due to errors in operations
Overall measure of quality (defects per million, weighted defects per unit, variability in units of standard deviation, etc.)

Quality of input to operations
Percentage of incoming material meeting specifications
Percentage of incoming data that is complete and error-free
Amount of downtime of manufacturing equipment, computer systems, and other support equipment
Percentage of critical operations with certified employees
Percentage of specifications or process instructions requiring changes after release

TABLE 22.17   Three Types of Action

Type of action to take	When to take action	Basic steps
Troubleshooting (part of quality control)	Performance indicator outside control limits	Identify problem
	Performance indicator in clear trend toward control limits	Diagnose problem
	Performance indicator normally meets target but does not now	Take remedial action
	Process indicator outside target range or control limits	
	Process indicator trend toward target or control limits	
Quality improvement	The control limits are so wide that it is possible for the process to be in control and still miss the targets	Identify project
		Establish project
	Performance indicator frequently misses its target	Diagnose the cause
		Remedy the cause
		Hold the gains
Quality planning	Many performance indicators for this process miss their targets frequently	Establish project
		Identify customers
	Customers have significant needs that the work product does not meet	Discover customer needs
		Develop product
		Develop process
		Design controls

Source: Adapted from Juran Institute (1995, pp. 5–7).

3. Pursue radical forms of improvement ("re-engineering") in addition to incremental forms (see Section 6, Process Management).

4. Effectively and quickly capture, share, and take action on experience-based information. A fast-food firm is creating an "intellectual network" of computer bulletin boards of information that include best practices information. Personnel will be able to use the system 24 hours a day, 7 days a week. Rethmeier (1995) explains how an alliance of 300 hospitals uses a "learning center" to create and transfer knowledge for continuous improvement. See also the *Harvard Business Review* (September–October 1994) for a special section on "regaining the lead in manufacturing." This section is based on a 4-year study of 20 development projects and identifies 7 elements of learning (core capabilities, guiding vision, organization and leadership, ownership and commitment, "pushing the envelope," prototypes, and integration).

5. Apply all the tools of improvement—technical and behavioral, simple and sophisticated. Increasingly, savings from improvement projects have "skimmed the cream off the top." The next round will require deeper analysis.

Many sections of this handbook present both methodologies and case examples on quality improvement; see particularly Section 5, The Quality Improvement Process, and Section 45, Statistical Process Control. The general literature is replete with examples from operations in virtually every industry. For examples in manufacturing see Kitamura et al. (1994) for a discussion of the Toyota production system; also, Hays and Gander (1993) explain how yield and turn-around time were improved on a printed circuit board production line. For examples in the service sector, see Aubrey and Gryna (1991) to learn about the experiences of over 1000 quality teams at a bank; also, Anderson et al. (1995) describe experiences (including advantages and disadvantages) in applying computer simulation to achieve improvement in claims processing and other activities in a health insurance firm.

QUALITY AND OPERATIONS CULTURE

For an organization to become superior in quality, it needs an unusual marriage:

1. Technologies to create products and processes that meet customer needs. Part of this is the design of individual jobs (that meet the criteria of self-control).

2. A culture throughout the organization that continually views quality as a primary goal. Quality culture is the pattern—the emotional scenery—of human habits, beliefs, and behavior concerning quality. Designing and maintaining jobs to meet the criteria of self-control are essential prerequisites to achieving a positive quality culture.

Some companies have a strong—but negative—quality ethic. Examples are legion: hide the nonconforming product, e.g., bury the rejected paint in the ground; and finesse the inspector, e.g., keep producing defective product and wait until an inspector discovers the situation. Such negative actions are often taken in order to achieve other objectives such as production quotas. To build a strong quality culture requires two steps: (1) collect information to determine the present quality culture and (2) take the steps necessary to change the culture.

Determining the Quality Culture.   Learning about the present quality culture in a firm can be accomplished by a carefully planned attitude survey on quality for various levels of operations supervision and the work force. However, be prepared for some sobering results. For a general discussion of quality culture, see Section 15. For a discussion of the results of a survey (20 questions) given to both American and Russian factory workers, see Pooley and Welsh (1994). Yavas and Burrows (1994) used 33 questions to compare the attitudes of American and Asian manufacturing managers on quality. Tabladillo and Canfield (1994) describe a 25-question survey employed at a hospital. Turner and Zipursky (1995) describe a survey of 20 questions that measure "employee commitment." The analysis of the results made use of several tools, including cause-and-effect diagrams, analysis of means on performance versus importance of factors, regression analysis, interrelation digraph, and quality function deployment. The road to developing a positive quality culture is lengthy and difficult—though essential for survival. The general approach of Section 15 to organizing for quality, the manager's role, and so on is germane to quality culture.

Changing the Quality Culture.   Developing a positive quality culture involves five key elements [for elaboration, see Juran and Gryna (1993, Chap. 8)]:

1. Create and maintain an awareness of quality. This means we must create and disseminate information on our current status of quality. The message must go to upper management, middle and lower management, and all other personnel—using languages that fit each territory.

2. Provide evidence of management leadership on quality. This is not only cheerleading but serving on a quality council, doing strategic planning for quality, providing resources for quality, and doing a host of other tasks to plan and deploy quality goals.

3. Provide for self-development and empowerment. This includes designing jobs for self-control, selection and training for jobs, organizing work using approaches for self-development such as self-managing teams, and encouraging personal commitment for quality.

4. Provide participation as a means of inspiring action. The forms of participation are almost endless: serve on a quality council, a quality circle, or an improvement team; be a process owner; take part in a product or process design review; or make presentations on quality.

5. Provide recognition and rewards. These expressions of esteem play an essential role in inspiring people on quality. Recognition takes the form of public acknowledgment for great performance on quality. Rewards are tangible benefits (salary increases, bonuses, promotions, etc.) for quality. Aside from these specifics, some countries that are moving toward democracy in the workplace

must address basic "quality of life" issues (e.g., clean bathrooms and other working conditions) before attempts at changing the quality culture will succeed.

Creating a positive culture is an important factor in building loyalty and retaining key personnel in operations. Reichheld (1993) explains the importance of loyal employees in achieving loyal customers.

ACKNOWLEDGMENTS

I extend my thanks to Charles Amaral and Jack Beatty of McDonald's Corporation, Erik Bredal and Kim O'Haver of Group Technologies Corporation, Derek S. Gryna of Mercantile/Bancorp, and Ron Rogge of J. S. Alberici Construction Company for their input and comments.

REFERENCES

American Management Association (1992). *Blueprints for Service Quality.* American Management Association, New York, pp. 51–64.

Anderson, DeAnn, Abetti, Frank, and Savage, Phillip (1995). "Process Improvement Utilizing Computer Simulation: Case Study." *ASQC Quality Congress Transactions 1995.* American Society for Quality Control, New York, pp. 713–724.

Aubrey, Charles A., II and Gryna, Derek S. (1991). "Revolution Through Effective Improvement Projects." *ASQC Quality Congress Transactions 1991.* American Society for Quality Control, Milwaukee, pp. 8–13.

Bemesderfer, John L. (1979). "Approving a Process for Production." *Journal of Quality Technology,* Vol. 11, No.1, pp. 1–12.

Berry, Leonard L., Parasuraman, A., and Zeithmal, Vllerie A. (1994). "Improving Service Quality in America: Lessons Learned." *Academy of Management Executive,* Vol. 8, No. 2, pp. 32–52.

Bettis, Richard, Bradley, Stephen P., and Hamel, Gary (1992). "Outsourcing and Industrial Decline." *Academy of Management Executive,* Vol. 6, No. 1, pp. 7–22.

Binroth, William. (1992). "Fail-Safe Manufacturing for Assembly Operations." *ASQC Quality Congress Transactions 1993.* American Society for Quality Control, Milwaukee, pp. 107–115.

Black, Sam P. (1993). "Internal Certification: The Key to Continuous Quality Success." *Quality Progress,* January, pp. 67–68.

Bothe, Davis R. (1997). *Measuring Process Capability. McGraw-Hill, New York..*

Bothe, Davis R. (1992). "A Capability Study for an Entire Product." *ASQC Quality Congress Transactions 1992.* American Society for Quality Control, Milwaukee, pp. 172–178.

Business Week. (1996). "Has Outsourcing Gone Too Far?" *Business Week,* April, 1, pp. 26–28.

Carlisle, Rodney (1981). "Shirt-Sleeve Quality." *Quality,* March, pp. 48–49.

Carpenter, Ben H. (1982). "Control of Copper Ore Roasting Exit Gas Quality." *ASQC Quality Congress Transactions 1982.* American Society for Quality Control, Milwaukee, pp. 748–755.

Chang, Tien-Chien, Wysk, Richard A., and Wang, Hus-Pin (1991). *Computer-Aided Manufacturing.* Prentice-Hall, Englewood Cliffs, NJ.

Clark, J. M., and Milligan, G. W. (1994). "How Sweet it Is—Quality Management in a Honey House: The Stickey Quality Problems of Honey." *Quality Engineering,* Vol. 6, No. 3, pp. 379–400.

Collins, William H., and Collins, Carol B. (1993). "Differentiating System and Execution Problems." *Quality Progress,* February, pp. 59–62.

Cooper, Robin and Slagmulder, Regine (1997). *Target Costing and Value Engineering.* Productivity Press, Portland, OR and Institute of Management Accountants, Montvale, NJ.

Craig, Robert J. (1993). "Six Sigma Quality: The Key to Customer Satisfaction." *ASQC Quality Congress Transactions 1993.* American Society for Quality Control, Milwaukee, pp. 206–212.

Craner, Barrett C. (1994). "Managers' Audit System: Managers Just Do It." *ASQC Quality Congress Transactions 1994.* American Society for Quality Control, Milwaukee, pp. 920–928.

Davis, Robert, Rosegrant, Susan, and Watkins, Michael (1995). "Managing the Link Between Measurement and Compensation." *Quality Progress,* February, pp. 101–106.

Dedhia, Navin S. (1985). "Process Audit System Effectiveness." *European Organization for Quality Control Annual Conference.* EOQC, Berne, Switzerland, pp. 159–173.

Dodson, B. L. (1993). "Determining the Optimal Target Value for a Process with Upper and Lower Specification Limits." *Quality Engineering,* Vol. 5, No. 3, pp. 393–402.

Early, John F. (1989). "Strategies for Measurement of Service Quality." *ASQC Quality Congress Transactions 1989.* American Society for Quality Control, Milwaukee, pp. 2–9.

Ford, J. I. and Leader, C. R. (1989). "Integrating Human Dynamics and Statistical Process Control." *Quality Engineering,* Vol. 1, No. 2, pp. 179–189.

Foster, Robert D., and Zirk, Wayne E.(1992). "Getting Operators to Really Use SPC: MCPS Can Help." *ASQC Quality Congress Transactions 1992.* American Society for Quality Control, Milwaukee, pp. 201–207.

Gass, K. C. (1994). "How to Make Procedures Work." *Quality Engineering,* Vol.7, No. 2, pp. 337–343.

Godfrey, A. Blanton (1995). "Critical Issues in Service Quality Management." Address for the Fourth Annual Service Quality Conference of the American Society for Quality Control, Baltimore.

Goldman, Steven L., Nagel, Roger N., and Preiss, Kenneth (1995). *Agile Competitors and Virtual Organizations.* Van Nostrand Reinhold, New York.

Hart, Christopher W. L. (1995). "The Power of Internal Guarantees." *Harvard Business Review,* January-February, pp. 64–74.

Harvard Business School (1994). *Case 9-694-076, Taco Bell.* Boston.

Hayes, Bob E. (1994). "How to Measure Empowerment." *Quality Progress,* February, pp. 41–46.

Hays, Thomas J., and Gander, Mary J. (1993). "Total Quality Transformation on a PCB Manufacturing Line." *ASQC Quality Congress Transactions 1993.* American Society for Quality Control, Milwaukee, pp. 112–118.

Herman, John T. (1989). "Capability Index—Enough for Process Industries?" *ASQC Quality Congress Transactions 1989.* American Society for Quality Control, Milwaukee, pp. 670–675.

Hestand, Randy (1991). "Measuring the Level of Service Quality." *Quality Progress,* September, pp. 55–60.

Hinckley, C. Martin, and Barkan, Philip (1995). "The Role of Variation, Mistakes, and Complexity in Producing Nonconformities." *Journal of Quality Technology,* Vol. 27, No. 3, pp. 242–249.

Jeffrey, Jaclyn R. (1995). "Preparing the Front Line." *Quality Progress,* February, pp. 79–82.

Juran Institute, Inc. (1995). *Work Team Excellence.* Wilton, CT.

Juran, J. M. (1992). *Juran on Quality by Design.* Free Press, New York.

Juran, J. M., and Gryna, Frank M. (1993). *Quality Planning and Analysis,* 3d ed. McGraw-Hill, New York.

Kane, Victor E. (1986). "Process Capability Indices." *Journal of Quality Technology,* Vol. 18, No. 1, pp. 41–52.

Katzenbach, J. R., and Smith, D. K. (1993). *Wisdom of Teams: Creating the High Performance Organization.* Harvard Business School Press, Boston.

Kearney, Francis J. (1984). "Management of Product Quality without a Quality Department." *ASQC Quality Congress Transactions 1984.* American Society for Quality Control, Milwaukee, pp. 249–252.

Keenan, Thomas M. (1995). "A System for Measuring Short-Term Producibility." *ASQC Quality Congress Transactions 1995.* American Society for Quality Control, Milwaukee, pp. 50–56.

Kegg, Richard L. (1985). "Quality and Productivity in Manufacturing System." *Annals of the CIRD International Association for Production Research,* Vol. 34, No. 2, pp. 531–534.

Kitamura, Toshiyuki, Hiller, Dennis E., and Ingram, Larry J. (1994). "Lean Production System Implementation at Supplier Base." *Impro94 Conference Proceedings.* Juran Institute, Wilton, CT, pp. 4C-1–4C-23.

Laffel, Glenn, and Plsek, Paul E. (1989). "Preliminary Results from a Quality Improvement Demonstration Program at Brigham and Women's Hospital." *Impro89 Conference Proceedings,* Juran Institute, Wilton, CT, pp. 8A-21–8A-27.

Lane, Patricia A. (1989). "Continuous Improvement—AT&T QA Audits." *ASQC Quality Congress Transactions 1989.* American Society for Quality Control, Milwaukee, pp. 772–775.

Latzko, William J. (1993). "A Bank Quality Model." *ASQC Quality Congress Transactions 1993.* American Society for Quality Control, Milwaukee, pp. 38–44.

Lee, Jay (1995). "Perspective and Overview of Manufacturing Initiatives in the United States." *International Journal of Reliability, Quality and Safety Engineering,* Vol. 2, No. 3, pp. 227–233.

Lehrman, Karl Henry (1991). "Capability Analyses and Index Interval Estimates: Quantitative Characteristics (variables)." *Quality Engineering,* Vol. 4, No. 1, pp. 93–130.

Lochner, Robert H. (1995). "Developing a Balanced Set of Measures in Healthcare." *ASQC Quality Congress Transactions 1995.* American Society for Quality Control, Milwaukee, pp. 293–300.

Madigan, James M. (1993). "Measures Matrix Chart: A Holistic Approach to Understanding Operations." *Quality Management Journal,* Vol. 1, No. 1, pp. 77–86.

Mann, David W. (1994). "Re-engineering the Manager's Role." *ASQC Quality Congress Transactions 1994.* American Society for Quality Control, Milwaukee, pp. 155–159.

McCoy, Paul F. (1991). "Using Performance Indexes to Monitor Production Processes." *Quality Progress,* February, pp. 49–55.

McDermott, Robin (1994). "The Human Dynamics of Total Quality." *ASQC Quality Congress Transactions 1994.* American Society for Quality Control, Milwaukee, pp. 225–233.

Melan, Eugene H. (1993). *Process Management.* McGraw-Hill, New York.

Mentch, C. C. (1980). "Manufacturing Process Optimization Studies." *Journal of Quality Technology,* Vol. 12, No. 3, pp. 119–129.

Middleton, David H. (1992). "Accrediting a Machine for a Lifetime of Quality." *ASQC Quality Congress Transactions 1992.* American Society for Quality Control, Milwaukee, pp. 151–157.

Myers, Dale H., and Heller, Jeffrey (1995). "The Dual Role of AT&T's Self-assessment Process." *Quality Progress,* Vol. 28. No. 1, pp. 79–83.

Nakajo, Takeshi, and Kume, Hitoshi (1985). "The Principles of Foolproofing and Their Application in Manufacturing." *Reports of Statistical Application Research,* Vol. 32, No. 2, pp. 10–29.

Newberg, Craig K., and Nielsen, James R. (1990). "The Pathway to Operator Control." *ASQC Quality Congress Transactions 1990.* American Society for Quality Control, Milwaukee, pp. 723–728.

Parker, H. V. (1981). "A Paper Mill Solves a Quality Control Problem with Process Control Data." *Quality Progress.* March, pp. 18–22.

Peña, Ed (1990). "Motorola's Secret to Total Quality Control." *Quality Progress,* October, pp. 43–45.

Perkins, Nancy S. (1994). "Can TQM Be Derailed by an Enemy from Within?" *ASQC Quality Congress Transactions 1994.* American Society for Quality Control, Milwaukee, pp. 731–738.

Pignatiello, Joseph H., Jr., and Ramberg, John S. (1993). "Process Capability Indices: Just Say No." *ASQC Quality Congress Transactions 1993.* American Society for Quality Control, Milwaukee, pp. 92–104.

Plsek, Paul E. (1989). "FMEA for Process Quality Planning." *ASQC Quality Congress Transactions 1989.* American Society for Quality Control, Milwaukee, pp. 484–489.

Pooley, John, and Welsh, Dianne H. B. (1994). "A Comparison of Russian and American Factory Quality Practices." *Quality Management Journal,* Vol. 1, No. 2, pp. 57–70.

Pyzdek, Thomas (1994). "Toward Service Systems Engineering." *Quality Management Journal,* Vol. 1, No. 3, pp. 26–42.

Raheja, Dev (1982). "Fault Tree Analysis—How Are We Doing?" *ASQC Quality Congress Transactions 1982.* American Society for Quality Control, Milwaukee, pp. 355–359.

Reichheld, Frederick F. (1993). "Loyalty-Based Management." *Harvard Business Review,* March-April, pp. 64–73.

Rethmeier, Kenneth A. (1995). "Creating the Learning Organizations." *Impro95 Conference Proceedings.* Juran Institute, Wilton, CT, pp. 3C.1-1–3C.1-11.

Rubinstein, Sidney P. (1993). "Democracy and Quality as an Integrated System." *Quality Progress.* September, pp. 51–55.

Scharlacken, John W. (1992). "The Advantages of Manufacturing Technology Planning." *Quality Progress.* July, pp. 57–62.

Schilling, Edward G. (1994). "The Importance of Sampling in Inspection." *ASQC Quality Congress Transactions 1994.* American Society for Quality Control, Milwaukee, pp. 809–812.

Schonberger, Richard J. (1996). *World Class Manufacturing: The Next Decade.* Free Press, New York.

Shay, Michael E., White, G. Randy, and Blackman, Paul (1991). "Team Work and Empowerment at the A. O. Smith Corporation." *ASQC Quality Congress Transactions 1991.* American Society for Quality Control, Milwaukee, pp. 801–807.

Shenoy, Muralidhar (1994). "Machine Monitoring for Quality Assurance." *ASQC Quality Congress Transactions 1994.* American Society for Quality Control, Milwaukee, pp. 439–445.

Shina, S. G. (1991). "The Successful Use of the Taguchi Method to Increase Manufacturing Process Capability." *Quality Engineering,* Vol. 3, No. 3, pp. 333–350.

Shirley, Britt M., and Gryna, Frank M. (1998). "Work Design for Self-Control in Financial Services: *Quality Progress,* May, pp. 67–71.

Shostack, G. Lynn (1984). "Designing Services that Deliver." *Harvard Business Review,* January-February, pp. 133–139.

Siff, Walter C. (1984). "The Strategic Plan of Control a Tool for Participative Management." *ASQC Quality Congress Transactions 1984.* American Society for Quality Control, Milwaukee, pp. 384–390.

Simons, Robert (1995). "Control in an Age of Empowerment." *Harvard Business Review,* March-April, pp. 80–88.

Skrabec, Q. R., Jr. (1991). "Using the Ishikawa Process Classification Diagram for Improved Process Control." *Quality Engineering,* Vol. 3, No. 4, pp. 517–528.

Snee, Ronald D. (1993). "Creating Robust Work Processes." *Quality Progress,* February, pp. 37–41.

Somerton, Diana G., and Mlinar, Sharon E. (1996). "What's Key? Tool Approaches for Determining Key Characteristics." *ASQC Quality Congress Transactions 1996.* American Society for Quality Control, Milwaukee, pp. 364–369.

Stalk, George Jr., and Hout, Thomas M. (1990). *Competing Against Time.* Free Press, New York.

Stampen, John O., and Stampen, Jacob O. (1995). "Training Function Deployment: A New Approach for Designing and Evaluating Employee Development Programs." *ASQC Quality Congress Transactions 1995.* American Society for Quality Control, Milwaukee, pp. 933–946.

Tabladillo, Mark F., and Canfield, Susan (1994). "Creation of Management Performance Measures from Employee Surveys." *Quality Management Journal 1994,* Vol. 1, No. 4, pp. 52–56.

Tarver, Mae G. (1984). "Multistation Process Capability-Filling Equipment." *ASQC Quality Congress Transactions 1984.* American Society for Quality Control, Milwaukee, pp. 281–288.

Turner, Robert B., and Zipursky, Lorne S. (1995). "Quality Tools Help Improve Employee Commitment." *ASQC Quality Congress Transactions 1995.* American Society for Quality Control, Milwaukee, pp. 770–776.

Whittingham, P. R. B. (1987). "Operator Self-Inspection." *ASQC Quality Congress Transactions 1987.* American Society for Quality Control, Milwaukee, pp. 278–286.

Willis, Roger G., and Sullivan, Kevin H. (1984). "CIMS in Perspective=Costs, Benefits, Timing, Payback Periods Are Outlined." *Industrial Engineering,* Vol. 16, No. 2, pp. 23–26.

Yavas, Burhan Fatih, and Burrows, Thomas M. (1994). "A Comparative Study of Attitudes of U.S. and Asian Managers Toward Product Quality." *Quality Management Journal,* Vol. 2, No. 1, pp. 41–56.

Yee, William, and Musselwhite, Ed (1993). "Living TQM with Workforce 2000." *ASQC Quality Congress Transactions 1993.* American Society for Quality Control, Milwaukee, pp. 141–146.

Ziegler, August H. (1995). "Self-Inspection Implementation: Beyond the Rhetoric." *ASQC Quality Congress Transactions 1995.* American Society for Quality Control, Milwaukee, pp. 618–621.

SECTION 23
INSPECTION AND TEST[1]

E. F. "Bud" Gookins

---

[1]In the Fourth Edition, the material on inspection and test was prepared by Joseph J. Zeccardi.

INTRODUCTION

Inspection and testing activities always involve the evaluation of a characteristic as it relates to a specific requirement. The requirement can be in the form of a standard, a drawing, a written instruction, a visual aid, or any other means of conveying the characteristic specification.

Inspection and testing functions can be done automatically, manually, or both in a sequential manner. The evaluation process consists of the following steps applied to each characteristic (Juran 1945, p. 23):

1. Interpretation of the specification

2. Measurement of the quality of the characteristic

3. Comparing 1 with 2

4. Judging conformance

5. Processing of conforming items

6. Disposition of nonconforming items

7. Recording of data obtained

These steps apply to both product and service items.

The inspection and testing evaluation can be determined by using the intrinsic senses of the human being (i.e., smell, taste, sight, hearing, and touch), or it can be made using a nonvariable gage, a nonvariable electronic or laser instrument, a nonvariable chemical or physical testing device, or any other method in which a decision is made based on simply an "accept" or "reject" determination. Such inspection is commonly referred to as *attribute inspection.*

The inspection and testing evaluation that is determined by using any measurement device, be it mechanical, electronic, laser, chemical, or any other method that will display data generated by physically measuring the characteristic, in which a decision is made based on actual value readout, is commonly referred to as *variable inspection.*

The primary purpose of inspection and testing is to determine whether products or services conform to specification. This purpose is often called *acceptance inspection* and *acceptance testing.* The components of the inspection and testing function can be broken down further into subclassifications. The most salient are listed in Table 23.1.

People engaged full time in inspection work commonly carry the title of *inspectors* but often are recognized as product appraisers, product auditors, and product verifiers for organizations that produce a manufactured item. For organizations engaged in nonmanufactured goods or services, the

TABLE 23.1 Subclassifications for Manufacturing Products and Service

Type	Prime function
Receiving (or incoming) inspection and testing	To ensure that incoming product is not used or processed until it has been inspected or tested and found to be conforming to specified requirements
In-process inspection and testing	To ensure that in-process product is not moved forward until it has been inspected or tested and found to be conforming to specified requirements
Final inspection and testing	To ensure that finished component and/or product is not dispatched until all the activities have been satisfactorily completed
Layout inspection and functional testing	To ensure that all customer engineering material and performance standards have been appraised prior to production
Shipping inspection and testing	To ensure that all shipped products are conforming to specified requirements
Qualification inspection and testing	To judge the service capability of the product and the possible extreme applications of the product
Dock inspection and auditing	To ensure that the product and its testing (product-related packaging, identification, information, etc.) are released to the customer conforming to all requirements
Service inspection (health inspector, environmental inspector, etc.)	To ensure that all specified requirements are met and to evaluate and measure nonconformancies found in the system
Nonproduction inspection and testing	To evaluate a specific task requested by quality assurance i.e., gage repeatability and reproducibility, process capability, 100% appraisal, etc.
Initial sample inspection request (ISIR)	To assure the customer that the first production run will be in conformance with all their designated characteristics and to submit those actual characteristic measurements and attributes to the customer for verification and approval
Production part approval process (PPAP)	A request by the customer indicating the level of inspection to conduct first production run checking all characteristics and indicating actual dimensions, gage repeat-ability and reproducibility studies, capability studies, material verification, or any other outside processing; e.g., heat-treating, plating, etc., and submit to the customer for approval (used primarily by the auto industry)

inspection work is identified pertaining to that particular function, e.g., safety inspectors, environmental inspectors, health inspectors, etc. People engaged in part-time inspection work are commonly referred to by the title of their major activities, e.g., machine operators, setup people, assemblers, welders, platers, foundry people, etc. The people engaged full time in the testing function commonly carry the title of *tester* but often are recognized as assaying technicians, laboratory technicians, chemists, metallurgists, etc.

Today, many manufacturing organizations have moved toward total automated inspection and testing or semiautomated inspection or testing augmented by productive operation appraisal. Within this movement of proactive product acceptance, the inspection and testing become integrated into the operational function and verified usually by a new breed of inspectors called *audit inspectors*.

PURPOSE OF INSPECTION AND TEST

The purpose of inspection and test is to determine the conformance of the product or service to the standard or specific requirements and to disposition the product or service based on the results of the evaluation. This determination involves three main decisions (Juran and Gryna 1980):

Conformance decision: To judge whether the product conforms to specification

Fitness-for-use decision: To decide whether nonconforming product is fit for use

Communication decision: To decide what to communicate to outsiders and insiders

The Conformance Decision.   Except in small companies, the number of conformance decisions made per year is simply huge. There is no possibility for the supervisory body to become involved in the details of so many decisions. Hence the work is organized in such a way that the inspectors or production workers can make these decisions. To this end, they are trained to understand the products, the standards, and the instruments. Once trained, they are given the job of making the inspections and of judging conformance. (In many cases the delegation is to automated instruments.)

Associated with the conformance decision is the disposition of conforming product; the inspector is authorized to identify the product ("stamp it up") as an acceptable product. This identification then serves to inform the packers, shippers, etc. that the product should proceed to its next destination (further processing, storeroom, customer). Strictly speaking, this decision to "ship" is made not by the inspectors but by management. With some exceptions, a product that conforms to specification is also fit for use. Hence the company procedures (which are established by the managers) provide that conforming products should be shipped as a regular practice.

The Fitness-for-Use Decision.   In the case of nonconforming product, a new question arises: Is this nonconforming product fit for use or unfit? In some cases the answer is obvious—the nonconformance is so severe as to make the product clearly unfit. Hence it is scrapped or, if economically repairable, brought to a state of conformance. However, in many cases the answer as to fitness for use is not obvious. In such cases, if enough is at stake, a study is made to determine fitness for use. This study involves securing inputs such as those shown in Table 23.2.

Once all the information has been collected and analyzed, the fitness-for-use decision can be made. If the amount at stake is small, this decision will be delegated to a staff specialist, to the quality manager, or to some continuing decision-making committee such as a material review board. If the amount at stake is large, the decision usually will be made by a team of upper managers.

The Communication Decision.   Inspection and test serve two purposes: to make decisions on the products and to generate data that provide essential information for a wide variety of uses, such as those listed in Table 23.1. The conformance and fitness-for-use decisions likewise are a source of essential information, although some of this is not well communicated.

Data on nonconforming products are usually communicated to the producing departments to aid them in preventing a recurrence. In more elaborate data-collection systems there may be periodic summaries to identify "repeaters" or the "top 10," which then become the subject of special studies.

When nonconforming products are sent out as fit for use, there arises the need for two additional categories of communication:

1. *Communication to "outsiders":*   They (usually customers) have a right and a need to know. All too often the manufacturing companies neglect to inform their customers when shipping nonconforming products. This may be as a result of bad experience; i.e., some customers will seize on such nonconformances to secure a price discount despite the fact that use of the product will not add to their costs. Usually, the neglect indicates a failure even to face the question of what to communicate. A major factor here is the design of the forms used to record the decisions. With rare excep-

TABLE 23.2   Inputs Required for Fitness-for-Use Decision

Input	Usual sources
Who will be the user?	Marketing
How will the nonconforming product be used?	Marketing, client
Are there risks to human safety or structural integrity?	Product research and design
What is the urgency?	Marketing, client
What are the company's and the users' economics?	All departments, client
What are the users' measures of fitness for use?	Market research, marketing, client

tions, these forms lack provisions that force those involved to make recommendations and decisions on (a) whether to inform the outsiders and (b) what to communicate to them.

2. *Communication to insiders:* When nonconforming goods are shipped as fit for use, the reasons are not always communicated to the inspectors and especially not to the production workers. The resulting vacuum of knowledge has been known to breed some bad practices. When the same type of nonconformance has been shipped several times, an inspector may conclude (in the absence of knowing why) that it is just a waste of time to report such nonconformances in the first place. Yet in some future case the special reasons (which were the basis of the decision to ship the nonconforming goods) may not be present. In like manner, a production worker may conclude that it is a waste of time to exert any effort to avoid some nonconformance that will be shipped anyway. Such reactions by well-meaning employees can be minimized if the company faces squarely the question: What shall we communicate to the insiders?

PREPRODUCTION EVALUATION

The inspection and testing functions are key elements of the production process. Without accurate and specific criteria for determining that the manufacturing or service product meets the customer's requirements, we expose the organization to uncontrolled, inefficient, and expensive processing as well as negative perceptions from customers. These resulting performances can be minimized—if not eliminated—by preproduction and service evaluations.

The approach to inspection and test planning follows closely the principles of quality planning as set out in Section 3, The Quality Planning Process. Application of these principles to the inspection job has been studied extensively, and good tools are available to facilitate inspection planning.

The Nature of "Lots" of Product.   It is useful here to define what is meant by *lot* and expand briefly on the term as applied to inspection and test. A *lot* is usually associated with physical product, especially in connection with sample inspection and test. Usually the product submitted for decision on conformance to standard consists of a lot. The true lot is an aggregation of product made under a common system of causes. When this ideal is met, the lot possesses an *inherent uniformity derived from the common system of causes.* The extent to which the lot conforms to this ideal greatly influences the approach to the product conformance decision and especially the kind and extent of sampling.

In its simplest form, the true lot emerges from one machine run by one operator processing one material batch, all under a state of statistical control, e.g., a single formulation of a drug product or a run of screw-machine parts turned from one piece of rod on one machine. A great deal of industrial production consists of true lots.

However, a great deal of other production consists of product mixtures that, in varying degrees, fall short of the ideal lot definition. Product made from several material batches, on several machines, or by several operators may be dumped into a common container. In shop language, this mixture is a "lot," but in more precise language it is only a "mixture." In continuous processes or in conveyor production, the process may well be common and constant, but the input materials may not be.

For precise and economic product conformance decisions, it is most helpful to *preserve the order.* This means that product is kept segregated in true lots or at least identified as to common cause. In addition, for those processes which exhibit a time-to-time variation or "drift" (e.g., the solution gradually becomes dilute. The tool gradually wears), preserving the order includes preserving the time sequence during which various portions of the lot were made. Any loss of order of manufacture also becomes a loss of some prior knowledge as to inherent uniformity. (See Section 22, Operations, under Process Capability: The Concept, Process Mixture, for a discussion of the effect of product mixture on process improvement, including application in services.)

Some products are naturally fluid and develop a homogeneity through this fluidity. Homogeneity from this new cause also can qualify the product as a true lot, with important implications for the sampling process.

When several true lots are combined for the purposes of acceptance, the combination is known as a *grand lot.* Such mixtures are very common, e.g., product from multiple cavities of molding operations or from multiple spindles of screw-machine operations. The two categories of single elements of product (e.g., discrete units and specimens) have their counterparts in two categories of lots.

The Lot as a Collection of Discrete Units.   Here the lot consists of numerous bolts, teacups, or refrigerators, each one of which is governed by the product specification. In batch production, the lot is usually determined by the obvious boundaries of the batch. In continuous production, the lot is usually defined as an arbitrary amount of production or as the amount produced during an arbitrary time span, e.g., a shift, a week.

The Lot as Coalesced Mass.   Here the lot also may consist of a batch, e.g., the melt of steel. In continuous production, the lot is again based on some arbitrary selection, e.g., 1 ton or a day's production.

The Inspection and Test Requirement Review.

Some organizations produce a standard or proprietary product that lends itself to very little change in configuration, materials, or processing, whereas some organizations are driven by ongoing changes or modifications to their products or services, and even some organizations are a "job shop" type, producing a product specifically to a customer's requirements and specifications. Regardless of the frequency or type of change to existing inspection and testing requirements, however, a review should be made prior to first production release or any subsequent revisions.

This review should examine any measurement parameter that would require special gaging and testing equipment that is different from that presently used as the method of measurement. It also should include any other input that would provide assurance that the customer's requirements will be met or that preproduction appraisal exceptions have been resolved.

The Inspection and Test Planner.

The planning can be done by anyone who understands the fitness for use of the product being inspected. Usually, however, the planning is done by a quality assurance staff planner, an inspection or test supervisor, and in some situations even the inspector or tester.

Where planning is done by a staff planner, it is recommended that his or her proposal be accepted by the inspection or testing supervisor before the plan becomes effective. The staff planner also is assigned a scope of responsibility within which to work. This scope determines which aspects of inspection or test planning are to be covered: inspection instructions, instrumentation, cost estimates, space and workplace design, documentation, etc. In large organizations, the planning is sometimes divided among specialists rather than being assigned by project. In smaller organizations, the planning may be done by the head of quality or assigned to an inspector or tester.

If the inspection or testing planning is service-oriented, it is usually conducted by the functioning individual conducting the inspection/testing or an immediate supervisor. If the inspection or testing planning is manufacturing-oriented, it is usually broken down into five categories:

1. Components completed within a single department, small series production—conducted by inspector/tester.

2. Components completed within a single department, large series production—conducted by inspection/testing supervision.

3. Simple components and services, purchased or in-house heat-treating, plating, casting—conducted by inspection/testing supervision.

4. Complex units, small series production (machine tools)—conducted by inspection/testing supervision.

5. Components produced by multiple department progression, subsystem test, or interdepartment units—conducted by quality planner.

Developing the Inspection and Test Plan.   For each inspection station, the planner lists the quality characteristics to be checked. To determine these, the planner considers the various sources of pertinent product information:

- The needs of fitness for use
- The product and process specifications as published by the engineers
- The customer's order, which references the product specification but may call for modifications
- The applicable industry standards and other general-use sources

For test stations, the planner must consider the functional and reliability parameters, such as

- The industrial standards
- Third-party requirements
- Application environments
- Customer expectations

For the service application, the planner must consider the characteristic criteria for the checklist method of evaluation, such as

- The needs of the customer
- Service industry standards
- The objectives and goals of the organization.

The specification information is seldom sufficient for the inspector/tester to meet the realities to be faced. The inspector/tester planner can help to bridge this gap in several ways:

1. *Clear up the meaning of the words used.*   Terminology for describing sensory qualities is often confusing. In one company, the term *beauty defects* was used generally to describe blemishes on the products. Some of these blemishes (scratches in the focal plan of an optical instrument) made the product unfit for service. Other blemishes, though nonfunctional, could be seen by the customers and were objectionable for aesthetic reasons. Still other nonfunctional blemishes could be seen by the company inspectors but not by consumers. However, because the multiple meanings of the term *beauty defect* had not been clarified, the inspectors rejected all blemishes. Data analysis showed that most of the blemishes were both nonfunctional and nonoffensive to customers. Hence new terminology was created to make the distinctions needed to describe the effect of blemishes. The clarification of terminology improved yields and opened the way to improvement in manufacturing processes as well. [Based on the consulting experience of J. M. Juran. For some added examples, see Juran (1952).]

2. *Provide supplemental information.*   Make it available on matters for which the specification is vague or silent, e.g., workmanship. Usually this can be done for entire commodity or component classes, with minimum individual analysis. The greatest needs for supplemental standards arise in new and rapidly changing technology; in such cases it is common to find that vague standards are provided to the inspectors. Vague standards create confusion among departments as well as among companies. Refer to Miller (1975) for a discussion on specifying test methods and specifications.

3. *Classify the characteristics for seriousness.*   This will help place the emphasis on the most important features of the product. (See Seriousness Classification, later in this section.) In the case of process characteristics, make use of the concept of dominance, as discussed in Section 22, Operations, under Planning Process Control.

4. *Provide samples, photographs, or other reference aids.*   This will help explain the meaning of the specification. The greatest single need is for visual standards (see below).

Inspection and Test Equipment.   Each product type requires a review of the gages and test equipment required prior to production. Many specifications can be satisfied in the inspection or testing appraisal using standard inspection and testing equipment. However, some characteristics require gaging and testing equipment that must be designed specially. The planner must make this decision and schedule the equipment prior to production. The preproduction assessment should be verified prior to production release, and any discrepancy should be corrected immediately, before first-piece acceptance.

Inspection and Test Locations.   Inspection and test stations usually are placed

At movement of goods between companies, usually called *supplier inspection or test*

Before starting a costly or irreversible operation, usually called *setup inspection*

At movement of goods between departments, usually called *process inspection* or *process testing*

As an integration of automatic inspection or testing within the process

On completion of the product, usually called *finished-goods inspection* or *final-product testing.*

For complex products, acceptance may require tests of mechanical compatibility, electrical mating, product performance under specified environmental conditions, and final configuration. These are usually called *systems tests.*

These general rules do not decide all questions of inspection and test stations. Complex supplier relations may require an inspection location at the supplier's plant. Some process operations may require a "station" from which the inspector patrols a large area. Other process operations may be sufficiently well in hand that no inspection stations are used between departments; instead, there is a station after completion of all operations. In assembly lines, inspection stations may be located on the line as well as at the end of the line. In still other situations there may be an added station after packing or at the customer's premises.

For each inspection location, instruct the inspector (or tester) what to inspect for and how to do it:

Just what the mission of that inspection or test station is, i.e., which qualities to check

How to determine whether a unit of product conforms to standard or not

How to determine whether a lot of product is acceptable or not (*lot criteria*)

What to do with conforming and nonconforming products

What records to make

While these categories of instruction are quite similar from one job to another, the degree of detail varies enormously.

In allocating the inspection work among the various inspection stations, the planner should be alert to the presence of "self-policing" operations. Some oversize parts will not enter tools or fixtures for further processing or cannot be assembled. Some parts are subjected to greater stresses during manufacture than during use. Some electrical circuit tests identify deficient components. Oil-pressure tests identify some undersize parts. [Refer to Trippi (1975) for a discussion on the optimal allocation of inspection resources; Ballou and Pazer (1982) for the optimal placement of inspection stations; and Eppen and Hurst (1974) for the optimal location of inspection stations in multistage production processes.]

INSPECTION AND TEST DOCUMENTED INFORMATION

As products have developed in complexity and technological advances, consistent and repetitive information properly documented is essential. The ability to appraise the product or service item the same exact way each and every time is imperative if acceptance or rejection criteria are to be cogently and enforceably judged.

Inputs into the Control Plan.   The final results of inspection and test planning are reduced to writing in one of several ways.

Inspection and Test Procedure.   This is a tailor-made plan for a specific component or product type. It always lists the characteristics to be checked, the method of check (e.g., visual, gage, etc.), and the instruments to be used. It may, in addition, include the seriousness classification of characteristics, tolerances and other piece criteria, list of applicable standards, sequence of inspection operations, frequency of inspection, sample size, allowable number of defects, and other lot criteria, as well as inspection stamps to be applied.

Inspection and test procedures are widely used in industry. In companies making complex systems or undergoing frequent design changes, these procedures become very numerous and consume extensive staff resources to prepare them.

The planner also should be alerted to the need for locating inspection stations at such operations as materials handling, storage, packing, and shipping. The fact that the departments doing these operations are not a part of production is of no consequence if product quality is affected.

Aspects that may require inspection planning include

- *Internal handling:*   Use of correct containers and other handling facilities; product protection against corrosion, contamination, etc.
- *Internal storage:*   Adequate identity and traceability
- *Packing:*   Product identification, lot numbers, traceability; protection against adverse environments; protection against damage due to handling, shipping, and unpacking; presence of incidental small parts and information circulars
- *Shipping:*   Care in loading; special markings required by customers

Once the planner has prepared the procedure, the interested departmental supervisors can be convened to reach agreement on who is to carry out which part of the inspection plan.

Inspection Data Planning.   The planner also determines the data-recording needs for each inspection station. In many cases the standard inspection report forms will meet the recording needs. For finished products, a special test document is usually provided. In addition, the planner makes provision for any special recording needed for frequency distribution, control charts, certification, traceability, etc.

The concept of separating inspection planning from execution has great value if properly applied. If planning is underapplied, there is increased risk of catastrophic product failure. If overapplied, the result is excess cost and much internal friction. Striking a sound balance requires periodic reappraisal of the major forces in contention as well as analysis of the conventional alarm signals, e.g., rising staff costs or abrasion between departments. In addition, the changing job situations influence the extent of formal planning needed, notably (1) the education, experience, and training of the work force, (2) the stability of the processes, and (3) the severity of the product requirements.

Error-Proofing.   The planner faces two responsibilities related to inspection error: (1) avoiding built-in sources of error and (2) providing positive means of foolproofing the inspection against error. See, for a detailed discussion, Inadvertent Inspector Errors, under Inspection Errors, below.

Overplanning.   In some companies, the writing of inspection plans is done extensively. New customer orders, new product designs, new process changes, new regulations, and so on, are all occasions for scrutiny by the quality engineers, who issue inspection plans accordingly. As this goes to extremes, the cost of planning rises, and the excess formality increases the training time for inspectors, the attention to trivia, the documentation, and the control effort generally. Error rates tend to increase, with adverse effects on inspection costs and inspector morale.

Dealing with excess planning costs takes several forms. One technique is to do the planning by computer or by other means of mechanizing much of what the engineers otherwise do manually. A second approach is to minimize the amount of tailor-made planning by extending the use of inspection and test manuals that have broad application. See Instruction Manual, below.

A third approach is to delegate some of the planning itself to the inspection supervisors and the inspectors. To do this usually requires preparation of a manual on inspection planning plus training the inspection force to do the planning for all except the vital few characteristics, which are reserved for the staff planners. Still another device is to agree, case by case, on the amount of detailed planning needed.

Human, Machine, and System.   A major decision in all planning is the extent to which tasks will be assigned to people or to machines and the related decision regarding delegation of tasks to people or to systems. Machines are superior for doing deeds that can be clearly defined and which require exacting attention to repetitive detail.

Table 23.3 is a list contrasting intellectual activities and proposes a division between person and machine as applied to inspection and test. (See also Thompson and Reynolds 1964.)

The study of the interrelationship of people, machines, and system masquerades under a variety of names: human factors, biomechanics, human engineering, ergonomics, and industrial psychology. Industrial managers, including quality managers, are commonly amateurs in the understanding of human capacities and especially human behavior. The behavioral scientists are the "professionals," but the subject is as yet hardly a science. In addition, communication between the practicing managers and the behavioral scientists is severely limited by differences in dialect and, especially, cultural background.

Procedure Manual (Includes Flow Diagram)

The Flow Diagram.   The more complex the product, the greater is the need to prepare a flow diagram showing the various materials, components, and processes that collectively or sequentially turn out the final product. To prepare the flow diagram, the planner visits the various locations, interviews

TABLE 23.3   Assignment to Machines Versus People

Lower intellectual activities	Higher intellectual activities
Things that can be expressed exactly	Things that cannot be expressed exactly
Decisions that can be made in advance	Decisions that cannot be made in advance
Arithmetic, algebraic, and and chesslike symbolic logic	Pattern recognition, judgment, creativity, foresight, leadership, and such thinking
Highly repetitive and, therefore menial	Random, having many degrees of freedom, never exactly the same
Can be reduced to logic and therefore programmed exactly into a machine	Cannot be programmed exactly but can use heuristic approximations as an aid
Those a small machine can handle completely, faster and more positively	A machine cannot handle completely and it becomes excessively large and uneconomical in attempting to do so
Design and programming require a high level of intelligence but, once done, the mental activity need not be repeated	This problem is never exactly the same and it must be reconsidered, that is, rethought out for each new decision
Involves decisions as to what is right or wrong; the person guesses and the machine monitors to prevent him or her from making a mistake; it does this positively enough for use in safety systems	People use the display which is driven by the machine and possibly a separate computer to assist them in making the choice type of decisions as to what is best, using the most advanced mathematical techniques
Requires a high degree of orderliness	Takes care of matters which cannot be arranged into any sort of orderly procedures
Includes the decisions that must be made rapidly by the machine in periods of congestions and in emergencies	Involves situations that develop more slowly, that will, sooner or later, require a considered decision.

the key people, observes the activities, and records findings. The planner simplifies the picture by good use of symbols. One common set of symbols is

○	Operation	D	Delay
⇨	Transportation	V	Storage
□	Inspection	⊡	Combined activity

(See also Section 3, The Quality Planning Process, for another view of flow diagraming.) In addition, the planner prepares proposals for improvement, sends copies of the diagram to all concerned, and then is ready to convene them for discussion of the diagram and the proposals.

Procedures for the inspection and testing activities to verify that specified requirements for the products are met are collected and organized in the inspection and testing procedures manual. A flow diagram should be incorporated into the procedure contents and should reflect the path the product takes and the types of inspection required along this path. Any changes to the procedures should be so noted by an ongoing document change system. All aspects of the throughput should be described in subsequent procedures with a clearly defined documentation trail back to the main or general procedure. Each organization should examine its flow diagram to determine if receiving, in-process (sometimes referred to as *patrol inspection*), final inspection (sometimes referred to as *finished-goods acceptance, dock auditing*), or special inspection (such as magnetic particle inspection, Zyglo, etc.) is applicable.

It is not uncommon to incorporate the in-process inspection function within the responsibility of the operators or even to integrate it into the machine or process. If so, this consolidation also should be spelled out as a procedure.

Instruction Manual.   The instruction manual elaborates the work of each discipline in the inspection and testing functions, including detailed instructions on how to do specific work. The document should spell out the proper method of inspecting and/or testing and should be detailed as to how to fill out an inspection or test log, report, and any other record of data retention. All detailed information should be clearly defined, with the documentation trail extending bilaterally between the quality systems manual and the inspection and test instructions.

CRITERIA FOR INSPECTION AND TEST DEVELOPMENT

The factors for determining the methods and evaluation functions of appraising products or services are based on knowledge available from multiple sources.

Prior Knowledge of Product or Service Performance.   In some cases, the concept of "audit of decisions" has been put to work so that suppliers, independent laboratories, workers, and so on have been qualified as able to give reliable product conformance decisions and in addition have accepted this very lot. In such cases, no further product inspection is necessary (beyond that inherent in "audit of decisions"). See, in this connection, Section 22, Operations, under Audit of Decisions.

Prior Knowledge of the Process.   To illustrate, a press operation stamps out 10,000 pieces. If the first and last pieces contain certain specified holes of correct size and location, it follows that the intervening 9998 pieces also carry holes of correct size and in the correct locations. Such is the inherent nature of press dies. In statistical language, the sample size is two pieces, and the number of allowable defects is zero. Yet despite the tiny sample size, this is a sound way to do the inspection for these characteristics in the example given.

The press example is rather simple. In more complex cases, there is need to measure process capability and to arrange specially to take the samples with knowledge of the order of production. One organized form of this is the conventional control chart method used for process control. For product conformance, the approach is less well organized.

Prior knowledge of the "process" as used here includes knowledge of the qualifications of the suppliers and workers who run the process. Workers who have qualified for licenses require less rigorous inspection of their work than operators who have not qualified. Suppliers who have established a record of good deliveries need not be checked as severely as suppliers who lack such a record.

Product Homogeneity.   When the product is a fluid, this fluidity contributes to homogeneity. The extent of this homogeneity can be established by taking multiple specimens and computing the dispersion (another form of study of process capability). The presence of uniformity through fluidity greatly reduces the need for random sampling and thereby greatly reduces the sample sizes.

Even when the product is a solid, the inspection planner should be alert to the possibility that it possesses homogeneity through former fluidity. For example, a centrifugal casting process was used to cast metal cylinders that were then destroyed during testing for strength. However, it was then found that the dispersion of several strength tests all made on one ring was not different from the ring-to-ring dispersion. This discovery made it possible to reduce the amount of product destroyed during test.

Economic Impact.   When it is important to allocate limited testing resources to those parts which cost the most to replace, unique testing strategies can be designed that use past test, line, and field history. [See Wambach and Raymond (1977) for a discussion of this application for reliability-critical parts.]

Input from Outside Inspection and Test Functions.   The "prior knowledge" does not automatically come to the inspection planner or the inspector. Some of this knowledge is already in existence as a byproduct of other activities and hence can be had for the procedural cost of retrieval. Other knowledge is not in existence and must be created by additional effort. However, this added effort is usually a one-time study, whereas the benefits then go on and on.

THE DEGREE OF INSPECTION AND TESTING NEEDED

It is evident that a determination of "How much inspection?" should be made only after there has been an evaluation of the other inputs to product knowledge. This evaluation can then dictate any of several levels of inspection.

No Inspection.   There is already adequate evidence that the product or service conforms, and hence no further inspection is needed.

Skip Lot.   There is already adequate evidence that the product or service conforms, but because of the nature of the characteristics being checked or the customer's requirement for some type of verification, a need to spot check the lots in batches every so often is conducted. This skip lot will remain effective until a nonconformance occurs; then the inspection or test reverts back to lot sampling.

Sampling Plans.   Where there is little or no prior knowledge and no product fluidity, the main source of product knowledge becomes product inspection through *random sampling*. The amount of this inspection can be determined "scientifically" once the tolerable level of defects in accepted product has been defined clearly. However, choice of these levels—using the sampling parameters AQL (acceptable quality level), AOQL (average outgoing quality level), etc.—is largely arbitrary and usually is determined by negotiation. In theory, the sampling parameters can be determined from economic considerations, i.e., the cost of detecting unsatisfactory lots versus the cost of failing to detect them. In practice, the "cost of detecting" is fairly easy to determine, but the cost of "failing to detect" is difficult to determine. For intangibles such as customer goodwill, there is no way known to make the determination with any useful precision.

One Hundred Percent Inspection and Test.   This alternative is usually used for final test of critical or complex products. In very critical cases, it is used to provide redundancy against the unreliability of 100 percent inspection. In these cases the amount may be 200 percent or over. In cases where "zero defects" constitute the objective or requirement, 100 percent automated inspection is required. [See Nygaard (1981) and the discussion on machine-vision systems later in this section.] In some cases, 100 percent inspection is required to satisfy legal or political requirements (Walsh 1974). In other cases, 100 percent inspection is the most cost-effective approach. [See Walsh et al. (1976) and (1978) for examples where 100 percent inspection is the cost-effective alternative in high-volume production and in testing hardness of finished steel wheels.]

One hundred percent inspection also may be used when process capability is inherently too poor to meet product specifications. Sampling is of no avail in such cases, since the accepted lots are usually no better than the rejected lots, i.e., the difference is merely the result of statistical variations in the respective samples. This does not apply in cases where the process is highly erratic so that some lots are truly conforming and others are not merely the result of statistical accidents. For such processes, sampling can be a useful way to separate the conforming lots from the nonconforming ones.

Whatever the motivation for selecting the 100 percent alternative, effective implementation goes beyond choosing the inspection methodology or test equipment. Physical arrangements and personnel procedures must be changed. In many cases, attitudes of in-house and vendor personnel must be changed through reeducation. [See Walsh et al. (1979b).]

With the advent of computer-based testing, 100 percent inspection is becoming more practical (Schweber 1982). However, it is not clear that it is cost-effective in all cases, even if the technology is readily available (Walsh et al. 1979a). This is especially true if the total costs of quality are considered (Gunter 1983).

Check List Inspection.   Where there is not a hardware sort of product that requires an assessment of the dimension, appearance, or testing parameters, the main source of evaluation is the check list type. Such service industries as hotels, restaurants, banks, and health care require only evidence of compliance or no compliance to a specific standard.

OTHER TYPES OF CONFORMANCE INSPECTIONS

Faced with the objective to minimize inspection costs and achieve maximum quality control, management must be sensitive to nontraditional types of conformance decision-making inspections.

Simulation.   With advancement of the digital computer, system-simulation techniques are used as an alternative to the experimental or analytical approach (Wang et al. 1981). Examples of application:

Receiving inspection simulation

Printed circuit board assembly inspection simulation

Camera subassembly and final inspection simulation

Automated Inspection and Test.   The first large-scale applications of automated testing were very likely done by the Western Electric Company during the 1920s. Current developments in microcomputers, artificial intelligence (AI), integrated computer-aided manufacturing (ICAM), robotics, and software are making automated inspection practical and cost-effective. [See "An Outlook for 3D Vision" (1984) and "ASD's Quality Assurance Program Rates in Top 10" (1984); also see the discussion on 100 percent inspection above and that on Machine Vision, in Human Factors in Inspection, below.]

Automated inspection and testing are used to reduce costs, improve precision, shorten time intervals, alleviate labor shortages, and avoid inspection monotony, among other advantages. In some industries, the labor problems now seem insoluble in the absence of automated inspection. Already in widespread use, automated inspection is still expanding, with no end in sight.

The economics of automation involve a substantial investment in special equipment to secure a reduction in operating costs. The crux of justifying the investment lies in the amount of repetitive work the equipment will be called on to perform. This estimate of the anticipated volume of testing therefore should be checked out with great care.

A common starting point in discovering opportunities for automated inspection is to make a Pareto analysis of the kinds of inspections and tests being conducted. The vital few types are identified. Estimates are then made of the personnel, costs, and other current problems associated with these tests. The economics of automation are then estimated, and the comparable figures are an aid in deciding on the feasibility of successful conversion. For complex equipment involving depot storage and field maintenance, the question of use of automated testing is itself highly complex and requires a tailor-made study of some magnitude.

Technologically, the "machine" poses many problems. It is less adaptable than the human being it replaces, so some changes may need to be made in the product to offset this rigidity. For example, the machine may hold the units to be tested by grasping certain surfaces whose dimensions were previously unimportant. Now these surfaces may need to be held to close dimensions because the machine is not as adaptable as the human inspector. Alternatively, the product design may need to be changed to provide for adequate location.

Beyond the work of original design, construction, and prove-in, the machine must be set up specially for each job. However, modular construction master test pieces and taped programs have considerably reduced setup time while improving reproducibility. Reliability generally has been high, and use of printed circuit cards and other modular components has so reduced the "mean time to repair" that downtime is generally below 5 percent for well-designed machines.

Automated gaging and testing are used extensively in the mechanical industries. They are also widely used in the electronics industries, especially for electronic components, where the problem of making connections to the automated test equipment is so severe that the original product design must provide especially for this.

In the chemical industries, the corresponding development has been the *autoanalyzer.* This already has made possible some extensive cost reductions and solutions of otherwise forbidding problems of recruitment of laboratory technicians. The autoanalyzer makes use of some equipment common to all tests—sensors, transducers, recorders, and computers. However, each type of analysis has its unique procedure for converting the material under test into a form suitable for sensing.

The rapid evolution from numerically controlled (NC) machines to computer numerical control (CNC) machines has brought with it related new technologies. Machines now available are capable of continuously informing the operator of the machine's state of adjustment. Specification data stored in the computer's database are compared with the continuous stream of workpiece measurements provided by the measurement sensors to warn the operator when an adjustment is needed and even, in some cases, make the adjustment automatically. This new control technology, with its programmable logic, simplifies the operator's task. In reducing the degree of nonconformance at the source and providing a richer stream of information for the inspector, it also holds the promise of reducing the volume of inspections necessary and making the inspector function much more reliable. Applying programmable logic to the control of tool cutter wear compensation makes possible longer tool life and products of higher grade and greater consistency.

These new CNC machines also can function in a "lights out" manufacturing setting, in which a single machine or machining cell can operate for long periods and many machine cycles without human intervention, like an airplane flying on automatic pilot. Technology now under development will incorporate setup approval as an integral part of the setup process and then retain in memory a record of the setup as a mode of objective evidence documentation. Only when the setup is approved does the software release the machine for production; if the setup is rejected, the software does not respond to a "run" command. This technology will allow manufacturing organizations to create a totally paperless system and still meet the requirements of ISO 9000 standards.

As we move into the next century, we will see machines that will be programmed to react to voice recognition with output able to make tool adjustments automatically, without operator intervention. All these technologies will contribute to increased consistency and productivity; equally important, they will help to improve product quality as well.

There are many recent and successful stories of how automated gaging and testing have improved defect detection and product reliability and even reduced the overall cost per unit of inspection and reject rate. Usually custom-built and highly engineered, these automated systems are best suited for application where certain conditions exist:

- Steady production output
- History of process control
- History of low reject rate
- Where products need individual appraisal

Automated inspection and test may be categorized into five main types: postprocess gaging, in-process gaging, testing, inspection, and assembly and test systems (Quinlan 1996). In all five types of inspection and testing, the automation or semiautomated concept provides high-confidence repeatability and a faster analytical mode that can be interfaced directly to a machine or process to display visually or audibly the statistical process variance above or below the control limits and signal the immediate need for process corrective action.

Strong advancements also have been made in high-speed visual inspection processes, where automated electronic inspection stations are either integrated or slightly off line from manufacturing operations. High-speed cameras capture the image displayed on a monitor, and in microseconds each part passing under the "eye" or segment of a continuous process automatically can undergo multi-characteristic checks. At the end of the inspection cycle, a computer monitor tells the operator if the part or strip passed. The monitor also can indicate why parts are rejected, while updating quality process statistics.

Prior inspection methods could spot only 85 to 90 percent of critical defects. With the new high speed visual automation, virtually all defects are caught. This advanced technology has resulted in increased productivity, reduced waste, and fewer defective parts passing through to the customer (Lincoln 1996).

Integrated Process Inspection and Test.   More and more companies are developing inspection and/or testing equipment that can be integrated into the process directly. Instead of having to physically transport the part to the inspection station, certain characteristics or testing parameters can be evaluated directly on machine or process equipment. For example, one company integrated a measurement system comprising a standard gage linked to a statistical analyzer and coupled with some specially made fixturing. This method allowed the company to make an appraisal decision without having to spend abundant amounts of time on transporting the product to the inspection station and inspection time for setup. This cost savings can be realized in productivity gains by closing the gap between manufacturing and inspection. The shorter the time required for appraisal, the more time is available for production and the greater is the impact of generating larger amounts of quality product.

Specialty gaging and control monitoring equipment has been developed to integrate directly into certain types of machinery and sold as an accessory or an attachment to an existing machine. See, for example, Figure 23.1, which shows a process-control monitor integrated into a coldforming machine for continuous monitoring of process variation. If a force load value falls outside specified parameters, the instrument signals a shutdown of the machine, and only after a quality evaluation is conducted and conformance is determined can the monitor be reactivated by the insertion of a key—usually under quality assurance ownership.

Computer-Aided Inspection.   The marriage of computer and inspector provides benefits to inspection in the form of (1) information or (2) assistance to enhance the conformance decision for almost any inspection situation, or both. The marriage of computers and inspectors takes the form of (1) providing information and/or (2) providing assistance to enhance the conformance decision for almost any inspection situation. This is especially applicable to inspecting precision machine parts and assembly inspection [see Holmes (1974) and Linn (1981) for further discussion]. The selective use of computer-aided inspection (CAI) techniques can minimize the more menial, repetitive inspection tasks and direct the human resource (the inspector) to preventive quality control.

Voice Entry.   Another unique enhancement of computer-aided technology is the voice data-entry system (VDES). VDES can be applied to incoming, in-process, and final inspection functions, as

FIGURE 23.1   Process control monitor.

well as some testing operations. Yet, somewhat embryonic in its applications, VDES will continue to develop in providing the inspection and testing desciplines with many practical benefits, such as

- Saving time by eliminating manual entry
- Reducing labor requirements
- Providing instantaneous response
- Accepting multiple languages and all speakers
- Allowing for remote control
- Permitting on-line, real-time control

Video Entry.   Video inspection techniques used to augment typical video applications called *Moiré contouring* are emerging as a viable method for inspection. Using an optical head with a projection system and camera, a monitor, and a computer equipped with an image-processing board, Moiré allows the operator to obtain a very dense collection of *X, Y, Z* measurements. This depth-information technology is used to find surface defects at various dimension and contours of the part.

The Moiré sensor has been used in several companies for visual enhancement of warping, to inspect hard-to-measure locations, and to examine other machining features with no sharp edges or lighting contrasts to define such shapes as bevels, tapers, etc.

Because Moiré fringe data are already separated into two dimensions (2D) and the depth information is encoded into a 2D line drawing, computer analysis is easier. Also, inspection applications let the manufacturer use the high-density *X, Y, Z* data to detect random part flaws and warpage, and the process can move from high-speed off-line inspection to high-speed on-line inspection and in-process control (see Kennedy 1996).

Optical Sensing.   Optical sensing also can be designed for some applications that may require greater sensing capabilities than would otherwise be possible. This optical scanning technology can measure dimensions by counting the number of beams blocked by an object passing through the scanned region.

Optical scanning systems typically consist of three components: a transmitter, a receiver, and a controller. The transmitter includes a series of light-emitting diodes (LEDs) in a linear array. The controller sequentially switches each LED to the "on" state in succession to generate a sequence of parallel beams of infrared light.

The receiver, with a corresponding linear arrangement of phototransistors, is modulated by the same controller (multiplexed) so that each phototransistor only detects light from its corresponding LED in the transmitter. When something changes the state of one of the receivers, the controller can generate an analog, parallel digital, or serial output (Strack 1996).

INSPECTION AND TESTING FUNCTIONS

Inspection functions are usually staffed by full-time inspectors or testers responsible to the Inspection Department. This is by no means universal. Some final inspection or test functions are staffed by full-time inspectors responsible to production. Many process inspection stations are staffed by production workers whose principal job is production.

Receiving (Incoming) Inspection and Testing.   The extent of inspection of products received from suppliers depends largely on the extent of prior planning for supplier quality control. In the extreme case of using surveillance and audit of decisions, there is virtually no incoming inspection except for identity. At the other extreme, many "conventional" products are bought under an arrangement that relies primarily on incoming inspection for control of supplier quality.

The inspectors and their facilities are housed in the receiving area to provide ready collaboration with other supplier-related activities, i.e., materials receiving, weighing, counting, and storage. Depending on the physical bulk and tonnage of product, entire shipments or just samples are brought to the inspection floor. The documentation routines provide the inspectors with copies of the purchase orders and specifications, which are filed by supplier name.

Inspection planning is conventional, as discussed under Inspection Data Planning above. However, there is usually a lack of prior knowledge of process capability, order of manufacture, etc. Consequently, the sampling plans involve random and (often) large samples, employing standard random inspection tables. Randomness becomes a severe problem in the case of large shipments, whether bulk or not. However, special arrangements can be made with the supplier. Setting acceptable quality levels (AQLs) has been a troublesome problem to such an extent that some industry standards have been worked up. In the absence of such standards, the AQLs are established based on precedent, past performance, or just arbitrarily. Then, as instances of rejection arise, the negotiations with vendors result in adjustment to the AQLs or other acceptable sampling criteria. Data feedback to vendors follows conventional feedback practice.

Process Inspection and Testing.   This commonly serves two purposes simultaneously:

1. *Provides data for making decisions on the product*; e.g., does the product conform to specification?

2. *Provides data for making decisions on the process;* e.g., should the process run or stop?

Because of the interrelation between process and product variables, process inspection involves observation of process variables as well as inspection of the product. These observations and inspections are made by both production and inspection personnel.

Product acceptance of work in process may be done in any of several stages or by a combination of them. These stages include the following.

Setup Inspection.   Some processes are inherently so stable that if the setup is correct, the entire lot will be correct, within certain limits of lot size. For such processes, the setup approval also can be used as the lot approval. Where a good deal is at stake, it is usual to formalize the setup inspection and to require that the process may not run until the inspector has formally approved the setup, e.g., by signing off, by stamping the first pieces, etc. (Garfinkel and Clodfelter 1984).

Patrol Inspection.   For processes that will not remain stable for the duration of the lot, it is usual to provide for periodic sampling to be conducted during the progression of the lot, making use of various techniques described in Section 24. The numerous plans in use consist mainly of variations of the following four types:

1. Preserve the order of manufacture under an arrangement such as is depicted in Figure 23.2. In this example, the machine discharges its production into a small container called a *pan*. The production operator periodically empties the pan into one of three larger containers:
   a. Into the junk box if the parts are junk.
   b. Into the reject box if the parts are questionable or are mixed good and bad.
   c. Into the tray if the parts are presumably good.
   The patrol inspector comes to the machine and checks the last few pieces being made. (He or she also may sample the tray.) Based on this check, the tray is disposed of in one of three ways:
   a. Into the junk box if the parts are junk.
   b. Into the reject box if the parts are questionable or are mixed good or bad.
   c. Into the good box if the parts are acceptable. The good box goes on to the next operation.
   Only the inspector may dispose of the tray, and only the inspector may place any product in the good box. The reject box is gone over by a sorter, who makes three dispositions:
   a. Junk to the Junk Department.
   b. Reoperate back to the Production Department.
   c. Good parts on to the next operation.

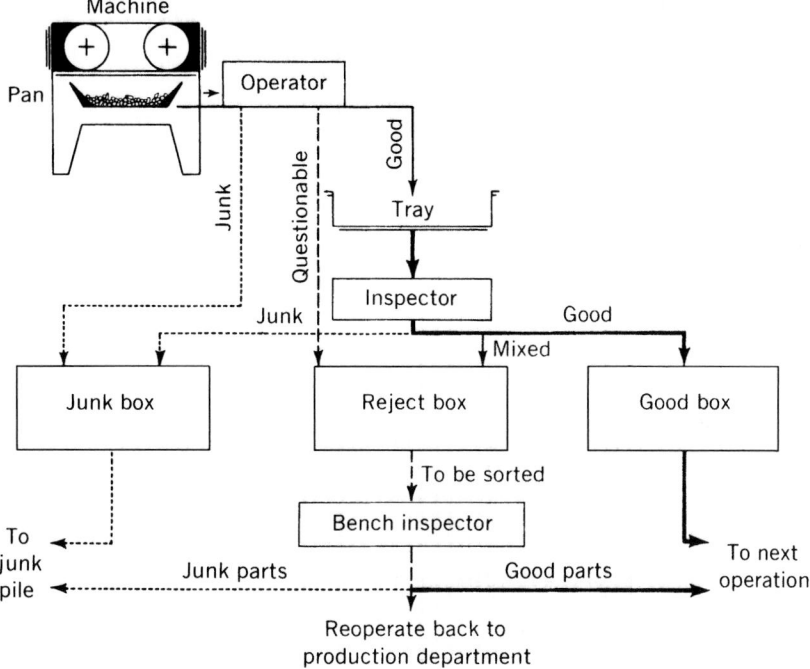

FIGURE 23.2   Patrol inspection plan based on preserving the order.

2. This method is similar to the first, but the inspection data from the last few pieces are posted to a control chart. If the process remains in control, all product made since the last check is accepted.

3. The accumulated product (e.g., in the tray of Figure 23.2) is sampled at random using some standard sampling plan, and acceptance is based on the sampling criteria.

4. The process variables are checked against a process specification, and the product is accepted provided the process conforms to specification. This method is usually restricted to cases in which there is to be a direct check on the product at later stages.

Tollgate Inspection.   This is a lot-by-lot product-acceptance procedure. Commonly it is done after the Production Department has concluded its operations. Sometimes the product is moved physically to an inspection area, where it waits its turn to be inspected. Sampling is at random, using standard sampling tables.

Tollgate inspection reduces congestion at the machines and clarifies departmental responsibility. The price paid is in added material handling, added floor space, loss of order of production, and greater difficulty in fixing individual responsibility.

Finished-Goods Inspection.   Most finished products are tested 100 percent for minimal simulation of use. Tests are often automated, as are the data recordings. Testing may be done either at inspection stations on the production line or in separate inspection areas.

Shipping Inspection.   For processes that are continuous and stable and/or do not require multiple operations before enclosure into a container, a shipping inspection and/or test will provide assurance of conformance before the containers are sealed.

Dock Audit.   This is an inspection of a random selection of product that is ready to be shipped. Such audits may be made daily, weekly, semiweekly, biweekly, etc. Whatever the frequency, it is important to check not only the product characteristics but also the labeling, shipping and handling methods, and many other requirement criteria.

Destructive Test.   A product that has been deemed stable and by its functional nature requires a lot or batch destruction test on its components or assemblies is subject to a random destructive evaluation. If any failure occurs, the lot or batch is automatically quarantined to future analysis.

QUALITY STANDARDS

The cornerstone of quality control is the specification. Specifications embody the minimum and maximum values (see also Bader 1980). However, they are usually *incomplete.* They tend to ignore *visual quality characteristics,* or they treat characteristics labeled as *workmanship* superficially. [Refer to the discussion of visual quality characteristics below; and see also Dodds (1967) and Alaimo (1969) for further discussion on workmanship standards.]

Seriousness Classification.   Some quality characteristics and defects are very important to fitness for use; others are not. The village artisan and the small-shop proprietor, with their first-hand knowledge of fitness for use, are able to concentrate their efforts on the most important qualities. In modern, large, complex organizations, the workers, inspectors, and many of the supervisors lack complete knowledge of fitness for use and thus are not fully clear on where to place their emphasis and how to make their decisions.

For example, one company studying the fabrication and inspection of machine parts divided the quality characteristics into four classes. Table 23.4 shows the effect of this classification on product tolerances and on the number of dimensions checked. The inspection time was reduced from 215 to 120 minutes. In addition, there were greater savings through lower rework costs, lower tooling costs, and lower engineering costs for disposition of nonconforming product (see also Allen 1959).

The seriousness classification takes into consideration information from a variety of sources:

The specification: This is the primary source. Because the specifications rarely reflect the most current input from customers and other sources, it is important that the team proactively seek current information from the sources listed here.

TABLE 23.4   Results of Seriousness Classification of Characteristics

Characteristic classification	Effect of classification on design tolerance	Effect of classification on amount of inspection	Number of dimensions checked — Before classification	Number of dimensions checked — After classification
Critical	None	None	154	154
Major	None	None	110	110
Minor A	Tolerance was increased by a specified amount (doubled, etc.) provided the part assembled satisfactorily	Inspection was made normally, but to wider tolerances	66	15
Minor B	Tolerance was ignored provided the part assembled satisfactorily	Inspection was eliminated	352	0
Total			682	279

Customer input: The classification is an even more useful guide to decision making when the team broadens the information base to include direct preproduction input from the whole range of customers—the purchaser, the end user, the various internal departments involved in manufacture, inspection, packaging, shipping, and so on. Then the classification can reflect more accurately the customer's needs and expectations of the product.

Manufacturing experience with the product: Problems and inadequacies, including evidence of nonconformance during manufacturing, help identify troublesome features.

Life testing and functional testing: When these functions are part of the production cycle, they should be included in the classification process and in structuring a formal classification model. The test results are useful input to classification decisions. Where there exists the possibility of product misuse—which, for a variety of unanticipated reasons, too frequently replaces the customer's originally intended use—the tests can help anticipate and prevent such misuse or at least help mitigate its effects. The proper classification ensures that the affected features will receive proper attention.

Failure during use: Field failure is the ultimate undesirable quality outcome. Any feature for which a causal link can be established to field failure is, by definition, a critical feature.

Seriousness classification is useful input both to the control plan and the overall quality plan. Some quality characteristics are multileveled in their seriousness criteria. For example, a shaft diameter specified as "1.000 ± 0.001" gives rise to two defects: oversize and undersize. These defects may be assigned different degrees of seriousness depending on the extent and effect of nonconformance. Some extensive defect lists, e.g., the list for glass bottles, have little resemblance to the list of characteristics set out in the specifications.

Some companies use the same system of classification for both characteristics and defects. However, there is enough uniqueness about each of the two lists to suggest that adoption of a single system should be preceded by a positive examination of the nature of the two lists. For example, the effect of seriousness classification on design decisions can be quite different from the effect on inspection decisions, as is evident from Table 23.4.

Formal systems of seriousness classification were evolved originally to serve specialized purposes. (The Bell System pioneered by developing a system to permit rating of quality of finished product. The U.S. Army developed systems to simplify the administration of acceptance of goods purchased from contractors.) However, as the systems came into being, they were found to have application to the entire progression of product from design through use: in quality specification, manufacturing planning, supplier relations, tooling, production, salvage, product auditing, and executing reporting. Vital qualities could now be identified with greater confidence, and it also became feasible to delegate class decisions and actions on a broad scale. For example, all class C defects could be assigned a common sampling plan, thereby avoiding the need for publishing numerous individual plans.

The multiple uses of seriousness classification systems make it desirable that the job of developing such a system be guided by an interdepartmental committee that has the responsibility for drafting a plan, modifying it, and recommending it for adoption. Such a committee has a series of tasks:

1. Determining the number of strata or classes of seriousness to use.
2. Defining each class.
3. Classifying each defect into one of the classes.

Number of Levels or Strata.   In theory, this number may be large; e.g., a defect may have any weight from 1000 down to 1. In practice, such a large number of weights is too complex to administer. The actual plans in use consist of only several classes. While choice of the actual number of classes is arbitrary, extensive experience has shown that three or four classes suffice for a wide variety of situations.

Definitions for the Classes.   These will differ with the nature of the product, process, etc. However, plans in existence tend to show striking similarity in definition, the result in part of the influence of the Bell System classification plan [see also Dodge (1928) and Dodge and Toreey (1956)]. Not only was this pioneering plan uncommonly well reasoned out; the men who devised it were later consultants to some of the U.S. armed services during World War II, and their thinking influenced the classification plans adopted by these services. These plans, in turn, influenced the plans adopted by the contractors to the armed services.

The standard definitions adopted by the Bell System [see also Dodge and Torrey (1956)] are shown in Table 23.5. Study of these definitions discloses that there is an inner pattern common to the basic definitions (Table 23.6).

A composite of definitions used in food industry companies is shown in Table 23.7. It is evident that there are industry-to-industry differences in products, markets, and so on that require a tailor-made wording for each industry. In addition, the lists are not static. The growth of government regulation has further influenced the definition, as has the problem of repairs and guarantees for long-life products.

It is also evident that the classifications must simultaneously take into account multiple considerations such as functional performance, user awareness, and financial loss. For example, the effects and awareness of a radio receiver's defects may be as follows:

Defect	Effect	User awareness
Open circuit in power supply	Set is inoperative	Fully aware
Short circuit in resistor	Excess power consumption	Seldom aware
Poor exterior finish	No effect	Usually aware
Poor dress or internal wiring	No effect	Seldom aware

Classifying the Defects.   This essential task is time-consuming, since there are always many defects to be classified. If the class definitions have been well drawn, the task becomes much easier.

During classifying, much confusion is cleared up. It is found that the seriousness of important visual defects depends not so much on whether the inspector can see them as on whether the consumer can see them. It is found that some words describing defects must be subdivided; i.e., a *stain* may be placed in two or three classes depending on severity and location. In many ways, the work of classifying defects is rewarding through clearing away misconceptions and giving a fresh view to all who participate.

Classification of Characteristics.   In some companies the formal "seriousness" classification is not of defects but of characteristics in the specifications. The classification may be in any of several alternatives:

1. *Functional or nonfunctional.*   Where a single set of drawings carries both functional ("end use") requirements and nonfunctional ("means to an end") requirements, it is important to make clear which is which. (This is not to be confused with mechanical, chemical, or electrical functioning. In products such as jewelry or textiles, the most important functional requirement is *appearance.*) The purposes served by these two classes are generally alike throughout industry:

Functional requirements are intended to	*Nonfunctional requirements are intended to*
Ensure performance for intended use	Inform the shop as to method of manufacture
Ensure long, useful life	Reduce cost of manufacture
Minimize accident hazards	Facilitate manufacture
Protect life or property	Provide interchangeability in the shop
Provide interchangeability in the field	Provide information to toolmakers
Provide competitive sales advantage	

TABLE 23.5 Serious Classification of Defects (Bell System)

Class A: Very serious (demerit value, 100)

1. Will surely cause an operating failure of the unit in service that cannot be readily corrected in the field, e.g., open relay winding, or
2. Will surely cause intermittent trouble, difficult to locate in the field, e.g., loose connection, or
3. Will render unit totally unfit for service, e.g., dial finger wheel does not return to normal after operation, or
4. Liable to cause personal injury or property damage under normal conditions of use, e.g., exposed part has sharp edges.

Class B: Serious (demerit value, 50)

1. Will probably cause an operating failure of the unit in service that cannot be readily corrected in the field, e.g., protective finish missing from coaxial plug, or
2. Will surely cause an operating failure of the unit in service that can be readily corrected in the field, e.g., relay contact does not make, or
3. Will surely cause trouble of a nature less serious than an operating failure, such as substandard performance, e.g., protector block does not operate at specified voltage, or
4. Will surely involve increased maintenance or decreased life, e.g., single contact disk missing, or
5. Will cause a major increase in installation effort by the customer, e.g., mounting holes in wrong location, or
6. Defects of appearance or finish that are extreme in intensity, e.g., finish does not match finish on other parts, requires refinishing.

Class C: Moderately serious (demerit value, 10)

1. May possibly cause an operating failure of the unit in service, e.g., contact follow less than minimum, or
2. Likely to cause trouble of a nature less than an operating failure, such as substandard performance, e.g., ringer does operate within specified limits, or
3. Likely to involve increased maintenance or decreased life, e.g., dirty contact, or
4. Will cause a minor increase in installation effort by the customer, e.g., mounting bracket distorted, or
5. Major defects of appearance, finish, or workmanship, e.g., finish conspicuously scratched, designation omitted or illegible.

Class D: Not serious (demerit value, 1)

1. Will not affect operation, maintenance, or life of the unit in service (including minor deviations from engineering requirements), e.g., sleeving too short, or
2. Minor defects of appearance, finish, or workmanship, e.g., slightly scratched finish.

TABLE 23.6 Inner Pattern: Seriousness Classification System

Defect class	Demerit weight	Cause personal injury	Cause operating failure	Cause intermittent operating trouble difficult to locate in field	Cause substandard performance	Involve increased maintenance or decreased life	Cause increase in installation effort by customer	Appearance, finish, or workmanship defects
A	100	Liable to	Will surely*	Will surely				
B	50		Will surely		Will surely	Will surely	Major increase	
C	10		May possibly		Likely to	Likely to	Minor increase	Major
D	1		Will not		Will not	Will not	Minor increase	Minor

*Not readily corrected in the field.

TABLE 23.7 Composite Definitions for Seriousness Classification in Food Industry

Defect	Effect on consumer safety	Effect on use	Consumer relations	Loss to company	Effect on conformance to government relations
Critical	Will surely cause personal injury or illness	Will render the product totally unfit for use	Will offend consumer's sensibilities because of odor, appearance, etc.	Will lose customers and will result in losses greater than value of product	Fails to conform to regulations for purity, toxicity, identification
Major A	Very unlikely to cause personal injury or illness	May render the product unfit for use and may cause rejection by the user	Will likely be noticed by consumer, and will likely reduce product salability	May lose customers and may result in losses greater than the value of the product; will substantially reduce production yields	Fails to conform to regulations on weight, volume, or batch control
Major B	Will not cause injury or illness	Will make the product more difficult to use, e.g., removal from package, or will require improvision by the user; affects appearance, neatness	May be noticed by some consumers, and may be an annoyance if noticed	Unlikely to lose customers; may require product replacement; may result in loss equal to product value	Minor nonconformance to regulations on weight, volume, or batch control, e.g., completeness of documentation
Minor	Will not cause injury or illness	Will not affect usability of the product, may affect appearance, neatness	Unlikely to be noticed by consumers, and of little concern if noticed	Unlikely to result in loss	Conforms full to regulations

"Which is which" becomes important because it directs the priorities of process design and many aspects of economics of manufacture, as well as the jurisdiction over waivers. When the engineers make this classification, they commonly add a designation such as *E* (for engineering) to the functional characteristics. All others are then assumed to be nonfunctional.

A comparable situation prevails in process specifications, where the need is to distinguish mandatory from advisory requirements, which correspond roughly to functional and nonfunctional requirements as applied to the product. However, the process specifications seldom make this distinction. (For further discussion, see Section 22, Operations, under Knowledge of "Supposed to Do.")

2. *Seriousness classification.*   When this method is used, it parallels closely the classification into critical, major, and minor as used for classification of defects. (The contention is often raised that the tolerances on the specifications are an automatic form of seriousness classification, i.e., anything with assigned tolerances must be met and is "therefore" critical. An alternative contention is that the closeness of the tolerances is a key to seriousness classification; i.e., the narrowest tolerances are assigned to the most critical characteristics. When these contentions are examined more closely, they are found to contain too many exceptions to serve as firm rules for classifications.)

The resulting classifications then become a supplement to the specification or are shown on the drawings themselves by some code designation, for example:

Critical   ⊕

Major A   ⊙

Major B   ○

Minor   Not marked

One large automotive company differentiates between regulatory and nonregulatory critical characteristics by the use of two different symbols.

3. *Segregation of functional requirements.*   Place in a separate document, such as an "engineering specification" or a "test specification."

4. *Shop practice tolerances versus special tolerances.*   This method is based on preparation of a shop practice manual that sets out general-use tolerances derived from the process capability of general-use machines and tools. Once published, these shop practice tolerances govern all characteristics not specially toleranced.

Who Classifies?   For defect classification, an interdepartmental committee is the ideal choice. This provides each department with the benefits derived from the process of active review, and it also produces a better final result (sometimes the committee goes further and establishes a plan for product rating, including demerit weights for each class; see Section 8, under Appraisal Costs). However, some companies assign a staff specialist to prepare a proposed classification, which is then reviewed by all interested departments. The specialist is usually a quality control engineer.

When the classification is limited to specified characteristics, e.g., functional versus nonfunctional, the designer usually prepares the draft.

SENSORY QUALITIES

Sensory qualities are those for which we lack technological measuring instruments and for which the senses of human beings must be used as measuring instruments. (For some special purposes, e.g., tests of toxicity, the test panel may consist of animals.) Sensory qualities may involve

- Technological performance of the product, e.g., adhesion of a protective coating, friction of a sliding fit
- Esthetic characteristics of consumer products, e.g., taste of food, odor of perfume, appearance of carpets, noise of room air conditioners

In common with other qualities, sensory qualities require:

1. Discovery of which characteristics are required and in what degree to meet the needs of fitness for use
2. Design of products that will possess these characteristics
3. Establishment of product and process standards and of tests that will simulate fitness for use
4. Judgment of conformance to the product and process standards

This multiplicity of tasks requires a corresponding multiplicity in type of sensory test panel used, choice of test design, and so on.

Customer Sensitivity Testing.   In this form of test, the purpose is to discover the "threshold" level at which customers can detect the presence of sensory qualities. The qualities under test may be "desirable." For example, if an expensive ingredient is used in a product blend, it is very useful to know the threshold concentration level that ensures customer recognition of the ingredient. The qualities under test may be "undesirable." For example, a product exhibits varying degrees of visual blemish. It is very useful to know the threshold degree of defectiveness that makes the customer respond negatively to the product.

In customer-sensitivity testing, a graduated set of samples is prepared, each exhibiting a progressively greater extent of the quality or deficiency under investigation. These samples are submitted to a customer panel as part of an organized study.

For example, in two companies—one making sterling silverware and the other making costume jewelry—studies were conducted to discover customer sensitivity to visual defects. In both companies, a committee of key people (from Marketing, Design, Manufacture, and Quality) structured a plan of study as follows:

1. An assortment of product was chosen to reflect the principal visual defects, the principal products in the product line, and the principal price levels.

2. These samples were inspected in the factory by the regular inspectors to determine the severity of the defects as judged by the frequency with which the inspectors rejected the various units of product.

3. The assortment of products was then shown to a number of customer panels chosen from those segments of the buying public which constituted important customer classes, e.g., suburban women, college students, etc. These panels reviewed the products under conditions that simulated use of the product, e.g., silverware in place settings on a dining room table. The customers were instructed (by printed card) somewhat as follows: "Assume you have previously bought these products and they have been delivered to you. Naturally, you will want to look them over to see that the merchandise is satisfactory. Will you be good enough to look it over, and if you see anything that is objectionable to you, will you please point it out to us?"

The resulting data showed that the customer panels were highly sensitive to some defects. For such defects, the strict visual standards were retained. For certain other defects, the customer sensitivity was far less than factory-inspector sensitivity. For such defects, the standards were relaxed. In still other instances, some operations had deliberately been omitted, but the customers proved to be insensitive to the effect of the omission. As a result, the operations were abolished.

Customer sensitivity testing is an extension of the principle that "the customer is right." This principle may be subdivided as follows:

1. The customer is right as to qualities he or she can sense. As to such qualities, the manufacturer is justified in taking action to make such qualities acceptable to the customer.

2. The customer is also right as to qualities he or she *cannot* sense. The manufacturer is not justified in adding costs to create an esthetic effect not sensed by the customer.

3. Where, for a given quality, the customer is sensitive to a limited level but not beyond, the manufacturer should take action to make the quality to that level but not beyond.

The intermediate marketing chain sometimes interferes with these principles. Sales clerks are proficient in emphasizing product differences, whether important or not. In turn, dealers are alert to seize on such differences to wring concessions out of competing manufacturers. A frequent result is that all manufacturers are driven to adopt wasteful standards, resulting in a needlessly high cost, e.g., finishes on nonworking or nonvisible surfaces. Elimination of such perfectionism commonly requires that the manufacturer secure data directly from consumers and then use the data to convince the distribution chain. These same data may be needed to convince other nonconsumers who exhibit perfectionist tendencies: upper management, designers, salespeople, inspectors, etc.

Visual Quality Characteristics.   These constitute a special category of sensory qualities. (Visual inspection remains the largest single form of inspection activity.) For these characteristics, the written specifications seldom describe completely what is wanted, and often inspectors are left to make their own interpretation. In such cases, inspectors are really making two judgments simultaneously:

1. What is the meaning of this visual characteristic of the specification, e.g., what is the standard?

2. Does this unit of product conform to the standard?

Where inspectors understand fitness for use, they are qualified to make both these judgments. If a particular inspector lacks this knowledge, he or she is qualified to make only judgment 2, no matter how long on the job. Extensive experience has shown that inspectors who lack this knowledge differ widely when setting standards and, in addition, do not remain consistent. (As an example, from the consulting experience of J. M. Juran, in an optical company, study of the methods used by 18 different inspectors, engineers, and so on disclosed the existence of six methods of counting the number of "fringes of irregularity.") Several methods are available to planners to clarify the standard for visual characteristics.

Visual Inspection and Test Standards.   The most elementary form of visual standard is the *limit sample*—a unit of product showing the worst condition acceptable. In using this standard, the inspector is aided in two ways:

1. The sample conveys a more precise meaning than does a written specification.
2. The inspection is now made by comparison, which is well known to give more consistent results than judgment in the absence of comparison.

A more elaborate form of visual standards involves preparation of an exhibit of samples of varying degrees of defects ranging from clearly defective to clearly acceptable. See, for an example involving solder connections, Leek (1975 and 1976), who describes how the companies Martin Marietta and Northrup supported series or "ranges" to provide manufacturing latitude in processing material while at the same time identifying minimum and maximum limits on visual attributes. This exhibit is used to secure the collective judgments of all who have a stake in the standard—consumers, supervisors, engineers, and inspectors. Based on these judgments, standards are agreed on, and limit samples are chosen. (It is also feasible to estimate, by sampling, what would be the yield of the process, and thereby the cost of defects, for any one of the various degrees of defectiveness.)

In products sold for esthetic appeal, appearance becomes a major element of fitness for use and commonly a major element of cost as well. In such cases, an exhibit of samples with varying degrees of defects intermingled with perfect units of product becomes a means of measuring consumer sensitivity (or insensitivity) to various defects. Use of consumer panels to judge such mixtures of product invariably confirms some previous concepts but also denies some long-standing beliefs held by managers as well as by the inspectors.

In the sterling silverware case mentioned earlier, consumers were quite sensitive to several types of defects—they held out 22 percent of the defects present. However, for the bulk of defects, the consumers were quite insensitive and found only 3 percent of such defects. The salespeople generally found twice as many defects as consumers but still considerably fewer than factory inspectors.

A further use of samples of various defects is to establish *grades* of defects. The concept of different grades is vital when a plant makes products that, while outwardly similar, are used in widely different applications, e.g., ball bearings used for precision instruments and those used for roller skates, lenses for precision apparatus and lenses for simple magnifiers, or sterling silverware and plated silverware. Unless the grades are well defined and spelled out in authoritative form, the risk is that the inspectors will apply one standard to all grades.

Once limit samples have been agreed to, there remains a problem of providing working standards to the inspection force. Sometimes it is feasible to select duplicates for inspection use while retaining the official standard sample in the laboratory. An alternative is to prepare photographs (sometimes stereoscopic) of the approved standards and to distribute these photographs instead.

Standardizing the Conditions of Inspection and Test.   Visual inspection results are greatly influenced by the type, color, and intensity of illumination, by the angle of viewing, by the viewing distance, and so on. Standardizing these conditions is a long step in the direction of securing uniform inspection results. In the case of esthetic visual qualities, the guiding rule for conditions of inspection is to simulate the conditions of use, but with a factor of safety.

Establish a *fading distance*. In some products the variety of visual defects is so great and the range of severity so wide that the creation of visual standards becomes prohibitively complex. An alternative approach is to standardize the conditions of inspection and then to establish a fading distance for each broad defect class. The definition for a defect becomes "anything that can be seen at the fading distance." (This technique appears to have been evolved in 1951 by N. O. Langenborg of St. Regis Paper Company. See also Riley 1979.)

Sensory Tests: Design and Analysis.   There are numerous designs of sensory tests, some of them quite complex. Some of the basic forms are described below.

Tests for Differences or Similarities.   These include

1. *The paired-comparison test.*   Product is submitted to members of a panel in pairs of samples. One sample is identified to each panelist as the standard or "control"; the other is the test sample. The panelist is asked to judge and record the difference on a scale of differences (such as no difference, slight difference, or pronounced difference). Some of the pairs have no difference; i.e., both are "controls."

2. *The triangle test.*   The panelist is asked to identify the odd sample in a group of three, two of which are alike. He or she also may be asked to estimate the degree of difference and to describe the difference between the odd sample and the two like samples.

3. *The duo-trio test.*   The panelist is asked to identify which of two samples is like the "control" to which she or he has been subjected previously. For example, in liquor manufacture, the aim is to make each batch indistinguishable in taste from past batches. The duo-trio test is used as a product-acceptance test. Each panelist tastes the "control," which he or she is told comes from previous product. Then each panelist tastes the two remaining samples, one of which is "control" and the other of which is the batch under test. However, the panelists are not told which is which. If the data make clear that the panel cannot distinguish the new batch from the control, the batch is accepted. Otherwise, it is reblended.

4. *Ranking test.*   Coded samples are submitted to each panelist, who is asked to rank them in the order of concentration.

Creating New Instruments to Measure Sensory Qualities.   Many sensory qualities formerly judged by human perception are now measured by instruments. This development of new instrumentation goes on apace using essentially the following approach (based on a procedure set out by Dr. Amihud Kramer 1952; see also Hains 1978):

1. Define precisely what is meant by the quality characteristic under discussion. This must be done with participation of all interested parties.

2. Discover, through analytical study, the subcharacteristics, and define them in a way that permits, in theory, measurement by some inanimate instrument.

3. Search the literature to become informed about methods already in existence or under the development for measuring these subcharacteristics. This search will disclose a number of such possible measurement methods.

4. Choose or create product samples that vary widely for the subcharacteristics. Test a limited number (10 to 50) of samples with each of the various measurement methods, and correlate these tests with evaluation by panels of human testers. The human evaluation here aims not to measure personal preferences but to rate the degree to which the samples possess the variable under study. Hence the main requirement of the panel is that it be able to discriminate the subcharacteristics under study. Discard those measurement methods which lack precision or which fail to reflect human evaluation.

5. For the remaining, more promising measurement methods, conduct tests on a larger number of samples (100 to 1000) also chosen to reflect the entire range of quality variation. In addition, conduct tests of duplicate samples using evaluation by human test panels.

6. Correlate the results of measurement against the human test panel evaluation; select the method that gives a high correlation. (Multiple correlation methods may be necessary.)

7. Improve and simplify the selected measurement method through further tests and correlations.

8. Establish a scale of grades through use of a human sensory test panel. At this stage, the prime purpose of the human test panel is to state preferences along the scale of measure. Hence the main requirement of this panel is that it be representative of the producers and users of the product.

9. Weigh the various subcharacteristics in accordance with their rated performance [see also Montville (1983) and Papadopoulos (1983)].

10. Develop the sampling procedures needed to apply the resulting method of measurement.

MEASUREMENT QUALITY: AN INTRODUCTION

Conduct of the quality function depends heavily on qualification of product and process characteristics. This quantification is done through a systematic approach involving

1. Definition of standardized units called *units of measure* that permit conversion of abstractions (e.g., length, mass) into a form capable of being quantified (e.g., meter, kilogram).

2. *Instruments* that are calibrated in terms of these standardized units of measure.

3. Use of these instruments to quantify or *measure* the extent to which the product or process possesses the characteristic under study. This process of quantification is called *measurement.*

The word *measurement* has multiple meanings, these being principally

1. The *process* of quantification; e.g., "The measurement was done in the laboratory."

2. The resulting number; e.g., "The measurement fell within the tolerances."

Measurement rests on a highly organized, scientific base called *metrology,* i.e., the science of measurement. This science underlies the entire systematic approach through which we quantify quality characteristics.

MEASUREMENT STANDARDS

The seven fundamental units of the International System (SI) of measurement are defined as shown in Table 23.8. It is seen that except for the kilogram, all units are defined in terms of natural phenomena. (The kilogram is defined as the mass of a specific object.)

Primary Reference Standards.   In all industrialized countries there exists a national bureau of standards whose functions include construction and maintenance of *primary reference standards.* These standards consist of copies of the international kilogram plus measuring systems that are responsive to the definitions of the fundamental units.

In addition, professional societies (e.g., the American Society for Testing and Materials) have evolved standardized test methods for measuring many hundreds of quality characteristics. These standard test methods describe the test conditions, equipment, procedure, and so on, to be followed. The various national bureaus of standards, as well as other laboratories, then develop primary reference standards that embody the units of measure corresponding to these standard test methods.

Primary reference standards have a distinct legal status, since commercial contracts usually require that "measuring and test equipment shall be calibrated . . . utilizing reference standards . . . whose calibration is certified as being traceable to the National Institute of Standards and

TABLE 23.8   Definitions of Fundamental Units of the SI System

Unit	Definition
Meter, m	1/650 763.73 wavelengths in vacuo of the unperturbed transition $2_{p10} \rightarrow 5d_5$ in ^{36}Kr
Kilogram, kg	Mass of the international kilogram at Sevres, France
Second, s	1/31 556 925 974 7 of the tropical year at 12^h ET, 0 January 1900, supplementarily defined in 1964 in terms of the cesium F, 4; M, 0 to F, 3; M, 0, transition, the frequency assigned being 9 192 631 770 Hz
Kelvin, K	Defined in the thermodynamic scale by assigning 273.16 K to the triple point of water (freezing point, 173.15 K = 0°C)
Ampere, A	The constant current which, if maintained in two straight parallel conductors of infinite length, of negligible circular sections, and placed 1 m apart in a vacuum, will produce between these conductors a force equal to 2×10^{-7} mks unit of force per meter of length.
Candela, cd	1/60 of the intensity of 1 cm^2 of a perfect radiator at the temperature of freezing platinum.
Mole, mol	An amount of substance whose weight in grams numerically equals the molecular formula weight.

Technologies." In practice, it is not feasible for the U.S. National Institute of Standards and Technologies to calibrate and certify the accuracy of the enormous volume of test equipment in use in shops and test laboratories. Instead, resort is made to a hierarchy of secondary standards and laboratories, along with a system of documented certification of accuracy.

Hierarchy of Standards.   The primary reference standards are the apex of an entire hierarchy of reference standards (Figure 23.3). At the base of the hierarchy there stands the huge array of *test equipment,* i.e., instruments used by laboratory technicians, workers, and inspectors to control processes and products. These instruments are calibrated against *working standards* that are used solely to calibrate these laboratory and shop instruments. In turn, the working standards are related to the primary reference standards through one or more intermediate secondary reference standards or *transfer standards.* Each of these levels in the hierarchy serves to "transfer" accuracy of measurement to the next lower level in the hierarchy.

Within the hierarchy of standards there are differences both in the physical construction of the standards and in their precision. The primary reference standards are used by a relatively few highly skilled metrologists, and their skills are a vital commitment to the high precision attained by these standards. As we progress down the hierarchy, the number of technicians increases with each level, until at the base there are millions of workers, inspectors, and technicians using test equipment to control product and process. Because of the wide variation in training, skills, and dedication among these millions, the design and construction of test equipment must feature ruggedness, stability, and foolproofing so as to minimize errors contributed by the human being using the equipment.

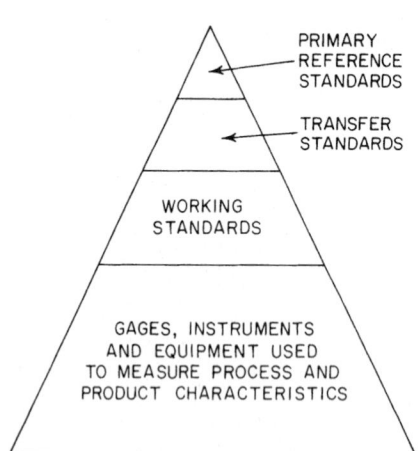

FIGURE 23.3   Hierarchy of standards.

Precision of measurement differs widely among the various levels of the hierarchy of standards. At the level of primary reference standards, the precision is determined by the state of the art. For example, Figure 23.4 shows the precision attained by the U.S. National Bureau of Standards (now the National Institute of Standards and Technology) when weighing loads across the spectrum of 10^{-10} to 10^6 (National Bureau of Standards 1965).

At the base of the hierarchy, the precision of measurement is determined by the needs of fitness for use as reflected by the product and process tolerances. While some specialists urge that the test equipment be able to "divide the tolerance into tenths," this ideal is by no means always attained in practice. However, the tolerances themselves have been tightened drastically over the centuries, and this tightening generally has paralleled the advances made in the state of the art of measurement. For example, accuracy of measurement of a meter of length has progressed from an error of 1000 per million (at the end of the fifteenth century) to an error of 0.0001 per million as we move into the twenty-first century.

Allocation of measurement errors among the working and transfer standards has been discussed widely but has not been well standardized. The precision gap between primary reference standards and product test equipment may be anywhere from one to several orders of magnitude. This gap must then be allocated among the number of levels of standards and laboratories (transfer plus working) prevailing in any given situation. (Some models have been worked out to show the interrelation among the cost of developing greater precision in the primary reference standard, cost of attaining precision at each level of transfer laboratory, and number of laboratories at each level. See, for example, Crow 1966.) When this problem of allocation was first faced, there was a tendency to conclude that each level should have a precision 10 times greater than the level it was checking. More recently, there has been growing awareness of how multiple levels of precision combine; i.e., their composite is better represented by the square root of the sum of the squares rather than by the arithmetic sum. This new awareness has caused many practitioners to accept a ratio of 5:1 rather than 10:1 for precision of working standards to product tolerances. This same ratio also has been tolerated among transfer standards as well.

ERROR OF MEASUREMENT

Product and process conformance is determined by measurements made by the test equipment at the bottom of the hierarchy of standards. Obviously, any error in these measurements has a direct bearing on the ability to judge conformance. On examination, the nature of measurement error is quite complex; even the terminology is confused. A clear understanding of the meaning of the

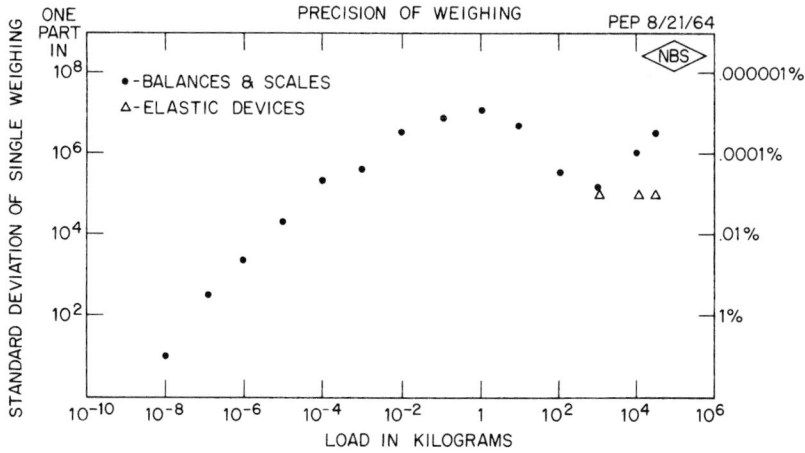

FIGURE 23.4   Precision of weighing.

measurements requires a minimal degree of understanding of the nature of measurement error. The starting point is to understand the nature of accuracy and precision. Figure 23.5 shows the meaning of these terms by example and by analogy. See, in this connection, ASTM 177-71 (listed in Appendix III) on use of the terms *precision* and *accuracy* as applied to measurement of the properties of materials. See also Mathur (1974) for a discussion on the influence of measurement variation in mass production of parts.

Accuracy.   Suppose that we make numerous measurements on a single unit of product and that we then compute the average of these measurements. The extent to which this average agrees with the "true" value of that unit of product is called the *accuracy* of the instrument or measurement system that was employed. The difference between the average and the true value is called the *error* (also *systematic error, bias,* or *inaccuracy*) and is the extent to which the instrument is out of calibration. The error can be positive or negative. The *correction* needed to put the instrument in calibration is of the same magnitude as the error but opposite in sign. The instrument is still considered *accurate* if the error is less than the *tolerance* or maximum error allowable for that grade of instrument.

Accuracy and *error* are quantified as a difference between (1) the average of multiple measurements and (2) the true value. As will be seen, each of these is surrounded by a fringe of doubt. Consequently, the expression of accuracy must show the extent of these doubts if the full meaning of the numbers is to be conveyed.

Precision.   Irrespective of accuracy of calibration, an instrument will not give identical readings even when making a series of measurements on one single unit of product. Instead, the measure-

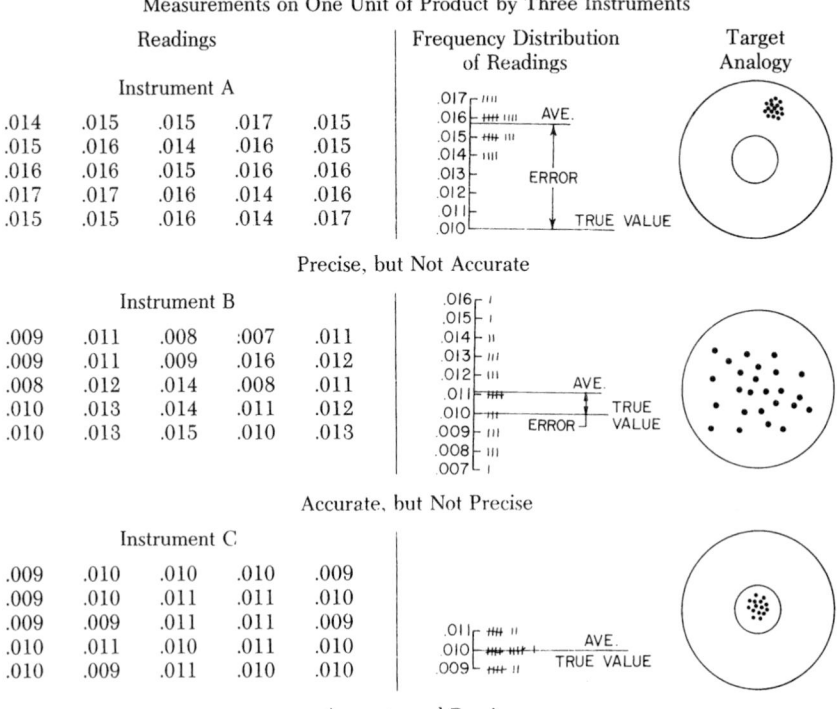

FIGURE 23.5   Accuracy and precision.

ments scatter about the average, as exemplified in Figure 23.5. The ability of the instrument to reproduce its own measurements is called its *precision,* and this varies inversely with the dispersion of the multiple (*replicated*) measurements.

Experience has shown that any measurement system has an inherent dispersion that is itself reproducible, measurable, and therefore (once known) predictable. This inherent precision of measurement parallels the inherent process capability of a machine tool. (The parallel extends to the requirement that the system be in a state of statistical control.) Quantification of precision is in terms of the standard deviation of replicated measurements and is expressed by σ (sigma), the statistical symbol for standard deviation of a population.

Normally, recalibration can improve the accuracy of an instrument by reducing its error. However, recalibration normally does not improve the precision of the instrument, since this precision remains relatively constant over the working range.

Sources of Error.
Systematic error and dispersion of measurements have their origin in several well-known components of measurement error. (In some industries, e.g., chemical processes, the measurement problems are so severe that development of valid test procedures is a major step in the launching of a new product or process.)

Within-Operator Variation.   The same operator, inspector, or technician, even when using the same measuring system on the same unit of product, nevertheless will come up with a dispersion of readings. This variation is usually referred to as *within-operator variation.*

Between-Operators Variation.   When two operators use the same measuring system on the same products, they usually will exhibit differences traceable to differences in operator technique. These differences are called *between-operators variation* and can be exhibited both as systematic error and as differences in dispersion.

Materials Variation.   In many cases it is not feasible to conduct replicated tests on the same "unit of product"; i.e., the product is changed or destroyed by testing. In other cases, the standard itself is consumable (e.g., hardness test blocks), so material variation affects the standard as well. In these cases, where replicate testing is not feasible, the variations due to operator, equipment, and test method are composited with the materials variation. Sometimes it is feasible to resolve these composites into their components, and sometimes it is not. A further complication is the case of perishable materials, which may require use of calibrations that relate time elapsed to degradation suffered.

Test Equipment Variation.   Instruments are subject to numerous sources of error, both within a single instrument and between instruments: nonlinearity, hysteresis (e.g., gear backlash), drift due to temperature rise, and sensitivity to extraneous "noise" (e.g., magnetic, thermal, electrical fields). Each technology is subject to its own unique array of troubles. These instrument troubles are multiplied by the *fixturing* troubles of connecting the instruments into the larger test equipment units and of connecting the test specimens for tests. These fixturing troubles include such problems as making good electrical connections, fastening mechanical linkage, locating probes precisely, and so on.

Test Procedure Variation.   In those cases where more than one test procedure is available to conduct measurement, it is essential to determine the relative variations, since these comprise one of the criteria for judging the adequacy of the procedure used.

Between-Laboratories Variation.   This is a major problem both within companies and between companies. Some major programs must await resolution of this problem before they can be concluded, e.g., industry standardization of materials, test equipment, and test procedures. In like manner, variation between vendor and buyer laboratories may be at the root of a major quality problem. So extensive is the need to reduce between-laboratories variation that standard procedures have been evolved for the purpose (see *Quality Control Handbook,* 4th ed., Section 18, under Interlaboratory Test Programs).

Composite Errors.   The observed measurements are, of course, a resultant of the contributing variations. Generally this resultant or composite is related to the component variables in accordance with the formula

$$\sigma_{obs}^2 = \sigma_w^2 + \sigma_b^2 + \sigma_m^2 + \sigma_e^2 + \sigma_f^2 + \cdots$$

where σ_{obs} is the standard deviation of the observed measurements and σ_w, σ_b, σ_m, σ_e, σ_f, etc. are the standard deviations reflecting the size of the variables that affect precision, i.e., within operator, between operators, material used, in test equipment, in test procedure, etc., respectively.

This relationship is valid provided the variables are independent of each other, which they often are. Where two or more of the variables are interrelated, then the equation must be modified. If, for example, variables A and B are interrelated, then

$$\sigma_T^2 = \sigma_A^2 + \sigma_B^2 + \rho_{AB}\sigma_A\sigma_B$$

where σ_T^2 = total variance
σ_A^2 = variance of A
σ_B^2 = variance of B
ρ_{AB} = the correlation coefficient (ρ) of A and B

In many cases it is feasible to quantify the effect of some component sources of variation by simple designs of experiment. When an instrument measures a series of different units of product, the resulting observations will have a scatter that is a composite of (1) the variation in the system of measurement and (2) the variation in the product itself. This relationship can be expressed as

$$\sigma_{obs} = \sqrt{\sigma_{prod}^2 + \sigma_{meas}^2}$$

where σ_{obs} = σ of the observed data
σ_{prod} = σ of the product
σ_{meas} = σ of the measuring method

Now, solving for σ_{prod},

$$\sigma_{prod} = \sqrt{\sigma_{obs}^2 - \sigma_{meas}^2}$$

It is readily seen that if σ_{meas} is less than one-tenth σ_{obs}, then the effect on σ_{prod} will be less than 1 percent. This is the basis of the rule of thumb that the instrument should be able to divide the tolerance into about 10 parts.

To illustrate, in one shop the validity of a new type of instrument was questioned on the ground that it lacked adequate precision. The observed variation σ_{obs} was 11 (coded). An experiment was conducted by having the instrument make replicate checks on the same units of product. The σ_{meas} was figured out to be 2. Then, since

$$\sigma_{prod}^2 = \sigma_{obs}^2 - \sigma_{meas}^2$$
$$\sigma_{prod} = \sqrt{121 - 4} = \sqrt{117} = 10.8$$

This was convincing proof that the instrument variation did not significantly inflate the product variation.

In another instance involving the efficiency of an air-cooling mechanism, the observed variation σ_{obs} was 23, and the variation on retests σ_{meas} was 16. Thereupon,

$$\sigma_{prod} = \sqrt{23^2 - 16^2} = \sqrt{529 = 256} = \sqrt{273} = 16$$

This showed that the measurement variation was as great as the product variation. Further study disclosed that the measurement variation could be resolved into

Variable	σ of that variable	σ^2
A	14	196
B	5	25
All other	7+	52
		273

It became clear that real progress could be made only by improving variable A, and the engineers took steps accordingly.

To quantify the individual components of variation requires still more elaborate analysis, usually through a special design of experiment (see Section 47, Design and Analysis of Experiments). [See also McCaslin and Grusko (1976) for a discussion on an attribute gage study procedure and Ezer (1979) for statistical models for testing vial-to-vial variation in medical laboratories.]

Statement of Error.   In publishing results, it is necessary to make clear the extent of error in those results. Lacking clear conventions or statements, those who review the results simply do not know how to evaluate the validity of the data presented. To make clear the extent of error present in the data, metrologists have adopted some guidelines that are ever more widely used.

Effect of Reference Standards.   Accuracy of an instrument is expressed as the difference between T, the "true" value, and $\overline{X}_m$, the average of the replicated measurements. The reference standard used is *assumed* to be the true value, but of course, this is not fully valid; i.e., the standards laboratory is able to make only a close approximation. In theory, the "true" value cannot be attained. However, the extent of error can be ascertained through the use of replication and other statistical devices. As long as the systematic error of the standard is small in relation to the error of the instrument under calibration (the usual situation), the error of the standard is ignored. If there is some need to refer to the error of the standard, the published measurements may include a reference to the standard in a form similar to "as maintained at the National Institute of Science and Technology."

Effect of Significant Systematic Error.   When the systematic error is large enough to require explanation, the approved forms of explanation consist of sentences appended to the data, stating (for example), "This value is accurate within $\pm x$ units," or "This value is accurate within $\pm y$ percent." [See National Bureau of Standards (1965).] If necessary, these statements may be further qualified by stating the conditions under which they are valid, e.g., temperature range.

It is a mistake to show a result in the form $a \pm b$ with no further explanation. Such a form fails to make clear whether b is a measure of systematic error, an expression of standard deviation of replicate measurements, or an expression of probable error, etc.

Effect of Imprecision.   The quantification of precision is through the standard error (standard deviation), which is the major method in use for measuring dispersion. In publishing the standard error of a set of data, care must be taken to clear up what are otherwise confusions in the interpretation.

1. Does the standard error apply to individual observations or to the average of the observations? Unless otherwise stated, it should be the practice to relate the published standard error to the published average, citing the number of observations in the average.

2. If uncertainty is expressed as a multiple of the standard error, how many multiples are used? An approved form of expression is "…with an overall uncertainty of ± 4.5 km/s derived from a standard error of 1.5 km/s."

3. Is the standard error based solely on the data presented or on a broader history of data? To clarify this requires still more intricate wording, since a dispersion based solely on the current data is itself uncertain [see Eisenhart (1968)].

Effect of Combined Systematic Error and Imprecision.   In these cases the expression of the published result must make clear that both types of error are present and significant. Eisenhart (1968) recommends a phraseology such as "…with an overall uncertainty of ±3 percent based on a standard error of 0.5 percent and an allowance of ±1.5 percent for systematic error."

Errors Negligible.   Results also may be published in such a way that the significant figures themselves reflect the extent of the uncertainty. For example, in the statement, "The resistance is 3942.1 Ω correct to five significant figures," the conventional meaning is that the "true" value lies between the stated value ±0.05 Ω.

CALIBRATION CONTROL

Measurement standards deteriorate in accuracy (and in precision) during use and, to a lesser degree, during storage. To maintain accuracy requires a continuing system of calibration control. The elements of this system are well known and are set out below.

(The terminology associated with *calibration control* is not yet standardized. To put an instrument into a state of accuracy requires first that it be tested to see if it is within its calibration limits. This test is often referred to as *checking* the instrument. If, on checking, the instrument is found to be out of calibration, then a *rectification* or *adjustment* must be made. This adjustment is called variously *calibration, recalibration,* or *reconditioning.* In some dialects, the word *calibration* is used to designate the combination of checking the instrument and adjusting it to bring it within its tolerances for accuracy.)

While the same system can be applied to all levels of standards, as well as to test equipment, there are some significant differences in detail of application. Transfer standards are exclusively under the control of standards laboratories staffed by technicians whose major interest is maintaining the accuracy of calibration. In contrast, test equipment and, to some extent, working standards are in the hands of those production, inspection, and test personnel whose major interest is product and process control. This difference in outlook affects the response of these people to the demands of the control system and requires appropriate safeguards in the design and administration of the system.

New-Equipment Control.   The control system regularly receives new elements in the form of additional standards, new units of test equipment, and expendable materials. These elements should be of proven accuracy before they are allowed to enter the system. The approach varies depending on the nature of the new item:

1. *Purchased precision standards.*   These include high-accuracy gage blocks, standard cells, etc. Control is based on the supplier's calibration data and on his or her certification that the calibration is traceable to the National Institute of Standards and Technology. Where such purchased standards represent the highest level of accuracy in the buyer's company, any subsequent recalibration must be performed by an outside laboratory, i.e., the supplier, an independent laboratory, or the National Institute of Standards and Technology.

2. *Purchased working standards.*   These are subjected to "incoming inspection" by the buying company unless the demonstrated performance of the supplier merits use of an audit of decisions (see Section 21).

3. *New test equipment.*   This equipment is intended for use in checking products and processes (Figure 23.6). However, it usually embodies measuring instruments of various sorts and may well include working standards as well, i.e., test pieces ("masters" for in-place check of calibration).

4. *Test materials.*   These include consumable standards as well as expandable supplies such as reagents or photographic film. Variability in such materials can affect the associated measurements and calibrations directly.

For example, a manufacturer of sandpaper needed a uniform material on which to test the abrasive qualities of the sandpaper. The engineer investigated the possibility of using plastic blocks and

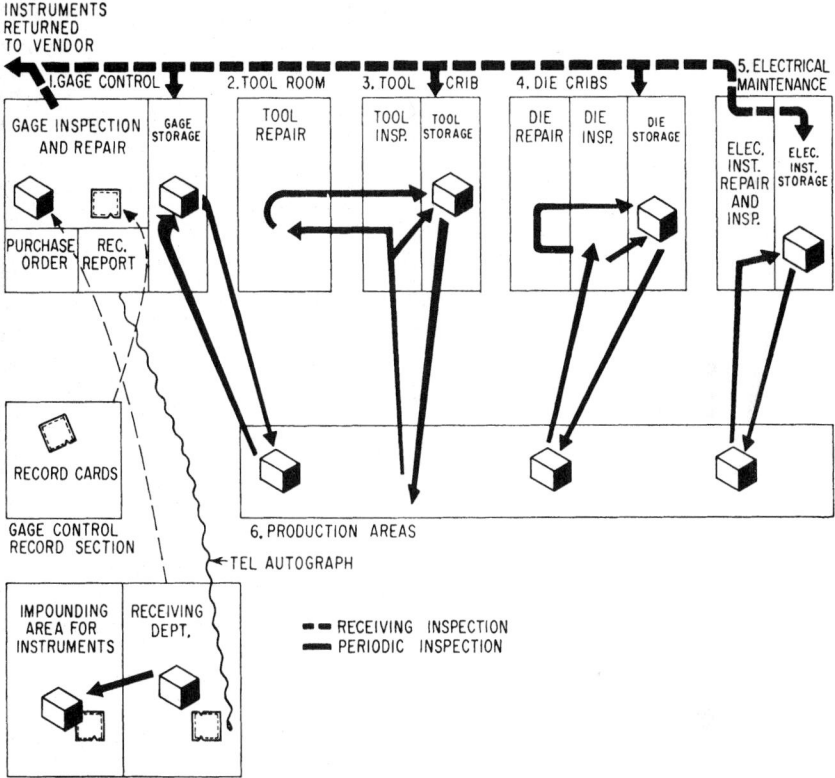

FIGURE 23.6 Flow diagram for gage control.

found that the plastics manufacturer was using the same sandpaper as a means of testing the toughness of the plastic. For some of these materials, the suppliers could provide data on variability. For the rest, it was necessary to discover the variability through analysis, as discussed earlier under Error of Measurement.

Inventory and Classification.   A systematic approach to calibration control starts with a physical inventory of all standards, instruments, and test equipment. (Where tooling is used as one of the means of product inspection, such tooling is commonly included in the list of items to be systematically controlled for accuracy.)

For each item that is to enter the system, a database record is created. This record contains the historical origin of the item, its assigned serial number, the checking schedule, and related information. The record is also designed to accommodate information on the results of checking and the repairs needed [see also Woods (1978)]. The physical test equipment is also marked with an assigned serial number for identification and traceability in the system.

Calibration Schedules.   These are established by class of equipment and are varied to reflect precision, nature, and extent of use and still other factors. At the outset, these schedules are established by judgment and bargaining. Later, as data become available on the results of checking, it becomes feasible to change the schedules in the interest of greater effectiveness and economy.

The broad intent of calibration schedules is to detect deterioration beyond tolerable levels of accuracy. This deterioration takes place primarily though use and secondarily through the passage of time. As a result, the calibration schedules describe the extent of use or of elapsed time in several ways:

1. *Elapsed calendar time.*   This method is in widest use. It establishes a fixed calendar time, e.g., 3 months, as a checking interval. At the end of the 3 months, steps are taken to check the equipment in accordance with schedule.

2. *Actual amount of use.*   This is based on counting the actual use, e.g., number of units of product checked by the equipment. The count may be made (*a*) manually, by the inspectors, (*b*) through automatic counters installed in the equipment, or (*c*) by programming the computers to show the amount of testing performed based on production schedules.

3. *Test accuracy ratio (TAR) control.*   This is a systems approach that analyzes the degree to which interrelated parameters are identified and controlled within a *traceability cone.* Minimum TARs for each traceable parameter are measured and controlled [see also Tobey (1979)].

Adherence to Schedule.   This vital detail makes or breaks the entire system of calibration control. Generally, the transfer standards and most working standards pose no problem of adherence to schedule, since they are in the custody of a few standards laboratories and a relatively few associated technicians. In contrast, the test equipment (and some working standards) are widely scattered over numerous locations and are in the custody of thousands of workers, inspectors, and testers. Some of these individuals can be relied on to see that the checking schedule is followed, but many cannot.

In part, the problem is one of lack of knowledge of when the recalibration is due. The shop personnel require the aid of a memory system if they are to know which piece of equipment is due to be checked that day. They may recall what the checking intervals for each class of equipment are, but they cannot recall what the date of the last calibration was.

Some systems for adherence to schedule make use of ingenious color codes or labels that mark on each unit of equipment the date it was put back into service. (These codes are often extended to identify the grades of the standards themselves, whether primary, secondary, etc.) For large units, the expiration date also may be entered on a maintenance card that is attached to the unit. Such dates are an aid to personnel for adhering to the checking schedule.

However, an added problem is that of motivation. The numerous users of test equipment are quite concerned with recalibration when trouble is encountered but less concerned when things seem to be going smoothly. In these latter cases, interruptions for calibration can even be a nuisance.

The solution is to give responsibility (for adherence to schedule) to the standards laboratory rather than to the production, inspection, and test personnel. When this proposal is made to practicing managers, they seldom accept it purely on grounds of theory of organization. However, when it is proposed that a sample of instruments be taken at random and checked for calibration (as a test of the existing "system" of calibration control), these same managers are quite willing to conduct such a test. The resulting disclosure of the actual state of calibration of the sample (of 25 to 100 instruments) is then decisive in convincing the managers of the need for a revision in the system of adherence to schedule. Under this assignment, thelaboratory organizes a plan of checking that will keep up with the scheduled load [see also Gebhardt (1982)].

In administering the checking plan in the time-interval system, the database is queried, using the calibration dates, to provide a list showing which standards and equipment are due to be checked in the forthcoming week.

Calibration Practice.   To ensure accuracy and to establish traceability, control laboratories have evolved some widely used procedures. Individual responsibility is established by requiring that all concerned sign for their actions. The equipment record, retained either in hardware or software, carries these signatures, as do the labels on the equipment. Dates are recorded for all actions in view of the role of elapsed time in the calibration procedures.

Manuals of practice are established, including tolerances for accuracy and methods to be used in calibration. In some types of test, these methods must be spelled out in detail, e.g., temperature or humidity controls, time cycle, human technique, etc. (Witness the detail of some of the ASTM standards on test method.)

Training programs are established for personnel, including (in some cases) formal qualification certificates to attest to proficiency. Equipment is tamperproofed through sealing the adjusting screws. (The seals are then imprinted with the stamp of the laboratory.) In like manner, panels and drawers of test equipment are lock-wired, and the wires are lead-sealed together. (The laboratory takes no responsibility when seals are broken, and the company takes stern measures against tampering with the seals.) As a means of assisting enforcement, quality assurance audits are conducted to review the calibration control procedures.

Record and Analysis of Results.   It is most useful to keep a record of the results of checking calibration and of the extent of work done to restore accuracy. Typically, such a record lists

Observed deficiencies in the equipment

Causes of out-of-calibration conditions

Repair time and recalibration time

Periodic analysis of these data then becomes the basis for

1. Reducing checking for equipment shown to be stable
2. Redesigning equipment to eliminate causes of repetitive failure

Organization for Calibration Control.   Ensuring that measuring devices are calibrated correctly is critical to ensure that product is conforming to the customer's requirements. The best inspector is only as good as his or her gages. If the gage is in error, then we could reject good parts or accept bad parts. Both results are costly to the company; both send a message to the customer that the company does not have a basic control system.

A technician trained in metrology—the science of gaging—is the determining factor in the accuracy of the gage by calibrating to a recognized master, traceable in the United States to the National Institute of Standards and Technology or similar international organization. The calibration process is performed by following a set of test procedures developed for scientific instruments.

The quality system can elect to calibrate all measuring devices in house, send them to a commercial calibration service, return the device to the original manufacturer, or a combination of all of the above. The determining factor is usually based on master calibration up-front cost, frequency of needed calibration control, and turnaround time of outside source.

Regardless of how the actual calibration function is allocated, the frequency of calibration of each measurement device must be specified by a sytematic calibration frequency program. This program defines the categories of inspection, measuring, and test equipment to be covered and assigns responsibility for operating the program and maintaining records.

There are no standard rules and no body of knowledge that dictate how often measuring instruments should be inspected. The determining factors of calibration are use and mishandling (Palumbo 1997).

Physical design of the laboratory workplace has been greatly complicated by the proliferation of many varieties of specialized testing: ultrasonic, x-ray, vibration, shock, acceleration, heat, humidity, etc. The details of these designs are beyond the scope of this handbook. The practitioner must consult with the available experts: equipment manufacturers, researchers, metrologists, and still others. This must be a continuing process, since there is continuing progress in development of new tests and standards.

HUMAN FACTORS IN INSPECTION

A myriad of factors can influence inspector behavior; they are summarized by Baker (1975), as shown in Table 23.9. One factor, visual acuity or sight, is the dominant sense in human beings, and great reliance is placed on it in inspection tasks. The effectiveness of the use of sight depends largely on eye movements that bring the images of significant features of the material being inspected to the most sensitive part of the retina. However, experience and studies have shown that the other factors in Table 23.9 have an interrelated effect on the effectiveness, productivity, reliability, and accuracy of the inspector. [See Megaw (1978) for related studies carried out in a textile factory.]

A detailed discussion of inspector errors below follows the discussion on human factors; however, it appears appropriate to first discuss the techniques and measurements developed to improve the reliability of the inspection function.

Machine Vision.   The term *machine vision,* or *noncontact inspection,* is applied to a wide range of electrooptical sensing techniques from relatively simple triangulation and profiling to three-dimensional object recognition and bin picking, techniques based on sophisticated computerized image-analysis routines. The applications are broad, ranging from relatively simple detection and measuring tasks to full-blown robot control. (See Table 23.10.)

The incentive to introduce machine-based systems, e.g., robots, is obvious—to eliminate human error. [See Spow (1984) for a discussion of robots in an automatic assembly application.] The key influences behind the growth of the machine-vision industry also include inspector capability, inspector productivity, and inspection costs. [See also Nelson (1984) for a review of machine-vision equipment, some designed to eliminate process contamination, in addition to human error.] Table 23.11 discusses these factors as they relate to inspecting printed circuit boards. [See Ken (1984) for a detailed discussion of automated optical inspection (AOI) of printed circuit boards (PCB); see also Denker (1984) for a detailed discussion on justifying investments in automatic visual PCB testing.]

Vision-based systems—human or machine—involve an inspection procedure: the examination of a scene. The examination, in turn, can lead to recognition of an object or feature, to a quality decision, or to the control of a complex mechanism (Schaffer 1984). When a human inspector is involved, human judgment and perception have an influence on the quality assessment process.

The most common application for vision systems involves measuring critical dimensions, detecting flaws, counting/sorting, assembly verification, position analysis, character or bar-code reading/verification, and determination of presence/absence of features on small parts.

TABLE 23.9   Variables Influencing Inspector Behavior

1. Individual abilities
    a. Visual activity
    b. General intelligence and comprehension
    c. Method of inspection
2. Task
    a. Defect probability
    b. Fault type
    c. Number of faults occurring simultaneously
    d. Time allowed for inspection
    e. Frequency of rest periods
    f. Illumination
    g. Time of day
    h. Objective of conformance standards
    i. Inspection station layout
3. Organizational and social
    a. Training
    b. Peer standards
    c. Management standards
    d. Knowledge of operator or group producing the item
    e. Proximity of inspectors
    f. Reinspection versus immediate shipping procedures

TABLE 23.10 Machine-Vision Applications

1. Inspection
   a. Dimensional accuracy
   b. Hole location and accuracy
   c. Component verification
   d. Component defects
   e. Surface flaws
   f. Surface-contour accuracy
2. Part identification
   a. Part sorting
   b. Palletizing
   c. Character recognition
   d. Inventory monitoring
   e. Conveyor picking (overlap, no overlap)
   f. Bin picking
3. Guidance and control
   a. Seam-weld tracking
   b. Part positioning
   c. Processing/machining
   d. Fastening/assembly
   e. Collision avoidance

TABLE 23.11 Selected Example of Factors and Effect of Automated Optical Inspection (AOI) on Printed Circuit Board

Manufacturing inspection costs	Inspection accounts for up to 30 percent of the manufacturing costs of complex double-sided and multilayer boards.
Inspection capability	Human inspection capability decreases well before 5 mil because of fatigue
Inspection productivity	AOI reduces the number of inspections from 30 to 7 per shift while obtaining a 1 percent yield improvement and fivefold increase in inspection speeds.

If parts are moving and not indexed, an electronic shutter camera may be needed. Many systems can handle randomly oriented parts, but throughput rates will be lower than for fixtured parts, and if the need is to inspect more than one side or area of the parts, separate cameras may be required.

Repetitive Function. In highly repetitive subjective inspection (e.g., inspecting parts on an automated paint line), the process of perceiving can become numbed or hypnotized by the sheer monotony of repetition. In some way the scanning process and the model become disconnected, and the observer sees only what he or she expects to see but does not see anything not actively expected. The situation often can result in bad parts passing through the process and should be evaluated closely for effectiveness before being initiated. The known remedies are

- Break up the benumbing rhythm with pauses.
- Introduce greater variety into the job.
- Increase the noticeability of the faults through enhancing.
- Provide background and contrast cues.
- Arrange for frequent job rotation.

Vigilance becomes a big problem when faults are obvious and therefore serious but infrequent and unpredictable. What traveler has not wondered about the protection provided by 100 percent x-ray and metal-detection inspection of airline passenger carryon baggage and personal effects? In 1993, an audit by the U.S. General Accounting Office of airport security measures in the United States revealed the ease with which the "secure" areas of America's airports could be penetrated. Of the hundreds of attempts to breach security, 75 percent were successful; some of these successes were attributable to a failure of 100 percent inspection. The auditing agent who succeeded in passing a live hand grenade through the screening system provided an especially troublesome example (Gleick 1996).

INSPECTION ERRORS

The inspector, as the human element in the inspection process, contributes importantly to inspection errors. Inspection errors due to the inspector, called *inspector errors,* are discussed here. The reliability of the human inspector was discussed previously under human factors (as opposed to noncontract machine inspection or the machine-vision system). Other sources of inspection error, e.g., vague specifications, lack of standards, inaccurate instruments, etc., are discussed elsewhere in this handbook. (*Note:* The problem of human error is common to operators, inspectors, and anyone else. For an extensive discussion of problems of worker error, see Section 22, under Concept of Controllability: Self-Control.)

Inspector errors are of several categories:

Technique errors

Inadvertent errors

Conscious errors

Each of these categories has its own unique causes and remedies. Collectively, these inspector errors result in a performance of about 80 percent accuracy in finding defects; i.e., inspectors find about 80 percent of the defects actually present in the product and miss the remaining 20 percent. [*Note by the editor (Juran*): Numerous studies in various countries have yielded the 80 to 20 ratio as a broad measure of quantified inspector accuracy. For example, Konz and coworkers (1981) found that inspectors were only catching about 80 percent of the defects in glass subassemblies.] However, little is known about the relative importance of each category of inspector error (e.g., lack of technique, inadvertence, conscious). For further discussion, see Tawara (1980).

Technique Errors.   Into this category are grouped several subcategories: lack of capacity for the job, e.g., color blindness; lack of knowledge due to insufficient education or job training; and lack of "skill," whether due to lack of natural aptitude or to ignorance of the knack for doing the job. Technique errors can be identified in any of several ways:

Check Inspection.   A check inspector reexamines work performed by the inspector, both the accepted and rejected product. Figure 23.7 shows an example of the results of such check inspection of the work of several inspectors. It is evident that inspectors C and F operate to loose standards, whereas inspector B operates to tight standards. Inspector E shows poor discrimination in both directions.

Round-Robin Inspections.   In this analysis, the same product is inspected independently by multiple inspectors. The resulting data, when arrayed in a matrix (usually with defect type along one axis and the inspectors along the other axis), shows the defects found by each inspector in relation to the inspectors as a group.

Repeat Inspections.   In this method, the inspector repeats his or her own inspection of the product without knowledge of his or her own prior results. The analysis of the resulting data discloses the extent of the consistency or lack of consistency of the inspector's judgments.

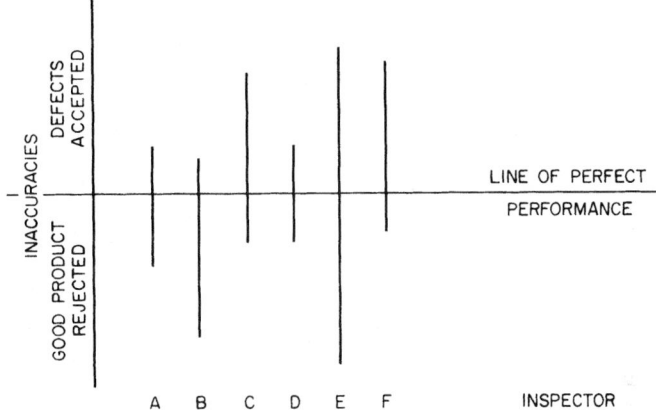

FIGURE 23.7   Analysis of inspection errors.

Standard Sample Array.   In this method, the inspector makes an "examination" by inspecting a prefabricated mixture of product consisting of good units plus various kinds of defects. The standard sample array is known by various names, including *job sample.* For added discussion, refer to Harris and Chaney (1969). All units were previously carefully graded by a team of experts and numbered for ready analysis of results. The inspector's score and his or her pattern of errors all point to the need, if any, for further training or other remedial steps. (In effect, the check inspection is conducted before the inspection.)

For example, in a company making glass bottles, an attempt was made to correlate process variables with the frequency and type of defects found by inspectors stationed at the cold end of the annealing lehr. The experiment failed because inspector variability from shift to shift exceeded product variability. This also threw suspicion on the accuracy of the inspection performed by the final product sorters at the end of the line. A standard sample array of 500 bottles was created and was used to examine the inspectors. (The examinations were conducted in the Training Department on a miniature lehr.) The suspicions turned out to be well founded (consulting experience of J. M. Juran).

Remedies for Technique Errors.   The need is to provide the missing skill or know-how and to answer the inspector's proper question, "What should I do differently from what I am doing now?" Unless the inspector is in a position to discover the answer for himself or herself, the answer must be provided by management. If no answer is provided, there will be no change in performance.

The various methods of analysis discussed earlier all can provide some clues that suggest the type of remedial action needed. In particular, use can be made of the concept of finding the knack. Under this concept, the data on inspector performance are analyzed to discover which inspectors give consistently superior performances and which are inferior. Next, a study is made of the work methods used by both types of inspectors to identify the differences in methods. Analysis of these differences often discovers what is the secret know-how (knack) being used to get the superior performance (or what is the secret ignorance that results in poor performance). Finally, the knack can be transferred to all inspectors through retraining (Czaja and Drury 1981 and Cooper 1980) or through being embodied in the technology (Kusch 1979).

Where the technique errors are the result of lack of job capacity, the foregoing may be of no avail, and the need may be to foolproof the operation (see below) or to reassign the inspector to a job for which he or she does have adequate job capacity.

Certification of Inspectors.   In critical inspections involving inspector judgment (e.g., interpreting x-rays of critical welds), it is increasingly the practice to require that the inspector be formally certified as qualified to do this job. [See Gibson (1983) for a further discussion of inspector qualifications in offshore industries.] The qualification process follows a well-standardized series of steps:

A formal training program on how to do the job

A formal examination, including a demonstration of successful performance of the job

A formal certificate attesting to the success in the examination

A license to do the job for some designated period of time

A program of audit to review performance and to serve as a basis for renewing the license

In some companies this concept of certification has been based on the "escape rate," i.e., the extent to which defects escape detection, as determined by subsequent check inspection. (See Measure of Inspector and Test Accuracy, below.) When this concept is used, a limited "license" (e.g., 2 months) is given to the inspector, subject to renewal if check inspection results continue to be favorable.

Inadvertent Inspector Errors. These errors are characterized by the fact that at the time the error is made, the inspector is not even aware he or she is making an error. Also, the best intentions are present—the inspector wants not to make any errors. The term *inadvertent* or *unavoidable* is used to connote the fact that the human being is simply unable to achieve perfection, no matter how good his or her intentions. (This topic is closely related to inadvertent worker errors. For added discussion, see Section 5, under Inadvertent Errors.)

The theory of inadvertence has wavered up and down. For years it was the sincere belief of many inspection supervisors that when product was inspected 100 percent, the inspectors would find all the defects. Numerous unpublished and published studies have since demonstrated that human inspectors do not find all defects present. By and large, the human inspector finds about 80 percent of the defects present and misses the remaining 20 percent.

While the 80 to 20 ratio is widely accepted, there are numerous aspects that are not fully researched, i.e., how this ratio changes with percentage defective in the product, with types of inspection (e.g., visual, mechanical gaging, electrical testing), with product complexity, with amount of time allotted for inspection, etc. For a discussion from the viewpoint of human factors plus some supporting data (e.g., that increased product complexity results in increased inspector error), see Harris and Chaney (1969, pp. 77–85).

Inspection fallibility can be demonstrated easily in the industrial classroom. The following sentence has been used thousands of times:

FEDERAL FUSES ARE THE RESULTS OF YEARS OF SCIENTIFIC STUDY COMBINED WITH THE EXPERIENCE OF YEARS

The sentence is flashed before the audience for 30 s or for a full minute. Each member is asked to count and record the number of times the letter *F* appears. When the record slips are collected and tallied, the result is invariable. Of the *F*'s present, only about 80 percent have actually been found.

The existence of so extensive an error rate has stimulated action on several fronts:

1. *To discover why inspectors make these errors.* To date, the research has not been adequate to provide conclusive answers, so industrial psychologists have not agreed on what the main causes are.

2. *To measure the extent of the errors.* Techniques for this are now available. See Measure of Inspector and Test Accuracy, below.

3. *To reduce the extent of these errors.* There is a wide assortment of remedies, as discussed below.

Remedies for Inadvertent Inspector Errors. In the absence of convincing knowledge of the causes of these errors, managers have resorted to a variety of remedies, all involving job changes in some form. These remedies include the following:

Error-Proofing. There are several forms of error-proofing that are widely applicable to inspection work: redundancy, countdown, and fail-safe methods. These are discussed in detail in Section 22, Operations, under Error-Proofing the Process. See also Inspection Planning, above.

Automation.   This is really a replacement of the repetitive inspection by an automation that makes no inadvertent (or other) errors once the setup is correct and stable. The economics of automation and the state of technology impose severe limits on the application of this remedy. See Automated Inspection, above.

Sense Multipliers.   Use can be made of optical magnifiers, sound amplifiers, and other devices to magnify the ability of the unaided human being to sense the defects. Development of a new instrument to do the sensing is the ultimate form of this multiplication. Evidently there is an optimum to the level of magnification, and this optimum can be discovered by experimentation. [For some industrial studies, see Harris and Chaney (1969, pp. 124–126, 137–142).]

Conversion to Comparison.   In many types of inspection, inspectors must judge products against their memories of the standard. When such inspectors are provided with a physical standard against which to make direct comparison, their accuracy improves noticeably. For example, in the optical industry, scratches are graded by width, and tolerances for scratches vary depending on the function of the product element (lens, prism, etc.) To aid the inspectors, plates are prepared exhibiting several scratches of different measured widths so that the inspectors can compare the product against a physical standard.

Standards for comparison are in wide use: colored plastic chips, textile swatches, forging specimens, units of product exemplifying pits and other visual blemishes, etc. Sometimes photographs are used in lieu of product. There are also special optical instruments that permit dividing the field of view to permit comparison of product with standard.

In some cases it is feasible to line up units of product in a way that makes any irregularities become highly conspicuous, e.g., lining up the holes in a row. (The childhood row of tin soldiers makes it obvious which one has the broken arm.) Some practitioners advocate inspecting units of product in pairs to utilize the comparison principle. See Shainin (1972) for further discussion.

Templates.   These are a combination gage, magnifier, and mask. An example is the cardboard template placed over terminal boards. Holes in the template mate with the projecting terminals and serve as a gage for size. Any extra or misplaced terminal will prevent the template from seating properly. Missing terminals become evident because the associated hole is empty.

Masks.   These are used to blot out the view of characteristics for which the inspector is not responsible and concentrate attention on the real responsibility. Some psychologists contend that when the number of characteristics to be inspected rises to large numbers, the inspector error rate also rises.

Overlays.   These are visual aids in the form of transparent sheets on which guidelines or tolerance lines are drawn. The inspector's task of judging the size or location of product elements is greatly simplified by such guidelines, since they present the inspector with an easy comparison for judging sizes and locations.

Checklists.   These may be as simple as a grocery shopping list used to verify that you purchased all the items you originally planned to buy. At the other extreme, a checklist may consist of the countdown for the lofting of a new space shuttle. [See Walsmann (1981) for a discussion of the purpose, advantages, and drawbacks as well as the development and implementation of various checklists.]

Reorganization of Work.   One of the theories of cause of inadvertent inspector errors is fatigue, due to inability to maintain concentration for long periods of time. Responses to this theory have been to break up these long periods in any of several ways: rest periods, rotation to other inspection operations several times a day, and job enlargement, e.g., a wider assortment of duties or greater responsibility. Some behavioral scientists urge reorganization of work on the broader ground of motivation theory, and they offer data to support this theory [see generally Harris and Chaney (1969, pp.

201–229)]. However, to date, there is no conclusive evidence that in Western culture reorganization of work (to provide greater participation, etc.) will give measurably superior results in work accomplishments. (See, in this connection, *Quality Control Handbook,* 4th ed., Section 10, under Processing System Design.)

Product Redesign.   In some instances the product design is such that inspection access is difficult or that needless burdens are placed on inspectors. In such cases, product redesign can help to reduce inspector errors as well as operator errors. For some examples, see Section 22, Operations, under Error-Proofing the Process. [See also Smith and Duvier (1984).]

Errorless Proofreading.   Beyond the techniques described in Section 22, Operations, under Error-Proofing the Process, there are special problems of error-proofing in inspection work. A major form of this is proofreading of text of a highly critical nature, i.e., critical to human safety and health. In such cases, the low tolerance for error has driven many companies to use redundant checking, despite which some errors still get through.

A closer look makes it clear that proofreading is of two very different kinds:

1. *Passive proofreading.*   Here, the proofreader takes no overt action. For example, he or she silently reads the copy while someone else reads the master aloud. Alternatively, the proofreader silently reads both documents and compares them. In such cases it is quite possible for extraneous matters to intrude and dominate his or her attention temporarily.

2. *Active proofreading.*   Here, the proofreader must take an overt action, e.g., he or she reads aloud, performs a calculation, etc. Such positive actions dominate the proofreader's attention and reduce the chance of error.

For a second example, in blood donor centers it was once usual to remove the whole blood from the donor, take it to a separate location, centrifuge it to remove a desired component, and then return the remaining fraction of the blood to the donor. This return demanded absolute assurance that the remaining fraction was being returned to that donor and to no one else. The system used involved piping both the donation and the return through tubing on which there were repeats of 10-digit numbers. The tubing was cut when the donation went to the centrifuge. Prior to return of the cells, two technicians checked to compare the two 10-digit numbers on the two ends of the cut tubing. One technician actively read the number on the tubing end attached to the donor. The other technician passively listened while comparing the number he or she heard with the number seen on the end of the tubing that was attached to the bag of blood cells.

It was sometimes feasible to use technology to make both technicians "active." For example, where the equipment was available, each was required to enter on a keyboard the number he or she saw. These signals went to a computer that compared the two numbers and signaled either a go-ahead or an alarm. This same principle of comparing two independent active sets of signals can be extended to any problem in proofreading.

Today another technology has eliminated altogether the need for this application of proofreading. The donor is now usually linked directly to the separating equipment. The blood is drawn from one arm, mixed with an anticoagulant, and passed through a separating machine that collects the specific component. The remaining blood components are returned to the donor in the opposite arm (Sataro 1997).

While the foregoing are listed as remedies for inadvertent inspector errors, most of them also can be used to reduce errors due to lack of skill or errors of a willful nature.

Procedural Errors.   Aside from inadvertent failures to find defects, there are inadvertent errors in shipment of uninspected product or even shipment of rejected product. These errors are usually the result of loose shipping procedures. For example, a container full of uninspected product may be moved inadvertently in with the inspected product; a container full of defectives may be moved inadvertently into the shipping area. Such errors can be reduced by error-proofing the identification and shipping routines:

1. The inspector must *mark the product* at the time of inspection. Sometimes the inspector places the good product in one box and the bad product in another or places the good product lengthwise and the bad product crosswise. Lacking the markings, there is always the risk that between shifts, during rest periods, etc., the unmarked product will go to the wrong destination.

2. The product markings should be so distinctive that the product "screams its identity" to packers and shippers. Bar codes should be used where appropriate.

3. The colors used for markers that identify good product should be used for no other purpose. These markers should be attached only when the inspector finds that there remains nothing to be done but ship the product.

4. Issuance of markers used to identify good product should be restricted to specially chosen personnel. (These markers are a form of company seal.) For products of substantial value, serial numbers may be used as a further control.

5. The markers should provide inspector identity. In some companies, the system of identification includes the operators and packers.

6. Shipping personnel should be held responsible for any shipment of goods failing to bear an inspector's approval.

Conscious Inspection and Test Errors

Management-Initiated.   The distinguishing features of the willful inspector error are that the inspector knows that he or she is committing the error and intends to keep it up. These willful errors may be initiated by management, by the inspector, or by a combination of both. However, with few exceptions, the major notorious quality errors and blunders have been traceable to the decisions of managers and engineers rather than to those of the inspectors at the bottom of the hierarchy. Management-initiated errors take several forms, all resulting in willful inspector errors.

Conflicting Management Priorities.   Management's priorities for its multiple standards (quality, cost, delivery, etc.) vary with the state of the economic cycle. When the state of management priorities is such that conformance to quality standards is subordinated to the need for meeting other standards, the inspectors' actions are inevitably affected, since they also are given multiple standards to meet.

Management Enforcement of Specifications.   When management fails to act on evidence of nonconformance and on cause of defects, the inspectors properly judge management's real interest in quality from these deeds rather than from the propaganda. For example, if the supervision or the material review board consistently accepts a chronic nonconformance condition as fit for use, the inspectors tend to quit reporting these defects, since they will be accepted anyhow.

Management Apathy.   When management makes no firm response to suggestions on quality or to inspector complaints about vague information, inadequate instruments, etc., the inspectors again conclude that management's real interests are elsewhere. Consequently, the inspectors do the best they can with information and facilities that they believe to be deficient.

Management Fraud.   Periodically a company manager attempts to deceive customers (or the regulators, etc.) through fictitious or deceitful records on quality. Seldom can a manager acting alone perpetrate such a fraud. The manager requires confederates who submit themselves to orders, usually in a way that makes clear to them the real character of what is going on. An inspector who is a willing accomplice (e.g., for a bribe) shares in the legal responsibility. However, the inspector also may be a most reluctant accomplice, e.g., an immediate superior gives orders to prepare nonfactual reports or to take actions clearly contrary to regulations. In such cases the inspector cannot escape taking some kind of risk, i.e., participation in a conspiracy versus the threat of reprisal if he or she fails to participate. These management-initiated errors parallel closely those associated with conscious worker errors. See Section 22, Operations, under Concept of Controllability: Self-Control.

Inspector-Initiated. These errors likewise take multiple forms, and some take place for "good" reasons. It is important to understand the distinctions among these forms, since any misunderstandings are a breeding ground for poor industrial relations.

Inspector Fraud. The inspector is subjected to a variety of pressures. The most rudimentary forms are those by production supervisors and operators pleading for a "break." Sometimes this extends to a collusion where piecework payments are involved, both for quality and for quantity certification. At higher levels are cases in which an inspector is exposed to suppliers who have a good deal at stake in the lot of product in question. Even a situation in which inspectors dealing directly with production supervisors who outrank them involves substantial pressures to which inspectors should not be subjected.

Another form of inspector fraud consists of reporting false results solely to improve the outward evidence of one's own efficiency or to make life more convenient. For example, Figure 23.8 shows the results reported by an inspector after taking a sample of *n* pieces from each of 49 lots. There is a large predominance of three defects per lot reported in the sample (exactly the maximum allowable number). The reason was found to be the inspector's reluctance to do the paperwork involved in a lot rejection.

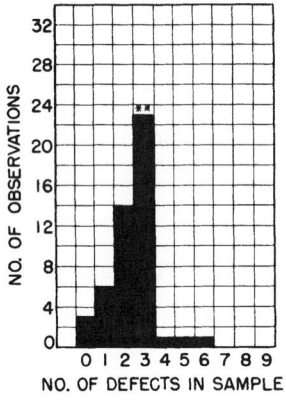

FIGURE 23.8 Inspector error made for personal convenience.

In the example in Figure 23.9, the inspector was to take a sample of 100 pieces from each lot, with no defects allowable. If one or more pieces were defective, an added sample of 165 pieces was to be taken, with a total of 3 defects allowed in the combined sample of 265. It is seen that the inspector reported defects in virtually every first sample of 100 pieces.

However, no defects were reported in most of the second (larger) samples. It was found that the inspector could improve personal efficiency by taking second samples, since the time allowance for taking the second samples was liberal. Inspector fraud can be minimized by

1. Filling inspection jobs only with persons of proved integrity
2. Restricting the down-the-line inspector to the job of fact finding, and reserving to the inspection supervision the job of negotiating and bargaining with other supervisors or executives
3. Rotating inspectors and workers to have both employee categories evaluate first hand the consequences of actions and decisions
4. Including inspectors in customer presentations or visits
5. Conducting regular check inspections and periodic independent audits to detect fraud
6. Taking prompt action where fraud is discovered

Inspector Shortcuts. These may be unauthorized omissions of operations that the inspector has reason to believe are of dubious usefulness; e.g., accidental omissions had failed to give evidence of trouble. In some cases there is a shared blame; i.e., management has imposed a highly disagreeable task. For example, in a company making "tin cans," one inspection involved cutting up a can with hand-held tin shears, submerging the pieces in chloroform to remove the enamel, and measuring the thickness of the bare pieces with a micrometer. The cutting process was tedious and the chloroform

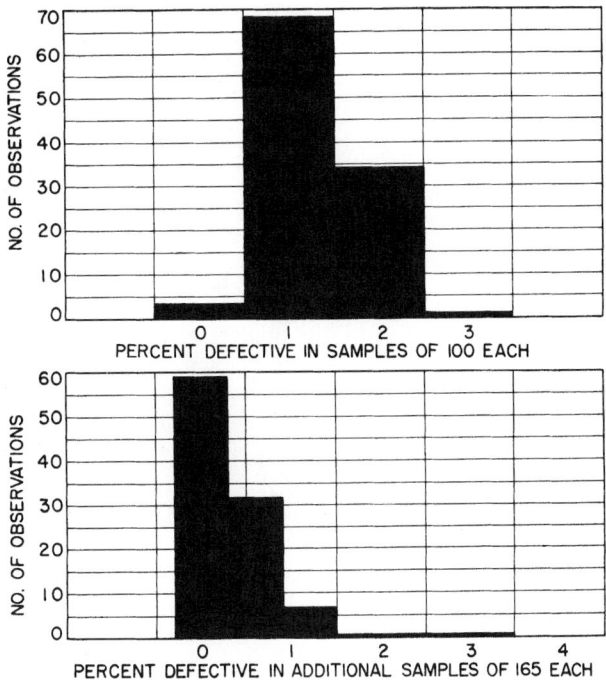

FIGURE 23.9   Inspector error made to improve personal efficiency.

was irritating to the skin, so the inspectors avoided the operation as much as possible. When better cutting tools and a different solvent were provided, the problem became minimal.

Flinching.   This is the tendency of inspectors to falsify the results of inspection of borderline product. Flinching is actually widespread among all persons who report on performance versus goals and especially when it is their own performance. Figure 23.10 shows a frequency distribution of measurement on volume efficiency of electronic receivers. There is an "excess" of readings at the specification maximum of 30, and there are no readings at all at 31, 32, or 33. Retest showed that the inspector recorded these "slightly over" units of product at 30. By this flinching, the inspector, in effect, changed the specification maximum from 30 to 33. This is a serious error (Juran 1935).

Flinching during variables measurements is easy to detect by check inspection, which is also conducted on a variables basis. Analysis of the inspector's variables data likewise will detect flinching (as in the preceding example).

The remedy for flinching is an atmosphere of respect for the facts as the ethical foundation of the Inspection Department. The main means for achieving this are examples set by the inspection supervisors.

One way *not* to deal with flinching is to criticize the inspector on the basis that the pattern of readings does not follow the laws of chance. Such criticism can be interpreted as being aimed at the symptom (the unnatural pattern of reading) rather than the disease (recording fictitious instead of factual readings). The risk is that inspectors will try to meet such criticisms by trying to make the false results look more natural, hence eliminating the symptom but not the disease.

Flinching also takes place during attribute inspection. Numerous studies have shown that inspector errors in rejecting good product outnumber the errors of accepting bad product. In part, this arises because the good product outnumbers the bad and hence affords greater opportunity for error. However, it also arises in part from the fact that acceptance of defects often comes dramatically to the attention of higher management, whereas rejection of good product seldom does so. These same

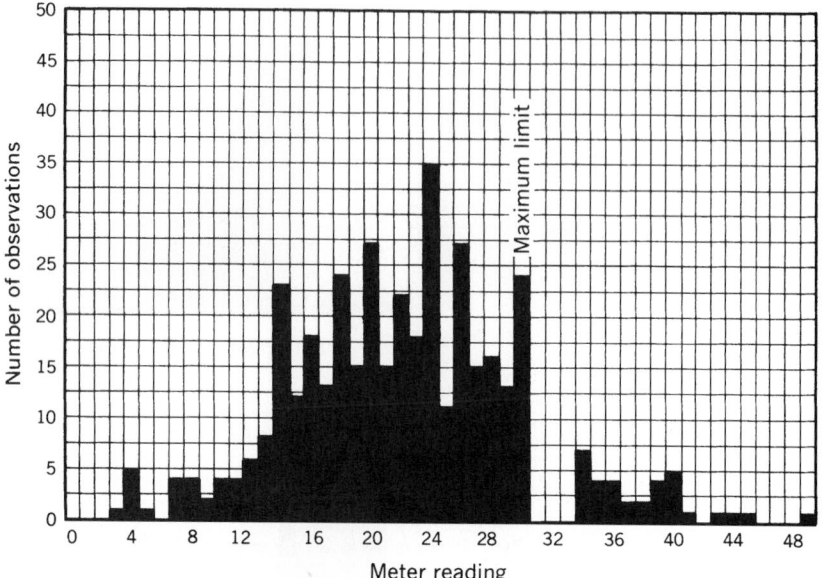

FIGURE 23.10 How inspector flinches at design limit.

studies show that when check inspection is introduced and check is made both of product rejected and of product accepted, the rejection of good product is reduced without affecting the acceptance of bad product.

Another form of flinching is to modify the inspection results to conform to the results that the inspector expects. For example, a visual inspection was being performed following a specified lapping operation. An experiment to omit the lapping operations, conducted without inspector awareness of omission of the operation, resulted in rejection of less than a third of unlapped product.

In some cases, flinching by inspectors is actually management-initiated through manager pressures that seem to the inspectors to leave no alternative. In one company, the inspectors making hardness tests were discovered to be flinching to an astonishing degree. This practice had been going on for years. It developed that the manufacturing vice president had designed this hardening process himself when he was the process engineer. At the time, he had deluded himself as to its capabilities and thereby had been the author of this long-standing practice (Juran, early consulting experience).

Rounding Off. The process of dispensing with unneeded accuracy is generally referred to as *rounding off.* Inspectors commonly round off their meter readings to the nearest scale division, as shown in Figure 23.11. The effect of rounding off is seen in the "picket fence" frequency distribution of Figure 23.10.

Rounding off is easy to detect from analysis of inspection data. A good analyst can, from the data alone, reconstruct the pattern of scale markings of an instrument without ever having seen the scale itself.

Rounding off is often a good thing, since it avoids undue attention to individual readings. Sometimes, however, the need for precision on individual readings is great enough that rounding off should not be practiced. The planner and inspection supervisor should be on the alert to identify situations in which rounding off is not tolerable, and they should provide accordingly.

Instruments and gages should be selected properly for the application. [Churchill (1956) gives a quantitative discussion on scale interval length and pointer clearance.] One practice is to require "readings to be recorded to the nearest . . . . "

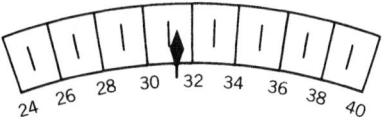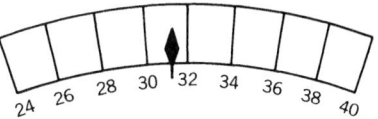

FIGURE 23.11   Rounding off.

Measure of Inspector and Test Accuracy.   The collective effect of inspector errors, from all causes, is so extensive that there is a need for measuring the extent of errors and for use of the data in controlling the effectiveness of inspectors. If this measurement is made only occasionally, use can be made of standard sample arrays (see above, under Technique Errors) and cross-check among inspectors, as well as check inspection. If the measurement is to be conducted regularly so as to discover trends in performance, then check inspection is necessary.

In conventional check inspection, a second inspector, i.e., a check inspector, reviews the decisions of the inspector by reexamining the product after it has been inspected. For an early example, refer to Taylor (1911). The best practice is to reexamine the rejected product as well as the accepted product.

Inspection errors may consist of accepting defective units of product or rejecting good units of product. If, in addition, the check inspection reviews the procedure followed by the inspector, other errors may be found, e.g., use of wrong issue of the specification, wrong instrument, improper filling out of documents, etc.

The convenient use of check inspection data to quantify inspector accuracy is to count the errors, to assign weights, and to use the composite of errors as an index of accuracy (of inaccuracy, usually). [See, for example, Gilman (1963).] In some schemes, the errors discovered in later operations or in customer complaints are included in the data. The composite of errors may be expressed in terms of percentage defective (found to exist in the inspected product) or in terms of demerits per unit. [Refer to Weaver (1975), which reports on a study of inspector accuracy during the production process, based on accept/reject decisions involving the product currently produced.] Either way, the scoring system is open to the objection that the inspector's accuracy depends, to an important degree, on the quality of the product submitted to him or her by the process; i.e., the more defects submitted, the greater is the chance of missing some.

A plan for measuring inspectors' accuracy in a way that is independent of incoming quality is that evolved in 1928 by J. M. Juran and C. A. Melsheimer. [See Juran (1935) for the original published description of this plan.] Under this plan, the check inspector, as usual, reexamines the inspected product, both the accepted and the rejected units.

In addition, the check inspector secures the inspector's own data on the original makeup of the lot, i.e., total units, total good, total defective. From these data, the following formulas emerge as applied to a single lot, which has been check inspected:

Accuracy of inspector = percent of defects correctly identified

$$= \frac{d - k}{d - k + b}$$

where d = defects reported by the inspector
     k = number of defects reported by the inspector but determined by the check inspector not to be defects
     $d - k$ = true defects found by the inspector
     b = defects missed by the inspector, as determined by check inspection
     $d - k + b$ = true defects originally in the product

Figure 23.12 illustrates how the percentage of accuracy is determined. The number of defects reported by the inspector, d, is 45. Of these, 5 were found by the check inspector to be good; that is,

$k = 5$. Hence $d - k$ is 40, the true number of defects found by the inspector. However, the inspector missed 10 defects; that is, $b = 10$. Hence the original number of defects, $d - k + b$, is 50, that is, the 40 found by the inspector plus the 10 missed. Hence

$$\text{Percentage of accuracy} = \frac{d - k}{d - k + b} = \frac{45 - 5}{45 - 5 + 10} = 80\%$$

In application of the plan, periodic check inspection is made of the inspector's work. Data on d, k, and b are accumulated over a period of months to summarize the inspector's accuracy, as for example:

Job no.	Total pieces	d,	b	k
3	1000	10	0	0
19	50	3	1	0
42	150	5	1	0
48	5000	10	4	0
Total		200	30	0

The totals give, for percentage accuracy:

$$\frac{d}{d + b} = \frac{200}{230} = 87\%$$

As is evident, the plan lends itself to simple cumulation of data. However, some compromise is made with theory to avoid undue emphasis on any one lot checked. Over a 6-month period, where the cumulative checks may reach 50 or more, the need for such compromise or weighting is diminished.

The check inspector also makes errors. However, these have only a secondary effect on the inspector's accuracy. In the preceding example, if the check inspector were only 90 percent accurate, only 27 of the 30 defects missed by the inspector would be found. The inspector's accuracy would become

$$\frac{200}{227} = 88.1\% \quad \text{instead of} \quad 87.0\%$$

In some situations, k is small and may be ignored. However, in other situations, notably for sensory qualities, the inspector may have a bias for rejecting borderline work. In such cases, it is feasible to use, as an added measure, the inspector's accuracy due to rejection of good pieces. This has been termed *waste*. Under the terminology used here:

$$\text{Waste} = \text{percentage of good pieces rejected} = \frac{k}{n - d - b + k}$$

where n is the total pieces inspected.

The *percent accuracy* is also equal to the percentage of material correctly inspected. This feature permits use of the plan in the pay formula of the inspector.

Some investigators have developed variations on the foregoing measures of inspector accuracy as applied to visual inspection. These include measures based on probability theory (Wang 1975) and on use of signal-detection theory in the analysis of industrial inspection (Ainsworth 1980).

The proportion of correct decisions made by the inspector is an intuitively good index of the inspector's efficiency when the costs of rejecting a good item and accepting a bad item are equal.

The two measures for evaluating inspector's efficiency, introduced by Wang (1975) are

$$N_\alpha = \frac{\text{number of true defects detected by the inspector}}{\text{number of true defects}}$$

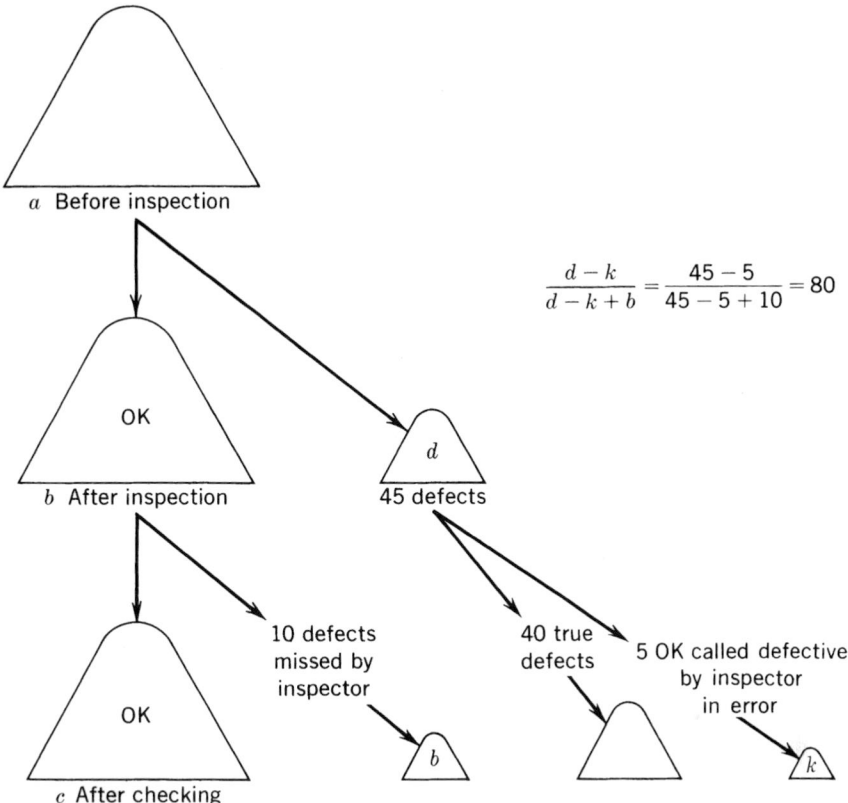

$$\frac{d-k}{d-k+b} = \frac{45-5}{45-5+10} = 80$$

FIGURE 23.12   Process for determining accuracy of inspectors.

$$N_\beta = \frac{\text{number of true defects detected by the inspector}}{\text{number of all defects detected by the inspector}}$$

If the two costs are not equal, the theory of signal detectability (TSD) provides a better measure for analyzing an inspector's performance. However, the use of TSD requires assumptions about normal and equal variant population distributions, and the probability density functions for the "good" and "defective" populations must be calculated. See Johnson and Funke (1980) for a discussion of the advantages and disadvantages of a number of human performance measures, including Wang's (1975) approach.

In the application of any plan to check the accuracy of inspectors, it is essential that the checks be at random. Neither the inspector nor the check inspector should know the schedule in advance. Random dice, cards from a pack, etc. should be used. It is also essential that the responsibility be clear. The inspector who has accepted defects under orders or through inaccurate instruments, etc. cannot be held responsible for the results.

INSPECTION AND TESTING SOFTWARE

Training.   As we move into the twenty-first century, we will be required to learn the concepts and application of computer-based technologies. Computer-based training has increasingly become a popular training tool for many organizations that do not require bilateral communications between instructor and trainee and can provide alternate training schedules without having to rely on facility and instructor availability. Software for computer-based programs range from PC disks and interactive CD/ROMs to Internet links and cover topics for everything from gage calibration to 100 percent on-time delivery analysis (Kennedy 1996).

Statistical Process Control Interface.
Software organizes collected data and tracks them to measure performance and standards in manufacturing processes. Many manufacturers have integrated the measurement system directly onto the machine to provide for the operator an immediate analysis of the process variation through charts and graphic data.

Data Collection.
Software collects data from peripheral devices and retrieves data from the collection devices and downloads them onto a host computer. These data can then be analyzed for trends or problem solving.

Gage Calibration.
Software is used for monitoring and recording of measurement device hardware accuracy, reportability, and gage performance. Features include gage type, gage due date, next calibration date, location, assignee, gage inspector (metrologist), and calibration history.

Simulation.
Software is used for preproduction simulation to check for machine tool collisions, verify material removal of a machine tool, tool life analysis, inspection and test station layouts, and many other inspection and test interfaces into production processing. See "Software for Manufacturing" (1996).

REFERENCES

Ainsworth, L. (1980). "The Use of Signal-Detection Theory in the Analysis of Industrial Inspection." *Quality Assurance,* vol. 6, no. 3, September, pp. 63–68.

Alaimo, A. P. (1969). "A Total Inspection System." *ASQC Technical Conference Transactions,* Los Angeles, pp. 71–78.

Allen, P. E. (1959). "Evaluating Inspection Cost." *ASQC Technical Conference Transactions,* Los Angeles, pp. 585–596.

"An Outlook for 3D Vision." (1984). *Assembly Engineering, Newsline,* August, p. 6.

"ASD's Quality Assurance Program Rates in Top 10." (1984). *Assembly Engineering, Newsline,* August, p. 6.

Bader, M. E. (1980). "Quality Assurance and Quality Control, Part 1: Specifications." *Chemical Engineering,* vol. 83, no. 3, Feb. 11, pp. 92–96.

Baker, E. M. (1975). "Signal Detection Theory Analysis of Quality Control Inspector Performance." *Journal of Quality Technology,* vol. 7, no. 2, April, pp. 62–71.

Ballou, D. P. and Pazer, H. L. (1982). "The Impact of Inspector Fallibility on the Inspection Policy in Serial Production Systems." *Management Science,* vol. 28, no. 4, April, pp. 387–399.

Churchill, A. V. (1956). "The Effect of Scale Interval Length and Pointer Clearance on Speed and Accuracy of Interpolation." *Journal of Applied Psychology,* vol 40, December, pp. 358–361.

Cooper, J. E. (1980). "The Care and Training of Your Inspectors." *ASQC Technical Conference Transactions,* Milwaukee, pp. 674–676.

Crow, E. L., (1966). "Optimum Allocation of Calibration Errors." *Industrial Quality Control,* November, pp. 215–219.

Czaja, S. J., and Drury, C. G. (1981). "Training Programs for Inspectors." *Human Factors,* vol. 23, no. 4, pp. 473–484.

Denker, S. P. (1984). "Justifying Your Investment in Automatic Visual PCB Testing." *Circuits Manufacturing,* October, pp. 56, 58, 60, 62.

Dodds, L. B. (1967). "Workmanships." *ASQC Technical Conference Transactions,* Chicago, pp. 249–252.

Dodge, H. F. (1928). *A Method of Rating Manufactured Product.* Reprint B-315, May, Bell Telephone Laboratories.

Dodge, H. F., and Torrey, M. N. (1956). "A Check Inspection and Demerit Rating Plan." *Industrial Quality Control,* vol. 13, no. 1, July, pp. 5–12.

Eisenhart, C. (1968). "Expression of the Uncertainties of Final Results." *Science,* June 14, pp. 1201–1204.

Eppen, G. D., and Hurst, E. G., Jr. (1974). "Optimal Location of Inspection Stations in a Multistage Production Process." *Management Science,* vol. 20, no. 8, April, pp. 1194–1200.

Ezer, S. (1979). "Statistical Models for Proficiency Testing." *ASQC Technical Conference Transactions,* Houston, pp. 448–457.

Garfinkel, D., and Clodfelter, S. (1984). "Contract Inspection Comes Into Its Own." *American Machinist,* October, pp. 90–92.

Gebhardt, C. (1982). "Color Me Calibrated." *Quality,* March, pp. 62–63.Gibson, J. D. (1983). "How Do You Recognize a Qualified Inspector?" *Quality Assurance for the Offshore Industry* (*London*), April, pp. 55–56.

Gilman, J. R. (1963). "Quality Reports to Management." *Industrial Quality Control,* May, pp. 15–17. [In this demerit scheme of check inspecting the work of inspectors (who are paid by piece work), the accuracy is expressed in a form equivalent to the number of demerits found per lot checked.]

Gleick, Elizabeth (1996). "No Barrier to Mayhem," *Time,* July 29, p. 42.

Gunter, B. (1983). "The Fallacy of 100% Inspection." *ASQC Statistical Division Newsletter,* vol. 5, no. 1, September, pp. 1–2.

Hains, R. W. (1978). "Measurement of Subjective Variables." *ASQC Technical Conference Transactions,* Chicago, pp. 237–244.

Harris, D. H., and Chaney, F. B. (1969). *Human Factors in Quality Assurance.* John Wiley & Sons, New York, pp. 107–113.

Holmes, H. (1974). "Computer Assisted Inspection." *The Quality Engineer,* vol, 38, no. 9, September, pp. 211–213.

Johnson, S. L. and Funke, D. J. (1980). "An Analysis of Human Reliability Measure in Visual Inspection." *Journal of Quality Technology,* vol. 12, no. 2, April, pp. 71–74.

Juran, J. M. (1935). "Inspector's Errors in Quality Control." *Mechanical Engineering,* vol. 59, no. 10, October, pp. 643–644.

Juran, J. M. (1945). *Management of Inspection and Quality Control.* Harper & Brothers, New York.

Juran, J. M. (1952). "Is Your Product Too Fussy?" *Factory Management and Maintenance,* vol. 110, no 8, August, pp. 125–128.

Juran, J. M., and Gryna, F. M., Jr. (1980). *Quality Planning and Analysis from Product Development Through Use.* McGraw-Hill, New York, pp. 357, 360–361.

Juran, J. M., and Gryna, F. M., Jr. (1988). *Juran's Quality Control Handbook,* 4th ed. McGraw-Hill, New York.

Karabatsos, N. (1983). "Serving Quality." *Quality,* vol. 22, no. 9, September, pp. 65–68.

Ken, J. (1984). "The (Artificial) Eyes Have It." *Electronic Business,* Sept. 1, pp. 154–162.

Kennedy, M. S. (1996). "Inspection-Focused for the Future." *Quality in Manufacturing,* vol. 7, no. 4, May-June, pp. 22–23.

Konz, S., Peterson, G., and Joshi, A. (1981). "Reducing Inspector Errors." *Quality Progress,* vol. 14, no. 7, July, pp. 24–26.

Kramer, Amihud. (1952). "The Problem of Developing Grades and Standards of Quality." *Food Drug Cosmetic Law Journal,* January, pp. 23–30.

Kusch, J. (1979). "Robots and Their Advantage in Inspection." *Proceedings, Society of Photo-Optical Instrumentation Engineers,* vol. 170: Optics in Quality Assurance, vol. II, pp. 40–42.

Leek, J. W. (1975). "See It as It Really Is." *ASQC Technical Conference Transactions,* San Diego, pp. 41–43.

Leek, J. W. (1976). "Benefits from Visual Standards." *Quality Progress,* December, pp. 16–18.

Lincoln, M. (1996). "Automated Inspection Boots Productivity, Quality," *Quality in Manufacturing,* vol. 7, no. 5, July/Aug., p. 18.

Linn, R. D. (1981). "Computer-Aided Inspection—Its Time Has Come." *ASQC Quality Congress Transactions,* San Francisco, pp. 599–602.

Mathur, C. P. (1974). "The Influence of Measurement Variation in Mass Production of Precision Parts." *Q.R. Journal,* January, pp. 1–5.

McCaslin, J. A., and Gruska, G. F. (1976). "Analysis of Attribute Gage Systems." *ASQC Technical Conference Transactions,* Toronto, pp. 392–400.

Meckley, D. G., III (1955). "How to Set Up a Gaging Policy and Procedure." *American Machinist,* vol. 99, no. 6, March 14, p. 133.

Megaw, E. D. (1978). "Eye Movements in Visual Inspection Tasks." *Quality Assurance,* vol. 4, no. 4, December, pp. 121–125.

Miller, E. M. (1975). "Test Methods and Specification Requirements." *ASQC Technical Conference Transactions,* San Diego, pp. 229–230.

Montville, V. L. (1983). "Color Control from Start to Finish." *Quality,* March, pp. 36–38.

National Bureau of Standards (1965). "Expression of the Uncertainties of Final Results." Chapter 23 in *Experimental Statistics,* NBS Handbook 91. U.S. Goverment Printing Office, Washington.

Nelson, A. V. (1984). "Machine Vision: Tomorrow's Inspections with Today's Equipment." *Evaluation Engineer,* vol. 23, no. 9, October, pp. 21, 24, 26, 29, 30, 34, 37.

Nygaard, G. M. (1981). "Why 100 Percent Inspection?" *Quality,* October, pp. 38–39.

Ohta, H., and Kase, S. (1980). "Evaluation of Inspectors in Sensory Tests—Qualification by Geometrical Methods and Classification by Bayesian Diagnosis Rule." *Journal of Quality Technology,* vol. 12, no. 1, January, pp. 19–24.

Palumbo, R. (1997). "Gage Calibration Is an Investment in Quality Production." *Quality Digest,* vol. 17, no. 2, February, p. 60.

Papadopoulos, N. (1983). "Instrumental Color Control—Where to Start." *Quality,* December, pp. 44–46.

"Portable Gage Lab Provides Flexibility and Minimizes Downtime." (1971). *Quality Management and Engineering,* November, p. 23.

Quinlan, J. C. (1996). "Automated Gaging, Test, and Inspection." *Quality in Manufacturing,* vol. 7, no. 3, April, pp. 8–9.

Riley, F. D. (1979). "Visual Inspection—Time and Distance Method." *ASQC Technical Conference Transactions,* Houston, pp. 483–490.

Sataro, Pat (1997). American Red Cross Blood Service, Farmington, Conn., in private correspondence with the editors.

Schaffer, G. (1984). "Machine Vision: A Sense for CIM." *American Machinist,* Special Report 767, June, pp. 101–120.

Schweber, W. (1982). "Programming for Control—Computerizing Measurement and Control." *Quality,* March, pp. 38–42.

Shainin, D. (1972). "Unusual Practices for Defect Control." *Quality Management and Engineering,* February, pp. 8, 9, 30.

Smith, J. R., and Duvier, H., III (1984). "Effect of Inspector Error on Inspection Strategies." *ASQC Quality Congress Transactions,* Chicago, pp. 146–151.

"Software for Manufacturing." (1996). *Quality in Manufacturing,* vol. 7, no. 3, April, p. 36.

Spow, E. E. (1984). "Automatic Assembly." *Tooling and Production,* October, pp. 46–47.

Strack, Charles M. (1996). *Scientific Technologies,* vol. 2, no. 3, November.

"Tape Controlled Machines at Sunstrand Aviation." (1970). *Quality Assurance,* June, pp. 30–34.

Tawara, N. (1980). "A Case Study on Measuring Inspection Performance for Inspection Job Design." *International Journal of Production Research,* vol. 18, no. 3, May-June, pp. 343–353.

Taylor, F. W. (1911). *Principles of Scientific Management.* Harper & Brothers, New York, pp. 80–96.

"The Vision Thing." (1996). *Quality in Manufacturing,* vol. 7, no. 4, May-June, p. 10.

Thompson, H. A., and Reynolds, E. A. (1964). "Inspection and Testing as a Problem in Man-Machine Systems Control Engineering." *Industrial Quality Control,* July, pp. 21–23.

Tobey, D. (1979). "Metrology = Calibration." *ASQC Technical Conference Transactions,* Houston, pp. 513–520.

Trippi, R. R. (1975). "The Warehouse Location Formulation as a Special Type of Inspection Problem." *Management Science,* vol. 21, no. 9, May, pp. 986–988.Walsh, L. (1974). "Back to One Hundred Percent Inspection??" Editorial comment. *Quality Management and Engineering,* March, p. 9.

Walsh, L., et al. (1976). "100% Inspection at Production Line Rates." *Quality,* November, pp. 30–31.

Walsh, L., et al. (1978). "Steel Wheel Maker Tests Hardness 100%" *Quality,* November, p. 37.

Walsh, L., et al. (1979a). "Can 100% Testing Be Eliminated?" *Quality,* May, pp. 42–43.

Walsh, L., et al. (1979*b*). "100% Inspection Plus." *Quality,* September, pp. 102, 104.

Walsmann, M. R. (1981). "The Check List—A Powerful Inspection Tool." *ASQC Technical Conference Transactions,* pp. 348–351.

Wambach, G. W., and Raymond, A. S. (1977). "The Optimum Sampling Plan." *ASQC Technical Conference Transactions,* Philadelphia, pp. 574–578.

Wang, S. C. (1975). "Human Reliability in Visual Inspection." *Quality,* September, pp. 24–25.

Wang, S. H. S., and Seppanen, M. S. (1981). "A System Simulation for Inspection Planning." *ASQC Quality Congress Transactions,* San Francisco, pp. 769–776.

Weaver, L. A. (1975). "Inspection Accuracy Sampling Plans." *ASQC Technical Conference Transactions,* San Diego, pp. 34–39.

Woods, D. G., and Zeiss, C. (1981). "Coordinate Measuring and Finite Metrology." *ASQC Quality Congress Transactions,* San Francisco, pp. 232–237.

Woods, K. C. (1978). "Calibration System for Measuring and Testing Equipment." *Quality Progress,* March, pp. 20–21.

SECTION 24
JOB SHOP INDUSTRIES

Leonard A. Seder

WHAT IS A JOB SHOP?

The terms "job shop" and "mass production shop," though widely used, are loosely defined. Managers who use these terms are well aware that industrial life as lived in the job shop differs considerably from that prevailing in the mass production shop. This difference extends to the problems of creating, controlling, and improving quality. This section undertakes to define the nature of the job shop and to explain the methods in use for dealing with job shop quality.

There is no single parameter which distinguishes the job shop from the mass production shop. Job shops vary in size from very small to very large. Some are captive; others are independent. Some serve sophisticated industrial customers; others serve relatively naive consumers. Their products range from one-of-a-kind nonrepeating items to large lots of frequently reordered stock items. Some make proprietary products of their own design. Others develop designs jointly with customers' designers. Many cannot even be classified neatly in the foregoing terms, since their product mix spreads across the whole spectrum of customer sophistication, design responsibility, lot sizes, repeat rate, etc.

Despite this difficulty of classification, it is possible to identify certain basic common types of job shops and to recognize among them differences and commonality that affect the fashioning of a quality control program to suit their individual needs. Table 24.1 identifies four common types of job shop, and shows some typical products or operations which exemplify each type.

Percent Repeat Jobs. The term "percent repeat jobs," which appears in the headings of Table 24.1, is one of the universal parameters of job shop operation. Percent repeat jobs is defined as the

24.1

TABLE 24.1 Types of Job Shops

		Typical products — or operations	
Type	Description	Percent repeat jobs low to moderate	Percent repeat jobs moderate to high
I	Large complex equipment	Locomotives Chemical plants Buildings Automated production equipment Radar sets	Farm equipment Aircraft Machine tools Printing presses
II	Small, simple end products and components	Fashion fabrics Industrial adhesives Circuit boards Fabricated metals Books	Tires Shoes Garments Wall covering Small appliances Metal shapes Automotive components Electronic components Private-label foods Furniture
III	Custom parts	Machined parts Forgings Weldments	Stampings Castings Molded plastics Screw-machine parts Molded rubber parts Extruded parts Containers
IV	Subcontracted services	Toolmaking Diemaking Moldmaking Printing Machining Testing	Heat treating Welding Plating Packaging Electropolishing

percentage of the total number of jobs in the factory in any one month that are identical repeats of job orders run previously.

The classification of low, moderate, and high percent repeat jobs have the following approximate values: "low" = under 35 percent; "moderate" = 35 to 80 percent; "high" = over 80 percent.

Large Complex Equipment. Companies in Type I (Table 24.1) produce large complex units, each made up of thousands of different parts and components, each of these in turn being defined by its own "drawing number." These companies call themselves job shops because an individual job order or contract usually calls for a very small number of such large units, often only one. Only if the percent repeat jobs is moderate to high are they able to justify manufacturing these units as stock items and making (or buying) the input parts in economic quantities. Lacking a stock of finished units or components, it is necessary to produce from "scratch," and the time pressure becomes severe.

Small End Products and Components. The Type II companies usually produce large quantities on any one order. However, they regard themselves as job shops because of the endless variations of size, shape, color, style, or configuration typically involved in their product lines. Even those with "standard" product lines frequently show hundreds of different "model numbers" in their catalogs. Those who make "specials" for the various customers' unique requirements have thousands of drawing numbers in their engineering files.

Custom Parts. These Type III companies are mainly in business as suppliers to Type I and II companies. They specialize in one or more of the processes listed and fill their shops with customer-designed parts of thousands of different configurations and compositions. Often they can satisfy a customer's annual requirements for a particular part number in just a few hours of production running time. The nature of the problems of such a job shop has been well portrayed by Furukawa, Kogure, and Ishizu (1981), as shown in Figure 24.1.

Subcontracted Services. Type IV companies differ from Type III only in that they tend to be small, independent shops specializing in particular operations, often working on customer-furnished material. Captive shops of this type are often in-house departments within a large Type I, II, or III company. Variety of jobs is again the rule, each job usually requiring only a few hours of production time.

Jobs per Worker per Week. All four types of job shops exhibit two recurring themes of commonality:

1. *Wide variety of designs* (due to a myriad of different configurations, options, colors, sizes, shapes, models)
2. *Short production time* for any individual production task on any one "job"

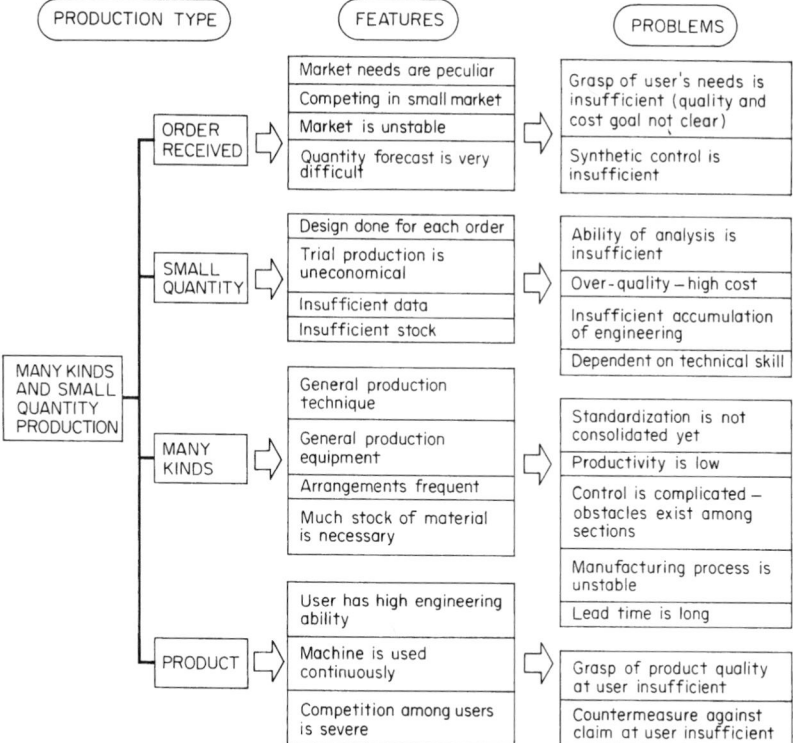

FIGURE 24.1 Problems of quality control at a plant producing to customer order.

These two factors may be conveniently combined into a single parameter of "jobs per worker per week," reflecting the average number of different orders, or different setups, or setup changes that will be handled by each worker over a week's time. (As more and more of the worker skills, memories, and decisions associated with setup changes are replaced by computers and automated setup changers, the name of this parameter will necessarily change to "jobs per machine per week.") Whatever the type of job shop, this number is generally much higher in the job shop than in the mass production shop. This fact has a direct bearing on the nature of the job shop quality program.

The percent repeat jobs (defined above) also varies in size among job shop types and within types. However, the percent repeat jobs is generally much lower for job shops as a class than for mass production shops. As implied in Table 24.1, the percent repeat jobs tends to be higher for some job shop types than others, e.g., Type II versus Type I. However, the rate varies over the whole range and for individual shops within one type.

Job Shop Grid.   When the two parameters of jobs per worker per week and percent repeat jobs are related to each other on the same diagram, there emerges a convenient way to quantify the distinction between the production shop and the job shop. The "job shop grid" in Figure 24.2 is designed to show this relationship.

The job shop grid opens the way to design and apply quality control methods which are keyed to the quantified parameters. At the outset it is evident that production shops are generally those with a small number of jobs per worker per week and a high percentage of repeat jobs. Above the level of 20 jobs per worker per week, we consider it a job shop regardless of the percentage of repeat jobs. Also, below a 50 percent repeat job rate, we consider it a job shop even though the number of jobs per worker per week is low.

THE JOB SHOP QUALITY PROGRAM

In a broad sense, the problems of job shop quality management are the same as for any other shop:

1. Planning of quality for new or modified products and processes
2. Controlling the quality during manufacture
3. Improving quality levels to reduce quality losses

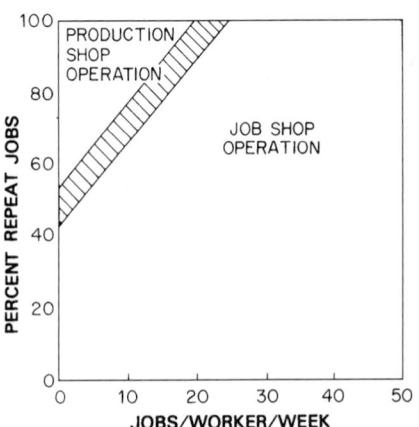

FIGURE 24.2   The job shop grid.

Similarly, the concepts and principles of solution of quality problems are those presented in Sections 3, 4, and 5, and related application sections. However, the numerous job orders (which create a high number of jobs per worker per week and a low percentage of repeat jobs) greatly influence the details of the job shop approach to quality.

The impact of these numerous job orders is not on the materials, processes, or people; these generally remain common to all jobs. Neither is the impact on the systems, practices, and procedures; these likewise remain common to all jobs. (These common "ingredients" may, however, contain the root cause of chronic quality problems.) Rather, the impact is similar to that involved in launching many "new products" every week. For each of these "new products" there is need to discover (1) what is "new," (2) how this affects product design, plan of manufacture, special tools, quality requirements, etc., and (3) what needs to be done to assure that the "newness" is correctly identified and complied with by all departments.

Stated another way, the impact of the numerous job orders is primarily on preproduction planning, and especially on *manufacturing planning*. This planning creates a considerable problem of communicating to all concerned what is "different" about each order so that responsive action can be taken. The amount of such communication can rise to enormous proportions because of the multiplying effect of (1) the number of job orders, (2) the number of ways in which each order is different, and (3) the number of processes, tools, etc., affected by each of these differences. A consequence of this great volume of communication is that the problem of quality control is a problem in *quality of communication* as much as a problem in conventional process and product quality control. As a corollary, when product nonconformance is detected, the correction to be made is very frequently in some detail of the job plan rather than in the product or manufacturing process.

In the light of the foregoing, the quality program for the job shop must include special provisions for:

Planning to communicate essential quality information to all concerned

Controlling the errors and inadequacies in this communication

Improving not merely the processes and products but also the planning and communication

THE JOB NUMERICS

As noted, preproduction planning is a major job shop activity, and involves every job. Since each job differs from all others in *design,* each requires its own *product specifications,* spelling out in detail the materials, formulation, configuration, end-product physical properties, quality and reliability requirements, and the rest. (Simplification is often possible in instances in which a single specification can be used to specify a whole "family" of items largely resembling each other but differing only in specific detail of size or color, etc.)

Since jobs also differ in the exact manufacturing process to be followed, each requires its own *manufacturing plan,* to communicate to Production and Inspection the necessary details of input materials, operation sequence, inspection or laboratory release points, special or unique tooling, in-process properties required, mandatory processing restrictions, and the like.

There appears to be no accepted generic term to represent, for a specific job order, all the details of product and manufacturing process. "Job documentation" comes close, but it sometimes is used to include the recorded quality data, which are not part of our definition. Hence the author has coined the term "primary job numerics" to serve as such a generic term. Table 24.2 summarizes and gives typical examples of these primary job numerics. Obviously, mass production shops must also have these same numerics. However, in the mass production shops the numerics are few in number, tend to become stabilized, and are easily remembered by shop personnel. In the job shop they are many in number, are frequently changed, and require constant reference to the written documents.

Not all job shops have responsibility for preparing the numerics to define *both* product and manufacturing plan. Types III and IV (Table 24.1) ordinarily receive product specifications from their

TABLE 24.2 Primary Job Numerics

Aspect of definition	Typical examples
To define the product	
Materials	Material specification numbers for metals, chemicals, agricultural products, etc.
Formulation	Specific proportions of various materials to be used
Configuration	Drawing or sketch showing dimensions, component parts, assembly details, etc.
End-product acceptability	Dimensional, physical, chemical, optical, metallurgical, electrical, visual, etc., tolerances Functional test requirements
Reliability	Maximum failure rate, or degree of degradation in specified endurance test
To define the manufacturing plan	
Input materials	Sources, subcontractors
Operation sequence	Exact order of primary, secondary, finishing, packaging, etc., operations Specific machines, baths, ovens (when restricted)
Inspection points	Location of inspection stations or laboratory release points
Unique tooling	Design of specific form tools, molds, dies, assembly fixtures, artwork, etc.
In-process properties	Dimensions, thicknesses, densities, colors, electrical outputs, chemical values, strengths, etc., needed at specific operations
Mandatory processing restrictions	Temperatures and times for bakes, heat treatments, reactions, drying, curing, pasteurizing, etc. Hold times between operations

customers and hence prepare only the manufacturing plan. Types I and II prepare both sets of numerics for their own products, but only product specifications for those materials and components which they purchase.

The primary job numerics have long been recognized as essential and have found expression in various types of "legitimate" documentation in the shop. The product specification, manufacturing drawing, material specification, formulation or batching sheet, tool drawings, exploded assembly view, route card, operation sheet, inspection detail sheet, test procedure, and job order card are the more common names for the various means of communicating the needed numerics to shop personnel.

In the evolving techniques of computer-aided manufacturing (CAM), the problem of communicating and *updating* the primary job numerics to the production floor has been solved. The engineers or technical staff persons generally specify the primary numerics with inputs from manufacturing or process engineers. These are communicated via computer to the production personnel. The once-frustrating problem of time delay between the decision to make a change and the making of that decision officially known by advancement of the "change letter" is thus now avoidable.

The primary job numerics outlined in Table 24.2 are the minimum details necessary to define and make the product. However, they are seldom sufficient to assure the quality of the end product or to attain economic operation. There are additional numerics which provide the added information needed to minimize product deficiencies, rejections, repairs, yield losses, and customer complaints.

Some of these supplementary job numerics are shown in Table 24.3. For each aspect of product and process definition, there are special details, unique to the individual job, that are of value to the shop personnel. Communicating these details to the shop personnel provides the advance knowledge that can often spell the difference between success and failure to meet end-product requirements or between high and low quality costs.

It is probably no exaggeration to say that the key to preventing product deficiencies in the job shop lies in recognizing the importance of *specifying* the supplementary numerics. Once this is understood, the same computerized method used for the primary numerics is extended to the supplementary. This virtually eliminates the many "debates" and misunderstandings that formerly took place. Not all of these supplementary numerics are needed for all jobs. Indeed, one of the real dilemmas faced by job shop managers is the decision of how far to go in this direction (see below).

JOB PLANNING

To generate, communicate, and comply with all these job numerics requires that a major element of the job shop quality program must be concerned with individual job planning. This involves

TABLE 24.3 Supplementary Job Numerics to Prevent Product Deficiencies and Losses

Aspect of definition	Typical examples
Materials	Special supplier requirements for process control Special gages or test methods Packaging requirements Classification of characteristics Acceptable quality levels Certifications required
Operation sequence	Special work instructions Exact details of important hand operations Permissible deviations from sequence
Inspection points	Special gages or test methods Classification of characteristics Acceptable quality levels
Unique tooling	Identification numbers Tool inspection details Permissible tool deviations
In-process properties	Optimum settings of process variables Special gages or test methods Plans of control for operations with setup approval criteria, running approval criteria Statistical control plans to be used
Mandatory processing	Tolerances for times, temperatures, etc. Certifications required
End-product acceptability	Special gages or test methods Special customer "idiosyncrasies" Classification of characteristics Acceptable quality levels Customer data submittals or certifications Visible defect acceptability limits Customer sampling plan impositions
Reliability	Testing details

(1) organizing for job planning, (2) detecting and correcting job planning errors, (3) improving the job planning.

Organizing for Job Planning.   A major question is how far to go in completeness of planning. It is usual to carry out planning of the *primary* job numerics in total or to leave only minor details to be worked out during the production run. However, the *supplementary* job numerics present a problem in striking the proper balance between overplanning and underplanning. Establishment of the numerics beforehand will work to avoid errors and misjudgments during the production run. However, the volume of detail and the lack of adequate information of some aspects (e.g., the expected rate of occurrence of specific defects, sequence deviations, or tool deviations; knowledge of the optimum settings of process variables) make it uneconomic or impossible to fill in all the details.

A major consideration in this decision is the "percent repeat jobs." A shop with a low percent repeat jobs must necessarily devote a major effort to planning the supplementary job numerics, since there is no "second chance." A shop with a high percent repeat jobs can place its major effort in control, at the sacrifice of planning, since the control activities will, over a period of time, influence the evolution of the correct numerics. Figure 24.3 shows this contrast diagrammatically.

A further consideration in extent of planning is the time schedule. Planning can be less than complete when the manufacturing cycle permits a trial lot to be piloted through ahead of the job order to pin down many of the supplementary job numerics or when the job running time is long enough to use data feedback and corrective action for the same purpose. (See Job Shop Control, below.)

Responsibility for job planning varies widely among job shops. In all but the smallest shops, the primary numerics are commonly developed and issued by a staff group, separate from line production. This group is variously designated as Research and Development, Engineering or Technical (especially when definition of product is part of the work), or as Manufacturing Engineering, Process Engineering, Production Engineering, Estimating, Planning or Industrial Engineering (when the planning is mostly limited to definition of manufacturing plan). However, there is no universal pattern of responsibility for generating the supplementary numerics. In some instances, staff quality engineers or process engineers have this responsibility. In other shops, line supervisors, inspection supervisors, and even workers and inspectors have the assignment. In still other cases, the supplementary inspection numerics are prepared by staff specialists, while the supervisor is left to his or her own devices to develop and convey information on tools, setups, settings of process variables, etc.

As is usual in matters of organizing, it is more important to be clear than to be logical or uniform. There is a need for providing the supplementary numerics, and the responsibility for doing

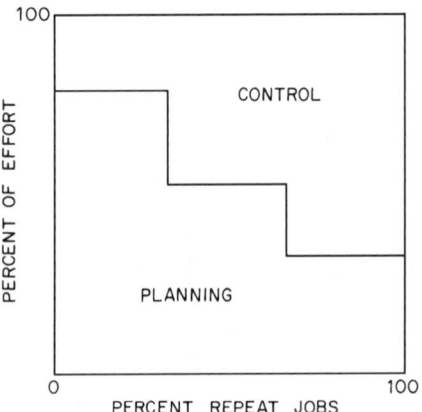

FIGURE 24.3   Effect of percent repeat jobs on allocation of planning effort.

so should be clear. The job shop which has left this question vague would do well to face it cleanly. As a general rule, if the generation and maintenance of any portion of the supplementary job numerics is to be delegated to Production, the responsibility and method must be made clear. Otherwise, in the author's experience, it tends to be neglected, poorly maintained, and ineffective.

Often the variety of work makes it necessary and possible to do the planning selectively on the basis of the size or complexity of the job. Jobs over a certain size, or involving new manufacturing techniques, or expected to become new standard products, are planned by the staff group, whereas small jobs, or those which involve only minor changes from "standard items" or previously run jobs, are planned by the line production people.

Detecting and Correcting Job Planning Errors.   The sheer number of details involved in the primary and supplementary numerics makes it inevitable that errors will occur in job planning. Some will be the inadvertent errors of misplaced decimal points, transposed digits, incorrect arithmetic, and the like. Others will arise from lack of sufficient knowledge, by the planner, of processes, economics of manufacture, capabilities, and reliabilities.

To minimize planning errors, it is useful to review the planning in some appropriate way. For large jobs, this review tends to be elaborate and formal. The planning documents are circulated to the key departments, after which there is a formal review meeting. This meeting not only goes over errors and refinements; it also identifies possible problem areas. In addition, it may determine whether there is need to provide "sample" or "trial lot" evaluation before full production. The review meeting also may establish the guidelines for delegating development of the supplementary numerics to lower levels of organization.

For smaller, low repeat rate jobs, the planning documents are likewise circulated to the key departments. However, the review and sign-off usually take place without formal review meetings.

Enlisting the participation of workers in identification of errors has been a goal of many quality programs. Schafer (1978) describes a successful program for error identification in a machine shop. "Job check" cards (see Figure 24.4) were developed for machining, welding, assembly, electrical, paint, packaging, and crating to aid workers in performing a systematic review. When errors were found, workers filled out a problem report, which was circulated to the planners for corrective action.

MACHINING JOB CHECK

PRINT/PROCESS REVIEW		MACHINE REVIEW	
• Blueprint Correct & Clear	• Information Adequate	• Correct Speed	• Correct Horizontal Setting
• Process Correct	• Unknowns / Problems	• Correct Feed	• Machine Functioning **OK**
• Process=Print	• Areas to Machine	• Correct Vertical Setting	• Machine Capable
TOOLS/FIXTURES/LOCATION REVIEW		**MATERIAL/PARTS REVIEW**	
• Tools Correct & Sharp	• Inspection Tools	• Correct Material	• Correct Prior Operations
• Correct Fixture	• Locating Surfaces **OK**	• Correct Size & Shape	• Part Complete
• Fixture Capable, Complete	• Holding Method for Machining	• Material Condition **OK**	• Appearance **OK**
1st PIECE SAMPLE CHECK		**IN PROCESS SAMPLE CHECK**	
• Part to Print	• Part Damage	• Part to Print	• Part Moving Location
• Part Usable	• Machining Acceptable	• Machine Drifting	• Machine Settings **OK**
• Off Specifications **OK'd**	• Appearance Acceptable	• Tool Dull	• Every 10th Piece **OK**

FIGURE 24.4   "Job check" card for operator detection of job planning errors.

For high percent repeat jobs, reliance is more heavily placed on the control system (see Job Shop Control, below).

Improving Job Planning.   Improvement of this planning, as applied to manufacture, involves preparation and use of machine and process capability knowledge in establishing tolerances, choosing processes, classifying characteristics, etc. The approach, which is generally conventional, is discussed in Section 22.

A concept known as "group technology" involves planning of jobbing work by identifying "families" of parts based on commonality of operations. This commonality then is used as a basis for standardization of drawings, tooling, etc., with an obvious residual effect on quality planning. The group technology idea extends beyond planning to machine layout, product flow, and cellular and flexible manufacturing systems. See also Section 22 under Group Technology.

JOB SHOP CONTROL

The "jobbing" nature of the job shop is derived from the diversity of products. However, the manufacturing processes which turn out these diverse products exhibit a high degree of commonality in materials, machines, instruments, and people. As a result, the job shop systems for quality control of manufacture closely parallel the systems in use in the mass production shops, but scaled to the size and needs of the particular job shop. In addition, the large number of jobs per worker per week and the mass of detail contained in the job numerics make it important to have special approaches to data feedback and corrective action.

Overall Control System.   Those minimal job shop systems that parallel the mass production systems are listed below, including references to the handbook sections that discuss conventional approaches:

Control system	Section reference
Supplier material	21
Identity and flow of "lots"	23
Process control decision making	45
Tool and equipment qualification and maintenance	23
Calibration and maintenance of measuring equipment	23
Disposition of rejected material	23
Analysis and followup of customer complaints	25

To formalize these systems, it is necessary to document them in a quality control manual which is then distributed to those concerned. Under ISO-9000 requirements, this is mandatory (see Section 11 under the Clauses of ISO 9001 and their Typical Structure for the ISO 9000 requirements of a quality manual).

In the job shop it is uncommonly important to provide a sound plan for making decisions on whether the process should run or stop, and to make clear delegation of responsibility for decision making on the factory floor. With limited workforce to spread over a multitude of jobs, it is also important that the job shop understand and make use of the concept of dominance (see Section 22 under Control Systems and the Concept of Dominance) in order to maximize the effectiveness of that laborpower. Setup dominance is the prevailing mode for most quality characteristics in the small-lot job shop, especially those of Types III and IV. The main reason is that the running time is usually so short that the "time-to-time" variation of the process is minimal. Hence "if the setup is right, the lot will be right." Accordingly, the job setup is a vital control station, and demands use of statistically valid plans for setup approval.

Earlier, simplified plans such as narrow-limit gaging and precontrol and control charts were used but were not widely understood. Computers have stepped into this breech. They have made statistical validity simple and automatic. There are numerous examples.

In the manufacture of expensive machined pieces, including multidimensional ones, special-purpose gages are used where needed to assure correctness of the *setup*. The data from these measurements are automatically fed into specially programmed computers to produce histograms, capabilities, control charts and other statistical evaluations to predict and thereby urge action, if needed, to assure lot conformance. (See Figure 24.5).

More generally, for any process, data from regular gages or visual inspection can be punched into hand-held computers by the person collecting the data. Instantly, the statistically valid "photograph" of the predicted lot quality from that setup flashes on the computer's miniscreen. The user doesn't have to understand the statistics; he or she sees the picture!

These developments have accelerated the trend toward establishment of a state of self-control by the operator, to whom the setup acceptance responsibility can now be delegated (See generally Section 22 under Self-Inspection).

Data Feedback and Corrective Action.   Many job shop managers have fallen into the trap of believing that once they have established an inspection system (even if, in a small shop, it means the hiring of the first inspector), they now have "quality control" and can relax. Now "quality control" will protect them against bad purchased materials, stop defects from being manufactured, and guard the outgoing product. It may well do these things, but an essential added need is to use the *information* gained from performing the inspection to *improve* quality. It can do this in several ways if appropriate feedback mechanisms are established:

1. For preventing defects in the unmanufactured portion of a job
2. For preventing defects in repeat orders of a job
3. For preventing defects in future orders for other jobs in the same "family"
4. For correcting problems in the "ingredients" (i.e., policies, systems, procedures, practices) common to all jobs

The extent of these benefits available to a particular job shop depends on the "jobs per worker per week," the "percent repeat jobs," and other factors. Consideration of these factors leads to the "current job approach" and the "repeat job approach" as two basic ways of achieving feedback and corrective action.

The Current Job Approach.   This is a means of preventing defects in the unmanufactured portion of a job through feedback of information from the manufactured portion of the same job. It can be used whenever the running time of the job is longer than the time necessary to give the feedback signal, diagnose the cause, and determine and implement the corrective action. See, schematically, the top part of Figure 24.6. The value of such prevention is so obvious as to provide an incentive to prompt feedback of data on quality troubles, and prompt corrective action on the feedback.

Figure 24.7 shows the mechanism used in one electronics assembly plant to secure such prompt feedback and corrective action. The results of subunit, unit, and systems test are recorded by serial number as to the item and the nature of the discrepancy. Copies of the test records are reviewed daily by a quality control engineer who determines the nature of each deficiency and initiates a "corrective-action request" (Figure 24.7) to the design engineer, manufacturing engineer, components engineer, test supervisor, test-set maintenance person, supplier liaison person, or other individual who can take the necessary action. The quality engineer follows up each of the requests and the associated replies until the matter is disposed of by action, or by decision that no action is necessary.

Corrective action that will benefit the unmanufactured product may either involve a change in the *job numerics* or correction of an error in *complying* with the job numerics.

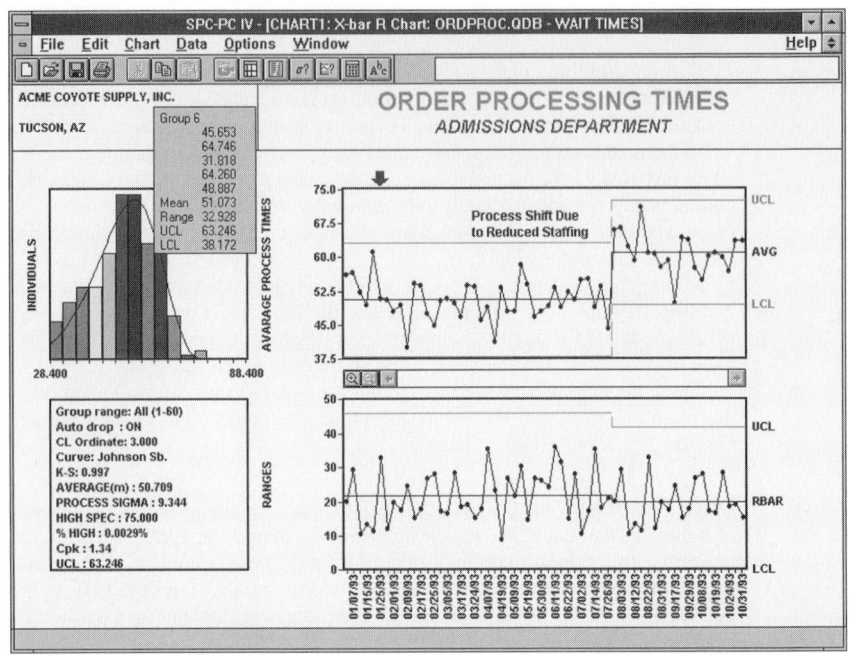

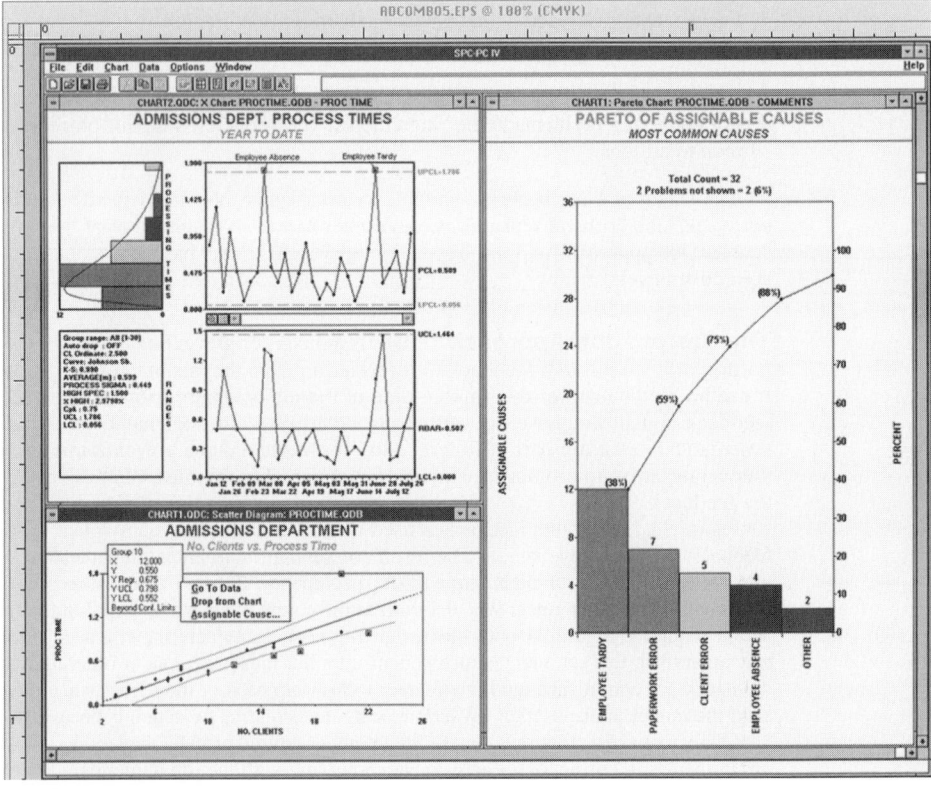

FIGURE 24.5 Computer software has enabled job shops to effectively utilize SQC tools. Shown are typical examples offered by Quality America, Inc.

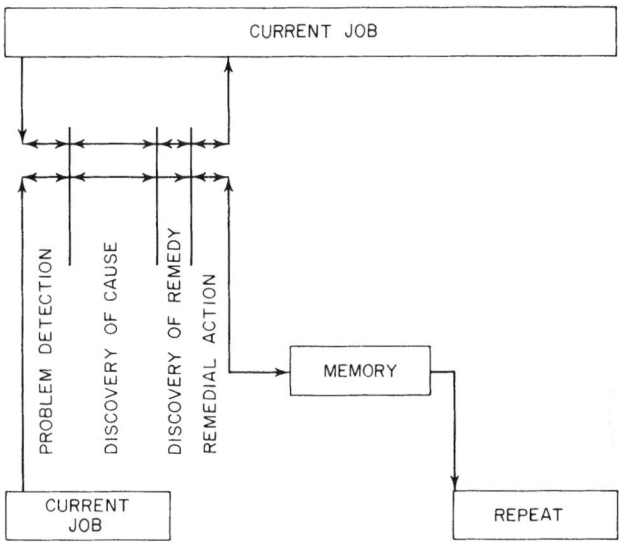

TIME →

FIGURE 24.6 Current job versus repeat job corrections.

This current job approach is most applicable when the following combination of circumstances is present:

1. The economic gain through preventing defects in the unmanufactured units is obviously greater than the cost of the feedback and prevention machinery; i.e., large lots or expensive items are involved.

2. The causes of the defects are obvious enough to permit prompt diagnosis.

3. The organization for feedback and followup can be kept simple, e.g., one employee with a clear assignment.

In the job shop with a low percentage repeat job rate it is often desirable to create deliberately the opportunity to use the current job approach even when the normal running time would be too short to use it. Instead of processing the entire order as a lot, a small "pilot" lot or single item precedes the main lot, and time is allowed for the feedback and correction to take place.

The pilot lot idea has long been used in job shops. However, in its early form the pilot lot was processed like any other lot. Then the final results were scanned, and if no trouble was reported in production, everything was assumed to be satisfactory, i.e., the tentative job numerics could be made permanent. Often the pilot lot was considered to be solely the means of making samples for customer approval. Experience has shown that there is much value in expending extra effort in planning the trial lot, collecting special data, analyzing the data, and using the results to modify the job numerics.

For example, when a mechanical piece part moves through a series of operations, all of which can influence a critical dimension, the fact that 9 of the 10 pieces in the pilot lot conform to final specifications might seem to be adequate to firm up the numerics; yet the full job order could easily run 30 percent defective as a result of inadequate process capability or wrong process centering. By requiring measurements of the 10 pieces after each operation, and treating the means and ranges statistically, the alert analyst would easily discover the problem and, moreover, would be able to identify the operation responsible, thus preventing a large loss when the remainder of the order is run.

| INSPECTION REPORT | INSTRUMENT DEVELOPMENT LABORATORIES, INC. 67 MECHANIC ST., ATTLEBORO, MASS. | | REPORT NUMBER |

FORM #164 REV.

DATE:

LOT QUANTITY	PART NUMBER:	DESCRIPTION:	VENDOR:
154	401185	Barrel	IDL

ACC.	REJ.	INSPECTOR	P.O.	R.R.	W.O.	A.O.	DATE REC'D IN INSP.
0	All	J. Couchie			0801B	0502	

CERTIFICATION REQUIRED

OPERATION NO. 150 – Complete

CLASS	A. Q. L.	SAMPLE SIZE	REJECTION NO.	NO. IN SAMPLE DEFECTIVE
MAJOR	1.5%	35	2	2
MINOR	4.0%	25	3	4
OTHER				

ACCEPTED PCS. RECEIVED BY:
M. Taylor – Stockroom
DATE:

REJECTED PCS. RECEIVED BY:
W. Pendergast – Prod. Control
DATE:

REJECTIONS

REASONS FOR: LIST BY ITEM:

1. 1.595 + 001 is 1.5947 to 1.5956
          −000

2. 0.125 ± 005 holes are 0.126 − 0.133

DISPOSITION

LIST BY ITEM:

1. Sort for undersize 1.595 diameter and
    return defectives for rework

2. Defective 0.125 hole sizes do not affect
    function or fit – accept as is.

Results of sorting: 143 accepted
                      11 rejected.

CORRECTIVE ACTION

Tooling correction promised by Industrial
Engineering.
File for follow - up on 12/2

MATERIAL REVIEW BOARD

QUALITY CONTROL:	DATE:
W. Wold	

PRODUCT ENGINEER:	DATE:
C. Logan	

GOVERNMENT:	DATE:
L. Thuotte - AFQCR	

FIGURE 24.7 Typical rejection, disposition, and corrective action scheme.

The Repeat Job Approach.   For many jobs the running time is so short that the job is completed before the sequence of "analysis, feedback, and corrective action" can be completed. In such cases, the knowledge gained from the analysis cannot be put to use on the "current" job. However, this same knowledge can be put to use on a repeat order *provided there is a memory system* which can:

1. Store the knowledge
2. Provide ready recall when repeat orders are received

The lower half of Figure 24.6 shows diagrammatically the time relationship that permits this "repeat job approach" to give to future orders the benefit of the knowledge gained from previous orders. Since this approach involves the costs of maintaining a memory system, it is most applicable when the percent repeat jobs is relatively high, of the order of 75 percent or more. It probably cannot be justified economically if the percent is low, say 25 percent or lower. In between, the economics of the specific situation must be examined to determine whether it is less costly to provide protection for all potential repeat orders or suffer the losses of repeating the error for that smaller number of jobs that will be reordered. In addition, there are special situations that may warrant a memory system, as in the case of small first runs of development work on complex equipment, for it is generally important to "debug" the job numerics before production orders are received.

Memory System for Job Numerics.   Before the computer came on the scene, an astonishing variety of memory systems had been invented to serve this need. In designing the computer system, the designer would do well to become familiar with some of these to understand what is needed, i.e., not just the *specifications,* but the *experience* gained on past job orders. For example, in a plant manufacturing custom aircraft parts (Type III), a "job history file" was maintained by the manufacturing staff group. Into this file went the customer drawing and/or the specification, the job numerics, in-process and final inspection data on each run, nonconformance record, problems encountered, corrective actions taken, results of troubleshooters' investigations, recommendations, and formal change requests issued. When a repeat order was received, the planner would refer to this file and review the data and notes before issuing the new manufacturing order. Similar planning memories were in use in other shops covering items purchased from suppliers and/or items subcontracted to Type IV job shops. The purpose of all this information was that, on a repeat order, the designer could revise the specification, if needed, and the planner, in turn, could revise the other numerics.

Experience with such memory systems led to an important conclusion. They were most useful when the mix of data and notes were "digested" and the planned changes for future lots prepared during or immediately after the completion of the current job order. They were least successful when the data and notes were left undigested until the repeat order was received. By then, the undigested information would have deteriorated badly and the time pressure to produce the repeat order would often stymie the effort to take advantage of the system.

"Digestion" requires establishing the discipline of corrective action investigation and follow-up, *even though the current job may already be completed.* The organization for investigation may, as in the case of the current job approach, be limited to one analyst when simple technology is involved. However, where the causes of defects and the needed corrections are not obvious, more talents are needed. In any event, responsibility for the investigation and decision should be clearly allocated. It may be a material review board, corrective action board, factory service group, quality engineering, or other specially designated team. The agreed-on corrective actions are ordinarily recorded, and the responsible department designated, together with the expected date of accomplishment (see Figure 24.7). Diligent follow-up by a systematic routine is then needed to assure that these intentions are executed during the interval between orders. Someone must therefore be given the job of "keeping a book" on pending corrective actions until completion.

Memory Systems for Manufacture.   These systems were created, usually by production departments, to alert personnel to the hazards of known prior errors of execution, and to evolve more optimum supplementary numerics. For worker-controllable defects, for example, special "warning" or "caution" slips are often attached to the blueprints or instructions in the job file maintained in the factory. The worker assigned to the repeat order is thereby "flagged" to exercise special care of a particular job. An example is seen in the "pitfall sheet" shown in Figure 24.8, which was used in a large job shop described by Fletcher and Novy (1972).

A further example is the job "setup card" file maintained for some processes. Each card is a record for a single job order. On the card are posted the conditions which prevailed in the process while that job was being run, as well as the results of inspection and test. For example, in calendering plastic film, the setup involves such numerics as roll speeds, roll temperatures, roll spacing, material feed rate, and many others. Often some of these are altered (and duly recorded) during the

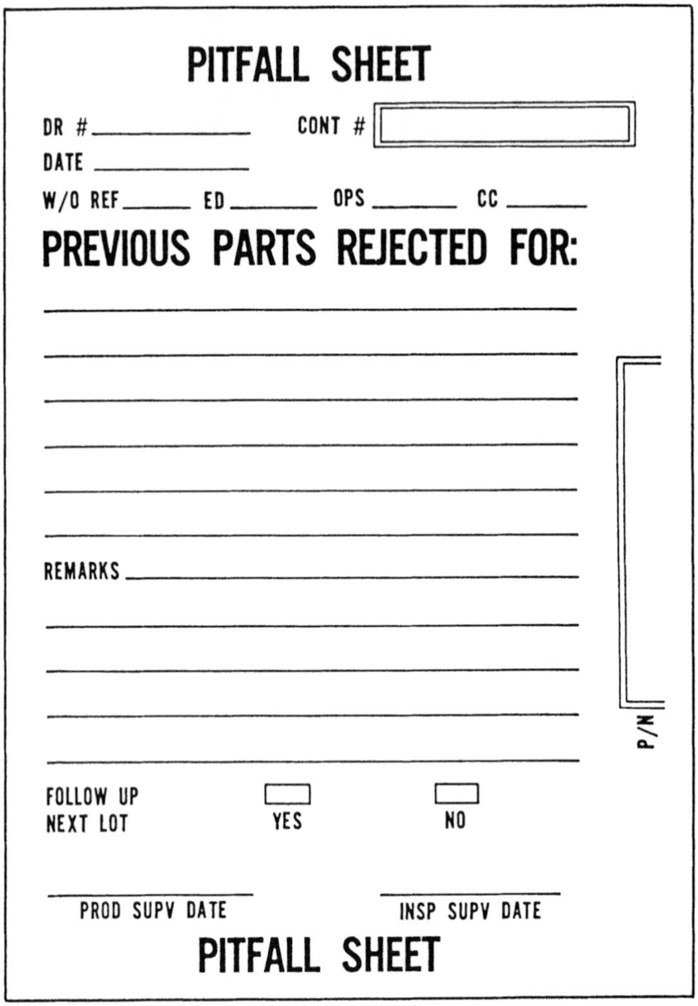

PITFALL SHEET

DR # _____   CONT # [   ]

DATE _____

W/O REF _____ ED _____ OPS _____ CC _____

PREVIOUS PARTS REJECTED FOR:

_____

_____

_____

_____

_____

_____

REMARKS _____

_____

_____

_____

P/N

FOLLOW UP
NEXT LOT   [   ]   [   ]
                  YES   NO

_____   _____
   PROD SUPV DATE   INSP SUPV DATE

PITFALL SHEET

FIGURE 24.8   Memory system "flag" for operators.

run to improve the quality of film being produced, based on the judgment of the supervisor. The subsequent inspection or lab test results are likewise posted to the card.

When this same plastic film is reordered, the setup person consults the card file to identify the lots which showed the best test results. He or she then tries to reproduce the process conditions which prevailed during the manufacture of this best product. As the card file builds up, a further step can be taken by analyzing the data through more sophisticated statistical methods, e.g., regression analysis (see Section 44). Computer programs offer an excellent opportunity to carry out such work "just in time" for the next order.

Memory Systems for Inspection.   Such memory systems have usually consisted of job history cards to which inspection results were posted. The resulting knowledge of job quality levels and frequencies of specific defects can be used for a variety of improvement purposes. It can warn of inspection errors, lead to revision of defect classifications and acceptable quality levels (AQLs), promote changes in inspection or test methods or gages, provide additional supplier instructions or notifications, furnish Pareto summaries of the vital few defects of each job, identify jobs where inspection or testing can be reduced, etc.

In addition, the memory system concept offers the job shop a way to diagnose the causes of "mysterious" defects, to determine process capabilities, to discover dominance, and to perform other statistical analyses. For example, in mass production, a few days or even hours may produce enough defects to provide the data needed for conclusive analysis. Job shop managers ordinarily are envious at these opportunities to collect and analyze data in such short order, and they often give up trying to apply such techniques to jobbing work. However, repeated small lots, plus a memory system, plus patience, will likewise furnish the data, analyses, and solutions. When the data are organized by machine center rather than by defect type, the economics of analysis may be more favorable.

Computer Memory Systems.   Computerization of such memory systems has provided much more effective and efficient accumulation of, and real-time access by production personnel to, the needed job numerics and useful information. At the same time, the stored data can be readily manipulated to prepare Pareto analyses and to perform the more sophisticated analyses referred to above. Rapid analyses made possible by the computer can go a long way in diagnosing job shop quality improvement problems, hitherto felt to be too complicated and time-consuming to pursue.

However, it is not obvious that computer programs will be dedicated to these memory systems unless the significance of the foregoing descriptions of the spontaneous and "unofficial" memory systems is recognized. They were the primitive responses to a real need; that need must now be satisfied by a well-planned computerized system.

SERVICE INDUSTRY JOB SHOPS

The thrust of this section is aimed at manufacturing industries where the term "job shop" originated and where the unique problems of such a shop tended to slow progress in applying quality control methods. However, many service industries, it turns out, are job shops. In hotels, hospitals, insurance agencies, and many others, everyday life is a variety of different events, patients, policies, or customers. The applicability of the methods of job planning and control discussed in this handbook section has been convincingly demonstrated by Patrick Mene, vice president of the Ritz-Carlton Hotel chain. He applied them to the planning and control of "events" (i.e., meetings and conferences). He defines the "primary event numerics" (his words) as "the minimum information to define event *products* and *work*." His numerics to define *products* include such items as

1. Types of guest rooms to be required
2. Configuration and seating capacities of function rooms
3. Particular audiovisual requirements
4. Food menus and specifications
5. Beverage brands and container sizes

The primary numerics for *work plans* include:

1. Required supplier input materials
2. Recipes and order of service for meals
3. Inspection points, rehearsals, food tasting
4. Exhibits, decorations, etc.

But, as in manufacturing, "the primary event numerics are seldom sufficient to assure the quality of the event…additional numerics are needed to minimize mistakes, rework, breakdowns, delays, inefficiencies, variation, customer complaints and rebates." The secondary event numerics to prevent defects and losses include:

1. Special supplier brand names, material requirements, test methods, AQLs
2. Sequence and exact details of important operations, special work instructions
3. Special test methods, control plans, AQLs, times, temperatures, staffing requirements, etc.
4. Special customer likes or dislikes

The Ritz-Carlton quality program includes the procedure for event *planning* to minimize errors. For large events, this includes formal review meetings and sign-off on the written plans. For smaller, simple, or repeat events, the written plan is developed without formal meetings. *Control* procedures involve ways of detection of errors in the event numerics or correction of errors in complying with the numerics. Efforts are concentrated on diagnosing the cause of an error "while the bits and pieces are fresh in mind" and, though the knowledge is too late to be used on the current event, it is put to use on a repeat order of the event or a future event in the same family. See also Sections 30 through 34 for quality in various service industries.

IMPACT OF ADVANCES IN METALWORKING TECHNOLOGY AND COMPUTERIZATION

Advances in metalworking production technology and computerization of process decision making have had a major impact on job shops engaged in the fabrication of metal parts. While most of these developments have been aimed at increasing productivity, they have dramatically altered the basic worker-machine relationships that previously characterized small batch production of metal parts.

New Developments.   The thrust of the new developments, from the quality standpoint, has been to improve process capability in a number of ways and reduce the dependence on worker judgment for process control decisions. Among the developments have been:

• Improved tooling, fixturing, and work movement methods, such as pneumatic and hydrostatic holding devices.
• Numerically controlled machine tools, in which punched tapes or computers establish and set machine operating speeds, feeds, and other cutting conditions.
• Automatic selection of proper tools, from a carousel holding as many as 60, and their insertion into the machine chucks, which sharply reduces job setup time and makes small lot production even smaller.
• Computer control of the motions of the machine to execute the desired functions from simple hole drilling to very complex contouring.
• Production of a variety of goods from a single set of tools and equipment by means of programmable automation (see Blumenthal and Dray 1985).
• Use of sophisticated robots for assembly operations and inspection of dimensions of parts (see Blumenthal and Dray 1985).
• Design of high speed, multidimensional gages which not only speed up difficult measurements, but do so simultaneously at a new level of accuracy and precision.
• Direct input of the results of such measurements into computers.
• Mind-boggling ability of the computer programs to instantly turn out all manner of data summaries, including process centering and capability, control charts, positional diagrams, multivariate charts, gage repeatability and reproducibility, and other useful information.

Effect of Developments.   The main effect of the metalworker developments has been to increase radically the amount of job planning and reduce, though by no means eliminate, the amount of job controlling in the traditional sense. In terms of the earlier discussion in this section, many of the supplementary numerics have had to be thought out and incorporated into the computer program. While such a move is to be applauded, it can bring on new needs for controlling:

> The programmer typically works in an office away from the factory floor. Only a few are experienced machinists....Programs often have mistakes....A misoriented tool can drill a hole in the wrong place or become chipped...and debugging takes anywhere from two hours to two weeks (Blumenthal and Dray 1985, p. 34).

> The machinist must still set up the workpiece to be cut, make adjustments to correct for tool wear and stop the machine if anything goes wrong (Shaiken 1985, p. 18).

> An operator may find that the rough casting to be machined is larger than the programmer expected and thus requires more cutting. An alloy may turn out to be harder than expected, in which case the part must be fed more slowly into the cutting tool (Shaiken 1985, p. 19).

Job Planning and Controlling.   Most of these developments make the need for detecting and correcting job planning errors even more of a necessity than before. Software development and quality assurance, discussed in Section 20, becomes a new routine operation in shops that adopt the new technology.

A second necessity is validation of the correctness of all the planning details, including the software, by thorough setup acceptance of the very first pieces from the automated process. Process centering and capability are determined using any one of a number of available software packages. In addition to providing a validation of the setup, this establishes a benchmark against which to judge later repeat runs.

Regular setup acceptance at the time of each subsequent job order can thus be reduced to a simple determination that no change has taken place since the validation, i.e., tool sharpness has been maintained, work placement is correct, feeds and speeds are optimum for *this* lot of material, etc.

QUALITY IMPROVEMENT

To remain competitive the job shop must constantly engage in improvement or "breakthrough," quite aside from its day-to-day problems of enforcing quality compliance. The mass production shop must likewise engage in breakthrough, and the conceptual approach to breakthrough is identical for these two forms of industrial organization. Where they differ is mainly in the nature of the improvement "project." Because of sheer volume, the mass production shop quality improvement project usually involves a specific defect on a specific product, e.g., the 3KL cylinders are out of round on the 2.500-in (63.5-mm) dimension. In contrast, the job shop quality improvement project is usually concerned with remedy of some common cause which cuts across a variety of jobs.

The preoccupation with individual jobs is often a detriment to organizing for job shop improvement. A given common cause may adversely affect, say 25 percent of the jobs (see, for example, the case of "inductance out of specification" below). Preoccupation with looking for "blame" or for the "corrective action" in each of the jobs affected may blind the managers to the existence of a common cause, and thereby to the opportunity for leveraged improvement.

Identification of logical job shop improvement projects is largely a matter of ingenious use of the Pareto principle (see Section 5 under Pareto Principle). The need is to identify those common causes which are at the root of the greatest amount of job shop trouble, and thereby will result in the greatest value of improvement for the least cost of analysis. Once a project has been chosen, the usual limitation to solution is more a matter of management than technology; someone must be liberated from the daily, job-to-job problems and given a license to diagnose the improvement projects.

Three approaches for project identification in job shops are presented below:

The chronic offenders approach

The product family approach

The non-job approach

The Chronic Offenders Approach.   In this approach, the use of the Pareto analysis is first one of identifying the few jobs that result in the bulk of the quality losses. The term "chronic offenders" refers to these few jobs.

For example, a manufacturer of a line of floor polishers found that punch-press scrap was an important quality cost. A Pareto analysis of this scrap by part number (Figure 24.9) established that six of the parts accounted for half of the value of the punch-press scrap. It became a logical project to reduce scrap on these chronic offenders, since the "percent repeat jobs" was high. (The detailed approach to analysis of causes and discovery of remedy is discussed in Section 5, The Quality Improvement Process.)

In the complex assembly job shop, interest centers on individual assembly defects rather than on jobs per se. Defects may be so numerous and varied on each job that corrective action must be concentrated on those recurring defects which account for the greatest dollar loss, or are the most serious to the customer, or both. A "chronic offender" chart is then made for each major job, indicating the predominant defects. When high losses and seriousness are both involved, the list can be a composite of "five top dollar-loss defects," plus all the serious ones.

Corrective action for chronic offenders follows the general methods discussed under the Repeat Job Approach, above. In addition, the list of chronic offenders is publicized for the attention and priority of all—managers, supervisors, analysts, production workers, inspectors, etc. (In one plant in which the defects were mainly worker-controllable, good results were achieved by posting the chronic offenders list in each production department.) Along with this, the plant manager received a weekly bulletin of progress in tackling these high dollar-loss items.

The chronic offenders list is never static. Some projects are removed from the list because they have been solved. New projects are added to the list as the result of new customer demands or com-

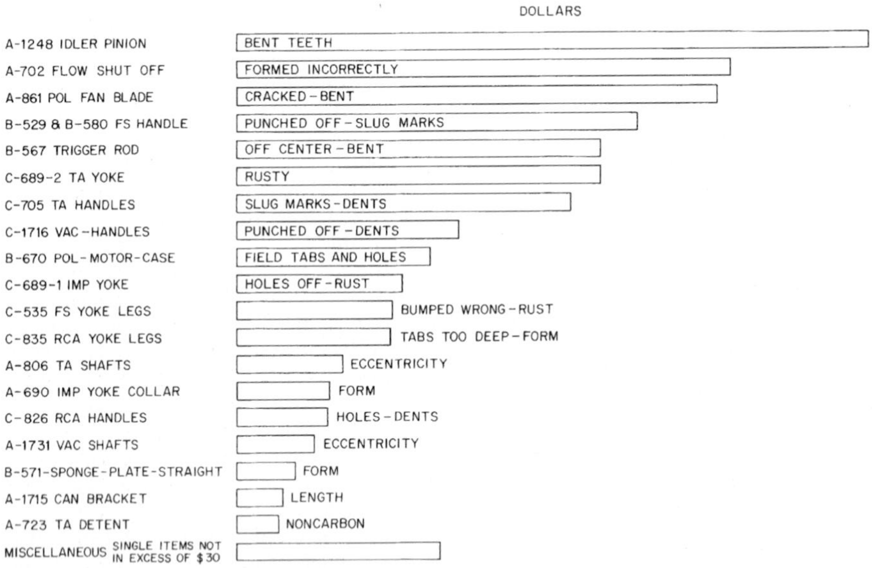

FIGURE 24.9   Pareto analysis by part number.

petitor practice. Accordingly, it is necessary to revise the list of the "worst" offenders periodically (in the same manner that law enforcement officers revise the list of the ten most wanted criminals).

The Product Family Approach. This approach utilizes the fact that all jobs in a product family have similar customer requirements, design specifications, sequence of operations, process variables, inspection instructions, or other job numerics. Under such conditions, any job is a repeat for any other job in the same product family. Through this relationship, the percent repeat jobs is greatly increased. In consequence, the effort of diagnosis and remedy is amortized over a greater number of jobs. This amortization tends to make this approach economic for moderate and even low percent repeat job rates.

The instances of "families" are legion. Tire manufacturers have hundreds of job specifications to cover all permutations of brand, size, fabric, construction, grade, wall color, tread design, etc. Yet, most of the numerics for, say, a four-ply nylon tire are alike (allowing for size scaling) for all members of that family. In the calendered vinyl plastics business, myriads of artistic printing and embossing patterns are applied to only a dozen or so basic families of laminated, unsupported, or coated films. Again, the job numerics for each of these families prior to printing and embossing are largely alike. In machine shops, the family concept has led to the group technology approach of identifying operations, sizes, materials, etc. that are common to a whole group of parts and restructuring the plan of manufacture.

The product family approach directs its efforts toward improving the job numerics for the entire family. This may come about in several ways:

1. By extending to all members of the family the knowledge gained when using the "current job" control plan. Once corrective action has become known for one job, such knowledge can be used to benefit other members of the family.

2. By utilizing the product family concept in setting up the memory system for the "repeat job" approach. Such usage supplies more data in a shorter time, and facilitates the identification of chronic offenders or defect concentrations. In turn, diagnosis of these "vital few" family problems leads to remedies which can be extended to the whole family.

3. By setting up a special project, in the absence of a memory system, to furnish the information in item 2 above and tackle the "vital few."

The Non-Job Approach. "Non-job" is used here in the sense of an approach to chronic defect reduction through discovery and elimination of common causes that are not job-related, i.e., the causes that cut across many jobs. Because the causes common to many jobs are quite numerous, the first step in the non-job approach is to use the Pareto analysis to identify those common causes that might warrant further analysis. The Pareto study is conducted in various ways—by defect type, failure mode, process, department, discrepancy, "basic cause," etc. Out of these studies emerges the most promising avenue for further study, usually that Pareto distribution in which the fewest number of defect types (or whatever) account for the greatest proportion of the trouble.

The usual starting point is to study the distribution of rejects or losses by *common symptoms,* on the normally valid premise that common symptoms will be found to have common causes when further analyzed. For example, an electronics assembly shop found (at subassembly) that a Pareto distribution by defect provided a good basis for further study, since one defect (solder) accounted for about 37 percent of all defects (Figure 24.10). Studies by failure mode, error type, discrepancy, etc., are similar in nature.

For many products, the study can usefully go one step further, even at the exploratory stage. Whenever the causes for the principal symptoms categories are "obvious" (i.e., all knowledgeable hands agree) from the nature of the symptoms, then an analysis can be made by *basic cause categories.* (Such obvious causes are so well recognized that they often find their way into the very name of the defect, such as "toolmarks," "incompletely lapped," "undercured," or "double-knurled.")

In an automobile tire plant, a Pareto distribution by defects was of no avail, since there was not sufficient concentration among the 85 identifiable defect types to justify study projects for each of

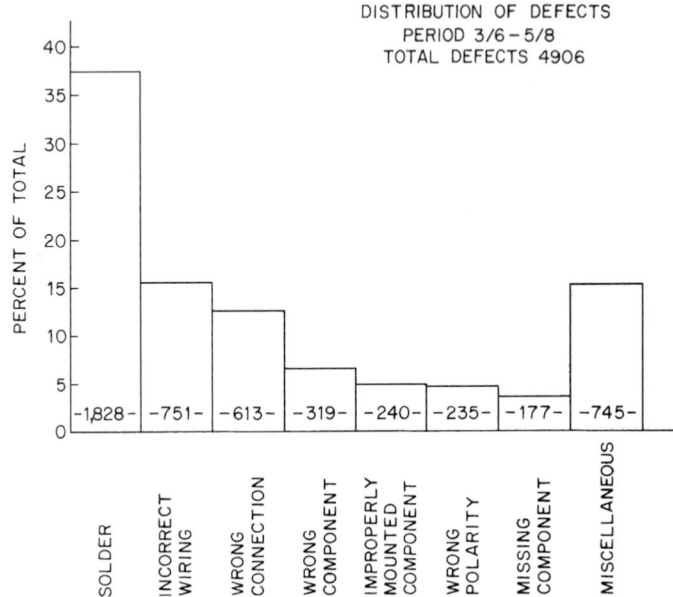

FIGURE 24.10   Pareto analysis by common symptoms.

about 20 principal defects. However, by regrouping the 85 defect types into 7 basic cause categories (Figure 24.11), it became evident that one operation (curing) required better control systems, while another operation (finishing) needed improved attention to setups. It also became evident that the relationship of ply angle to width required a more complete planning of the supplementary job numerics by Engineering.

In still other cases in which causes are not obvious and defects might be the result of any of several possible causes, it is nevertheless instructive to attempt to classify by basic cause categories in the job shop. This can be done by setting up a special study for a limited period (e.g., a week or a month), during which time each rejection or error is carefully traced to its origin by a task force representing Engineering, Production, and Inspection. Based on the facts unearthed, they try to agree on the cause classification.

For example, a sheet metal fabricating shop studied its rejections in this way and obtained the Pareto distributions shown in Figure 24.12. The most promising direction for study was the basic cause category "operator error," since operator errors (acknowledged as such by the operator in each case) were by a wide margin the biggest single class.

One machine tool builder prepared a check sheet to assist supervisors in analysis of causes of defects. The check sheet required each foreman to:

Describe the defect in terms of the specification, and of the effect on assembly or customer

Identify the source of process dominance (see Section 22 under Planning Process Controls)

Identify the plan in use for detecting nonconformance

Identify where the defect occurred and where it was found

Determine the extent to which worker self-control was present (see Section 22 under Concept of Controllability)

Determine basic cause for the defect (Figure 24.13) (see Section 5 under Diagnostic Journey)

The resulting data offered many possible useful Pareto analyses.

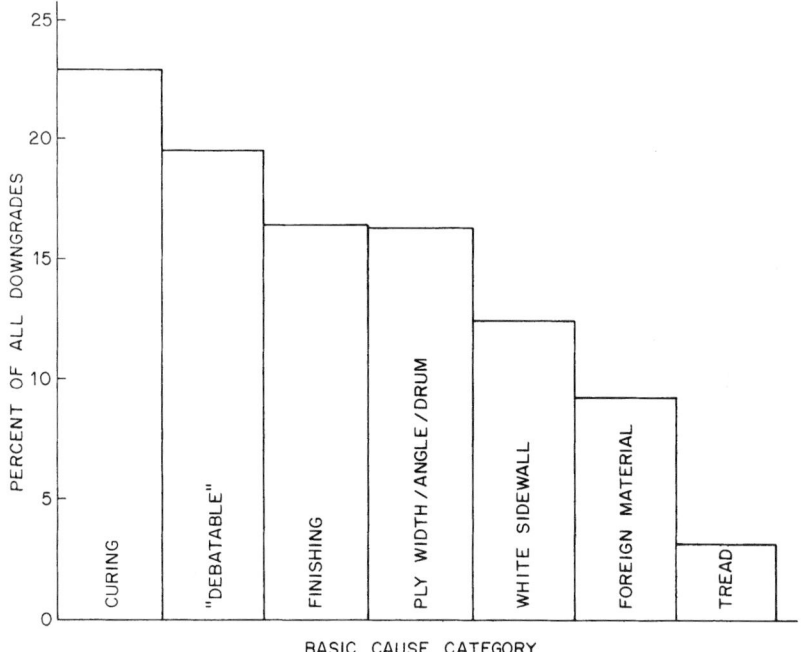

FIGURE 24.11 Pareto analysis of auto tire downgrades by basic causes.

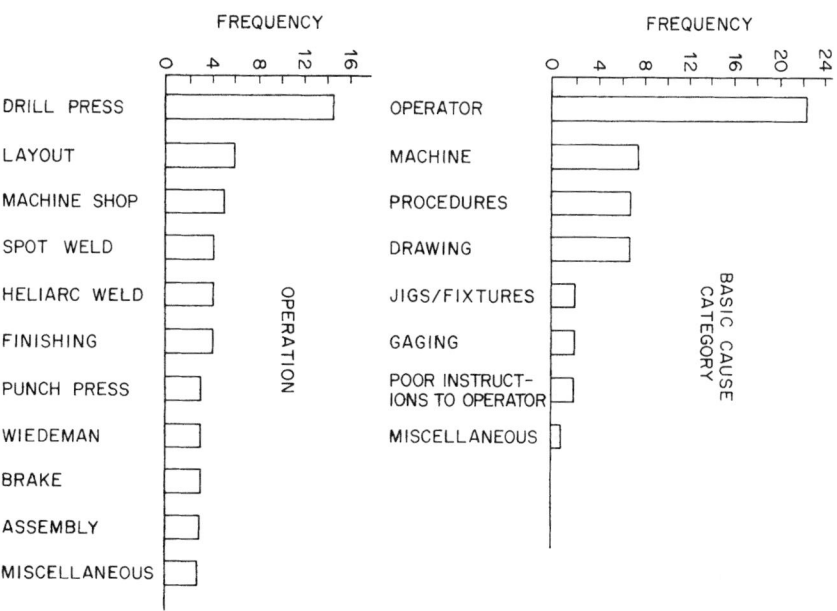

FIGURE 24.12 Two-way Pareto analysis by operation and basic cause category.

WHAT WAS THE BASIC CAUSE FOR THE DEFECT?

____Specification not clear	____Tool wear
____Spec misinterpreted by operator	____Operator loading wrong
____Routing error	____Operator repeatability
____Machining sketch error	____Machine repeatability
____Tape error	____Measurement method
____Tooling error	____Measurement error
____Fixture wrong	____Deviation from routing sequence
____Operator setup error	____Supplier material defective
____Layout error	____Machine malfunction
____Prior operation wrong	____Wrong feeds and speeds
____Other (describe below)	____Operator inattention

FIGURE 24.13   Check sheet for basic cause analysis in a machine shop.

In a company making electrical inductors, "inductance out of specification" was by far the most frequent defect. Frequency distributions of a dozen high-reject lots showed two different symptoms patterns:

1. Some lots had adequate within-lot variability, but had poor centering of the mean of the distribution, resulting in rejects all beyond one of the limits.

2. Other lots had excessive within-lot variability, resulting in defects outside of both limits.

The common cause in this case was in the Engineering department "sample shop," which established product tolerances (e.g., inductance, resistance) as well as manufacturing instructions for wire size and number of turns. The practice of the sample shop was to make a single trial coil to meet the customer's electrical requirements. Then, the wire size and number of turns that had been used to make this trial coil were incorporated in the manufacturing instructions. The system was defective because it failed to consider the effect of production variability from coil to coil. The results of one trial coil provided no information on natural coil-to-coil variability from which to establish the tolerance width, nor was the sample coil necessarily at the center of the tolerance range, as assumed by the Engineering department—hence the observed symptoms when the Production department followed the manufacturing instructions for wire size and number of turns. When the sample shop procedure was changed to making three coils, with calculated consideration for expected variability, there soon was a substantial reduction in the rejections.

It is evident from such cases, especially the last one, that job shop quality improvement is not confined to changing the job numerics or to providing a warning to Production the next time the job is run. The "common cause" often interacts randomly with jobs; it can affect any job at any time. Once tracking of such a common cause gets under way, the diagnostic trail frequently leads back to the fundamentals of the very system of preparing specifications, of assigning tolerances, of bidding or estimating, of controlling processes, etc.

Diagnostic Techniques.   The techniques for diagnosis and remedy of the chronic problems in the job shop are similar to those in any shop; they are covered in detail in Section 5. However, the job shop diagnostician is seldom able to collect large quantities of data at will; lot sizes are too small. This calls for ingenuity by the diagnostician in collecting the needed data from repeat orders of a job or from several members of a family. Often a variant of the "memory system" (see under The Repeat Job Approach, above) provides the answer by accumulaing the data gradually over a period of time.

For example, machine capabilities in a job shop can be determined from an accumulation of measurements of 5 to 10 pieces per job over a series of jobs. As the data accumulate, a statistically adequate basis for estimating machine capability emerges from the "within-job" range. (See Section 22, under Process Capability.)

Similarly, dominance can be identified by recording 5 or 10 "first piece" and "last piece" measurements on a series of jobs in the memory system. If, for a given machine or process, the series of jobs shows no significant change between the two sets of measurements, setup dominance is indicated; otherwise, time dominance. In the latter event, the accumulated data give important quantified information on whether the time-to-time variation is steady or erratic from run to run. Recording of worker identity is the key to identifying worker dominance.

Even "within-piece" or positional concentration of defects can be discovered by accumulation of data from small lots. In a plant making large castings for pumps and air compressors, a condition of leaks in castings was remedied only after patient recording of the location of the leaks, month by month. In this case the castings were made a few at a time, repaired, and shipped out. The memory device was a drawing copy on which the diagnostician accumulated all leak locations.

REMEDIES FOR JOB SHOP PROBLEMS

When the identified problems are job-related, so are the remedies. In the chronic offenders and product family approaches, the remedies indicated by the diagnoses are usually changes in the job numerics. When the identified problems are not job- related, as in the non-job approach, the remedies must go deeper, e.g., modification of the overall system of specification, planning, or control.

Challenging the Basic Premises.   The most difficult remedies are those for which it is necessary to question the basic premises or axioms underlying management thinking. These premises are often of such long standing that little effort is being devoted toward changing them or even questioning them. These premises are further entrenched because their effect is interdepartmental, i.e., several major company departments are involved. In consequence, a change requires acquiescence or formal approval from the upper management of the company.

Some widespread examples of the need for challenging basic premises are listed below. In studying these examples, it is well to keep in mind that these premises were very likely well founded in years gone by but have meanwhile become obsolete by the slow, undetected movement of events.

Unrealistic Specifications.   The large number of jobs per worker per week (so usual in the job shop) exposes the production and inspection personnel to a very large number of quality characteristics. Under such conditions, systems of "unrealistic specifications loosely enforced" become unmanageable because the shop people must carry in their heads so much detail of how loosely to enforce the specifications.

For job shops engaged in custom work, the way to avoid recurring violation of specifications (by workers who conclude that the tolerance is unrealistic or unimportant) seems, on the face of it, to give binding force to the specifications. However, no amount of criticizing, threatening, or pleading will assure compliance if the process capability is inadequate. The trouble is that the system is founded on a defective premise. The remedy lies not in more intense use of the present system; the remedy lies in change of the basic system.

This is not as easy as it sounds, since the system is logical once the basic premise is accepted. It is common to discuss enforcement of tolerances without questioning the basic premise itself. Such discussions may settle the specific instance without settling the broad question. It is only when the question "tight tolerances loosely enforced versus realistic tolerances rigidly enforced" appears on some important agenda as a topic in its own right that the question has been brought out in the open.

Informal Communication.   In very small model shops and specialty shops the communication from designer and planner to mechanic is highly informal and includes much oral communication. As the shop grows, this close relationship is gradually eroded by sheer size and complexity, and the need is for greater formality and greater reliance on written communication. However, in some job shops this communication retains many aspects of the informality of the model shop or speciality shop despite the fact that these have long since been outgrown. As discussed under The Job Numerics, above, the communication of the supplementary job numerics is a necessary response in the modern industrial world of multiplicity of requirements.

Quotation Review.   The prevailing practice in quoting prices to customers is to base them on cost estimates prepared by an estimator who makes use of cost standards based on historical data.

In some types of product, the precision demanded has become such that the decisive factor in meeting cost and delivery standards is the ability to hold tolerances. Yet seldom is the estimator provided with adequate standard data (on the cost of precision) to come up with quotations which reflect the realities of holding the precision demanded.

Of course, adequate standard data should be prepared and made available to the estimators, who in turn should be trained in how to use them. Until this is done, the job shop is well advised to bring the Quality department into the quotation procedure so that available quality capability knowledge is utilized. Use of this knowledge can aid in identifying unrealistic tolerances, predicting costs, anticipating gaging problems, defining vague characteristics, and improving inspection planning.

The basic premise here is that the estimator should be able to prepare the quotation. The premise is sound only if the estimator is equipped with the data on which a sound quotation can be built up.

Quality Planning.   The "basic premise" question here is primarily one of separation of planning from execution. However, the question extends to numerous facets—choice of methods, tool control, gage control, definition of responsibilities, feedback systems, etc. The really decisive question is whether to formalize or not. Once there is a decision to formalize, the people involved can usually find ways appropriate to their needs.

An example of combined quality planning and execution is the Engineering department model shop. These shops are staffed by skilled model makers working directly with the engineering designers. The atmosphere is highly informal, with little reliance on drawings, tolerances, methods sheets, or other written communication. The model maker is expected to make the model by utilizing general-use machinery, to create ingenious setups so as to avoid expenditure for tools, to consult freely with the designer on open questions, and even to contribute ideas to the design itself. In such an organization form, reliance for quality is on the model maker rather than on some formal system.

As the job shop grows, the need arises for a greater degree of separation of planning from execution. This need is met by the creation of separate planners and a Planning department. (The quality counterpart of this is separate quality engineers and a Quality Engineering department.) The planners soon find (as did the model makers before them) that the planning should not be uniformly applied to all jobs or functions. Some jobs are more defect-prone, more expensive, more unstable than others.Hence there arises the need for a rationale or logic to determine which part of the planning is to be done by the planners and which is to remain with the shop personnel.

Outdated Factory Organization.   Many job shops organize their machinery on the colony plan, e.g., all lathes are in one room, all presses in another, etc. The intention is to reduce investment in machinery and to develop skills in the respective processes. This colony form of machine organization multiplies greatly the problem of preparing the job numerics and increases the opportunities for error. (It also increases process inventories, overall manufacturing intervals, and the complexities of process control.)

Here the basic premise is that as the job shop grows, the colony organization must be retained. This premise has been questioned on the grounds that growth should be through creation of "cellular" groupings, which use special machine designs to minimize the preparation of extensive job numerics, increase machine utilization, improve coordination, etc. Highly automated forms of such cellular groupings, known as "flexible manufacturing systems," are in use, which can significantly lower production and quality costs. A good description has been given by Black (1983).

For example, a manufacturer of cigarette making machinery embarked on a program of:

1. Use of light alloys to increase speed of metal cutting
2. Design of special machines on the Numerical Control principle to perform multiple operations during a single setup
3. Design of special inspection machines to verify the setups
4. Organization of the shop into small, compact crews (see the discussion of "Group Technology" under Improving Job Planning, above)

Multiple Suppliers.   For the small job shop, material usage is so modest that when an adequate source of supply has been established for any specific material, there is little point looking for a second source. As the shop grows, material usage grows with it, and there may be a need to shift from single to multiple suppliers for some materials. However, the basic premise of single suppliers may meanwhile have become so rooted that it blocks consideration of multiple suppliers.

Worker Motivation.   The economics of job shop planning favor a higher degree of delegation to the work force than is readily feasible in the mass production shop. This delegation reduces the prevalence of worker monotony and boredom, but also increases the extent of worker controllability of defects. (See generally Section 15 and especially under Open Communications. See also Section 22 under Concept of Controllability.) As these defects are brought to light, the managers conclude that since worker inattention, blunder, etc., created these defects (which is often true), it follows that better worker attention, etc., will eliminate all defects. An extension of this logic is that the way to improve quality is to penalize workers for defects. However, the logic is based on a defective premise, since even if worker errors can be eliminated, there still remain the management-controllable defects (which usually are about 80 percent of all defects).

Actually, the wide delegation of duties to shop personnel, so prevalent in job shops, creates a favorable climate for new approaches to increasing job interest and improving worker motivation. The job shop is thereby a good laboratory for testing out some of the modern ways being evolved to improve motivation (see Section 15 under Design Principles of Work and Organization).

THE SMALL JOB SHOP

The approaches discussed in this section for planning, controlling, and improving quality require much technique and effort beyond that needed for the basic "line" activities of designing and producing the product. In the large job shop this additional work is mostly performed by staff specialists in a "Quality Engineering" department. However, the small job shop seldom can justify use of such full-time specialists. Neither can this small shop endure high quality losses. The answer to this dilemma is universal for all small enterprises: everyone wears several hats. The necessary quality "staff" activities do get carried out, but as part-time tasks for someone who is busy with many other part-time tasks.

For example, in one small job shop a wide array of quality tasks was assigned as shown in Table 24.4.

REFERENCES

Black, J. T. (1983). "Cellular Manufacturing Systems Reduce Setup Time, Make Small Lot Production Economical." *Industrial Engineering,* November, pp. 36–48.

Blumenthal, Marjory, and Dray, Jim (1985). "The Automated Factory: Vision and Reality." *Technology Review,* vol. 88, no. 1, January, pp. 30–37.

Fletcher, O. L., and Novy, E. (1972). "Application of Hypergeometric Sampling Plans in a Large Job Shop."

TABLE 24.4   Assignment of Quality Tasks in a Small Job Shop

Tasks	Assigned to
Receiving inspection, in process inspection and test	Line inspectors and testers
Gage control and gage procurement	One full-time technician
Reliability test and evaluation, inspection planning, test equipment design, statistical methods	One quality engineer
Quality laboratory, special process controls	One laboratory technician
Supplier control, troubleshooting	Laboratory technician and quality manager
New design review	Quality manager
Command of the department	Quality manager

ASQC Technical Conference Transactions, Milwaukee, pp. 489–500.

Furukawa, O., Kogure, M., and Ishizu, S. (1981). "Systems Approach to QC System of Job Production." *ASQC Quality Congress Transactions,* Milwaukee, pp. 225–262.

Groover, M. P., and Zimmers, E. W. (1980). "Energy Constraints and Computer Power Will Greatly Impact Automated Factories in the Year 2000." *Industrial Engineering,* November, pp. 34–43.

Schafer, Reed (1978). "Error Reduction in the Job Shop." *ASQC Technical Conference Transactions,* Milwaukee, pp. 162–166.

Shaiken, H. (1985). "The Automated Factory: The View from the Shop Floor." *Technology Review,* vol. 88, no. 1, January, pp. 17–24.

Swaton, Lawrence E., and Green, Carl P., Jr. (1973). "Automated Process Audit and Certification Program." *Quality Management and Engineering,* vol. 12, no. 3, March, pp. 24–28.

SECTION 25
CUSTOMER SERVICE[1]

Edward Fuchs

INTRODUCTION

During the 1980s, customer service emerged as a critical success driver in many companies that pursue the management approaches of leading-edge enterprise. The compelling logic behind the importance of customer service is simple (Grant and Schlesinger 1995). As rapid cycle times diminish the ability of providers to differentiate their products, customer service is one of only very few ways to motivate customers to enhance behaviors that affect the enterprise's sales and profitability. Customer service includes transactions with customers and relationships with customers that occur before and after purchase of product or service. These transactions and relationships differ for different types of products and services. It is convenient to view possible customer service elements as innovative or traditional, as is illustrated in Table 25.1.

It should be understood that the list in Table 25.1 is not a complete listing of possible customer service elements. In fact, leading-edge companies always strive to find new elements that will provide a competitive edge. I call these elements *key customer satisfiers* (KCSs). In some cases, KCSs involve the transfer of value to customers. In other cases, they involve relationships with customers. Assignments to the traditional and innovative categories are not always the same for different markets. What may be innovative in one market may be traditional in another.

The *value proposition* of a product is the totality of the service and tangible product elements that are perceived by customers as valuable, leading to decisions to purchase the products offered for sale in preference to competitors' products. The weight of evidence (e.g., see Grant and Schlesinger 1995) is that the marketing trend for many products and services continues to be the extension of the value proposition to include more transactional (value-transfer) items and more and more relational

---

[1] In the Fourth Edition, material for the section on customer service was supplied by Frank M. Gryna.

TABLE 25.1   Traditional and Innovative Elements of Customer Service

Traditional presale elements	Innovative presale elements	Traditional postsale elements	Innovative postsale elements
Sales information Ordering	Market research for key customer satisifiers	Packaging Transport Delivery Installation Maintenance Complaint handling Problem solving	Report cards Customer care Value-added information access

services. This is so because the trend is continuing for the products in many industries to become undifferentiated and commoditized. This trend is especially true in mature industries but is increasingly the case in industries such as consumer electronics and personal computers. Consequently, the traditional criteria used to distinguish between the product (manufacturing) and service sectors are breaking down. Products often include tangible goods bundled together with associated real-time services and ongoing information exchange. Anyone who has purchased and installed a personal computer software package will recognize this extended value proposition.

This section describes the quality considerations of customer service, the use of customer service as a strategic differentiator, and the design, structural, and operational considerations of high-quality customer service performance. Elements of behavioral science, organizational design, and systems engineering all are brought to bear on these topics (Pyzdek 1994). Reference to other sections of this handbook is made where appropriate.

CONSIDERATIONS OF QUALITY IN CUSTOMER SERVICE

There are four important areas where quality must be considered in customer service: strategic intent, design, organizational structure, and operations. The first is *strategic intent.* As part of a strategy formulation, the overall approach to quality is determined for the product or service. A subset of the overall approach is the use of customer service elements as strategic differentiators. Examples of marketplace advantages that may be achieved by providing value-added customer service are reduced customer turnover, increased repurchasing performance, increased product or service use, and customer referrals. Value-adding customer service also may yield higher profit margins and growth in market share. The customer service strategy is determined as part of the annual business planning cycle or as part of the product planning cycle. Customer service elements are hypothesized based on insights from the available body of knowledge about the industry and its marketplace. The insights are then verified using systematic methods. The results of the verification step may be confirmation that the proposed approaches to customer service indeed are opportunities to achieve marketplace advantages or that modified or alternative approaches yield more value-adding capability. Iteration may be required until there is closure and consensus on the strategy to be undertaken.

The second important area where quality must be considered is the *design area.* Companies that emphasize customer service design customer service capabilities as part of the product/service life cycle. Structures and processes are devised during the design phase to implement the strategy for customer service that was determined in the strategy formulation. The approach to design that is presented here is patterned after the service systems engineering framework of Pyzdek (1994). Pyzdek's framework combines concepts and methods from the systems engineering, organizational design, and behavioral science fields. All these components are important to achieve an effective design of customer service delivery capabilities.

The third area where quality must be considered is *organizational structure.* This is where quality is built into the organization's customer service delivery structure according to the requirements of the product or service life-cycle design. Important considerations are enabling technologies and training, process engineering and reengineering, legacy migration, and outsourcing. It is common to find that process models for customer service capabilities involve almost every business function and, therefore, almost every part of the organization. This creates an opportunity to use customer service as an integrating force that knits together a company's quality thrust. However, in many firms this pervasive characteristic also creates a problem of distributed responsibility and accountability for effective customer service.

The last important area where quality must be considered is the *operational area.* This is where results are achieved, measured, and improved. The ongoing performance of operating customer service processes is important here and includes such considerations as human performance, process control, process capability, process improvement, and performance metrics. Three areas of consideration dominate an effective end-to-end life-cycle design. The first includes continuous communication, reward, recognition, and development of customer-facing associates; second is continuous oversight of processes in search of stress cracks that may be caused by saturated processes, by changes in customer needs, or by other considerations not congruent with the design intent and implementation criteria; and the third is continual environmental scans in search of early warning signs of changing markets, of new opportunities, and of new competitive threats. Common to all these considerations is the need for instrumentation for collection and analysis of information.

Here I present a model that features customer service as a strategic differentiator. To illustrate the use of the model, data and examples are provided, taken from the literature, from published benchmarks, and from Baldrige Award–winning companies. The reader may wish to review the contents of Sections 3, 13, and 18 as preparation for this topic.

The model presented in Figure 25.1 helps us visualize customer service as a vehicle to achieve strategic goals and objectives for a product or a service or for a family of products or services (Innis 1994). During the planning period for a specific product, or during the annual business planning interval for the forthcoming year, customer service is proposed as a competitive differentiator. The elements of customer service that may be appropriate for the particular type of product or service are selected. These elements are a subset of the assumed KCSs for the product or service. Insight for this inductive step may come from experience with previous products, from customer feedback such as complaint or suggestion data, from competitive analyses and environmental scans, or from new technology or business capabilities. A structured approach for this step, called *service blueprinting,* is described below, under Strategic Intent, AT&T Universal Card Services.

At this point in the strategy development, these insights should be verified. This verification step often is omitted, leaving the ultimate verification to the marketplace. This sometimes turns out to be an expensive shortcut. Therefore, every effort should be made to obtain verification. KCS data for the industry or for the specific product or service must be acquired. Identifying and addressing the key customer satisfiers constitute an essential step toward achieving customer satisfaction. KCS data are acquired from market research studies conducted in appropriately defined market segments. The KCS data provide insight into the specific product and service attributes that are important to customers in each market segment and the degree of importance the customers attach to each attribute. Depending on the industry, on its marketplace characteristics, and on the particular type of the product or service, one or both of two types of market research are commonly used for this purpose (see Section 18).

In many cases, it is possible to verify the proposed KCSs by using such preimplementation techniques as focus groups or market research surveys or by limited prototyping experiences. In other cases, it may be necessary to use a rapid cycle-time implementation of initial production to get feedback. This is called a *beta test* in some industries. A more traditional term is *field trial.* These alternatives are illustrated in an expansion of the Lisrel model as shown in Figure 25.2.

The well-known quality function deployment technique often is used as the vehicle to integrate two important steps, the understanding of what is important to customers in market segments (e.g., KCS data) and the correlation of the important attributes, stated in terms meaningful to the customer, with specific product and service attributes, stated in terms that are meaningful to the designer (see Section 3).

When the strategic approach to customer service is verified, improved and updated, and documented, it is usually included in a preliminary business case.

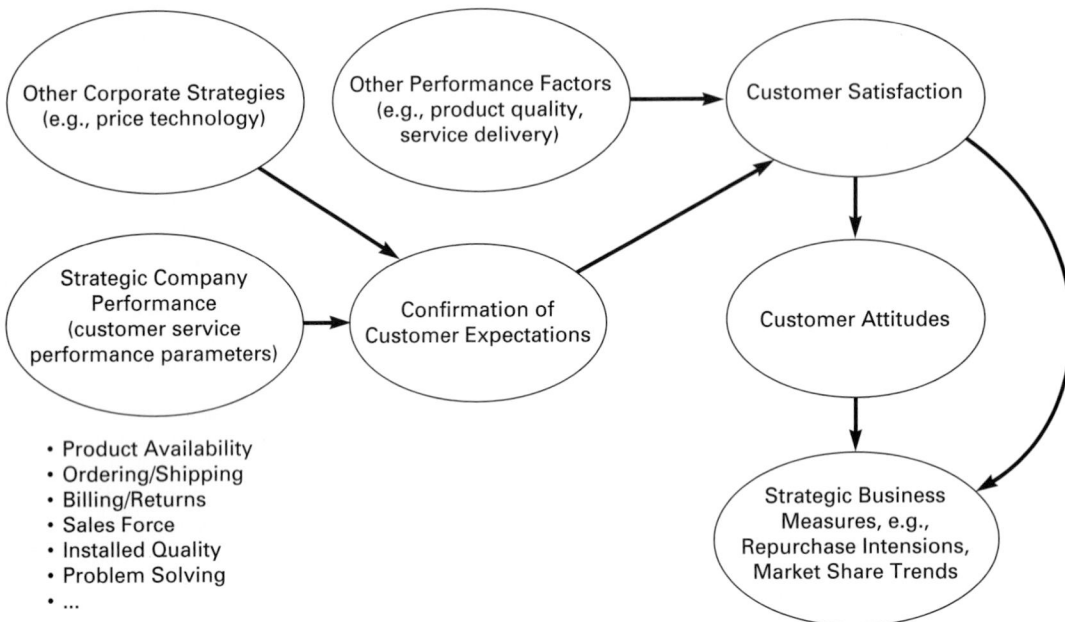

FIGURE 25.1   Combined Lisrel-Dresner model of customer service.

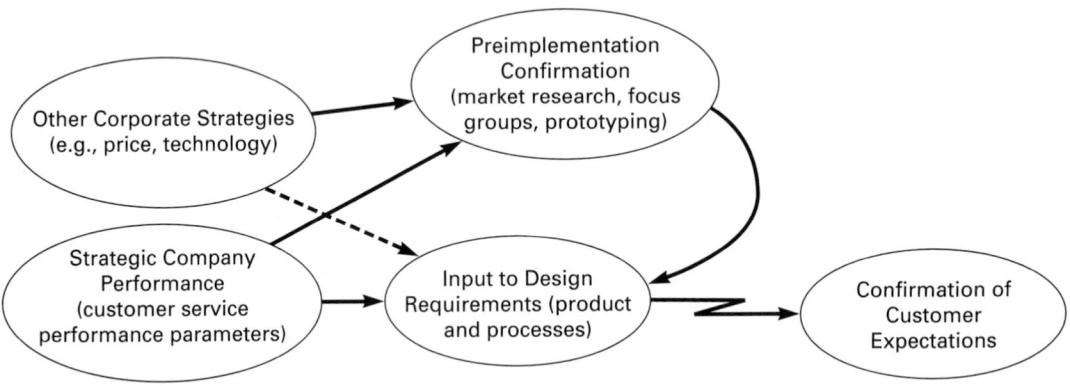

FIGURE 25.2   Expanded Lisrel model of customer service.

STRATEGIC INTENT

AT&T Universal Card Services.   The Universal Card Services business unit of AT&T provides an impressive example. From the time the business was first created in March 1990, it took only 30 months to become the second largest credit card issuer in the United States. How did AT&T differentiate what had been a commodity product and service? What prompted AT&T at the outset to rush in and try to establish a winning credit card business? AT&T was looking for ways to reinforce brand loyalty for long-distance calling services. A credit card that is also a calling card seemed like a natural marriage. However, there were hurdles to overcome. Customers were not satisfied with the credit cards that they had. Card users did not like the high fees, and they did not generally like the service that they received. They complained angrily about surly service representatives who took weeks to process changes and who embroiled anxious customers who had lost their cards in tangles of paperwork and phone calls. AT&T asserted that it could do better. Using the service blueprinting process, described below, AT&T chose a differentiating capability that had never been provided to credit card owners in the past: world-class customer service. To differentiate the customer service significantly, several important moments of truth in the customer service blueprints were selected. Market research studies showed clearly that one moment of truth would not have been enough.

The blueprinting process is illustrated and summarized in Figure 25.3. The blueprint is usually recorded as a flowchart of the particular process under study. The flowchart is segmented to clearly illustrate the lines of interaction between the service and the customer and the so-called line of invisibility. The line of invisibility is the boundary between the directly customer-facing subprocesses and the "back-office" subprocesses that support the customer-facing elements. Using the blueprint diagram, it becomes possible to answer the questions summarized in the bullet list in Figure 25.3. These answers provide the basis for selecting key moments of truth and for making other implementation decisions.

The service blueprint, which was implemented as a process, included fast, accurate, supportive, and helpful customer service representatives. The service representatives are empowered and rewarded for helping customers. First, a rigorous *daily* customer satisfaction measurement process was put in

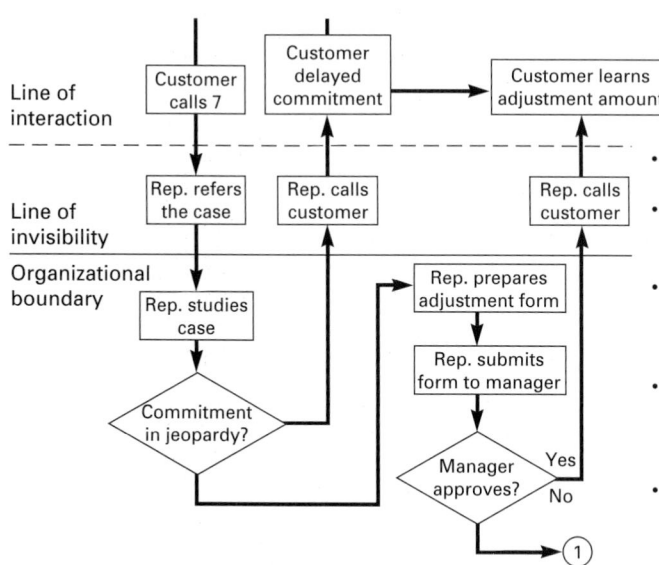

FIGURE 25.3   Example of the customer service blueprint (see Section 3).

	Quarter or Period	Number of quality days as percent of total	Bonus as percent of salary
Associate Quality Days and Bonus	1990-4th quarter	76.1%	6.4%
	1991-1st quarter	87.8%	11.4%
	1991-2nd quarter	92.3%	9.9%
	1991-3rd quarter	96.7%	12.0%
	1991-4th quarter	95.7%	11.6%
	1992-1st quarter	60.0%	10.6%
	1992-2nd quarter	75.8%	7.5%
	1992-3rd quarter	76.1%	7.9%
	1992-4th quarter	95.7%	10.8%
	1993-1st quarter	84.4%	9.4%
Management Quality Days and Bonus	1991	87.9%	5.6%
	1992	66.1%	4.7%
	1993-1st quarter	76.6%	5.6%

FIGURE 25.4   UCS quality days: performance and bonuses. [*From AT&T Operations (1998).*]

place, coupled with a *daily* bonus plan. Figure 25.4 offers an example of bonuses actually received during a period of time. On those days when the service representatives do not receive their bonus, no manager receives a bonus. Second, the customer service processes are supported by leading-edge technology. This gives the service representatives easy and rapid access to all the information that could possibly be needed in support of the customer. Third, the customers are given differentiating value in the form of lower interest rates than are provided by any other vendor. And fourth, the company undertook the role of the agent of the customer in relationships with credit bureaus and other users of credit information.

In recent years, AT&T Universal Card Services has initiated more than 300,000 new accounts per month. This growth stems in part from the sales story that is part of the front-end, customer relationship-building process. It also comes from finely tuned customer service operations that also were part of the strategic design of the business and which have been improved and enhanced over time. This customer service strategy was part of AT&T's Baldrige Award–winning approach.

New England Telephone.   The Interexchange Customer Service Center (ICSC) of New England Telephone Company knew in 1988 that it had to make changes. ICSC is the interface between the local telephone company and the interexchange carriers. Its customers are AT&T, MCI, Sprint, mobile, radio, and television common carriers. The customers did not like the job that the center was doing, nor did management or the employees (Clarke 1992). Morale was low. The organization conducted a customer assessment and created a service blueprint. It was discovered that customers thought that New England Telephone was reactive rather than proactive. It found the company unable to relate individual jobs to customer needs. The service blueprint revealed that the organization design was that of a job shop. Employees each took a piece of paper in a hand-off and processed it. Everyone did the job according to the job description, but accountability for the results occurred only at executive levels. It was not possible to relate individual tasks to customer needs, nor to coordinate and track customer service requests and results. ICSC developed a strategic vision, which was to "…outperform competitors by providing our customers the highest level of service quality. We will be the 'best of the best' within our industry.…" The solution was to reengineer the operation, establishing a customer-focused team structure responsible for all processing and service support for a customer segment. The changes affected not only ICSC but also all the company organizations that provide technical and operational services, such as network and accounting.

Hewlett-Packard Company.   Hewlett-Packard summarizes its overall strategy for quality with two acronyms (Spechler 1988). The first acronym is *FURPS*. The letters in the acronym have the following definitions:

- F = functionality. The feature set, capabilities, comparability, and security.
- U = usability. The human factors, consistency and documentation of the product.
- R = reliability. The frequency and severity of failures. The predictability and accuracy of the product.
- P = performance. The speed and efficiency of the product, as well as resource consumption.
- S = supportability. Maintainability and servicability of the product, along with its ability to be installed.

The second category of quality attributes uses the acronym *AART*. AART refers to the relationship with customers. The letters in the acronym have the following meanings:

- A = anticipation. The ability to identify, understand, and help solve customer needs before they become problems.
- A = availability. The degree to which products and services provide for uninterrupted use at full functionality.
- R = responsiveness. The ability to provide timely, accurate, and complete information and/or solutions to customer-initiated requests for help.
- T = transitions. The ease of initial startup and ongoing changes as individual products and services evolve and conform to new needs and technologies.

Many of these attributes are strongly related to customer service. Two areas of implementation where Hewlett-Packard sought to differentiate itself with exceptional customer service are the field customer engineers (CEs) and the response center engineers. The field customer engineers, who provide on-site hardware maintenance, have been equipped with hand-held portable terminals. Through these terminals, they receive notification of service calls they need to fulfill. They use the terminals to upload symptoms and to download diagnoses. At the completion of the call, the CE enters repair details, such as revised failure symptoms, parts used, and labor effort. This information enters a relational database, where it is used for parts inventory management, staff capacity management, determining training needs, and to produce the quality metrics that are part of the strategic goals and objectives.

The methodology used by Hewlett-Packard to develop, deploy, and track strategic plans, goals, and objectives for these attributes is called the *hoshin kanri planning process. Hoshin kanri* was introduced within Hewlett-Packard in the Yokogawa Hewlett-Packard subsidiary in Japan. *Hoshin* has been loosely translated into English and is generally referred to in Western countries as *policy deployment* (see Sections 3 and 13).

Motorola. Motorola's strategic customer service thrust was prompted by findings from the CEO's personal visits with key customers (Motorola 1989). The CEO visited the top 10 corporate customers during a 5-month interval in 1988 and then presented the findings to the Motorola Corporate Quality Council. Quality was the battleground of the 1980s, but the council concluded that service was evolving as the next competitive challenge. In response, one business sector established a service strategy task force to focus on the issues of customer service and to develop recommendations. The task force comprised eight senior managers, including seven officers. The eighth member was the chairperson, who was appointed to facilitate and coordinate the task. They left their current job responsibilities and moved to an off-site location for 3 months. The task force began its work with customer and associate assessments using formal survey instruments. They surveyed 54 customers worldwide and 220 Motorola business-sector employees. The highlight of the customer survey was that responsiveness emerged as the competitive issue; customers took for granted perfect quality and on-time delivery. By contrast, the employee assessment revealed that employees believed on-time delivery to be the most important issue (and, at that time, the area in which their performance was weakest). Further, the employees believed that service and responsiveness were areas of company strength and needed only improvements in teamwork, communications, and consistent goals to support them.

The task force grouped the issues revealed by the data into six categories. Then, using Pareto analysis, the task force identified issues that warranted action. (See Section 5, The Quality Improvement Process, under Quality Improvement Goals in the Business Plan.) The task force came up with 52 action items, of which the top 10 were considered urgent. One of the six categories was on-time delivery. Examples of key action items in this area included the following three statements (Motorola 1989):

1. One hundred percent on-time delivery will be the acceptable level of performance on which compensation will be based. (This replaced an incremental improvement goal.)
2. There will be disciplined use of the on-time delivery index. This means no recutting of orders, no early shipments unless approved by the customer, and no partial shipments unless approved by the customer.
3. Training in and application of the methods of short-cycle-time production and order realization will be required for all organizations, with applied measurements and goals.

Another catagory of issue from among the top six was responsiveness. Here the task force decided on the following key action items (Motorola 1989):

1. A schedule of formal visits with all key customers
2. A no-questions-asked return policy
3. Development of systems and procedures to take an order anywhere in the world and ship it anywhere in the world
4. Elimination of the "will advise," "will schedule," and "Motorola to advise" from the backlog lexicon

The consequences of these actions and comparable action on the other issues were reflected by a number of Motorola quality awards and in independent third-party customer satisfaction assessments that revealed improved perceived performance.

DESIGN

Many of the design approaches that apply to customer service processes apply to the design of any business process and are well covered in Section 6, Process Quality Management, and in the reengineering literature, the human factors literature, the organization design literature, and elsewhere. Pyzdek integrates the systems engineering considerations, the organization, and the human factors considerations into an integrated design paradigm that he calls *service systems engineering.* The term *service systems engineering* actually has been used for decades in industries such as the transportation industry and the telecommunications industry to size traffic-carrying systems and facilities. Techniques employing approaches from the fields of applied probability and statistics are found in the systems engineering and operations research literature. What distinguishes Pyzdek's approach is the integration of these approaches with the very important behavioral and organization considerations (Pyzdek 1994).

Pyzdek defines a *service delivery system* as a systematic arrangement of human resources and technology designed to create successful service encounters. A *service encounter* is the direct interaction between a retail or service firm and a client. As described in the preceding section, success is measured in terms of a set of business performance criteria.

Systems Engineering.   Table 25.2 identifies the major inputs, activities, and outputs of the service delivery systems engineering process. Managing these steps for quality involves knowing how the precise and measurable quality objectives for the final customer service translates into specific, measurable performance standards for the inputs, outputs, and activities.

TABLE 25.2 Major Inputs, Activities, and Outputs of Service Delivery Systems

Inputs	Activities	Outputs
Information from business case, including • Market segments • General service definition • Demand or volume forecast • Quality, reliability, other performance goals	Analyze/clarify user needs and service objectives	Create service requirements document, including • Functions • Performance • Quality and reliability standards
Service objectives and customer needs	Create a functional framework	High-level service architecture
Service constraints	Create functional and cost boundaries	High-level cost objectives
Technology roadmap	Analyze available technologies	High-level technology architecture
Feasibility analyses and exploratory service development results	Plan service development	Service design and implementation documentation

Organizational Structure.   The planning and management controls associated with the specific customer services are intended to guide and support managers and associates in all parts of the organization in their definition of meaningful tasks and in the specification of procedures to monitor their effective completion. From this perspective, it is easy to recognize the central importance of organizational structure. Responsibility has to do with the nature of the tasks entrusted to each individual. The design process is a primary vehicle to identify, in a coordinated manner, the major tasks faced by the enterprise and to organize those tasks in the most effective way (Hax and Maluf 1984).

The organizational structure to perform the customer service functions may be defined as "the relatively enduring allocation of work roles and administrative mechanisms that create a pattern of interrelated work activities and allows the organization to conduct, coordinate and control its work activities" (Jackson and Morgan 1978). There are three accepted basic types of organization for managing customer service work and two newer, emerging approaches. The accepted organizational types are functional, divisional, and matrix. They are important design baselines because these organizational structures have been tested and studied extensively, and their advantages and disadvantages are well known. The matrix structure is a hybrid combination of the functional and divisional archetypes. The newer, emerging organizational designs are process and network organizations.

The design attributes associated with division managers, functional managers, process managers, and network managers of customer services are summarized in Table 25.3. There is emerging evidence that divisional and functional organizations may not have the flexibility to adapt to rapidly changing marketplace or technological changes. This is especially true of larger customer service organizations associated with large companies. For this reason, there is a trend in large companies to outsource the customer service function, a topic that is addressed below.

Human Resources.   Behavioral science helps us to understand human capabilities, capacities, and needs for the design of customer service systems. This understanding is becoming increasingly important for several reasons. First, modern technology, especially information technology, is providing the foundation for customer service systems. For these systems to be successful, it is essential to design effective human-machine coupling. Second, modern technology is increasingly embodied in products and services, making new and unusual demands on customers. For these products and services to be successful, user-friendly human-machine interfaces again are essential. Third, to the extent that the user-friendliness of products and services is less than perfect, customers will make use of customer services to help them through the rough spots.

TABLE 25.3   Design Attributes of Various Roles in a Customer Service Organization

	Division manager	Function manager	Process manager	Network leader
Strategic orientation	Entrepreneurial	Professional	Cross-functional	Dynamic
Focus	Customer	Internal	Customer	Variable
Objectives	Adaptability	Efficiency	Effectiveness	Adaptability, speed
Operational responsibility	Cross-functional	Narrow, parochial	Broad, pan-organizational	Flexible
Authority	Less than responsibility	Equal to responsibility	Equal to responsibility	Ad hoc, based on leadership
Interdependence	May be high	Usually high	High	Very high
Personal style	Initiator	Reactor	Active	Proactive
Ambiguity of task	Moderate	Low	Variable	Can be high

Sources: The first two columns are adapted from the work of Financial Executive Research Foundation, Morristown, N.J. The last two columns represent the work of the author.

Because new customer service systems involve more interactions between users and supporting systems than ever before, making these systems easy to use and error-free is a major goal of their design. Properly designed information input arrangements, easy-to-read screens, announcements, timings, and instructions increase user acceptance, minimize errors, and promote effective use. Training programs for customer service representatives and easy-to-understand and easy-to-use instructions for customers are very important.

The design of new customer service systems can be evaluated for user-friendliness by coordinated studies that include (Bell 1983)

- Analysis of present customer service systems
- Benchmarking of best-in-class customer service systems
- User interviews
- Laboratory studies of protocols for user-system interactions
- Field tests of customer services
- In-service follow-up studies for continuous improvement

Investigation of best-in-class systems is particularly useful. In the area of customer service, certain companies are well known for the efficacy of the approaches, e.g., L.L. Bean for order entry, Fidelity Investments for account inquiries, and GE for broad customer support.

It is clear that the customer service representatives as well as all other employees whose work affects the customers' perception of service are critical elements of the customer service design. The design of the roles of individuals in the enterprise, including those whose jobs involve customer service, must be within the frame of a human resource architecture for the enterprise. An example of such an architecture is that in Figure 25.5. In order for the customer service design to be successful, the design must address more than the functional processes and systems associated with the service. Each of the people considerations identified in Figure 25.5 also must be included, either directly or by reference to the overall human resource architecture design documents.

Electronic Data Systems (EDS).   In 1992, EDS put in place a matrix organizational structure radically different from those in most successful U.S. companies (EDS Homepage Internet www). The matrix structure is a hybrid that has elements of the divisional and network designs. The structure involves 35 to 40 strategic business units (SBUs). The number changes frequently, as it would in a network design, due to business demands. There are approximately the same number of strategic support centers (SSCs), which support SBU requirements. The organizational relationships between the SBUs and the SSCs form a matrix. Think of the various SSCs as the entries in the columns of the matrix and the SBUs as the entries in the rows of the matrix. Stanford University's business school has included in its core curriculum a case study of this organization design. In this design, key customer service teams reside organizationally in the SBU, which is the profit center, while commodity production services (e.g., EDS information-processing centers) are outside the profit centers, in the support organizations. The SSCs have the people with the subject matter expertise to perform key functions for the SBUs, and the SBU managers have the budget to pay them for their support services. EDS's approach to human resources is based on customer focus and rewards. These are the key elements of the HR architecture.

- *Customer focus:*   The company keeps its people focused on what is important: quality delivery. The company practices a severe approach to employee alignment—"If you can't change the people, change the people!"
- *Reward system:*   Since 1993, EDS has changed its reward system to emphasize team awards. This appears to be comparable to the approach practiced by AT&T's Universal Card Services, described earlier.

STRUCTURE

Presales customer service systems support such services as sales information, ordering, and market research (see Table 25.1). These are both outbound and inbound services, since they involve service representative-originated calls to customers and customer–originated calls for service. In the diverse universe of differing products and services that are addressed by this section, the term *call* should be understood broadly as referring to both personal calls, such as calls on a customer by a salesperson, and remote calls, such as telephone calls by a customer to place an order or to request sales information. Postsales customer service systems support such services as complaint handling and problem solving; again, refer to Table 25.1. These systems are usually primarily inbound services, involving calls from the customer or from installers or maintenance personnel to the support system.

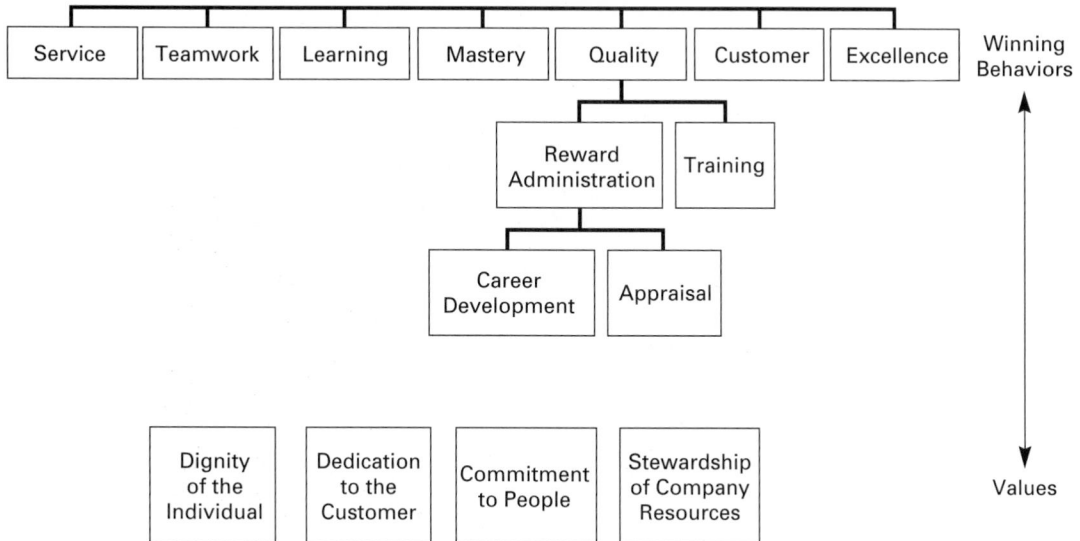

FIGURE 25.5   Human resource architecture.

Examples of configurations of such systems are included later in this section. A wide variety of tools is in use to help the customer service personnel and to help the customers. Examples of tools, organized by category, are shown in Table 25.4.

The configuration of customer service tools to execute the customer service process can take a variety of forms. Two of these are illustrated in Figure 25.6. The configuration in Figure 25.6*a* is a traditional one, where each function is supported by its own work center. The configuration in Figure 25.6*b* is a more advanced arrangement, one that typically emerges from a TQM or a re-engineering initiative.

TABLE 25.4   Some Important Tools in Customer Service Systems

Access tools	Process tools	Enabling tools
Automatic call distributors	Customer information resource databases	Intelligent agent software within the product
Voice-response units	Remote access tools	On-line support services to log requests, check status, etc.
On-line bulletin boards	Outsourcing vendors	User access to knowledge bases
Fax servers		Integrated voice recognition/response

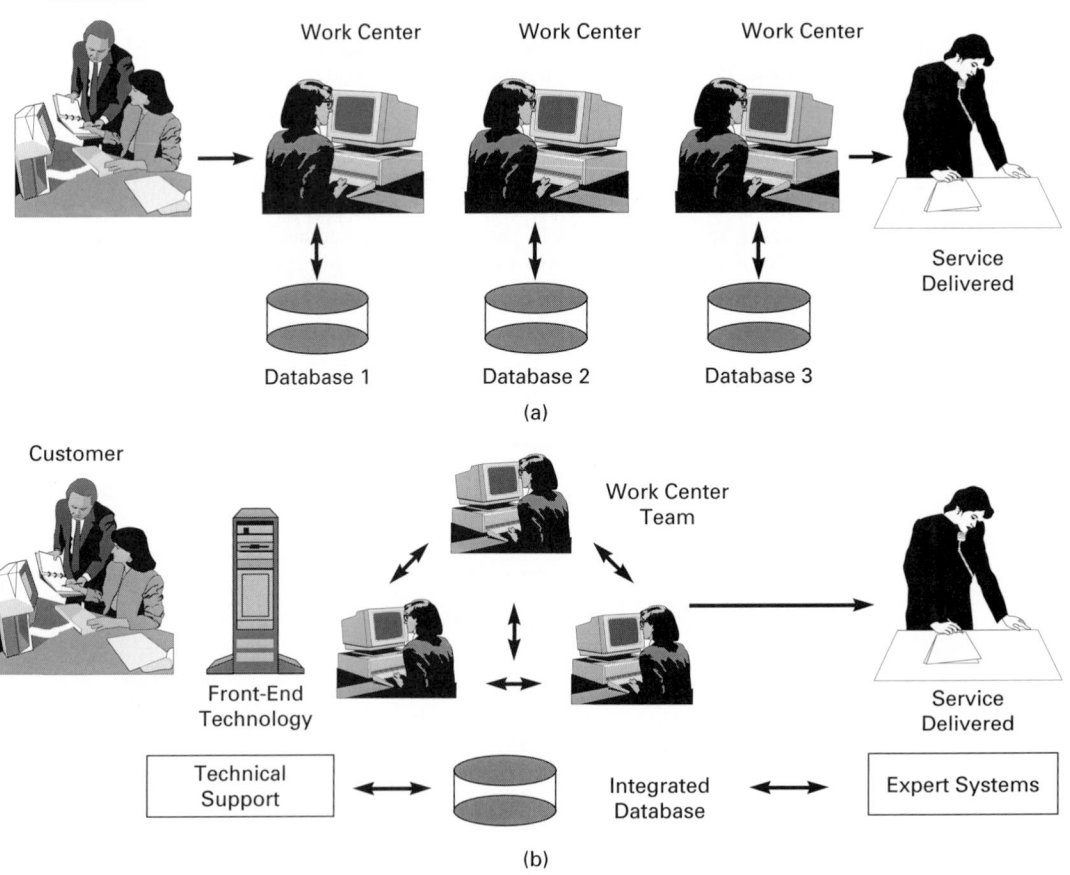

FIGURE 25.6   (*a*) A traditional organization with functional work centers and separate databases. (*b*) A more advanced organization with team work center and integrated database.

The Work Center team is a colocated group consisting of the major work functions required for delivery of customer service from the initiation of a request for service through the delivery of the service. The attributes of the design of the Work Center team for high-quality performance are

- The team includes all the necessary work functions.

- The team bears complete responsibility for satisfying the customer. Performance of the team is measured accordingly.

- The team uses a set of work processes that are formulated in the design step that eliminate rework and review and that meet service delivery quality objectives.

- The team is provided with the technology and systems that enable its members to meet the objectives.

- The team is sized and trained according to the criteria included in the design process.

Structurally, we may think of these teams and associated systems and tools as defining an enterprise in terms of its processes, as shown in Figure 25.7. As we have seen, there are numerous ways in which these processes may be organized to work together.

Consider the customer service process for an enterprise that provides "help desk" or "customer care" types of service to customers who have problems with a purchased product. The product could be an unsatisfactory software package for a PC, or a malfunctioning dishwasher, or a shirt purchased by mail order that is the wrong size. Let's see what the customer service process of Figure 25.7 might look like, in terms of the tools described in Table 25.4, organized in configurations such as those in Figure 25.6. Such a configuration is illustrated in Figure 25.8.

Incoming calls encounter a voice response unit (VRU), which engages the caller in a dialogue to determine the nature and the priority of the call. Based on this information, the call is routed to the first available analyst. The analyst examines the symptoms provided by the call and either resolves the matter or directs the call to a subject matter expert at the help desk. If the problem is complex, the analyst may obtain technical assistance. If the problem is very complicated or unusual, it may require the support of the product vendor for resolution.

Customer Care Center Example.   Quality function deployment (QFD) is an effective tool to provide the coupling between the customer satisfiers and the design process. One of the methods used in QFD is the "house of quality." Figure 25.9 is an example of the house of quality developed by a re-engineering team for an enterprise that provides desktop computing services to business customers. The customer requirements data, in the language of the users, were gathered from customer interviews. The data were transferred to the house of quality, and design attributes and performance targets were determined by the re-engineering team. The customer needs were in the categories Knowledge and Responsiveness. Under the latter heading, the customers identified many subcategories: Promptness, Completion Estimates, Status, Access, and Communications. For each performance category, the customers and the re-engineering team agreed on one or more performance measures that appropriately capture the customer need. Then the re-engineering team developed a set of design attributes and priorities that formed the house of quality. Notice that the attributes that comprise the rows in Figure 25.9 fall into three design categories: systems, human resources, and organizational structure.

The design attributes of Figure 25.9 then were translated by the re-engineering team into a design concept, portrayed in Figure 25.10. The specific design concept looks similar to the hypothetical design of Figure 25.8. The key needs that were identified in the QFD exercise, which are emphasized in the design concept, are

- Increased knowledge on the part of the Customer Care Center associates

- Systems and resources that provide a high degree of responsiveness

- Systems that provide increased accessibility to the Customer Care Center

- Accurate and complete responses to customer questions

Corporate Management Process
Product Management Process
Sales and Marketing Process
Product Development Process
Order Management Process
Customer Service Process

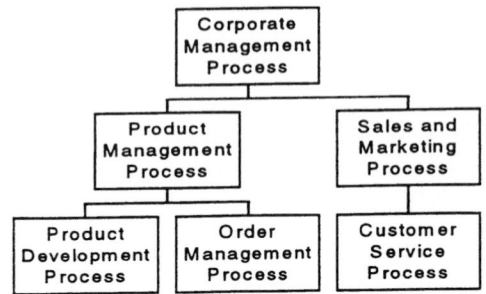

FIGURE 25.7 An enterprise described in terms of its major processes.

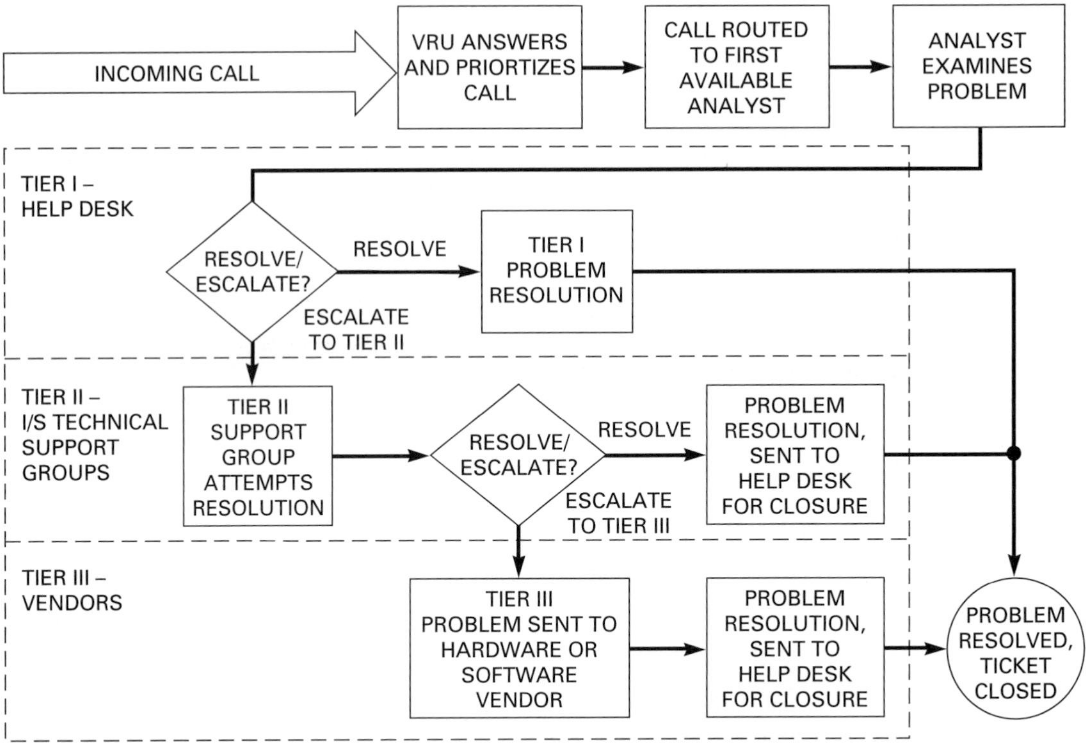

INCOMING CALL
VRU ANSWERS AND PRIORTIZES CALL
CALL ROUTED TO FIRST AVAILABLE ANALYST
ANALYST EXAMINES PROBLEM
TIER I – HELP DESK
RESOLVE/ ESCALATE?
RESOLVE
TIER I PROBLEM RESOLUTION
ESCALATE TO TIER II
TIER II – I/S TECHNICAL SUPPORT GROUPS
TIER II SUPPORT GROUP ATTEMPTS RESOLUTION
RESOLVE/ ESCALATE?
RESOLVE
PROBLEM RESOLUTION, SENT TO HELP DESK FOR CLOSURE
ESCALATE TO TIER III
TIER III – VENDORS
TIER III PROBLEM SENT TO HARDWARE OR SOFTWARE VENDOR
PROBLEM RESOLUTION, SENT TO HELP DESK FOR CLOSURE
PROBLEM RESOLVED, TICKET CLOSED

FIGURE 25.8 The "customer care" or "help desk" organization. [*From AT&T Operations (1998).*]

The team also prepared performance objectives for the design concept, which could be used to test the design for compliance with customer needs. These are examples of the objectives for the Call Center:

- The Call Center will be capable of resolving 60 percent of customer service requests without the need for second- or third-tier support.
- An average call to the Call Center will take about 10 minutes to resolve.

WHATs vs. HOWs Legend

Strong	○	9
Moderate	●	3
Weak	△	1

Columns (HOWs):

- KNOWLEDGE
  - (1) % basic commands and locks answered on initial call
  - (2) % printer problems resolved on initial call
  - (3) % e-mail questions answered on initial call
  - (4) % networking questions answered on initial call
  - (5) % backup/restore problems completed on initial call
  - (6) % desktop software questions answered on initial call
- RESPONSIVENESS / PROMPTNESS
  - (7) time for response
- COMPLETION ESTIMATES
  - (8) % of applicable problems for which estimate is provided
- STATUSING
  - (9) % customers given intermediate status
  - (10) % customers notified after problem is resolved (if applicable)
- ACCESS
  - (11) % of calls that receive live person (8 AM-8 PM)
- COMMUNICATION DURING OUTAGE
  - (12) interval between updates on system status

WHATs	(1)	(2)	(3)	(4)	(5)	(6)	(7)	(8)	(9)	(10)	(11)	(12)
NEED KNOWLEDGEABLE PERSONNEL (28)												
knowledgeable about editors and other tools (4)				●								
knowledgeable about computing platforms (1)	●			●								
knowledgeable about e-mail (3)			●									
knowledgeable about IMS interface (1)				●								
knowledgeable about PC (5)	○	○	●		●							
knowledgeable about equipment (2)					●							
knowledgeable about UNISON (2)		●										
knowledgeable about access methods (1)			○	●								
NEED RESPONSIVE SERVICES (26)												
need quick response (7)							●					
need estimate of time to fix problem (3)								●				
need intermediate status report (4)									●	●		
customer needs to know or establish priority (5)									●			
need to know that job is complete (2)												
need to know what was done to remove problem (2)										●		
NEED QUICK ACCESS TO HELP DESK (24)												
need fast call answers (3)							●					
need short or nonexistent queue (11)											●	
need to talk to live person (4)											●	
need line coverage early/late (3)												
need to lnow length of queue (1)											●	
NEED COMMUNICATION ABOUT OUTAGES (12)												●
NEED SINGLE POINT OF CONTACT (10)	○	○	○	○	○	○						
NEED SHORT CYCLE TIME (6)												
short cycle time for reboot (1)		●										
short cycle time for restore (1)				●								
NEED ERROR-FREE TROUBLE TICKET CREATION (6)	△	△	△	△	△	△						
Design targets	>75%	>90%	>90%	>80%	>95%	>80%	Immediate	100	100	100	>90%	<30 min.

FIGURE 25.9 The house of quality. [*From AT&T Operations (1998).*]

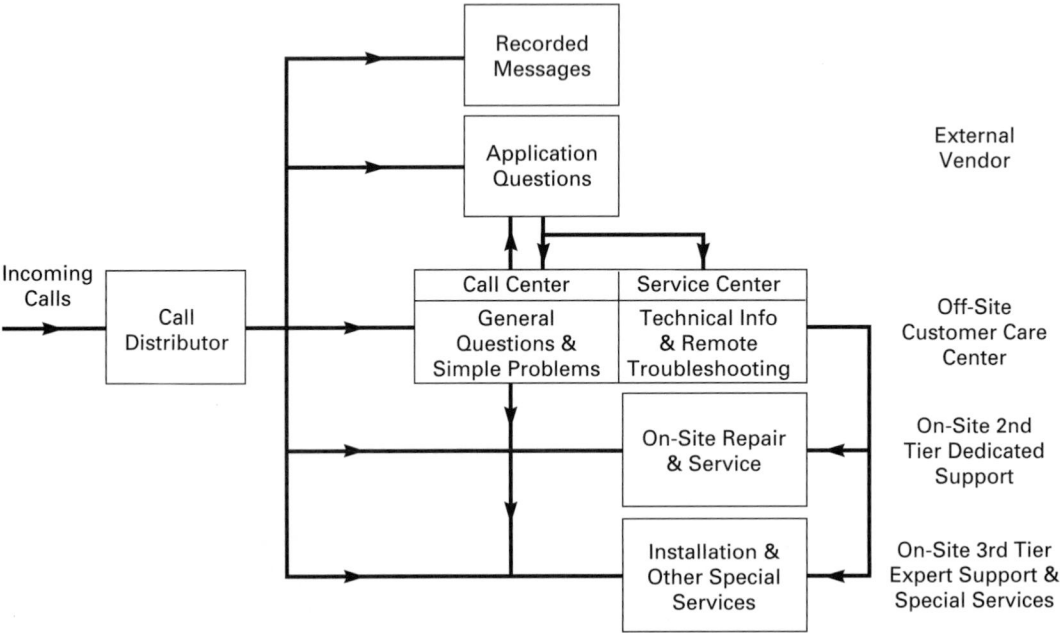

FIGURE 25.10   Process-design concept model. [*From AT&T Operations (1998).*]

- The Call Center will function as a learning organization.
- The Call Center will comprise both permanently assigned employees and employees assigned on a rotating basis from the functional service organizations supported by the Call Center.
- Permanent employees will provide general customer service and also will be responsible for quality and training functions.
- Employees on rotating assignment will provide specialized skills to the Call Center but also will support general customer service to gain familiarity with the "voice of the customer."
- All Call Center agents will be trained in customer contact skills and in the skills required for solution of common problems.
- All solutions to customer problems will be entered into the system knowledge database.
- All transactions with customers will be recorded for analysis to determine root causes and to support quality improvement.

OPERATIONS

The last important area where quality must be considered is the operational area. This is where results are achieved, measured, and improved. The ongoing performance of operating customer service processes includes such considerations as human performance, process control, process capability, and process improvement. Performance metrics are important here. These considerations for customer service share quality methods in common with most other business functions. Therefore, interested readers should refer to earlier sections to understand the generic methods. In particular, the following sections are important: Section 4, The Quality Control Process; Section 5, The

Quality Improvement Process; Section 6, Process Management; Section 9, Measurement, Information, and Decision Making; Section 15, Human Resources and Quality; and Section 16, Training for Quality. This section will focus mainly on examples of the operations of well-designed, well-structured customer service capabilities that employ the methods described in these earlier sections.

PROCESSES AND SYSTEMS

The first requirement for successful management, control, and improvement of customer service processes is to have well-defined, well-understood processes. Often, steps to manage and improve quality are undertaken without process definitions because no one has taken the time to think through and document, especially in process diagrams, the way work is done. Ideally, the initial versions of such documents are produced as part of the service design. An example of process diagram for the process described in Figure 25.10 is shown in Figure 25.11.

The process diagram uses a particular flowcharting convention (*AT&T Operations Engineering Workbook*). There are many available process flowcharting conventions and tools. The particular conventions used here are summarized in Figure 25.12.

A detailed understanding of the conventions used in Figure 25.11 is not important here; the words and arrows convey enough meaning for our purposes. It is important to remember that it is not possible to control and improve customer service operations without process descriptions and associated metrics. Attempts at control and improvement without process descriptions and metrics are only guesswork. And metrics without defined processes only make for busywork, with no real value. Worse yet, lacking processes, it is tempting to measure the performance of the people who provide customer service. The consequences of so doing are severe. The people will do their work to make the metrics look good rather than providing customer service that maximizes customer satisfaction. And the people will lobby strongly for weak metrics that may very well be irrelevant to customer needs or business requirements.

In addition to process control and improvement, it is important to recognize that as an enterprise grows or as customer services are extended and augmented, process capability must be monitored. This involves continuous oversight of customer service processes in search of stress cracks that may be caused by saturated processes, by changes in customer needs, or by other considerations not congruent with the design intent and implementation criteria. Process capability studies are addressed in Section 4. When the demands of the process exceed the process capability, the process must be reengineered. Process reengineering also may yield improved efficiency or effectiveness, especially when it incorporates new or improved technology.

OUTSOURCING

In many enterprises, business process re-engineering (BPR) initiatives introduce issues relating to legacy migration and outsourcing. *Legacy migration* is a term used to connote the migration of customer service systems from their traditional (legacy) processes and associated systems to modern, leading-edge processes and systems. Traditional approaches have legacy hardware, software, training, and cultures that must be overcome to replace them effectively with modern approaches that provide competitive advantages. The corollary issue is whether the customer service functions should be done at all inside the enterprise, or whether the enterprise should *outsource* the functions to companies that specialize in performing them. Companies that provide customer service functions on an outsourced basis are prominent in the customer service arena, performing many of the associated functions at superior levels. The conventional wisdom on this question is that if the customer service functions are strategic to the enterprise's business strategy, as defined earlier, under Strategic Intent, then the functions should be performed internally. This

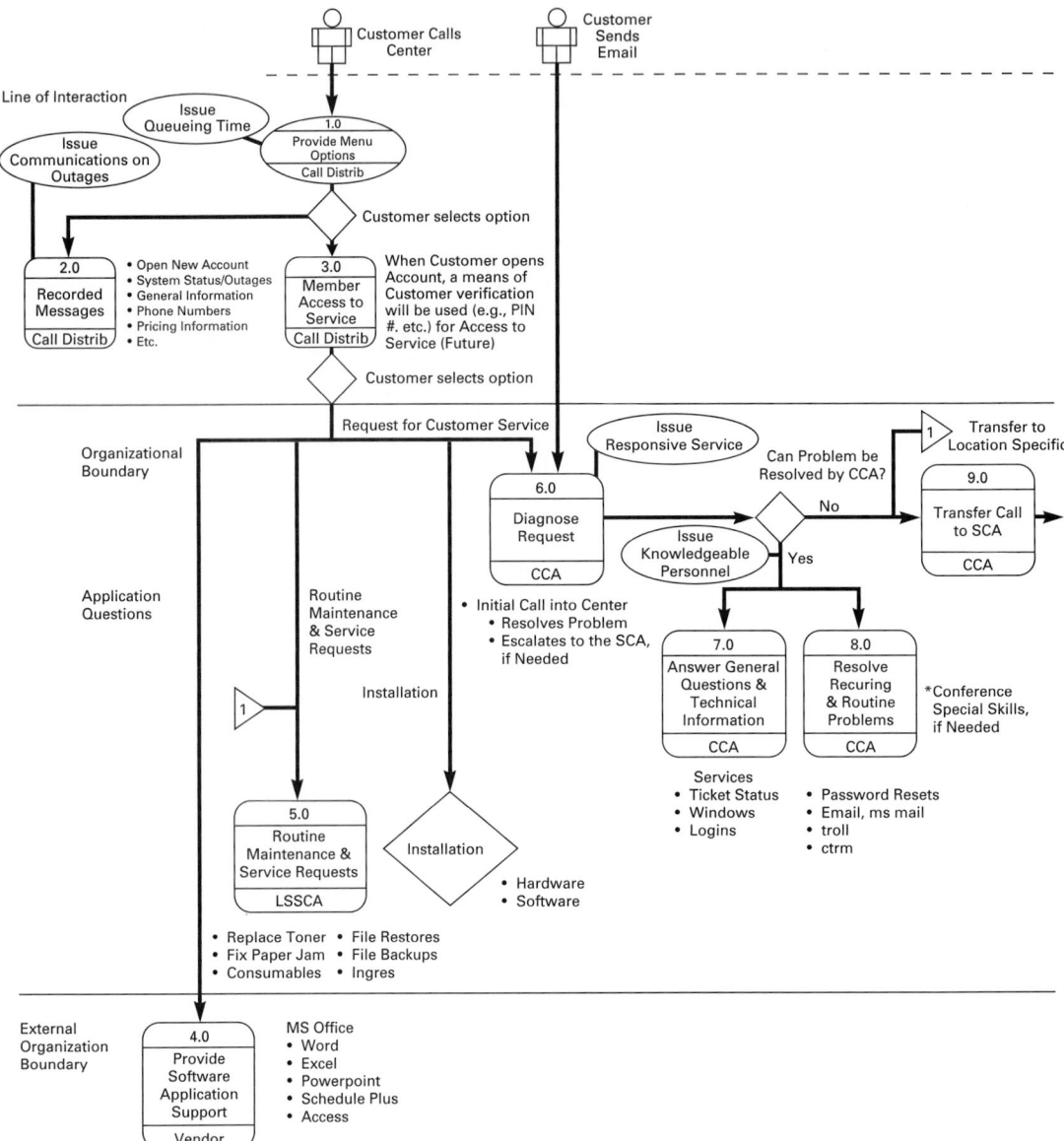

FIGURE 25.11   The customer care center: high-level overview. [*From AT&T Operations (1998).*]

is easy to understand, since the functions are part of a strategy to be better than the best competitor in some critical dimension of service. On the other hand, if the functions are not strategic, then the question of whether they can be performed at levels of quality and at costs that are comparable with those of the *outside* companies is a valid one. Customer service functions of many quality-award-winning companies are provided by outside companies unbeknownst to the customers.

MEASUREMENT AND METRICS

Common to all approaches designed to achieve excellence in customer service operations is instrumentation for the collection and analysis of information. We must begin with some basic definitions.

- *Metric:* Unit of measurement. Common metrics in everyday life are miles (or kilometers) to measure distance and pounds (or kilograms) to measure weight.

- *Statistical metric:* The result of a calculation procedure. For instance, the most commonly used statistical metric is the arithmetic average of a set of numerical values. One could easily calculate the average weight of people riding in the elevator with you by asking them their individual weights, adding the numbers, together with your weight, and dividing the sum by the total number of people in the elevator. Of course, the people in the elevator may not know their weights exactly, so the calculated average would be an estimate.

- *Data:* A set of individual measurements. The data are usually the input to a statistical metric. For instance, the weights of the individuals in the elevator comprise a set of data.

- *Analysis:* Structured ways of comparing data and statistical metrics. For instance, Tom weighs more than the average of the people in the elevator.

With these definitions in mind, we are ready to view Figure 25.13, a process diagram for defining, collecting, and analyzing metrics (I am indebted to the AT&T OE Process Management Consultants for most of this material on this topic.). In the following explanations, it is assumed that

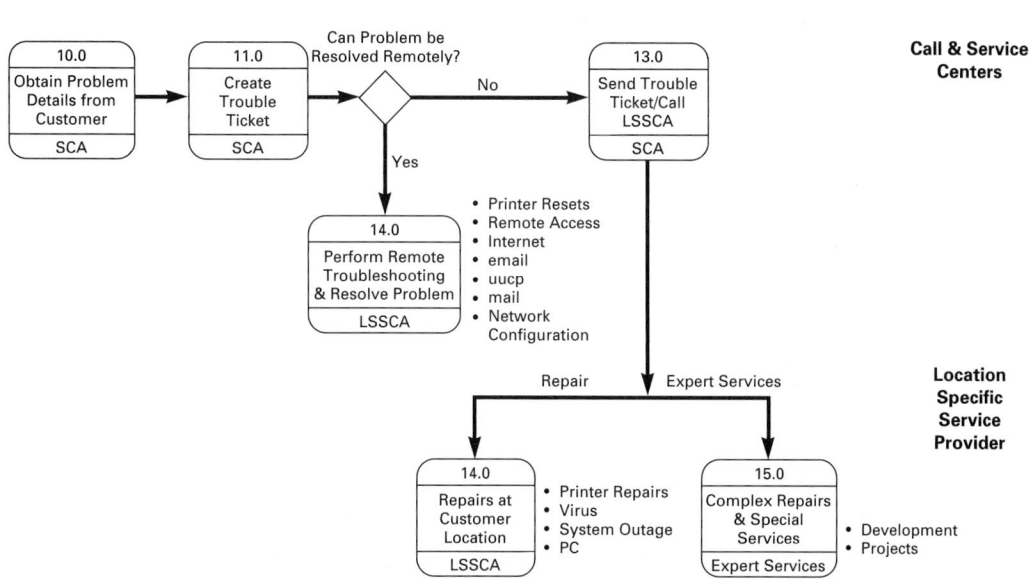

FIGURE 25.11 *(Continued)*

SYMBOLS

OEW flow-charting symbols are designed to capture a maximum amount of information and still maintain an uncluttered appearance for the resulting flowcharts. They also ensure a consistency and uniformity between process descriptions.

Processes are represented by large diamonds with the name of the process written inside.

Processes external to the process to be improved are depicted as diamonds with "cut corners". The name of the external process is written inside the diamond.

The names of organizations external to the process to be improved (e.g., Customer or Vendor), are shown as a aquare with "cut corners".

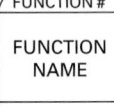

Functions performed by organizations or work centers are symbolized by a vertical rectangular, with rounded corners. The rectangle is divided into three areas. The top area contains the acronym and function number. The middle area contains the name given to the function. The bottom area identifies the organization performing the function.

A function which has been improved will have a asterisk (*) before the function number.

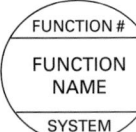

Functions performed solely by computer systems are depicted by a circle divided into three areas, The top two areas are as described above; the bottom area contains the name of the computer system.

INFORMATION NAME

MATERIAL NAME

There are two types of interfaces (inputs/outputs). A solid line is used for information. A dashed line is used for material. The name of the information or material is written on the line.

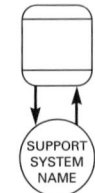

Support systems are depicted by a circle containing the name of the information system. This symbol is used when the system is supporting a function. Two parallel, vertical lines show input and output flowing between the system and function.

FIGURE 25.12   Flowcharting: symbols and conventions. [*From AT&T Operations (1998).*]

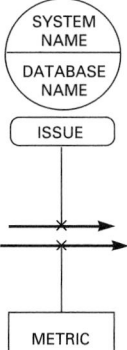

A small circle divided into two sections depicts a specific database that is being accessed by a system. The top half contains the name of the computer support system, and the bottom half contains the name of the database.

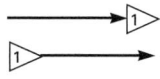

Any issue identified within a process is shown in an oval above or below the point at which it occurs. The point is marked by an X and is connected to the oval by a line.

Any metric associated with part of a process is shown in a rectangle above or below the point at which the metric is measured. The measured point is marked by an X and is connected to the rectangle by a line.

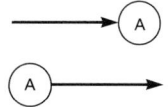

Feedback symbols are used to prevent flow-charts from becoming too cluttered. Feedbacks are always labelled numerically. Identical labels must be given to both ends of the feedback flow.

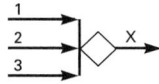

Connector symbols are used when the splitting of a flow must be divided between pages. The symbol is a small circle with a capital letter. The identical letter is given to both sides of the division.

DECISION POINTS: Decision points are shown as small diamonds. These are points at which choices are made based on pre-specified conditions called decision rules. They are generally given as supporting function descriptions (see p. 8).

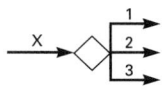

In this case, a decision rule in an upstream sending function, based on outputs 1, 2 and 3 dictates what occurs at X. For example, the decision rule may be: if 1 and 3 are both present, then send 2 to X, otherwise take no action.

In this case, a decision rule in an upstream sending function, based on the output x, dictates what occurs at 1, 2 and 3. For example, the decision rule may be, if X is greater than 10 send X through to 1, if it is between 5 and 10 send through to 2, otherwise send through to 3.

OE Flow-charting conventions assure that complex flow-charts are interpreted in only one way. They also help to ensure consistency and uniformity between projects.

DIRECTION: Process flow-charts show the time sequence of the functions performed. The progression is from left to right.

The following situation could arise when constructing process flow diagrams: Five functions; A, B, C, D, and E, are considered. Functions A and B receive materials from external suppliers and perform separate operations on them. Their products go to a "storage" function C where they are kept for some time before being sent to D for further processing. When the product of function D is ready, it goes to C for storage for additional time, before being sent to E for final test and shipment to an external customer.

FIGURE 25.12   *(Continued)*

the activities identified on the far left end of Figure 25.13 have taken place; process descriptions exist, and business needs, in the form of goals and objectives, have been defined as part of the strategy formulation and service definition.

The first steps leading to the deployment of customer service metrics are to identify the requirements for metrics and to define the specific metrics that satisfy the requirements. Usually, metrics address some or all of three process considerations:

- *Customer service process effectiveness:* Does the output of the process step meet customer needs? One approach to this metric is to quantify the defects that escape the process.
- *Customer service process efficiency:* Does the process make good use of resources? One approach to this metric is to quantify defects internal to the process, such as rework.
- *Supplier effectiveness:* Do the inputs to the process meet its needs? An approach to this metric is to quantify incoming defects.

The metrics recommended in the preceding paragraph all focus on defects. In the sense intended for these metrics, a *defect* is anything that does not meet or exceed the requirements of the customer, the business, or the process. Since the focus on defects is primarily on internal defects, rather than on defects that affect the customer, the emphasis on defects is a preventive approach, eliminating problems before they have any effect on the service provided to the customer. Therefore, it is important to have a realistic threshold for what is called a *defect.*

When defining the metrics, it is important to ensure that the data required to calculate them can be collected readily and reliably. The metrics team should think through the sources of the data, the likely methods of collection, how defects will be identified, the amount of effort required, and the cycle time for data collection. All the key characteristics should be represented with metrics, even though all of them may not be implemented initially. The metrics team should think through how each method may affect behavior and be modified if required to ensure that the metrics will drive the desired behavior.

IBM Corporation.   Large companies that emphasize customer service, such as IBM, approach the measurement process in a comprehensive way (Edosomwan 1988). Customer service quality measurements at IBM address four areas that affect customers:

- Customer requirements

  1. Customer partnership
  2. Quality assurance
  3. Reliability
  4. Empathy
  5. Durability
  6. Responsiveness

- Task requirements

  1. Supplier options
  2. Vendor options
  3. Operational options
  4. Department options
  5. Interfunctional options
  6. Cross-functional options
  7. Production options
  8. Delivery options
  9. Consumption options

- Organizational requirements

  1. Management commitment
  2. Education and training

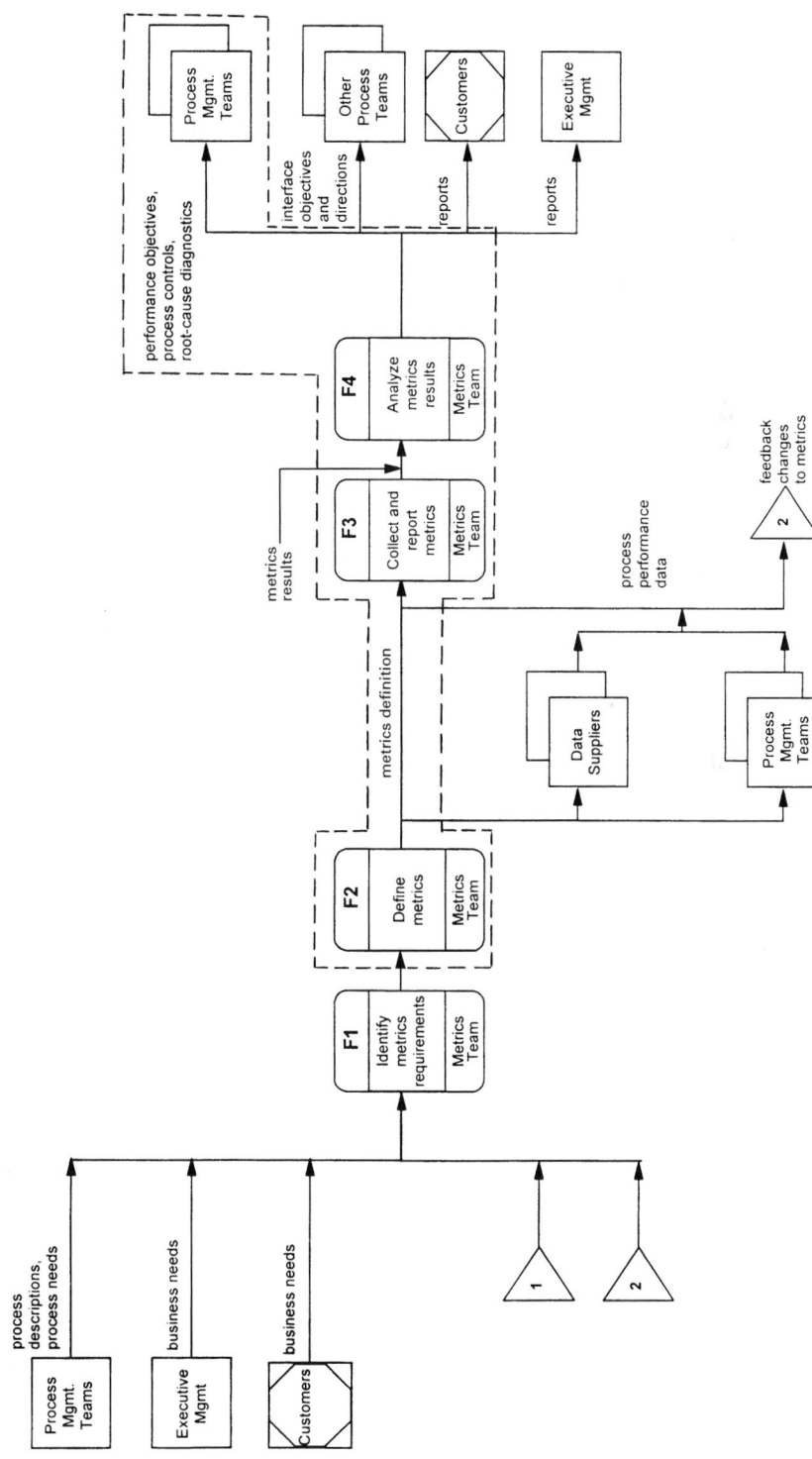

FIGURE 25.13 Defining, collecting, and analyzing metrics.

3. Defined roles and responsibilities
4. Recognition and reward
5. Communications
6. Goal-setting process

- Process control requirements

  1. Measurement tools
  2. Evaluation tools
  3. Improvement tools
  4. Information
  5. Resource allocations
  6. Planning
  7. Feedback mechanisms
  8. Environment-monitoring tools

Each of the four areas is conceptually addressed at five levels, as shown in Table 25.5. Examples of specific types of metrics that IBM uses to implement the preceding structured approach are summarized in Table 25.6 (Edosomwan 1988).

The examples include objective metrics, such as defect levels and error rates, and subjective measures, such as competitive quality ratings. In both cases, the usefulness of the data are determined by the effectiveness of the definitions of what constitutes a defect or an error, how the customer questionnaires and surveys are designed and implemented, and how the data are analyzed and used to drive control, improvement, and planning efforts. Therefore, education and training of all levels of personnel are critically important. According to IBM, the education and training requirements summarized in Table 25.7 are appropriate for a medium-sized division. Recall that the content, allotted times, and budgets are those which were given in 1988.

The education and training summarized in Table 25.7 do not include the specialized training that is given to the customer-facing customer service personnel. The training for the customer service personnel is tailored by business unit to the particular approaches, systems, and strategies employed for the products and services sold and supported by the business unit.

TABLE 25.5   Impacts of Quality in Customer Service

Level of service quality	Potential areas of impact
Fitness for use	Customer
Fitness for standard	Operational processes
Fitness for demand	Cycle times
Fitness for planning	Planning service quality variables
Fitness for control	Protecting customer requirements

Source: Edosomwan (1988).

TABLE 25.6   Some Types of Quality Metrics in IBM Customer Service

Types of metrics	Sources of data
System in-house performance defects	Manufacturing, engineering, and quality assurance departments
System field performance defects	Customer service departments and marketing offices
Field replacements due to in-service failures	Spare parts logs and marketing office data
Competitive quality ratings	Customer report cards
Engineering changes and updates	Development, manufacturing, and engineering quality reports
Performance error rates in-plant	Plant control charts
Field performance error rates	System performance control charts

Source: Edosomwan (1988).

TABLE 25.7   Education and Training for Customer Service at IBM

Level	Training highlights	Training time required	Training budget allocation

Upper management	Customer partnership
	Customer requirements
	Goal setting
	Key customer service
	  quality programs
		One-week initial
		Three-hour periodic
		  follow-ups
			9% of total

Middle management	Customer partnership
	Resource allocation
	Customer service program
	  development
		$1^1/_2$-week initial
		Five-hour periodic
		  follow-ups
		Program oversight
			23% of total

Supervisors	Program oversight Quality management tools Feedback processes	Two weeks initial One-day periodic updates	23% of total

Technical staff
  and operators
	Tools for reliability, durability,
	  and customer satisfaction
		Three weeks initial
		Two-days periodic
		  follow-up
			45% of total

Source: Edosomwan (1988).

HUMAN RESOURCES

The human resource (HR) issues for enterprises that focus on excellent customer service involve both the customer-facing staff and all the associates in the enterprise who support the customer-facing staff and the customer service strategies and processes.

The key issues concerning the customer service staff are summarized in Table 25.8. The operational framework for the customer service issues summarized in Table 25.8 should include goals, metrics, and improvement approaches for each HR issue. Examples of such a framework are summarized in Table 25.9.

EMPLOYMENT CONSIDERATIONS

Quality Progress reported a study by Abt Associates of 50,000 customer service employees. The study found that customer satisfaction depends on the degree to which customer service employees believe that they have the capability, the tools, and the organization's support to provide for the customers' needs (Struebing 1996). The same article reported an expert assessment that customer service initiatives can be headed for success or failure at the very first step, the recruiting of customer-facing employees. Here, personality is the critical characteristic.

CONTINUOUS IMPROVEMENT

The philosophy, methods, and tools for the continuous improvement of customer service are similar to those for other functions and are generally covered in Section 6. It is important to remember that continuous improvement involves incremental improvement when appropriate and discontinuous improvement when required. Part of the improvement process involves continual environmental scans in search of early warning signs of changing markets, of new opportunities, and of new competitive threats. In some cases the improvement needs or opportunities will be customer-driven, in some cases they will be competitor-driven, and in some cases they will be driven be emerging technological capabilities. In any of these cases, improvement may be emphasized by including it as part of the strategic planning process discussed earlier in this section.

TABLE 25.8 Human Resources: Quality Initiatives for a Customer Service Organization

Human resource issues	Key operational initiatives (examples)	Key strategic initiatives (examples)
Recruiting	Active relationships with suppliers of customer service recruits	Identify competencies and skills required for customer service job functions
Education and training	Conduct skill assessments and gap analyses	Develop and deliver functional skills and people skills
Tools and technology	Instrument the information technology and tools to provide metrics for improvements and rewards	Acquire and deploy information technology and tools in support of customer service functions
Involvement	Implement feedback and incentive building processes	Develop team-centered customer service organization structures
Empowerment	Provide leadership, role models, and examples for institutionalizing empowerment	Create stakeholder relationships between customer service associates and the enterprise
Performance and recognition	Provide immediate, visible feedback and reward for exemplary performance	Ensure that recognition and reward processes emphasize desired behaviors and business results

TABLE 25.9 Human Resources: Operational Framework for Quality Initiatives

HR issue	Goal	Metric	Improvement approach
Recruiting	100% follow-up within 48 h	Cycle time to yes/no decision	Process improvement for cycle-time reduction
Education and training	Fully qualified front-line staff	% Customer service staff with up-to-date education and training milestones	Cascaded goals and objectives for policy deployment
Technology and tools	Provision systems to permit answering 100% of incoming calls within 15 s	% calls with answer times greater than 15 s	Associates and managers review performance daily at start-of-shift meeting
Involvement	Cascaded personal and team goals and objectives for everyone	Suggestions for improvement per associate	Customer service teams
Empowerment	Every associate has the authority and the tools to satisfy the customers	Leadership survey scores	360-degree assessments
Reward and recognition	Every associate has documented SMART goals (Specific, Measurable, Agreed-to, Realizable, Time-bound)	Associate and management performance assessment scores on goal setting and realization	SMART goals include shared team goals and individual goals; some goals are SMAST (S=Stretch)

L.L. Bean.   The first step by L.L. Bean to initiate an environment of continuous improvement in its customer service operations was to assign a full-time facilitator to help the customer service manager in the effort (*Customer Service Newsletter* 1995). A transition team was formed to re-engineer the customer service process to include a continuous improvement capability. The team was formed based on expertise in four essential competencies.

- *Customer focus:*   Are team members committed to meeting the needs of external and internal customers, and can they build effective relationships with all customers?
- *Team play:*   Are team members willing to share knowledge and information with others? "We don't want people who are going to try to build job security by hoarding information," noted the team facilitator.

- *Good communications skills.*
- *Flexibility and adaptability* to changing work conditions.

The team began by understanding the current customer service process. The team discovered that it was based on hand-offs, many of which were unnecessary. The hand-offs were especially problematic whenever there was a transaction that a front-line employee could not handle. In those cases, the employee transferred the customer to another department. The transfer took place off-line. The receiving department did the necessary work and then got back to the customer with the resolution. In most cases, this took 2 or 3 days to complete.

The team developed a pilot prototype program before trying to initiate changes throughout L.L. Bean. It decided that it was too risky to apply untested process improvements to 1200 people at one time. The changes involved a new front-line customer service organization structure that possessed the following characteristics:

- *External-customer focus:*   It maximized convenience for customers by eliminating unnecessary hand-offs, moving the customers quickly and efficiently to the person that can meet their needs.

- *Process-oriented structure:*   The prior organizational structure was functionally oriented. In the new structure, the customer service representative is accountable for the entire transaction from the initial customer interaction to the final resolution.

- *Learning organizational structure:*   The minimum hand-off, end-to-end accountability design emphasized learning opportunities. The service representatives were now enabled to build reservoirs of knowledge sufficient to identify the root causes of customer service problems and ways to prevent their occurrence at the sources. The Customer Service Department became a forum for discussing the day's activities, what went wrong and what went well, with a focus on preventing problems from recurring, rather than getting better at resolving them.

Two months after initiation of the pilot program, the productivity of the pilot customer service team increased by 38 percent. This was a surprise to the team leader, who had budgeted for an initial decline in productivity. The most important discovery made during the pilot effort was that productivity, good service, and job satisfaction are complementary results.

REFERENCES

AT&T Operations. "An Action-Oriented Approach to Implementing PQMI." *AT&T Operations Engineering Workbook.* (Available in *Process Quality Management*, Lucent Technologies Publications, Indianapolis, IN, 1998).

Bell System (1983). "Human Factors and Behavioral Science." *The Bell System Technical Journal,* vol. 62, no. 6, part 3, July-August.

Clarke, J. Barry, Eileen F. Mahoney, and Susan E. Robishaw (1993). "New England Telephone Opens Customer Service Lines to Change." *National Productivity Review,* Winter. The authors describe this effort as a reengineering project and discuss in some detail the problems encountered and the countermeasures that led to ultimate success. Although the term *service blueprinting* is not used, the approach is described as part of the overall solution.

Customer Service Newsletter (1995). "You're Not Giving `World-Class' Customer Service Until You Can Prevent Problems," vol. 23, no. 5, May.

Edosomwan, Johnson Aimie (1988). "A Program for Managing Service Quality," in Jay W. Spechler (ed.): *When America Does it Right: Case Studies in Service Quality.* Industrial Engineering and Management Press, Norcross, Ga.

EDS Homepage Internet (www). "A Yankee Group Profile: Outsourcing Leadership Series. In Electronic Data Systems Corp."

Grant, Alan W. H., and Leonard A. Schlesinger. (1995). "Realize Your Customers' Full Profit Potential." *Harvard Business Review,* September-October, pp. 59–72.

Hax, Arnaldo C., and Maluf, Nicolas S. (1984). *Strategic Management.* Prentice-Hall, Englewood Cliffs, N.J.

Innis, Daniel E. (1994). "Modelling the Effects of Customer Service Performance on Purchase Intentions in the Channel." *Journal of Marketing Theory and Practice.* Spring. Combined Lisrel model. Note that this model is a particular application of the "service satisfaction model" presented as Figure 3 in "Toward Service Systems Engineering," by Thomas Pyzdek, in *Quality Management Journal,* April 1994. Pyzdek attributes the model to Wirtz and Bateson.

Jackson, J. H., and Morgan, C. P. (1978). *Organization Theory: A Macro Perspective for Management,* Prentice-Hall, Englewood Cliffs, N.J.

Motorola, Inc. (1989). Benchmarking visit to AT&T, August 16.

Pyzdek, Thomas (1994). "Toward Service Systems Engineering." *Quality Management Journal,* April.

Spechler, Jay W. (1988). *When America Does it Right: Case Studies in Service Quality.* Industrial Engineering and Management Press, Norcross, Ga.

Struebing, Laura (1996). "Customer Loyalty: Playing for Keeps." *Quality Progress,* February, pp. 25–30.

SECTION 26
ADMINISTRATIVE AND SUPPORT OPERATIONS

James A. F. Stoner
Charles B. Wankel
Frank M. Werner

INTRODUCTION

Exceeding customer expectations involves numerous activities throughout the organization. In the early years of the quality "movement," attention was devoted predominantly to those activities which most directly influenced the nature of the product or service (e.g., design, purchase of materials, fabrication, inspection). However, there are many activities that, though indirectly influencing quality of product or service, have a major impact on customer satisfaction. As dramatic improvements in product and service quality have become more widespread, increasing attention has been devoted to improving the quality of these less directly visible contributors to customer satisfaction or dissatisfaction.

A broadly accepted generic term for these activities has not yet emerged in the literature or in organizational practice. In this handbook, the term *administrative and support operations* will be used to designate these activities. Responsibility for these activities is normally located in "staff" or administrative functions, such as Finance, Human Resources, or Corporate Law. Examples include the Human Resources Department's handling of the payroll so that employees get paid and the Legal Department's support to researchers in the Research and Development (R&D) Department in preparing and filing patent applications. These operations are defined more extensively later in this section, but we will not emphasize the distinctions between *administrative activities* and *support activities* because, by their nature, the role of all administrative and support activities is to support the rest of the organization in creating and delivering products and services that exceed customers' expectations.

Summary and Conclusions.   In organizing to achieve and sustain revolutionary rates of improvement in quality and customer satisfaction, companies find that organizational structures, systems, relationships, and all parts of the organizational culture change greatly (see also Sections 5 and 15). These changes are no less profound in administrative and support departments—such as Finance, Human Resources, and Legal Services—than they are anywhere else in the organization. This section of the handbook looks at how administrative and support operations within organizations can be and are being improved as part of the global quality revolution and how the functional departments that are the traditional homes of these operations are changing.

Introducing quality into administrative and support operations produces three broad types of change. Some change relates to the processes of administrative and support work, some to the content of that work, and some to the context within which that work is performed.

Process: Well-established ("traditional") administrative and support operations are examined using the tools of quality management and are often significantly improved.

Content: The work of administrative and support functions changes as new operations are undertaken and old operations are eliminated or transferred to other organizational units.

Context: The roles of administrative and support personnel and their relationships to the rest of the organization change—and new roles and relationships emerge—as administrative and support areas become more concerned about serving both external and internal customers, as the nature of their work changes, and as they increase their rate of learning from the rest of the organization.

The major conclusions of this section are

Administrative and support functions can and do use essentially the same quality improvement tools in essentially the same ways with essentially the same results as other parts of the organization.

Increases in customer satisfaction arise from improvements in traditional services—with the improvements showing up as defect reduction, cycle-time reduction, and cost reduction—and the development of new services.

Beneficiaries of these quality improvements include both internal and external customers.

Role and relationship changes that departments and their members undergo are particularly important.

In general, the functional departments that provide these administrative and support operations follow rather than lead the production and operations functions, sometimes with a considerable lag.

The kinds of changes in functional departments described in this section are likely to continue to evolve in new ways as organizations invent and discover new structures and relationships more consistent with a total quality culture (see also Section 15, Human Resources and Quality).

These conclusions suggest that all organizational members should expect and should plan for revolutionary rates of quality improvement in all administrative and support operations, just as they can and should expect and plan for such improvements in the products and services they provide to their external customers.

Defining and Identifying Administrative and Support Processes. The processes that make up the administrative and support operations of an organization can be looked at as occurring in three related areas: administering the internal activities of the organization, supporting members of the organization in serving external customers, and supporting external customers—defined broadly to include suppliers and other outside stakeholders—in doing their business with the organization.

The administrative side of these activities encompasses operations required for the organization to sustain itself as an ongoing entity so that it can serve its customers. Administrative activities involve processes that affect primarily or exclusively internal customers. They may affect internal customers in their roles as employees of the organization or in their roles as producers serving external customers. An example of an administrative activity with an impact on employees as internal customers is human resources' handling of the payroll so that employees get paid. Other examples include managing pension funds and retirement programs, career systems, compensation programs, training systems, health care programs, and so on.

Administrative processes that affect employees in their roles as producers involve developing, maintaining, and using systems and activities—such as management information systems and plant maintenance—that enable the organization's members to do work that serves the ultimate customer. An example is the contribution facilities management groups make in maintaining clean and comfortable working conditions.

The support side of these activities encompasses operations that contribute more directly to the work of other departments as they serve the organization's external customers or do other necessary activities to ensure competitive health. An example would be the Legal Department's support to researchers in the R&D Department in preparing and filing patent applications. Support activities also assist customers and other stakeholders in dealing with the organization; examples include invoicing and processing payments from external customers.

Some activities that we have treated as administrative and support operations may not fit easily into this category because their connection with external customers—or at least stakeholders—is quite strong and direct. For example, investor relations functions, frequently located within the corporate finance staff, prepare annual reports and deal with securities analysts. We have treated these activities as support activities because they assist external customers in dealing with the organization.

The Need for Improvement. In the preceding edition of this handbook, Gryna (1988) suggested that significant improvement opportunities exist for the reduction of error rates and cycle time in the execution of useful and effective administrative and support systems. He reported a study in an electronics firm that showed that "50 percent of time cards were in error," "40 percent of all travel reservations made in one month were changed" and two were changed nine times, and 10 percent of performance appraisals sampled in "one month were returned because of lack of signatures." He noted that "little work has been done to quantify the extent and cost of errors in support operations or administrative operations. It is the author's belief that the cost of poor quality for these activities is as great as the cost of poor quality associated with the product."

As Gryna suggests, the costs of defects and of slow cycle time in useful and valuable administrative systems may be quite large. However, these costs may be lower than the costs arising from ineffective and unnecessary administrative processes. One of the costs of poor quality that has not been quantified is the interference in the organization's work that arises from the existence of outdated, unnecessary, and ineffective administrative and support operations. Administrative and support procedures that do not add value can easily arise and then become encrusted parts of organizational systems and programs. Ineffective or useless systems waste not only the resources of the administrative and support personnel who administer and execute them but also the time and energy of those who collect the data demanded, fill out the required forms, wait for the required approvals, and so on.

Beyond opportunities to reduce the costs of poor quality, there is a need to improve administrative and support operations to make them consistent with the organization's emerging quality management systems and culture. Administrative and support processes inconsistent with the rest of the organization can generate significant drag on many activities; processes aligned with other organizational units and processes can enable superior performance.

Relation to Other Sections of This Handbook.   Because administrative and support operations employ the same types of quality tools and implementation processes used in other parts of the organization, Sections 3, 4, 5, 6, 12, 13, 14, and 45, on Quality Planning, Quality Control, Quality Improvement, Process Quality Management, Benchmarking, Strategic Quality Planning, Total Quality Management, and Statistical Process Control, are particularly relevant to this section. Because the changes in the work, roles, and relationships of administrative and support members can be quite large and are part of broad cultural changes in organizations, Sections 15 and 16, Human Resources and quality and Training for Quality, are also quite relevant to this section.

PROGRESS IN ACHIEVING QUALITY IMPROVEMENTS IN ADMINISTRATIVE AND SUPPORT SERVICES

Overview.   Although progress in improving administrative and support operations historically has been slower than in many production operations, progress can be dramatic once the changes are started. Recent evidence suggests that organizations can expect rates of quality improvement in administrative and support operations comparable with what they achieve in production. Errors in issuing paychecks can be reduced from hundreds a month to one or two, cycle time to complete credit analyses or to draft contracts can be reduced from weeks to minutes, and the cost of closing the corporate books of large multinational companies can be reduced by tens of millions of dollars per year—with fewer errors and greatly reduced cycle time. This part of this section begins with a historical perspective on improvements in administrative and support operations. It then discusses reasons why quality progress in administrative and support functions has tended to be slower than in areas more directly in touch with the final customer and how these historical reasons are changing, yielding the current situation in which rates of improvement in administrative and support services can be similar to those in other activities.

Historical Perspective.   The manufacturing or service company's mission is to produce a continuing flow of products and services that exceed customer expectations, to market and support these products and services effectively and efficiently, and thereby to generate income that sustains the organization and rewards its various stakeholders.

An early awareness of the effect of product and service quality on income led companies to establish quality control departments within their traditional operations. These activities focused initially on final inspection of completed products and then quality assurance during production. These quality control operations developed most rapidly and extensively in manufacturing companies. Moreover, within manufacturing companies, they were most highly developed on those aspects of product progression which are most obvious to users and producers, such as materials, manufacturing operations, and final performance features. The effect of administrative and support operations on product quality and customer satisfaction was less obvious, and this less obvious connection may have contributed to the slower development of quality controls in these "indirect" activities.

Sullivan (1986) has described the development of companywide quality control as a seven-stage process. In his model, administrative and support functions are likely to become actively involved in the third stage, but not before. The first stage is "product-focused," involving "inspection after production, audits of finished products, and problem-solving activities." The second stage is "process-oriented," involving "quality assurance during production including SPC and foolproofing." In the third stage, attention shifts to the rest of the organization. In this stage, which he refers to as "systems-oriented," all departments are involved. Although he cites only design, manufacturing, sales, and service, an increasing systems orientation would logically include administrative and support operations.

Sullivan's remaining four stages are (4) "humanistic," changing "the thinking of all employees through education and training," (5) "society-oriented," "product and process design optimization for more robust function at lower costs," (6) "cost-oriented," which involves integrating Taguchi's societal loss function concept into the design process and into the total management system to produce products and services that are fully competitive on a total cost-to-society basis, and (7) "consumer-oriented," involving "quality function deployment to define the `voice of the customer' in operational terms." [All quotations are from Sullivan (1986).]

The ways quality management has evolved in companies in the late 1980s and 1990s may be considerably different from the process Sullivan described in his 1986 article, but his framework does suggest many of the changes occurring in administrative and support functions. Sullivan's framework seems particularly accurate in describing a follower rather than a leader role for administrative and support operations in adopting quality approaches. A number of historical barriers to quality improvement in administrative and support operations have contributed to this follower role.

Historical Barriers to Quality Progress. Perhaps the most important factor contributing to this follower orientation has been the historically noncompetitive nature of support and administrative operations in most companies. Administrative and support activities primarily served captive internal customers who often were without access to other suppliers. These internal suppliers were not exposed to a clear competitive wakeup call like the ones that occurred in many industries as their products competed unsuccessfully against the dramatically improved quality of foreign competitors, most notably the Japanese. Administrative and support activities were shielded from the competitive forces that drove the quality revolution in its spread across companies, industries, sectors, and nations.

Other factors that contributed to this follower role include a lack of information about the quality dimensions of administrative and support processes, a lack of alertness to the profit impacts of administrative and support operations, a lack of success stories about improving these processes, a lack of experienced leaders for making the improvements, and a lack of process awareness.

The connection between ineffective administrative and support activities and company competitiveness and profitability has been much less clear than the connections between product quality and cost and organizational profitability and competitiveness. The early success stories that energized quality improvements in many companies occurred overwhelmingly on the plant floors of close competitors and led to attempts to achieve similar gains. The experienced quality experts who spread quality approaches from one company to another usually were production experts; rarely were they experienced in improving administrative and support processes. In fact, a widely held belief was that companies could not "do quality" in finance, in marketing, or in legal activities. Many administrators also have been slow to recognize that administrative and support operations accomplish their missions through productive processes, just as do all other parts of an organization. With its strong emphasis on improving processes, the quality approach seemed less relevant to people who did not recognize that they were, in fact, involved in processes.

This lack of awareness of processes in administrative and support functions is particularly interesting and may have existed for a variety of reasons, including

Lower levels of training in subjects that relate to systems and processes, such as engineering and other science-based disciplines

Processes that are not highly visible because they are episodic and infrequently repeated (e.g., drafting legal contracts, analyzing major investment projects, dealing with turnover of team members on a project team)

The belief that administrative and support activities are not processes at all but are "an art rather than a science," depending primarily on creative or professional skills (e.g., preparing an advertising campaign, preparing a patent application, managing an investment portfolio)

The existence of traditional functional silos that interrupt the natural flow of processes and obscure their interconnections across functional activities, supported by the belief that functional silos are desirable or inevitable (e.g., assuming that it is appropriate for all market research activities to be concentrated in a department of expert market researchers)

Falling Barriers and Rising Expectations and Pressures. Many of these barriers to quality improvement are falling rapidly while expectations and pressures for quality improvements are rising. Successes in using quality management approaches to make improvements and to change to more effective roles have shown that progress is possible and profitable. Training and consulting resources for improvement inside and outside the organization have increased greatly. Many individuals in administrative and support activities now recognize the existence of processes and the importance of serving their internal and external customers.

In addition, two related changes in the 1980s and 1990s—increased competitive pressures on companies and the rise of outsourcing—may be reducing the "follower gap" in administrative and support operations and contributing to the much more aggressive adoption of quality approaches that has been occurring recently. As companies have faced progressively more competitive situations, the pressure to "do more with less" has fallen heavily on administrative and support operations. At the same time, increased attention to the potential value of outsourcing has reduced the protected nature of these operations. Members of administrative and support functions increasingly face the choice of making dramatic improvements in quality, cycle time, and cost-effectiveness or allowing their operations to be supplied by contracts with outside organizations that can provide their services at lower cost and higher quality [see, for example, Alexander and Young (1996), Harkins et al. (1996), and Lacity et al., (1996)]. Because of these changes, quality improvements in administrative and support operations are increasing at a rapid pace.

CHANGES FROM QUALITY IMPROVEMENTS IN ADMINISTRATIVE AND SUPPORT SERVICES

Overview.   To a large extent, the types of quality improvements achieved in administrative and support operations are the same as those achieved in other parts of the organization. However, there is one type of improvement that is moderately "special" to administrative and support operations in the sense that it is more likely to occur in those operations than in other organizational operations. In this part of this section we discuss the first three broad categories of changes and illustrate them with examples drawn from the various administrative and support operations. We then discuss the more unusual improvement area. We also identify the opportunity for quality champions elsewhere in the organization to contribute to quality in administrative and support areas.

Three Types of Quality-Based Improvements.
Quality-based improvements in administrative and support operations may be grouped into three broad and well-recognized categories: (1) improvements arising from examining traditional activities and doing them better, (2) improvements arising from doing different things, and (3) improvements arising from changes in roles and relationships. (See also Section 15, Human Resources and Quality.)

Figure 26.1 is a model of these three types of improvements. The organization's administrative and support operations are shown in transition, from operating within a system of traditional management (TM), represented by the steep pyramid, to functioning within systematic quality management (SQM), shown as a flattened and inverted "plate." The plate, flattened to represent the elimination of unnecessary organizational layers and inverted to symbolize the changing role of managers from commanding to supporting the rest of the organization, represents the current state of the art of quality management—the best practice of companies such as the Deming and Baldrige Prize winners of the late 1990s.

The figure also shows a dotted arrow continuing upward to the right and leading toward an amorphous object. This part of the diagram suggests the continually evolving nature of the quality management paradigm toward a state of management practice not yet reached by any existing organization. The path of that dotted arrow might be the evolution of organizations from total quality to learning to world class discussed by Hodgetts et al. (1994).

In the diagram, two arrows indicate improvements from examining traditional work through the quality management lens; some work (arrow 1a) is left unchanged, whereas other work (arrow 1b) is redesigned. The other two arrows represent improvements from doing different things,

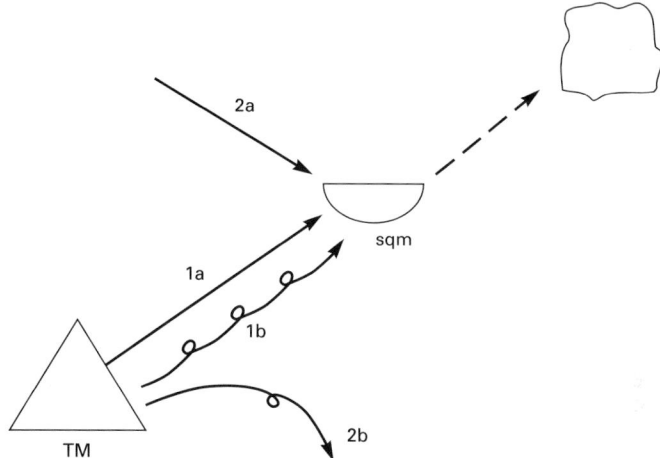

FIGURE 26.1   Types of quality-based improvements.

either by undertaking new value-adding operations (arrow 2a) or by eliminating non-value-adding activities or transferring work better performed elsewhere to other organizational units (arrow 2b). Surrounding the diagram is the rest of the organization, symbolizing the third type of improvement, changes in the roles and relationships of administrative and support personnel (Stoner and Werner 1992*b*).

Improvements from Doing the Same Things Better.   The first type of change begins with the study of what the organizational unit is presently doing to locate opportunities for improvement. While some activities in the unit already may be consistent with best practice and systematic quality management, more often there are numerous opportunities to improve quality and customer service and to reduce costs. This is done both through small-scale continuous improvements [what Imai (1986) calls *kaizen*] and through large-scale, discontinuous improvements (what Imai calls *innovation* or others have called *re-engineering*.

Kaizen involves continuously improving existing operations in incremental steps. For example, many organizations have used quality improvement teams to improve their handling of reimbursement for travel expenses. Federal Express redesigned its expense reporting system by replacing manual forms with a computer screen that eliminates missing data and mathematical errors. As a result, it now reimburses employees within 1 day and in the process has eliminated the need for most cash advances (Stoner and Werner 1993, p. 44). Of course, over time, a series of small changes may aggregate into very large changes.

Imai's innovation involves dramatic, discontinuous, large-scale changes in what the function does or how it does it. As Section 6 discusses, innovation—or re-engineering—is the redesign of existing processes to accomplish the same (or different) tasks in significantly different ways. When IBM Credit Corporation studied its credit approval process to understand why it took a full week to process a customer request, it found that much of the week was wasted while the documents waited for the next specialist's time. IBM Credit threw out the existing process and designed a completely new system built around a single customer representative responsible for each part of the process. Now, with one person handling the documents, the loan request is processed within a few hours (Hammer and Champy 1993, pp. 36–39).

Improvements from Doing Different Things.   The second type of improvement involves doing different things: adopting new activities and eliminating those which are found to add little or no value. When these changes are quite large in scope and impact, they also may be called *re-engineering*. These changes increase the effectiveness of the organization by reducing the inefficiencies, frustrations, and waste of time and effort that occur when administrative and support groups undertake work better done elsewhere in the organization (or not at all!) or do not perform operations that they are best suited for. For example, one major form of inefficiency occurs when administrative and support groups perform inspections, sign-offs, and policing functions that create delays and costs without adding commensurate value.

As administrative and support personnel begin to question the value added by their activities, they often discover new products and services they can provide to other parts of the organization (or perhaps to external customers). American Standard Companies discovered an opportunity to improve its internal business processes by modeling them after the quality processes that had been adopted successfully in their manufacturing activities. The company created a transformation team that conducts 5-day workshops at its administrative offices around the world to teach and install these new methods.

In administrative and support functions, one of the largest changes involves stopping activities that have outlived their usefulness. For example, early in their quality improvement initiatives, many finance functions find that they are producing periodic reports—often at considerable expense—that simply are not used elsewhere in the organization. Suspecting this to be true, the CFO of Stanley Tools once stopped sending out all financial reports; his suspicions were confirmed when few recipients complained. Stanley eliminated many of the reports and redesigned the others to provide recipients with the data they truly needed (seminar presentation at Fordham University, October 9, 1990).

Improvements from Changing Roles and Relationships.   The third type of improvement involves changes in the roles of administrative and support personnel and in their relationships with others in the organization, transformations that alter the ways individuals and groups work together. These changes come about as administrative personnel begin to see the organizational units they support as internal customers and increase their rate of learning (or begin to learn) from the rest of the organization.

Keating and Jablonsky (1990) and Stoner and Werner (1993, 1995) have reported large changes in the ways corporate finance members work with other organizational members in some companies. Finance members have abandoned the role of corporate police, no longer seeing themselves as "the only adults in the company." Instead, they act like "business partners" who work collaboratively with other organizational members. Keating and Jablonsky make this point using the analogy of a successful sports team. It is the players on the field who win the games, but a winning team also requires effective coaches and a strong franchise to provide supporting resources. They observe that

> …the unique challenge facing the financial professional…is learning how to become a player and a coach. This is not an insignificant task when individuals have been trained in school and [have] continued professionally to be scorekeepers or independent commentators. Getting close to the business means getting on the field as part of the management team [Keating and Jablonsky 1990, p. 15].

A Bonus from Eliminating Non-Value-Adding Administrative Work.   There is a special aspect of administrative and support operations that has enabled quality initiatives to make unusual—and often major—contributions to the competitiveness of organizations, to their ability to improve customer satisfaction, and to their ability to provide an enjoyable working environment (what W. Edwards Deming called "joy in work"). This unusual type of improvement arises from the elimination of non-value-adding work by administrative and support functions. It is similar to the results achieved when any part of an organization eliminates an unimportant or useless activity, but frequently it is also a bit different.

When the parts of an organization that serve customers directly—the front line—eliminate non-value-adding work, the major beneficiary is usually the unit itself that eliminates the non-value-adding work, although the external customer also may benefit. When administrative and support operations eliminate non-value-adding work, the major beneficiaries are usually others: the customers of the administrative support functions and the entire organization.

This difference in likely beneficiaries arises because administrative and support operations frequently require work to be performed by other parts of the organization to meet government regulations, to conform to procedures and policies, to provide data intended to be useful to others inside and outside the organization, and so on. Administrative and support functions do not simply serve internal customers; they also impose on other organizational units demands that can get in the way of doing business: reports that require data to be provided by other units, internal audit inspections and requirements for changes that can be onerous or not worth the effort to make, delays in hiring that prevent work from being done, lengthy legal documentation and other steps to prevent or defend law suits, and so on. When ineffectively designed and implemented, these administrative requirements create very real burdens for the rest of the organization and are of questionable value in improving customer service.

Ford discovered the magnitude of the costs imposed by non-value-adding administrative work when it conducted its now well-known study of Mazda in the early 1980s to learn how Mazda could produce higher-quality cars at significantly lower cost. One key conclusion was that "…Mazda's efficiency gains resulted from dispersing many so-called accounting activities to nonfinance functions.…Quoting from the study: 'This environment helps eliminate duplication of effort. Mazda activities handle a broad range of responsibilities related to their operation, and it is not necessary for detailed information to be passed to accounting for processing, data entry and reporting'" (Keating and Jablonsky 1990, p. 79).

Administratively imposed burdens often have a second, perhaps even more perverse effect: They can deprive people of the ability and motivation to improve the organization. One Ford employee related, "People got to the point where they said…it's not my role to control costs, it's what those finance people are for" (Keating and Jablonsky 1990, p. 87). AT&T's Chief Accountant Bernie Ragland, referring to the company's regulated history and associated internal conformance procedures, "argued that the single most negative cost associated with regulation is not the specific dollar outlay required to maintain the complex fiscal accountability system; the most negative cost is the 'conformance mentality that regulation breeds.'" This effect "is particularly insidious because it is pervasive, and over time, becomes an habitual pattern of behavior that, even when recognized, is not easily broken…you cause people not to think about doing things'" (Keating and Jablonsky 1990, p. 168).

Administratively imposed requirements, such as unnecessary reports to staff organizations, are particularly vexing to other organizational members because the administrative departments are nonpaying customers that cannot be avoided. External customers frequently can protect themselves from being forced to do non-value-adding work by taking their business elsewhere, but internal customers rarely have that opportunity. One of the major contributions of the quality approach has been the removal of many non-value-adding requirements from administrative and support functions. Although the savings in time and money within administrative and support budgets may be quite significant, the savings in the other organizational units that are freed of the administrative burdens are frequently much greater. In this sense, this type of change deserves special attention when assessing the value of quality-based improvements in administrative and support operations.

Assisting Administrative and Support Areas. Administrative and support operations offer particularly fertile ground for contributions by the traditional quality champions from production and quality departments. Quality tools and approaches that have become part of the cultures of an organization's production and operations areas can be deployed in administrative and support areas with results comparable with those achieved elsewhere in the organization.

The quality successes achieved in other parts of the organization provide at least three stimuli that encourage rapid progress in administrative and support areas. The stimuli relate to expectations about rates of quality improvement, acceptability of quality initiatives, and resources for making improvements:

Expectations: As other parts of the organization achieve quality improvements, their expectations of high-quality performance from administrative and support functions rise, and they start encouraging or (putting pressure on) those functions to achieve similar improvements.

Acceptability: The successes achieved elsewhere in the organization demonstrate to administrative and support personnel that the techniques actually work in their own company (reducing the "not invented here" syndrome).

Resources: The individuals and teams who achieved progress elsewhere in the organization are available as resources—a pool of internal quality consultants within the company, familiar with the organization and its internal operations.

INTERNAL AUDIT: AN EXAMPLE OF PROFOUND CHANGES IN ADMINISTRATIVE AND SUPPORT OPERATIONS

Overview.   The three types of quality-based improvements (doing things better, doing different things, changing roles and relationships) and the value of eliminating non-value-adding work can be illustrated in all administrative and support activities. However, one of the most impressive illustrations of the scope and impact of these changes is occurring in the internal audit function. For a variety of reasons, it is a particularly interesting example of the ways administrative and support functions are changing as companies progress on their quality journeys.

First, internal audit can be looked at as the quality control unit of the management system, with every bit as much potential for improvement as quality control activities in manufacturing operations. In many companies, however, the audit function has been performed in ways quite inconsistent with the organizational culture, roles, and ways of work necessary for achieving revolutionary rates of quality improvement and increases in customer satisfaction. In these companies, internal audit has been a corporate police force, increasing stress and tension as it proved its value—and superior influence—by finding errors and weaknesses in the departments being audited. Rather than building collaboration and breaking down barriers, such departments traditionally did the opposite. Changes in some internal audit groups represent a virtual reversal of past practices—dramatic changes in ways of contributing to an organization's success.

Second, the changes in internal audit are significantly increasing the value added by that function while reducing the cost of performing it. Third, these changes are permitting other parts of the organization to modify the ways they work, resulting in marked additional cost reduction and work simplification. And fourth, the changes in internal audit serve as models of the changes occurring elsewhere in administrative and support operations.

Examples of Improvements.   The examples that follow are representative of the exciting changes taking place in internal audit groups and illustrate the four types of quality-based improvements discussed earlier.

Improvements from Doing Things Better—Improving Well-Established Processes.   American Standard is one of many companies that have redesigned their audit reports to reduce cycle time. The new process has eliminated drafts and even the traditional audit recommendations. Computer-assisted audit programs have been developed and adopted—and shared with the operating groups—to permit the examination of large data sets without reliance on nonstatistical small samples.

Motorola's internal audit group has successfully applied the company's 6σ and $10\times$ cycle-time reduction initiatives to its activities, redesigning and simplifying its audit processes dramatically. Although the company has grown considerably in recent years, Internal Audit has remained the same size while increasing the scope and depth of its work. In one initiative, Motorola is building internal control processes into its operations to allow its units to audit themselves and to permit Internal Audit to shift its focus from auditing transactions to auditing the self-audit processes. In another initiative, Internal Audit's improvement approaches have been used by Motorola's Public Contracts Compliance Office to collaborate with a key customer, the U.S. Department of Defense. The collaboration has reduced the cycle time of government audits from over 1 year to 2 to 3 weeks while producing audits of higher quality at significantly lower cost to both parties.

Improvements from Doing Things Better—Achieving Traditional Goals in Very Different Ways.
The primary goal for internal audit departments frequently has been stated in terms of improving the internal control environment. At Gulf Canada Resources, a process called *control self-assessment* (CSA) has replaced some traditional auditing activities and increased the effectiveness of the ones that remain. In CSA, the unit being audited identifies its own control issues in a team-based meeting using computerized survey feedback methods. Control becomes internal to the operating unit and

intrinsically motivated rather than arriving in the form of an external police force providing extrinsic motivation. The CSA workshops reveal more and different control concerns than tradition-al audits and provide a broad picture of the corporate culture. CSA produces more audit findings and larger numbers of high-quality audit findings than are found with traditional audits (Makosz and McCuaig 1990*a*) and detects serious, hidden companywide problems ("cultural rogue elephants") that appear as ambiguous problems throughout the company in ways that disguise their seriousness and overall cumulative impact (Stoner and Werner 1995).

Improvements from Doing Different Things—Undertaking New Operations.   American Standard's audit group responded to the company's adoption of a just-in-time (JIT) manufacturing system by changing the framing and content of its work. Training of auditors was extended to include the just-in-time concepts, and auditors now spread JIT concepts throughout the company and assist the operating units with the related measurement systems.

An early supporter and adopter of the company's quality initiatives, Baxter International's Internal Audit Group quickly discovered that a key benefit of quality processes is the time and energies freed up to add additional value to its audit customers and the organization. One of the more striking ways it has invested these resources is in undertaking "special projects" in areas as diverse as electronic data interchange, treasury processes, managing overlapping responsibility and authority, and assisting Baxter's ultimate customers. Projects such as these have helped Internal Audit integrate its traditional audit responsibilities with an internal consulting role that has significant cost and knowledge advan-tages over the use of external consultants. Now the demand for Internal Audit's services far exceeds the group's resources, a significant change from the days when Internal Audit was seen as an adversary.

Improvements from Changing Roles and Relationships—Increasing Organizational Learning and Becoming an Organizational Change Agent.   At Raychem, Internal Audit views auditing as a process containing many "continuous learning loops" and interprets its activities as building and strengthening those loops (Stoner and Werner, 1995). At one level, these activities seek to make Internal Audit a vehicle for creating a high-trust, self-regulating, continuously improving control environment. At another level, Internal Audit is working to make itself and the rest of the company superior learning organizations. The continuous learning loops seek at least five types of changes in Raychem's audit process:

- *Anticipatory change:*   Responding in advance to audit customer needs
- *Self-control:*   Building auditee self-control directly into audit processes
- *Trust:*   Increasing trust in the audit process by making it more predictable and win-win
- *Added value:*   "Migrating up the value-added food chain" by replacing lower-value-added audit activities with higher-value-added activities
- *Diversity:*   Increasing Internal Audit's professional diversity through the use of guest auditors and audit staff members with nontraditional backgrounds

Perhaps Internal Audit's most exciting new role is that of organizational change agent. American Standard has created an innovative analogue to JIT in manufacturing that is moving the company toward a just-in-time office environment. It is aggressively spreading best practices throughout the company—for example, establishing a *kanban* system for corporate office supplies after observing its successful use in manufacturing. And with the luxury of fewer deadlines than other parts of the organization, it is becoming a transformational thinker, playing an increasing and ongoing role in the evolution of the company as a whole.

Improvements from Eliminating Non-Value-Adding Work—Recognizing When Organizational Changes Render Traditional Activities Obsolete.   When the American Standard Companies adopted their JIT manufacturing system, its audit group realized that much of its traditional work—such as veri-fying inventory balances—added little, if any, value to the company or its customers and would naturally disappear as inventory balances were dramatically reduced. Internal Audit responded by eagerly shedding these activities so that it could devote its resources to tasks seen to be truly value-adding.

Improvements from Eliminating Non-Value-Adding Work—Ceasing Activities That Can Be Done Better Elsewhere.   One traditional component of an internal audit report is the auditors' "recommendations," the auditors' instructions to the management of the audited unit for improving areas of managerial weakness or inadequate control identified in the audit's findings. Traditionally, the findings and the parallel recommendations have been an important way auditors have demonstrated they were doing their job well. An estimate of the savings achieved from implementing the auditors' recommendations is a metric used by some companies to calculate the value of the audit process. For example, some audit groups have a goal of saving three times the cost of the audit and measure their success in doing so by calculating the value of recommendations that actually get implemented.

Other audit groups, including those at Raychem and American Standard, have concluded that there is significant cost and little added value in these recommendations; they no longer make them. Instead, they issue their reports with the findings from the audit, the actions the unit's management has committed to take to correct the control or managerial concerns raised in the findings, the names of the responsible individuals, and the timetable for completing the actions.

The reasoning behind such changes is interesting. Recommendations are frequently subject to negotiation between auditors and auditees and cause delay both in issuing the audit report and in fixing the problem. For example, when American Standard was studying its audit process, it found that one of the major time consumers in the process was the time it took to reach agreement on recommendations with the auditee. When the auditor and the audited unit differed in preferred ways to resolve control problems, coming to an agreed wording on recommendations often was tedious and time-consuming. Even when there was agreement on the recommendations, it was not always clear who first developed the ideas for improving a situation. The audited unit could easily feel that the auditors were claiming credit for ideas the unit had already formulated or had even already started implementing. Furthermore, audit recommendations that were unnecessarily costly to implement or were at odds with management's insights often were damaging to the relationship between auditor and auditee. American Standard concluded that the long-established practice of including recommendations in the audit report was not actually adding value but was instead creating unnecessary work and delays—a major and very valuable insight of their study of its audit processes (Stoner and Werner 1995, pp. 7–36).

Companies that have eliminated audit recommendations from their reports have concluded that since it is typically the unit's managers and not the auditors who most deeply understand the nature of their business, it is the managers who are in the best position to design improved controls, especially since they ultimately will be held accountable for the success of the changes.

ENHANCING THE QUALITY OF ADMINISTRATIVE AND SUPPORT ACTIVITIES

Overview.   The well-established quality planning, control, and improvement processes are as effective in improving and maintaining administrative and support processes as they are in improving other organizational processes. In this part of this section we note some of the aspects of planning, control, and improvement in administrative and support operations that are worthy of special comment.

Quality Planning.   The quality planning process for administrative and support operations does not differ inherently in content from the quality planning process for manufacturing, but it does often differ in context and motivation. The basic quality planning steps are described in Section 3 and are summarized by Juran in many places (such as Juran 1988 (Planning), p. 14). In very brief form, they are "Identify the customers and their needs. Develop a product that meets those needs. Develop a process capable of producing the product." These steps are as appropriate for administrative and support services as they are for designing and producing an automobile or for performing open heart surgery.

Although the quality planning process for administrative and support operations may be essentially the same as for manufacturing, contextual differences include the low frequency with which brand-new processes are designed from scratch, the comparatively low power of customers to demand quality improvements, and the relative ease with which information on key elements of customer satisfaction can be gathered from internal customers.

The infrequency of establishing new systems and the lower power of internal customers to demand quality improvements from their internal suppliers (since internal customers often have difficulty taking their business elsewhere) both tend to reduce the likelihood or ease of systematic planning for quality improvement. On the other hand, the closeness of internal customers simplifies gathering data about their needs, and this makes quality planning easier and perhaps even more effective.

Whereas each new physical product or new customer service offers a clear opportunity to design in quality from the very beginning, administrative and support services tend to experience undisciplined growth and are typically already well entrenched throughout the organization before they receive systematic quality management attention. For example, when a quality improvement team at Aid Association for Lutherans (AAL) studied the organization's billing and collection process in the early 1990s, it discovered many opportunities for improvement. (The AAL experience is discussed further below, under Other Examples of Quality Improvement in Administrative and Support Activities, Finance.) The process had been designed in 1960 when AAL offered only one product, whole-life insurance. As new products were added, each received its own bill with its own billing cycle. "[Billing procedures resulting from] the piecemeal addition of new products proved cumbersome; manual procedures were often required to make payments `fit' into the existing whole-life system. The system made no one happy. AAL found it financially inefficient, and customers got multiple bills and few payment options" (Sharman 1995, p. 30).

A well-known exception to the tendency of organizations to allow administrative and support operations to emerge and grow in an unplanned manner is the AT&T Universal Credit Services (UCS) company that was formed in 1990 and won the Baldrige Prize in 1992. In UCS's case, the company's original business plan was an integrated strategic plan for business and quality that designed high quality internal administrative and support processes from the very beginning (Kordupleski et al. 1993). A year after the card was introduced, senior management used input from associates to revise its values, mission, and vision (Kahn 1995). One UCS administrative process designed from the beginning to be consistent with quality management principles and to support high-quality performance is UCS's compensation and reward system (Davis et al. 1995). This system is discussed later in this section.

Three aspects of quality planning for administrative and support operations that are important to note are the role of breakthrough goals, the use of benchmarking, and the participation of administrative and support personnel on cross-functional process re-engineering projects.

Setting Breakthrough Goals. Companies that set breakthrough improvement goals to challenge themselves to rethink the way they perform every activity provide an environment in which dramatic improvements in administrative and support operations are expected. Motorola's series of $10\times$ and $100\times$ defect-reduction goals in the 1980s and its $10\times$ cycle-time reduction goals in the 1990s are well known examples. When these goals are companywide, they involve administrative and support operations. For example, Motorola's defect-reduction goals prompted the finance function's first major quality project—reducing the time required to close the corporate books—and also was the initial stimulus for its Internal Audit function to measure its performance in new and creative ways and to demand the same level of quality of itself as the manufacturing side of the company (Stoner and Werner 1994, pp. 153–154). Subsequently, Internal Audit responded to the company's cycle-time challenge by working to reduce the time spent on each audit activity to one-tenth its prior amount (Stoner and Werner 1995, p. 122).

Competitive Benchmarking to Identify Improvement Opportunities. Competitive benchmarking has become a common practice within administrative and support functions to identify and prioritize operations for continuous and large-scale improvement. In the early 1980s, before competitive benchmarking was a widely used quality tool, Motorola began inviting guests—leaders in their fields—to speak to its people. Motorola gives credit to Westinghouse forteaching Motorola, during an invited visit, how to use quality-improvement task forces (Stoner and Werner 1994, p. 152). One of the conspicuous successes of the task forces was the introduction of competitive benchmarking into Motorola's quality efforts throughout the company. Early administrative and support functions benchmarked were the processing of accounts receivable, accounts payable, and invoices.

When Southern Pacific Transportation Company began its quality journey in the fall of 1990, it made innovative use of the Interstate Commerce Commission's R1 reports, annual submissions from all railroads detailing operating and cost data. The company was emerging from a period of ill-fated expansion that had culminated in a failed merger attempt and 5 years in trust. The R1 reports showed its costs to be higher than those of its competitors in most categories. Southern Pacific's Finance Department responded by comparing the operating and financial information of its competitors with its own R1 data, line item by line item. Included were two calculations of savings—first, the savings if Southern Pacific matched the average performance of its competitors on each line item and, second, the savings if it matched the performance of its best competitor in each activity. Southern Pacific concluded that at least $400 million could be saved by bringing current operations in each activity up to the standards of the competitors most similar on important dimensions to Southern Pacific (Stoner and Werner 1995, p. 213). The comparison, which is redone each year as new data become available, is used as Southern Pacific prioritizes its quality efforts and has been a significant factor in breaking down the belief, previously held by many in the company, that order-in-magnitude improvements in quality and costs were not possible (Stoner and Werner 1995, pp. 205–229).

Cross-functional Process Re-engineering Projects.   Many cross-functional process re-engineering teams include members from administrative and support areas. For example, a team working to improve the order-fulfillment process logically would include members from finance and accounting. This provides an opportunity for administrative and support personnel to learn about and experience quality activities, even if there are no quality initiatives being pursued within their own functions.

By participating on these teams, administrative and support staff are exposed to many of the skills of quality improvement. They experience first hand successful quality applications. They learn to appreciate better the role that administrative functions play in supporting the rest of the organization. And they are often prompted to initiate similar quality efforts in their own processes.

Quality Control.

The classic tools of quality control and the "seven new tools" (Mizuno 1988) are just as useful in administrative and support operations as they are in other operations. Gryna (1988, pp. 8–20) provided a series of examples of the use of the seven classic tools of quality control (flow diagrams, histograms, Pareto charts, run charts, cause-and-effect diagrams, control charts, and scatter diagrams) in administrative and support operations. Figure 26.2 is a flowchart for payroll changes provided by International Paper Company. Figure 26.3 provides a control chart for the time to process a freight invoice.

Quality Improvement.

The quality improvement process challenges the notion that quality problems represent immutable fate (see Section 5). It addresses performance deficiencies in stable processes that have become accepted as inevitable by organizational members, even when those deficiencies are unacceptable. The objective of the quality improvement process is to examine the deficient process for root causes of the deficiencies and remove or mitigate the effects of the root causes—permanently improving process performance and reducing the costs of poor quality associated with the deficiencies. Such improvement is carried out project by project. Quality improvement processes provide a convenient vehicle for accomplishing many goals concurrently, including improving quality, reducing costs, involving and developing people, training in managing for quality, and contributing toward the total organizational transformation required for competing in a global economy based on ever-improving quality.

As noted earlier, systematic quality improvement approaches include both the small-scale, continuous improvement efforts sometimes called *kaizen* and the large-scale, discontinuous improvements usually called *re-engineering*. There has been considerable confusion about re-engineering and its role in managing for quality. Cole (1994) provides a valuable discussion of this confusion and suggests how re-engineering fits into the tools of managing for quality.

Both these approaches, continuous and discontinuous improvement efforts, are team-oriented and seek to improve process performance. The discontinuous efforts are usually biased toward radical gain through radical change—the "clean slate" approach. Continuous improvement efforts aim at important but less ambitious gains in performance. They are generally applied to the process as it is

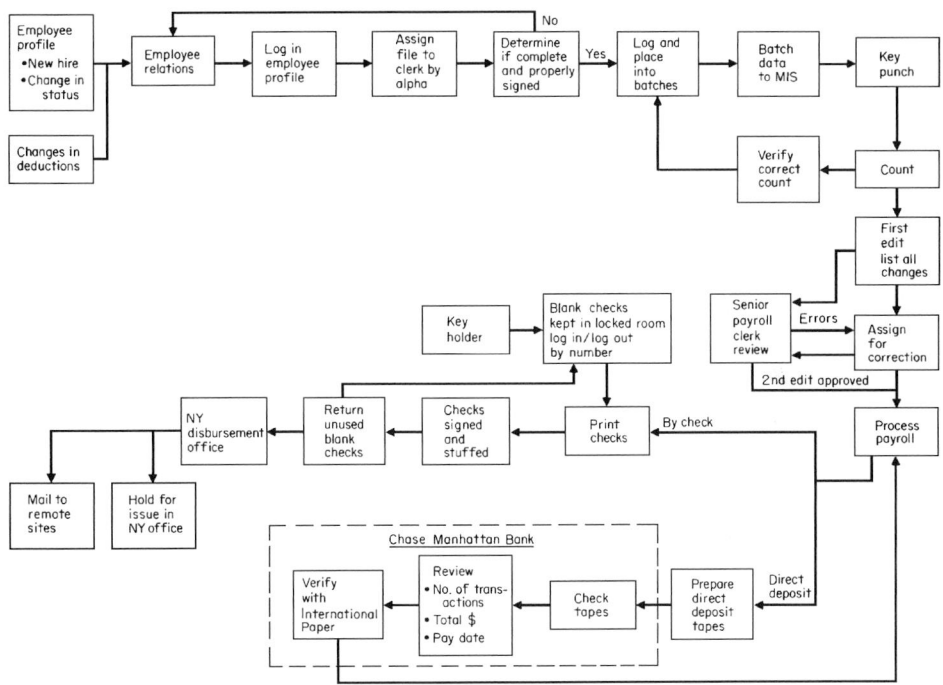

FIGURE 26.2   Flowchart for payroll changes. (*Courtesy of International Paper Company.*)

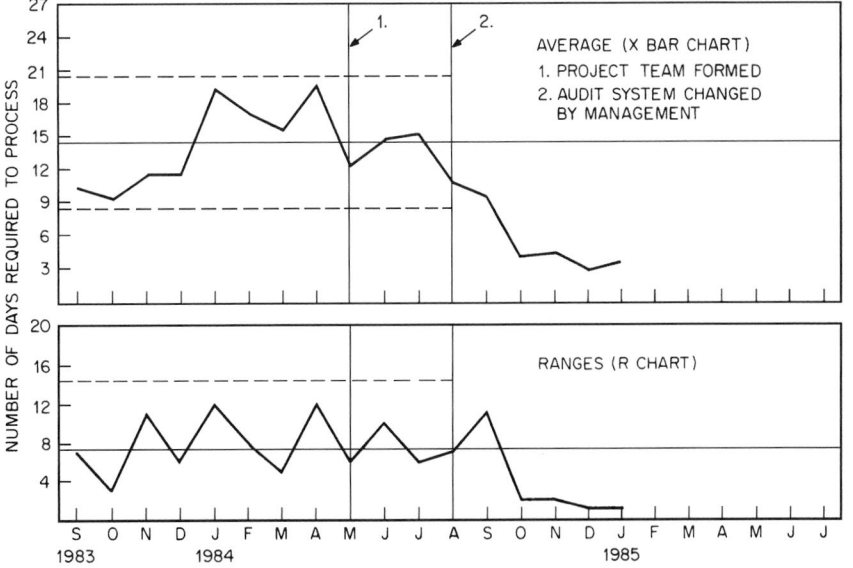

FIGURE 26.3   Control charts for time required to process bills. (*From Baker and Artinian 1985.*)

found and forego the temptation to overhaul the entire process before thorough study. The continuous approach tends to be less disruptive in two ways. First, the changes are usually incremental, leaving in place an improved but still familiar process. Second, the organizational changes required to execute the improved process are likely to be small by comparison with the discontinuous approach. Consequently, there is less near-term uncertainty as to whether the improved process will work. The more far-reaching organizational changes associated with discontinuous approaches bring with them near-term turbulence in the organization and the uncertainty of the new and unfamiliar. Some of the difficulties involved in using the discontinuous approach are discussed in Hall et al. (1993) Hammer and Stanton (1995).

OTHER EXAMPLES OF QUALITY IMPROVEMENT IN ADMINISTRA-TIVE AND SUPPORT ACTIVITIES

Overview. In an earlier part of this section we illustrated quality improvement by profiling innovative changes taking place in Internal Audit. In this part of this section we look at other examples of quality improvement throughout administrative and support units.

The examples are organized by functional area: accounting, controllership, and internal audit, finance, human resources, information systems, legal services, and marketing. However, few take place solely within the boundaries of traditional functional silos. To the contrary, many illustrate the power of quality improvements to break down these structural barriers to improved performance.

Accounting, Controllership, Internal Audit.
Accounting's traditional role of measuring and reporting business performance places it naturally at the center of many quality improvement efforts. Since collecting organizational data involves numerous repetitive steps, accounting is a particularly fertile area for process improvement. And since, as many argue, "what gets measured gets managed," accounting plays a key role in communicating what is important within an organization.

Closing the Books. Many companies have significantly reduced the length of time required to close their corporate books. In doing so, they also have reduced the cost of this repetitive corporate activity and freed up resources for use in activities that add greater value.

One of the earliest and best-known examples is the reduction in time required to close Motorola's corporate books each month. As the project began in the early 1980s, the monthly close averaged 9 working days. As a result, the updated forecast for the coming month was not available until 11 or 12 working days of that month had passed, and the month typically was two-thirds over before the operating committee meeting to review the prior month's activity and the current month's plan.

Motorola found that much of the cycle time was spent correcting erroneous journal entries, waiting for data from overseas units, and entering and correcting data in a headquarters computer. "In the past, the accuracy of journal entries—greater than 98.6 percent—had been considered acceptable. However, the company made some 600,000 entries per month worldwide, and even with 98.6 percent accuracy, 8000 entries each month were wrong. Correcting and reconciling those entries at month's end were enormously time consuming." Improvement efforts led to a decrease in monthly journal entry errors from roughly 8000 to 2. Overseas data started being forwarded directly to headquarters rather than through three intermediary locations that routinely approved the information after considerable delay. The number of local data-entry steps was reduced by bringing back in-house previously outsourced data-entry activity. These and other changes reduced the length of time required to close the corporate books to 4 days in 1990 and to 2 days in 1992. Motorola estimated that the reduction from 6 to 4 days saved $20 million per year, with another $10 million saved with the subsequent reduction from 4 to 2 days (Stoner and Werner 1994, pp. 154–155). Many other companies have initiated similar book-closing projects, e.g., Raychem's Five-Day RACE—Raychem Accounting Close Engineering (Stoner and Werner 1995, p. 158).

Reporting Nontraditional Data.   As the focus of systematic quality management on customers and processes has become more widely understood, organizations have expanded their traditional accounting-based measurement systems to capture and disseminate these new performance dimensions. Customer satisfaction data are gathered through direct questioning and surveys and are summarized in many ways, including Solectron's grading system (described below, under Finance) and AT&T's customer value-added (CVA) indexes, which compare the satisfaction of AT&T's customers with that of its competitors (Kordupleski et al. 1993). Process functioning is typically captured by measuring total errors, often weighted by importance to customers, as Federal Express does with its service quality indicators (SQIs); by measuring failures in relationship to opportunities to fail, as is done by Motorola's widely copied 6σ measurement system; and by computing process cycle time (Stoner and Werner 1993).

> Recognizing the importance of these new measures and of making them visible, the chief financial officer of IBM Credit Corporation asked each business unit to support the accounting function in adding one customer satisfaction measure and one process measure to the monthly reports of the unit's performance (interview with Stoner and Werner, November 22, 1991).

Improving Internal Audit.   Changes in corporate Internal Audit in many companies have been so numerous and so dramatic that we used Internal Audit as a comprehensive example earlier. These changes have moved these audit groups from their traditional roles as corporate police to new supportive roles that add greater value to the organization at the same time as they yield greater compliance with corporate control systems.

Finance.   The following examples, typical of quality progress in finance, illustrate successes in improving financial processes, distributing finance's work to place it closer to customers, integrating financial analysis with other operations, and improving the alignment between the organization's financial goals and those of its shareholders.

Improving the Accounts Payable Process.   Early in the 1980s, the Ford Motor Company investigated ways of improving its accounts payable process. Plans to introduce automation were developed that promised an impressive 20 percent reduction in the 500 workers then required to pay the suppliers of Ford's North American operations. A benchmarking study, however, revealed that Mazda required only 5 employees to pay its suppliers worldwide and forced Ford to rethink its improvement plans (Hammer 1990).

A subsequent study of the payables process revealed that department members spent most of their time resolving discrepancies among the three documents required for each buy: purchase order, receiving report, and invoice. If the three documents matched, a check was issued; if not, an investigation was begun to resolve the differences. Since there were 14 data items on the documents, mismatches were common, and investigations could be quite time-consuming.

Ford redesigned its accounts payable process to eliminate the receiving report and invoice and to reduce the number of items needing matching to three: supplier code, part number, and quantity. While a purchase order is still sent to the supplier, it is not copied to accounts payable; instead, the data are entered into a database accessible at all receiving docks. When goods arrive, the receiving clerk checks the computer to see if they match the order. If not, they are rejected and returned to the supplier. If they match, the shipment is accepted, and the arrival information is entered into the database, which automatically triggers payment to the supplier at the appropriate time. The new process requires only about 150 workers and saves millions of dollars every year.

Improving the Billing and Collections Process.   The simple billing and collection process Aid Association for Lutherans (AAL) designed in the 1960s for its single financial product—whole-life insurance—was appropriate for the company's and its customers' needs at that time. However, by the early 1990s, AAL, the largest fraternal benefit society in the United States, was offering to its members life and disability insurance, annuities and retirement products, mutual funds, long-term care and Medicare supplemental coverage, and savings and loan products through a credit union.

Over the years, the company had adapted the old system for each new product and product feature. However, many of the new products and features had billing patterns different from whole-life insurance, leading to unwieldy manual adjustments and software patches. For example, a customer purchasing more than one product received separate bills on different billing cycles for each.

AAL formed a team of managers and specialists in billing and collections that studied and diagramed the process, identified and surveyed customers of the process to identify their requirements, and identified, mapped to customer requirements, and prioritized some 50 product or process features. Customers asked for a single invoice for all their accounts with certain characteristics: clearly identifiable as a bill, accurate, arrives on time, easy to read and understand, and permits a variety of payment options. A parallel analysis identified as major problems various elements of cost of poor quality—internal failure costs (rework within AAL) and external failure costs (rework caught by or affecting external customers).

Over a 15-month period, the team began to re-engineer major parts of the process and also made incremental changes in existing parts. Among the incremental savings: Increased use of electronic funds transfer is expected to save $75,000 per year, and providing greater payment flexibility and improving the posting of retirement plan billing are expected to save another $50,000 to $80,000 annually. AAL expects the re-engineered process to include a single billing account for each customer, electronic transmission of new business applications from the field staff to home office processing systems, and the use of image-processing technology to minimize and automate application processing, all providing significant additional cost savings and customer satisfaction (Hooyman and Harshbarger 1994).

Checking Customer Credit-Worthiness.   Solectron, a recipient of the Malcolm Baldrige National Quality Award in 1991, is a contract manufacturer of electronic components and systems. In the company's early days, there was no need for a formal procedure for checking a customer's credit-worthiness. The company was small relative to its customers and could not afford to extend much credit. Its customers were either large, well-established companies or small firms personally known to Solectron's management. However, by the mid-1980s, Solectron and its customer base had grown to a point where extending credit was becoming increasingly important and appropriate.

Finance responded initially by developing an elaborate and traditional credit analysis process similar to commercial bank systems. The system proved cumbersome, however. Few members of the sales force had the requisite finance skills or wanted to take time from their sales activities, so credit analysis took place within the Finance Department. Salespeople lost considerable time as they contended for the attention of the company's sole credit analyst, lobbied for credit approval, and were forced to wait for the credit decision. It was difficult to be responsive to prospects. Worse, a negative decision meant that the time invested to cultivate a potential customer was wasted, leaving the sales staff and the rejected prospect frustrated and very likely angry.

Solectron simplified the process. Now Finance divides potential customers into three groups: "A customers," those which are large and financially healthy and for which no credit check is required; "B customers," the middle-sized prospects for which it is important to check credit-worthiness; and "C customers," those which are financially weak and are not to be pursued unless a special strategic rationale exists. Salespeople are free to pursue A customers without further financial review. For B customers, Finance designed a credit scoring sheet that uses easily available data, is simple for the salespeople to fill out early in the prospecting process, and sums to a number that translates directly into the credit line Solectron is willing to extend.

The new process has eliminated the frustrations of the prior system. Salespeople no longer waste time pursuing unacceptable credit risks and now know exactly how Finance will respond to a new customer. Because it is so easy to understand, the form itself teaches the sales staff how and why Finance makes the credit decision. Finance staffers are now seen as a support system and no longer as the "bad guys" who interfere with sales. And by distributing finance skills to the sales force, the company's credit specialist has eliminated much routine work and can devote more time to supporting strategic marketing decisions (Stoner and Werner 1994, pp. 186–187).

Capital Budgeting.   Several companies now view their capital budgeting activities in process terms. In 1989, Alcoa began a project to improve the company's capital expenditure decision process in response to benchmarking and other data that indicated that the company lagged behind its competitors in efficiency of capital utilization. A 16-person, cross-functional, cross-divisional, high-level team, including 7 vice presidents, reviewed past data and studied potential causes of suboptimal capital expenditure decisions. The team concluded that the company's request for authorization process was seriously flawed because it did not consistently provide the right technical and business information to make the decision, reveal the underlying quality problem-solving process used to arrive at the recommendation, provide quantitative data to allow project results to be verified, or provide appropriate accountability for project success. Among the improvements was the development of a decision analysis summary to provide project reviewers with more detailed analyses and strategic, market, and manufacturing process context (Rosenfeld 1990).

The *business case approach* (BCA) is a project analysis process used by Federal Express that emphasizes a broad companywide focus rather than a narrow financial or departmental one and the building of a team bringing diverse perspectives into the analysis at early stages. Most of the financial analysis takes place during repeated iterations as the project concept and design evolve rather than at the end, when the project design has been completed. Along the way, financial models of the project and its implications are assembled, so the repeated iterations may be analyzed as alternative approaches. The BCA process does not end when a completed proposal is presented to senior management nor when senior management makes its decision. Rather, the modeling and documentation are retained so that implementation alternatives can be evaluated at key decision points if the project is accepted, or the project can be reconsidered as new technological or market data emerge if the project is rejected (Stoner and Werner 1994, pp. 131–137).

Targeting Shareholders. The search for superior service for shareholders has led to innovative approaches in the investor-relations function. A growing activity within these functions involves creating a better match between a company's financial goals and breadth of public ownership and the investment needs of the company's shareholders.

Founded as a long-distance telephone carrier, MCI Corporation initially attracted investors looking for that industry's historic pattern of high dividends and low earnings volatility. As the company diversified into local telephone service, media services, and communications outsourcing, however, its earnings and dividend patterns could no longer be predicted to adhere to those patterns. MCI's investor-relations group responded by creating and using a database of institutional shareholder preferences to locate potential new investors (Brenner 1996).

The stock of EMC Corporation, a manufacturer of technologically innovative computer data storage systems, initially attracted a large number of short-term-oriented technical traders when the company was growing at a very rapid rate. As EMC's growth rate settled down, the company made a concerted effort to pursue shareholders with a long-term investment horizon (Brenner 1996, pp. 60–61).

In 1992, GATX Corporation began a project to create additional liquidity for its investors by expanding its shareholder base. At that time, five institutional shareholders owned 51 percent of the company's outstanding stock, and their trading activities often caused significant price fluctuations. By 1986, that 51 percent was owned by 28 institutions, another 20 percent was owned by retail shareholders, daily trading volume had increased significantly, and large institutional trades were being made without an appreciable effect on the company's stock price (Brenner 1996).

Human Resources. In Section 15 of this handbook, the term *human resources* (HR) is used to refer to the culture, relationships, and behaviors within organizations that are necessary to achieve high-quality performance on a sustained basis. This part of this section builds on Section 15 by addressing how corporate- and divisional-level human resources functions contribute to achieving these types of culture, relationships, and behaviors.

Human resources functions play key roles in enabling organizations to achieve the types of skills, attitudes, values, and other cultural changes that are required to move from traditional command and control management systems to modern quality-based management systems. The deep cultural change involved in making this transition requires that a great amount of "people change" must occur. Human resources functions are especially well placed to contribute to this transformation because the function is actively involved in designing and administering many of the levers of organizational change—such as recruiting, selecting, training, developing, rewarding, compensating, and promoting. The function is also well placed to contribute because the values that have long been "preached" by HR professionals—teamwork, collaboration, power equalization, courage in risk-taking, integrity, respecting people, seeking excellence, etc.—are far more consistent with the emerging quality-based management systems than with traditional command and control management systems.

Although some well-established human resources activities will continue to be performed much as they have been performed in the past, a large amount of change in HR practices and philosophy is needed in virtually all organizations if these functions are to play important roles in the organization's quality transformation. In some companies, these changes are taking place. Schonberger (1996) discusses how HR premises and practices are changing in organizations that are approaching world-class manufacturing status and provides many examples. Blackburn and Rosen (1993) studied the HR practices of eight Baldrige Award winners and reported "evidence of a paradigm shift in the HR policies by those organizations" (p. 50). They found that the companies had developed "portfolios" of human resource management policies to complement their strategic quality management objectives. Other companies seem to be lagging far behind. For example, a survey of human resources functions in 245 companies from a wide range of industries and regions of the United States indicated that most HR managers "talked the talk" well but were very slow in "walking their talk" (Blackburn and Rosen 1995).

The discussion below addresses ways in which traditional human resources operations are being improved; how HR's philosophy, ways of doing its traditional work, and traditional roles are changing; and a new role that is emerging for human resources functions.

Improving Traditional Human Resource Operations.   Well-established human resource operations have provided rich grounds for improvement using classic quality improvement approaches. Leonard (1986) reports how a cross-functional team developed flow diagrams to analyze and improve the process for recruiting exempt-salary personnel at Rogers Corporation. The quality improvement process also has been used to reduce employee dissatisfaction with savings and investment plans and to reduce the costs of administering these types of plans. Since very early in its existence AT&T Universal Card Services has used quality teams on an ongoing basis to improve quality process measures such as how quickly Human Resources responds to job résumés and issues employee paychecks (Hall et al. 1993).

Changing HR's Philosophy and Ways of Doing Its Traditional Work.   A number of HR functions have been wrestling with the changes necessary to bring their philosophy and practices more into alignment with the needs of quality-driven companies and the requirements of a quality-driven global competitive environment. Haddock et al. (1995), for example, used previous empirical and conceptual studies to develop a list of characteristics of the HR philosophy of organizations deeply committed to quality management. These characteristics, listed in Figure 26.4, include treating employees as valuable resources and as partners, taking a long view of the ongoing personal fits of employees to the organization, involving employees actively, and encouraging employee growth and contributions that go well beyond traditional job descriptions (Cardy and Dobbins 1996).

The differences between the HR philosophy and practices appropriate for the traditional command-and-control, Tayloristic management system and those appropriate for quality-driven organizations have been discussed by several writers, including Petrick and Furr (1995). Compared with traditional management systems, information is shared more readily and widely, customers and employees are attended to much more, and education and training go beyond a narrow job focus.

Blackburn and Rosen (1993), Haddock et al. (1995), Petrick and Furr (1995), Schonberger (1996), and other authors have suggested many HR practices consistent with managing for quality. These include

- *Staffing (recruitment, selection, and utilization):*   Going beyond screening largely or exclusively on technical job skills to looking for recruits with interest and skills in teamwork, problem solving, and quality improvement; looking at employees' fit with the larger organization rather than with a particular job; emphasizing the ongoing development and effective utilization of the organization's human resources

- Employees and professionals are treated as valuable resources, and their central role in improving quality is emphasized.
- The psychological contract between employee and employer becomes more of a partnership in which employees are recognized as key sources of competitive advantage.
- The personal fit of employees with the organization and the taking of a long-term perspective are consistently emphasized.
- Ideas and techniques of employee involvement that fit the specifics of an organization's situation are integrated or coordinated with continuous quality improvement efforts.
- Employees are encouraged to develop and to accomplish as much as they can, both qualitatively and quantitatively, within the framework of organizational needs rather than to focus on what the specific job requirements are.

FIGURE 26.4   Human resources philosophy consistent with managing for quality. (*Adapted from Haddock et al. 1995, p. 144.*).

- *Performance measurement and evaluation:*   Emphasizing developmental feedback rather than judgments of current performance, using customer and peer evaluations, placing greater emphasis on team and organizational goals than on individual ones

- *Compensation:*   Providing a "basket of values" tailored to each individual rather than only base pay and benefits (Schonberger 1996), emphasizing team and organizational incentives rather than individual incentives, emphasizing nonfinancial rewards including opportunities for self-management and recognition, paying for skills mastered rather than basing pay on job titles, experimenting with ways to avoid the problems of linking performance appraisal to compensation, experimenting with ways to avoid the problems of weakening intrinsic motivation by focusing on extrinsic rewards (Kohn 1994)

- *Training and development:*   Providing comprehensive and continuous education and training, setting progressively more aggressive goals for the amount of training provided to all employees every year, preparing for cross- and interfunctional work, providing training in group and quality improvement skills

- *Management development:*   Preparing managers for roles as facilitators and coaches; training in team building, strategy, and vision development techniques; training in leadership, self-leadership, and self-management (DiPietro 1993)

- *Relationship to organized labor:*   Seeking collaborative rather than adversarial relations with unions, particularly in activities related to quality and continuous improvement (Redman and Mathews 1998, Wilkinson et al. 1998).

- *The role of the human resource function in management:*   Emphasizing HR's key strategic role in acquiring and developing human resources, deemphasizing HR's traditional preoccupation with routine personnel activities, training employees to carry out many traditional human resource management activities on their own, seeking ways to enable employees to carry out newly emerging HR activities.

Changing Roles and Relationships.   Schonberger has noted many changes in roles and relationships in HR departments of companies becoming world-class manufacturers. HR is more involved in high-level and strategic decisions while simultaneously shrinking in size. The more mundane bureaucratic orientation of HR in traditional organizations, centered around activities such as creating job descriptions and classifications, becomes less of a priority. Instead, an increasing emphasis is placed on selection and development. HR professionals shift their emphasis to team-based and self-directed training and education rather than making HR the main "trainer" of other employees and managers. The facilitator role of HR professionals also becomes more significant (Schonberger 1996). One approach used by organizations such as IBM has been to diffuse the HR function over each major unit to enable greater customization of services rather than a "one-size fits all" approach (Caudron 1993).

On the basis of a study of HR functions and quality management in the United Kingdom, Wilkinson et al. (1993) identified five phases of an organization's transformation to quality and suggested important roles the HR function can play in each. At the beginning of the transformation—the "formulation or developmental" phase—HR can help shape the quality management approach the entire organization chooses to follow. When the organization has committed to a quality management approach—the "introductory" phase—HR's role shifts to training managers and facilitators in quality management techniques and developing vehicles to communicate the move into quality management. When significant progress on a variety of quality initiatives has been achieved, the "maintenance and reinforcement" phase has been reached, and HR's role involves maintaining momentum and sustaining a high profile for existing and new quality initiatives. One important measure to do so would involve providing incentive-compensation structures to reward progress in quality management and to evaluate and improve those structures over time. Two other phases—a companywide review and a review of the HR function's activities—overlap or are concurrent with the first three phases. In the companywide "review" phase, HR's contribution involves ongoing evaluation of the quality management infrastructure through such techniques as internal surveys and external benchmarking. Finally, in the review of HR's operations phase, HR assesses and improves both its standard HR operations and its support of its internal customers in their implementation of quality management.

Undertaking a (Almost) New Role.   The variety of roles HR can play in the various phases Wilkinson et al. (1993) describe suggests that HR functions are ideally placed to play major roles in moving their organizations from the traditional command and control management system to a modern quality-based system. Wilkinson et al. note that HR can play a "change agent" role by operating with a "high profile" at the strategic decision-making level in the organization. This potential change agent role can be similar in impact to the role designed for the Organizational Development (OD) Departments and teams from the 1960s onward—departments that frequently were located within human resources functions or closely related to them in the organizational structure. Although some "OD change" efforts were dramatic successes, many were not. Part of the difficulties OD Departments experienced arose from lack of sustained top-level support and commitment (similar to Deming's call for "constancy of purpose"). However, another major factor was surely the lack of a powerful, integrated management "technology" like the one that has created the global quality revolution (Mooney 1986, Stoner and Wankel 1990, 1991). With this new global management technology now becoming progressively better understood and with its competitive power progressively more obvious, HR functions are now very well placed to play the organizational transformation role originally intended for their organizational development departments.

Information Systems.   Jurison (1994) points out that for continuous improvement, information systems' role is "that of support, providing data collection, analysis and decision support functions" (p. 13). "Considering the fundamental importance of statistical data to quality improvement initiatives, it follows that information technology, with its ability to capture, process, and disseminate data, must play a key role in [continuous quality improvement]" (Jurison 1994, p. 4).

In making significant process change, information technology is often "an essential enabler without which the process could not be re-engineered. It has the potential to reduce organizational complexity, eliminate unnecessary work, simplify and streamline communications and coordination, and facilitate teamwork. Many re-engineered processes make use of [information systems] in replacing outmoded processes that originated before the advent of modern computer and telecommunications technology" (Jurison 1994, p. 13).

In addition to the important role of information systems in improving processes, they also are critically important in many other aspects of managing for quality. These include gathering customer satisfaction data and disseminating them throughout the organization; establishing effective communications processes such as electronic data interchange between the organization and its customers, suppliers, and other stakeholders; and enabling effective internal communications among individuals and teams with e-mail, team-based groupware, and other computer networking vehicles. For example, many quality organizations have linked their information systems with those of their customers and suppliers. Now, when a customer places an order, the company and its suppliers immediately see and can respond to the implications for deliveries of materials and scheduling of work. This type of seamless system reduces the delays and errors that might occur if requirements analysis, ordering, materials delivery, and production planning had to await other, more complex, and more error-prone flows of information. Many other examples of the use of information systems and technologies to achieve quality goals appear throughout this handbook.

Legal Services. One of the more exciting areas of quality improvement in support operations is the corporate legal activity. For some corporate pessimists, the term *legal services* is an oxymoron, and efforts to achieve high quality in legal services seem doomed to failure because of the attitudes of many attorneys. In an article on the Malcolm Baldrige National Quality Award, David Garvin (1991, p. 86) noted that "examiners use various techniques to assess horizontal deployment" (the spread of quality practices across an organization). "One examiner noted that, on site visits, he heads immediately for a company's legal or maintenance group to see how it has responded to the quality effort."

In some companies the examiners would find dramatic improvements led by attorneys who are deeply committed to the quality ethic. Lyondell Petrochemical Company, Motorola, and AT&T are three companies whose Legal Departments have achieved solid successes in using quality approaches.

Jeffrey Pendergraft, general counsel and vice president of Lyondell, described the evolution of quality management in Lyondell's Legal Department as having occurred in successive stages: customer focus, teamwork, empowerment, process analysis and problem solving, benchmarking, and supplier partnerships (Stoner et al. 1993, pp. 9–10). When the company applied for the Baldrige Award, the Legal Department "got involved in the Baldrige process and…began to understand what process analysis was all about."

> As far as process analysis is concerned, we are using what we call the "potato analysis" (the Pareto analysis is too sophisticated for us lawyer types). Everybody in the department is keeping checklists of time consuming functions. Those functions with the most checks get analyzed. The difficult thing was forcing the staff to take the time to think about the process because they were so busy putting out fires. When they did that, they found tremendous opportunities for improving productivity (Stoner et al. 1993, p. 9).

The Legal Department greatly simplified the approval of feed stock purchase contracts (which frequently were being reviewed and approved by attorneys on both sides of the transaction months after the transaction had been completed), sponsored a task force of the American Corporate Counsel Association to prepare a draft crude oil purchase agreement that would be standardized and available for use by all companies in the industry, and shortened the time to negotiate long-term contracts for petrochemical sales dramatically by turning the contracts around in 2 days. Negotiating these contracts often had taken so long that sales had been taking place for a year before the attorneys had formed an agreement.

> We were able to [turn the contracts around in 2 days] by standardizing the terms and documentation of the business transaction. We were able to set up the documentation on a document assembly system so that every contract looked like an original—not a standard form. The document assembly system was developed by a legal assistant, and it has become a key competitive advantage in the industry (Stoner et al. 1993, p. 20).

Richard H. Weise, former senior vice president, general counsel and secretary of Motorola, has written that when Motorola embarked on its quality journey:

> Quality became a religion and the law department was expected to follow. We started by going through the motions of creating law department initiatives which seemed to line up with those of the corporation (constant respect for people, uncompromising integrity, development of criteria by survey and benchmarking, six sigma quality, measurement, cycle-time reduction, process design, etc.).
>
> To my surprise, I found that each corporate initiative and imperative actually applied to the legal function very well. We began to design systems and develop processes which changed the way we were doing our work. We saw and measured increases in productivity and cost effectiveness and we witnessed marked increases in the quality of our work product, the quality of our morale and the quality of our thinking (Weise 1993–1996, vol. I, p. xxvii).

Improving the Patent Protection Process.   In 1990, Vincent Rauner of Motorola reported on the Legal Department's "first trial" of Motorola's quality improvement process—the preparation of patent applications (Rauner 1990). The importance of patents to Motorola's global competitiveness would be hard to exaggerate, and Rauner noted that the company "has dozens of attorneys preparing hundreds of applications per year," a major effort for the company and for its Intellectual Property Department. Rauner described how the project began initially with a focus on what happened within the Legal Department and then expanded—like so many quality improvement initiatives—to encompass a much broader recognition of the determinants of quality in the process. The department

> …mapped the activity in the Intellectual Property Department, initially starting with the receipt of an invention disclosure, then an optional search of the prior art, next the start of the patent drafting job, then obtaining the patent drawings, getting further information from the inventor, finalizing the documents, getting inventor signatures, and finally filing the documents in the U.S. Patent and Trademark Office. We looked at the times involved in these various steps and the particular steps where quality improvement would be significant. Then the thinking began to expand and the real issues became clearer
>
> First, we saw that the entire process from start to finish must be mapped. The starting point really is the conception of the idea by the engineer, then his initial record making, his testing of the idea, or experimentation, the decision to pursue or not to pursue for patent, all of which must occur before the project even reaches our department. Then after we file the patent application there is, of course, prosecution or arguing with the Patent Office Examiner, usually followed by patent issuance, all before the real test is encountered. The final proof of the pudding is whether the patent stands up in infringement negotiations or in a court action.
>
> Second, in our preliminary analysis, we noted the times for the various steps and the fact that such times were unnecessarily and harmfully long in many cases.
>
> Third, the customer situation was apparent. One customer is the inventor, i.e., the patent write-up must satisfy him. Another customer is the Patent Office Examiner who must be satisfied or the patent will never issue. Finally, an infringer must accept, and not find defects in the patent or it will not be protecting technology. If litigation is necessary, the judge must not find defects either (Rauner 1990).

This first patent process initiative yielded a series of improvements in quality and cycle time. The map of the process led to development of a patent-filing template that standardized "the format and the manner of writing up the application document." The template saved time and increased the likelihood that the application would be acceptable, with a minimum of modification, in most countries. The improved process allowed "better focus and more time for the difficult part of the preparation which is the drafting of claims and analyzing an optimum definition of the invention"—the place where the "'art' of the patent attorney comes into play."

In less than a year, the length of time between the date of conception of an invention and the decision to pursue a patent was reduced by about 30 percent, and the metric used by the company to indicate how rapidly the Legal Department acts on the invention idea once it is received fell by 74 percent, even with additional time taken in the critical parts of drafting of claims and defining the invention (Rauner 1990). These improvements in quality and cycle time have continued during the 1990s.

One of AT&T's many quality initiatives in its legal function also involved large investments in intellectual property. With a portfolio of approximately 25,000 active patents worldwide, the income from patent licensing was large and important to the corporation. In 1994 an Intellectual Property Process Quality Improvement Team was formed. The team was chartered to conduct a comprehensive review of AT&T's intellectual property–related processes and to determine whether those processes were maximizing the value of the corporation's patent and technology assets (Greene 1995). One of the team's conclusions was that "AT&T could earn more than five times the economic return on its intellectual property assets than was being achieved" (Greene 1995, pp. 4–34).

The original team led to seven process QITs dealing with such processes as patent assertion (defining and improving the process for identifying patent users and enforcing AT&T patent rights), licensing process (defining licensing process improvements to guide Intellectual Property Teams and the newly organized cross-functional Intellectual Property Department in licensing efforts), and collection process (monitoring and ensuring compliance with Intellectual Property Agreements). Greene concluded "that aggressive application of quality principles in a corporate headquarters division can yield dramatic improvements, even though headquarters divisions typically face shifting stakeholder expectations and budgetary uncertainties" (Greene 1995, pp. 4–30).

Reusing Knowledge to Improve Contract Preparation Processes. In the early 1990s, Motorola's Legal Department sought to achieve a "step function improvement in the computer and communications abilities of the department." In doing so, it simultaneously addressed the tendency for attorneys to keep reinventing the wheel in repetitive processes, such as drafting contracts.

Two major initiatives to improve the contract-drafting process—the development of a set of model contracts (the "forms freezer") and an accessible inventory of best-practice contractual clauses (the "clause closet")—are described in Figure 26.5. The figure is taken from an end-of-1990 report on improvement activities in the Legal Department. Both the forms freezer and the clause closet continued to be extended and improved in the 1990s. And recently, the department has developed mechanisms for capturing detailed analyses of substantive legal issues (Substantive Advice Memoranda) in computer databanks and making them more readily accessible for future use.

Preventing Quality Problems. In addition to improving many of their legal processes, Motorola's Legal Department has placed a strong emphasis on preventing quality problems in at least four major ways: moving from a reactive stance to a proactive stance, moving from dispute resolution to dispute avoidance, establishing a matrix management system, and improving the management development process in the Legal Department. Suggested procedures and the philosophy underlying each of these approaches are discussed in Weise (1993–1996).

The Motorola Legal Department's many initiatives in customer satisfaction, quality improvement, defect reduction, and cycle-time reduction all contribute to the shift from a reactive to a proactive stance. Two initiatives that require special mention are in the areas of crisis management and conducting annual client review meetings to manage the department's relationships with its internal customers.

The Legal Department has taken the lead in working with other parts of the organization to anticipate and plan for areas in which crises are likely to occur. When potentially disruptive events can be anticipated and prepared for, they are no longer considered crises, since they can be avoided or since plans and training for effective response can be in place when they do occur. In the same mode, preplanning and generalized crisis-management training reduce the damage from crises that cannot be anticipated. A key factor in such preparation involves training corporate officers and line personnel not trained in the legal issues and implications of crisis events well before those surprises occur (Weise 1993–1996, vol. III, chap. 22).

The Legal Department also has taken a more proactive stance by holding client review meetings with its major corporate customers, the operating divisions. These meetings are used to assess customer

---

LAW DEPARTMENT
1991 OMDR
Key Focus Areas for Accelerated Organizational
Change and Improved Performance

The Law Department is continuing to implement a fundamental change in the manner of carrying out its function. The focal point is a step-function improvement in the computer and communications abilities of the department.

A major part of the Law Department's function is the drafting of contracts. Recognizing the cycle-time reduction and quality improvement possibilities of computer-aided drafting, we made excellent progress during 1990 on two separate computer-aided contract drafting projects.

The first project was the development of a "forms freezer." This consists of more than 100 "fill in the blank" contracts on floppy disks available to all Motorola attorneys with PCs. The "forms freezer" works best for relatively simple standardized contracts.

For more complex, customized contracts, the Law Department developed a "clause closet." This is a computerized drafting system that uses the smallest reusable components of the Law Department's best contracts indexed on a computer so that they can be quickly retrieved and used interchangeably. These components as such are reusable in many more instances than when they were hidden in a specific contract form. We expect to continue increasing the "clause closet" database and thus continue to increase the quality and shorten the cycle time in preparation of contracts.

FIGURE 26.5 Creating a "forms freezer" and "clause closet" at Motorola. (*From Weise 1993, pp. 11–27, exhibit 2B.*)

satisfaction, report on costs charged for the services supplied, assess the effectiveness of the services, jointly develop plans for improved service at lower cost, and anticipate and plan for the coming years' legal work for the division (Weise, private communication, June 11, 1996; Weise 1993–1996, vol. I, chap. 8; Stoner et al. 1993, pp. 30–33).

One of the greatest payoffs from the Legal Department's investments in quality management has arisen from the department's success in building the alternative dispute resolution (ADR) philosophy and mechanisms into Motorola's corporate culture. ADR involves two major approaches to resolving disagreements: one involves adjudicatory procedures—such as arbitrators, referees, or private judges—that replace the traditional binding decision of a judge or jury. The second involves facilitated or structured negotiations, such as settlement conferences, summary jury trials, minitrials, and mediation (Weise 1993, vol. I, pp. 6–23). ADR offers many advantages over the slow, costly, and relationship-damaging traditional litigation process. In addition to large savings in time and costs, the ADR process is also more conducive to effective intraorganizational and interorganizational learning, improving the chances that individual or collaborative steps are taken to avoid similar disagreements in the future (Weise 1993, vol. I, chap. 6).

Redesigning the Legal Department. Motorola's Legal Department is using matrix management approaches to reduce the dangers of bureaucracy in the department. Weise believes that

> …matrix management needs to be addressed in all corporate legal departments. Lawyers don't need traditional management, and corporate law departments are very likely to be over-managed. Law firms have traditionally been filled with legal entrepreneurs, and have been vertically strong, but horizontally weak. As corporate law departments grow, they tend to become bureaucratic and this tendency has to be avoided (Weise, private communication, June 11, 1996).

Efforts to develop a more fluid organizational structure and less hierarchical and formal working climate include strong support for participative management practices—long a part of Motorola's management approach—the use of many quality improvement teams and task forces, and training and empowering nonattorneys to take over work traditionally done by attorneys—when that work cannot be eliminated altogether. Initiatives like these at Motorola are consistent with Samborn's (1994) description of ways paralegals contribute to and benefit from efforts that empower nonattorneys to perform work normally reserved for attorneys and to participate in quality improvement initiatives. Samborn reports that many paralegals embrace systematic quality improvement approaches, and some have become quality leaders in legal departments and law firms.

One of Motorola's recent major initiatives involves improving the professional and management development process in the Legal Department. In 1993, the department undertook a major review and redesign of those processes when survey results of the Legal Department's members showed considerable concern about the extent to which they had understandable, meaningful, and satisfying career plans and paths. The review of existing "personnel-type" programs yielded a surprisingly large number of programs with significant amounts of overlap in intents and purposes but without clear job descriptions and comprehensive job performance criteria. A flowchart of legal organization personnel-related activities was developed and helped the department conclude that a complete overhaul of the processes was required. A management task force was commissioned that developed a comprehensive professional performance system (Weise 1993–1996).

Marketing. Hurley (1994, p. 45) suggests that marketing organizations are often slow to embrace systematic quality efforts, due, in part, to a belief that marketing is already "doing quality" given its traditional focus on the external customer. However, there are many areas in which substantial progress is being made.

Hearing "the Voice of the Customer." Companies adept at systematic quality management place few barriers between the voice of the customer and all parts of their organization. This principle played an important role as Toyota Motor Sales USA constructed its marketing function. All quality information is kept in a central database accessible to all parts of the organization, and customized reports and graphs from this database can be generated at workstations in all field locations. Data are collected through hundreds of thousands of contacts—surveys, telephone calls, etc.—with customers annually. The data are broken down by dealer and downloaded to each dealership to spur local corrective action in response to customer dissatisfaction. A customer satisfaction committee meets monthly to assess satisfaction levels, review plans to improve satisfaction, and communicate status and progress to top management. Based on these findings, senior management sets priorities and allocates resources to improve customer satisfaction (Hurley 1994, pp. 47–48).

Xerox USA surveys every customer after installation of one of its machines, conducts random telephone interviews with customers throughout the year, and benchmarks its service against its competitors on key dimensions every other year. Senior executives also are assigned to take customer complaint calls on a rotating basis (Hurley 1994, p. 48).

Solectron asks its customers to complete and fax back a weekly report card asking about the company's performance on the dimensions of product quality, on-time delivery, communication, service responsiveness, and overall company performance. Customers score the company on a letter-grade scale with a matching point score for each letter: A (100 points), A− (90), B (80), B− (75), C (0), and D (−100). The only acceptable score is a straight A. Solectron holds weekly early-morning meetings, open to all employees, at which the data are discussed, responsibility for investigating and remedying low scores is taken, and improvements are reported and shared (Stoner and Werner 1993, p. 98).

Defining Quality in Customers' Terms.   Companies that explore their customers' needs in-depth often learn how to expand their definition of their own products and services to their competitive advantage. Federal Express learned to see its service as going beyond package delivery to include package tracking and sophisticated ordering and billing options. Texas Instruments Defense Systems and Electronics discovered that its customers want a total-value solution and not always the one that is most technologically elegant. Toyota USA realized that its customers are as interested in the quality of their purchase and service experiences as they are in the car itself (Hurley 1994, p. 49).

Marketing research in quality organizations helps define the relative importance to customers of various quality dimensions. For example, Xerox discovered that there were three segments to the market for copiers, each with a different definition of quality. Customers in the low-volume segment (<5000 copies per month) primarily wanted reliability. Customers in the high-volume segment (>100,000 copies per month) valued reliability but also wanted service features such as fast response time in the event of a breakdown (Hurley and Laitamaki 1995, p. 65).

Federal Express conducts surveys to compile its "Hierarchy of Horrors," a list of the service failures of most concern to its customers. These are weighted according to their importance as reported by the customers and are then used to create a daily service quality indicator (SQI), a measure of companywide service quality that is posted throughout the organization. Federal Express's goal is to drive the SQI toward zero by systematically eliminating failures in meeting customer expectations, with an emphasis on those failures most distressing to its customers. (Hurley 1994, p. 48; Stoner and Werner 1993, pp. 42–43).

AT&T conducts surveys to discover the relative importance to its customers of each of its business processes and their characteristics and then uses these data to prioritize its quality-improvement efforts (Hurley and Laitamaki 1995, p. 66). Texas Instruments, Toyota USA, and Xerox are among the companies that use forms of quality function deployment (see also Section 3) to align their goals and plans with customer needs (Hurley 1994, p. 49).

Nurturing Long-Term Customer Relationships.   One change in marketing practice driven by systematic quality is a move away from "short-term conquest marketing, in which potential customers are reacquired each period, to a long-term approach that creates loyal buyers over the company's life span by consistently delivering quality and value to them." Federal Express identified its most profitable customers and organized around understanding and meeting their business needs. Globe Metallurgical creates cross-organizational teams composed of both customer and Globe staff to align its operations with its customers' strategic plans. Marriott's computer system tracks the preferences and purchases of its best customers so that employees can personalize their conversations with these customers and anticipate their requests, creating a stronger relationship (Hurley 1994, p. 46).

Improving Marketing Processes.   Zytec Corporation redesigned its order process to reduce cycle time and cost when it discovered that it took longer for a salesperson to process an order than for the company to manufacture the product (Hurley 1994, p. 44).

After reducing the manufacturing cycle time for its pagers to well under 1 day, Motorola discovered that it had created an opportunity to redesign its order process. In the new process, a salesperson with a portable computer can take an order in the morning, transmit it by modem to the factory by midday, and promise delivery by air express by the next morning.

Competing via Quality.   For some companies, having a well-functioning quality program is becoming necessary to compete successfully. Many companies, including Federal Express, Ford, and Xerox now expect, and sometimes demand, that their suppliers meet increasingly higher quality standards. These same companies report that their own, well-recognized quality programs provide greater access to potential customers by increasing their credibility (Hurley 1994, p. 47).

CONCLUSION

Systematic quality improvement in administrative and support operations historically has lagged behind quality improvement in production and operations activities. However, an increasing number of organizations are discovering that the opportunities for quality gains in administrative and support areas are often as great as opportunities elsewhere in the organization. Administrative and support functions are learning to use the same quality improvement tools in the same ways as other parts of the organization. The results are the same as well: increased quality, customer satisfaction, and revenues and decreased costs through defect reduction, cycle-time reduction, the development of new services, and the elimination of burdens placed on customers and suppliers. The beneficiaries of these quality improvements include both internal and external customers.

In implementing systematic quality management, administrative and support personnel find that both they and their departments undergo significant changes in roles and relationships with others. Perhaps the most fundamental change is from a traditional role of scorekeeper and corporate police to team member and facilitator.

The kinds of changes described in this section are likely to continue to evolve as organizations invent and discover new systems, structures, and relationships more and more consistent with a total quality culture. All organizational members should expect and plan for revolutionary rates of quality improvement throughout administrative and support operations, just as they should expect and plan for such improvements in the products and services they provide to their external customers.

REFERENCES

Alexander, M., and Young, D. (1996). "Strategic Outsourcing." *Long Range Planning,* vol. 29, no. 1, pp. 116–119.

Baker, Edward M., and Artinian, Harry L. (1985). "The Deming Philosophy of Continuing Improvement in a Service Organization: The Case of Windsor Export Supply." *Quality Progress,* June, pp. 61–69.

Blackburn, Richard, and Rosen, Benson (1993). "Total Quality and Human Resource Management: Lessons Learned from Baldrige Award–Winning Companies." *Academy of Management Executive,* vol. 7, no. 3, pp. 49–66.

Blackburn, Richard, and Rosen, Benson (1995). "Does HRM Walk the TQM Talk?" *HRMagazine,* vol. 40, no. 7, pp. 69–72.

Brenner, Lynn (1996). "Shareholder Targeting." *CFO,* February, pp. 57–61.

Cardy, Robert L., and Dobbins, Gregory H. (1996). "Human Resource Management in a Total Quality Organizational Environment: Shifting from a Traditional to a TQHRM Approach." *Journal of Quality Management,* vol. 1, no. 1, pp. 5–20.

Caudron, Shari (1993). "How HR Drives TQM," *Personnel Journal,* vol. 72, no. 4, pp. B48ff.

Champy, James (1995). *Re-engineering Management.* Harper Business, New York.

Cole, Robert E. (1994) "Reengineering the Corporation: A Review Essay." *Quality Management Journal,* vol 1, no. 4, July, pp. 77–85.

Davis, Robert, Rosegrant, Susan, and Watkins, Michael (1995). "Managing the Link Between Measurement and Compensation." *Quality Progress,* February, pp. 101–106.

DiPietro, R. A. (1993). "TQM: Evolution, Scope and Strategic Significance for Management Development." *Journal of Management Development,* vol. 12, no. 78, pp. 11–18.

Garvin, David A. (1991). "How the Baldrige Award Really Works." *Harvard Business Review,* November-December, pp. 80–95.

Glasser, Gerald J. (1995). "Quality Audits of Paperwork Operations—The First Step Toward Quality Control." *Journal of Quality Technology,* vol. 17, no. 2, pp. 100–107.

Greene, R. Michael (1995). "Intellectual Property Management in a Dynamic Organizational Context." *1995 Management for Quality Research and Development Symposium.* Juran Institute, Wilton, CT, pp. 4-29–4-38.

Gryna, Frank M. (1988). "Administrative and Support Operations," in Juran, J. M. (ed.). *Juran's Quality Control Handbook,* 4th ed. McGraw-Hill, New York, pp. 21.1–21.23.

Haddock, Cynthia Carter, et al. (1995). "The Impact of CQI on Human Resources Management." *Hospital and Health Services Administration,* vol. 40, no. 1, pp. 138–153.

Hall, Gene, Rosenthal, Jim, and Wade, Judy (1993). "How to Make Reengineering *Really* Work." *Harvard Business Review,* November-December, pp. 119–131.

Hammer, Michael (1990). "Reengineering Work: Don't Automate, Obliterate." *Harvard Business Review,* July-August, pp. 104–112.

Hammer, Michael, and Champy, James (1993). *Reengineering the Corporation.* Harper Business, New York.

Hammer, Michael, and Stanton, Steven A. (1995). *The Reengineering Handbook.* Harper Business, New York, pp. 14–33.

Harkins, Philip J., Brown, Stephen M., and Sullivan, Russell (1996). *Outsourcing and Human Resources: Trends, Models, and Guidelines.* LER Press, Lexington, MA.

Higgins, Brian K., and Dice, Christopher M. (1984). "Quantifying White Collar Functions." *National Productivity Review,* Summer, pp. 288–302.

Hodgetts, Richard M., Luthans, Fred, and Lee, Sang M. (1994). "New Paradigm Organizations: From Total Quality to Learning to World-Class." *Organizational Dynamics,* vol. 22, no. 3, pp. 5–19.

Hooyman, Judith A., and Harshbarger, Richard W. (1994). "Re-engineering the Billing and Collections Process." *IMPRO-94 Conference Proceedings.* Juran Institute, Wilton, CT, pp. 7A-3–7A-18.

Hurley, Robert F. (1994). "TQM and Marketing: How Marketing Operates in Quality Companies." *Quality Management Journal,* July, pp. 42–52.

Hurley, Robert F., and Laitamaki, Jukka M. (1995). "Total Quality Research: Integrating Markets and the Organization." *California Management Review,* vol. 38, no. 1, pp. 59–78.

Imai, Masaaki (1986). *KAIZEN: The Key to Japan's Competitive Success.* McGraw-Hill, New York.

International Paper Company (1985). "CTQ Flow Charting," in *Quality Management Concepts.* IPCO, New York.

Juran, J. M. (1988). *Juran on Planning for Quality.* Free Press, New York.

Juran, J. M. (1989). *Juran on Leadership for Quality: An Executive Handbook.* Free Press, New York.

Jurison, Jaak (1994). "The Role of Information Systems in Total Quality Management." *Knowledge and Policy,* vol. 7, no. 2, pp. 3–16.

Kahn, Paul G. (1995). "Pacing the Quality Race." *Credit World,* vol. 83, no. 5, pp. 24–26.

Keating, Patrick J., and Jablonsky, Stephen F. (1990). *Changing Roles of Financial Management—Getting Close to the Business.* Financial Executives Research Foundation, Morristown, NJ.

Kohn, Alfie (1994). *Punished by Rewards.* Houghton Mifflin, Boston.

Kordupleski, Raymond E., Rust, Roland T., and Zahorik, Anthony J. (1993). "Why Improving Quality Doesn't Improve Quality (Or Whatever Happened to Marketing?)." *California Management Review,* vol. 35, no. 3, pp. 82–95.

Lacity, Mary C., Willcocks, Leslie P., and Feeny, David F. (1996). "The Value of Selective IT Sourcing." *Sloan Management Review,* vol. 37, no. 3, pp. 13–25.

Latzko, William J. (1985). "Process Capability in Service and Administrative Operations." *ASQC Quality Congress Transactions.* Milwaukee, pp. 168–173.

Leonard, Joseph W. (1986) "Why MBO Fails So Often." *Training and Development Journal,* vol. 40, no. 6, June, pp. 38–39.Makosz, P. G., and McCuaig, B. W. (1990*a*). "Is Everything under Control? A New Approach to Corporate Governance." *Financial Executive,* January-February, pp. 24–29.

Makosz, P. G., and McCuaig, B. W. (1990b). "Internal Audit—Ripe for a Renaissance." *Internal Auditor,* December, pp. 43–49.

Mizuno, Shigeru (ed.) (1988). *Management for Quality Improvement: The 7 New QC Tools.* Productivity Press, Portland, OR.

Mooney, Marta (1986). "Process Management Technology." *National Productivity Review,* vol. 5, pp. 386–391.

Petrick, Joseph A., and Furr, Diana S. (1995). *Total Quality in Managing Human Resources.* St. Lucie Press, Delray Beach, FL.

Rauner, Vincent J. (1990). "Pursuing Quality in Patent Applications." *IMPRO-90 Conference Proceedings.* Juran Institute, Wilton, CT, pp. 2C-1–2C-9.

Redman, Tom, Mathews, Brian (1998). "Service Quality and Human Resources Management: A Review and Research Agenda." *Personnel Review,* vol. 27, no. 1.

Redman, Tom, Mathews, Brian, Wilkinson, Adrian, and Snape, Ed (1995). "Quality Management in Services: Is the Public Sector Keeping Pace?" *International Journal of Public Sector Management,* vol. 8, no. 7, pp. 21–34.

Rosenfeld, Manny (1990). "A Quality-Based Capital Decision Process." *IMPRO-90 Conference Proceedings.* Juran Institute, Wilton, CT, pp. 7D-24–7D-36.

Samborn, Hope Viner (1994). Total Quality Management." *Legal Assistant Today,* November-December, pp. 40–43.

Schonberger, Richard J. (1996). *World Class Manufacturing: The Next Decade.* Free Press, New York.

Sharman, G. K. (1995). "Breaking Through in the Finance Function." *The Total Quality Review,* March-April, pp. 29–33.

Stoner, James A. F., and Wankel, Charles B. (1990). "World Class Managing: Two Pages at a Time," Book I. Graduate School of Business, Fordham University, New York, unpublished manuscript.

Stoner, James A. F., and Wankel, Charles B. (1991). "Teaching the New Global Management Paradigm: Five Years' Experience." *Academy of Management Best Papers Proceedings.* Miami Beach, pp. 126–130.

Stoner, James A. F., and Werner, Frank M. (1992*a*). *Remaking Corporate Finance—The New Corporate Finance Emerging in High-Quality Companies.* McGraw-Hill, New York.

Stoner, James A. F., and Werner, Frank M. (1992*b*). "Changing Roles of Finance: Strategies for Bringing Finance into Quality Management." CFRI Conference, Financial Executives Institute, New York, November 16.

Stoner, James A. F., and Werner, Frank M. (1993). *Finance in the Quality Revolution—Adding Value by Integrating Financial and Total Quality Management.* Financial Executives Research Foundation, Morristown, NJ.

Stoner, James A. F., and Werner, Frank M. (1994). *Managing Finance for Quality: Bottom-Line Results from Top-Level Commitment.* ASQC Quality Press, Milwaukee, WI, and Financial Executives Research Foundation, Morristown, NJ.

Stoner, James A. F., and Werner, Frank M. (1995). *Internal Audit and Innovation.* Financial Executives Research Foundation, Morristown, NJ.

Stoner, James A. F., et al. (1993). "The Quality Revolution in Legal Practice: Removing the Oxymoron from 'Legal Service.'" Graduate School of Business, Fordham University, New York, unpublished manuscript.

Sullivan, L. P. (1986). "The Seven Stages in Company-Wide Quality Control." *Quality Progress,* May, pp. 77–83.

Weise, Richard H. (1991–1996). *Representing the Corporate Client: Designs for Quality,* vols. I, II, III, and periodic supplements. Prentice-Hall Law and Business and Aspen Law and Business, Aspen Publishers, Englewood Cliffs, NJ.

Wilkinson, Adrian, Marchington, Mick, and Dale, Barrie (1993). "Enhancing the Contribution of the Human Resource Function to Quality Improvement." *Quality Management Journal,* October, pp. 35–46.

Wilsinson, A., Redman, T., Snape, E., and Marchington, M. (1998). *Managing with Total Quality Management.* Macmillan, New York.

---

[1]In the Fourth Edition, material for the section on administrative and support operations was supplied by Frank M. Gryna.

SECTION 27
PROCESS INDUSTRIES

Donald W. Marquardt

QUALITY MANAGEMENT IN THE PROCESS INDUSTRIES

In the process industries, as in other sectors of the economy, Quality Management is the umbrella framework for managing the quality of a product. The philosophy, managing procedures, and technology should provide an operational system in which Marketing, Research and Development, Production, and Support personnel can work together to meet increasingly stringent customer requirements.

The system must deal with all facets of a product's life span from the product's conception through commercialization and subsequent improvements, as shown in Figure 27.1. When a product initially is developed, the emphasis is on designing quality into the product through optimizing functionality and producibility. For established products, the emphases are on maintaining and continually improving product conformance to quality requirements.

The most significant quality improvements are accomplished in those businesses that broadly implement a quality management system. The "system" feature:

- Provides an implementation process
- Interconnects the operational techniques
- Requires and expedites communications in the organization.

The system feature is the vehicle that drives quality improvements as a business strategy. The system approach is particularly important in the process industries.

CHARACTERISTICS OF THE PROCESS INDUSTRIES

The "process industries" have special needs in the technology for quality management. The process industries typically have continuous processes, or batch processes with many batches per year of a given product type.

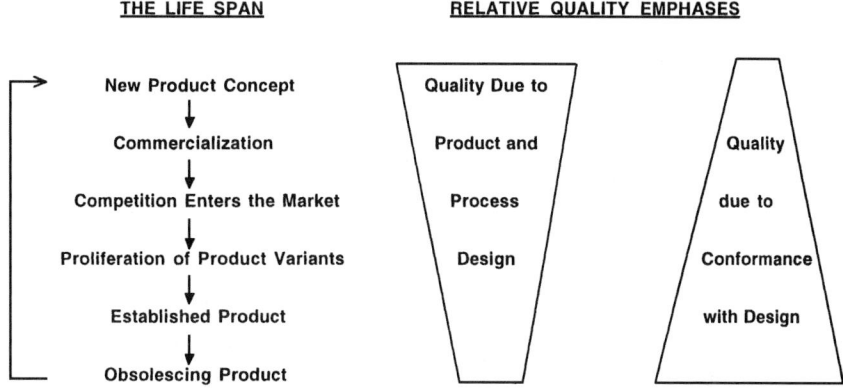

FIGURE 27.1 The product life span. [*From Marquardt (1991).*]

Typical products of the process industries are:

- Solid materials:

  Pieces, particles, powders

  Discrete or continuous sheets

  Chopped or continuous filaments

- Liquid materials:

  High and low viscosity

- Gaseous materials:

Such products can be sampled in specified volumes or weights at specified production points to measure a quality property. The property value may vary throughout a volume or quantity of product, and may change with time. Often the relationship between a measured property and the functionality of the end-use product is not fully understood.

Frequently, the process industry product is an input material for the customer's process. The product must meet this customer's processing requirements as well as requirements of the end user of the ultimate product. Such products are examples of the generic product category called "processed materials."

The foregoing features distinguish the process industries from the mechanical industries, where the emphasis is on the making of parts or the assembly of parts, that is, products of the generic product category called "hardware."

Most existing quality control methodology was developed for hardware products of the mechanical industries, so it may not be too surprising to find that emphases and methodologies in this section are often different from the traditional quality control literature. However, small parts made in large quantities often benefit from the statistical methodologies for processed materials.

DEALING WITH LARGE MEASUREMENT VARIABILITY

A primary difference between the process industries and the mechanical industries is the amount of variability associated with measurement processes. In the mechanical industries, measurement variability often is, or is perceived to be, small or negligible, usually less than 10 percent of total variability. Many measurements in the mechanical industries are based on properties that have absolute or near absolute reference standards, such as dimension, weight, or electrical or optical properties.

These measurements often are of a nondestructive nature, so that the same sample of material can be measured multiple times.

In the process industries, measurement variability is typically larger, often about half, and occasionally as high as 80 percent of the total variability. Most process industry measurements are complex, highly specialized, and not traceable to absolute standards. Examples are measurements of:

- Relative viscosity of a polymer
- Dyeability of a textile yarn
- Speed of a photographic film
- Strength of a plastic film

Many of these measurements involve destructive testing as well, so that local product nonuniformities cannot be disentangled from the measurement variability. In the process industries, measurement variability must therefore be taken explicitly into account in virtually all quality management activities. These include setting product specifications, control of the production process, product characterization and release, and planning experiments to seek improvements.

In recent years, products of the mechanical industries increasingly require tighter manufacturing tolerances, and incorporate components or mechanisms that have process industry characteristics. The technology in this section should be particularly attractive to mechanical industry producers facing such trends.

FUNDAMENTAL CONCEPTS

Generic Product Categories. International standardized terminology (e.g., ISO 9000-1:1994) recognizes that the term "product" encompasses four "generic product categories:"

- Hardware
- Software
- Processed materials
- Services

(See Section 11, Table 11.4, for further detail.)

Every Product Is the Result of a Process. The term "production process" has wide applicability. For example, the product units from a production process may be:

- "Processed materials" from a process industry company
- "Widgets" from a hardware manufacturer
- "Test results" from a laboratory
- "Confirmed reservations" from a travel agency
- "Sales made" from a sales organization
- "Deliveries completed" from a distribution organization

Each of these situations involves a production process. Each process creates product units that have measurable properties relating to quality as perceived by the customer. For example, the customers of a distribution organization may perceive quality in terms of the timeliness of deliveries, delivery to the proper destination, and physical condition of the delivered item.

It is important to approach the subject of quality from this viewpoint that *all work is accomplished by a process,* whether in Marketing, Manufacturing, Delivery, Research and Development, Personnel, or other functions. See ISO 9000 (1994).

Facets of Quality.   Discussion and communications about quality have often been unproductive because of failure to distinguish among the four facets of quality (see ISO 9000 1994). The four facets are described in Section 11. Two facets (quality of product design and quality of conformance with product design) receive explicit discussion in this section because of specific needs in the process industries.

Quality of Product Design.   To meet customer needs a product design has to provide the intended characteristics and functionality. For example, when different chemical forms of processed materials are designed for the same markets, we may judge a particular product to have superior quality of product design because it offers additional or enhanced features that improve functionality. Quality of product design can be quantitatively measurable characteristics such as strength, speed, chemical resistance, or subjective characteristics like styling, texture, or odor.

There is an important link between quality of product design and quality of conformance with design. If a product consistently provides its intended functionality despite typical variations in the environments in which the product is produced and used, the product design is said to be "robust." The attainment of robustness in product design has received a great deal of emphasis in recent years.

Quality of Conformance with Product Design.   Quality of conformance with design (or, more simply, quality of conformance) refers to the uniformity of the characteristics and the consistency of functionality of all product units produced day after day. Good quality of conformance means that the characteristics, properties, features, and functionality consistently satisfy their intended specifications.

A BALANCED APPROACH TO ACHIEVE EXCELLENCE IN QUALITY

The Quality Management Success Triad.   Quality management requires three coequal and interrelated facets (Marquardt 1984):

- Quality philosophy and policy
- Quality management systems
- Quality technology systems

These three facets form the quality management success triad (Figure 27.2). The achievement of continued excellence in managing the quality of products requires that balanced attention be given to all three facets and the linkages among them. Too often, quality management systems have emphasized only one or two of these facets.

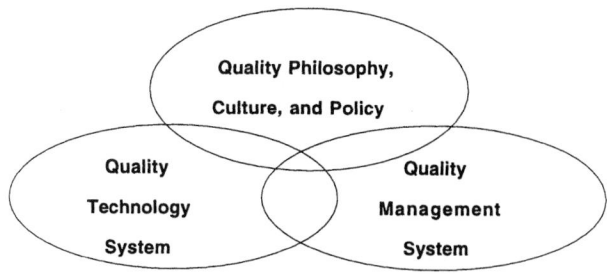

FIGURE 27.2   The quality management success triad. [*From Marquardt (1991)*.]

Quality Philosophy, Culture, and Policy.   The operation of any enterprise reflects the underlying philosophy of those who lead it. Thus, philosophy is the first facet. It affects the operation of financial systems, the choice of product types, the attitudes on external social issues, the approach to employee safety and health, and many other aspects of the business. In short, the management philosophy shapes the culture of the organization.

The process that is most central to the role of top management is the strategic planning process. If quality management principles are effectively applied in top management's strategic planning, that process can be a mechanism to drive quality leadership throughout the enterprise. (See Section 14, Total Quality Management.)

Quality Management System.   The second facet is the management procedures that are used to achieve, monitor, control, and improve the functional, financial, and human relations performance of the enterprise. Management procedures include mechanisms for allocating, organizing, and improving resources.

Management procedures should incorporate quantitative metrics and other criteria ("report cards") to monitor and evaluate the performance of the organizational units and personnel. Metrics which have exclusive focus on costs, yields, and output, provide disincentives toward achieving high product quality. Thus, report-card design is a key element of quality management. This is an important but difficult area because many disincentives are subtle and not easily foreseen during quality systems design (Marquardt 1994).

Quality Technology System.   Quality Management strategy is complete only when we include the third facet—technology elements that are used to achieve, monitor, control, and improve the quality of the products. To appreciate the importance of this third aspect, we need only look at the effect that new technology has had throughout world history. Repeatedly, new technology has been the driving force behind great changes in the arts, in philosophy, in life style. New technology gives birth to new tools, which affect economic costs and opportunities, and are the driving force behind changes in the strategy of competition between individuals, businesses, or nations.

QUALITY PROBLEMS IN THE PROCESS INDUSTRIES

Systems Problems Versus Worker Problems.   Many quality professionals of wide experience (Deming 1967, 1972, 1982; Juran 1974), have observed that about 85 percent of quality problems are management or systems problems and only 15 percent are worker problems. This experience applies to the chemical process industries. A management or systems problem is one that the individual production worker did not create, has no influence over, and usually does not have the proper information and tools to diagnose. Diagnosis and correction of such systems problems are management responsibilities, although production workers often play a crucial role.

Chronic Quality Problems.   In processed material manufacturing it is particularly important to distinguish between chronic problems and rare-event quality problems. For an established product that has been in production for some time, a good quality management system can dramatically reduce or eliminate "chronic" quality problems. (See Section 5, The Quality Improvement Process.) Some symptoms of chronic quality problems in the process industries are

- A large proportion of total production is affected.
- A large proportion of product is downgraded from first quality, using euphemistic labels such as "special lots," "subcodes," etc.
- A regular practice exists of segregating product for shipment to specific customers on the basis of their specific quality needs.

- First-pass yields to first-grade product are chronically low.
- Complex pricing arrangements persist for specific customers and/or lots, where the prices do not stem from actual differences in mill costs or marketing costs.

The most severe symptom (and consequence) of chronic quality problems is

- Erosion of market share, specifically due to better quality of competitive product

True Rare-Event Quality Problems.   In sharp contrast to chronic problems are the true rare-event problems. A true rare-event quality problem is a source of product nonconformity presumed to be entirely absent in all normal product. In processed materials these typically are nonconformities that occur because of occasional equipment malfunctions, occasional bad supplies of raw material, and the like. These rare-event problems are often called sporadic problems. (See Section 5, The Quality Improvement Process). The key symptoms are

- A very small proportion of total production is affected, typically only a small fraction of 1 percent of product units over a year's time. True rare-event problems are not representative of the usual production population.
- Each instance of quality breakdown can be presumed to be due to a specific unusual malfunction. These malfunctions should be sought out and corrected. Often the cause can only be diagnosed by analysis of patterns of rare-event failures whose occurrences have been recorded over an extended period of time.

Routine sampling of products will never be frequent enough to detect and weed out a majority of the "true rare events" that do occur. When detection is necessary because of a high economic stake for a particular type of rare event, it is usually better to monitor the appropriate production process conditions continuously and to trigger an alarm (or better still, initiate an automatic corrective action) when a process breakdown occurs, rather than attempt to discover the "needle in a haystack" by extensive testing of final product. The responsibility for elimination of rare events should be focused toward personnel in product design, production process design, and maintenance, rather than toward the process control and product release personnel, who can neither detect nor correct most true rare events in the normal course of their work.

Chronic Quality Problems That Appear to Be Rare Events.   Often in the process industries quality problems appear to be rare events, but really are chronic quality problems. These "apparent rare events" arise as follows:

- *Slippage of the process average:*   As illustrated in Figure 27.3a, the output from an "on-aim" process will have a distribution of values for any product characteristic (i.e., property). If, as in Figure 27.3b, slippage occurs causing the process average to deviate from its aim point, then a small fraction of the product coming from a tail of the distribution may be outside acceptable limits. If the quality management system does not alert production personnel to the slippage of the average, these apparent rare events may only be detected by the customer, and may result in customer complaints. These apparent rare events are representative of the usual production population when such slippage occurs, and the occurrence of such problems can be detected early and can be eliminated by standard quality management procedures using regular sampling methods.
- *Low-count defects:*   Many important product characteristics, such as defects or nonconformities, are quantifiable only by counting their frequency of occurrence in a sample of the product. A product characteristic that gives a low count (perhaps fewer than one nonconformity per sample on the average) may, nevertheless, be a phenomenon that is continuously or usually present in normal product, even though only infrequently counted in a typical routine sample. Such counts are representative of the usual production population and when their frequency increases (a form of process slippage), the problems can be detected and can be eliminated by quality management procedures.

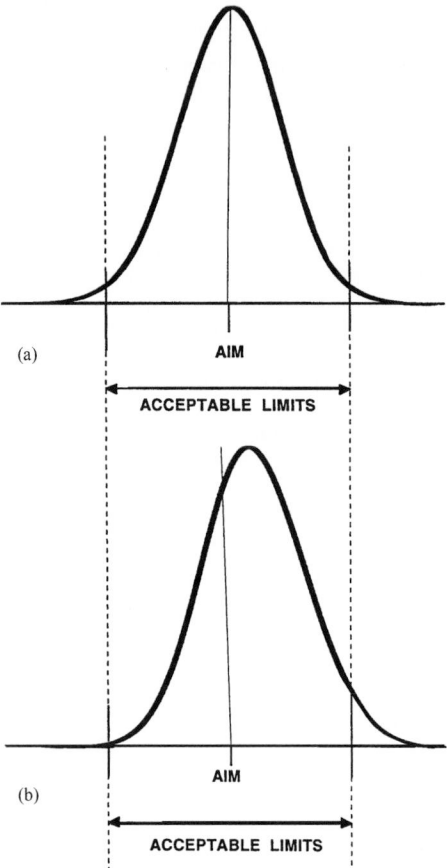

FIGURE 27.3   Effect of slippage of process average.
(*a*) Process average is on-aim; negligible proportion of
product is outside acceptable limits. (*b*) Process average
is off-aim; larger proportion of product is outside acceptable limits. [*From Marquardt (1991).*]

In many production organizations, the "firefighting" effort spent chasing after both true and
apparent rare-event problems exceeds the effort spent on actually solving the chronic problems.
Some firefighting is almost inevitable, but the bigger danger is that the chronic problems often are
not recognized as problems to be solved, but come to be viewed as a necessary way of life.
Nevertheless, chronic problems usually carry the larger economic stake, by far.

On-Aim Control Contrasted with In-Limits Control.   The implicit process control
concept that has traditionally been employed in many process industry production sites is "in-limits"
control. Under this concept, process measurements are taken periodically. If the measured data fall anywhere within the acceptable limits for the product, the process status is considered to be acceptable.
Commonly, the "within acceptable limits" decision is made on the basis of each successive
single measured test result, or the average of a small subgroup of test results. This procedure disregards
the inherent variability, which gives rise to the distribution shown in Figure 27.3 even when the process
average is held constant. Under the "in-limits" control paradigm the acceptable limits usually are equated with the product specifications and are usually wider than the inherent process variability. The

process is allowed to run indefinitely in the condition shown in Figure 27.3*b*. Customers of such producers of processed materials may learn to accommodate to the tail of the distribution that is above the upper limit. For example, customers may adjust the conditions of their own production process to compensate for the actual location of the producer's process average. Subsequently, a change in the producer's raw material or process procedures may cause the process average to shift to a value below the aim point—but still within the acceptable limits. Now the customers must contend with a distribution having a tail that goes outside the opposite limit, or with a transition-period mixture of the two distributions. Such shifts are a sure way to cause customer complaints and loss of business.

The "on-aim" control concept, which is central to good quality management, does not allow indefinite operation with the process average displaced from its aim point. As soon as the accumulated evidence from the statistical process control scheme shows that the process is off-aim, action is taken to bring the process back to its aim point. This is based on the understanding that customer satisfaction and marketplace performance are better when the product is maintained close to its aim point.

DESIGNING THE PRODUCTION PROCESS

Strategic Aspects of Economy of Scale.   In the process industries, the design of the production process has an enormous impact on the ability to achieve consistently high product quality. Both the intended characteristics and the uniformity of the product are affected by production process design. Achievement of the desired average characteristics of the product is a major goal in production process design. Likewise, the achievement of product uniformity during continuous process operation with a single product variety is also a major goal in production process design. Unfortunately, achievement of product uniformity under discontinuous operation due to multiple product varieties has not often been adequately incorporated as a major goal in production process design. The resulting quality problems and their penalty on long-term financial performance have often been severely underestimated.

The economies of scale in production process design are always obvious to the designer. They lead to process designs with large single-line continuous processes, or batch processes with very large batches. Typically, at the time of process design for a new product, or for expansion of capacity for an existing product, the new capacity requirement appears to focus on only one or a few product varieties. Hence, the tendency is to design a large single-line continuous process or a batch process with very large batches. The potential for good product uniformity may appear to be high for such process designs due to inherent blending of the large process holdup volumes and the economic feasibility of sophisticated instrumentation and control. However, as a product line matures, it is typical for the marketplace to demand a larger number of product varieties. Good marketing strategy will hold proliferation of product varieties to a minimum. Nevertheless, a mature market usually involves many important segments requiring different product varieties.

Building Flexibility into the Production Process.   The only fully satisfactory answer is to build flexibility into the original production process design. To "design it right the first time" means recognizing the inevitability of proliferation of product types. It means designing the process with

- Small in-process volumes
- Short lag times
- No blending vessels
- Quick, reliable transition procedures

One promising strategy is to construct a series of small-volume plants, bringing smaller increments of capacity on-line only when needed by the current product volume, and retaining design flexibility to incorporate features needed by the evolving mix of product types demanded by the market. Each continuous process production line, or each batch facility, would then be small enough to allow (for each product variety it produces) a production run length that involves sufficient sampling intervals for effective process control. Multiplant production strategy can then encompass both a flexible multivariety product mix and consistently good quality of conformance. Long-term financial performance should be favorably impacted.

The concept of Continuous Flow Manufacturing (CFM) can enhance this strategy. Under this concept the equipment and facilities in each smaller-volume production line are dedicated full time to that product. If multiple product types are simultaneously produced, each has it own full-time dedicated equipment. This eliminates unnecessary variability introduced into the product by using different combinations of equipment each time the product is run. The dedicated equipment strategy (and often personnel, too) facilitates prompt diagnosis and correction of off-aim results due to specific pieces of production facilities.

Process controllability is fundamental. A continuous process that runs year-round on a single product variety is ideal from a controllability viewpoint. A batch process that produces hundreds of batches of a single product variety year-round is comparable. In both cases, the regular, frequent opportunities to sample and measure intermediate and final product make it feasible to employ control strategies that detect small quality deviations or trends before they become serious, and to feed back corrective action to prevent serious deviations. Production run lengths for controllability purposes must be measured in numbers of sampling intervals. For good controllability, a production run must be much longer than the combined effect of all lags in the system. These include inherent process lags caused by physical holdup volume or chemical reaction time, lags due to sampling and measuring the product properties, and lags due to the response time of the control procedure. Usually this means a minimum production run of several weeks for good controllability in large-scale production lines. Small-scale production lines with fully automated computer-controlled sampling, measurement, and feedback may achieve good controllability in shorter production runs.

Another promising strategy is to construct a higher-volume plant whose design focuses upon quick, reliable transition procedures. Production facilities capable of such transitions between product styles have been devised and implemented in a number of mechanical industries. There are examples in automobile assembly operations. Modular process design is employed, permitting quick, reliable exchange of modules for the differing product styles. Computerized changes in control settings may be involved. The key requirement is that the process will reliably start up on-aim for all product properties immediately after changeover.

In the process industries, continuous process designs with small in-process volumes, short lag times, no blending vessels, and quick reliable transition procedures are a direct analog of the flexible manufacturing systems (FMSs) that have been successful in the mechanical industries. FMS supports the goal of just-in-time (JIT) inventory management. However, just-in-time cannot function properly unless the supplier has a flexible production process. Without a flexible production process, combined with an effective quality management system, efforts to comply with JIT for customers will merely result in the customer's raw materials inventory being carried by the supplier as a final product inventory. In that event, the general economy is in no better position than before.

The process industries, by and large, have not yet implemented process designs with this degree of modularity so as to achieve quick, reliable changeover.

THE ROLE OF PRODUCTION SCHEDULING AND INVENTORY CONTROL

An existing production facility built with large continuous processing equipment, or large batch equipment, will be forced to adapt to the evolving multivariety product mix. Production planning personnel tend to adopt production strategies that optimize only the direct cost of inventory, customer response time, and the like. They often ignore the indirect cost of inadequate product quality. The inevitable result in a continuous process is a production strategy calling for frequent rotation among all product varieties. This requires many short production runs. Serious quality problems are certain to result because each production run may be nearly over before it is possible to attain and maintain on-aim operation. The large process holdup time may inherently defeat all process control strategies From such a situation, the "transition" product made at the beginning of each run (in a continuous process) is a large fraction of the total production. Various approaches have been used in attempting to cope with these problems:

• Large mixing vessels can be used to blend final product for the purpose of smoothing out property variations, especially those due to transitions. This adds to the mill cost of the product due to increased investment, larger in-process inventory, and increased manufacturing labor. Moreover, the functionality of processed material blended to a nominal average property level usually is not the same as the functionality of product made uniformly at the intended level.

• The transition material can be reworked or recycled to bring it within specifications after further processing. This adds to mill cost and decreases net production capacity. Moreover, reworked processed material usually does not have the same functionality as material originally made to specifications.

• A portion of transition or recycled material can be included with, or blended into, regular shipments of on-aim product. This increases inventory and handling costs and increases the risk of nonfunctionality in customer use.

• The transition or recycled material can be discarded or sold at a reduced price. This has obvious economic penalties.

Analogous problems occur in a large-batch facility that must produce many product varieties. Each variety may require only one or a few batches per year. This degrades the ability to keep production personnel and procedures tuned to a constant state of "standard process" operation. Every batch is handled like an experimental run. The typical result is large batch-to-batch variability.

Whether continuous or batch, the symptoms of chronic quality problems are then encountered. Obvious partial answers are to reduce the number of product varieties being marketed, to revise production planning to require longer production run lengths (continuous processes) or increased number of like batches run each year, and to improve start-up control strategies to get the process on-aim as promptly as possible. The principal part of the answer must come from better production process design strategy.

THE PRODUCTION PROCESSES CRITICAL FOR QUALITY MANAGEMENT

Especially in the process industries, several processes within the "production" operation are critical in accomplishing quality of conformance. Critical production processes generate and use data. The information flows in various paths throughout the network of processes.

In Figure 27.4 the critical production processes are diagrammed. The production process itself is one of five types of processes that operate together:

• Production process
• Sampling processes
• Measurement processes
• Decision/control processes
• Computing processes

The shipped product cannot have consistently good quality of conformance unless all of these critical processes work together properly. In addition to the production process, each of the other processes also must be considered.

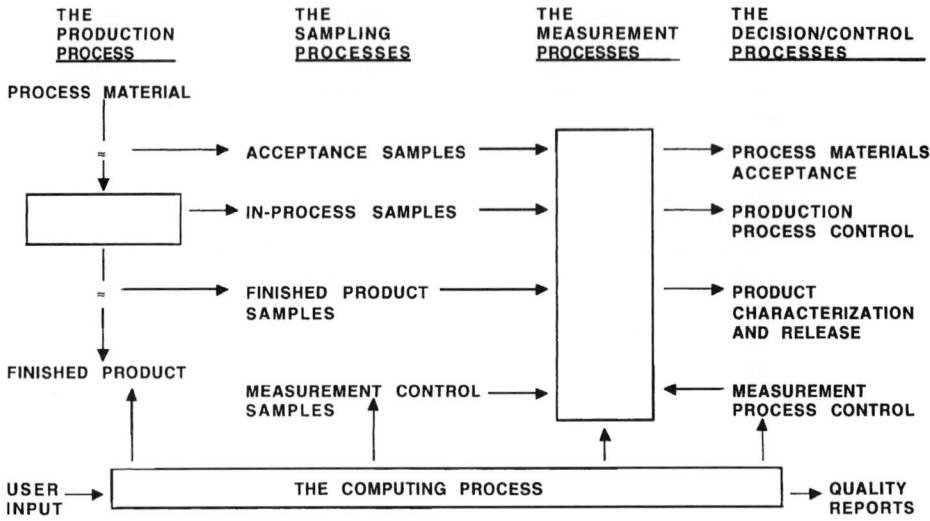

FIGURE 27.4   Critical production processes. [*From Marquardt (1991)*.]

The Sampling Processes.   When samples are taken for any of the purposes shown in Figure 27.4, the required structure and size of the sample must be determined quantitatively. This sampling plan will depend on where the major sources of product and measurement variability are present. Moreover, the sample must be taken in a way that makes it truly typical of the material which it must represent, so that:

- "Acceptance samples" of incoming materials may be required to verify conformity to their required specifications. In a well-developed quality system, suppliers' measurements can be relied upon, which minimizes the amount of acceptance sampling required, thus reducing redundant costs in the value-adding chain from supplier to producer.
- "In-process samples" are needed for production process control.
- "Finished product samples" are needed for production process control, for product characterization, and for product release.
- "Measurement control samples" are needed to keep the measurement processes accurate.

The samples that result from these processes must be tagged with identifications telling where they came from and how and when they were obtained.

Most quality management procedures are applied to a single property at a time. Although a single physical test sample is, in practice, frequently used to obtain test results for multiple product properties, the term "sample" in this section normally refers to a single product property sample.

The Measurement Processes.   The output of the measurement processes is numbers (the measured property values for each sample submitted). Some measurement processes are conducted in the laboratory, while others are done on the production floor. Some are highly automated, while others are labor intensive. The measurement processes often are as complex as the production process, and are just as capable of going out of control. If reported test results have large measurement error, they may cause improper actions to be taken in the decision processes. This can cause the product uniformity to be poorer than if the process were left alone.

The Decision/Control Processes.   The decision processes are of two types:

- "Process Control" decisions leading to actions to maintain the production process and the measurement processes on-aim.
- "Accept and Release" decisions leading to actions regarding product disposition to ensure that only conforming process materials are used and only conforming product is released.

The Computing Processes.   Computer hardware and software typically implement parts of all of the preceding processes. Data from the measurement processes may be entered and stored in a computer database, either manually or by automatic instrument data entry. Automated instrumentation may be used to produce sample measurements. Data collected from these processes are then used as input to the control and decision processes. Diagnostic data analysis software may be used to produce graphs, reports, and other quantitative summaries to detect quality problems and monitor quality status.

The computer hardware and software processes involved in these tasks may be complex. The reliability and adequacy of the hardware and software used in these computing systems need to be given attention.

CONCEPTS OF STANDARD PROCESS AND STANDARD PRODUCT

Production Process Specifications.   To ensure the production of a product that consistently meets customer requirements, the production process must be consistent, for example, from day to day and month to month. The set of product properties listed on the product specifications should, to the best of the producer's knowledge, encompass all the characteristics of the product that affect its ability to satisfy the needs of customers. However, it is not enough to rely upon the consistent satisfaction of established product specification limits. There are two reasons for this:

- The product properties for which specification limits have been established may not capture all the important characteristics of the product that affect its functionality in anticipated applications.
- Some customers may use the product in end uses or environments not anticipated by the producer, and these may be suitable end uses, but the producer's tests do not directly ensure consistent functionality in these applications. In such situations, there may be failures in quality of conformance despite the producer's best intentions. An "insurance policy" is needed.

The Standard Process as an Insurance Policy.   Insistence upon a uniform production process is the principal mechanism by which a producer can ensure against the unknown risks just described. A uniform production process from one point in time to another, and from one production facility to another, tends to ensure the consistency of the product properties the producer does not measure, in the same way that it tends to ensure consistency of the properties and characteristics the producer does measure. Thus, there is need for "production process specifications" which define a "standard process." These complement the product specifications. The "standard process" is an essential tool for achieving conformance capability. Standard process protocols should be documented and readily available to all production personnel. In an organization that operates in compliance with the International Standards ISO 9001 or ISO 9002 (ASQC 1996), these protocols are among the "procedures" and "work instructions" available to all relevant personnel. If the standard process is treated as a secret, there can be no standard process.

For each process and each product type, "standard process" conditions should be specified, together with fixed "process limits" for each process setting or processing procedure. The standard process and process limits should be documented formally as "standard operating procedures."

Moves of manipulatable control variables within the documented "standard operating conditions" are permitted for process control, but limited as to allowable magnitude and authorization level required for the move. If more than one process line is producing the same product type, the lines may be operated at different conditions within the standard process if this is required to maintain on-aim control of product properties on each process line.

Definition of the process limits must sometimes be a matter of judgment; but wherever possible, the process control variable limits should be determined by conducting a statistically designed process calibration experiment surrounding the standard process conditions as a center point. (See Section 47, Design and Analysis of Experiments.) When limits for process control variables are specified on the basis of judgment, a rule of thumb is to set limits that exclude the most extreme 2 percent of process conditions used during extended past periods during which a consistently acceptable product was produced.

The concept of standard process also includes the following:

- No change may be made in the operating conditions of any production process for "testing" or "experimental" purposes without prior notification of all potentially affected organizational functions. By definition, product made under test or experimental conditions is not made under the standard process.

- No change may be made in:

  design of the product itself (or introduction of a new product)

  production process equipment

  raw materials or incoming parts (except routine replacements-in-kind of expendable items or materials)

without prior notification of all potentially affected organizational functions. By definition, product made after such changes is not standard product until the changes have been qualified, that is, it has been verified that the product meets all customer requirements and meets product specifications for all regular finished product characteristics for an extended period of time.

Control of Both Production Process and Measurement Process.   Continual on-aim control of the production process is a part of the standard process and is a fundamental concept of good quality management. The production process control scheme is designed to promptly detect a shift of specified magnitude by the process average away from the aim point. System design procedures should ensure that the values of the specified shift and the product specification limits are mutually compatible.

Continual control of the accuracy of the measurement processes also is an essential part of the standard process concept. System design procedures should ensure that a measurement drift of specified size will be detected more promptly than is required for detecting a comparable drift in the production process. In that manner, measurement problems are discovered and fixed before they can cause problems in production process control or in product release decisions.

Every Process in a Quality Management System Must Be Audited.   A quality management system enables objectivity and integrity in assessing the true status of product functionality. Its role is parallel to the financial system which enables objectivity and integrity in assessing the true economic status of a product.

Periodic financial audits are universally accepted as a means to ensure integrity of a producer's financial management system. Likewise, periodic audits of every process in a quality management system are essential to maintain continued system integrity.

Conforming Product.   "Conforming product" is, by definition, product that is in conformance with product design specifications and is produced from standard materials in conformance with the standard process. It is the ultimate purpose of a quality management system to

produce only conforming product. It is important to provide tools to assess product conformance and to deal with situations where existing process performance does not consistently meet performance goals.

QUALITY EVALUATION BASED ON TRUE PRODUCT VARIABILITY FOR CRITICAL PROPERTIES

Each product property can be classified according to how it will be administered within the producer's quality management system. Critical properties are those properties for which the made-product total variability is so wide as to cause persistent difficulty in meeting current or near-term anticipated customer's needs. The lower and upper limits between which a critical property must fall in order to conform to specifications can be called "unit tolerance limits." These are limits on the scale of true product values with which shipped product units are to comply. The product specification format in Figure 27.5 shows the information needed to state adequately the quantitative true product value unit tolerance limits.

Both the customer and the producer evaluate product quality on the basis of final product characteristics. The customer needs to determine how well the product received will meet user needs. The producer needs an accurate evaluation of the quality of the product shipped. Variability in critical product characteristics is the enemy of quality of conformance. Variability in the measurement of critical product properties can make attainment of quality extremely difficult. Every quantitative value observed, however, is subject to variability introduced by the process of sampling and measuring the product characteristics. That is, every observation has the following structure:

Observed value = true value + measurement error

The statistical term "measurement error" refers to the net effect of all sources of sampling or measurement variability that cause an observed value to deviate from the true value; the term does not imply that an error or mistake has been made. Thus, the true value of a product characteristic is a value that does not contain any sampling or measurement error. A fundamental quality management system concept is the notion of quality evaluation based on true value for critical properties.

The customer and the producer both seek a common basis for discussion and evaluation of the product quality, a quantitative "true value" basis free of any sampling or measurement error. Straightforward statistical methods provide procedures for quantifying separately variability due to true product and variability due to measurement. These procedures require some extra effort. They are cost-effective for those critical, hard-to-control properties where quality of conformance really counts in the marketplace. Use of procedures to characterize product quality of conformance on a true value basis is valuable for both producer and customer

The inevitable presence of measurement variability is one reason for the necessity of using statistical methods in product quality work. The other reason is that the true property values of product units also vary from unit to unit. Thus, we are always faced with the necessity of making decisions about a variable product in the face of noisy measurements on that product. Especially in the process

Property	Product unit	Intended value	Unit tolerance limits		Test method
			Lower	Upper	
1					
2					
3					
...					

FIGURE 27.5 Format for true product value quantitative specifications. [*From Marquardt (1991).*]

industries, it is not uncommon to find that the measurement variability is as large as, or larger than, the true product variability. Statistical analysis of variance techniques are used to break up the total variance observed for a product property into useful "variance components." It is then possible to quantify true product variability separately from measurement variability and to estimate the proportion each variance component contributes to the total variance. For critical properties it is desirable to express the product specifications on a true-value basis.

IMPORTANCE OF OBJECTIVELY DESCRIBED PRODUCT SPECIFICATIONS

To design and produce a product that will meet customer requirements, we must be specific about what those requirements are. The necessary content and format of a product specification form a primary vehicle by which a producer and a customer can communicate about the satisfaction of customer requirements. The product specifications document the best approximation of the translation of customer needs and requirements into measurable product properties which can be used by the producer to manage the production process. Especially in the process industries, a small set of defined measurable product properties cannot always fully describe the needs and requirements of all end users. Consequently, the product specifications must be seen as an ever-evolving document reflecting the current understanding of the producer and customers as partners in the translation process.

Three crucial matters have often been ignored by producers in setting specification limits:

- Definition of the product unit
- Definition of the test method to be used
- Appreciation of the role of component sources of variability

Meaningful, technically sound product specification limits can be developed only when the product unit, the test method, and the variance components are all taken into account.

Product Specifications — Content and Format.   A product specification serves both as a documentation of the (specified or implied) agreement between the customer and a producer and is a compendium of information on the product (ASQC 1996b). Ideally, the customer is directly involved in setting product specifications. In many situations, the agreement is between the producer's Marketing and Production organizations, with Marketing representing many customers' needs. The specification contains two types of information:

- *Descriptive information on the product:*   Name, identification code, chemical composition, engineering designs and drawings, uses and functionality, units of measurement, delivery units and conditions, and other qualitative characteristics of the product as well as proper, safe handling and storage information.
- *Quantitative specifications for measurable product properties:*   Numerical values of intended levels of properties and ranges or limits. Most products have from 10 to 100 such measured properties listed on their product specifications. Many products have one, or several properties in their "critical few." These quantitative specifications (intended values and limits):

  Document the best current definition of the product, in measurable terms, that is expected to meet the needs of customers and that can be supplied commercially by the producer with current technology and facilities

  Are for a prescribed measurement method

  Can apply only to properties that can be measured on shipped product

The specifications are incomplete unless all of these items are provided. The entry under "test method" usually is a code reference to a document describing the standard test method (including test equipment, materials, and protocol). For some critical properties, such as impurity levels, only a lower or an upper one-sided tolerance specification may be needed.

Three terms may be interchangeable in some contexts, but may be numerically different in other contexts (Figure 27.6). For clarity of meaning, use each term where appropriate.

The value of the aim point X_0 used in production process control usually is set equal to the intended value.

For noncritical properties, the unit tolerance limits on the product specification format are replaced by observed value limits, usually 3 to 4 standard deviations of a test result above and below the intended value. The format shown in Figure 27.7 can be used.

Definitions of Product Unit Terms.   A product may be gathered into various unit quantities for specific purposes, for example, a warehousing unit, a shipping unit, or a customer-use unit. It is important to define precisely a "product unit" of finished product to which product specifications will apply. This is essential for quantitatively defining product quality and applying the correct sampling system and statistical procedures. The term "product unit" applies equally to a physical product, a software product, or a service. For processed materials selection of the appropriate unit of finished product in a given instance should take into account two features:

- The physical units (bags, rolls, bottles, pallets, etc.) in which the finished product is handled by the producer.
- The physical units in which the customer will use the product.

The product unit for product specifications is best selected as the smallest conveniently handled quantity of product within which a customer will likely detect a significant departure from intended functionality, if such departure exists. Considerable discretion often is available in selecting a product unit for quality management purposes. The product unit may be as small as the "unit

Term	Usage
Intended value	The value of a property which specifies the product functionality *intended by the product design* (and documented in the product specification) for every product unit.
Aim value X_0	The value of a property which specifies the value *intended to be achieved by the production process* for every product unit made.
Average value	The value of a property which specifies the *average* value actually achieved by the production process, averaged over many product units.

FIGURE 27.6   Definitions of product specification terms. [*From Marquardt (1991).*]

Property	Intended value	Observed value limits		Test method	Product Unit
		Lower	Upper		
1					
2					
3					
...					

FIGURE 27.7   Format for observed value quantitative specifications. [*From Marquardt (1991).*]

quantity of sampled material" from which a single "test result" is obtained, or it may be many "adjacent" multiples of that unit quantity. Although the product unit choice could vary from property to property, it is operationally helpful to select the same product unit for all product styles and properties in a given product line.

Generic Concepts of Product Unit and Unit Tolerance Limits.   The terms "generic product unit" and "generic unit tolerance limits" (GTLs) may be used in this discussion for true product value quantitative specifications without reference to any specific form of product unit. In most of the illustrative discussions, the product is a processed material that can be sampled and measured at various points in the product unit. The GTLs apply to the true average of the property over an entire product unit. If the product is an item whose property is only realizable and measurable at one point in the product unit, then the GTLs apply to the true value at that point. Accordingly, these product-unit-concepts apply equally to process industry products and mechanical industry products.

Definition of product specifications for product properties that are subject to aging requires special clarification. The term "shipped product" is not usually meant to include aging shifts, except for the aging that occurs in the normal time lag from production until sampling for product characterization and release, usually less than a few days. In each such instance where aging is involved, the time lags should be part of the product specifications.

Unit Package—the Common Case.   In the most common case, the appropriate choice of product unit is "unit package." The unit tolerance limits for critical properties then apply to the true average of the shipped product property for the entire unit package. The unit package should be a unit of product such that within-package variability is not a predominant source of variability. It can be noted that product from the process industries often undergoes physical blending during normal use by customers, resulting in some degree of averaging of property values that may vary throughout each product unit.

Examples of such products and their typical unit package definitions are presented in Figure 27.8.

The continuous filament yarn example needs special discussion. It is an example where the end-use product (a textile fabric) is constructed from multiple unit packages by a process (knitting or weaving). Some "blending" occurs because of the random allocation of yarn tubes to adjacent yarns in a fabric, but the continuity of each threadline is maintained. The resulting blending is not as effective in reducing visual fabric defects as is the blending of staple fiber where each filament is short and is separately blended. Nevertheless, extensive experience indicates that a tube of yarn is an operationally desirable product unit for quality management purposes.

Unit tolerance limits are, in this case, unit package tolerance limits. If within-package variability must be controlled, it can itself be treated as a property of the package. To simplify presentation of concepts and terminology in this section, the discussion is restricted to the package structure. Other product unit definitions should be selected only for special needs.

Variance Components for the Unit Package Case.   The terms "variance" and "variance components," respectively, describe the overall, total variability and the meaningful component sources of variability of product properties and measurement methods. These terms have precise mathematical definitions that are used in computations.

Product	Unit package
Staple fiber	Bale of staple
Powdered pigment	Bag of powder
Pelletized elastomer	Bag of pellets or pallet of bags
Paint	Can of fluid or carton of cans
Insecticide	Metal drum of fluid
Continuous filament yarn	Tube of yarn
Photographic film	Box of film sheets or roll of sheet material

FIGURE 27.8   Typical products and their unit packages. [*From Marquardt (1991).*]

When the product unit is a package, the product variance components are defined as shown in Figure 27.9. These components are always present, although their magnitudes will differ from one product and property to another. For any choice of product unit, measurement variance components are always present.

The time interval corresponding to the "same nominal test time" must be defined in context. In many cases, it should be one laboratory shift. Typically, a "nominal test time" is the time interval over which a group of samples, submitted at the same time, would normally be tested. VST, defined in this manner, includes all short-term sources of measurement variability that affect a test result, including differences among multiple instruments or multiple operators that may typically be assigned to analyze the several samples. If the measurement is a destructive one, the VST component inextricably includes the local product variability associated with the quantity of product used to measure a test result, as well as the short-term measurement variability itself. VLT includes all long-term sources of measurement variability.

A "test result" is a single numerical value that is the end result of carrying out a test method for a specific property. A "test method" is a specified (documented) set of test equipment, test materials, and test protocol, whose input is a specified unit quantity of sampled material. The test protocol may require more than one test specimen to produce one test result. For example, the protocol may require averaging measurements from several test specimens.

The five variance components for the package unit can conveniently be visualized in the nested structure shown in Figure 27.10.

Variance component	Symbol
Lot-to-lot variance about the true process average	VLL
Package-to-package variance about the true lot average	VPP
Within-package variance about the true unit package average	VWP
Short-term measurement variance of nominally identical, disguised samples tested at the same nominal test time	VST
Long-term measurement variance; all measurement variance in excess of short term	VLT

FIGURE 27.9   Definitions of variance components for the unit package. [*From Marquardt (1991).*]

Total variance	Total true product property variance	External to lot (VLL)	
		Within lot	Package-to-package within lot (VPP)
			Within package (VWP)
	Total measurement variance	Short-term (VST)	
		Long-term (VLT)	

FIGURE 27.10   Components of variance structure for the unit package. [*From Marquardt (1991).*]

Product Specifications and Market Requirements.   Historical practice in setting product specification limits has varied from one situation to another. In the process industries, defining both customer requirements and requirements for incoming materials has been difficult because the requirements are rarely known precisely, and have been based on experience and judgment.

The problem of basing specifications on incomplete information is particularly acute when one organizational unit gains a monopoly on input to specification setting. Where specifications have been provided initially by Research and Development, they may reflect the technical judgment of the Research and Development group based on the limited experience of the laboratory and pilot projects. Where specifications have been provided by Marketing, they may reflect Marketing's desire to meet customer requirements, without regard to technical capability to do so in Production. And where Manufacturing dominates the setting of specifications, those specifications are likely to be easily met in production but ignore the needs of the customer, or fail to take into account the conditions imposed by a new technology. None of these unilateral procedures can ever be totally satisfactory because none of the involved parties has all the information necessary to make the proper judgment. That is why, in everyday production and marketing practice, the product specification limits often have become a source of friction or have been ignored altogether.

Whenever customer needs for a critical property are known, the specifications should be based on these needs. If customer needs are not precisely known, the specifications should result from a consensus-forming process among Production, Marketing, and the customer. Research and Development should play an explicit supporting role in this process. Production should provide data describing the inherent magnitude of variability in the finished product and in the measurement procedures, based not upon idealized process or measurement capability, but upon the actual historical performance. All parties should understand that wider specifications could increase the producer's yield and conformance and reduce production costs. Narrower specifications, on the other hand, may result in a more competitive product and larger market share and should reduce customer complaints and claims paid to customers for shipped product that does not perform as expected.

In practice, there is not usually a sharp point of demarcation between good product and bad product. Product having actual product levels not far outside the specifications may perform adequately for its intended use, in most instances. Hence, companies should use the terms "conforming" and "nonconforming" to specifications and avoid, in this context, terms such as good, bad, and defective. This usage is consistent with current terminology in quality systems standards (ISO 8402:1994).

As process improvements are implemented, the product variability should decrease and become clearly narrower than the specification limits. Some customers may develop uses for the product that depend upon the "de facto specification limits" that are narrower than the documented specification limits. It will then become necessary to agree on narrower specification limits to recognize the new trade needs. Alternatively, the property might be removed from the "critical property" list because the process performance is now consistently better than trade needs.

Often, in the process industries, the trade need is not well known, or there are so many different end uses, each with its own specification limits requirement, that it is impractical to develop quantitative data on trade need for all segments of the market. In this case, the producer's process performance can be used as input to discussions between the producer and customer (or between the producer's Production and Marketing organizations so long as the customer is properly represented), where the intent is to arrive at mutually agreeable specification limits.

MEASURING QUALITY OF CONFORMANCE WITH PRODUCT DESIGN

Quantitative Definition of Conformance.   "Conformance" is defined quantitatively as the percent of product units which meet product specifications, assessed over a suitable period of time. This quantitative definition of conformance applies to critical properties whose product specifications are on a true product value basis. In various contexts, we need to distinguish:

- "Conformance of product made," which refers to the percent of all product units made which meet true product value specification limits
- "Conformance of product shipped," which applies to the product units released for shipment and refers to the percent of released product units which meet true product value specification limits. When a formal lot release procedure is in place, the product units shipped may not include all product units made.

In this section, the term "conformance" always means conformance of product made, unless explicitly stated to mean conformance of product shipped. Both conformance of product made and conformance of product shipped can be estimated using appropriate computing methods. The definition of "conformance of product shipped" carries with it an assumption: the product units shipped are the same as the product units released.

Goal Conformance as a Fixed Reference for Product Specifications. An appropriate "goal conformance" level should be selected. Once selected, the goal conformance becomes the fixed reference point against which product quality of conformance is thereafter reckoned.

Typical values of goal conformance are 99 percent or 99.7 percent, but the appropriate value depends upon product characteristics and other factors. For marketing and internal administrative convenience the goal conformance level should be the same for all products within a product line. As uniformity improves for a given product, the width of the product specification range can decrease, with goal conformance held constant.

The approach recommended here differs from the "6σ" approach used widely in the mechanical industries. The "6σ" approach keeps the specifications constant, and strives to continually improve the percent conformance to become virtually 100 percent. The approach recommended here keeps the goal conformance constant at a value (99 percent or 99.7 percent) that can be quantitatively assessed with moderate sample size, and strives to continually narrow the width of the product specification range. Both approaches have the same ultimate intent.

Experience Curves as Measures of Continual Improvement. To display progress (or lack thereof) in continual improvement, a useful tool is a plot of the standard deviation for a single product unit (package) versus calendar time. Such plots are examples of "experience curves."

Experience curves can be plotted for properties whose specifications are on a true-product value basis or an observed value basis. In the latter case, the plotted standard deviation contains both true product variability and measurement variability. Various forms of experience curves can be useful.

For example, for properties having true product value specification limits, the width of the product specification range provides a formal measure of quality of conformance. The width is a specified multiple of the standard deviation of true product unit values for any specific property. The width of the specification range should become smaller periodically as the product uniformity is improved.

REFERENCE BASES FOR MEASUREMENT CALIBRATION AND PRODUCT CONTROL

Hierarchy of Reference Bases. Each routine test measurement method is itself an on-going process subject to all the ills that befall production processes. Each such process needs a reference basis. The measurement process is the reference basis for the production process. "Recognized standards" and/or "control samples" are used as reference bases for the measurement process. Reference bases are used for the purpose of maintaining measurement procedures and equipment in satisfactory condition to run routine analyses.

Recognized Standards as Reference Bases.   Many test methods are direct measurements of dimension, weight, time, temperature, electrical quantities, and the like. Such measurements usually are the most important properties for many products of the "mechanical industries." These measurement methods periodically must be calibrated within each manufacturing plant and each test laboratory against secondary standards of high accuracy, stability, and uniformity. The secondary standards are, in turn, calibrated against primary standards of ultimate accuracy, for example, at the National Institute of Standards and Technology (NIST), in a traceable sequence of steps. The end result is that such measurements can have small variance and small bias if suitable attention is given to

- Periodic calibration against recognized standards
- Test equipment maintenance
- Standardized test procedures
- Operator training

Some products require high absolute accuracy of such direct measurement; this may present serious problems. Elaborate metrology programs may be required to ensure adequate calibration, standardized procedures, maintenance, and training.

Control Samples as Reference Bases.   In the process industries, many test methods are incapable of direct traceability to recognized standards. Since they are simulations of customer-use conditions, they are indirect, multistep, unique to a product, or otherwise nonstandard.

These test methods usually involve some direct measurements that can be accurately calibrated individually but not collectively as a total test procedure. Adequately maintaining these test methods requires more than just a good metrology program, more than just periodic calibration, equipment maintenance, standardized procedures, and operator training. For these test methods "control samples" must be used as reference bases, and control procedures must be used to maintain control.

A control supply is a quantity of regular first-grade product, properly characterized and validated, that is retained for use in testing subsequently manufactured product. Many issues in measurement control center on the proper choice and use of control samples and the control supplies from which they are taken.

CONTROL OF THE MEASUREMENT PROCESS

Test Method Administration.   Good test method administration, whether physically located in a laboratory or elsewhere, requires regular calibration of test instruments against recognized standards. Examples are

- A standard weight to calibrate the zero setting of a weighing scale
- A standard white reflecting plate to calibrate a colorimeter
- A standard solution to titrate a chemical test instrument

It must be clearly understood, however, that these calibration procedures are not subject to many of the actual sources of variability and bias in taking, preparing, and conducting measurements of routine production samples. Consequently, virtually all test methods need statistical process control such as Twin Metric control or Cumulative Sum (CUSUM) control of the full measurement process. (See Section 45 for a discussion of CUSUM control.)

Twin Metric control, a form of CUSUM control, offers exceptionally favorable balance among control scheme performance, simplicity, and intuitive user interface; this section refers to Twin Metric control (Marquardt 1993, 1997; Marquardt and Ulery 1991, 1992) as a prototype for state-of-the-art statistical process control in today's computerized working environments. (See Section 45, Statistical Process Control for additional discussion of statistical process control.)

Availability of Means of Calibration.   By employing a suitable control supply for a measurement process, procedures such as Twin Metric control can be used very effectively to detect drifts, level changes, and sensitivity changes, so that appropriate corrective action may be taken. Quite analogous to production control applications, Twin Metric or CUSUM for measurement control are equally valid for situations where a predetermined calibration knob exists and for situations where no such designated control variable is available.

Specification of Aim point X_0.   The measurement process control, such as Twin Metric, will normally use the nominal value of a control supply as the aim point, X_0. Often, the true average is not as important as the uniformity of measurements by different instruments using that control supply or by a single instrument using that control supply at different times. Thus, it is important that the control supply property level remain stable with time, or at least that any changes of the control be adequately characterized. The measurement control aim point, X_{0M}, should be near the production process aim point, X_0. Multiple measurement control supplies and statistical process control schemes may be necessary to satisfy this objective when the measurement is used for several products which have different property levels.

Specification of Other Statistical Process Control Parameters.   The relationship between the design of the measurement process control scheme and the design of the production process control scheme is important.

The acronym ARL (for *a*verage *r*un *l*ength) is commonly used (Bissell 1986; Champ and Woodall 1987; Kemp 1961; Lucas 1973, 1976, 1982, 1985*a,* 1985*b*; Marquardt 1997, Page 1961) for the average number of process control sampling intervals before a control scheme will produce a signal. ARL(0), the value of ARL when the process average is on-aim, should be large. ARL(1), the value of ARL when the process average is off-aim by one multiple of the process standard deviation (SPROC), should be small. For example, the classic Shewhart chart with control limits at ±3 SPROC has ARL(0)=370, ARL(1)=44. Twin Metric control and CUSUM control provide combinations of ARL(0) and ARL(1) that are far better than Shewhart charts. ARL, measured in this (dimensionless) way in number of sampling intervals, can be converted to units of elapsed time or units of production volume if desired. To ensure product quality in routine production the relationship between production process control and measurement process control should be an explicit element of design. Figure 27.11 displays the correspondence between parameters of production process control and parameters of measurement process control.

The standard Twin Metric design procedure described by Marquardt (1997) is used in selecting the parameters for both production process control and measurement control. However, different criteria are applied.

Production process control parameter	Translation	Measurement process control parameter
X_0	Process aim point	X_{0M}
ARL(Δ)	Average number of $ARL_M(\Delta)$ measurements to produce a signal when process is off-aim by (Δ) (SPROC)	
SPROC	Process standard deviation	$SPROC_M$

FIGURE 27.11   Parameters for production and measurement control. [*From Marquardt (1991).*]

A measurement process control scheme should be designed so that a measurement level change from X_{0M} will be signaled and corrective action is taken before the change results in a production process control signal and an unnecessary adjustment in the production process.

Let D be a fixed magnitude of production process shift that is important to detect promptly in the measured product property level.

Further, let $\Delta = D/SPROC$.

$ARL_M(\Delta)$ should be smaller than $ARL(\Delta)$. A practical place to begin is to set $D = SPROC$ (i.e., $\Delta = 1$), and try to have $ARL_M(\Delta) \leq \frac{1}{2} ARL(\Delta)$, where ARL and ARL_M are expressed in the same units of time. The intent is to allow time to correct the measurement problem after its detection before a production process signal occurs.

To satisfy the desired inequality, the following design variables are available:

- *SPROC and SPROC$_M$:* SPROC$_M$ is likely to be smaller than SPROC for the same sampling interval, sample size, and structure because the measurement control supply is likely to be more uniform than routine product made, even during periods of good production process control.

- *ARL(Δ) and ARL$_M$(D):* Since the ARLs are to be compared in the same units of elapsed time, the inequality can be satisfied more easily if the sample interval for measurement control is shorter than the sample interval for production process control.

- *ARL(0) and ARLM(0):* It is usually possible to have a value of $ARL_M(0)$ (for the on-aim measurement process) which is smaller than the value of $ARL(0)$ (for the on-aim production process). This is because false alarm signals on the measurement process can usually be tolerated more readily than false alarm signals on the production process. Having $ARL_M(0) < ARL(0)$ pushes the relationship between $ARL_M(\Delta)$ and $ARL(\Delta)$ toward satisfaction of the inequality.

This discussion of the effect of design variables on ARLs applies to all forms of statistical process control (e.g., Twin Metric, CUSUM, or even Shewhart charts).

SPROCM Estimation.   A simple, robust estimate of the measurement process standard deviation SPROCM can be determined by the mean square successive difference (MSSD) method. Marquardt (1993) discusses the MSSD method and useful generalizations.

Sampling Frequency and Test Protocol.   As a general rule, samples for Twin Metric or other statistical process control of a measurement process should be tested each day (or each shift), preferably at a random time. An exception would be infrequently measured properties for which control samples need to be taken only when routine production samples are being processed through the measurement process. Another exception would be a test method that tends to require frequent recalibrations. In that case more frequent measuring of the control supply may be needed.

Test results from the measurement control supply should be obtained using the same test protocol as is used for test results from production samples.

SELECTION AND PREPARATION OF CONTROL SAMPLES

Control Supply Selection.   Control supplies for measurement control should come from regular product typical of routine production. Candidate control packages should be "validated" before final selection as the control supply. The validation process should include several criteria:

- The production process should be qualified as within "standard process" limits and as operating "on-aim" during the period when the control packages were produced.

- Test results from the candidate control packages should fall within the product specification limits for all properties.

- Samples used to estimate the values of X_0 and SPROC_M of the control supply should span many nominal test times to ensure that long-term measurement variability is well represented. Normally, the value of X_0 must be estimated by the average of a series of measurements on the control supply itself, covering at least 30 calendar days. This value of X_0 should be based on at least 60 sets of measurement data covering at least 90 days.

- Where the control supply is to be used at more than one production site, validation may be appropriate on an interlaboratory basis.

Control Sample Preparation.   The control supplies should be stored in an environment that will prevent damage and minimize degradation. If the product is a discrete, particulate, or liquid material, the control samples for measurement on a given day should each represent a random sample from the original package(s), either by virtue of being taken from a random location in the package(s) each time, or by virtue of thorough stirring or blending of the control package(s). If the product is a continuous sheet, filament, or the like, then control samples used for the actual measurement should be spaced widely enough to be a representative sample of the whole package. When this is not practical, use of multiple or composite control samples is especially important.

Control samples should be identified in a manner that does not call special attention to them as distinct from regular production samples. The goal is to avoid special treatment (hence distortion of level or variability) by measurement operators.

Composite, Multiple, and Staggered Control Samples

Composite Control Samples.   Some measurement methods allow sample quantities to be obtained from several control packages and then plied, blended, or composited into a single control sample to give a single test result. When plying, blending, or compositing is feasible, it provides a useful procedure to reduce the contribution of within-control-package variability and (provided two or more packages are composited) the contribution of between-package variability to SPROCM. Compositing does not reduce the contribution to SPROCM from short- and long-term measurement variability.

Multiple Control Samples.   Using the average of separate test results from multiple control samples is more effective for reducing SPROCM than a single test from a composite control sample. The average of separate test results from multiple control samples reduces the contribution of within- and between-control-sample variability to SPROCM, to the maximum degree feasible, and also reduces the contribution from measurement error, especially the short-term measurement variance component. Typically, this greater effectiveness of multiple control samples outweighs the cost savings from composite control samples.

Staggered Control Samples.   Where the control supply uses multiple control packages, they should be replaced preferably on a staggered schedule rather than all at once. For example, if eight control packages are maintained (whether for compositing or for multiple control samples), a practical scheme would be to replace (the oldest) two at a time. Such staggering is helpful to maintain a stable control supply average age, hence a stable control supply average level, when consistent control level changes due to aging are present.

Control Samples at More Than One Level.   Where multiple products have a wide range of levels on a measurement or where two families of products are produced—a type that runs high in the data range and a type that runs low in the data range—then two separate measurement control procedures, such as Twin metric with control supplies at the two levels, may be advisable.

USING CONTROL SUPPLIES

Control Samples as Computational References.   For some test procedures there is large variability from one test time to another, but these large systematic errors at any one test time affect equally all samples processed at the same time. The best approach to such problems is tighter control of the test method operating conditions and procedures to reduce the time-to-time variability that inflates the long-term measurement variance component. In situations where a practical route has not yet been found to reduce the time-to-time variability, it is common to calibrate all the routine samples run at a given time by simultaneously analyzing some control samples. Then the reported test result is determined by referencing the raw test results computationally to the control sample results from the same test time. In the simplest instance the referencing is done by simple differencing:

$$\text{Reported test result} = OV_{PS} - OV_{CS}$$

where OV_{PS} = process sample observed value and OV_{CS} = control sample observed value.

The consequence of this approach is that the short-term measurement variance component for the reported test result is double that for the process sample observed value. This comes about because the variance of a difference is the sum of the variances, assuming the observed values are statistically independent with respect to short-term sources of variability. Under the assumption that the routine samples and the control samples have the same variance, the variance of their difference is double the variance of one observation. Hence, the use of a control sample as a computational reference is beneficial only if the reduction in the long-term measurement variance is substantially greater that twice the value of the short-term measurement variance. The short-term measurement variance inflation can be made less than a doubling by using, as computational references, the average of several simultaneously analyzed control samples.

Sometimes an observed value is referenced to a control by ratio rather than difference. Propagation of error theory shows that the inflation of the short-term measurement variance component due to a ratio is quantitatively similar to the inflation due to a differencing operation, so the same guideline is appropriate.

Multiple Measurement Process Configurations.   When any one of many measurement process configurations, such as multiple instruments or multiple operators, may be used in routine testing of the same product property, it is necessary that all such measurement process configurations be maintained at the same value of X_0. To maintain the same value of X_0, it is desirable to use the same controls for all. This enables expeditious detection of biases among instruments or operators. For example, where there are multiple instruments, Twin Metric schemes could be maintained for each instrument, the average of all instruments, or differences between each instrument and the average of all instruments. Obviously, not all of these should be used for any one application.

Confirming the Effectiveness of a Measurement Adjustment or Calibration.   When a measurement Twin Metric or CUSUM signal is received and proper action is taken, it is advisable to carry out additional testing with controls to confirm that the action has in fact returned the measurement process to the aim point. Such a confirmation step is particularly advisable when the adjustment or calibration of the measurement is known to be inaccurate or has a history of poor behavior.

TESTING PRODUCT WHEN MEASUREMENT PROCESS IS OFF-AIM

When the measurement process is detected to be off-aim and reaction to this knowledge returns the measurement process to the aim point quickly, no special steps need be considered in dealing with the product or the production process. If the measurement process is not returned promptly to the aim point, the measurement process output should not be reported or used to control the production process or to characterize the product. Simply stated, when the measurement process is off-aim too long, then the product is not being produced under standard operating procedures and should be given special marketing treatment.

A backup measurement method (often less automated, slower, and more costly per test result) may be available for emergency use. This backup method must itself be periodically validated to maintain its accuracy for when it is needed.

VALIDATING NEW MEASUREMENT METHODS

If we cannot measure a problem adequately, we cannot correct it either. For this reason, periodic adoption of new/better measurement methods is a principal route to quality improvement.

Whenever it is proposed to substitute a new measurement method for an existing method to measure the same characteristic of the product, it is necessary to establish that the new method is adequate for the needs. It is not enough to demonstrate that the new method is cheaper, faster, less labor intensive, and the like. It must have three performance characteristics:

- The slope (sensitivity) of the average response of the measurement process must be sufficiently large versus the property measured to provide the needed average responsiveness.

- The variability of the measurement process must be small enough in comparison to the slope of the average response so that an adequate signal-to-noise ratio is obtained.

- The measurement process, with its documented test protocol and its procedures for periodic calibration, must be demonstrated to be stable and reliable in actual use.

The first two performance characteristics can only be demonstrated by a designed experiment. (See Section 47, Design and Analysis of Experiments.)

The third performance characteristic can be demonstrated by an extended simultaneous overlap period during which both the old and the new methods are used, allowing the measurement variability sources for the new measurement method (i.e., SPROCM and/or VST, VLT) to be estimated under conditions of actual use. During the time when data are collected for purposes of estimating the value of X_0 and SPROCM for the new method, the old test method should remain in place for purposes of process control and product characterization.

The relationship between the old and new methods may often be demonstrated by scatter plots and calculations. However, it must be understood that the observed numerical correlation coefficient between two measurement methods will be lower than the correlation between the true average responses due to the measurement variances for both methods (Hald 1952, p. 615). Moreover, the new measurement method may not measure precisely the same product characteristic as the old method. This is a vexing problem for many processed material products. This points up a fourth important performance characteristic that any measurement method applied to final product must have:

- Measurements from the new method must correlate with customer use requirements.

Here again, experimental design methods are the tools to establish adequacy of the measurement method.

ESTIMATING AND MAINTAINING VARIANCE COMPONENTS FOR THE PRODUCT AND MEASUREMENTS

Previously in this section the "package" form of product unit was defined, and with it, the five variance components that are needed. The tool called "analysis of variance" (ANOVA) is the statistical method for separating the total observed variability in a measured product property into sources of variation, and for estimating the magnitude of variation that can be attributed to each source. (See Section 47, Design and Analysis of Experiments.)

Analysis of variance should be used in many ways, including

- To analyze "routine production process data" that are obtained to monitor the process average and variability, and to provide information to periodically update the quality system design.
- To analyze on-going "maintenance data" that are obtained to maintain current estimates of the measurement process and within-package variance components. When the maintenance data are combined with the routine production process data, all five variance components can be estimated separately and updated regularly.

Analysis of Variance Using Production Process Data.   One-way analysis of variance, also known as "among and within groups ANOVA," is a statistical method for isolating two sources of variation in a measured product property and for estimating the magnitude of variation that can be attributed to each of the two sources. (See Section 47, Design and Analysis of Experiments.) This discussion is intended to highlight the relevant concepts of ANOVA and to illustrate the calculations required when the product unit is a package. In practice, the ANOVA computations should be computerized as part of the software system.

Guideline for Obtaining Variance Components Data.   The variance components estimates should be based on data that:

- Provide at least 60 degrees of freedom for the estimate of each component
- Cover at least 60 production days
- Cover at least 90 calendar days

When both routine production data and maintenance data are being collected regularly (e.g., daily), this guideline is straightforward to implement. There are strong reasons for this guideline, both theoretical and practical.

In theoretical terms, the guideline is important to ensure enough data so that variance component estimates are adequately close to their true values. The theoretical effect of degrees of freedom in the precision of estimating a simple variance (single variance component) for a random sample is shown in Table 27.1.

Thus, 60 to 90 degrees of freedom are required to get a variance estimate that will be within 25 to 30 percent of the true value with reasonable confidence. Small variance component estimates derived from ANOVA calculations may have somewhat greater statistical variability. This level of accuracy is necessary and is sufficient for the various uses of the variance components.

TABLE 27.1   Confidence Limits on Estimate of Variance.

Degrees of freedom	90% Confidence limits on the ratio: estimated variance/true variance
4	0.18, 2.4
10	0.39, 1.8
60	0.72, 1.32
90	0.77, 1.25

Source: Donald W. Marquardt (1991). *Product Quality Management,* DuPont Engineering, Wilmington DE.

In practical terms, the guideline is also important. From experience with a wide variety of applications, it has been found that adequate stability of process and measurement variances can only be obtained when sampling covers the elapsed time periods prescribed in the guideline.

Variance Component Estimates from Production Process Data. The two mean squares calculated in the one-way ANOVA are not themselves the variance components (i.e., VLL, VPP, VWP, VST, and VLT). The within-lot mean square is only influenced by sources of variability occurring within a lot. The lot-to-lot sum of squares, on the other hand, is influenced not only by lot-to-lot sources of variability but within-lot sources as well.

Call the within-lot variance component VPPU, for the package product unit case, where the U stands for "uncorrected." VPPU is the source from which VPP is calculated, but it must be "corrected" for the effects of other within-lot variance components. Call the lot-to-lot variance component VLLU. If three packages were sampled per lot, the lot-to-lot mean square is an estimate of:

$$VPPU + 3 \times VLLU$$

The one-way ANOVA table can now be completed as in Section 47, Design and Analysis of Experiments, and expressions for expected mean squares can be used to solve for estimates of VPPU and VLLU.

These expressions for the expected mean squares are exact for the true (population) values of the mean squares and variance components. The estimates of the variance components are obtained by substituting the estimates for the population values, and solving for VPPU and VLLU. The expected mean square expressions in Figure 27.12 show that VPPU = MSWL directly; then VLLU = (MSLL − VPPU)/a.

Neither VPPU nor VLLU is an estimate of any of the five desired package product unit variance components, VLL, VPP, VWP, VST, or VLT. Both VPPU and VLLU are estimates of combinations of the variance components. As mentioned earlier, VPPU includes all sources of variability occurring within the lot, namely, VPP, VWP, and VST, while VLLU includes all sources of variability occurring external to lots, namely, VLL and VLT. In fact:

VPPU estimates VPP + VWP + VST

VLLU estimates VLL + VLT

Estimating VWP, VST, and VLT From the Maintenance Sampling Plan. It is important to estimate and update all variance components on a regular basis, using data accumulated during on-going operation of the system. The routine production process data normally are used for production process control and sometimes for product release, in addition to their use in estimating and updating variance components. As described earlier, estimates of

$$VPPU = VPP + VWP + VST$$
$$VLLU = VLL + VLT$$

come from the routine production process data.

Source	Sum of squares	Degrees of freedom	Mean square	Expected mean square
Lot-to-lot	SSLL	DFLL	MSLL	VPPU + a × VLLU
Within-lot	SSWL	DFWL	MSWL	VPPU
Total	SSTOT	DFTOT	MSTOT	
Note: a is the number of packages sampled per lot.				

FIGURE 27.12   ANOVA table with expected mean squares. [*From Marquardt (1991)*.]

To be able to estimate all five variance components, extra production samples—the "maintenance samples"—are required at regular time intervals. Typically, a group of maintenance samples is submitted to the measurement facility daily whenever the process is making the product. The maintenance data are used to estimate and, periodically, to update ("maintain") the values of VWP, VST, and VLT.

Maintenance Sampling for Measurement Components of Variance.

The minimum adequate extra sampling for maintenance of VWP, VLT, and VST requires a group of four samples to be submitted routinely at scheduled intervals (e.g., daily) to be measured for each sampled property, a so-called ABCD plan. The four maintenance samples for the ABCD plan are designated A, B, C, and D. A fixed sampling strategy should be used in selecting the A, B, C, and D samples. For example:

- Sample A might always be taken from near the "top" or "outside" of the package.
- If so, then sample B should always be taken in a fixed adjacent relationship to A, for example, immediately "after" or "below."
- If A is taken near the "top" or "outside," then C should be near the "bottom" or "inside."
- Sample D should always be taken immediately "after" or "below" sample C.

A and C are tested "today"; B and D are tested at a later time, for example, "tomorrow."

The lag between the times of testing the (A, C) and the (B, D) samples ideally should be long enough for all long-term sources of measurement variability to come into play. Sometimes a 1-day lag is not enough, and VLT may be underestimated if a longer lag time is not used.

Note: Care should be taken in defining what is meant by a "test time." If the routine production process samples are taken and tested daily, then the test time for maintenance sampling should be 1 day as well; if the routine samples are tested during each shift, especially if the process is controlled and product is released by shift, the maintenance sampling test time could be a shift. In any case, A and C are tested right away (within the test time definition) and B and D are tested later (often, the next shift or tomorrow).

One degree of freedom is sacrificed to remove any fixed effects when estimating each of VWP, VST, and VLT according to the procedure outlined in the following paragraphs.

Contrast Method of Variance Components Estimation.

The data from the ABCD plan can be analyzed by general ANOVA procedures similar to those outlined in Section 47. The simple computing procedure recommended here for the special ABCD structure, the "contrast method," gives results identical to the general ANOVA, uses convenient numerical procedures, and supplies useful diagnostic information.

For the ith maintenance set, $i = 1, 2, \ldots, R$, where R is the number of maintenance sets of data, compute three quantities (contrasts):

$$WP_i = (-A_i - B_i + C_i + D_i)/2$$

$$LT_i = (-A_i + B_i - C_i + D_i)/2$$

$$ST_i = (+A_i - B_i - C_i + D_i)/2$$

When the samples are taken in the recommended fixed sampling pattern, it becomes possible to:

- Test for the existence, or monitor the magnitude, of a consistent average slope within packages (top to bottom or outside to inside, as the case may be)
- Test for the existence, or monitor the magnitude, of a consistent average degradation or bias between the samples analyzed "today" and those analyzed "tomorrow"
- Test for the existence, or monitor the magnitude, of a consistent average change of slope (interaction) within a package

The presence of consistent profiles or biases of these types will not inflate the variance component estimates. However, such consistent profiles or biases represent possibly serious deficiencies in the product or the measurement processes. Technical programs should be initiated to eliminate the biases if their magnitudes are large enough to be important in practice.

The next step in the contrast method determination of the variance components is to compute the following three averages:

$$\overline{WP} = \frac{1}{R} \sum_{i=1}^{R} WP_i$$

$$\overline{LT} = \frac{1}{R} \sum_{i=1}^{R} LT_i$$

$$\overline{ST} = \frac{1}{R} \sum_{i=1}^{R} ST_i$$

Next, compute the three mean squares:

$$MSWP = \frac{1}{R-1} \sum_{i=1}^{R} (WP_i - \overline{WP})^2$$

$$MSLT = \frac{1}{R-1} \sum_{i=1}^{R} (LT_i - \overline{LT})^2$$

$$MSST = \frac{1}{R-1} \sum_{i=1}^{R} (ST_i - \overline{ST})^2$$

Finally, the variance components are calculated as:

$$VWP = \frac{MSWP - MSST}{2}$$

$$VLT = \frac{MSLT - MSST}{2}$$

$$VST = MSST$$

and each has $R-1$ degrees of freedom.

The ABCD ANOVA calculations correspond exactly to those for a two-way crossed ANOVA, where both the within-package and long-term factors are crossed with each other and only one replicate is included in the data. The expected mean squares apply to this two-way crossed ANOVA model. The short-term "factor" is really the interaction plus short-term measurement. In much collective experience, this factor effect has never been statistically significant.

Estimating and Maintaining VLL and VPP.   Estimates of VLLU and VPPU and estimates of VWP, VST, and VLT are combined as follows:

$$VPP = VPPU - VWP - VST$$

$$VLL = VLLU - VLT$$

This completes the calculations of all five package product unit variance components.

In practice, the variance component updating procedures (including calculating initial estimates of all components) should be done by computer. The data are entered into the computer database lot by lot (daily, by shift, or whatever) as they become available. The ANOVA software, both routine ANOVA and maintenance ANOVA, should contain diagnostic procedures to warn of any outlying test results, outlying averages, flinching, or unusual patterns in the data. Useful diagnostic procedures include sample sequence plots and histograms, CUSUM sequence plots, and sequence plots of lot averages and of maintenance contrasts.

If routine production process data have missing observations, the routine ANOVA formulas must be modified. It may be best simply to discard incomplete routine or maintenance data sets, so long as at least 60 complete sets are available in both cases for analysis. The causes of the missing data should be investigated periodically, using techniques such as Pareto analysis of the circumstances when data are missing.

The procedure that has been described here for collecting and analyzing routine production data and maintenance data will produce values of VLL, VPP, VWP, VST, and VLT that estimate the variability encountered with the way the plant processes and the measurement facilities are being operated.

PRODUCT CHARACTERIZATION

Having determined quantitatively the five variance components as described in the foregoing paragraphs, it is straightforward to calculate the implied conformance of product made. If the process is operating on-aim, first calculate

SPROD = true product standard deviation, the standard deviation of the true product property
from product unit to product unit

$$= \frac{\text{VLL} = + \text{VPP}}{2}$$

Then calculate

$$Z = \frac{\text{UTL (High)} - \text{UTL(Low)}}{2 \text{ (SPROD)}}$$

The UTLs are the unit tolerance limits (Figure 27.8). Then find the conformance of product made using a two-sided table of the cumulative normal distribution.

Some representative table values are presented in Table 27.2.

For example, if the UTLs are 2.5 multiples of SPROD above and below X_0, then the Z value is 2.5 and the conformance of product made is 98.8 percent.

If the process is operating off-aim, two values of Z must be calculated, one for each tail of the distribution.

$$Z_1 = \frac{\text{ProcessAverage} - \text{UTL(Low)}}{\text{SPROD}}$$

$$Z_2 = \frac{\text{UTL(High)} - \text{ProcessAverage}}{\text{SPROD}}$$

Then find the conformance corresponding to each of Z_1 and Z_2 using a one-sided table of the cumulative normal distribution. Some representative values are presented in Table 27.3.

Then the conformance of product made is obtained by adding the two conformances and subtracting 100. The conformance of product made decreases steadily as the process average moves off-aim.

For example, suppose the actual process average is one multiple of SPROD above X_0. If the UTLs are 2.5 multiples of SPROD above and below X_0 then the value of Z_1 is 3.5 and the value of Z_2 is 1.5. The conformance of product made is then $99.98 + 93.32 - 100 = 93.30\%$.

If an effective process control scheme is in place, it is unlikely that the process average would be off-aim as much as one multiple of SPROD for the entire period of producing a shipment of product

TABLE 27.2 Percent Conforming Product for Representative Z Values—On-Aim Process

Z	Conformance, percent
1.00	68.3
1.50	86.6
2.00	95.5
2.50	98.8
2.58	99.0
3.00	99.7

Source: Donald W. Marquardt (1991). *Product Quality Management,* DuPont Engineering, Wilmington DE.

TABLE 27.3 Percent Conforming Product for Representative Z Values—Off-Aim Process

Z	Conformance, percent
1.0	84.13
1.5	93.32
2.0	97.72
2.5	99.38
3.0	99.86
3.5	99.98

Source: Donald W. Marquardt (1991). *Product Quality Management,* DuPont Engineering, Wilmington DE.

for a customer. However, it also is unlikely that the process average would be precisely on-aim for the entire period. Thus, for this hypothetical example, the effective conformance of product made will likely be somewhere between 98.8 and 93.3 percent on a true-product-value basis.

It is well to remind ourselves at this point that only critical properties are candidates for treatment on a true-product-value basis. Often, with such properties the measurement variance (VST + VLT) is a substantial fraction of the total variance (VLL + VPP + VWP + VST + VLT). Then the interval UTL(high) − UTL(low) will be substantially smaller than the typical range of routine measurements from the process. Statistical process control becomes more difficult, and any attempt to improve conformance by releasing only product that is within "release limits" becomes inherently ineffective. This is a quite common circumstance in the process industries for such critical properties.

The prevalence of this circumstance is one reason why the most effective statistical control procedures, such as Twin Metric or CUSUM, are worthwhile in the process industries.

USE OF THE VARIANCE COMPONENTS FOR IMPROVEMENT OF THE QUALITY MANAGEMENT SYSTEM

Variance components are to be developed and maintained only for properties that are "critical," that is, properties that require careful control and are most in need of improvement.

The five variance components are invaluable information for purposes of quality management system improvement. A simple Pareto analysis, that is, ranking of the five variance components from largest to smallest is a first step. Suppose the VPP component is the largest, being, say, half of the total of the five variance components. It is then clear that the biggest quality management problem is with the product itself; in particular, the predominant source of product variability is among pack-ages within a lot. That knowledge immediately directs improvement attention to those features of the production process that can contribute to variability within a lot. On the other hand, suppose VST is the largest, being half or more of the total of the five variance components. This indicates that the biggest quality management problem is with the measurement system, and comes from the short-term sources of measurement variability. Improvement attention should then be directed to those aspects of the measurement process that can contribute to short-term variability in the measurement.

In the absence of such quantitative information about the magnitude of key sources of variability, many companies have worked for years on the wrong part of the system and have failed to achieve the improvement they sought. Periodic updates of the variance components provide quantitative evidence of the degree of success of improvement efforts.

USE OF THE VARIANCE COMPONENTS FOR ROUTINE SAMPLE DESIGN

The MSSD (mean square successive difference) method was recommended in a foregoing paragraph as the means to estimate the standard deviation, SPROC, for process control purposes. The MSSD method is recommended for SPC on both critical and noncritical properties.

For those few properties on which variance components are developed and maintained, the vari-ance components provide guidance in designing the structure of the Twin Metric or CUSUM sam-pling plan. For that purpose the following formula can be used to take account of the expected relationship between the sampling plan structure and the value of SPROC, where

NPP = total number of packages sampled per lot

NWP = total number of distinct samples per lot

NST = total number of test results per lot

NLT = total number of test times per lot

For example, if there is one observation for each of four packages sampled in each lot, and these are submitted to the laboratory for analysis in two groups at different times, the numbers are NPP=4, NWP=4, NST=4, and NLT=2.

The goal is to make SPROC as small as practically feasible. The routine SPC sampling plan is described by the denominators in the formula. The formula shows that, for most purposes, one obser-vation per package is optimum unless VST is by far the largest variance component. Similarly, NLT is often set to 1 unless VLT is a large variance component. In situations where a formal product release system is deemed necessary, the variance components are essential to effective release system design and evaluation. (See Li and Owen 1979; NBS 1959; Owen and Boddie 1976; Owen and Wiesen 1959.)

QUALITY MANAGEMENT SYSTEM UPDATES

Intent, Definition, and Timing of an Update.   The quality management system will reli-ably produce the claimed product quality levels if all functions of the system are properly designed, and carried out as designed. To guarantee continued compliance with the quality management sys-tem design, all system elements must be updated on a periodic basis. The activities necessary to accomplish the update serve as part of an internal audit of the system. It is important to document the administrative procedures and the organizational responsibilities to ensure that the necessary preparations are completed correctly on a timely basis.

The activities that take place during the update often identify specific data problems, quality man-agement system design problems, or the like. Fixing any of these can lead to an improvement in the system or its utilization, and lead to actual product improvements. Used in this manner, the update is an excellent source of ideas for continual improvement.

An update refers to:

- Assembly of data
- Calculation of variance components, SPROCs, etc., using the most recent data
- Calculations of system performance criteria (conformance, yield, and cost) for the period since the last update
- Calculations of predicted system performance criteria
- Recording of this information
- Discussions among Production, Marketing, and other appropriate personnel to decide whether any changes are needed in any component of the quality management system
- Documentation of the decisions taken

In preparation for an update, Production personnel analyze the accumulated data since the last update. For critical properties of high-volume products, the variance components and other statistics should be updated, say, quarterly. Updates for lower-volume products may only be practicable at longer intervals.

AUDITS

Every process in a quality management system should be audited periodically to ensure the objectivity and integrity of system performance. (See Section 11.) In many situations, such internal quality audits serve an additional role as precursors to external audits for external quality assurance requirements.

When audits and their resulting reports are wisely administered, audits are perceived by all participants as a mutually helpful vehicle to improve the quality system and the product quality.

Preparation of an audit plan should include a detailed review of everything that has been agreed to be done in implementing the quality management system.

Obviously, not everything can be audited at frequent intervals. The audit plans should establish audit priorities and audit schedules that reflect the priorities. High-priority items will be audited frequently, low-priority items infrequently.

Lower-Level and Higher-Level Audits.   At each organizational level the audit team should consist of persons who collectively have knowledge of the activities being audited and their proper procedures. To ensure objectivity, at each audit level the team should not be composed entirely of local personnel. The nonlocal personnel can include: higher-level management; technical personnel or supervision from the corresponding organization at another site; technical personnel or supervision from a distinct, but related, organization at the same site; and personnel from staff specialties such as Research and Development, Statistics, Marketing, and Quality Management Systems.

Numerical Audits and Procedures Audits.   In discussing periodic system updates, it was noted that the activities necessary to accomplish an update serve as key parts of the numerical aspect of an internal audit of the quality management system.

When developing routine system updates, the integrity of the numerical data should be questioned by graphical examination of data. In particular, the histograms of maintenance data and of routine product data should be inspected to detect flinching, outliers, or other anomalies. Time plots of these data are also informative, especially for revealing any long-term trends, cycles, or other patterns. The accuracy of data entry, storage, and manipulation should be spot-checked.

Procedures audits must include on-site observations of actual quality procedures. The objective is to assist the audited site in complying with quality management principles. Inspection details will differ from situation to situation, but should include:

- Verification that original data are accurately and promptly entered into the database system (computer or manual) with no form of flinching (Marquardt 1994).
- Verification that Twin Metric or CUSUM signals are followed by prompt and effective action.
- Verification that product release decisions are being made and followed properly.
- Verification that process materials acceptance decisions are being made and followed correctly.
- Verification that designs for Twin Metric or CUSUM and for product release are updated correctly and whenever required.
- Verification that standard operating conditions and standard operating procedures are being followed.
- Verification that standard procedures are properly documented and current versions (only) are readily available to those who need them.
- Verification that accurate records are kept.
- Verification that Production operators, inspectors, laboratory technicians, and others are following proper procedures, including:

Production equipment checkout and control

Process instrumentation calibration

Sample taking

Sample preparation and handling

Laboratory instrument calibration

Laboratory control sample validation and handling

Product packaging and handling

Product labeling

- Verification that action has been taken to correct quality system deficiencies identified in previous audits.

A written report should be prepared after each audit, and follow-up procedures should be established to verify that any quality management system deficiencies uncovered by the audit are corrected.

ACKNOWLEDGMENTS

The contents of this section are adapted and updated from portions of a comprehensive book on product quality management produced in 10 successive editions by the corporate group which I managed within the DuPont Company from 1964 through 1991. In all, 16 members of the group were co-authors of that book. All of them are skilled statisticians with a great deal of experience in dealing with process industry problems in practice.

I served as overall editor, and was the principal author of most of the portions I have chosen to adapt for this section of Juran's Handbook. Some of the chosen topics derive from important early technical initiatives of several of the other 15 co-authors. In particular, K. A. Chatto and R. E. Scruby pioneered in the 1960s in the use of variance components, the package structure for the variance components, the use of true-value specifications, and the ABCD maintenance plan. W. H. Fellner and others extended that work. The discussion in this section describes only the most straightforward practical uses of that body of work. J. M. Lucas pioneered in the 1970s in the use of CUSUM and important CUSUM enhancements for process control in a computer-friendly computational form with practical design methods. This section incorporates the CUSUM approach, which is so important in the process industries, only by reference to key papers in the literature. My intent in preparing this section has been to focus upon key quality management issues in the process industries and, in doing so, to build upon other sections of Juran's Handbook.

REFERENCES

ASQC (1987). *Quality Assurance for the Chemical and Process Industries: A Manual of Good Practices,* American Society for Quality Control, Milwaukee, WI.

ASQC (1996*a*). *ISO 9000 Guidelines for the Chemical and Process Industries,* 2d ed., American Society for Quality Control, Chemical and Process Industries Division, Chemical Interest Committee, Milwaukee, WI.

ASQC (1996*b*). *Specifications for the Chemical and Process Industries,* American Society for Quality Control, Chemical and Process Industries Division, Chemical Interest Committee, Milwaukee, WI.

Bissell, A. F. (1986). "The Performance of Control Charts and CUSUMs Under Linear Trend." *Journal of the Royal Statistical Society,* series C, vol. 35, p. 214; corrigendum to ibid., vol. 33, pp. 145–151.

Champ, C. W. and Woodall, W. H. (1987). "Exact Results for Shewhart Control Charts with Supplementary Runs Rules." *Technometrics,* vol. 29, pp. 393–399.

Deming, W. E. (1967). "What Happened in Japan." *Industrial Quality Control,* vol. 24, pp. 89–93.

Deming, W. E. (1972). "Report to Management." *Quality Progress,* vols. 5, 2, 3, and 41.

Deming, W. E. (1982). *Quality, Productivity, and Competitive Position.* Massachusetts Institute of Technology, Center for Advanced Engineering Study, Cambridge, MA.

Hald, A. (1952). *Statistical Theory with Engineering Applications.* John Wiley & Sons, New York.

ISO 9000 (1994). International Standard. *Quality Management and Quality Assurance Standards—Guidelines for Selection and Use,* International Organization for Standardization, Geneva, Switzerland. Available through American National Standards Institute, New York; also available as ANSI/ASQC Q9000:1994 from American Society for Quality Control, Milwaukee, WI.

Juran, J. M., ed. (1974). *Quality Control Handbook,* 3rd ed., McGraw-Hill, New York, p. 18-3

Kemp, K. W. (1961). "The Average Run Length of the Cumulative Sum Chart When a `V' Mask is Used." *Journal of the Royal Statistical Society,* series B, vol. 23, pp. 149–153.

Li, L. and Owen, D. B. (1979). "Two-Sided Screening Procedures in the Bivariate Case." *Technometrics,* vol. 21, pp. 79–85.

Lucas, J. M. (1973). "A Modified V-Mask Control Scheme." *Technometrics,* vol. 13, pp. 833–847.

Lucas, J. M. (1976). "The Design and Use of V-Mask Control Schemes." *Journal of Quality Technology,* vol. 8, pp. 1–12.

Lucas, J. M. (1982). "Combined Shewhart-CUSUM Quality Control Schemes." *Journal of Quality Technology,* vol. 14, pp. 51–59.

Lucas, J. M. (1985*a*). "Counted Data CUSUMs." *Technometrics,* vol. 27, pp. 129–144.

Lucas, J. M. (1985*b*). "Cumulative Sum (CUSUM) Control Schemes." *Communications in Statistics, Theoretical Methods,* vol. 14, no. 11, pp. 2689–2704.

Marquardt, D. W. (1984). "New Technical and Educational Directions for Managing Product Quality." *The American Statistician,* vol. 38, pp. 8–14.

Marquardt, D. W. (1991). *Product Quality management,* DuPont Engineering, Wilmington, DE.

Marquardt, D. W. (1993). "Estimating the Standard Deviation for Statistical Process Control." *International Journal of Quality and Reliability Management.* vol. 10, no. 8, pp. 60–67.

Marquardt, D. W. (1994). "Report Card Issues in Quality Management." *Quality Management Journal,* vol. 1, no. 3, pp. 16–25.

Marquardt, D. W. (1997). "Twin Metric Control-CUSUM Simplified in a Shewhart Framework." *International Journal of Quality and Reliability Management,* vol. 14, no. 3, pp. 220–233.

Marquardt, D. W. and Ulery, D. L. (1991). "Twin Metric-A Strategy for Improved SPC." *IMPRO 1991 Conference Proceedings,* Juran Institute, Inc., Wilton CT.

Marquardt, D. W. and Ulery, D. L. (1992). "Twin Metric Control-Improving SPC to Meet Supply Chain Realities of the 1990s." *Transactions of the 46th Annual Quality Congress,* American Society for Quality Control, pp. 367–373.

NBS (1959). "Tables of the Bivariate Normal Distribution Function and Related Functions." *Applied Mathematics Series 50,* National Bureau of Standards, U.S. Government Printing Office, Washington, DC.

Owen, D. B. and Boddie, J. W. (1976). "A Screening Method for Increasing Acceptable Product with Some Parameters Unknown." *Technometrics,* vol. 18, pp. 195–199.

Owen, D. B. and Wiesen, J. M. (1959). "A Method of Computing Bivariate Normal Probabilities." *The Bell System Technical Journal,* pp. 553–572.

Page, E. S. (1961). "Cumulative Sum Charts." *Technometrics,* vol. 3, pp. 1–9.

SECTION 28
QUALITY IN A HIGH-TECH INDUSTRY

M. K. Detrano

INTRODUCTION

In order to understand the role of quality in a high-tech industry, it is important first to define what both "quality" and "high tech" mean in this section. The term "quality" is interpreted very broadly and can be thought of as a synonym for "business excellence", i.e., that attribute of business systems and processes which contributes to successful and sustainable business results, encompassing a balance among employee, customer, shareholder, and other stakeholder (community, environment, etc.) values.

The term "high tech" is used to refer to any business that produces a product that has "high tech" attributes, uses processes that rely on "high tech" methods of production, or addresses a marketplace that has "high tech" product or service needs. What, then, are these "high tech" product, process, or market needs attributes? From a product point of view, "high tech" can be defined as *a product that is propelled by a powerful technology learning curve and that has a complex architecture with high feature density, speed, and/or bandwidth requirements.*

While products from the electronics and communications industries are obvious examples of high tech, the definition can be interpreted more expansively, especially when examining process and marketplace attributes. For instance, high tech processes have many of these challenges:

- Rapid time to market with shorter development and life cycles
- Very well-educated, technically excellent, multicultural work force
- A growing body of de facto global standards
- A significant reliance on "qualified" suppliers and complex manufacturing systems

The preceding characteristics represent internal aspects of high tech. The external or marketplace needs have these attributes:

- A customer base that may have little or no experience with the product or technology
- A marketplace that is willing to accept innovative solutions to problems they may not yet know they have

There are probably no aspects of total quality management (TQM) principles in "any tech" industries that do not pertain to high-tech industries as well. Therefore, no attempt will be made in this section to outline all TQM activities pertaining to industry. Only selected aspects of quality management principles that are particularly relevant to high-tech industries will be discussed.

Finally, the discussion will roughly track the 1997 U.S. Malcolm Baldrige National Quality Award Criteria, highlighting topics as they pertain to each of the following Baldrige categories:

- Category 1: Leadership
- Category 2: Strategic Planning
- Category 3: Customer and Market Focus
- Category 4: Information and Analysis
- Category 5: Human Resource Development and Management
- Category 6: Process Management
- Category 7: Business Results

This is in keeping with the intent to define "quality" as "business excellence." As the understanding of the impact of quality continues to mature and expand, it is appropriate to broadly address all areas of business excellence. This is especially true in high-tech industries, where there is a growing intolerance for mediocrity in terms of time to market, quality of products and services, competitive value, and flawless execution.

BALDRIGE CATEGORY 1: LEADERSHIP

Speed is clearly a major differentiator in a high-tech industry. This has important implications on the leadership and management systems.

Speed manifests itself in various important ways. These include

- The organization's and the leader's ability to learn
- The process and length of time it takes to reach organizational alignment
- How long it takes to channel technical breakthroughs into innovative products
- How quickly informed decisions are made

Each of these areas will be addressed in turn below.

Ability to Learn.   First, good leaders are not only able to learn but they are also *willing to be seen learning*. They are in a continuous learning mode with shorter and shorter cycles. They create an environment open to learning and change, and they expect the people in their organizations to behave similarly. This enables both the leaders and the work force to continually and quickly develop the better answers that are demanded in their high-tech world.

Organizational Alignment.   Second, the leaders must set direction but not overmanage. Constancy of purpose with repetitive attention to clearly articulated goals is important. The use of a policy deployment (or similar) process with mutually negotiated goals, metrics, cascaded objectives, and frequent review sessions is very beneficial. Such a process can reach all levels in the organization quickly and is helpful in a high-tech environment because technical innovation frequently can create apparently conflicting goals. In addition, it helps leaders conduct a systematic analysis for near- and longer-term goals. It helps communicate the critical few breakthrough projects needed to make the organization successful. And it helps employees know what is important and how they personally connect to the organization's goals.

Figure 28.1 provides a sample for the structure of an organizational policy deployment matrix. Typically, these are layered downward through the organization, with a referencing scheme (numerical or otherwise) that allows lower levels, including individuals, to relate their goals to higher-level ones. Each organization, in turn, uses the matrix/matrices of the organizations above it *and beside it* as guidance and insight into the development of its own matrix, negotiating with higher-level organizations as necessary. The essence of the matrix-development task, therefore, is to develop a set of goals and objectives that have these characteristics:

- They specifically address the scope and area of responsibility of that particular organization.
- They are aligned with the overall business goals and objectives.
- They are limited to a reasonable number of clearly understood and mutually agreed-to projects and/or metrics.
- They are synergistic but not conflicting with neighboring organizations.

Each individual employee, with the supervisor's input, now develops his or her own set of personal goals from these organizational matrices. These employee goals, representing a mutually negotiated agreement between the employee and his or her supervisor, should have the same four key characteristics as outlined for organizations above. *In some cases, the identical "individual" goals are applied uniformly to a team of employees, where the team has agreed to work as a unit, i.e., they succeed or fail together.* Figure 28.2a presents a possible standard template for an individual employee goal based on the higher-level matrix. Figure 28.2b uses this template to derive a sample goal from the higher-level organizational goals shown in the example in Figure 28.1.

An important point to note is that policy deployment can be used to define and track incremental business goals as well as breakthrough goals. It is helpful to keep the number of breakthrough goals very small, e.g., two to three, because too many large projects will dilute the benefit of concentrated effort. Having a clear understanding of which goals are breakthrough versus those which are incremental will help ensure that appropriate effort, staff, reporting timeframes, expectations, and rewards are instituted commensurate with the type of goal.

The use of the traditional plan-do-check-act model here helps to show the importance of linking strategic planning to policy deployment. The entire goal-setting and cascading portion of the organization's policy deployment process should be completed before the beginning of the fiscal year and should be integrated with the strategic planning process for that year. This can be thought of as the "plan-do" steps, i.e., "plan" through the development of the strategic plan and the more tactical policy deployment matrices, and then "do" through the cascading of the policy deployment matrices throughout the organizations. The review portion of the policy deployment process, the "check" and "act" steps, should be arranged so that every organizational metric is reviewed daily, monthly, and/or quarterly, i.e., checked, as appropriate. Figure 28.3 is a sample of a monthly metric reporting sheet. Trending data are required and corrective action, i.e., the "act" portion, instituted against root causes in cases where the trends are in the wrong direction or behave outside expected control limits. (See Baldrige Category 2: Strategic Planning, below, and Section 13 for a broader discussion of strategic planning.)

These monthly metric reports, or a condensed version of them, can be "made visible" to all employees as a continuous reminder of priorities and goals, i.e., constancy of purpose. A simple example of a visibility method is the use of video monitors mounted in high-traffic hallways. Information is refreshed daily and has a simple choreography designed to catch the eye. Such a display can summarize policy deployment metric status as well as community announcement activities, e.g., new employee arrivals, upcoming ISO audits, etc.

One note of caution is necessary regarding the reporting portion of policy deployment. Most often policy deployment fails because the entire process is allowed to degenerate into the creation of a stack of arcane monthly reports that no one looks at, responds to, or is held accountable for. This could be indicative of a number of root causes:

Corporate goal	Subproject	Project owner	Metric	Metric reporter	Target (current year)	Long-term target
1.0 Increase customer satisfaction	1.1 Institute rigorous survey method	J. Smith	1.1a Monthly survey results	K. Andersen	10 months	1 every month
			1.1b Quarterly survey analysis	B. Baker	3 of 4 quarters	Every quarter
	1.2 Create customer response team	R. Johnson	1.2a Percent customer complaints resolved	V. Jones	60%	95%
	1.3 Achieve ISO registration	D. White	1.3a Successful external audit	O. Thompson	During 3rd quarter	N.A. (Need long-term goal for ISO)
2.0 Enhance employee engagement	2.1 Pilot self-directed team concept	E. Edwards	2.1a Choose and train pilot group	E. Edwards	End of 1st quarter	N.A.
			2.1b Percent favorable response to questionnaire	P. Brown	80%	95%
3.0 Create additional share owner value	3.1 Decrease accounts receivable	W. Green	3.1a Percent product DOA	Q. Director	75%	99%
		R. Johnson	1.2a See above			
	3.2 Reduce COGS	C. I. Owe	3.2a Manufacturing cost reduction	T. Bender	10%	35%

FIGURE 28.1 Policy deployment matrix.

28.4

GOAL DEPLOYMENT TEMPLATE

Goal: (State what the goal is in one sentence.)

Goal Reference Number: (In order to keep the numbers consistent, use a reference number that refers back to the policy deployment matrix.)

Plans/Steps/Process to Accomplish Goal: (Outline briefly what are the steps that are needed to accomplish this goal. Note: This is not a replacement for a detailed work plan but a vehicle to clarify expected work effort.)

Possible Barriers to Overcome: (Identify any issues or situations that could impede progress. Note: This helps to visualize areas for special attention for both the employee and his/her support manager.)

Probability of Meeting Goal (High, Medium, Low): (Identify probability of success. Note: Most goals have a high probability of success. However, if the goal is a "stretch" goal that is perceived to be very difficult, a medium or low probability may be more realistic. This does not mean that the employee will not work hard to accomplish it, but it does give the employee the sense of support from his/her manager that the manager acknowledges the difficulty of the goal. In this case, there should be an accompanying discussion in the "Steps" and "Barriers" paragraphs above about ways to increase the probability of success.)

To Whom Is Goal Deployed: (Note: This is normally the employee. It also frequently includes other people that the employee is expected to influence or work closely with.)

Metric(s) to Measure Progress:
  Name: (Identify some metric that is measurable that will show progress or completion.)
  Source: (Identifying the source of the metric helps eliminate vague metrics.)
  Target: (Helps to realistically assess the usefulness of the goal and the metric.)

Goal Owner: (Person responsible for the goal. It may be the employee, the employee's manager, etc.)

Signatures for Concurrence:

Employee: **Date:**

Goal Owner: **Date:**

Manager/Support Person: **Date:**

FIGURE 28.2a Goal deployment template description.

EMPLOYEE GOAL DEPLOYMENT

Goal: Become a productive team member for the new customer response program.

Goal Reference Number: 1.1

Plans/Steps/Process to Accomplish Goal: Take training to become certified in product knowledge. Complete workshop on "Talking with Unhappy Customers." Develop spreadsheet to help analyze customer call statistics.

Possible Barriers to Overcome: Training material may not be available in time. May need to find more informal ways to learn about products. (My manager can help negotiate training schedules, etc. on behalf of the new team.)

Probability of Meeting Goal (High, Medium, Low): High

To Whom Is Goal Deployed: I. M. Worker

Metric(s) to Measure Progress:
  Name: Course completion rate
  Source: Training records
  Target: 2 courses, one each quarter, completed by end of 2Q.

  Name: Spreadsheet usage
  Source: Monthly reports
  Target: All team reports produced by this software from 3Q on.

Goal Owner: I. M. Worker

Signatures for Concurrence:

Employee: *I. M. Worker*   **Date:**

Goal Owner: (*Same*)   **Date:**

Manager/Support Person: *U. R. Boss*   **Date:**

FIGURE 28.2b Employment goal deployment sample.

Metric: *% Customer Complaints Resolved*   **Sponsor:** *R. Johnson*

Cascade Reference Number: *1.2a*

Policy Deploy.Objectives: **Level 1:** *To Increase Customer Satisfaction*

                                     Level 2: *To Create Customer Response Team*

                                     Level 3:

QI Story	**Step 1:**	*Need systematic way to respond to customer complaints.*
-or-	**Step 2:**	*Data of current situation shows no repeatable "system".*
Improvement	**Step 3:**	*Root cause analysis shows need to formalize responsibility, et al.*
Plan	**Step 4:**	*Currently instituting needed countermeasures.*
	Step 5:	*Some early results are beginning to show improvement.*
	Step 6:	*Will formalize new approach next year if continued successful.*
	Step 7:	*Will start next level of work toward BIC of 95%.*

Trend Chart: (Show this year's goal and B-I-C or competitive benchmark.)

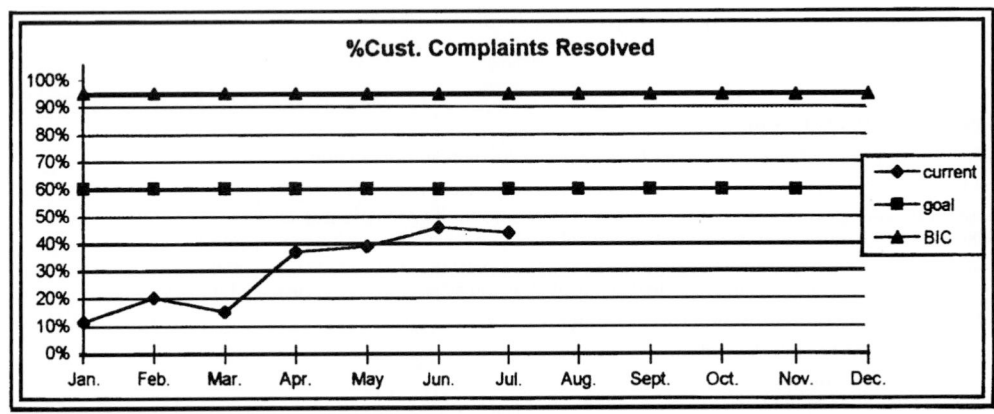

Person Tracking This Data: *V. Jones*

Explanation of Spikes/Dips: *Begun to implement Q.I. Story Countermeasures*
                                      in March. Results beginning to show improvement
                                      in April. Final Countermeasures will be in place
                                      by August.

FIGURE 28.3   Monthly metric report.

- Wrong choice of metrics
- Lack of a suitable venue for review and action
- Waning attention to the need for constancy of purpose

The first two causes can be addressed through the use of problem-solving techniques. The third cause, however, may be a symptom of waning leadership attention. It requires direct leadership intervention.

Finally, at the individual employee level, performance evaluation against goals is a major tool in the overall human resources performance management process. (See Baldrige Category 5: Human Resource Development and Management.) Individual goals should be reviewed at least quarterly and updated as necessary because individual assignments and goals are likely to change more frequently than overall organizational goals.

Product Innovation.   A third leadership challenge is to channel technical excellence into innovative products rapidly. Policy deployment, coupled with a systematic approach to ensuring that customer needs are well understood by the R&D community, can help this. (See Baldrige Category 3: Customer and Market Focus, Reaching the R&D Work Force.) A dual-reporting organization that allows a program manager (one who manages a new product or service) to operate as the link between research and development (R&D) and the customer, while offloading administrative overhead to a functional manager, has shown success in some areas. This can be augmented by a strong focus on teams and teamwork for that product or service, including team members from business, marketing, and sales. All team members must be fully conversant in the technology. Lack of understanding on anyone's part will slow down or misguide that information flow between technology capability and customer needs. (See more in Baldrige Category 5: Human Resource Development and Management.)

Decision Making.   Lastly, the leadership and management system must be able to make informed decisions quickly, particularly where new products are concerned. A well-followed and well-cared-for "gate process" has the potential to move product decisions along expeditiously. While the basic purpose of a gate process (see process flow in Figure 28.4) is to ensure that the right things are done at the right time via checklists at each gate review (everything from funding approvals to documentation available at time of product ship), it has the added benefit of creating a rhythm for moving the product along through its various stages or "gates." If the process is followed, with aggressive target dates set for passage through each gate, it is expected that new products will be introduced sooner, more smoothly, and with a better sense of cost of development, while at the same time weeding out false starts early.

BALDRIGE CATEGORY 2: STRATEGIC PLANNING

This area touches on two aspects of strategic planning: (1) a comprehensive and thorough plan-development process that includes feedback loops and (2) a deployment and execution process. If either of these areas is incomplete or performed outside the normal flow of business, the effort will fall short of its potential. It is important to note that the planning process needs to include an update-cycle capability. In high-tech industries, the rate of change demands continuous evaluation and improvement.

Plan Development.   The process that develops the plan must include many constituencies, but in a high-tech industry it is essential to include technology (i.e., where the technology is going, what breakthroughs are needed, and how fast the technology will change) as a major component. To accomplish this, real diversity of ideas is needed. In high-tech businesses, not all strategy ideas are "found" at the top. Therefore, it is important to include the "diversities" of the organization regardless of hierarchy. [A *Harvard Business Review* article (Hamel 1996), addresses this issue in detail.]

Each project, before it is approved to move to the next stage, i.e., through a "Gate", must undergo a "Gate Review". (Note: this requires that a detailed project schedule has been defined, including the projected dates for each Gate Review at the time of initiation of the project.)

A "Gate Review" is a formal meeting, conducted by a moderator with the assistance of a scribe, that systematically reviews a checklist of items that are required in order to allow the project to move through the gate. The review is conducted by a set of "gatekeepers", i.e., a cross-functional group of people representing each organization that has responsibility for some part of the project. Consensus is usually required.

The "Gate Process" has a formal process owner who works with the organization to define the five checklists and other process details, keeps track of gate review statistics (projects, schedules, etc.), and applies process improvement suggestions, as needed.

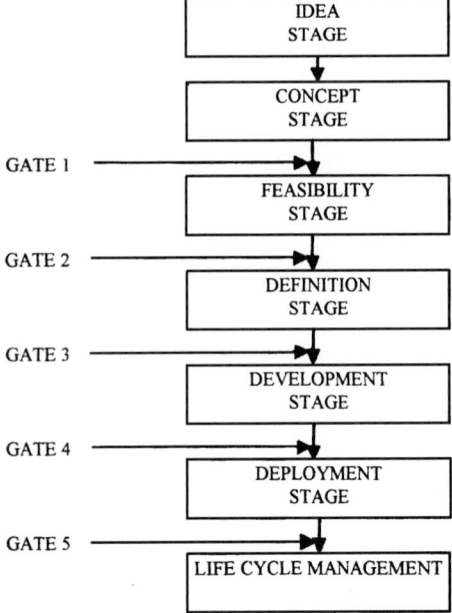

FIGURE 28.4   Gate process.

A second impact of high tech, and an outgrowth of this diversity, is the need to ensure that R&D breakthrough project plans and foundational technology work that is clearly linked to the goals of the business are included in the strategic plan.

A third consideration for plan development in a high-tech business is integration of contingency planning. When speed to market is important and the product is based on new technology, the product-development process, as well as the product delivery and customer expectation management processes, must account for the risk implied from the unknown.

This leads to the last, albeit most important, consideration: that of input from the customer, together with input from the market and the sales channels. This feedback loop is essential. As more and more high-tech business becomes "customized," the ability to see, hear, and accommodate the "customer of one," i.e., the individual needs of every customer, is critical. Also, when designing a new product concept, it is important to understand the impact on the customer/market that was created by a previous product. This is due to the ability of new high-tech products to actually define new markets or cannibalize old ones. A comprehensive "voice of the customer" system can fulfill this need for systematic and comprehensive customer input. (See more on this topic in Baldrige Category 3: Customer and Market Focus.)

Deployment and Execution.   The deployment and execution phase of strategic planning can be linked directly to and accomplished by policy deployment and human resources performance management practices. The objectives set out in the policy deployment matrix can be captured directly from the strategic plan. As discussed earlier, regular reviews of progress toward these objectives is essential to maintaining the relevancy of the strategic plan, i.e., constancy of purpose.

BALDRIGE CATEGORY 3: CUSTOMER AND MARKET FOCUS

Customers are just as important in a high-tech environment as in any other environment. However, since there is a temptation in high-tech industries to focus more on the technology of the product or the process, some aspects of customer satisfaction should be highlighted. These include

- Achieving balance between a customer and a product focus
- Systematic surveying
- Reaching the R&D work force
- Managing customer expectations

Note that the term "voice of the customer" is used informally several times. It implies a recognition that input from the customer comes in many forms and through many different channels and that these sources should be analyzed and integrated systematically.

Balancing Product and Customer Focus.   In high-tech industries, there is an obvious need to focus on technology leadership. However, attention to the voice of the customer is a critical balance if the new product is to fit a useful application. Being mindful that the introduction of new technology can create discontinuities in the market, it is vital, therefore, to listen both to existing customers/markets and to those who would become users of the newer technology. (Wireless telephony that allows someone to carry a phone with them wherever they go rather than being tethered to fixed locations is an obvious example of this discontinuity. Computing technology that moved from mainframes and timeshare to PCs and client/server is another.)

While much emphasis is placed on the chain of traceability from customer requirements through product delivery, it may be useful for high-tech companies to initiate this chain even before the customer is necessarily aware of what is desired or wanted. Gale (1994) discusses the concept of product attributes and their seven-stage life cycle, from "basic" attributes, i.e., required with no competitive edge, to "latent," i.e., not yet visible or apparent in the marketplace. Table 28.1 describes each of these seven stages.

When searching for and listening to the voice of the customer, high-tech businesses that are mindful of this hierarchy can now probe for clues about new product possibilities based on what customers portray as "unique," "desired," or "latent" attributes, i.e., needs they have that are not yet fulfilled and hence are new product opportunities.

TABLE 28.1   Stages of the Attribute Life Cycle

Stage	Description
1. Latent	Not yet visible or apparent
2. Desired	Known but not currently supplied
3. Unique	Only the pioneer scores well
4. Pacing	One supplier is already ahead
5. Key	Differences in performance determine competitiveness
6. Fading	"Catch-up" moves begin to take away the top performer's edge
7. Basic	All suppliers perform well—no competitive edge

Source: Gale (1994).

Systematic Surveying.   Speed and the acceleration of change in a high-tech industry also influence when and to whom to listen in the customer/market environment. Systematic surveying processes, showing trends in the market and the competition, are particularly useful if executed at regular intervals that keep tempo with market and technology changes. Surveys and surveying techniques that crosscut the management hierarchy of the customer base, including key decision makers, technology directors, and line technicians, fill in a more complete picture of how the customer needs are being perceived and met. Next, feedback to the customer to clarify what was heard and what should be expected helps to manage the risk of wasting ill-affordable time on bad assumptions. Finally, an aggressive action plan to assimilate and act on what was heard reassures customers that their time was used productively.

Reaching the R&D Work Force.   High-tech industries need to have a wide, heavily used, and effective channel by which the voice of the customer can be clearly heard throughout the ranks of the technical community. While this may be difficult to create because of the distance in the value chain from R&D to the customer (customer to sales to marketing to product management to product planning to design to development to research), user group programs, technical seminars, formalized customer-visit processes, and creative partnering are examples of such channels. High-tech leaders who demonstrate equally high respect for the customer and the technical community and recognize the necessity to bring these groups together lay a good foundation for effective channel implementation.

Customer Expectations.   Customer focus is *bidirectional*. Since uncertainties and unfamiliarities are intrinsic to new technology, it is just as important for the company to *speak* to its customer/market as it is expected to *listen* to its customers. Educating the customer/market about the new technology and helping to align expectations regarding the capabilities of the new technology are responsibilities that a high-tech company or industry must assume.

BALDRIGE CATEGORY 4: INFORMATION AND ANALYSIS

Topics to be addressed here fall into four broad areas:

- Finding and using the right metrics
- Using information technology
- The role of benchmarking
- The methods and benefits of overall organizational assessment

Metrics.   It is not easy to find the right metrics. Relating back to policy deployment, any metric that is chosen for tracking should pertain to the objectives on the policy deployment matrix in some way and should support the greater good intended by the highest-level matrix. An example is useful here. Suppose one of the highest-level goals of an organization is to increase customer satisfaction. Further suppose that root-cause analysis of customer feedback data indicates that one of the issues for many of the customers is that installation time is too long. In response, a tools group within a product support division in the organization decides to invent tools to ease the installation burden for existing products. (Product re-engineering is surely one of the objectives for the design group, but that is longer term.) A metric for the tools group, therefore, might have to do with the speed with which they develop tools. This metric of development speed would pertain to the higher-level metrics of customer satisfaction and installation time and, at the same time, support the intended greater good. If, on the other hand, the tools group chose the metric "number of tools built," it could potentially build many tools that were not very useful even though their metric performance would be good. This attention to avoiding local optimization is what is meant by supporting the greater good. One therefore can see that while a metric might be far too detailed and obscure at the highest levels, it can nevertheless be a valid metric at lower levels if it can affect directly the performance of a higher-level metric.

In addressing an additional point that metrics also need to be truly measurable and quantifiable, some metrics must be broken down into constituent parts to be measurable. For example, in an R&D community, technical excellence, while very important, is in itself quite vague. It could, however, actually be decomposed into individual excellence, team excellence, and technical leadership excellence. Now how does an organization measure something like individual excellence? Individual excellence could be further broken down into attributes that address hiring, continuing education, and retention. Specific measures for these attributes can now be enumerated, such as hiring statistics from target schools and markets, attrition rates, participation in training programs, and so on.

Product-development indicators are especially important for high-tech products. One category of product-development indicator covers the overall development process, such as the frequency with which new products are being introduced or how effective vis-à-vis revenue is to the development process. A second category that is especially important for high-tech products pertains to the product at different stages of its development cycle, such as a metric having to do with testing and performance uncertainty. For instance, in many cases, hardware test data results are heavily dependent on the resolution of available test instrumentation, which, in an area of high technology, may approach the tolerances of the product. Hence empirical and testing data from field trials, prototyping, etc. are essential from the *earliest stages* because of the higher degree of uncertainty.

In summary, every level or function in the organization will likely have its own set of metrics, some of which are common across the organization and others of which relate directly to locally derived objectives. Identifying the right metrics, especially to ensure traceability to the customer, may take analysis of the task and insight into what behavior or outcome is sought. Typical metrics categories might include revenue, customer value, cycle time, productivity, cost (including cost of quality), reliability, defect rate, on-time delivery, technical excellence, product-development indicators, and manufacturing statistics.

Information Technology.   Due to the combination of a global work force and a need for speed, the technology of the information systems is important. The systems must be integrated and able to gather, process, distribute, and share data quickly. This helps to close the geographic gap, make the employees more knowledgeable, and enhance the sense of awareness of the current situation. All these aspects help to bring a global work force together faster. The example of a hardware-development process illustrates the point here. Traditionally, the R&D group would design a circuit board, for example, and then create a working model, test it, and finally send it to the factory for manufacture. In turn, the factory engineers would redesign aspects of the circuit board to better accommodate supplier and production issues that had not been fully considered by the R&D staff. Today, a highly integrated design database coupled with a "concurrent" engineering process, i.e., R&D and factory working from the same design database at essentially the same time, can bring the two work groups together from the very beginning of the design phase even if not colocated. Significant interval reductions can be achieved, as well as side benefits such as better overall design and a stronger sense of teamwork.

Benchmarking.   Benchmarking of product attributes is sometimes very difficult because new product performance characteristics are generally a closely held and well-controlled body of knowledge. However, high-tech companies can benchmark process capabilities. Benchmarking can help an organization to quickly study, assimilate, and improve processes based on the experience of others. High-tech companies in particular that are driven by speed can benefit from this because it helps them avoid the loss of time due to a more traditional trial-and-error method.

The areas for benchmarking include almost any process that is part of the day-to-day business. An obvious example of this is the 6σ improvement process that has been adopted by many companies interested in defect reduction (Harry n.d.).

Management System Assessment.   Finally, a systematic management assessment process is essential to maintaining leading-edge performance. Because of the rate of innovation and change in a high-tech industry, it is important to guard against the tendency to fall behind if the organization is not growing, learning, and improving continuously. An efficient method of regular and periodic assessment with improvement planning built into the policy deployment process is a good countermeasure.

There are many different models against which an organization may audit itself for this purpose. The U.S. Malcolm Baldrige National Quality Award and the European Quality Award are two such models.

These models can be modified and used outside the intended formal award structure to bring the assessment process directly into the management and governance structure of the business. For a high-tech company whose survival depends on speed and quality, an integrated assessment process can serve as a constant barometer of its health. In this environment, the assessment process can be stream-lined in time (e.g., on a regular basis, the leadership team can perform a self-assessment under the guidance of a trained examiner) and/or focused on a specific area (e.g., the R&D community might perform an assessment focusing on its ability to innovate.) Figure 28.5 is an example of an assessment process that falls between the formality of a national quality award application process and a simple leadership brainstorming session. It serves to suggest the existence of useful alternatives.

BALDRIGE CATEGORY 5: HUMAN RESOURCE DEVELOPMENT AND MANAGEMENT

The most obvious challenge in a high-tech business is to ensure that the organization has motivated and correctly trained people when and where it needs them in the global business environment. This applies not only to the R&D work force but also to the factory, business, sales, and customer service employee base. This requires lead time and is an important consideration in the strategic planning process and the policy deployment process.

In addition, highly trained technical work forces create the need for HR-related processes that specifically address the following needs:

- Nourishment of a feeder pool through trade school, university, and research partnerships
- Expectation of obsolescence and how to avoid it through a constant learning mentality
- Motivation to excel and incentive to innovate
- Tolerance for failure and risk-taking
- Retention of a scarce resource

Innovative processes that bring together technical expertise, corporate education, and HR can produce insights into core competency skills. They also can design and provide the best and quickest ways to acquire, develop, and maintain these needed skills.

Step 1: Using Baldrige criteria, the coach prepares a set of questions. The coach then conducts an initial round of sessions with focus group(s) (a group of subject matter experts from the organization, preferably the leadership team). Done in two adjacent half-day sessions, using a timekeeper:

- Each question is displayed on a Vugraph. 2 minutes to read.
- 5–10 minute brainstorm, "What are we doing?"
- 5 minutes to highlight strengths.

Step 2: The coach prepares summaries of focus group results for each question.

Step 3: The coach reconvenes original focus group(s) to

- Review and validate the findings
- Score the question
- Identify gaps based on the findings
- Propose possible countermeasures

The coach guides and helps calibrate scoring. An average score is developed.

Step 4: The coach prepares a summary by question, identifying strengths, gaps, and score for each question.

Step 5: [Optional] The coach leads an independent assessment team through the focus group findings. They develop an independent score, compare it to the focus group score, and resolve differences.

Step 6: Coach prepares executive summary and shares results with leadership team.

FIGURE 28.5   Simplified Baldrige assessment process.

In addition, to quickly and efficiently tap the potential of the large, skilled, distributed work force, it is important to avoid creating "linear systems." (Linear systems here generally imply processes that serialize events between and across groups of people.) The use of "parallel processes", e.g., policy deployment or concurrent engineering, and distributed information networks, serves to facilitate parallel yet coordinated activity. Here again, as mentioned in Baldrige Category 1: Leadership, a management approach that sets clear direction, expectations, and accountability but does not serialize or micromanage, i.e., one that has "loose hands on the reins," is essential to the success of parallel systems. One would not like to envision an organization where its manager inserts himself or herself into every decision, thereby throttling back decision frequency to a point that is commensurate with the "bandwidth" of the manager!

No amount of process, leadership direction, training strategies, and so on will succeed, however, if the employee body is not motivated to contribute to the success of the business. One technique that tends to couple approaches for motivation, resource retention, and business goal alignment is the use of a broad recognition system. Assuming that business values and goals have been communicated clearly through leadership initiatives and policy deployment processes, a follow-up system that formally and informally rewards behaviors aligned with those values and goals can be motivational in itself. For a highly technical work force in which recognition traditionally has been realized through intellectual and frequently individual achievement, a recognition program can highlight the value of team participation, speed, and innovation. Regardless of the monetary value of a recognition "event" (anything from a simple thank-you note to an expensive gift or free vacation), such an overall program serves to reinforce across the work force the power and the value of its individual and team contributions.

An even more powerful approach starts with the selection of the employees through a testing system that measures individual tendencies with respect to the core values of the business. Simply stated: Those who have the values that align with the organization's values are invited to join the team. For instance, if the need for collaboration among team members is identified as an essential element for success, then potential employees are interviewed to ascertain their affinity for collaboration before they are hired or invited to join the team. *Only those who demonstrate the desired attitudes, in addition to having the needed technical skills, are acceptable for that position.* A prerequisite to this approach is the existence of a set of principles or core values against which to measure candidates.

While this approach is probably part of every hiring practice to some degree, a thorough and systematic application uniformly across the organization can quickly set up an environment that has a very high probability of operating according to its stated principles and core values. For a high-tech company, the values of teamwork, collaboration, the need to be respectful of others, the drive to demand respect for every individual, a predisposition to excitement about new things, and the need to achieve and celebrate are the types of values that can positively influence the productivity of an organization. The other two essential ingredients in this formula are that there is a "structure" that supports and nurtures this environment—e.g., a separate location, perhaps, or no privileged parking (other than customers)—and that everyone *lives* the principles. If a transgression is sensed, any employee can and will speak to the transgressor, regardless of level. A related requirement that aligns with this approach is the use of process. Any process that has been adopted is rigorously followed with substance and conviction.

BALDRIGE CATEGORY 6: PROCESS MANAGEMENT

In the past, "process" has been considered anathema by many in the technical community. In practice, however, processes, whether formal or informal, are the repositories of organizational learning. The better the management of process (not to imply unnecessary complexity or overhead), the better is the ability to systematically produce quality results repeatedly and, hopefully, faster and faster. Therefore, a discussion of the management of process in a high-tech business is an important first step before addressing the issue of engaging the technical community in process.

The Management of Process.   To meet the high-tech challenges of speed, global work force demands, and other issues, processes need to be well documented, effective, speedy, tolerant of very little variation in output, easy to use, easy to change, and integrated with each other. In addition, processes need a supportive environment around them, including tools, owners, coaches, and related "care and feeding," such as management support and recognition programs. ISO 9000 registration, when accompanied by a healthy corrective action process, invites consideration of the preceding aspects and has helped organizations introduce process discipline into their activities.

ISO 9000 requires that processes are monitored for trends that indicate that sloppiness is creeping in. (*Sloppiness* here means things like missing quality records, processes not being followed, and dwindling leadership involvement.) When and if this is detected, root-cause analysis and corrective action are performed. If, however, the "corrective action" clause is not invoked frequently in order to maintain processes that are continually refined, the tendency is to tolerate or, worse, circumvent a system of processes that have become stagnant and obstructive.

It has been said that "if you do anything the same way twice, you are not learning to do it better." The point here is an interesting one. This idea, which on the surface may appear to be in conflict with the requirement of repeatability, is actually quite aligned with it, as can be seen in the simple process management flowchart in Table 28.2.

Note that "cycle" in the flowchart is loosely defined to imply some time period of homogeneous work, e.g., a software release, a major contract development, or a yearly performance review cycle. In some sense, the shorter the cycle and the more meaningful the improvement, the better the organization is at learning, innovating, and changing. Care needs to be taken, however, not to introduce a feeling of chaos. Further, quantum changes in technology could render parts or all of a process obsolete, or competition could demand significantly better performance. In these cases, aggressive process re-engineering is required.

In addition to ISO 9000 and the specifics of process management, supporting tools and environments that help build a healthy process management system include the following:

- *Leadership's constancy of purpose:*   Continually reinforcing and rewarding process improvement. Leaders themselves should "own" breakthrough process changes.
- *Policy deployment:*   This is a useful process to gather attention and promote momentum for breakthrough process re-engineering efforts.
- *Quality improvement story* (Qualtec 1991):   This is a useful seven-step tool for tackling process-improvement activities. See Table 28.3 for a simplified chart of the seven steps.

Engagement of the Technical Work Force.   And now to the question of engaging the technical community in process. The scientific method is in itself a process but is not labeled as such. In software, there is a growing body of evidence that careful application of process can substantially improve productivity and quality. Well-known processes such as quality function deployment (QFD, a.k.a. "house of quality") clearly provide R&D with a credible (to them) understanding of customer needs. These observations explicitly demonstrate to the technical world the benefits of process as a daily approach to product development. Leadership reinforcement, together with a responsive corrective action system, signals to this community that process is the acceptable approach. Process audit practices such as the Capers Jones CHECKPOINT software measurement method (Jones 1991) and the Software Engineering Institute's Capability Maturity Model (Paulk 1995), in addition to ISO audits, are

TABLE 28.2   Simplified Process Management Flowchart

Step	Action	Effect
Step 1	Define and deploy the process.	Stabilize the process and define the baseline.
Step 2	Use it for a cycle.	Measure it and gain experience with how the process operates.
Step 3	Change it via the corrective action process.	Improve the process per the experience gained by using it.
Step 4	Deploy the changes.	Systematically track the state of the process.
Step 5	Go to step 2.	Create an opportunity to improve the process again.

TABLE 28.3   Quality Improvement Story Simplified Flowchart

Step	Description
Step 1: Reason for improvement	Identify a theme (problem area) and reason for working on it. Identify team and schedule.
Step 2: Current situation	Select a problem (problem statement), set a target, and collect data on all aspects of the problem.
Step 3: Analysis	Identify and verify the root causes of the problem via cause and effect analysis.
Step 4: Countermeasures	Plan and implement countermeasures to the selected root causes of the problem.
Step 5: Results	Confirm that the problem and its root causes have been decreased and that the target has been met.
Step 6: Standardization	Systematically prevent the problem and its root causes from reoccurring. Replicate solutions, as appropriate, in other areas.
Step 7: Future plans	Plan how to tackle remaining problems. Evaluate Team's effectiveness.

Source: Qualtec Quality Services, Inc. (1991).

practices that help maintain alignment with process improvement. Finally, when connected to the customer, process has the added benefit of ensuring that customer input is systematically made available to the technical community, something generally sought but not always available.

Regarding this connection to the customer, a strong relationship between the sales/marketing organizations and the R&D organization provides a foundation for productive customer-R&D interactions, i.e., forums where R&D people discuss technology visions with customer groups and receive feedback on the visions. For more in-depth product understanding, process tools such as QFD can provide systematic support to the development process from the point of view of product definition. ISO 9000 then plays a strong role because of the discipline it implies in the development and manufacturing environment for the creation and deployment of the product with traceability to the customer.

Fundamentally and in summary, all this comes back to the organization's ability to accept and improve processes. The true measure of process-management capability then becomes the *speed at which small process improvements or major breakthroughs can be transferred into the mainstream operation* without losing efficiency, integration, and other aspects of the processes.

BALDRIGE CATEGORY 7: BUSINESS RESULTS

Results are really the ultimate metrics that show how well or how poorly a process or set of processes is producing sustained performance. It is important that the metrics represent not just financial results but a balance among shareholder, employee, customer, and other stakeholder values. For high-tech businesses, indicators of innovation, work force effectiveness, product quality, supplier quality, process management (especially speed), and customer satisfaction are essential. The major question to answer, then, is, "Are results meeting the goals defined by the strategy and policy deployment processes, and are the goals aggressive enough, i.e., best in class, breakthrough, or incremental, to sustain the desired trajectory?"

SUMMARY

Quality in a high-tech business has to do with a well-balanced mix of customer focus, process management and improvement, and engagement and alignment of skilled employees in support of clearly understood and cascaded goals. Particular tools and methods help to achieve the desired results. The major tools and methods that have been highlighted here include

- Policy deployment
- Process management and ISO 9000 certification
- Management system assessment against Baldrige criteria
- Customer interaction programs
- Employee engagement processes, e.g., training, recognition, etc.

The leadership of the business or industry, through its behavior, defines the approaches that will be used for each of these components. The customers of the business or industry ultimately decide the effectiveness of those approaches.

ACKNOWLEDGMENTS

I wish to acknowledge the thoughtful and insightful contributions of the following Lucent Technologies colleagues who freely shared their ideas and experiences with me: Augie Corsico, Product Realization Director; Peggy Dellinger, Manager, Corporate Quality and Customer Satisfaction; Martha Huss, Product Realization Customer Satisfaction Manager; Bob Martin, Bell Laboratories Technical Officer; Lex McCusker, Quality Director, Consumer Products; Lou Monteforte, North American Region Quality Director (and former Quality Director, AT&T Transmissions Systems, Baldrige Award recipient); Mike Pennotti, Business Communications Systems Quality Director (and Baldrige site visit recipient); John Pittman, Vice President, Chief Quality and Customer Satisfaction Officer; Diana Risell, Customer Satisfaction Director; Bill Robinson, Bell Laboratories and Network Systems Quality Director; Bill Skeens, Network Systems Wireless Product Development Vice President; Ruth Spaulding, Advanced Technology Systems Manager of Quality and Strategy; and George Zysman, Wireless Chief Technical Officer.

REFERENCES

Gale, Bradley T. (1994). *Managing Customer Value.* The Free Press, Old Tappan, NJ.

Hamel, Gary (1996). "Strategy as Revolution." *Harvard Business Review,* July-August, pp. 69–82.

Harry, Mikel J. (n.d.). *The Nature of Six Sigma Quality.* Government Electronics Group, Motorola, Inc., Schaumburg, IL.

Jones, Capers (1991). *Applied Software Measurement.* McGraw-Hill, New York.

Paulk, Mark, et al. (1995) *Capability Maturity Model.* Addison-Wesley, Reading, MA.

Qualtec Quality Services, Inc. (1991). *Total Quality Management, Q I Story: Tools and Techniques*, Marshall Qualtec, Scottsdale, AZ.

SECTION 29
AUTOMOTIVE INDUSTRY[1]

Yoshio Ishizaka

INTRODUCTION

The word "quality" has taken on new meaning in the automotive industry over the past two decades. No longer is quality simply a statistical scorecard on freedom from defects or the measurement of fit and finish. Today, quality has a much broader meaning that involves a customer's inner feelings about a product and the company that offers it.

The new definition of quality takes in the basics of performance, comfort, environmental suitability, and affordability, but it adds certain elements of what is known as "production quality", related to the maker's ability to perform consistently better, and "ownership quality", which deals with customer satisfaction.

---

[1]In the Fourth Edition, material for this section on the automotive industry was supplied by J. Douglas Ekings. In the Third Edition, material for the section on the automotive industry was supplied by Soichiro Toyoda.

Two other significant shifts have occurred in recent years to raise the level of quality and quality consciousness in the automotive industry. One is the industry's move to *anticipating* customer requirements rather than responding to them, as in the past. Increasingly, automakers have employed sophisticated demographic and other studies to learn more about tomorrow's consumer, whereas in the past the industry relied on comments from customers as information was gathered for product development.

The other important change is the increase in closeness between automakers and their key suppliers. In the past, suppliers tended to be treated as vendors and were selected mainly on the basis of price and delivery capabilities. Today, the industry's key component suppliers are nearly full partners with the major companies they supply, with both automaker and supplier reaching into one another's designs, plans, and quality-improvement mechanisms. This increased confidence, trust, and reliance serve to form shared-destiny relationships that are crucial to the improvement of quality in the automotive industry.

Historical Perspective of Automotive Quality.

The automotive industry has grown from the days of hand craftsmanship into a complex infrastructure of the globally competitive automotive giants of today. Figure 29.1 reveals five eras in the growth of the automotive industry.

The earliest automobiles were built largely by hand, with each bearing as much quality as was put into it by a single craftsperson or a small group of them. Assembly was careful, time-consuming, and costly. The resulting quality was high, but expensive and inconsistent. Only the wealthy could afford a motorcar.

Henry Ford is widely regarded as the father of mass-production quality. He designed reliable cars that could be built rapidly, consistently, and inexpensively by people with less than master skills. Ford's vehicles met the consumer satisfaction requirements of the day, and America took to the road as ordinary families became able to afford automobiles.

Alfred Sloan advanced the concept of mass production but noted that consumers were raising the standard for customer satisfaction. He recognized that quality had begun to mean that a product had to meet customer expectations, not just a manufacturer's standards. Thus he created a great variety of motor vehicles, and his General Motors Corporation grew by offering cars for every purse and purpose.

Later came the rise of consumerism, followed by environmentalism, as will be explained.

Quality in the Automobile Business.

There are three dimensions to quality in the automotive industry: quality in product, quality in production, and quality in ownership.

Quality in product is the product's overall ability to perform required functions. In the case of an automobile, this means certain performance capabilities, such as the ability to accelerate to 60 mi/h in a certain time; comfort and entertainment features, such as a quiet, smooth ride, controlled climate, and ergonomically designed components such as audio equipment; environmental acceptability, which means the vehicle is fuel-efficient, clean, and safe; and affordability, or meeting the customer's ability to pay.

Quality in production is the ability to produce consistent quality as designed while still meeting volume and cost targets. Within this important dimension are four functions. The first is production of a quality product, measured by defects per hundred. The second ensures operational quality, which

1. Traditional quality	Craftsman
2. Mass-production quality	Henry Ford
3. Customer satisfaction	Alfred Sloan
4. Rise of consumerism	Safety
5. Green movement	Conservation and clean air

FIGURE 29.1   Quality, historical perspective

is the plant's ability to introduce new models, remain flexible, and still maintain consistency. Efficiency is third and is the key to producing even higher quality as volume increases. The fourth production function is cost. The plant must be able to produce an affordably priced product at a profit.

Quality in ownership is the overall ability to satisfy customers throughout their ownership life cycle. This is a critical dimension of quality, but it is the one least understood. Ownership itself is the first function. In it, our progress is measured from the sales experience through each phase of initial ownership including trade-in and repurchase for all subsequent owners. It involves the quality of the purchase experience, the everyday use of the vehicle, service, repairs, and trade-in. Cost of ownership is another function, taking in everything from down payment and monthly payments through operating expenses, maintenance costs, and insurance, finally coming down to resale value. The third function involves the intangibles—the psychological value of this ownership experience. It is centered in pride of ownership, the owner's self-image, and other special feelings created by this relationship with a vehicle.

Industry Structure. The automotive industry has undergone a concentration that accelerated greatly in the postwar years. Today, the world has 65 international automakers, of which 22 compete in the United States. Under the pressure of cost and currency differentials and government urgings to balance trade, numerous international companies have opened manufacturing and assembly plants in the United States over the past two decades. Today, the United States is home to 61 passenger-vehicle assembly plants.

Design and Production. The cost of designing and manufacturing motor vehicles is very high, owing to the complexity of the vehicles, the rigid quality and safety standards involved in making them, plus the rigorous testing required and the frequency of design change. Thus design and manufacture are carried out in a very few very large companies. Increasingly in recent years, portions of the product design function have been shared by major automobile makers with their supplier companies. This sharing has added a dimension to the responsibility for quality assurance.

Marketing. Most manufacturers design and produce a variety of models in order to attract customers of varying means and tastes. Vehicles are sold through nationwide networks of franchised dealerships, some of which offer more than one manufacturer's products. Vehicle dealerships, most of which also offer used vehicles and service and repair facilities, numbered 22,650 in the United States as of January 1, 1998.

The External Climate. External factors began to affect the industry more and more in the postwar years. These include growing consumer activism, safety issues, product liability, government regulation, and the rise of environmental concerns.

Consumerism. As the automobile industry's annual-model cycle matured in the 1950s and 1960s, the opportunity for product improvement often was overlooked in the race to create change for its own sake—change more likely to involve fashion than function. Consumers began to resist this trend, demanding that vehicles meet quality standards not only at the time of delivery but throughout their lives and even at the time of trade-in. In this era of consumerism, automotive quality began to be measured over time.

Safety. Along with consumerism came new standards for safety. Many years after safety glass became standard in American automobiles, a great rush of safety-related improvements came to market, including advanced passive-restraint devices, antilock brakes, energy-absorbing steering columns, crumple zones, child safety seating, and stronger side-impact protection, among many others. Dealing as they do with the preservation of human life and limb, these improvements further raised the standard of quality required in automotive design and production.

Improvements that aimed to promote visibility, accident avoidance, and survivability in crashes were accompanied by some intended to ease the job of collision repair and to lower its cost. Certain safety improvements, notably passenger seat belts and other restraint devices, have proven to be successful in reducing injuries, and their installation is now mandated by federal law in the United States.

Product Liability.   Manufacturers have been required to consider product liability, which has grown to become a serious concern, especially in the United States. It plays a significant role in product design and manufacturing. As consumers become more aware of manufacturers' legal obligations and the legal system encourages legal action when faulty product is the suspected cause of personal injury, automobile companies have had to become more cautious not only in product design and manufacturing but also in documentation communicated to consumers in association with their products.

Government Regulation.   Government regulation has had a profound effect on the automobile industry. In the United States, the National Highway Traffic Safety Administration (NHTSA) regulates federal safety requirements that can affect design, cost, and consumer expectations of vehicles. Government regulations also mandate certain fuel-efficiency standards for motor vehicles sold in the United States. Although it is difficult to estimate the cost of compliance, increasing NHTSA mandates and other controls have tended to increase manufacturers' costs and vehicle prices significantly.

Environmental Issues.   Two factors combined to create the green movement: One was the development of uncertainties in the world's fuel supplies; the other was worldwide recognition of the problem of air pollution. Following the oil crises of the midseventies and early eighties and the rise of air pollution to unhealthful levels in many of America's largest cities, the industry no longer could overlook fuel efficiency and exhaust emissions in its measurement of quality. All manufacturers of motor vehicles sold in the United States now operate under federal laws mandating increases in fuel efficiency and decreases in emissions.

Concept of Quality Assurance.
The concept of quality goes beyond just meeting customer expectations and fitness for use, as has been explained. This concept extends to serviceability, costs, and emotional issues and remains until the eventual disposal of the vehicle at the end of its useful life. Toyota employs a system, illustrated in Figure 29.2, that also is characteristic of those in most automobile manufacturing companies.

This system for quality assurance was developed in Toyota Motor Corporation along with the company's renowned Toyota production system, to which it is closely tied. As standardization of product and process became the keys to consistent, high-quality production, certain communication disciplines became the keys to quality assurance. These were centered in the steady feedback of information from audits, inspections, testing, and analysis throughout the process of creating and producing vehicles.

Flexible manufacturing based on consumer preferences became the norm. Prior to this, the industry built huge banks of vehicles not to customer requests but on a push system of factory capacities. These changes, and inventory reductions that accompanied the lean-production, just-in-time philosophy, have radically altered production activity.

Lean Production.
Several aspects of the Toyota production system combine to create what is known as "lean production", a concept now widely emulated by American automobile makers. Lean production involves the removal of waste from every step in the production chain—waste of energy, motion, time, and resources. It is based on the pull-through method of inventory control that Toyota's Taiichi Ohno devised in the 1950s, now known as "just-in-time".

Through training and partnering with suppliers, extensive cost reductions have been experienced by both manufacturers and suppliers as they order, build, and ship materials and vehicles on a pull-through or customer-demand system. With inventory reduction, hidden quality problems surface quickly and must be addressed immediately. This is often referred to as "lowering the water level," which will expose rocks (or problems) previously unknown or concealed. In this way, lean production yields important quality-assurance benefits.

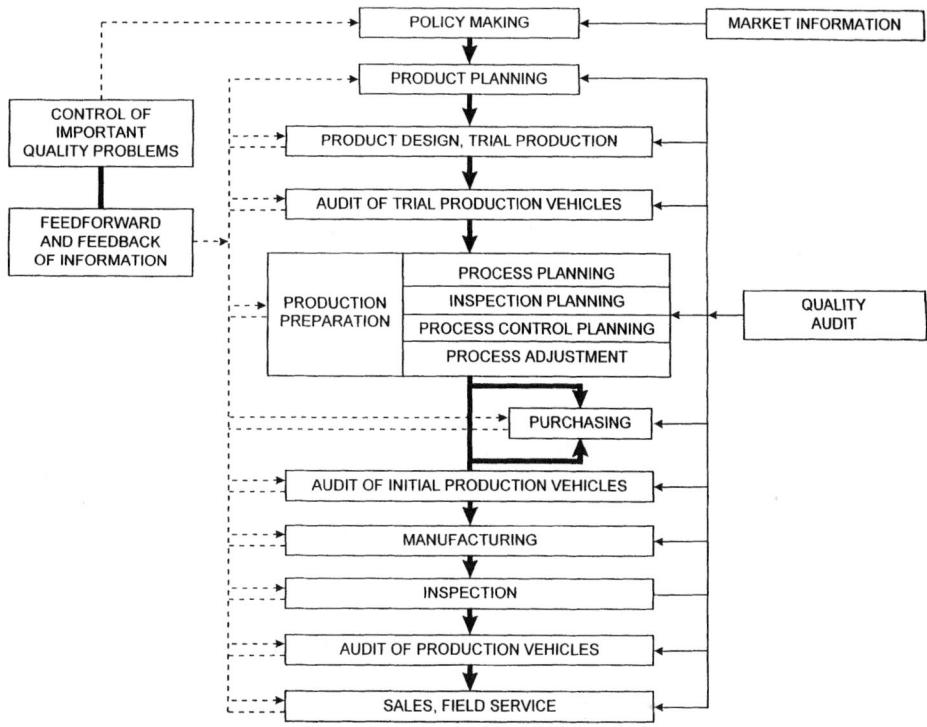

FIGURE 29.2   Toyota concept of quality. (*Toyoda, Soichiro. Automotive Industry 3d ed.*)

PRODUCT PLANNING

As increased consumer awareness and demands added fuel to the competition between growing automobile companies, product planning needed to follow the improvements in mass-production systems. As production systems improved and frequent major or minor model changes were required for competitiveness, the time required for new model planning and launch had to decrease. Thus the concepts of chief engineer, product platforms, and simultaneous engineering became the norm for all companies.

In Toyota, the chief engineer concept meant one person led a team of experts from various disciplines to plan and launch a vehicle in an efficient and effective manner. As the leader, the chief engineer must organize human resources and facilitate discussions to solve complex issues and ultimately is responsible for the success of the vehicle. To design and produce vehicles efficiently, the chief engineer is encouraged to use common parts and to look at class distinction in regard to vehicle content. Focus is also placed on engineering capability and assembly simplicity.

In the United States, many companies employ a platform concept that has radically improved both the quality and efficiency of the development process. Further steps are being taken to streamline timing and personnel efficiency to improve competitiveness.

The once-normal 36-month lead time for product planning and launch is no longer acceptable. Using the chief engineer and platform concepts, Chrysler and Toyota have achieved 23- to 28-month launches. The new target at Toyota is 18 months—half the historically acceptable norm. Shortened lead times also support customer-satisfaction improvements because they permit the timely response to economic circumstances and customer preferences that can change over periods of less than 3 years.

Figure 29.3 shows a typical approach to new model planning based on the 3-year term that was once normal in the industry.

Time-Phased Planning.   The initial step in long-range planning is market research. To illustrate consumer buying trends, Figures 29.4a and 29.4b show research for customer vehicle purchase patterns. New model proposals set parameters to further define vehicles for the detailed planning phase. Parameters include the vehicle type, such as sedan, coupe, or station wagon; vehicle specifications for function, performance, reliability, service, and maintenance; timing for development; and cost and finance items, including selling price and profit.

Critical Quality Problems.   An important step in the new model proposal is defining critical quality problems that must be addressed. This step is illustrated in Figure 29.5. The design group must provide data to support this activity and track its timely completion.

Design Planning.   The design planning phase includes development of detailed specifications of components and their interface with major body/frame dimensions that are concurrently being developed. Drawings are developed, and consideration is given to historical data on like-vehicle development/manufacturing to "design in" foolproofing or elimination of such concerns.

Design Review.   The design review phase is a critical step in the design process. A cross-functional team representing all disciplines reviews the design in detail. The team verifies and identifies safety and critical characteristics to ensure that specifications are clearly defined and that

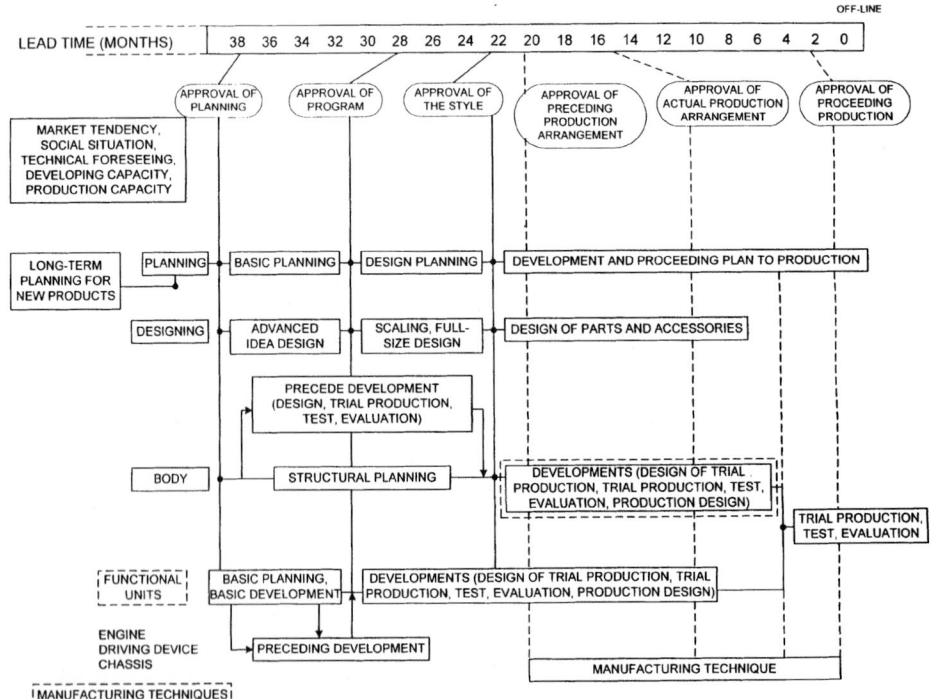

FIGURE 29.3   Product planning activity timing. (*Toyoda, Soichiro. Automotive Industry.*)

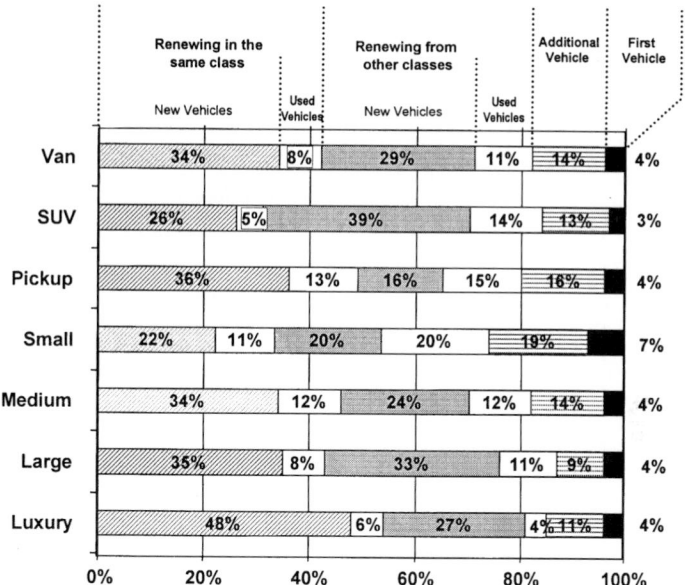

FIGURE 29.4a Customer vehicle purchase patterns. (*1996 Toyota New Buyer Survey.*)

Type	Replaced Vehicle				Additional Vehicle	First Vehicle	Total
	Same Class		Different Class				
	Bought New	Bought Used	Bought New	Bought Used			
Van	30	10	32	13	14	1	100
SUV	20	5	43	17	14	2	100
Pickup	36	15	17	16	15	2	100
Small	23	11	21	23	17	5	100
Medium	30	8	30	15	13	3	100
Large	34	6	38	10	10	2	100
Luxury	48	7	28	4	12	2	100

Note: Owing to rounding errors, numbers in row may fail to total 100.

FIGURE 29.4b Customer patterns of vehicle purchase (in percent). (*1996 Toyota New Buyer Survey.*)

all supporting functions comply with specifications. Marketing and practical aspects are also evaluated. Consensus must be reached to ensure that all activities and input are in line with the overall plan guidelines. The effort required to identify and address issues early in the design phase can save tremendous costs and delays later in the program.

Laboratory Testing.   Laboratory testing during the design and prototype phase is important for all new products, materials, new or changed tooling, or new suppliers. With supplier participation in such testing, critical characteristics pertaining to function, durability, environmental conditions, simulated driving conditions, and crashworthiness can be measured prior to part finalization.

Evaluation under Actual Conditions.   Prototype parts can be tested on current vehicles, either under test-track or actual road conditions, to obtain data that do not depend solely on laboratory

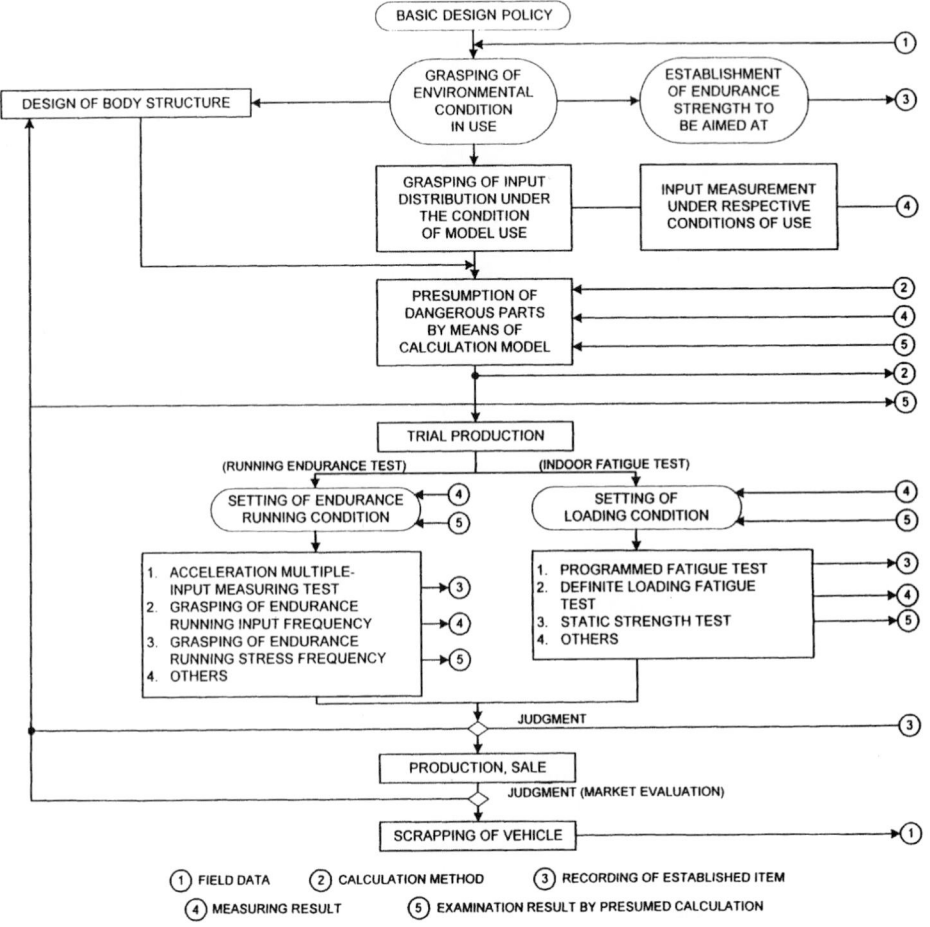

FIGURE 29.5   Idea model for body strength. (*Toyoda, Soichiro. Automotive Industry.*)

tests. This activity is important to warranty or long-term durability design activities. By incorporating this activity into customer-fleet or employee-fleet use, companies can collect extensive data on real-time use that may include extremely severe conditions and use, as with taxi or delivery-vehicle fleets. This testing may occur prior to release of the production part to all customers or in preparation for a product enhancement in the future.

Safety Evaluation.   Individual components and completed vehicles are tested exhaustively to ensure the greatest possible protection for occupants. Various means are used to test components in the laboratory. Vehicle testing will be done to measure the effects of head-on, barrier-impact, rollover, and side-impact crashes, as well as the latest testing that includes variations of head-on impact at differing angles. Much of this information is used to enable insurance companies to calculate the cost of repair of a damaged vehicle and to establish facts on which to base safety-oriented advertisements.

Overall Evaluation.   Constant feedback and communication are critical to the success of the design review, laboratory, and usage testing activities. Figure 29.6 depicts the steps of the feedback process that supports such activity. In this process, the more communication, the better.

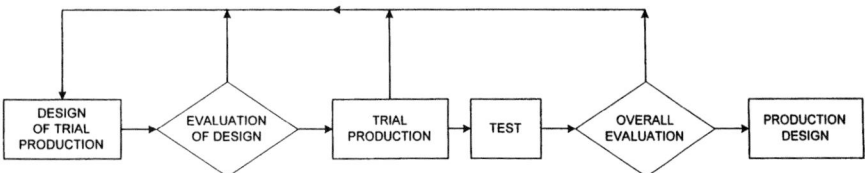

FIGURE 29.6   Design evaluation communication system. (*Toyoda, Soichiro. Automotive Industry.*)

Preparation for Production.   As in the planning phase, production personnel are represented on the cross-functional team to support preparation for production. This phase directly supports the vehicle launch activity. Production personnel will help coordinate pilot production either at a separate facility or at the production line. As pilots are run, information is gathered to assist with the ramp-up to full scale. The participants may be the people who will train production-line workers. Others who will later be assigned as trainers may participate as well. The point of this activity is verification of issues such as the production process, tooling, and quality-control parameters.

Organization for New Model Planning.   The organization supporting this activity varies not only by company but by model and complexity. The larger the company and more complex the change, the larger is the group and the number of departments. Normally, coordination of such activities can be provided by

1. Planners assigned full time to specific models
2. Task forces or committees with assigned representatives to facilitate communication
3. Formalized plans written to encompass all groups and required support
4. Process and procedures such as sign-offs

SUPPLIER MANAGEMENT

Supplier support and supplier partnerships become increasingly important in a lean-production environment. (See Section 21: Supplier Relations.) Gone is the day of the adversarial relationship between automobile manufacturers and suppliers. While the Japanese companies have practiced supplier partnerships for years, Chrysler and Ford are fast approaching the same practices. By developing trust between manufacturer and supplier, many mutual benefits can be achieved.

Supplier Relations.   The component-supplier community can be immensely helpful to the automobile manufacturer in providing design expertise. If the manufacturer provides performance specifications to the supplier, leaving the component company to proceed with design, costs to the automaker can be sharply reduced. To support this activity, the manufacturer must allow supplier representation in the early phase of vehicle development—a big step in the partnering philosophy. This also requires agreements to allow suppliers to invest in development costs with some assurance of being awarded the business. The other element of trust involves suppliers opening their books and sharing costing information to support the manufacturer's objectives.

This is not a short-term activity, and it requires a great deal of interface between the management teams of both the supplier and the manufacturer to develop the required trust and support. This cooperation is essential for long-term competitiveness in the automotive arena.

Partnerships have long been a strength of the Toyota system. Through its close relationship with Toyota, a supplier can become stronger and more efficient. Various Toyota groups share the Toyota production system (TPS) expertise with their suppliers. Toyota groups in the United States have been receiving TPS training from the experts at Toyota Motor Corporation in Japan, and the sharing of this knowledge with suppliers has proven beneficial, since Toyota has experienced outstanding supplier improvements. As an example, the following facts show one company's improvement experience after 2 years of working with Toyota on TPS implementation:

Finished goods inventory reduction	82%
Work-in-process reduction	59%
Plantwide productivity increases	32%
Quality index improvements	58%
Production lead-time reduction	74%

The overall improvement depends on the supplier's system at the beginning of the project, but experience has shown that results like those in the preceding example are not uncommon. This activity takes a great deal of time and requires a large commitment on the supplier's part. Actual monetary investments are normally small; training, research, and reorganizing efforts can involve substantial investments of time. All suppliers have benefited many times over from these investments.

Often taken for granted by manufacturers is the openness of two-way communications between supplier and manufacturer. In many cases, however, the supplier will not share thoughts in order to appear not to question the customer's decisions. In order for the partnership to work properly, the customer must depend on suppliers for honest feedback on design, processes, and any other issues that may affect the output of their partnership. For their part, suppliers are entitled to a fair hearing of their honest feedback.

Supplier Base Reduction.   If suppliers are to believe that they will be rewarded for the commitment of resources to a customer's product development, they must be able to see an opportunity to capture a significant share of the customer's business. Often this means that the number of suppliers must be reduced. By reducing the number of suppliers, the manufacturer's staffing and support also can be reduced, thus lowering costs. The supplier list can be reduced in a number of ways.

1. *Tiering.* Using large, tier-one suppliers to supply completed assemblies, rather than just their components, improves production efficiency, material handling costs, and supplier interface expenses. The tier-one supplier establishes relationships with the other component suppliers and becomes the final design, development, and assembly point for the completed assembly.

2. *Exclusive contracts.* As partnerships grow, it becomes normal for some suppliers to develop specific parts, as long as they uphold their portion of the partnership. Communication lines, development expertise, and overall system understanding are critical to these relationships. In some cases, supplier representatives have offices and full-time employees located in the manufacturer's facility to support that activity.

3. *Outsourcing.* In many cases it can be more expensive and cumbersome for automobile manufacturers to continue producing various components in-house than to rely on external suppliers. If a manufacturer has aging facilities, outdated production methods, or dwindling expertise, it may make sense to join forces with a supplier who is an expert in production of a required component.

Supplier Selection.   As new products, processes, and increased volumes begin to demand the addition of supply resources, certain criteria must be established for their selection. Supplier surveys may be performed by individual departments or, preferably, by a cross-functional team representing all disciplines to ensure that all aspects of the company's capability are considered. Many factors may be considered, such as

1. Previous experience with the supplier
2. Affiliated company experience (parent, sister, joint venture, etc.)
3. Reputation or recommendations of other companies

Quality Assurance by the Supplier.   As with other supplier activities, communication is critical for both supplier and manufacturer to understand the supplier quality assurance activities. This begins with early supplier involvement in the development process. Once the product is agreed on, designed, developed, and prototyped, four basic steps are key to ensuring customer satisfaction with the product.

1. *Product approval*
    a. Verification of understanding *and agreement on specifications.* The supplier must understand the manufacturer's expectations and be able to deliver timely production quantities consistently on that level. The supplier also must understand the broader system and know how each component fits into it.
    b. Initial sample approval. Samples provided for testing must come from production lots, not from laboratory or prototype production, if production capability is to be measured properly. All systems require first-production-piece verification. The additional requirements for sample size, test requirements, and reporting format may vary, but the intent is that the supplier must use this process to verify that expectations for continuous mass production can be met. On completion of this activity, the supplier will submit the part and information to the manufacturer for review. If all criteria and expectations are met, the supplier is given formal, written approval for production to begin. Should a concern arise, it must be addressed immediately to ensure that the part meets expectations and that timing is not delayed.
    c. Comparison of test methods. The approval process should include a check of the compatibility of test methods. The supplier should include the test data with the test certificate attached to samples, including any information about accelerated life testing and destructive testing. The supplier should be given the automobile company's test results in order to discover and correct any differences in testing techniques.
2. *Process review.* A common practice with new suppliers is to audit the production process prior to awarding business. Process reviews of current suppliers also are conducted periodically. During the audits, discussions are held on improvements based on a comparison with other suppliers and the experience of the auditor. Also, suppliers may have process improvements or changes that require approval or on-site review by the manufacturer. On-site supplier visits can be useful, since continuous improvement activities will strengthen the process, partnership, and product quality. It cannot be stressed enough that manufacturers should view this as an opportunity and a small investment in the future of the relationship. This activity also strengthens both companies and builds trust through face-to-face meetings. Should the supplier experience difficulty or need assistance with a process, the door should be open to contact the manufacturer and ask for help. Within the manufacturer's ranks, experts should be available to be dispatched to support the supplier. The supplier also may seek the assistance of outside consultants known to the manufacturer to be capable of supporting both process and management improvement activity.
3. *Feedback.* Communications between the automobile company and the supplier must be open, whatever the topic. The supplier should expect honest, fact-based feedback from the manufacturer. The better the information provided, the quicker and more responsive will be the supplier's reaction. This is a true test of both partners' ability to give detailed information and respond quickly with a detailed, concrete action plan. No production system is ever completely foolproof, so when a concern does arise, responsiveness and support are of the utmost importance.
4. *Continuous improvement, or kaizen.* The supplier's work force should be made to understand that simply maintaining the present level of quality is not enough and that constant effort must be applied in the pursuit of improvement. The manufacturer's quality assurance group can help spearhead continuous improvement throughout the supplier firm by making available trainers and training materials; fostering quality-improvement competition between work shifts, departments, and teams; meeting with supplier teams to discuss quality issues; and recognizing continuous improvement by means of on-site supplier award presentations. If managements of both the manufacturer and the supplier regularly demonstrate their wholehearted support of *kaizen* in these ways, associates are more likely to buy into team problem solving, quality circles, and other techniques for ensuring continuous improvement.

Controls on the Supplier.   Automobile companies exercise control over suppliers in several ways. Some of these, such as the process of supplier selection, joint quality planning, and supplier quality assurance, take place before production. Others take effect after deliveries begin.

Acceptance Inspection.   Initial quantities of components from mass-produced lots are subjected to random sampling for the first 1 to 3 months of production. If defects are found, sampling quantities are increased and action taken as needed. If random sampling indicates quality within agreed standards, sampling is reduced and finally ceases, with the automaker relying thereafter on the supplier's test certificates and periodic quality audits. At this point in the relationship, dual sampling is avoided, and the supplier takes on greater responsibility for both product quality and its proof.

Supplier Quality Audits.   An annual review of a supplier's facilities is normally conducted to ensure that the supplier is adhering to federal requirements and quality control plans in system testing and document control activities. By contract, suppliers must notify manufacturers of process changes, so documentation and the actual process can be reviewed during the visit. The on-site visit will include many subjects, some of which are listed in Figure 29.7.

Supplier Ratings.   A rating system encompassing various supplier activities can be used for quality-improvement activities and awards. The system may compare the supplier against expectations, previous performance, and other suppliers. As an example, a summary page from the Toyota quality alliance rating system, administered by Toyota Motor Sales USA, Inc., is shown in Figures 29.8*a* and 29.8*b*.

MANUFACTURING

Development of Manufacturing Methods.   Over the years, substantial growth in motor vehicle production volume led to the automation of numerous production-line activities, with the result that the economics of production became more favorable. A corollary benefit—greater quality—was created when automated machines and associated transfer devices demanded improved uniformity of input components to maintain continuous production. Thus automation yielded not only greater productivity and better economic use of human resources but also improved quality.

This benefit continues to follow the increasing use of computers to guide design and manufacturing processes, since computers make possible far greater reproducibility and consistency—essential to quality assurance—than is possible with purely mechanical means.

Production Preparation.   Automobile companies maintain departments that carry out the preparatory steps through which the production force can maintain the desired level of quality: facilities and process planning, and trial mass production.

Organization	Incoming inspection procedure
Quality control policies	Process deviations
Product identification	Supplier performance records
Lot ID and traceability	Cosmetic acceptance standards
Test procedures	Defective material handling
Test results	Supplier defect notification
Process flow diagrams	Drawing control
Engineering change procedures	Specification documents
Inspection standards	Specification deviations

FIGURE 29.7   Supplier audit subjects. (*Note:* This is not al all-inclusive list.)

Product quality
   Quality Assurance   20
   Warranty   15
Operational quality
   Deliver   15
   Mispack/damage claims   5
   Purchasing   20
Product support
   Marketing   12
   Program management   5
   Technical development   8
Total points   100

(a)

Category	Group	Department
Product quality		
1. Quality assurance	U.S. products	Supplier development
2. Warranty	Service/U.S. products	Warranty administration
Operational quality		
1. Delivery	Parts	Parts supply
2. Mispack/damage claims	Parts	Parts invoicing
3. Purchasing	U.S. products	Procurement
Product support		
1. Marketing	parts	Parts marketing
2. Program management	U.S. products	Program management
3. Technical development	U.S. products	Parts and accessory engineering

Note: Suppliers should direct all inquiries regarding any portion of the evaluation to their respective buyers.

(b)

FIGURE 29.8 (*a*) Supplier evaluation scoring categories. (*b*) TMS responsibility by scoring category. (*Toyota Quality Alliance Rating System, Toyota Motor Sales, USA, Inc.*)

Facilities and Process Planning.   The process-planning step specifies operations to be performed, the tools and machines to be used, and other aspects of production technique. Included in this step is an examination of the quality capability of the process to see if it meets the tolerances and dependability specified in the design. For the production of components related to basic vehicle functions or safety, the most stable and foolproofed production and inspection systems should be used.

Trial Mass Production.   When all is in readiness for production, trial mass production is carried out to confirm that quality standards continue to be met as volume production proceeds. The variability or dispersion of the product at all stages, from manufacture of parts through final assembly and testing as automobiles, is measured closely. Dispersion data can be used to identify any processes requiring improved uniformity. Trial mass production also helps to identify any required changes in tooling.

Process Control.   The control of manufacturing processes becomes simpler with each addition of computerization in the workplace. Relying as it does on rapid feedback of information, process control is well served by computer processing of quality data. Some production functions rely on older, conventional processes subject to human variability and provide little feedback. In these, worker training and motivation become ever more important as production volumes increase.

The concept of process capability is in widespread use throughout the automotive industry. Because process capability can vary with process conditions, it can be useful to maintain a record of process capability and to investigate any significant variances, especially deterioration.

Operator process control and product acceptance empowerment also became popular in recent years. Equipped with information on quality standards, machine operators are being empowered to take greater control in regulating the process to meet standards. Operators also are being empowered to accept or reject product under the operator self-inspection concept. The operator decides whether the process should continue or stop and whether the product meets quality specifications.

Automatic process regulation takes the place of human judgment in many applications, relying on automatic measurement and feedback for regulation of processes without human intervention. Maintenance of such systems does require periodic human intervention, and inspections of the process capability lead to necessary repairs, adjustments, and replacements.

Process Improvement.   Historically, industry has relied on engineers and supervisors to contribute changes to industrial processes. Worker contributions were limited to suggestion-box ideas and received relatively little recognition.

During the 1960s, Japanese industries launched extensive educational programs for workers at all levels on ways to control and improve processes. At Toyota Motor Corporation, such programs began in 1961 and included such matters as quality control, Pareto diagrams, characteristic factor diagrams, histograms, control charts, and correlation diagrams. Quality circles were initiated to foster improvement in intradepartmental processes, with remarkable results. A later campaign to eliminate defects and claims due to process deficiencies and operator error also was responsible for great improvements in quality.

Kanban.   *Kanban* is one of the primary tools of the just-in-time system that is used to facilitate an even flow of production and an even distribution of work among the various stages of manufacturing and transportation. With thousands of items such as engine components, drive train parts, sheet metal pieces, seats, and other interior parts, *kanban* maintains an orderly and efficient flow of goods, material, and information throughout the entire manufacturing process.

Kanban is usually a printed card in a clear plastic cover that contains specific information such as part description, part number, and quantity or lot size. This card is affixed to the various containers, bins, and racks that hold the parts. These *kanban* cards are used to withdraw additional parts from the preceding process to replace the ones that are used. It is the concept of "sell one, buy one." In so doing, only the right parts are used, in the right quantities and at the right time. You might look at it as a waste-free means of producing and conveying materials.

INSPECTION

In-line and final inspectors have been an important part of automobile manufacturing for many years. The inspection operators review material, component, and assembly quality to ensure conformance to standards. Data provided by the inspection functions can facilitate product and process improvements to foolproof product quality. The information is used primarily for immediate operator feedback and machine adjustments. This information is used for reports to management for comparison and tracking purposes to support cost justification for machine, product, or process improvements. Management support can be directed to areas requiring improvement based on inspection feedback. This focus can greatly assist management, and the overall process will benefit. To a considerable extent, the role of inspector in some automobile companies has been taken by production-line operators as a part of the operator process control and product acceptance empowerment mentioned earlier.

Classification of Defects by Seriousness.   The automobile industry has developed a widely accepted classification system, with three major groupings for all defects.

A. Safety or critical functions that can endanger operators or passengers or render the vehicle functionally inoperative, such as brake function, electrical operations, or steering.

B. Operations that affect primary functions of the vehicle or major appearance items that most customers would not accept, such as inoperative locking mechanisms; faded, chipped, or peeling paint; or noisy operation of engine or brakes.

C. A third category includes items that do not affect vehicle functions or appearance items not leading to customer complaints, such as crooked labels or stripes, underbody rust, or an inoperative glove box light.

Concerning customer satisfaction, *A* defects will be returned for repair, *B* defects normally will be returned, and *C* defects are almost never returned if they are the only issue found. This severity rating helps automobile companies focus on major issues.

Organization for Inspection.   Incoming, in-process, and final inspections are the three types commonly used in automobile plants. Inspection methods are usually developed by each group to support customer satisfaction by focusing inspections based on product, process, and operator concerns. In many cases, formalized procedures requiring documentation must be tracked [e.g., Federal Motor Vehicle Safety Standard (FMVSS) requirements, or other safety items]. Many of these issues can be machine-verified, but issues that cannot be verified are 100 percent checked by inspection personnel. Other inspectors may do random audits of machine processes to verify that machine results and readings are accurate.

- *Incoming* checks are performed on raw materials or purchased components. Different testing requirements based on different raw materials must be developed based on hardness, strength, clarity, and otherfactors.
- *In-process* checks include items such as dimensional checks (stampings, machining, molding); equipment temperature, pressure, and timing (casting, forging, molding); fit verifications; and functional verifications. Major functional components such as axles, transmissions, and engines may be operated prior to final assembly to save repair time and costs should a defect be found.
- *Final vehicle inspection* is performed after all components are assembled and the vehicle is complete. Many functional checks, adjustments, and verifications are required. Based on vehicle complexity, some additional functional issues may need to be verified. A few of these are

  Water test: Ensures leak-free vehicles

  Front wheel alignment: Verifies toe, caster, and camber to ensure best handling and long tire wear

  Brake function: Ensuring no leaks and that all components functioning

  Headlight aim

  Complete functional check

  Prior to shipping, a complete appearance check is done to ensure that no damage, paint, or fit concerns reach the customer. Although most inspections are redundant or additional overhead activities, they are generally accepted as a requirement by automobile manufacturers to promote customer satisfaction.

FIELD SERVICE

Of concern to all automotive manufacturers are the sale and service of vehicles they produce. Since most vehicles are sold through independently owned dealers, manufacturers do not have direct control of this activity.

Role of the Dealer.   It is the obligation of the dealer to deliver the vehicle in good, clean condition, with emphasis on satisfying the customer during the sale and beyond. Manufacturers spend a great deal of time, effort, and money to develop procedures to support dealer sales and service. Recognizing that these activities may be the only experience the customer has with the manufacturer other than vehicle performance, automobile companies provide extensive training for sales and delivery activities. This assists dealers to improve initial reaction and retain the customer for future sales. Training provided by the manufacturers for technical service of the vehicles is even more extensive than sales training. Certification programs have been developed for service technicians to create levels and competition among technicians. Manufacturers support sales and technician training with the hope and expectation of increased customer satisfaction as dealership personnel become better and stronger assets.

Vehicle Service System.   During the warranty period, dealers are almost guaranteed that customers will return to the dealership for maintenance and warranty repairs, if required. Most of these charges are paid directly to the dealer by the manufacturer. Dealerships must satisfy customer needs during the warranty period in the hope that the customer will return to them for maintenance and repairs beyond the warranty period. By satisfying customers and creating ongoing business, the service area can be an excellent profit center for the dealership.

As the automobile manufacturers increase the warranty period and coverage of their vehicles, dealerships are guaranteed more service activity that can lead to increased after-warranty service. Figure 29.9 shows the initial maintenance log for a typical manufacturer. Dealerships have many sources of parts to use after the warranty period, but during the warranty, manufacturers require the use of original equipment (OE) parts. All manufacturers have extensive parts distribution systems to support dealer requirements for these parts in a timely manner. Toyota, for example, established its North American Parts Logistics Division to improve local parts sourcing and to serve as a parts distribution network supplying all North American Toyota dealers, export markets, and certain General Motors vehicles.

Once the warranty period has expired, the parts-purchasing option is open to the dealership. Since the OE parts have satisfied the dealer's needs and are competitively priced, the manufacturer urges the dealer to continue to use them. This supports the manufacturer by ensuring that

1. High-quality parts are used.
2. Specifications including fit and functions are correct.
3. Manufacturer's distribution systems are supported for ongoing activity.

The use of factory parts also will help support high customer satisfaction.

Feedback of Information on Field Quality.   The performance of vehicles and parts in the field is information essential to the manufacturer. Not only is this information pertinent to continuous-improvement activity on current parts and future designs, it is vital to ensuring customer satisfaction. Safety concerns are normally monitored by government regulatory agencies such as NHTSA in the United States, and the manufacturer must communicate to the agency all pertinent activities to support the issue. On nonsafety issues, customer satisfaction is a major concern, and the way the recall campaign is handled can be critical to satisfying customers. Figure 29.10 shows a typical feedback system for manufacturers.

Field Service Beyond the Warranty Period.   As repairs to customer vehicles are made during the warranty period, extensive information is required to support dealer payment. This information is extremely helpful for improvements to parts, decisions to launch a recall campaign to repair or replace defective parts, or setting realistic parts performance expectations based on actual field data. Figure 29.11 shows many of the inputs required to complete a warranty claim. The information is very similar for all manufacturers.

Manufacturers require that many of the parts being replaced be returned for teardown and analysis. Root-cause analysis using the "plan-do-check-action" (PDCA) model (Figure 29.12) can be performed by the responsible engineer and/or supplier using these returned parts. This information is much more useful than a written description of the technician's opinion.

Fleet Feedback.   Vehicles in customer or automaker-employee fleets also are sources of information. With customer fleets, high mileage or hard use may accelerate the feedback obtained.

5,000 Miles or 4 Months*

☐ Replace engine oil and oil filter
Additional Maintenance Items for Special Operating Conditions:
Please refer to page 30 of this supplement to determine if your vehicle requires the additional maintenance items.

Inspect the Following:
☐ Air filter   ☐ Ball joints and dust covers
☐ Brake: linings, discs/drums   ☐ Drive shaft boots (re-torque flange bolts)
☐ Steering linkages   ☐ Body/chassis nuts and bolts

Dealer Service Verification

Date: _____

Mileage: _____

10,000 Miles or 8 Months*

☐ Replace engine oil and oil filter
Additional Maintenance Items for Special Operating Conditions:
Please refer to page 30 of this supplement to determine if your vehicle requires the additional maintenance items.

Inspect the Following:
☐ Air filter   ☐ Ball joints and dust covers
☐ Brake: linings, discs/drums   ☐ Drive shaft boots (re-torque flange bolts)
☐ Steering linkages   ☐ Body/chassis nuts and bolts

Dealer Service Verification

Date: _____

Mileage: _____

7,500 Miles or 6 Months*

☐ Replace engine oil and oil filter

Dealer Service Verification

Date: _____

Mileage: _____

---

* Use the white background boxes to follow 5,000 mile oil change intervals or the shaded background boxes to follow 7,500 mile oil change intervals. Please refer to page 29 of this supplement for further information and to determine which interval is right for your driving circumstances.

5,000 Mile Oil Change Intervals	7,500 Mile Oil Change Intervals

FIGURE 29.9 Maintenance log.

29.17

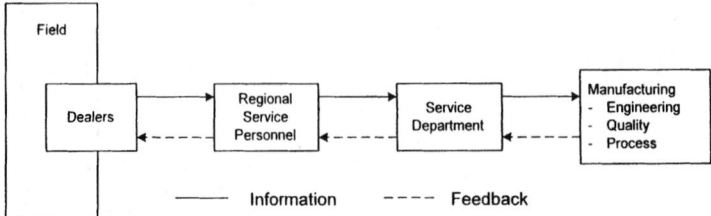

FIGURE 29.10 Feedback.

Car line	Delivery date
Model year	Sales district
Assembly plant	Repair date
Serial number	Mileage at repair
Body style	Part number
Engine type	Defect code
Transmission type	Vendor code
Axle ratio	Cost of repair
Month of production	

FIGURE 29.11 Warranty input information.

With employee vehicles, prototype or improved parts can be tested to verify effectiveness of the changes.

Customer Complaints. All manufacturers have toll-free telephone service for customers to gain information and assistance or to voice complaints. In many cases, this information can be used to support product or process improvements. Also, quick reactions by the appropriate individuals can ensure the maintenance of high customer satisfaction. The overall goals of this activity are improved product quality and increased customer satisfaction.

QUALITY AUDIT

As an additional check prior to shipping vehicles to customers, manufacturers do random-sample vehicle audits. Although many audits take place in-plant, analysis of the final assembled vehicle can verify product and process quality throughout the entire production system. The final vehicle audit is broken up into various areas.

Specification. This includes torque tests of various fasteners, verifying component correctness, function tests for setting within specified ranges such as parking brake or shift adjustments, air-conditioning function, and a leak test.

Customer Acceptance. This includes fit, finish, and function.

Road Test. Driving vehicles measures noise, vibration, harshness (NVH) issues and functional or dynamic problems such as front end or wheel balance.

Water Test. Ensures that all sealing areas are complete. Issues found in these areas are immediately referred to root-cause analysis and countermeasure development. The final vehicle audit

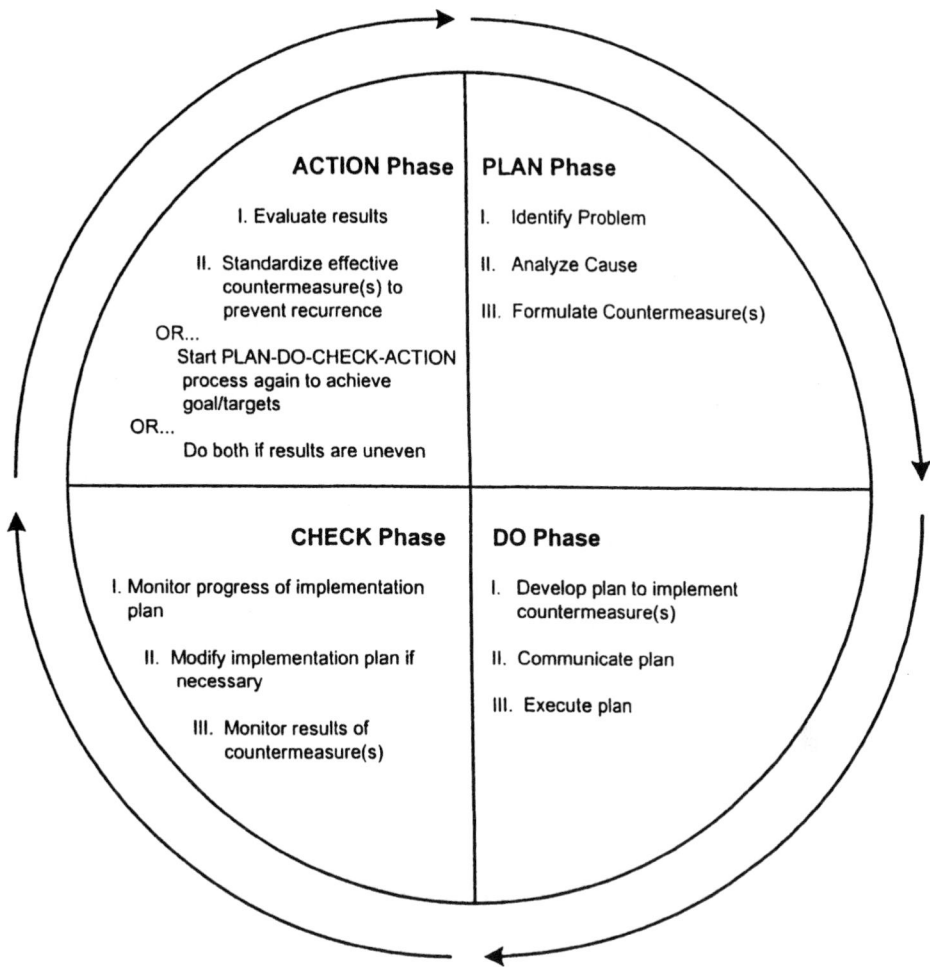

ACTION Phase

I. Evaluate results

II. Standardize effective
countermeasure(s) to
prevent recurrence

OR...

Start PLAN-DO-CHECK-ACTION
process again to achieve
goal/targets

OR...

Do both if results are uneven

PLAN Phase

I.   Identify Problem

II. Analyze Cause

III. Formulate Countermeasure(s)

CHECK Phase

I. Monitor progress of implementation
plan

II. Modify implementation plan if
necessary

III. Monitor results of
countermeasure(s)

DO Phase

I.   Develop plan to implement
countermeasure(s)

II. Communicate plan

III. Execute plan

FIGURE 29.12   PDCA cycle.

receives a high priority within the manufacturing facility, since the vehicles selected were approved throughout the system and are ready to be sent to the customer.

Additional in-system checks may be performed by the personnel of production, tooling, and other departments to verify that their areas are meeting requirements. All have the same goals of defect-free quality and increased customer satisfaction.

THE OUTLOOK FOR AUTOMOTIVE QUALITY

Steady Gains.   The world's automotive industry met the challenge of converting from handcraft to mass production and went on to produce tremendous improvements in product quality. Better processes, materials, equipment, and worker training yielded great gains in consistency as production lines moved faster and faster in mass production's early years. The industry also responded well to the growth of consumerism and environmental awareness with substantial improvements in product quality and safety and in the reduction of emissions and effluents both from vehicles and from the plants that produce them.

Next Steps.   In the final years of the twentieth century and into the next, the industry will consolidate and reinforce these gains in product quality, reliability, and environmental acceptability. Further quality gains will come from team production methods, from the use of robotics, and from just-in-time production control and the other techniques that in Toyota are known as the "Toyota production system". Supplier relationships, already very close in many parts of the industry, will grow closer still, with manufacturers and suppliers sharing even more of their product plans, production methods, and quality disciplines.

The growth of intelligent transportation systems, which will have intelligent vehicles traveling intelligent highways, will bring about new quality concerns and opportunities. Distance sensors, "smart" cruise control, onboard navigation systems, built-in diagnostic devices and other electronic equipment will add complexities to the measurement of quality and require new approaches to in-plant process control and quality assurance methods. Many of the suppliers of these components may be companies not now involved with automakers and unfamiliar with automobile industry quality initiatives.

The New Battleground.   The primary quality battleground, however, will shift from the production plant to the retail sale. More and more surveys, studies, and focus-group comments show that automobile buyers associate quality not only with the product but also with the purchase and service experiences that accompany it—the components of customer satisfaction.

Success in the retail battleground may be the automobile industry's greatest challenge, for it is here that the automobile manufacturer exercises the least control over any aspect of the design/production/distribution/sale process. Although the manufacturer influences this part of the process, the conversion of prospects into customers rests with independent dealers.

If the automaker is to meet the buyer's needs, it needs to know much more about changes in buyer demographics and about the changing ways motor vehicles fit into new lifestyles. Manufacturers must consider the new and greater pressures on the buyer's time and financial resources, changes in commuting habits, the growth of entrepreneurship, and the work-at-home phenomenon, among other factors.

Some opportunities are available to the automobile manufacturer interested in improving the sale and service experience, however. Dealer training may be the greatest of these opportunities, since manufacturers can do much to raise the level of knowledge and professionalism among dealers' sales and service employees. Most manufacturers have programs in place for this kind of training. All are likely to be accelerated as competition drives the need for more and better customer relations.

Another area for improvement is emphasis on "lean distribution", the field equivalent of lean production. Manufacturers will be working on ways to improve their responsiveness to vehicle orders, which would have the combined effects of pleasing customers with prompt delivery as it shrinks costly dealer inventories.

Finally, manufacturers will be able to improve the quality of the vehicle-purchase experience through customer education and communication, mainly by way of the Internet. Growing numbers of automobile buyers are gathering information about motor vehicles electronically from their homes as they begin the search. This puts customer and manufacturer into direct, two-way communication in ways never before available. New capabilities lent by Internet communications range from simple matters such as showing the automobile buyer how a certain vehicle would look in his or her choice of colors and appearance options to detailed answers to highly technical questions about performance, materials, warranty details, and other matters.

The automotive industry is being challenged to change from quality considerations based on product to those which include all aspects of the ownership experience. How we meet this challenge will test each of us about responsiveness, adaptability, and genuine customer-service attitudes.

REFERENCES

Five sources of specific information, charts, graphs, and data were used. They are

Hiller, Dennis, Ingram, Larry, and Kitamura, Toshi (1994). "Production System Implementation at Supplier Base." *IMPRO 1994 Conference Proceedings,* Juran Institute, Wilton, CT.

Sakai, Shinji (1994). "Rediscovering Quality—The Toyota Way." *IMPRO 1994 Conference Proceedings,* Juran Institute, Wilton, CT.

Toyoda, Soichiro (1979). "Automotive Industry—Section 42." *Quality Control Handbook,* 3d ed. McGraw-Hill, New York.

Waltz, Robert (1996). "Managing Customer Loyalty." *IMPRO 1996 Conference Proceedings,* Juran Institute, Wilton, CT.

Womack, James, Roos, Daniel, and Jones, Daniel (1990). *The Machine That Changed The World.* Harper Collins Publishers, New York.

I acknowledge with thanks several contributions to this work from the experience of David L. Pearce of Toyota Technical Services.

SECTION 30
TRAVEL AND HOSPITALITY INDUSTRIES[1]

Patrick Mene

---

[1] In the Fourth Edition, material for the section on the service industries was supplied by Charles D. Zimmerman III and John W. Enell.

MISSION OF THE TRAVEL AND HOSPITALITY INDUSTRIES

While the travel and hospitality industries are different from each other, they are strongly related. The differences are apparent in their respective missions. The travel industry provides travel advice to travel planners as well as transportation and travel-related services for travelers. The hospitality industry provides food, lodging, meeting places, and hospitality-related services to travelers.

The two industries are related by their interdependence—the hospitality industry depends on the travel industry to deliver its customers; the travel industry depends on destinations, including resorts and conference sites, to justify its existence as a travel provider. For this reason, companies in the two industries are also frequently related through joint marketing. (For example, it is common for conference registration material to make provision for special ticket prices with a preferred air carrier and discounted rooms rates with a preferred hotel at or near the conference site.)

In this section, quality as practiced in the travel and hospitality industries is presented primarily through the experience of airline and hotel companies, respectively. There are also some examples from the experience of other participants in the industries, such as travel planners and cruise lines.

The section covers these broad areas:

- The definitions of *quality*.
- Quality challenges and barriers.
- A brief history of the industry, including the way in which societal and legislative changes affected the approach to managing for quality. The emphasis is on the years since 1950.
- The quality management processes—quality planning, quality control, and quality improvement—as practiced in these industries.
- The importance and management of supplier quality.
- Benchmarks—the industries' best performances.
- A prognosis of quality in these industries.

DEFINITION OF QUALITY IN TRAVEL AND HOSPITALITY PRODUCTS

Within these industries, quality has long been defined in terms of product features or product grade. Three basic grades—luxury grade, deluxe grade, and economy grade—are frequent in a product line and serve to illustrate the approach. These three grades are based on the number and kinds of features, benefits, and advantages expected by the customer. Product grade also establishes the expected level of user satisfaction. Luxury is the highest level, and economy the lowest. Although the grade names for a given product may be different and the grade structure may contain more or fewer divisions, customers are generally made aware, through advertising, sales prospectuses, and the like, that a grade structure does exist. This grade structure may be signaled through brand name or class of service within a brand. Table 30.1 depicts the three product grades in several dimensions.

TYPICAL QUALITY CHALLENGES IN THE TRAVEL AND HOSPITALITY INDUSTRIES

Importance of Nonfunctional Expectations.
Most customers of airlines, hotels, and cruise ships have expectations that extend beyond functional requirements of the product or service being used, especially when the travel is for long periods or a special occasion. Their ultimate desire

TABLE 30.1 Product Grades

Grade	Luxury	Deluxe	Economy
Features (what the product promises to the user)	Beautiful, pristine, comfortable surroundings	Attractive, clean, comfortable surroundings	Pleasant, clean, comfortable surroundings
	Highly personal service	Courteous service	Courteous service Attractive price levels
	Timely delivery	Timely delivery	Timely delivery
	Immediate and complete problem resolution	Problem resolution	Problem resolution
Ultimate benefit (what the product does for the user)	A positive, memorable experience	Freedom from a negative experience	Lowest cost
Personal advantage (what the user can do because of the product)	Be more productive, comfortable, and prestigious	Be productive and comfortable	Justify the cost to themselves or to others

is a "memorable experience" that makes them feel well (Cleveland 1988). The characteristics of the experience are accumulated matters of a psychological nature—the look and feel of the surroundings, the warmth of the greeting, the feeling of comfort and importance. These sensory matters create personal advantages for customers that have a major effect on their loyalty and willingness to pay premium prices.

The sense of feeling can be easily damaged by an indifferent attitude, an unclean carpet, a distracting noise, an insincere smile, or any inappropriate appearance such as an unsightly "out of order" sign. The damage is compounded because humans have a tendency to judge what they cannot see by what they can see. In the words of Tom Peters, "Coffee stains on the airliner food tray suggest poor engine maintenance." The magnitude of the damage multiplies rapidly because the news (of the offensive sensory qualities) spreads quickly.

The importance of nonfunctional requirements is explained by Rapaille, a cultural anthropologist who makes customer loyalty a subject of his research. In Rapaille's view, the decision-making process customers go through when selecting any product or service is based on emotion or what he terms the "logic of emotion." For a brief summary of his theories, see Bernowski (1996). For a data-based approach to nonfunction requirements, see below, under Quality Improvement: Services, Special Jobs, and Events, for the example from KLM Royal Dutch Airline.

Overdesign.   While customers often harbor unrealistic expectations, developers and operators of high-grade products often set nonfunctional standards to levels beyond those sensed by the customers, adding cost rather than value. One example is the former practice of a leading New York restaurant. Their menus were printed on an imported paper known as "elephant hide" that had the appearance of parchment paper with an ultrasmooth feeling to the touch. While the cost of the paper was more than double typical graphic products, the tactile perception of it by the diner was very ordinary. This product merely added cost, not value.

Knowledge of Travel Planners.   The purchase of travel and hospitality products takes place from a variety of intermediate travel planners (although some purchases are made directly by the traveler). These travel planners include administrative support people in an organization, travel agents, and meeting planners. Airlines, hotels, and cruise ships are faced with the challenge of establishing the superiority of their product (however minor it may be) not only to the traveler but also to all the intermediate planners throughout the travel planning chain. Each of these intermediaries varies in ability to evaluate quality and influence travelers. The following summarizes the knowledge and economic strength of these planners. (This part of this section will not attempt to address the entire complex distribution system of the industries.)

Meeting planners can be viewed as the "vital few" of travel planners. They usually represent a small number of organizations that hold many, often large meetings at various destinations throughout the year. Their economic strength cannot be ignored. Their technical knowledge is very high for several reasons: (1) industry training and certification, (2) extensive experience planning and managing repeat events, and (3) collection and analysis of event data.

Corporate travel planners have the same economic strength as meeting planners. Their role is to develop travel policies (usually with the advice of a travel organization expert). It is noteworthy that travel and entertainment expenditures are typically the third largest in an organization, preceded by payroll and data-processing costs. Some quality experts and travel planners regard corporate travel policies that solely attempt to minimize travel expenditures as antagonistic to employee productivity. They contend that travel which is selected on price alone causes delays, complexity, stress, and fatigue. The ultimate effect on the traveling employee is reduced productivity.

Travel agents have been gaining economic strength since the Airline Deregulation Act of 1978 (see below under Government Influence). Most independent travel agents have become members of various franchise or industry association brands, generally known as consortiums, to compete with leading global brands. Like meeting planners, travel agents, owing to their extensive travel experience, have the capability to evaluate quality.

As of 1995, the travel agent industry, especially in the United States, faced several threats. With the introduction of the Internet (i.e., the computer information highway), travel customers have direct access to travel products and instantaneous verification of travel arrangements. In addition, most major U.S. carriers have begun to limit travel agent commissions on domestic flight tickets sold. The extension of this policy to international flights is also under study. To compensate for this changing environment, large travel companies have begun bundled prices for their corporate customers in the form of fee-based arrangements.

Administrative support people in an organization typically plan travel for people in the organization. This may include the inbound visitor or the outbound travel of a senior leader. Collectively, these intermediaries are significant. Their technical knowledge is limited because they do not routinely evaluate product quality.

In all cases, airlines, hotels, and cruise ships provide product information and personable salespersons to assist the travel planner in evaluating product quality and demonstrating the noticeable differences to their travelers. Where product superiority is minimal, marketing skills can make a remarkable difference in sales income. For more on this topic, see the example in Section 7 on Consumer Preference and Share of Market.

Product Perishability.   Airline seats, hotel guest rooms, and cruise ship cabins are the most perishable product of travel and hospitality enterprises, more so than food. (One-half an apple not sold today can be used tomorrow, but an airline seat or a guest room not sold today is lost forever.) Inventory management practices must be planned and managed carefully to avoid costly spoilage.

The decor of cabins and guest rooms also will progressively age over time and ultimately will perish unless scheduled process checks and corrective action are conducted (i.e., scheduled housekeeping and maintenance).

Limited Awareness of Process Management.   Traditionally, companies in the industries have brought most of their objectives and resources to bear on attracting and retaining customers by emphasizing the features and benefits of the products offered. There has been little awareness of the processes necessary to create and deliver these features and benefits. In fact, for years, the prevailing mindset has denied that the concept of "process" had application in the industry. ("That's manufacturing. We're different.") Consequently, processes were typically some combination of inefficient, obsolete, excessively bureaucratic, or even nonexistent. To have no concept of process is to deprive an organization of a powerful arsenal of process-related logic and tools to prevent or correct quality and related productivity problems. Without such preventive or corrective action, a whole class of chronic, severe problems can plague the complex organizational processes that cut across multiple departments. Accepted as fate—"the way things are"—these attitudes have caused organizations to accommodate them. This accommodation is typically a many-level management hierarchy in which it is not unusual to have organizational units with a ratio of workers to managers as small as 7 to 1. This ratio imposes substantial salary costs as well as a mindset that is further damaging to quality and human relations. This bureaucratic approach fosters a social system where easily replaced unskilled workers in narrowly defined jobs merely carry out the plans of managers. This low-quality work life transforms itself into low-quality products of cross-departmental processes as well as high organizational costs. For more on this subject, see Section 6, Process Management.

DYNAMIC NATURE OF THE INDUSTRIES

Typically, 30 percent of travel orders placed by customers will change or be canceled. This high rate of change causes scheduling and information problems as well as extra work and costs. This information-dominant condition provides strong justification for (1) building flexibility into the reservation and production systems and (2) a control system that places the emphasis on the accuracy and up-to-dateness of information provided to all concerned. These industry challenges have a major effect on sales income and costs. The effects of these challenges are summarized by grade in Table 30.2.

BARRIERS TO PROBLEM ANALYSIS

Multiple Types of Processes. Airline and hotel products employ three basic types of processes: (1) continuous production, (2) special job, and (3) service delivery. Each of these processes exists for different reasons and has distinguishing features. They are not mutually exclusive.

TABLE 30.2 Typical Quality Challenges

	Grade		
Challenge	Luxury	Deluxe	Economy
Psychological features and benefits are inadequate.	S	S	S
Appearance standards are set to levels beyond those sensed by the customer.	C	C	C
Product superiority is not established with travel planners.	S	S	S
Airline seats, guest rooms, and cabins perish when not sold.	C	C	C
Furnishings and decor "wear out."	C-S	C-S	C-S
Processes are ineffective, inefficient, obsolete, bureaucratic, or nonexistent.	C-S	C-S	C-S
Process planning and control depends on an elaborate management hierarchy.	C-S	C-S	C-S
Unskilled workers in narrowly defined jobs merely carry out the plans of managers.	C-S	C-S	C-S
Information about customer orders undergoes frequent change.	C	C	C

Key: Challenge has a major effect on C= costs; S = sales income.

1. *Continuous production* processes exist when the product has a standard design that must be produced repeatedly. These processes share many similarities with production and assembly work. Some examples are

- Fulfilling airline or hotel reservation requests
- Providing airline seat assignment
- Assigning hotel guest rooms
- Cleaning and resupplying of aircraft
- Cleaning and resupplying of hotel guest rooms
- Preparing airline or hotel food
- Handling airline baggage
- Providing hotel valet parking

Although this type of work usually provides tangible products (e.g., reservation information cards, restocked guest rooms or a plate of food), service delivery processes are involved as well.

2. *Special job* processes exist when there are multiple designs for a product requiring something new or different from each production run. Some examples are

- Generating a cruise ship itinerary
- Providing a chartered aircraft flight
- Staging a meeting, convention, or banquet

This type of process relies heavily on communicating essential information to all concerned, and the production run is typically short. Again, service delivery processes are involved.

3. *A service delivery* process exists when the supplier meets the customer to conduct a transaction face to face. It typically involves delivery of one or more of the following: (a) personal effort to add comfort or well-being, (b) information to solve a problem, and (c) tangible products. Some examples in the airline and hotel industry are

- Welcoming a traveler
- Providing directions
- Relieving a traveler of baggage
- Serving a beverage or a meal
- Providing forgotten items

It is important to note that service delivery processes may be embedded in both the continuous production and the special job processes. It is useful to understand service delivery as a continuum. At one end is the *individualized service* of a bank machine. This mechanized process can identify an individual by bank card number and process the *individual request*. At the other end of the continuum is the warm, comfortable, *personal service* one would provide at home to guests. Since service delivery is central to hospitality organizations, especially high-grade ones, the service delivery processes are highly personalized and carefully planned and managed.

The Logistics Are Complex.   Service delivery is involved with virtually all work in a travel/hospitality enterprise. Since service is consumed promptly by the customer (at the time of delivery), three logistical problems exist when analyzing trouble:

1. Analysis can easily interfere with customer satisfaction.

2. It is not feasible to stop delivery to identify a problem cause.

3. The information collected is usually in the form of recollections, not real-time observations.

The time elapsed in service delivery is usually so short that most problems (such as providing incorrect directions to a traveler) are not even detected until the service delivery is complete.

Short-Term Economic Pressure. Airlines and hotels are capital-intensive enterprises whose performance is closely monitored by external financial analysts. Financial performance reviews of these investments are conducted on at least a quarterly basis. Accordingly, exhibiting positive quarterly financial results is often more of a priority than discovering the causes of good results and generating competitiveness. In the words of one well-known consultant, "an imbalance exists between counting the golden eggs and making the goose healthier." Table 30.3 summarizes process characteristics and barriers to problem analysis.

THE RISE OF CURRENT QUALITY PRACTICES

Much has been done within specific organizations to create a historical perspective of the industries (Abbey 1995, Forsyeth 1995). An explanation of this work follows.

Few National Carriers or Hotel Companies. Throughout the 1930s and 1940s, there were few national carriers or hotel companies in the United States. Airline carriers had a tiny share of the travel market and were primarily regional; most hotels were small and independently owned. Airlines and hotels were located in population and trade centers, and sea travel was international aboard transatlantic ocean liners.

No Standardization of Product. Although there were airlines and hotels with common ownership, even the most prominent of them lacked standard policies, procedures, product design, or referral networks.

Changes in Society. Beginning in the 1950s, however, society began to experience dramatic change in demographics and in habits and practices relating to travel and lodging.

1. *Population growth.* Following World War II, the rate of household formation jumped, and the population began growing significantly (at a rate of 1.35 percent compounded annually), especially in the South and the Mountain and Pacific regions.

2. *Population shift.* In addition to this growth, the population began shifting; the Sunbelt (especially Florida and Texas) and the western states (Colorado, Arizona, and California in particular) experienced a tremendous influx of people.

3. *Greater life expectancy.* The population grew also because life expectancy increased, a reflection of advances in medicine and public health measures.

4. *Improved incomes.* Individual real incomes improved in the postwar economy, and two-income families became more common. After the belt-tightening war years, families suddenly had more money to spend on travel and leisure.

TABLE 30.3 Airline and Hotel Processes and Barriers to Analysis of Chronic Trouble

Process type	Product design	Production/delivery run length	Barrier
Continuous production	Standard design	Long	Hundreds of functional and sensory variables to control
Special job or event	Multiple designs	Short	Most analysis cannot be put to use on the current job
Service delivery	Customer-employee interface	Brief	Measurement can only be conducted after the delivery (i.e., the data are retrospective recollections)

5. *Increased leisure time.* Leisure time increased in the early postwar years, when the 40-hour work week became commonplace and additional legal holidays were given to workers. Job market factors such as part-time work and job sharing contributed to the increased amount of leisure time available to workers.

6. *Proliferation of convention centers.* The 1950s and 1960s experienced a booming U.S. economy. As businesses and business and fraternal organizations grew, people needed facilities for conventions and meetings. Some cities had civic centers or auditoriums that could accommodate groups. As business expanded into the suburbs or outgrew the limited facilities of the smaller city-based convention centers, there was a surge in the demand (and supply) of convention-related travel and hospitality products. This demand occurred in cities and in regional and resort destinations.

7. *Expansion of the highway system.* Construction of the Interstate highway system began in 1956. Soon, the 42,000-mile system became an important factor in the number of Americans traveling, both for business and for leisure. Vehicle registrations grew dramatically, and Americans took to the roads in great numbers.

8. *Increased air travel and growth in airport infrastructure.* Air travel became a commonplace part of the American business and leisure scene. By the early 1980s, more than 700 airports were certified for passenger service, including 23 large hub airports (in Chicago, New York City, Dallas, Atlanta, and so on). The large hubs were destinations in their own right, as well as connection points for an increasing number of domestic and international flights. In addition, 35 medium hubs served regional areas such as the Southwest or Northeast, and 62 small hubs provided statewide connections for a growing number of business and leisure travelers.

9. *Expanded connoisseurship.* Consumers became increasingly informed, astute, and discriminating in making travel and hospitality decisions. Travelers were increasingly unwilling to accept limited comforts, whatever the product grade.

10. *Decline of international sea travel.* Rising costs and the increasing attractiveness of air travel caused international sea travel to decline and finally to virtually disappear.

INDUSTRY RESPONSES TO SOCIETAL CHANGES

Consolidations: National Alliances and Hotel Chains.   The industries evolved rapidly to meet the challenges posed by a changing society. Independent and regional airline and hotel enterprises began to seek affiliations to compete more effectively. The basic motivation of joining a national alliance was the added prestige (i.e., the ability to trade on the quality reputation of the alliance) as well as the increased purchasing power.

Common Management.   Not until the 1950s did national carriers or hotel companies adopt common management to coordinate all commercial activities. During this period of coordination among the affiliated business units of an airline or hotel company, each consolidated organization developed its own policies, procedures, product designs, amenities, and referral networks.

To meet the demands of road travelers in the 1950s, the lodging industry responded with the development of chain properties. Holiday Inn, Ramada Inn, and Howard Johnson's were among the lodging pioneers along interstate highways. Hilton Hotels chose to capitalize on the expanded demands for convention centers by developing or purchasing facilities that were fit for large conventions.

By the 1960s, giant air carriers attempted to capitalize on their quality reputations by extending their product lines to include hotels. United Airlines, TWA, and American Airlines each developed or acquired a lodging brand. While this approach eventually was abandoned in the United States, the airlines of other countries later adopted it. Some examples are

Airline	Lodging product
Air France	Meridien
Japan Airlines	Nikko
KLM Royal Dutch Airlines	Golden Tulip
SwissAir	Swissotel

The growth of affiliations and information technology created the conditions for the first toll-free reservation systems in the 1960s. These automated systems provided instantaneous verification of travel and lodging arrangements.

In retrospect, early quality efforts were characterized by standardization of product design to distinguish products from those of competitors, a response to the requirements of management. It would be some time before managers would recognize that response to customer requirements was an even more effective competitive strategy and vital to long-term business health.

GOVERNMENT INFLUENCE

Two acts of legislation have greatly shaped the economic and competitive situation of today's hospitality/travel industries in the United States: The Airline Deregulation Act of 1978 and the Economic Recovery Act of 1981.

Airline Deregulation Act of 1978.   With the first legislation, airlines ceased to be regulated as public utilities. The gradual removal of economic controls was designed to encourage competition and improve service quality among U.S. air carriers. Some effects have been predictable and some surprising.

Predictable effects

1. *Passenger traffic increased.* In 1977, U.S. airlines carried 240 million passengers. By 1994, the number totaled 605 million.
2. *The price of air travel decreased.* As of 1995, about 90 percent of all air passengers paid at a discount from standard fare. Passenger traffic on low-fare airlines grew from 3 percent of the total in 1980 to 12 percent of the market in 1993, a trend that continues with no end in sight.
3. *Competition grew.* Immediately following deregulation, a flurry of new airlines entered the market. Although most of these new entrants failed or were merged into other carriers, the competition is still very stiff.
4. *Large carriers have abandoned small communities.* Major airlines are leaving the service of smaller cities to commuter carriers that feed traffic to their hubs for connecting service. In 1977, commuter airlines carried 9.2 million passengers. By 1994, the number had risen to 56.5 million.

Surprising effects

1. While traffic growth has been favorable for large, full-service airlines, most of the growth experienced by the top three airlines happened soon after deregulation was initiated in 1978. Demand for full-service airlines seems to have flattened, owing to adverse economic conditions and the attractiveness of low-fare airlines.
2. Chapter 11 reorganization has affected pricing. Since deregulation, there have been 117 bankruptcy filings in the airline industry, including 18 major U.S. carriers. Carriers operating under Chapter 11 are protected from creditors. This enables them to offer discounted fares, forcing competitors to match their prices at unprofitable levels.
3. The airline industry is evolving into two major segments; one segment will feature convenient schedules; and the other, attractive pricing.

Economic Recovery Act of 1981.   To stimulate capital investments in the United States, Congress allowed for investment tax credits as well as shorter depreciation and amortization schedules. As a result, hotels were built without regard for market demand. This condition changed with the Tax Reform Act of 1986. There remains an average surplus of hotel rooms amounting to 40 percent (as of 1995), a surplus that is expected to persist until the turn of the twenty-first century.

QUALITY PLANNING IN THE TRAVEL AND HOSPITALITY INDUSTRIES

How to Think About Quality Planning in the Travel and Hospitality Industries.
The launch of a new aircraft, a new hotel, or a new cruise ship is a capital-intensive, high-risk under-taking. Inadequate quality planning creates unnecessary costs, possibly even catastrophe.

Long airport check-in lines, uncomfortable surroundings in a hotel, runways crumbling at a new airport, food poisoning, and a ship that overturns at sea are all potential results of inadequate quality planning. [For further examples, see Juran Institute Quality Minute Video Education Series (1994)].

A well-documented example of the cost of poor quality planning occurred in a major resort hotel in the United States. This property featured a magnificent outdoor pool and sunbathing area, accommodating several hundred guests. Among the most popular products served there was the frozen yogurt and the yogurt-based cream drinks made by machine at the poolside service area. A chronic problem was the unreliability of the yogurt machine. The failures were frequent, causing lost sales, rework, added steps, accidents, employee fatigue and turnover, and customer dissatis-faction. Management of the property decided that the purchase of a new yogurt machine would remedy the trouble. The choice of a $12,000 unit was influenced by a $5000 discount on the pur-chase price.

Management was surprised when the workers at the pool area complained about the new equip-ment. Some managers resented the protest, even claiming the work force was "insatiable." An inves-tigation by the hotel purchasing agent uncovered the following facts:

1. The capacity of the new machine was inadequate for production needs.
2. The work force was not trained in an important particular: to store the yogurt mix between 30 and 45°F. The use of improperly stored mix rendered the unit inoperable.
3. The workers were substituting expensive ingredients.
4. Procuring the substitute ingredients caused extra work paid at the overtime rate.
5. Customers of the pool bar were dissatisfied with the sporadic unavailability of yogurt during the summer.

In the words of one worker, "Twelve thousand dollars was spent; things are worse; yet management thinks they saved $5000." Worker input was a vital missing element in the original planning and in the replanning of yogurt production. Sound quality planning can produce processes and products that are robust against failure and avoid attendant costs of failure.

Quality Planning Facilities and Equipment.
Aircraft, hotels, and ships are built to order by capable suppliers not made by the purchaser. Planning for such an investment typically involves a well-organized joint planning effort. It requires assembling a team of experts from several disci-plines: marketing, finance, architecture, engineering, construction, operations, and so on. The team usually draws architectural and other discipline expertise from outside firms that specialize in the discipline. The planning team draws marketing and operational expertise from inside the purchaser's organization. The team's mission is to identify and incorporate the necessary product features and to identify and avert potential sources of failure in the finished facility. In addition to providing their market and operational knowledge, the representatives of the purchaser's organization play two addi-tional roles that are vital to the success of the completed project. One role is to keep before the team the economic constraints of the project; the second is to provide a consistent management process within which the team can effectively carry out the project.

The process of planning a new hotel facility will illustrate some proven approaches. Typically, the team uses one of two basic approaches to develop a new facility: (1) "last model built" or (2) bench-marking "the best."

The last model built approach uses the last hotel as a starting point for the next step in an evolu-tionary continuous improvement. The benchmarking approach is used when the hotel company is seeking a fresh approach to facility design. The team of professionals identifies the best-in-industry facilities or even best-in-class outside the industry. The airline industry has benchmarked hotels for equipment, cruise ships have engaged hotel interior architects, and hotels have benchmarked cruise ships for facility design and equipment. British Airways engaged an industrial design firm known for its work on expensive boats to redesign its first class cabins (Goldsmith 1995). From these benchmarking experiences, the team of professionals identifies the most important facility features of its new product, whether they be functional or sensory qualities. For further discussion, see Section 12, Benchmarking.

It is typical of the industries to use existing products or competitive products as a source of inspiration to answer the question: Which product features best respond to customer needs? This approach is in contrast to the approach of home appliance makers, for example, who are more likely to use extensive market research into customer needs as a basis for new product design. The last model built approach fosters consistency of product design as well as a low degree of difficulty in planning and execution. Conversely, this approach not only invites carryover failures, it also can lead to product obsolescence. The benchmarking approach can generate breakthrough improvements, yet it is more difficult to manage the new and untried.

Some travel products and travel-product specifications became so closely identified with quality levels that their names and terms became descriptors in everyday language. *Ritzy* is synonymous with elegance and refinement, based on the luxury hotels of César Ritz. The term *posh* is an acronym for port outbound starboard home, which specifies cabins that are free from the glare of the sun on both legs of a round-trip ocean liner journey between the East Coast of the United States and Europe.

Once the basic concept is developed, the performance evaluation of the new facility or equipment is measured in a series of thorough reviews and tests. These critical examinations are centered on validating four characteristics: (1) life safety, (2) appearance, (3) feasibility, and (4) reliability (i.e., time-oriented performance.)

The life-safety validation is typically carried out under the supervision of the appropriate regulatory agency. Matters of appearance are validated by the expert sensory evaluation of architects, whereas economic feasibility is monitored by the expert financial analyst. Validating reliability is a more complex issue than the preceding items. Since facilities and equipment in the industries are purchased products, reliability largely depends on (1) supplier reliability data (today, most suppliers use their internal reliability data as marketing evidence regarding product performance and (2) interviewing customers who use the product.

Another form of design assurance is the construction of product models. Airlines, hotels, and cruise lines build and test cabins and guest rooms. Some are laboratory simulations, whereas others are small-scale tests in existing facilities. The Hong Kong and Shanghai Hotel Company, owners and operators of the Peninsula Group of hotels (located predominately in the Far East), use "home testing" for small, untried equipment. New or untried guest room equipment must be installed in the home of an upper manager for 1 year's practical use prior to implementation. This approach overcomes the pressure to use the latest technology as well as gaining insights into the intended versus real use of equipment by customers.

Since the ultimate test of new facilities is customer use, most enterprises in the industries conduct some form of postproject evaluation of the new product. These evaluations usually include some form of postcustomer reaction and cost analysis.

Unfortunately, in my experience, the industries have not yet seen the value of continuously assessing and improving the product development process itself. This form of 1-day review by upper managers (made popular by Japanese manufacturers) would benefit the entire organization. This audit in some organizations receives the personal attention of the president of the enterprise. Not surprising, this review is often termed the *presidential audit*. This examination makes new products more salable, eliminates waste, reduces capital costs, and improves the return on investment.

Once new facilities clear the design validation phase, preparations for regular operations begin. Since construction and installation schedules and costs can gain the upper hand, it is important that the startup process be reliable. Accordingly, most enterprises use readiness systems such as countdowns or checklists (as practiced in hospital operating rooms and spacecraft launches) to get new products off to a good start.

Products in the hospitality and travel industries are subject to cyclic demand patterns. New products are typically launched in advance of a peak demand period, allowing time to identify and correct deficiencies before the peak occurs.

As mentioned earlier in this section, the management of facilities/equipment inventory (e.g., aircraft seats, hotel guest rooms, and cruise ship cabins) is a complex subject. The complexity is caused by two compounding factors: (1) the inventory is more perishable than food and when lost is lost forever, and (2) 30 percent of travel plans either change or are canceled. Today, inventory management in the industries is very scientific. The object is clear: Sell every item (i.e., seat, guest room, cabin) *today* at the best possible price. The airline industry has been at the frontier of this special form of inventory management. It was the first to use automated inventory management systems to provide decision support in controlling inventory flow and price levels. It would be no exaggeration to say that improving the yield of facilities and equipment has a major effect on the profitability of an enterprise in the hospitality and travel industries.

Quality Planning for Service.   Unlike facilities/equipment planning, which is dominated by consulting specialists, planning for new service is monopolized by managers of the enterprise. Most frequently, their qualification for this role is their experience in customer service. (The training and experience of industrial engineers, much valued in planning of services within industrial companies, is not seen as relevant to service planning within the hospitality and travel industries, with the exception of a few major airlines.)

At the Ritz-Carlton Hotel Company, a Malcolm Baldrige National Quality Award recipient, upper managers are personally involved in planning service for each new hotel. The foundation of their approach is the "Gold Standards" (see Figure 30.1). These standard specifications and processes guide the activities and decisions of the entire organization, especially in the ambiguous situations commonly encountered in personal service delivery. When new hotel development occurs, experienced managers migrate to the new location to select scientifically a work force that shares the organization's values. This scientific selection process, developed by a supplier-partner, uses an empirical-based instrument to "screen in" promising managers and workers. The screening instrument compares the candidate's character traits (i.e., spontaneous behaviors) to those of consistent superior performers of the Ritz-Carlton.

During the 7 days prior to initial customer occupancy of the hotel, upper managers play an active role. In the first hour of employment, the President/C.O.O. assertively instills and models the "Gold Standards" for all new recruits. These new recruits then spend the next 2 days in classroom sessions with trainers. During these sessions, the new recruits gain a better understanding of the "Gold Standards." In essence, every new recruit of the Ritz-Carlton Hotel Company must master the employee-customer interface process shown in Figure 30.2. This process requires and allows anyone in the entire work force to (1) break away from routine duties when the customer is dissatisfied or requires something that is not in the standard array of products or (2) provide immediate positive action, (3) document the incident for future analysis, and (4) "snap back" to routine duties. The preceding relationship-management process training is a strictly enforced prerequisite for new recruits (at any level of the organization) before they can begin functional training or have any customer contact.

British Airways provides another example of personal involvement of upper managers in service-delivery planning. After World War II, this national airline of England made a practice of hiring former military personnel. As a result, the culture and organization of the airline was decidedly militaristic. Their core beliefs centered around a traditional hierarchy, conformance to regulations, and safety. Despite the airline's core technical competencies, its losses required substantial subsidies from the government.

The company surveyed 45,000 customers for opinions and attitudes regarding the airline's products. The findings were not entirely surprising: "Your people, while technically competent, are cold, aloof, uncaring and bureaucratic." These findings set in motion cultural and organizational countermeasures that have improved the financial performance of the airline dramatically (Geojiales 1995).

Quality Planning for Special Jobs.   Once aircraft, hotels, or cruise ships are in regular operations, special jobs occur that require special planning and coordination. These jobs may take

THREE STEPS OF SERVICE

1
A warm and sincere greeting. Use the guest name, if and when possible.

2
Anticipation and compliance with guest needs.

3
Fond farewell. Give them a warm good-bye and use their names, if and when possible.

"We Are Ladies and Gentlemen Serving Ladies and Gentlemen"

THE RITZ-CARLTON®

CREDO

The Ritz-Carlton Hotel is a place where the genuine care and comfort of our guests is our highest mission.

We pledge to provide the finest personal service and facilities for our guests who will always enjoy a warm, relaxed yet refined ambience.

The Ritz-Carlton experience enlivens the senses, instills well-being, and fulfills even the unexpressed wishes and needs of our guests.

THE RITZ-CARLTON® BASICS

1 The Credo will be known, owned and energized by all employees.

2 Our motto is: "We are Ladies and Gentlemen serving Ladies and Gentlemen". Practice teamwork and "lateral service" to create a positive work environment.

3 The three steps of service shall be practiced by all employees.

4 All employees will successfully complete Training Certification to ensure they understand how to perform to The Ritz-Carlton standards in their position.

5 Each employee will understand their work area and Hotel goals as established in each strategic plan.

6 All employees will know the needs of their internal and external customers (guests and employees) so that we may deliver the products and services they expect. Use guest preference pads to record specific needs.

7 Each employee will continuously identify defects (Mr. BIV) throughout the Hotel.

8 Any employee who receives a customer complaint "owns" the complaint.

9 Instant guest pacification will be ensured by all. React quickly to correct the problem immediately. Follow-up with a telephone call within twenty minutes to verify the problem has been resolved to the customer's satisfaction. Do everything you possibly can to never lose a guest.

10 Guest incident action forms are used to record and communicate every incident of guest dissatisfaction. Every employee is empowered to resolve the problem and to prevent a repeat occurrence.

11 Uncompromising levels of cleanliness are the responsibility of every employee.

12 "Smile – We are on stage." Always maintain positive eye contact. Use the proper vocabulary with our guests. (Use words like – "Good Morning," "Certainly," "I'll be happy to" and "My pleasure").

13 Be an ambassador of your Hotel in and outside of the work place. Always talk positively. No negative comments.

14 Escort guests rather than pointing out directions to another area of the Hotel.

15 Be knowledgeable of Hotel information (hours of operation, etc.) to answer guest inquiries. Always recommend the Hotel's retail and food and beverage outlets prior to outside facilities.

16 Use proper telephone etiquette. Answer within three rings and with a "smile." When necessary, ask the caller, "May I place you on hold." Do not screen calls. Eliminate call transfers when possible.

17 Uniforms are to be immaculate; Wear proper and safe footwear (clean and polished), and your correct name tag. Take pride and care in your personal appearance (adhering to all grooming standards).

18 Ensure all employees know their roles during emergency situations and are aware of fire and life safety response processes.

19 Notify your supervisor immediately of hazards, injuries, equipment or assistance that you need. Practice energy conservation and proper maintenance and repair of Hotel property and equipment.

20 Protecting the assets of a Ritz-Carlton Hotel is the responsibility of every employee.

FIGURE 30.1   The Ritz-Carlton Hotel Company "Gold Standards."

the form of chartered air travel, a special cruise itinerary, or a special event such as a convention or banquet. Such jobs usually involve the requirements of multiple customers traveling and/or convening together. These special jobs often account for 50 percent or more of an enterprise's revenue. Virtually every organization in the industries has a planner for special jobs or events who knows the capabilities of the organization as well as the economics of special jobs or events. This internal planner determines the requirements of the customer, usually through personal interview and historical

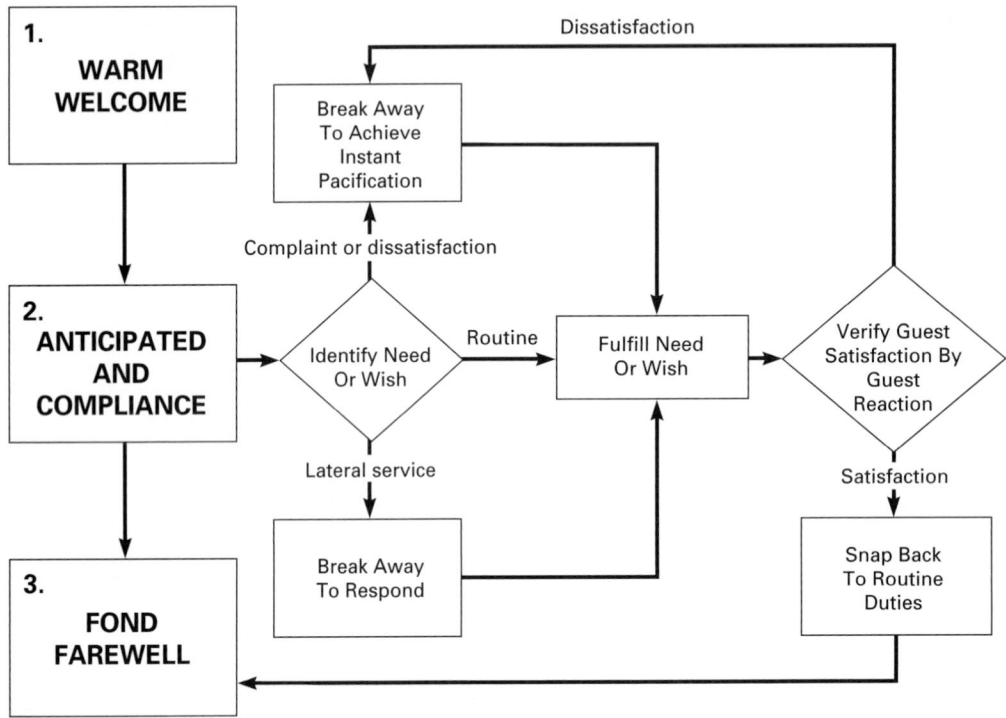

FIGURE 30.2   The Ritz-Carlton Hotel Company "Three Steps of Service."

information. The planner then creates and circulates documents to all concerned for execution of the job. It appears easy! However, even for a knowledgeable and experienced internal event planner, the entire process is filled with potential difficulties for many reasons. The event planner, although a high-level coordinator in the organization, has no hierarchical authority. The work usually passes through multiple divisions and/or departments and is subject to the preexisting difficulties between and within these work jurisdictions. The internal planner can easily underplan or overplan the job or event.

A case of overplanning occurred in a special job where the planner established a 150-point checklist for one aspect of the job. Not only did this list require many detailed inspections and tests, the minimal results of this extra attention to the job output were undetectable by the customers. Taken to the extreme, overplanning a special event or job increases the cost of planning, the attention to trivia, documentation, and the control effort in general.

Conversely, underplanning increases the likelihood the job or event fails to respond to the requirements of the customer and the work becomes more costly than necessary. An example of underplanning resulted in costly litigation. A hotel failed to determine the seating requirements of a special event. The operating forces, functioning without specifications, were left to decide the number of seats per table. Their decision was different from the seating plan of the host organization. The effect of the underplanning damaged the reputation and aim of both the customer and the hotel while adding extra costs to both.

The amount of event communication can rise to enormous proportions, fostering inaccurate and missing information. Furthermore, problems of a job or event usually receive only immediate action, not longer-term corrective action. As a result, the entire event or job can be filled with mistakes, rework, breakdowns, delays, inefficiencies, variations, rebates, and massive inspection efforts. To eliminate or reduce these problems and their extra costs, upper management must create a seamless interface between the planners and those who execute the plan. This organization should develop event specifications so that internal event planners understand the minimum information needed to define and produce an event as well as special event details to prevent defects and extra costs. Only the people who perform the work can provide this vital information to the internal planner of the event or job.

The seamless organization also must have standing and special meetings to review event or job plans. The main purpose of the standard meeting is to provide redundant verification of circulated document information. The special meeting is designed to focus on a single event or job to identify potential failures. The special review meeting is especially important when the job or event is new and untried. When potential failures are identified, the organization works to correct the problem before it reaches the customer and notifies the customer of the situation.

On the surface, alerting the external customer may appear to be an unnecessary, even harmful disclosure. However, research has demonstrated that special event planners expect to be informed on all matters of progress (Power 1994).

Finally, when problems do occur, the organization must first work to solve the problems event by event, because most problems have their origins in the planning of the event itself. In doing so, the organization can prevent a recurrence in repeat jobs or like jobs.

Ritz-Carlton has developed a simple but powerful aid to event planning. The setup sequence (Figure 30.3) and the operations sequence (Figure 30.4) provide a clear conceptual framework that applies to all special events. The *setup sequence,* borrowed from the logistics of the industrial job shop, shows the four categories of elements that constitute a special event, an abbreviation of the generic planning steps for the elements in each category, and the planning goal for each element by category. The grand goal of planning for any special event is a state of readiness, achieved when the goal for each individual planning element is achieved.

The *operations sequence* presents in simple terms the key interfaces between the hotel organization and the major external suppliers and the responsibilities of the major contributors in executing the plan for a special event. The operations sequence also provides a graphic (and perhaps comforting) reminder that much of the work of managing a "special" event consists of coordinating tasks that are part of the hotel's everyday routine, such as laundry, billing, housekeeping, and the like.

At Ritz-Carlton, managers who have had to master the planning and execution of the special event through repetitive involvement have welcomed the introduction of the setup and operations sequences. Managers find these tools useful in discussing an event plan with others, training inexperienced planners to plan and execute events effectively, and reducing the length of time it takes for inexperienced planners to achieve full productivity in event planning and execution. As they have come to appreciate the power of these tools as applied to special events, the managers also have begun to apply the concepts to all work flow generally. Furthermore, these tools provide an ice-breaking example of cross-disciplinary learning. Managers acknowledge that these graphic tools, inspired by the experience of the industrial manufacturing job shop, are very useful in application to hospitality issues. Such acknowledgment opens the door to discovery of other useful concepts from other disciplines outside the hospitality industry. (For a full discussion of special job or event quality, see Section 24, Job Shop Industries.)

Another application of special event or job planning can be useful to restaurants. In this case, each occupied table should be considered a special job. In order to fully understand what's new or different about each table's order and execute the order without failure, the dining room function and the kitchen function must be horizontally integrated. They must agree on the following things:

• The minimum information needed to define and produce a food order

• The additional information needed to prevent errors on a food order (e.g., degree of doneness on meat orders, the seat location for each item to be delivered, etc.)

• A system to coordinate the final food preparation of each order with the start of food delivery

Restaurants have made great strides in reconciling freshly prepared food and timely delivery. This success is a result of high-quality planning, not high levels of personnel or automation. Today, most restaurant kitchens use some variation of the quality plan shown in Figure 30.5.

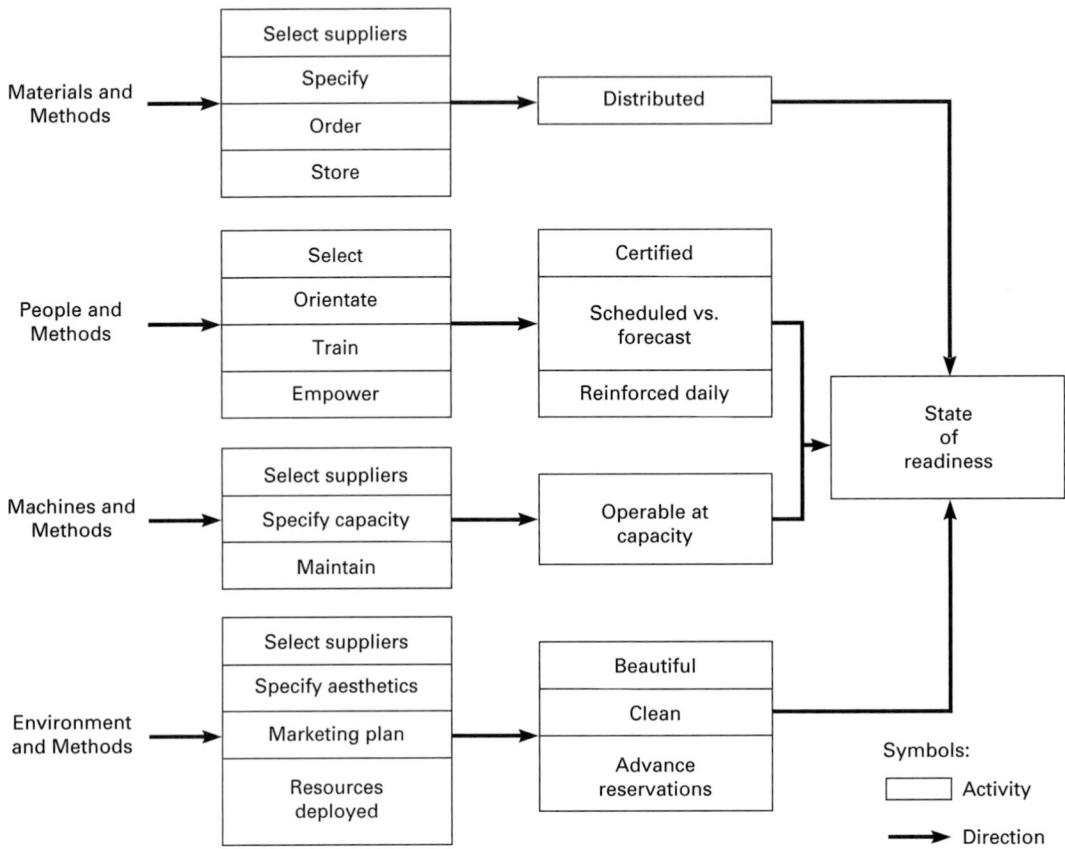

FIGURE 30.3 Setup sequence for special event. (*Courtesy of The Ritz-Carlton Hotel Company.*)

The Role of *Savoir Faire* in Travel and Hospitality Planning. It simply is not feasible to detect the perferences of every customer. Accordingly, many decisions about ambiance, protocol, and general appropriateness are left to the travel or hospitality organization. For this very reason, most high-grade products in the industries employ leaders who have an ability to say or do the right thing in any social situation. This knack (which is often called *savoir faire* from the French "to know what to do") is learned behavior, usually acquired through emulation of a role model early in one's career.

CONTROLLING QUALITY IN TRAVEL AND HOSPITALITY INDUSTRIES

Although the cardinal rules of controlling quality generally apply to the travel and hospitality industries (Juran 1988), it would be fair to say quality performance in these industries does not receive the same degree of control as other internal measures such as schedules or finance. The appropriate controls to be used are most easily determined by understanding the characteristics of the processes employed in the industries.

Worker-dominant processes: When the process is dominated by people (such as service delivery), the control system relies heavily on the talent, skill, and knack of the individual worker.

Material-dominant processes: When the process is dominated by materials (such as uniform purchases or some forms of food production), the control system depends on supplier relations and inspection of delivered lots to identify the unacceptable.

Time-dominant processes: When the process changes with time (such as unsold cabins or guest rooms losing value, machines or decor exhibiting wear and tear), the control system relies on frequent process checks.

Information-dominant processes: When the process undergoes frequent information changes (such as travel reservations, staging a special event or job), the control system addresses the accuracy and the up-to-dateness of the information. (For more on information-dominant processes, see above under Quality Planning for Special Jobs.)

Setup-dominant processes: When the process is dominated by a setup (such as arrangements for a special event or banquet or the placement of baggage-delivery equipment), the control system relies on verification of the setup (i.e., the design).

Facilities and Equipment Quality Control.   When facilities and equipment have reached operating conditions, quality is monitored with emphasis on four dimensions: (1) life safety, (2) continuity of electrical/mechanical service, (3) inventory management yield, and (4) appearance of

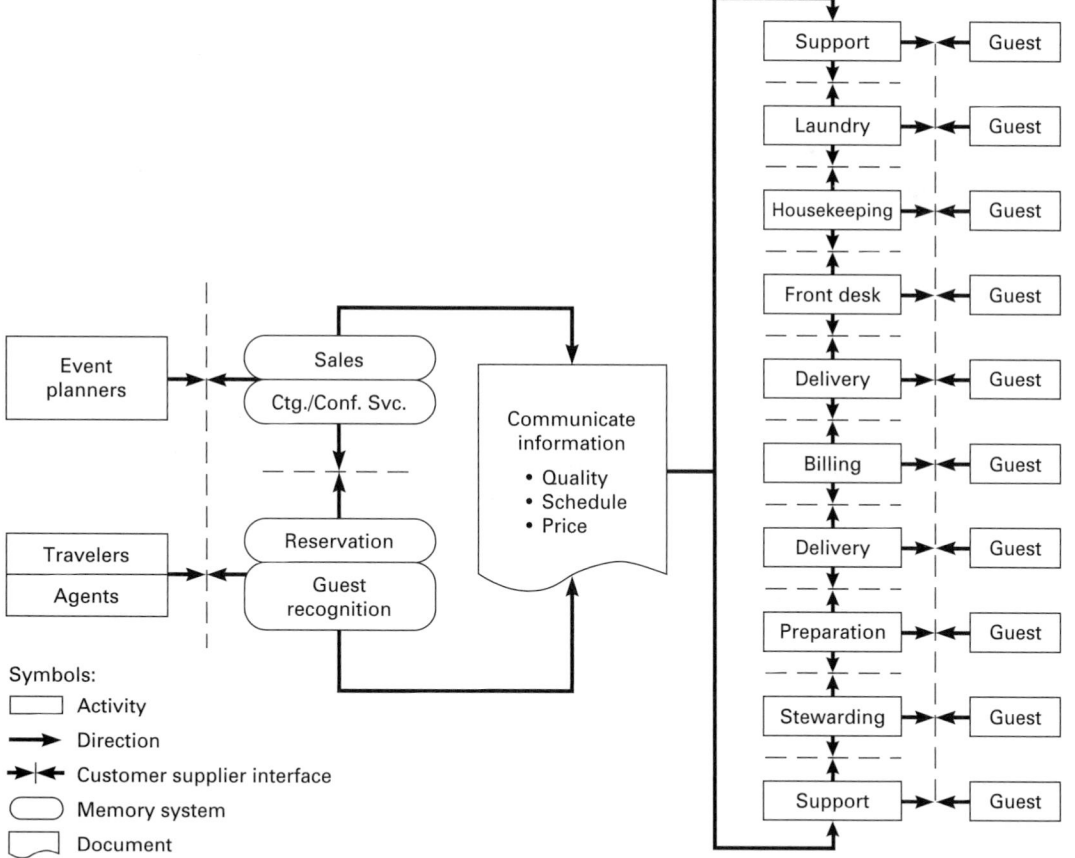

FIGURE 30.4   Operations sequence for special event. (*Courtesy of The Ritz-Carlton Hotel Company.*)

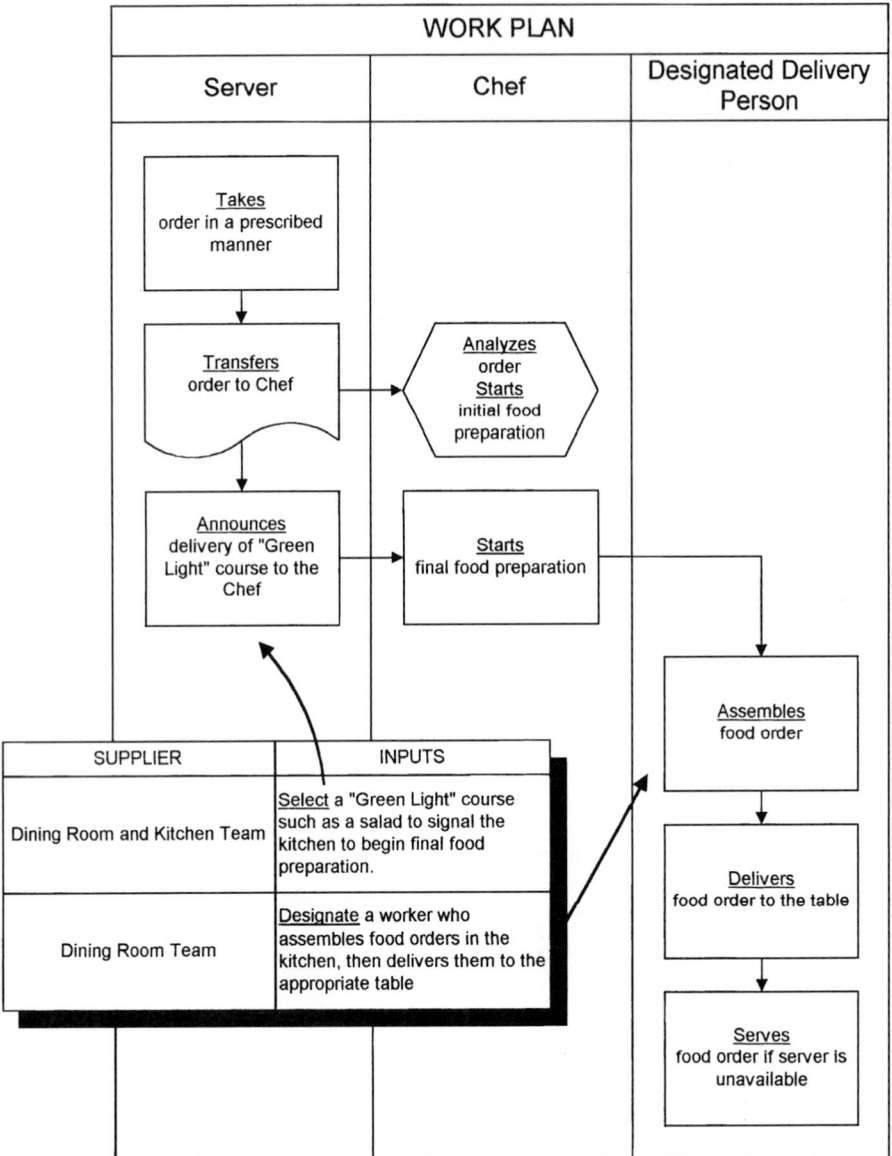

FIGURE 30.5 Quality food delivery plan.

surroundings. Since many of the things measured change with time (e.g., unsold cabins and guest room age), frequent appraisals may be necessary. These appraisals usually take the form of scheduled process checks. Table 30.4 is a matrix of facilities/equipment quality control.

For the matter of aircraft maintenance, fundamental improvements have been developed. The airlines have abandoned scheduled overhauls in favor of reliability-centered maintenance (also called *on-condition maintenance,* since 89 percent of the failure patterns were not age-related). This subject is addressed in Section 7, Quality and Income, and Section 25, Customer Service. For more on the topic of sensory inspections, see Section 23, Inspection and Test.

Service Quality Control. The activities to control service delivery in the travel and hospitality industries should be centered on controlling the behavior, skills, and knack of the workers. For example, service employees should be empathetic and have a high degree of skill in resolving problems. (Unempathetic workers will not carry out customer-oriented policies and procedures adequately.) If there is no process to (1) control or screen out unempathetic candidates and (2) develop the necessary skills and knack to carry out the services required by customers, significant inspection and corrective action will be necessary. The inspection usually takes the form of management review, both daily inspection and performance appraisal systems. This approach has inherent shortcomings.

Regarding inspection, experience has demonstrated that inspection activities detect only 80 percent of the deficiencies, release the worker from the responsibility for quality, are an added cost, and often blame the workers for problems that are beyond their control. A more enlightened approach is to verify the designs of the service delivery and then put the work force in a state of self-inspection or self-control. This alternative works to recruit people who share the organization's values of courtesy, timely delivery, and problem resolution. Through daily reinforcement of these values, the worker spontaneously controls his or her attitude, makes final conformance assessments before service delivery, and extends himself or herself to solve the problems of customers. Periodically, the workers and managers evaluate the customers' assessment of service delivery to identify chronic, severe trouble. Progressive organizations also educate their work force about quality logic, tools, and processes to prevent or correct difficulties.

Generally, the basic quality training program in the travel and hospitality industries covers a number of tools and topics:

1. Service quality logic and processes (i.e., face-to-face interactions between employee and customers)

2. A problem-solving process that includes a diagnostic journey to identify problem causes and a remedial journey to eliminate the causes

3. Statistical tools embedded in the problem-solving process to collect information, analyze data, display data, and plan actions

TABLE 30.4 Facilities and Equipment Quality Control

Timing	Appraisal	Appraiser	Purpose	Example
Life safety dimension				
Scheduled	Process tests and checkouts	Management	Prevent potential failures	Inspection of fire alarm system
Continuity of electrical/mechanical service				
Scheduled	Process overhaul	Equipment operators	Prevent potential failures	Overhaul of the central plant air-conditioning unit
Scheduled	Process tests and checkouts	Patrol operator	Prevent potential failures	Monthly filter examinations of ventilation units
Unscheduled	Process failure identification	Entire work force	Immediate corrective action	Reporting an inoperable air-conditioning unit
Cabin/guest room inventory management yield				
Scheduled	Plan review	Management	Corrective action	Reducing prices when the inventory sold is less than planned
Appearance of surroundings				
Scheduled	Sensory evaluation	Management	Prevent aging of decor	Determining which furnishings must be renovated or replaced over the next 12 months
Unscheduled	Sensory evaluation	Entire work force	Correct unacceptable appearances	Reporting a stained carpet

More advanced components also may include some form of quality planning process, i.e., providing the logic to do something for the first time or to improve a process. These processes and tools are a powerful arsenal to be used in planning, controlling, or improving any aspect of an enterprise in the industries.

Special Job or Event Quality Control.

Special job or event control is distinguished in two ways. Since the process undergoes frequent information changes, the control system addresses the accuracy and the up-to-dateness of the information. The second distinction is the manner in which problems are addressed. Whereas most continuous production and service delivery processes concentrate on the chronic, severe problems first, special event or job control concentrates on the problems within each job or event. An effective way to manage event-by-event quality involves holding a pre- and postevent meeting with the customer. During these meetings, most problems will be identified from the current event. The enterprise then documents each problem in an event log. Each problem is assigned to the appropriate person in the enterprise to diagnose and remedy. All data relating to the problems of the events (i.e., theories, tests, potential solutions, and the like) are kept with the master data file of the special event or job. Since most problems with special jobs or events originate with the planning of that event or job, remedies typically involve redefining what is essential to communicate and to whom. This approach prevents problems in repeat events and like jobs.

Quality Control in Food Production.

In view of the advanced technological equipment used to control the many sensory aspects of production in the commercial kitchens of packaged food manufacturers, it may seem surprising that travel and hospitality organizations still rely on the sensory evaluation of the human beings preparing the food (e.g., professional chefs). The evaluation used by the professional chef is strongly oriented to relations with suppliers, sorting of inferior materials, verification of the process setup, and frequent process checks. The basis of the chef's quality evaluation is either conformance to standards (i.e., recipes handed down by revered teachers, personal specialties, etc.) or food that conforms to the requirements of customers. Typically, it requires the leadership of the organization to convince the professional chef that the marketplace is the final judge of quality, not the chef.

One of the most sophisticated applications of statistical process control in the industries is in a large airline catering enterprise. The worldwide flight kitchens of this progressive organization have undergone extensive process analysis (with the aid of a special consultant) to improve "first-pass yields" (i.e., the percentage of work done right within the time frame). As a result of this analysis, *the work* is able to tell the organization what the kitchen is actually capable of achieving. Should a kitchen perform above or below the norms of "like" operations, the kitchen receives special attention by upper managers. To make sure their flight kitchens continuously achieve quality and related productivity targets, this enterprise conducts internal process audits. As a result, this organization has more than doubled its profitability in an 18-month period.

Quality Monitoring in the U.S. Airline Industry.

Customer dissatisfaction in the U.S. airline industry is monitored by the U.S. Department of Transportation. Each month the office of Aviation Enforcement and Proceedings develops the *Air Travel Consumer Report*. The report is designed to assist consumers with information on the quality of services provided by airlines. The report is divided into four sections. These sections deal with flight delays, mishandled baggage, oversales, and consumer complaints. The report normally is released by the end of the first week of each month. To obtain a single copy, write to the Office of Aviation Enforcement and Proceedings, U.S. Department ofTransportation, 400 7th Street, C-75/Room 4l07, Washington, DC 20590.

QUALITY IMPROVEMENT IN THE TRAVEL AND HOSPITALITY INDUSTRIES

Overcoming Resistance to Change.

Quality improvement activity was largely confined to technical training of the work force and the renovation or replacement of facilities and equipment until the 1980s. The industry's management focus was on business results. This was the classic business school approach: Sell hard, raise prices, force production to make it cheaper, and finance creatively. The beliefs in this approach began to change for several reasons. First, many leaders were learning about the stunning improvements of companies using total quality management (TQM). Second, they began to recognize that their enterprises were producing tangible, repeatable products, not just services. Third, they realized that every product (functional or sensory) was created by some process. Fourth, they simply had run out of ideas on how to make quantum improvements with a changing environment. The old beliefs were not driven out easily. Many of these beliefs had been handed down for nearly a century, and the pride of industry experts was considerable. The stimulus for the eventual change was strong leaders. These leaders put stress on their organizations to improve while reaffirming the ability of the enterprise to meet the new challenge. Finally, these leaders illustrated how the change would benefit the work force. This approach of stress, potential, and payoff was effective at overcoming cultural resistance to change. Currently, improvement projects are more diverse, reflecting the growing diversity of tools to realize the principles of modern quality management: customer as focal point, detailed planning, re-engineered processes, practice, and teamwork.

Quality Improvement: New Product Development.   Because of changing traveler preferences and emerging market segments, the industries have redefined target customers and repositioned themselves to be more competitive. This has led them to develop beyond the traditional broad grades or segments: luxury, deluxe, and economy. Now products (especially facilities and equipment) are designed, developed, and marketed for specific target customers. These targets are new segments within existing segments, and the products are new as well as the images, generating improved competitiveness. This product segmentation usually takes the form of tier or niche segmentation (Abbey 1995). For example, in the cruise line industry, increased availability and economy of air travel and rising costs of marine operations caused the decline of international sea travel aboard luxury ocean liners. Many companies in the industry reinvented themselves by designing and developing new facilities and equipment for specific target customers. There emerged many cruise products that occupy particular niches in the marketplace: (1) sun and shopping itinerary—1 week or less; (2) seasonal itinerary—1 to 2 weeks; (3) exotic itinerary of multiple ports with indepth shore programs—10 days to 3 weeks (Kaznova 1995). As of 1995, demand for cruise products exceeded the supply.

One example of tier and niche segmentation hospitality occurred in West Los Angeles in the early 1980s. A series of six "all-suite" hotels were designed and developed by a single company at six product grades ranging from economy to luxury. Each suite hotel was specially designed for people in the creative industries (i.e., motion picture, music, and advertising travelers). Since creative-industry travelers in Los Angeles have extended stayover requirements, these "artistic havens" were favored over traditional hotel products.

The new Boeing 777 was designed to anticipate the needs of the people who use and operate the plane. It features more comfortable seating, a quiet rest room, reading lights that can be replaced by the cabin crew, as well as mechanical systems that are more repairable.

Not all product development segmentation is capital-intensive. Customer-focused travel agent companies have identified new market segments and developed new products for them. Once, as a motivation to superior employee performance, organizations offered houseware, appliances, electronics, and like products. Over time, high achievers accumulated every conceivable reward and lost interest in more. Travel agencies, realizing the changing need, designed and developed comprehensive incentive travel products, creating a new industry.

Quality Improvement: Services, Special Jobs, and Events.   There was a time when quality service was perceived as unattainable, except in high-grade, high-priced products. Now, enterprises at all product grades are striving to improve service quality to remain competitive.

There are two distinct approaches used in the industries to improve service and special job quality. The first approach is to focus on the development of positive relationships between staff and guests. This leadership approach requires the selection of staff with a demonstrated talent for establishing positive interpersonal relationships and further developing that talent through training in such subjects as relationship management. Figure 30.6 gives some examples of proven *inside-out improvement strategies* in the travel and hospitality industries.

The second approach is to redesign the service delivery processes to provide more customer focus. Success at this approach requires teamwork and worker involvement in planning and execution. These practices can improve worker pride and commitment, management productivity, and customer loyalty. These *organizational and management practices* are referred to as the *outside-in approach* (i.e., focusing on the capability of the organizational processes). The following are some examples of proven outside-in improvement strategies in the industries.

Marriott Hotels and Resorts standardized the best practice of one of its individual hotels. The quality objective was to reduce the elapsed time of the registration process. The means to reach this objective is the integration of the front desk function with the function of the bellpersons (i.e., the people who escort newly registered guests to their room and carry their luggage). To the extent possible, the front desk now preassigns a suitable guest room for arriving guests. Registration information and keys are assembled and placed in an area easily accessible to bellpersons. Arriving guests are greeted by a bellperson, who collects the appropriate key packet and escorts the guests (and their luggage) directly to their room (Hadley 1995).

The Ritz-Carlton Hotel in Dearborn, Michigan (Ritz-Carlton 1994) set out to reduce the number of defects in its guest rooms as well as increase the productivity of room cleaning and resupply. A cross-functional team, with the guidance of an advisor, re-engineered room preparation processes based on the following data: (1) cleaning and resupply defects, (2) cycle time, and (3) worker travel time. The improvements in the new work processes are described in Table 30.5.

KLM Royal Dutch Airlines has translated qualitative service issues into quantitative management tools. Their approach involves summarizing traveler survey data into 10 items referred to as the *service decathlon* (Figure 30.7).

Group A items determine how a customer feels treated as an individual. Group B items all deal with the human body's physical comfort. Group C contains items that can make flying fun.

While Singapore Airlines is praised for having outstanding cabin crews and Virgin is a top performer for entertainment, KLM strives for overall excellence. Like an Olympic decathlon, one does not have to achieve a number 1 position on all items to get the gold medal. KLM turns this service perception data into product improvements (Kaznova 1995).

SUPPLIER QUALITY

Background.   Because the travel and hospitality industries use large quantities of purchased products, supply quality is a topic that deserves upper management attention. Major purchases have an impact on the organization's image and on its financial resources. Major purchases include facilities, furniture, fixtures, and equipment. They also include automated systems and creative services such as advertising and public relations. In most enterprises, food and alcoholic and nonalcoholic beverages are major purchases. Minor purchases generally involve operating supplies, such as uniforms, paper products, chemicals, and cleaning supplies. Minor purchases also include services that are outsourced, such as preventive maintenance, landscaping, copy machine leasing, and pest con-

Inside-out strategy	Purpose
"Character trait" employee selection	Select new recruits whose spontaneous behavior matches the organization's values
Personal improvement education	Rescript the personal and interpersonal behavior of the work force
Coaching, individual recognition, and rewards	Positive (or corrective) reinforcement of behaviors

FIGURE 30.6   Examples of inside-out strategies. (*Source: Frequent Flyer/J. D. Power and Associates Airline Customer Satisfaction Study—U.S. Flights, J. D. Power, Agoura Hills, CA.*)

TABLE 30.5 Ritz-Carlton Dearborn: Housekeeping Process Improvements

Process feature	Formerly	Currently	Benefit
Clean linen and terry delivery to guest floors	Clean linen was "pushed" from the laundry to the guest floor storage areas	Clean linen and terry are "pulled" from the laundry based on forecasts	Less inventory needed No stock-outs Fewer delays
Sequence of guest room cleaning and resupply	Option of the individual worker	Master cleaning schedule based on customer requirements	Less delay at the front desk Fewer room assignment deficiencies
	Separate cleaning and minibar resupply functions	Minibar resupply is combined with room cleaning	Lower labor cost Less intrusion on the guest
	Progressive work by one person	Concurrent team approach	Job enrichment Fewer inadvertent errors
Quality control	100% inspection	Self-inspection Random audits	Increased attention and responsibility by the work force
Quality improvement	Informal	Formal process involving workers	50% reduction of defects 10% reduction in worker travel time Productivity increased 25% Improved employee satisfaction from 75% to 91%

Group	Service decathlon item	Contribution, Overall Score
A	1. Friendly crew	25%
A	2. Efficient crew	18%
B	3. Ground services	9%
B	4. Meal service	9%
B	5. Drink service	9%
B	6. Seat comfort	9%
B	7. Cabin environment*	9%
C	8. Entertainment	4%
C	9. Punctuality	4%
C	10. Information	4%
	Overall score	100%

* Cleanliness, climate, interior

FIGURE 30.7 KLM traveler survey data. [*Source: Frank Schaper (1995). Presentation: "KLM Service Decathalon," National Quality Roundtable VI, March 15–17, Ritz Carlton Hotel Rancho Mirage.*]

trol. In the airline industry there is a growing trend to outsource "under the wing" support services such as fueling, baggage handling, maintenance, and the like.

How Suppliers Are Selected.   Suppliers to enterprises in the travel and hospitality industries are generally selected on three criteria: (1) capability to meet customer requirements on time, every time, (2) knowledge of how their products are used by the enterprise, and (3) cost of the product. Supplier evaluation against these three parameters is usually determined by (1) visiting the supplier's facility and assessing the capabilities by reviewing data, interviewing people, and making observations, or (2) interviewing customers of the supplier. Less frequently suppliers are also selected on their willingness and ability to improve quality and related productivity.

Although most enterprises in the travel and hospitality industries seek a "partnership" with their suppliers, some organizations maintain an adversarial relationship to maximize short-term gains. Most organizations understand that the real cost of a purchased product must include the cost of inspection and corrective action needed. However, some enterprises continue to buy on original price alone. (The case study earlier involving the hotel resort yogurt machine is a good example of relying solely on original price.)

Dependence on Specifications and Brand Names.

Organizations in the industries purchase hundreds, even thousands, of products. The organization defines product requirements in specifications or by accepting the characteristics of a trusted brand name product. Usually, there are separate approaches to developing specifications depending on whether the product is a current product or a new product. The current-product approach begins with selecting a sample that is fit for use and then asking the supplier to describe or specify the product. The new-product approach begins with a description of the organization's requirements, which the supplier then uses to develop samples for testing. The organization selects the product that is fit for use and then has the supplier describe or specify the product. It is important to note that specification development is an example of quality by design. Only through extensive testing and real-world use can the enterprise determine the supplier's capability or management of quality.

Dependence on Receiving Inspection.

Manufacturing organizations understand that quality control of purchased products can be achieved with various practices. These practices involve reviewing the process of the supplier rather than inspecting the final product. These process reviews include process audits (by the supplier or the customer) and reviews of process data that accompany delivery of the product. Unfortunately, most travel and hospitality organizations depend on 100 percent inspection by specialists. The dependence on this ineffective method can be attributed to several conditions. The first condition is that for many products there is a limited number of suppliers. With more demand than supply, there is no motivation for the supplier to improve. The second condition is that many suppliers and customers are willing to endure an adversarial relationship. The third condition is that few suppliers and customers have an understanding of the cost of poor quality and related productivity penalty.

Considerations in Determining the Number of Suppliers to Select.

Usually a large number of suppliers can be identified for every product need, but organizations typically prefer to have as few suppliers as possible. There are arguments for and against single-source supply. Figure 30.8 displays these arguments.

Each travel and hospitality organization must relate these factors to its own situation. Single-source supply has its place in the travel and hospitality industries, and its use is increasing.

Arguments *for* single-source supply	Arguments *against* single-source supply
• Reduces time to negotiate with suppliers, process orders, and receive incoming goods • Builds trust between supplier and customer • The arrangement reflects the synergies of a partnership effort • Opportunity for economies of scale • Supplier can offer full benefit of expertise	• There is a lack of specialized or expert information for each product within the organization • There may be fewer products available within a product line • Customer may give up price leverage

FIGURE 30.8 Analysis of single-source supply.

Benchmarks in the Travel and Hospitality Industries.   The U.S. airline industry has the most readily available and comprehensive data on customer satisfaction in the travel and hospitality industries. These data are the result of an annual study (since 1993) by *Frequent Flyer/J. D. Power and Associates Airline Customer Satisfaction Study—U.S. Flights.* The study is based on a survey of frequent travelers who make, on average, more than 25 trips by airline per year. The study measures the relative importance of each of 10 components of customer satisfaction and the performance of nine major U.S. carriers in each of the component categories. The components, ranked by relative importance, and the leader in each category for l995 are presented in Table 30.6.

The International Air Transport Association (IATA) conducts customer satisfaction performance surveys among transatlantic long-haul passengers on the major airlines between North America and Europe. Although individual airline scores are confidential, it is possible to find interesting aspects of overall performance from aggregate data IATA provides.

PROGNOSIS OF QUALITY IN THESE INDUSTRIES

Dramatic Changes in the Organization of Work.   In the years ahead, enterprises in the travel and hospitality industries will dramatically adjust the way they work to achieve quality, productivity, and profit. Rather than putting the needs of the organization first and then motivate the work force to adapt, they will first address the social needs of the work force to increase their energy, attention, and commitment to the company mission. The social needs are described as the six critical human requirements (Cabana 1995): (1) a sense that people are their own boss, (2) opportunity to learn on the job and keep on learning, (3) an optimal level of variety, (4) mutual support and respect, (5) a sense that one's work is meaningful, and (6) a desirable future. These needs are best addressed collectively by self-directed work design and control, i.e., by self-directed work teams. A *self-directed work team* is a small group with a common goal. It is enabled and allowed to make decisions, with the understanding that the decisions must be by team consensus. Only the group can succeed or fail. Rather than relying on one person, a manager, for leadership, each team member masters some management functions, and team members learn to rely on each other. The responsibilities of self-directed work teams may include identifying customer needs, designing and improving the way work is done, setting goals and monitoring progress, scheduling and planning the work, managing budgets and profitability, and designing, producing, selling, and distributing products and services. The self-directed approach is of benefit to the workers and to the organization in terms of the quality of the work performed. Furthermore, managers have more time than ever to work on new initiatives. Self-directed work teams also bring on another shift in thinking. The work force becomes a group of multiskilled individuals performing broad, meaningful duties rather than unskilled, easily

TABLE 30.6   Customer Satisfaction for Air Carriers in the United States

Category	Importance, %	Carrier
On-time performance	l7	Southwest
Aircraft/attendants	l6	American/Delta (tied)
Schedule flight accommodations	14	American/United (tied)
Airport check-in	l3	Delta
Frequent flyer program	9	Northwest
Food service	9	American
Seating comfort	8	TWA
Gate location	6	Delta
Postflight services	5	Southwest
In-flight amenities	3	Delta
TOTAL	l00	

Source: J. D. Power and Associates 1995.

replaceable people in narrowly defined jobs. Not surprisingly, the more skills workers learn, the more desirable is their future.

Several organizations in the industries have begun to experiment with self-directed work teams. Marriott Hotels, Resorts and Suites uses a self-directed work team approach in its New Jersey Reservation Office. As a result, the performance of the office has improved, especially in reducing the number of abandoned calls.

The Ritz-Carlton Hotel Company has allowed its hotels to reorganize for improved quality productivity and profit. Their hotel in Tysons Corner, Virginia, converted each and every work area from a traditional hierarchy to a self-directed work team over a 2-year period. Since self-directed work teams are vulnerable to overemphasizing their part in the organization, the hotel connected the self-directed teams into five seamless process teams. These process teams were based on the way customers came in contact with the hotel: (1) prearrival team, (2) arrival, stayover, and departure team, (3) dining services team, (4) banquet services team, and (5) engineering and security team. As a result of these efforts, the hotel has made significant improvements in quality, productivity, and profitability.

Emphasis on Process Management.   The needs of the organization also will change. Travel and hospitality organizations will place more emphasis on process management as part of a strategy to adapt to change and to achieve better quality, productivity, salability, and higher profits.

Threats of an Electronic World Where Efficiency Rules.   A business scenario exists, in the view of many futurists, where the growth of communications technology and the overriding concern for efficiency create the conditions for a virtual corporate office space (Cleveland 1996). The travel and hospitality industries would be permanently affected should this scenario dominate the future. The reasons for this are explained as follows: New technology such as the hologram will be available by the turn of the century. A *hologram* is a real-time, three-dimensional laser image that can be projected to another location. This means that people can be in contact with each other electronically but are not in physical proximity. People will be able to choose where they do their work, such as their home. With the virtual office at home and the ability to contact coworkers or customers virtually, travel will decrease, hotel nights will drop considerably, and people will decrease the number of miles traveled each year.

The likelihood of this radical change is great because the unrelenting demands for cost reduction, high efficiency, innovation, and constant change are universal. With people working at their homes, organizations will close their offices. This can occur because your digital screen does not just connect you to the other person's office, it takes you down the hall of the virtual office so that you make the kinds of contacts you did in 1996.

The same futurists contend that the realization of this efficiency scenario may be limited because people will feel disintegrated and distressed. The stress and strains of this efficient electronic world could cause the return of a relatively familiar office in which the "human factor" dominates. In any event, the long-term business strategy of the travel and hospitality industries should address these powerful future forces.

REFERENCES

Abbey, James R. (1995). *Sales and Marketing for Hospitality Operations.* American Hotel and Motel Association, Lansing, MI, pp. 3–7, 8, 11.

Bernowski, Karen (1996). "America's Nostalgic Affair with Loyalty." *Quality Progress Magazine,* February.

Cabana, Steven (1995). "Participative Design: Effective, Flexible and Successful, Now!" *Journal for Quality and Productivity,* January-February.

Cleveland, Charles (1988). *Decision-making: Super Luxury Hotels 1988.* Communications Development, Inc., Des Moines, IA.

Cleveland, Charles (1995). *The Future 1995*. Communications Development, Inc., Des Moines, IA.

Forsyeth, Steven E. (1995). "Effects of Airline Deregulation in the Airline Industry." Presentation to Edison Electric Institute, Delta Airlines, Atlanta, GA.

Geojiales, Nick (1995). "Quality Management at British Airways." Presentation to Coopers Lybrand, Venice, Italy.

Goldsmith, Charles (1995). "Jet Ahoy! First Class Flyers Go `Yachting.'" *Wall Street Journal,* December 4.

Hadley, Helena (1995). Light, private communication to Patrick Mene, Marriott Hotels and Resorts, Washington.

Juran Institute (1994). *The Alligator Hatchery.* Quality Minute Video Education Series. Center for Video Education, North White Plains, NY.

Juran, J. M. (1988). *Quality Control Handbook,* 4th ed., McGraw-Hill, NY.

Kaznova, Brian (1995). Private communications to Patrick Mene. Kaznova Consulting, Atlanta, GA.

Nichols, Robert (1995). Private communication to Patrick Mene: "Quality Practices in the Airline Industry." Sky West Airlines, Phoenix, AZ.

Power, J. D., & Associates (1994, 1995). *Meeting Planner Market Research,* 1994; *Annual Frequent Flyer Airline Research,* 1995. Aurora, CA.

The Ritz-Carlton Hotel Company (1993, 1994). "Malcolm Baldrige National Quality Award Application Summary 1993," Atlanta, GA; Dearborn Housekeeping Project 1994," Dearborn, MI.

SECTION 31

GOVERNMENT SERVICES: REINVENTION OF THE FEDERAL GOVERNMENT

Al Gore

THE COMPELLING NEED TO TRANSFORM GOVERNMENT

As President Clinton has told the nation, he and his team have worked hard since 1993 to create a leaner, but not meaner, federal government, one that works hand in hand with states, localities, businesses, and community and civic associations to manage resources wisely while helping those Americans who cannot help themselves. In January 1996, the President stated, "We know big government does not have all the answers. We know there's not a program for every problem. We know, and we have worked to give the American people a smaller, less bureaucratic government in Washington. And we have to give the American people one that lives within its means. The era of big government is over. But we cannot go back to the time when our citizens were left to fend for themselves."

For the first time in a generation, we live in a balanced-budget age, and although we may disagree on the route of achieving this objective, we share a common destination. The budget will be balanced. This new environment, together with the birth of the information economy, the death of the Cold War, and an assortment of end-of-the-century jitters, has raised an old question with a new sense of urgency: How should the federal government operate?

President Clinton identified this question early in his first term of office, and 3 years ago he asked me to begin working on it. I'm proud of what our federal employees have done to reinvent government. With their help, their ideas, and their leadership, we are eliminating 16,000 pages of regulations. We are implementing the suggestions that federal employees who work on the front lines have been providing to us, suggestions that have never before been heard clearly on a sustained basis, and indeed have sometimes not even been offered because of the fear that those who suggested changes might somehow be subject to retaliation. However, by hearing and implementing their suggestions during Phase I of our reinvention efforts, we have created the smallest government since the administration of President John F. Kennedy. Indeed, we have reduced the size of the federal work force by more than 350,000 positions from 1993 to the present. Because of our efforts and our partnership with federal employees, the government work force as a percentage of the civilian work force is now smaller than it has been since 1933.

We haven't just shrunk the size of the government. Again in partnership with federal employees, we are actually making it work better. We've got a long way to go, and we understand that very well, but we have made progress, and we're beginning to make even more rapid progress. For example, the Federal Emergency Management Agency no longer has the reputation for being slow and bureaucratic—it is now renowned for its assistance in response to disasters, and is praised on a bipartisan basis every time it is called upon to respond. The Social Security Administration now gives world-class service to our senior citizens. The Small Business Administration has reduced its size, cut its paperwork dramatically, and increased its loans. Compared to the steady growth of the bureaucracy year-in and year-out, to which our country was formerly accustomed, these recent achievements since 1993 represent a new pattern and are remarkable.

ROLE OF THE PRESIDENT'S MANAGEMENT COUNCIL

Agency leadership in implementing our efforts to transform government has primarily been provided by the President's Management Council (PMC), a new organization established by President Clinton in 1993, when he began his first term of office. The President asked me to serve as Chairman of the PMC, which is composed of the Chief Operating Officer of every Cabinet department and several other agencies. They, in turn, have directly advised the agency head regarding the agency's overall management needs, and have directed the formulation of specific agency plans for transformation. Members of the PMC have been responsible as well for implementation of the administration's general management reforms associated with the reinvention of government. These general reform efforts are designed to improve customer service, streamline the personnel, procurement, and budget systems, analyze field office structures, and reduce the size of the federal work force, while ensuring more efficient and effective operation of the large systems which carry out the agency's primary mission.

The PMC has worked closely with employee representatives and associations of government managers to make labor-management partnerships a reality. They have also worked closely with senior federal management officials, as well as with members of Congress, to analyze agency plans, systems, operations, and outcomes associated with these reforms. On behalf of the President's Management Council, the Federal Quality Consulting Group[1] was asked to develop guidance and training materials in order to provide consistency in agency quality management efforts with the criteria specified in the Malcolm Baldrige National Quality Award used in the private sector, and similar criteria used in the Presidential Award for Quality. Senior federal executives on loan to the Federal Quality Consulting Group continue to provide this guidance to agencies and assist them in their reinvention efforts.

THE FIRST PHASE OF REINVENTING GOVERNMENT

In past years, debates about government programs were usually dominated by discussions over how much the government should spend, rather than on what the spending would accomplish. For most

---

[1] The Federal Quality Consulting Group may be reached at 1700 G Street, N.W., Washington, DC 20552, telephone (202) 632-6068.

Americans, however, the debates were largely academic, since for well over a decade the public has been saying that government simply is not working. Clearly, dramatic steps had to be taken to restore public confidence in government.

The first phase of reinventing the federal government occurred during the President's first term of office. It was characterized by agency zeal as voluminous rules and regulations were read, consolidated, or eliminated. Agencies tried a wide variety of service quality initiatives, and conducted training and skills development on the part of managers and supervisors primarily, and new quality initiatives began to take hold.

Initial steps in quality management were designed to make government smaller, better managed, and more efficient. The administration announced our vision for the federal government, "To create a government that works better and costs less." We called the first phase of reinventing government "Creating a Customer-Driven Government," and used the following model for agency quality improvement initiatives.

This model is flexible and allows agencies to choose from a variety of service quality approaches to the establishment and evolution of a quality organization. As indicated by the "Learning and Improvement Cycle" of the model in Figure 31.1, we realize our efforts at improvement must be ongoing.

THE SECOND PHASE OF REINVENTING GOVERNMENT

Americans expect and deserve commonsense government—a government that performs well, uses their tax dollars wisely, views them as valued customers, does not impose excessive burdens, and makes a positive impact on their lives when it addresses such problems as crime and poverty and the challenges of employment and education.

Consequently, during the second phase of reinventing government we initiated five basic steps to achieve reform and decide how the federal government will operate in the future:

1. Improve customer service dramatically

2. Increase the use of regulatory partnerships

3. Create performance-based partnership grants

4. Establish single points of contact for communities

5. Transform the federal work force

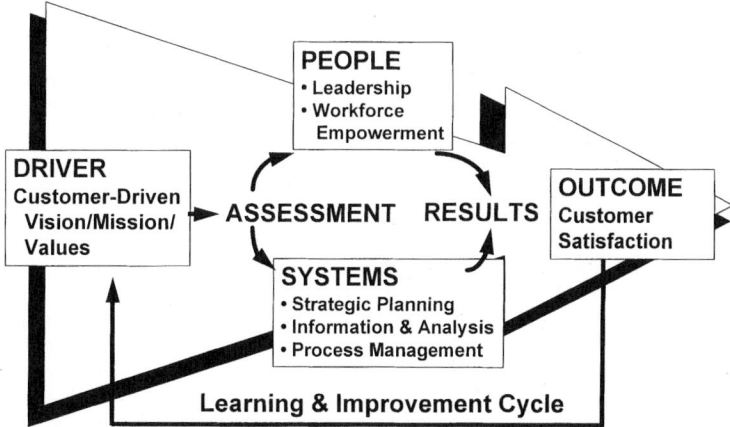

FIGURE 31.1 Creating a customer-driven government—transforming both culture and systems.

Improve Customer Service Dramatically. The first step in reforming government and achieving quality organizations in a balanced-budget age was critical. It was designed to restore America's confidence in the federal government. On March 4, 1996, I pledged to the American public on behalf of President Clinton that the federal government would immediately begin to deliver visible, recognizable, and measurable improvements in customer service. For example, a commitment made by the Passport Office in the U.S. Department of State resulted in shrinking the long lines for service.

Agency commitments to transform their operations and services will be posted in the Internet to make it easier for taxpayers to hold us accountable. We want to hear from taxpayers about their experiences in receiving services from federal agencies and departments, and specifically with regard to how we're doing in keeping the commitments that are posted on the Internet. Our address is www.whitehouse.gov. Look in the "What's New" section, and you'll see the commitments.

During Phase I of reinventing government, when we first started talking to government agencies about customer service, very few of them really understood what we meant by the phrase "customer service." Frontline employees viewed customer service as keeping their supervisors happy, and the culture of the organization has tended to reinforce that view in the past. Customer service for many heads of agencies meant pleasing Congress and particularly the appropriations subcommittees of Congress. Entire federal programs were designed and implemented without ever really finding out what the customers wanted in the first place. And since government had little idea of who its customers were, the idea of setting measurable standards for customer service was really a strange concept.

Working with federal employees, we have changed that. We went out and looked at how the best in business delivered first-rate customer service, and we've started doing the same for our customers. For the first time, most government agencies have established and published customer service standards, and they have engineered their processes to figure out how they can meet those standards and steadily improve on them. Now, just as Federal Express customers know that they're guaranteed to have their package delivered overnight, taxpayers going to a Social Security office will know that they will be seen within 10 minutes. Students calling for information on direct loans will know that they will get through to a live human being within 35 seconds.

These and over a thousand other measurable standards mark a dramatic change in the way government views and treats its customers. We're seeing the signs of success already. For example, an independent survey concerning the provision of 800-line telephone service rated a regional office of the Social Security Administration higher than any American business competing in that category.

We have recently established one-stop Internet access to the federal government through the U.S. Business Advisor. We're redesigning the telephone book blue pages so that people will be able to look under "P" for passports, not "S" for State Department. We've challenged agencies by setting a goal that "everyone in America will know" that government service has improved and continues to get better.

Increase the Use of Regulatory Partnerships. The second way we're transforming government is to make partnerships with the private sector the rule and not the exception. So far, such partnerships have mainly been in the form of pilot projects scattered through the regulatory agencies. We are bringing these partnerships into the mainstream of our regulatory philosophy.

The vast majority of Americans believe that their government has an important role in ensuring their safety and protecting their environment. Government oversight has helped make our workplaces safer and kept our food safe and healthy. Strong enforcement of environmental statutes has helped to restore the health of rivers and has brought back the bald eagle. Our challenge is not only to maintain that progress, but actually to do better.

We know that regulatory agencies can do better by focusing on results as well as process. We've seen it work. In Maine, the Occupational Safety and Health Administration (OSHA) realized that all of its fines and penalties on businesses that were not complying with workplace health and safety rules were producing a lot of income, but they weren't producing any improvement in workplace safety. So they abandoned the old way of doing business and formed alliances with industry management and labor, and focused on results. They have identified and corrected 14 times as many hazards as they did in the old way. Productivity is up, and most importantly, injuries in the workplace are down by one-third in their jurisdiction.

In Miami, the Customs Service used to force companies to spend as much time on government paperwork as on their business. One company had to fill out 700,000 forms each year. Then, Customs sat down with these firms and ascertained what was really needed, and what wasn't. Now that company files one form per month. By streamlining work processes and seeking better results, Customs has been able to confiscate more illegal drugs—seizures are way up—while at the same time helping legitimate businesses grow. Meanwhile, the people who were waiting in lines, after getting off international airline flights, are moving through much faster than before. As a matter of fact, their complaint now is that they're having to wait to get their baggage, and so the airlines are getting their baggage handlers together and challenging them to reinvent the way they handle that job so they can be good enough for government work!

Through programs such as Project XL, Green Light, and 33/50, we have seen the Environmental Protection Agency and hundreds of companies team up to keep a billion pounds of pollutants from being emitted into the air. We've said, "If you can get the job done cleaner and cheaper, then go to it. And throw the book away. Give us a way to measure your progress toward exceeding the goal you've committed to exceed."

Why are these partnerships working? Because they are focused on results. The goal is to reduce injuries, stop drugs, and cut emissions, not to make sure that businesses are penciling in the proper lines on the proper forms. These partnerships are working because government and industry are joining hands, not locking horns. The government recognizes that many corporate leaders share the same goals and are interested in working cooperatively to achieve them. And because they know their businesses far better than the government ever could, they also know best how to attack their own problems, using all of the innovations characteristic of the private sector. Treating them as adversaries wastes those positive inclinations, and it stretches federal resources thin instead of focusing them on the biggest problems.

These partnerships will be new for many of us in government and in industry. Old habits are sometimes hard to break, and this new approach will not be without some difficulties. However, we've seen it work, and we're going to make it happen.

Create Performance-Based Partnership Grants.   The third step in reinventing government in a balanced-budget age was to forge new relationships with communities. On March 4, 1996, I announced the administration's decision to require that every time a grant program comes up for reauthorization, we will ask Congress to turn it into a performance partnership and, if necessary, consolidate it with other related programs.

So far, more than 600 separate federal grant programs, each with its own rules and requirements, have participated. We're going to shift the focus again from process to performance. Together, federal, state, and local governments will set the goals, and the communities will decide how best to meet them.

The goals are to produce better results, to increase accountability to the public for outcomes, to reduce red tape and micromanagement, and to provide greater flexibility in how services are designed and delivered. For example, among these 600-plus grant programs are a number relating to child immunization. Some provide funds to help get children to the clinic, so they can receive the vaccines that their parents have learned about. However, the goal of the child immunization program should be to increase the percentage of two-year-olds who have all their shots.

The President has called for the consolidation of 271 diverse grant programs into 27 performance partnership grants. We have created 105 "Empowerment Zones" and "Enterprise Communities" to focus on local needs, and we've started signing agreements with states to create these new partnerships. We entered into such a new partnership with Oregon several years ago, to promote healthier children, more stable families, and a more highly skilled work force. I have recently visited Connecticut to enter into a similar partnership to improve that state's poorest communities through economic development and neighborhood revitalization approached in a brand-new partnership context. We are now in the process of expanding these partnerships across the country.

Establish Single Points of Contact for Communities.   Several years ago I announced that for every single community of more than 150,000 residents, the Department of Housing and Urban Development will select a person to serve as the single point of contact. For each of these communities we're going to give the legendary "nameless, faceless bureaucrat" a name and a face recognizable in that community—an individual who can facilitate the solution of problems that these communities have which relate to the federal government.

Communities have interactions with the federal government on scores of different issues, ranging from "Head Start" to highways. To get their work with the federal government done, communities had to go door to door. Why? Because these interactions are dictated now by the way the federal government is structured, and not by what the communities need. There is no focal point for dealing with the communities' issues in the executive branch of government. We are consequently now establishing more of these community focal points.

Transform the Federal Work Force.   The fifth step in transforming the federal government is to transform the federal work force itself. Our administration's reinvention efforts have saved $137 billion and cut the federal work force by over 350,000.

To achieve this, we have submitted to Congress legislation to vastly expand the demonstration authority in civil service law. This expansion of authority will allow large numbers of government agencies to design personnel systems suited to their mission. Again, I do not anticipate any partisan opposition to this particular step. As any CEO can tell you, reinvention will be only as successful as the partnership with workers. Let me emphasize that our federal work force is one that any private sector executive would be proud of. However the personnel system they toil under is not.

We have as diverse a set of missions as any conglomerate—more so—but we use a personnel system which by law applies a single set of rules to all federal employees, from patent attorneys to park rangers. This "one-size-fits-all" approach, designed in the last century, simply will not meet the challenges of the next century. No corporation would operate this way. For example, General Electric makes light bulbs, secures mortgage loans, and leases cars. G.E. would not try to squeeze such a diverse cadre of workers into a single personnel system, and neither should the federal government.

We need a new model, decentralized and focused on the mission of each organization. It should hold federal workers to the highest degree of accountability, and give line managers more authority over personnel decisions. This is why we want to change the law and liberate people to build a government that does work better and cost less.

THE THIRD PHASE OF REINVENTING GOVERNMENT

Today the federal government is proud of having the longest-running and most successful government reform effort in history. This has been accomplished through partnerships in and out of government, and with the application of new technologies. The key to our success has been the initiatives and applications of reinvention policies by the people on the front line, providing service directly to the American people. For the first time in history the federal government has measurable standards for serving the public. We have safer workplaces, less crime, easier buying practices, faster drug approvals, and faster delivery of mail. We have cut unnecessary regulations and outdated programs, trimmed our own work force, and saved taxpayers billions of dollars. Our reinvention efforts have been renamed the National Partnership for Reinventing Government, to reflect our commitment to our vision of "America at Our Best."

I have had the privilege of seeing the workings of government through the lenses of both Congress and the White House. As the executive branch moves into a new balanced-budget world, the Congress must join us. This may mean conducting some reinventing inside Congress itself.

Members of both parties have talked about reforming the Congressional committee structure to align congressional oversight more squarely with executive branch organization. A private sector company that reported to multiple boards of directors would have real difficulty in defining its mission and improving its efficiency. Yet many of our executive agencies report to multiple committees and subcommittees of Congress. While we recognize that some overlap is inevitable, the current situation is ridiculous. For example, the Environmental Protection Agency now reports on a regular basis to 28 committees and 43 subcommittees. There are 19 congressional committees and 33 subcommittees that have jurisdiction over federal programs for children and families. Surely there is room for significant streamlining and improvement here.

Now that we set multiyear goals for overall budgeting, do we really need to stick to single-year appropriations? Can we move to more multiyear appropriations so that government managers have a more stable environment in which to plan and to invest, and so that all of the local and state governments that wait each year for the federal government to conclude its annual appropriations can also have more flexibility to plan ahead? Surely the answer must be yes, even though I know these are contentious issues.

Both parties in Congress have put forward some interesting and sound reform proposals. However, the point I want to make is that Congress must join us in the reinvention effort. In 1994 Congress passed and President Clinton signed the "Government Performance and Results Act," which provides the framework for strategic planning for almost all federal agencies. This historic piece of legislation, which had bipartisan support, sets us on a path to a performance-based government. What Congress has passed for the executive branch should be matched by its own internal re-examination. Only by working together can Congress and the executive branch, along with the states and local governments, achieve the reinvention of government in a balanced-budget era, and the provision of public services necessary for the American people.

Increase the Use of Information Technology. Today's information technology is the great enabler for reinvention. It allows us to rethink, in fundamental ways, how people work and how we serve customers. For example, in Miami, compliance officers of the U.S. Customs Service can use their computers to compare passenger manifests with up-to-the-minute risk lists. Before the planes land, officers identify passengers who will be the focus of their inspections. Similarly, on the World Wide Web, the Social Security Administration lets you order your personal benefits estimate by filling out a form on screen, taking only 5 minutes to do so.

Not all technology applications require high-end components. More Americans contact the government by phone than any other way, so the Internal Revenue Service built Telefile to allow 20 million filers to submit their 1040EZ tax forms using touch-tone phones. It takes 8 minutes, is paperless, and has an error rate a fraction of that found on paper returns.

The potential payoff from technology is huge, but not automatic. Workable solutions can be had at reasonable costs by following these principles, derived from the best practices of the private and public sectors:

- Don't automate the old process, reengineer. The new technologies bring new possibilities, like putting services on the Web and letting customers get them when they want.

- Buy off the shelf. Commercial products provide big variety and capability, and new products are added every day. It almost always makes sense to give up a few performance features to get something that costs less and has been thoroughly tested.

- Check on investments. Tracking not just costs and schedules, but whether the new technology is paying off as promised, is critical in spending the taxpayers' money and in applying good management principles to large and small government programs.

- Integrate information. Many agencies duplicate data collected by others. For example, 40 agencies gather trade information. We are constantly looking for chances for agencies to share workload and data, thereby reducing the number of federal programs and activities, while providing outstanding service to our customers.

Focus Regulators on Compliance, Not Enforcement.   The President has issued directives which told agency heads to follow the following five strategies:

Cut obsolete regulations

Reward results, not red tape

Visit grassroots partnerships

Negotiate, don't dictate

Reduce regulatory reporting

The first strategy deals with what is in the rulebooks. We are taking steps to get rid of what's outdated, and rewrite the rest in plain English. We believe that better compliance occurs when people understand what's expected of them.

The remaining strategies deal with what is most important in reinventing regulation—improving the relationships between regulators and the regulated community. Experience shows that most businesses and communities do want to comply with regulations and will do so if they can figure out what they're supposed to do. Agencies are proving that, working with new partners, agreeing on the goals, allowing room for innovation, and providing all the help possible to those who want to comply. We are finding that because regulatory time is no longer being wasted on the good guys, agencies can better focus their attention on the few cheaters.

CREATING A MORE BUSINESSLIKE GOVERNMENT

We have asked agencies to take steps to make their agency programs and operations more businesslike. In order to help them do so, I have approached many of our country's finest companies, and have received guidance from their CEOs as to how a more businesslike government may be achieved. Help from such companies as the Cadillac Division of General Motors, Disney, Federal Express, Harley- love to my horse is an hour awa and Davidson, Ritz-Carlton, Wal-Mart, and Xerox enabled us to take first steps in creating a businesslike environment.

All this help from America's best companies is paying off for the American people:

- Social Security Administration 1-800 telephone service has been ranked the "best in business."
- General Services Administration is managing inventory like General Electric.
- National Aeronautics and Space Administration's Martian explorer phones home through an off-the-shelf Motorola modem.
- U.S. Department of Agriculture's Rural Development Agency runs its mortgage portfolio just like Citicorp.
- Food and Drug Administration is partnering with medical researchers to get safe, effective treatment to the people who need it really fast.

We believe that Americans are already receiving good service and good value from top American companies. These same Americans deserve and have the right to expect no less from government. We are grateful that America's best companies are helping us to deliver in the same outstanding manner.

IN CONCLUSION

In conclusion, as we continue to reinvent the federal government, we know that the challenge facing us is very large but not impossible to conquer. We have singled out a few agencies for special attention, those mentioned in this report, to provide an idea of the range and type of services currently being reengineered. The challenge is great, but the desire of federal employees and agency constituents is even greater: to achieve a more customer-friendly, streamlined, and effective government. We must work together with leaders from business, state and local governments, and the university community if we are to succeed in the challenge of reinventing our government.

COPYRIGHT DISCLAIMER

All or any part of this section may be reproduced without permission. Reproduction expenses will be borne by the user of this document.

SECTION 32
HEALTH CARE SERVICES

Donald M. Berwick
Maureen Bisognano

INTRODUCTION

Few industries have changed so radically in structure, focus, and process as health care in the United States has in recent years. Integrated delivery systems have replaced hospitals as the powerful force in health care delivery. The aim of the system has moved from caring for patients with disease and injury to improving the health of entire communities. Process redesign has produced changes in the kinds of care provided, the site in which care is received, and the extent to which the patient is an active participant in the plan for care.

Health care is America's second largest industry, surpassed in total expenditures only by education. A trillion dollars changes hands each year in the purchase of medical products and services, with care being given in over 6300 hospitals, 1000 health maintenance organizations, 720,000 nursing homes, and 200,000 medical offices. Health care involves the work of over 600,000 physicians in the United States, 1,900,000 nurses, 155,000 dentists, and hundreds of thousands of others in a myriad of allied health professions and support services.

The complexities of the system are evident in a comparison with other industries; a manufacturing company employing 4000 people will categorize staff in about 50 job titles; a typical health care organization of 4000 will utilize 500 titles. This specialization, originally designed to improve quality, now creates multiple handoffs for any patient procedure and contributes to a breakdown in quality processes.

The economic scale of American health care dwarfs that in other countries. On a per capita basis, Americans spend almost 40 percent more on health care than in the next most costly health care system (Canada's), and over twice as much as in many other systems in the Western world. Over 15 percent of America's Gross Domestic Product goes to health care, with the comparable figures being 11 percent in Canada, 8 percent in the United Kingdom, and 6 percent in Denmark. Throughout most of the last half of the twentieth century, American health care costs grew at a much faster rate than did the economy as a whole. In total dollars, the bill for health care doubled in the decade of the 1970s, and then doubled again between 1980 and 1989. On average, the medical price index, the price of a "market basket" of medical products and services, has risen 2 to 3 times faster than the consumer price index during the last third of this century. One health care economist in 1970 wrote of his dismay and disbelief that health care costs might soon rise to over 7 percent of the GDP in the United States, little imagining that, by the end of the century, that figure would more than double.

Yet despite its immense scale, American health care falls short of the social need it aims to fulfill. Because of anomalies in the financing system for U.S. health care, as of the late 1990s, over 30 million Americans lacked insurance as a means to pay for their care, and simple indicators of population health place Americans surprisingly lower in health status than many other populations in the developed world. The United States ranked twenty-first in the world in infant mortality rate (deaths in the first 30 days of life) in 1995, and American life expectancy is 3.4 years shorter for men and 2.8 years shorter for women than that of the best in the world, Japan. Surveyed about satisfaction with their health care, Americans routinely give the system a dichotomous rating: They rate *their own* doctors highly, but express strong dissatisfaction with the quality and cost of American health care *in general.*

INITIATIVES FOR CHANGE

Discontent with the high cost and variable outcomes of health care has led, in the last quarter of the 20th century, to a number of initiatives in public policy, finance, and organization of care—all in an effort to measure, control, and improve the value of care, and to increase the accountability of health care providers to the public and to insurers. Important examples include the following:

Prepaid Financing of Care.   Provider organizations are given a prospective annual budget to meet the needs of enrolled populations of patients, instead of simply charging insurers without limit for care as it is consumed. Such prepaid financing is supposed to change the incentives for providers of care from "doing more" (under fee-for-service payment) to "conserving resources" (under prepayment), and to shift the risk for the increases in cost from the payor to those providing care.

Health Maintenance Organizations.   "Managed care" systems are designed as systems of health care linked to provide health care services to members for a periodic fee, regardless of the extent of services required. In managed care systems there is often an emphasis on improving the health status of the member population to decrease the cost of care for preventable chronic disease or injury.

"Gatekeeping" and Other Forms of Utilization Review.   Providers of care must justify to insurers their choices to use expensive tests, surgery, specialty referrals, and so forth, especially if their patterns of use deviate from prevailing patterns among their peers. The controls on resource consumption are placed in the hands of a clinician who assumes the role of care manager and who orchestrates the level of testing and service delivered to the patient.

Innovations in Programs of Care.   For example, a major shift in *site of care* has occurred in the past several decades, as procedures and tests, formerly done only in inpatient settings, are now performed safely and at lower cost in outpatient offices and clinics. Only 15 years ago, cataract surgery required a week-long hospitalization and a quiet recovery period at home. Today, advances in technology and process changes permit such surgery to take place in physician offices or surgery centers, requiring only a 2-hour total visit/procedure time. Even more exciting innovations have occurred in outreach programs extending technical care into patients' homes, and in new communicationmethods to involve patients more directly in decision making about their own care, such as choosing between medical and surgical management of prostate disease and breast cancer. Telemedicine links specialists with rural physicians in remote areas and allows for the latest in diagnostic capabilities to reach patients who would otherwise need to travel for such access.

"Report Cards" and Measurement Systems.   The performance of health care organizations is assessed by standardized instruments, and often the results of measurement are made public to help inform the choices of payers and patients. Business coalitions in key cities across the nation have accumulated comparative cost and outcomes data to aid business leaders in the decisions on which providers will care for employees. In many companies, the cost of health insurance for employees is the number one increasing cost and these data are considered a key tool in controlling what was considered an "uncontrollable" cost.

Health care leaders are also using these comparative data and internal "balanced score cards" to define organizational priorities for change and to demonstrate progress to key stakeholders.

FOUR ARENAS OF IMPROVED PROCESS

Health services research has demonstrated major opportunities for improvement of the performance of health care processes in at least four arenas: health status outcomes, service characteristics (measures of satisfaction and ease of use of the system), breadth of access (including equity among racial and socioeconomic groups), and levels of waste.

Health Status.   With respect to health status, Americans are not nearly as healthy as they could be, given available scientific knowledge. Unintentional injuries are the major cause of premature death, and many are preventable through simple steps such as wearing automobile seatbelts, using helmets when riding bicycles or motorcycles, and fencing off swimming pools. Simple counseling by health care providers can lead many people to change their life-style choices in a more healthful direction. Almost half of all deaths annually in the U.S. are caused by alterable choices in life-style and behavior. Tobacco use, alone, accounts for 19 percent of deaths in this country, by inducing heart disease, cancer, and respiratory disease (Table 32.1).

Technical medicine also could be safer and more effective than it is. Adverse drug events—complications from medication use—occur in over 6 percent of all hospital admissions, and many are due to avoidable system errors. Unnecessary surgery and testing add hazards without benefit for

TABLE 32.1   Actual Causes of Death in the United States in 1980

Causes	Deaths	
	Estimated no.	Percentage of total deaths
Tobacco	400,000	19
Diet/activity patterns	300,000	14
Alcohol	100,000	5
Microbial agents	90,000	4
Toxic agents	60,000	3
Firearms	35,000	2
Sexual behavior	30,000	1
Motor vehicle	25,000	1
Illicit use of drugs	20,000	<1
Total	1,060,000	50

many patients. In one study, for example, more than half of the operations done on carotid arteries were unnecessary according to scientific literature. Rates of "inappropriate" care (care that, on scientific grounds, cannot help the patient) range between 10 and 50 percent for many frequently performed procedures. Cesarean section rates in the United States rose from 5 percent in 1970 to over 23 percent in 1995, without any strong evidence of clinical benefit to mothers or infants. Such inappropriate care both raises costs and introduces risks.

Service Characteristics.   Service characteristics of health care have also lagged behind those in other industries. Waiting times in health care systems—both waits for appointments and waits on-site at the time of appointment—are often very long, and many care-giving institutions experience frequent complaints from patients and families because of incomplete communication, impersonal encounters, and lapses in continuity of care. An average physician office visit takes hours and involves care by several staff members and 23 different procedural steps. The information systems in health care organizations often add to the complexity. They have not been utilized, as they have in other service industries, to produce a sense of confidence and friendliness by assuring that all appropriate information is cascaded to key staff in anticipation of patient needs.

Breadth of Access, Levels of Waste.   Access to care in the United States is generally good, but not for all portions of the population. The health care insurance system allows over 30 million Americans to remain without health insurance. Some are very healthy, and do not notice. But others are at high risk of suffering devastating health care bills, or delay their own care imprudently because of inability to pay. For lower income groups, the gaps in access may be severe. Almost one in four pregnant women in America's inner cities lacks adequate prenatal care.

Above all, the *costs* of health care in America are far from optimal. We have already noted above how great is the difference in per capita health care expenses comparing the United States with other developed nations. Some of that difference may relate to features of the U.S. system that are not available in others—for example, we have more "high technology" imaging machines and generally newer hospital facilities—but, to a large extent, the higher costs of health care in the United States represent higher levels of waste—costs of poor quality. Compared to other nations, we in the United States do more laboratory tests, use more hospital bed-days, perform more surgery, and use more minor visits to doctors. We tend to use our capital (expensive diagnostic machines, costly hospital space, and equipment) less fully, and we use many more futile interventions in the final stages of life. Unlike other markets, the health care economy is largely "supply-driven." That is, the availability of hospital beds, specialist services, and imaging equipment, for example, appears to be the greatest single determinant of the rate of use of those resources. For historical reasons, the United States has accumulated a larger supply of specialized health care than other countries, and many health economists believe that this is a *cause,* not a *consequence,* of our high rates of use and therefore a cause of our high costs. Most important, this increased use of high-cost care has apparently not led to better health outcomes for Americans.

FURTHER POTENTIAL IMPROVEMENT

The following simple list of 11 major areas for potential improvement of American health care is based on author Berwick's 1994 health service research findings. The list is only a sample; many additional entries would also be supported by available research.

Aim 1: Increase appropriateness of practice.   Reduce the use of inappropriate surgery, admissions, and tests. Important initial targets may include: management of stage I and II breast cancer, prostatectomy, carotid endarterectomy, coronary artery bypass surgery, low back pain management, hysterectomy, endoscopy, blood transfusion, chest x-rays, and prenatal ultrasound.

Aim 2: Increase effective preventive practices.   Improve health status through reduction in "upstream" causes of illness, including especially: smoking, handgun violence, preventable injuries in children, and alcohol and cocaine abuse.

Aim 3: Reduce cesarean section rates.   Reduce cesarean section rates to below 10 percent, without compromise in maternal or fetal outcomes.

Aim 4: Reduce unwanted care at the end of life.   Reduce the use of unwanted and ineffective medical technologies at the end of life.

Aim 5: Rationalize pharmaceutical use.   Adopt simplified formularies, and streamline pharmaceutical use, especially for antibiotics and for drug prescriptions for the elderly.

Aim 6: Involve patients in decisions.   Increase the frequency with which patients participate actively in decision making about therapeutic options.

Aim 7: Reduce wait states.   Decrease uninformative waiting in all its forms.

Aim 8: Reduce, consolidate, and regionalize high-technology services.   Reduce the total supply of high-technology medical and surgical care. Consolidate high-technology services into regional and community-wide centers.

Aim 9: Reduce wasteful and duplicative recording.   Reduce the frequency of duplicate data entry and of recording of information never used in medical record and administrative systems.

Aim 10: Reduce inventory costs.   Reduce inventory levels.

Aim 11: Reduce racial and economic health status inequities.   Reduce the racial gap in infant mortality and low birthweight.

American health care has enormous opportunities for improvement. Despite this, we must acknowledge that American health care is, nonetheless, in many ways the very best in the world. Patients who can afford it come from all over the world to receive their care in American facilities, and the American biomedical research community is the largest wellspring in the world of new knowledge used to improve the effectiveness of treatments. Many leaders in health care systems throughout the world have received their training in U.S. facilities, and many nations, both developed and developing, base their system designs and approaches to the prevention and treatment of disease on American models.

APPROACHES TO QUALITY CONTROL IN HEALTH CARE: HISTORY AND PREVAILING METHODS

Quality of care has been a concern of health care leaders for as long as we have written records of medicine. Quality scholar John Williamson reports the following "quality control" text in Hammurabi's code, dating from 2000 B.C.: "If a physician should operate on a man for a severe wound with a bronze lancet and cause the man's death; or open an abscess (in the eye) of a man…and destroy the eye, they shall cut off his (the physician's) fingers." Standards of ethical conduct were established in the Hippocratic Oath in the fourth century B.C. In the Middle Ages, regulations in regions of France, Germany, Italy, and elsewhere defined who was and was not permitted to practice surgery, obstetrics, and drug prescribing. In America, licensing of physicians to practice began in New York in 1760, and the first American specialty society, the College of Physicians, was founded in 1787.

Arguably, the modern era of concern for quality of medical care in America began at about the turn of the twentieth century. Academic medicine took a major step forward with the publication in 1910 of the Flexner report, which both documented the deficiencies of medical training at that time and set out new standards for the education of physicians based upon a scientific view of the practice of medicine. As a result of the Flexner report, half of the medical schools then active in the United States closed their doors, and medical training thereafter became subject to stringent accreditation procedures.

At about the same time, medical professional organizations began to dominate the landscape of professional certification. Prime among them was the American College of Surgeons (ACS), which undertook, in 1916, an extended project to study quality of care and to develop standards for American hospitals in areas such as medical staff organization, record keeping, and availability of diagnostic and therapeutic facilities. The initial findings in 1919 from the so-called Hospital

Standardization Program were disturbing; only 89 of the 692 surveyed hospitals with over 100 beds had met the new American College of Surgeons Standards.

The work of the ACS, along with other professionally driven efforts at self-inspection, led, in the 1950s, to the formation of the Joint Commission on Accreditation of Hospitals [later renamed the Joint Commission on Accreditation of Health Care Organizations (JCAHO)], which rapidly became the major accrediting body for hospitals throughout the nation. The JCAHO purpose, largely unchanged throughout the years, is to establish standards of care for hospitals and to conduct surveys that teach and promote improved systems of care and safety. The scope of work of the Commission extends from physical plant inspection to medical documentation review to interviews with hospital staff and physicians ascertaining their level of organizational involvement and their capacity to perform effectively and to improve their work. Still today, the triennial accreditation site visit from a JCAHO team is a recurring milestone for most American hospitals. Other prevailing forms of outside inspection include those from state and local health departments to review and enforce regulations bearing on the operations of clinics and hospitals, and regular procedures for licensure and relicensure of doctors and other health professionals, as administered by the state licensing board.

While the Joint Commission, state health departments, and professional boards of registration were developing as the major forms of external quality inspection for hospitals, a set of conventional forms of internal inspection gradually became routine, even traditional, inside the organization of hospitals themselves. These internal surveillance systems, many required by the Joint Commission, consist mainly of committees drawn from the medical staff, each with jurisdiction to review specific components of care process and outcome. Tissue Committees, for example, review organs and tissues removed in surgical procedures. Pharmacy and Therapeutics Committees review medical use and set pharmacy policy. Morbidity and Mortality Conferences, especially in surgical departments, review cases of operative death or complication. To staff these quality control groups, and often also to help prepare for Joint Commission surveys, most hospitals maintain Quality Assurance departments, whose members collect necessary data, prepare documents, and conduct special studies of quality-related issues. Membership is traditionally drawn mainly from nursing backgrounds.

An extensive research literature on quality of care has developed during the century in parallel with these accreditation and inspection activities. Researchers have explored methods of inspection— such as "explicit review" and "implicit review," in the jargon of health care quality assurance. Explicit review consists of reviews of care processes against written criteria. For example, such a review may ask: Did patients with anemia on screening tests receive the correct follow-up laboratory tests according to a preexisting protocol? Implicit review consists of summative judgments by recognized clinical experts rating the adequacy of care without reference to specific, preexisting criteria.

One of the seminal authors in the history of health care quality research, Professor Avedis Donabedian, offered in 1966 what remains the dominant categorical framework defining possible objects of inspection, whether by explicit or implicit means: "structure, process, and outcome." Donabedian claimed that quality assessment could study the resources and organizational architecture of care (structure), the sequences of diagnostic and therapeutic activity (process), or the health status, mortality rate, and functional results of care (outcomes), and that each object of study could shed light of a different type on the overall pattern of quality of care (Donabedian 1966).

The inspection-oriented foundations of so-called quality assurance in health care took a significant step of maturation in the 1970s and 1980s due largely to research at the Rand Corporation, which was employing an experimental design to assess the effects of various forms of health care insurance on the processes and outcomes of care (Brooke et al. 1979; Lohr and Brooke 1984). As an important byproduct of their main research plan, Rand's investigators developed a carefully crafted set of surveys and measures to assess quality. They employed a very broad definition of "quality of care," encompassing patient satisfaction, ease of access, and appropriateness, as well as the more traditional definitions of health outcomes. They broadened the definition of "health outcome" itself, by describing dimensions such as emotional well-being, social and role functioning, and physical comfort, rather than stopping with simplistic, unidimensional measures of health. For example, the instrument analyzes days lost to work and the degree of effective work ability 6 weeks after a surgical procedure in addition to a more introspective measure such as complications at the time of discharge after surgery.

Most important of all, these researchers showed that simple questionnaires and record abstracting forms could have excellent properties—validity, reliability, and sensitivity to changes over time. Colleagues in other research centers built upon this work, producing, by the end of the century, a relatively well-developed tool kit of sound measures of the effect of health care on its customers.

Meanwhile, researchers concerned primarily with the processes of care (in Donabedian's sense of the word) had become aware of large variation among health care practitioners in their approach to diagnosis and therapy. Pathfinding research by Dartmouth Professor John Wennberg (Wennberg and Gittelsohn 1973) showed that, adjusted for the "case-mix" (age, gender, etc.) of the populations treated, rates of use of laboratory tests, surgical operations, and hospital bed-days varied greatly, depending on which doctor, which hospital, and what geographical region, both within the United States and among countries. In one study in the mid-1980s, for example, Wennberg found a variation in rates of hysterectomy in women before the age of 70 years of 350 percent between two cities in Maine less than 100 miles apart.

Wennberg's work, along with the many confirmatory studies that followed, raised interest in the need for *standards of practice* to reduce this unexplained variation. In the last two decades of the twentieth century, a virtual subindustry developed in the United States of both public and private groups who developed and promulgated guidelines and protocols for care, some allegedly based on scientific literature, others on expert opinion, and still others on prevailing practice patterns. Some regulatory agencies seized upon such guidelines as another component for their systems of external surveillance of quality of care, and payers (insurance companies and corporations covering their employees' health care needs) began widespread use of the guidelines to help them decide what care to pay for, and what not. National guidelines for the care of common conditions, such as cardiac chest pain, low back pain, and diabetes, emerged from the work of several federal agencies, as well (see Figure 32.1 for an example).

Using more modern information systems and claims databases, many health care systems have recently begun offering direct feedback reports to physicians as individuals and groups, showing their rates of utilization of resources or, sometimes, their degree of adherence to protocols for care. Reactions to such feedback vary. In some cultures, physicians show fear, wariness, or anger at this apparent invasion of their professional autonomy. Doctors in other systems appear to welcome the feedback, and use it to engage each other in discussion or to set in place personal learning plans. One of the most dramatic examples of successful arrangements for feedback of performance data to physicians has been the work of the Maine Medical Assessment Foundation (Wennberg and Keller 1994), which sponsors peer group collaborations of specialists who receive such data confidentially, and who then use it to explore the causes of variation among the individual members, often with prompt results in reduced variation and correction of outlier patterns.

In summary, the system's familiarity with issues of quality management comes traditionally in its use of various forms of inspection—internal and external—applied both to its work procedures (in guidelines, accreditation, and certification) and, to a smaller degree, to its outcomes. Senior health care leaders mainly seek to assure quality through accreditation of facilities and people, through periodic external and internal reviews against standards, and through the study of unusual events and complications. For the most part, more formal statistical methods for quality control are not in evidence even decades after they have become conventional in many industrial settings. Exceptions include a few technical areas in care, where statistical process control theory makes its appearance in supervisory use of run charts and control charts on a local basis, such as in clinical laboratories (where control charts are used to maintain machine calibration) and radiology units (where graphical methods are used routinely to control the temperature of x-ray film developer solutions).

QUALITY IMPROVEMENT

Health care's traditional reliance on external and internal inspection to maintain quality has had its expected effects on performance. The extensive inspections are very costly, but are agreed to, by providers and outsiders alike, as the best they can do. Reviews do tend to call attention to serious

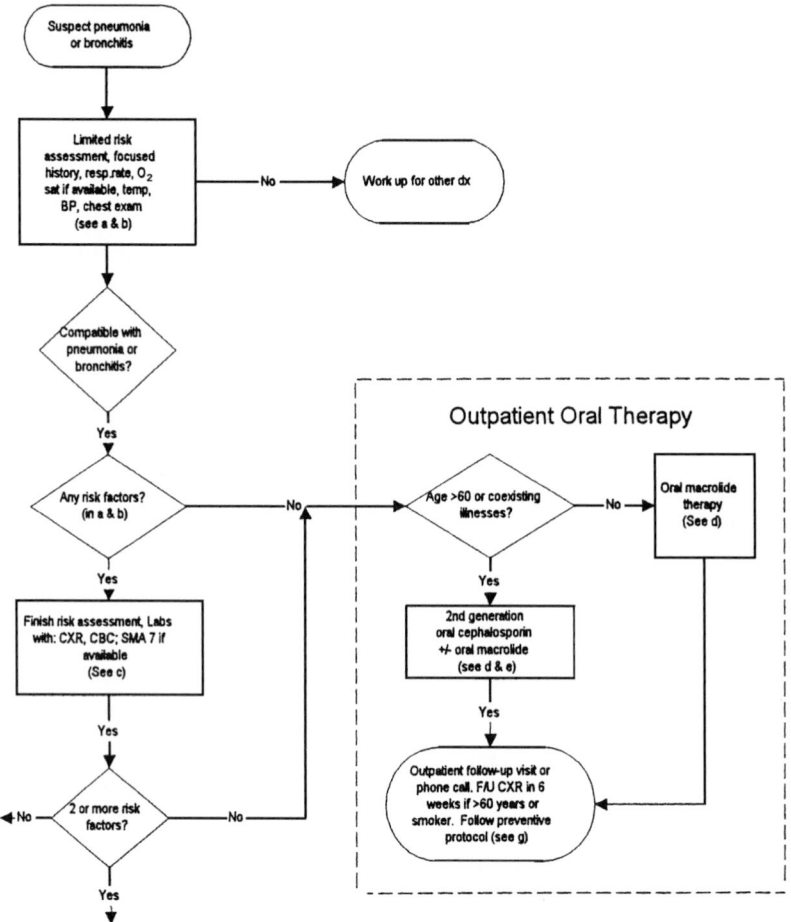

FIGURE 32.1   Guideline for care of patient with pneumonia.

problems, some of which are remedied correctly, and others which invite overreactions that have probably added to the cost and complexity of care without much gain in quality or efficiency. One good example of waste of this type is the accretion of requirements and guidelines for medical record keeping, which have resulted today in a medical record whose size, complexity, and format confound accurate and efficient use. No modern industry outside health care keeps records as wasteful as those in medicine. On the whole, quality assurance in health care is viewed as cumbersome, occasionally revealing, and a necessary evil.

At a deeper level, however, quality improvement concepts, though they have had a second seat to inspection, are not so new at all in medicine. The Joint Commission itself developed and refined an improvement procedure, which it recommended to hospitals for their internal quality assurance procedures, and which incorporates many of the aspects of modern approaches to "plan-do-check-act" (PDCA) cycles (Figure 32.2).

In another example of going beyond the inspection model, John Williamson, the dean of American health care quality researchers, drew heavily upon an adult learning model for his work, and proposed eloquently that quality assurance should involve a "cybernetic" (i.e., feedback) process in which data on performance were systematically used to identify opportunities for further improvement. Williamson's

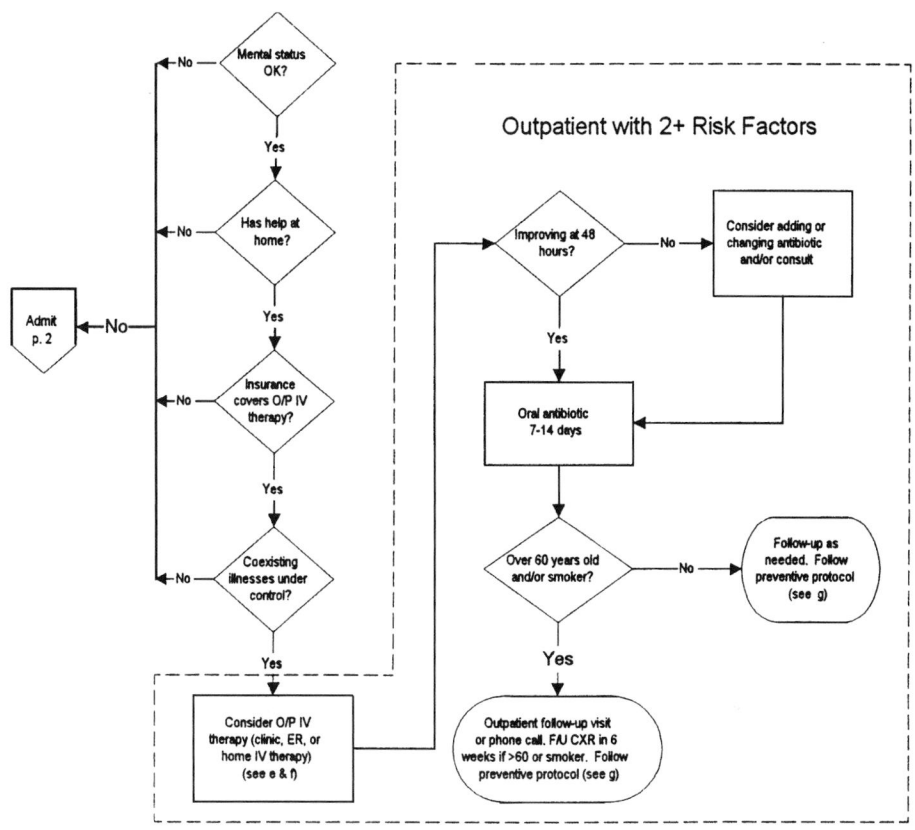

Risk Factors for the Immunocompetent Patient

(a) <u>History</u>
   Coexisting Illnesses
   - COPD
   - CHF
   - Diabetes
   - Chronic renal failure
   - Chronic liver failure
   - Hospitalization prior 12 months
   - Postsplenectomy
   - ETOH or malnutrition
   - Tuberculosis
   - Bronchitis
   Suspicion of Aspiration
   Age >65

(b) <u>Physical</u>
   Vital signs
     Adult: Respiratory rate >30
     Blood pressure <90/50
     Temperature >101
     SaO_2 <66 or PaO_2 <50
   Acute altered mental status
   Extrapulmonary site of infection (e.g. septic
     arthritis, meningitis, etc.)
   Coexisting illness findings

(c) <u>Laboratory</u>
   Chest film complicated
   - multilobar
   - cavitation
   - effusion
   CBC with differential
   - WBC < 4,000 > 20,000
   - Hgb < 9 or Hci < 30
   BUN & creatnine
   - BUN >20 or Creatinine > 1.2
   Coexisting illness lab abnormalities

Therapies
(d) <u>Macrolides</u>
   Erythromycin 500 mg po qld x 10-14 days
   Blaxin (Clarithromycin) 250 mg po bid x
     10-14 days
   Zithromax (Azithromycin) 500 mg po 1st day,
     250 mg qd x 4 days
   If mycrolide intolerant, use Doxycycline 100
     mg po bid x 10-14 days

(e) <u>Cephalosporins</u>
   Oral Therapy: Cefin (Cafuroxime) 250-500
     mg bid po x 10-14 days

If allergic, use Blaxin (Clarithromycin) 250 mg
   PO bid x10-14 days or Cleocin
   (Clindamycin) 300 mg PO every 8 hours
IV Therapy:
   2nd Generation:
     - Kefurox (Cefuroxime) 1.5 grams IV
       every 8 hours
   3rd Generation:
     - Claforan (Cefotaxime) 2 grams IV
       every 8 hours
     - Rocephin (Cefriaxone) 1-2 grams IV
       every 24 hours
If penicillin allergic, give Cleocin (Clindamycin)
   900 mg IV every 8 hours + Nebcin
   (Tobramycin) IV every 8 hours (dose by
   patients)

Notes
(f)   Medicare and Medicaid coverage is complex;
      verify insurance coverage with home IV
therapy agency

(g)   Evaluate for pneumococcal and/or flu
      vaccination with follow-up visit if needed.

FIGURE 32.1   *(Continued)*

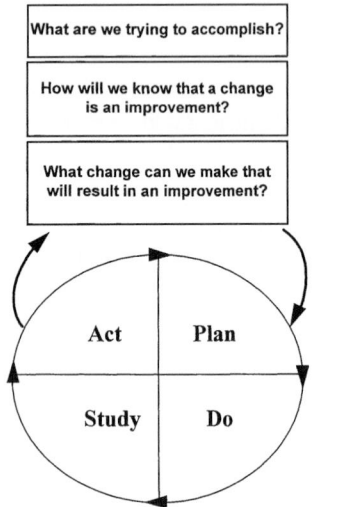

FIGURE 32.2   Model for improvement. (*Langley et al. 1996.*)

recommendations, as early as the 1960s, were clear harbingers of the modern approach to quality improvement and design (Williamson 1971). Nonetheless, through most of the twentieth century, health care quality activities remained those of inspection, with its primary aim of stabilization.

A new emphasis on the opportunity for improvement began to blossom in the mid-1980s. Due in part to the exploratory work of the National Demonstration Project on Quality Improvement in Health Care (Berwick et al. 1990), and in part to the entry of industrial quality professionals into positions of influence on hospital boards and in health care management, health care organizations became gradually more receptive to the ideas that stabilization was not enough, that important improvements in cost and quality could be achieved, and that new managerial methods—quality management—might help in health care, even though these methods had first appeared in other industries.

The early efforts of this sort, such as those documented in the monograph report on the National Demonstration Project, *Curing Health Care,* concentrated largely on classical business processes that appeared also in health care organizations—processes like scheduling, equipment maintenance, and transportation. In one highly successful project in 1987, for example, the University of Michigan Hospitals, working with the help of a quality professional from Corning, Inc., reduced waiting times in its ambulatory care clinic by 89 percent in a few months. Efforts in improvement at Intermountain Health Care's LDS Hospital in Salt Lake City reduced perioperative wound infection rates from 1.9 percent to 0.4 percent, compared with the national average of 4 percent, and a standardization project to reduce the number of different prostheses used in total hip replacement saved almost $1 million annually, while achieving better functional outcomes for patients (Pestotnik et al. 1996; James 1993; Morrissey 1996).

On the basis of these initial successes, quality improvement methods rapidly spread among dozens, then hundreds, of American hospitals and other health care organizations. Senior management groups organized Quality Councils; formal Quality Improvement teams became commonplace, and more fundamental redesigns of care began to make systems more patient-friendly. In general, the health care models for managing these improvements closely paralleled those in other industries. Perhaps for this reason, the models worked more smoothly in segments of health care that, from the start, looked more "corporate" in structure.

Less-well-developed managerial environments, such as medical staffs in hospitals, office-based medical practices, nursing homes, and interprofessional processes (such as those involving both doctors and nurses) proved less susceptible to repackaged industrial quality improvement methods. Not being employees, and tending to see themselves as customers of hospitals, rather than as partners or employees, doctors, for example, exhibited difficulty understanding and buying into coordinated, corporate objectives. They objected to attending improvement team meetings regularly as contributors, and redesigning their own work to fit better into the system as a whole. The various special languages and turf boundaries that have developed in health care (and that professional certification processes perversely reinforce in an effort to protect quality) have stood in the way of whole-hearted collaboration on systemic improvement. Hospital records still often maintain separate "nursing diagnoses" and "medical diagnoses." Professions sometimes do not even share common lounge areas, cafeterias, or meeting times, and it is still common in hospitals to find "nursing notes" and "doctors' notes" in separate sections of the same medical records.

Thus medical care, perhaps even more than other industries, finds itself susceptible to forms of fragmented efforts that impede systemic vision and optimization of the whole. Even more, old habits of work die hard in medicine. Physicians, nurses, and others are trained in highly conservative modes of work; they often regard changes as dangerous until proven to a standard far more stringent than in other industries. The learning cycles (plan-do-check-act) so characteristic of robust quality improvement can therefore feel especially threatening to health care professionals trained, first of all, "to do no harm."

Like other industries that came new to improvement methods, health care organizations often simply do not seem to believe that significant improvement is possible. Many tend to regard disease outcomes as biologically predetermined, patient expectations for comfort and service as "unrealistic," and excessive health care costs as inevitable. (All these constraints are, of course, quite real, unless the processes of work can be systematically changed and improved on the basis of data—unless, that is, quality is managed actively.)

Yet, despite the cultural barriers, the promise of quality improvement in health care remains great. A deeper look at two improvement projects shows how it can work under the best of circumstances.

Case I: The Northern New England Cardiovascular Disease Study Group.

In the mid-1980s, the five hospital centers and 17 cardiothoracic surgeons performing open heart operations in the three northern New England states (Maine, New Hampshire, and Vermont) began receiving some unwelcome news from the Health Care Financing Administration (HCFA), the federal agency that administers health care payment for Americans over 65 years of age who are covered by Medicare. HCFA's leaders had begun an annual program of feedback to hospitals of their mortality rates for Medicare patients in specific diagnostic categories. Initial findings for the northern New England centers showed wide variation in death rates associated with coronary artery bypass graft surgery—from 4 percent to 9 percent among the five institutions, and from 2 percent to 11 percent among the individual surgeons (O'Connor et al. 1996).

The surgeons doubted the adequacy of HCFA's "case-mix adjustment" procedures, which used variables like age, gender, and comorbidity in a statistical model to "level the playing field" for outcomes of care among the hospitals and surgeons. If Surgeon A's patients tended, on the whole, to be older than Surgeon B's, then it seemed only fair to adjust for the increased risk of surgery in older people before declaring any differences in the two surgeons' mortality rates to be associated with the "quality" of their care. When HCFA published mortality rates, it was natural for the surgeons to suspect that any differences among them were due to unmeasured, extraneous differences in the incoming mix of patients. In defense, the surgeons began a collaboration with a Dartmouth epidemiologist to develop a better case-mix adjustment model using a prospective, comprehensive database.

A strong case-mix adjustment model emerged, able to account for some of the observed variation in death rates, but the residual differences among centers and surgeons remained large even after the new adjustment. By 1989, the cardiovascular surgeons of northern New England faced a problem: *By their own measures* the mortality rate in coronary surgery still varied by over 300 percent among hospitals and more than 500 percent among surgeons.

With remarkable courage and honesty, the surgeons decided to continue their collaboration, but not now for the purpose of playing defense against HCFA's data release. They decided to use their data to support improvement efforts, and they began with comparisons of process. Interdisciplinary teams of surgeons, nurses, open-heart pump technicians, and others made site visits among the institutions (even though those institutions competed with each other for patient referrals) with the aim of studying variations in technical approaches to surgery and surrounding support systems—patient selection, preparation, surgical technique, postoperative care, rehabilitation, and so on. The variation in processes that they observed astounded them, and reinforced their own intent to discover and document "best practices" in care. They found important innovations within their own group in approaches to control of bleeding, reduction of time on the heart bypass machine, use of medications, patient education, and much more.

Within a year, the payoff for this collaborative improvement effort began, and, thanks to their careful system of prospective data collection, the surgeons soon documented the improvements on their own charts. Between 1989, when the first improvement efforts began, and 1991, the Northern

New England Cardiovascular Disease Study Group documented a 40 percent decrease in operative mortality rates for coronary bypass patients in the region, adjusted for case-mix by their own statistical model. What is more, the decrease occurred not only among the hospitals and surgeons originally at the high end of the variation first reported by HCFA, but among all of the surgeons and centers combined. Even the best got better.

Although they eschewed the jargon and formalisms of classical industrial quality improvement, the Northern New England Group used many of the basic, driving principles: benchmarking, measuring process and outcome variables over time, breaking down barriers among centers and departments, setting ambitious improvement aims, maintaining control charts to classify variation correctly, innovating in small plan-do-check-act cycles, and maintaining a system for reflection and learning from their own experiences. (The group met regularly as a whole at least four times a year throughout the project period—and, as of 1996, they are still meeting regularly.)

Case II: Collaborating to Reduce Delays.   Late in 1995, 28 organizations from across the United States and Canada joined in a collaborative improvement effort to reduce waiting times and delays. Frustrated by delays in operating room starts and turnaround times, emergency room waiting, and delayed access to office appointments, these organizations, ranging from community hospitals to major integrated delivery systems, began to test a focused improvement model (Figure 32.3) based on a common aim, to reduce the delay by 50 percent over the course of 1 year.

Teams from each of the organizations met in at a learning session organized by the Institute for Healthcare Improvement. With a specific organizational goal for their own organization in mind, the teams learned from experts in the field about lean process methods and principles for improvement in handoffs and queueing. Each team worked to learn and to apply the concepts to tests of change back in their organization's area of study. Sewickly Valley Hospital tested the following changes in improving operating room flow:

- Scheduling unpredictable cases at the end of the day or in a separate room
- Working to optimize surgery team utilization rather than operating room utilization
- Doing tasks in parallel and converting internal tasks to external to reduce turnover time between cases
- Staggering start times for the first cases of the day (Nolan et al. 1997)

Team members learned from the rapid cycles they tested at their own sites and from the other participant teams as well (Figure 32.4). Two subsequent learning sessions reinforced skills and encouraged many cycles of testing changes that accumulated to substantial progress for a number of the organizations (Nolan et al. 1997).

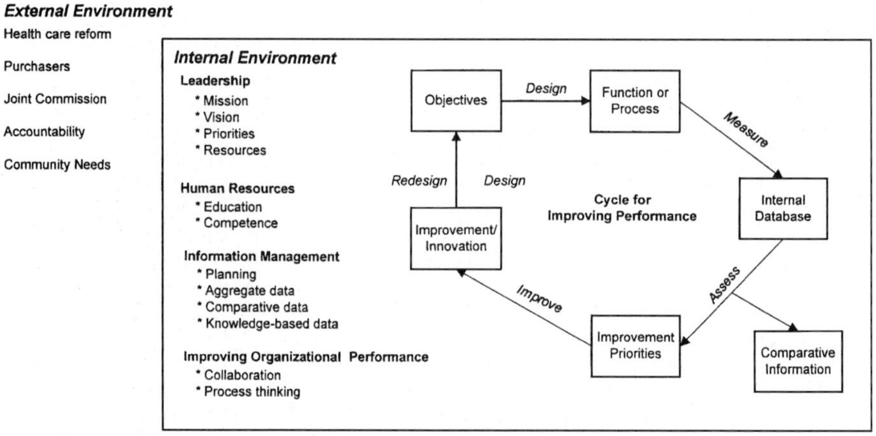

FIGURE 32.3   Framework for improving performance. (*Joint Commission Accreditation of Health Care Organizations.*)

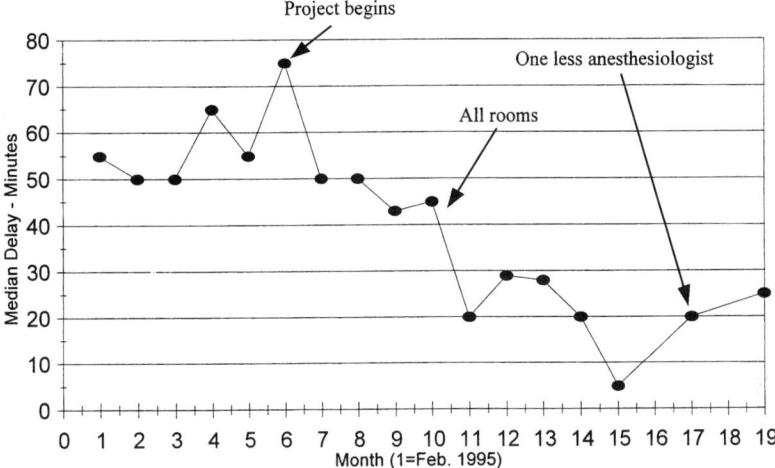

FIGURE 32.4 Sewickley Hospital: median delay 11 a.m. and 2 p.m. (*Nolan and Schall 1996.*)

Delays have long been a chronic symptom within health care systems, and these organizations have demonstrated results now that are being shared internally to widen the reach of the projects in each. The results provided some energy to an industry struggling with substantial cost reductions; these teams proved that customer satisfaction improvement and waste reduction can happen at the same time. These cases and hundreds of other early successful (and unsuccessful) experiences in applications of quality improvement methods in health care settings underline lessons such as the following.

1. *In health care, as in all other known examples of successful improvement, the role of leaders is crucial:* Improvement must be a led and managed undertaking; it does not happen by accident. This leadership challenge is especially acute in health care organizations, which often have traditions of divided leadership (among nurses, doctors, and administrators, for example) or dysfunctional barriers among staff areas.

2. *Breaking down barriers among functional areas is necessary for effective system changes:* Medical care has a strong tradition of suboptimization of functions, especially where professions (medicine, nursing, pharmacy, physical therapy, etc.) have taken exclusive control over activities within their own boundaries. The advantages of this professional autonomy have been great, as each profession has refined its technical skills and sense of discipline. But the price has been high, as well, as each profession's special vocabulary, technical priorities, physical space, and prerogatives have combined to decrease anyone's ability to view the system of care as an integrated whole. Members of successful health care teams have crossed disciplinary boundaries, and cut new windows on shared processes. Dr. Robert Master, a Boston-based physician, has redesigned care for a challenging population of patients by eliminating these disciplinary boundaries. The patients, severely challenged with chronic diseases such as end-stage neuromuscular disorders, cerebral palsy, and quadriplegia, historically received costly care in hospitals. Their care was directed by specialists, and many lived in medical centers. The pattern of their care was not directed toward independence. Dr. Master changed all that, working with a theory that these patients should be at home, not in hospitals. He created a team of providers, with nurse practitioners taking the lead role in coordination and management. The effect of the role reversal has been dramatic improvement in both cost and quality. Today 87 percent of all medical contacts for these patients occur in the home. Chair cars, ambulances, clinic fees, and specialist charges are almost things of the past. The use of specialists has dropped to $26 per member per month (pmpm), which is about the same as a normal HMO spends per person for speciality care for a normal, healthy enrolled population ($22 pmpm). Whereas Medicaid claims data show that normally 55 percent to 70 percent of all dollars for patients like this go to the hospital sector, in Dr. Master's program, this has fallen to 30 percent (Master 1997). His program spends $140 pmpm for nurse practitioner services, $60 pmpm for primary care physicians, and only $26 pmpm for specialty care. Quality reviews show better outcomes and higher satisfaction than ever before, and there has been literally no voluntary disenrollment. (Berwick 1996).

3. *Sound data, soundly analyzed, are as important in improving health care as in any other industry:*   With voluminous medical records, constant physiological monitoring, response to extensive regulatory reporting requirements, and increasing levels of computerization, health care is awash in data. But improvement teams often find that the data lie unused, unrefined, and displayed, if at all, in uninformative lists and charts, rather than in graphs, run charts, and control charts that "tell a story" on the basis of which theories for improvement can be developed and tested. Culturally, doctors seem especially receptive to data-based approaches to improvement. Physicians left cold by philosophical explanations of the principles of quality improvement can come alive in a simple presentation of evidence of variation or of tests of a hypothesis about the cause of a recurrent defect. More than a few health care improvement efforts have accelerated once the potential benefits of measurement and display of information were demonstrated.

4. *"Customer focus" is as meaningful in health care as in any other industry:*   When the concept of "customer" was first introduced by quality consultants into health care organizations, an allergic reaction occurred. "I don't have 'customers,' " was an angry, recurrent refrain from doctors and nurses, "I have 'patients.' " The reaction seemed to come from a misunderstanding that "focusing on customers" somehow demeaned the importance of the doctor's technical expertise or commitment to professionalism.

But the idea of customer focus was, in fact, entirely convergent with the trends of quality research in health care throughout the last quarter of the twentieth century. The investigators in the Rand health insurance experiments and others in the 1970s and 1980s developed an important focus on "patient-based measurements" of quality of care, including the insight that patient self-reports of health status and function were among the most reliable and valid forms of assessment of the outcomes of care. (Until then, most "outcome" measures were of variables observed by the doctor—like blood pressure, pulse, and fever—rather than variables reported directly by patients.)

The concept of "patient-centered care" (health care that continually views itself from the patient's perspective and adjusts processes to meet needs) also made inroads into health care, most impressively in places like the experimental "Planetree Units" established in a few American hospitals to test entirely new forms of interaction with patients. In Planetree settings, patients wear street clothes, instead of hospital smocks, and are invited to read and to write in their own medical records.

Becoming more receptive to the concept, health care leaders now understand that they serve multiple "customers," including patients, families, payers, communities, and referring institutions, and that each has legitimate and important needs. Each also defines "quality" somewhat differently. For patients and families, "quality" includes dimensions of health status outcome, accessibility, communication, comfort, dignity, cleanliness, convenience, problem resolution, and respect for the individual's time, among others. Payers value all of these (as representatives of the insured population), but also insist strongly upon cost control based on rational parsimony and waste reduction, and upon the timely provision of accurate accounting information and benefits management. Increasingly, communities as a whole demand from health care new forms of participation in collaborative health improvement, respect for the environment, and coordinated services, especially for the elderly and people with multiple problems.

As the flow of patients among institutions becomes more complex, demands increase for smooth coordination of referral patterns, "seamlessness" as patients move among sources of primary care (where they may first receive a diagnosis of, say, breast cancer), secondary care (where initial tests and treatment may begin), tertiary care (where high-technology procedures should be concentrated), and community services (for rehabilitation, employment support, social service, home care, and so forth). The watchword for redesign and improvement of health care in the 1990s has been "integration," as new amalgams of providers—hospitals merged with each other, physician-hospital organizations, vertically integrated systems with hospitals, offices, laboratories, and nursing homes—seek to create this seamless system of former fragments.

5. *Cost containment through improvement is attainable in health care:*   But the consciousness about the favorable relationship between cost and quality is only dawning. For many decades, a "more-is-better" attitude dominated thinking about health care quality. America overbuilt health care, providing technology, hospital beds, specialty services, and laboratory capacity that, as mentioned above, not only vastly exceeded the supply of these services in other developed countries, but also acted as drivers of demand and cost. In a fee-for-service financing environment, it was easy for health care organizations to maintain wasteful levels of inventory, inefficient programs for use of capital, high levels of scrap, and, in comparison to analogous processes in industry, low productivity.

Increasingly familiar with process improvement methods, many health care organizations defined cost reduction as Improvement Priority 1. No one knows yet what levels of savings can be achieved while improving quality through a thorough war on waste in health care systems, but the evidence abounds of inappropriate practice, redundant and complex processes, excessive inventory, and wasteful waiting. Table 32.2 lists some examples of cost savings achieved by health care improvement teams through relatively simple process changes.

Deaconess Hospital in Boston used an improvement method from General Motors to reduce excess inventory from operating room shelves and saved $200,000. Standardizing operative stapling devices and prostheses helped Mayo Clinic to hold costs down. Changing the way that nurses, therapists and nutritionists team up to provide care has dramatically reduced complications such as infections and skin breakdown at several hospitals and resulted in cost savings with better patient outcomes.

6. *Benchmarking is useful:*   Despite their professional collegiality, health care organizations and professionals traditionally work in remarkable isolation from each other. Few doctors routinely visit with colleagues to discover differences and innovations, and few methods exist by which hospitals and health maintenance organizations can discover best practices and copy each other. That is changing. Health care organizations are consolidating into integrated systems whose components can have easier access to each other, and students of quality management in health care have discovered the power of benchmarking through visits, trade associations, and collaboratives. Sometimes, the first "wake up" call has come from the payer community, as in Cincinnati, San Francisco, and Chicago, for example, where payer collaboratives have collected and published performance data on managed care plans. In one case, payer data first showed the care planorganiza-

TABLE 32.2   Quality-Improvement Steps at Intermountain Health Care

Samples of projects large and small that reduced operating costs in 1994

Project	Expense reduction
System project:	
Standardizing hematology procedures	$259,000 in supplies,
Standardizing chemistry reagents	$58,000 in equipments
Decrease in costs of coronary artery	$209,000 in supplies
bypass surgery	$245,000 (since 1993)
Facility-specific project:	
Emergency room lab turnaround time	$18,000 in salary and supplies
Fast-track extubation	$575,500 in shorter length of stay
Decreased retake rate for x-rays	$11,200 in film costs
Long-term clinical care improvement efforts	

- 50% decrease in adverse drug events since 1991, with much of that coming in the first year. A computerized alert system was a key factor in avoiding prescriptions that could pose a risk to patient.
- 80% decrease in postoperative wound infection rates, reducing an incidence of nearly 2% in 1990 to a fraction of a percent in 1994.

Source: Intermountain Health Care application for the National Quality Health Care Award.

tions that their rates of hysterectomy varied eightfold from the highest plan to the lowest—a fact they did not know until the payers documented it.

Once alerted to the variation, curious health care organizations can make good use of benchmarking to discover new ways to approach their own work. Sometimes the best benchmarking is outside health care entirely. The Mayo Clinic leadership team has made fruitful visits to winners of the Malcolm Baldrige National Quality Award—companies like AT&T, IBM, Eastman Chemical, and Milliken—and has directly grafted process ideas from those pioneers into the work of the Clinic.

7. *Involvement of all staff in improvement is powerful:*   The dedication to excellence of health care workers at all levels is no surprise, but until recently most organizations have had difficulty tapping that energy to support systemic improvements. As a result, improvement in health care has tended to move slowly, lacking the momentum of total involvement. But counter examples are emerging. At Wesley Medical Center in Wichita, Kansas, a single project on reducing waiting times in the operating room, which began as part of Wesley's participation in a national collaborative benchmarking effort, was replicated internally within a few months as the hospital's leaders built upon what they had learned to support delay-reduction projects in 85 departments.

8. *Financing systems can be a major barrier to improvement:*   The suboptimization of professions and functions in health care has its mirror image in traditional payment systems. As one example, consider the perverse incentives in the compensation of hospitals in a capitated system (in which a strong incentive exists to shorten a patient's length of stay to save money) where doctors are still paid separately for each inpatient visit (so that shortening length of stay reduces physician incomes).

With the hospital aiming to reduce resource consumption, and the physicians (who control the ordering of all tests and the length of a patient's stay) reimbursed for each day of stay, and in some cases for the analysis of each test ordered, conflicts arise quickly in what improvements should be made. In another recent case, a hospital successfully reduced the frequency of unnecessary admission of patients with chest pain by better identifying in the emergency room those who did (or did not) have a heart attack. Their success led to a safe reduction in the total cost of chest pain treatment because people without heart attacks were more likely to avoid admission, but naturally caused an increase in the cost-per-admitted-case for that smaller number of patients who actually entered the hospital. (After all, a larger proportion of admitted patients were now turning out actually to have had a heart attack, and required the full resources of the hospital for diagnosis and treatment.) An unwise, but very important, health care insurance company, noticing that the *cost per case* was rising, and failing to notice the savings on a population base, caused a major crisis in the hospital by threatening to cut off reimbursement because of "excessive cost."

A third, and even more compelling, example of perverse financing affected the Magic Valley Medical Center in Twin Falls, ID (Roessner 1993). The leadership of the hospital, having made a commitment to help Magic Valley become "the healthiest community in America," made a major investment in leading the reduction of bicycle head injuries in the community by supporting the use of bicycle helmets. Helmet use did increase, head injuries in children fell by almost 50 percent in three months, and the hospital harvested both the satisfaction of saving lives and the severe economic problem of a shortfall in Emergency Department revenues as those well-paying cases of head injury decreased dramatically. Only a firm sense of mission allowed them to stay the course (Roessner 1993).

Clever managers have been able to maintain momentum for improvement despite these absurd payment paradoxes, but, for the longer term, payers and providers of care are coming to realize that the basic structure of finance in American health care is less favorable to rapid, strategic improvement than it may be in other industries. Innovations that consolidate payment and make rational systemic improvement more attractive are now beginning. The most significant of these changes is the expansion of *capitated* payment, in which provider systems are paid for the total costs of care and illness for populations, and in which, therefore, rational savings at one point in the system can be reallocated elsewhere, and in which the incentives of professionals, institutions, and consumers of care are better aligned.

THE FUTURE

Major trends toward better quality of care are now well underway in health care. Integrated delivery systems, more sophisticated information systems, and capitated payment are all helping to focus energies on improvement—especially on waste reduction. Better informed about the performance of health care organizations, consumers and payers are beginning to make their choices about where to receive care based on technical and service quality characteristics, as well as on price. The new century will witness the maturing of a quality-driven health care system, probably with at least the following characteristics:

1. *Much improved service quality:*   The time is ending when health care customers will tolerate—or must endure—waiting times, communication lapses, and process failures that have long ago become intolerable in other industries. Rental agencies, hotels, airlines, or retail stores that had the same waiting times as the average hospital or doctor's office today would be out of business quickly. Quality improvement in health care will drive new standards of service in health care.

2. *Decreasing total costs:*   With a refined understanding of the nature of waste, it is probable that leading health care organizations will achieve completely unprecedented production efficiencies within the next decade. We have at least the international comparisons to sustain confidence that excellent care is achievable at far lower cost than prevails in America today. The early success of individual project teams, able to return the same 6-to-1 or 10-to-1 return on investment while improving the experience of customers, needs only to be rolled up into strategic, deployed, systemwide improvement efforts to yield enormous savings for the system as a whole. Health care leaders are smart enough to build on this success over time.

3. *Health care will downsize, especially in high-technology services:*   Because of health care's unusual characteristic of "supply-driven costs," effective cost containment will require a smaller, tighter industry than we have as the 1990s close. We can expect fewer specialists, hospital beds, and sites of high-technology service per capita within the next decade. This should not spell a decrease in quality or service. On the contrary, research suggests that a "volume-outcome" relationship exists for many forms of technical care. Hospitals performing a small number of sophisticated procedures each year tend to have worse outcomes than those doing more. On two counts then—cost and outcome—the American public will be well-served by a consolidation of sources of advanced technical care into fewer, larger specialized centers.

4. *Prevention will take on new energy:*   The "upstream" causes of disease are in many cases controllable, but not within the classical boundaries of health care delivery. A rational public will maintain and increase investments in injury prevention, smoking cessation, reduction of alcohol and drug abuse, increasing physical exercise, and wise diet. At this moment, health care financing favors none of these, but the quality of outcomes depends on shifting the pattern of investment.

5. *Participative decision making will become more widespread:*   Important research now shows that, when patients become involved in making decisions about their own care (for example in choosing between medical and surgical treatment for prostate disease, in selecting from among therapeutic options in breast cancer, and in self-monitoring in diabetes and asthma), satisfaction, outcomes, and costs all tend to improve. The trend is so strong that it will make good business sense for health care organizations to build upon this notion of partnership as a core concept in the design of care systems of the future.

6. *Information systems and remote communication will advance:*   The greatest technical advances of all in the next phase of development of health care may be more in the realm of information management than in diagnosis and treatment. By the early twenty-first century, we can expect common use of automated patient records, computerized order entry, regular monitoring of both processes and outcomes of care, and remote forms of "telemedicine." It is already possible for the world's best interpreters of MRI scans or CT images to receive on-line images good enough to interpret locally from almost anywhere in the world. The well-trained nurse who answers a phone call from an anxious mother in the middle of the night can be next door or a thousand miles away and be equally effective in coordinating care. We are only a few steps away from remotely guided procedures, in which technical experts can not only talk, but also act, from a remote site. The potential for improvements in cost, outcomes, and service is extraordinary.

7. *Job boundaries will change:*   The traditional professional classifications of medical care—doctors, nurses, respiratory therapists, laboratory technicians, and dozens of others—reflect the historical configuration of care systems. As the configuration changes, so will the jobs. This type of change is progressing much more slowly than logic or need would dictate, due largely to the well-established forms of professional self-regulation. But the boundaries are fraying, and fundamental change will sooner or later arrive. Many tasks and procedures currently done by doctors could easily be reallocated. Repetitive technical tasks may be done more competently and less expensively by technicians; just as intravenous lines formerly inserted only by doctors are put in today far more comfortably by IV teams, so may the colonoscopies, hernia repairs, fracture setting, and angiographies of the future be done well by people with highly focused training. Managed care systems and well-run clinics are already assigning to physicians' assistants and nurse practitioners tasks previously only given to doctors. Expert computer systems, which are already helping doctors to choose antibiotics or to make difficult diagnoses, will become better complements to traditional approaches to diagnosis. [The visionary Dr. Larry Weed suggests that expecting doctors to remember all pertinent diagnostic options is like expecting travel agents to remember airline schedules; to him, neither makes sense in the computer age (Weed 1968).]

8. *Organizational boundaries will change:*   As professions evolve, so will organizations. In fact, the heart and core of the American health care system of the twentieth century—the hospital—may become a dinosaur in the twenty-first. Gradually, it is becoming the case that patients not sick enough to be in intensive care units may not be sick enough to require a hospital bed at all. Better alternatives will exist. In the future, the whole hospital may be an intensive care unit; other patients will be at home or in a new kind of step-down setting. Sick asthmatic patients already skillfully use in their own homes devices and drugs that 10 years earlier would have been found only in emergency rooms.

9. *"More is better" will give way to "First, prove it works."*   A fundamental shift in the burden of proof for health care practices will take firm hold in the coming years. Scientific evidence and public consciousness are converging into a realization that health care does not generally provide what people want most: health; and that excessive care produces excessive risks. Combined with the problem of high cost, this consciousness should make providers and patients both more wary of technology than they have been in the past and more inclined to question the need for a test or treatment than to request it. One example of this new, more prudent attitude is in the work of the U.S. Preventive Services Task Force, which in 1989 and then again in 1995 produced a *Guide to Clinical Preventive Services,* reviewing the scientific evidence for and against over 200 preventive services—like screening tests, counseling on risk factors, and immunizations (U.S. Preventive Services Task Force 1996). The Task Force report documents strong evidence in favor of some preventive practices, but, even more impressively, shows that little or no evidence exists to support many others, some of which are in common use. Table 32.3 lists some of the recommendations in the *1995 Guide.* The general trend is toward what some call "evidence-based practice" in medical care; and it will mark a new, more scientific era in clinical work.

10. *Breakthrough performance will emerge:*   In the end, the momentum behind quality improvement in other industries has not come from theory, it has come from evidence, especially evidence owned by the competition. Great cars from competitors, not great ideas about cars, caused the American automotive industry to change. So it will be in health care. It is only a matter of when. The crucial turning point will have come when there exists for health care what Toyota was for the automobile—a breakthrough example, operating at an unprecedented level of performance and built for less than anyone had theretofore imagined possible. We do not yet have such a model in medicine. We have breakthroughs in process, superb cost reductions, and exciting new designs. But all of these achievements remain at the level of individual process, product, or service. We have improved parts, in some cases dramatically, but no one has yet fundamentally improved the whole—an entire system of care.

It is, at last, just within the reach of health care leaders to do so early in the twenty-first century. The rewards will be thrilling, and, as a consequence of that achievement, both health care and health itself in the early twenty-first century will be forever transformed.

TABLE 32.3   Guide to Clinical Preventive Services

Screening for hypertension		
Intervention	Level of evidence	Strength of recommendation
Periodic blood pressure measurement in persons aged ≥21 years	I	A
Measurement of blood pressure in children and adolescents during office visits	II-2, II-3, III	B
Screening for breast cancer		
Routine mammogram every 1–2 years with or without annual clinical breast exam		
Women aged 40–49	I	C
50–69	I, II-2	A
70–74	I, II-3	C
≥75	III	C
Annual clinical breast exam without periodic mammograms		
Women aged 40–49	III	C
50–59	I	C
≥60	III	C
Routine breast self-exam	I, II-2, III	C
Screening for cervical cancer		
Regular Pap testing in women who are or have been sexually active and who have a cervix	II-2, II3	A
Discontinuation of regular Pap testing in women aged >65	III	C
Routine cervicography or colposcopy	III	C
Routine testing for HPV infection	III	C
Screening for prostate cancer		
Routine digital rectal exam	II-2	D
Routine prostate-specific antigen or other serum tumor markers	I, II-2, III	D
Routine transrectal ultrasound	II-2, III	D
Screening for lung cancer		
Routine chest x-ray or sputum cytology	I, II-1, II-2	D

Strength of recommendations:

A. There is good evidence to support the recommendation that the condition be specifically considered in a periodic health examination.

B. There is fair evidence to support the recommendation that the condition be specifically considered in a periodic health examination.

C. There is insufficient evidence to recommend for or against the inclusion of the condition in a periodic health examination, but recommendations may be made on other grounds.

D. There is fair evidence to support the recommendation that the condition be excluded from consideration in a periodic health examination.

Level of evidence:

I. Evidence obtained from at least one properly randomized controlled trial.

II-1. Evidence obtained from well-designed controlled trials without randomization.

II-2. Evidence obtained from well-designed cohort or case-control analytic studies, preferably from more than one center or research group.

II-3. Evidence obtained from multiple time series with or without the intervention. Dramatic results in uncontrolled experiments (such as the results of the introduction of penicillin treatment in the 1940s) could also be regarded as this type of evidence.

III. Opinions of respected authorities, based on clinical experience; descriptive studies and case reports; or reports of expert committees.

REFERENCES

Berwick, D. M., Godfrey, A. B., Roessner, J. (1990). *Curing Health Care: New Strategies for Quality Improvement,* Jossey-Bass, San Francisco.

Berwick, D. M. (1996). "Changing the Boundaries of Health Care." *Quality Connection,* Summer; vol. 5, no. 3, pp. 1–3.

Brooke, R. H., Ware, J. E., Davies-Avery, A., Stewart, A. L., Donald, C. A., Rogers, W. H., Williams, K. N., and Johnston, S. A. (1979). "Overview of Adult Health Status Measures Fielded in Rand's Health Insurance Study." *Medical Care,* vol. 17, no. 97 (suppl.), pp. 1–131.

Donabedian, A. (1966). "Evaluating the Quality of Medical Care." *Milbank Memorial Fund Quarterly,* vol. 44, no. 3 (suppl.), July, pp. 166–206.

James, B. C. (1993). "Implementing Practice Guidelines Through Clinical Quality Improvement." *Frontiers of Health Services Management,* vol. 10, pp. 3–37.

Langley, G. J., Nolan, K. M., Nolan, T. W., Norman, C. L., Provost, L. P. (1996). *The Improvement Guide: A Practical Approach to Enhancing Organizational Performance.* Jossey-Bass, Inc., San Francisco.

Lohr, K. N., and Brooke, R. H. (1984). "Quality Assurance in Medicine." *American Behavioral Scientist,* vol. 27, pp. 583–607.

Master, Robert (1997). Personal communication with the author.

Morrissey, J. (1996). "IHC Sets Pace On Quality Improvement." *Modern Health,* January 29, p. 42.

Nolan, T., and Schall, M. W. (1996). *Waits and Delays Throughout The Healthcare System.* Institute for Healthcare Improvement, Boston, MA.

Nolan, T. W., Schall, M. W., Roessner, J. (1997). *Reducing Delays and Waiting Times Throughout the Healthcare System,* Institute for Healthcare Improvement, Boston.

O'Connor, G. T., Plume, S. K., Olmstead, E. M. (1996). "A Regional Intervention to Improve Hospital Mortality Associated With Coronary Artery Bypass Graft Surgery." *Journal of American Medical Association,* vol 272, pp. 841–846.

Pestotnik, S. L., Classen, D. C., Evans, R. S., Burke, J. P. (1996). "Implementing Antibiotic Practice Guidelines Through Computer-Assisted Decision Support: Clinical and Financial Outcomes." *Annals of Internal Medicine,* vol. 124, pp. 884–890.

Roessner, J. (1993). "The Healthiest Place in America." *Quality Connections,* vol. 2, pp. 10–11.

U.S. Preventive Services Task Force (1996). *Guide to Clinical Preventive Services,* 2nd ed. Williams & Wilkins, Baltimore.

Weed, L. L. (1968). "Medical Records That Guide and Teach," *New England Journal of Medicine,* vol. 278, pp. 593–600, 652–657.

Wennberg, J. E., Gittelsohn, A. (1973). "Small Area Variations in Health Care Delivery." *Science,* vol. 1823, pp. 1102–1108.

Wennberg, J. E., Keller, R. (1994). "Regional Professional Foundations." *Health Affairs,* vol. 13, pp. 257–263.

Williamson, J. W. (1971). "Evaluating Quality of Patient Care: A Strategy Relating Outcome and Process Assessment." Journal of American Medical Association, vol. 218, pp. 564–569.

SECTION 33
FINANCIAL SERVICE INDUSTRIES

Charles A. Aubrey II
Robert E. Hoogstoel

INTRODUCTION

What is financial service? Financial services are services related to cash and other financial assets. There are seven types of service that financial services institutions may perform. The first four are distinctly *financial* services:

1. Safekeeping of customers' currency or other assets and fiduciary responsibility for them.
2. Conversion, transfer, and transformation of currency and other customer assets.
3. Adding value to customers' assets by means which may include:

- *Return on deposits:*   Providing return on capital or assets which the customer places, deposits, trusts, or invests with the financial institution.
- *Investments and loans:*   Providing the customer with capital or other assets to generate value.
- *Insurance:*   Providing protection against possible financial loss in the event of certain predefined risks.

4. Underwriting or guaranteeing customer projects through some combination of advice, marketing of customer instruments (stock and bonds, for example), and guarantee of minimum financial support (providing line of credit, buying an amount of stock, or insuring against a loss).

The last three services are common to many service institutions and are not unique to financial services.

5. Interacting with customers in a variety of ways, including sales presentations, responding to customer service inquiries, and providing advice.
6. Accounting and reporting to detail and summarize the transactions related to the customers' relationship, including billing and tax information.
7. Providing a profit to the owners of the company—those who provide the operating capital (common to all kinds of companies).

THE FINANCIAL SERVICE SECTOR

The financial service sector comprises companies engaged in the creation and delivery of financial services. Traditionally, these companies are organized by industry, where each industry is concentrated on a common line of business, such as commercial lending, credit card services, or casualty insurance. Commercial banks, savings and loan institutions, brokerage firms, investment companies, investment banking institutions, insurance companies, credit card companies, and credit companies (including pawn shops) are all part of the financial service sector.

This tidy description of the organization of the financial service sector provides a convenient point of departure. It should be kept in mind, as discussed below (Sector Organization; Boundaries; History), that the sector is undergoing rapid change and this description does not always square with the growing complexity of the sector as it is emerging.

Sector Organization; Boundaries; History.   A special word should be said about the boundaries which divide the various industries within the financial services sector, their history, their present status, and the implications these boundaries have for the activities within each of the industries. Most commercial and retail customers are in the habit of labeling financial institutions on the basis of a memory of financial institutions of the recent past. Labels like "bank," "savings and loan" and "credit union" are generic terms which were once precisely defined in the United States by the regulations which governed the financial services sector. U.S. consumers associated these generic institutions with their products: loans, deposits, and transactions related to them. Similarly, insurance companies were associated with the insurance products they created and sold, and brokerage and investment firms were associated with packaging, promoting, and selling investment in stocks, bonds, and other assets, such as real estate.

There is good historical reason for the structure of the financial services industries. In 1933, at the depth of the Great Depression, which was precipitated by the stock market crash of 1929 and subsequent bank failures, the U.S. Congress reacted to the economic devastation by passing the Glass-Steagall Act (Werner and Stoner 1995). Because Congress saw the root cause of the Great Depression to be the involvement of banks in the promotion of securities, Glass-Steagall limited this activity. This act and other regulatory measures drew boundaries around various segments of the financial services sector and prohibited the companies within each of these industries from participating in business outside the limits of their respective industry territories. Thus, banks were prohibited from dealing in securities, brokerage firms were prohibited from performing certain banking functions, and so on. The resulting model of distinct financial businesses was, to a great degree, adopted and enforced internationally by a great majority of countries.

Today, various legislation in the United States has modified portions of the Glass-Steagall Act. The pattern began to change in 1980, with the passage of the Depository Institutions Deregulation and Monetary Control Act, which took the first steps to reduce differences between commercial banks and thrift institutions. By 1994, with the passage of the Riegle-Neal Interstate Banking and Branching Efficiency Act, which removed restrictions on interstate bank branch operations, the system of regulations which had maintained the pattern of financial services for over half a century was virtually gone. The effort to redefine and restructure the industry which began in 1980 is clearly evident in the mergers, acquisitions, and restructuring throughout the financial services industry in the late 1980s and 1990s.

As a result of this deregulation, the lines of demarcation are becoming blurred. For example, today banks can earn up to 10 percent of their revenue through underwriting activity. They can sell securities and the insurance of other companies (though they cannot underwrite them). They are forbidden to make the related investment decisions or to hold the resulting portfolio.

Another example of the blurring lines is seen when a bank sells mutual funds or allows sales agents of an insurance company to operate in a bank branch. Such cooperative arrangements enable the bank to offer a more complete line of financial services in order to keep and attract customers. Such intra-sectoral cooperation has extended competition beyond the competition that existed between organizations within a sector; the competition now takes place between sectors. Table 33.1 summarizes the activity of the financial services companies, by industry, represented in the 1996 Fortune 500.

Customers now face a wide choice of providers for services that once resided securely in companies of a single industry. For example, checking accounts, once the exclusive domain of the commercial bank, can now also be obtained from a savings and loan, a savings bank, even from a stock brokerage or mutual fund provider.

DEFINITION OF QUALITY IN FINANCIAL SERVICES

Juran (Section 2, How to Think About Quality) proposes a definition of quality which we will use here. Quality is the combination of: (1) features which attract customers and satisfy their needs and (2) freedom from deficiencies, which avoids dissatisfying customers.

TABLE 33.1 Profile of Financial Services Sector

Industry (no. of organizations)	Revenue*	Profit*	Profit as a percent of:			Employment	Revenue per employee
			Revenue	Assets	Equity		
Commercial banks (55)	$271,691	$32,728	13	1	15	1,041,623	$260,834
Insurance: life and health (mutual) (21)	$180,286	$4,438	3	1	11	255,396	$705,908
Diversified financial (15)	$105,264	$9,660	10	1	18	232,804	$452,157
Insurance: property and casualty (stock) (27)	$100,947	$13,953	8	2	13	296,814	$340,102
Insurance: life and health (stock) (20)	$89,053	$4,426	7	1	11	209,567	$424,938
Insurance: property and casualty (mutual) (9)	$60,337	$2,657	8	3	11	116,187	$519,309
Brokerage (7)	$53,957	$2,452	5	0	10	105,561	$511,145
Savings institutions (8)	$16,373	$1,226	8	1	10	41,983	$389,991

*000,000 omitted.
Source: *Fortune.* April 28, 1996.

Features.   Features are those attributes or characteristics that combine to constitute a product.

The thing sold to customers is some bundle of services, generally referred to as "product." Usually, the product includes some tangible elements (a report, a receipt, cash in hand) and much that is intangible (a feeling of safety and security; speed of information access by 24-hour telephone line; speed, accuracy, and pleasantness of response to an inquiry). In many cases the intangible aspects are more important to the customer than the tangible ones.

Salable services require features that customers want and need. Sales and revenue depend on an understanding of those wants and needs throughout the duration of the supplier's relationship with the customer—from initial inquiry, through sales presentation, purchase decision, and the customer's subsequent decisions, conscious or unconscious, to maintain the relationship or make further purchases.

The variety of features which financial products and services offer is growing each year. The growth is propelled by the ever-growing capabilities of available computer hardware and software. One result is increasing customization of financial products to the needs of individual customers and classes of customers.

In addition to features associated with services generally, there are three features that are especially important to financial services: fiduciary responsibility; return on an investor's investment (ROI) (from the borrower's viewpoint this is interest paid on a loan); and risk.

The suitable combination of fiduciary behavior, return on investment, and risk is an important feature that customers use to: (1) decide what company to do business with, (2) select which products and services to utilize for each company the customers choose, and (3) how much of the product or service the customers will utilize or purchase.

Features: Fiduciary Responsibility.   Fiduciary responsibility is the exercise of prudent judgment and acting in the best interests of the customer. It includes keeping the customer's assets safe, as well as those of the company, and earning the best return possible given the amount of risk the customer and company have agreed to accept. These responsibilities are clearly written in company policy and procedures. In the case of larger transactions, such as underwriting stocks or bonds for issue, commercial loans, trusts, arranging large customer investment portfolios, or managing large company insurance portfolios, the fiduciary responsibilities are also written in the contract.

Fiduciary responsibility is an important feature of any financial transaction. As important as its actual exercise is the customer's perception that it is being exercised. Once the customer loses faith in the fiduciary responsibility of an institution, that trust may be impossible to repair. When many customers lose faith simultaneously the result can be calamitous for the institution. In the early 1980s, Continental Illinois National Bank was forced to "write off" over $1 billion of loans. (Writing off is a recognition by the managers that there is no hope of recovering any of the principal or interest due on the loan.) The effect of the disclosure of this massive write-off was to stimulate a "run" on the bank, in which many depositors, fearing the loss of their assets, sought simultaneously to withdraw funds from the bank. They withdrew so much money the bank became illiquid. The Federal Reserve stepped in and loaned the bank what was necessary to meet depositor demands, but the damage was done. The bank's stock value dropped from over $20 per share to under $2, where it remained until dramatic steps were taken, including rechartering. Many customers never returned. Half of the 16,000 employees lost their jobs.

Features: Return; Interest paid.   Return on investment (ROI) and interest rate paid (I) on a loan are computations of the same quantity (the annual cost of money, in percent) from two different viewpoints—that of the investor and that of the borrower. Where the loan (or investment) period is 1 year, the computation is straightforward:

$$I = \text{ROI} = \frac{S - P}{P}$$

where P = the amount borrowed (or the amount invested) and S = the amount paid back (or the amount collected) at the end of one year.

Complexity increases as the number of interest periods increases, and when fees, compound interest, currency translations, taxes, and the like are introduced. As the computation becomes more complex, it also becomes more difficult for the customer to compare competing financial products.

The largest and most sophisticated customers have complex models to make the calculation and comparison. The small retail customers must do the best they can from what they read and are able to calculate. For example, it is very difficult to compare the interest cost of an automobile purchase loan with that of a lease, despite the fact that many laws exist in retail consumer financial services to simplify the products and allow comparison by the consumer. Because of these difficulties, some consumers depend on their financial institution and accept on faith its advice on investment return and interest rate.

High-trust fiduciary responsibility, linked with other features of very high importance, such as ease of doing business, computer access, and personalized service may reduce the importance of return on investment or interest rates paid. As long as these rates are in the "ballpark" compared to competitors, they may become less important features.

Features: Risk. Risk is a correlate of return. Greater risk generally accompanies greater return. All financial transactions are accompanied by risk—conditions which threaten the success of the transaction and realization of the return.

The risk associated with a transaction is generally expressed in terms of the uncertainty of success, in the language of probability and statistics. For a given transaction, the best data available are used to create a statistical model as a prediction of the outcomes of the transaction. Risk is expressed in terms of the dispersion or spread of those outcomes.

The quantification of risk is illustrated by the risk associated with investing in a hypothetical stock index fund. (An index fund is a mutual fund comprising a package of common stocks selected to emulate the price behavior of some stock index such as Standard & Poor's 500.) This hypothetical fund will emulate the Dow-Jones Industrial Average ("the Dow"), chosen because among stock indices representing the behavior of the U.S. stock market, the Dow has the longest history, having been established over 100 years ago. Though an imperfect representative of the broad market, the Dow will serve to illustrate the relationship of income and risk.

Although every investment primer and the prospectus of every mutual fund will tell the prospective investor that past performance is no guarantee of future performance, the fact remains that past performance is important and often is the only guide to future performance. Table 33.2 is a portion of the data table based on the annual closing value of the Dow for the years 1914 to 1990, inclusive. We assume that dividend income from our stock is small compared to the price appreciation over time. Likewise, we assume the costs of trades to keep our portfolio composition identical to the Dow are negligible. Thus, we define the income from the fund for a given year as that year's capital appreciation. From these values it is possible to compute the effective annual rate of return on investment for 76 periods of 1 year, 75 periods of 2 years, 74 periods of 3 years, and so on to the one period of the entire 76 years. For each of those groups of periods, we can compute the sample standard deviation s.

The assembled results of these computations provide a statistical description of the behavior of the Dow over the 76-year period. Those with confidence that the Dow will continue to behave like this over the long term can use this model to predict the statistical picture of future behavior. In any case, the numbers give a sense of the magnitude of risk faced by an investor in this hypothetical fund.

The distribution of annual interest rates for each of the 76 years, is assumed to be reasonably approximated by the standard normal distribution. On this basis, we use the sample standard deviation s as an estimate of the standard deviation σ (sigma), and estimate the probability of any given range of return for an investment in the "fund." For example, for any given year, absent any other knowledge of the market conditions, we can say that with about 95 percent probability the ROI will fall between $\pm 1.645s$ of the average annual ROI. The computation yields values of these limits of -39.56 and $+49.92$ percent.

Before despairing that the range is too large to be of use in decision making, it is well to remember that in predicting next year's performance from the vantage point of this year, we have much more information on which to base the prediction than is suggested by these data; further, the data have more to tell under further analysis (see Figure 33.1).

The data tell us that over the period covered the expected (average) annual return is 9.5 percent; 5 percent of the individual years experienced a return of -4.0 percent or less; 95 percent of the individual years experienced a return of 29.0 percent or less; thus, 90 percent of the time the annual

TABLE 33.2   Dow-Jones Industrial Average: Annual Closing Prices, 1918 to 1990

Year	Close	Annualized percent return for varying investment periods				
		1 year	2 year	3 year	4 year	5 year
1914	56.76					
1915	99.15	74.68				
1916	95.00	−4.19	29.37			
1917	74.38	−21.71	−13.39	9.43		
1918	82.20	10.51	−6.98	−6.06	9.70	
1919	107.23	30.45	20.07	4.12	1.98	13.57
1920	71.95	−32.90	−6.44	−1.10	−6.71	−6.21
1921	81.10	12.72	−13.03	−0.45	2.19	−3.11
1922	98.73	21.74	17.14	−2.72	4.69	5.83
	...	...	...	...	...	...
	...	...	...	...	...	...
1984	1211.57	−3.74	7.60	11.46	5.88	7.63
1985	1546.67	27.66	10.85	13.91	15.30	9.92
1986	1895.95	22.58	25.09	14.63	16.02	16.72
1987	1938.83	2.26	11.96	16.97	11.41	13.12
1988	2168.57	11.85	6.95	11.92	15.67	11.49
1989	2753.20	26.96	19.17	13.24	15.51	17.84
1990	2633.66	−4.34	10.20	10.75	8.56	11.23
Average		5.18	5.18	5.18	5.18	5.18
Min.		−61.69	−44.01	−37.75	−33.15	−21.61
Max.		74.68	38.14	33.98	31.63	25.72
s		22.37	15.24	11.64	10.14	8.97
99th percentile		57.24	40.64	32.26	28.78	26.05
95th percentile		41.98	30.25	24.33	21.86	19.93
90th percentile		33.86	24.72	20.10	18.18	16.68
50th percentile		5.18	5.18	5.18	5.18	5.18
10th percentile		−23.50	−14.36	−9.74	−7.82	−6.32
5th percentile		−31.62	−19.89	−13.97	−11.51	−9.58
1st percentile		−46.88	−30.28	−21.91	−18.42	−15.70

Source: *Dow-Jones Averages 1885–1990, Phyllis S. Pierce, ed, Business One Irwin, Homewood, IL, 1991.*

return lay between −4.0 percent and 29.0 percent. These values become, for lack of any better basis for prediction, our estimates of probability for a time horizon of 1 year. Similarly, we estimate the 90 percent envelope of probabilities for the time period (in years) N of 2, 3, 4, etc.

The data, plotted in Figure 33.2, help us visualize and quantify the uncertainty suggested by these numbers and the way in which time reduces uncertainty. The long-term average return of 9.5 percent, while attractive, is by no means certain in an investment of just 1 year. The range from 4.0 to 29.0 may be unacceptably wide. The prospect of such uncertainty may drive us to invest in a bank certificate of deposit with a rate of return of only 5 percent, but with far narrower limits of uncertainty.

Credit application data can be compared with statistical models of data on repayment behavior of loan holders to assess creditworthiness of the loan applicant, based on the estimated risk of loan default. For life insurance products, mortality data are statistically modeled to create and price life insurance products on the basis of estimated risk of death of a beneficiary while the insurance is in force.

Risk and Risk Assessment.   A key to profitability is accurate risk assessment. When a financial institution underwrites a project, it is investing funds in the completion of that project in expectation of a competitive return on that investment. Some level of risk is always associated with profitability, irrespective of the product, service, or the industry which offers them. In financial services the risk is the more apparent because the product itself directly involves money.

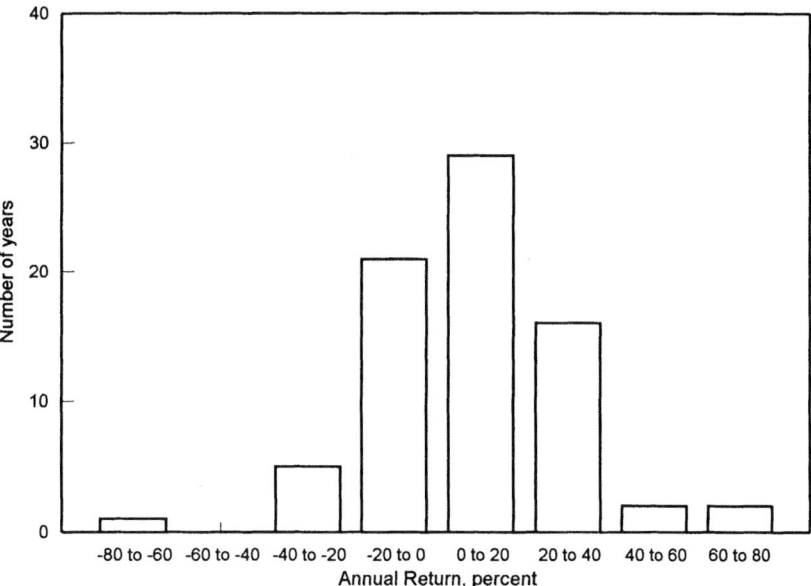

FIGURE 33.1   Distribution of annual ROI values for the years 1918 to 1990 of a hypothetical fund of the Dow Jones Industrial Average. (*Source: Adapted from Dow-Jones Averages 1885–1990, Phyllis S. Pierce, ed., Business One Irwin, Homewood, IL, 1991.*)

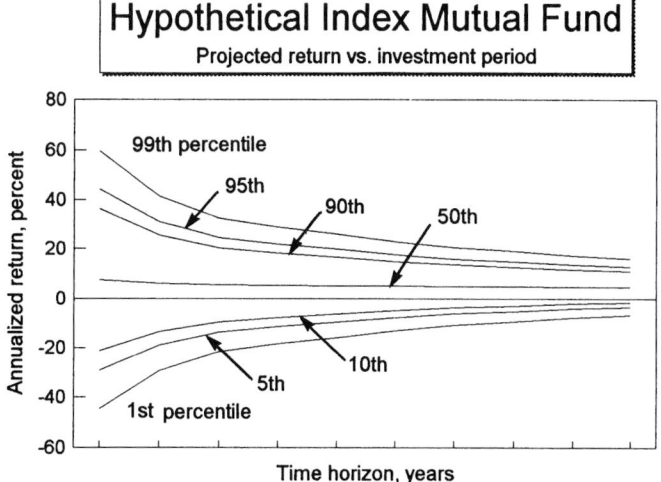

FIGURE 33.2   Risk versus investment time horizon. (*Source: Adapted from Dow-Jones Averages 1885–1990, Phyllis S. Pierce, ed., Business One Irwin, Homewood, IL, 1991.*)

It is important for both parties to a financial transaction to be able to assess the comparative risk of the transaction; the penalty for mistake may in some cases exceed the amount of the investment. Both parties will seek to make the ratio of cost of risk to reward acceptably low. The ratio is an important measure on which a customer bases such decisions as selection of financial-service provider, selection from among competing financial products, and allocation of investment among selected products. In other words, the estimated ratio of cost of risk to reward is an important measure of quality in financial services.

Defects or Deficiencies.   A defect or deficiency occurs whenever a product fails to perform as designed or as promised. Defects can be tangible (e.g., incorrect terms on a loan document, a deposit or delivery of currency in incorrect amount, missing pages in an insurance policy, delivery of stock or bond certificates too late to complete a trade, unavailability of an automatic teller machine) or intangible (e.g., surly behavior of a bank teller or customer service inquiry clerk, misquote of loan application fee, failure to open the office or switchboard at the advertised time, chronically busy telephone lines, failure to answer an incoming telephone call).

Many of the features and deficiencies mentioned above are common to services in general. Significant features that are most usually associated with financial services are fiduciary responsibility, risk, and financial return (or interest paid).

Challenges and Obstacles for Quality in Financial Services.   There are a few quality issues which stand out in the financial services sector: creating uniformity of service, comparison with leaders in other service sectors, quality-initiatives fatigue, downsizing and consolidation, and automation.

Creating Uniformity of Service.   Many financial service organizations have branch operations and extensive telephone service networks. In this environment, creating and maintaining uniform service of the highest level is extremely difficult. A customer who is overserviced at one location and underserviced at another will probably be disappointed. In most financial service companies it is primarily the actions and attitudes of people that constitute the service. Trust in individual employees is as important as trust in the institution and in the branch office. Therefore, hiring, training, developing, motivating, and rewarding are extremely important.

Comparison with Leaders in Other Service Sectors.   There is a carryover effect from other service companies, even companies in other sectors. Fast, competent, friendly and high-value service from a fast-food restaurant, a dry cleaner, or an auto repair shop creates expectations for the same from bank, broker, and insurance agent. Many surveys show that financial services lag in this respect. [See the discussion under The American Customer Satisfaction Index (ASCI), below.]

Quality-Initiatives Fatigue.   Many financial service organizations went through "Quality" in the past three decades: "smile training" in the 1960s, quality circles in the 1970s, and culture modification in the 1980s. Having tried all these, they either feel they have done it all or are just tired of all the quality initiatives which have come and gone without much effect.

Downsizing and Consolidation.   Financial services, like many other sectors, is experiencing corporate downsizing and consolidation. For example, between 1980 and 1990 the number of commercial banks in the United States fell from 14,000 to 12,000, and by 1996 to 9800—a reduction of 30 percent in just 16 years. Although the pattern in other western countries like Canada and Great Britain is to consolidate into five or six giant institutions, experts in the United States have predicted that several thousand commercial banks will remain in this country in the long run. In fact, in banking, dozens of banks start up successfully each year to serve customers that larger consolidated banks are not satisfying.

Consolidation is also the prevailing trend in other financial services—brokerage, savings bank, insurance, credit card, mortgage, and trust.

The reduction of personnel has been even greater than the reduction in the number of institutions. When a merger occurs, a reduction in combined staff of from 10 to 20 percent is common. In addition, ongoing efforts at productivity improvement and cost reduction have created a sectorwide reduction in work force of between 5 and 8 percent. These changes have placed a significant burden on the remaining staff, and in many cases have been reflected in reduction of service to customers.

Automation.   Much increase of efficiency and productivity has come from automation in financial services. In banks, for example, computerizing customer interfaces and transaction activity cuts cost significantly. Unfortunately, it also keeps customers out of the branches, where sales of new service traditionally occurs—person to person. Pressure from shareholders to increase profits has caused many companies to reduce costs by automating operations and eliminating staff.

PROOF OF THE NEED—EVIDENCE OF QUALITY TROUBLE IN FINANCIAL SERVICES

There is much evidence of customer dissatisfaction with the quality of financial services. Some of the evidence follows:

The American Customer Satisfaction Index (ACSI).

The index is a cross-industry measure of the satisfaction of U.S. household customers with the quality of the goods and services available to them. Included are goods and services produced within the United States and those provided as imports by foreign firms that have substantial market share or dollar sales. The index provides an objective economic measurement of the improvement needed in financial services as well as in other industries. The ACSI is cosponsored by the University of Michigan Business School, where its methodology was developed, and the American Society for Quality (ASQ, formerly the American Society for Quality Control). Results of the first ACSI assessment, based on data collected May 10 to July 22, 1994, were released in October 1994.

The ACSI measures customer satisfaction on a uniform scale scale from 0 to 100, where 100 is best. The first value of the national ACSI was 74.5. This baseline measure is the benchmark for comparing customer satisfaction over time. The ACSI covers seven sectors of the economy and key industries within each sector. These sectors are (1) Manufacturing/Nondurables, (2) Manufacturing/Durables, (3) Transportation/Communications/Utilities, (4) Retail, (5) Finance/Insurance, (6) Services, and (7) Public Administration/Government. These sectors account for 75 percent of the Gross Domestic Product. (For a detailed description of the ACSI, contact the American Society for Quality, 611 East Wisconsin Avenue, Milwaukee, WI 53201-3005; phone 414-272-8575.)

Table 33.3 shows those financial service companies large enough and receiving enough national responses to receive individual indices. The index values may be interpreted as relative ratings. Of special interest is the fact that the 3-year trend is slightly downward, suggesting an opportunity for companies applying quality management methodology to gain a competitive edge.

Quality Costs.

Section 8, Quality and Costs, proposes four categories of quality-related cost: prevention, appraisal, internal failure, and external failure. Examples of these costs from the financial service sector follow:

Prevention: New product review, quality design or quality planning activities, training programs, creation of written policies and procedures, analysis of quality information, and conduct of quality improvement projects.

Appraisal: Inspection of incoming work, supplies, and material; periodic inspection of work in process; checking, balancing, verifying, and final inspection of account transactions; conducting customer surveys; and analysis of customer correspondence/complaints.

TABLE 33.3   ACSI Scores for Representative Companies in Financial Services

Commercial bank		Insurance: life		Insurance: property and casualty	
Organization	ASCI score	Organization	ASCI score	Organization	ASCI score
Banc One Corporation	74	Aetna Life & Casualty	73	Allstate Insurance Group	73
BankAmerica Corporation	67	Metropolitan Life Insurance Company	73	Farmers Group, Inc.	71
Chase Manhattan	70	The Prudential Insurance Company of America	74	State Farm Insurance	79
CitiCorp	70	Travelers, Inc.	NA	Other companies	75
First Interstate Bancorp	71	Other companies	74		
First Union Corporation	73				
Key Corp.	76				
Nations Bank Corporation	73				
Norwest Corporation	76				
Wells Fargo & Company	71				
Other banks	77				

*NA = not available.
Source: American Society for Quality, Milwaukee.

Internal failure: Machine downtime, scrap and waste due to improperly processed forms or reports, and rework of incorrectly processed work.

External failure: Investigation time for complaints, payment of interest penalties, reprocessing of an item, scrap due to improperly processed or incorrect forms or reports, time spent with disgruntled customers, and lost business due to providing poor service.

The sum of failure costs plus the portion of appraisal cost necessitated by a high failure rate is referred to as "cost of poor quality." These are costs which would disappear if the product or service were defect-free.

The sum of all four categories of quality costs constitutes the total cost of quality. Experience shows that the latter three categories, called the "cost of poor quality," range from 10 to 30 percent of sales income or 25 to 40 percent of operating expense, not including the cost of funds. This magnitude, frequently verified, represents a large opportunity for improvement. Organizations such as Banc One Corporation, American Express, and Prudential Insurance Corporation use cost of poor quality to drive quality improvement. The data in Table 33.4 give examples of this opportunity in one large bank organization.

Customer Loyalty and Retention.   Customer loyalty is the purchasing behavior exhibited toward a financial services provider by a customer who places a major percentage of his or her financial services business with that provider. Customer retention means keeping a customer as a buyer of services.

In financial service companies the rates of customer loyalty and customer retention are often so low that as much as 30 percent of the customer base is lost each year, equivalent to a complete turnover in a little over 3 years. The typical American household uses six different financial service

TABLE 33.4   Quality Costs by Activity

Department/service	Quality cost, % of total expense	Percent of total quality cost Prev/App	Failure	Measured defect rate, %	Individual improvement opportunity, $
Bank card services	94	10	90	25	119,000
Credit card processing	NA	36	64	10	93,240
Word processing	25	43	57	0.69	77,480
Commercial credit	48	56	44	NA	40,810
General accounting	49	45	55	5	29,200
Automatic teller machine	37	32	68	0.02	25,200
Customer inquiry center	26	64	36	NA	18,000
Commercial loan operation	50	42	58	NA	16,660
Consumer credit collections	14	77	23	2	11,280
Commercial lending	17	64	36	NA	8,620
Installment loan credit	NA	66	33	NA	5,400
Training	NA	89	11	NA	4,800
Deposit services	NA	30	70	0.01	NA

NA = not available.

organizations, where one to three could provide the services required. The principal reason for this seemingly unnecessary spreading around of work is poor quality of service. Financial service companies too frequently treat their products as commodities. They often seem unwilling or unable to distinguish their products from those of their competitors, and fail to develop in their customers a feeling of loyalty to the product or brand or to the organization. In that circumstance, the customer's cure for dissatisfaction is to quit and take the business to another provider. At the same time, while companies may be unable to retain customers, they recognize that their competitors share the same problem. This gives rise to marketing efforts aimed at capturing the business of the disaffected customers of other companies by means such as advertising the ease of switching accounts, provision of free checks, instant replacement of existing insurance policy, or waiving of transfer fees in a new brokerage account.

When poor quality service translates to a lack of customer loyalty and poor customer retention, the costs are almost always larger than they appear at first. Casual analysis of a lost account may suggest that the account was small anyway, with little loss to the company. Deeper analysis frequently reveals that in earlier times—say 6 to 12 months before—the customer had multiple accounts, all with healthy balances. The customer departed a little at a time, not all at once, and in that time took substantial business away.

Further, the cost of replacing a customer is demonstrably greater than the cost of retaining one. The least costly means of gaining sales and market share is to treat the existing customer well, rather than trying to create a replacement.

For more on customer loyalty and customer retention, see Section 7, Quality and Income.

DESIGNING WORLD-CLASS FINANCIAL SERVICES

Customer-Driven or Technology-Driven Product Development?   New products and features come to market in two essentially different ways. The first depends on a product based on the "great idea," a novel idea of great potential benefits around which a marketing effort is developed. The automatic teller machine (ATM) is an example of such a product. The customer can, by introducing a magnetically coded identification card, withdraw cash from any one of a network of machines at any time. For banks, the principal attraction is the reduction of the substantial labor cost of the withdrawal transaction (the cost of a teller to handle the face-to-face transaction, and the costs of processing, transporting, and storing the resulting paper records); for the customer, the advertised benefits are speed and convenience of withdrawing cash at any time of the day or night.

One of the drawbacks of this product development strategy is its reliance on the great idea. Great ideas are difficult to come by and the investment required to promote them can be prohibitive.

The second way to develop product is to begin with a clearly researched profile of customer needs, and design the product with features that address the most important of those needs. While the resulting products may not create dramatic competitive breakthroughs, the approach enables the company to maintain for the product line an acceptable rate of improvement and, for each improvement, acceptable odds of successfully launching and sustaining it.

There is a third way to develop new products which bears mentioning. It is to copy competitors in introducing new products rather than to take the initiative in developing new product. Some refer to this as the "lemming" process, after the legendary behavior of lemmings, which follow the crowd regardless of consequences. Such a process assures that its practitioners will never know the success that the big idea can bring, or even the success which is based on a thorough understanding of customer needs. The lemmings will always be late to benefit from emerging trends. When a trend becomes a boom, and the boom leads to overcapacity, the lemmings are frequently the first victims of the bust that follows.

Developing World-Class Products and Services. The product development process starts with identifying current and potential customers. Next, "listening posts," the mechanisms by which to discover the needs of these customers, are established. These listening posts may be strategically placed individuals (such as sales personnel or customer service personnel), customer surveys, telephone listening/recording, complaints, customer interviews, focus groups, field service personnel, advisory councils, and so on.

In some organizations, sales personnel and customer transaction personnel use every interaction as an opportunity to probe and ask specific questions. See Table 33.5.

When the listening posts are well designed and the information they generate is well coordinated, their contribution can help to develop new products and to sell existing product that is already developed of which the customer is not aware.

USAA and American Express are companies which collect and analyze information generated in all types of customer contact. All written and telephone complaints, compliments, and inquiries are collected, classified, and analyzed. The resulting information provides a clear indication of what is going well, what needs to be improved and what new needs are emerging from the customers.

Information from the listening posts helps to identify needs and to assure that they are associated with the appropriate set of customers, in order to improve service, add features, and develop new products.

For example, funds transfer between accounts is a product useful to an individual who makes transfers among multiple accounts in one bank, as well as to a company with far-flung operations making transfers among hundreds of accounts. While the basic transaction is the same for both customers, the required documentation differs greatly. The individual may be perfectly well served by a monthly statement of all transactions and balances. The company will likely need daily end-of-day statements delivered to multiple locations.

TABLE 33.5 Questions to Discover Customer Needs

- Is there anything else I can do for you?
- Are there any problems or needs today you have in managing your financial affairs?
- Are there any products or services you wish we offered?
- Are you aware of any products on the market that we don't offer?

Needs, once identified, must be translated into measurable terms. This is especially true in the service environment, in which there may exist a belief that "intangible" is a synonym for "immeasurable." Thus, a "fast" transaction must be expressed in days, minutes, or seconds. "Easy" portfolio liquidation must be described in terms which make clear who is authorized to do the liquidating, when they may do it, under what controls, and so on.

Once needs are so translated, the translation must be validated or modified by the customer. Table 33.6 is an example of customer needs and translation for terms associated with one product: check truncation. (Check truncation is a feature associated with the checking account statement. For an account with the check truncation feature, the account statement contains the vital information for each check cleared, but the actual written and processed checks are not returned.)

Each of the needs, clearly stated and measurable, is addressed by appropriate product features. When all the product features are developed, the product emerges and the delivery process can be developed. World-class companies have found that the customers should be consulted at each step of this development process to be sure that the product being developed continues to meet the need. Table 33.7 extends the example of check truncation to product features and process features.

TABLE 33.6   Check Truncation: Needs and Translation

Primary	Secondary	Tertiary	Translation	Units of measure	Sensor
\[Needs\]					
Security	Account insurance		Institution, account insured by FDIC	Insurance in place, advertised; Y/N	Consumer
	Financial soundness		Sound financial statement	Expert judgment	Expert
Utility	Acceptability		All recipients will accept checks	Y/N	Marketing survey
	Flexibility		Pay by mail, in person; get cash; deposit money	Y/N	Product design
Convenience	Account information access	By telephone	Account information accessible by telephone	Average call wait time, seconds	Telephone log
	Funds access	By ATM	24-hour deposit/withdraw anywhere	% ATM monthly downtime (minutes $\times$ 100/43,200)	ATM log
	Documentation	Payment receipt	Proof of payment available when needed	% items requested but not found	Quality performance report
	Problem resolution	Knowledgeable staff	Staff perceived as capable of solving customers' problems	Survey response	Consumer survey
		By telephone	Problem resolution available by phone	Number of complaints	Complaint log
				% of problem calls abandoned	Phone log
	Overdraft protection	Covers reasonable operation	Covers reasonable overdraft	Survey response	Consumer survey
				Number of complaints	Complaint log
		Overline credit available	Overline credit available	Y/N	Product design

TABLE 33.7 Check Truncation: Product and Process Features

Product feature	Product goal	Process feature	Process goal
Statement	Accuracy: <0.01% error rate	Check encoding	<0.005% miscoded
		Check read/post	<0.005% misread
		Statement design change	Error free
			Customer driven
	Timeliness: >99% mailed on day of printing	Statement printing	On schedule
		Statement envelope mailing	No delay
			On schedule
	Readability	Design of layout	Readable
		Words	No complaints
		Statement printing	Clean: readable

CONTROLLING QUALITY AND CUSTOMER SERVICE

Measuring from the Customer Perspective—What They Think Is Important.
Controlling quality of service depends on the measurement of performance in terms of the features that matter to the customer— primarily the external customers, but the internal customers, too.

The time to develop measures and performance levels that meet customer needs is during the design of the product. For existing products that have already been developed and are being delivered today, performance measures that do exist are too frequently based on the needs of internal customers.

Developing customer-oriented performance measures to assure products is a task that requires the help of listening posts, discussed above. Most useful is the customer satisfaction survey process.

The customer satisfaction survey begins with the study of the complete service cycle for the product for which measures and performance standards are to be developed. Figure 33.3 shows, from the customers' perspective, all of the features of the product through its entire life. Then each feature is written into a "satisfaction" and an "importance" question (Figure 33.4). Next, an actual performance question is written for each feature. Figure 33.5 offers an example.

In the example of Figure 33.5, suppose a disbursement time of 48 to 72 hours correlates with "satisfied," and 24 to 48 hours correlates with "very satisfied." The goal for funds transfer should be to achieve "very satisfied," and the corresponding numeric goal is 24 hours. Goals should always be set to create very satisfied customers. Customer loyalty research shows that customers who are merely satisfied are not loyal. A satisfied customer is 51 percent likely to buy again, almost literally as likely as not, whereas very satisfied customers exhibit an 85 percent probability of buying again. These probabilities explain why so many individuals and companies have multiple suppliers of financial services. It is difficult for an organization or its products to sustain a customer at "very satisfied," so the customer buys from multiple organizations.

It is instructive, using survey data from questions of the sort in Figure 33.4, to assess the importance of each feature on a simple scale (1 = low to 4 = high), and to use the same scale to measure satisfaction performance. Table 33.8 shows a 2 × 2 matrix into which the pairs of values—importance and performance—can be plotted to guide the measurement effort. Each quadrant suggests an action. The features whose importance ranks high (3 or 4) fall in the right-hand column. All of these must be measured. For any which have no measure, one must be devised. The performance standards can be determined by correlating the numerical performance reported by the customer to the customer's reported level of satisfaction.

Through the administration and analysis of periodic surveys, and through the continuous improvement of service that survey results stimulate, it can be expected that the list of questions can

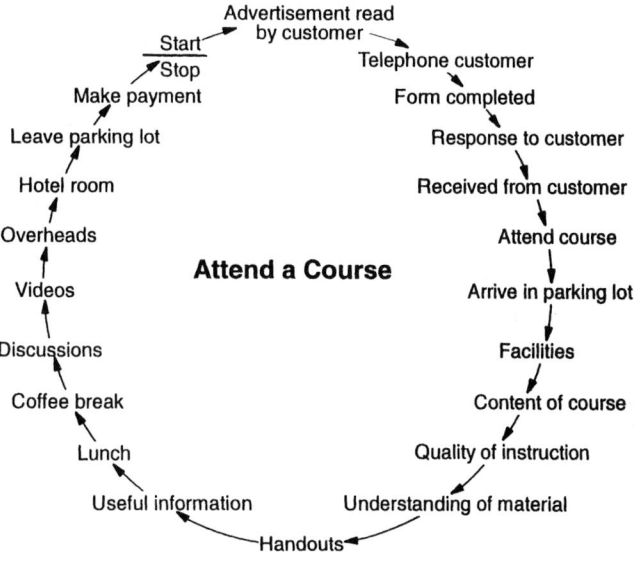

FIGURE 33.3   Service cycle (attend a course).

A. How satisfied are you with following branch staff service features?

	Very satisfied	Somewhat satisfied	Somewhat dissatisfied	Very dissatisfied
1. Greeting you with a smile	1	2	3	4
2. Concerned and caring	1	2	3	4
3. Helpfulness	1	2	3	4
4. Friendliness	1	2	3	4
5. Using you name	1	2	3	4
6. Handling phone transactions efficiently	1	2	3	4
7. Answering questions and resolving problems	1	2	3	4
8. Processing transactions without error	1	2	3	4
9. Knowledge of bank products and services	1	2	3	4
10. Thanking you for your business	1	2	3	4
11. Helpful in identifying your financial needs and solutions	1	2	3	4
12. Ability to help plan, find alternatives, and solve financial problems	1	2	3	4

B. Which FIVE features of the services listed below are MOST IMPORTANT to you in judging performance? Read through the items below and circle the five most important.

1. Greeting you with a smile
2. Concerned and caring
3. Helpfulness
4. Friendliness
5. Using your name
6. Handling phone transactions efficiently
7. Answering questions and resolving problems
8. Processing transactions without error
9. Knowledge of bank products and services
10. Thanking you for your business
11. Helpful in identifying your financial needs and solutions
12. Ability to help plan, find alternatives and solve financial problems

FIGURE 33.4   Customer satisfaction survey.

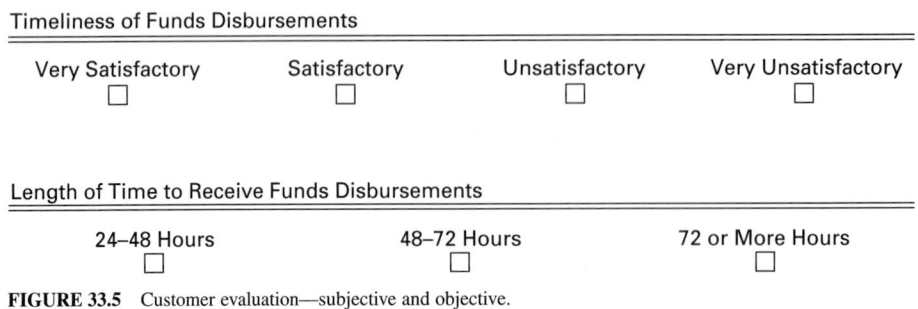

FIGURE 33.5   Customer evaluation—subjective and objective.

TABLE 33.8   Service Improvement Measurement Matrix

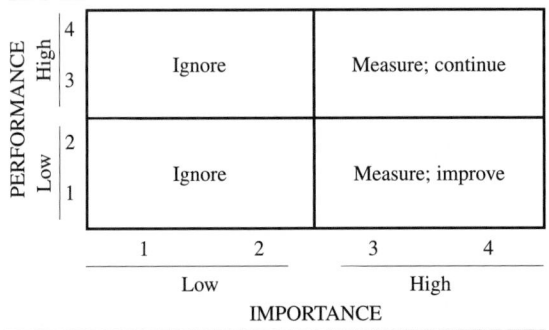

be narrowed to a few which capture the essence of customers' assessment of performance. An ongoing monthly survey of a small random sample will assure that service remains at "very satisfied" or is moving toward it.

Internally, proxy measures are needed to track those features that are important to the customer. In the financial services environment, with such extensive use of information technology, many measures can be collected automatically, for 100 percent of transactions and in real time. The time saved in automatic measurement can be used for improvement. Where manual measures are required, sampling should be used. (For more on sampling, see Section 46, Acceptance Sampling.)

Performance measures are processed in a feedback loop. As the data are collected, either by sample or through the automated system, the actual performance is compared to the standard, and action is taken on the difference.

The detailed plan for this measurement—who does it, how often it should be done, with how large a sample—is documented in a control matrix or control plan. Table 33.9 is an example. The plan provides the procedures for sampling and comparing to a standard and specifies the actions to be taken if performance does not meet standard. A graph of the actual data against the standard displays the quality level for easy comprehension and action. The tool that is used most often is a control chart. (See Section 45, Statistical Process Control.)

In successful organizations, performance reporting is designed to facilitate continuous improvement. For example, organizations with duplicate operations want to discover best practices within the organization so as to make them the performance standard throughout. Table 33.10 is an example of a performance report from a multibranch bank. Examples of measures and their standards for a full service, multibranch commercial bank are shown in Table 33.11. Examples of measurement feedback reports for a credit card company are provided in Tables 33.12 and 33.13.

TABLE 33.9   Control Matrix/Plan

Feature	Measure	Standard	How	Sample	Parameter	Action	Who
Accuracy	In balance	Within $2.00	Clerk calculation	100%	<$2.00	Nothing	NA
					>$2.00	Review to find error	Balance clerk
					More than two per shift	Notify supervisor and improvement team	Balance clerk

THE CUSTOMER AS A SUPPLIER

In service organizations, the customer is often a supplier, too. The brokerage customer, in placing an order, must supply the detailed transaction data: buy 100 shares of General Motors stock at a price not to exceed $150 a share. The bank customer that prepares a shipment of coin and currency to be deposited is supplying material input to the bank's sorting/counting process. If the customer, as supplier, errs in providing input, the service provider will be unable to deliver output that will satisfy the customer's needs.

The financial service provider's best defense against customer blame for error in output is to prevent customers from making errors in input. Three approaches used by financial service providers are customer education, errorproofing customer input, and monitoring the quality of customer input.

Customer Education to Be a Good Supplier.   At a large bank, a study of a sample of out-of-balance deposit transactions (in which the deposit claimed by the customer does not match the bank's count) revealed that a major cause of the condition was the occurrence of unsegregated currency and coins in the deposit. Further, in addition to the extra work required to balance the deposit, the need to sort the coin and currency imposed its own cost on the transaction. Analysis showed that the total effect was to double the cost of the deposit transaction, a cost which all customers bore, though the increase in cost was caused by only 10 percent of the customers.

Those few customers were targeted for training. The bank's deposit department sent a delegation to the companies' premises to train the personnel responsible for preparing the deposits. The result was to reduce processing time at the bank by more than half, which, in turn, reduced costs and allowed a price reduction. The deposit error rate was reduced from 15 percent to $1/10$ of 1 percent. The customers reported a reduction of over 30 percent in time spent in deposit preparation. Bank and customers both benefited from this customer education.

Errorproofing.   There are a number of ways to errorproof customer input. Paper forms for input by mail, fax, or in person serve as a checklist to assure complete input. Menus—on-screen or by voice—guide input by telephone, kiosk, or computer terminal. "Help" buttons can further assist electronic entry. Requiring the customer to verify a playback of the entered data is another form of errorproofing. Another technique is to have the customer enter the data or information twice and validate the corresponding data entries that don't match. Errorproofing as a part of training makes a powerful combination for reducing customer input error.

Monitoring and Measuring Customer Input.   In the application of training and errorproofing, it is important to monitor customer input to verify effectiveness and expose any need for additional improvement. Monitoring should be done as close to the input transaction as possible—to correct errors quickly and to keep error costs down. The ideal is to monitor input as the customer is making the entry. The customer can be given immediate corrective feedback and training in how to prepare and conduct the transaction properly the next time. Training can also include spot tips on errorproofing the transaction, such as double checking, checklisting, and the like.

TABLE 33.10 Performance Report from a Large Multibranch Bank

Project name/affiliate bank	Description	Plan or actual			Improvement					Annual savings or revenue, $
		Start date	End date	Implant date	Customer	Quality	Service	Environment	Product	
Savings bonds/Dallas Contact: Name/#	Streamline processing of corporate savings bond program for corporate customers	11/96	8/97	10/97	X	X			X	4,125
NSF and OD/Houston Contact: Name/#	Review overdraft policies with the intent of: maximizing fee income, minimizing bank loss, providing consistent level of service	3/96	1/97	1/97	X	X	X			100,000
Additional cards/Dayton Contact: Name/#	To find a more efficient method of processing requests from card-holders for additional cards of existing accounts	1/97	6/97	7/97	X	X	X			7,280
⋮		⋮	⋮		⋮	⋮	⋮		⋮	⋮
Home equity service, indirect QIT/Columbus Contact: Name/#	Develop a marketing program to obtain home equity loans through indirect dealers when consumers prefer to use their home as collateral to buy other consumer goods.	9/97	12/97	2/98	X	X	X			54,000
Credit approvals/Cleveland Contact: Name/#	Streamline credit procedures to enhance customer service, minimize delays, improve productivity, document new procedures.	3/97	10/97	11/98	X	X	X		X	142,740
Total contribution					18	20	21	13	19	1,696,131

TABLE 33.11 Quality Measures in a Multibranch Commercial Bank

Operation	Measure	Standard, %
Retail deposits	Number of incorrect new DDA account documents/total number of new DDA accounts	0.50
Retail deposits	Number of incorrect new IRA documents/total number of new IRA accounts	0.50
Retail branches	Number of transaction errors/total number of transactions	0.10
Retail branches	Total dollars over and short/total dollars of cash in and out	0.01
Mortgage loans	Number of applications not committed on time/number of applications underwritten	0.30
Mortgage loans	Number of applications not closed on time/number of applications closed	0.20
Commercial loans	Loan payments posted incorrectly/total loan payments	1.00
Commercial loans	Number of payments backlogged/number of payments received	0.04
Retail installment loans	Number of loans not closed in 45 days/total number of loans closed	0.10
Retail installment loans	Number of complete credit applications not answered within standard	1.00
Accounting/financial	Number of GL entry errors/total number of GL entries	0.05
Accounting/financial	Number of GL reports delivered late/total number of GL reports	0.15
Human resources	Number of payroll errors/total number of employee payments	1.50
Human resources	Number of days of formal training/total number of employees	0.05
Human resources	Number of performance appraisals late/total number of performance appraisals due	0.02

QUALITY IMPROVEMENT—THE NEVER-ENDING JOURNEY

Continuous Customer Input—More on Customer Listening Posts. In addition to their importance in product development and in quality control, customer listening posts are also useful in helping identify what needs to be improved. The customer-complaint desk is a particularly effective listening post for improvement ideas. Customer satisfaction studies and customer loyalty studies also provide useful input for quality improvement efforts.

Customer Complaints and Recovery. For several reasons it is useful to centralize customer complaint and inquiry management. First, it is useful to create a cadre of experts to deal with customer problems. When customers call, it is best that they are greeted by knowledgeable personnel who have available all the tools they need and know how to use them. Availability and capability are extremely important in dealing with customer problems. The speed with which the problem is solved will have a significant impact on the customer's final satisfaction. It is a worthy goal to solve a problem at the first call, if possible, and in the same day in any case. Research shows that each time it is necessary to recontact the customer or transfer an inquiry, customer satisfaction drops by 8 to 10 percent.

The process of fixing the customer problem is called "recovery." The centralized customer complaint group uses a number of strategies for recovery to minimize the customer's dissatisfaction. These include making an apology, replacing the product (including service) quickly, conveying a

TABLE 33.12 Monthly Quality Report

Quality measure	Standard	September	August	July	June	May	April	Avg. perf. to std.
Average time to answer, seconds	20.00	36.00	40.00	25.00	21.00	20.00	18.00	
Performance to standard, %		55.6	50.0	80.0	95.2	100.0	100.0	82.0
Number of abandoned calls/total number of calls, %	2.00	4.00	3.40	2.60	3.10	3.20	3.10	
Performance to standard, %		50.0	58.8	76.9	64.5	62.5	64.5	62.9
Number days to handle cancellation request	5.00	3.00	4.00	6.00	6.00	4.00	3.00	
Performance to standard, %		100.0	100.0	83.3	83.3	100.0	100.0	100.0
Number days to handle written investigation	5.00	6.00	5.00	3.00	4.00	4.00	4.00	
Performance to standard, %	03.3	33.0	100.0	100.0	100.0	100.0	100.0	
Number of unresolved investigations to regulatory agencies	0.00	26.00	41.00	37.00	42.00	40.00	57.00	
Performance to standard, %		.0	.0	.0	.0	.0	.0	0.0
Number of state insertion errors	0.00	.05	.02	1.00	.02	1.00	8.00	
Performance to standard, %		.0	.0	.0	.0	.0	.0	0.0
Number of plastic insertion errors	0.00	.00	.00	.00	.00	.00	.00	
Performance to standard, %		100.0	100.0	100.0	100.0	100.0	100.0	100.0

sense of understanding and empathy, making recompense for value lost, offering a token compensation, and following up later with the customer to be sure that all reasonable effort has been made to put the matter right.

Customers left unsatisfied by a recovery effort represent a loss in at least five ways: (1) they will likely stop giving business to the offending organization; (2) they will probably give future business to a competing one; (3) they may give the offending organization negative publicity by reciting their story of dissatisfaction to business colleagues who are potential customers; (4) they will likely never offer good word-of-mouth publicity, as they might have had the recovery experience been good; and (5) their tie to the offending organization, measured by intent to repurchase, could actually have been strengthened by the combination of initial problem and subsequent recovery, properly handled.

Table 33.14 compares the repurchase intent of customers of a financial services firm who had an unsatisfactory recovery experience to customers who had a very satisfactory one.

This repurchase behavior demonstrates that effective recovery is important as a part of the effort to retain present customers. Courting customers through good recovery practice amounts to an effective and inexpensive alternative to creating a new customer to replace each one driven away by poor recovery. That this makes good business sense becomes clear upon study of the costs associated with developing a new customer, closing the account of a departing customer, and replacing the lost income stream. Additionally, financial service organizations find that immediately after a problem has been fixed the customer is ripe to be cross-sold a service (10 to 20 percent of the time the customer will buy). When an attempted sale follows the fixing of a problem customer satisfaction scores go up 3 to 5 percent more than they do as a result of only fixing the problem, and 5 to 10 percent if the sale is successful.

It should also be recognized that as a customer's relationship with a financial service provider develops, the contribution of that business to the firm's profit increases. The longer-term customer has a larger portfolio, a higher balance, uses more services, and therefore incurs higher fees and provides economies of scale with consolidated accounts and statements.

As important as it is in the short run to solve the immediate customer's problem, it is equally important in the long run to capture information about the problem, code it, and consolidate all such infor-

TABLE 33.13 Quality Trend Report

Meas. no.	Quality measure	Total defects	Total volume	Defect rate or defects	Standard	Performance to standard, %	Weight	Weighted performance
148	No. abandoned calls/total calls	8,594	214,862	4.00%	2.00%	50.0	1	50.0
454	Average time to call pickup, seconds	36	NA	36	20	55.6	1	55.6
455	No. days to handle cancel request	3	NA	3	5	166.7	1	166.7
456	No. days to handle written investigation	6	NA	6	5	83.3	1	83.3
457	No. of unresolved investigations to regulatory agencies	26	NA	26	0	0.0	1	0.0
458	No. of statement insertion errors	11	209	0	0	0.0	1	0.0
459	No. of days statement production not met standard	1	NA	1	0	0.0	1	0.0
460	No. of applications not processed within standard	0	NA	0	0	200.0	1	200.0
462	No. of credit bureau correction requests	1,725	NA	1,725	0	0.0	1	0.0
463	No. of customer complaints about collections	5	NA	5	0	0.0	1	0.0
464	No. of credit inquiries not processed to standard	2	NA	2	0	0.0	1	0.0
465	No. of hours system down	2	NA	2	0	0.0	1	0.0
466	No. of hours credit card system down	1	NA	1	0	0.0	1	0.0
467	No. of payments not posted	1,154	NA	1,154	0	0.0	1	0.0
468	Payments not posted, in thousands of dollars	10,787	NA	10,787	0	0.0	1	0.0
478	No. of plastic insertion errors	0	NA	0	0	200.0	1	200.0
479	No. of other insertion errors	2	211	0.01	0	0.0	1	0.0
		8,594	214,862				17	755.6
	Overall total defect rate, %		4.00			Overall performance, %		44.4

NA=not applicable.

mation as a database to help identify what processes need to be improved. An example of a complaint and inquiry management process which features such a consolidated database is shown in Figure 33.6.

Continuous Improvement—Project by Project. In Section 5, The Quality Improvement Process, the general principles of quality improvement are laid out in detail. The present section highlights some examples of the application of these principles in the financial services sector and resulting lessons.

To focus the improvement effort on quality of customer service can provide high payback for an organization.

TABLE 33.14   Repurchase Intent

Type of customer	Percent who intend to repurchase	Percent who will recommend purchase to others
Typical satisfied customer with no problem	51	65
Customer with problem; unsatisfied with recovery	30	46
Customer with problem; very satisfied with recovery	79	88

Source: TARP.

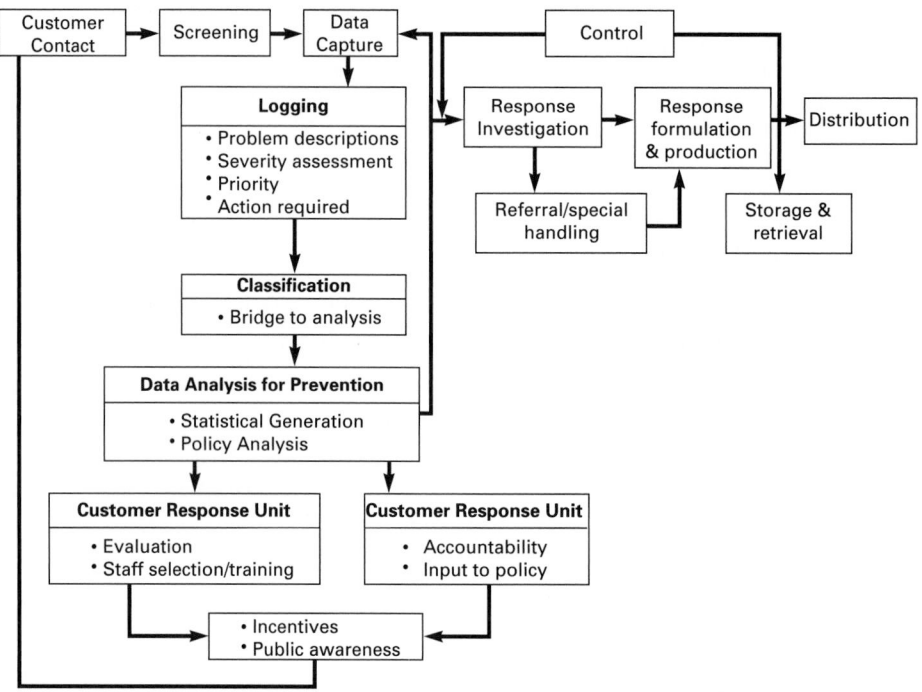

FIGURE 33.6   Customer-response framework: complaint/inquiry management.

In the first year of a quality improvement process, Banc One realized a net savings and revenue enhancement of three-quarters of a million dollars; in the sixth year, net savings and revenue enhancement exceeded 16 million dollars.

In 1991, a study was conducted of over 1000 improvement teams at 35 financial institutions (Aubrey and Gryna 1991). The most successful teams, measured in terms of saving and revenue enhancement, defect reduction, customer satisfaction improvement, and increase in employee satisfaction, shared some important characteristics:

- *Team makeup:*   Mix—75 percent officer/manager level, 25 percent nonexempt; average team size—7; "ideal" team size—4 or 5
- *Team member selection:*   By management
- *Training:*   Two days minimum
- *Project selection:*   By management or the quality council

- *Projection Duration:* Three to four months, meeting weekly for 90 minutes.
- *Improvement tools used:* Brainstorming, Pareto analysis, surveys, cause and effect diagram, data collection, flowcharting, work simplification, and cost/benefit analysis.

Table 33.15 is a partial list of projects completed by a large interstate bank during a single year.

In medium- to large-size organizations, tracking projects makes it possible to manage their progress, share essential information about them with other parts of the organization, and to replicate solutions where appropriate. A quality improvement tracking report system is illustrated in Table 33.15.

Project reporting is most effective when it is on line with search mode. In addition to general publicity through press and house organs, it is useful to publicize projects with potential for replication directly to the areas which would benefit from it, to exploit all possible opportunities for replication.

IMPLEMENTING AND GETTING STARTED WITH TQM IN FINANCIAL SERVICE

Section 14, Total Quality Management, covers in detail the process of planning and introducing TQM. The present section offers a few illustrative examples from financial services.

Training and Resource Commitments. Training must be conducted and carried out first with the chief executive officer and the chief operating officer down through all levels of management within the organization. As an example, training for quality was of the utmost importance to NatWest USA. A total of 2500 employees—including the chairman, president, and every senior and executive vice president received training.

TABLE 33.15 Quality Improvement Projects at a Large Bank

Project no.	Department/branch	Project title	Identified cost reduction/revenue enhancement, $
1	Accounting services	Computer output team	466,000
2	Consumer credit collections	Retail collections job descriptions	366,421
3	Financial administration	Employee expense reimbursement	341,040
4	Bank card	Late fee review	288,388
5	Commercial banking	Account profitability/private banking	220,303
6	Operations	Cash and due team	220,124
7	Commercial loan	Commercial loan services	212,610
8	Interbank/cross-function	Cash management	168,364
9	Corporate banking	Small business lending	161,000
10	Branches	Vault cash	153,138
11	Operations	MICR reject	146,656
12	Commercial loans	Commercial loan standardization	145,446
13	Bank card associates	Credit procedure review	142,740
14	Special industries group	Credit services	138,000
...	...	...	
61	Asset/accounting	Collateral protection insurance	27,400
62	Bank card/collection	FCS/CMS nondelinquent accounts	26,926
63	Commercial banking	Commercial loan documentation	25,650
64	Operations analysis	PC maintenance	25,640
Total cost-reduction opportunity			6,054,895

All of NatWest USA's 4700 employees attended a 1- or 2-day training sessions. This training ensured a common language and approach and provided a standard set of tools and techniques to address the opportunities at hand.

Each course included a broad-based discussion of quality, creating a common definition of quality, customer/supplier relationships, identifying and meeting customer needs (for both internal and external customers), the cost of quality, quality measurement and reporting, and tools and techniques for data collection and analysis. Case studies utilizing current bank data and processes were key elements of the training.

Different kinds of training with different objectives must be made available to different types of employees. Figure 33.7 demonstrates the array of courses offered by a Fortune 500 financial service company that is relatively mature in the quality management process.

Some of the best-performing financial service companies report 10 to 15 days of training per year per employee.

Strategic Planning.   An extremely important element to sustain and drive a total quality effort is strategic planning. Quality must be integrated into the business plans of financial services companies or any company for that matter. This keeps quality from becoming an adjunct activity and elevates it in importance, as it should be, with operations, finance, marketing, etc.

It can become a strategy of its own or can be integrated with other strategies. Quality activities such as measures, goals, standards, projects, training, and teams make the strategic plan a reality by executing the strategies. See Section 13, Strategic Deployment, for more details regarding strategic planning.

Management Involvement—Professional and Personal.   Effective leadership of the quality improvement process requires senior management to set an example and tone for quality within their organizations. In some companies, this includes the designation of a chief quality officer, with responsibilities analogous to those of the chief financial officer. These responsibilities include planning and assuring that the organization's plans for quality are developed and are being carried out, assuring that quality is being managed and controlled to plan, and assuring that quality improvement is continuous. The chief quality officer is most able to be effective when the position reports to the CEO, president, or the chief operating officer of the company.

The Dawn of Revolution.   More and more financial services are becoming automated every day; this is especially true in retail services. In banking, there is a fundamental shift underway—from brick-and-mortar, people-intensive service to service which is totally electronic. According to the American Bankers Association, as of 1996 over 40 percent of transactions were electronic; the figure is projected to be 60 percent by 1998. Soon, it will not matter whether the bank, broker or insurance company is in Boston, Beijing, or Barcelona.

Managing the quality of customer service
Leading and facilitating teams
Customer satisfaction measurement
Quality measurement
Quality cost measurement
First impressions
Beyond customer satisfaction
Customer-first teams
Customer inquiry handling
Market damage measurement

FIGURE 33.7   Content of financial service quality training.

- In obtaining a loan, the original signed note is the only physical document required; it can be transferred by overnight courier. The money and all of the other information can be sent by fax, on the Internet, and by wire transfer.

- A deposit and investment account can be set up and transactions conducted by fax, Internet, and wire transfer.

- Life insurance application and payment can also be done by fax, Internet, and wire transfer. Only the medical examination resists automation as of this writing. The results, though, can be transmitted by fax to the insurance company. A nurse can visit the applicant at home to administer a blood test, take blood pressure and pulse, and so on.

PC/modem services	58%
Internet services	53%
Screen phone services	47%
PC/third-party network	42%
Interactive TV services	26%
Other automation	25%

FIGURE 33.8   U.S. banks' service plans for 1998.

As electronic capability increases, the accompanying reduction in costs, cycle time, defects, and errors increases profit.

In the 1980s home banking ventures were a failure. Today, over 40 percent of American homes have personal computers; an even higher percentage of businesses have them. This equipment base creates a far more fertile climate for electronic financial services than existed just 10 years ago. Figure 33.8 shows what U.S. bankers planned for 1998, according to the American Bankers Association.

An electronic innovation of great potential for change in financial services is the "smart" card. The smart card resembles the credit card in appearance, but differs radically in that it contains electronic circuitry that enables it to be electronically loaded with monetary value and be relatively independent, except for a simple device to electronically read and transfer value from it. A person can receive salary or incentive payment on the card, make payments and transfers, pay for services, and get cash. This expansion of services is potentially very attractive to customers. Additionally, the card promises services that are faster, more reliable, more accurate, in a wider variety, and at lower cost.

Dozens of smart card trials are being carried out around the globe. Smart cards are being tested in France and Spain and were also tested in Atlanta at the 1996 Summer Olympics. Dutch banks, including ING Bank, are issuing the cards now and expect them to be distributed nationwide by the end of 1998. American Express pays bonus incentives to its employees with smart cards it developed.

The smart card and other elements of the electronic revolution portend great change in financial services. Improved services for customers will be accompanied by a great change in the nature and number of jobs in financial services. There will, no doubt, be more jobs requiring higher skill levels, such as customer service, programming, computer operation, and so on. In proportion, there will be fewer clerical jobs.

Companies in the financial services sector will find that the tools of quality management will provide welcome help in mastering this rush of innovation and change, and may prove to be the competitive edge.

REFERENCES

Aubrey, Charles A., II (1988). "Effective Use of Quality Cost Applied To Service." *ASQC Quality Congress Transactions,* Dallas.

Aubrey, Charles A., II (1989). "Continuous Improvement To Meet Customer Needs." *ASQC Quality Congress Transactions,* Toronto.

Aubrey, Charles A., II (1988). *Quality Management in Financial Services,* 2nd printing. Hitchcock, Wheaton, IL.

Aubrey, Charles A., II, and Felkins, Patricia K. (1988). *Teamwork—Involving People in Quality and Productivity Improvement.* Quality Resources, White Plains, NY.

Aubrey, Charles A., II and Gryna, Derek S. (1991). "Revolution Through Effective Improvement Projects." *ASQC Quality Congress Transactions,* Milwaukee.

Berg, Eric N. (1987). "Improving Banking Machines." *The New York Times,* September 9.

Berg, Sanford V., and Lynch, John G., Jr. (1995). "Regulatory Measurement and Evaluation of Telephone Service Quality." *Quality Control and Applied Statistics,* March-April.

Berry, Thomas H., et al. (1995). "Quality In Financial Services: The Vanguard Story." *IMPRO 1995 Conference Proceedings,* Juran Institute, Inc. Wilton, CT.

Brennan, John J. (1991). "Investing in Quality at the Vanguard Group. If It's Not Broken, Improve It Everyday Anyway." *IMPRO 1991 Conference Proceedings,* Juran Institute, Inc. Wilton, CT.

Bulkeley, William M. (1987). "News Flash! Firm Has Bad Year and Tells Shareholders about It!" *The Wall Street Journal,* May 4.

Butterfield, Fox (1987). "North's $10 Million Mistake: Sultan's Gift Lost in a Mixup." *The York Time News,* May 13.

Calian, Sara (1995). "Firms Unveil Prospectuses Readers Can Digest." *The Wall Street Journal,* July 31.

Deutsch, Howard, and Metviner, Neil J. (1992). "Quality in Banking: The Competitive Edge." *Bank Administration.*

Feder, Barnaby J. (1988). "Securities Firms Streamlining Computer Systems Despite the Crash." *The New York Times,* March 2.

Frigo, Robert, and Janson, Robert (1994). "GE's Financial Services Operation Achieves Quality Results through `Work-out Process'." *Quality Control and Applied Statistics,* vol. 39, no. 6, November-December.

Gryna, Frank M. (1995). *Exploratory Research on Quality Assessment in Financial Services.* The University of Tampa.

Gwin, John M., and Lindgren, John H., Jr. (1993). "Reaching the Service-Sensitive Retail Consumer." *Journal of Retail Banking.*

Henderson, G. L. (1992). "GIROBANK: The First Bank to Win a British Quality Award." *Quality Forum,* vol. 18, no. 2, June.

Hestand, Randy (1991). "Measuring the Level of Service Quality." *Quality Progress,* September.

Johnson, Randall (1991). "Centralized Customer Service Functions Boost Service, Create Cost Efficiencies." *The Service Edge,* vol. 4, no. 11, Lakewood Publication, November.

Klein, Jerilyn, and Wiseman, Paul (1992). "NationsBank, Dean Witter Form Venture." *USA Today,* October 27.

Kleinfield, N. R. (1993). "The Check-Cashing Industry is Growing—And Not Just in Poor Neighborhoods." *The New York Times.*

McClure, Beverley J. (1991). "USAA's Culture of Quality: Keys to Making Quality a Way of Life." *IMPRO 1991 Conference Proceedings,* Juran Institute, Inc. Wilton, CT.

Myerson, Allen R. (1992). "NationsBank Will Create a Firm with Dean Witter." *The New York Times,* October 27.

Oxman, Jeffrey A. (1992). "The Global Service Quality Measurement Program at American Express Bank." *National Productivity Review,* Summer.

Smith, Priscilla Ann (1987). "Americus Philip Morris Trust Faces Risk of Extinction as Stock Price Surges." *The Wall Street Journal,* August 28.

Stankard, Martin (1992). "Critical Success Factors for Bank Improvement Projects." PDG, Inc. *Productivity Views,* January-February.

Stankard, Martin (1993). "Sponsoring Employee Teams—23 Tips." Productivity Development Group, Inc., vol.10, no.2.

Stewart, Steve R. (1987). "First National Bank of Chicago: Commitment to Quality Cuts Costs, Adds Customers." *American Productivity Center,* October.

Stoner, James A. F., and Werner, Frank M. (1995). "Managing Finance for Quality: Bottom-Line Results from Top-Level Commitment." *American Society for Quality Control,* ASQC Quality Press, Milwaukee.

Weinstein, Michael (1989). "Consumers Give Their Banks Lower Grades Than Last Year." *The Daily Financial Services News,* October 5.

Welch, James F. (1992). "Service Quality Measurement at American Express Traveler's Cheque Group." *National Productivity Review,* Autumn.

Werner, Frank M., and Stoner, James A. F. (1995). *Modern Financial Managing,* preliminary edition, Harper Collins College Publishers, New York.

SECTION 34
SECOND-GENERATION DATA QUALITY SYSTEMS

Thomas C. Redman

INTRODUCTION

This section discusses the ubiquity and poor quality of data. More specifically, it explains why the manager must be concerned about poor data quality and discusses the approaches to improvement. Most people recognize that if there are errors in data, they should be corrected. And indeed, good software packages can help find certain errors in even the largest databases. But new data are created at enormous rates, so the job of error detection and correction never ends. Leading enterprises have achieved superior results with so-called second-generation data quality systems, which are those focused on preventing future errors. This section is a guide to the leader who wishes to upgrade to a second-generation data quality system and reap the benefits of improved customer satisfaction, lowered costs, and more confident decision making.

This section begins with an outline of the steps taken by one executive to do so. It then summarizes critical differences between data and other resources and the complexities they engender for data quality. Next, it sketches the business case for second-generation data quality systems. Lastly, it describes good practice in defining and implementing them.

THIS REPORT IS WRONG!

To illustrate one way that implementation of a second-generation data quality system could proceed, consider a fairly large service enterprise, organized in several business units and staff functions. The enterprise is profitable, but its industry is extremely competitive and becoming more so. In this example, the head of one business unit is frustrated that summaries of financial performance are late, incomplete, and inaccurate. Important decisions are delayed, then made in haste. Poorer decisions that are harder to implement are the direct result.

This executive decides to start a small program to explore the benefits of a second-generation data quality system. He or she selects an important data-driven business problem. (Experience suggests that areas and problems involving money, such as billing, revenue, customer accounts, and marketing, are especially impacted by poor data and can produce case studies with bottom-line results rather quickly.) A *customer need analysis* is conducted to determine who is concerned about this problem and to understand their data needs. Next, primary *sources of* (*internal*) *data are identified*. A *planning project* identifies two or three important gaps, and *improvement projects* follow. Upon completion, *controls* are implemented to hold the gains.

The people responsible for these projects share their successes with this executive's staff. The executive urges further projects and each staff member agrees to conduct a test project or two. In conducting these tests, several staff members recognize that the data cross their functions and they begin to explore *process management* as a means of working together. Another acquires important data from a commercial source, and decides to explore *supplier management.*

Not all projects succeed. But enough do and the reports that originally frustrated the executive are now much better. The staff decides to devote a portion of each bi-weekly staff meeting to data quality, effectively forming itself as a *Data Council.* And it, at the vocal urgings of the executive, decides to be more vigorous in its efforts. It agrees to aggressive improvement targets after a high-level planning effort and decides it must formalize its approach to meet them. Council members also recognize the need for *vision* and *policy,* and devote time over several months crafting them.

In parallel, our executive decides it is time to inform the CEO of his/her unit's data quality program. He or she naturally wants *recognition,* both personally and for those who led improvement efforts. More importantly, projects increasingly need the cooperation of other units, which are reluctant to provide it (importantly, few data quality systems impact beyond the span of control of their most senior sponsor). And the cycle repeats itself across the enterprise.

In time, the enterprise and executive discover the techniques described herein. They will also learn some important features of data.

PROPERTIES OF DATA

In many respects, the techniques used by the executive and enterprise mentioned previously are as described throughout this handbook. But data differ from other resources in some critical ways, and these differences can have important, yet subtle, impact on the data quality program. These differences (Levitin and Redman 1998) include the following:

Consumability: Unlike other resources, data are consumed with use.

Copyability: Data records can be copied for a fraction of the cost of the original. You simply can't do this with other resources.

Computer Storage: Data, unlike other resources, can be stored cheaply and easily in almost unimaginable quantities on computers.

Depreciability: Data, can, in principle, be immortal. While their value does not diminish with use, the utility of most data deteriorates rapidly in time.

Fragility: Paper data records are occasionally accidentally destroyed, but they are more apt to be lost. Data stored in computers are much more fragile.

Intangibility: Perhaps the most obvious difference between data and other resources is data's intangibility. While data recordings can be seen and touched, data themselves are intangible.

Nonfungibility: Fungibility means that one unit of a resource can easily be exchanged for another, assuming another unit is available. But data "units" are inherently unique. You simply cannot substitute one person's date of birth for another's.

Renewal: Whenever pertinent features of the real-world change, data values change and/or new ones are created. New data result from everyday business—each customer transaction, each shipment, indeed, practically every activity leads to new data. This happens at astounding rates. This property of data, called "renewal," does not really apply to other resources. In most cases, there is an inherent lag time until all databases are updated.

Shareability: To a larger degree than any other resource, data may be shared.

Source: In contrast to other resources, data are generated by a tremendous number of sources. In many cases, the original sources of many data sets are undocumented or even unknown. The Internet is exacerbating this problem.

Transportability: Data are also unlike any other resource in the degree, ease, and speed with which they can be transported over long distances.

Valuation: Neither markets nor standard accounting practices exist for most data.

Versatility: Data collected and used for one purpose are often used in other applications. Data-driven marketing is one such example. But some alternate uses of data are illegitimate. Data about a person's age, for example, cannot be used in a hiring decision.

Implications for Data Quality.   The following are some of the more obvious ways that properties of data influence the data quality program. First, that which is "out of sight is often out of mind." As data are stored neatly away in computers, there seems to be a tendency to pay them less attention, particularly given the other compelling and highly visible problems facing the enterprise. So many enterprises are not even aware of their data quality issues. This problem is exacerbated by the lack of accepted methods of valuation. It is much easier to spur management action when there are clear monetary costs or benefits.

Second, more than anything else, the high rates of data creation (renewal) ensure that error detection and correction won't work well.

Third, since data are not tangible, they have no physical properties. This complicates measurement. Managers and technicians know how to measure physical properties such as length, viscosity, time, and impedance. But all important data quality dimensions are abstract and so are difficult to measure. For example, you can't tell by direct examination whether most data are correct.

A number of properties help create difficult political situations [see Davenport et al. (1993), Strassman (1994)]. It is interesting that while data are shareable, this is the exception rather than the rule in most enterprises. Instead, since data are relatively inexpensive to store, copy, and transport, organizations within an enterprise often acquire, store, and manage their own. This immediately raises issues of ownership—issues that are among the most brutal in many enterprises. It also makes any kind of centralized planning for data and standards difficult to establish and enforce.

A number of properties (copyability, transportability, nonfungibility, fragility, and cost of storage) conspire to increase redundancy and contribute to a "save everything" mentality, including data that are no longer useful. Yet all redundancy is not bad. For example, redundancy allows users to work with data in the environments of their choosing (for example, it helps make the office-at-home feasible).

BUSINESS CASE FOR SECOND-GENERATION DATA QUALITY SYSTEMS

Second-generation data quality systems make good business sense. The case is summarized as follows:

- Data are used by every activity conducted by the enterprise.
- Most data are of poor quality.
- Current efforts to find and correct errors don't work well.

- Left alone, the problem will become more critical as data become even more important. The Internet exacerbates this problem.
- Second-generation data quality systems cost less and produce better results.

The following subsections explain these points in more detail.

Data Are Ubiquitous.   In the previous example, the "call to action" stemmed from poor decisions. It is axiomatic that decisions will be no better than the data on which they are based. Nor should one expect any other activity that takes poor data as inputs to yield superior results. And indeed, data are ubiquitous—virtually every activity in which the enterprise engages requires data. Consider the following:

- Data are both critical inputs to and outputs of almost all "work" performed by an enterprise. Data are used to serve customers, manufacture products, manage inventory, and so forth.
- Data support managerial and professional work and, as in the previous example, are critical inputs to almost all decision making at all levels of the enterprise. Data are the means by which the enterprise knows about its other resources—financial, human, and so forth.
- Data may be combined in almost unlimited ways in the search for new opportunities, market niches, process improvements, and new products and services.
- Because definitions of common terms like "customer" and "service" are captured in data, they (data) contribute to the enterprise's culture. They "fill the white space" in the organization's chart.
- Enterprises strive to convert tacit "knowledge" into data. For example, a company's salespeople may have warm personal relationships with the company's most important customers. But for the company as a whole to serve these customers, important aspects of the relationships must be specified in data. (Some authors have noted that "data" are the raw material for "information," which in turn are the raw materials for "knowledge." The reverse direction is even more important. Specifically, knowledge created or developed by an individual or group must eventually become structured data so others can apply that knowledge.)

It is interesting to note that these activities can be taking place simultaneously, using the same data. Figure 34.1 depicts a typical scenario. The end-to-end process of information supply, new data creation, processing, and use is called an "information chain." "Information products" are the outputs along the way—the data recorded in databases, reports, analyses, and so forth.

Most Data Are of Poor Quality.   At the time of this writing, the best known data quality problem is the so-called "year 2000 (Y2K) problem." Examples of other common data quality problems include:

- *Low Accuracy:*   Accuracy is probably the most carefully studied data quality issue. Numerous studies yield error rates ranging from 1 to 75 percent. Direct mailers find that up to 20 percent of their flyers are returned as undeliverable; customers find billing errors; scanners report incorrect prices; etc.
- *Inconsistency:*   Data values in two databases, "owned" by two organizations within a bank, disagree. The two organizations cannot determine who their common customers are.
- *Difficulties in interpretation:*   A clothier may provide shirts in four sizes: small, medium, large, and extra large. In the computer system supporting manufacturing, these sizes are coded "1," "2," "3," and "4," respectively. Later, operators can't recall whether "1" means small or extra large.
- *Unmanageable redundancy:*   Many enterprises have any number of copies of the same data. L. P. English (1998) cites a company that had 43 such redundant databases. In addition to adding complication and expense, redundancy makes it more difficult to maintain consistency.

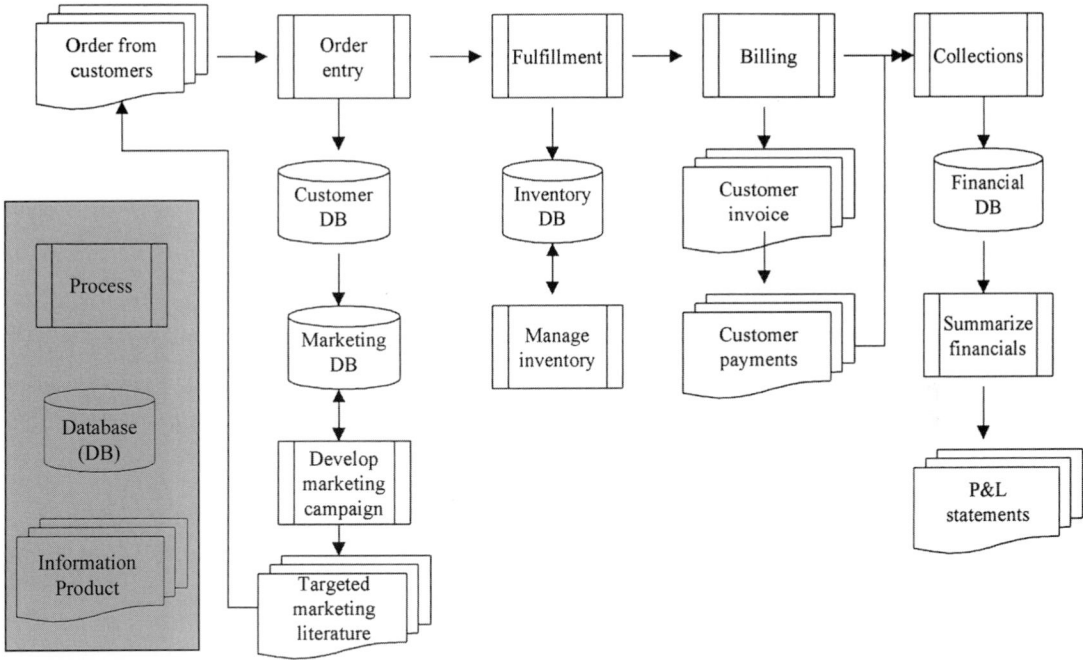

FIGURE 34.1   An example of an information chain (or chains depending on how the enterprise chooses to manage) and the numerous recodings in databases and information products produced.

Poor quality data seem to attack information chains like viruses—there is no way to predict exactly where they will strike next or the impact they will have.

Error Detection and Correction Yield Poor Results.   Naturally many enterprises are not blind to poor data quality. So they undertake efforts to detect and correct errors. Direct inspection is one way of doing so. Many organizations review the data supplied to them by upstream organizations to identify data they "know are wrong" because the values are outside accepted domains. Others search for inconsistencies in a collection (or collections) of data. Thus, if a customer's telephone area code = 999, the value cannot be correct. Or if the telephone area code = 212 (New York) and address zip code = 90210 (California), then at least one value must be in error. The search for inconsistencies can be quite sophisticated, involving numerous fields, several databases, and sophisticated error logic. These searches may be computerized and a number of good software programs are available. Once errors have been found, they must be corrected and in some cases, error correction logic helps do so. But correcting errors is often more difficult than finding inconsistencies. In some cases the company must make reference to the real world to determine correct values.

Even conscientiously applied efforts to detect and correct errors do not yield satisfactory results. The litigation alone from noncompliance in the Y2K problem may cost $1 trillion in the United States (Thompson 1997). Of course, "everyday" problems have costs also. The total costs across the typical enterprise are summarized in Table 34.1.

Some costs are stated in quantitative financial terms. Those affecting customers or employees are not so easily quantified but may be even more important. First, is the impact of poor data on customer satisfaction. Customers receive much data as a byproduct of the product or service they receive. The invoice for product sent is a good example. Many customers are remarkably

Table 34.1 The Costs of Poor Data Quality to the Typical Enterprise

Typical problems
Inaccurate data
Inconsistencies across databases
Inappropriate formats
Difficulties in interpretation
Unmanaged redundancy

Typical costs*
Operational costs:
Lowered customer satisfaction
Increased cost: 8–12% of revenue in the few, carefully studied cases; for service organizations, 40–60% of expense
Lowered employee satisfaction
Tactical costs:
Poorer decision making: Poorer decisions that take longer to make
More difficult to implement data warehouses
More difficult to "mine" data and re-engineer
Increased organizational mistrust
Strategic costs:
More difficult to set and execute strategy
Contribute to issues of data ownership
Compromise ability to align organizations
Divert management attention

*Operations, tactics, and strategy represent a loose hierarchy of work performed. Operations are the day-to-day tasks such as order entry, customer support, and billing. Tactics are the decisions of short- and mid-term consequence and work to support them usually made by middle managers. Strategy involves long-term direction.

Source: After Redman (1998).

unforgiving of billing (and other data) errors, reasoning that any company that can't bill them properly simply can't provide a good product or service.

Second is the impact on employees. One cannot expect the hotel clerk, dealing with irate travelers whose reservations have been lost, to have high job satisfaction. And many enterprises have dozens, even hundreds, of such "customer care" jobs. This issue occurs at the organizational level as well. When one organization depends on another for input data and those data are frequently wrong, it is natural for the former to develop a poor opinion of the latter. And their ability to work together in the future is compromised.

The Importance of Data Will Continue to Grow.
Data have always been important to commerce and their importance continues to grow rapidly. They are the fuels for economic growth in the Information Age.

Increasingly, important data are computerized. This trend has accelerated for several decades, driven by impressive advances in database, networking, and communications technologies. It has made data available to new, unsophisticated users and exacerbated issues of poor data quality. Many people have a tendency to assume that "if it is in the computer, it must be right." Such people are more easily victimized by poor quality data. Even sophisticated users cannot be expected to have familiarity with the nuances of all the data they encounter. The Internet is taking these phenomena in directions that are not fully understood. Specifically, companies are making their data, heretofore closely held, directly available to Internet users, a remarkably diverse group with disparate needs that are likely to be quite different from those of internal users. Thus, data quality issues too will become both more critical and difficult as a result.

Second-Generation Data Quality Systems.   In recent years superior approaches that focus not on individual errors, but rather on identifying and eliminating root causes of entire categories of errors have been developed. Companies find they can make order-of-magnitude improvements, often by eliminating some rather simple (once they have been identified) problems. After Ishikawa (1990), we call techniques aimed at error detection and correction "first-generation techniques" and those aimed at eliminating root causes "second-generation techniques."

By a *"data quality system"* we mean the totality of an enterprise's efforts to manage, control, and improve data quality. The system includes:

• Activities aimed at understanding customer needs
• Activities aimed at detecting and correcting errors and activities aimed at preventing future errors (including control and improvement)
• Activities to build management infrastructure to do so effectively and efficiently

Second-generation data quality systems are those that emphasize prevention of future errors over error detection and correction and build management infrastructure to do so effectively and efficiently.

Practitioners have learned that second-generation data quality systems cost far less and produce much better results. Consciously applied, these systems have helped many organizations reduce error rates by factors of 10 to 100, cut cycle time in half, and reduce costs by up to 75 percent. Customer service is better and employees feel much greater involvement in their jobs, so morale improves. These results, summarized in Table 34.2, stand in marked contrast to the baseline of Table 34.1.

DATA DEFINED AND DIMENSIONS OF DATA QUALITY

Data Defined.   As used here, "data" per se (or a "data collection," "data set," etc.) consist of two interrelated components, "data models" and "data values" (Fox and Redman 1994). Data models are the definitions of entities, attributes, and relationships among them that enterprises use to structure their view of the real world. Enterprises often model a given portion of the world

Table 34.2   Typical Results from Implemenation of a Second-Generation Data Quality System

Typical improvements
Accuracy improved by 1 to 2 orders of magnitude
Easy-to-read formats
Redundancy and inconsistency minimized

Typical benefits
Operational benefits:
A primary source of customer dissatisfaction eliminated
Decreased cost: Two-thirds to three-quarters of cost of error detection and correction eliminated
Important cycle times cut in half
Employees feel empowered
Tactical benefits:
More confident decision making
Re-engineering opportunities suggested
New technology implementation easier
Organizations build trust by working together
Strategic benefits:
Issues of data ownership eased
Easier to set and execute strategy

differently. For example, an employer may be interested in a person as an *employee,* the IRS as a *taxpayer.* The person is the same real-world entity in both cases, the employer and IRS are interested in different attributes. Data values are the specific realizations of an attribute of the data model for specified entities. The "123-45-6789" in taxpayer social security number = 123456789 is a data value.

"Data records," as distinct from data per se, are the physical realizations of data stored in paper files, in spreadsheets, in databases, and so forth and presented to users in ways that (one hopes) make them easy to store and use. Note that data (i.e., the models and values) are abstract, while data records are their tangible realizations.

It is evident that data are not just random facts and figures, but rather are quite structured. Especially for computerized data, several steps must be completed before they can be used. These steps, summarized in Figure 34.2, include model development, acquisition of the values, storage, selection of what the user wants to see, and presentation to the user. All can impact customer satisfaction, though some steps are more pertinent to the computer systems, than to data. And many of them are carried out by separate organizations. Thus data models are the responsibility of those who develop databases, variously called Information Technology, Information Systems, or Information Management Systems (all abbreviated IT herein). Data values are created within or obtained from external suppliers by line organizations or business units in the course of everyday business. IT is also usually responsible for the manner in which data are recorded and presented to users.

The most important common dimensions of data quality are discussed below.

Dimensions of Data Quality.
"Dimensions" of data quality stem from user needs. The "properties" described previously are intrinsic to data. Customers naturally have needs or requirements that bear on each major constituent of data (model, values, and records). It is important to distinguish closely related needs, such as accessibility, that bear more on supporting technology, from those directly pertinent to data. Table 34.3 lists the most important "dimensions of data quality" [The definitions of many of these dimensions are quite technical and the interested readers are referred to Fox and Redman (1994) and Levitin and Redman (1995).]

To summarize, high-quality data are data that are fit for use in their intended operational, decision-making, planning, and strategic roles. As Figure 34.3 depicts, data models and presentation define the features of high-quality data. And data values must be free of defects. Supporting technologies must make data secure yet accessible and ensure privacy.

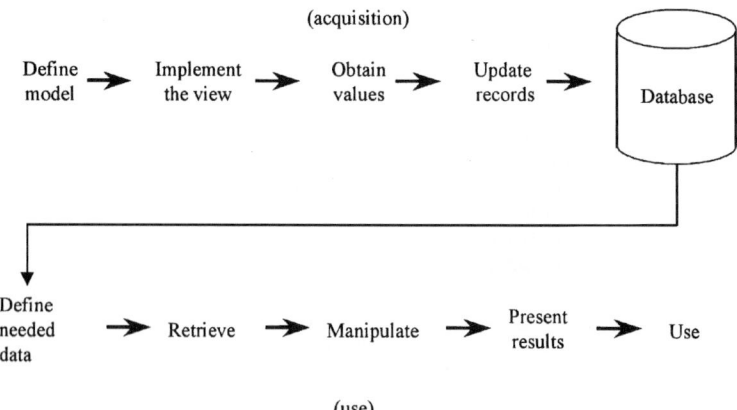

FIGURE 34.2   The data life-cycle model features activities needed to define a data model, obtain values, store and process data, and present the user with what he/she wants.

Table 34.3   Dimensions of Data Quality*

Quality dimensions of a data model		
Scope	*Comprehensiveness*	*Essentialness*
Level of detail	*Attribute granularity*	*Precision of domains*
Composition	*Naturalness*	*Identifiability*
	Homogeneity	*Simplicity*
Content	*Relevance*	*Obtainability*
	Clarity of definition	
View consistency	*Semantic consistency*	*Structural consistency*
Reaction to change	*Robustness*	*Flexibility*
Quality dimensions of data values		
	Accuracy	*Completeness (entities and attributes)*
	Consistency	*Currency (cycle time)*
Quality dimensions of data records and presentation		
Formats	*Appropriateness*	*Format precision*
	Use of storage	*Correct interpretation*
	Flexibility	*Portability*
	Represent null values	
Physical instances	*Representation consistency*	
Other dimensions often associated with data		
	Accessibility	*Appropriate use*
	Privacy	*Redundancy*
	Security	

Source: Fox et al. (1994) and Levitin and Redman (1995). For an alternative formulation, see Wang and Strong (1996).

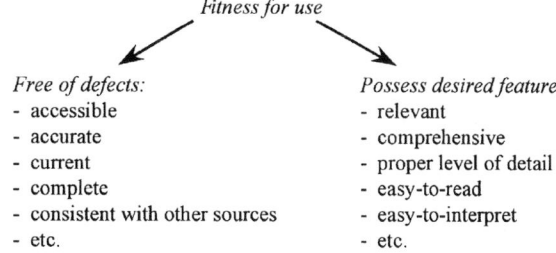

Fitness for use

Free of defects:
- accessible
- accurate
- current
- complete
- consistent with other sources
- etc.

Possess desired features:
- relevant
- comprehensive
- proper level of detail
- easy-to-read
- easy-to-interpret
- etc.

FIGURE 34.3   Data are of high quality if they are "fit for use" in their intended operational, decision-making, and other roles. Fitness implies both freedom from defects and possession of desired features. Most users associate dimensions associated with data models as desired features and the lack of other features as defects.

SECOND-GENERATION DATA QUALITY SYSTEMS

The thrust of second-generation quality systems is to consistently identify and prevent the most important root causes of future errors. This requires both proper technique and management infrastructure. This section describes each element of a second-generation data quality system. (See Table 34.4 for a list.) Each enterprise must craft and evolve its own data quality system. Four elements, senior leadership and support, quality planning, quality control, and quality improvement are required. Other elements are selected based on the enterprise's opportunities and challenges, its culture and organization, and the origin of its most important data.

Table 34.4 Elements of Second Generation Data Quality Systems

Management infrastructure	Process description
Senior leadership and support	Measurement
Data quality vision	Quality planning
Data quality policy	Quality control
Supplier management	Quality improvement
Process management	Process (re-)design
Change management	Inspection and test (data editing)
Database of record	Quality assurance
Strategic data quality management	Document assurance
Training and education	Rewards and recognition
Technical capabilities	Domain knowledge
Identification of information chains	Standards
Customer needs analysis	Quality handbook

Management Infrastructure

Senior Leadership and Support. As we have observed, data cross organizational boundaries in the blink of an eye and the politics associated with data can be brutal. So senior leadership, support, and intervention in the conflicts that are sure to arise are essential. These may be provided by a Data Council (or a Quality Council). Senior management is ultimately responsible for crafting the data quality system and leading its implementation. It ensures that data quality efforts are directed at the most pressing business problems and opportunities. To do so, it develops and deploys a vision and policy (see the following two paragraphs), sets quality goals, selects planning, control, and improvement "projects," and provides cross-functional coordination for projects that require it. It allocates funds for data quality efforts and ensures that the enterprise is properly trained.

Data Quality Vision. People need to know where the enterprise "is going." A vision is a "picture" of the enterprise's desired future state with respect to data and information, including a rationale for people to work to create that future state. Developing the vision forces the Data Council to think about the enterprise's long-term data and information needs to ensure business success. A vision, supported by broad communication, helps align the enterprise and motivates people to work together to achieve the desired future state.

Data Quality Policy. A data quality policy (or simply data policy) is management's statement of intent regarding long-term data and information quality. It serves as a "guide for managerial action." It should recognize data as "business assets" and, accordingly, specify improvement objectives and management accountabilities for achieving them. The best second-generation policies delineate accountabilities along information chains from those that create data models to creators of data values to data users (Redman 1995, 1996). Like the vision, formulating policy forces the Council to think broadly and deeply about management accountability for data. Widely communicated policies help align the enterprise and provide a basis for decision making by lower-level managers.

Supplier Management. Almost all enterprises receive data from outside. Some is purchased, while other data, such as an invoice, are the byproduct of other goods and services. The quality of these data is extremely important. Supplier management is the overall program for managing suppliers, including selecting them, ensuring that they understand what is expected, measuring performance against expectations, and negotiating improvements to close gaps.

Process Management. Well-managed information chains ensure that data values created throughout will be of high quality. Process management (see Section 6) provides the infrastructure and technique needed to do so. Experience confirms that most individual functions within a single organization work fairly well. But "problems" and/or opportunities occur when organizational boundaries are crossed (and as noted in Figure 34.1, data cross organizational lines in the blink of an eye). Process management focuses specifically on organizational interfaces. It provides a structured framework for utilizing many of the techniques presented in the next subsection and is a proven method for making and sustaining improvements to data.

Change Management. Most enterprises/organizations have first-generation data quality systems. Second-generation systems require them to think and act differently and change is always difficult. Experience shows that when change is managed, the risks can be reduced (Kotter 1996). Enterprises are advised to be conscious of these issues and actively address them.

Database of Record. Most enterprises have too much redundant data. But how to manage and reduce redundancy? It is natural to try to do so by recognizing an official "standard" data source to be used throughout the enterprise. But people have resisted what they perceive to be authoritarian directives about data, particularly when the designated source is difficult to use, inaccessible, out-of-date, or otherwise of low quality. A database of record addresses this issue directly. It provides a set of quality standards (for example, accuracy greater than 99.5 percent) and a manager ("custodian," "keeper," or "steward") charged with ensuring that the source satisfies them. Only then is the database designated as the "approved master source" for that data.

Strategic Data Management. In most enterprises, today's data do not meet the enterprise's current needs. And all expectations are that the typical enterprise's data needs will grow exponentially in the future. Strategic data management aims to ensure that the enterprise's top-line business strategy is "data-enabled," that the enterprise has the data and information assets (especially data sources, information chains, and the ability to exploit them) to effect its strategy.

Training and Education. Management must ensure that all involved have the knowledge, skills, and tools needed to improve data quality.

Technical Capabilities. We now turn our attention to the technical capabilities needed to focus improvement efforts, make and sustain gains, and ensure continuity.

Identification of Information Chains. Not all data are created equal. The enterprise should define and execute a process (either formal or informal) of identifying the data and information chains most critical to the enterprise. The data quality program should be focused on these assets.

Customer Needs Analysis. The customer is the final arbiter of quality, so understanding who they are and their needs, prioritizing those needs, and communicating to those who need to know is essential. Good second-generation systems focus incessantly on users of data, formally document and keep customer needs, and keep them current.

Process Description. Although data cross organizational boundaries in remarkably confused paths, managers need to understand these paths. Process description is the means of acquiring and documenting this understanding. There are many ways to conduct this process. Good descriptions of information chains include suppliers, the steps taken to produce information products, customers, and all other essential aspects (including data, organization, supporting technologies, etc.) that may impact performance. The simple act of describing what is actually happening often brings incongruities to the surface, and these incongruities become opportunities for improvement.

Measurement. Management of data quality proceeds on the basis of fact (see also Section 9). And facts are needed at many levels of the enterprise. At a low level, measurement is the process of quantifying information chain performance, including aligning measures to customer needs, specifying the measurement protocol, collecting data, and presenting results. Good second-generation practice emphasizes measurement *within information chains* at the points of new data creation. "Data tracking" is one method to do so (see Redman 1995, 1996). It also helps solve the intangibility issue noted previously. At the enterprise level, measurement also refers to the overall system or collection of measurement processes and management summaries to track overall progress. One important overall measurement may be the cost of poor data quality.

Quality Planning.   At the enterprise level, planning is the regular (e.g., annual) process of setting quality goals or targets and/or improvement and putting in place the means to achieve those goals. At the "project" level, planning is a team-oriented process that creates or replans new information products, information chains, or controls to meet specific customer needs. The first steps of a re-engineering project also constitute quality planning.

Quality Control.   Quality control is the process of evaluating (quality) performance, comparing that performance with standards or goals, and acting on the difference (see Section 4). Establishing and maintaining control is essential because it provides the basis for predicting that errors will not occur in the future.

Quality Improvement.   Good second-generation practice calls for use of a structured team process for reducing errors and other deficiencies in information chains and information products. This process involves identifying and selecting improvement opportunities (projects), chartering teams to make improvements, completing those projects, and "holding the gains" (see Section 5). Like the other elements of the Juran trilogy, quality improvement is essential to second-generation data quality systems. Participating in and completing projects is the vehicle by which most people are inculcated into the "quality culture."

It is important to recognize the proper role of computer, networking, and database technologies in data quality improvement. These technologies have been quite effective in enabling well-established and well-managed information chains to perform faster and cheaper. But technology alone [see also Landauer (1995) and Strassman (1997)] is not the key to improving data quality. Indeed an overreliance on technology appears to exacerbate inadequacies. Information chains must be put into reasonable working order before applying the newest technology. For years, quality gurus have advised against automating ineffective factories. So this prescription simply extends that time-tested maxim to data.

Process (Re-)design.   Some information chains are fated to yield poor results because they are poorly designed. A chain in which raw data must be manually rekeyed into several computers is a good example. It simply won't work effectively or efficiently. The planned "blueprint" of an information chain (including suppliers, the sequence of work activities, interfaces, supporting technologies, management accountabilities, and information products delivered to customers) should incorporate principles of good design. For example, the tools needed by the process owner (i.e., measurement and control) should be incorporated into a process design. Other best practices are given in Redman (1996, Chapter 14).

Inspection and Test (Data Editing).   Data editing is the process of determining if data values satisfy consistency criteria. Editing is usually applied in data clean-ups and, as such, is more properly a first-generation technique. Edits may be employed in second-generation systems at the points of data creation and entry to prevent defective data from proceeding further. Failures (and corrections) must be counted and classified and used in control and improvement projects. Editing may also be employed as part of a supplier program and is sometimes necessary for complex information products.

Quality Assurance.   A quality assurance program consists of audits to determine the degree to which the data quality system, as designed, is deployed and functioning.

Document Assurance.   Procedures are needed to control documents and data/information that relate to requirements of the data quality system. In particular, a master list of all current policies, procedures, and results should be accessible to those that need it.

Rewards and Recognition.   As noted previously, second-generation data quality systems require people to think and act differently. Rewards and recognition that provide the reinforcement and feedback of superior performance should be part of the enterprise's merit/compensation rating system. Specifically, compensation increases and promotion decisions should take employees' contributions to data quality into account.

Domain Knowledge.   The enterprise should take care to learn more about data as a resource (see Levitin and Redman 1998), the data they use, the way data create value, and techniques to manage them. As we have noted, data are different from other resources. Intimate knowledge of their properties is critical.

Standards.   Standards are accepted definitions, rules, and bases of comparison, usually developed and agreed-upon by a body with authority to do so. Standards can advance a data quality program. For example, a standard definition of common terms such as "customer" could help reduce the number of partially redundant and disparate databases. But standards have been very difficult to effect in most enterprises. The various organizations just can't seem to agree on a common definition. The root cause of the issue seems to be that the various organizations think differently about their customers and their thoughts are captured in their data models. They fear that a standard definition would compromise their relationships with customers.

There is no easy resolution to this issue. As a practical matter, it is usually best to postpone work on standards until other elements of the enterprise's second-generation data quality system are in place. And the Data Council should first gain some experience with relatively less contentious standards.

Quality Handbook.   As a quality system matures, it is appropriate to codify it into a published "book" containing an enterprise's quality policy, important concepts and definitions, and procedures. Ideally the handbook is customized to the enterprise, general enough for widespread use, yet specific enough to help focus the enterprise's efforts.

CONCLUSIONS

An important analogy likens a database to a lake. Water represent the data and the streams feeding the lake represent information chains. Animals who drink from the lake and others who enjoy the lake represent users. Presented with a polluted lake, the community has three choices:

- It can suffer the consequences.
- It can filter the lake water.
- It can eliminate sources of pollution upstream.

Experience confirms that the best results are obtained with the third approach. For data, this analogy explains precisely why second-generation data quality systems are so successful.

ACKNOWLEDGEMENTS

The author wishes to thank Bill Barnard of the Juran Institute, Bob Pautke of AT&T Labs, and Charles and Joan Redman of Nashville, Tennessee, for their insights, suggestions, and improvements to this section. The author also wishes to thank his wife Nancy for carefully proofreading drafts.

REFERENCES

Davenport, T. H., Eccles, R. G., and Prusak, L. (1992). "Information Politics." *Sloan Management Review,* vol. 34, no.1, pp. 53–65.

English, L. P. (1998). "The High Costs of Low Quality Data." *DM Review,* January.

Fox, C., Levitin, A., and Redman, T. (1994). "The Notion of Data and Its Quality Dimensions." *Information Processing and Management,* vol. 30, no. 1, pp. 9–19.

Ishikawa, K. (1990). *Introduction to Quality Control.* 3A Corporation, Tokyo.

Kotter, J. P. (1996). *Leading Change.* Harvard Business School Press, Cambridge, MA.

Landauer, T. K. (1995). *The Trouble with Computers: Usefulness, Usability, and Productivity.* The MIT Press, Cambridge, MA.

Levitin, A. V., and Redman, T. C. (1993, 1995). "A Model of Data (Life) Cycles with Application to Quality." *Information and Software Technology,* vol. 35, no. 4, April, pp. 217–223.

Levitin, A. V., and Redman, T. C. (1995). "Quality Dimensions of a Conceptual View." *Information Processing and Management,* vol. 31, no. 1, January, pp. 81–88.

Levitin, A. V., and Redman, T. C. (1998). "Data vs. Traditional Resources: Properties, Implications, and Prescriptions for Management." submitted to *Sloan Management Review.*

Redman, T. C. (1995). "Opinion: Improve Data Quality for Competitive Advantage." *Sloan Management Review,* vol. 36, no. 2, winter, pp. 99–107.

Redman, T. C. (1996). *Data Quality for the Information Age.* Artech House, Norwood, MA.

Redman, T. C. (1998). "The Impact of Poor Data Quality on the Typical Enterprise." *Communications of the ACM,* vol. 41, no. 2, February, pp. 79–82.

Strassman, P. (1994). *The Politics of Information Management.* The Information Economics Press, New Canaan, CT.

Strassman, P. (1997). *The Squandered Computer: Evaluating the Business Alignment of Information Technologies.* The Information Economics Press, New Canaan, CT.

Thompson, R. J. (1997). "The Unspeakably High Cost of Noncompliance." *InformationWeek,* June 30.

Wang, R. Y., and Strong, D. M. (1996). "Beyond Accuracy: What Data Quality Means to Data Consumers." *Journal of Management Information Systems,* vol. 14, no. 4, spring, pp. 5–34.

SECTION 35
QUALITY AND SOCIETY

Joseph M. Juran

THE BACKGROUND

Human society has depended on quality since the dawn of history. In primitive societies this dependence is on the quality of natural goods and "services." Human life can exist only within rather narrow limits of climatic temperature, air quality, food quality, and so on. For most primitive societies, life even within these narrow limits is marginal, and human beings in most primitive societies live precariously. Hours of work are often long and exhausting. Life spans are shortened by malnutrition, disease, natural disasters, and so on. To reduce such risks primitive societies created nonnatural aids to their mental and physical capabilities, aids such as:

- Division of labor.
- Community forms of society, such as villages.
- Artificial shelter, e.g., houses.
- Processing of natural materials to produce nonnatural goods such as pottery, textiles, tools, weapons.
- Lessons learned. The experience of the past—when to plant crops, which berries are poisonous— is handed down from generation to generation.

The subsequent growth of commerce and of science and technology greatly expanded the extent and variety of nonnatural goods and services. As a result human beings in many modern industrial societies live longer and safer lives. They are largely shielded from the perils which their ancestors faced. However, all those nonnatural goods and services have created a new dependence, and therefore new risks.

Life behind the Quality Dikes. Years ago the author coined the phrase "life behind the quality dikes" to designate these new risks (Juran 1969). In industrial societies, great masses of human beings place their safety, health, and even their daily well-being behind numerous protective "dikes" of quality control. For example, the daily safety and health of the citizenry now depend absolutely on the quality of manufactured products: drugs, food, aircraft, automobiles, elevators, tunnels, bridges, and so on. In addition, the very continuity of our life style is built around the continuity of numerous vital services: power, transport, communication, water, waste removal, and many others. A major power failure paralyzes the lives of millions.

There are numerous minor breaks in the quality dikes—occasional failures of goods and services. These are annoying as well as costly. Far more serious are the terrifying major breaks such as Chernobyl, Bhopal, Three Mile Island.

Not only individuals but also nations and their economies live dangerously behind the dikes of quality control. National productivity relies on the quality of product and process design. National defense relies on the quality of complex weaponry. The growth of the national economy is keyed to the reliability of its systems for energy, communication, transport, and so on.

So while technology confers wonderful benefits on society, it also makes society dependent on the continuing performance and good behavior of technological goods and services. This is life behind the quality dikes—a form of securing benefits but living dangerously. Like the Dutch who have reclaimed much land from the sea, we secure benefits from technology. However we need good dikes—good quality—to protect us against the numerous service interruptions and occasional disasters. These same risks have also led to legislation which at the outset was bitterly opposed by industrial companies. Since then it has become clear that the public is serious about its concerns. What is encouraging is that users (whether individuals or nations) are willing to pay for good dikes.

The ability to cope with breaks in the quality dikes varies remarkably among users. Large organizations (industrial companies, governments) employ technologists or otherwise use their economic and political strengths to plan, control, and improve quality. In contrast, the individuals (consumers, the citizenry) find themselves pitted against forces which to them seem as mysterious and overpowering as natural forces seemed to their primitive ancestors.

Any one individual has only a very limited capacity to deal with these forces. However, these individuals are very numerous. Collectively their economic and political powers are formidable. These powers have emerged as a movement generally called *consumerism*. This movement, though loosely organized, has become influential in providing individual members of society with protection and recourse relative to breaks in the quality dikes.

THE GROWTH OF CONSUMERISM[1]

Consumerism is a popular name for the movement to help consumers solve their problems through collective action.

No one knows whether the *rate* of consumer grievances has grown over the centuries. However we know that the *volume* of grievances has grown to enormous numbers due to the growth in volume of goods and services. By the mid-twentieth century, consumer frustrations had reached levels which stimulated attacks on industrial companies for their alleged responsibility for consumers' problems. Then, when most companies failed to take appropriate action, the resulting vacuum attracted numerous contenders for leadership of a consumerism movement: government agencies, politicians, social reformers, consumer advocates, consumer associations, standardization organizations, independent test laboratories, and still others. A risk arose that a bargaining agent would emerge to intervene between industrial companies and their customers.

[1]*Note:* The text for this topic includes some extracts and paraphrases of material from Juran (1995, chap. 17).

Consumer Problems and Perceptions.   Starting in the 1970s researchers began to identify the dominant consumer problems as well as the perceptions of the groups in interest: consumers, consumer organizations, government, business, insurance companies, and so on. Table 35.1 lists the major quality-oriented consumer problems as derived from one such study (Sentry 1976, p. 5).

Consumer expectations sometimes rise faster than the market rate of improvement (Sentry 1976, p. 30.) In addition, consumer perceptions can differ from the realities. For example, many consumers believed that quality of product was getting worse; that "products don't last as long as they used to." Yet the author's studies of specific product lines have almost always found that quality has kept improving.

Consumers are generally more negative than positive on the attitudes of business toward problems of consumers. They strongly favor competition as a means of ensuring higher quality, safer products, and better prices. They also feel that most advertising is misleading, and that much is seriously misleading (Sentry 1976, pp. 7, 8, 12).

During the 1970s, consumer perceptions of the job done by specific industries varied widely. The favorable perceptions included banks, department stores, small shopkeepers, telephone companies, supermarkets and food stores, and airlines. At the other end of the spectrum, consumers had poor perceptions of car manufacturers, the advertising industry, the oil industry, garages and auto mechanics, used car dealers (Sentry 1976, p. 13).

While consumer perceptions are sometimes in error, the perceptions are important in their own right. People act on their perceptions, so it is important to understand what are the perceptions of consumers. For additional findings on consumer perceptions, see the study sponsored by the American Society for Quality Control (ASQC 1980).

Consumers generally felt that there was much they could do to help themselves relative to quality. They felt that the necessary product information was available but that the information was not being used by consumers. They had similar views with respect to product safety. They generally felt that most products were safe if used properly; also that many product safety problems arose because of failure to read the instructions properly (Sentry 1976, pp. 9, 10).

Remedial Proposals.   There are a number of these, amid much difference of opinion. The differences arise in part because of the impact on costs and prices (see below). In addition there are differences due to a contest for power. The various consumer organizations and government departments all feel that they should play larger roles, and that certain traditional powers of business should be restricted.

Ideally, the remedies should eliminate the causes of consumer problems at their source. The consumerism movement has been skeptical that such prevention will take place at the initiative of the industrial companies. Hence the main proposals have related to establishing ways to enable consumers to judge beforehand whether they are about to buy trouble.

Access to Information before Purchase.   Consumers could make better buying decisions if they had access to information on competitive product test data, field performance, and so on. Many

TABLE 35.1   Major Consumer Quality-Oriented Problems

Poor quality of many products
Failure to live up to advertising claims
Poor quality of after-sales service and repairs
Misleading packaging or labeling
Futility of making complaints: nothing substantial will be done
Inadequate guarantees or warranties
Failure of companies to handle complaints properly
Too many dangerous products
The absence of reliable information about the different goods and services
Not knowing what to do when something goes wrong with a purchased product
The difficulty of choosing which of the competing products to buy

Source: Sentry (1976), p. 5.

industrial companies possess such information but will not disclose it—they regard it as proprietary. They do disclose selected portions, but mainly to aid in sale of the product. The risk of bias is obvious.

Consumers' needs for information extend also to after-sale service, response to complaints, and so on. Here again, the companies regard such information as proprietary.

The lack of information from industrial companies has created a vacuum which has attracted alternative sources of product information to help consumers judge which products to buy and which to avoid. One such source is test laboratories which are independent of the companies that make and sell the products.

Under this concept a competent laboratory makes an expert, independent evaluation of product quality so that consumers can obtain the unbiased information needed to make sound purchasing judgments. Adequate consumer test services require professionals and skilled technicians, well-equipped test laboratories, acquisition of products for test, and dissemination of the resulting information. Financing of all these needs is so severe a problem that the method of financing determines the organization form and the policies of the test service.

Product Testing: Consumer-Financed.   In this form the test laboratory derives its income by publishing its test results, usually in a monthly journal plus an annual summary. Consumers are urged to subscribe to the journal on the ground that they will save money by acquiring the information needed to make better purchasing decisions. Advertisements of such test laboratories raise questions such as "Would you pay $100 for an appliance when independent tests show that a $75 appliance is just as good?"

In their operation, these consumer-financed test laboratories buy and test competitive products, evaluate their performance and failures, compare these evaluations with the product prices, and rate the products according to some scale of relative value. The ratings, test result summaries, descriptions of tests conducted, and so on, are published in the laboratory's journals. The industrial companies play no role in the testing and evaluation. In addition, the companies are not permitted to quote the ratings, test results, or other material published in the journals.

It is seen that the service offered to consumers consists of:

1. The laboratory's test results, which are mainly objective and unbiased.
2. Judgments of values which are subjective and carry a risk of bias, i.e., the stress of the advertising (showing the consumer that some lower-priced products are as good as higher-priced products) creates a bias against higher-priced products. More importantly, the judgments are not necessarily typical of consumers' judgments.

Despite the obvious problems of financing a test service out of numerous consumer subscriptions, there are many such services in existence in affluent and even in developing countries. In the United States, the most widely known source of such tests is Consumers Union. The test results are published in the journal *Consumer Reports.*

Product Testing: Government-Financed.   Governments have long been involved in matters of product quality, originally to protect the safety and health of the citizenry, and later, the environment. For elaboration, see below, under Government Regulation of Quality.

The most recent extension has been in the area of consumer economics. Some of this has been stimulated by the consumerism movement. A byproduct has been the availability of some product test results and other quality-related information. This information is made available to the public, whether in published form, or on request.

Government-Subsidized Tests.   In some countries, the government subsidizes test laboratories to test consumer products and to publish the results as an aid to consumers. The rationale is that there is a public need for this information, and that hence the costs should be borne by the public generally.

Mandated Government Certification.   Under this concept, products are required by law to be independently approved for adequacy before they may be sold to the public. This concept is applied in many countries to consumer products for which human safety is critical (e.g., pharmaceuticals, foods). For other products, there has been a sharp division in practice. Generally, the market-based economies have rejected mandated government certification for (noncritical) consumer products, and have relied on the forces of the competitive marketplace to achieve quality. In contrast, the planned economies, as exemplified by the former Soviet Union, went heavily into the setting of standards for consumer products and the use of government laboratories to enforce compliance to these standards (which had the force of law).

Product Testing: Company-Financed. In this form, industrial companies buy test services from independent test laboratories in order to secure the mark (certificate, seal, label) of the laboratory for their products. In some product categories it is unlawful to market the products without the mark of a qualified testing service. In other cases it is lawful, but the mark is needed for economic reasons—the insurance companies will demand extraordinarily high premiums or will not provide insurance at all.

An example of a sought-after mark is that of Underwriters Laboratories, Inc. (UL). Originally created by the National Board of Fire Underwriters to aid in fire prevention, UL (now independent) is involved in the general field of fire protection, burglary protection, hazardous chemicals, and still other matters of safety. Its activities include:

- Developing and publishing standards for materials, products, and systems.
- Testing manufacturers' products for compliance with these standards (or with other recognized standards).
- Awarding the UL mark to products which comply. This is known as "listing" the products.

Numerous other laboratories are similarly involved in safety matters, e.g., steam boilers and marine safety. Some of these laboratories have attained a status in their specialty which confers a virtual monopoly on performing the tests.

Another purpose of securing a mark from an independent test service is to help market the product. Companies vary in their views of the value of such "voluntary" marks. Strong companies tend to feel that their own brand or mark carries greater prestige than that of the test laboratory, and that the latter has value only for weak companies. The test services which offer this category of marks vary widely in their purpose and in their objectivity.

In some countries the voluntary mark is offered by the national standardization bodies, such as Japan Standards Association or the French AFNOR. The mark is awarded to products that meet their respective product standards. Companies that wish to use the JIS mark (Japan Industrial Standards) or the NF mark (Normale Francais) must submit their products for test and must pay for the tests. If the products qualify, the companies are granted the right to use the marks.

Data Banks on Business Practices. Many consumer grievances are traceable to company business practices, such as evasiveness in meeting the provisions of the guarantee. The Pareto principle applies—a comparative few companies are named in the bulk of the grievances. In this way, a data bank on company business practices can help to identify the vital few "bad guys" and aid in reducing their influence.

The organizations known as Better Business Bureaus (BBB) created one such data bank. A description of BBBs and what they do is given in the following quotation from *Consumer's Resource Handbook* (1980):

> BBBs are non-profit organizations sponsored by private businesses. There are 147 BBB locations across the U.S. today, sponsored by local and national business. While BBBs vary from place to place, most offer a variety of basic services. These include: general information on products or services, reliability reports, background information on local businesses and organizations, and records of companies' complaint handling performances. Depending on the policy of the individual BBB, it may or may not tell you the nature of the complaints registered against a business. BBBs accept written complaints, and will contact a firm on your behalf.

The BBB receives complaints from consumers (among others) on unethical business practices, and endeavors as an ombudsman (see below) to get these practices changed. When citizens call the BBB, they are able to learn whether the company under inquiry has a record of complaints lodged against it.

BBBs also are active in helping consumers and local business firms to settle consumer complaints. (See below).

Consumer Education.　Beyond product tests and data banks on business practices, still other forms of before-purchase information are available to consumers. Some government departments publish information describing the merits (or lack of merits) of products and product features in general. However, the most often used source of product information is advice received from relatives and friends who have experience to share. Consumers regard such advice as reliable (Sentry 1976, p. 55).

The limiting factors in consumer use of the available independent data sources are the consumers themselves. Fewer than 10 percent of families in the United States subscribe to *Consumer Reports.* The lack of use of other helpful information (much of which is available free to consumers) may well have its origin in a school system that makes little provision to educate children in one of the major roles they will play as adults—the role of consumer.

Some observers explain "unwise" consumer behavior on grounds other than lack of education. They note that many consumers spend money on narcotic drugs, alcohol, or tobacco; kill themselves (and others) by driving too fast or in a drunken state; eat "junk" food; gamble their money away. It is understandable that some skeptics conclude that consumers who are gullible or stupid will learn only from their mistakes.

Consumer organizations are quite aware that "most consumers do not use the information available about different products in order to decide to buy one of them." (Sentry 1976, p. 10). However, consumer organizations never characterize consumers (their clientele) as being gullible or stupid. (*Consumer Reports* 1977).

The Standards Organizations.　There are many of these. For example, in the United States, those of importance to consumers include:

- Leading manufacturers and merchants, whose standards exert wide influence on their suppliers and competitors
- Industry bodies such as the American Gas Association or the Association of Home Appliance Manufacturers
- Professional organizations such as the American Society for Testing and Materials
- Independent agencies such as Underwriters Laboratories, Inc.
- The American National Standards Institute (ANSI), which is a recognized clearinghouse for committees engaged in setting national standards and is the official publisher of the approved standards
- The National Institute of Standards and Technology, formerly the National Bureau of Standards, the government agency which establishes and maintains standards for metrology

Standards for Consumer Products.　Awarding a mark presupposes the existence of some standard against which the product can be tested on an objective basis. Providing such standards for consumer products has not received the priorities given to standards for metrology, basic materials, and other technological and industrial needs. However, the consumerism movement has very likely stimulated the pace of developing these standards. Industry associations especially have been stimulated to undertake more of this type of activity.

A serious limitation on creating standards for consumer products is the pace of product obsolescence versus the time required to set standards. Usually, it takes years to evolve a standard due to the need for securing a consensus among the numerous parties in interest. For subject matter such as metrology or basic materials, the standards, once approved, can have a very long life. However, for consumer products the life is limited by the rate of obsolescence, and for many products the life of the standard is so short as to raise serious questions about the economics of doing it at all.

In some cases the obsolescence is traceable to the zeal of the marketers. For example, one measure of the quality of mechanical watches has been the number of jewels. Then some manufacturers began to include nonfunctional jewels to provide a basis for claiming higher quality. It became necessary to redefine the word "jewel."

A further problem in standards for consumer products is that the traditional emphasis of the standardization bodies has been on "time zero"—the condition of the product when tested prior to use. However, many products, especially the most costly, are intended to give service for years. Many consumer problems are traceable to field failures during service, yet most consumer product standards do not adequately address the "abilities"—reliability, maintainability, and so on. (See generally Juran 1970.)

Objectivity of Test Services. Unless the testing service is objective, consumers may be misled by the very organization on which they thought they could rely. The criteria for objectivity include:

- *Financial independence:* The income of the test service should have no influence on the test results. This independence is at its best when the income is derived from sources other than the company whose products are under test. Failing this, the payments by the company should be solely for the testing service and in no way contingent on the test results.

  One example of failure to meet this criterion is any test service which carries on the dual activities of (1) offering a mark based on product test and approval and (2) publishing a journal of general circulation in which companies that receive the mark are required to place advertisements. In such cases the risk of conflict of interest is very high, so consumers should be cautious about giving credence to such marks.

- *Organizational independence:* The personnel of the test service should not be subordinate to the companies whose products are undergoing test.

- *Technological capability:* This obvious need includes a qualified professional staff, adequate test equipment and competent management. Whether the managers should be the sole judges of such capabilities is open to question.

So important is the question of objectivity that in cases of government controls on quality it is usual to write into the statute the need for defining criteria for what constitutes a qualified test laboratory. The administrator of the act then becomes responsible for certifying laboratories against these criteria.

The Resulting Information. Consumer test services offer consumers a wide range of information. The principal forms include:

- Comparative data on competitive products for (1) price and (2) fitness for use, plus judgments of comparative values. In this form, the information is also a recommendation for action.

- Data on product conformance to the standards. In this form, consumers are thrown on their own to discover competitive prices and to make a judgment on comparative fitness for use. For many consumers, it is a burden to provide this added information.

- Evidence of product conformance to the standard (through the mark.) Here the consumer is largely asked to equate the standard with fitness for use and to use other means to discover competitive differences and competitive prices.

Information on conformance to standard is quite useful to industrial buyers, but less so to consumers. For consumers the optimal information consists of comparative data on fitness for use, plus comparative data on cost of usage.

Traditional test services do not provide adequate information as to certain important quality problems faced by consumers: products arrive in defective condition; products fail during use; response to consumer complaints is poor.

- *Products defective on arrival:* Test services typically conduct their tests on a small sample—one or a few units of product. These nevertheless enable the test service to judge whether the product design can provide fitness for use. However, the sample is too small to provide information on how often units will be defective on arrival.

- *Products fail during use:* Traditionally, test services have evaluated consumer products at "time zero"—prior to use. For long-life products this is no longer good enough–there is need for information on field failure rates. Some test services now do conduct a degree of life testing, but the number of units tested is too small to predict field failure rates. There are some efforts to secure such information through questionnaires sent to consumers. An alternative source is to secure information from the repair shops.

- *Poor response to customer complaints:* Here the situation is at its worst. The test laboratory and its instruments are irrelevant, since the needed information relates to the competence, promptness, and integrity of the service organizations.

Remedies after Purchase.   Consumers who encounter product quality problems during the warranty period have a choice of alternatives. They may be able to resolve the problem unaided; i.e., they study the product information and then apply their skills and ingenuity. More usually they must turn to one of the companies directly in interest: the merchant who sold them the product; the manufacturer who made the product. If none of these provides satisfaction, the consumers have still other alternatives for assistance (see below).

Warranties.   Quality warranties are a major after-purchase aid to consumers. However, many consumers feel that warranties are not understandable. In addition, most feel that warranties are written mainly to protect manufacturers rather than consumers. Nevertheless, consumers are increasingly making the warranty an input to their buying decisions. This means also that warranties are increasingly important as marketing tools (Sentry 1976, pp. 14, 15).

By a wide margin, consumers complain to the merchant (store, dealer) rather than to the manufacturer. A third choice is to complain to the Better Business Bureau. (Sentry 1976, p. 15).

Better Business Bureaus (BBB).   The following is quoted from *Consumer's Resource Handbook* (1980).

> BBBs attempt to settle consumer complaints against local business firms. A BBB considers a consumer complaint settled when:
>
> **1.** The customer receives satisfaction.
> **2.** The customer receives a reasonable adjustment—in other words, gets what was paid for.
> **3.** The company provides proof that the customer's demands are unreasonable or unwarranted.
>
> The BBB does not: judge individual products or brands, handle complaints concerning the prices of goods or services, or give legal advice.
>
> More than 100 of the 147 BBBs offer binding arbitration to those who ask for it, and others are beginning programs. Arbitration is a way for people to settle a dispute by having an impartial person or board (people who have nothing to gain or lose from the decision) decide the outcome of the dispute. In arbitration, parties are bound by the decision, and it can be enforced by the courts. Do not enter arbitration lightly since you must follow the decision that is made.
>
> BBBs also handle false advertising cases. Your local BBB looks into local advertising, while the BBBs' National Advertising Division (NAD) checks out complaints about national advertising.
>
> How to Reach Them: To find a BBB, check your local phone book, local consumer office, or library.

The Ombudsman.   Ombudsman is a Swedish word used to designate an official whose job is to receive citizens' complaints and to help them secure action from the government bureaucracy. The ombudsman is familiar with government organization channels and is able to find the government official who has the authority or the duty to act. The ombudsman has no authority to compel action, but has the power to publicize failures to act.

The concept of the ombudsman has been applied to problems in product quality. Some companies have created an in-house ombudsman and have publicized the name and telephone number. Consumers can phone (free of charge) to air grievances and to secure information. In the United States a more usual title is Manager (Director), Consumer Affairs (Relations). Such a manager usually carries added responsibilities for stimulating changes to improve relations with consumers on a broad basis. In one company these efforts resulted in programs to effectuate a consumer "bill of rights," which includes rights to safety, to be informed, to choose, to be heard, and to redress. For elaboration, see Peterson (1974).

Another form is the industry ombudsman. An example is the Major Appliance Consumer Action Panel (a group of independent consumer experts) created by the Association of Home Appliance Manufacturers to receive complaints from consumers who have not been able to secure satisfaction locally.

Still another form is the Joint Industry-Consumer Complaint Board. Examples are the government-funded boards which mediate and adjudicate consumer disputes in some Scandinavian industries. The boards have no power to enforce their awards other than through publicity given to unsatisfied awards. Yet they have met with wide acceptance by and cooperation from the business people.

The concept of the ombudsman is fundamentally sound. It is widely supported by consumers and regulators as well as by a strong minority of business managers (Sentry 1976, p. 77). Some newspapers provide an ombudsman service as part of their department of Letters to the Editor.

Mediation. Under the mediation concept, a third party—the mediator—helps the contestants to work out a settlement. The mediator lacks the power of enforcement—there is no binding agreement to abide by the opinion of the mediator. Nevertheless mediation stimulates settlements. Best (1981) reports that the New York City Department of Human Affairs achieved a 60 percent settlement rate during 1977 and 1978.

The mediation process helps to open up the channels of communication and thereby to clear up misunderstandings. In addition, an experienced mediator exerts a moderating influence which encourages a search for a solution.

Arbitration. Under arbitration the parties agree to be bound by the decision of a third party. Arbitration is an attractive form of resolving differences because it avoids the high costs and long delays inherent in most lawsuits. In the great majority of consumer claims, the cost of a lawsuit is far greater than the amount of the claim. Nevertheless there are obstacles to use of the arbitration process. Both parties must agree to binding arbitration. There is need to establish local, low-cost arbitration centers and to secure the services of volunteer arbitrators at nominal fees or no fees. These obstacles have limited the growth of use of arbitration for consumer complaints.

Consumer Organizations. There are many forms of consumer organizations. Some are focused on specific products or services such as automotive safety or truth in lending. Others are adjuncts of broader organizations such as labor unions or farm cooperatives. Still others are organized to deal broadly with consumer problems. In addition, there are broad consumer federations, national and international, which try to improve the collective strength of all local and specialized consumer groups.

Government Agencies. These exist at national, state, and local levels of government. All invite consumers to bring unresolved complaints to them as well as to report instances of business malpractice. These complaints aid the agency in identifying widespread problems, which, in turn, become the basis for:

1. Conducting investigations in depth
2. Proposing new legislation
3. Issuing new administrative regulations

The agencies also try to help complaining consumers, either in an ombudsman role or by threat of legal action. However, in practice, broad government agencies are unable to become involved in specific consumer grievances due to the sheer numbers. See below, under Government Regulation of Quality; The Enforcement Process.

No Remedy.   Under the prevailing free-enterprise, competitive market system, many valid consumer complaints result in no satisfaction to the consumer. Nevertheless the system includes some built-in stabilizers. Companies which fail to provide such satisfaction also fail to attract repeat business. In due course they mend their ways or lose out to companies who have a better record of providing satisfaction. In the experience of the author, every other system is worse.

Perceptions of the Consumer Movement.   There is wide agreement, including among business managers, that "the consumer movement has kept industry and business on their toes." There is also wide agreement that the consumer movement's demands have "resulted in higher prices." Despite this, most of the public feels that the "changes are generally worth the extra cost." Consumers feel strongly that the consumer advocates should consider the costs of their proposals. However, a significant minority of the consumers believe that the advocates do not consider the costs involved (Sentry 1976, pp. 39, 40, 42, 47).

GOVERNMENT REGULATION OF QUALITY

From time immemorial, "governments" have established and enforced standards of quality. Some of these governments have been political—national, regional, local. Others have been nonpolitical: guilds, trade associations, standardization organizations, and so on. Whether through delegation of political power or through long custom, these governing bodies have attained a status which enables them to carry out programs of regulation as discussed below.

Standardization.   With the evolution of technology came the need for standardizing certain concepts and practices.

- *Metrology:* One early application of standardization was to units of measure for time, mass, and other fundamental constants. So basic are these standards that they are now international in scope.
- *Interchangeability:* This level of standardization has brought order out of chaos in such day-to-day matters as household voltages and interchangeability of myriads of the bits and pieces of an industrial society. Compliance is an economic necessity.
- *Technological definition:* A further application of standardization has been to define numerous materials, processes, products, tests, and so on. These standards are developed by committees drawn from the various interested segments of society. While compliance is usually voluntary, the economic imperatives result in a high degree of acceptance and use of these standards.

The foregoing areas of regulation are all related to standardization, and have encountered minimal resistance to compliance. Other areas do encounter resistance, in varying degrees.

Safety and Health of the Citizenry.   A major segment of political government regulation has been to protect the safety and health of its citizens. At the outset the focus was on punishment "after the fact"—the laws provided punishment for those whose poor quality had caused death or injury. Over the centuries there emerged a trend to regulation "before the fact"—to become preventive in nature.

For example, in the United States there are laws which prescribe and enforce safety standards for building construction, oceangoing ships, mines, aircraft, bridges, and many other structures. Other laws aim at hazards which have their origins in fire, foods, pharmaceuticals, dangerous chemicals, and so on. Still other laws relate to the qualifications needed to perform certain activities essential to public safety and health, such as licensing of physicians, professional engineers, airline pilots. Most recently these laws have proliferated extensively into areas such as consumer product safety, highway safety, environmental protection, and occupational safety and health.

Safety and Economic Health of the State.   Governments have always given high priority to national defense: recruitment and training of the armed forces; quality of the weaponry. With the growth of commerce, laws were enacted to protect the economic health of the state. An example is laws to regulate the quality of exported goods in order to protect the quality reputation of the state. Another example is laws to protect the integrity of the coinage. (Only governments have the right to debase the currency.) In those cases where the government is a purchaser (defense weapons, public utility facilities), government regulation includes the normal rights of a purchaser to assure quality.

Economics of the Citizenry.   Government regulation relative to the economics of the citizenry is highly controversial in market economies. Some of the resistance is on ideological grounds—the competitive marketplace is asserted to be a far better regulator than a government bureau. Other resistance is based on the known deficiencies of the administration of government regulation (see below). Some of the growth of this category of regulation has been stimulated by the consumerism movement.

The Volume of Legislation.   Collectively, the volume of quality-related legislation has grown to formidable proportions. A desk reference book (Kolb and Ross 1980) includes lists (in fine print) of appendixes as follows:

- 21 pages of exposure limits for toxic substances
- 93 pages of hazardous materials and the associated criteria for transportation
- 24 pages of American National Standards for safety and health
- 36 pages of Federal record-retention requirements
- 38 pages of standards-setting organizations

In the United States, much of this legislation is within the scope of the Federal Trade Commission, which exercises a degree of oversight relative to "unfair or deceptive practices in commerce." That scope has led to specific legislation or administrative action relative to product warranties, packaging and labeling, truth in lending, and so on.

In a sense these actions all relate to representations made to consumers by industrial companies. In its oversight the Federal Trade Commission stresses two major requirements:

1. The advertising, labeling, and other product information must be clear and unequivocal as to what is meant by the seller's representation.
2. The product must comply with the representation.

These forms of government regulation are a sharp break from the centuries-old rule of *caveat emptor* (let the buyer beware). That rule was (and is) quite sensible as applied to conditions in the village marketplaces of developing countries. However it is not appropriate for the conditions prevailing in industrialized, developed countries. For elaboration, see Juran (1970).

The Plan of Regulation.

Once it has been determined to regulate quality in some new area, the approach follows a well-beaten path. The sequence of events listed below, while described in the language of regulation by political government, applies to nonpolitical government as well.

The Statute.   The enabling act defines the purpose of the regulation and especially the subject matter to be regulated. It establishes the "rules of the game" and creates an agency to administer the act.

The Administrator.   The post of administrator is created and given powers to establish standards and to see that they are enforced. To this end he or she is armed with the means for making awards and applying sanctions on matters of great importance to the regulated industries.

The Standards.   The administrator has the power to set standards and may exercise this power by adopting existing industry standards. These standards are not limited to products; they may deal with materials, processes, tests, descriptive literature, advertising, qualifications of personnel, and so on.

Test Laboratories.   The administrator is given power to establish criteria for judging the qualifications of "independent" test laboratories. Once these criteria are established, he or she also may have the power to issue certificates of qualification to laboratories which meet the criteria. In some cases administrators have the power to establish their own test laboratories.

Test and Evaluation.   Here there is great variation. In some regulated areas, agency approval is a prerequisite to going to market, e.g., new drug applications or plans for the operation and maintenance of a new fleet of airplanes. Some agencies put much stress on surveillance, i.e., review of the companies' control plans and adherence to those plans. Other agencies emphasize final product sampling and test.

The Seal or Mark.   Regulated products are frequently required to display a seal or mark to attest to the fact of compliance with the regulations. Where the regulating agency does the actual testing, it affixes this mark; e.g., government meat inspectors physically stamp the carcasses.

More usually, the agency does not test and stamp the product. Instead, it determines, by test, that the product *design* is adequate. It also determines, by surveillance, that the companies' systems of control are adequate. Any company whose system is adequate is then authorized to affix the seal or mark. The statutes always provide penalties for unauthorized use of the mark.

Sanctions.   The regulatory agency has wide powers of enforcement, such as the right to:

- Investigate product failures and user complaints
- Inspect companies' processes and system of controls
- Test products in all stages of distribution
- Recall products already sold to users
- Revoke companies' right to sell, or to apply the mark
- Inform users of deficiencies
- Issue cease-and-desist orders

Effectiveness of Regulation.
Regulators face the difficult problem of balance—protecting consumer interests while avoiding creation of burdens which in the end are damaging to consumer interests. In part the difficulty is inherent because of the conflicting interests of the parties. However, much of the difficulty is traceable to unwise agency policies and practices in carrying out the regulatory process. These relate mainly to the conceptual approach, setting standards, the enforcement process, and cost of regulation.

The Conceptual Approach.   An example is seen in the policies employed by the National Highway Traffic Safety Administration (NHTSA) for administering two laws enacted in 1966:

1. The National Traffic and Motor Vehicle Safety Act, directed primarily at the vehicle
2. The Highway Safety Act, directed primarily at the motorist and the driving environment

Even prior to 1966, the automobile makers, road builders, and so on had improved technology to an extent which provided the motorist with the means of avoiding the "first crash," i.e., accidents due to collisions, running off the road, etc. The availability of seat belts then provided the motorist with greatly improved means of protection against the "second crash." This crash takes place when the sudden deceleration of a collision hurls the occupants against the steering wheel, windshield, and so on.

At the time NHTSA was created, the United States' traffic fatality rate was the lowest among all industrial countries. It was also known, from overwhelming arrays of data, that the motorist was the limiting factor in traffic safety:

- Alcohol was involved in about half of all fatal accidents.

- Young drivers (under age 24) constituted 22 percent of the driver population but were involved in 39 percent of the accidents.

- Excessive speed and other forms of "improper" driving were reported as factors in about 75 percent of the accidents. (During the oil crisis of 1974 the mandated reduction of highway speeds resulted in a 15 percent reduction in traffic fatalities, without any change in vehicles.)

- Most motorists did not buy safety belts when they were optional, and most did not wear them when they were provided as standard equipment.

In the face of this overwhelming evidence NHTSA paid little attention to the main problem—improving the performance of the motorists. Instead, NHTSA concentrated on setting numerous standards for vehicle design. These standards did provide some gains in safety with respect to the second crash. However the gains were minor, while the added costs ran to billions of dollars—to be paid for by consumers in the form of higher prices for vehicles.

The policy is seen to have been one of dealing strictly with a highly visible political target—the automobile makers, while avoiding any confrontation with a large body of voters. It was safe politically but it did little for safety. For elaboration, see Juran (1977).

Setting Standards.   A major regulatory question is whether to establish design standards or performance standards.

- Design standards consist of precise definitions, but they have serious disadvantages. Their nature and numbers are such that they often: lack flexibility, are difficult to understand, become very numerous, become prohibitive to keep up to date.

- Performance standards are generally free from the above disadvantages. However, they place on the employer the burden of determining how to meet the performance standard, i.e., the burden of creating or acquiring a design. Performance standards also demand level of compliance officers who have the education, experience, and training needed to make the subjective judgments of whether the standard has been met.

These alternatives were examined by a presidential task force assigned to review the safety regulations of the Occupational Safety and Health Administration. The task force recommended a "performance/hazard" concept. Under that concept, the standard would "codify into a requirement the fact that a safe workplace can be achieved only by ensuring that employees are not exposed to the hazards associated with the use of machines. Under this standard, the employer would be free to determine the most appropriate manner in which to guard against any hazard which is presented, but his compliance with the requirement is objectively measurable by determining whether or not an employee is exposed to the hazard." For elaboration, see OSHA Safety Regulation 1977. For added discussion, see Tye (1988).

The Enforcement Process.   A major deficiency in the regulatory process is failure to concentrate on the vital few problems. Regulatory agencies receive a barrage of grievances: consumer complaints, reports of injuries, accusations directed at specific products, and so on. Collectively the numbers are overwhelming. There is no possibility of dealing thoroughly with each and every case. Agencies which try to do so become hopelessly bogged down. The resulting paralysis then becomes a target for critics, with associated threats to the tenure of the administrator, and even to the continued existence of the agency.

In the United States the Occupational Safety and Health Administration faced just such a threat in the mid-1970s. In response it undertook to establish a classification for its cases based on the seriousness of the threats to safety and health. It also recalled about 1000 safety regulations which were under attack for adding much to industry costs and little to worker safety.

With experience, the agencies tend to adopt the Pareto principle of vital few and useful many. This enables them to concentrate their resources and to produce tangible results.

Choice of the vital few is often based on quantitative data such as frequency of injuries or frequency of consumer complaints. However, subjective judgment plays an important role, and this enables influential special pleaders to secure high priority for cases which do not qualify as being among the vital few.

How to deal with the "useful many" needs for assistance has been a perplexing problem for all agencies. The most practical solution seems to have been to make clear that the agency is in no position to resolve such problems. Instead the agency provides consumers with information and educational material of a self-help nature: where to apply for assistance; how to apply for assistance; what are the rights of the consumer; what to do and not to do. See, for example, *Consumer's Resource Handbook* (1980).

The failure of regulators to deal forthrightly with such consumer problems has no doubt contributed to the mediocre status given to regulators by the public, in response to the question: which (of four options) would you like to be primarily responsible for the job of seeing that consumers get a fair deal? For elaboration, see Sentry 1976, p. 70.

A Rule for Choosing the Vital Few.   In 1972 the author proposed the following as a quantitative basis for separating the vital few from the rest, on matters of safety:

Any hour of human life should be as safe as any other hour.

To effectuate such a policy it is first necessary to quantify safety nationally, on some common basis such as injuries per million worker-hours of exposure. In general, the data for such quantification are already available, though some conversions are needed to arrive at a common unit of measure. For example, statistics on safety at school are computed on the basis of injuries per 100,000 student-days, motor vehicle statistics are on a per 100 million miles of travel basis, and so on.

The resulting national average will contain a relative few situations which are well above the average and a great many which are below. Those above the average would automatically be nominated to membership in the vital few. Those below the average would not be so nominated; the burden of proof would be on any special pleader to show why something below the national average should take priority ahead of the obvious vital few. For elaboration, see Juran (1972).

The Costs and Values of Regulation.   The costs of regulation consist largely of two major components:

1. *The costs of running the regulatory agencies:*   These are known with precision. In the United States they have risen to many billions of dollars per year. These costs are paid for by consumers in the form of taxes that are then used to fund the regulatory agencies.

2. *The costs of complying with the regulations:*   These costs are not known with precision, but they are reliably estimated to be many times the costs of running the regulatory agencies. These costs are in the first instance paid for by the industrial companies, and ultimately by consumers in the form of higher prices. For an example of a study of industry costs, see The Business Roundtable (1978).

The value of all this regulation is difficult to estimate. (There is no agreement on what is the value of a human life.) Safety, health, and a clean environment are widely believed to be enormously valuable. Providing consumers with honest information and prompt redress is likewise regarded as enormously valuable. However such general agreements provide no guidelines for what to do in specific instances. Ideally, each instance should be examined as to its cost-value relationship. Yet the statutes have not required the regulators to do so. The regulators have generally avoided facing up to the idea of quantifying the cost-value relationship.

Until 1994 the support for studying the cost-value relationships came mainly from the industrial companies. For example, a study of mandated vehicle safety systems found that:

"...states which employ mandatory periodic inspection programs do not have lower accident rates than those states without such requirements."

"…only a relatively small portion of highway accidents—some 2 to 6 percent—are conclusively attributable to mechanical defects."

"…human factors (such as excess speeds) are far more important causes of highway accidents than vehicle condition." For elaboration see Crain 1980.

The indifference of regulators to costs inevitably creates some regulations and rigid enforcements so absurd that in due course they become the means for securing a change in policy. The companies call such absurdities to the attention of the media, who relish publicizing them. (The media have little interest in scholarly studies.) The resulting publicity then puts the regulators on the defensive while stimulating the legislators to hold hearings. During such hearings (and depending on the political climate) the way is open to securing a better cost-value balance.

The political climate is an important variable in securing attention to cost-versus-value considerations. During the 1960s and 1970s the political climate in the United States was generally favorable to regulatory legislation. Then, during the 1980s the climate changed, and with it a trend toward requiring cost justifications. This trend then accelerated in late 1994, when the elections enabled the opponents of regulation to gain majority status in the national legislatures. For added discussion, see Dowd 1994.

PRODUCT SAFETY AND PRODUCT LIABILITY

Growth of the Problem. Until the early twentieth century, lawsuits based on injuries from use of products (goods and services) were rarely filed. When filed, they were often unsuccessful. Even if successful, the damages awarded were modest in size.

During the twentieth century these lawsuits have, in the United States, grown remarkably in numbers. By the mid-1960s they were estimated to have reached over 60,000 annually and by the 1970s to over 100,000 per year. (Most are settled out of court.) This growth in numbers of lawsuits has been accompanied by an equally remarkable growth in the sizes of individual claims and damages. From figures measured in thousands of dollars, individual damages have grown to a point where awards in excess of $100,000 are frequent. Damages in excess of $1,000,000 are no longer a rarity.

In some fields the costs of product liability have forced companies to abandon specific product lines or go out of business altogether.

> Twenty years ago, 20 companies manufactured football helmets in the United States. Since that time, 18 of these companies have discontinued making this product because of high product liability costs. (Grant 1994).

Several factors have combined to bring about this growth in number of lawsuits and in size of awards. The chief factors include:

- The "population explosion" of products. The industrial society has placed large numbers of technological products into the hands of amateurs. Some of these products are inherently dangerous. Others are misused. The injury *rate* (injuries per million hours of usage) has probably been declining, but the total *number* of injuries has been rising, resulting in a rise in total number of lawsuits.

- The erosion of company defenses. As these lawsuits came to trial, the courts proceeded to erode the former legal defenses available to companies.

Formerly, a plaintiff's right to sue a manufacturer rested on one of two main grounds:

A *contract* for sale of the product, with an actual or implied warranty of freedom from hazards. Given the contract relationship, the plaintiff had to establish "privity," i.e., that he or she was a party to the contract. The courts have in effect abolished the need for privity by taking the position that the implied warranty follows the product around, irrespective of who is the user.

Negligence by the company. Formerly the burden of proof was on the plaintiff to show that the company was negligent. The courts have tended to adopt the principle of "strict liability" on the ground that the costs of injuries resulting from defective products should be borne "by the manufacturers that put such products on the market rather than by the injured persons who are powerless to protect themselves." In effect, if an injury results from use of a product that is unreasonably dangerous, the manufacturer can be held liable even in the absence of negligence. (Sometimes the injured persons are not powerless. Some contribute to their injuries. However, juries are notoriously sympathetic to injured plaintiffs).

The literature on growth of the problem is extensive, as is the literature on proposals for remedy. See especially Harrington and Litan 1988 and references cited. See also Grant 1994; Smith 1991; Egington 1989; McGuire 1988; Wargo 1987.

Defensive Actions.   The best defense against lawsuits is to eliminate the causes of injuries at their source. All company functions and levels can contribute to making products safer and to improving company defenses in the event of lawsuits. The respective contributions include the following:

- *Top management.*   Formulate a policy on product safety; organize product safety committees and formal action programs; demand product dating and product traceability; establish periodic audits of the entire program; support industry programs which go beyond the capacity of the unaided company. To this list should be added a scoreboard—a measure of the injury rate of the company's products relative to an appropriate benchmark. A useful unit of measure is the number of injuries per million hours of usage, since most major data banks on injuries are already expressed in this form or are convertible to this form.

- *Product design:*   Adopt product safety as a design parameter; adopt a fail-safe philosophy of design; organize formal design reviews; follow the established codes; secure listings from the established laboratories; publish the ratings; utilize modern design technique.

- *Manufacture:*   Establish sound quality controls, include means for errorproofing matters of product safety; train supervisors and workers in use of the product as part of the motivation plan; stimulate suggestion on product safety; set up the documentation needed to provide traceability and historical evidence.

- *Marketing:*   Provide product labeling for warnings, dangers, antidotes; train the field force in the contract provisions; supply safety information to distributors and dealers; set up exhibits on safety procedures; conduct tests after installation, and train users in safety; publish a list of dos and don'ts relative to safety; establish a customer relations climate which minimizes animosity and claims. Contracts should avoid unrealistic commitments and unrealistic warranties. Judicious disclaimers should be included to discourage unjustified claims.

- *Advertising:*   Require technological and legal review of copy; propagandize product safety through education and warnings. Avoid "puffing"—it can backfire in liability suits, e.g, if a product is advertised as "absolutely safe." During advertising review, one of the questions should be "How would this phrase sound in a courtroom?"

- *Customer Service:*   Observe use of the product to discover the hazards inherent during use (and misuse); feed the information back to all concerned; provide training and warnings to users.

- *Documentation:*   The growth of safety legislation and of product liability has enormously increased the need for documentation. A great deal of this documentation is mandated by legislation, along with retention periods. (For a compilation, see Kolb and Ross 1980, pp. 547–584.)

Consumers exhibit a wide range of product knowledge, including the lowest. In consequence, actual use of the product can differ significantly from intended use. For example, some stepladders include a light platform which is intended to hold tools or materials (e.g., paint) but is not intended to carry the weight of the user. However, some users nevertheless do stand on these platforms with resulting injury to themselves.

Most modern policy is to design products to stand up under actual usage rather than intended usage. For added discussion, see Farrow 1991; see also Scofield 1986 relative to product labeling.

Defense against Lawsuits.   The growth of product liability lawsuits has led to reexamination of how best to defend against lawsuits once they are filed. Experience has shown the need for special preparation for such defense, including:

- Reconstruction of the events which led up to the injury
- Study of relevant documents—specifications, manuals, procedures, correspondence, reports
- Analysis of internal performance records for the pertinent products and associated processes
- Analysis of field performance information
- Physical examinations of pertinent facilities
- Analysis of the failed hardware

All this should be done promptly, by qualified experts, and with early notification to the insurance company. For elaboration, see Gray et al. (1975, pp. 67–93); also Kolb and Ross (1980, pp. 275–286).

Whether and how to go to trial involves a great deal of special knowledge and experience. See generally, Gray et al (1975); also Kolb and Ross (1980, pp. 275–286).

Defense through Insurance.   Insurance is widely used as a defense against product liability. But the costs have escalated sharply, again because of the growth in number of lawsuits and size of awards. In some fields insurance has become a major factor in the cost of operations. (Soaring insurance rates have forced some surgeons to take early retirement.)

Some comprehensive studies have examined the problem of insurance as applied to product liability. See McGuire 1988; Harrington and Litan 1988. See also Interagency Task Force Final Report on Product Liability (1978, chap. III), and Kolb and Ross (1980, pp. 287–327).

Prognosis.   As of the mid-1990s there remained some formidable unsolved problems in product liability. To many observers the United States' legal system contained some serious deficiencies:

- Lay juries lack the technological literacy needed to determine liability on technological matters. In most other developed countries, judges make such decisions.
- Lay juries are too easily swayed emotionally when determining the size of awards.
- In the United States, "punitive damages" may be awarded along with compensatory damages and damages for "pain and suffering." Punitive damages contribute greatly to inflated awards.
- In the United States, lawyers are permitted to work on a contingency fee basis—a concept that assertedly stimulates lawsuits. This arrangement is illegal in most countries.
- The adversary system of conducting trials places the emphasis on winning rather than on doing justice.
- Only a minority of the award money goes to the injured parties. The majority goes to lawyers and to pay administrative expenses.

By the mid-1990s some elements of this legal system were under active review in the national Congress. However, the system which has endured these deficiencies is deeply rooted in the United States culture, so it is speculative whether it will undergo dramatic change. A major obstacle has been the lawyers. They have strong financial interests in the system, and they are very influential in the legislative process—many legislators are lawyers.

In most developed countries the legal system for dealing with product liability is generally free from the above asserted deficiencies. Those same countries are also largely free from the extensive damage which product liability is doing to the United States economy.

(For some incisive comments on the deficiencies in the United States' legal system, see Grant 1994.)

Personal Liability.   An overwhelming majority of product liability lawsuits have been aimed at the industrial companies; they and their insurers have the greatest capacity to pay. As a corollary, such civil lawsuits are rarely aimed at individuals, e.g., design managers or quality managers. These individuals have little cause for concern with respect to civil liability. They are not immune from lawsuits but they are essentially immune from payment of damages.

Criminal liability is something else. Now the offense (if any) is against the state, and the state is the plaintiff. Until the 1960s, prosecution for criminal liability in product injury cases was directed almost exclusively at the corporations rather than the managers. During the 1960s and the 1970s the public prosecutors became more aggressive with respect to the persons involved. The specific targets were usually the heads of the companies but sometimes included selected subordinate managers such as for product development or for quality.

A contributing factor has been an earlier provision of the Food, Drug and Cosmetic Act making it a crime to ship out adulterated or misbranded drugs. This provision was interpreted by the United States Supreme Court to be applicable to the head of a company despite the fact that he or she had not participated in the events and even had no knowledge of the goings-on. For an analysis, see O'Keefe and Shapiro (1975). Also O'Keefe and Isley (1976).

For the great majority of industrial managers the threat of criminal liability is remote. Before there can be such liability, the manager must be found guilty of (1) having *knowingly* carried out illegal actions or (2) having been grossly negligent. These things must be proved to a jury beyond a reasonable doubt. It is a difficult proof. (Many guilty criminals escape conviction because of this difficulty.)

ENVIRONMENTAL PROTECTION

A special category of government regulation is environmental protection (EP). On the face of it, EP is a twentieth-century phenomenon. However, there is a school of thought suggesting that EP originated in a conservation movement to preserve lands that were being exploited by European colonists during the seventeenth and eighteenth centuries. For elaboration, see Grove 1992.

The Industrial Revolution of the mid-eighteenth century opened the way to mass production and consumption of manufactured goods at rates that grew exponentially. To support this growth required a corresponding growth in production of energy and materials. The resulting goods conferred many benefits on the societies that accepted industrialization. However, there were unwelcome by-products, and these also grew at exponential rates.

Generating the needed energy produced emissions that polluted the air and water. Nuclear power created the problem of nuclear waste disposal as well as the risk of radiation leaks. Mining for raw materials damaged the land, as did disposition of toxic wastes. Ominous threats were posed by ozone depletion and the risk of global warming. Disposition of worn-out and obsolete products grew to problems of massive proportions. All this was in addition to problems posed by the numerous inconveniences and occasional disasters caused by product failures during service. (See above under Life behind the Quality Dikes).

Industrial companies were generally aware that they were creating these problems, but their priorities were elsewhere. Public awareness lagged, but by the mid-twentieth century the evidence had become overwhelming. Responding to public pressures, governments enacted much legislation to avoid worsening the problem, and provided funds to undo some of the damage.

The new legislation was at first strongly resisted by industrial companies because of the added costs it imposed. Then as it became clear that EP was here to stay, the ingenuity of industry began to find ways to deal with the problem at the source—to use technology to avoid further damage to the environment. A striking example is Japan's achievement in energy conservation. During the period 1973–1990, despite continuing growth in industrial production, there was no increase in energy consumption (Watanabe 1993).

Public and media preoccupation with specific instances of environmental damage has tended to stimulate allocation of funds to undo such damage. However the long-range trend seems to be toward prevention at the source. A recent survey estimated that during 1991, of the research and development budgets of United States industrial companies, about 13 percent was directed at technology to minimize environmental damage (Rushton 1993).

Recognition of the importance of EP is now evident in many ways. For example:

- Many countries have created new ministries to deal with the problem of EP.

- Many industrial companies have created high-level posts for the same purpose.

- Numerous conferences are being held, including at the international level, with participation from government, industry and academia (Strong 1993).

- An extensive and growing body of literature has emerged. Some of this is quite specific. See for example on asbestos, Mossman et al. 1990; on design for recycling, Bylinsky 1995; also Penev and de Ron 1994.

- Companies have also evolved specific processes for addressing the problems of EP. These generally consist of:

  Establishment of policies and goals with respect to EP.

  Establishment of specific action plans to be carried out by the various company functions.

  Audits to assure that the action plans are carried out.

In addition, the ingenuity of companies has begun to find ways to reduce the costs of providing solutions. Table 35.2 lists some of the identified problems and the associated opportunities for solution (Rushton 1993).

TABLE 35.2 Environmental Problems and Opportunities for U.S. Industry

Problem	Opportunity
Performance chemicals and materials	
Air pollution	Emission control catalysts, clean motor fuels
Hazardous substances	Asbestos substitutes, chlorine substitutes, PCB substitutes
Land pollution	Lagoon liners
Oil spills	Oil absorbents, surfactants
Ozone layer depletion	CFC substitutes, UV hazard reduction technologies
VOC emissions	Power coatings, radiation-cured coatings, water-based coatings
Water pollution	Water treatment chemicals
Food	
Caffeine	Supercritical fluid extraction
Disposable packaging pollution	Recyclable or degradable packaging
Fat	Fat substitutes, "lite" foods
Short shelf life	Antioxidants, aseptic packaging, controlled/modified atmosphere packaging
Sugar	Low-calorie sweeteners
Environmental management	
Hazardous substance treatment	Asbestos removal, waste remediation
Hazardous substance prevention	Process redesign, spent oil recycling, waste prevention and minimization
Pollutant detection and monitoring	Analytic laboratory services, sensors
Solid waste storage and disposal	Waste recycling, incineration
Water supply	Low water consumption processes, water purification, water recycling
Health care and safety	
Automotive safety	Air bags, antilock brake systems
Bioincompatibility, rejection	Thromboresistant biomaterials, biodegradable implants
Disease diagnosis and treatment	Diagnostic reagents, instruments, services

TABLE 35.2   (Continued)

Problem	Opportunity
	Health care and safety
Equipment safety	Inspection and testing services
Medical waste pollution	Medical waste minimization, incineration.
Personal safety	Flame-retardant materials, protective clothing
Product tampering	Tamper-evident packaging
Side effects of drugs	Controlled-release drug delivery systems, biosensors

Source: Rushton (1993).

The emerging consensus is that the best solution lies in industrial efficiency (Strong 1993).

REFERENCES

ASQC (1980), "Consumer Attitudes on Quality in the United States." American Society for Quality Control, Milwaukee.

Best, Arthur (1981). *When Consumers Complain.* Columbia University Press, New York.

The Business Roundtable (1978). *Cost of Government Regulation Study for the Business Roundtable.* This is a study of the direct incremental costs incurred by 48 companies in complying with the regulations of six federal agencies during 1977. The Business Roundtable, New York.

Bylinsky, Gene (1995). "Manufacturing for Reuse." *Fortune,* February 6, pp. 102–112.

Consumer Reports (1977). "Laetrile, the Political Success of a Scientific Failure." *Consumer Reports,* August, pp. 444–447.

Consumer's Resource Handbook (1980). A publication of the United States Office of Consumer Affairs. Includes: a "complaint handling primer"; where to go for assistance; functions, services and information available from Federal offices; directories of federal, state, and local offices. Available from Consumer Information Center, Dept. 532G, Pueblo, CO 81009.

Crain, W. Mark (1980). "Vehicle Safety Inspection Systems. How Effective?" American Enterprise Institute for Public Policy Research, Washington, DC.

Dowd, Ann Reilly (1994). "Environmentalists Are on the Run." *Fortune,* September 19, pp. 91–104.

Eginton, Warren W. (1989). "An Overview of Federal and State Legislative Developments in Torts and Product Liability." *1989 ASQC Quality Congress Transactions,* Toronto, pp. 922–931.

Farrow, John H.(1991). "Product Liability." *ASQC Quality Congress Transactions,* Milwaukee, pp. 51–56.

Grant, Eugene L. (1994). "Why Product-Liability and Medical-Malpractice Lawsuits Are So Numerous in the United States." *Quality Progress,* December, pp. 63–65.

Gray, I., A. L. Bases, C. R. Martin, and A. Sternberg (1975). *Product Liability: A Management Response.* (Emphasis is on defense against product liability suits.) American Management Association, New York.

Grove, Richard H.(1992). "Origins of Western Environmentalism." *Scientific American,* July, pp. 42–47.

Harrington, Scott, and Robert E. Litan (1988). "Causes of the Liability Insurance Crisis." *Science,* February 12, pp. 737–741.

Interagency Task Force on Product Liability; Final Report (undated; about 1978.) (A comprehensive study on product liability, including the impact, the legal implications, the insurance problems, and especially, the merits of various proposals for remedy.) Distributed by National Technical Information Service, Springfield, VA 22161.

Juran, J. M. (1969). "Mobilizing for the 1970s." *Quality Progress,* August, pp. 8–17.

Juran, J. M. (1970). "Consumerism and Product Quality." *Quality Progress,* July 1970, pp. 18–27.

Juran, J. M. (1972). "Product Safety." *Quality Progress,* July 1972, p. 30–32.

Juran, J. M. (1977). "Auto Safety, A Decade Later." *Quality,* October 1977, pp. 26–32; November, pp. 54–60; December, pp. 18–21. Originally presented at the 1976 Conference of the European Organization for Quality, Copenhagen.

Juran, J. M. (1995). *A History of Managing for Quality.* Quality Press, 1995.

Kolb, John, and Steven S. Ross (1980). *Product Safety and Liability—A Desk Reference.* (A comprehensive reference treatise: attaining safety throughout the product life cycle; defenses against lawsuits; insurance; reference tables.) McGraw-Hill, New York.

McGuire, E. Patrick (1988). "The Impact of Product Liability." Report 908, The Conference Board, New York.

Mossman, B. T., J. Bigbon, M. Corn, A. Seaton, and J. B. L. Gee (1990). "Asbestos: Scientific Developments and Implications for Public Policy." *Science,* January 19, pp. 294–301.

National Business Council for Consumer Affairs. (1973). *Safety in the Market Place.* (Makes 14 recommendations, along with details for implementation.) Superintendent of Documents, Washington DC, Stock No. 5274-00009.

O'Keefe, Daniel F., Jr., and M. H. Shapiro (1975). "Personal Criminal Liability Under the Federal Food, Drug and Cosmetic Act—The Dotterweich Doctrine." *30 Food-Drug-Cosmetic Law Journal 5,* January.

O'Keefe, Daniel F., Jr., and C. Willard Isley (1976). "Dotterweich Revisited—Criminal Liability Under the Federal Food, Drug and Cosmetic Act." *31 Food-Drug-Cosmetic Law Journal 2,* February.

OSHA Safety Regulation (1977). This is the report of a presidential task force assigned to review the safety regulatory practices of the Occupational Safety and Health Administration. American Enterprise Institute for Public Policy Research, Washington, DC, 1977.

Penev, Kiril D., and Ad J. deRon (1974). "Development of Disassembly Line for Refrigerators." *Industrial Engineering,* November, pp. 50–53.

Peterson, Esther (1974). "Consumerism as a Retailer's Asset." *Harvard Business Review,* May–June, pp. 91–101.

Rushton, Brian M. (1993). "How Environmental Protection Affects U.S. R&D." *The Bridge,* National Academy of Engineering, Washington, DC, Summer, pp. 16–21.

Scofield, Eugene L. (1986). "A Quality System for Labeling Excellence." *ASQC Quality Congress Transactions,* Anaheim, pp. 159–161.

Sentry (1976). *Consumerism at the Crossroads.* Results of a national opinion research survey on the subject. Sentry Insurance Co., Stevens Points, WI.

Smith, Lee (1991). "Trial Lawyers Face a New Charge." *Fortune,* August 26, pp. 85–89.

Strong, Maurice F. (1993). "The Road from Rio." *The Bridge,* Summer, pp. 3–7.

Travelers Insurance Company (1972). *The Act and Its Principal Features.* Explanation of the Consumer Product Safety Act. The Travelers Insurance Companies.

Travelers Insurance Company (1973). *A Management Guide to Product Quality & Safety.* Sets out contributions and responsibilities of various company functions, relative to product safety. The Travelers Insurance Companies.

Tye, Josh B. (1988). "The Anatomy of a Voluntary Safety Standard." *1988 ASQC Quality Congress Transactions,* Dallas, pp. 238–244.

Wargo, John J. (1987). "Product Safety and Product Liability—An Overview." *1987 ASQC Quality Congress Transactions,* Minneapolis, pp. 77–83.

Watanabe, Chihiro (1993). "Energy and Environmental Technologies in Sustainable Development: A View from Japan." *The Bridge,* Summer, pp. 8–15.

SECTION 36
QUALITY AND THE NATIONAL CULTURE

J. M. Juran

INTRODUCTION

The goal of high quality is common to all countries. This common goal must compete with other national goals amid the massive forces—political, economic, and social—which determine the national priorities. This section examines these forces and their effect on the problems of attaining quality.

The growth of international trade and of multinational companies has required that attention be directed to understanding the impact of national culture on managing for quality. To aid in this understanding, the subject is organized under the following general subdivisions:

Developing economies: The special problems of managing for quality in such economies are discussed in Section 37, Quality in Developing Countries.

Other economies: Other sections discuss the problems of managing for quality in specific economies:

- 38:   Quality in Western Europe
- 39:   Quality in Central and Eastern Europe
- 40:   Quality in The United States
- 41:   Quality in Japan
- 42:   Quality in the People's Republic of China
- 43:   Quality in Latin America

In all types of national economy, there are natural resources and limitations which influence the priority of goals. However, an even greater force is that of human leadership and determination. Historically, these human forces have been more significant than natural resources in determining whether goals are attained.

The words "capitalistic," "socialistic," and "developing" are simple labels for some very complex concepts. The broad definition of "capitalism" is private ownership of the means of production and distribution, as contrasted with state ownership under socialism. Yet all self-styled capitalistic countries include a degree of state ownership, e.g., in matters of health, education, transport, and communication. Similarly, the self-styled socialistic countries contain, in varying degrees, some private ownership of enterprises for production of goods and services. In like manner, countries which are "developing" in the industrial sense may be highly developed in terms of other aspects of national maturity, e.g., political or social. The reader is urged to keep in mind that the words "capitalistic," "socialistic," and "developing" are used in a relative sense and cannot be considered as absolutes.

The subject matter of this section and of the companion Sections 37 through 43 are of obvious interest and importance to those engaged (or contemplating engagement) in operations of an international nature. Such operations are becoming ever more extensive as trade barriers are progressively removed. However, removal of governmental barriers has little effect on cultural barriers. These remain as a continuing problem until the cultural patterns (and the reasons behind them) are understood, appreciated, and taken into account.

In the economic sense, the capitalistic developed countries are the "vital few." The developing countries are the most numerous, occupy most of the land surface, and include most of the human population. However, it is the capitalistic developed countries which produce the bulk of the world's goods and services. This great importance (in the economic sense) suggests that those who engage in international trade should acquire a working knowledge of the cultures which prevail in the respective countries.

QUALITY IN CAPITALISTIC ECONOMIES

All capitalistic economies exhibit some basic similarities which influence the importance of quality in relation to other goals in the economy.

Competition in Quality.   Capitalistic societies permit and even encourage competition among enterprises, including competition in quality. This competition in quality takes multiple forms.

Creation of New Enterprises.   A frequent reason for the birth of new enterprises is poor quality of goods or services. For example, a neighborhood has outgrown the capacity of the local food shop or restaurant, so the clients must wait in long queues before they can receive service. In such cases, entrepreneurs will sense a market opportunity and will create a new enterprise which attracts clients by offering superior service.

The ease of creating new enterprises is a far greater force in quality improvement than is generally realized. All economies, whether capitalistic or socialistic, suffer poor quality during shortages of goods. Creation of new enterprises is one means of alleviating shortages, and thereby of eliminating an invariable cause of poor quality.

Product Improvement.   A common form of competition in quality is through improving products so that they have more appeal to the users and can therefore be sold successfully in the face of competition from existing products. These product improvements come mainly from internal product development carried on by existing companies. In addition, some product improvements are designed by independents who either launch new enterprises or sell their ideas to existing companies.

New Products.   These may be "products" or even new systems approaches, e.g., designs which minimize user maintenance. The industrial giants of today include many members founded on new systems concepts. As with product improvements, the new products may originate through development from within or through acquisition from the outside.

Competition in quality results in duplication of products and facilities. Such duplication is regarded as wasteful by some economists. However, the general effect has been to stimulate producers to outdo each other, with resulting benefit to users.

Direct Access to Marketplace Feedback.   In the capitalistic economies, the income of the enterprise is determined by its ability to sell its products, whether directly to users or through an intermediate merchant chain. If poor quality results in excessive returns, claims, or inability to sell the product, the manufacturers are provided with the warning signals which are a prerequisite to remedial action.

This severe and direct impact of poor quality on the manufacturers' income has the useful by-product of forcing manufacturers to keep improving their market research and early warning signals, so as to be able to respond promptly in case of trouble.

Direct access to the marketplace is not merely a matter of receiving complaints and other information about bad quality, important though that is. Even more important is the access to the marketplace *before* products are launched and sales programs are prepared. In the capitalistic economies, the autonomous companies all make their own forecasts on how much they expect to sell. Their ability to thrive depends on how well they are able to realize their forecasts. The potential benefits and detriments force the companies to pay attention to the needs of the marketplace, since it provides their income.

Protection of Society.   The autonomy of capitalist enterprises enables them to misrepresent their products, sell unsafe products, damage the environment, fail to live up to their warranties, and so on. The extent of such practices has been large enough to generate extensive preventive legislation. In this connection, see Section 35, Quality and Society.

CULTURAL DIFFERENCES

There are many of these, including:

Language: Many countries harbor multiple languages and numerous dialects. These are a serious barrier to communication.

Customs and traditions: These and related elements of the culture provide the precedents and premises which are guides to decisions and actions.

Ownership of the companies: The pattern of ownership determines the strategy of short-term versus long-term results, as well as the motivations of owners versus nonowners.

The methods used for managing operations: These are determined by numerous factors such as reliance on system versus people; extent of professional training for managers; extent of separation of planning from execution; careers within a single company versus mobile careers.

Suspicions: In some countries, there is a prior history of hostilities resulting from ancient wars, religious differences, membership in different clans, and so on. The resulting mutual suspicions are then passed down from generation to generation.

It is clearly important to learn about the nature of a culture before negotiating with members of that culture. Increasingly, companies have provided special training to employees before sending them abroad. Similarly, when companies establish foreign subsidiaries, they usually train local nationals to qualify for the senior posts.

MULTINATIONAL COLLABORATION

Collaboration across cultures is a many-faceted problem. For example, a system may be designed in country A but the subsystem designs may come from other countries. In like manner, companies from multiple countries may supply components, carry out manufacture, marketing, installation, maintenance, and so on.

Numerous methodologies have been evolved to help coordinate such multinational activities. Those widely used include:

Standardization: This is accomplished through organizations such as the International Organization for Standardization (ISO) and the International Electrotechnical Commission (IEC). A special application is the Allied Quality Assurance Publication (AQAP) standards widely used by the North Atlantic Treaty Organization (NATO) countries for multinational contracting. Some of the standards used by these and other organizations are listed in Appendix III, Selected Quality Standards, Specifications, and Related Documents.

The consortium: This form involves creating an association of companies from various countries. The consortium is usually dedicated to a specific project, e.g., the Airbus. [See Debout (1978) for elaboration.]

Contract Management: In many cases, the prime contractor provides a coordinating service for the subcontractors (who may include a consortium). [See McClure (1979) relative to the F16 aircraft; see also McClure (1976).]

Technology Transfer: This is carried out in numerous well-known ways: international professional societies and their committees; conferences; exchange visits; training courses and seminars. In large, multinational companies, such activities are carried out within the companies as well. [For an example involving the use of multinational quality councils, see Groocock (1978).]

REFERENCES

Debout, E. (1978). "European Aerospace Cooperation and Quality." *International Conference on Quality Control,* Tokyo, pp. A1-11 to A1-16.

Groocock, J. M. (1978). "Quality Councils—A Means for International Cooperation." *International Conference on Quality Control,* Tokyo, pp. A1-17 to A1-22.

McClure, J. Y. (1976). "Quality—A Common International Goal." *ASQC Technical Conference Transactions,* Milwaukee, pp. 459–466.

McClure, J. Y. (1979). "Procurement Quality Control Within the International Environment." *ASQC Technical Conference Transactions,* Milwaukee, pp. 643–649.

SECTION 37
QUALITY IN DEVELOPING COUNTRIES

Lennart Sandholm

A HETEROGENEOUS GROUP OF COUNTRIES

Countries are often classified as "developing" or "developed." This terminology is misleading, as countries that are classified as developing could in many respects be more advanced than some so-called developed countries; for example, when such facets of human life as morale, culture, social relations, democratic rights, and equal opportunities are taken into consideration. A clearer grouping would be into "industrialized" versus "less industrialized" countries or, alternatively, into "economically developed" versus "less economically developed" countries. This is, in fact, what is meant by the developed versus developing classification.

Neither developing nor developed countries can be regarded as forming a homogeneous group; they show differences in terms of industrial development, natural resources, size, economic strength, access to markets, national policy, human resources, and other criteria.

The heterogeneity of developing countries is also reflected in other classification systems. The term "least developed countries" is used to denote some 50 countries in the context of United Nations discussions. These countries are the weakest partners in the international community, with the most formidable structural problems. The term "newly industrialized countries," commonly abbreviated NIC, implies countries which have recently advanced from a developing status to a developed status. This group includes some countries in Southeast Asia referred to as the "Asian tigers." As a consequence of the collapse of the Soviet Union there are "countries in transition to a market economy." This group previously had a centrally planned economy. Now they are converting to a market economy.

TRADE GLOBALIZATION AND LIBERALIZATION

The Uruguay Round of Multilateral Trade Negotiations was held under the General Agreement on Tariffs and Trade (GATT) from 1986 to 1993. The final act of these negotiations was signed at a meeting in Marrakesh, Morocco in April 1994. The "Marrakesh Declaration" affirmed that the results of the Uruguay Round would "strengthen the world economy and lead to more trade, investment, employment and income growth throughout the world." Although tariff reductions had continuously been undertaken under the GATT framework earlier, liberalization of trade had been threatened by the increased use of protectionist measures such as export restraints and import quotas. The Uruguay Round of negotiations achieved substantial results toward the reversal of protectionist measures.

The developing countries took a very active and influential part in the Uruguay Round of negotiations. These countries will be affected to a higher degree by the new agreement than by earlier agreements which involved industrialized countries only.

The new trade situation includes reductions of subsidies and tariff and nontariff barriers, more open and fair trade without protection, and rules for trade in goods, services, and "intellectual properties." This will entail greater access to new markets, as well as increased competition from imports on the domestic market. In the new competitive situation, quality will become more important to developing countries than ever before.

The contribution made by developing countries to global output is increasing rapidly. According to the World Bank (*Global Economic Prospects and the Developing Countries,* 1995, p. 65), these countries will account for 38 percent of the growth in global output in 1995–2010, compared to 22 percent in the 1980s. Their share of global output will rise from 21 percent in 1994 to 27 percent in 2010. The report states that "the increasing integration of developing countries into the global economy represents a major—perhaps the most important—opportunity for raising the welfare of both developing and industrial countries in the long term."

TECHNOLOGY IN DEVELOPING COUNTRIES

In many developing countries the manufacturing sector may be divided into a modern component, a modernizing component, and a nonmodern component, which utilize different technology.

Modern component: This component consists of the largest industrial enterprises, which are located mainly in urban areas where infrastructure and an adequate work force, including skilled workers, are available, and which use modern technology. This group of enterprises also includes subsidiaries of multinational companies. The modern component is mainly found in the most industrialized developing countries.

Modernizing component: This component includes mainly small to medium-sized industrial enterprises, located mostly in urban but also in some rural areas, in which various intermediate levels of technology are used. Enterprises in this component might be suppliers to larger enterprises.

Nonmodern component: This consists of small industrial enterprises and artisan workshops, located largely in rural but also in urban areas, which use traditional and upgraded traditional technologies.

Discussions of the technological basis for industrial development in developing countries are often focused on the choice of appropriate technologies and their development, transfer, and implementation.

The term "appropriate technology" for developing countries is often taken to mean simple and labor-intensive technology. The concept is based on the observations that industrial technologies are designed in the developed countries and that conditions in these countries are quite different from those in the developing countries. Technologies that are more suitable to conditions in developing countries are needed.

The concept of appropriate technology includes factors other than economic efficiency and growth; for example, employment, working conditions, and provision of basic needs. This means that appropriate technologies will differ from country to country depending on the significance of these various factors.

FACTORS IMPEDING QUALITY IMPROVEMENT

Developing countries face several problems with regard to quality. The nature of these problems differs depending on the phase of development the country is in. Consequently, the solutions to the problems also differ.

An increasing number of developing countries are liberalizing their economies and adopting export-oriented policies. These changes lead to an increased awareness of the importance of quality.

Discussions that the author has had with many representatives of developing countries show that there are several factors impeding improvement of quality in developing countries, of which the major ones seem to be.

Low Purchasing Power. The vast majority of people are poor. Their purchase decisions are based on price consideration only. The manufacturers consequently aim at low prices, using cheap and low-quality materials.

Shortage of Goods and Absence of Competition. The shortage of goods provides some guarantee to the manufacturers that everything produced will be sold; as a result, they show very little interest in quality. Restrictions on the importation of goods, along with high custom barriers, protect locally produced goods against competition from goods produced in more industrialized countries.

Foreign Exchange Constraints. Most developing countries have a shortage of foreign exchange, and the industrial sector of the economy has to compete with other sectors for the insufficient amount available. This leads to obsolete technology, inadequate machinery and poor material, all of which have an adverse effect on quality.

Incomplete Infrastructure. The infrastructure is not satisfactory. In most developing countries there are shortcomings in areas such as power supply, transport, communication, and education. In addition, specific services in areas important to quality development, e.g., standardization, testing, training, and consulting, are not adequate for the needs of the enterprises.

Inadequate Leadership.   There is a short-term view on the business, which leads to a quantity-oriented management culture. Business leaders rely on a few key members of personnel. The need for an overall coordination of activities is overlooked. Quality is regarded as a technical issue only, managed by technicians. There is no proper awareness of the strategic importance of quality to the enterprise among owners and top managers.

Inadequate Knowledge.   The managerial as well as technical knowledge of personnel in industry is generally limited. In some developing countries the problem of limited knowledge is compounded by the transient nature of the work force. (It is not uncommon to find that 20 to 50 percent of work force is replaced within 3 to 6 months). Under these circumstances, it is difficult to achieve a skilled work force. The high illiteracy rate in many developing countries adds to the problem.

INDUSTRIAL DEVELOPMENT AND QUALITY

Industrial development usually progresses in identifiable phases, from a primitive, agricultural subsistence economy to a sophisticated economy producing manufactured goods for export. Five phases of development can be defined (Juran 1975), as described below.

Phase I. Subsistence Economy.   Economic activity consists mainly of the production of subsistence goods for local consumption (agriculture, fishing, etc.). Quality is low—there is a lack of quality standards, technology, test facilities, etc. Quality control takes place mainly by consumer inspection of products in the village marketplace.

Phase II. Export of Natural Materials.   In this phase the economy undertakes export of natural materials such as fruits, fibers, and minerals. Selling these goods in the international market requires adherence to international quality standards, which are usually higher than domestic standards. Quality therefore has to be improved. The contracts for export normally include the quality specifications to be met, the tests to be used, and the sampling procedures to be followed, which requires test laboratories, instruments, and appropriate knowledge. In order to provide for the necessary services, a national standards body is set up.

Phase III. Export of Processed Materials.   Local processing of materials is started, and the economy shifts to export of processed rather than raw materials, e.g., metals instead of ore, plywood instead of logs, canned instead of fresh fruit. The economy must now include the acquisition, operating, and maintenance of technological processes. International quality standards for processed products have to be met, and process controls have to be introduced. Supplier relationships concerning quality have to be developed, since packaging materials, raw materials, etc., are supplied from external sources within the country. The traditional work of the standards body is expanded. In addition, new needs arise as tools from the quality management profession (statistical methodology, quality planning, supplier quality activities, organization for quality, etc.) are introduced. This requires training and consulting services.

Phase IV. Integrated Manufacture for Domestic Use.   In this phase the economy undertakes integrated manufacture of modern industrial and consumer products for domestic use. The industries now have to manage quality in all stages of industrial production, from determining the market need through product development, design, manufacture, and marketing. This requires not only training and consulting services but also professional development through research work, conferences and seminars, publications, quality society activities, exchange of views with colleagues, etc.

Phase V. Export of Manufactured Products.   Finally, the manufactured products are sold on the international market, where they have to compete with products from other countries that have fully developed industrial economies. Foreign buyers are increasingly requesting suppliers to conform to the international series of standards on quality management systems (ISO 9000). The industries now have to develop, document, and introduce such systems. They must also, in many cases, ensure that these systems are certified by an accredited institution. This means that there is a need for training in ISO 9000 standards and auditing of quality management systems. Certification bodies have to obtain accreditation, which implies registration of certified auditors.

QUALITY ACTIVITIES

The different phases of industrial development require various activities to attain, improve, and control quality. In a subsistence economy these consist primarily of inspection by consumers.

In the early stages of industrialization there is an increasing need for standardization activities, such as the preparation of standards covering specifications, testing methods, sampling methods, etc. It is also necessary to develop applied and legal metrology, as well as a national testing capability. A certification scheme for selected products is sometimes introduced. All these tasks are usually referred to the national standards body.

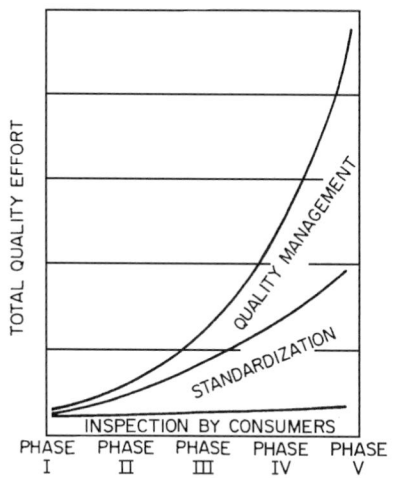

FIGURE 37.1   The growth of total quality effort in industrial development. (*Based on Juran 1975.*)

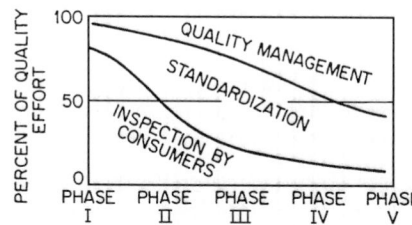

FIGURE 37.2   The relative importance of the quality activities. (*Based on Juran 1975.*)

As industrialization continues, new needs arise, which cannot be satisfied entirely by traditional standardization activities on a national level. These new needs have to be fulfilled primarily by in-house quality activities, e.g., quality planning, design reviews, new-product testing, supplier surveys and controls, process capability analyses, inspection planning, quality audits, quality evaluation, quality data feedback, poor quality cost analyses, benchmarking, quality improvement programs, quality policies and objectives, quality systems and organizational development. Knowledge about these activities and the appropriate tools (under the designation of quality management) is required, and this creates a need for education and training, development of quality specialists and managers, consulting assistance, etc. These needs are accommodated with measures that may include training courses, conferences and seminars, research committees, publications, and/or quality societies.

Consequently, there are a wide range of quality activities, the importance of which depends on the stage of industrial development. Basically, the activities can be grouped as follows:

Inspection by consumers: Inspection of products in the marketplace.

Standardization: National standards on terminology, sampling methods, testing methods, specifications, codes of practice, and quality management systems; applied and legal metrology; national testing facilities; certification; legislation.

Quality management: Application of managerial tools to plan for quality, attain and control quality, follow up and improve quality, as well as to organize for quality and develop competence in the field.

Figure 37.1 shows the development of total quality measures in the industrial development. Figure 37.2 shows the importance of the activities on a relative basis. It is clear that the dominance is shifting from inspection by consumers to standardization, and from standardization to quality management.

QUALITY MANAGEMENT

Phases of Development. There is an increasing interest in quality worldwide. Over the past 10 years, more and more companies in industrialized countries have become involved in quality activities of various kinds. Management journals include quite a lot about quality. National and international conferences highlighting quality are held frequently.

This development is also influencing developing countries. In the more industrially advanced of these countries, the interest in quality is equal to that in industrialized countries. In other countries, interest is emerging.

Five phases can be identified in the development of quality in manufacturing enterprises as well as in service organizations (Sandholm 1996).

Dormant phase: Companies do not feel any threat in the marketplace. They earn an acceptable income. Executives are satisfied with the business results. They experience no need to give any special consideration to quality.

Awakening phase: The situation is dramatically changed. Market shares are lost. Income drops. Profit turns into loss. Executives awake and feel that they are facing a crisis.

Groping phase: Upon awakening, executives realize that they have to do something in the field of quality. But what? Trendy tools and methods are there as a possibility, highlighted very much in business literature and at management seminars and conferences. Lacking any sound knowledge in how to manage for quality, executives just select whatever presents itself. The groping phase is a period of trial and error.

Action phase: Some companies discover that the trendy tools and methods do not lead to excellent results. They then start to carry out an effective program for changing the situation. Such a program includes a change of the internal culture, as well as improvements of products and processes.

Maturity phase: A real sign of maturity is when quality is no longer talked about in the enterprise. Full customer satisfaction is achieved through perfect processes in all areas of the organization. The concept of quality applies not only to products, i.e., the goods and services produced and supplied, but also to all supporting activities. A total quality approach is applied, which includes all processes and functions, as well as the involvement of everyone in the organization. Quality is just a natural aspect of the work, permeating the entire organization. Executives regard quality in the same natural way as they regard finances. The maturity phase has been reached by successful Japanese companies. In the West, the list of winners of the Malcolm Baldrige National Quality Award in the United States and the European Quality Award in Europe contains companies that can be fairly considered to have reached the maturity phase.

In Western industrialized countries, the dormant phase continued up to the end of the 1970s, when the awakening came as a result of Japanese competition. The awakening phase in Japan had taken place 30 years earlier. The groping phase in the West is going on right now. The maturity phase is still to come in most Western companies.

The dormant phase has prevailed in most developing countries due to governmental policies of protecting local manufacturers from competition from abroad. However, the governments in many developing countries introduced policies implying liberalization of the economy and opening up the market for foreign competition early in the 1990s. For local enterprises this meant that they entered into the awakening phase, which was rapidly succeeded by the groping phase.

Developments in Western industrialized countries have a strong impact on what is done in the field of quality in developing countries, particularly those countries which are more industrialized. The methods and tools which become trendy in the groping phase in the West will be applied (somewhat later, though) in developing countries as well.

A list of trendy methods and tools includes "Zero Defects," quality circles, statistical process control (SPC), quality function deployment (QFD), seven tools, "TQM," ISO 9000, benchmarking, process re-engineering, etc. There is a phenomenon of methods and tools of this kind emerging, being highlighted for some time and then fading away.

A veritable explosion took place in this field in the 1980s, when it dawned on many Western companies that they were being driven out of business by Japanese companies. Companies in the West found that they had to do something and were willing to try anything.

There is nothing wrong with these methods and tools as such. The fault lies in how they are implemented. They are used as general methods and tools for quality improvement and, used in this manner, will only lead to marginal improvement. They ought to be used only when an analysis indicates that they are the appropriate measures to eliminate specific problems or to better meet the needs of customers. The same is true for all other methods and tools which are part of the quality profession and which have not yet been widely publicized, even though they may have much greater effects on the results achieved.

Areas of Interest.   During the 1990s there has been an increasing interest in three areas: total quality management, ISO 9000, and national quality awards. In this respect, the more industrialized developing countries do not very differ significantly from countries with a developed market economy.

Total Quality Management.   The concept of quality has formerly been discussed exclusively in relation to products, i.e., the goods and services which an organization produces and supplies. Nowadays, quality is increasingly discussed in a broader context. The concept most commonly referred to is total quality, which includes the quality of all internal processes and functions as well as the involvement of everyone in the organization.

In this context, reference is often made to the term "total quality management." An increasing number of conference papers referring to total quality management in the title are being presented. Many of these papers, however, do not actually deal with total quality management. They propagate certain narrow concepts or tools. Presentations on the experiences of companies in developing countries really practicing total quality management are, in fact, rare.

ISO 9000.   The international series of standards for quality systems, ISO 9000, has had an immense impact. A growing number of enterprises in developing countries are taking ideas from these standards as a basis for developing and introducing procedures for their own quality activities, primarily relating to product quality. More and more quality activities are tending to focus on ISO 9000.

For most enterprises, the driving force behind the development, documentation, and implementation of a quality system based on the requirements in the ISO 9000 series of standards is commercial. Enterprises in developing countries find that they are no longer accepted as suppliers by customers in industrialized countries if they do not apply a documented quality system. Or they may find that they are losing market shares to competitors who do have a quality system based on the ISO 9000 standards.

National standards bodies are very active in promoting the use of ISO 9000. Promotion material is distributed frequently. ISO 9000 receives extensive publicity in the media through these bodies. Conferences and seminars are organized to highlight the merits of ISO 9000.

A certain infrastructure on a national level is a prerequisite for using the ISO 9000 approach to quality (see below, under Institutional Infrastructure). The enterprises need services in areas such as consulting, training, auditing and certification. In addition, the infrastructure shall provide for the accreditation of certification bodies, as well as for the registration of certified auditors for quality systems (see below, under Certification).

The United Nations Industrial Development Organization (see below, under External Assistance) has conducted a survey of developing countries and emerging economies in Latin America and the Caribbean, Africa, Asia, and Eastern Europe on the subject of the ISO 9000 series of standards and the implications of these standards upon the trade of those countries (United Nations Industrial Development Organization 1996). Special attention was paid to the situation of small and medium-sized enterprises. The survey showed that awareness of ISO 9000 was highest in Latin America and lowest in African countries. The highest awareness was among multinational companies and large national enterprises, while small and medium-sized enterprises showed a low awareness. ISO 9000 was perceived to be of most importance for exporters, particularly in Latin America and Asia. This was not the case in Africa. It had hardly any importance for importers and little relevance for producers for the domestic market.

The International Trade Centre UNCTAD/WTO (see below, under External Assistance) has published two books on the implementation of the ISO 9000 standards in small and medium-sized enterprises in developing countries. One is a guide (International Trade Centre UNCTAD/WTO and International Organization for Standardization 1996), published jointly with the International Organization for Standardization (see below, under External Assistance), that provides guidance on methodology for implementing the elements of ISO 9000 quality management systems as well as guidance on preparing enterprises for third-party certification. The second book is a handbook (International Trade Centre UNCTAD/WTO 1996) dealing with basic concepts of quality management. The book explains the structure and content of the ISO 9000 standards and discusses the establishment, implementation, auditing, and third-party certification of quality management systems.

National Quality Awards.   In order to promote quality, national quality award schemes are set up in an increasing number of developing countries. These schemes started to emerge early in the 1990s in the more industrialized developing countries. The schemes are based on either the Malcolm Baldrige National Quality Award (in the United States) or the European Quality Award.

The quality awards have become an important ingredient of national programs to promote an awareness of quality among manufacturers and service providers (see below, under National Promotion).

UNIDO Quality Program.   The United Nations Industrial Development Organization (UNIDO) has designed a program for quality development in enterprises in developing countries (Maizza-Neto et al. 1994). The program, called the UNIDO Quality Program, is based on a systemwide approach to continuously improving every aspect of an organization's production process to achieve higher levels of quality—while simultaneously holding down costs. The program recognizes that poor product quality and low productivity are major impediments to the viability of many enterprises in developing countries.

The system has two major parts: an inner quality loop and an outer management loop (Figure 37.3). The inner quality loop involves the main activities that are responsible for compliance with specifications and the creation of consumer satisfaction. The outer management loop has to do with the operational performance of the enterprise. Operational indicators are developed. These will show the entrepreneurs that the effect of an important strategic decision can be detected and corrected. The implementation starts with a survey to identify subjects which the entrepreneurs themselves consider to be important. Once the priorities for changes have been identified, the technical assistance of UNIDO, in the form of the expert advice of national and international consultants, is available to support the modernization of interested pilot enterprises. The results are continuously evaluated through the operational indicators. For the evaluation of these indicators, UNIDO has developed two computer softwares: BEST (Business Environment Strategic Toolkit) and FIT (Financial Improvement Toolkit). BEST assists the entrepreneurs in the operation and management of the enterprise, while FIT assists in financial decisions concerning strategy.

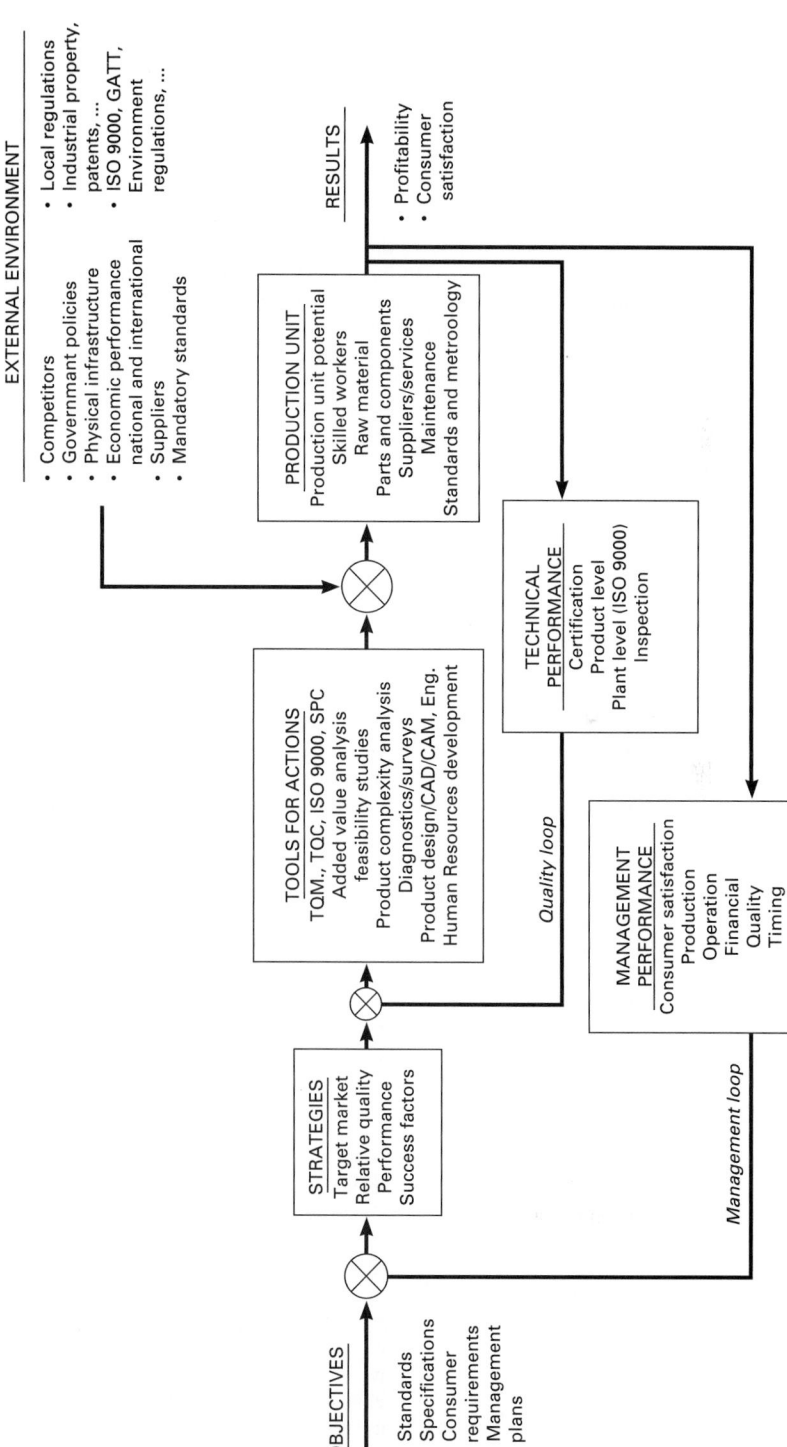

FIGURE 37.3 The inner quality loop and the outer management loop in the UNIDO Quality Program. *(Maizza-Neto et al. 1994.)*

The approach described above has been successfully implemented in 95 companies, in the capital goods sector, in 10 Latin American countries. The same approach is planned for the food processing sector of seven African countries.

NATIONAL EFFORTS FOR QUALITY

The problems related to quality in developing countries are of such a nature that they cannot be solved on a company level only. In fact, efforts on a national level are necessary, and these depend largely on the national policy of the country. In some countries, the government is ahead in developing means for an effective quality program. National efforts include.

Standardization: Preparation of national standards covering terminology, sampling methods, testing methods, specifications, quality management systems, and codes of practice

Certification: To attest that products comply with standards, and that quality-related activities are carried out in accordance with certain standards on quality management systems

Export inspection: Ensuring the quality of certain products for export through preshipment inspection

Legislation: Enforcement of standardization, certification, accreditation, export inspection, and other requirements through acts of parliament and legislature

National promotion: National programs to promote a general awareness of quality

Education and training: Development of the necessary knowledge and skills, as well as exertion of influence on attitudes

External Assistance: Assistance from international organizations, bilateral aid programs, transnational corporations, and other sources to shortcut the development process

Institutional infrastructure: Services offered by institutions in the areas of standardization, certification, accreditation, testing, metrology, quality consulting, and training

Professional societies: To develop the competence of quality professionals and practitioners

The steps to be taken on a national level are determined mainly by the governments of the respective developing countries. The way this is done depends on the policy of the government. In some developing countries there is a centrally planned economy, and in these we will often find a strong reliance on governmental institutions and legislation. In other countries only some basic needs (e.g., standardization, education) are met by the government. There is a growing trend among developing countries toward less government intervention and more deregulation.

STANDARDIZATION

Standardization plays a major role in promoting industrial and economic development of a country. Many developing countries have realized this and as a consequence have set up national institutions to handle the standardization activities.

Standardization refers to the process of formulating and applying standards to a specific activity. There are several types of standards:

Terminology standards: Terms used in technical and legal documents have to be clearly defined. Terminology and symbols are a means of communication.

Basic standards: Standards on units of measure are a prerequisite for trade, engineering, health care, etc. There are also other basic standards in fields like civil engineering, electrical engineering, and mechanical engineering.

Dimensional standards: By making components to certain standardized dimensions, interchangeability will be achieved.

Performance standards: Criteria reflecting the fitness for use of the products are important for consumers. These, as well as safety-related requirements, are given in performance standards.

Testing and inspection standards: Test and inspection data can differ depending on the methods used. Therefore, standards describing the equipment to be used, the procedures to be adopted and the evaluation of the data obtained are important. This also includes the sampling procedures to be used.

Quality systems standards: Industrial buyers are increasingly referring to the international series of standards for quality systems (ISO 9000) in the procurement of equipment and materials. An increasing number of developing countries have as a consequence adopted these standards as national standards.

National Standardization. Standardization activities on a national level are dealt with by a national standards body. This institution, carrying full government recognition (in many developing countries through legislation), is responsible for the development and publication of national standards, as well as for keeping them up to date. In preparing standards, the national standards body calls upon the knowledge and experience of manufacturers, users, government departments, universities, etc. This is normally done by setting up technical committees with this wide representation.

The following procedure is applicable in relation to the national standards bodies activities on preparing national standards (International Organization for Standardization 1994a, pp. 32–42):

1. Identification of standardization subjects
2. Justification of projects and assignment of priorities
3. Approval of projects and their inclusion in the work program of the national standards body
4. Development of the standards
5. Approval of the standards by the national standards body and their publications

As an example of standards preparation, the procedure used by the Standards Association of Zimbabwe (SAZ) is shown in Figure 37.4.

A checklist for identifying subjects and assigning priorities for national standardization is given in Figure 37.5.

National standards can be mandatory or voluntary. Mandatory standards are found in countries with a centrally controlled economy, while countries with a free-enterprise economy normally have voluntary standards. In most countries, however, there is a mixed or selective approach in the enforcement of national standards. This means that standards are voluntary, except those dealing with safety and health. Such standards are mandatory.

This mixed or selective approach would fit the needs of many developing countries where the situation is as follows (International Organization for Standardization 1991, p. 21):

1. Low level of standards consciousness in circles involved in production and distribution of goods and services, combined with a low literacy rate in the country as a whole
2. Coexistence of modern and traditional industries
3. Small or unrecognized industries that, by virtue of their number, contribute to substantial production
4. Wide variation in manufacturing and processing capabilities
5. Uncertainty about the availability of raw materials of assured quality
6. Sporadic or low consumer demand for standard quality
7. Difficulty in ensuring an adequate enforcement mechanism
8. Need for ensuring essential health and safety protection of the consumers

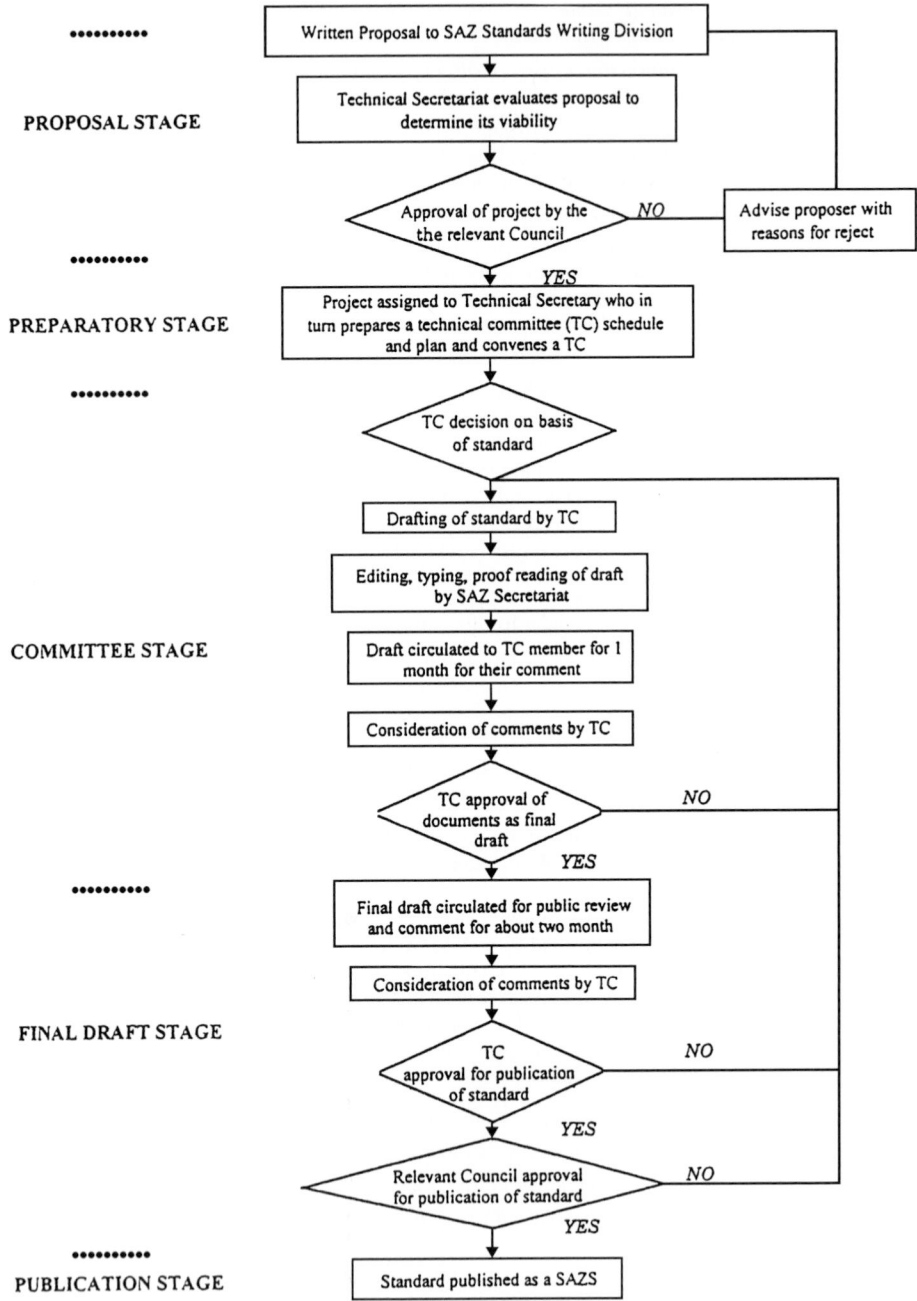

FIGURE 37.4   Standardization procedure used by the Standards Association of Zimbabwe (SAZ). (*Mhlanga 1995, p. 56.*)

1. Which of the following objectives should be included in the standards?

	YES	NO

Mutual understanding
Health, safety, protection of environment
Interface, interchangeability
Fitness for purpose
Variety control
Others (please specify)

2. Which of the following aspects are to be standardized?

a. Terminology symbols/signs
b. Characteristics
 • Dimensional
 • Mechanical
 • Chemical (including microbiological)
 • Thermal
 • Electrical
c. Marketing, labelling, packaging, transport
d. Sampling
e. Testing
f. Others (please specify)

3. What is (are) the specific aim(s) of the standard? E.g.,

a. Government requirements in support of
 • Public sector procurement
 • Legislation
b. International/regional harmonization
c. Commercial factors associated with
 • Exports
 • Imports
 • Internal trade
d. Certification

4. What technical expertise is available?

a. Research institutions/universities
b. Manufacturers
c. Users (consumers)
d. Independent personnel
e. Government departments
f. Others (please specify)

5. Is it envisaged that these aspects can be covered

In a single publication?
Or in parts?

6. Are any of the aspects listed in 4 or 5 above covered by existing national legislation in the country? _____

Please specify such legislation. _____

7. Are any of these aspects covered by existing

a. International standards
b. National standards
c. Other specifications of written requirements?

8. Are any of the requirements included in standards considered to be of outstanding importance by your organization?

9. What is the estimated time needed to complete the project? _____

10. What liaison will be necessary with other technical committees or outside bodies? _____

11. Who will be the ultimate users of the standard (number and/or names)? _____

FIGURE 37.5 Checklist for identifying fields for the establishment of national standards. (*Source: International Organization for Standardization 1994b*, pp. 93, 94.)

Regional Standardization.   Developing countries in the same region have similarities in climate, culture, governmental policies, consumption, industrial production, etc. Therefore, there might be a need for common standards. In order to deal with these regional issues in the field of standardization, regional standards organizations have been formed.

Such regional standards organizations exist in Africa (African Regional Organization for Standardization, ARSO), in Arab Countries (Arab Organization for Standardization and Metrology, ASMO), in the Caribbean (Caribbean Common Market Standards Council, CARICOM), and in Latin America (Comisión Panamericana de Normas Técnicas, COPANT).

International Standardization.   International standards are an important means of communication in international business and trade. With the globalizing of trade, the need for international standardization increases.

There are three international standardization bodies which are most important to developing countries:

The International Organization for Standardization (*ISO*):   This a worldwide federation of the national standards bodies of some 110 countries, many of them developing countries. The mission of ISO is "to promote the development of standardization and related activities in the world, with a view to facilitating the international exchange of goods and services, and to developing cooperation in the spheres of intellectual, scientific, technological, and economic activity." This work results in international agreements which are published as international standards.

The International Electrotechnical Commission (*IEC*):   This is a body preparing international standards in the electrotechnical field. IEC and ISO have a close cooperation.

The Codex Alimentarius Commission (*CAC*):   This is a body jointly set up by the Food and Agricultural Organization of the United Nations (FAO) and the World Health Organization (WHO) in order to prepare international standards in the food sector.

Developing countries can gain by adopting or adapting international standards as national standards, as the process of standards development is time-consuming and costly. Many developing countries do not have the professional resources necessary to prepare standards in certain areas.

Adopting and adapting international standards as national standards has several advantages (International Organization for Standardization 1986, p. 5):

1. It supports export promotion
2. It creates flexibility in imports and ensures the quality of imported products, services, etc.
3. It facilitates the transfer of technology
4. It supplements the production of national standards
5. It gives national authority to the use of international standards

As international standards are providing the key to international markets, it is important to developing countries to be able to see such standards internationally established for products that are vital to their national economies. In addition, they are concerned for the safety and the health of their citizens.

A developing country may find a need for an international standard to serve one or more of the following purposes (International Organization for Standardization 1994b, p. 3):

1. For an important export commodity: to ensure equitable prices that correspond to the delivered quantity and quality and to avoid rejects due to nonconformance to different conditions of export markets.
2. For an imported product that may affect the health and safety of users or have environmental implications: to have a technically sound and universally accepted basis for government regulations to protect the consumer and the ecology.

3. For a service that is being offered in the country to local and foreign customers: to maintain an acceptable level that would ensure safety (such as in building construction and health services) and/or competitiveness with other countries that offer similar services (such as tourist services).

4. For products and components traded with other countries: to ensure ability to interact and interchangeability.

5. For information, commercial, cultural, and other types of exchange: to facilitate communication between nations.

It may be difficult for developing countries to launch and to get acceptance for proposals on international standards in fields of interest to them. In order to help developing countries to prepare their proposals in a way that will have a good chance of acceptance by other countries, ISO has published some guidelines on how to launch a standards initiative (International Organization for Standardization 1994b).

CERTIFICATION

Certification means to verify conformance with certain requirements. This can take many forms, ranging from a simple statement by the manufacturer that the product conforms to the specification to a third party certification in which an independent body verifies that the manufacturer operates in accordance with a recognized standard on quality management systems.

Certification on a national level is very often combined with standardization. This means that national standards bodies, in addition to dealing with standardization on a national level, may also perform third-party certification on a national level.

Product Certification. This involves checking and certifying that products comply with the standards. The purpose of a product certification program is to give the buyer confidence that the product is of a certain quality or that it meets quality requirements. Such certification goes beyond the seller's assurances that the product conforms to the requirements and beyond the buyer's own verification—instead, systems operated by impartial bodies are used (third-party certification). The impartial certification body can be a governmental or nongovernmental organization. In developing countries the national standards body normally assumes this responsibility.

In developing countries there are various reasons for having a third-party certification program. One is to upgrade quality in the domestic market. Because of the shortage of goods and the absence of competition which very often prevail in developing economies, product quality is likely to be poor, and a mandatory certification system can provide a minimum quality level. A second reason is to promote exports. A certification system can be an important factor in enabling developing countries to secure access to foreign markets. A third reason is to prevent importation of products of inferior quality. Some developing countries have had the misfortune to be used as dumping grounds for unscrupulous foreign manufacturers.

A third-party certification can take various forms:

Type testing: A sample of the product is tested according to a test specification in order to verify conformance with certain specified quality requirements. The testing is carried out by the certification body or by a recognized testing institute or laboratory.

Audit testing: In order to provide for a subsequent assurance, the type testing can be followed by an audit test of samples purchased in the market place or selected from the manufacturer's production before shipment.

Batch testing: Manufactured batches are tested. This could be done on a sampling basis or 100 percent.

Assessment: Specialists from the certification body visit the manufacturer's factory in order to find out how the quality control activities are performed.

These forms can be combined into certification systems (Figure 37.6). Products that come under a certification program and have been found to comply with the standards are marked with a certification mark upon the granting of a license issued by the certification body (in most cases the national standards body). The manufacturer has to support the application with the following documents:

1. Articles of incorporation
2. Organizational chart
3. Quality control staff complement, their designations, qualifications, training courses attended, and number of hours thereof

---

System No. 1—Type testing

Type testing is a method under which a sample of the product is tested according to a prescribed test method in order to verify the compliance of a model with a specification. It is the simplest and most limited form of independent certification of a product both from the point of view of the manufacturer and the approval authority.

System No.2—Type testing followed by subsequent surveillance through audit testing of samples purchased on the open market

A system based on type testing (see System No. 1) but with some follow-up action to check that subsequent production is in conformity. Open market audit testing means a random audit testing of the type tested model from distributors' or retailers' stock.

System No.3—Type testing followed by subsequent surveillance through audit testing of factory samples

A system based on type testing (System No. 1) but with some follow-up action to check that subsequent production conforms. Audit testing of factory samples involves a regular check of samples of the type-tested models selected from the manufacturer's production before dispatch.

System No.4—Type testing followed by subsequent surveillance through audit testing of samples from both the open market and the factory

A system based on type testing (System No. 1) but with follow-up action to check that subsequent production conforms. Audit testing both of factory samples and open market samples.

System No.5—Type testing and assessment of factory quality control and its acceptance followed by surveillance that takes into account the audit of factory quality control and the testing of samples from the factory and the open market

A system based on type testing (System No. 1), with assessment and approval of the manufacturer's quality control arrangement followed by regular surveillance through inspection of factory quality control and audit testing of samples from both the open market and the factory.

System No.6—Factory quality control assessment and its acceptance only

Sometimes known as the approved firm or approved manufacturing method of certification. A system under which the manufacturer's capability to produce a product in accordance with the required specification, including the manufacturing methods, quality control organizations, and type and routine testing facilities are assessed and approved, in respect of a discrete technology. This system can be applied particularly where the specification covers a type of manufacture, possibly a material, but where the end product may take a variety of forms for which there are no particular specifications.

System No.7—Batch testing

Batch testing is a system under which a batch of a product is sample tested and from which a verdict on the conformity with the specification is issued.

System No.8—100% testing

100% testing is a system under which each and every item certified is tested to the requirements of the technical specification.

FIGURE 37.6 Description of third-party certification systems. (*Source: International Organization for Standardization and International Electrotechnical Commission 1992, pp. 156–171.*)

4. List of training courses regularly conducted by the company or other outside training courses attended by company personnel

5. Flow process diagram indicating the inspection points, frequency of inspection and key quality characteristics to be inspected at each control point

6. Brief description of manufacturing process

7. Copy or summary of management policies

8. Copy or description of test plan and specifications used by the company

9. Copy of sampling plans

10. Copy or a sample of work instructions or other related documents

11. Sample copies of frequency tables, histograms and/or statistical control charts and records of inspection/test results

12. Brief description of action taken on defectives

13. Brief description of the system for the preservation, segregation, and handling of all items

14. List of measuring and testing equipment with nominal capacities at each inspection point and final product testing together with their evidence of ownership

15. Brief description of calibration program (including frequency of calibration)

16. Brief description of equipment maintenance program

System Certification.   There is a trend to move from product certification to system certification. This means that the supplier's credibility is demonstrated by the assessment and registration of his quality system.

The assessment of the supplier's quality system is carried out by an independent certification body against an applicable quality system standard in order to establish whether the system conforms to the standard. Standards usually referred to are the international standards on quality management systems ISO 9001 and ISO 9002 (or the equivalent national standards). If the assessment shows that the supplier's quality system conforms to the relevant quality system standard, the supplier is entered into a register of certified companies. The idea is that this will indicate to the user of the register that the supplier is capable of producing a certain product or range of products. The registration is maintained by surveillance through periodic audits of the supplier's quality system, performed by the certification body. This will ensure continued conformity of the quality system with the standard in question.

In some developing countries, the national standards body has been assigned the quality system certification task. As soon as the auditing capability of the staff is developed and the necessary management structure is set up, quality system certification will be one of the major tasks of the national standard bodies in developing countries. As there is a great demand for quality system certification, many foreign (mainly European) certification agencies are operating in developing countries on a commercial basis. The certification has also become a business opportunity for local consultants. The less scrupulous ones form two organizations: one to set up the quality systems and to write the quality manuals for their clients, and one to certify that these quality systems and manuals conform to the quality system standards ISO 9001 or ISO 9002. The certification agency itself might have achieved accreditation by a body in Europe on a rather loose ground, making any of the certificates it issues of doubtful validity.

There are doubts about the validity of quality system certification, as this mainly relies on documentation in the form of quality manuals, work procedures, instructions, and records generated during the operation of the quality system. Product quality, and the customers' perception of quality, are not considered. Such doubts are more valid in developing countries compared to industrialized countries (Lal 1996a). In developing countries top management more often pays only lip-service to quality. They see the ISO 9000 certification as "some kind of magic key" which will open doors to export markets for their products. For them it is most important to obtain the formal certificate. There will be no basic change in the quality culture of the enterprise, nor will any real product quality improvements take place. The limited experience in quality management among assessors and auditors in developing countries will also create doubts about the validity of the certification.

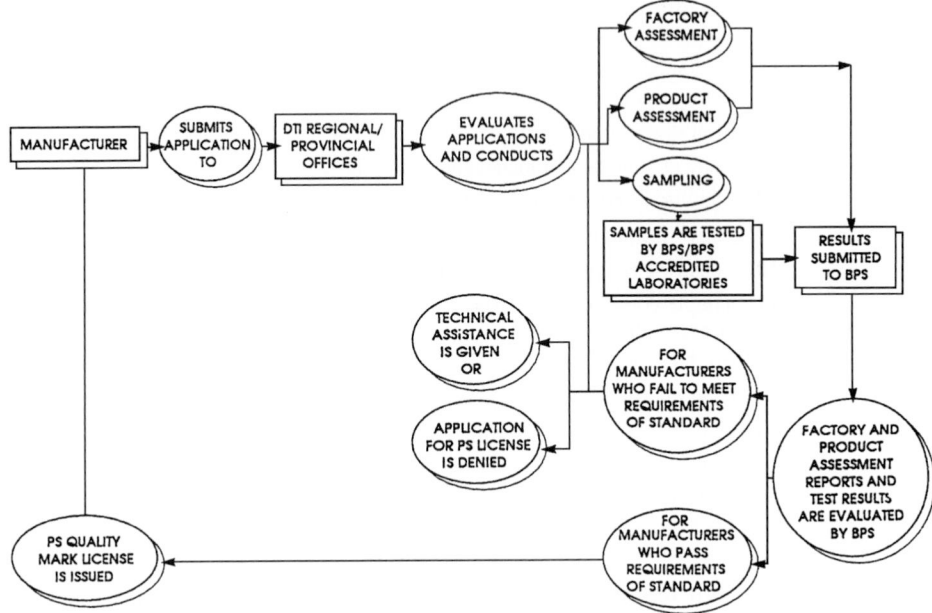

FIGURE 37.7 Product certification procedure used in the Philippines. (*Courtesy Bureau of Product Standards, Philippines.*)

Worldwide Recognition of Certificates. In order to encourage good practice in conformity assessment activities worldwide among the organizations that perform them, the International Organization for Standardization and the International Electrotechnical Commission in 1985 set up the Committee on Conformity Assessment (CASCO). CASCO has three basic objectives: (1) to study the means of assessing the conformity of products, processes, services, and quality systems to appropriate standards or other technical specifications; (2) to prepare international guides relating to the testing, inspection, and certification of products, processes, and services, and to the assessment of quality systems, testing laboratories, inspection bodies, and certification bodies and their operation and acceptance; (3) to promote mutual recognition and acceptance of national and regional conformity assessment systems and the appropriate use of international standards for testing, inspection, certification, assessment, and related purposes (Donaldson 1996). The guides are available from the International Organization for Standardization (see below, under External Assistance).

The worldwide growth in the certification of enterprises to ISO 9000 has led to a desire to establish a global system of mutual recognition of certificates. Therefore, on the basis of a proposal from CASCO, ISO and IEC initiated a voluntary worldwide system called Quality System Assessment Recognition (QSAR). The objective of QSAR is to ensure that ISO 9000 certificates issued to any enterprise, anywhere in the world, by any quality system certification body recognized under the QSAR program, are accepted everywhere. When QSAR has established itself, developing countries could benefit from such a worldwide recognition system in their efforts to export goods and services.

EXPORT INSPECTION

In assuring foreign buyers goods of an acceptable quality, plans for preshipment testing and inspection of products for export are used. Some developing countries have legislation dealing with export inspection. The legislation generally applies to specific commodities that are of key importance for the national economy (coconut, coffee, cocoa, fruit, jute, rice, rubber, etc.).

As a consequence of the deregulation of the markets, the trend is to get away from compulsory systems for export inspection enforced by law. Instead the export inspection is a matter for the market forces, which means that the trading partners (the exporters and the importers) set up and agree on how to ensure the quality of the deliveries. Third-party certification is then increasingly applied.

An example of an export inspection program which has been successful in promoting export is the fish inspection program in Thailand (Suwanrangsi 1995). The Fish Inspection and Quality Control Division (FIQD), established in 1992 within the Department of Fisheries, has gained recognition in major importing countries such as Australia, Japan, Canada, and the European Union for its active role in controlling the performance of the seafood export industry.

The Thai fish inspection program involves plant inspection and preshipment inspection. The plant inspection, carried out two to four times a year, covers the condition and maintenance of construction, equipment, processing operations, plant hygiene, and personnel. The preshipment inspection includes sampling of batches to be delivered. Sensory and microbiological assessment, as well as testing for contamination, toxicology, etc., are performed according to the requirements of the importing country. FIQD also provides consultancy service and training to the seafood industry in matters related to quality.

LEGISLATION

National standards bodies in developing countries have been established by acts of parliament or legislature. Such acts, normally called "standards acts," stipulate the role of the national standards institute in promoting standardization throughout the country. Provisions related to the enforcement of standards are sometimes incorporated into some of these laws; for example, foods acts or certification mark acts.

The standards act may also deal with weights and measures. In some countries such standards are covered by a separate weights and measures act, which also deals with the testing of weighing and measuring equipment to ensure that it is fit for use in trade.

There may also be legislation on export inspection, which is the case for certain items in some developing countries.

NATIONAL PROMOTION

In an increasing number of developing countries, national programs for promoting a general awareness of quality have been launched with the involvement of government agencies and trade and consumer organizations. The national programs usually have the following components:

High-level recognition: Support from ministries and important national organizations is granted. Even the head of state may be involved.

Publicity: Public media such as newspapers, magazines, radio, and television are used.

Conferences, seminars, and other meetings: Speeches are given by political leaders, industrialists, quality professionals, etc.

Slogans: Slogans such as "Quality first" are disseminated through posters, pamphlets, stickers, badges, etc.

Logotype: The campaign has a common emblem, displayed on posters, flags, pamphlets, etc.

Awards: Deserving companies and individuals are recognized through awards, which are presented with great publicity.

Quality month: The promotional activities may be concentrated in a particular month.

National Quality Awards.   Promoting quality among enterprises by means of a national quality award program has in recent years become an important factor to improve the competitiveness of the industrial sector in many industrialized countries. Experience shows that these programs stimulate improvements of quality and productivity to an extent that goes beyond the effects of ISO 9000. The awards criteria have become a yardstick for enterprises that allows them to assess their own situation and to guide their improvement efforts.

National quality awards have also been launched in some developing countries, e.g., Argentina, Brazil, Colombia, India, Republic of Korea, Malaysia, and Philippines (Stephens 1995). In most cases, either the Malcolm Baldrige National Quality Award or the European Quality Award has served as a model in the development of the setup and the awards criteria.

In some developing countries, the responsibility for the national quality award is assumed by a governmental organization, in others by a foundation financially supported by the private sector. The latter is the case in Argentina and Brazil. The National Quality Award in Argentina was established by a legislation passed by the Argentinean parliament. The administration of the award is delegated to a private foundation (FUNDAPRE, the National Quality Award Foundation), funded by member companies and by the government (Bertin 1996).

In India, the Rajiv Gandhi National Quality Award, named for the late prime minister of India, was launched in 1991. The award is promoted by the Indian Government and administered by the Bureau of Indian Standards. The presentations are made by the prime minister. A list of the award criteria and the point allocations is shown in Table 37.1.

There are four awards, consisting of one for large-scale manufacturing units, one for small-scale manufacturing units, one for service sector organizations, and one for best overall. In addition, there are six commendation certificates each for large-scale and small-scale manufacturing units: metallurgical, electrical and electronic, chemical, food and drug, textile, and engineering industry and others. The objectives of the award are stated as follows:

- Encouraging Indian companies to make significant improvements in quality for maximizing consumer satisfaction and for successfully facing competition in the global market as well.

- Recognizing the achievements of those companies which have improved the quality of their products and services and thereby set an example for others.

- Establishing guidelines and criteria that can be used by industry in evaluating their own quality improvement efforts.

- Providing specific guidance to other organizations which wish to learn how to achieve excellence in quality, by making available detailed information on the "Quality Management Approach" adopted by award winning organizations to change their cultures and achieve eminence.

TABLE 37.1 Criteria of the Rajiv Gandhi National Quality Award in India

Criteria	Marks
Leadership	100
Policies and strategies	100
Human resource management	50
Resources	100
Processes	150
Customer satisfaction	200
Employees satisfaction	50
Impact on society	100
Business results	150
Total	1000

Source: Rajiv Gandhi National Quality Award. Procedure and Application Form, Bureau of Indian Standards.

EDUCATION AND TRAINING

A key to quality upgrading is education and training, which involves developing necessary knowledge and skills, as well as influencing attitudes. In developing countries, it is necessary to direct activities of this kind not only toward manufacturers and service providers but also toward consumers; consequently, the activities are very far-reaching. The difficulties are aggravated by the normally high illiteracy rate in these countries.

National Level.   Education and training in the quality field can be dealt with in different ways within a developing country:

Educational Institutions.   Quality management topics are to an increasing extent included in the curricula of engineering universities and institutes, particularly in the more industrialized developing countries. The courses offered deal mainly with the more trendy tools, such as statistical process control, quality function deployment, benchmarking, and ISO 9000. Topics related to top management are rare. The courses are taught by professors who have limited experience in industrial work and management. There is a limited supply of locally developed textbooks.

Under a project of the European Union and with the involvement of three European universities, an undergraduate course on total quality management has been developed at the Indian Institute of Technology in New Delhi (van der Wiele 1996). The course has a modular basis with a content according to Figure 37.8. Each module includes case studies from Indian industry. The course was designed to meet the following objectives:

- Present an overall view of quality fundamentals for engineering graduates of all disciplines
- To present the international developments in the field of total quality management

---

Module I. Fundamentals   (6 lectures)

General introduction (definitions, evolution etc.). Contributions from quality gurus. Major conclusions and a general framework for understanding TQM. Examples of companies. Tools and techniques. Costs of quality.

Module II. Human Resource Management   (8 lectures)

General framework. Attitude, value system, and behavioral pattern. Motivation models. Team concept. Human resouce development. Quality circle. Quality education and training. Employee empowerment.

Module III. Tools and Techniques   (12 lectures)

Significance. Seven QC tools. Pareto analysis. Process control chart. Seven management tools. Just-in-time. Quality function deployment. Statistical process control. Process capability. Tolerances and interference. Quality and maintenance.

Module IV. Systems and Procedures   (8 lectures)

Quality system structure. Quality-related standards. Certification, testing, registration, and accreditation. Quality manual. Quality information systems. Quality audit. ISO 9000.

Module V. Implementation   (7 lectures)

Quality strategy as a competitive tool. Customer-supplier chain. Quality policy deployment. System strategies and tactics. Leadership. Motivational elements. Breakthrough improvement. Continuous quality improvement. Quality awards and self-assessment. Benchmarking. Employee empowerment. Implementation barriers. Impact on society.

FIGURE 37.8   Content of undergraduate course on total quality management given at the Indian Institute of Technology. (*Courtesy of Indian Institute of Technology, New Delhi, India.*)

- To create an awareness of the major factors in evolution on the quality front
- To expose the students to the important tools and techniques of quality management
- To emphasize the role of standards in maintaining and improving quality
- To appreciate the role played by human factors like motivation, leadership, and teamwork in sustaining quality
- To imbibe the principles of quality and apply them to achieve personal improvement

Courses and Seminars Offered by Associations, Institutes, and Other Organizations.   In countries that have reached a higher level of industrialization, courses and seminars are offered by national institutions for standardization, productivity, etc. and by professional associations (e.g., manufacturers' associations, national quality societies), as well as by consultants. Some countries are active in inviting foreign lecturers. As a consequence of the great interest in ISO 9000 certification, the majority of courses offered are in this field (e.g., implementation of ISO 9000, preparation for ISO 9000 certification, quality system auditing). International consulting and registration firms (mainly European-based) offer lead assessor training that can lead to registration by the International Register of Certified Assessors (IRCA).

Meetings and Conferences.   An important activity of a national quality organization is to hold meetings and conferences at which practitioners can exchange ideas and experiences. Countries with a more developed industrial sector generally have a quality association that is active in this way. Such organizations are the Chilean Association for Quality (ASCAL), the China Quality Control Association, and the Philippine Society for Quality Control.

Self-Instruction.   Independent study of books and journals can provide considerable knowledge in the quality management field, particularly for managers and engineers. In many developing countries there is, however, a shortage of such literature, owing to the lack of foreign exchange, a national language with limited readership, etc. Some national quality organizations publish a journal or newsletter as a means of promoting the professional development of their members.

In-House Training Programs.   Successful enterprises in industrialized countries have adopted a strategy that includes a massive in-house training program with the objectives of changing everyone's attitudes and giving new skills and knowledge. In these enterprises training in quality is provided for everyone regardless of function and level. Such training starts with top management and then works its way down the organization, level by level. This kind of massive training is rare in developing countries. In some larger enterprises, training in various narrow concepts and tools is provided.

On-the-Job Training.   On-the-job training is the principal method of training workers and inspectors, even in developing countries. Instructions are given by the supervisor or a more experienced inspector. The result depends on both the technical ability and the ability of the instructor to instruct and motivate the participants. In general, these abilities vary more in countries that have a limited industrial tradition.

International Level.   Developing countries also have opportunities for training on the international level by sending trainees to more industrialized countries. International organizations such as the United Nations Industrial Development Organization (UNIDO), the International Organization for Standardization (ISO), and the Asian Productivity Organization (APO) organize quality training programs for developing countries. (For information on these programs consult the International Trade Centre and the International Organization for Standardization at the addresses given under References.)

   The United Nations Development Organization, in cooperation with the Government of Japan (Ministry of International Trade and Industry, MITI) and the Japanese Association for

Overseas Technical Scholarships (AOTS), organizes a 5-week training program on Quality Improvement of Industrial Products every 2 years. The structure of the program is given in Figure 37.9. The program, intended for quality managers and production managers, focuses on problem-solving.

Since 1973, the government of Sweden has been sponsoring training programs in the field of quality for developing countries. A 7-week course entitled "Total Quality Management" is held in Sweden twice annually, and a 3-week top management seminar entitled "Quality Leadership" is held once annually. The training programs include lectures, case studies and study visits. The content of the theoretical training sessions of the "Total Quality Management" course is listed in Figure 37.10. The training programs in Sweden have been attended by quality professionals and managers from some 80 countries in Africa, Asia, Europe, Latin America, and the Caribbean. (Address information for the Swedish International Development Agency is given under References.)

EXTERNAL ASSISTANCE

External assistance plays a significant part in the industrial and economic growth of developing countries by making it possible to shorten the process of development. There are various forms of assistance.

Assistance from International Organizations.   There are international organizations providing assistance to developing countries in the field of quality. Some of them work worldwide (mainly within the United Nations system), others work on a regional basis. Significant organizations of this kind are:

European Union (EU).   EU offers financial support to development projects in developing countries. Address:

Rue de la Loi 200   Telephone: +32 2 2991111
B-1049 Brussels   Telefax: +32 2 2950138
Belgium   Internet: http://europa.eu.Int

Food and Agriculture Organization of the United Nations (FAO).   FAO has a mandate to raise levels of nutrition and standards of living, to improve agricultural productivity and the conditions for rural populations. The organization is involved in land and water development, plant and animal production, forestry, fisheries, economic and social policy, investment, nutrition, food standards and commodities and trade. Address:

Viele delle Terme di Caracalla   Telephone: +39 6 52251
00100 Rome   Telefax: +39 6 52253152
Italy   E-mail: webmaster@fao.org

International Organization for Standardization (ISO).   ISO is a worldwide federation of national standards bodies in more than 100 countries, with the objective to promote standardization and related activities worldwide. Within ISO there is a committee on developing country matters (DEVCO) with the aim of assisting developing countries and being a forum for the discussion of all aspects of standardization in this group of countries. In the program for developing countries (ISO 1995), ISO/DEVCO has identified six elements: identification and accommodation of standardization needs in developing countries, preparation and publication of development manuals, training, participation in ISO standards committee meetings, development of international standards needed by developing

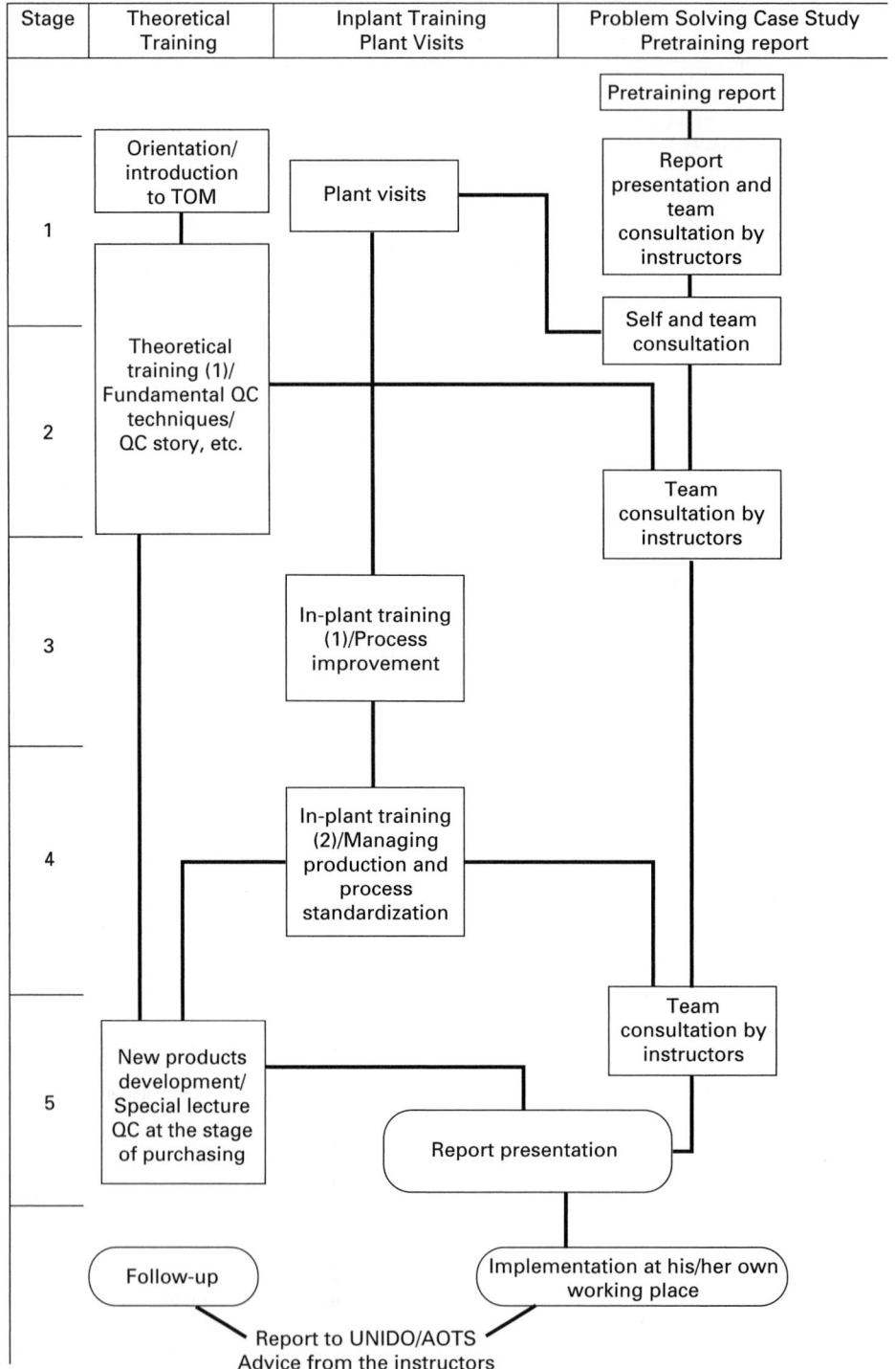

FIGURE 37.9   Structure of training program on Quality Improvement of Industrial Products held in Japan. (Courtesy UNIDO, MITI, and AOTS.)

Introduction: Quality; the quality function; quality management; role of upper management; quality planning; the concept of total quality; developments in the quality profession; profitability and quality; quality strategies; terminology

Statistical tools: Basic concepts, the statistical tool kit; methods of summarizing data; probability distributions; process capability; tools for process studies; control charts; acceptance sampling; point and internal estimates; tests of hypotheses; transformations of data; regression and correlation analysis; design of experiments

Metrology: Measurement technology; error of measurement; calibration control

Inspection: The nature of inspection; statistical tools in inspection; automated inspection; inspection accuracy; inspection workplace; inspection feedback; inspection by operator; inspection planning; planning of incoming inspection; planning of process inspection; planning of final inspection

Reliability and dependability: Dependability concepts; reliability requirements; failures and failure mechanisms; reliability prediction; fault tree analysis (FTA) and safety analysis; failure mode and effects analysis (FMEA); reliability determination by testing; reliability growth and screening; reliability verification; maintainability requirements, prediction and testing; availability requirements and testing; maintenance support conditions; environmental testing in general

New-product quality: Product phases and development programs; forms of early warnings, design reviews; the concept and definition phase; the preliminary design and development phase, the final design phase; the pilot production phase; series production, marketing and installation; use and maintenance phase; project management for new product quality; new-product planning

Quality specifications: Purpose and content; setting requirements; revisions; standards

Supplier relations: Supplier relations; purchase documents; supplier selection; joint quality planning; activities in the supplier's plant; incoming inspection; supplier surveillance; supplier certification; complaints; supplier quality improvement; single-source suppliers

Manufacture of quality: Manufacturing planning; nature of process; providing capable process; providing capable instruments; process control; self-inspection; inspection feedback to production; process improvement; traceability; quality responsibilities on the factory floor; automated manufacturing

Customer relations: Customer; field intelligence; complaints; marketing; product safety and liability; consumerism; government regulation of quality

Quality assessment: Why assessment?; standing in the marketplace; customer-focused quality analysis; quality culture; assessment of quality activities; using quality awards criteria; quality auditing; costs of poor quality; quality data, benchmarking

Quality audit: Quality audit concept; product quality audit; process quality audit; systems quality audit; total quality audit (quality surveys)

Economics of quality: Quality and economics; quality and income; quality costs; user's costs; life cycle costing

Quality data: Basic concepts; in-plant quality data; usage data; quality cost data; reports on quality

Improving quality: Control versus breakthrough; need for quality improvement; quality improvement sequence; obtaining information; project identification; diagnosis; remedies; follow-up and feedback; organizing for improvement; resistance to change; training in quality improvement; seven tools

Human factors in quality control: Controllability; theories of motivation; human resource management practices; role of management; role of work force; role of teams; quality motivation for managers; training; inspection accuracy

Quality leadership: Recognized champions versus potential champions; strategic quality management; quality strategies; role of upper management; quality policy; quality objectives (goals); implementing total quality

Quality system: Quality systems terminology; the systems concept; systems requirements and standards (ISO 9000); the quality manual; importance of ISO 9000

Organization for quality: Basic concept; quality work elements; evolution of the quality organization; organization for acceptance; organization for prevention; organization for improvement; organization for coordination; organization for assurance; role of the quality department

Quality consulting: Efficiency of the quality function; studies of the efficiency of the quality function; evaluation; report; who carries out the evaluation?

Developing countries and quality: Situation in developing countries; industrial development and quality; national efforts for quality; standardization; certification; export inspection; legislation; national promotion; education and training; external assistance; institutional infrastructure; national quality society

FIGURE 37.10 Content of theoretical training sessions of the total quality management course given in Sweden.

countries, documentation and information systems. ISO also organizes regional courses and seminars related to the ISO 9000 standards. Address:

P.O. Box 56   Telephone: +41 22 749 01 11
CH-1211 Geneva 20   Telefax: +41 22 733 34 30
Switzerland   E-mail: central@isocs.iso.ch

International Trade Centre UNCTAD/WTO (ITC).   ITC works with developing countries and economies in transition to set up effective trade promotion programs for expanding their exports and improving their import operations. This includes assistance in the quality management field in order to improve product quality and competitiveness in international markets. Services given are advisory missions, consultancy, development programs, research, publications, etc. ITC operates computer databases, including "Qualidata," which covers information on export quality management, and "Qualicontacts," containing worldwide data on quality-related institutions. Address:

Palais des Nations   Telephone: +41 22 730 01 11
CH-1211 Geneva 10   Telefax: +41 22 733 44 39
Switzerland   E-mail: itcreg@intracen.org
                                                 Internet: http://www.unicc.org/itc/

United Nations Development Programme (UNDP).   UNDP normally plays the chief coordinating role for operational development activities undertaken by the whole United Nations system. The UNDP activities focus on six priority themes: poverty elimination and grass roots development; environment and natural resources; management development; technical cooperation among developing countries; transfer and adaptation of technology; and women in development. Address:

One United Nations Plaza   Telephone: +1 212 9065000
New York, NY 10017   Telefax: +1 212 9065001
USA

United Nations Industrial Development Organization (UNIDO).   UNIDO is the specialized agency of the United Nations aimed at supporting and promoting industrial development in developing countries in order to overcome these countries social and economic difficulties and to achieve a greater stake in the global market. UNIDO's quality, standardization, and metrology program provides multidisciplinary services in areas such as total quality management, just-in-time manufacture, statistical quality control, and business strategies. Address:

Vienna International Centre   Telephone: +43 1 211310
P.O. Box 300   Telefax: +43 1 235156
A-1400 Vienna   E-mail: omaizza-neto@unido.org
Austria   Internet: http://www.unido.org

World Bank.   The World Bank Group comprises five organizations. Two of them, the International Bank for Reconstruction and Development (IBRD) and the International Development Association (IDA), provide loans to developing countries for development projects. Address:

1818 H Street, NW   Telephone: +1 202 4771234
Washington, DC 20433   Telefax: +1 202 4776391
USA   Internet: http://www.worldbank.org

World Health Organization (WHO).   WHO is the United Nations directing and coordinating authority on international health work. The organization has a wide range of functions including promotion of technical cooperation; improving nutrition and environmental hygiene; improving standards of teaching and training in the health, medical, and selected professions; establishing standards for biological and pharmaceutical products. Address:

20, Avenue Appia	Telephone: +41 22 7912111
CH-1211 Geneva 27	Telefax: +41 22 7910746
Switzerland	E-mail: postmaster@who.ch

Bilateral Assistance.   A great deal of the external assistance to developing countries is through bilateral aid from industrialized countries. In this way, quality experts have assisted in various developing and training programs.

An example of this kind of assistance is a 3-year program founded by the U.S. Agency for International Development (USAID) to help enterprises in Egypt with ISO 9000 implementation and certification. A similar program for the countries in Southern Africa is funded by the Norwegian government. The Swedish government funds training programs in quality management organized in developing countries. The Japanese MITI (Ministry of International Trade and Industry) is active in assisting developing countries in Southeast Asia in total quality management development.

Assistance from Transnational Companies.   External assistance may also take the form of collaboration with foreign manufacturers to obtain benefits such as technical know-how in joint ventures, import of plants and equipment, or consultant service. This kind of assistance is vital to developing countries.

INSTITUTIONAL INFRASTRUCTURE

Industrial enterprises require access to an infrastructure of institutions able to render a wide range of services for instance, in the areas of standardization, certification, accreditation, testing, metrology, quality consulting, and training. Developing countries that are in the process of industrializing must also provide for development of such an institutional infrastructure.

National Standards Body.   In most developing countries, a national standards body is in operation primarily to provide services in standardization, certification, testing, and metrology. Unlike standards bodies in the West, national standards bodies in developing countries are usually governmental agencies established by law and founded by their respective governments. Because of the importance of standardization to both the public and private sector, some developing countries restructure their national standards bodies as joint public/private sector bodies. In the early stages of industrial development some form of government support is needed. A balanced representation of all groups interested in standardization should be considered: industry, trade, consumers, professional associations, the government.

If an integrated approach is used, the list of tasks of national standards bodies includes (International Organization for Standardization 1994a, p. 11):

- The preparation and promulgation of national standards

- The promotion of the adoption and application of national standards at all levels in the country

- The promotion of standardization as a technical activity and an integral yet distinct function of management in the country

- The promotion and, if necessary, provision of third-party guarantee of the conformity of products through a national product certification program

- The promotion of quality in industry and services and, eventually, the provision of third-party certification of quality management systems implemented by industrial firms and service operators

- The provision of means for collecting and disseminating information on standards and related technical matters both nationally and internationally
- Supporting export promotion programs by providing information on the standards and regulations of export markets, promoting quality by testing products, and certifying their conformity to export specifications
- Protecting the consumer, particularly for safety and against health hazards
- Supporting through standards the efforts undertaken to safeguard the environment from hazards related to products and processes
- Safeguarding the country against the dumping of inferior quality goods as well as unsafe and ecologically hazardous goods
- The promotion of metrology as a necessary adjunct to standardization
- The coordination, for the benefit of the country, of standardization and related activities carried out at various levels: company, national, regional, and international
- The promotion and, if necessary, operation of an accreditation system for testing and calibration laboratories, to promote the precision of their results and enhance confidence in their work
- Undertaking tests for industry
- Offering technological advice to both government and industry
- Organizing training in standardization and related matters
- Conducting investigations to quantify the benefits of standardization with subsequent dissemination of information

The national standards bodies have not traditionally provided any services to industry in the fields of quality consulting and training. In consequence of the great interest for ISO 9000 certification and registration, an increasing number of national standards bodies in developing countries have started offering training in areas related to ISO 9000, such as quality systems implementation and auditing.

The mission of a national standards body is given in Figure 37.11.

In most countries the standards bodies are governed by a council which is responsible for working out policy guidelines, as well as for approving standards. The drafting of standards is supervised by technical committees representing manufacturers, users, university-affiliated institutions, research centers, etc. Usually, there are technical committees in particular fields, such as electrical, mechanical, and civil engineering, chemicals, and textiles.

The staff of a national standards body is headed by a chief executive officer or director, who is usually in charge of departments for standardization, certification (sometimes called quality control or quality assurance), metrology, laboratory services, information, and administration. The organizational structure of a fairly well-developed national standards body is given in Figure 37.12.

National Quality Council.   With the increasing awareness of the fact that quality is an important element behind the economic growth of a country, the emerging necessity of having a national institutional infrastructure for the following activities is growing more and more apparent (Lal 1996b, p. 167):

Creating awareness about quality and its economic benefits including administering the national quality award

Promoting new concepts in quality management

Training in quality-related subjects

BPS MISSION STATEMENT

We at BPS, in partnership with other organizations, are committed to contribute through standardization to national economic and social development by helping industries raise the quality and competitiveness of their products and by fostering consumer and environmental protection, and to serve our clients with the quality of service that is attainable with our resources.

RENATO V. NAVARRETE

Director

22 March 1993

FIGURE 37.11   Mission statement of a national standards body. (*Courtesy Bureau of Product Standards, Philippines.*)

Accreditation of certification bodies

Registration of certified auditors for quality systems

These activities should not be assigned to governmental agencies only, although governments have to take a major initiative in these areas. They should be autonomous, free from bureaucratic control, and set up in close cooperation with the private sector.

The Quality Council of India is an example of such a council. It is an autonomous body composed of about 10 members from governmental departments and about 20 members from industrial and trade organizations, academic institutions, consumer associations, certification bodies, quality award-winning companies, etc. The chairman is nominated by the prime minister. The structure is shown in Figure 37.13.

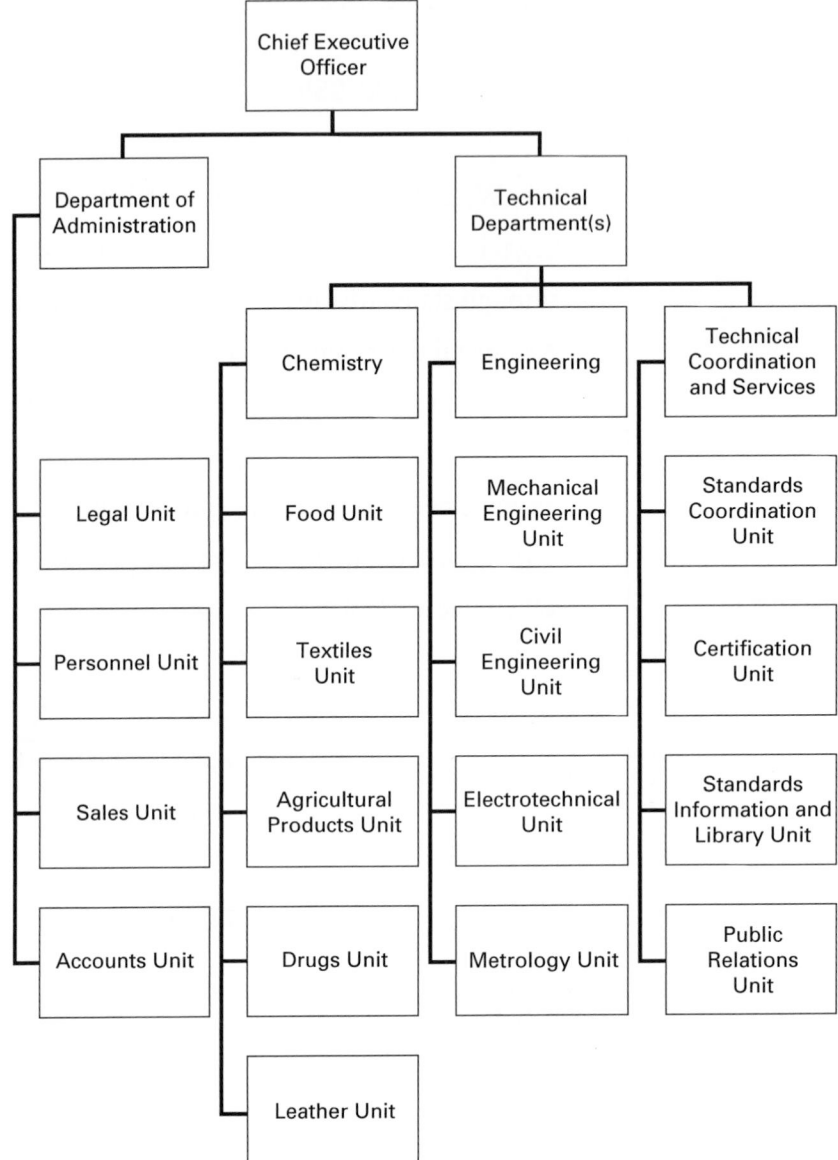

FIGURE 37.12   Organization of a national standards body. (*Source: International Organization for Standardization 1994a , p. 22.*)

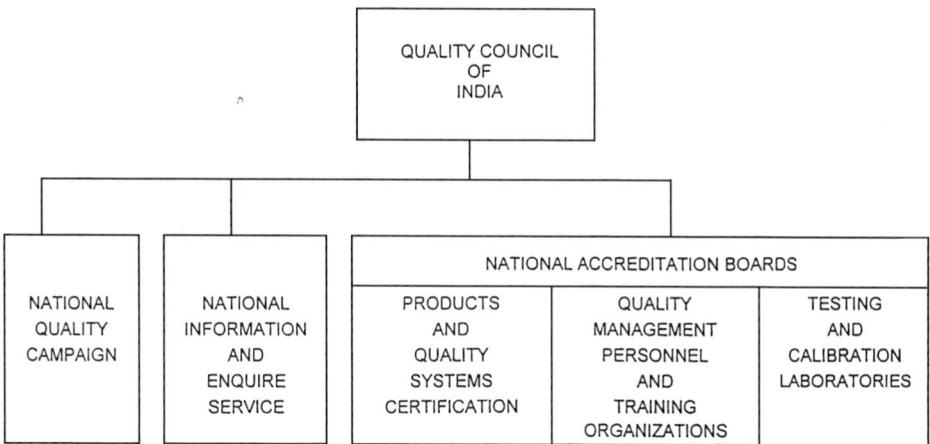

FIGURE 37.13   Quality Council of India. (*Source: Lal 1994, p. 21.*)

PROFESSIONAL SOCIETIES

A national quality society can play an important role in promoting quality nationwide, both in developing and in industrially more developed countries. Quality societies or associations are being formed in an increasing number of countries. Some of the national societies offer individual membership only and some offer institutional membership only, but most offer both.

Some societies are very active, for example, the Argentine Institute of Quality Control (IACC), the Brazilian Association of Quality Control (ABCQ), the Chilean Association for Quality (ASCAL), the China Quality Control Association, the Institute of Quality Control Malaysia, and the Philippine Society for Quality Control (PSQC). These societies organize conferences and seminars, conduct training programs, and distribute information (newsletters, journals, etc.).

A case in point is the Philippine Society for Quality Control, founded in 1969. At the end of 1996 it had 300 institutional members and 100 individual members. The mission of PSQC is phrased as follows: "We are an internationally recognized organization highly committed to promote quality as a culture and a way of life in our industries, government, academe, non-government organizations and communities. We lead and influence the transformation of the Philippines into a world class country by being proactive in the practice of total quality management." The values are phrased: "The organization is continuously striving for excellence. This is to be achieved through: respect and concern for people, continuous improvement, customer orientation and teamwork."

PSQC's objectives are

- To promote quality awareness and practices in both the private and public sectors through seminars, conferences, publications, and awareness campaigns
- To promote the development of members in quality leadership and technology
- To link with international organizations for updates in recent trends and the latest technologies

PSQC's programs and activities are

- To organize luncheon dialogues and symposia, seminars, and training programs
- To organize a National Quality Forum during the National Quality Month in October
- To disseminate quality information and updates about PSQC through quarterly issues of the society's newsletter, "Q...Point"

- To establish links with government agencies in promoting quality practices
- To coordinate "ugnayan" (the Filipino word for "linking") with host member companies for plant tours and sharing of their quality programs and success stories
- To form Youth Quality Chapters in universities to increase youth involvement and participation in making quality a way of life
- To administer the selection of the Philippine Quality Award winner every year

The term of office is 2 years, but an election is held every year for nine and six members alternately, to allow for smooth transitions and prevent a complete change in the composition of the board. The trustees elect officers from among themselves every year. They subsequently form working committees from the rest of the membership to help carry out the programs for the year. The immediate past-president is the 16th member of the board as an advisor.

PSQC is a member of the regional organization Asia-Pacific Quality Control Organization (APQCO). Another regional organization of national quality societies is the Latin American Organization for Quality (OLAC). These regional organizations regularly organize regional conferences.

CONCLUSION

The so-called developing countries are in various phases of economic and industrial development—some have just started with manufacture, whereas others are almost fully industrialized. No matter which phase a country is in, the quality of the goods and services provided is important to its economic growth. Quality management in a broad sense will ensure that goods and services are fit for domestic use, reduce the waste of scarce resources, and facilitate the export of national products.

It is of great importance that developing countries use appropriate methods in quality efforts. Simple basic techniques will in most cases give a better result than currently fashionable methods (to which too much attention is often paid). Before methods are selected, the actual needs have to be determined. This seemingly obvious point, however, is often overlooked. The focus should be shifted from a method-oriented to a more result-oriented approach, with increased emphasis placed on studies of established facts. Such an approach will provide a better basis for successful development.

There is an obvious need for extensive studies on how to deal with quality in developing countries on both a national and a company level and on the relations between the two levels. Models and plans for different levels of industrial development, different industrial structures, different branches of industry, etc. should be worked out. This is an important international undertaking.

ACKNOWLEDGMENT

Thanks are due to Dr. Noriaki Kano (Professor at the Science University of Tokyo, Japan), Gen. H. Lal (Director General, FICCI QUALITY FORUM, Federation of Indian Chambers of Commerce and Industry, New Delhi, India), Dr. Octavio Maizza-Neto (Coordinator, Industrial Quality Group, United Nations Industrial Development Organization, Vienna, Austria), Mr. Enrique Sierra (Senior Advisor on Export Quality Control, International Trade Centre UNCTAD/WTO, Geneva, Switzerland), and Dr. Anwar El-Tawil (Director, Program for Developing Countries, International Organization for Standardization, Geneva, Switzerland) who have given the author useful information by sharing with him their great experience of quality management in developing countries. In his work in developing countries, the author has had the privilege of meeting many dedicated people, both as colleagues and as students, who are too numerous to be mentioned by name. Their contributions are highly appreciated.

REFERENCES

Donaldson, John L. (1996). "Profile of ISO/CASCO Committee on Conformity Assessment." *ISO 9000 News,* vol. 5, no. 2, pp. 3–6.

Global Economic Prospects and the Developing Countries. (1995). The World Bank, Washington, DC.

International Organization for Standardization (1986). *Guidelines for Selecting and Adopting/Adapting International Standards for National Use.* International Organization for Standardization, Geneva.

International Organization for Standardization (1991). *Development Manual 6. Application of Standards.* International Organization for Standardization, Geneva.

International Organization for Standardization (1994a). *Development Manual 1. Establishment and Management of a National Standards Body.* International Organization for Standardization, Geneva.

International Organization for Standardization (1994b). *Launching a Standards Initiative. Guidelines for Developing Countries.* International Organization for Standardization, Geneva.

International Organization for Standardization (1995). *ISO Programme for Developing Countries 1995–1997.* International Organization for Standardization, Geneva.

International Organization for Standardization and International Electrotechnical Commission (1992). *Certification and Related Activities.* International Organization for Standardization and International Electrotechnical Commission, Geneva.

International Trade Centre UNCTAD/WTO (1996). *Applying the ISO 9000 Quality Management System.* International Trade Centre UNCTAD/WTO, Geneva.

International Trade Centre UNCTAD/WTO and International Organization for Standardization (1996). *ISO 9000 Quality Management Systems. Guidelines for Enterprises in Developing Countries,* 2nd ed. International Trade Centre UNCTAD/WTO and International Organization for Standardization, Geneva.

Juran, J. M. (1975). "Standardization and Quality." *Quality Progress,* vol. 8, no. 2, pp. 4, 5.

Lal, H. (1994). "National Quality Council." *Quality Quest,* vol. 3, no. 2, pp. 15–22.

Lal, H. (1996a). "Is certification at risk from lack of standardization?" *ISO 9000 News,* vol. 5, no. 1, pp. 19–22.

Lal, H. (1996b). "Quality Strategy for Globalising the Economy of Developing Countries." In *Quality Without Borders, Silver Jubilee Book.* Sandholm Associates, Djursholm, Sweden, pp. 163–169.

Maizza-Neto, Octavio, et al. (1994). *Quality, Standardization, and Metrology.* UNIDO, Vienna.

Mhlanga, Kinnios En (1995). *Quality in Zimbabwe.* Standards Association of Zimbabwe, Harare.

"National Quality Award, Argentina." In *Quality Without Borders, Silver Jubilee Book.* Sandholm Associates, Djursholm, Sweden, pp. 37–41.

Sandholm, Lennart (1996). "Maturity in Quality—Still to Come?" In *Quality Without Borders, Silver Jubilee Book.* Sandholm Associates, Djursholm, Sweden, pp. 187–195.

Stephens, Kenneth S. (1995). "National Quality Awards: A Developing Country Perspective." In *The Best on Quality.* Book series of International Academy for Quality, vol. 5. ASQC Quality Press, Milwaukee, pp. 179–210.

Suwanrangsi, Sirilak (1995). "Fish Inspection and Quality Control in Thailand—A Model for Developing Countries." *ITC Export Quality,* no. 47, October.

United Nations Industrial Development Organization (1996). *Trade Implications of International Standards for Quality and Environmental Management Systems. Survey Results.* United Nations Industrial Development Organization, Vienna.

van der Wiele, Ton (1996). "TQM Initiatives at the Institute of Technology in Delhi." *Institute of Directors, Conference Proceedings,* New Delhi.

Lists on quality training programs offered for developing countries can be obtained from:

International Trade Centre UNCTAD/WTO
Functional Advisory Services Section, Division of Trade Services
Palais des Nations
CH-1211 Geneva 10
Switzerland
Telefax + 41 22 733 44 39

International Organization for Standardization
Director, Programme for Developing Countries
ISO Central Secretariat
P.O. Box 56
CH-1211 Geneva 20
Switzerland
Telefax: +41 22 733 34 30

The governmental agency responsible is the Swedish International Development Agency (Sida). The program director is the author. Information is available from Swedish embassies and consulates, as well as from:

Sandholm Associates
P.O. Box 28
S-182 05 Djursholm
Sweden
Fax: + 46 8 755 19 51
e-mail: quality@sandholm.se

SECTION 38
QUALITY IN WESTERN EUROPE

Tito Conti

INTRODUCTION

Europe is composed of many different nations, cultures and languages. From whatever point of view you consider it, its differences are easier to pinpoint than its similarities. The current move toward economic and political integration—a long and difficult process—should eliminate the remaining pockets of economic and social backwardness and boost the continent's political and economic strength, while maintaining legitimate differences and the specific cultural identity of each nation. Quality, which is a cultural issue first and a technical issue second, is no exception as far as diversity is concerned. There is no such thing as European quality. Significant differences exist among and even within the various countries. The quality profiles of the nations vary enormously, too. Germany, for example, has a solid reputation for product quality; many Northern European countries provide excellent service quality. Other countries offer outstanding quality in specific sectors: France and Italy are typical examples. France is famous for its *haute couture* and its food and wines, while Italy has a reputation for design, fashion, and goldwork. But despite these great differences, over the last 10 years, Europe has made up a great deal of ground in relation to the United States of America and Japan. In the second half of the 1980s, large corporations still employed the services of non-European specialists to plan their Total Quality Management (TQM) strategies. Today, halfway through the 1990s, both the corporate sector and the leading quality consultants—a small minority, but this is the situation everywhere—stand up well to comparison with the United States and even with Japan.

In discussing quality in Western Europe, this section provides a brief overview of the main development trends and then looks at a specific and highly important issue: the impact of the Single Market on the quality culture (the Single Market is the term normally used in Europe to refer to the market formed by the 12 original members of the European Community, which has gradually extended to the Free Exchange Zone and continues to expand as new member states join what is now called the European Union). At the level of the individual nations, only Germany, France, and Great Britain are analyzed separately, since they represent three typical examples of the evolution of quality in Europe after the Second World War and thus cover the main trends in the other nations. The country-by-country analysis ends with a brief summary of distinctive developments in other West European countries. This section then looks at the European quality organizations, which have played an important part in recent quality developments, and discusses Europe's quality awards. The final portion of this section discusses environmental quality.

QUALITY DEVELOPMENT TRENDS IN EUROPE AFTER THE SECOND WORLD WAR

In Western Europe as in other areas of the western world, from the 1950s to the 1970s, quality was considered a priority issue mainly in defense, aerospace, telecommunications, electronuclear energy and energy in general, chemicals, and other high-technology sectors. An indication of the level of interest in quality disciplines (specifically, in statistical quality control) was the formation of national quality associations in many European countries during the early 1950s. In the area of consumer durables and consumer goods in general, which were enjoying a period of high demand, a capacity for innovation and the ability to produce large volumes at low costs were the main priorities. Healthy market performance meant staying more or less in line with the typical (and often mediocre) quality standards of the relevant price/performance class, which, given the lack of specific competitive stimuli, tended to remain stable. Certain niches, brands, or even entire geographical areas stood out for the high quality and reliability of their products, but as a rule, higher quality positioned a product in a higher price class compared with products offering similar performance. Countries like Germany, for instance, had a reputation for superior product quality, especially in niche markets (e.g., luxury automobiles, electric home appliances), but since this quality stemmed from high professional skills and craftsmanship, product prices were usually higher. (See Quality in Germany below.)

For the strategic sectors in the first group, quality, reliability, and safety were essential. Price was a secondary variable. Many suppliers financed R&D work for their commercial productions with the margins on their military sales. On the other hand, these sectors occupied a position of strategic importance for Europe during the period concerned: the defence sector in relation to the cold war, the others in relation to postwar reconstruction and creation of the infrastructures needed for new growth. Taken together, these considerations explain the development of a quality culture and quality practices predominantly based on the specific characteristics of this group of sectors.

The first characteristic is the contractual relationship between the supplier and the customer. The supplier provides goods for a specific customer, who stipulates requirements in a contract, in as precise and as clear a form as possible. The supplier undertakes to provide what the contract requests, and therefore regards quality as conformity with predefined specifications and standards. This produced a quality culture based on the concept of standards conformity and fulfillment of specifications. The focus was on execution (engineering and production) rather than on goals, which had already been set. This was a "reactive" approach to quality: the customer orders; the supplier provides what has been ordered—and (if possible) nothing more, since any extra is not paid for.

The second characteristic of this first group of sectors is that quality and reliability requirements are generally very high. Preventing failures is essential. The focus on ensuring reliability meant that engineering techniques improved. Progress was slower in production, where the quality and early-life reliability levels requested were often achieved not through process improvements but through final-product testing and screening (prolonged testing, run-ins, burn-ins), which raised costs significantly.

As product quality requirements rose, producers acquired greater skills in technical quality disciplines geared to technology and the product. Among customers, the 1960s witnessed a continuous increase in the number of inspectors appointed to approve products and plants. (See Quality in Great Britain, which discusses the role of the British Ministry of Defence).

Quality took a major step forward with the move from inspection to preventive assessment of suppliers' quality systems. Once again, the development originated in the defense field, following the introduction of NATO's APAQ standards (1968). In similar fashion, the U.K. BS 5750 standard (1979) was drawn up on the recommendation of the Ministry of Defence, to replace the endless series of second-party audits with third-party certification (Hutchins 1995).

At the beginning of the 1980s, therefore, the quality culture and quality practices in Europe were closely associated with the concept of conformity with standards or previously defined specifications. Table 38.1 summarizes the situation of quality standards during 1980.

In the second half of the 1970s, however, word had begun to spread about the success achieved by Japan with a "proactive" approach that perceived quality as a competitive factor. Doctor Juran's article "Japanese and Western Quality: a Contrast" (Juran 1979) was a revelation in many quarters. Large companies operating in highly competitive international markets were a particularly receptive audience. In the early 1980s, these companies began to experiment with total quality, often with little success in the first instance. For a number of years, the standards-based quality culture and the new concept of continuous improvement, customer focus, and process control evolved separately: two converging tributaries, the temperature, speed, and composition of whose waters are so different that they flow side by side without mixing. At this point, the EC Council began examining the question of the technical barriers inside the Single Market.

THE CREATION OF THE EUROPEAN SINGLE MARKET AND ITS IMPACT ON THE EVOLUTION OF THE QUALITY CULTURE

The European Community (EC) Council Resolution Of May 1985. On May 7, 1985, the EC Council adopted a resolution regarding technical harmonization and standardization which was designed to bring about the rapid elimination (by the end of 1992) of the technical barriers that, together with physical and fiscal barriers, were obstructing the creation of the Single Market. The EC strategy, known as the *nouvelle approche,* was based on the following principles:

1. Limitation of compulsory technical regulations to the essential issues of security and protection of health and the environment.

2. Harmonization of member states' voluntary technical standards, through discussions among the national and European standardization bodies. Implementation of specific measures by the Council to provide guidance for these bodies and strengthen their role.

3. Rapid definition of a testing and certification strategy to permit mutual recognition of certifications.

4. Acceptance of a manufacturer's declared standards conformity once that manufacturer's quality system is certified as compliant with European quality assurance standards.

TABLE 38.1 Quality Standards

	Market sectors	
Type of standard	Strategic (defense, etc.)	Commercial
Product	Defined by customer	Defined by a market which was sometimes indifferent to quality
Quality system and processes	Developed through cooperation within the sector; widely available and widely applied	Not generally available or applied

At this point, however, work on the ISO 9000 standards (published in 1987) had reached an advanced stage.

With the approach of January 1, 1993, and the official launch of the Single Market, the European political system made the pleasant discovery that a regulatory tool intended for worldwide application was about to be introduced, paving the way to elimination of technical barriers in Europe (McMillan 1994). The European Standardization Committee (CEN) immediately incorporated the ISO 9000 standard into the European standards with the name EN 29000 (now EN 9000) and initiated an emergency procedure to complete work on the remaining standards (EN 45000). For national governments, which apart from a few notable exceptions had until then lent a very half-hearted ear to the messages on total quality coming from various sectors of private industry, the ISO 9000 standards presented quality in a formulation that was much closer to their way of thinking. The simplicity of certification and mutual recognition had enormous appeal, especially for those with no previous experience of quality, who tended to overestimate the potential of the certification approach, hailing it as a sort of magic wand for the development of quality in Europe.

In this period, the European Foundation for Quality Management (see below) was set up, for the specific purpose of focusing the attention of senior executives from business and state administrations on the question of total quality. Despite the EFQM's high-level sponsors (the chairmen of 14 leading European companies), the interest it attracted in political circles and the media was small in comparison with the interest fueled by certification. The EC allocated abundant funds to the national standardization bodies, to be used, rightly, for the rapid harmonization of standards across Europe. Less justifiably perhaps, these funds helped to spread a standards-based quality culture, which was clearly out of step with the times.

Ratified by the 1987 Single Market Act, the 1985 Resolution of the EC Council would have unforeseen effects on the evolution of quality both in and outside Europe.

Fears of "Fortress Europe."   In 1987, the Malcolm Baldrige Award was launched. For the next few years, the TQM model used by the award was the center of U.S. attention, and many companies adopted it as a self-assessment guideline, even though they were not competing for the award itself. The introduction of the ISO 9000 standards therefore went almost unnoticed in the United States. Nevertheless, the Malcolm Baldrige model provided for product and service quality—the key issue of the ISO 9000 standards—in category 5. It was not until 1990 that the excitement in Europe about the ISO 9000 standards began to arouse the interest of the United States. At this time, Europe's sights were firmly set on the 1993 deadline, fueling suspicions in non-European political and economic circles that construction of a "Fortress Europe" was the ultimate goal. The ISO 9000 standards and the growing demand for preventive certification of the conformity of suppliers' quality systems with those standards resembled nothing if not a bulwark of that fortress.

In May 1990, the ASQC invited European quality specialists to attend its congress, as did the Juran Institute and a number of other bodies, in an attempt to find out what lay behind the European craze for ISO 9000. This was followed by a general rush to the ISO 9000 camp in North America, too, a trend that could be justified from a commercial point of view. It is less easy to explain when one considers the content of the standards. After a decade of total quality and continuous improvement, it is strange, to say the least, that quality assurance—an essential issue, but only one aspect, indeed a preliminary condition, of the wider theme of total quality—should have attracted such enormous attention.

The reaction in Japan was less extreme. The ISO 9000 standards were analyzed in detail (Kume 1990) and compared with current practice. Without doubt, Japan grasped the commercial significance of certification, as well as the value of adopting the criterion of ISO 9000 conformity in contractual relations, especially in geographical areas where partnership between customer and supplier is difficult to achieve. Its verdict on the ISO 9000 standards as such was favorable, with the proviso that the concept of continuous improvement should be integrated into the standards approach.

Midterm Effects of the Emphasis on ISO 9000 and Certification.   The political focus on the ISO 9000 standards and certification has had positive and negative repercussions for quality in Europe. The effects on the process of market unification, which was the main goal, are positive. As far as the effects on the development of quality itself is concerned, it is too soon to draw conclusions, but a brief analysis is possible.

The main benefit is the rapid spread of basic quality assurance know-how. Huge numbers of small to medium enterprises, operating mainly in domestic markets or in fields that are still unexposed to international competition, have had their attention drawn to quality system certification by industrial customers, distributors, and clients in the public sector. Other operators have been involved in an indirect way, often through sectoral associations. Without doubt, had the need to remove technical barriers within the Single Market not been such an urgent priority, the number of companies reached by the quality message would have been on a far smaller scale.

The reverse of the picture is that the emphasis on ISO 9000 and certification has distracted attention away from competitive quality for too long; certification is an important requirement, but it is a precompetitive requirement. Moreover, this "distraction" appeared at a critical time, just when total quality concepts were beginning to make headway in corporate culture. A groundless antagonism frequently developed between the two approaches, which can only be explained as the result of political intervention in quality issues (in many European countries, state economic intervention was a significant factor during the period considered) and the appetizing business potential of certification.

Unfortunately, the quality organizations jumped on the bandwagon, too. Certification was the sole focus of interest of many national quality associations for years. The EOQ congresses at the end of the 1980s were dedicated almost entirely to standards-based quality [see The European Organization for Quality (EOQ) below]. No decisive change occurred until 1992, when the EOQ took active steps to redress the balance with TQM. The European Foundation for Quality Management, which was set up in 1988 to promote TQM, has played an important role in re-establishing an equilibrium, most notably with the launch in 1991 of the European Quality Award.

Table 38.2 illustrates the singular attention to certification in Europe.

TABLE 38.2   ISO 9000 Certifications Awarded in Europe

Country/region	1989	January 93*	March 95*	December 95*
Austria		101	667	1,133
Belgium		180	1,226	1,716
Denmark		326	1,183	1,314
Finland		185	646	772
France		1,049	4,277	5,535
Germany		790	5,875	10,236
Greece		18	162	248
Ireland		100	1,410	1,617
Italy		188	3,146	4,814
The Netherlands	250 (KEMA)	716	4,198	5,284
Norway		91	679	890
Portugal		48	257	389
Spain		43	942	1,492
Sweden		229	871	1,095
Switzerland	150 (SQS)	410	1,520	2,065
United Kingdom	10,000 (BS5750)	18,577	44,107	52,591
Other Europe		41	751	1,419
Total Europe		23,092	71,917	92,610
Europe excluding U.K.		4,515	27,810	40,019
Rest of world		4,829	23,559	34,779
World total		27,921	95,476	127,389

*__Source:__ *The Mobil Survey,* 6 August 1996

The 1990s: Setting a New Course.   Around 1994, European and national government bodies began to set a new course. After extensive consultations, in 1994 the Directorate General III (DGIII) of the European Commission (responsible for internal market and industry) published a working paper entitled "A European Quality Promotion Policy." This document stated that, with the completion of the regulatory phase designed to create the correct operating conditions for the Single Market, the Commission intended to launch a new phase of promotion and support to enable business enterprises in any part of Europe to pursue their goals within a balanced European competitive environment. The paper listed the following strategic objectives:

- Satisfaction of the expectations of consumers and of the community in general
- Development of human potential
- Respect for the environment
- Prudent use of resources
- Efficient and effective enterprise management
- The ability to create jobs, in particular through creativity and innovation

The strategies outlined in the paper are largely derived from TQM strategies. Three main groups of measures are identified—development of human resources, improvement of production structures, and development of a European quality image—which are intended to support TQM strategies in business and in state administrations.

The European Quality Promotion Policy document was endorsed by the European Union's Industry Council in November 1996.

In 1995, the European Union promoted the following initiatives: a European Quality Award for small to medium enterprises to flank the current Award, which is open to companies with more than 250 employees, and an annual European Quality Week (to be held the second week of November). These initiatives are organized by the two European quality organizations, the EOQ and the EFQM, under the umbrella of the European Quality Platform (see below).

COUNTRY-BY-COUNTRY ANALYSIS

Quality in France.   Many large French companies were nationalized in the immediate postwar period (Renault in 1945 for example); others were nationalized during the early 1980s. The trend has been reversed since the mid-1990s. The decline in quality generally associated with nationalization—which many European countries have experienced—has not been seen in France. Many state-controlled French companies have been leaders in quality in Europe during the 1980s and 1990s: Renault is a case in point, its leadership exemplified by the success of the Espace automobile and the Formula I engine. This is probably a result of the high quality of managers trained in France's *grandes écoles,* which provide the country's major corporations with approximately 80 percent of their management recruits (the management schools such as the Ecoles des Hautes Etudes Commerciales, and the scientific schools, e.g., the Ecoles Polytechniques) (Gogue 1988).

The key elements of the French approach to quality from the mid-1940s to the mid-1990s are as follows:

1. The central role of the State in promoting quality, as a result of its dominant presence in the economy (the above-mentioned nationalizations are the most extreme manifestation of this). State intervention appears to have been beneficial, again very probably due to the high quality of management (in this case, the quality of managers in the state administration, which is superior to that of other European countries).
2. The deliberate focus on the regions and, in particular, on small to medium enterprises (SMEs). This has stimulated dynamic territorial activity and created a flourishing bottom-up movement.

3. The particular characteristics of the French Quality Award, which was introduced in 1993: The award is for SMEs (companies with fewer than 500 employees) and is based on a two-tier bottom-up selection procedure, from regional to national level.

This is the background to the development of quality in France following the formation in 1957 of the French Quality Association (AFCIQ), a particularly significant event (Dragomir and Halais 1994).

In 1975, the Ministry of Industry set up an Office for Industrial Product Quality and Standardization (SQUALPI). In 1978, the Loi Scrivener consumer protection law was passed. In 1980, the French Association for Standardization (AFNOR) published the NF X 50-110 quality management and assurance standard, which was abandoned in 1987 with the introduction of the international ISO 9000 standards. Also in 1980, the AFCIQ and a number of other groups instituted the "Industrie et Qualité" Award.

1981 was proclaimed Quality Year by the Ministry of Industry; the following year, together with the Ministry for Higher Education, the Ministry of Industry created the "Club Enseignement-Qualité" (Quality Training Club).

In France as in other European countries, the 1980s witnessed a proliferation of quality associations. However—and this is another positive element that distinguishes France from other nations—in 1991 these groups merged into the French Quality Movement (MFQ). Jean-René Fourtou, the chairman of a major French company, Rhône Poulenc, was appointed president of the MFQ.

In 1984 the French government drew up a first program of measures to improve quality in French industry. A far more complete, wider-ranging second program covering quality, certification, safety, and training was launched in 1993. Initiatives included the introduction of an environment mark, a food mark and a new Quality Award.

Measures taken to involve the Regions included the Quality Train in 1985 and the "Tour de France" in 1987. A second, more incisive Tour was organized in 1993 to mark the introduction of the new Quality Award, with an average of 2000 businesses taking part in each stretch.

The Prix Français de la Qualité (PFQ, French Quality Award) was launched in 1993 by the Ministry of Industry and the MFQ (which runs the award). A particularly interesting feature of the French scheme is that although it is related conceptually to the European Quality Award (EQA), it is geared to small- and medium-sized businesses and operates at two levels, regional and national (*Prix Français de la Qualité* 1994). Companies wishing to apply for the award must first compete successfully at the regional level in order to qualify at the national level. (In the first year alone, approximately 1000 companies competed for the 22 regional awards. Thirty-six regional winners then went on to compete for thenational prize, which was awarded to one company); a further three won a special mention (Qualité en Mouvement 1993).

The high level of participation in the regional awards reflects the strong French focus on SMEs. These companies usually find it difficult to take part in national awards, let alone European awards: For them, the regional level is the natural starting point. Companies that win regional awards have already acquired a certain degree of experience, self-confidence, and determination and find it easier to qualify at national level. In 1995, the French model was taken as the reference by the two European organizations, the EOQ and the EFQM, when they were asked by the EU Commission to plan a European Award for SMEs. And in fact, this Award is organized at a third level, after the regional and national prizes.

To conclude, within the general European quality scenario, the French situation is notable for the stronger, more dynamic presence of the State, at both central and regional levels. With a few rare exceptions, most notably in the automobile sector, the response of the business community has been less forceful. Progress in product and service quality does not as yet appear to match the efforts of central and local government.

Quality in Germany. Although the concepts of scientific management and Taylorism came to Germany, as to other countries, with widespread industrialization, they developed through the filters of the German manufacturing culture, which was strongly rooted in the guilds and in the skills training and apprenticeship systems. The guilds had established precise quality standards for their members and for the training of artisans and shopkeepers. These standards, typically based on a 3- to 4-year apprenticeship, still constitute the basic model today. Awareness of the merits of this system and the consequent determination to defend its values meant that German industrialization developed on the basis of the skills of the country's workers. This is the reason for the "craftsmanship" of much of German industry, even in medium to large organizations; routine tasks are performed by low-skilled labor (often temporary immigrant workers), while highly skilled experts always play a key role in guaranteeing quality.

While Germany's large and, above all, its medium-sized companies retain a highly skilled, craftsmanlike approach, its small businesses (with fewer than 10 employees), which represent approximately 12 percent of gross domestic product (GDP) and employ about 16 percent of the workforce, are still heavily influenced by the imprint of the guilds. They account for more than 40 percent of apprentice training (Schlötel 1988).

Another significant factor in the rationalization of the German industrial system, which has had considerable benefits for quality, is the great emphasis on standardization. The Deutscher Normenausschuss (DNA), the German standards committee which drew up the Deutsche Industrienorm (DIN) standards, was formed in 1926. The DNA has always been a positive embodiment of the typical German industrial policy, in which the intervention of the State has tended to converge with the interests of the private sector, producing great competitive advantages for the economic system as a whole. The DIN standards apply German rationality to the industrial system in order to optimize overall benefits for consumers and manufacturers. They have played a leading role in the creation of standards, not only in Germany, but also in Europe and worldwide.

Another important organization is the VDE Pruf-und-Zertifierungsinstitut (Test and Certification Institute), which was formed by the Verband Deutscher Elektrotechnicker (VDE) in 1971 from a test laboratory that had existed since 1920. The VDE quality and safety mark for electrical products is granted in accordance with rules that for the most part have become DIN standards. The VDE mark is regarded as a guarantee of high quality and safety for the consumer, in and outside Germany.

A further step forward in the rationalization of the industrial system was the creation of the Rationalisierungs-Kuratorium der Deutschen Wirtschaft (RKW, Board for the Rationalization of the German Economy), which coordinates the activities of all the various rationalization bodies. The RKW is another example of broad cooperation among social and economic groups for the common good (Lerner 1995).

The traditional high quality of German products is also linked to another distinctive feature of the system, its trademarks and quality marks. The use of the trademark to guarantee product quality stems from the guilds. On the German market, Fugger, Welser, and Henkel's Zwilling Factory in Solingen were familiar names to everyone, and their products were automatically accepted without any form of inspection. Paradoxically, it was a defensive move by the British Parliament (the Merchandise Marks Act of 1887) that fueled awareness of the quality of German products outside Germany. People began to associate the "Made in Germany" mark with high quality, specifically quality in terms of above-average reliability and solidity (and often weight!).

Frequently, this higher quality and reliability was and still is coupled with a premium price, which the purchaser looking for a superior product is willing to pay. The premium price is usually the consequence of higher costs, as a result of overdimensioning and higher skilled labor content. As a result, German products tend to be regarded as niche products, offering craftsmanlike quality and reliability. Many of these niche markets are highly sophisticated, covered by a large number of small and medium high-tech enterprises, which are often worldwide market leaders (Simon 1996).

As far as quality marks are concerned, the Reichs-Ausschuss für Lieferbedingungen (RAL) was set up in 1925. Another broadly held private-public concern, the RAL ensures that product and service quality marks are properly used. Today, it also supervises use of the environmental mark created by the Federal Ministry of the Interior, to guarantee product environment-friendliness (Schlötel 1988).

Another guarantee of product quality is the Stiftung Warentest, a product testing foundation founded in 1961, which conducts systematic tests on different competitors' products and publishes the results in *Test* magazine. Its findings are usually reported by the media and used in advertising and are therefore an extremely significant factor in consumers' purchasing decisions.

In Germany, therefore, the emphasis on product quality predominates. Quality has been the main focus of German industry to date, and will probably continue to be so for the foreseeable future. Nevertheless, the risk of problems emerging in relation to other industrial systems that have had greater success in combining quality with efficiency and resolving the apparent conflict between quality and large-scale production should not be underestimated.

In this connection, German industry has been slow to grasp the importance of Quality Management, first at the level of quality assurance and later at the level of Total Quality Management. Its leadership in technical quality seems somehow to have overshadowed the development of a systemic, integrated vision of quality. As far as the ISO 9000 standards are concerned, Germany has made up rapidly for lost time: by 1995 approximately 10,000 companies had obtained certification. The absence of a strong military production sector and the corresponding focus on the market could be the reason for the relatively low importance previously attached to quality system standards, which are essential factors in the customer-supplier relationship in the military, nuclear, and space industries. Comparison with the situation in Britain supports this theory.

Germany has only just moved into the age of TQM; it remains to be seen whether its recovery here will be as rapid as it was in the ISO 9000 area. Greater problems could well emerge, given the enormous importance that corporate Germany attaches to professional skills. A strong sense of profession creates a strong functional identity (people who work in the R&D field—or in production, etc.—take pride in their specific skills). This makes it more difficult to achieve the interfunctional integration at process level that is vital to optimize costs and times together with quality. Similar difficulties could arise in attempting to obtain significant improvements, which always require interfunctional groups. Combining efficiency and quality is the challenge that many German companies will be facing in the next few years.

The Deutsche Gesellschaft für Qualitat (DGQ) has had an important role in spreading the quality culture in Germany. It was founded in 1952 and has been a full member organization of the EOQ since the latter was formed. In 1995 it had more than 6000 members, of whom about 25 percent were corporate members and 75 percent individual members. The DGQ is based in Frankfurt and has approximately 60 full-time employees. Its main activity is education and training, with approximately 1500 courses per year, attended by more than 30,000 people. These figures refer to 1993, when 7500 diplomas were awarded to participants at courses that have become *de facto* qualifications on the labor market. The DGQ also has a research unit, which conducts applied research jointly with business and universities.

This brief review of the development and current situation of quality in Germany cannot end without mention of Walter Masing, entrepreneur, lecturer, and quality pioneer in Europe, who was a cofounder and the first president of the European Organization for Quality. In addition to the many honors received from national associations and universities, in 1996 Dr. Masing was awarded the EOQ gold medal, a tribute that had previously been given only to Dr. J. M. Juran.

Quality in Great Britain. The main characteristic of the British approach to quality in the period from the early 1970s to the mid-1990s is certainly the great emphasis on standards-based quality and certification. This is not to say that more advanced aspects of quality have been neglected; indeed, Britain was one of the first European countries to introduce the new TQM concepts, especially among large companies operating in highly competitive fields. But in Britain's divaricate vision of quality, certification has certainly predominated, at least at the quantitative level and in terms of image. With more certifications than the rest of Europe put together (see Table 38.2), during the period in question the United Kingdom has been the symbol and reference model for supporters of the standards-based approach to quality. The divergence between this view and the TQM view, a typical trend in Europe from 1987 to 1995, has assumed fairly extreme proportions in the United Kingdom, which can be taken to exemplify the phenomenon. The excessive emphasis on standards-based quality in Britain is probably the reason why the TQM vision has evolved as a counterreaction rather than as a complementary development. The emphasis on standards is itself the outcome of the crucial economic weight of large customers and, in particular, of the Ministry of Defence, the great champion of standards, which, through its procurement contracts, controlled approximately 10 percent of the products manufactured in the United Kingdom when the Defence Standards for Quality Assurance were introduced in 1973 (Hutchins 1995).

It was toward the end of the 1970s that an approach to quality based on the experience of companies operating on highly competitive open markets, and geared to customer satisfaction and continuous improvement (as opposed to the customer-driven view geared to obtaining products compliant with specifications), began to emerge. This new vision sprang both from the direct experience of British industry and, indirectly, from the experience of U.S. business, which at that time was beginning to come under severe competitive pressure from Japan. This antagonism has created a situation in which, on one hand, the standards-based view has focused so single-mindedly on quality control and assurance that it has often lost sight of the customer; on the other, the TQM vision has sometimes been taken to such an extreme that it has lost sight of the fundamental need to ensure product and service quality through rigorous monitoring of every phase in the life cycle. Although this divergence obviously has no legitimate basis, it has developed nonetheless and has had a negative impact on quality development in every European country.

The phenomenon can be analyzed when discussing the development of quality in Great Britain, since it has acquired such significant proportions there. The split was formalized in 1992, when the British Quality Association (BQA, formed in 1981 to handle quality in the corporate sector) separated from the Institute of Quality Assurance (IQA). Founded in 1919 and granted its present title in 1972, the IQA was a Full Member Organization of the EOQ for both bodies. Officially, the purpose of the separation was to create a new body with the characteristics required by the government to run the new British Quality Award, which is based on the European award and therefore has a strong TQM content. The organization of the award was thus assigned to the BQA, which changed its name to the British Quality Foundation. But although, formally speaking, the division appears to have come about for contingent reasons connected with the award, it was in fact the last act in a deliberate distancing of the new groups that support the TQM approach from the historical nucleus that had promoted quality assurance for so many years. It is a false ideological distinction, which is bound to be a not insignificant obstacle to the correct growth of the quality culture in Europe.

With the loss of its branch that handled quality in the corporate sector, the IQA has abolished the clause that limited membership to professionals and now also welcomes corporate members. To date, quality assurance remains its chief mission, but the association has explicitly acknowledged the need to cover the entire spectrum of issues from control to quality assurance and total quality management.

The BS 5750 Standards and Certification Fever.   A comparative analysis of the development of quality concepts in the United Kingdom and in Germany in the postwar period reveals similarities between the United Kingdom and the United States on one side, and Germany and Japan on the other. Here is a possible explanation, based on a plausible interpretation of the facts, which at the very least merits further historical investigation. Like the United States, Britain was one of the victors of the Second World War; like the United States, it regarded defense as a top priority, partly in response to the Cold War. Like the United States at world level, the United Kingdom at European level had the strongest military defense industry, with the Ministry of Defence playing a key role. The strong emphasis on conformity with specifications, an essential requirement in this type of contractual relationship, could hardly fail to have a decisive impact on the development of the quality culture. Whereas the military commitments of Germany, like those of Japan, were no longer of a level to make the respective ministries of defense key "accounts" of industry. In Germany and Japan, industry was driven, first, by the needs of postwar reconstruction and later by the desire for expansion on the international marketplace, and therefore tended to regard quality as competitive added value for customers rather than as conformity with standards. These similarities should not be taken too far, however: Japan has tended to set its sights on high-volume markets and therefore on process-based quality, at minimum costs; Germany has concentrated on niche markets with high added value, where quality may involve higher costs. But certainly in both countries the approach to quality has been driven by the need to boost market share rather than to comply with standards and specifications.

The United Kingdom's particular focus on standards-based quality dates back to 1968 and the Raby report. The Ministry of Defence formed the Raby Committee to resolve the problem of inspections, which had grown out of all proportion, revealing all their intrinsic limitations in assuring the quality of purchased products. The Raby report led to the creation and approval in 1973 of the Defence Standards for Quality Assurance, based on NATO's AQAP standards (Allied Quality Assurance Publications).

During the same period, a number of other large organizations in both the public and the private sectors, including the Central Electricity Generating Board and a number of automobile manufacturers, had developed their own standards for auditing the quality systems and processes of their suppliers. Multiple audits thus began to be a serious problem for many companies. In 1979, the BSI published the BS 5750 standard, which was immediately adopted by the Ministry of Defence (Hutchins 1995) and, after some initial difficulties, by an increasing number of organizations and businesses. This was the start of the certification phenomenon in the United Kingdom, which would have significant repercussions on the history of quality, not just in Britain, but throughout Europe.

In 1987, when the ISO 9000 standards (which were based on the BS 5750 standards) were introduced, the number of certified operations was approximately 7000; in 1993, it had risen to approximately 19,000, and by the end of 1995 to approximately 53,000 (see Table 38.2).

It was not until 1994 that the first detailed critical reviews of Britain's attitude to certification were conducted (as opposed to the criticisms regularly made by the opposing group). The SEPSU report (SEPSU 1994) commissioned by the Royal Society and the Royal Academy of Engineering and, in the same year, a report by the BFQ put the spotlight on the following problems:

- In many cases, BS 5750 (or ISO 9000) certification is wrongly seen as an almost miraculous event, which makes companies automatically capable of generating quality, instead of as a first step toward competitive quality.

- Certification is unrelated to customer satisfaction; it has effectively shifted attention away from product quality and customer satisfaction.

- The certification business has led to the creation of a surplus of certifying bodies (more than 40 in 1995), and this has certainly lowered average quality. Significant inconsistencies exist among the approaches of these bodies, because of their number and the shortcomings of the National Accreditation Board (NAB).

- The economic support of certification provided by the State has encouraged a marketing-type approach by ISO 9000 consultants.

- High demand for consultants to help prepare companies for certification has led to a decrease in the quality of the service offered and to a tendency to sell standard solutions for situations and needs that vary widely.

- There is a general tendency to neglect specific technological and sector characteristics.

The rapid growth of the certification business in Britain has been amplified by the creation of extensive consultancy opportunities abroad as certification fever has spread to Europe and the rest of the world. These opportunities have been fueled by Britain's image as a leader in certification, which is leveraged, at times deservedly, at other times less so, by British certification bodies and consultancies, and by the fact that English is so widely used around the world. Consultancy work and certification of both companies and assessors outside the United Kingdom has reached huge proportions, generating a captive market and stimulating a desire to exploit demand by extending standards and certification to pathological levels (the creation of TQM standards, for example).

The most significant step taken by the British government to promote standards and certification was the publication of the "Standards, Quality and International Competitiveness" white paper and the simultaneous launch of the National Quality Campaign, to promote the BS 5750 quality assurance system standards in British industry. From 1983 to 1989, the Department of Trade and Industry spent around £19 million in a series of quality initiatives (Lascelles and Dale 1989). The fact that in 1983, when the need for research into the competitive dimension of quality as introduced by the Japanese had already become evident, a major national campaign focused on promotion of standards and certification is symptomatic. To quote David Hutchins: "It seems that rather than discover why the approach did not provide a remedy for a worsening of Britain's competitive position with regard to the Far East, the advocates [*of the campaign*] were saying that all that was necessary was the same medicine but with a larger spoon." (Hutchins 1995.)

Today, halfway through the 1990s, it is still too soon to draw any conclusions about the effectiveness of such a massive drive toward standards and certification. The impression is that the results in terms of improvements in products and services and increased productivity are well below expectations. Britain is certainly the most important test bench as far as certification is concerned, since no other nation has achieved a similar scale. We can only hope that in the United Kingdom and the rest of Europe, the artificial separation between quality assurance and approaches that put the emphasis on the customer and continuous improvement can be resolved, in the interest of European industrial competitiveness, with a global vision based on a correct balance of all the various elements. (Signs of a move in this direction are already beginning to emerge.)

The Citizen's Charter.   In the area of government services for citizens, the British Government was one of the first in Europe to introduce systematic quality promotion schemes, as part of its program of privatization of government services. It has since been imitated by other European governments with the wave of privatizations in the 1990s.

In the 1980s, the U.K. government introduced significant reforms in the way public services were organized and run. These reforms addressed the problems of efficiency and effectiveness in central government, local government, and the National Health Service.

The first objective was to abolish bodies that filled no useful function. Those remaining were divided into two categories: those to be privatized, and those to remain in the public sector. In 1994 almost 1 million jobs passed to the private sector. Moreover, in areas where the public sector continued to be responsible for the provision of a function, an assessment was made of whether government should provide the function directly or through the private sector. The purpose was evidently to improve efficiency through competition.

The Citizen's Charter is designed to ensure that citizens receive adequate standards of service, irrespective of the provider. Basically, the idea is to create a competitive situation wherever possible, as a precondition for achieving quality. And since the focus is on services for the public, the government wishes to guarantee that the quality made possible by competition is actually produced. In cases where competition cannot be created, action is being taken—including incentive schemes— to guarantee the required quality.

The Citizen's Charter was launched in 1991. It is a 10-year rolling program of improvements to public services, based on the following principles (Hilton 1994):

- To set and monitor standards for services and to publish performance results
- To provide openness and full information about how services are run and what they cost
- To provide choice wherever possible
- To make services available equally to all who are entitled to them
- To provide courtesy and helpfulness
- To put things right
- To improve value for money

Special charters are planned for each of the main public services. By 1994, 38 sectors had published charters, including central and local government, schools, universities, hospitals, police services, rail services, prisons, tax collection, and the regulated utilities: gas, electricity, water, telecommunications.

A charter mark scheme has been devised as recognition for organizations that, in the opinion of users, deliver outstanding service. The earnings of those who run the services are in part performance-related.

The Citizen's Charter has been running for a number of years, and assessments of its success are many and varied. Once again, the results have been lower than expected: It is difficult to change people's attitudes, especially in the state administration. Nevertheless, some admirable improvements have been achieved, for example in the Post Office: in 1989–90, 78.1 percent of first-class letters were delivered on the next working day after mailing. This figure rose to 89.8 percent in 1991–92 and to 92 percent in 1993–94.

One significant merit of Britain's Citizen's Charter is that it has led to the introduction of similar schemes by other European governments, giving fresh stimulus to a sector that is traditionally extremely static. This is why the Charter is discussed at some length here, despite the fact that, in terms of real quality of public services, a number of northern European countries could justifiably claim superiority.

Notes on Other Countries.   Although detailed descriptions of every West European country are not possible here, a series of comments and clarifications are provided to complete the picture.

We can begin with Switzerland, which has a well-deserved reputation for the quality of its manufactured goods (e.g., wristwatches, large electromechanical equipment, chemicals) and its services (e.g., banking and tourism). Switzerland was one of the first countries in Europe to institute a national quality system certification organization (the SQS, formed in 1983) and it has one of the highest certification rates. Many of its manufacturing companies and service providers pursue TQM strategies.

The Netherlands also has a long-standing quality tradition and was one of the first countries to introduce certification (1982). The Dutch Quality Foundation dates back to 1953 and was a founder member of the EOQ. In 1984, the Dutch government launched the first national quality program, intended primarily for SMEs, which form the backbone of the country's economy. A second program, introduced in 1988, provided 50 percent government funding for more than 100 projects designed to promote quality principles and methodologies within the corporate sector. Dutch universities also play a significant role. Finally, mention should be made of the Dutch Quality Award (which is based on the European Award but introduces a number of interesting differences). The Award is run by the Dutch Quality Institute, which was set up in 1992.

Scandinavia, too, woke up to quality issues in the 1980s. In 1984, Sweden instituted the first TQM chair, at the University of Linköping. In 1990, the Swedish Quality Institute was formed to run the Swedish Quality Award (1992), which is based on the Malcolm Baldrige model. In the business world, the wave of privatization at the beginning of the 1990s encouraged the spread of TQM. Certification is also strongly promoted, especially in Denmark. All three Scandinavian nations pay particular attention to the quality of services, most notably in health care, and to environmental quality. Sweden takes a special interest in customer satisfaction issues, at both the academic level and in practical applications. It was the first country to introduce a national customer satisfaction observatory, the Swedish Customer Satisfaction Barometer (SCSB), which was created in 1989 and was the model for the American Customer Satisfaction Index formed in 1994. The SCSB monitors customer satisfaction in a number of industries and individual corporations within those industries. It has five objectives:

1. To compare industries
2. To compare individual firms with the industry average
3. To make comparisons over time
4. To predict long-term performance
5. To answer specific questions

Germany also has a Customer Satisfaction Index, which was created in 1992. Research is currently underway to assess the possibility of introducing a European customer satisfaction index.

Certification also has an important place in Ireland and Italy. The situation in Italy (whose national certification body, the AICQ, was one of the five founder members of the EOQ) merits a brief comment. In the corporate world, the various sectoral associations (for the mechanics industry, electronics, chemicals, etc.) effectively control certification in their respective sectors, even though the market is open to anyone wishing to enter the certification business. It is obviously in these associations' interests to keep certification quality high; moreover, since assessors are frequently recruited from the ranks of those who have spent their working life in the particular sector, they are able to assess the factors that really count, and avoid the bureaucratic approach adopted by assessors who are familiar with the standards but not with specific processes. This combination of factors has tended to keep the level of certification relatively high in Italy, compared with countries where the business aspect of certification tends to predominate.

THE EUROPEAN QUALITY ORGANIZATIONS

The European quality organizations have been a major contributing factor to the rapid growth of interest in quality over the years. So a specific section is dedicated to them here. Readers wishing to examine in greater detail the issues discussed here, which are necessarily touched on only very briefly, should refer directly to these bodies. Two major organizations operate at the European level, the EOQ and the EFQM; the bodies active at the national level are usually the member associations of the EOQ and in some cases the organizations that run the awards.

The European Organization for Quality (EOQ).   The European Organization for Quality is the federation of the European national quality organizations. Its aims are specified in the following mission statement:

> To facilitate the exchange of information and experience of quality theories and best practices across Europe, in order to enhance the competitiveness of the European economic system, with special attention to small and medium enterprises; to promote the growth of quality in public services and the educational system.

To accomplish these missions the EOQ:

- Promotes and coordinates the activities of its member organizations and all working units
- Organizes annual congresses, seminars, and forums
- Publishes the journal *European Quality*
- Participates in and contributes to projects at European level

The EOQ was founded in 1957 by the national quality organizations of France, the Federal Republic of Germany, Italy, The Netherlands, and the United Kingdom as the European Organization for Quality Control (EOQC). Since then, membership has expanded to other national quality organizations, as shown below. By 1996, thirty-one national quality associations had become Full Member Organizations (FMOs) of the EOQ.

Date	Country
1961	Denmark, Sweden
1962	Norway
1963	Czechoslovakia
1966	Yugoslavia
1970	German Democratic Republic
1971	Portugal
1972	Hungary
1974	Belgium
1976	Turkey

Date	*Country*
1979	Austria, Greece
1980	Ireland
1989	Iceland
1991	The former Union of Soviet Socialist Republics becomes Russia and the Federal Republic of Germany incorporates the former German Democratic Republic
1992	The former Yugoslavia splits into separate states; Slovenia is the first of the new states to enter the EOQ
1993	Croatia and Estonia; the former Czechoslovakia splits into the Czech Republic and the Slovak Republic
1994	The former Yugoslavian Republic of Macedonia
1995	Latvia
1996	The Ukraine

In 1989, the EOQC changed its name to European Organization for Quality (EOQ), to take account of the new developments in quality concepts after quality control: quality assurance, quality management, Total Quality Management.

One of the outstanding characteristics of the EOQ in its early years was its policy of accepting organizations from all over Europe, West and East, in a period when the continent was deeply divided. Promoting exchanges of experience between quality specialists within the two extremely different economies was considered of greater importance than the political divisions on the two sides of the Iron Curtain. This was demonstrated to the world when an Annual Congress was organized in Czechoslovakia as early as 1969 and when a Soviet citizen was elected President in 1978.

The EOQ is governed by Swiss law. Its General Secretariat is located in Bern, Switzerland, and is currently composed of three people. This low number is a significant indication of the difference between the EOQ and other quality organizations, such as the ASQC. Whereas the latter is a national association (covering the entire U.S. territory, although it is a highly decentralized organization consisting of territorial divisions), the EOQ is a federation of national associations (the Full Member Organizations). The EOQ reflects the political organization of Europe, a continent composed of separate sovereign states, which are gradually moving toward economic integration and partial political unification (among the 15 member states that make up the European Union and other states that will gradually join the EU), without to date at least overriding the absolute preeminence of each country's national identity. Similarly, within the EOQ, the national quality associations tend to maintain their specific prerogatives and dedicate most of their attention and resources to national initiatives. The limited proportion of resources that go to the EOQ (to cover membership fees and the participation of their representatives at the meetings of the statutory bodies, committees/sections, and workgroups) enables the federation to provide harmonization, coordination, and leadership on special strategic initiatives of European significance or on operations for which a centralized organizational approach improves efficiency by enhancing synergies among the various bodies.

The prevalence of the FMOs' national activities over the European activities conducted by the EOQ is confirmed by staff figures. If the staffs of the 31 national organizations are added together, the total is approximately 250 people, compared with 3 at the EOQ General Secretariat. The national bodies also rely on the work of many volunteers. The EOQ executive organs feel, however, that the central structure should be strengthened to take account of the expansion of activities at the European level fueled by the growing economic and political integration of the EU.

Table 38.3 lists the FMOs of the EOQ, showing their individual and corporate members.

Structure and Activities.   The decision-making organ of the EOQ is the General Assembly. It consists of the representatives of the Full Member Organizations, the members of the Executive, and the

TABLE 38.3   EOQ Full Member Organization (FMO) Membership, January 1996

FMO	Individual	Corporate
Austria	40	430
Belgium		1,900
Bulgaria	N.A.	N.A.
Croatia	240	60
Czech Republic	1,350	25
Denmark	1,011	744
Estonia	86	23
Finland	2,400	300
France	1,000	2,000
Germany	6,000	1,500
Greece	3,980	382
Hungary	850	260
Iceland	188	282
Ireland		2,200
Italy	2,000	800
Latvia	176	52
Macedonia (except Yugoslavia)	60	55
The Netherlands		350
Norway	1,910	1,010
Poland	N.A.	N.A.
Portugal	1,487	928
Romania	15	265
Russia	320	—
Slovakia	115	151
Slovenia	650	—
Spain	3,211	1,113
Sweden	3,500	160
Switzerland	30	1,820
Turkey	1,701	3,661
United Kingdom	13,000	600
Total EOQ	45,320	21,071

N.A.=not available.

General Secretariat. The General Assembly meets twice a year: at a summer session during the Annual Congress and at a winter session. The Executive is composed of the President, the Immediate Past President and four Vice Presidents. The activities of the EOQ are conducted through:

- The Annual Congress
- The official EOQ journal, the *European Quality Journal*
- The Committees and Sections
- The Unit for the Registration and Qualification of Quality Professionals
- The Evolution of Quality Working Group

The Annual Congress.   Held every year since 1957, the EOQ Congress is hosted by the member countries on a rotating basis. The first Congress, chaired by Walter Masing, was held in Paris in 1957. The Congress calendar has already been drawn up until 2005 (1997: Trondheim, Norway; 1998: Paris; 1999: Spain; 2000: Hungary; 2001: Turkey; 2002: United Kingdom; 2003: The Netherlands; 2004: Bulgaria; 2005: Italy).

The Congress is usually held in the month of June; every other year it ends with the election of a new president and Executive.

The European Quality Journal. The *European Quality Journal* is the EOQ's official journal. A bimonthly magazine, it was introduced in January 1994, when it replaced the *EOQ Journal.* The journal is owned by the EOQ and published by European Quality Publications, J. & M. Kelly. Its editorial position is decided by an Editorial Board consisting of the Editor, the Secretary General, and another EOQ representative. The journal also has the support of an Advisory Board of United States and Japanese as well as European experts. The journal is a managerial rather than technical publication; it is intended for managers in general, in every type of organization, public or private, profit-making or not, and for the academic world.

The Committees and Sections. The number of committees and sections has varied over the years, depending on requirements and on individual FMO availability to chair new groups. In 1995, the following sections were operational: Automotive, Pharmaceutical Industry, Construction Industry, Consultancy, Healthcare; the following committees were operational: Education and Training, Dependability, Glossary, Statistical Methods, Software, Human Factors in Quality Management, Environmental Quality, Service Quality.

The Unit for the Registration and Qualification of Quality Personnel. The Unit runs the Harmonized Scheme for the Qualification and Registration of Quality Personnel, which was introduced by the EOQ in January 1994 as a means of establishing high standards for the training and qualification of quality personnel throughout Europe and paving the way toward mutual recognition of all the relevant professional figures. As of 1996, the EOQ Scheme identifies three categories of quality personnel: the Quality Professional, the Quality System Manager, and the Quality System Auditor. The Scheme is based on the training experience of a number of FMOs, who in 1991 decided to launch a process of harmonization in order to create a common scheme. It has also been adopted by the FMOs who were not involved in its planning and has thus become the official EOQ Scheme. If it is accepted by the relevant bodies, it will also become the European Scheme. Today, the Scheme is run by the EOQ in line with the requirements of the EN 45013 and ISO 10011 standards. The EOQ therefore acts as a *de facto* accrediting body and will officially be recognized as such if the Scheme is integrated into the European system. The EOQ's FMOs act as certifying bodies for professional figures in their respective countries. Under EU rules, this function is not confined to the FMOs but may also be performed by other bodies that meet the requirements specified in the relevant EOQ document. Nevertheless, the FMOs are the EOQ agents in their respective countries and guarantee correct application. Figure 38.1 illustrates entry requirements and the certification process.

The Evolution of Quality Working Group. This group monitors and, when possible, anticipates the evolution of quality theories and practices, in order to help keep Europe at the forefront of developments. One of the reasons behind the formation of the Group in 1994 was the realization that, over the years, quality had developed in a one-sided manner in Europe and within the EOQ itself, with the result that the organization was in danger of losing the leadership envisaged by its vision and mission statements, and confirmed by its history. After 1987, quality system certification based on the ISO 9000 standards had become the key issue in Europe and in the EOQ Congresses. High demand for certification had created new opportunities for the national quality associations, which were able to offer their experience in quality training and launch new activities in the area of assessor certification. This increase in activity not only boosted the importance and visibility of the national organizations, it also led them to enlarge their structures, with the result that technical structures gradually came to prevail over professional skills. More and more time was spent resolving problems related to budgets, expansion, internal organization, and competition with other groups attracted by the new business opportunities. Specialist, professional issues were neglected or, often, not even perceived. Developments in the field of Total Quality Management were relegated to the sidelines in the FMOs and consequently in the EOQ, because of the overwhelming priority given to certification.

The Evolution of Quality Working Group was set up to provide the EOQ with an observatory that would identify and analyze major trends in quality as distinct from specific issues—albeit legitimate ones—of contingent concern, business opportunities of interest to certain groups or the latest fads.

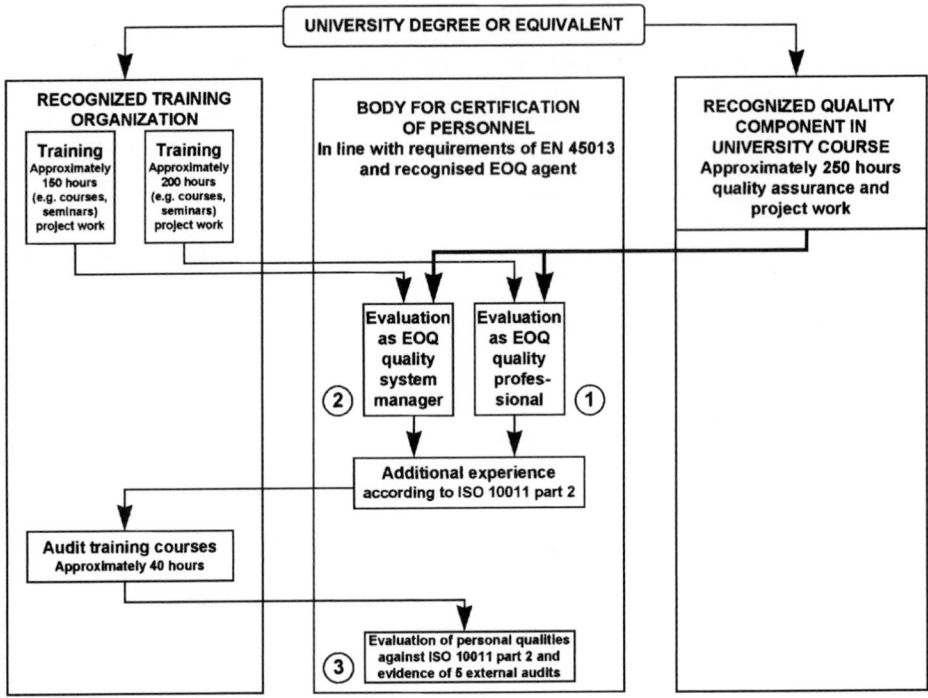

FIGURE 38.1   The EOQ harmonized scheme for the registration and certification of quality personnel (1 = quality professional, 2 = quality system manager, 3 = quality auditor).

The Group is therefore a point of reference for the EOQ and the FMOs, providing support in making the correct decisions and avoiding the pitfalls of fashion. It meets about three times a year and organizes an annual forum to discuss a theme of particular current interest. It also provides support for the FMO that hosts the EOQ Annual Congress, helping to choose the main theme for debate, select speakers, and assess the articles received in response to the call for papers. It also examines the key themes in the papers, before discussion by the Executive.

The European Foundation for Quality Management (EFQM).   By the 1980s, the new quality messages had spread to many European companies and many quality managers had experienced their benefits at firsthand. The main difficulty was how to involve top management and line management in general. Sensitive to this problem, the quality associations attempted to draw the attention of corporate chairmen and chief executives on one hand and politicians on the other to the new quality issues, but with little success.

In 1987, the chairman of N.V. Philips Gloeilampenfabrieken, the Dutch electrotechnical and electronics multinational, contacted 13 other major European companies (14 EFQM founder companies) with an idea for the creation of a European body to stimulate corporate interest in Total Quality Management, which was seen as a major strategic factor in the recovery of Europe's industrial competitiveness. The 14 chairmen assigned their quality strategy managers to analyze the project. The result was a proposal to set up an organization that would be complementary to those already operating in the quality field. If the strong point of the existing associations was the know-how and experience of quality managers and quality experts and their weak point was the lack of top management participation, then a highly visible, direct top management presence would be the goal of the new organization. On September 15, 1988, the chairmen of the 14 companies met in Brussels to sign a letter of intent for the formation of the European Foundation for Quality Management. The Foundation was formally constituted the following year, on October 19, in Montreux (Switzerland), again in the presence of the fourteen chairmen. The Policy Document approved in Montreux set out the vision, missions, and objectives of the EFQM. Specifically, the Foundation's missions were as follows:

- First, to support the management of Western European companies in accelerating the process of making quality a decisive influence in achieving a global competitive advantage
- Second, to stimulate and, where necessary, support the participation of all segments of the Western European community in quality improvement activities and to enhance the quality culture.

For the first five years, the EFQM was directed by a Governing Committee consisting of the chairmen of the founder companies. The policies drawn up by the Committee were transformed into initiatives by an Executive Committee composed of the chairmen's representatives and implemented by a staff headed by a Secretary General. An Advisory Board of representatives of the Executive flanked the Secretary General during the intervals between the meetings of the Executive Committee. In 1994, EFQM membership conditions were changed, with the only distinction now being between Regular Members (public or private companies, or governmental bodies) and Associated Members (nonprofit organizations). Correspondingly, the Governing Committee and the Executive Committee are elected from among the members. The Foundation's official headquarters are in Eindhoven (Netherlands), but since 1992 the 20-strong staff headed by the Secretary General has operated from the EFQM Representative Office in Brussels.

In 1996, the EFQM had approximately 500 members.

EFQM Activities. The main activities of the EFQM are as follows:

- *Recognition:* Covers management of the European Quality Awards (by far the most important activity).
- *Services for managers:* The Annual EFQM Representatives Meeting, the Quality Management Open Days, seminars, common interest days, working groups, etc.
- *Communication:* Publication of the *Quality Link* newsletter and monographs.
- *Education, training, and research:* Organization of the annual Learning Edge Conference and other activities.
- *Winners' conferences:* The conferences at which the winners of the awards illustrate their paths toward excellence for the benefit of other European companies. The conferences are organized by the local EFQM members.
- *Award-related training activities:* Training for award assessors and company internal assessors.

In the main countries in which the EFQM operates, local activity groups have been formed to translate documents and coordinate the national group of award assessors.

The Annual Forum. This is the EFQM's main public event. The first Forum was held in 1989 in Montreux, the second in 1990 in London, and the third in 1991 in Paris. Since 1992 (Madrid), the Forum has also included the European Quality Award presentation ceremony. The Forum has been held in Turin (1993), Amsterdam (1994), Berlin (1995), Edinburgh (1996), Stockholm (1997), and Paris (1998).

Although the presentation of the Award is a key highlight, the Forum offers an intensive schedule of events of great interest to large numbers of European managers: papers by European corporate chairmen and CEOs and workshops on carefully selected issues. The Forum is also the venue for the annual meeting of the Foundation's Governing Committee.

The EFQM 2000 Program. In 1996, the EFQM reviewed its vision and missions and launched a long-term program.

The Foundation's new *vision* is

> To be a leading organization, recognized on a global basis, for the development and promotion of a consistent approach to Total Quality Management as the vehicle for the achievement of business excellence in European organizations.

Its *missions* are

- To stimulate and support the participation of organizations throughout Europe in improvement activities leading to excellence in customer satisfaction, impact on society and business results.
- To support the managers of European organizations in enhancing the role of Total Quality as a decisive factor in achieving a global competitive advantage.

Under the EFQM 2000 program, the Foundation's *objectives* are as follows:

1. For the European Model for Business Excellence to be recognized as the key strategic framework for managing an organization and identifying improvement opportunities, regardless of the nature and size of that organization.
2. For the European Quality Award and Prizes to be recognized internationally as major achievements, and the winners to be acknowledged as role models of business excellence.
3. To provide membership satisfaction, through value-for-money services.
4. For the philosophy, methods, tools, and techniques of Total Quality to be accepted as a key element of education and training in Europe, at every level.
5. To establish constructive relationships with the national quality organizations, the European Organization for Quality, and the European Union.
6. To operate on a sound financial basis.

The EFQM has set itself a membership target of 800 by the year 2000.

The European Quality Platform.   Given the complementarity between the EOQ and the EFQM, on September 1, 1994, the two organizations created the European Quality Platform as a means of optimizing the overall benefits for Europe of their activities. The Platform's vision reads as follows:

> The European Quality Platform is the leading body in relation to communication and coordination in the field of quality matters throughout Europe.

> The European Quality Platform serves and supports the European Commission and all private and governmental organizations in Europe in their attempt to promote quality at all levels and in all sectors.

The Platform comprises a Board with three representatives from each of the two organizations, which meets at regular intervals, a permanent link between the EOQ and EFQM Secretaries General and, primarily, a range of joint activities. Leadership of these activities is always handled by one or another of the organizations, but planning and supervision are conducted jointly by special Steering Committees. From 1994 to 1996, the Platform's joint activities were as follows:

- Creation of the new European Quality Award for Small to Medium Enterprises
- European Quality Week
- Preparation of a book illustrating case studies of TQM strategies developed by European companies.

These three activities were part of the programs of the European Commission (DGIII) and therefore received Commission funding. The Platform's future plans involve activities in education and training, publication of information about quality, and promotion of Professional Quality Figures.

THE EUROPEAN QUALITY AWARDS

The European Quality Prizes and the European Quality Award (EQA) were launched in 1991 by the EFQM, with the support of the EOQ and the European Commission. The EFQM is responsible for their management and funding. To receive a prize, an organization must "…demonstrate that its approach to Total Quality Management has contributed significantly to satisfying the expectations of its customers, employees, and others who have an interest in it, over a number of years.…" Since the candidates do not compete against one another, but are assessed separately against a predefined level of excellence, a number of prizes can be awarded in the same year. The "best of the best" receives the European Quality Award. Originally, the awards were intended for large or medium-large business organizations. In 1996, a separate award was introduced for public service providers. An award for small to medium enterprises (SMEs) will be introduced in 1997 (see notice of brochures about the awards under References).

If the applicant is an independent company, the sole condition for eligibility is that at least 50 percent of its activities over the previous 5 years must have been conducted in Europe. If the applicant is part of a larger organization, the following conditions must also be met: (1) the applicant must have a unique company name and its own unique brands; (2) the applicant must have more than 250 employees (from 1997 on, this will no longer be a condition; 250 employees will be the dividing line between the awards for large organizations and SMEs); (3) output supplied to the parent company must be less than 50 percent of total output; (4) the applicant must perform a sufficiently broad range of business functions (European Quality Award).

The following companies won European Quality Awards from 1992 to 1995 (the first company in the list for each year is the Award winner, the other companies are Prize winners):

- 1992: Rank Xerox Ltd.; BOC Ltd., Special Gases, Industria del Ubierna SA, UBISA, Milliken European Division.
- 1993: Milliken European Division; ICL Manufacturing Division (now D2D).
- 1994: D2D (Design to Distribution); Ericsson SA, IBM SEMEA.
- 1995: Texas Instruments Europe; TNT Express (UK) Ltd.
- 1996: Brisa-Bridgestone Sabanci Tire Co. SA; British Telecommunications plc; Netas, Northern Electric Telekomunikayson AS.
- 1997: SGS Thomson; British Telecommuncations plc; TNT UK Ltd.; Netas, Northern Electric Telekommunikayson AS. 1997 was also the first year of the European quality award for small/medium enterprises: Award winner: Beksa. Prize winners: Gasnalsa, Gas Natural de Alava; DiEU; ABB Semiconductors; DD Williamson; Prec-Cast Foundry Ltd.; Landhotel Schindlerhof.

The EQA timetable is approximately as follows: submission of preliminary application data (information about the company and fulfillment of eligibility criteria), end of January; submission of final Application Document, beginning of March; site visits, June; announcement of the award winners, October/November.

A team of five to seven assessors, headed by a senior assessor, is assigned to each application. After the application is examined separately by each member of the team, the senior assessor calls for a "consensus meeting" to reach qualitative agreement (on strengths and areas for improvement) and quantitative agreement (on scores). The resulting "consensus report" is then presented by the senior assessor to the EQA Jury, which examines all the consensus reports and selects the applicants which will receive a site visit. After the visits have been completed, the EQA Jury meets to reach its final verdict. The number of assessors involved in the process varies, depending on the number of applicants.

The EQA is based on the model in Figure 38.2. The model expresses the concept that *customer satisfaction, people (employee) satisfaction* and a positive *impact on society* can be achieved through *leadership* as the driving force behind *policy and strategy, people management,* and management of *resources* and *processes,* and that this leads ultimately to excellence in *business results.* The clear division between *enablers* and *results,* with a 50 percent weighting attributed to each of the two blocks, is the distinguishing characteristic of the EQA.

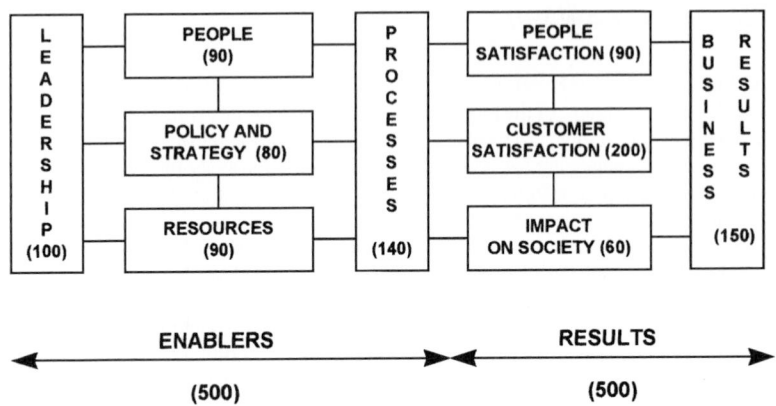

FIGURE 38.2   The European Quality Model, with weights.

THE EUROPEAN ENVIRONMENTAL MANAGEMENT SYSTEM

The world's first set of environmental management system standards, BS 7750, was drawn up by the British Standards Institute (BSI) in 1992. Clearly, in becoming the first body to establish an environmental standard, the BSI was able to capitalize on its years of experience in quality system standards (BS 5750 and, later, ISO 9000).

At the European level, the first Resolution of the EC Council in this field, passed on December 3, 1992, expressed the view that a preventive guideline geared to the market economy in the area of environmental protection could benefit industrial competitiveness and should be adopted wherever possible. The following year saw the publication of EC Ruling 1836/93 on the voluntary compliance of industrial companies with a community eco management and auditing scheme designed to raise the environmental efficiency of industrial operations. In preparation for the introduction of the scheme, the Ruling provided a set of legislative, technical, and managerial guidelines covering the development of "environmental management systems" within companies, the implementation of the necessary audits, and information for the public. The Eco Management and Audit Scheme (EMAS) became operational in 1995. By 1996, therefore, three systems existed: the BS 7750 standard, the EMAS, and the ISO 14000 standard, the final draft of which was approved in Oslo in June 1995. The formal standard is due to be published by the ISO in the second half of 1996 (Kiesgen 1995).

Though similar in many ways, the three systems contain basic differences in emphasis. The ISO standard is considered to be the least prescriptive, because of the compromises introduced to make it internationally acceptable. The EMAS standard is regarded as the most demanding. Many observers in Europe believe that EMAS registration should be the ultimate goal and that both the BS 7750 and the ISO 14000 standards should be regarded as staging posts. However, the CEN (the European standardization body) is examining the possibility of narrowing the gap between the EMAS and ISO 14001, leading ultimately to the adoption of the ISO standard.

The approach of the environmental efficiency certification system for industrial operations outlined by the EU Ruling is similar to that adopted for quality system certification in relation to the free circulation of goods within the Union. The EMAS registration roadmap is as follows (Mullins 1996):

1. Adopt an environmental policy. This must be a corporate policy designed to ensure:

   • Compliance with the relevant environmental regulations
   • A commitment to continuous improvement

2. Conduct a site environmental review. The aim is to identify the site's environmental impact and relevant issues.

3. Develop an environmental program. The program should be set out in accordance with the company's environmental policy and the outcome of the site environmental review.

4. Set up an environmental management system.

5. Perform an environmental audit using EMAS, BS 7750, or ISO 14001 criteria. The frequency of the audit depends on the specific business, but it must be performed at least once every 3 years.

6. Prepare an environmental statement for publication. The aim is to ensure that all interested parties are aware of the environmental impact of the site and the company's environmental management policy.

7. Validation. An accredited environmental verifier scrutinizes the company's environmental policy, program, management and audit procedures, to ensure that they meet EMAS requirements.

REFERENCES

Brochures about the awards can be requested from the EFQM, European Foundation for Quality Management, Avenue des Pléiades 15, 1200 Brussels, Belgium.

Dragomir, Rado, and Halais, Bernard (1994). *Quelques Points de Repère sur l'Histoire de la Qualité en France.* Ministère de l'Industrie des Postes et Télécommunications et du Commerce Extérieur, Paris.

Gogue, Jean-Marie (1988). "Quality in France." *Quality Control Handbook,* 4th ed. J. M. Juran, editor. McGraw-Hill, New York.

Hilton, Brian (1994). "The Citizen's Charter." Paper presented at 38th EOQ Quality Congress, Lisbon.

Hutchins, David (1995). "History of Managing for Quality in the UK." *History of Managing for Quality,* J. M. Juran, ed. Quality Press, Milwaukee.

Juran, J. M. (1979). "Japanese and Western Quality: a Contrast." *Quality,* January and February.

Kiesgen, Gerrit (1995). "The Marriage of Environment to Quality." *European Quality,* vol. 2, no. 2.

Kume, H. May (1990). "Quality Management: Japan and the West." *Quality,* May, p.16.

Lascelles D. M., and Dale, B. G. (1989). "The UK Department of Trade & Industry National Quality Campaign: 1983 to January 1989." *International Journal of Operations and Production Management,* vol. 9, no. 6.

Lerner, Franz (1995). "History of Quality Assurance in Germany." *History of Managing for Quality,* J. M. Juran, ed. Quality Press, Milwaukee.

McMillan, J. (1994). "The Usage of the EN 29000/ISO 9000 Series of Standards in Community Legislation." Paper presented at the 38th EOQ Quality Congress, Lisbon.

Mullins, Sandra (1996). Choice of Clean-Up Acts. *European Quality Publications,* vol. 3, no. 2.

Prix Français de la Qualité (1994). Mouvement Français pour la Qualité, Paris, April.

Qualité en Mouvement (1993). Paris. Special Dossier on the Prix Français de la Qualité, March–April.

Schlötel, Ernst (1988). "Quality in the Federal Republic of Germany." *Quality Control Handbook,* 4th ed. J. M. Juran, editor. McGraw-Hill, New York.

SEPSU (1994).

SEPSU Report on Quality Management. Royal Society, London.

Simon, Herman (1996). *Hidden Champions: Lessons from 500 of the World's Best Unknown Companies.* Harvard Business School Press, Boston, MA.

The European Quality Award, Application Brochure, European Foundation for Quality Management, Eindhoven, The Netherlands.

SECTION 39
QUALITY IN CENTRAL AND EASTERN EUROPE

Robert E. Hoogstoel

OBJECTIVES OF THIS SECTION

This brief survey of quality as experienced in central and eastern Europe has two principal objectives:

1. To review the history of the region, especially with respect to how quality has been managed

2. To identify the forces now at work which may affect the way quality is managed in the future

RECENT HISTORY OF CENTRAL AND EASTERN EUROPE

Central and Eastern Europe Defined.   In this section, "central and eastern Europe" signifies the area of Europe whose countries, in the period beginning in the years following World War II proclaimed themselves socialist or communist under single-party rule. This includes the countries which were allied with the Soviet Union (U.S.S.R.)—Bulgaria, Czech Republic, the German Democratic Republic, Hungary, Poland, Romania, and Slovakia—or were actually incorporated into it as socialist republics—Belorus, Estonia, Latvia, Lithuania, Ukraine. It also includes socialist countries that stood apart from the U.S.S.R.—Yugoslavia and Albania. And, because of its decisive influence in quality as in other matters in this region, we include Russia, then at the center of the U.S.S.R. This is not a complete list, nor have we tried to anticipate the emergence of new countries in the region as the result of negotiation or civil war. For a discussion of quality the list will suffice; it contains the most industrialized and therefore the most influential countries in the region regarding quality.

Recent History.   It is good to remember that the recent history of this area, both economic and political, is a turbulent one. The experiences of self-rule before 1989 were brief and scattered. In the countries that were the remnants of the Austro-Hungarian Empire or Prussia, self-rule was limited to the period following the end of World War I, in 1918. Those newly independent countries were Austria, Hungary, Poland, Czechoslovakia, Yugoslavia, and the Baltic states of Lithuania, Latvia, and Estonia. Self-rule for them effectively ended with the rise of Nazi Germany, beginning around 1935. After that there was occupation, World War II, and more occupation. Soon after the end of World War II in 1945, with the exception of Austria, all of them, plus eastern Germany, became communist, under the rule of various national Communist Parties, most dominated by the Soviet Union through economic and military agreements enforced by military power. All told, there was little over a decade to develop democratic forms, too short a time to establish democratic traditions. Further, during the 40 and more years before 1989, the experience with market-based economics in central and eastern Europe was limited to black markets, transactions outside of state control, and condemned and punished as criminal.

It is this common history and the challenges which that history poses that bind these countries with respect to quality. Each of the countries, whatever the new form of organization toward which it is tending, aspires to create products—goods and services—which can compete in the world marketplace. And each must find a way to create the environment in which that aspiration can be fulfilled.

Political and Economic Organization under Communism.

This part of Europe is in transition. Rather than speculate where each will be in the future, it is perhaps more fruitful to look backward and examine the common factors which today affect the state of quality in this region.

For over 40 years (longer in the case of Russia), the region was governed under a political-economic model dramatically different from that of the Western countries. The model and the consequences of its implementation created the climate for industrial quality. The features of the model profoundly affected the management of quality. The habits of management and thought left behind still affect quality management. Some of the features and their effects were

- *The economy was centrally planned.* Smith (1976) discusses some of the effects in Russia of commitment to plans issuing from the top of the political hierarchy and promulgated downward. There was pressure on managers to "meet the plan." Fearing retribution for failing to make the plan, and motivated by a system of bonuses for meeting the plan, managers throughout the economy resorted to all manner of shady schemes to either meet the plan or be certain that blame for failure to do so was attributable elsewhere. These distortions led to unrealistic plans, which in turn led to increasing shortages of materials throughout the economy, with which managers had to deal with increasingly creative dishonesty.

- *In effect, the state was the only customer.* Regarding consumer goods, this meant that huge central agencies represented the needs and wishes of the general population. There was no direct customer input to the design of goods or delivery of services, and no effective feedback of product experience in the field. The consumer had to accept what was available and suffer with the quality (Smith 1976).

- *Orientation was toward production volume. The central plan required commitment to numerical production quotas.* The most obvious shortcut to meeting the plan (or to exceeding it and achieving production stardom) was to shortcut quality. Khrushchev (1970) describes his investigation of an epidemic of failing automobile tires. He found that the factory had, at the request of a visiting commissar of transport, reduced the amount of reinforcing wire at the edges and of cording under the tread. This action was motivated by a wish to meet the plan and save material. Unfortunately, it was uninformed by the technical requirements of an automobile tire. Although the incident took place in 1939, it was typical of the problems that continued to plague industry under the communist system.

- *Much of the industrial production was committed to military purposes.* The military, as the preferred customer, always enjoyed direct input in the factories, and had either their own "closed" plants or plants for mixed civilian-military production. High priority for the military meant consequent low priority for the remainder of the economy. The practical effect was a two-tier system of quality. The first tier was occupied by military goods; the second tier by everything else.

- *The plan committed the state to heavy-industrial development.* In the earliest days of communist rule in Russia, this policy was based on the objective to convert an agrarian economy into an industrial one. The policy outlasted the need and became an end itself, promoted for propaganda purposes (Smith 1976). This has left the entire region with gigantic factories unable to react to rapidly changing market demands and rapidly changing technology.

- *There was little or no private ownership of property.* The most apparent effect of public ownership was the disregard for maintenance evident in decaying public works of all sorts—factories, apartment blocks, sewer systems, and so on. One expert in the region (Behrman 1998) has observed that "where the owner is 'everyone,'...no one feels responsible for maintenance." A related effect was the perfecting of a notoriously poor work ethic, summed up in the workers' joke: "We pretend to work and they pretend to pay us."

Effects of the Political and Economic Organization. Implementation of this system and its plans and goals involved cultural changes on an unprecedented scale. Motivation was presumably based on the common good, not the good of the individual, but this motivation was never sufficient. At least one Western student of the Soviet Union (Figes 1997) has observed that communism was bound to fail because it sought to achieve the impossible: the transformation of human nature.

From the start, state coercion played a large role in moving plans forward. There was state control of information and cultural affairs, reflected in censorship of publications unfriendly to the state and punishment of dissident behavior. At the factory level, authoritarian management prevailed.

Information control had many adverse implications for quality. The reactor explosion at Chernobyl in 1986 dramatically revealed some of these. Before the accident, it was government policy to conceal from the general public accidents at nuclear power stations and to forbid publication of "information about the unfavorable ecological impact of energy-related facilities on operating personnel, the population, and the environment." This contributed to widespread complacency when the accident occurred. Further, in 1985, a newly appointed minister of energy further impeded exchange of information and reduced the ability to anticipate and prepare for reactor accidents by abolishing the central coordinating body of the ministry. Most obviously, when the accident occurred, the authorities failed to alert the immediate population and the people who lived downwind of the accident, resulting in extensive unnecessary exposure to radiation (Medvedev 1991).

By 1995, among children in the affected area, the incidence of thyroid cancer was reported to be 285 times the preaccident rate (Crossette 1995). Medvedev (1991), a Russian expert in nuclear power, condemns generally the government's control of information with these words in the preface to his account of the Chernobyl disaster: "...Chernobyl demonstrated the ignominious failure and the sheer insanity of the administrative-command system."

On a less frightening dimension, the concentration of economic information and decision making in the hands of a few resulted in widespread ignorance of microeconomic and financial concepts, contributing also to poor preparation for life in a market economy.

Quality Management—1945 to 1990: State Quality Control

Council for Mutual Economic Assistance (CMEA). During the communist period, most countries in the area, under the leadership of the Soviet Union, formed a trading alliance—the Council for Mutual Economic Assistance. Among other matters of mutual concern, the alliance cooperated on the approach to industrial product quality. Between 1971 and 1980, the CMEA introduced programs to promote economic integration among CMEA countries. One requirement for success was the development and adoption of a wide variety of uniform quality standards.

Unified System of State Quality Control (USSQC). The approach to quality was based on a Russian model established in 1978, known as the Unified System of State Quality Control. Under USSQC, quality control was a function of the central government, involving the coordination of economic planning (including planning for quality control at all levels of the economy). USSQC is described in detail by Egermayer in the fourth edition of this Handbook (Egermayer 1988).

In practice, the most widely adopted function of these state QC systems was product certification, according to three quality categories or grades: highest—products whose quality reaches or surpasses the highest quality level of comparable domestic or foreign products; first—products whose characteristics satisfy the requirements of currently valid technical standards; and second—products whose characteristics do not satisfy current requirements and are of poor quality. [These categories probably correspond to those described earlier by a metalworker in a Russian steel factory, as reported by Smith (1976): first, for the military; second, for export; and third, for "common" domestic use.]

Details varied from country to country, but "highest" and "first" grade products generally carried an approval stamp of certification from a national laboratory.

Although there were official reports of general improvement in product quality as a result of this system, Egermayer (1988) reported that implementation of product certification in the CMEA countries was "not uniform, and certification results [were] not internationally valid."

Price premiums and penalties were awarded to factories, according to the outcome of product certification tests of the factory's product. These incentives were supposed to provide strong motivation and promote the interests of manufacturers in the achieving product certification and improving product quality. In fact, the effects were slight.

USSQC included direct quality control at the worker level, based on three principles: (1) The majority of errors committed by an employee on the job can be prevented, (2) every employee is responsible for the quality of work done and therefore is obliged to control it, and (3) every employee should know the exact job duties and must be given all facilities to accomplish them. As implemented, the system evolved to include worker certification for self-inspection, financial bonuses for attaining such certification, and measures of workers' quality performance.

This system was adapted to various national cultures, appearing as "Fehlerfrei Arbeit" (failure-free work) in the German Democratic Republic, "DO-RO" (shorthand for "good work" in Polish) in Poland, and so on.

The USSQC approach fit comfortably with the Soviet ideals of centralized planning and control. However, evidence is scarce of competitive results attributable to the system itself.

More on the Two-Tier System.   Some military goods were—and continue to be—successful, even in the export market. Russian rockets are an example. The rockets are now in great demand for launching commercial communications satellites. The rockets appeal to commercial customers not because of leading-edge technology (which they don't have), but because of high reliability in service, low initial cost, and ease of maintenance. These characteristics are traced to simple and rugged design, typical of many Russian military products (Broad 1996).

The story is different for consumer goods, whose reception in Western markets has been cool at best. For example, automotive products, such as Lada (U.S.S.R.), Yugo (Yugoslavia) and Daccia (Romania), and farm tractors, such as Belorus (U.S.S.R.) and Ursus (Poland), were unable, despite dramatic price advantage, to overcome their quality deficiencies in competition with Japanese and Western products.

USSQC could not achieve the stated objective of a high rate of quality improvement. The structure of incentives always favored military goods, never consumer goods. Furthermore, the system described was at complete odds with the social conditions existing in the factories. First, standards for consumer goods were technical standards, established centrally, at a great social distance and with little input from the consumers who would actually use the products involved. Second, the contribution of metrology depended largely on the appropriateness of the standards used. If the standard were not appropriate—for example, for lack of input from the consuming public—it mattered little how precisely and accurately the product was measured or how high the product scored against standard. Beyond that, the environment of chronic shortage of material and parts meant that managerial energy was concentrated on making production quotas, not on quality. Finally, emphasis on inspection was in sharp contrast to the Western experience in recent years. In the West, industrial managers have learned that it is futile to "inspect in" quality. Dependence on product inspection as a means for quality attainment has been generally abandoned in favor of control of the process by which the product is made.

Evidence of poor quality results is plentiful in technical literature from former CMEA countries, in periodicals and general literature, and in accounts of contemporary travelers. Hedrick Smith's "The Russians" (1976) and "The New Russians" (1990) treat the matter of productivity and quality as part of his studies of the Russian people.

How It Worked.   Juran (1994) summarized the Soviet approach to quality management in consumer goods with an example:

Consider a product such as shoes. A planning ministry decided what shoes to produce (styles, sizes, quantities, schedules, etc.), a production ministry made plans to produce the shoes, dividing the work among the various production facilities, a marketing ministry planned for the distribution and sale of the shoes, and the consumer bought the shoes at a retail outlet—perhaps a shop or a store at his or her place of work. The standardization ministry was responsible to create appropriate quality standards. (In fact that ministry, faced with the need for large numbers of standards, itself had a quota for the production of standards.) The enforcement branch of the standards ministry provided inspection at the factory. The factory's income depended in part on the quality level as determined by the inspectors. It was illegal to ship substandard goods.

Under this system, if a consumer discovered a defect in a pair of shoes, he or she complained to the manager of the shop where they were purchased. That complaint had to work its way up through the hierarchy of the marketing ministry—consumer to store manager to local office of the marketing ministry, then up the chain in that ministry to a minister, then into the production ministry hierarchy— say "production" to "light industry" to "shoes" and finally to the factory where the shoes were made. The process took months, and by that time the complaint might have undergone some amazing distortion along the way (see also Kamm 1995). Whether relief of the problem resulted, indeed, whether the information was of any use at all by that time, is questionable. Thus was the system in practice deprived of two vital elements of effective quality control: rapid and direct feedback on deviation from standard; and the direct input of the customer.

One characteristic of consumer commerce in the Soviet era, noted by virtually all writers on the subject of the Soviet Union's consumer marketplace, was the absence of customer orientation on the part of sellers—shop personnel and retail service providers—and the powerlessness of consumers to stimulate a change. In fact, the very word "customer" within the communist system had a whiff of the counterrevolutionary about it and was better not mentioned. The famously stoic Russian shoppers, and their counterparts throughout the region, simply bore up—stood in long lines, suffered shabby treatment at the hands of shop clerks (who, in turn, had to go out and suffer the same when they did *their* shopping), withstood the disappointment of limited selection of goods and poorly stocked store shelves—and carried on as best they could.

USSQC, its predecessor systems, and the ideology which brought them forth had 70 years in Russia and more than 40 years in the other countries of CMEA to pervade the economic systems in which they were embedded. A system so pervasive over so long a period of time cannot be easily abandoned, even now that the economies have begun a transition to something other than communism. Remnants of the system remain in the mind. The leaders of the former CMEA countries will have to assess the system, identify the parts which contribute to competitive quality, and identify and discard the parts that do not.

QUALITY MANAGEMENT TODAY AND TOMORROW

The Forces at Work.   Today, there are numerous forces at work which are sharply changing the course of quality management in central and eastern Europe.

- *Local appearance of Japanese and Western goods:* In newly opened markets, Western products—manufactured goods, foodstuffs, even services—began to appear. They were at first too expensive for most people to afford. But their presence often created a new standard of comparison with local products.

- *Availability for purchase of Japanese and Western goods in the local market place:*   With the shift to a market economy, quality standards were determined in the marketplace, not in the committee rooms and laboratories of the standards organizations. Survival of local products required action which would bring quality to levels which are competitive in the world market. The introduction of Western automobiles produced a rapid decline in marketability of those made in the region. For example, when the German Democratic Republic (East Germany) and the Federal Republic of Germany (West Germany) joined, the West German government agreed to exchange East German marks for West German at the rate of one for one. The citizens in the east who had substantial savings were suddenly enriched. Many, attracted by the superior quality of the newly available Western automobiles, bought one. The Trabant and the Wartburg, two mainstays of the East German roads, went out of business.

- *The need and desire to compete at home against Western imports:*   Governments in the region know that high import tariffs to protect local industry against these imports can only be a temporary remedy. In the long term, local industry must make competitive products or suffer loss of business.

- *Need and desire to sell high-value goods on the international market:*   Governments know that the key to long-term economic health is vigorous participation in the world market. World competition requires products of world-class quality.

- *Joint ventures with international corporations:*   Joint ventures are proving to be effective vehicles of technology transfer in market-oriented management, including quality management.

Many successful international corporations, attracted by the market potential and the availability of a well-educated work force at low cost, have been rushing into the region for some years. They have formed local partnerships, joint ventures, and other commercial relationships which involve local people directly with foreign (Western and Asian) quality approaches, standards, and procedures, as well as with foreign goods and services. Such companies as ABB, General Electric, General Motors, Hewlett-Packard, Honda, and Toyota bring more than investment in modern physical plant; they bring business systems, including quality systems, which have been developed over decades, and which have enabled them to create products whose quality ranks among the best in the world.

An example is the introduction of Coca-Cola into Romania, described by Nash (1995). At the end of 1991, 2 years after the overthrow of the Ceacescu regime, Coke had no presence in the country. By 1994, as many as 25,000 kiosks and other small retail shops started or maintained their business because of Coca-Cola. In an interview, an economic advisor to the president of Romania said he viewed the arrival of multinational firms such as Coca-Cola as vehicles for transferring organizational and managerial skills. Coke's decision to develop local suppliers has meant that the manufacturers of bottles, plastic cases, and labels have all begun to experience Western-style quality standards and concepts.

In Moscow, within 5 years of opening in 1990, the McDonald's restaurant on Pushkin Square had become the company's busiest in the world. To achieve this, McDonald's had to develop extensive local supply of ingredients to their standards, and to train an army of servers to the company's service standards. Perhaps more significant in the long run is the emulation of the McDonald's model by another Russian restaurant, Russkoye Bistro, which presents traditional Russian food items in the same format and employing the same quality approach—fresh ingredients and good service at relatively low prices (Specter 1995).

Viewed from the local perspective, involvement in these management systems may be viewed in two ways: as an encroachment on local culture and autonomy to be resisted, or as first steps on the road to competitive quality, an opportunity to be embraced. In fact, there is evidence that each view has its adherents. The most reliable predictor of this is age, older people being likely to resist, younger ones to embrace. The older one was at the time of the shift from the old system to the new, the more difficult was the transition, hence the more likely the case for resistance. (This phenomenon is nothing new or strange, nor is it unique to the case of the change of system being discussed here. Millions of words have been written on the subject of a variety of cultural changes and the individual behaviors they stimulated, from agricultural economy to industrial, from manual labor to machine-assisted labor, from manufacturing economy to service, and so on.)

Lorber (1993) tells of being offered advice when he took the post as manager of Hewlett Packard's new office in Prague. He was told to "forget the older people; work with the young ones." "How old is old?" he asked. His friend responded that after the age of 25 it's too late. That seemed a bleak viewpoint, but it is true that younger employees had a far easier time of adjusting to the conditions of the marketplace than those who had made their careers accommodating themselves to a different economic and political structure. The older employees require much help to learn and apply the new concepts.

Total Quality Management, Customers, and Democratic Traditions. The concept of Total Quality Management (TQM) developed in the context of market economics and democratic tradition. In a market-based economy, the identity of the customer can be made clear, and the producer is free to use all means to ascertain (and even to influence) the customer's needs. In a centrally planned economy, decisions as to who is the customer and what are thecustomer's needs are hidden from view. The customer being the very foundation stone of TQM, the formerly planned economies seeking to establish TQM will need to restore respectability to market transactions and the word "customer," and train participants in market business basics.

Democratic tradition affects the flow of information. TQM depends on the free exchange of information within an organization. Lacking an environment which encourages such free exchange, TQM is not achievable. Where the political system discouraged free exchange, its participants must be trained to overcome the habits of guarded exchange.

TQM Criteria—The Ideal. The principal goal of TQM is to attain competitive (or world-class) quality. The criteria of the Malcolm Baldrige National Quality Award provide a widely accepted definition of comprehensive quality management. The Baldrige criteria support a model which includes:

1. Success measured in terms of delighted customers, empowered employees, higher revenue, and lower operating costs

2. Attainment of success through management processes which include quality planning, quality control, and quality improvement

3. Organizational infrastructure on which these processes depend, including a documented quality system, customer-supplier collaboration in partnerships, involvement of everyone in the organization in the quality effort, measurement and information systems for key business variables, and education and training of all in the organization, when and as required

4. A foundation of the quality effort which includes strategic planning and management to the plan, hands-on leadership by top executives, and a focus on customers, their wants, and their needs.

These elements of the Baldrige model provide a framework for assessing an organization's ability to progress toward TQM.

TQM Criteria in Practice

Delighted Customers. In the old regime, the concept of customer was alien, even subversive. Likewise, "consumer" had no currency. In Poland's Popular Encyclopedia of 1982, consumer was defined only in terms of the biological food chain (Lanigan and Bielska 1994). Habits of thought and behavior toward customers (or consumers), leftover from times of the single supplier and chronic shortages, still often reveal themselves in brusque customer treatment.

To delight customers, one must first recognize that customers exist and that commercial success depends on how well the customers are served. Where employees have operated in state monopolies for all of their careers, there is no reason to expect them to behave appropriately when they are suddenly thrust into the open market. This is not so different from the situation in the West, where it is not uncommon for workers who never face an external customer directly, let alone employees who have no experience with the concept of customer, to believe that they have no customers themselves. This state of belief is one of the first hurdles for organizations to overcome in establishing an approach to quality.

It will be necessary to educate all employees—managers and workers alike—whether they deal directly with external customers or exclusively with internal ones, to view those to whom they supply product and service as customers in the commercial sense, as a means to encourage a sense of partnership in the interest of the ultimate external customer, the one who buys the finished product or service.

In Poland, the author encountered an example that typifies the changes that will come with the shift to market orientation. An assembler of automobile components, a division of a major automobile maker, provides more product for other auto makers than to its parent company. The managing director has gone so far as to separate assembly operations by floor according to customer (which also reduces the probability of confusing one customer's parts for another's), and to identify these areas by prominently displaying each customer's logo on the exterior of the building and in its designated assembly area within the building.

Empowered Employees. In 1994, the director of a Polish factory reported that in the past, each fax message from his plant had required an authorizing management signature. Although he had announced a policy change which eliminated that signature requirement, some employees still climbed the stairs to his fourth floor office to ask for his signature. They were too accustomed to the old rules to feel comfortable ignoring them and adapting themselves to the new ones. Those employees lacked a sense of empowerment.

Employee empowerment means three things. First, that the employee is in a state of self-control, that is, able to know what is expected, know current performance as measured against those expectations, and adjust the process to bring actual performance into line with expectations. Second, the employee is able to make recovery, i.e., has freedom, within limits, to make things right when things go wrong. Third, the employee has the opportunity to participate in planning and improving the work. Key conditions of empowerment are (1) information which is both timely and accurate, (2) trust among the workers and managers, and (3) motivation to act in the interest of quality.

In a visit to a multiplant Polish manufacturing company, in plant visits and classroom meetings, it was common to be asked how to get accurate and timely information—from customers, from suppliers, from workers, from managers. Some groups asked how to get people to tell the truth. From the managers, this had the flavor of getting people to admit error. The workers seemed to suggest they were blamed for things they couldn't control. Questions like these are consistent with the formality of relationship between workers and managers and between layers of management which is, to Western eyes, extreme, and which could only impede the free flow of information within the company. (It is worth noting here that when a group of Polish supplier executives visited a progressively managed factory in Connecticut, they remarked favorably on the informality and apparent ease of communication between managers and workers.)

The questions are also consistent with the history within the company of not sharing information. This failure to share information dates from the communist period, when information was charged with political significance. Where financial information is involved, the situation is compounded by the fact that financial and cost information have never been available in a form that is useful in making decisions in the market setting. Further, there is also a history of punishing people who provided unwelcome information—a practice known in the West as "shooting the messenger." This was common during the time when the survival and promotion of managers was tied to meeting production goals, and when reporting bad news likely implicated the messenger as a barrier to meeting the goals.

As to trust, it has been frequently noted that the history of the past 50 years has seriously eroded trust in the former communist societies. Often enough, this lack of trust is reinforced by the fact that some former Party functionaries still hold positions of influence in companies that are still state-owned.

As to motivation, judging from the frequent questions on the subject directed to the author by employees and managers in Poland, there is a belief that (1) the quality effort will require more work from employees and (2) the employees should receive payment for that extra work. (Our view is that quality is an important part of every employee's job. To separate "quality work" from "normal work" is artificial and will damage the quality effort.)

Developing a remedy for these conditions will take time and patience. The management of such a company (and this one is probably typical) must first be aware of the conditions, recognize the conditions as a problem, recognize and agree on the causes of the problem, and begin setting a strategy to remove the causes. It will take education and, above all, continuous demonstration by the management at all levels that it is not only safe to participate and to share information, but that it is important for the future of the organization and the job security of everyone to do so.

For many companies enduring this transition, employee participation does not fit easily with the traditional management style. But participation is vital if they are to become as effective as they can be. This will require skilled leadership and facilitation as teams are set in motion. There are a number of basic skills in which managers will need to be trained, including finance, motivating people, identifying and valuing customers, sharing information, etc.

Employee empowerment also requires capable processes. In factories, this brings into question the ability of the existing manufacturing technologies to consistently meet the quality goals which they are supposed to meet. The problem of capability became acute in factories which were run for years without updating or even proper maintenance. This was typical throughout the automobile industry in the East.

Reduced Costs and Increased Profits. There is a more fundamental negative effect of inadequate process capability than the effect on potential employee empowerment. It is the effect on the unit cost of product—especially as the result of rework. For example, in the paint shop of one automobile factory, to paint a body in a popular metallic finish required three passes through the paint booth, with an enormous amount of hand rubbing between passes.

In an extreme case, an automobile factory in Yugoslavia was in such disarray that, as a visiting consultant reported to the author, one finished car on the assembly line had wheels loose to casual touch, another had a blue front seat and a brown back one, and all the cars emerging from the paint booth had so much airborne debris in the paint that they "looked like Brillo pads." A shortage of funds—for power lug-nut wrenches, for sufficient inventory of seats, for air filters in the paint drying booth—was offered as the common cause for all of these and other production woes. But the consultant noted that a major contributor was the absence of consistent management and supervision in the plant. In a perverse application of "worker democracy," a work group was always at liberty to replace by majority vote any supervisor with which it disagreed. Any supervisor attempting to improve production in his area became an immediate candidate for replacement. (For these and a variety of other reasons, the company was unable to market a minimally acceptable product. The company slid quietly into industrial history not long after the consultant's visit.)

Getting Started. Providing executive and managerial leadership in this quality effort is a challenge which should not be underestimated. When the transition to a market economy began, there was little experience within these countries in managing for other than production quotas. It is an encouraging sign that many managers show interest in learning what they should do to manage for quality.

If executives and managers agree on the goals on which the Baldrige criteria are based, other elements of the Baldrige model provide guidance as to *what* the organization requires to accomplish those goals—quality processes and adherence to them, infrastructure elements, and the foundation of planning, leadership, and customer focus. The matter of *how* to provide these supporting elements is more difficult. All of these elements are likely to be very different from anything the managers have experienced before, and the foundation of planning, leadership, and customer focus is probably equally novel in their experience. It must be remembered, after all, that even in the Western setting, examples of the companies conducting themselves according to the Baldrige criteria are not so common.

When the transition to a market economy began, these countries had no tradition of relationship with or dependence on a customer. The concept of customer delight and the rest must have seemed meaningless, if not absurd. There was in the West a fair amount of skepticism, if not outright pessimism, about the prospect of rapid change.

Today, there are many hopeful signs that the changes can be made. The experience of Coca Cola and McDonald's, mentioned above, are examples. Milbank (1994) reports rapidly rising exports from the region, especially from Poland, the Czech Republic, and Hungary. This rise in exports is enabled by dramatic improvements in quality, made possible, in turn, by financial and technical help from Western partners. Three Hungarian companies are presented as examples—a printing and packaging company, a liquor distiller, and a frozen food company—all competing successfully outside their home country.

We have stated, from the customer's viewpoint, some of the things that must change to improve quality and its management in the region. There is growing number of companies that are making those changes happen. Only time will tell how fast the changes proceed and when they will be pervasive enough to put the region as a whole on a secure path to full participation in the world marketplace.

REFERENCES

Behrman, Jack (1998). "In Praise of Maintenance." *The Chapel Hill News,* May 3.

Broad, William J. (1996). "Russian Rockets Get Lift in U.S. From Cautious and Clever Design." *The New York Times,* October 29.

Crossette, Barbara (1995). "Chernobyl Trust Fund Depleted as Problems of Victims Grow." *The New York Times,* November 29.

Egermayer, F. (1988). "Quality in Socialist Countries." *Quality Control Handbook,* 4th ed., J. M. Juran, editor. McGraw-Hill, New York.

Figes, Orlando (1997). *A People's Tragedy: A History of the Russian Revolution.* Viking, New York.

Juran, J. M. (1994). In response to a question at a seminar, December 1994, based on his discussions with industrial managers in Russia during the Soviet period.

Kamm, Henry (1995). "Poland Reawakens to Its History as Communism's Mirror Shatters." *The New York Times,* January 26, 1995.

Khrushchev, Nikita S. (1970). *Khrushchev Remembers.* Little, Brown and Company, Boston.

Lanigan, Edward P., and Bielska, Ewa (1994). "Quality Renaissance in Poland: A True Ethic Shines Through the Tarnished Image." *Proceedings, 38th EOQ Annual Congress,* Lisbon, Portugal. Portuguese Association for Quality, Lisbon.

Lorber, Franz (1993). Remarks during the question period following presentation of "How to Successfully Establish Quality Management in a Start-up Operation." *IMPRO 1993 Conference Proceedings.* Juran Institute, Wilton, CT.

Medvedev, Grigori (1991). *The Truth About Chernobyl* (English translation by Evelyn Rossiter). Basic Books (no city indicated).

Milbank, Dana (1994). "New Competitor: East Europe's Industry Is Raising Its Quality and Taking on West." *The Wall Street Journal,* September 24.

Nash, Nathaniel C. (1995). "Coke's Great Romanian Adventure." *The New York Times,* February 26.

Smith, Hedrick (1976). *The Russians.* Quadrangle/The New York Times Book Co., New York.

Smith, Hedrick (1990). *The New Russians.* Random House, New York.

Specter, Michael (1995). "Borscht and Blini to Go: From Russia's Capitalists, an Answer to McDonald's." *The New York Times,* August 9.

SECTION 40
QUALITY IN THE UNITED STATES

J. M. Juran

THE BACKGROUND

The economy of the United States of America (U.S.) rests mainly on a base of numerous autonomous producers and marketers of goods and services. These autonomous companies are characterized by:

1. A high concentration of industry in relatively few companies. The number of companies runs to over a million, but the top 1000 companies account for most of the goods and services produced.

2. A high degree of private ownership of these large companies. Normally, a large company will have thousands of owners, no one of whom owns more than a few percent of the company.

3. A "professional" management. The companies are run by professional managers—persons who consider their lifetime career to be that of managing. These managers become the real power in the company, since the owners are too numerous. In addition, under the prevailing legal system of boards of directors, the managers usually dominate the board. The managers and the concept of professional management are among the main strengths of the U.S. economy.

    The features of autonomous companies and professional managers to run them have a considerable impact on how quality is managed. Within the flexibility permitted by the "anarchy of the marketplace," each company determines which products it will make or stop making, what quality policies it will employ, and so on. Innovation plays an important role throughout, owing to the rather unusual industrial history of the country.

The early European colonists faced the problems and opportunities associated with exploiting the natural resources of a huge land mass. An innovative spirit was developed in the early agricultural days, and carried over when the nation industrialized. Self-reliance and risk-taking emerged as respected traditions. These traditions then raised entrepreneurship and individualism to a state of respect. The resulting companies tended to organize in ways which assigned responsibility to individuals rather than to teams. The tradition of self-reliance also stimulated job mobility. In the U.S., workers, engineers, and managers tend to change jobs more often than their counterparts in other countries. The concept of a lifetime career has usually been viewed as being associated with a trade, a union, or a profession rather than with a specific company.

Early Systems of Managing for Quality.   Late in the eighteenth century, the colonists broke with their European rulers and established an independent United States. The domestic economy was unified by the laws governing movement of goods in interstate commerce. These laws avoided the obstacles inherent in the national boundaries then prevailing in Western Europe—passports, customs offices, import duties, and so on—that have plagued the countries of Europe for centuries. The absence of such barriers enabled the United States to become a unified common market and contributed to the speed with which the country emerged as an economic superpower.

As the colonies began to industrialize, they generally followed the craftsmanship concept which prevailed in their European country of origin. Apprentices learned a trade and qualified to become craftsmen. Achievement of quality was one of the essential skills learned by the apprentice. A major force for assuring quality of product was the village form of society in which the craftsman met face-to-face with the users. In a shop of any size, the master carried out a form of product inspection and process audit which provided added quality assurance. Alternatively, the master delegated this function to an inspector.

When the Industrial Revolution of the mid-eighteenth century was exported from Europe to the United States, the colonists again followed European practice. Many craftsmen became factory workers, and many masters became factory foremen. Quality was assured as before—by the skills of the craftsmen supplemented by supervisory audit or by departmental inspection.

The Taylor System and Its Impact.   Late in the nineteenth century many American companies broke sharply with European tradition by adopting the Taylor system of "Scientific Management." The basic concept was the separation of planning from execution. This separation made possible a considerable increase in productivity and was a major contributor to making the United States the world leader in productivity. (For elaboration, see Juran 1973.)

The Taylor system also included some adverse side effects. It dealt a crippling blow to the concept of craftsmanship. In addition, the new emphasis on productivity had a negative effect on quality. To restore the balance, the factory managers created a central Inspection department headed by a chief inspector. The various departmental inspectors were transferred to the new department over the bitter opposition of production supervisors. In due course, the inspection departments grew into broad-based organizations called variously Quality Control, Quality Assurance, Quality Management, and so on. These organizations evolved quality-oriented specialties such as quality engineering and reliability engineering.

The central activity of these quality-oriented departments remained that of inspection and test—separating good product from bad. The prime benefit of this activity was to reduce the risk that defective products would be shipped to customers. However, there were serious detriments:

This central activity of the quality department helped to foster a widespread belief that achievement of quality was the responsibility of the quality department.

In turn, this belief hampered efforts at eliminating the causes of defective products–the responsibilities were confused.

As a result, failure-prone products and incapable processes remained in force and continued to generate high costs of poor quality.

What emerged *de facto* was a concept of managing for quality somewhat as follows:

Each functional department carried out its assigned function and then delivered the result to the next function in the sequence of events.

At the end, the quality department separated the good product from the bad.

For defective product which escaped to the customer, redress was to be provided through customer service based on warranties.

By the standards of later decades, this concept of prime reliance on inspection and test was unsound. However, it was not a handicap if competitors employed the same concept, and such was usually the case. Despite the deficiencies inherent in this "concept of detection," American goods came to be well regarded as to quality. In some product lines, American companies became quality leaders. In addition, the American economy grew to superpower size. Some of this growth was achieved in ways which had implications for quality:

Entrepreneurs were on the alert to create sales in various ways: e.g., bring new, improved products to market; create additional production capacity to eliminate shortages. (Elimination of shortages also eliminates an inevitable cause of poor quality.)

Managers were willing to invest in facilities to improve productivity. Some of those investments (e.g., in machines, tools, instruments) improved quality as well.

The United States became a leader in the concept of a "professional" approach to management, involving extensive training for managers and specialists.

The growing number of quality specialists developed numerous new methods and tools specifically oriented to managing for quality. However, use of these methods was limited by the prevailing functional organization forms and, especially, by upper management's limited understanding of how to manage for quality.

WORLD WAR II AND ITS IMPACT

During World War II, American industry was faced with the added burden of producing enormous quantities of military products, many of which made use of new, sophisticated technology. However, the basic system of managing for quality remained unchanged. Each function carried out its responsibility and delivered the result to the next function in the sequence. At the end, inspection and test separated the good from the bad.

The military clients secured their quality assurance largely by additional inspection and test. Not until well after World War II did they evolve a concept of mandating the quality system to be followed by contractors.

A part of the American grand strategy during World War II was to shut off production of many civilian products: automobiles, household appliances, entertainment products, and so on. A massive shortage of goods developed amid a huge buildup of purchasing power. It took the rest of that decade (the 1940s) for supply to catch up with demand. In the interim, the manufacturing companies gave top priority to meeting delivery dates, so that quality of product went down. (Quality always goes down during shortages.) The habit of giving top priority to delivery dates then persisted long after the shortages were gone.

During this progression of events, the priority given to quality declined significantly. In addition, the leadership of the quality function became vague and confused. What emerged was a concept in which upper management became detached from the process of managing for quality. (For elaboration on the impact of World War II, see AT&T 1989, Grant 1991, Juran 1991, Wareham and Stratton 1991.)

THE JAPANESE QUALITY REVOLUTION AND ITS IMPACT

Following World War II, the Japanese embarked on a course of reaching national goals by trade rather than by military means. The major manufacturers, who had been largely involved in military production, were faced with converting to civilian products. The chief obstacle to selling these products in international markets was a national reputation for shoddy goods, created by export of poor-quality goods prior to World War II.

The Japanese adopted a variety of strategies for improving their quality. (See generally, Section 41, Quality in Japan.) In the judgment of the author, several of those strategies were decisive in creating a successful revolution in quality.

The upper managers personally took charge of leading the revolution.

They trained all levels and functions of the hierarchy in how to manage for quality.

They trained the specialists in statistical process control.

They undertook quality improvement at a continuing, revolutionary pace.

They provided means for the work force to participate in control and improvement of quality.

In the early postwar period, the affected American companies logically considered Japanese competition to be in price rather than in quality. Their response was to shift the manufacture of labor-intensive products to low-cost areas, often abroad. Then, as the years unfolded, price competition declined while quality competition increased. However, the American companies generally failed to recognize these trends or to heed the warning signals. In 1966, the author sounded an alarm at the annual conference of the European Organization for Quality Control:

The Japanese are headed for world quality leadership and will attain it in the next two decades because no one else is moving there at the same pace. (Juran 1967)

During the 1970s and 1980s, numerous Japanese manufacturers greatly increased their share of the American market. A major reason was superior quality. Many industries were affected: automobiles, consumer electronics, steel, machine tools, and so on. Some research quantified the quality differences. [See Juran 1979 (color television sets); also Garvin 1983 (room air conditioners).]

RESPONSES TO THE JAPANESE QUALITY REVOLUTION

The most obvious effect of the Japanese quality revolution was a massive export of goods to the United States. These goods were welcomed by consumers because of their superior quality along with their competitive and even lower prices. However, these same goods did much damage to other sectors of the American economy:

The affected manufacturing companies were damaged by the resulting loss of sales.

Workers and their unions were damaged by the resulting "export of jobs."

The national economy was damaged by the resulting unfavorable trade balance.

Some of the American companies' responses to the Japanese invasion had no relation to improving American competitiveness in quality.

Block the Imports.   Some of the affected companies tried to respond by reducing or eliminating the imports. They urged legislators to establish restrictive import quotas and tariffs. They urged criminal prosecutions on the grounds of violation of laws against "dumping," (selling below cost, or at "less than fair value"). They filed civil lawsuits on the grounds of unfair trade practices. They appealed to the citizenry to "Buy American."

These responses did not arouse broad sympathy among the buying public. Influential journalists, economists, legislators, and others pointed out that restriction of imports generates serious side effects: Buyers are deprived of better values; restriction invites retaliatory restriction; companies have no incentive to become more competitive; and so on. (For some case examples in which import restrictions damaged the very industries they were intended to protect, see Levinson 1987.)

Reduce Costs.   Some companies viewed the problem as one of price competition, arising from the low wage rates then prevailing in Japan. Such companies responded by moving their production to low wage areas, including locations overseas. These actions often did reduce labor costs but did not solve the main problem which was competition in quality.

Give Up.   Still other companies concluded that to become competitive in quality required expenditures (in product design and process facilities) which would not yield adequate return on the investment. These companies either sold out or otherwise went out of business.

> During the 1960s, there were over 30 American-owned companies making color television sets. By the early 1990s there was only one.

INITIATIVES TO IMPROVE QUALITY

By the end of the 1970s, the American quality crisis had reached major proportions. It attracted the attention of the national legislators and administrators. It was featured prominently in the media—it was regularly "on the front page." It forced many chief executive officers (CEOs) to become involved in managing for quality.

During the 1980s, a great many American companies undertook initiatives to deal with the quality crisis. These initiatives were largely focused on three strategies:

Exhortation.   Some consultants proposed a sweeping solution by exhorting the work force to make no mistakes—to "do it right the first time." This simplistic approach was persuasive to those managers who, at the time, believed that the primary cause of their company's quality problems was the carelessness and indifference of the work force. The facts were that the bulk of the quality problems had their origin in managerial and technological processes. In due course, this approach was abandoned but not before it generated a lot of divisiveness.

Training in Statistical Methods.   During the 1980s, many American companies undertook to train company personnel in application of statistical methods to quality problems. The term "Statistical Process Control" (SPC) became the popular label for this training.

While SPC is a useful tool, most companies assumed it to be the panacea claimed by its advocates. The companies lost precious years before learning that leadership in quality comes from multiple strategies, no one of which is a panacea. To make matters worse, the training was done before the companies had identified their quality problems and defined their quality goals. In a sense, the personnel were trained in remedies before the diseases were known.

> Eastman Chemical Company, when relating its approach to managing for quality (it became a 1993 winner of the Malcolm Baldrige National Quality Award), stated that it had trained 10,000 of its personnel in SPC. However, many trainees lacked the opportunity to apply the training, so much was forgotten. (Eastman Chemical Co. 1994)

Quality Improvement, Project by Project.   One of the consulting companies, Juran Institute, Inc., created and published a series of videocassettes titled "Juran on Quality Improvement," (Juran 1980). These were tested by many companies. Some achieved notable quality improvements while others did not. The decisive variable was the extent of personal leadership provided by the upper managers. By the end of the 1980s, the improvement process described in those videocassettes had become the basic model for the process of continuous quality improvement adopted by most companies.

Results of the Initiatives of the 1980s.   In retrospect, the quality initiatives of the 1980s were deeply disappointing. Most fell well short of their goals. Some produced negative results—the companies lost several years of potential progress. The poor results were due mainly to poor choice of strategies and to poor execution of valid strategies. In turn, these were largely traceable to the limitations of leadership by upper managers who lacked training and experience in managing for quality. In the minds of some observers, the lessons learned during the 1980s were chiefly *lessons in what not to do.*

THE ROLE MODELS

During that same disappointing decade of the 1980s, a relatively few company initiatives achieved stunning results. Such companies attained quality leadership—"world-class quality"—and thereby became the role models for the rest of the American economy.

The role models were few in number. They included the winners of the Malcolm Baldrige National Quality Award plus other companies that had achieved similar results. Together, they made up only a tiny part of the economy. Yet there were enough of these companies to prove that world-class quality is attainable within the American culture—"They did it, so it must be doable."

The successes achieved by the role-model companies stimulated great interest among upper managers and others who sought to learn how such stunning results had been achieved. The role models were quite willing to share information about the strategies they had used to achieve those results. In addition, steps were taken to share the lessons learned through company visits, conferences, publications, and so on.

LESSONS LEARNED—THE CORE STRATEGIES

Each role-model company is different. In groping for ways to attain world-class quality, each serves as a laboratory, testing out various strategies, adopting some, modifying others, rejecting still others. In this sense each role-model company is unique. Yet despite differences among the role-model companies, analysis shows that their strategies have much in common. There is *a core list of strategies* which were widely adopted by most of the role models. These core strategies, some of which are listed below, deserve careful study—they are a body of *lessons learned,* a list of the key strategies which enabled the role models to achieve those stunning results.

Customer Focus.   All role models adopted the concept that the customer has the last word on quality. Adoption of this concept then led to intensified action to identify: who are the customers, internal as well as external; what are the needs of customers; what product features are required to meet those needs; how do customers decide which of the competing products to buy; and so on. (For elaboration, see Section 3, The Quality Planning Process.)

Many quality problems of the past have been traced to failure to meet the needs of *internal* customers. As a result, customer focus is increasingly being extended to include internal as well as external customers.

A widespread example has been product designs which designers "threw over the wall" to be produced, sold, and serviced by internal customers—other company departments.

The concept of customer focus has led to broader use of the concept of participation. This minimizes the damage done when planners are unaware of (or indifferent to) the problems their plans will create for their customers. It provides early warning—those affected are able to point out, "If you plan it that way, here are the problems we will face." The participation concept is also being extended to supplier relations, a popular label being "partnering."

Upper Managers in Charge.   One element present in all successes, and absent in most failures, was the personal involvement of the upper managers. In effect, the upper managers took charge of quality by accepting responsibility for certain roles, including:

Serve on the Quality Council

Establish the strategic quality goals

Provide the needed resources

Provide quality-oriented training

Stimulate quality improvement

Review progress

Give recognition

Revise the reward system

Many upper managers resisted such additions to their own workload. Their preference was to establish broad goals and then to urge their subordinates to meet the goals. However, the lessons learned from the role models are that *the above roles are not delegable*—they must be carried out by the upper managers, personally.

Strategic Quality Planning.   The role models recognized that the new priority given to quality requires enlarging the business plan to include quality-related goals. These goals are then "deployed" to identify the needed actions and resources, to establish responsibility for taking the actions, and so on. The resulting plans parallel those long used to meet goals for sales and profit. A common name for this concept is Strategic Quality Planning. (For elaboration, see Section 13, Strategic Deployment.)

The Concept of "Big Q."   The role models grasped the concept that managing for quality should not be limited to manufacturing companies and manufacturing processes. It should also include service companies and business processes. This concept broadens the area under the "quality umbrella." (For details, see Table 2.1 of Section 2.) Some companies call this broader concept "Big Q," in contrast to the traditional concept which they call "Little Q."

Cost of Poor Quality.   Cost of poor quality (COPQ) consists of those costs which would disappear if everything were perfect—if there were no errors, no waste, no field failures, and so on.

As upper managers were drawn into managing for quality, they learned a good deal about COPQ. Some of what they learned came as surprises.

Relation to Big Q: Many upper managers had assumed that COPQ consisted of the cost of running the quality department, or alternatively, the costs of deficient factory goods and processes. It is now widely accepted that COPQ should include costs traceable to deficiencies anywhere—deficiencies in Big Q.

COPQ is huge: In the early 1980s, the author estimated that in the United States, close to a third of the work done consisted of redoing what had been done before. Depending on the nature of the industry, COPQ consumed between 20 and 40 percent of the total effort. Translated into financial terms, the sums are staggering. Translated into other terms, the effects are equally staggering: delays in getting new products to market or in providing service, damage done to customer relations, damage to internal morale, and so on.

Higher quality costs less: Many upper managers have long believed that to attain higher quality requires increasing the costs. This belief is often valid as applied to "higher quality" in the sense of better product features to increase sales. The belief is seldom valid as applied to "higher quality" in the sense of less errors, less redoing, fewer field failures, and so on. In these latter cases, higher quality almost always costs less, and often a lot less.

Upper managers who have become deeply involved in managing for quality have gained new insights relative to COPQ. They came to realize that high COPQ presented an opportunity for cost reductions, at a higher return on investment than virtually any other managerial activity. (For additional discussion, see Section 8, Quality and Costs.)

Quality Improvement.   Without exception, the role models went extensively into quality improvement—most of the stunning results came from projects to improve quality. These projects extended to all activities under the Big Q umbrella. They reduced costs, raised productivity, shortened cycle times, improved customer service, and so on.

Quality improvement required special organization. The vital few projects were carried out by multifunctional teams of managers and specialists. The useful many projects were carried out at lower levels, including members of the work force.

The role models also adopted the concept that quality improvement must go on year after year—it must be woven into the company culture. To this end, they mandated that goals for quality improvement be included in the annual business plans. They also redesigned the systems of recognition and reward to give added weight to performance on quality improvement. (For elaboration, see Section 5, The Quality Improvement Process.)

Business Process Quality Management ("Re-engineering").   A major extension of quality improvement was to the area of business processes. This extension resulted from fresh thinking relative to the multifunctional processes prevalent in functional organizations.

Figure 40.1 shows the interrelation between the typical "vertical" functional organization and the "horizontal macroprocesses" through which things get done.

Each horizontal macroprocess consists of numerous steps or "microprocesses" which thread their way through multiple functions. Every microprocess has an "owner," but there is no clear "ownership" of the macroprocess.

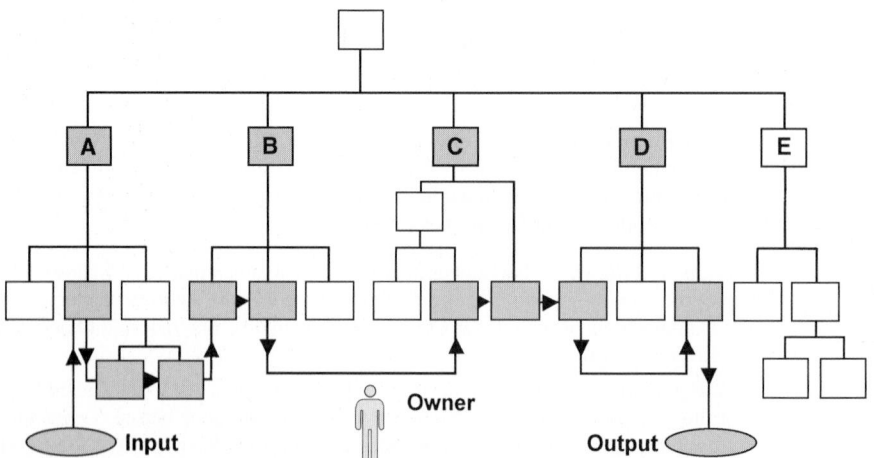

FIGURE 40.1   Interrelation between "vertical" functions and "horizontal" processes. [*Adapted from Re-engineering Processes for Competitive Advantage: Business Process Quality Management, 2nd ed. (1994). Juran Institute, Inc., Wilton, CT, pp. 3–35.*]

The role models concluded that each key macroprocess should have an owner, and they took action to create such owners (individuals or teams). They also defined the responsibilities of an owner, including responsibility for improving the macroprocess. An important part of the "stunning results" achieved by the role models came from improvements made in the business processes. (For elaboration, see Section 6, Process Management.)

Training for Quality.   The earliest training for quality was the system of apprenticeship through which young boys became certified as craftsmen. More recently, with the proliferation of inspectors and inspection departments, there emerged training for inspectors—interpreting specifications, use of measuring instruments, and so on.

During World War II, the American government sponsored courses in statistical quality control, based on the Shewhart control chart and other tools developed by AT&T during the 1920s, (Working 1945). Following the end of the war, these courses continued to be offered by some colleges (as extension courses), by societies such as American Management Association (AMA) and the American Society for Quality Control (ASQC), and by consultants. Then, as quality departments broadened their scope, there emerged courses oriented to the functional needs of those departments: inspection and test, quality engineering, reliability engineering, and advanced courses in statistical methodology such as design of experiments and analysis of variance.

During the 1970s, there emerged the quality crisis resulting from the Japanese quality revolution. As companies tried to respond, it became clear that training for quality should not be confined to the quality department—it should be extended to all functions, and to all levels of the hierarchy. It also became clear that this extension required design of new courses, especially courses in managing for quality.

For example, when planning a new product or process, it is usual to assign a project planner to plan for the numerous parameters: technology, finance, schedule, quality, and so on. Such planners are often experts in the technology, but seldom in planning for quality. This is known as "quality planning by amateurs." One of the remedies is to train the planners in how to plan for quality, as set out in Section 3, The Quality Planning Process.

(Some large companies extended the concept of training in managing for quality to their suppliers. They urged and even demanded that their suppliers provide training in managing for quality to the appropriate personnel).

Much of the training done during the 1980s failed to produce tangible results. The chief reasons included:

The line managers (the customers) often did not participate in the planning of the courses.

Training through exhortation (banners and slogans) was frequently counterproductive.

Training in "awareness" failed to provide employees with answers to the question "What should I do that is different from what I have been doing?"

There was overemphasis on changing attitudes and underemphasis on changing behavior.

Training in the use of tools was usually done before identifying the quality problems or setting quality goals.

Training in quality improvement consisted of training in use of statistical tools rather than by being assigned to a quality improvement team.

By the 1990s numerous course designs were available for training in managing for quality. No consensus had been reached, but three designs were in wide use:

One was based on the criteria for the Malcolm Baldrige National Quality Award.

Another was based on Deming's lectures in statistics plus his "14 points" (Deming 1986).

A third was based on the Juran Trilogy, which organizes the subject matter into three fundamental processes: quality planning, quality control, and quality improvement. (For elaboration, see Section 2, How to Think About Quality, under How to Manage for Quality: The Juran Trilogy.)

There is also a growing feeling among industrial companies that, while quality has risen greatly in importance, the national educational system has not kept up with this trend. As a result, the graduates lack knowledge of the subject, forcing the companies to fill the gap through training. Schools at all levels have begun to address this problem. In addition, some companies have set up alliances with selected schools to help redesign the curricula, provide training materials, train faculty members, and otherwise support the alliances. (For elaboration on training for quality, see Section 16, Training for Quality.)

Measurement of Quality.   Measurement of quality has long been used at the technological level. What is new is measuring quality at the business level: customer satisfaction, competitors' quality, performance of key business processes, and so on. To meet such needs may require inventing new measures as well as new methods of analysis and presentation. The need for measurement may also require creating a National Quality Index to parallel indexes already in use, such as for consumer prices, unemployment, and productivity. (For elaboration, see Section 9, Measurement, Information, and Decision-Making.)

Benchmarking.   The concept of benchmarking grew out of the need to establish quality goals based on factual analysis rather than empiricism. The approach is to discover, for the process under study, what is the best performance, whether within one's own company, or in a competitor's company, or in a completely different industry.

For example, in one company, the best warehouse takes an average of five working days to fill customers' orders. The leading competitor takes an average of four days. A company in a different industry takes only three days. The benchmarked goal then becomes three days. There may well be a reaction "It can't be done," and this may be true as applied to the present process. However, the response is "It's being done now." So the problem is then to create (or re-create) a process which can meet the benchmark. The concept of benchmarking has been widely accepted in the United States. Progress is being made to develop data banks on what are the best known performances, and on the methods used to achieve them. (For elaboration, see Section 12, Benchmarking.)

Empowerment.   As of the early 1990s, many American companies still retained the separation of planning from execution inherent in the Taylor system of Scientific Management. Such companies were failing to make use of a huge underemployed asset—the education, experience, and creativity of the work force. It was generally agreed that the Taylor system was obsolete and should be replaced, but there was no consensus on what should replace it.

Replacing the Taylor system requires transfer of tasks from specialists and supervisors to non-supervisory workers. The word "empowerment" has become a label for such transfer. Empowerment takes various forms, all of which have been undergoing test. The more usual forms of empowerment have included:

Establish worker self-control: This requires providing workers with all the essentials for doing good work: means of knowing what are the quality goals; means of knowing what is the actual process performance; and means for adjusting the process in the event that quality does not conform to goals. A state of self-control empowers workers to make decisions on the *process,* decisions such as: Is the process in conformance? Should the process continue to run or should it stop? Ideally, such decisions should be made by the work force. There is no shorter feedback loop.

Establish worker self-inspection: This empowers workers to make decisions on whether the *product* conforms to the quality goals. Such empowerment shortens the feedback loop, confers a greater sense of job ownership, and removes the police atmosphere created by use of inspectors.

Enlarge workers' jobs: The enlargement may be horizontal—assigning a greater assortment of tasks within the same function to reduce the monotony of short-cycle work. It may also be vertical—assigning multiple functions around the core task. A widespread example has been the training and empowerment of workers who answer telephones, to enable them to provide "one-stop shopping" to customers who call in.

Establish self-directed teams of workers: Under this concept, teams of workers are trained to conduct operations which consist of multiple functions as well as multiple tasks. The empowerment may include process planning, establishing work schedules, deciding who is to perform which tasks, recruiting new team members, maintaining discipline, and still other responsibilities formerly carried out by specialists and supervisors.

The concept of self-directed teams has been widely tested. The published results indicate that quality and productivity improve significantly. The ratio of workers to managers rises sharply. Jobs cross functional lines and become team jobs. Workers become team members. All this requires extensive training.

Because empowerment involves extensive transfer of work from supervisors and specialists to the work force, it is meeting much cultural resistance. There is also some resistance from labor unions. They sense that empowerment establishes a new communication link between management and the work force which may weaken the linkage between workers and the union.

In the view of the author, replacing the Taylor system is an idea whose time has come. It is also his view that all of the above options will grow, and that the major successor to the Taylor system will be self-directed teams of workers.

Motivation; Recognition; Reward.   To meet the new competition in quality has required company personnel to adapt to numerous changes such as:

Quality is to receive top priority.

Personnel are to accept training in various quality-related disciplines.

A new responsibility—quality improvement—is added to the traditional list of responsibilities.

The use of teams requires the personnel to learn how to behave as team members.

Generally, American companies have recognized that for such changes to be accepted, it is necessary to make revisions with respect to motivation. The companies responded by increasing the use of recognition and, to a lesser degree, by revising the reward systems.

Recognition is public acknowledgment of superior performance. The companies expanded their use of prizes, plaques, ceremonial dinners, publicity, and so on. Generally, they did this with skill and in good taste.

While recognition relates to voluntary action, the reward system relates to the mandated actions which define the job description. Here, the company responses were less sure-footed—there was no precedent on how to make the needed changes. Mostly, the companies expanded the list of parameters used annually to judge employee performance by adding a new parameter such as "performance on quality improvement." (Some companies even failed to realize that there was need for changing the reward system.)

Total Quality Management (TQM).   By the late 1980s, it was becoming clear to upper managers that competitiveness and quality leadership could not be achieved by pecking away—by bringing in this or that tool or technique. Instead, it was necessary to apply the lessons learned (from the role models) to all functions and all levels, and to do so in a coordinated way. The popular label adopted to designate that collection of lessons learned was "Total Quality Management," or TQM. (The usual Japanese term is Company Wide Quality Control.)

There has been no agreed standard definition for TQM, so communication has been confused within companies, in training courses, and in the general literature. This confusion has since been reduced by publication of the criteria used by the National Institute for Standards and Technology (NIST) to judge the applications for the United States' Malcolm Baldrige National Quality Award (Baldrige Award). NIST, Gaithersburg, MD, administers the United States Malcolm Baldrige National Quality Award.

Those criteria have been widely disseminated, NIST has filled over a million requests for application forms. While there have been relatively few applications for the award, many companies have conducted self-audits against the criteria. In addition, as state and local quality awards have prolif-erated, many have used the Baldrige Award criteria as major inputs to their own list of criteria. By the early 1990s, this wide exposure had, in the opinion of the author, made the Baldrige Award cri-teria the most widely accepted definition of what is included in TQM.

PROGNOSIS

Until the 1980s, the prognosis for the United States was gloomy. Japanese companies had success-fully invaded the American market with products which offered superior quality and value. The resulting public perception then became a force in its own right, continuing to damage those American companies who had been slow to respond.

Emergence of Role Models.   During the 1980s, the quality crisis deepened despite quality initiatives launched by many companies—most of those initiatives fell far short of their goals. The good news was the emergence of role models, discussed above, and identification of the strategies they used to become quality leaders. The job ahead then became one of scaling up—of applying those lessons learned across the entire economy, including the giant service industries—health, edu-cation, and government.

The Urge to Scale Up.   By the early 1990s, some powerful forces had converged to stim-ulate scaling up. The growing quality crisis had raised awareness of the subject, as did the growth of awards for quality, notably the Malcolm Baldrige National Quality Award. Self-assessment against the Baldrige criteria helped many companies to identify their strengths and weaknesses. The results achieved by the role-model companies stimulated a desire to secure similar results. The publicized "lessons learned" showed the way to get such results. The role models demanded better quality from their suppliers, who in turn transmitted those demands through the entire supplier chain. The supplier base shrank, and a major test for supplier survival was to attain world-class quality.

An additional and growing force is the increasing sophistication of American upper managers. For decades, they had been detached from the quality function—they had delegated the job of man-aging for quality to their quality managers. As a result, the upper managers were, in the true sense of the word, ignorant of how to manage for quality.

When the quality crisis deepened, the upper managers were forced to move in. At first their ignorance led to poor choice of strategies. But once upper managers moved in they did not remain ignorant, they learned from their mistakes as well as from the role models.

A further powerful force is waiting to emerge—the urge to "buy American." Most Americans do prefer to buy American, all other things being equal. During the 1960s and 1970s, other things were not equal, so the urge to buy American was overcome by the superior quality and value of Japanese products. We can expect the quality gap to narrow in the coming century. That will translate into growth in market share for American companies, once customer perception catches up with the realities. Some of this has happened already.

Yet another force which urges scaling up is the rise of quality to a position of prominence in the public mind. Quality has moved to center stage due to a convergence of multiple trends:

Growing public awareness of the role of quality not only in competitive trade, but also in other fields such as national defense

Pressure from consumer organizations for better quality and more responsive redress if products fail

Fear of major disasters and near disasters arising from quality failures

Growing concerns about damage to the environment

Action by the courts to impose strict liability (For elaboration, see Section 35, Quality and Society)

Limitations to Scaling Up.   Despite the forces urging scaling up, progress will be slow. The American economy is huge and, like a huge aircraft carrier, it has a great turning radius. Most companies have a sizable backlog of needed quality improvements; it will take them years to work this off. It will take additional years to improve the quality planning processes so as to minimize creating new chronic wastes.

A preview of the pace of scaling up is seen in the time required for companies to attain the status of role models. No company known to the author became a role model in less than 6 years; more usually it took 8 to 10 years. This length of time was consumed by several common steps:

Conduct a pilot test of selected strategies

Analyze the results

Make the needed mid-course corrections

Scale up

The Upcoming Century of Quality.   The twentieth century can rightly be called the Century of Productivity. During that century, the United States became the most productive country on earth, thanks in part to adoption of the Taylor system. Productivity is still an important element of competition, but meanwhile, quality has moved to center stage. The twenty-first century will probably be known as the Century of Quality.

The lessons learned from the role models have identified the principal sources of quality leadership, such as:

Upper managers take charge of quality by carrying out certain nondelegable roles. (See above, under LESSONS LEARNED—THE CORE STRATEGIES; Upper Managers in Charge.)

Quality improvement is carried out at a revolutionary rate, year after year.

In the view of the author, the United States is now well poised to share world quality leadership during the next century. The failures of the 1980s provided lessons learned about what *not* to do. The role models provided lessons learned about what *to* do. Scaling up is under way, and the pace seems to be accelerating.

Nevertheless, for this scaling up to permeate the massive size of the American economy will take decades (as it did in Japan). Moreover, public perception lags behind events, sometimes for years. It may well take another two or three decades before "Made in USA" is widely accepted as a universal symbol of world-class quality.

ACKNOWLEDGMENT

This section includes some extracts from Juran, J. M., ed., *A History of Managing for Quality,* sponsored by Juran Foundation Inc., and published by Quality Press, Milwaukee, WI, 1995. The author is grateful to the Juran Foundation for permission to use those extracts.

REFERENCES

AT&T (1989). "A History of Quality Control and Assurance at AT&T, 1920–1970." This series of 14 videocassettes contains interviews with W. Edwards Deming, J. M. Juran, Ralph Wareham, and others who played significant roles in quality control during the 50-year span 1920–1970. Available from AT&T Customer Information Center, phone 1-800-432-6600, select code 500-721.

Deming, W. Edwards (1986). *Out of the Crisis.* Massachusetts Institute of Technology, Center for Applied Engineering Study, Cambridge, MA.

Eastman Chemical Co. (1994). Papers presented at Quest for Excellence Conference. National Institute for Standards and Technology, Washington, DC.

Garvin, David A. (1983). "Quality on the Line." *Harvard Business Review,* September-October, pp. 64–75.

Grant, E. L. (1991). "Statistical Quality Control in the World War II Years." *Quality Progress,* December, pp. 31–36.

Juran, J. M. (1967). "The QC Circle Phenomenon." *Industrial Quality Control,* January, pp. 329–336.

Juran, J. M. (1973). "The Taylor System and Quality Control." A series of articles in *Quality Progress,* May through December (listed under "Management Interface").

Juran, J. M. (1979). "Japanese and Western Quality—A Contrast." *Quality,* January and February. Also published under the same title, in *Proceedings of the International Conference on Quality Control,* Japanese Union of Scientists and Engineers, Tokyo, 1978, pp. A3-11 to A3-25.

Juran, J. M. (1980). "Juran on Quality Improvement." A series of 16 videocassettes plus associated manuals. Published by Juran Institute, Inc., Wilton, CT.

Juran, J. M. (1991). "World War II and the Quality Movement." *Quality Progress,* December, pp. 19–24.

Juran, J. M. (1995). ed., *A History of Managing for Quality.* Sponsored by Juran Foundation Inc., Quality Press, Milwaukee, WI.

Levinson, Marc (1987). "Asking for Protection Is Asking for Trouble." *Harvard Business Review,* July–August, pp. 42–47.

Wareham, Ralph E., and Stratton, Brad (1991). "Standards, Sampling and Schooling." *Quality Progress,* December, pp. 38–42.

Working, Holbrook (1945). "Statistical Quality Control in War Production." *Journal of the American Statistical Association,* December, pp. 425–447.

SECTION 41
QUALITY IN JAPAN

Yoshi Kondo
Noriaki Kano

DEVELOPMENT OF MODERN QUALITY CONTROL IN JAPAN

Introduction of Quality Control from the United States.   Prior to World War II, Japanese research on and application of modern quality control were limited. Japanese product quality was poor relative to international levels. These poor products were sold only at ridiculously low prices, and it was difficult to secure repeat sales. Among the exceptions were the high-technology products of some Japanese companies, primarily for military use, which were manufactured without successful application of mass production techniques.

The concepts and techniques of modern quality control were introduced from the United States immediately after World War II. The General Headquarters (GHQ) of the Allied Occupation Forces in Japan was experiencing difficulties with the poor state of the country's communication systems and the defective quality and late delivery of communication equipment and components ordered from Japanese manufacturers. GHQ's Civil Communication Section (CCS) was instructed to provide communication equipment manufacturers with business management guidance, including advice on quality control. Many Japanese manufacturing companies received help from the members of the Section such as Frank Polkinghorn, Charles Protzman, W. G. Magil, and Homer Sarasohn.

CCS Course.   To bring all these efforts together, Protzman and Sarasohn ran a seminar in the autumn of 1949 called the CCS Course, mainly for the top executives of communication equipment manufacturers. This course was designed to elevate the quality of management, and quality control was a part of it (Ikezawa et al. 1990, Hopper 1985).

It was thought then that the statistical methods used in the quality control activities were very helpful, indeed, indispensable, for the reconstruction and development of Japanese industries. It should be noted that during the War, Japanese industries were almost completely destroyed. Since Japan lacks abundant natural resources and has virtually the highest population density in the world, it became an overriding national priority to design and manufacture industrial products of superior quality and export them to foreign countries. Modern quality control is the most important and indispensable tool for improving and maintaining the quality of manufactured products.

Establishment of the Japanese Union of Scientists and Engineers.   The Japanese Union of Scientists and Engineers (JUSE) was established in April 1946 and has been at the core of quality control activities in Japan (Kondo 1978). This nonprofit organization is not financially supported by, or controlled by, the government. It was established with the aim of "contributing to the development of culture and industry through the comprehensive promotion of various projects and activities needed for the advancement of science and technology." To achieve this aim, close cooperation between scientists and engineers has been emphasized, as is evident in the name of the organization.

Among JUSE's early activities was the formation of the Quality Control Research Group in 1949, which was composed of people from industry, academic institutions, and government. The same year, the Basic Quality Control Course (the first one lasting 12 months, and subsequent ones lasting 6 months) was inaugurated with the aim of reporting the Quality Control Research Group's findings to industry. The course has since been held 89 times through April 1996 and has been attended by 29,741 engineers, who went on to provide the nucleus of quality control activities in their respective companies.

A famous American, W. Edwards Deming, accepted the invitation of JUSE to visit Japan in 1950. He lectured at JUSE's 8-day quality control courses for engineers and quality control seminars for top management held in several large cities in Japan. His lectures at these seminars helped the Japanese participants to understand the importance of statistical quality control (SQC) in manufacturing industries (Deming 1986).

Deming Prizes.   In recognition of Deming's friendship and contributions to Japan, at JUSE's suggestion the Deming Prize was established in 1951 to encourage the development of quality control in Japan. JUSE serves as secretariat to the Deming Prize Committee. The Deming prizes include the Deming Prize itself and the Deming Application Prize. The former is awarded every year to a person whose contribution is judged outstanding in theoretical research work and in the practical application of statistical methods. Those who promote increased use of statistical methods in the industries are also eligible.

The Deming Application Prize is awarded every year to the companies (including public institutions) or divisions that have achieved the most distinctive improvement of performance through the implementation of company-wide quality control based on SQC. The Deming Application Prize provides a powerful incentive for Japanese companies to promote and achieve their quality control activities.

Since their inception, the prizes have been awarded to 140 companies and 5 divisions. JUSE's managing director is responsible for the procedures of awarding the Deming prizes.

Development of the Quality Control Concept into Company-Wide Quality Control. Statistical methods were found by Japanese engineers to be very effective for assigning causes of variation in manufacturing processes, clarifying the correlation between manufacturing conditions and product quality, and reducing the work force needed for inspection by introducing sampling inspection techniques, among other benefits. However, during the first decade after their introduction from the United States in the late 1940s, the application of these methods was limited only to the fields of manufacturing and inspection.

Although the application to these processes yielded remarkable results, it became clear that it was not a sufficient condition for achieving the main objective of quality control, that is, customer satisfaction. To achieve this objective, it is of course necessary not only to place more emphasis on the processes that take place before manufacturing (e.g., market surveys, research, planning, development, design, and purchasing), but also to apply the quality control approach to those that take place after inspection (e.g., packaging, storage, transportation, distribution, sales, and after-sales service).

For example, it is well known that design flaws often occupy the top position in Pareto diagrams of complaints relating to household electrical appliances. Eliminating such shortcomings, elucidating their causes and taking steps to prevent them from recurring in the design of new products are important actions not only for eliminating customer dissatisfaction, but also for improving the health and character of the company itself. More and more people began to recognize the significance of this in the mid-1950s, as trade liberalization became imminent. The importance of company-wide quality control (CWQC) was emphasized and began to be understood by manufacturers. For Japan, with its scarcity of natural resources and need to pay for them by trading in highly competitive international markets, improving product quality to levels acceptable for export was indispensable.

In 1954, Joseph M. Juran visited Japan at JUSE's invitation to hold quality control courses for top and middle managers. These courses had an immeasurably large impact on Japanese quality control in the sense that they extended the quality control philosophy to almost every area of corporate activity and clearly positioned quality control as a management tool. Taking its lead from these courses, JUSE initiated the Middle-Management Quality Control Course in 1955 and the Special Quality Control Course for Executives in 1957. These courses are improved and still held today (Ishikawa and Kondo 1969, Mizuno and Kume 1978).

Teijin Co. (a synthetic fiber manufacturer) and Sumitomo Electric Industries Co. won the Deming Application Prize in 1961 and 1962, respectively. In these companies, the quality control activities were defined broadly to include marketing, design, manufacturing, inspection, sales, and administration departments and subsidiaries, and the companies achieved outstanding—even epochal—results, for which the prizes were awarded. Their success stimulated other Japanese companies, providing them with a very powerful incentive to broaden their quality control activities.

The internal quality control audit by top management, started in many companies in the latter part of the 1950s, was also found to be very effective for promoting and improving CWQC activities. The type of CWQC practiced in Japan has two principal features. The first is the wide span of coverage of the quality control activities practiced, and the second principal feature of CWQC is total employee participation in quality control activities and ancillary activities. They are explained in more detail later.

PDCA Cycle. A usual definition of control is checking and directing action. This means comparing the actual results of an action with a standard or target, monitoring the disparity between the two, and adopting corrective measures if that disparity becomes abnormally large. This process is the familiar plan-do-check-act (PDCA) cycle shown in Figure 41.1a.

Following this PDCA cycle is more effective than adopting the perfectionist approach of concentrating exclusively on developing flawless plans. In addition to the factors which can be controlled accurately, usually there are many other extraneous factors likely to influence the results, and it is almost impossible to establish standards over all such factors. For this reason, even though the plan may be nearly perfect, we still need to keep on checking and taking corrective actions in this way.

The Crosby-style exhortation to "do things right the first time" is criticized (Tsuda 1990), for if this were all that was necessary to obtain good quality, we would all have an easy time of it. But it seldom is possible to judge immediately whether things have, in fact, been done right. The PDCA

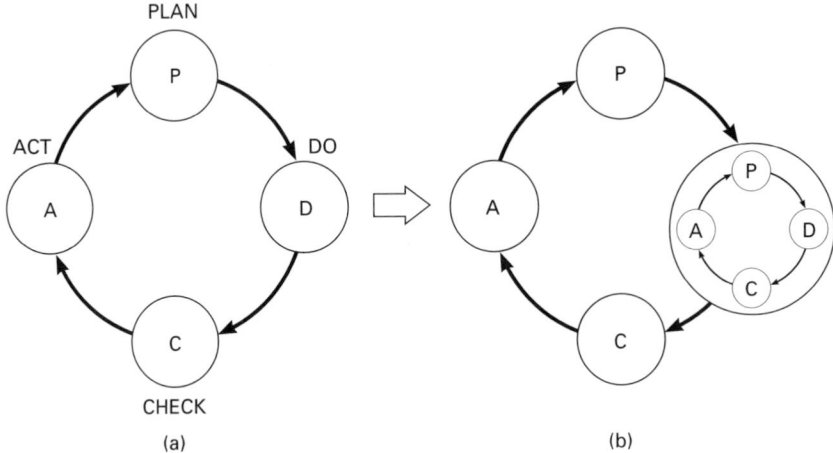

FIGURE 41.1   PDCA cycle. (*Source: Kondo 1977.*)

cycle means continually looking for a better method of doing things. By following this PDCA cycle, it is expected that the results will be obtained, and also that the process itself will be improved in an upward spiral. This leads to improvement and strengthening of the company structure.

In some forms of manufacturing planning, the quality standard and operations manual are established by the engineering staff and management, and the workers are requested only to perform their jobs in accordance with the established manual. Thus the planning and execution are separated.

In such cases, if manufactured products are found to be defective, the supervisor may seek the causes and reproach the worker. The worker may then reply, "I am not responsible for the defect. I honestly followed the operation manual that you gave me. You are responsible for the result." It is clear that when workers are responsible only for following the established manual, their responsibility for quality becomes obscure. Such vague responsibility is detrimental to high quality of conformance, which is achieved only if the workers are conscious of quality and have a keen sense of responsibility.

It is true that the workers are assigned to perform the manufacturing job. However, this job performance is also composed of a plan-do-check-act cycle, as shown in Figure 41.1*b*. The extent to which the PDCA cycle is followed in this portion of the overall job is considered to reflect the self-control of workers. Everyone is in a state of self-control to a greater or lesser extent. Thanks to this ability of self-control, we humans can enjoy our lives, including sports and leisure. Of course, education and training are to a certain extent necessary to cultivate the self-control capacity of the workers.

Regarding checking, we tend to stress identifying the deficiencies in a process. We advocate quantifying deficiencies relating to quality, quantity, and cost, and expressing them in the form of hard data, since this is the first step in improving the process. This is correct and should be strongly emphasized. However, it should be underlined that as long as processes have deficiencies they are also bound to have their opposite, that is, strengths. While it is important to identify deficiencies and prevent them from recurring by eliminating their causes, it is equally important to identify strengths and ensure their recurrence by standardizing their causes.

There are two types of corrective action—temporary and permanent. The former is aimed at results, while the latter is targeted at processes. Because temporary corrective action consists of adjusting or reworking the results of a process, it can be implemented without knowing why the abnormality occurred in the first place. Permanent corrective action, on the other hand, consists of investigating the abnormality, identifying the reasons for its occurrence, and either preventing or institutionalizing its recurrence by eliminating or standardizing its causes. This makes it essential to understand the causes. Permanent corrective action thus focuses on the process rather than on the results. The reason companies introduce and promote quality control is said to be the desire to improve company health. In order for this to happen, it is not good enough for the company simply to produce acceptable outputs; the internal processes giving rise to those outputs must also pass muster. The permanent corrective actions we undertake every day in our workplaces may be individually insignificant, but together these small improvements can result in major improvements to the health of an entire company. In light of this, permanent corrective action is obviously far more important than temporary corrective action.

Education and Training of the First-Line Workers and the Birth of the Quality Control Circle.   The importance of the role of first-line workers has been recognized along with the progress of company-wide quality control. Without the daily efforts of those workers, the quality of conformance of manufactured products cannot be achieved. To offer training to the workers, a 13-week shortwave radio series, "Quality Control for the First-Line Supervisors," was planned and broadcast from October to December 1956, and it was continued by NHK (the Japan Broadcasting Corp.) until 1962. During the first year about 100,000 transcripts of the radio broadcast text were sold. In 1959 a weekly television series on quality control was initiated. *Quality Control Text for Foremen,* edited by K. Ishikawa, was published by JUSE in 1960, and 200,000 copies were sold before the end of 1967. Thus the education and training of supervisors and first-line workers were carried out very enthusiastically. This provided the most important basis for the birth of the quality control circle movement.

In 1962, JUSE began publishing the monthly magazine *Gemba-to-QC* (*Quality Control for Foremen*) as a sister magazine of *Hinshitsu Kanri* (*Statistical Quality Control*), which began publication in 1950. This new publication aimed at

1. Education and training of supervisors and workers and dissemination of statistical methods among them
2. Formation of quality control circles
3. Application of the workers' knowledge to their own daily jobs, attainment of the set target, and elevation of their own capability

A quality control circle consists of a group of workers and a foreman who voluntarily meet to solve job-oriented quality problems. These activities are intended to be tightly linked with company-wide quality control activities. The quality control circle activities are the quality control activities of the first-line workers on the shop floor, who are responsible for attainment of the design quality of manufactured products. Quality control circle members study activities on the shop floor, using the *Hinshitsu Kanri* magazine as a textbook, and they thus become the core of the activities.

In May 1962, the first quality control circle was registered at the Quality Control Circle Headquarters of JUSE in Tokyo, and thereafter the number of quality control circles and the number of members have been increasing year after year. The number of registered quality control circles at the end of May 1996 was 397,216; registered members numbered 3,038,038.

Recognition of the daily efforts of quality control circle members is important, and JUSE undertook to organize quality control circle conferences for presentation of the members' case studies. The first Foremen's Quality Control Conference was held in Tokyo in November 1962, which was the third nationwide Quality Month; 12 papers were presented and 235 foremen attended. In parallel with this annual conference, the first nationwide Quality Control Circle Conference was held in May 1963, where 19 papers were read and 193 people attended. These conferences and local quality control circle meetings have continued with great success. Mutual visits and discussions among quality control circle members from different companies were also revealed to be very effective in motivating the workers and broadening their viewpoints. These visits were started in March 1963, and JUSE assists in arranging them when its services are requested.

According to the Quality Control Circle *Koryo* (*General Principles of the Quality Control Circles*), published by JUSE in 1980 and revised in 1990 and 1995, the major purposes of the quality control circle movement are

1. To improve the leadership and management abilities of the foremen and first-line supervisors in the workshop and to encourage improvement by self-development
2. To increase employee morale and simultaneously create an environment in which everyone is more conscious of quality problems and the need for improvement
3. To function as a nucleus for company-wide quality control (CWQC) at the workshop level

The basic goals behind these immediate purposes are

1. To display human capabilities fully and eventually to draw out unlimited possibilities
2. To develop respect for humanity and build a happy, bright workshop, which is meaningful to work in
3. To contribute to the improvement and development for the enterprise

As recommended by Juran, JUSE has been sending foremen's quality control teams composed of quality control circle leaders and members abroad since 1968. They visit various plants in foreign countries and present papers on their own case studies in the workshop at various conferences and meetings. The Quality Control Circle Cruising Seminar was also started in 1971; the members spend two weeks on a boat trip and attend seminars and discussion meetings held on the boat. They visit various countries in Southeast Asia during the trip.

Regarding the major effects of quality control circle activities, it can be said in the first place that the willingness, creativity, and viewpoints of first-line workers are enhanced or broadened, which results in the formation of centers of quality control activities in the manufacturing process. The elevation of morale and the improvement of human relations among workers are clear. The reduction of manufacturing defects is always evident, and this brings out the elevated level of quality assurance. Thus the engineers and staff personnel can, without any worry, entrust to the foremen and workers the greatest part of their daily duties as troubleshooters in the manufacturing line, and they can concentrate their efforts on their own proper duties (development of new products and techniques, etc.)

The recent quality control circle activities are further marked by the following features (Kondo 1976):

1. *Division of a quality control circle into subcircles and minicircles:*   As study among workers progresses and their capabilities are enhanced, the quality control circles are often divided into smaller circles. Some workers become the leaders of these smaller circles, to which the foremen, who were the former leaders of the larger quality control circles, serve as advisors and promoters.

2. *Formation of joint quality control circles:*   Joint quality control circles are also often organized by combining quality control circles along manufacturing lines or quality control circles of manufacturing and inspection, etc. These joint quality control circles are effective in finding and solving new quality problems.

3. *Service of workers as quality control circle leaders:*   Formerly, it was the custom for the foremen to be elected as the quality control circle leader. However, as the quality control circle members progressed in their studies, many of them assumed prominent leadership roles in quality control circles. In many Japanese companies it became customary to rotate the leadership among the quality control circle members.

4. *Establishment of autonomous administrative systems in quality control circle activities:* Conferences and meetings of quality control circles are always conducted by the attending members who were assigned as the session moderators. The company staff members and university professors serve only as the advisors.

5. *Expansion of quality control circle themes:*   The themes being chosen include not only the reduction of defects, elevation of productivity, and reduction of manufacturing and inspection costs, but also improvements in preventive maintenance jobs, the manufacturing schedule, and other aspects of production.

6. *Improvement of techniques employed:*   In quality control circle activities the elementary statistical tools are called the "seven tools." These are: stratification of data, Pareto diagram, checksheet, histogram, cause-and-effect diagram, graphs such as control charts, and scatter diagram. In addition, techniques such as regression analysis, process capability studies, analysis of variance, and value analysis are being used.

7. *Quality control circle activities in supporting operations:*   Quality control circles are being formed, and their activities are becoming broader, not only in manufacturing and inspection workshops but also in supporting services such as warehousing, transportation, purchasing,

administration, reception, and telephone operation. The above-mentioned statistical methods are helpful and effective in quality control circles in these areas.

8. *Quality control circle activities in subsidiary companies:*   Quality control circle activities in subsidiaries that make parts and components or carry out an intermediate manufacturing process are recognized to be very important, in fact indispensable, for ensuring product quality, and parent companies are encouraging the formation of quality control circles in their subsidiaries. Mutual visits and discussion among quality control circle members from the parent company and from its subsidiaries are always effective in identifying and solving quality problems.

Along with the expansion of company-wide quality control activities in the service industries in Japan, quality control circles are becoming active and effective in hospitals and in companies that operate hotels, restaurants, banks, department stores, supermarkets, retail stores, etc.

Quality Revolution in Japanese Industry—Breakthrough.   At the end of World War II, the former Japanese military and political leaders were no longer in power, having been replaced in large part by relatively young industrialists who wanted Japan to advance as an industrialized country and not to fall back into the old agricultural economy of the type prevalent in some parts of developing countries. After this decision was made, however, they faced a difficult road. Poor product quality was a principal obstacle; no one wanted to repeatedly buy such low-quality goods. For a country so lacking in raw materials, the inability to sell finished goods for export also meant an inability to earn foreign currency and hence an inability to buy the materials needed to create an upward spiral of industrial development. Thus a revolution in product quality became essential (Juran 1981). This quality revolution has been taking place in Japan since the early 1950s as the result of efforts to apply the concepts and techniques of statistical quality control on a company-wide scale. Three of the principal reasons often cited as to why this has been relatively easy to accomplish in Japan are as follows.

1. Japan is a uniform society composed of people of the same race who speak the same language.
2. Japan has a comprehensive system of compulsory education, and the general education level is high.
3. Japanese companies practice the custom of lifetime employment and the turnover rate of employees is therefore low.

Of these three, the second is likely to continue in the future and education levels may well improve even further. However, labor shortages are becoming more and more severe and it is impossible to say whether the first will continue to hold, although this will depend on how the labor shortages are dealt with. Even the third may not apply so strongly in the future as more young people become reluctant to take up employment in industry and those who do so tend to change jobs more frequently. It is important to keep a watchful eye on these trends. Various other reasons are given in addition to the above three.

In Japan, apart from a limited number of occupations such as medicine and the law, professionalism is not so strongly entrenched as it is in the Western countries. Each of the unique features of these countries has its own particular advantages and disadvantages, but their existence gives rise to some noteworthy differences.

For example, in the United States, people improve their status by moving from company to company without changing their specialty to any great extent. In contrast, in Japan, although there may be exceptions, people typically get promoted by changing jobs within the same company. This process of promotion makes it necessary for them to learn a series of different jobs, so in one way it may be rather hard on them. However, the richer experience within the company and the deeper understanding they gain of the relationships among different departments are great advantages. The fact that the custom of lifetime employment continues in Japan even today, whether or not it will in the future, may be because this in-company promotion process gives people the opportunity to experience a wide range of different jobs and helps them remain interested in their work over a long period of time.

The quality revolution was indispensable for the survival of Japanese industries. However, the following "trade-off argument" is often commonplace:

Improving quality is a good thing, but it raises costs, so it needs to be done judiciously.

There is an optimum to quality of conformance with regard to the manufacturing cost. As shown by the solid line in Figure 41.2, increased conformance reduces the losses incurred by defects, but the cost of quality improvement needed for greater conformance rises sharply as quality approaches the perfect state. Thus the optimum should always fall short of perfection because the total cost soars as the percent defective approaches zero.

However, the above optimum is doubtful. Both basic manufacturing cost and losses incurred by defects are easily defined, but the cost of quality improvement is usually indefinable. When creative ideas by which we can increase the conformance with less additional cost are introduced, the curve of quality improvement cost is shifted down as shown with broken lines in the figure. Then the total cost can be reduced, and the optimum moves toward zero defect. Thus the optimum is indefinable and movable. We must really search for the ways and means by which we can reduce the defects with minimum cost, instead of trying to find the indefinable optimum. It may be said that this approach is a "breakthrough," which is quite different from the superficial optimization. It is clear that the successful breakthrough is always accompanied by the creative idea and strong will of the people concerned (Kondo 1977).

Dividing quality into two categories of must-be quality and attractive quality is an important step to attain customer satisfaction (Kano et al. 1984). As described later, the former quality can be related to the reduction of cost and the latter concerns the enlargement of market size and share and the increase of profit. Thorough investigation of both qualities is indispensable for the quality revolution.

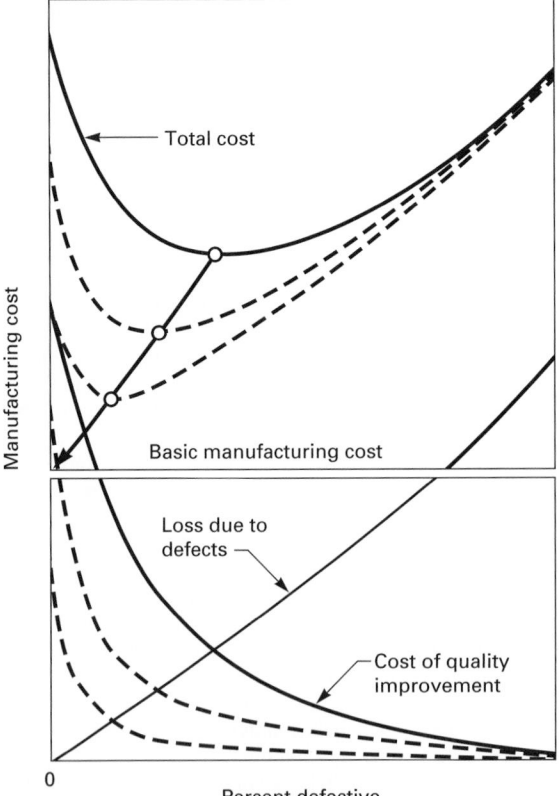

FIGURE 41.2   Plausible optimum of manufacturing cost. (*Source: Kondo 1977.*)

Juran (1981) summarized three features of quality control activities in Japanese industries that created the revolution in quality:

1. A massive quality-related training program
2. Annual programs of quality improvement
3. Upper management leadership of the quality function

Company-Wide Quality Control after the Energy Crises.

Although there were variations among countries and industries, the world economy in general showed significant growth from the latter half of the 1960s to the early 1970s. This rapid growth was checked by the first and second energy crises, in 1973 and 1979, respectively. Japan is particularly vulnerable economically, since it depends on other countries for its supplies of crude oil and other natural resources. During the energy crises, there were even attempts in some quarters to corner the market in goods such as detergents and toilet paper, but the Japanese government and individual companies took various steps to deal with the situation.

The energy-saving measures introduced during that time can be broadly divided into energy saving in manufacturing processes and the like, and the creation of energy-saving products. In addition, companies had to institute a wide range of countermeasures on an organizationwide scale. These measures included providing education and training for the surplus personnel they now had in their production and engineering departments (as a result of the drop in manufacturing volumes) and assigning them to sales departments to provide extra support. Various benefits accrued from giving a higher priority to these kinds of countermeasures. For example, engineering staffs dispatched to sales departments were able to obtain a better understanding of the market, and the education and training programs carried out by companies enabled them to speed up production extremely rapidly when the economic good times returned.

The Japanese economy was subsequently rocked by the rapid appreciation in the value of the yen. However, many Japanese companies were able to emerge stronger each time they came through one of these difficult experiences. In this sense, the energy crises in the 1970s could be regarded as heralding the new dawn of the 1980s and 1990s.

According to Juran (1981), by the mid-1970s the Japanese had caught up with their Western counterparts in terms of creating quality products. The successful efforts made by Japanese industry to cope with the energy crises and to create quality products could probably be considered early indicators of the industrial restructuring that was to take place in Japan beginning in the early 1980s. In addition, these efforts and the restructurings are evidence of the realization by the Japanese that they were in the process of surpassing other countries and becoming a world leader in terms of product quality, and point to Japan's new sense of mission and responsibility.

Since the early 1980s, diversification of business has been a common feature of Japanese companies. However, expansion into a new area of business is accomplished in a distinctly Japanese way. While a company may be extremely enthusiastic about the move into a new line of business, and while this move may mean reducing involvement in an old, mature business, it is only in exceptional circumstances that the company abandons the old business completely. This is because Japanese companies tend to feel that they ought to continue with a business as long as some customers still require them to, even if the business has matured and demand has shrunk. Many companies even continue to actively develop new products in these mature fields with considerable success.

A notable difference between Japan and Western attitudes toward companies is that in Japan, even privately owned firms are regarded more as public institutions than as their owners' possessions. Japanese executives appear to feel more strongly that they are stewarding their companies for the benefit of society.

An interesting feature of new product development in Japanese companies is the active development of applications, particularly nonmilitary ones, for new products. Compared with new products for military use, products for the civilian sector are not subject to as rigorous usage conditions and generally have lower quality and reliability requirements. However, the demand for products in the civilian sector is far greater than that for the military, making it easier for industries to become established in this sector. There are also more opportunities for producing improved versions of products, and it is easy to obtain quality information under a variety of usage conditions. The development of carbon fibers for use in fishing rods and tennis rackets is a good example.

Because of intense competition in global and domestic markets, a company's ability to develop new products and technology became an important and indispensable factor determining its competitive advantage. It became extremely difficult for a company without this kind of ability to ensure its continued growth through diversification and entry into new business fields. In the hope of making some important gains, more and more Japanese companies are establishing research and development facilities in Japan and overseas.

Japan's direct overseas investment began to increase in the latter half of the 1960s, and the rate of increase has been particularly high since the late 1980s. This overseas activity builds on a long and successful experience of joint ventures between Japanese and foreign corporations going back to the 1970s. It is interesting to see that there are many examples in which the excellent quality management activities of a Japanese subsidiary have had beneficial effects on their overseas parent companies and have contributed greatly to the promotion of quality management in those companies. In fact, the joint ventures of Aisin Warner, Fuji Xerox, Yokogawa-Hewlett-Packard, and the Bipolar Department of Texas Instruments, Japan, have won the Deming Application Prize, and they have accelerated the progress of quality management in their parent companies.

After being invited in 1961 by the government of Thailand to manufacture dry cells there, Matsushita Electric Industries faced up to and overcame many difficulties in making the business a success (Tsutsumi 1988, 1991). The company used this valuable experience to formulate the following basic policy regarding its overseas operations:

1. We will conduct the sort of business operations that will be welcomed by our host country.

2. We will execute our business activities in accordance with the policies of the host nation's government. At the same time, we will make constant daily efforts to ensure that the host country's government fully understands Matsushita's philosophy.

3. We will promote thorough localization of people, materials, finance, and know-how in the local community in the spirit of responsible self-management.

For this purpose, Matsushita intends to foster a positive participative outlook in a desire for improvement on the part of local employees. It ensures that all employees enjoy their daily work and find it worthwhile and engenders a sense of responsibility toward their manufacturing activities. Matsushita encourages them to see that they are contributing to their own country's development through those activities. These basic policies are backed up by specific strategies actively promoted in each of Matsushita's overseas companies.

In this way, more and more Japanese companies are setting up manufacturing operations abroad, and more and more of the components and materials used in Japan are being purchased overseas. How to use the CWQC expertise effectively in this kind of global industry, particularly in Japanese overseas manufacturing operations, is an important subject that will probably require continued investigation.

Deming Application Prize for Overseas Companies.   Originally the Deming Application Prize was for Japanese companies only. However, companies from countries outside Japan expressed a strong interest in it. In response, the Deming Prize Committee drafted new regulations for the operation of the Deming Application Prize in 1984, making it possible for companies from countries outside Japan to apply. This opened the door for overseas companies passing the examination to receive the Deming Application Prize for Overseas Companies. The first company to win this prize was the American electricity utility Florida Power and Light Company, in 1989. The next was Philips Taiwan, in 1991, and the third was AT&T Power Systems in the United States, in 1994. These successes, which indicated Japanese management systems can be applied to any kind of industry anywhere in the world, attracted worldwide interest.

COMPANY-WIDE QUALITY CONTROL IN JAPAN

Two Basic Features of Japanese Company-Wide Quality Control.   Japanese quality control activities have been gradually broadened from the narrow fields of manufacturing and inspection to almost all company branches. As mentioned earlier, it became widely known that achievement of "fitness for use and environment" is important to ensure product quality and to secure customer satisfaction and that it is realized by improving not only the quality of conformance but also the quality of design.

The Deming Prize Committee for example, defined the company-wide quality control as follows:

> It is the activity of economically designing, producing, and supplying products and services of the quality demanded by customers, based on customer-focused principles and with full consideration of the public welfare. It achieves corporate objectives through the efficient repetition of the PDCA cycle of planning, implementation, evaluation, and corrective action, doing so by means of the understanding and application of the statistical approach and statistical methods by all employees to all activities for assuring quality, where such activities include the chain of activities comprising survey, research, development, design, purchasing, production, inspection, and marketing, together with all other related activities both inside and outside the company.

Quality control activities in Japanese industries were expanded in the 1960s (Ishikawa 1965) to:

1. Establishing the top management policy on quality and the long-term quality control plan of the entire company to realize the policy
2. Introducing the quality control concept and techniques into new-product development
3. Establishing the quality assurance system, which covers the whole company
4. Conducting quality control audits
5. Expanding quality control activities to include the sales and marketing activities of the agents, the trading firms, the stores and shops, etc.

A second feature of Japanese quality control is the willingness of employees to participate in the quality control activities of the company. For example, the quality control circle movement, discussed before, came from this idea. In Japanese companies quality control activities are not restricted to quality control staff but include all personnel of the company, from the president to the factory workers and the sales persons. Among them, the leadership of top management is indispensable for launching and continuing the activities. Thus the quality control activities in Japanese companies, which we call "company-wide quality control," is a movement that involves the entire company.

Company-Wide Quality Control Education and Training.   Study of and education in modern quality control were started in Japan in 1949. It was a prevailing thought then among the members of the Quality Control Research Group in JUSE that a Japanese model of quality control should be established because of differences in background and cultural pattern between Japan and Western countries (Ishikawa 1972). For example, company-wide quality control, in which all company employees participate, is a specifically Japanese approach. The concept of professionalism, which is rather widespread in Western countries, had not yet been established in Japan as of 1949.

Introduction and promotion of company-wide quality control led to a revolution in management philosophy, which required lengthy, persevering efforts in education and training. Thus, since the early 1950s education and training in quality control have been continued for everyone from top management to first-line workers in each and every department, including research and development, designing, manufacturing, inspection, purchasing, marketing, sales, and administration. As an example, training courses that have been held by JUSE are summarized in Table 41.1. Currently there are more than 60 quality control training courses of different kinds, which are held regularly by

TABLE 41.1   Education and Training Courses of JUSE

Title	Frequency per year	Year established
Quality control [QC]		
QC Top Management Course (5 days)	3	1957
QC Executive Course (5 days)	5	1962
QC Introductory Course for Executives and Management (3 days)	5	1981
QC Middle Management Course (12 days)	7	1955
QC Basic Course for Assistant to Section Chief (6 days)	4	1992
QC Basic Course (30 days)	5	1949
QC Introductory Course (8 days)	8	1957
QC Introductory Course for Quality Function Deployment (4 days)	5	1989
QC Basic Course for Foremen (6 days)	16	1967
QC Basic Course for Group Leaders (4 days)	12	1974
TQC Instructor Course (6 days)	3	1976
QC Course for Purchasing Department (10 days)	1	1971
QC Introductory Course for Purchasing Department (4 days)	1	1983
QC Course for Sales Department (10 days)	1	1968
QC Introductory Course for Sales Department (4 days)	2	1983
QC Middle Management Course for Seven Management Tools (4 days)	2	1993
Introductory Course for Seven Management Tools for QC (3 days)	15	1984
Introductory Course for Seven Management Tools for QC in Sales Department (4 days)	2	1989
Policy Management Seminar for TQC (3 days)	4	1989
Product Planning Seven Tools Seminar (4 days)	1	1995
QC Course for GNP (Pharmaceutical) (3 days)	1	1977
QC Introductory Course for GNP (Pharmaceutical) (2 days)	1	1977
QC circle [QCC]		
QC Circle Executive Course (1 day)	1	1994
QC Circle Middle Management Course (2 days)	11	1980
QC Circle Instructor Course (6 days)	11	1972
QC Circle Leader Course (3 days)	42	1977
QC Circle Leader Course for Service-Sales Industries (3 days)	2	1990
QC Circle Cruising Seminar (13 days)	1	1971
Reliability [RE]		
RE Management Course (3 days)	1	1966
RE Course (15 days)	1	1960
Basic Course (4 days)	5	1960
RE Six-Day Course (6 days)	1	1980
RE Seminar on Electronics and Machinery Systems (3 days)	1	1995
RE Course on FMEA-FTA (2 days)	13	1976
RE Course on Design Review (2 or 3 days)	12	1977
RE Course on Checklists (3 days)	1	1989
RE Course on Test (3 days)	2	1983
RE Course on Failure Analysis (3 days)	2	1985
RE Course on Computer Aided Reliability Engineering (3 days)	1	1991
Design of experiment [DE]		
DE Tokyo Course (12 days)	1	1955
DE Osaka Course (20 days)	1	1962
DE Introductory Course (8 days)	6	1963

TABLE 41.1 Education and Training Courses of JUSE (*Continued*)

Title	Frequency per year	Year established
Multivariate analysis [MA]		
MA Seminar (7 days)	1	1984
MA Advanced Course (4 days)		1970
MA Basic Course (4 days)		1984
Operations research [OR]		
Corporate Strategy Managers Course (6 days)		1962
OR Introductory Course (5 days)		1987
Industrial engineering [IE]		
IE Seminar (16 days)		1963
IE Basic Course for Foreman (6 days)		1971
Marketing research [MR]		
MR Seminar (16 days)		1963
Software production control [SPC]		
SPC Course for Managers (6 days)	1	1988
SPC Course for Engineers (8 days)	2	1980
SPC Course on Design Review (3 days)	1	1994
Sensory inspection [SI]		
Sensory Inspection Seminar (11 days)	1	1957
Introductory Courses for Sensory Inspection Seminar (3 days)	1	1995
Product liability [PL]		
PL Prevention Introductory Course (3 days)	5	1973
Product Safety Advanced Course for Engineers (2 days)	2	1994
Product Safety Advanced Course for Promoters (3 days)	2	1994
Other management techniques		
Statistical Application Seminar for Clinical Test (7 days)	1	1972
New Finite Element Method Introductory Seminar (3 days)	1	1987
Finite Element Method Seminar for Fluid Mechanics (3 days)	1	1977
Cost Reduction Seminar (6 days)	2	1981
VE Basic Course for Foremen (5 days)	1	1984
Analytic Hierarchy Process Seminar (2 days)	1	1992
Logistics System Design Seminar (3 days)	1	1993
Other management techniques		
Data Envelopment Analysis Seminar (2 days)	1	1994
Biostatistical Application Seminar for Pharmaceutical Data (24 days)	1	1989
ISO 9000 (JIS Z 9900)		
JAB Accredited ISO 9000 Assessor Training Course (5 days)	5	1995
ISO 9000s Internal Auditor Training Course (4 days)	10	1994
ISO 9000s Promoter Course (2 days)	4	1994
ISO 9000s Introductory Course (1 day)	3	1994

Source: Ishikawa 1969.

nonprofit organizations such as JUSE, Japanese Standards Association (JSA), and the Association for Overseas Technical Scholarship (AOTS).

Furthermore, many Japanese companies are enthusiastic about quality control education for the employees of both the parent company and the subsidiaries. Many companies have their own education and training programs. Education and training in company-wide quality control usually start with top management and are then extended to middle management, supervisors, and first-line workers. It is often emphasized that the progress of company-wide quality control exactly reflects the leadership of the company's top management.

On-the-job training is also emphasized. One example is the training that results from internal quality control audits by top management, which are effective in obtaining the facts within the company and lead to appropriate corrective actions. Another example is the on-the-job training of engineers. Concerning the training of engineers who design color television sets, Juran (1978) commented as follows:

> A second aspect of training of designers is "practical experience." One major area for such experience is in the production shop, to give the designers an awareness of some of the realities faced by the production personnel and thereby to give them a better understanding of how to design for "producibility." A second major area for designer experience acquisition is in field service work. Through such experience, the designers learn much about the conditions of use, the problems of diagnosing field failures, the difficulties of making repairs, etc. As a result, they understand better how to design for reliability and maintainability.
>
> With respect to such "training by practical experience," there are wide differences between Japan and the West. It is common, though not invariable, for Japanese companies to require that designers acquire such shop and field experiences before being assigned to key responsibilities in product design. In the West the requirement for such experience is unusual.

For the above-mentioned purposes of broadening the viewpoint of designers and engineers, job rotation of employees is also emphasized in Japanese companies. The lack of established professionalism in Japanese society is thought favorable for this rotation.

Policy Management ("Hoisin Kanri") in Japanese Companies.   Quality control activity in Japanese industries is company-wide, and top management personnel are in the position of leading and promoting the quality control activities of their companies. They are responsible for deciding top policy on quality of manufactured products and service and for establishing the long-term plan of company-wide quality control in order to realize that policy. In addition, they evaluate whether the policy and the plan are being realized on schedule and whether any corrective actions need to be taken by top management. These activities are a form of company-wide PDCA cycle and are called "policy management" ("hoisin kanri") in Japanese industries. This concept of policy management is followed in many but not all Japanese companies. Internal quality control audits by top management, which will be described later, are an effective way to evaluate the results as a basis for appropriate corrective actions.

Recently, many Japanese companies have undertaken to investigate and decide at the start of every fiscal year what their long-term plans and targets will be for the coming 3 to 5 years, the plan and target for the first year being set to coincide with the present fiscal year. In this way, long-term plans are taken into consideration in establishing each annual or semiannual plan and target.

A company's basic business philosophy is of fundamental importance, since it underpins the enterprise's annual and long-term policies and provides its employees with a standard by which to measure their behavior. In the future, it will probably become of even greater importance for every company to work out a philosophy that can be accepted and bought into by all its employees, is regarded as an attractive feature by its customers, and forms the basis of goals shared by its entire work force.

The company's basic philosophy has a close bearing on the quality created by the company. Quality has had a far longer history in the life of humankind than either cost of productivity, and is the only one of the three that is a common concern of both company and customers (Kondo 1988). It is for reasons such as these that quality is regarded as a more human concept than cost or productivity. From now on, while quality may not be everything, it will almost certainly be an essential attribute that can attract customers and act as the focus of a shared commitment on the part of all employees. As described before, quality improvements effected by creative methods result in lower cost and higher productivity and the development of new products with attractive qualities that meet customers' true needs, expand markets, and increase corporate profitability.

It is extremely important for the upper managers to tell everyone in the company their basic philosophy regarding the quality of the products and services their company offers, and the quality management activities the company is undertaking to ensure that quality. This quality policy forms the guidelines for establishing specific quality targets for new products and quality improvement plans for existing products in the course of policy management.

It is widely understood that planning should be results-oriented rather than procedure-oriented. For example, a plan to extend standardization might at the end of the year have produced 100 new procedure manuals. If all this effort failed to improve operating results, this would be evidence that the plan had been procedure-oriented. At the investigation and discussion stage of the draft annual plan, the persons concerned are encouraged to offer many alternative proposals. It is of great importance at this stage to discuss thoroughly the true aim of the proposals and to clarify the "resultant present problem." The above-mentioned extension of standardization is merely a procedure. The high percentage of defects and rework, low productivity and yield, etc., are the resultant present problems. After the resultant problems become clear, data are collected and analyzed. The "vital few" problems can be further determined with a Pareto diagram. Two-stage Pareto analysis always makes the problems and the corrective actions clearer. Once a procedure-oriented policy such as that described above has been converted to a results-oriented policy, the resultant policy consists of the following three items:

1. *Aims:*   The reasons why the policy should be implemented
2. *Goals:*   The direction to move in, how far to go, and the deadline
3. *Priority procedure:*   The methods to be used

Goals should be inspiring; if not, people do not give them serious attention, and they are not achieved. Two Japanese examples are given here.

At the Car Radio Division of Matsushita Electric Industries Co., engineers and staff personnel up to the division manager were discussing an unattained goal of 10 percent price reduction for a car radio as requested by an automobile company. When former President Konosuke Matsushita visited the division, he was told of the situation. He said, "You should consider 15 percent cost reduction when you are requested by a customer to reduce the price by 10 percent." Because he was the founder of the company and was esteemed as a "godfather" by all employees, they started to investigate cost reduction possibilities more thoroughly and finally achieved a 13 percent reduction. After hearing of this success, Matsushita visited the automobile company and expressed his appreciation by saying, "Thanks to your request for a 10 percent price reduction, we have succeeded in achieving a 13 percent cost reduction. Thank you very much for that."

At the Nankai Plant of Bando Chemical Co. near Osaka, where V-shaped belts are manufactured, it was previously the custom for the plant manager to announce the monthly production target and to urge the work force to achieve this target. However, the target was never achieved even though the employees exerted all possible efforts, especially toward the end of every month. On the advice of an outside consultant, the plant management changed the procedure for determining the monthly target by having the plant manager prepare a draft target and ask the work force in every workshop to thoroughly investigate the possibility and the ways and means of attaining the target. In the early stage of this revised procedure, the sum of the individual targets proposed by the work force was more demanding than the draft target indicated by the manager. The plant manager, however, established the proposals of the work force as the targets of the month because they were the result of their thorough investigations. Interestingly, the work force always achieved their targets. Moreover, a few months later the overall target proposed by the work force started to rise month after month, and in half a year or so it even exceeded the previously unmet target of the manager.

There are two ways of determining the target, from the top down and from the bottom up. The example of Matsushita is top-down, and the case of Bando is bottom-up. Both of these were very successful. Top management is always concerned about the future of the company, and the top-down target is usually determined from the company's needs. On the other hand, the employees usually investigate the draft target indicated by the top management from the viewpoint of feasibility. If the draft target is not investigated thoroughly by the employees, they easily find their own good excuses when it is not achieved.

Many Japanese companies, during annual or semiannual planning, adopt a combination of top-down and bottom-up approaches. Top management prepares a draft plan, which then is discussed, for example, among the top managers and the division managers. From this discussion the draft plans of each division are made. The draft plan of each division is then further discussed among the divisional and lower managers and finally among the supervisors and quality control circle leaders. After these investigations and discussions, the detailed draft plans are formulated. They are brought to the top management, which decides on the final annual plan of the company. This procedure of deploying the annual or semiannual planning throughout the company is called "playing catch" in Japanese companies. The procedure is believed to be effective in establishing the annual plan, although it is somewhat time-consuming. Through this procedure, what was a norm enforced by top management is revised to become the voluntary target of each employee. This revision is extremely important for motivation.

The establishment of results-oriented plans makes evaluation easier, and the characteristics used in the targets become the basis for review by the respective managers. Corrective actions are cooperative. When it becomes clear that a target will not be achieved, for example, companies often assign additional budget, work force, etc., in order to attain the target within the time limit. Although this may be called a type of corrective action, it is actually a superficial countermeasure, or adjustment. A matter of greater importance is to detect the assignable causes by which failures, defects, rework, delays, etc., are created and to remove them from the process. This is the action of "cause removal," which is essential for improving the basic process.

An example of the breakdown of a broad goal into subgoals is that of Komatsu Ltd., a manufacturer of construction machinery. This company originated the "flag diagram" by skillfully combining the Pareto and Ishikawa diagrams (see Kondo 1977). An example is shown in Figure 41.3.

This diagram shows machining time classified into several major items according to the Pareto principle. Each item is further broken down into the respective secondary items. The target line is drawn in the diagram of each item. The subsequent performance is also plotted in the same diagram and compared with the respective target. Each diagram becomes an item for review by the responsible manager and for appropriate corrective action in the event of significant deviations in performance. Such diagrams make it easier for all employees involved to understand their own situations, the roles of their colleagues, and the interrelationship among them. Such understanding contributes to the common interest. New and worthwhile ideas are easily born from discussions among employees who have common interests and yet can see the problems from different viewpoints.

Internal Quality Control Audit by Top Management.   Internal quality control audit by top management is one of the outstanding features of Japanese company-wide quality control (Kondo 1969). It is carried out in ways similar to the activities of checking and taking corrective actions in the policy management discussed above. The aim of this internal quality control audit differs from that of the external audit—its purpose is not merely for the employees "to pass the examination" but to stimulate mutual discussion between the auditors and the people involved in order to find ways and means to improve the present situation. Corrective actions on both sides are required. Thus the internal audit is educational in character, involving on-the-job training as well as the survey itself. It may be defined as follows: The internal quality control audit is a study of the present situation regarding the system of company-wide quality control and the quality functions of whole processes in order to find and take the necessary corrective actions. This is the reason why many Japanese companies prefer to call the internal quality control audit the "quality control diagnosis" or the "discussion meeting of [those performing] quality control activities with top management."

The procedure of carrying out the internal audit is not fixed but is flexible according to the situation of the division, the department, the plant, or the branch office audited. It is also flexible according to the kind of internal audit being undertaken in the company; usually a company has a few kinds of audit at different levels, as will be mentioned below.

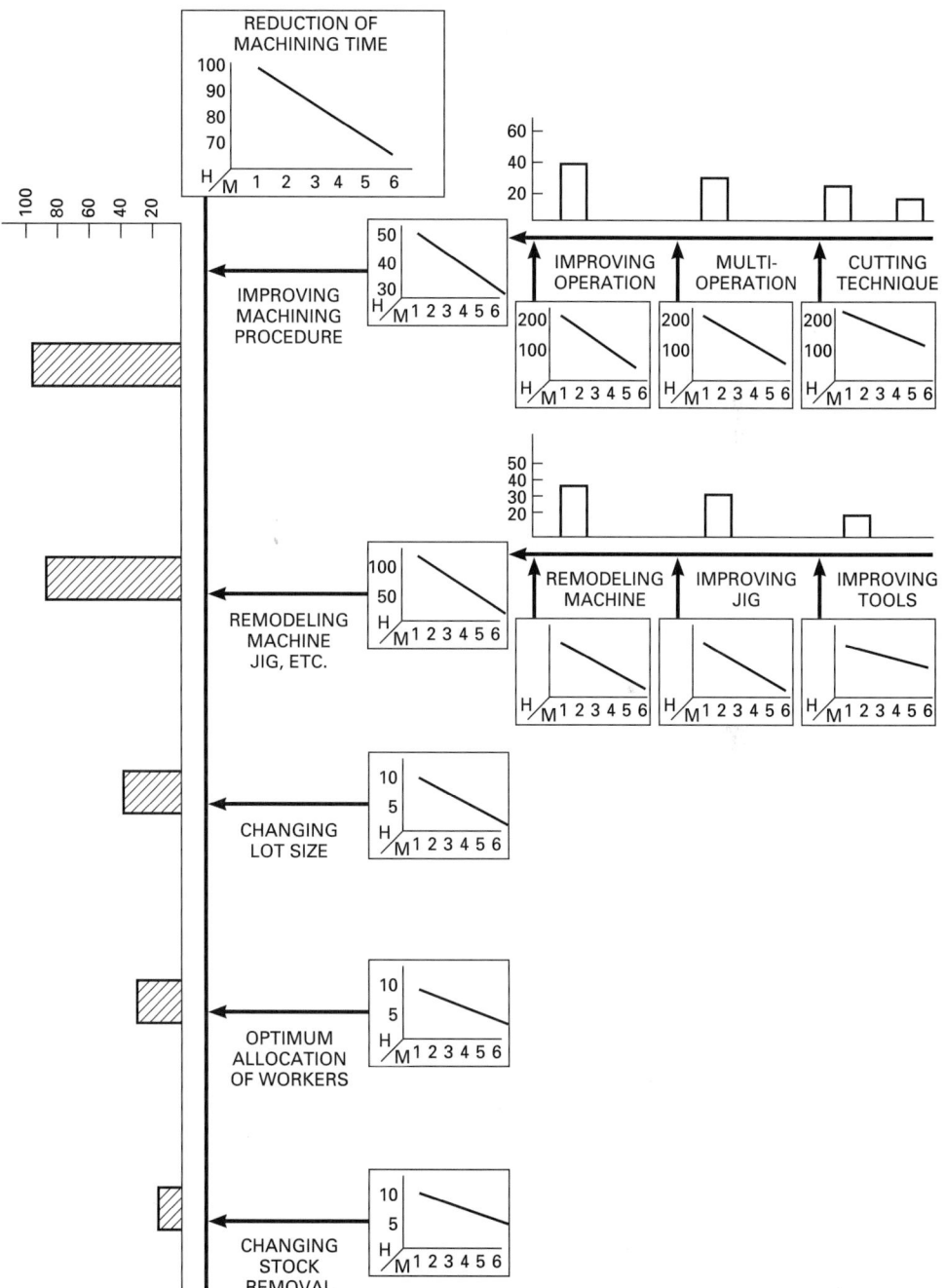

FIGURE 41.3 An example of "flag diagram." (*Source: Kondo 1977.*)

The top managers of a Japanese company are in the position of leading and promoting the company's quality control activities. They are responsible for determining the top policy on the quality of products and services and for establishing the long-term quality control plan in order to realize that policy. The aim of the internal audit is to determine whether the policy and the plan are being realized and attained on schedule and whether any corrective actions by top management are needed.

Mizuno (1967) explained the aim of the audit as follows:

> It is essentially desirable in the quality control function that effective daily checks be made and that the important quality problems in the company be pinpointed in order to take corrective actions. However, it is embarrassing for a company to remain unaware of the problems involved. The aim of the internal audit is to investigate and analyze systematically the hidden causes of those quality problems that cannot be detected by daily checking. Because the audit in any form is an energy-consuming job, it is a must for the auditors and the coordinators to pursue resultant benefits that are worthy of the input energy.

The educational character of the audit is considerable. The audit offers the best chance for top management to grasp systematically those facts that may reflect on themselves. The employees audited are also given opportunities to examine and to rearrange their daily work. Moreover, the internal audit contributes to the improvement of mutual understanding and human relations among the employees. Such an opportunity can hardly be obtained through the daily meetings and reports. For these reasons, it is often effective to announce beforehand the audit theme and the checklist being used in the audit.

The audit is usually carried out in either of the following ways: In some companies the audit is done by the president and by the members of the board of directors separately; in other companies it is done by a top management team that includes the president. In the audit by the directors, the emphasis is usually on the general management of the company. In the audit by the management team, there is usually more emphasis on specifics than in the audit by the directors. The two audits are planned to be correlated with each other. Usually, the auditors are accompanied by quality control staff members and sometimes by a university professor as a third party. The audit is done either with or without a predetermined theme. When the theme of the audit is not determined beforehand, it usually covers a wide range of items.

At Toshiba Corp., for example, it was reported (Sugimoto 1968) that the following items were checked by the president during the plant tour:

1. Putting the shop in order—its cleanliness and working environment
2. Layout of the shop and the machines
3. Flow of materials and the line balance
4. Material handling
5. Operation standards
6. Efficient use of jigs and tools, mechanization, and automation
7. Maintenance of machines and measuring instruments
8. Inventories of on-line stocks, materials, and parts
9. Attitude and motivation of workers
10. Administration of storehouse
11. Reducing the number of slips and chits
12. Content and worker-hours of indirect work
13. Production control
14. Quality assurance
15. Supervision of subcontractors
16. Maintenance of buildings, roads, and incidental facilities

When experienced auditors are not available, an audit covering such a wide range of topics tends to become a loose and formalistic one, which cannot be effective. This is why the audit with a predetermined theme is preferred in many companies; these audits become more intensive. There is also a prevailing tendency for the audit to develop from a departmental one into a functional one, with emphasis being given to the functional interrelationship among departments.

Usually the internal quality control audit is done once or twice a year. This is because it is related to annual planning of the company ("policy management"); the schedule of the audit parallels that of the planning.

The auditors often request a short report explaining the current situation of the department that is to undergo audit. This report enables the reporters ("auditees") to rearrange their ideas in a proper form. Although a checklist is sometimes prepared, the audit items do not always follow it. Since this checklist summarizes the ideas and the points of the audit, it is often effective to announce it before the audit.

Although the way of carrying out the audit is flexible, it usually includes a plant tour and a round-table discussion, which take place after presentation of the report about the current situation of the department. It is essential during the plant tour for the members of the auditing team to be very observant in order not to miss any details. To walk through the entire manufacturing process, from the storehouse to the final inspection, for example, is an effective way to locate and "shoot" the troubles. It is also important for the top managers to talk to the foremen and workers to encourage and motivate them; the audit offers a golden opportunity for conversation between them, which can hardly occur in the course of daily work.

Concerning the short report that is read before the discussion, it is advisable to review the actions taken by the department on comments and recommendations made at the preceding audits. The discussion based on the report and on the facts observed during the plant tour is the most important part of the audit.

After the discussion, comments and recommendations are summarized and announced by the leader of the auditing team. The consultant (often a university professor) also contributes suggestions. In addition to the recommendations for the department to take corrective actions, the statement should include the recommendations made for actions by the auditors. It is also important to include not only the faults but also the merits in the performance of the department under audit. Of course, the recommendations made should be seriously considered by the department audited, and the results should be checked at subsequent audits (Itoh 1974).

QUALITY ASSURANCE AND NEW-PRODUCT DEVELOPMENT

Customer-Oriented Concept.   Quality assurance is the most important company-wide quality control activity in Japanese companies. Several features of the Japanese approach are discussed below.

The first Japanese emphasis is on the requirement that the activities of quality assurance be customer-oriented. While this concept is well known, it is quite common for companies to judge the adequacy of their quality assurance by the amount paid out as a result of customer complaints. In effect, that consists of judging the adequacy of quality assurance by the cost to the company rather than the cost to the customer. The "user's quality cost," as discussed by Gryna (1977), takes account of customer demand and is based on a customer-oriented approach. However, although this user's cost concept is useful, it fails to take account of certain other aspects of customer satisfaction which are not quantifiable in monetary terms but are nevertheless influential in creating customer confidence.

"Must-Be" Quality and "Attractive" Quality.   Ishikawa (1978) prefers to classify product quality as "backward-looking" and "forward-looking" qualities. Later Kano et al. (1984) gave a detailed consideration on this bidirectional way of perceiving quality, calling the former type "must-be" quality and the latter "attractive" quality. Thus they have a dualistic relationship with each other: some products sell well, even though they are the subject of many complaints, because they are highly attractive to customers, while others that receive few complaints do not sell at all because they lack appeal.

Thus, to obtain positive customer satisfaction, we must not only achieve must-be quality by eliminating defects and customer complaints but we must also give our products attractive qualities. Must-be quality is expressed in terms of indicators such as the defect rate, rework rate, and number of customer complaints. It therefore has the property of universality and may be expressed similarly for different types of products. Attractive qualities, on the other hand, are usually highly individual and are consequently different for different products. Thus, while statistical tools such as control charts are effective in controlling the former type of quality, the latter type is best achieved by learning as much as possible from individual examples of failure and success.

Takenaka Komuten, a Japanese building construction company, made extensive process improvements and succeeded in greatly reducing defects and reworks such as cracks in concrete or water leaks. Takenaka designers raised the question: Is this enough to satisfy our customers? To learn the answer, they visited and surveyed employees in the hospitals they had designed and built. During these interviews, they learned for the first time, for example, that nurses encountered daily troubles with the hospital facilities; that patients on the window side complained of the traffic noise and wished to be moved to quieter rooms, and that serious problems with maintenance of air conditioners were found early every summer. (All this had been going on despite the fact that the hospital director and some physicians reported that they were very satisfied with the building and facilities.) The findings were promptly reported to the design department, which took action to revise the design manuals and checklists. This, in turn, resulted in a remarkable increase in the company's share of the hospital construction market.

Quality Costs versus the Manufacturer's Conscience.   Quality costs are frequently discussed in the context of corporate quality assurance activities. Among the many proposals made concerning the classification of quality costs, those made by Feigenbaum (1983) are the most significant. He classified quality costs into the following four categories:

1. Prevention costs
2. Appraisal costs
3. Internal failure costs
4. External failure costs

It is desirable to reduce such quality costs to as low a level as possible and to maintain them at that level. However, as implied by Juran's description of such costs as "costs of poor quality" (Juran and Gryna 1988), they relate mainly to must-be quality, and activities to reduce them have little to do with actively furnishing products with attractive qualities. Such activities are therefore useful for reducing costs but are insufficient in themselves to expand the market size and share and to increase profitability. Both of these activities are important because minimum quality costs do not necessarily mean maximum profit.

There are two main problems with taking quality costs as indicators of quality assurance. The first of these is that, while quality costs address the issue of must-be quality, they take no account of attractive quality. It is therefore impossible for this approach alone to shed light on the conditions needed to produce customer satisfaction.

The second problem is that quality costs represent the amount spent by a manufacturer for maintaining and improving quality; they do not represent the losses or quality maintenance costs borne by customers as a result of poor quality. Although quality is a common concern of both manufacturers and customers, these two parties often view quality from very different perspectives; for example, when complaints are made about quality, the financial losses sustained by customers often far exceed the amount spent by manufacturers for dealing with the complaint. Consequently, while these kinds of quality costs may be useful indicators of a manufacturer's cost-reduction performance, it is doubtful whether they are suitable for measuring the efficacy of its customer quality assurance.

Some attempts have been made to consider quality costs from the user's standpoint as opposed to the costs that must be borne by the manufacturer, mainly in relation to must-be quality. For example, Gryna lists the following seven items as user's quality costs relating to industrial products (Gryna 1977):

1. Cost of repair

2. Cost of effectiveness loss

3. Cost of maintaining extra capacity because of expected failures

4. Cost of damages caused by a failed item

5. Lost income

6. Extra investment cost compared to competing products

7. Extra operating and maintenance costs compared to competing products

These user's quality costs are probably more suitable quality assurance indicators than the manufacturer's quality costs described earlier. Furthermore, because of the recent importance given to environmental issues, it is also necessary to take into account the cost of protecting the environment during manufacturing and use of the product and after the product has reached the end of its useful life (e.g., the cost of scrapping it or recycling it).

Like manufacturer's quality costs, these user's quality costs also focus principally on must-be quality. Since attractive qualities are more product-specific and less immediately obvious than must-be qualities, as described before, it is difficult to assess them in the form of quality costs. However, user's attractive qualities are important product qualities that can create larger markets, bigger market shares, and greater profits for manufacturers, so it is important to take positive steps to develop products that feature them.

More generally, reducing users' quality costs is both an important and necessary aspect of customer quality assurance. However, this alone is insufficient. Ensuring customers' positive satisfaction (including providing them with attractive quality) depends on securing their trust and confidence in the manufacturer, and this is something that cannot be bought with money alone. The first thing that a manufacturer must do to win this confidence is to act conscientiously when it comes to customer quality assurance.

"Fitness for Use and Environment" and "Surplus Quality."   In Japan the concept of fitness for use is enlarged to include fitness for the environment. For example, room air conditioners are used extensively in the densely populated Japanese cities. Beyond the need for comfortable temperature and humidity are the requirements of quiet operation and low consumption of electricity. At the initial stage of product introduction, some complaints came from neighbors who were disturbed by the noise of their next-door-neighbor's air conditioner.

In the context of manufacturers' and customers' different perspectives regarding quality, a few points about surplus quality are discussed. Surplus quality is divided into the following two categories:

1. Quality that clearly appears excessive to both the manufacturer and the customer.

2. Quality that tends to appear excessive to the manufacturer but that is strongly demanded by the customer.

As far as the first of these is concerned, it is important for the manufacturer and customer to work together in establishing limit samples and controlling these rigorously.

Acceptable items may be found among those that have failed inspection when there are stringent requirements concerning the quality of parts used in a customer's automated production line in order to maintain the productivity of that line. An extremely low defect rate (1 ppm or less) of electronic components was requested when large numbers of parts were being used for television sets in order to maintain the defect rate of the completed sets at an acceptably low level. Although it is also important in such a case to try to reduce the number of components used (e.g., by combining or modularizing them) and to make attempts to automate the inspection process, the most important thing is to minimize the defect rate of the components and, if possible, reduce it to zero.

Particularly today, when the lifetimes of products are decreasing due to the increasingly rapid appearance of new products on the market, it is becoming more and more important to cope with the competition in quality by achieving zero defects right from the start of new production.

Achieving extremely low or zero defect rates, which may appear at first sight to represent excessive quality, demonstrates the success of manufacturers in developing superior technology and high-quality products. Since they have gone to the trouble of developing these new technologies and products, it surely is extremely important for companies to make effective use of them by actively developing new fields of application for them.

Autonomous Inspection.   Needless to say, the task of manufacturing is to manufacture products whose quality conforms to the quality of design: it is the production of conforming products. This definition of manufacturing includes autonomous inspection, which consists of workers checking their own products to see whether or not they conform. This is the basic premise of autonomous inspection.

Various conditions must be satisfied in order to get autonomous inspection up and running. The most important of these is to decide on the methods by which autonomous inspection is to be carried out (including sampling), clarify the division of responsibility between autonomous inspection and proxy inspection (described below), include the labor-hours used for autonomous inspection in the manufacturing labor-hour figures, provide the measuring equipment needed, and give people the necessary training.

The practice of autonomous inspection does not mean that the people working in the manufacturing process are required to check for every single quality that they are responsible for building into the product. Some qualities will only be checked for the first time when the product is processed or assembled in subsequent processes, and others would require too much time or skill for the people working in the manufacturing process to check them. Some qualities will be inspected by people in subsequent processes or by special inspection personnel rather than by the people who actually make the product. Since such inspections are carried out on behalf of the people who make the product, they are referred to as "proxy inspections." The results of proxy inspections should immediately be fed back to the people manufacturing the product as a motivational tool and to use in the manufacturing process in a way that confirms the controlled state of the process. At the same time, it is hoped that a spirit of cooperation and teamwork will be fostered between the people of the manufacturing process and the people who inspect it on their behalf.

Autonomous inspections, including proxy inspections, have the following additional features: People are instantly aware of the quality of the products they have made. They feel happy if their products are satisfactory, and will wonder what has gone wrong if their products do not conform or an abnormality is discovered in the manufacturing process. Then, because the results are known immediately after a product has been manufactured, autonomous inspections will definitely stimulate people's curiosity and desire for improvement and will motivate them to try to improve the situation.

Process Capability and Control Charts.   Needless to say, the following two conditions must be satisfied in order to make products whose quality conforms to their quality of design, that is, for manufacturing conforming products:

1. Create no defects.
2. Permit no abnormalities.

In order to see whether these two conditions are satisfied, process capability indices and control charts are used.

Recent advances in factory automation, on the other hand, have led to the appearance of many automated and mechanized processes with far better process capability indices than ever before. However, this means that there is a danger of people mistakenly believing that because process capabilities have improved and no more defects are occurring, control charts are no longer needed. It is strongly emphasized once again that defects and abnormalities are different.

Cause and effect relationship should also be clarified to identify the possible relationships between product qualities and the various factors thought most likely to influence them. It is necessary to confirm the principal factors affecting product quality by checking correlation in the case of variables and by stratifying the data in the case of attributes. Rigorously controlling any factors ascertained by these techniques to have a significant effect on quality enables us to reduce product quality variation and decrease the incidence of abnormalities. The skillful execution of this kind of upstream control is extremely important (Ishikawa 1962).

Creativity and Work Standardization.   Clearly, the desire to work is closely connected to creativity and is in fact inseparable from it. On the other hand, necessity and importance of work standardization is often emphasized from the standpoint of improving work efficiency and assuring quality. Concerning work standardization, however, the following two types of problems are pointed out:

1. Work standardization conflicts with motivation, since it restricts the creativity and ingenuity of the people engaged in the work and reduces their opportunities to exercise those faculties.

2. Even after a lot of time and efforts have been put into standardizing work methods, the standards are actually not often complied with. One survey showed that, although most Japanese companies stipulate in abstract terms that their in-house standards are to be obeyed, more than half of them do not have any definite procedures for ensuring that they are in fact enforced. Preparing standards is known to be a difficult and time-consuming job. Is it so difficult for workers to follow the standards for the work they have been allocated?

To discuss the question of creativity and standardization in more concrete terms, let us examine work standards in a manufacturing process. They usually include the following three items:

1. The aims of the work. In a manufacturing process, this corresponds to the quality standards for the products that the process must produce.

2. Constraints on carrying out the work. The most important restrictions are those designed to ensure employee safety and preserve the quality created in upstream processes.

3. The means and methods to be employed in carrying out the work.

Of these three items, item 1 must always be achieved and item 2 must be scrupulously obeyed by whoever is responsible for doing the work. Clearly, everyone must make conforming products and work safely. But, must item 3 be obeyed in the same way as item 2 regardless of who is responsible for the work? As emphasized previously, establishing and enforcing prescribed means and methods encourages people to avoid responsibility for failure and claim that the failure was not their fault because they followed the stipulated methods. This must be strenuously guarded against.

One of the grounds for asserting that item 3 must be obeyed is that the standardized means and methods are the most productive and efficient, regardless of who uses them. At least the people who drew up the standards think so. However, in view of people's different characteristics and habits, it is highly unlikely that any single standard could be the most efficient for everyone, no matter how carefully it was formulated.

We know that this kind of standardization of action is missing from sports. To excel at a sport, we must first master the basic actions by reading textbooks and taking lessons, but this alone will never allow us to produce a new world record right away. The only way to keep improving our personal best is to build on these basic actions through hard work, that is, by continually practicing and exerting great ingenuity to discover the method that suits us best.

In light of this, item 3 should be regarded as the training manual for beginners, equivalent to the basic actions in sports discussed above, and not as rigid instructions to be obeyed without fail. Item 3 is for helping people understand the basic actions and making the process of learning the job more efficient.

In using these manuals, it is also important to make clear to all trainees at the end of their basic training that the working methods they have learned so far are no more than the standard actions that are useful hints for improvement, and that, having mastered them, they should actively try to further develop methods of working that really suit themselves as individuals. They should be told that this will help them to improve their skills, and that the managers actively support and encourage them to do so. Conversely, forcing novices to perform standard actions exactly as they have been taught is an absurd way to proceed, since it not only leads to shirking responsibility but also prevents them from improving their skills.

If workers are encouraged to improve their skills, they are requested to use their own initiative to develop standard actions into practical working methods, and discover the secrets of performing the work efficiently. Creativity and standardization are thus not mutually exclusive but mutually complementary. By encouraging and promoting the standardization described here, managers will help the people engaged in the work make full use of their creativity, and discover methods of doing the work even better.

Teamwork Relationship among Departments.   In order to achieve the goal of quality assurance, it is important, even indispensable, first to improve the quality of design to meet the target of fitness for use and environment. Second, the quality of manufactured products should be in conformance with this quality of design.

Although this procedure for assuring product quality is accepted, another general rule emphasized in Japan is teamwork among all departments. In other words, the cycle of plan-do-check-act should be rotated to involve marketing, design, manufacture, inspection, and sales. Quality assurance is one of the most important company-wide functions. It is also important to remember that quality assurance, cost and profit control, and control of factors such as delivery performance, production volume, and productivity are closely interrelated These activities are often regarded as cross-functional management, which consists of the following steps:

1. Making plans for each important corporate function
2. Passing the plans down to the specific line divisions that will implement them
3. Using policy management and daily management within these line divisions to implement the plans, check the results, and take any necessary corrective action
4. Further reviewing the results of implementing the plans from the company-wide standpoint and then taking any necessary corrective actions

The key point of cross-functional management is that the action plans for each of the functions formulated from the company-wide standpoint should not be implemented just by the supervisory head office or operating division department in charge of those functions. They should be passed down to all of the relevant organizational units within the company and be reflected in the duties of each, and they should be implemented, evaluated, and subjected to corrective action through teamwork among the different units. The PDCA cycle is rotated for each function on a company-wide basis by consolidating individual units' implementation results for each function and comparing these with the company-wide cross-functional objectives. Thus the head office or operating division supervising department does not bear the sole responsibility for implementing cross-functional management; it is shared by all of the company's line divisions in accordance with their respective duties.

A company usually sets up supervisory departments such as a quality assurance department within its head office or operating divisions to plan these activities, support their implementation, and evaluate them from the company-wide standpoint. However, these supervising departments usually are not very large.

It is also usual practice to set up a quality assurance committee for each function in order to effectively promote cross-functional management while liaisoning between the different departments. Each of these committees is chaired by the director responsible for the particular function, and the head office or operating division supervisory department responsible for each function acts as the committee's secretariat.

Teamwork Relationship with Relatively Few Suppliers.   The quality and reliability of modern complex products depend heavily on the quality and reliability of components and parts. This product quality cannot be assured solely by 100 percent inspection. Japanese manufacturers have been trying to establish and maintain a cooperative relationship with their suppliers in order to improve and assure component quality. This requirement will probably become more stringent as the number of new products entering the market increases and their life cycles shrink.

Such strict quality requirements cannot be met by means of suppliers' final inspections or manufacturers' incoming inspections based on the usual types of sampling inspection plans. In addition to implementing automated "screening inspections," manufacturers are placing more and more importance on promoting the type of quality control that extends to the quality assurance activities of suppliers themselves.

As a result, increased emphasis is being placed on establishing cooperative relationships and good teamwork between the suppliers making the components and the manufacturers using them. The establishment of teamwork relationships starts with the selection of suppliers. Questionnaires are distributed and visits made, and the suppliers are screened on the basis of financial status, facilities, basic ideas and policies of top management, enthusiasm for quality improvement, level of education and training of employees, etc. Large Japanese companies usually have an established procedure for this selection and a standardized checklist.

Good teamwork relationships with suppliers are based on mutual trust and confidence and cannot usually be established in the short term; it takes at least a few years. Importance is attached to the basic philosophy and the enthusiasm of the top management personnel. Because of the smaller size of the supplier companies, quality control activities and product quality and reliability are rather rapidly improved under the positive leadership of their top management. Independent ideas and attitudes of the suppliers are emphasized, rather than a dependent relationship to the purchasing company. Positive guidance and quality control educational services are made unsparingly available by the company on the understanding that voluntary efforts will be made by the suppliers. The importance of education and training—which become more effective in the relatively long term—is emphasized.

There is a recent trend in Japan for smaller manufacturers of components and parts to have their own test laboratories for these components so as to ensure that their quality and reliability are adequate before permitting the design department of the purchaser to specify them for use in the final manufactured products. Such test laboratory activities by the parts manufacturers are also effective in the development of new products of their own. It is fairly common in Japan for data on the process capability index and manufacturing process control charts in supplier organizations to be sent to the purchasing company and utilized in the manufacturing, inspection, and design departments in addition to the quality control department. Internal quality control audit by top management is voluntarily carried out by the suppliers. In this audit, appropriate representatives of the purchasing company are sometimes invited to attend the audit and express comments as outside authorities.

"Ten Principles for Vendee-Vendor Relations from the Standpoint of Quality Control" were established in 1966 and presented by Ishikawa (1969) at the International Conference on Quality Control 1969, Tokyo, as follows:

Preface: Both vendee and vendor should have mutual confidence, cooperation, and high resolve of "live and let live" based on the responsibilities of enterprises for the public. Following the above spirit, both parties should sincerely practice the following "Ten Principles."

Principle 1: Both vendee and vendor are fully responsible for quality control application with mutual understanding and cooperation on their quality control systems.

Principle 2: Both vendee and vendor should be independent of each other, and esteem the independence of the other party.

Principle 3: Vendee is responsible to bring clear and adequate information and requirements to the vendor, so that the vendor can understand what should be manufactured.

Principle 4: Both vendee and vendor, before entering into business transactions, should conclude a rational contract between them in respect of quality, quantity, price, delivery term, and terms of payment.

Principle 5: Vendor is responsible for the assurance of quality that will give full satisfaction to vendee. Vendor is also responsible for submitting necessary and actual data upon the vendee's request, if necessary.

Principle 6: Both vendee and vendor should decide the evaluation method of various items beforehand, which will be admitted as satisfactory to both parties.

Principle 7:   Both vendee and vendor should establish, in their contract, the systems and procedures through which they can reach amicable settlement of disputes whenever any problems occur.

Principle 8:   Both vendee and vendor, taking into consideration the other party's standing, should exchange information necessary to carry out better quality control.

Principle 9:   Both vendee and vendor should always perform business control activities sufficiently, such as in ordering, production and inventory planning, clerical work, and systems, so that their relationship is maintained upon an amicable and satisfactory basis.

Principle 10:   Both vendee and vendor, when dealing with business transactions, should always take full account of the consumer's interest.

These principles were formulated through enthusiastic cooperation and thorough discussions among purchasers and suppliers. Their meaning is obvious, although they have not always been adhered to. It is hoped that they will further be utilized as much as possible as a basic guideline for improving the manufacturer-supplier relationship.

Quality Information—Quality Complaints versus Systematic and Positive Collection of Quality Information.   In a paper on quality design, Aiba (1966) emphasized the importance of securing external quality information from customers in the marketplace. Such raw information is then converted into measurable quality characteristics.

In this connection, external quality complaints, while demanding prompt and remedial action, are collectively a poor index of quality. They reflect must-be quality—the poor quality of product dissatisfaction. In addition, quality complaints are subject to the following kinds of problems as a measure of quality:

1. Complaints are influenced by the unit price of the product.
2. Any kind of information, not only that on complaints, can very easily become distorted when transmitted orally.
3. Complaints are also affected by the type of customer and the economic climate.
4. Some complaints vanish en route to the manufacturer.

Quality complaints might be expressed as sporadic and passive information on must-be quality. However, they are still a valuable source of information that we should use to improve the quality of our products.

In tandem with this process, the systematic and positive collection and utilization of external quality information on the following are also important:

1. Customer demands, i.e., how the commodities are used by the customers and the conditions of use
2. Quality of similar commodities manufactured by competitors
3. Actual conditions of transportation and storage by distributors' channels and by retailers
4. Present and future market trends

At the Cotton Underwear Division of Gunze Co., Japan, the division director received a letter from an elderly male customer expressing his hearty appreciation for the excellent quality and durability of Gunze cotton underwear, stating he had been wearing the same articles for more than 10 years. This customer also urged the director to develop new products of such high durability. The director requested a consultant's opinion about the development of such new products. The consultant asked, "How long do customers usually wear your underwear?" Since the division had not investigated the life span of the underwear, it conducted a market survey. When the results were summarized in the form of a histogram, the distribution of life span clearly was well approximated by a normal distribution curve, with a mean of about 3 years and a standard deviation of about 0.5 years. It became evident from this histogram that the number of customers who continue to wear the same underwear for longer than 4.5 years was 1 to 2 per 1000. The data suggested that the market for a new product of much higher durability was very limited.

To avoid possible adverse consequences of reacting to sporadic customer information, it is necessary to collect market quality information on a systematic basis. Such systematic collection is comparatively easy for products of high unit price, as in the case of the Takenaka Komuten construction company, mentioned earlier. With products of low unit price, some ingenious methods are necessary. A questionnaire card is often enclosed with the product, asking customers their reasons for purchasing, the name of the retailer, impressions of product performance, conditions of use, etc. In Japan customers often mail this card immediately after their purchase and before using the product. Hence their comments about quality and reliability are meaningless. Furthermore, only a small proportion of the cards are usually sent back to the manufacturer. A better way is to utilize the returned cards to design follow-up surveys and then to apply sampling techniques based on types of customers and conditions prevailing during use. Interviews with the customers are usually far more fruitful than mailed questionnaires.

Apart from special situations, the external quality information is usually collected by regular company sales or marketing personnel. However, although the collection and feedback of external quality information is an important part of their quality assurance task, it is not their only job, and they usually have to squeeze it into the gaps in their busy and complicated daily sales and service schedules.

Consequently, the enthusiasm of these people for collecting external quality information at the request of design and engineering departments easily tends to evaporate if the purpose for which it is being collected is unclear, if hard won information is left idle and not utilized for new-product development or quality improvement, or, if after having worked so hard to collect the information, they are not informed of the benefits achieved through its utilization.

As users of this external quality information, which has taken so much effort to collect, the engineering and design departments should therefore provide positive feedback to the sales people responsible for collecting it. They should tell them how effective it has been and how they would like it to be improved. If, through poor communication, the sales personnel collecting the external quality information receive no reaction from the departments requesting it no matter how much information they pass on, they will feel that their efforts are not valued. Conversely, if the quality information that they have spent so much time and trouble gathering is used effectively within the company and they are given information on how it can be improved, this will increase their enthusiasm for collecting it. This is very important both for motivation and for the smooth operation of a quality information system.

New-Product Development.   Since Japan has few natural resources of its own, the only way it can survive and grow is by obtaining the necessary foreign currency by profitably developing, manufacturing, and exporting high-value-added products that satisfy customers' actual and potential needs and that they are delighted to use. It is impossible to achieve this in regard to product quality simply by reducing user's quality costs relating to must-be quality. It also requires action to satisfy customers in positive ways, including the development of new products. Through this, we aim not just to minimize quality costs but to maximize corporate profits.

Quality function deployment (QFD) (Kogure and Akao 1983; Akao et al. 1987), a "design approach" based on deductive reasoning, has recently been widely adopted as an effective technique for new-product development, with eye-opening results. At the same time, however, we must not forget that the analytical approach using the quality control story is also highly effective. It is important to make positive use of the valuable lessons that past successes and failures have taught us, and not just discard them because the present project happens to be different from previous ones.

We know that we can increase the likelihood of a project's success if we appoint as team leader someone who has already succeeded with previous projects, even if they were different from the present project. This is because such people not only have confidence in their ability to succeed but also know how to make use of their past experiences in the present project—in other words, they follow the PDCA cycle.

Many companies try to prevent the repetition of new-product development failures by collecting examples of past failures and having those involved in current projects study them. This is also an attempt to utilize the analytical approach based on the quality control story. However, as long as they are collecting examples of failures, it is equally important to try collecting examples of past successes and make use of these. The reason for the success of a project is not usually the opposite of the cause of its failure; the two are often different. Publishing compendiums of failure case studies certainly reduces the chance of making similar mistakes, but it is doubtful whether it increases the possibility of success.

Companies that are good at new-product development and enjoy a high success rate usually appoint their deputy CEO or a senior director as their new-product development manager. It is often said that the enemies of new-product development are to be found within the company. When the work of developing a new product or technology begins, some people will always come out with negative comments like, "There's no point in doing that," or "We already know it can't be done." A new idea is like a flower bud—it is easily squashed, but we do not know how good it really is until it has been allowed to develop to some extent. The most important duty of the new-product development manager is to foster these ideas he or she believes are important and are likely to prove useful in the future, shield them from criticism within the company, and make something of them.

In many Japanese companies new products are classified into the following groups on the basis of their novelty:

1. New in the world
2. New in Japan
3. New in the company
4. Model change of a current product, etc.

Such a classification is effective for determining the steps and stresses in new-product development. For a new-product of class 1, the quality, including safety, and the reliability should be most strictly assured even though time and money are needed. The new users of these products should also be broadly pursued and developed.

It is frequently stressed that the development of new applications must not be ignored when thinking about new-product development. In view of the many new materials that have been developed recently, it seems that the development of the material precedes the development of uses for it. In short, the development is "seeds-led", with the development of needs lagging behind. An example of this is the recent development of a brassiere employing shape-memory alloys that is selling well, particularly in the American market. This illustrates how important it is, in developing new applications, to be thoroughly familiar with the quality characteristics of newly developed materials or technology, to identify latent customer needs that existing technology has been unable to satisfy, and to pool ideas to come up with ways of resolving the situation.

In contrast, for new products of classes 3 and 4, the emphasis is on promptness, so that product development cycle time can be shortened in order to cope with competition in the market.

During quality design, there is a need for conversion of true quality characteristics requested by customers into measurable substitute characteristics (Aiba 1966). The QFD technique mentioned above is being widely used for this purpose. After converting the true quality characteristics into the measurable substitute characteristics, it becomes easier to determine the quality specifications for each item.

The measurable substitute characteristics thus evolved become the focus of various checkpoints in the manufacturing and inspection processes. During the design and pilot stages of the manufacturing process, a "quality control process chart" is established for assurance of quality in the manufacturing process. Figure 41.4 shows an example of such a process chart, which, as may be seen, uses substitute characteristics. Thus the quality function is implemented throughout the whole company (Kogure and Akao 1983).

The steps in new-product development are roughly as follows:

1. Market research
2. Conceptualization
3. Experiments, market research

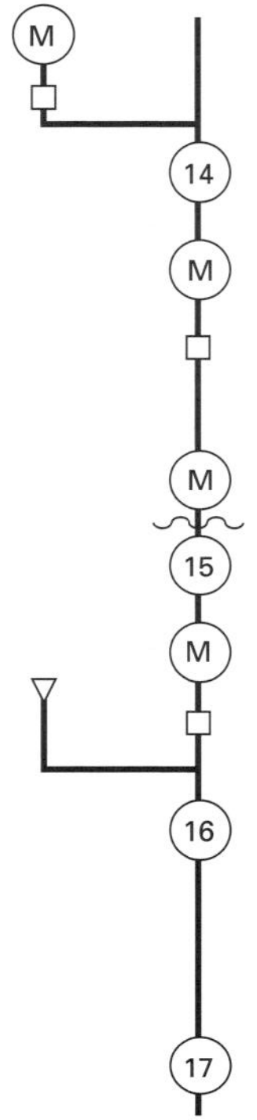

NAME OF PART	FLOW CHART MATERIAL PROCESS	FLOW CHART MAIN PROCESS	NAME OF PROCESS	OPERATION MANUAL	ITEM	METHOD	PERSON	SAMPLING	MEASUREMENT
Lead chip			Preforming	32-2-RC-1	T.W.	X-R Chart	Mr. F	Every lot, random n = 5	Autobalance
					T. dia.	Check sheet	Mr. F	Every lot, random n = 10	Gauge
					Appearance	p-Chart	Mr. F	Every lot, total n = 300	Eyes
					Temperature		Mr. F	Twice a day	Drum
					Speed constancy			Twice a day	
			Preheat	22-RCG -005 -006	Temperature	Check sheet	Mr. S	Twice a day	
			Forming	222-RCG	L-dimension	X-R chart	Mr. S	Every lot, random n = 5	Slide calipers
					Temperature	Check sheet	Mr. S	Once a day	
					Pressure	Check sheet	Mr. S	Once a day	
			Aging	222-RCG 6-1	Temperature	Check sheet	Mr. S	Once a day	

FIGURE 41.4 A QC process chart for a solid resistor. (*Source: Ishiwara 1975.*)

4. Application for patent

5. Quality design

6. Pilot-scale manufacturing and qualification testing

7. Process design

8. Full-scale production testing and sales

9. Mass production

10. Termination of initial warning system

11. Termination of mass production

These steps and the company departments involved may be summarized by a chart of the quality assurance system, an example of which is shown in Figure 41.5. This chart will vary with the novelty of the products.

Development of new products is usually a time-consuming undertaking, involving many trial-and-error procedures. However, the procedures of evaluation, qualification, review, etc. carried out during the various steps can be standardized, which becomes the first step in introducing quality control to new product development.

Market Research and Hypothesis Testing.   Two stages of market research are shown in the above-listed steps of new product development (steps 1 and 3). The preliminary market research is rather general in character and may lead to an idea for a new product. The later market research is to confirm the effectiveness of this idea.

For example, in the late 1960s (Yoneyama 1969), the half-size camera was becoming popular in Japan. The first manufacturer was company A. At company B, there was discussion among top managers about developing a new half-size camera to compete with that of company A. Mr. Y. was appointed a senior member of the development team. First, he collected statistics on monthly camera sales from the relevant government report and classified them into sales of half-size and full-size cameras. He found that total sales were rising mainly because of the increased sales of half-size cameras. When fitting the increase in demand for half-size cameras to a logistic curve, it became clear that this increase was not likely to slow down. From this study, top management decided to develop a new camera.

The second market research program (step 3) was then launched. It happened that company B was also selling color film, which was ultimately developed at the company's central laboratory. It was found from the analysis of developed film that the pictures taken indoors by half-size cameras were often blurred. It was decided on the basis of this fact that the new camera should be equipped with a better lens.

Mr. Y went out every Sunday with two counters in his pockets. He strolled in downtown Tokyo and then in suburban resorts and counted the number of people who carried a camera; he pushed one counter for each full-size camera and pushed the other counter for each half-size camera. He found that the ratio of half-size to full-size cameras was lower downtown than in the suburbs. It became clear that the customers were buying the carrying convenience and light weight of a half-size camera. It was then further decided that the new camera should be lighter than the camera of company A.

These two features of company B's new half-size camera became very profitable sales points. It should be noted that the ideas conceived in step 2 above need to be examined and confirmed in the market research of step 3. Systematic and positive collection of quality information from the market is also important for determining test conditions in qualification and for improving the items subject to design evaluation and review in the intermediate stages.

Design Reviews—Corrective Action on Physical Objects and Process.   Design review, evaluation, and qualification done at proper stages of development, such as steps 5, 6, and 8, are prevalent in Japanese companies. These evaluations aim at determining whether

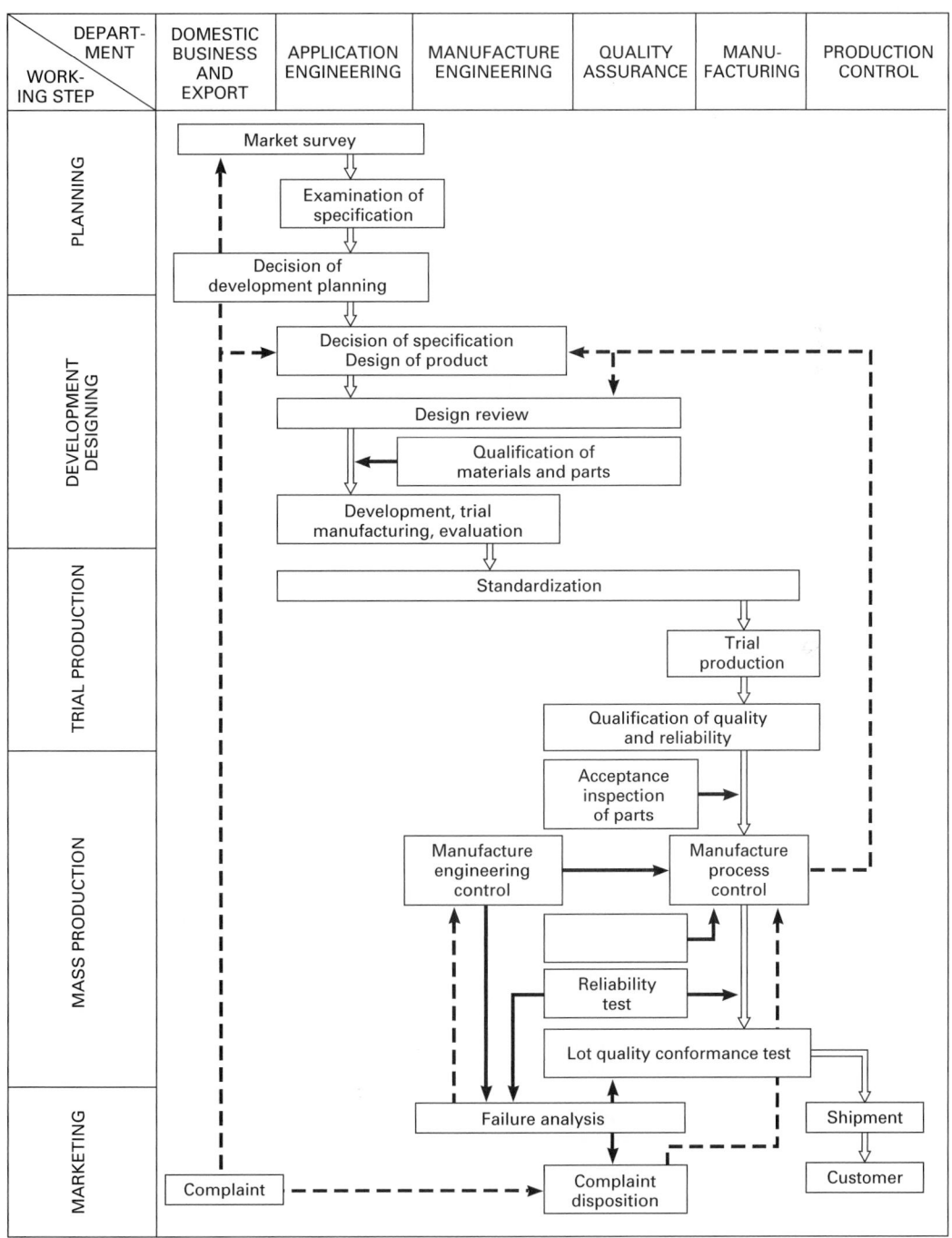

FIGURE 41.5   Chart of quality assurance system. (*Source: Takanashi 1978.*)

1. The quality of design was appropriate

2. The design of the foregoing procedures was adequate

With regard to exported goods, quality information from foreign markets is extremely important because environmental conditions (temperature, humidity, dust, etc.), laws, weights and measures, customs, conditions of use, physical constitution of customers, etc. are quite different compared to the domestic market. Although these data in foreign markets cannot be anticipated in the earlier stages of foreign trade, records of previous failures of various kinds are accumulated and utilized as much as possible to prevent recurrences. Those corrective actions that improve the design, testing, and evaluation procedures are most effective.

Full-Scale Production Testing.   The aim of the full-scale test (or trial), which is the final stage of new product development, is not always clear. It is often carried out by skilled workers, with quite satisfactory results. When mass production starts, however, the manufacturing department is always worried about problems in the manufacturing process. The reasons are clear: They reflect the differences in the skill of the workers engaged in the full-scale test run and in mass production. The distribution of the quality of components and parts is also sometimes different. The true aim of the full-scale test should then be clearly defined and achieved. One of its most important purposes is to thoroughly eliminate weaknesses in manufacturing and take corrective actions before the start of mass production.

Taking this approach even further, we arrive at the concept of upstream management discussed previously. Not only should we eliminate all remaining problems before the start of mass production by performing full-scale production testing under mass production conditions, but we should also investigate those problems closely to determine whether or not they could have been discovered only at the full-scale production testing stage or whether they could, in fact, have been discovered in earlier processes and prevented from occurring in the first place. We should also think about how the development process could be improved in order for this to happen. We should put these improvements into effect, check the results, and identify any remaining problems. Thus we will continue to rotate the PDCA cycle.

REFERENCES

Aiba, K. (1966). "Significance of Quality Designing." *Hinshitsu Kanri* (*Statistical Quality Control*), vol. 17, pp. 88–89 (Japanese).

Akao, Y., Ofuji, T., and Naoi, T. (1987). "Survey and Reviews on Quality Function Deployment in Japan," *Proceedings of International Conference on Quality Control '87,* Union of Japanese Scientists and Engineers, Tokyo, pp. 171–182.

Deming, W. E. (1986). *Out of the Crisis,* Massachusetts Institute of Technology, Center for Advanced Engineering Study, Cambridge, MA.

Feigenbaum, A. V. (1983). *Total Quality Control,* 3rd ed. McGraw-Hill, New York, p. 109.

Gryna, Frank, M. (1977). "Quality Costs: User vs. Manufacturer." *Quality Progress,* vol. 10, no. 6, pp. 10–13.

Hopper, K. (1985). "Quality, Japan and the U.S.: The First Chapter." *Quality Progress,* vol. 18, no. 9, pp. 34–41.

Ikezawa, T., et al. (1990). "The Origin of Japanese Quality Control." *Hinshitsu Kanri* (*Total Quality Control*), vol. 41, no. 1, pp. 173–78; no. 2, pp. 171–177; no. 3, pp. 263–270; no. 4, pp. 362–368; no. 5, pp. 457–464; no. 6, pp. 547–551; no. 7, pp. 1064–1071 (Japanese).

Ishikawa, K. (1962). *Kanrizu-ho* (*Control Chart*). Japanese Union of Scientists and Engineers, Tokyo (in Japanese).

Ishikawa, K. (1965). "Recent Trend of Quality Control." *Reports of Statistical Applications and Research, Japanese Union of Scientists and Engineers,* vol. 12, no. 1, pp. 1–17.

Ishikawa, K., and Kondo, Y. (1969). "Education and Training for Quality Control in Japanese Industry." *Quality,* vol. 8, no. 4, pp. 90–96.

Ishikawa, K. (1969). "Ten Principles of Vendee-Vendor Relations from the Standpoint of Quality Control." *Proceedings of International Conference on Quality Control.* Japanese Union of Scientists and Engineers, Tokyo, pp. 333–336.

Ishikawa, K. (1972). "Quality Control Starts and Ends with Education." *Quality Progress,* vol. 5, no. 8, p. 18.

Ishikawa, K. (1978). "Quality Control Specialists and Standardization." *Proceedings, International Conference on Quality Control.* Japanese Union of Scientists and Engineers, Tokyo, pp. A6-5–A6-10.

Ishiwara, K. (1975). "Process Flow Chart for Management." *Hinshitsu Kanri (Statistical Quality Control),* vol. 26, pp. 332–333 (Japanese).

Itoh, S. (1974). "Executive Reports on Quality: Audit Reports and Quality Reports." *Reports on Statistical Applied Research,* vol. 21, no. 3, pp. 65–77.

Juran, J. M. (1978). "Japanese and Western Quality—A Contrast." *Quality Progress,* vol. 11, no. 12, pp. 10–18.

Juran, J. M. (1981). "Product Quality—A Prescription for the West." *Proceedings, 25th Conference European Organization for Quality Control,* Paris, June, vol. 3, pp. 221–242.

Juran, J. M., and Gryna, F. M. (1988). *Quality Control Handbook,* 4th ed. McGraw-Hill, New York, p. 4.18.

Kano, N., Seraku, N., Takahashi, F., and Tsuji, S. (1984). "Attractive Qualities and Must-Be Qualities." *Hinshitsu (Quality),* vol. 14. no. 2, pp. 147–156 (Japanese).

Kogure, M., and Akao, Y. (1983). "Quality Function Development and CWQuality Control in Japan." *Quality Progress,* vol. 16, no. 10, pp. 25–29.

Kondo, Y. (1969). "Internal Quality Control Audit in Japanese Companies." *Quality,* no. 4, pp. 97–103.

Kondo, Y. (1976). "The Roles of Manager in Quality Control Circle Movement." *Reports of Statistical Applications and Research, Japanese Union of Scientists and Engineers,* vol. 23, pp. 71–81.

Kondo, Y. (1977). "Creativity in Daily Work." *ASQC Technical Conference Transactions,* Philadelphia, pp. 430–439.

Kondo, Y. (1978). "JUSE—A Center of Quality Control in Japan." *Quality Progress,* vol. 11, no. 8, pp. 14–15.

Kondo, Y. (1988). "Quality Through Millennia." *Quality Progress,* vol. 21, no. 12, p. 83.

Mizuno, S. (1967). "Execution of Internal Quality Control Audit." *Hinshitsu Kanri (Statistical Quality Control),* vol. 18, pp. 835–839 (Japanese).

Mizuno, S., and Kume, H. (1978). "Development of Education and Training in Quality Control." *Reports of Statistical Applications and Research, Japanese Union of Scientists and Engineers,* vol. 25, pp. 36–60.

Quality Control Circle Koryo (1980). Quality Control Circle Headquarters, Japanese Union of Scientists and Engineers, Tokyo.

Sugimoto, T. (1968). "Quality Control Audit by the Top Management." *Hinshitsu Kanri (Statistical Quality Control),* vol. 19, pp. 1136–1140 (Japanese).

Takanashi, M. (1978). "Quality Assurance System for the Integrated Circuit." *Proceedings of International Conference on Quality Control.* Japanese Union of Scientists and Engineers, Tokyo, pp. C1-51–C1-56.

Tsuda, Y. (1990). "The Quality Situation in Europe—A Communication from Belgium." *Hinshitsu Kanri (Total Quality Control),* vol. 41, pp. 241–251 (Japanese).

Tsutsumi, S. (1988). "Quality Control Training for Local Staff of Japanese Companies Abroad, Including Top Management." *Proceedings of 46th JUSE Quality Control Symposium,* pp. 35–47 (Japanese).

Tsutsumi, S. (1991). "Basic Business Principle of Our Company and Quality Activities at the Overseas Plants." *Proceedings of Asian Quality Symposium,* Tokyo, pp. 27–32.

Yoneyama, T. (1969). *Hinshitsu Kanri no Hanashi (Some Topics on Quality Control).* Japanese Union of Scientists and Engineers, Tokyo, pp. 115–118 (Japanese).

SECTION 42

QUALITY IN THE PEOPLE'S REPUBLIC OF CHINA

Yuanzhang Liu

THREE THOUSAND YEARS OF QUALITY—AN INTRODUCTION

Historical Background.   China's long recorded history can be traced back to the twenty-first century B.C., more than 4000 years ago. Its political history began taking recognizable shape during the period when the Yangtze River basin was dominated in succession by two great feudal dynasties, first the Shang, whose kings ruled from around 1750 B.C. until 1125 B.C., then the Zhou, who conquered the Shang people around 1125 and ruled until 250 B.C.

The last 2 centuries of Zhou rule were marked by warfare among the city-states of the Middle Kingdom, as the Yangtze region was then called by its inhabitants. The larger city-states began to conquer and absorb the smaller, weaker ones. The Zhou dynasty collapsed late in the third century B.C., bringing a temporary end to political order. In 221 B.C., Shi Huangdi, king of the western city-state of Qin, unified the Middle Kingdom politically for the first time, becoming its first emperor and the founder of the short-lived Qin dynasty (221–206 B.C.). (Though the dynasty ended with the death of its founder, its name lives on, in altered form, as "China.")

Over the next two millennia, China was ruled by a succession of imperial dynasties. When the Qing dynasty (1644–1911 A.D.) came to an end, relinquishing rule in 1911, the 4000-year era of feudal/dynastic rule in China ended with it.

For purposes of tracing the history of quality management, it isn't necessary to trace the history of the rise and fall of all of the dynasties which came to rule China. It is only necessary to point out the fact of this succession. The emergence of a dynasty generally signaled a period of relative absence of threat, either from internal factions or external invaders, when social and technological development could take place. The eclipse of a dynasty, involving the transfer of great power, was usually accompanied by military action and social disorder which threatened all development.

The Shang and Zhou dynasties are associated with the earliest expressions of high civilization in China, especially their rich religious and ceremonial life. The production of bronze ceremonial artifacts was an important part of that tradition. By the early years of the Shang dynasty, the handicraft industry which produced these objects was already established. Its outstanding achievement is recorded in classical Chinese writings and is evidenced by archaeological relics and by the remaining items of handicraft. This industry was the foundation for the manufacturing tradition that survived the long succession of dynasties that followed.

All of this evidence leaves no doubt that the management of product quality has long been understood and practiced in China. However, the West has had access to few details of quality management as practiced by the ancient Chinese. The work of Jin et al. (1995) offers a rare exception.

It may be appropriate to review briefly what is known of the history of quality management, to better understand quality management as it is practiced in China today.

The Nature.   In ancient times, quality management was applied only to the handicraft industry, which at that time included many trades, such as metallurgy, vehicles, ships, textile and leather, pottery and woodworking, weapons, musical instruments, and architecture. As written in *Zuozhuan, the Thirteenth Year of Cheng Gong* (Annals of the Zhou dynasty), "Sacred ritual and war are major matters of prime importance to the State." Strict quality control began first on those products to be used in offering ritual sacrifice and in making war. The fine quality of bronze vessels, sets of bells, and swords made in the Shang and Zhou dynasties are well-known and provides tangible evidence of attention paid to their quality.

China's ancient handicraft industry was dominated by workshops owned by feudal lords, especially by the emperor. (For simplicity, we will call all such workshops "officially owned workshops.") This does not mean that there were no civilian handicrafts made during the long history of China. But no matter whether in scale, or in sophistication of technology or of management, civilian industry was no match for officially owned industry. The civilian handicraft industry was mainly a collection of small family workshops. To create such an enterprise, a family spent a lot of energy and undertook a great deal of risk. Risk was especially high during the chaos which often accompanied a change of dynasties. Lacking official protection, an individual workshop was often too weak to survive. On the other hand, the officially owned workshops, recognized and valued by rulers in every dynasty, had the power to survive and develop. As an important element of the surviving officially owned workshops, the official quality control and quality management systems survived virtually intact, with only incremental change over time. This helps account for the continuity of the quality management system throughout China's 3000 years of feudal history.

There are three fundamental aspects of this ancient system of quality management which are of interest: its content, its limitations, and its lessons for the future.

Content.   Of the content of the Chinese quality management system, four main points can be made.

Concept.   From the earliest times, it was based on a clear concept of quality. In 403 B.C., in *Kao Gong Ji* (Records in Inspecting the Works), it was written that "Heaven having time, earth having energy, material having beauty, work having skill, add these four and the result is quality." The recognition that quality is the result of many contributing factors was reflected in the practices of later generations of workshops, including the way they applied the concept of division and cooperation of labor as well as management.

Training.   The system paid full attention to training and caring for skilled labor. China's workers as a class were never treated well, nor did they attain high status in the social hierarchy of the feudal system. In fact, during the early dynasties, craftsmen were slaves. Nevertheless, the emperor used to visit their workshops to inspect the quality of their product and to inquire about their techniques and skill. As slavery was gradually abolished, the officially owned workshops began to use conscription to recruit craftsmen. After the Qin dynasty, many generations of succeeding dynasties practiced the same methods of recruitment. As in the times of slave labor, the craftsmen working for the officially owned workshops were forced to live together in the workshop, to make it more convenient for their bosses to manage and train them. The craftsmen thus conscripted were of relatively high technical capability to begin with; living collectively with fellow craftsmen made them more skillful through mutual teaching and learning. In addition, the collective arrangement made it easier to enforce official standards and rules for production and quality, assuring the product's conformity and superiority.

Standards. From the time the first emperor of Qin unified the metrological system in China, succeeding dynasties all promulgated laws and decrees to enforce the adoption of a unified standard of measures and weights, which helped greatly in the uniform practice of quality control in all industries throughout the country. For instance, according to *Tang Lu Shu Yi, Za Lu Men* (Introduction to the Laws of the Tang Dynasty, Miscellaneous Categories), compiled in 635–640 A.D., a law stipulated that measuring tools were to be checked every August, and were to be used only after the seals were affixed. Moreover, the concept of standardization went beyond its application to measuring tools. Application extended to the industrial products themselves, as well as to production practice, with the introduction of interchangeability of parts. The famous terra cotta army buried in the tomb of the first emperor of Qin was actually assembled from parts.

Table 42.1 presents a sample of surviving writings which bear on the subject of managing quality over the period from 403 B.C. (during the Zhou dynasty) until late in the Qing (last) dynasty. All of these books contain compilations of industrial standards and specifications. *Tian Gong Kai Wu*, the last reference in Table 42.1, is an especially important example of writings on Chinese technology, and was praised by Joseph Needham, the world's pre-eminent authority on the history of Chinese science and technology.

TABLE 42.1 A Sample of Ancient Chinese Writings on Quality Management

Kao Gong Ji (Records in Inspecting the Works)	403 B.C.	A recognition of quality as the combined result of "the time of heaven, the energy of earth, the beauty of material, and the skill of the workman"
Tang Lu Shu Yi, Za Lu Men (Introduction to the Laws of the Tang Dynasty, Miscellaneous Categories)	Compiled in 635–640 A.D.	A law stipulated that measuring tools were to be checked every August, and were to be used only after the seals were affixed.
Wu Jing Zong Yao (Compendium of the Most Important Military Techniques)	650–950 A.D.	Subject: weapons manufacture
Ying Zao Fa Shi (Architecture Rules and Methods)	Song dynasty (960–1219 A.D.)	Subject: architecture
Zi Ren Yi Xun (Teachings of the Deceased)	Yuan dynasty (1279–1368 A.D.)	Subject: textiles
Long Jiang Chuan Chang Zhi (Records of the Long Jiang Shipyard)	Ming dynasty (1368–1644 A.D.)	Subject: shipbuilding
Cong Cheng Zuo Fa Gui Ze (Regulations in Engineering Projects)	Qing dynasty (1644–1911 A.D.)	Subject: construction
Tian Gong Kai Wu (Technology and Manufacture)	1637 A.D. by Sung Yingxing	Subject: manufacturing

Responsibility.   The system provided for strict responsibility. From the time of the Zhou dynasty onward, the centralized autocratic state had a centralized system of quality control over the whole process of handicraft production. Special officials were appointed to manage specific organizations in charge of various production matters, from the administrative ministries down to the local workshops. Those officials, together with the craftsmen, were to be responsible for the quality of the product. For this purpose, a unique measure, entitled "Articles to be inscribed with the names of the craftsmen and the officials in charge," was enacted as early as the Zhou dynasty, and was continued in force by the governments of later dynasties. If someone made product of inferior quality, he could be traced and was to be punished properly. For the sake of justice and fairness, a system of product examination was devised, including in-process mutual, patrol, and final inspection. A method of sampling inspection was invented and used as part of the system. China's ancient quality management, though rather primitive in its early days, became quite systematic and efficient in its later development.

Limitations.   This system had limitations. In the 3000-year period from the Zhou dynasty to the Qing dynasty, the basic political system changed little; so also did the basic organizational structure for control of industry, which, by and large, was a collection of officially owned and bureaucratically managed business. Despite the fact that these businesses faced no competition, and despite the fact that they were protected by state authority, their quality control and management were carried out with strict discipline. However, because all the products were demanded either for the luxury of the royal court or for the needs of the state, production cost was totally ignored. Furthermore, a state policy of "stressing agriculture and suppressing commerce" was adopted at the time of the Zhou dynasty, and maintained for the 2 millennia which followed. There may have been a valid argument for such an emphasis on agriculture at the expense of commerce 2000 years ago, but it goes without saying that clinging to this policy posed a severe hindrance to the development of a national manufacturing industry in China's feudal era.

As a consequence of these factors, the development of quality management stagnated—no further innovation was believed to be necessary. Another more serious consequence was that the science and technology of China also began to stagnate and to fall behind. Whereas until the sixteenth century China was among the most advanced nations of the world, it began a period of decline, the victim, perhaps, of its feudalism and its self-sufficient small-scale peasant economy.

Lessons for the Future.   Without the felt need to innovate, and without the help of advancing science and technology, quality management could not evolve further. When we look at the China of today, we can find in its state-owned industries, in their administration by various ministries, and quality management by governmental regulations, a considerable resemblance to the ancient system. Yet this ought not to be surprising, as China is, after all, a country of tradition.

While tradition can sometimes stifle progress, it can stimulate it, too. The history of China's ancient quality management system reveals brilliant achievements associated with its application to the production of handicrafts. Today, China faces a new environment—political, economic, and scientific. One challenge of the new environment is to create a new quality management system. Should not the study of this history also provide useful guidance for the men and women of China who face this challenge?

QUALITY IN POSTREVOLUTION CHINA

Quality in the Early Years of the People's Republic (1949–1952).   When the People's Republic was first established in 1949, the predominant task was to fully utilize the meager industry that was left following the Civil War to produce as much as possible to meet the needs of the country and people. All privately owned factories were turned over to the state and rebuilt into state-run businesses. How to run such reborn factories became a problem. There was no experience to follow, no lessons to learn from, and no knowledge base to refer to. What could be relied upon were only the factory workers who had been just liberated and were full of enthusiasm for production. A new form of organization had to be worked out. Some new concepts for quality and productivity had to be put forward.

Worker Terms.   First of all, a democratic management campaign was carried out in the whole country to abolish the feudal gangmaster system which had so cruelly exploited the workers. For example, the Ministry of Fuel Industry issued such an order in 1950 to all mines over the country. Then a new form of organization, the "Administrative Committee," was set up in every factory which was composed of workers of the factory and cadres dispatched by the government to discuss and decide important matters. Needless to say, the quantity and quality of products were the first items to be placed on the agenda. Consequently, laborers enjoyed a new status; they were workers of course, but also, in a sense, managers. A very harmonious relation between management staff and line workers was thus formed.

In such a background and environment workers started to create their own teams. They began to be aware of their responsibility and improved their labor discipline and skill through team activities. A rationalization suggestion campaign spread spontaneously across the country. In August 1950 the Government Administration Council of the People's Republic passed a resolution to give awards for invention, technical innovation, and rationalization, putting the campaign on a more effective path. The consciousness of the labor did raise the industrial productivity and on the whole product quality reached a suitable level for civilian goods but was not satisfactory for more stringent requirements, particularly in military weapons. There had been a report saying that the point of fall of rocket shells produced in 1950 scattered very inaccurately due to the unevenness of charges used in different factories. The official view revealed the fact that inspections and quality control had not been taken seriously and properly. According to a statistical report of September 1950, which covered 29 weapons factories, only 21 had a quality inspection department. Of these, only 5 reported to the factory director. The remaining 8 factories had no inspection system of any sort at all. The sole dependence on the worker's individual consciousness of quality and productivity proved to be insufficient without a coordinated system of quality management.

Corrective Measures.   In April 1951 the Central Government began to establish an independent inspection department within every factory. Meanwhile, the central industrial ministries and their local agencies were also asked to have a quality supervision department. In the years 1951 and 1952, a quality management and supervision system was completed, comprising three levels—central government, local authorities, and factory management. The first task was to unify the test and inspection procedures and standards for most of the products which had lacked interchangeability and maintainability. Efforts for this purpose could be seen from the decision promulgated by the central government in October 1951, which made clear the duties and rights of inspection and demanded test equipment to be consolidated, in-process-inspection to be implemented, and inspection specifications to be documented in detail.

Inspection and standardization on one hand, operations improvement on the other—these were the two hands on which China was relying to exercise her quality management in the period of the Korean War. Workers concentrated on the improvement of their operations, because they knew that the only thing they could do to guarantee their product quality and quantity, with machinery equipment being so old and raw material so inferior, was to operate more efficiently and more effectively. They understood that this was the only way to meet the needs both in the front and rear. In cotton mills, coal mines, iron and steel works, machine-building factories and in railroad transportation, there appeared heroic workers who not only created advanced operations methods but also inspired the morale of their group by their devoted labor. As a matter of course, a call from the laboring masses and then a campaign organized by the State to learn from those model workers followed. This kind of learning campaign has since become a tradition which formed an important part of quality management.

Quality Management Introduced from the Soviet Union (1953–1960).   China started her first 5-year plan in 1953 with the ending of the Korean War. The basic task was to build

up an initial foundation of industrialization centering around the 156 major projects helped by the Soviet Union. All technology and management necessary were brought to China by the thousands of Soviet experts who came as advisers at every industrial construction site. By the end of 1957, industries that China had never before had, such as automobile, airplane, machine tool, electricity-generation equipment, and high-alloy steel, were successfully established. It was in this period that the rough outlines of quality management as a systematic and scientific activity was first proposed. Since then, quality management has come to have tremendous influence on the later stages in China's economic development.

Organization for Quality Management

Centralized Leadership.   A sound quality management system is an institutional guarantee of product quality for which commitment of the leadership is most vital. Following the Soviet experience, the chief of each level in the industrial management hierarchy, governmental and factory, was supposed to bear the entire responsibility and absolute leadership in what was called the "one-boss system." The factory director was in such a position for the factory management, including quality, and the director of the inspection department was in the same position as far as the department was concerned. A circular promulgated in this connection by the Central Committee of the Communist Party of China demanded party organizations in every factory to help strengthen the system. Everyone in the factory had to obey any order from the chief of the group, section, department, and factory office to which he or she belonged. Though the "one-boss system" gave full attention to the accomplishment of every production task, it was criticized as neglecting democratic management and hence inhibiting the workers' initiative. The Soviet-style centralized leadership was abandoned finally in 1961; it was replaced by leadership on the basis of consensus.

The Organizational Hierarchy of Quality Management.   During the early years of the People's Republic, China was divided into six large administrative areas. Each area had had its own relatively independent area government. In the years 1952 to 1954, the central government gradually eliminated the large administrative area and tightened up its integrated industrial management in order to push the 5-year plan more vigorously. Industrial Ministries were rearranged and charged with full power and responsibility of administration respectively. State-owned factories were put under the direct and strict administrative management of those ministries, each of which had a department of quality or technical supervision. At the level of provincial and municipal government, there were also established local industrial bureaus and corresponding departments to execute the administrative supervision delegated by the central ministries. Aside from the central and local authorities, the State Economic Commission was created to establish quality management principles and policies and to coordinate the quality management of different ministries and localities. In factories, quality management was carried out at different levels: factory, workshop, and group. The inspection department director had a private staff and posted inspectors in the workshops and on the line, and in worker groups. This is how quality management was deployed, through local agencies, from the central government down to the factory floor. So it remained for the whole period of China's planned economy.

Various Functions of Quality Management.   Production units in China's factories were unique and simple, in that such important matters as purchasing and sales were not their business. Each factory was categorized by product and size and reported to a central ministry and local bureau appropriate to that category. Those administrative authorities planned and assigned production quotas to factories. Other commercial ministries provided materials to and acquired products from factories at prices set by the government. There were neither a market nor a commodity exchange, but only products delivered through official circulation channels. Factories could not be regarded as enterprises. Instead, in fact, they were only workshops, since there was no risk to take on the part of the factory, profit, if any, being turned over to the State, and loss, no matter how much, being made up by the State. Regarding quality management, quality standards (and sometimes even product specifications and drawings) were given to the factory by the department in charge of each ministry. The factory was concerned mainly with process specification and control, equipment control and maintenance, operation and work instructions, inspections (incoming, in-process, and outgoing), and disposition of nonconforming products.

In every factory the chief engineer was responsible for all the technical aspects of quality management, while the director of the inspection department was in charge only of the inspections. The independent position given to the inspection department assured the authority of inspection with the side effect that it lost contact with the engineering department and line workers. Consequently, the responsibility of product quality was always an issue of dispute. The inspection department, particularly its director, was supposed to take the final responsibility, but had no authority or responsibility regarding product design or process control. Quality should be the comprehensive result of the efforts of all departments. However, the quality management system of that time was a hindrance to the realization of that ideal.

Shifting the Emphasis from the Product to the Process

Manufacturing Process Design.   In 1953, the Ministry of Machinery Industry issued the "Regulations for the Work of Inspection Department" and "Regulations for the Trial Production of New Product," which together laid the foundation of quality management on both the industrial ministry and factory levels. Though these regulations put inspection first in quality management, they also incorporated Soviet-style technical supervision. This supervision began with the specification and design of product and then proceeded to the specification and design of the process. A trial production committee was set up and the inspection department was asked to participate to study the methods and results of test and analysis. The inspection department assisted in drafting technical standards and making modifications to correct deficiencies identified during the qualification test programs. So it could be said that in this way the inspection department was engaged in preventive actions. But it must be added that factories also received concrete help and guidance in their quality management from the ministries, where well-educated and talented technical personnel were concentrated. Therefore, in those days Chinese quality management was not carried out solely by factories, but was done in cooperation between factories and the ministries to which they reported.

Manufacturing Specification.   It was gradually recognized through the implementation of trial production that product quality could not be guaranteed by inspection procedures and standards alone and was determined by good workmanship and proper process control, and that both of them should be stipulated by written standards, instructions, and other illustrations. Industries started to systematically work out process specifications and operation instructions. In every factory the inspection department organized its members to learn the process specifications and workmanship criteria and supervise their implementation in collaboration with process technicians and work-group leaders. A task was thus added to the inspection department: to check the integrity and applicability of the process specifications drawn up by other engineering departments. It was intended that the inspection department would reduce its degree of separation from other departments and its isolation in the factory. But things went in a contrary direction. So long as the problem of product quality responsibility had not been solved to the understanding of all parties concerned, disputes between manufacturing and inspection, which had always existed, now intensified. A complete solution had to wait for many years.

Establishing the Metrological System.   Invasion by big powers brought different metrological systems to old China. One of the problems that the early industrialization of new China faced was the confusing state of metrological work. With industries and factories being built one after another, the administrative authorities felt it important to unify the system. Beginning in 1954, they set up special departments for metrology and standardization in factories. Measuring and test equipment was installed, qualified personnel were trained, and by 1957 a metrology system began to take shape in state-owned factories. The State Council established the State Bureau of Metrology in 1955 to unify and administer the metrological work nationwide, and issued a decree in 1959 which formally stipulated the adoption of the metric system.

The Disastrous "Great Leap Forward."   Everything in quality management seemed to be developing fairly well. Management by the Soviet model, in spite of its shortcomings, had finally put the quality management of the newly built factories on a regular basis. At the same time, modern quality management developed in the United States was introduced to China. In 1957, an operations

research group was founded within the Chinese Academy of Sciences. They began their work with research in statistical quality control, offering courses, training staff, and conducting application experiments in factories. In August 1968, dizzied by the success of industrialization (though the success was in fact very preliminary), the politburo of the Communist Party ordered that a "Great Leap Forward" movement be launched. The backyard steel furnace and the people's commune were the two "indigenous" inventions that were promoted in this movement. They were meant to speed up the development of China's economy and to help China catch up with advanced countries. But they disregarded objective conditions and scientific reason. As a result, the national economy was badly hurt. Industrial management was discarded as a nuisance. In factories, inspection departments were all dismantled, and quality management was forced to disappear. The farce continued for 2 years. And who could have known at the time that when it was over an even madder one would take its place.

Quality in Self-Reliant China (1961–1965)

The Break with the Soviet Union and Its Management. In 1960, relations between China and the Soviet Union were suddenly ended. All Soviet experts were ordered to leave China, almost overnight, taking with them drawings and technical documents of unfinished industrial construction projects. This aroused in the Chinese people a spirit of self-reliance. From the beginning, the Soviet "one-boss system," the core of Soviet-style management, had been unpopular among Chinese workers. By 1959, some factories had begun quietly changing to a management style based on consensus. So, in March 1960, as an endorsement of this movement, the Control Committee of the Communist Party circulated an instruction generally known as the "Charter of the Anshan Iron and Steel Company," which stressed harmonious leadership through close collaboration of labor and management. It was so named because there had previously been passed to China the "Charter of the Magnetogorsk Iron and Steel Company" representing the stiff and rigid Soviet management. The Chinese edition of the "Charter" was apparently the antithesis of the Soviet one. It was intended to emancipate people's minds from blind worship of untested teachings and proved to be correct for conditions as they existed in China. It encouraged a mass movement of technological innovation and managerial reform.

The Emerging Chinese Model. In September 1961 the Central Committee of the Communist Party promulgated the "Seventy Regulations in Industry." Drawn from both the positive and negative experiences of economic development and administrative practices of the previous years, the regulations outlined the principles of industrial management to be carried out. It was actually the first comprehensive summary of the exploration of socialist economic management and it had far-reaching influence in later years.

Factory Director's Responsibility. The regulations clearly stipulated that the factory director took the management responsibility under the leadership of the Party committee in the factory. The decision left a question unanswered from beginning to end. It was expected that the Party committee could motivate the workers and establish good morale, and that the director could manage deputies and department heads in professional discipline. In fact, if the secretary of the Party committee and the factory director consulted and cooperated with each other, then the system would have worked well. If, on the contrary, they did not cooperate for any reason, the system would have led to an unstable or even a harmful situation.

In any case, the responsibility system was built on delegation by the director and regular division of management, and it cleared away the confusion and obstacles from various sources. As to quality management, inspection still played the leading role, but within a better framework than before. Beside the usual first-piece, patrol, and final inspection, there was implemented a combination of self-inspection, mutual inspection, and specialized inspections. A strong quality consciousness was to be consolidated through the new system of inspection.

Democratic Management. The gist of the "Charter of Anshan Iron and Steel Company" was the so-called two participation, one reform and three-in-one combination, i.e., management staff participate in line work, workers participate in management; reform through mass movement; management, technicians, and workers collaborate in combination to solve technological and managerial problems. The "Regulations in Industry" also promulgated a democratic management system which embodied the principle that the factory director was responsible for duties under the guidance of the Party committee on the one hand and under the supervision of the Worker Congress on the other. Meanwhile, a campaign of "Learning from Daqing" took place all over the country. Daqing was the first big oil field, discovered in 1960. Within 3 years a huge oil refinery had been constructed there. Needless to say, Daqing contributed enormously to China's economy, but above all was its democratic management, of which the main point was its managerial training of all employees "to be honest, to be strict and to be same." That is, to be an honest person, speaking and acting honestly; to be strict in organization, requirement, attitude, and discipline; and to behave in the same way, no matter whether day or night, in good or bad weather, inspected or not inspected, whether the leader is on site or not on site.

Essentially, the Daqing experience was the revolutionary and scientific spirit which kept the workers' production enthusiasm alive and the enterprise management efficient and effective. With this spirit in daily work, quality management began to put prevention first. For instance, because line inspectors were chosen from quality-conscious and experienced line workers, they were asked to fill a triple role: first, to be a propagandist to explain to the line worker why quality comes first; second, to be an instructor, to tell the line workers how they should operate to protect their product from nonconformity; and last, to be the inspector—to decide whether the products meet the standards. This was a useful measure to prevent nonconforming products, but also proved to be a wise method to avoid the disputes between production and inspection which had often occurred in the past.

A National Campaign. From the beginning of the People's Republic, democratic management in factories was sought. In the best circumstances, democratic management is easier said than done; widespread distrust of the then-dominant autocratic Soviet management model made it even more difficult to achieve. By early in 1957, statistical quality control (SQC) had already been introduced to China by scholars returned from study in the United States. Courses had been offered, technical staff had been trained, and experiments had been done in some factories. However, while SQC remained a useful tool in the hands of the technicians and engineers in charge of inspection, it would be rejected by the management and labor as well. This lesson had already been learned at that time. When the call to "integrate theory with practice" sounded over the country, SQC people, research scientists and college professors, went down to the grass-roots units and realized that SQC had to connect itself with the democratic management movement then in progress if it was to become a useful element of quality management. Data collection, for example, would come to nothing without the wholehearted cooperation of the line workers. For another example, when the concept of the process capability study was introduced to factory people by lectures and experiments, the workers spontaneously grasped the meaning and intuitively recognized that this was the common language by which engineering, production, and inspection could communicate in order to guarantee product quality and was the practical way to implement the "three-in-one combination" and the "two participation" democratic management.

Advancing from product inspection to process control, management, technical staff, and workers got together to discuss quality problems and related matters, such as product specifications, equipment maintenance, work standards, and production costs. Here staff were able to get first-hand data and material in the workshop, and workers were able to work in concert with management. Appreciation of the new quality control thus grew gradually. In 1964, the Ministry of Machine Building decided to promote process control through process capability study in all its factories.

The Cultural Revolution (1966–1976).
The so-called Cultural Revolution started, developed, and ended beyond rational imagination. It was a farce, a tragedy. It was a nightmare, a disaster. Daily life— political, economic, and social—fell totally into turmoil. Economic management systems—from the central government to local authorities—were gravely damaged. Order in work and production was disrupted completely. Every principle of the "Seventy Regulations in Industry," such as the proper relationship between the State's centralized administration and the enterprise's independent management, and the management responsibility system, were criticized as evil institutions which "bother, block, and suppress" the working class. Instead the "three-nil factory" was

recommended, which had no administration from above, no management of itself, no rules and regulations at all. Under these circumstances, quality management, which had been implemented and improved so much, became a taboo. In 1972, industrial product quality had deteriorated so dangerously that a proposal was raised by the National Planning Conference to re-emphasize "quality first"; it was overruled. The "Cultural Revolution" brought disorder, irresponsibility, and low morale to factories and put China's economy to the verge of collapse. What is more, intellectuals, engineers, and technicians were vilified as "stinking rascals," their endeavor to absorb knowledge from advanced countries abroad and to apply it to the economic construction of their own country was taken as "blind worship and faith in things foreign." Be it SQC or TQC, it was forbidden in those sad years.

Reform and Open Door (1977 to Present)

Cultural Revolution.   The "Cultural Revolution" was ended in October 1976. Various efforts were undertaken to bring order out of the chaos created in those 10 years. Government organizations were gradually restored and replenished. Factories resumed regular production order and proper leadership, step by step. In April 1977, the State Council convened a national industrial conference which, among other things, decided to promote a nationwide campaign: "Trust me with quality." The campaign aimed to arouse the quality consciousness and sense of responsibility of workers, asking them to inscribe their names on their own products, a measure very like the old one carried out in the dynasties. Though the request was liable to make workers overintense on the processing production line, and could not continue for long, it did educate the workers that quality was a serious matter. Since then the campaign has evolved to an ordinary team-management of factories, where a group or team of good workmanship and quality product would be given the title "Team can be trusted." In April 1978 a new "Thirty Regulations in Industry" was issued, the main points of which were to rectify the management leadership of factories and to put product quality at the first place for factories in fulfilling the state-planned production assignment. A responsibility system was once more put forward and it was stipulated this time that a deputy director of the factory was to be responsible for product quality and related matters. A number of factories with difficult and long-standing problems thus recovered within a short time after the release of the Regulations.

Qinghe Woolen Mill.   As a matter of fact, Total Quality Control began independently at Qinghe Woolen Mill in 1976 and at the Beijing Internal Combustion Engine Factory in 1977 with help from China's own experts, and the success achieved had drawn the attention of the country. But the dispatch of QC experts from Komatsu, a world-famous manufacturing company of Japan, working in the Beijing Internal Combustion Engine Factory under the guidance of Dr. Kaoru Ishikawa in the summer of 1978, gave a tremendous boost to TQC in China. All of a sudden, factory people were astonished to hear about a system of quality control in which all factory members had to participate, and rushed to the Beijing Internal Combustion Engine Factory from all over the country to learn what it was. Next year a group of technicians and engineers of the Factory was kindly invited by the President of Komatsu, Mr. Ryoichi Kawai, to come to its Koyama Engine Factory to practice TQC in the TQC atmosphere of Komatsu. The book written by members of the group, recording their experience in Koyama and published after their return, was circulated so widely that TQC in the early days in China was always connected with the names of Komatsu and Beijing Internal Combustion Engine Factory.

Reform and Opening.   In December 1978 China adopted a policy of "reform and opening to the outside world." It proposed emancipating the mind and at the same time seeking truth from facts. It demanded that issues in management methods and management institutions as well as economic policies be studied and solved carefully. A common acknowledgment was reached that management and technology were the two wheels of the vehicle of production in which China's technology lagged behind the times and management was even further behind. The new policy greatly encouraged Chinese people to strive for a better understanding of the outside world and a better way to rebuild their country. High official delegations were sent to Japan, the United States, and Western European countries, one after another, to learn business management. They brought back new knowledge and fresh excitement. The China Enterprise Management Association and the China Quality Control Association (CQCA) were set up separately in 1978 and 1979 at the suggestion of one of these delegations. The mission of CQCA with its local and trade branches was to promote TQC among enterprises and to provide consultation and advisory services to governments of different levels. When the State Economic Commission began to take the responsibility of promoting TQC in state-owned factories within the conditions of the planned economic system, CQCA actually became the acting body of the Commission in this respect.

In 1978, the State Council approved the proposal from the quality professionals to make September "Quality Month," during which prestigious awards were given to the enterprises with outstanding quality product and quality management. Propaganda on TQC was conducted on a national scale. Mass media were mobilized. In 1980, a TV program on TQC was broadcast for the first time and repeated in different versions in succeeding years. The TQC TV program was specially featured as an educational course.

Students were recruited through industrial ministries and were qualified if they passed the examination directed by CQCA and other authorities. It was reported that more than 10 million people of different occupations have watched the TQC TV program in the past 15 years. By the time the "Provisional Regulations for the Implementation of TQC" was issued by the State Economic Commission in 1980, TQC had been disseminated broadly throughout the country. Beginning in 1978, QC circles were organized in state-owned factories, and they have held their local and national conventions every year since. In 1983 the "Provisional Regulations for QC Circles" was issued by the State Economic Commission which put the QC circle activities on a healthy footing and in a more influential position. In 1980, the number of QC circles was estimated to be 40,000; by 1995, the officially registered QC circles had increased in number to 1,360,000, with an economic benefit to the enterprises of more than 20 billion yuan ($US 1=8.31 yuan). By the end of 1985, 38,000 state-owned factories in a variety of industries had implemented TQC with special departments in charge of the effort. On the basis of the Seventh Five-Year Plan (1986–1990), which stressed repeatedly the importance and necessity of quality management, the State Economic Commission made a corresponding resolution to examine and reinforce the TQC of 8200 medium- and large-scale factories within the 5-year plan period. Though small in number, these factories account for 60 percent of the entire annual industrial output of China. The heavy promotional task fell on CQCA. A new set of criteria was designed for this purpose which focused on the establishment of a quality assurance system.

New Era.   Thus a new era of quality management arrived, a result of the Reform and Open Door policy. The foremost breakthrough was the exchange of ideas and experiences with foreign quality professionals and organizations. As mentioned above, Dr. Ishikawa came first in 1978 and came almost every year thereafter until his death in 1988. Dr. Lennart Sandholm, Dr. Genichi Taguchi, and Dr. Yoshio Kondo were among the earliest experts to visit China. In 1982 in Beijing, Dr. J. M. Juran delivered a week-long series of lectures. Despite Dr. Juran's advanced age, the lectures were moving. Dr. H. James Harrington's enthusiastic help was welcomed. Dr. Hitoshi Kume was also a frequent visitor. Their personal lectures and advices had been most valuable to the development of TQC in China. Doctor A. V. Feigenbaum, though he did not come in person, consented warmly to be an honorary advisor to CQCA. There were so many foreign experts who came to help that it would be impossible to give all their names here. In addition, quality organizations such as the Japanese Union of Scientists and Engineers (JUSE) and the American Society for Quality Control (ASQC) were not reluctant to render their aid and hospitality to China. Numerous delegations and study groups from China also visited many foreign countries, and it must be noted that the first official QC delegation to the United States was invited by the American Association for the Advancement of Science in 1983, and had the honor of meeting prominent American scientists and entrepreneurs on the occasion arranged by the American National Academy of Sciences and the American National Academy of Engineering. Yet the big event would be the First Congress of the Asia-Pacific Quality Organization convened at Beijing in 1985, where Chinese quality professionals were able to become acquainted with so many colleagues from abroad at home. The exchanges of knowledge and experiences not only enriched the QC expertise on the China side but also enhanced mutual understandings among experts of different countries, the latter being by far the most beneficial and valuable outcome.

Readjustment.   In the years since the Reform of 1977, China's economy has undergone several periods of readjustment, restructuring, consolidation, and improvement. In each period it has been emphasized that enterprises should strengthen themselves by improving quality rather than expanding in quantity. This message undoubtedly supported the promotion of TQC. Nevertheless, it has not always been clear for TQC. For instance, in 1988 the "Law on the Industrial Enterprises Owned by All the People" was promulgated, which made a clear separation between ownership and management, assigning more power and responsibility to the enterprise manager.

In the process of enforcing the law, the administrative functions of the government department shifted from exerting tight control on the management of enterprises to creating a better environment for the development of enterprises. This shift should be very helpful to the implementation of TQC in enterprises, but enterprise managers widely misjudged the intent of the Law and took advantage of their enlarged power to pursue quick profit by expanding quantity at the expense of quality. Under these circumstances, quality management was weakened and the inspection department was even abolished in some cases. From this bitter experience several conclusions were drawn. First, Total Quality Control must be "Top's Quality Control" meaning that the top management must learn first and be committed in person if there is to be a really effective and sustainable TQC. Second, product quality must have a veto over other production performance. In the computation of workers' wages and bonuses, their work quality must have priority over quantity and other matters. Third, in the final analysis quality is in the hands of workers. Therefore, their quality consciousness must be first motivated before the training of skill. Equally important was the legislation for product quality and quality management. "The Law of Standardization," effective on April 1, 1989, encourages adoption of international standards. In December 1993, the State Economic and Trade Commission, State Planning Commission, State Science and Technology Commission, and State Technical Supervision Bureau jointly issued the "Regulations of Adoption of International Standards and Foreign Advanced Standards," which supplies preferential merits to enterprises which do so.

As early as December 1988 China adopted the ISO 9000 series of 1987 for national standards in quality management, adding a few technical complements and changing the coding system. The converted standards posed some problems in communication and cooperation with foreign experts because of their nonconformity with the original. In December 1989, a National Technical Committee, as the counterpart of ISO TC 176, was formed to be the technical authority over the standardization of Quality Management and Quality Assurance in China. The "Provisions of Quality Product Certification," issued by the State Council in May 1991, made existence of a quality system a necessary condition for an enterprise to apply for the certificate. As of the end of May 1996, 721 certificates of accreditation had been issued by state-recognized authorities to enterprises in China. Overseas organizations, such as Underwriters Laboratories Inc. in the United States, Canadian Standards Association, British Standards Institute, and others are also involved in quality certification in China. The whole certification system is helping China's enterprises to produce commodities of better quality and safety, in conformity with international standards. In August 1992, the State Council issued the "Decision on Further Strengthening Quality Management". The document summarized the achievements and shortcomings of quality management since the beginning of the Reform and Open Door Policy, and stressed the crucial meaning of quality and quality management. It required governments and enterprises at all levels to have a sense of urgency and crisis regarding quality.

The spirit of the 10 articles of the State Council's decision was to emphasize three elements: the full utilization of the market mechanism to force enterprises to improve their product quality, provision of a framework of laws to guide and regulate enterprises' quality management, and the education of the people to exercise their legal rights regarding product quality. Accordingly, in September 1993, the National People's Congress (NPC) passed the "Laws of Product Quality" stipulating the rights and duties of the producer, and introducing for the first time in China the idea of product liability. A month later, in October 1993, NPC passed the "Laws of Consumer Rights Protection," which made clear the consumer's right to complain and to be compensated for inferior quality.

Though not perfect, these two laws have already begun to put pressure on producers to pay due attention to their product quality, and encourage consumers to seek compensation instead of accepting inferior quality goods in silence.

Quality Long March, a TV Program which began in 1992, is a lively example of educational work in consumers' rights. TV reporters travel around the country to collect and broadcast consumers' opinions on product quality and at the same time secure responses to their complaints from the producers concerned. The TV program did a good job in making known to the consumers their legal right to protect themselves from inferior-quality products and in warning producers of their duties to produce quality goods. Quality Long March has now become a regular national program every year during Quality Month, and has invited quality experts to tour-lecture on quality since 1995. Another event of Quality Month, the China High Level Forum on Quality, also began in 1992. High officials of the State, business representatives, quality experts, and scholars meet at the Forum to discuss quality issues and make pertinent proposals to the parties concerned. Vice premier Zhu Rongji told the Forum in 1992 that quality should be the life of China's economy. In 1993, another vice premier, Li Langing, proposed to the Forum that China could prosper only through superior quality. Their words greatly encouraged China's quality professionals to strive for better quality of products, services, and life.

QUALITY AND THE TRANSITION TO THE MARKET ECONOMY

China's quality and quality management cannot be separated from the economic system of China. A brief introduction to the changes of the economic system is necessary for a clear discussion of quality and quality management. From the first 5-year plan period (1953–1957) to that of the fifth (1976–1980) China had practiced the centralized planned economic system. The Reform and Open Door Policy was proclaimed in December 1978, halfway through the fifth 5-year plan period.

In 1980, the first test of a different economic system was implemented at Shenzhen, a small area near Hong Kong. A system of "mainly planned economy supplemented by adjustment of market" was proposed favorably in the Twelfth National Congress of the Communist Party of 1982. This was equal to admitting the law and function of the market. In the Thirteenth National Congress of 1987 the wording changed to the "planned commodity economy."

How much difference there is between these two phrases is a question which has puzzled many people. In any case, one thing was felt for sure: The economic system of China was going to change. The Fourteenth National Congress of 1992 made the proposal of the "Socialist Market Economy," and the resolution passed by the National People's Congress in 1993 determined officially the new economic system. During the whole process of change of economic system, administration of enterprises by the government and enterprise management itself changed too. So did the management of quality.

The Challenge for China in the World Market. The transition from the centralized planned economy to the socialist market economy is a challenge as well as an opportunity. The challenge comes from the market. If China meets the challenge with success, then China will prosper. In this way the market offers the opportunity as well. To meet the challenge China must deepen the Reform and widen the Opening.

Nature and Status of State-Owned Enterprises. The first important and urgent task is to restructure and rejuvenate the state-owned enterprises, which lost their vitality during the almost 30 years of planned economy. The separation of ownership and management power stipulated by the State Industrial Enterprises Law of 1988 has to be accomplished by restructuring the state-owned enterprises to corporate organizations which make their own managerial decisions, take full responsibility for their own profit and loss, develop themselves by their own efforts and resources, and restrain themselves by observing laws and regulations. In 1992, the government issued laws and regulations for State enterprises to change their management mechanism and defined the modern corporation system as the goal of enterprise reform. In 1994, from the tens of thousands of State-owned enterprises, 100 large- and medium-sized ones were chosen to take part in the experimental introduction of the modern corporation structure. The crucial and difficult task in this experiment was transforming the existing administrative relationship between the government and the enterprise into an economic one; that is, the government is the owner of the enterprises' property, which is operated by the enterprise manager. The results were satisfactory and the reform began to spread nationwide in 1996. Though there are many problems, such as property rights, huge debts, and inactive employees, the reform of State-owned enterprises will be achieved with the successful macro economic control.

Unification of Democratic Management and the Legal System.   According to the State Industrial Enterprises Law of 1988, the state is supposed to ensure that staff and workers enjoy the status of the masters of their enterprises, and the enterprise should, through the staff and workers' congress, practice democratic management. The trade union in the enterprise should represent and safeguard the interests of the staff and workers, and should organize them for participation in democratic management and democratic supervision. It is the first time in history that democratic management has been thus stipulated by law. The establishment of a modern corporation system is combined with democratic management. The "Law on Corporations," effective on July 1, 1994, further stipulates the relationship among owner, manager, staff, and workers and the right of the staff and workers' congress and trade union to send representatives to the meeting of the Board of Directors in case there is discussion of matters concerning the interests of the staff and workers. The "Labor Law," effective on January 1, 1995, stipulates the rights and obligations of labor. Under this law, democratic management not only assures the right of employees to participate but also imposes a duty to obey the rules and regulations of the enterprise.

International Business Environment.   Foreign trade has increased rapidly since the Reform and Opening, particularly after the change to the market economy. China has become a country relying heavily on foreign trade; in 1995, the combined total value of exports and imports was more than 40 percent of Gross Domestic Product (GDP). In the same year, foreign investment in China amounted to 150 billion U.S. dollars. Competition in both the domestic and foreign marketplace is becoming more and more intensified. In recent years, the sharp cut in China's import tariffs has especially accelerated the import of foreign-made commodities, which in turn, has put pressure on State-owned enterprises. Competition is good, but Chinese producers of commodities lack competitiveness. This is precisely one of the reasons that the State has speeded up the reform of State-owned enterprises in order to improve product quality and cost. The ISO 9000 family of quality-management and quality-assurance standards is currently popular for the same reason.

An Assessment of Quality in China Today.
Much has been said about quality management at different stages of China's economic development. Let us turn to the question of the status of quality today and the impact on quality of the changes described in the economic system and in quality management.

Quality in Manufacturing Industries.   In the early period of the planned economy, product quality as a whole was guaranteed, though the quality standards were low and the variety of products was limited. But, in parallel with the progress of economic reform, collective and private enterprises mushroomed, suddenly enlarging the workforce without prior and proper training and without resources for management and supervision of product quality. State-owned enterprises expanded their production in pursuit of quick profit by emphasizing quantity at the expense of quality. Even shoddies and counterfeits appeared in the market. A sampling survey of product quality, conducted quarterly by the State Technical Supervision Bureau, shows the percentage of the product tested which met the applicable standards. The summary results for some representative years from 1985 through 1995 are

1985	65.4
1987	77.0

1991	80.0
1992	70.1
1993	70.9
1994	69.8
1995	75.4

The kinds of products and producers inspected vary in each survey; hence, strictly speaking, the figures are not comparable. Particularly, the survey is intended to inspect enterprises whose products caused consumer's complaints. Therefore, the results are not fully indicative of the general situation but rather reflect the product quality of those enterprises whose quality management was poorer and often ignored the national standards. The State-owned large-scale enterprises are usually much better. For instance, the survey results for the first half of 1996 reveal that an average of only 80 percent of products inspected met quality standards, 10 percent higher than the worst years. The large-sized state-owned enterprises in that survey achieved 91 percent.

Quality in Service Industries. It is perhaps a remnant of the "Cultural Revolution" that the service provided by State-owned enterprises is notoriously poor. A TV program on TQC jointly sponsored by the Ministries of Commerce, Post and Telecommunications, Railways, and others was broadcast for the service industry of China over and over again from 1987 to 1992. It introduced the activities of outstanding QC circles in the service industry and some very simple QC techniques presumably helpful in improving service quality. The response from society was unexpectedly favorable. The Beijing municipal government asked the TV network to broadcast it for the citizens of Beijing on the eve of the Beijing Asian Olympic Games. In 1995 another TV program introduced ISO 9004-2. In it, the Director of the State Technical Supervision Bureau, Li Chuanqing, presented the opening address.

Quality in Government Service. The 40 years of economic construction of China can be divided into three stages: In the first stage (1949–1952) government took over the economy; in the second stage (1953–1978) government ran the economy, in the third stage (1979–present) government has undertaken to reform the economy. Now economic reform has come to a crossroad which demands reform of government itself. As mentioned above, the key to the establishment of a modern corporation system lies in the separation of government from enterprises. A 3-year plan which began in 1993 has been implemented to reform the administrative management system and the functions of government departments. It set out first to reduce administrative personnel in the departments affiliated with the State Council by 20 percent and local government departments by 25 percent. (The first target was achieved by the middle of 1995, and the outlook for the second is good, according to the report sent to the central government by local institutions.)

It then began to readjust government department functions and the relationships between different government departments to avoid duplicating their functions and to improve their efficiency. Readjustment on the central level has basically been completed. Internal branches and personnel have been established. Government departments thus strengthened their supervision and macro guidance of economic development through prices, taxation and other financial means without touching concrete matters of the management of enterprises. Government services probably is the only area where TQC has not entered yet, though most of the ministries have a quality control association under their administration as a peripheral organization through which the ministry provides guidance of quality management to enterprises and engages in international exchanges of information and experience in quality management.

EPILOGUE

Quality versus Speed. In retrospect China's product quality has not met expectations, and the reasons are many. However, the dominant one lies in the "go-for-speed" policy explicitly or implicitly cherished by government officials and followed by enterprise managers. In years past, when officials in charge of a department or a locality were evaluated on their administrative performance, economic growth was almost the sole factor on which they were judged. It was not unusual that stock or products unsold or even useless were calculated as output value. The result was a large output value but small economic benefit.

The "Outline to the Ninth Five-Year Plan (1996–2000) for National Economic and Social Development and the Long-Term Target for the Year 2010," passed by the National People's Congress in April 1996, pointed out that "high speed" is contingent on efficiency and quality, and put forward two fundamental changes necessary for maintaining the sustained, rapid, and healthy development of the national economy: first is the shift from the traditional planned economy to the socialist market economy; the second is the shift of the economic growth mode from extensive to intensive. For this end, it is necessary to bring the role of market mechanism into full play and to reform China's State-owned enterprises from extensive management to intensive management so that they can compete in the market to their advantage. Total Quality Management as an effective means of quality improvement and as a philosophy of management can definitely serve well the demand of an intensive-growth mode.

"Outline of Rejuvenation of Quality."   On September 3, 1996, the fiftieth meeting of the State Council discussed and passed a national program, "Outline of Rejuvenation of Quality." The document was drafted by quality professionals of many fields under the leadership of the State Technical Supervision Bureau in July 1993. The draft had been sent to all ministries to solicit opinions, and the final version was once again discussed and revised by members of the State Economic and Trade Commission in August 1995. The long duration of drafting, discussing, revising, and approving illustrates the complex nature of quality. Everybody knows it, everybody can say something of it, and in the end nobody knows truly what it is.

The outline points out that quality is a matter of strategy in the development of the national economy; the economy will pay heavily if quality is neglected. It set goals for product quality and quality management to be achieved by the year 2010. To promote TQM continuously and to seriously implement the ISO 9000 family of standards, "Quality Management and Quality assurance," are the two important items among others in the national program. Introduction of foreign advanced technology and management is also stressed.

Since this day, September 3, 1996, China finally has had a national program of quality rejuvenation.

REFERENCES

Jin Quipeng, Chen Meidong, and Lin Wenzhao (1995). "Ancient China's History of Managing for Quality." *A History of Managing for Quality,* J. M. Juran, ed. Quality Press, Milwaukee.

Schafer, Edward H., and the editors of Time-Life Books (1967). *Ancient China,* 6th printing, Time-Life Books, New York, 1976.

SECTION 43
QUALITY IN LATIN AMERICA

Marcos E. J. Bertin

FROM ECONOMY TO QUALITY

The industrialization process began to accelerate in Latin America in the late 1950s. At that time, business leaders responsible for results (company presidents, general managers, owners) became involved in management processes as never before. They were very interested and eager to learn about the elements of business management—marketing, finances, costs, and personnel. They had only a minor interest in quality control. Most of their time was devoted to lobbying the government on matters of trade and to negotiating on labor issues with very strong unions. The degree of their interest in any subject varied according to the then-current level of inflation, economic activity, price and salary controls, import duties, taxes on exports, foreign exchange controls, interest rates, and the probability of devaluation of the local currencies. The leaders also had to deal with the huge government-owned monopolies (all financed with big government deficits) on which their businesses relied for various vital products and services. The most important of these were oil (extraction, production, and distribution), utilities (electricity and water supply), communications, transportation, and steel. With so many monopolies under its control, government, also a major buyer, had many suppliers, thus exerting another form of influence on the economies of the region. Therefore, it is not surprising that business leaders were totally occupied with the effects of government regulations and controls.

At any given moment, a reliable indicator of which business factors were the critical ones was to be found in the categories of experience sought in the help-wanted advertisements in the newspapers. One day the advertisements might stress commercial experience; the next day, financial; the day after, government relations.

Latin America has a long history of able entrepreneurs who knew how to ensure the growth, profitability, and, even in difficult times, survival of their companies. Many of them did it with success, despite an uncertain and even hostile business environment.

Quality and excellence were not critical factors in the business environment of the period from 1950 to 1980. In that economic environment, industrial growth happened largely because of government protection; soft loans were available at terms that amounted to a subsidy to promote investments in local production during periods of high inflation and high import duties. With some exceptions, quality, both in products and services, was below international standards.

43.1

Quality control was a subject of interest only within the factory. The people who became the quality pioneers in Latin America were engineers and other technical professionals working in local factories. They became very interested in the manufacturing improvements being reported, mainly in the United States of America, and were eager to learn about the new techniques being credited for those improvements.

As early as 1959, the American Society for Quality Control (ASQC) and, later, the European Organization for Quality (EOQ) and the Japanese Union of Scientists and Engineers (JUSE) helped to develop new professional societies in Argentina, Brazil, Chile, Colombia, and Mexico. During the 1960s and 1970s, experts from the United States, Europe, and Japan belonging to these organizations and to the International Academy of Quality (IAQ) gave courses and participated in conferences that were supported by leading companies. But presidents and general managers did not attend. They did not regard quality as a critical business factor; indeed, it was difficult to dispute this opinion in a time of financial crisis.

In the years 1989 to 1992, when annual inflation reached 2500 percent in Brazil and 1700 percent in Argentina (where it peaked at around 96 percent per month in March 1992), the entire attention and energy of chief executives had to be focused on financial measures to preserve their enterprises. The price of inattention was bankruptcy. The legacy of the prevailing economic system was enormous deficits in government budgets, deficits in the balance of trade, and ongoing crippling inflation.

In the 1980s, Latin America went through dramatic changes. On the political side, military governments were replaced by democratic governments. On the economic side, governments were forced by the severe monetary crisis and the requirements of a highly competitive global economy to carry out economic reforms based on free-market principles.

The most important changes were

- Privatization of government monopolies
- Improvement of the national infrastructure, both in quality and productivity, in response to the needs of business
- Reduction of regulations
- Reduction of government budget deficits (to zero in some countries)
- Control of inflation at low levels
- Reduction of customs duties for finished products, raw materials, and equipment
- Opportunity for businesses to increase participation in international trade, due to trade agreements such as NAFTA (North American Free Trade Agreement, the trade agreement of 1993, which reduced trade restrictions among Canada, Mexico, and the United States) and MERCOSUR (the common market between Argentina, Brazil, Paraguay, and Uruguay), and to lower customs import duties on equipment and raw materials
- A significant increase in private-sector capital investments
- Introduction of government policies conducive to competitiveness.

Throughout the region, the fact that free elections brought democratic candidates to office was widely interpreted as a social consensus of support for these many changes.

The economic environment changed dramatically, and with it changed many rules of the game for business. Many companies and institutions did not survive the crisis; either their cultures changed dramatically or they were succeeded by altogether new companies. As the environment changed, the interest of most CEOs shifted from survival issues to strategies for improving business competitiveness. Quality, productivity, and cost became the key strategic issues. For the first time, presidents and general managers in many companies became directly involved in quality programs—within the company and at the national level. However, there are still many owners and CEOs to convince, especially in the millions of small- and medium-size companies (SMCs) that account for about 40 percent of the gross industrial product of Latin America and which employ around 60 percent of its work force.

According to a 1995 study: "In a nutshell…Chile and Argentina led the revival of Latin America" (*The World Competitiveness Report* 1995). Brazil, Colombia, Peru, and Mexico also showed significant economic improvements over the period 1990 to 1995. (Mexico suffered political and economic setbacks in 1994 and 1995, and is working its way back to health in these areas.)

THE EVOLUTION OF QUALITY: A COUNTRY-BY-COUNTRY REVIEW

The recent history of quality in Latin America is a history of four critical factors:

1. Business competitiveness
2. The economic climate
3. CEO involvement
4. Participation of quality professionals

ARGENTINA

The Business Foundation for Quality and Excellence (FUNDECE).   In the early 1980s, some CEOs foresaw the changes that were approaching in the economy. In 1987, after meeting informally for a few years, they founded the Business Foundation for Quality and Excellence. Its mission is

1. To promote the improvement of the competitiveness of business through the implementation of formal quality efforts. FUNDECE emphasizes the importance of the commitment and involvement of CEOs.
2. To improve the quality of life through education, health care, and government services.

Membership in FUNDECE is open only to CEOs. Experience suggests that CEOs listen to and share ideas easily with other CEOs. Only CEOs can attend FUNDECE meetings. When nonmember organizations participate by special invitation, it is only the CEO who is invited. Sometimes high government officials are also invited.

Participants in the meetings value this policy of including only CEOs, as it reinforces the need to maintain the focus of discussion at the highest policy level. There have been a few occasions when second-level managers asked for permission to participate. The requests were always diplomatically but firmly refused. Some of those who were refused admission as lower managers later became presidents and, therefore, members of FUNDECE. One of them was elected to the Board of Directors. Those who had endured refusal as lower managers acknowledged that even at the time of refusal they felt the policy to be sound. In the author's opinion, anyone wishing to create an organization with a mission similar to that of FUNDECE should consider the implications of this restriction on membership.

As of 1995, the membership of FUNDECE comprised the presidents and general managers of 130 enterprises. Of the enterprises represented, about two-thirds are large (companies having 200 employees or more and sales of at least $US40 million for industrial companies, or sales of at least $US24 million for service companies) and one-third are SMCs (companies that do not meet the definition for "large"). The breakdown by company type is roughly 60 percent industrial, 40 percent service. Among large companies, there are two industrial companies for each service company; among the SMCs, the service companies are slightly more numerous than industrial companies. The industrial companies include Argentine affiliates of multinational companies such as IBM, General Motors, Hewlett-Packard, and Reckitt & Colman. There are also Argentine companies not necessarily well known outside the country. Service companies include such categories as private universities, banks, insurance companies, office equipment providers, health care providers, advertising agencies, and accounting firms.

The Strategic Committee of FUNDECE—a group of presidents and general managers of major companies in Argentina—created the FUNDECE Strategic Model (Figure 43.1). This model serves as a basis for the yearly FUNDECE action plan.

The most important activities of FUNDECE are

- Monthly breakfast meetings for the sharing of experiences and ideas. During visits sponsored by FUNDECE, foreign experts [from IAQ (International Academy for Quality), ASQ (American Society for Quality), EOQ (European Organization for Quality), and JUSE (Japanese Union of Scientists and Engineers)] frequently participate in these meetings. The breakfasts are the dominant and most effective activity of FUNDECE.

- Special projects. Two projects currently under development are the creation of a consultant data bank and the establishment of a library to include a collection of reference cases. Part of FUNDECE's sponsorship of such projects is to assign and fund a project leader. Special projects is the second most effective activity of FUNDECE.

- Seminars for professionals, middle managers, and academia.

- Teleconferences, including the satellite broadcast of the National Quality Forum—the industry quality event cosponsored each year by ASQC and *Fortune* magazine.

- Management in Argentina of the ASQC certification system.

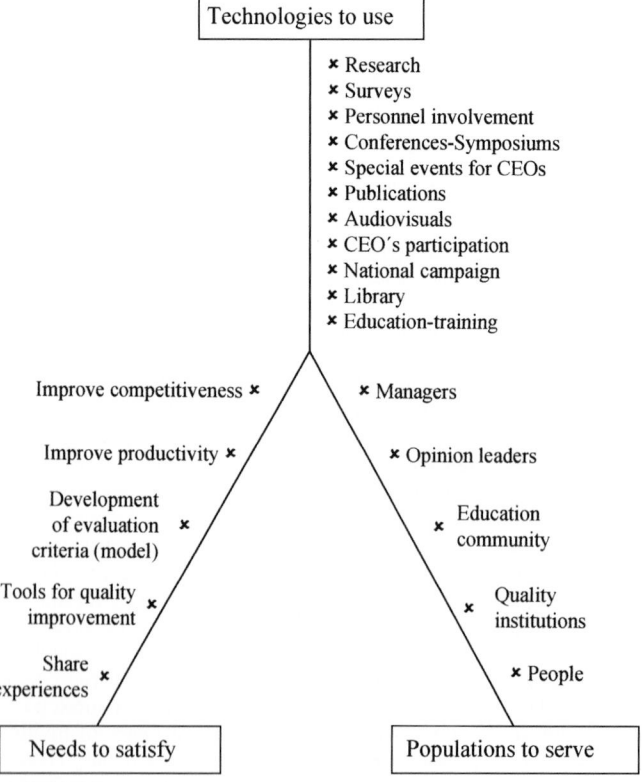

FIGURE 43.1   FUNDECE strategic model. (*FUNDECE, 1995.*)

- Publication of a newsfax with quality information, both local and international.

- Publishing of videos. These are videos developed by committees chaired by CEOs that include members from an advertising agency and other organizations. The videos feature such topics as quality implementation and problem diagnosis, and are for general business use, both for member and nonmember organizations. Also, in cooperation with the Argentine Advertising Council (an association of advertising agencies and other companies engaged in advertising) and the Argentine government, FUNDECE helped create two TV spot commercials on the subject of quality, for showing on all the country's TV channels.

- Close relations are maintained with international quality organizations—ASQC, IAQ, EOQ, EFQM (European Foundation for Quality Management), JUSE, and with national quality organizations, both in Latin America and the rest of the world.

Three FUNDECE Projects

The National Quality Award.   In 1992, the government called upon FUNDECE to take a leading role in the promotion of quality in industry, services, education, and health care. The objective was to develop an efficient national infrastructure to carry out this promotion. Working with a commission in the legislative house of representatives, the first step was to create a national quality award. The award criteria would give business and other organizations the quality model. A FUNDECE project leader guided the efforts of a team that included members of the House of Representatives commission; Wayne Cassatt, then associate director of the Malcolm Baldrige National Quality Award in the United States' National Institute of Science and Technology (NIST); and members of FUNDECE affiliated with North American companies that had won the award in the United States (e.g., Xerox, IBM). That project leader became the managing director of the National Quality Award Foundation (FUNDAPRE), a private institution created by national law to manage the award for the private sector.

A separate award was established for the government sector, administered by the Secretariat of Public Function. The award criteria provide the quality model preferred by FUNDECE members and the government. Used in self-assessment, the criteria are effective tools, particularly if the scoring system allows for giving more weight to those items that are critical to the particular business assessing itself.

The Argentina National Quality Award began in 1994 with strong support from the government, FUNDECE, and the Malcolm Baldrige National Quality Award administration.

The major difference between the Baldrige and the Argentina Award was that FUNDECE insisted that the scoring of the Argentina Award stress business results. CEOs played a major role in establishing the criteria and weighting points. For some criteria, the points are closer to those of the Baldrige (for example, for customers and planning), and in others they are closer to Europe's (for example, for leadership, personnel satisfaction, social responsibility). All the comparisons are to 1992 data. Since then, weightings in all the national awards have been changed in response to suggestions and experiences of businesses involved in the awards. As of this writing, there is not yet much sharing of experience and ideas among the different national awards administrations. In 1996, at a meeting organized in Lima by Wayne Cassatt, the managers of the Award from Argentina, Brazil, Colombia, Chile, and Peru decided to hold an annual International Quest for Excellence meeting with the managers of awards and companies that won the award. The purpose of the meeting is to share experiences and results, as well as discuss new awards for education and health care. The first took place in Brazil in September 1997.

The Professional Institute for Quality.   The CEOs identified the need for an organization in which their quality managers, their second-level managers of key areas (marketing, finance, etc.), and university professors could be trained, kept up to date, interchange ideas, do research, interact with similar organizations in other countries, and provide support to the National Quality Award and other future projects. To address this need, FUNDECE established a project to organize the technical arm of FUNDECE—a professional society to provide the "how to" of quality. The Argentina Professional

Institute for Quality and Excellence (IPACE) was founded for this purpose. The leader of the IPACE project is today its Managing Director.

The National Quality Movement.   The 1995 FUNDECE project related to the need to focus the national effort for quality and excellence in key areas. Many institutions, both government and private, are working on this. In the sense that more people are becoming aware of the importance of quality, this is good. Still, there are areas where coordination is needed to help people and institutions move in the same direction.

Business leaders should assume the responsibility of leading the process of change for two reasons: (1) it is necessary to convince many more CEOs (CEOs will listen better to other CEOs) and (2) the CEOs' training and know-how is precisely to lead organizations and projects effectively. In the past, national quality programs failed because they lacked such leadership; professionals and government specialists led these programs in an economic environment where quality was not a critical factor for business.

In December 1994, FUNDECE initiated the creation of an integrated National Quality Movement (see Figure 43.2). It is composed of six key organizations: (1) FUNDECE; (2) FUNDAPRE; (3) IPACE; (4) The National Institute for Industrial Technology (INTI)—a government institution dedicated to giving assistance to industries, in technology and in metrology; (5) The Argentinean Institute for Materials Rationalization—the standards institution that also works on ISO 9000 certification; and (6) the General Secretariat of the President—dedicated to distributing information on quality, motivating the involvement and cooperation of major institutions interested in quality, and optimizing the effect of government actions.

The mission of the National Quality Movement is to coordinate major quality activities, to improve the effectiveness of the major programs of various government and private institutions, and to ensure that all organizations and businesses view quality as a national cultural value. To accomplish this mission requires a common vision and culture. It requires a suitable structure, sufficient resources, and coordinated strategies.

In Argentina, as in many other countries of the world, debate continues on the question of the correct national quality model. Some argue for adoption of the criteria of the national quality award; others support the ISO 9000 series of quality standards as a more suitable model for promotion. FUNDECE's view is that although ISO 9000 certification is sometimes necessary for commercial reasons and a good introduction to quality for newcomers, receiving ISO registration does not

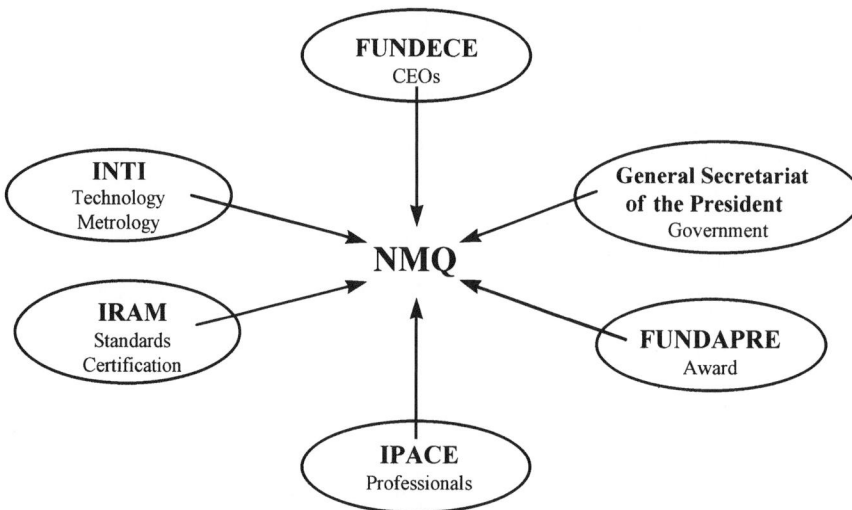

FIGURE 43.2   National movement for quality. (*FUNDECE, 1995.*)

necessarily result in acceptable quality levels in today's highly competitive world. Argentina has very few ISO 9000 certifications. However, in the period from March to November 1995 the number of certifications rose from 37 to 69, indicating that these certifications are on the rise, particularly for SMCs.

In the field of education, universities have shown interest in the subject and application of quality. Three university presidents are already members of FUNDECE. The Buenos Aires Institute of Technology offers a master's degree in quality; several other universities offer postgraduate courses in quality management. The Minister of Education is studying with FUNDECE and FUNDAPRE the recent developments in quality awards for educational institutions in the United States, such as those conferred by the NIST and the City of New York.

BRAZIL

There has been a significant increase in the participation of CEOs in quality programs since 1990, the year the government launched the Brazilian Program for Quality and Productivity (PBQP) as part of its program to modernize the Brazilian economy along the same lines as the other major countries in the region.

The PBQP is led by the government, with the participation of the following sectors and institutions:

- *Business:*   The National Confederations of Industry and Commerce (CNI) and press, advertising, and radio and TV associations.
- *Professional societies:*   The Brazilian Association for Quality Control and other societies representing consultants and specialists in fields that include human resources, research, standards, and industrial engineering.
- *Government institutions:*   The National Institute for Metrology Standards and Industrial Quality, the Brazilian Service for Support of Micro- and Small Companies, the National Service for Industrial Training, and several others in the areas of deregulation, science and technology, and health.
- *Government companies:*   Telebras, Electrobras, Usiminas, and Petrobras—huge monopolies in communications, electricity, mining, and oil.
- *Government offices:*   The ministries of Economics, Education, Labor, Agriculture, and Health, and other groups dedicated to science and technology, education, and quality. The coordination structure of the PBQP, Figure 43.3, is a matrix of general subprograms and sector subprograms.

The general subprograms are

- Awareness and motivation—to promote quality in all sectors of society
- Development and promotion of modern management methods—to stimulate the adoption, by private and government business and by government services, of modern methods for managing quality and productivity
- Training of personnel concerning quality and productivity
- Technological services for quality—to improve the infrastructure of services such as standards, quality certification, metrology, tests, and technical information.
- Institutional coordination—to promote the coordination of activities for quality and productivity among government, industry, commerce, services, academia, and science and technology organizations

The sector subprograms are individual programs designed for specific sectors, such as industry, public administration, state government agencies, and so on. Each sector prepares a reference plan that includes:

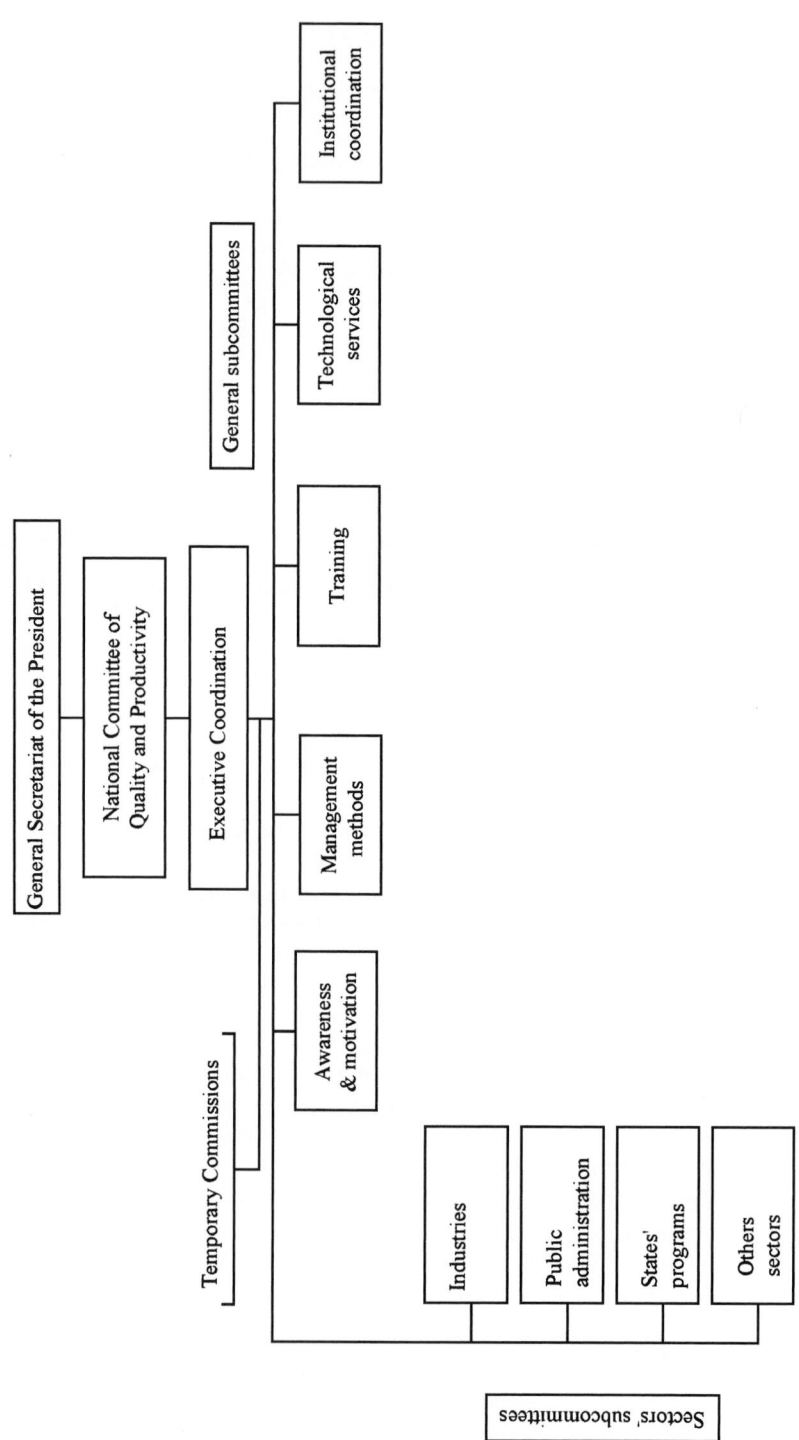

FIGURE 43.3 Brazilian program of quality and productivity structure. (*Industrial Policy Secretary, Ministry of Industry, Commerce and Tourism of Brazil, 1990.*)

- Quality and productivity diagnosis
- National and international trends in quality and productivity
- Objectives
- Strategies and action plans
- Proposed projects
- A management plan

The PBQP, through the National Quality and Productivity Committee (comprising four government officials and three representatives of private business designated by the government) provides a review of each sector subprogram as well as guidance in the creation of the sector support structure and in carrying out the subprogram. The PBQP was revised in 1995, Figure 43.4, adding the voices of consumers and workers to the coordination body. Development and publication of indicators was also included as a separate function. Supplier development and financial support were added to the general subprograms and tourism, agriculture and livestock, and services and commerce were added to the sector subprograms.

The most effective parts of the PBQP, as reported by the CNI and several CEOs, are the actions taken by the private sector, led by industries through their chambers, where each industry (e.g., chemicals, steel, leather) is represented by its own chamber. Each industry chamber has quality and productivity committees working for industry improvement and collaborating with the PBQP. Today 240 business and technical institutions are working on these programs under the industry sector sub-committees. CEOs are becoming increasingly involved in these activities.

The PBQP has made another contribution whose significance may be difficult for readers from outside the region to understand. Historically, national development programs throughout Latin America have focused on population centers—Buenos Aires, Rio, Mexico, etc. It has been a natural consequence of taking programs where there is a base of industry to build on. A frequent effect has been neglect of other geographic areas, especially rural ones. The PBQP program, however, has been effective throughout the country, with the involvement of the different states.

The major contributions of the PBQP to quality and productivity in Brazil are

- It increased collaboration between government and private institutions.
- It motivated the adoption of quality and productivity programs in many private companies.
- It provided a timely and reliable source of information to help business adapt to the new, highly competitive environment.
- It revealed a new role for business chambers, accustomed as they were to focusing on lobbying.
- It reinforced the importance of standards, metrology, and certification.
- It reinforced the importance of customer service.
- It promoted closer relations between customers and suppliers.
- It increased training efforts, and contributed to a change in management mentality.

Some weaknesses remain:

- Government leadership—the actions in place have little support in the various government departments
- A low rate of workers' union participation
- Poor coordination of program activities
- Insufficient infrastructure—privatization of government enterprises is necessary to improve the efficiency of the economic system in transportation, energy, and communications and to revise the role of government.
- A low rate of participation by universities and technology centers—These organizations have great potential, not yet realized, to support the efforts to improve quality and productivity.

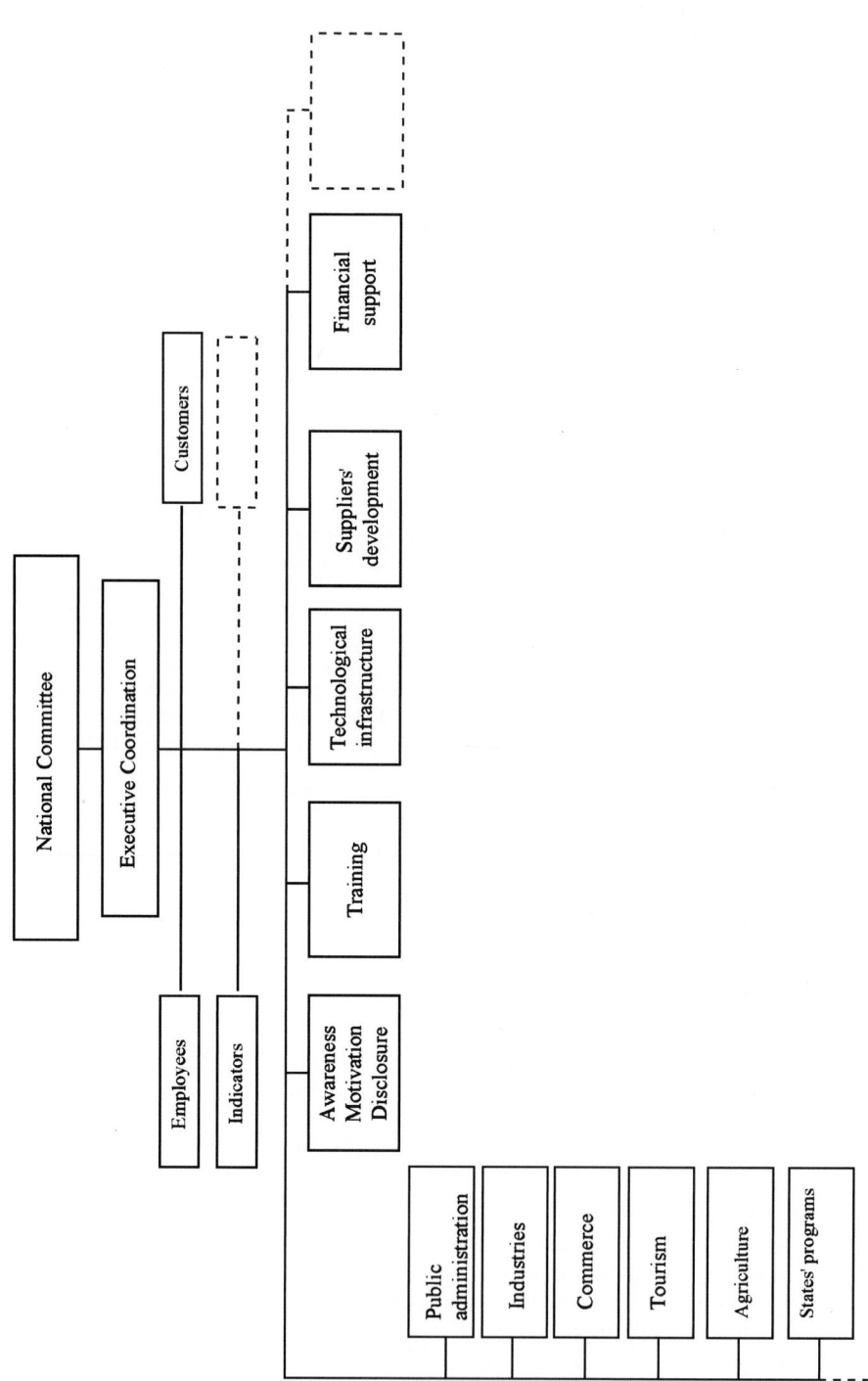

FIGURE 43.4 Brazilian program of quality and productivity structure PBQP 2000. (*Industrial Policy Secretary, Ministry of Industry, Commerce and Tourism of Brazil, 1995.*)

- Absence of design and innovation as topics in the programs
- Numbers of unconvinced executives, particularly within the SMCs
- The reduced resources available from the government

These efforts are reflected in increasing acceptance of Brazilian goods in international trade. Furthermore, in the period 1992 to 1994, industrial goods exported from Brazil accounted for a rising proportion of total value of goods exported—from 73.4 percent of total exports to 76.6 percent. (See Figure 43.5.)

This increase was especially impressive, considering that in the same period the world prices of the traditional commodities exported rose significantly. In the same period, Argentina showed a comparable increase—from 62.6 percent to 65.8 percent. The MERCOSUR agreement also had an influence on these results. With the elimination of tariff barriers between the member countries (Brazil, Argentina, Uruguay, and Paraguay), trading among these countries more than doubled, and many new investments have come into the area, induced by the incentive of a larger market potential.

Of all the countries in Latin America, Brazil has the most ISO 9000 certifications—over 1000 and increasing as of 1995 (63 percent of the total for the region). Brazilian companies see certification as an effective first step toward a quality program, or toward fulfilling a commercial need to export or to sell, in the local market, to certain major local companies.

As in the rest of the world, the emphasis is switching from satisfying the requirements of ISO 9000 to performance according to the criteria of the National Quality Award.

The Brazilian National Quality Award was initiated in 1992. As in other countries in the region, the award criteria derived, with minor adaptations, from the Baldrige Award model. The NIST supplied all the needed information regarding such matters as the award criteria, selection and training of examiners, selection of judges, code of ethics, evaluation process, etc. The Brazilian Quality Award Foundation, which administers the award, is supported mainly by private business. To date, the Foundation has trained about 1000 examiners and contributed improvements to the award process, such as a computerized system to simplify the evaluation of examiners. The Award is now presented by the president of Brazil.

CHILE

Chile was the first country in the region to carry out economic reforms, beginning in the early 1970s. These growing reforms have provided a very favorable economic environment for business, reflected in

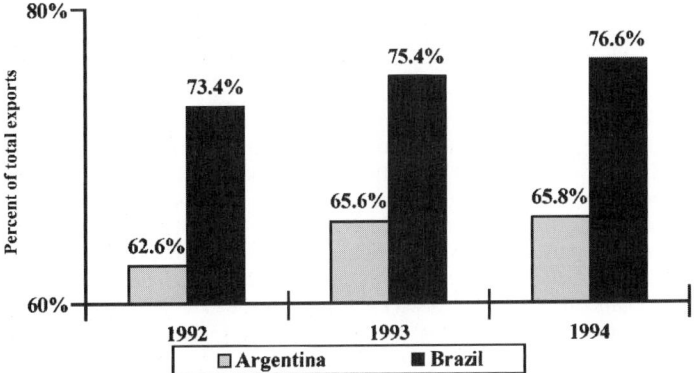

FIGURE 43.5 Industrial exports share of the total. (*Brazil: Secretary of Industrial Policy, Ministry of Industry, Commerce and Tourism, 1995; Argentina: National Institute of Statistics and Census, 1995.*)

renewed growth and prosperity. CEOs were able, for the first time in years, to concentrate their energies on managing their companies and increasing sales, particularly of exports to the United States, Europe, Asia, and elsewhere in Latin America. Some larger companies began investing in new factories in Argentina, Peru, and Brazil. Customer demands forced companies engaged in international trade—export of goods and investment in foreign plants—to incorporate quality management into their operations. As in the other countries of the region, many heads of smaller companies remain unconvinced of the relevance of quality to their operations. Even in those smaller companies which are growing rapidly toward international prominence, the CEOs have been reluctant to put energy into developing a national infrastructure to support quality; they judge it to be more important to spend the energy growing their own businesses. As of 1994, just a few pieces of quality infrastructure were in place:

- The Chilean Association for Quality, the organization for quality professionals, is very active in conferences, courses, and publications. There is very little participation by CEOs.

- Several universities offer postgraduate studies in quality management (Chile and Santiago). The Catholic University has a very active center for productivity and quality.

- The government provides both an excellent economic environment for SMCs and, through an industrial promotion institution, financial help for quality improvement programs. The Industrial Productivity Center, a private organization representing 3000 SMCs, administers funds for this promotional activity.

In 1995, a collaboration of government and some interested CEOs led to the founding of the National Center for Quality and Productivity. The Center is chaired by the Minister of Economics and comprises the heads of business organizations (CEOs in industry, commerce, and services), universities, several government departments, and representatives of workers' unions. Some leaders had seen a dangerous trend in the complacency and boastfulness with which some businesses reacted to small early successes. The reaction, they felt, revealed industry's failure to grasp the need for continuous improvement in the face of rapidly changing and highly competitive global markets. The Center was a response to what they thought was the need to develop further the infrastructure for quality, to create a model such as the Baldrige Award criteria, and to conduct research to identify the most successful quality tools for Chile.

The Center's first action (in 1996) was to present its first National Quality Award. The NIST collaborated in the development of the Award.

The organization, strategies, and plans of the National Center for Quality and Productivity are not available. The Center will probably be led by business and dedicated initially to improving the national infrastructure for education and training on quality, helping the introduction and implementation of ISO 9000 certification, and generating quality awareness, especially in SMCs.

COLOMBIA

National Quality Week began in Colombia in 1975; a national quality award has been presented annually since then. The leading private organization for quality is the Colombian Institute for Technical Standards (ICONTEC). The 15-member Board of ICONTEC contains 10 members from business organizations and 5 from government. Besides providing consulting and publication services and conducting training courses, ICONTEC serves as a registrar for ISO 9000. It is the interface between the government and private business for quality programs. Historically, business participation has largely been among second-level managers, but CEOs are becoming increasingly involved.

Other institutions focused on quality are the Colombian Association for Quality, the organization for quality professionals; the Colombian Association for Quality Circles; and several business and government organizations, including two in health care—Hospital Management (GEHOS) and Companies for Promotion of Health (EPS).

Future needs for quality promotion include involving SMCs and increasing opportunities for education and training.

MEXICO

In 1988, a group of major industrial and service companies came together to create the Mexican Foundation for Total Quality (FUNDAMECA) just a year after the need had been identified. Today, FUNDAMECA has a membership of more than 500 individuals and organizations. Membership is not limited to CEOs.

In 1989, in close cooperation with the Secretary of Commerce and Industrial Promotion, FUNDAMECA announced the creation of the National Quality Award. The first award was presented by the president of Mexico in 1990. FUNDAMECA has administered the award since its inception. In the same year, FUNDAMECA created the Mexican Association for Total Quality Managers, an organization for quality professionals.

In Mexico, CEOs mobilized for quality for the same reasons that had moved their southern neighbors to action: their companies needed to become competitive in the free market economy and within newly emerging common markets, especially NAFTA. Because of its proximity and participation in NAFTA, the United States participated in this mobilization through institutions such as the NIST, the Quality and Productivity Management Association, and the Interamerican Network of Quality and Productivity Centers.

The member companies of FUNDAMECA made significant progress during the early years, with many successes. There remains much to do to gain the involvement of the many companies not yet a part of FUNDAMECA.

Several Mexican universities, such as the Tecnológico de Monterrey, have postgraduate studies in total quality management. Years of intensive training of professionals both in Mexico and in the United States have resulted in the development of many capable professionals in the field of quality.

The economic crisis of 1994 and 1995 was a strong incentive for change, and Mexico has in place the necessary institutions for this process.

PERU

The National Society for Industry (SNI), founded in 1896, claims a membership today of 2000 Peruvian companies. The society began working on quality programs as early as 1980, but with no success, due to the economic environment at that time. When the economic changes affecting the region began to happen in Peru in 1991, SNI organized quality workshops and created a quality award. In 1992, the government made the award official. In that same year, the government and SNI founded the Center for Industrial Development, a private institution with strong ties to the government, to administer the award. In 1995, a joint team of government and SNI members worked with the NIST to incorporate lessons from the Malcolm Baldrige Award into Peru's award. CEOs of member companies of SNI provide the driving force of leadership in these organizations.

URUGUAY

Facing the new challenges of Uruguayan membership in MERCOSUR, Uruguayan CEOs founded the Uruguayan Association of Companies for Total Quality (AUECE) in 1993. By 1995, more than 50 CEOs were members of AUECE. FUNDECE (of Argentina) worked very closely with AUECE and helped stimulate its organization.

The National Committee for Quality, chaired by the Secretariat of the President of Uruguay, developed and manages the National Quality Award. The first award was presented in 1993 by the president of Uruguay.

AUECE, like FUNDECE in Argentina, is leading the Uruguayan quality programs.

CONCLUSIONS—FOUR CRITICAL FACTORS

Four critical factors underlie the emergence of quality in the countries of Latin America: business competitiveness; a favorable economic climate; the leadership and involvement of business leaders, especially the CEOs of business organizations; and the availability of capable quality professionals.

The first critical factor, competitiveness of business, is measured in comparative terms. In today's global economy, every company in the world with similar capabilities must be considered a potential competitor. Competitiveness is a measure of a company's ability to make and sell product against all their potential competitors in the global marketplace. Quality is an important determinant of that ability. The opportunities and challenges offered by common markets, such as NAFTA and MERCOSUR, combined with the importance of quality, are powerful motivators for individual businesses to improve quality performance. In turn, improved quality performance in these businesses exerts a positive effect on the country's aggregate quality performance.

The second critical factor for quality success is a favorable economic environment. Most of the countries in Latin America have been carrying out similar economic reforms since 1990. The economic environment is changing very rapidly toward open, free, and unregulated markets. Governments are aware that their policies must encourage competitiveness. Creating those policies means a new role for government, one that requires government to work with business rather than dominate it.

The third critical factor is the participation of CEOs, both in leading the quality improvement efforts within their own companies and in organizing and leading the larger effort within the country.

The fourth critical factor is to have the participation of capable professionals to support the implementation of the quality efforts.

To date, the major accomplishments in Latin America are the establishment of the national awards, a growing trend toward ISO 9000 certifications, and the collaboration of CEOs in establishing quality institutions at the national level. What remains to be done?

- Advance further in economic reform
- Convince the leaders of smaller companies to implement quality programs
- Improve the quality of education at all levels
- Include the quality sciences in the curriculum of universities
- Intensify training in quality at all levels and in all sectors
- Introduce the application of quality concepts and methodologies into sectors not yet affected

In Argentina, FUNDECE remains at the center of the quality movement, and serves as a model for other countries in Latin America. FUNDECE's major activities are designed to provide information to its members, to professionals (through IPACE), and to other sectors of society (through the National Quality Movement) about what is being done in quality in Argentina and in other parts of the world. FUNDECE maintains a sustaining membership in the ASQC as a means of helping its members participate in ASQC activities and of keeping them up to date in quality subjects.

An example of the impact of FUNDECE is the case of health care in Argentina, a sector that has not changed its traditional approach to quality in many years. There are today several health care organizations that participate actively in FUNDECE. These groups are gathering information and support to launch a quality initiative.

In other countries, organizations such as FUNDECE continue to provide the means to bring the latest developments in quality to various sector leaders, offering hope that the list of work yet to be done is more than a "wish list."

ACKNOWLEDGMENTS

Mario Casellini, Managing Director, and Diego Hollweck, Project Leader of FUNDECE. José Di Fabrizio, President of IPACE and Quality Manager for Latin America of Hewlett Packard. Mario

Mariscotti, Managing Director of FUNDAPRE and Wayne Cassatt, Former Associate Director for Malcolm Baldrige Program managed by NIST. Enrique Sierra, Senior Quality Management of Exports Consultant of the International Trade Centre (CCI). Rodolfo Cruz Venegas, Secretary General of ASCAL. Claudio Viggiani and Luis Alberto Frías, General Managers of Firmenich SA in Brazil and Mexico respectively.

REFERENCE

The World Competitiveness Report (*1995*). World Economic Forum and IMD, p. 16.

SECTION 44
BASIC STATISTICAL METHODS

Edward J. Dudewicz

Department of Mathematics,
Syracuse University, Syracuse, New York

THE STATISTICAL TOOL KIT

Most decision making in quality control, as in most other areas of modern human endeavor (e.g., evaluation of new medical treatments and scanning machines, planning of scientific polling, and marketing and investment strategies, to name a few), rests on a base of *statistics*—defined narrowly as the collection, analysis, and interpretation of data or, more broadly, as "the science of decision making under uncertainty." For the practitioner, *statistics* can be thought of as a kit of tools that helps to solve problems. The statistical tool kit shown in Table 44.1 lists problems to be solved, applicable statistical tools, and where in this handbook the tool is to be found.

Examples of actual practice will be used as extensively as the space allocated to it allows to provide the reader with both a model for solution and a data set with correct analysis that can be used to verify the accuracy of local computer software. Annotated computer program output from such packages as SAS and BMDP will be used in many of these examples.

In addition to the basic statistical methods discussed in this section, four other sections cover specific areas—Section 45, Statistical Process Control; Section 46, Acceptance Sampling; Section 47, Design and Analysis of Experiments; and Section 48, Reliability Data Analysis. Many other sections include additional applications. Also, Appendix III, Selected Quality Standards, Specifications, and Related Documents, includes documents on statistical techniques and procedures.

SOURCES AND SUMMARIZATION OF DATA

The source of a set of data that we desire to analyze to solve a problem is a very important consideration. The two sources we will address and that are the most common are historical data and data from planned experimentation. Investigators using historical data are like blind people probing an elephant, for reasons discussed below under Historical Data, Their Uses, and Caveats, and under Data from Planned Experimentation. Note that all data need careful review, as discussed under Data Screening, below.

Planning for Collection and Analysis of Data. The tools cited in Table 44.1 must be used in an effective manner to yield a return appropriate for the cost of using them. To achieve this return, it is not sufficient to plug numbers into formulas. The full process must include careful planning of data collection, analysis of the data to draw statistical conclusions, and making the transition to answer the original technical problem. A checklist of some of the key steps in achieving this is as follows:

1. Collect sufficient background information to translate the engineering problem statement into a specific statement that can be evaluated by statistical methods.
2. Plan the collection of data.
   a. Determine the type of data needed. Variables data (readings on a scale of measurement) may be more expensive than attributes data (go or no-go data), but the information is much more useful.
   b. Determine if any past data are available that are applicable to the present problem; however, bear in mind the hazards of historical data sets.
   c. If the problem requires an evaluation of several alternative decisions, obtain information on the economic consequences of a wrong decision.
   d. If the problem requires the estimation of a parameter, define the precision needed for the estimate.
   e. Determine if the error of measurement is large enough to influence the sample size or the method of data analysis; laboratory error often can dwarf experimental variability.
   f. Define the assumptions needed to calculate the required sample size.
   g. Calculate the required sample size considering the desired precision of the result, statistical risk, variability of the data, measurement error, and other factors.

TABLE 44.1   The Statistical Tool Kit

Problem	Statistical tool	Reference pages or sections
Planning a statistical investigation	Planning and analyzing data for solving specific problems	44.2–44.4
Summarizing data	Frequency distributions, histograms, and indices	44.7–44.17
Predicting future results from a sample	Probability distributions	44.23–44.41
Determining a probability involving several events	Basic theorems of probability	44.17–44.23
Determining the significance of difference between two sets of data or between a set of data and a standard value	Tests of hypotheses	44.58–44.81
Determining the sample size required for testing a hypothesis	Sample size determination for hypothesis testing	44.78–44.79
Determining the ability of a sample result to estimate a true value	Confidence limits	44.41–44.46
Determining the sample size required to estimate a true value	Sample size determination for estimation	44.45–44.46
Determining tolerance limits on single characteristics	Statistical tolerance limits	44.47–44.51
Determining tolerance limits for interacting dimensions	Tolerance limits for interacting dimensions	44.50–44.54
Incorporating past information in predicting future events	Bayes' Theorem	44.21–44.22 44.54–44.58
Incorporating economic consequences in defining decision rules	Statistical decision theory	44.22
Converting data to meet statistical assumptions	Transformations of data	44.81–44.84
Predicting system performance	Monte Carlo sampling methods	44.84–44.85
Determining group membership	Clustering and discrimination	44.86–44.87
Determining which is the best	Selection of the best	44.87–44.88
Evaluating the relationship between two or more variables by determining an equation to estimate one variable from knowledge of the other variables	Regression analysis	44.88–44.108
Controlling process quality by early detection of process changes:		
1. Using measurements data	Variables control charts	Sec. 45
2. Using attributes data	Attributes control charts	Sec. 45
Evaluating quality of lots to a previously defined quality level		
1. Quality measured on an attributes basis	Attributes sampling plans	Sec. 46
2. Quality measured on a variables basis	Variables sampling plans	Sec. 46
3. Sampling to determine reliability	Reliability sampling plans	Sec. 46
4. Bulk product	Bulk sampling plans	Sec. 46
Planning and analyzing experiments:		
1. Investigating the effect of varying one factor	One-factor experiment	Sec. 47
2. Investigating the effect of varying two or more factors	Designs for two or more factors	Sec. 47
3. Investigating the variability of laboratory measurements	Interlaboratory tests	Sec. 47
4. Experimenting under process conditions to determine optimum settings of variables	Evolutionary operation (EVOP)	Sec. 47
5. Determining the optimum set of values of a group of variables that affect a response variable	Response surface methodology (RSM)	Sec. 47
Predicting performance without failure (reliability)	Reliability prediction and analysis	Sec. 48

h. Define any requirements for preserving the order of measurements when time is a key parameter.

i. Determine any requirements for collecting data in groups defined so as to reflect the different conditions that are to be evaluated.

j. Define the method of data analysis and any assumptions required.

k. Define requirements for any computer programs that will be needed.

3. Collect the data.

a. Use methods to ensure that the sample is selected in a random manner.

b. Record the data and also all conditions present at the time of each observation.

c. Examine the sample data to ensure that the process shows sufficient stability to make predictions valid for the future.

4. Analyze the data.

a. Screen the data.

b. Evaluate the assumptions previously stated for determining the sample size and for analyzing the data. Take corrective steps (including additional observations) if required.

c. Apply statistical techniques to evaluate the original problem.

d. Determine if further data and analysis are needed.

e. Conduct sensitivity analyses by varying key sample estimates and other factors in the analysis and noting the effect on final conclusions.

5. Review the conclusions of the data analysis to determine if the original technical problem has been evaluated or if it has been changed to fit the statistical methods.

6. Present the results.

a. Write a report, including an executive summary.

b. State the conclusions in meaningful form by emphasizing results in terms of the original problem rather than the statistical indices used in the analysis.

c. Present the results in graphic form where appropriate. Use simple statistical methods in the body of the report, and place complicated analyses in an appendix.

7. Determine if the conclusions of the specific problem apply to other problems or if the data and calculations could be a useful input to other problems.

Historical Data, Their Uses, and Caveats.   *Historical data* are data that we already have and which may seem to be relevant to a question or problem that has arisen. Such data are sometimes also called *existing data sets*. Often data are saved during the production process, for example. If a satisfactory process goes out of control after some years of operation, it is often suggested that it would save both time and expense to statistically analyze the historical data rather than perform a planned experiment to obtain data that could lead to process correction. Thus we have available data that may consist of measurements Y (such as a process yield, e.g., the strength of a material produced) and associated process variables $x_1, x_2,\ldots, x_k$ (such as x_1 = pressure and x_2 = acid concentration, with $k = 2$).

This situation is extremely different from that where experiments have been run at each of a number of settings of $x_1,\ldots, x_k$ that were selected in advance by statistical design criteria, and often little can be learned from such data even with the most thorough statistical analysis. Some of the reasons for this are

The x's may be highly correlated with each other; hence it may not be possible to separate an effect as due to (for example) x_1 or x_2.

The x's may have been manipulated to try to control the output Y of the process (some of them perhaps even in directions that move the output in directions that are not desired), hence giving spurious indications of directions of effects when analyzed.

The x's may cover a very small part of the possible operating range, so small that any indications of changes in Y attributable to changes in the x's may be overwhelmed by the size of the variability of the process (measured by "standard deviation," discussed below).

Other variables that affect the output of the process (e.g., time of day, atmospheric conditions, operator running the process, etc.) may not have been held constant and may in fact be the real causes of changes observed in the process (while an analysis conducted based only on the x's may erroneously yield a model that has no basis in reality).

For these and other reasons, much more information generally can be obtained from a carefully designed experiment than from extensive analysis of historical data sets collected in uncontrolled circumstances. The best one usually can hope for from such historical data set analysis is an indication of the most important variables to include in the designed experiment.

As an example of one of the failings described above, suppose that a historical data set consists of a yield Y at each of the five (x_1, x_2) pairs given in Table 44.2. We see that there is what appears to be a good spread on the x_1 (acid concentration) values, from 80 to 110, and also a good spread on the x_2 (pressure) values, from 105 to 144. However, a plot of the data points (x_1, x_2), as shown in Figure 44.1, shows that in fact all five of the data points lie on a straight line. Thus there is no hope of any analysis of these data telling us whether any effect we see is due to x_1 or to x_2. The points do not "cover" the space of x_1 between 80 and 110 and x_2 between 105 and 144 well at all. Also, to solve the production problem, it may be desirable to explore outside this space of historical operation, and this is not allowed for in the historical data set.

A problem such as that shown in Figure 44.1 is called *multicollinearity* of the data points, i.e., of the sets of $(x_1 \dots, x_k)$ that we have available for analysis. With $k = 2$, such a problem is easy to detect by a graph such as that in Figure 44.1, and such a graph should always be made. With $k = 3$ it may be possible to detect such a situation graphically using computer graphics packages. Often, however, k is much larger than 2 or 3, and then statistical analysis is needed to detect multicollinearity. Near multicollinearity (i.e., when the points almost fall on a line when $k = 2$, on a plane when $k = 3$, or on what is called a $k - 1$ dimensional hyperplane when k is larger than 3) is just as much of a problem and is much harder to detect. For an extensive bibliography and comments on computer routines for this problem, see Hoerl and Kennard (1981). For some more recent results, see Huh and Olkin (1995), where numerical illustrations using the classic data of Longley (1967) are included.

Data from Planned Experimentation.   *Data from planned experimentation* are data gathered in an attempt to study a problem that has arisen or is contemplated. Such data are gathered at various settings of the variables felt to be of importance $(x_1, \dots, x_k)$, while holding constant (if possible, and recording the values of in any case) all other variables that could conceivably have an effect on the output $(x_{k+1}, \dots, x_m)$—for example, atmospheric pressure may not be able to be controlled in most circumstances but should be monitored and recorded in every experiment if it is felt beforehand that it might have an effect on the output. (If it is irrelevant, it can be disregarded later; if it is relevant but not recorded, we will not be able to detect that relevance.) Details of designs to use, i.e., how to choose the values of $(x_1, \dots, x_k)$ for the experiments once the variables to be varied have been chosen, are considered in Section 47, Design and Analysis of Experiments, and in Dudewicz and Karian (1985).

Data Screening.   Once the data specified in the preceding checklist for collection and analysis of data, step 2, has been collected (step 3), the first step in its analysis (step 4a) is to screen the data.

TABLE 44.2   Historical Data Set

Data point	x_1 (Acid concentration)	x_2 (Pressure)
1	100	131.0
2	90	118.0
3	105	137.5
4	110	144.0
5	80	105.0

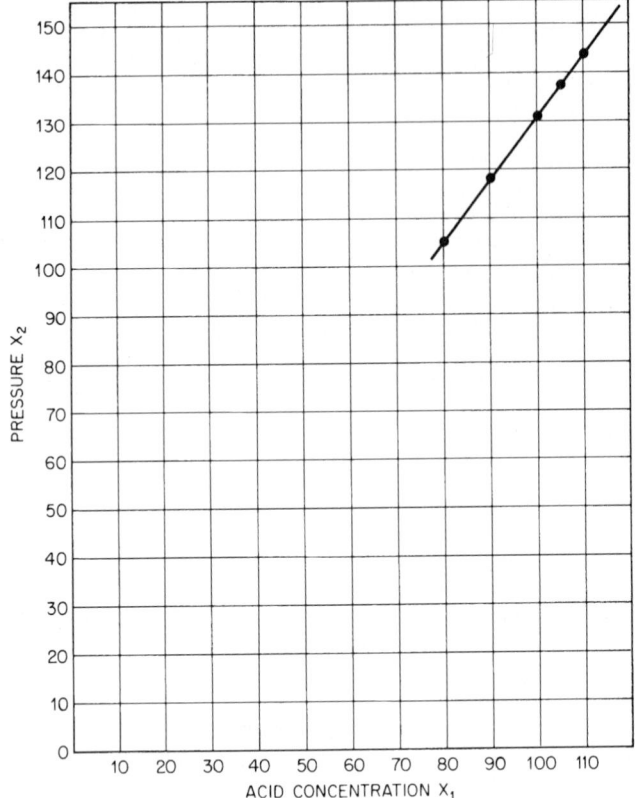

FIGURE 44.1   Plot of (x_1, x_2) in a historical data set. Five data points have been plotted, and a straight line has been drawn through them.

We will now discuss the need for this critical (and often omitted) step, some common methods of screening, and a very commonly used method that has some great dangers.

The Need for Data Screening.   A data set that contains no instances of incorrectly transcribed values, contains no values that are technically correct but where the experiment went awry for some reason (such as equipment malfunction), and where the basic model does not change its form over the region of experimentation is called a *clean* data set. Such data sets are commonly expected to be obtained by experimenters who exercise care in their experimental conduct and recording. However, contrary to this expectation of the experimenters, few statisticians have ever seen a clean data set (despite many years of studying many data sets arising in many areas). It follows, then, that all data sets need, as a first stage of analysis, to be examined for values that may cause invalid inferences to be made if those values are left in the data set. Procedures for performing this examination are called *data screening methods*. Among the most powerful such methods are those which are used on the results of a regression analysis (studied later in this section).

There are also a number of methods that can and should be used at the outset, before any regression analyses are performed, with the goal of detecting *outliers,* that is, observations (or groups of observations) that deviate markedly from the other available data. Numerous tests are available for detecting outliers [see Sheesley (1977) for some of these]. One simple rule calls an observation an outlier if it lies 2.5, 3, or 4 standard deviations or further from the mean (Draper and Smith 1981). This is discussed further below.

Methods of Data Screening.   One of the most common methods of data screening is to classify observations as outliers if they are outside an interval of L multiples of the standard deviation about the mean. (Standard deviation is discussed below under Sample Characteristics.) The number L is commonly taken to be 2.5, 3, or 4. The larger L is, the less likely it is that outliers will be detected, while the smaller L is, the more good observations one will wrongly detect as potential outliers. For example, from Table B in Appendix II, we see that if $L = 3$, then $(100)(.0027) = 0.27$ percent of the observations will be further than 3 standard deviations from the mean even if there are no outliers in the data set; this assumes a normal distribution for the observations. Thus, if one uses $L = 3$, one expects to find about 3 possible outliers in a data set of 1000 data points (since 0.27 percent of 1000 is 2.7, i.e., roughly 3). As the data set being considered becomes larger, the more possible outliers one will identify even if there are no problems with the data (which is quite unlikely). For this reason,

> Outliers should be deleted from the analysis only if they can be traced to specific causes (such as recording errors, experimental errors, and the like).

> Typically, one takes L to depend on the size of the data set to be screened; with $n = 1000$ points, $L = 3$ is reasonable; with $n = 100$, $L = 2.5$ can be used and only $(100)(0.0124) = 1.24$ outliers will be expected to be found if the data have no problems.

After bad data are deleted or replaced (this is desirable if the experiment can be rerun under comparable conditions to those specified in the experimental plan), the data should be screened again: With the "worst" points removed/corrected, less extreme cases may come to be identified as possible outliers. Another commonly used method, that of crossplots, is discussed below.

Crossplots and Their Hazards.   In *crossplots* (also called *scatter plots*), one simply plots each pair of variables in the data set on a set of axes. For example, in the example of Table 44.2, one would plot Y versus x_1, Y versus x_2, and x_1 versus x_2. The last of these plots was given in Figure 44.1 (and showed some problems with the data that have already been discussed). Note that when using this technique, one uses all variables that were measured (and not just the variables that are thought to be of primary interest; see Nelson 1979, Section 4).

Such plots must be used with great caution. While points that seem odd (e.g., away from the majority of the data points) should be subjected to examination to see if there are problems with them, one should not use such plots to conclude relationships of the yield with x_1 and/or x_2. For example, suppose one has the data set of Table 44.3. Then from crossplots of Y versus x_1 and Y versus x_2 (see Figure 44.2), one would be tempted to conclude that Y is a decreasing function of x_1 and an increasing function of x_2 (and so, in attempting to maximize yield, might set x_1 as low as possible and x_2 as high as possible). However, as can easily be verified, the data in Table 44.3 came exactly from the relationship (model) $Y = 10 + x_1 + 2x_2$. Thus, in fact, Y is an increasing function of x_1 (not a decreasing function as the crossplot had suggested). Daniel (1977) has suggested that the data set of Table 44.4 may give additional insight here. Crossplots for that data set are given in Figure 44.3. Thus one sees that even with a strong true relationship (here $Y = 10 + x_1 + 2x_2$ is used again), the crossplots may yield no insight at all. Even worse, clearly by choosing the points (x_1, x_2) one may give the Y versus x_1 relationship any slope (negative, as with Table 44.3 and Figure 44.2; zero; or positive), or the relationship can be smoothly curved in any direction or degree of complexity. (Note that the regression methods given later in this section would not be fooled by the relationships in the data of Table 44.4. They would give the true relationships.) Thus crossplots are useful for detection of possible outliers; however, they are not a substitute for regression and can easily be misused.

Descriptive Statistics for Summarizing Data.

Many of the most practical methods of summarizing data are quite simple in concept. Depending on the goals of the data summarization, sometimes one method will provide a useful and complete summarization. More often, two or more methods will be used to attain the clarity of description that is desired. Several key methods are plots versus time order of data, frequency distributions and histograms, sample characteristics (mean, median, mode, variance, standard deviation, and percentiles), measures of central tendency/location, and measures of dispersion.

TABLE 44.3   Data Set *A* for Crossplots Example

Data point	x_1	x_2	Y
1	−2	1	10
2	−1	−2	5
3	0	−5	0

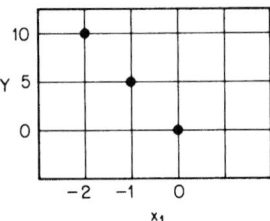 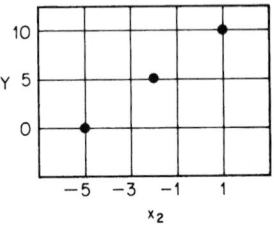

FIGURE 44.2   Crossplots of Y versus x_1 and of Y versus x_2 for data set *A* of Table 44.3.

TABLE 44.4   Data Set *B* for Crossplots Example

Data point	x_1	x_2	Y
1	−2	1	10
2	−1	−2	5
3	0	0	10
4	1	2	15
5	2	−1	10

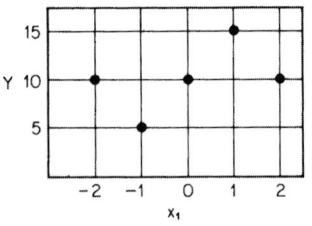 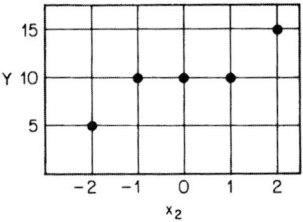

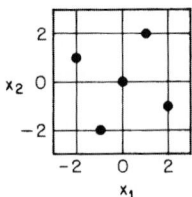

FIGURE 44.3   Crossplots for data set *B* of Table 44.4.

Plots versus Time Order of Data.   After a data set has been obtained, it is very instructive in many situations to plot the output Y against the time order in which the experiments were run (which is essentially a crossplot of Y versus time t). Among the possible phenomena that may come to our attention from such a plot are these:

A few observations, often at the start of the experimentation, are far from the others; this often represents a learning curve of the experimenters with the experimental situation, and those experiments should be repeated if possible.

There are trends within each day (or within each week, etc.); this may represent such phenomena as warming of a machine or process, operator fatigue, or similar time-related trends.

Variability decreases (or increases) with time; this may be due to a learning curve or raw material characteristics (as when one lot of material is used up and the next lot has less or greater heterogeneity).

While the preceding trends may be apparent even in a plot of the original observations Y versus time, they are often more easily spotted in plots of the residuals of the observations (difference between the observed and predicted value) after a regression analysis. See later in this section under Regression Analysis.

Frequency Histograms.   The *frequency histogram* (or distribution) is a statistical tool for presenting numerous data in a form that makes clearer the central tendency and the dispersion along the scale of measurement, as well as the relative frequency of occurrence of the various values.

Table 44.5 shows "raw data" representing measurements of electrical resistance of 100 coils. A practitioner scanning these 100 numbers has difficulty in grasping their meaning.

Table 44.6 shows the same data after tabulation. Note how the analyst's tallies in the column "Tabulation" make more evident where the central tendency is and what the dispersion is. The column "Frequency" is merely a recorded count of these same tallies. The column "Cumulative frequency" shows the number of coils with resistance equal to or greater than the associated resistance value.

Table 44.6 exhibits a range of values from 3.44 to 3.27, or 17 intervals of 0.01 Ω each. When it is desired to reduce the number of such intervals, the data are grouped into "cells." Table 44.7 shows the same data grouped into a frequency distribution of only six cells, each 0.03 Ω wide. Grouping the data into cells simplifies presentation and study of the distribution but loses some of the detail. (However, one can always go back to the original data if necessary.)

The following are the steps taken to construct a frequency distribution:

1. Decide on the number of cells. Table 44.8 provides guidelines that are adequate for most cases encountered. These guidelines are not rigid and should be adjusted when necessary; their aim is not only to provide a clear data summary but also to reveal any underlying pattern in the data.

TABLE 44.5   Resistance (Ohms) of 100 Coils

3.37	3.34	3.38	3.32	3.33	3.28	3.34	3.31	3.33	3.34
3.29	3.36	3.30	3.31	3.33	3.34	3.34	3.36	3.39	3.34
3.35	3.36	3.30	3.32	3.33	3.35	3.35	3.34	3.32	3.38
3.32	3.37	3.34	3.38	3.36	3.37	3.36	3.31	3.33	3.30
3.35	3.33	3.38	3.37	3.44	3.32	3.36	3.32	3.29	3.35
3.38	3.39	3.34	3.32	3.30	3.39	3.36	3.40	3.32	3.33
3.29	3.41	3.27	3.36	3.41	3.37	3.36	3.37	3.33	3.36
3.31	3.33	3.35	3.34	3.35	3.34	3.31	3.36	3.37	3.35
3.40	3.35	3.37	3.32	3.35	3.36	3.38	3.35	3.31	3.34
3.35	3.36	3.39	3.31	3.31	3.30	3.35	3.33	3.35	3.31

TABLE 44.6   Tally of Resistance Values of 100 Coils

Resistance, ohms	Tabulation	Frequency	Cumulative frequency
3.45			
3.44	I	1	1
3.43			
3.42			
3.41	II	2	3
3.40	II	2	5
3.39	IIII	4	9
3.38	⊥⊦⊦ I	6	15
3.37	⊥⊦⊦ III	8	23
3.36	⊥⊦⊦ ⊥⊦⊦ III	13	36
3.35	⊥⊦⊦ ⊥⊦⊦ IIII	14	50
3.34	⊥⊦⊦ ⊥⊦⊦ II	12	62
3.33	⊥⊦⊦ ⊥⊦⊦	10	72
3.32	⊥⊦⊦ IIII	9	81
3.31	⊥⊦⊦ IIII	9	90
3.30	⊥⊦⊦	5	95
3.29	III	3	98
3.28	I	1	99
3.27	I	1	100
3.26			
Total		100	

TABLE 44.7   Frequency Distribution of Resistance Values

Resistance, ohms		Frequency	Cumulative frequency
Boundaries	Midpoints		
3.415–3.445	3.43	1	1
3.385–3.415	3.40	8	9
3.355–3.385	3.37	27	36
3.325–3.355	3.34	36	72
3.295–3.325	3.31	23	95
3.265–3.295	3.28	5	100
		100	

2. Calculate the approximate cell interval i. The cell interval equals the largest observation minus the smallest observation divided by the number of cells. Round this result to some convenient number.
3. Construct the cells by listing cell boundaries. As an aid to later calculation:
   a. The cell boundaries should be to one more decimal place than the actual data and should end in a 5.
   b. The cell interval should be constant throughout the entire frequency distribution.
4. Tally each observation into the appropriate cell and then list the total frequency f for each cell.

There are several ways of showing a frequency distribution in graphic form. The most popular is the frequency histogram. Figure 44.4 shows the electrical resistance data of Table 44.7 depicted in histogram form. The diagram is so easy to construct and interpret that it is widely used in elementary analysis of data.

One example of wide, effective use of frequency histograms is comparison of process capabilities with tolerance limits. The histogram of Figure 44.5 shows a process that is inherently capable of holding the tolerances drawn on the same figure. The high degree of defectives being produced is the result of running this process at a setting that does not locate its central tendency near the midpoint of the tolerance range. (See Section 22, under Operations Analysis, for other examples.)

Analyses of histograms to draw conclusions beyond the sample data customarily should be based on at least 50 measurements.

Sample Characteristics: Mean, Median, Mode, Variance, Standard Deviation, Percentiles.   Faced with a large data set, *descriptive statistics* furnish a simple method of extracting information from what often seems at first glance to be a mass of numbers without rhyme or reason to it. These characteristics of the data may relate to a "typical (or central) value" (mean, median, mode), a measure of how much variability is present (variance, standard deviation), or a measure of frequency (percentiles). The first two types of characteristics (typical value and variability) will be discussed below, but first we will present the concept of percentiles.

TABLE 44.8   Number of Cells in Frequency Distribution

Number of observations	Recommended number of cells
20–50	6
51–100	7
101–200	8
201–500	9
501–1000	10
Over 1000	11–20

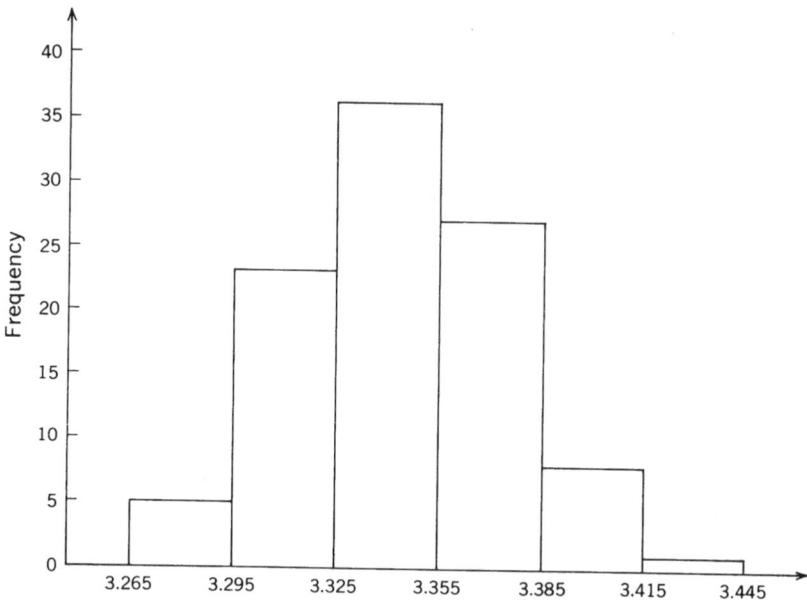

FIGURE 44.4   Histogram of resistance.

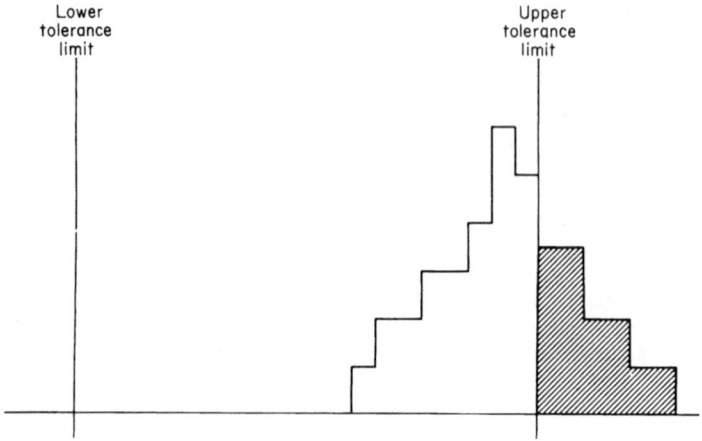

FIGURE 44.5   Histogram of a process.

A *percentile curve* is a plot of the percentile rank of the data against the data values. For example, for the data of resistance of 100 coils given in Table 44.6, 1 percent are at or below resistance 3.27, 2 percent are at or below 3.28, 5 percent are at or below 3.29, and so on, as given in Table 44.9. The percentile rank plot (or percentile curve) for these data is given in Figure 44.6. Note that while the data will result in a "rough" curve (since the percentile curve of the data will jump at each data point and remain constant between actual data points), a smooth curve drawn through the data will be a better representation of reality; this is the dashed curve drawn in Figure 44.6. (Thus while values 3.42 and 3.43 did not occur in our sample, we expect they would in a larger sample; hence the smooth curve is the one we would utilize to assess their chances of occurring.) Note that no data are discarded (or grouped) in making the percentile curve of a data set; in this sense, this is a more precise process than construction of a histogram of the data (since no information is "lost" in the process). Most statistical work uses the percentile curve, under the name *distribution function* of the data. (In early papers this was called the *cumulative* distribution function.) Since the distribution gives the proportion of the data falling at or below each value, the graph and curve are the same; only the scale on the vertical axis is changed to read from 0 to 1 (since 0 to 100, divided by 100, runs from 0 to 1).

Measures of Central Tendency/Location.   Most frequency distributions exhibit a *central tendency*, i.e., a shape such that the bulk of the observations pile up in the area between the two extremes. Central tendency is one of the two most fundamental concepts in all statistical analysis.

There are three principal measures of central tendency: arithmetic mean, median, and mode. The *arithmetic mean* (the ordinary "average") is used for symmetric or near symmetric distributions or for distributions that lack a clearly dominant single peak. The arithmetic mean $\overline{X}$ is the most generally used measure in quality work. It is employed so often to report average size, average yield, average percent defective, etc. that control charts have been devised to analyze and keep track of it. Such control charts can give early warning of significant changes in the central value (see Section 45, Statistical Process Control).

The *mean* is calculated by adding the observations and dividing by the number of observations. A short method for calculating the mean is given in a subsequent example under Measures of Dispersion.

The *median* (the middle value when the figures are arranged according to size) is used for reducing the effects of extreme values or for data that can be ranked but are not economically measurable (shades of color, visual appearance, odors) or for special testing situations. If, for example, the average of five parts tested is used to decide whether a life-test requirement has been met, then the lifetime of the third part to fail can sometimes serve to predict the average of all five, and thereby the decision of the test can be made much sooner. As shown on Figure 44.6, the median is simply the horizontal scale value where the percentile curve reaches the height 50 percent.

The *mode* (the value that occurs most often in data) is used for severely skewed distributions, describing an irregular situation where two peaks are found, or for eliminating the effects of extreme values. The statistical "efficiency" of these measures varies. See Dixon and Massey (1969, chap. 9) or Dudewicz (1976, pp. 221–222) for elaboration.

Measures of Dispersion.   Data are always scattered around the zone of central tendency, and the extent of this scatter is called *dispersion* or *variation*. A measure of dispersion is the second of the two most fundamental measures in all statistical analysis.

TABLE 44.9   Percentile Rank of Resistance Values in Table 44.6

Resistance (x)	3.27	3.28	3.29	3.30	3.31	3.32	3.33	3.34
Percentile rank (y)	1	2	5	10	19	28	38	50
	3.35	3.36	3.37	3.38	3.39	3.40	3.41	3.42
	64	77	85	91	95	97	99	99
	3.43	3.44						
	99	100						

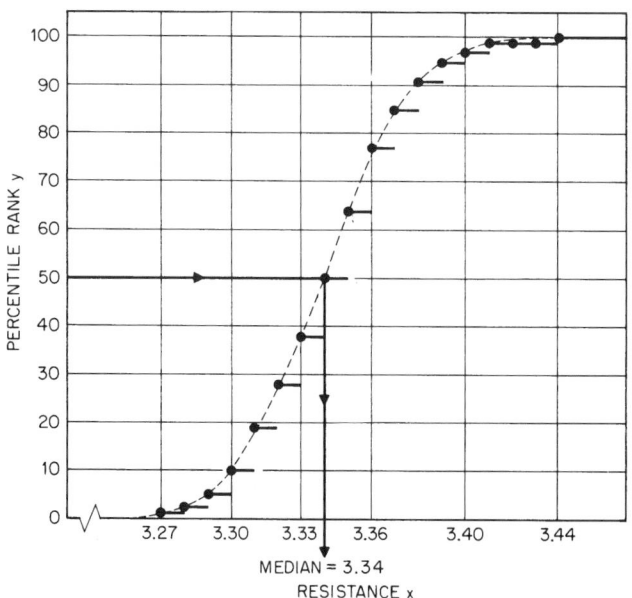

FIGURE 44.6   Percentile curve of a data set.

There are several measures of dispersion. The simplest is the *range,* which is the difference between the maximum and minimum values in the data. Since the range is based on only two values, it is most useful when the number of observations is small (about 10 or fewer).

The most important measure of variation is the *standard deviation.* The definition of the sample standard deviation is

$$s = \sqrt{\frac{\Sigma(X - \overline{X})^2}{n - 1}}$$

where s = sample standard deviation
Σ = "sum of"
X = observed values
$\overline{X}$ = arithmetic mean
n = number of observations

For calculation purposes, an equivalent formula is

$$s = \sqrt{\frac{n\Sigma(X^2) - (\Sigma X)^2}{n(n - 1)}}$$

The square of the standard deviation is called the *variance.* There is also a measure called *covariance,* which gives information on the relationship between pairs of observations on characteristics X and Y. This is defined as

$$s_{XY} = \frac{\Sigma[(X - \overline{X})(Y - \overline{Y})]}{n - 1}$$

Further discussion of the relationship between two or more variables is given later in this section under Regression and Correlation Analysis.

With data in frequency distribution form, shortcut calculations can simplify finding the average and the standard deviation. This is illustrated in Table 44.10. To start, an arbitrary origin A is assumed as 3.37. A zero is arbitrarily placed on this line in the d' column. The other figures in this column indicate how many cells the entry is above or below the arbitrary zero. Minus signs are attached when the entry is smaller than the assumed value, 3.37. The fd' values in column (4) are found by multiplying together the entries in columns (2) and (3). Similarly $f(d')^2$ is found by multiplying the figures in columns (3) and (4). Note that the totals in the last two columns are identified in the formulas as $\Sigma fd'$ and $\Sigma f(d')^2$, respectively, and i is the cell interval. Since the multiplications are small enough to be carried out mentally, the complete table can be made quickly.

$$\overline{X} = A + \left(\frac{\Sigma fd'}{n}\right) i = 3.37 + \left(\frac{-87}{100}\right) 0.03 = 3.344$$

$$s = i \sqrt{\frac{n\Sigma f(d')^2 - (\Sigma fd')^2}{n(n - 1)}}$$

$$s = 0.03 \sqrt{\frac{100(185) - (-87)^2}{100(99)}} = 0.031$$

For sample sizes of about 10 or fewer observations, the standard deviation can be approximated from the range by calculating R/d_2, where d_2 is a factor in Appendix II, Table A. For example, suppose the first column of values in Table 44.5 represents a sample of 10. The range is 3.40−3.29, or 0.11. From Table A in Appendix II, d_2 = 3.078. The estimate of the standard deviation is therefore 0.11/3.078 = 0.036. This is much simpler than calculating the standard deviation directly. Subsequent

TABLE 44.10 Calculation of Average and Standard Deviation

Midpoint (1)	Frequency f (2)	d' (3)	fd' (4)	$f(d')^2$ (5)
3.43	1	$+2$	2	4
3.40	8	$+1$	8	8
3.37	27	0	0	0
3.34	36	-1	-36	36
3.31	23	-2	-46	92
3.28	5	-3	-15	45
	$\Sigma = 100$		$\Sigma = -87$	$\Sigma = 185$

topics in this section further illustrate this feature of the range. Dixon and Massey (1969, pp. 136–140) furnish procedures and tables for a variety of applications of the range.

A final measure of variation is the *coefficient of variation.* This is defined as the standard deviation divided by the mean and is thus a relative measure of variation. It can be helpful in comparing several sets of similar data that differ in mean value but may have some commonality in *relative* variation.

The methods of summarizing data covered in the previous paragraphs can be performed on a computer, as discussed below. (Also see Section 10 in general for additional information on computer programs for quality control.)

Stem-and-Leaf Plots, Boxplots, and Statistical Graphics.   Looking at data to find patterns and other characteristics is very important. In fact, it is often said that "the first rule of statistics is: Look at the data." This does not mean to pass the data on to be fed into some computer program without a careful examination of them. It does mean to visually scan the data, to calculate classic measures of important characteristics, to check to see if known measures of validity are satisfied, and to make graphic representations of some data characteristics. There are many methods and approaches to this initial stage of data analysis, some favored by one or another school of statisticians, all of some use, and the total group being of significant size and utility. Some widely valued methods (in addition to those covered in detail earlier) include stem-and-leaf plots and boxplots, for which one can refer to many introductory statistics texts. The area as a whole is part of what is called *statistical graphics,* and computer programs to perform such calculations are widely available. However, to achieve their full potential (and rise above mediocrity or inferiority), these methods need to be used carefully and intelligently (as we saw earlier with crossplots, for example). An excellent reference in this regard (one that is filled with examples, and when the examples are of bad graphics shows how to use a good graphic on the same data) is Schmid (1983).

Accurate Calculation of Descriptive Statistics.   Accurate calculation of even simple descriptive statistics is not as easy a task as it might seem at first, as the example in the paragraph on coding of data below shows. In particular,

Calculation by hand is an error-prone process, especially if there are 10 or more data points.

Construction of computer programs to perform the calculations is a task that requires knowledge of both statistics and numerical analysis as well as computer programming; hence it is inadvisable to "roll your own" in most cases.

Many of the computer software routines that come with computers (especially microcomputers) are not of high quality.

In light of the preceding, it is recommended that a high-quality package of computer routines for statistics be obtained and used for all such analyses. In particular, I suggest packages called SAS,

BMDP, STATPRO, and LABONE. These are of high quality and among them cover a wide variety of computers on which they will run (including IBM and Data General mainframe computers, APPLE and IBM PC microcomputers, and a variety of minicomputers). Additional packages that are in wide use and widely valued include MINITAB and SPSS. There are many other packages, some specially designed for particular areas of statistics (such as design of experiments). It is good to bear in mind that quality varies widely; SAS is felt by many to be the "gold standard" of such packages.

For details of SAS, see SAS Institute, Inc. (1982). For BMDP, see Dixon (1983). For STATPRO, see Pinsky (1983). For LABONE, see Levy and Dumoulin (1984). In each case, a telephone call will bring current information regarding which computers the package is available for, as well as licensing information.

Section 10, under Computer Applications and Quality Systems, presents additional information on statistical packages. The graphics available with these packages (such as SAS/GRAPH with SAS) are of such a quality that they would often justify the expenditure for the package to management. The American Society for Quality annually publishes a "Directory of Software for Quality Assurance and Quality Control" (see, for example, ASQC 1996).

Coding of Data. Suppose that we have five observations, $X_1, ..., X_5$ as in Table 44.11 and wish to compute the mean and standard deviation. Then on a pocket calculator (such as the Texas Instruments TI-55) or on a mainframe computer, we find

$$\sum_{i=1}^{5} X_1 = 49345 \qquad \sum_{i=1}^{5} X_i^2 = 4.8698581 \times 10^8$$

$$\overline{X} = 9869 \qquad s^2 = 0.985 \qquad s = 0.992$$

The only problem with these answers is that they are *wrong*. For example, 9869 is less than all the observations; hence it cannot be the mean. And the standard deviation is a measure of the spread in the data; that spread from largest to smallest is 0.00008; hence 0.992 is much too large. The problem that has led to this inaccuracy is that the computers used keep only about eight decimal places of accuracy, and this results in discarding digits that are needed for accurate calculation. If it can be so troublesome to calculate the mean and standard deviation of five numbers accurately, clearly, problems of meaningful size require careful analysis of numerical inaccuracy, as discussed above, and for this reason, software should not be trusted without a careful analysis of its capabilities. The four packages suggested are of high quality.

If one must use untested software, it is recommended that one *code* the data; that is, calculate using $Y_1, ..., Y_5$, where $Y_i = a(X_i + b)$ for some a and b. For example, if we choose $a = 10^5$ and $b = -9869$, we will have $Y_1 = 13$, $Y_2 = 7$, $Y_3 = 15$, $Y_4 = 8$, and $Y_5 = 9$. For these we calculate (on the same computers)

$$\sum_{i=1}^{5} Y_i = 52 \qquad \sum_{i=1}^{5} Y_i^2 = 588 \qquad \overline{Y} = 10.4 \qquad s_Y^2 = 11.8 \qquad s_Y = 3.4351$$

which are exactly correct. Now it can be shown that the relation between $\overline{X}$, s_X^2 and $\overline{Y}$, s_Y^2 is

$$\overline{X} = \frac{1}{a}\overline{Y} - b \qquad s_X^2 = \frac{1}{a^2}s_Y^2$$

TABLE 44.11   Five Data Points

$X_1 = 9869.00013$
$X_2 = 9869.00007$
$X_3 = 9869.00015$
$X_4 = 9869.00008$
$X_5 = 9869.00009$

Hence we find (exactly again)

$$\overline{X} = 10.4 \times 10^{-5} + 9869 = 9869.000104$$

$$s_X^2 = 11.8 \times 10^{-10} \text{ (hence } s_X = 0.000034)$$

This method of *coding the data* will preserve accuracy in many other, more complex statistical calculations (such as regression) and is in fact done internally by many of the high-quality statistical computer packages. Thus it should be used whenever you are using a package other than one of the high-quality ones listed earlier. In addition, many absurd results from inaccurate software will be detected early if one observes the cardinal rule of statistics, namely, "Look at the data." (It is all too common to have the data gathered and analyzed via computer without a careful look at them with the measures recommended in this section, and this has led to many costly problems for many companies.)

In terms of the choice of a and b for the coding, the best values to use are $a = 1/s_X$ and $b = -\overline{X}$. If these values cannot be calculated in a trustworthy manner by the software, the simple estimates of the next paragraph may be used instead.

Simple Estimates of Location and Dispersion. Simple estimates of the center and the spread of a set of data are often desired (e.g., for use in coding of data, as discussed earlier, and also for rapid analysis of a data set under time pressure). Two simple measures for the center are the median ($\widetilde{X}$) and the midrange, expressed as

$$\frac{\text{Maximum } (X) + \text{minimum } (X)}{2}$$

A simple measure for the variability is

$$\frac{\text{Maximum } (X) - \text{minimum } (X)}{4}$$

For the data of Table 44.11, the true value was $\overline{X} = 9860.000104$. The two simple methods yield 9869.00009 and 9869.000110, respectively. Similarly, the true value of the standard deviation was $s_X = 0.000034$, and the simple estimate yields 0.000020. (If we use Table A in Appendix II and take the range divided by d_2 instead of the range divided by 4, we will have a better estimate. With five data points, $d_2 = 2.326$, so our estimate of s_X is then $0.000080/2.326 = 0.000034$.)

PROBABILITY MODELS FOR EXPERIMENTS

A distinction is made between a sample and a population (or *universe*). A *sample* is a limited number of measurements taken from a large source. A *population* is a large source of measurements from which the sample is taken. (Note that a population may physically exist, such as all stereo sets in a certain lot. It also may be conceptual, as all experiments that might be run.)

A *probability distribution* is a mathematical formula that relates the values of the characteristic with their probability of occurrence in the population. Figure 44.7 summarizes some distributions. When the characteristic being measured can take on any value (subject to the fineness of the measuring process), its probability distribution is called a *continuous* probability distribution. For example, the probability distribution for the resistance data of Table 44.7 is an example of a continuous probability distribution because the resistance could have any value, limited only by the fineness of the measuring instrument. Experience has shown that most continuous characteristics either follow one of several common probability distributions, the *normal* distribution, the *exponential* distribution, and the *Weibull* distribution, or can be fitted with an empirical estimate, as discussed later in this section. These distributions find the probabilities associated with occurrences of the actual values of the characteristic. Other continuous distributions (e.g., t, F, and chi square) are important in data analysis but do not provide probabilities of occurrence of actual values.

When the characteristic being measured can take on only certain specific values (e.g., integers 0, 1, 2, 3, etc.), its probability distribution is called a *discrete* probability distribution. For example, the distribution for the number of defectives r in a sample of five items is a discrete probability distribution because r can only be 0, 1, 2, 3, 4, or 5. The common discrete distributions are the Poisson, binomial, negative binomial, and hypergeometric (see Figure 44.7).

The following paragraphs explain how probability distributions can be used with a sample of observations to make predictions about the larger population.

Sample Space.   Statistics deals with the *outcomes of experiments*. When an experiment is performed, some outcome results; let us denote a typical outcome by the symbol e. Such an outcome is called a *simple event*. If we list all the possible outcomes of the experiment of interest to us, that set is called the *sample space* of the experiment.

As an example, if we perform the experiment of tossing three coins and observing for each coin whether it lands heads (H) or tails (T), the sample space will contain the eight possible outcomes

$$\text{HHH} \quad \text{HHT} \quad \text{HTH} \quad \text{THH} \quad \text{HTT} \quad \text{THT} \quad \text{TTH} \quad \text{TTT}$$

For simplicity of notation, let us denote these simple outcomes, respectively, by e_1, e_2, e_3, e_4, e_5, e_6, e_7, and e_8.

We associate a number called *probability* with each of the simple events. We think of this number as representing the proportion of times each simple event would occur in a very large number of experiments of this type. For example, the probability of HHH in our experiment of tossing three coins is usually taken to be $1/8 = 0.125$ because it typically occurs in about one-eighth of a large number of experiments where three coins are tossed. We denote the probability of a simple event e by $P(e)$; thus we usually have $P(\text{HHH}) = 1/8$.

Since some outcome occurs in each experiment, when we add up the proportion of times that each e in the sample space occurred, we must obtain a sum of 1. Since probabilities represent what those proportions would be in a large number of experiments, we also must have *probabilities that sum to 1 when all outcomes are accounted for.* For example, in our example with three coins,

$$P(e_1) + P(e_2) + P(e_3) + P(e_4) + P(e_5) + P(e_6) + P(e_7) + P(e_8) = 1$$

Events.   Very often we are interested not in a simple event but in a combination of them, called a *composite event*. For example, the event "more heads than tails" occurs if and only if one of the simple events e_1, e_2, e_3, e_4 (i.e., the simple events HHH, HHT, HTH, THH) occurs in our example of tossing three coins. The frequency with which we find "more heads than tails" will be the sum of the relative frequencies of e_1, e_2, e_3, and e_4. Thus we say the probability of the event "more heads than tails" is the sum of the probabilities of the simple events that comprise the event "more heads than tails":

$$P \text{ (more heads than tails)} = P(e_1) + P(e_2) + P(e_3) + P(e_4)$$

To make this simpler to write, we often denote the event of interest by a symbol, such as A for the event "more heads than tails." Then

$$P(A) = P(e_1) + P(e_2) + P(e_3) + P(e_4)$$

Thus *the probability of a composite event is the sum of the probabilities of all the simple events that comprise it.* (Note that simple events are always mutually exclusive.) Since in the example with three coins we have $P(e_1) = P(e_2) = \ldots = P(e_8) = 1/8$, we find

$$P(A) = 1/8 + 1/8 + 1/8 + 1/8 = 1/2$$

DISTRIBUTION	FORM	PROBABILITY FUNCTION	COMMENTS ON APPLICATION
NORMAL		$y = \dfrac{1}{\sigma\sqrt{2\pi}}\, e^{-\frac{(X-\mu)^2}{2\sigma^2}}$ μ = Mean σ = Standard deviation	Applicable when there is a concentration of observations about the average and it is equally likely that observations will occur above and below the average. Variation in observations is usually the result of many small causes.
EXPONENTIAL		$y = \dfrac{1}{\mu}\, e^{-\frac{x}{\mu}}$	Applicable when it is likely that more observations will occur below the average than above.
WEIBULL		$y = \alpha\beta\,(X-\gamma)^{\beta-1} e^{-\alpha(X-\gamma)^{\beta}}$ α = Scale parameter β = Shape parameter γ = Location parameter	Applicable in describing a wide variety of patterns of variation, including departures from the normal and exponential.
POISSON*		$y = \dfrac{(np)^r\, e^{-np}}{r!}$ n = Number of trials r = Number of occurrences p = Probability of occurrence	Same as binomial but particularly applicable when there are many opportunities for occurrence of an event, but a low probability (less than 0.10) on each trial.
BINOMIAL*		$y = \dfrac{n!}{r!(n-r)!}\, p^r q^{n-r}$ n = Number of trials r = Number of occurrences p = Probability of occurrence $q = 1-p$	Applicable in defining the probability of r occurrences in n trials of an event which has a probability of occurrence of p on each trial.
NEGATIVE BINOMIAL*		$y = \dfrac{(r+s-1)!}{(r-1)!(s!)}\, p^r q^s$ r = Number of occurrences s = Difference between number of trials and number of occurrences p = probability of occurrence $q = 1-p$	Applicable in defining the probability that r occurrences will require a total of $r+s$ trials of an event which has a probability of occurrence of p on each trial. (Note that the total number of trials n is $r+s$.)
HYPERGEOMETRIC*		$y = \dfrac{\dbinom{d}{r}\dbinom{N-d}{n-r}}{\dbinom{N}{n}}$	Applicable in defining the probability of r occurrences in n trials of an event when there are a total of d occurrences in a population of N.

FIGURE 44.7   Summary of common univariate probability distributions. (Asterisks indicate that these are discrete distributions, but the curves are shown as continuous for ease of comparison with the continuous distributions.)

i.e., we expect to find more heads than tails (when three coins are tossed) in about 50 percent of such experiments.

In the example with three coins, we have *equally likely simple events,* i.e., $P(e_i) = P(e_j)$ for all i, j. When this is true, it follows that for any composite event A we have

$$P(A) = \frac{\text{number of simple events in } A}{\text{number of points in the sample space}}$$

In the case of the three coins, this yields the same answer obtained before, namely, $P(A) = 4/8 = 1/2$.

We say two composite events A_1 and A_2 are *mutually exclusive* if no e_i is in both A_1 and A_2. For example, if A_1 is the event "2 heads" and A_2 is the event "more tails than heads," then A_1 and A_2 are mutually exclusive because $A_1 = \{e_2, e_3, e_4\}$ and $A_2 = \{e_5, e_6, e_7, e_8\}$ have no point in common. We often express the fact that A_1 and A_2 are mutually exclusive in shorthand by writing

$$A_1 A_2 = \phi$$

If A_1 and A_2 are mutually exclusive, then for the event "A_1 or A_2" (which occurs if and only if at least one of A_1, A_2 occurs), *we have*

$$P(A_1 \text{ or } A_2) = P(A_1) + P(A_2)$$

This follows because $P(A_1 \text{ or } A_2)$ equals the number of e_i in either A_1 or A_2 divided by the total number of simple events; since there are no points in both A_1 and A_2, this is the same as taking the number of points in A_1 and adding to it the number in A_2 and then dividing by the total number of simple events. In our example, $P(A_1 \text{ or } A_2) = P(A_1) + P(A_2) = 3/8 + 4/8 = 7/8$.

In our example so far, we have discussed the events

A: "more heads than tails"

A_1: "2 heads"

A_2: "more tails than heads"

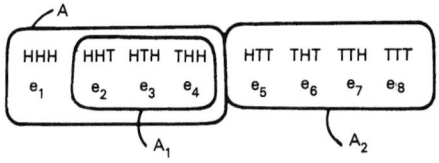

FIGURE 44.8   Events and sample space, experiment of tossing 3 coins.

These are shown on the sample space in Figure 44.8. Here A and A_1 are not mutually exclusive, so the simple addition of the probabilities does not hold for them, since $P(A) + P(A_1) = 4/8 + 3/8 = 7/8$ counts the points e_2, e_3, e_4 twice as to their probabilities. Thus we see that a correct equation for $P(A \text{ or } A_1)$ will need to subtract this overcounting part, which is $P(A \text{ and } A_1)$, i.e., the probability that we have an experimental outcome where both A *and* A_1 occur, which is $P(A$ and $A_1) = 3/8$ in this example. Thus we have reasoned to the fact that *for any events A and A_1 the addition rule is*

$$P(A \text{ or } A_1) = P(A) + P(A_1) - P(A \text{ and } A_1)$$

For mutually exclusive events, the $P(A$ and $A_1)$ would be 0 so that this would reduce to simple addition of $P(A)$ and $P(A_1)$ in that case.

Rules of Probability, Combinatorics.

Probability theory underlies all decisions that are based on sampling. As we have seen, probability is expressed as a number that lies between 1.0 (certainty that an event will occur) and 0.0 (impossibility of occurrence), and the most intuitive definition of probability is one based on a frequency interpretation. In the simple case when an event A can occur in s cases out of a total of n possible and equally probable cases, then the probability that the event will occur is

$$P(A) = \frac{s}{n} = \frac{\text{number of successful cases}}{\text{total number of possible cases}}$$

Counting s and n can be complex, in which case it is called a problem of *combinatorics.*

Example: A lot consists of 100 parts. A single part is selected *at random,* which means that each of the 100 parts has an equal chance of being selected. Suppose a lot contains a total of 8 nonconforming parts. Then the probability of drawing a single part that is non-conforming is 8/100, or 0.08.

The following theorems are useful in solving problems involving probability:

Theorem 1: If $P(A)$ is the probability that an event A will occur, then the probability that A will not occur is $1 - P(A)$.

Theorem 2: If A and B are two events, then the probability that either A or B occurs is

$$P(A \text{ or } B) = P(A) + P(B) - P(A \text{ and } B)$$

A special case of this theorem occurs when A and B cannot occur simultaneously (i.e., A and B are "mutually exclusive"). Then the probability that either A or B occurs is

$$P(A \text{ or } B) = P(A) + P(B)$$

Example: The probabilities of r defectives in a sample of six units from a 5 percent defective lot are found below by the binomial. The probability of zero defectives is 0.7351; the probability of one defective is 0.2321. The probability of zero or one defective is then $0.7351 + 0.2321$, or 0.9672.

Theorem 3: If A and B are two events, then the probability of the joint occurrence of both A and B is

$$P(A \text{ and } B) = P(A) \times P(B|A)$$

where $P(B|A) =$ probability that B will occur assuming A has already occurred.

A special case of this theorem occurs when the two events are *independent*; i.e., when the occurrence of one event has no influence on the probability of the other event. If A and B are independent, then the probability of both A and B occurring is

$$P(A \text{ and } B) = P(A) \times P(B)$$

Example: A complex system consists of two major independent subsystems. The probability of successful performance of the first subsystem is 0.95; the corresponding probability for the second subsystem is 0.90. Both subsystems must operate successfully to achieve total system success. The probability of the successful operation of the total system is therefore $0.95 \times 0.90 = 0.855$.

The preceding theorems have been stated in terms of two events but can be expanded for any number of events.

Conditional Probability; Bayes' Theorem.

In *conditional probability,* we seek an answer to such questions as, "If I know that A_2 has occurred, then on those trials of the experiment where A_2 has occurred, how often does A_1 occur?" We use a special shorthand symbol for this conditional probability:

$$P(A_1|A_2)$$

which is read as "the probability of A_1, given that A_2 is known to have occurred" and is calculated from the formula

$$P(A_1|A_2) = \frac{P(A_1 \text{ and } A_2)}{P(A_2)}$$

If $P(A_1|A_2) > P(A_1)$, we say A_2 carries "positive information" about A_1.

If $P(A_1|A_2) < P(A_1)$, we say A_2 carries "negative information" about A_1.

If $P(A_1|A_2) = P(A_1)$, we say A_2 carries no information about A_1 or that A_1 and A_2 are "independent events." In this last case, knowing that A_2 has (or has not) occurred does not change the chances of A_1 occurring.

The powerful *conditional probability reversal formula,* namely,

$$P(A_1|A_2)P(A_2) = P(A_1 \text{ and } A_2) = P(A_2 \text{ and } A_1) = P(A_2|A_1)P(A_1)$$

is the basis of Bayes' theorem. In the simplest setting, this states that

$$P(A_2|A_1) = \frac{P(A_1|A_2)P(A_2)}{P(A_1|A_2)P(A_2) + P(A_1|\text{not } A_2)P(\text{not } A_2)}$$

(Here, "not A_2," is the event that A_2 does not occur.) For worked solutions of problems in probability, see Dudewicz (1993).

The techniques presented under Tests of Hypotheses in this section consist of analyzing a sample of observations and reaching a conclusion (with defined sampling risks) to accept or reject a hypothesis. The experimenter considers the consequences of drawing incorrect conclusions and, to a lesser degree, the likelihood that extreme values of the population parameter will occur. However, this is usually done on a qualitative basis and involves judgment. In practice, a sample size is limited by economics, and the experimenter defines the type I error (usually 0.05 or 0.01) in numerical terms and then must accept the type II error that results with the sample size fixed by economics. There is a methodical way of defining the consequences of the type I and type II errors and the likelihood of extreme values. The approach involves Bayes' theorem and statistical decision theory.

Statistical Decision Theory.   This concept requires two items of information not formally used in classic analysis:

1. The economic consequences of making type I or type II errors.
2. The probabilities that different values of the population parameter will occur. (The classical approach has no assumption concerning different values of the population parameter.)

Statistical tables and sampling plans based on Bayes' theorem or statistical decision concepts are not common, but the concepts can have a significant effect, and therefore development work seems imminent and worthwhile. Oliver and Springer (1972) give an example of Bayesian acceptance sampling tables. Hadley (1967) provides background material including examples on tests of hypotheses, confidence limits, and acceptance sampling plans. Lenz and Rendtel (1984) compare the performance of MIL-STD-105D, Skip Lot, and Bayesian sampling plans. See also the discussion in Section 46, under Bayesian Procedures.

Simpson's Paradox.   What is called *Simpson's paradox* was discovered in 1951 by E. H. Simpson. Since then, it has been found to have many important implications in industry, medicine, and many other fields; a discussion is given in Dudewicz and Mishra (1988, pp. 55–57) and examples are given in Wardrop (1995) and its references. I will give an example to illustrate the problem and discuss how to avoid it.

Suppose that a company has two plants in different parts of the world. Both make a state-of-the-art product that is not yet fully understood (and hence much rework or scrap results). A new process modification is to be tried in both plants, head to head with the present method. The results of the trial come back and are summarized for management as follows: 46 percent successes (on 11,000 units made) with the new method but only 11 percent successes (on 10,100 units made) with the present method. Do you have enough information, and if so, what decision do you make on selection of a method to use in the future?

It is tempting to decide to use the new process; after all, it had over four times the success rate of the present process with a substantial number of trials. However, suppose we ask for more than the executive summary report, namely, for results in both the plants. In fact, it may be the case that the present method was substantially superior in *both* the plants, as illustrated in Table 44.12. There we see that the new method decreased the success rate in Plant 1 from 10 to 5 percent and in Plant 2 from 95 percent with the present method to 50 percent with the new method. Thus the present method was substantially superior in both plants (both Plant 1, which operates in adverse conditions and has a correspondingly lower success rate, and Plant 2, which has a higher success rate and thus received 10,000 of the new materials to test with).

This is paradoxical because the present method does better in both plants, but when one sums over plants, it does worse. In fact, people often think that this is impossible before studying a numerical example such as we have provided. To avoid erroneous decisions (such as selection of the inferior method), there are a number of actions one can take. First, in the design of trials that will take place in different facilities, assign the same number of test cases to each facility. (This is not always possible, since facilities may vary in size, prior commitments, etc.) Second, when assessing results that come from several (two or more) locations, machines, technicians, states, etc., insist that more than the overall summary of successes by method be provided (namely, one also needs to see the successes by location, machine, technician, state, etc.). This should allow one to avoid errors due to Simpson's paradox (and in addition will allow one to detect possibly important nonuniformity in success rates at different locations, machines, etc. that might otherwise go undetected).

DISCRETE PROBABILITY DISTRIBUTIONS

Discrete probability distributions are used to model situations where the outcome of interest can take on only a few discrete values (such as 0 or 1 for failure or success or 0, 1, 2, 3,…as a number of occurrences of some event of interest). Below I give the model leading to the most commonly occurring such distributions and consider where one can obtain numerical values of their probabilities, and their uses in quality control.

The Discrete Uniform Distribution.   If each of the values $x_1,\ldots, x_n$ is equally likely to occur as the result of an experiment, then we say the value obtained has the uniform distribution on the set of values $x_1,\ldots, x_n$. In this case the probability of x_i is $1/n$. Since the probabilities are so simple, no special tables are needed.

Model Leading to a Uniform Distribution.   The model leading to a uniform distribution is random selection from a finite population in which each value occurs the same number of times. (This makes values equally likely to occur in the sample.)

Uses of Random Choices.   Random choices are often used in sampling inspection. For example, suppose that a lot of 1000 items is sequentially numbered from 500 through 1499. Then the chance

TABLE 44.12 Simpson's Paradox with Two Methods in Two Locations

		Plant 1	Plant 2
Present method:	Successes	1000	95
	Failures	9000	5
New method:	Successes	50	5000
	Failures	950	5000

that an item selected at random from the lot will have number i (for any i between 500 and 1499) is $1/1000 = 0.001$. The probability that such an item will have a serial number at least 1400 is $100/1000 = 0.10$.

The Binomial Distribution.

If the probability of occurrence p of an event is constant on each of n independent trials of the event, then the probability of r occurrences in n trials is

$$\frac{n!}{r!(n-r)!}\, p^r q^{n-r}$$

where $q = 1 - p$.

In practice, the assumption of a constant probability of occurrence is considered reasonable when the population size is at least 10 times the sample size (under this circumstance, the change in probability from one trial to the next due to depletion of the population is negligible).

Table F in Appendix II provides partial tables for the binomial and gives references for more complete tables. King (1971, chaps. 20 through 22) discusses binomial probability.

Model Leading to a Binomial Distribution. When n independent trials of an experiment each have a constant probability p of occurrence of an event of interest (commonly termed a *success*), then the number of occurrences follows a binomial distribution. The name comes from the fact that the factor

$$\frac{n!}{r!(n-r)!}$$

in the probabilities is called a *binomial coefficient* in mathematics.

Binomial Probabilities and Uses. A lot of 100 units of product is submitted by a vendor whose past quality has been about 5 percent nonconforming. A random sample of six units is selected from the lot. The probabilities of various sample results are given in Table 44.13.

In using the formula, note that $0! = 1$. Table F in Appendix II lists binomial probabilities in cumulative form, i.e., the probability of r or fewer occurrences in n trials. For the preceding example, the probability of 1 or fewer nonconforming items in a sample of 6 can be read from the table as 0.9672. Note that this is the sum of the probabilities for $r = 0$ and $r = 1$, i.e., $0.7351 + 0.2321 = 0.9672$.

The Hypergeometric Distribution.

Occasionally, the assumptions of the Poisson (see below) or binomial cannot be met even approximately. Subject only to the assumption of a random

TABLE 44.13 Table of Binomial Probabilities

r	P (exactly r defectives in 6) $=$ $[6!/r!\,(6-r)!](0.05)^r(0.95)^{6-r}$
0	0.7351
1	0.2321
2	0.0306
3	0.0021
4	0.0001
5	0.0000
6	0.0000

sample, the hypergeometric gives the probability of exactly r occurrences in n trials from a lot of N items having d defectives as

$$\frac{\binom{d}{r} \binom{N-d}{n-r}}{\binom{N}{n}}$$

where $\binom{N}{n}$ is the "combinations" of N items taken n at a time and is equal to $N!/[n!(N-n)!]$, where $N! = [N(N-1)(N-2) \cdots 1]$ and $0! = 1$. The calculations can be avoided by using tables such as those prepared by Lieberman and Owen (1961). Duncan (1974) compares the results of Poisson, binomial, and hypergeometric distributions.

Model Leading to a Hypergeometric Distribution.   The hypergeometric distribution is appropriate when independent trials are conducted, but the probability of occurrence of the event of interest changes from trial to trial because of depletion of a finite population. Because of their simpler form, in this situation one often uses the binomial or Poisson distributions if their assumptions are approximately met.

Hypergeometric Probabilities and Uses.   A lot of 100 units is submitted by a vendor whose past quality has been about 5 percent nonconforming. A random sample of 20 units is selected from the lot. To calculate the probability of 0 nonconforming in 20, note that the lot has 5 nonconforming items and 95 conforming. Then

$$P(0 \text{ in } 20) = \frac{\binom{5}{0} \binom{95}{20}}{\binom{100}{20}} = \frac{\left[\dfrac{5!}{0!(5-0)!}\right]\left[\dfrac{95!}{20!(95-20)!}\right]}{\dfrac{100!}{20!(100-20)!}} = 0.319$$

Repeat substitutions into the formula are made to find $P(r \text{ in } 20)$, where r in this example is 0, 1, 2, 3, 4, and 5.

The Poisson Distribution.

In practice, the most important discrete distribution is the Poisson. It is an approximation to more exact distributions (the hypergeometric and the binomial) and applies when the sample size is at least 16, the population size is at least 10 times the sample size, and the probability of occurrence p on each trial is less than 0.1. (These conditions are often met.)

Figure 44.7 states the Poisson probability function, but the real work is done by cumulative probability tables.

Model Leading to a Poisson Distribution.   As well as being an approximation to more exact distributions, the Poisson is the exact distribution when certain assumptions are met. These assumptions are that events occur at random (in time, or in space, or in location, for example) with a probability of occurrence roughly proportional to the length of time (or volume of space, or area) and that there is no "clumping." [For details, see Dudewicz (1976, Section 3.2).] For example, if a target 0.1 mi^2 in size is known to be contained in an area 10 mi^2 in size, and this area is shelled at random (one shell at a time so there will be no clumping), the probability of a hit will be $0.1/10 = 0.01$. The number of hits will follow a Poisson distribution, and the number of shells fired may be set so that the probability of eight or more hits on the target will be at least 0.95; this is often done in practice when it is known that eight or more hits will effectively destroy the target and a 95 percent kill probability is desired.

Poisson Probabilities and Uses.   A lot of 300 units of product is submitted by a supplier whose past quality has been about 2 percent nonconforming. A random sample of 40 units is selected from the lot. Table E in Appendix II provides the probability of r or fewer defectives in a sample of n units. (The application of these probabilities is explained in Section 46, under Operating Characteristic, OC, Curve.) Entering the table with a value of np equal to $40(0.02)$, or 0.8, gives Table 44.14. Individual probabilities can be found by subtracting cumulative probabilities. Thus the probability of exactly two defectives is $0.953 - 0.809$, or 0.144.

The Negative Binomial Distribution.

The negative binomial distribution is one of the most commonly occurring distributions in situations where the sample size is not set in advance but rather is determined as the experiment proceeds.

Model Leading to a Negative Binomial Distribution.   If the probability of occurrence of an event is constant from trial to trial and we make trials until we find m occurrences, then the probability that r trials will be needed is

$$\frac{(r-1)!}{(m-1)!(r-m)!} \, p^m (1-p)^{r-m}$$

where r can be m, $m+1$, $m+2$,.... This equation is equivalent to the more complex one listed in Figure 44.7. Other situations leading to a negative binomial distribution are discussed in Chapter 5 of Johnson and Kotz (1969). Tables are available in Williamson and Bretherton (1963) in case direct calculation is burdensome; Johnson and Kotz (1969) give references to additional tables.

Negative Binomial Probabilities and Uses.   A large lot is inspected until the first defective ($m = 1$) is found; if this occurs in the first five trials, the lot is rejected. Hence the lot will be accepted if no defective is found in the first five trials (and thus trials 6, 7,... need not be performed—we will never inspect more than five items with this scheme). If the lot is 10 percent nonconforming, then from Table

TABLE 44.14   Table of Poisson Probabilities

r	Probability of r or fewer in sample
0	0.449
1	0.809
2	0.953
3	0.991
4	0.999
5	1.000

TABLE 44.15   Table of Negative Binomial Probabilities ($m=1, p=0.10$)

r	Probability r trials are needed to find $m = 1$ defectives
1	0.10
2	0.09
3	0.081
4	0.0729
5	0.06561
6	0.059049
7	0.053144
8	0.047830

44.15 we see that the probability the lot is accepted is $1 - (0.10 + 0.09 + 0.081 + 0.0729 + 0.06561)$ $= 0.59049$.

The Multinomial Distribution.   The discrete probability distributions discussed up to this point (the uniform, binomial, hypergeometric, Poisson, and negative binomial) all relate to situations that are *univariate,* that is, where the outcome of interest relates to one variable's value (such as the number of defectives in a sample of size n in the binomial case). However, there are important cases where the outcome is *multivariate.* One such example is that where in a sample of size n one observes both the number needing rework and the number to be scrapped, since there are two quantities, this is called a *bivariate* situation. The multinomial distribution can be used when there are any number of categories into which the items may be classified.

Model Leading to a Multinomial Distribution.   If exactly one of the events $E_1, \dots, E_k$ occurs on each of n independent trials and the probabilities of occurrence of the events are respectively $p_1, \dots,$ p_k (with $p_1 + \dots + p_k = 1$ so that one of them must occur), then the probability that E_1 occurs x_1 times and E_2 occurs x_2 times,..., and E_k occurs x_k times (where $x_1 + x_2 + \dots + x_k = n$ since there are n trials and one of $E_1, \dots, E_k$ must occur on each trial) is

$$\frac{n!}{x_1! x_2! \dots x_k!} \, p_1^{x_1} p_2^{x_2} \cdots p_k^{x_k}$$

Tables are not widely available, and calculations are usually done directly from the probability formula, or (if n is large) the multivariate normal distribution (a continuous multivariate distribution discussed later in this section) is used as an approximation.

Multinomial Probabilities and Uses.   Suppose that $n = 5$ large assemblies are manufactured and inspected each day. The results of each inspection are either pass, rework, or scrap. Past results have shown that 80 percent pass, 15 percent need rework, and 5 percent need to be scrapped. What are the probabilities of the various possible outcomes of one day's output?

Figure 44.9 shows the probabilities of the outcomes, calculated directly from the basic formula. Note that once the number to be reworked (e.g., 1) and the number to be scrapped (e.g., 0) are specified, the number passed is determined (e.g., $5 - 1 - 0 = 4$). We have had to use the multinomial probability distribution because the categories are not independent (an item cannot be both passed and scrapped).

Selecting a Discrete Distribution.   Selection of which discrete distribution to use is usually made either by knowledge of the underlying situation or by fitting a model from relative frequency probability. In either case, a test of the model selected is desirable to check its validity.

Selection from a Model of Reality.   In many cases one will know (or assume) that the model that leads to one of the distributions we have discussed underlies the practical situation. For example, if one draws 50 items at random from a large lot with $100p$ percent defective, one will assume the binomial model with $n = 50$ and probability p of a defective on each trial.

(Chi Square) Test of Model Validity.   To test the validity of an assumed discrete model, where cell i should occur with probability p_i, one compares the observed cell totals with those predicted by the model using the chi square test discussed later (Test 12*b*). Such a confirmatory test can be omitted only if one is willing to run the risk of assuming a model that in fact may have little basis in reality. [See Section 9.12 of Dudewicz and Mishra (1988) and especially Problem 9.12.2 on p. 532.]

Empirical Models via Relative Frequency Probability.   If there is little or no reason to lead to the adoption of one of the specific models discussed, a model can be fitted to the data using the relative frequencies observed in the past. For example, if one has observed that in 100 items produced in the

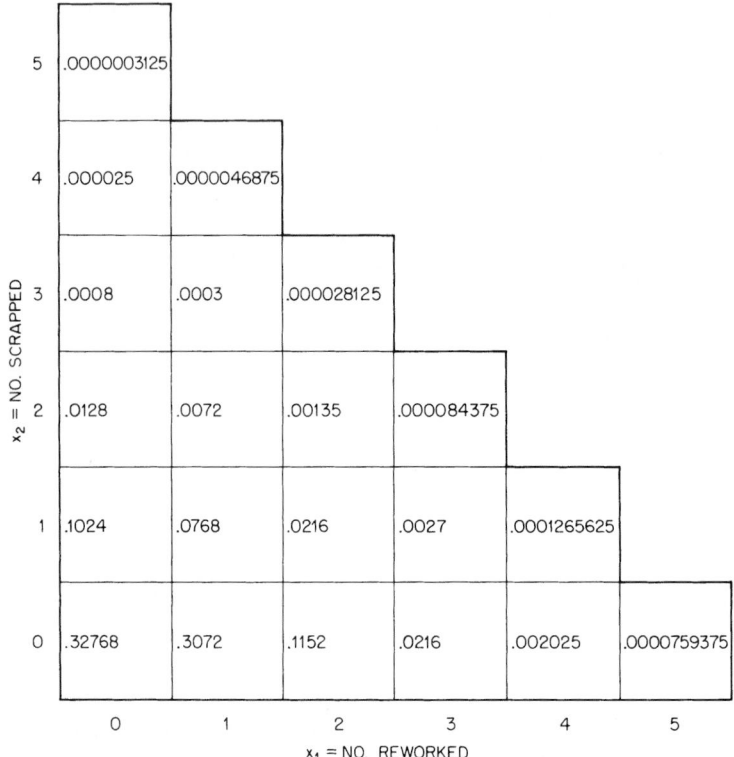

FIGURE 44.9   Table of multinomial probabilities, $k = 3$ Categories, $n = 5$, $p_1 = P(\text{Rework}) = 0.15$, $p_2 = P(\text{Scrap}) = 0.05$, $p_3 = P(\text{Pass}) = 0.80$.

past there have been blemishes present in the numbers given in Table 44.16, one would estimate the probability of r blemishes as its relative frequency in the past work, i.e., as in Table 44.17. Such a model should be tested by taking a new sample and performing a chi square test as previously discussed. See also Bootstrap Methods later in this section.

CONTINUOUS PROBABILITY DISTRIBUTIONS

Continuous probability distributions are used to model situations where the outcome of interest can take on values in a continuous range (such as all values greater than zero for the failure time of a motor that is run continuously). Below I give the model leading to the most commonly occurring such distributions and consider where one can obtain numerical values of their probabilities and their uses in quality control.

The Continuous Uniform Distribution.   If all values between a and b ($a < b$) are possible, and the chances of the value being in a subinterval are proportional to its length, then the uniform distribution is appropriate. The probability function is flat over the interval (a,b), where $y = 1/(b - a)$. Thus the probability the value is in a subinterval of length c is $c/(b - a)$. Since the probabilities are so simple, no special tables are needed.

Model Leading to a Uniform Distribution. The model leading to a continuous uniform distribution over the range (*a*,*b*) is random selection of a value between *a* and *b*. For example, if a valve on a water line is spun at random between pressure 0 lb/in² (closed) and 100 lb/in² (fully open), then the resulting pressure will be a uniform random variable on (0, 100).

Uses of Random Numbers. Uniform random variables on the range (0, 1) are called *random numbers*. Such variables are often used to drive digital computer simulation models and are of great importance in simulation studies of quality systems. Full details on sources of high-quality random numbers may be found in Dudewicz and Ralley (1981). Special considerations for simulation uses on microcomputers are given in Dudewicz et al. (1985).

The Exponential Distribution.
The exponential probability function is

$$y = \frac{1}{\mu}\, e^{-X/\mu}$$

Figure 44.7 shows the shape of an exponential distribution curve. Note that the normal and exponential distributions have distinctly different shapes. An examination of the tables of areas shows that 50 percent of a normally distributed population occurs above the mean value and 50 percent below it. In an exponential population, 36.8 percent are above the mean and 63.2 percent below the mean. (This refutes the intuitive idea that the average is always the 50 percent point.) The property of a higher percentage below the average sometimes helps to indicate applications of the exponential. For example, the exponential describes the loading pattern for some structural members because smaller loads are more numerous than larger loads. The exponential is also useful in describing the distribution of failure times of certain complex equipment.

Model Leading to an Exponential Distribution. It can be shown that the exponential distribution of failure times arises when failures occur "at random" (and are not due to wearout but to such items as random shocks). In fact, the exponential distribution is characterized as the only continuous distribution with the "lack of memory property" that the chances of the item living an additional t_0 time units depend only on the length t_0 and not on how long the item has already been in use [see Dudewicz (1976), pp. 88, 106, for details].

Predictions with Exponential Distributions. Predictions based on an exponentially distributed population require only an estimate of the population mean. For example, the time between successive failures of a complex piece of equipment is measured, and the resulting histogram is found to

TABLE 44.16 Blemishes per Item in Past Work

No. of blemishes	No. of occurrences
0	50
1	25
2	25

TABLE 44.17 Blemish Probabilities via Relative Frequency Estimation

No. of blemishes	Probability
0	.50
1	.25
2	.25

resemble the exponential probability curve. The results of a sample of measurements indicate that the average time between failures (commonly called *MTBF*, or *mean time between failures*) is 100 h. What is the probability that the time between two successive failures of this equipment will be at least 20 h?

The problem is one of finding the area under the curve beyond 20 h (Figure 44.10). Table C in Appendix II gives the area under the curve beyond any particular ratio X/μ. In this problem,

$$\frac{X}{\mu} = \frac{20}{100} = 0.20$$

and from Table C in Appendix II the area under the curve beyond 20 h is thus 0.8187. The probability that the time between two successive failures is greater than 20 h is 0.8187; i.e., there is about an 82 percent chance that the equipment will operate without failure continuously for 20 or more hours. Similar calculations would give a probability of 0.9048 for 10 or more hours. In Section 48, Reliability Concepts and Data Analysis, this probability is calculated for the specified mission time of a product, and the result is called *reliability*. These analyses also could be made using exponential probability paper.

Example: The Relationship Between Part and System Reliability: It is often assumed that the probability of survival P_s (the system reliability) is the product of the individual reliabilities of the *n* parts within the system: $P_s = P_1 P_2 \dots P_n$. This is known as the *product rule*. The formula assumes (1) that the failures of any part will cause failure of the system and (2) that the reliabilities of the parts are independent of each other, i.e., that the reliability of one part is not dependent on the reliability of another part. [Evans (1966) gives a good discussion of this and other assumptions in reliability calculations.] A set of lights in series on a Christmas tree demonstrates the product rule. These assumptions are usually not 100 percent correct. However, the formula is a convenient approximation that should be refined as information becomes available on the interrelationships of parts and their relationship to the system. [The redundancy formula (see below under Example: Redundancy) is an example of this.] I will now illustrate the product rule.

Suppose that the following reliability requirements have been set on the subsystems of a communications system:

Subsystem	Reliability (for a 4-h period), %
Receiver	0.970
Control system	0.989
Power supply	0.995
Antenna	0.996

What is the expected reliability of the overall system if the preceding requirements are met?

$$P_s = (0.970)(0.989)(0.995)(0.996) = 0.951$$

The chance that the overall system will perform its function without failure for a 4-h period is 95 percent.

If it can be assumed that each part follows the exponential distribution, then

$$P_s = e^{-t_1 \lambda_1} e^{-t_2 \lambda_2} \dots e^{-t_n \lambda_n}$$

Further, if *t* is the same for each part,

$$P_s = e^{-t \Sigma \lambda}$$

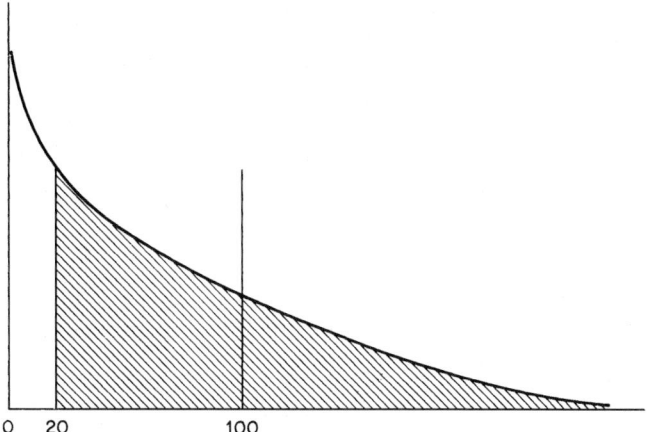

FIGURE 44.10   Distribution of time between failures.

Thus, when the failure rate is constant (and therefore the exponential distribution applies), a "reliability prediction" of a system can be made based on the addition of the part failure rates. This is illustrated in Section 19, under Designing for Reliability, Maintainability, and Availability.

Example: Redundancy: A number of system designs have been devised that attempt to increase system reliability by introducing redundancy. The simplest such system is the *parallel system,* which operates if at least one of its components operates. If each component has reliability $R(t_0)$, then the parallel system consisting of *m* components has reliability equal to [see Dudewicz (1976, p. 39)]:

$$1 - [1 - R(t_0)]^m$$

which is greater than $R(t_0)$.

Many other designs have been introduced, with a view both to reliability and to cost, such as the *k*-out-of-*n* systems. For details on these and other aspects of reliability, see Zacks (1983), Ireson (1982), and Lloyd and Lipow (1982).

The Weibull Distribution.   The Weibull distribution is a family of distributions having the general density function

$$y = \alpha\beta(X - \gamma)^{\beta-1} e^{-\alpha(X-\gamma)^{\beta}}$$

where α = scale parameter, β = shape parameter, and γ = location parameter.

The curve of the function (Figure 44.7) varies greatly depending on the numerical values of the parameters. Most important is the shape parameter β, which reflects the pattern of the curve. Note that when β is 1.0, the Weibull function reduces to the exponential and that when β is about 3.5 (and $\alpha = 1$ and $\gamma = 0$), the Weibull closely approximates the normal distribution. In practice, β varies from about $^1/_3$ to 5. The scale parameter α is related to the peakedness of the curve; i.e., as α changes, the curve becomes flatter or more peaked.

The location parameter γ is the smallest possible value of X. This is often assumed to be zero, thereby simplifying the equation. It is often unnecessary to determine the values of these parameters because predictions are made directly from Weibull probability paper. King (1971, pp. 136–140) gives procedures for graphically finding α, β, and γ.

The Weibull covers many shapes of distributions. This makes it popular in practice because it reduces the problem of examining a set of data and deciding which of the common distributions (e.g., normal or exponential) fits best. In particular, both IFR (increasing failure rate) and DFR (decreasing failure rate) cases are included, respectively, with $\beta > 1$ and $\beta < 1$ [see Dudewicz (1976, pp. 88–89)].

Model Leading to a Weibull Distribution.   It can be shown that a Weibull distribution arises if an exponential variable is raised to a power; i.e., if Y is exponential, then $Y^{1/\beta}$ has a Weibull distribution (Dudewicz 1976, p. 89).

Predictions with Weibull Distributions.   An analytical approach for the Weibull distribution (even with tables) is cumbersome, and the predictions are usually made with Weibull probability paper. For example, five heat-treated shafts were stress tested until each of them failed. The fatigue life (in terms of number of cycles to failure) is

10,263

12,187

16,908

18,042

23,271

The problem is to predict the percentage failure of the population for various values of fatigue life. The solution is to plot the data on Weibull paper, observe if the points fall approximately in a straight line, and if so, read the probability predictions (percentage failure) from the graph.

Although Weibull plotting can follow the mean rank procedure of normal probability paper (see below under Predictions with Normal Distributions), much of the literature on Weibull applications uses "median ranks." Table D in Appendix II gives, for various sample sizes, the values of the median rank. (Note that the mean rank procedure does not require a table.) The median ranks necessary for this particular example are based on a sample size of five failures and are as shown in Table 44.18. (The mean rank estimates are shown for comparison.) The cycles to failure are now plotted on the Weibull graph paper against the corresponding values of the median rank (see Figure 44.11). These points fall approximately in a straight line [King (1971, pp. 126–128) describes how to modify a plot to help obtain a straight line], so it is assumed that the Weibull distribution applies. The vertical axis gives the cumulative percentage of failures in the population corresponding to the fatigue life shown on the horizontal axis. For example, about 50 percent of the population of shafts will fail in less than 17,000 cycles. About 90 percent of the population will fail in less than 24,000 cycles. By appropriate subtractions, predictions can be made of the percentage of failures between any two fatigue life limits.

It is tempting to extrapolate on probability paper, particularly to predict life. For example, suppose the minimum fatigue life were specified as 8000 cycles and the five measurements above were from tests conducted to evaluate the ability of the design to meet 8000 cycles. Since all five tests

TABLE 44.18   Table of Median and Mean Ranks

Failure number i	Median rank	Mean rank $= \dfrac{i}{5 + 1}$
1	0.1294	0.1667
2	0.3147	0.3333
3	0.5000	0.5000
4	0.6853	0.6667
5	0.8706	0.8333

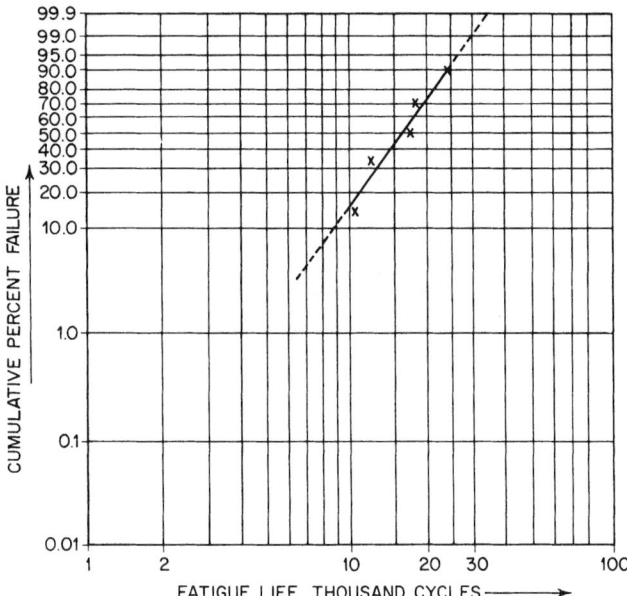

FIGURE 44.11   Distribution of fatigue life on Weibull probability paper.

exceeded 8000 cycles, the design seems adequate and therefore should be released for production. However, extrapolation on the Weibull paper predicts that about 8 percent of the *population* of shafts would fail in less than 8000 cycles. This suggests a review of the design before release to production. Thus the small *sample* (all *within* specifications) gave a deceiving result.

Extrapolation can go in the other direction. Note that a probability plot of life-test data does not require that all tests be completed before the plotting starts. As each unit fails, the failure time can be plotted against the median rank. If the early points appear to be following a straight line, then it is tempting to draw in the line *before* all tests are finished. The line can then be extrapolated beyond the actual test data, and life predictions can be made without accumulating a large amount of test time. The approach has been applied to predicting, *early in a warranty period,* the "vital few" components of a complex product that will be most troublesome. However, extrapolation has dangers. It requires the judicious melding of statistical theory and engineeirng experience and judgment.

Moult (1963) describes the use of a Weibull plot in comparing the suitability of two types of steel for use in bearings. The plot is shown in Figure 44.12. Nelson (1982) discusses Weibull paper. Probability graph paper is available for the normal, exponential, Weibull, and other probability distributions. (A source is TEAM, Technical and Engineering Aids for Management, Box 25, Tamworth, NH 03886.) Although the mathematical functions and tables provide the same information, the graph paper reveals *relationships* between probabilities and values of X that are not readily apparent from the calculations. For example, the reduction in percentage defective in a population as a function of wider and wider tolerance limits can be easily portrayed by the graph paper.

The Normal Distribution.   Many engineering characteristics can be approximated by the normal distribution:

$$y = \frac{1}{\sigma\sqrt{2\pi}}\, e^{-(X-\mu)^2/2\sigma^2}$$

where $e = 2.718$, $\pi = 3.141$, μ = population average, σ = population standard deviation.

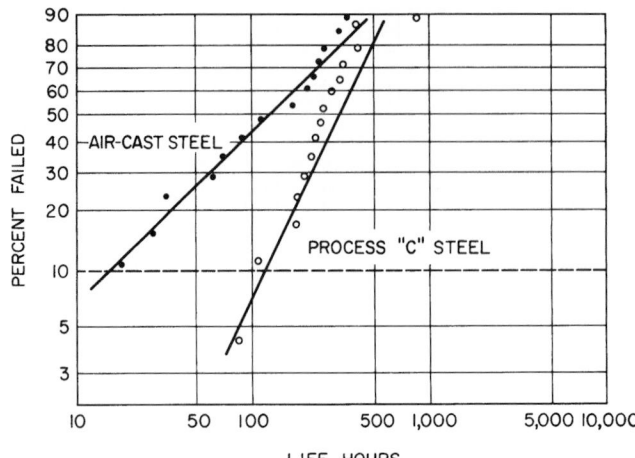

FIGURE 44.12   Composite fatigue endurance—process "C" vacuum degassing
versus air cast AISI 8620.

Problems are solved with a table, but note that the distribution requires estimates of only the average μ and standard deviation σ of the population (unless otherwise indicated, Greek symbols will be
used for population values and Roman symbols for sample values) in order to make predictions about
the population. The curve for the normal probability distribution is related to a frequency distribution
and its histogram. As the sample becomes larger and larger and the width of each cell becomes smaller and smaller, the historgram approaches a smooth curve. If the entire population (in practice, the population is usually considered infinite, e.g., the potential production from a process) were measured, and
if it were normally distributed, the result would be as shown in Figure 44.7. Thus the shape of a histogram of sample data provides some indication of the probability distribution for the population. If the
histogram resembles the "bell" shape shown in Figure 44.7, this is a basis for assuming that the population follows a normal probability distribution. (It is not necessary that the sample histogram be perfectly normal—the assumption of normality is applied only to the population, and small deviations
from normality are expected in random samples.) Hahn (1971) gives a practical discussion of assuming normality. (The name *normal* distribution dates back to a time when all other distributions were
erroneously thought to be abnormal. Today, some prefer the name *Gaussian* distribution.)

Model Leading to a Normal Distribution (Additive Errors, Central Limit Theorem).   It can be
shown that if a variable Y is the result of adding many other variables and those variables are not
highly dependent on each other, then Y will have approximately a normal distribution. [This result is
called the *central limit theorem*; see Dudewicz (1976, p. 149).] Statisticians usually recommend that
10 or more terms be added before this result is relied on to produce normality; however, a number
of applied studies have shown good results with as few as three terms being added.

Predictions with Normal Distributions.   Predictions require just two estimates and a table. The
estimates are

$$\text{Estimate of } \mu = \overline{X} \quad \text{and} \quad \text{estimate of } \sigma = s$$

The calculations of the sample $\overline{X}$ and s are made by one of the methods previously discussed.
    For example, from past experience, a manufacturer concludes that the burnout time of a particu-
lar light bulb it manufactures is normally distributed. A sample of 50 bulbs has been tested and the
average life found to be 60 days, with a standard deviation of 20 days. How many bulbs in the entire
population of light bulbs can be expected to be still working after 100 days of life?

The problem is to find the area under the curve beyond 100 days (see Figure 44.13). The area under a distribution curve between two stated limits represents the probability of occurrence. Therefore, the area beyond 100 days is the probability that a bulb will last more than 100 days. To find the area, calculate the difference between a particular value X and the average of the curve in units of standard deviation:

$$K = \frac{X - \mu}{\sigma}$$

In this problem, $K = (100 - 60) \div 20 = +2.0$. Table B in Appendix II shows, for $K = 2$, a probability of 0.9773. Applied to this problem, the probability that a bulb will last 100 days or less is 0.9773. The normal curve is symmetrical about the average, and the total area is 1.000. The probability of a bulb's lasting more than 100 days then is $1.0000 - 0.9773$, or 0.0227, or 2.27 percent of the bulbs in the population will still be working after 100 days.

Similarly, if a characteristic is normally distributed, and if estimates of the average and standard deviation of the population are obtained, this method can estimate the total percentage of production that will fall within engineering specification limits.

Figure 44.14 shows representative areas under the normal distribution curve (these can be derived from Table B in Appendix II). Thus 68.26 percent of the *population* will fall between the average of the population plus or minus 1 standard deviation of the population, 95.46 percent of the population will fall between the average $\pm 2\sigma$, and finally, $\pm 3\sigma$ will include 99.73 percent of the population. The percentage of a *sample* within a set of limits can be quite different from the percentage within the same limits in the population. This important fact is crucial in testing hypotheses (covered later in this section).

Another way of making predictions based on a normal distribution employs probability paper. Probability paper is so constructed that data from a particular kind of distribution plots as a straight line; i.e., a sample of data from a normally distributed population plots approximately as a straight line on normal probability paper. (Small deviations from a straight line are expected because the data represent a *sample* of the population.) The following are the steps taken to plot a set of individual observations on probability paper:

1. Arrange the observations in ascending values. The smallest value is given a rank i of 1 and the largest value a rank of n.

2. For each value, calculate the cumulative frequency.

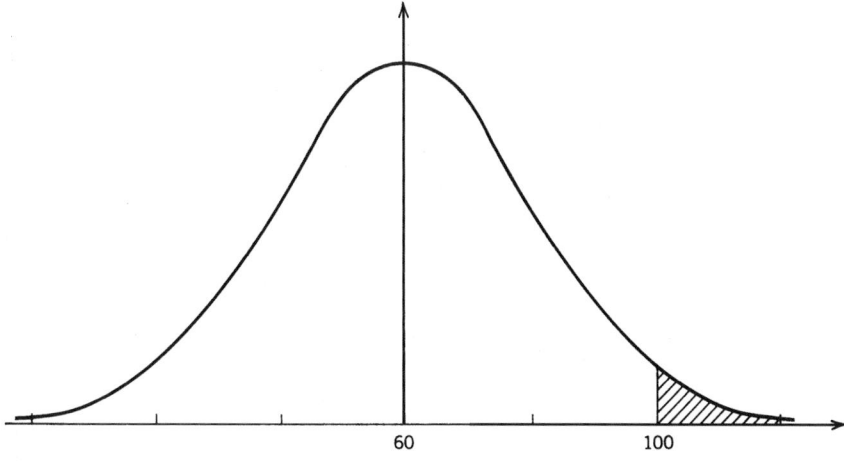

FIGURE 44.13   Distribution of light bulb life.

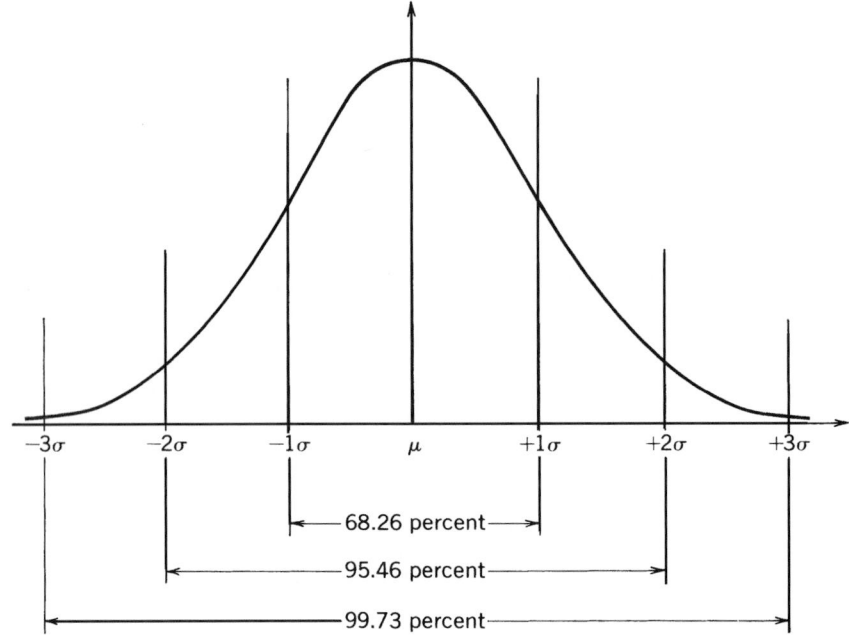

FIGURE 44.14   Areas of a normal curve.

3. For each value, calculate

$$\frac{\text{Cumulative frequency}}{n+1} \times 100$$

This provides the mean rank probability estimate, in percent, for plotting the data.

4. Plot the observed values against their mean rank probability estimate.

If the observations are in frequency distribution form, the procedure is the same, except that instead of using the observed values, the probability estimates are plotted against the cell boundaries. This is illustrated for the resistance data (see Table 44.19).

The plot is shown in Figure 44.15. Lower cell boundaries are plotted against the last column of Table 44.19 using the upper (Percent Over) scale. The line has been drawn in by eye, and the fit appears reasonable. This line represents an estimate of the population, and predictions like those obtained from the normal probability table can be read directly from the graph. For example, 5 percent of the population of coils will have resistance values greater than about 3.39. Also, 95 percent will have values greater than about 3.29. (Therefore, 95 − 5, or 90 percent, will have values between 3.29 and 3.39.)

Figure 44.16 shows a form that incorporates probability paper plotting with further analysis such as confidence limits and control limits. (This type of form was originally developed by E. F. Taylor.)

King (1971) gives a practical description of probability paper procedures for the normal and other important distributions. While fit is often evaluated by eye when we have clearly good fit (as in Figure 44.15), statistical tests are available and should be used in cases that are not so clearcut; see Iman (1982) for details and graphs on which this analysis can be performed.

The Lognormal Distribution.   If $Y = e^Z$, where Z has a normal distribution, Y is said to have a *lognormal distribution* (since the logarithm of Y has a normal distribution).

TABLE 44.19   Resistance Data

Cell boundaries	Frequency	Cumulative frequency	$\dfrac{\text{Cumulative frequency}}{100+1}(100)$
3.415–3.445	1	1	0.99%
3.385–3.415	8	9	8.90
3.355–3.385	27	36	35.60
3.325–3.355	36	72	71.30
3.295–3.325	23	95	94.10
3.265–3.295	5	100	99.00
	100		

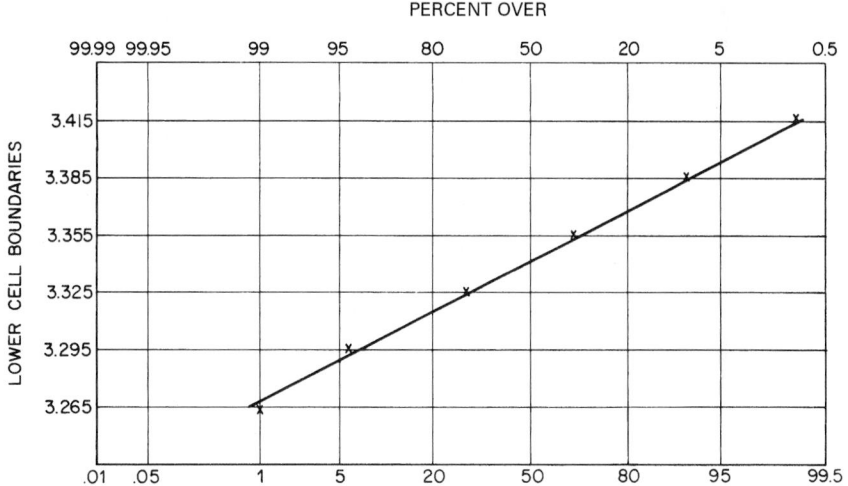

FIGURE 44.15   Cumulative probability plot of Table 44.19.

Model Leading to a Lognormal Distribution (Multiplicative or Percentage Effects).   Lognormal variables arise when effects are percentages or multiplicative, which is common in biological and many other applications. For example, if a random percentage of a stock of items goes bad each time period, the percentage still good after a large number (10 or more) of time periods will be lognormally distributed. (This result follows from the central limit theorem already discussed when considering the normal distribution, since the logarithm of a product is the sum of the logarithms of the items in the product.)

Predictions with Lognormal Distributions.   As with the normal distribution, predictions require two estimates. For details, see Cohen and Whitten (1981). Note that the lognormal distribution is positively skewed and is widely employed as a model for distribution of life spans, reaction times, income distributions, and other economic data.

Often the mean, standard deviation, and probabilities of the lognormal variable Y itself are of basic interest; thus, while one can easily estimate these for Z (the logarithm of $Y = e^{Z}$), this does not answer the real problem. For example, if Y is the lifetime of some system, one may want to estimate the average life of the system—not the mean of the logarithm of Y. This is why the special methods referred to above have been developed for this distribution.

Mixture Distributions. Y is said to have a *mixture distribution* if Y results from source i a percentage $100p_i$ of the time ($i = 1, 2,\dots$).

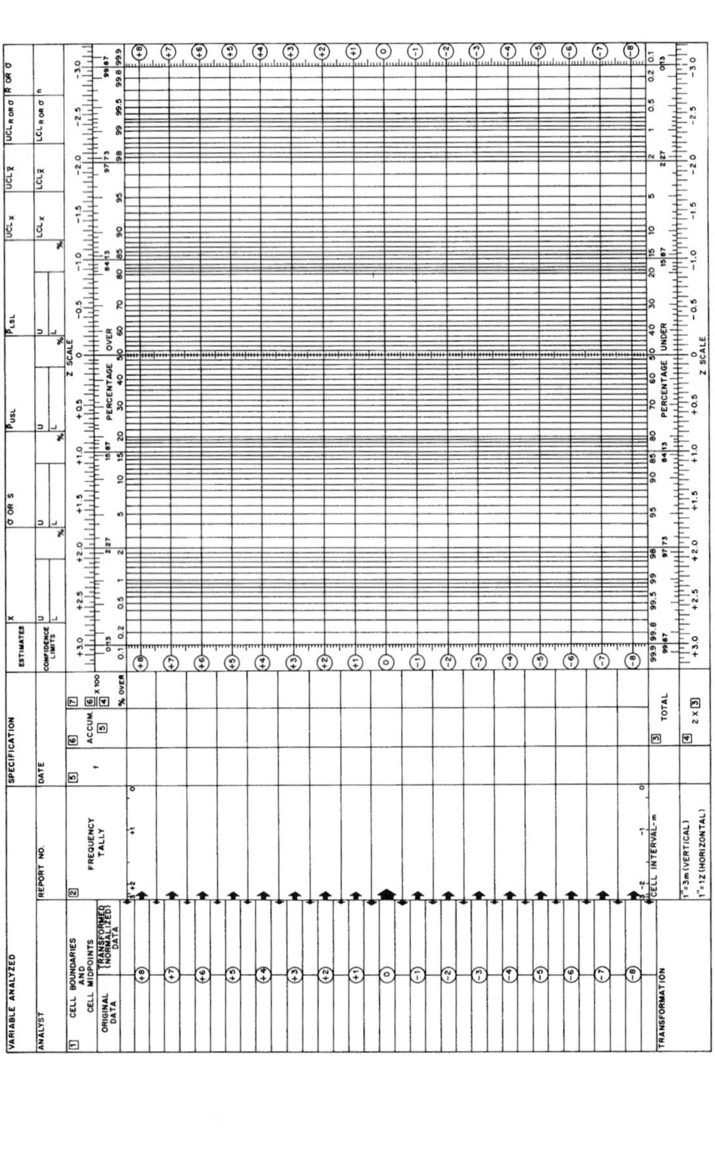

FIGURE 44.16 Normal distribution analysis sheet.

Model Leading to a Mixture Distribution. When output from several sources is mixed (e.g., output from several suppliers, several plants, several machines, several workers, and so on), the quality characteristics of the resulting mix have mixture distributions. If each of the components coming into the mix has *exactly* the same distribution, then the mix also will have that distribution. However, if the components coming into the mix have different distributions, then the mix will have a *mixture distribution.*

Fitting of Mixtures. As can be seen from Figure 44.7, the distributions we have considered so far are unimodal (have one peak). When in practice one sees two or more peaks in the histogram, one suspects that a mixture underlies the data. In some cases, this itself leads to a study of the items coming into the mix, often to find a problem in one of the streams of what should be homogeneous product.

In other cases, the streams coming into the mix cannot be separated (or it is not desired to separate them), but rather one wishes to fit the density of Y as $p_1 f(Z_1) + p_2 f(Z_2) + \ldots + p_k f(Z_k)$ for some k (2 or more). Here the p_i's add to 1 (100 percent of the mix), and often the Z_i's are known to be normal. To fit the distribution of the Y, one must then select $k - 1$ p's (since they add to 1, the last one is then determined), k means, and k variances. This process requires use of modern computer software such as LABONE. As an example, consider the data of Table 44.20. A histogram of these data shows that a mixture (of two terms, since there are two peaks) may be involved. The distributions seem to the eye to be normal. Using LABONE Expert Statistical Programs (ESP), we are able to easily fit a mixture of normal distributions.

The Multinormal Distribution.
The continuous probability distributions discussed up to this point (the uniform, exponential, Weibull, normal, lognormal, and mixture) all relate to situations that are *univariate,* that is, where the outcome of interest has one component (such as lifetime). If there are additional components of interest (such as weight and height), then the outcome is *multivariate* (in this case of three, *trivariate*). The multinormal distribution is appropriate when each of the components has a normal distribution and is the most widely used continuous multivariate distribution.

Model Leading to a Multinormal Distribution. The same sort of additive process that leads to a univariate normal distribution leads to a multivariate normal distribution when more than one component is being measured.

Predictions with Multinormal Distributions. Predictions with multinormal distributions require computer packages in most cases. The details, with computer code and examples, are discussed by Siotani et al. (1985).

The Extended Generalized Lambda Distribution.
A one-parameter lambda distribution was proposed in 1960 by J. Tukey, generalized in 1972 and 1974 by J. S. Ramberg and B. Schmeiser,

TABLE 44.20 Data for Mixture Analysis

3.37	3.34	3.48	3.32	3.33	3.38	3.34	3.31	3.43	3.34
3.29	3.46	3.30	3.31	3.43	3.34	3.34	3.46	3.39	3.34
3.45	3.36	3.30	3.42	3.33	3.35	3.45	3.34	3.32	3.48
3.32	3.37	3.44	3.38	3.36	3.47	3.36	3.31	3.43	3.30
3.35	3.43	3.38	3.37	3.54	3.32	3.36	3.42	3.29	3.35
3.48	3.39	3.34	3.42	3.30	3.39	3.46	3.40	3.32	3.43
3.29	3.41	3.37	3.36	3.41	3.47	3.36	3.37	3.43	3.36
3.31	3.43	3.35	3.34	3.45	3.34	3.31	3.46	3.37	3.35
3.50	3.35	3.37	3.42	3.35	3.36	3.48	3.35	3.31	3.44
3.35	3.36	3.49	3.31	3.31	3.40	3.35	3.33	3.45	3.31

and developed and tabled by Ramberg et al. (1979). While it fits a wide variety of curve shapes and can fit any mean and any variance, it cannot fit all combinations of skewness and kurtosis. Thus it is useful when it fits, but for some data sets it does not fit. Recently, Karian et al. (1996) gave an extension called the *extended generalized lambda distribution* (EGLD), which can fit any and all values of the mean, variance, skewness, and kurtosis; tables needed in applications were given in Dudewicz and Karian (1996). Thus now the benefits of this empirical family are available in all univariate data sets. These methods are extended to bivariate GLD cases in Karian and Dudewicz (1999).

Selecting a Continuous Distribution.   Selection of which continuous distribution to use is usually made either by knowledge of the underlying situation or by fitting a model to the histogram (often via plots on probability papers of the most usual distributions). In either case, a test of the model selected is desirable to check its validity.

Selection from a Model of Reality.   In many cases one will know (or assume) that the model that leads to one of the distributions we have discussed underlies the practical situation. For example, if the life distribution of the equipment under study has, in the past, always been adequately fitted by a Weibull model (though with parameters that change from application to application), one will usually start with a Weibull assumption.

Testing Distributional Assumptions (Probability Plotting, Tests for Specific Distributions).   In practice, a distribution is assumed by evaluating a sample of data. Often it is sufficient to evaluate the shape of the histogram or the degree to which a plot on probability paper follows a straight line. These convenient methods do require judgment (e.g., how "straight" must the line be?) because the sample is never a perfect fit; quantitative tests for probability plots should be used (see Iman 1982). Be suspicious of the data if the fit is "perfect." "Goodness of fit" tests (see Tests of Hypotheses later in this section) evaluate any distribution assumption using quantitative criteria.

Fitting Empirical Probability Distributions.   If there is little or no reason to suggest one of the specific models discussed, or if they are rejected (e.g., because of a poor probability paper fit), an alternative is to fit an empirical model. Such models can adapt to a wide range of distributional shapes, including many of those of the specific models discussed above. One of the most widely used empirical families has been the *generalized lambda distribution* (GLD) family. However, as discussed earlier, the new EGLD enhances its capabilities and should be used instead; often the two give the same result, but when the GLD has difficulty, the EGLD can still match the population's sample moments.

As an example, Table 44.21 presents data (from p. 219 of Hahn and Shapiro, 1967) on the coefficient of friction of a metal in 250 samples. Using procedures and tables given in Ramberg et al. (1979), a probability function can be developed.

TABLE 44.21   Coefficient-of-Friction Data

Range	Frequency
Less than .015	1
0.015–0.020	9
0.020–0.025	30
0.025–0.030	44
0.030–0.035	58
0.035–0.040	45
0.040–0.045	29
0.045–0.050	17
0.050–0.055	9
0.055–0.060	4
More than 0.060	4
Total	250

As a check on the goodness of the fit, it is recommended that the probability function always be plotted on the same graph as the histogram for a visual assessment of the fit (which can be supplemented by a chi square test if desired). This is done in Figure 44.17, and we see that the fit is excellent (a chi square test comes to the same conclusion).

Bootstrap Methods.   Bootstrap methods suggest that one fit a model such as the EGLD to the data, and then, assuming that the fit passes testing, use that model in a "bootstrapping" analysis. (See The Generalized Bootstrap and Bootstrap Method later in this section.)

STATISTICAL ESTIMATION

In statistical estimation, we make inferences about parameters of a population from data on a sample. For example, if we have a random sample of 100 items from a large lot, information on the sample can be used to infer information about the proportion of defectives p in the lot. This inference takes the form of either a single number (a *point estimate*) or a pair of numbers (an *interval estimate*); there are several types of interval estimates, depending on our goals. If we found that 15 of the 100 items in the sample were defective, we would estimate p as being $15/100 = 0.15$ (point estimate); a typical interval would be to state that we are 95 percent confident that p is between 0.08 and 0.22 (confidence interval estimate). Thus a *confidence interval* sets limits on the unknown parameter, here the proportion p. Two other types of intervals often needed are *prediction intervals* and *tolerance intervals*. Let X denote the number of defectives in a future sample; a prediction interval sets limits on X, such as

$$P[L_1 \leq X \leq U_1] = 0.95$$

In this example, a *prediction interval* would state that X would be between 5 and 25; these are limits within which one could be 95 percent sure the number of defectives in a future sample of 100 items would lie. A *tolerance interval* sets limits (L, U) such that one can be 95 percent sure that at

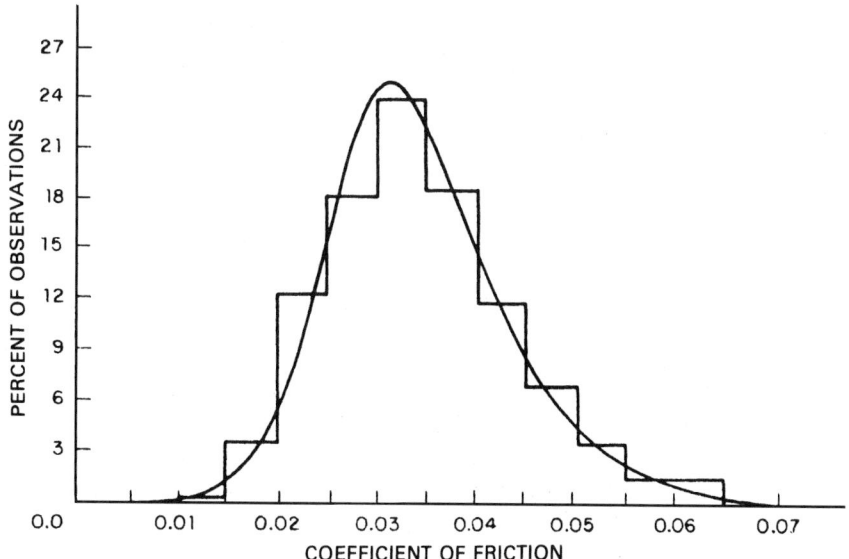

FIGURE 44.17   Coefficient of friction relative frequency histogram and the fitted distribution.

least 99 percent of the population will be included within these limits. In this example, the population is all lots of 100 items drawn from the same process; at least 99 percent of the lots will have a proportion of defectives between $L = 0.05$ and $U = 0.25$. These items are discussed in more detail below.

Point Estimates.   Point estimates are customarily the points at which the interval estimates are centered. In many cases, it is preferable to give an interval estimate, since that estimate tells us how much uncertainty is associated with the estimate. For example, if we observe that 15 of 100 items chosen at random from a very large lot are defective, then we will estimate the proportion of defectives in the lot as $15/100 = 0.15$. Similarly, if we observe three defectives in a sample of 20 items, we will estimate the proportion of defectives as $3/20 = 0.15$. However, in the first case the interval estimate (at 95 percent confidence) will be that p is between 0.08 and 0.22, while in the latter case the interval estimate (at 95 percent confidence) will be that p is between 0.00 and 0.31. In either case, if forced to estimate proportion defective in the lot by a single number, that number would be 0.15; however, the uncertainty in that estimate is much greater with the smaller sample size (where as much as 31 percent of the population might be defective) than with the larger sample size (where as much as 22 percent of the population might be defective). The typical point estimates are covered below, as the centers of the respective intervals.

Confidence Interval Estimates.   *Estimation* is the process of analyzing a sample result in order to predict the corresponding value of the population parameter. For example, a sample of 12 insulators has an average impact strength of 4.952 ft·lb (6.7149 N·m). If this is a representative sample from the process, what estimate can be made of the true average impact strength of the entire population of insulators?

1. The *point estimate* is a single value used to estimate the population parameter. For example, 4.952 ft·lb (6.7149 N·m) is the point estimate of the average strength of the population.

2. The *confidence interval* is a range of values that includes (with a preassigned probability called *confidence level*) the true value of a population parameter. *Confidence limits* are the upper and lower boundaries of the confidence interval. Confidence level is the proportion of times in the long run that an assertion about the value of a population parameter is correct.

Duncan (1974) provides a thorough discussion of confidence limits. The explanation here indicates the concept behind the calculations.

If the population mean is μ, the probability that the sample mean will be between

$$\mu \pm 1.96 \, \frac{\sigma}{\sqrt{n}}$$

is equal to 0.95:

$$P\left(\mu - 1.96 \, \frac{\sigma}{\sqrt{n}} \leq \overline{X} \leq \mu + 1.96 \, \frac{\sigma}{\sqrt{n}}\right) = 0.95$$

This is algebraically equivalent to saying that the sample mean plus 1.96 standard deviations of means lies above μ and the sample mean minus 1.96 standard deviations of means lies below μ:

$$P\left(\mu \leq \overline{X} + 1.96 \, \frac{\sigma}{\sqrt{n}} \text{ and } \overline{X} - 1.96 \, \frac{\sigma}{\sqrt{n}} \leq \mu\right) = 0.95$$

or

$$P\left(\overline{X} - 1.96 \, \frac{\sigma}{\sqrt{n}} \leq \mu \leq \overline{X} + 1.96 \, \frac{\sigma}{\sqrt{n}}\right) = 0.95$$

Or the 95 percent confidence interval on μ is $\overline{X} \pm 1.96 \; (\sigma/\sqrt{n})$. Before the sample is taken, this interval has a 0.95 probability of including the population value: 95 percent of the set of such intervals would include the population value. In practice, this is interpreted to mean that there is a 95 percent "confidence" that the limits based on one sample will include the true value.

For the sample of 12 insulators, suppose that $\sigma = 0.25$. Then, the 95 percent confidence limits are

$$\overline{X} \pm 1.96 \; \frac{\sigma}{\sqrt{n}} = 4.952 \pm 1.96 \; \frac{(0.25)}{\sqrt{12}} = 4.811 \text{ and } 5.093$$

This is interpreted to mean that there is 95 percent confidence that μ is between 4.811 and 5.093. The 95 percent is the confidence level (confidence levels of 90, 95, or 99 percent are usually assumed in practice, and some statisticians call these the "holy" numbers), and 4.811 and 5.093 are the limits of the confidence interval. A confidence level is associated with an assertion based on actual measurements and measures the proportion of times that the assertion will be true in the long run. Confidence limits are limits that include the true value with a preassigned degree of confidence (the confidence level).

Table 44.22 summarizes confidence limit formulas for common parameters. The following examples illustrate some of these formulas.

Example: Sixty-one specimens of brass have a mean hardness of 54.62 and an estimated standard deviation of 5.34. Determine the 95 percent confidence limits on the mean.

Solution:

$$\text{Confidence limits} = \overline{X} \pm t \; \frac{s}{\sqrt{n}}$$

$$= 54.62 \pm 2.00 \; \frac{5.34}{\sqrt{61}}$$

$$= 53.25 \text{ and } 55.99$$

There is 95 percent confidence that the true mean hardness of the brass is between 53.25 and 55.99.

Example: A radar system has been operated for 1200 h, during which time eight failures occurred. What are the 90 percent confidence limits on the mean time between failures for the system?

Solution:

$$\text{Estimated } m = \frac{1200}{8} = 150 \text{ h}$$

$$\text{Upper confidence limit} = \frac{2(1200)}{7.962} = 301.4$$

$$\text{Lower confidence limit} = \frac{2(1200)}{26.296} = 91.3$$

There is 90 percent confidence that the true mean time between failures is between 91.3 and 301.4 h. [Epstein (1960) discusses several cases of making estimates from life test data.]

Confusion has arisen on the application of the term *confidence level* to a reliability index such as mean time between failures. Using a different example, suppose the numerical portion of a reliability requirement reads as follows: "The MTBF shall be at least 100 h at the 90 percent confidence level." This means that

1. The minimum MTBF must be 100 h.

TABLE 44.22 Summary of Confidence Limit Formulas and Graphs

Parameters	Formulas
1. Mean of a normal population (standard deviation known)	$\overline{X} \pm K_{\alpha/2} \dfrac{\sigma}{\sqrt{n}}$ where $\overline{X}$ = sample average K = normal distribution coefficient (see Appendix II, Table B) σ = standard deviation of population n = sample size
2. Mean of a normal population (standard deviation unknown)	$\overline{X} \pm t_{\alpha/2} \dfrac{s}{\sqrt{n}}$ where t = Student's distribution coefficient (with $n-1$ degrees of freedom) (see Appendix II, Table G) s = estimated σ
3. Standard deviation of a normal population *a.* Using sample standard deviation *b.* Using sample range	Upper confidence limit = $B_U s$ Lower confidence limit = $B_L s$ where B_U and B_L are numerical factors given in Natrella (1963, p. T-34), Dixon and Massey (1969, p. 140)
4. Population fraction defective based on attribute data (fraction defective in sample)	See Appendix II, Chart N
5. Population fraction defective based on variables data ($\overline{X}$ and s in sample)	See Kirkpatrick (1970)
6. Difference between the means of two normal populations (standard deviations σ_1 and σ_2 known)	$(\overline{X}_1 - \overline{X}_2) \pm K_{\alpha/2} \sqrt{\dfrac{\sigma_1^2}{n_1} + \dfrac{\sigma_2^2}{n_2}}$ where K = normal distribution coefficient (see Appendix II, Table B)
7. Difference between the means of two normal populations ($\sigma_1 = \sigma_2$ but unknown)	$(\overline{X}_1 - \overline{X}_2) \pm t_{\alpha/2} \sqrt{\dfrac{1}{n_1} + \dfrac{1}{n_2}}$ $\times \sqrt{\dfrac{\sum (X - \overline{X}_1)^2 + \sum (X - \overline{X}_2)^2}{n_1 + n_2 - 2}}$ where t = Student's distribution coefficient (with degrees of freedom = $n_1 + n_2 - 2$)(see Appendix II, Table G)
8. Difference between the means of two normal populations (σ_1 and σ_2 both unknown)	$(\overline{X}_1 - \overline{X}_2) \pm t^*_{\alpha/2} \sqrt{\dfrac{\sum (X - \overline{X}_1)^2}{n_1(n_1 - 1)} + \dfrac{\sum (X - \overline{X}_2)^2}{n_2(n_2 - 1)}}$ where t^* = Student's distribution coefficient (with the smaller of $n_1 - 1$ and $n_2 - 1$ as degrees of freedom)

TABLE 44.22 Summary of Confidence Limit Formulas and Graphs (*Continued*)

Parameters	Formulas
9. Mean time between failures (based on an exponential population of time between failures)	Upper confidence limit $= 2T/\chi^2_{\alpha/2}$ Lower confidence limit $= 2T/\chi^2_{1-\alpha/2}$ where $T = $ total test time on all units and DF $= 2r$, where r is a preassigned number of failures and where $\chi^2 = $ chi square distribution coefficient (see Appendix II, Table L)
10. Reliability (based on a Weibull population)	See Thoman, Bain, and Antle (1970)
11. Availability (based on an exponential population of time between failures and log normal population of repair time)	See Gray and Lewis (1967)

2. Actual tests shall be conducted on the product to demonstrate with 90 percent confidence that the 100-h MTBF has been met.

3. The test data shall be analyzed by calculating the observed MTBF and the lower one-sided 90 percent confidence limit on MTBF.

4. The lower one-sided confidence limit must be ≥ 100 h.

The term *confidence level,* from a statistical viewpoint, has great implications on a test program. Note that the observed MTBF must be *greater than* 100 if the lower confidence limit is to be ≥ 100. Confidence level means that sufficient tests must be conducted to demonstrate, with statistical validity, that a requirement has been met. Confidence level does *not* refer to the qualitative opinion about meeting a requirement. Also, confidence level does *not* lower a requirement; i.e., a 100-h MTBF at a 90 percent confidence level does not mean that 100 h is desired but that 0.90×100, or 90 h, is acceptable. Such serious misunderstandings have occurred. When the term is used, a clear understanding should be verified and not assumed.

Determination of the Sample Size Required to Achieve a Specified Precision in an Estimate. Additional tests will increase the precision of the estimates obtained from a test program. The increase in precision usually does not vary linearly with the number of tests—doubling the number of tests usually does not double the precision (even approximately). Further, if the sample is selected randomly and if the sample size is less than 10 percent of the population size, then precision depends primarily on the absolute size of the sample rather than on the sample size expressed as a percentage of the population size. Thus a sample size that is 1 percent of a population of 100,000 may be more precise than a 10 percent sample from a population of 1000 (see Hahn 1972).

The cost of additional tests must be evaluated against the value of the additional precision. Confidence limits can help to determine the size of a test program required to estimate a product characteristic within a specified precision. Suppose it is desired to estimate the true mean life of a battery. The estimate must be within 2.0 h of the true mean if the estimate is to be of any value. The variability is known as $\sigma = 10.0$. A 95 percent confidence level is desired on the confidence statement. The 2.0 h is the desired confidence interval half-width, so

$$2.0 = \frac{(1.96)(10)}{\sqrt{n}} \qquad n = 96$$

A sample of 96 batteries will provide a mean that is within 2.0 h of the true mean (with 95 percent confidence). Notice the type of information required: (1) desired width of the confidence interval (the precision desired in the estimate), (2) confidence level desired, and (3) variability of the characteristic under investigation. The number of tests required cannot be determined until the engineer furnishes these items of information.

Table 44.23 summarizes formulas and graphs useful in determining the sample size required to estimate a population parameter with a specified precision. The following examples illustrate some of the formulas.

Example: A sample must be selected to estimate the population mean length of a part. It appears reasonable to assume that length is normally distributed. An estimate of the standard deviation is not available, but process knowledge suggests that "almost all" production falls between 2.009 and 2.027 in. As a first approximation, the standard deviation is estimated as $(2.027 - 2.009)$ divided by 6, or 0.003 in. It is desired that the estimate of μ be within 0.001 in of the true μ and that the estimation statement be made at the 95 percent confidence level. Referring to Appendix II, Chart S, $E/s = 0.001/0.003 = 0.33$, and the required sample size is about 37. It is instructive to calculate n for other values of E and s (see Table 44.24). Such a *sensitivity analysis* is helpful in evaluating the cost of extra tests against the value of extra precision.

Example: It is desired to estimate the standard deviation σ of a population within 20 percent of the true value at the 95 percent confidence level. Referring to Appendix II, Chart T, the required degrees of freedom is about 46 and, therefore, the sample size is $46 + 1$, or 47.

TABLE 44.23   Summary of Sample Size Formulas and Graphs

Parameters	Formulas
1. Mean of a normal population (σ known)	$n = \dfrac{K_{\alpha/2}^2 \sigma^2}{E^2}$ where K = normal distribution coefficient E = maximum allowable error in estimate (desired precision)
2. Mean of a normal population (σ estimated)	See Appendix II, Chart S
3. Standard deviation of a normal population	See Appendix II, Chart T
4. Fraction defective of a population	$n \doteq p\,(1 - p) \left(\dfrac{K_{\alpha/2}}{E}\right)^2$ where p = estimate of the population fraction defective. If no estimate of p is available, assume "worst case" of $p = 0.5$

TABLE 44.24   Effect of E and s on n

Maximum error E	Standard deviation s		
	0.002	0.003	0.004
0.0008	27	56	98
0.0010	18	37	64
0.0020	7	12	18

Prediction Intervals.   A prediction interval is used when the desire is not an estimate of population characteristics directly but rather a prediction of what we will find when we take a future item from the population. For example, in the example of $n = 61$ specimens of brass with a mean hardness of 54.62 and an estimated standard deviation of 5.34, we previously found that 95 percent confidence limits on the mean were

$$\overline{X} + t\,\frac{s}{\sqrt{n}} = 53.25 \text{ and } 55.99$$

Now we ask: What limits can we be 95 percent sure the next item sampled will have its hardness within? The appropriate interval here is

$$\overline{X} + ts\,\sqrt{1 + \frac{1}{n}} = 54.62 \pm (2.00)(5.34)(1.0082)$$

$$= 43.85 \text{ and } 65.39$$

For further considerations and tables, see Hahn (1970a, 1970b).

Tolerance Intervals.   *Statistical tolerance limits* are similar to *process capability*; i.e., they show the practical boundaries of process variability (see Section 22, under Operations) and therefore can be a valuable input in the determination of engineering tolerance limits (which specify the allowable limits for product acceptance). Methods for calculating statistical tolerance limits are of two types—those which assume a normal distribution and those which do not require any distributional assumption. Table 44.25 summarizes these methods.

Table 44.26 shows data I will use to illustrate these methods. Five samples of four each were taken and an outside dimension of a cathode pole recorded. A confidence level of 95 percent and a population percentage of 99 percent have been chosen.

Using method 1 and the standard deviation *s*, the statistical tolerance limits are

$$\overline{X} \pm ks = 1.00287 \pm 3.615(0.00034) = 1.00164 \text{ and } 1.00410$$

Using method 2 and the overall range *R* of the combined data, the limits are

$$\overline{X} \pm K_1 R = 1.00287 \pm 1.005(0.00134) = 1.00152 \text{ and } 1.00422$$

Using method 3 and the average of the ranges *R*, the limits are

$$\overline{X} \pm K_2 \overline{R} = 1.00287 \pm 1.783(0.00078) = 1.00148 \text{ and } 1.00426$$

These methods assume that the characteristic is normally distributed. Method 4 is "distribution-free" and assumes only that the distribution is continuous and the sample is a random one (these assumptions apply to all methods). The statistical tolerance limits by this method are simply the extreme observations in the combined sample, i.e., 1.00231 and 1.00365. Appendix II, Table W indicates that at least 78.4 percent of the population will be included within these limits. (Note that Appendix II, Table X provides the sample size required to include 99 percent of the population; i.e., a sample of 473 is needed to be 95 percent confident that the sample extremes would include 99 percent of the population.)

When it is feasible to assume a normal distribution, method 1 is preferred because it usually provides the narrowest set of limits while recognizing the variation in the sample. Methods 2 and 3 are good approximations. If normality cannot be assumed, then method 4 is appropriate but at the cost of a larger sample size. In practice, a partial sample can first be obtained to evaluate the assumption of normality. If normality can be assumed, the partial sample is then used to determine statistical tolerance limits. Otherwise, the full sample should be taken and the distribution-free approach (method 4) applied to determine the limits.

Tolerance intervals also have been developed for other cases, such as where the distribution is exponential. Ranganathan and Kale (1983) give such intervals that are resistant to the presence of an

TABLE 44.25 Methods for Calculating Statistical Tolerance Limits

Method	Distribution assumption	Formula for limits	Source of factor
1. Measure a sample of n items and calculate the average $\overline{X}$ and standard deviation s	Normal	Two-sided limits: $\overline{X} \pm ks$ One-sided limit: $\overline{X} + ks$ or $\overline{X} - ks$	Appendix II, Table V Appendix II, Table V
2. Measure a sample of n items and calculate the average $\overline{X}$ and range R	Normal	Two-sided limits: $\overline{X} \pm K_1 R$	Appendix II, Table U
3. Measure N samples of n items each and calculate the grand average $\overline{\overline{X}}$ and average range $\overline{R}$	Normal	Two-sided limits: $\overline{\overline{X}} \pm K_2 \overline{R}$	Bingham (1962, p. 37)
4. Define the population percentage P which must be included between the tolerance limits. Measure a sample of n and observe the largest and smallest values	None	Two-sided limits: The probability is γ that at least P % of the population will be between the sample extremes. One-sided limit: The probability is γ that at least P % of the population will be less than the largest value in the sample (or greater than the smallest value).	Appendix II, Table W Natrella (1963, p. T-76)

TABLE 44.26 Data on Cathode Pole Dimension

Sample 1	Sample 2	Sample 3	Sample 4	Sample 5	
1.00263	1.00306	1.00293	1.00291	1.00310	
1.00298	1.00328	1.00343	1.00247	1.00281	$\bar{X} = 1.00287$
1.00293	1.00274	1.00239	1.00268	1.00256	$s = 0.00034$ (standard deviation of the
1.00285	1.00303	1.00274	1.00365	1.00231	20 observations about $\bar{X}$)
$\bar{X}$ 1.00285	1.00303	1.00287	1.00293	1.00269	$\bar{\bar{R}} = 0.00078$
R 0.00035	0.00054	0.00104	0.00118	0.00079	$\bar{R} = 0.00134$

44.49

outlier in the observations. Their Section 5 includes an example with data on reliability of air conditioning on a Boeing 727 jet aircraft.

All the preceding methods involve two probabilities, i.e., a confidence level γ and the probability P of falling within limits. This is confusing, but these two probabilities are needed to obtain a mathematically correct statement concerning the limits. An approximation uses the sample average $\overline{X}$ and standard deviation s and regards these as highly reliable estimates of μ and σ. If normality is assumed, then 99 percent statistical tolerance limits are calculated as

$$\overline{X} \pm 2.58s$$

where the value of 2.58 is obtained from the normal distribution table (Appendix II, Table B):

$$\overline{X} \pm 2.58 = 1.00287 \pm 2.58\,(0.000274)$$

$$= 1.00216 \text{ and } 1.00358$$

These limits are then interpreted to mean that 99 percent of the population is within 1.00217 and 1.00357. Another approach sets the limits at simply $\overline{X} \pm 3s$. At best, these are only approximate because $\overline{X}$ and s are not exactly equal to μ and σ. Bingham (1962) discusses the $X \pm 3s$ approximation. (Confidence limit calculations could indicate the size of the possible error.)

Another approach to simplify the probability statement uses the Chebyshev inequality theorem, which holds for any continuous distribution. The theorem states that the probability of obtaining a value that deviates from μ by more than k standard deviations is less than $1/k^2$. For the limits to include 99 percent,

$$0.01 = \frac{1}{k^2}$$

or

$$k = 10$$

The 99 percent limits would be calculated at $\overline{X} \pm 10s$. These limits are distribution-free, and the prediction statement is simple; i.e., at least 99 percent of the population is within $\overline{X} \pm 10s$. However, the multiple of 10 is highly conservative and results in limits much wider than any of the other methods.

For the methods listed in Table 44.25, no provision is made for the division of the remaining $100(1 - P)$ percent between the upper and lower tails of the distribution. Owen and Frawley (1971) give a procedure and tables for setting limits that do provide for controlling the percentage outside each of the two limits.

Statistical tolerance limits are sometimes confused with other limits used in engineering and statistics. Table 44.27 summarizes the distinctions among five types of limits. Hahn (1970a, 1970b) gives an excellent discussion with examples and tables to illustrate the differences among several types of limits. Also see Harter (1983), under "Tolerance Limits."

Tolerance Limits for Interacting Dimensions.   *Interacting dimensions* are those which mate or merge with other dimensions to create a final result. Setting tolerance limits on such dimensions is discussed in the following paragraphs. Setting tolerance limits on noninteracting dimensions makes use of the methods presented under Statistical Tolerance Limits in this section.

Conventional Method Relating Tolerances on Interacting Dimensions.   Consider the simple mechanical assembly shown in Figure 44.18. The lengths of components *A, B,* and *C* are interacting dimensions because they determine the overall assembly length.

The conventional method of relating interacting dimensions is simple addition. For the example of Figure 44.18,

$$\text{Nominal value of the result} = \text{nominal value}_A + \text{nominal value}_B + \text{nominal value}_C$$

TABLE 44.27   Distinction among Limits

Name of limits	Meaning
Tolerance limits	Set by the engineering design function to define the minimum and maximum values allowable for the product to work properly.
Statistical tolerance limits	Calculated from process data to define the amount of variation that the process exhibits. These limits will contain a specified proportion of the total population.
Prediction limits	Calculated from process data to define the limits which will contain all of k future observations.
Confidence limits	Calculated from data to define an interval within which a population parameter lies.
Control limits	Calculated from process data to define the limits of chance (random) variation around some central value.

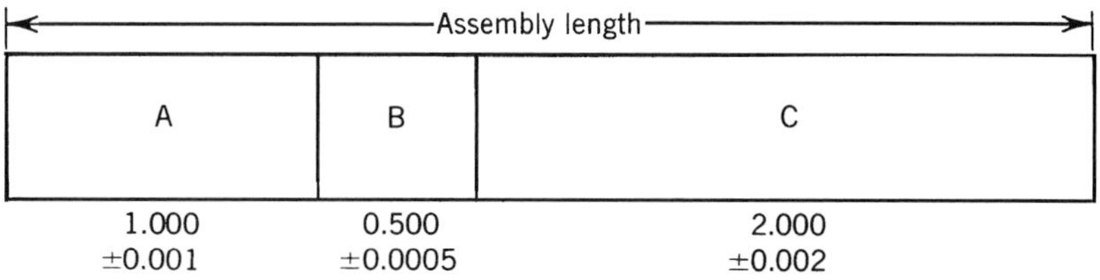

FIGURE 44.18   Mechanical assembly.

Tolerance T of the result $= T_A + T_B + T_C$

Nominal value of assembly length $= 1.000 + 0.500 + 2.000 = 3.500$

Tolerance of assembly length $= 0.0010 + 0.0005 + 0.0020 = \pm 0.0035$

This method assumes 100 percent interchangeability of components making up the assembly. If the component tolerances are met, then all assemblies will meet the assembly tolerance determined by the simple arithmetic addition.

The approach of adding component tolerances is mathematically correct but often too conservative. Suppose that about 1 percent of the pieces of component A are expected to be below the lower tolerance limit for component A, and suppose the same for components B and C. If a component A is selected at random, there is, on average, 1 chance in 100 that it will be on the low side, and similarly for components B and C. The key point is this: If assemblies are made at random, and if the components are manufactured independently, then the chance that an assembly will have all three components simultaneously below the lower tolerance limit is

$$\frac{1}{100} \times \frac{1}{100} \times \frac{1}{100} = \frac{1}{1,000,000}$$

There is only about one chance in a million that all three components will be too small, resulting in a small assembly. Thus, setting component and assembly tolerances based on the simple addition formula is conservative in that it fails to recognize the extremely low probability of an assembly containing all low (or all high) components.

Statistical Method of Relating Tolerances on Interacting Dimensions.   This method states for the example Figure 44.18:

$$\text{Nominal value of the result} = \text{nominal value}_A + \text{nominal value}_B + \text{nominal value}_C$$

$$\text{Tolerance of the result} = \sqrt{T_A^2 + T_B^2 + T_C^2}$$

Then:

$$\text{Nominal value of the assembly} = 1.000 + 0.500 + 2.000 = 3.500$$

$$T \text{ of the assembly} = \sqrt{(0.001)^2 + (0.0005)^2 + (0.002)^2} = \pm0.0023$$

Practically all (but not 100 percent) of the assemblies will fall within 3.500 ± 0.0023. This is narrower than 3.500 ± 0.0035 (the result by the arithmetic method).

In practice, the problem often is to start with a defined end result (e.g., assembly length specification) and set tolerances on the parts. Suppose the assembly tolerance was desired to be ±0.0035. Listed in Table 44.28 are two possible sets of component tolerances that when used with the quadratic formula will yield an assembly tolerance equal to ±0.0035. The tolerance set using the conventional formula is also shown.

The advantage of the statistical formula is larger component tolerances. With alternative 1, the tolerance for component A has been doubled, the tolerance for component B has been quadrupled, and the tolerance for component C has been kept the same as the original component based on the simple addition approach. If alternative 2 is chosen, similar significant increases in the component tolerances may be achieved. This formula, then, may result in a larger component tolerance with no change in the manufacturing processes and no change in the assembly tolerance. Note that the *largest single* tolerance has the greatest effect on the overall result.

The disadvantage of the quadratic formula is that it involves several assumptions that, even if met, will still result in a small percent (theoretically 0.27 percent) of results not conforming to the limits set by the formula. The assumptions are

1. The component dimensions are independent and the components are assembled randomly. This assumption is usually met in practice.

2. Each component dimension should be normally distributed. Some departure from this assumption is permissible.

3. The actual average for each component is equal to the nominal value stated in the specification. For the original assembly example, the actual averages for components A, B, and C must be 1.000, 0.500, and 2.000, respectively. Otherwise, the nominal value of 3.500 will not be achieved for the assembly, and tolerance limits set about 3.500 will not be realistic. Thus it is important to control the *average* value for interacting dimensions. This means that process control techniques are needed using variables measurement rather than go no-go measurement.

A summary of the two methods of tolerance is given in Table 44.29.

The statistical tolerancing formula applies both to assemblies made up of physically separate components and to a chain of several interacting dimensions within one physical item. Further, the result of the interacting dimensions can be an outside dimension (assembly length) or an internal result (clearance between a shaft and hole).

TABLE 44.28 Comparison of Statistical and Conventional Methods

Component	Statistical		Conventional
	Alternative 1	Alternative 2	
A	±0.002	±0.001	±0.0010
B	±0.002	±0.001	±0.0005
C	±0.002	±0.003	±0.0020

TABLE 44.29 Comparison of Conventional and Statistical Tolerancing

Factor	Conventional	Statistical
Risk of items not interacting properly	No risk; 100% interchangeability of items	Small percent of final results will fall outside limits (but these can sometimes be corrected with selective assembly)
Utilization of full tolerance range	Method is conservative; tolerances on interacting dimensions are smaller than necessary	Permits larger tolerances on interacting dimensions
Special process control techniques	None	Average of each interacting dimension must be controlled using variables measurement
Statistical assumptions	None	Interacting dimensions must be independent and each must be normally distributed
Lot size for components	Any size	Lot size should be moderately large (to assure balancing effect on extreme interacting dimensions)

Further Applications of Statistical Tolerancing. It is easy to be deceived into concluding that the statistical method of tolerancing is merely a change from an expression of tolerances in the form of limits on each component to a form of

1. Upper and lower limits on the average $\overline{X}$ of the mass of components

2. An upper limit to the scatter σ of the components

The change is much more profound than mere form of the specification. It affects the entire cycle of manufacturing planning, production, inspection, quality control, service, etc. It is, in effect, *a new philosophy of manufacture.*

The first published example of a *large-scale application* of statistical tolerancing appears to be that of the L-3 coaxial system (a broad-band transmission system for multiple telephone or television channels). Dodge et al. (1953) discuss the application.

The general plan was

1. Discovery of the key quality characteristics of each component element of the system.

2. Determination of the precision of measurement to separate measurement variability from process and product variability.

3. Collection of data on process capability for the key qualities, to aid in establishing realistic tolerances.

The foregoing were preliminary to

4. Establishment of tolerances for the key quality characteristics in the dual form of a maximum on the standard deviation σ and limits on the average $\overline{X}$. The limits on X were established as $\pm(^1/_3)\sigma$ around the nominal.

5. Establishment of control procedures.

It was recognized that the limits on σ and $\overline{X}$ required further interpretation if the intent of the designers was to be carried out by the manufacturers. To this end, three forms of product acceptance were established:

1. *Control charts.* Shewhart control charts for $\overline{X}$ and σ could be used for product acceptance, provided "eligibility" was established (seven consecutive subgroups, of five pieces each, all met the control limits for $\overline{X}$ and σ) and provided subsequent statistical control was maintained (based on chart results plus absence of major changes in process).

2. *Batch control.* This was based on examination of a sample of (normally) 50 pieces by variables measurements, with limits on $\overline{X}$ and σ appropriate to the sample size of 50. Each batch stood or fell on its own measurements.

3. *Detailed classification.* Product that did not qualify under 1 or 2 was measured in detail. The resulting conforming units were classified into one of three variable classes. The packaging was then done by selecting classified units in such a way that each package contained an assortment of product which conformed to the intent of the design as to $\overline{X}$ and σ.

Grant and Leavenworth (1980) discuss statistical tolerancing, including an application to shafts and holes. The Western Electric Company, Inc. (1982), in its *Statistical Quality Control Handbook* (pp. 122–127), presents examples and discusses the assumptions. Peters (1970) discusses statistical tolerancing, including a method for recognizing cost differences among components. Choksi (1971) discusses the use of computer simulation to determine optimum tolerances.

The concept may be applied to several interacting variables in an engineering relationship. The nature of the relationship need *not* be additive (assembly example) or subtractive (shaft and hole example). The formula can be adapted to predict the variation of results that are the product and/or the division of several variables. Mouradian (1966) discusses these applications.

Bayesian Estimates.   Bayesian estimation can be used when the parameter to be estimated can be considered to be a random variable for which we know the distribution. For example, in sampling inspection, the proportion of defectives p in a lot may be a random variable about which we can fit a distribution by our sampling inspection over time. If so, then in the future that information can be used to provide quality assurance with less sampling (see Lenz and Rendtel 1984).

When it is not possible to cumulate information about a stable process, some have proposed that we use our "feelings" about the parameter to choose a statistical distribution for it and then proceed as if the parameter were a random variable with that distribution. This is called the *personal probability approach,* and those who use it are called *Bayesians.* Some of the proponents of this approach say that it is the only method that any sensible person should use, and this has been cause for bitter debates and ill-will. In my view, while a person in a management position might find this a reasonable way to express his or her insights quantitatively, in most cases this will be an unscientific way of simply incorporating prejudices into the decision process, resulting in costly errors. This approach is of some use in general statistical decision theory (see Chapter 12 of Dudewicz 1976), but there it is used to generate a set of decision rules that contains all good rules, not just one rule based on one's "feelings." The Bayesian approach should be considered whenever information can be gathered over time on a stable process.

Intervals with Fixed Width and Precision.   The intervals considered up to now typically either had a random width (e.g., parts 7 and 8 of Table 44.22) or required that one know such parameters as variances (e.g., part 6 of Table 44.22). If one can take observations in two stages, then one can control both the width and the confidence. Let me illustrate for two normal means when we do not know the variances (and do not know that they are equal). Here we can proceed in two stages as follows:

Sample n_0 observations from each of $k = 2$ populations (n_0 at least 10 is desirable).

Determine the total sample size for population i as

$$n_i = max\,[n_0 + 1,\,(ws_i)^2] \qquad \text{with } w = h_{n_0}\,(2,\,(1 + P^*)/2)/d$$

where h is from Table 44.30, d is the desired half-width, $k = 2$ populations, and P^* is the confidence desired.

TABLE 44.30 Multipliers $h=h_{n_0}(k, P^*)$ Needed for Solving Two-Stage Testing, Confidence Interval, and Selection of the Best Problems*

k	P^*	n_0												
		2	3	4	5	6	7	8	9	10	15	20	25	30
2	.75	2.00	1.37	1.21	1.14	1.10	1.07	1.05	1.04	1.03	1.00	0.99	0.98	0.98
	.80	2.75	1.76	1.54	1.44	1.38	1.35	1.32	1.30	1.29	1.25	1.24	1.23	1.22
	.85	3.93	2.27	1.94	1.80	1.72	1.68	1.64	1.62	1.60	1.55	1.53	1.51	1.51
	.90	6.16	3.04	2.50	2.29	2.18	2.11	2.06	2.02	2.00	1.93	1.90	1.88	1.87
	.95	12.63	4.57	3.50	3.11	2.91	2.79	2.71	2.66	2.61	2.50	2.45	2.42	2.41
	.975	25.42	6.54	4.59	3.94	3.63	3.45	3.33	3.24	3.18	3.02	2.95	2.91	2.88
	.99	63.7	10.28	6.31	5.14	4.60	4.30	4.11	3.98	3.89	3.64	3.54	3.48	3.45
3	.75	3.52	2.15	1.86	1.74	1.67	1.63	1.60	1.57	1.56	1.51	1.49	1.48	1.47
	.80	4.59	2.59	2.20	2.04	1.95	1.89	1.85	1.83	1.80	1.75	1.72	1.71	1.70
	.85	6.31	3.17	2.62	2.40	2.28	2.21	2.16	2.13	2.10	2.03	1.99	1.98	1.96
	.90	9.64	4.05	3.22	2.90	2.73	2.63	2.57	2.52	2.48	2.39	2.34	2.32	2.30
	.95	19.40	5.86	4.29	3.75	3.48	3.32	3.21	3.14	3.08	2.94	2.87	2.84	2.81
	.975	38.7	8.25	5.50	4.63	4.22	3.98	3.82	3.72	3.64	3.43	3.35	3.30	3.27
	.99	96.2	12.83	7.44	5.91	5.23	4.86	4.62	4.46	4.34	4.04	3.92	3.85	3.81
4	.75	4.77	2.66	2.25	2.08	1.99	1.93	1.89	1.86	1.84	1.78	1.75	1.74	1.73
	.80	6.16	3.13	2.59	2.38	2.26	2.19	2.14	2.11	2.08	2.01	1.98	1.96	1.95
	.85	8.41	3.77	3.03	2.75	2.60	2.51	2.45	2.41	2.37	2.28	2.24	2.22	2.21
	.90	12.80	4.75	3.66	3.26	3.06	2.93	2.85	2.80	2.75	2.63	2.58	2.55	2.54
	.95	25.76	6.80	4.80	4.14	3.81	3.62	3.50	3.41	3.34	3.17	3.10	3.06	3.03
	.975	51.4	9.53	6.10	5.05	4.57	4.29	4.11	3.99	3.90	3.67	3.57	3.51	3.48
	.99	128	14.79	8.21	6.40	5.62	5.19	4.92	4.74	4.60	4.27	4.13	4.05	4.01
5	.75	5.95	3.05	2.53	2.32	2.21	2.14	2.09	2.06	2.03	1.96	1.93	1.91	1.90
	.80	7.65	3.56	2.89	2.63	2.49	2.40	2.35	2.30	2.27	2.19	2.15	2.13	2.12
	.85	10.43	4.25	3.34	3.00	2.83	2.72	2.65	2.60	2.56	2.46	2.41	2.39	2.37
	.90	15.90	5.32	4.00	3.53	3.29	3.15	3.05	2.99	2.94	2.81	2.75	2.72	2.69
	.95	32.04	7.58	5.20	4.42	4.05	3.84	3.70	3.60	3.53	3.34	3.26	3.21	3.18
	.975	64.1	10.61	6.58	5.37	4.83	4.52	4.32	4.18	4.08	3.83	3.72	3.66	3.62
	.99	160	16.47	8.82	6.78	5.90	5.43	5.13	4.93	4.79	4.43	4.28	4.20	4.14
6	.75	7.10	3.39	2.76	2.52	2.38	2.30	2.25	2.21	2.18	2.10	2.06	2.04	2.03
	.80	9.12	3.93	3.13	2.82	2.66	2.57	2.50	2.45	2.42	2.32	2.28	2.26	2.24
	.85	12.43	4.66	3.60	3.21	3.01	2.89	2.81	2.75	2.70	2.59	2.54	2.51	2.49
	.90	18.96	5.82	4.28	3.74	3.47	3.31	3.21	3.13	3.08	2.93	2.87	2.84	2.81
	.95	38.29	8.26	5.53	4.66	4.25	4.01	3.86	3.75	3.67	3.46	3.38	3.33	3.30
	.975	76.7	11.56	6.97	5.64	5.03	4.69	4.48	4.33	4.22	3.95	3.83	3.77	3.73
	.99	192	17.97	9.34	7.09	6.13	5.62	5.30	5.09	4.93	4.55	4.39	4.30	4.25
7	.75	8.23	3.68	2.96	2.67	2.53	2.44	2.37	2.33	2.30	2.21	2.17	2.14	2.13
	.80	10.58	4.25	3.33	2.99	2.81	2.70	2.63	2.58	2.54	2.43	2.38	2.36	2.34
	.85	14.42	5.03	3.81	3.37	3.15	3.02	2.93	2.87	2.82	2.70	2.64	2.61	2.59
	.90	22.01	6.27	4.51	3.92	3.62	3.45	3.33	3.25	3.19	3.04	2.97	2.93	2.91
	.95	44.53	8.88	5.81	4.85	4.41	4.15	3.98	3.87	3.79	3.57	3.47	3.42	3.39
	.975	89.3	12.43	7.32	5.86	5.20	4.84	4.61	4.45	4.34	4.05	3.93	3.86	3.82
	.99	223	19.34	9.79	7.35	6.32	5.78	5.44	5.21	5.05	4.64	4.48	4.39	4.33
8	.75	9.36	3.95	3.13	2.81	2.65	2.55	2.48	2.43	2.40	2.30	2.25	2.23	2.21
	.80	12.02	4.55	3.51	3.13	2.93	2.81	2.73	2.68	2.63	2.52	2.47	2.44	2.42
	.85	16.40	5.37	4.01	3.52	3.28	3.13	3.04	2.97	2.92	2.78	2.72	2.69	2.67
	.90	25.05	6.68	4.72	4.07	3.75	3.56	3.44	3.35	3.29	3.12	3.05	3.01	2.98
	.95	50.76	9.45	6.06	5.03	4.54	4.27	4.09	3.97	3.88	3.65	3.55	3.50	3.46
	.975	101.9	13.24	7.63	6.05	5.35	4.97	4.72	4.56	4.44	4.13	4.00	3.93	3.89
	.99	256	20.62	10.20	7.58	6.49	5.91	5.56	5.32	5.15	4.73	4.55	4.46	4.40
9	.75	10.49	4.20	3.28	2.93	2.75	2.64	2.57	2.52	2.48	2.37	2.33	2.30	2.28
	.80	13.47	4.82	3.67	3.25	3.04	2.91	2.82	2.76	2.72	2.60	2.54	2.51	2.49
	.85	18.37	5.68	4.18	3.65	3.39	3.23	3.13	3.06	3.00	2.86	2.80	2.76	2.74
	.90	28.08	7.06	4.91	4.21	3.86	3.66	3.53	3.44	3.37	3.20	3.12	3.08	3.05
	.95	57.0	9.99	6.29	5.18	4.66	4.37	4.19	4.06	3.96	3.72	3.62	3.56	3.53
	.975	114.5	13.99	7.91	6.22	5.48	5.08	4.82	4.65	4.52	4.20	4.07	4.00	3.95
	.99	287	21.8	10.58	7.79	6.64	6.03	5.66	5.41	5.23	4.80	4.62	4.53	4.46

TABLE 44.30   Multipliers $h = h_{n_0}(k, P^*)$ Needed for Solving Two-Stage Testing, Confidence Interval, and Selection of the Best Problems* *(Continued)*

k	P*	2	3	4	5	6	7	8	9	10	15	20	25	30
								n_0						
10	.75	11.60	4.43	3.42	3.04	2.85	2.73	2.65	2.60	2.55	2.44	2.39	2.36	2.35
	.80	14.90	5.08	3.82	3.36	3.13	3.00	2.90	2.84	2.79	2.66	2.60	2.57	2.55
	.85	20.34	5.98	4.33	3.77	3.48	3.32	3.21	3.13	3.08	2.92	2.86	2.82	2.80
	.90	31.12	7.41	5.09	4.33	3.96	3.75	3.62	3.52	3.45	3.26	3.18	3.14	3.11
	.95	63.2	10.49	6.50	5.32	4.77	4.47	4.27	4.14	4.04	3.79	3.68	3.62	3.58
	.975	127.1	14.70	8.17	6.38	5.60	5.17	4.91	4.73	4.60	4.26	4.13	4.05	4.01
	.99	318	22.9	10.92	7.98	6.77	6.14	5.75	5.49	5.31	4.86	4.68	4.58	4.51
11	.75	12.72	4.64	3.54	3.14	2.93	2.81	2.72	2.66	2.62	2.50	2.45	2.42	2.40
	.80	16.34	5.32	3.95	3.46	3.22	3.07	2.98	2.91	2.86	2.72	2.66	2.63	2.61
	.85	22.31	6.25	4.48	3.87	3.57	3.40	3.28	3.20	3.14	2.98	2.91	2.87	2.85
	.90	34.15	7.75	5.25	4.44	4.06	3.83	3.69	3.59	3.51	3.32	3.24	3.19	3.16
	.95	69.4	10.97	6.70	5.44	4.87	4.55	4.35	4.21	4.10	3.84	3.73	3.67	3.63
	.975	139.7	15.38	8.41	6.52	5.71	5.26	4.99	4.80	4.66	4.32	4.18	4.10	4.05
	.99	350	24.0	11.25	8.16	6.90	6.24	5.84	5.57	5.38	4.91	4.73	4.62	4.56
12	.75	13.84	4.85	3.66	3.23	3.01	2.88	2.79	2.73	2.68	2.55	2.50	2.47	2.45
	.80	17.77	5.54	4.07	3.56	3.30	3.14	3.04	2.97	2.92	2.77	2.71	2.68	2.66
	.85	24.27	6.52	4.61	3.97	3.65	3.47	3.35	3.26	3.20	3.03	2.96	2.92	2.90
	.90	37.17	8.07	5.40	4.54	4.14	3.91	3.76	3.65	3.57	3.37	3.28	3.24	3.21
	.95	75.6	11.42	6.88	5.56	4.96	4.63	4.42	4.27	4.16	3.89	3.78	3.72	3.68
	.975	152.3	16.02	8.63	6.66	5.08	5.34	5.06	4.86	4.72	4.37	4.23	4.15	4.10
	.99	382	25.0	11.55	8.32	7.01	6.32	5.91	5.64	5.44	4.96	4.77	4.67	4.60
13	.75	14.95	5.04	3.77	3.31	3.08	2.94	2.85	2.78	2.73	2.60	2.54	2.51	2.49
	.80	19.21	5.76	4.19	3.64	3.37	3.21	3.10	3.03	2.97	2.82	2.76	2.72	2.70
	.85	26.24	6.77	4.74	4.06	4.73	3.53	3.41	3.32	3.25	3.08	3.01	2.97	2.94
	.90	40.20	8.38	5.54	4.64	4.22	3.97	3.82	3.71	3.63	3.42	3.33	3.28	3.25
	.95	81.8	11.86	7.05	5.67	5.05	4.70	4.48	4.33	4.22	3.94	3.82	3.76	3.72
	.975	164.8	16.64	8.85	6.78	5.90	5.42	5.12	4.92	4.78	4.42	4.27	4.19	4.14
	.99	413	26.0	11.84	8.47	7.11	6.41	5.98	5.70	5.50	5.01	4.81	4.71	4.64
14	.75	16.06	5.23	3.87	3.39	3.15	3.00	2.90	2.83	2.78	2.65	2.59	2.55	2.53
	.80	20.64	5.97	4.30	3.72	3.44	3.27	3.16	3.08	3.02	2.86	2.80	2.76	2.74
	.85	28.20	7.01	4.86	4.14	3.80	3.60	3.46	3.37	3.30	3.12	3.05	3.01	2.98
	.90	43.23	8.67	5.67	4.73	4.29	4.04	3.87	3.76	3.68	3.46	3.37	3.32	3.29
	.95	88.1	12.27	7.22	5.77	5.12	4.76	4.54	4.38	4.27	3.98	3.86	3.80	3.75
	.975	177.4	17.24	9.05	6.90	5.98	5.49	5.18	4.98	4.83	4.46	4.31	4.22	4.17
	.99	445	27.0	12.11	8.61	7.21	6.48	6.05	5.76	5.55	5.05	4.85	4.74	4.68
15	.75	17.17	5.41	3.97	3.46	3.21	3.05	2.95	2.88	2.83	2.69	2.63	2.59	2.57
	.80	22.07	6.17	4.40	3.80	3.50	3.32	3.21	3.13	3.07	2.91	2.84	2.80	2.77
	.85	30.17	7.24	4.97	4.22	3.86	3.65	3.52	3.42	3.35	3.16	3.09	3.01	3.01
	.90	46.25	8.96	5.80	4.82	4.36	4.09	3.93	3.81	3.72	3.50	3.41	3.35	3.32
	.95	94.3	12.68	7.37	5.87	5.20	4.83	4.59	4.43	4.32	4.02	3.90	3.83	3.79
	.975	190.0	17.81	9.24	7.01	6.06	5.55	5.24	5.03	4.88	4.50	4.34	4.26	4.20
	.99	476	27.9	12.37	8.75	7.30	6.55	6.11	5.81	5.60	5.09	4.88	4.78	4.71
16	.75	18.28	5.58	4.06	3.53	3.26	3.11	3.00	2.93	2.87	2.72	2.66	2.63	2.60
	.80	23.50	6.36	4.50	3.87	3.56	3.37	3.26	3.17	3.11	2.94	2.87	2.83	2.81
	.85	32.13	7.46	5.07	4.29	3.92	3.70	3.56	3.47	3.39	3.20	3.12	3.07	3.05
	.90	49.28	9.23	5.92	4.90	4.42	4.15	3.97	3.85	3.77	3.54	3.44	3.39	3.35
	.95	100.5	13.07	7.52	5.96	5.27	4.88	4.64	4.48	4.36	4.06	3.93	3.86	3.82
	.975	202.6	18.37	9.42	7.11	6.14	5.62	5.29	5.08	4.92	4.53	4.38	4.29	4.24
	.99	508	28.8	12.62	8.88	7.39	6.62	6.16	5.86	5.65	5.12	4.92	4.81	4.74
17	.75	19.40	5.74	4.15	3.59	3.32	3.15	3.05	2.97	2.91	2.76	2.69	2.66	2.63
	.80	24.93	6.55	4.59	3.94	3.61	3.42	3.30	3.21	3.15	2.98	2.90	2.86	2.84
	.85	34.09	7.68	5.17	4.37	3.98	3.75	3.61	3.51	3.43	3.24	3.15	3.11	3.08
	.90	52.30	9.50	6.03	4.97	4.48	4.20	4.02	3.90	3.81	3.57	3.47	3.42	3.38
	.95	106.7	13.45	7.66	6.04	5.33	4.94	4.69	4.52	4.40	4.09	3.96	3.89	3.85
	.975	215.3	18.90	9.60	7.21	6.21	5.67	5.34	5.12	4.96	4.57	4.41	4.32	4.27
	.99	542	29.6	12.86	9.00	7.47	6.68	6.22	5.91	5.69	5.16	4.95	4.84	4.76

TABLE 44.30 Multipliers $h=h_{n_0}$ $(k,\ P^*)$ Needed for Solving Two-Stage Testing, Confidence Interval, and Selection of the Best Problems* *(Continued)*

k	P*	n_0												
		2	3	4	5	6	7	8	9	10	15	20	25	30
18	.75	20.51	5.90	4.23	3.66	3.37	3.20	3.09	3.01	2.95	2.79	2.72	2.69	2.66
	.80	26.36	6.73	4.68	4.00	3.67	3.47	3.34	3.25	3.19	3.01	2.93	2.89	2.87
	.85	36.05	7.89	5.27	4.43	4.03	3.80	3.65	3.55	3.47	3.27	3.18	3.14	3.10
	.90	55.3	9.76	6.14	5.04	4.54	4.25	4.06	3.94	3.84	3.60	3.50	3.45	3.41
	.95	112.9	13.81	7.80	6.13	5.39	4.99	4.73	4.56	4.44	4.12	3.99	3.92	3.87
	.975	227.7	19.43	9.77	7.31	6.28	5.73	5.39	5.16	5.00	4.60	4.43	4.35	4.29
	.99	571	30.5	13.09	9.12	7.55	6.74	6.27	5.95	5.73	5.19	4.98	4.86	4.79
19	.75	21.61	6.06	4.32	3.71	3.42	3.24	3.13	3.04	2.98	2.82	2.75	2.71	2.69
	.80	27.79	6.91	4.77	4.06	3.72	3.51	3.38	3.29	3.22	3.04	2.96	2.92	2.89
	.85	38.01	8.09	5.37	4.50	4.08	3.85	3.69	3.59	3.51	3.30	3.21	3.16	3.13
	.90	58.4	10.01	6.25	5.11	4.59	4.29	4.10	3.97	3.88	3.63	3.53	3.47	3.44
	.95	119.1	14.17	7.93	6.20	5.45	5.04	4.78	4.60	4.47	4.15	4.02	3.95	3.90
	.975	240.2	19.94	9.93	7.40	6.34	5.78	5.43	5.20	5.04	4.63	4.46	4.37	4.32
	.99	604	31.3	13.31	9.23	7.62	6.80	6.31	5.99	5.77	5.22	5.00	4.89	4.81
20	.75	22.72	6.21	4.39	3.77	3.46	3.28	3.16	3.08	3.02	2.85	2.78	2.74	2.72
	.80	29.22	7.08	4.85	4.12	3.76	3.56	3.42	3.32	3.26	3.07	2.99	2.95	2.92
	.85	39.98	8.29	5.46	4.56	4.13	3.89	3.73	3.62	3.54	3.33	3.24	3.19	3.16
	.90	61.4	10.25	6.35	5.18	4.64	4.34	4.14	4.01	3.91	3.66	3.56	3.50	3.46
	.95	125.3	14.52	8.05	6.28	5.51	5.08	4.82	4.64	4.51	4.18	4.04	3.97	3.92
	.975	252.9	20.43	10.09	7.49	6.40	5.83	5.48	5.24	5.07	4.66	4.49	4.40	4.34
	.99	635	32.1	13.52	9.34	7.69	6.86	6.36	6.03	5.80	5.25	5.03	4.91	4.84
21	.75	23.83	6.36	4.47	3.82	3.51	3.32	3.20	3.11	3.05	2.88	2.81	2.77	2.74
	.80	30.65	7.24	4.93	4.17	3.81	3.59	3.45	3.36	3.29	3.10	3.02	2.97	2.94
	.85	41.94	8.48	5.54	4.62	4.18	3.93	3.77	3.65	3.57	3.35	3.26	3.21	3.18
	.90	64.4	10.49	6.44	5.24	4.69	4.38	4.18	4.04	3.94	3.69	3.58	3.52	3.48
	.95	131.5	14.86	8.17	6.35	5.56	5.13	4.86	4.67	4.54	4.21	4.07	3.99	3.95
	.975	265.7	20.92	10.24	7.57	6.46	5.88	5.52	5.28	5.11	4.68	4.51	4.42	4.36
	.99	667	32.8	13.72	9.44	7.76	6.91	6.40	6.07	5.84	5.27	5.05	4.93	4.86
22	.75	24.94	6.50	4.54	3.88	3.55	3.36	3.23	3.15	3.08	2.91	2.83	2.79	2.76
	.80	32.08	7.40	5.01	4.23	3.85	3.63	3.49	3.39	3.32	3.12	3.04	2.99	2.96
	.85	43.90	8.67	5.63	4.67	4.22	3.97	3.80	3.69	3.60	3.38	3.29	3.24	3.20
	.90	67.4	10.72	6.54	5.30	4.74	4.42	4.22	4.08	3.98	3.71	3.60	3.54	3.51
	.95	137.7	15.19	8.29	6.42	5.61	5.17	4.89	4.71	4.57	4.23	4.09	4.02	3.97
	.975	278	21.39	10.39	7.65	6.52	5.92	5.56	5.31	5.14	4.71	4.53	4.44	4.38
	.99	700	33.6	13.93	9.54	7.83	6.96	6.44	6.11	5.87	5.30	5.07	4.95	4.88
23	.75	26.05	6.64	4.61	3.92	3.59	3.39	3.27	3.18	3.11	2.93	2.85	2.81	2.78
	.80	33.51	7.56	5.08	4.28	3.89	3.67	3.52	3.42	3.34	3.15	3.06	3.02	2.99
	.85	45.86	8.85	5.71	4.73	4.27	4.00	3.83	3.72	3.63	3.41	3.31	3.26	3.22
	.90	70.4	10.95	6.63	5.36	4.78	4.46	4.25	4.11	4.00	3.74	3.63	3.57	3.53
	.95	143.9	15.51	8.40	6.49	5.66	5.21	4.93	4.74	4.60	4.26	4.11	4.04	3.99
	.975	291	21.85	10.53	7.73	6.57	5.96	5.59	5.35	5.17	4.73	4.56	4.46	4.40
	.99	730	34.3	14.12	9.63	7.89	7.01	6.48	6.14	5.90	5.32	5.10	4.97	4.90
24	.75	27.16	6.78	4.68	3.97	3.63	3.43	3.30	3.20	3.14	2.95	2.88	2.83	2.81
	.80	34.93	7.72	5.15	4.33	3.93	3.70	3.55	3.45	3.37	3.17	3.08	3.04	3.01
	.85	47.82	9.03	5.78	4.78	4.31	4.04	3.87	3.75	3.66	3.43	3.33	3.28	3.24
	.90	73.5	11.17	6.72	5.42	4.82	4.49	4.28	4.14	4.03	3.76	3.65	3.59	3.55
	.95	150.1	15.83	8.51	6.55	5.71	5.25	4.96	4.77	4.63	4.28	4.14	4.06	4.01
	.975	303	22.30	10.67	7.80	6.63	6.01	5.63	5.38	5.20	4.76	4.58	4.48	4.42
	.99	762	35.1	14.31	9.73	7.95	7.05	6.52	6.18	5.93	5.37	5.12	4.99	4.92
25	.75	28.27	6.91	4.74	4.02	3.67	3.46	3.33	3.23	3.16	2.98	2.90	2.85	2.83
	.80	36.36	7.87	5.22	4.38	3.97	3.74	3.58	3.48	3.40	3.19	3.11	3.06	3.03
	.85	49.78	9.21	5.86	4.83	4.35	4.07	3.90	3.77	3.68	3.45	3.35	3.30	3.26
	.90	76.5	11.38	6.80	5.47	4.87	4.53	4.31	4.17	4.06	3.78	3.67	3.61	3.57
	.95	156.3	16.14	8.62	6.62	5.75	5.28	4.99	4.80	4.66	4.30	4.16	4.08	4.03
	.975	316	22.74	10.80	7.88	6.68	6.05	5.66	5.41	5.23	4.78	4.60	4.50	4.44
	.99	794	35.8	14.49	9.82	8.01	7.10	6.56	6.21	5.96	5.37	5.14	5.01	4.94

*The table entries are from Table 4 on pp. 17–23 of "New Tables for Multiple Comparisons with a Control (Unknown Variances)," by E. J. Dudewicz, J. S. Ramberg, and H. J. Chen, *Biometrische Zeitschrift*, vol. 17 (1975), pp. 13–26. Reprinted with the permission of Akademie-Verlag, Berlin.

Take $n_1 - n_0$ more observations on population 1 and $n_2 - n_0$ more on population 2.

Compute the sample means $\overline{X}_1$ of n_1 observations, and $\overline{X}_2$ of n_2 observations and the interval is $\overline{X}_1 - \overline{X}_2 \pm d$ (i.e., we are $100P^*$ percent sure that the difference of the population means is within d units of the difference of the sample means). For an example of the calculations involved, see Example of Selection of the Best below.

STATISTICAL TESTS OF HYPOTHESES

A *statistical hypothesis* is an assertion about a population, often about some parameter of a population. *Tests of hypotheses* (also called *tests of significance*) were designed so that experimenters would not ascribe causes to variations in data that were in fact due simply to random variation (and thus did not need a cause to explain them). Thus statistical hypothesis testing is a modern-day version of the medieval Occam's razor principle that "one should not multiply causes without reason." (William of Occam was an English Franciscan philosopher who died about 1349.) For example, if a process has a mean weight of 14.90 lb per item produced, a change is made with a view to increasing the weight per item produced, and a sample of 10 items (taken after the process change) has a mean weight of 15.10 lb, this *does not* necessarily mean that the process mean has been shifted up: It could be that it has remained the same (or has even decreased) and that we are simply seeing the results of the randomness of the process. Making correct inferences in the face of such possibilities is the gist of the area of hypothesis testing.

Basic Concepts, Types of Errors.   *Hypothesis* as used here is an assertion made about a population. Usually the assertion concerns the numerical value of some parameter of the population. For example, a hypothesis might state that the mean life of a population of batteries equals 30.0, written as H:μ_0 = 30.0. This assertion may or may not be correct. A *hypothesis test* is a test of the validity of the assertion and is carried out by analysis of a sample of data.

There are two reasons why sample results must be evaluated carefully. First, there are many other samples that, by chance alone, could be drawn from the population. Second, the numerical results in the sample actually selected can easily be compatible with several different hypotheses. These points are handled by recognizing two types of error which can be made in evaluating a hypothesis:

1. *Reject* the hypothesis when it is *true.* This is called the *type 1 error*; its probability is called the *level of significance* and is denoted by α.

2. *Accept* the hypothesis when it is *false.* This is called the *type II error*; its probability is usually denoted by β (though some authors call it $1-\beta$).

These error probabilities can be controlled to desired values.

The type I error is shown graphically in Figure 44.19 for the hypothesis H:μ_0 = 30.0. The area between the vertical lines represents the *acceptance region* for the hypothesis test: If the sample result falls within the acceptance region, the hypothesis is accepted. Otherwise, it is rejected. Notice that there is a small portion of the curve that falls outside the acceptance region. This portion (α) represents the probability of obtaining a sample result outside the acceptance region, even though the hypothesis is correct.

Suppose it has been decided that the type I error must not exceed 5 percent. This is the probability of rejecting the hypothesis when, in truth, the true average life is 30.0. The acceptance region can be obtained by locating values of average life that have only a 5 percent chance of being exceeded when the true average life is 30.0. Further, suppose a sample n of four measurements is taken and $\sigma = 10.0$.

Remember that the curve represents a population of sample averages because the decision will be made on the basis of a sample average. Sample averages vary less than individual measurements according to the relationship $\sigma_{\overline{X}} = \sigma/\sqrt{n}$.

Further, the distribution of sample averages is approximately normal even if the distribution of the individual measurements (going into the averages) is not normal [see Grant and Leavenworth

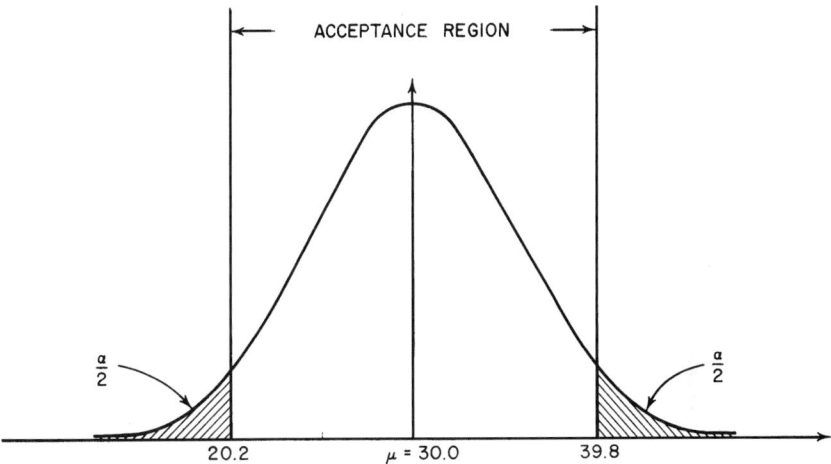

FIGURE 44.19   Acceptance region for H:μ_0 = 30.0.

(1972, pp. 69–71)]. The approximation holds best for large values of *n* but is adequate for *n* as low as 4.

Table B in Appendix II shows that a 2.5 percent area in each tail is at a limit that is 1.96 standard deviations from 30.0. Then under the hypothesis that μ_0 = 30.0, 95 percent of sample averages will fall within ±1.96$\sigma_{\bar{x}}$ of 30.0, or

$$\text{Upper limit} = 30.0 + 1.96\,\frac{10}{\sqrt{4}} = 39.8$$

$$\text{Lower limit} = 30.0 - 1.96\,\frac{10}{\sqrt{4}} = 20.2$$

The acceptance region is thereby defined as 20.2 to 39.8. If the average of a random sample of four batteries is within this acceptance region, the hypothesis is accepted. If the average falls outside the acceptance region, the hypothesis is rejected. This decision rule provides a type I error of 0.05.

The type II or β error, the probability of accepting a hypothesis when it is false, is shown in Figure 44.20 as the shaded area. Notice that it is possible to obtain a sample result within the acceptance region, even though the population has a true average that is not equal to the average stated in the hypothesis. The numerical value of β depends on the true value of the population average (and also on *n*, σ, and α). This is depicted by an *operating characteristic (OC) curve.*

The problem now is to construct an operating characteristic curve to assess the magnitude of the type II (β) error. Since β is the probability of *accepting* the original hypothesis (μ_0 = 30.0) when it is *false,* the probability that a sample average will fall between 20.2 and 39.8 must be found when the true average of the population is something other than 30.0. This has been done for many values of the true average, and the result is shown in Figure 44.21. [This curve should not be confused with that of a normal distribution of measurements. In some cases the shape is similar, but the meanings of an OC curve and a distribution curve are entirely different; Juran and Gryna (1980, pp. 410–412) give the detailed calculations; also see Dudewicz (1976, pp. 272–275).] Thus the OC curve is a plot of the probability of accepting the original hypothesis as a function of the true value of the population parameter (and the given values of *n*, σ, and α).

Use of the Operating Characteristic Curve in Selecting an Acceptance Region.

The acceptance region was determined by dividing the 5 percent allowable α error into equal parts

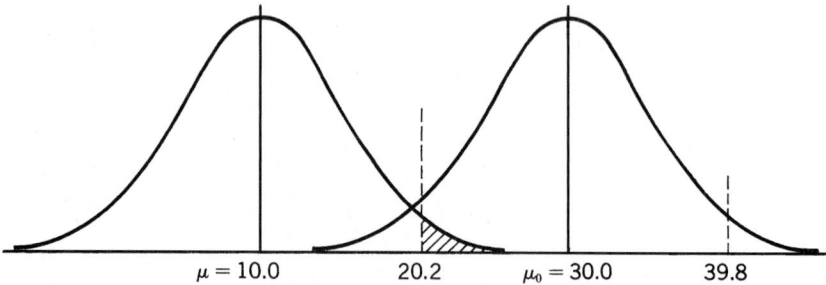

FIGURE 44.20   Type II or β error.

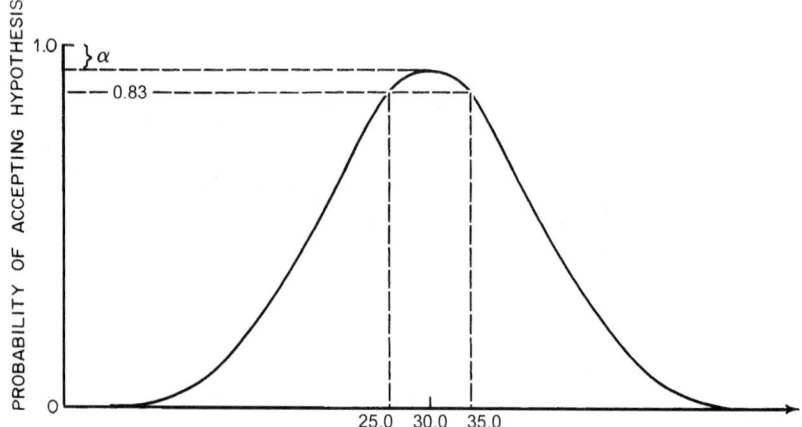

FIGURE 44.21   Operating characteristic curve.

(see Figure 44.19). This is called a *two-tailed test.* The entire 5 percent also could be placed at either the left or the right tail of the distribution curve (Figure 44.22). These are *one-tailed tests.*

Operating characteristic curves for tests having these one-tailed acceptance regions can be developed following the approach used for the two-tailed region. Although the α error is the same, the β error varies for the three tests. (See Figure 9.2-2 on p. 275 of Dudewicz 1976.)

In some problems, knowledge is available to indicate that if the true average of the population is *not* equal to the hypothesis value, then it is on one side of the hypothesis value. For example, a new material of supposedly higher average strength will have an average equal to or *greater than* that of the present material. Such information will help select a one-tailed or two-tailed test to make the β error as small as possible. The following guidelines are based on the analysis of OC curves:

Use a one-tailed test with the entire α risk in the right tail if (1) it is known that (if μ_0 is not true) the true mean is $>\mu_0$ or (2) values of the population mean $<\mu_0$ are acceptable and we are interested only in detecting a population mean $>\mu_0$. [Use a one-tailed test with the entire α risk in the left tail if (1) it is known that (if μ_0 is not true) the true mean is $<\mu_0$ or (2) values of the population mean $>\mu_0$ are acceptable and we are interested only in detecting a population mean $<\mu_0$.]

Use a two-tailed test if (1) there is no prior knowledge on the location of the true population mean or (2) we are interested in detecting a true population mean $<$ or $>$ the μ_0 stated in the original hypothesis. (With a two-tailed test, the hypothesis is sometimes stated as the original hypothesis $H_0:\mu_0 = 30.0$ against the alternative hypothesis $H_1:\mu_0 \neq 30.0$. With a one-tailed test, $H_0:\mu_0 = 30.0$ against the alternative $H_1:\mu_1 < 30.0$ if α is placed in the left tail or $H_1:\mu_1 > 30.0$ if α is placed in the right tail.)

Every test of hypothesis has an OC curve. Duncan (1974) and Natrella (1963) are good sources of OC curves. [Some references present "power" curves, but power is simply $1 -$ (the probability of acceptance) $= 1 - \beta$.]

With this background, our discussion now proceeds to the steps for testing a hypothesis.

Testing a Hypothesis When the Sample Size Is Fixed in Advance.   Ideally, desired values for the type I and type II errors are defined in advance and the required sample size determined (see later discussion on Determining the Sample Size Required for Testing a Hypothesis). If the sample size is fixed in advance because of cost or time limitations, then usually the desired type I error is defined and the following procedure is followed:

1. State the hypothesis.
2. Choose the type I error. Common values are 0.01, 0.05, or 0.10.
3. Choose the test statistic for testing the hypothesis.
4. Determine the acceptance region for the test, i.e., the range of values of the test statistic that result in a decision to accept the hypothesis.
5. Obtain the sample of observations, compute the test statistic, and compare the value to the acceptance region to make a decision to accept or reject the hypothesis.
6. Draw an engineering conclusion.

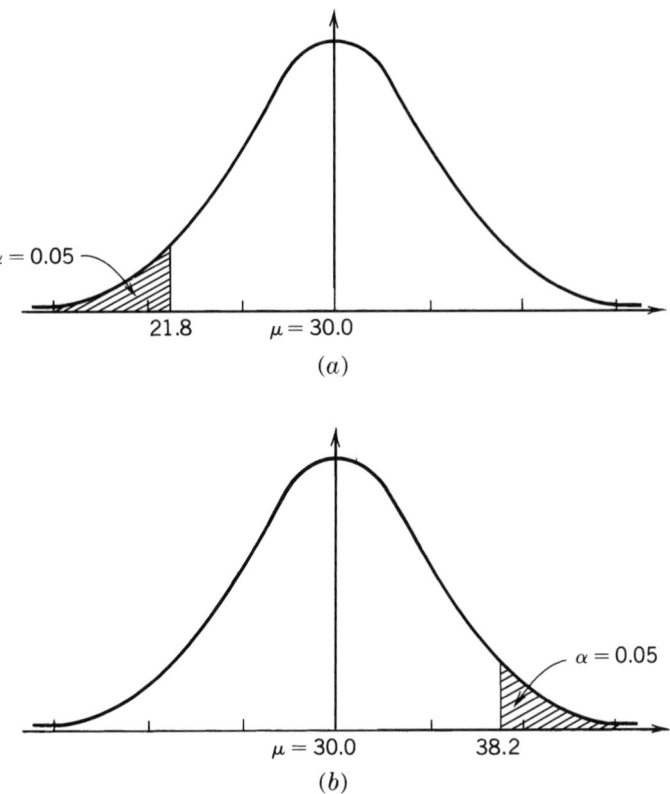

FIGURE 44.22   (*a*) Entire 5 percent error on left tail. (*b*) Entire 5 percent on right tail.

In the case of fixed sample size, the hypothesis is said to be "accepted" in the sense that there is insufficient evidence for the hypothesis to be rejected. However, this does not mean that the hypothesis is true. (See the introductory paragraph of Statistical Tests of Hypotheses above and Drawing Conclusions from Tests of Hypotheses below.) For this reason, one often uses the terminology "fail to reject" rather than "accept."

Table 44.31 summarizes some common tests of hypotheses. [These are tests commonly mentioned in the literature and used in practice. A number of them assume a normal distribution, Harter and Dubey (1967) define tests for the mean and variance but assuming a Weibull distribution—which really covers a family of distributions.] The procedure is illustrated through the following examples. Further examples and elaboration of the procedure are provided in Dixon and Massey (1969), Johnson and Leone (1964), Duncan (1974), Dudewicz (1976), and Natrella (1963). Table 44.31 lists a few unique or additional references for certain tests. For nonparametric tests, see especially Gibbons (1997).

Example: Tests on eight units of an experimental engine showed that they operated, respectively, for 28, 26, 31, 29, 25, 27, 28, and 27 min with 1 liter of a certain kind of fuel. A proposed specification states that the engine must operate for an average of at least 30 min. Does the engine meet the requirement? Assume a 5 percent significance level.

Solution: Using Test 1*b* of Table 44.31,

$$H_0: \mu = 30.0$$

$$H_1: \mu < 30.0$$

Test statistic:

$$t_1 = \frac{\overline{X} - \mu_0}{s/\sqrt{n}}$$

Acceptance region:

$$\text{Degrees of freedom DF} = 8 - 1 = 7$$

$$t \geq -1.895$$

A mathematical derivation of *degrees of freedom* is beyond the scope of this handbook, but the underlying concept can be stated. *Degrees of freedom* is a measure of the assurance involved when a sample standard deviation is used to estimate the true standard deviation of a universe. When the true standard deviation is known, DF $= \infty$. More generally, DF equals the number of measurements used to determine the sample standard deviation minus the number of constants estimated from the data in order to compute the standard deviation. In this example, it was necessary to estimate only one constant (the sample average) in order to compute the standard deviation, therefore DF $= 8 - 1$.

Analysis:

$$\overline{X} = 27.6 \qquad s = 1.86$$

$$t = \frac{27.6 - 30.0}{\dfrac{1.86}{\sqrt{8}}} = -3.65$$

Conclusion: Reject the hypothesis. There is sufficient evidence to conclude that the engine does not meet the requirement.

TABLE 44.31 Summary Table of Tests of Hypotheses

Test statistic and its distribution	Assumptions	Remarks
Test 1. The mean of a population is equal to μ_0 (H:$\mu = \mu_0$).		
$(a)\ U = \dfrac{\overline{X} - \mu_0}{\sigma/\sqrt{n}}$ Normal distribution (Appendix II, Table B)	X is distribution-free but should be continuous and have only one mode.	Standard deviation of population is known.
$(b)\ t = \dfrac{\overline{X} - \mu_0}{s/\sqrt{n}}$ t distribution with DF $= n - 1$ (Appendix II, Table G)	X is normally distributed.	Standard deviation of population is estimated by sample s.
$(c)\ \tau_1 = \dfrac{\overline{X} - \mu_0}{R}$ τ_1 distribution (Appendix II, Table H)	X is normally distributed.	Dispersion of the population is estimated from sample range.
$(d)\ r =$ number of occurrences of less frequent sign for a two-tail test (or number of positive or negative signs for a one-tail test) Distribution of r (Appendix II, Table I)	Distribution-free but population should be continuous and symmetrical.	Data are analyzed by evaluating signs of $(X_i - \mu_0)$. (This is a "sign" test; see Gibbons, 1997, p. 112.)

TABLE 44.31 Summary Table of Tests of Hypotheses (*Continued*)

Test statistic and its distribution	Assumptions	Remarks
Test 2. The means of two populations are equal ($H: \mu_1 = \mu_2$).		
(a) $U = \dfrac{\overline{X}_1 - \overline{X}_2}{\sqrt{\sigma_1^2/n_1 + \sigma_2^2/n_2}}$ Normal distribution (Appendix II, Table B)	X_1 and X_2 are distribution-free but should be continuous and have only one mode. If the populations are not normally distributed, sample sizes n_1 and n_2 should be large so that sampling distribution of U is approximately normal.	Standard deviations of populations are known.
(b) $t = \dfrac{\overline{X}_1 - \overline{X}_2}{\sqrt{\dfrac{1}{n_1} + \dfrac{1}{n_2}}\sqrt{\dfrac{[(n_1 - 1)s_1^2 + (n_2 - 1)s_2^2]}{n_1 + n_2 - 2}}}$ t distribution with DF $= n_1 + n_2 - 2$ (Appendix II, Table G)	$\sigma_1 = \sigma_2$. X_1 and X_2 are normally distributed.	Standard deviations of populations are estimated by sample s_1 and s_2.
(c) $\tau_d = \dfrac{\overline{X}_1 - \overline{X}_2}{0.5(R_1 + R_2)}$ Distribution of τ_d (Appendix II, Table J)	X_1 and X_2 are normally distributed.	Dispersions of populations are estimated by sample ranges.

(d) $t' = \dfrac{\bar{X}_1 - \bar{X}_2}{\sqrt{s_1^2/n_1 + s_2^2/n_2}}$ t distribution with DF = $\min(n_1 - 1, n_2 - 1)$ (Appendix II, Table G)	X_1 and X_2 are normally distributed.	Standard deviations of populations are estimated by sample s_1 and s_2 (no assumption that $\sigma_1 = \sigma_2$).
(e) $t = \dfrac{\bar{d}}{s_d/\sqrt{n}}$ t distribution with DF = number of pairs $- 1$ (Appendix II, Table G)	Populations are normally distributed.	Data are taken in n pairs, and difference d within each pair is calculated.
(f) r = number of occurrences of less frequent sign for a two-tail test (or number of positive or negative signs for a one-tail test) Distribution of r (Appendix II, Table I)	n_1 and n_2 each ≥ 6. Populations must be continuous and symmetrical; each of the two observations in a pair has been obtained under similar conditions.	Data are analyzed by evaluating signs of $(X_1 - X_2)$. (This is a "sign" test.) Distribution of r (Appendix II, Table I)

Test 3. Two characteristics on one product have specified population means (H:$\mu_X = \mu_{X0}$ and $\mu_Y = \mu_{Y0}$) (Johnson and Leone, 1964).

$T^2 = \dfrac{s_Y^2}{s_X^2 s_Y^2 - s_{XY}}(X - \mu_{X0})^2 - \dfrac{2s_{XY}}{s_X^2 s_Y^2 - s_{XY}} \cdot$ $(X - \mu_{X0})(Y - \mu_{Y0}) + \dfrac{s_X^2}{s_X^2 s_Y^2 - s_{XY}}(Y - \mu_{Y0})^2$ where $s_{XY} = \dfrac{n\Sigma(XY) - \Sigma X \Sigma Y}{n(n - 1)}$ where critical value for $T^2 = \dfrac{2(n - 1)F}{n - 2}$ and where the F distribution has $DF_1 = 2$; $DF_2 = n - 2$ (Appendix II, Table K)	Bivariate normal distribution.	Population standard deviations are estimated by s_X and s_Y (T^2 is called Hotelling's T^2).

TABLE 44.31 Summary Table of Tests of Hypotheses (*Continued*)

Test statistic and its distribution	Assumptions	Remarks
Test 4. The standard deviation of a population is equal to σ_0 (H:$\sigma = \sigma_0$).		
$\chi^2 = \dfrac{(n-1)s^2}{\sigma_0^2}$ Chi-square distribution with DF = $n - 1$ (Appendix II, Table L)	Population is normally distributed.	Standard deviation of population is estimated by sample s.
Test 5. The standard deviations of two populations are equal (H:$\sigma_1^2 = \sigma_2^2$).		
(a) $F = \dfrac{s_1^2}{s_2^2}$ F distribution with DF$_1 = n_1 - 1$ and DF$_2 = n_2 - 1$ (Appendix II, Table K)	Populations are normally distributed.	Standard deviations of populations are estimated by sample s_1 and s_2.
(b) $F' = \dfrac{R_1}{R_2}$ Distribution of F' (Appendix II, Table M)	Distribution-free but X_1 and X_2 should be continuous.	Dispersions of populations are estimated by sample ranges.
Test 6. The proportion of a population exhibiting a certain characteristic is p_0 (H:$p = p_0$).		
(a) $U = \dfrac{X - np_0}{\sqrt{np_0(1 - p_0)}}$ Normal distribution (Appendix II, Table B)	$n \geq 100$. Only for large sample sizes.	Proportion of population is estimated by sample proportion.
(b) Determine confidence limits (Appendix II, Table N) and observe if p_0 falls within the limits		Proportion of population is estimated by sample proportion (useful when $n < 100$).

Test 7. The proportions in two populations are equal ($H{:}p_1 = p_2$).

$U = \dfrac{X_1/n_1 - X_2/n_2}{\sqrt{\hat{p}(1-\hat{p})(1/n_1 + 1/n_2)}}$ where $\hat{p} = \dfrac{X_1 + X_2}{n_1 + n_2}$ Normal distribution (Appendix II, Table B)	$np > 5$ for each population. Sample sizes n_1 and n_2 must be large so that sampling distribution of U is approximately normal.	Proportions in populations are estimated by sample proportions.

Test 8. Proportion of correct decisions on a sensory evaluation is p_0 ($H{:}p = p_0$).

(a) $U = \dfrac{X/n - 0.50}{\sqrt{0.25/n}}$ Normal distribution (Appendix II, Table B)	$p = 0.5$, n should be $>$ 30.	Judge is asked to identify which of two specimens is the same as a control specimen originally given to him or her. (This is a "duo-trio" test.)
(b) $U = \dfrac{X/n - 0.33}{\sqrt{0.22/n}}$ Normal distribution (Appendix II, Table B)	$p = 0.33$, n should be $>$ 30.	Judge is asked to identify which of three specimens is different from the other two. (This is a "triangle" test.)

TABLE 44.31 Summary Table of Tests of Hypotheses *(Continued)*

Test statistic and its distribution	Assumptions	Remarks

Test 9. Samples are from identically distributed populations [H:$F(X_1) = F(X_2)$ where $F(X)$ = distribution function] [Johnson and Leone, 1964].

Test statistic and its distribution	Assumptions	Remarks
(*a*) For evaluation of means T' = sum of ranks in smaller sample Distribution of T' (Appendix II, Table O)	Distribution-free; if $n_1 >$ 8 and $n_2 > 8$, the distribution of statistic T' can be closely approximated by normal.	Data are evaluated by ranking the combined observations from the two samples (1 for the smallest, 2 for the next smallest, etc.). Then calculate the sum of the ranks in the smaller sample. A rejected hypothesis leads to the conclusion that the means are different. (This is a "rank sum" test.)
(*b*) For evaluation of standard deviations T' = sum of ranks in smaller sample Distribution of T' (Appendix II, Table O)	Distribution-free; if $n_1 >$ 8, and $n_2 > 8$, the distribution of statistic T' can be closely approximated by normal.	Data are evaluated by ranking the combined observations but ranking assigns 1 to smallest observation, 2 to largest observation, 3 to next smallest, etc. The sum of the ranks for the smaller sample is calculated. A rejected hypothesis means that the standard deviations are different. (This is a "rank sum" test.)

Test 10. The observations in a sample have been randomly drawn from a single population (Bennett and Franklin, 1954, and Dixon and Massey, 1969).

$n_1 \geq 10,\ n_2 \geq 10$.

(a) u = number of runs
Distribution of runs (Appendix II, Table P)

Data are evaluated in terms of number of sequences or "runs" above and below the median. (This is a "runs" test.)

(b) $M = \dfrac{\sum\limits_{i=1}^{n-1}(X_{i+1} - \overline{X}_i)^2}{\Sigma(X_i - \overline{X})^2}$

$U = \dfrac{1 - M/2}{\sqrt{(n-2)/[(n-1)(n+1)]}}$

Normal distribution (Appendix II, Table B)

1. $n \geq 4$.
2. Normal population.

Data are evaluated in terms of differences between successive observations in a sequence $(X_{i+1} - X_i)$. (This is the "mean square successive difference, M, test.")

Test 11. An observation does belong to the same population as the other observations in a sample (Dixon and Massey, 1969; Grubbs, 1969).

Both X_1 and X_2 are to be evaluated if extreme observations in either direction are undesirable.

X is normally distributed. Population mean and standard deviation are unknown.

Data are evaluated by arranging data in order of magnitude and comparing the distance of one extreme observation from other observations with a measure of variability. (This is an outlier test.)

n	r
$3 \leq n \leq 7$	$r_{10} = \dfrac{X_2 - X_1}{X_n - X_1}$
$8 \leq n \leq 10$	$r_{11} = \dfrac{X_2 - X_1}{X_{n-1} - X_1}$
$11 \leq n \leq 13$	$r_{21} = \dfrac{X_3 - X_1}{X_{n-1} - X_1}$
$14 \leq n \leq 25$	$r_{22} = \dfrac{X_3 - X_1}{X_{n-2} - X_1}$

Distribution of r (Appendix II, Table Q)

TABLE 44.31 Summary Table of Tests of Hypotheses (*Continued*)

Test statistic and its distribution	Assumptions	Remarks
Test 12. A sample of data comes from a population with the specified probability function.		
(*a*) D = largest deviation of actual % cumulative frequency from theoretical % cumulative frequency Distribution of D (Gibbons, 1997, Table C)	Distribution should be continuous.	Data are evaluated by first plotting on probability paper. The largest deviation of a plotted point from a straight line is then evaluated. This test applies particularly where $n < 30$. (This is the "Kolmogorov-Smirnov goodness-of-fit test.")
(*b*) $\chi^2 = \Sigma \dfrac{(f_a - f_e)^2}{f_e}$	$n > 30$ and preferably > 100.	Data are evaluated by first constructing a frequency distribution. Theoretical frequencies (based on the distribution assumption) are calculated for each cell. The actual (f_a) and theoretical (f_e) frequencies are then compared. This test can be used for continuous and discrete distributions. If any theoretical frequency is less than 5, the cell involved should be combined with one or more adjacent cells. (This is the χ^2 goodness-of-fit test.)

Distribution	DF
Normal	Number of cells minus 3
Exponential	Number of cells minus 2
Weibull	Number of cells minus 4
Poisson	Number of cells minus 2
Binomial	Number of cells minus 2

Test statistics for all distributions follow the chi square distribution (Appendix II, Table L)

Example: *Solve the previous example using the range instead of the standard deviation.*

Solution: *Using Test 1c of Table 44.31,*

$$H_0: \quad \mu = 30.0$$

$$H_1: \quad \mu < 30.0$$

Test statistic:

$$\tau_1 = \frac{\overline{X} - \mu_0}{R}$$

Acceptance region:

$$\tau_1 \geq -0.230$$

Analysis:

$$\tau_1 = \frac{27.6 - 30.0}{6} = -0.40$$

Conclusion: Reject the hypothesis. There is sufficient evidence to conclude that the engine does not meet the requirement.

Example: Solve the previous example using the sign test.

Solution: Using Test 1d of Table 44.31,

$$H_0: \quad \mu = 30$$

$$H_1: \quad \mu < 30$$

Test statistic: Number of positive signs r.

Acceptance region:

$$r > 1 \text{ (one-tailed test)}$$

Analysis:

X	$X - \mu_0$	
28	−	
26	−	
31	+	
29	−	$r = 1$
25	−	
27	−	
28	−	
27	−	

Conclusion: Reject the hypothesis. There is sufficient evidence to conclude that the engine does not meet the requirement.

Example: Five batches of rubber were made by each of two recipes and tested for tensile strength with the following results:

Recipe 1	Recipe 2
3067	3200
2730	2777
2840	2623
2913	3044
2789	2834

Test the hypothesis that average strength is the same for the two recipes. Assume a 5 percent significance level.

Solution: First, Test 5a of Table 44.31 tests the assumption of equal variances. The outcome of this is used to decide whether to use Test 2b or 2d to evaluate the question about average strength.

$$H_0: \quad \sigma_1^2 = \sigma_2^2$$
$$H_1: \quad \sigma_1^2 \neq \sigma_2^2$$

Test statistic:

$$DF_1 = 5 - 1 = 4 \qquad DF_2 = 5 - 1 = 4$$
$$F = \frac{(s_1)^2}{(s_2)^2}$$

Acceptance region:

$$\frac{1}{9.60} \leq F \leq 9.60$$

Analysis:

$$s_1^2 = 16{,}923.7$$
$$s_2^2 = 51{,}713.3$$
$$F = \frac{16{,}923.7}{51{,}713.3} = 0.33$$

Conclusion: Accept the hypothesis. This is used to satisfy the assumption of equal variances in the following test of hypothesis. Now, using Test 2b,

$$H_0: \quad \mu_1 = \mu_2$$
$$H_1: \quad \mu_1 \neq \mu_2$$

Test statistic:

$$t = \frac{\overline{X}_1 - \overline{X}_2}{\sqrt{\frac{1}{n_1} + \frac{1}{n_2}} \sqrt{\frac{[(n_1 - 1)s_1^2 + (n_2 - 1)s_2^2]}{n_1 + n_2 - 2}}}$$

Acceptance region:

$$DF = 5 + 5 - 2 = 8$$

$$-2.306 \le t \le +2.306$$

Analysis:

$$t = \frac{2867.8 - 2895.6}{\sqrt{\frac{1}{5} + \frac{1}{5}}\sqrt{\frac{(5-1)16{,}923.7 + (5-1)51{,}713.3}{5 + 5 - 2}}} = -0.2373$$

Conclusion: Accept the hypothesis. There is insufficient evidence to conclude that the recipes differ in average strength.

Example: Solve the previous example using ranges instead of standard deviations.

Solution: First, Test 5b of Table 44.31 tests the assumption of equal variances.

$$H_0: \quad \sigma_1^2 = \sigma_2^2$$

$$H_1: \quad \sigma_1^2 \neq \sigma_2^2$$

Test statistic:

$$F' = \frac{R_1}{R_2}$$

Acceptance region:

$$0.32 < F' < 3.2$$

Analysis:

$$R_1 = 3067 - 2730 = 337$$

$$R_2 = 3200 = 2623 = 577$$

$$F' = \frac{337}{577} = 0.58$$

Conclusion: Accept the hypothesis. This is used to satisfy the assumption of equal variances in the following test of hypothesis. Now, using Test 2c,

$$H_0: \quad \mu_1 = \mu_2 \quad H_1: \quad \mu_1 \neq \mu_2$$

Test statistic:

$$\tau_d = \frac{\overline{X}_1 - \overline{X}_2}{0.5(R_1 + R_2)}$$

Acceptance region:

$$-0.493 \le \tau_d \le +0.493$$

$$R_1 = 3{,}067 - 2{,}730 = 337$$

$$R_2 = 3{,}200 - 2{,}623 = 577$$

Analysis:

$$\tau_d = \frac{2867.8 - 2895.6}{0.5(337 + 577)} = -0.061$$

Conclusion: Accept the hypothesis. There is insufficient evidence to conclude that the recipes differ in average strength.

Testing a Hypothesis When the Sample Size Is Not Fixed in Advance.   As noted earlier under Testing a Hypothesis When the Sample Size Is Fixed in Advance, ideally the desired values of type I and type II errors are defined in advance and the required sample size determined. However, if the sample size is fixed in advance, this cannot be done in most cases, and one can control only the type I error (not the type II error). One can control both when the sample size is not fixed in advance, as illustrated now for the case of testing if the means of two populations are equal ($H : \mu_1 = \mu_2$).

For type I error equal to α and type II error equal to β when the means differ by δ^* (a positive number) in absolute value, one proceeds as follows:

Sample n_0 observations from each of the ($k = 2$) populations (it is desirable that n_0 be at least 10 if possible).

Determine the total sample size for population i as

$$n_i = max[n_0 + 1, (ws_i)^2]$$

where w solves the equation (solve by trial and error)

$$P_{n_0}(-h - \delta^* w) + P_{n_0}(-h + \delta^* w) = \beta$$

where $h = h_{n_0}(2, 1 - \alpha/2)$ is found from Table 44.30, and $P_{n_0}(t)$ is tabled in Table 44.32.

Take $n_1 - n_0$ more observations from population 1 and $n_2 - n_0$ more observations on population 2. Reject the hypothesis that the means are equal if the sample means (based on all the data) differ by more than h/w.

As an example, suppose we have an initial sample size of $n_0 = 15$ and desire type I error $\alpha = .05$ and type II error $\beta = .10$ when the means differ by $\delta^* = 4.0$. Then from Table 44.30, $h = h_{n_0}(2, 1 - \alpha/2) = h_{15}(2, .975) = 3.02$. We find the w that solves

$$P_{15}(-3.02 - 4w) + P_{15}(-3.02 + 4w) = .10$$

in two steps. First, we find w approximately by solving $P_{15}(-3.02 + 4w) = .10$ using Table 44.32. Since Table 44.32 has entries of .4999 and larger, we first convert the equation using the fact that $P_{n_0}(-v) = 1 - P_{n_0}(v)$ for all v, obtaining

$$1 - P_{15}(3.02 - 4w) = .10 \quad \text{or} \quad P_{15}(3.02 - 4w) = .90$$

Since $P_{15}(1.9) = .8968$ and $P_{15}(2.0) = .9079$, w will (to a first approximation) be in the range of the solutions of

$$3.02 - 4w = 1.9 \quad \text{and} \quad 3.02 - 4w = 2.0$$

namely, $w = (3.02 - 1.9)/4 = 0.28$ and $w = (3.02 - 2.0)/4 = 0.255$. Thus let us try a value of $w = 0.255$. For this value,

$$P_{15}[-3.02 - (4)(0.255)] + P_{15}[-3.02 + (4)(0.255)] = P_{15}(-4.04) + P_{15}(-2.0) = 1 - P_{15}(4.04)$$
$$+ 1 - P_{15}(2.0) = 1 - .9946 + 1 - .9079 = .0054 + .0921 = .0975 = .10$$

TABLE 44.32 Probabilities $P_{n_0}(t)$ Needed for Testing if Two Means Are Equal in Two Stages*

t	2	3	4	5	6	7	8	9
0.0	.4999	.4999	.4999	.4999	.4999	.4999	.4999	.4999
0.1	.5158	.5208	.5229	.5241	.5248	.5254	.5257	.5260
0.2	.5317	.5415	.5457	.5481	.5496	.5506	.5514	.5520
0.3	.5473	.5620	.5684	.5719	.5742	.5757	.5768	.5777
0.4	.5628	.5822	.5907	.5954	.5984	.6004	.6019	.6030
0.5	.5779	.6021	.6126	.6184	.6221	.6246	.6265	.6279
0.6	.5927	.6215	.6340	.6410	.6453	.6483	.6505	.6521
0.7	.6071	.6404	.6549	.6629	.6679	.6713	.6738	.6757
0.8	.6211	.6588	.6751	.6841	.6897	.6936	.6964	.6986
0.9	.6345	.6765	.6946	.7046	.7108	.7151	.7182	.7206
1.0	.6475	.6935	.7134	.7242	.7311	.7357	.7391	.7417
1.1	.6600	.7099	.7313	.7431	.7504	.7554	.7591	.7618
1.2	.6720	.7255	.7485	.7610	.7688	.7742	.7780	.7810
1.3	.6834	.7405	.7648	.7781	.7863	.7919	.7960	.7991
1.4	.6943	.7547	.7803	.7942	.8029	.8087	.8130	.8162
1.5	.7048	.7681	.7950	.8094	.8184	.8245	.8289	.8322
1.6	.7147	.7809	.8088	.8238	.8331	.8393	.8439	.8473
1.7	.7242	.7930	.8218	.8372	.8467	.8532	.8578	.8613
1.8	.7332	.8044	.8340	.8498	.8595	.8660	.8707	.8742
1.9	.7418	.8151	.8454	.8615	.8714	.8780	.8827	.8863
2.0	.7499	.8252	.8561	.8725	.8824	.8891	.8938	.8974
2.1	.7577	.8347	.8661	.8826	.8926	.8993	.9040	.9075
2.2	.7651	.8436	.8754	.8920	.9020	.9087	.9134	.9169
2.3	.7721	.8520	.8841	.9007	.9107	.9173	.9220	.9254
2.4	.7788	.8599	.8922	.9087	.9187	.9252	.9298	.9332
2.5	.7852	.8673	.8996	.9162	.9260	.9324	.9369	.9402
2.6	.7912	.8742	.9066	.9230	.9327	.9390	.9434	.9466
2.7	.7970	.8807	.9130	.9293	.9388	.9450	.9493	.9524
2.8	.8025	.8868	.9190	.9350	.9444	.9504	.9546	.9576
2.9	.8078	.8925	.9245	.9403	.9495	.9553	.9594	.9623
3.0	.8128	.8979	.9297	.9452	.9541	.9598	.9637	.9665
3.1	.8176	.9029	.9344	.9497	.9583	.9638	.9676	.9703
3.2	.8221	.9076	.9388	.9537	.9622	.9675	.9711	.9736
3.3	.8265	.9121	.9429	.9575	.9657	.9708	.9742	.9766
3.4	.8307	.9162	.9466	.9609	.9688	.9737	.9770	.9793
3.5	.8347	.9201	.9501	.9640	.9717	.9764	.9795	.9817
3.6	.8385	.9238	.9533	.9669	.9743	.9788	.9817	.9838
3.7	.8422	.9273	.9563	.9695	.9766	.9809	.9837	.9857
3.8	.8457	.9305	.9591	.9719	.9788	.9829	.9855	.9874
3.9	.8491	.9336	.9617	.9741	.9807	.9846	.9871	.9889
4.0	.8524	.9365	.9640	.9761	.9824	.9861	.9885	.9902
4.1	.8555	.9392	.9662	.9779	.9840	.9875	.9898	.9913
4.2	.8585	.9417	.9683	.9796	.9854	.9888	.9909	.9923
4.3	.8614	.9442	.9702	.9812	.9867	.9899	.9919	.9932
4.4	.8641	.9464	.9720	.9826	.9879	.9909	.9928	.9940
4.5	.8668	.9486	.9736	.9839	.9890	.9918	.9936	.9947
4.6	.8694	.9506	.9751	.9851	.9899	.9926	.9943	.9953
4.7	.8719	.9526	.9765	.9862	.9908	.9933	.9949	.9959
4.8	.8743	.9544	.9779	.9872	.9916	.9940	.9954	.9964
4.9	.8766	.9561	.9791	.9881	.9923	.9946	.9959	.9968
5.0	.8788	.9578	.9803	.9889	.9930	.9951	.9964	.9972
5.1	.8810	.9593	.9813	.9897	.9935	.9956	.9967	.9975

TABLE 44.32 Probabilities $P_{n_0}(t)$ Needed for Testing if Two Means Are Equal in Two Stages* *(Continued)*

t.	n_0								
	10	11	12	13	14	15	20	25	30
0.0	.4999	.4999	.4999	.4999	.4999	.4999	.4999	.4999	.4999
0.1	.5262	.5264	.5266	.5267	.5268	.5269	.5272	.5274	.5275
0.2	.5524	.5528	.5531	.5533	.5535	.5537	.5544	.5547	.5555
0.3	.5783	.5789	.5793	.5797	.5800	.5803	.5812	.5818	.5822
0.4	.6039	.6046	.6052	.6057	.6061	.6065	.6077	.6085	.6089
0.5	.6290	.6299	.6306	.6312	.6317	.6322	.6337	.6346	.6352
0.6	.6534	.6545	.6554	.6561	.6567	.6572	.6591	.6601	.6608
0.7	.6772	.6785	.6794	.6803	.6810	.6816	.6837	.6849	.6857
0.8	.7003	.7016	.7027	.7037	.7045	.7051	.7075	.7089	.7098
0.9	.7224	.7239	.7252	.7262	.7271	.7278	.7304	.7319	.7329
1.0	.7437	.7453	.7466	.7478	.7487	.7495	.7523	.7539	.7550
1.1	.7640	.7657	.7671	.7683	.7693	.7702	.7732	.7749	.7761
1.2	.7833	.7851	.7866	.7879	.7889	.7899	.7930	.7948	.7961
1.3	.8015	.8034	.8050	.8063	.8075	.8084	.8117	.8136	.8149
1.4	.8187	.8207	.8223	.8237	.8249	.8259	.8293	.8313	.8326
1.5	.8348	.8369	.8386	.8400	.8412	.8422	.8457	.8478	.8491
1.6	.8499	.8520	.8538	.8552	.8564	.8575	.8611	.8631	.8645
1.7	.8639	.8661	.8679	.8693	.8706	.8716	.8753	.8774	.8788
1.8	.8770	.8792	.8809	.8824	.8837	.8847	.8884	.8905	.8919
1.9	.8890	.8912	.8930	.8945	.8957	.8968	.9004	.9025	.9039
2.0	.9001	.9023	.9041	.9056	.9068	.9079	.9115	.9136	.9149
2.1	.9103	.9124	.9142	.9157	.9169	.9180	.9215	.9236	.9249
2.2	.9196	.9217	.9235	.9249	.9261	.9272	.9307	.9327	.9340
2.3	.9281	.9302	.9319	.9333	.9345	.9355	.9389	.9409	.9421
2.4	.9358	.9379	.9395	.9409	.9421	.9430	.9464	.9482	.9495
2.5	.9428	.9448	.9464	.9478	.9489	.9498	.9530	.9548	.9560
2.6	.9491	.9511	.9526	.9539	.9550	.9559	.9590	.9607	.9618
2.7	.9548	.9567	.9582	.9594	.9605	.9614	.9643	.9659	.9670
2.8	.9599	.9617	.9632	.9644	.9654	.9662	.9690	.9706	.9716
2.9	.9645	.9663	.9676	.9688	.9697	.9705	.9732	.9746	.9756
3.0	.9686	.9703	.9716	.9727	.9736	.9743	.9768	.9782	.9791
3.1	.9723	.9739	.9751	.9761	.9770	.9777	.9800	.9813	.9821
3.2	.9756	.9771	.9782	.9792	.9800	.9807	.9828	.9841	.9848
3.3	.9785	.9799	.9810	.9819	.9826	.9833	.9853	.9864	.9871
3.4	.9811	.9824	.9834	.9843	.9850	.9855	.9874	.9885	.9891
3.5	.9833	.9846	.9856	.9864	.9870	.9875	.9893	.9902	.9908
3.6	.9854	.9865	.9874	.9882	.9888	.9893	.9909	.9917	.9923
3.7	.9871	.9882	.9891	.9898	.9903	.9908	.9923	.9930	.9935
3.8	.9887	.9897	.9905	.9912	.9917	.9921	.9934	.9941	.9946
3.9	.9901	.9911	.9918	.9924	.9928	.9932	.9944	.9951	.9955
4.0	.9913	.9922	.9929	.9934	.9938	.9942	.9953	.9959	.9962
4.1	.9924	.9932	.9938	.9943	.9947	.9950	.9960	.9966	.9969
4.2	.9933	.9941	.9947	.9951	.9955	.9958	.9967	.9971	.9974
4.3	.9942	.9949	.9954	.9958	.9961	.9964	.9972	.9976	.9979
4.4	.9949	.9955	.9960	.9964	.9967	.9969	.9976	.9980	.9982
4.5	.9955	.9961	.9966	.9969	.9972	.9974	.9980	.9983	.9985
4.6	.9961	.9966	.9970	.9973	.9976	.9978	.9983	.9986	.9988
4.7	.9966	.9971	.9974	.9977	.9979	.9981	.9986	.9989	.9990
4.8	.9970	.9974	.9978	.9980	.9982	.9984	.9988	.9991	.9992
4.9	.9974	.9978	.9981	.9983	.9985	.9986	.9990	.9992	.9993
5.0	.9977	.9981	.9983	.9985	.9987	.9988	.9992	.9993	.9994
5.1	.9980	.9983	.9986	.9987	.9989	.9990	.9993	.9995	.9995

*The table entries are from p.52 of E.J. Dudewicz and S.R. Dalal (1975), "Allocation of Observations in Ranking and Selection With Unequal Variances." *Sankhyā*, vol. 73B, pp.28-78. Acknowledgment is made to the Indian Statistical Institute for permission to reproduce these tables.

(The approximation process could be carried further, but for most practical purposes this w will suffice.) Then, if $s_1 = 17.3$, we will need a total sample of size

$$n_1 = \max\{15 + 1, [(0.255)(17.3)]^2\} = \max(16, 19.46) = 20$$

(since sample sizes must be integers, we round up). Since we already have 15 observations, $20 - 15 = 5$ more will be required from population 1. Similarly, if $s_2 = 10.4$, we will need a total sample of size

$$n_2 = \max\{15 + 1, [(0.255)(10.4)]^2\} = \max(16, 7.03) = 16$$

Since we already have 15 observations from population 2, $16 - 15 = 1$ more will be required. If the sample means based on all the data are $\overline{X}_1$ and $\overline{X}_2$, we reject the hypothesis that $\mu_1 = \mu_2$ if the sample means differ by more than

$$\frac{h}{w} = \frac{3.02}{0.255} = 11.84$$

For example, if $\overline{X}_1 = 38.3$ and $\overline{X}_2 = 50.2$, the sample means differ by 11.9, and we reject the null hypothesis that the means are equal.

Drawing Conclusions from Tests of Hypotheses.   The payoff for these tests of hypotheses comes from reaching useful conclusions. The meaning of "Reject the hypothesis" or "Accept the hypothesis" is shown in Table 44.33 along with some analogies to explain subtleties of the meanings.

When a hypothesis is rejected, the practical conclusion is that "the parameter value specified in the hypothesis is wrong." This conclusion is made with strong conviction—roughly speaking at a confidence level of $100 (1 - \alpha)$ percent. The key question then is: Just what is a good estimate of the value of the parameter for the population? Help can be provided on this question by calculating the "confidence limits" for the parameter discussed under Statistical Estimation: Confidence Interval Estimates.

TABLE 44.33   The Meaning of a Conclusion from Tests of Hypotheses

	If hypothesis is rejected	If hypothesis is accepted
Adequacy of evidence in the sample of observations	Sufficient to conclude that hypothesis is false	Not sufficient to conclude that hypothesis is false; hypothesis is a reasonable one but has *not* been proved to be true
Difference between sample result (e.g., $\overline{X}$) and hypothesis value (e.g., μ_0)	Unlikely that difference was due to chance (sampling) variation	Difference could easily have been due to chance (sampling) variation
Analogy of guilt or innocence in a court of law	Guilt has been established beyond a reasonable doubt	Have not established guilt beyond a reasonable doubt
Analogy of a batting average in baseball	If player got 300 base hits out of 1000 times at bat, this is sufficient to conclude that the overall batting average is about 0.300	If player got 3 hits in 10 times, this is not sufficient to conclude that the overall average is about 0.300

When a hypothesis is accepted, the numerical value of the parameter stated in the hypothesis has not been proved, but it has not been disproved. It is *not* correct to say that the hypothesis has been proved as correct at the $100 \ (1 - \alpha)$ percent confidence level. Many other hypotheses could be accepted for the given sample of observations and yet only one hypothesis can be true. Therefore, an acceptance does *not* mean a high probability of proof that a specific hypothesis is correct. (All other factors being equal, the smaller the sample size, the more likely it is that the hypothesis will be accepted. Less evidence certainly does not imply proof.) For this reason, often today the wording used is "the hypothesis was not rejected at level of significance α" rather than "the hypothesis was accepted at level α." Only when the sample size is not fixed in advance do we have an indication that the true value does not differ from the hypothesized value by more than $\delta*$ with risk β.

With an acceptance of a hypothesis, a key question then is: What conclusion, if any, can be drawn about the parameter value in the hypothesis? Two approaches are suggested:

1. Calculate confidence limits on the sample result (see the previous topic of Statistical Estimation). These confidence limits define an interval within which the true population parameter lies. If this interval is small, then an acceptance decision on the test of hypothesis means that the true population value is either equal to or close to the value stated in the hypothesis. Then it is reasonable to act as if the parameter value specified in the hypothesis is in fact correct. If the confidence interval is relatively wide, then this is a stern warning that the value stated in the hypothesis has not been proved and that the true value of the population might be far different from that specified in the hypothesis.

2. Construct and review the operating characteristic curve for the test of hypothesis. This defines the probability that other possible values of the population parameter could have been accepted by the test. Knowing these probabilities for values relatively close to the original hypothesis can help draw further conclusions about the acceptance of the original hypothesis. For example, Figure 44.21 shows the OC curve for a hypothesis that specified that the population mean is 30.0. Note that the probability of accepting the hypothesis when the population mean μ is 30.0 is 0.95 (or $1 - \alpha$). Also note that if μ really is 35.0, then the probability of accepting $\mu = 30.0$ is still high (about 0.83). If μ really is 42.0, the probability of accepting $\mu = 30.0$ is only about 0.33.

Care must always be taken in drawing engineering conclusions from the statistical conclusions, particularly when a hypothesis is accepted. [Rutherford (1971) discusses a procedure for drawing conclusions which requires that a choice be made between two policies for drawing conclusions, i.e., conservative and liberal.]

Determining the Sample Size Required for Testing a Hypothesis.

The previous subsections assumed that the sample size was fixed by nonstatistical reasons and that the type I error only was predefined for the test. The ideal procedure is to predefine the desired type I and type II errors and calculate the sample size required to cover both types of errors.

The sample size required will depend on (1) the sampling risks desired (α and β), (2) the size of the smallest true difference that is to be detected, and (3) the variation in the characteristic being measured. The sample size can be determined by using the "operating characteristic" curve for the test. Table 44.34 summarizes methods useful in determining the sample size required for two-sided tests of certain hypotheses. [Further sources of OC curves are Duncan (1974) and Natrella (1963).]

Suppose it were important to detect the fact that the average life of the batteries cited previously was 35.0. Specifically, be 80 percent sure of detecting this change ($\beta = 0.2$). Further, if the true average was 30.0 (as stated in the hypothesis), there should be only a 5 percent risk of rejecting the hypothesis ($\alpha = 0.05$). In using Appendix II, Chart R, d is defined as

$$d = \frac{\mu - \mu_0}{\sigma} = \frac{35.0 - 30.0}{10} = 0.5$$

Entering with $d = 0.5$ and $P_a = 0.2$ (the β risk), the curves indicate that a sample size of about 30 is required.

TABLE 44.34 Summary of Sample Size Graphs and Tables

Hypothesis	Graph or table
1. Mean of a population $= \mu_0$ (σ known)	Appendix II, Chart R
2. Mean of a population $= \mu_0$ (σ estimated by s)	Duncan (1974, p. 539)
3. Means of two populations are equal (σ_1 and σ_2 known)	Natrella (1963, pp. T-16, T-17)
4. Means of two populations are equal ($\sigma_1 = \sigma_2$ but estimated by s_1 and s_2)	Natrella (1963, pp. T-16, T-17)
5. Standard deviation of a population $= \sigma_0$	Duncan (1974, p. 324)
6. Standard deviations of two populations are equal	Duncan (1974, p. 572)

Duncan (1974) discusses the calculation of the sample size required to meet the type I and II errors. In practice, however, one is often not sure of desired values of these errors. Reviewing the operating characteristic curves for various sample sizes can help to arrive at a decision on the sample size required to reflect the relative importance of both risks. It is especially important to consider β as well as α, lest meaningless results be obtained. (Note that randomizing so as to reject H_0 100α percent of the time yields a test with level of significance α. That in itself, without consideration of β, is trivial.)

Relation to Confidence Intervals. Confidence limits provide a set of limits within which a population parameter lies (with specified probability). Tests of hypotheses evaluate a specific statement about a population parameter. These procedures are related, and most hypothesis tests also can be made using confidence limit calculations.

Example: A sample of 12 insulators has an average strength of 4.95 ft·lb (6.7149 N·m). The standard deviation of the population is known to be 0.25 ft·lb (0.34 N·m). It is desired to test the hypothesis that the population mean is 5.15 ft·lb (6.9834N·m).

Solution using tests of hypotheses: Table 44.31 defines the test statistic 1a and $U = (\overline{X} - \mu_0)/(\sigma/\sqrt{n})$, and U is normally distributed. If $\alpha = 0.05$, the acceptance region is a U between ± 1.96. Then

$$H_0: \quad \mu = \mu_0 = 5.15$$

$$H_1: \quad \mu \neq \mu_0$$

$$U = \frac{4.95 - 5.15}{0.25/\sqrt{12}} = -2.75$$

Since the sample index is outside the acceptance region, the hypothesis is rejected. The procedure using confidence limits is:

1. State the hypothesis concerning the value of a population parameter.
2. Obtain a sample of data and calculate confidence limits for the population parameter.
3. If the hypothesis value falls within the confidence limits, accept the hypothesis. If the hypothesis value falls outside the confidence limits, reject the hypothesis.

Solution using confidence limits: From Table 44.22, parameter 1, the confidence limits are

$$\overline{X} \pm K_{\alpha/2} \frac{\sigma}{\sqrt{n}}$$

The 95 percent confidence limits are $4.95 \pm 1.96 \, (0.25/ \sqrt{12}) = 4.81$ and 5.09. As the hypothesis value falls outside the confidence limits, the hypothesis is rejected. This is the same conclusion reached by using the hypothesis testing procedure.

Confidence limit concepts and tests of hypotheses are therefore alternative approaches to evaluating a hypothesis. [For certain hypotheses, these two approaches will result in slightly different type I errors (see Barr 1969).] As discussed under Drawing Conclusions from Tests of Hypotheses, confidence limits are a valuable supplement to the test of hypothesis procedure. For example, in the preceding example, not only do confidence limits tell us that μ is not 5.15, they also tell us that μ is between 4.81 and 5.09.

Standard Cases.   Some of the most important practical cases have been covered in Table 44.31, namely:

Binomial proportion	Tests 6, 8
Two binomial proportions	Test 7
Normal mean	Test 1
Two means	Test 2
Bivariate normal mean	Test 3
One normal standard deviation	Test 4
Two standard deviations	Test 5
Two distributions are equal	Test 9
Random order of observations	Test 10
Test for outliers	Test 11

Many of these problems also can be solved using sequential tests. There, the sample size is not set in advance, but based on the data we decide how many observations are needed. For example, if one decided to inspect a fixed number of items such as 100 items and reject the lot if 15 or more defectives were found, clearly one could stop sampling as soon as the fifteenth defective were found. Similar ideas allow savings in numbers of observations in most of the standard cases listed above and are especially important when sampling is costly or time-consuming. For details, see Govindarajulu (1981).

Paired versus Unpaired Data.   In Test 2*e* of Table 44.31 a test is given that is appropriate when data are taken in pairs and the difference within each pair is used as the basic data. This procedure is often used in order to "wash out" the effects of variables that are believed to have effects but whose effects we do not wish to study.

For example, suppose that there is an effect of the operator of the machine, and we wish to compare two types of operation on that machine—but do not wish to evaluate the size of the operator effect. Then by letting each operator perform both operations and taking the difference, we wash out the effect of the operator. (Whereas if one operator performed all of one procedure and another operator performed all of the other procedure, differences observed might be due not to the procedures but to the operators.)

As another example, in testing of mailing lists to evaluate competing advertising copy, often an "A/B split" is used. That is, one type of copy goes to names 1, 3, 5, 7,…on the list, while the other type goes to names 2, 4, 6, …. Since (on ZIP-code-ordered lists) adjacent listings may be expected to be more similar than entries far apart, this is an appropriate pairing.

This technique of pairing is *not* used when there is no advance pairing of the data. For example, it is an error to pair items by their sequence in a data listing (where often they may be sorted by some other characteristic).

Statistical Significance versus Practical Significance.   Suppose we are using Test 1*a* of Table 44.31 to test the hypothesis that the mean is 30, and wish a two-tailed test with level of significance 0.05. Then we will reject H if U is outside the interval $(-1.96, 1.96)$. If we find $U = 3.15$, one way of reporting the result of the test is to state that H was rejected at level 0.05.

Another way of reporting is to find for which level of significance the acceptance interval would be $(-3.15, 3.15)$. From Table B in Appendix II we see that level is $(2)(0.00082) = 0.00164$. This is called the *significance probability* of the test we have conducted. It is the smallest level of significance at which we would reject H for the data we observed.

One advantage of the significance probability is that if we use it, then we can report the significance probability as 0.00164 without choosing a level of significance. Anyone reading our report can use the level of significance he or she believes is appropriate: If theirs is less than 0.00164, they do not reject, while if theirs is equal to or greater than 0.00164, they do reject the hypothesis.

One disadvantage of the significance probability is that it can be very small (indicating, one would think, a "very significant" result) even when the true mean is close to 30. For example, $U = 3.15$ when $\overline{X} = 33.15$ and $\sigma/\sqrt{n} = 1.00$; $U = 3.15$ also when $\overline{X} = 30.0315$ and $\sigma/\sqrt{n} = 0.01$. In each case the significance probability is 0.00164 (i.e., 0.164 percent). In the first case, the confidence interval on the mean at 95 percent confidence runs over (31.19, 35.11), while in the second, the interval runs over (30.0119, 30.0511). In terms of *practical significance,* the latter is much more likely to be a trivial difference to the practitioner than is the former. However, there is no way to tell these two situations apart by using the significance probability. For this reason, it is recommended that instead confidence intervals be computed and presented.

ADDITIONAL STATISTICAL TOOLS

Today, a large number of statistical tools are used in quality control. The statistical tool kit (see Table 44.1) stresses the statistical base of collection, analysis, and interpretation of data. *Transformations,* discussed below, are a method often used to ensure that data will meet the assumptions of statistical procedures, while *Monte Carlo sampling methods* and *clustering and discrimination procedures* are powerful methods whose use in quality control is now growing. They allow analysis with minimal assumptions and analysis of multivariate characteristics, respectively. *Bootstrap methods* are a relatively recent attempt to simplify modeling and analysis; in their *generalized bootstrap* form they achieve this with minimal drawbacks. *Selection of the best* is an alternative goal (versus hypothesis testing or confidence intervals) that should be used when one's experiment has a goal of selection.

Transformations of Data. Most of the statistical methodology presented in this section assumes that the quality characteristic follows a known probability distribution. The analysis and conclusions that result are, of course, strictly valid only to the extent that the distribution assumption is correct. Under Tests of Hypotheses, a "goodness-of-fit" test was presented for quantitatively evaluating a set of data to judge the validity of a distributional assumption. Moderate deviations of a sample of observations from a theoretical population assumption are to be expected because of sampling variation. The goodness-of-fit test determines whether the deviation of the sample from a theoretical assumption is likely to have been due to sampling variation. If it turns out as unlikely, then it is concluded that the assumption is wrong.

Sometimes a set of data does not fit one of the standard distributions such as the normal distribution. One approach uses "distribution-free" statistical methods for further analysis. Some of these were listed under Tests of Hypotheses, and Natrella (1963) and Gibbons (1997) present further material. However, these methods often require larger sample sizes than conventional methods for equivalent statistical risks. Some other approaches to analysis are

1. Examine the data to see if there is a nonstatistical explanation for the unusual distributional pattern. For example, the output of each of several supposedly identical machines may be normally distributed. If the machines have different means or standard deviations, then the combined output probably has an unusual distribution pattern such as the mixture distribution already discussed in this section. In this case, separate analyses could be made for each machine.

2. Analyze the data in terms of averages instead of individual values. As stated under Basic Concepts, Types of Error, sample averages closely follow a normal probability distribution even if the population of individual values from which the sample averages came is not normally distributed. If it is sufficient to draw a final conclusion on a characteristic in terms of the average value, the normal distribution assumption can be applied. However, the conclusions apply only to the average value and not to the individual values in the population. (Predicting the percentage of a population falling outside engineering limits illustrates the situation where analysis in terms of the average would not be sufficient because engineering limits refer to individual values rather than averages.)

3. Use the Weibull probability distribution. The Weibull distribution is really a group of many continuous distributions with each distribution uniquely defined by numerical values of the parameters of the Weibull probability function (e.g., a beta value of 1.0 indicates an exponential distribution). If a set of data yields an approximate straight-line plot on Weibull paper, the straight line then directly provides estimates of the probabilities for the population. Whether the exact form of the probability distribution is normal, or exponential, or another distribution becomes somewhat secondary because the straight-line plot provides the needed probability estimates.

4. Make a transformation of the original characteristic to a new characteristic that is normally distributed. Figure 44.23 summarizes several of these mathematical transformations. These transformations are useful for (*a*) achieving normality of measured results, (*b*) satisfying the assumption of equal population variances required in certain tests, and (*c*) satisfying the assumption of additivity of effects required in certain tests. Natrella (1963) discusses transformations for all these uses. Romeu and Ozturk (1996) provide graphic tests of normality (even in multivariate cases).

The most common transformations for achieving normality are

$$\xi_1(X_1) = \sqrt{X_1 - a}$$

$$\xi_2(X_1) = X_1^{1/3}$$

$$\xi_3(X_1) = \log_{10}(X_1)$$

$$\xi_4(X_1) = \arcsin \sqrt{X_1}$$

$$\xi_5(X_1) = \sinh^{-1} \sqrt{X_1}$$

If one of these, say, $\xi(X_1)$, is normally distributed, the mean and variance of $Y_i = \xi(X_i)$ may be estimated by

$$\bar{Y} = \sum_{j=1}^{n} \frac{Y_j}{n} \qquad S_Y^2 = \sum_{j=1}^{n} \frac{(Y_j - \bar{Y})^2}{(n-1)}$$

However, interest in many cases is not in the expected value $E\xi(X_1)$ and the variance $\text{Var}\,\xi(X_1)$ but in the original problem units EX_1 and $\text{Var}(X_1)$. Simply using the inverse transformation—for example, to estimate EX_1 by $\bar{Y}^2 + a$ in the case of ξ_1—results in a biased estimate. Good estimators for the mean of the X's are given in Table 44.35. Good estimators of the variance of the X's allow us to find approximate 95 percent confidence intervals for the mean of the X's; such estimates are given in Table 44.36. For example, when using $\sqrt{X_1 - a}$, a 95 percent confidence interval for the mean of the X's is

$$\bar{Y}^2 + a + \left(1 - \frac{1}{n}\right) s_Y^2 \pm 2\sqrt{\lambda}$$

where

$$\lambda = \frac{4}{n} s_Y^2 \bar{Y}^2 + s_Y^4 \left\{ \left(1 - \frac{1}{n}\right)^2 - \frac{n-1}{n+1}\left[1 - 2\left(1 - \frac{1}{n}\right)^2 + 3\left(1 - \frac{1}{n}\right)^4\right]\right\}$$

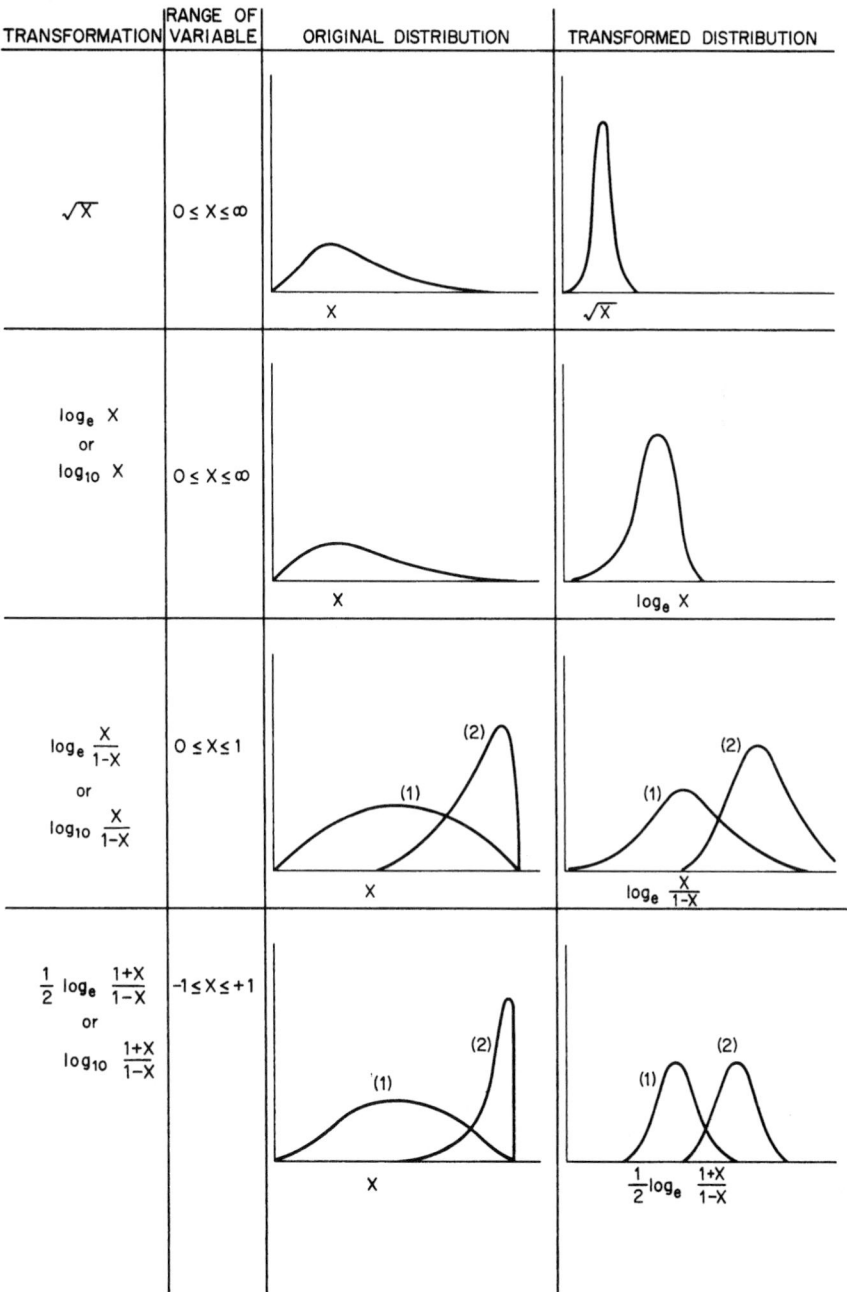

FIGURE 44.23   Summary of some transformations.

TABLE 44.35   Transformations and Estimators of $E(X_1)$†

$\xi(X_1)$	Estimator of $E(X_1)$
$\sqrt{X_1 - a}$	$\overline{Y}^2 + a + \left(1 - \dfrac{1}{n}\right) s_Y^2$

†For estimators in the cases $\log_{10}(X_1)$, arcsin $\sqrt{X_1}$, and $\sinh^{-1}\sqrt{X_1}$, see Dudewicz (1983).

TABLE 44.36   Estimators of Variances of Estimators of $E(X_1)$†

$\xi(X_1)$	Estimator of variance of estimator of $E(X_1)$
$\sqrt{X_1 - a}$	$\dfrac{4}{n} s_Y^2 \, \overline{Y}^2 + s_Y^4 \left\{ \left(1 - \dfrac{1}{n}\right)^2 - \dfrac{n-1}{n+1}\left[1 - 2\left(1 - \dfrac{1}{n}\right)^2 + 3\left(1 - \dfrac{1}{n}\right)^4\right]\right\}$

†For estimators in the cases $\log_{10}(X_1)$, arcsin $\sqrt{X_1}$, and $\sinh^{-1}\sqrt{X_1}$, see Dudewicz (1983).

Full references are given by Dudewicz (1983), who also covers procedures for dealing with cases where variances are unequal and data are normal. It is standard to recommend that equality of variability should be investigated, even when data are normal. Procedures for dealing with variance inequality when it is found had not been available until recent years. Recently, Dudewicz and Dalal (1983) showed how to compare several new processes with a standard process in this setting; their paper also includes consideration of nonnormality and a data set (their Section 5) with numerical details and normal probability plots for an example arising with solvents.

Monte Carlo Sampling Methods.   Monte Carlo sampling methods are finding increasing and important uses in quality control. For example, Gutt and Gruska (1977) use them to predict quality problems that may result from variation in manufacturing and assembly operations. This method is based on fitting distributions to data (such as the GLD distribution discussed above under Continuous Probability Distributions) and sampling from them using random numbers (also discussed previously), employing the resulting data to assess the performance of the simulated system (which allows optimization of the system before it is built or modified). Other uses occur in optimization (Golden et al. 1984), location modeling (Golden and Eiselt 1992), vehicle routing (Golden 1993), and inventory management (Dudewicz 1997). For example, a recent optimization method called ARSTI that uses random intervals (Edissonov 1994) compares very well with previous methods. Since one can choose to apply more than one method and then take the better result, one can only gain by incorporating such new methods into one's work [which is facilitated by the fact that a FORTRAN computer program for ARSTI is given by Edissonov (1994)].

Bootstrap Methods.   In many quality control problems, full solutions have been developed under the assumption that one knows the underlying probability distribution. Since often one does not have this knowledge in practice, one needs to estimate the probability distribution (see Selecting a Discrete Distribution and Selecting a Continuous Distribution earlier in this section). In the *bootstrap method,* one does not estimate the probability distribution from a set of data like the tensile strengths of five batches of rubber made with recipe 1 used in the illustration of testing average strength: 3067, 2730, 2840, 2913, and 2789. That is, one does not test (for example) normality of the data and (if the test fails to reject) use procedures that assume a normal probability distribution. Instead, the bootstrap method takes samples at random with replacement from the data we have and uses them to try to answer the question of interest. If one is interested, for example, in a 90 percent confidence interval for the mean tensile strength with recipe 1, then one proceeds as follows: Draw samples of size $n = 5$ repeatedly from the basic data points until one has N such samples (with $N = 500$ being a popular choice—see note 1 below); for each sample, calculate the sample mean, thus obtaining $N = 500$ sample means, say, $\overline{X}_1, \overline{X}_2, \ldots, \overline{X}_{500}$; delete the smallest 25 (5 percent) and largest 25 (5 percent) of these 500 sample means [since $(.10)(500)/2 = 25$]; and state that the mean tensile strength is between the smallest and largest of the remaining 450 sample means.

The bootstrap method is simple and attractive; it seems to yield (without any complicated statistics or mathematics) solutions to difficult problems, i.e., to give us "something for nothing." Since there is no free lunch, we should be suspicious. In fact, the method can behave badly if the sample is not large (and most of the theory is developed as the sample size becomes infinite). For example, suppose that one desires to study the maximum rainfall over 100 years and has data on 5 years. In resampling those 5 years, one will never observe a higher rainfall than the largest of the 5 measurements with the bootstrap method. Clearly, the study will be greatly misled by this (e.g., we may recommend an inadequately sized dam or levee). Thus while the bootstrap method can be useful in some settings, it is fraught with danger. A more robust version (which yields about the same results when the sample sizes are large but is not so fragile when the sample sizes are small) is the generalized bootstrap method discussed below.

The bootstrap type of method was used as early as 1967 but did not gain wide acceptance until it was given the name *bootstrap method* by B. Efron in 1979, after which it experienced an explosion of interest. It in some ways generalizes the method given by Quenouille in 1949, which gained wide acceptance when christened the *jackknife method* by J. Tukey in 1958. History, references, and examples of its flaws (and how to remedy them) are given by Dudewicz (1992) [also see Section 15.6 of Dudewicz and Mishra (1988)].

The Generalized Bootstrap.

The generalized bootstrap was introduced by Dudewicz (1992) as a generalization of the bootstrap method with superior properties in the small-sample (few observations) case. [For a textbook discussion, see Section 6.6 of Karian and Dudewicz (1991).] Basically, with the generalized bootstrap, one takes the observations and fits an appropriate probability distribution from a broad class such as the extended generalized lambda distribution discussed earlier in this section. One's random samples are then taken from the fitted distribution. Thus, in the context of the five rubber batches discussed under Bootstrap Methods earlier, one fits an EGLD to the $n = 5$ data points. Then $N = 500$ random samples of size 5 are drawn from the fitted EGLD (not from the basic 5 data points), and analysis proceeds as in the bootstrap method. This method has been shown to do better when one has few data points (but to do as well when one has many) in recent studies [e.g., Sun and Dietland-Müller (1996) have an excellent exposition with real-data examples].

Bootstrap methods, especially in the form of the generalized bootstrap, would suggest that one fit a model to the data (e.g., a Poisson or other model) and then (assuming the fit passes testing; see Test of Model Validity above) use that model in one's analysis, which can proceed by bootstrapping. (The generalized bootstrap method allows for there being a possibility one could observe three or more blemishes sometime in the future, while the bootstrap method—which is not recommended—always assumes the probability of three or more is zero just because we did not observe three or more in any of the set of data we have.)

Note 1:   The statistical literature recommends the use of at least $N = 200$ replications. Recent work by Sun and Müller-Schwarze (1996, pp. 482–483) suggests that considerable gains in accuracy can be had by requiring at least $N = 500$ replications. The number of replications made with one's data set should not be confused with one's actual data set's size (see Note 2).

Note 2:   Bootstrap methods are widely used for sample sizes that are quite small [such as 9 in Sun and Müller-Schwarze (1996)] due to the need to draw reliable conclusions from small data sets (often gathered over a considerable period at considerable effort, such as 10 years for the Sun and Müller-Schwarze data). I would strongly recommend the generalized bootstrap for cases with fewer than 100 data points.

Clustering and Discrimination.   *Clustering and discrimination* methods are a part of the area of statistics called *multivariate analysis* (Siotani et al. 1985). A typical type of problem where these methods are used in quality control is when several different kinds of malfunctions within a production facility cause product to fall outside engineering limits. It is often difficult to determine the causes of the malfunction in any one case. Then clustering a number of cases may reveal causal links via common factors over the clusters. (That is, this method allows one to ask, "What do the cases with malfunctions of each type have in common?")

As an example of the power of the "discrimination" method, Fisher (1936) gave the data excerpted in Table 44.37. [A convenient source of the full data set is Dixon (1983, p. 520).]This consists of two length and two width measurements on each of three distinct varieties that might be found in the same location. We wish to know: How well can the varieties (which can be classified by a more involved analysis without error) be classified by just use of the two length and two width measurements? After these data are entered into a computer, program 7M of the BMDP set of programs (Dixon 1983) may be used to answer this question. The program code is given in Figure 44.24. From the resulting output, of key interest are the so-called *canonical variables,* which are the linear combinations of L_1, L_2, W_1, and W_2 that best discriminate among the three groups. In this example, these turn out to be

$$V_1 = 2.10510 + 0.82938L_1 + 1.53447W_1 - 2.20121L_2 - 2.81046W_2$$

$$V_2 = -6.66147 + 0.02410L_1 + 2.16452W_1 - 0.93192L_2 + 2.83919W_2$$

A plot of the (V_1, V_2) values for the 150 data points is given in Figure 44.25, with $T = 1, 2, 3$ cases labeled *A, B, C,* respectively, and shows the excellent results obtained. These results are deemed excellent because they allow us to classify a future observation into the correct group with high probability of being correct. For example, if we find $V_1 = 7.20$ and $V_2 = 1.00$, we are virtually certain that group *A* is involved. (This plot is produced by program 7M.)

Heteroscedastic Discrimination.   Traditional discrimination methods have assumed that one knows that the variances are equal and have not been able to specify that the misclassification proba-

TABLE 44.37   Two Length (L_1, L_2) and Two Width (W_1, W_2) Measurements on 150 Individuals (50 from Each of Types $T = 1, 2, 3$), in 0.01-cm Units*

L_1	W_1	L_2	W_2	T
50	33	14	02	1
64	28	56	22	3
65	28	46	15	2
67	31	56	24	3
63	28	51	15	3
46	34	14	03	1
69	31	51	23	3
62	22	45	15	2
59	32	48	18	2
46	36	10	02	1
61	30	46	14	2
60	27	51	16	2
65	30	52	20	3
56	25	39	11	2
65	30	55	18	3
⋮	⋮	⋮	⋮	⋮

*To save space, only the first 15 individuals' measurements are shown.

```
//   EXEC    BIMED, PROG =BMDP7M
/problem title is 'fisher data'.
/input variables are 5. format is '(4f3.1, f3.0)'.
/variable names are L1, w1, L2, w2, t.
grouping is t.
/group codes(5) are 1 to 3.
names(5) are set, ver, vir.
/end
```

FIGURE 44.24   BMDP 7M program for dataset analysis.

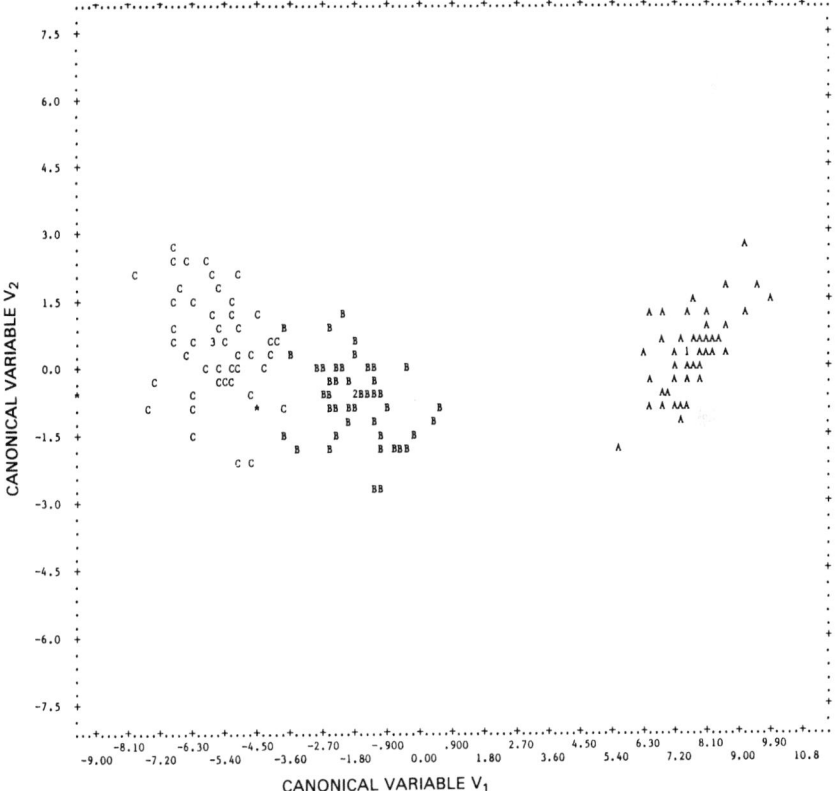

FIGURE 44.25   Plot showing groups (*A, B, C*) discriminated by "canonical variables" V_1 and V_2. Overlap of different groups is indicated by *.

bility will be estimated within .01 (say). However, owing to development of the heteroscedastic method [see Dudewicz (1995) for a review], this is now possible in some cases and expected to become more available in the next few years. In other cases, the goal may be to estimate the overlap (rather than directly discriminate the populations); for some methods here, see Mulekar and Mishra (1997).

Selection of the Best versus Testing Hypotheses.   In hypothesis testing, one may (e.g., in the setting of Test 2 in Table 44.31) assess whether one can reject the assertion that the means of two populations are equal. While this is an appropriate question in some settings, there are other settings where one knows a priori that one must select one or the other of the two populations and wants to select the one with the larger mean (called the "best" population). Procedures for this setting were first published in 1954 by R. E. Bechhofer, and since that time, over 500 publications and several books have been devoted to the problem. [For a categorized list and reviews, see Dudewicz and Koo (1982).]

Example of Selection of the Best: Suppose we are considering two recipes for rubber and want to select the one with the larger tensile strength. We make $n_0 = 5$ batches with each recipe and test for tensile strength with the following results:

Recipe 1	Recipe 2
3067	3200
2730	2777
2840	2623
2913	3044
2789	2834

We desire to be at least 95 percent sure that the one we select has a true mean tensile strength no further than $\delta^* = 120$ units from the best one. We then calculate the $s_1^2 = 16{,}923.7$ and $s_2^2 = 51{,}713.3$ in the first samples and take more observations (since we do not know the variances in advance of the first samples, we are not in a position to know how many total samples we will need). The total sample sizes are to be

$$n_1 = \max[n_0 + 1, (s_1 h/\delta^*)^2] = \max\{5 + 1, [(130.09)(3.11)/120]^2\}$$

$$= \max(6, 11.4) = 12$$

$$n_2 = \max[n_0 + 1, (s_2 h/\delta^*)^2] = max\{5 + 1, [(227.41)(3.11)/120]^2\}$$

$$= \max(6, 34.7) = 35$$

where $h = h_{n_0}(k, P^*) = h_5(2, 0.95)$ comes from Table 44.30 (with $k = 2$ because we are seeking the best of two populations, $P^* = 0.95$ because we desire 95 percent certainty, and $n_0 = 5$ because the first samples were of five observations). We then take $n_1 - 5 = 12 - 5 = 7$ more observations from recipe 1 and $n_2 - 5 = 35 - 5 = 30$ more observations from recipe 2. We then compute the sample mean $\overline{X}_1$ of all 12 observations from recipe 1 and the sample mean $\overline{X}_2$ of all 35 observations from recipe 2, selecting the recipe that produces the larger of $\overline{X}_1$ and $\overline{X}_2$, asserting that that recipe has the larger mean (or a mean no further than 120 units of tensile strength from the best recipe). For example, if we find $\overline{X}_1 = 3067$ and $\overline{X}_2 = 2895$, we will select recipe 1. We will state that we are at least 95 percent sure that recipe 1 is either the best recipe or (in any case) has a mean tensile strength no further than 120 strength units from the best (if it is not the best).

Procedures are also available for selection of the best of more than two populations [see Section 6.3 of Karian and Dudewicz (1991)]; one can essentially use the same procedure with the appropriate k being used when h is looked up in Table 44.30.

REGRESSION AND CORRELATION ANALYSIS

Many quality control problems require estimation of the relationship between two or more variables. Often interest centers on finding an equation relating one particular variable to a set of one or more variables. For example, how does the life of a tool vary with cutting speed? Or how does the octane number of a gasoline vary with its percentage purity?

Regression analysis is a statistical technique for estimating the parameters of an equation relating a particular variable to a set of variables. (Some authors refer to this as *least squares* or *curve fitting.*) The resulting equation is called a *regression equation.*

Some experimental data for the tool life example are given in Table 44.38 [from Johnson and Leone (1964, p. 380)] and plotted in Figure 44.26. Tool life is the *response variable* (also called the *dependent variable* or the *predictand*), and cutting speed is the *independent variable* (also called the *predictor variable*). In this case, the independent variable is controllable; i.e., it is fixed by the experimenter or the operator of the machine. In the second example, both the octane number and the percentage purity are random. The data for this example (from Volk 1956) are given in Table 44.39 and plotted in Figure 44.27. Since the goal is to predict the octane number, it is regarded as the dependent variable, and the percentage purity is considered as the independent variable. (In many problems there are a number of independent variables, and in some cases this set of independent variables includes both random and controllable variables.)

The computations for two-variable regression problems can be done quite easily on a calculator, but when there are many variables, the number of computations becomes overwhelming. (Even in the two-variable case, many of the computer programs available fail to provide numerically accurate calculations. It is strongly recommended that one *not* write one's own regression program and that only major tested software packages such as SAS, BMDP, and the like be utilized.) With modern digital computer multiple regression programs, the number of variables is not a restriction. To under-

TABLE 44.38   Tool Life (Y in Minutes) versus Cutting Speed (X in Feet per Minute)

Y	X
41	90
43	90
35	90
32	90
22	100
35	100
29	100
18	100
21	105
13	105
18	105
20	105
15	110
11	110
6	110
10	110

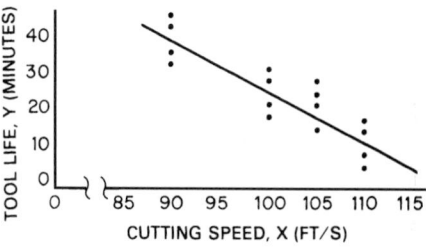

FIGURE 44.26   Tool life Y versus cutting speed X.

TABLE 44.39 Octane Number Y versus Percentage Purity X

Y	X
88.6	99.8
86.4	99.7
87.2	99.6
88.4	99.5
87.2	99.4
86.8	99.3
86.1	99.2
87.3	99.1
86.4	99.0
86.6	98.9
87.1	98.8

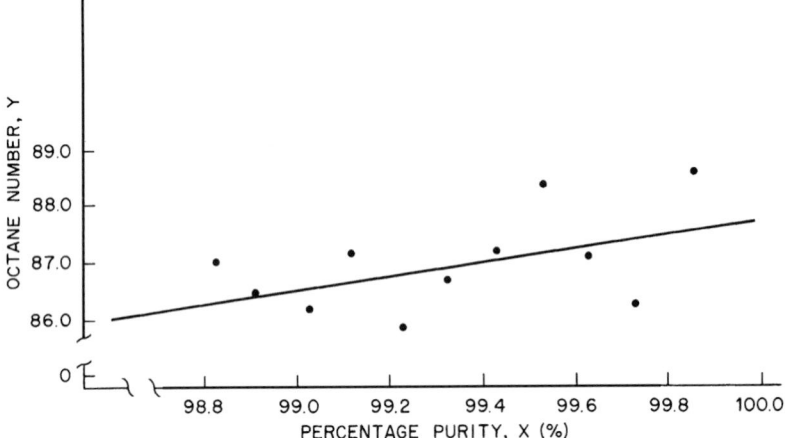

FIGURE 44.27 Octane number versus percentage purity.

stand and interpret the results of multidimensional problems, a thorough knowledge of the two- and three-variable cases is necessary.

There are many reasons for constructing regression equations. Although the motives do not affect the calculations, they do affect the interpretation of the results. In some cases, regression analysis is used to describe the nature of a relationship in a quantitative manner. Often the *goals* are more specific. In the first example, where the cutting speed is controllable, the objective might be to find the particular value of cutting speed which minimizes tool wear or some cost function based on tool wear. Least squares regression also can be used to determine the important independent variables in a process, e.g., whether process variables such as moisture, pressure, or temperature affect a quality characteristic of the product such as strength.

In other problems, where the independent variable is not controllable, the goal may be to predict the value of the dependent or response variable. This might be done because the independent variable is easier to measure than the dependent variable. Or the independent variable may be available before the dependent variable, and hence it would be desirable to forecast the value of the dependent variable before it occurs. In still other cases it might allow a destructive test to be replaced by a nondestructive test.

The following list includes a number of *uses of regression equations:*

1. Forecasting and prediction
2. Quantitatively describing the relationship between a particular variable and another set of variables
3. Interpolating between values of a function
4. Determining the important independent variables
5. Locating the optimum operating conditions
6. Discriminating between alternative models
7. Estimating particular regression coefficients

For any of these goals stated, the *basic steps in a regression study* are those of the checklist for planned experimentation given at the beginning of this section. A summary, specifically relating the steps to regression, is

1. Obtain a clear statement of the objectives of the study. Determine which variable is to be the response variable and which variables can be included as independent variables. In addition, obtain some measure of the precision of the results required—not necessarily in statistical terminology. (It is important to have a thorough understanding of what use will be made of the regression equation, since this may preclude the use of certain variables in the equation and will also help to give an understanding as to how much effort and money should be devoted to the project.)

2, 3. Specify collection procedures for the data. Collect the data. (The end results can only be as good as the data on which they are based. Careful planning at this stage is of considerable importance and can also simplify the analysis of the data.)

4. Prepare crossplots (plots of one variable versus another) of the data to obtain information about the relationships between the variables; screen the data; calculate the regression equation; and evaluate how well it fits the data (including looking at transformations of variables for a better fit, or the removal of variables from an equation if they do not improve the prediction). Give measures of the precision of the equation and any procedure for using the equation. Also specify procedures for updating the equation and checks to determine whether it is still applicable, including control charts for the residuals (observed value-predicted value). (Section 24, Statistical Process Control, discusses control charts.)

5, 6, 7. As in the checklist for planned experimentation.

A number of texts have been written on regression, including Daniel and Wood (1971) and Draper and Smith (1981). These include computer programs and output. Dudewicz and Karian (1985) also discuss design questions in detail. In addition to regression, other techniques have been devised for the analysis of multivariate data; see Kramer and Jensen (1969, 1970) and Siotani et al. (1985) for details. (One of these techniques is discriminant and cluster analysis, already discussed in this section.)

Our discussion of regression begins with a single predictor problem and then proceeds to problems with more than one predictor variable and a discussion of computer programs and outputs, with their interpretation. While many texts emphasize advanced mathematical aspects of regression, this is not needed for a practical understanding now that high-quality software is available; hence we find no need for such mathematics. This makes this important subject accessible to most quality practitioners.

Simple Linear Regression. Many problems involve only a single predictor variable X. (The dependent variable Y is often related to other predictor variables, which have either been held constant during the experiment or their effects judged to be much smaller than that of X.) These problems are often referred to as ones of *simple linear regression.*

Graphing the Data. A first step in any study of relationships between variables is to plot a graph of the data (often called a *scatter diagram*). The convention is to plot the response variable on the vertical axis and the independent variable on the horizontal axis. A graph can provide a great deal of information concerning the relationship between variables and often suggests possible models for the data. The data plotted in Figure 44.26 suggest that Y is linearly related to X over the range of this experiment. (If this were not the case, various transformations of the data as well as curvilinear relationships could also be considered. Often the relationship can be "linearized" by taking the logarithm of one or both of the variables.)

A graph also can indicate whether any of the observations are outliers, i.e., observations that deviate substantially from the rest of the data. (Outliers may be due to measurement errors or recording errors, in which case they should be corrected or deleted. They may be due to process changes or other causes, and the investigation of these changes or causes may provide more information than the analysis of the rest of the data.) No outliers are apparent in Figure 44.26.

A closer inspection of the graph can give an indication of the variability of Y for fixed X. In addition, it may show that this variability remains constant over all X or that it changes with X. In the latter case the method of weighted least squares (see Draper and Smith 1981) may be preferred to the standard least squares technique discussed here.

The Model.   After graphing the data, we want to obtain an equation relating Y to X. To do this, a model for the data must be postulated. (I emphasize that the proposed model may be modified during the course of the study and is just a starting point.)

A possible model for the data in Figure 44.26 is

$$Y = \beta_0 + \beta_1 X + \epsilon$$

where β_0 and β_1 are the unknown intercept and slope, respectively, of the regression line. The model assumes that Y is a linear function of X plus a random error term, denoted by ϵ. This random error may be due to errors in the measurement of Y and/or to the effects of variables not included in the model, which is called *equation error.* The X's are assumed to be measured with negligible error. For the data in Figure 44.26, the X's are fixed; however, the same model can be used when the X's are random as in Figure 44.27.

Estimating the Prediction Equation.   The objective is to find estimates (b_0, b_1) of the unknown parameters (β_0, β_1) and thus obtain a prediction equation

$$\hat{Y} = b_0 + b_1 X$$

where $\hat{Y}$ is the predicted value of Y for a given value of X.

Least squares provides a method for finding estimates of these parameters from a set of N observations $(Y_1, X_1), \ldots, (Y_N, X_N)$. The estimates are called *least squares estimates* because they minimize the sum of the squared deviations between the observed and predicted values of the response variable $\Sigma(Y_m - \hat{Y}_m)^2 = \Sigma(Y_m - b_0 - b_1 X_m)^2$. These ideas are illustrated in Figure 44.28. (For a mathematical derivation of the estimates, see any of the texts mentioned in the introduction.)

If (1) the observations are independent, (2) the variance of the errors is constant over these observations, and (3) the linear model postulated is correct, the least squares estimates are the "best linear unbiased estimates": In the class of linear unbiased estimates of the parameters, the least squares estimates have the smallest variance. (Even if these conditions are not satisfied, the least squares technique can be used, although modifications or other methods may provide better estimates.) Note that no assumption has been made concerning the distribution of the random error, and in particular a normal distribution is not assumed. No assumption on this error term will be required until confidence intervals and tests of hypotheses are constructed.

The least squares estimates for the parameters of the linear model $Y = \beta_0 + \beta_1 X + \epsilon$ are

$$b_1 = \frac{\Sigma(X_m - \overline{X})(Y_m - \overline{Y})}{\Sigma(X_m - \overline{X})^2}$$

$$b_0 = \overline{Y} - b_1 \overline{X}$$

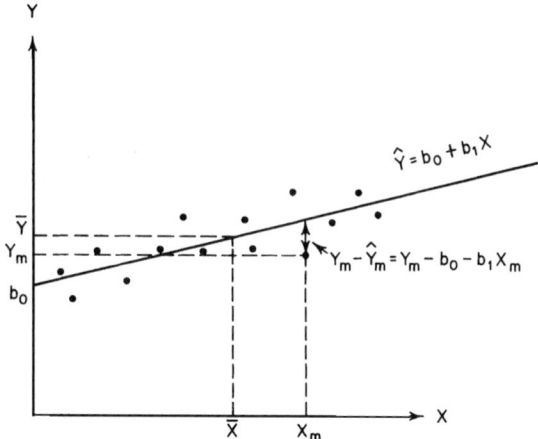

FIGURE 44.28   Least squares.

where $\overline{X} = \sum X_m/N$ and $\overline{Y} = \sum Y_m/N$ are sample averages. All these summations range from $m = 1$ to $m = N$. (Except where needed, the additional notation will be omitted for typographic simplicity.)

As can be seen, b_1 is related to the sample correlation coefficient

$$r = \frac{\sum(X_m - \overline{X})(Y_m - \overline{Y})}{\sqrt{\sum(X_m - \overline{X})^2 \sum(Y_m - \overline{Y})^2}}$$

by

$$b_1 = r\sqrt{\frac{\sum(Y_m - \overline{Y})^2}{\sum(X_m - \overline{X})^2}}$$

However, the concept of correlation is meaningful only when both the variables are random, whereas b_1, the least squares estimate of the rate of change of Y per unit change in X, has meaning for both the case of random X and controllable or fixed X.

The sums, sums of squares, and sum of crossproducts for the data ($N = 16$) given in Table 44.38 are

$$\sum X_m = 90 + 90 + \ldots + 110 = 1620$$

$$\sum Y_m = 41 + 43 + \ldots + 10 = 369$$

$$\sum X_m^2 = 8100 + 8100 + \ldots + 12,000 = 164,900$$

$$\sum Y_m^2 = 1681 + 1849 + \ldots + 100 = 10,469$$

$$\sum X_m Y_m = 3690 + 3870 + \ldots + 1100 = 36,170$$

(Note that calculations for regression are very susceptible to both human and numerical error because of their complexity. Hence good software should be used. The numerical examples in this section can be used both to test one's understanding of that software and to test its accuracy.)

The summary statistics are computed using the following computational formulas:

$$\overline{X} = \sum X_m/N = 101.25 \qquad \overline{Y} = \sum Y_m/N = 23.06$$

$$\sum(X_m - \overline{X})^2 = \sum X_m^2 - \frac{(\sum X_m)^2}{N} = 164{,}900 - \frac{(1{,}620)^2}{16} = 875.00$$

$$\sum(Y_m - \overline{Y})^2 = \sum Y_m^2 - \frac{(\sum Y_m)^2}{N} = 10{,}469 - \frac{(369)^2}{16} = 1958.94$$

$$\sum(X_m - \overline{X})(Y_m - \overline{Y}) = \sum X_m Y_m - \frac{\sum X_m \sum Y_m}{N}$$

$$= 36{,}170 - \frac{(1620)(369)}{16}$$

$$= -1191.25$$

From these results the least squares estimates can be calculated as

$$b_1 = \frac{-1191.25}{875} = -1.3614$$

$$b_0 = 23.06 - (-1.3614)(101.25) = 160.9018$$

and hence the prediction equation is

$$\hat{Y} = 160.90 - 1.3614X$$

The prediction equation is sometimes written in terms of deviations from averages, i.e., $\hat{Y} = \overline{Y} + b_1(X - \overline{X})$, which for this example becomes

$$\hat{Y} = 23.06 - 1.3614(X - 101.25)$$

Examining the Prediction Equation.   After estimating the coefficients of the prediction equation, the equation should be plotted over the data to check for gross calculation errors. Roughly half the data points should be above the line and half below it. In addition, the equation should pass exactly through the point $(\overline{X}, \overline{Y})$.

A number of criteria exist for judging the adequacy of the prediction equation. One common measure of the adequacy of the prediction equation is the proportion of variation R^2 explained. To compute R^2, the sum of the squared deviations of the Y_m about $\overline{Y}$ is partitioned into two parts, the sum of squares due to regression and the residual sum of squares:

$$\sum(Y_m - \overline{Y})^2 = SS(REG) + SS(RES)$$

$$= \sum(\hat{Y}_m - \overline{Y})^2 + \sum(Y_m - \hat{Y}_m)^2$$

$$= b_1\sum(X_m - \overline{X})(Y_m - \overline{Y}) + \sum(Y_m - \hat{Y}_m)^2$$

From this, the proportion of the variation $\sum(Y_m - \overline{Y})^2$ explained by the regression is computed as

$$R^2 = \frac{SS(REG)}{\sum(Y_m - \overline{Y})^2}$$

$$= \frac{b_1\sum(X_m - \overline{X})(Y_m - \overline{Y})}{\sum(Y_m - \overline{Y})^2}$$

$$= \frac{(-1.3614)(-1191.25)}{1958.94} = 0.828$$

Thus in this example the prediction equation explains 82.8 percent of the variation of the tool life.

Another interpretation of R^2 (when both the independent and the dependent variables are random) is as the square of the sample multiple correlation coefficient. When there is only one independent variable, this reduces to the square of the sample correlation coefficient r defined earlier.

Although R^2 is a useful measure of the adequacy of the prediction equation, an estimate of the variability of the Y's about the regression equation is usually more important. Either the sample variance s_e^2 or its square root s_e, called the *standard error of the estimate,* can be used. The latter is often preferred because it is measured in the same units as Y. Both of these, as well as other results, can be obtained from the analysis of variance (ANOVA) given in Table 44.40.

The corrected total sum of squares and the regression sum of squares are calculated from the summary statistics and the estimate of the regression coefficient [some authors include the total sum of squares, uncorrected, in the ANOVA table, partitioning it into two parts—the corrected sum of squares and the sum of squares due to $\overline{Y}$ (or b_0) (see Draper and Smith 1981)]. Although the residual sum of squares can be calculated directly, it is more easily obtained as the difference between the corrected total sum of squares and the sum of squares due to regression. Each of these sums of squares has an associated degrees of freedom (see Testing a Hypothesis When the Sample Size Is Fixed in Advance). The corrected total sum of squares has $N - 1$ degrees of freedom, since one degree of freedom is used in estimating the mean. For this *one*-variable model, there is *one* degree of freedom associated with the regression sum of squares, leaving $(N - 1) - 1 = N - 2$ degrees of freedom associated with the residual sum of squares. The mean squares (MS) are calculated by dividing the sum of squares by their associated degrees of freedom. The estimate of the variance of Y about the regression line is $s_e^2 = $ MS(RES); hence the standard error of the estimate is $s_e = \sqrt{\text{MS(RES)}}$.

From the mean squares an F statistic can be calculated as

$$F_{\text{CALC}} = \frac{\text{MS(REG)}}{\text{MS(RES)}}$$

If (1) the ϵ's in the original model are normally distributed with a common variance, (2) the observations are independent, and (3) the postulated linear model is correct, then the regression can be tested for significance, i.e., the statistical hypothesis

$$H_0: \quad \beta_1 = 0$$

can be tested against the alternative hypothesis

$$H_1: \quad \beta_1 \neq 0$$

by comparing F_{CALC} with the tabulated F_{TAB} at an appropriate level of significance α. If $F_{\text{CALC}} > F_{\text{TAB}}$, we conclude that the regression is significant and that the prediction equation is a better predictor of Y than $\overline{Y}$. Although it is difficult to check the assumptions stated above, the test is not extremely sen-

TABLE 44.40 ANOVA Table (Linear Model)

Source	Sum of squares	Degrees of freedom	Mean square
1. Due to regression (b_1)	$b_1\Sigma(X_m - \overline{X})(Y_m - \overline{Y})$	1	MS (REG) = SS(REG)/1
2. Residual	$\Sigma(Y_m - \hat{Y}_m)^2$†	$N - 2$	MS (RES) = SS(RES)/$(N - 2)$
3. Total corrected for the mean	$\Sigma(Y_m - \overline{Y})^2$	$N - 1$	

†Obtained by subtracting (1) from (3).

sitive to departures in the distribution of ϵ from normality if the number of observations is relatively large. If the X's are random, this test must be interpreted in a conditional sense, i.e., given the values of the X's.

For the example, the analysis of variance (ANOVA) table is given in Table 44.41. (See Section 26, under Completely Randomized Design: One Factor, k Levels.) The regression is significant at an $\alpha = 0.01$ level ($F_{TAB} = 8.86$) and $s_e = \sqrt{24.08} = 4.91$.

It is important to note that even when the regression is significant, the unexplained variability can still be large, and the prediction equation may not be of any value.

Residuals, Outliers, Confidence and Prediction Bands, Extrapolation; Lack of Fit—Replicated Observations.   If it is feasible to replicate, i.e., take more than one observation of Y at one or more values of X, the adequacy of the model also can be tested. (An estimate of the pure error sometimes may be available from sources outside the immediate experiment.) In this case, the SS(RES) can be partitioned into two parts—that due to pure error, SS(PE), and that due to lack of fit, SS(LF).

Suppose that there are N_m readings $Y_{m1}, Y_{m2}, \ldots, Y_{mN_m}$ at x_m, where $m = 1, 2, \ldots, k$. The contribution to the sum of squares due to pure error for X_m is

$$\sum_{j=1}^{N_m} (\hat{Y}_{mj} - \overline{Y}_m)^2 = \sum_{j=1}^{N_m} Y_{mj}^2 - \frac{\left(\sum_{j=1}^{N_m} Y_{mj} \right)^2}{N_m}$$

and the associated degrees of freedom is $N_m - 1$. The SS(PE) is just the sum of these k contributions, and the associated degrees of freedom (DF) is

$$\sum_{m=1}^{k} (N_m - 1) = \sum_{m=1}^{k} N_m - k$$

For the example given in Table 44.38,

X_m		SS(PE)	DF
90	$41^2 + 43^2 + 35^2 + 32^2 - (151)^2/4 =$	78.75	3
100	$22^2 + 35^2 + 29^2 + 18^2 - (104)^2/4 =$	170.00	3
105	$21^2 + 13^2 + 18^2 + 20^2 - (72)^2/4 =$	38.00	3
110	$15^2 + 16^2 + 6^2 + 10^2 - (42)^2/4 =$	41.00	3
	Total	327.75	12

The SS(LF) is found by subtraction as

$$\text{SS(LF)} = \text{SS(RES)} - \text{SS(PE)} = 337.14 - 327.75 = 9.39$$

TABLE 44.41   ANOVA Example

Source	Sum of squares	Degrees of freedom	Mean square	F_{CALC}
1. Due to regression	1621.80	1	1621.80	67.35
2. Residual	337.14	14	24.08	
3. Total corrected for the mean	1958.94	15		

and the lack of fit degrees of freedom is obtained in a similar manner as $14 - 12 = 2$. The mean squares are then found by dividing the sum of squares by the appropriate degrees of freedom and

$$F_{\text{CALC}} = \frac{\text{MS(LF)}}{\text{MS(PE)}}$$

If the F_{CALC} is greater than the tabled F, the lack of fit is "significant," and a better or more complete model is needed (e.g., $Y = \beta_0 + \beta_1 X + \beta_2 X^2$). Plots of the residuals ($Y_m - \hat{Y}_m$) versus X_m are particularly helpful in suggesting alternative models. Some examples are given in Figure 44.29. [Daniel and Wood (1971, pp. 19–24) present graphs of a number of nonlinear functions and give transformations that "linearize" them.] In each case the model $Y = \beta_0 + \beta_1 X$ was postulated and the plots are of the resulting residuals.

If F_{CALC} is less than the tabled F, the model is accepted. This does not mean that other variables should not be considered in the model, but only that the form of X in the model is adequate.

The calculations for our example are summarized in Table 44.42. The lack of fit is judged not significant at an α level of 0.05 ($F_{\text{TAB}} = 3.89$). Hence the postulated model is accepted, and the residual mean square is used as the estimate of the variance.

If replication is not possible, e.g., X is random rather than controllable, the Y values corresponding to X values that are close together can be used to obtain an estimate of the variability and hence judge the lack of fit. [See pp. 123–125 of Daniel and Wood (1971).]

Confidence Intervals.   Both R^2 and s_e^2 provide measures of the reliability or adequacy of a prediction equation. Confidence intervals provide another measure of the reliability of the various

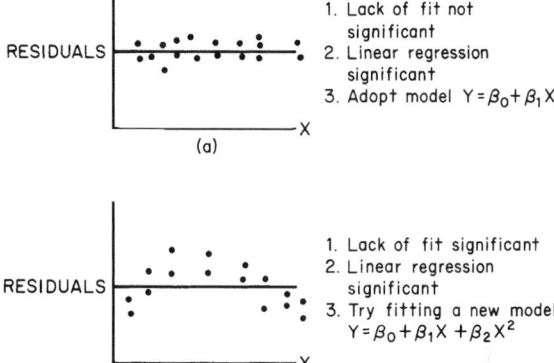

FIGURE 44.29   Residuals and lack of fit.

TABLE 44.42   ANOVA Lack of Fit Example

Source	Sum of squares	Degrees of freedom	Mean square	F_{CALC}
Lack of fit	9.39*	2	4.695	0.172
Pure error	327.75	12	27.3125	

*Obtained by subtracting SS(PE) and SS(RES).

estimates. All these confidence intervals are based on the square root of the residual mean square. A $(1 - \alpha)$ two-sided confidence interval for the slope β_1 is given by

$$b_1 \pm \frac{ts_e}{\sqrt{\Sigma(X_m - \overline{X})^2}}$$

where the value t is obtained from Appendix II, Table G, with $N-2$ degrees of freedom. The term in the denominator plays the role that $n^{1/2}$ plays in confidence intervals on population means. For the example, the 0.95 confidence interval on β_1 is

$$-1.36 \pm \frac{(2.145)(4.91)}{\sqrt{875}} = -1.36 \pm 0.356$$

The term $s_e/\sqrt{\Sigma(X_m - \overline{X})^2}$ is often called the *standard error of the regression coefficient.*

In addition to the confidence interval on β_1, more importantly, *confidence intervals* also can be constructed *for the mean of Y at a given value of X.* The $(1 - \alpha)$ confidence interval on the mean of Y at X (or equivalently on $\beta_0 + \beta_1 X$) is

$$b_0 + b_1 X \pm ts_e \sqrt{\frac{1}{N} + \frac{(X - \overline{X})^2}{\Sigma(X_m - \overline{X})^2}}$$

where X is the value at which the confidence interval is being constructed and t again has $N - 2$ degrees of freedom. (By letting $X=0$, a confidence level for β_0 is obtained.)

In addition to the assumptions previously stated, these confidence intervals also require (1) that the independent variable is fixed rather than random and (2) that the errors are normally distributed. However, if the X's are random, confidence intervals can still be calculated, but they must be interpreted in a conditional sense. Confidence intervals are not sensitive to departures from normality if the sample size is reasonably large. This is not the case for the following interval, which is very sensitive to the normality assumption. *Least absolute value* (LAV) and *Chebyshey estimation* are two possible alternatives to least squares estimation, which are less sensitive to model departures than is least squares. For sources of efficient computer algorithms, with a detailed numerical example, see Dielman and Pfaffenberger (1984).

In addition to a confidence interval on the expected value of Y at a given X, there may be a need for *an interval estimate for a future individual observation on Y at X.* [A more complete discussion of confidence intervals is given in Draper and Smith (1981). See Daniel and Wood (1971) for a confidence interval that simultaneously includes the whole line. See Dudewicz (1976, p. 427) for a plot of the interval for all X, called a *prediction band,* and its uses.] In this case the interval also must take into account the variability of Y about $\beta_0 + \beta_1 X$, and the result is

$$b_0 + b_1 X \pm ts_e \sqrt{1 + \frac{1}{N} + \frac{(X - \overline{X})^2}{\Sigma(X_m - \overline{X})^2}}$$

where t has $N - 2$ degrees of freedom. Computations of these intervals for various values of X are given in Table 44.43.

Multiple Regression.   Although there are many problems involving single predictor variables, more often there are many predictor variables. A generalization of the least squares technique, previously discussed, can be used to estimate the coefficients of the multivariable prediction equation. This problem is called *multiple regression.*

The General Model.   For a problem with k predictor variables, the model can be written as

$$Y = \beta_0 + \beta_1 X_1 + \ldots + \beta_k X_k + \epsilon$$

TABLE 44.43 Computation of Confidence Limits

X (1)	$b_0 + b_1 X$ (2)	$\dfrac{(X - \overline{X})^2}{\Sigma(\overline{X}_m - \overline{X})^2}$ (3)	$\sqrt{\dfrac{1}{N} + (3)}$ (4)	$ts \times (4)$ (5)	90% confidence limits on the mean of Y at X		$\sqrt{1 + \dfrac{1}{N} + (3)}$ (8)	$ts \times (8)$ (9)	90% prediction limits on Y at X	
					Lower (2) − (5) (6)	Upper (2) + (5) (7)			Lower (2) − (9) (10)	Upper (2) + (9) (11)
90	38.37	0.14464	0.4551	3.94	34.43	42.31	1.0987	9.50	28.87	47.87
100	24.76	0.00179	0.2535	2.20	22.56	26.96	1.0316	8.92	15.84	33.68
($\overline{X}$) 101.5	($\overline{Y}$) 23.06	0.0	0.2550	2.17	20.89	25.23	1.0308	8.91	14.15	31.97
110	11.15	0.08750	0.3870	3.36	7.79	14.51	1.0724	9.27	1.88	20.42

Note: Numbers in parentheses are column numbers. $t_{14,0.95} = 1.761$.

where the β's are unknown parameters and ϵ is the random error. These variables may be transformations of the original data. For example, in predicting gasoline yields from data on the specific gravity and vapor pressure of crude oil, Y may be the log of the gasoline yield, X_1 the crude oil specific gravity, X_2 the crude oil vapor pressure, and X_3 the product of the crude oil specific gravity with its vapor pressure.

The general model includes polynomial models in one or more variables such as

$$Y = \beta_0 + \beta_1 X_1 + \beta_2 X_2 + \beta_3 X_1^2 + \beta_4 X_2^2 + \beta_5 X_1 X_2 + \epsilon$$

[which is called the *full quadratic model* of Y on X_1 and X_2 and is of great use in designed experiments (see Dudewicz and Karian 1985)]. This is still a linear model, since the term *linear model* means that the model is linear in the β's. [See Draper and Smith (1981) for a discussion of models which are nonlinear.]

Estimating the Prediction Equation.   The objective now is to find the least squares estimates $(b_0, b_1,..., b_k)$ of the unknown parameters $(\beta_0, \beta_1,..., \beta_k)$ and obtain a prediction equation

$$\hat{Y} = b_0 + b_1 X_1 + ... + b_k X_k$$

where $\hat{Y}$ is the predicted value of Y for the given values of $X_1,..., X_k$. Letting $x_i = X_i - \bar{X}_i$ and using the fact that

$$b_0 = \bar{Y} - b_1 \bar{X}_1 - ... - b_k \bar{X}_k$$

this prediction equation can be expressed in the alternative form

$$\hat{Y} = \bar{Y} + b_1 x_1 + ... + b_k x_k$$

To simplify the formulas, the observations also can be expressed as deviations from their sample averages; i.e., for the mth observation $x_{im} = X_{im} - \bar{X}_i$ and $y_m = Y_m - \bar{Y}$. Then the least squares estimates of the $k + 1$ parameters of the multivariable linear model $Y = \beta_0 + \beta_1 X_1 + ... + \beta_k X_k$ can be obtained by solving the set of $k + 1$ linear equations:

$$b_1 \sum x_{1m}^2 + b_2 \sum x_{1m} x_{2m} + ... + b_k \sum x_{1m} x_{km} = \sum x_{1m} y_m$$
$$b_1 \sum x_{1m} x_{2m} + b_2 \sum x_{2m}^2 + ... + b_k \sum x_{2m} x_{km} = \sum x_{2m} y_m$$
$$\vdots \qquad\qquad\qquad\qquad\qquad\qquad \vdots$$
$$b_1 \sum x_{1m} x_{km} + b_2 \sum x_{2m} x_{km} + ... + b_k \sum x_{km}^2 = \sum x_{km} y_m$$
$$b_0 = \bar{Y} - b_1 \bar{X}_1 - b_2 \bar{X}_2 - ... - b_k \bar{X}_k$$

(All the above summations are on m and range from 1 to N.)

Solving these "reduced normal" equations simultaneously can be tedious, is error prone, and involves matrix algebra. Many reference works emphasize how to perform these calculations accurately. To the user of modern accurate statistical software, these calculations are of no direct importance: That user can trust that they are being done accurately and concentrate on statistical aspects of model adequacy, interpretation, and use.

Examining the Prediction Equation.   After obtaining $(b_0, b_1,..., b_k)$, an ANOVA table can be constructed and the adequacy of the prediction equation evaluated by a number of criteria. The ANOVA table, which is a generalization of that derived for the single predictor variable, is given in Table 44.44. The third row in Table 44.44 is the same as in Table 44.40. Note that the expressions in the first row reduce to those in the first row of Table 44.40 when $k=1$.

Since there are k variables in the model, the sum of squares due to regression has k degrees of freedom associated with it. In addition, since k coefficients and one intercept have been estimated, the residual sum of squares has $N - (k + 1) = N - k - 1$ degrees of freedom associated with it. The F

TABLE 44.44   ANOVA Table (Linear Model)

Source	Sum of squares	Degrees of freedom	Mean square
1. Due to regression	$b'a$†	k	MS(REG) = SS(REG)/k
2. Residual	$\Sigma(\overline{Y}_m - \hat{Y}_m)^2$‡	$N - k - 1$	MS(RES) = SS(RES)/ $(N - k - 1)$
3. Total corrected for the mean	$\Sigma(Y_m - \overline{Y}_m)^2$	$N - 1$	

†The sum of squares due to regression can be written as $b'a = b_1\Sigma x_{1m}y_m + b_2\Sigma x_{2m}y_m + \cdots + b_k\Sigma x_{km}y_m$. [Note that b' is the transpose of the b vector, i.e., $b' = (b_1\ b_2 \cdots b_k)$. Those not familiar with vector notation can regard $b'a$ as a shorthand for the sum given in this footnote. Since computations are usually done with computer software, details of vector notation are not given and will not be needed by most readers.]
‡Obtained by subtracting (1) from (3).

statistic is calculated as before—i.e., F = MS(REG)/MS(RES)—and $s_e = \sqrt{SS(RES)/(N-k-1)}$. In this case, the F statistic can be used to test the statistical hypothesis

$$H_0: \quad \beta_i = 0 \qquad (i = 1, 2,..., k)$$

against the alternative statistical hypothesis

$$H_1: \quad \text{Some } \beta_i \neq 0 \qquad (i = 1, 2,..., k)$$

The only change from simple linear regression is in the degrees of freedom used to look up the F_{TAB} ($k, n - k - 1$ versus $1, n - 2$).

The proportion of variation explained by the equation (R^2) can be obtained from the ANOVA table. Note that R^2 does not depend on the number of variables in the equation, but s_e^2 does. In fact, if a new variable is added to the model (and the least squares estimates and ANOVA table are recomputed), the value of R^2 *cannot decrease*. However, s_e^2 can either increase or decrease, since it depends on the residual degrees of freedom in addition to the residual sum of squares, which decreases by one when a new variable is added.

Confidence Intervals.   Confidence intervals for individual β's can be developed as

$$b_i \pm ts_e \sqrt{c_{ii}}$$

where t has $(N - k - 1)$ degrees of freedom, and c_{ii} is defined below. However, since the b_i ($i = 1,..., k$) have a joint distribution and are in general not uncorrelated, care must be taken in the interpretation of sets of these confidence intervals [see, for example, Draper and Smith (1981)].

More usefully, a confidence interval on the regression equation at a point $x = (x_1,..., x_k)$, where $x_i = X_i - \overline{X}_i$ is given by

$$\overline{Y} + b_1x_1 + ... + b_kx_k \pm ts_e \left(\frac{1}{N} + x'Cx\right)^{1/2}$$

where t has $N - k - 1$ degrees of freedom, and $x'Cx$ is a quadratic form that takes into account the covariances and variances of the b's. (Here C is as defined in the next paragraph.)

In a similar manner, an interval for a future Y at X is given by

$$\overline{Y} + b_1x_1 + ... + b_kx_k \pm ts_e \left(1 + \frac{1}{N} + x'Cx\right)^{1/2}$$

where t has $N - k - 1$ degrees of freedom. (Here $C = S^{-1}$, where S is the $k \times k$ matrix whose entry in row i and column j is $\Sigma x_{im}x_{jm}$. In practice, as we will see, these matrix calculations are done by the computer, and the user need not bother with them—or even understand the concept of a *matrix* and its *inverse.*)

An example: The methods discussed can now be illustrated by an example furnished by Mason E. Wescott (example 5 from *Mimeo Notes,* Mason E. Wescott, Rochester Institute of Technology, Rochester, NY) with $k = 2$ predictor variables. The problem is to relate the green strength (flexural strength before baking) of electric circuit breaker arc chutes to the hydraulic pressure used in forming them and the acid concentration. The data are given in Table 44.45, with hydraulic pressure and green strength given in units of 10 lb/in^2 and the acid concentration given as a percentage of the nominal rate for 20 observations. Two-variable plots of the data are given in Figures 44.30, 44.31, and 44.32. Summary statistics including sums, sums of squares, and crossproducts, both raw and corrected, as well as the sample means, are given in Table 44.46.

The estimates are

$$b_2 = 4.162940$$

$$b_1 = 1.571779$$

$$b_0 = \bar{Y} - b_1\bar{X}_1 - b_2\bar{X}_2 = 16.27475$$

The C matrix is

$$c_{22} = 0.00048916$$

$$c_{12}(=c_{21}) = -0.000058099$$

$$c_{11} = 0.00029787$$

TABLE 44.45   Arc Chute Data

Green strength Y in units of 10 lb/in^2	Hydraulic pressure X_1 in units of 10 lb/in^2	Acid concentration X_2, as % of nominal rate
665	110	116
618	119	104
620	138	94
578	130	86
682	143	110
594	133	87
722	147	114
700	142	106
681	125	107
695	135	106
664	152	98
548	118	86
620	155	87
595	128	96
740	146	120
670	132	108
640	130	104
590	112	91
570	113	92
640	120	100

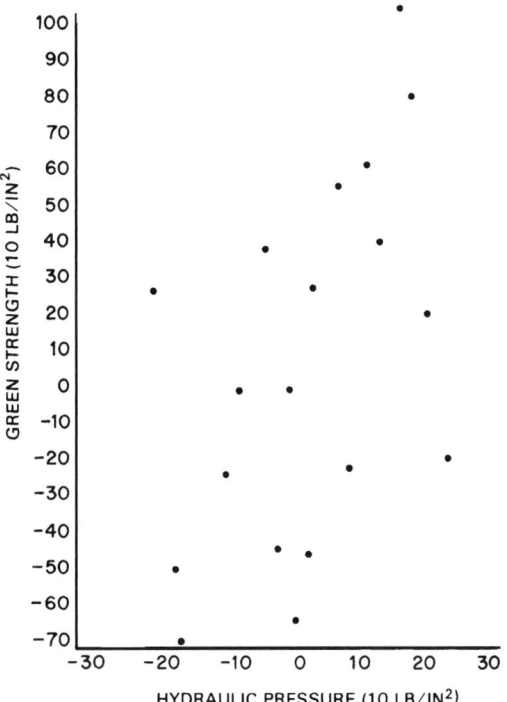

FIGURE 44.30   Green strength versus hydraulic pressure
(deviations from averages).

and hence

$$C = \begin{vmatrix} 297.869 & -58.099 \\ -58.099 & 489.160 \end{vmatrix} \times 10^{-6}$$

The prediction equations can be calculated as

$$\hat{Y} = 16.277 + 1.572X_1 + 4.163X_2$$

$$\hat{Y} = 641.600 + 1.572(X_1 - 131.400) + 4.163(X_2 - 100.600)$$

(The latter form will be used in the remaining computations.)

The ANOVA table (Table 44.47) for the example follows directly from the summary statistics of Table 44.46 and from Table 44.44. The residual mean square error is 228.0, and hence the standard deviation s_e is 15.100.

The 95 percent confidence intervals on β_1 and β_2 are obtained as

$$1.572 \pm 2.110 \times 15.100 \times \sqrt{297.869} \times 10^{-3}$$

$$= 1.572 \pm 0.550 = 1.02 \text{ and } 2.12$$

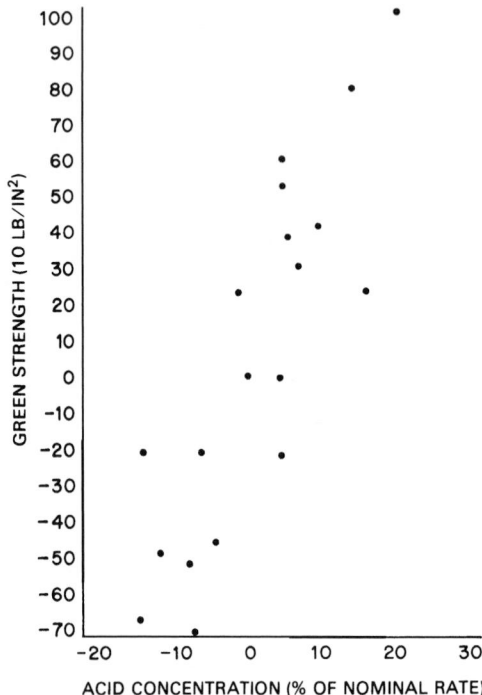

FIGURE 44.31   Green strength versus acid concentration (deviations from averages).

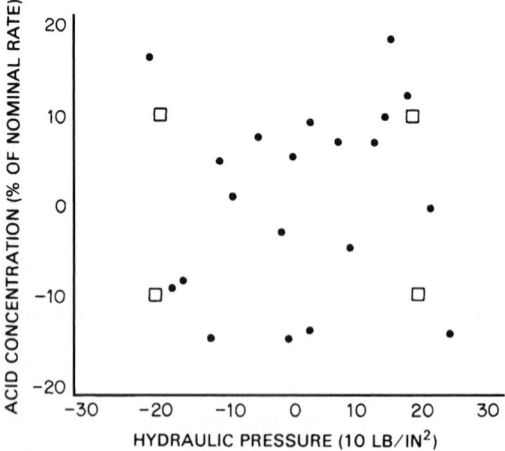

FIGURE 44.32   Acid concentration versus hydraulic pressure (deviations from averages). Rectangles indicate points at which confidence intervals have been calculated.

TABLE 44.46 Summary Statistics

$N = 20$		
$\Sigma Y_m = 12{,}832$	$\Sigma X_{1m} = 2628$	$\Sigma X_{2m} = 2012$
$\Sigma Y_m^2 = 8{,}286{,}988$	$\Sigma X_{1m}^2 = 348{,}756$	$\Sigma X_{2m}^2 = 204{,}500$
$\Sigma X_{1m}Y_m = 1{,}693{,}226$	$\Sigma X_{2m}Y_m = 1{,}300{,}253$	$\Sigma X_{1m}X_{2m} = 264{,}785$
$\overline{Y} = 641.6$	$\overline{X}_1 = 131.4$	$\overline{X}_2 = 100.6$

S		a
$s_{11} = 3436.8$	$s_{12} = 408.2*$	$a_1 = 7{,}101.2\dagger$
$s_{21} = 408.2$	$s_{22} = 2092.8$	$a_2 = 9{,}353.8$

$*s_{12} = \Sigma x_{1m}x_{2m} = \Sigma X_{1m}X_{2m} - (\Sigma X_{1m}\Sigma X_{2m}/N) = 264{,}785 - [(2628)(2012)/20] = 408.2.$
$\dagger a_1 = \Sigma x_{1m}y_m = \Sigma X_{1m}Y_m - (\Sigma X_{1m}\Sigma Y_m/N) = 1{,}693{,}226 - [(2628)(12{,}832)/20] = 7101.2.$

TABLE 44.47 ANOVA Example

Source	Sum of squares	Degrees of freedom	Mean square
1. Due to regression	50,100.8*	2	25,050.40
2. Residual	3,876.0†	17	228.0
3. Total corrected for the mean	53,976.8	19	

$*(1.571779)(7101.2) + (4.16294)(9353.8) = 50{,}100.825.$
†By subtraction.

and

$$4.163 \pm 2.110 \times 15.0997 \times \sqrt{489.16} \times 10^{-3}$$

$$= 4.163 \pm 0.705 = 3.46 \text{ and } 4.87$$

Confidence intervals for $\beta_0 + \beta_1 X_1 + \beta_2 X_2$ and for a future observation of Y, at five combinations of X_1 and X_2, are given in Table 44.48. (The five points at which the confidence intervals are computed are also indicated in Figure 44.32.)

An additional excellent example is provided in Golden and Wasil (1992, pp. 227–245), where 37,000 observations gathered from 34 stations in the Chesapeake Bay are used to develop 10 regression models for salinity dynamics. Data quality, model building, model results, and model validation are all discussed.

Computer Programs. Because of the widespread popularity of regression, almost every computer facility has at least one and most have many regression programs. These programs may have been locally written, or they may have been obtained from other sources. In the latter case, the program usually has been modified in some manner so that it can be run on the local computer system and satisfy the needs of the local users.

Studies by Longley (1967) and Wampler (1970) indicate that the user should *strongly* prefer software such as that of SAS and BMDP. One should not presume that the program has been checked just because a sample data problem is given in the program manual. Unfortunately, a number of the algorithms used in these programs are often taken directly from desk calculator instructions. These algorithms are often not good and can produce numerically inaccurate results, even in

TABLE 44.48 Computation of Confidence Intervals

X_1 (1)	X_2 (2)	$\hat{Y}$ (3)	$(x'Cx)$ (4)	$\sqrt{\dfrac{1}{N}+(4)}$ (5)	$ts \times (5)$† (6)	95% confidence limits on the mean of Y at X_1,X_2		$\sqrt{1+\dfrac{1}{N}+(4)}$ (9)	$ts \times (9)$ (10)	95% prediction limits on Y at X_1,X_2	
						Lower (3) − (6) (7)	Upper (3) + (6) (8)			Lower (3) − (10) (11)	Upper (3) + (10) (12)
111.4	90.6	568.53	0.1457*	0.4424	14.10	554.43	582.63	1.0935	34.84	533.69	603.37
111.4	110.6	651.79	0.1922	0.4921	15.68	636.11	667.47	1.1145	35.51	616.28	687.30
131.4	100.6	641.60	0.0	0.2236	7.12	634.48	648.72	1.0247	32.65	608.95	674.25
151.4	90.6	631.41	0.1922	0.4921	15.68	615.73	647.09	1.1145	35.51	595.90	666.92
151.4	110.6	714.67	0.1457	0.4424	14.10	700.57	728.77	1.0935	34.85	679.82	749.52

Note: Numbers in parentheses are column numbers.

*Since $x' = (111.4 - 131.4, 90.6 - 100.6) = (-20, -10)$, then

$$x'Cx = (-20,-10)\begin{bmatrix} 297.869 & -58.099 \\ -58.099 & 498.160 \end{bmatrix} \times 10^{-6} \begin{bmatrix} -20 \\ -10 \end{bmatrix} = (-5,376.39,\ -3,819.62) \times 10^{-6} \begin{bmatrix} -20 \\ -10 \end{bmatrix} = 0.1457.$$

†t_{TAB} with 17 degrees of freedom is 2.110, $s = 15.100$.

double precision. For example, Longley found that many of the programs computed the squared deviations of a variable about its mean by the computational formula $\Sigma X_m^2 - (\Sigma X_m)^2/N$, rather than by $\Sigma(X_m - \bar{X})^2$. Since these quantities are the base for most of the regression calculations, numerical errors may be present in all the results.

Since the user may not be able (or simply does not want to invest the time) to check a program, a few simple checks for this purpose are given below. These can be used to gain some idea of the limitations of a program; however, I again strongly recommend against "rolling your own" when SAS, BMDP, and (perhaps) other excellent software is available at reasonable prices and is kept up to date with advances in statistics.

1. Add the residuals about the regression line. They should sum to zero, within rounding error.

2. If the residuals sum to zero, make additional runs of the problem after adding 10, 100, 1000, 10,000, etc. to each variable. The coefficients will begin to change at a point where round-off error occurs.

3. As a check on the accuracy of the inversion routine, run a problem with two variables X_1 and X_2. Then make another run with the same response variable but two new independent variables $X_1^* = X_1 + X_2$ and $X_2^* = X_1 - X_2$. The following results should hold: $b_1 = b_1^* + b_2^*$ and $b_2 = b_1^* - b_2^*$.

Longley (1967) gives some additional checks, and Wampler (1970) lists results on many regression programs.

A write-up is usually available with regression programs; it should include a complete discussion of the *input* required and an explanation of the *output* and *options* available, as well as a complete statement of the calculation formulas and a sample problem. (Although documenting a program is a difficult task, poorly written documentation is often a warning of a poorly or improperly written computer program.)

While the input formats of regression programs vary, most programs have an option that allows the user to specify a variety of transformations of the data, such as logs, powers, and crossproducts. Typical regression outputs include ANOVA tables, residuals plots, and other statistics, in addition to the estimates of the regression coefficients. Although often omitted, an *echo-check,* i.e., a printout of the original and transformed data, is essential. Often "strange" regression results can be traced to a misplaced decimal point in an observation, the wrong variables being read in, or incorrect use of the transformation option. (If this is not available in a program, ask the computer center to modify the program so that it is automatically printed out unless the user deletes it.)

R^2 and C_p Criteria for Model Choice.

In many situations there are a large number of possible variables for a model, and the problem is to select the "vital few" from these "useful many," instead of obtaining the complete regression equation. There are many reasons for not using all the variables. For example, a subset of variables can provide a better prediction equation than the full set, even though the full set has a higher R, since the full set also will include more variability. In addition, equations with fewer variables are easier to understand and hence more likely to gain acceptance and be used.

Unless the data come from a properly designed experiment, there is no simple test for significant variables. Since there are $2^k - 1$ possible prediction equations to evaluate, for large k it is obvious that a brute-force approach is not feasible (e.g., if $k = 20$, then $2^k - 1 = 1,048,575$).

Stepwise regression is a heuristic technique for avoiding this computational problem. It begins by selecting the single independent variable that is the "best" predictor in the sense that it maximizes R^2. Then it adds variables to the equation in a sequential manner, in order of importance. At each step the variable added is the one that increases the regression sum of squares (and hence R^2) or equivalently reduces the residual sum of squares by the largest amount. This procedure not only selects variables but deletes variables previously selected, if at some point they no longer appear important.

Stepwise regression does not guarantee that the "best" set of variables will be included in the final equation. However, it does provide an efficient method for reducing the number of variables k to a manageable size; e.g., if $k = 100$, then stepwise regression can be used to select the best 25 or fewer for more exhaustive study. Stepwise regression programs are widely available; see Draper and Smith (1981).

In addition to stepwise regression, numerous other techniques have been developed. One that seems to have great potential was developed by Hocking and Leslie (1967) and improved by LaMotte and Hocking (1970). Their algorithm finds the "best" subset of variables of size 1, 2,..., k, where k is the total number of variables submitted. Although the computations require more time than the stepwise procedure, this procedure guarantees the best subset in the R^2 sense and in addition gives a number of the "contending" subsets. (Of course, if the number of variables is small, we could compute all regressions.) These "best" regression algorithms are available in BMDP and can handle k up to 20 or 25 without computer time problems.

The set of possible variables should be selected on the basis of preliminary investigations of the factors that influence the response variables. The indiscriminate use of regression analysis to "find" relationships, where no facts suggest the existence of a relationship, often leads to nonsensical results. Unfortunately, this is usually discovered after the prediction equation fails miserably in predicting future observations.

If a large amount of data is available, one portion of it can be used for selecting variables and estimating coefficients, saving the remainder of the data for testing the derived equations. In any case, the equation should be periodically reviewed as new data become available.

A powerful tool in modern regression analysis is the C_p statistic [see, e.g., Dudewicz and Karian (1985, pp. 236, 413)]. While R^2 measures the goodness of the regression equation in predicting the data points in the data set one is using to develop a model, C_p estimates the variance of future predictions made using the model. While adding a variable will increase the R^2 (even if that variable is totally unrelated to what we are trying to predict), C_p typically decreases as variables are added to the model, then increases. Thus, searching for the minimal C_p statistic in all possible regressions is a reasonable approach—though if a large gain in R^2 can be obtained with a modest increase in C_p, it should be taken.

REFERENCES

ASQC (1996). "QA/QC Software Directory." *Quality Progress,* April, pp. 31–59.

Barr, Donald R. (1969). "Using Confidence Intervals to Test Hypotheses." *Journal of Quality Technology,* vol. 1, no. 4, pp. 256–258.

Bennett, Carl A., and Franklin, Norman L. (1954). *Statistical Analysis in Chemistry and the Chemical Industry.* John Wiley & Sons, New York, chap. 11.

Bingham, R. S., Jr. (1962). "Tolerance Limits and Process Capability Studies." *Industrial Quality Control,* vol. XIX, no. 1, pp. 36–40.

Choksi, Suresh (1971). "Computer Can Optimize Manufacturing Tolerances for an Economical Design." *ASQC Technical Conference Transactions.* Milwaukee, pp. 323–330.

Cohen, A. Clifford, and Whitten, Betty Jones (1981). "Estimation of Lognormal Distributions." *American Journal of Mathematical and Management Sciences,* vol. 1, no. 2, pp. 139–153.

Daniel, Cuthbert (1977). Personal communication, August.

———— and Wood, Fred S. (1971). *Fitting Equations to Data.* John Wiley & Sons, New York.

Dielman, Terry E., and Pfaffenberger, Roger C. (1984). "Computational Algorithms for Calculating Least Absolute Value and Chebyshev Estimates for Multiple Regression." *American Journal of Mathematical and Management Sciences,* vol. 4, nos. 1 and 2, pp. 169–197.

Dixon, W. J. (1983). *BMDP Statistical Software, 1983 Printing with Additions.* University of California Press, Berkeley, CA.

———— and Massey, Frank J., Jr. (1969). *Introduction to Statistical Analysis,* 3rd ed. McGraw-Hill, New York.

Dodge, H. F., Kinsburg, B. J., and Kruger, M. K. (1953). "The L3 Coaxial System—Quality Control Requirements." *Bell System Technical Journal,* vol. 32, pp. 943–1005.

Draper, N. R., and Smith, H. (1981). *Applied Regression Analysis,* 2nd ed. John Wiley & Sons, New York.

Dudewicz, Edward J. (1976). *Introduction to Statistics and Probability.* Holt, Rinehart and Winston, New York. (Currently published by American Sciences Press, Columbus, OH. Address: P.O. Box 21161, Columbus 43221-0161.)

——— (1983). "Heteroscedasticity." In N. L. Johnson, S. Kotz, and C. B. Read (eds.), *Encyclopedia of Statistical Sciences,* vol. 3. John Wiley & Sons, New York, pp. 611–619.

——— (1992). "The Generalized Bootstrap." In Jöckel, K.-H., Rothe, G., and Sendler, W. (eds.), *Bootstrapping and Related Techniques,* vol. 376 of Lecture Notes in Economics and Mathematical Systems. Springer-Verlag, Berlin, pp. 31–37.

——— (1993). *Solutions in Statistics and Probability.* 2nd ed. American Sciences Press, Columbus, OH.

——— (1995). "The Heteroscedastic Method: Fifty+ Years of Progress 1945–2000." In Hayakawa, T., Aoshima, M., and Shimizu, K. (eds.), *MSI-2000: Multivariate Statistical Analysis,* vol. I. American Sciences Press, Columbus, OH, pp. 179–197.

——— (ed.) (1997). *Modern Digital Simulation Methodology, III: Advances in Theory, Application, and Design—Electric Power Systems, Spare Parts Inventory, Purchase Interval and Incidence Modeling, Automobile Insurance Bonus-Malus Systems, Genetic Algorithms—DNA Sequence Assembly, Education, & Water Resources Case Studies.* American Sciences Press, Columbus, OH.

——— and Dalal, Siddhartha R. (1983). "Multiple-Comparisons with a Control when Variances Are Unknown and Unequal." *American Journal of Mathematical and Management Sciences,* vol. 3, no. 4, pp. 275–295.

——— and Karian, Zaven A. (1985). *Modern Design and Analysis of Discrete-Event Computer Simulations.* IEEE Computer Society Press, Los Angeles, CA. (Address is P.O. Box 80452, Worldway Postal Center, Los Angeles 90080; Order No. CN597.)

——— and ——— (1996). "The EGLD (Extended Generalized Lambda Distribution) System for Fitting Distributions to Data with Moments: II. Tables." In Dudewicz, E. J. (ed.), *Modern Digital Simulation Methodology, II: Univariate and Bivariate Distribution Fitting, Bootstrap Methods, and Applications to CensusPES and CensusPlus of the U.S. Bureau of the Census, Bootstrap Sample Size, and Biology and Environment Case Studies.* American Sciences Press, Columbus, OH, pp. 271–332.

——— and Koo, J. O. (1982). *The Complete Categorized Guide to Statistical Selection and Ranking Procedures.* American Sciences Press, Columbus, OH.

——— and Mishra, S. N. (1988). *Modern Mathematical Statistics.* John Wiley & Sons, New York.

——— and Ralley, Thomas G. (1981). *The Handbook of Random Number Generation and Testing with TES-TRAND Computer Code.* American Sciences Press, Columbus, OH.

———, Karian, Zaven A., and Marshall, Rudolph James, III (1985). "Random Number Generation on Microcomputers." *Proceedings of the Conference on Simulation on Microcomputers.* The Society for Computer Simulation (Simulation Councils, Inc.), La Jolla, CA. (Address: P.O. Box 2228, La Jolla 92038.)

Duncan, Acheson J. (1974). *Quality Control and Industrial Statistics,* 4th ed. Richard D. Irwin, Homewood, IL.

Edissonov, I. (1994). "The New ARSTI Optimization Method: Adaptive Random Search with Translating Intervals." *American Journal of Mathematical and Management Sciences,* vol. 14, pp. 143–166.

Epstein, Benjamin (1960). "Estimation from Life Test Data." *Technometrics,* vol. 2, no. 4, pp. 447–454.

Evans, Ralph A. (1966). "Problems in Probability." *Proceedings of the Annual Symposium on Reliability.* IEEE, New York, pp. 347–353.

Fisher, R. A. (1936). "The Use of Multiple Measurements in Taxonomic Problems." *Annals of Eugenics,* vol. 7, pp. 179–188.

Gibbons, Jean Dickinson (1997). *Nonparametric Methods for Quantitative Analysis,* 3rd ed. American Sciences Press, Columbus, OH.

Golden, Bruce L. (ed.) (1993). *Vehicle Routing 2000: Advances in Time-Windows, Optimality, Fast Bounds, and Multi-Depot Routing.* American Sciences Press, Columbus, OH.

——— and Eiselt, H. A. (eds.) (1992). *Location Modeling in Practice: Applications (Site Location, Oil Field Generators, Emergency Facilities, Postal Boxes), Theory, and History.* American Sciences Press, Columbus, OH.

——— and Wasil, Edward A. (eds.) (1992). *Fisheries: Control and Management via Application of Management Sciences (with a Comprehensive Annotated Bibliography of Applications).* American Sciences Press, Columbus, OH.

———, Assad, Arjang A., and Zanakis, Stelios H. (eds.) (1984). *Statistics and Optimization: The Interface.* American Sciences Press, Columbus, OH.

Govindarajulu, Z. (1981). *The Sequential Statistical Analysis of Hypothesis Testing: Point and Interval Estimation, and Decision Theory.* American Sciences Press, Columbus, OH.

Grant, E. L., and Leavenworth, R. S. (1980). *Statistical Quality Control,* 5th ed., McGraw-Hill, New York.

Gray, H. L., and Lewis, Truman (1967). "A Confidence Interval for the Availability Ratio." *Technometrics,* vol. 9, no. 3, pp. 465–471.

Grubbs, Frank E. (1969). "Procedures for Detecting Outlying Observations in Samples." *Technometrics,* vol. 11, no. 1, pp. 1–21.

Gutt, J. D., and Gruska, G. F. (1977). "Variation Simulation." *ASQC Technical Conference Transactions.* Milwaukee, pp. 557–563.

Hadley, G. (1967). *Introduction to Probability and Statistical Decision Theory.* Holden-Day, San Francisco.

Hahn, G. J. (1970*a*). "Additional Factors for Calculating Prediction Intervals for Samples from a Normal Distribution." *Journal of the American Statistical Association,* vol. 65, no. 332, pp. 1668–1676.

―――― (1970*b*). "Statistical Intervals for a Normal Population: 1. Tables, Examples, and Applications; II. Formulas, Assumptions, Some Derivations." *Journal of Quality Technology,* vol. 2, no. 3, pp. 115–125; vol. 2, no. 4, pp. 195–206.

―――― (1971). "How Abnormal is Normality?" *Journal of Quality Technology,* vol. 3, no. 1, pp. 18–22.

―――― (1972). "The Absolute Sample Size Is What Counts." *Quality Progress,* vol. V, no. 5, pp. 18–19.

―――― and Shapiro, S. S. (1967). *Statistical Models in Engineering.* John Wiley & Sons, New York.

Harter, H. Leon (1983). *The Chronological Annotated Bibliography of Order Statistics,* vol. II: *1950–1959.* American Sciences Press, Columbus, OH.

―――― and Dubey, Satya D. (1967). *Theory and Tables for Tests of Hypotheses Concerning the Mean and the Variance of a Weibull Population.* Document AD 653593, Clearinghouse for Federal Scientific and Technical Information, U.S. Department of Commerce, Washington.

Hocking, R. R., and Leslie, R. N. (1967). "Selection of the Best Subset in Regression Analysis." *Technometrics,* vol. 9, no. 4, pp. 531–540.

Hoerl, Arthur E., and Kennard, Robert W. (1981). "Ridge Regression—1980. Advances, Algorithms, and Applications." *American Journal of Mathematical and Management Sciences,* vol. 1, no. 1, pp. 5–83.

Huh, M.-H., and Olkin, I. (1995). "Asymptotic Aspects of Ordinary Ridge Regression." In Hayakawa, T., Aoshima, M., and Shimizu, K. (eds.), *MSI-2000: Multivariate Statistical Analysis,* vol. I. American Sciences Press, Columbus, OH, pp. 239–254.

Iman, Ronald L. (1982). "Graphs for Use with the Lillefors Test for Normal and Exponential Distributions." *The American Statistician,* vol. 36, no. 2, pp. 109–112.

Ireson, W. Grant (ed.) (1982). *Reliability Handbook.* McGraw-Hill, New York.

Johnson, Norman L., and Kotz, Samuel (1969). *Discrete Distributions.* Houghton Mifflin, Boston.

―――― and Leone, Fred C. (1964). *Statistics and Experimental Design,* vol. I. John Wiley & Sons, New York.

Juran, J. M., and Gryna, Frank M., Jr. (1980). *Quality Planning and Analysis,* 2nd ed. McGraw-Hill, New York.

Karian, Z., and Dudewicz, E. J. (1991). *Modern Statistical, Systems, and GPSS Simulation.* Computer Science Press/W.H. Freeman and Company, Book Publishing Division of Scientific American, New York. Currently published by American Sciences Press, Columbus, OH.

――――, Dudewicz, E. J., and McDonald, P. (1996). "The Extended Generalized Lambda Distribution System for Fitting Distributions to Data: History, Completion of Theory, Tables, Applications, the `Final Word' on Moment Fits." *Communications in Statistics: Simulation and Computation,* vol. 25, pp. 611–642.

―――― and Dudewicz, Edward J. (1999). *Fitting Statistical Distributions to Data: The Generalized Lambda Distribution (GLD) and the Generalized Bootstrap (GB) Methods.* CRC Press, Boca Raton, FL.

Kececioglu, D., and Cormier, D. (1964). "Designing a Specified Reliability into a Component." *Proceedings of the Third Annual Aerospace Reliability and Maintainability Conference.* Society of Automotive Engineers, Washington, p. 546.

King, James R. (1971). *Probability Charts for Decision Making.* The Industrial Press, New York.

Kirkpatrick, R. L. (1970). "Confidence Limits on a Percent Defective Characterized by Two Specification Limits." *Journal of Quality Technology,* vol. 2, no. 3, pp. 150–155.

Kramer, C. Y., and Jensen, D. R. (1969, 1970). "Fundamentals of Multivariate Analysis." *Journal of Quality Technology,* vol. 1, no. 2, pp. 120–133, no. 3, pp. 189–204, no. 4, pp. 264–276; vol. 2, no. 1, pp. 32–40.

LaMotte, L. R., and Hocking, R. R. (1970). "Computational Efficiency in the Selection of Regression Variables." *Technometrics,* vol. 12, no. 1, pp. 83–93.

Lenz, H.-J., and Rendtel, U. (1984). "Performance Evaluation of the MIL-STD-105 D, Skip-Lot Sampling Plans, and Bayesian Single Sampling Plans." In Lenz, H.-J., Wetherill, G. B., and Wilrich, P.-Th. (eds.), *Frontiers in Statistical Quality Control 2.* Physica-Verlag, Rudolf Liebing GmbH & Co., Würzburg, Germany, pp. 92–106.

Levy, George C., and Dumoulin, Charles L. (1984). *The LAB ONE NMRI Spectroscopic Data Analysis Software System: Revision 2.00 User's Manual.* Department of Chemistry, Syracuse University, Syracuse, NY (Address: Dept. of Chemistry, Browne Hall, Syracuse University, Syracuse 13244).

Lieberman, G. J., and Owen, D. B. (1961). *Tables of the Hypergeometric Probability Distribution.* Stanford University Press, Stanford, CA.

Lloyd, David K., and Lipow, Myron (1982). *Reliability: Management, Methods and Mathematics,* 2nd ed. Prentice-Hall, Englewood Cliffs, NJ.

Longley, J. W. (1967). "An Appraisal of Least Squares Programs for the Electronic Computer from the Point of View of the User." *Journal of the American Statistical Association,* vol. 62, no. 319, pp. 819–841.

Lusser, R. (1958). *Reliability Through Safety Margins.* United States Army Ordnance Missile Command, Redstone Arsenal, AL.

Martin Marietta Corp. (1966). *Reliability for the Engineer,* book 5: *Testing for Reliability.* Martin Marietta, Orlando, FL, pp. 29–31.

Moult, John F. (1963). "Critical Agents in Bearing Fatigue Testing." *Lubrication Engineering,* December, pp. 503–511.

Mouradian, G. (1966). "Tolerance Limits for Assemblies and Engineering Relationships." *ASQC Technical Conference Transactions.* Milwaukee, pp. 598–606.

Mulekar, Madhuri S., and Mishra, Satya N. (1997). "Inference on Measures of Niche Overlap for Heteroscedastic Normal Populations." In Hayakawa, T., Aoshima, M., and Shimizu, K. (eds.), *MSI-2000: Multivariate Statistical Analysis,* vol. III. American Sciences Press, Columbus, OH, pp.163-185.

Natrella, Mary G. (1963). *Experimental Statistics, National Bureau of Standards Handbook 91.* U.S. Government Printing Office, Washington, DC.

Nelson, Wayne (1979). *How to Analyze Data with Simple Plots.* Volume 1 of Dudewicz, E. J. (ed.). *The ASQC Basic References in Quality Control: Statistical Techniques.* American Society for Quality Control, Milwaukee.

———— (1982). *Applied Life Data Analysis.* John Wiley & Sons, New York.

Oliver, Larry R., and Springer, Melvin D. (1972). "A General Set of Bayesian Attribute Acceptance Sampling Plans." *Technical Papers of the 1972 Conference, American Institute of Industrial Engineers,* pp. 443–455.

Owen, D. B., and Frawley, W. H. (1971). "Factors for Tolerance Limits Which Control Both Tails of the Normal Distribution." *Journal of Quality Technology,* vol. 3, no. 2, pp. 69–79. See also Frawley, W. H., Kapadia, C. H., Rao, J. N. K., and Owen, D. B. (1971). "Tolerance Limits Based on Range and Mean Range." *Technometrics,* vol. 13, no. 3, pp. 651–656.

Peters, J. (1970). "Tolerancing the Components of an Assembly for Minimum Cost." ASME Paper 70-Prod.-9. *Transactions of the ASME Journal of Engineering for Industry.* American Society of Mechanical Engineers, New York.

Pinsky, Paul D. (1983). *Statpro: The Statistics and Graphics Database Workstation, Statistics User's Guide.* Wadsworth Electronic Publishing Company, Boston (Address: Statler Office Building, 20 Park Plaza, Boston 02116).

Ramberg, John S., Dudewicz, Edward J., Tadikamalla, Pandu R., and Mykytka, Edward F. (1979). "A Probability Distribution and its Uses in Fitting Data." *Technometrics,* vol. 21, no. 2, pp. 201–214.

Ranganathan, J., and Kale, B. K. (1983). "Outlier-Resistant Tolerance Intervals for Exponential Distributions." *American Journal of Mathematical and Management Sciences,* vol. 3, no. 1, pp. 5–25.

Rich, Barrett G., Smith, O. A., and Korte, Lee (1967). *Experience with a Formal Reliability Program.* SAE Paper 670731, Farm, Construction and Industrial Machinery Meeting, Society of Automotive Engineers, Warrendale, PA.

Romeu, J. L., and Ozturk, A. (1996). "A New Graphical Test for Multivariate Normality." In Hayakawa, T., Aoshima, M., and Shimizu, K. (eds.), *MSI-2000: Multivariate Statistical Analysis,* vol. II. American Sciences Press, Columbus, OH, pp. 5–48.

Rutherford, John R. (1971). "A Logic Structure for Experimental Development Programs." *Chemical Technology,* March, pp. 159–164.

SAS Institute, Inc. (1982). *SAS User's Guide: Basics,* 1982 Edition. SAS Institute, Cary, NC (Address: P.O. Box 8000, Cary 27511).

Schmid, C. F. (1983). *Statistical Graphics, Design Principles and Practices.* John Wiley & Sons, New York.

Sheesley, J. H. (1977). "Tests for Outlying Observations." *Journal of Quality Technology,* vol. 9, no. 1, pp. 38–41.

Siotani, Minoru, Hayakawa, T., and Fujikoshi, Y. (1985). *Modern Multivariate Statistical Analysis: A Graduate Course and Handbook.* American Sciences Press, Columbus, OH.

Sun, L., and Müller-Schwarze, D. (1996). "Statistical Resampling Methods in Biology: A Case Study of Beaver Dispersal Patterns." *American Journal of Mathematical and Management Sciences,* vol. 16, pp. 463–502.

Thoman, D. R., Bain, L. J., and Antle, C. E. (1970). "Maximum Likelihood Estimation. Exact Confidence Intervals for Reliability and Tolerance Limits in the Weibull Distribution." *Technometrics,* vol. 12, no. 2, pp. 363–371.

Volk, William (1956). "Industrial Statistics." *Chemical Engineering,* March, pp. 165–190.

Wampler, R. H. (1970). "On the Accuracy of Least Squares Computer Programs." *Journal of the American Statistical Association,* vol. 65, no. 330, pp. 549–565.

Wardrop, R. L. (1995). "Simpson's Paradox and the Hot Hand in Basketball." *The American Statistician,* vol. 49, no. 1, pp. 24–28.

Wescott, M. E. *Mimeo Notes.* Rochester Institute of Technology, Rochester, NY.

Western Electric Company, Inc. (1982). *Statistical Quality Control Handbook.* American Society for Quality Control, Milwaukee.

Williamson, E., and Bretherton, M. H. (1963). *Tables of the Negative Binomial Probability Distribution.* John Wiley & Sons, New York.

Zacks, Shelemyahu (1983). *Workshop on Statistical Methods of Reliability Analysis for Engineers.* Center for Statistics, Quality Control, and Design, State University of New York, Binghamton, NY.

SECTION 45
STATISTICAL PROCESS CONTROL[1]

Harrison M. Wadsworth

INTRODUCTION

This section presents the use of statistical techniques to control processes. Historically, the processes controlled statistically were manufacturing processes; however, in recent years, other processes such as those used by organizations dealing primarily in services have recognized the power of these techniques. The section starts with a few definitions and comments on notation. Where possible, the definitions used are those found in the most common terminology standards.

---

[1] In the Fourth Edition, material for the section on statistical process control was supplied by Dorian Shainin and Peter D. Shainin.

These standards are identified here as American National Standards; however, there are also international standards with the same numerical designations. The standards used are

- ANSI/ISO/ASQC A8402-1994, *Quality Management and Quality Assurance—Vocabulary*
- ANSI/ISO/ASQC A3534-1993, *Statistics—Vocabulary and Symbols*

    Part 1: *Probability and General Statistical Terms*

    Part 2: *Statistical Quality Control*

The reader is referred to these standards for definitions of additional terms. ANSI/ISO/ASQC A8402-1994 has replaced ANSI/ASQC A3 and ANSI/ISO/ASQC A3534-2 has replaced ANSI/ASQC A1 and A2 that were referenced in earlier editions of this handbook.

Definitions

- *Process:*   Set of interrelated resources and activities that transform inputs into outputs. [Note: Resources may include personnel, finance, facilities, equipment, techniques, and methods (ANSI/ISO/ASQC A8402-1994, clause 1.2).]
- *State of statistical control:*   State in which the variations among the observed sampling results can be attributed to a system of chance causes that does not appear to change with time. [Note: Such a system of chance causes generally will behave as though the results are simple random samples from the same population) (ANSI/ISO/ASQC A3534-2-1993, clause 3.1.5).]
- *Process in control, stable process:*   Process in which each of the quality measures (e.g., the average and variability or fraction nonconforming or average number of nonconformities of the product or service) is in a state of statistical control (ANSI/ISO/ASQC A3534-2-1993, clause 3.1.6).
- *Chance causes:*   Factors, generally many in number but each of relatively small importance, contributing to variation that have not necessarily been identified. [Note: Chance causes are sometimes referred to as *common causes* of variation (ANSI/ISO/ASQC A3534-2-1993, clause 3.1.9).]
- *Assignable causes:*   Factors (usually systematic) that can be detected and identified as contributing to a change in a quality characteristic or process level. [Notes: (1) Assignable causes are sometimes referred to as *special causes* of variation; (2) many small causes of change are assignable, but it may be uneconomic to consider or control them; in this case they should be treated as chance causes (ANSI/ISO/ASQC A3534-2-1993, clause 3.1.8).]
- *Control chart:*   Chart with upper and/or lower control limits on which values of some statistical measure for a series of samples or subgroups are plotted, usually in time or sample number order. The chart frequently shows a central line to assist detection of a trend of plotted values toward either control limit. [Note: In some control charts, the control limits are based on the within-sample or within-subgroup data plotted on the chart; in others, the control limits are based on adopted standard or specified values applicable to the statistical measures being plotted on the chart (ANSI/ISO/ASQC A3534-2-1993, clause 3.3.1).]

Notation.   The standard statistical notation will be used wherever possible. That is, a measurement of a quality characteristic will be denoted by an x, parameters will be denoted by Greek letters, and statistics by Roman letters. An overbar will denote an average, and double overbars will indicate an average of averages. Other symbols will be defined as they are used.

THEORY AND BACKGROUND OF STATISTICAL PROCESS CONTROL

Concern over variation in manufactured products produced by the Western Electric Company and studies of sampling results led Dr. Walter A. Shewhart of the Bell Laboratories to the development of the control chart as early as 1924. An unpublished memorandum by Shewhart dated May 16, 1924 contains the first known control chart. In subsequent papers, Shewhart developed the concept more completely. See Shewhart (1926a, 1926b, 1927, and 1931) for more details about these early efforts. In 1931 Shewhart published his classic book, *Economic Control of Quality of Manufactured Product.* The first applications of the control chart by Shewhart were on fuses, heat controls, and station apparatus at the Hawthorne Works of the Western Electric Company. By 1927–1928, many other applications throughout the Western Electric Company were initiated by another member of the Quality Assurance Department of Bell Laboratories, Harold F. Dodge. These applications were as an adjunct to sampling inspection plans and as a basic device for demerits-per-unit charts for periodic quality assurance ratings.

In 1935, Dodge prepared the *ASTM Manual on Presentation of Data,* which addresses many types of control charts. This document is currently known as ASTM MNL 7 (1990). It has been used extensively by industry over the years. In 1941–1942, Dodge served as chairman of the American Standards Association Committee Z1, which published ASA standards Z1.1, Z1.2, and Z1.3 on control charts. Widespread use of control charts followed the initial publication of these standards. Revised versions of these three standards are now available from ANSI and ASQ (see ANSI/ASQC 1985a, 1985b, 1985c). Training courses on quality control principles (including the use of control charts and acceptance sampling plans) were sponsored by the Office of Production Research and Development of the War Production Board during World War II as a means for maintaining quality and promoting continual improvement. There have been many extensions and modifications of the basic control charts of Shewhart over the years. Control charts are now widely used in every industry. They are the principal tools of statistical process control (SPC).

In his studies, Shewhart observed that variation occurs in all things in nature as well as in manufactured goods. The study of this variation and its reduction are the principal vehicles of quality improvement. The control chart is the most important tool available to do this. Even though no two items are alike (variation), groups of observations form either predictable patterns or no patterns at all. The latter indicate that process improvement is necessary. This led Shewhart to conclude two important principles: (1) variation is inevitable, and (2) single observations form little or no basis for objective decision making. In order to determine patterns of observations, they may be plotted in several ways. One way is to form a histogram of the observations that presents a picture of the distribution. Another way is to plot the observations in the order in which they were obtained. This forms a line chart and is useful for observing trends or cycles in the data.

Shewhart recommended the use of a line chart to observe the data. He further indicated that there are two causes of variation. One he called *chance causes.* These are events that cause relatively minor fluctuations in the data. Each may be so small that its occurrence is not important or may be uneconomic to correct. Even though each contributes relatively minor fluctuations, together they form a pattern. Shewhart concluded from the central limit theorem and from empirical observations that they often formed approximately a normal distribution. The second cause of variation he called *assignable*; that is, a cause can be assigned for the fluctuations observed. These are sources of variation that cause a significant departure of the data from the pattern formed by the chance causes. The control chart developed by Shewhart contains a set of limits around the hypothesized normal distribution of chance causes. Any observation falling outside these limits indicates the presence of an assignable cause. In addition, because the observations are plotted in order of occurrence, trends or other unnatural patterns may be readily observed.

A control chart, then, is a graphic representation of the variation in the computed statistics being produced by the process. It has a decided advantage over presentation of the data in the form of a histogram in that it shows the sequence in which the data were produced. It reveals the amount and nature of variation by time, indicates statistical control or lack of it, and enables pattern interpretation and detection of changes in the process. The basic charts developed by Dr. Shewhart consisted of charts for averages and standard deviations. Other charts were developed for sample ranges, percentage nonconforming, and number of nonconformities per item or hundred items. Further improvements and modifications to the basic charts also have been made over the years. Such charts as cumulative sum charts and exponentially weighted moving average charts have allowed for quicker detection of small shifts in the parameter being followed. All these types of control charts will be reviewed briefly in this section.

STEPS TO START A CONTROL CHART

1. Choose the quality characteristic to be charted. In making this choice, there are several things to consider:
   a. Choose a characteristic that is currently experiencing a high number of nonconformities or items that do not conform. A Pareto analysis is useful to assist the process of making this choice.
   b. Identify the process variables contributing to the end-product characteristics to identify potential charting possibilities.
   c. Choose characteristics that will provide appropriate data to identify and diagnose problems. In choosing characteristics, it is important to remember that attributes provide summary data and may be used for any number of characteristics. On the other hand, variables data are used for only one characteristic on each chart but are necessary to diagnose problems and propose action on the characteristic.
   d. Determine a convenient point in the production process to locate the chart. This point should be early enough to prevent nonconformities and to guard against additional work on nonconforming items.
2. Choose the type of control chart.
   a. The first decision is whether to use a variables chart or an attributes chart. A variables chart is used to control individual measurable characteristics, whereas an attributes chart may be used with go no-go type of inspection. An attributes chart is used to control percentage or number of nonconforming items or number of nonconformities per item. A variables chart provides the maximum amount of information per item inspected. It is used to control both the level of the process and the variability of the process. An attributes chart often provides summary data that can be used to improve the process by then controlling individual characteristics.
   b. Choose the specific type of chart to be used. If a variables chart is to be used, decide whether the average and range or the average and standard deviation are to be charted. If small shifts in the mean are important, a cumulative sum or exponentially weighted moving average chart may be used. The disadvantage of these two latter charts is that they are more difficult for the practitioner to use and understand. If subgroups are not possible, individual readings may be used, but these are to be avoided if possible. For attributes charts, the percentage nonconforming or number of nonconforming items may be charted. In some cases, the number of nonconformities per inspection item may be preferable. All these charts will be discussed later.
3. Choose the center line of the chart and the basis for calculating the control limits. The center line may be the average of past data, the average of data yet to be collected, or a desired (standard) value. The limits are usually set at ±3 standard deviations, but other multiples of the standard deviation may be used for other risk factors. The use of 3 standard deviations results in a negligible risk of looking for problems that do not exist, i.e., false alarms. However, this multiple may result in an appreciable risk of failing to detect a small shift in the parameter being studied. Smaller multiples increase the risk of looking for a false alarm but reduce the risk of failing to detect a small shift. The fact that it is usually much more expensive to look for problems that do not exist than to miss some small problems is the reason that the ±3σ limits are usually chosen.
4. Choose the rational subgroup or sample. It should be pointed out that the term *sample* is usually used, but *sample* could mean an individual value, and samples of more than one are desirable for control charts if feasible. For variables charts, samples of size 4 or 5 are usually used, whereas for attributes charts, samples of 50 to 100 are often used. Attributes charts in fact may be used with 100 percent inspection as a reflection of the underlying process involved. In addition to the size of the sample, the samples should be selected in such a way that the chance of a shift in the process is minimized during the taking of the sample (thus a small sample should be used); whereas the chance of a shift, if it is going to occur, is at a maximum between samples. This is the concept of *rational subgrouping*. Thus it is better to take small samples periodically than to take a single large sample. Experience is usually the best method for deciding on the frequency of taking samples. That is, the known rate of a chemical change or the known rate of tool wear

should be considered when making these decisions. If such experience is not available, samples should be taken frequently until such experience is gained.

5. Provide a system for collecting the data. If control charts are to become a shop tool, the collection of data must be an easy task. Measurement must be made simple and relatively free of error. Measuring instruments must give quick and reliable readings. If possible, the measuring instrument actually should record the data, since this will eliminate a common source of errors. Data sheets should be designed carefully to make the data readily available. The data sheets must be kept in a safe and secure place, free from dirt or oil.

6. Calculate the control limits and provide adequate instruction to all concerned on the meaning and interpretation of the results. Production personnel must be knowledgeable and capable of performing corrective action when the charts indicate it.

CONSTRUCTING A CONTROL CHART FOR VARIABLES FOR ATTAINING A STATE OF CONTROL (NO STANDARD GIVEN CHARTS)

In this case we assume that we know nothing about the process, but we wish to determine if it is in a state of statistical control, i.e., to see if there are only chance causes of variation present. The procedure for all types of charts is

1. Take a series of 20 to 30 samples from the process.

2. During the taking of these samples, keep accurate records of any changes in the process such as a change in operators, machines, or materials.

3. Compute trial control limits from these data.

4. Plot the data on a chart with the trial limits to determine if any of the samples were out of control, i.e., if any plotted points are outside the control limits.

If none of the plotted points are outside the trial control limits, we can say the process is "in control," and these limits may be used for maintaining control. If, on the other hand, some of the plotted points are outside the trial control limits, we say the process is "not in control." That is, there are assignable causes of variation present. In such a case we must determine, from the records in step 2 above if possible, the cause of each out-of-control point, eliminate these samples from the data, and recalculate the trial control limits. If some points are outside these new limits, this step must be repeated until no points are outside the trial control limits. These final limits may then be used for future control.

$\overline{X}$ **and** R **Charts.**   This set of two charts is the most commonly used statistical process control procedure. It is used whenever we have a particular quality characteristic that we wish to control, since we can use the charts with only one characteristic at a time. In addition, the data must be of a measurement or variables type. Most users of process control are interested in individual items of product and the values of a few quality characteristics on these items. Averages and ranges computed from small samples or subgroups of individual items provide very good measures of the nature of the underlying universe. They permit us to control and otherwise make decisions about the process from which the items came. The chart for averages is used to control the mean or central tendency of the process, whereas the chart for ranges is used to control the variability. In place of the range, the sample standard deviation is sometimes used, but the range (the largest minus the smallest values in the sample) is easier to calculate and is easier to understand by the operators.

Using the convention of control limits set at ±3σ, the control limits for the $\overline{X}$ chart will be set at

$$\mu \pm 3\sigma_{\overline{X}} = \mu \pm 3\sigma / \sqrt{n}$$

where μ = the process mean

σ = the process standard deviation

n = the sample size

Since the parameters μ and σ are unknown, we must estimate them from the data. The best estimate of μ is $\bar{\bar{X}}$, where $\bar{\bar{X}}$ is the average of the sample averages $\bar{X}$, and $\bar{\bar{X}}$ is calculated as

$$\bar{\bar{X}} = \sum x/n$$

The process standard deviation may be estimated by a function of the average of the sample ranges $\bar{R}$. The average range provides a good estimate of the standard deviation for small samples, say, less than 12. The estimate of the standard deviation is the average range divided by a constant d_2, where d_2 is a function of the sample size. Thus the control limits for the $\bar{X}$ chart are calculated as

$$\text{UCL}_{\bar{X}} = \bar{\bar{X}} + 3\,\bar{R}/(d_2\sqrt{n}) = \bar{\bar{X}} + A_2\,\bar{R}$$

$$\text{LCL}_{\bar{X}} = \bar{\bar{X}} - 3\bar{R}/(d_2\sqrt{n}) = \bar{\bar{X}} - A_2\,\bar{R}$$

where $A_2 = 3/(d_2\sqrt{n})$.

The range chart will be calculated similarly as $\mu_R \pm 3\sigma_R$, where μ_R and σ_R are the mean and standard deviation of the distribution of the sample ranges, respectively. The mean range is estimated by $\bar{R}$, and the standard deviation of the range is $\sigma_R = d_3\sigma$ · This may be estimated by $d_3\bar{R}/d_2$. Therefore, we calculate the control limits for the range chart as

$$\text{UCL}_R = \bar{R} + 3d_3(\bar{R}/d_2) = (1 + 3d_3/d_2)\bar{R} = D_4\,\bar{R}$$

$$\text{LCL}_R = \bar{R} - 3d_3(\bar{R}/d_2) = (1 - 3d_3/d_2)\bar{R} = D_3\,\bar{R}$$

where the factors A_2, D_3, and D_4 are functions of the sample size n and are tabulated in Table 45.1 along with d_2, d_3, and some other factors that will be used later in this section. A more extensive table will be found in Appendix II, Table A.

To use these two charts, 20 to 25 samples of the same size n are taken from the process. Samples of 4 or 5 are usually used for these charts. The average and range are calculated for each sample. The 20 to 25 averages and ranges are then averaged to find the grand average and the average range. The upper and lower control limits may then be calculated for each of the charts. It will be noticed in Table 45.1 that for $n = 6$ or less, the value for D_3 is 0. This is so because if the lower control limit were to be calculated, it would be negative. This does not make sense because the range is always positive. The fact that the value for this factor is 0 does not mean that the lower control limit for the range is 0. It means that there is no lower control limit for the range for these small samples.

There are many examples of these charts in the literature. The book *Statistical Quality Control Handbook* (AT&T 1984) contains many illustrations of control charts used by the Western Electric

TABLE 45.1 Control Chart Factors

n	d_2	d_3	c_4	c_5	A	A_2	A_3	B_3	B_4	B_5	B_6	D_3	D_4
2	1.13	0.85	0.80	0.60	2.12	1.88	2.66	0	3.27	0	2.61	0	3.27
3	1.69	0.89	0.89	0.46	1.73	1.02	1.95	0	2.57	0	2.28	0	2.58
4	2.06	0.88	0.92	0.39	1.50	0.73	1.63	0	2.27	0	2.09	0	2.28
5	2.33	0.86	0.94	0.34	1.34	0.58	1.43	0	2.09	0	1.96	0	2.11
6	2.53	0.85	0.95	0.30	1.23	0.48	1.29	0.03	1.97	0.03	1.87	0	2.00
7	2.70	0.83	0.96	0.28	1.13	0.42	1.18	0.12	1.88	0.11	1.81	0.08	1.92
8	2.85	0.82	0.97	0.26	1.06	0.37	1.10	0.19	1.82	0.18	1.75	0.14	1.86
9	2.97	0.81	0.97	0.25	1.00	0.34	1.03	0.24	1.76	0.23	1.71	0.18	1.82
10	3.08	0.80	0.97	0.23	0.95	0.31	0.98	0.28	1.72	0.28	1.67	0.22	1.78

Company. Other books with excellent descriptions of control charts and examples of their use are Grant and Leavenworth (1996), Ott and Schilling (1990), and Wadsworth et al. (1986).

Example 45.1:   Figures 45.1 and 45.2 illustrate the use of $\overline{X}$ and R charts. The figures are self-explanatory in that they follow all the procedures outlined earlier. It may be observed that the range chart is in control, but the sequence of points in the $\overline{X}$ chart starts low and after an adjustment exhibits good control for awhile. However, the last eight points are above the center line, and three of these points are above the upper control limit.

***s* Charts.**   The sample standard deviation s may be used with the $\overline{X}$ chart in place of the sample range to measure the process dispersion. In such cases we would calculate the standard deviation of each sample, find the average of the sample standard deviations, and calculate the control limits from the equations below:

$$UCL_s = B_4\overline{s}$$

$$LCL_s = B_3\overline{s}$$

where values of B_3 and B_4 may be found in Table 45.1 and in Appendix II, Table A.

The standard deviation has superior mathematical properties over the range, and with present hand calculators and computers, it is relatively simple to calculate. However, the range is much easier for the operator with limited statistical training to understand, and for this reason, the range chart is the most often used chart to control process variability.

INTERPRETATION OF CONTROL CHARTS

It is customary to place the $\overline{X}$ chart directly above the range (or standard deviation) chart. Thus the two statistics computed from each sample are easily located, and the relationship between them is obvious. We look for any unusual points or patterns in the plotted data for either chart. In order to consider this, let us first understand natural patterns for control chart data. A stable process, i.e., one under statistical control, generally will not produce a discernible unnatural pattern. It will produce a random array of data that possesses several underlying characteristics. We have observed previously that the control limits are established at the extremities of the underlying distribution of the statistic being studied. Hence the data from a controlled process should have the following characteristics when plotted on a control chart:

- Most of the plotted points occur near the centerline.
- A few of the points occur near the control limits.
- Only an occasional rare point occurs beyond the control limits.
- The plotted points occur in a random manner with no clustering, trending, or other departure from a random distribution.

Having considered natural patterns, we now consider unnatural patterns of plotted points on a control chart. Certainly, a pattern lacking one or more of the preceding characteristics would be an unnatural pattern, but let us consider each of them in turn and consider possible reasons for data to not behave according to these rules.

Most of the data on an $\overline{X}$ chart may not occur near the centerline for several reasons. If the mean changes from time to time, the data may cluster around two lines above and below the centerline. This might occur when there are, for example, two machines or two persons operating at slightly different levels feeding into the production stream.

The absence of the second of the preceding characteristics of a controlled process, the occurrence of a few points near the control limits of an $\overline{X}$ chart, may be caused by two processes having

CONTROL CHART DATA SHEET

PART NUMBER: 102J		DESCRIPTION: SHAFT		
LOT NO.: 195		MACHINE NO.: SBL-20		DEPT.: M
OPERATION: FACING TO LENGTH	ORDER NO.: 7-18	SHIFT: 1	DATE: 12-17-48	INSPECTOR: SMITH
	OPERATOR: BROWN			

SAMPLE #	1	2	3	4	5	6	7	8	9	10	11	12	13
	.9382	.9382	.9385	.9379	.9384	.9385	.9387	.9387	.9388	.9381	.9386	.9385	.9386
	.9378	.9380	.9382	.9380	.9385	.9385	.9385	.9386	.9382	.9385	.9387	.9384	.9386
	.9385	.9380	.9383	.9384	.9387	.9385	.9385	.9385	.9386	.9383	.9385	.9384	.9386
	.9375	.9382	.9379	.9384	.9386	.9385	.9387	.9382	.9386	.9386	.9385	.9384	.9381
TOTAL	3.7520	3.7524	3.7529	3.7527	3.7542	3.7540	3.7544	3.7540	3.7542	3.7535	3.7545	3.7538	3.7539
$\bar{X}$	.93800	.93810	.93822	.93818	.93855	.93850	.93860	.93850	.93855	.93838	.93862	.93845	.93848
R	.0010	.0002	.0006	.0005	.0003	.0000	.0002	.0005	.0006	.0005	.0002	.0001	.0005

SAMPLE #	14	15	16	17	18	19	20	21	22	23	24	25
	.9382	.9386	.9388	.9385	.9387	.9388	.9385	.9386	.9384	.9387	.9387	.9390
	.9387	.9387	.9387	.9387	.9387	.9385	.9387	.9390	.9386	.9388	.9387	.9389
	.9388	.9387	.9388	.9385	.9383	.9386	.9386	.9390	.9386	.9388	.9389	.9389
	.9388	.9386	.9386	.9384	.9389	.9366	.9385	.9389	.9388	.9386	.9390	.9390
TOTAL	3.7545	3.7546	3.7549	3.7541	3.7538	3.7545	3.7538	3.7483	3.7544	3.7549	3.7553	3.7558
$\bar{X}$	.93862	.93865	.93872	.93852	.93865	.93862	.93865	.93888	.93860	.93872	.93882	.93895
R	.0006	.0001	.0002	.0003	.0006	.0003	.0003	.0004	.0004	.0002	.0003	.0001

NO.	$\bar{X}$	R
1	.93800	.0010
2	.93810	.0002
3	.93822	.0006
4	.93818	.0005
5	.93855	.0003
6	.93850	.0000
7	.93860	.0002
8	.93850	.0005
9	.93855	.0006
10	.93838	.0005
11	.93862	.0002
12	.93845	.0001
13	.93848	.0005
14	.93862	.0006
15	.93865	.0001
16	.93872	.0002
17	.93852	.0003
18	.93865	.0006
19	.93838	.0003
20	.93865	.0003
21	.93888	.0004
22	.93860	.0004
23	.93872	.0002
24	.93882	.0003
25	.93895	.0001
TOTAL	23.46353	0.0090

$$\bar{X} = \frac{23.46353}{25} \qquad \bar{R} = \frac{0.0090}{25}$$

$$\bar{X} = .938541 \qquad \bar{R} = .00036$$

LIMITS FOR AVERAGES CHART

UPPER CONTROL LIMIT = $\bar{X} + A_2\bar{R}$
LOWER CONTROL LIMIT = $\bar{X} - A_2\bar{R}$

$\bar{X} = .938541$ } COMPUTED FROM INSPECTION DATA. SEE
$\bar{R} = .00036$ } LAST TWO COLUMNS ON THIS SHEET.

CONSTANT A_2 IS OBTAINED FROM TABLES (SEE
FIGURE 22). FOR A SAMPLE SIZE = 4, A_2 = .729

$A_2\bar{R}$ = .729 x .00036 = .000262
U.C.L. = .938541 + .000262 = .938803
L.C.L. = .938541 - .000262 = .938279

LIMITS FOR RANGES CHART

UPPER CONTROL LIMIT = $D_4\bar{R}$
LOWER CONTROL LIMIT = $D_3\bar{R}$

$\bar{R}$ = .00036 (SEE LAST COLUMN ON THIS SHEET)

CONSTANTS D_3, D_4 ARE OBTAINED FROM TABLES (SEE
FIGURE 22). FOR A SAMPLE SIZE = 4, D_3 = 0.000, D_4 = 2.282
U.C.L. = 2.282 x .00036 = .00082
L.C.L. = 0 x .00036 = 0

FIGURE 45.1 Data sheet for determining if a process is in control.

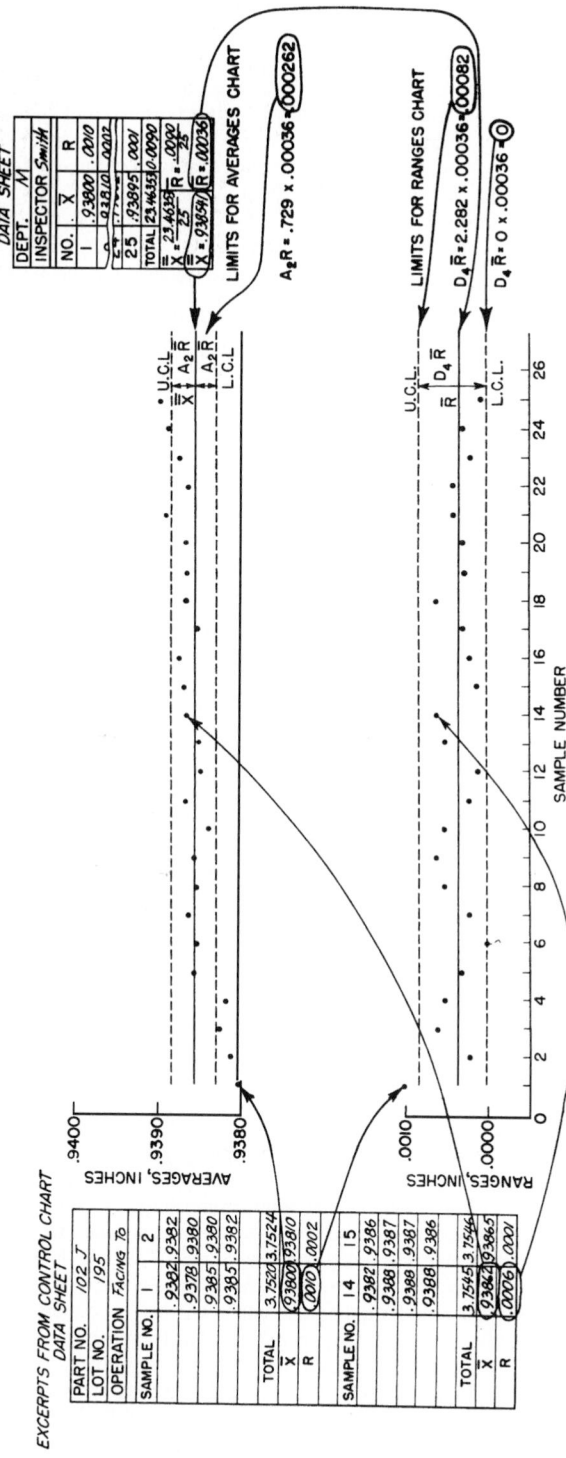

FIGURE 45.2 How to construct the Shewhart control chart.

different means being mixed together in the same sample. The sample averages will then tend to fluctuate about the centerline, but the range of each sample will be inflated, causing the control limits to be spread apart. In this case, the chart for averages might seem to be in very good control, but the range chart will show excessive variation. Data patterns lacking the third of the preceding characteristics, an occasional, rare point beyond the control limits, may indicate either frequent changes in the process average or an increased dispersion.

For the fourth characteristic of natural patterns, examples of unnatural patterns are sudden shifts in level, trends, bunching, or clustering. A detailed discussion of this topic may be found in the AT&T *Statistical Quality Control Handbook* (1984), where a dictionary of control chart patterns is developed.

If the process is "in control," there is approximately a risk of 0.0013, or 0.13 percent, of investigating a change in the process when there is none because a point falls outside one of the control limits. There is an equal risk of a point falling outside the other control limit, making a total risk of 0.0026, or 0.26 percent, of searching for an assignable cause that does not exist. This is often called the *risk of a type I error* or an α risk. It is also called the *producer's risk*. If the process is really out of control, the occurrence of a point outside a control limit is much more likely. How likely depends on how far the process is out of control. The risk of a point not falling outside the control limits and thus failing to detect a shift of a given size is the *risk of a type II error* or a β risk. This risk is also called the *consumer's risk*.

The control limits are usually set for a low α risk. Limits much wider than 3σ would save only a few needless investigations and might cause a delay in investigating out-of-control situations. Limits much narrower would increase the incidence of needless investigations while not appreciably reducing delay in the detection of out-of-control situations. The rule of investigating each point that falls outside the 3σ control limits is therefore a compromise.

Once the control limits have been set, there is information that may be obtained from the chart that is not exploited by the preceding rule. It is rare and about equally likely to have the following events occur when the process is in control:

1. One point outside one of the 3 standard deviation control limits
2. Two of three successive points outside 2 standard deviations
3. Four of five successive points outside 1 standard deviation
4. Eight successive points on the same side of the centerline

Any one of these should be regarded as a signal of an out-of-control condition that requires attention.

The use of all four of the preceding rules is optional and has proved practical in many cases. They were advocated in 1956 in the Western Electric Company's quality control program. The last three apply only to control charts with symmetric limits. Collectively, their use increases the α risk when the process is in control but decreases the β risk when it is not in control.

CONTROL CHARTS FOR INDIVIDUALS

Sometimes it is not practical to take a subgroup from a process. This is particularly true for chemical or other continuous processes or when studying variables such as temperature or pressure. Other situations for which a single observation makes the most sense are accounting data, efficiency, ratios, expenditures, or quality costs. In such a situation, if we took several samples from the process at the same time, we would really only be checking our measuring device and not the manufacturing process itself. In this instance we would only take one observation each time we sample. A control chart based on one observation is called a *chart for individuals* or an X chart. Since we are not using sample averages, we do not have the benefit of the central limit theorem, which tells us that the distribution of averages is approximately normal regardless of the underlying distribution. Therefore, an X chart is much more sensitive to a lack of normality of the underlying distribution than is an $\bar{X}$ chart.

Since we are only taking one observation, we do not have an obvious source of an estimate of the standard deviation that we can use to determine the control limits for an X chart. We have two possible choices: We can use a moving range, e.g., the difference between two successive observations, or we can calculate the standard deviation of 20 or more successive observations. If the process does not shift, both statistics will give about the same answer. However, if the process does shift, the moving range will minimize the effect of the shift. Therefore, that method is usually used.

The basic procedure for developing X charts is as follows:

1. Select the measurable characteristic to be studied.

2. Collect enough observations (20 or more) for a trial study. The observations should be far enough apart to allow the process to be potentially able to shift.

3. Calculate control limits and the centerline for the trial study using the formulas given later.

4. Set up the trial control chart using the centerline and limits, and plot the observations obtained in step 2. If all points are within the control limits and there are no unnatural patterns, extend the limits for future control.

5. Revise the control limits and centerline as needed (by removing out-of-control points or observing trends, etc.) to assist in improving the process.

6. Periodically assess the effectiveness of the chart, revising it as needed or discontinuing it.

Computing the Control Limits.   Control limits for the X chart are calculated similarly to those for the $\overline{X}$ chart. That is, we set the control limits at the centerline ±3 standard deviations. They would then be set at

$$\mu \pm 3\sigma$$

The average of the observations $\overline{X}$ is commonly used to estimate the process mean μ. For the standards-given case, the known standard deviation would be used in the preceding equation. For the no-standards-given case, we must estimate the standard deviation. As mentioned earlier, we can use either the moving range or the standard deviation for this estimate.

Moving Range.   The moving range is the difference between the largest and smallest of two successive observations. Thus for a total of n observations there will be $n - 1$ moving ranges. We would average the moving ranges to get an average moving range $\overline{R}$. The control limits would thus be set in a similar manner to those for the charts using subgroups as

$$UCL_x = \overline{X} + 3\overline{R}/d_2 = \overline{X} + E_2\overline{R}$$

$$LCL_x = \overline{X} - 3\overline{R}/d_2 = \overline{X} - E_2\overline{R}$$

where values of $E_2 = 3/d_2$ will be found in Appendix II, Table Y. If we employ the common rule of using moving ranges of size 2, E_2 from Appendix II, Table Y, is 3 divided by 1.13, or 2.66. The control limits for the individuals chart would thus be set at

$$UCL_x = \overline{X} + 2.66\overline{R}$$

$$LCL_x = \overline{X} - 2.66\overline{R}$$

Standard Deviation.   As an alternative to the moving range method, the standard deviation s of all the trial observations is sometimes calculated. In this case, the control limits would be set at

$$\overline{X} \pm 3s$$

If the process average shifts, e.g., in the case of a trend, the standard deviation method will tend to overstate the variability. If the process average remains relatively constant, both methods will result in approximately the same control limits.

CONSTRUCTING CONTROL CHARTS FOR VARIABLES WHEN A STANDARD IS GIVEN

If standard values of the parameters are given, we have what is commonly called a *standards-given chart*. In this case, the standard for the mean is denoted as $\bar{X}_0$, and the standard for the standard deviation is denoted as σ_0. The control limits for the $\bar{X}$ chart will be at

$$\bar{X}_0 \pm 3\sigma_0/\sqrt{n} = \bar{X}_0 \pm A\sigma_0$$

where A is a function of the sample size n and is tabulated in Table 45.1 and Appendix II, Table A. In the case of control charts for individuals, the X chart, the control limits would be the same, except that since the sample size is 1, the limits would be merely $\bar{X}_0 \pm 3\sigma_0$.

CONTROL CHARTS FOR ATTRIBUTES

Control charts for variables require actual measurements, such as length, weight, tensile strength, etc. Thus go no-go data cannot be used for such charts. Charts for attributes, on the other hand, can be used in situations where we only wish to count the number of nonconforming items or the number of nonconformities in a sample. There are several advantages of attributes charts over variables charts.

1. Attributes charts can be used to cover many different nonconformities at the same time, whereas a separate chart must be used for each quality characteristic with variables charts.

2. The inspection required for attributes charts may be much easier than that for variables charts. We merely need to know if the item being inspected meets the specified requirements.

3. Attributes charts may be used for visual inspections for such attributes as cleanliness, correct labeling, correct color, and so on.

4. Attributes charts do not depend on an underlying statistical distribution.

On the other hand, variables charts need a much smaller sample size. In the case of the charts discussed earlier, we only used sample sizes of 4 or 5, whereas attributes charts would require sample sizes of at least 50. Attributes control charts are often used for 100 percent inspection, whereas this would be difficult for variables charts. The most common control charts for attributes are the p chart for percentage nonconforming, the np chart for number of nonconforming items, the c chart for number of nonconformities, and the u chart for number of nonconformities per item. We will discuss each of these charts in order.

Control Charts for Percentage Nonconforming (p).   The variable to be controlled here is the percentage or fraction of each sample that is nonconforming to the quality requirements. Thus the number of inspected items containing one or more nonconformities is divided by the number of items inspected. This is the fraction nonconforming. Sometimes this ratio is multiplied by 100, and the variable plotted is the percentage nonconforming. Assuming that the process is constant, the underlying distribution would be the binomial distribution. The details of this distribution are covered in Section 44. For relatively large samples, the binomial distribution can be approximated adequately by the normal distribution, and as with the control charts for variables, virtually all the data should then fall within 3 standard deviations of the mean. If data fall outside these 3 standard deviation limits, this would indicate a lack of statistical control. Therefore, the control limits are set at these values. Recall that the standard deviation of a binomial variable is

$$\sigma_p = \sqrt{\frac{p(1-p)}{n}}$$

Therefore, the upper and lower control limits for a *p* chart will be at

$$\bar{p} \pm 3 \sqrt{\frac{\bar{p}(1 - \bar{p})}{n}}$$

Although the *p* chart is usually used for go no-go types of inspection, it also may be used for measurement inspection. Here a piece being inspected is considered nonconforming if its measurements are outside a set of specified limits. However, this use of the *p* chart is not recommended because it is less able to diagnose causes of nonconformities. A chart for determining control limits for a *p* chart will be found in Appendix II, Chart Z.

A *p* chart may be used when the sample size is constant or not constant. Since the *n* in the preceding expression for the control limits is the sample size, if *n* varies from subgroup to subgroup, as is often the case when the chart is used to plot 100 percent inspection data, the control limits will vary. They will be wider for small subgroups than for large ones. If the subgroup size varies, we have three possibilities.

1. We can calculate the average size of the subgroups. This is appropriate when the sizes are similar or all data lie near the centerline.
2. We can calculate separate control limits for each subgroup. This might lead to a rather confusing appearing chart.
3. We can find the average subgroup size and use the resulting control limits, but when a point falls near the limits, calculate the actual limits using the actual subgroup size.

The third approach is the recommended one. In using it, we must remember that a subgroup size larger than the average will mean the limits move in toward the centerline, so if a point lies outside the limits based on the average *n,* there is no point in calculating a new limit.

In setting up a *p* chart, we would, as for variables charts, collect 20 to 25 samples over enough time to allow the process to change. If the sample sizes are equal, we would determine the number of nonconforming items in each sample, divide each by the subgroup size, and average them. This is the $\bar{p}$ in the preceding expression for the control limits. In finding this average *p,* we first add all the numbers of nonconforming items and the numbers of items inspected in each subgroup. Then we divide the first total by the second to get the average. This procedure is essential if the subgroup size changes. As an example to show the importance of this procedure, suppose that we have five lots of a finished product as shown in Table 45.2. *N* is the lot size, *x* is the number of nonconforming items in each lot, and *p* is the fraction nonconforming. We are inspecting all items in each lot.

If we simply average the fraction nonconforming values in the right hand column, we would get a value for $\bar{p}$ of 0.590/5 = 0.118. This would give us upper control limits, using the actual lot sizes of 0.161, 0.255, 0.152, 0.215, and 0.197. None of the lots are out of control. However, if we instead find $\bar{p}$ as 90/1600 = 0.056, we get upper control limits of 0.087, 0.154, 0.081, 0.125, and 0.113. In this case, three of the five lots (lots 2, 4, and 5) are above the upper control limit, indicating that this does not represent a stable process. The true average fraction nonconforming is 0.056 not 0.118. The latter gives equal representation to each lot despite their widely differing sizes.

TABLE 45.2 Inspection Results of 5 Lots

Lot	*N*	*x*	*p*
1	500	32	0.064
2	50	10	0.200
3	800	10	0.013
4	100	18	0.180
5	150	20	0.133
Totals	1600	90	0.590

For all types of control charts for attributes, the lower control limit, when calculated using the appropriate expressions, may turn out to be negative. This, of course, makes no sense, so we simply do not have a lower control limit in these cases. In the preceding example, this applies to lots 2, 4, and 5. Note that lot 3 is actually below its lower control limit, which is, in this case, 0.032. This means that the quality, measured in terms of fraction nonconforming, is better than the average of the other lots. Some people use zero as a lower control limit when the calculated limit is negative. This might lead to a mistaken notion that a sample value of zero means that the subgroup is not in control.

Example 45.2:   Table 45.3 and Figure 45.3 illustrate a p chart. Table 45.3 contains the result of final testing of permanent magnets used in electrical relays. There were a total of 14,091 magnets tested, and 1030 were found to be nonconforming. The average number of magnets tested per week was $14{,}091/19 = 741.6$. The average fraction nonconforming was $1030/14{,}091 = 0.073$. Figure 45.3 illustrates the control chart with limits based on these values of $\bar{n}$ and $\bar{p}$. Using the preceding formulas, the upper and lower control limits are set at 0.073 ± 0.0287, or 0.102 and 0.044.

Control Charts for Number of Nonconforming Items (*np*).

In this case we plot the number of nonconforming items in each subgroup. Since p is x/n, x is *equal to np,* where x is the number of nonconforming items in a sample, and the chart is often called an *np chart.* For this type of chart, we must have a constant subgroup size. In this case the control limits are set at

$$n\bar{p} \pm 3 \sqrt{n\bar{p}(1 - \bar{p})}$$

TABLE 45.3   *p* Chart Data

Subgroup: Week No.	Production during week Week ending	No. magnets inspected	No. defective magnets	Fraction defective p
1	12/3	724	48	0.067
2	12/10	763	83	0.109
3	12/17	748	70	0.094
4	12/31	748	85	0.114
5	1/7	724	45	0.062
6	1/14	727	56	0.077
7	1/21	726	48	0.066
8	1/28	719	67	0.093
9	2/4	759	37	0.049
10	2/11	745	52	0.070
11	2/18	736	47	0.064
12	2/25	739	50	0.068
13	3/4	723	47	0.065
14	3/11	748	57	0.076
15	3/18	770	51	0.066
16	3/25	756	71	0.094
17	4/1	719	53	0.074
18	4/8	757	34	0.045
19	4/15	760	29	0.038
Totals		14,091	1030	
Averages		741.6	54.2	0.073

The heading note: Inspected at: Final assembly and test

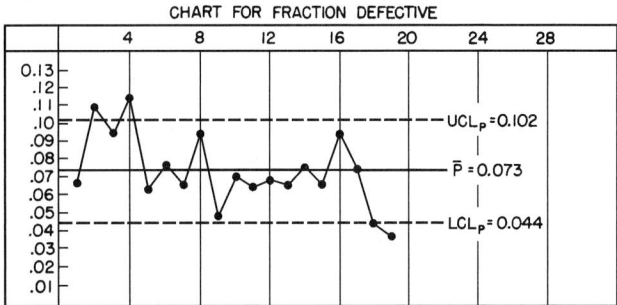

FIGURE 45.3   *p* chart for permanent magnets.

or, using $\overline{X}$ as the average number of nonconformities per subgroup, the control limits may be stated equivalently as

$$\overline{X} \pm 3 \ \sqrt{\overline{X}(1 - \overline{X}/n)}$$

Control Charts for Number of Nonconformities (*c*).

If we wish to plot the number of nonconformities, where each item inspected may have several nonconformities and each nonconformity is counted, we have a *c* chart, where *c* is the number of nonconformities in each sample. In this case, the underlying distribution is the Poisson distribution, discussed in Section 44. Recall that in this distribution the standard deviation is the positive square root of the mean, so if we again take 20 to 25 samples and calculate the average number of nonconformities, $\overline{c}$, the control limits will be set at

$$\overline{c} \pm 3\sqrt{\overline{c}}$$

A chart for determining control limits for a *c* chart will be found in Appendix II, Chart AA.

A *c* chart requires an *equally large* number of opportunities for a nonconformity to occur in each subgroup inspected. Thus, for example, if we are inspecting the number of defective solder connections in each circuit board, they must all have the same number of connections. If not, we must use the *u* chart, to be discussed next.

Example 45.3:   A control chart for *c* is used to control nonconformities in sheeted material. Table 45.4 shows the results of a series of pinhole tests of paper intended to be impervious to oils. Specimen sheets 11 by 17 inches in size were taken from production, and colored ink was applied to one side of each sheet. Each inkblot that appeared on the other side of the sheet within 5 minutes was counted as a nonconformity.

The centerline of the chart is set at $\overline{c} = 200/35 = 8.0$ nonconformities per sheet. Control limits are set at $8.0 \pm 3 \ \sqrt{8} = 8.0 \pm 8.5$, or 0 and 16.5. That is, there is only an upper limit set at 16.5. The resulting *c* chart is found in Figure 45.4.

Control Chart for Number of Nonconformities per Item (*u*).

This chart is sometimes called a *standardized c chart*. It is used when more than one item makes up a sample but each item may have more than one nonconformity. The variable plotted on the chart is the number of nonconformities per item. Thus the number of items in a sample does not need to remain constant, as it does for the *c* chart. The control limits are calculated similarly to those for a *c* chart except that the variable is *c/n*, where *n* is the number of items in the sample. For example, if we are inspecting for defective solder joints on printed circuit boards where the boards have different numbers of solder connections, we would divide the number of nonconforming solder connections on each board by

TABLE 45.4   *c* Chart Data

Sheet number	Number of pinholes	Sheet number	Number of pinholes
1	8	14	6
2	9	15	14
3	5	16	6
4	8	17	4
5	5	18	11
6	9	19	7
7	9	20	8
8	11	21	18
9	8	22	6
10	7	23	9
11	6	24	10
12	4	25	5
13	7	Total	200

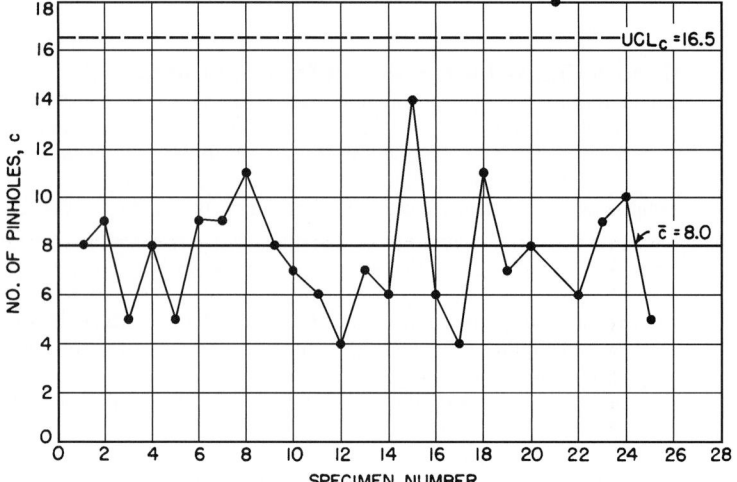

FIGURE 45.4   c chart for pinholes in paper.

the number of connections on that board.

The control limits for a *u* chart are set at

$$\bar{u} \pm 3 \sqrt{\bar{u}/n}$$

Example 45.4:   As an example of the use of a *u* chart, the data in Table 45.5 are the results of the inspection of 10 samples of cloth for imperfections introduced during processing. From the table, we determine that $\bar{u} = 59/1360 = 0.043$. We might then calculate the average sample size as $1360/10 = 136$. Using this figure for *n* in the preceding expression for the control limits, we get

$$UCL_u = 0.043 + 3\sqrt{0.043/136} = 0.043 + 0.054 = 0.097$$

$$LCL_u = 0.043 - 3\sqrt{0.043/136} = 0.043 - 0.054 = <0$$

TABLE 45.5 Inspection Results of Cloth

Square meters of cloth inspected	Nonconformities found	u
200	5	0.025
80	7	0.088
100	3	0.030
300	15	0.050
120	4	0.033
90	6	0.067
250	10	0.040
50	1	0.020
100	6	0.060
70	2	0.029
Totals 1360	59	

There is no lower control limit, and only the second sample is close to the upper control limit; however, since the size of this sample is 80, less than the average sample size of 136, the correct upper control limit will be greater than 0.097, and this sample is in control.

CUMULATIVE SUM (CUSUM) CONTROL CHARTS

Another type of control chart is the *cumulative sum control chart* (often abbreviated as *CUSUM chart*). This chart is particularly useful for detecting small changes (between 0.5σ and 2.5σ) in the parameter being studied. For shifts larger than approximately 2.5σ, the Shewhart-type charts discussed previously are just as good or somewhat better and are easier to understand and use. A CUSUM chart is a plot of the cumulative sum of the deviations between each data point, e.g., a sample average, and a reference value T. Thus this type of chart has a memory feature not found in the previous types of charts. It is usually used to plot the sample average $\overline{X}$, although any of the other statistics, such as s or p, may be used. It is also often used for individual readings, particularly for chemical processes. This section will only discuss the CUSUM chart for averages. For a discussion of other types of CUSUM charts, the reader is referred to Wadsworth et al. (1986). For CUSUM charts, the slope of the plotted line is the important thing, whereas for the previous types of charts it is the distance between a plotted point and the centerline.

CUSUM charts, like other control charts, are interpreted by comparing the plotted points to critical limits. However, the critical limits for a CUSUM control chart are neither fixed nor parallel. A mask in the shape of a V is often constructed. It is laid over the chart with its origin over the last plotted point. If any previously plotted point is covered by the mask, it is an indication that the process has shifted.

Construction of a CUSUM Control Chart for Averages. The following steps may be followed to develop a CUSUM control chart for averages:

1. Obtain an estimate of the standard error of the statistic being plotted; e.g., $\sigma_{\overline{x}}$ may be obtained from a range chart or from some other appropriate estimator. If a range chart is used, the estimate is $\overline{R}/(d_2\sqrt{n})$ or $A_2\overline{R}/3$.

2. Determine the smallest amount of shift in the mean D for which detection is desired. Calculate $\delta = D/\sigma_{\overline{x}}$.

3. Determine the probability level at which decisions are to be made. For limits equivalent to the standard 3σ limits, this is $\alpha = 0.00135$.

4. Determine the scale factor k. This is the change in the value of the statistic to be plotted (vertical scale) per unit change in the horizontal scale (sample number). Ewan (1963) recommends that k be a convenient value between $1\sigma_{\bar{x}}$ and $2\sigma_{\bar{x}}$, preferably closer to $2\sigma_{\bar{x}}$.

5. Obtain the lead distance d from Table BB in Appendix II using the value of δ obtained in step 2.

6. Obtain the mask angle θ from Table BB in Appendix II by setting D/k equal to δ in the table and reading θ from the table. Straight-line interpolation may be used if necessary.

7. Use d and θ to construct the V mask.

8. The sample size for a CUSUM chart for averages is usually the same as for the $\bar{X}$ chart. However, Ewan (1963) suggests, for best results, that one use

$$n = 2.25 s^2 / D$$

where s is an estimate of the process standard deviation.

Operation of the CUSUM V Mask.   The mask is placed over the last point plotted. If any of the previously plotted points are covered by the mask, a shift has occurred. Points covered by the top of the mask indicate a decrease in the process average, whereas those covered by the bottom of the mask indicate an increase. The first point covered by the mask indicates the approximate time at which the shift occurred. If no previous points are covered by the mask, the process is remaining in control.

Some Cautions

1. Periodically check the process variability with an R or s chart before drawing conclusions about the average level of the process.

2. Watch for gradual changes (trends) or changes that come and go in a short period of time. Such changes are not as apparent on a CUSUM chart as they are on an $\bar{X}$ chart.

3. The individual measurements are assumed to follow the normal distribution.

Example 45.5:   The data in Table 45.6 summarize measurements of 20 samples of 4 each taken on the percentage of water absorption in common building brick. A reference value of $T = 10.0$ was used. To illustrate the CUSUM chart, the original data were modified to introduce a decrease in the average level of 2.0 percent, starting with subgroup 11. A range chart shows a value of $\bar{R}$ of 8.08.

1. $\sigma_{\bar{x}}$ is estimated as $\bar{R}/d_2\sqrt{n} = 8.08/2.059\sqrt{4} = 1.96$.

2. We wish to detect a shift in the process mean of $D = 1\sigma_{\bar{x}}$; therefore, $\delta = 1.96/1.96 = 1.0$.

3. We wish to use $\alpha = 0.00135$, the value corresponding to the standard control chart.

4. The scale factor k is calculated as $2\sigma_{\bar{x}} = 2(1.96) = 3.92 \approx 4$.

5. We enter Table BB in Appendix II to get the lead distance of the mask as $d = 13.2$.

6. We substitute $D/k = 2/4 = 0.5$ for δ in Table BB in Appendix II to get $\theta = 14°$.

Figure 45.5 shows the CUSUM chart with the mask at sample number 17. The mask was moved from left to right with the zero point on the mask over each plotted point. For the first 16 points, the mask did not cover any of the previously plotted points. At point 17 the shift was detected, and the mask indicated that the shift occurred at point 10 or 11.

Figure 45.5 also shows the same data plotted on a standard $\bar{X}$ chart. Such a chart would not have detected the shift based on a single point outside the control limits. However, it would have detected the shift at point 17 based on the fact that four of five successive points were outside the 1σ limits. A comparison of the CUSUM chart and the standard $\bar{X}$ chart usually involves the average run length (ARL). This is the average number of sample points plotted at a specified quality level before the chart detects a shift from a previous level. Ewan (1963) compares the CUSUM and $\bar{X}$ charts for various

TABLE 45.6 Data on Percentage Water Absorption

Sample no.	$\overline{X}$	$\overline{X} - 10.0$	$\sum (\overline{X} - 10.0)$
1	15.1	5.1	5.1
2	12.3	2.3	7.4
3	7.4	−2.6	4.8
4	8.7	−1.3	3.5
5	8.8	−1.2	2.3
6	11.7	1.7	4.0
7	10.2	0.2	4.2
8	11.5	1.5	5.7
9	11.2	1.2	6.9
10	10.2	0.2	7.1
11	7.6	−2.4	4.7
12	6.2	−3.8	0.9
13	8.2	−1.8	−0.9
14	7.8	−2.2	−3.1
15	6.8	−3.2	−6.3
16	6.1	−3.9	−10.2
17	4.3	−5.7	−15.9
18	8.5	−1.5	−17.4
19	7.7	−2.3	−19.7
20	9.7	−0.3	−20.0

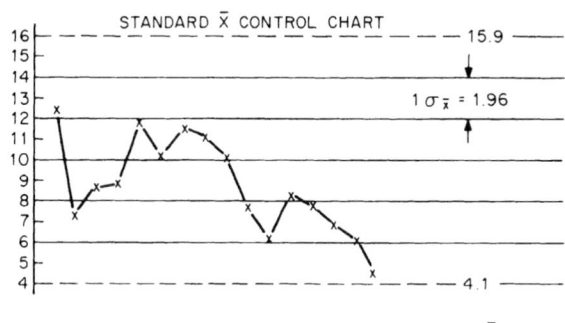

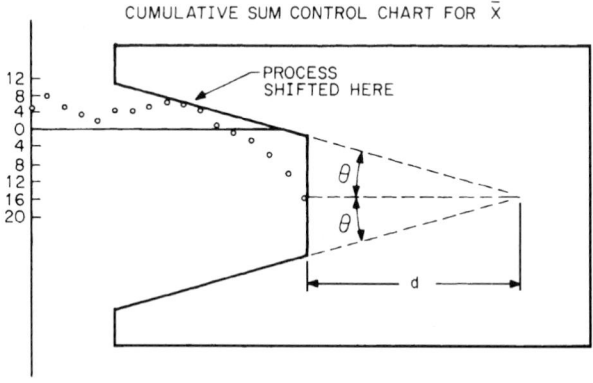

FIGURE 45.5 Comparison of cumulative sum and standard control charts.

amounts of shift in the process mean. For shifts between $0.5\sigma_{\bar{x}}$ and $2.0\sigma_{\bar{x}}$ the CUSUM chart detects the shift with many fewer samples than are needed by the $\bar{X}$ chart. For larger shifts there is no longer any advantage to the CUSUM chart. The preceding statement assumes that the $\bar{X}$ chart uses only the one point outside the control limits rule. As indicated in the preceding example, if the other rules for detecting out-of-control conditions are used, the advantage of the CUSUM chart is diminished.

CUSUM Limits Using the Tabulation Method.   For some processes it may not be convenient to use a V mask. An alternative tabulation method may be used that is particularly well suited for computer applications. This method is equivalent to the charting method with the mask. The procedure is as follows:

1. Form the CUSUM as $C_1 = \sum (\bar{X}_i - K_1)$, where $K_1 = T + D/2$, to detect a shift upward.
2. Form the CUSUM as $C_2 = \sum (\bar{X}_i - K_2)$, where $K_2 = T - D/2$, to detect a shift downward.
3. Tabulate these quantities sequentially with $\bar{X}_i$ ignoring negative values of C_1 and positive values of C_2. That is, reset the upper CUSUM to zero when it is negative and the lower CUSUM to zero when it is positive.
4. Watch the progress of the C_1 and C_2 values. When either value equals or exceeds $Dd/2$ in absolute value, a signal is produced.

Example 45.6:   We will use the same data as in Example 45.5 to illustrate this technique. Recall that for that example, we had $T = 10.0$, $D = 2$, and $d = 13.2$. Then we have for the preceding steps,

1. $K_1 = T + D/2 = 10.0 + 2/2 = 11.0$ and $C_1 = \sum (\bar{X}_i - 11.0)$
2. $K_2 = T - D/2 = 10.0 - 2/2 = 9.0$ and $C_2 = \sum (\bar{X}_i - 9.0)$
3. Decision limit $= Dd/2 = 13.2$

This procedure is illustrated by Table 45.7.

THE EXPONENTIALLY WEIGHTED MOVING AVERAGE CONTROL CHART

Another type of control chart is the exponentially weighted moving average (EWMA) control chart. It was first introduced by Roberts (1959) and later by Wortham and Ringer (1971), who proposed it for applications in the process industries as well as for applications in financial and management control systems for which subgroups are not practical. Like the CUSUM charts, it is useful for detecting small shifts in the mean. Single observations are usually used for this type of chart. The single observations may be averages (when the individual readings making up the average are not available), individual readings, ratios, proportions, or similar measurements. A brief discussion of the chart is presented here. The interested reader should consult the preceding references or other more recent publications such as Hunter (1986), Lowry et al. (1992), Lucas and Saccucci (1990), Ng and Case (1989), Crowder (1989), and Albin et al. (1997).

The plotted statistic is the weighted average of the current observation and all previous observations, with the previous average receiving the most weight, that is,

$$Z_t = \lambda x_t + (1 - \lambda)Z_{t-1} \qquad 0 < \lambda < 1$$

where $Z_0 = \mu$

Z_t = the exponentially weighted moving average at the present time t

Z_{t-1} = the exponentially weighted moving average at the immediately preceding time

TABLE 45.7   Tabulation of Data on Water Absorption

Sample no.	$\overline{X}$	$\overline{X} - 11.0$	C_1	$\overline{X} - 9.0$	C_2	Remarks
1	15.1	4.1	4.1	6.1	>0	
2	12.3	1.3	5.4	3.3	>0	
3	7.4	−3.6	1.8	−1.6	−1.6	
4	8.7	−2.3	<0	−0.3	−1.9	
5	8.8	−2.2	<0	−0.2	−2.1	
6	11.7	0.7	0.7	2.7	>0	
7	10.2	−0.8	<0	1.2	>0	
8	11.5	0.5	0.5	2.5	>0	
9	11.2	0.2	0.7	2.2	>0	
10	10.2	−0.8	<0	1.2	>0	
11	7.6	−3.4	<0	−1.4	−1.4	
12	6.2	−4.8	<0	−2.8	−4.2	
13	8.2	−2.8	<0	−0.8	−5.0	
14	7.8	−3.2	<0	−1.2	−6.2	
15	6.8	−4.2	<0	−2.2	−8.4	
16	6.1	−4.9	<0	−2.9	−11.3	
17	4.3	−6.7	<0	−4.7	−16.0	Lower signal
18	8.5	−2.5	<0	−0.5	−16.5	
19	7.7	−3.3	<0	−1.3	−17.8	
20	9.7	−1.3	<0	0.7	−17.1	

x_t = the present observation
λ = the weighting factor for the present observation

The x_t are assumed to be independent, but the sample statistics Z_t are autocorrelated. However, Wortham and Ringer (1971) demonstrated that for large t, the sample statistic is normally distributed when the x_t are normally distributed with mean μ and variance σ^2. That is,

$$E(Z_t) = \mu$$

and

$$\text{Var}(Z_t) = \sigma^2[\lambda/(2-\lambda)][1 - (1 - \lambda)^{2t}]$$

As t increases, the last term bracketed on the right-hand side converges rapidly to one, and the corresponding expression for the variance becomes

$$\text{Var}(Z_t) \approx \sigma^2 [\lambda/(2 - \lambda)]$$

By choosing $\lambda = 2/(t + 1)$, the variance approximation becomes

$$\text{Var}(Z_t) = \sigma^2/t \quad \text{(the variance of averages of sample size } t\text{)}$$

Under these conditions, the control limits become $\hat{\mu} \pm 3 \sqrt{\hat{\sigma}^2/t}$. For other values of λ, the control limits are

$$\text{UCL} = \hat{\mu} + 3\hat{\sigma} \sqrt{\lambda/(2 - \lambda)}$$

$$\text{LCL} = \hat{\mu} - 3\hat{\sigma} \sqrt{\lambda/(2-\lambda)}$$

For the first few observations, the first equation for the variance should be used. This can be illustrated by the following example. If a good estimate of σ is not available, a range chart should be used with $\hat{\sigma}$ estimated by $\overline{R}/d_2$. In the case of individuals, the average moving range can be used as with the control chart for individuals.

Example 45.7:   To illustrate the EWMA control chart, we will use the same data as used in Example 45.5. Recall from that example that we have results of 20 samples of size 4 each taken on the percentage of water absorption in common building brick. The averages are shown in Table 45.6 and again in Table 45.8. The range chart gave us an estimate of $\sigma_{\bar{x}}$ of 1.96. For the first few samples we will use the complete formula for the standard deviation, and we will use the target value of 10 for our estimate of μ. For this example, a λ of 0.2 was used.

The resulting Minitab output with the data plotted is shown in Figure 45.6 It may be observed from the table and the chart that the control limits have stabilized after the eighth or ninth sample. The plot drops below the lower control limit on the sixteenth sample and stays there for the rest of the run.

The design parameters of this chart are the multiple of σ and the value of λ used. It is possible to choose these parameters to closely approximate the performance of the CUSUM chart. There have been several theoretical studies of the average run length of EWMA charts; see, for example, Lucas and Saccucci (1990). They provide average run length tables for a large range of values of λ and control chart widths. In general, Montgomery (1991) recommends values of λ between 0.05 and 0.25, with $\lambda = 0.08$, $\lambda = 0.10$, and $\lambda = 0.15$ being popular choices. Small values of λ should be used to detect small shifts. Three standard deviation limits, as with the Shewhart charts, seem to work well in most cases; however, if λ is less than 0.10, a value of 2.75σ works somewhat better.

Like the CUSUM chart, the EWMA chart is more effective than the $\bar{X}$ chart in detecting small shifts (less than 2.5σ) in the mean; however, both charts perform worse than the $\bar{X}$ chart for larger shifts. In order to overcome this difficulty, some authors have suggested plotting both the Shewhart limits and the EWMA limits on the same chart (see, for example, Albin et al. 1997).

TABLE 45.8   Data and Calculations for EWMA Chart

Sample t	$\bar{X}_t$	Z_t	Control limits for Z_t	
			LCL	UCL
0		10.00		
1	15.1	11.02	8.8	11.2
2	12.3	11.28	8.5	11.5
3	7.4	10.50	8.3	11.7
4	8.7	10.14	8.2	11.8
5	8.8	9.87	8.2	11.9
6	11.7	10.24	8.1	11.9
7	10.2	10.23	8.1	11.9
8	11.5	10.48	8.1	11.9
9	11.2	10.62	8.1	11.9
10	10.2	10.54	8.1	12.0
11	7.6	9.95	8.1	12.0
12	6.2	9.20	8.0	12.0
13	8.2	9.00		
14	7.8	8.76		
15	6.8	8.37		
16	6.1	7.92		
17	4.3	7.20		
18	8.5	7.46		
19	7.7	7.51		
20	9.7	7.95		

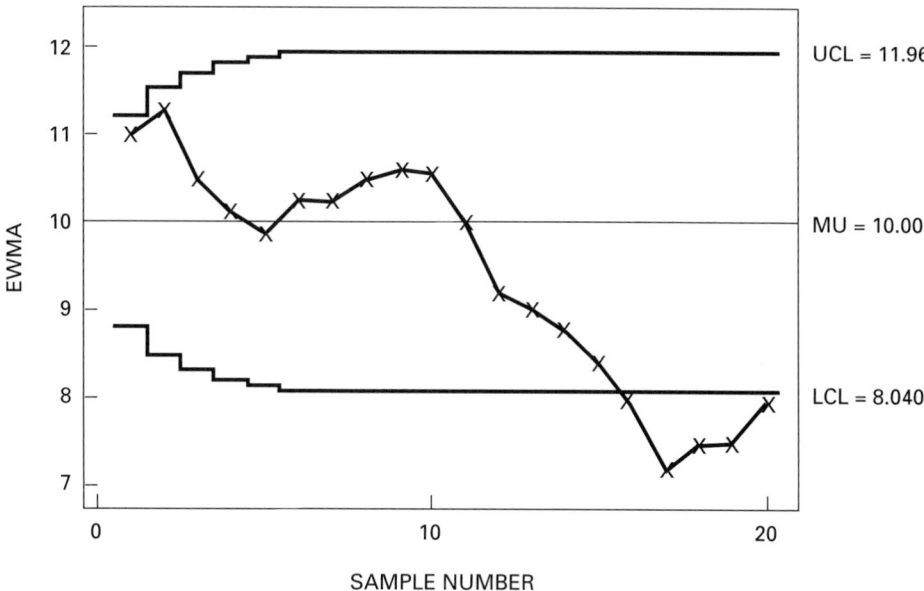

FIGURE 45.6   EWMA chart.

SHORT-RUN CONTROL CHARTS

Some processes are carried out in such short runs that the usual procedure of collecting 20 to 30 samples to establish a control chart is not feasible. Sometimes these short runs are caused by previously known assignable causes that take place at predetermined times. Hough and Pond (1995) discuss four ways to construct control charts for such situations:

1. Ignore the systematic variability, and plot on a single chart.
2. Stratify the data, and plot them on a single chart.
3. Use regression analysis to model the data, and plot the residuals on a chart.
4. Standardize the data, and plot the standardized data on a chart.

The last option has received the most consideration. It involves the use of the transformation

$$Z = \frac{X - \mu}{\sigma}$$

to remove any systematic changes in level and variability. This standardization of Shewhart charts has been discussed by Nelson (1989), Wheeler (1991), and Griffith (1996). Control charts of this form have control limits of ±3.0. These charts are sometimes called *Z charts.*
There are several variations of these charts, some of which are as follows:

1. *Difference charts,* in which a constant such as a known mean (μ), an average from past data ($\overline{X}$), a target value (T), or the specification limit is subtracted from each observation. Burr (1989) discussed this chart and included its use to determine process capability.
2. *Standardized charts,* in which a constant as above is subtracted from each observation and the result is divided by a second constant. When the divisor is the standard deviation, the resulting Z values have a standard deviation of 1, and the control limits are ±3 for $\alpha = 0.0027$. The divisor

can be a known standard deviation (σ) or an estimate from past data, such as $\overline{R}/d_2$. Sometimes the divisor is not a standard deviation but some other value. For example, Burr (1989) discussed a measure used at Kodak in which the nominal was subtracted from the observation and the result divided by half the tolerance. This had the advantage of indicating the fraction of the tolerance used by the process as well as the closeness to the nominal.

3. *Short-run $\overline{X}$, R charts,* in which the statistic

$$Z = (\overline{X} - \overline{\overline{X}})/\overline{R}$$

is plotted against control limits for the mean set at $\pm A_2$. The statistic $R/\overline{R}$ is plotted on a range chart with control limits set at D_3 and D_4. The result is a set of dimensionless charts that are suitable for plotting different parts or runs on the same chart.

Another chart for short runs is the Q chart discussed by Quesenberry (1991, 1995a, 1995b). This technique allows the chart to be constructed from initial data without the need of previous estimates of the mean or variance. It allows charting to begin at the start of a production run. Furthermore, the probability integral transformation is used to achieve normality for Q from nonnormal data such as the range or standard deviation. Quesenberry explains how the method may be used for charts for the mean or standard deviation when those parameters are either known or unknown.

To illustrate the Q chart, consider the case where normally distributed measurements are to be plotted when the mean is unknown and the variance is known. The chart for the process location is constructed as follows:

1. Collect individual measurements: $x_1, x_2, x_3, \dots, x_r$.

2. For the rth point to be plotted, compute

$$Q_r(X_r) = [(r-1)/r]^{1/2} [(x_r - \overline{X}_{r-1})/\sigma] \qquad r = 1, 2, 3,\dots$$

where $\overline{X}_{r-1}$ is the average of the previous $r - 1$ points.

3. Plot $Q_r(X_r)$ for each of the data points against control limits of ± 3.0.

The chart for process variation is constructed similarly:

1. Use the same measurements: $x_1, x_2,\dots, x_r$.
2. For the rth data point to be plotted (plot for only even r)
   a. Compute

$$[(x_r - x_{r-1})^2/2\sigma^2], \qquad r = 2, 4, 6,\dots$$

   b. Find the percentile of the χ^2 distribution with 1 degree of freedom for the value computed in *a*.
   c. Find the normal deviate value for the percentile determined in *b* and set it equal to $Q(R_r)$.
3. Plot $Q(R_r)$ for even data points against control limits of ± 3.0.

BOX-JENKINS MANUAL ADJUSTMENT CHART

J. S. Hunter has suggested an important addition to the charting tools discussed earlier (Hunter 1997). Whereas the Shewhart, CUSUM, and EWMA charts for variables data *monitor* processes, the charts Hunter calls the *Box-Jenkins Manual Adjustment Charts* may be used to *regulate* them. These charts are based on the early work of Box and Jenkins (1962, 1970). Box and Luceño (1997) have recently published a text with a thorough explanation of these techniques. Hunter (1997) includes a worked example accompanied by many graphs.

The Shewhart, CUSUM, and EWMA charts discussed earlier require an operator to plot the time history of the statistic of interest and to leave the process alone as long as the plotted points fall within the control limits and satisfy the run rules. This assumes that the process mean μ is constant and that departures of the data from the mean are independent and normally distributed with a constant variance σ^2. The charts thus *monitor* the process. The objective is to reduce variability by the elimination of special (assignable) cause events identified by the charts.

The Box-Jenkins manual adjustment charts provide a mechanism to forecast and *regulate* the process after each observation. The procedure assumes the process to be *nonstationary*; i.e., the process level is changing constantly, and the variance is increasing. The objective of the procedure is to keep the variance as small as possible.

It is important to note that if the observed deviations from the mean are independent, this method should not be used, and if it is used, it may tend to inflate the process variance. However, if the deviations are not independent, a Box-Jenkins manual adjustment chart can provide both a forecast and the adjustments necessary to force the process to be on target with minimum variance. For a complete discussion of this technique, the reader is referred to the paper by Hunter (1997).

MULTIVARIATE CONTROL CHARTS

When there are two or more correlated quality characteristics that must be controlled simultaneously, such as both the length and weight of a part, the use of multivariate control charts is necessary. If we construct both the dimensions with separate control charts, the region of control will be rectangular, whereas the region using multivariate methods will be elliptical. This means that some observations will be in control in one case and out of control in the other.

Multivariate control charts are beyond the scope of this section, but the reader is referred to Alt et al. (1998), Jackson (1985), or Woodall and Neube (1985) for a detailed discussion of this topic. Multivariate versions of Shewhart, CUSUM, and EWMA charts are available in the literature. They are all discussed with references in Alt et al. (1998).

PRE-CONTROL

The control charts discussed so far in this section provide limits based solely on the observed variation of the process and therefore provide a means to detect process changes. In some situations it is important to detect the presence of only those changes which might incur the presence of nonconforming items. For such cases, control limits may be derived using a combination of the observed frequency and the product specifications. The process is then permitted to change as long as nonconforming items do not become imminent. The chart gives alarm signals only when nonconforming items threaten. Modified control limits [see Wadsworth et al. (1986, pp. 256–266)] and acceptance control charts are two examples of such charts. Acceptance control charts are described in Section 46. Another example of this approach is the PRE-control technique.

PRE-Control starts with a process centered between the specification limits and detects shifts that might result in making some parts outside of the specification limits. PRE-Control requires no plotting and no computations from the sample data. It only requires the inspection of two items to give control information. It uses the normal distribution to determine significant changes in the process that might result in nonconforming items. The principle of PRE-control assumes that the process uses up the entire tolerance spread. That is, the difference between the upper and lower specification limits is 6σ with the process exactly centered.

Two PRE-control (PC) limits are set one-fourth of the way in from each specification limit, as in Figure 45.7. This results in five regions that are often colored for practical applications. The two regions outside the specification limits are colored red. The two regions between the PC limits and the specification limits are colored yellow, and the middle region between the PC limits is colored green. If the distribution were as indicated in Figure 45.7a, 86 percent of the parts will be in the green zone, whereas 7 percent or one part in 14 will be in each yellow zone. The probability that two successive parts will fall in the same yellow zone will be 1/14 times 1/14, or 1/196 (approximately 1 in 200). Therefore, we can say that if two parts in a row fall in the same yellow zone, there is a much greater chance (195/196) that the process mean has shifted than not. It is advisable then to adjust the process toward the center.

It is equally unlikely to get one part in one yellow zone and the next in the other yellow zone. This would not indicate that the process has shifted but that some new factor has been introduced that has caused the variability to increase to an extent that nonconforming pieces are inevitable. An immediate study of the process and some corrective action must be made before it is safe to continue its operation.

These principles lead to the following set of rules that summarize the PRE-control procedure:

1. Divide the specification band with PC lines located one-fourth of the way in from each specification limit. If desired, color the zones appropriately, as indicated above, red, yellow, and green.

2. Start the process.

3. If the first piece is in the red (nonconforming) zone, adjust the process.

4. If the first piece is in the yellow (caution) zone, check the next piece.

5. If the second piece is in the same yellow zone, adjust the process toward the center.

6. If the second piece is in the green (good) zone, continue the process, and adjust the process only when two pieces in a row are in the same yellow zone.

7. If two successive pieces are in opposite yellow zones, stop the process and take action to reduce the variability.

8. When five successive pieces fall in the green zone, *frequency gaging* may start and continue as long as the average number of checks to an adjustment is 25. While waiting for five pieces in the green zone, if a piece falls in the yellow zone, restart the count.

9. During frequency gaging, make no process adjustments until a piece exceeds the PC line (yellow zone). If this occurs, check the next piece and continue as in step 6.

10. When the process is adjusted, five successive pieces in the green zone must again be made before returning to frequency gaging.

11. If the operator checks more than 25 pieces without having to adjust the process, the frequency of checking may be reduced so that more pieces are produced between checks. If, on the other hand, adjustment is needed before 25 pieces are checked, the frequency of checking should be increased. An average of 25 checks between adjustments is an indication that the frequency is correct.

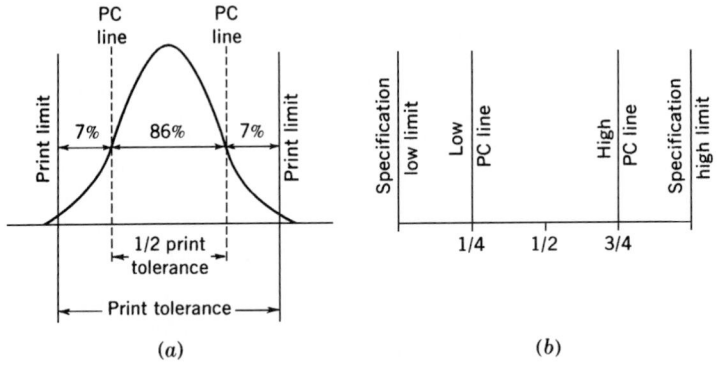

FIGURE 45.7 (*a*) Assumptions underlying PRE-control. (*b*) Location of PRE-control lines.

For one-sided specification limits, a single PC limit is established halfway between the tolerance and zero (in the case of flatness or concentricity) or one-fourth the distance between the specification limit and the best piece (in the case of a variable like yield or strength). In these cases there is but one yellow zone, one red zone, and one green zone.

The PRE-control technique indicates changes in the process mean or variability. The technique is simple to use and understand, it can make use of go no-go gages, and it can guarantee a certain percentage nonconforming if adjustments are made when indicated. On the other hand, process control techniques such as this, which make use of narrow limit gaging, must be introduced to the shop with care. Unless fully explained, the limits appear to be tightening up the tolerances.

Further discussion of this technique and the statistics behind it may be found in Brown (1966) and Shainin (1984). A comparison of PRE-control with $\overline{X}$ and R charts may be found in Sinibaldi (1985). Ledolter and Swersey (1997) evaluate the PRE-control procedure and conclude that PRE-control "is not an adequate substitute for control charts." They go on to explain that control charts are useful for identifying assignable causes of variation and distinguishing them from chance causes. This is particularly important in the early stages of a process when the process capability is likely to be low ($C_p < 1$). PRE-Control is poorly suited for this situation, and its use is likely to lead to excessive tampering with the process.

STATISTICAL CONTROL OF AUTOMATED PROCESSES

Other sections in this handbook have discussed automated manufacturing processes and the role of computers in both manufacturing and nonmanufacturing processes. The march toward automation has sparked some important innovations in data analysis for the control of processes:

1. Continuous-reading instrumentation that yields large amounts of data on process variables and quality characteristics.
2. Automation of statistical analyses (e.g., calculation of averages, standard deviations, and control limits). This is discussed in the next paragraph.
3. Comparison of process results with preset numerical standards. This comparison may result in
   a. The generation of a document giving pertinent information on a nonstandard condition.
   b. The generation of an error signal that automatically makes a process adjustment.
4. In recent years a great improvement in size control devices has been realized. While an item is being ground, a measuring pair of contacts monitors the change in size and electronically instructs the grinding wheel to prepare to stop as the desired size is being approached.

Software for Statistical Process Control.   Many personal computer programs are available for statistical analysis of data, including all types of control charts. Gages with direct digital output into hand-held data-collection devices are available. These devices will store the gage readings collected on the shop floor and download them into a personal computer in the office. As an alternative, gage readings may be entered into the collection device with its keyboard.

The software will calculate the sample statistics, initial control limits, and the control chart. Control limits may be easily recalculated periodically, and ±2σ limits can be calculated as an additional guide. Most software will provide additional summaries and analyses such as listing of the raw data, out-of-specification values, histograms, checks for runs and other patterns within control limits, tests for normality, process capability calculations, Pareto charts, and trend analyses.

The relatively low cost of personal computers and the availability of software have contributed substantially to the renewed interest in $\overline{X}$ and R and other control charts. This computer software combination also has made it practical to collect large quantities of data and subject them to complex analyses. However, process improvement still requires the identification of the vital few and often of unexpected variables causing excess variation. Computer analysis of previous data are no substitute for such tools as design of experiments. Computers are most helpful when the diagnostician has created a template for a spreadsheet program and least helpful when a packaged statistical analysis program is used. Users of such programs are rarely familiar with the detailed logic of the prepackaged programs and can be led to erroneous conclusions.

The American Society for Quality annually publishes a directory of software for process control and other statistical uses (ASQC 1996). In addition, the *Journal of Quality Technology* (ASQ) and *Quality* (Hitchcock Publishing Company) have computer columns describing programs for process control.

REFERENCES

Albin, S. L., Kang, L., and Shea, G. (1997). "An X and EWMA Chart for Individual Observations." *Journal of Quality Technology,* vol. 28, no. 1, pp. 41–48.

Alt, F. B., Smith, N. D., and Jain, K.(1998). "Multivariate Quality Control," in Wadsworth, H. M., (ed.): *Handbook of Statistical Methods for Engineers and Scientists,* 2d ed. McGraw-Hill, New York, chap. 21.

ANSI/ISO/ASQC A3534 (1993). *Statistics—Vocabulary and Symbols,* Part 1: *Probability and General Statistical Terms*; Part 2: *Statistical Quality Control.* American Society for Quality, Milwaukee.

ANSI/ISO/ASQC A8402 (1994). *Quality Management and Quality Assurance—Vocabulary.* American Society for Quality, Milwaukee.

ANSI/ASQC B1 (1985*a*). *Guide for Quality Control.* American Society for Quality, Milwaukee.

ANSI/ASQC B2 (1985*b*). *Control Chart Method of Analyzing Data.* American Society for Quality, Milwaukee.

ANSI/ASQC B3 (1985*c*). *Control Chart Method of Controlling Quality during Production.* American Society for Quality, Milwaukee.

ASQC (1996). "Quality Progress' 13th Annual QA/QC Software Directory." *Quality Progress,* vol. 29, no. 4, pp. 32–59.

ASTM (1990). *ASTM Manual on Presentation of Data.* American Society for Testing Materials, Philadelphia.

AT&T (1984). *Statistical Quality Control Handbook.* AT&T Technologies, Indianapolis.

Box, G. E. P., and Jenkins. G. M. (1962). "Some Statistical Aspects of Adaptive Optimization and Control." *Journal of the Royal Statistical Society* [B], vol. 24, pp. 297–343.

Box, G. E. P., and Jenkins. G. M. (1970). *Time Series Analysis, Forecasting and Control.* Holden Day, San Francisco.

Box, G. E. P., and Luceño, A. (1997). *Statistical Control by Monitoring and Adjustment.* Wiley, New York.

Brown, N. R. (1966). "Zero Defects the Easy Way with Target Area Control." *Modern Machine Shop,* July, pp. 96–100.

Burr, J. T. (1989). "SPC in the Short Run." *Transactions: ASQC Quality Congress.* Milwaukee, pp. 778–780.

Crowder, S. V. (1989). "Design of Exponentially Weighted Moving Average Schemes." *Journal of Quality Technology,* vol. 21, no. 2, pp. 155–162.

Ewan, W. D. (1963). "When and How to Use Cu-SUM Charts." *Technometrics,* vol. 5, no. 1, February, pp. 1–22.

Grant, E. L., and Leavenworth, R. S., (1996). *Statistical Quality Control,* 7th ed., McGraw-Hill, New York.

Griffith, G. K. (1996). *Statistical Process Control Methods for Long and Short Runs,* 2d ed., ASQC Quality Press, Milwaukee.

Hough, L. D., and Pond, A. D. (1995). "Adjustable Individual Control Charts for Short Runs." *Proceedings: 40th ASQC Annual Quality Congress.* Milwaukee, pp. 1117–1125.

Hunter, J. S. (1986). "The Exponentially Weighted Moving Average." *Journal of Quality Technology,* vol. 18, no. 2, pp. 203–210.

Hunter, J. S. (1997). "The Box-Jenkins Manual Adjustment Chart," *Proceedings: 51st Annual Quality Congress.* American Society for Quality, Milwaukee, pp. 158–169.

Jackson, J. E. (1985). "Multivariate Quality Control." *Communications in Statistical Theory Methods,* vol. 14, pp. 2657–2688.

Ledolter, J., and Swersey, A. (1997). "An Evaluation of Pre-Control." *Journal of Quality Technology,* vol. 29, no. 2, pp. 163–171.

Lowry, C. A., Woodall, W. H., Champ, C. W., and Rigdon, S. E. (1992). "Mutivariate Exponentially Weighted Moving Average Control Charts." *Technometrics,* vol. 34, no. 1, pp. 46–53.

Lucas, J. M., and Saccucci, M. S. (1990). "Exponentially Weighted Moving Average Control Schemes: Properties and Enhancements." *Technometrics,* vol. 32, no. 1, pp. 1–12.

Montgomery, D. C. (1991). *Introduction to Statistical Quality Control.* Wiley, New York.

Nelson, L. S. (1989). "Standardization of Shewhart Control Charts." *Journal of Quality Technology,* vol. 21, no. 4, pp. 287–289.

Ng, C. H., and Case, K. E. (1989). "Development and Evaluation of Control Charts Using Exponentially Weighted Moving Averages." *Journal of Quality Technology,* vol. 21, no. 3, pp. 242–250.

Ott, E. R., and Schilling, E. G. (1990). *Process Quality Control,* 2d ed., McGraw-Hill, New York.

Quesenberry, C. P. (1991). "SPC Q Charts for Start-Up Processes and Short or Long Runs." *Journal of Quality Technology,* vol. 23, no. 3, pp. 213–224.

Quesenberry, C. P. (1995a). "On Properties of Binomial Q Charts for Attributes." *Journal of Quality Technology,* vol. 27, no. 3, pp. 204–213.

Quesenberry, C. P., (1995b). "On Properties of Poisson Q Charts for Attributes." *Journal of Quality Technology,* vol. 27, no. 4, pp. 293–303.

Roberts, S. W. (1959). "Control Chart Tests Based on Geometric Moving Averages," *Technometrics,* vol. 1, no. 3, pp. 239–250.

Shainin, D. (1984). "Better Than Good Old $\bar{X}$ and R Charts Asked by Vendors." *ASQC Quality Congress Transactions,* pp. 302–307.

Shewhart, W. A. (1926a). "Quality Control Charts." *Bell System Technical Journal,* pp. 593–603.

Shewhart, W. A. (1926b). "Finding Causes of Quality Variations." *Manufacturing Industries,* pp. 125–128.

Shewhart, W. A. (1927). "Quality Control." *Bell System Technical Journal,* pp. 722–735.

Shewhart, W. A. (1931). *Economic Control of Quality of Manufactured Product.* Van Nostrand-Reinhold, New York. A reprint of this classic work has been published and is available from the American Society for Quality, Milwaukee.

Sinibaldi, F. J. (1985). "PRE-Control, Does It Really Work with Non-Normality." *ASQC Quality Congress Transactions,* pp. 428–433.

Wadsworth, H. M., Stephens, K. S., and Godfrey, A. B. (1986). *Modern Methods for Quality Control and Improvement.* Wiley, New York.

Wheeler, D. J. (1991). *Short Run SPC.* SPC Press, Inc., Knoxville, TN.

Woodall, W. H., and Neube, M. M. (1985). "Multivariate CUSUM Quality Control Procedures." *Technometrics,* vol. 27, pp. 285–292.

Wortham, A. W., and Ringer, L. J. (1971). "Control Via Exponential Smoothing." *The Logistic Review,* vol. 7, no. 3, pp. 33–40.

SECTION 46
ACCEPTANCE SAMPLING[1]

Edward G. Schilling

---

[1]In the Fourth Edition, the section on Acceptance Sampling was prepared by E. G. Schilling and Dan J. Sommers. The present author gratefully acknowledges the work of his previous co-author, and the authors of previous editions, and takes full responsibility for the present manuscript.

INTRODUCTION

Acceptance Sampling—General.   Acceptance sampling refers to the application of specific sampling plans to a designated lot or sequence of lots. Acceptance sampling procedures can, however, be used in a program of *acceptance control* to achieve better quality at lower cost, improved control, and increased productivity. This involves the selection of sampling procedures to continually match operating conditions in terms of quality history and sampling results. In this way the plans and procedures of acceptance sampling can be used in an evolutionary manner to supplement each other in a continuing program of acceptance control for quality improvement with reduced inspection. It is the objective of acceptance control in any application to eventually phase out acceptance sampling in favor of supplier certification and process control. After explaining a variety of specific sampling procedures, this section concludes with suggestions on how and when to progress from sampling inspection toward reliance on process control and check inspection and eventually to no inspection at all, depending on the stage of the life cycle of the product and the state of control.

The disposition of a lot can be determined by inspecting every unit ("100 percent inspection") or by inspecting a sample or portion of the lot. Economy is the key advantage of acceptance sampling as compared with 100 percent inspection. However, sampling has additional advantages:

1. Economy due to inspecting only part of the product
2. Less handling damage during inspection
3. Fewer inspectors, thereby simplifying the recruiting and training problem
4. Upgrading the inspection job from piece-by-piece decisions to lot-by-lot decisions
5. Applicability to destructive testing, with a quantified level of assurance of lot quality
6. Rejections of entire lots rather than mere return of the defectives, thereby providing stronger motivation for improvement

Sampling also has some inherent disadvantages:

1. There are risks of accepting "bad" lots and of rejecting "good" lots.
2. There is added planning and documentation.
3. The sample usually provides less information about the product than does 100 percent inspection.

Acceptance sampling under a system of acceptance control provides assurance that lots of product are acceptable for use. It is assumed that acceptable product is normally being submitted by honest, conscientious, capable suppliers. The plans provide warnings of discrepancies as a means of forestalling use of the product while providing signals of the presence of assignable causes for both the producer and the consumer. If this is not so, other means of dealing with the supplier should be used—such as discontinuance or 100 percent inspection. This is why it is called acceptance sampling.

Why Is Sampling Valid?   The broad scheme of use of sampling is shown in Figure 46.1. It is universally realized that inspection of a sample gives information then and there about the quality of the pieces in the lot. But that is only the beginning. Beyond this, a vast area of knowledge is unfolded because *the product can tell on the process.*

The sample, being the result of the variables present in the process at the time of manufacture, can give evidence as to those variables. Thereby it is possible to draw conclusions as to whether the process was doing good work or bad at the time it produced the samples.

From these conclusions about the process, it is then possible to reverse the reasoning, so that the known *process now can tell on the product.* We already know about the inspected product, but the knowledge of the process gives information about the uninspected pieces. In this way acceptance sampling of a lot is valid not only because the uninspected pieces are neighbors of the inspected

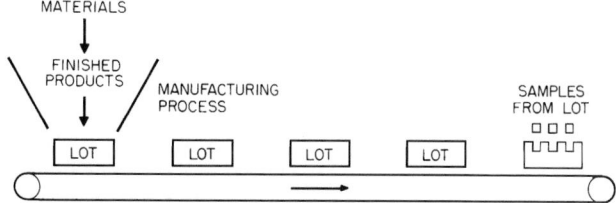

FIGURE 46.1   Measurement of a sample tells (1) whether the pieces in the sample are good or bad; (2) whether the process, at the time the sample was made, was doing good work or bad; (3) whether the uninspected pieces made at the same time by the same process are good or bad (principle of sampling inspection); (4) whether the process is stable; (5) whether the unmanufactured pieces are going to be good or bad (control inspection).

pieces; acceptance sampling is also valid because the uninspected pieces may be derived from the same process which the inspected pieces have labeled as "good."

When we go a step further and examine a series of samples, we learn whether the process is stable or not. Once the series of samples tells on the process by certifying stability, we can use this knowledge of stability to predict the quality of unmanufactured product.

Cowden (1957) has summarized the characteristics of a good acceptance plan. It will:

1. Protect the producer against having lots rejected when the product is in a state of control, and satisfactory as to level and uniformity
2. Protect the consumer against the acceptance of bad lots
3. Provide long-run protection to the consumer
4. Encourage the producer to keep the process in control
5. Minimize sampling costs, inspection, and administration
6. Provide information concerning product quality

Types of Sampling.   Any acceptance sampling application must distinguish whether the purpose is to accumulate information on the immediate *product* being sampled or on the *process* which produced the immediate lot at hand. Accordingly, two types of sampling have been distinguished:

Type A: Sampling to accept or reject the immediate lot of product at hand

Type B: Sampling to determine if the process which produced the product at hand was within acceptable limits

The type of sampling will determine the appropriate probability distribution to be used in characterizing the performance of the plan. In addition, the type of data generated will also play a role. In acceptance sampling, data can be of the following types:

• Attributes—go no-go information

Defectives—usually measured in proportion or percent defective. This refers to the acceptability of units of product for a wide range of characteristics.

Defects—usually measured by actual count or as a ratio of defects per unit. This refers to the number of defects found in the units inspected, and hence can be more than the number of units inspected.

• Variables—measurement information

Variables—usually measured by the mean and standard deviation. This refers to the distribution of a specific measurable characteristic of the product inspected.

Terminology of Acceptance Sampling.   The terminology of acceptance sampling has evolved over the years into a precise and well-directed set of terms defining various properties of acceptance sampling plans and procedures. These are clearly described in the international standard ANSI/ISO/ASQC A3534 (1993), "Statistics—Vocabulary and Symbols."

It is important to distinguish between product which is not fit for use and product which does not meet specification requirements. This is done in ANSI/ISO/ASQC A3534 (1993), Part 2, which uses the term "defect" in relation to the former and "nonconformity" in relation to the latter. These are defined as

> *Defect:*   The nonfulfillment of an intended usage requirement.
>
> *Nonconformity:*   The nonfulfillment of a specified requirement.

Clearly a unit of product containing one or more defects or nonconformities is a "defective" or "nonconforming unit."

Since acceptance sampling plans usually relate to requirements imposed by the customer (internal or external), the terms "defect" and "defective" will be generally used here. Furthermore, in referring to the literature of acceptance sampling, the terms "defect" and "defective" will be commonly found. However, in appropriate instances in this section such as variables sampling plans which compare measurements to specifications, the terms "nonconformity" and "nonconforming unit" will be used.

Acceptance Sampling Procedures

Forms of Sampling.   Sampling plans can be classified in two categories: attributes plans and variables plans.

Attributes Plans.   In these plans, a sample is taken from the lot and each unit classified as conforming or nonconforming. The number nonconforming is then compared with the acceptance number stated in the plan and a decision is made to accept or reject the lot. Attributes plans can further be classified by one of the two basic criteria:

1. Plans which meet specified sampling risks provide protection on a lot-by-lot basis. Such risks are
   a. A specified quality level for each lot (in terms of percent defective) having a selected risk (say 0.10) of being accepted by the consumer. The specified quality level is known as the lot tolerance percent defective or limiting quality (p_2); the selected risk is known as the consumer's risk (β).
   b. A specified quality level for each lot such that the sampling plan will accept a stated percentage (say 95 percent) of submitted lots having this quality level. This specified quality level is termed the acceptable quality level (AQL). The risk of rejecting a lot of AQL quality (p_1) is known as the producer's risk (α).
2. Plans which provide a limiting average percentage of defective items for the long run. This is known as the average outgoing quality level (AOQL).

Variables Plans.   In these plans, a sample is taken and a measurement of a specified quality characteristic is made on each unit. These measurements are then summarized into a simple statistic (e.g., sample mean) and the observed value compared with an allowable value defined in the plan. A decision is then made to accept or reject the lot. When applicable, variables plans provide the same degree of consumer protection as attributes plans while using considerably smaller samples.

Attributes plans are generally applied on a percent defective basis. That is, the plan is instituted to control the proportion of accepted product which is defective or out of specification. Variables plans for percent defective are also used in this way. Such plans provide a sensitivity greater than attributes but require that the shape of the distribution of individual measurements must be known and stable. The shape of the distribution is used to translate the proportion defective into specific values of process parameters (mean, standard deviation) which are then controlled.

Variables plans can also be used to control process parameters to given levels when specifications are directed toward the process average or process variability and not specifically to percent defective. These variables plans for process parameter do not necessarily require detailed knowledge of the shape of the underlying distribution of individual measurements.

Sampling plans used in reliability and in the sampling of bulk product are generally of this type. Published plans in the reliability area, however, usually require detailed knowledge of the shape of the distribution of lifetimes. Some of the important features of attributes and variables plans for percent defective are compared in Table 46.1.

The principal advantage of variables plans for percent defective over corresponding attributes plans is a reduction in the sample size needed to obtain a given degree of protection. Table 46.2 shows a comparison of variables sample sizes necessary to achieve the same protection as the attributes plan: $n = 125$, $c = 3$ (sample size of 125 units, allowable number of defectives of 3).

Types of Sampling Plans.   In single-sampling plans, the decision to accept or reject a lot is based on the results of inspection of a single group of units drawn from the lot. In double-sampling plans, a smaller initial sampling is usually drawn, and a decision to accept or reject is reached on the basis of this smaller first sample if the number of defectives is either quite large or quite small. A second sample is taken if the results of the first are not decisive. Since it is necessary to draw and inspect the second sample only in borderline cases, the average number of pieces inspected per lot is generally smaller with double sampling. In multiple-sampling plans, one, or two, or several still smaller samples of n individual items are taken (usually truncated after some number of samples) until a decision to accept or reject is obtained. The term "sequential-sampling plan" is generally used when

TABLE 46.1   Comparison of Attributes and Variables Sampling Plans for Percent Defective

Feature	Attributes	Variables
Inspection	Each item classified as defective or nondefective. Go no-go gages may be employed.	Each item measured. Inspection more sophisticated. Higher inspection and clerical cost.
Distribution of individual measurements	Need not be known.	*Must* be known (normal usually assumed).
Type of defect	Any number of defect types can be assessed by one plan.	Separate plan required for each type of defect.
Sample size	Depends on protection required.	Smaller sample size for same protection as attributes (at least 30% smaller*).
Process information	Percent defective.	Percent defective plus valuable information on process average and variability for corrective action.
Severity	Weights all defectives of a given kind equally.	Weights each unit inspected by its proximity to specifications.
Evidence to supplier	Defectives available as evidence.	Possible for lot to be rejected on sample containing no defectives.
Measurement errors	Measurements not recorded.	Measurements available for review.
Screened lots	No effect on performance of plan.	Screened lots may be rejected in error even though they contain no defectives.

*Bowker and Goode (1952), pp. 32–33. Assumes single sample of one characteristic.

TABLE 46.2 Comparison of Variables and Attributes Sample Sizes*

$p_1 = 0.0109; \alpha = 0.05; p_2 = 0.0535; \beta = 0.10$

Plan	Sample size
Single-sampling attributes	125
Variables:	
σ known	19
σ unknown (s)	52
σ unknown ($\overline{R}$ of groups of 5)	75
Sequential sampling, σ known (ASN at p_1)	10.3

*Specifications assumed to be $> 6\sigma$ apart if two-sided.

the number of samples is unlimited and a decision is possible after each individual unit has been inspected.

Sampling Schemes and Systems.   While simple sampling plans are often employed solely in sentencing individual lots, sampling schemes and systems are generally used in acceptance control applications involving a steady flow of product from the producer. The ANSI/ISO/ASQC A3534 (1993) standard defines a sampling plan as "a specific plan which states the sample size(s) to be used and the associated criteria for accepting the lot," and a sampling scheme as "a combination of sampling plans with rules for changing from one plan to another." Finally, it defines an acceptance sampling system as, "a collection of sampling schemes, each with its own rules for changing plans, together with criteria by which appropriate schemes may be chosen." Thus, $n = 134$, $c = 3$ is a sampling plan; Code J, 1.0 percent AQL is a sampling scheme; and ANSI/ASQC Z1.4 (1993) is a sampling system.

Published Tables and Procedures.   From the point of view of ease of negotiation between the producer and the consumer, it is usually best to use the published procedures and standards. This avoids problems of credibility created when one party or the other generates its own sampling plan. Also, the legal implications of using plans which appear in the literature (such as Dodge-Romig plans) or have been subjected to national consensus review (such as ANSI/ASQC Z1.4, 1993) are obvious. Unique plans, specifically generated for a given application, are probably best used internally.

Screening.   In view of advances in automatic inspection equipment and computer integrated manufacturing, 100 percent inspection, or screening, has become more attractive. This is particularly true in the short run; however, dependence on screening may inhibit efforts for continual improvement as a long-term strategy. Screening procedures have been designed to address important considerations such as the selection of a screening variable, prior information on the population, cost, losses due to erroneous decisions, variation in product quality, and environmental considerations. Methods include: statistical models, such as Deming's all-or-none rule, Taguchi's model for tolerance design, and economic and statistical models using correlated variables. Other issues include the possibility of burn-in, group testing, allocation of inspection effort, selection of process parameters, and selective assembly. An excellent systematic literature review will be found in Tang and Tang (1994). An example of the use of a correlated variable to improve 100 percent inspection is given by Mee (1990). Jaraiedi, Kochhar, and Jaisingh (1987) discuss a model for determining the average outgoing quality for product which has been subjected to multiple 100 percent inspections. A computer program for application of their results will be found in Nelson and Jaraiedi (1987).

SAMPLING RISKS AND PARAMETERS

Risks.   When acceptance sampling is conducted, the real parties of interest are the producer (supplier or Production department) and the consumer, i.e., the company buying from the supplier or the department to use the product. Since sampling carries the risk of rejecting "good" lots and of accepting "bad" lots, with associated serious consequences, producers and consumers have attempted to standardize the concepts of what constitutes good and bad lots, and to standardize also the risks associated with sampling. These risks are stated in conjunction with one or more parameters, i.e., quality indices for the plan. These indices are as follows.

Producer's Risk.   The producer's risk α is the probability that a "good" lot will be rejected by the sampling plan. In some plans, this risk is fixed at 0.05; in other plans, it varies from about 0.01 to 0.10. The risk is stated in conjunction with a numerical definition of the maximum quality level that may pass through the plan, often called the acceptable quality level.

Acceptable Quality Level.   Acceptable quality level (AQL) is defined by ANSI/ISO/ASQC A3534 as follows: "When a continuing series of lots is considered, a quality level which for purposes of sampling inspection is the limit of satisfactory process average." A sampling plan should have a low producer's risk for quality which is equal to or better than the AQL.

Consumer's Risk.   The consumer's risk β is the probability that a "bad" lot will be accepted by the sampling plan. The risk is stated in conjunction with a numerical definition of rejectable quality such as lot tolerance percent defective.

Lot Tolerance Percent Defective.   The lot tolerance percent defective (LTPD) is the level of quality that is unsatisfactory and therefore should be rejected by the sampling plan. A consumer's risk of 0.10 is common and LTPD has been defined as the lot quality for which the probability of acceptance is 0.10; i.e., only 10 percent of such lots will be accepted. (LTPD is a special case of the concept of limiting quality, LQ, or rejectable quality level, RQL. The latter terms are used in tables that provide plans for several values of the consumer's risk as contrasted to a value of 0.10 for the LTPD.) Limiting quality (LQ) plans will be found in the international standard ISO 2859-2 (1985) and in the U.S. national standard Q3 as well as the LTPD plans in Dodge and Romig (1959).

A third type of quality index, average outgoing quality limit, is used with 100 percent inspection of rejected lots and will be discussed later in this section.

The producer's and consumer's risks and associated AQL and LTPD are summarized graphically by an operating characteristic curve.

Operating Characteristic Curve.   The operating characteristic (OC) curve is a graph of lot fraction defective versus the probability that the sampling plan will accept the lot.

Figure 46.2 shows an ideal OC curve for a case where it is desired to accept all lots 3 percent defective or less and reject all lots having a quality level greater than 3 percent defective. Note that all lots less than 3 percent defective have a probability of acceptance of 1.0 (certainty); all lots greater than 3 percent defective have a probability of acceptance of zero. Actually, however, no sampling plan exists that can discriminate perfectly; there always remains some risk that a "good" lot will be rejected or

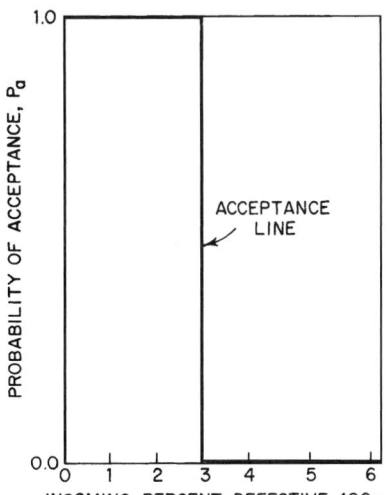

FIGURE 46.2   An ideal sampling plan performance.

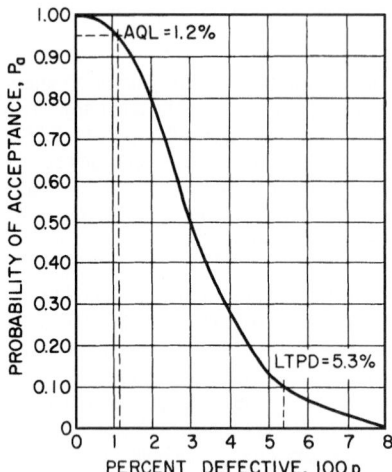

FIGURE 46.3   An actual sampling plan performance.

that a "bad" lot will be accepted. The best that can be achieved in practice is to control these risks.

Figure 46.3 shows the curve of behavior that would be obtained if an operator were instructed to take a sample of 150 pieces from a large lot and to accept the lot if no more than four defective pieces were found. Such curves can be constructed from the appropriate probability distribution. For example, using the Poisson table (Appendix II, Table E), we find that for a sample size of 150 ($n = 150$) and 2 percent defective ($p = 0.02$) we would expect $np = 3.0$ defectives in the sample. The table shows for $r = 4$ and $np = 3.0$, the probability of acceptance (for four or fewer defectives) is 0.815.

It is seen from this curve that a lot 3 percent defective has one chance in two of being accepted. However, a lot 3.5 percent defective, though technically a "bad" lot, has 39 chances in 100 of being accepted. In like manner, a lot 2.5 percent defective, though technically a good lot, has 34 chances in 100 of not being accepted.

The effect of parameters of the sampling plan on the shape of the OC curve is demonstrated in Figure 46.4, where the curve for perfect discrimination is given along with the curves for three particular acceptance sampling plans. The following statements summarize the effects:

1. When the sample size approaches the lot size or, in fact, approaches a large percentage of the lot size, and the acceptance number is chosen appropriately, the OC curve approaches the perfect OC curve (the rectangle at p_1).

2. When the acceptance number is zero, the OC curve is exponential in shape, concave upward (see curves 2 and 3).

3. As the acceptance number increases, the OC curve is pushed up, so to speak, for low values of p, and the probability of acceptance for these quality levels is increased, with a point of inflection at some larger value of p (see curve 1).

4. Increasing the sample size and the acceptance number together gives the closest approach to the perfect discrimination OC curve (see curve 1).

It is sometimes useful to distinguish between type A and type B OC curves (see Dodge and Romig, 1959, pp. 56–59). Type A curves give the probability of acceptance for an individual lot that comes from finite production conditions that cannot be assumed to continue in the future. Type B curves assume that each lot is one of an infinite number of lots produced under essentially the same production conditions. In practice, most OC curves are viewed as type B. With the few exceptions noted, this section assumes type B OC curves.

Average Sample Number Curve.   The average sample number (ASN) is the average number of units inspected per lot in sampling inspection, ignoring any 100 percent inspection of rejected lots. In single-sampling inspection the ASN is equal to n, the sample size. However, in double- and multiple-sampling plans the probability of not reaching a decision on the initial sample, and consequently being forced to inspect a second, third, etc., sample must be considered. For a double-sampling plan with sample sizes n_1 and n_2, the average sample number, ASN is simply

$$\text{ASN} = n_1 + (1 - P_{a1})\, n_2$$

where P_{a1} is the probability of acceptance on the first sample. The ASN for multiple sampling is more complicated to calculate; see Schilling (1982).

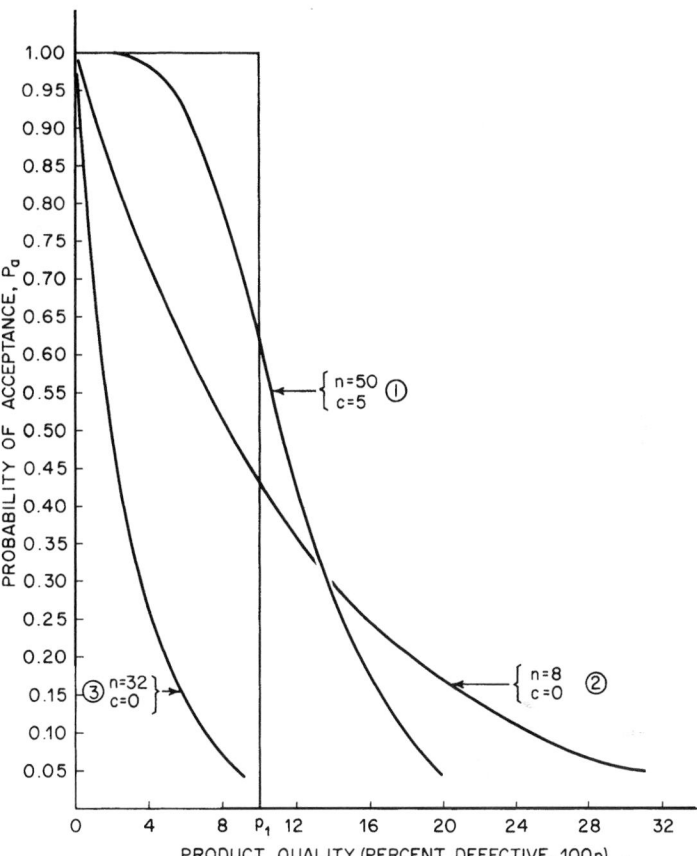

FIGURE 46.4 Shape of the OC curve.

A comparison of ASN for some specific plans, roughly matching $n = 125$, $c = 3$, is shown in Figure 46.5. (The calculations assume that all samples selected are completely inspected.) As indicated, double and multiple sampling generally lead to economies over single sampling when quality is either very good or very poor.

Average Outgoing Quality Limit.

The acceptable quality level (AQL) and lot tolerance percent defective (LTPD) have been cited as two common quality indices for sampling plans. A third commonly used index is the average outgoing quality limit (AOQL). The AOQL is the upper limit of average quality of outgoing product from accepted lots and from rejected lots which have been screened.

The AOQL concept stems from the relationship between the fraction defective before inspection (incoming quality) and the fraction defective after inspection (outgoing quality) when inspection is nondestructive and rejected lots are screened. When incoming quality is perfect, outgoing quality must likewise be perfect. However, when incoming quality is very bad, outgoing quality will also be near perfect, because the sampling plan will cause most lots to be rejected and 100 percent inspected. Thus, at either extreme—incoming quality very good or very bad—the outgoing quality will tend to be very good. It follows that between these extremes is the point at which the percent defective in the outgoing material will reach its maximum. This point is known as the AOQL.

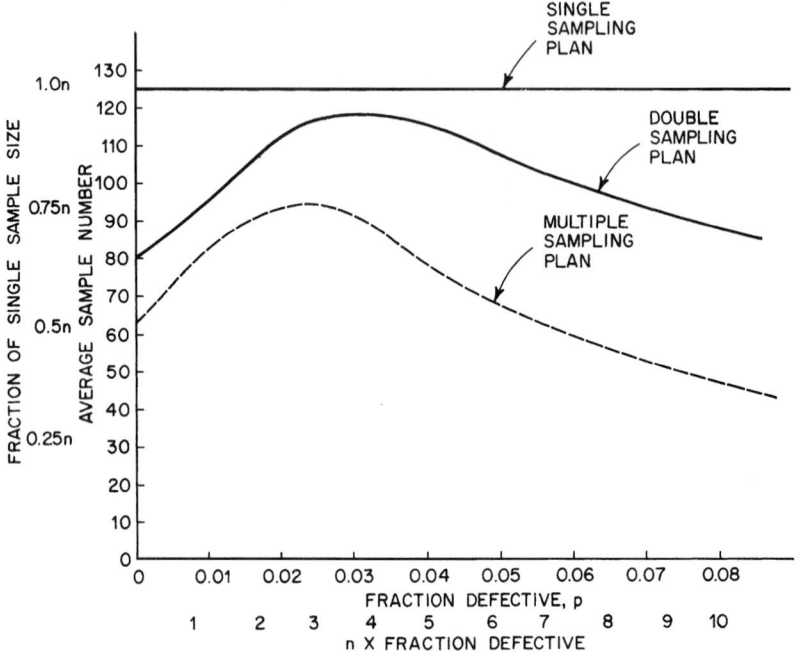

FIGURE 46.5   Average sample number versus fraction defective.

If p is incoming quality, P_r is the probability of lot rejection, and all rejected lots are screened and made free of defects (i.e., 0 percent), then

$$\text{AOQ} = (p)P_a + (0)P_r = (p)P_a$$

This calculation is approximate in that it does not account for the small effect of 100 percent inspection of the units in the samples from accepted lots. Taking this into account:

$$\text{AOQ} = pP_a \left(1 - \frac{n}{N} \right)$$

for sample size n and lot size N.

Average Outgoing Quality Curve.   An example of the calculation for the AOQL and a plot of the average outgoing quality (AOQ) curve is shown in Table 46.3 and Figure 46.6. Schilling (1982) has indicated that for acceptance numbers of five or less

$$\text{AOQL} \simeq \frac{0.4}{n} (1.25c + 1)$$

Average Total Inspection.   The average total inspection (ATI) takes into account the likelihood of 100 percent inspection of rejected lots when this is possible; i.e., inspection is nondestructive. The lot size N must now be taken into account. Again with single-sampling plans with sample size n, lot size N, probability of acceptance P_a, and probability of rejection P_r,

$$\text{ATI} = P_a n + P_r N$$

TABLE 46.3 Computations for Average Outgoing Quality Limit (AOQL); for This Example, $n = 78$, $c = 1$

Incoming quality fraction defective $= p$	np	Probability of acceptance $= P_a$	Average outgoing quality (AOQ) $= p \times P_a$
0.005	0.39	0.940	0.00470
0.010	0.78	0.820	0.00820
0.015	1.17	0.680	0.01020
0.020	1.56	0.550	0.01100*
0.025	1.95	0.430	0.01075
0.030	2.34	0.330	0.00990
0.035	2.73	0.250	0.00875
0.040	3.12	0.190	0.00760
0.045	3.51	0.140	0.00630
0.050	3.90	0.100	0.00500
0.055	4.29	0.075	0.00402
0.060	4.68	0.050	0.00300

*AOQL $\cong$ maximum AOQ $\cong 0.01100 = 1.1\%$.

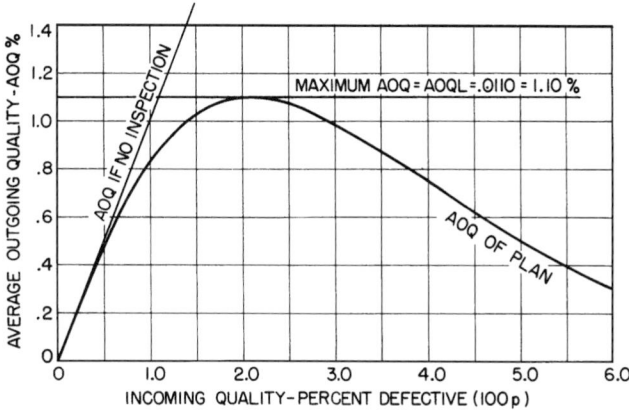

FIGURE 46.6 AOQ curve and AOQL for a typical sampling plan.

An extension to double sampling is simply

$$ATI = n_1 + (1 - P_{a1})n_2 + (N - n_1 - n_2)(1 - P_a)$$

where P_{a1}, is the probability of acceptance on the first sample. Extension to multiple-sampling plans of k levels is straightforward and will not be derived here. See Schilling (1982).

Minimum Total Inspection.
Minimum inspection per lot, for a given type of protection, can be illustrated by the following example:

Assume that a consumer establishes acceptance criteria as an LTPD of 5 percent. A great many sampling plans meet this criterion. Some of these plans are

Take from each lot a sample n of	Accept the lot if the number of defectives does not exceed the maximum acceptance number c
46	0
78	1
106	2
134	3
160	4

Figure 46.7 shows the corresponding operating characteristic curves. Each plan has an LTPD of 5 percent; i.e., the probability of acceptance of a submitted inspection lot with fraction defective of 0.05 is 0.10. Which plan should be used? One logical basis for choosing among these plans is to use that one which gives the least inspection per lot.

The total number of units inspected consists of (1) the sample which is inspected for each lot and (2) the remaining units which must be inspected in those lots which are rejected by the sampling inspection. The number of lots rejected in turn depends on the normal level of defectives in the product so that minimum inspection is a function of incoming quality.

A sample computation of minimum inspection per lot is shown in Table 46.4. It is assumed that rejected lots are detail (100 percent) inspected. For small acceptance numbers the total inspection is high because many lots need to be detailed. For large acceptance numbers the total is again high, this time because of the large size of samples. The minimum sum occurs at a point between these extremes.

From the foregoing it is seen that for any specified conditions of

Lot tolerance percent defective

Consumer's risk

Lot size

Process average

it becomes possible to derive the values of

$$n = \text{sample size}$$

$$c = \text{allowable number of defectives in the sample}$$

to obtain

$$I_m = \text{minimum inspection per lot}$$

Similar reasoning may be applied in making a selection from a group of sampling plans designed to give the same average outgoing quality limit (AOQL).

Complete tables for sampling have been derived for minimum inspection per lot for a variety of $\bar{p}$, LTPD, and AOQL values (e.g., Dodge and Romig 1959). These will be described later.

Selection of a Numerical Value of the Quality Index.   There are three alternatives for evaluating lots: no inspection, inspect a sample, 100 percent inspection (or more). An economic evaluation of these alternatives requires a comparison of *total* costs under each of the alternatives.

Enell (1954) has suggested that a break-even point be used in the selection of quality index:

$$P_b = \frac{I}{A}$$

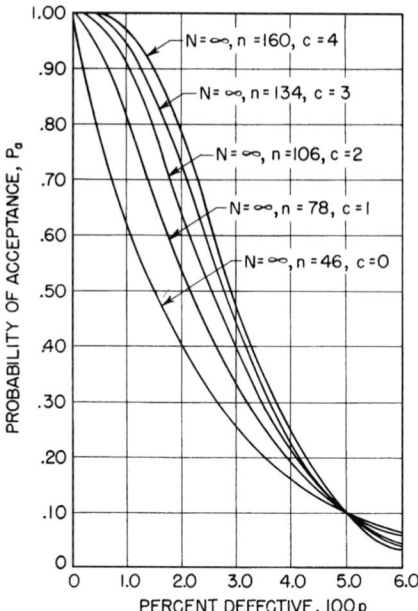

FIGURE 46.7   Family of sampling plans each having LTPD = 0.05.

TABLE 46.4   Computation of Minimum Inspection per Lot for 5 Alternative n-c Combinations Appropriate to Lots of $N=1000$ Articles

All plans have the same lot tolerance percent defective. Incoming material has a process average percent defective, $\bar{p} = 0.5\%$.

				Average no. pieces inspected		
Sample size n	Allowable number of defects c	Probability of acceptance by sampling P_a	Probability of inspecting residue of lot $1 - P_a$	In sample* n	In rest of lot† $(N - n) \times (1 - P_a)$	Total inspected per lot
46	0	0.795	0.205	46	196	242
78	1	0.940	0.060	78	55	133
106	2	0.983	0.017	106	15	121‡
134	3	0.995	0.005	134	4	138
160	4	0.998	0.002	160	2	162

*The sample size indicates the number inspected from each lot.
†The size of the uninspected residue of the lot, multiplied by the probability that it will have to be inspected because of rejection of the sample.
‡This is the minimum sought.

where I = inspection cost per item and A = damage cost incurred if a defective slips through inspection. If it is thought that the lot quality p is less than p_b, the total cost will be lowest with sampling inspection or no inspection. If p is greater than p_b, 100 percent inspection is best. If p is unknown, it may be best to sample using an appropriate sampling plan such as an AOQL scheme.

For example, a microcomputer device costs $0.50 per unit to inspect. A damage cost of $10.00 is incurred if a defective device is installed in the larger system. Therefore,

$$p_b = \frac{0.50}{10.00} = 0.05 = 5.0\%$$

If it is expected that the percent defective will be greater than 5 percent, then 100 percent inspection should be used. Otherwise, use sampling or no inspection.

The formula assumes that the sample size is small compared to the lot size, the cost to replace a defective found in inspection is borne by the producer or is small compared to the damage or inconvenience caused by a defective, and no inspection errors occur.

As a 5 percent defective quality level is the break-even point between sorting and sampling, the appropriate sampling plan should provide for a lot to have a 50 percent probability of being sorted or sampled; i.e., the probability of acceptance for the plan should be 0.50 at a 5 percent defective quality level. The operating characteristic curves in a set of sampling tables such as ANSI/ASQC Z1.4 can now be examined to determine an AQL. For example, suppose that the device is inspected in lots of 3000 pieces. The operating characteristic curves for this case (code letter K) are shown in ANSI/ASQC Z1.4. The plan closest to having a P_a of 0.50 for a 5 percent level is the plan for an AQL of 1.5 percent. Therefore, this is the plan to adopt. Other economic models have been developed. For example, see Tagaras (1994) for the case of variables sampling.

In practice, the quantification of the quality index is a matter of judgment based on the following factors:

1. Past performance on quality
2. Effect of nonconforming product on later production steps
3. Effect of nonconforming product on fitness for use
4. Urgency of delivery requirements
5. Cost to achieve the specified quality level

IMPLEMENTATION OF AN ACCEPTANCE SAMPLING PROCEDURE

Assumptions Made in Sampling Plans

Inspection Error.   In implementing acceptance sampling plans, it is commonly assumed that:

1. The inspectors follow the prescribed sampling plan.
2. The inspection is made without error; i.e., no human or equipment mistakes are made in measurement or in judging conformance.

In practice, these assumptions are not fully valid. See Section 23, under Inspection Errors.

The effect of inspection error has received considerable attention in the literature of quality. For example, see Johnson, Kotz, and Rodriguez (1985, 1986, 1988, 1990) and Suich (1990).

Units of Product.   These may consist of (1) discrete pieces or (2) specimens from bulk material. The criteria used for judging the conformance of a single unit of product to standard are somewhat different for these two categories. The criteria used for judging lot conformance differ even more widely.

Seriousness Classification of Defects. Many sampling plans set up their criteria for judging lot conformance in terms of an allowable number of defects in the sample. Since defects differ greatly in seriousness, the sampling plans must somehow take these differences into account.

Where there exists a formal plan for seriousness classification of defects, the sampling plans may be structured so that:

1. A separate sampling plan is used for each seriousness class, e.g., large sample sizes for critical defects, small sample sizes for minor defects.
2. A common sampling plan is used, but the allowable number of defects varies for each class; e.g., no critical defects are allowed, but some minor defects are allowed.
3. The criteria may be established in terms of defects per hundred units, the allowable number being different for each class.
4. The criteria may be based on demerits per unit; i.e., all defects found are converted to a scale of demerits based on the classification system.

In the absence of a formal system of seriousness classification, all defects are considered equally important during the sampling inspection. However, when nonconforming lots are subsequently reviewed to judge fitness for use, the review board gives consideration to the seriousness of the defects.

Lot Formation.
The general approach to lot formation is discussed in Section 23, under Degree of Inspection and Test Needed. While most acceptance sampling plans can be validly applied regardless of how lots are formed (skip-lot plans are an exception), the economics of inspection and the quality of the accepted product are greatly influenced by the manner of lot formation.

The interrelation of lot formation to economics of inspection is discussed in Section 23 and will not be elaborated here. The interrelation of lot formation to quality of accepted product can be seen from a single example.

Ten machines are producing the same product. Nine of these produce perfect product. The tenth machine produces 100 percent defectives. If lots consist of product from single machines, the defective product from the tenth machine will always be detected by sampling. If, however, the lots are formed by mixing up the work from all machines, then it is inevitable that some defects will get through the sampling plan.

The fact that lot formation so strongly influences outgoing quality and inspection economics has led to some guidelines for lot formation:

1. Do not mix product from different sources (processes, production shifts, input materials, etc.) unless there is evidence that the lot-to-lot variation is small enough to be ignored.
2. Do not accumulate product over extensive periods of time (for lot formation).
3. Do make use of extraneous information (process capability, prior inspections, etc.) in lot formation. Such extraneous information is especially useful when product is submitted in isolated lots, or in very small lots. In such cases the extraneous information may provide better knowledge on which to base an acceptance decision than the sampling data.
4. Do make lots as large as possible consistent with the above to take advantage of low proportionate sampling costs. (Sample sizes do not increase greatly despite large increases in lot sizes.)

When production is continuous (e.g., the assembly line) so that the "lot" is necessarily arbitrary, the sampling plans used are themselves designed to be of a "continuous" nature. These continuous sampling plans are discussed later in this section.

Selecting the Sample. The results of sampling are greatly influenced by the method of selecting the sample. In acceptance sampling, the sample should be representative of the lot. In those cases where the inspector has knowledge of how the lot was formed, this knowledge can be used in selecting the sample by stratification (see below). Lacking this knowledge, the correct approach is to use random sampling.

Random Sampling.   All published sampling tables are prepared on the assumption that samples are drawn at random; i.e., at any one time each of the remaining uninspected units of product has an equal chance of being the next unit selected for the sample. To conduct random sampling requires that (1) random numbers be generated and (2) these random numbers be applied to the product at hand.

Random numbers are available in prefabricated form in tables of random numbers (Appendix II, Table CC). One uses such a table by entering it at random (without "looking") and then proceeding in some chosen direction (up, down, right, left, etc.) to obtain random numbers for use. Numbers which cannot be applied to the product arrangement are discarded.

Random numbers may also be generated by various devices. These include:

1. *Calculators or computers:*   Many calculators are available with random-number routines built in. Computers are, of course, an excellent source of random numbers. Statistical software often has random numbers built in.

2. *A bowl of numbered chips or marbles:*   After mixing, one is withdrawn and its number recorded. It is then replaced and the bowl is again mixed before the next number is withdrawn.

3. *Random number dice:*   One form is icosahedron (20-sided) dice. There are three of these, each of a different color, one for units, one for tens, and one for hundreds. (Each die has the numbers from 0 to 9 appearing twice.) Hence one throw of the three dice displays a random number within the interval 000 to 999.

Once the random numbers are available, they must be adapted to the form in which the product is submitted. For systematically packed material, the container system can be numbered to correspond to the system of random numbers. For example, a lot might be submitted in 8 trays, each of which has 10 rows and 7 columns. In such a case, the trays might be numbered from 0 to 7, the rows from 0 to 9, and the columns from 0 to 6. Then, using three-digit random numbers, the digits are assigned to trays, rows, and columns, respectively.

For bulk packed materials, other practical procedures may be used. In the case of small parts, they may be strewn onto a flat surface which is marked with grid lines in a 10×10 arrangement. Based on two-digit random numbers, the cell at the intersection of these digits is identified. Within the cell, further positional identity can be determined with a third digit.

For fluid or well-mixed bulk products, the fluidity obviates the need for random numbers, and the samples may be taken from "here and there."

Stratified Sampling.   When the "lots" are known to come from different machines, production shifts, operators, etc., the product is actually multiple lots which have been arbitrarily combined. In such cases, the sampling is deliberately stratified; i.e., an attempt is made to draw the sample proportionately from each true lot. However, within each lot, randomness is still the appropriate basis for sampling.

A further departure from randomness may be due to the economics of opening containers, i.e., whether to open few containers and examine a few pieces from each.

Sampling Bias.   Unless rigorous procedures are set up for sampling at random and/or by stratification, the sampling can deteriorate into a variety of biases which are detrimental to good decision making. The more usual biases consist of:

1. Sampling from the same location in all containers, racks, or bins

2. Previewing the product and selecting only those units which appear to be defective (or nondefective)

3. Ignoring those portions of the lot which are inconvenient to sample

4. Deciding on a pattern of stratification in the absence of knowledge of how the lot was made up

The classic example is the legendary inspector who always took samples from the four corners and center of each tray and the legendary production worker who very carefully filled these same spots with perfect product.

Because the structured sampling plans do assume randomness, and because some forms of sampling bias can significantly distort the product acceptance decisions, all concerned should be alert to plan the sampling to minimize these biases. Thereafter, supervision and auditing should be alert to assure that the actual sampling conforms to these plans.

ATTRIBUTES SAMPLING

Overview of Single, Double, Multiple, and Sequential Plans. In general, single, double, multiple, and sequential sampling plans can be planned to give lots of specified qualities nearly the same chance of being accepted; i.e., the operating-characteristic curves can be made quite similar (matched) if desired. However, the best type of plan for one producer or product is not necessarily best for another. The suitability of a plan can be judged by considering the following factors:

1. Average number of parts inspected
2. Cost of administration
3. Information obtained as to lot quality
4. Acceptability of plan to producers

The advantages and disadvantages of the four forms of sampling plans are tabulated in Table 46.5. The average number of parts that need to be inspected to arrive at a decision varies according to the plan and the quality of the material submitted. In cases where the cost of inspection of each piece is substantial, the reduction in number of pieces inspected may justify use of sequential sampling despite its greater complexity and higher administrative costs. On the other hand, where it is not practicable to hold the entire lot of parts while sampling and inspection are going on, it becomes necessary to set aside the full number of items that may need to be inspected before inspection even begins. In these circumstances single sampling may be preferable if the cost of selecting, unpacking, and handling parts is appreciable. It is of course simplest to train personnel, set up records, and administer a single-sampling plan. A crew of inspectors hastily thrown together cannot easily be taught all the intricacies of the more elaborate plans. However, double-sampling plans have been demonstrated to be simple to use in a wide variety of conditions, economical in total cost, and acceptable psychologically to both producer and consumer.

Lot-by-Lot Inspection. When product is submitted in a series of lots (termed "lot-by-lot"), the acceptance sampling plans are defined in terms of

N = lot size

n = sample size

c = acceptance number, i.e., the allowable number of defects in the sample

r = rejection number

When more than one sample per lot is specified, the successive sample sizes are designated as n_1, n_2, n_3, etc. The successive acceptance numbers are c_1, c_2, c_3, etc. The successive rejection numbers are r_1, r_2, r_3, etc.

Selecting Sampling Plans. Sampling plans are often specified by choosing two points on the OC curve: the AQL denoted by p_1 and the LTPD symbolized by p_2. Tables have been developed to facilitate the selection process by using unity (np) values from the Poisson distribution. See Cameron (1952), Duncan (1974), and Schilling and Johnson (1980). Such tables use the operating

TABLE 46.5 Comparative Advantages and Disadvantages of Single, Double, and Multiple Sampling

Feature	Single sampling	Double sampling	Multiple sampling	Sequential
Acceptability to producer	Psychologically poor to give only one chance of passing the lot	Psychologically adequate	Psychologically open to criticism as being indecisive	Psychologically open to criticism as being more indecisive than multiple
Number of pieces inspected per lot*	Generally greatest	Usually (but not always) 10 to 50% less than single sampling	Generally (but not always) less than double sampling by amounts of the order of 30%	Minimum over all attributes plans
Administration cost in training, personnel, records, drawing and identifying samples, etc.	Lowest	Greater than single sample	Greater	Greatest
Information about prevailing level of quality in each lot	Most	Less than single sample	Less than double	Least

*This is not to be confused with total cost of inspection, which includes administration cost of the plan.

ratio, $R = p_2/p_1$, in conjunction with the unity value of (np_1) or (np_2) to obtain the sample size and acceptance number(s) for the plan whose OC curve passes through the points specified. Often other unity values are provided which assess other properties of the plan. Such tables are illustrated below by the Schilling-Johnson (1980) table, reproduced in part as Table 46.6, which uses producer risk $\alpha = 0.05$ and consumer risk $\beta = 0.10$ with equal sample sizes at each stage of a double or multiple sampling plan. The table provides matched sets of single, double, and multiple sampling plans, such that their OC curves are essentially equivalent. For example, if $p_1 = 0.012$ and $p_2 = 0.053$ then $R = 4.4$. The closest value given under the column for R is 4.058, corresponding to an acceptance number of 4. The sample size is then found to be $n = 7.994 \div 0.053 = 150.8$, which would usually be taken to be 150. The operating characteristic curve for the plan can be seen in Figure 46.3.

Single Sampling.   In single sampling by attributes, the decision to accept or reject the lot is based on the results of inspection of a single sample selected from the lot. The operation of a single-sampling plan by attributes is given in Figure 46.8. The characteristics of an attributes sampling plan are given in Table 46.7.

Double Sampling.   In double sampling by attributes, an initial sample is taken, and a decision to accept or reject the lot is reached on the basis of this first sample if the number of nonconforming units is either quite small or quite large. A second sample is taken if the results of the first sample are not decisive. The operation of an attributes double-sampling plan is given in Figure 46.9. The characteristics of an attributes double-sampling plan are given in Table 46.8.

Multiple Sampling.   In multiple sampling by attributes, more than two samples can be taken in order to reach a decision to accept or reject the lot. The chief advantage of multiple sampling plans is a reduction in sample size for the same protection. The operation of multiple sampling by attributes is given in Figure 46.10. The characteristics of a multiple sampling plan are given in Table 46.9.

Sequential Sampling.   In sequential sampling, each item is treated as a sample of size 1, and a determination to accept, reject, or continue sampling is made after inspection of each item. The major advantage of sequential sampling plans is that they offer the opportunity for achieving the minimum sample size for a given protection. The operation of sequential sampling by attributes is given in Figure 46.11. The characteristics of a sequential sampling plan are given in Table 46.10. Attributes sequential plans will be found tabulated by producer's and consumer's risk in the international standard ISO 8422 (1991).

Rectification Schemes.   Rectification schemes are used when it is desired to ensure that the average outgoing quality level of a series of lots will not exceed specified levels. Such schemes employ 100 percent inspection (screening), with nonconforming items replaced by conforming items.

There are two basic types of rectification schemes: LTPD schemes and AOQL schemes. LTPD schemes ensure consumer quality level protection for each lot, while AOQL schemes ensure AOQL protection for a series of lots. Both types of schemes minimize ATI at the projected process average percent nonconforming. See Dodge and Romig (1959).

Continuous Sampling.   Some production processes deliver product in a continuous stream rather than on a lot-by-lot basis. Separate plans have been developed for such continuous production based on the AOQL concept. These plans generally start with 100 percent inspection until some consecutive number of units free of defects is found and then provide for inspection on a sampling basis until a specified number of defective units is found. One hundred percent inspection is then instituted again until a specified number of consecutive good pieces is found, at which time sampling is reinstituted. Continuous sampling plans have been proposed by Harold F. Dodge and modifications

TABLE 46.6 Unity Values for Construction and Evaluation of Single, Double, and Multiple Sampling Plans

$(n_1 = n_2 = \cdots = n_k$; # indicates acceptance not allowed at a given stage)

Plan	Acceptance numbers	$R = p_2/p_1$	np_2		.99	.95	.90	.75	.50	.25	.10	.05	.01	.005	.001	.0005	.0001
													Probability of acceptance				
0S	Ac = 0 Re = 1	44.893	2.303	np ASN n_1	.0101 1	.0513 1	.105 1	.288 1	.693 1	1.386 1	2.303 1	2.996 1	4.605 1	5.298 1	6.908 1	7.601 1	9.206 1
1S	Ac = 1 Re = 2	10.958	3.890	np ASN n_1	.149 1	.355 1	.532 1	.961 1	1.678 1	2.693 1	3.890 1	4.744 1	6.638 1	7.430 1	9.234 1	10.000 1	11.759 1
1D	Ac = 0 1 Re = 2 2	12.029	2.490	np ASN n_1	.0860 1.079	.207 1.168	.310 1.228	.566 1.321	1.006 1.368	1.661 1.316	2.490 1.206	3.124 1.137	4.649 1.045	5.324 1.026	6.914 1.007	7.604 1.004	9.209 1.001
1M	Ac = # # 0 0 1 1 2 Re = 2 2 2 3 3 3	8.903	.917	np ASN n_1	.0459 3.254	.103 3.501	.148 3.637	.252 3.774	.416 3.640	.643 3.169	.917 2.601	1.121 2.270	1.602 1.761	1.815 1.618	2.325 1.388	2.549 1.319	3.075 1.205
2S	Ac = 2 Re = 3	6.506	5.322	np ASN n_1	.436 1	.818 1	1.102 1	1.727 1	2.674 1	3.920 1	5.322 1	6.296 1	8.406 1	9.274 1	11.230 1	12.053 1	13.934 1
2D	Ac = 0 3 Re = 3 4	5.357	3.402	np ASN n_1	.363 1.298	.635 1.443	.827 1.511	1.231 1.581	1.816 1.564	2.566 1.450	3.402 1.306	3.986 1.222	5.290 1.097	5.852 1.066	7.201 1.025	7.810 1.016	9.295 1.005
2M	Ac = # 0 0 1 2 3 4 Re = 2 3 3 4 4 5 5	6.244	1.355	np ASN n_1	.111 2.432	.217 2.789	.293 2.983	.451 3.207	.683 3.165	.988 2.776	1.355 2.261	1.635 1.950	2.343 1.470	2.671 1.344	3.458 1.167	3.803 1.122	4.602 1.060

Plan	Ac	Re															
3S	3	4	np	4.891	.823	1.366	1.745	2.535	3.672	5.109	6.681	7.754	10.045	10.978	13.062	13.935	15.922
			$\dfrac{\text{ASN}}{n_1}$	6.681	1	1	1	1	1	1	1	1	1	1	1	1	1
3D	1 4	4 5	np	4.398	.635	1.000	1.246	1.750	2.465	3.373	4.398	5.130	6.808	7.542	9.270	10.019	11.757
			$\dfrac{\text{ASN}}{n_1}$	4.398	1.130	1.245	1.316	1.421	1.470	1.414	1.293	1.211	1.084	1.053	1.017	1.010	1.003
3M	#012346	3345667	np	4.672	.200	.348	.446	.642	.910	1.246	1.626	1.901	2.553	2.848	3.566	3.887	4.650
			$\dfrac{\text{ASN}}{n_1}$	1.626	2.461	2.820	3.026	3.286	3.288	2.935	2.450	2.156	1.693	1.559	1.340	1.274	1.163
4S	4	5	np	4.058	1.279	1.970	2.433	3.369	4.671	6.274	7.994	9.154	11.605	12.594	14.795	15.711	17.792
			$\dfrac{\text{ASN}}{n_1}$	7.994	1	1	1	1	1	1	1	1	1	1	1	1	1
4D	3 5	6 6	np	4.102	1.099	1.633	1.992	2.728	3.789	5.162	6.699	7.762	10.047	10.978	13.062	13.933	15.909
			$\dfrac{\text{ASN}}{n_1}$	6.699	1.025	1.077	1.125	1.233	1.341	1.345	1.242	1.164	1.055	1.033	1.009	1.005	1.001
4M	#123456	3446677	np	4.814	.266	.440	.558	.798	1.141	1.591	2.118	2.502	3.385	3.763	4.640	5.016	5.884
			$\dfrac{\text{ASN}}{n_1}$	2.118	2.128	2.300	2.417	2.590	2.618	2.384	2.021	1.792	1.427	1.326	1.174	1.132	1.070
5S	5	6	np	3.550	1.785	2.613	3.152	4.219	5.670	7.423	9.275	10.513	13.109	14.150	16.455	17.411	19.578
			$\dfrac{\text{ASN}}{n_1}$	9.275	1	1	1	1	1	1	1	1	1	1	1	1	1
5D	2 6	5 7	np	3.547	1.116	1.630	1.959	2.607	3.490	4.579	5.781	6.627	8.537	9.357	11.253	12.066	13.928
			$\dfrac{\text{ASN}}{n_1}$	5.781	1.097	1.199	1.263	1.360	1.405	1.352	1.243	1.171	1.064	1.039	1.012	1.007	1.002
5M	#123579	45678910	np	3.243	.490	.700	.830	1.079	1.410	1.814	2.270	2.604	3.411	3.776	4.642	5.017	5.884
			$\dfrac{\text{ASN}}{n_1}$	2.270	2.496	2.906	3.143	3.459	3.516	3.188	2.677	2.347	1.791	1.628	1.367	1.292	1.171

Source: Schilling and Johnson, 1980, p. 221.

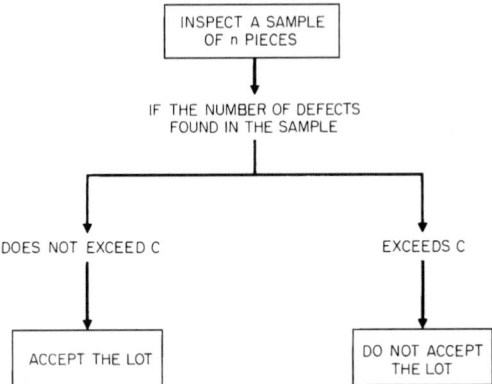

SCHEMATIC OPERATION OF SINGLE SAMPLING

INSPECT A SAMPLE
OF n PIECES

IF THE NUMBER OF DEFECTS
FOUND IN THE SAMPLE

DOES NOT EXCEED C

EXCEEDS C

ACCEPT THE LOT

DO NOT ACCEPT
THE LOT

FIGURE 46.8   Schematic of single sampling. In practice, the lot not to be accepted may be repaired, junked, etc. Sampling tables usually assume that the lot is detail-inspected and the defective pieces are all repaired or replaced by good pieces.

TABLE 46.7   Single Sampling by Attributes

Example: Determine a plan which will have AQL = 100 p_1 = 1.0% with producer risk α = 0.05 and LTPD = 100p_2 = 5.0% with consumer risk β = 0.10. Suppose a sample is taken from a lot of N = 500 and four nonconforming units are found.

Summary of plan	Calculations
I. Restrictions: Random sample of dichotomous data	

II. Necessary information
  A. Producer quality level, p_1
  B. Producer risk, α
  C. Consumer quality level, p_2
  D. Consumer risk, β
	II.
	  A. p_1 = 0.01
	  B. α = 0.05
	  C. p_2 = 0.05
	  D. β = 0.10

III. Selection of plan (using Table 46.6)
  A. Calculate operating ratio
 	
$$R = p_2 / p_1$$
 	
  B. Choose single sampling plan (S) from Table 46.6 which shows $R \leq$ the value calculated
  C. Acceptance and rejection numbers are given
  D. Determine sample size from value of (np_2) shown for the plan as $n = (np_2)/p_2$; round up
	III.
	  A. R = 0.05/0.01
	    = 5
	
	  B. Choose Plan 3S since R shown is 4.891
	
	  C. Ac = 3
	    Re = 4
	  D. $n = \dfrac{6.681}{0.05}$

TABLE 46.7 Single Sampling by Attributes (*Continued*)

Summary of plan	Calculations

IV. Elements
    A. Sample size: See III above
    B. Statistic: d = number nonconforming in sample
    C. Decision criteria:
        1. Accept if $d \leq c$
        2. Reject if $d > c$
	IV.
	    A. n = 134
	    B. d = 4
	
	    C. $4 > 3$, reject

V. Action: Dispose of lot as indicated by decision rules.	V. Reject

VI. Measures (using Table 46.6)
    A. Probability of acceptance (OC curve)
        1. Under each value of probability of acceptance listed across top of table, there corresponds a value of np shown for the plan
        2. Divide the values of np by sample size n to get p
 	
        3. Draw OC curve from corresponding values of p and P_a
    B. Average sample number (ASN curve)
        1. Calculate p from P_a as above
 	
        2. Under the value np from which p was calculated is listed a value of ASN/n_1
        3. Multiply ASN/n_1 by n to obtain ASN corresponding to p
        4. Draw ASN curve from corresponding values of p and P_a
    C. Average outgoing quality
        1. Use formula
$$AOQ \simeq p\,P_a$$
 	
    D. Average outgoing quality limit
 	
$$AOQL \simeq \frac{0.4}{n}(1.25c + 1) \text{ for } c \leq 5$$
 	
    E. Average total inspection
$$ATI = n\,P_a + N(1 - P_a)$$
	VI.
	    A.
	        1. $P_a = 0.50$ corresponds to $np = 3.672$
	
	        2. $p = \dfrac{3.672}{134}$
	           = 0.027
	        3. $p = 0.027$
	           $P_a = 0.50$
	    B.
	        1. $P_a = 0.50$
	           $p = 0.027$
	        2. $ASN/n_1 = 1$
	
	        3. $(1)(134) = 134$
	
	        4. $p = 0.027$
	          $ASN = 134$
	    C.
	        1. at $p = 0.027$,
	          $AOQ = 0.027\,(0.50)$
	               = 0.014
	    D.
	        1. $AOQL \simeq$
	$$\frac{0.4}{134}[1.25(3)+1] \simeq 0.019$$
	    E. at $p = 0.027$
	        $ATI = 134(0.5) + 500(0.5)$
	           = 317

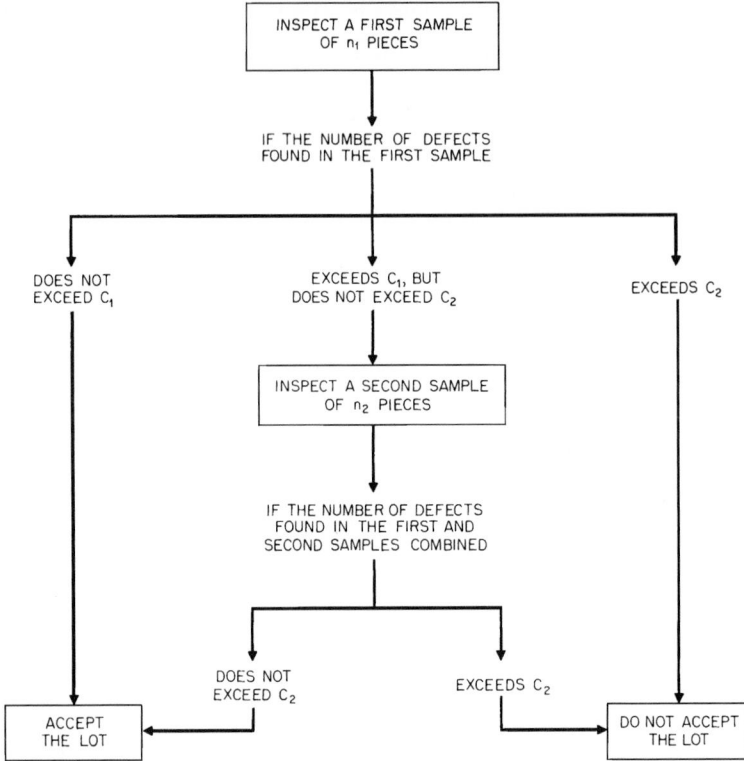

SCHEMATIC OPERATION OF DOUBLE SAMPLING

INSPECT A FIRST SAMPLE
OF n_1 PIECES

IF THE NUMBER OF DEFECTS
FOUND IN THE FIRST SAMPLE

DOES NOT EXCEED C_1

EXCEEDS C_1, BUT DOES NOT EXCEED C_2

EXCEEDS C_2

INSPECT A SECOND SAMPLE
OF n_2 PIECES

IF THE NUMBER OF DEFECTS
FOUND IN THE FIRST AND
SECOND SAMPLES COMBINED

DOES NOT EXCEED C_2

EXCEEDS C_2

ACCEPT THE LOT

DO NOT ACCEPT THE LOT

FIGURE 46.9   Schematic of double sampling. In regard to the lot not to be accepted, inspect the remainder of the pieces, replacing or repairing those defective.

TABLE 46.8   Double Sampling by Attributes

Example: Determine a double sampling plan which will have AQL = $100p_1$ = 1.0% with producer risk α = 0.05 and LTPD = $100p_2$ = 5.0% with consumer risk β = 0.10. Suppose a sample is taken from a lot of N = 500 and four nonconforming units are found.

Summary of plan	Calculations
I. Restrictions: Random sample of dichotomous data	
II. Necessary information	II.
A. Producer quality level, p_1	*A.* p_1 = 0.01
B. Producer risk, α	*B.* α = 0.05
C. Consumer quality level, p_2	*C.* p_2 = 0.05
D. Consumer risk, β	*D.* β = 0.10
III. Selection of plan (using Table 46.6)	III.
A. Calculate operating ratio	*A.* R = 0.05/0.01
$R = p_2 / p_1$	= 5

TABLE 46.8 Double Sampling by Attributes (*Continued*)

Summary of plan	Calculations
B. Choose double sampling plan (*D*) from Table 46.6 which shows $R \leq$ the value calculated	*B.* Choose plan 3D since R shown is 4.398
C. Acceptance (*a*) and rejection (*r*) numbers are given as Ac and Re	*C.* First sample $a_1 = 1, r_1 = 4$ Second sample $a_2 = 4, r_2 = 5$
D. Determine first sample size from value of (np_2) shown for the plan as $n = (np_2)/p_2$ This is sample size of each of the double samples	*D.* $n = 4.398/0.05$ $= 88$

IV. Elements
   A. Sample size: See III above
	IV.
	
	  A. $n_1 = 88$
	     $n_2 = 88$

B. Statistic: $d_1 =$ number nonconforming in first sample; $d_2 =$ number nonconforming in second sample	*B.* $d_1 = 4$

C. Decision criteria
     1. Accept on first sample if $d_1 \leq a_1$
     2. Reject on first sample if $d_1 \geq r_1$
     3. Take a second sample if $a_1 < d_1 < r_1$
     4. Accept on second sample if $d_1 + d_2 \leq a_2$
     5. Reject on second sample if $d_1 + d_2 \geq r_2$
	C.
	
	    1. 4 not ≤ 1
	    2. $4 \geq 4$, reject

V. Action:
  Dispose of lot as indicated by decision rules.
	V.
	
	  Reject

VI. Measures (using Table 46.6)
  A. Probability of acceptance (OC curve)
    1. Under each value of probability of acceptance listed across the top of table, there corresponds a value of np shown for the plan
	VI.
	  A.
	    1. $P_a = 0.50$ corresponds to $np = 2.465$

2. Divide the values of np by sample size n to get p	2. $p = \dfrac{2.465}{88}$ $= 0.028$

3. Draw OC curve from corresponding values of p and P_a
  B. Average sample number (ASN curve)
    1. Calculate p from P_a as above
	3. $p = 0.028$
	      $P_a = 0.50$
	  B.
	    1. $P_a = 0.50$
	      $p = 0.028$

2. Under the value np from which p was calculated is listed a value of ASN/n_1	2. $ASN/n_1 = 1.470$
3. Multiply ASN/n_1 by n_1 to obtain ASN corresponding to p	3. 1.470(88) $= 129.4$
4. Draw ASN curve from corresponding values of p and P_a	4. $p = 0.028$ ASN $= 129.4$

C. Average outgoing quality
    1. Use formula
 	
         $AOQ \simeq p\, P_a$
	C.
	    1. at $p = 0.028$
	
	      $AOQ = 0.028(.5)$
	           $= 0.014$

TABLE 46.8 Double Sampling by Attributes (*Continued*)

Summary of plan	Calculations

D. Average outgoing quality limit
   1. AOQL = maximum of AOQ curve
	D.
	   1. AOQL = 0.0151 at
	        p = 0.022

E. Average total inspection ATI = $n_1 + n_2(1 - P_{a_1}) + (N - n_1 - n_2)(1 - P_a)$ where
   P_{a_1} is probability of acceptance on first sample
	E. P_{a_1} = probability no more than one defective in 88
	   P_{a_1} = 0.29
	   ATI = 88 + 88 (0.71) + 324(0.50)
	       = 312.5

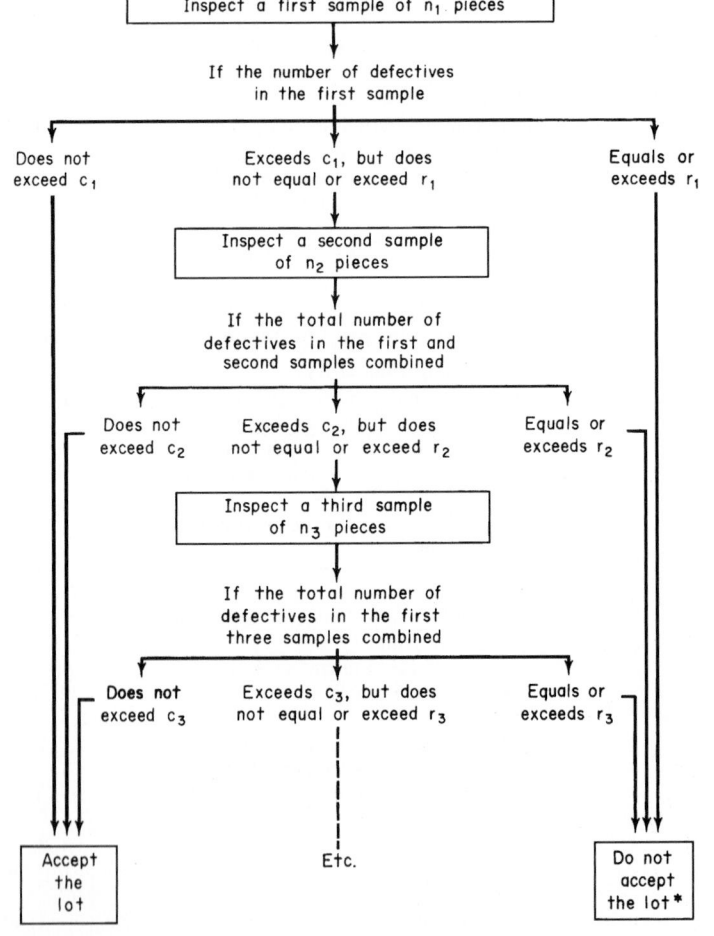

FIGURE 46.10 Schematic operation of multiple sampling. The asterisk means some of these plans continue to the "bitter end"; i.e., the taking of samples continues if necessary until the lot is fully inspected, unless the plan has meanwhile "made up its mind." Other plans, described as "truncated," are designed to force a decision after a certain number of inconclusive samples have been examined.

TABLE 46.9 Multiple Sampling by Attributes

Example: Determine a multiple sampling plan which will have AQL = $100p_1$ = 1.0% with producer risk $\alpha = 0.05$ and LTPD = $100p_2$ = 5.0% with consumer risk $\beta = 0.10$. Suppose a sample is taken from a lot of $N = 500$ and four nonconforming units are found.

Summary of plan	Calculations
I. Restrictions: Random sample from dichotomous population	

II. Necessary information
  A. Producer quality level, p_1
  B. Producer risk, α
  C. Consumer quality level, p_2
  D. Consumer risk, β
	II.
	  A. $p_1 = 0.01$
	  B. $\alpha = 0.05$
	  C. $p_2 = 0.05$
	  D. $\beta = 0.10$

III. Selection of plan (Table 46.6)
  A. Calculate operating ratio
      $R = p_2/p_1$
  B. Choose multiple sampling plan (M) from Table 46.6 which shows $R \le$ the value calculated
  C. Acceptance (a) and rejection (r) numbers are given as Ac and Re; note # indicates acceptance is not possible at that sample
  D. Determine first sample size from value of (np_2) shown for the plan as
      $n = (np_2)/p_2$
This is sample size of each of the multiple samples
	III.
	  A. $R = 0.05/0.01 = 5$
	
	  B. Choose plan 3M since R shows 4.672
	
	  C.
	    Ac = #, 0, 1, 2, 3, 4, 6
	    Re = 3, 3, 4, 5, 6, 6, 7
	
	  D. $n = 1.626/0.05$
	    = 33

IV. Elements
  A. Sample size: See III above
  B. Statistic: $\sum_{i=1}^{k} d_i$ = total number nonconforming up to kth sample
  C. Decision Criteria
    1. Accept on kth sample if
      $\sum_{i=1}^{k} d_i \le a_k$
    2. Reject on kth sample if
      $\sum_{i=1}^{k} d_i \ge r_k$
    3. Continue sampling if
      $a_k < \sum_{i=1}^{k} d_i < r_k$
	IV.
	  A. $n = 33$
	  B. $\sum_{i=1}^{1} d_i = 4$
	
	  C.
	    1. # indicates cannot accept
	
	    2. $4 \ge 3$, reject on first sample

V. Action Dispose of lot as indicated by decision rules	V. Reject

VI. Measures (using Table 46.6)
  A. Probability of acceptance (OC curve)
    1. Under each value of probability of acceptance listed across the top of the table, there corresponds a value of np shown for the plan
	VI.
	  A.
	    1. $P_a = 0.50$ corresponds to $np = 0.910$

TABLE 46.9   Multiple Sampling by Attributes (*Continued*)

Summary of plan	Calculations
2. Divide the values of np by sample size n to get p	2. $p = \dfrac{0.910}{33} = 0.028$
3. Draw OC curve from corresponding values of p and P_a	3. $p = 0.028$ $P_a = 0.50$
B. Average sample number (ASN curve)	B.
1. Calculate p from P_a as above	1. $P_a = 0.50$ $p = 0.028$
2. Under the value np from which p was calculated is listed a value of ASN/n_1	2. ASN/n_1 = 3.288
3. Multiply ASN/n_1 by n to obtain ASN corresponding to p	3. ASN = 3.288(33) = 108.5
4. Draw ASN curve from corresponding values of p and P_a	4. p = 0.028 ASN = 108.5
C. Average outgoing quality	C.

1. Use formula AOQ $\simeq p\,P_a$	1. at p = 0.028
	     AOQ $\simeq$ 0.028(.5)
	       = 0.014

D. Average outgoing quality limit
    AOQL = maximum of AOQ curve
	D.
	   AOQL = 0.0148 at
	       p = 0.022

E. Average total inspection (approximate)
    ATI $\approx$ ASN $(P_a) + N(1 - P_a)$
	E.
	   ATI $\simeq$ 108.5(0.5) + 500(0.5)
	      $\simeq$ 304

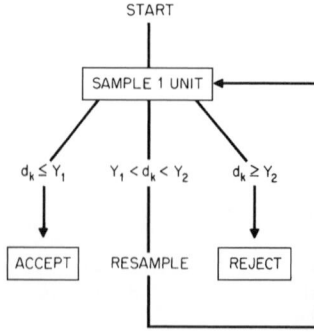

FIGURE 46.11   Schematic operation of sequential sampling. (*From Schilling, 1982, p. 156.*)

TABLE 46.10   Sequential Sampling by Attributes

Example: Determine a sequential sampling plan which will have AQL = $100p_1$ = 1% with producer risk α = 0.05 and LTPD = $100p_2$ = 5% with consumer risk β = 0.10. Suppose all the samples taken from a lot of 500 are defective.

Summary of plan	Calculations
I. Restrictions Random sample of dichotomous data	
II. Necessary information A. Producer quality level, p_1 B. Producer risk, α C. Consumer quality level, p_2 D. Consumer risk, β	II. A. p_1 = 0.01 B. α = 0.05 C. p_2 = 0.05 D. β = 0.10

III. Selection of plan
A. Parameters
1. $h_1 = \log[(1 - \alpha)/\beta]\,/\,\{\log(p_2/p_1) + \log[(1 - p_1)/(1 - p_2)]\}$
2. $h_2 = \log[(1 - \beta)/\alpha]\,/\,\{\log(p_2/p_1) + \log[(1 - p_1)/(1 - p_2)]\}$
3. $s = \log[(1 - p_1)/(1 - p_2)]\,/\,\{\log(p_2/p_1) + \log[(1 - p_1)/(1 - p_2)]\}$
B. Acceptance criteria
1. Acceptance line
$y_1 = sk - h_1$
2. Rejection line
$y_2 = sk + h_2$
 	
3. k = number of samples taken
	III.
	A.
	1. h_1 = 1.3639
	
	2. h_2 = 1.7510
	
	3. s = 0.02499
	
	B.
	1. y_1 = 0.02499k
	   − 1.3639
	2. y_2 = 0.02499k
	   + 1.7510
	
	3. Take k = 2

IV. Elements A. Sample size Not definite since samples are taken one at a time B. Statistic d_k = cumulative number nonconforming by the kth sample C. Decision criteria 1. Accept if $d_k \le y_1 = sk - h_1$ 2. Reject if $d_k \ge y_2 = sk + h_2$ 3. Continue sampling if $y_1 < d_k < y_2$	IV. A. B. At k = 2 d_2= 2 C. 1. 2 not $\le$ −1.31, cannot accept 2. 2 $\ge$ 1.8, reject 3. Discontinue sampling
V. Action: Dispose of lot as indicated by the decision rule	V. Reject
VI. Measures A. Probability of acceptance (OC curve) 1. Pick arbitrary value of h 2. Then $$p = \frac{1 - [(1 - p_2)/(1 - p_1)]^h}{(p_2/p_1)^h - [(1 - p_2)/(1 - p_1)]^h}$$ and $$P_a = \frac{[(1 - \beta)/\alpha]^h - 1}{[(1 - \beta)/\alpha]^h - [\beta/(1 - \alpha)]^h}$$	VI. A. 1. h = 1 2. p = 0.01 P_a = 0.95

TABLE 46.10   Sequential Sampling by Attributes (*Continued*)

Summary of plan	Calculations

B. Average sample number (ASN curve)
  1. Calculate p and P_a as above
 	
  2. Then
	B.
	  1. p = 0.01
	     P_a = 0.95
	  2. ASN = 81

$$\text{ASN} = \frac{\left\{ \begin{array}{l} P_a \log [(\beta/(1 - \alpha)] + \\ (1 - P_a) \log [(1 - \beta)/\alpha] \end{array} \right\}}{\left\{ \begin{array}{l} p \log (p_2/p_1) + (1 - p) \\ \log [(1 - p_2)/(1 - p_1)] \end{array} \right\}}$$

Summary of plan	Calculations

C. Average outgoing quality (AOQ curve)
  1. Use approximate formula
     $\text{AOQ} \simeq p\, P_a$
	C.
	  1. At p = 0.01
	     $\text{AOQ} \simeq 0.01(0.95)$
	     = 0.0095

D. Average outgoing quality limit
     AOQL = maximum of AOQ curve
	D. AOQL = 0.0145 at
	     p = 0.021

E. Average total inspection (approximate)
     $\text{ATI} \approx \text{ASN}\,(P_a) + N(1 - P_a)$
	E. At $p = 0.01$
	     $\text{ATI} \approx 81(.95) + 500$
	     (0.05)
	     $\simeq 102$

developed by Dodge and Torrey (1951). Dodge (1970) recounts these and other developments in the evolution of continuous sampling plans. MIL-STD-1235B (1981) provides a tabulation of continuous sampling plans.

The following are prerequisites for application of the single-level continuous sampling plans:

1. The inspection must involve "moving product," i.e., product which is flowing past the inspection station, e.g., on a conveyer belt.

2. Rapid 100 percent inspection must be feasible.

3. The inspection must be relatively easy.

4. The product must be homogeneous.

CSP-1 Plans.   CSP-1 plans use 100 percent inspection at the start. When i successive units are found to be acceptable and when there is assurance that the process is producing homogeneous product, 100 percent inspection is discontinued and sampling is instituted to the extent of a fraction f of the units. The sampling is continued until a defective unit is found. One hundred percent inspection is then reinstated and the procedure is repeated. Dodge (1943) has provided graphs which can be used to select such plans based on the AOQL desired. See also Schilling (1982).

For example, consider the plan $i = 35, f = \frac{1}{10}$. One hundred percent inspection is used until 35 successive units are found nondefective. Sampling, at the rate of 1 unit in 10, is then instituted. If a defective is found, 100 percent inspection is reinstituted and continued until 35 successive units are found nondefective, at which time sampling is reinstituted.

CSP-2 Plans.   CSP-2 plans were developed to permit sampling to continue even if a single defective is found. Again, 100 percent inspection is used at the start until i successive units are found free of defects. When sampling is in effect and a defective is found, 100 percent inspection is instituted only if a second defective occurs in the next i or fewer sample units inspected. Details on the selection of such plans will be found in Dodge and Torrey (1951) or in Schilling (1982).

Stopping Rules.   Murphy (1959) has investigated four different "stopping rules" for use with CSP-1 plans. He concludes that a useful rule is to stop when a specified number of defectives is found during any one sequence of 100 percent inspection.

Multilevel Continuous Sampling.   Continuous sampling plans have been developed by Lieberman and Solomon (1955) and others which reduce the sampling rate beyond that of other continuous plans when the quality level is better than a defined AOQL. In addition, sudden changes in the amount of inspection are avoided by providing for several "levels" of inspection. Figure 46.12 outlines the procedure for using the plan. The procedure provides for reducing the sampling rate each time i successive units are found to be free of defects. The first sampling rate after leaving 100 percent inspection is f, and each succeeding sampling rate is f raised to one larger power. The number of sampling levels is k. Thus, if $f = \frac{1}{2}$ and $k = 3$, the successive sampling rates are $\frac{1}{2}$, $\frac{1}{4}$, and $\frac{1}{8}$.

Notice that all the continuous plans discussed so far are based on the AOQL concept, i.e., requiring periods of 100 percent inspection during which only nondefective product is accepted. This action controls the average defectiveness of accepted product at some predetermined level. In all these cases, inspection must be nondestructive to employ the AOQL principle.

Additional detail on continuous sampling plans will be found in Stephens (1979). A sequential approach to continuous sampling is given in Connolly (1991).

Skip-Lot Schemes.

Skip-lot sampling plans are used when there is a strong desire to reduce the total amount of inspection. The approach is to require some initial criterion be satisfied, such as 10 or so consecutively accepted lots, and then determine what fraction of lots will be inspected. Given this fraction, the lots to be inspected are chosen using some random selection procedure. The assumptions on which the use of skip-lot plans are based are much like those for chain-sampling plans (see below). OC curves for the entire skip-lot sampling scheme can be derived, and an AOQL concept does apply. Strong dependence is made on the constancy of production and consistency of process quality as well as faith in the producer and inspector. The skip-lot sampling scheme was devised by Dodge (1955a) and is, in essence, an extension of continuous sampling acceptance plans with parameters i and f. He designated the initial scheme as SkSP-1. Here, a single determination of acceptability was made from a sample of size $n = 1$ from each lot. Lots found defective under sampling were 100 percent inspected or replaced with good lots. The AOQL then was in terms of the long-term average proportion of defective *lots* that would reach the customer. This is called AOQL-2. Later, Dodge and Perry (1971) incorporated the use of a "reference" sampling plan to determine whether a lot is acceptable or rejectable. Such skip-lot plans are designated SkSP-2 and require 100 percent inspection of rejected lots. The resulting AOQL is in terms of the long-term average proportion of individual product *units* that would reach the customer. This is called AOQL-1.

The application of skip-lot sampling should be confined to those instances where a continuous supply of product is obtained from a reasonably stable and continuous process.

The parameters of the SkSP-2 plan (in terms of continuous sampling) are:

i = number of successive lots to be found conforming to qualify for skipping lots either at the start or after detecting a nonconforming lot

f = fraction of lots to be inspected after the initial criteria have been satisfied

n = sample size per inspected lot

c = acceptance number for each inspected lot

For example, if $i = 15$, $f = \frac{1}{3}$, $n = 10$, $c = 0$, the plan is to inspect 10 units from each lot submitted until 15 consecutive lots are found conforming, i.e., have no defective in the sample. Then one-third of the submitted lots are chosen at random for inspection and inspected to the same n and c. As long as all lots are found to be conforming, the f-rate applies. When a nonconforming lot is identified, reversion to inspection of all lots occurs until again 15 consecutive lots are found to be conforming, and then skipping is permissible again.

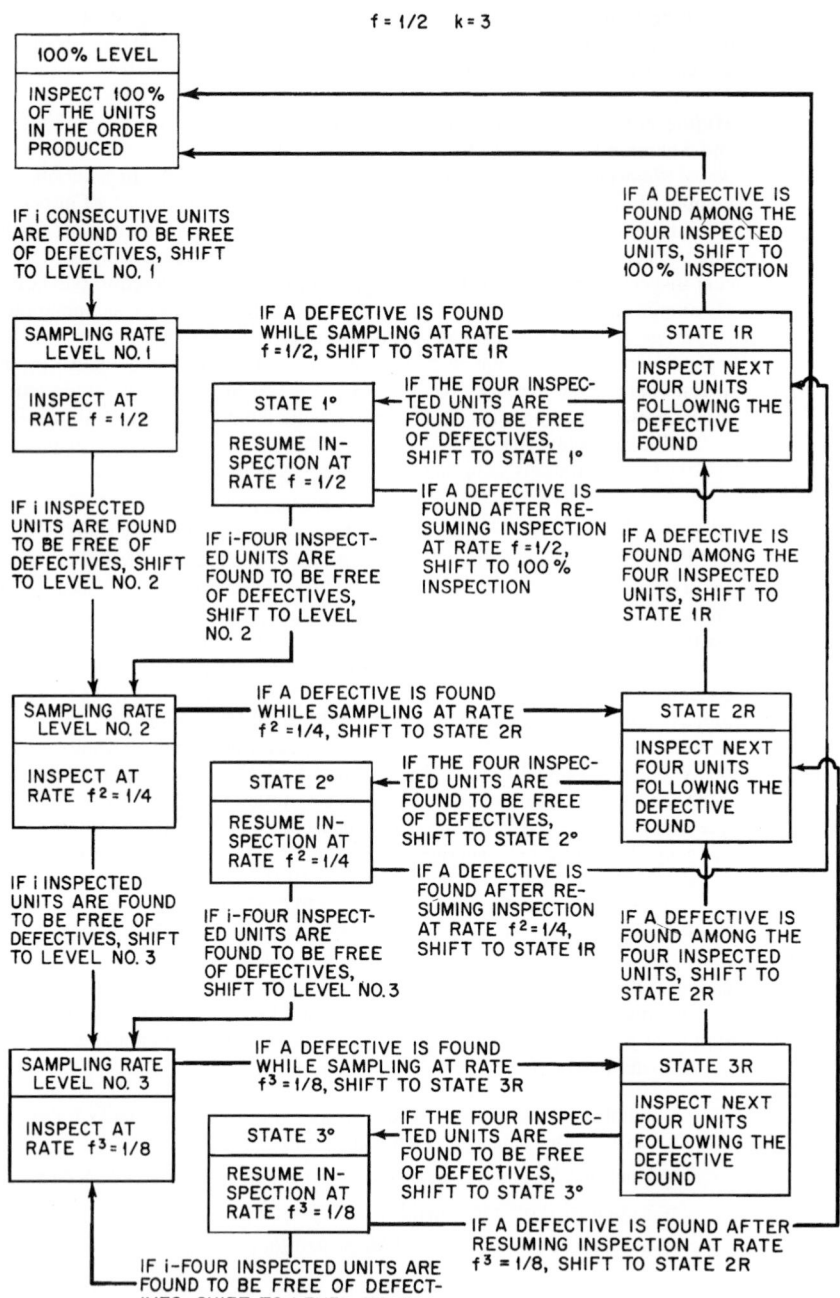

FIGURE 46.12 Procedure for multilevel continuous sampling.

Many values of f and i are possible. For SkSP-1 these may be obtained from the Basic Curves for Plan CSP-1. See Dodge (1955a). In using SkSP-2, unity values for selection of a plan having a designated operating ratio will be found in Dodge and Perry (1971). See also Schilling (1982).

One further consideration is whether to apply the skip-lot procedure to only one characteristic of a product, to all characteristics simultaneously, or just to some. The plan is obvious and straightforward when applied only to one product characteristic, perhaps a particularly expensive one in time or dollars. If applied to several characteristics, it might be best to inspect the f fraction for these characteristics on each lot. For example, if 6 characteristics are candidates for skip-lot sampling and $f = \frac{1}{3}$, perhaps 2 of these would be examined on one lot, two on another, and two on the third, again in some random fashion, so that all lots receive some inspection. Of course, if the most expensive part of the sampling scheme were forming the lot, sampling from it, and keeping records, the choice might be to inspect all characteristics on the same sample units. This argument may be extended to the case where all inspection characteristics are on a skip-lot basis; but if inspection is very complex, this is difficult to visualize.

Perry (1973) has extended skip-lot plans in many ways including the incorporation of multi-level plans. See Perry (1973a). Skip-lot plans are described in detail by Stephens (1982).

Chain Sampling.   Chain-sampling plans utilize information over a series of lots. The original plans by Dodge (1955), called ChSP-1, utilized single sampling on an attributes basis with n small and $c = 0$. The distinguishing feature is that the current lot under inspection can also be accepted if one defective unit is observed in the sample provided that no other defective units were found in the samples from the immediately preceding i lots, i.e., the chain. Dodge and Stephens (1965, 1966) derived some two-stage chain-sampling plans which make use of cumulative inspection results from several samples and are generalizations and extensions of ChSP-1. Conversely, ChSP-1 is a subset of the new plans, and the discussion here will be in terms of the original plan.

Before discussing the details of the plans, the general characteristics of chain sampling are described. Chain-sampling plans, in comparison with single-sampling plans, have the characteristic of "bowing up" the OC curve for small fractions defective while having little effect on the end of the curve associated with higher fractions defective. The effectiveness of chain-sampling plans strongly depends upon the assumptions on which the plans are based, namely:

1. Production is steady, as a continuing process.
2. Lot submittal is in the order of production.
3. Attributes sampling is done where the fraction defective p is binomially distributed.
4. A fixed sample size from each lot is assumed.
5. There is confidence in the supplier to the extent that lots are expected to be of essentially the same quality.

Chain-sampling plans are particularly useful where inspection is costly and sample sizes are relatively small. However, they may also be found useful with large sample sizes. The advantage over double-sampling plans is the fixed sample size. The disadvantage is that moderate changes in quality are not easily detected. However, major changes in quality are detected as easily with chain-sampling plans having much smaller sample sizes as with single-sampling with the same AQL.

The original ChSP-1 plan is to inspect a sample of n units from a lot, accept the lot if zero defectives were found in the n units, or accept the lot if one defective were found in the sample and no defectives were found in the samples from the immediately preceding i lots. See Figure 46.13.

Formulas for the OC curves for ChSP-1 are given by Dodge (1955) and for the two-stage procedure will be found in Dodge and Stephens (1965, 1966) while Raju (1991) has developed three-stage chain plans. All are based on the type B sampling situation. Soundararajan (1978) has prepared tables of unity values for selection of chain-sampling plans from various criteria. Stephens (1982) provides additional detail on implementing chain plans. Govindaraju (1990) gives tables for determining ChSP-1 plans given acceptable quality level and limiting quality level.

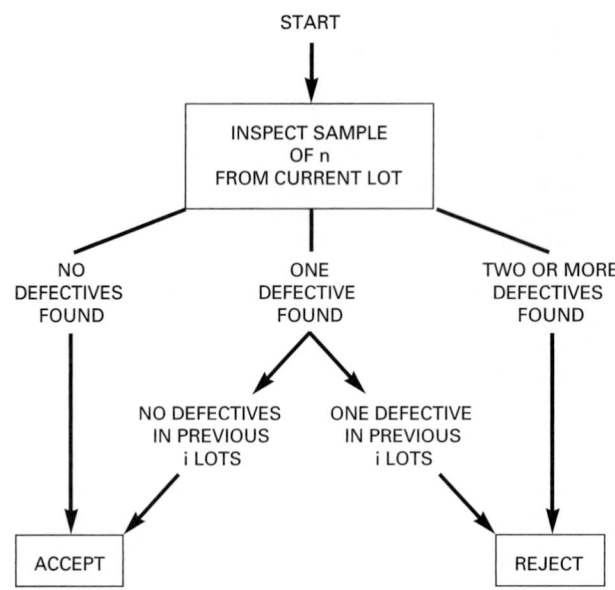

FIGURE 46.13   Flowchart of ChSP-1 chain sampling scheme. (*From Schilling, 1982, p. 452.*)

Cumulative Sum Sampling Plans.

A scheme has been devised by Beattie (1962) whereby the continuous inspection approach may be employed when inspection is destructive. Acceptance or rejection of the continuously produced product is based on the cumulation of the observed number of defectives. In addition, it is possible to discriminate between two levels of quality, an acceptable quality level (AQL) and rejectable quality level (RQL).

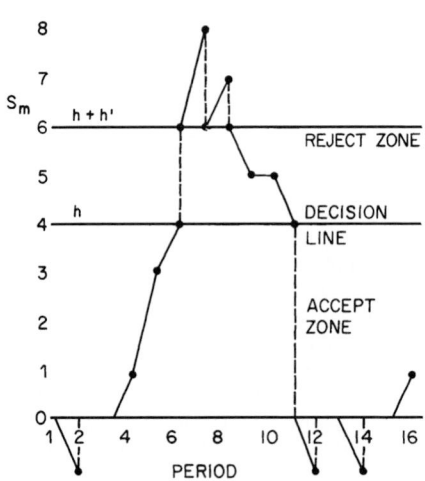

FIGURE 46.14   Form of chart for acceptance under cumulative sum sampling plans.

Prior to the development of continuous and cumulative sum plans, the following situation frequently occurred when inspection was destructive. Since production was continuous, usually there was no technically rational procedure for defining lots. A common procedure was to accumulate product until a lot was formed, but then the time delay in making a decision on the product was often too long, and a few short periods of bad production could cause a large quantity of good product to be rejected. Various approaches to lessening this jeopardy were used, all with some logical or practical drawback.

The procedure of Beattie (1962) establishes two zones, an accept zone and a reject zone, and product is accepted or rejected according to the cumulative sum of defectives observed. Figure 46.14 illustrates the plan. To implement the procedure, a sample of size n is chosen at regular intervals from production. The units are inspected and the number of defective units d_i in the ith sample recorded. Then,

$$S_m = \sum_{i=1}^{m} (d_i - k)$$

is accumulated, where k is a parameter of the scheme and S_m is computed and plotted according to the following rules:

1. Start the cumulation, S_m, at zero.
2. Accept product as long as $S_m < h$, where h is a second parameter of the scheme. When $S_m < 0$, return the cumulation to zero.
3. When h is crossed or reached from below, reject product and restart the cumulation at $h + h'$ (h' is a third parameter of the plan).
4. Continue rejecting product as long as $S_m > h$. When $S_m > h + h'$, return calculation to $h + h'$ and continue rejecting product until $S_m < h$.
5. When h is crossed or reached from above, accept product and restart cumulation at zero.

The discriminatory capability of the plans is controlled through the choice of $k, h,$ and h' and their combinations. In familiar terms, k is akin to an acceptance number and h defines how far one is willing to deviate from this number over a long run and still accept product. The quantity h' is essentially how much evidence is required to be assured the process has been corrected.

The protection offered by this type of plan, which is similar to a type B acceptance situation, is defined by an OC curve which is determined by the ratio of the average run length (ARL) in the accept zone to the sum of the ARL in the accept and reject zones.

Published Tables and Procedures. The published plans cover primarily three quality indices: AQL, LTPD, and AOQL.

AQL Plans. Sampling schemes indexed by AQL are designed to give high assurance for type B situations (that is, sampling from a process) of lot acceptance when the process fraction defective is less than or equal to that AQL. These plans are devised for producer's protection or to keep the producer's risk small. Consideration is given to what might be called the other end of the plan, the consumer's risk, through the switching rules. If there is interest in controlling both producer's risk and consumer's risk with the use of schemes indexed by AQL values, the best approach is to study the scheme OC curves and choose the plan which provides adequate protection against both types of risks. See Schilling and Sheesley (1978); also Schilling and Johnson (1980).

MIL-STD-105 is the best-known sampling scheme indexed by AQL. The AQLs presented in this standard form a sequence which is evident in the standard. The sequence chosen, while arbitrary, does provide a sensible grouping and indexing of plans for use. The steps to be applied in AQL sampling acceptance procedures are summarized in Table 46.11.

LTPD Plans. Sampling schemes indexed by LTPD or other values of limiting quality (LQ) are designed for type A situations that are essentially the mirror image of plans indexed by AQL. That is, the plans are chosen such that the consumer's risk of accepting a submitted lot of product with quality equal to or worse than the LQ is equal to or less than some specified value. Specifically, LTPD plans provide a 10 percent probability of accepting lots of the listed LTPD quality. The Dodge-Romig sampling scheme is the best-known set of plans indexed by LTPD. The derivative standards of MIL-STD-105 also provide some LQ plans. Schilling (1978) has developed a set of type A lot sensitive plans with $c = 0$ which are selected directly from the lot size and the LTPD using a simple table. Properties of these plans under curtailment have been developed by Govindaraju (1990).

AOQL Plans. Sampling schemes indexed by AOQL are derived to provide assurance that the long-run average of accepted quality, given screening of rejected lots, will be no worse than the indexed AOQL value. The Dodge-Romig (1959) plans provide the best available tables of sampling plans indexed by AOQL.

TABLE 46.11 Summary of Lot-by-Lot Sampling Procedure

Establishment of standards (by inspection executive):
  Decide what shall be a unit of product.
  Classify the quality characteristics for seriousness.
  Fix an acceptable quality level for each class.
  Fix an inspection level for the product.
Installation of procedure (by inspection supervisor):
  Arrange for formation of inspection lots.
  Decide what type of sampling shall be used (single, double, multiple).
  Choose sampling plan from tables.
Operation of procedure (by line inspector):
  Draw sample units from each inspection lot.
  Inspect each sample unit.
  Determine whether to accept or reject the inspection lot (if sampling for acceptance) or
    whether to urge action on the process (if sampling for control).
Review of past results (by inspection supervisor):
  Maintain a record of lot acceptance experience and cumulative defects by successive lots.
  Determine whether to tighten or reduce inspection on future lots.
Feedback
  Insure that the results of inspection are fed back to all components and especially the
    producer to motivate action for quality improvement.

AQL Attributes Sampling System. Most attributes AQL sampling systems are derivatives of the 105 series of acceptance sampling procedures, issued by the United States Department of Defense. This series was initiated with MIL-STD-105A in 1950 and concluded with MIL-STD-105E issued 10 May 1989. This series was canceled 27 February 1995. The best known 105 derivatives are the U.S. commercial version ANSI/ASQC Z1.4 (1993) and the international standard ISO-2859-1 (1989). With the cancellation of MIL-STD-105E, ANSI/ASQC Z1.4 was adopted by the U.S. military for future acquisitions.

ISO 2859-1 (1989) is part of the international ISO 2859 series of integrated standards to be used with the ISO 2859-1 AQL system. ISO 2859-0 (1995) provides an introduction to acceptance sampling as well as a guide to the use of ISO 2859-1, which is the central AQL sampling system in the series. ISO 2859-2 (1985) gives Limiting Quality (LQ) sampling plans for isolated lots to be used individually or in cases where the assumptions implicit in AQL schemes do not apply. ISO 2859-3 (1991) presents skip-lot sampling procedures which may be used as a substitute for Reduced Sampling in ISO 2859-1. Related U.S. national standards are ANSI/ASQC S2 (1995) which is an introduction to the ANSI/ASQC Z1.4 (1993) AQL scheme; ANSI/ASQC Q3 (1988) for Limiting Quality; and ANSI/ASQC S1 (1996) for skip-lot inspection. The international standard ISO 8422 (1991) also provides attribute sequential plans which may be used with ISO 2859-1.

The quality index in ANSI/ASQC Z1.4 (1993) and its international counterpart ISO 2859-1 (1989) is the acceptable quality level (AQL):

- A choice of 26 AQL values is available ranging from 0.010 to 1000.0. (Values of 10.0 or less may be interpreted as percent defective or defects per hundred units. Values above 10.0 must be interpreted as defects per hundred units.)
- The probability of accepting at AQL quality varies from 89 to 99.5 percent.
- Nonconformances are classified as A, B, C according to degree of seriousness.
- The purchaser may, at its option, specify separate AQLs for each class or specify an AQL for each kind of defect which a product may show.

The user also specifies the relative amount of inspection or inspection level to be used. For general applications there are three levels, involving inspection in amounts roughly in the ratio of 1 to 2.5 to 4. Level II is generally used unless factors such as the simplicity and cost of the item, inspection cost, destructiveness of inspection, quality, consistency between lots, or other factors make it appropriate to use another level. The standard also contains special procedures for "small-sample inspection" where small sample sizes are either desirable or necessitated by some aspects of inspection. Four additional inspection levels (S1 through S4) are provided in these special procedures.

The procedure for choice of plan from the tables is outlined below.

1. The following information must be known:
   a. Acceptable quality level.
   b. Lot size.
   c. Type of sampling (single, double, or multiple).
   d. Inspection level (usually level II).
2. Knowing the lot size and inspection level, obtain a code letter from Table 46.12.
3. Knowing the code letter, AQL, and type of sampling, read the sampling plan from one of the nine master tables (Table 46.13 is for single-sampling normal inspection; the standard also provides tables for double and multiple sampling).

For example, suppose that a purchasing agency has need for a 1 percent AQL for a certain characteristic. Suppose also that the parts are bought in lots of 1500 pieces. From the table of sample size code letters (Table 46.12), it is found that letter K plans are required for inspection level II, the one generally used. Then the plan to be used initially with normal inspection would be found in Table 46.13 in row K. The sample size is 125. For AQL = 1.0, the acceptance number is given as 3 and the rejection number is given as 4. This means that the entire lot of 1500 units may be accepted if 3 or fewer defective units are found in the sample of 125, but must be rejected if 4 or more are found. Where an AQL is expressed in terms of "nonconformities per hundred units," this term may be substituted for "nonconforming units" throughout. Corresponding tables are provided in the standard for tightened and reduced inspections.

Inspection Severity—Definitions and General Rules for Changing Levels.   The commonly used AQL attributes acceptance sampling plans make provisions for shifting the amount of inspection

TABLE 46.12   Sample Size Code Letters*

Lot or batch size		General inspection levels		
		I	II	III
2 to	8	A	A	B
9 to	15	A	B	C
16 to	25	B	C	D
26 to	50	C	D	E
51 to	90	C	E	F
91 to	150	D	F	G
151 to	280	E	G	H
281 to	500	F	H	J
501 to	1,200	G	J	K
1,201 to	3,200	H	K	L
3,201 to	10,000	J	L	M
10,001 to	35,000	K	M	N
35,001 to	150,000	L	N	P
150,001 to	500,000	M	P	Q
500,001 and over		N	Q	R

*Sample size code letters given in body of table are applicable when the indicated inspection levels are to be used. The Standard includes an added table of code letters for small-sample inspection.

TABLE 46.13 Master Table for Normal Inspection (Single Sampling)

Acceptable Quality Levels (normal inspection)

Sample size code letter	Sample size	0.010		0.015		0.025		0.040		0.065		0.10		0.15		0.25		0.40		0.65		1.0		1.5		2.5		4.0		6.5		10		15		25		40		65		100		150		250		400		650		1000	
		Ac	Re	Ac	Re	Ac	Re	Ac	Re	Ac	Re	Ac	Re	Ac	Re	Ac	Re	Ac	Re	Ac	Re	Ac	Re	Ac	Re	Ac	Re	Ac	Re	Ac	Re	Ac	Re	Ac	Re	Ac	Re	Ac	Re	Ac	Re	Ac	Re	Ac	Re	Ac	Re	Ac	Re	Ac	Re	Ac	Re
A	2	↓	↓	↓	↓	↓	↓	↓	↓	↓	↓	↓	↓	↓	↓	↓	↓	↓	↓	↓	↓	↓	↓	↓	↓	↓	↓	↓	↓	↓	↓	↓	↓	0	1	1	2	2	3	3	4	5	6	7	8	10	11	14	15	21	22	30	31
B	3	↓	↓	↓	↓	↓	↓	↓	↓	↓	↓	↓	↓	↓	↓	↓	↓	↓	↓	↓	↓	↓	↓	↓	↓	↓	↓	↓	↓	↓	↓	0	1	1	2	2	3	3	4	5	6	7	8	10	11	14	15	21	22	30	31	44	45
C	5	↓	↓	↓	↓	↓	↓	↓	↓	↓	↓	↓	↓	↓	↓	↓	↓	↓	↓	↓	↓	↓	↓	↓	↓	↓	↓	↓	↓	0	1	1	2	2	3	3	4	5	6	7	8	10	11	14	15	21	22	30	31	44	45	↑	↑
D	8	↓	↓	↓	↓	↓	↓	↓	↓	↓	↓	↓	↓	↓	↓	↓	↓	↓	↓	↓	↓	↓	↓	↓	↓	↓	↓	0	1	1	2	2	3	3	4	5	6	7	8	10	11	14	15	21	22	30	31	44	45	↑	↑	↑	↑
E	13	↓	↓	↓	↓	↓	↓	↓	↓	↓	↓	↓	↓	↓	↓	↓	↓	↓	↓	↓	↓	↓	↓	↓	↓	0	1	1	2	2	3	3	4	5	6	7	8	10	11	14	15	21	22	30	31	44	45	↑	↑	↑	↑	↑	↑
F	20	↓	↓	↓	↓	↓	↓	↓	↓	↓	↓	↓	↓	↓	↓	↓	↓	↓	↓	↓	↓	↓	↓	0	1	1	2	2	3	3	4	5	6	7	8	10	11	14	15	21	22	30	31	44	45	↑	↑	↑	↑	↑	↑	↑	↑
G	32	↓	↓	↓	↓	↓	↓	↓	↓	↓	↓	↓	↓	↓	↓	↓	↓	↓	↓	↓	↓	0	1	1	2	2	3	3	4	5	6	7	8	10	11	14	15	21	22	30	31	44	45	↑	↑	↑	↑	↑	↑	↑	↑	↑	↑
H	50	↓	↓	↓	↓	↓	↓	↓	↓	↓	↓	↓	↓	↓	↓	↓	↓	↓	↓	0	1	1	2	2	3	3	4	5	6	7	8	10	11	14	15	21	22	30	31	44	45	↑	↑	↑	↑	↑	↑	↑	↑	↑	↑	↑	↑
J	80	↓	↓	↓	↓	↓	↓	↓	↓	↓	↓	↓	↓	↓	↓	↓	↓	0	1	1	2	2	3	3	4	5	6	7	8	10	11	14	15	21	22	30	31	44	45	↑	↑	↑	↑	↑	↑	↑	↑	↑	↑	↑	↑	↑	↑
K	125	↓	↓	↓	↓	↓	↓	↓	↓	↓	↓	↓	↓	↓	↓	0	1	1	2	2	3	3	4	5	6	7	8	10	11	14	15	21	22	30	31	44	45	↑	↑	↑	↑	↑	↑	↑	↑	↑	↑	↑	↑	↑	↑	↑	↑
L	200	↓	↓	↓	↓	↓	↓	↓	↓	↓	↓	↓	↓	0	1	1	2	2	3	3	4	5	6	7	8	10	11	14	15	21	22	30	31	44	45	↑	↑	↑	↑	↑	↑	↑	↑	↑	↑	↑	↑	↑	↑	↑	↑	↑	↑
M	315	↓	↓	↓	↓	↓	↓	↓	↓	↓	↓	0	1	1	2	2	3	3	4	5	6	7	8	10	11	14	15	21	22	30	31	44	45	↑	↑	↑	↑	↑	↑	↑	↑	↑	↑	↑	↑	↑	↑	↑	↑	↑	↑	↑	↑
N	500	↓	↓	↓	↓	↓	↓	↓	↓	0	1	1	2	2	3	3	4	5	6	7	8	10	11	14	15	21	22	30	31	44	45	↑	↑	↑	↑	↑	↑	↑	↑	↑	↑	↑	↑	↑	↑	↑	↑	↑	↑	↑	↑	↑	↑
P	800	↓	↓	↓	↓	↓	↓	0	1	1	2	2	3	3	4	5	6	7	8	10	11	14	15	21	22	30	31	44	45	↑	↑	↑	↑	↑	↑	↑	↑	↑	↑	↑	↑	↑	↑	↑	↑	↑	↑	↑	↑	↑	↑	↑	↑
Q	1250	↓	↓	↓	↓	0	1	1	2	2	3	3	4	5	6	7	8	10	11	14	15	21	22	30	31	44	45	↑	↑	↑	↑	↑	↑	↑	↑	↑	↑	↑	↑	↑	↑	↑	↑	↑	↑	↑	↑	↑	↑	↑	↑	↑	↑
R	2000	↑	↑	0	1	1	2	2	3	3	4	5	6	7	8	10	11	14	15	21	22	30	31	44	45	↑	↑	↑	↑	↑	↑	↑	↑	↑	↑	↑	↑	↑	↑	↑	↑	↑	↑	↑	↑	↑	↑	↑	↑	↑	↑	↑	↑

⇩ = Use first sampling plan below arrow. If sample size equals, or exceeds, lot or batch size, do 100 percent inspection.

⇧ = Use first sampling plan above arrow.

Ac = Acceptance number.

Re = Rejection number.

and/or the acceptance number as experience indicates. If many consecutive lots of submitted product are accepted by an existing sampling plan, the quality of submitted product must exceed that specified as necessary for acceptance. This makes it desirable to reduce the amount and cost of inspection (with a subsequent higher risk of accepting an occasional lot of lesser quality) simply because the quality level is good. On the other hand, if more than an occasional lot is rejected by the existing sampling plan, the quality level is either consistently lower than desired or the quality level fluctuates excessively among submitted lots. In either case it is desirable to increase the sampling rate and/or reduce the acceptance number to provide greater discrimination between lots of adequate and inadequate quality.

Three severities of inspection are provided: normal, tightened, and reduced. All changes between severities are governed by rules associated with the sampling scheme. Normal inspection is adopted at the beginning of a sampling procedure and continued until evidence of either lower or higher quality than that specified exists. Schematically, the rules for switching severities specified in MIL-STD-105D and E, ANSI/ASQC Z1.4 (1993) and ISO 2859-1 (1989) are given in Figure 46.15.

From the schematic it can be seen that the criteria for change from normal to tightened and back to normal are simple and straightforward. (This is a vast improvement over early issues of MIL-STD-105.) Again, the change from reduced to normal is straightforward and occurs with the first indication that quality has slipped. Considerably more evidence is required to change from normal to reduced.

In using schemes like MIL-STD-105E, ANSI/ASQC Z1.4 (1993), or ISO 2859-1 (1989), the customer assumes that the changes from the normal to tightened and reduced to normal are a necessary part of the scheme. (Recall that such AQL plans provide primarily producer's assurance that quality at the AQL level will be accepted.) The change from normal to tightened or reduced to normal occurs when evidence exists that the quality level has deteriorated. In this way, the consumer's protection is maintained.

On the other hand, the supplier is interested in keeping inspection costs as low as possible consistent with the demands placed upon it. Certainly it is desirable to change from tightened to normal inspection when conditions warrant. Generally, it would be economic to change from normal to reduced except in those cases where record-keeping costs or bother exceed the saving in reduced sampling effort. The initiative for these types of change generally rests with the supplier; only in those instances where an economic advantage exists would a customer insist on reduced inspection. This might occur, e.g., when the customer is using these sampling plans, where inspection is destructive or degrading, or when the psychological impact on the supplier makes reduced inspection a motivational force.

The case could arise where a complex item is being inspected for many characteristics, some of which are classified as critical (A), some major (B), and some minor (C). A different sampling plan could be in effect for each class. It is also possible on this same product to have reached the situation where, for example, reduced inspection could be in effect for A defects, tightened inspection for B defects, and normal inspection for C defects. The bookkeeping involved with complete flexibility of sampling plan choice by classification of characteristics and inspection severity can be enormous. On the other hand, the savings involved with lower inspection rates, especially with destructive inspection, could be sizable. Inspection cost and convenience, along with the consequences of the error of applying an improper sampling plan, are the main determinants in any decision to use anything but the highest sampling rate associated with all defect classifications and whether to take advantage of reduced inspection for one or more defect classes when allowed. Of course, when tightened inspection is dictated by sampling experience, there is no recourse but to adopt it for at least that classification or characteristic, other than to discontinue acceptance inspection of submitted product.

The standard also provides limiting quality (LQ) single-sampling plans with a consumer's risk of 10 percent (LTPD) and 5 percent for use in isolated lot acceptance inspection. If other levels of LQ are desired, the individual OC curves can be examined to adopt an appropriate plan.

Finally, a table of AOQL values is included for each of the single-sampling plans for normal and tightened inspection. These may also be used as rough guides for corresponding double- and multiple-sampling plans.

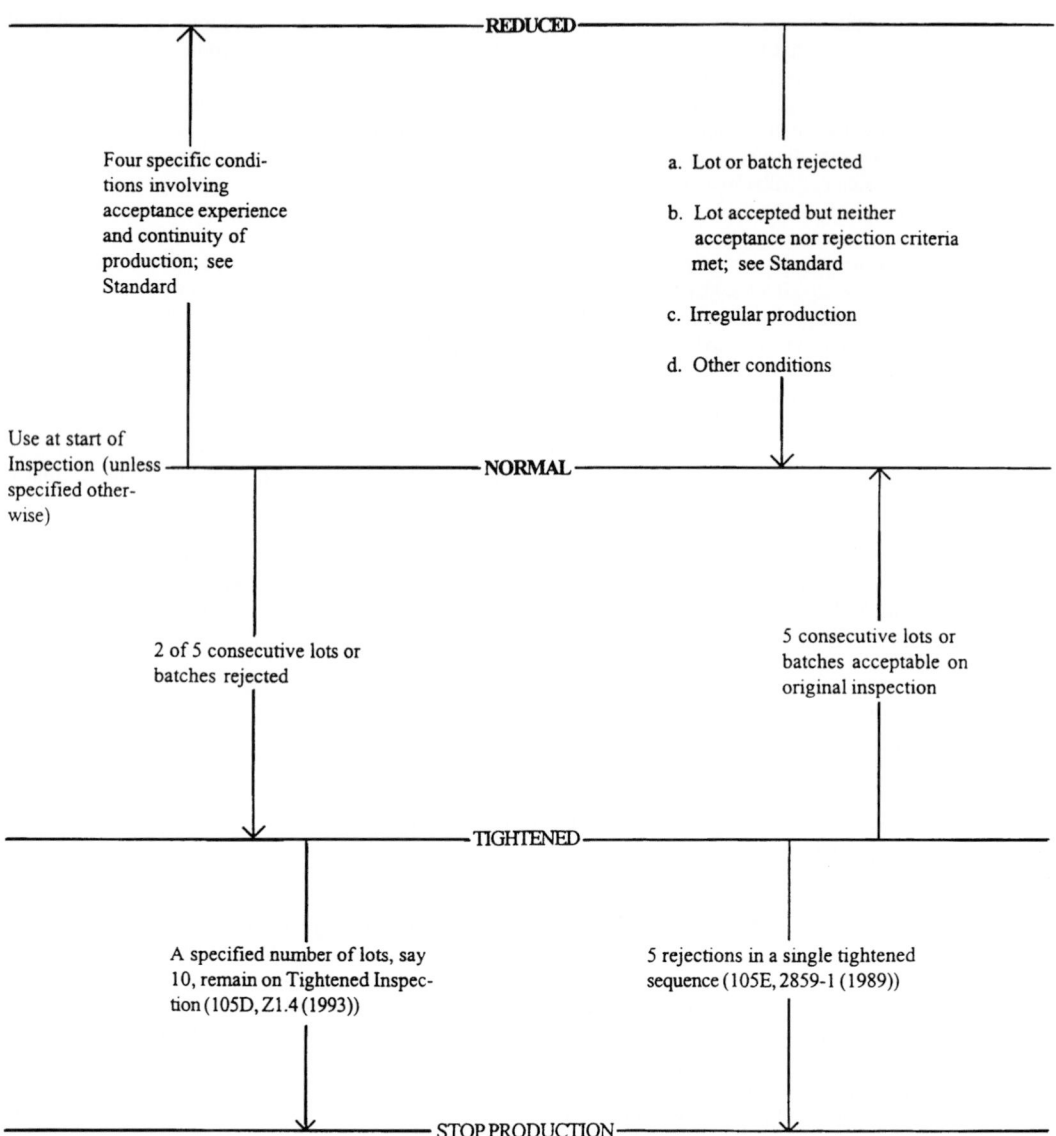

FIGURE 46.15   Rules for switching inspection severity, MIL-STD-105D and E, ANSI/ASQC Z1.4 (1993) and ISO 2859-1 (1989).

The sampling plans in MIL-STD-105E, ANSI/ASQC Z1.4, and ISO 2859-1 are sufficiently varied in type (single, double, multiple), amount of inspection, etc., to be useful in a great number of situations. The inclusion of operating characteristic curves and average sample size curves for most of the plans is a noteworthy advantage of the standards.

Dodge-Romig Sampling Tables.   Dodge and Romig (1959) provide four different sets of tables:

1. Single-sampling lot tolerance tables (SL)
2. Double-sampling lot tolerance tables (DL)

3. Single-sampling AOQL tables (SA)

4. Double-sampling AOQL tables (DA)

All four types of plans were constructed to give minimum total inspection for product of a given process average. All lots rejected are assumed to be screened, and both the sampling and the expected amount of 100 percent inspection were considered in deriving the plan which would give minimum inspection per lot. This is a particularly appropriate approach for a manufacturer's inspection of its own product as when the product of one department is examined prior to use in another, especially when processes are out of control. Practically, it may be reasonable to use the same theory even when the sampling is done by the purchasing company and the detailing of rejected material is done by the supplying company, since in the long run all the supplier's costs to provide material of a specified quality are reflected in the price.

The first and second sets of tables are classified according to lot tolerance percent defective at a constant consumer's risk of 0.10. Available lot tolerance plans range from 0.5 to 10.0 percent defective. In contrast, the third and fourth sets of tables are classified according to the average outgoing quality limit (AOQL) which they assure. Available AOQL values range from 0.1 to 10.0 percent. Lot tolerance plans emphasize a constant low consumer's risk (with varying AOQLs). In other words, they are intended to give considerable assurance that individual lots of poor material will seldom be accepted. The AOQL plans emphasize the limit on poor quality in the long run, but do not attempt to offer uniform assurance that individual lots of low quality will not get through. The relative importance of these two objectives will guide the choice of types of plan.

Table 46.14 shows a representative Dodge-Romig table for single sampling on the lot tolerance basis. All the plans listed in this table have the same risk (0.10) of accepting submitted lots that contain exactly 5 percent of defective units. The table has six columns. Each of these lists a set of plans appropriate to a specified average value of incoming quality. For example, if the estimated process average percent defective is between 2.01 and 2.50 percent, the last column at the right gives the plans that will provide the minimum inspection per lot. However, the assurance that a lot of quality 5 percent defective will be rejected is the same for all columns, so an initial incorrect estimate of the process average would have little effect except to increase somewhat the total number of pieces inspected per lot. The selection of a plan from this table thus requires only two items of information: the size of the lot to be sampled and the prevailing average quality of the supplier for the product in question. If the process average is unknown, the table is entered at the highest value of process average shown.

Zero Acceptance Number Plans.

In an increasingly litigious society, zero acceptance number plans have enjoyed increasing popularity. This is due, in part, from the notion that no defectives in the sample implies there are no defectives in the rest of the lot; and that defectives found in the sample indicate that the rest of the lot is equally contaminated. This is in line with a zero defects approach to quality. Continual improvement, however, implies lack of perfection and a constant effort to reveal and appropriately deal with quality levels that are not zero. If the state-of-the-art quality levels are indeed not zero, plans with nonzero acceptance numbers and correspondingly larger sample sizes will better discriminate between quality levels, thus facilitating corrective action when needed. Sample size is important as well as the acceptance number! A lot 13 percent defective has 50:50 odds of rejection by the plan $n = 5$, $c = 0$ or by the plan $n = 13$, $c = 1$. Thus, $c = 0$ and $c = 1$ can give equal protection at this point. However, the $c = 1$ plan can discriminate between AQL and LTPD levels that are closer together.

The zero acceptance number plans are useful for emphasizing zero defects and in product liability prevention. The $c = 0$ requirement, however, forces the consumer quality level to be 41 times higher than the producer quality level, since this is a mathematically unavoidable characteristic of $c = 0$ plans. If the consumer quality level is to be closer to the producer quality level, other acceptance numbers must be used. This is why lawyers like and statisticians dislike exclusive use of $c = 0$ plans. A simple estimator of the number of nonconformances outgoing from a $c = 0$ plan with rectification has been given by Greenberg and Stokes (1992).

TABLE 46.14 Dodge-Romig Table* for Lot Tolerance Single Sampling

Lot tolerance percent defective = 5.0%

Lot size	Process average % 0–0.5			0.06–0.50			0.51–1.00			1.01–1.50			1.51–2.00			2.01–2.50		
	n^*	c^*	AOQL %	n	c	AOQL %	n	c	AOQL %	n	c	AOQL %	n	c	AOQL %	n	c	AOQL %
1–30	All	0	0	All	0	0	All	0	0	All	0	0	All	0	0	All	0	0
31–50	30	0	0.49	30	0	0.49	30	0	0.49	30	0	0.49	30	0	0.49	30	0	0.49
51–100	37	0	0.63	37	0	0.63	37	0	0.63	37	0	0.63	37	0	0.63	37	0	0.63
101–200	40	0	0.74	40	0	0.74	40	0	0.74	40	0	0.74	40	0	0.74	40	0	0.74
201–300	43	0	0.74	43	0	0.74	70	1	0.92	70	1	0.92	95	2	0.99	95	2	0.99
301–400	44	0	0.74	44	0	0.74	70	1	0.99	100	2	1.0	120	3	1.1	145	4	1.1
401–500	45	0	0.75	75	1	0.95	100	2	1.1	100	2	1.1	125	3	1.2	150	4	1.2
501–600	45	0	0.76	75	1	0.98	100	2	1.1	125	3	1.2	150	4	1.3	175	5	1.3
601–800	45	0	0.77	75	1	1.0	100	2	1.2	130	3	1.2	175	5	1.4	200	6	1.4
801–1,000	45	0	0.78	75	1	1.0	105	2	1.2	155	4	1.4	180	5	1.4	225	7	1.5
1,001–2,000	45	0	0.80	75	1	1.0	130	3	1.4	180	5	1.6	230	7	1.7	280	9	1.8
2,001–3,000	75	1	1.1	105	2	1.3	135	3	1.4	210	6	1.7	280	9	1.9	370	13	2.1
3,001–4,000	75	1	1.1	105	2	1.3	160	4	1.5	210	6	1.7	305	10	2.0	420	15	2.2
4,001–5,000	75	1	1.1	105	2	1.3	160	4	1.5	235	7	1.8	330	11	2.0	440	16	2.2
5,001–7,000	75	1	1.1	105	2	1.3	185	5	1.7	260	8	1.9	350	12	2.2	490	18	2.4
7,001–10,000	75	1	1.1	105	2	1.3	185	5	1.7	260	8	1.9	380	13	2.2	535	20	2.5
10,001–20,000	75	1	1.1	135	3	1.4	210	6	1.8	285	9	2.0	425	15	2.3	610	23	2.6
20,001–50,000	75	1	1.1	135	3	1.4	235	7	1.9	305	10	2.1	470	17	2.4	700	27	2.7
50,001–100,000	75	1	1.1	160	4	1.6	235	7	1.9	355	12	2.2	515	19	2.5	770	30	2.8

Note: n = size of sample; entry of "All" indicates that each piece in lot is to be inspected.
 c = allowable defect number for sample.
 AOQL = average outgoing quality limit.

Source: Reproduced from Dodge and Romig (1959) by permission of the publisher and of Bell Telephone Laboratories, Inc.

Several plans have been developed to incorporate the $c = 0$ requirement. These are

- *Lot sensitive plan:*   Presents a simple table for Type A sampling that can be used to determine the sample size given the lot size for a desired LTPD. See Schilling (1978).
- *Parts per million AOQL sampling plans:*   Give Type B plans to achieve stated AOQL in parts per million given lot size. Also shows LTPD of plans presented. See Cross (1984).
- *AQL sampling plans:*   Codified Type B plans indexed by AQL and lot size compatible with MIL-STD-105E Normal Inspection (only). See Squeglia (1994).
- *AQL sampling scheme:*   Switches sample sizes between two $c = 0$ plans to increase discrimination over single sampling plans. Called TNT plans, they utilize switching rules in the manner of MIL-STD-105E. See Calvin (1977).

The tightened-normal-tightened (TNT) plan is of special interest in that it improves upon the operating characteristics of a single sample $c = 0$ plan. The switching rules of 105 or its derivatives may be used to switch between normal and tightened inspection. Two plans are used, each with $c = 0$: a normal plan with a given sample size n_N and a tightened plan with a larger sample size n_T. Switching between them will build a shoulder on the OC curve of the normal plan, thus increasing discrimination.

The normal and tightened plan sample sizes n_N and n_T can be found using a method developed by Schilling (1982) as follows:

$$n_T = \frac{230.3}{\text{LTPD}(\%)} \qquad n_N = \frac{5.13}{\text{AQL}(\%)}$$

Thus, if it is desired to achieve AQL = 1 percent and LTPD = 5 percent, $n_T = 230.3/5 = 46.1 \simeq 46$ and $n_N = 5.1/1 \simeq 5$. Then, using Z1.4 switching rules to switch between the plans $n = 46$, $c = 0$ for tightened inspection and $n = 5$, $c = 0$ for normal inspection, the desired values of AQL and LTPD can be obtained. Procedures and tables for the selection of TNT plans are given in Soundararajan and Vijayaraghavan (1990).

VARIABLES SAMPLING

Overview.   In using variables plans, a sample is taken and a measurement of a specified quality characteristic is made on each unit. These measurements are then summarized into a simple statistic (e.g., sample mean) and the observed value is compared with an allowable value defined in the plan. A decision is then made to accept or reject the lot. When applicable, variables plans provide the same degree of consumer protection as attributes plans while using considerably smaller samples.

Table 46.15 summarizes various types of variables plans showing, for each, the assumed distribution, the criteria specified, and special features.

Sampling for Percent Nonconforming.   When interest is centered on the proportion of product outside measurement specifications, and when the underlying distribution of individual measurements is known, variables plans for percent nonconforming may be used. These plans relate the proportion of individual units outside specification limits to the population mean through appropriate probability theory. The sample mean, usually converted to a test statistic, is then used to test for the position of the population mean.

Assume the distribution of individual measurements is known to be normal and a plan is desired such that the OC curve will pass through the two points $(p_1, 1 - \alpha)$ and (p_2, β) where

$$p_1 = \text{acceptable quality level}$$
$$1 - \alpha = \text{probability of acceptance at } p_1$$

TABLE 46.15 Summary of Variables Sampling Plans

Type of plan	Plan	Assumed distribution	Criteria specified	Features
Percent nonconforming	Single-sampling variables plan	Normal	Acceptable and rejectable percent nonconforming	Formulas for determining sample size and acceptance criteria to meet defined risks.
	Double-sampling variables plan	Normal	Acceptable and rejectable percent nonconforming	Tables for determining sample size and acceptance criteria to meet defined risks.
	Narrow-limit gauging	Normal	Acceptable and rejectable percent nonconforming	Tables for determining sample size and acceptance criteria to meet defined risks.
	Lot plot	None	Allowable percent nonconforming	Requires 50 measurements. Simple calculations and graphical procedure used to evaluate lot.

	Method	Distribution/applicability	Parameter	Description
Lot or process parameter—general	Test of hypothesis	Appropriate for test	Mean or standard deviation	Formulas for determining sample size and acceptance criteria to meet defined risks.
	Acceptance control chart	Appropriate for determining acceptable and rejectable values of mean	Mean	Graphical procedure to determine if mean falls within defined limits.
	Sequential sampling	Normal	Mean or standard deviation	Procedures for evaluating one measurement at a time to determine if mean falls within defined limits. Complex but total sample size lower than with other plans.
Published plans for percent nonconforming	MIL-STD-414 ANSI/ASQC Z1.9 ISO 3951	Normal	Acceptable quality level (percent defective)	Tables and procedures for lot evaluation to a specified AQL. Requires tightened and reduced inspection. OC curves given. Includes mixed plans for use when lots have been screened before submission to sampling inspection.
Lot or process parameter—bulk sampling	Specific bulk sampling models	Appropriate for test	Mean	Formulas for determining sample size to estimate mean with specified confidence interval. Applicable for gaseous, liquid, or solid products which occur in nondiscrete units.

p_2 = rejectable quality level

β = probability of acceptance at p_2

The OC curve should appear as indicated in Figure 46.16. Some important plans of this type are described below.

Single Sampling.

The rationale for variables sampling plans for percent non-conforming is illustrated in Figure 46.17, which assumes the underlying distribution of measurements to be normal with standard deviation σ known.

Suppose the following sampling plan is used to test against an upper specification limit U:

1. Sample n items from the lot and determine the sample mean $\overline{X}$.
2. Test against an acceptance limit $(U - k\sigma)$, k standard deviation units inside the specification.
3. If $\overline{X} \le (U - k\sigma)$, accept the product; otherwise reject the product.

The situation is analogous, but reversed, for a lower specification limit. If the distribution of individual measurements is normal, as shown, a proportion p of the product above the specification limit U implies the mean of the distribution must be fixed at the position indicated by μ. Means of samples of size n are, then, distributed about μ, as shown; so the probability of obtaining an $\overline{X}$ not greater than $(U - k\sigma)$ is indicated by the shaded area of the distribution of sample means. This shaded area is the probability of acceptance when the fraction nonconforming in the process is p. Note that the normal shape supplies the necessary connection between the distribution of the sample means and the proportion of product nonconforming. While, for reasonable sample sizes, the distribution of sample means will be normal, regardless of the shape of the underlying distribution of individual measurements, it is the underlying distribution of measurements itself that determines the relationship of μ and p. Hence, the plan will be quite sensitive to departures from normality.

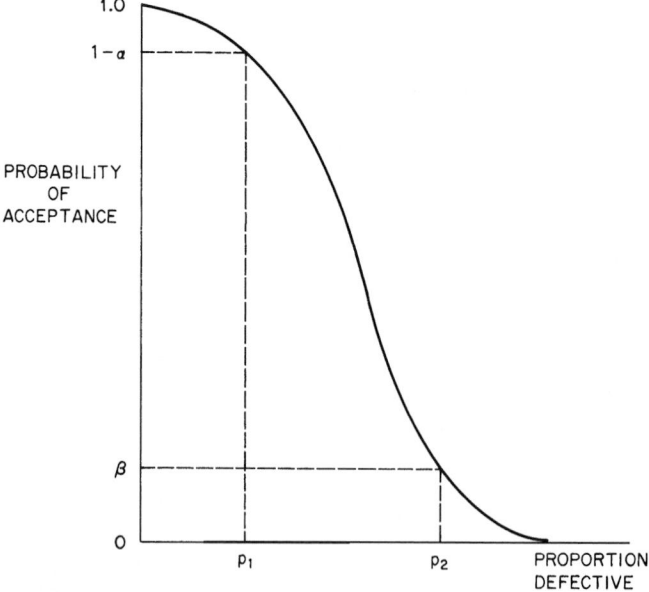

FIGURE 46.16   Operating characteristic curve.

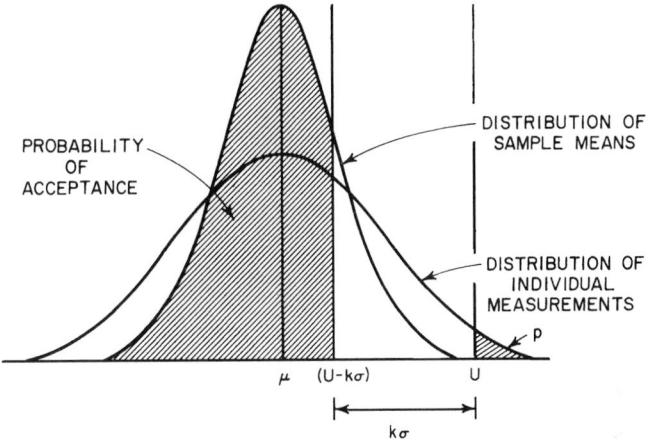

FIGURE 46.17 $(U-k\sigma)$ method.

Since $\overline{X} \leq U - k\sigma$ is equivalent to $(U - \overline{X})/\sigma \geq k$, the above sampling plan may be expressed as follows:

1. Sample n items from the lot and determine the sample mean $\overline{X}$.
2. If $(U - \overline{X})/\sigma \geq k$, accept the product; otherwise reject the product. Note that for a lower specification limit L, the inequality becomes $(\overline{X} - L)/\sigma \geq k$.

This is the method used to specify variables sampling plans in MIL-STD-414 (1957) and its civilian versions, ANSI/ASQC Z1.9 (1993) and ISO 3951 (1989).

Using Figure 46.17 and normal probability theory, it is possible to calculate the probability of acceptance P_a for various possible values of p, the proportion defective. For example, from Figure 46.17 we see that using an upper tail z value, $z_{1-P_a} = (z_p - k)\sqrt{n}$. So the probability of exceeding z_{1-P_a} is $1 - P_a$ from which we obtain P_a. By symmetry, this relationship can be used with a lower specification limit as well.

A graph of P_a versus p traces the operating characteristic curve of the acceptance sampling plan. Figure 46.18 shows the operating characteristic curve of the variables plan $n = 19$, $k = 1.908$, testing against a single-sided specification limit with known standard deviation. For comparative purposes, the OC curve of the attributes plan $n = 125$, $c = 3$ is also given. Note that the OC curves intersect at about $p = 0.01$ and $p = 0.05$, indicating roughly equivalent protection at these fractions defective.

Probability theory appropriate to other methods of specifying variables plans (e.g., standard deviation unknown, double specification limits) can be used to give the OC curves and other properties of these procedures. Formulas for determining plans to meet specific prescribed conditions can also be derived. Note that the OC curves of variables plans are generally considered to be type B. That is, they are regarded as sampling from the process producing the items inspected, rather than the immediate lot of material involved.

Two-Point Variables Plans for Percent Nonconforming.

Two-point single-sampling plans for variables inspection can be readily obtained using an approximation derived by Wallis. For example, a two-point plan testing against a single specification limit, incorporating producer's quality level p_1 and consumer's quality level p_2 with producer's and consumer's risks fixed at $\alpha = 0.05$ and $\beta = 0.10$ may be found using

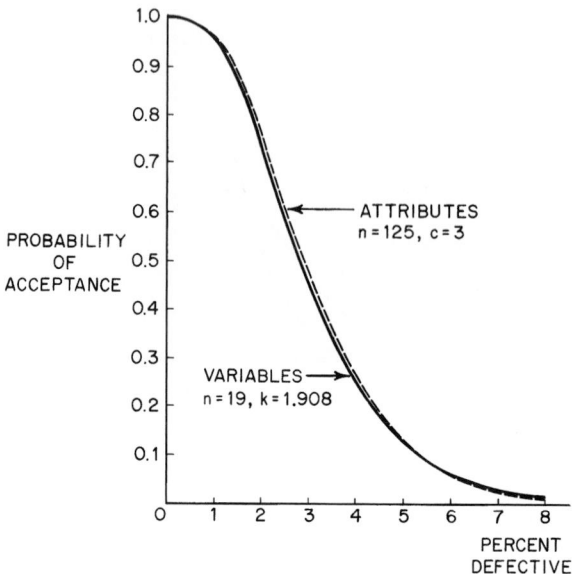

FIGURE 46.18 Operating characteristic curve.

$$k = 0.438Z_1 + 0.562Z_2$$

$$n_\sigma = \frac{8.567}{(Z_1 - Z_2)^2}$$

or

$$n_s = \left(\frac{1 + k^2}{2}\right) n_\sigma$$

Here Z_1 is the standard normal deviate corresponding to an upper tail area of p_1, and Z_2 is the standard of normal deviate corresponding to an upper tail area of p_2. Use n_σ or n_s depending on whether the standard deviation is known or unknown. For other risks or double specification limits more detailed procedures are necessary. See Duncan (1974).

As an example of the use of the procedure, consider $p_1 = 0.01$, $p_2 = 0.05$, $\alpha = 0.05$, $\beta = 0.10$. Application of the formulas with $Z_1 = 2.326$ and $Z_2 = 1.645$ gives $k = 1.94$, $n_\sigma = 18.5 \approx 19$ and $n_s = 53.3 \approx 54$. This plan closely matches the attributes plan $n = 125$, $c = 3$. Using the formula given above we see that for $p = p_1 = 0.01$, $Z_{1 - P_a} = (2.326 - 2.94) \sqrt{19} = 1.68$ so $1 - P_a = 0.046$ and $P_a = 0.954$.

A procedure for applying double-sampling variables plans was first presented by Bowker and Goode (1952). Tables to expedite the selection and application of these plans are given by Sommers (1981). Double sampling by variables can account for about a 20 percent reduction in average sample number from single sampling. For example, the plan matching the single-sampling plan derived above would have average sample size of 14.9 for the known standard deviation plan and 41.4 when the standard deviation is unknown. The application of double-sampling by variables is analogous to double sampling by attributes.

The relative efficiency of two-point variables plans over the matched attributes plans varies with k. Hamaker (1979) has shown that

$$\frac{n_a}{n_\sigma} = 2\pi p_k (1 - p_k) e^{k^2}$$

where p_k is the upper tail normal area corresponding to the standard normal deviate $Z = k$, and n_a is the attributes sample size. For the example above, $k = 1.94$ so $p_k = .0262$ and $n_a/n_\sigma = 6.9$ and we observe that $131/19 = 6.9$.

Narrow-Limit Gaging for Percent Nonconforming.   Narrow- (or compressed-) limit gaging plans bridge the gap between variables and attributes inspection by combining the ease of attributes with the power of variables inspection to reduce sample size. An artificial specification limit is set inside the specification limit. Samples are selected and gaged against this artificial "narrow limit." The narrow limit is set using the properties of the normal distribution, to which the product is assumed to conform. A standard attributes sampling plan is applied to the results of gaging to the narrow limit and the lot is sentenced accordingly.

The criteria for the narrow-limit gaging plan are, then, as follows:

n = sample size

c = acceptance number

t = number of standard deviation units the narrow-limit gage is set inside the specification limit.

The standard deviation σ is assumed to be known.

A set of two-point plans having a minimum sample size for selected values of producer's and consumer's quality levels with $\alpha = 0.05$ and $\beta = 0.10$ has been given by Schilling and Sommers (1981). They have also compiled a complete set of narrow-limit plans matching MIL-STD-105E, ANSI/ASQC Z1.4, and ISO 2859-1.

Consider the plan having producer's and consumer's quality levels $p_1 = 0.01$ and $p_2 = 0.05$ with $\alpha = 0.05$ and $\beta = 0.10$. The appropriate plan given in the Schilling-Sommers tables is $n = 28$, $c = 14$, $t = 1.99$. A narrow-limit gage is set 1.99σ inside the specification limit. A sample of 28 is taken and gaged against each unit. If 15 or more units fail the narrow limit, the lot is rejected. Otherwise, it is accepted.

A good approximation to the optimum narrow limit plan can be constructed from the corresponding known standard deviation variables plan (n_σ, k) as follows:

$$n = \frac{3n_\sigma}{2} \qquad t = k \qquad c = \frac{3n_\sigma}{4} - \frac{2}{3}$$

For the example of the two-point variables plan given above, we have seen that $n_\sigma = 18.5$ and $k = 1.94$

$$n = 27.8 \simeq 28 \qquad t = 1.94 \qquad c = 13.2 \simeq 14$$

Narrow-limit plans have had many successful applications and are readily accepted by inspectors because of the ease with which they can be applied.

Lot Plot.   Probably no variables acceptance sampling plan matches the natural inclination of the inspector better than the lot plot method developed by Dorian Shainin (1950) at the Hamilton Standard Division of United Aircraft Co. The procedure employs a histogram and rough estimates of the extremes of the distribution of product to determine lot acceptance or rejection. A standard sample size of 50 observations is maintained.

The method is useful as a tool for acceptance sampling in situations where more sophisticated methods may be inappropriate or not well received by the parties involved. The lot plot plan is especially useful in introducing statistical methods. The subjective aspects of the plan (classification of frequency distributions, their construction, etc.) and its fixed sample size suggest the use of more objective procedures in critical applications. The wide initial acceptance of Shainin's (1952) work attests to its appeal to inspection personnel. The lot plot method is outlined in Table 46.16. See Grant and Leavenworth (1979) for details. For a critical review, see Moses (1956).

TABLE 46.16   Variables Plans for Percent Defective Lot Plot

Example: A lot plot is to be used in inspecting the width of caps. A sample of 50 is taken in 10 subgroups of 5 with the following results:

1	2	3	4	5	6	7	8	9	10
0.2538	0.2581	0.2556	0.2531	0.2501	0.2521	0.2541	0.2555	0.2489	0.2529
0.2519	0.2571	0.2542	0.2566	0.2506	0.2557	0.2499	0.2569	0.2557	0.2579
0.2508	0.2521	0.2521	0.2534	0.2534	0.2569	0.2514	0.2553	0.2542	0.2565
0.2537	0.2545	0.2521	0.2557	0.2516	0.2541	0.2536	0.2496	0.2529	0.2577
0.2529	0.2563	0.2518	0.2519	0.2559	0.2524	0.2492	0.2512	0.2546	0.2541

These data are shown analyzed in Fig. 46.19. Should the lot be accepted?

Summary of plan	Calculations

I.   Restrictions: None.

II. Necessary information: Specification limits.

III. Selection of plan.
  · *A.* Plan is constant lot to lot.
    B. A special form (Fig. 46.19) is used to apply plan.

IV. Elements
    A. Sample size: A random sample of 50 pieces is taken from the lot in 10 subsamples of 5 each. Subsample identification is maintained.

    B. Statistic (for symmetric distribution).
      1. Statistic is $\overline{\overline{X}} \pm 3\hat{\sigma}$ calculated as follows:
        a. Construct cells for frequency distribution on chart.
          (1) Determine mean $\overline{X}_1$ and range R_1 for the first subgroup
          (2) Position $\overline{X}_1$ at line number 0 on lot plot form
          (3) Set cell width w so that $w \simeq R_1/4$
          (4) Fill in lot plot form with cell midpoints

        b. Tally measurements for frequency distribution using subsample number as tally mark. Tally marks form a histogram.

        c. Record range of each subsample on form in terms of number of cells between lowest and highest tally mark for each subgroup.

        d. Calculate grand mean $\overline{\overline{X}}$ from frequency distribution in terms of line numbers above (+) and below (−) arbitrary origin taken as the zero cell.

B.

1*a.* $\overline{X}_1 = 0.2526$, $R_1 = 0.003$

$\overline{X}_1 \sim 0.253$ at 0

$w \simeq \dfrac{0.003}{4} = 0.00075$

take $w = 0.001$

1*b.* See Fig. 46.19

1*c.* See Fig. 46.19 under "range" on right side

1*d.* Zero cell shown as arrow in Fig. 46.19. $\overline{\overline{X}} = +0.14$

TABLE 46.16 Variables Plans for Percent Defective Lot Plot (*Continued*)

Summary of plan	Calculations
e. Draw $\overline{\overline{X}}$ on chart in terms of line numbers.	1*e.* $\overline{\overline{X}}$ drawn 0.14 cell widths above middle of zero cell
f. Estimate 3σ of line numbers from average of subsample ranges $3\hat{\sigma} = 3\overline{R}/d_2 = 1.29\ \overline{R}$	1*f.* $3\hat{\sigma} = 1.29\left(\dfrac{51}{10}\right) = 6.6$
g. Label $\overline{\overline{X}} \pm 3\hat{\sigma}$ in terms of line numbers as (1) ULL = Upper Lot Limit ($\overline{\overline{X}} + 3\hat{\sigma}$) (2) LLL = Lower Lot Limit ($\overline{\overline{X}} - 3\hat{\sigma}$)	1*g.* See Fig. 46.19 marked ULL and LLL
h. Draw specification limits on chart.	1*h.* See Fig. 46.19 marked SPEC

C. Decision criteria.
1. Acceptance criterion.
   a. Symmetric distribution well within specification limits—accept automatically.
   b. Symmetric distribution other than above:
     (1) Lot limits within specification—accept.
     (2) Lot limits outside specification—estimate proportion of product outside specification with special technique using code strip. If less than allowable value—accept. See reference below.
   c. Nonsymmetric, bimodal distributions, etc. Special technique provided for estimating proportion out of specification using code strip. See reference below. If less than allowable value—accept.
2. Rejection criterion: Reject otherwise.	

V. Action: Dispose of lot as indicated. The Lot plot form provides a useful communication device with vendor.	V. Reject the lot
VI. Characteristics: See source reference below.	
VII. Reference: Shainin, 1950.	

A special lot plot card is helpful in simplifying some of the calculations. Figure 46.19 shows the form filled out for the example given in Table 46.16.

Grand Lot Schemes.   Acceptance inspection and compliance testing often require levels of protection for both the consumer and the producer that make for large sample size relative to lot size. A given sample size can, however, be made to apply to several lots jointly if the lots can be shown to be homogeneous. This reduces the economic impact of a necessarily large sample size. Grand lot schemes, as introduced by L. E. Simon (1941), can be used to effect such a reduction. Application of the grand lot scheme has been greatly simplified by incorporating graphical analysis of means procedures in verifying the homogeneity of a grand lot. In this way individual sublots are subjected to control chart analysis for uniformity in level and variation before they are combined into a grand lot. The resulting approach can be applied to attributes or variables data, is easy to use, provides high levels of protection economically, and can reduce sample size by as much as 80 percent. It may be applied to unique "one-off" lots, isolated lots from a continuing series, an isolated sequence of lots, or to a continuing series of lots. The procedure has been described in depth by Schilling (1979).

Sampling for Process Parameter.   When process parameters are specified, sampling plans can be developed from analogous tests of significance with corresponding OC curves. These plans do not require percent nonconforming to be related to the process mean, since the specifications to which they are applied are not in terms of percent nonconforming. This means that assumptions of process distribution may not need to be as rigorously held as in variables plans for percent nonconforming.

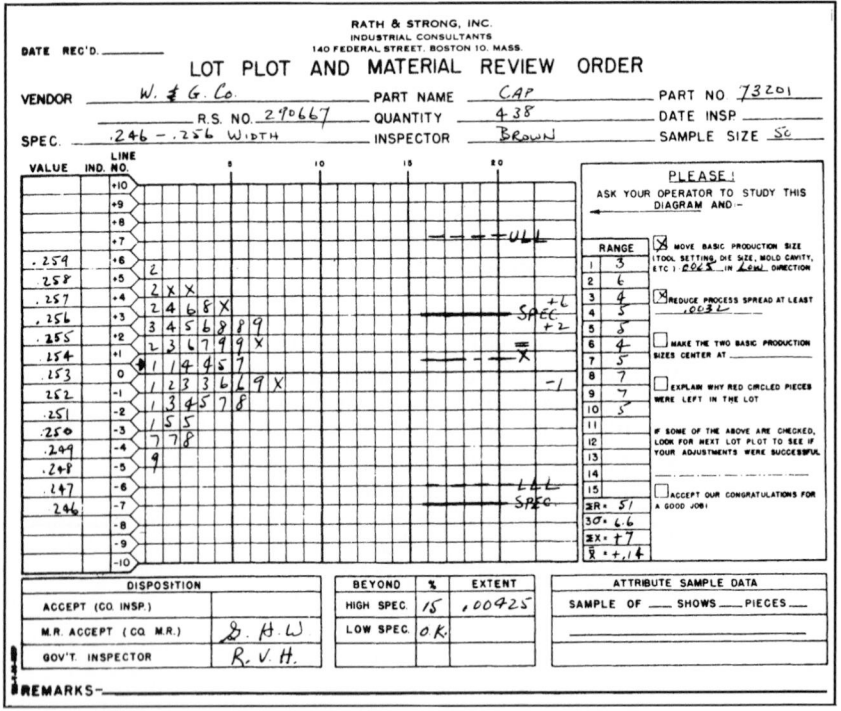

FIGURE 46.19   Illustration of lot plot method.

With specifications stated in terms of process location or variability, as measured by specific values of the mean μ or the process standard deviation σ, interest is centered not on fraction defective, but rather on controlling the parameters of the distribution of product to specified levels. From specifications of this type, it is usually possible to distinguish two process levels which may be used as bench marks as conceived by Freund (1957):

1. APL = acceptable process level—a process level which is acceptable and should be accepted most of the time by the plan.
2. RPL = rejectable process level—a process level which is rejectable and should be rejected most of the time by the plan.

The probability of acceptance for each of these process levels is usually specified as:

$$1-\alpha = \text{probability of acceptance at the APL}$$
$$\beta = \text{probability of acceptance at the RPL}$$

where α = producer's risk, β = consumer's risk.

Variables plans appropriate for this type of specification can be derived from the operating characteristic curves of appropriate tests of hypotheses. This is the case for single-sampling plans for process parameter which are, simply, appropriately constructed tests of hypotheses, e.g., testing the hypothesis that μ equals a specific value, against a one- or two-sided alternative. Thus, the statistical tests presented in Section 44 under Statistical Tests of Hypotheses can be used for this type of acceptance sampling plan.

Sequential sampling procedures have been developed which are particularly useful when levels of process parameter are specified. They usually offer a substantial decrease in sample size over competing procedures, although they may be difficult to administer. To use sequential sampling:

1. Take a sample of one measurement at a time.
2. Plot the cumulative sum T of an appropriate statistic against the sample number n.
3. Draw two lines

$$T_2 = h_2 + sn$$
$$T_1 = -h_1 + sn$$

where the intercepts h_1 and h_2 are values associated with the plan used and the symbol s is not a standard deviation but is a constant computed from the values of the acceptable process level (APL) and the rejectable process level (RPL). The use of s here corresponds to its use in the literature of sequential sampling plans.

4. Continue to sample if the cumulative sum lies between these lines, and take the appropriate action indicated if the plot moves outside the lines.

Procedures for constructing such plans and determining appropriate values of h_1, h_2, and s are given in detail in Duncan (1974) and Schilling (1982).

Acceptance control charts offer a unique answer to the problem of sampling for process parameter and can be used to implement such plans when an acceptable process level and rejectable process level are defined in terms of the mean value. They satisfy the natural desire of inspection personnel to observe quality trends and to look upon sampling as a continuing process.

These charts incorporate predetermined values of consumer and producer risk in the limits and so provide the balanced protection for the interested parties that is often lacking in the use of a conventional control chart for product acceptance.

It is not necessary that the population of individual measurements be normally distributed. The distribution must be known so that acceptable and rejectable values of the mean can be calculated. The procedure then uses the normal distribution in the analysis of the sample mean because the distribution of sample means of samples of reasonable size may be regarded as normal for any distribution of individual measurements.

The procedure for implementing this technique is shown in Table 46.17. The acceptance control chart concept is shown in Figure 46.20, and an acceptance control chart example is shown in Figure 46.21. See Freund (1957) for additional details.

TABLE 46.17   Variables Plans for Process Parameter—Acceptance Control Charts

Example: The specification limits for electrical resistance are 620 and 680, the AQL 2.5%, and the standard deviation 13. Assuming a normal distribution of individual measurements, the mean may be as low as $620 + 1.96 (13)$, or 646, or as high as $680 - 1.96 (13)$, or 654. This pair of values represents the range of the acceptable process level (APL). It was decided that the rejectable process level would occur when 14% was beyond a specification limit. Thus, the range of RPL was $620 + 1.08 (13)$ and $680 - 1.08 (13)$, or 634 and 666. Should the lot be accepted if $\overline{X} = 647$?

Summary of plan	Calculations

I.   Restrictions: None

II.   Necessary information (single-sided specification)
  A. σ = known standard deviation
  B. μ_1 = APL (acceptable process level) with $P_a = 1 - a$
  C. μ_2 = RPL (rejectable process level) with $P_a = \beta$

III.   Selection of plan: See below

IV.   Elements
  A. Sample size
  $$n = \left[\frac{(z_\alpha + z_\beta)\,\sigma}{\mu_2 - \mu_1}\right]^2$$
  where z_p cuts off upper tail area of p in standard normal curve
  B. Statistic: $\overline{X}$ = mean of sample of n
  C. Decision criteria
    1. Compute:
    $$d = \frac{z_\alpha}{z_\alpha + z_\beta}\,|\mu_2 - \mu_1|$$
    and set the acceptance control limit, ACL, a distance d from APL in the direction of the RPL. Sign of $|\mu_2 - \mu_1|$ ignored.
    2. Construct an acceptance control chart (Fig. 46.20) and accept if $\overline{X}$ falls within acceptance control limits; reject otherwise. Double-sided specification chart shown (see remarks below). Use appropriate half of chart for single-sided specification.

II.
  A. $\sigma = 13$
  B. $\mu_1 = 654$, $P_a = 0.95$
  C. $\mu_2 = 666$, $P_a = 0.10$

IV.
  A. $n =$
  $$\left[\frac{(1.645 + 1.282)(13)}{12}\right]^2$$
  $n = 10.06 \sim 10$

  B. $\overline{X} = 647$
  C. $d =$
  $$\left(\frac{1.645}{1.645 + 1.282}\right)(12)$$
  $= 6.74$
  Upper ACL $= 654 + 6.74$
  $= 660.74$

  By symmetry
  Lower ACL $= 646 - 6.74$
  $= 639.26$

  Plot as in Fig. 46.21

TABLE 46.17 Variables Plans for Process Parameter—Acceptance Control Charts (*Continued*)

Summary of plan	Calculations
V. Action: Single lot disposed of as indicated by chart	V. Accept the lot
VI. Characteristics: Two points originally specified give indication of OC curve	
VII. Reference:	Freund, 1957.
VIII. Remarks	VIII.
A. Above formulas are for single upper or single lower process limits, or for both if (Upper ACL − Lower ACL) $\geq k\sigma / \sqrt{n}$ where:	*A.* Can use both upper and lower limits since $(660.74 - 639.26) > \dfrac{(5)\,(13)}{\sqrt{10}}$ $21.48 > 20.56$

$$\begin{array}{c|c} \alpha & k \\ \hline 0.05 & 5 \\ 0.01 & 6 \\ 0.001 & 7 \end{array}$$

If (Upper ACL − Lower ACL) $< k\sigma / \sqrt{n}$, see above reference for appropriate factors

B. If standard deviation is estimated from control chart, see reference above for appropriate limits

C. Advisable to run range chart with acceptance control chart to ensure stability of variation

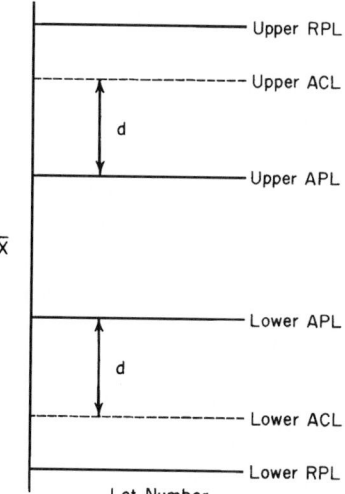

Upper RPL

Upper ACL

d

Upper APL

$\bar{x}$

Lower APL

d

Lower ACL

Lower RPL

Lot Number

FIGURE 46.20 Acceptance control chart concept.

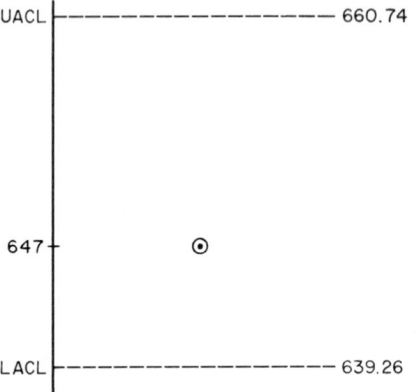

FIGURE 46.21   Acceptance control chart example.

Published Tables and Procedures.   There is often much more involved in acceptance sampling than simple tests of hypotheses. Sampling plans applied individually to guard against an occasional discrepant lot can be reduced to hypothesis tests. Sampling plans, however, may be combined into sampling schemes, intended to achieve a predetermined objective. Sampling schemes, as overall strategies using one or more sampling plans, have their own measures such as AOQ (average outgoing quality) or ATI (average total inspection), not to be found in hypothesis testing. Thus, MIL-STD-105E (1989) and its civilian versions ANSI/ASQC Z1.4 (1993) and ISO 2859-1 (1989) together with their variables counterpart MIL- STD-414 (1957) and its modified civilian versions ANSI/ASQC Z1.9 (1993) and ISO 3951 (1989) are sampling schemes which specify the use of various sampling plans under well-defined rules. The latter assume the individual measurements to which they are applied to be normally distributed.

The variables standards allow for the use of three alternative measures of variability: known standard deviation σ, estimated standard deviation s, or average range $\bar{R}$. If the variability of the process producing the product is known and stable as verified by a control chart, it is profitable to use σ. The choice between s and $\bar{R}$, when σ is unknown, is an economic one. The range requires larger sample sizes but is easier to understand and compute. The operating characteristic curves given in the standards are based on the use of s, the σ and $\bar{R}$ plans having been matched, as closely as possible, to those using s.

MIL-STD-414 and ANSI/ASQC Z1.9 (1993) offer two alternative procedures. In addition to the method using an acceptance constant k, each standard also presents a procedure for estimating the proportion of defective in the lot from the variables evidence. The former method is called Form 1; the latter is called Form 2. Form 2 is the preferred procedure in MIL-STD-414 since the switching rules for reduced and tightened inspection cannot be applied unless the fraction defective of each lot is estimated from the sample. The switching rules of ANSI/ASQC Z1.9 were patterned after MIL-STD-105D (1964) and can be used with Form 1 or Form 2. ISO 3951 presents a graphical alternative to the Form 2 procedure to be used with switching rules also patterned after MIL-STD-105D.

Application of these variables schemes follows the pattern of MIL-STD-105E, which was also an AQL sampling scheme. Sample sizes are determined from lot size, and after choosing the measure of variability to be used and the Form of the acceptance procedure, appropriate acceptance limits are obtained from the standard. As in MIL-STD-105E, operating characteristic curves are included in the variables standards. These should be consulted before a specific plan is instituted. Note that the plans contained in MIL-STD-414 and MIL-STD-105E did not match; however the plans in ANSI Z1.9 are matched to MIL-STD-105E, ANSI/ASQC Z1.4 and ISO 2859-1 as are those in ISO 3951.

Since they are AQL schemes, these standards are based on an overall strategy which incorporates switching rules to move from normal to tightened or reduced inspection and return depending on the quality observed. These switching rules are indicated in Figure 46.22 for ANSI/ASQC Z1.9 (1993) and should be used if the standard is to be properly applied. The switching rules for ANSI/ASQC Z1.9 (1993) are the same as those of ANSI/ASQC Z1.4 (1993) except that the limit numbers for reduced inspection have been eliminated. This allows these standards to be readily interchanged. ISO 3951 switching rules differ slightly but are analogous to 105, while MIL-STD-414 rules are unique.

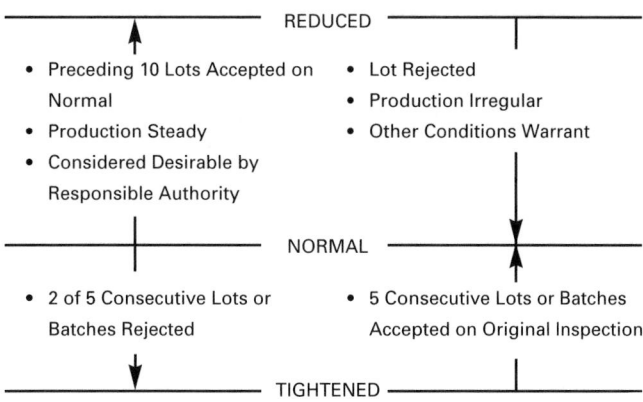

FIGURE 46.22 Switching rules for ANSI/ASQC Z1.9 (1993).

A check sequence for application of MIL-STD-414, ANSI/ASQC Z1.9 and ISO 3951 is given in Figure 46.23. Table 46.18 and Table 46.21 show the specific steps involved in application of the two forms using the sample standard deviation as a measure of variability. The graphical method of ISO 3951 is shown in Table 46.24 and Figure 46.24. Although the basic procedures of ANSI/ASQC Z1.9 (1993) and the graphical methods of ISO 3951 are analogous to MIL-STD-414, it should be noted that ranges, inspection levels, and tables of plans are different in these standards, since Z1.9 and ISO 3951 were revised to match MIL-STD-105D while 414 remained matched to MIL-STD-105A (1950). Procedures for upper, lower, and double specification limits are indicated in these tables together with appropriate references to the standard and an illustrative example. Modifications to the procedure, necessary when variability is measured by average range or a known standard deviation, are described. Table 46.25 shows the relationship of the statistics and procedures used under the various measures of variability allowed, for each of the forms.

These standards have a liberal supply of excellent examples. The reader is referred to the standards for more detailed examples of their applications.

Variables sequential plans for use with ISO 3951 (1989), ISO 2859-1 (1989), and their counterparts are provided in ISO 8423 (1991).

Mixed Plans.   One disadvantage of variables is the fact that screened lots may at times be rejected by a sample $\bar{X}$ or s indicating percent defective to be high when, actually, the discrepant material has been eliminated. To prevent rejection of screened lots, double-sampling plans have been developed which use a variables criterion on the first sample and attributes on the second sample. Lots are accepted if they pass variables inspection; however, if they do not pass, a second sample is taken and the results are judged by an attributes criterion. In this way screened lots will not be rejected, since rejections are made only under the attributes part of the plan. These plans provide a more discriminating alternative when it is necessary to use $c = 0$. This type of plan was proposed by Dodge (1932). The plan has been discussed in some detail by Bowker and Goode (1952), Gregory and Resnikoff (1955), and Schilling and Dodge (1969). MIL-STD-414 and ANSI/ASQC Z1.9 (1993) allow for the use of such procedures, although only ANSI/ASQC Z1.9 (1993) is properly matched to MIL-STD-105E or ANSI/ASQC Z1.4 (1993) to allow for proper use of mixed plans. The steps involved are shown in Table 46.26.

RELIABILITY SAMPLING

Overview.   Sampling plans for life and reliability testing are similar in concept and operation to the variables plans previously described. They differ to the extent that, when units are not all

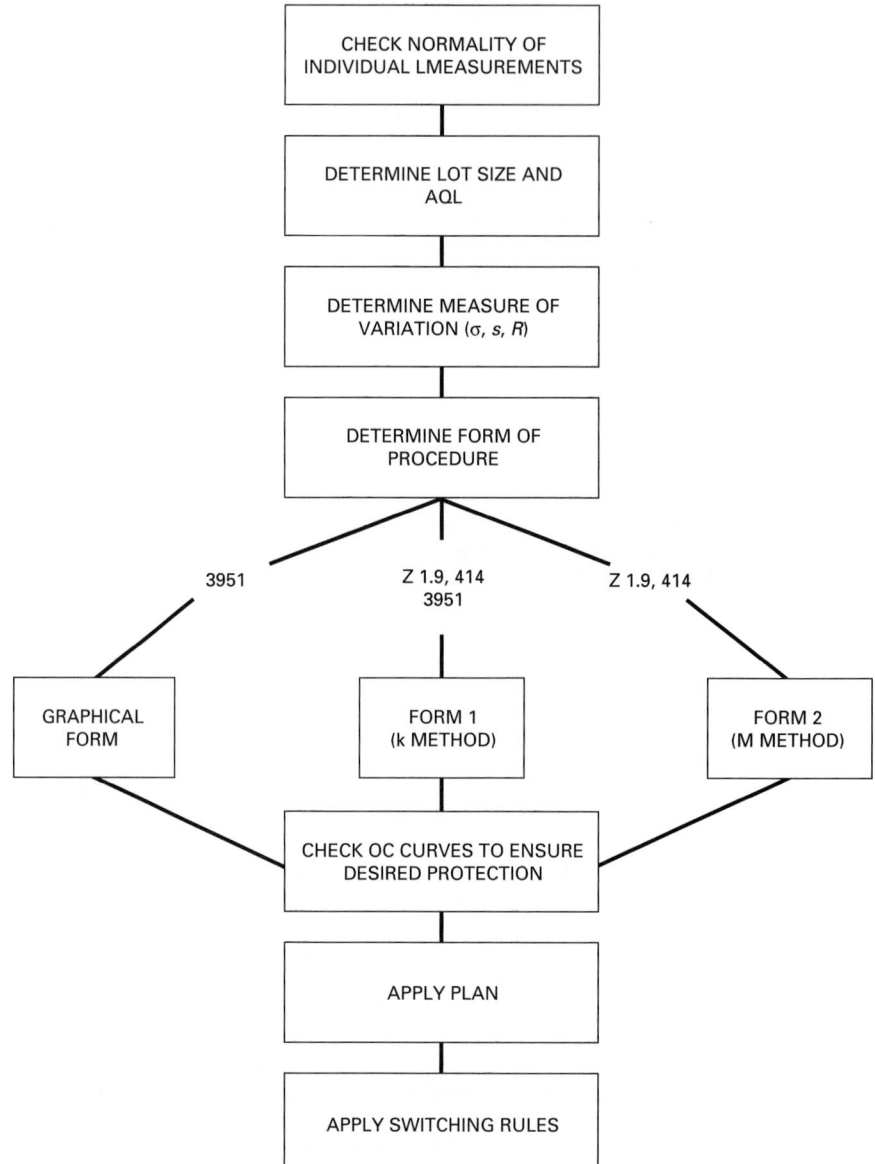

FIGURE 46.23   Check sequence for MIL-STD-414, ANSI/ASQC Z1.9, and ISO 3951.

TABLE 46.18 MIL-STD-414 (1957), ANSI/ASQC Z1.9 (1993) and ISO 3951 (1989), Variability Unknown, Standard Deviation Method, Form 1

Example: The specification for electrical resistance of a certain component is 620 to 680 ohms. A lot of 100 is submitted for inspection, normal inspection, with AQL = 2.5 percent. Should the lot be accepted if $\bar{X} = 647$ and $s = 17.22$?

Summary of plan	Calculations
I. Restrictions: Individual measurements normally distributed	MIL-STD-414 Example
II. Necessary information	II.
A. Lot size	A. Lot size=100
B. AQL	B. AQL=2.5%
C. Severity of inspection: normal, tightened, reduced	C. Normal inspection
III. Selection of plan	III.
A. Determine code letter (Table 46.19 for MIL-STD-414 only) from lot size and inspection level [normally, inspection level IV is used in MIL-STD-414 and inspection level II in ANSI/ASQC Z1.9 (1993) and ISO 3951 unless otherwise specified]	A. Code F
B. From code letter and AQL, determine (Table 46.20 for MIL-STD-414 only)	B.
1. Sample size = n	$n=10$
2. Acceptance constant = k	$k=1.41$
C. Double specification limits: obtain MSD = $F(U-L)$, where F is obtained from appropriate table in the Standard	C. MSD = 0.298(680 − 620) = 17.88
IV. Elements	IV.
A. Sample size: See above	A. $n = 10$
B. Statistic	B. $T_U = (680 − 647)/17.22 = 1.92$
1. Upper specification: $T_U = (U − \bar{X})/s$	$T_L = (647 − 620)/17.22 = 1.57$
2. Lower specification: $T_L = (\bar{X} − L)/s$	
3. Double specification: T_U and T_L	
C. Decision Criteria	C.
1. Acceptance criterion	
a. Upper specification: $T_U \geq k$	1.92 > 1.41
b. Lower specification: $T_L \geq k$	1.57 > 1.41
c. Double specification: $T_U \geq k$, $T_L \geq k$, and $s \leq$ MSD	17.22 < 17.88
2. Rejection criterion: Reject otherwise	
V. Action: Dispose of lot as indicated and refer to switching rules for next lot	V. Accept the lot
VI. Characteristics: OC curves given.	
VII. Reference: 1957 version of MIL-STD-414, ANSI/ASQC Z1.9 (1992) and ISO 3951 (1989)	
VIII. Remarks (use appropriate tables from Standard)	
A. Range method	
1. Use $\bar{R}$ of subsamples of 5 if $n \geq 10$; use R if $n < 10$	
2. Substitute $\bar{R}$ for s in statistics	
3. Double specifications—use values of f (for MAR) in place of F (for MSD), where MAR is the maximum allowable range	
B. Variability known: Substitute σ for s in statistics	
C. ISO 3951 uses a special procedure for separate AQLs	

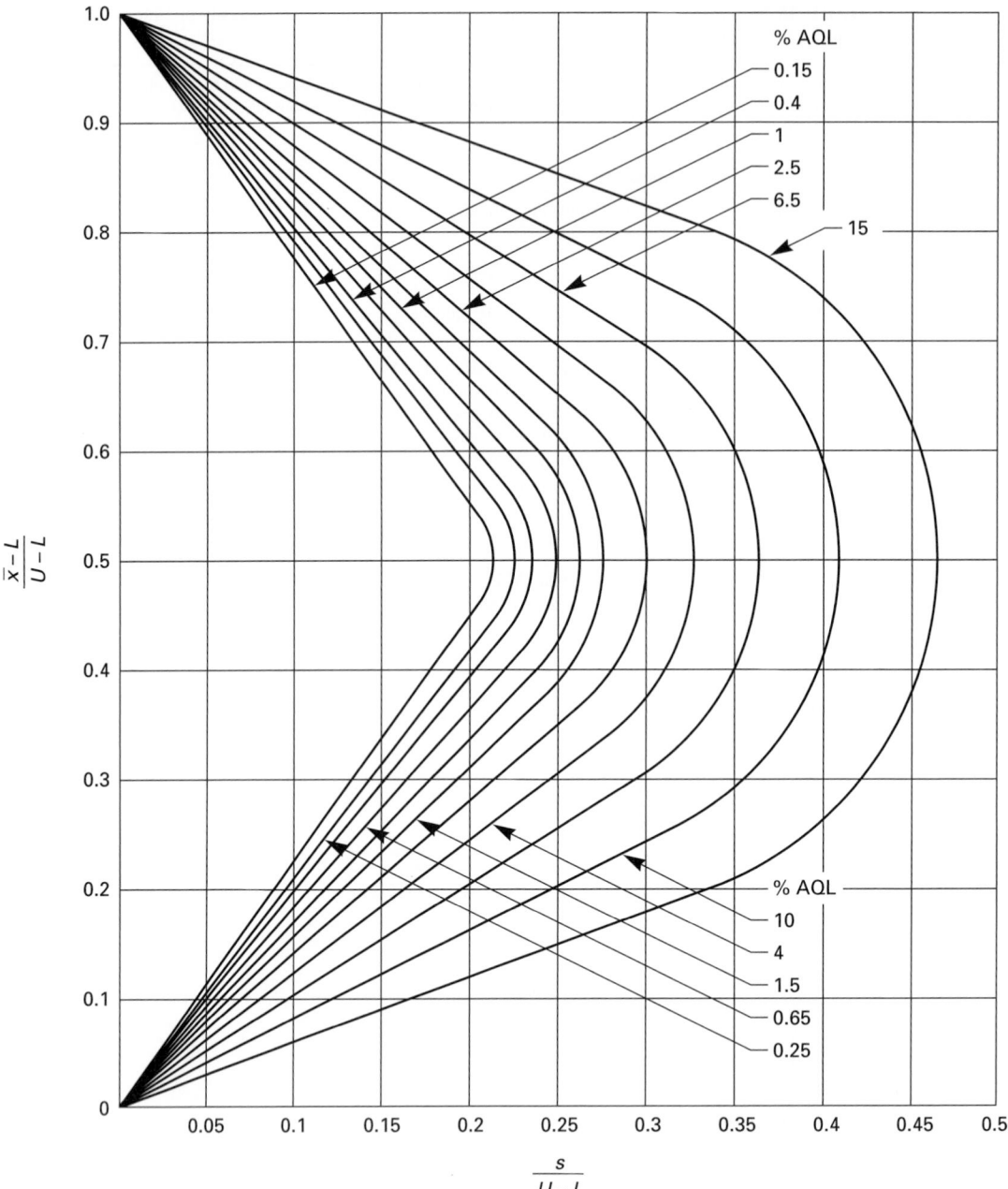

FIGURE 46.24   Acceptance curves for combined double specification limits: standard deviation method—sample size code letter F (sample size 10).

TABLE 46.19   MIL-STD-414 Sample Size Code Letters

Lot size	Inspection levels				
	I	II	III	IV	V
3–8	B	B	B	B	C
9–15	B	B	B	B	D
16–25	B	B	B	C	E
26–40	B	B	B	D	F
41–65	B	B	C	E	G
66–110	B	B	D	F	H
111–180	B	C	E	G	I
181–300	B	D	F	H	J
301–500	C	E	G	I	K
501–800	D	F	H	J	L
801–1,300	E	G	I	K	L
1,301–3,200	F	H	J	L	M
3,201–8,000	G	I	L	M	N
8,001–22,000	H	J	M	N	O
22,001–110,000	I	K	N	O	P
110,001–550,000	I	K	O	P	Q
550,001 and over	I	K	P	Q	Q

*Sample size code letters given in body of table are applicable when the indicated inspection levels are to be used.

run to failure, the length of the test becomes an important parameter determining the characteristics of the procedure. Further, time to failure tends to conform naturally to skewed distributions such as the exponential or as approximated by the Weibull. Accordingly, many life test plans are based on these distributions. When time to failure is normally distributed and all units tested are run to failure, the variables plans assuming normality, discussed above, apply; attributes plans such as ANSI/ASQC Z1.4 or ISO 2859 may also be used. Life tests, terminated before all units have failed, may be

1. Failure terminated—a given sample size n is tested until the rth failure occurs. The test is then terminated.

2. Time terminated—a given sample size n is tested until a preassigned termination time T is reached. The test is then terminated.

Furthermore, these tests may be based upon specifications written in terms of one of the following characteristics:

1. *Mean life:*   The mean life of the product
2. *Hazard rate:*   Instantaneous failure rate at some specified time t
3. *Reliable life:*   The life beyond which some specified proportion of items in the lot or population will survive
4. *Failure rate* (FR or λ):   The percentage of failures per unit time (say 1000 hours of test)

Relation of Life Characteristics.   Specification and test of various life characteristics are intimately related. Tables 46.27 and 46.28 will be found useful in converting life test characteristics. Formulas for various characteristics are shown in terms of mean life μ. Thus, using the tables, it will be found that a specification of mean life $\mu = 1000$ hours for a Weibull distribution with $\beta = 2$ is

TABLE 46.20 MIL-STD-414 Master Table for Normal and Tightened Inspection for Plans Based on Variability Unknown, Standard Deviation Method

(Single specification limit, form 1)

All *k* values shown under Acceptable quality levels (normal inspection).

Sample size code letter	Sample size	0.04	0.065	0.10	0.15	0.25	0.40	0.65	1.00	1.50	2.50	4.00	6.50	10.00	15.00
B	3	↓	↓	↓	↓	↓	↓	↓	↓	↓	1.12	0.958	0.765	0.566	0.341
C	4	↓	↓	↓	↓	↓	↓	↓	1.45	1.34	1.17	1.01	0.814	0.617	0.393
D	5	↓	↓	↓	↓	↓	↓	1.65	1.53	1.40	1.24	1.07	0.874	0.675	0.455
E	7	↓	↓	↓	↓	2.00	1.88	1.75	1.62	1.50	1.33	1.15	0.955	0.755	0.536
F	10	↓	↓	↓	2.24	2.11	1.98	1.84	1.72	1.58	1.41	1.23	1.03	0.828	0.611
G	15	2.64	2.53	2.42	2.32	2.20	2.06	1.91	1.79	1.65	1.47	1.30	1.09	0.886	0.664
H	20	2.69	2.58	2.47	2.36	2.24	2.11	1.96	1.82	1.69	1.51	1.33	1.12	0.917	0.695
I	25	2.72	2.61	2.50	2.40	2.26	2.14	1.98	1.85	1.72	1.53	1.35	1.14	0.936	0.712
J	30	2.73	2.61	2.51	2.41	2.28	2.15	2.00	1.86	1.73	1.55	1.36	1.15	0.946	0.723
K	35	2.77	2.65	2.54	2.45	2.31	2.18	2.03	1.89	1.76	1.57	1.39	1.18	0.969	0.745
L	40	2.77	2.66	2.55	2.44	2.31	2.18	2.03	1.89	1.76	1.58	1.39	1.18	0.971	0.746
M	50	2.83	2.71	2.60	2.50	2.35	2.22	2.08	1.93	1.80	1.61	1.42	1.21	1.00	0.774
N	75	2.90	2.77	2.66	2.55	2.41	2.27	2.12	1.98	1.84	1.65	1.46	1.24	1.03	0.804
O	100	2.92	2.80	2.69	2.58	2.43	2.29	2.14	2.00	1.86	1.67	1.48	1.26	1.05	0.819
P	150	2.96	2.84	2.73	2.61	2.47	2.33	2.18	2.03	1.89	1.70	1.51	1.29	1.07	0.841
Q	200	2.97	2.85	2.73	2.62	2.47	2.33	2.18	2.04	1.89	1.70	1.51	1.29	1.07	0.845
Tightened inspection AQL		0.065	0.10	0.15	0.25	0.40	0.65	1.00	1.50	2.50	4.00	6.50	10.00	15.00	

Acceptable quality levels (tightened inspection)

Note: All AQL values are in percent defective.
↓ Use first sampling plan below arrow, that is, both sample size as well as *k* value. When sample size equals or exceeds lot size, every item in the lot must be inspected.
Source: 1957 version of MIL-STD-414.

TABLE 46.21 MIL-STD-414 and ANSI/ASQC Z1.9 (1993), Variability Unknown, Standard Deviation Method, Form 2

Example: The specification for electrical resistance of a certain electrical component is 620 to 680 ohms. A lot of 100 is submitted for inspection, normal inspection, with AQL = 2.5 percent. Should the lot be accepted if $\overline{X}$ = 647 and s = 17.22?

Summary of plan	Calculations
I. Restrictions: Individual measurements normally distributed	MIL-STD-414 Example
II. Necessary information 　*A.* Lot size 　*B.* AQL 　*C.* Severity of inspection: normal, tightened, reduced	II. 　*A.* Lot size = 100 　*B.* AQL = 2.5% 　*C.* Normal inspection
III. Selection of plan 　*A.* Determine (Table 46.19 for MIL-STD-414 only) code letter from lot size and inspection level (usually, inspection level IV is used in MIL-STD-414 and inspection level II in ANSI/ASQC Z1.9) 　*B.* From code letter and AQL, determine (Table 46.22 for MIL-STD-414 only) 　　1. Sample size = n 　　2. Value of M	III. 　*A.* Code F 　*B.* 　　$n = 10$ 　　$M = 7.29$
IV. Elements 　*A.* Sample size: See above 　*B.* Statistic 　　1. Upper specification: $Q_U = (U - \overline{X})/s$ 　　2. Lower specification: $Q_L = (\overline{X} - L)/s$ 　　3. Double specification: Q_U and Q_L 　*C.* Estimate Percent Defective from Table 46.23 　　1. Upper specification: estimate p_U (%) from Q_U and n 　　2. Lower specification: estimate p_L (%) from Q_L and n 　　3. Double specification: estimate p (%) = p_U (%) + p_L(%) 　*D.* Decision criteria 　　1. Acceptance criterion 　　　*a.* Upper specification: p_U (%) < M 　　　*b.* Lower specification: p_L(%) < M 　　　*c.* Double specification: p(%) < M 　　Note: if AQLs not equal on upper and lower specifications, obtain M for each and apply *a, b, c,* above, using larger of two M values in *c*	IV. 　*A.* n = 10 　*B.* 　　$Q_U = (680 - 647)/17.22$ 　　　$= 1.92$ 　　$Q_L = (647 - 620)/17.22$ 　　　$= 1.57$ 　*C.* 　　p_U (%) = 1.68 　　p_L(%) = 4.92 　　p(%) = 6.60 　*D.* 　　6.60 < 7.29

TABLE 46.21 MIL-STD-414 and ANSI/ASQC Z1.9 (1993), Variability Unknown, Standard Deviation Method, Form 2 (*Continued*)

Summary of plan	Calculations
2. Rejection criterion: Reject otherwise	
V. Action: Dispose of lot as indicated and refer to switching rules for next lot.	V. Accept the lot
VI. Characteristics: OC curves given	
VII. Reference: 1957 version of MIL-STD-414 and ANSI/ASQC Z1.9 (1993).	
VII. Remarks (use appropriate tables from Standard) *A.* Range method—similar except: 1. Use $\overline{R}$ of subsamples of 5 if $n \geq$ 10; use R if $n < 10$ 2. Substitute $\overline{R}/c$ for s in statistics, where c is a scale factor given with n and M in the Standard *B.* Variability known—similar except: 1. Substitute σ/v for s in statistics, where v factor is given with n and M in the Standard	

equivalent to a hazard rate of 0.000157 at 100 hours or to a reliable life of 99.22 percent surviving at 100 hours.

Exponential Distribution: H108. *Quality Control and Reliability Handbook* MIL-HDBK-108 (1960) is widely used and presents a set of life test and reliability plans based on the exponential model for time to failure. The plans contained therein are intended for use when mean time to failure β is specified[2] in terms of acceptable mean life β_0 and unacceptable mean life β_1. Testing may be conducted:

With replacement: Units replaced when failure occurs. Test time continues to be accumulated on replacement unit.

Without replacement: Units not replaced upon failure.

The handbook contains three types of plans:

1. *Life tests terminated upon occurrence of a preassigned number of failures:* Here, n units are tested until r failures occur. The average life is calculated and compared with an acceptable value defined by the plan, and a decision is made.

2. *Life tests terminated at a preassigned time:* Here, n units are tested for a specified length of time T. If T is reached before r failures occur, the test is stopped and the lot accepted. If r failures occur before T is reached, the test is stopped and the lot rejected.

3. *Sequential life testing plans:* Here n units are placed on test, and time and failures are recorded until sufficient data are accumulated to reach a decision at specified risk levels. Periodically

---

[2]Note that elsewhere in this handbook the mean value is described by the symbol μ.

TABLE 46.22 MIL-STD-414, Master Table for Normal and Tightened Inspection for Plans Based on Variability Unknown, Standard Deviation Method

(Double specification limit and form 2, single specification limit)

Sample size code letter	Sample size	_	_	_	_	_	_	Acceptable quality levels (normal inspection)							
		0.04	0.065	0.10	0.15	0.25	0.40	0.65	1.00	1.50	2.50	4.00	6.50	10.00	15.00
		M	M	M	M	M	M	M	M	M	M	M	M	M	M
B	3	↓	↓	↓	↓	↓	↓	↓	↓	↓	7.59	18.86	26.94	33.69	40.47
C	4	↓	↓	↓	↓	↓	↓	↓	1.53	5.50	10.92	16.45	22.86	29.45	36.90
D	5	↓	↓	↓	↓	↓	↓	1.33	3.32	5.83	9.80	14.39	20.19	26.56	33.99
E	7	↓	↓	↓	↓	0.422	1.06	2.14	3.55	5.35	8.40	12.20	17.35	23.29	30.50
F	10	↓	↓	↓	0.349	0.716	1.30	2.17	3.26	4.77	7.29	10.54	15.17	20.74	27.57
G	15	0.099	0.186	0.312	0.503	0.818	1.31	2.11	3.05	4.31	6.56	9.46	13.71	18.94	25.61
H	20	0.135	0.228	0.365	0.544	0.846	1.29	2.05	2.95	4.09	6.17	8.92	12.99	18.03	24.53
I	25	0.155	0.250	0.380	0.551	0.877	1.29	2.00	2.86	3.97	5.97	8.63	12.57	17.51	23.97
J	30	0.179	0.280	0.413	0.581	0.879	1.29	1.98	2.83	3.91	5.86	8.47	12.36	17.24	23.58
K	35	0.170	0.264	0.388	0.535	0.847	1.23	1.87	2.68	3.70	5.57	8.10	11.87	16.65	22.91
L	40	0.179	0.275	0.401	0.566	0.873	1.26	1.88	2.71	3.72	5.58	8.09	11.85	16.61	22.86
M	50	0.163	0.250	0.363	0.503	0.789	1.17	1.71	2.49	3.45	5.20	7.61	11.23	15.87	22.00
N	75	0.147	0.228	0.330	0.467	0.720	1.07	1.60	2.29	3.20	4.87	7.15	10.63	15.13	21.11
O	100	0.145	0.220	0.317	0.447	0.689	1.02	1.53	2.20	3.07	4.69	6.91	10.32	14.75	20.66
P	150	0.134	0.203	0.293	0.413	0.638	0.949	1.43	2.05	2.89	4.43	6.57	9.88	14.20	20.02
Q	200	0.135	0.204	0.294	0.414	0.637	0.945	1.42	2.04	2.87	4.40	6.53	9.81	14.12	19.92
		0.065	0.10	0.15	0.25	0.40	0.65	1.00	1.50	2.50	4.00	6.50	10.00	15.00	
								Acceptable quality levels (tightened inspection)							

Note: All AQL and table values are in percent defective.

↓ Use first sampling plan below arrow, that is, both sample size as well as M value. When sample size equals or exceeds lot size, every item in the lot must be inspected.

Source: 1957 version of MIL-STD-414.

TABLE 46.23 Table for Estimating the Lot Percent Defective Using Standard Deviation Method

Sample sizes

Q_U or Q_L	3	4	5	7	10	15	20	25	30	35	40	50	75	100	150	200
0.1	47.2	46.7	46.4	46.3	46.2	46.1	46.1	46.1	46.0	46.0	46.0	46.0	46.0	46.0	46.0	46.0
0.2	44.5	43.3	42.9	42.5	42.4	42.2	42.2	42.2	42.2	42.1	42.1	42.1	42.1	42.1	42.1	42.1
0.3	41.6	40.0	39.4	38.9	38.6	38.4	38.4	38.3	38.3	38.3	38.3	38.3	38.2	38.2	38.2	38.2
0.4	38.7	36.7	35.9	35.3	34.9	34.7	34.6	34.6	34.6	34.6	34.5	34.5	34.5	34.5	34.5	34.5
0.5	38.8	33.3	32.4	31.7	31.4	31.2	31.1	31.0	31.0	31.0	31.0	30.9	30.9	30.9	30.9	30.9
0.6	32.6	30.0	29.0	28.3	27.9	27.7	27.6	27.6	27.6	27.5	27.5	27.5	27.5	27.5	27.4	27.4
0.7	29.3	26.7	25.7	25.0	24.7	24.5	24.4	24.3	24.3	24.3	24.3	24.3	24.2	24.2	24.2	24.2
0.8	25.6	23.3	22.5	21.9	21.6	21.4	21.3	21.3	21.3	21.3	21.2	21.2	21.2	21.2	21.2	21.2
0.9	21.6	20.0	19.4	18.9	18.7	18.5	18.5	18.5	18.5	18.4	18.4	18.4	18.4	18.4	18.4	18.4
1.0	16.7	16.7	16.4	16.1	16.0	15.9	15.9	15.9	15.9	15.9	15.9	15.9	15.9	15.9	15.9	15.9
1.1	9.8	13.3	13.5	13.5	13.5	13.5	13.5	13.5	13.5	13.5	13.5	13.5	13.6	13.6	13.6	13.6
1.2	0	10.0	10.8	11.1	11.2	11.3	11.4	11.4	11.4	11.4	11.4	11.5	11.5	11.5	11.5	11.5
1.3	0	6.7	8.2	8.9	9.2	9.4	9.5	9.5	9.6	9.6	9.6	9.6	9.6	9.6	9.6	9.7
1.4	0	3.3	5.9	7.0	7.4	7.7	7.8	7.9	7.9	7.9	7.9	8.0	8.0	8.0	8.0	8.0
1.5	0	0	3.8	5.3	5.9	6.2	6.3	6.4	6.5	6.5	6.5	6.6	6.6	6.6	6.6	6.6

Sample sizes

Q_U or Q_L	3	4	5	7	10	15	20	25	30	35	40	50	75	100	150	200
1.6	0	0	2.0	3.8	4.5	4.9	5.1	5.2	5.2	5.3	5.3	5.3	5.4	5.4	5.4	5.4
1.7	0	0	0.7	2.6	3.4	3.8	4.0	4.1	4.2	4.2	4.2	4.3	4.4	4.4	4.4	4.4
1.8	0	0	0	1.6	2.5	2.9	3.1	3.2	3.3	3.4	3.4	3.4	3.5	3.5	3.5	3.6
1.9	0	0	0	0.9	1.8	2.2	2.4	2.5	2.6	2.6	2.6	2.7	2.8	2.8	2.8	2.8
2.0	0	0	0	0.4	1.2	1.6	1.8	1.9	2.0	2.0	2.1	2.1	2.2	2.2	2.2	2.2
2.1	0	0	0	0.1	0.7	1.2	1.3	1.4	1.5	1.5	1.6	1.6	1.7	1.7	1.7	1.8
2.2	0	0	0	0	0.4	0.8	1.0	1.1	1.1	1.2	1.2	1.2	1.3	1.3	1.3	1.4
2.4	0	0	0	0	0.1	0.3	0.5	0.5	0.6	0.6	0.7	0.7	0.7	0.8	0.8	0.7
2.6	0	0	0	0	0	0.1	0.2	0.3	0.3	0.3	0.3	0.4	0.4	0.4	0.4	0.4
2.8	0	0	0	0	0	0	0.1	0.1	0.1	0.1	0.2	0.2	0.2	0.2	0.2	0.2
3.0	0	0	0	0	0	0	0	0	0.1	0.1	0.1	0.1	0.1	0.1	0.1	0.1

TABLE 46.24   ISO 3951 (1989) Variability Unknown, Standard Deviation Method, Double Specifications with Combined AQL, Graphical Procedures

Example: The specification for electrical resistance of a certain component is 620 to 680 ohms. A lot of 100 is submitted for inspection, inspection level II, normal inspection, with AQL = 2.5 percent. Should the lot be accepted if $\bar{X} = 647$ and $s = 17.227$?

Summary of plan	Calculations
I. Restrictions: Individual measurements, normally distributed	
II. Necessary information	II.
A. Lot size	A. Lot size = 100
B. AQL (combined for both specification limits)	B. AQL = 2.5 percent
C. Severity of inspection	C. Normal inspection
III. Selection of plan	III.
A. Determine code letter from lot size and inspection level; usually inspection level II is used	A. Code F
B. From code letter and AQL, determine sample size	B. $n = 10$
IV. Elements	IV.
A. Sample size: See above	A. $n = 10$
B. Statistic: $\dfrac{s}{U-L}$ and $\dfrac{\bar{X}-L}{U-L}$	B. 0.287 and 0.450
C. Decision criteria	C.
1. If $s > \mathrm{MSSD} = f_s(U-L)$ reject outright	1. $17.22 < .298(60) = 17.88$
2. Select acceptance chart corresponding to code letter and AQL and plot above statistics on chart	2. See Figure 46.24
3. Acceptance criterion: Point plots inside accept zone	3. Point plots inside accept zone. Accept the lot
4. Rejection criterion: Point plots outside accept zone	
V. Action: Dispose of lot as indicated and refer to switching rules for next lot; see above	
VI. Characteristics: OC curves given	
VII. Reference: ISO 3951 (1989)	
VIII. Remarks: Analogous methods are available for average range $\bar{R}$ or known standard deviation σ, by substituting $\bar{R}$ or σ in the above.	

throughout the test, the time accumulated on all units is calculated and compared with the acceptable amount of time for the total number of failures accumulated up to the time of observation. If the total time exceeds the limit for acceptance, the lot is accepted; if the total time is less than the limit for rejection, the lot is rejected. If the total time falls between the two limits, the test is continued.

Plans are given for various values of the consumer and producer risks, and operating characteristic curves are provided for life tests terminated at a preassigned number of failures or preassigned time. Special tables are also included showing the expected saving in test time by increasing the sample size or by testing with replacement of failed units.

TABLE 46.25 Application of MIL-STD-414, ANSI/ASQC Z1.9 (1993), and ISO 3951

Step	Section	Form 1	Form 2	Graphical
Preparatory		Obtain k and n from appropriate tables	Obtain M and n from appropriate tables	Select appropriate acceptance chart
Determine criteria	Section B (s)	$T_U = \dfrac{U - \overline{X}}{s}$ $T_L = \dfrac{\overline{X} - L}{s}$	$Q_U = \dfrac{U - \overline{X}}{s}$ $Q_L = \dfrac{\overline{X} - L}{s}$	$A = \dfrac{\overline{X} - L}{U - L}$ $V = \dfrac{s}{U - L}$
	Section C $(\overline{R})$	$T_U = \dfrac{U - \overline{X}}{\overline{R}}$ $T_L = \dfrac{\overline{X} - L}{\overline{R}}$	$Q_U = \dfrac{(U - \overline{X})c}{\overline{R}}$ $Q_L = \dfrac{(\overline{X} - L)c}{\overline{R}}$	$A = \dfrac{\overline{X} - L}{U - L}$ $V = \dfrac{\overline{R}}{U - L}$
	Section D (σ)	$T_U = \dfrac{U - \overline{X}}{\sigma}$ $T_L = \dfrac{\overline{X} - L}{\sigma}$	$Q_U = \dfrac{(U - \overline{X})v}{\sigma}$ $Q_L = \dfrac{(\overline{X} - L)v}{\sigma}$	$A = \dfrac{\overline{X} - L}{U - L}$ $V = \dfrac{\sigma}{U - L}$
Estimation			Enter table with n and Q_U or Q_L to get p_U or p_L	
Action	Single specification	Accept if $T_U \geq k$ or $T_L \geq k$	Accept if $p_U \leq M$ or $p_L \leq M$	
	Double specification	Accept if* $T_U \geq k$, $T_L \geq k$ and $s <$ MSD or $\overline{R} <$ MAR	Accept if $p_U + p_L \leq M$	Accept if (A, V) plots inside acceptance curve
Standard/ Specification		414, 1.9/single, double* 3951/single, double with separate AQL's	414, 1.9/single, double	3951 double with combined AQL

Note: c = scale factor; $v = \sqrt{\dfrac{n}{n-1}}$

* Not official procedure.
Source: ANSI/ASQC Z1.9 (1993), ISO 3951 (1989).

Weibull Distribution: TR-3, TR-4, TR-6, TR-7. United States Defense Department quality control and reliability technical reports TR-3 (1961), TR-4 (1962), TR-6 (1963), and TR-7 (1965) have presented sampling plans based on an underlying Weibull distribution of individual measurement t.

Plots on probability paper or goodness-of-fit tests must be used to assure that individual measurements are distributed according to the Weibull model. See Wadsworth, Stephens, and Godfrey (1982). When this distribution is found to be an appropriate approximation to the failure distribution, methods are available to characterize a product or a process in terms of the three parameters of the Weibull distribution (see Sections 44 and 48 under the Weibull Distribution). These include probability plots and also point and interval estimates. Sampling plans are available for use with the Weibull approximation, which assumes shape and location parameters to be known. The plans are given in the technical reports mentioned above and are based in the following criteria:

TABLE 46.26 Variables Plans for Percent Defective, Mixed Variables—Attributes, ANSI/ASQC Z1.9 (1993)

Example: The specification for electrical resistance of a certain component is 620 to 680 ohms. A lot of 100 is submitted for inspection, inspection level II, normal inspection, with AQL = 2.5 percent. Should the lot be accepted if $\bar{X} = 647$ and $s = 17.22$?

Summary of plan	Calculations
I. Restrictions: Measurements normally distributed	

II. Necessary information
    A. Lot size
    B. AQL
	II.
	    A. Lot size = 100
	    B. AQL = 2.5%

III. Selection of plan
    A. Using AQL and Lot Size, select appropriate variables plan from ANSI/ASQC Z1.9 (1993), Normal Inspection
    B. Using AQL and Lot Size, select single-sampling attributes plan from ANSI/ASQC Z1.4 (1993), using Tightened Inspection
	III.
	    A. $n = 10$
	       $M = 7.26$
	
	    B. ANSI/ASQC Z1.4 gives Code *F*
	       $n = 32$
	       $c = 1$

IV. Elements
    A. Sample size: See above; use items drawn in first sample as part of second sample
    B. Statistic: Use appropriate statistics from ANSI/ASQC Z1.9 (1993) and ANSI/ASQC Z1.4 (1993), as indicated in each standard
    C. Decision criteria
       1. Apply ANSI/ASQC Z1.9 (1993) plan
         a. Accept lot if plan accepts
         b. Otherwise, apply ANSI/ASQC Z1.4 (1993) plan, taking additional samples to satisfy sample size requirements
       2. Apply ANSI/ASQC Z1.4 (1993) plan if necessary
         a. Accept lot if plan accepts
         b. Otherwise reject lot
	IV.
	    A. First sample: $n = 10$
	       Second sample:
	       $n = (32 - 10)$
	       $n = 22$
	
	    C. Table 46.21 indicates ANSI/ASQC Z1.9 (1993) plan accepts; if the plan rejected the lot, an additional 22 samples would be drawn and the ANSI/ASQC Z1.4 (1993) plan applied to the number of defectives in the combined sample of 32

V. Action *A.* Dispose of lot as indicated	V. Accept the lot

VI. Characteristics
    A. Since the procedure outlined is a "dependent" mixed plan, see the following references:
       1. σ known: Schilling and Dodge (1969)
       2. σ unknown: Gregory and Resnikoff (1955)
       3. Approximation for σ unknown: Bowker and Goode (1952)	

Source: ANSI/ASQC Z1.9 (1993).

TABLE 46.27   Life Characteristics for Two Failure Distributions

$$Exponential\ f(t) = \frac{1}{\mu} e^{-t/\mu}$$

$$Weibull*f(t) = \frac{\beta t^{\beta-1}}{\eta^\beta} e^{-(t/\eta)^\beta}\ where\ \mu = \eta\Gamma\left(1 + \frac{1}{\beta}\right)$$

Life characteristic	Exponential	Weibull
Proportion $F(t)$ failing before time t	$F(t) = 1 - e^{-t/\mu}$	$F(t) = 1 - e^{-g(t/\mu)^\beta}$
Proportion $R(t)$ of population surviving to time t	$R(t) = e^{-t/\mu}$	$R(t) = e^{-g(t/\mu)^\beta}$
Mean life, ML or mean time between failures	μ	μ
Hazard rate, $Z(t)$, instantaneous failure rate at time t	$Z(t) = \frac{1}{\mu}$	$Z(t) = \frac{\beta g t^{\beta-1}}{\mu^\beta}$
Cumulative hazard rate $M(t)$ for period 0 to t	$M(t) = \frac{t}{\mu}$	$M(t) = \frac{g t^\beta}{\mu^\beta}$
Failure rate λ or average hazard rate period 0 to t, $m(t)$	$\lambda = \frac{1}{\mu}$	$m(t) = \frac{g t^{\beta-1}}{\mu^\beta}$

*Weibull parameters explained in Section 48. The formulas given here are those of H108 (exponential) and TR-3 (Weiball).

TABLE 46.28   Values of $g=[\Gamma(1+1/\beta)]^\beta$ for Weibull Distribution*

β	0.0	0.1	0.2	0.3	0.4	0.5	0.6	0.7	0.8	0.9
0.0	$\cdots$	4.5287	2.6052	1.9498	1.6167	1.4142	1.2778	1.1794	1.1051	1.0468
1.0	1.0000	0.9615	0.9292	0.9018	0.8782	0.8577	0.8397	0.8238	0.8096	0.7969
2.0	0.7854	0.7750	0.7655	0.7568	0.7489	0.7415	0.7348	0.7285	0.7226	0.7172
3.0	0.7121	0.7073	0.7028	0.6986	0.6947	0.6909	0.6874	0.6840	0.6809	0.6778

β	0.33	0.67	1.33	1.67	3.33	4.00	5.00
g	1.8171	1.2090	0.8936	0.8289	0.6973	0.6750	0.6525

*The columns of this table are subdivisions of the rows. Thus when $\beta = 1.2$, the value of g is 0.9292.

1. Mean life criterion (TR-3)
2. Hazard rate criterion (TR-4)
3. Reliable life criterion (TR-6)
4. All three (TR-7)

The tables cover a wide range of the family of Weibull distributions by providing plans for shape parameter β from $1/3$ to 5. The technical reports abound in excellent examples and detailed descriptions of the methods involved.

Technical report TR-7 (1965) provides factors and procedures for adapting MIL-STD-105D plans to life and reliability testing when a Weibull distribution of failure times can be assumed. Tables of the appropriate conversion factors are provided for the following criteria:

Table	Criterion	Conversion factor
1	Mean life	$(t/\mu) \times 100$
2	Hazard rate	$tZ(t) \times 100$
3	Reliable life ($r = 0.90$)	$(t/p) \times 100$
4	Reliable life ($r = 0.99$)	$(t/p) \times 100$

Each table is presented in three parts, each of which is indexed by 10 values of $\beta (\beta = {}^1\!/3, {}^1\!/2, {}^2\!/3, 1, 1^1\!/3, 1^2\!/3, 2, 2^1\!/2, 3^1\!/2, 4)$. TR-7 is used in a manner analogous to the other three technical reports.

BULK SAMPLING[3]

Bulk material may be of gaseous, liquid, or solid form. Usually it is sampled by taking increments of the material, blending these increments into a single composite sample, and then, if necessary, reducing this gross sample to a size suitable for laboratory testing.

If bulk material is packaged or comes in clearly demarked segments, if it is for all practical purposes uniform within the packages, but varying between packages, and if the quality of each package in the sample is measured, then the sampling theory developed for discrete units may be employed.

A special theoretical discussion is necessary for the sampling of bulk material:

1. If the packages are uniform but the increments from individual packages are not tested separately; instead they are physically composited, in part at least, to form one or more composite samples that are tested separately.

2. If the contents of the packages are not uniform so that the question of sampling error arises with respect to the increments taken from the packages.

3. If the bulk material is not packaged and sample increments have to be taken from a pile, a truck, a railroad car, or a conveyer belt.

In the above circumstances, the special aspects that make bulk sampling different from the sampling of discrete indivisible units are:

1. The possibility of physical compositing and the subsequent physical reduction (or subsampling) that is generally necessary.

2. The need in many cases to use a mechanical sampling device to attain the increments that are taken into the sample. In this case the increments are likely to be "created" by the use of the sampling device and cannot be viewed as preexisting.

Objectives of Bulk Sampling. In most cases the objective of sampling bulk material is to determine its mean quality. This may be for the purpose of pricing the material or for levying custom duties or other taxes, or for controlling a manufacturing process in which the bulk material may be used. It is conceivable that interest in bulk material may also at times center on the variability of the material; or, if it is packaged, on the percent defective; or on the extreme value attained by a segment or package, as described in ASTM (1968). In view of the limited space that is available, the discussion will be restricted to estimation of the mean quality of a material.

---

[3]This section is condensed from Section 25A of the third edition of this handbook, prepared by Acheson J. Duncan.

Special Terms and Concepts. A number of special terms and concepts are used in the sampling of bulk material. These are

1. *Lot:* The mass of bulk material the quality of which is under study—not to be confused with a statistical population.

2. *Segment:* Any specifically demarked portion of the lot, actual or hypothetical.

3. *Strata:* Segments of the lot that are likely to be differentiated with respect to the quality characteristic under study.

4. *Increment:* Any portion of the lot, generally smaller than a segment.

5. *Sample increments:* Those portions of the lot initially taken into the sample.

6. *Gross sample:* The totality of sample increments taken from the lot.

7. *Composite sample:* A mixture of two or more sample increments.

8. *Laboratory sample:* That part of a larger sample which is sent to the laboratory for test.

9. *Reduction:* The process by which the laboratory sample is obtained from a composite sample. It is a method of sampling the composite sample. It may take the form of hand-quartering or riffling or the like.

10. *Test-unit:* That quantity of the material which is of just sufficient size to make a measurement of the given quality characteristic.

11. *Quality of a test-unit:* The expected value of the hypothetically infinite number of given measurements that might be made on the test-unit. Any single measurement is a random sample of one from this infinite set. The analytical variance is the variance of such measurements on the infinite set.

12. *Mean of a lot:* If a lot is exhaustively divided into a set of M test-units, the mean of the qualities of these M test-units is designated the mean of the lot. It is postulated that this mean will be the same no matter how the M test-units are obtained. This assumes that there is no physical interaction between the quality of test-units and the method of division. See item 16 below.

13. *Mean of a segment (stratum, increment, composite sample, or laboratory sample):* Defined in a manner similar to that used to define the mean of a lot. It is assumed that the segment is so large relative to the size of a test-unit that any excess over the integral number of test-units contained in the segment can be theoretically ignored. If this is not true, then the quality of the fraction of a test-unit remaining is arbitrarily taken to be the quality of the mean of the segment minus this fraction.

14. *Uniformity:* A segment of bulk material will be said to be uniform if there is no variation in the segment. If, for example, every cubic centimeter of a material contains exactly the same number of "foreign particles," the density of these particles would be said to be uniform throughout the segment. See the note under item 15, however.

15. *Homogeneous:* A segment of bulk material will be said to be homogeneous with respect to a given quality characteristic if that characteristic is randomly distributed throughout the segment.
   Note: The character of being uniform or homogeneous is not independent of the size of the units considered. The number of foreign particles may be the same for every cubic meter of a material, and with respect to this size unit the material will be said to be uniform. For units of size 1 cubic centimeter, however, there may be considerable variation in the number of foreign particles, and for this size of unit the material would not be judged to be uniform. The same considerations are involved in the definition of homogeneity. The number of foreign particles per cubic meter could vary randomly from one cubic meter to the next, but within each cubic meter there might be considerable (intraclass) correlation between the number of foreign particles in the cubic centimeters that make up the cubic meter.

16. *Systematic physical bias:* If the property of the material is physically affected by the sampling device or method of sampling employed, the results will have a systematic bias. A boring or cutting device, for example, might generate sufficient heat to cause loss of moisture.

17. *Physical selection bias:* If a bulk material is a mixture of particles of different size, the sampling device may tend to select more of one size particle than another. This means that if a segment was exhaustively sampled by such a device, early samples would tend to have relatively more of certain size particles than later samples.

18. *Statistical bias:* A function of the observations that is used to estimate a characteristic of a lot, e.g., its mean, is termed a "statistic." A statistic is statistically biased if in many samples its mean value is not equal to the lot characteristic it is used to estimate.

Determination of the Amount of Sampling. Since the variance of a sample mean $\sigma_{\bar{X}}^2$ is a function of the amount of sampling, say the number of increments taken, then once a model has been adopted and a formula obtained for $\sigma_{\bar{X}}^2$, it becomes possible to determine the amount of sampling required to attain a confidence interval of a given width or to attain a specified probability of making a correct decision.

A similar approach is involved in determining the amount of sampling to get a desired set of risks for lot acceptance and rejection.

Models and Their Use. Sampling plans for discrete product have been cataloged in a number of tables. This has not yet been possible for bulk sampling, and instead, a "sampling model" must be created for each type of bulk material and the model used to determine the sample size and acceptance criteria for specific applications.

A bulk sampling model consists of a set of assumptions regarding the statistical properties of the material to be sampled plus a prescribed procedure for carrying out the sampling. A very simple model, for example, would be one in which it is assumed that the quality characteristics of the test-units in a lot are normally distributed and simple random sampling is used.

With the establishment of a model, a formula can generally be derived for the sampling variance of an estimate of the mean of a given lot. This must be uniquely derived based on the type of product, lot formation, and other factors. The reader is urged to consult the references for specific formulas (see especially Duncan, 1962). From an estimate of this sampling variance, confidence limits can be established for the lot mean, and/or a decision with given risk can be made about the acceptability of the lot.

Let the variance of a sample mean be denoted as $\sigma_{\bar{X}}^2$ and its estimate as $s_{\bar{X}}^2$. Then "0.95 confidence limits" for the mean of the lot will be given by

$$0.95 \text{ confidence limits for } \mu = \overline{X} \pm t_{0.025} s_{\bar{X}}$$

where μ is the mean of the lot, $\overline{X}$ is the sample mean, and $t_{0.025}$ is the 0.025 point of a t-distribution for the degrees of freedom involved in the determination of $s_{\bar{X}}^2$.

If a decision is to be made on the acceptability of a lot, a criterion for acceptability will take some such form as

$$\text{Accept if } t = \frac{\overline{X} - L}{s_{\bar{X}}} \text{ is positive or if it has a negative value numerically less than } t_{0.025}$$

where L is the lower specification limit on the product and $t_{0.025}$ is the 0.025 point of a t-distribution for the degrees of freedom involved in the determination of $s_{\bar{X}}^2$. See Grubbs and Coon (1954).

Models for Distinctly Segmented Bulk Material ("Within and Between" Models). Much bulk material comes in distinctly segmented form. It may be packaged in bags, bales, or cans, for example, or may come in carloads or truckloads.

For distinctly segmented material it can be established that the overall variance of individual test-units is, for a large number of segments each with a large number of test-units, approximately equal to the sum of the variance between segments and the average variance within segments. In what follows, the variance within segments is assumed the same for all segments.

Model 1A. Isolated Lots, Nonstratified Segments.   For an isolated lot of distinctly segmented material, the sampling procedure will be to take an increment of m test-units from each of n segments, reduce each increment to a laboratory sample, and measure its quality X. The mean of the n test-units is taken as an estimate of the mean of the lot.

Model 1B. One of a Series of Lots.   Suppose the current lot is one of a series of lots of distinctly segmented material and that estimates of the between and within variances have been made in a pilot study, together with estimates of the reduction variance and test variance. It will be assumed that the reduction variance yielded by the pilot study is valid for larger amounts than that used in the study.

With the given prior information, an estimate of the sampling variance for the current sample estimate of mean lot quality can be based on the pilot study, and there is need only for a current check on the continued validity of this study. Consequently, composite samples can be used requiring only a few measurements, and the cost of inspection of the current lot may be considerably curtailed.

The "Within and Between" Models: Stratified Segments.   In some situations the quality characteristic of the bulk material may be stratified in that in each segment it may vary from layer to layer and is not randomly distributed throughout the segment. Difficulties in formulating a model for stratified segments of this kind can be overcome if the strata are reasonably parallel and if the increments taken from the sample segments are taken perpendicular to the strata and penetrate all strata. In taking a sample from a bale of wool, for example, a thief could cut a sample running vertically from top to bottom of the bale.

Models for Bulk Material Moving in a Stream.
In many instances the bulk material to be sampled is moving in a stream, say on a conveyer belt. In such instances it is the common practice to take increments systematically from the stream, the increment being taken across the full width of the stream.

Isolated Lots.   If increments are taken at random from the stream, we would have a simple random sample from the lot, and we could proceed much as indicated for isolated lots of distinctly segmented material.

Model 2A. A Stream of Lots: A Segregation Model.   When a stream of bulk material persists for some time, with possible interruptions, determinations of quality and/or action decisions may have to be made for a number of lots. Here, a pilot study might be profitable.

The kind of pilot study needed will depend on the assumptions about the statistical properties of the material in the stream. If the quality of the material varies randomly in the stream, then a pilot study based on a number of randomly or systematically taken increments would be sufficient to determine the variance of the material, and this could be used to set up confidence limits for the means of subsequent lots or make decisions as to their acceptability, even though in each of these lots a single composite sample was taken.

Obtaining the Test-Units.   Three factors are important in obtaining the test-units.

1. *The models discussed above assume random sampling:*   Either the increments are picked at random (using, preferably, random numbers if the units can be identified) or the increments are selected systematically from material that is itself random. Random sampling may have to be undertaken while the material is in motion or being moved. If random sampling cannot be used, special efforts should be made to get a representative picture of a lot, noting strata and the like. Some element of randomness must be present in a sampling procedure to yield a formula for sampling variance.

2. *Grinding and mixing:*   In the processing of bulk material and in the formation of composite samples, grinding and/or mixing may be employed in an attempt to attain homogeneity or at least to reduce the variability of the material.

If a material can be made homogeneous for the size increment that is to be used in subsequent sampling, then an increment of that size can be viewed as a random sample of the material. The attainment of homogeneity of bulk material in bags, cartons, barrels, etc., is thus a worthy objective when the contents of these containers are to be sampled by a thief or other sampling instruments. However, random sampling does not guarantee minimum sampling variation, and grinding and mixing may lead to a reduction in overall sampling variation without attaining homogeneity. See Cochran (1977).

Although grinding and mixing are aimed at reducing variability, these operations may in some circumstances cause segregation and thus increase variability.

3. *Reduction of a sample to test-units:*   Measurements often may be made directly on the sample itself. However, sometimes a portion of the sample must be carefully reduced in either particle size or physical quantity to facilitate laboratory testing. The unreduced portion of the sample may be retained for subsequent reference for legal purposes or verification of results. An example of a technique is coning and quartering. The material is first crushed and placed in a conical pile, which is then flattened. The material is then separated into quarters and opposite quarters selected for further quartering or as the final test-units.

Bicking (1967) and Bicking et al. (1967) provide unique details on obtaining test-units for a large variety of specific bulk products.

Tests of Homogeneity.   The homogeneity of bulk material may be tested by $\bar{X}$-charts and c-charts. For mixtures of particles that can be identified, local homogeneity may be tested by running a χ^2 short-distance test. See Shinner and Naor (1961).

BAYESIAN PROCEDURES

The concept that experience or analytical studies can yield prior frequency distributions of the quality of submitted lots and that these "prior" distributions can in turn be used to derive lot-by-lot sampling plans has gained some popularity in recent years. This is generally termed the "Bayesian approach." Bayes' theorem and the Bayesian decision theory concept are discussed in Section 44, Basic Statistical Methods. The general concept of the use of prior information is discussed in Section 23, under Degree of Inspection and Testing Needed.

Considerable literature exists on Bayesian sampling. However, a very limited number of sampling tables based on particular prior distributions of lot quality are available. Calvin (1984) provides procedures and tables for implementing Bayesian plans. Another example is the set of tables by Oliver and Springer (1972), which are based on the assumption of a Beta prior distribution with specific posterior risk to achieve minimum sample size. This avoids the problem of estimating cost parameters. It is generally true that a Bayesian plan requires a smaller sample size than does a conventional sampling plan with the same consumer's and producer's risk. Among others, Schaefer (1967) discusses single sampling plans by attributes using three prior distributions of lot quality. Given specified risks α and β, sampling plans which satisfied these risks and which minimized sample size were determined. For example, with a particular prior distribution, $n = 6$ and $c = 0$ gave protection equivalent to a conventional sampling plan with $n = 34$, $c = 0$.

Advantages similar to those quoted above are usually cited. However, one prime factor is frequently ignored: How good is the assumption on the prior distribution? Hald (1960) gives an extensive account of sampling plans based on discrete prior distributions of product quality. Hald also employs a simple economic model along with the discrete prior distribution and the hypergeometric sampling distribution to answer a number of basic questions on costs, relationship between sample size and lot size, etc. Schaefer (1964) also considers the Bayesian operating characteristic curve. Hald (1981) has also provided an excellent comparison of classical and Bayesian theory and methodology for attributes acceptance sampling. The Bayesian approach to sequential and nonsequential acceptance sampling has been described by Grimlin and Breipohl (1972). Most such plans incorporate cost data reflecting losses involved in the decision making process to which the plan is to be applied. An excellent review of Bayesian and non-Bayesian plans has been given by Wetherill and Chin (1975).

While the explicit specification of a prior distribution is characteristic of the classical Bayes approach, procedures have been developed to incorporate much of the philosophy and approach of Bayes without explicit specification of a prior distribution. These methods are based on the incorporation of past data into an empirical estimate of the prior, and hence the approach is called "empirical Bayes." An excellent description of this method of estimation has been presented by Krutchkoff (1972). Application of the empirical Bayes methodology to attributes sampling has been given by Martz (1975). Craig and Bland (1981) show how it can be used in variables sampling.

These ideas have been further extended by the application of shrinkage estimators to the empirical Bayes problem as described by Morris (1983). One outstanding application of such procedures is in the so-called "universal sampling plan" which has been utilized in the quality measurement plan (QMP) of American Telephone and Telegraph. Audit sample sizes are based on historical process control, economics, quality standards, and the heterogeneity of audit clusters as described by Hoadley (1981). Strictly speaking, the weights for the Stein shrinkage estimators utilized in the procedure are developed through a classical Bayes approach, and so this application is more properly viewed as Bayes empirical Bayes methodology. A method for estimating quality using data from both accepted and rejected lots based on QMP is given in Brush and Hoadley (1990).

Application of Bayesian methods has been hindered by the difficulty involved in correct assessment of the necessary prior distributions and in collecting and keeping current the required cost information. Clearly, greater reliance on prior empirical information and the potential of the computer for generating cost information should be helpful in this regard. Pitfalls in the selection of a prior distribution have been pointed out by Case and Keats (1982).

While studies have indicated that Bayesian schemes may be quite robust to errors in the prior distributions and loss functions, they nevertheless assume the prior to be stationary in the long-term sense (i.e., a process in control). Classical methods do not make this assumption. Bayesian plans are, in fact, quite application-specific, requiring extensive information and update for proper application, and like variables sampling, they are applied one characteristic at a time. Nevertheless, where appropriate, they provide yet another tool for economic sampling.

HOW TO SELECT THE PROPER SAMPLING PROCEDURES

The methods of acceptance sampling are many and varied. It is essential to select a sampling procedure appropriate to the acceptance sampling situation to which it is to be applied. This will depend upon the nature of the application itself, on quality history, and the extent of knowledge of the process and the producer. Indeed, according to Dodge (1950, p. 8):

> A product with a history of consistently good quality requires less inspection than one with no history or a history of erratic quality. Accordingly, it is good practice to include in inspection procedures provisions for reducing or increasing the amount of inspection, depending on the character and quantity of evidence at hand regarding the level of quality and the degree of control shown.

This will be discussed further in the Conclusion, below.

The steps involved in the selection and application of a sampling procedure are shown in Figure 46.25 taken from Schilling (1982). Emphasis here is on the feedback of information necessary for the proper application, modification, and evolution of sampling. As Ott and Schilling (1990) have pointed out:

> There are two standard procedures that, though often good in themselves, can serve to postpone careful analysis of the production process:

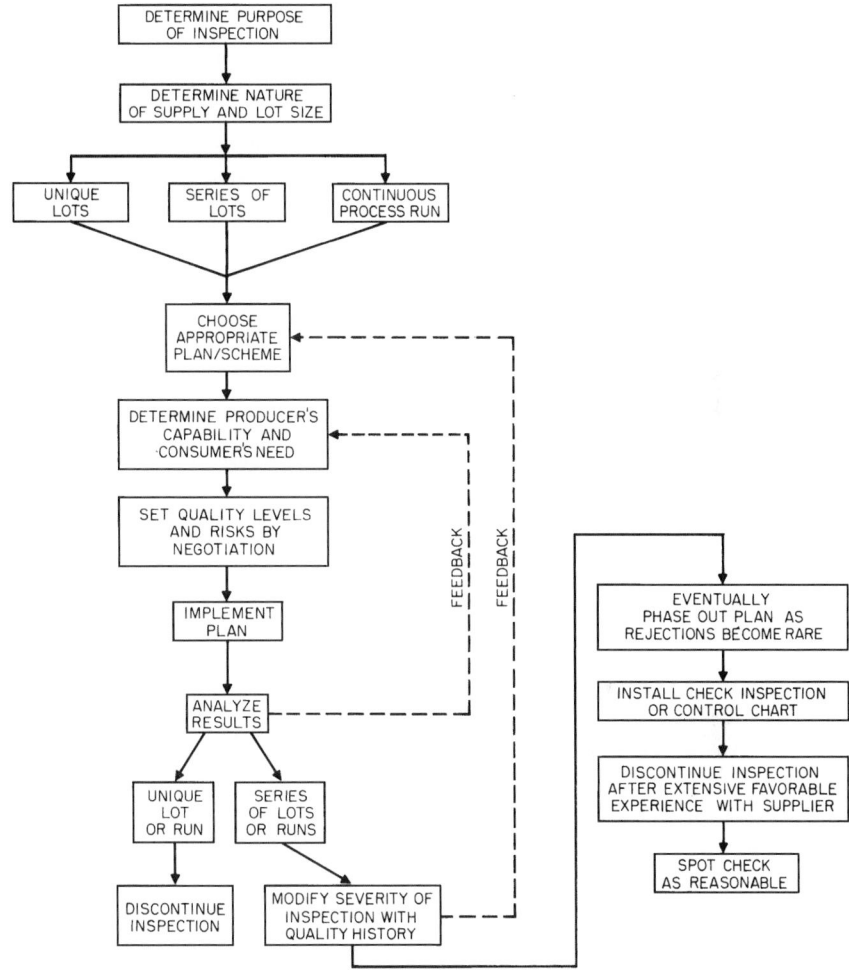

FIGURE 46.25   Check sequence for implementation of sampling procedure.

1. On-line inspection stations (100% screening). These can become a way of life.
2. On-line acceptance sampling plans which prevent excessively defective lots from proceeding on down the production line, but have no feedback procedure included.

These procedures become bad when they allow or encourage carelessness in production. It gets easy for production to shrug off responsibility for quality and criticize inspection for letting bad quality proceed.

It is therefore necessary for sampling to be constantly subject to modification in a system of acceptance control. Each plan has a specific purpose for which it is to be applied. Table 46.29 shows the relation of plan and purpose for some of the plans which have been discussed. Those not discussed here will be found described in detail in Schilling (1982). These plans then are the elements involved in a system of acceptance control. Their effective use in a continuing effort to provide protection while reducing inspection and moving to process control is a function of the ingenuity, integrity, and industry of the user.

TABLE 46.29   Selection of Plan

Purpose	Supply	Attributes	Variables
Simple guarantee of producer's and consumer's quality levels at stated risks	Unique lot	Two-point plan (Type A)	Two-point plan (type B)
	Series of lots	Dodge-Roming LTPD Two-point plan (Type B)	Two-point plan (type B)
Maintain level of submitted quality at AQL or better	Series of lots	MIL-STD-105E ANSI/ASQC 1.4 ISO 2859	MIL-STD-414 ANSI/ASQC Z1.9 ISO 3951
Rectification guaranteeing AOQL to consumer	Series of lots	Dodge-Roming AOQL	Use measurements as go-no go
	Flow of individual units	CSP-1, 2, 3 Multilevel plan MIL-STD-1235B	
Reduced inspection after good history	Series of lots	Skip-lot Chain	Lot plot Mixed variable-attributes Narrow-limit gaging
Check inspection	Series of lots	Demerit rating	Acceptance control chart
Compliance to mandatory standards	Unique lot	Lot-sensitive plan	Mixed variables-attributes with $c = 0$
	Series of lots	TNT plan	
Reliability sampling	Unique lot	Two-point plan (type B)	MIL-HDBK-108 TR-7
	Series of lots	LTPD plan	TR-7 using MIL-STD-105D switching rules

Source: Schilling, (1982) p.569.

COMPUTER PROGRAMS FOR ACCEPTANCE SAMPLING

A variety of computer programs are available for application of specific sampling plans. They have become increasingly important, particularly in the application of selected plans and in the development of plans for specific applications. Some early fundamental programs are listed in Table 46.30. It should be noted, however, that it is questionable whether these programs will ever replace hard copy procedures and standards which play a vital role in the negotiations between producer and consumer and which are immediately available for in-plant meetings and discussions at any location.

A number of commercial packages intended for quality control applications contain acceptance sampling options. In addition, various programs have been presented in the literature. For example, Guenther (1984) presents a program for determining rectifying single sampling plans while Garrison and Hickey (1984) give sequential plans, both by attributes. A program for determining the ASN of curtailed attributes plans is given by Rutemiller and Schaefer (1985). McWilliams (1990) has provided a program which deals with various aspects of hypergeometric, Type A, plans including double sampling. McShane and Tumbull (1992) present a program for analysis of CSP-1 plans with finite run length. Goldberg (1992) gives a PC program for rapid application of the sequential approach to continuous sampling. A program for selecting quick switching systems will be found in Taylor (1996). These programs supply code. Implementation of plans on spread sheets is presented in Cox (1992–1993). Expert systems are discussed in Fard and Sabuncuoglu (1990) and Seongin (1993). Thus, the computer can be expected to enhance and diversify application of sampling plans.

TABLE 46.30 Elementary Computer Programs for Acceptance Sampling (Optional Input/Output in Brackets)

Purpose	Input	Output	Reference
Single sampling—derivation	p_1, p_2, α, β, (lot size, p)	Plan (ATI at p, AOQL)	Snyder and Storer, 1972
Double sampling—derivation	AQL, LTPD (lot size) (assumes $\alpha = 0.05$, $\beta = 0.10$)	Plans for $n_2 = n_1$ and $n_2 = 2n_1$ with AOQL (p, P_a, AOQ, ATI, ASN)	Chow, Dickinson, and Hughes, 1973 and 1975
Multiple sampling—derivation	AQL, LTPD (lot size) (assumes $\alpha = 0.05$, $\beta = 0.10$)	Plans for equal n with AOQL (p, P_a, AOQ, ATI, ASN)	Hughes, Dickinson, and Chow, 1973
Single, double, multiple sampling—evaluation (P_a given p or p given P_a) (hypergeometric, binomial, Poisson, normal)	Distribution, sample sizes, acceptance/rejection numbers, fractions defective to be evaluated, probabilities of acceptance to be determined (lot size)	Plan control table, p, P_a, ASM, AOQ, ATI	Schilling, Sheesley, and Nelson, 1978

TABLE 46.30 Elementary Computer Programs for Acceptance Sampling (Optional Input/Output in Brackets) (*Continued*)

Purpose	Input	Output	Reference
Dodge continuous sampling—evaluation CSP-1, 2, 3	Fractions defective, index of plan (1, 2, 3), f, i, k	p, F, P_a, AOQ	Sheesley, 1975
MIL-STD-414—implementation acceptance/rejection Decision given data	Number lots, type inspection (T, N, R), measure of variability, lot size, specification limits, AQL, data	Plan $\hat{p}$, M, accept or reject	Nelson, 1977

Snyder, D. C. and Storer, R. F. (1972). "Single Sampling Plans Given an AQL, LTPD, Producer and Consumer Risks." *Journal of Quality Technology,* vol. 4, no. 3, July 1972, pp. 168–171.

Chow, B., Dickinson, P. C., and Hughes, H. (1973). "A Computer Program for the Solution of Double Sampling Plans." *Journal of Quality Technology,* vol. 4, no. 4, October 1973, pp. 205–209; also, *Journal of Quality Technology,* vol. 5, no. 4, October 1975, p. 166.

Hughes, H., Dickinson, P. C., and Chow, B. (1973). "A Computer Program for the Solutions of Multiple Sampling Plans." *Journal of Quality Technology,* vol. 5, no. 1, January 1973, pp. 39–42.

Schilling, E. G., Sheesley, J. H., and Nelson, P. R. (1978). "GRASP: A General Routine for Attribute Sampling Plan Evaluation." *Journal of Quality Technology,* vol. 10, no. 3, July 1978, pp. 125–130.

Sheesley, J. H. (1975). "A Computer Program to Evaluate Dodge's Continuous Sampling Plans." *Journal of Quality Technology,* vol. 7, no. 1, January 1975, pp. 43–45.

Nelson, P. R. (1977). "A Computer Program for Military Standard 414: Sampling Procedures and Inspection by Variables for Percent Defective." *Journal of Quality Technology.* vol. 9, no. 2, April 1977, pp. 82–86.

Source: Schilling, 1982, p. 587.

CONCLUSION—MOVING FROM ACCEPTANCE SAMPLING TO ACCEPTANCE CONTROL

There is little control of quality in the application of a sampling plan to an individual lot. Such an occurrence is static, whereas control implies movement and direction. When used in a system of *acceptance control,* over the life of the product, however, the plan can provide:

Protection for the consumer

Protection for the producer

Accumulation of quality history

Feedback for process control

Pressure on the producer to improve the process

Acceptance control involves adjusting the acceptance sampling procedure to match existing conditions with the objective of eventually phasing out the inspection altogether. This is much as the old-time inspector varied the inspection depending on quality history. Thus, acceptance control is "a continuing strategy of selection, application and modification of acceptance sampling procedures to a changing inspection environment" [see Schilling (1982, p. 546)]. This involves a progression of sampling procedures applied as shown in Table 46.31.

These procedures are applied as appropriate over the lifetime of the product in a manner consistent with the improvement of quality and the reduction and elimination of inspection. This can be seen in Table 46.32. For a detailed discussion and example of moving to audit sampling, see Doganaksoy and Hahn (1994).

Modern manufacturing is not static. It involves the development, manufacture, and marketing of new products as well as an atmosphere of continuing cost reduction, process modification, and other forms of quality improvement. Changes are continually introduced into the production process which are deliberate, or unexpected from an unknown source. When this happens, acceptance sampling is necessary to provide protection for the producer and the consumer until the process can be brought into control. This may take days, months, or years. Under these circumstances, the inspection involved should be systematically varied in a system of acceptance control which will reduce inspection reciprocally with the increase of the learning curve in the specific application. This can provide the protection necessary for the implementation of proper process control in a manner portrayed roughly in Figure 46.26. Acceptance control and process control are synergistic in the sense that one supports proper use of the other. As Dodge (1969, p. 156) has pointed out:

The "acceptance quality control system"…encompassed the concept of protecting the consumer from getting unacceptable defective material and encouraging the producer in the use of process quality control

TABLE 46.31 Progression of Attribute Sampling Procedures

Past results	Quality history			Criterion
	Little	Moderate	Extensive	
Excellent	AQL plan	Chain	Demerit rating or remove inspection	Almost no (<1%) lots rejected
Average	Rectification or LTPD plan	AQL plan	Chain	Few (<10%) lots rejected
Poor	100% inspection	Rectification or LTPD plan	Discontinue acceptance	Many (≥10%) lots rejected
Amount	Fewer than 10 lots	10–50 lots	More than 50 lots	

TABLE 46.32   Life Cycle of Acceptance Control Application

Stage	Step	Method
Preparatory	Choose plan appropriate to purpose	Analysis of quality system to define the exact need for the procedure
	Determine producer capability	Process performance evaluation using control charts
	Determine consumer needs	Process capability study using control charts
	Set quality levels and risks	Economic analysis and negotiation
	Determine plan	Standard procedures if possible
Initiation	Train inspector	Include plan, procedure, records, and action
	Apply plan properly	Ensure random sampling
	Analyze results	Keep records and control charts
Operational	Assess protection	Periodically check quality history and OC curves
	Adjust plan	When possible change severity to reflect quality history and cost
	Decrease sample size if warranted	Modify to use appropriate sampling plans taking advantage of credibility of supplier with cumulative results
Phase out	Eliminate inspection effort where possible	Use demerit rating or check inspection procedures when quality is consistently good. Keep control charts
Elimination	Spot check only	Remove all inspection when warranted by extensive favorable history

Source: Schilling, 1982, p. 566.

by: varying the quantity and severity of acceptance inspections in direct relation to the importance of the characteristics inspected, and in inverse relation to the goodness of the quality level as indicated by these inspections.

It is in this sense that acceptance control is an essential part of the quality system.

REFERENCES

ANSI/ASQC Q3 (1988). *Sampling Procedures and Tables for Inspection of Isolated Lots by Attributes,* American Society for Quality, Milwaukee.
ANSI/ASQC S1 (1996). *An Attribute Skip-Lot Sampling Program,* American Society for Quality, Milwaukee.

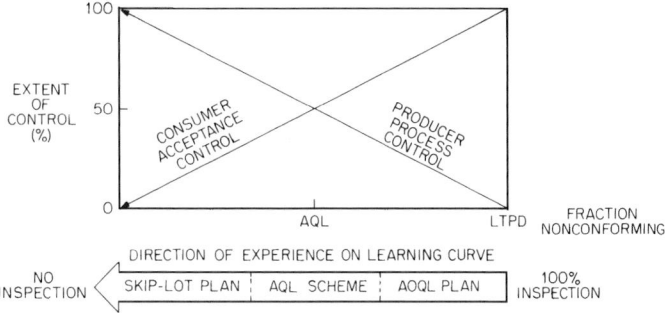

FIGURE 46.26   Synergy between acceptance control and process control. (*From Schilling, 1982, p. 570.*)

ANSI/ASQC S2 (1995). *Introduction to Attribute Sampling,* American Society for Quality, Milwaukee.

ANSI/ASQC Z1.4 (1993). *Sampling Procedures and Tables for Inspection by Attributes.* American Society for Quality, Milwaukee.

ANSI/ASQC Z1.9 (1993). *Sampling Procedures and Tables for Inspection by Variables for Percent Nonconforming.* American Society for Quality, Milwaukee.

ANSI/ISO/ASQC A3534 (1993). *Statistics—Vocabulary and Symbols, Part I: Probability and General Statistical Terms; Part II: Statistical Quality Control,* American Society for Quality, Milwaukee.

ASTM (1968). *Standard Methods of Test for Sampling Coal (D2234—68).* American Society for Testing and Materials, Philadelphia.

ASTM (1981). *Recommended Practice for Sampling Industrial Chemicals (E300—73).* American Society for Testing and Materials, Philadelphia.

Beattie, D. W. (1962). "Continuous Acceptance Sampling Procedure Based Upon a Cumulative Sum Chart for the Number of Defectives." *Applied Statistics,* vol. 11, no. 3, November, pp. 137–147.

Bicking, C. A. (1967). "The Sampling of Bulk Materials." *Materials Research and Standards,* March, pp. 95–116.

Bicking, C. A., Donovan, T. A., Sosnowski, T. S., and Bicking, C. M. (1967). *Bibliography on Precision Bulk Sampling and Related Applications of Statistics—Supplement.* CA Report 10, Technical Association of the Pulp and Paper Industry.

Bowker, A. H., and Goode, H. P. (1952). *Sampling Inspection by Variables.* McGraw-Hill, New York.

Brush, G. G., and Hoadley, B. (1990). "Estimating Outgoing Quality Using the Quality Measurement Plan." *Technometrics,* vol. 32, no. 1, February, pp. 31–41.

Burington, R. S., and May, D. C. (1970). *Handbook of Probability and Statistics with Tables,* 2nd ed. McGraw-Hill, New York.

Calvin, T. W. (1977). "TNT Zero Acceptance Number Sampling." *ASQC Technical Conference Transactions,* Milwaukee, pp. 35–39.

Calvin, T. W. (1984). *How and When to Perform Bayesian Acceptance Sampling.* American Society for Quality, Milwaukee.

Cameron, J. M. (1952). "Tables for Constructing and for Computing the Operating Characteristics of Single Sampling Plans." *Industrial Quality Control,* vol. 9, no. 1, July, pp. 37–39.

Case, K., and Keats, J. B. (1982). "On the Selection of a Prior Distribution in Bayesian Acceptance Sampling." *Journal of Quality Technology,* vol. 14, no. 1, January, pp. 10–18.

Cochran, W. G. (1977). *Sampling Techniques.* 3rd ed. John Wiley & Sons, New York.

Connolly, C. (1991). "A Sequential Approach to Continuous Sampling." *Quality Engineering,* vol. 3, no. 4, pp. 529–535.

Cowden, D. J. (1957). *Statistical Methods in Quality Control.* Prentice-Hall, Englewood Cliffs, NJ, pp. 489–490.

Cox, M. A. A. (1992–1993). "The Implementation on Spreadsheets of Sampling Curves Associated with Acceptance/Rectification Plans." *Quality Engineering,* vol. 5, no. 2, pp. 191–212.

Craig, J. A., Jr., and Bland, R. P. (1981). "An Empirical Bayes Approach to a Variables Acceptance Sampling Plan Problem." *Communications in Statistics—Theory, Methods,* vol. A10, no. 23, pp. 2399–2410.

Cross, R. (1984). "Parts Per Million AOQL Sampling Plans." *Quality Progress,* vol. 17, no. 11, November, pp. 28–34.

Dodge, H. F. (1932). "Statistical Control in Sampling Inspection." *American Machinist,* October, pp. 1085–1088.

Dodge, H. F. (1943). "A Sampling Plan for Continuous Production." *Annals of Mathematical Statistics,* vol. 14, no. 3, pp. 264–279.

Dodge, H. F. (1950). "Inspection for Quality Assurance." *Industrial Quality Control,* vol. 7, no. 1, June, p. 8.

Dodge, H. F. (1955). "Chain Sampling Inspection Plan." *Industrial Quality Control,* vol. 11, no. 4, January, pp. 10–13.

Dodge, H. F. (1955a). "Skip-lot Sampling Plans." *Industrial Quality Control,* vol. 11, no. 5, February, pp. 3–5.

Dodge, H. F. (1969). "Notes on the Evolution of Acceptance Sampling Plans, Part II." *Journal of Quality Technology,* vol. 1, no. 3, July, p. 156.

Dodge, H. F. (1970). "Notes on the Evolution of Acceptance Sampling Plans, Part IV." *Journal of Quality Technology,* vol. 2, no. 1, January, pp. 1–8.

Dodge, H. F., and Perry, R. L. (1971). "A System of Skip-Lot Plans for Lot-by-Lot Inspection." *ASQC Technical Conference Transactions,* Milwaukee, pp. 469–477.

Dodge, H. F., and Romig, H. G. (1959). *Sampling Inspection Tables,* 2nd ed. John Wiley & Sons, New York (republished in Wiley Classics Edition, 1998).

Dodge, H. F., and Stephens, K. S. (1965). "Some New Chain Sampling Inspection Plans." *ASQC Technical Conference Transactions,* Milwaukee, pp. 8–17.

Dodge, H. F., and Stephens, K. S. (1966). "Some New Chain Sampling Inspection Plans." *Industrial Quality Control,* vol. 23, no. 2, August, pp. 61–67.

Dodge, H. F., and Torrey, M. N. (1951). "Additional Continuous Sampling Inspection Plans." *Industrial Quality Control,* vol. 7, no. 5, March, pp. 7–12.

Doganaksoy, N., and Hahn, G. J. (1994). "Moving from Every Lot Inspection to Audit Sampling." *Journal of Quality Technology,* vol. 26, no. 4, October, pp. 261–273.

Duncan, A. J. (1962). "Bulk Sampling Problems and Lines of Attack." *Technometrics,* vol. 4, no. 3, August, pp. 319–344.

Duncan, A. J. (1974). *Quality Control and Industrial Statistics.* Richard D. Irwin, Homewood, IL.

Enell, J. W. (1954). "What Sampling Plan Shall I Choose?" *Industrial Quality Control,* vol. 10, no. 6, May, pp. 96–100.

Fard, N. S., and Sabuncuoglu, I. (1990). "An Expert System for Selecting Attribute Sampling Plans." *International Journal of Computer Integrated Manufacturing,* vol. 3, no. 6, pp. 364–372.

Freund, R. A. (1957). "Acceptance Control Charts." *Industrial Quality Control,* vol. 14, no. 4, October, pp. 13–23.

Garrison, D. R., and Hickey, J. J. (1984). "Wald Sequential Sampling for Attribute Inspection." *Journal of Quality Technology,* vol. 16, no. 3, July, pp. 172–174.

Goldberg, G. F. (1992). "Program for Sequential Approach to Continuous Sampling." *Quality Engineering,* vol. 5, no. 1, pp. 85–93.

Govindaraju, K. (1990). "Certain Observations on Lot Sensitive Sampling Plan." *Communications in Statistics—Theory and Methods,* vol. 19, no. 2, pp. 617–627.

Govindaraju, K. (1990). "Selection of ChSP-1 Sampling Plans for Given Acceptable Quality Level and Limiting Quality Level." *Communications in Statistics—Theory and Methods,* vol. 19, no. 6, pp. 2179–2190.

Grant, E. L., and Leavenworth, R. S. (1979). *Statistical Quality Control,* 5th ed. McGraw-Hill, New York.

Greenberg, B. S., and Stokes, S. L. (1992). "Estimating Nonconformance Rates After Zero-Defect Sampling with Rectification." *Technometrics,* vol. 34, no. 2, May, pp. 203–213.

Gregory, G., and Resnikoff, G. J. (1955). *Some Notes on Mixed Variables and Attributes Sampling Plans.* Technical Report 10, Applied Mathematics and Statistics Laboratory, Stanford University, Stanford, CA.

Grimlin, D. R., and Breipohl, A. M. (1972). "Bayesian Acceptance Sampling." *IEEE Transactions on Reliability,* vol. R-21, no. 3, August, pp. 176–180.

Grubbs, F. E., and Coon, H. J. (1954). "On Setting Testing Limits Relative to Specification Limits." *Industrial Quality Control,* vol. 10, no. 5, March, pp. 15–20.

Grubman, S., Martin, C. A., and Pabst, W. R., Jr. (1969). "MIL-STD-690B Failure Rate Sampling Plans and Procedures." *Journal of Quality Technology,* vol. 1, no. 3, July, pp. 205–216.

Guenther, W. C. (1984). "Determination of Rectifying Inspection Plans for Single Sampling by Attributes." *Journal of Quality Technology,* vol. 16, no. 1, January, pp. 56–63.

Hald, A. (1960). "The Compound Hypergeometric Distribution and a System of Single Sampling Inspection Plans Based on Prior Distributions and Costs." *Technometrics,* vol. 2, no. 3, August, pp. 275–340.

Hald, A. (1981). *Statistical Theory of Sampling Inspection by Attributes.* Academic Press, New York, pp. 532.

Hamaker, H. C. (1979). "Acceptance Sampling for Percent Defective by Variables and by Attributes." *Journal of Quality Technology,* vol. 11, no. 3, July, pp. 139–148.

Hillier, F. S. (1969). "$\overline{X}$ and R-Chart Control Limits Based on a Small Number of Subgroups." *Journal of Quality Technology,* vol. 1, no. 1, January, pp. 17–26.

Hoadley, B. (1981). "The Universal Sampling Plan." *ASQC Quality Congress Transactions,* Milwaukee, pp. 80–87.

ISO-2859-0 (1995). *Sampling Procedures for Inspection by Attributes—Part 0: Introduction to Acceptance Sampling.* ISO, Geneva.

ISO-2859-1 (1989). *Sampling Procedures for Inspection by Attributes—Part 1: Sampling Plans Indexed by Acceptable Quality Level.* ISO, Geneva.

ISO-2859-2 (1985). *Sampling Procedures for Inspection by Attributes—Part 2: Sampling Plans Indexed by Limiting Quality for Isolated Lot Inspection.* ISO, Geneva.

ISO 2859-3 (1991). *Sampling Procedures for Inspection by Attributes—Part 3: Skip-Lot Sampling Procedures.* ISO, Geneva.

ISO 3951 (1989). *Sampling Procedures and Charts for Inspection by Variables for Percent Nonconforming.* ISO, Geneva.

ISO 8422 (1991). *Sequential Sampling Plans for Inspection by Attributes.* ISO, Geneva.

ISO 8423 (1991). *Sequential Sampling Plans for Inspection by Variables for Percent Nonconforming (Known Standard Deviation).* ISO, Geneva.

Jaraiedi, M., Kochhar, D. S., and Jaisingh, S. C. (1987). "Multiple Inspections to Meet Desired Outgoing Quality." *Journal of Quality Technology,* vol. 19, no. 1, January, pp. 46–51.

Johnson, N. L., Kotz, S., and Rodriguez, R. N. (1985). "Statistical Effects of Imperfect Inspection Sampling: I. Some Basic Distributions." *Journal of Quality Technology,* vol. 17, no. 1, January, pp. 1–31.

Johnson, N. L., Kotz, S., and Rodriguez, R. N. (1986). "Statistical Effects of Imperfect Inspection Sampling: II. Double Sampling and Link Sampling." *Journal of Quality Technology,* vol. 18, no. 2, April, pp. 116–138.

Johnson, N. L., Kotz, S., and Rodriguez, R. N. (1988). "Statistical Effects of Imperfect Inspection Sampling: III. Screening (Group Testing)." *Journal of Quality Technology,* vol. 20, no. 2, April, pp. 98–124.

Johnson, N. L., Kotz, S., and Rodriguez, R. N. (1990). "Statistical Effects of Imperfect Inspection Sampling: IV. Modified Dorfman Sampling." *Journal of Quality Technology,* vol. 22, no. 2, April, pp. 128–137.

Krutchkoff, R. G. (1972). "Empirical Bayes Estimation." *The American Statistician,* vol. 26, no. 5, December, pp. 14–16.

Lieberman, G. J., and Solomon, H. (1955). "Multi-Level Continuous Sampling Plans." *Annals of Mathematical Statistics,* vol. 26, no. 4, pp. 686–704.

McShane, L. M., and Turnbull, B. W. (1992). "New Performance Measures for Continuous Sampling Plans Applied to Finite Production Runs." *Journal of Quality Technology,* vol. 24, no. 3, July, pp. 153–161.

McWilliams, T. P. (1990). "Acceptance Sampling Plans Based on the Hypergeometric Distribution." *Journal of Quality Technology,* vol. 22, no. 4, October, pp. 319–327.

Martz, H. F. (1975). "Empirical Bayes Single Sampling Plans for Specified Posterior Consumer and Producer Risks." *Naval Research Logistics Quarterly,* vol. 22, no. 4, December, pp. 651–665.

Mee, R. W. (1990). "An Improved Procedure for Screening Based on a Correlated, Normally Distributed Variable." *Technometrics,* vol. 32, no. 3, August, pp. 331–337.

MIL-HDBK-53 (1965). *Guide for Sampling Inspection.* Government Printing Office, Washington, DC, June 30.

MIL-HDBK-108 (1960). *Sampling Procedures and Tables for Life and Reliability Testing (Based on Exponential Distribution).* Quality Control and Reliability Handbook, U.S. Department of Defense, Government Printing Office, Washington, DC.

MIL-STD-105A (1950). *Sampling Procedures and Tables for Inspection by Attributes.* Government Printing Office, Washington, DC.

MIL-STD-105B (1958). *Sampling Procedures and Tables for Inspection by Attributes.* Government Printing Office, Washington, DC.

MIL-STD-105C (1961). *Sampling Procedures and Tables for Inspection by Attributes.* Government Printing Office, Washington, DC.

MIL-STD-105D (1964). *Sampling Procedures and Tables for Inspection by Attributes.* Government Printing Office, Washington, DC.

MIL-STD-105E (1989). *Sampling Procedures and Tables for Inspection by Attributes.* Government Printing Office, Washington, DC.

MIL-STD-414 (1957). *Sampling Procedures and Tables for Inspection by Variables for Percent Defective.* Military Standard, U.S. Department of Defense, Government Printing Office, Washington, DC.

MIL-STD-690B (1960). *Failure Rate Sampling Plans and Procedures.* Military Standard, U.S. Department of Defense, Government Printing Office, Washington, DC.

MIL-STD-781C (1977). *Reliability Tests: Exponential Distribution.* Military Standard, U.S. Department of Defense, Government Printing Office, Washington, DC.

MIL-STD-1235B (1981). *Continuous Sampling Procedures and Tables for Inspection by Attributes.* Government Printing Office, Washington, DC.

Morris, C. N. (1983). "Parametric Empirical Bayes Inference: Theory and Applications." *Journal of the American Statistical Association,* vol. 78, no. 131, March, pp. 47–55.

Moses, L. E. (1956). "Some Theoretical Aspects of the Lot Plot Sampling Inspection Plan." *Journal of the American Statistical Association,* vol. 51, no. 273, pp. 84–107.

Murphy, R. B. (1959). "Stopping Rules with CSP-1 Sampling Inspection Plans in Continuous Production." *Industrial Quality Control,* vol. 16, no. 5, November, pp. 10–16.

Nelson, P., and Jaraiedi, M. (1987). "Computing the Average Outgoing Quality after Multiple Inspections." *Journal of Quality Technology,* vol. 19, no. 1, January, pp. 52–54.

Nelson, W. (1983). *How to Analyze Reliability Data.* American Society for Quality, Milwaukee.

Oliver, L. R., and Springer, M. D. (1972). *A General Set of Bayesian Attribute Acceptance Plans.* American Institute of Industrial Engineers, Norcross, GA.

Ott, E. R., and Schilling, E. G. (1990). *Process Quality Control,* 2nd ed. McGraw-Hill, New York, pp. 133.

Pabst, W. R., Jr. (1963). "MIL-STD-105D." *Industrial Quality Control,* vol. 20, no. 5, November, pp. 4–9.

Perry, R. L. (1973). "Skip-Lot Sampling Plans." *Journal of Quality Technology,* vol. 5, no. 3, July, pp. 123–130.

Perry, R. L. (1973a). "Two-Level Skip-Lot Sampling Plans—Operating Characteristic Properties." *Journal of Quality Technology,* vol. 5, no. 4, October, pp. 160–166.

Raju, C. (1991). "Three-Stage Chain Sampling Plans." *Communications in Statistics—Theory and Methods,* vol. 20, no. 5/6, pp. 1777–1801.

Rutemiller, H. C., and Schaefer, R. E. (1985). "A Computer Program for the ASN of Curtailed Attributes Sampling Plans." *Journal of Quality Technology,* vol. 17, no. 2, April, pp. 108–113.

Schaefer, R. E. (1964). "Bayesian Operating Characteristic Curves for Reliability and Quality Sampling Plans." *Industrial Quality Control,* vol. 21, no. 3, September, pp. 118–122.

Schaefer, R. E. (1967). "Bayes Single Sampling Plans by Attributes Based on the Posterior Risk." *Naval Research Logistics Quarterly,* vol. A, no. 1, March, pp. 81–88.

Schilling, E. G. (1970). "Variables Sampling and MIL-STD-414." *Transactions of 26th Quality Control Conference of the Rochester Society for Quality Control,* March 30, pp. 175–188.

Schilling, E. G. (1978). "A Lot Sensitive Sampling Plan for Compliance Testing and Acceptance Inspection." *Journal of Quality Technology,* vol. 10, no. 2, April, pp. 47–51.

Schilling, E. G. (1979). A Simplified Graphical Grand Lot Acceptance Sampling Procedure." *Journal of Quality Technology,* vol. 11, no. 3, July, pp. 116–127.

Schilling, E. G. (1982). *Acceptance Sampling in Quality Control.* Marcel Dekker, New York and Basel.

Schilling, E. G., and Dodge, H. F. (1969). "Procedures and Tables for Evaluating Dependent Mixed Sampling Plans." *Technometrics,* vol. 11, no. 2, May, pp. 341–372.

Schilling, E. G., and Johnson, L. I. (1980). "Tables for Construction of Matched Single, Double and Multiple Sampling Plans with Applications to MIL-STD-105D." *Journal of Quality Technology,* vol. 12, no. 4, October, pp. 220–229.

Schilling, E. G., and Sheesley, J. H. (1978). "The Performance of MIL-STD-105D under the Switching Rules." *Journal of Quality Technology,* Part 1: vol. 10, no. 2, April, pp. 76–83; Part 2: vol. 10, no. 3, July, pp. 104–124.

Schilling, E. G., and Sommers, D. J. (1981). "Two-Point Optimal Narrow Limit Plans with Applications to MIL-STD-105D." *Journal of Quality Technology,* vol. 13, no. 2, April, pp. 83–92.

Schmee, J. (1980). "MIL-STD-781C and Confidence Intervals." *Journal of Quality Technology,* vol. 12, no. 3, April, pp. 98–105.

Seongin, K., Lee, C. S., Yang, J. R., and Wang, H. C. (1993). "An Expert System Approach to Administer Acceptance Control." *Industrial Engineering,* vol. 25, no. 6, June, pp. 57–59.

Shainin, D. (1950). "The Hamilton Standard Lot Plot Method of Acceptance Sampling by Variables." *Industrial Quality Control,* vol. 7, no. 1, July, pp. 15–34.

Shainin, D. (1952). "Recent Lot Plot Experiences Around the Country." *Industrial Quality Control,* vol. 8, no. 5, March, pp. 20–29.

Shinner, R., and Naor, P. (1961). "A Test of Randomness for Solid-Solid Mixtures." *Chemical Engineering Science,* vol. 15, no. 3/4, pp. 220–229.

Simon, L. E. (1941). *An Engineer's Manual of Statistical Methods.* John Wiley & Sons, New York.

Sommers, D. J. (1981). "Two-Point Double Variables Sampling Plans." *Journal of Quality Technology,* vol. 13, no. 1, January, pp. 25–30.

Soundararajan, V. (1978). "Procedures and Tables for Construction and Selection of Chain Sampling Plans (ChSP-1)." *Journal of Quality Technology,* Part 1: vol. 10, no. 2, April, pp. 56–60; Part 2: vol. 10, no. 3, July, pp. 99–103.

Soundararajan, V., and Vijayaraghavan, R. (1990). "Construction and Selection of Tightened-Normal-Tightened (TNT) Plans." *Journal of Quality Technology,* vol. 22, no. 2, April, pp. 146–153.

Squeglia, N. L. (1994). *Zero Acceptance Number Sampling Plans,* 4th ed. ASQC Quality Press, Milwaukee.

Stephens, K. S. (1979). *How to Perform Continuous Sampling (CSP).* American Society for Quality, Milwaukee.

Stephens, K. S. (1982). *How to Perform Skip-Lot and Chain Sampling.* American Society for Quality, Milwaukee.

Suich, R. (1990). "The Effects of Inspection Errors on Acceptance Sampling for Nonconformities." *Journal of Quality Technology,* vol. 22, no. 4, October, pp. 314–318.

Tagaras, G. (1994). "Economic Acceptance Sampling by Variables with Quadratic Quality Costs." *IIE Transactions,* vol. 26, no. 6, pp. 29–36.

Tang, K., and Tang, J. (1994). "Design of Screening Procedures: A Review." *Journal of Quality Technology,* vol. 26, no. 3, July, pp. 209–226.

Taylor, W. A. (1996). "A Program for Selecting Quick Switching Systems." *Journal of Quality Technology,* vol. 28, no. 4, October, pp. 473–479.

TR-3 (1961). *Sampling Procedures and Tables for Life and Reliability Testing Based on the Weibull Distribution (Mean Life Criterion).* Quality Control and Reliability Technical Report, U.S. Department of Defense, Government Printing Office, Washington, DC.

TR-4 (1962). *Sampling Procedures and Tables for Life and Reliability Testing Based on the Weibull Distribution (Hazard Rate Criterion).* Quality Control and Reliability Technical Report, U.S. Department of Defense, Government Printing Office, Washington, DC.

TR-6 (1963). *Sampling Procedures and Tables for Life and Reliability Testing Based on the Weibull Distribution (Reliable Life Criterion).* Quality Control and Reliability Technical Report, U.S. Department of Defense, Government Printing Office, Washington, DC.

TR-7 (1965). *Factors and Procedures for Applying MIL-STD-105D Sampling Plans to Life and Reliability Testing.* Quality Control and Reliability Technical Report, U.S. Department of Defense, Government Printing Office, Washington, DC.

Wadsworth, H. M., Stephens, K. S., and Godfrey, A. B. (1986). *Modern Methods of Quality Control and Improvement.* John Wiley and Sons, New York, pp. 635–646.

"Wescom, Inc.—A Study in Telecommunications Quality." (1978). *Quality,* August, p. 30.

Wetherill, G. B., and Chin, W. K. (1975). "A Review of Acceptance Sampling Schemes with Emphasis on the Economic Input." *International Statistical Review,* vol. 43, no. 2, pp. 191–209.

SECTION 47
DESIGN AND ANALYSIS OF EXPERIMENTS[1]

J. Stuart Hunter

*with Mary G. Natrella, E. Harvey Barnett,
William G. Hunter, and Truman L. Koehler*

[1]This section borrows extensively from the Third and Fourth Editions, in which Mary Natrella helped prepare the section on design and analysis of experiments; E. Harvey Barnett helped prepare the section on evolutionary operation; and William G. Hunter and Truman L. Koehler helped prepare the section on response surface methodology. The present author, J. Stuart Hunter, gratefully acknowledges this earlier work and takes full responsibility for all changes in organization and emphasis.

INTRODUCTION

Experiments, statistically designed or not, are a component of the *learning process.* We experiment to learn. Learning through experimentation is a complex mechanism combining one's hopes, needs, knowledge, and resources. How well one succeeds will be a function of adherence to the *scientific method,* the most rapid means for speeding the learning process.

The scientific method is an iterative process. Ideas based upon one's current state of knowledge lead to experiments designed to answer questions. The experiments in turn lead to data that serve to confirm and modify these initial ideas. An iterative loop is established between ideas (*hypotheses*), the design of the experiment, the production of data, and the subsequent analysis (*inference*) leading on to new ideas, new experiments, new data, and newer inferences. This iterative process continues until an adequate state of knowledge is acquired and confirmed. The entire process is always slowed by "noise," the results of "errors" of measurement and experimental "variability." We "see through the glass darkly" and thus data analyses must be conditioned by statements of *uncertainty,* by considerations of probability. Statistics is the science, the provider of the language and logic, that combines the roles of hypotheses, data, inference, and uncertainty within the scientific method.

Two famous statisticians, G. E. P. Box and W. Edwards Deming, encapsulate the fundamentals of the scientific method into four segments: Box's "conjecture, design, experiment, analysis" and Deming's "plan, do, check, act." Both statements illustrate the role of statistics as an integral part of the scientific method; see Deming (1982), Box (1976), Ishikawa (1985).

In applying the scientific method, a worker's intelligence, experience and resources dominate the speed and success of any problem-solving program. The art of statistics comes into formal play in two places: in helping design the experiments to produce data heavily laden with information and in the data analysis wherein the statistician's responsibility is to uncover all the information of value. Simply stated: The application of the statistical design and analysis of experiments accelerates the learning process.

An experiment has been defined as a "considered course of action aimed at answering one or more carefully framed questions." Framing the question and planning for action are best accomplished through a team effort. An isolated experimenter working alone has become an increasingly rare phenomenon: the resources of information germane to any problem-solving and learning activity are too vast and varied to be left to a single individual.

Computing and Analysis.   Almost 10 years have elapsed since the publication of the fourth edition of *Juran's Quality Control Handbook* (1988). This brief period has witnessed a spectacular growth in the use of the personal computer. Today the personal computer and associated software not only reduce the burdens of computations but can directly assist in the selection of an experimental design. Computer-constructed graphical displays of data quickly assist in data analysis. The users manual accompanying many software programs is often equivalent to a reference book on the construction and analysis of statistical designs, offering help in the pathways of analysis and providing illustrative examples. Many statistical software programs go far beyond the subject of statistical design of experiments and include elements of industrial quality control, linear and nonlinear regression, reliability and nonparametric studies, and multivariate and time series analyses. The reader is encouraged to secure the use of a personal computer and software program capable of performing the analyses associated with the statistical design of experiments.

The challenge, of course, is to be wise in the use of the powerful tools of statistics and modern computation. This Handbook well serves as an introduction to the application of good statistical practices devoted to the pursuit of quality. As a further aid the Statistics Division of the American Society for Quality has produced a series of *How To* booklets (see ASQ in the references) designed to assist beginners in the applications of statistics. The best advice for those just starting out in the statistical design of experiments is KISS: Keep it simple statistician! One should first experience the design and analysis of a small program with limited amounts of data, perhaps a replicated balanced block to compare a few treatments, or a simple factorial to study the simultaneous influences of factors upon a response. Further, the beginner must recognize that there has never been a signal in the absence of "noise" and then learn to measure the noise: to estimate the variance of the observations. Beginners often confuse noise (errors) due to measurement and instrumentation with the more serious source of noise (error) caused by the failure of repeated experiments to provide identical results. Experimental error encompasses measurement error and is far more serious. A beginner will also quickly learn the importance of always plotting the data (Chambers et al. 1993; Cleveland 1993).

The methods of analysis and computation given here, and those found in most textbooks, are intended for hand or desk calculation. The modern personal computer and associated software dramatically change this environment, with some computer software programs providing levels of assistance that resemble artificial intelligence; the experimenter is asked questions and is guided through almost every stage of experimental design and analysis. Two general cautions are in order: (1) the hand-desk computing methods given here should not be literally translated into a computer program—there are alternative powerful computer-based methods of data reduction and analysis; and (2) the user of packaged statistical programs should be as critical a consumer of this as of everything else.

The journal *Applied Statistics* (The Royal Statistical Society, London) has in each issue a *Statistical Software Review* section providing careful and critical commentary on the utility of various software programs along with, usually, the reply of the producer of the software. The quarterly journal *The American Statistician,* of the American Statistical Association, has a section titled *Statistical Computing Software Reviews.* The reviews in both these journals can be of great value in judging the value of software. *Applied Statistics* also provides an additional section on *Statistical Algorithms* for those requiring special programs not commonly found in ordinary software. More general information on software for statistical purposes is provided in Section 44, Basic Statistical Methods. See also Wadsworth (1990).

Basic Definitions.   Several fundamental terms are widely used throughout this section. They may be defined as follows:

Factor.   A "factor" is one of the controlled or uncontrolled variables whose influence upon a response is being studied in the experiment. A factor may be quantitative, e.g., temperature in degrees, time in seconds. A factor may also be qualitative, e.g., different machines, different operators, switch on or off.

Level (Version).   The "levels" ("versions") of a factor are the values of the factor being examined in the experiment. For quantitative factors, each chosen value becomes a level, e.g., if the experiment is to be conducted at four different temperatures, then the factor *temperature* has four *levels*. In the case of qualitative factors, *switch on or off* becomes two levels (versions) for the switch factor; if six machines are run by three operators, the factor *machine* has six levels (versions) while the factor *operator* has three levels (versions).

Treatment.   A "treatment" is a single level (version) assigned to a single factor during an experimental run, e.g., temperature at 800 degrees. A "treatment combination" is the set of levels for all factors in a given experimental run. For example, an experimental run using an 800-degree temperature, machine 3, operator A, and switch off would constitute one treatment combination.

Experimental Units.   The "experimental units" consist of the objects, materials, or units to which treatments are being applied. They may be biological entities, natural materials, fabricated products, etc.

Experimental Environment.   The "experimental environment" comprises the surrounding conditions that may influence the results of the experiment in known or unknown ways.

Block.   A factor in an experimental program that has influence as a source of variability is called a "block." The word is derived from early agricultural usage, in which blocks of land were the sources of variability. A block is a portion of the experimental material or of the experimental environment that is likely to be more homogeneous within itself than between different portions. For example, specimens from a single batch of material are likely to be more uniform than specimens from different batches. A group of specimens from such a single batch would be regarded as a block. Observations taken within a day are likely to be more homogeneous (to have smaller variance) than observations taken across days. *Days* then becomes a block factor.

Experimental Design.   The formal plan for conducting the experiment is called the "experimental design" (also the "experimental pattern"). It includes the choice of the responses, factors, levels, blocks, and treatments and the use of certain tools called planned grouping, randomization, and replication.

Models.
Experimental designs are created to help explain the association between a response variable η (η=eta; sometimes the symbol μ=mu is used) and other factors $x = x_1, x_2, x_3, \ldots, x_k$ (sometimes called "variables") thought to influence η. We say, "η (eta) is a function of x," that is, $\eta = f(x)$. Of course, *observing* the true response η entrains "noise" or "error" and produces observations $y = \eta + \epsilon$. The "expected value" of y is η. We are now faced with a "two model" problem: every observation y requires a model for η and another for ϵ.

The Model for the Error ϵ.   The model for the ϵ usually assumed is that the errors are independent, normally distributed with an expected value of zero and a constant variance σ^2; that is, the errors are white Gaussian noise. One's ability to discern associations between a response of η and factors x and to make inferences about future performance are profoundly influenced by the ϵ, the "errors" (noise). The errors (noise) attending a series of experiments have two primary components, those due to measurement and those attending the repetition of the experiment. Considerations of *measurement* error; the precision, traceability, specificity, calibration, and cost of the measurements; and variability due to sampling are always important. However, measurement and sampling errors alone will underestimate the contribution of errors associated with running and then repeating an experiment. *Experimental* error, the failure of agreement between two or more separate experiments run under the same conditions, is of crucial importance. Experimental error includes measurement and sampling errors as components. Special experimental designs (the hierarchical designs, see below) can provide experimenters with the ability to measure separately all error components.

The assumption that the errors ϵ are normally distributed with zero expectation and constant variance is important. Transformations of experimental data are often required to accomplish these attributes. However, the assumption of independent errors is especially crucial whenever probability statements (hypothesis tests or confidence intervals) are made. To guarantee independence, acts of randomization should be part of every experimental design protocol. Randomization also provides support for the constant variance assumption.

The Model for the Response η.   Implicit in every experiment is a *response model* $\eta = f(x)$ descriptive of how η changes as the factors x are changed. The specifics of this model are often not clear in the mind of the experimenter; one object of the experiments is to reduce this ambiguity. Most experimental designs are constructed to provide sufficient data to estimate the parameters in a very general model, thus allowing the data to identify appropriate subsets of models for the experimenter's appraisal.

The Model for the Observations **y.**   The *observation model* is $y = \eta + \epsilon$. Thus the experimenter must separate motion among observations into two parts: one part assignable to η and ultimately to the changes in the factors x, and a second part assignable to ϵ, the noise. This separation of the roles of η and ϵ is formally recognized in an analysis of variance (ANOVA) table, an easy computation provided experiments are properly planned.

Tools for Good Experimentation.   Good experimentation is an art and depends heavily upon the prior knowledge and abilities of the experimenter. Some tools of importance in the statistical planning and analysis of the design of experiments follow.

Blocking (Planned Grouping).   Beyond the factors x_1, x_2, ..., x_k selected for study, there are often other "background" variables that may also influence the outcome of the experimental program, variables such as raw material batches, operators, machines, or days. The influences of these variables upon the response are not under the control of the experimenter. These variables are commonly called "blocks," a legacy of the day when different blocks of land were used in agricultural experimentation. When an experimenter is aware of blocking variables it is often possible to plan experimental programs to reduce their influence. In designing experiments, wide use is made of the reduced variability occurring *within* blocks to accentuate the influences of the studied factors. Designs that make use of this uniformity within blocks are called "blocked" designs and the process is called "planned grouping."

Randomization.   The sequence of experiments and/or the assignment of specimens to various treatment combinations in a purely chance manner is called "randomization." Such assignment increases the likelihood that the effect of uncontrolled variables will balance out. It also improves the validity of estimates of experimental error variance and makes possible the application of statistical tests of significance and the construction of confidence intervals. Whenever possible, randomization is part of the experimental program.

Replication.   "Replication" is the repetition, the rerunning, of an experiment or measurement in order to increase precision or to provide the means for measuring precision. A single replicate consists of a single observation or experimental run. Replication provides an opportunity for the effects of uncontrolled factors or factors unknown to the experimenter to balance out and thus, through randomization, acts as a bias-decreasing tool. Replication also helps to detect gross errors in the measurements. In replications of groups of experiments, different randomizations should apply to each group.

Rerun experiments are commonly called "replicates." However, a sequence of observations made under a single set of experimental conditions, under a single replicate, are simply called "repeated observations."

Reproducibility and Repeatability.   In manufacturing, reproducibility measures the variability between items manufactured on different days or on different machines. Repeatability measures sources of variability that are more local or immediate, assignable to item measurements or to the variability occurring between adjacent items manufactured in sequence.

Requisites and Tools.   Table 47.1 lists some of the requisites for sound experimentation and shows the way in which these tools contribute to meeting these objectives. A checklist that can be helpful in all phases of an experiment is given in Table 47.2. Good references are Bicking (1954), Hahn (1977, 1984), Bisgaard (1992), Bishop et al. (1982), and Hoadley and Kettenring (1990), and Coleman and Montgomery (1993).

TABLE 47.1   Some Requisites and Tools for Sound Experimentation

Requisites	Tools
1. The experiment should have carefully defined objectives. See Table 26.2.	1. The definition of objectives requires all the specialized subject matter knowledge of the experimenter, and involves such things as: 　*a.* Choice of factors, including their range 　*b.* Choice of experimental materials, procedures, and equipment 　*c.* Choice of the metric for the factors (e.g., temperature or log temperature) and method of measurement
2. As far as possible, effects of factors should not be obscured by other variables.	2. The use of an appropriate experimental pattern helps to free the comparisons of interest from the effects of uncontrolled variables and simplifies the analysis of results.
3. As far as possible, the experiment should be free from bias, conscious or unconscious.	3. Some variables may be taken into account by planned grouping. Use randomization. Replication helps randomization to do a better job.
4. Experiments should provide a measure of experimental error variance (precision).	4. Replication provides the measure of variance and randomization ensures its validity.
5. Precision of experiment should be sufficient to meet objectives set forth in requisite 1.	5. Greater precision may be achieved by: refinements of measurement and experimental technique, experimental pattern (including planned grouping), replication.

CLASSIFICATION OF EXPERIMENTAL DESIGNS

Statisticians by themselves do not design experiments, but they have developed a number of structured schedules called "experimental designs," which they recommend for the taking of measurements. These designs have certain rational relationships to the purposes, needs, and physical limitations of experiments. Designs also offer certain advantages in economy of experimentation and provide straightforward estimates of experimental effects and valid estimates of variance. There are a number of ways in which experiment designs might be classified, for example, the following:

1. By the number of experimental factors to be investigated (e.g., single-factor versus multifactor designs)

2. By the structure of the experimental design (e.g., blocked, factorial, nested, or response-surface design)

3. By the kind of information the experiment is primarily intended to provide (e.g., estimates of effects, estimates of variance, or empirical mappings)

Some of the common statistical experimental designs are summarized in Table 47.3. Basic features of the designs are summarized in terms of these criteria of classification, and the details of design and analysis are given under the topics that follow. The analysis for observed responses is always based on a statistical model unique to the specific design.

TABLE 47.2   Checklist for Planning Test Programs

A. Obtain a clear statement of the problem.
1. Identify the problem area in quantitative terms.
2. Identify the response(s) to be measured, the factors that may be varied, the factors to be held constant, and the factors that cannot be controlled.
3. Identify the ranges or limitations of the measurements and of the experimental factors.
B. Collect available background information.
1. Investigate all available sources of information.
2. Tabulate data pertinent to planning the experimental program.
3. Be quantitative.
C. Design the experimental program.
1. Hold a conference of all parties concerned.
   a. State the propositions to be explored.
   b. Agree on magnitude of differences in the response considered worthwhile.
   c. Outline possible alternative outcomes.
   d. Choose the factors to be studied.
   e. Determine practical range of factors and specify levels.
   f. Choose the measurements and methods of measurement.
   g. Consider the effect of sampling variability and of precision of the measurement methods.
   h. Consider possible interrelationships (interactions) of the factors.
   j. Determine influences of time, cost, materials, manpower, instrumentation, and other facilities and of extraneous conditions such as weather.
   k. Consider personnel and human relations requirements of the program.
2. Design the experimental program in preliminary form.
   a. Prepare a systematic and inclusive schedule, which includes the randomization pattern.
   b. Provide for stepwise performance or adaptation of schedule if necessary.
   c. Eliminate effect of variables not under study by controlling, balancing, or randomization.
   d. Minimize the number of experimental runs consistent with objectives.
   e. Choose the method of statistical analysis.
   f. Arrange for orderly accumulation of data.
3. Review the experimental design program with all concerned.
   a. Adjust the program as required.
   b. Spell out the steps to be followed in unmistakable terms.
D. Plan and carry out the experimental work.
1. Develop methods, materials, and equipment.
2. Carry out the experimental design in some random order.
3. Record ancillary data.
4. Record any modifications of the experimental design.
5. Take precautions in the collection and recording of data, especially data from extra experiments and missing experiments.
6. Record progress of the program by date, run number, and other ancillary data.
E. Analyze the data.
1. Review the data with attention to recording errors, omissions, etc.
2. Use graphics: plot the data, plot averages, plot simple graphs.
3. Apply appropriate statistical techniques.

TABLE 47.2   Checklist for Planning Test Programs (*Continued*)

F. Interpret the results.
   1. Consider all the observed data.
   2. Confine initial conclusions to strict deductions from the experimental evidence at hand.
   3. Elucidate the analysis in both graphical and numerical terms.
   4. State results in terms of verifiable probabilities.
   5. Arrive at conclusions as to the technical meaning of results as well as their statistical significance.
   6. Point out implications of the findings for application and for further work.
   7. Account for any limitations imposed by the data or by the methods of analysis used.
G. Prepare the report.
   1. Describe work clearly, giving background, pertinence of problems, meaning of results.
   2. Use tabular and graphic methods of presenting data, and consider their possible future use.
   3. Supply sufficient information to permit readers to verify results and to draw their own conclusions.
   4. Limit conclusions to objective summary of evidence.

Split Plot.   When certain (major) factors are difficult to change, other (minor) factors are run in an experimental design within each of the settings of the major factors. The Taguchi inner and outer array designs are of this genre. Many split-plot designs confound minor factor interactions with major factor main effects.

Major factors are not blocks. Major factors are studied for their main effects and interactions; blocks identify random variables. Further, blocks are assumed to have no interactions with the factors under study. When both the major and minor factors are random variables, the designs are identified as "nested."

COMPLETELY RANDOMIZED DESIGN: ONE FACTOR, k LEVELS

The completely randomized design is appropriate when a total of N experimental units are available for the experiment and there are k treatments (or levels of the factor) to be investigated. Of the total number N, it is usual to assign *randomly* an equal number of trials n to each of the k treatments.

Example.   A study was made to investigate the effect of three different conditioning methods on the breaking strength T (in pounds per square inch) of cement briquettes. Fifteen briquettes were available from one batch and were assigned at random to the three methods. The results are summarized in Table 47.4. The purpose of the experiment was to investigate the effect of conditioning methods on breaking strength, and the analysis was designed to answer the question: Do the mean breaking strengths differ for the different conditioning methods?

This is an example of a randomized one-factor experiment. Only one experimental factor (method of conditioning) is under study. There are three methods; i.e., the number of treatments k equals 3. The number of units n assigned at random to each treatment is 5. The total number of experimental units N is 15.

Analysis.   Almost everyone today possesses a computer and the software capable of performing the arithmetic associated with the analysis of most experimental designs. Once the data for the fully

TABLE 47.3 Classification of Designs

Design	Type of application	Structure	Information sought
Completely randomized	Appropriate when only one experimental factor is being investigated.	Basic: One factor is investigated by allocating experimental units at random to treatments (levels of the factor). Blocking: none.	1. Estimate and compare treatment effects. 2. Estimate variance.
Factorial	Appropriate when several factors are to be investigated at two or more levels and interaction of factors may be important.	Basic: Several factors are investigated at several levels by running all combinations of factors and levels. Blocking: none.	1. Estimate and compare effects of several factors. 2. Estimate possible interaction effects. 3. Estimate variance.
Blocked factorial	Appropriate when number of runs required for factorial is too large to be carried out under homogeneous conditions.	Basic: Full set of combinations of factors and levels is divided into subsets so that some high-order interactions are equated to blocks. Each subset constitutes a block. All blocks are run. Blocking: Blocks are usually units in space or time. Estimates of certain interactions are sacrificed to provide blocking.	1. Same as factorial except that certain high-order interactions cannot be estimated.

TABLE 47.3 Classification of Designs *(Continued)*

Design	Type of application	Structure	Information sought
Fractional factorial	Appropriate when there are many factors and levels and it is impractical to run all combinations.	Basic: Several factors are investigated at several levels but only a subset of the full factorial is run. Blocking: Sometimes possible.	1. Estimate and compare effects of several factors. 2. Estimate certain interaction effects (some may not be estimable). 3. Certain small fractional factorial designs may not provide sufficient information for estimating the variance.
Randomized block	Appropriate when one factor is being investigated and experimental material or environment can be divided into blocks or homogeneous groups.	Basic: Each treatment or level of factor is run in each block. Blocking: Usually with respect to only one variable.	1. Estimate and compare effects of treatments free of block effects. 2. Estimate block effects. 3. Estimate variance.
Balanced incomplete block	Appropriate when all the treatments cannot be accommodated in a block.	Basic: Prescribed assignments of treatments to blocks are made. Every pair of treatments will appear at least once in the experimental design, but each block will contain only a subset of pairs.	1. Same as randomized block design. All treatment effects are estimated with equal precision. Treatment averages must be adjusted for blocks.
Partially balanced incomplete block	Appropriate if a balanced incomplete block requires a larger number of blocks than is practical.	Basic: Prescribed assignments of treatments to blocks are made.	1. Same as randomized block design but all treatments are not estimated with equal precision.

TABLE 47.3 Classification of Designs *(Continued)*

Design	Type of application	Structure	Information sought
Latin square	Appropriate when one primary factor is under investigation and results may be affected by two other experimental variables or by two sources of nonhomogeneity. It is assumed that no interactions exist.	Basic: Two cross groupings of the experimental units are made corresponding to the columns and rows of a square. Each treatment occurs once in every row and once in every column. Number of treatments must equal number of rows and number of columns Blocking: With respect to two other variables in a two-way layout.	1. Estimate and compare treatment effects, free of effects of the two blocked variables. 2. Estimate and compare effects of the two blocked variables. 3. Estimate variance.
Youden square	Same as Latin square but number of rows, columns, and treatments need not be the same.	Basic: Each treatment occurs once in every row. Number of treatments must equal number of columns. Blocking: With respect to other variables in a two-way layout.	1. Same as Latin square.
Nested	Appropriate when objective is to study relative variability instead of mean effect of sources of variation (e.g., variance of tests on the same sample and variance of different samples).	Basic: Factors are strata in some hierarchical structure; units are tested from each stratum.	1. Relative variation in various strata, components of variance.

TABLE 47.3 Classification of Designs *(Continued)*

Design	Type of application	Structure	Information sought
Response surface	Objective is to provide empirical maps (contour diagrams) illustrative of how factors under the experimenter's control influence the response.	Factor settings are viewed as defining points in the factor space (may be multidimensional) at which the response will be recorded	Maps illustrating the nature of the response surface
Mixture designs	Same as factorial designs.	Many unique arrays. Factor settings are constrained. Factor levels are often percentages that must sum to 100%. Other factor level constraints are possible.	Same as factorial

TABLE 47.4 Breaking Strength T of Cement Briquettes, lb/in^2

	Method 1	Method 2	Method 3
	553	553	492
	550	599	530
	568	579	528
	541	545	510
	537	540	571
Total T	2749	2816	2631
n	5	5	5
Average $\bar{y}$	549.8	563.2	526.2
Estimate of variance s^2	145.7	626.2	864.2
Degrees of freedom	4	4	4

randomized design displayed in Table 47.4 have been placed into a computer with an appropriate statistical design of experiments software program, then graphics similar to Figure 47.1 and an analysis of variance table such as that displayed in Table 47.5 become available. Nevertheless, we include here the details of the computations for the reader. The ability to perform one's own hands-on calculations is often of great value in reducing the "black-box" approach to data analysis.

The analysis of these data begins with a plot of the three treatment averages as shown in Figure 47.1. The "reference distribution" for these averages will be explained shortly. The average responses

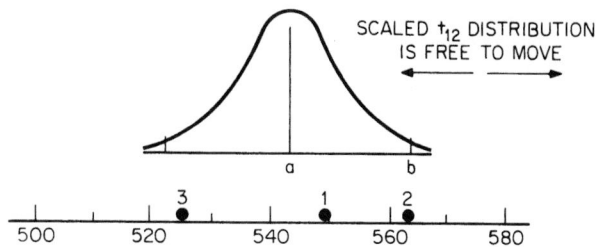

FIGURE 47.1   Plot of treatment averages and their reference t-distribution. (Distance from a to $b = t_{\nu,\,\alpha/2}\sqrt{s^2/n} = t_{12,\,0.025}\sqrt{s^2/n} = 2.179\sqrt{545.4/5} = 22.8$.)

TABLE 47.5   Analysis of Variance for a One-Factor Experiment

Source of variation	Sum of squares	Degrees of freedom	Mean square
Between treatments	SSB = 3,509.2	$(k-1) = 2$	MSB = SSB/$(k-1)$ MSB = 3509.2/2 = 1754.6
Within treatments	SSW = 6,544.4	$k(n-1) = 12$	MSW = SSW/$k(n-1)$ MSW = 6544.4/12 = 545.4
Total	TSS = 10,053.6	$(N-1) = 14$	

are obviously different. The key question is whether the observed differences are due solely to the inherent variability of the observations or caused by this variability plus real differences between the treatment means. (In this section on experimental design the word "mean" is used to connote the *expected value* of an average, that is, the value an average would take if an infinite number of observations were made.) The analysis of variance (ANOVA) is a basic statistical technique in the analysis of such data and is illustrated for the data in Table 47.5.

Analysis of Variance.   Referring to Table 47.4, the total T is

$$T = 8196 \qquad N = 15 \qquad n_i = 5 \qquad i = 1, 2, 3$$

Calculate the following:

The *uncorrected* total sum of squares $\Sigma y^2 = 4{,}488{,}348$

$$C = \text{correction factor (a special constant)} = \frac{T^2}{N} = \frac{(8196)^2}{15} = 4{,}478{,}294.4$$

TSS=*corrected* total sum of squares
$$= \Sigma y^2 - C = 4{,}488{,}348 - 4{,}478{,}294.4 = 10{,}053.6$$

SSB = between-treatments sum of squares

$$= \sum_i \frac{T_i^2}{n} - C = \frac{22{,}409{,}018}{5} - C = 4{,}481{,}803.6 - 4{,}478{,}294.4 = 3509.2$$

SSW = within-treatments sum of squares = TSS $-$ SSB = $10{,}053.6 - 3509.2 = 6544.4$

(SSW is here obtained by subtraction. It may be obtained directly for the special case of the completely randomized design by calculating $\Sigma y^2 - [(\Sigma y)^2/n]$ for each treatment and summing for all treatments.)

The corrected total sum of squares TSS has $N - 1$ degrees of freedom (DF), the between-treatments sum of squares SSB has $(k - 1)$ degrees of freedom, and the within-treatments sum of squares SSW has $(N - k)$ degrees of freedom.

The mean-square column in the table is computed as follows:

$$\text{Mean square between treatments MSB} = \frac{\text{SSB}}{k - 1}$$

$$\text{Mean square within treatments MSW} = \frac{\text{SSW}}{N - k}$$

These calculated quantities are inserted in Table 47.5.

Calculate $F = \text{MSB/MSW} = 1754.6/545.5 = 3.22$. Choose α, the significance level of the test. If the calculated F exceeds $F_{1-\alpha}$ from Appendix II, Table K, for $(k - 1)$ and $k (n - 1)$ degrees of freedom, conclude that there are differences among treatment means. For example, $\alpha = 0.05$ level, $F_{0.95}$ for (2,12) degrees of freedom $= 3.89$; the calculated F does not exceed this value, and we do not have sufficient evidence to conclude that the mean breaking strength is different for the different conditioning methods. The differences between the treatment averages are thus assumed to be due to the error variance (the noise). Had the hypothesis that there are no treatment effects been rejected by the F test, then one could conclude that at last one of the mean breaking strengths differs from the others. The formal issue of comparing treatment means, the problem of "multiple comparisons," is discussed below.

In this example, the n_i are all equal. Designs that have an equal number of observations in each treatment are generally to be preferred. Such designs provide each treatment with an equal opportunity for comparison against all other treatments. On rare occasions, as when one of the treatments is a standard against which all other treatments are to be compared, more observations are placed in the standard treatment than in the alternatives. When the n_i are not all equal, use the following formula for the between-treatments sum of squares:

$$\text{SSB} = \frac{T_1^2}{n_1} + \frac{T_2^2}{n_2} + \ldots + \frac{T_k^2}{n_k} - C$$

Here $\text{SSW} = \text{TSS} - \text{SSB}$ and $\text{MSW} = \text{SSW}/(N - k)$ as before. The MSW (the overall estimate of variance) can also be obtained by estimating the variance within each treatment and pooling these estimates. The pooled estimate has $N - k$ degrees of freedom.

Graphical Analysis.   An approximate graphical analysis of these data is provided by sketching an appropriate "reference" distribution for the average, as illustrated in Figure 47.1. When σ^2 (the population variance) or, equivalently, σ (the population standard deviation) is known, averages may be referred to a normal distribution with standard error $\sigma/\sqrt{n}$. Here σ^2 is unknown, and its estimate s^2 must be determined from the data. Averages must be referred to a Student's t-distribution, suitably scaled by $s/\sqrt{n}$. The number of degrees of freedom in t is equal to the number of degrees of freedom in s^2. The estimate of σ^2 is obtained from the analysis of variance table ($s^2 = \text{MSW}$) or by pooling the separate estimates of variance obtained from within each treatment classification. The pooled estimate of variance is given by the weighted average of the individual estimates, the weights being their degrees of freedom. Thus

$$s^2 = \sum_i \frac{(n_i - 1)s_i^2}{N - k}$$

$$= \frac{4(145.7) + 4(626.2) + 4(864.2)}{16} = 545.5$$

$$s = 23.34 \text{ with 12 degrees of freedom}$$

A sketch of the appropriately scaled t-distribution appears in Figure 47.1. The distance from the center of the t-distribution to each extremal point equals $t_{12,\,0.025}s/\sqrt{n} = 2.18(23.35)/\sqrt{5} = 22.8$, and 95 percent of the distribution falls within the range $2(22.8) = 45.6$. The distribution is easily sketched; there is no need for great precision in its shape save that it look reasonably bell-shaped. The distribution can be moved back and forth, and has been located in this instance so that it just fits over the three averages. Thus, based on this graphical evidence, the suggestion that all three averages could have come from the same parent distribution and hence could be estimates of the identical mean seems reasonable.

This graphical interpretation is confirmed by using the F test in the analysis of variance table; that is, the computed F ratio was not statistically significant. When averages do not fit reasonably under the distribution, the graphical analysis suggests that the differences between the treatment averages reflect real differences between the treatment means. Such graphical evidence could, of course, be verified by an F test. The scaled reference distribution can be of great value in interpreting treatment averages. Often, although it is technically possible to place the reference distribution over all the averages and simultaneously to obtain a nonsignificant F test, interesting differences between groups of the averages may become clear. The F test does not consider patterns among the averages, a factor that can be of great importance to the analyst (see Box, Hunter, and Hunter 1978). Alternative approximate analysis techniques, in particular those based upon the use of the range statistic in place of the estimated standard deviation, are available [see the papers by Kurtz et al. (1965) and by Sheesley (1980)].

GENERALIZED COMMENTS ON A COMPLETELY RANDOMIZED DESIGN

The completely randomized design is simple to organize and analyze and may be the best choice when the experimental material is homogeneous and when background conditions can be well controlled during the experiment.

The advantages of the design are

1. Complete flexibility in terms of number of treatments and number of units assigned to a treatment
2. Simple analysis
3. No difficulty with lost or missing data

In planning the experiment, n units are assigned at random to each of the k treatments. When the data have been taken, the results are set out as in Table 47.6.

TABLE 47.6   Completely Randomized Design

Observations within treatments*	Treatments				
	1	2	3	$\cdots$	k
1	y_{11}	y_{12}	y_{13}	$\cdots$	y_{1k}
2	y_{21}	y_{22}	y_{23}	$\cdots$	y_{2k}
3	y_{31}	y_{32}	y_{33}	$\cdots$	y_{3k}
.	.	.	.		.
.	.	.	.		.
.	.	.	.		.
n	y_{n1}	y_{n2}	y_{n3}	$\cdots$	y_{nk}

*The entire nk observations are recorded in random order *without* regard to the treatment classifications.

Displayed this way, the results of experiments are indistinguishable from a situation in which there has been no design and no allocation at all but in which several different samples have been tested from each of several different sources of material or several observations have been made under each of several different conditions. Whether the observations come from units randomly allocated to several different treatments or from units obtained from several different sources, the data table looks the same, and in fact the analysis will be essentially the same.

This simple one-factor design is called "completely randomized" to distinguish it from other experiment designs in which either randomization is constrained or the principle of "blocking," or planned grouping, has been made part of the structure.

One-Way Analysis of Variance—Models.   The results of an experiment run according to a completely randomized design are summarized in a one-way table such as Table 47.6. The completely randomized design is called a "one-way" classification of data, whether or not the data came from a designed experiment. To discuss the associated analysis of variance, statisticians require "models" for the data. In the case of a one-way classification analysis of variance, the most appropriate model is determined by answering the question: Do the several groups (into which the data are classified) represent *unique* groups of interest to the experimenter? If they do, the model is called "Model I," the "Fixed Effects Model." If, on the other hand, the groups are considered to be a random sample from some population made up of many such groups, the model is called "Model II," the "Random Effects Model." For example, suppose that the data in Table 47.4 were not from a completely randomized design in which 15 briquettes were allocated at random to three unique conditioning treatments of interest to the experimenter. Suppose instead the column headings were "Batch 1," "Batch 2," "Batch 3," where the "batches" represented some convenient grouping of briquettes so that five briquettes were tested per batch. In the original experimental program, in which the three conditioning treatments of the designed experiment were the unique treatments of interest to the experimenter, the data may be represented by the Fixed Effects Model (Model I), whereas in the second program the three batches presumably were a random sample of batches, and hence these data are represented by the Random Effects Model (Model II).

For both Model I and Model II, the experimenter is trying to determine whether the three groups are different in mean value. For Model II the experimenter may also be interested in knowing about the "components of variance"; that is, the variance between samples from the same batch and the variance existing between batches. Knowledge of the variability of different samples within a batch and between different batches is helpful in planning how many samples to test in future experiments.

Data obtained from a designed experiment, as described for this completely randomized design, are usually considered to be represented by Model I, since presumably the experimenter includes the treatments of interest. If the data correspond to Model II, the analysis of variance table and F test are used with one extra step, which requires adding an extra column labeled "Expected Mean Square," to Table 47.5.

	Expected mean square
Between groups	$\sigma_w^2 + n\sigma_b^2$
Within groups	σ_w^2

For the data in Table 47.5, we have:

	Mean square	Expected mean square
Between groups	MSB $= 1754.6$	$\sigma_w^2 + 5\sigma_b^2$
Within groups	MSW $= 545.4$	σ_w^2

The quantity σ_b^2 is called the "*between* component of variance," and σ_w^2 is called the "*within* component of variance"; MSB is an estimate of the "Expected Between-Groups Mean Square"; and MSW is an estimate of the "Expected Within-Groups Mean Square." Estimates of s_b^2 and s_w^2 or σ_b^2 and σ_w^2 can be obtained as follows:

MSB is set equal to $s_w^2 + n s_b^2$ and MSW is set equal to s_w^2:

$$s_w^2 = 545.4$$

$$s_b^2 = \frac{1754.6 - 545.4}{5} = \frac{1209.2}{5} = 241.8$$

The total variance assignable to a single observation from one randomly chosen batch and a single briquette is estimated as

$$s^2 = s_b^2 + s_w^2$$

BLOCKED DESIGNS

The several levels, or versions, of a studied factor or group of studied factors are called "treatments," and the major objective of an experimenter is to study the influences of these different treatment levels upon some response. Often all the levels of the studied factors are repeated each day or with a different operator, machine, supply of raw materials, etc. Each complete replication of the set of treatments is called a "block." The experimenter should plan the treatments so as to prevent differences between the blocks from influencing the comparisons between the treatments. For example, if the blocks in the experiment are days, the first aim of the experiment is to evaluate the effects of the studied factors free of the effects of day differences. A secondary aim might actually be to measure the effects of the days to help in planning future experiments. In blocked designs it is generally assumed that blocking factors do not interact with studied factors. In the simplest block designs the data, when taken, can be summarized in a two-way table, as illustrated in Table 47.7. Note that this design is *not* a factorial design. In a factorial design all the factors (here rows and columns) are at predetermined levels, whereas in the design under consideration here, the blocking factor is not under the control of the experimenter. The blocking factor is, however, recognized as capable of influencing the response. The experimenter's objective is to remove from the influences of the studied factors any possible contributions to the response that are provided by the blocking factors. Blocking factors are commonly environmental phenomena outside the control of the experimenter.

The interest in the factor called blocks has several objectives. Some of these are:

TABLE 47.7 Schematic for a Simple Block Design

	Treatments			
	1	2	$\cdots$	k
Block 1	y_{11}	y_{12}	$\cdots$	y_{1k}
Block 2	y_{21}	y_{22}	$\cdots$	y_{2k}
.	.	.	.	.
.	.	.	.	.
.	.	.	.	.
Block b	y_{b1}	y_{b2}	$\cdots$	y_{bk}

1. The aim of the experimenter is to estimate effects of treatments free of block effects; numerical estimates of block effects are not particularly needed. For example, if blocks are days, day-to-day differences should be eliminated as sources of variability and are of no particular interest in themselves.

2. The primary aim is to estimate effects of the treatments (the studied factors) and secondarily to have estimates of block effects.

3. Sometimes the treatment effects and the block effects are of almost equal interest. In this case "block design" is analogous to a "two-factor experiment," but the experimenter must be sure that the studied and blocking factors do not interact before using a block design data analysis. If interaction between factors exists or is suspected, the design and analysis for a factorial experiment must then be used.

The simplest design with one-way blocking is the "randomized block design."

RANDOMIZED BLOCK DESIGN

In comparing a number of treatments, it is clearly desirable that all other conditions be kept as nearly constant as possible. Unfortunately, the required number of tests is often too large to be carried out under similar conditions. In such cases, the experimenter may be able to divide the experiment into blocks, or planned homogeneous groups. When each such group in the experiment contains exactly one observation on every treatment, the experimental plan is called a randomized block plan. The treatments are run *in a random order* within the blocks.

There are many situations in which a randomized block plan can be profitably utilized. For example, a comparison of several levels of some factor may take several days to complete. If we anticipate that the different days may also have an influence upon the response, then we might plan to observe all of the factor levels on each day. A day would then represent a block. In another situation, several operators may be conducting the tests, and differences between operators may be expected. The tests or observations made by a given operator can be considered to represent a block. The size of a block—that is, the number of tests contained within the block—may be restricted by physical considerations. In general, a randomized block plan is one in which each of the treatments appears exactly once in every block. The treatments are allocated to experimental units at random within a given block. The results of a randomized block experiment can be exhibited in a two-way table such as Table 47.8, in which we have $b = 4$ blocks and $k = 6$ treatments. Since each treatment occurs exactly once in every block, the treatment totals or averages are directly comparable without adjustment.

Example.   The data in Table 47.8 represent the conversion gain of four resistors measured under six different conditions. The response, conversion gain, is defined as the ratio of available current-noise power to applied dc power expressed in decibel units and is a measure of the efficiency with which a resistor converts dc power to available current-noise power. Each test condition involves the same four resistors. The experimenter is interested in comparing differences between conditions (the studied factor) clear of possible influences due to the resistors (the blocking factor). A quick review of Table 47.8 indicates large differences between the resistors, i.e., between the block averages. The key question is whether, with this resistor variability eliminated, the experimenter can now detect real differences between the test conditions since the differences between the observed condition averages are small and may merely reflect experimental error.

Analysis.   Some computer software programs may title the analysis of variance associated with the randomized block experiment as a "two-way" analysis of variance. In a randomized block experiment primary interest rests in the treatment averages and their standard errors. Block averages are always of interest, but since blocks cannot be controlled by the experimenter (they represent environ-

TABLE 47.8 Randomized Block Design Response: Conversion Gain of Resistors

Resistor (blocks)	Test set (treatments)						Row total	Row average
	1	2	3	4	5	6		
1	138.0	141.6	137.5	141.8	138.6	139.6	$B_1 = 837.1$	$b_1 = 139.52$
2	152.2	152.2	152.1	152.2	152.0	152.8	$B_2 = 913.5$	$b_2 = 152.25$
3	153.6	154.0	153.8	153.6	153.2	153.6	$B_3 = 921.8$	$b_3 = 153.63$
4	141.4	141.5	142.6	142.2	141.1	141.9	$B_4 = 850.7$	$b_4 = 141.78$

Column totals

$T_1 = 585.2$ $T_2 = 589.3$ $T_3 = 586.0$ $T_4 = 589.8$ $T_5 = 584.9$ $T_6 = 587.9$ Grand total $G = 3523.1$

Column averages

$\bar{y}_1 = 146.30$ $\bar{y}_2 = 147.32$ $\bar{y}_3 = 146.50$ $\bar{y}_4 = 147.45$ $\bar{y}_5 = 146.22$ $\bar{y}_6 = 146.98$

ments within which the treatments have been randomly run) their importance is commonly secondary to that of the treatments. Of course, comparisons between averages requires an estimate of the variance σ^2. The primary reason for the analysis of variance computations is to get an estimate of the variance (to quantify the noise) clear of all assignable causes. Plots of the averages are always required.

The analysis of a randomized block experiment depends on a number of assumptions. We assume that each of the observations is the sum of four components. If we let y_{ij} be the observation on the ith treatment in the jth block, then

$$y_{ij} = \eta + \phi_i + \beta_j + \epsilon_{ij}$$

The term η is the grand mean, ϕ_i is the effect of treatment i, β_j the effect of block j, and ϵ_{ij} the experimental error associated with the measurement y_{ij}. (The subscripts $i = 1, 2, ..., k$ and $j = 1, 2, ..., b$.) The mean for the ith treatment equals $\eta + \phi_i$ and the mean for the jth block equals $\eta + \beta_j$. The terms ϕ_i and β_j represent, respectively, the unique contributions (effects) of treatments and blocks. The estimate of the mean η is given by $\bar{y}$, the grand average. Letting $\bar{y}_i$ equal the average for the ith treatment, the estimate of treatment effect ϕ_i is $\bar{y}_i - \bar{y}$. Similarly, $\bar{y}_j - \bar{y}$ estimates the block effect β_j.

In order to make interval estimates for or tests of hypotheses on the treatment or block contributions, we assume that the values of the experimental error ϵ_{ij} are independently and normally distributed with constant variance. If the experiment is randomized properly, failure of these assumptions will, in general, not cause serious difficulty.

A more serious difficulty occurs when count data are recorded. Count data are frequently Poisson-distributed, and hence the variance of the observations is linked directly to their mean. In such circumstances, it is best first to take the square roots of the count data and then to proceed with the estimation of effects and the analysis of variance.

Reference Distribution for Treatment Averages. The plot of the $k = 6$ treatment averages is displayed in Figure 47.2

To construct the appropriate reference distribution to judge these averages and to test hypotheses, an estimate of the experimental error variance σ^2 is required. Using the model, the associated analysis of variance table can now be constructed.

Randomized Block Analysis of Variance. The analysis of variance table for this randomized block experiment data with $k = 6$ treatments and $b = 4$ blocks is given in Table 47.9.

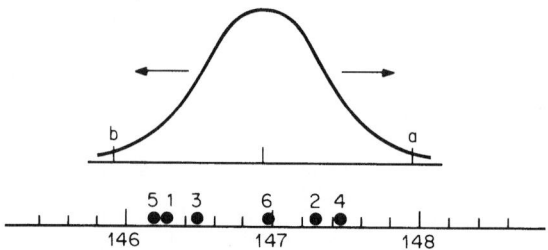

FIGURE 47.2 Plot of $k=6$ treatment averages and their reference t-distribution. (Distance from a to $b = 2ts/\sqrt{b} = 2.02$.) Numbers above scale are treatment numbers from Table 47.8.

TABLE 47.9 Analysis of Variance Table: Randomized Block Design

	Sum of squares (SSq)	Degrees of freedom (DF)		Mean square
Blocks: SSB	927.665	$(b-1)$	3	309.222
Treatments SST	5.598	$(k-1)$	5	1.120
Residual SSR	13.467	$(b-1)(k-1)$	15	$0.898 = s^2$
Total corrected SSq	946.730	$(bk-1)$	23	

The details of these computations are as follows:

Let $N = bk$ = total number of observations. Here $N = (4)(6) = 24$.

Let G = grand total of all the observations. $G = 3523.1$.

Let Σy_{ij}^2 = sum of the $y_{ij}^2 = 518,123.13$.

Total corrected SSq$=\Sigma y_{ij}^2 - G^2/N = 518,123.13 - (3523.1)^2/24$.

Let B_i = total for block i, $i = 1, 2, ..., b$. Here $b = 4$.

Blocks: SSB $= \Sigma_i B_i^2/k - G^2/N = 518,104.065 - G^2/N = 927.665$.

Let T_j = total for treatment j, $j = 1, 2, ..., k$. Here $k = 6$.

Treatments SST $= \Sigma_j T_j^2/b - G^2/N = 517,181.998 - G^2/N = 5.598$.

The residual SSR is obtained by subtraction.

The "mean squares" = (sum of squares)/(degrees of freedom).

An excellent explanation of these computations can be found in Box, Hunter, and Hunter (1978).

To test the hypothesis that all the treatment means are equal select an α risk level (commonly 0.05) and perform an $F_{\alpha,v1,v2}$ test. Thus

$$F_{v1,v2} = F_{5,15} = \frac{SST/(k-1)}{s^2} = \frac{1.120}{0.898} = 1.247$$

The critical $\alpha = 0.05$ value of $F_{0.05,5,15} = 2.90$. (Critical values of the F ratio are found in Table K, Appendix II.) The computed F is less than the critical F and we therefore declare there is insufficient evidence to lead to the rejection of the hypothesis that all treatment effects are zero. We may not actually believe this hypothesis, but we cannot reject it. A similar test that all block effects are zero gives:

$$F_{v1,v2} = F_{3,15} = \frac{SSB/(b-1)}{s^2} = \frac{309.222}{0.898} = 344.345$$

The computed F is far greater than the critical $F_{0.05,3,15} = 3.29$, and the hypothesis that there are zero block effects is rejected.

Figure 47.2 shows the plot of the six treatment averages and their associated "reference distribu-tion." The reference distribution for averages is a normal distribution scaled by $\sqrt{\sigma^2/n}$ when σ^2 is known. When, as in this case, only $s^2 = 0.898$ with $v = 15$ degrees of freedom is known, the refer-

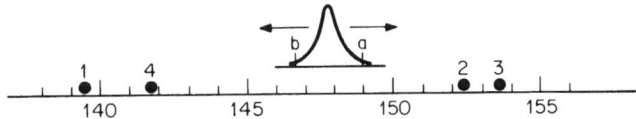

FIGURE 47.3   Plot of $b = 4$ block averages from Table 47.8 and their reference t-distribution. (Distance from a to $b = 2ts/\sqrt{k} = 1.65$.)

ence distribution for averages becomes a t-distribution scaled by $\sqrt{s^2/n}$. The distribution displayed in Figure 47.2 is the bell-shaped t curve, 95 percent of its area contained within the interval

$$2t_{\alpha,\nu}\sqrt{s^2/n} = 2(2.131)\sqrt{(0.898)/4} = 2.02$$

(The number of observations in a treatment average is $n = b = 4$.) The curve can be sketched in by hand; great precision is not required. One imagines this bell-shaped t-distribution to move back and forth horizontally. If two or more averages appear to be nested under a single location of the distribution, the inference is that differences between these averages may be due solely to the variance of the observations, a graphical test analogous to a nonsignificant F test. If some of the averages, individually or in clusters, seem to aggregate beyond the bell curve, this is taken as a signal that real differences exist, analogous to a significant F test.

Figure 47.3 displays the $b = 4$ block averages and their reference t-distribution. Once again, 95 percent of the distribution contained in the interval is given by

$$2t_{\alpha,\nu}\sqrt{s^2/n} = 2(2.131)\sqrt{(0.898)/6} = 1.65$$

(The number of observations in a block average is $n = k = 6$.) It is obvious from viewing Figure 47.3 that the reference distribution, wherever horizontally positioned, cannot reasonably account for the four averages. Block averages 1 and 4 are distinctly different from 2 and 3 and from one another. There is some indication, though slight, that averages 2 and 3 may be from a single distribution and hence merely reflections of the variability of averages about their expected mean value.

MULTIPLE COMPARISONS

An alternative graphical device for comparing treatment averages is the *analysis of means,* originated by Ott (1967). A plot similar to the Shewhart control chart is constructed. The control limits for the charts are easily obtained using special tables that adjust for the multiple comparisons that are possible. See Ott (1967), Schilling (1973), Nelson, L. (1983), Ott and Schilling (1990), Nelson, P. (1993), and, for good examples, Ramig (1983).

The graphical technique of a sliding reference distribution is subjective, the experimenter's eye and good judgment called into play. Many alternative formal techniques are available to demonstrate whether differences exist between the treatment means. Commonly, special intervals are computed such that averages appearing together within an interval may not be declared statistically significantly different. One approach is to compute the *least significance difference* (LSD) where

$$\text{LSD}_\alpha = t_{\nu,\alpha/2}\sqrt{\left(\frac{1}{n_i} + \frac{1}{n_j}\right)s^2}$$

If the absolute value of the difference between any two averages is greater than the LSD, the two treatment averages may be declared statistically significantly different. Of course, in comparing k averages there are $k(k - 1)/2$ possible pairs, and many other comparisons (contrasts) are possible when k is modestly large. In LSD comparisons a fixed α risk (usually $\alpha = 0.05$) is maintained for each comparison regardless of the number of comparisons made. The overall α risk is considerably increased when multiple tests are performed. For a fixed overall α risk for all possible *pairs* of treatments, one may wish to use the Tukey Studentized range statistic (Tukey 1949). Dunnett's method (Dunnett 1964; Bechofer and Tamhane 1983) is used for all $(k - 1)$ differences from a standard. The Scheffé test is used to make *all* possible comparisons (Scheffé 1953). Other multiple comparison methods are the Bonferroni interval (Dunn and Clark 1987); and the Duncan multiple range (Duncan, D. B., 1955). The Bonferroni and Duncan multiple range procedures can be found in most computer software programs. A good overall reference to the problems of multiple comparisons is Miller (1981). For multiple comparison of variances see Spurrier (1992).

BALANCED INCOMPLETE BLOCK DESIGNS

In an incomplete block design, all the treatments cannot be accommodated within a single block. To illustrate, consider a production manager who wishes to study the differences between the products supplied by six different suppliers. Unfortunately, personnel and equipment limit the number of suppliers that can be studied to three a day. The production manager is concerned that day-to-day differences might upset comparisons between suppliers and wishes to block the contributions of days, but the individual blocks are not large enough to encompass all six treatments. The appropriate experimental design to use is a "balanced incomplete block design," as illustrated in Table 47.10. The six suppliers (treatments) labeled A, B, C, D, E, and F are then studied in groups of three within each day (block). The blocks and the sequence of trials within the blocks are to be chosen in some random order.

Note that in the design displayed in Table 47.10 every letter supplier is tested the same number of times and every pair of letters appears within a block the same number of times. Another design appropriate to the case of six treatments constrained to be studied three at a time is the "combinatoric" design, that is, a design consisting of all combinations of six things taken three at a time. The combinatoric balanced incomplete block would have required 20 blocks; the design illustrated in Table 47.10 requires only 10.

One consequence of using an incomplete block design is that each treatment average must be adjusted for the blocks in which it appears and the differences between the *adjusted* treatment averages appraised. The computations are straightforward but go beyond what can be accommodated in this handbook. Interested readers are referred to Cochran and Cox (1957), Natrella (1963), and Box, Hunter, and Hunter (1978).

To enumerate the situations in which it is possible to construct a balanced incomplete block design, the quantities $r, b, t, k, L, E,$ and N are defined as follows:

r = number of replications (the number of times a treatment appears)

b = number of blocks in the plan

TABLE 47.10   A Balanced Incomplete Block Design

Days	Before randomization	After randomization	Six treatments
			A B C D E F
1	ABE	DBC	t = 6 treatments
2	ABF	CDA	b = 10 blocks
3	ACD	FBD	k = 3 treatments/block
4	ACF	CEB	r = 5 replicates/treatment
5	ADE	DEF	L = 2, i.e., each
6	BCD	EBA	treatment pair
7	BCE	AFB	appears twice
8	BDF	DEA	
9	CEF	FCE	
10	DEF	CFA	

t = number of treatments

k = block size, i.e., the number of treatments that can appear in each block

L = number of blocks in which a given treatment pair appears: $L = r(k - 1)/(t - 1)$

E = a constant used in the analysis: $E = tL/rk$

N = total number of observations: $N = tr = bk$

Plans are indexed in Table 47.11 for $4 < t < 10$ and $r < 10$. For an extensive listing of the designs and many worked examples see Cochran and Cox (1957) and Natrella (1963).

General Comments on Block Designs. In the simplest type of block design, Randomized Blocks, each block is large enough to accommodate all the treatments one wishes to test. In Incomplete Block Designs, the block size is not large enough for all treatments to be tested in every

TABLE 47.11   Balanced Incomplete Block Plans ($4 < t < 10$ and $r < 10$)

t	k	r	b	L	E^*	Plan†
4	2	3	6	1	2/3	1
	3	3	4	2	8/9	Comb‡
5	2	4	10	1	5/8	2
	3	6	10	3	5/6	Comb
	4	4	5	3	15/16	Comb
6	2	5	15	1	3/5	3
	3	5	10	2	4/5	4
	3	10	20	4	4/6	5
	4	10	15	6	9/10	6
	5	5	6	4	24/25	Comb
7	2	6	21	1	7/12	Comb
	3	3	7	1	7/9	7
	4	4	7	2	7/8	8
	6	6	7	5	35/36	Comb
8	2	7	28	1	4/7	9
	4	7	14	3	6/7	10
	7	7	8	6	48/49	Comb
9	2	8	36	1	9/16	Comb
	3	4	12	1	3/4	11
	4	8	18	3	27/32	12
	5	10	18	5	9/10	13
	6	8	12	15	15/16	14
	8	8	9	7	63/64	Comb
10	2	9	45	1	5/9	15
	3	9	30	2	20/27	16
	4	6	15	2	5/6	17
	5	9	18	4	8/9	18
	6	9	15	5	25/27	19
	9	9	10	18	80/81	Comb

*The constant $E = tL/rk$ is used in this analysis.
†For the plan classification number see Clatworthy (1973).
‡"Comb" indicates plans constructed by taking all possible combinations of t treatments in groups (block size) of k.

block. In Balanced Incomplete Block Designs, treatments are assigned to blocks that lead to equal precision in the estimation of differences between treatments.

If Randomized Block and Balanced Incomplete Block Designs do not meet the needs of the experimenter with regard to number of blocks, size of blocks, number of treatments, etc., other kinds of plans are available, for example, *partially* balanced incomplete block designs (Clatworthy 1973), and "chain block" designs, which are useful whenever observations are expensive and the experimental error is small. [See Natrella (1963) and Fleiss (1986) for the structure and details of analysis.]

General-purpose software programs written initially for statisticians employing mainframe computers rapidly perform the computations associated with the incomplete block designs. Only the more advanced personal computer design of experiments software programs offer lists of balanced incomplete block designs and assistance in their analysis. Despite their obvious importance and value, the application of these designs within industry is slight.

LATIN SQUARE DESIGNS

A Latin square design (or a Youden square plan, described later) is useful when it is necessary to investigate the effects of different levels of a studied factor while simultaneously allowing for two specific sources of variability or nonhomogeneity, i.e., two different *blocking* variables. Such designs were originally applied in agricultural experimentation when the sources of nonhomogeneity in fertility were simply the two directions on the field, and the "square" was literally a square plot of ground. Its usage has been extended to many other applications in which there are two sources of nonhomogeneity (two blocking variables) that may affect experimental results—for example, machines and positions or operators and days. The studied variable, the experimental treatment, is then associated with the two blocking variables in a prescribed fashion. The use of Latin squares is restricted by two conditions:

1. The numbers of rows, columns, and treatments must all be equal.

2. There must be *no* interactions between the row, the column, and the studied factors (see Factorial Experiments—General, for discussion of interaction).

As an example of a Latin square, suppose we wish to compare four materials with regard to their wearing qualities. Suppose further that we have a wear-testing machine that can handle four samples simultaneously. The two blocking variables might be the variations from run to run and the variations among the four positions on the wear machine. A 4 × 4 Latin square will allow for both sources of homogeneity. The Latin square plan is shown in Table 47.12 (the four materials are labeled A, B, C, and D). Note that every letter (treatment) appears once in every row and once in every column. Examples of Latin squares from size 3 × 3 to 7 × 7 are given in Table 47.13.

Strictly speaking, every time we use a Latin square we should choose a square at random from the set of all possible squares of its size. The tables of Fisher and Yates (1964) give complete collections of all the squares from 3 × 3 up to 12 × 12. Once a given square is chosen, permute the

TABLE 47.12   A 4 × 4 Latin Square

	Position number			
Run	1	2	3	4
1	*A*	*B*	*C*	*D*
2	*B*	*C*	*D*	*A*
3	*C*	*D*	*A*	*B*
4	*D*	*A*	*B*	*C*

TABLE 47.13 Selected Latin Squares

3 × 3	4 × 4			
	1	2	3	4
A B C	A B C D	A B C D	A B C D	A B C D
B C A	B A D C	B C D A	B D A C	B A D C
C A B	C D B A	C D A B	C A D B	C D A B
	D C A B	D A B C	D C B A	D C B A

5 × 5	6 × 6	7 × 7
A B C D E	A B C D E F	A B C D E F G
B A E C D	B F D C A E	B C D E F G A
C D A E B	C D E F B A	C D E F G A B
D E B A C	D A F E C B	D E F G A B C
E C D B A	E C A B F D	E F G A B C D
	F E B A D C	F G A B C D E
		G A B C D E F

columns at random, permute the rows at random, and assign the letters randomly to the treatments to provide a completely randomized design.

The analysis of the Latin square design is discussed in most textbooks and is available on most design of experiments personal computer software programs. The analysis of variance table is a simple extension of the randomized block table. Plots of the treatment averages and their reference distribution and/or multiple comparison hypothesis tests are the major objective of the analysis. Plots of the block averages and estimates of block effects are always informative, but since blocks cannot be controlled by the experimenter, their importance is only tangential to the analysis of the treatments. The reader is warned, firmly, that the mathematical model underlying the analysis of a Latin square (or any of its associated designs, the Graeco-Latin and the hyper Graeco-Latin square) assumes that *no interactions* exist between rows, columns, or any of the treatment classifications. Failure to meet this "no interactions" requirement leads to biased estimates of the treatment effects and the row and column (block) effects, and also biases the estimate of the variance σ^2. Unbiased estimates of σ^2 can be obtained by repeating the entire design or by partial replication [see Youden and Hunter (1955)].

The Latin square is *not* a factorial design, i.e., a design that allows for interactions between the separate factors composing the design. If there are interactions likely, the experimenter is advised to use a factorial or fractional factorial design and associated mode of data analysis. The 3×3 Latin square design is sometimes called the L9 orthogonal array. See Hunter (1989) for an example of the dangers that can arise from the misuse of the Latin squares.

YOUDEN SQUARE DESIGNS

The Youden square, like the Latin square, allows for two experimental sources of inhomogeneity. The conditions for the use of the Youden square, however, are less restrictive than those for the Latin square. The use of Latin square plans is restricted by the fact that the number of rows, columns, and treatments must all be the same. Youden squares have the same number of columns and treatments, but a fairly wide choice in the number of rows is possible. We use the following notation:

t = number of treatments to be compared

b = number of levels of one blocking variable (columns)

k = number of levels of another blocking variable (rows)

r = number of replications of each treatment

L = number of times that two treatments occur in the same block

In a Youden square, $t = b$ and $k = r$.

Some Youden square plans are given in Table 47.14. The analysis of the Youden squares must be carefully handled; in particular, the treatment averages must be adjusted for the rows in which they appear *before* they can be compared. Further, the standard error of the adjusted averages requires special computation. Reference should be made to the textbooks listed at the end of this section for the numerical details.

PLANNING INTERLABORATORY TESTS

We present here only a few simple techniques found useful in the planning and analysis of interlaboratory (round-robin) tests. The very early article by Wernimont (1951) remains an excellent introduction to the general problem. Other early contributors to the field are Youden (1967) and Mandel (1964), both members of the early National Bureau of Standards. An overall view of the importance of interlaboratory comparisons can be found in Hunter (1980). The best source of detailed information on round-robin procedures can be found in the publications of committee E11 of the American Society for Testing and Materials (ASTM). The text by Moen, Nolan, and Provost (1991) has several worked examples including a graphical method of analysis due to Snee (1983). Repeatability and Reproducibility (R&R) studies are often part of an interlaboratory testing program; see Barrentine (1991), Montgomery (1991), and Automotive Industry Action Group (1990).

A Rank Sum Test for Laboratories.   In almost any set of interlaboratory test data, some of the reported results fall so far out from the main body of results that there is a real question as to whether these data should be omitted in order to avoid distortion of the true picture. It is always a difficult problem to decide whether or not outlying results should be screened. One does not wish to discard a laboratory's results without good reason; on the other hand, if a laboratory is careless or not competent, one does not wish to "punish" the test method. A ranking test for laboratories due to

TABLE 47.14   Youden Square Arrangements

$t = 3$	$t = 4$		$t = 5$
A B	A B C		A B C D
B C	B A D		B A E C
C A	C D B		C D A E
	D C A		D E B A
			E C D B

$t = 6$	$t = 7$	or	$t = 7$
A B C D E	A B D		A C D E
B F D C A	B C E		B D E F
C D E F B	C D F		C E F G
D A F E C	D E G		D F G A
E C A B F	E F A		E G A B
F E B A D	F G B		F A B C
	G A C		G B C D

Youden (1963) is described here. This is only one of several nonparametric ranking procedures that may be of interest to the reader. Excellent references on these nonparametric approaches are the texts by Hollander and Wolfe (1973), Conover (1980), Gibbons and Chakraborti (1992), and Iman (1994).

An interlaboratory test usually involves sending several materials containing some particular chemical element or compound or possessing some physical quality to each of several laboratories. The ranking test for laboratories uses the recorded measured responses of the materials to rank the laboratories. The data from the interlaboratory test are summarized in a two-way table with materials as rows and laboratories as columns (or vice versa).

For each material, the laboratory having the largest result is given rank 1, the next largest rank 2, etc. (Tied values are treated as is usual in ranking procedures, each tied value being given the average of those ranks that would have been assigned if the values had differed.)

For each laboratory, the assigned ranks are summed over all materials. A laboratory that is consistently high in its ability to measure the response will show a lower rank sum, and a laboratory that is consistently low will show a higher rank sum than the average or expected rank sum. The question is whether such rank sums are excessively high or excessively low. To decide this, tables have been provided (see Table 47.15).

A Ruggedness Test for Use by the Initiating Laboratory. Very often a test method is judged to have acceptable precision by the original laboratory, but when the test is performed by several laboratories, the results are disappointing. The reason is usually that the original laboratory has carefully controlled conditions and equipment and that the operating conditions in other laboratories are slightly different. (There are always slight deviations, which are permissible within the instructions contained in the standard procedure for the test method.) Youden (1967) proposed that the initiating laboratory investigate the effects of such deviations by deliberately introducing small variations in

TABLE 47.15 Approximate 5% Limits for Ranking Scores

Number of laboratories participating	Numbers of materials									
	3	4	5	6	7	8	9	10	11	12
3	· · ·	4	5	7	8	10	12	13	15	17
	· · ·	12	15	17	20	22	24	27	29	31
4	· · ·	4	6	8	10	12	14	16	18	20
	· · ·	16	19	22	25	28	31	34	37	40
5	· · ·	5	7	9	11	13	16	18	21	23
	· · ·	19	23	27	31	35	38	42	45	49
6	3	5	7	10	12	15	18	21	23	26
	18	23	28	32	37	41	45	49	54	58
7	3	5	8	11	14	17	20	23	26	29
	21	27	32	37	42	47	52	57	62	67
8	3	6	9	12	15	18	22	25	29	32
	24	30	36	42	48	54	59	65	70	76
9	3	6	9	13	16	20	24	27	31	35
	27	34	41	47	54	60	73	79	85	91
10	4	7	10	14	17	21	26	30	34	38
	29	37	45	52	60	67	73	80	87	94
11	4	7	11	15	19	23	27	32	36	41
	32	41	49	57	65	73	81	88	96	103
12	4	7	11	15	20	24	29	34	39	44
	35	45	54	63	71	80	88	96	104	112

Note: Let L laboratories test each of M materials. Assign ranks 1 to L for each material. Sum the ranks to get the score for each laboratory. The mean score is $M(L + 1)/2$. The entries are lower and upper limits that are included in the approximate 5% critical region.

the method, a "ruggedness test," so as to be prepared for the variations resulting when the test is used by other laboratories. In order to minimize the extra work required for the original laboratory, he proposed that the Plackett-Burman designs for 7, 11, 15 factors be used to detect such effects. If significant effects result from such variations of conditions in a single laboratory, the method needs further refinement before interlaboratory tests are run. (See Thomas and Kiwanga 1993.)

Youden Two-Sample Plan.   A simple plan to investigate the performance of laboratories and of the test procedure itself was suggested by Youden (1959) and reprinted in Ku (1969). Samples of two materials (*A* and *B*) are sent to each laboratory in the program. The two materials should be similar in kind and in the value of the property to be measured. The laboratories should have the same internal precision. The pairs of results are used to plot a graph on which the *x* and *y* scales are equal and each laboratory is represented by one point. A laboratory's result on sample *A* is the *x* coordinate and its result on sample *B* is the *y* coordinate of that point. There will be as many points as there are laboratories. For graphical diagnosis, a vertical line is drawn through the median of all points in the *x* direction and a horizontal line through the median of all points in the *y* direction. The lines could be drawn through the *x* and *y* averages just as well, but the medians are convenient for quick graphical analysis.

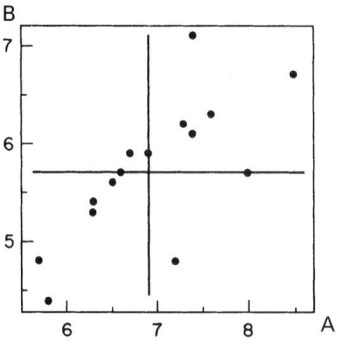

FIGURE 47.4   Percent phthalic anhydride in two paint samples—Youden plot showing systematic differences.

Individual points that are very far removed from the main body of the results indicate laboratories that should probably be screened from the analysis. The two intersecting median lines divide the space into four quadrants, and the first (and often revealing) step in the analysis is to look at the distribution of points among the quadrants. If only random errors of measurement were operating, there would be a circular scatter of points with roughly equal numbers in each quadrant. The plots of most real-life interlaboratory data, however, show concentrations in the upper right and lower left quadrants (see Figure 47.4). If a laboratory is high on both samples, its point will lie in the upper right; if a laboratory is low on both samples, its point will lie in the lower left. Being high (or low) on both samples is an indication that a laboratory has somehow put its own stamp on the procedure, i.e., that there are systematic differences between the laboratories. Where these systematic differences exist, the points will tend to lie along a long, narrow ellipse. Assuming that the two materials are similar in kind and in value of the property measured, as prescribed, and that the scatter in results for sample *A* does turn out to be approximately the same as the scatter for sample *B,* we can calculate an estimate of the standard deviation of a single result as follows:

1. Calculate the "signed differences" $d = A - B$ for each laboratory; that is, compute the difference and keep the sign (for the *i*th laboratory $d_i = A_i - B_i$).
2. Calculate $\bar{d}$, the algebraic average of the *d*'s.
3. Calculate $d'_i = d_i - \bar{d}$.
4. Take the absolute d' values and calculate their average; that is, drop the signs before averaging.
5. Multiply this value by 0.886 to get an estimate *s* of the standard deviation of a single result. (The value 0.886 is $1/d_2$, Appendix II, Table A, for $n = 2$.)

A circle can now be drawn that is expected to contain any stated percentage of the points. The circle is centered at the median point and its radius (for the stated percentage to be contained within it) is obtained by multiplying *s* (from Step 5) by the factor given in Table 47.16.

TABLE 47.16 Radius of Circle on Youden Plot in Terms of Multiples of the Standard Deviation

Percent of points within circle	Multiple of the standard deviation
90	2.146
95	2.448
99	3.035

Points lying outside the circle usually indicate laboratories with systematic differences. Further deductions are possible from such plots (see Youden, 1959); they have been used in a wide variety of applications, including chemical and engineering tests and standards comparisons.

NESTED (COMPONENTS OF VARIANCE) DESIGNS

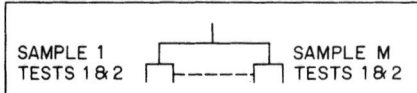

FIGURE 47.5 A two-stage balanced nested design. (*Reprinted with permission from Bainbridge 1965.*)

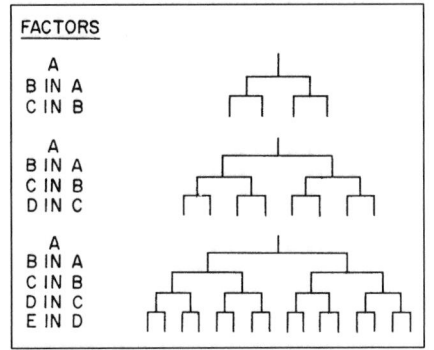

FIGURE 47.6 Balanced nested designs for three, four, and five factors. (*Reprinted with permission from Bainbridge 1965.*)

Most experimental designs are primarily intended to provide estimates of the means and differences or other comparisons between the means of experimental treatments. However, in investigations associated with interlaboratory comparisons, or the repeatability and reproducibility of measuring instruments, sources of variability become the studied factors, and the primary knowledge sought concerns the relative importance of these sources. Such investigations are called "components of variance" studies. Designs intended to provide estimates of the components of variance arising from various sources are called "nested" or "hierarchical" designs. A simple nested design is shown in Figure 47.5 wherein M samples are taken and then within each sample duplicate tests are made. The variance of all the observations is then partitioned into two components: that assignable uniquely to the samples and that uniquely to the duplicates. This is a two-stage balanced nested design. Three or more classifications can be nested, as illustrated in Figure. 47.6.

The analysis of nested designs and estimation of components of variance is based upon a "random effects" model for the observations in which, except for a single constant term, all elements are random variables. (See further discussion under the sections One-Way Analysis of Variance—Models and Split-Plot Factorial Experiments.) Needed are the expected values of the mean squares obtained from an analysis of variance table; see Snee (1974). The designs and their analysis is described in most textbooks on experimental design [for example, Dunn and Clark (1987); Neter, Wasserman, and Kutner (1990); Hogg and Ledolter (1992); Montgomery (1991); and Burdick and Graybill (1992)]. An example of the estimation of components of variance that does not evoke the analysis of variance is given in Box, Hunter, and Hunter (1978). A Bayesian method for estimating variance components is given in Box and Tiao

(1973). An entire textbook devoted to the subject is Searle, Casella, and McCulloch (1992). Most design of experiments software programs will provide the expected values of the mean squares needed to determine the components of variance for the simple balanced designs.

In Figure 47.6, at each stage only two subunits of each unit are taken, but the total number of tests multiplies rapidly as the number of stages increases. Because of the rapidly increasing total number of tests, only a few units are usually used at the top levels. In other words, balanced nested designs tend to provide too little information on the upper levels (the initial stages, or factors *A* and *B*) and often provide more than enough information at the bottom levels (factors *E,* for example). Bainbridge (1965) has considered alternative "unbalanced" nested designs with a fixed number of tests. He prefers a design, which he calls a "staggered nested design" that is easy to administer and provides about the same number of degrees of freedom for each factor. Bainbridge shows staggered nested designs for three, four, five, and six factors (see Figure 47.7 for the designs and their analysis).

PLANNING THE SIZE OF THE EXPERIMENT

Methods for determining the number of observations required for estimating the mean and variance with certain precision, or for comparing two sets of data with regard to mean and variance with certain risks of error, are given in Section 44, Basic Statistical Methods. A method for determining the number of observations required for comparing several groups is given here.

For example, the analysis of variance *F* test (see Completely Randomized Design: One Factor, *k* Levels) is designed to test the hypothesis that all group means are the same, i.e., $\eta_1 = \eta_2 \ldots \eta_k = \eta$. The corresponding averages $\bar{y}_1, \bar{y}_2, \ldots, \bar{y}_k$ computed from the recorded data will, of course, be different. The outcome of the test of hypothesis depends on the significance level α at which the test is performed, the true variability of individual observations, the number of observations per average, and the size of the true difference (if any) between group means. When planning experiments, if there are no restrictions on the number of observations that can be made, one should specify the size of those differences in means that are considered important from a practical standpoint. When the significance level at which the test is to be made is also specified, existing tables or charts can be used to determine the necessary sample size (number of observations per average) for achieving a stated probability $(1 - \beta)$ of detecting differences between the means of the required size. To use such tables, we compute a quantity

$$\phi^2 = \frac{n \sum_i (\eta_i - \eta)^2}{k\sigma^2} = \frac{n\Sigma\delta_i^2}{k\sigma^2}$$

where n = number of observations per group (to be determined)
    k = number of groups
    σ^2 = true value of within-group variance (assumed same for all groups; can be estimated from previous similar work)
    η_i = mean for *i*th group
    η = grand mean

Let $(\eta_i - \eta) = \delta_i$. The sum of the δ_i values must equal zero.

Appendix II, Table DD gives values ϕ^2 for $\alpha = 0.01$ and $\beta = 0.2$, $(1 - \beta = 0.80)$, and DF_1 and DF_2 degrees of freedom. In the simple case used in the example, $DF_1 = k - 1$ and $DF_2 = k(n - 1)$. Other charts and tables are available in slightly different form and for additional values of α and β. See, for example, Dixon and Massey (1969), Owens (1962), or Odeh and Fox (1991).

Example.   Consider the experiment shown in Table 47.4. Suppose that another experiment is to be run and that we wish to determine beforehand how many briquettes to test using each method in

Sources of Variance	Sums of Squares	Degrees of Freedom	Expectations of Mean Squares	Format of A-Units
A	(5) – CF	m–1	$\sigma_c^2 + 1\frac{2}{3}\sigma_b^2 + 3\sigma_a^2$	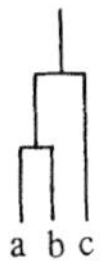
B in A	(3)+(4)–(5)	m	$\sigma_c^2 + 1\frac{1}{3}\sigma_b^2$	
C in B	(1)+(2)–(3)	m	σ_c^2	
Total	(1)+(2)+(4)–CF	3m–1	4A Three Factors	a b c
A	(7) – CF	m–1	$\sigma_d^2 + 1\frac{1}{2}\sigma_c^2 + 2\frac{1}{2}\sigma_b^2 + 4\sigma_a^2$	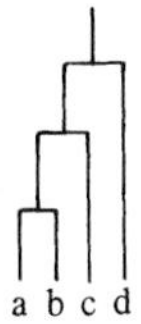
B in A	(5)+(6)–(7)	m	$\sigma_d^2 + 1\frac{1}{6}\sigma_c^2 + 1\frac{1}{2}\sigma_b^2$	
C in B	(3)+(4)–(5)	m	$\sigma_d^2 + 1\frac{1}{3}\sigma_c^2$	
D in C	(1)+(2)–(3)	m	σ_d^2	
Total	(1)+(2)+(4)+(6)–CF	4m–1	4B Four Factors	a b c d
A	(9) – CF	m–1	$\sigma_e^2 + 1\frac{2}{5}\sigma_d^2 + 2\frac{1}{5}\sigma_c^2 + 3\frac{2}{5}\sigma_b^2 + 5\sigma_a^2$	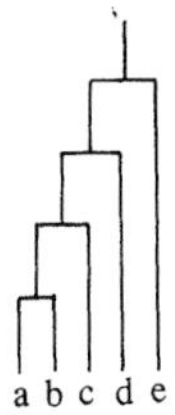
B in A	(7)+(8)–(9)	m	$\sigma_e^2 + 1\frac{1}{10}\sigma_d^2 + 1\frac{3}{10}\sigma_c^2 + 1\frac{3}{5}\sigma_b^2$	
C in B	(5)+(6)–(7)	m	$\sigma_e^2 + 1\frac{1}{6}\sigma_d^2 + 1\frac{1}{2}\sigma_c^2$	
D in C	(3)+(4)–(5)	m	$\sigma_e^2 + 1\frac{1}{3}\sigma_d^2$	
E in D	(1)+(2)–(3)	m	σ_e^2	
Total	(1)+(2)+(4)+(6)+(8)–CF	5m–1	4C Five Factors	a b c d e
A	(11) – CF	m–1	$\sigma_f^2 + 1\frac{1}{3}\sigma_e^2 + 2\sigma_d^2 + 3\sigma_c^2 + 4\frac{1}{3}\sigma_b^2 + 6\sigma_a^2$	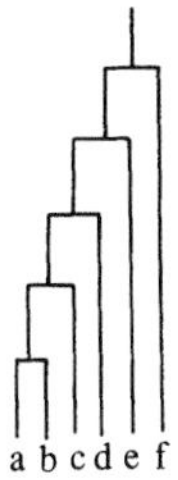
B in A	(9)+(10)–(11)	m	$\sigma_f^2 + 1\frac{1}{15}\sigma_e^2 + 1\frac{1}{5}\sigma_d^2 + 1\frac{2}{5}\sigma_c^2 + 1\frac{2}{3}\sigma_b^2$	
C in B	(7)+(8)–(9)	m	$\sigma_f^2 + 1\frac{1}{10}\sigma_e^2 + 1\frac{3}{10}\sigma_d^2 + 1\frac{3}{5}\sigma_c^2$	
D in C	(5)+(6)–(7)	m	$\sigma_f^2 + 1\frac{1}{6}\sigma_e^2 + 1\frac{1}{2}\sigma_d^2$	
E in D	(3)+(4)–(5)	m	$\sigma_f^2 + 1\frac{1}{3}\sigma_e^2$	
F in E	(1)+(2) – (3)	m	σ_f^2	
Total	(1)+(2)+(4)+(6)+(8)+(10)–CF	6m–1	4D Six Factors	a b c d e f

TOTALS NEEDED TO GET SUMS OF SQUARES

(1) $= \Sigma a^2$

(2) $= \Sigma b^2$

(3) $= \dfrac{\Sigma (a+b)^2}{2}$

(4) $= \Sigma c^2$

(5) $= \dfrac{\Sigma (a+b+c)^2}{3}$

(6) $= \Sigma d^2$

(7) $= \dfrac{\Sigma (a+b+c+d)^2}{4}$

(8) $= \Sigma e^2$

(9) $= \dfrac{\Sigma (a+b+c+d+e)^2}{5}$

(10) $= \Sigma f^2$

(11) $= \dfrac{\Sigma (a+b+c+d+e+f)^2}{6}$

$CF = \dfrac{(\text{Grand Total})^2}{\text{Total No. of Tests}}$

FIGURE 47.7 Staggered nested designs for three, four, five, and six factors. (*Reprinted with permission from Bainbridge 1965.*)

order to achieve a certain discrimination between the means for the three methods. If the statistical test is to be done at the $\alpha = 0.01$ level and if we want the probability of detecting the postulated differences to be at least 0.8, we can use Appendix II, Table DD. Assume $\sigma^2 = 545$, an estimate of the variance determined from the previous experiment. Suppose the following differences between the means are considered practically important:

$$\delta_1 = \eta_1 - \eta = -30$$

$$\delta_2 = \eta_2 - \eta = +20$$

$$\delta_3 = \eta_3 - \eta = +10$$

(Obviously, many different values for the δ's will yield the same value for $\Sigma \delta^2$ and therefore the same ϕ^2.)

The δ's chosen should be meaningful for each experimental situation. Here we have postulated three particular differences; in other situations the pattern of the differences might take on special meanings. For example, if the groups were increasing levels of a quantitative variable such as temperature, a meaningful pattern for the δ's might be a constant change in mean from one level to the next higher one. (Remember that the δ_i's must sum to zero.)

$$\phi^2 = \frac{n \Sigma \delta_i^2}{k \sigma^2}$$

$$\phi^2 = \frac{n(1400)}{3(545)} = \frac{1400n}{1635} = 0.86n$$

$$DF_1 = k - 1 = 3 - 1 = 2$$

$$DF_2 = k(n - 1) = 3n - 3$$

Using Appendix II, Table DD, we must find two values of n, one that gives ϕ^2 larger than required and one that gives a smaller value than required:

n	$DF_2 = 3n - 3$	Tabled ϕ^2	Desired $\phi^2 = 0.86n$
7	18	6.05	6.02
8	21	5.83	6.88

The "tabled ϕ^2" for $n = 8$ was obtained by linear interpolation. The solution lies between $n = 7$ and $n = 8$, and we take the larger n. Eight observations per group will give us an 80 percent chance of detecting the postulated differences when we do an F test at the $\alpha = 0.01$ level.

This method may be used for multifactor experiments provided the proper values for DF_1 and DF_2 are used. It is used when the purpose of the experiment is to compare group averages, and it works for any number of groups provided the number of observations per group is large enough. In this case and in the case described below, equal numbers of observations should be taken in each group.

For another kind of experiment, in which the purpose is to compare the between-group variance with the within-group variance (see discussion of Model II, Random Effects Model, under One-Way Analysis of Variance—Models), a *minimum* number of *groups* is required to achieve desired discrimination in terms of the relative variability. For example, see Table 47.17, where α and β are the risks of the two kinds of error, δ_0 is an "acceptably small" value of the ratio σ_b/σ_w (large enough to achieve a significant result), and δ_1 is an unacceptably large value for σ_b/σ_w.

TABLE 47.17 Minimum Number of Groups—Random Effects Model

$\alpha = \beta = 0.05$		$\alpha = \beta = 0.01$	
δ_1/δ_0	Minimum number of groups	δ_1/δ_0	Minimum number of groups
1.5	35	1.5	68
2.0	14	2.0	25
2.5	9	2.5	16
3.0	7	3.0	12

Useful discussions on determining the number of observations are given in the texts by Cochran and Cox (1957), Cox (1958), and Odeh and Fox (1991). Extensive tables are given in the papers by Kastenbaum, Hoel, and Bowman (1970).

FACTORIAL EXPERIMENTS—GENERAL

Factorial designs are most frequently employed in engineering and manufacturing experiments. In a factorial experiment several factors are controlled at two or more levels, and their effects upon some response are investigated. The experimental plan consists of taking an observation at each of all possible combinations of levels that can be formed from the different factors. Each different combination of factor levels is called a "treatment combination."

Suppose that an experimenter is interested in investigating the effect of two factors, amperage (current) level and force, upon the response y, the measured resistivity of silicon wafers. In the past, one common experimental approach has been the so-called one-factor-at-a-time approach. This experimental strategy studies the effect of first varying amperage levels at some constant force and then applying different force levels at some constant level of amperage. The two factors would thus be varied one at a time with all other conceivable factors held as constant as possible. The results of such an experiment are fragmentary in the sense that we learn about the effect of different amperage levels only at one force level and the effect of different force levels at only one amperage level. The effects of one factor are conditional on the chosen level of the second factor. The measured resistivity of the wafer at different current levels may, of course, be different when a different force level has been chosen. Similarly, any observed relation of resistivity to force level might be quite different at other amperage levels. In statistical language, there may be an "interaction effect" between the two factors over the range of interest, and the one-at-a-time procedure does not enable the experimenter to detect the interaction.

In a factorial experiment, the levels of each factor are chosen, and a measurement is made at each of all possible combinations of levels of the factors. Suppose that five levels of amperage and four levels of force are chosen. There would thus be 20 possible combinations of amperage and force, and the factorial experiment would consist of 20 trials. In this example, the term "level" is used in connection with quantitative factors, but the same term is also used when the factors are qualitative.

In the analysis of factorial experiments, one speaks of "main effects" and "interaction effects" (or simply "interactions"). Estimated main effects of a given factor are always functions of the average yield response at the various levels of the factor. When a factor has two levels, the estimated main effect is the difference between the average responses at the two levels, i.e., the averages computed over all levels of the other factors. In the case in which the factor has more than two levels, there are several main effect components (linear, quadratic, cubic, etc.), the number of estimable main effect components being one less than the number of levels. Other comparisons, called treatment "contrasts," are possible. If the difference in the expected response between two levels of factor A remains constant over the levels of factor B (except for experimental error), there is no interaction between

A and *B*; that is, the *AB* interaction is zero. Figure 47.8 shows two examples of response, or yield, curves; one example shows the presence of an interaction and the other shows no interaction. If there are two levels each of the factors *A* and *B,* then the *AB* interaction (neglecting experimental error) is the difference in the average yields of *A* at the second level of *B* minus the difference in the average yields of *A* at the first level of *B.* If there are more than two levels of either *A* or *B,* then the *AB* inter-action can be composed of more than one component. If we have *a* levels of the factor *A* and *b* levels of the factor *B,* then the *AB* interaction has $(a - 1)(b - 1)$ independent components. A two-factor interaction (e.g., *AB*) is also called a "second-order" effect or "coupled" effect.

For factorial experiments with three or more factors, interactions can also be defined. For example, the *ABC* interaction is the interaction between the factor *C* and the *AB* interaction (or, equivalently, between the factor *B* and the *AC* interaction or between *A* and the *BC* interaction). A three-factor interaction (e.g., *ABC*) is a "third-order" effect.

FACTORIAL EXPERIMENT WITH TWO FACTORS

A two-factor experiment is the simplest kind of multifactor experiment; i.e., all possible combinations of the levels of the two factors are run. For example, measurements of the response resistivity of a silicon wafer are usually made at a standard amperage level while using 150 g of force. Let us consider an investigation in progress to see what happens when other values of force and amperage are employed. Four values of force are to be investigated (25, 50, 100, 150 g), along with five levels of amperage (levels 1, 2, 3, 4, and 5, where level 3 is the standard level). An experimental trial is made at each of the $4 \times 5 = 20$ possible combinations. The data can be displayed in a two-way array, as in Table 47.18. This is an unreplicated two-factor multilevel factorial experiment. Some textbooks will describe it as having two "crossed" factors.

Analysis.   The data for this 4×5 factorial are displayed in Table 47.18. The first stage of the analysis is to compute, plot, and review both the column and row averages shown in Table 47.19.

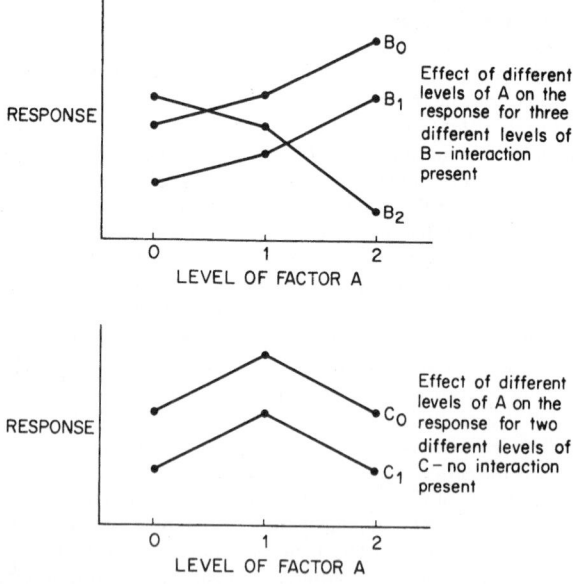

FIGURE 47.8   Response curves showing presence or absence of interaction.

TABLE 47.18 Resistivity Measurements

Force, g	Current level 1	2	3	4	5	Row totals R_i
25	11.84	11.83	11.84	11.81	11.96	$R_1 = 59.28$
50	11.84	11.88	11.88	11.87	11.90	$R_2 = 59.37$
100	11.77	11.80	11.80	11.81	11.88	$R_3 = 59.06$
150	11.79	11.80	11.80	11.80	11.87	$R_4 = 59.06$
Column totals C_j	47.24	47.31	47.32	47.29	47.61	$T = 236.77$

TABLE 47.19 Table of Averages and Effects

Row number	Row average	Row effect
1	11.8560	0.0175
2	11.8740	0.0355
3	11.8120	−0.0265
4	11.8120	−0.0265

Column number	Column average	Column effect
1	11.8100	−0.0285
2	11.8275	−0.0110
3	11.8300	−0.0085
4	11.8225	−0.0160
5	11.9025	0.0640

Plots of the response versus the factor levels are always useful; in this case they would point up the noticeably higher resistivity values at amperage level 5 and the apparent changes in the resistivity with increasing force, particularly at the outer values of amperage. When the factors are quantitative and the levels equally spaced, there are simple methods to check on various types of trends (linear, quadratic, etc.) in the response measurements as a function of varying levels of a factor (see Hicks 1982).

The analysis of variance can be used to test two hypotheses: (1) the mean resistivity at all levels of force is the same and (2) the mean resistivity at all levels of amperage is the same.

The construction of the analysis of variance table would proceed as follows:

$$r = \text{number of rows} = 4$$

$$c = \text{number of columns} = 5$$

$$T = \text{grand total} = \sum_j C_j = \sum_i R_i = 236.77$$

$$N = \text{total number of observations} = r \times c = 20$$

$$C = \text{the correction factor} = \frac{T^2}{N} = 2803.001645$$

$$\text{SSR} = \text{row sum of squares} = \frac{\sum_i R_i^2}{c} - C = 0.014855$$

$$\text{SSC} = \text{column sum of squares} = \frac{\sum_j C_j^2}{r} - C = 0.021430$$

$$\text{TSS} = \text{total corrected sum of squares}$$

$$= (\text{all observations squared}) - C$$

$$= 2803.043500 - 2803.001645 = 0.041855$$

$$\text{SSE} = \text{error (or residual) sum of squares}$$

$$= \text{TSS} - \text{SSR} - \text{SSC} = 0.005570$$

The analysis of variance table for this (unreplicated) two-factor crossed factorial is shown in Table 47.20. Tests of hypotheses that no treatment effects exist are now possible. For rows conclude that statistically significant effects exist between rows if $F =$ (row mean square)/(error mean square) is greater than $F_{1-\alpha}$ for $(r - 1)$ and $(r - 1)(c - 1)$ degrees of freedom (from Appendix II, Table K). Similarly, for columns conclude that statistically significant effects exist if $F =$ (column mean square)/(error mean square) is greater than the tabulated value of F for $(c - 1)$ and $(r - 1)(c - 1)$ degrees of freedom. In this example, both F tests reject the hypothesis that no effects exist.

The analysis of variance table and its associated tests of hypotheses are only one part of the analysis of data from a factorial design. The model postulated for the data assumes that the force and amperage factor effects are additive. Thus there can be no interactions; that is, the effects of force upon resistivity remain unchanged whatever the levels of the second factor. One way to check this assumption is to compute Tukey's one-degree-of-freedom test for nonadditivity (see Snedecor and Cochran 1967 or Box, Hunter, and Hunter 1978). The test for nonadditivity here proved to be nonsignificant.

Table 47.19 has been extended to include the estimated row and column *effects* (an estimated row effect is the row average minus the grand average; an estimated column effect is the column average minus grand average). The row and column effects can then be used to make a two-way table of "residuals" in which the residual for the cell in the ith row and jth column is equal to the observation in that cell minus the sum of the grand average, the ith row effect, and the jth column effect. The table of residuals, Table 47.21, is examined for individual large values (indicating a possibly erroneous observation) and for unusual patterns in sign and size (indicating possible interaction effects). The residuals should be plotted in the time order in which the treatment combinations were run; any indication of trends indicative of other factors disturbing the response. The residuals can also be plotted on normal probability paper as a check on the normality assumption.

The discussion thus far assumes that only one determination per cell was made. To obtain a truly valid estimate of the error variance from this experiment, the cell's observations must be replicated. When experiments are replicated (ideally with each cell containing the same number of observations), it is useful to have a table similar to Table 47.18, where now in each cell both the average and the estimate of variance are recorded. Homogeneity of variance tests may be made (see Snedecor and Cochran 1967; Duncan 1974; Dyer and Keating 1980), although it is well to remember that such

TABLE 47.20   Analysis of Variance of Resistivity Measurements Given in Table 47.18

Source of variation	Sum of squares (SS)	Degrees of freedom (DF)	Mean square = SS/DF	F*
Rows (force)	0.014855	$(r - 1) = 3$	0.004952	10.67†
Columns (current)	0.021430	$(c - 1) = 4$	0.005358	11.54†
Error	0.005570	$(r - 1)(c - 1) = 12$	0.000464	
Total	0.041855	$(rc - 1) = 19$		

*F = mean square (source)/mean square (error).
†Significant at the 1% level.

TABLE 47.21   Residuals, Row and Column Effects, Grand Average

Force, g	Amperage, A					Force effects
	Level 1	Level 2	Level 3	Level 4	Level 5	
25	0.0125	−0.0150	−0.0075	−0.0300	0.0400	0.0175
50	−0.0055	0.0170	0.0145	0.0120	−0.0380	0.0355
100	−0.0135	−0.0010	−0.0035	0.0140	0.0040	−0.0265
150	0.0065	−0.0010	−0.0035	0.0040	−0.0060	−0.0265
Current effects	−0.0285	−0.0110	−0.0085	−0.0160	0.0640	11. 8385 = $\bar{y}$

tests require near exact normality if they are to be useful. The analysis of variance assumes these variance estimates to be homogeneous and pools them.

A plot of each cell average versus cell estimate of variance (or standard deviation) is often revealing. Individual outlying points or a pattern of dependence of variability on average value should be looked for. (In the latter case, the need for a transformation of the data should be considered; see Box, Hunter, and Hunter 1978.)

In this experiment, replicate measurements were made in each cell of the table since the investigator was interested in finding out about possible interactions and whether the variance of the resistivity measurements would be constant at the extreme values of the factors. In this original experiment, no significant interactions were found, nor was the variance nonhomogeneous.

The analysis of variance of a replicated two-factor crossed factorial design experiment is easily modified from that of the unreplicated one. The procedure is to calculate the following quantities and insert them in Table 47.22:

$$r = \text{number of rows}$$

$$c = \text{number of columns}$$

$$k = \text{number of determinations per cell}$$

$$N = \text{total number of observations} = krc$$

$$T = \text{grand total}$$

$$C = T^2/N$$

$$\text{SSR} = \text{row sum of squares} = \frac{\sum_i R_i^2}{kc} - C$$

$$\text{SSC} = \text{column sum of squares} = \frac{\sum_j C_j^2}{kr} - C$$

$$\text{SSI} = \text{Interaction sum of squares}$$

$$= \frac{\sum_{ij} (\text{cell}_{ij} \text{ total})^2}{k} - \text{SSR} - \text{SSC} - C$$

$$\text{TSS} = \text{total sum of squares} = \Sigma y^2 - C$$

$$\text{SSE} = \text{error sum of squares} = \text{TSS} - \text{SSR} - \text{SSC} - \text{SSI}$$

TABLE 47.22   Analysis of Variance Table for Two-Factor Factorial (k Replicates per Cell)

Source of variation	Sum of squares	Degrees of freedom	Mean square		F
Rows	SSR	$r - 1$	$SSR/(r - 1)$	$= MSR$	MSR/MSE
Columns	SSC	$c - 1$	$SSR/(c - 1)$	$= MSC$	MSC/MSE
Interaction	SSI	$(r - 1)(c - 1)$	$SSI/(r - 1)(c - 1)$	$= MSI$	MSI/MSE
Error	SSE	$rc(k - 1)$	$SSE/rc(k - 1)$	$= MSE$	
Total	SST	$krc - 1$			

The above instructions will fill in all the cells in the "Sum of squares" column of the analysis of variance table (Table 47.22), but the similarities to Table 47.5 should be noted. The value of the mean square error (MSE) in Table 47.22 can also be obtained by pooling the ($r \times c$) estimates of the within-cell variance.

SPLIT-PLOT FACTORIAL EXPERIMENTS

In general a factorial design is a program of experiments consisting of all possible combinations of levels (versions) of k different factors (variables). There can be m_1 levels of factor 1 combined with m_2 levels of factor 2,..., combined with m_k levels of factor k to give a total of $N = m_1 \times m_2 \times ... \times m_k$ experimental trials.

Care must be taken that the N experimental trials composing a factorial design are all run in a random sequence. Consider for example an experimental program for studying temperature and pressure each at three levels along with four different versions of catalyst, that is, a $3 \times 3 \times 4$ factorial design in $N = 36$ trials. To be a standard factorial all 36 runs would be performed in random order. However, it is very likely that the experimental design will be run in a split-plot (nested) arrangement; that is, the nine temperature-pressure runs will be randomly performed for each catalyst separately. The analysis of a split-plot factorial design, most particularly the computation of correct confidence intervals and tests of hypotheses concerning treatment means, usually requires the attention of a professional statistician. In this handbook only fully randomized designs are discussed.

Split-plot designs are alike in structure to the "nested" designs discussed earlier. Commonly, the designs are called "split-plot" when the objective is the study of treatment means and called "nested" when the objective of the experimenter is to estimate components of variance. Statisticians, and most computer software programs, distinguish between these two forms of experimental design with their different objectives. The factorial split-plot arrangement requires a "Type I" analysis of variance while that for the nested designs requires a "Type II." Occasionally, experimental designs are employed in which estimates of both means and components of variance are required. The subtleties of analysis become important and one should seek the advice of a professional statistician.

FACTORIAL EXPERIMENTS WITH k FACTORS (EACH FACTOR AT TWO LEVELS)

The 2^k factorial designs have widespread industrial applicability. The designs permit the separate estimation of the individual effects and the interaction effects of the k factors in an experimental program in which all k factors are varied simultaneously in a carefully organized pattern of trials.

Symbols.   A factorial experiment with k factors, each at two levels, is known as a 2^k factorial experiment. The experiment consists of 2^k trials, one trial at each combination of levels of the fac-

TABLE 47.23   Different Notations for the 2^3 Factorial Design

Run no.	Geometric notation			Alternative notation			Japanese notation			Classical notation		
	A	B	C	A	B	C	A	B	C	A	B	C
1	−	−	−	0	0	0	1	1	1	1		
2	+	−	−	1	0	0	2	1	1	a		
3	−	+	−	0	1	0	1	2	1	b		
4	+	+	−	1	1	0	2	2	1	ab		
5	−	−	+	0	0	1	1	1	2	c		
6	+	−	+	1	0	1	2	1	2	ac		
7	−	+	+	0	1	1	1	2	2	bc		
8	+	+	+	1	1	1	2	2	2	abc		

tors. To identify the individual trials, different notations are used, as illustrated in Table 47.23. One convention is to label each factor by a letter (or numeral) and then to denote the two levels (versions) of each factor by a plus (+) and a minus (−) sign. Commonly the minus sign refers to the lower level, the standard conditions, or the absence of the factor. Thus, if there are three factors labeled A, B, and C, the eight trials comprising the 2^3 factorial design are as shown in Table 47.23. The (+, −) notation is sometimes referred to as "geometric." For example, the eight (±, ±, ±) factor settings for the 2^3 design may be interpreted as giving the (±1, ±1, ±1) coordinates of the eight vertices of a cube. Alternative notations are to employ 0 and 1, respectively, or, following the Japanese tradition earlier established by Taguchi, 1 and 2 for the two versions of each factor. The classical convention is to denote the two versions of each lettered factor by the presence and absence of its corresponding lowercase letter, as is also illustrated in Table 47.23. Here the trial in which all factors are at their "low" level is denoted by a 1. The sequence of trials in Table 47.23 is written in standard or "Yates" order. The trials would, of course, be run in random order.

Example.   The data in Table 47.24 are taken from a larger experiment on fire-retardant treatments for fabrics. The excerpted data are intended only to provide an example for demonstrating the technique of analysis. The experiment has four factors, each at two levels, i.e., it is a 2^4 factorial. Note that all factors are qualitative in this experiment. The experimental factors and levels (versions) are

Factors	Levels
A—Fabric	−Sateen
	+Monk's cloth
B—Treatment	−Treatment x
	+Treatment y
C—Laundering	−Before laundering
	+After one laundering
D—Direction	−Warp
	+Fill

The observations reported in Table 47.24 are inches burned, measured on a standard-sized sample after a flame test. For convenience, alternative design notations representing the treatment combinations appear beside the resulting observation.

Estimation of Main Effects and Interactions.   Obtaining the estimates of main effects and interactions from a 2^k factorial design (and 2^{k-p} fractional factorial designs) is available on

TABLE 47.24 Results of Flame Test of Fire-Retardant Treatments (a 2^4 Factorial Experiment)

Geometric notation				Classical notation	Response yield, inches burned
A	B	C	D		
−	−	−	−	1	4.2
+	−	−	−	a	3.1
−	+	−	−	b	4.5
+	+	−	−	ab	2.9
−	−	+	−	c	3.9
+	−	+	−	ac	2.8
−	+	+	−	bc	4.6
+	+	+	−	abc	3.2
−	−	−	+	d	4.0
+	−	−	+	ad	3.0
−	+	−	+	bd	5.0
+	+	−	+	abd	2.5
−	−	+	+	cd	4.0
+	−	+	+	acd	2.5
−	+	+	+	bcd	5.0
+	+	+	+	abcd	2.3

Treatment combinations

almost all design of experiments software programs. The computations are simple to do by hand but become quickly tedious and hence ideal work for a computer (see Bisgaard 1993a). Further, the associated analysis of variance table is similarly easy and hence widely available. Nevertheless, we include here a description of the hand computations for the interested reader, since they are also easily done on a spread sheet.

The 2^k factorial designs permit the estimation of all k main effects (first-order effects), all $k(k − 1)/2$ two-factor interactions, all $k(k − 1)(k − 2)/3!$ three-factor interactions, etc. Each estimated effect is a statistic of the form $\bar{y}_+ − \bar{y}_-$; that is, it is expressed by the difference between two averages, each containing $2^{k−1}$ observations. For a 2^4 design the analyst would thus be able to estimate, in addition to the grand average, four main effects, six two-factor interactions, four three-factor interactions, and a single four-factor interaction, giving a total of 16 statistics. Remarkably, all these statistics are "clear" (orthogonal) of one another; that is, the magnitudes and signs of each statistic are in no manner influenced by the magnitude and sign of any other.

The question as to which observations go into which average for each estimated effect is determined from the k columns of + and − signs that together form the experimental design (the design column "vectors"). Additional column vectors of + and − signs are then constructed for each interaction, as illustrated in Table 47.25. For example, the vector of signs labeled AB is obtained by algebraically multiplying, for each row, the + or − sign found in column A by the + or − sign found in column B.

Table 47.25 also contains the column of observations. To estimate the AB interaction effect, all the observations carrying a + sign in the AB column are placed in $\bar{y}_+$ and those with a minus sign in $\bar{y}_-$. The estimated AB interaction effect $(\bar{y}_+ − \bar{y}_-)$ is therefore:

$$\frac{4.2 + 2.9 + \ldots + 2.3}{8} − \frac{3.1 + 4.5 + \ldots + 5.0}{8} = \frac{27.0}{8} − \frac{30.5}{8} = \frac{-3.5}{8} = -0.4375$$

TABLE 47.25 Table of Signs for Calculating Effects for a 2^4 Factorial

A	B	C	D	AB	AC	AD	BC	BD	CD	ABC	ABD	ACD	BCD	ABCD	Obs.*
−	−	−	−	+	+	+	+	+	+	−	−	−	−	+	4.2
+	−	−	−	−	−	−	+	+	+	+	+	+	−	−	3.1
−	+	−	−	−	+	+	−	−	+	+	+	−	+	−	4.5
+	+	−	−	+	−	−	−	−	+	−	−	+	+	+	2.9
−	−	+	−	+	−	+	−	+	−	+	−	+	+	−	3.9
+	−	+	−	−	+	−	−	+	−	−	+	−	+	+	2.8
−	+	+	−	−	−	+	+	−	−	−	+	+	−	+	4.6
+	+	+	−	+	+	−	+	−	−	+	−	−	−	−	3.2
−	−	−	+	+	+	−	+	−	−	−	+	+	+	−	4.0
+	−	−	+	−	−	+	+	−	−	+	−	−	+	+	3.0
−	+	−	+	−	+	−	−	+	−	+	−	+	−	+	5.0
+	+	−	+	+	−	+	−	+	−	−	+	−	−	−	2.5
−	−	+	+	+	−	−	−	−	+	+	+	−	−	+	4.0
+	−	+	+	−	+	+	−	−	+	−	−	+	−	−	2.5
−	+	+	+	−	−	−	+	+	+	−	−	−	+	−	5.0
+	+	+	+	+	+	+	+	+	+	+	+	+	+	+	2.3

*Obs. = observations.

47.41

Yates' Algorithm. An alternative and more rapid method for obtaining estimates of main effects and interactions for two-level factorials, called "Yates' algorithm," applies to all two-level factorials and fractional factorials. The first step in Yates' algorithm is to list the observed data in Yates order, as illustrated in Table 47.26. The generation of the values in Table 47.26 proceeds as follows:

1. A two-level factorial with r replicates contains $N = r2^k$ runs. The associated Yates' algorithm table will have $k + 2$ columns, the first of which contains the experimental design, i.e., the 2^k treatment combinations in standard (Yates) order.

2. In column 2, enter the observed yield corresponding to each treatment combination listed in column 1. If the design is replicated, enter the total for each treatment combination.

3. In the top half of column 3 enter, in order, the sums of consecutive *pairs,* of entries in column 2, i.e., the first plus the second, the third plus the fourth, and so on. In the bottom half of column 3 enter, in order, the differences between the same consecutive pairs of entries, i.e., second entry minus first entry, fourth entry minus third entry, etc. Change the sign of the top (first of the pair) and algebraically add.

4. Obtain columns 4, 5, ..., $k+2$, in the same manner as column 3, i.e., by obtaining in each case the sums and differences of the pairs in the preceding column in the manner described in step 3.

5. The entries in the last column (column $k + 2$) are labeled $g(T)$, $g(A)$, $g(B)$, $g(AB)$, etc. The letters in the parentheses correspond to the $+$ signs in the geometric notation. The first value $g(T)$ is divided by N to give the grand average. Estimates of the remaining main effects and interactions are obtained by dividing each $g(...)$ by $N/2$. (*Note:* The remaining steps of this procedure are checks on the computations.)

6. The sum of all the individual responses (column 2) should equal the total given in the first entry of column 6, i.e., $g(T)$ must equal the grand total.

7. The sum of the squares of the quantities in column 2 should equal the sum of the squares of the entries in column $(k + 2)$ divided by 2^k.

8. Each $g(...)$ in the last column equals the sum of observations carrying a $+$ sign minus the sum of observations carrying a $-$ sign when the columns of signs displayed in Table 47.25 are employed.

TABLE 47.26 Yates Method of Analysis Using Data of Table 47.24

A B C D	2	3	4	5	6		Estimated effects	
$-\ -\ -\ -$	4.2	7.3	14.7	29.2	57.5 $=$	$g(T)$	Average $=$	3.5938
$+\ -\ -\ -$	3.1	7.4	14.5	28.3	$-12.9 =$	$g(A)$	$A =$	-1.6125
$-\ +\ -\ -$	4.5	6.7	14.5	-5.2	2.5 $=$	$g(B)$	$B =$	0.3125
$+\ +\ -\ -$	2.9	7.8	13.8	-7.7	$-\ 3.5 =$	$g(AB)$	$AB =$	-0.4375
$-\ -\ +\ -$	3.9	7.0	-2.7	1.2	$-\ 0.9 =$	$g(C)$	$C =$	-0.1125
$+\ -\ +\ -$	2.8	7.5	-2.5	1.3	$-\ 0.5 =$	$g(AC)$	$AC =$	-0.0625
$-\ +\ +\ -$	4.6	6.5	-3.5	-0.8	1.3 $=$	$g(BC)$	$BC =$	0.1625
$+\ +\ +\ -$	3.2	7.3	-4.2	-2.7	0.5 $=$	$g(ABC)$	$ABC =$	0.0625
$-\ -\ -\ +$	4.0	-1.1	0.1	-0.2	$-\ 0.9 =$	$g(D)$	$D =$	-0.1125
$+\ -\ -\ +$	3.0	-1.6	1.1	-0.7	$-\ 2.5 =$	$g(AD)$	$AD =$	-0.3125
$-\ +\ -\ +$	5.0	-1.1	0.5	0.2	0.1 $=$	$g(BD)$	$BD =$	0.0125
$+\ +\ -\ +$	2.5	-1.4	0.8	-0.7	$-\ 1.9 =$	$g(ABD)$	$ABD =$	-0.2375
$-\ -\ +\ +$	4.0	-1.0	-0.5	1.0	$-\ 0.5 =$	$g(CD)$	$CD =$	-0.0625
$+\ -\ +\ +$	2.5	-2.5	-0.3	0.3	$-\ 0.9 =$	$g(ACD)$	$ACD =$	-0.1125
$-\ +\ +\ +$	5.0	-1.5	-1.5	0.2	$-\ 0.7 =$	$g(BCD)$	$BCD =$	-0.0875
$+\ +\ +\ +$	2.3	-2.7	-1.2	0.3	0.1 $=$	$g(ABCD)$	$ABCD =$	0.0125
Total	57.5							

Sum of squares 219.15   3506.40/16=219.15

The corresponding estimated effects are given by $g(\ldots)/(N/2)$. The algorithm is best explained with an example.

Example. The example shown in Table 47.24 has 2^4 runs. Thus the associated Yates algorithm will have six columns, as shown in Table 47.26. The grand average is $\bar{y} = 57.5/16 = 3.5938$. The next entry in column 6 is $g(A) = -12.9$. The estimated main effect of factor A is then

$$A \text{ effect} = \frac{-12.9}{8} = -1.6125$$

The estimate of the main effect of A can be checked by taking the average of the responses recorded on the high $(+)$ side of the factor A and subtracting the average response on the low $(-)$ side to give $\bar{y}_+ - \bar{y}_- = 22.3/8 - 35.2/8 = -12.9/8 = -1.6125$.

The remaining effects are similarly computed. Thus, the estimated AD interaction effect $= -2.5/8 = -0.3125$.

The following steps are checks on the computations in Table 47.26:

6. The sum of column 2 equals $g(T)$.

7. The sum of squares of the entries in column 2 equals 219.5. The sum of squares in column 6 divided by $2^4 = 3506.4/16 = 219.15$.

Testing Main Effects and Interactions. The grand average and 15 estimated effects obtained from the 2^4 design appear in the right-hand column of Table 47.26. The standard error (SE) of each estimated effect is given by

$$\text{SE(effect)} = \text{SE}(\bar{y}_+ - \bar{y}_-) = 2s/\sqrt{N}$$

where $N=$total number of observations. The $100(1-\alpha)$ percent confidence limits are given by

$$\text{Effect} \pm t_{\alpha/2}\,[\text{SE(effect)}]$$

Needed is s^2, the estimate of the experimental error variance σ^2. An estimate of σ^2 can always be obtained from truly replicated trials, each set of replicates providing a single estimate of variance and then all the estimates pooled. However, in this example each trial was performed only once and some alternative procedure for securing an estimate of σ^2 is needed. We turn now to the analysis of variance table for the 2^k factorial design.

The Analysis of Variance for the 2^k Factorial. It is a rare design of experiments computer software program that omits the computations for a factorial design analysis of variance table. The computations are easy, though lengthy, when done by hand. We use our 2^4 example to demonstrate these computations for the interested reader. They are easily done with a computer spread sheet program.

The total variability of the observations is measured by $\Sigma(y_i - \bar{y})^2 = \Sigma y_i^2 - (\Sigma y_i)^2/N$, $i = 1, 2, \ldots, N$. In this example $\Sigma(y_i - \bar{y})^2 = 219.15 - (57.5)^2/16 = 12.51$. If this variability could be completely assignable to random errors, then the estimate of the variance σ^2 is given by $s^2 = \Sigma(y_i - \bar{y})^2/(N-1)$ with $(N-1)$ degrees of freedom. However, some of the movement amongst the observation y_i is likely caused by the influences of the controlled factors which make up the experimental design. The contribution of each factorial effect is given by its "sum of squares" $= N(\text{effect})^2/4$, labeled SSq, with one degree of freedom, as illustrated in Table 47.27.

The residual sum of squares represents variability remaining after all assignable causes have been subtracted. In Table 47.27 all 15 degrees of freedom with their associated SSq "sum of squares" are present, and thus the residual sum of squares and degrees of freedom are both zero.

To estimate the variance σ^2 in this unreplicated factorial design, we must declare some of the estimated effects to be manifestations of noise, i.e., not real effects and likely equal to zero. The most

TABLE 47.27   Analysis of Variance Table: 2^4 Factorial

Source of variation	SSq	DF
Total $= \Sigma(y_i - \bar{y})^2$	12.509375	15
A effect $= 16\,(-1.6125)^2/4$	10.400625	1
B effect $= 16\,(+0.3125)^2/4$	0.390625	1
C effect $= 16\,(-0.1125)^2/4$	0.050625	1
D effect $= 16\,(-0.1125)^2/4$	0.050625	1
AB effect $= 16\,(-0.4375)^2/4$	0.765625	1
AC effect $= 16\,(-0.0625)^2/4$	0.015625	1
AD effect $= 16\,(-0.3125)^2/4$	0.390625	1
BC effect $= 16\,(+0.1625)^2/4$	0.105625	1
BD effect $= 16\,(+0.0125)^2/4$	0.000625	1
CD effect $= 16\,(-0.0625)^2/4$	0.015625	1
ABC effect $= 16\,(+0.0625)^2/4$	0.015625	1
ABD effect $= 16\,(-0.2375)^2/4$	0.225625	1
ACD effect $= 16\,(-0.1125)^2/4$	0.050625	1
BCD effect $= 16\,(-0.0875)^2/4$	0.030625	1
$ABCD$ effect $= 16\,(+0.0125)^2/4$	0.000625	1
Residual sum of squares	0	0

reasonable collection of such effects is the three and four factor interactions. We thus sum their sum of squares and degrees of freedom and compose an estimate of variance $s^2 = 0.323125/5 = 0.064625$ with $\nu = 5$ degrees of freedom. The estimated standard deviation is $s = 0.254$. The standard error of an effect is then SE(effect) $= 2s/\sqrt{N} = 0.127$, and the $100(1 - \alpha)$ percent confidence limits for an effect are

$$\pm t_{\nu\alpha/2}\text{SE(effect)} = \pm t_{5,0.025}(0.127) = \pm 2.571(0.127) = \pm 0.326$$

Thus, an (approximate) 95 percent confidence interval for all remaining effects is given by effect $\pm$ 0.326. Any estimated effect whose confidence interval "effect $\pm$ 0.3265" includes zero may be declared not statistically significant. Of the estimated effects in this example only A and AB can be declared statistically significant and the effects B and AD nearly so. Computer software programs will perform analysis of variance F tests (or equivalent t tests) and compute their probabilities. These tests provide identical inferences. The analyst is advised to look at the interval statements rather than try to judge the differences between small probabilities. It is a rare individual who can really appreciate the true meaning of probabilities such as 0.07 versus 0.035.

A less formal analysis is to plot the absolute values of the estimated effects into a Pareto diagram. Here a Pareto diagram of the estimated effects indicates that factors A, AB, AD, and B are the "vital few." A Pareto diagram will identify "significant" effects almost as well as a collection of confidence intervals or a set of F tests in an ANOVA table (Hunter, W.G. 1977; Lenth 1989; Schmidt and Launsby 1991).

Collapsing the 2^k Factorials.   Collapsing 2^k designs into lower "dimensionality" is an important strategy in the application of factorial designs. In practice it is very unlikely that all the possible main effects and interactions are real when the number of factors $k \geq 3$. In fact, particularly in screening situations where many factors are being co-studied, several individual factors may have no detectable main effect or interaction influences upon a response. Factorial designs for k modestly large are thus said to have "hidden replication," that is, an excess of degrees of freedom over and

TABLE 47.28   Collapsed 2^4 Giving a Replicated 2^3

A B D	Responses	Difference	
− − −	4.2, 3.9	0.3	
+ − −	3.1, 2.8	0.3	
− + −	4.5, 4.6	−0.1	
+ + −	2.9, 3.2	−0.3	$s^2 = \Sigma d^2/2n_d$
− − +	4.0, 4.0	0	
+ − +	3.0, 2.5	0.5	$s^2 = 0.57/2(8)$
− + +	5.0, 5.0	0	$= 0.03560$
+ + +	2.5, 2.3	0.2	

beyond those necessary to explain the response. Thus, when a factorial design is unreplicated, many estimated effects and their associated sums of squares and degrees of freedom may become available to employ in the estimate of variance. In this 2^4 example it appears that factor C makes no large contribution to the response, either as a main effect or as an interaction. The factor's effects are so small over its region of exploration as to be indistinguishable from noise. The 2^4 factorial now collapses into a replicated 2^3 design in the effective factors A, B, and D. The collapsed design is displayed in Table 47.28.

The estimated effects for factors A, B, and D remain unchanged. An estimate of variance can now be obtained from the repeated runs. When there are pairs of observations, a shortcut computation for s^2 is given by:

$$s^2 = \frac{\sum_i d_i^2}{2n_d}$$

where the d_i are the differences between the $i = 1, 2, \ldots, d$ pairs of observations and n_d is the number of differences. This estimate of variance has $\nu = n_d$ degrees of freedom. Thus, for this example:

$$s^2 = \frac{0.57}{(2)(8)} = 0.0356$$

with $\nu = 8$ degrees of freedom and $s = 0.1887$. The standard error of each effect (excluding all those carrying the label C, of course), is

$$\text{SE(effect)} = \frac{2s}{\sqrt{N}} = \frac{2(0.1887)}{\sqrt{16}} = 0.0943$$

Using t with $\nu = 8$ degrees of freedom, the 95 percent confidence limits for the estimated effects are given by

$$(t_{\alpha/2})\text{SE(effect)} = \pm 2.306(0.0943) = \pm 0.2175$$

Half-Normal Plots.   Daniel (1959) proposed a simple and effective technique for use in the interpretation of data from the two-level factorial designs. This technique consists of plotting the absolute values of the estimated effects on normal probability paper (dropping the effect's signs), leading to "half-normal" plots. Effects indistinguishable from noise will fall along a straight line; effects that are statistically significant will fall well off the line. Full-normal plots consisting of the estimated effects *with* their signs are also possible; see Box, Hunter, and Hunter (1978). Normal probability plot routines are found in most statistical software programs. Half-normal plots are a particularly useful diagnostic tool offering evidence of wild observations and other conditions that violate the assumptions of homogeneous variance, normality, and randomization. "Guardrails" can be used to judge the divergence of the plotted effects from the line; see Zahn (1975) and Taylor (1994). The Lilliefors (1967) test for normality can be helpful in judging all normal plots. For this and other nonparametric methods see Iman (1994).

EVOP: EVOLUTIONARY OPERATION

An important application of experimental design in the production environment was proposed by Box (1957). Essentially, a simple experimental design, run repeatedly, provides a routine of small systematic changes in a production process. The objective is to force the process to produce information about itself while simultaneously producing product to standards. Only small changes in the process factors are allowed, and the consequences of these changes must be detected in the presence of the many natural variabilities that surround the process. The repetition of an experimental design, commonly a 2^2 factorial with center point, permits the blocking of many of the disturbances that commonly influence production. Through the process of replication, the design provides steadily improving estimates of the main effects and interactions of the studied factors.

Response Surface.   A response surface (see Box 1954) is a graphical representation of the connection between important independent variables, controlled factors, and a dependent variable. (An independent variable is a factor that is, or conceivably could be, controlled. Examples are flow rate and temperature. The value of a dependent variable is the result of the settings of one or more independent variables.) Most processes have several dependent variables, such as yield, impurities, and pounds per hour of a byproduct. These responses are usually smooth and may be graduated approximately by simple contours such as a family of lines or arcs. We ordinarily work on processes that have unknown response surfaces—if they were known, the work would not be necessary. See Response Surface Designs below.

A response surface for a process might look like the one in Figure 47.9, which shows the yield of a catalytic oxidation as a function of temperature and feed rate of hydrocarbon. If this information were known, the pounds per hour of product could be determined and better operating conditions selected for any desired production rate. The response surface is initially unknown, but improvement can be made if we only find out which way is up. Multiple regression can be used to approximate the response contours (see Section 44, under Multiple Regression).

EVOP Technique.   The problem, then, is to increase profit in an operating plant with minimum work and risk and without upsetting the plant. These are the steps:

1. Survey company reports and open literature on the process. Study cost, yield, and production records.
2. Study this section on EVOP and preferably the definitive text (Box and Draper 1969).
3. Obtain agreement and support from production management. Organize a team and hold training sessions.
4. Select two or three controllable factors that are likely to influence the most important response.
5. Change these factors in repeated small steps according to a plan.
6. After the second repetition of the plan (Cycle 2) and each succeeding cycle, estimate the effects.
7. When one or more of the effects is significant, move to the indicated better operating conditions and start a new EVOP program, perhaps with new ranges or new factors.
8. After eight cycles, if no factor has been shown to be effective, change the ranges or select new variables (Box and Draper 1969).
9. Continue moving the midpoint of the EVOP plan and adjust the ranges as necessary.

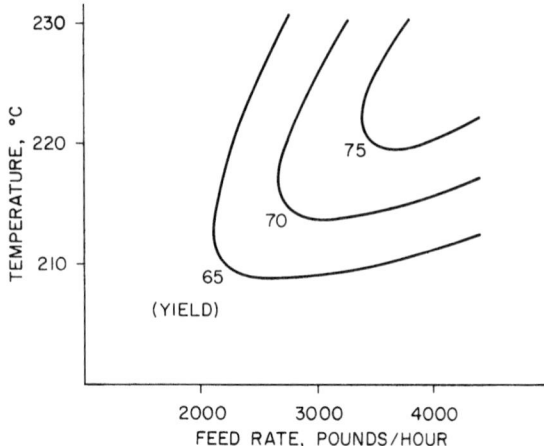

FIGURE 47.9   Typical response surface: yield as a function of feed rate and temperature for a catalytic reactor.

10. When a maximum has been obtained, or the rate of gain is too slow, drop the current factors from the plan and run a new plan with different factors.

The following topics explain these steps in detail.

Literature Search.   Sources of information include the process instructions, company reports, manufacturers' literature, patents, textbooks, and encyclopedias of technology. Do not neglect people. Company personnel, consultants, and operators can all contribute. Search for information on

1. Important independent variables

2. Test methods for intermediate and final results

3. Recommended procedures

4. Records of good and bad results and their causes

5. Long-term history of results; plant production rate and yield by week or month and similar data; effect of past changes in equipment and conditions

When information is contradictory, an EVOP program is an ideal strategy for resolving the conflict. Always consider the physical and chemical principles that apply.

The EVOP Design.   EVOP uses planned runs that are repeated over and over (replicated). One plan in wide use is the two-level complete factorial. There are important reasons to maintain observations on a known set of conditions, called a "reference point." For simplicity in the present discussion, let this point be the center of the square formed by the vertices of the factorial.

Example.   This example shows coded data from an actual EVOP program (Barnett 1960). The problem involved a batch organic reaction, and after the steps above were followed, the two factors and their ranges were selected as shown in Figure 47.10. The response Y is a coded yield in pounds per batch and should be maximized. In this diagram, the reference run (batch) was made at 130°C for $3^{1}/_{2}$ h. The next batch was made at 120°C for 3 h, and so on. The first *cycle* contains five runs, one at each of the conditions. Samples were taken from each batch and analyses were obtained. If the process were continuous, it would be allowed to stabilize after each change of conditions.

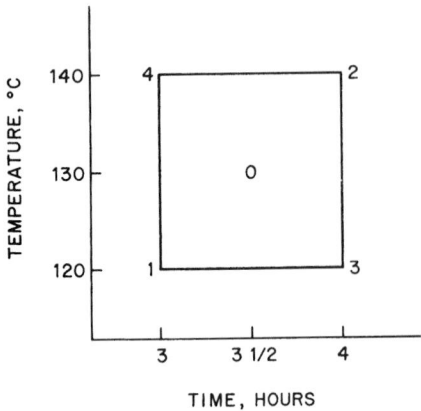

FIGURE 47.10 An EVOP plan. Numbers are in run order; 0 is the reference run.

Warning: It is not unusual to have difficulty in obtaining a representative sample. Procedures and tests are discussed in Sections 44 and 46.

Effects can be estimated at the end of Cycle 1, but in the absence of repeated trials, no estimate of error variance is available. At the conclusion of Cycle 2 both estimated effects and error variance are possible. The estimated effects and their confidence intervals may be computed on a form such as Figure 47.11, which shows this example. The form helps reduce the work and minimizes mistakes. Instructions should be printed on the reverse side (see Figure 47.12).

The error term that puts the magnitude of effects into perspective is obtained from the range by use of factor *K* first derived by Box and Hunter (1959). The uncertainty in estimating the effects is stated as a confidence interval and the Change-in-mean effect (CIM) is calculated by comparing the results at the outer four corners with the result in the center. The use of this last information is discussed below.

A "phase" is defined as all the cycles that use the same settings of the same factors. The phase average shows the general level of the response and can be used to compare different phases.

The left part of the form is used to record data and estimate effects. It also has the scaled diagram of the phase. The right third is used to determine the error limits of effects and corner aver-

J-227-x

CALCULATION OF AVERAGES

OPERATING CONDITIONS	0	1	2	3	4
SUM FROM PREVIOUS CYCLE	938	666	1104	904	948
AVERAGE FROM PREVIOUS CYCLE	938	666	1104	904	948
NEW OBSERVATIONS	792	840	883	825	996
DIFFERENCES (WATCH SIGNS)	+146	−174	+221	+79	−48
NEW SUMS (N.S.)	1730	1506	1987	1729	1944
NEW AVERAGES (N. S./N)	A) 865	B) 753	C) 994	D) 864	E) 972

CALCULATION OF EFFECTS

A EFFECT: C 994, D 864, F 1858; B 753, E 972, G 1725; 1858 − 1725; 2) 133; + 66

B EFFECT: C 994, E 972, 1966; B 753, D 864, 1617; 1966 − 1617; 2) 349; + 174

AB EFFECT: B 753, C 994, 1747; D 864, E 972, 1836; 1836 − 1747; 2) 89; − 44

CHANGE-IN-MEAN EFFECT: F 1858, G 1725, H 3583; A 865 × 4 = 3460; 3583 − 3460; 5) 123; + 25

PHASE MEAN: H 3583, A 865, 5) 4448, 890

DESIGN: B, TEMP, °C, 140, 120, 4, 2, 0, 1, 3, 3, 4, TIME, HRS., A

CALCULATION OF STANDARD DEVIATION

PREVIOUS SUM S ———

NEW S = RANGE × K = 395 × 0.30 = 118

NEW SUM S = 118

NEW AVERAGE S A = NEW SUM S/(N−1) = 118 / 1 = 118

PREVIOUS AVERAGE S A = ———

CALCULATION OF 2 S. E. ERROR LIMITS

FOR NEW AVERAGES AND NEW EFFECTS: L 1.41 × S A 118 = ± 166

FOR CHANGE-IN-MEAN EFFECT: M 1.26 × S A 118 = ± 149

FACTORS

N	K	L	M
2	0.30	1.41	1.26
3	0.35	1.15	1.03
4	0.37	1.00	0.89
5	0.38	0.89	0.80
6	0.39	0.82	0.73
7	0.40	0.76	0.68
8	0.40	0.71	0.63

CORNING GLASS WORKS
EVOLUTIONARY OPERATIONS
TWO VARIABLE WORK SHEET

PRODUCT EXAMPLE
RESPONSE YIELD, POUNDS
PHASE I CYCLE (N) 2
BY EHB DATE 1-11-71

REMARKS:
TEMPERATURE EFFECT IS BORDERLINE
RUN ANOTHER CYCLE

MANUFACTURING AND ENGINEERING DIVISION

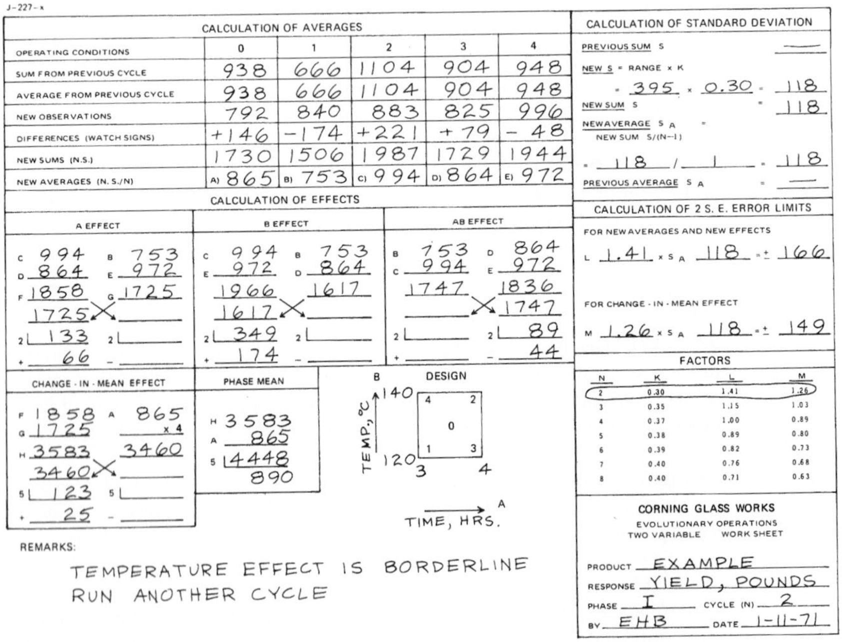

FIGURE 47.11 Calculations form at the end of Cycle 2. (*Modified from Barnett 1960.*)

Differences	Subtract new observations from old averages. Note the algebraic sign of the difference. (938 − 792 = 146)
New Sums	Add the new observations to the old sums. (938 + 792 = 1,730)
New Averages	Divide the new sums by N, the number of the cycle. (1730/2 = 865)
Calculation of Effects	(For example, the A effect)
	Write the new averages for operating conditions 2 and 3 opposite (C) and (D). Add these two to get the number in space (F). Carry out corresponding operations to get the number for space (G).
	The next operation is subtraction. If (F) is larger than (G), recopy (G) under (F) and subtract (G) from (F). If (G) is larger than (F), recopy (F) under (G) and subtract (F) from (G). In either case divide by 2 and the sign of quotient is shown on the form. (The A effect is +66.)
Change-in-mean effect	Copy (F) and (G) as shown and add these two. (A) is multiplied by 4. The next operation is subtraction as above. Divide by 5. (The Change-in-mean effect is +25.)
Average phase	Copy (H) and (A) from Change-in-mean box, add them, and divide by 5. (The phase mean is 890.)
Calculation of Standard Deviation	Range. The range is the algebraic difference between the most positive and most negative differences. The range is always positive. (The range of +146, −174, +221, +79, and −48 is 395.) The standard deviation is estimated by taking the range times K.
Constants	Read K, L, and M factors from the table.
Note:	The numerical illustrations apply to the data of Fig. 47.11.

FIGURE 47.12   Instructions for EVOP form.

ages. The error limits are called "2 S.E." for "two standard errors." The estimated effect ±2 S.E. covers the usual 95 percent confidence region. Caution should be exercised in claiming statistical significance until after two or three cycles.

For the present example, at the end of Cycle 2 the A effect (time) is estimated as 66 ± 166; it lies somewhere between −100 and +232. Thus, the true value of this effect could be negative, positive, or nil. The B (temperature) effect, however, is estimated to be 174 ± 166, or in the range of +8 to 340. Technically, it is likely to be a positive real effect. The interaction AB is small, and so is the change in mean. Since the confidence region for B is so close to zero and following the advice above to be cautious at Cycle 2, another cycle is run. Its results are shown in Figure 47.13.

After Cycle 3, the B effect (temperature) was declared statistically significant, since its likely values fall in the range of 169 ± 92, or +77 to +261. It does not appear that data from more cycles would change the conclusion that temperature should be increased to increase the response y.

When statistical significance is found for one factor but not for the other, *move* the plan in the desirable direction for the "discovered" important factor and increase the range for the second (nonsignificant) one. Possible old and new phases of an EVOP program are shown in Figure 47.14. When significance is found for both variables, the center of the plan is moved in both directions in proportion to the size of the effects. This is the direction of steepest ascent (see Determine Direction of Steepest Ascent below, under First-Order Strategy). Whenever the plan is changed, a new *phase* is started. During the second and later phases, the previous estimate of standard deviation is often used since it was obtained under the current operating method.

Moves.   To be conservative, as EVOP should be, moves are contiguous; i.e., one or more of the points in the old phase and new phase coincide. This limits the moves to those types shown in Figure 47.14.

There is nothing "magic" about drawing these plans as squares—1 h does not equal 20°C anyway. A particularly strong signal may justify a move to a plan that does not adjoin the previous one.

Change-in-Mean.   The Change-in-mean effect is the difference between the results at the center point and the average of the other four peripheral points. It is therefore a signal of curvature as shown

J-227-x'

CALCULATION OF AVERAGES						CALCULATION OF STANDARD DEVIATION	
OPERATING CONDITIONS	0	1	2	3	4	PREVIOUS SUM S	*118*
SUM FROM PREVIOUS CYCLE	*1730*	*1506*	*1987*	*1729*	*1944*	NEW S = RANGE × K	
AVERAGE FROM PREVIOUS CYCLE	*865*	*753*	*994*	*864*	*972*	= *119* × *0.35* = *42*	
NEW OBSERVATIONS	*934*	*853*	*1066*	*852*	*953*	NEW SUM S = *160*	
DIFFERENCES (WATCH SIGNS)	*−69*	*−100*	*−72*	*+12*	*+19*	NEW AVERAGE S_A =	
NEW SUMS (N.S.)	*2664*	*2359*	*3053*	*2581*	*2897*	NEW SUM S/(N−1)	
NEW AVERAGES (N. S./N)	A) *888*	B) *786*	C) *1018*	D) *860*	E) *966*	= *160* / *2* = *80*	

PREVIOUS AVERAGE S_A ___ *118* ___

CALCULATION OF EFFECTS

A EFFECT	B EFFECT	AB EFFECT
C *1018* B *786*	C *1018* B *786*	B *786* D *860*
D *860* E *966*	E *966* D *860*	C *1018* E *966*
F *1878* G *1752*	*1984* *1646*	*1804* *1826*
1752 ✕	*1646* ✕	*1804* ✕
2 ⌐ *126* 2 ⌐	2 ⌐ *338* 2 ⌐	2 ⌐ 2 ⌐ *22*
+ *63* −	+ *169* −	+ − *11*

CALCULATION OF 2 S. E. ERROR LIMITS

FOR NEW AVERAGES AND NEW EFFECTS

L *1.15* × S_A *80* = ± *92*

FOR CHANGE - IN - MEAN EFFECT

M *1.03* × S_A *80* = ± *82*

FACTORS

N	K	L	M
2	0.30	1.41	1.26
3	0.35	1.15	1.03
4	0.37	1.00	0.89
5	0.38	0.89	0.80
6	0.39	0.82	0.73
7	0.40	0.76	0.68
8	0.40	0.71	0.63

CHANGE - IN - MEAN EFFECT	PHASE MEAN
F *1878* A *888*	H *3630*
G *1752* × 4	A *888*
H *3630* *3552*	S *4518*
3552 ✕	*904*
S ⌐ *78* S ⌐	
+ *16* −	

DESIGN
TEMP, °C
140 — 4 — 2
    — 0 —
120 — 1 — 3
    — 3 — 4
TIME, HRS. → A

CORNING GLASS WORKS
EVOLUTIONARY OPERATIONS
TWO VARIABLE WORK SHEET

PRODUCT *EXAMPLE*
RESPONSE *YIELD, POUNDS*
PHASE *I* CYCLE (N) *3*
BY *EHB* DATE *1-12-71*

REMARKS:

TEMPERATURE EFFECT IS +77 ⟶ +261

INCREASE TEMPERATURE & START PHASE II

MANUFACTURING AND ENGINEERING DIVISION

FIGURE 47.13 Calculation form after Cycle 3.

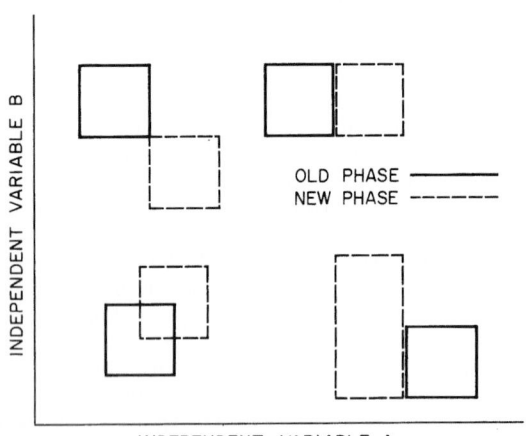

OLD PHASE ———
NEW PHASE -------

INDEPENDENT VARIABLE B

INDEPENDENT VARIABLE A

FIGURE 47.14 Possible relations of old phase and new phase.

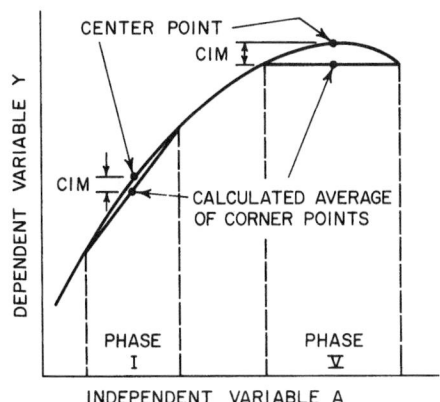

FIGURE 47.15   Cross section through a response surface. CIM indicates curvature.

in Figure 47.15. It is used in conjunction with the effects to indicate when a maximum (or a minimum) has been reached and to indicate the sensitivity of the response to changes in the independent factors. In the Taguchi literature the response would be termed "robust" to changes in the independent variable (Taguchi 1978). In rare cases it may happen that the first phase is located symmetrically about the maximum with respect to the two independent factors chosen. In this case the factors should be nonsignificant but the Change-in-mean may be significant.

Blocking.   A process response ordinarily changes slightly with time, reflecting changes in sources of raw material, changes in air temperature from day to night, and so on.

Runs made close together in time are expected to be more nearly alike than those over a longer interval. Blocking is used to minimize the trouble caused by temporal changes of this type. For the EVOP calculations shown here, a block is one cycle. Changes in average level that occur between cycles are completely eliminated from the estimated effects, as can be seen by adding a constant to the five runs of Cycle 3, and recalculating the effects. (The phase average is changed, of course.)

Multiple Dependent Responses.   So far the explanation has been in terms of a single dependent response. This is rather unrealistic, except for the profit variable. Most processes have several dependent responses that must be measured or calculated, such as yield, production rate, percent impurity, or pounds of byproduct. A calculation sheet is made for each dependent response, and statistical significance may be noted on one dependent sheet but not on others. In this case, it may be well to run another cycle or two to get more information on the other dependent responses before a move is made.

The most troublesome case occurs when the indicated directions for improvement of two responses (say production rate and percent impurity) do not agree. The EVOP program has brought information from the production process. Decisions as to what to do next now rest upon information supplied by the EVOP program coupled with information to be supplied by the subject matter experts.

BLOCKING THE 2^k FACTORIALS

Experimenters often find difficulty in maintaining a homogeneous experimental environment for all the experiments required in a 2^k factorial. For example, an experimenter might need 2 days to run the eight trials required in a 2^3 factorial. The question is how to choose the trials to be run each day so as not to disturb the estimates of the major effects of the three factors, i.e., how to "block" the design into two blocks of four runs each. Here blocking is accomplished by sacrificing the interaction estimate of least concern, i.e., the three-factor interaction. The procedure to be followed is illustrated in Table 47.29, part *a,* for a 2^3 factorial. First, the plus and minus signs of the 2^3 design are written down. Next, the columns of plus and minus signs commonly used to estimate the *ABC* interaction is constructed and labeled the block "generator." Those runs carrying a plus sign in the block generator column form the first block; those carrying a minus sign form the second block.

When this design is employed, the estimate of the three-factor interaction (abbreviated 3fi) cannot be distinguished from the block effect; the block effect and 3fi effect are "confounded." All other estimated effects are clear of the block effect.

TABLE 47.29 Partitioning the 2^3 Factorial Design

a. Partitioning the 2^3 into two blocks of four runs

2^3			ABC = block generator	2^3 in two blocks			
				(+) Block		(−) Block	
A	B	C		A B C		A B C	
−	−	−	−	+ − −		− − −	
+	−	−	+	− + −		+ + −	
−	+	−	+	− − +		+ − +	
+	+	−	−	+ + +		− + +	
−	−	+	+				
+	−	+	−				
−	+	+	−				
+	+	+	+				

b. Partitioning the 2^3 into four blocks of two runs

2^3			Block generators	
A	B	C	AB	BC
−	−	−	+	+
+	−	−	−	+
−	+	−	−	−
+	+	−	+	−
−	−	+	+	−
+	−	+	−	−
−	+	+	−	+
+	+	+	+	+

+ + Block	− + Block	+ − Block	− − Block
A B C	A B C	A B C	A B C
− − −	+ − −	+ + −	− + −
+ + +	− + +	− − +	+ − +

To partition the design into four blocks of two runs each, the proper block generators are provided by the two columns of plus and minus signs associated with the interactions AB and BC as illustrated in Table 47.29, part *b.* Note that the generators produce four combinations of minus and plus signs, each combination identifying a block of two runs. In this particular design all 2fi (two-factor interactions) are confounded with blocks.

The block generators must be carefully chosen. The blocking arrangements for the 2^3, 2^4, and 2^5 designs appear in Table 47.30. A more complete table and description of factorial design blocking appears in Box, Hunter, and Hunter (1978).

FRACTIONAL FACTORIAL EXPERIMENTS (EACH FACTOR AT TWO LEVELS)

If there are many factors, a complete factorial experiment, requiring all possible combinations of the levels of the factors, involves a large number of tests—even when only two levels of each fac-

TABLE 47.30   Blocking Arrangements for the 2^k Factorials

k = number of factors	Block size	Block generators	Interactions confounded with blocks
3	4	*ABC*	*ABC*
	2	*AB,BC*	*AB,BC, AC*
4	8	*ABCD*	*ABCD*
	4	*ABC, ACD*	*ABC, ACD,BD*
	2	*AB,BC,CD*	all 2fi and 4fi
5	16	*ABCDE*	*ABCDE*
	8	*ABC,CDE*	*ABC,CDE, ABDE*
	4	*ABC,BCD,CDE*	*ABC,BCD,CDE, AD, ABDE, BE, ACE*
	2	*AB,BC,CD,DE*	all 2fi and 4fi

tor are being investigated. In these cases, it is useful to have a plan that requires fewer tests than the complete factorial experiment. The fraction is a carefully prescribed subset of all possible combinations. The analysis of fractional factorials is relatively straightforward, and the use of a fractional factorial does not preclude the possibility of later completion of the full factorial experiment.

Confounding (Aliasing, Biasing).   In a complete factorial experiment we have 2^k experimental trials. The 2^k experiments can be used to give independent estimates of all 2^k effects. In a fractional factorial (say the fraction $1/2^p$) there will be only 2^{k-p} experiments, and therefore only 2^{k-p} independent estimates are possible. In designing the fractional plans (i.e., in selecting an optimum subset of the 2^k total combinations), the goal is to keep each of the 2^{k-p} estimates as unbiased or "clear" as possible, i.e., to keep the estimates of main effects and if possible second-order interactions mutually unbiased, or nearly so.

To explain, consider the following 2^{3-1} fractional (the one-half, 2^{-1}, of the 2^3 factorial):

A B C	Observed
− − +	$y_1 = 8$
+ − −	$y_2 = 11$
− + −	$y_3 = 9$
+ + +	$y_4 = 14$

The main effects are given by the statistics $\bar{y}_+ - \bar{y}_-$, where once again the plus and minus subscripts of each letter in the design identify the observations entering each average. Thus, the main effect of *A* is estimated to be $(11 + 14)/2 - (8 + 9)/2 = 4.0$. The main effects of *B* and *C* are, respectively, $(9 + 14)/2 - (8 + 11)/2 = 2$ and $(8 + 14)/2 - (11 + 9)/2 = 1.0$. Now consider the estimate of the two-factor interaction *AB*. The analyst will find that the signs required to estimate the *AB* interaction are identical to those already employed to estimate the main effect of *C*. The main effect of *C* and the two-factor interaction *AB* are *confounded.* Said another way, the statistic $\bar{y}_+ - \bar{y}_- = 1.0$ has an "alias" structure; that is, the statistic may be identified as either *C* or *AB*. In fact, the expected value of the statistic equals *C*+*AB,* the sum of the two effects, and in the absence of clear information on the main effect of *C,* we cannot tell whether the *AB* effect is plus, minus, large, or small. The reader will note that estimate *A* is confounded with *BC,* as is *B* with *AC.*

When some or all main effects are confounded with two-factor interactions, the fractional factorial design is said to be of "Resolution III." When one or more of the main effects are confounded with (at least) three-factor interactions, the fractional is said to be a "Resolution IV" design. Fractionals with main effects confounded with (at least) four-factor interactions are of "Resolution V," etc. (See Box and Hunter 1959).

DESIGNING A FRACTIONAL FACTORIAL DESIGN

Let N equal the number of runs and k the number of factors to be investigated. When $N = 2^k$, we have a full factorial design. When $N = 2^{k-p}$, we have a $(1/2)^p$ replicate of the 2^k factorial; for example, a 2^{7-3} is a one-eighth replicate of a 2^7 factorial and contains 16 runs.

To design a one-half replicate design in N runs, first write down (Yates order is best) the full factorial design in $(k - 1)$ factors. Next write down the column of signs associated with the highest-order interaction. These signs are now used to define the versions of the kth factor. For example, to construct the 2^{4-1} design, begin with a 2^3 factorial in factors A, B, and C as illustrated in Table 47.31. Next to the columns for A, B, and C write down the column of signs associated with the ABC interaction. Use these signs to identify the two versions of factor D. (The other one-half fraction is obtained by reversing the signs of the column ABC.)

TABLE 47.31   Constructing the 2^{4-1} Fractional Factorial*

Generator				Principal design				Alternative design†			
A	B	C	ABC = D	A	B	C	D	A	B	C	D
−	−	−	−	−	−	−	−	−	−	−	+
+	−	−	+	+	−	−	+	+	−	−	−
−	+	−	+	−	+	−	+	−	+	−	−
+	+	−	−	+	+	−	−	+	+	−	+
−	−	+	+	−	−	+	+	−	−	+	−
+	−	+	−	+	−	+	−	+	−	+	+
−	+	+	−	−	+	+	−	−	+	+	+
+	+	+	+	+	+	+	+	+	+	+	−

*Example run: Run no. 2 requires the experimenter to hold factor A at +, factor B at −, factor C at −, and factor D at +.
†The alternative fraction is obtained by reversing the signs of the ABC vector, that is, by setting $D = -ABC$.

To construct a one-quarter replicate design, two columns of signs are required in addition to the standard factorial in N runs; the one-eighth replicate design requires three additional columns, etc. The columns of signs to be used must be carefully chosen; they are listed in Table 47.32 for designs up to $k = 7$ factors. Table 47.32 is an adaptation of a much more extensive table given in Box, Hunter, and Hunter (1978). Extensive listings of fractional factorial designs can also be found in Diamond (1989).

Most design of experiment software programs provide two-level fractional factorial designs of any desired resolution along with the alias-confounding patterns associated with each estimated effect. Given below is an example of the construction of a 2^{4-1} and design and analysis of a 2^{6-2} factorial.

Example.   To construct a fractional factorial design for $k = 6$ factors in $N = 16$ runs, first write down the full factorial 2^4 design in factors A, B, C, and D. Consulting Table 47.32, the vectors of plus and minus signs associated with the interaction ABC are now used to define the versions of factor E. The signs of the BCD interaction are similarly used to define the versions of factor F. The completed 2^{6-2} design is displayed in Table 47.33 along with observed responses, Yates' algorithm, and identified effects.

TABLE 47.32 Vectors Used for the Construction of Fractionals

Number of runs N	Number of factors k				
	3	4	5	6	7
4	2^{3-1} $\pm AB = C$	NA	NA	NA	NA
8		2^{4-1} $\pm ABC = D$	2^{5-2} $\pm AB = D$ $\pm AC = E$	2^{6-3} $\pm AB = D$ $\pm AC = E$ $\pm BC = F$	2^{7-4} $\pm AB = D$ $\pm AC = E$ $\pm BC = F$ $\pm ABC = G$
16			2^{5-1} $\pm ABCD = E$	2^{6-2} $\pm ABC = E$ $\pm BCD = F$	2^{7-3} $\pm ABC = E$ $\pm BCD = F$ $\pm ACD = G$
32				2^{6-1} $\pm ABCDE = F$	2^{7-2} $\pm ABCD = F$ $\pm ABDE = G$
64					2^{7-1} $\pm ABCDEF = G$

TABLE 47.33 A 2^{6-2} Resolution IV Fractional Factorial and Associated Yates Analysis

Generators: $E=ABC$ and $F=BCD$
Defining relation: $I=ABCE=BCDF=ADEF$

$A\,B\,C\,D\,E\,F$	Obs.*	Yates algorithm				Effects	Identification†
$-\;-\;-\;-\;-\;-$	124	271	541	1137	2405	150.3125	Average
$+\;-\;-\;-\;+\;-$	147	270	596	1268	11	1.375	$A + BCE + ABCDF + DEF$
$-\;+\;-\;-\;+\;+$	145	284	615	-1	35	4.375	$B + ACE + CDF + ABDEF$
$+\;+\;-\;-\;-\;+$	125	312	653	12	-139	-17.375	$AB + CE + ACDF + BDEF$
$-\;-\;+\;-\;+\;+$	138	307	3	27	93	11.625	$C + ABE + BDF + ACDEF$
$+\;-\;+\;-\;-\;+$	146	308	-4	8	-1	-0.125	$AC + BE + ABDF + CDEF$
$-\;+\;+\;-\;-\;-$	162	323	3	-63	35	4.375	$BC + AE + DF + ABCDEF$
$+\;+\;+\;-\;+\;-$	150	330	9	-76	169	21.125	$ABC + E + ADF + BCDEF$
$-\;-\;-\;+\;-\;+$	125	23	-1	55	131	16.375	$D + ABCDE + BCF + AEF$
$+\;-\;-\;+\;+\;+$	182	-20	28	38	13	1.625	$AD + BCDE + ABCF + EF$
$-\;+\;-\;+\;+\;-$	181	8	1	-7	-19	-2.375	$BD + ACDE + CF + ABEF$
$+\;+\;-\;+\;-\;-$	127	-12	7	6	-13	-1.625	$ABD + CDE + ACF + BEF$
$-\;-\;+\;+\;+\;-$	168	57	-43	29	-17	-2.125	$CD + ABDE + BF + ACEF$
$+\;-\;+\;+\;-\;-$	155	-54	-20	6	13	1.625	$ACD + BDE + ABF + CEF$
$-\;+\;+\;+\;-\;+$	154	-13	-111	23	-23	-2.875	$BCD + ADE + F + ABCEF$
$+\;+\;+\;+\;+\;+$	176	22	35	146	123	15.375	$ABCD + DE + AF + BCEF$

*Obs.=observations.

†Expected value of the effect from the defining relation.

Identifying the Estimates. The 2^{6-2} design was generated by setting $E = ABC$ and $F = BCD$. A simple procedure for identifying the biases (aliases) of the effects estimable from this design is as follows. Multiply the expression $E=ABC$ by E and the expression $F=BCD$ by F. This gives $E^2 = ABCE$ and $F^2 = BCDF$. Now adopt the rule that whenever a symbol appears squared, it is replaced by an I, the "identity," a symbol equivalent to the numeral 1. We now have for the design "generators" $I = ABCE$ and $I = BCDF$. Multiplying the generators together gives the *defining relation* $I = ABCE = BCDF = AB^2C^2DEF$, which reduces to $I = ABCE = BCDF = ADEF$.

When Yates' algorithm is applied to the 16 runs of the 2^{6-2}, the algorithm estimates 15 effects and provides each with its initial name, as illustrated in Table 47.33. The defining relation is now employed to determine the additional names (aliases or biases) of each of these statistics. Thus, the statistic labeled the "main effect" of A actually equals $A = BCE = ABCDF = DEF$, an expression obtained by multiplying through the defining relation by the symbol A. Similarly, the statistic initially called the "ABC interaction" actually estimates $ABC = E = ADF = BCDEF$. The estimates and their full names are given in Table 47.33.

Five of the estimates appear unusually large and are good candidates for measured phenomena distinguishable from natural variability (noise). Using only their first- and second-order names we have: -17.375 estimates $AB + CE$, 11.625 estimates C, 21.125 estimates E, 16.375 estimates D, and 15.375 estimates $DE + AF$. A reasonable interpretation of these statistics is that factors C, D, and E have detectable important influences upon the response over their studied ranges, while factors A and B do not. This conclusion obviously needs confirmation, but it represents a good first guess. The 2^{6-2} design now collapses into a 2^3 factorial repeated in factors C, D, and E.

OTHER FRACTIONAL FACTORIALS

Although the 2^{k-p} fractional factorial designs discussed here are the most frequently used, many other fractionals exist. For example the Plackett and Burman (1966) designs are two-level fractional factorials whose number of runs N is not a power of 2 but a multiple of 4. The $N = 12$ design for $k \leq 11$ factor design is displayed in Table 47.34. The templates for producing the designs for $N = 20, 24, 28,$ and 36 can be found in Box and Draper (1987) and Myers and Montgomery (1995).

All Plackett and Burman designs are Resolution III. The alias structure associating main effects and two factors interactions is not as readily available as those of the regular 2_{III}^{k-p} designs. However, *every* Resolution III design can be made into a Resolution IV design by the principle of "fold-over."

TABLE 47.34 A Plackett and Burman Design
($k = 11$ factors, $N = 12$ runs)

A	B	C	D	E	F	G	H	I	J	K
+	−	+	−	−	−	+	+	+	−	+
+	+	−	+	−	−	−	+	+	+	−
−	+	+	−	+	−	−	−	+	+	+
+	−	+	+	−	+	−	−	−	+	+
+	+	−	+	+	−	+	−	−	−	+
+	+	+	−	+	+	−	+	−	−	−
−	+	+	+	−	+	+	−	+	−	−
−	−	+	+	+	−	+	+	−	+	−
−	−	−	+	+	+	−	+	+	−	+
+	−	−	−	+	+	+	−	+	+	−
−	+	−	−	−	+	+	+	−	+	+
−	−	−	−	−	−	−	−	−	−	−

To "fold over" a design one merely writes it down again with all signs reversed. A fold-over design combined together with its original Resolution III design forms a design of Resolution IV.

Fractional factorials are not limited to the 2^{k-p} designs. The 3^3 factorial design can be reduced into a variety of fractions including the Latin Square, see Hunter (1985a) for graphical displays. Mixed level fractional factorials are also available; see Addelman (1962). Many novel fractionals have been published: Hahn and Shapiro (1966), Webb (1968a, b), Margolin (1969), Anderson and Thomas (1978), and Rechtschaffner (1967). Fractional factorials when the number of runs is less than the number of variables, "supersaturated" designs, are also possible; see Booth and Cox (1962) and Lin (1993, 1995).

Several design of experiments computer software programs will construct unique fractional factorial designs, allowing the experimenter to not only choose the number of runs, but also the number of factors and levels. The reader can only be warned that the application of designs with small numbers of runs and many factors assumes great simplicity in the mathematical model for the response function under study, most particularly that only main effects exist and that all interactions are either zero or truly near zero. One is also well advised to leave some redundancy in one's design, i.e., extra runs to provide degrees of freedom that can be employed in estimating the experimental error variance σ^2 [see Snee (1985) and Berk and Picard (1991)]. (There has never been a signal in the absence of noise, and one should plan on measuring the noise as well as possible signals.

Screening Experiments.   The saturated 2_{III}^{k-p} designs for studying k factors in $N = k + 1$ runs can be of value in early screening efforts to detect important factors among many candidates. The possible biasing influences of interactions is very serious in these applications and should always be kept in mind. Certain software programs can construct fractional factorial designs to provide estimates of all k main effects and certain prechosen two-factor interactions. However, the assumption that all interactions of importance can be announced prior to the experimental program being designed and run can be naive when one considers the number of such interactions that may be possible. Furthermore, the discovery of interactions can easily be as important as the identification of main effects. Conservative practice in the use of fractional factorials generally requires designs of Resolution IV, i.e., the ability to separate main effects from two-factor interactions. See Hurley (1994); Tang and Tang (1994); and Haaland and O'Connell (1995).

Orthogonal Arrays.   The terminology "orthogonal array," used by the earliest creators of balanced block experimental designs, has been popularized by Taguchi (1987); see below. All the 2^k, 2^{k-p}, Plackett and Burman, and Latin square type designs can be called "orthogonal arrays." The number of runs associated with each orthogonal array is often identified by the notation $L(N)$, as for example the $L8$ orthogonal array is the 2^3 (or 2^{7-4}) two-level factorial and the $L9$ and the $L27$ are the 3^2 and 3^3 factorials. The $L36$ can be viewed either as a three-level design or as a 6×6 Latin square.

The classification "orthogonal array" is appropriate to any experimental design that can provide estimates of effects having zero correlations. The designs are sometimes described as "main effect clear" designs, although they are often adapted to take into account certain interactions. The 2^{k-p} and 3^{k-p} fractionals, the Latin square designs (and the Graeco-Latin and Hyper-Graeco-Latin square designs) thus qualify, as do the mixed-level factorials and fractional factorials and the balanced block designs. The Box and Behnken (1960) designs form novel fractions of the three-level orthogonal arrays. The terminology "orthogonal arrays" recognizes the geometric multidimensional nature of all these designs; that is, in the N-dimensional space of the observations, the vectors representing the effects to be estimated are all mutually perpendicular. One orthogonal array design popularized by Taguchi is the $L9$ design, the 3^{4-2} for studying four factors each at three levels in nine runs. A critique of the application of this and other three-level orthogonal arrays is found in Hunter (1985).

In listing orthogonal array designs the Taguchi literature uses the notation (1, 2) and (1, 2, 3) to identify the levels (versions) of each variable instead of the geometric notation (-1, $+1$) and (-1, 0, $+1$). Examples of the analysis of orthogonal arrays employing the Taguchi terminology and methodology are provided by Barker (1990, 1994), Kacker (1985), Phadke et al. (1983), Phadke (1989), and Taguchi (1978, 1987). These authors pay particular attention to the use of "inner" and "outer" orthogonal arrays, or in a parallel terminology, to "design" and "noise" matrices. The designs are similar in structure to the classical split-plot designs.

To construct unique designs for estimating main effects and *selected* interactions, Taguchi employs "linear graphs" associated with each orthogonal array; see Kacker and Tsui (1990), Wu and Chen (1991), and Wu, Mao, and Ma (1990). Linear graphs provide a geometric analogue to the use of fractional factorial defining relations (see Identifying the Estimates under Designing a Fractional Factorial Design, above). Almost always, classical methods employing defining relations can provide experimental designs identical to or better than those provided by the application of linear graphs.

TAGUCHI OFF-LINE QUALITY CONTROL

The Japanese engineer–quality expert Genechi Taguchi must be credited with much of today's interest in the use of factorial and fractional factorial designs on the part of the automotive, communication, and assembly industries; see Taguchi (1978). Within these industrial environments experiments are run to identify the settings of both product design parameters and process variables that will simultaneously provide a manufactured item whose response is robust to process variability while meeting the customer's product expectations and possible environmental challenges. The adaptation of statistical experimental design to these objectives has its origins in Taguchi's early work in the communications industries in Japan in the 1950s. The strategy is called "parameter design" or "robust design." It is important to note that the word "design" takes differing connotations: product design, process design, and statistical design.

Taguchi requires manufactured products be created to meet the following criteria:

1. To protect the product from sources of variability occurring within the manufacturing process
2. To have minimum variability about the customer's target values
3. To be robust to environmental factors encountered by the customer

More formally, a product's response y is considered to be a function of "controllable" factors x and "noise" factors z. The objective is to choose settings of x that will make the product's response y insensitive to variability associated with both x and z and still meet target specifications with least variability.

Inner and Outer Arrays.   The statistical designs associated with the Taguchi approach to product and process design usually contain both an "inner" and "outer" array, or "design matrix" and "noise matrix," each constructed from the orthogonal arrays. (See previous section.) The inner array consists of a statistical experimental design employing the controllable factors x, while the outer array is a statistical experimental design in the noise factors z which are now intentionally varied. (Occasional x factors may also be included as "noise" factors in an outer array.) The entire design forms a split-plot-like experimental array with the z outer array repeated within each of the settings of the x inner array. For example, if the inner array were a $L16 = 2^{8-4}$ and the outer array a $L9 = 3^{4-2}$ there would be a total of $16 \times 9 = 144$ experiments, each of the 16 runs of the $L16$ containing its own 9-run $L9$ design. Experimental designs providing for inner and outer arrays that employ fractional factorial arrangements are also possible; see Shoemaker, Tsui, and Wu (1991) and Montgomery (1991).

At each setting of the inner array Taguchi now determines a "signal to noise" statistic composed from the outer array, noise matrix, observations. Taguchi focuses on a quadratic loss function $Q = (\eta - \theta)^2 + \sigma^2$, where η is the expected product response, θ the target value (the quantity $\eta - \theta$ is bias), and σ^2 the variance of the observed responses. To aid in minimizing the loss function, Taguchi defines the "signal to noise" ratio SN, where commonly SN $= 10 \log_{10} (\eta/\sigma)^2$, $\eta = E(y)$, and $\sigma^2 = \mathrm{Var}\ (y)$. At each setting of the inner array the statistic SN $= 10 \log_{10} (\bar{y}/s)^2$ is computed using the n observations from the outer array occurring only at that setting. Other definitions for

SN are also suggested. For example, when a higher response is preferred, Taguchi proposes $SN = 10 \log_{10}[(y_1^2 + y_2^2 + \cdots + y_n^2)/n]$ and, for lower desired response, the statistic $SN = 10 \log_{10}[(1/y_1^2 + 1/y_2^2 + \cdots + 1/y_n^2)/n]$, where $y_1, y_2, \ldots, y_n$ are the n observations from the outer array unique to each setting of the inner array.

Most statisticians recommend that the averages and estimated variances obtained at each of the points of the inner array be separately analyzed. Members of the Taguchi school continue to recommend the analysis of the various signal to noise statistics. No closure to the debate seems imminent. One thing is clear. The fraternity of quality engineers and statisticians is indebted to Prof. Taguchi for proposing the concept of the design of robust products (parameter design) and for adapting the arts of statistical design of experiments to that end use.

A large body of literature exists describing and offering examples of the Taguchi approach. An excellent summary and critique of the methodology identified with Prof. Taguchi, and possessing an extensive bibliography, appears in a discussion organized by Nair (1992). Major authors identified with the Taguchi approach are Taguchi (1978, 1986, 1987); Kacker (1985); Kacker and Shoemaker (1986); Kacker and Tsui (1990); Barker (1990, 1994); Phadke (1989); Phadke et al. (1983); Leon, Shoemaker, and Kacker (1987); and their various coworkers. Authors who have discussed the Taguchi work include Bisgaard (1993b), Box (1988), Box and Jones (1986), Box and Myers (1986), Easterling (1985), Goh (1993), Grove and Davis (1991b), Hunter (1985a, 1989), Hurley (1994), Lucas (1985), Miller et al. (1993), Montgomery (1991), Nair and Shoemaker (1990), Stephens (1994), Tribus and Sconyi (1989), and Vining and Myers (1990).

RESPONSE SURFACE DESIGNS

Response Surface Methodology (RSM) has been successfully used to optimize many different kinds of industrial units, processes, and systems. It is an experimental approach and has been applied in research and development laboratories and sometimes on actual plant equipment itself. In the latter situation, however, Evolutionary Operation is often more appropriate. Evolutionary Operation is an alternative form of RSM that is useful for both objectives of screening and optimizing.

RSM experimental designs require that important factors influencing a process, identified perhaps by a screening experiment, be varied in a carefully chosen pattern of experiments. Commonly two controlled variables and a single response variable are studied. The data obtained are then analyzed with the primary objective of providing a rough map (usually a contour representation) of the response surface over the region of the controlled variables investigated. The mathematical models employed are the first-order and second-order polynomials. Thus, the fitted response surface may be planar (a first-order approximation to the "true" surface) or nonplanar or curved (a second-order approximation). The fitted models are obtained using ordinary least squares estimation procedures (regression analysis). Often a fitted response surface will suggest alternative levels of a factor to provide better yields. Thus, a program of RSM may go through several stages of mapping before "best" conditions are identified. When more than two controlled variables are studied, contour surfaces are employed. See Box and Draper (1987). Nor is it necessary to map only a single response. Two or more responses can be separately mapped and their maps superimposed to identify regions of "optimum" operability. Finding "optimum" operating conditions does not always mean finding the factor settings that give the biggest or smallest response. Suggestions for further reading include Box, Hunter, and Hunter (1978); Box and Draper (1987); Khuri and Cornell (1987); Mason, Gunst, and Hess (1989); Haaland (1989); and Myers and Montgomery (1995). A history of RSM appears in Myers, Khuri, and Carter (1989).

Modern computers and software programs have made RSM a most useful and valuable statistical tool. Not only are the burdens of computation minimized, but the ability of computers to display maps of the fitted response surfaces provides the analyst with vivid insights into the nature of the responses and factors under investigation. Most software programs will allow an experimenter to obtain a first- or second-order mapping of an unknown response surface employing almost any collection of data; all that is needed is a good least-squares regression program. However, good experimentation requires a careful selection of points. A poorly designed response surface program is analogous to viewing a scene through an astigmatic lens. The consequence is a warped view. In most circumstances a first- or second-order rotatable (nonastigmatic) response surface design offers the best strategy. Fitting response surfaces to haphazardly acquired data is called PARC analysis by some statisticians (PARC: Practical Accumulated Records Computations, or Planning After Research is Completed.) The use of the standard first- and second-order RSM designs for $k = 2$ factors and a single response is described here.

In implementing RSM, a number of statistical procedures discussed in other sections of this Handbook are used. The concept of RSM was first developed and described by Box and Wilson (1951). At first RSM was used primarily as an experimental optimization technique in the chemical industry. Since then, however, it has found application in many other fields (see Hill and Hunter 1966).

RSM can be usefully regarded as consisting of two stages:

1. First-order stage, in which a first-order mathematical model is contemplated, a factorial or other first-order design performed, the data fitted, the contours of the response drawn, and the direction of steepest ascent determined and pursued

2. Second-order stage, in which a second-order mathematical model is contemplated, a central-composite or other second-order design performed, the data fitted, the contours drawn, a canonical analysis performed, and an optimum located

Response Surface Methodology is actually more flexible than these brief definitions indicate. A skeletal outline, which shows some of the possible paths through an RSM study, is given in flow-diagram form in Figure 47.16.

Weakness of One-Variable-at-a-Time Approach.
A popular method of experimentation is the one-factor-at-a-time approach. Each factor, in turn, is varied while all the rest of the factors are held at some fixed, constant levels. One trouble with this approach is that a false optimum can be reached. Consider the following hypothetical illustration.

Example.
Under study is a chemical reaction in which there are two factors of interest, the concentration of one of the reactants and the reaction time. What settings for these two factors will maximize the yield? The best known settings, at the outset of the investigation, are a concentration of 25 percent and a time of 1 h (see Figure 47.17).

Following a one-factor-at-a-time approach, the engineer first runs a series of experiments by varying the time, while holding the concentration at 25 percent. The results show that a maximum yield of about 65 percent is obtained when the time is 1.9 h (position E on the line A to B in Figure 47.17). Holding the time fixed at this value, varying concentration along the line C to D, and obtaining a maximum at 25 percent, the engineer reaches the conclusion that the maximum yield (65 percent) is achieved when the concentration is 25 percent and the time is 1.9 h. This conclusion, however, is incorrect.

Response Surface Approach.
The actual situation, unknown to the experimenter, is shown in Figure 47.17. Here the yield is shown as a function of both concentration and time. The solid curved lines in the figure are contour lines of constant yield. For example, there is an entire set of conditions of concentration and time that give an 80 percent yield. The contour surface can be viewed as a mountain; the peak of the mountain is the point P. The contours of 90, 80, and so forth, can be viewed as altitudes. These numbers represent the percentage yields.

The engineer's objective was to find those settings for the concentration and time that would give the maximum yield. Viewed geometrically, what the engineer was trying to do was to climb to the highest point on the mountain. The attempt failed for a fairly simple reason.

Figuratively speaking, by varying time, the engineer first traversed the hill going along a path from point A to point B (see Figure 47.17). Between A and E the path led up the mountain, but then

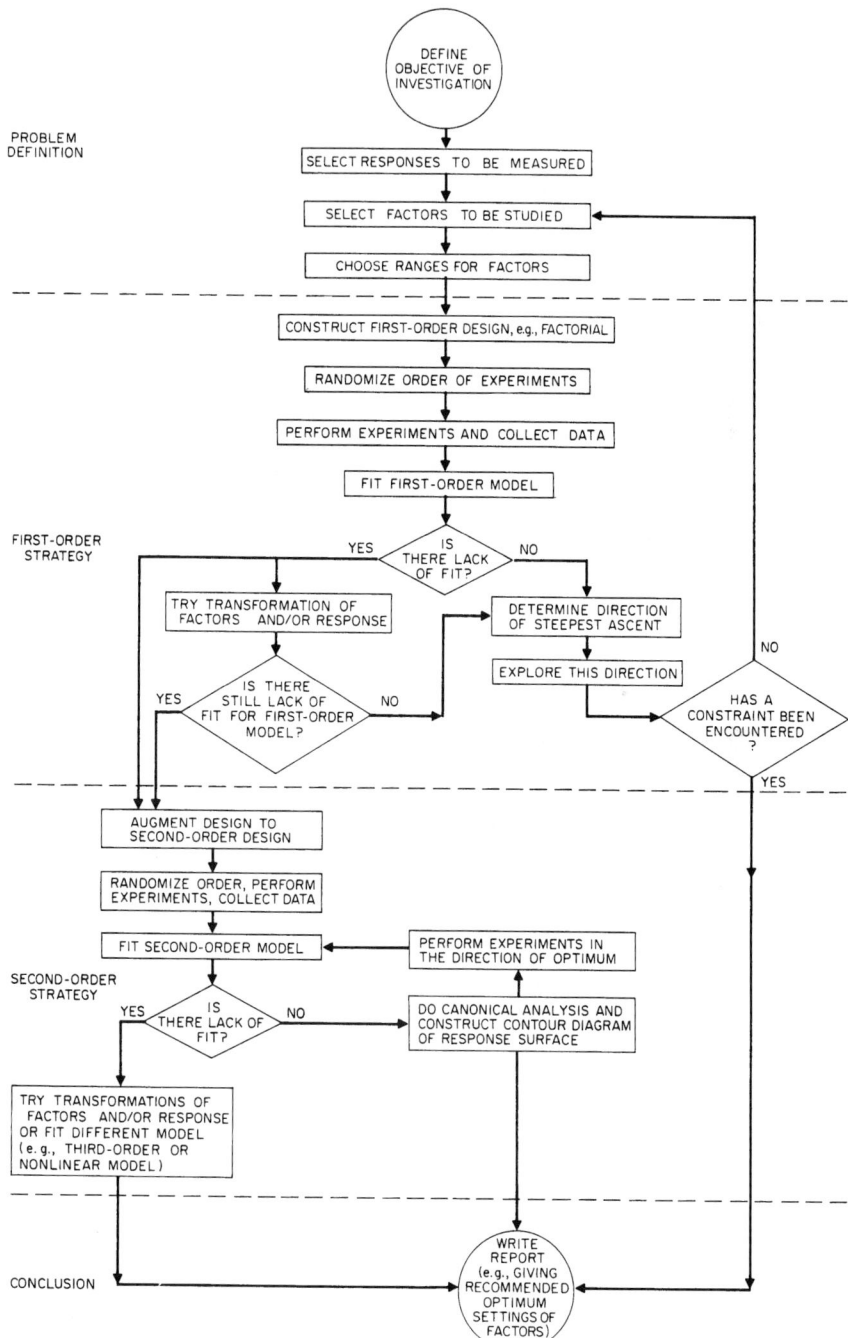

FIGURE 47.16   Outline of main ideas of Response Surface Methodology.

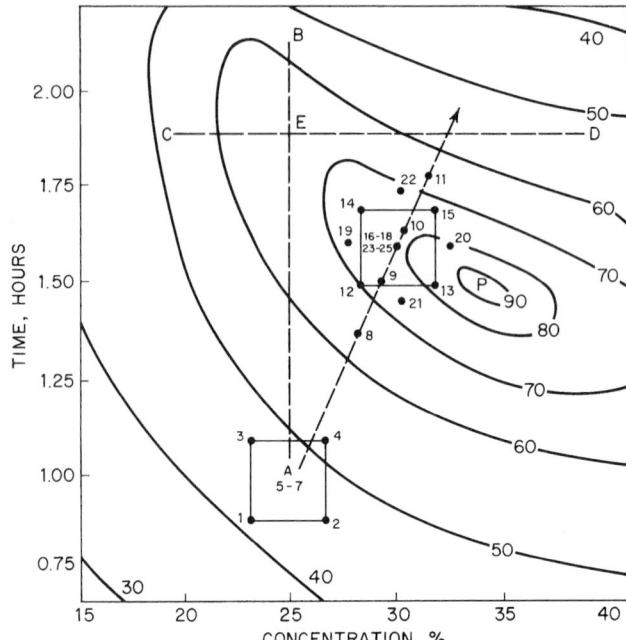

FIGURE 47.17 Response surface showing yield of a chemical reaction as a function of concentration and time.

at point E it started to go down the other side. From point E to point B the engineer was walking down the other side of the hill. The traverse for varying concentration (C to E to D) is shown.

The experimenter has achieved a yield of only 65 percent (at E), whereas a yield in excess of 90 percent (at P) is possible. This higher yield can be achieved by *simultaneously* increasing concentration and decreasing time from the experimenter's reported "optimum" values.

If the contours of the hill were circular and there were no experimental error, this one-at-a-time-procedure would have taken the engineer to the highest point on the hill. In general, the contours of real response surfaces are not circular nor experimental error (noise) absent, and thus what is needed is a more sophisticated experimental strategy such as RSM.

Beginning of Program. The RSM approach (see Figure 47.16) will now be applied to the example of maximizing the yield.

Define Objective of Investigation. It is of the utmost importance to define clearly the objective of the study to be undertaken. It is surprising how often in practice this step is either ignored or not given the careful attention it deserves. This often leads to difficulties later on. In the present example the objective is to maximize the yield. The objective, in general, may involve multiple criteria, that is, to maximize yield while simultaneously meeting other objectives such as minimizing impurity and obtaining an acceptable range of viscosity.

Select Factors and Ranges. The next step is to select the factors to be studied together with the ranges over which they are to be studied. It is necessary to understand the technical aspects of the experimental situation for this to be done intelligently. The specific *scale* over which each factor is to be studied must also be chosen. For example, instead of varying time linearly in units of hours, the experimenter might choose the basic scale to be the logarithm of the number of hours. In the present example, the variables concentration and time are selected. Initially, it is decided to vary concentration from 23 to 27 percent and time from 0.9 to 1.1.

First-Order Strategy

Construct Design and Collect Data. The 2^2 factorial design with three center points, shown in Tables 47.35, is constructed. [Further discussion of the number of center points and other matters on setting up the design is given in Cochran and Cox (1957), Hunter (1959), Box and Draper (1987), and Myers and Montgomery (1995).] The order of the seven runs is randomized, the experiments are performed, and the results shown in Table 47.35 are obtained. The results are displayed in Figure 47.18.

Fit First-Order Model and Check for Lack of Fit. The analysis of these results can be carried out in either one of two equivalent ways. The effects and interaction of the factorial design can be calculated with their associated 95 percent confidence intervals, as is also shown in Table 47.35.

TABLE 47.35 Results of First-Order Design

Run number	X_1 = concentration		X_2 = time		Y = yield
	%	Coded units	Hours	Coded units	%
1	23	-1	0.9	-1	43.7
2	27	$+1$	0.9	-1	44.5
3	23	-1	1.1	$+1$	47.2
4	27	$+1$	1.1	$+1$	51.8
5	25	0	1.0	0	46.8
6	25	0	1.0	0	45.9
7	25	0	1.0	0	45.3

Calculation of main effects

Concentration: $(-Y_1 + Y_2 - Y_3 + Y_4)/2 = (-43.7 + 44.5 - 47.2 + 51.8)/2 = 2.7$
Time: $(-Y_1 - Y_2 + Y_3 + Y_4)/2 = (-43.7 - 44.5 + 47.2 + 51.8)/2 = 5.4$
Interaction: $(+Y_1 - Y_2 - Y_3 + Y_4)/2 = (43.7 - 44.5 - 47.2 + 51.8)/2 = 1.9$
Curvature: $(+Y_1 + Y_2 + Y_3 + Y_4)/4 - (Y_5 + Y_6 + Y_7)/3 = 46.8 - 46.0 = 0.8$

Calculation of confidence intervals

Concentration: $\pm 2ts/\sqrt{n} = \pm 2(4.30)(0.755)/\sqrt{4} = \pm 3.25$
Time: $\pm 2ts/\sqrt{n} = \pm 2(4.30)(0.755)/\sqrt{4} = \pm 3.25$
Interaction: $\pm 2ts/\sqrt{n} = \pm 2(4.30)(0.755)/\sqrt{4} = \pm 3.25$
Curvature: $\pm ts\sqrt{(1/n) + (1/n_0)} \equiv \pm(4.30)(0.755)\sqrt{(¼) + (⅓)}$
$= \pm 2.48$

Effects	Calculated 95% confidence interval, % yield
Concentration	2.7 ± 3.25
Time	5.4 ± 3.25
Interaction	1.9 ± 3.25
Curvature	0.8 ± 2.48

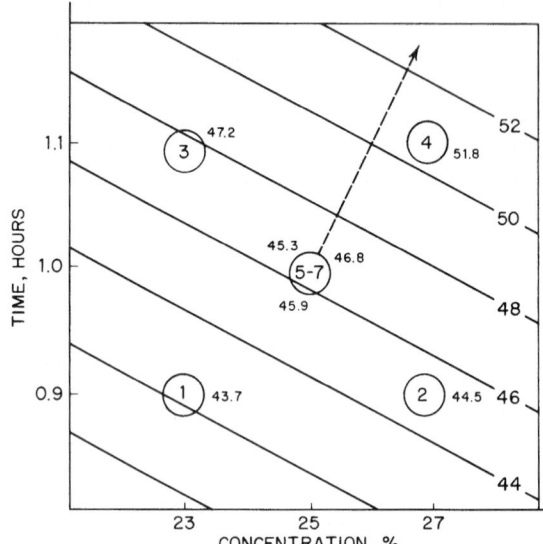

FIGURE 47.18   Results of first-order design with fitted first-order (planar) response surface.

As the center-point conditions have been repeated three times, an estimate of the variance can be readily obtained. (If repeat runs have not been performed, it might be possible to obtain an appropriate estimate in some other way, for example, from some external source, past experience, or from a technique like half-normal plots; see Daniel 1959.) With the three values 46.8, 45.9, and 45.3, $s^2 = 0.57$ is calculated as an estimate of the variance of an individual observation with two degrees of freedom.

Employing the estimated standard deviation $s = 0.755$ with two degrees of freedom, the confidence intervals suggest that the main effects of concentration and time are significant while the estimates of second-order effects (interaction and curvature) are indistinguishable from zero. We are thus able to employ as an approximation to the unknown response surface the first-order model

$$Y = \beta_0 + \beta_1 X_1 + \beta_2 X_2 + \epsilon$$

where Y is the observed response and the β's are coefficients to be estimated from the data. The quantities X_1 and X_2 are independent variables representing the experimental factors concentration and time, where

$$X_1 = \frac{\text{concentration(\%)} - 25}{2}$$

$$X_2 = \frac{\text{hours} - 1.0}{0.1}$$

These expressions for X_1 and X_2 code the original settings of concentration and time to match those of the 2^2 factorial design with center point given in Table 47.37 and displayed in Figure 47.18. For example, when concentration = 23 and time = 0.9 hours, $X_1 = -1$ and $X_2 = -1$. The quantity ϵ in the model is assumed to be a random error normally distributed, independent, with constant variance σ^2. Although standard regression techniques (see Section 44) can be used to fit the first-order model, estimates of the β's are readily obtained as a consequence of the 2^2 factorial design with center point. The estimate of β_0 is is the average, the estimate of β_1 one half the concentration effect, and the estimate of β_2 one half the time effect. Thus the fitted equation becomes

$$\hat{Y} = 46.46 + 1.35X_1 + 2.70X_2$$

The equation in terms of the original variables is

$$\hat{Y} = 28.23 + 0.675[\text{conc}\%] + 1.35[\text{hours}]$$

Setting $\hat{Y} = 44.0$ will produce the straight line contour labelled 44 in Figure 47.18. The fitted contours suggest the response surface to be well represented by the plane.

$$\hat{Y} = B_0 + B_1X_1 + B_2X_2 = 46.46 + 1.35X_1 + 2.70X_2$$

in the X_1, X_2 coordinate system.

A second method for evaluating the fit is to use the analysis of variance. The resulting ANOVA table (Table 47.36) indicates that the first-order model (above equation) adequately fits the data. (See Section 44; also Draper and Smith 1981.) The ratio of the lack of fit mean square divided by the pure error mean square is 4.13, and since this value is less than $F_{2,2}(0.95) = 19.0$, there is no evidence of lack of fit of the first-order model. Since there is no evident lack of fit, it is reasonable to study the implications of the fitted first-order model (above equation). The plane described by this equation is represented in Figure 47.18 by the straight contour lines.

Determine Direction of Steepest Ascent. The direction of steepest ascent is indicated in Figure 47.18. (For further details on direction of steepest ascent, see Cochran and Cox 1957, p. 357, and Box and Draper 1987.) It is perpendicular to the contour lines. Four experiments (numbers 8 to 11) in this direction indicate that the center of a second design should be approximately at a concentration of 31 percent and a time of 1.6 h. The design employed and the data obtained after performing the runs in random order are shown in Table 47.37 as runs 12 to 18. An analysis of the data shows apparent lack of fit (Table 47.38). The ratio of the lack of fit mean square divided by the pure error mean square is 26.8, and since this value is greater than $F_{2,2}(0.95) = 19.0$, there is evidence of lack of fit of the first-order model.

Second-Order Strategy

Construct Design and Collect Data. Since lack of fit is detected, the design is augmented by adding runs 19 to 25 to form the second-order (central composite) design shown in Table 47.37. In general,

TABLE 47.36 ANOVA Table: First-Order Model, First Design*

Source	Sum of squares	Degrees of freedom	Mean square
Mean b_0	15,107.86	1	
b_1	7.29	1	7.29
b_2	29.16	1	29.16
Lack of fit	4.71	2	2.36
Pure error	1.14	2	0.57
Total	15,150.16	7	

*In the literature of response surface methodology, it is customary that the ANOVA table include a term for the sum of squares for the mean. In other uses of ANOVA, some authors and computer software programs exclude the sum of squares for the mean.

TABLE 47.37   Results of Second-Order Design

Run number	Concentration (coded units)X_1	Time (coded units)X_2	Yield Y, %
First-order design			
12	−1	−1	69.3
13	+1	−1	85.1
14	−1	+1	72.8
15	+1	+1	73.6
16	0	0	80.9
17	0	0	78.4
18	0	0	80.4
Augmenting runs			
19	$-\sqrt{2}$	0	71.4
20	$+\sqrt{2}$	0	78.9
21	0	$-\sqrt{2}$	73.9
22	0	$+\sqrt{2}$	69.1
23	0	0	76.4
24	0	0	78.5
25	0	0	76.3

TABLE 47.38   ANOVA Table: First-Order Model, Second Design

Source	Sum of squares	Degrees of freedom	Mean square
Mean b_0	41,734.32	1	
b_1	68.89	1	68.89
b_2	16.00	1	16.00
Lack of fit	94.12	2	47.06
Pure error	3.50	2	1.75
Total	41,916.83	7	

if a model does not fit, it may be advantageous, instead of immediately considering a higher-order model, to consider transformations of the factors and/or the responses. See Box and Cox (1964), Box and Tidwell (1962), and Draper and Hunter (1967).

Fit Second-Order Model and Check for Lack of Fit.   The fitted second-order equation obtained by least squares is

$$\hat{Y} = 78.50 + 3.40X_1 - 1.85X_2 - 3.75X_1X_2 - 1.21X_1^2 - 3.03X_2^2$$

The contours of this equation are shown in Figure 47.19 with the second-order design results. No lack of fit is evident from either visual inspection or statistical calculation (see Table 47.39). The form of the above equation can be simplified so the shape of the response surface can be better appreciated. It is difficult to visualize the surface from the equation because it contains six constants. A canonical analysis, which involves a translation and rotation of the coordinates from the original (X_1, X_2) axes to the new (Z_1, Z_2) axes, gives an equation containing only three constants:

$$Y - 173.83 = -0.0332Z_1^2 - 8.4075Z_2^2$$

This equation indicates that because of the negative coefficients for Z_1^2 and Z_2^2, the fitted response surface has a maximum point. A direction in which to proceed at the next stage to search for the maximum is indicated by the arrow in Figure 47.19. The arrow points toward the "top of the mountain." The investigation might terminate after experimenting in this direction, perhaps with a few added points in the vicinity of the maximum. In some situations it may be useful to perform a full

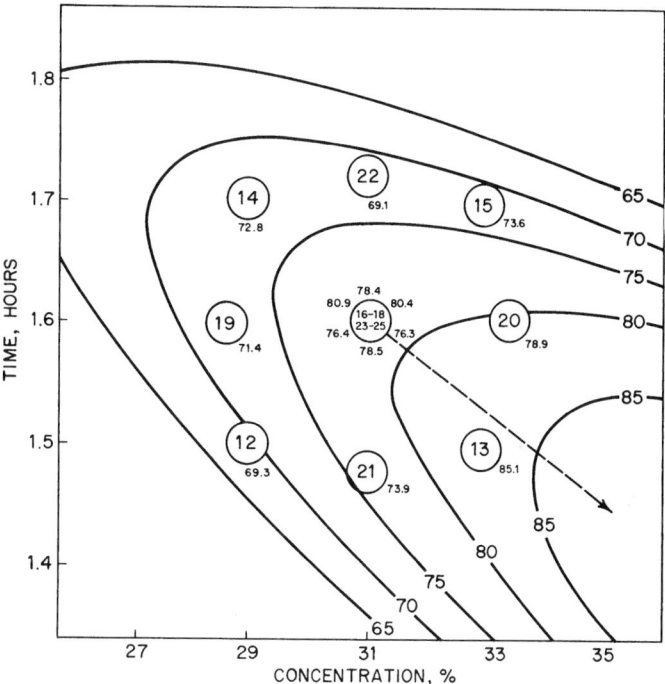

FIGURE 47.19 Results of second-order design with fitted second-order (nonplanar) response surface.

TABLE 47.39 ANOVA Table: Second-Order Model

Source	Sum of squares	Degrees of freedom	Mean square
Mean b_0	81,016.07	1	
First-order b_i	119.87	2	59.94
Pure second-order b_{ii}	80.73	2	40.37
Mixed second-order b_{ij}	56.25	1	56.25
Lack of fit	2.23	3	0.74
Pure error	21.77	5	4.35
Total	81,296.92	14	

Note: The ratio of the lack of fit mean square divided by the pure error mean square is 0.17, and since this value is less than $F_{0.95}$ for 3, 5 degrees of freedom (5.41), there is no evidence of lack of fit of the second-order model.

second-order design near the final optimum. [For further details on canonical analysis and RSM in general, see Box and Draper (1987) and Myers and Montgomery (1995).]

Many response surface experimental designs are available, in particular the three-level factorials, the central composite, the rotatable designs, and Box-Behnken designs. For response surface designs using a minimum number of runs while preserving many of the qualities of the larger designs see Draper (1985) and Draper and Lin (1990). Some computer software programs can provide unique response surface designs (commonly D-optimal) constructed to match special constraints provided by the experimenter.

MIXTURE DESIGNS

In some experiments with mixtures, the property of interest depends on the proportions of the mixture components and not on the amounts (volume or weight) of the individual components. For example, stainless steel is a mixture of different metals, and its tensile strength depends on the proportions of the metallic elements present; gasoline is ordinarily a blend of various stocks, and the octane rating of the final blend depends on the proportions going into the blend. The proportions of the components of a mixture must add up to unity, and in the most general case the proportion of any component may range from zero to unity. An important reference text is Cornell (1990).

In the design of mixtures, the factor space available for experimentation is constrained, since the proportions used must sum to unity. It has been shown that if the number of components in the mixture is q, the available factor space becomes a regular $(q - 1)$-dimensional simplex (e.g., a triangle for $q = 3$), a tetrahedron for $q = 4$).

A natural approach would be to take a uniformly spaced distribution of experimental points over the available factor space. This results in the simplex lattice designs proposed by Scheffé (1958). A (q, m) lattice, for example, is a lattice for q components, where the proportions for each component have $m+1$ equally spaced values from 0 to 1, i.e., the values 0, $1/m$, $2/m$, etc. For three components, the proportions of each component would be 0, $1/2$, 1 when $m=2$; and 0, $1/2$, $2/3$, 1 when $m = 3$. The lattice resulting when $m = 2$ is called the quadratic lattice, the lattice resulting when $m = 3$ is called the cubic lattice, etc. (see Figure 47.20).

In addition, modified lattices can be made by adding center points to the two-dimensional face or faces of the quadratic lattice. This provides a useful design called the "special cubic lattice."

The number of points k required for any lattice except the special cubic is found by using the formula

$$k = \frac{(m + q - 1)!}{m!(q - 1)!}$$

The number of points required for the special cubic is

$$k = \frac{q(q + 1)}{2} + \frac{q(q - 1)(q - 2)}{6}$$

The number of points required for several values of m and q is given in Table 47.40.

The property of interest is measured at each of the design points (corresponding to mixtures of different proportions). Simplified polynomials are used to relate the response variable y to the various mixture proportions used.

Another useful design called the "special cubic" by Scheffé (1958) requires seven points for three-component mixtures—the six points of a $(q = 3, m = 2)$ lattice plus a seventh point at $X_1 = 1/3$, $X_2 = 1/3$, $X_3 = 1/3$.

The seven mixtures are the three pure components, the three binary mixtures, and the ternary mixture, as shown in Table 47.41.

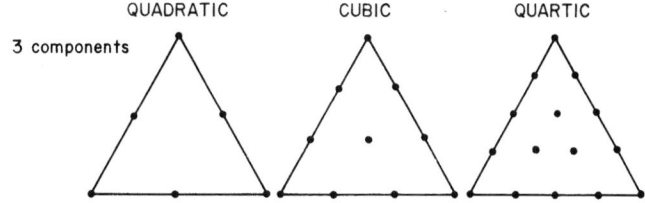

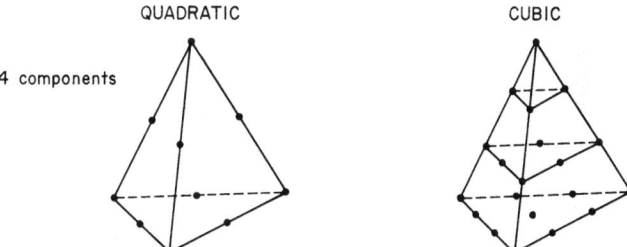

FIGURE 47.20 Lattice designs for three- and four-component mixtures. (*Reprinted with permission from Gorman and Hinman 1962.*)

TABLE 47.40 Number of Points Required for Lattice Designs

Number of components q	Type of lattice			
	Quadratic, $m = 2$	Special cubic, $m = 2$	Cubic, $m = 3$	Quartic, $m = 4$
3	6	7	10	15
4	10	14	20	35
5	15	25	35	70
6	21	41	56	126
8	36	92	120	330

TABLE 47.41 Design Points for Special Cubic (Three-Component Mixture)

Point number	x_1	x_2	x_3	Response
1	1	0	0	y_1
2	0	1	0	y_2
3	0	0	1	y_3
4	½	½	0	y_4
5	½	0	½	y_5
6	0	½	½	y_6
7	⅓	⅓	⅓	y_7

The "special cubic" corresponds to the equation:

$$y = B_1X_1 + B_2X_2 + B_3X_3 + B_{12}X_1X_2 + B_{13}X_1X_3 + B_{23}X_2X_3 + B_{123}X_1X_2X_3$$

The computed coefficients are

$$b_1 = y_1 \qquad b_2 = y_2 \qquad b_3 = y_3$$

$$b_{12} = 4y_4 - 2(y_1 + y_2)$$

$$b_{13} = 4y_5 - 2(y_1 + y_3)$$

$$b_{23} = 4y_6 - 2(y_2 + y_3)$$

$$b_{123} = 27y_7 - 12(y_4 + y_5 + y_6) + 3(y_1 + y_2 + y_3)$$

The subject of mixture designs now has a vast body of literature, and many computer software programs devote attention to both their design and analysis. Computer printouts of contour representations are particularly valuable. Factors composing mixture designs are often constrained to fall within narrow ranges, thus forming isolated mixture regions and requiring novel experimental designs difficult to obtain without a computer. Constrained factors are sometimes combined with factors not constrained, once again leading to unique designs and analyses. The designs may also be blocked and run sequentially. The textbook by Cornell (1990a) is devoted entirely to the topic of mixture experiments. See also Myers and Montgomery (1995). Important papers are Gorman and Hinman (1962), Thompson and Myers (1968), Snee (1973, 1979), Crosier (1984), Piepel and Cornell (1994), and Draper et al. (1993).

GROUP SCREENING DESIGNS

Novel experimental designs for finding the few effective factors out of a very large number of possible factors have been called "group screening designs." These designs have the following structure: groups are formed, each containing several factors; the groups are tested; and individual factors of the groups that prove to contain significant factors are then separately tested. Such designs, proposed by Connor (1961) and further studied by Watson (1961), are intended to minimize the amount of experimentation required.

The experimental variables are divided into groups, and each group is treated as a single variable until an effect on the response variable is shown.

The following assumptions are made:

1. All factors initially have the same probability of being effective.
2. The factors do not interact.
3. The directions of effects, if they exist, are known.

The number of factors is $f = gk$, where g = number of groups and k = number of factors per group. For example, consider an experiment with nine factors, which are divided into three groups of three factors each (i.e., $g = 3$, $k = 3$). The upper and lower levels of the groups are defined as follows:

1. Group factor X consists of factors A, B, C.

   Level 1: All three factors at lower level (0, 0, 0)

   Level x: All three factors at upper level (1, 1, 1)

2. Group factor Y consists of factors D, E, F.

   Level 1: All three factors at lower level (0, 0, 0)

Level y: All three factors at upper level (1, 1, 1)

3. Group factor *Z* consists of factors *G, H, I.*

Level 1:   All factors at lower level (0, 0, 0)

Level z: All factors at upper level (1, 1, 1)

The first-stage design studies the *group* factors, for example by using a half-replicate of a 2^3 factorial. This requires the four group treatment combinations *x, y, z,* and *xyz* corresponding to treatment combinations for the nine factors as follows:

$$x(1,1,1\ 0,0,0,0,0,0)$$

$$y(0,0,0\ 1,1,1\ 0,0,0)$$

$$z(0,0,0\ 0,0,0\ 1,1,1)$$

$$xyz(1,1,1\ 1,1,1\ 1,1,1)$$

The results of the first-stage experiment will indicate which group factors contain at least one effective factor. A second-stage experiment, which may consist of a half-replicate of a 2^3, will then be run on each effective group factor to determine which of the individual factors are effective. For further details, see Watson (1961). Patel (1962) gives detailed procedures for two-, three-, and four-stage screening tests.

The application of group screening designs that has been discussed here is to the identification of effective experimental factors, but there is extensive literature relating to the screening of effective responses, e.g., to compounds and drugs, and to the group testing of individuals. Papers of interest are by Ehrenfeld (1972), Pocock (1983), Mundel (1984), Hwang (1984), and Hayre (1985). An excellent review of the entire problem of group screening is provided by Tang and Tang (1994).

REFERENCES

Addelman, S. (1962). "Symmetrical and asymmetrical fractional factorial plans." *Technometrics,* vol. 4, pp. 47–58.

American Society for Quality, Statistics Division, ASQ Quality Press, Milwaukee.

 1: *How To Analyze Data with Simple Plots,* Wayne Nelson

 2: *How To Perform Continuous Sampling,* 2nd ed., K. S. Stephens

 3: *How To Test for Normality and Other Distribution Assumptions,* S. S. Shapiro

 4: *How To Perform Skip-lot and Chain Sampling,* 2nd ed., K. S. Stephens

 5: *How To Run Mixture Experiments for Product Quality,* J. A. Cornell

 6: *How To Analyze Reliability Data,* W. Nelson

 7: *How To and When to Perform Bayesian Acceptance Sampling,* T. W. Calvin

 8: *How To Apply Response Surface Methods,* J. A. Cornell

 9: *How To Use Regression Analysis in Quality Control,* D. C. Crocker

10: *How To Plan an Accelerated Life Test—Some Practical Guidelines,* W. Meeker and G. Hahn

11: *How To Perform Statistical Tolerance Analysis,* N. Cox

12: *How To Choose a Proper Sample Size,* G. G. Brush

13: *How To Use Sequential Statistical Methods,* T. P. McWilliams

14: *How To Construct Fractional Factorial Experiments,* R. F. Gunst and R. L. Mason

15: *How To Determine Sample Size and Estimate Failure Rate in Life Testing,* E. C. Moura

16: *How To Detect and Handle Outliers,* B. Iglewicz and D. C. Hoaglin

Anderson, D. A., and Thomas, A. M. (1978). "Resolution IV fractional factorial designs for the general asymmetric factorial." *Communications in Statistics,* vol. A8, pp. 931–943.

Automotive Industry Action Group (1990). *Measurement System Analysis Reference Manual.* Chrysler, General Motors, and Ford, Detroit.

Asao, M. (1992). "A simulation game of experimental design." *Quality Engineering,* vol. 4, pp. 497–517.

Bainbridge, T. R. (1965). "Staggered, nested designs for estimating variance components." *Industrial Quality Control,* vol. 22, pp. 12–20.

Barker, T. B. (1990). *Engineering Quality by Design: Interpreting the Taguchi Approach.* Marcel Dekker, New York.

Barker, T. B. (1994). *Quality by Experimental Design,* 2nd ed., Marcel Dekker, New York.

Barnett, E. H. (1960). "Introduction to Evolutionary Operation." *Industrial Engineering Chemistry,* vol. 52, pp. 500–503.

Barrentine, L. B. (1991). *Concepts for R&R Studies.* ASQC Quality Press, Milwaukee.

Bechhofer, R. E., and Tamhane, H. E. (1983). "Design of experiments for comparing treatments with a control: Tables of optimal allocation of observations." *Technometrics,* vol. 25, pp. 87–95.

Beckman, R. J., and Cook, R. D. (1983). "Outlier...s." *Technometrics,* vol. 25, pp. 119–149.

Berk, K. N., and Picard, R. R. (1991). "Significance tests for saturated orthogonal arrays." *Journal of Quality Technology,* vol. 23, pp. 73–89.

Bicking, C. A. (1954). "Some uses of statistics in the planning of experiments." *Industrial Quality Control,* vol. 10, pp. 23–31.

Bisgaard, S. (1992). "Industrial use of statistically designed experiments: Case study references and some historical anecdotes." *Quality Engineering,* vol. 4, pp. 547-562.

Bisgaard, S. (1993*a*). "Spreadsheets for the analysis of two-level factorials." *Quality Engineering,* vol. 6, pp. 149–157.

Bisgaard, S. (1993*b*). "Iterative analysis of data from two-level factorials." *Quality Engineering,* vol. 6, pp. 319–330.

Bishop, T., Peterson, B., and Trayser, D. (1982). "Another look at the statistician's role in experimental planning and design." *The American Statistician,* vol. 36, pp. 387–389.

Booth, K. H. V., and Cox, D. R. (1962). "Some systematic supersaturated experimental designs." *Technometrics,* vol. 4, pp. 489–495.

Bowman, K. O., Hopp, T. H., Kacker, R. N., and Lundegard, R. J. (1991). "Statistical quality control techniques in Japan." *Chance,* vol. 4, pp. 15–21.

Box, G. E. P. (1954). "The exploration and exploitation of response surfaces: some general considerations and examples." *Biometrics,* vol. 10, pp. 16–36.

Box, G. E. P. (1957). "Evolutionary Operation: A method for increasing industrial productivity." *Applied Statistics,* vol. 6, pp. 81–101.

Box, G. E. P. (1976). "Science and Statistics." *Journal of the American Statistical Association,* vol. 71, pp. 791–799.

Box, G. E. P. (1988). "Signal to noise ratios, performance criteria and transformations" (with discussion). *Technometrics,* vol. 30, pp. 1–40.

Box, G. E. P. (1990). "Do interactions matter?" *Quality Engineering,* vol. 2, pp. 365–369.

Box, G. E. P. (1991). "Finding bad values in factorial designs." *Quality Engineering,* vol. 3, pp. 405–410.

Box, G. E. P. (1992). "Sequential experimentation and sequential assembly of designs." *Quality Engineering,* vol. 5, pp. 321–330.

Box, G. E. P. (1992). "What can we find out from eight experimental runs?" *Quality Engineering,* vol. 4, pp. 619–627.

Box, G. E. P. (1992). "What can we find out from sixteen experimental runs?" *Quality Engineering,* vol. 5, pp. 167–181.

Box, G. E. P., and Behnken, D. W. (1960). "Some three level designs for the study of quantitative variables." *Technometrics,* vol. 2, pp. 477–482.

Box, G. E. P., Bisgaard, S., and Fung, C. A. (1988). "An explanation and critique of Taguchi's contributions to quality engineering." *Quality and Reliability Engineering International,* vol. 4, pp. 123–131.

Box, G. E. P., and Cox, D. R. (1964). "The analysis of transformations." *Journal of the Royal Statistical Society, Series B,* vol. 26, p. 211.

Box, G. E. P., and Draper, N. R. (1969). *Evolutionary Operation.* John Wiley, New York.

Box, G. E. P., and Draper, N. R. (1987). *Empirical Model Building and Response Surfaces.* John Wiley, New York.

Box, G. E. P., and Hunter, J. S. (1959). "Condensed calculations for evolutionary operation programs". *Technometrics,* vol. 1, pp. 77–95.

Box, G. E. P., Hunter, W. G., and Hunter, J. S. (1978). *Statistics for Experimenters,* John Wiley, New York.

Box, G. E. P., and Jones, S. (1986). "Split plot designs for robust product experimentation." *Journal of Applied Statistics,* vol. 19, pp. 2–36.

Box, G. E. P., and Myers, R. D. (1986). "An analysis for unreplicated fractional factorials." *Technometrics,* vol. 28, pp. 11–18.

Box, G. E. P., and Myers, R. D. (1986). "Dispersion effects from fractional designs." *Technometrics,* vol. 28, pp. 19–27.

Box, G. E. P., and Tiao, G. C. (1973). *Bayesian Inference in Statistical Analysis.* Addison-Wesley, Reading, MA.

Box, G. E. P., and Tidwell, P. W. (1962). "Transformations of the independent variable." *Technometrics,* vol. 4, p. 531.

Box, G. E. P., and Wilson, K. B. (1951). "On the experimental attainment of optimum conditions." *Journal of the Royal Statistical Society, Series B,* vol. 13, p. 1.

Burdick, R. K., and Graybill, F. A. (1992). *Confidence Intervals on Variance Components,* Marcel Dekker, New York.

Chambers, J. M., and Hastie, T. J. (1992). *Statistical methods in S,* Wadsworth, Pacific Grove, CA.

Chambers, J. M., Cleveland, W. S., Kleiner, B., and Tukey, P. A. (1993). *Graphical Methods for Data Analysis,* Chapman and Hall, New York.

Clatworthy, W. H. (1973). *Tables of the Two-Associate-Class Partially Balanced Designs.* National Bureau of Standards Applied Mathematics Services Publication 63, U.S. Government Printing Office, Washington, DC.

Cleveland, W. S. (1993). *Visualizing Data,* Hobart Press, Summmit, NJ.

Cochran, W. G., and Cox, G. M. (1957). *Experimental Designs,* 2nd ed., John Wiley, New York.

Coleman, D. E., and Montgomery, D. C. (1993). "A systematic approach to planning for a designed industrial experiment." *Technometrics,* vol. 35 (with discussion), pp. 1–27.

Connor, W. S. (1961). "Group screening designs." *Industrial and Engineering Chemistry,* vol. 53, pp. 69A–70A.

Conover, W. J. (1980). *Practical Non-Parametric Statistics,* John Wiley, New York.

Cornell, J. A. (1990). *Experiments with Mixtures: Designs, Models, and the Analysis of Mixture Data,* 2nd ed., John Wiley, New York.

Cornell, J. A. (1995). "Fitting models to data from mixture experiments containing ordered factors." *Journal of Quality Technology,* vol. 27, pp. 13–33.

Cornell, J. A. (1990*a*). *How to Run Mixture Experiments for Product Quality* (rev. ed.). American Society for Quality Control, Milwaukee.

Cornell, J. A. (1990*b*). *How to Apply Response Surface Methodology* (rev. ed.). American Society for Quality Control. Milwaukee.

Cox, D. R. (1958). *Planning of Experiments.* John Wiley, New York.

Crocker, D. C. (1990). *How to Use Regression Analysis in Quality Control* (rev. ed.). American Society for Quality Control, Milwaukee.

Crosier, R. B. (1984). "Mixture experiments: geometry and pseudo-components." *Technometrics,* vol. 26, pp. 209–216.

Crowder, M. J., and Hand, D. J. (1990). *Analysis of Repeated Measures,* Chapman and Hall, London.

Daniel, C. (1959). "Use of half-normal plots for interpreting factorial two-level experiments." *Technometrics,* vol. 1, pp. 311–341.

Deming, W. E. (1982). *Quality, Productivity and Competitive Position.* MIT Center for Advanced Engineering Study, Cambridge, MA.

Diamond, W. (1989). *Practical Experimental Designs.* Van Nostrand Reinhold, New York.

Dixon, W. J., and Massey, F. J. (1969). *Introduction to Statistical Analysis.* McGraw-Hill, New York.

Draper, N. R. (1985). "Small composite designs." *Technometrics,* vol. 27, pp. 173–180.

Draper, N. R., and Hunter, W. G. (1967). "Transformations: Some examples revisited." *Technometrics,* vol. 11, p. 53.

Draper, N. R., and John, J. A. (1988). "Response surface designs for quantitative and qualitative variables." *Technometrics,* vol. 30, pp. 423–428.

Draper, N. R., and Lin, D. K. J. (1990). "Small response surface designs." *Technometrics,* vol. 32, pp. 187–194.

Draper, N. R., Lewis, S. M., John, P. W. M., Prescott, P., Dean, A. M., and Tuck, M. G. (1993). "Mixture designs for four components in orthogonal blocks." *Technometrics,* vol. 35, pp. 268–276.

Draper, N. R., and Smith, H., Jr. (1981). *Applied Regression Analysis,* 2nd ed., John Wiley, New York.

Duncan, A. J. (1974). *Quality Control and Industrial Statistics,* 4th ed. Richard D. Irwin, Homewood, IL.

Duncan, D. B. (1955). "Multiple range and multiple F tests." *Biometrics,* vol. 11, p. 1.

Dunn, O., and Clark, V. (1987). *Applied Statistics: Analysis of Variance and Regression,* 2nd ed., John Wiley, New York.

Dunnett, C. W. (1964). "New tables for multiple comparison with a control." *Biometrics,* vol. 11, p. 1.

Dyer, D. D., and Keating, J. P. (1980). "On the determination of critical values of Bartlett's test." *Journal of the American Statistical Association,* vol. 75, pp. 313–319.

Easterling, R. G. (1985). "Discussion of Kacker's paper." *Journal of Quality Technology,* vol. 17, pp. 191–192.

Ehrenfeld, S. (1972). "On group sequential sampling." *Technometrics,* vol. 14, pp. 167–174.

Everitt, B. S. (1994). *A Handbook of Statistical Analyses Using S-Plus,* Chapman and Hall, London.

Fisher, R. A., and Yates, F. (1964). *Statistical Tables for Biological, Agricultural, and Medical Research,* 6th ed. Stechert-Hafner, New York.

Fleiss, J. L. (1986). *The Design and Analysis of Clinical Experiments,* John Wiley, New York.

Fries, A., and Hunter, W. G. (1980). "Minimum aberration 2^{k-p} designs." *Technometrics,* vol. 22, pp. 601–608.

Gibbons, J., and Chakraborti, S. (1992). *Nonparametric Statistical Inference,* 3rd ed. Marcel Dekker, New York.

Goh, T. N. (1993). "Taguchi methods: some technical, cultural and pedagogical perspectives." *Quality & Reliability International,* vol. 9, pp. 185–202.

Gorman, J. W., and Hinman, J. E. (1962). "Simplex lattice designs for multicomponent systems." *Technometrics,* vol. 4, pp. 463–487.

Grove, D. M., and Davis, T. P. (1991a). *Engineering Quality and Experimental Design,* John Wiley, New York.

Grove, D. M., and Davis, T. P. (1991b). "Taguchi's idle column method." *Technometrics,* vol. 33, pp. 349–354.

Gunst, R. F., and Mason, R. L. (1991). *How to Construct Fractional Factorial Experiments,* ASQC Quality Press, Milwaukee.

Haaland, P. D. (1989). *Experimental Design in Biotechnology.* Marcel Dekker, New York.

Haaland, P. D., and O'Connell, M. R. (1995). "Inference for effect-saturated fractional factorials." *Technometrics,* vol. 37, pp. 82–93.

Hahn, G. J. (1977). "Some things engineers should know about experimental design." *Journal of Quality Technology,* vol. 9, pp. 13–20.

Hahn, G. J. (1984). "Experimental design in a complex world." *Technometrics,* vol. 26, pp. 19–31.

Hahn, G. J., and Shapiro, S. S. (1966). "A catalog and computer program for the design and analysis of orthogonal symmetric and asymmetric fractional factorial experiments," Technical Report 66-C-165, G.E. Research & Development Center, Schenectady, NY.

Hayre, L. S. (1985). "Group sequential sampling with variable group sizes." *Journal of the Royal Statistical Society, Series B,* vol. 47, pp. 463–487.

Hicks, C. R. (1982). *Fundamental Concepts in the Design of Experiments,* 3rd ed., Holt, Reinhart & Winston, New York.

Hill, W. J., and Hunter, W. G. (1966). "A review of response surface methodology: a literature survey." *Technometrics,* vol. 8, pp. 571–590.

Hoadley, A., and Kettenring, J. (1990). "Communications between statisticians and engineers/physical scientists." *Technometrics,* vol. 32, pp. 243–274.

Hogg, R., and Ledolter, J. (1992). *Applied Statistics for Engineers and Physical Scientists,* 2nd ed. Macmillan, New York.

Hollander, M., and Wolfe, D. (1973). *Nonparametric Statistical Methods.* John Wiley, New York.

Hunter, J. S. (1959). "Determination of optimum operating conditions by experimental methods," Parts I, II, and III. *Industrial Quality Control,* December 1958, January, February 1959.

Hunter, J. S. (1980). "The national system of scientific measurement." *Science,* vol. 210, pp. 869–874.

Hunter, J. S. (1985a). "Statistical design applied to product design." *Journal of Quality Technology,* vol. 17, pp. 210–221.

Hunter, J. S. (1989). "Let's all beware the Latin Square." *Quality Engineering,* vol. 4, pp. 453–466.

Hunter, W. G. (1977). "Some ideas about teaching design of experiments with 2^5 examples of experiments conducted by students." *The American Statistician,* vol. 31, pp. 12–17.

Hurley, P. (1994). "Interactions: Ignore them at your own risk." *Journal of Quality Technology,* vol. 21, pp. 174–178.

Hwang, F. K. (1984). "Robust group testing." *Journal of Quality Technology,* vol. 16, pp. 189–195.

Iglewicz, B., and Hoaglin, D. C. (1993). *How to Detect and Handle Outliers.* ASQC Quality Press, Milwaukee.

Iman, Ronald L. (1994). *A Data Based Approach to Statistics.* Duxbury Press, Belmont, CA.

Ishikawa, K. (1985). *What is Total Quality Control?* Prentice Hall, Englewood Cliffs, NJ.

Jones, M. C., and Rice, J. A. (1992). "Displaying the important features of a large collection of similar curves." *The American Statistician,* vol. 46, pp. 140–145.

Kacker, R. N. (1985). "Off-line quality control, parameter design and the Taguchi method" (with discussion). *Journal of Quality Technology,* vol. 17, pp. 176–209.

Kacker, R. N., and Shoemaker, A. C. (1986). "Robust design: a cost effective method for improving manufacturing process." *AT&T Technical Journal,* vol. 65, pp. 39–50.

Kacker, R. N., and Tsui, K. L. (1990). "Interaction graphs: graphical aids for planning experiments." *Journal of Quality Technology,* vol. 22, pp. 1–14.

Kastembaum, M. A., Hoel, D. G., and Bowman, K. O. (1970). "Sample size requirements; one-way analysis of variances, randomized block designs." *Biometrika,* vol. 57, pp. 421–430, 573–578.

Khuri, A. I., and Cornell, J. A. (1987). *Response Surfaces: Designs and Analyses,* Marcel Dekker, New York.

Ku, H. H. (ed.) (1969). *Precision Measurement and Calibration—Statistical Concepts and Procedures,* National Bureau of Standards Special Publication 300, Vol. 1. U.S. Government Printing Office, Washington, DC.

Kurtz, T. E., Link, B. F., Tukey, J. W., and Wallace, D. L. (1965). "Short-cut multiple comparisons for balanced single and double classifications." *Technometrics,* vol. 7, pp. 95–165.

Lenth, R. V. (1989). "Quick and easy analysis of unreplicated factorials." *Technometrics,* vol. 31, pp. 469–473.

Leon, R. V., Shoemaker, A. C., and Kacker, R. N. (1987). "Performance measure independent of adjustment: an explanation and extension of Taguchi's signal to noise ratio." *Technometrics,* vol. 29, pp. 253–285.

Lilliefors, H. W. (1967). "On the Kolmogorov-Smirnov test for normality with mean and variance unknown." *Journal of the American Statistical Association,* vol. 62, pp. 399–402.

Lin, D. K. J. (1993). "A new class of supersaturated designs." *Technometrics,* vol. 35, pp. 28–31.

Lin, D. K. J. (1995). "Generating systematic supersaturated designs." *Technometrics,* vol. 37, pp. 213–225.

Lucas, J. M. (1985). "Comment on `Off line quality control, parameter design and the Taguchi method' by K. N. Kacker." *Journal of Quality Technology,* vol. 17, pp. 195–197.

Mandel, J. (1964). *Statistical Analysis of Experimental Data.* John Wiley, New York.

Margolin, B. H. (1969). "Results on factorial designs of resolution IV for the 2^n and $2^n 3^m$ series." *Technometrics,* vol. 11, pp. 431–444.

Mason, R. L., Gunst, R. F., and Hess, J. L. (1989). *Statistical Design and Analysis of Experiments with Applications to Engineering and Science.* John Wiley, New York.

Miller, R. G., Jr. (1981). *Simultaneous Statistical Inference,* 2nd ed. Springer-Verlag, New York.

Miller, A., Sitter, R. R., Wu, C. F. J., and Long, D. (1993). "Are large Taguchi-style experiments necessary? A reanalysis of gear and pinion data." *Quality Engineering,* vol. 6, pp. 21–37.

Moen, R. D., Nolan, T. W., and Provost, L. P. (1991). *Improving Quality Through Planned Experimentation.* McGraw-Hill, New York.

Montgomery, D. C. (1991). *Introduction to Statistical Quality Control,* 2nd ed. John Wiley, New York.

Montgomery, D. C. (1991). *Design and Analysis of Experiments.* John Wiley, New York.

Montgomery, D. C. (1991). "Using fractional factorial designs for robust design process development." *Quality Engineering,* vol. 3, pp. 193–205.

Montgomery, D. C., and Voth, S. R. (1994). "Multicollinearity and leverage in mixture experiments." *Journal of Quality Technology,* vol. 26, pp. 96–108.

Mundel, A. (1984). "Group testing." *Journal of Quality Technology,* vol. 16, pp. 181–188.

Myers, R. H., Khuri, A. I., and Carter, W. H. (1989). "Response surface methodology: 1966–1988." *Technometrics,* vol. 31, pp. 137–157.

Myers, R. H., and Montgomery, D. C. (1995). *Response Surface Methodology.* John Wiley, New York.

Nachtscheim, C. J. (1987). "Tools for computer aided design of experiments." *Journal of Quality Technology*, vol. 19, pp. 132–160.

Nair, V. J. (1988). "Analyzing dispersion effects from replicated factorial experiments." *Technometrics*, vol. 30, pp. 247–257.

Nair, V. J., ed. (1992). "Taguchi's parameter design: A panel discussion." *Technometrics*, vol. 34, pp. 127–161.

Nair, V. J., and Shoemaker, A. C. (1990). "The role of experimentation in quality engineering: a review of Taguchi's contributions," pp. 247–277 in *Statistical Design and Analysis of Experiments*, Ghosh, G., ed. Marcel Dekker, New York.

Natrella, M. C. (1963). *Experimental Statistics*, National Bureau of Standards Handbook 91, U.S. Government Printing Office, Washington, DC.

Nelson, L. S. (1983). "Exact critical values for use with the analysis of means." *Journal of Quality Technology*, vol. 15, pp. 40–44.

Nelson, P. R. (1993). "Additional tables for the analysis of means and extended tables of critical values." *Technometrics*, vol. 35, pp. 61–71.

Neter, J., Wasserman, W., and Kutner, M. H. (1990). *Applied Linear Statistical Models: Regression, Analysis of Variance and Experimental Designs*, 3rd ed. Richard D. Irwin, Homewood, IL.

Noether, G. E. *Introduction to Statistics: The Nonparametric Way*. Springer-Verlag, New York.

Odeh, R. E., and Fox, M. (1991). *Sample Size Choice: Charts for Experimenters with Linear Models*. Marcel Dekker, New York.

Oehlert, G. W. (1994). "Isolating one cell interactions." *Technometrics*, vol. 36, pp. 403–408.

Ott, E. R. (1967). "Analysis of means—a graphical procedure." *Industrial Quality Control*, vol. 24, pp. 101–109.

Ott, E. R., and Schilling, E. G. (1990). *Process Quality Control*, 2nd ed. McGraw-Hill, New York.

Owens, D. B. (1962). *Handbook of Statistical Tables*. Addison-Wesley, Reading, MA.

Patel, M. S. (1962). "Group screening with more than two stages." *Technometrics*, vol. 4, pp. 209–217.

Peace, G. S. *Taguchi Methods: A Hands-on Approach*. Addison-Wesley, Reading, MA.

Phadke, M. S. (1989). *Quality Engineering Using Robust Design*. Prentice Hall, Englewood Cliffs, NJ.

Phadke, M. S., Kacker, R. N., Speeney, D. V., and Greico, M. J. (1983). "Off-line quality control in integrated circuit fabrication using experimental design." *Bell System Technical Journal*, vol. 62, pp. 1273–1309.

Piepel, G. F., and Cornell, J. A. (1994). "Mixture experiment approaches: Examples, discussion and recommendations." *Journal of Quality Technology*, vol. 26, pp. 177–196.

Plackett, R. L., and Burman, J. P. (1946). "The Design of Optimum Multi-Factorial Experiments." *Biometrica* vol. 33, pp. 305–325.

Pocock, S. (1983). *Clinical Trials*. John Wiley, New York.

Ramig, P. F. (1983). "Applications of the analysis of means." *Journal of Quality Technology*, vol. 15, pp. 19–25.

Rechtshaffner, R. (1967). "Saturated fractions of the 2^n and 3^m factorial designs." *Technometrics*, vol. 9, pp. 569–575.

Scheffé, H. (1953). "A method for judging all contrasts in the analysis of variance," *Biometrika*, vol. 40, p. 87.

Scheffé, H. (1958). "Experiments with mixtures." *Journal of the Royal Statistical Society, Series B*, vol. 20, pp. 344–360.

Schilling, E. G. (1973). "A systematic approach to the analysis of means." *Journal of Quality Technology*, vol. 5, pp. 93–108, 147–159.

Schmidt, S. R., and Launsby, R. G. (1991). *Understanding Industrial Designed Experiments*, Air Academy Press, Colorado Springs.

Shoemaker, A., Tsui, K.-L., Wu, J. C. F. (1991). "Economical Experimentation Methods for Robust Design." *Technometrics*, vol. 33, pp. 415–428.

Searle, S. R., Casella, G., and McCulloch, C. E. (1992). *Variance Components*. John Wiley, New York.

Sheesley, J. H. (1980). "Attributes comparison of k samples involving variables or data using the analysis of means." *Journal of Quality Technology*, vol. 12, pp. 47–52.

Snedecor, G. W., and Cochran, W. G. (1980). *Statistical Methods*. The Iowa University Press, Ames, IA.

Snee, R. D. (1973). "Design and analysis of mixture experiments." *Journal of Quality Technology*, vol. 3, pp. 159–169.

Snee, R. D. (1974). "Computation and Use of Expected Mean Squares in Analysis of Variance." *Journal of Quality Technology*, vol. 6, pp. 128–137.

Snee, R. D. (1979). "Experimental designs for mixture systems with multicomponent constraints." *Communications in Statistics,* vol. A8, pp. 303–326.

Snee, R. D. (1983). "Graphical analysis of process variation studies." *Journal of Quality Technology,* vol. 15, pp. 76–88.

Snee, R. D. (1985). "Computer-aided design of experiments: Some practical experiences." *Journal of Quality Technology,* vol. 17, pp. 222–236.

Snell, E. J., and Simpson, H. R. (1991). *Applied Statistics: A Handbook of GENSTAT Analyses.* Chapman and Hall, London.

Spector, P. (1994). *An Introduction to S and S-Plus.* Duxbury Press, Belmont, CA.

Spurrier, J. D. (1992). "Optimal designs for comparing the variances of several treatments with that of a standard treatment." *Technometrics,* vol. 34, pp. 332–339.

Stephens, Matthew P. (1994). "Comparison of robustness of Taguchi's method with classical ANOVA under conditions of homogeneous variances." *Quality Engineering,* vol. 7, pp. 147–167.

Taguchi, G. (1978). "Off-line and on-line quality control systems." *Proceedings of International Conference on Quality,* Tokyo.

Taguchi, G. (1986). *Introduction to Quality Engineering.* Asian Productivity Association, Tokyo.

Taguchi, G. (1987). *Systems of Experimental Design,* Vol. 1. Unipub, Kraus International Publications, New York.

Tang, K., and Tang, J. (1994). "Design of screening procedures: A review." *Journal of Quality Technology,* vol. 26, pp. 209–226.

Taylor, G. A. R. (1994). "Analysis of experiments by using half normal plots." *The Statistician,* vol. 43, pp. 529–536.

Thomas, G. E., and Kiwanga, S. S. (1993). "Use of ranking and scoring methods in the analysis of ordered categorical data from factorial experiments." *The Statistician,* vol. 42, pp. 55–67.

Thompson, W. O., and Myers, R. H. (1968). "Response surface designs for mixture problems." *Technometrics,* vol. 10, pp. 739–756.

Tribus, M., and Sconyi, G. (1989). "An alternative view of the Taguchi approach." *Quality Progress,* vol. 22, pp. 46–48.

Tukey, J. W. (1949). "Comparing individual means in the analysis of variance." *Biometrics,* vol. 5, p. 99.

Vining, G. C., and Myers, R. H. (1990). "Combining Taguchi and response surface philosophies: A dual response approach." *Journal of Quality Technology,* vol. 22, pp. 38–45.

Wadsworth, H. (ed.) (1990). *Handbook of Statistical Methods for Engineers and Scientists,* McGraw-Hill, New York.

Wang, J. C., and Wu, C. F. J. (1991). "An approach to the construction of asymmetrical orthogonal arrays." *Journal of the American Statistical Association,* vol. 86, pp. 450–456.

Watson, G. S. (1961). "A study of the group screening method." *Technometrics,* vol. 3, pp. 371–388.

Webb, S. R. (1968a). "Non-orthogonal designs of even resolution." *Technometrics,* vol. 10, pp. 291–299.

Webb, S. R. (1968b). "Saturated sequential factorial design." *Technometrics,* vol. 10, pp. 535–550.

Wernimont, G. (1951). "Design and Analysis of Interlaboratory Studies of Test Methods." *Analytical Chemistry,* vol. 23, p. 1572.

Welch, W. J., Yu, T. K., Kang, K. S., and Sacks, J. (1990). "Computer experiments for quality control by parameter design." *Journal of Quality Technology,* vol. 22, pp. 15–22.

Winer, B. J. (1971). *Statistical Principles in Experimental Design,* 2nd ed. McGraw-Hill, New York.

Wu, C. F. J., and Chen, Y. Y. (1992). "A graph aided method for planning two-level experiments when certain interactions are important." *Technometrics,* vol. 34, pp. 162–175.

Wu, C. F. J., Mao, S. S., and Ma, F. S. (1990). "Sequential elimination of levels—An investigation method using orthogonal arrays" (Chinese). *Chinese Journal of Applied Probability and Statistics,* vol. 6, pp. 185–204.

Youden, W. J. (1959). "Graphical diagnosis of interlaboratory tests." *Industrial Quality Control,* vol. 15, pp. 1–5.

Youden, W. J. (1963). "Ranking laboratories by round-robin tests." *Materials Research and Standards,* vol. 3, pp. 9–13.

Youden, W. J. (1967). *Statistical Techniques for Collaborative Tests.* Association of Official Analytical Chemists, Washington, DC.

Youden, W. J., and Hunter, J. S. (1955). "Partially replicated Latin squares." *Biometrics,* vol. 5, pp. 99–104.

Zahn, D. A. (1975). "Modifications of and revised critical values for the half-normal plot." *Technometrics,* vol. 17, pp. 189–200.

SECTION 48

RELIABILITY CONCEPTS AND DATA ANALYSIS

William Q. Meeker
Luis A. Escobar
Necip Doganaksoy
Gerald J. Hahn

INTRODUCTION

Relationship between Quality and Reliability. Rapid advances in technology, development of highly sophisticated products, intense global competition, and increasing customer expectations have put new pressures on manufacturers to produce high-quality products. While producers often think of high quality in terms of minimizing scrap and rework, customers are concerned with functionality, reliability, and overall product lifetime. Customers expect a purchased product to meet or exceed life expectations and to be safe. Technically, reliability is often defined as the probability that a system, vehicle, machine, device, or other product will perform its intended function under some specified operating conditions, for a specified period of time. Improving reliability is an important part of the larger overall goal of improving product quality. There are many definitions of quality, but there is general agreement that an unreliable product is *not* a high-quality product. Condra (1993) emphasizes that "reliability is quality over time."

Modern programs for improving reliability of existing products and for assuring continued high reliability for the next generation of products require quantitative methods for predicting and assessing product reliability and for providing early signals and information on root causes of failure. In many cases this will involve the collection and analysis of reliability data from studies such as laboratory tests (or designed experiments) of materials, devices, and components, tests on early prototype units, careful monitoring of early-production units in the field, analysis of warranty data, and systematic longer-term tracking of product in the field. This frequently also involves the careful planning of such programs so as to ensure that the most meaningful information is obtained. Reliability evaluations often present a challenge beyond that normally encountered in quality evaluations because there is usually an elapsed time between when the product is built and when the reliability information is forthcoming.

Applications of Reliability Data Analysis.   Some major goals in obtaining reliability data include:

1. Obtaining early identification of failure modes and understanding and removing their root causes— and thereby improving reliability.
2. Obtaining field failure information to help improve the product design, perhaps for the next product release.
3. Determining how long each unit should be run prior to shipment (and at what conditions) in order to avoid premature field failures. This is sometimes referred to as "burn-in."
4. Quantifying reliability to determine whether or not a product is ready for release; i.e., does the product meet the specified level of reliability?
5. Predicting warranty costs.
6. Deciding on the need for recall of a product in the field (and, if so, the extent of the recall).

Although the last four objectives are the ones that analysts are asked to respond to most often, concentrating on them is sometimes premature. The first objective is the most fundamental since one cannot demonstrate a high level of reliability if, in fact, it does not exist, and clearly it is desirable to avoid expensive burn-in and field recalls. However, much of the discussion in this section addresses the last four goals. Section 3 (The Quality Planning Process) and Section 19 (Quality in Research and Development) consider the entire process of assuring high reliability.

Sources and Types of Reliability Data.   It is important to distinguish between the following types of reliability data:

- A sequence of reported system repair times (or the times of other system-specific events) for a repairable system.
- The time of failure (or other clearly specified event) for nonrepairable units or components (including *nonrepairable* components within a *repairable* system).

When reliability tests are conducted on larger systems and subsystems (even those that may be repaired), it is essential that detailed component-level information on cause of failure be obtained. This is especially important if the purpose of the data collection is improvement of system reliability, as opposed to mere assessment of overall system reliability. Subsequent subsections describe methods for data analysis for nonrepairable units or components as well as for analyzing system repair data.

Reliability data arise from many different kinds of situations. Examples include:

- Laboratory tests to study durability, wear, or other lifetime properties of particular materials, components, or subsystems.
- Operational life tests on complete systems or subsystems, conducted before a product is released to customers.
- Field operation by customers.

Reliability data are typically censored, i.e., some units remain unfailed, and (at best) only their survival times, but not their failure times, are known. The most common reason for censoring is the need to analyze data before all units fail. The analysis of censored data is more complicated when the censoring times of unfailed units differ. This would happen when different units of the product are placed on test or enter into the field at different times, as is usually the case in analyzing field failure data. It may also be the case when units have different degrees of exposure over time or when one is evaluating failures due to a particular failure mode (in which case failures from other independent modes are treated as censored observations in the data analysis). An important assumption needed for standard analysis of censored data is that the censoring time for a unit is independent of when that unit would have failed. For example, if a unit were removed from the field because it is about to fail, treating it as a censored observation would bias the analysis.

Computer Software. In this section we focus on the methods and applications of reliability data analysis. We provide formulas only when they provide insight and will help readers to understand how to use the method.

Although it is possible to do some of the simplest reliability data analyses by hand, for maximally effective analyses, computer processing along with display using modern high-resolution graphics should be used.

Unfortunately, only a few of the best-known standard data analysis computer packages have adequate capabilities for doing reliability data analysis. As software vendors become more aware of their customers' needs, capabilities in commercial packages can be expected to improve.

SAS PROC RELIABILITY (SAS Institute 1996), SAS JMP (SAS Institute 1995), S-Plus (Statistical Sciences 1996), Minitab 12 (Minitab 1997), and a specialized program called WinSMITH (Abernethy 1996), can do product life data analysis for the simple (single-distribution) situation for which all life data are assumed to be in a single common environment. Ease of use and the character of the user interface vary widely over these packages. In addition, each of these programs, except WinSMITH, can do regression and accelerated life test analyses. SAS JMP can also analyze data with more than one failure mode. SAS PROC RELIABILITY can, in addition, do the nonparametric repairable systems analyses described in this section. Meeker and Escobar (1998) describe S-Plus functions that can be used to do all of the analyses described in this section.

LIFE DATA MODELS

This section deals mainly with nonrepairable components or other products that are replaced rather than repaired upon failure (or time to first failure on repairable products). The following are described:

1. Statistical models for representing time to failure of nonrepairable products.
2. Methods for estimating quantities of interest (e.g., failure probabilities, distribution quantiles) from time-to-failure data, with and without making any distributional assumptions.

We discuss initially models that deal with time to failure in a single-use environment. In our subsequent discussion of accelerated life tests we will describe corresponding regression models that explain the effect that factors like temperature have on life.

Time-to-Failure Model Concepts.   The distribution of time to failure T of a product can be characterized by a cumulative distribution function (cdf), a probability density function (pdf), a survival function (sf), or a hazard function (hf). The cdf and pdf are common ways of describing a statistical model in many applications (and are treated in elementary statistics courses). The hf, on the other hand, is not as well known, but it has particular applicability to product life data analysis. These functions are described below and illustrated (for one possible time-to-failure distribution) in Figure 48.1. The choice of which function or functions to use depends on the specific practical problem that is being addressed. All of these functions are important for one purpose or another.

The cdf for T, $F(t) = \mathrm{Pr}\,(T \le t)$, gives the *probability* that a unit, selected at random from the population or process, will fail before time t. Alternatively, $F(t)$ can be interpreted as the *proportion* of units in the population (or from some stationary process) that will fail before time t. (Here a stationary process is defined as one that generates units that have a $F(t)$ that does not change over time.)

The sf (also known as the *reliability function*) is the complement of the cdf, $S(t) = \mathrm{Pr}\,(T > t) = 1 - F(t) = \int_t^\infty f(x)\,dx$, and gives the probability of surviving until time t.

The pdf for a continuous random variable T is defined as the derivative of $F(t)$ with respect to t: $f(t) = dF(t)/dt$. Thus, for a positive random variable, $F(t) = \int_0^t f(x)\,dx$. The pdf can be used to represent relative frequency of failure times as a function of time and can be thought of as a smoothed histogram of a large number of observed failure times.

The hf (also known as the hazard rate, the instantaneous failure rate function, and by various other names) is defined by:

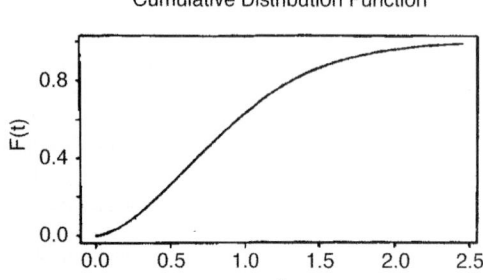

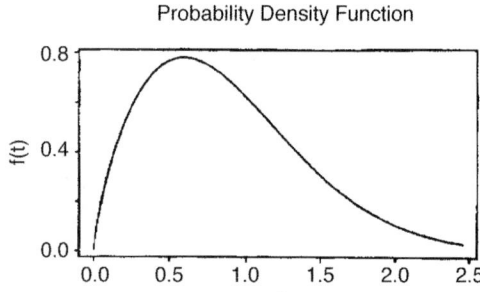

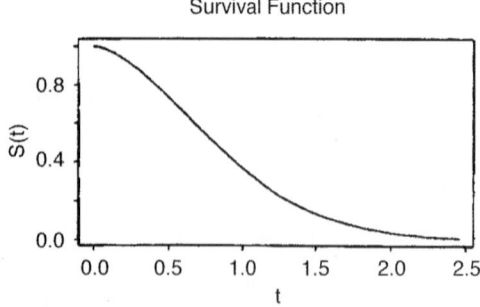

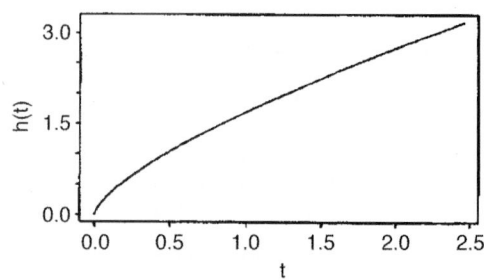

FIGURE 48.1   Typical time-to-failure cdf, pdf, sf, and hf.

$$h(t) = \lim_{\Delta t \to 0} \frac{\Pr(t < T \le t + \Delta t | T > t)}{\Delta t} = \frac{f(t)}{1 - F(t)}$$

The hf expresses the propensity to fail in the next small interval of time, given survival to time t. That is, for small Δt, $h(t) \times \Delta t \approx \Pr(t < T \le t + \Delta t | T > t)$. The hf can be interpreted as a failure rate in the following sense. If there is a large number of items [say $n(t)$] in operation at time t then $n(t) \times h(t)$ is approximately equal to the number of failures per unit time [or $h(t)$ is approximately equal to the number of failures per unit time per unit at risk]. Because of its close relationship with failure processes and maintenance strategies, reliability engineers often model time to failure in terms of $h(t)$. The "bathtub curve" shown in Figure 48.2 provides a useful conceptual model for the hazard of some product populations. There may be early failures of units with quality-related defects (infant mortality). During much of the useful life of a product, the hazard may be approximately constant because failures are caused by external shocks that occur at random. Late-life failures are due to wear-out. Many reliability tests focus on one side or the other of this curve.

Other Quantities of Interest in Reliability Analysis. The mean (also known as the "expectation" or "first moment") of a positive random variable T is a measure of central tendency or "average" time, and it is defined by:

$$E(T) = \int_0^\infty t f(t)\, dt = \int_0^\infty [1 - F(t)]\, dt$$

When the probability density function $f(t)$ is highly skewed, e.g., has a long right tail (as is common in many life data applications), the mean time to failure may differ appreciably from other measures of central tendency like the median time to failure (the time by which exactly half of the population of units will fail). The mean time to failure $E(T)$ is sometimes abbreviated to MTTF.

The traditional parameters of a statistical model (mean and standard deviation) are often *not* of primary interest in reliability studies. Instead, design engineers, reliability engineers, managers, and customers are interested in specific measures of product reliability or particular characteristics of the failure-time distribution, e.g., distribution quantiles. In particular, the quantile t_p is the time at which a specified proportion p of the population will fail. Also, $F(t_p) = p$. For example, $t_{.20}$ is the *time* by which 20 percent of the population will fail. Alternatively, frequently one would like to know $F(t)$, the *probability* of failure associated with a particular number of hours, days, weeks, months, or years of usage, e.g., the probability of a product failing (or not failing) during the first 5 years in the field.

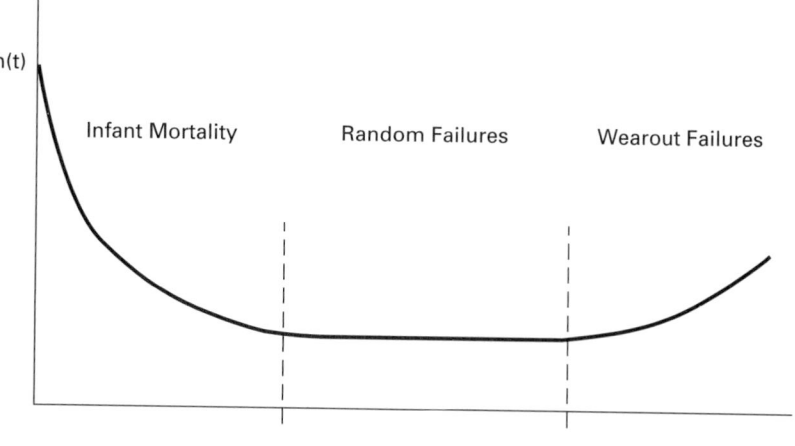

FIGURE 48.2 Bathtub curve hazard function.

Common Parametric Time-to-Failure Distributions

The Exponential Distribution.   The exponential distribution cdf is

$$F(t; \theta) = 1 - \exp\left(-\frac{t}{\theta}\right) \qquad t > 0$$

where the distribution's single parameter $\theta > 0$ is a scale parameter. The exponential distribution has no shape parameter and, thus, has only a single unique shape. The exponential cdf, pdf, and hazard function are graphed in Figure 48.3 for $\theta = .5$, 1, and 2. The p quantile of the exponential distribution is $t_p = -\theta \log (1 - p)$. The exponential distribution has the important characteristic that its hazard function $h(t) = 1/\theta$ is constant (does not depend on time t). A constant hazard implies that, for an unfailed unit, the probability of failing in the next interval of time is independent of the unit's age. Physically, a constant hazard implies that units in the population are not subject to an infant mortality failure mode and also are not wearing out or otherwise aging. Thus, this model may be appropriate when failures are induced by an external phenomenon that is independent of product life and past history. An example might be the life in service of a dish in a restaurant, whose end of life is the result of breakage. The exponential distribution is commonly, and sometimes incorrectly, used because of its simplicity. It would *not* be appropriate for modeling the life of mechanical components (e.g., bearings) subject to some combination of fatigue, corrosion, or wear or for electronic components that exhibit wear-out properties over their life (e.g., lasers and filament devices).

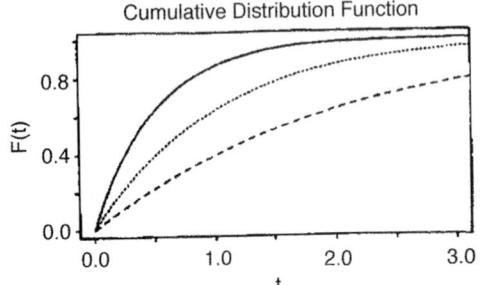

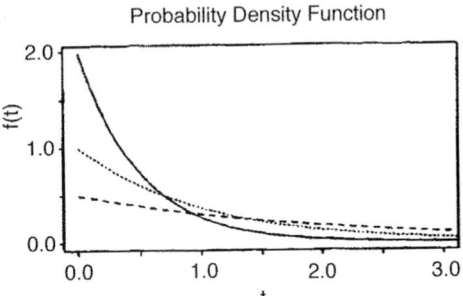

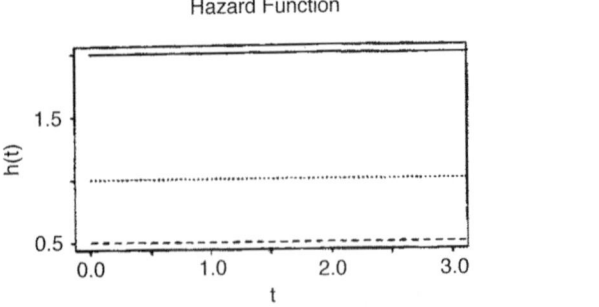

FIGURE 48.3   Exponential distribution cdf, pdf, and hf for $\theta = .5$, 1, and 2.

The Weibull Distribution.   The Weibull distribution cdf is

$$F(t; \eta, \beta) = 1 - \exp\left[-\left(\frac{t}{\eta}\right)^{\beta}\right] \qquad t > 0$$

where $\beta > 0$ is a shape parameter and $\eta > 0$ is a scale parameter (which is also approximately the .63 quantile of this distribution). The p quantile of the Weibull distribution is $t_p = \eta[-\log(1-p)]^{1/\beta}$. For example, the Weibull distribution median is $t_{.50} = \eta[-\log(1-.5)]^{1/\beta}$. The Weibull hazard function is $h(t) = (\beta/\eta)(t/\eta)^{\beta-1}$. The exponential distribution is a special case of the Weibull with $\beta = 1$. The practical importance of the Weibull distribution stems from its ability to describe failure distributions with many different commonly occurring shapes. As illustrated in Figure 48.4 with $0 < \beta < 1$, the Weibull hazard function is decreasing and with $\beta > 1$, the Weibull hazard function is increasing. Thus the Weibull distribution can be used to describe either decreasing hazard caused by infant mortality or increasing hazard due to wear-out, but *not* both. The Weibull distribution is often used as a model for the life of insulation and many other products as well as a model for strength of materials. For some applications there is theoretical justification for its use.

The logarithm of a Weibull random variable, $Y = \log(T)$, follows a smallest extreme value distribution so that

$$\Pr[\log(T) \le \log(t)] = F(t; \mu, \sigma) = \Phi_{sev}\left(\frac{\log(t) - \mu}{\sigma}\right)$$

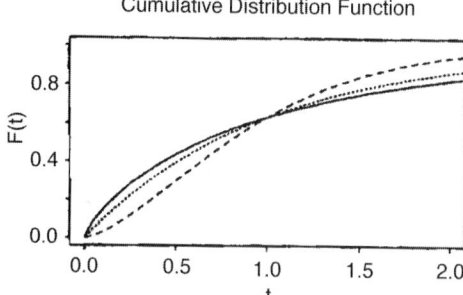

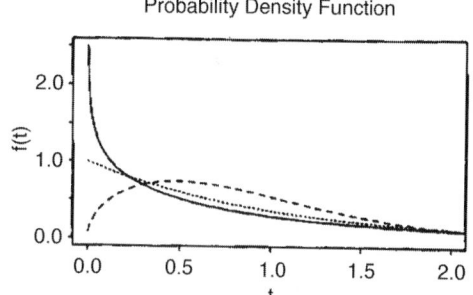

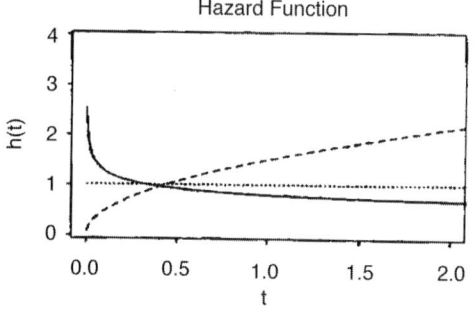

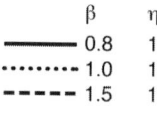

FIGURE 48.4   Weibull distribution cdf, pdf, and hf for scale parameter $\eta=1$ and shape parameter $\beta=.8$, 1.0, and 1.5.

where $y = \log(t)$, $\Phi_{sev}(z) = 1 - \exp[-\exp(z)]$, $\mu = \log(\eta)$ is a location parameter with $-\infty < \mu < \infty$, and $\sigma = 1/\beta > 0$ is a scale parameter. The relationship between the Weibull and the smallest extreme value distribution is similar to the relationship between the normal and lognormal distributions and the relationship will be used later in this section in explaining methods for failure-time regression analysis and accelerated test modeling.

The Lognormal Distribution.   The normal distribution, though playing a central role in other statistical applications, is infrequently appropriate as a time-to-failure model. This is because failure can occur at times shortly after 0 hours, but they cannot be negative. Logarithms of failure times, however, are often described well by a normal distribution. This is equivalent to fitting a lognormal distribution. The lognormal cdf is

$$F(t; \mu, \sigma) = \Phi_{nor}\left[\frac{\log(t) - \mu}{\sigma}\right] \qquad t > 0$$

where Φ_{nor} is the cdf for the standard normal distribution (see Section 44). The parameter $\exp(\mu)$ is a scale parameter and the median of the distribution, $\sigma > 0$ is a shape parameter. The p quantile of the lognormal distribution is $t_p = \exp[\mu + \Phi_{nor}^{-1}(p)\sigma]$, where Φ_{nor}^{-1} is the inverse of the standard normal distribution. The relationship between the lognormal and normal distributions is often used to simplify the use of the lognormal distribution.

The lognormal distribution is commonly used as a model for the distribution of failure times. As shown in Figure 48.5, the lognormal hazard function always starts at 0, increases to a maximum, and then approaches 0 for large t.

Following from the central limit theorem (described in Section 44), application of the lognormal distribution could be justified for a random variable that arises from the product of a number of identically distributed independent positive random effects. For example, it has been suggested that the lognormal is an appropriate model for time to failure caused by a degradation process with combinations of random rate constants that combine multiplicatively. Correspondingly, the lognormal distribution is widely used to describe failure times due to fatigue and in microelectronic applications.

Other Life Distributions.   For more details for these and other distributions that can be used to describe life distributions, see Chapter 8 of Hahn and Shapiro (1967) and Chapters 4 and 5 of Meeker and Escobar (1998).

Multiple Failure Modes.   In more complex situations (e.g., products that have *both* infant mortality and wear-out modes, as in the bathtub curve situation, which was described earlier) the data cannot be fit by any of the preceding models. This is often the case when there is more than one failure mode. In such cases, if the individual failure modes can be identified, it is usually better to do the evaluations separately for each of these modes and then combine the results to get total system probabilities. See Chapter 5 of Nelson (1982) or Chapter 15 of Meeker and Escobar (1998) for examples.

ANALYSIS OF CENSORED LIFE DATA

This subsection illustrates simple, but powerful, methods for analyzing life data. The first example uses data from a laboratory life test. The second example illustrates the analysis of multiply censored data from a field tracking study. In both cases we proceed as follows:

1. Analyze the data graphically, using a nonparametric procedure (one that does not require any assumptions about the form of the underlying distribution) plotted on special probability paper. Such plots also allow us to explore the adequacy of possible distributional models.

2. If a parametric distribution provides an adequate description of the available data, estimate the distribution parameters.

Cumulative Distribution Function

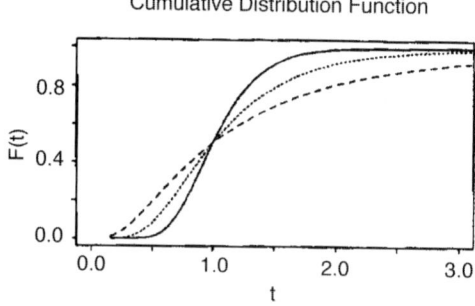

Probability Density Function

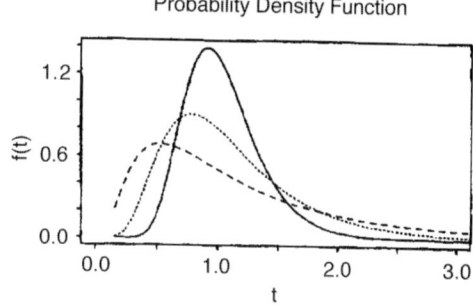

Hazard Function

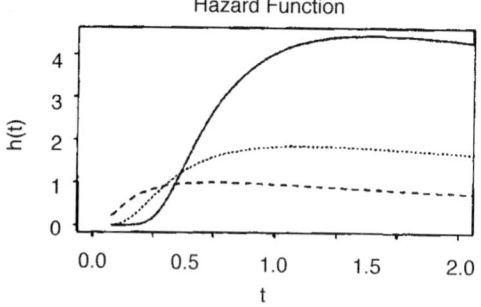

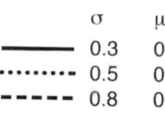

FIGURE 48.5   Lognormal distribution cdf, pdf, and hf for scale parameter $\exp(\mu) = 1$ and for shape parameter $\sigma = .3, .5,$ and $.8$.

3. Using these parameter estimates, estimate the distribution quantiles of interest, population proportion failing at specified times, hazard function values, and appropriate confidence intervals on some or all of these estimates.

Simple Life Test Data (with Single Censoring).

For sample life test data with n observations, continuous or very frequent monitoring for failures, and all censoring at the end of the observation period (typical for some laboratory life tests), a nonparametric estimate of $F(t)$ can be computed as

$$\hat{F}(t) = \frac{\text{number of failures up to time } t}{n} \tag{48.1}$$

This is a step function that jumps by an amount $1/n$ at each failure time (unless there are ties, in which case the estimate jumps by the number of tied failures divided by n). This estimate requires no assumption about the form of the underlying distribution. The estimate can be computed only up to the censoring time, and not beyond.

Example: Chain Link Fatigue Life.   Parida (1991) gives the results of a load-controlled high-cycle fatigue test conducted on 130 chain links. The 130 links were randomly selected from a population of different heats used to manufacture the links. Each link was tested until failure or until it had run for 80 thousand cycles, which ever came first. There were 10 failures—one each at 33, 46, 50, 59,

62, 71, 74, and 75 thousand cycles and two at 78 thousand cycles. The remaining 120 links had not failed by 80 thousand cycles. Figure 48.6 shows the step-function nonparametric estimate of $F(t)$ for these data.

Probability Plotting.   Comparing data in plots like Figure 48.6 with theoretical cumulative distributions like those shown in Figures 48.3, 48.4, and 48.5 is difficult, because the human eye cannot easily compare nonlinear curves. Probability plots display data on special *probability scales* such that a particular theoretical cumulative distribution is a straight line when plotted on such a scale. For example, taking the log of the Weibull quantile gives the straight line $\log(t_p) = \log(\eta) + (1/\beta)\log[-\log(1-p)]$. Thus, appropriately plotting the data on probability paper, and assessing whether the plot reasonably approximates a straight line, provides a simple assessment of the adequacy of an assumed model. In addition, probability plots are commonly used to estimate

- Distribution quantiles
- Probabilities of surviving to specified lifetimes

Figure 48.7 is a Weibull probability plot of the chain link failure data. One common method of making a probability plot is to plot the ith largest time (or number of cycles at failure) value on the time axis versus $(i - .5)/n$ on the probability axis for $i = 1,2,\dots$. This is equivalent to plotting, at each jump in the step function of Figure 48.6, the time of the step versus the point halfway between steps, as shown in Figure 48.7. Drawing a curve through these points allows one to estimate distribution quantiles and survival probabilities, without assuming any distributional model. However, the points plotted on Figure 48.7 fall nearly along a straight line, indicating that the Weibull distribution provides a good fit to the data. Figure 48.7 contains a special "β" scale that allows one to obtain a graphical estimate of the Weibull shape parameter. This is done by drawing a line through the cross mark on the top of the plot, parallel to the line that seems to best fit the data points. Then the graphical estimate of β is read off the "β" scale. In this case, the graphical estimate of the Weibull shape para-

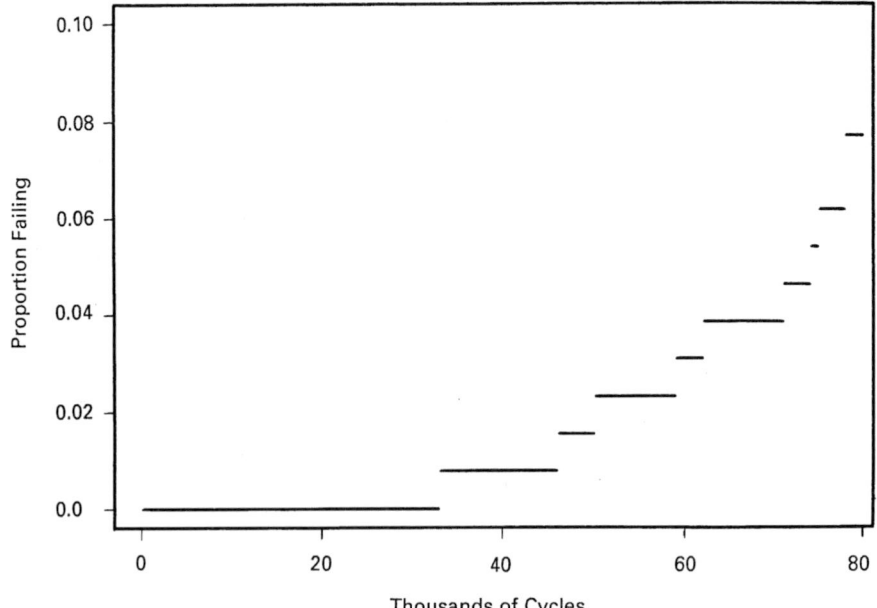

FIGURE 48.6   Nonparametric estimate of $F(t)$ for the chain link failure data.

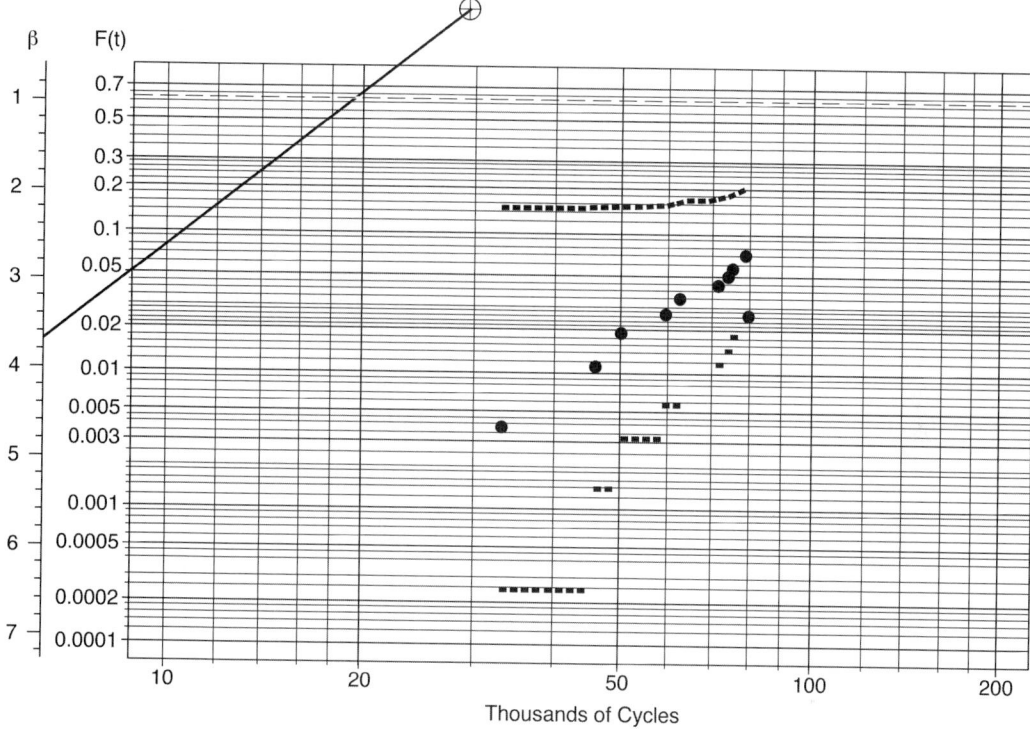

FIGURE 48.7   Weibull probability plot for the chain link failure data with 95 percent simultaneous confidence bands.

meter is $\hat{\beta} \approx 3.62$. A graphical estimate of the Weibull scale parameter η can be obtained by drawing a line to fit the plotted points and reading off an estimate of the .63 quantile, giving $\hat{\eta} \approx 185$ in the example. The grid lines on Figure 48.7 are useful for plotting by hand or for reading numbers from the plot. It can be argued that the grid lines interfere with the interpretation of the graph, and their inclusion should be a software option.

The light dotted points on this probability plot are 95 percent simultaneous nonparametric (i.e., without any parametric distribution assumption) confidence bands that help judge the sampling uncertainty in the estimate of $F(t)$. These confidence bands are obtained by using the method described in Nair (1984) and are defined by the inversion of a distributional goodness of fit test. If one can draw a straight line through the confidence bands, as is the case here, then the distribution implied by the probability paper (Weibull for Figure 48.7) cannot be ruled out as a model that might have generated the data.

A similar lognormal plot (not shown here) suggested the data could also have come from a lognormal distribution. The lognormal distribution assumption provides somewhat similar estimates for, say, quantiles within the range of the data, but appreciable differences outside the range.

Maximum Likelihood Estimation.   The maximum likelihood (ML) method provides a formal, objective, statistical approach for fitting a life distribution, using an assumed model, and estimating various properties, such as distribution quantiles and failure probabilities. A probability plot can also be used to show the ML estimate of $F(t)$. In fact, fitting a Weibull distribution using ML is an objective way of fitting a straight line through data points on a probability plot. Computer programs are convenient for this task. Least squares (using standard linear regression methods) can also be used to fit such a line to the plotted points. ML is, however, the preferred method, especially with censored data, because of its desirable statistical properties and other theoretical justifications. Figure 48.8 is similar to Figure 48.7, but gives a computer-generated Weibull probability plot for the chain link data with the fitted ML line superimposed. The dotted lines are drawn through a set of 95 percent parametric *pointwise* normal-approximation confidence intervals for $F(t)$ (computed as described in Chapter 8 of Meeker and Escobar 1998). For some purposes, the uncertainty implied by the intervals in Figure 48.8 would suggest that more information is needed. The next subsection discusses methods for choosing the sample size to control the width of confidence intervals for Weibull quantiles.

The probability plot in Figure 48.8 might invite extrapolation beyond the range of the observed times to failure in either direction. It is important to note, however, that such extrapolation is dangerous; the data alone do not tell us about the shape of $F(t)$ outside the range of 30 to 90 thousand cycles. Moreover, the confidence intervals, even though they are wider outside the data range, reflect only the uncertainty due to limited data, and assume that the fitted Weibull model holds beyond, as well as within, the range of the data.

Guidelines for Choosing the Sample Size in Life Tests.

Generally, larger samples lead to more precise estimates. This subsection shows how to choose the sample size n to control precision in a simple life test. Certain input information is required, as illustrated in the following example.

Planning Values for Sample Size Determination.   For purposes of illustration assume that investigators intend to run a life test for a new mechanical device. The test will run for 30 days and the experimenters, from past experience, expect that about 7 percent of the units will fail in that time. From past experience with similar products, they expect the data to fit a Weibull distribution with a shape parameter near $\beta = 3$.

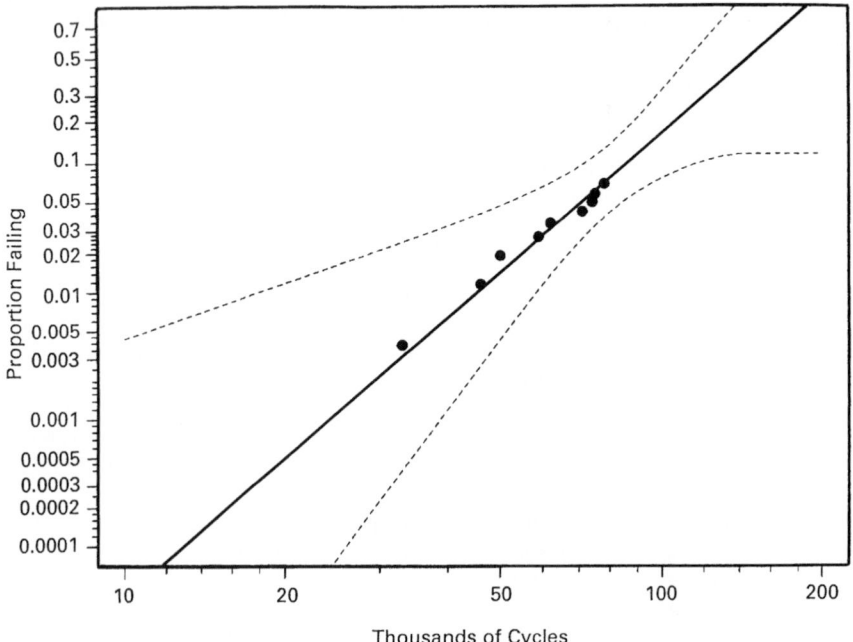

FIGURE 48.8   Weibull probability plot with the Weibull ML estimate and a set of pointwise approximate 95 percent confidence intervals for $F(t)$ for the chain link failure data.

Using Simulation to Evaluate Precision.   The plots in Figure 48.9 compare simulations of ML cdf estimates for the above testing situation (Weibull distribution with $\eta = 30[-\log(1-.07)]^{-1/3} = 71.92$ days and $\beta = 3$) with a fixed censoring time of 30 days for sample sizes $n = 40, 160, 640,$ and 2560; each simulation is based on 30 random samples of size n. In each plot, the thicker, longer line shows the cdf of the assumed underlying "true" population distribution, from which the simulated data were generated. The other lines show the cdf, as estimated from each of the 30 simulations. A comparison of the plots shows that with a small sample size, the estimated line can depart appreciably from the true line. Looking at the horizontal dashed line allows us to see the improvement in precision for estimating the time at which 1 percent of the population will fail as the sample size increases. For example, for $n = 40$, the 30 estimates of $t_{.01}$ range from 5.9 to 28.0 days, while for $n = 2560$, the range is 11.2 to 14.2 days. Thus, such simulations allow us to assess the precision that can be expected in the estimates from a planned study and from this choose an appropriate sample size.

Analytical Formula for Sample Size.   If one prefers an analytic approach to find the sample size needed to estimate the quantile t_p of a Weibull distribution with a specified degree of precision, one can use a simple approximate formula, supplemented by a set of curves. Thus, in our example, we might wish to estimate the .01 quantile within 40 percent of its true value with 95 percent confidence. The approximate formula is

$$n = \frac{z_{(1-\alpha/2)}^2 V^{\square}}{[\log(R_T)]^2}$$

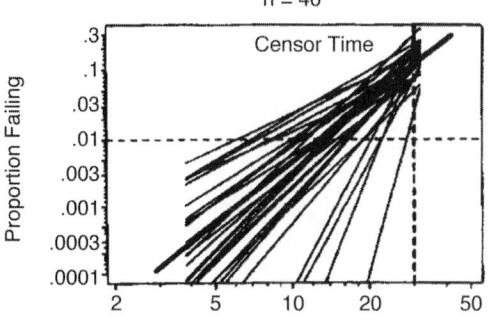

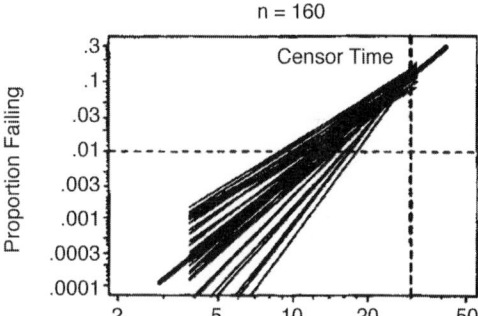

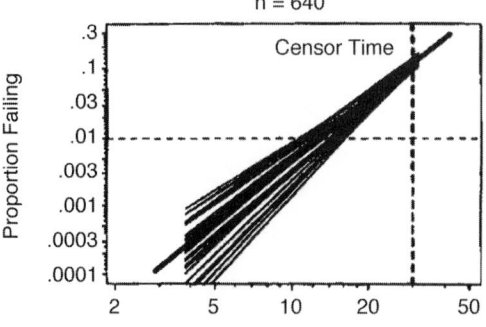

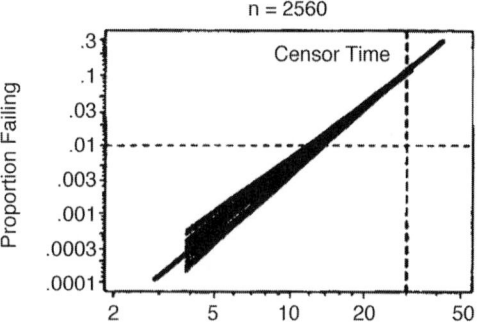

FIGURE 48.9   Weibull probability plots showing the cdf, and the estimated .01 quantile as estimated by ML, for 30 simulated samples of size n=40, 160, 640, and 2560 from a Weibull distribution with η=71.92 and β=3.

where $z_{(1-\alpha/2)}$ is the $1 - \alpha/2$ quantile of the standard normal distribution, $1 - \alpha$ is the desired level of confidence, $R_T > 1$ is the target precision, and $V^\square$ is the value read from Figure 48.10, multiplied by $(\beta^\square)^2$ where $\beta^\square$ is a "planning value" for the Weibull distribution parameter β. In our example $\beta^\square = 3$. For a 95 percent confidence interval $\alpha = .05$, $1 - \alpha/2 = .975$ so $z_{.975} = 1.96$. Typical values of R_T range from 1.2 to 2, where the desire is to have confidence interval endpoints not more than $100(R_T - 1)$ percent above and below the point estimate for t_p. In our example $R_T = 1.4$. Figure 48.10 gives a plot of the large-sample approximate variance factor $\beta^2 \times V$ for estimating t_p. A smaller variance implies better estimation precision. Each line gives $\beta^2 \times V$ as a function of the quantile p whose value is to be estimated from the life test; in our example $p = .01$. There are different lines for different values of p_c, the expected proportion failing during the life test. In our example, $p_c = .07$. The curves in Figure 48.10 are then used to read $\beta^2 \times V$ using p and p_c and then solve for the value $V^\square$ of V, using $\beta^\square$. If the model and planning values are correct, the actual precision achieved will be smaller than the target precision R_T with probability approximately .5.

For the example, entering Figure 48.10 with $p_c = .07$ and $p = .01$, we read the ordinate $V^\square \times (\beta^\square)^2 \approx 70$, so $V^\square \approx 70/(3)^2 = 7.778$. Thus, with $R_T = 1.4$,

$$n = \frac{z_{(1-\alpha/2)}^2 V^\square}{[\log (R_T)]^2} = \frac{(1.96)^2(7.778)}{[\log (1.4)]^2} \approx 264$$

is the number of items that should be tested. In summary, in order to estimate the 0.01 quantile within 40 percent of its true value with 95 percent confidence, we need to test approximately 264 units up to 30 days (under the assumption of a Weibull distribution with shape parameter 3 and an expected 7 percent failures after 30 days).

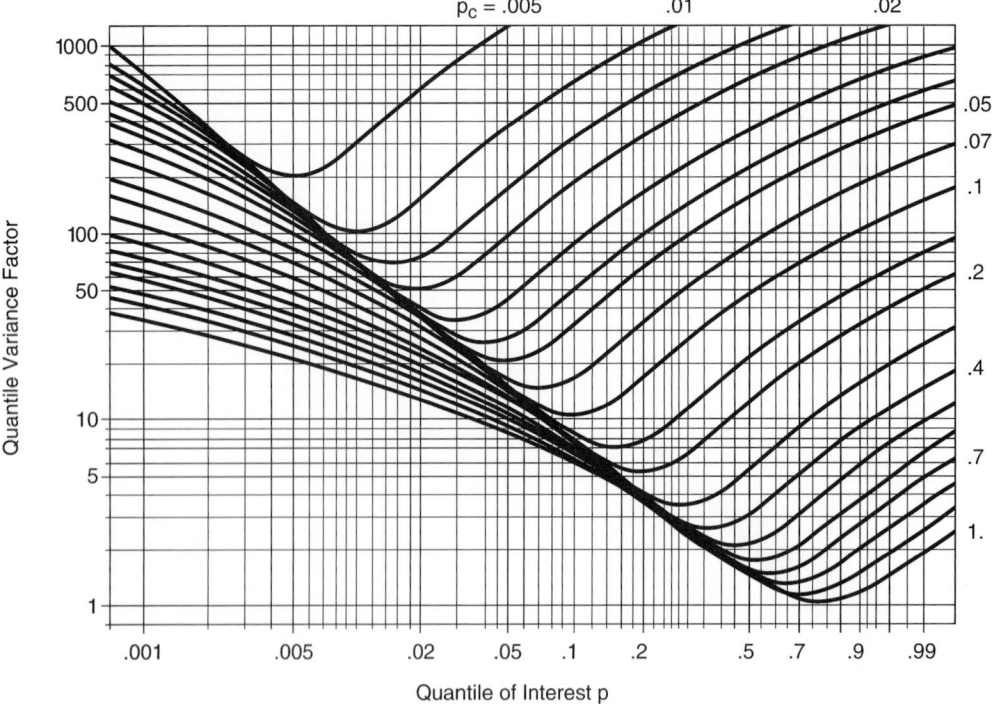

FIGURE 48.10 Large-sample approximate variance factor $\beta^2 \times V$ for ML estimation of Weibull distribution quantiles as a function of p_c, the population proportion failing by censoring time t_c and p, the quantile of interest.

Analysis of Multiply Censored Data. Multiple censoring can arise when units go into service at different times or units accumulate varying running times over a particular elapsed time period (likely in the field), or in a variety of other situations (such as the analysis of a particular failure mode, for which units that fail independently due to other modes are taken as still running with regard to the mode under evaluation). In this case, often some running times for unfailed units are less than some failure times. Multiple censoring is encountered frequently in practice, especially in dealing with field life data. For such data, the cumulative failure probability $F(t)$ cannot be estimated directly using (48.1). Instead an alternative estimator, to be described below, is used. The ML estimates and approximate confidence intervals that were previously mentioned still do apply.

Example: Shock Absorber Failure Data. O'Connor (1985) gives failure data for shock absorbers. At the time of analysis, failures had been reported at 6700, 9120, 12,200, 13,150, 14,300, 17,520, 20,100, 20,900, 22,700, 26,510, and 27,490 km. There were 27 units in service that had not failed. The running times for these units were 6950, 7820, 8790, 9660, 9820, 11,310, 11,690, 11,850, 11,880, 12,140, 12,870, 13,330, 13,470, 14,040, 17,540, 17,890, 18,450, 18,960, 18,980, 19,410, 20,100, 20,150, 20,320, 23,490, 27,410, 27,890, and 28,100 km.

To estimate $F(t)$ with multiply censored data, we use the following nonparametric procedure (i.e., no distribution model assumed). Let $t_1, t_2, \ldots, t_r$ denote the times at which failures occurred, let d_i denote the number of units that failed at t_i, and let n_i denote the number units that are unfailed just before t_i. The estimator of $F(t)$ for all values of t between t_i and t_{i+1} is

$$F(t_i) = 1 - \prod_{j=1}^{i} \left(1 - \frac{d_j}{n_j}\right) \qquad i = 1, \ldots, r \qquad (48.2)$$

This is the well-known product-limit or Kaplan-Meier estimator. It can be used only within the range of the failure data. The justification for this estimator is based on the relationship between the cdf and the discrete-time hazard function and is described more completely in Chapter 3 of Meeker and Escobar (1998). Calculations of the estimated cumulative failure probability $F(t_i)$ and the estimated survival probability $S(t_i)$ to 12,200 km for the shock absorber data are illustrated in Table 48.1. This table could have gone up to 27,490 km—the largest failure time.

Figure 48.11 shows the shock absorber data plotted on Weibull probability paper, with the points plotted at each of the failure times at a height halfway between the jumps in the nonparametric cdf estimate in Table 48.1 (extended to 27,490 km). Figure 48.11, like Figure 48.8, also shows the Weibull ML cdf estimate and a set of approximate 95 percent pointwise confidence intervals for $F(t)$. The plot shows that the Weibull distribution fits the data well. A similar lognormal probability plot

TABLE 48.1 Nonparametric Estimates of $F(t)$ for the Shock Absorber Data up to 12,200 km

t_i, km	d_j	r_j	n_j	d_j/n_j	$1-d_j/n_j$	$\hat{S}(t_i)$	$\hat{F}(t_i)$
6,700	1	0	38	1/38	37/38	.97368	.0263
6,950	0	1	37				
7,820	0	1	36				
8,790	0	1	35				
9,120	1	0	34	1/34	33/34	.94505	.0549
9,660	0	1	33				
9,820	0	1	32				
11,310	0	1	31				
11,690	0	1	30				
11,850	0	0	29				
11,880	0	1	28				
12,140	0	1	27				
12,200	1	0	26	1/26	25/26	.90870	.0913

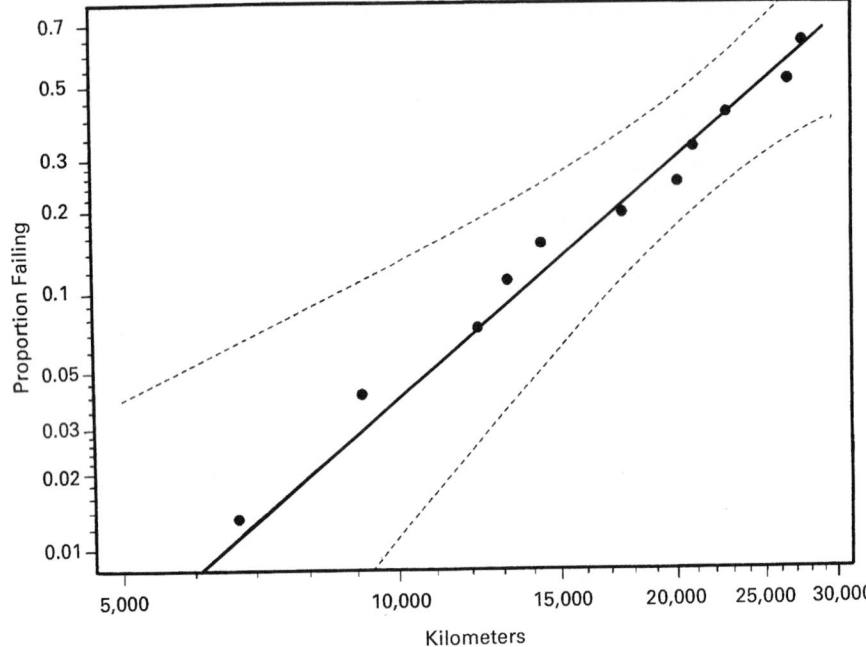

FIGURE 48.11 Weibull probability plot of shock absorber failure times with maximum likelihood estimates and approximate 95 percent pointwise confidence intervals for $F(t)$.

(not shown here) did not fit nearly as well. Although the plotted points represent a nonparametric estimate that is *not* based on any assumed distribution, the straight line drawn through the points on Weibull probability paper, by ML, freehand, or some other method is a parametric estimate that, unlike the nonparametric estimate, allows extrapolation in time. Extrapolation outside the range of the data is, however, dangerous because the assumed distribution model, though providing a reasonable fit near the center of the data, may no longer apply beyond the extremes.

Chapter 4 of Nelson (1982) describes hazard plotting, an alternative graphical method of nonparametric estimation for multiply censored data. The two methods are very similar and generally provide similar estimates.

ACCELERATED LIFE TEST MODELS AND DATA ANALYSIS

Estimating the time-to-failure distribution or long-term performance of components of *high-reliability* products is particularly difficult. Most modern products are designed to operate without failure for years, decades, or longer. Thus, we might expect that few units will fail in a test of practical length at normal use conditions. For example, the design and construction of a new communications satellite or a new appliance may allow only 8 months to test components that are expected to be in service for 15 or 20 years. For such applications, Accelerated Life Tests (ALTs) are used widely, particularly to obtain timely information on the reliability of simple product components, materials, and to provide early identification (and removal) of failure modes, thus improving reliability. There are difficult practical and statistical issues involved in accelerating life, especially for a complicated product that can fail in different ways. Here we describe accelerated life tests for components, materials, or products with a single failure mode. Extensions are described in Chapter 7 of Nelson (1990). As previously suggested, the analysis of life data with various independent failure modes often calls for considering failure modes one at a time, treating failures from other modes as if they are unfailed, and then combining the results.

Often, information from tests at high levels of stress (e.g., use rate, temperature, voltage, or pressure) is extrapolated, through a physically reasonable statistical model, to obtain estimates of life or long-term performance at lower, normal levels of stress. In some cases, stress is increased or otherwise changed during the course of a test (step-stress and progressive-stress ALTs). ALT results are used in the reliability-design process to assess or demonstrate component and subsystem reliability, certify components, detect failure modes, compare different manufacturers, and so forth as *speedily* as possible. ALTs have become increasingly important because of rapidly changing technologies, more complicated products with more components, the pressures for rapid new product introduction, and higher customer expectations for better reliability (in place of a traditional "ship and fix later" approach).

Methods of Acceleration. Three different methods of accelerating a reliability test are

- Increase the use rate (or cycling rate) of the product. A typical toaster is designed for a median lifetime of 20 years with a usage rate of twice each day. If, instead, we test the toaster 365 times each day, we could reduce the median lifetime (for many potential failure modes) to about 40 days. Also, because it is not necessary to have all units fail in a life test, useful reliability information might be obtained in a matter of days instead of months. This form of acceleration is the simplest to handle, since the methods for a single distribution can generally be applied directly (after making an appropriate time transformation). Thus, this situation will not be discussed further. The key assumption is that the increased use rate will not introduce new failure modes or change the damage per cycle rate for existing failure modes (e.g., due to overheating of cycled components).

- Increase the aging rate of the product. For example, increasing temperature or humidity can accelerate the chemical processes of certain failure modes such as chemical degradation (resulting in eventual weakening) of an adhesive mechanical bond or the growth of a conducting filament across an insulator (eventually causing a short circuit).

- Increase the level of stress (e.g., voltage or pressure) under which test units operate. A unit will fail when its *strength* drops below applied stress. Thus, a unit with degrading strength at a high level of stress will generally fail more rapidly than it would have failed at a low level of stress.

Combinations of these methods of acceleration can also be employed. A factor like voltage will, for some failure modes, both increase the rate of an electrochemical reaction (thus accelerating the aging rate) and increase stress relative to strength. In such situations, when the effect of stress is complicated, there may not be enough knowledge to provide an adequate physical model for acceleration. Empirical models may or may not be useful. See Chapter 2 of Nelson (1990) or Chapter 18 of Meeker and Escobar (1998) for more discussion.

Acceleration Models. Analysis of accelerated life tests, other than those that just increase the product use rate, requires a physically reasonable model relating acceleration variables like temperature, voltage, pressure, and size to time to failure. One then fits the model to the data to describe the effect that the factors have on failure.

Physical Acceleration Models. For well-understood failure modes, it may be possible to use a model based on physical/chemical theory that will be capable not only of describing the failure-causing process over the range of the data, but also allow extrapolation to use conditions based on accelerated testing. Usually the actual relationship between acceleration variables and the actual failure mechanism is extremely complicated. Often, however, it is possible to find a physical simplification that will adequately capture and describe the dominant aspect of the process. For example, failure may be affected by a complicated chemical process with many steps, but there may be one rate-limiting (or dominant) step, and a good understanding of this part of the process may provide a model that is adequate for extrapolation. See Meeker and LuValle (1995) and Chapter 2 of Nelson (1990) for examples.

Empirical Acceleration Models.   When there is little understanding of the chemical or physical processes leading to failure, it may be impossible to develop a model based on physical/chemical theory. An empirical model may be the only alternative. However, such a model might provide an excellent fit to the available data, but may be highly incorrect when extrapolated to the use conditions of real interest. In some situations there may be extensive experience with an empirical model and this may provide the needed basis for extrapolation to use conditions.

Some guidelines for the use of acceleration models include:

- Use physical theory as much as possible, and choose accelerating variables (e.g., temperature, voltage, or humidity) that correspond to factors that cause actual failures.

- Investigate previous attempts to accelerate failure modes similar to the ones of interest. There are, for example, many research reports and papers on physics of failure. Chapter 2 of Nelson (1990) or Chapter 18 of Meeker and Escobar (1998) provide examples and references.

- Design accelerated tests to minimize, as much as possible, the degree of extrapolation required [see Meeker and Hahn (1985), Chapter 6 of Nelson (1990), and Chapter 20 of Meeker and Escobar (1998)]. Highly accelerated testing can cause failure modes that would not occur at use conditions. If extraneous failures are not recognized they can lead to seriously incorrect conclusions. Note that product life data can be analyzed, using maximum likelihood methods, even though at some test conditions the majority, or even all, of the test units remain unfailed. This allows us to be much more conservative in establishing testing conditions than would otherwise be the case, (i.e., include tests at or close to the use conditions). See Meeker and Hahn (1985) or Nelson (1990) for details.

- Most accelerated tests are used to obtain information about a single, relatively simple, failure mode (or corresponding degradation measure). If there is more than one failure mode, it is likely that the different failure modes will be accelerated at different rates and, unless this is accounted for in the modeling, testing, and analysis, the resulting inferences could be seriously incorrect.

- In practice, acceleration relationships are difficult to verify in their entirety. The accelerated life test data should be used to look for departures from the assumed acceleration model. It is important to recognize, however, that the available data will generally provide very little power to detect anything but the most serious departures, and typically there is no useful diagnostic information close to use conditions.

- Simple models with the right shape have generally proven to be more useful for extrapolation to use conditions than elaborate multiparameter models. For example, even if there is some evidence of curvature in ALT data, a simple linear model relating the accelerated stress variable to time to failure will generally provide more accurate extrapolation than a fitted quadratic model.

- Sensitivity analyses should be used to assess the effect of perturbing uncertain inputs (e.g., inputs related to model assumptions). For example, it is useful to compare the extrapolations to use conditions with different assumed time-to-failure distributions, or different physically reasonable models relating stress to life.

- Include some test units at conditions at which no failures would be expected during the duration of the test (e.g., at or near use conditions). If failures do, indeed, take place at these conditions it would raise serious concerns about reliability (and suggest that the assumed model is inadequate).

- Accelerated life test programs should be planned and conducted by teams including individuals knowledgeable about the product and its use environment, the physical/chemical/mechanical aspects of the failure mode, and the statistical aspects of the design and analysis of reliability experiments.

The next subsection describes a simple temperature-acceleration model and illustrates the analysis of data from an ALT.

Elevated Temperature Acceleration.

High temperature has been said to be the enemy of reliability. Increasing temperature is one of the most commonly used methods to accelerate failures.

RELIABILITY CONCEPTS AND DATA ANALYSIS

The Arrhenius relationship is a well-known model describing the effect that temperature has on R, the rate of a simple chemical reaction. This model for reaction rate can be written as

$$R(\text{temp}) = \gamma_0 \exp\left(\frac{-E_a}{k_B \times \text{temp K}}\right) = \gamma_0 \exp\left(\frac{-E_a \times 11{,}605}{\text{temp K}}\right)$$

where temp K = temp °C + 273.15 is temperature in the absolute Kelvin scale, $k_B = 1/11{,}605$ is Boltzmann's constant in units of electron volts per degrees Celsius and E_a is the reaction activation energy in electron-volts. The parameters E_a and γ_0 are generally unknown product or material characteristics. In some simple situations, the failure-causing process can be modeled adequately by such a simple chemical reaction. In such cases, if log (T) follows a distribution with a location and scale parameter (so that the time to failure T could, for example, be described by a lognormal distribution or a Weibull distribution) with parameters μ and σ then the life distribution can be written as a function of temperature:

$$\Pr\,(T \leq t;\, \text{temp}) = \Phi\left[\frac{\log\,(t) - \mu(\text{temp})}{\sigma}\right]$$

where $\Phi = \Phi_{\text{nor}}$, and $\Phi = \Phi_{\text{sev}}$ (see earlier discussion) for the lognormal and Weibull distributions, respectively, $\mu(\text{temp}) = \beta_0 + \beta_1 x$, $x = 11{,}605/(\text{temp K})$, and $\beta_1 = E_a$. This is known as the Arrhenius failure-time model. One would then use life data at various accelerated temperatures and the resulting observed times to failure to estimate β_0, β_1, and σ.

Although the Arrhenius model does not apply to all temperature-acceleration problems and will be adequate over only a limited range of temperatures (depending on the particular application), it is used widely in many areas of application. Nelson (1990, p. 76) comments that "…in certain applications (e.g., motor insulation), if the Arrhenius relationship…does not fit the data, the data are suspect rather than the relationship."

Similar life-stress regression relationships have been used for other accelerating variables like voltage, pressure, humidity, cycling rate, and specimen size. See Chapter 2 of Nelson (1990) and Chapter 18 of Meeker and Escobar (1998) for examples and further references.

Planning an ALT.   Usually accelerated life tests need to be conducted within stringent cost and time constraints. Careful planning is essential. Resources need to be used efficiently, and the degree of extrapolation minimized, to the greatest degree possible. Meeker and Hahn (1985), Chapter 6 of Nelson (1990), and Chapter 20 of Meeker and Escobar (1998) describe methods for planning statistically efficient ALTs that meet practical constraints. During the test planning phase of a study, experimenters should try to explore the kind of results that they might obtain as a function of the assumed model and proposed test plan. Simulation methods, such as described in discussion of the guidelines for choosing sample size, also provide useful insights in this process.

Strategy for Analyzing ALT Data.   This subsection outlines a useful procedure for analyzing ALT data when multiple units are life tested at three or more conditions (e.g., three or more temperatures). See, for example, the data in Table 48.2.

The basic idea is to start by analyzing the data at each test condition separately and then to fit a model that ties together the data at the different conditions. Briefly, the strategy, as illustrated in the example that follows, is to:

1. Construct probability plots of the data at each test condition (level of the accelerating factor) separately, to suggest and explore the adequacy of possible distributional models.

2. At each test condition with two or more failures, fit models, as suggested by the previous step, individually to the data using the ML method. Plot the ML lines on a multiple probability plot for each of these conditions. Use the plotted points and fitted lines to assess the reasonableness of assumptions (such as a constant standard deviation at each test condition, e.g., Figure 48.12).

TABLE 48.2 Failure Intervals or Censoring Times and Testing Temperatures from an ALT Experiment on a New-Technology Integrated Circuit Device

Temperature °C	Number of devices	Hours Lower	Hours Upper	Status
150	50		1536	Censored
175	50		1536	Censored
200	50		96	Censored
250	1	384	788	Failed
250	3	788	1536	Failed
250	5	1536	2304	Failed
250	41		2304	Censored
300	4	192	384	Failed
300	27	384	788	Failed
300	16	788	1536	Failed
300	3		1536	Censored

3. Fit a model to the assumed relationship between life and the accelerating variable. Plot the fitted model on probability paper, together with a plot of the data (e.g., Figure 48.13).

4. Perform residual analyses and other diagnostic checks of the model assumptions (as in standard regression analysis). See Chapter 5 of Nelson (1990). (This is not done for the example here.)

5. Obtain the desired estimates and confidence intervals, and assess their reasonableness (e.g., Figure 48.14).

For further examples and further discussion of methods for analyzing ALT data, see Nelson (1990) or Meeker and Escobar (1998).

Example.   Table 48.2 gives the results of an ALT on a new-technology integrated circuit (IC) device. The device inspection process involved an electrical diagnostic test conducted periodically to determine whether or not failure had occurred. This test was expensive because it was manual and required much time on a special machine. Thus, only a few inspections could be conducted on each device. One common method of planning the times for such inspections is to choose a first inspection time and then space the further inspections to be equally spaced on a log scale. In this case, the first inspection was after 1 day with subsequent inspections at 2 days, 4 days, and so on (except for a day when the person doing the inspection had to leave early). Tests were run at 150, 175, 200, 250, and 300°C. Life tests in which failures are recorded only at inspection times lead to interval-censored data (sometimes called "interval" or "read out" data). When a unit fails the test, all that is known is that there has been a failure in the elapsed time since the previous inspection. Interval-censored data require special statistical methods that are described in Chapter 9 of Nelson (1982), Chapter 3 of Nelson (1990), and Chapters 3, 7, and 19 of Meeker and Escobar (1998). Table 48.2 gives the interval in which failure occurred for the devices that failed. Correspondingly, for devices that did not fail, Table 48.2 gives the running time at the last inspection.

At the time of analysis, failures had been found only at the two highest temperatures. After early failures had been observed at 250 and 300°C there was some concern that no failures would be observed at 175°C before the time at which decisions would have to be made. Thus the 200°C test was started later than the others and only limited running time was accumulated by the time of the analysis.

The developers were interested in estimating the activation energy of the suspected failure mode and the long-life reliability of the components. In particular, they wanted to estimate the proportion of devices in the product population that would fail by 100 thousand hours (about 11 years) at the use condition of 100°C.

Figure 48.12 is a lognormal probability plot and estimated ML cdf of the failures at 250 and 300°C along with the ML estimates of the individual lognormal cdfs. The differing slopes in the plot suggest the possibility that the lognormal shape parameter sigma changes from 250 to 300°C. Such a change could be caused by a change in failure mode. Failure modes with a higher activation energy, that might never be seen at low levels of temperature, can appear at higher levels of temperature (or other acceleration factors). An approximate 95 percent confidence interval on $\sigma_{250}/\sigma_{300}$ is [1.01, 3.53] (calculations not shown here), suggests that the difference could be real. These results also suggested that detailed physical failure mode analysis should be done for at least some of the failed units and that, perhaps, the accelerated test should be extended until some failures are observed at lower levels of temperature. However, in the subsequent analyses we shall assume that σ is constant at the different temperatures.

Table 48.3 gives Arrhenius-lognormal model ML estimation results and confidence bounds for the new-technology IC device under the assumed model. Figure 48.13 is a lognormal probability plot of the data at 250°C and 300°C showing the Arrhenius-lognormal model cdf fit to the data. This figure also shows lognormal cdf estimates from the fitted model for the other test temperatures as well as the estimated cdf, and an approximate 95 percent confidence interval at the use condition of 100°C. We note that the upper bound of the 95 percent confidence interval on the proportion failing by 100,000 hours at this temperature is less than .0001 (0.01 percent).

Figure 48.14 is an Arrhenius plot of the lognormal model fit to the new-technology IC device ALT data, showing the times at which 1, 10, and 50 percent of the units are expected to fail as a function of time. This plot shows the rather extreme extrapolation needed to make inferences at the use conditions of 100°C. If the projections are close to reality, it appears unlikely that there will be any failures below 200°C during the remaining 3000 hours of testing, and, as mentioned before, this was the reason for testing additional units at 200°C.

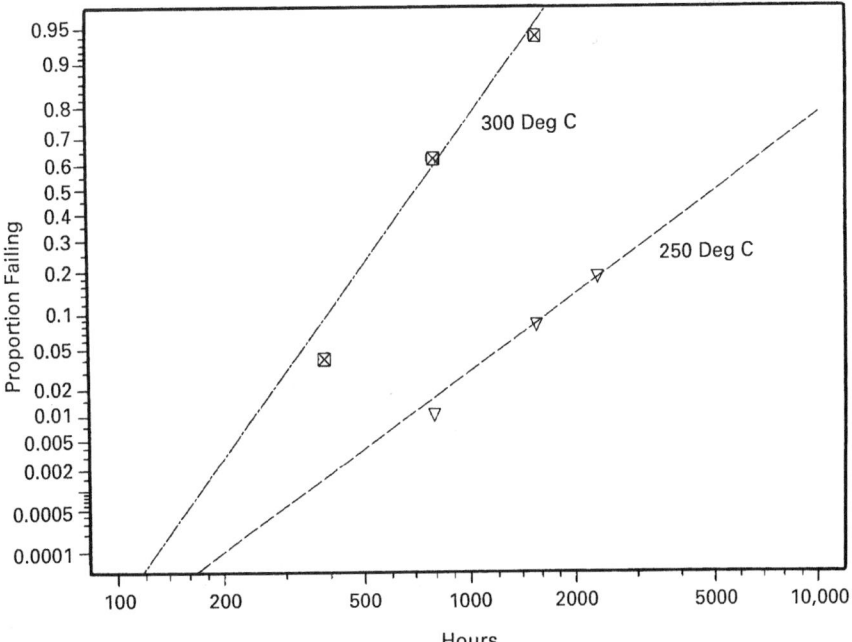

FIGURE 48.12   Lognormal probability plot and estimated ML cdf of the failures at 250 and 300°C for the new-technology integrated circuit device ALT experiment.

TABLE 48.3 Individual Lognormal ML Estimation Results for the New-Technology IC Device*

Parameter	ML estimate	Standard error	Approximate 95% confidence intervals	
			Lower	Upper
β_0	−10.2	1.5	−13.2	−7.2
β_1	.83	.07	.68	.97
σ	.52	.06	.42	.64

*The loglikelihood is $\mathscr{L} = -88.36$. The confidence intervals are based on the normal approximation method.

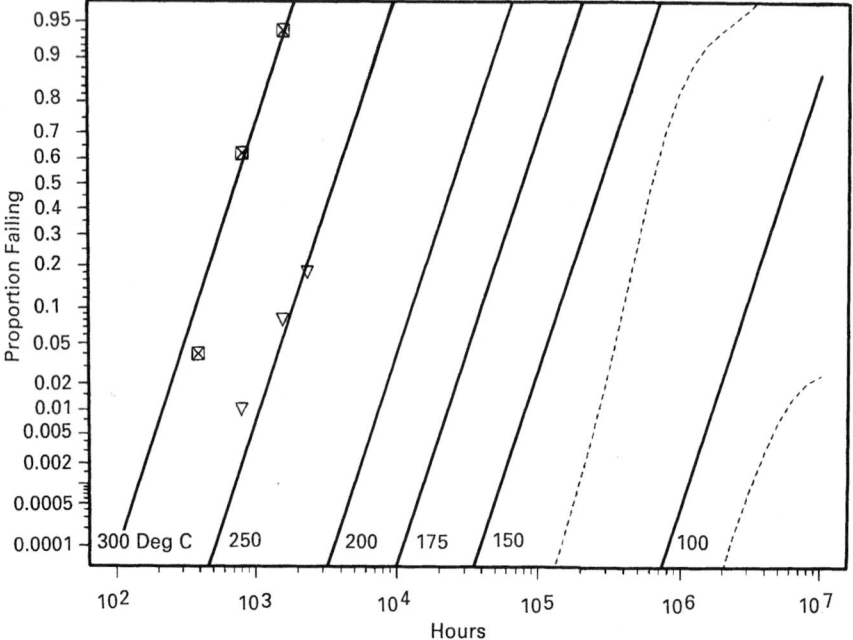

FIGURE 48.13 Lognormal probability plot showing the Arrhenius-lognormal model ML estimation results for the new-technology IC device at the test temperatures and (with the approximate 95 percent confidence interval) at 100°C.

SYSTEM RELIABILITY CONCEPTS

The system failure probability $F_T(t)$ is the probability that the system fails before time t. The failure probability of the system is a function of time in operation t (or other measure of use), the operating environment(s), the system structure, and the reliability of system components, interconnections, and interfaces (including, for example, human operators).

This subsection describes several simple system structures. Not all systems fall into one of these categories, but the examples provide building blocks to illustrate the basics of using system structure to compute system reliability. Complicated system structures can generally be decomposed into collections of the simpler structures presented here. Thus the methods for evaluation of system relia-

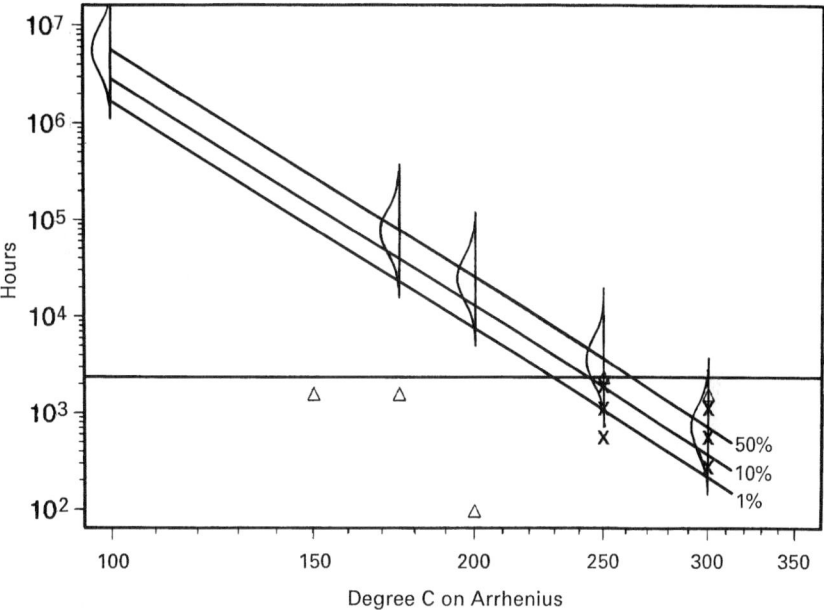

FIGURE 48.14   Arrhenius plot showing the Arrhenius-lognormal model ML estimates for the new-technology IC device. Failure and censored observation times are indicated by × and Δ respectively.

bility can be adapted to more complicated structures. For more information, see texts such as O'Connor (1985), Høyland and Rausand (1994), or Meeker and Escobar (1998).

Terminology for System Reliability.   Consider the time to failure of a system (i.e., all components starting a time 0) with s independent components. The cdf for component i is denoted by $F_i(t)$. The corresponding reliability (or survival probability) is $S_i(t) = 1 - F_i(t)$. The cdf for the system is denoted by F_T. This cdf is determined by the F_i's and the system structure, that is, $F_T(t) = g[F_1(t), \ldots, F_s(t)]$. To simplify the presentation, this function will be expressed as $F_T = g(F_1, \ldots, F_s)$.

Systems with Components in Series.   A series structure with s components works if and only if all the components work. Examples of systems with components in series include chains, high-voltage multicell batteries, inexpensive computer systems, and inexpensive decorative tree lights using low-voltage bulbs. For a system with two independent components in a series, illustrated in Figure 48.15, the cdf is

$$F_T(t) = \Pr(T \leq t) = 1 - \Pr(T > t) = 1 - \Pr(T_1 > t \text{ and } T_2 > t)$$

$$= 1 - [\Pr(T_1 > t) \, \Pr(T_2 > t)] = 1 - (1 - F_1)(1 - F_2) \qquad (48.3)$$

FIGURE 48.15   A system with two components in series.

For a system with s independent components $F_T(t) = 1 - \prod_{i=1}^{s} (1 - F_i)$ and for a system with s independent components, each of which has the same time to failure distribution [for example, multicell batteries $(F = F_i, i = 1, ..., s)$], $F_T(t) = 1 - (1 - F)^s$. The system hazard function, for a series system of s independent components, is the sum of the component hazard functions:

$$h_T(t) = \sum_{i=1}^{s} h_i(t)$$

Figure 48.16 shows the relationship between system reliability $1 - F_T(t)$ and individual component reliability $1 - F_T(t)$ for systems with different numbers of identical independent components in series. This figure shows that extremely high component reliability is needed to maintain high system reliability if the system has many components in series. If the system components are not independent, then the first line of (48.3) still gives $F_T(t)$, but the evaluation has to be done with respect to the bivariate distribution of T_1 and T_2 with a similar generalization to a multivariate distribution for more than two components.

Importance of Part Count in Product Design.   An important rule of thumb in reliability engineering design practice is "keep the part count small," meaning keep the number of individual parts (or components in series) in a system to a minimum. Besides the cost of purchasing and handling of additional individual parts, there is also an important reliability motivation for having a smaller number of parts in a product. For example, the design of a new-technology computer modem uses a higher level of microelectronic integration and requires only 20 discrete parts instead of the 40 parts required in the previous generation. As a specific example, the system hazard function $h_T(t)$ becomes particularly simple if a constant hazard rate (or equivalently, an exponential time-to-failure distribution) provides an adequate model for time to failure for the system components or parts. In this case, as a rough approximation, assuming that all failures are due to part failures, and that the parts have the same hazard function, the new design with only 20 parts will experience only half of the failures of the old design. Allowing that failures can occur at interfaces and interconnections between parts with the same frequency in the new and old designs would further increase the reliability improvement because of the larger number of such interfaces with a higher number of parts. With nonconstant hazard function (more common in practice), and parts with different hazard functions, the idea is similar.

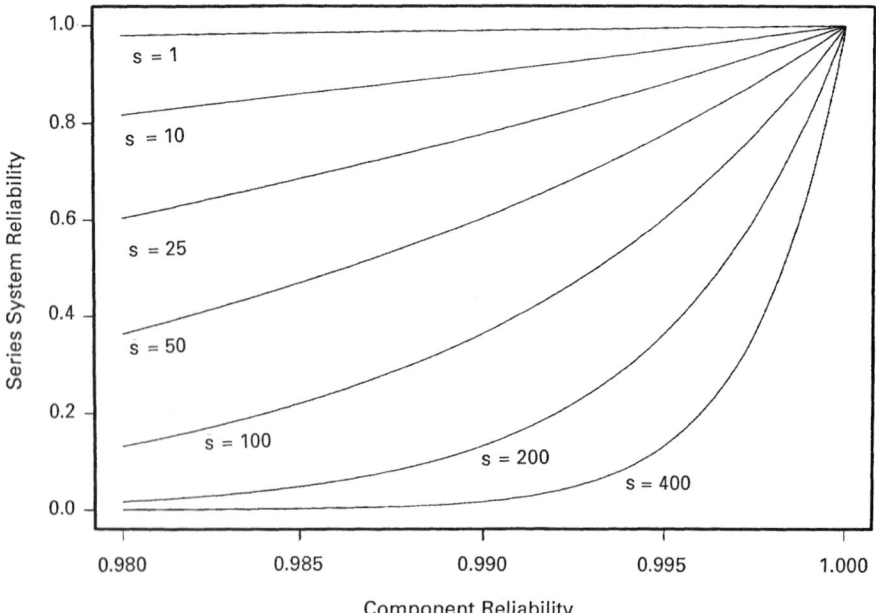

FIGURE 48.16   Reliability of a system with s identical independent components in series.

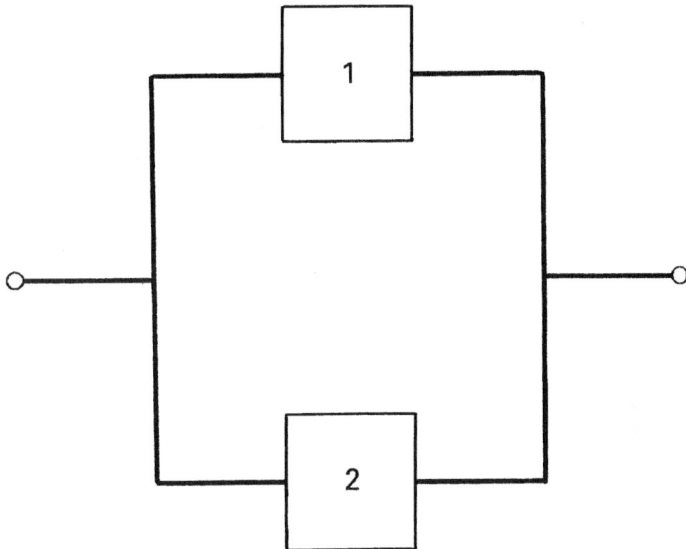

FIGURE 48.17   A system with two components in parallel.

This illustration assumes, of course, that the reliability of the parts in the new system will be the same (or at least similar to) the reliability of the individual parts of the old system, and that the stress in operation on each part remains the same. If the new system uses parts from an immature production process with low part reliability, or the operating stress on individual components is increased, the new system could have lower reliability.

Systems with Components in Parallel.   Parallel redundancy is often used to improve the reliability of weak links or critical parts of larger systems. A parallel structure with s components works if at least one of the components works. Examples of systems with s components in parallel include automobile headlights, RAID computer disk array systems, stairwells with emergency lighting, overhead projectors with a backup bulb, and multiple light banks in a classroom. For two independent parallel components, illustrated in Figure 48.17,

$$F_T(t) = \Pr(T \leq t) = \Pr(T_1 \leq t \text{ and } T_2 \leq t)$$

$$= \Pr(T_1 \leq t)\Pr(T_2 \leq t) = F_1 F_2 \qquad (48.4)$$

For s independent components $F_T(t) = \Pi_{i=1}^{s} F_i$ and for s independent identically distributed components $(F_i = F, i = 1, ..., s)$, $F_T(t) = F^s$. Figure 48.18 shows the relationship between system reliability $1 - F_T(t)$ and individual component reliability $1 - F(t)$ for different numbers of identical independent components in parallel. The figure shows the dramatic improvement in reliability that parallel redundancy can provide. If the components are not independent, then the first line of (48.4) still gives $F_T(t)$, but the evaluation has to be done with respect to the bivariate distribution of T_1 and T_2. A similar generalization applies for more than two nonindependent components.

REPAIRABLE SYSTEM DATA

The previous discussion dealt with reliability data analysis for nonrepairable components (or devices). Since a nonrepairable component can fail only once, time to failure data from a sample of

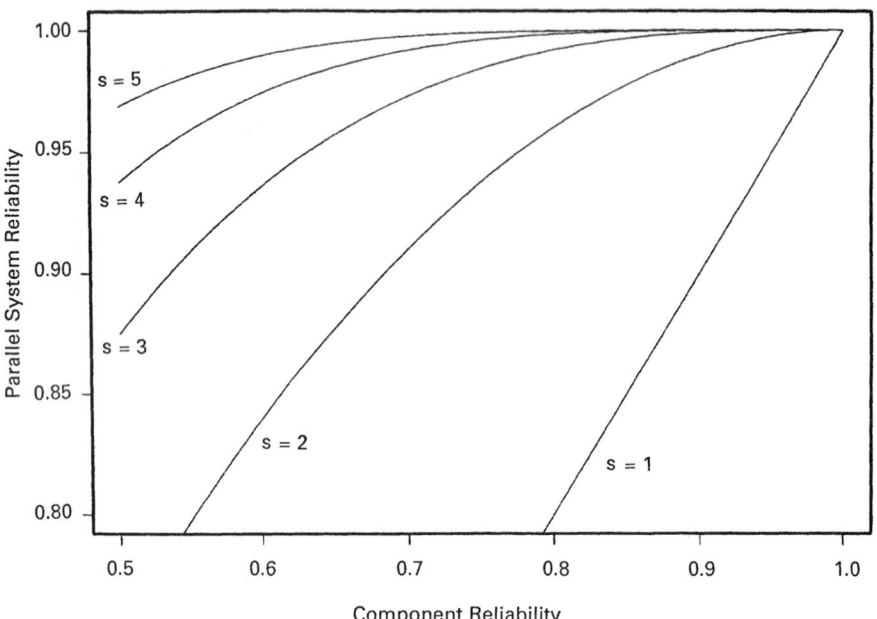

FIGURE 48.18   Reliability of a system with *s* independent components with identical time to failure distributions, in parallel.

nonrepairable components consist of the times to first failure for each component. In most instances involving nonrepairable components, the assumption of independent and identically distributed failure times is a reasonable one and suitable lifetime distributions (such as the Weibull or lognormal) are used to describe the distribution of failure times. In contrast, repairable system data typically consist of multiple repair times (or times between repairs) on the same system since a repairable system can be placed back in service after repair.

The purpose of many reliability studies is to describe the trends and patterns of repairs on failures over time for an *overall system* or collection of systems. Data consist of a sequence of system repair times for similar systems. When a single component or subsystem in a larger system is repaired or replaced after a failure, the distribution of the time to the next system repair will depend on the overall state of the system at the time just before the repair and the nature of the repair. Thus, repairable system data, in many situations, should be described with models that allow for changes in the state of the system over time or for dependencies between repairs over time.

Repairable system data can be viewed as sequence of repair times $T_1, T_2, \ldots$. The model for such data is sometimes called a "point process." Some applications have repair data on only one system. In most applications there are data from a collection of systems, typically monitored over a fixed observation period (t_0, t_a), where, often, $t_0 = 0$. The observation period often differs from system to system. In some cases, exact repair times are recorded. In other cases, the numbers of repairs within time intervals are reported.

From the repair data, one would typically like to estimate:

- The distribution of the times between repairs, $\tau_j = T_j - T_{j-1} (j = 1, 2, \ldots)$ where $T_0 = 0$
- The number of repairs in the interval $(0, t)$ as a function of t
- The expected number of repairs in the interval $(0, t)$ as a function of t
- The recurrence rate of replacements as a function of time t

These questions lead to the analyses in the following example.

Example: Times of Replacement of Diesel Engine Valve Seats.   The following data will be used to illustrate the methods. Repair records for a fleet of 41 diesel engines were kept over time. Table 48.4 gives the times of replacement (in number of days of service) of the engine's valve seats. This is an example of data on a group of systems. The data were given originally in Nelson and Doganaksoy (1989) and also appear in Nelson (1995). Questions to be answered from these data (not all of which are discussed here) include:

- How many replacements can the fleet owner expect, on the average, during the first year of operation of a locomotive?
- Does the replacement rate increase or decrease with diesel engine age and at what rate?
- How many replacement valves will be needed in the next two calendar years for this fleet?
- Can the valve replacement times be modeled as a renewal process (so that simple methods for independent observations can be used for further analysis and prediction)?

Simple data plots provide a good starting point for analysis of system repair data. Figure 48.19 is an event plot of the valve seat repair data showing the observation period and the reported replacement times. Note that the length of the observation period differed from engine to engine.

  Initial evaluations of the data suggested some differences in repair vulnerability among locomotives. This is a separate subject of study, and is ignored in the following evaluations.

Nonparametric Model for Point Process Data.

For data on a single system, the cumulative number of repairs up to time t is denoted as $N(t)$. The corresponding model, used to describe a population of systems from the same population, is based on the mean cumulative function (MCF) at time t. The MCF is defined as the average or expected number of repairs per system before time

TABLE 48.4   Diesel Engine Age at Time of Replacement of Valve Seats

Engine ID	Number of days observed	Engine age at replacement time, days	Engine ID	Number of days observed	Engine age at replacement time, days
251	761		403	593	
252	759		404	589	573
327	667	98	405	606	165 408 604
328	667	326 653 653	406	594	249
329	665		407	613	344 497
330	667	84	408	595	265 586
331	663	87	409	389	166 206 348
389	653	646	410	601	
390	653	92	411	601	410 581
391	651		412	611	
392	650	258 328 377 621	413	608	
393	648	61 539	414	587	
394	644	254 276 298 640	415	603	367
395	642	76 538	416	585	202 563 570
396	641	635	417	587	
397	649	349 404 561	418	578	
398	631		419	578	
399	596		420	586	
400	614	120 479	421	585	
401	582	323 449	422	582	
402	589	139 139			

Source: Nelson and Doganaksoy (1989).

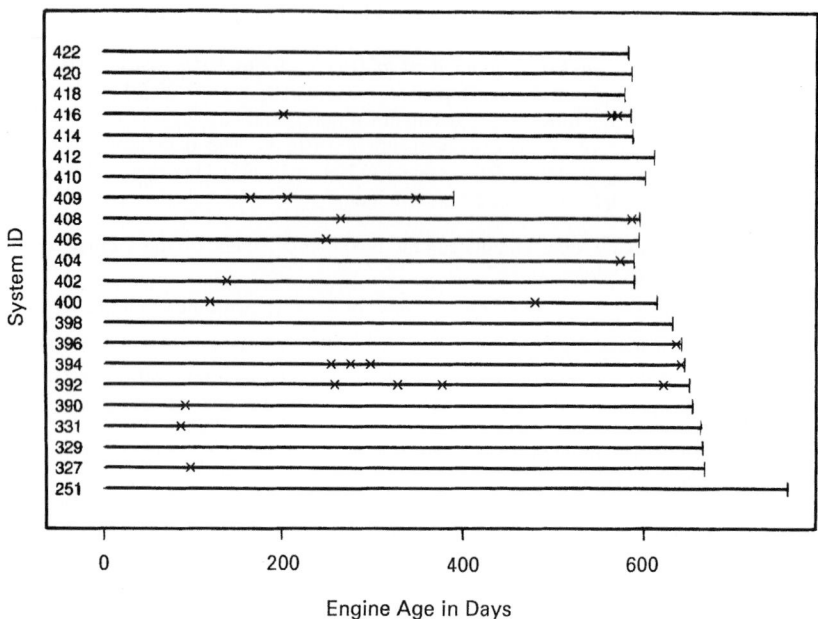

FIGURE 48.19 Valve seat replacement event plot for a subset of 22 of the 43 diesel engines.

t, i.e., $\mu(t) = E[N(t)]$, where the expectation is over the entire population of systems. Assuming that $\mu(t)$ is differentiable,

$$v(t) = \frac{dE[N(t)]}{dt} = \frac{d\mu(t)}{dt}$$

defines the recurrence rate of repairs per system for the system population. This can also be interpreted as an *average* rate of repair occurrence for individual systems.

Although data on number of repairs (or other specific events related to reliability) are encountered frequently, the methods given here can be used to model other quantities accumulating in time, including continuous variables like cost. Then, for example, if $C(t)$ is the cumulative repair cost for a system up to time t, then $\mu(t) = E[C(t)]$ is the average cumulative cost per system in the time interval $(0, t)$.

Nonparametric Estimation of the MCF. Given $n \geq 1$ repairable systems, the following method can be used to estimate the MCF. The method is nonparametric in the sense that it does not require specification of a parametric model for the repair-time point process. The method assumes that the sample data are taken randomly from a population of MCF functions. It is also assumed that the time at which we stop observing a system does not depend on the process. Thus, it is important that the time at which a unit is censored is not systematically related to any factor related to the repair time distribution. Biased estimators will, for example, result if units follow a staggered entry into service (e.g., one unit put into service each month) and if there has been a design change that has increased the repair probability of the more recent systems introduced into service. Then newer systems have a more stressful life and will fail earlier, causing an overly optimistic trend over time on the estimated recurrence rate $v(t)$. In such cases, data from different production periods must be analyzed separately or the change in the recurrence rate needs to be modeled as a function of system age and calendar time.

Let $N_i(t)$ denote the cumulative number of system repairs for system i at time t and let $t_{ij}, j = 1, ..., m_i$ be the failure (or repair, or other event) times for system i. A simple estimator of the MCF at time t would be the sample mean of the available $N_i(t)$ values for the systems still operating at time t. This estimator is simple, but appropriate only if all systems are still operating at time t. Thus, this method can be used in the diesel engine example up to $t = 389$ days, or up to 578 days if we ignore the data on engine 409. A more appropriate estimator, allowing for multiple censoring, and providing an unbiased estimate of the MCF is described in Nelson (1988), Nelson (1995), Lawless and Nadeau (1995), and Meeker and Escobar (1998). A plot of the MCF estimate versus age indicates whether the reliability of the system is increasing, decreasing, or unchanging over time.

MCF Estimate for the Valve Seat Replacements.   Figure 48.20 shows the estimate of the valve seat MCF as a function of engine age in days. The estimate increases sharply after 650 days, but this is based on only a small number (i.e., 10) of systems that had a total operating period exceeding 650 days. The uncertainty in the estimate for longer times is reflected in the width of the confidence intervals [the computation of the estimate beyond 389 days and the confidence limits is explained in Nelson (1995) and Chapter 16 of Meeker and Escobar (1998)].

Nonparametric Comparison of Two Samples of Repair Data.   Decisions often need to be made on the basis of a comparison between two manufacturers, product designs, environments, etc. Doganaksoy and Nelson (1998) describe methods for comparing samples from two different groups of systems.

OTHER TOPICS IN RELIABILITY

Sources of Reliability Data and Information.   Component and subsystem reliability information is needed as input to reliability models. Such information comes from a number of different sources. For example:

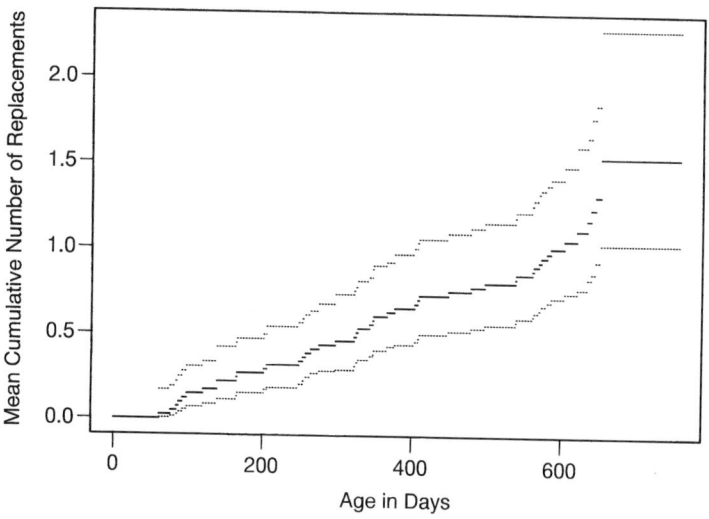

FIGURE 48.20   Estimate of the mean cumulative number of diesel engine valve seat repairs, with a 95 percent confidence interval.

- Laboratory tests are used widely, especially to test new materials and components, when there is little past experience. Such testing is generally expensive and may have limited applicability to product field reliability, but it may provide early warning of reliability problems. Care must be taken to assure that failure modes relevant to field use are obtained and that, if the goal is reliability prediction, the test conditions can be accurately related to actual field conditions. As described earlier in this section, laboratory tests are often accelerated with the goal of getting component reliability information more quickly.

- Carefully collected field data, when available, may provide the most accurate information about how components and systems behave in the field. It is important, however, to ensure that field data are consistently and unambiguously recorded to allow speedy and easy analysis. Field warranty data often contain no information on units that do not fail and may be biased, for example, because units in the harshest environments tend to fail sooner. Also, field data generally come too late to help avoid costly reliability problems.

- Reliability handbooks and data banks can be useful [e.g., Klinger, Nakada, and Menendez (1990) and MIL-HDBK 217F (1991)]. One common complaint about such handbooks, however, is that data become obsolete by the time they are published, or shortly after.

- Expert knowledge is often used when no other source of information is available.

Unless data are collected from carefully planned and conducted studies and properly maintained and analyzed, obtaining unbiased estimates, and quantifying uncertainty may be impossible.

FMEA and FMECA.   Products and systems often have complicated designs that are the result of the efforts of several design teams. Management for system reliability requires a global process to assure up front that the product/system reliability will meet customer requirements.

Failure Modes and Effect Analysis (FMEA) is a systematic, structured method for identifying system failure modes and assessing the effects or consequences of the identified failure modes. Failure Modes and Effect Criticality Analysis (FMECA) considers, in addition, the importance of identified failure modes, with respect to safety, successful completion of system mission, or other criteria. The goal of FMEA/FMECA is to identify and possibly remove possible failure modes at a specified level of system architecture. These methods are typically used in product/system design review processes. The use of FMEA/FMECA usually begins in the early stages of product/system conceptualization and design. Then the FMEA/FMECA evolves over time along with changes in the product/system design and accumulation of information about product/system performance in preproduction testing and field experience. FMEA/FMECA is used during the product/system design phase to help guide decision making. It is also used to develop product/system guidelines for system repair and maintenance procedures, to improve system safety, and to provide direction for reliability improvement efforts.

Operationally, FMEA/FMECA begins by defining the scope of the analysis, specified by the system level at which failures are to be considered. FMEA/FMECA can be conducted at various levels in a product or a system. It might be done initially for individual subsystems. Then the results can be integrated to provide an FMEA/FMECA for an entire system comprising many subsystems. For example, an FMEA to study the reliability of a telecommunications relay repeater might consider, as basic components, each discrete device in the electronic circuit (e.g., ICs, capacitors, resistors, diodes). At another level, an FMEA for a large telecommunications network might consider as components all of the network nodes and node interconnections (ignoring the electronic detail within each node).

The next step in the FMEA/FMECA process is the identification of all components that may fail during the life of the product. This is followed by identification of all component interfaces or connections that might fail. In many applications, environmental and human factors–related failures are considered in defining failure modes. Finally the effects of the identified failure modes are delineated. Determining the effect of failure modes and combinations of failure modes uses the detailed specification of the relationship among the product/system components (system structure). Special worksheets and/or computer software can be used to organize all of the information.

MIL-STD-1629A (1980) and books like Høyland and Rausand (1994) and Sundararajan (1991) outline in more detail and provide examples of the procedures for performing an FMEA/FMECA analysis. Høyland and Rausand (1994) also list several computer programs designed to facilitate such analyses.

Fault Trees.

The FMEA/FMECA process described above is sometimes referred to as a "bottom-up" approach to reliability modeling. Fault tree analysis, on the other hand, quantifies system failures using a "top-down" approach. First, one or more critical "top events" (such as loss of system functionality) are defined. Then in a systematic manner, the combination (or combinations) of conditions required for that event to occur are delineated. Generally this is done by identifying how failure-related events at a higher level are caused by lower-level "primary events" (e.g., failure of an individual component) and "intermediate events" (e.g., failure of a subsystem). Information from an FMEA analysis might be used as input to this step. The information is organized in the form of a "fault tree diagram" with the top event at the top of the diagram. Events at different levels of the tree are connected by logic gates defined by Boolean logic (e.g., AND, OR, Exclusive OR gates).

A complete fault tree can be used to model the probability of critical system events. Additional inputs required for this analysis include the probabilities or the conditional probabilities of the primary events. With this information and the detailed system structure specification provided by the fault tree, it is possible to compute critical event probabilities. See O'Connor (1985), Sundararajan (1991), Høyland and Rausand (1994), or Lewis (1996) for examples.

Fault tree diagrams are, in one sense, similar to the reliability block diagrams presented earlier in this subsection. It is generally possible to translate from one to the other. Fault tree analysis, however, differs in its basic approach to system reliability. Reliability block diagrams are structured around the event that the system does *not* fail. Fault tree analysis focuses on the critical top events like loss of system functionality or other safety-critical events. The tree shows, directly, the root causes of these top events, and other contributing events, at all levels within the scope of the analysis. The structure and logic of the fault tree itself provides not only a mechanism for quantitative reliability assessment, but also clearer insight into possible approaches for reliability improvement.

Designed Experiments to Improve Reliability.

Experimental design can be an important tool for reliability improvement. Different kinds of experiments can obtain information for assessing and improving reliability. These include life tests and accelerated life tests, such as those described earlier in this section, to assess the durability or lifetime distribution of materials and components, for which there is one or a small number of identifiable failure modes. Information from such tests is used to make design decisions that will result in acceptable (or better) reliability for the tested components.

At another level one may conduct experiments to define system configurations or operating factors to enhance reliability by making components more robust to varying manufacturing conditions or operating/environmental conditions that might be encountered in service. Generally such experiments also focus on a single failure mode or a few failure modes for a single component or subsystem. Multifactor robust-design experiments (RDE) provide methods for systematic and efficient reliability improvement. These are often conducted on prototype units and subsystems and focus on failure modes involving interfaces and interactions among components and subsystems. Among many possible product design factors that may impact a system's reliability, RDEs empirically identify the important ones and find levels of the product design factors that yield consistent high quality and reliability. Graves and Menten (1996) provide an excellent description of experimental strategies that can be used to help design products with higher reliability. Byrne and Quinlan (1993) describe an experiment to design a flexible push cable that will have long life under varying operating conditions. Tseng, Hamada, and Chiao (1995) describe the use of a factorial experiment to choose manufacturing variables that will extend the life of fluorescent light bulbs. Phadke (1989) describes an experiment to optimize performance, and thus reliability, of a copying machine subsystem.

The need to quickly develop new high-reliability products has motivated the development of other new-product testing methods. The purpose of these testing methods is to rapidly identify and eliminate potential reliability problems (failure modes or other system weaknesses) early in the design stage of product development. One such testing method is known as STRIFE (*stress-life*) testing. The basic idea of STRIFE testing is to aggressively stress and test prototype or early production units (using, for example, temperature cycling, power cycling, and vibration) to force failures. Although it may be useful to test only one or two units, appropriately selected units can provide important additional information on unit-to-unit variation. Bailey and Gilbert (1981) describe an example in which the complete STRIFE test and improvement program was successfully completed in three weeks. Kececioglu and Sun (1995) also describe STRIFE testing and present several interesting case studies. Schinner (1996) provides a detailed description of preproduction "accelerated tests" that can be used to discover or assess the potential impact of product failure modes.

Reliability Demonstration.   Reliability demonstration is used to assess whether product reliability meets specifications. The reliability specification is often stated in terms of the mean or a specified percentile of the life distribution of the product. For example, an industrial purchaser of a component may require that the percentage of failing parts during the first 5 years of service must not exceed 0.5 percent.

Demonstration of product reliability usually involves some type of a life test. The basic considerations in planning a reliability demonstration test are similar to those of acceptance sampling (see Section 46), although some special aspects arise as noted below. A reliability demonstration test plan specifies the number of units to be placed on test, test time, and criteria for successful demonstration. Acceptance test plans involving attribute data [such as MIL-STD-105E (1989) or the corresponding civilian standard ANSI/ASQC (1993a) Z1.4] are applicable sometimes to reliability demonstration when the outcome of a test is either a success or a failure, such as in demonstrating start-up reliability of a system. Likewise, standard acceptance test plans for the normal distribution [MIL-STD-414 (1957) or the corresponding civilian standard ANSI/ASQC (1993b) Z1.9] can sometimes be used for reliability demonstration when all units are tested to failure, provided distributional assumptions of the plan are satisfied (e.g., logs of failure times are often modeled with a normal distribution). However, special test plans for reliability demonstration are needed in situations where testing is to be terminated before all units fail. Abernethy (1996) describes small-sample Weibull demonstration plans. Special test plans for reliability are also needed to accommodate life distributions other than the normal (such as the Weibull) which are widely used in reliability data analysis.

Reliability demonstration test plans for a wide range of situations have been documented in various handbooks and standards published by the U.S. Defense Department. For example, U.S. Department of Defense (1960), *Quality Control and Reliability Handbook H-108,* provides failure-terminated (test is stopped upon occurrence of preassigned number of failures), time-terminated (test duration is predetermined) and sequential test plans for the exponential distribution. Similar test plans for the Weibull distribution are contained in U.S. Department of Defense (1965) *Handbook TR 7.* Schilling (1982) provides a thorough overview of common test plans used in reliability demonstration.

Screening and Burn-in.   Many products, ranging from electronic components to electronic systems to automobiles, experience reliability problems in the early break-in part of life. Such "infant mortality" failures often occur in a small proportion of units with manufacturing defects. After all or most of these defective units have failed, population failure rates typically reach a low level until wear-out failures begin to occur (see earlier discussion of "bathtub curve"). For some products (e.g., desktop computers), the wear-out phase of life is well beyond the technological life of the product. However, in many applications involving critical systems or where repair is impossible or highly expensive (e.g., satellite systems, undersea systems, and critical aircraft systems), infant mortality failures present an important obstacle to achieving the needed level of reliability.

Manufacturers, in their quality and reliability improvement programs, strive to eliminate or at least reduce the rate of occurrence of infant mortality failures. In such applications, screening or burn-in techniques are often used to identify and remove defective components or to remove products containing defects before they are released to customers.

Burn-in tests can be viewed as a type of 100 percent inspection or screening of the product population. All units are run for a period of time before shipment or installation. To accelerate the process, components such as integrated circuits may be run at high levels of temperature and/or other stresses. The ability to use acceleration is much more limited for systems and subsystems. See Jensen and Petersen (1982) for an engineering approach to this subject. Also, see Nelson (1990, p. 43) for more information and references.

Environmental Stress Screening (ESS) was developed as a means of accelerated burn-in for units at the system or subsystem (e.g., circuit pack) level. ESS uses mild, but complicated, stressing such as combinations of temperature cycling, physical vibration, and perhaps stressful operational regimes (e.g., running computer chips at higher than usual clock speeds and lower than usual voltages) to help identify the weak units. Tustin (1990) gives a motivational description of the methodology and several references. MIL-STD-2164 (1985) provides a useful overview of ESS procedures. Kececioglu and Sun (1995) provide a comprehensive treatment of ESS. Nelson (1990, p. 39) gives additional references, including military standards.

Burn-in and ESS are inspection/screening schemes. In line with the modern quality goal of eliminating reliance on mass inspection, most manufacturers prefer not to do burn-in or ESS. Such methods are also expensive and may not be totally effective. By improving reliability through continuous improvement of the product design and the manufacturing process, it may be possible to reduce or eliminate reliance on screening tests except, perhaps, in the most critical applications.

Reliability Growth. Reliability of a system continually evolves during its design, development, testing, production, and field use. This ongoing change is referred to as "reliability growth" (although, in fact, reliability may actually improve or deteriorate). Growth usually results from efforts to discover design (or manufacturing) flaws and implement fixes to affect all future manufactured units. Thus, later-generation systems should have better reliability than their predecessors. Sometimes this process is referred to as "test, analyze, and fix" (TAAF). The basic idea is to find and fix reliability problems that had not been found earlier. Reliability growth models and data-fitting methods allow predictions of product reliability due to such improvements. Reliability growth analysis usually involves fitting a reliability growth model (such as the popular Duane plot) to failure data on systems built over time. These empirical models are then used to predict the reliability of future generations of systems. For further information and references on reliability growth, see MIL-HDBK-189 (1981), Klion (1992), Ascher and Feingold (1984), and Chapter 11 of Pecht (1995).

Software Reliability. State-of-the-art reservation, banking, billing, accounting, and other financial and business systems depend on complicated software systems. Additionally, modern hardware systems of all kinds, from automobiles and televisions to communications networks and spacecraft, contain complicated circuitry. Most of these electronic systems depend heavily on software for important functionality and flexibility. For many systems, software reliability has become an important limiting factor in system reliability.

The Institute of Electrical and Electronic Engineers defines software reliability as "the probability that software will not cause a system failure for a specified time under specified conditions." Software reliability differs from hardware reliability in at least one important way. Hardware failures can often be traced to some combination of a physical fault and/or physical/chemical degradation that progresses over time, perhaps accelerated by stress, shocks, or other environmental or operating conditions. Software failures, on the other hand, are generally caused by inherent faults in the software that are usually present all along. Actual failure may not occur until a particular set of inputs is used or until a particular system state or level of system load is reached. The state of the software itself does not change without intervention.

Software errors differ in their criticality. Those who work with personal computers know that from time to time the system will stop functioning for unknown reasons. The cause is often software-related (i.e., it would not have occurred if the software had been designed to anticipate the conditions that caused the problem). Restarting the computer and the application will seem to make the problem disappear. Important data in the application being used at the time of the failure may or may not have been lost. Future versions of the operating system or the application software may correct such problems. In safety-critical systems (e.g., medical, air traffic control, or military systems) software failures can have much more serious (e.g., life-threatening) consequences.

For some purposes, statistical methods for software reliability are similar to those used in predicting reliability growth of a system or analyzing data from a repairable system, treated earlier in this subsection. Software data often consist of a sequence of times of failures (or some other specific event of interest) in the operation of the software system. Software reliability data are collected for various reasons, including assessment of the distribution of times between failures, tracking the effect of continuing efforts to find and correct software errors, making decisions on when to release a software product, assessing the effect of changes to improve the software development process, etc.

Numerous models have been suggested and developed to model software reliability data. The simplest of these describe the software failure rate as a smooth function of time in service and other factors, such as system load and amount of testing or use to which the system has been exposed. In an attempt to be more mechanistic and to incorporate information from the fix process directly into the software reliability model, many of these models have a parameter corresponding to the number of faults remaining in the system. In some models, the failure rate would be proportional to the number of faults. When a "repair" is made, there is some probability that the fault is fixed and, perhaps, a probability that a new fault is introduced.

For more information on software reliability and software reliability models, see Musa, Iannino, and Okumoto (1987), Shooman (1983), Neufelder (1993), Chapter 6 of Pecht (1995), or Azem (1995).

ACKNOWLEDGMENTS

Parts of this section were taken from Meeker and Escobar (1998) with permission from John Wiley & Sons, Inc. We would like to thank Elaine Miller and Denise Riker for their help in editing and typing parts of the manuscript.

REFERENCES

Abernethy, R. B. (1996). *The New Weibull Handbook.* Self-published, 536 Oyster Road, North Palm Beach, FL 33408-4328.

ANSI/ASQC (1993a). Z1.4-1993: *Sampling Procedures and Tables for Inspection by Attributes.* Available from American National Standards Institute, Customer Service, 11 West 42nd St., New York, NY 10036.

ANSI/ASQC (1993b). Z1.9-1993: *Sampling Procedures and Tables for Inspection by Variables for Percent Nonconforming.* Available from American National Standards Institute, Customer Service, 11 West 42nd St., New York, NY 10036.

Ascher, H., and Feingold, H. (1984). *Repairable Systems Reliability.* Marcel Dekker, New York.

Azem, A. (1995). *Software Reliability Determination for Conventional and Logic Programming.* Walter de Gruyter, New York.

Bailey, R. A., and Gilbert, R. A. (1981). "STRIFE Testing for Reliability Improvement." *Proceedings of the Institute of Environmental Sciences,* pp. 119–121.

Byrne, D., and Quinlan, J. (1993). "Robust Function for Attaining High Reliability at Low Cost." *1993 Proceedings Annual Reliability and Maintainability Symposium,* pp. 183–191.

Condra, L. W. (1993). *Reliability Improvement with Design of Experiments.* Marcel Dekker, New York.

Doganaksoy, N., and Nelson, W. (1998). "A Method to Compare Two Samples of Recurrence Data." *Life Data Analysis,* vol 4, pp. 51-63 .

Graves, S., and Menten, T. (1996). "Designing Experiments to Measure and Improve Reliability." *Handbook of Reliability Engineering and Management,* 2nd ed. W. G. Ireson, C. F. Coombs, and R. Y. Moss, eds. McGraw-Hill, New York, Chapter 11.

Hahn, G. J., and Shapiro, S. S. (1967). *Statistical Models in Engineering.* John Wiley & Sons, New York.

Høyland, A., and Rausand, M. (1994). *System Reliability Theory: Models and Statistics Methods.* John Wiley & Sons, New York.

Jensen, F., and Petersen, N. E. (1982). *Burn-in, An Engineering Approach to Design and Analysis of Burn-in Procedures.* John Wiley & Sons, New York.

Kececioglu, D., and Sun, F. (1995). *Environmental Stress Screening: Its Quantification, Optimization and Management.* PTR Prentice Hall, Englewood Cliffs, NJ.Klinger, D. J., Nakada, Y., and Menendez, M. A. (1990). *AT&T Reliability Manual.* Van Nostrand Reinhold, New York.

Klion, J. (1992). *Practical Electronic Reliability Engineering.* Van Nostrand Reinhold, New York.

Lawless, J. F., and Nadeau, C. (1995). "Some Simple Robust Methods for the Analysis of Recurrent Events." *Technometrics,* vol 37, pp. 158–168.

Lewis, E. E. (1996). *Introduction to Reliability Engineering.* John Wiley & Sons, New York.

Meeker, W. Q., and Escobar, L. A. (1998). *Statistical Methods for Reliability Data.* John Wiley & Sons, New York.

Meeker, W. Q., and Hahn, G. J. (1985). "How to Plan Accelerated Life Tests: Some Practical Guidelines." *ASQC Basic References in Quality Control: Statistical Techniques.* Available from the American Society for Quality, 310 W. Wisconsin Ave., Milwaukee, WI 53203, vol. 10.

Meeker, W. Q., and LuValle, M. J. (1995). "An Accelerated Life Test Model Based on Reliability Kinetics." *Technometrics,* vol. 37, pp. 133–146.

MIL-HDBK-189 (1981). *Reliability Growth Management.* Available from Naval Publications and Forms Center, 5801 Tabor Ave., Philadelphia, PA 19120.

MIL-HDBK-217F (1991). *Reliability Prediction for Electronic Equipment.* Available from Naval Publications and Forms Center, 5801 Tabor Ave., Philadelphia, PA 19120.

MIL-STD-105E (1989). *Military Standard, Sampling Procedures and Tables for Inspection by Attributes.* Available from Naval Publications and Forms Center, 5801 Tabor Ave., Philadelphia, PA 19120.

MIL-STD-414 (1957). *Military Standard, Sampling Procedures and Tables for Inspection by Variables for Percent Defective.* Available from Naval Publications and Forms Center, 5801 Tabor Ave., Philadelphia, PA 19120.

MIL-STD-1629A (1980). *Failure Modes and Effects Analysis.* Available from Naval Publications and Forms Center, 5801 Tabor Ave., Philadelphia, PA 19120.

MIL-STD-2164 (1985). *Environmental Stress Screening Process for Electronic Equipment.* Available from Naval Publications and Forms Center, 5801 Tabor Ave., Philadelphia, PA 19120.

Minitab (1997). *Version 12 User's Guide,* Minitab Inc., State College, PA.

Musa, J. D., Iannino, A., and Okumoto, K. (1987). *Software Reliability: Measurement, Prediction, Application.* McGraw-Hill, New York.

Nair, V. N. (1984). "Confidence Bands for Survival Functions with Censored Data: a Comparative Study." *Technometrics,* vol. 26, pp. 265–275.

Nelson, W. (1982). *Applied Life Data Analysis.* John Wiley & Sons, New York.

Nelson, W. (1988). "Graphical Analysis of System Repair Data." *Journal of Quality Technology,* vol. 20, pp. 24–35.

Nelson, W. (1990). *Accelerated Testing: Statistical Models, Test Plans, and Data Analyses.* John Wiley & Sons, New York.

Nelson, W. (1995). "Confidence Limits for Recurrence Data—Applied to Cost or Number of Product Repairs." *Technometrics,* vol. 37, pp. 147–157.

Nelson, W., and Doganaksoy, N. (1989). "A Computer Program for an Estimate and Confidence Limits for the Mean Cumulative Function for Cost or Number of Repairs of Repairable Products." TIS Report 89CRD239, General Electric Company Research and Development, Schenectady, NY.

Neufelder, A. (1993). *Ensuring Software Reliability.* Marcel Dekker, New York.

O'Connor, P. D. T. (1985). *Practical Reliability Engineering,* 2nd ed., John Wiley & Sons, New York.

Parida, N. (1991). "Reliability and Life Estimation from Component Fatigue Failures below the Go-No-Go Fatigue Limit." *Journal of Testing and Evaluation,* vol. 19, pp. 450–453.

Pecht, M., ed. (1995). *Product Reliability, Maintainability, and Supportability Handbook.* CRC Press, Boca Raton, FL.

Phadke, M. S. (1989). *Quality Engineering Using Robust Design.* Prentice Hall, Englewood Cliffs, NJ.

SAS Institute (1995). *JMP User's Guide,* Version 3.1. SAS Institute Inc., Cary, NC.

SAS Institute (1996). *SAS/STAT™ Software: Changes and Enhancements through Release 6.11.* SAS Institute Inc., Cary, NC. (Proc RELIABILITY is part of the SAS/QC software.)

Schilling, E. G. (1982). *Acceptance Sampling in Quality Control.* Marcel Dekker, New York.

Schinner, C. (1996). "Accelerated Testing." *Handbook of Reliability Engineering and Management,* 2nd ed., W. G. Ireson, C. F. Coombs, and R. Y. Moss, eds. McGraw-Hill, New York, Chapter 12.

Shooman, M. L. (1983). *Software Engineering: Design, Reliability, and Management.* McGraw-Hill, New York.

Statistical Sciences, (1996). *S-PLUS User's Manual,* Volumes 1, 2, Version 3.4. Statistical Sciences, Inc., Seattle.

Sundararajan, C. (1991). *Guide to Reliability Engineering.* Van Nostrand Reinhold, New York.

Tseng, T. S., Hamada, M., and Chiao, C. H. (1995). "Using Degradation Data from a Factorial Experiment to Improve Fluorescent Lamp Reliability." *Journal of Quality Technology,* vol. 27, pp. 363–369.

Tustin, W. (1990). "Shake and Bake the Bugs Out." *Quality Progress,* September 1990, pp. 61–64.

U.S. Department of Defense (1960). "Sampling Procedures and Tables for Life and Reliability Testing." *Quality Control and Reliability (Interim) Handbook (H-108).* Office of the Assistant Secretary of Defense (Supply and Logistics), Washington, DC.

U.S. Department of Defense (1965). "Factors and Procedures for Applying MIL-STD-105D Sampling Plans to Life and Reliability Testing." *Quality Control and Reliability Assurance Technical Report (TR 7).* Office of the Assistant Secretary of Defense (Installations and Logistics), Washington, DC.

APPENDIX I
GLOSSARY OF SYMBOLS AND ABBREVIATIONS

This list is meant to provide a single reference source for symbols and abbreviations commonly used in the literature of the quality sciences, especially in the sections of this handbook. It represents a major update and revision of this appendix for this edition.

a = acceptance number in a sampling plan.

a = combination of factors A, B, C, ..., n in a 2^n experiment in which only A occurs at the high level; similarly for $b, c, ..., n$.

A = arbitrary origin for grouped data.

A = factor in graphical application of variables sampling plan.

A = unit cost of acceptance (damage done by a defective piece which slips through inspection).

A = a multiplier of σ_0 used to locate the 3-sigma control limits above and below the central line on an X chart.

A_3 = a multiplier of $\bar{s}$ used to locate the 3-sigma control limits above and below the central line on an $\bar{X}$ chart.

A_2 = a multiplier of $\bar{R}$ used to locate the 3-sigma control limits above and below the central line of an $\bar{X}$ chart.

A_c = in sampling acceptance schemes, the acceptance number, i.e., the maximum allowable number of defective pieces in a sample of size n.

ACL = acceptance control limit, i.e., a distance d from APL in the direction of the RPL.

ANOVA = analysis of variance.

AOQ = average outgoing quality, i.e., the long-term average quality of product leaving the inspection department after acceptance sampling and any detailing found necessary is performed.

AOQL = average outgoing quality limit, i.e., an upper limit on the AOQ.

APL = acceptable process level, i.e., a process level which is acceptable and should be accepted most of the time by the plan.

AQL = acceptable quality level, i.e., a specified quality level for each lot such that the sampling plan will accept a high percentage (say 95 percent) of submitted lots having this quality level.

ARL = average run length.

ASN = average sample number, i.e., the average number of units inspected per lot in sampling inspection.

ATI = average total inspection, i.e., average number of items inspected in a lot under a specified acceptance procedure.

b = number of blocks in a randomized or balanced incomplete block experimental design.

$b_1, b_2, \ldots$ = estimates of regression coefficients $\beta_1, \beta_2, \ldots$.

B_3 = a multiplier of $\bar{s}$ used to locate the 3-sigma lower control limit on a chart for σ.

B_4 = a multiplier of $\bar{s}$ used to locate the 3-sigma upper control limit on a chart for σ.

B_5 = a multiplier of σ_0 used to locate the 3-sigma lower control limit on a chart for σ.

B_6 = a multiplier of σ_0 used to locate the 3-sigma upper control limit on a chart for σ.

B_U, B_L = factors for confidence limits on σ using s.

c = number of defects, usually in a sample of stated size.

c = in sampling acceptance plans, the acceptance number, i.e., the maximum allowable number of defective pieces in a sample of size n. (In some sampling tables the symbol Ac is used.)

c = a scale factor given in MIL-STD-414.

c = acceptance number in a sampling plan, also A_c.

$\bar{c}$ = average number of defects per sample in a series of samples.

c_0 = standard or aimed-at average number of defects in a sample of stated size. c_0 may also refer to the population average number of defects per sample.

c_2 = ratio of the expected value of $\hat{\sigma}$ in a long series of samples to the σ of the population from which they were drawn.

c_4 = factor for estimating the standard deviation from $\hat{\sigma} = \bar{s}/c_4$.

c_5 = factor for estimating the standard deviation of s, $\sigma_s = c_5\sigma$, so $\hat{\sigma}_s = c_5\bar{s}/c_4$.

c_{ij} = element in the ith row and jth column of a matrix C.

C = cost of repairing or replacing a defective once found.

C = extra capacity required.

C = correction factor, constant.

C_p = variance of regression predictions.

C_p = capability index for process centered between specification limits.

C_{pk} = performance index.

$\hat{C}_{pk}$ = estimate of C_{pk}.

$C(t)$ = cumulative repair cost for a system up to time t.

C_1, C_2 = adjusted cumulative sum for tabulation method.

ChSP = chain sampling plan.

CIM = change-in-mean-effect in Evolutionary Operations.

CSP = continuous sampling plan.

d = the difference to be detected in a test divided by the measure of variability.

d = the difference in readings in a paired sample.

d = the number of defectives.

d = lead distance in cumulative sum charts.

d_i = number of defective units in ith sample.

d_i = signed differences.

$\bar{d}$ = the average difference in readings in a paired sample.

d' = deviation, in cells, from the assumed origin of a frequency distribution.

d_2 = ratio of the expected value of $\bar{R}$ (in samples of size n) to the σ of the population, i.e., $\bar{R} = d_2\sigma$.

D = largest deviation of actual percent cumulative frequency from theoretical percent cumulative frequency.

D = in a cumulative sum control chart, the least amount of change in the average that it is desired to detect.

D_1 = a multiplier of σ_0 used to locate the 3-sigma lower control limit on a chart for R.

D_2 = a multiplier of σ_0 used to locate the 3-sigma upper control limit on a chart for R.

D_3 = a multiplier of $\bar{R}$ used to locate the 3-sigma lower control limit on a chart for R.

D_4 = a multiplier of $\bar{R}$ used to locate the 3-sigma upper control limit on a chart for R.

D_i = cumulative number of defective units at the ith sample.

df = degrees of freedom, the number of independent comparisons possible with a given set of observations (also called DF).

DF = degrees of freedom, the number of independent comparisons possible with a given set of observations (also called df).

DPU = defects per unit.

e = the base of the natural logarithm, a constant whose value is 2.71828+.

e = simple event.

e_{ij} = the experimental error associated with the measurement Y_{ij}.

E = maximum allowable error in estimate (desired precision).

$E(T)$ = expectation of T; also mean time to failure.

E_2 = a multiplier of $\bar{R}$ to determine the 3-sigma control limits on a chart for individuals.

EVOP = evolutionary operation.

f = frequency; generally, the number of observed values in a given cell of a frequency distribution.

f = sampling rate in continuous sampling.

f = in Skip Lot sampling, the fraction of lots to be inspected after the initial criteria have been satisfied.

$f(t)$ = probability density function (pdf).

F = ratio of two estimates of variance or the distribution of this ratio.

F_{CALC} = calculated value of F.

F_{TAB} = tabulated value of F.

F' = ratio of two sample ranges or the distribution of this ratio.

$F(t)=Pr(T \le t)$ = cumulative distribution function (cdf).

$F_T(t)$ = system failure probability.

$\hat{F}(t)$ = empirical estimate of distribution function.

$F(t, \theta)$ = exponential distribution function with mean θ.

$F(t, \eta, \beta)$ = Weibull distribution function with shape parameter η and scale parameter β.

$F(t, \mu, \sigma)$ = normal distribution function with location parameter μ and scale parameter σ; $F(t, \theta, \sigma) = \Phi_{NOR}[(t - \mu)/(\sigma)]$.

g = a numerical factor used in calculations for the Weibull distribution.

G = grand total of all observations.

GLD = generalized lambda distribution.

h = multiplier for two-stage testing, confidence interval, and testing of the best.

h = a parameter of cumulative sum sampling plans.

$h_T(t)$ = system hazard function

$h(t)$ = hazard function; $h(t) = f(t) / (1 - F(t))$

h' = a parameter of cumulative sum sampling plans.

h_1, h_2 = intercept values in a sequential sampling plan for process parameter.

H: = hypothesis.

i = cell interval; for grouped data, the distance from a point in one cell to a similar point in the next cell.

i = in Skip Lot sampling, the number of successive lots to be found conforming to qualify for skipping lots either at the start or after detecting a nonconforming lot.

i = number of successive acceptable units in continuous sampling.

i = number of lots having samples with no defectives in chain sampling.

I = cost of inspecting one piece.

I_m = minimum inspection per lot.

I = the square identity matrix.

I = identity symbol.

k = number of sampling levels in continuous sampling plans.

k = a parameter of cumulative sum sampling plans.

k = number of treatments or levels of the factor to be investigated.

k = acceptance constant for variables plan.

k = block size in a balanced incomplete block or Youden square design.

k = number of factors in a $2k$ design.

k = number of points required for a lattice design for $m + 1$ equally spaced values and q components.

k = number of populations for intervals of fixed width and precision.

k = number of predictor variables in regression.

k = sample number.

k = tolerance interval factor for the normal distribution using s.

K = the difference between a particular value and the average in units of standard deviation. $K = (X - \mu)/\sigma$. Also called z.

K = in Evolutionary Operation, a factor used in determining the error term. Converts range into estimated standard deviation.

K_1, K_2 = factors for tabulation method cumulative sum chart.

K_1, K_2 = tolerance interval factors for the normal distribution using the range.

$\mathscr{L}$ = likelihood.

L = a numerical factor used in calculations for Evolutionary Operations.

L = loss function.

L = multiples of the standard deviation.

L = lower specification limit for a quality characteristic.

L, E = constants in a balanced incomplete block or Youden square design.

$L(N)$ = notation for orthogonal array with N runs.

LACL = in variables sampling plans, the lower acceptance control limit.

LCL = lower control limit on a control chart.

LQ = limiting quality.

LTPD = lot tolerance percent defective, i.e., the level of defectiveness that is unsatisfactory and therefore should be rejected by the sampling plan.

LSD = least significant difference.

m = number of occurrences.

m = number of levels in the primary (A) factor in a staggered nested design.

m = number of levels of a factor.

m_R = true mean of ranges.

M = in MIL-STD-414, an acceptance limit.

M = a numerical factor used in calculations for Evolutionary Operation.

MAR = maximum allowable range.

MCF = mean cumulative function, the expected number of repairs per system before time t.

ML = maximum likelihood.

MS = mean square.

MSB = mean square between treatments.

MSD = maximum standard deviation for variables sampling plan.

MSE = error mean square.

MSI = mean square for interaction.

MS (RES) = residual mean square.

MSW = mean square within treatments.

MTBF = mean time between failures.

n = number of articles or observed values in a sample or subgroup. Also, the number of trials of some event.

n_i = number assigned to each treatment.

n_0 = in a response surface experimental design, the number of center points.

n_0 = initial sample size.

n_1 = in double sampling, the number of pieces in the first sample.

n_2 = in double sampling, the number of pieces in the second sample.

n_a = attributes sample size.

n_0, n_s = sample sizes for variables plans with known and unknown standard deviation.

n_T, n_N = sample sizes for a TNT sampling plan.

$\bar{n}$ = average sample size.

$n!$ = $n(n-1)(n-2)\cdots(1)$, where n is a positive integer; $n! = 1$ for $n = 1$ and for $n = 0$.

np = number of defective articles in a sample of size n.

$n\bar{p}$ = average value of np in a set of sample size n.

N = number of articles in a lot or population.

N = number of experimental units.

N = number of data pairs in regression.

OC = operating characteristic, a plot describing the risks in a sampling plan.

p = fraction nonconforming, i.e., the ratio of the number of nonconforming units to the total number of nonconforming and conforming units.

p = probability of occurrence of an event.

p = fractionalization element in a 2^{k-p} experiment design.

$P*$ = confidence level.

$p(\%)$ = in variables sampling plans, estimate of two-sided percent defective.

p_1 = in sampling plans, the acceptable fraction nonconforming.

p_2 = in sampling plans, the rejectable fraction nonconforming.

p_0 = aimed-at or standard values of the fraction of nonconforming articles; also, the true value of p in a lot or population being sampled.

$\bar{P}$ = average fraction nonconforming, i.e., the total number of nonconforming units found in a set of samples divided by the total number of units in the samples.

p_b = break-even value of fraction nonconforming for which cost of inspection of $1/p_b$ units is equal to cost of damage done by one nonconforming.

p_c = expected proportion failing.

p_k = upper tail p value corresponding to $z = k$.

$P_{n_0}(t)$ = probability for testing if two measures are equal in two stages.

$p_L(\%)$ = in variables sampling plans, estimate of percent defective below the lower specification limit.

$p_U(\%)$ = in variables sampling plans, estimate of percent defective.

P = the population percentage included between statistical tolerance limits.

P_a = probability of accepting a given lot. Also, the probability of accepting a hypothesis.

$P(A)$ = probability of occurrence of event A.

$P(B|A)$ = conditional probability of B, given A has occurred.

P_r = the probability of lot rejection.

PC = PRE-Control.

PE = pure error.

P_s = probability of survival. The probability of failure-free operation for a time period equal to or greater than t. This is identical with R, reliability.)

q = $1 - p$, the probability that a particular event will not happen in a single trial.

q = in experimental design, number of components in a mixture.

Q = quadratic loss function.

$Q_r(X_r)$ = statistic for Quisenberry Q chart for the mean with known σ.

$Q(R_r)$ = statistic for Quisenberry Q chart for process variation with known σ.

Q_L, Q_U = quality indices used in MIL-STD-414.

r = the number of occurrences of some event, e.g., the number of defectives in a sample, the number of occurences of the less frequent sign in a test of hypothesis. Also, the distribution of this statistic.

r = the sample correlation coefficient.

r = number of replications.

r = rejection number in a sampling plan.

$r_{10}, r_{11}, r_{21}, r_{22}$ = criteria for testing outliers.

R = range of a set of n numbers, i.e., the difference between the largest number and the smallest number.

R = operating ratio (p_2/p_1) of a sampling plan.

$\overline{R}$ = mean of several ranges.

R_e = the rejection number, i.e., the number of defective pieces in a sample of size n which causes rejection of the lot.

$R(t)$ = reliability; proportion of population surviving to time t; also called R.

REG = regression.

RES = residual.

RPL = rejectable process level, i.e., a process level which is rejectable and should be rejected most of the time by the plan.

RQL = rejectable quality level.

RSM = response surface methodology, an experimental approach used to optimize many different kinds of industrial unit processes and systems.

R^2 = the proportion of variation explained by a regression model. Also the square of the sample multiple correlation coefficient.

s = sample estimate of σ (standard deviation of population); e.g., s is the sample estimate of the standard deviation of individual values, $s_{\overline{X}}$ is the sample estimate of the standard deviation of sample means.

s = $\sqrt{\Sigma(x - \overline{x})^2/(n - 1)}$.

s = in sequential sampling plans, a constant computed from the values of the APL and the RPL.

s^2 = sample estimate of σ^2 (variance of a population); e.g., s^2 is the sample estimate of the variance of individual values, $s_{\overline{X}}^2$ is the sample estimate of the variance of sample means.

s_0 = standard deviation for a standards given control chart.

s_d = standard deviation of differences.

s_{XY} = covariance between X and Y.

S_m = test statistic in cumulative sum sampling plans.

$S(t)$ = $\text{Pr}(T > t)$ = survival function; $S(t) = P_s = 1 - F(t)$, also known as reliability function.

SE = standard error.

SKSP = Skip Lot sampling plan.

SN = signal-to-noise ratio.

SPC = statistical process control.

SS = sum of squares; e.g., SSE is sum of squares for error.

SS(LF) = sum of squares due to lack of fit.

SS(PE) = sum of squares due to pure error.

SSB = between-treatments sum of squares.

SSC = column sum of squares.

SSE = error sum of squares.

SSI = interaction sum of squares.

SSR = row sum of squares.

SSR = residual sum of squares.

SST = total sum of squares, also TSS.

SSW = within-treatments sum of squares.

t = a specified period of failure-free operation.

t = number of treatments in a randomized block design.

t = statistic used to compare sample means or the distribution of the statistic.

t' = statistic used to compare sample means when the population standard deviations cannot be assumed equal. $t' = (\bar{X}_1 - \bar{X}_2)/ \sqrt{s_1^2/n_1 + s_2^2/n_2}$

t = number of standard deviation units.

t_c = censoring time.

t_p = pth quantile of T.

T = in sequential sampling, the cumulative sum of an appropriate statistic against the sample number n.

T = in cumulative sum control charts, the reference value.

T = tolerance.

T = observation total.

T = total test time.

T' = sum of ranks for two-sample Mann-Whitney test.

T_L = $(\bar{X} - L)/s$, a statistic used in MIL-STD-414.

T_U = $(U - \bar{X})/s$, a statistic used in MIL-STD-414.

T^2 = a statistic used to test population means on two characteristics (called Hotelling's T^2).

TSS = corrected total sum of squares.

u = defects per unit.

$\bar{u}$ = total number of nonconformities in all samples divided by the total number of units in all samples, i.e., the average number of nonconformities per unit.

U = symbol used for statistics (for testing hypotheses) that follow a normal distribution.

U = upper specification limit for a quality characteristic X.

UCL = upper control limit on a control chart.

ULL, LLL = parameters for lot plot sampling.

v = factor in graphical application of variables sampling plan.

v = factor in variables sampling plans for known variability.

$\text{var}(Z_t)$ = variance of exponentially weighted moving average.

w = cell width.

w = the ranges of the n observations on each treatment.

W = importance weights.

x = a vector of elements.

x' = the transpose of the x vector.

$x_1, x_2,...$ = observed value of some variable, usually a quality characteristic.

$\overline{X}$ = arithmetic mean, the average of a set of numbers $X_1, X_2, X_3, ..., X_n$ is the sum of the numbers divided by n.

$\overline{X}_0$ = an aimed-at or standard value of a quality characteristic. Also used to represent the true but often unknown mean of a universe being sampled.

$\overline{\overline{X}}$ = mean of several $\overline{X}$ values. Often called the grand average.

X = a matrix of elements.

$\tilde{X}$ = median value of X.

y = density or frequency function.

y_1, y_2 = acceptance and rejection values in sequential sampling.

Y = response characteristic.

Y = observed value of a variable.

Y_i = the yield for run number i.

Y_{ij} = the observation on the ith treatment in the jth block.

$\overline{Y}$ = average of values of Y variable.

$\overline{Y}_0$ = in response surface methodology, the average of the center points.

$\hat{Y}$ = the predicted value for Y.

z_{1-Pa} = normal z value for caluclating OC curve.

z_p = z value for fraction p in the upper tail of the normal distribution.

Z = normal distribution coefficient.

Z_t = statistic for exponentially weighted moving average plot.

α (alpha) = probability of rejecting the hypothesis under test when it is true. (Called the type I error or level of significance.) In acceptance sampling, α = the producer's risk.

α = scaling parameter of the Weibull distribution.

β (beta) = probability of accepting the hypothesis under test when it is false. (Called the type II error.) In acceptance sampling, β is the consumer's risk.

β_0, β_1 = acceptable and unacceptable mean life.

$\beta_1, \beta_2,...$ = in regression, the unknown parameters of the model.

β = shape parameter of the Weibull distribution.

$\beta^2 V$ = large sample approximate variance factor for Weibull estimation.

β_i = a term peculiar to a given block. It is the amount by which the response of a given treatment in the ith block differs from the response of the same treatment averaged over all blocks, assuming no experimental error.

γ (gamma) = confidence level.

γ = location parameter of the Weibull distribution.

Γ (gamma) = the gamma function.

δ (delta) = width of confidence interval.

δ = in a cumulative sum control chart, standardized shift in mean it is desired to detect, $\delta = D/\sigma$.

δ^* = distance between true value and hypothesized value.

ϵ (epsilon) = a random error term of a linear function.

η (eta) = scale parameter of the Weibull distribution.

η = group means, grand mean.

θ (theta) = mask angle for cumulative sum chart (one-half total angle of mask).

θ = in reliability, mean life.

θ = in cumulative sum control charts, an angle on the mask used in the chart.

λ (lambda) = failure rate, i.e., the proportion of failures per unit time.

λ = weighting factor for exponentially weighted moving average.

μ (mu) = the population mean (average), e.g., the mean of a lot, the mean time between failures, or the mean life. (μ_0 is the acceptable value of the mean and μ_1 is the unacceptable value of the mean.)

μ_0 = a standard value for μ.

ν (nu) = degrees of freedom.

$\nu(t)$ = recurrence rate of repairs per system.

ξ (xi) = transformation value.

π (pi) = the constant 3.14159+.

σ (sigma) = the population standard deviation; e.g., σ is the standard deviation of indiviual values, $\sigma_{\bar{x}}$ is the standard deviation of sample means. (Some literature uses σ' in place of σ.)

$\bar{\sigma}$ = mean of several standard deviations.

$\hat{\sigma}$ (sigma hat) = estimate of population standard deviation computed from a sample of size n; $\hat{\sigma} = \sqrt{\Sigma(x - \bar{x})^2 / n}$.

σ_0 = a standard value for σ.

σ_p = standard deviation of proportions.

σ_R = standard deviation of sample ranges.

σ_w^2 = within component of variance, estimated by s_w^2.

σ_b^2 = between component of variances, estimated by s_b^2.

σ^2 = population variance, error variance.

Σ (sigma) = a mathematical sign meaning "take the algebraic sum of the quantities which follow."

τ_d (tau$_d$) $= (\bar{X}_1 - \bar{X}_2/[^1\!/2(R_1 + R_2)]$, a statistic used to test the hypothesis about μ_1 and μ_2. Also, the distribution of the statistic.

τ_1 $= (\bar{X} - \mu_0)/R$, a statistic used to test the hypothesis about μ_0. Also the distribution of the statistic.

τ_j $=$ time between repairs.

ϕ_i (phi) $=$ in a randomized block experimental design, a term peculiar to the ith treatment, constant for all blocks regardless of the block in which the treatment occurs.

ϕ^2 $=$ quantity for determining sample size in analysis of variance.

$\Phi^{-1}(p)$ $=$ inverse of distribution function giving value of random variable having cumulative probability p.

Φ_{SEV} $=$ smallest extreme value distribution function.

Φ_{NOR} $=$ normal distribution function.

$\Phi(p)$ $=$ distribution function.

χ^2 (chi^2) $= (n - 1)s^2/\sigma_0^2$ or the distribution of this ratio.

∞ $=$ infinity.

See Appendix III for a listing of standards covering symbols and other matters.

APPENDIX II
TABLES AND CHARTS

For more extensive tables see Owen, D. B. (1962). *Handbook of Statistical Tables.* Addison-Wesley, Reading, MA. 1962

Table A Factors for Computing Control Chart Lines*

Observations in Sample, n	Chart for averages — Factors for control limits			Chart for standard deviations — Factors for central line		Factors for control limits				Chart for ranges — Factors for central line			Factors for control limits			
	A	A_2	A_3	c_4	$1/c_4$	B_3	B_4	B_5	B_6	d_2	$1/d_2$	d_3	D_1	D_2	D_3	D_4
2	2.121	1.880	2.659	0.7979	1.2533	0	3.267	0	2.606	1.128	0.8865	0.853	0	3.686	0	3.267
3	1.732	1.023	1.954	0.8862	1.1284	0	2.568	0	2.276	1.693	0.5907	0.888	0	4.358	0	2.574
4	1.500	0.729	1.628	0.9213	1.0854	0	2.266	0	2.088	2.059	0.4857	0.880	0	4.698	0	2.282
5	1.342	0.577	1.427	0.9400	1.0638	0	2.089	0	1.964	2.326	0.4299	0.864	0	4.918	0	2.114
6	1.225	0.483	1.287	0.9515	1.0510	0.030	1.970	0.029	1.874	2.534	0.3946	0.848	0	5.078	0	2.004
7	1.134	0.419	1.182	0.9594	1.0423	0.118	1.882	0.113	1.806	2.704	0.3698	0.833	0.204	5.204	0.076	1.924
8	1.061	0.373	1.099	0.9650	1.0363	0.185	1.815	0.179	1.751	2.847	0.3512	0.820	0.388	5.306	0.136	1.864
9	1.000	0.337	1.032	0.9693	1.0317	0.239	1.761	0.232	1.707	2.970	0.3367	0.808	0.547	5.393	0.184	1.816
10	0.949	0.308	0.975	0.9727	1.0281	0.284	1.716	0.276	1.669	3.078	0.3249	0.797	0.687	5.469	0.223	1.777
11	0.905	0.285	0.927	0.9754	1.0252	0.321	1.679	0.313	1.637	3.173	0.3152	0.787	0.811	5.535	0.256	1.744
12	0.866	0.266	0.886	0.9776	1.0229	0.354	1.646	0.346	1.610	3.258	0.3069	0.778	0.922	5.594	0.283	1.717
13	0.832	0.249	0.850	0.9794	1.0210	0.382	1.618	0.374	1.585	3.336	0.2998	0.770	1.025	5.647	0.307	1.693
14	0.802	0.235	0.817	0.9810	1.0194	0.406	1.594	0.399	1.563	3.407	0.2935	0.763	1.118	5.696	0.328	1.672
15	0.775	0.223	0.789	0.9823	1.0180	0.428	1.572	0.421	1.544	3.472	0.2880	0.756	1.203	5.741	0.347	1.653
16	0.750	0.212	0.763	0.9835	1.0168	0.448	1.552	0.440	1.526	3.532	0.2831	0.750	1.282	5.782	0.363	1.637
17	0.728	0.203	0.739	0.9845	1.0157	0.466	1.534	0.458	1.511	3.588	0.2787	0.744	1.356	5.820	0.378	1.622
18	0.707	0.194	0.718	0.9854	1.0148	0.482	1.518	0.475	1.496	3.640	0.2747	0.739	1.424	5.856	0.391	1.608
19	0.688	0.187	0.698	0.9862	1.0140	0.497	1.503	0.490	1.483	3.689	0.2711	0.734	1.487	5.891	0.403	1.597
20	0.671	0.180	0.680	0.9869	1.0133	0.510	1.490	0.504	1.470	3.735	0.2677	0.729	1.549	5.921	0.415	1.585
21	0.655	0.173	0.663	0.9876	1.0126	0.523	1.477	0.516	1.459	3.778	0.2647	0.724	1.605	5.951	0.425	1.575
22	0.640	0.167	0.647	0.9882	1.0119	0.534	1.466	0.528	1.448	3.819	0.2618	0.720	1.659	5.979	0.434	1.566
23	0.626	0.162	0.633	0.9887	1.0114	0.545	1.455	0.539	1.438	3.858	0.2592	0.716	1.710	6.006	0.443	1.557
24	0.612	0.157	0.619	0.9892	1.0109	0.555	1.445	0.549	1.429	3.895	0.2567	0.712	1.759	6.031	0.451	1.548
25	0.600	0.153	0.606	0.9896	1.0105	0.565	1.435	0.559	1.420	3.931	0.2544	0.708	1.806	6.056	0.459	1.541

*The above table is a copy of Table 27 in *ASTM Manual on Presentation of Data and Control Chart Analysis.* (1976). ASTM Publication STP15D. American Society for Testing and Materials, Philadelphia, pp. 134–135. Used with permission.

Notes: For $n > 25$, $A = 3/\sqrt{n}$, $A_3 = 3/c_4\sqrt{n}$, $c_4 \cong 4(n-1)/(4n-3)$; $B_3 = 1 - 3/c_4\sqrt{2(n-1)}$, $B_4 = 1 + 3/c_4\sqrt{2(n-1)}$,

$$B_5 = c_4 - 3/\sqrt{2(n-1)}, B_6 = c_4 + 3/\sqrt{2(n-1)}$$

FORMULAS

Purpose of chart	Chart for	Central line	3-Sigma control limits
For analyzing past inspection data for control ($\overline{\overline{X}}$, $\overline{s}$, $\overline{R}$ are average values for the data being analyzed)	Averages	$\overline{\overline{X}}$	$\overline{\overline{X}} \pm A_3\overline{s}$, or $\overline{\overline{X}} \pm A_2\overline{R}$
	Standard deviations	$\overline{s}$	$B_3\overline{s}$ and $B_4\overline{s}$
	Ranges	$\overline{R}$	$D_3\overline{R}$ and $D_4\overline{R}$
For controlling quality during production ($\overline{X}_0$, σ_0, R_0, are selected standard values; $R_0 = d_2\sigma_0$ for samples of size n)	Averages	$\overline{X}_0$	$\overline{X}_0 \pm A\sigma_0$ or $\overline{X}_0 \pm A_2R_0$
	Standard deviations	s_0 or $c_4\sigma_0$	$B_5\sigma_0$ and $B_6\sigma_0$
	Ranges	R_0 or $d_2\sigma_0$	$D_1\sigma_0$ and $D_2\sigma_0$

TABLE B Normal Distribution*

Proportion of total area under the curve from $-\infty$ to $K = \dfrac{X - \mu}{\sigma}$. To illustrate: when $K = + 2.0$, the probability is 0.9773 of obtaining a value equal to or less than X.

K	0.00	0.01	0.02	0.03	0.04	0.05	0.06	0.07	0.08	0.09
−3.5	0.00023	0.00022	0.00022	0.00021	0.00020	0.00019	0.00019	0.00018	0.00017	0.00017
−3.4	0.00034	0.00033	0.00031	0.00030	0.00029	0.00028	0.00027	0.00026	0.00025	0.00024
−3.3	0.00048	0.00047	0.00045	0.00043	0.00042	0.00040	0.00039	0.00038	0.00036	0.00035
−3.2	0.00069	0.00066	0.00064	0.00062	0.00060	0.00058	0.00056	0.00054	0.00052	0.00050
−3.1	0.00097	0.00094	0.00090	0.00087	0.00085	0.00082	0.00079	0.00076	0.00074	0.00071
−3.0	0.00135	0.00131	0.00126	0.00122	0.00118	0.00114	0.00111	0.00107	0.00104	0.00100
−2.9	0.0019	0.0018	0.0017	0.0017	0.0016	0.0016	0.0015	0.0015	0.0014	0.0014
−2.8	0.0026	0.0025	0.0024	0.0023	0.0023	0.0022	0.0021	0.0021	0.0020	0.0019
−2.7	0.0035	0.0034	0.0033	0.0032	0.0031	0.0030	0.0029	0.0028	0.0027	0.0026
−2.6	0.0047	0.0045	0.0044	0.0043	0.0041	0.0040	0.0039	0.0038	0.0037	0.0036
−2.5	0.0062	0.0060	0.0059	0.0057	0.0055	0.0054	0.0052	0.0051	0.0049	0.0048
−2.4	0.0082	0.0080	0.0078	0.0075	0.0073	0.0071	0.0069	0.0068	0.0066	0.0064
−2.3	0.0107	0.0104	0.0102	0.0099	0.0096	0.0094	0.0091	0.0089	0.0087	0.0084
−2.2	0.0139	0.0136	0.0132	0.0129	0.0125	0.0122	0.0119	0.0116	0.0113	0.0110
−2.1	0.0179	0.0174	0.0170	0.0166	0.0162	0.0158	0.0154	0.0150	0.0146	0.0143
−2.0	0.0228	0.0222	0.0217	0.0212	0.0207	0.0202	0.0197	0.0192	0.0188	0.0183
−1.9	0.0287	0.0281	0.0274	0.0268	0.0262	0.0256	0.0250	0.0244	0.0239	0.0233
−1.8	0.0359	0.0351	0.0344	0.0336	0.0329	0.0322	0.0314	0.0307	0.0301	0.0294
−1.7	0.0446	0.0436	0.0427	0.0418	0.0409	0.0401	0.0392	0.0384	0.0375	0.0367
−1.6	0.0548	0.0537	0.0526	0.0516	0.0505	0.0495	0.0485	0.0475	0.0465	0.0455
−1.5	0.0668	0.0655	0.0643	0.0630	0.0618	0.0606	0.0594	0.0582	0.0571	0.0559
−1.4	0.0808	0.0793	0.0778	0.0764	0.0749	0.0735	0.0721	0.0708	0.0694	0.0681
−1.3	0.0968	0.0951	0.0934	0.0918	0.0901	0.0885	0.0869	0.0853	0.0838	0.0823
−1.2	0.1151	0.1131	0.1112	0.1093	0.1075	0.1057	0.1038	0.1020	0.1003	0.0985
−1.1	0.1357	0.1335	0.1314	0.1292	0.1271	0.1251	0.1230	0.1210	0.1190	0.1170

TABLE B (*Continued*)

K	0.00	0.01	0.02	0.03	0.04	0.05	0.06	0.07	0.08	0.09
−1.0	0.1587	0.1562	0.1539	0.1515	0.1492	0.1469	0.1446	0.1423	0.1401	0.1379
−0.9	0.1841	0.1814	0.1788	0.1762	0.1736	0.1711	0.1685	0.1660	0.1635	0.1611
−0.8	0.2119	0.2090	0.2061	0.2033	0.2005	0.1977	0.1949	0.1922	0.1894	0.1867
−0.7	0.2420	0.2389	0.2358	0.2327	0.2297	0.2266	0.2236	0.2207	0.2177	0.2148
−0.6	0.2743	0.2709	0.2676	0.2643	0.2611	0.2578	0.2546	0.2514	0.2483	0.2451
−0.5	0.3085	0.3050	0.3015	0.2981	0.2946	0.2912	0.2877	0.2843	0.2810	0.2776
−0.4	0.3446	0.3409	0.3372	0.3336	0.3300	0.3264	0.3228	0.3192	0.3156	0.3121
−0.3	0.3821	0.3783	0.3745	0.3707	0.3669	0.3632	0.3594	0.3557	0.3520	0.3483
−0.2	0.4207	0.4168	0.4129	0.4090	0.4052	0.4013	0.3974	0.3936	0.3897	0.3859
−0.1	0.4602	0.4562	0.4522	0.4483	0.4443	0.4404	0.4364	0.4325	0.4286	0.4247
−0.0	0.5000	0.4960	0.4920	0.4880	0.4840	0.4801	0.4761	0.4721	0.4681	0.4641

K	0.09	0.08	0.07	0.06	0.05	0.04	0.03	0.02	0.01	0.00
+0.0	0.5359	0.5319	0.5279	0.5239	0.5199	0.5160	0.5120	0.5080	0.5040	0.5000
+0.1	0.5753	0.5714	0.5675	0.5636	0.5596	0.5557	0.5517	0.5478	0.5438	0.5398
+0.2	0.6141	0.6103	0.6064	0.6026	0.5987	0.5948	0.5910	0.5871	0.5832	0.5793
+0.3	0.6517	0.6480	0.6443	0.6406	0.6368	0.6331	0.6293	0.6255	0.6217	0.6179
+0.4	0.6879	0.6844	0.6808	0.6772	0.6736	0.6700	0.6664	0.6628	0.6591	0.6554
+0.5	0.7224	0.7190	0.7157	0.7123	0.7088	0.7054	0.7019	0.6985	0.6950	0.6915
+0.6	0.7549	0.7517	0.7486	0.7454	0.7422	0.7389	0.7357	0.7324	0.7291	0.7257
+0.7	0.7852	0.7823	0.7794	0.7764	0.7734	0.7704	0.7673	0.7642	0.7611	0.7580
+0.8	0.8133	0.8106	0.8079	0.8051	0.8023	0.7995	0.7967	0.7939	0.7910	0.7881
+0.9	0.8389	0.8365	0.8340	0.8315	0.8289	0.8264	0.8238	0.8212	0.8186	0.8159
+1.0	0.8621	0.8599	0.8577	0.8554	0.8531	0.8508	0.8485	0.8461	0.8438	0.8413
+1.1	0.8830	0.8810	0.8790	0.8770	0.8749	0.8729	0.8708	0.8686	0.8665	0.8643
+1.2	0.9015	0.8997	0.8980	0.8962	0.8944	0.8925	0.8907	0.8888	0.8869	0.8849
+1.3	0.9177	0.9162	0.9147	0.9131	0.9115	0.9099	0.9082	0.9066	0.9049	0.9032
+1.4	0.9319	0.9306	0.9292	0.9279	0.9265	0.9251	0.9236	0.9222	0.9207	0.9192
+1.5	0.9441	0.9429	0.9418	0.9406	0.9394	0.9382	0.9370	0.9357	0.9345	0.9332

z	0.00	0.01	0.02	0.03	0.04	0.05	0.06	0.07	0.08	0.09
+1.6	0.9452	0.9463	0.9474	0.9484	0.9495	0.9505	0.9515	0.9525	0.9535	0.9545
+1.7	0.9554	0.9564	0.9573	0.9582	0.9591	0.9599	0.9608	0.9616	0.9625	0.9633
+1.8	0.9641	0.9649	0.9656	0.9664	0.9671	0.9678	0.9686	0.9693	0.9699	0.9706
+1.9	0.9713	0.9719	0.9726	0.9732	0.9738	0.9744	0.9750	0.9756	0.9761	0.9767
+2.0	0.9773	0.9778	0.9783	0.9788	0.9793	0.9798	0.9803	0.9808	0.9812	0.9817
+2.1	0.9821	0.9826	0.9830	0.9834	0.9838	0.9842	0.9846	0.9850	0.9854	0.9857
+2.2	0.9861	0.9864	0.9868	0.9871	0.9875	0.9878	0.9881	0.9884	0.9887	0.9890
+2.3	0.9893	0.9896	0.9898	0.9901	0.9904	0.9906	0.9909	0.9911	0.9913	0.9916
+2.4	0.9918	0.9920	0.9922	0.9925	0.9927	0.9929	0.9931	0.9932	0.9934	0.9936
+2.5	0.9938	0.9940	0.9941	0.9943	0.9945	0.9946	0.9948	0.9949	0.9951	0.9952
+2.6	0.9953	0.9955	0.9956	0.9957	0.9959	0.9960	0.9961	0.9962	0.9963	0.9964
+2.7	0.9965	0.9966	0.9967	0.9968	0.9969	0.9970	0.9971	0.9972	0.9973	0.9974
+2.8	0.9974	0.9975	0.9976	0.9977	0.9977	0.9978	0.9979	0.9979	0.9980	0.9981
+2.9	0.9981	0.9982	0.9983	0.9983	0.9984	0.9984	0.9985	0.9985	0.9986	0.9986
+3.0	0.99865	0.99869	0.99874	0.99878	0.99882	0.99886	0.99889	0.99893	0.99896	0.99900
+3.1	0.99903	0.99906	0.99910	0.99913	0.99915	0.99918	0.99921	0.99924	0.99926	0.99929
+3.2	0.99931	0.99934	0.99936	0.99938	0.99940	0.99942	0.99944	0.99946	0.99948	0.99950
+3.3	0.99952	0.99953	0.99955	0.99957	0.99958	0.99960	0.99961	0.99962	0.99964	0.99965
+3.4	0.99966	0.99967	0.99969	0.99970	0.99971	0.99972	0.99973	0.99974	0.99975	0.99976
+3.5	0.99977	0.99978	0.99978	0.99979	0.99980	0.99981	0.99981	0.99982	0.99983	0.99983

*Adapted with permission from Grant, Eugene L. and Leavenworth, Richard S. (1972). *Statistical Quality Control*, 4th ed. McGraw-Hill, New York, pp. 642–643.

TABLE C   Exponential Distribution*

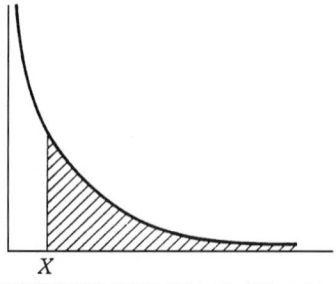

$\dfrac{X}{\mu}$	0.00	0.01	0.02	0.03	0.04	0.05	0.06	0.07	0.08	0.09
0.0	1.000	0.9900	0.9802	0.9704	0.9608	0.9512	0.9418	0.9324	0.9231	0.9139
0.1	0.9048	0.8958	0.8860	0.8781	0.8694	0.8607	0.8521	0.8437	0.8353	0.8270
0.2	0.8187	0.8106	0.8025	0.7945	0.7866	0.7788	0.7711	0.7634	0.7758	0.7483
0.3	0.7408	0.7334	0.7261	0.7189	0.7118	0.7047	0.6977	0.6907	0.6839	0.6771
0.4	0.6703	0.6637	0.6570	0.6505	0.6440	0.6376	0.6313	0.6250	0.6188	0.6126
0.5	0.6065	0.6005	0.5945	0.5886	0.5827	0.5769	0.5712	0.5655	0.5599	0.5543
0.6	0.5488	0.5434	0.5379	0.5326	0.5273	0.5220	0.5169	0.5117	0.5066	0.5016
0.7	0.4966	0.4916	0.4868	0.4819	0.4771	0.4724	0.4677	0.4630	0.4584	0.4538
0.8	0.4493	0.4449	0.4404	0.4360	0.4317	0.4274	0.4232	0.4190	0.4148	0.4107
0.9	0.4066	0.4025	0.3985	0.3946	0.3906	0.3867	0.3829	0.3791	0.3753	0.3716
	0.0	0.1	0.2	0.3	0.4	0.5	0.6	0.7	0.8	0.9
1.0	0.3679	0.3329	0.3012	0.2725	0.2466	0.2231	0.2019	0.1827	0.1653	0.1496
2.0	0.1353	0.1225	0.1108	0.1003	0.0907	0.0821	0.0743	0.0672	0.0608	0.0550
3.0	0.0498	0.0450	0.0408	0.0369	0.0334	0.0302	0.0273	0.0247	0.0224	0.0202
4.0	0.0183	0.0166	0.0150	0.0130	0.0123	0.0111	0.0101	0.0091	0.0082	0.0074
5.0	0.0067	0.0061	0.0055	0.0050	0.0045	0.0041	0.0037	0.0033	0.0030	0.0027
6.0	0.0025	0.0022	0.0020	0.0018	0.0017	0.0015	0.0014	0.0012	0.0011	0.0010

*Adapted with permission from Selby, S. M. (ed.) (1969). *CRC Standard Mathematical Tables*, 17th ed. The Chemical Rubber Co., pp. 201–207.

TABLE D Median Ranks*

Sample size = n

	1	2	3	4	5	6	7	8	9	10	11	12	13	14	15	16	17	18	19	20
1	.5000	.2929	.2063	.1591	.1294	.1091	.0943	.0830	.0741	.0670	.0611	.0561	.0519	.0483	.0452	.0424	.0400	.0378	.0358	.0341
2		.7071	.5000	.3864	.3147	.2655	.2295	.2021	.1806	.1632	.1489	.1368	.1266	.1178	.1101	.1034	.0975	.0922	.0874	.0831
3			.7937	.6136	.5000	.4218	.3648	.3213	.2871	.2594	.2366	.2175	.2013	.1873	.1751	.1644	.1550	.1465	.1390	.1322
4				.8409	.6853	.5782	.5000	.4404	.3935	.3557	.3244	.2982	.2760	.2568	.2401	.2254	.2125	.2009	.1905	.1812
5					.8706	.7345	.6352	.5596	.5000	.4519	.4122	.3789	.3506	.3263	.3051	.2865	.2700	.2553	.2421	.2302
6						.8909	.7705	.6787	.6065	.5481	.5000	.4596	.4253	.3958	.3700	.3475	.3275	.3097	.2937	.2793
7							.9057	.7979	.7129	.6443	.5878	.5404	.5000	.4653	.4350	.4085	.3850	.3641	.3453	.3283
8								.9170	.8194	.7406	.6756	.6211	.5747	.5347	.5000	.4695	.4425	.4184	.3968	.3774
9									.9259	.8368	.7634	.7018	.6494	.6042	.5650	.5305	.5000	.4728	.4484	.4264
10										.9330	.8511	.7825	.7240	.6737	.6300	.5915	.5575	.5272	.5000	.4755
11											.9389	.8632	.7987	.7432	.6949	.6525	.6150	.5816	.5516	.5245
12												.9439	.8734	.8127	.7599	.7135	.6725	.6359	.6032	.5736
13													.9481	.8822	.8249	.7746	.7300	.6903	.6547	.6226
14														.9517	.8899	.8356	.7875	.7447	.7063	.6717
15															.9548	.8966	.8450	.7991	.7579	.7207
16																.9576	.9025	.8535	.8095	.7698
17																	.9600	.9078	.8610	.8188
18																		.9622	.9126	.8678
19																			.9642	.9169
20																				.9659

*Adapted with permission from "The Table of Median Ranks of Sample Values on Their Population with an Application to Certain Fatigue Studies." (1951). *Industrial Mathematics*, no. 2, p. 7.

TABLE E Poisson Distribution*

1000 × probability of r or fewer occurrences of event that has average number of occurrences equal to np.

np \ r	0	1	2	3	4	5	6	7	8	9
0.02	980	1,000								
0.04	961	999	1,000							
0.06	942	998	1,000							
0.08	923	997	1,000							
0.10	905	995	1,000							
0.15	861	990	999	1,000						
0.20	819	982	999	1,000						
0.25	779	974	998	1,000						
0.30	741	963	996	1,000						
0.35	705	951	994	1,000						
0.40	670	938	992	999	1,000					
0.45	638	925	989	999	1,000					
0.50	607	910	986	998	1,000					
0.55	577	894	982	998	1,000					
0.60	549	878	977	997	1,000					
0.65	522	861	972	996	999	1,000				
0.70	497	844	966	994	999	1,000				
0.75	472	827	959	993	999	1,000				
0.80	449	809	953	991	999	1,000				
0.85	427	791	945	989	998	1,000				
0.90	407	772	937	987	998	1,000				
0.95	387	754	929	984	997	1,000				
1.00	368	736	920	981	996	999	1,000			
1.1	333	699	900	974	995	999	1,000			
1.2	301	663	879	966	992	998	1,000			
1.3	273	627	857	957	989	998	1,000			
1.4	247	592	833	946	986	997	999	1,000		
1.5	223	558	809	934	981	996	999	1,000		
1.6	202	525	783	921	976	994	999	1,000		
1.7	183	493	757	907	970	992	998	1,000		
1.8	165	463	731	891	964	990	997	999	1,000	
1.9	150	434	704	875	956	987	997	999	1,000	
2.0	135	406	677	857	947	983	995	999	1,000	

np \ r	0	1	2	3	4	5	6	7	8	9
2.2	111	355	623	819	928	975	993	998	1,000	
2.4	091	308	570	779	904	964	988	997	999	1,000
2.6	074	267	518	736	877	951	983	995	999	1,000
2.8	061	231	469	692	848	935	976	992	998	999
3.0	050	199	423	647	815	916	966	988	996	999
3.2	041	171	380	603	781	895	955	983	994	998
3.4	033	147	340	558	744	871	942	977	992	997
3.6	027	126	303	515	706	844	927	969	988	996
3.8	022	107	269	473	668	816	909	960	984	994
4.0	018	092	238	433	629	785	889	949	979	992
4.2	015	078	210	395	590	753	867	936	972	989
4.4	012	066	185	359	551	720	844	921	964	985
4.6	010	056	163	326	513	686	818	905	955	980
4.8	008	048	143	294	476	651	791	887	944	975
5.0	007	040	125	265	440	616	762	867	932	968
5.2	006	034	109	238	406	581	732	845	918	960
5.4	005	029	095	213	373	546	702	822	903	951
5.6	004	024	082	191	342	512	670	797	886	941
5.8	003	021	072	170	313	478	638	771	867	929
6.0	002	017	062	151	285	446	606	744	847	916

np \ r	10	11	12	13	14	15	16
2.8	1,000						
3.0	1,000						
3.2	1,000						
3.4	999	1,000					
3.6	999	1,000					
3.8	998	999	1,000				
4.0	997	999	1,000				
4.2	996	999	1,000				
4.4	994	998	999	1,000			
4.6	992	997	999	1,000			
4.8	990	996	999	1,000			
5.0	986	995	998	999	1,000		
5.2	982	993	997	999	1,000		
5.4	977	990	996	999	1,000		
5.6	972	988	995	998	999	1,000	
5.8	965	984	993	997	999	1,000	
6.0	957	980	991	996	999	999	1,000

TABLE E   (*Continued*)

np \ r	0	1	2	3	4	5	6	7	8	9
6.2	002	015	054	134	259	414	574	716	826	902
6.4	002	012	046	119	235	384	542	687	803	886
6.6	001	010	040	105	213	355	511	658	780	869
6.8	001	009	034	093	192	327	480	628	755	850
7.0	001	007	030	082	173	301	450	599	729	830
7.2	001	006	025	072	156	276	420	569	703	810
7.4	001	005	022	063	140	253	392	539	676	788
7.6	001	004	019	055	125	231	365	510	648	765
7.8	000	004	016	048	112	210	338	481	620	741
8.0	000	003	014	042	100	191	313	453	593	717
8.5	000	002	009	030	074	150	256	386	523	653
9.0	000	001	006	021	055	116	207	324	456	587
9.5	000	001	004	015	040	089	165	269	392	522
10.0	000	000	003	010	029	067	130	220	333	458

np \ r	10	11	12	13	14	15	16	17	18	19
6.2	949	975	989	995	998	999	1,000			
6.4	939	969	986	994	997	999	1,000			
6.6	927	963	982	992	997	999	999	1,000		
6.8	915	955	978	990	996	998	999	1,000		
7.0	901	947	973	987	994	998	999	1,000		
7.2	887	937	967	984	993	997	999	999	1,000	
7.4	871	926	961	980	991	996	998	999	1,000	
7.6	854	915	954	976	989	995	998	999	1,000	
7.8	835	902	945	971	986	993	997	999	1,000	
8.0	816	888	936	966	983	992	996	998	999	1,000
8.5	763	849	909	949	973	986	993	997	999	999
9.0	706	803	876	926	959	978	989	995	998	999
9.5	645	752	836	898	940	967	982	991	996	998
10.0	583	697	792	864	917	951	973	986	993	997

np \ r	20	21	22
8.5	1,000		
9.0	1,000		
9.5	999	1,000	
10.0	998	999	1,000

np \ r	0	1	2	3	4	5	6	7	8	9
10.5	000	000	002	007	021	050	102	179	279	397
11.0	000	000	001	005	015	038	079	143	232	341
11.5	000	000	001	003	011	028	060	114	191	289
12.0	000	000	001	002	008	020	046	090	155	242
12.5	000	000	000	002	005	015	035	070	125	201
13.0	000	000	000	001	004	011	026	054	100	166
13.5	000	000	000	001	003	008	019	041	079	135
14.0	000	000	000	000	002	006	014	032	062	109
14.5	000	000	000	000	001	004	010	024	048	088
15.0	000	000	000	000	001	003	008	018	037	070

np \ r	10	11	12	13	14	15	16	17	18	19
10.5	521	639	742	825	888	932	960	978	988	994
11.0	460	579	689	781	854	907	944	968	982	991
11.5	402	520	633	733	815	878	924	954	974	986
12.0	347	462	576	682	772	844	899	937	963	979
12.5	297	406	519	628	725	806	869	916	948	969
13.0	252	353	463	573	675	764	835	890	930	957
13.5	211	304	409	518	623	718	798	861	908	942
14.0	176	260	358	464	570	669	756	827	883	923
14.5	145	220	311	413	518	619	711	790	853	901
15.0	118	185	268	363	466	568	664	749	819	875

np \ r	20	21	22	23	24	25	26	27	28	29
10.5	997	999	999	1,000						
11.0	995	998	999	1,000						
11.5	992	996	998	999	1,000					
12.0	988	994	997	999	999	1,000				
12.5	983	991	995	998	999	999	1,000			
13.0	975	986	992	996	998	999	1,000			
13.5	965	980	989	994	997	998	999	1,000		
14.0	952	971	983	991	995	997	999	999	1,000	
14.5	936	960	976	986	992	996	998	999	999	1,000
15.0	917	947	967	981	989	994	997	998	999	1,000

*Adapted with permission from Grant, E. L. and Leavenworth, Richard S. (1972). *Statistical Quality Control,* 4th ed. McGraw-Hill, New York.

TABLE F   Binomial Distribution*

Probability of r or fewer occurrences of an event in n trials, where p is the probability of occurrence on each trial.

n	r	0.05	0.10	0.15	0.20	0.25	0.30	0.35	0.40	0.45	0.50
2	0	0.9025	0.8100	0.7225	0.6400	0.5625	0.4900	0.4225	0.3600	0.3025	0.2500
	1	0.9975	0.9900	0.9775	0.9600	0.9375	0.9100	0.8775	0.8400	0.7975	0.7500
3	0	0.8574	0.7290	0.6141	0.5120	0.4219	0.3430	0.2746	0.2160	0.1664	0.1250
	1	0.9928	0.9720	0.9392	0.8960	0.8438	0.7840	0.7182	0.6480	0.5748	0.5000
	2	0.9999	0.9990	0.9966	0.9920	0.9844	0.9730	0.9571	0.9360	0.9089	0.8750
4	0	0.8145	0.6561	0.5220	0.4096	0.3164	0.2401	0.1785	0.1296	0.0915	0.0625
	1	0.9860	0.9477	0.8905	0.8192	0.7383	0.6517	0.5630	0.4752	0.3910	0.3125
	2	0.9995	0.9963	0.9880	0.9728	0.9492	0.9163	0.8735	0.8208	0.7585	0.6875
	3	1.0000	0.9999	0.9995	0.9984	0.9961	0.9919	0.9850	0.9744	0.9590	0.9375
5	0	0.7738	0.5905	0.4437	0.3277	0.2373	0.1681	0.1160	0.0778	0.0503	0.0312
	1	0.9774	0.9185	0.8352	0.7373	0.6328	0.5282	0.4284	0.3370	0.2562	0.1875
	2	0.9988	0.9914	0.9734	0.9421	0.8965	0.8369	0.7648	0.6826	0.5931	0.5000
	3	1.0000	0.9995	0.9978	0.9933	0.9844	0.9692	0.9460	0.9130	0.8688	0.8125
	4	1.0000	1.0000	0.9999	0.9997	0.9990	0.9976	0.9947	0.9898	0.9815	0.9688
6	0	0.7351	0.5314	0.3771	0.2621	0.1780	0.1176	0.0754	0.0467	0.0277	0.0156
	1	0.9672	0.8857	0.7765	0.6554	0.5339	0.4202	0.3191	0.2333	0.1636	0.1094
	2	0.9978	0.9842	0.9527	0.9011	0.8306	0.7443	0.6471	0.5443	0.4415	0.3438
	3	0.9999	0.9987	0.9941	0.9830	0.9624	0.9295	0.8826	0.8208	0.7447	0.6562
	4	1.0000	0.9999	0.9996	0.9984	0.9954	0.9891	0.9777	0.9590	0.9308	0.8906
	5	1.0000	1.0000	1.0000	0.9999	0.9998	0.9993	0.9982	0.9959	0.9917	0.9844
7	0	0.6983	0.4783	0.3206	0.2097	0.1335	0.0824	0.0490	0.0280	0.0152	0.0078
	1	0.9556	0.8503	0.7166	0.5767	0.4449	0.3294	0.2338	0.1586	0.1024	0.0625
	2	0.9962	0.9743	0.9262	0.8520	0.7564	0.6471	0.5323	0.4199	0.3164	0.2266
	3	0.9998	0.9973	0.9879	0.9667	0.9294	0.8740	0.8002	0.7102	0.6083	0.5000
	4	1.0000	0.9998	0.9988	0.9953	0.9871	0.9712	0.9444	0.9037	0.8471	0.7734
	5	1.0000	1.0000	0.9999	0.9996	0.9987	0.9962	0.9910	0.9812	0.9643	0.9375

n	x										
6		0.9922	0.9963	0.9984	0.9994	0.9998	0.9999	1.0000	1.0000	1.0000	1.0000
8	0	0.0039	0.0084	0.0168	0.0319	0.0576	0.1001	0.1678	0.2725	0.4305	0.6634
	1	0.0352	0.0632	0.1064	0.1691	0.2553	0.3671	0.5033	0.6572	0.8131	0.9428
	2	0.1445	0.2201	0.3154	0.4278	0.5518	0.6785	0.7969	0.8948	0.9619	0.9942
	3	0.3633	0.4770	0.5941	0.7064	0.8059	0.8862	0.9437	0.9786	0.9950	0.9996
	4	0.6367	0.7396	0.8263	0.8939	0.9420	0.9727	0.9896	0.9971	0.9996	1.0000
	5	0.8555	0.9115	0.9502	0.9747	0.9887	0.9958	0.9988	0.9998	1.0000	1.0000
	6	0.9648	0.9819	0.9915	0.9964	0.9987	0.9996	0.9999	1.0000	1.0000	1.0000
	7	0.9961	0.9983	0.9993	0.9998	0.9999	1.0000	1.0000	1.0000	1.0000	1.0000
9	0	0.0020	0.0046	0.0101	0.0207	0.0404	0.0751	0.1342	0.2316	0.3874	0.6302
	1	0.0195	0.0385	0.0705	0.1211	0.1960	0.3003	0.4362	0.5995	0.7748	0.9288
	2	0.0898	0.1495	0.2318	0.3373	0.4628	0.6007	0.7382	0.8591	0.9470	0.9916
	3	0.2539	0.3614	0.4826	0.6089	0.7297	0.8343	0.9144	0.9661	0.9917	0.9994
	4	0.5000	0.6214	0.7334	0.8283	0.9012	0.9511	0.9804	0.9944	0.9991	1.0000
	5	0.7461	0.8342	0.9006	0.9464	0.9747	0.9900	0.9969	0.9994	0.9999	1.0000
	6	0.9102	0.9502	0.9750	0.9888	0.9957	0.9987	0.9997	1.0000	1.0000	1.0000
	7	0.9805	0.9909	0.9962	0.9986	0.9996	0.9999	1.0000	1.0000	1.0000	1.0000
	8	0.9980	0.9992	0.9997	0.9999	1.0000	1.0000	1.0000	1.0000	1.0000	1.0000
10	0	0.0010	0.0025	0.0060	0.0135	0.0282	0.0563	0.1074	0.1969	0.3487	0.5987
	1	0.0107	0.0232	0.0464	0.0860	0.1493	0.2440	0.3758	0.5443	0.7361	0.9139
	2	0.0547	0.0996	0.1673	0.2616	0.3828	0.5256	0.6778	0.8202	0.9298	0.9885
	3	0.1719	0.2660	0.3823	0.5138	0.6496	0.7759	0.8791	0.9500	0.9872	0.9990
	4	0.3770	0.5044	0.6331	0.7515	0.8497	0.9219	0.9672	0.9901	0.9984	0.9999
	5	0.6230	0.7384	0.8338	0.9051	0.9527	0.9803	0.9936	0.9986	0.9999	1.0000
	6	0.8281	0.8980	0.9452	0.9740	0.9894	0.9965	0.9991	0.9999	1.0000	1.0000
	7	0.9453	0.9726	0.9877	0.9952	0.9984	0.9996	0.9999	1.0000	1.0000	1.0000
	8	0.9893	0.9955	0.9983	0.9995	0.9999	1.0000	1.0000	1.0000	1.0000	1.0000
	9	0.9990	0.9997	0.9999	1.0000	1.0000	1.0000	1.0000	1.0000	1.0000	1.0000

*Adapted with permission from Miller, Irwin and Freund, John E. (1965). *Probability and Statistics for Engineers*. Prentice-Hall, Englewood Cliffs, NJ, pp. 388–389.

For more extensive tables see The Staff of Harvard University Computation Laboratory (1955). *Tables of Cumulative Binomial Probability Distribution*. Harvard University Press, Cambridge, MA. See also Robertson, W. H. (1960). *Tables of the Binomial Distribution Function for Small Values of p*. Sandia Corp. Monograph, available from the Office of Technical Services, Department of Commerce, Washington, DC.

TABLE G* Distribution of *t*

Value of t corresponding to certain selected probabilities (i.e. tail areas under the curve). To illustrate: The probability is 0.975 that a sample with 20 degrees of freedom would have t = +2.086 or smaller.

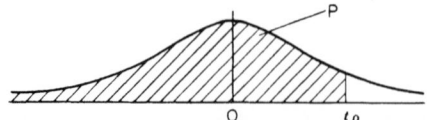

DF	$t_{.60}$	$t_{.70}$	$t_{.80}$	$t_{.90}$	$t_{.95}$	$t_{.975}$	$t_{.99}$	$t_{.995}$
1	0.325	0.727	1.376	3.078	6.314	12.706	31.821	63.657
2	0.289	0.617	1.061	1.886	2.920	4.303	6.965	9.925
3	0.277	0.584	0.978	1.638	2.353	3.182	4.541	5.841
4	0.271	0.569	0.941	1.533	2.132	2.776	3.747	4.604
5	0.267	0.559	0.920	1.476	2.015	2.571	3.365	4.032
6	0.265	0.553	0.906	1.440	1.943	2.447	3.143	3.707
7	0.263	0.549	0.896	1.415	1.895	2.365	2.998	3.499
8	0.262	0.546	0.889	1.397	1.860	2.306	2.896	3.355
9	0.261	0.543	0.883	1.383	1.833	2.262	2.821	3.250
10	0.260	0.542	0.879	1.372	1.812	2.228	2.764	3.169
11	0.260	0.540	0.876	1.363	1.796	2.201	2.718	3.106
12	0.259	0.539	0.873	1.356	1.782	2.179	2.681	3.055
13	0.259	0.538	0.870	1.350	1.771	2.160	2.650	3.012
14	0.258	0.537	0.868	1.345	1.761	2.145	2.624	2.977
15	0.258	0.536	0.866	1.341	1.753	2.131	2.602	2.947
16	0.258	0.535	0.865	1.337	1.746	2.120	2.583	2.921
17	0.257	0.534	0.863	1.333	1.740	2.110	2.567	2.898
18	0.257	0.534	0.862	1.330	1.734	2.101	2.552	2.878
19	0.257	0.533	0.861	1.328	1.729	2.093	2.539	2.861
20	0.257	0.533	0.860	1.325	1.725	2.086	2.528	2.845
21	0.257	0.532	0.859	1.323	1.721	2.080	2.518	2.831
22	0.256	0.532	0.858	1.321	1.717	2.074	2.508	2.819
23	0.256	0.532	0.858	1.319	1.714	2.069	2.500	2.807
24	0.256	0.531	0.857	1.318	1.711	2.064	2.492	2.797
25	0.256	0.531	0.856	1.316	1.708	2.060	2.485	2.787
26	0.256	0.531	0.856	1.315	1.706	2.056	2.479	2.779
27	0.256	0.531	0.855	1.314	1.703	2.052	2.473	2.771
28	0.256	0.530	0.855	1.313	1.701	2.048	2.467	2.763
29	0.256	0.530	0.854	1.311	1.699	2.045	2.462	2.756
30	0.256	0.530	0.854	1.310	1.697	2.042	2.457	2.750
40	0.255	0.529	0.851	1.303	1.684	2.021	2.423	2.704
60	0.254	0.527	0.848	1.296	1.671	2.000	2.390	2.660
120	0.254	0.526	0.845	1.289	1.658	1.980	2.358	2.617
∞	0.253	0.524	0.842	1.282	1.645	1.960	2.326	2.576

*Adapted with permission from Dixon, W. J. and Massey, F. J., Jr. (1969). *Introduction to Statistical Analysis,* 3rd ed. McGraw-Hill, New York. Entries originally from Fisher, R. A. and Yates, F. *Statistical Tables.* Oliver & Boyd, London, Table III.

TABLE H* Percentile for $\tau_d = \dfrac{\overline{X} - \mu_0}{R}$

Sample size	$\phi_{.95}$	$\phi_{.975}$	$\phi_{.99}$
2	3.175	6.353	15.910
3	0.885	1.304	2.111
4	0.529	0.717	1.023
5	0.388	0.507	0.685
6	0.312	0.399	0.523
7	0.263	0.333	0.429
8	0.230	0.288	0.366
9	0.205	0.255	0.322
10	0.186	0.230	0.288
11	0.170	0.210	0.262
12	0.158	0.194	0.241
13	0.147	0.181	0.224
14	0.138	0.170	0.209
15	0.131	0.160	0.197
16	0.124	0.151	0.186
17	0.118	0.144	0.177
18	0.113	0.137	0.168
19	0.108	0.131	0.161
20	0.104	0.126	0.154

*Adapted with permission from Lord, E. (1957). "The Use of the Range in Place of the Standard Deviation in the *t* Test." *Biometrika,* vol. 34.

TABLE I Critical Values of *r* for the Sign Test*

Percentages are values for α for a two-tail test. (Two-tail percentage points are given for the binomial for p = 0.05.)

N	1%	5%	10%	25%
1				
2				
3				0
4				0
5			0	0
6		0	0	1
7		0	0	1
8	0	0	1	1
9	0	1	1	2
10	0	1	1	2
11	0	1	2	3
12	1	2	2	3
13	1	2	3	3
14	1	2	3	4
15	2	3	3	4
16	2	3	4	5
17	2	4	4	5
18	3	4	5	6
19	3	4	5	6
20	3	5	5	6
21	4	5	6	7
22	4	5	6	7
23	4	6	7	8
24	5	6	7	8
25	5	7	7	9
26	6	7	8	9
27	6	7	8	10
28	6	8	9	10
29	7	8	9	10
30	7	9	10	11
31	7	9	10	11
32	8	9	10	12
33	8	10	11	12
34	9	10	11	13
35	9	11	12	13
36	9	11	12	14
37	10	12	13	14
38	10	12	13	14
39	11	12	13	15
40	11	13	14	15
41	11	13	14	16
42	12	14	15	16
43	12	14	15	17
44	13	15	16	17
45	13	15	16	18
46	13	15	16	18
47	14	16	17	19
48	14	16	17	19
49	15	17	18	19
50	15	17	18	20

*Adapted with permission from Dixon, W. J. and Massey, F. J., Jr.(1969). *Introduction to Statistical Analysis*, 3rd ed. McGraw-Hill, New York.

TABLE J* Percentiles for $\tau_d = \dfrac{\overline{X}_1 - \overline{X}_2}{\frac{1}{2}(R_1 + R_2)}$

$n = n_A = n_B$	$\phi'_{.95}$	$\phi'_{.975}$	$\phi'_{.99}$
2	2.322	3.427	5.553
3	0.974	1.272	1.715
4	0.644	0.813	1.047
5	0.493	0.613	0.772
6	0.405	0.499	0.621
7	0.347	0.426	0.525
8	0.306	0.373	0.459
9	0.275	0.334	0.409
10	0.250	0.304	0.371
11	0.233	0.280	0.340
12	0.214	0.260	0.315
13	0.201	0.243	0.294
14	0.189	0.189	0.276
15	0.179	0.216	0.261
16	0.170	0.205	0.247
17	0.162	0.195	0.236
18	0.155	0.187	0.225
19	0.149	0.179	0.216
20	0.143	0.172	0.207

*Adapted with permission from Lord, E. (1947). "The Use of the Range in Place of the Standard Deviation in the *t* Test. " *Biometrika,* vol. 34.

TABLE K* Distribution of F

Values of F corresponding to certain selected probabilities (i.e., tail areas under the curve). To illustrate: The probability is 0.05 that the ratio of two sample variances obtained with 20 and 10 degrees of freedom in numerator and denominator, respectively, would have F = 2.77 or larger. For a two-sided test, a lower limit is found by taking the reciprocal of the tabulated F value for the degrees of freedom in reverse. For the above example, with 10 and 20 degrees of freedom in numerator and denominator, respectively, F is 2.35 and 1/F is 1/2.35, or 0.43. The probability is 0.10 that F is 0.43 or smaller or 2.77 or larger.

n_2 \ n_1	1	2	3	4	5	6	7	8	9
					$F_{.95}(n_1, n_2)$				
1	161.4	199.5	215.7	224.6	230.2	234.0	236.8	238.9	240.5
2	18.51	19.00	19.16	19.25	19.30	19.33	19.35	19.37	19.38
3	10.13	9.55	9.28	9.12	9.01	8.94	8.89	8.85	8.81
4	7.71	6.94	6.59	6.39	6.26	6.16	6.09	6.04	6.00
5	6.61	5.79	5.41	5.19	5.05	4.95	4.88	4.82	4.77
6	5.99	5.14	4.76	4.53	4.39	4.28	4.21	4.15	4.10
7	5.59	4.74	4.35	4.12	3.97	3.87	3.79	3.73	3.68
8	5.32	4.46	4.07	3.84	3.69	3.58	3.50	3.44	3.39
9	5.12	4.26	3.86	3.63	3.48	3.37	3.29	3.23	3.18
10	4.96	4.10	3.71	3.48	3.33	3.22	3.14	3.07	3.02
11	4.84	3.98	3.59	3.36	3.20	3.09	3.01	2.95	2.90
12	4.75	3.89	3.49	3.26	3.11	3.00	2.91	2.85	2.80
13	4.67	3.81	3.41	3.18	3.03	2.92	2.83	2.77	2.71
14	4.60	3.74	3.34	3.11	2.96	2.85	2.76	2.70	2.65
15	4.54	3.68	3.29	3.06	2.90	2.79	2.71	2.64	2.59
16	4.49	3.63	3.24	3.01	2.85	2.74	2.66	2.59	2.54
17	4.45	3.59	3.20	2.96	2.81	2.70	2.61	2.55	2.49
18	4.41	3.55	3.16	2.93	2.77	2.66	2.58	2.51	2.46
19	4.38	3.52	3.13	2.90	2.74	2.63	2.54	2.48	2.42
20	4.35	3.49	3.10	2.87	2.71	2.60	2.51	2.45	2.39
21	4.32	3.47	3.07	2.84	2.68	2.57	2.49	2.42	2.37
22	4.30	3.44	3.05	2.82	2.66	2.55	2.46	2.40	2.34
23	4.28	3.42	3.03	2.80	2.64	2.53	2.44	2.37	2.32
24	4.26	3.40	3.01	2.78	2.62	2.51	2.42	2.36	2.30
25	4.24	3.39	2.99	2.76	2.60	2.49	2.40	2.34	2.28
26	4.23	3.37	2.98	2.74	2.59	2.47	2.39	2.32	2.27
27	4.21	3.35	2.96	2.73	2.57	2.46	2.37	2.31	2.25
28	4.20	3.34	2.95	2.71	2.56	2.45	2.36	2.29	2.24
29	4.18	3.33	2.93	2.70	2.55	2.43	2.35	2.28	2.22
30	4.17	3.32	2.92	2.69	2.53	2.42	2.33	2.27	2.21
40	4.08	3.23	2.84	2.61	2.45	2.34	2.25	2.18	2.12
60	4.00	3.15	2.76	2.53	2.37	2.25	2.17	2.10	2.04
120	3.92	3.07	2.68	2.45	2.29	2.17	2.09	2.02	1.96
∞	3.84	3.00	2.60	2.37	2.21	2.10	2.01	1.94	1.88

*Adapted with permission from Pearson, E. S. and Hartley, H. O. (eds.) (1958). *Biometrika Tables for Statisticians,* 2nd ed. Cambridge University Press, New York, vol. I.

Note: n_1 = degrees of freedom for numerator. n_2 = degrees of freedom for denominator.

10	12	15	20	24	30	40	60	120	∞
				$F_{.95}\,(n_1,\,n_2)$					
241.9	243.9	245.9	248.0	249.1	250.1	251.1	252.2	253.3	254.3
19.40	19.41	19.43	19.45	19.45	19.46	19.47	19.48	19.49	19.50
8.79	8.74	8.70	8.66	8.64	8.62	8.59	8.57	8.55	8.53
5.96	5.91	5.86	5.80	5.77	5.75	5.72	5.69	5.66	5.63
4.74	4.68	4.62	4.56	4.53	4.50	4.46	4.43	4.40	4.36
4.06	4.00	3.94	3.87	3.84	3.81	3.77	3.74	3.70	3.67
3.64	3.57	3.51	3.44	3.41	3.38	3.34	3.30	3.27	3.23
3.35	3.28	3.22	3.15	3.12	3.08	3.04	3.01	2.97	2.93
3.14	3.07	3.01	2.94	2.90	2.86	2.83	2.79	2.75	2.71
2.98	2.91	2.85	2.77	2.74	2.70	2.66	2.62	2.58	2.54
2.85	2.79	2.72	2.65	2.61	2.57	2.53	2.49	2.45	2.40
2.75	2.69	2.62	2.54	2.51	2.47	2.43	2.38	2.34	2.30
2.67	2.60	2.53	2.46	2.42	2.38	2.34	2.30	2.25	2.21
2.60	2.53	2.46	2.39	2.35	2.31	2.27	2.22	2.18	2.13
2.54	2.48	2.40	2.33	2.29	2.25	2.20	2.16	2.11	2.07
2.49	2.42	2.35	2.28	2.24	2.19	2.15	2.11	2.06	2.01
2.45	2.38	2.31	2.23	2.19	2.15	2.10	2.06	2.01	1.96
2.41	2.34	2.27	2.19	2.15	2.11	2.06	2.02	1.97	1.92
2.38	2.31	2.23	2.16	2.11	2.07	2.03	1.98	1.93	1.88
2.35	2.28	2.20	2.12	2.08	2.04	1.99	1.95	1.90	1.84
2.32	2.25	2.18	2.10	2.05	2.01	1.96	1.92	1.87	1.81
2.30	2.23	2.15	2.07	2.03	1.98	1.94	1.89	1.84	1.78
2.27	2.20	2.13	2.05	2.01	1.96	1.91	1.86	1.81	1.76
2.25	2.18	2.11	2.03	1.98	1.94	1.89	1.84	1.79	1.73
2.24	2.16	2.09	2.01	1.96	1.92	1.87	1.82	1.77	1.71
2.22	2.15	2.07	1.99	1.95	1.90	1.85	1.80	1.75	1.69
2.20	2.13	2.06	1.97	1.93	1.88	1.84	1.79	1.73	1.67
2.19	2.12	2.04	1.96	1.91	1.87	1.82	1.77	1.71	1.65
2.18	2.10	2.03	1.94	1.90	1.85	1.81	1.75	1.70	1.64
2.16	2.09	2.01	1.93	1.89	1.84	1.79	1.74	1.68	1.62
2.08	2.00	1.92	1.84	1.79	1.74	1.69	1.64	1.58	1.51
1.99	1.92	1.84	1.75	1.70	1.65	1.59	1.53	1.47	1.39
1.91	1.83	1.75	1.66	1.61	1.55	1.50	1.43	1.35	1.25
1.83	1.75	1.67	1.57	1.52	1.46	1.39	1.32	1.22	1.00

TABLE K (*Continued*)

n_2 \ n_1	1	2	3	4	5	6	7	8	9
				$F_{.975}$ (n_1, n_2)					
1	647.8	799.5	864.2	899.6	921.8	937.1	948.2	956.7	963.3
2	38.51	39.00	39.17	39.25	39.30	39.33	39.36	39.37	39.39
3	17.44	16.04	15.44	15.10	14.88	14.73	14.62	14.54	14.47
4	12.22	10.65	9.98	9.60	9.36	9.20	9.07	8.98	8.90
5	10.01	8.43	7.76	7.39	7.15	6.98	6.85	6.76	6.68
6	8.81	7.26	6.60	6.23	5.99	5.82	5.70	5.60	5.52
7	8.07	6.54	5.89	5.52	5.29	5.12	4.99	4.90	4.82
8	7.57	6.06	5.42	5.05	4.82	4.65	4.53	4.43	4.36
9	7.21	5.71	5.08	4.72	4.48	4.32	4.20	4.10	4.03
10	6.94	5.46	4.83	4.47	4.24	4.07	3.95	3.85	3.78
11	6.72	5.26	4.63	4.28	4.04	3.88	3.76	3.66	3.59
12	6.55	5.10	4.47	4.12	3.89	3.73	3.61	3.51	3.44
13	6.41	4.97	4.35	4.00	3.77	3.60	3.48	3.39	3.31
14	6.30	4.86	4.24	3.89	3.66	3.50	3.38	3.29	3.21
15	6.20	4.77	4.15	3.80	3.58	3.41	3.29	3.20	3.12
16	6.12	4.69	4.08	3.73	3.50	3.34	3.22	3.12	3.05
17	6.04	4.62	4.01	3.66	3.44	3.28	3.16	3.06	2.98
18	5.98	4.56	3.95	3.61	3.38	3.22	3.10	3.01	2.93
19	5.92	4.51	3.90	3.56	3.33	3.17	3.05	2.96	2.88
20	5.87	4.46	3.86	3.51	3.29	3.13	3.01	2.91	2.84
21	5.83	4.42	3.82	3.48	3.25	3.09	2.97	2.87	2.80
22	5.79	4.38	3.78	3.44	3.22	3.05	2.93	2.84	2.76
23	5.75	4.35	3.75	3.41	3.18	3.02	2.90	2.81	2.73
24	5.72	4.32	3.72	3.38	3.15	2.99	2.87	2.78	2.70
25	5.69	4.29	3.69	3.35	3.13	2.97	2.85	2.75	2.68
26	5.66	4.27	3.67	3.33	3.10	2.94	2.82	2.73	2.65
27	5.63	4.24	3.65	3.31	3.08	2.92	2.80	2.71	2.63
28	5.61	4.22	3.63	3.29	3.06	2.90	2.78	2.69	2.61
29	5.59	4.20	3.61	3.27	3.04	2.88	2.76	2.67	2.59
30	5.57	4.18	3.59	3.25	3.03	2.87	2.75	2.65	2.57
40	5.42	4.05	3.46	3.13	2.90	2.74	2.62	2.53	2.45
60	5.29	3.93	3.34	3.01	2.79	2.63	2.51	2.41	2.33
120	5.15	3.80	3.23	2.89	2.67	2.52	2.39	2.30	2.22
∞	5.02	3.69	3.12	2.79	2.57	2.41	2.29	2.19	2.11

10	12	15	20	24	30	40	60	120	∞
				$F_{.975}$ (n_1, n_2)					
968.6	976.7	984.9	993.1	997.2	1,001	1,006	1,010	1,014	1,018
39.40	39.41	39.43	39.45	39.46	39.46	39.47	39.48	39.49	39.50
14.42	14.34	14.25	14.17	14.12	14.08	14.04	13.99	13.95	13.90
8.84	8.75	8.66	8.56	8.51	8.46	8.41	8.36	8.31	8.26
6.62	6.52	6.43	6.33	6.28	6.23	6.18	6.12	6.07	6.02
5.46	5.37	5.27	5.17	5.12	5.07	5.01	4.96	4.90	4.85
4.76	4.67	4.57	4.47	4.42	4.36	4.31	4.25	4.20	4.14
4.30	4.20	4.10	4.00	3.95	3.89	3.84	3.78	3.73	3.67
3.96	3.87	3.77	3.67	3.61	3.56	3.51	3.45	3.39	3.33
3.72	3.62	3.52	3.42	3.37	3.31	3.26	3.20	3.14	3.08
3.53	3.43	3.33	3.23	3.17	3.12	3.06	3.00	2.94	2.88
3.37	3.28	3.18	3.07	3.02	2.96	2.91	2.85	2.79	2.72
3.25	3.15	3.05	2.95	2.89	2.84	2.78	2.72	2.66	2.60
3.15	3.05	2.95	2.84	2.79	2.73	2.67	2.61	2.55	2.49
3.06	2.96	2.86	2.76	2.70	2.64	2.59	2.52	2.46	2.40
2.99	2.89	2.79	2.68	2.63	2.57	2.51	2.45	2.38	2.32
2.92	2.82	2.72	2.62	2.56	2.50	2.44	2.38	2.32	2.25
2.87	2.77	2.67	2.56	2.50	2.44	2.38	2.32	2.26	2.19
2.82	2.72	2.62	2.51	2.45	2.39	2.33	2.27	2.20	2.13
2.77	2.68	2.57	2.46	2.41	2.35	2.29	2.22	2.16	2.09
2.73	2.64	2.53	2.42	2.37	2.31	2.25	2.18	2.11	2.04
2.70	2.60	2.50	2.39	2.33	2.27	2.21	2.14	2.08	2.00
2.67	2.57	2.47	2.36	2.30	2.24	2.18	2.11	2.04	1.97
2.64	2.54	2.44	2.33	2.27	2.21	2.15	2.08	2.01	1.94
2.61	2.51	2.41	2.30	2.24	2.18	2.12	2.05	1.98	1.91
2.59	2.49	2.39	2.28	2.22	2.16	2.09	2.03	1.95	1.88
2.57	2.47	2.36	2.25	2.19	2.13	2.07	2.00	1.93	1.85
2.55	2.45	2.34	2.23	2.17	2.11	2.05	1.98	1.91	1.83
2.53	2.43	2.32	2.21	2.15	2.09	2.03	1.96	1.89	1.81
2.51	2.41	2.31	2.20	2.14	2.07	2.01	1.94	1.87	1.79
2.39	2.29	2.18	2.07	2.01	1.94	1.88	1.80	1.72	1.64
2.27	2.17	2.06	1.94	1.88	1.82	1.74	1.67	1.58	1.48
2.16	2.05	1.94	1.82	1.76	1.69	1.61	1.53	1.43	1.31
2.05	1.94	1.83	1.71	1.64	1.57	1.48	1.39	1.27	1.00

TABLE K (*Continued*)

n_2 \ n_1	1	2	3	4	5	6	7	8	9
					$F_{.99}(n_1, n_2)$				
1	4,052	4,999.5	5,403	5,625	5,764	5,859	5,928	5,982	6,022
2	98.50	99.00	99.17	99.25	99.30	99.33	99.36	99.37	99.39
3	34.12	30.82	29.46	28.71	28.24	27.91	27.67	27.49	27.35
4	21.20	18.00	16.69	15.98	15.52	15.21	14.98	14.80	14.66
5	16.26	13.27	12.06	11.39	10.97	10.67	10.46	10.29	10.16
6	13.75	10.92	9.78	9.15	8.75	8.47	8.26	8.10	7.98
7	12.25	9.55	8.45	7.85	7.46	7.19	6.99	6.84	6.72
8	11.26	8.65	7.59	7.01	6.63	6.37	6.18	6.03	5.91
9	10.56	8.02	6.99	6.42	6.06	5.80	5.61	5.47	5.35
10	10.04	7.56	6.55	5.99	5.64	5.39	5.20	5.06	4.94
11	9.65	7.21	6.22	5.67	5.32	5.07	4.89	4.74	4.63
12	9.33	6.93	5.95	5.41	5.06	4.82	4.64	4.50	4.39
13	9.07	6.70	5.74	5.21	4.86	4.62	4.44	4.30	4.19
14	8.86	6.51	5.56	5.04	4.69	4.46	4.28	4.14	4.03
15	8.68	6.36	5.42	4.89	4.56	4.32	4.14	4.00	3.89
16	8.53	6.23	5.29	4.77	4.44	4.20	4.03	3.89	3.78
17	8.40	6.11	5.18	4.67	4.34	4.10	3.93	3.79	3.68
18	8.29	6.01	5.09	4.58	4.25	4.01	3.84	3.71	3.60
19	8.18	5.93	5.01	4.50	4.17	3.94	3.77	3.63	3.52
20	8.10	5.85	4.94	4.43	4.10	3.87	3.70	3.56	3.46
21	8.02	5.78	4.87	4.37	4.04	3.81	3.64	3.51	3.40
22	7.95	5.72	4.82	4.31	3.99	3.76	3.59	3.45	3.35
23	7.88	5.66	4.76	4.26	3.94	3.71	3.54	3.41	3.30
24	7.82	5.61	4.72	4.22	3.90	3.67	3.50	3.36	3.26
25	7.77	5.57	4.68	4.18	3.85	3.63	3.46	3.32	3.22
26	7.72	5.53	4.64	4.14	3.82	3.59	3.42	3.29	3.18
27	7.68	5.49	4.60	4.11	3.78	3.56	3.39	3.26	3.15
28	7.64	5.45	4.57	4.07	3.75	3.53	3.36	3.23	3.12
29	7.60	5.42	4.54	4.04	3.73	3.50	3.33	3.20	3.09
30	7.56	5.39	4.51	4.02	3.70	3.47	3.30	3.17	3.07
40	7.31	5.18	4.31	3.83	3.51	3.29	3.12	2.99	2.89
60	7.08	4.98	4.13	3.65	3.34	3.12	2.95	2.82	2.72
120	6.85	4.79	3.95	3.48	3.17	2.96	2.79	2.66	2.56
∞	6.63	4.61	3.78	3.32	3.02	2.80	2.64	2.51	2.41

10	12	15	20	24	30	40	60	120	∞
				$F_{.99}\ (n_1,\ n_2)$					
6,056	6,106	6,157	6,209	6,235	6,261	6,287	6,313	6,339	6,366
99.40	99.42	99.43	99.45	99.46	99.47	99.47	99.48	99.49	99.50
27.23	27.05	26.87	26.69	26.60	26.50	26.41	26.32	26.22	26.13
14.55	14.37	14.20	14.02	13.93	13.84	13.75	13.65	13.56	13.46
10.05	9.89	9.72	9.55	9.47	9.38	9.29	9.20	9.11	9.02
7.87	7.72	7.56	7.40	7.31	7.23	7.14	7.06	6.97	6.88
6.62	6.47	6.31	6.16	6.07	5.99	5.91	5.82	5.74	5.65
5.81	5.67	5.52	5.36	5.28	5.20	5.12	5.03	4.95	4.86
5.26	5.11	4.96	4.81	4.73	4.65	4.57	4.48	4.40	4.31
4.85	4.71	4.56	4.41	4.33	4.25	4.17	4.08	4.00	3.91
4.54	4.40	4.25	4.10	4.02	3.94	3.86	3.78	3.69	3.60
4.30	4.16	4.01	3.86	3.78	3.70	3.62	3.54	3.45	3.36
4.10	3.96	3.82	3.66	3.59	3.51	3.43	3.34	3.25	3.17
3.94	3.80	3.66	3.51	3.43	3.35	3.27	3.18	3.09	3.00
3.80	3.67	3.52	3.37	3.29	3.21	3.13	3.05	2.96	2.87
3.69	3.55	3.41	3.26	3.18	3.10	3.02	2.93	2.84	2.75
3.59	3.46	3.31	3.16	3.08	3.00	2.92	2.83	2.75	2.65
3.51	3.37	3.23	3.08	3.00	2.92	2.84	2.75	2.66	2.57
3.43	3.30	3.15	3.00	2.92	2.84	2.76	2.67	2.58	2.49
3.37	3.23	3.09	2.94	2.86	2.78	2.69	2.61	2.52	2.42
3.31	3.17	3.03	2.88	2.80	2.72	2.64	2.55	2.46	2.36
3.26	3.12	2.98	2.83	2.75	2.67	2.58	2.50	2.40	2.31
3.21	3.07	2.93	2.78	2.70	2.62	2.54	2.45	2.35	2.26
3.17	3.03	2.89	2.74	2.66	2.58	2.49	2.40	2.31	2.21
3.13	2.99	2.85	2.70	2.62	2.54	2.45	2.36	2.27	2.17
3.09	2.96	2.81	2.66	2.58	2.50	2.42	2.33	2.23	2.13
3.06	2.93	2.78	2.63	2.55	2.47	2.38	2.29	2.20	2.10
3.03	2.90	2.75	2.60	2.52	2.44	2.35	2.26	2.17	2.06
3.00	2.87	2.73	2.57	2.49	2.41	2.33	2.23	2.14	2.03
2.98	2.84	2.70	2.55	2.47	2.39	2.30	2.21	2.11	2.01
2.80	2.66	2.52	2.37	2.29	2.20	2.11	2.02	1.92	1.80
2.63	2.50	2.35	2.20	2.12	2.03	1.94	1.84	1.73	1.60
2.47	2.34	2.19	2.03	1.95	1.86	1.76	1.66	1.53	1.38
2.32	2.18	2.04	1.88	1.79	1.70	1.59	1.47	1.32	1.00

TABLE L* Distribution of χ^2

Values of χ^2 corresponding to certain selected probabilities (i.e. tail areas under the curve). To illustrate: The probability is 0.95 that a sample with 20 degrees of freedom, taken from a normal distribution, would have $\chi^2 = 31.41$ or smaller.

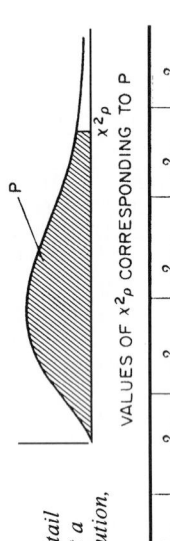

VALUES OF χ^2_P CORRESPONDING TO P

DF	$\chi^2_{.005}$	$\chi^2_{.01}$	$\chi^2_{.025}$	$\chi^2_{.05}$	$\chi^2_{.10}$	$\chi^2_{.90}$	$\chi^2_{.95}$	$\chi^2_{.975}$	$\chi^2_{.99}$	$\chi^2_{.995}$
1	0.000039	0.00016	0.00098	0.0039	0.0158	2.71	3.84	5.02	6.63	7.88
2	0.0100	0.0201	0.0506	0.1026	0.2107	4.61	5.99	7.38	9.21	10.60
3	0.0717	0.115	0.216	0.352	0.584	6.25	7.81	9.35	11.34	12.84
4	0.207	0.297	0.484	0.711	1.064	7.78	9.49	11.14	13.28	14.86
5	0.412	0.554	0.831	1.15	1.61	9.24	11.07	12.83	15.09	16.75
6	0.676	0.872	1.24	1.64	2.20	10.64	12.59	14.45	16.81	18.55
7	0.989	1.24	1.69	2.17	2.83	12.02	14.07	16.01	18.48	20.28
8	1.34	1.65	2.18	2.73	3.49	13.36	15.51	17.53	20.09	21.96
9	1.73	2.09	2.70	3.33	4.17	14.68	16.92	19.02	21.67	23.59
10	2.16	2.56	3.25	3.94	4.87	15.99	18.31	20.48	23.21	25.19
11	2.60	3.05	3.82	4.57	5.58	17.28	19.68	21.92	24.73	26.76
12	3.07	3.57	4.40	5.23	6.30	18.55	21.03	23.34	26.22	28.30
13	3.57	4.11	5.01	5.89	7.04	19.81	22.36	24.74	27.69	29.82
14	4.07	4.66	5.63	6.57	7.79	21.06	23.68	26.12	29.14	31.32
15	4.60	5.23	6.26	7.26	8.55	22.31	25.00	27.49	30.58	32.80
16	5.14	5.81	6.91	7.96	9.31	23.54	26.30	28.85	32.00	34.27
18	6.26	7.01	8.23	9.39	10.86	25.99	28.87	31.53	34.81	37.16
20	7.43	8.26	9.59	10.85	12.44	28.41	31.41	34.17	37.57	40.00
24	9.89	10.86	12.40	13.85	15.66	33.20	36.42	39.36	42.98	45.56
30	13.79	14.95	16.79	18.49	20.60	40.26	43.77	46.98	50.89	53.67
40	20.71	22.16	24.43	26.51	29.05	51.81	55.76	59.34	63.69	66.77
60	35.53	37.48	40.48	43.19	46.46	74.40	79.08	83.30	88.38	91.95
120	83.85	86.92	91.58	95.70	100.62	140.23	146.57	152.21	158.95	163.64

*Adapted with permission from Dixon, W. J. and Massey, F. J., Jr. (1969). *Introduction to Statistical Analysis*, 3rd ed. McGraw-Hill, New York.

TABLE M* Percentiles of $F' = \dfrac{R_1}{R_2}$

Values of F' corresponding to certain selected cumulative probabilities. To illustrate: The probability is 0.95 that the ratio of sample ranges R_1/R_2 is 2.6 or less when $n_1 = n_2 = 5$.

n_2	Cumulative probability	n_1 2	3	4	5	6	7	8	9	10
2	0.025	0.039	0.217	0.37	0.50	0.60	0.68	0.74	0.79	0.83
	0.05	0.079	0.31	0.50	0.62	0.74	0.80	0.86	0.91	0.95
	0.95	12.7	19.1	23	26	29	30	32	34	35
	0.975	25.5	38.2	52	57	60	62	64	67	68
3	0.025	0.026	0.160	0.28	0.39	0.47	0.54	0.59	0.64	0.68
	0.05	0.052	0.23	0.37	0.49	0.57	0.64	0.70	0.75	0.80
	0.95	3.19	4.4	5.0	5.7	6.2	6.6	6.9	7.2	7.4
	0.975	4.61	6.3	7.3	8.0	8.7	9.3	9.8	10.2	10.5
4	0.025	0.019	0.137	0.25	0.34	0.42	0.48	0.53	0.57	0.61
	0.05	0.043	0.20	0.32	0.42	0.50	0.57	0.62	0.67	0.70
	0.95	2.02	2.7	3.1	3.4	3.6	3.8	4.0	4.2	4.4
	0.975	2.72	3.5	4.0	4.4	4.7	5.0	5.2	5.4	5.6
5	0.025	0.018	0.124	0.23	0.32	0.38	0.44	0.49	0.53	0.57
	0.05	0.038	0.18	0.29	0.40	0.46	0.52	0.57	0.61	0.65
	0.95	1.61	2.1	2.4	2.6	2.8	2.9	3.0	3.1	3.2
	0.975	2.01	2.6	2.9	3.2	3.4	3.6	3.7	3.8	3.9
6	0.025	0.017	0.115	0.21	0.30	0.36	0.42	0.46	0.50	0.54
	0.05	0.035	0.16	0.27	0.36	0.43	0.49	0.54	0.58	0.61
	0.95	1.36	1.8	2.0	2.2	2.3	2.4	2.5	2.6	2.7
	0.975	1.67	2.1	2.4	2.6	2.8	2.9	3.0	3.1	3.2
7	0.025	0.016	0.107	0.20	0.28	0.34	0.40	0.44	0.48	0.52
	0.05	0.032	0.15	0.26	0.35	0.41	0.47	0.51	0.55	0.59
	0.95	1.26	1.6	1.8	1.9	2.0	2.1	2.2	2.3	2.4
	0.975	1.48	1.9	2.1	2.3	2.4	2.5	2.6	2.7	2.8
8	0.025	0.016	0.102	0.19	0.27	0.33	0.38	0.43	0.47	0.50
	0.05	0.031	0.14	0.25	0.33	0.40	0.45	0.50	0.53	0.57
	0.95	1.17	1.4	1.6	1.8	1.9	1.9	2.0	2.1	2.1
	0.975	1.36	1.7	1.9	2.0	2.2	2.3	2.3	2.4	2.5
9	0.025	0.015	0.098	0.18	0.26	0.32	0.37	0.42	0.46	0.49
	0.05	0.030	0.14	0.24	0.32	0.38	0.44	0.48	0.52	0.55
	0.95	1.10	1.3	1.5	1.6	1.7	1.8	1.9	1.9	2.0
	0.975	1.27	1.6	1.8	1.9	2.0	2.1	2.1	2.2	2.3
10	0.025	0.015	0.095	0.18	0.25	0.31	0.36	0.41	0.44	0.48
	0.05	0.029	0.13	0.23	0.31	0.37	0.43	0.47	0.51	0.54
	0.95	1.05	1.3	1.4	1.5	1.6	1.7	1.8	1.8	1.9
	0.975	1.21	1.5	1.6	1.8	1.9	1.9	2.0	2.0	2.1

*Adapted with permission from Dixon, W. J. and Massey, F. J., Jr. (1969). *Introduction to Statistical Analysis,* 3rd ed. McGraw-Hill, New York.

CHART N Confidence Limits for Fraction Defective*

Enter the horizontal scale with the sample fraction defective. Rise vertically to the upper and lower curves for the stated sample size. Read the corresponding upper and lower confidence limits on the vertical scale. To illustrate: If a sample of 50 is 20% defective, the 95% confidence limits on the population fraction defective are 10 and 35%.

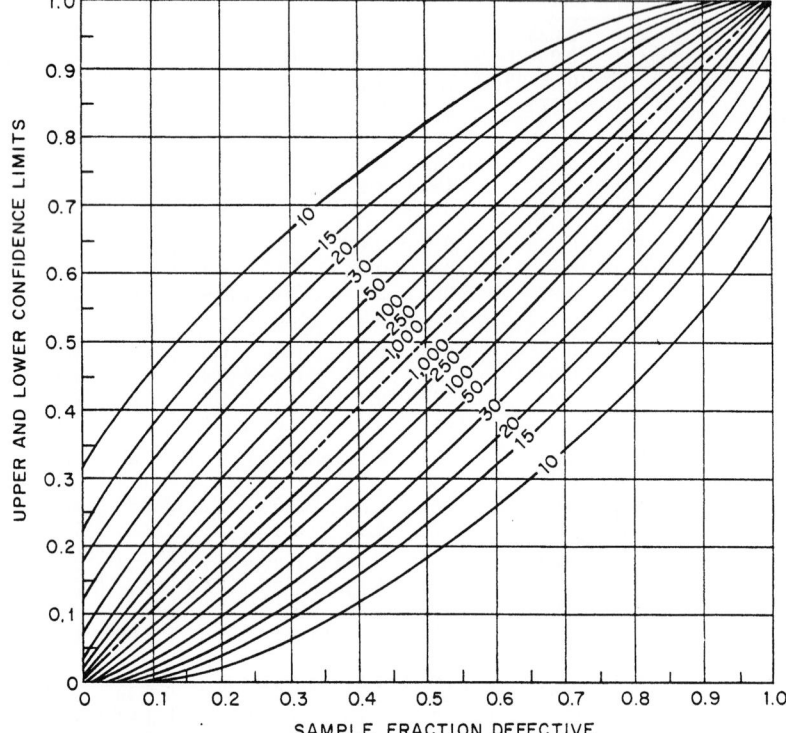

*By permission of Prof. E. S. Pearson from Clopper, C. J. and Pearson, E. S. (1934). "The Use of Confidence or Fiducial Limits Illustrated in the Case of the Binomial." *Biometrika,* vol. 26, p. 404.

TABLE O Critical Values of Smaller Rank Sum for the Wilcoxon-Mann-Whitney Test*

n_2	α for 2-sided test	α for 1-sided test	n_1 (smaller sample)											
			1	2	3	4	5	6	7	8	9	10	11	12
3	0.20	0.10		3	7									
	0.10	0.05			6									
	0.05	0.025												
	0.01	0.005												
4	0.20	0.10		3	7	13								
	0.10	0.05			6	11								
	0.05	0.025				10								
	0.01	0.005												
5	0.20	0.10		4	8	14	20							
	0.10	0.05		3	7	12	19							
	0.05	0.025			6	11	17							
	0.01	0.005					15							
6	0.20	0.10		4	9	15	22	30						
	0.10	0.05		3	8	13	20	28						
	0.05	0.025			7	12	18	26						
	0.01	0.005				10	16	23						
7	0.20	0.10		4	10	16	23	32	41					
	0.10	0.05		3	8	14	21	29	39					
	0.05	0.025			7	13	20	27	36					
	0.01	0.005				10	16	24	32					
8	0.20	0.10		5	11	17	25	34	44	55				
	0.10	0.05		4	9	15	23	31	41	51				
	0.05	0.025		3	8	14	21	29	38	49				
	0.01	0.005				11	17	25	34	43				
9	0.20	0.10	1	5	11	19	27	36	46	58	70			
	0.10	0.05		4	9	16	24	33	43	54	66			
	0.05	0.025		3	8	14	22	31	40	51	62			
	0.01	0.005			6	11	18	26	35	45	56			
10	0.20	0.10	1	6	12	20	28	38	49	60	73	87		
	0.10	0.05		4	10	17	26	35	45	56	69	82		
	0.05	0.025		3	9	15	23	32	42	53	65	78		
	0.01	0.005			6	12	19	27	37	47	58	71		
11	0.20	0.10	1	6	13	21	30	40	51	63	76	91	106	
	0.10	0.05		4	11	18	27	37	47	59	72	86	100	
	0.05	0.025		3	9	16	24	34	44	55	68	81	96	
	0.01	0.005			6	12	20	28	38	49	61	73	87	
12	0.20	0.10	1	7	14	22	32	42	54	66	80	94	110	127
	0.10	0.05		5	11	19	28	38	49	62	75	89	104	120
	0.05	0.025		4	10	17	26	35	46	58	71	84	99	115
	0.01	0.005			7	13	21	30	40	51	63	76	90	105

*Reproduced with permission from Tate, M. W. and Clelland, R. C. (1957). *Non-parametric and Shortcut Statistics*. The Interstate Printers & Publishers, Danville, IL.

TABLE P   Limiting Values for Number of Runs above and below the Median of a Set of Values*

n_1 = number of values above the median and n_2 = number of values below the median.

$m = n_1 = n_2$	Probability of an equal or smaller number of runs		Probability of an equal or larger number of runs	
	$\alpha = 0.05$	$\alpha = 0.01$	$\alpha = 0.05$	$\alpha = 0.01$
5	3	2	9	10
6	3	2	11	12
7	4	3	12	13
8	5	4	13	14
9	6	4	14	16
10	6	5	16	17
11	7	6	17	18
12	8	7	18	19
13	9	7	19	21
14	10	8	20	22
15	11	9	21	23
16	11	10	23	24
17	12	10	24	26
18	13	11	25	27
19	14	12	26	28
20	15	13	27	29
21	16	14	28	30
22	17	14	29	32
23	17	15	31	33
24	18	16	32	34
25	19	17	33	35
26	20	18	34	36
27	21	19	35	37
28	22	19	36	39
29	23	20	37	40
30	24	21	38	41

*Reproduced with permission from Swed, Freda S. and Eisenhart, C. (1943). "Tables for Testing Randomness of Grouping in a Sequence of Alternatives." *Annals of Mathematical Statistics,* vol. XIV, pp. 66 and 87, Tables II and III.

TABLE Q Criteria for Testing for Extreme Mean*

Statistic	No. of observations	$P_{.90}$	$P_{.95}$	$P_{.98}$	$P_{.99}$
$r_{10} = \dfrac{X_2 - X_1}{X_n - X_1}$	3	0.886	0.941	0.976	0.988
	4	0.679	0.765	0.846	0.889
	5	0.557	0.642	0.729	0.780
	6	0.482	0.560	0.644	0.698
	7	0.434	0.507	0.586	0.637
$r_{11} = \dfrac{X_2 - X_1}{X_{n-1} - X_1}$	8	0.479	0.554	0.631	0.683
	9	0.441	0.512	0.587	0.635
	10	0.409	0.477	0.551	0.597
$r_{21} = \dfrac{X_3 - X_1}{X_{n-1} - X_1}$	11	0.517	0.576	0.638	0.679
	12	0.490	0.546	0.605	0.642
	13	0.467	0.521	0.578	0.615
$r_{22} = \dfrac{X_3 - X_1}{X_{n-2} - X_1}$	14	0.492	0.546	0.602	0.641
	15	0.472	0.525	0.579	0.616
	16	0.454	0.507	0.559	0.595
	17	0.438	0.490	0.542	0.577
	18	0.424	0.475	0.527	0.561
	19	0.412	0.462	0.514	0.547
	20	0.401	0.450	0.502	0.535
	21	0.391	0.440	0.491	0.524
	22	0.382	0.430	0.481	0.514
	23	0.374	0.421	0.472	0.505
	24	0.367	0.413	0.464	0.497
	25	0.360	0.406	0.457	0.489

*Adapted with persmission from Dixon, W. J. and Massey, F. J., Jr. (1969). *Introduction to Statistical Analysis,* 3rd ed. McGraw-Hill, New York.

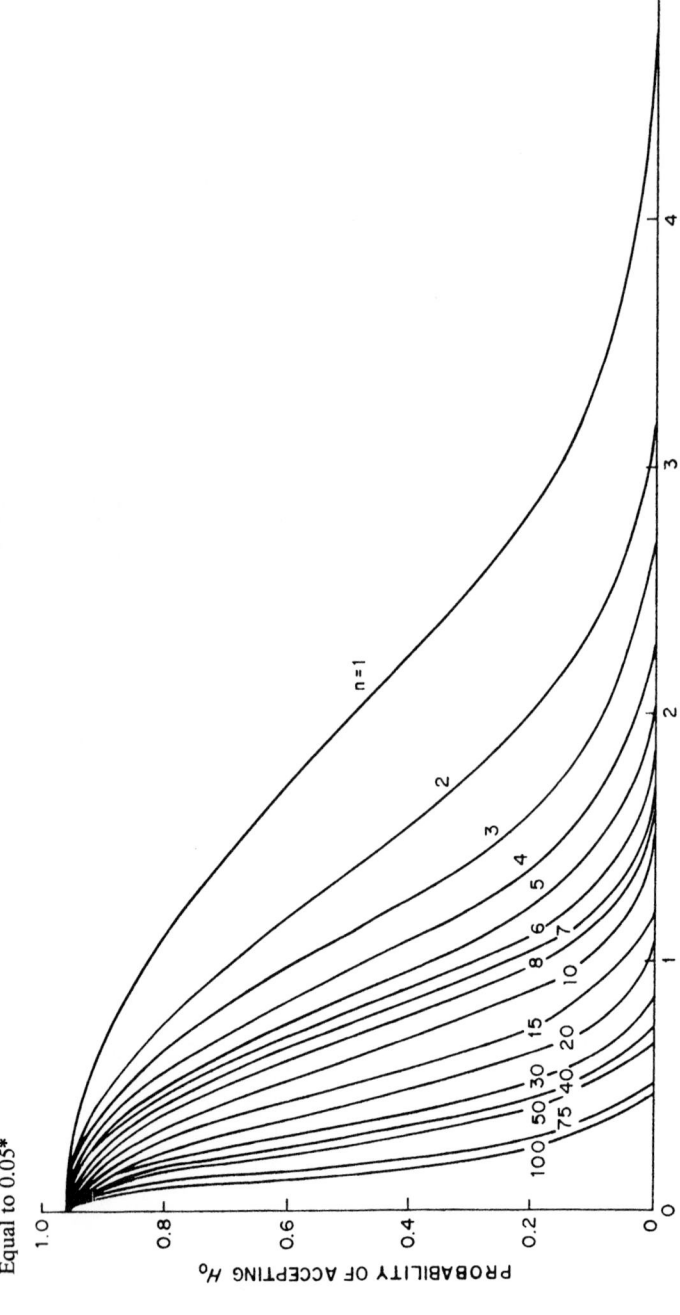

CHART R Operating Characteristics of the Two-Sided Normal Test for a Level of Significance Equal to 0.05*

PROBABILITY OF ACCEPTING H_0

n=1
2
3
4
5
6
7
8
10
15
20
50 30
75 40
100

*Adapted with permission from Ferris, Charles D., Grubbs, Frank E., and Weaver, Chalmers L. (1946). "Operating Characteristics for the Common Statistical Tests of Significance." *Annals of Mathematical Statistics*, June.

CHART S   Size of Sample for Arithmetic Mean When σ is Unknown*

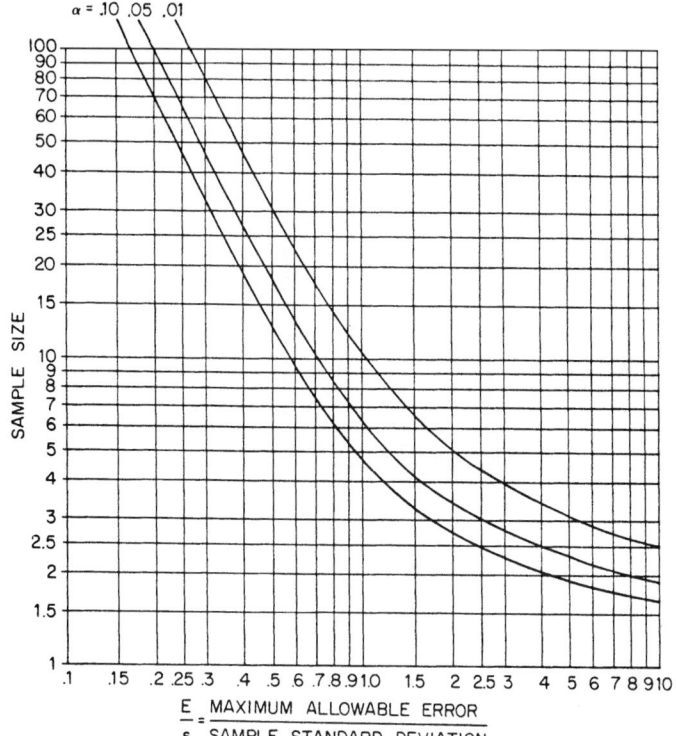

$$\frac{E}{s} = \frac{\text{MAXIMUM ALLOWABLE ERROR}}{\text{SAMPLE STANDARD DEVIATION}}$$

*Reproduced with permission from Weida, Frank M. and Lum, Mary D. (1953). *Statistical Inference, Reliability, and Significance.* WADC Technical Report 53-149, U.S. Air Force.

CHART T   Number of Degrees of Freedom Required to Estimate the Standard Deviation within $P\%$ of Its True Value with Confidence Coefficient γ

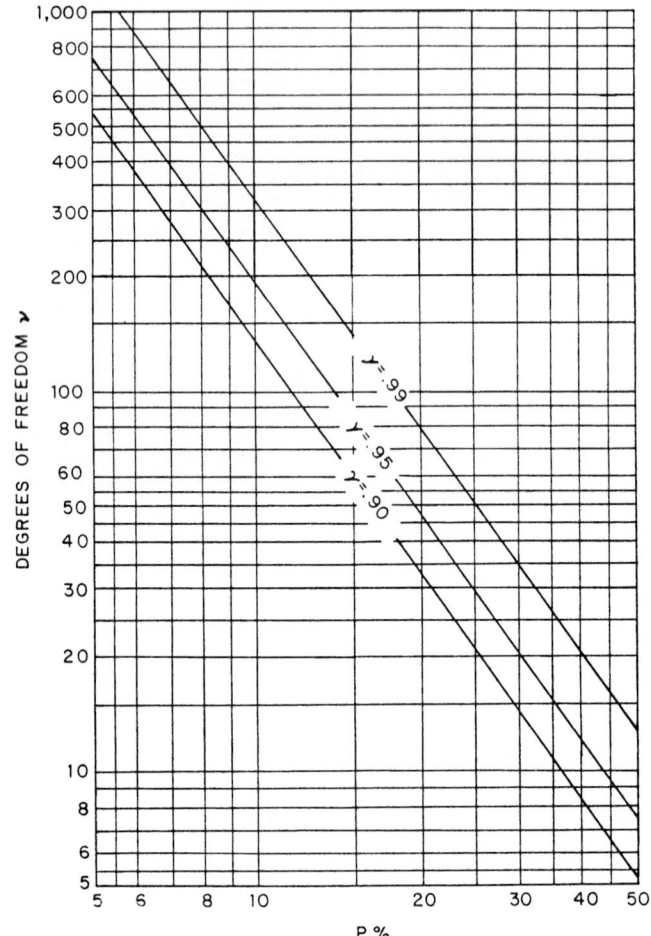

*Adapted with permission from Greenwood, J. A. and Sandomire, M. M. (1950). "Statistics Manual, Sample Size Required for Estimating the Standard Deviation as a Percent of Its True Value." *Journal of the American Statistical Association,* vol. 45, p. 258. The manner of graphing is adapted with permission from Crow, E. L., Davis, F. A., and Maxfield, M. W. (1955). *NAVORD Report 3369.* NOTS 948, U.S. Naval Ordnance Test Station, China Lake, CA. (Reprinted by Dover Publications, New York, 1960.)

TABLE U Tolerance Factors for Normal Distribution*

Factors K_1 such that the probability is γ that at least a proportion P of the distribution will be included between $\overline{X} \pm K_1 R$ where $\overline{X}$ is the mean and R is the range in a sample of size n.

	$\gamma = 0.90$				$\gamma = 0.95$				$\gamma = 0.99$			
P / n	0.90	0.95	0.99	0.999	0.90	0.95	0.99	0.999	0.90	0.95	0.99	0.999
2	11.298	13.294	17.090	21.374	22.635	26.634	34.238	42.821	113.429	133.469	171.576	214.588
3	3.069	3.631	4.711	5.936	4.399	5.206	6.752	8.509	9.951	11.776	15.275	19.249
4	1.877	2.227	2.902	3.672	2.422	2.873	3.744	4.737	4.233	5.021	6.543	8.279
5	1.428	1.697	2.216	2.812	1.749	2.078	2.715	3.444	2.709	3.219	4.205	5.335
6	1.194	1.420	1.857	2.360	1.418	1.686	2.206	2.803	2.042	2.429	3.178	4.038
7	1.050	1.248	1.635	2.080	1.222	1.453	1.903	2.420	1.678	1.996	2.615	3.325
8	0.951	1.131	1.483	1.888	1.090	1.297	1.700	2.165	1.449	1.724	2.261	2.878
9	0.879	1.046	1.372	1.747	0.997	1.187	1.556	1.981	1.290	1.536	2.014	2.565
10	0.824	0.981	1.286	1.639	0.926	1.103	1.446	1.843	1.176	1.400	1.836	2.340
11	0.780	0.929	1.219	1.554	0.871	1.037	1.361	1.735	1.088	1.296	1.701	2.168
12	0.745	0.887	1.164	1.484	0.827	0.985	1.292	1.648	1.020	1.215	1.594	2.033
13	0.715	0.852	1.118	1.426	0.790	0.940	1.235	1.575	0.964	1.148	1.507	1.922
14	0.690	0.822	1.079	1.377	0.759	0.904	1.187	1.514	0.917	1.093	1.435	1.830
15	0.669	0.797	1.046	1.334	0.733	0.873	1.146	1.462	0.878	1.046	1.373	1.753
16	0.650	0.774	1.016	1.297	0.710	0.845	1.110	1.417	0.845	1.007	1.322	1.687
17	0.633	0.755-	0.991	1.265	0.690	0.822	1.109	1.377	0.816	0.972	1.277	1.630
18	0.619	0.737	0.968	1.235	0.672	0.801	1.051	1.342	0.790	0.941	1.236	1.578
19	0.605	0.721	0.947	1.209	0.656	0.782	1.027	1.311	0.768	•0.916	1.203	1.535
20	0.594	0.707	0.929	1.186	0.642	0.765	1.005	1.282	0.748	0.892	1.171	1.495

*Adapted with permission from Mitra, S. K. (1957). "Tables for Tolerance Limits for a Normal Population Based on Sample Mean and Range on Mean Range." *Journal of the American Statistical Association,* vol. 52, no. 277, March, p. 92.

TABLE V One-Sided and Two-sided Statistical Tolerance Limit Factors k for a Normal Distribution*

Factors k such that the probability is γ that at least a proportion P of the distribution will be less than $\overline{X} + ks$ (or greater than $\overline{X} - ks$) where $\overline{X}$ and s are estimates of the mean and standard deviation computed from a sample size of n. Two-sided factors cover $\overline{X} \pm ks$.

		Two-sided Factors				One-sided Factors					
		γ = 0.95				**γ = 0.95**					
P		0.90	0.95	0.99	0.999	0.90	0.95	0.99	0.999		
n											
3						4.258	5.310	7.340	9.651		
4						3.187	3.957	5.437	7.128		
5						2.742	3.400	4.666	6.112		
6		4.408	5.409	7.334	9.540	2.494	3.091	4.242	5.556		
7		3.856	4.730	6.411	8.348	2.333	2.894	3.972	5.201		
8		3.496	4.287	5.811	7.566	2.219	2.755	3.783	4.955		
9		3.242	3.971	5.389	7.014	2.133	2.649	3.641	4.772		
10		3.048	3.739	5.075	6.603	2.065	2.568	3.532	4.629		
11		2.897	3.557	4.828	6.284	2.012	2.503	3.444	4.515		
12		2.773	3.410	4.633	6.032	1.966	2.448	3.371	4.420		
13		2.677	3.290	4.472	5.826	1.928	2.403	3.310	4.341		
14		2.592	3.189	4.336	5.651	1.895	2.363	3.257	4.274		
15		2.521	3.102	4.224	5.507	1.866	2.329	3.212	4.215		
16		2.458	3.028	4.124	5.374	1.842	2.299	3.172	4.164		
17		2.405	2.962	4.038	5.268	1.820	2.272	3.136	4.118		
18		2.357	2.906	3.961	5.167	1.800	2.249	3.106	4.078		
19		2.315	2.855	3.893	5.078	1.781	2.228	3.078	4.041		
20		2.275	2.807	3.832	5.003	1.765	2.208	3.052	4.009		
21		2.241	2.768	3.776	4.932	1.750	2.190	3.028	3.979		
22		2.208	2.729	3.727	4.866	1.736	2.174	3.007	3.952		
23		2.179	2.693	3.680	4.806	1.724	2.159	2.987	3.927		
24		2.154	2.663	3.638	4.755	1.712	2.145	2.969	3.904		
25		2.129	2.632	3.601	4.706	1.702	2.132	2.952	3.882		
30		2.029	2.516	3.446	4.508	1.657	2.080	2.884	3.794		
35		1.957	2.431	3.334	4.364	1.623	2.041	2.833	3.730		
40		1.902	2.365	3.250	4.255	1.598	2.010	2.793	3.679		
45		1.857	2.313	3.181	4.168	1.577	1.986	2.762	3.638		
50		1.821	2.296	3.124	4.096	1.560	1.965	2.735	3.604		

One-sided Factors, **γ = 0.99** (P = 0.90, 0.95, 0.99, 0.999):

n	0.90	0.95	0.99	0.999
3	6.158	7.655	10.552	13.857
4	4.163	5.145	7.042	9.215
5	3.407	4.202	5.741	7.501
6	3.006	3.707	5.062	6.612
7	2.755	3.399	4.641	6.061
8	2.582	3.188	4.353	5.686
9	2.454	3.031	4.143	5.414
10	2.355	2.911	3.981	5.203
11	2.275	2.815	3.852	5.036
12	2.210	2.736	3.747	4.900
13	2.155	2.670	3.659	4.787
14	2.108	2.614	3.585	4.690
15	2.068	2.566	3.520	4.607
16	2.032	2.523	3.463	4.534
17	2.001	2.486	3.415	4.471
18	1.974	2.453	3.370	4.415
19	1.949	2.423	3.331	4.364
20	1.926	2.396	3.295	4.319
21	1.905	2.371	3.262	4.276
22	1.887	2.350	3.233	4.238
23	1.869	2.329	3.206	4.204
24	1.853	2.309	3.181	4.171
25	1.838	2.292	3.158	4.143
30	1.778	2.220	3.064	4.022
35	1.732	2.166	2.994	3.934
40	1.697	2.126	2.941	3.866
45	1.669	2.092	2.897	3.811
50	1.646	2.065	2.863	3.766

Two-sided Factors†

2	15.978	18.800	24.167	30.227	32.019	37.674	48.430	60.573	160.193	188.491	242.300	303.054
3	5.847	6.919	8.974	11.309	8.380	9.916	12.861	16.208	18.930	22.401	29.055	36.616
4	4.166	4.943	6.440	8.149	5.369	6.370	8.299	10.502	9.398	11.150	14.527	18.383
5	3.494	4.152	5.423	6.879	4.275	5.079	6.634	8.415	6.612	7.855	10.260	13.015
6	3.131	3.723	4.870	6.188	3.712	4.414	5.775	7.337	5.337	6.345	8.301	10.548
7	2.902	3.452	4.521	5.750	3.369	4.007	5.248	6.676	4.613	5.488	7.187	9.142
8	2.743	3.264	4.278	5.446	3.136	3.732	4.891	6.226	4.147	4.936	6.468	8.234
9	2.626	3.125	4.098	5.220	2.967	3.532	4.631	5.899	3.822	4.550	5.966	7.600
10	2.535	3.018	3.959	5.046	2.839	3.379	4.433	5.649	3.582	4.265	5.594	7.129
11	2.463	2.933	3.849	4.906	2.737	3.259	4.277	5.452	3.397	4.045	5.308	6.766
12	2.404	2.863	3.758	4.792	2.655	3.162	4.150	5.291	3.250	3.870	5.079	6.477
13	2.355	2.805	3.682	4.697	2.587	3.081	4.044	5.158	3.130	3.727	4.893	6.240
14	2.314	2.756	3.618	4.615	2.529	3.012	3.955	5.045	3.029	3.608	4.737	6.043
15	2.278	2.713	3.562	4.545	2.480	2.954	3.878	4.949	2.945	3.507	4.605	5.876
16	2.246	2.676	3.514	4.484	2.437	2.903	3.812	4.865	2.872	3.421	4.492	5.732
17	2.219	2.643	3.471	4.430	2.400	2.858	3.754	4.791	2.808	3.345	4.393	5.607
18	2.194	2.614	3.433	4.382	2.366	2.819	3.702	4.725	2.753	3.279	4.307	5.497
19	2.172	2.588	3.399	4.339	2.337	2.784	3.656	4.667	2.703	3.221	4.230	5.399
20	2.152	2.564	3.368	4.300	2.310	2.752	3.615	4.614	2.659	3.168	4.161	5.312
21	2.135	2.543	3.340	4.264	2.286	2.723	3.577	4.567	2.620	3.121	4.100	5.234
22	2.118	2.524	3.315	4.232	2.264	2.697	3.543	4.523	2.584	3.078	4.044	5.163
23	2.103	2.506	3.292	4.203	2.244	2.673	3.512	4.484	2.551	3.040	3.993	5.098
24	2.089	2.480	3.270	4.176	2.225	2.651	3.483	4.447	2.522	3.004	3.947	5.039
25	2.077	2.474	3.251	4.151	2.208	2.631	3.457	4.413	2.494	2.972	3.904	4.985
26	2.065	2.460	3.232	4.127	2.193	2.612	3.432	4.382	2.460	2.914	3.865	4.935
27	2.054	2.447	3.215	4.106	2.178	2.595	3.409	4.353	2.446	2.914	3.828	4.888
30	2.025	2.413	3.170	4.049	2.140	2.549	3.350	4.278	2.385	2.841	3.733	4.768
35	1.988	2.368	3.112	3.974	2.090	2.490	3.272	4.179	2.306	2.748	3.611	4.611
40	1.959	2.334	3.066	3.917	2.052	2.445	3.213	4.104	2.247	2.677	3.518	4.493
45	1.935	2.306	3.030	3.871	2.021	2.408	3.165	4.042	2.200	2.621	3.444	4.399
50	1.916	2.284	3.001	3.833	1.996	2.379	3.126	3.993	2.162	2.576	3.385	4.323

*Adapted from Lieberman, Gerald J. (1958). "Tables for One-Sided Tolerance Limits." *Industrial Quality Control*, vol. XIV, no. 10, April, p. 8. Adapted with permission of the American Society for Quality Control.
†Adapted with permission from Eisenhart, C., Hastay, M. W., and Wallis, W. A. (1947). *Techniques of Statistical Analysis*. McGraw-Hill, New York.

TABLE W *P for Interval between Sample Extremes**

γ is the probability that an interval will cover a proportion P of the population with a random sample of size N.

N \ γ	0.5	0.7	0.9	0.95	0.99	0.995
2	0.293	0.164	0.052	0.026	0.006	0.003
4	0.615	0.492	0.321	0.249	0.141	0.111
6	0.736	0.640	0.490	0.419	0.295	0.254
10	0.838	0.774	0.664	0.606	0.496	0.456
20	0.918	0.883	0.820	0.784	0.712	0.683
40	0.959	0.941	0.907	0.887	0.846	0.829
60	0.973	0.960	0.937	0.924	0.895	0.883
80	0.980	0.970	0.953	0.943	0.920	0.911
100	0.984	0.976	0.962	0.954	0.936	0.929
150	0.990	0.984	0.975	0.969	0.957	0.952
200	0.992	0.988	0.981	0.977	0.968	0.961
500	0.997	0.996	0.993	0.991	0.987	0.986
1,000	0.999	0.998	0.997	0.996	0.994	0.993

*Adapted with permission from Dixon, W. J. and Massey, F. J., Jr. (1969). *Introduction to Statistical Analysis,* 3rd ed. McGraw-Hill, New York.

TABLE X *N for Interval between Sample Extremes**

P \ γ	0.50	0.70	0.90	0.95	0.99	0.995
0.995	336	488	777	947	1,325	1,483
0.99	168	244	388	473	662	740
0.95	34	49	77	93	130	146
0.90	17	24	38	46	64	72
0.85	11	16	25	30	42	47
0.80	9	12	18	22	31	34
0.75	7	10	15	18	24	27
0.70	6	8	12	14	20	22
0.60	4	6	9	10	14	16
0.50	3	5	7	8	11	12

*Adapted with permission from Dixon, W. J. and Massey, F. J., Jr. (1969). *Introduction to Statistical Analysis,* 3rd ed. McGraw-Hill, New York.

TABLE Y E_2 Factors for Control Charts

Number of observations in subgroup	E_2
2	2.660
3	1.772
4	1.457
5	1.290
6	1.184
7	1.109
8	1.054
9	1.010
10	0.975
11	0.946
12	0.921
13	0.899
14	0.881
15	0.864

CHART Z Control Limits of p Charts

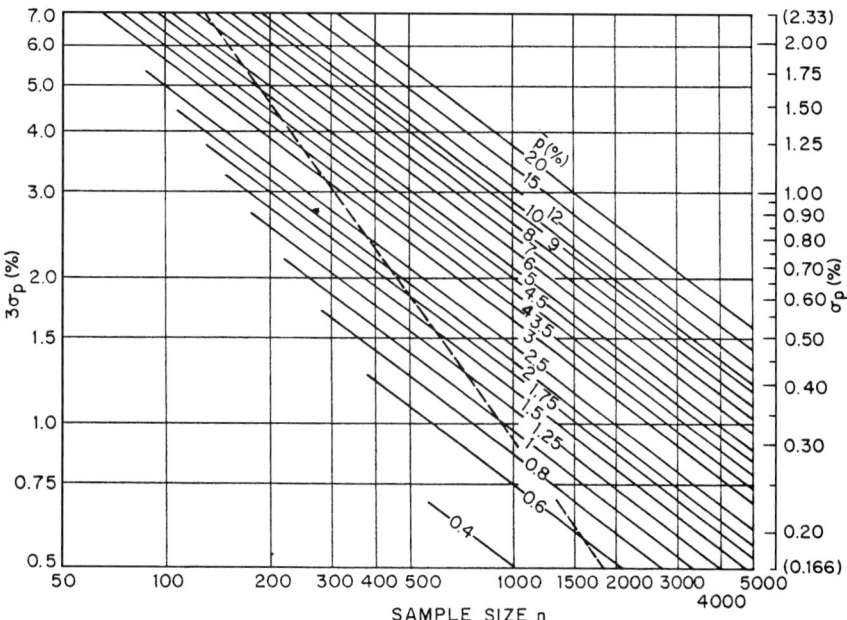

CHART AA Control Limits for c, Number of Defects per Sample*

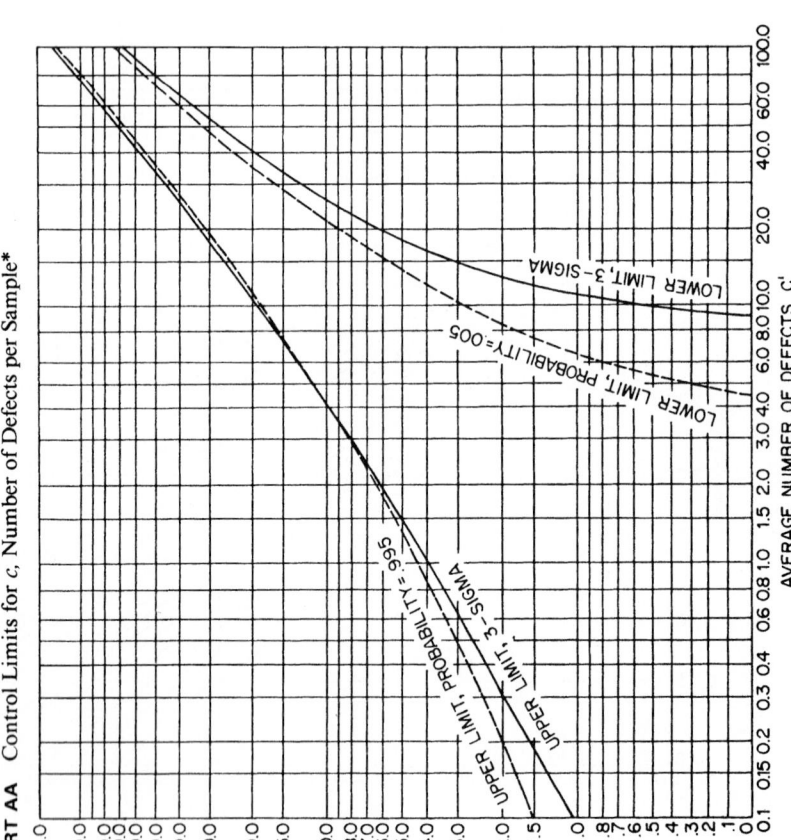

*Reproduced by permission from American War Standard Z1.3-1942, American Standards Association, New York.

TABLE BB Factors for Cumulative Sum Control Chart*

	$2\,\alpha_0 = 0.0027$ $\alpha_0 = 0.00135$†	
δ	θ	d
0.2	5°43'	330.4
0.4	11°19'	82.6
0.5	14°00'	52.9
0.6	16°42'	36.7
0.8	21°48'	20.6
1.0	26°34'	13.2
1.2	30°58'	9.2
1.3	32°59'	7.8
1.4	35°00'	6.7
1.6	38°40'	5.2
1.8	41°59'	4.1
2.0	45°00'	3.3
2.2	47°44'	2.7
2.4	50°12'	2.3
2.6	52°26'	2.0
2.8	54°28'	1.7
3.0	56°19'	1.5

*Adapted with permission from Johnson, Norman L. and Leone, Fred C. (1964). *Statistics and Experimental Design in Engineering and Physical Sciences.* John Wiley & Sons, New York, vol. I, p. 322.

†For limits comparable with the 3-sigma limits used in the Shewhart control chart.

TABLE CC   Random Numbers*

1306	1189	5731	3968	5606	5084	8947	3897	1636	7810
0422	2431	0649	8085	5053	4722	6598	5044	9040	5121
6597	2022	6168	5060	8656	6733	6364	7649	1871	4328
7965	6541	5645	6243	7658	6903	9911	5740	7824	8520
7695	6937	0406	8894	0441	8135	9797	7285	5905	9539
5160	7851	8464	6789	3938	4197	6511	0407	9239	2232
2961	0551	0539	8288	7478	7565	5581	5771	5442	8761
1428	4183	4312	5445	4854	9157	9158	5218	1464	3634
3666	5642	4539	1561	7849	7520	2547	0756	1206	2033
6543	6799	7454	9052	6689	1946	2574	9386	0304	7945
9975	6080	7423	3175	9377	6951	6519	8287	8994	5532
4866	0956	7545	7723	8085	4948	2228	9583	4415	7065
8239	7068	6694	5168	3117	1586	0237	6160	9585	1133
8722	9191	3386	3443	0434	4586	4150	1224	6204	0937
1330	9120	8785	8382	2929	7089	3109	6742	2468	7025
2296	2952	4764	9070	6356	9192	4012	0618	2219	1109
3582	7052	3132	4519	9250	2486	0830	8472	2160	7046
5872	9207	7222	6494	8973	3545	6967	8490	5264	9821
1134	6324	6201	3792	5651	0538	4676	2064	0584	7996
1403	4497	7390	8503	8239	4236	8022	2914	4368	4529
3393	7025	3381	3553	2128	1021	8353	6413	5161	8583
1137	7896	3602	0060	7850	7626	0854	6565	4260	6220
7437	5198	8772	6927	8527	6851	2709	5992	7383	1071
8414	8820	3917	7238	9821	6073	6658	1280	9643	7761
8398	5224	2749	7311	5740	9771	7826	9533	3800	4553
0995	8935	2939	3092	2496	0359	0318	4697	7181	4035
6657	0755	9685	4017	6581	7292	5643	5064	1142	1297
8875	8369	7868	0190	9278	1709	4253	9346	4335	3769
8399	6702	0586	6428	7985	2979	4513	1970	1989	3105
6703	1024	2064	0393	6815	8502	1375	4171	6970	1201
4730	1653	9032	9855	0957	7366	0325	5178	7959	5371
8400	6834	3187	8688	1079	1480	6776	9888	7585	9998
3647	8002	6726	0877	4552	3238	7542	7804	3933	9475
6789	5197	8037	2354	9262	5497	0005	3986	1767	7981
2630	2721	2810	2185	6323	5679	4931	8336	6662	3566
1374	8625	1644	3342	1587	0762	6057	8011	2666	3759
1572	7625	9110	4409	0239	7059	3415	5537	2250	7292
9678	2877	7579	4935	0449	8119	6969	5383	1717	6719
0882	6781	3538	4090	3092	2365	6001	3446	9985	6007
0006	4205	2389	4365	1981	8158	7784	6256	3842	5603
4611	9861	7916	9305	2074	9462	0254	4827	9198	3974
1093	3784	4190	6332	1175	8599	9735	8584	6581	7194
3374	3545	6865	8819	3342	1676	2264	6014	5012	2458
3650	9676	1436	4374	4716	5548	8276	6235	6742	2154
7292	5749	7977	7602	9205	3599	3880	9537	4423	2330
2353	8319	2850	4026	3027	1708	3518	7034	7132	6903
1094	2009	8919	5676	7283	4982	9642	7235	8167	3366
0568	4002	0587	7165	1094	2006	7471	0940	4366	9554
5606	4070	5233	4339	6543	6695	5799	5821	3953	9458
8285	7537	1181	2300	5294	6892	1627	3372	1952	3028

*Adapted with permission from Owen, Donald B. (1962). *Handbook of Statistical Tables.* Addison-Wesley, Reading, MA. Courtesy U.S. Atomic Energy Commission.

TABLE DD   Values of ϕ^2 for Determining Sample Size in Analysis of Variance*

$\alpha = 0.01; \beta = 0.2$

DF₂＼DF₁	1	2	3	4	5	6	7	8	9
2	80.37	106.63	119.75	127.62	132.87	136.63	139.45	141.63	143.38
4	17.28	18.58	18.95	19.11	19.18	19.21	19.23	19.24	19.24
6	11.36	11.12	10.77	10.49	10.27	10.11	9.97	9.86	9.77
8	9.41	8.76	8.21	7.83	7.54	7.32	7.15	7.01	6.89
10	8.47	7.63	7.02	6.58	6.26	6.03	5.84	5.68	5.56
12	7.91	6.98	6.33	5.87	5.54	5.29	5.09	4.93	4.80
14	7.55	6.56	5.88	5.41	5.07	4.81	4.61	4.45	4.31
16	7.30	6.26	5.56	5.09	4.75	4.49	4.28	4.11	3.98
18	7.11	6.05	5.35	4.86	4.51	4.24	4.04	3.87	3.73
20	6.96	5.89	5.17	4.68	4.33	4.06	3.85	3.68	3.54
24	6.76	5.66	4.93	4.41	4.08	3.80	3.57	3.42	3.28
30	6.55	5.42	4.68	4.19	3.82	3.55	3.33	3.16	3.02
40	6.35	5.20	4.45	3.96	3.57	3.31	3.10	2.92	2.79
60	6.18	5.00	4.25	3.74	3.37	3.10	2.88	2.70	2.55
80	6.10	4.88	4.16	3.65	3.28	2.99	2.76	2.59	2.43
120	6.00	4.80	4.04	3.53	3.17	2.89	2.66	2.50	2.34
240	5.90	4.71	3.96	3.46	3.06	2.79	2.56	2.40	2.25
∞	5.84	4.62	3.87	3.35	2.98	2.70	2.47	2.29	2.14

*These tables are computed from Lehmer, Emma (1944). "Inverse Tables of Probabilities of Errors of Second Kind." *Annals of Mathematical Statistics,* vol. 15, p. 390. Reproduced from Dixon, W. J. and Massey, F. J., Jr. *Introduction to Statistical Analysis,* 1st ed. McGraw-Hill, New York, p. 330.

10	12	15	20	24	30	40	60	120	∞
144.82	147.02	149.30	151.63	152.84	154.06	155.30	156.55	157.83	159.09
19.24	19.24	19.24	19.22	19.21	19.21	19.19	19.18	19.18	19.17
9.69	9.57	9.44	9.30	9.22	9.14	9.07	8.99	8.90	8.81
6.80	6.64	6.48	6.31	6.21	6.12	6.02	5.91	5.81	5.70
5.45	5.29	5.11	4.92	4.82	4.71	4.61	4.49	4.38	4.26
4.69	4.52	4.33	4.13	4.02	3.91	3.80	3.68	3.56	3.43
4.20	4.02	3.83	3.63	3.52	3.40	3.28	3.16	3.03	2.89
3.86	3.68	3.48	3.27	3.16	3.04	2.92	2.80	2.66	2.52
3.61	3.43	3.23	3.01	2.90	2.78	2.66	2.53	2.39	2.24
3.42	3.23	3.03	2.82	2.70	2.58	2.46	2.32	2.18	2.03
3.13	2.96	2.76	2.53	2.43	2.31	2.16	2.02	1.88	1.72
2.90	2.70	2.50	2.27	2.16	2.02	1.88	1.74	1.59	1.42
2.66	2.46	2.25	2.02	1.90	1.77	1.61	1.46	1.30	1.13
2.43	2.23	2.02	1.78	1.66	1.52	1.37	1.21	1.04	0.841
2.31	2.13	1.90	1.66	1.54	1.39	1.25	1.08	0.902	0.689
2.22	2.02	1.80	1.56	1.44	1.28	1.12	0.960	0.766	0.528
2.13	1.90	1.69	1.44	1.32	1.17	1.00	0.828	0.624	0.345
2.02	1.81	1.58	1.34	1.21	1.05	0.884	0.704	0.472	0.000

APPENDIX III

SELECTED QUALITY STANDARDS, SPECIFICATIONS, AND RELATED DOCUMENTS

This list is limited to documents likely to be of general interest. August Mundel collected, organized, and assembled the material to bring this appendix up to date. There are many other standards that refer to the control and measurement of specific quality, reliability, etc., of specific products. In general, these have not been listed.

The U.S. Department of Defense has, for the most part, given up the maintenance and use of its own standards and is letting other organizations inherit, revise, and republish the standards. There is still a vast amount of information in the U.S. MIL standards not available elsewhere. These standards can, for the most part, be obtained and are therefore listed.

The list is based on the original summary by B. A. MacDonald and M. V. Petty entitled "List of Quality Standards, Specifications and Related Documents," (February 10, 1987, unpublished). Much of the other material is the result of the Standards Column of *Quality Engineering* magazine, edited by August B. Mundel, where the data from MacDonald and Petty has also been published.

The following list is divided into three parts as follows:

System and program requirements	Statistical techniques and procedures	Inspection and test methods
Definitions	Definitions, symbols, formulas	Definitions, units of measurement
Quality	Analysis of data	Inspection and test methods and
Reliability	Control charts	requirements
Maintainability	Acceptance sampling	Evaluation of inspection and test
Software		systems
Safety		Precision and accuracy
Certification programs		
Other		

In some cases, the official document designation includes abbreviations or other notations:

Amd.: Amendment

Chg not: Change notice

Supp: Supplement

'79: 1979

Following the complete listing, addresses are given for the organizations issuing the documents.

SYSTEM AND PROGRAM REQUIREMENTS

Definitions

ANSI/ISO/ASQC 8402-1994	Quality Management and Quality Assurance
ANSI/IEEE 610-12-1990	*Software Engineering Terminology, Standard Glossary of*
ANSI/ISO/ASQC A3534-1 1993	*Statistics—Vocabulary and Symbols Probability and Statistical Terms*
ANSI/ISO/ASQC A3534-2 1993	*Statistics—Vocabulary and Symbols Statistical Quality Control*
ANSI/ISO/ASQC A3534-3 1998	*Statistics—Vocabulary and Symbols: Part 3, Design of Experiments*
IEEE/ASTM SI-1	*Standard for Use of the International System of Units (SI)—The Modern Metric System*
IEC 271A (first, second, third supplements) 1984	*List of Basic Terms, Definitions, and Related Mathematics for Reliability*
MIL-STD-1093 4 April 69	*Quality Assurance Terms and Definitions*
MIL-STD-280A 7 July 69	*Definitions of Item Levels, Exchangeability, Models, and Related Items*
MIL-STD 721C	*Definition of Terms for Reliability, Maintainability, Human Factors and Safety*

Quality

ANSI 10011	*Guidelines for Auditing Quality Systems*
ANSI/ASQC C1 1996 (ANSI Z1.8 1971)	*Specifications of General Requirements for a Quality Program*
ANSI/ISO/ASQC 9000-1-1994	*Quality Management and Quality Assurance Standards—Guidelines for Selection and Use*
ANSI/ISO/ASQC 9001-1994	*Quality Systems—Model for Quality Assurance in Design, Development, Production, Installation and Servicing*
ANSI/ISO/ASQC 9002-1994	*Quality Systems—Model for Quality Assurance in Production, Installation and Servicing*
ANSI/ISO/ASQC 9003-1994	*Quality Systems—Model for Quality Assurance in Final Inspection and Test*
ANSI/ISO/ASQC Q 10014-97	*Guidelines for Managing the Economics of Quality*
ANSI/IEC/ASQC 1160	*Formal Design Review*
AQAP-1	*NATO Requirements for an Industrial Quality Control System*
AQAP-2	*Guide for the Evaluation of a Contractor's Quality Control System for Compliance with AQAP-1*
AQAP-3	*List of Sampling Schemes used in NATO Countries*
AQAP-4	*NATO Inspection Systems Requirements for Industry*
AQAP-5	*Guide for the Evaluation of a Contractor's Inspection System for Compliance with AQAP-4*
AQAP-6	*NATO Measurement and Calibration System Requirements of Industry*
AQAP-7	*Guide for the Evaluation of a Contractor's Measurements and Calibration System for Compliance with AQAP-6*

AQAP-8	*NATO Guide to the Preparation of Specifications for the Procurement of Defence Materiel*
AQAP-9	*NATO Basic Inspection Requirements for Industry*
AQAP-10	*NATO Guide for a Government Quality Assurance Programme*
AQAP-11	*NATO Guideline for the Specification of Technical Publications*
AQAP-15	*Glossary of Terms Used in QA STANAGs and AQAPs*
ASQC E1	*Quality Program Guidelines for Project Phase of Non-nuclear Power Generating Facilities*
BSI HDBK 22:1981	*Quality Assurance* (contains 15 publications)
CAN 3 Z299-1—CSA	*Quality Assurance Program—Category 1*
CAN 3 Z299-2—CSA	*Quality Assurance Program—Category 2*
CAN 3 Z299-3—CSA	*Quality Assurance Program—Category 3*
IEC Guide 102 (1996-03)	*Specification Structures for Quality*
ISO Guide 33	*Use of Certified Reference Materials*
ISO Guide 35	*Certification of Reference Materials—General and Statistical Principles*
ISO/IEC Guide 43	*Proficiency Testing by Interlaboratory Comparisons*
ISO 2602	*Statistical Interpretation of Test Results—Estimation of the Mean–Confidence Interval*
ISO 2854	*Statistical Interpretation of Data—Techniques of Estimation and Tests Relating to Means and Variances*
MIL-HDBK-50	*Evaluation of a Contractor's Quality Program*
MIL-Q-9858A	*Quality Program Requirements*
MIL-S-19500G	*General Specification for Semiconductor Devices*
MIL-STD 454K	*Standard General Requirements—Electronic Equipment*
MIL-STD-1521B	*Technical Reviews and Audits of System, Equipments, and Computer Software*
MIL-STD-1535A	*Supplier Quality Assurance Program Requirements*
MIL-STD-2164	*Failure Reporting, Analysis and Corrective Action Systems*
MIL-T-50301	*Quality Control System Requirements for Technical Data*
NHB 5300.4 (1B) (NASA)	*Quality Program Provisions for Aeronautical and Space System Contractors*
NHB 5300.4 (1D-2) (NASA)	*Safety, Reliability, Maintainability and Quality Provisions for the Space Shuttle Program*
NHB 5300.4 (2B-1) (NASA)	*Quality Assurance Provisions for Government Agencies*
NQA-2 (ANSI/ASME)	*Quality Assurance Requirements for Nuclear Power Plants*
STP 616 (ASTM)	*Quality Systems in the Nuclear Industry*

Reliability

ANSI/IEC/ASQC 300-3-1	*Dependability Management Part 3: Application Guide Section 1: Analysis Techniques for Dependability: Guide on Methodology*
ANSI/IEC/ASQC D1025-1997	*Fault Tree Analysis (FTA)*

ANSI/IEC/ASQC D1070-1997	*Compliance Test Procedures for Steady State Availability*
ANSI/IEC/ASQC D1078-1997	*Analysis Techniques of Dependability—Reliability Block Diagram Method*
ANSI/IEC/ASQC D1123-1997	*Reliability Testing—Compliance Test Plans for Success Ratio*
ANSI/IEC/ASQC D1164-1997	*Reliability Growth Statistical Tests and Estimation Methods*
ANSI/IEC/ASQC D1165-1997	*Application of Markov Techniques*
IEC 50 Chap 191	*International Electro-technical Vocabulary,* Chapter 191, "Dependability and Quality of Service"
IEC 60272 (1968)	*Preliminary Reliability Considerations*
IEC 60300	*Reliability and Maintainability Management*
IEC 60300-1 (1993-04)	*Dependability Management—Part 1: Dependability Programme Management*
IEC 60300-2 (1995-12)	*Dependability Management—Part 2: Dependability Programme Elements and Tasks*
IEC 60300-3-2 (1993-10)	*Dependability Management Part 3: Application Guide Section 2: Collection of Dependability Data from the Field*
IEC 60300-3-9 (1995-12)	*Dependability Management—Part 3: Application Guide— Section 9: Risk Analysis of Technological Systems*
IEC 60319 (1978-01)	*Presentation of Reliability Data on Electronic Components (or Parts)*
IEC 60362 (1971)	*Guide for the Collection of Reliability, Availability, and Maintainability Data from Field Performance of Electronic Items*
IEC 60409	*Guide for the Inclusion of Reliability Clauses into Specifications for Components (or Parts) for Electronic Equipment*
IEC 60571-3	*Electronic Equipment Used on Rail Vehicles Part 3: Components, Programmable Electronic Equipment for Electronic System Reliability*
IEC 60605-1 (1978-01)	*Equipment Reliability Testing—Part 1: General Requirements*
IEC 60605-2 (1994-10)	*Equipment Reliability Testing—Part 2: Design of Test Cycles*
IEC 60605-3-1 (1986-09)	*Equipment Reliability Testing—Part 3: Preferred Test Conditions—Sec.1: Indoor Portable Equipment—Low Degree of Simulation*
IEC 60605-3-2 (1986-09)	*Equipment Reliability Testing—Part 3: Preferred Test Conditions—Sec.2: Equipment for Stationary Use in Weatherprotected Locations—High Degree of Simulation*
IEC 60605-3-3 (1992-11)	*Equipment Reliability Testing—Part 3: Preferred Test Conditions—Section 3: Test Cycle 3 Equipment for Stationary Use in Partially Weatherprotected Locations— Low Degree of Simulation*
IEC 60605-3-4 (1992-07)	*Equipment Reliability Testing—Part 3: Preferred Test Conditions—Section 4: Test Cycle 4: Equipment for Portable and Non-stationary Use—Low Degree of Simulation*

IEC 60605-4 (1986-07)	*Equipment Reliability Testing—Part 4: Procedure for Determining Point Estimates and Confidence Limits for Equipment Reliability Determination Tests*
IEC 60605-6 (1997-04)	*Equipment Reliability Testing—Part 6: Tests for the Validity of the Constant Failure Rate or Constant Failure Intensity Assumptions*
IEC 60605-7	*Equipment Reliability Testing—Part 7: Compliance Test Plans for Failure Rate and Mean Time between Failures Assuming Constant Failure Rate*
IEC 60812 (1985-07)	*Analysis Techniques of System Reliability—Procedure for Failure Mode and Effects Analysis (FMEA)*
IEC 60863 (1986-05)	*Presentation of Reliability, Maintainability, and Availability Predictions*
IEC 61014 (1989-11)	*Programmes for Reliability Growth*
IEC 61025 (1990-10)	*Fault Tree Analysis*
IEC 61069-5 (1994-12)	*Industrial-Process Measurement and Control—Evaluation of System Properties for the Purpose of System Assessment—Part 5: Assessment of System Dependability*
IEC 61070 (1991-11)	*Compliance Test Procedures for Steady-State Availability*
IEC 61078 (1991-11)	*Analysis Techniques for Dependability—Reliability Block Diagram Method*
IEC 61123 (1991-12)	*Reliability Testing Compliance Test Plans for Success Ratio*
IEC 61163-1	*Reliability Stress Screening—Part 1: Repairable Items Manufactured in Lots*
IEC 61164 (1995-06)	*Reliability Growth—Statistical Test and Estimation Methods*
IEC 61165 (1995-01)	*Application of Markov Techniques*
MIL-STD-781C (Chg Notice 1)	*Reliability Design, Qualification and Production Acceptance Test: Exponential Distribution*
MIL-STD-785B Notice 1	*Reliability Program for Systems and Equipment Development and Production*
MIL-STD-790C Notice 1	*Reliability Assurance Program for Electronic Parts Specifications*
MIL-STD-1543A	*Reliability Program Requirements of Space and Missile Systems*
MIL-STD-1629A (Chg Notice 2)	*Procedure for Performing a Failure Mode, Effects and Criticality Analysis*
MIL-STD 1635	*Reliability Growth Testing*
MIL-STD 2068	*Reliability Development Tests*
MIL-STD-2101	*Failure Classification for Reliability Testing*
MIL-HDBK-189	*Reliability Growth Management*
MIL-HDBK-217E	*Reliability Prediction for Electronic Equipment*
MIL-HDBK-251	*Reliability/Design Thermal Applications*
MIL-HDBK-338	*Electronic Reliability Design Handbook, Vols. 1 and 2*
NHB 5300.4(1A) NASA	*Reliability Program Provisions for Aeronautical and Space Systems Contractors*
NHB 5300.4(1D-2)	*Safety, Reliability, Maintainability and Quality Provisions for the Space Shuttle Program*

Maintainability

DOD-HDBK-472	*Maintainability Prediction*
IEC 60706-1 (1982-01)	*Guide on the Maintainability of Equipment Part 1: Sections 1, 2, and 3 Introduction, Requirements, Maintainability Program*
MIL-STD-470A	*Requirements for Systems and Equipment*
MIL-STD-471A	*Maintainability, Verification/Demonstration/Evaluation*
NHB-5300.4 (1D2) NASA	*Safety, Reliability, Maintainability, and Quality Provisions for the Space Shuttle Program*

Safety

ANSI/NFPA 70	*National Electrical Code* (National Fire Protection Association)
MIL-STD-882B	*System Safety Program Requirements*
NHB-5300.4 (1D2) NASA	*Safety, Reliability, Maintainability, and Quality Provisions for the Space Shuttle Program*

Software

AQAP 13 (NATO)	*Software Quality Control System Requirements*
AQAP 14 (NATO)	*Guide for the Evaluation of a Contractor's Software Quality Control System for Compliance*
DI-R-3521-1982 (DoD)	*Software Quality Assurance Plan*
IEEE 610.12-1990	*Standard Glossary of Software Engineering Terminology*
IEEE 730-1989	*Software Engineering Quality Assurance Plans, Standard for*
IEEE 730.1 1995	*Software Quality Assurance Planning, Guide for*
IEEE 828-1990	*Software Configuration Management Plans, Standard for*
IEEE 829-1983	*Software Test Documentation, Standard for*
IEEE 830-1993	*Software Requirements Specification, Recommended Practice for*
IEEE 982.1-1988	*Standard Dictionary of Measures to Produce Reliable Software*
IEEE 982.2-1988	*Guide for the Use of IEEE Standard Dictionary of Measures to Produce Reliable Software*
IEEE 1012-1986	*Standard for Software Verification and Validation Plans*
IEEE 1059-1993	*Software Verification and Validation Plans, Guide for*
IEEE 1061-1992	*Software Quality Metrics Methodology, Standard for*

Certification Programs

ANSI 34.1	*Certification Procedures, Practice for*
ANSI 34.2	*Self-Certification by Producer or Supplier*
ANSI/IEC 599	*National Electric Process Certification Standard*
ANSI/ISO/ASQC 9000-1-1994	*Quality Management and Quality Assurance Standards— Guidelines for Selection and Use of 9001, 9002 and 9003*

ANSI/ISO/ASQC 9000-2-1993	*Quality Management and Quality Assurance Standards— Generic Guidelines of the Application of 9001, 9002 and 9003*
ANSI/ISO/ASQC 9000-3-1991	*Quality Management and Quality Assurance Standards— Generic Guidelines of the Application of 9001, to the Development, Supply and Maintenance of Software*
ANSI/ISO/ASQC 9001-1994	*Quality Systems—Model for Quality Assurance in Design, Development, Production, Installation and Servicing*
ANSI/ISO/ASQC 9002-1994	*Quality Systems—Model for Quality Assurance in Production, Installation and Servicing*
ANSI/ISO/ASQC 9003-1994	*Quality Systems—Model for Quality Assurance in Final Inspection and Test*
ANSI/ISO/ASQC A8402-1994	*Quality Management and Quality Assurance—Vocabulary*
ANSI/ISO/ASQC 9004-2-1991	*Quality Management and Quality System Elements— Guidelines for Services*
ANSI/ISO/ASQC 9004-3-1993	*Quality Management and Quality System Elements— Guidelines for Processed Materials*
ANSI/ISO/ASQC 9004-4-1993	*Quality Management and Quality System Elements— Guidelines for Quality Improvement*
ANSI/ISO/ASQC 10007-1995	*Quality Management—Guidelines for Configuration Management*
ISO/IEC Guide 7	*Requirements for Standards Suitable for Product Certification*
ISO/IEC Guide 16	*Code of Principles on Third-Party Certification Systems and Related Standards*
ISO/IEC Guide 22	*Information on Manufacturer's Declaration of Conformity and Standards or Other Technical Specifications*
ISO/IEC Guide 23	*Method of Indicating Conformity with Standards for Third-Party Certification Systems*
ISO/IEC Guide 24	*Guidelines for the Acceptance of Testing and Inspection Agencies by Certification Bodies*
ISO/IEC Guide 27	*Guidelines for Corrective Action to Be Taken by a Certification Body in the Event of Either Misapplication of Its Mark of Conformity to a Product, or Products which Bear the Mark of the Certification Body Being Found to Subject Persons or Property to Risk (Excludes Certificates of Conformity)*
ISO/IEC Guide 28	*General Rules for a Model Third-Party Certification System for Products*
ISO/IEC Guide 31	*Contents of Certificates of Reference Materials*
ISO/IEC Guide 40	*General Requirements for the Acceptance of Certification Bodies*
ISO/IEC Guide 42	*Guidelines for a Step-by-Step Approach to an International Certification System*
ISO/IEC Guide 44	*General Rules for ISO or IEC International Third-Party Certification Schemes for Products*
ISO 2859-4	*Sampling Procedures for Inspection by Attributes—Part 4: Procedures for Assessment of Stated Quality Levels*

Other

ANSI/ISO/ASQC Q10007-1995	*Quality Management—Guidelines for Configuration Management*
ASME Y14.5M 1994	*Dimensioning and Tolerancing*
ASME Y14.5.1M 1994	*Mathematical Definition of Dimensioning and Tolerancing*
DOD-STD-480A	*Configuration Control—Engineering Changes, Deviations and Waivers*
DOD-STD-2101	*Classification of Characteristics*
FED-STD 209B	*Clean Room and Work Station Requirements, Controlled Environment*
ISO/IEC Guide 26	*Justification of Proposals for the Establishment of Standards*
ISO/IEC Guide 37	*Instructions for Use of Products of Consumer Interest*
MIL-H-46855B	*Human Engineering Requirements for Equipment and Facilities*
MIL-HDBK-727	*Design Guidance for Producibility*
MIL-STD-449A	*Engineering Management*
MIL-STD-481A	*Configuration Control—Engineering Changes, Deviations and Waivers (short form)*
MIL-STD 482A	*Configuration Status Accounting Data Elements and Related Forms*
MIL-STD-483A	*Configuration Management Practices for Systems, Equipment, Munitions and Computer Programs*
MIL-STD-680A	*Contractor Standardization Program Requirements*
MIL-STD-1246A	*Product Cleanliness Levels and Contamination Control Program*
MIL-STD-1472C	*Human Engineering Design Criteria for Military Systems, Equipment and Facilities*
MIL-STD 1546	*Parts, Materials and Process Standardization Control and Management Programs for Spacecraft and Launch Vehicles*

STATISTICAL TECHNIQUES AND PROCEDURES

Definitions, Symbols, Formulas

ANSI/ISO/ASQC A3534-1 1993	*Statistics—Vocabulary and Symbols Probability and Statistical Terms (Replace AI A2)*
ANSI/ISO/ASQC A3534-2 1993	*Statistics—Vocabulary and Symbols Statistical Quality Control*
ANSI/ISO/ASQC A3534-3 1998	*Statistics—Vocabulary and Symbols Part 3, Design of Experiments*
IEEE/ASTM SI-10	*Standard for Use of the International System of Units (SI), The Modern Metric System*
ISO 3207	*Statistical Interpretation of Data—Determination of Statistical Tolerance Interval*

ISO 3301	*Statistical Interpretation of Data—Comparison of Two Means in the Case of Paired Observations*
ISO 3494	*Statistical Interpretation of Data—Power of Tests Relating to Means and Variances*
ISO 5479	*Tests for Departure from the Normal Distribution*

Analysis of Data

ASTM E29-93	*Using Significant Digits in Test Data to Determine Conformance with Specifications, Standard Practice for*
ASTM E178-94	*Dealing with Outlying Observations, Standard Practice for*
ASTM E678-90	*Evaluation of Technical Data, Standard Practice for*
ASTM Z7240Z-98	*Applying Statistical Quality Assurance Techniques to Evaluate Analytical System Performance, Practice for*
ASTM E122	*Choice of Sample Size to Estimate a Measure of Quality for a Lot or Process, Practice for*
IEC 60493	*Statistical Analysis of Aging Test Data, Guide for*
IEC 60493-1 (1974-01)	*Part 1 Methods Based on Mean Values of Normally Distributed Test Results*
ISO 2602	*Statistical Interpretation of Test Results—Estimation of the Mean-Confidence Interval*
ISO 2854	*Statistical Interpretation of Data—Techniques of Estimation and Test Results Relating to Means and Variances*
ISO 3207	*Statistical Interpretation of Data: Determination of Statistical Tolerance Interval (also Addendum 1)*
ISO 3301	*Statistical Interpretation of Data—Comparison of Two Means in the Case of Paired Observations*
ISO 3494	*Statistical Interpretation of Data—Power of Tests Relating to Means and Variances*
ISO Hndbk 3 1981	*Statistical Methods*
ISO/IEC Guide 45	*Guidelines for the Presentation of Test Data*
ISO 8595	*Interpretation of Statistical Data—Estimation of a Median*
ISO 11462-1	*Guidelines for Implementation of Statistical Process Control (SPC)*
STP 468/A (ASTM)	*Evaluation of Technical Data, Standard Practice for (Manual on Methods for Retrieving and Correlating Technical Data)*

Control Charts

ANSI/ASQC B1-B3 1991	*Quality Control Chart Methodologies (Contains B1, B2, B3)*
ASTM MNL 7 1970	*Manual on the Presentation of Data and Control Chart Analysis*
ISO 7870	*Control Charts—General Guide and Introduction*
ISO/TR 7871	*Cumulative Sum Charts—Guide to Quality Control and Data Analysis Using CUSUM Techniques*

ISO 7873	*Control Charts for Arithmetic Average with Warning Limits*
ISO 8258	*Shewhart Control Charts*
ISO 7966	*Acceptance Control Charts*

Acceptance Sampling

ANSI/ASQC S1 1996 & ISO 2859-3	*Sampling Procedures for Inspection by Attributes—Part 3 Skip-Lot Sampling Procedures*
ANSI/ASQC S2 1995	*Introduction to Attribute Sampling*
ANSI/ASQC Z1.4	*Sampling Procedures and Tables for Inspection by Attributes*
ANSI/ASQC Z1.9	*Sampling Procedures and Tables for Inspection by Variables for Percent Nonconforming*
ANSI/ASQC Q3 1988	*Sampling Procedures and Tables for Inspection of Isolated Lots by Attributes*
ANSI/EIA 585	*Zero Acceptance Number Samplings Procedures and Tables for Inspection by Attributes of Isolated Lots*
ANSI/EIA 584	*Zero Acceptance Number Sampling Procedures and Tables for Inspection by Attributes of a Continuous Manufacturing Process*
ANSI/EIA 591	*Assessment of Quality Levels in PPM Using Variables Test Data*
FED-STD-358	*Sampling Procedures*
IEC 60410 (1973-01)	*Sampling Plans and Procedures for Inspection by Attributes (Use ISO 2859 Part 1)*
IEC 60419	*Guide for the Inclusion of Lot-by-Lot and Periodic Inspection Procedures in Specifications for Electronic Components (or parts) (Use ISO 2859, Part 1)*
ISO 2859-0	*Sampling Procedures for Inspection by Attributes— Part 0 Introduction to the ISO 2859 Attribute Sampling System*
ISO 2859-1	*Sampling Procedures for Inspection by Attributes—Part 1: Sampling Plans Indexed by Acceptable Quality Level (AQL) for Lot-by-Lot Inspection*
ISO 2859-2	*Sampling Procedures for Inspection by Attributes—Part 2: Sampling Plans Indexed by Limiting Quality (LQ) for Isolated Lot Inspection*
ISO 2859-4	*Sampling Procedures for Inspection by Attributes—Part 4: Procedures for Assessment of Stated Quality Levels*
ISO 3951	*Sampling Procedures and Charts for Inspection by Variables for Percent Nonconforming*
ISO 8422	*Sequential Sampling Plans for Inspection by Attributes*
ISO 8423	*Sequential Sampling Plans for Inspection by Variable for Percent Nonconforming (known standard deviation)*
ISO/TR 8550	*Guide for Selection of an Acceptance Sampling System, Scheme, or Plan for Inspection of Discrete Items in Lots*
ISO 10725	*Acceptance Sampling Plans and Procedures for the Inspection of Bulk Materials*

ISO 11648-1	*Statistical Aspects of Sampling from Bulk Materials—Part 1: General Introduction*
MIL-STD-105	*Sampling Procedures and Tables for Inspection by Attributes (Use ANSI/ASQC Z1.4 or ISO 2859-1)*
MIL-STD-414	*Sampling Procedures and Tables for Inspection by Variables for Percent Defective*
MIL-STD-690B (Chg Notice 2)	*Failure Rate Sampling Plan and Procedures*
MIL-STD-1235B	*Continuous Sampling Procedures and Tables for Inspection by Attributes*
MIL-HDBK-53-1A	*Guide for Attribute Lot Sampling Inspection*
MIL-HDBK-106	*Multi-Level Continuous Sampling Procedures and Table for Inspection by Attributes*
MIL-HDBK-107	*Single Level Continuous Sampling Procedures and Table for Inspection by Attributes*
MIL-HDBK-108	*Sampling Procedures and Tables for Life and Reliability Testing (Based on the Exponential Distribution)*
MIL-HDBK-109	*Statistical Procedures for Determining Validity of Supplier Attribute Inspection*
TR-3 (DoD)	*Sampling Procedures and Tables for Life and Reliability Testing—Based on the Weibull Distribution (Mean Life Criterion)*
TR-4 (DoD)	*Sampling Procedures and Tables for Life and Reliability Testing—Based on the Weibull Distribution (Hazard Rate Criterion)*
TR-6 (DoD)	*Sampling Procedures and Tables for Life and Reliability Testing—Based on the Weibull Distribution (Reliable Life Criterion)*
TR-7 (DoD)	*Factors and Procedures for Applying MIL-STD-105 Sampling Plans to Life and Reliability Testing* (see note at MIL-STD-105)

INSPECTION AND TEST METHODS

Definitions, Units of Measurement

ASTM E548-94	*General Criteria Used for Evaluating Laboratory Competence, Standard Guide for*
ASTM E 994-95	*Calibration and Testings Laboratory Accreditation Systems, General Requirements for Operation and Recognition, Standard Guide for*
ASTM 177-90a (R1996)	*Use of the Terms Precision and Bias in ASTM Test Methods, Standard Practice for*
IEEE/ASTM SI 10-1997	*Standard for the Use of the International System of Units (SI)*
ISO Guide 2	*General Terms and Their Definitions Concerning Standardization, Certification, and Testing Laboratory Accreditation*

| ISO/IEC Guide 30 | *Terms and Definitions Used in Connection with Reference Materials* |
| MIL-STD-1309C | *Definitions of Terms for Test, Measurement and Diagnostic Equipment* |

Inspection and Test Methods and Requirements

ANSI/ASQC E2-1996	*Guide to Inspection Planning*
ANSI/ASQC M1	*Calibration Systems, American National Standard for Calibration System*
ANSI/ASQC E4-1994	*Specification and Guidelines for Quality Systems for Environmental Data Collection and Environmental Technology Programs*
ANSI/ISO 14001-1996	*Environmental Management Systems—Specifications with Guidance for Use*
ANSI/ISO 14004-1996	*Environmental Management Systems—General Guidelines on Principles, Systems and Supporting Techniques*
ANSI/ISO 14010-1996	*Guidelines for Environmental Auditing—General Principles of Environmental Auditing*
ANSI/ISO 14011-1996	*Guidelines for Environmental Auditing—Audit Procedures—Auditing of Environmental Systems*
ANSI/ISO 14012-1996	*Guidelines for Environmental Auditing—Qualification Criteria for Environmental Auditors*
IEC 60605-1 (1978-01)	*Equipment Reliability Testing—Part 1: General Requirements*
IEC 60605-2 (1994-10)	*Equipment Reliability Testing—Part 2: Design of Test Cycles*
IEC 60605-3-1 (1986-09)	*Equipment Reliability Testing—Part 3: Preferred Test Conditions Indoor Portable Equipment—Low Degree of Simulation*
IEC 60605-3-2 (1986-09)	*Equipment Reliability Testing—Part 3: Preferred Test Conditions, Equipment for Stationary Use in Weather-protected Locations—High Degree of Simulation*
IEC 60605-3-3 (1992-11)	*Equipment Reliability Testing—Part 3: Preferred Test Conditions—Section 3: Test Cycle 3 Equipment for Stationary Use in Partially Weather Protected Locations—Low Degree of Simulation*
IEC 60605-3-4 (1992-07)	*Equipment Reliability Testing—Part 3: Preferred Test Conditions—Section 4: Test Cycle 4: Equipment for Portable and Non-stationary Use—Low Degree of Simulation*
IEC 60605-4 (1986-07)	*Equipment Reliability Testing—Part 4: Procedure for Determining Point Estimates and Confidence Limits for Equipment Reliability Determination Tests*
IEC 60605-6 (1997-04)	*Equipment Reliability Testing—Part 6: Tests for the Validity of the Constant Failure Rate or Constant Failure Intensity Assumption*
IEC 60605-7	*Equipment Reliability Testing—Part 7: Compliance Test Plans for Failure Rate and Mean Time Between Failures Assuming Constant Failure Rate*

ISO 1	*Standard Reference Temperature for Industrial Length Measurements*
ISO R286	*ISO System of Limits and Fits—Part 1: General Tolerances and Deviations*
ISO R 1938	*ISO System of Limits and Fits—Part II: Inspection of Plain Workpieces*
ISO Guide 12	*Comparative Testing of Consumer Products*
ISO/IEC Guide 36	*Preparation of Standard Methods of Measuring Performance (SMMP) of Consumer Goods*
ISO/IEC Guide 39	*General Requirements for the Acceptance of Inspection Bodies*
ISO/IEC Guide 43	*Development and Operation of Laboratory Proficiency Testing*
MIL-STD-120	*Gage Inspection*
MIL-STD-202F	*Test Methods for Electronic and Electrical Component Parts*
MIL-STD-252B	*Classification of Visual and Mechanical Defects for Equipment, Electronic, Wires and Other Devices*
MIL-STD-271E	*Nondestructive Testing Requirements for Metals*
MIL-STD-415D	*Test Provisions for Electronic Systems and Associated Equipment, Design Criteria for*
MIL-STD-810D	*Environmental Test Methods and Engineering Guidelines*
MIL-STD-1520C	*Corrective Action and Disposition System for Nonconforming Material*
MIL-STD-1540C	*Test Requirements for Space Vehicles*
MIL-STD-2164	*Environmental Stress Screening Process for Electronic Equipment*
MIL-STD-2165	*Testability Program for Electronic Systems and Equipment*
MIL-STD-45662	*Calibration System Requirement*
MIL-T-5422F	*Testing Environmental, Airborne Electronic and Associated Equipment*
MIL-I-6970E	*Inspection Program Requirements, Nondestructive Testing for Aircraft and Missile Materials and Parts*
MIL-M-38793	*Manuals, Technical, Calibration Procedures, Preparation of*
MIL-I-45208	*Inspection System Requirements*
MIL-HDBK-204	*Inspection Equipment Design*
MIL-HDBK-333	*Handbook for Standardization of Nondestructive test Methods Vol I and II*
NHB 5300.4 (1C)NASA	*Inspection System Provisions for Aeronautical and Space System Materials, Parts, Components and Services*
STP 335 (1963)	*Manual for Conducting and Interlaboratory Study of a Test Method*
STP 540 ASTM	*Sampling Standards and Homogeneity*
STP 624 ASTM	*Nondestructive Testing Standards—A Review*

Evaluation of Inspection and Test Systems

ANSI/ASQC E2-1996	*Guide to Inspection Planning*

ISO/IEC Guide 25	*General Requirements for the Competence of Calibration and Testing Laboratories*
ISO/IEC Guide 38	*General Requirements of the Acceptance of Testing Laboratories*
MIL-HDBK-51	*Evaluation of a Contractor's Inspection System*
MIL-HDBK-52A	*Evaluation of a Contractor's Calibration System*
MIL-STD-410D Notice 2	*Nondestructive Testing Personnel Qualifications and Certification (Eddy Current, Liquid Penetrant, Magnetic Particle, Radiographic and Ultrasonic)*
STP 814 1983 ASTM	*Evaluation and Accreditation of Inspection and Test Activities*

Precision and Accuracy

ASTM E 691-92	*Conducting an Interlaboratory Test Program to Determine the Precision of Test Methods, Standard Practice for*
ASTM E 1267-88	*ASTM Standard Specification Quality Statements, Standard Guide for*
ASTM E 177-R96	*Use of Terms Precision and Bias in ASTM Test Methods, Standard Practice for*
"GUM" BIPP/IEC/IFCC/ISO/ IUPAC/OIML	*Guide to the Expression of Uncertainty in Measurement* (published by ISO)
ISO 5725-1	*Accuracy (Trueness and Precision) of Measurement Methods and Results—Part 1: General Principles and Definitions*
ISO 5725-2	*Accuracy (Trueness and Precision) of Measurement Methods and Results—Part 2: Basic Methods for the Determination of Repeatability and Reproducibility of a Standard Measurement Method*
ISO 5725-3	*Accuracy (Trueness and Precision) of Measurement Methods and Results—Part 3: Intermediate Measures of the Precision of a Standard Measurement Method*
ISO 5725-4	*Accuracy (Trueness and Precision) of Measurement Methods and Results—Part 4: Basic Methods for the Determination of the Trueness of a Standard Measurement Method*
ISO 5725-5	*Accuracy (Trueness and Precision) of Measurement Methods and Results—Part 5: Alternative Methods for the Determination of the Precision of a Standard Measurement Method*
ISO 5725-6	*Accuracy (Trueness and Precision) of Measurement Methods and Results—Part 6: Use in Practice of Accuracy Values*
ISO 11095	*Linear Calibration Using Reference Materials*
ISO 11453	*Statistical Interpretation of Data—Tests and Confidence Intervals Relating to Proportions*
ISO 11843-1	*Decision Limit—Detection Limit—Capability of Detection—Part 1: Terms and Definitions*
ISO/TR 13425	*Guide for the Selection of Statistical Methods in Standardization and Specification*
"VIM" BIPP/IEC/IFCC/ISO/ IUPAC/OIML	*International Vocabulary of Basic and General Terms in Metrology* (published by ISO)

Copies of the preceding standards and specifications can be ordered by writing to the organizations indicated in the following:

ANSI standards	American National Standards Institute 11 West 42nd Street New York, NY 10066 www.ansi.org
	American Society for Quality 611 East Wisconsin Avenue Milwaukee, WI 53201-3005 www.asq.org
ASTM publications	American Society for Testing and Materials 100 Barr Harbor Drive West Conshohocken, PA 19428-2959 www.astm.org
AQAP	North Atlantic Treaty Organization Autoroute de Zaventem 1110 NATO (Brussels), Belgium www.nato.int
	In the United States, contact: Defense Automated Printing Services 700 Robbins Avenue Bldg. 4, Section D Philadelphia PA 19111-5094 www.dodssp.daps.mil
BSI	British Standards Institution 389 Chiswick High Road London W4 4AL, United Kingdom www.bsi.org.uk
CSA	Canadian Standards Association 178 Rexdale Blvd. Rexdale (Toronto), Ontario Canada M9W 1R3 www.csa.ca
DOD	The Department of Defense The Pentagon Washington, DC 20301-1155 www.dodssp.daps.mil
Federal standards	General Services Administration GSA/FSSP Specification Section, Room 6654 78 D Street S.W. Washington, DC 20407 http://pub.fss.gsa.gov/pub/fed-specs.html
IEC publications	American National Standards Institute 11 West 42nd Street New York, NY 10036 www.iec.ch
ISO publications	American National Standards Institute 7 West 42nd Street New York, NY 10026 www.ansi.org

Military handbooks, military specifications, and military standards	Defense Automated Printing Services 700 Robbins Avenue Bldg. 4, Section D Philadelphia PA 19111-5094 www.dodsp.daps.mil
NASA reliability, maintainability, and assurance publications	Superintendent of Documents U.S. Government Printing Office Washington, DC 20402 www.nasa.gov

APPENDIX IV
QUALITY SYSTEMS TERMINOLOGY

The following terms and definitions are selected from those given in the American National Standard ANSI/ASQC A8402-1994, *Quality Management and Quality Assurance—Vocabulary.* The reader is urged to consult the full standard for the full vocabulary list and for valuable notes and comments associated with individual terms. The document is available from American Society for Quality, 611 East Wisconsin Avenue, Milwaukee, WI 53202. In many cases, the authors of specific sections have used slightly different or expanded definitions of some terms. The following terms and definitions are provided for general understanding and use.

Conformity: Fulfillment of specific requirements.

Customer: Recipient of a product provided by the supplier.

Dependability: Collective term used to describe the availability performance and its influencing factors: reliability performance, maintainability performance, and maintenance-support performance.

Design review: Documented, comprehensive, and systematic examination of a design to evaluate its capability to fulfill the requirements for quality, identify problems, if any, and propose the development of solutions.

Deviation permit: See *Production permit.*

Entity: Item which can be individually described and considered.

Grade: Category or rank given to entities having the same functional use but different requirements for quality.

Inspection: Activity such as measuring, examining, testing, or gauging one or more characteristics of an entity and comparing the results with specified requirements in order to establish whether conformity is achieved for each characteristic.

Objective evidence: Information which can be proved true, based on facts obtained through observation, measurement, test, or other means.

Procedure: Specified way to perform an activity.

Process: Set of interrelated resources and activities which transform inputs into outputs.

Process quality audit: See *Quality audit.*

Product: Results of activities or processes.

Product quality audit: See *Quality audit.*

Production permit (Deviation permit):   Written authorization to depart from the originally specified requirements for a product prior to its production.

Quality: Totality of characteristics of an entity that bear on its ability to satisfy stated and implied needs.

Quality assurance: All the planned and systematic activities implemented within the quality system, and demonstrated as needed, to provide adequate confidence that an entity will fulfill requirements for quality.

Quality audit: Systematic and independent examination to determine whether quality activities and related results comply with planned arrangements and whether these arrangements are implemented effectively and are suitable to achieve objectives. *Note:* The quality audit typically applies to, but is not limited to, a quality system or elements thereof, to processes, to products, and to services. Such audits are often called "quality system audit," "process quality audit," product quality audit," "service quality audit."

Quality control: Operational techniques and activities that are used to fulfill requirements for quality.

Quality loop/quality spiral: Conceptual model of interacting activities that influence quality as well as the losses incurred when satisfactory quality is not achieved.

Quality management: All activities of the overall management function that determine the quality policy, objectives, and responsibilities, and implement them by means such as quality planning, quality control, and quality improvement within the quality system.

Quality plan: Document setting out the specific quality practices, resources, and sequence of activities relevant to a particular product, project, or contract.

Quality planning: Activities that establish the objectives and requirements for quality and for the application of quality system elements.

Quality policy: Overall intentions and direction of an organization with regard to quality, as formally expressed by top management.

Quality surveillance: Continual monitoring and verification of the status of an entity and analysis of records to ensure that specified requirements are being fulfilled.

Quality system: Organizational structure, procedures, processes, and resources needed to implement quality management.

Quality system audit: See *Quality audit.*

Requirements for quality: Expression of the needs or their translation into a set of quantitatively or qualitatively stated requirements for the characteristics of an entity to enable its realization and examination.

Requirements of society: Obligations resulting from laws, regulations, rules, codes, statutes, and other considerations.

Service: Result generated by activities at the interface between the supplier and the customer and by supplier internal activities to meet the customer needs.

Specification: Document stating requirements.

Supplier: Organization that provides a product to the customer.

Traceability: Ability to trace the history, application, or location of an entity by means of recorded identifications.

Verification: Confirmation by examination and provision of objective evidence that specified requirements have been fulfilled.

Waiver: Written authorization to use or release a product which does not conform to the specified requirements.

APPENDIX V
QUALITY IMPROVEMENT TOOLS*

INTRODUCTION

Ishikawa's classic *Guide to Quality Control* (1972) is generally credited as the first training manual of problem-solving tools specifically presented for use in quality improvement. In the Japanese original, the book was a training reference for factory workers who were members of quality control (QC) circles. To the "seven tools" of that volume (flow diagrams, brainstorming, cause-effect diagrams, Pareto analysis, histograms, and scatter diagrams), we here add three more. Two are implied but not dealt with in detail in the Ishikawa work—data collection and graphs and charts. The third, box plots, is of more recent origin (Tukey 1977), but has proved itself a worthy complement or substitute for histograms under certain conditions. There are many other useful tools; this list is not exhaustive, nor could it or any list be so. But the list here is a useful starting place for quality improvement activity. Quality improvement teams that have mastered these tools are well prepared for most of the problems they are likely to face.

As presented by Juran Institute in the course *Quality Improvement Tools,* the tools are integrated with a structured quality improvement process (Figure V.1). Each tool is described and instructions for the tool's use are presented. In addition, the process steps in which the tool is used are identified. This information is captured in an application map in matrix form (Figure V.1), with each column corresponding to a tool, and each row corresponding to a process step. At each intersection is a symbol indicating the frequency of use of that tool at that process step (frequent, infrequent, and very rarely).

The process map is a valuable guide to problem-solving teams in these ways:

1. It reminds the team that there is a structured order to the problem-solving process and helps keep the team on track.

2. At a given step, if the team is at a loss what to do next, one of the frequently used tools may suggest the next action.

3. At a given step, using a tool indicated as rarely used is a signal to the team to reconsider its course of action. A convenient example is the use of brainstorming (which is an effective way to develop a list of theories, ideas, and opinions of group members) to test theories (which always requires data, not the opinions of the team members).

The tools are briefly described and illustrated here. A more thorough treatment is available from a number of texts, including the course notes for *Quality Improvement Tools* or from *Modern Methods for Quality Control and Improvement* by Wadsworth, Stephens, and Godfrey (1986).

---

*In response to inquiries from users of previous editions of this handbook, the editors include this introduction to some of the more important and popularly used quality improvement tools. Material in this appendix is adapted from material provided in the course *Quality Improvement Tools* presented by Juran Institute, Inc., Wilton, CT.

Quality Improvement Steps and Activities — Quality Tools

Quality Improvement Steps and Activities	Box Plot	Brainstorming	Cause-Effect Diagram	Data Collection	Flow Diagrams	Graphs and Charts	Histogram	Pareto Analysis	Scatter Diagram	Stratification
1. Identify a Project										
a. Nominate projects		■		■	■					
b. Evaluate projects		×	×	■	□	□		■		
c. Select a project				■				■		□
d. Ask: Is it quality improvement?	×		×	■	□		×	■	×	□
2. Establish the Project										
a. Prepare a mission statement					□	□		□		□
b. Select a team					□					
c. Verify the mission				□	□	□				
3. Diagnose the Cause										
a. Analyze symptoms	□	×	×	■	■	□	□	■		■
b. Confirm or modify the mission		×	×	■	■			■		□
c. Formulate theories		■	■		□					□
d. Test theories	■	×	×	■	■		■	■	■	■
e. Identify root cause(s)	■	×	×	■	■	■	■	■	■	■
4. Remedy the Cause										
a. Evaluate alternatives		■	□	■	■		□	□	□	□
b. Design remedy	□	×	×	□	■	■	□	□	□	□
c. Design controls	□	×	×		■			□	□	□
d. Design for culture	□	×	×							
e. Prove effectiveness		■								
f. Implement				■				□	□	□
5. Hold the Gains										
a. Design effective quality controls										
b. Foolproof the remedy	□	×	×		■			□	□	□
c. Audit the controls					■					
6. Replicate Results and Nominate New Projects										
a. Replicate the project results				■		■		■		
b. Nominate new projects				■	■			□	□	□
					■			■		

Legend: ■ Frequently used   □ Occasionally used   (dots) Rarely used   × Never used

FIGURE V.1   Applications for quality improvement tools. (*Juran Institute.*)

THE TOOLS

Flow Diagram.   A graphic representation of the sequence of steps needed to produce some output. The output may be a physical product, a service, information, or a combination of the three. The symbols of a flow diagram are specific to function and are explained in Figure V.2. Figure V.3 shows the use of these symbols in the flow diagram of an airline ticketing process.

Brainstorming.   A group technique for generating constructive and creative ideas from all participants. Use of this tool should provide new ideas, or new applications and novel use of existing ideas. The technique is outlined in Figure V.4.

Cause-Effect ("Fishbone") Diagram.   Developed by Kaoru Ishikawa, this tool is frequently called the Ishikawa diagram in his honor. Its purpose is to organize and display the interrelationships of various theories of root cause of a problem. By focusing attention on the possible causes of a specific

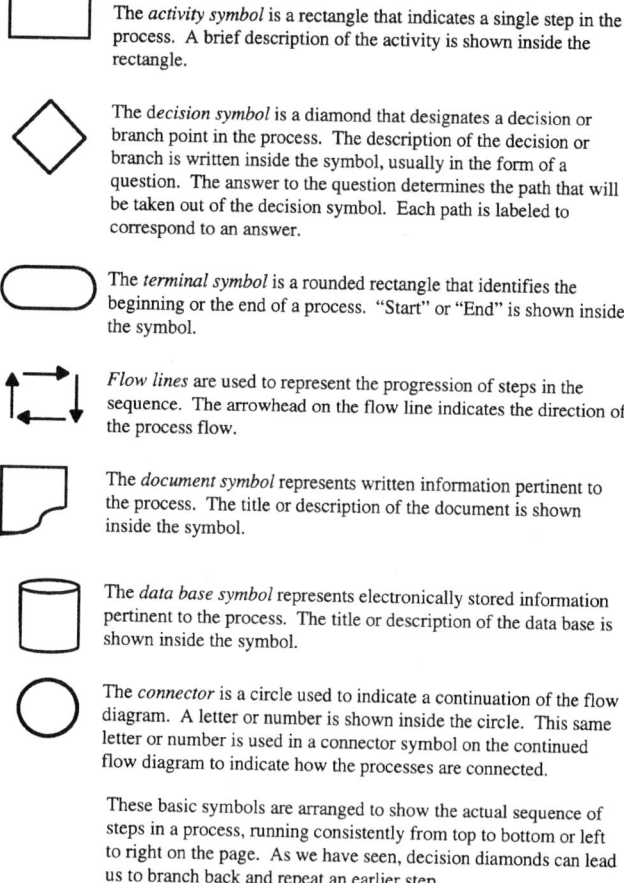

The *activity symbol* is a rectangle that indicates a single step in the process. A brief description of the activity is shown inside the rectangle.

The *decision symbol* is a diamond that designates a decision or branch point in the process. The description of the decision or branch is written inside the symbol, usually in the form of a question. The answer to the question determines the path that will be taken out of the decision symbol. Each path is labeled to correspond to an answer.

The *terminal symbol* is a rounded rectangle that identifies the beginning or the end of a process. "Start" or "End" is shown inside the symbol.

Flow lines are used to represent the progression of steps in the sequence. The arrowhead on the flow line indicates the direction of the process flow.

The *document symbol* represents written information pertinent to the process. The title or description of the document is shown inside the symbol.

The *data base symbol* represents electronically stored information pertinent to the process. The title or description of the data base is shown inside the symbol.

The *connector* is a circle used to indicate a continuation of the flow diagram. A letter or number is shown inside the circle. This same letter or number is used in a connector symbol on the continued flow diagram to indicate how the processes are connected.

These basic symbols are arranged to show the actual sequence of steps in a process, running consistently from top to bottom or left to right on the page. As we have seen, decision diamonds can lead us to branch back and repeat an earlier step.

FIGURE V.2   Symbols used in flow diagramming.

AV.4 APPENDIX V

Customer Arrives at Airport Needing to Purchase Ticket or Check Bag

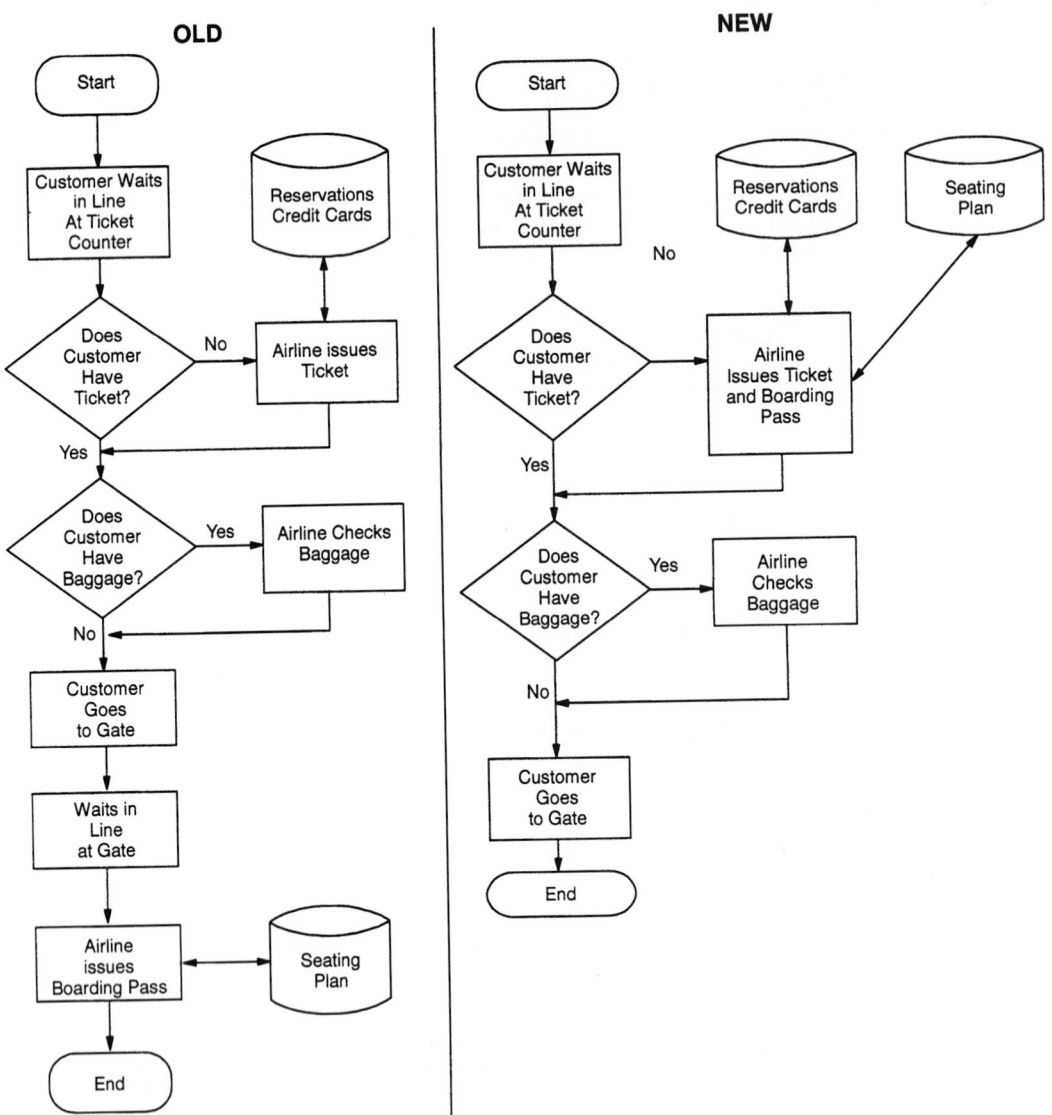

FIGURE V.3 Before and after flow diagrams of an airline ticketing process. (*Juran Institute.*)

- Good ideas are not praised or endorsed. All judgment is suspended initially in preference to generating ideas.
- Thinking must be unconventional, imaginative, or even outrageous. Self-criticism and self-judgment are suspended.
- To discourage analytical or critical thinking, team members are instructed to aim for a large number of new ideas in the shortest possible time.
- Team members should "hitchhike" on other ideas, by expanding them, modifying them, or producing new ones by association.

FIGURE V.4 Brainstorming technique. (*Juran Institute.*)

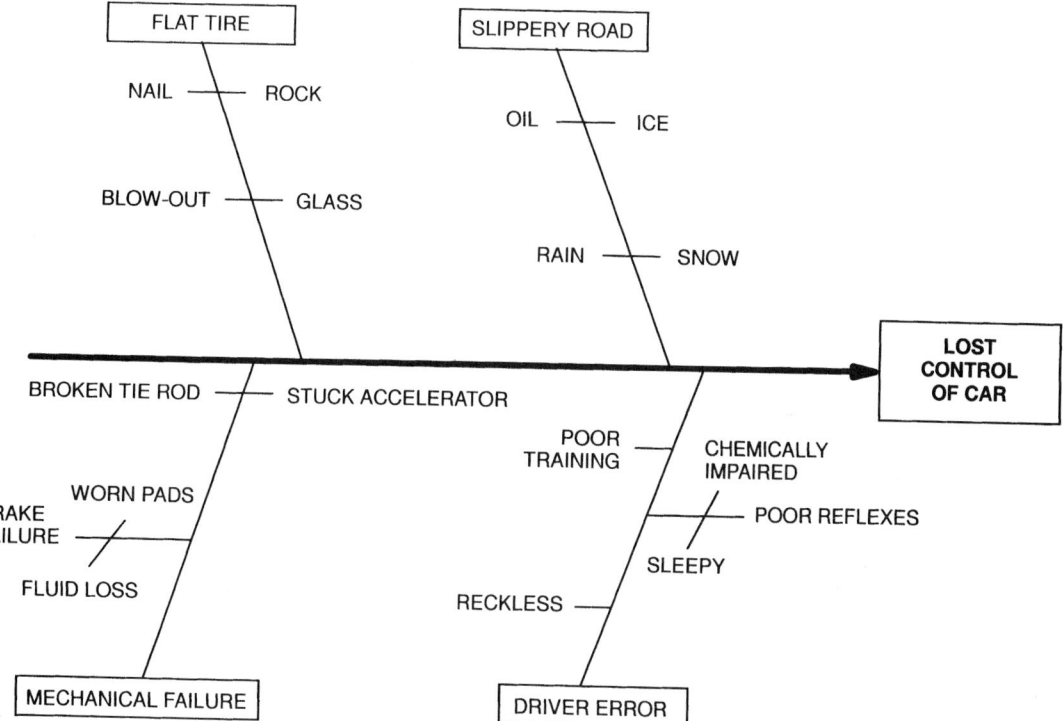

FIGURE V.5 Cause-effect diagram for lost control of car. (*Juran Institute.*)

problem in a structured, systematic way, the diagram enables a problem-solving team to clarify its thinking about those potential causes, and enables the team to work more productively toward discovering the true root cause or causes. Figure V.5 is an example.

Data Collection. The gathering of the objective data needed to shed light on the problem at hand, and in a form appropriate for the tool selected for the analysis of the data. Types of data collection include check sheets (providing data and trends), data sheets (simple tabular or columnar format), and checklists (simple listing of steps to be taken). Figures V.6 and V.7 are examples.

Graphs and Charts. A broad class of tools used to summarize quantitative data in pictorial representations. Three types of graphs and charts that prove especially useful in quality improvement are line graphs, bar graphs, and pie charts. A *line graph* connects points which represent pairs of numeric

COMPONENTS REPLACED BY LAB

Enter a mark for each component replaced. Mark like
the following: /   //   ///   ////   ~~/////~~

Time Period: 22 Feb to 27 Feb 1988
Repair Technician: Bob

TV SET MODEL 1013

Integrated circuits	~~/////~~
Capacitors	~~/////~~ ~~/////~~ ~~/////~~ ~~/////~~ ~~/////~~ //
Resistors	//
Transformers	////
Commands	
CRT	/

TV SET MODEL 1017

Integrated circuits	///
Capacitors	~~/////~~ ~~/////~~ ~~/////~~ ~~/////~~ ~~/////~~ //
Resistors	/
Transformers	//
Commands	~~/////~~ ~~/////~~ ~~/////~~ / //
CRT	/

TV SET MODEL 1019

Integrated circuits	/
Capacitors	~~/////~~ ~~/////~~ ~~/////~~ ~~/////~~ ///
Resistors	/
Transformers	//
Commands	
CRT	/

FIGURE V.6   Check sheet for TV component failures. (*Juran Institute.*)

data, to display the relationship between two continuous numerical variables (e.g., cost and time). In Figure V.8 each pair of data consists of a year and the cost of poor quality recorded in that year. The cost of poor quality is shown as a function of the year. A *bar graph* also portrays the relationship between pairs of variables, but only one of the variables need be numeric (Figure V.9). A *pie chart* (Figure V.10) shows the proportions of the various classes of a phenomenon being studied that make up the whole.

Stratification.   This is the separation of data into categories. It is used to identify which categories contribute to the problem being solved and which categories are worthy of further investigation. It is

- Read temperature to nearest degree off meter number 5. Date: _____6-7-88_____

- Record the temperature in the table below.

  Line #: _____13_____

- Reading should be taken on the hour (± 5 minutes).

- Use the "Notes" section to record anything unusual. Inspector: _____Ginny Smith_____

* Question? Contact Larry Fine x2222

Time of Day	Temperature (°F)
0800	240
0900	242
1000	236
1100	236
1200	236

Time of Day	Temperature (°F)
1300	227
1400	230
1500	224
1600	220
1700	220

Notes: • *1100 hours reading taken at 1112*
  • *The line was stopped between 1310 and 1330*

FIGURE V.7   A data sheet showing solder bath temperature. (*Juran Institute.*)

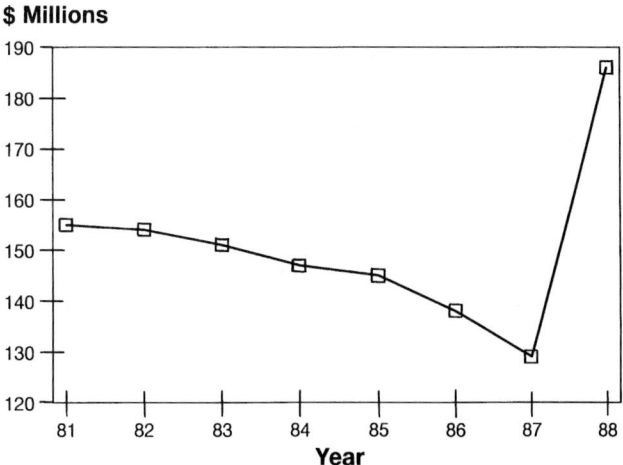

$ Millions

FIGURE V.8   A deceptive vertical scale, intended to show the cost of poor quality. (*Juran Institute.*)

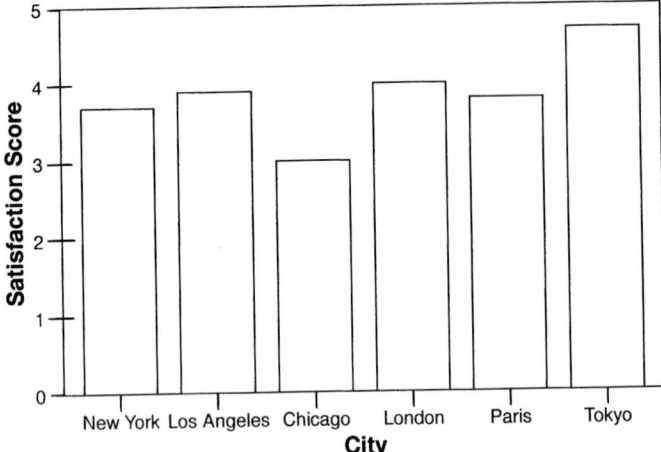

FIGURE V.9   A bar graph showing average customer satisfaction scores by city. (*Juran Institute.*)

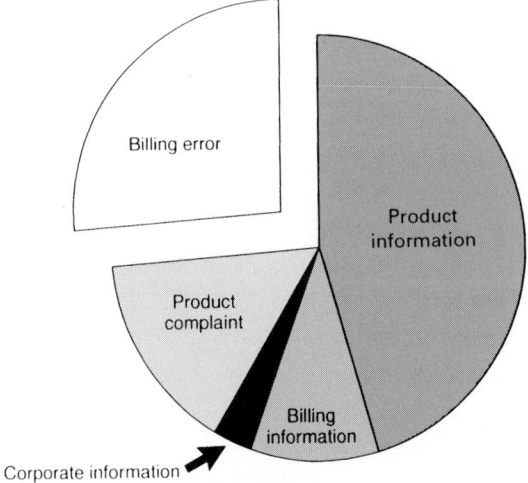

FIGURE V.10   A pie chart showing phone calls received by customer service in one month. (*Juran Institute.*)

an analysis technique that helps pinpoint the location or source of a quality problem. It may be necessary to stratify the data in many different ways. Figures V.11 and V.12 illustrate stratification of data related to field failure of an electronic component. The first stratification is by supplier, the second by shipper. The analysis reveals that the problem relates to shippers (specifically a new shipper), not suppliers. Stratification is the basis for the application of other tools, such as Pareto analysis and scatter diagrams.

Pareto Analysis.   This is a tool used to establish priorities, dividing contributing effects into the "vital few" and "useful many." A Pareto diagram includes three basic elements: (1) the contributors

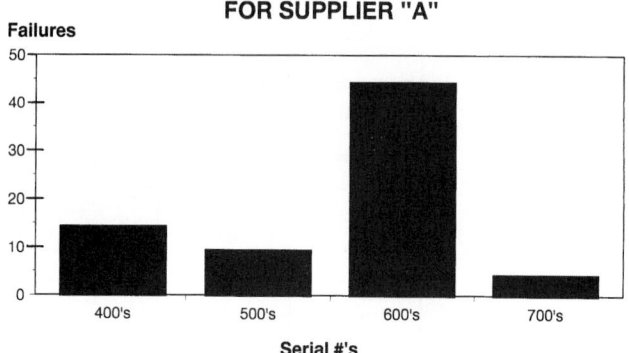

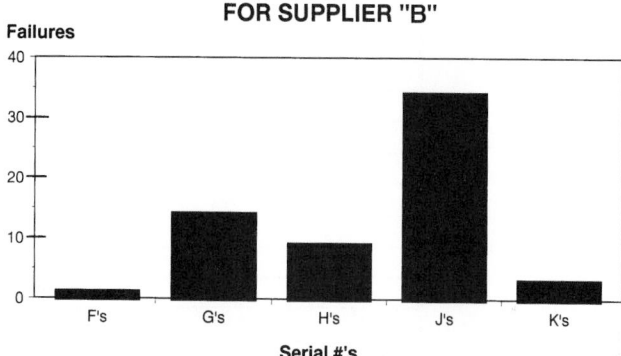

FIGURE V.11   Second-stage stratification of failure. Shows RF driver failure by serial number. (*Juran Institute.*)

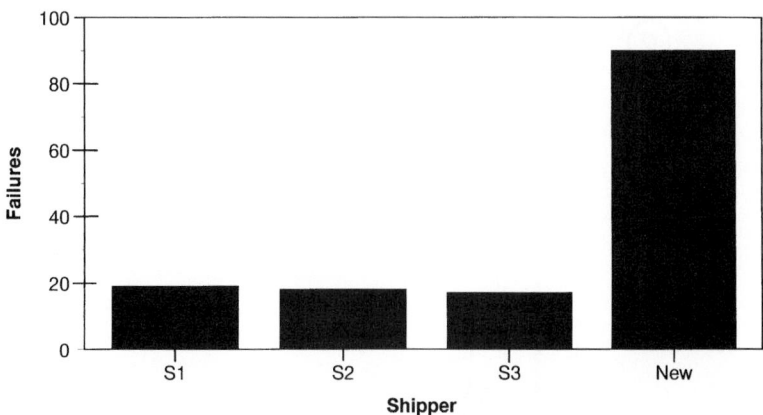

FIGURE V.12   Stratification by new variable. Shows RF driver failures by shipper. (*Juran Institute.*)

① Order-Form Item	② Number of Errors	③ Percent of Total	Cumulative-Percent of Total
G	44	29	
J	38	25	29
M	31	21	54
Q	16	11	75
B	8	5	86
D	5	3	91
C	3	2	95
A	1	0.67	97
O	1	0.67	98
R	1	0.67	98
N	1	0.67	99
L	1	0.66	99
I	0	0	100
E	0	0	100
H	0	0	100
K	0	0	100
F	0	0	100
P	0	0	100
TOTAL	150	100	100

FIGURE V.13   Pareto table of errors on order forms. (*Juran Institute.*)

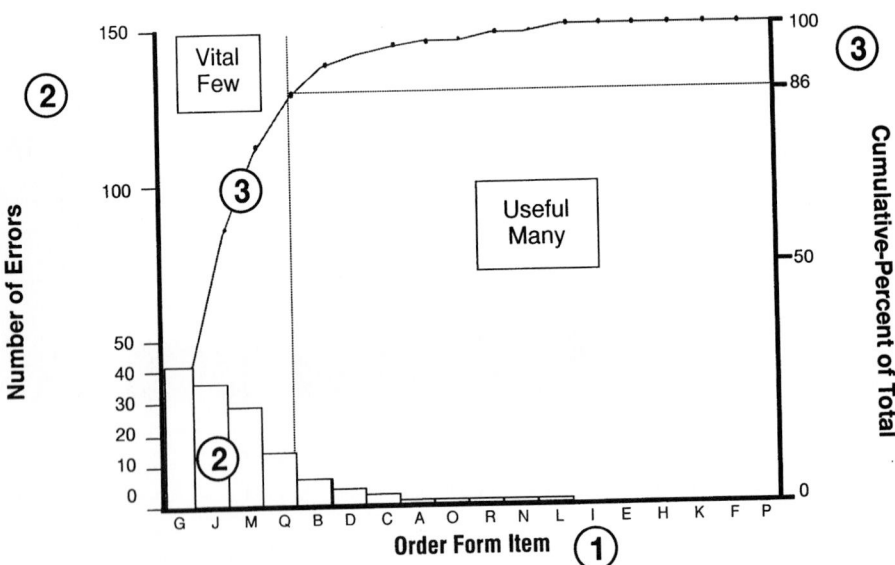

FIGURE V.14   Pareto diagram of errors on order forms. (*Juran Institute.*)

to the total effect, ranked by the magnitude of contribution; (2) the magnitude of the contribution of each expressed numerically; and (3) the cumulative-percent-of-total effect of the ranked contributors. Figure V.13 shows a Pareto table. Figure V.14 shows the same data in a Pareto diagram. Note the three basic elements as reflected in each figure.

Histogram. This is a picture of the distribution of a set of measurements. A histogram is a graphic summary of variation in a set of data. Four concepts related to variation in a set of data underlie the usefulness of the histogram: (1) values in a set of data almost always show variation, (2) variation displays a pattern, (3) patterns of variation are difficult to see in simple numerical tables, and (4) patterns of variation are easier to see when the data are summarized pictorially in a histogram. Analysis consists of identifying and classifying the pattern of variation displayed by the histogram, then relating what is known about the characteristic pattern to the physical conditions under which the data were created to explain what in those conditions might have given rise to the pattern. The data in Figure V.15a shows days elapsed between an interdepartmental request for an interview and the actual interview. In Figure V.15b, the histogram helped a team recognize the unacceptable range of time elapsed from request to interview. It also provided the team with a vivid demonstration of a human-created phenomenon—the rush at day 15 to get as many requests completed within the 15-day goal. The histogram directed the team's attention toward steps to reduce the duration (and with it the spread) of the process.

Scatter Diagram. This is a tool for charting the relationship between two variables to determine whether there is a correlation between the two which might indicate a cause-effect relationship (or indicate that no cause-effect relationship exists). Figure V.16 shows scatter diagrams used to explore three potential causes for clerical errors. Potential causes, chosen from a cause-effect diagram are: fatigue, working with a mixture of alphabetic and numeric keyboard characters, and distracting background noise. The diagrams indicate the alphanumeric mixture appears to be the most likely root cause of the clerical errors.

Box Plot. This is a graphic, five-number summary of variation in a data set. The data are summarized by: the smallest value, second quartile, median, third quartile, and largest value. The box plot can be used to display the variation in a small sample of data or for comparing the variation in

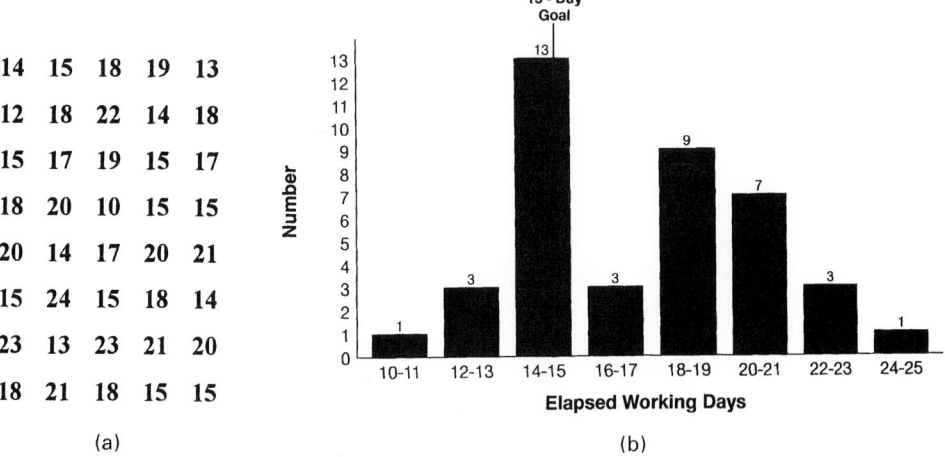

FIGURE V.15 Response to job evaluation requests. (*a*) Data showing elapsed time (in working days) from receipt of request to preliminary review and contact with manager. (*b*) Histogram. (*Juran Institute.*)

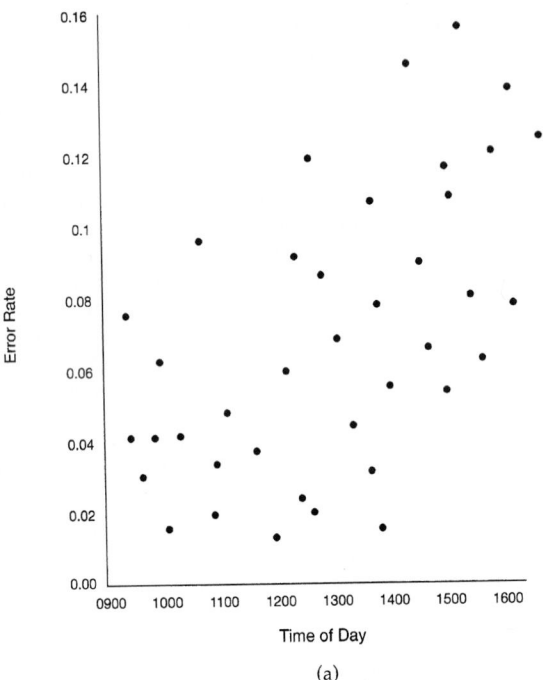

Time of Day

(a)

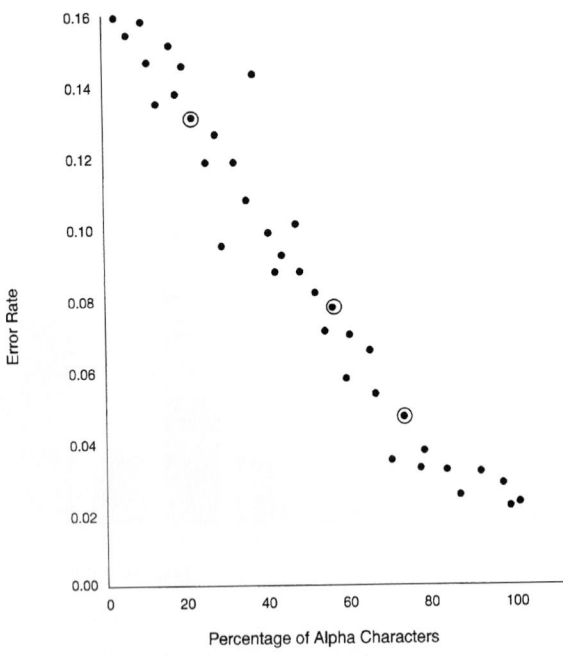

Percentage of Alpha Characters

(b)

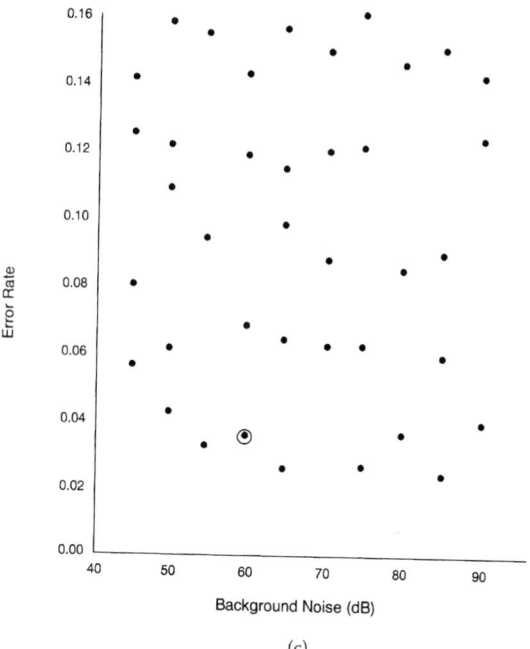

(c)

FIGURE V.16 Scatter diagrams used to explore potential causes of error in clerical data-entry errors. (*a*) Time of day; (*b*) the percentage of alphabetical characters; and (*c*) background noise. (*Juran Institute.*)

A		B	
243	345	192	251
207	268	156	102
272	290	228	215
45	251	125	145
226	183	279	188

(a)

Contractor	1	2	3	4	5	6	7	8	9	10
A	45	183	207	226	243	251	268	272	290	345
B	102	125	145	156	188	192	215	228	251	279

(b)

FIGURE V.17 Box plots used to compare response times (in minutes) of two copier repair contractors, A and B. (*a*) Raw data; (*b*) ordered data set; and (*c*) basic box plots. (*Juran Institute.*)

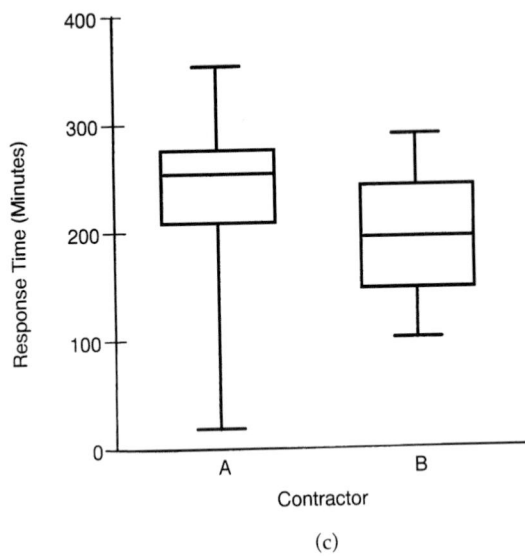

(c)

FIGURE V.17   Box plots used to compare response times (in minutes) of two copier repair contractors, A and B. (*a*) Raw data; (*b*) ordered data set; and (*c*) basic box plots. (*Juran Institute.*)

a large number of related distributions. The box plot (sometimes called a "box and whisker chart") can be useful even when the number of data points is too small to produce a meaningful histogram. In analyzing the performance of two copy-repair contractors, a potential customer created the analysis in Figure V.17. It is based on 10 repair calls each to contractors A and B. Figure V.17*a* and *b* shows the response time of two copy machine contractors as raw data and as "ordered" (sequenced) data. Figure 17*c* shows the same data as a box plots. The comparison of the two plots favors contractor B: B's median response time is less than A's and exhibits less variation.

REFERENCE

Tukey, J. W. (1977). *Exploratory Data Analysis.* Addison-Wesley, Reading, MA.

Wadsworth, Harrison M., Stephens, Kenneth S., and Godfrey, A. Blanton (1986). *Modern Methods for Quality Control and Improvement.* John Wiley and Sons, New York.

NAME INDEX

See also Glossary of Symbols and Abbreviations, Appendix I; Subject Index.

SUBJECT INDEX

Collapsing 2^k factorial experiments, **47**.44, **47**.45
Collection and analysis of data, **44**.2–**44**.4
College of Physicians, **32**.5
Colombian Institute for Technical Standards
   (ICONTEC), **43**.12
Comisión Panamericana de Normas Técnicas
   (COPANT), **37**.14
Commitment:
   defined, **15**.3
   employee empowerment and, **15**.3
   work design for, **15**.11–**15**.14
Committee on Conformity Assessment (CASCO),
   37.18
Common causes of variation, **45**.2
Communication errors:
   defined, **5**.63, **5**.64
   diagnosis and remedy, **5**.63–**5**.65
Communism's quality legacy in central and eastern
   Europe, **39**.1–**39**.5
Companywide quality management in Japan,
   41.9–**41**.19
Competitive evaluation:
   by field studies, **18**.15–**18**.18
   in product development, **3**.32
Complaints:
   in automotive industry, **29**.18
   effect on income, **7**.8–**7**.11
   effect on sales of handling, **3**.15
   as field intelligence, **18**.7
   as indicator of potential customer defection,
    18.21
   vs. systematic collection of quality data, **41**.26,
    41.27
Completely randomized design of experiments,
   47.8–**47**.17
Completeness, defined, **19**.16
Composite sample, in bulk sampling, **46**.72
Computer programs:
   for acceptance sampling, **46**.78–**46**.80
   BEST (Business Environment Strategic Toolkit),
    37.8
   FIT (Financial Improvement Toolkit), **37**.8
   for regression, **44**.105, **44**.107
   for reliability data analysis, **48**.3
   for statistical process control, **45**.27, **45**.28
Computer software industry, adoptions and
   extensions of ISO 9000 standards,
   11.22–**11**.24
Computer storage, of data, **34**.2
Computer systems:
   applications to quality systems, **10**.1–**10**.21
   and communications, **10**.1, **10**.2
   and computer-aided inspection, **10**.10–**10**.13
   and design control, **10**.2–**10**.5

Computer systems (*Cont.*):
   future trends in applications to quality, **10**.19,
    10.20
   and reliability and safety, **10**.15–**10**.19
   software attributes in, **10**.7–**10**.9
   and statistics, **10**.9, **10**.10
   and testing and validation, **10**.5–**10**.7
   virus protection in, **10**.2
   (*See also* Microprocessor)
Computer-aided design (CAD), defined, **22**.35
Computer-aided inspection, **10**.10–**10**.14
Computer-aided manufacturing (CAM), defined,
   22.35
Computer-integrated manufacturing (CIM):
   defined, **22**.35
   functions of, **22**.35, **22**.36
Computing and analysis in design of experiments,
   47.2, **47**.3
Computing processes in process industries,
   27.13
Concentration diagram (*see* Defect concentration
   diagram)
Conditional probability, **44**.21, **44**.22
Conference Board, **8**.10
Confidence interval estimates, **44**.42–**44**.46
   confidence interval, defined, **44**.42
   confidence level, defined, **44**.42
   confidence limits, defined, **44**.42
   point estimate, defined, **44**.42
   in regression, **44**.97–**44**.99, **44**.101-44.105
Confidence level, **44**.42
Confidence limits, for fraction defective, chart of,
   AII.28
Configuration management, **19**.31, **19**.32
Conformance:
   decision, **23**.3, **23**.4
   of process, **4**.16–**4**.20
   of product, **4**.20–**4**.24
   of product vs. fitness for use, **4**.20–**4**.24
Conformance to design:
   defined, **27**.20, **27**.21
   experience curves as measures of continual
    improvement, **27**.21
   goal conformance as fixed reference for product
    specifications, **27**.21
Conformance to specification, **2**.2, **2**.3
   vs. fitness for use, **7**.4
   in process industries, **27**.14
Confounding, in design of experiments, **47**.53,
   47.54
Conscious errors:
   defined, **5**.62
   diagnosis and remedy, **5**.62, **5**.63
   in inspection, **23**.47

ABOUT THE EDITORS

JOSEPH M. JURAN, Co-Editor-in-Chief, has published the leading reference and training materials on managing for quality, a field in which he has been the international leader for over 70 years. A holder of degrees in engineering and law, Dr. Juran has pursued a varied career in management as an engineer, industrial executive, government administrator, university professor, corporate director, and management consultant. As a member of the Board of Overseers, he helped to create the U.S. Malcolm Baldrige National Quality Award and has received over 50 medals and awards from 14 countries, including The Order of the Sacred Treasure from the Emperor of Japan for ". . . the development of Quality Control in Japan and the facilitation of U.S. and Japanese friendship"; and the National Medal of Technology from the President of the United States for "his lifetime work of providing the key principles and methods by which enterprises manage the quality of their products and processes, enhancing their ability to compete in the global marketplace." He is also founder of the consulting firm of Juran Institute, Inc., and founder of Juran Foundation, Inc. The latter is now a part of the Juran Center for Leadership in Quality at the Carlson School of Management in the University of Minnesota. Among his 20 books, the Handbook is the international reference standard.

A. BLANTON GODFREY, PH.D., Co-Editor-in-Chief, is Chairman and Chief Executive Officer of Juran Institute, Inc.. Under his leadership Juran Institute has expanded its consulting, training, and research services to over 50 countries. Previously, Dr. Godfrey was head of the Quality Theory and Technology Department of AT&T Bell Laboratories. He is a Fellow of the American Statistical Association, a Fellow of the American Society for Quality, a Fellow of the World Academy of Productivity Science, and an Academician of the International Academy for Quality. He is a co-author of *Modern Methods for Quality Control and Improvement* and *Curing Health Care*. He contributed to the creation of the Malcolm Baldrige National Quality Award and served as a judge from 1988 to 1990. In 1992 he received the Edwards Medal from ASQ for his outstanding contributions to the science and practice of quality management. Dr. Godfrey holds an M.S. and a Ph.D. in Statistics from Florida State University and a B.S. in Physics from Virginia Tech.

ROBERT E. HOOGSTOEL is Associate Editor for all but the statistical sections of the Handbook. As Manager of Technical Education at Dana Corporation, he conducted training and consulting in quality management throughout the corporation. For ten years he was a consultant with Juran Institute, working with clients worldwide in the manufacturing and process industries. He co-authored with Dr. Juran the Institute's training course in quality planning, and appears in the Institute's video training series, "Juran on Quality Planning." He has engineering degrees from Cornell University and the University of California at Berkeley. He is a registered professional engineer in New York State.

EDWARD G. SCHILLING, PH.D., is Associate Editor for the statistical sections of the Handbook. He is Professor of Statistics in the Center for Quality and Applied Statistics at Rochester Institute of Technology, where he has held the position of Director of the Center and Chair of the Graduate Statistics Department. Prior to joining R.I.T. he was manager of the Lighting Quality Operation for the Lighting Business Group of the General Electric Company. He received his B.A. and M.B.A. degrees from SUNY Buffalo, and his M.S. and Ph.D. degrees in statistics from Rutgers University, New Jersey. Dr. Schilling is a Fellow of the American Statistical Association and the American Society for Quality, and a member of the Institute of Mathematical Statistics, the American Economic Association, and the American Society for Testing and Materials. He received the ASQ Shewhart Medal in 1983 and its Brumbaugh Award four times. He is registered as a professional engineer in California and is certified by ASQ as a quality and reliability engineer. His two books, *Acceptance Sampling in Quality Control* and *Process Quality Control* (with E. R. Ott), are among the leading texts in the field.